A Member of the International Code Fa

INTERNATIONAL CODE COUNCIL®

INTERNATIONAL BUILDING CODE®

COMMENTARY - Vol. II

2003

First Printing: February 2004
Second Printing: September 2004
Third Printing: February 2005

ISBN # 1-58001-128-4

PRINTED IN THE U.S.A.

PREFACE

The principal purpose of the Commentary is to provide a basic volume of knowledge and facts relating to building construction as it pertains to the regulations set forth in the 2003 *International Building Code*. The person who is serious about effectively designing, constructing and regulating buildings and structures will find the Commentary to be a reliable data source and reference to almost all components of the built environment

As a follow-up to the *International Building Code*, we offer a companion document, the *International Building Code Commentary—Volume II.* Volume II covers Chapters 16 through 35 of the 2003 *International Building Code*. The basic appeal of the Commentary is thus: it provides in a small package and at reasonable cost thorough coverage of many issues likely to be dealt with when using the *International Building Code* — and then supplements that coverage with historical and technical background. Reference lists, information sources and bibliographies are also included.

Throughout all of this, strenuous effort has been made to keep the vast quantity of material accessible and its method of presentation useful. With a comprehensive yet concise summary of each section, the Commentary provides a convenient reference for regulations applicable to the construction of buildings and structures. In the chapters that follow, discussions focus on the full meaning and implications of the code text. Guidelines suggest the most effective method of application, and the consequences of not adhering to the code text. Illustrations are provided to aid understanding; they do not necessarily illustrate the only methods of achieving code compliance.

The format of the Commentary includes the full text of each section, table and figure in the code, followed immediately by the commentary applicable to that text. At the time of printing, the Commentary reflects the most up-to-date text of the 2003 *International Building Code*. As stated in the preface to the *International Building Code,* the content of sections in the code which begin with a letter designation (i.e., Section 3401) are maintained by another code development committee. Each section's narrative includes a statement of its objective and intent, and usually includes a discussion about why the requirement commands the conditions set forth. Code text and commentary text are easily distinguished from each other. All code text is shown as it appears in the *International Building Code*, and all commentary is indented below the code text and begins with the symbol ❖.

Readers should note that the Commentary is to be used in conjunction with the *International Building Code* and not as a substitute for the code. **The Commentary is advisory only;** the code official alone possesses the authority and responsibility for interpreting the code.

Comments and recommendations are encouraged, for through your input, we can improve future editions. Please direct your comments to the Codes and Standards Development Department at the Chicago District Office.

TABLE OF CONTENTS

Chapter 16: Structural Design

General Comments

This chapter contains the commentary for the following structural topics: definitions of structural terms, construction document requirements, load combinations, dead loads, live loads, snow loads, wind loads, soil lateral loads, rain loads, flood loads and earthquake loads.This chapter provides minimum design requirements so that all buildings and structures are proportioned to resist the loads and forces that are likely to be encountered. The loads specified herein have been established through research and service performance of buildings and structures. The application of these loads and adherence to the serviceability criteria will enhance the protection of life and property. The wind loads and snow loads in this chapter are generally based on the 2000 edition of ASCE 7. The code's earthquake load requirements are based on the 2000 edition of ASCE 7 as well as the National Earthquake Hazards Reduction Program's (NEHRP) *Recommended Provisions for the Development of Seismic Regulations for New Buildings* (FEMA 368). The NEHRP Provisions were prepared by the Building Seismic Safety Council (BSSC) for the Federal Emergency Management Agency (FEMA).

Purpose

The purpose of this chapter is to prescribe minimum structural loading requirements for use in the design and construction of buildings and structures with the intent to minimize hazard to life and improve the occupancy capability of essential facilities after a design level event or occurrence.

SECTION 1601 GENERAL

1601.1 Scope. The provisions of this chapter shall govern the structural design of buildings, structures and portions thereof regulated by this code.

❖ Unless stated otherwise, the load criteria found in this chapter are applicable to all buildings and structures, or portions thereof, including new buildings, alterations, additions and repairs. See Chapter 34 for the application of the requirements in this chapter relating to existing structures.

SECTION 1602 DEFINITIONS

1602.1 Definitions. The following words and terms shall, for the purposes of this chapter, have the meanings shown herein.

❖ Definitions of terms can help in the understanding and application of the code requirements. The purpose for including these definitions within this chapter is to provide more convenient access to them without having to refer back to Chapter 2. For convenience, these terms are also listed in Chapter 2 with a cross reference to this section. The use and application of all defined terms, including those defined herein, are set forth in Section 201.

ALLOWABLE STRESS DESIGN. A method of proportioning structural members, such that elastically computed stresses produced in the members by nominal loads do not exceed specified allowable stresses (also called "working stress design").

❖ This definition establishes the difference between the allowable stress design (ASD) method and the strength design method. In this methodology, the elastically computed stresses determined from unfactored loads cannot exceed the specified allowable stresses, which will contain a factor of safety.

BALCONY, EXTERIOR. An exterior floor projecting from and supported by a structure without additional independent supports.

❖ This definition is provided to differentiate live loads between a balcony and a deck, based on the likelihood of different live loads being applied.

BASE SHEAR. Total design lateral force or shear at the base.

❖ This definition is needed for the understanding and application of Section 1616. For earthquake loads, base shear is the result of the building weight being accelerated by ground motion during an earthquake.

BASIC SEISMIC-FORCE-RESISTING SYSTEMS.

❖ These definitions are needed to apply the system limitations on basic seismic-force-resisting systems in Section 1617.6. (e.g., see Table 1617.6.2). They provide the distinctions between the various types of systems that the code recognizes.

Bearing wall system. A structural system without a complete vertical load-carrying space frame. Bearing walls or bracing elements provide support for substantial vertical loads. Seismic lateral force resistance is provided by shear walls or braced frames.

❖ This is a system in which bearing walls support most of the gravity loads. With no moment-resisting frame available, seismic forces are resisted either by the bearing walls acting as shear walls or by braced frames.

Building frame system. A structural system with an essentially complete space frame providing support for vertical loads. Seismic lateral force resistance is provided by shear walls or braced frames.

❖ In contrast to the bearing wall system, this is a system in which a space frame supports most of the gravity loads. A building frame system includes shear walls or braced frames for lateral (seismic) load resistance.

Dual system. A structural system with an essentially complete space frame providing support for vertical loads. Seismic lateral force resistance is provided by a moment frame and shear walls or braced frames.

❖ Like a building frame system, the dual system utilizes a space frame to support most of the gravity loads. It resists lateral loads using a combination of moment frames and either braced frames or shear walls. The two structural systems are connected to share the seismic loads.

Inverted pendulum system. A structure with a large portion of its mass concentrated at the top; therefore, having essentially one degree of freedom in horizontal translation. Seismic lateral force resistance is provided by the columns acting as cantilevers.

❖ Inverted pendulum structures exhibit little redundancy or overstrength. Inelastic behavior is concentrated at the base. Compared to other systems, inverted pendulum systems have less energy dissipation capacity.

Moment-resisting frame system. A structural system with an essentially complete space frame providing support for vertical loads. Seismic lateral force resistance is provided by moment frames.

❖ This is a system in which moment frames support most of the gravity loads as well as providing resistance to seismic forces.

Shear wall-frame interactive system. A structural system which uses combinations of shear walls and frames designed to resist seismic lateral forces in proportion to their rigidities, considering interaction between shear walls and frames on all levels. Support of vertical loads is provided by the same shear walls and frames.

❖ This system consists of reinforced concrete moment frames and shear walls designed to resist seismic loads based on relative rigidities. It is limited in use to areas of low seismicity.

BOUNDARY MEMBERS. Strengthened portions along shear wall and diaphragm edges (also called "boundary elements").

❖ Boundary members transfer wind and seismic loads from diaphragms to adjoining structural members.

Boundary element. In light-frame construction, diaphragms and shear wall boundary members to which sheathing transfers forces. Boundary elements include chords and drag struts at diaphragm and shear wall perimeters, interior openings, discontinuities and reentrant corners.

❖ The boundary element in wood diaphragms and shear walls is typically the rim joists (diaphragms) and the studs and plates surrounding the sheathing (shear walls). The boundary elements are important in transferring the diaphragm or shear wall loads into the structure below.

CANTILEVERED COLUMN SYSTEM. A structural system relying on column elements that cantilever from a fixed base and have minimal rotational resistance capacity at the top with lateral forces applied essentially at the top and are used for lateral resistance.

❖ Cantilevered column systems are commonly used to resist seismic and wind lateral loads.

COLLECTOR ELEMENTS. Members that serve to transfer forces between floor diaphragms and members of the lateral-force-resisting system.

❖ These elements are named collectors since they "collect" lateral loads and transfer them to the lateral-force-resisting system.

CONFINED REGION. The portion of a reinforced concrete component in which the concrete is confined by closely spaced special transverse reinforcement restraining the concrete in directions perpendicular to the applied stress.

❖ This definition is needed to apply the reinforced concrete provisions for resistance to seismic effects.

DEAD LOADS. The weight of materials of construction incorporated into the building, including but not limited to walls, floors, roofs, ceilings, stairways, built-in partitions, finishes, cladding and other similarly incorporated architectural and structural items, and fixed service equipment, including the weight of cranes. All dead loads are considered permanent loads.

❖ This definition is necessary to distinguish certain loads from live loads, and for use in load combinations, as specified by Section 1605. The design dead load is to be determined in accordance with Section 1606.

DECK. An exterior floor supported on at least two opposing sides by an adjacent structure, and/or posts, piers or other independent supports.

❖ This definition is provided to differentiate live loads between a balcony and a deck, based on the different live loads specified in Table 1607.1.

DEFORMABILITY. The ratio of the ultimate deformation to the limit deformation.

High deformability element. An element whose deformability is not less than 3.5 when subjected to four fully reversed cycles at the limit deformation.

Limited deformability element. An element that is neither a low deformability or a high deformability element.

Low deformability element. An element whose deformability is 1.5 or less.

❖ This is a measure of how much a material is able to deform and still resist a load. These definitions are needed to apply the earthquake provisions in this chapter.

DEFORMATION.

Limit deformation. Two times the initial deformation that occurs at a load equal to 40 percent of the maximum strength.

Ultimate deformation. The deformation at which failure occurs and which shall be deemed to occur if the sustainable load reduces to 80 percent or less of the maximum strength.

❖ Deformation is similar to deflection as a result of resisting an applied load. These definitions are needed to apply the earthquake provisions in this chapter.

DESIGN STRENGTH. The product of the nominal strength and a resistance factor (or strength reduction factor).

❖ This definition is needed to apply the strength design requirements in the code. The design strength is the nominal strength that is multiplied by a resistance or strength reduction factor that is less than one.

DIAPHRAGM. A horizontal or sloped system acting to transmit lateral forces to the vertical-resisting elements. When the term "diaphragm" is used, it shall include horizontal bracing systems.

❖ Floor and roof diaphragms act to transfer the lateral forces due to wind or seismic loads to the vertical-resisting elements, such as shear walls or moment frames supporting them at their perimeter or intermittent locations.

Diaphragm, blocked. In light-frame construction, a diaphragm in which all sheathing edges not occurring on a framing member are supported on and fastened to blocking.

❖ Blocked diaphragms are horizontal or nearly horizontal assemblies designed to resist high shear forces in light-frame construction. Diaphragm sheathing may be applied with its long dimension either perpendicular or parallel to the main framing members. When the edge of the sheathing is not supported by the main framing member, it is considered to be unblocked. Blocking is accomplished by installing a framing member parallel to the otherwise unsupported edge of the sheathing. In some cases, the code may require blocking, while in other cases, the capacity of a diaphragm assembly will vary based on whether it is blocked or unblocked.

Diaphragm boundary. In light-frame construction, a location where shear is transferred into or out of the diaphragm sheathing. Transfer is either to a boundary element or to another force-resisting element.

❖ Diaphragm boundary is typically the connection between the floor or roof sheathing and the band board surrounding the diaphragm.

Diaphragm chord. A diaphragm boundary element perpendicular to the applied load that is assumed to take axial stresses due to the diaphragm moment.

❖ A diaphragm acts as a deep horizontal beam. The chords of the beam are the elements at the boundary of the diaphragm that are perpendicular to the direction of the applied load.

Diaphragm, flexible. A diaphragm is flexible for the purpose of distribution of story shear and torsional moment when the computed maximum in-plane deflection of the diaphragm itself under lateral load is more than two times the average drift of adjoining vertical elements of the lateral-force-resisting system of the associated story under equivalent tributary lateral load (see Section 1617.5.3).

❖ A flexible diaphragm deforms more than a rigid diaphragm when subjected to the same load. Load distribution from a flexible diaphragm is usually based on the tributary area associated with vertical-resisting elements. This definition is needed to apply the earthquake and wind load provisions in this chapter. Figure 1602.1(1) illustrates the criteria for determining whether a diaphragm is flexible or not. Note that the reference to "average drift of adjoining vertical elements" is actually the average (relative) deflection of those elements for the story being evaluated.

Diaphragm, rigid. A diaphragm is rigid for the purpose of distribution of story shear and torsional moment when the lateral deformation of the diaphragm is less than or equal to two times the average story drift.

❖ This definition is needed to apply the earthquake and wind load provisions in this chapter. A rigid diaphragm deforms less than a flexible diaphragm when subjected to the same load. Load distribution from a rigid diaphragm is based on the stiffness of the vertical-resisting elements.

DURATION OF LOAD. The period of continuous application of a given load, or the aggregate of periods of intermittent applications of the same load.

❖ An understanding of duration of load is necessary since the allowable design stresses for wood members and fasteners are a function of the length of time that the wood member has to resist the applied load.

ELEMENT.

Ductile element. An element capable of sustaining large cyclic deformations beyond the attainment of its nominal strength without any significant loss of strength.

Limited ductile element. An element that is capable of sustaining moderate cyclic deformations beyond the attainment of nominal strength without significant loss of strength.

Nonductile element. An element having a mode of failure that results in an abrupt loss of resistance when the element is deformed beyond the deformation corresponding to the development of its nominal strength. Nonductile elements cannot reliably sustain significant deformation beyond that attained at their nominal strength.

❖ These definitions are needed for the proper applications of the earthquake provisions in the code. Ductile elements are effective in resisting dynamic earthquake effects.

EQUIPMENT SUPPORT. Those structural members or assemblies of members or manufactured elements, including braces, frames, lugs, snuggers, hangers or saddles, that transmit gravity load and operating load between the equipment and the structure.

❖ This definition is needed to properly apply the structural loads in this chapter.

ESSENTIAL FACILITIES. Buildings and other structures that are intended to remain operational in the event of extreme environmental loading from flood, wind, snow or earthquakes.

❖ This definition is needed to properly determine the structural loads to be applied to the facility. The design loads are higher for an essential versus a nonessential facility.

FACTORED LOAD. The product of a nominal load and a load factor.

❖ This definition is needed to determine the proper structural loads to be applied according to the code. Factored loads are specified for the strength design [or load and resistance factor design (LRFD)] methodology.

FLEXIBLE EQUIPMENT CONNECTIONS. Those connections between equipment components that permit rotational and/or translational movement without degradation of performance.

❖ This definition is needed to properly account for the behavior of the connections when subjected to load.

FRAME.

❖ All of the terms defined under the general term "frame" are needed to determine the proper response modification coefficient (R) and the deflection amplification factor (C_d) in Section 1617.6 for a particular seismic-resisting system.

Braced frame. An essentially vertical truss, or its equivalent, of the concentric or eccentric type that is provided in a building frame system or dual system to resist lateral forces.

❖ Examples of braced frames are shown in Figure 1602.1(2).

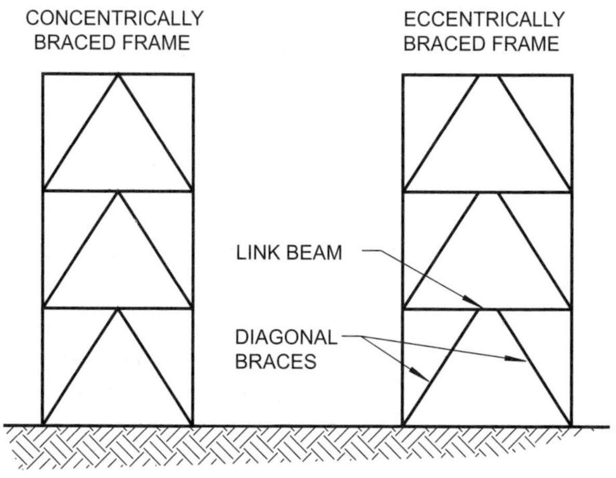

Figure 1602.1(2)
EXAMPLE OF BRACED FRAMES

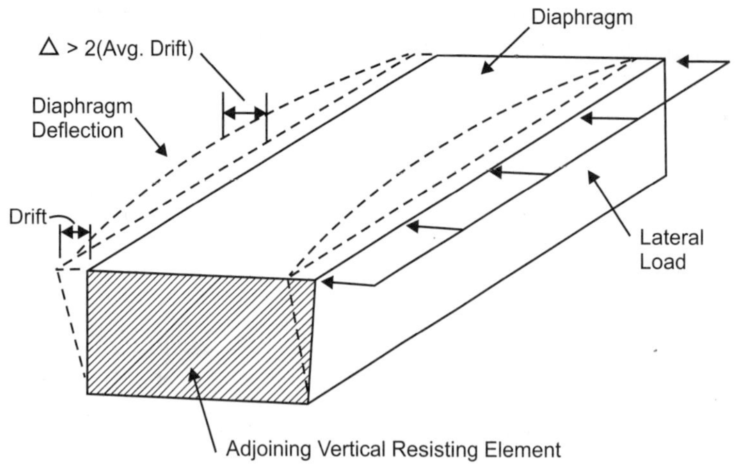

Figure 1602.1(1)
FLEXIBLE DIAPHRAGM DETERMINATION

Concentrically braced frame (CBF). A braced frame in which the members are subjected primarily to axial forces.

❖ An example of a concentrically braced frame is shown in Figure 1602.1(2).

Eccentrically braced frame (EBF). A diagonally braced frame in which at least one end of each brace frames into a beam a short distance from a beam-column or from another diagonal brace.

❖ An example of an eccentrically braced frame is shown in Figure 1602.1(2). All of the defined terms listed under "Eccentrically braced frames" are needed to apply the structural steel requirements of Section 2205. The detailed design provisions for seismic design of structural steel eccentrically braced frames are in the AISC *Seismic Provisions for Structural Steel Buildings*.

Ordinary concentrically braced frame (OCBF). A steel concentrically braced frame in which members and connections are designed in accordance with the provisions of AISC Seismic without modification.

❖ This definition is needed for the proper application of the earthquake provisions in this chapter and the steel design provisions in Chapter 22. The difference between OCBF and SCBF (referenced below) pertains to the level of detailing, which affects response modification coefficients and deflection amplification factors.

Special concentrically braced frame (SCBF). A steel or composite steel and concrete concentrically braced frame in which members and connections are designed for ductile behavior.

❖ This definition is needed for the proper application of the earthquake provisions in this chapter and the steel design provisions in Chapter 22.

Moment frame. A frame in which members and joints resist lateral forces by flexure as well as along the axis of the members. Moment frames are categorized as "intermediate moment frames" (IMF), "ordinary moment frames" (OMF), and "special moment frames" (SMF).

❖ This definition is needed for the proper application of the earthquake provisions in this chapter.

An intermediate moment frame of reinforced concrete has higher ductility than a reinforced concrete ordinary moment frame. The higher ductility is due to the reinforcing requirements of Chapter 19. The specific requirements for reinforcement are in ACI 318.

An ordinary moment frame design may be of reinforced concrete or steel framing. Ordinary moment frames of steel and concrete have use limitations for buildings assigned to high seismic design categories (see Section 1617.6 for the limitations). These limitations are caused by a lower ductility for this class of moment frames than for intermediate or special moment frames.

Special moment frames may be of either steel or reinforced concrete. They are called "special" since the design includes unique detailing of the frame such that large deformations can occur without the loss of seismic lateral load resistance. The specific design requirements for these frames are in ACI 318 for reinforced concrete and the AISC *Seismic Provisions for Structural Steel Buildings*.

GUARD. See Section 1002.1.

❖ This is simply a cross reference to the section where the complete definition is located. It is listed here since guard loads are addressed in this chapter.

IMPACT LOAD. The load resulting from moving machinery, elevators, craneways, vehicles and other similar forces and kinetic loads, pressure and possible surcharge from fixed or moving loads.

❖ This definition identifies the scope of the type of loading addressed in Section 1607.8. The effect of impact loads on a structure can be significantly higher than the weight of the impacting elements because of their movement or vibration.

JOINT. A portion of a column bounded by the highest and lowest surfaces of the other members framing into it.

❖ This definition is needed for the application of the earthquake provisions and the concrete reinforcement requirements in Chapter 19.

LIMIT STATE. A condition beyond which a structure or member becomes unfit for service and is judged to be no longer useful for its intended function (serviceability limit state) or to be unsafe (strength limit state).

❖ This definition is needed for the application of the earthquake requirements in this chapter.

LIVE LOADS. Those loads produced by the use and occupancy of the building or other structure and do not include construction or environmental loads such as wind load, snow load, rain load, earthquake load, flood load or dead load.

❖ This definition identifies the scope of the type of loading included in Section 1607. Generally, live loads are not environmental loads or dead loads, but are transient in nature and will vary in magnitude over the life of a structure.

LIVE LOADS (ROOF). Those loads produced (1) during maintenance by workers, equipment and materials; and (2) during the life of the structure by movable objects such as planters and by people.

❖ This definition is needed for the proper application of the load combinations in this chapter. This definition clarifies that roof loads, such as snow loads, are not live loads.

LOAD AND RESISTANCE FACTOR DESIGN (LRFD). A method of proportioning structural members and their connections using load and resistance factors such that no applicable limit state is reached when the structure is subjected to appropriate load combinations. The term "LRFD" is used in the design of steel and wood structures.

❖ This definition is needed for the proper application of the steel design requirements in Chapter 22 and the wood design requirements in Chapter 23.

LOAD FACTOR. A factor that accounts for deviations of the actual load from the nominal load, for uncertainties in the analysis that transforms the load into a load effect, and for the probability that more than one extreme load will occur simultaneously.

❖ This definition is needed for the application of the load combinations in this chapter.

LOADS. Forces or other actions that result from the weight of building materials, occupants and their possessions, environmental effects, differential movement and restrained dimensional changes. Permanent loads are those loads in which variations over time are rare or of small magnitude, such as dead loads. All other loads are variable loads (see also "Nominal loads").

❖ This definition is needed for the proper application of the structural load requirements in this chapter.

LOADS EFFECTS. Forces and deformations produced in structural members by the applied loads.

❖ This definition is needed to properly apply the structural load requirements in this chapter.

NOMINAL LOADS. The magnitudes of the loads specified in this chapter (dead, live, soil, wind, snow, rain, flood and earthquake).

❖ This definition is needed for the proper application of the structural loads in this chapter.

NOTATIONS.

D = Dead load.

E = Combined effect of horizontal and vertical earthquake-induced forces as defined in Sections 1616.4.1 and 1617.1.

E_m = Maximum seismic load effect of horizontal and vertical seismic forces as set forth in Sections 1616.4.1 and 1617.1.

F = Load due to fluids.

F_a = Flood load.

H = Load due to lateral pressure of soil and water in soil.

L = Live load, except roof live load, including any permitted live load reduction.

L_r = Roof live load including any permitted live load reduction.

P = Ponding load.

R = Rain load.

S = Snow load.

T = Self-straining force arising from contraction or expansion resulting from temperature change, shrinkage, moisture change, creep in component materials, movement due to differential settlement or combinations thereof.

W = Load due to wind pressure.

❖ These notations are used in this chapter, primarily in Section 1605. Generally, the definition of notations is included in the section where they are used in formulas. Thus, it is not necessary to refer back to Section 1602 to understand the formulas.

OTHER STRUCTURES. Structures, other than buildings, for which loads are specified in this chapter.

❖ This definition is needed for the application of earthquake design requirements for other structures included in this chapter.

P-DELTA EFFECT. The second order effect on shears, axial forces and moments of frame members induced by axial loads on a laterally displaced building frame.

❖ This definition is needed to apply properly the earthquake requirements referenced in Section 1617.4. The P-delta effect has a significant impact on the stresses for structural systems that have high story drifts. As the story drift increases, the resulting member forces and moments increase. These structural systems are usually steel moment frames that resist all of the lateral loads from an earthquake without assistance from shear walls or braced frames.

PANEL (PART OF A STRUCTURE). The section of a floor, wall or roof comprised between the supporting frame of two adjacent rows of columns and girders or column bands of floor or roof construction.

❖ This definition is needed to apply the structural load requirements in this chapter.

RESISTANCE FACTOR. A factor that accounts for deviations of the actual strength from the nominal strength and the manner and consequences of failure (also called "strength reduction factor").

❖ This definition is needed to apply the material strength adjustments by way of the resistance factor that is specified in the structural material chapters and the referenced structural standards.

SHALLOW ANCHORS. Shallow anchors are those with embedment length-to-diameter ratios of less than eight.

❖ This definition is needed to apply the anchor design requirements in Chapter 19.

SHEAR PANEL. A floor, roof or wall component sheathed to act as a shear wall or diaphragm.

❖ This definition is needed for the application of the structural design requirements for wind and earthquake loads in this chapter.

SHEAR WALL. A wall designed to resist lateral forces parallel to the plane of the wall.

❖ A shear wall carries primarily in-plane (rather than out-of-plane) lateral loads in addition to gravity loads. A shear wall relies on the in-plane properties (height and length) of the wall, coupled with the structural materials properties of the wall, to resist lateral loads. The lateral loads can be due to wind, earthquake or other external forces. A partition wall that is not designed to resist in-plane loads is not a shear wall. The term "shear wall" is used primarily in the earthquake provisions (see Sections 1613 through 1623) as part of the seismic-force-resisting system.

SPACE FRAME. A structure composed of interconnected members, other than bearing walls, that is capable of supporting vertical loads and that also may provide resistance to seismic lateral forces.

❖ A space frame refers to a three-dimensional moment frame that resists gravity loads. It does not include bearing walls, and it may or may not be used to resist lateral forces.

SPECIAL TRANSVERSE REINFORCEMENT. Reinforcement composed of spirals, closed stirrups or hoops and supplementary cross ties provided to restrain the concrete and qualify the portion of the component, where used, as a confined region.

❖ This definition is needed for the application of the earthquake loads in this chapter and the reinforced concrete design requirements in Chapter 19.

STRENGTH, NOMINAL. The capacity of a structure or member to resist the effects of loads, as determined by computations using specified material strengths and dimensions and equations derived from accepted principles of structural mechanics or by field tests or laboratory tests of scaled models, allowing for modeling effects and differences between laboratory and field conditions.

❖ This definition is needed to apply the structural analysis and structural material design requirements in this chapter and the structural materials requirements in Chapters 19 through 23.

STRENGTH, REQUIRED. Strength of a member, cross section or connection required to resist factored loads or related internal moments and forces in such combinations as stipulated by these provisions.

❖ This definition is needed to apply the structural analysis requirements in this chapter and the structural material design requirements in Chapters 19 through 23.

STRENGTH DESIGN. A method of proportioning structural members such that the computed forces produced in the members by factored loads do not exceed the member design strength [also called "load and resistance factor design" (LRFD)]. The term "strength design" is used in the design of concrete and masonry structural elements.

❖ This definition is needed to apply the structural analysis requirements in this chapter and the structural material design requirements in Chapters 19 through 23.

WALL, LOAD BEARING. Any wall meeting either of the following classifications:

1. Any metal or wood stud wall that supports more than 100 pounds per linear foot (plf) (1459 N/m) of vertical load in addition to its own weight.
2. Any masonry or concrete wall that supports more than 200 plf (2919 N/m) of vertical load in addition to its own weight.

❖ This definition is necessary since the structural requirements in this chapter vary for load-bearing versus nonload-bearing walls. The term is intended to apply to walls that support part of the structural framework of the building.

WALL, NONLOAD BEARING. Any wall that is not a load-bearing wall.

❖ This definition identifies the scope of the code provisions that apply to this type of wall. Nonload-bearing walls also commonly include partition walls. Nonload-bearing walls do not support any other portion of the building and only support the weight of the wall itself.

SECTION 1603
CONSTRUCTION DOCUMENTS

1603.1 General. Construction documents shall show the size, section and relative locations of structural members with floor levels, column centers and offsets fully dimensioned. The design loads and other information pertinent to the structural design required by Sections 1603.1.1 through 1603.1.8 shall be clearly indicated on the construction documents for parts of the building or structure.

Exception: Construction documents for buildings constructed in accordance with the conventional light-frame construction provisions of Section 2308 shall indicate the following structural design information:

1. Floor and roof live loads.
2. Ground snow load, P_g.
3. Basic wind speed (3-second gust), miles per hour (mph) (km/hr) and wind exposure.
4. Seismic design category and site class.

❖ The purpose of this section is to specifically require the design professional to provide the building official with the appropriate structural details and design load data for determination of compliance with this chapter.

The construction documents are required to contain sufficient detail for the building official to perform plan review and field inspection, as well as for construction activity. Dimensions indicated on architectural drawings are not required to be duplicated on the structural drawings and vice versa. The design loads, to be indicated by the design professional on the construction documents, are to be consistent with the loads used in the structural calculations. Note that the loads are not required to be on the construction drawings but must be included within the construction documents in a manner such that the design loads are clear for all parts of the structure (see Chapter 2 for the definition of "Construction documents"). The building official is to compare the loads on the construction documents with the applicable minimum required loads as specified by this chapter. The inclusion of the load design information is assurance by the design professional that the structure has been designed for the indicated loads, which are required to be equal to or greater than the loads required by the code. It should be emphasized that these requirements for construction documents are applicable regardless of the involvement of a registered design professional, which is regulated by the applicable state's licensing laws.The exception provides a less extensive list of structural data to be indicated for buildings constructed in accordance with the conventional wood-frame provisions of Section 2308. This is appropriate in view of the prescriptive wood-frame requirements.

1603.1.1 Floor live load. The uniformly distributed, concentrated and impact floor live load used in the design shall be indicated for floor areas. Live load reduction of the uniformly distributed floor live loads, if used in the design, shall be indicated.

❖ The purpose of the requirement in this section is to provide information for the building official to facilitate the plan review process. The floor live loads, which are indicated on the construction documents by the design professional, are required to meet or exceed the loads in Section 1607. Any live load reductions taken are also to be indicated.

1603.1.2 Roof live load. The roof live load used in the design shall be indicated for roof areas (Section 1607.11).

❖ This section provides information allowing the building official to facilitate the plan review process. The roof live loads, indicated on the construction documents by the design professional, are required to meet or exceed the loads in Section 1607.11.

1603.1.3 Roof snow load. The ground snow load, P_g, shall be indicated. In areas where the ground snow load, P_g, exceeds 10 pounds per square foot (psf) (0.479 kN/m^2), the following additional information shall also be provided, regardless of whether snow loads govern the design of the roof:

1. Flat-roof snow load, P_f.
2. Snow exposure factor, C_e.
3. Snow load importance factor, I_s.
4. Thermal factor, C_t.

❖ The roof snow load design basis, indicated on the construction documents (design drawings or specifications) by the design professional, provides information allowing the building official to facilitate the plan review process. The flat-roof snow load, snow exposure factor, snow load importance factor, I_s, and roof thermal factor are not to be less than required by Section 1608.

1603.1.4 Wind design data. The following information related to wind loads shall be shown, regardless of whether wind loads govern the design of the lateral-force-resisting system of the building:

1. Basic wind speed (3-second gust), miles per hour (km/hr).
2. Wind importance factor, I_W, and building category.
3. Wind exposure, if more than one wind exposure is utilized, the wind exposure and applicable wind direction shall be indicated.
4. The applicable internal pressure coefficient.
5. Components and cladding. The design wind pressures in terms of psf (kN/m^2) to be used for the design of exterior component and cladding materials not specifically designed by the registered design professional.

❖ The wind load design basis, indicated on the construction documents (design drawings or specifications) by the design professional, provides information allowing the building official to facilitate the plan review process. All five of the indicated items are to be on the submitted construction documents. Each of the indicated items is an important parameter in the determination of the wind resistance that is required in the building framework. The building official should verify that the information is on the construction documents during the plan review process. The correctness of the listed items is the responsibility of the owner or the owner's design professional.

1603.1.5 Earthquake design data. The following information related to seismic loads shall be shown, regardless of whether seismic loads govern the design of the lateral-force-resisting system of the building:

1. Seismic importance factor, I_E, and seismic use group.
2. Mapped spectral response accelerations S_S and S_1.
3. Site class.
4. Spectral response coefficients S_{DS} and S_{D1}.
5. Seismic design category.
6. Basic seismic-force-resisting system(s).
7. Design base shear.
8. Seismic response coefficient(s), C_S.
9. Response modification factor(s), R.
10. Analysis procedure used.

❖ The earthquake load design basis, indicated on the construction documents by the design professional, provides information allowing the building official to facilitate the plan review process. All buildings, except those indicated in the exceptions to Section 1614.1, are to be designed for earthquake effects.The earthquake design data for a specific building are required to meet or exceed the requirements in Sections 1613 through 1623 (see Sections 1613 through 1623 for an explanation of terms used in the specified list).

1603.1.6 Flood load. For buildings located in flood hazard areas as established in Section 1612.3, the following information, referenced to the datum on the community's Flood Insurance Rate Map (FIRM), shall be shown, regardless of whether flood loads govern the design of the building:

1. In flood hazard areas not subject to high-velocity wave action, the elevation of proposed lowest floor, including basement.
2. In flood hazard areas not subject to high-velocity wave action, the elevation to which any nonresidential building will be dry floodproofed.
3. In flood hazard areas subject to high-velocity wave action, the proposed elevation of the bottom of the lowest horizontal structural member of the lowest floor, including basement.

❖ The flood hazard elevation information to be shown on the construction documents by the design professional provides information allowing the building official to facilitate the plan review process. By citing the specified flood information, the design professional is indicating that the building was designed in accordance with the flood hazard requirements in Section 1612. Depending on the nature of the designated flood hazard area, certain elevation requirements are to be met. In flood hazard areas not subject to high-velocity wave action (commonly called A Zones), the lowest floor of all buildings and structures, or the elevation to which a nonresidential building is dry floodproofed, is to be located at or above the design flood elevation (DFE). In flood hazard areas subject to high-velocity wave action (commonly called V Zones), the bottom of the lowest horizontal member of buildings and structures is to be located at or above the DFE. This is one of the most important requirements to provide resistance to flood damage for buildings and structures located in designated flood hazard areas. It is also the main factor used in determining flood insurance premium rates for buildings and structures. Constructing a building or structure with its lowest floor (or extent of dry floodproofing) below DFE will result in significantly higher flood insurance premiums for the building owner. It is important to note that, while not a requirement of the code, the National Flood Insurance Program (NFIP) requires dry floodproofing to extend to 1 foot (305 mm) above the DFE in order for the building or structure to receive a favorable flood insurance premium rate.

1603.1.7 Special loads. Special loads that are applicable to the design of the building, structure or portions thereof shall be indicated along with the specified section of this code that addresses the special loading condition.

❖ Indication of special loads on the construction documents by the design professional provides information allowing the building official to facilitate the plan review process. The design professional is expected to determine the special loads that the occupancy will impose on the structure.

1603.1.8 System and components requiring special inspections for seismic resistance. Construction documents or specifications shall be prepared for those systems and components requiring special inspection for seismic resistance as specified in Section 1707.1 by the registered design professional responsible for their design and shall be submitted for approval in accordance with Section 106.1. Reference to seismic standards in lieu of detailed drawings is acceptable.

❖ This section provides for construction documents for the systems and components specified in Section 1707.1. Generally, the systems and components are the seismic-force-resisting systems, certain mechanical and electrical equipment and systems and architectural components for buildings where the seismic performance category is high. Construction documents are needed for these important items to verify that they comply with code requirements.

1603.2 Restrictions on loading. It shall be unlawful to place, or cause or permit to be placed, on any floor or roof of a building, structure or portion thereof, a load greater than is permitted by these requirements.

❖ The loads that this section is referring to are the various structural loads specified in this chapter. For example, Table 1607.1 includes the minimum live loads for building design. Unless the building is designed for higher loads than specified in Table 1607.1, those values are not to be exceeded. Note that the loads in Table 1607.1 are minimum live loads. A building is permitted to be designed for higher loads, in which case the higher loads would be the limit of the actual applied loads.

1603.3 Live loads posted. Where the live loads for which each floor or portion thereof of a commercial or industrial building is or has been designed to exceed 50 psf (2.40 kN/m^2), such design live loads shall be conspicuously posted by the owner in that part of each story in which they apply, using durable signs. It shall be unlawful to remove or deface such notices.

❖ This section requires that live loads be posted for most occupancies, since many of the live loads specified in Table 1607.1 exceed 50 pounds per square foot (psf) (2.40 kN/m^2). Where part of the floor is designed for 50 psf (2.40 kN/m^2) or less and part for more than 50 psf (2.40 kN/m^2), the live loads are required to be posted for those portions more than 50 psf (2.40 kN/m^2). The code requires that the posting be done in the part where it applies. For example, an assembly area such as a restaurant would need to have the live load posted in the dining room.

1603.4 Occupancy permits for changed loads. Construction documents for other than residential buildings filed with the building official with applications for permits shall show on each drawing the live loads per square foot (m^2) of area covered for which the building is designed. Occupancy permits for buildings hereafter erected shall not be issued until the floor load signs, required by Section 1603.3, have been installed.

❖ The live load is required to be on the construction documents for other than residential buildings to facilitate the plan review process. It also serves as a record of the building design for future reference when the occupancy of the building could change. The design live load signs required by Section 1603.3 need to be in place prior to the occupancy of the building for reference. The posting gives the building department easy access for field verification.

SECTION 1604 GENERAL DESIGN REQUIREMENTS

1604.1 General. Building, structures and parts thereof shall be designed and constructed in accordance with strength design, load and resistance factor design, allowable stress design, empirical design or conventional construction methods, as permitted by the applicable material chapters.

❖ This section identifies the various design methods that are permitted by the code and referenced design standards. The details for these design methods are either explicitly specified in the code or are located in the design standards that are referenced in the structural material design chapters for concrete, aluminum, masonry, steel and wood. For example, empirical design of masonry is addressed in Section 2109. The design of masonry using working stress design is required to be in accordance with ACI 530/ASCE 5/TMS 402.

1604.2 Strength. Buildings and other structures, and parts thereof, shall be designed and constructed to support safely the factored loads in load combinations defined in this code without exceeding the appropriate strength limit states for the materials of construction. Alternatively, buildings and other structures, and parts thereof, shall be designed and constructed to support safely the nominal loads in load combinations defined in this code without exceeding the appropriate specified allowable stresses for the materials of construction.

Loads and forces for occupancies or uses not covered in this chapter shall be subject to the approval of the building official.

❖ This section describes the strength and allowable stress design methods in the code. It also gives the building official approval authority for structural loads in buildings used for occupancies that are not addressed by the code.

1604.3 Serviceability. Structural systems and members thereof shall be designed to have adequate stiffness to limit deflections and lateral drift. See Section 1617.3 for drift limits applicable to earthquake loading.

❖ The deflection limits for structural members must be in accordance with the requirements in this and subsequent sections. This section also provides a reference to the story drift limitations in the earthquake portion of this chapter. Generally, deflection limits are needed for the comfort of the building occupants and so that the structural member's deflection does not cause damage to supported construction. Excessive deflection can also contribute to excessive vibration, which is discomforting to the occupants of the building.

1604.3.1 Deflections. The deflections of structural members shall not exceed the more restrictive of the limitations of Sections 1604.3.2 through 1604.3.5 or that permitted by Table 1604.3.

❖ The deflection of structural members is limited by the code and some material standards for damage control of supported construction and human comfort. Generally, the public equates visible deflection, or even detectable vibration, with an unsafe condition (which in many cases is not true). The intent of this section is that deflection is not to exceed either the limitations in the applicable material design standard or the applicable specified requirements.

TABLE 1604.3
DEFLECTION LIMITS[a, b, c, h, i]

CONSTRUCTION	L	S or W^f	$D + L^{d,g}$
Roof members:[e]			
Supporting plaster ceiling	$l/360$	$l/360$	$l/240$
Supporting nonplaster ceiling	$l/240$	$l/240$	$l/180$
Not supporting ceiling	$l/180$	$l/180$	$l/120$
Floor members	$l/360$	—	$l/240$
Exterior walls and interior partitions:			
With brittle finishes	—	$l/240$	—
With flexible finishes	—	$l/120$	—
Farm buildings	—	—	$l/180$
Greenhouses	—	—	$l/120$

For SI: 1 foot = 304.8 mm.

a. For structural roofing and siding made of formed metal sheets, the total load deflection shall not exceed $l/60$. For secondary roof structural members supporting formed metal roofing, the live load deflection shall not exceed $l/150$. For secondary wall members supporting formed metal siding, the design wind load deflection shall not exceed $l/90$. For roofs, this exception only applies when the metal sheets have no roof covering.

b. Interior partitions not exceeding 6 feet in height and flexible, folding and portable partitions are not governed by the provisions of this section. The deflection criterion for interior partitions is based on the horizontal load defined in Section 1607.13.

c. See Section 2403 for glass supports.

d. For wood structural members having a moisture content of less than 16 percent at time of installation and used under dry conditions, the deflection resulting from $L + 0.5D$ is permitted to be substituted for the deflection resulting from $L + D$.

e. The above deflections do not ensure against ponding. Roofs that do not have sufficient slope or camber to assure adequate drainage shall be investigated for ponding. See Section 1611 for rain and ponding requirements and Section 1503.4 for roof drainage requirements.

f. The wind load is permitted to be taken as 0.7 times the "component and cladding" loads for the purpose of determining deflection limits herein.

g. For steel structural members, the dead load shall be taken as zero.

h. For aluminum structural members or aluminum panels used in roofs or walls of sunroom additions or patio covers, not supporting edge of glass or aluminum sandwich panels, the total load deflection shall not exceed $l/60$. For aluminum sandwich panels used in roofs or walls of sunroom additions or patio covers, the total load deflection shall not exceed $l/120$.

i. For cantilever members, l shall be taken as twice the length of the cantilever.

❖ The deflection limits in this table apply when they are more restrictive than those in the structural design standards that are indicated in Sections 1604.3.2 through 1604.3.5. Note that the deflection limits for exterior walls and interior finishes vary with flexible or brittle finishes. A flexible finish is intended to be one that has been designed to accommodate the higher deflection indicated and remain serviceable. A brittle finish is any finish that has not been designed to accommodate the deflection allowed for a flexible finish.The deflection limit for a roof member supporting a plaster ceiling is intended to apply only for a plaster ceiling. The limit for a roof member supporting a gypsum board ceiling is that listed in the table for "supporting a nonplaster ceiling."

1604.3.2 Reinforced concrete. The deflection of reinforced concrete structural members shall not exceed that permitted by ACI 318.

❖ The deflection limitations in ACI 318 are not to be exceeded for reinforced (and prestressed) concrete (see ACI 318 for detailed deflection requirements).

1604.3.3 Steel. The deflection of steel structural members shall not exceed that permitted by AISC LRFD, AISC HSS, AISC 335, AISI -NASPEC, AISI-General, AISI-Truss, ASCE 3, ASCE 8-SSD-LRFD/ASD, and the standard specifications of SJI Standard Specifications, Load Tables and Weight Tables for Steel Joists and Joist Girders as applicable.

❖ The design standard to be met depends on the type of steel structural member. For example, steel joists are to meet the deflection limitations in the Steel Joist Institute's (SJI) standard and rolled steel members are to meet the deflection criteria in AISC-ASD and AISC-LRFD.

1604.3.4 Masonry. The deflection of masonry structural members shall not exceed that permitted by ACI 530/ASCE 5/TMS 402.

❖ The deflection of masonry beams and lintels is limited by ACI 530/ASCE 5/TMS 402 (see ACI 530/ASCE 5/TMS 402 for deflection requirement details).

1604.3.5 Aluminum. The deflection of aluminum structural members shall not exceed that permitted by AA-94.

❖ The deflection of aluminum structural members is limited to that in AA-94 for aluminum or that permitted by Table 1604.3 of the code.

1604.3.6 Limits. Deflection of structural members over span, *l*, shall not exceed that permitted by Table 1604.3.

❖ The limits specified in Table 1604.3 apply to the indicated members for any of the structural materials. As indicated in Section 1604.3.1, the deflection limits in the structural material standards apply when they are more restrictive than indicated in the table.

1604.4 Analysis. Load effects on structural members and their connections shall be determined by methods of structural analysis that take into account equilibrium, general stability, geometric compatibility and both short- and long-term material properties.

Members that tend to accumulate residual deformations under repeated service loads shall have included in their analysis the added eccentricities expected to occur during their service life.

Any system or method of construction to be used shall be based on a rational analysis in accordance with well-established principles of mechanics. Such analysis shall result in a system that provides a complete load path capable of transferring loads from their point of origin to the load-resisting elements.

The total lateral force shall be distributed to the various vertical elements of the lateral-force-resisting system in proportion to their rigidities considering the rigidity of the horizontal bracing system or diaphragm. Rigid elements that are assumed not to be a part of the lateral-force-resisting system shall be permitted to be incorporated into buildings provided that their effect on the action of the system is considered and provided for in design. Provisions shall be made for the increased forces induced on resisting elements of the structural system resulting from torsion due to eccentricity between the center of application of the lateral forces and the center of rigidity of the lateral-force-resisting system.

Every structure shall be designed to resist the overturning effects caused by the lateral forces specified in this chapter. See Section 1609 for wind, Section 1610 for lateral soil loads and Sections 1613 through 1623 for earthquake.

❖ This section includes the general requirements for structural analysis. The principles stated in this section are those commonly found in structural engineering textbooks.The requirement that structural analysis be capable of demonstrating a complete load path is essential to the adequate resistance of the structural system to wind loads or earthquake effects. The load path is to be capable of transferring all of the loads from their point of application onto the structure to the foundation. It is also important that nonstructural rigid elements be properly accounted for in the design. For example, a partial-height rigid masonry wall placed between steel columns in a steel frame will resist the horizontal shear load and cause bending in the column unless a flexible joint is provided between the wall and the column.

1604.5 Importance factors. The value for snow load, wind load and seismic load importance factors shall be determined in accordance with Table 1604.5.

❖ The table of importance factors is used in the calculation of wind, snow loads and seismic forces (see commentary, Table 1604.5).

TABLE 1604.5. See page 16-12.

❖ The importance factors in this table are used in the calculation of snow and wind loads and seismic effects. The value of the loads varies in proportion to the importance factor value. This result is a higher factor of safety for the facilities that have higher importance factors. The value of the importance factor generally increases with the importance of the facility. For example, the types of facilities listed as Category III have the indicated characteristics described in the table such that they are assigned importance factors greater than 1. The facilities listed as Category IV are indicated to be essential facilities. They are assigned the highest importance factors. The facilities are essential in that their continuous use is needed. For example, fire, rescue and police stations and emergency vehicle garages are needed to be in service during and immediately after a major windstorm or earthquake event.

1604.6 In-situ load tests. The building official is authorized to require an engineering analysis or a load test, or both, of any construction whenever there is reason to question the safety of the construction for the intended occupancy. Engineering analysis and load tests shall be conducted in accordance with Section 1713.

❖ The building official has the option of requiring either a structural analysis, an on-site in-situ load test or both, in accordance with Section 1713, on an existing structure, building or portion thereof if there is reasonable doubt as to structural integrity. The building official should document his or her reasons for the testing requirement. Whenever possible, the concern should be addressed by structural analysis since load testing a structure is very expensive. One example would be an analysis by a third-party engineering firm acceptable to both the building official and the owner.The structural integrity may be examined for items such as visible signs of excessive settlement, or lateral deflection, such as cracks in concrete foundation walls or exces-

TABLE 1604.5
CLASSIFICATION OF BUILDINGS AND OTHER STRUCTURES FOR IMPORTANCE FACTORS

CATEGORY[a]	NATURE OF OCCUPANCY	SEISMIC FACTOR I_E	SNOW FACTOR I_S	WIND FACTOR I_W
I	Buildings and other structures that represent a low hazard to human life in the event of failure including, but not limited to: • Agricultural facilities • Certain temporary facilities • Minor storage facilities	1.00	0.8	0.87[b]
II	Buildings and other structures except those listed in Categories I, III and IV	1.00	1.0	1.00
III	Buildings and other structures that represent a substantial hazard to human life in the event of failure including, but not limited to: • Buildings and other structures where more than 300 people congregate in one area • Buildings and other structures with elementary school, secondary school or day care facilities with an occupant load greater than 250 • Buildings and other structures with an occupant load greater than 500 for colleges or adult education facilities • Health care facilities with an occupant load of 50 or more resident patients but not having surgery or emergency treatment facilities • Jails and detention facilities • Any other occupancy with an occupant load greater than 5,000 • Power-generating stations, water treatment for potable water, waste water treatment facilities and other public utility facilities not included in Category IV • Buildings and other structures not included in Category IV containing sufficient quantities of toxic or explosive substances to be dangerous to the public if released	1.25	1.1	1.15
IV	Buildings and other structures designed as essential facilities including, but not limited to: • Hospitals and other health care facilities having surgery or emergency treatment facilities • Fire, rescue and police stations and emergency vehicle garages • Designated earthquake, hurricane or other emergency shelters • Designated emergency preparedness, communication, and operation centers and other facilities required for emergency response • Power-generating stations and other public utility facilities required as emergency backup facilities for Category IV structures • Structures containing highly toxic materials as defined by Section 307 where the quantity of the material exceeds the maximum allowable quantities of Table 307.7(2) • Aviation control towers, air traffic control centers and emergency aircraft hangars • Buildings and other structures having critical national defense functions • Water treatment facilities required to maintain water pressure for fire suppression	1.50	1.2	1.15

a. For the purpose of Section 1616.2, Categories I and II are considered Seismic Use Group I, Category III is considered Seismic Use Group II and Category IV is equivalent to Seismic Use Group III.

b. In hurricane-prone regions with $V > 100$ miles per hour, I_w shall be 0.77.

sive vibration when the assembly is loaded. The procedure must simulate the actual load conditions to which the structure is subjected during normal use.

1604.7 Preconstruction load tests. Materials and methods of construction that are not capable of being designed by approved engineering analysis or that do not comply with the applicable material design standards listed in Chapter 35, or alternative test procedures in accordance with Section 1711, shall be load tested in accordance with Section 1714.

❖ The alternative test procedure described in Section 1711 and the preconstruction load test procedure described in Section 1714 are intended to apply to materials or an assembly of structural materials that do not have an accepted analysis technique; thus, they are approved for use by way of the alternative test procedure. The preconstruction test procedure in Section 1714 includes the determination of the allowable superimposed design load (see Section 1714 for details).

1604.8 Anchorage.

1604.8.1 General. Anchorage of the roof to walls and columns, and of walls and columns to foundations, shall be provided to resist the uplift and sliding forces that result from the application of the prescribed loads.

❖ This section includes a specific load path requirement that is generally required by Section 1604.4. An example of the condition described in this section is roof uplift as a result of high winds. Table 2308.10.1 requires roof uplift connectors for wood-frame construction. Note that net roof uplift occurs at all wind speeds indicated in the table; thus, uplift connectors are required for roof wood framing for all of the indicated wind speed locations.

1604.8.2 Concrete and masonry walls. Concrete and masonry walls shall be anchored to floors, roofs and other structural elements that provide lateral support for the wall. Such anchorage shall provide a positive direct connection capable of resisting the horizontal forces specified in this chapter but not less than a minimum strength design horizontal force of 280 plf (4.10 kN/m) of wall, substituted for "*E*" in the load combinations of Section 1605.2 or 1605.3. Walls shall be designed to resist bending between anchors where the anchor spacing exceeds 4 feet (1219 mm). Required anchors in masonry walls of hollow units or cavity walls shall be embedded in a reinforced grouted structural element of the wall. See Sections 1609.6.2.2 and 1620 for wind and earthquake design requirements.

❖ Because the response of building construction elements to earthquake ground motion is a function of their mass, heavy elements such as concrete or masonry walls will be laterally loaded during a seismic event. One of the major problems associated with buildings during a severe earthquake is the pulling away of heavy concrete or masonry walls from their attachments to floors or roofs, resulting in the loss of vertical-resisting elements designed to transmit the lateral forces from the diaphragms (floors and roofs) down to the foundations. Such detachments are hazardous to life and property. This section prescribes a minimum strength design load for anchorages between concrete or masonry walls and horizontal diaphragms. The connection requirements apply to both bearing walls and nonstructural walls. The wall anchorage must be a positive connection that does not rely on friction for strength.

The reference to the "*E*" load in this section is the earthquake effect "*E*" included in the load combinations in Section 1605.

1604.8.3 Decks. Where supported by attachment to an exterior wall, decks shall be positively anchored to the primary structure and designed for both vertical and lateral loads as applicable. Such attachment shall not be accomplished by the use of toenails or nails subject to withdrawal. Where positive connection to the primary building structure cannot be verified during inspection, decks shall be self-supporting. For decks with cantilevered framing members, connections to exterior walls or other framing members shall be designed and constructed to resist uplift resulting from the full live load specified in Table 1607.1 acting on the cantilevered portion of the deck.

❖ This requirement for the positive anchorage of decks is a result of failures that have occurred primarily on nonengineered residential decks. Toenail connections are very weak since many of the field connections split the wood framing member. Nails that are installed in line with applied tension forces pull out easily. Wood deck framing could be attached to a supporting ledger board by joist hangers that provide nails loaded in shear for the vertical and horizontal loads. Uplift occurs at the point of connection between the wood framing and the exterior wall for cantilevered deck framing. This section highlights the need for resistance to uplift by a positive connection at the exterior wall.

SECTION 1605
LOAD COMBINATIONS

1605.1 General. Buildings and other structures and portions thereof shall be designed to resist the load combinations specified in Section 1605.2 or 1605.3 and Chapters 18 through 23, and the special seismic load combinations of Section 1605.4 where required by Section 1620.2.6, 1620.2.9 or 1620.4.4 or Section 9.5.2.6.2.11 or 9.5.2.6.3.1 of ASCE 7. Applicable loads shall be considered, including both earthquake and wind, in accordance with the specified load combinations. Each load combination shall also be investigated with one or more of the variable loads set to zero.

❖ Generally, there are two types of load combinations specified in the code: those to be used with a strength design or LRFD and those to be used with ASD. Where additional load combinations are specified in the structural material chapters, they apply also. Note that this section also requires that engineers consider additional load cases where variable loads are not acting concurrently with other loads. It is necessary to explore such possibilities, since they can at times result in the most critical load condition for some parts of a structure.

1605.2 Load combinations using strength design or load and resistance factor design.

1605.2.1 Basic load combinations. Where strength design or load and resistance factor design is used, structures and portions thereof shall resist the most critical effects from the following combinations of factored loads:

$1.4D$ **(Equation 16-1)**

$1.2D + 1.6L + 0.5(L_r \text{ or } S \text{ or } R)$ **(Equation 16-2)**

$1.2D + 1.6(L_r \text{ or } S \text{ or } R) + (f_1 L \text{ or } 0.8W)$ **(Equation 16-3)**

$1.2D + 1.6W + f_1 L + 0.5(L_r \text{ or } S \text{ or } R)$ **(Equation 16-4)**

$1.2D + 1.0E + f_1 L + f_2 S$ **(Equation 16-5)**

$0.9D + (1.0E \text{ or } 1.6W)$ **(Equation 16-6)**

where:

f_1 = 1.0 for floors in places of public assembly, for live loads in excess of 100 pounds per square foot (4.79 kN/m^2), and for parking garage live load.

f_1 = 0.5 for other live loads.

f_2 = 0.7 for roof configurations (such as saw tooth) that do not shed snow off the structure.

f_2 = 0.2 for other roof configurations.

Exception: Where other factored load combinations are specifically required by the provisions of this code, such combinations shall take precedence.

❖ This section lists the load combinations for strength design or LRFD methods. See the definitions in Section 1602 for an explanation of the notations. The basis for these load combinations is Section 2.3.2 of ASCE 7.

1605.2.2 Other loads. Where *F, H, P* or *T* is to be considered in design, each applicable load shall be added to the above combinations in accordance with Section 2.3.2 of ASCE 7. Where F_a is to be considered in design, the load combinations of Section 2.3.3 of ASCE 7 shall be used.

❖ This section addresses cases where the indicated load effects are to be included in the design (see Section 1602 for an explanation of the notations). Where a building is located in a flood zone, the additional load combinations in ASCE 7, Section 2.3.3, are to be included.

Under a cooperative agreement with FEMA, ASCE completed an extensive analysis of flood loads and flood load combinations. The results of this study were included in the 1998 edition of ASCE 7. For all buildings and structures located in designated flood hazard areas, Section 1605.2.2 specifies that flood load and flood load combinations using the strength design method are to be determined in accordance with Section 2.3.3 of ASCE 7, which states:

Load Combinations Including Flood Load. When a structure is located in a flood zone (Section 5.3.1), the following load combinations shall be considered:

1. In V Zones or Coastal A Zones, 1.6W in combinations (4) and (6) shall be replaced by 1.6W + 2.0F_a.
2. In noncoastal A Zones, 1.6W in combinations (4) and (6) shall be replaced with .8W + 1.0F_a.

1605.3 Load combinations using allowable stress design.

1605.3.1 Basic load combinations. Where allowable stress design (working stress design), as permitted by this code, is used, structures and portions thereof shall resist the most critical effects resulting from the following combinations of loads:

D **(Equation 16-7)**

$D + L$ **(Equation 16-8)**

$D + L + (L_r \text{ or } S \text{ or } R)$ **(Equation 16-9)**

$D + (W \text{ or } 0.7E) + L + (L_r \text{ or } S \text{ or } R)$ **(Equation 16-10)**

$0.6D + W$ **(Equation 16-11)**

$0.6D + 0.7E$ **(Equation 16-12)**

Exceptions:

1. Crane hook loads need not be combined with roof live load or with more than three-fourths of the snow load or one-half of the wind load.
2. Flat roof snow loads of 30 psf (1.44 kN/m^2) or less need not be combined with seismic loads. Where flat roof snow loads exceed 30 psf (1.44 kN/m^2), 20 percent shall be combined with seismic loads.

❖ See Section 1602 for an explanation of the notations. These basic load combinations for ASD are based on Section 2.4.1 of ASCE 7.

1605.3.1.1 Load reduction. It is permitted to multiply the combined effect of two or more variable loads by 0.75 and add to the effect of dead load. The combined load used in design shall not be less than the sum of the effects of dead load and any one of the variable loads. The 0.7 factor on *E* does not apply for this provision.

Increases in allowable stresses specified in the appropriate materials section of this code or referenced standard shall not be used with the load combinations of Section 1605.3.1 except that a duration of load increase shall be permitted in accordance with Chapter 23.

❖ The load reductions in this section are based on the lower probability that two or more variable loads will be at a maximum level at the same point in time. These reductions are similar to those of Section 2.4.1 of ASCE 7.

1605.3.1.2 Other loads. Where *F, H, P* or *T* are to be considered in design, the load combinations of Section 2.4.1 of ASCE 7 shall be used. Where F_a is to be considered in design, the load combinations of Section 2.4.2 of ASCE 7 shall be used.

❖ See the definitions in Section 1602 for an explanation of the notations used in this section. ASCE 7 provides load combinations in the indicated sections for cases where the indicated additional loads are to be included. The study discussed in the commentary to Section 1605.2.2 determined flood load and flood load combination factors for use with the ASD method. This section specifies that flood load and flood load combinations using the ASD method are to be determined in accordance with Section 2.4.2 of ASCE 7, which states:

Load Combinations Including Flood Load. When a structure is located in a flood zone, the following load combinations shall be considered:

1. In V Zones or Coastal A Zones (Section 5.3.1), $1.5F_a$ shall be added to other load combinations (5), (6) and (7), and *E* shall be set equal to zero in (5) and (6).
2. In noncoastal A Zones, $0.75F_a$ shall be added to the combinations (5), (6) and (7), and *E* shall be set at zero in (5) and (6).

1605.3.2 Alternative basic load combinations. In lieu of the basic load combinations specified in Section 1605.3.1, structures and portions thereof shall be permitted to be designed for the most critical effects resulting from the following combinations. When using these alternate basic load combinations that include wind or seismic loads, allowable stresses are permitted to be increased or load combinations reduced, where permitted by the material section of this code or referenced standard. Where wind loads are calculated in accordance with Section 1609.6 or Section 6 of ASCE 7, the coefficient ω in the following equations shall be taken as 1.3. For other wind loads ω shall be taken as 1.0.

$D + L + (L_r \text{ or } S \text{ or } R)$ **(Equation 16-13)**

$D + L + (\omega W)$ **(Equation 16-14)**

$D + L + \omega W + S/2$ **(Equation 16-15)**

$D + L + S + \omega W/2$ **(Equation 16-16)**

$D + L + S + E/1.4$ **(Equation 16-17)**

$0.9D + E/1.4$ **(Equation 16-18)**

Exceptions:

1. Crane hook loads need not be combined with roof live load or with more than three-fourths of the snow load or one-half of the wind load.
2. Flat roof snow loads of 30 pounds per square foot (1.44 kN/m^2) or less need not be combined with seismic loads. Where flat roof snow loads exceed 30 psf (1.44 kN/m^2), 20 percent shall be combined with seismic loads.

❖ These are alternate load combinations to those in Section 1605.3.1 for use with ASD. They are based on the allowable stress load combinations in the *Uniform Building Code*™ (UBC™) (see the definitions in Section 1602 for an explanation of the notations in this section).

1605.3.2.1 Other loads. Where *F, H, P* or *T* are to be considered in design, 1.0 times each applicable load shall be added to the combinations specified in Section 1605.3.2.

❖ See the definitions in Section 1602 for an explanation of the notation used in this section. As indicated, the applicable loads are to be added to each of the load combinations in Section 1605.3.2.

1605.4 Special seismic load combinations. For both allowable stress design and strength design methods, where specifically required by Sections 1613 through 1622 or by Chapters 18 through 23, elements and components shall be designed to resist the forces calculated using Equation 16-19 when the effects of the seismic ground motion are additive to gravity forces and those calculated using Equation 16-20 when the effects of the seismic ground motion counteract gravity forces.

$1.2D + f_1 L + E_m$ **(Equation 16-19)**

$0.9D + E_m$ **(Equation 16-20)**

where:

E_m = The maximum effect of horizontal and vertical forces as set forth in Section 1617.1.

f_1 = 1.0 for floors in places of public assembly, for live loads in excess of 100 psf (4.79 kN/m^2) and for parking garage live load.

f_1 = 0.5 for other live loads.

❖ The purpose of these load combinations is to find the most critical load combination forces when seismic forces are included. The first formula is for the maximum compressive force and the second formula is for the maximum tension force. Note that these are not general load cases, but are only to be applied where specified in this section by the indicated seismic sections of this chapter and Chapters 19 through 23 for structural materials requirements.

1605.5 Heliports and helistops. Heliport and helistop landing or touchdown areas shall be designed for the following loads, combined in accordance with Section 1605:

1. Dead load, D, plus the gross weight of the helicopter, D_h, plus snow load, S.
2. Dead load, D, plus two single concentrated impact loads, L, approximately 8 feet (2438 mm) apart applied anywhere on the touchdown pad (representing each of the helicopter's two main landing gear, whether skid type or wheeled type), having a magnitude of 0.75 times the gross weight of the helicopter. Both loads acting together total 1.5 times the gross weight of the helicopter.
3. Dead load, D, plus a uniform live load, L, of 100 psf (4.79 kN/m^2).

❖ Detailed requirements for helistops and heliports, including definitions, can be found in Section 412.5 of the code. These structural design requirements serve two purposes. First, they establish the minimum dead load and live load criteria that are specific (unique) to the design of these facilities. Secondly, they specify how the dead and live loads are to be utilized in the load combinations of Section 1605.2 or 1605.3.

The first load case considers the helicopter weight as a dead load on the pad to be combined with other dead loads, as well as the design snow load. The second load case considers dead load along with concentrated live loads that are based on the weight of the helicopter plus an impact load from landing. The third load case includes all other dead loads combined with an assembly-type uniform live load.

SECTION 1606
DEAD LOADS

1606.1 Weights of materials and construction. In determining dead loads for purposes of design, the actual weights of materials and construction shall be used. In the absence of definite information, values used shall be subject to the approval of the building official.

❖ When determining the design dead loads, the actual weights or a conservative estimate of the actual weights of all materials and construction are to be used to calculate the required design values for the structural design. As a design guideline, see the unit weights of common construction materials and assemblies in Tables C3-1 and C3-2 of ASCE 7.

The unit dead loads listed in the tables for assembled elements are usually given in units of pounds per square foot of surface area (i.e., floor areas, wall areas, ceiling areas, etc.). Unit dead loads for materials used in construction are given in terms of density (pounds per cubic foot). The unit weights given in the tables are generally single values, even though a range of weights may actually exist. The average unit weights given are generally suitable for design purposes; however, where there is reason to believe that the actual weights of assembled elements or construction materials may substantially exceed the tabular values, then the situation should be investigated and the highest values used.

1606.2 Weights of fixed service equipment. In determining dead loads for purposes of design, the weight of fixed service equipment, such as plumbing stacks and risers, electrical feeders, heating, ventilating and air-conditioning systems (HVAC) and fire sprinkler systems, shall be included.

❖ The weights of service equipment, such as plumbing stacks and risers; heating, ventilating and air-conditioning (HVAC) equipment; elevators and elevator machinery; fire protection systems and similar fixed equipment are to be determined and applied as dead loads. The operating loads of fixed service equipment required for design purposes are generally furnished by the manufacturers or suppliers. The actual weight of equipment installed as part of tenant occupancies is generally unknown during the design phase of a building project, but should be anticipated and accounted for as a unit weight in estimating design live loads (see Section 1607). Wherever heavy concentrated equipment loads are encountered because of tenant use, the load-carrying capacity of the supporting structural elements is to be investigated and, if necessary, reinforced.

SECTION 1607
LIVE LOADS

1607.1 General. Live loads are those loads defined in Section 1602.1.

❖ The live load requirements for the design of buildings and structures are based on the type of occupancy. Live loads are transient loads that vary with time. Generally, the design live load is that which is believed to be near the maximum transient load for a given occupancy.

TABLE 1607.1. See page 16-17.

❖ The design values of live loads for both uniform and concentrated loads are shown in the table as a function of occupancy. The values given are conservative and include both the sustained and variable portions of the live load. It should be noted that the "occupancy" category listed is not necessarily group specific. For example, an office building may be classified as Group B, but still contain incidental storage areas. Depending on the type of storage, the areas may warrant storage live loads of either 125 or 250 psf (5.98 or 11.9 kN/m^2) to be applied to the space in question.

TABLE 1607.1
MINIMUM UNIFORMLY DISTRIBUTED LIVE LOADS AND MINIMUM CONCENTRATED LIVE LOADS[g]

OCCUPANCY OR USE	UNIFORM (psf)	CONCENTRATED (lbs.)
1. Apartments (see residential)	—	—
2. Access floor systems Office use Computer use	 50 100	 2,000 2,000
3. Armories and drill rooms	150	—
4. Assembly areas and theaters Fixed seats (fastened to floor) Lobbies Movable seats Stages and platforms Follow spot, projections and control rooms Catwalks	 60 100 100 125 50 40	—
5. Balconies (exterior) On one- and two-family residences only, and not exceeding 100 ft.^{2}	100 60	—
6. Decks	Same as occupancy served[h]	—
7. Bowling alleys	75	—
8. Cornices	60	—
9. Corridors, except as otherwise indicated	100	—
10. Dance halls and ballrooms	100	—
11. Dining rooms and restaurants	100	—
12. Dwellings (see residential)	—	—
13. Elevator machine room grating (on area of 4 in.^{2})	—	300
14. Finish light floor plate construction (on area of 1 in.^{2})	—	200
15. Fire escapes On single-family dwellings only	100 40	—
16. Garages (passenger vehicles only) Trucks and buses	40 Note a See Section 1607.6	
17. Grandstands (see stadium and arena bleachers)	—	—
18. Gymnasiums, main floors and balconies	100	—
19. Handrails, guards and grab bars	See Section 1607.7	
20. Hospitals Operating rooms, laboratories Private rooms Wards Corridors above first floor	 60 40 40 80	 1,000 1,000 1,000 1,000
21. Hotels (see residential)	—	—
22. Libraries Reading rooms Stack rooms Corridors above first floor	 60 150[b] 80	 1,000 1,000 1,000
23. Manufacturing Light Heavy	 125 250	 2,000 3,000
24. Marquees	75	—
25. Office buildings File and computer rooms shall be designed for heavier loads based on anticipated occupancy Lobbies and first-floor corridors Offices Corridors above first floor	 100 50 80	 2,000 2,000 2,000
26. Penal institutions Cell blocks Corridors	 40 100	—
27. Residential One- and two-family dwellings Uninhabitable attics without storage Uninhabitable attics with storage Habitable attics and sleeping areas All other areas except balconies and decks Hotels and multifamily dwellings Private rooms and corridors serving them Public rooms and corridors serving them	 10 20 30 40 40 100	—
28. Reviewing stands, grandstands and bleachers	Note c	—
29. Roofs	See Section 1607.11	
30. Schools Classrooms Corridors above first floor First-floor corridors	 40 80 100	 1,000 1,000 1,000
31. Scuttles, skylight ribs and accessible ceilings	—	200
32. Sidewalks, vehicular driveways and yards, subject to trucking	250[d]	8,000[e]
33. Skating rinks	100	—
34. Stadiums and arenas Bleachers Fixed seats (fastened to floor)	 100[c] 60[c]	—
35. Stairs and exits One- and two-family dwellings All other	100 40 100	Note f
36. Storage warehouses (shall be designed for heavier loads if required for anticipated storage) Light Heavy	 125 250	—
37. Stores Retail First floor Upper floors Wholesale, all floors	 100 75 125	 1,000 1,000 1,000
38. Vehicle barriers	See Section 1607.7	
39. Walkways and elevated platforms (other than exitways)	60	—
40. Yards and terraces, pedestrians	100	—

(continued)

Notes to Table 1607.1

For SI: 1 inch = 25.4 mm, 1 square inch = 645.16 mm^2, 1 pound per square foot = 0.0479 kN/m^2, 1 pound = 0.004448 kN.
1 pound per cubic foot = 16 kg/m^3

a. Floors in garages or portions of buildings used for the storage of motor vehicles shall be designed for the uniformly distributed live loads of Table 1607.1 or the following concentrated loads: (1) for garages restricted to vehicles accommodating not more than nine passengers, 3,000 pounds acting on an area of 4.5 inches by 4.5 inches; (2) for mechanical parking structures without slab or deck which are used for storing passenger vehicles only, 2,250 pounds per wheel.

b. The loading applies to stack room floors that support nonmobile, double-faced library bookstacks, subject to the following limitations:
   1. The nominal bookstack unit height shall not exceed 90 inches;
   2. The nominal shelf depth shall not exceed 12 inches for each face; and
   3. Parallel rows of double-faced bookstacks shall be separated by aisles not less than 36 inches wide.

c. Design in accordance with the ICC *Standard on Bleachers, Folding and Telescopic Seating and Grandstands*.

d. Other uniform loads in accordance with an approved method which contains provisions for truck loadings shall also be considered where appropriate.

e. The concentrated wheel load shall be applied on an area of 20 square inches.

f. Minimum concentrated load on stair treads (on area of 4 square inches) is 300 pounds.

g. Where snow loads occur that are in excess of the design conditions, the structure shall be designed to support the loads due to the increased loads caused by drift buildup or a greater snow design determined by the building official (see Section 1608). For special-purpose roofs, see Section 1607.11.2.2.

h. See Section 1604.8.3 for decks attached to exterior walls.

1607.2 Loads not specified. For occupancies or uses not designated in Table 1607.1, the live load shall be determined in accordance with a method approved by the building official.

❖ Whenever an occupancy or use of a structure cannot be identified with the listing shown in Table 1607.1, then the live load values used for design are required to be determined by the design professional and subject to the approval of the building official. Aside from the obvious intent of this requirement, however, which is to prescribe a minimum design load value, some caution needs to be exercised by the design professional in determining the appropriate design live load value. For example, the table shows that heavy storage areas must be designed for a uniform live load of 250 psf (11.9 kN/m^2). This is a minimum value. Storage warehouses or storage areas within manufacturing facilities containing items such as automobile parts, electrical goods, coiled steel, plumbing supplies and bulk building materials generally have live loads ranging between 300 and 400 psf (14.4 and 19 kN/m^2). Similarly, storage facilities containing dry goods, paints, oil, groceries or liquor often have loadings that range between 200 to 300 psf (9.6 to 14.4 kN/m^2). Another example is a heavy manufacturing facility that makes generators for the electric power industry. Some of the production areas in this type of facility require structural floors that support loads of 1,000 psf (47.9 kN/m^2) or more, which is about seven times the live load specified in Table 1607.1.

1607.3 Uniform live loads. The live loads used in the design of buildings and other structures shall be the maximum loads expected by the intended use or occupancy but shall in no case be less than the minimum uniformly distributed unit loads required by Table 1607.1.

❖ Studies have shown that building live loads consist of a sustained portion based on the day-to-day use of the facilities, and a variable portion created by unusual events such as remodeling, temporary storage of materials, the extraordinary assemblage of people for an occasional business meeting or social function (i.e., holiday party) and similar events. The sustained portion of the live load will likely vary during the life of a building because of tenant changes, rearrangement of office space and furnishings, changes in the nature of the occupancy (i.e., number of people or type of business), traffic patterns and so on. In light of this variability of loadings that are apt to be imposed on a building, the code provisions simplify the design procedure by expressing the applicable load as either a uniformly distributed live load or a concentrated live load on the floor area. It should be pointed out that this section does not require the concurrent application of uniform live load and concentrated live load. In other words, this section requires that either the uniform load or the concentrated load be applied, so long as the type of load that produces the greater stress in the structural element under consideration is utilized.

1607.4 Concentrated loads. Floors and other similar surfaces shall be designed to support the uniformly distributed live loads prescribed in Section 1607.3 or the concentrated load, in pounds (kilonewtons), given in Table 1607.1, whichever produces the greater load effects. Unless otherwise specified, the indicated concentration shall be assumed to be uniformly distributed over an area 2.5 feet by 2.5 feet [6.25 ft^2 (0.58 m^2)] and shall be located so as to produce the maximum load effects in the structural members.

❖ A building or portion thereof is subjected to concentrated floor loads commensurate with the use of the facility. For example, in Group B, a law office may have stacks of books and files that impose large concentrations of loads on the supporting structural elements. An industrial facility may have a tank full of liquid material on a mezzanine that feeds a machine on the floor below. The structural floor of a stockroom may support heavy bins containing metal parts and so on. The exact locations or nature of such concentrated loadings is not usually known at the time of the design of the building. Furthermore, new sources of concentrated loadings will be added during the life of the structure, while some or all of the existing sources will be relocated; therefore, because of the uncertainties of the sources of concentrated loads as well as their weights and locations, the code provides typical loads to be used in the design of structural floors consistent with the type of use of the facility. The minimum concentrated loads to be used for design are contained in Table 1607.1. Concentrated loads are not required to be applied simultaneously, with the uniform live loads also specified in Table 1607.1. Concentrated loads are to be applied as an independent load condition at the location on the floor that produces the greatest stress in the structural members being designed. The single concentrated load is to be placed at any location on the floor. For example, in an office use area, the floor system is to be designed for either a 2,000-pound (8897 N) concentrated load (unless the anticipated actual concentrated load is higher) applied at any location in the office area, or the 50 psf (2.40 kN/m²) live load specified in Table 1607.1, whichever results in the greater stress in the supporting structural member.

1607.5 Partition loads. In office buildings and in other buildings where partition locations are subject to change, provision for partition weight shall be made, whether or not partitions are shown on the construction documents, unless the specified live load exceeds 80 psf (3.83 kN/m²). Such partition load shall not be less than a uniformly distributed live load of 20 psf (0.96kN/m²).

❖ The weights of all partitions are to be considered as dead loads and are to be estimated in accordance with the partition layouts shown in the building construction documents. In the absence of definitive partition layouts, a unit value of 20 psf (0.96 kN/m²) of floor area must be used in the design when the design floor live load is 80 psf (3.83 kN/m²) or less. This is a conservative value and is used in addition to the required uniformly distributed live load given in Table 1607.1. In all cases, provisions for the weight of partitions must be made in the structural design, except where the specified minimum uniformly distributed live load exceeds 80 psf (3.83 kN/m²).

1607.6 Truck and bus garages. Minimum live loads for garages having trucks or buses shall be as specified in Table 1607.6, but shall not be less than 50 psf (2.40 kN/m²), unless other loads are specifically justified and approved by the building official. Actual loads shall be used where they are greater than the loads specified in the table.

❖ The uniform load specified in this section is to be applied to the garage floor in accordance with Section 1607.6.1 (also see Table 1607.6 for the specified concentrated loads that are to be included in the design). The uniform and concentrated loads are to be applied as separate load cases and not at the same time.

TABLE 1607.6
UNIFORM AND CONCENTRATED LOADS

LOADING CLASS[a]	UNIFORM LOAD (pounds/linear foot of lane)	CONCENTRATED LOAD (pounds)[b]	
		For moment design	For shear design
H20-44 and HS20-44	640	18,000	26,000
H15-44 and HS15-44	480	13,500	19,500

For SI: 1 pound per linear foot = 0.01459 kN/m, 1 pound = 0.004448 kN, 1 ton = 8.90 kN.

a. An H loading class designates a two-axle truck with a semitrailer. An HS loading class designates a tractor truck with a semitrailer. The numbers following the letter classification indicate the gross weight in tons of the standard truck and the year the loadings were instituted.

b. See Section 1607.6.1 for the loading of multiple spans.

❖ These are the uniform and concentrated loads that have been established by the American Association of State Highway Transportation Officials (AASHTO) by truck live loads (see Section 1607.6.1 for directions regarding the application of these loads). The uniform load in the table is in pounds per linear foot (plf) of lane. The load is to be divided by 10 feet (3048 mm) to get the area loading of 64 or 48 psf (3.1 or 2.3 kN/m²), depending on the truck loading used.

1607.6.1 Truck and bus garage live load application. The concentrated load and uniform load shall be uniformly distributed over a 10-foot (3048 mm) width on a line normal to the centerline of the lane placed within a 12-foot-wide (3658 mm) lane. The loads shall be placed within their individual lanes so as to produce the maximum stress in each structural member. Single spans shall be designed for the uniform load in Table 1607.6 and one simultaneous concentrated load positioned to produce the maximum effect. Multiple spans shall be designed for the uniform load in Table 1607.6 on the spans and two simultaneous concentrated loads in two spans positioned to produce the maximum negative moment effect. Multiple span design loads, for other effects, shall be the same as for single spans.

❖ This section specifies the application of live loads so as to produce the maximum stresses. This provision is the truck live loading that has been established by AASHTO.

1607.7 Loads on handrails, guards, grab bars and vehicle barriers. Handrails, guards, grab bars as designed in ICC A117.1 and vehicle barriers shall be designed and constructed to the structural loading conditions set forth in this section.

❖ The requirements of this section are intended to provide an adequate degree of structural strength and stability to handrails, guards, grab bars and vehicle barriers.

1607.7.1 Handrails and guards. Handrail assemblies and guards shall be designed to resist a load of 50 plf (0.73 kN/m) applied in any direction at the top and to transfer this load through the supports to the structure.

Exceptions:

1. For one- and two-family dwellings, only the single, concentrated load required by Section 1607.7.1.1 shall be applied.
2. In Group I-3, F, H and S occupancies, for areas that are not accessible to the general public and that have an occupant load no greater than 50, the minimum load shall be 20 pounds per foot (0.29 kN/m).

❖ The loading in this section is the maximum anticipated load from use of the handrail by a crowd of people on the stairway. The exceptions provide lower loads for circumstances where the public does not use the handrail. The loading in this section is not to be applied with any other design load; it is a separate load case.

1607.7.1.1 Concentrated load. Handrail assemblies and guards shall be able to resist a single concentrated load of 200 pounds (0.89 kN), applied in any direction at any point along the top, and have attachment devices and supporting structure to transfer this loading to appropriate structural elements of the building.

This load need not be assumed to act concurrently with the loads specified in the preceding paragraph.

❖ The concentrated loading in this section is not to be applied with any other design load; it is a separate load case. The load simulates the maximum anticipated load from a person grabbing or falling into the handrail or guard.

1607.7.1.2 Components. Intermediate rails (all those except the handrail), balusters and panel fillers shall be designed to withstand a horizontally applied normal load of 50 pounds (0.22 kN) on an area equal to 1 square foot ($0.093 m^2$), including openings and space between rails. Reactions due to this loading are not required to be superimposed with those of Section 1607.7.1 or 1607.7.1.1.

❖ This is a localized design load for the guard members and is not to be applied with any other loads. It is to be applied horizontally at a 90-degree (1.57 rad) angle with the guard members. The number of balusters that would resist this load are those within the 1 square foot ($0.093 m^2$) area in the plane of the guard.

1607.7.1.3 Stress increase. Where handrails and guards are designed in accordance with the provisions for allowable stress design (working stress design) exclusively for the loads specified in Section 1607.7.1, the allowable stress for the members and their attachments are permitted to be increased by one-third.

❖ This section clarifies that the ASD stresses are permitted to be increased by one-third when the handrail or guard is subjected to any one of the design loads specified in Sections 1607.7.1 through 1607.7.1.2. This is appropriate since the design loads are the maximum anticipated loads.

1607.7.2 Grab bars, shower seats and dressing room bench seats. Grab bars, shower seats and dressing room bench seat systems shall be designed to resist a single concentrated load of 250 pounds (1.11 kN) applied in any direction at any point.

❖ These live loads provide for the normal anticipated loads from the use of the grab bars, shower seats and dressing-room bench seats.

1607.7.3 Vehicle barriers. Vehicle barrier systems for passenger cars shall be designed to resist a single load of 6,000 pounds (26.70 kN) applied horizontally in any direction to the barrier system and shall have anchorage or attachment capable of transmitting this load to the structure. For design of the system, the load shall be assumed to act at a minimum height of 1 foot, 6 inches (457 mm) above the floor or ramp surface on an area not to exceed 1 square foot ($305 mm^2$), and is not required to be assumed to act concurrently with any handrail or guard loadings specified in the preceding paragraphs of Section 1607.7.1. Garages accommodating trucks and buses shall be designed in accordance with an approved method that contains provision for traffic railings.

❖ This section is needed for the design of passenger car and light truck vehicle barriers. For bus and heavy truck vehicle barrier design criteria, the state department of transportation should be contacted. The 6,000-pound (26.70 kN) load includes the impact effect. The loads are intended for use with a one-third increase in allowable stresses when using the ASD method.

1607.8 Impact loads. The live loads specified in Section 1607.2 include allowance for impact conditions. Provisions shall be made in the structural design for uses and loads that involve unusual vibration and impact forces.

❖ In cases where "unordinary" live loads are likely to occur in a building that impose impact or vibratory forces on structural elements (i.e., elevators, machinery, craneways, etc.), additional stresses and deflections are produced in the structural system. Where unusual vibration (dynamic) and impact loads are likely to occur, the code requires that the structural design take these effects into account. Typically, the dynamic effects are approximated through the application of an equivalent static load equal to the dynamic load effects. In most cases, an equivalent static load is sufficient. A dynamic analysis is usually not required.

1607.8.1 Elevators. Elevator loads shall be increased by 100 percent for impact and the structural supports shall be designed within the limits of deflection prescribed by ASME A17.1.

❖ The static load of an elevator must be increased to account for the effect of elevator motion. For example, when an elevator comes to a stop, the load on the elevator supports is significantly higher than the weight of the elevator and the occupants. This effect increases with the acceleration and deceleration rate of the elevator.

1607.8.2 Machinery. For the purpose of design, the weight of machinery and moving loads shall be increased as follows to allow for impact: (1) elevator machinery, 100 percent; (2) light machinery, shaft- or motor-driven, 20 percent; (3) reciprocating machinery or power-driven units, 50 percent; (4) hangers for floors or balconies, 33 percent. Percentages shall be increased where specified by the manufacturer.

❖ The specified increases for machinery loads include the vibration of the equipment, which increases the effective load. The load increase for reciprocating machinery versus rotating shaft-driven machinery is to account for the higher vibration. The increase for floor or balcony hangers recognizes that these structural systems are subject to vibration as a result of the way they are supported.

1607.9 Reduction in live loads. The minimum uniformly distributed live loads, L_o, in Table 1607.1 are permitted to be reduced according to the following provisions.

❖ The basis for the live load reduction in Sections 1607.9 through 1607.9.1.4 is ASCE 7. This section provides for reductions in the minimum uniformly distributed live loads specified in Table 1607.1. Section 1607.3 directs the designer to utilize the greater live loads produced by the intended occupancy, but not less than the minimum uniformly distributed live loads listed in Table 1607.1. Thus, where actual live loads will or are likely to occur based on the intended occupancy, live load reductions are not applicable to those members supporting such loads.

1607.9.1 General. Subject to the limitations of Sections 1607.9.1.1 through 1607.9.1.4, members for which a value of $K_{LL}A_T$ is 400 square feet (37.16 m²) or more are permitted to be designed for a reduced live load in accordance with the following equation:

$$L = L_o\left(0.25 + \frac{15}{\sqrt{K_{LL}A_T}}\right) \quad \textbf{(Equation 16-21)}$$

For SI: $$L = L_o\left(0.25 + \frac{4.57}{\sqrt{K_{LL}A_T}}\right)$$

where:

L = Reduced design live load per square foot (meter) of area supported by the member.

L_o = Unreduced design live load per square foot (meter) of area supported by the member (see Table 1607.1).

K_{LL} = Live load element factor (see Table 1607.9.1).

A_T = Tributary area, in square feet (square meters). L shall not be less than $0.50L_o$ for members supporting one floor and L shall not be less than $0.40L_o$ for members supporting two or more floors.

❖ This section provides formulas for calculating the reduced design live loads based on the method in ASCE 7. The concept is that where the design live load is governed by the minimum live loads in Table 1607.1, the actual load on a large area of the floor is very likely to be less than the nominal live load in the table. Thus, the allowable reduction increases with the tributary area of the floor that is supported by a structural member; therefore, a girder that supports a large tributary area would be allowed to be designed for somewhat less live load per square foot than a floor beam that supports a smaller total floor area.

TABLE 1607.9.1
LIVE LOAD ELEMENT FACTOR, K_{LL}

ELEMENT	K_{LL}
Interior columns Exterior columns without cantilever slabs	4 4
Edge columns with cantilever slabs	3
Corner columns with cantilever slabs Edge beams without cantilever slabs Interior beams	2 2 2
All other members not identified above including: Edge beams with cantilever slabs Cantilever beams Two-way slabs Members without provisions for continuous shear transfer normal to their span	1

❖ The purpose of this table is to provide the coefficients for the reduced live load formulas that are in Section 1607.9.1. The value of the coefficients converts the tributary area of the member listed to the influence area of the supported area. The influence area of a structural member is the floor area where a portion of any applied load is supported by the member.

1607.9.1.1 Heavy live loads. Live loads that exceed 100 psf (4.79 kN/m²) shall not be reduced except the live loads for members supporting two or more floors are permitted to be reduced by a maximum of 20 percent, but the live load shall not be less than L as calculated in Section 1607.9.1.

❖ The purpose of this section is to limit live load reductions. It is likely that the full live load will occur at a given floor level for occupancies where the live load is required to be greater than 100 psf (4.79 kN/m²). Thus, reduced live loads are not allowed for these conditions except as described in this section.The loads on structural members, such as columns and bearing walls that support two or more floors, are allowed to be reduced by 20 percent. This reduction for elements supporting more than one floor applies to heavy floor design live loadings that exceed 100 psf (4.79 kN/m²). The basis for this allowance is that the floor areas on all floor levels that contribute to the loading of a supporting member are unlikely to all be fully loaded simultaneously.

1607.9.1.2 Passenger vehicle garages. The live loads shall not be reduced in passenger vehicle garages except the live loads for members supporting two or more floors are permitted to be reduced by a maximum of 20 percent, but the live load shall not be less than L as calculated in Section 1607.9.1.

❖ This section limits the live load reduction for passenger vehicles to only those members that support more than two floors. Thus, floor framing members that support only a part of one floor are not allowed to reduce the live load from the 40 psf (1.92 kN/m^2) that is specified in Table 1607.1. This is consistent with Section 4.8.3 of ASCE 7.

1607.9.1.3 Special occupancies. Live loads of 100 psf (4.79 kN/m^2) or less shall not be reduced in public assembly occupancies.

❖ This section does not allow the live load to be reduced for an assembly occupancy, since it is very likely that the load will be imposed over a large floor area.

1607.9.1.4 Special structural elements. Live loads shall not be reduced for one-way slabs except as permitted in Section 1607.9.1.1. Live loads of 100 psf (4.79 kN/m^2) or less shall not be reduced for roof members except as specified in Section 1607.11.2.

❖ One-way slabs are permitted to have the live load reduced only if they support more than one floor, in accordance with Section 1607.9.1.1.The uncertainties of roof loads, including snow loads, preclude live load reductions. Roof minimum load reductions are allowed based on the provisions of Section 1607.11.2. Roof minimum loads are allowances for imposed loads as a result of maintenance and repair.

1607.9.2 Alternate floor live load reduction. As an alternative to Section 1607.9.1, floor live loads are permitted to be reduced in accordance with the following provisions. Such reductions shall apply to slab systems, beams, girders, columns, piers, walls and foundations.

1. A reduction shall not be permitted in Group A occupancies.
2. A reduction shall not be permitted when the live load exceeds 100 psf (4.79 kN/m^2) except that the design live load for columns may be reduced by 20 percent.
3. For live loads not exceeding 100 psf (4.79 kN/m^2), the design live load for any structural member supporting 150 square feet (13.94 m^2) or more is permitted to be reduced in accordance with the following equation:

$$R = r(A - 150) \qquad \textbf{(Equation 16-22)}$$

For SI: $R = r(A - 13.94)$

Such reduction shall not exceed 40 percent for horizontal members, 60 percent for vertical members, nor R as determined by the following equation:

$$R = 23.1\,(1 + D/L_o) \qquad \textbf{(Equation 16-23)}$$

where:

A = Area of floor or roof supported by the member, square feet (m^2).

D = Dead load per square foot (m^2) of area supported.

L_o = Unreduced live load per square foot (m^2) of area supported.

R = Reduction in percent.

r = Rate of reduction equal to 0.08 percent for floors.

❖ This section includes an alternative floor live load method that is permitted to be used instead of the method indicated in Sections 1607.9 through 1607.9.1.4. The basis for this section is the 1997 *Uniform Building Code*.

1607.10 Distribution of floor loads. Where uniform floor live loads are involved in the design of structural members arranged so as to create continuity, the minimum applied loads shall be the full dead loads on all spans in combination with the floor live loads on spans selected to produce the greatest effect at each location under consideration. It shall be permitted to reduce floor live loads in accordance with Section 1607.9.

❖ For continuous floor members loaded such that the live loads of a building are distributed in some bays and not in others, some of the structural elements will be subjected to greater stresses because of partial loading conditions as compared to full loading on all spans. This code section requires the engineer to consider partial loadings that produce the greatest design forces for any location in the design of continuous floor elements.

For example, Figure 1607.10 shows a continuous multispan girder with partial loading. The Type I loading condition shows that only the alternate spans have uniform live loads, which produces:

- Maximum positive moments at the centers of the loaded spans (A-B, C-D, E-F) and
- Maximum negative moments at the centers of the unloaded spans (B-C, D-E, F-G).

The Type II live load distribution shows two loaded adjacent spans with alternate spans loaded beyond these, which produces:

- Maximum negative moment at Support D and
- Maximum girder shears.

To obtain the maximum total stresses imposed on the girder, the dead load moments and shears must be added to those produced by the partial live loadings.

1607.11 Roof loads. The structural supports of roofs and marquees shall be designed to resist wind and, where applicable, snow and earthquake loads, in addition to the dead load of construction and the appropriate live loads as prescribed in this section, or as set forth in Table 1607.1. The live loads acting on a sloping surface shall be assumed to act vertically on the horizontal projection of that surface.

❖ In addition to dead and live loads (typically during construction), the roof's structural system is to be designed and constructed to resist environmental loads caused by wind, snow and earthquakes.

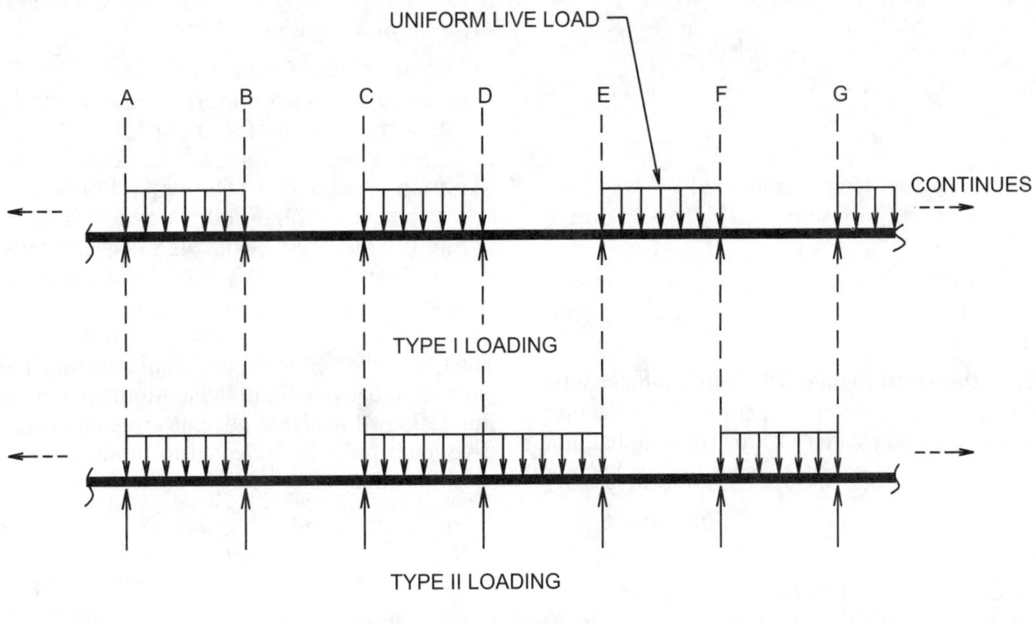

Figure 1607.10
PARTIAL LOADING

1607.11.1 Distribution of roof loads. Where uniform roof live loads are involved in the design of structural members arranged so as to create continuity, the minimum applied loads shall be the full dead loads on all spans in combination with the roof live loads on adjacent spans or on alternate spans, whichever produces the greatest effect. See Section 1607.11.2 for minimum roof live loads and Section 1608.5 for partial snow loading.

❖ For continuous roof construction, partial loadings must be included in the design of structural elements to determine the governing loading situation. For example, Figure 1607.10 shows a continuous multispan girder with partial loading. The Type I loading condition shows that only the alternate spans have uniform live loads, which produces:

- Maximum positive moments at the centers of the loaded spans (A-B, C-D, E-F) and
- Maximum negative moments at the centers of the unloaded spans (B-C, D-E, F-G).

The Type II live load distribution shows two loaded adjacent spans with alternate spans loaded beyond these, which produces:

- Maximum negative moment at Support D and
- Maximum girder shears.

To obtain the maximum total stresses imposed on the girder, the dead load moments and shears must be added to those produced by the partial live loadings.

1607.11.2 Minimum roof live loads. Minimum roof loads shall be determined for the specific conditions in accordance with Sections 1607.11.2.1 through 1607.11.2.4.

❖ The minimum roof live loads typically occur during roof maintenance, construction or repair. This section also refers to loads for occupied roofs.

1607.11.2.1 Flat, pitched and curved roofs. Ordinary flat, pitched and curved roofs shall be designed for the live loads specified in the following equation or other controlling combinations of loads in Section 1605, whichever produces the greater load. In structures where special scaffolding is used as a work surface for workers and materials during maintenance and repair operations, a lower roof load than specified in the following equation shall not be used unless approved by the building official. Greenhouses shall be designed for a minimum roof live load of 10 psf (0.479 kN/m²).

$L_r = 20R_1R_2$ **(Equation 16-24)**

where: $12 \leq L_r \leq 20$

For SI: $L_r = 0.96\, R_1R_2$

where: $0.58 \leq L_r \leq 0.96$

L_r = Roof live load per square foot (m²) of horizontal projection in pounds per square foot (kN/m²).

The reduction factors R_1 and R_2 shall be determined as follows:

$R_1 = 1$ for $A_t \leq 200$ square feet (18.58 m²)

(Equation 16-25)

$R_1 = 1.2 - 0.001A_t$ for 200 square feet $< A_t <$ 600 square feet

(Equation 16-26)

For SI: $1.2 - 0.011A_t$ for 18.58 square meters $< A_t <$ 55.74 square meters

$R_1 = 0.6$ for $A_t \geq 600$ square feet (55.74 m^2)

(Equation 16-27)

where:

A_t = Tributary area (span length multiplied by effective width) in square feet (m^2) supported by any structural member, and

F = for a sloped roof, the number of inches of rise per foot (for SI: $F = 0.12 \times$ slope, with slope expressed in percentage points), and

F = for an arch or dome, rise-to-span ratio multiplied by 32, and

$R_2 = 1$ for $F \leq 4$ **(Equation 16-28)**

$R_2 = 1.2 - 0.05\ F$ for $4 < F < 12$ **(Equation 16-29)**

$R_2 = 0.6$ for $F \geq 12$ **(Equation 16-30)**

❖ This section provides formulas for the determination of roof load. The concept is to provide enough strength to account for the maintenance, construction or repair of the roof. Generally, the roof live load in this section is 24 psf (1.15 kN/m^2) for roof members that support small tributary areas of less than 200 square feet (18.58 m^2) and less than 24 psf (1.15 kN/m^2) as the tributary area increases.

This section also provides for a lower roof live load for a greenhouse, since it is not likely that loads from maintenance or repair will exceed the specified 10 psf (0.479 kN/m^2).

1607.11.2.2 Special-purpose roofs. Roofs used for promenade purposes shall be designed for a minimum live load of 60 psf (2.87 kN/m^2). Roofs used for roof gardens or assembly purposes shall be designed for a minimum live load of 100 psf (4.79 kN/m^2). Roofs used for other special purposes shall be designed for appropriate loads, as directed or approved by the building official.

❖ Roofs that are to be occupied during social events incidental to the principal use of the facility are to be designed for a minimum uniform live load of 60 psf (2.87 kN/m^2). The promenade deck of a residential penthouse located on the main roof of an apartment building is an example of this type of use. Where roofs are designed to be used as roof gardens or to support large gatherings of people as a function accompanying the educational or assembly uses of a facility, the roofs are required to be designed to a minimum live load of 100 psf (4.79 kN/m^2). The minimum live loads specified in Table 1607.1 are not required to be added to the design load requirements for special-purpose roofs specified in this section.

1607.11.2.3 Landscaped roofs. Where roofs are to be landscaped, the uniform design live load in the landscaped area shall be 20 psf (0.958 kN/m^2). The weight of the landscaping materials shall be considered as dead load and shall be computed on the basis of saturation of the soil.

❖ Those areas of a special-purpose roof that are to be landscaped are required to be designed for a minimum uniform live load of 20 psf (0.958 kN/m^2) to accommodate the occasional loads associated with the maintenance of plantings. The weight of landscaping materials is to be considered as dead loads in the design of the roof structure, which is to be combined with the live load (see Section 1605 for load combinations to be included).

1607.11.2.4 Awnings and canopies. Awnings and canopies shall be designed for a uniform live load of 5 psf (0.240 kN/m^2) as well as for snow loads and wind loads as specified in Sections 1608 and 1609.

❖ Awning and canopy structures covered with fabric materials are to be designed to sustain a live load of 5 psf (0.240 kN/m^2) as well as the specified snow and wind loads of Sections 1608 and 1609. The live load, snow load and wind load are to be combined according to Section 1605.

1607.12 Crane loads. The crane live load shall be the rated capacity of the crane. De sign loads for the runway beams, including connections and support brackets, of moving bridge cranes and monorail cranes shall include the maximum wheel loads of the crane and the vertical impact, lateral and longitudinal forces induced by the moving crane.

❖ This section provides a general description of the crane loads that are required to be included in the design. The supporting structure for the crane is to be designed for a combination of the maximum wheel load, vertical impact and horizontal load as a simultaneous load combination. The typical arrangement for a top-running bridge crane is shown in Figure 1607.12.

1607.12.1 Maximum wheel load. The maximum wheel loads shall be the wheel loads produced by the weight of the bridge, as applicable, plus the sum of the rated capacity and the weight of the trolley with the trolley positioned on its runway at the location where the resulting load effect is maximum.

❖ The maximum vertical wheel load occurs when the trolley is moved as close as possible to the supporting beams under consideration. This results in the greatest portion of the crane weight, the design weight lifted load and the wheel vertical impact load on the supporting beams.

1607.12.2 Vertical impact force. The maximum wheel loads of the crane shall be increased by the percentages shown below to determine the induced vertical impact or vibration force:

Monorail cranes (powered) . . . . 25 percent

Cab-operated or remotely operated bridge cranes (powered) . . . . 25 percent

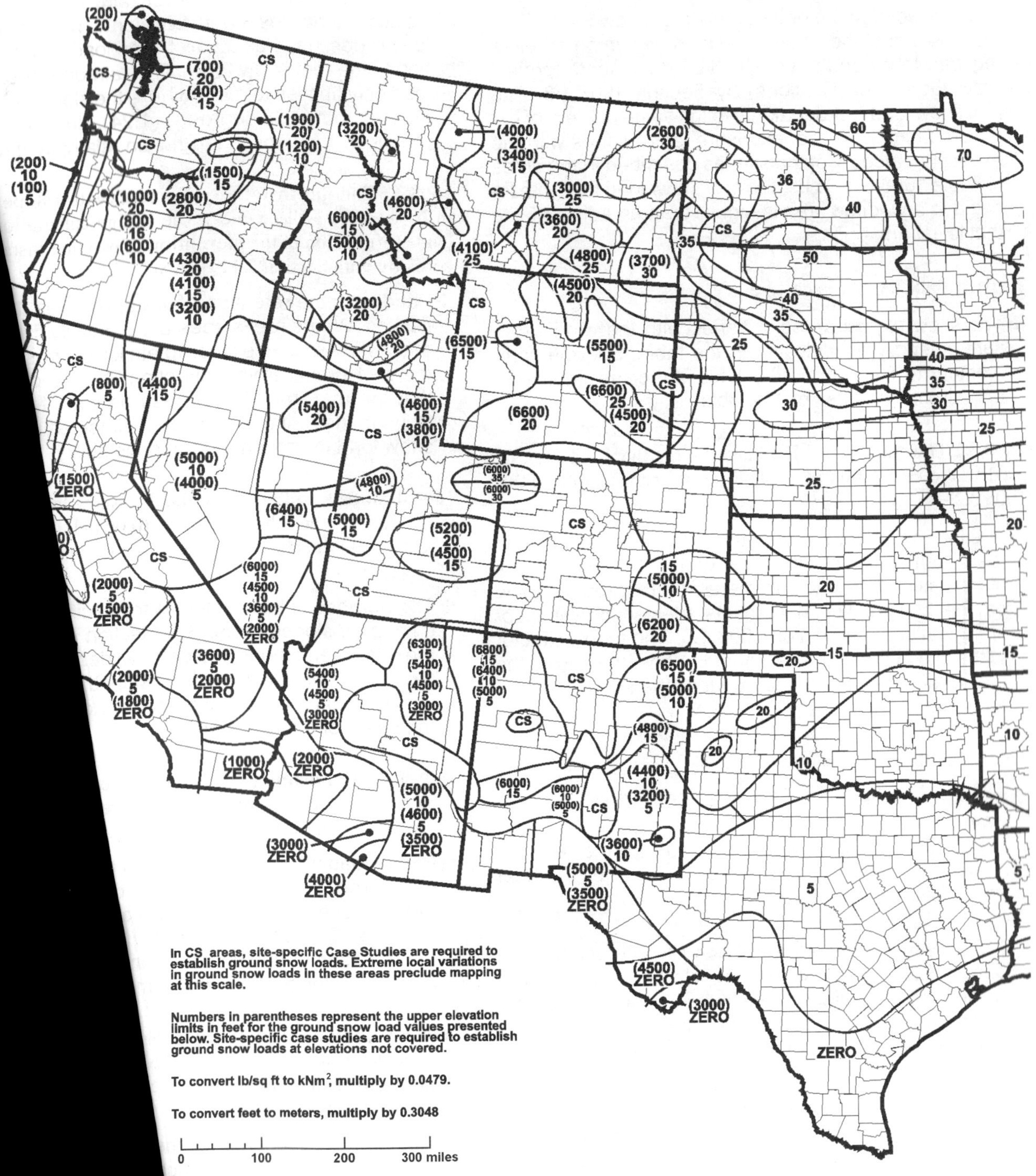
In CS areas, site-specific Case Studies are required to establish ground snow loads. Extreme local variations in ground snow loads in these areas preclude mapping at this scale.
Numbers in parentheses represent the upper elevation limits in feet for the ground snow load values presented below. Site-specific case studies are required to establish ground snow loads at elevations not covered.
To convert lb/sq ft to kN/m², multiply by 0.0479.
To convert feet to meters, multiply by 0.3048
0
100
200
300 miles

FIGURE 1608.2
GROUND SNOW LOADS, p_g, FOR THE UNITED STATES (psf)

Pendant-operated bridge cranes (powered) . . . . 10 percent

Bridge cranes or monorail cranes with hand-geared bridge, trolley and hoist . . . . . . 0 percent

❖ A vertical impact force is necessary to account for the impact from the starting and stopping movement of the suspended weight from the crane. Vertical impact is also created by the movement of the crane along the rails.

1607.12.3 Lateral force. The lateral force on crane runway beams with electrically powered trolleys shall be calculated as 20 percent of the sum of the rated capacity of the crane and the weight of the hoist and trolley. The lateral force shall be assumed to act horizontally at the traction surface of a runway beam, in either direction perpendicular to the beam, and shall be distributed according to the lateral stiffness of the runway beam and supporting structure.

❖ This section is necessary to define the design lateral force on the crane supports. Lateral force at the right angle to the crane rail is caused by the lateral movement of the lifted load and from the frame action of the crane.

1607.12.4 Longitudinal force. The longitudinal force on crane runway beams, except for bridge cranes with hand-geared bridges, shall be calculated as 10 percent of the maximum wheel loads of the crane. The longitudinal force shall be assumed to act horizontally at the traction surface of a runway beam, in either direction parallel to the beam.

❖ This section is needed to define the longitudinal force on the crane supports, which is caused from the longitudinal motion of the crane with the lifted load.

1607.13 Interior walls and partitions. Interior walls and partitions that exceed 6 feet (1829 mm) in height, including their finish materials, shall have adequate strength to resist the loads to which they are subjected but not less than a horizontal load of 5 psf (0.240 kN/m²).

❖ The requirements of this section are intended to provide sufficient strength and durability of the wall framing and of the finished construction to provide a minimum level of resistance to nominal impact loads that commonly occur in the use of a facility and also to resist HVAC pressurization.

SECTION 1608
SNOW LOADS

1608.1 General. Design snow loads shall be determined in accordance with Section 7 of ASCE 7, but the design roof load shall not be less than that determined by Section 1607.

❖ The intent of this section is that the code requirements are the same as the technical requirements in Section 7 of ASCE 7. The snow load provisions in ASCE 7 are based on over 40 years of ground snow load data, and include requirements for thermal resistance of the roof structure, a rain-on-snow surcharge, partial loading on continuous beam systems and ponding instability from melting snow or rain on snow. The text of the snow load section contains many of the variables that must be determined for the calculation of snow loads: ground snow load, exposure factor, importance factor and thermal factor. The maps and tables necessary for determining each of these is in the code, consistent with the information provided in ASCE 7 so that the building official is able to verify these fundamental criteria as well as de-

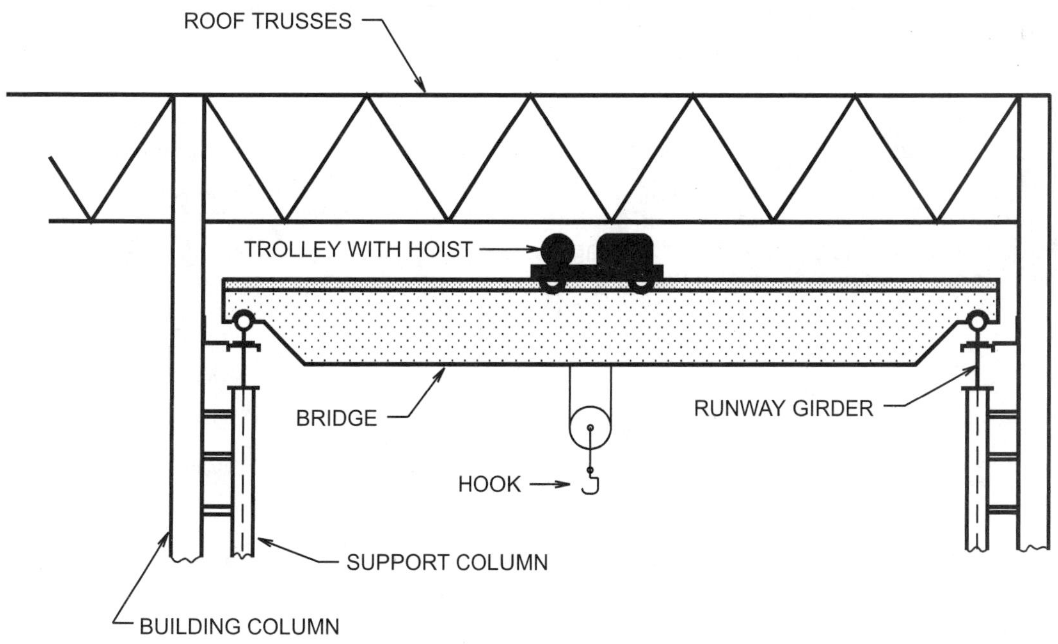
ROOF TRUSSES
TROLLEY WITH HOIST
BRIDGE
RUNWAY GIRDER
HOOK
SUPPORT COLUMN
BUILDING COLUMN

Figure 1607.12
TOP-RUNNING BRIDGE CRANE

termine the conditions where other critical factors (such as rain-on-snow surcharge) must be included in the design. Section 1608 allows design professionals to fully utilize the provisions of the leading national consensus standard for the calculation of snow loads, while simultaneously providing building officials with key information in the code to verify the appropriate snow load criteria.

1608.2 Ground snow loads. The ground snow loads to be used in determining the design snow loads for roofs are given in Figure 1608.2 for the contiguous United States and Table 1608.2 for Alaska. Site-specific case studies shall be made in areas designated CS in Figure 1608.2. Ground snow loads for sites at elevations above the limits indicated in Figure 1608.2 and for all sites within the CS areas shall be approved. Ground snow load determination for such sites shall be based on an extreme value statistical analysis of data available in the vicinity of the site using a value with a 2-percent annual probability of being exceeded (50-year mean recurrence interval). Snow loads are zero for Hawaii, except in mountainous regions as approved by the building official.

❖ The ground snow loads on the maps in Figure 1608.2 of the code are generally based on over 40 years of snow depth records. The snow loads on the maps are those that have a 2-percent annual probability of being exceeded (a 50-year mean recurrence interval). The maps were generated from data through the winter of 1991-92, and from data through the winter of 1993-94 where the snows were heavy. The mapped snow loads are not increased much from a single snowy winter, since most reporting stations have more than 20 years of snow data. The map values indicate the ground snow load in pounds per square foot. In mountainous areas, the map indicates the highest elevation that is appropriate for the use of the associated snow load. For elevations higher than indicated on the map, a site-specific case study is required to determine the appropriate snow load. Some of the mountainous areas in the western United States require site-specific case studies due to high elevations. Assistance in the determination of an appropriate ground snow load for these areas may be obtained from the U.S. Department of Army Cold Regions Research and Engineering Laboratory in Hanover, New Hampshire.

TABLE 1608.2. See below.

❖ The purpose of this table is to provide the ground snow loads, P_g, for Alaskan locations. These values are needed to determine the appropriate roof snow loads for the indicated locations. The roof snow load is to be determined according to Section 1608.3 for flat roofs and Section 1608.4 for sloped roofs.

FIGURE 1608.2. See page 16-28.

❖ See the commentary to Section 1608.2 for an overview of the snow load map. See Section 7.2 of the commentary of ASCE 7 for a complete description of methodology used in developing the contour lines shown on the figure.

1608.3 Flat roof snow loads. The flat roof snow load, p_f, on a roof with a slope equal to or less than 5 degrees (0.09 rad) (1 inch per foot = 4.76 degrees) shall be calculated in accordance with Section 7.3 of ASCE 7.

❖ The purpose of this section is to specify the appropriate section of ASCE 7 for the determination of snow loads on roofs where the slope does not exceed 5 degrees (0.09 rad) [approximately 1 inch per foot (4.76 degrees)]. See Section 7.3 of ASCE 7 for the design equations and details.

TABLE 1608.2
GROUND SNOW LOADS, p_g, FOR ALASKAN LOCATIONS

LOCATION	POUNDS PER SQUARE FOOT	LOCATION	POUNDS PER SQUARE FOOT	LOCATION	POUNDS PER SQUARE FOOT
Adak	30	Galena	60	Petersburg	150
Anchorage	50	Gulkana	70	St. Paul Islands	40
Angoon	70	Homer	40	Seward	50
Barrow	25	Juneau	60	Shemya	25
Barter Island	35	Kenai	70	Sitka	50
Bethel	40	Kodiak	30	Talkeetna	120
Big Delta	50	Kotzebue	60	Unalakleet	50
Cold Bay	25	McGrath	70	Valdez	160
Cordova	100	Nenana	80	Whittier	300
Fairbanks	60	Nome	70	Wrangell	60
Fort Yukon	60	Palmer	50	Yakutat	150

For SI: 1 pound per square foot = 0.0479 kN/m^2.

1608.3.1 Exposure factor. The value for the snow exposure factor, C_e, used in the calculation of p_f shall be determined from Table 1608.3.1.

❖ The snow load exposure factor, C_e, is one of the parameters included in the determination of snow loads. This factor is one of the basic snow load design parameters required to be on the construction documents (project drawings or specifications) by Section 1603.1.3. The roof snow load is proportional to the value of the roof exposure factor. The roof exposure factor varies with the exposure category of the site, as described in the notes to Table 1608.3.1. Thus, the roof snow load for a site located in a wooded area is higher than an adjacent site that is flat and open.

TABLE 1608.3.1. See page 16-30.

❖ The snow exposure factor varies with the wind exposure category as indicated in the table. See the commentary to Section 1608.3.1 for further description of the purpose and use of this table.

1608.3.2 Thermal factor. The value for the thermal factor, C_t, used in the calculation of p_f shall be determined from Table 1608.3.2.

❖ The thermal factor is one of the parameters included in the determination of roof snow loads. This factor is one of the basic snow load parameters that is required to be on the construction documents (design drawings or specifications) by Section 1603.1.3. The thermal factor takes into account the heat that is transmitted from the interior of the structure that reduces the snow depth on the roof.

TABLE 1608.3.2. See page 16-30.

❖ See the commentary to Section 1608.3.2 for an overview of the thermal factor, C_t. The table entry for continuously heated greenhouses accounts for the reduced snow depths on greenhouses as a result of the heat transmission of a glass roof.

1608.3.3 Snow load importance factor. The value for the snow load importance factor, I_s, used in the calculation of p_f shall be determined in accordance with Table 1604.5. Greenhouses that are occupied for growing plants on production or research basis, without public access, shall be included in Importance Category I.

❖ The snow load importance factor is one of the basic roof snow load parameters and thus is required by Section 1603.1.3 to be on the construction documents (design drawings or specifications). The snow load importance factor takes into account the relative importance of structure from a human safety standpoint.

1608.3.4 Rain-on-snow surcharge load. Roofs with a s less than $^1/_2$ inch per foot (2.38 degrees) shall be designed

rain-on-snow surcharge load determined i
Section 7.10 of ASCE 7.

❖ The rain-on-snow surcharge load
creased weight of snow after rain.
details are addressed in Section
load is described as a surcharge
tion to the roof snow load. The r
load only applies to roofs where
inch per foot (2.38 degrees) (1
remain in snow much longer

1608.3.5 Ponding instability. Fo
inch per foot (1.19 degrees), th
clude verification of the preven
cordance with Section 7.11 of

❖ Ponding is the retention
flection. Where roof slo
of snow could cause a
the roof drains from
that the roof framinc
slope to the roof dra
snow load to preve

1608.4 Sloped roof s
with a slope greater
4.76 degrees) shall
of ASCE 7.

❖ The purpose
criteria for sl
the design e
Generally,
load for a
the effect

1608.5 Pa
snow loac
accordan

❖ Part
on
ev
ex
fr

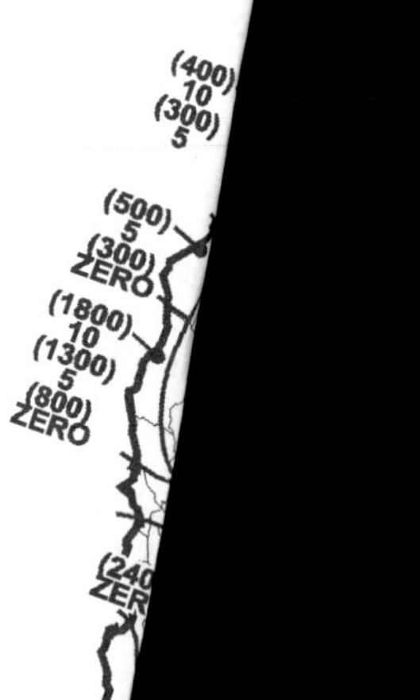

FIGURE 1608.2–continued
GROUND SNOW LOADS, p_g, FOR THE UNITED STATES (psf)

TABLE 1608.3.1
SNOW EXPOSURE FACTOR, C_e

TERRAIN CATEGORY[a]	EXPOSURE OF ROOF[a,b]		
	Fully exposed[c]	Partially exposed	Sheltered
A (see Section 1609.4)	N/A	1.1	1.3
B (see Section 1609.4)	0.9	1.0	1.2
C (see Section 1609.4)	0.9	1.0	1.1
D (see Section 1609.4)	0.8	0.9	1.0
Above the treeline in windswept mountainous areas	0.7	0.8	N/A
In Alaska, in areas where trees do not exist within a 2-mile radius of the site	0.7	0.8	N/A

For SI: 1 mile = 1609 m.

a. The terrain category and roof exposure condition chosen shall be representative of the anticipated conditions during the life of the structure. An exposure factor shall be determined for each roof of a structure.

b. Definitions of roof exposure are as follows:
1. Fully exposed shall mean roofs exposed on all sides with no shelter afforded by terrain, higher structures or trees. Roofs that contain several large pieces of mechanical equipment, parapets which extend above the height of the balanced snow load, h_b, or other obstructions are not in this category.
2. Partially exposed shall include all roofs except those designated as "fully exposed" or "sheltered."
3. Sheltered roofs shall mean those roofs located tight in among conifers that qualify as "obstructions."

c. Obstructions within a distance of 10 h_o provide "shelter," where h_o is the height of the obstruction above the roof level. If the only obstructions are a few deciduous trees that are leafless in winter, the "fully exposed" category shall be used except for terrain category "A." Note that these are heights above the roof. Heights used to establish the terrain category in Section 1609.4 are heights above the ground.

TABLE 1608.3.2
THERMAL FACTOR, C_t

THERMAL CONDITION[a]	C_t
All structures except as indicated below	1.0
Structures kept just above freezing and others with cold, ventilated roofs in which the thermal resistance (*R*-value) between the ventilated space and the heated space exceeds 25h · ft² · °F/Btu	1.1
Unheated structures	1.2
Continuously heated greenhouses[b] with a roof having a thermal resistance (*R*-value) less than 2.0h · ft² ·°F/Btu	0.85

For SI: 1 h · ft² · °F/Btu = 0.176 m² · K/W.

a. The thermal condition shall be representative of the anticipated conditions during winters for the life of the structure.

b. A continuously heated greenhouse shall mean a greenhouse with a constantly maintained interior temperature of 50°F or more during winter months. Such greenhouse shall also have a maintenance attendant on duty at all times or a temperature alarm system to provide warning in the event of a heating system failure.

1608.7 Drifts on lower roofs. In areas where the ground snow load, p_g, as determined by Section 1608.2, is equal to or greater than 5 psf (0.240 kN/m²), roofs shall be designed to sustain localized loads from snow drifts in accordance with Section 7.7 of ASCE 7.

❖ Drifts onto lower roofs are a common cause of roof failures after a heavy snow. This section of the code specifies the criteria for the design for drifts; Section 7.7 of ASCE 7 addresses windward and leeward drifts. A windward drift is one that is formed on the windward side of a high bay wall area. A leeward drift is formed on the leeward wall of a high bay wall area. ASCE 7 contains the equations and design details that are required to be used for this loading.

1608.8 Roof projections. Drift loads due to mechanical equipment, penthouses, parapets and other projections above the roof shall be determined in accordance with Section 7.8 of ASCE 7.

❖ Roof projections of the type described in the code result in roof snow drifts after a heavy snow. This section of the code requires both a windward and leeward drift load to be included in the design of the roof. The drift loads are in addition to the snow load on the roof (see Section 7.8 of ASCE 7 for the detailed design requirements).

1608.9 Sliding snow. The extra load caused by snow sliding off a sloped roof onto a lower roof shall be determined in accordance with Section 7.9 of ASCE 7.

❖ The purpose of this section is to include the effect of snow from a sloped roof sliding onto an adjacent lower roof. The snow that slides from a higher sloped roof is in addition to the snow load already on the lower roof (see Section 7.9 of ASCE 7 for detailed requirements).

SECTION 1609
WIND LOADS

FIGURE 1609. See page 16-32.

❖ Figure 1609 is a duplication of Figure 6-1 of ASCE 7 and provides basic wind speeds based on 3-second-gusts at 33 feet (10 058 mm) above ground for Exposure Category C (see Section 1609.4). The reference to the 50-year mean recurrence interval (MRI), used in previous maps, has been removed, reflecting the fact that the MRI is greater than 50 years along the hurricane coastline. Nonhurricane wind speeds are still based on a 50-year MRI.

Earlier maps (earlier editions of ASCE 7 and the model codes preceding the code) incorporated fastest-mile wind speed maps. "Fastest mile" is defined as the average speed of one mile of air that passes a specific reference point. It is important to recognize the differences in averaging times between fastest-mile and the new 3-second gust maps. The averaging time for a 90-mph fastest-mile wind speed is (t = 3,600/V) 40 seconds. Obviously, due to greater averaging time, for a given location, the fastest-mile wind speed will be less than the 3-second-gust wind speed. The fact that the wind speed values are higher does not necessarily indicate higher wind loads. Buildings and structures resist wind loads, not wind speeds. Wind speed, although a significant contributor, is only one of several variables and factors that affect wind forces. Wind loads are affected by atmosphere and aerodynamics. Other elements that affect actual wind forces as wind flows across a bluff body include shape factors (C_p), gust effect factors (G) and the velocity pressure that is a function of wind speed, exposure and topography, among others.

The change from the fastest-mile wind speed map to a 3-second-gust map was necessary for the following reasons. First, weather stations across the United States no longer collect fastest-mile wind speed data. Additionally, the perception of the general public will be more favorable where the code wind speeds are higher, although the design wind pressures were not changed significantly. The map includes a more complete analysis of hurricane wind speeds than previous maps, since more data was available for sites away from the coast. In western states, the 85- to 90-mph contour boundary follows along the Washington, Oregon and California eastern state lines. This is because inland wind data was such that there was no statistical basis to place them elsewhere. The design wind speeds do not include the effects of tornadoes. The probability of a tornadic event at any particular location is so low that tornadoes do not appear in the 50-year statistical storm data used to formulate the inland portion of the map. See the commentary to Section 6.5.4 in ASCE 7 for more information.

1609.1 Applications. Buildings, structures and parts thereof shall be designed to withstand the minimum wind loads prescribed herein. Decreases in wind loads shall not be made for the effect of shielding by other structures.

❖ The intent of this section is to provide minimum criteria for the design and construction of buildings and other structures to resist wind loads. These regulations serve to reduce the potential for damage to property caused by windstorms and to provide an acceptable level of protection to building occupants. The objective also includes the prevention of damage to adjacent properties because of the possible detachment of major building components (e.g., walls, roofs, etc.), structural collapse or flying debris and for the safety of people in the immediate vicinity.

The requirements for wind design given in this section of the code generally reflect the wind load provisions of ASCE 7. For a better understanding of the wind load provisions of this section, it is important to have a fundamental knowledge of the effects of high-velocity wind forces on buildings and other structures. Wind/structure interactions can be characterized as follows:

When wind encounters a stationary object such as a building, the airflow changes direction and produces several effects on the building. Exterior walls and other vertical surfaces facing the wind (windward side) and perpendicular to its path are subjected to inward (positive) pressures; however, wind does not stop on contact with a facing surface, but flows around and over the building. Since wind cannot negotiate sharp corners, such as corners of walls or eaves, or over ridges and roof corners, the airflow separates from the downwind surfaces due to high turbulence and localized pressures, which results in outward (negative) pressures. This phenomena produces suction or outward pressures (negative) on the sidewalls, leeward wall and, depending on geometry, the roof. The basic external wind-flow effects are illustrated in Figure 1609.1.

Figure 1609.1 shows a flat roof and the resulting negative pressures caused by external wind; however, pressures may differ on sloping roofs in the direction perpendicular to the ridge. Roof surfaces (on the windward side) with shallow slopes are generally subjected to outward (negative) pressures—the same as flat roofs. Moderately sloping roofs [about 30 degrees (0.5235 rad)] may be subjected to either inward (positive) pressures, outward (negative) pressures or both—negative pressure in the lower part of the roof and positive pressure in the area of the ridge; however, the code does not require consideration of this scenario. High-sloping roofs (windward side) respond similar to walls and sustain positive wind pressures. Sloped roof sections on the leeward side are subjected to negative pressures, regardless of the degree of slope. For sloped roofs where the wind direction is parallel to the ridge, the wind pressures act similarly to a flat roof where the roof is subject to outward (negative) pressures.

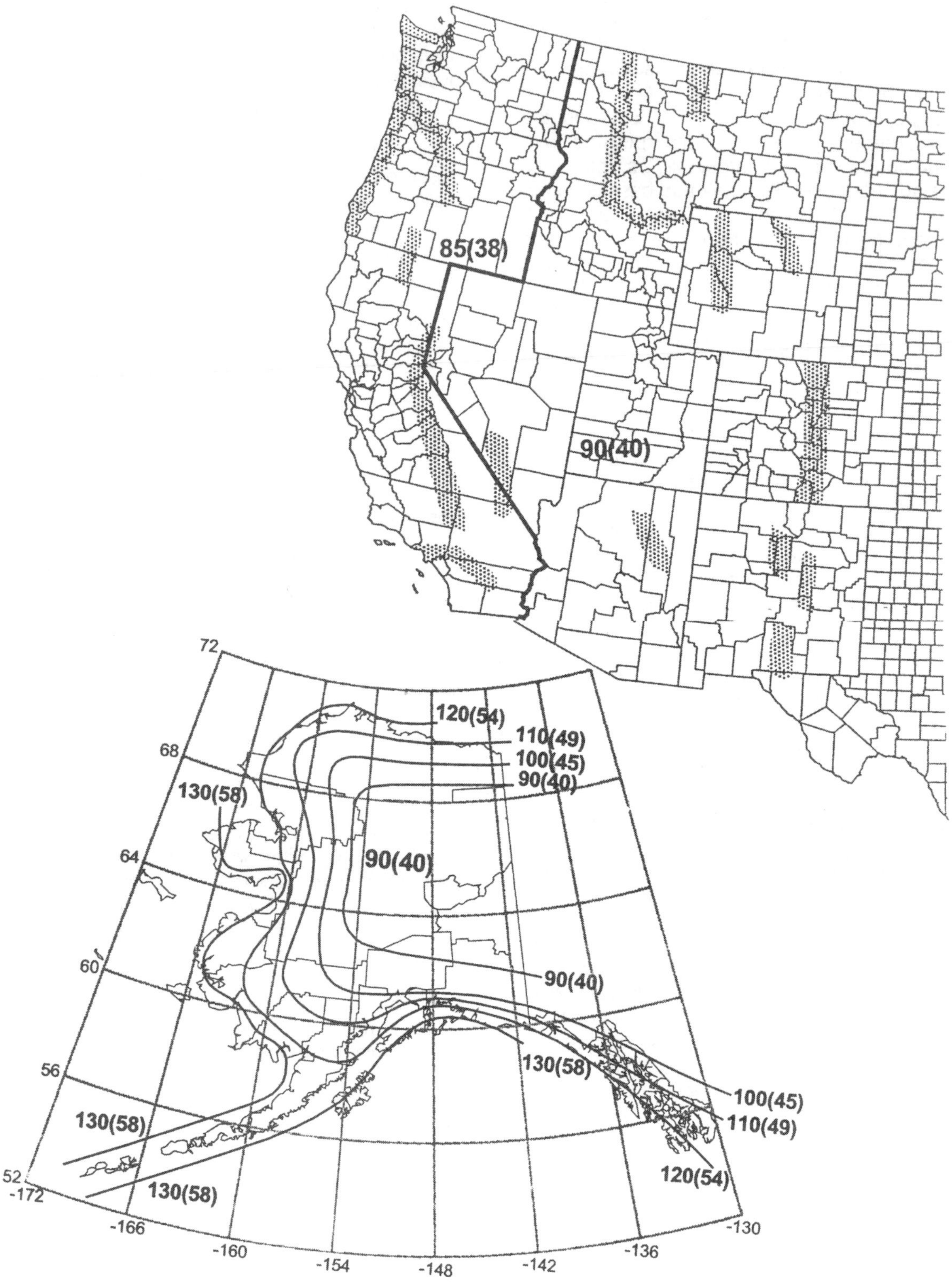

FIGURE 1609
BASIC WIND SPEED (3-SECOND GUST)

Pendant-operated bridge cranes (powered) . . . . 10 percent

Bridge cranes or monorail cranes with hand-geared bridge, trolley and hoist . . . . . . 0 percent

❖ A vertical impact force is necessary to account for the impact from the starting and stopping movement of the suspended weight from the crane. Vertical impact is also created by the movement of the crane along the rails.

1607.12.3 Lateral force. The lateral force on crane runway beams with electrically powered trolleys shall be calculated as 20 percent of the sum of the rated capacity of the crane and the weight of the hoist and trolley. The lateral force shall be assumed to act horizontally at the traction surface of a runway beam, in either direction perpendicular to the beam, and shall be distributed according to the lateral stiffness of the runway beam and supporting structure.

❖ This section is necessary to define the design lateral force on the crane supports. Lateral force at the right angle to the crane rail is caused by the lateral movement of the lifted load and from the frame action of the crane.

1607.12.4 Longitudinal force. The longitudinal force on crane runway beams, except for bridge cranes with hand-geared bridges, shall be calculated as 10 percent of the maximum wheel loads of the crane. The longitudinal force shall be assumed to act horizontally at the traction surface of a runway beam, in either direction parallel to the beam.

❖ This section is needed to define the longitudinal force on the crane supports, which is caused from the longitudinal motion of the crane with the lifted load.

1607.13 Interior walls and partitions. Interior walls and partitions that exceed 6 feet (1829 mm) in height, including their finish materials, shall have adequate strength to resist the loads to which they are subjected but not less than a horizontal load of 5 psf (0.240 kN/m^2).

❖ The requirements of this section are intended to provide sufficient strength and durability of the wall framing and of the finished construction to provide a minimum level of resistance to nominal impact loads that commonly occur in the use of a facility and also to resist HVAC pressurization.

SECTION 1608
SNOW LOADS

1608.1 General. Design snow loads shall be determined in accordance with Section 7 of ASCE 7, but the design roof load shall not be less than that determined by Section 1607.

❖ The intent of this section is that the code requirements are the same as the technical requirements in Section 7 of ASCE 7. The snow load provisions in ASCE 7 are based on over 40 years of ground snow load data, and include requirements for thermal resistance of the roof structure, a rain-on-snow surcharge, partial loading on continuous beam systems and ponding instability from melting snow or rain on snow. The text of the snow load section contains many of the variables that must be determined for the calculation of snow loads: ground snow load, exposure factor, importance factor and thermal factor. The maps and tables necessary for determining each of these is in the code, consistent with the information provided in ASCE 7 so that the building official is able to verify these fundamental criteria as well as de-

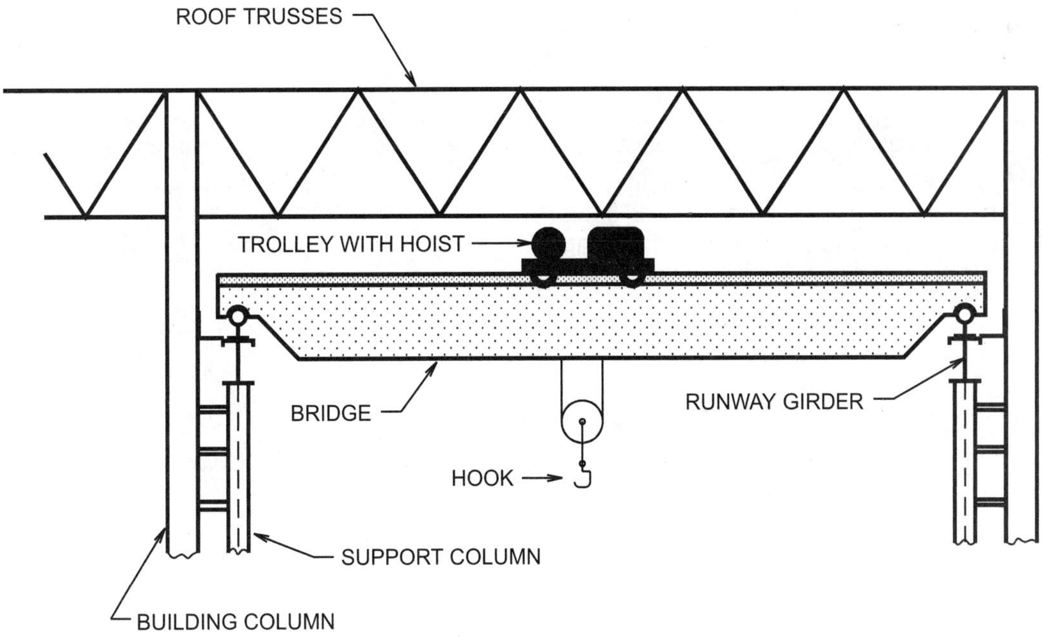

Figure 1607.12
TOP-RUNNING BRIDGE CRANE

termine the conditions where other critical factors (such as rain-on-snow surcharge) must be included in the design. Section 1608 allows design professionals to fully utilize the provisions of the leading national consensus standard for the calculation of snow loads, while simultaneously providing building officials with key information in the code to verify the appropriate snow load criteria.

1608.2 Ground snow loads. The ground snow loads to be used in determining the design snow loads for roofs are given in Figure 1608.2 for the contiguous United States and Table 1608.2 for Alaska. Site-specific case studies shall be made in areas designated CS in Figure 1608.2. Ground snow loads for sites at elevations above the limits indicated in Figure 1608.2 and for all sites within the CS areas shall be approved. Ground snow load determination for such sites shall be based on an extreme value statistical analysis of data available in the vicinity of the site using a value with a 2-percent annual probability of being exceeded (50-year mean recurrence interval). Snow loads are zero for Hawaii, except in mountainous regions as approved by the building official.

❖ The ground snow loads on the maps in Figure 1608.2 of the code are generally based on over 40 years of snow depth records. The snow loads on the maps are those that have a 2-percent annual probability of being exceeded (a 50-year mean recurrence interval). The maps were generated from data through the winter of 1991-92, and from data through the winter of 1993-94 where the snows were heavy. The mapped snow loads are not increased much from a single snowy winter, since most reporting stations have more than 20 years of snow data. The map values indicate the ground snow load in pounds per square foot. In mountainous areas, the map indicates the highest elevation that is appropriate for the use of the associated snow load. For elevations higher than indicated on the map, a site-specific case study is required to determine the appropriate snow load. Some of the mountainous areas in the western United States require site-specific case studies due to high elevations. Assistance in the determination of an appropriate ground snow load for these areas may be obtained from the U.S. Department of Army Cold Regions Research and Engineering Laboratory in Hanover, New Hampshire.

TABLE 1608.2. See below.

❖ The purpose of this table is to provide the ground snow loads, P_g, for Alaskan locations. These values are needed to determine the appropriate roof snow loads for the indicated locations. The roof snow load is to be determined according to Section 1608.3 for flat roofs and Section 1608.4 for sloped roofs.

FIGURE 1608.2. See page 16-28.

❖ See the commentary to Section 1608.2 for an overview of the snow load map. See Section 7.2 of the commentary of ASCE 7 for a complete description of methodology used in developing the contour lines shown on the figure.

1608.3 Flat roof snow loads. The flat roof snow load, p_f, on a roof with a slope equal to or less than 5 degrees (0.09 rad) (1 inch per foot = 4.76 degrees) shall be calculated in accordance with Section 7.3 of ASCE 7.

❖ The purpose of this section is to specify the appropriate section of ASCE 7 for the determination of snow loads on roofs where the slope does not exceed 5 degrees (0.09 rad) [approximately 1 inch per foot (4.76 degrees)]. See Section 7.3 of ASCE 7 for the design equations and details.

TABLE 1608.2
GROUND SNOW LOADS, p_g, FOR ALASKAN LOCATIONS

LOCATION	POUNDS PER SQUARE FOOT	LOCATION	POUNDS PER SQUARE FOOT	LOCATION	POUNDS PER SQUARE FOOT
Adak	30	Galena	60	Petersburg	150
Anchorage	50	Gulkana	70	St. Paul Islands	40
Angoon	70	Homer	40	Seward	50
Barrow	25	Juneau	60	Shemya	25
Barter Island	35	Kenai	70	Sitka	50
Bethel	40	Kodiak	30	Talkeetna	120
Big Delta	50	Kotzebue	60	Unalakleet	50
Cold Bay	25	McGrath	70	Valdez	160
Cordova	100	Nenana	80	Whittier	300
Fairbanks	60	Nome	70	Wrangell	60
Fort Yukon	60	Palmer	50	Yakutat	150

For SI: 1 pound per square foot = 0.0479 kN/m^2.

1608.3.1 Exposure factor. The value for the snow exposure factor, C_e, used in the calculation of p_f shall be determined from Table 1608.3.1.

❖ The snow load exposure factor, C_e, is one of the parameters included in the determination of snow loads. This factor is one of the basic snow load design parameters required to be on the construction documents (project drawings or specifications) by Section 1603.1.3. The roof snow load is proportional to the value of the roof exposure factor. The roof exposure factor varies with the exposure category of the site, as described in the notes to Table 1608.3.1. Thus, the roof snow load for a site located in a wooded area is higher than an adjacent site that is flat and open.

TABLE 1608.3.1. See page 16-30.

❖ The snow exposure factor varies with the wind exposure category as indicated in the table. See the commentary to Section 1608.3.1 for further description of the purpose and use of this table.

1608.3.2 Thermal factor. The value for the thermal factor, C_t, used in the calculation of p_f shall be determined from Table 1608.3.2.

❖ The thermal factor is one of the parameters included in the determination of roof snow loads. This factor is one of the basic snow load parameters that is required to be on the construction documents (design drawings or specifications) by Section 1603.1.3. The thermal factor takes into account the heat that is transmitted from the interior of the structure that reduces the snow depth on the roof.

TABLE 1608.3.2. See page 16-30.

❖ See the commentary to Section 1608.3.2 for an overview of the thermal factor, C_t. The table entry for continuously heated greenhouses accounts for the reduced snow depths on greenhouses as a result of the heat transmission of a glass roof.

1608.3.3 Snow load importance factor. The value for the snow load importance factor, I_s, used in the calculation of p_f shall be determined in accordance with Table 1604.5. Greenhouses that are occupied for growing plants on production or research basis, without public access, shall be included in Importance Category I.

❖ The snow load importance factor is one of the basic roof snow load parameters and thus is required by Section 1603.1.3 to be on the construction documents (design drawings or specifications). The snow load importance factor takes into account the relative importance of a structure from a human safety standpoint.

1608.3.4 Rain-on-snow surcharge load. Roofs with a slope less than $^1/_2$ inch per foot (2.38 degrees) shall be designed for a rain-on-snow surcharge load determined in accordance with Section 7.10 of ASCE 7.

❖ The rain-on-snow surcharge load accounts for the increased weight of snow after rain. The design method details are addressed in Section 7.10 of ASCE 7. The load is described as a surcharge load since it is in addition to the roof snow load. The rain-on-snow surcharge load only applies to roofs where the slope is less than $^1/_2$ inch per foot (2.38 degrees) (1:24), since water tends to remain in snow much longer on relatively flat roofs.

1608.3.5 Ponding instability. For roofs with a slope less than $^1/_4$ inch per foot (1.19 degrees), the design calculations shall include verification of the prevention of ponding instability in accordance with Section 7.11 of ASCE 7.

❖ Ponding is the retention of water as a result of roof deflection. Where roof slopes are relatively flat, the weight of snow could cause a complete loss of positive slope to the roof drains from deflection. This section requires that the roof framing stiffness be such that a positive slope to the roof drain will be maintained under the roof snow load to prevent ponding instability.

1608.4 Sloped roof snow loads. The snow load, p_s, on a roof with a slope greater than 5 degrees (0.09 rad) (1 inch per foot = 4.76 degrees) shall be calculated in accordance with Section 7.4 of ASCE 7.

❖ The purpose of this section is to specify the snow loads criteria for sloped roofs. See Section 7.4 of ASCE 7 for the design equations and detailed design requirements. Generally, the snow load for sloped roofs is the snow load for a flat roof multiplied by a factor that accounts for the effect of the roof slope.

1608.5 Partial loading. The effect of not having the balanced snow load over the entire loaded roof area shall be analyzed in accordance with Section 7.5 of ASCE 7.

❖ Partial loading is the requirement to vary the snow load on the roof such that the roof is designed for the likely event of unequal depth of snow on adjacent spans. For example, Section 7.4 of ASCE 7 requires that three different cases of snow load arrangement be included in the design for continuous beam systems.

1608.6 Unbalanced snow loads. Unbalanced roof snow loads shall be determined in accordance with Section 7.6 of ASCE 7. Winds from all directions shall be accounted for when establishing unbalanced snow loads.

❖ This section accounts for the unbalanced snow loads that result from wind blowing part of the snow off the windward side of a roof. The snow is increased on one side of the roof and decreased on the other. This loading condition is sometimes the controlling load for the design of roof members for hip, gable, folded plate, sawtooth, barrel vault and dome roofs (see Section 7.6 of ASCE 7 for the detailed design requirements).

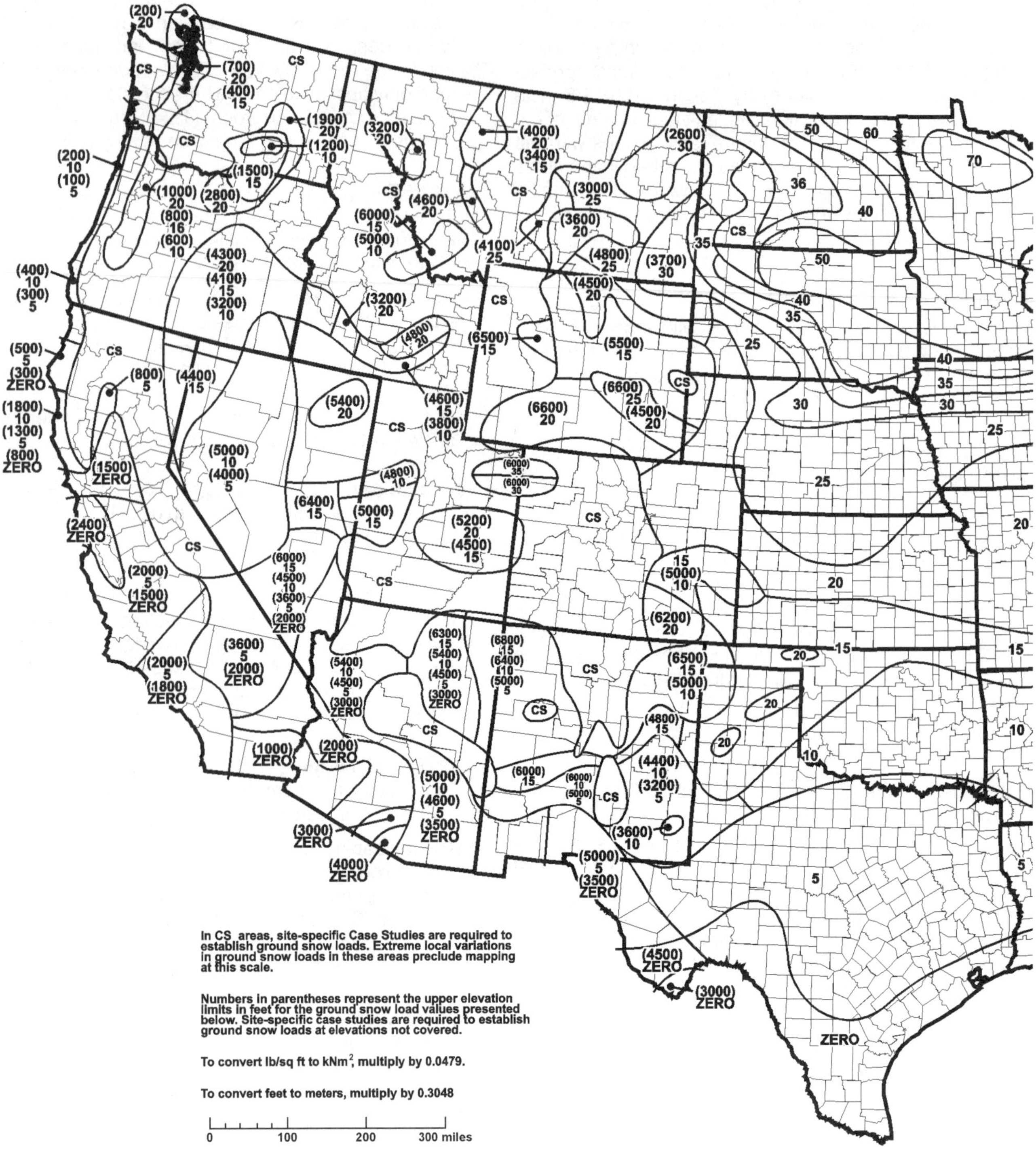
In CS areas, site-specific Case Studies are required to establish ground snow loads. Extreme local variations in ground snow loads in these areas preclude mapping at this scale.
Numbers in parentheses represent the upper elevation limits in feet for the ground snow load values presented below. Site-specific case studies are required to establish ground snow loads at elevations not covered.
To convert lb/sq ft to kNm², multiply by 0.0479.
To convert feet to meters, multiply by 0.3048
0 100 200 300 miles

FIGURE 1608.2
GROUND SNOW LOADS, p_g, FOR THE UNITED STATES (psf)

FIGURE 1608.2–continued
GROUND SNOW LOADS, p_g, FOR THE UNITED STATES (psf)

TABLE 1608.3.1
SNOW EXPOSURE FACTOR, C_e

TERRAIN CATEGORY[a]	EXPOSURE OF ROOF[a,b]		
	Fully exposed[c]	Partially exposed	Sheltered
A (see Section 1609.4)	N/A	1.1	1.3
B (see Section 1609.4)	0.9	1.0	1.2
C (see Section 1609.4)	0.9	1.0	1.1
D (see Section 1609.4)	0.8	0.9	1.0
Above the treeline in windswept mountainous areas	0.7	0.8	N/A
In Alaska, in areas where trees do not exist within a 2-mile radius of the site	0.7	0.8	N/A

For SI: 1 mile = 1609 m.

a. The terrain category and roof exposure condition chosen shall be representative of the anticipated conditions during the life of the structure. An exposure factor shall be determined for each roof of a structure.

b. Definitions of roof exposure are as follows:
1. Fully exposed shall mean roofs exposed on all sides with no shelter afforded by terrain, higher structures or trees. Roofs that contain several large pieces of mechanical equipment, parapets which extend above the height of the balanced snow load, h_b, or other obstructions are not in this category.
2. Partially exposed shall include all roofs except those designated as "fully exposed" or "sheltered."
3. Sheltered roofs shall mean those roofs located tight in among conifers that qualify as "obstructions."

c. Obstructions within a distance of 10 h_o provide "shelter," where h_o is the height of the obstruction above the roof level. If the only obstructions are a few deciduous trees that are leafless in winter, the "fully exposed" category shall be used except for terrain category "A." Note that these are heights above the roof. Heights used to establish the terrain category in Section 1609.4 are heights above the ground.

TABLE 1608.3.2
THERMAL FACTOR, C_t

THERMAL CONDITION[a]	C_t
All structures except as indicated below	1.0
Structures kept just above freezing and others with cold, ventilated roofs in which the thermal resistance (*R*-value) between the ventilated space and the heated space exceeds 25h · ft^2 · °F/Btu	1.1
Unheated structures	1.2
Continuously heated greenhouses[b] with a roof having a thermal resistance (*R*-value) less than 2.0h · ft^2 ·°F/Btu	0.85

For SI: 1 h · ft^2 · °F/Btu = 0.176 m^2 · K/W.

a. The thermal condition shall be representative of the anticipated conditions during winters for the life of the structure.

b. A continuously heated greenhouse shall mean a greenhouse with a constantly maintained interior temperature of 50°F or more during winter months. Such greenhouse shall also have a maintenance attendant on duty at all times or a temperature alarm system to provide warning in the event of a heating system failure.

1608.7 Drifts on lower roofs. In areas where the ground snow load, p_g, as determined by Section 1608.2, is equal to or greater than 5 psf (0.240 kN/m^2), roofs shall be designed to sustain localized loads from snow drifts in accordance with Section 7.7 of ASCE 7.

❖ Drifts onto lower roofs are a common cause of roof failures after a heavy snow. This section of the code specifies the criteria for the design for drifts; Section 7.7 of ASCE 7 addresses windward and leeward drifts. A windward drift is one that is formed on the windward side of a high bay wall area. A leeward drift is formed on the leeward wall of a high bay wall area. ASCE 7 contains the equations and design details that are required to be used for this loading.

1608.8 Roof projections. Drift loads due to mechanical equipment, penthouses, parapets and other projections above the roof shall be determined in accordance with Section 7.8 of ASCE 7.

❖ Roof projections of the type described in the code result in roof snow drifts after a heavy snow. This section of the code requires both a windward and leeward drift load to be included in the design of the roof. The drift loads are in addition to the snow load on the roof (see Section 7.8 of ASCE 7 for the detailed design requirements).

1608.9 Sliding snow. The extra load caused by snow sliding off a sloped roof onto a lower roof shall be determined in accordance with Section 7.9 of ASCE 7.

❖ The purpose of this section is to include the effect of snow from a sloped roof sliding onto an adjacent lower roof. The snow that slides from a higher sloped roof is in addition to the snow load already on the lower roof (see Section 7.9 of ASCE 7 for detailed requirements).

SECTION 1609
WIND LOADS

FIGURE 1609. See page 16-32.

❖ Figure 1609 is a duplication of Figure 6-1 of ASCE 7 and provides basic wind speeds based on 3-second-gusts at 33 feet (10 058 mm) above ground for Exposure Category C (see Section 1609.4). The reference to the 50-year mean recurrence interval (MRI), used in previous maps, has been removed, reflecting the fact that the MRI is greater than 50 years along the hurricane coastline. Nonhurricane wind speeds are still based on a 50-year MRI.

Earlier maps (earlier editions of ASCE 7 and the model codes preceding the code) incorporated fastest-mile wind speed maps. "Fastest mile" is defined as the average speed of one mile of air that passes a specific reference point. It is important to recognize the differences in averaging times between fastest-mile and the new 3-second gust maps. The averaging time for a 90-mph fastest-mile wind speed is (t = 3,600/V) 40 seconds. Obviously, due to greater averaging time, for a given location, the fastest-mile wind speed will be less than the 3-second-gust wind speed. The fact that the wind speed values are higher does not necessarily indicate higher wind loads. Buildings and structures resist wind loads, not wind speeds. Wind speed, although a significant contributor, is only one of several variables and factors that affect wind forces. Wind loads are affected by atmosphere and aerodynamics. Other elements that affect actual wind forces as wind flows across a bluff body include shape factors (C_p), gust effect factors (G) and the velocity pressure that is a function of wind speed, exposure and topography, among others.

The change from the fastest-mile wind speed map to a 3-second-gust map was necessary for the following reasons. First, weather stations across the United States no longer collect fastest-mile wind speed data. Additionally, the perception of the general public will be more favorable where the code wind speeds are higher, although the design wind pressures were not changed significantly. The map includes a more complete analysis of hurricane wind speeds than previous maps, since more data was available for sites away from the coast. In western states, the 85- to 90-mph contour boundary follows along the Washington, Oregon and California eastern state lines. This is because inland wind data was such that there was no statistical basis to place them elsewhere. The design wind speeds do not include the effects of tornadoes. The probability of a tornadic event at any particular location is so low that tornadoes do not appear in the 50-year statistical storm data used to formulate the inland portion of the map. See the commentary to Section 6.5.4 in ASCE 7 for more information.

1609.1 Applications. Buildings, structures and parts thereof shall be designed to withstand the minimum wind loads prescribed herein. Decreases in wind loads shall not be made for the effect of shielding by other structures.

❖ The intent of this section is to provide minimum criteria for the design and construction of buildings and other structures to resist wind loads. These regulations serve to reduce the potential for damage to property caused by windstorms and to provide an acceptable level of protection to building occupants. The objective also includes the prevention of damage to adjacent properties because of the possible detachment of major building components (e.g., walls, roofs, etc.), structural collapse or flying debris and for the safety of people in the immediate vicinity.

The requirements for wind design given in this section of the code generally reflect the wind load provisions of ASCE 7. For a better understanding of the wind load provisions of this section, it is important to have a fundamental knowledge of the effects of high-velocity wind forces on buildings and other structures. Wind/structure interactions can be characterized as follows:

When wind encounters a stationary object such as a building, the airflow changes direction and produces several effects on the building. Exterior walls and other vertical surfaces facing the wind (windward side) and perpendicular to its path are subjected to inward (positive) pressures; however, wind does not stop on contact with a facing surface, but flows around and over the building. Since wind cannot negotiate sharp corners, such as corners of walls or eaves, or over ridges and roof corners, the airflow separates from the downwind surfaces due to high turbulence and localized pressures, which results in outward (negative) pressures. This phenomena produces suction or outward pressures (negative) on the sidewalls, leeward wall and, depending on geometry, the roof. The basic external wind-flow effects are illustrated in Figure 1609.1.

Figure 1609.1 shows a flat roof and the resulting negative pressures caused by external wind; however, pressures may differ on sloping roofs in the direction perpendicular to the ridge. Roof surfaces (on the windward side) with shallow slopes are generally subjected to outward (negative) pressures—the same as flat roofs. Moderately sloping roofs [about 30 degrees (0.5235 rad)] may be subjected to either inward (positive) pressures, outward (negative) pressures or both—negative pressure in the lower part of the roof and positive pressure in the area of the ridge; however, the code does not require consideration of this scenario. High-sloping roofs (windward side) respond similar to walls and sustain positive wind pressures. Sloped roof sections on the leeward side are subjected to negative pressures, regardless of the degree of slope. For sloped roofs where the wind direction is parallel to the ridge, the wind pressures act similarly to a flat roof where the roof is subject to outward (negative) pressures.

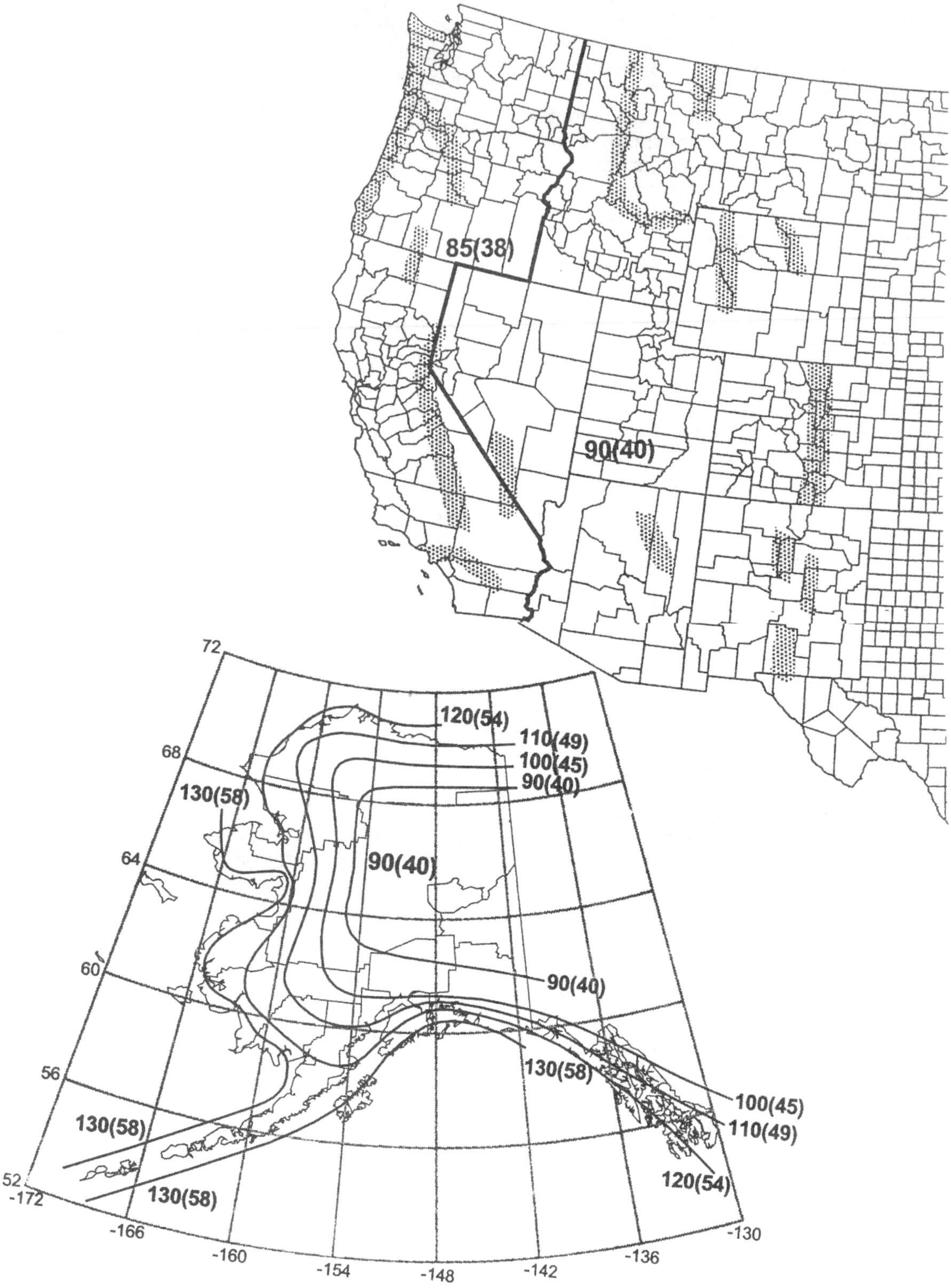

FIGURE 1609
BASIC WIND SPEED (3-SECOND GUST)

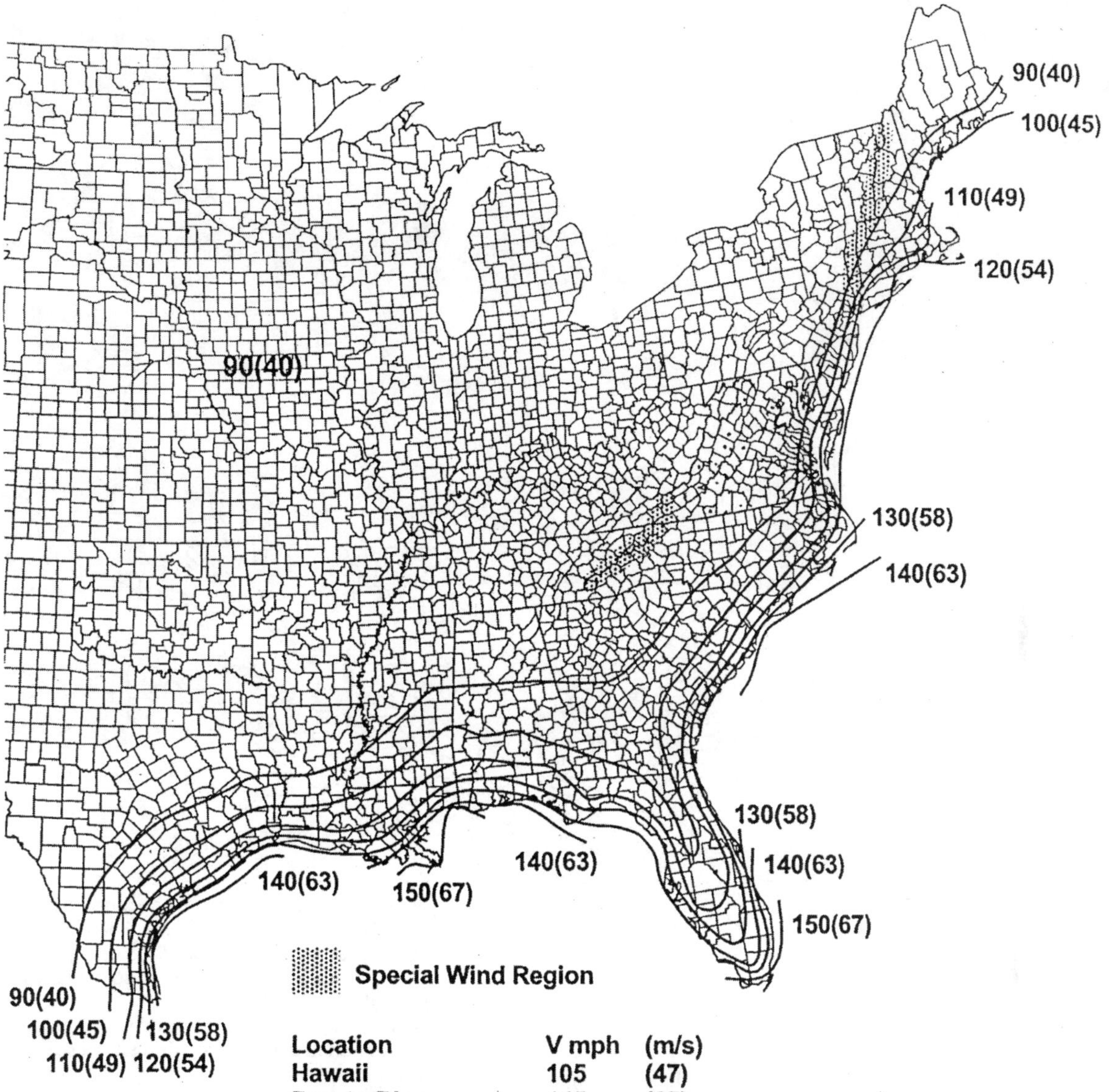

Location	V mph	(m/s)
Hawaii	105	(47)
Puerto Rico	145	(65)
Guam	170	(76)
Virgin Islands	145	(65)
American Samoa	125	(56)

FIGURE 1609—continued
BASIC WIND SPEED (3-SECOND GUST)

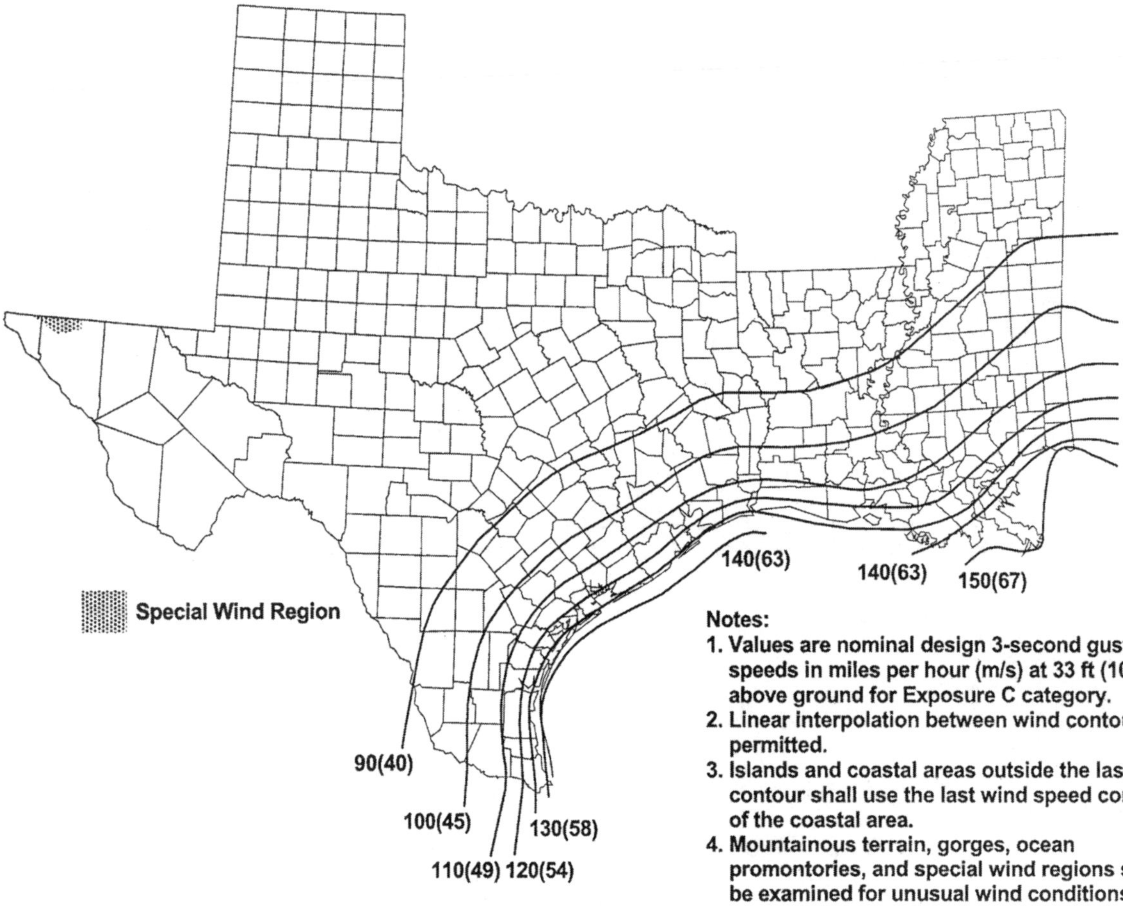

Notes:

1. Values are nominal design 3-second gust wind speeds in miles per hour (m/s) at 33 ft (10 m) above ground for Exposure C category.
2. Linear interpolation between wind contours is permitted.
3. Islands and coastal areas outside the last contour shall use the last wind speed contour of the coastal area.
4. Mountainous terrain, gorges, ocean promontories, and special wind regions shall be examined for unusual wind conditions.

FIGURE 1609–continued
BASIC WIND SPEED (3-SECOND GUST)
WESTERN GULF OF MEXICO HURRICANE COASTLINE

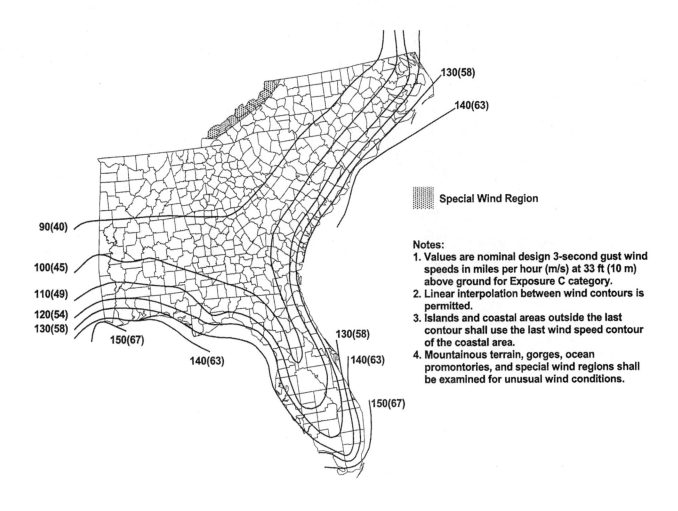

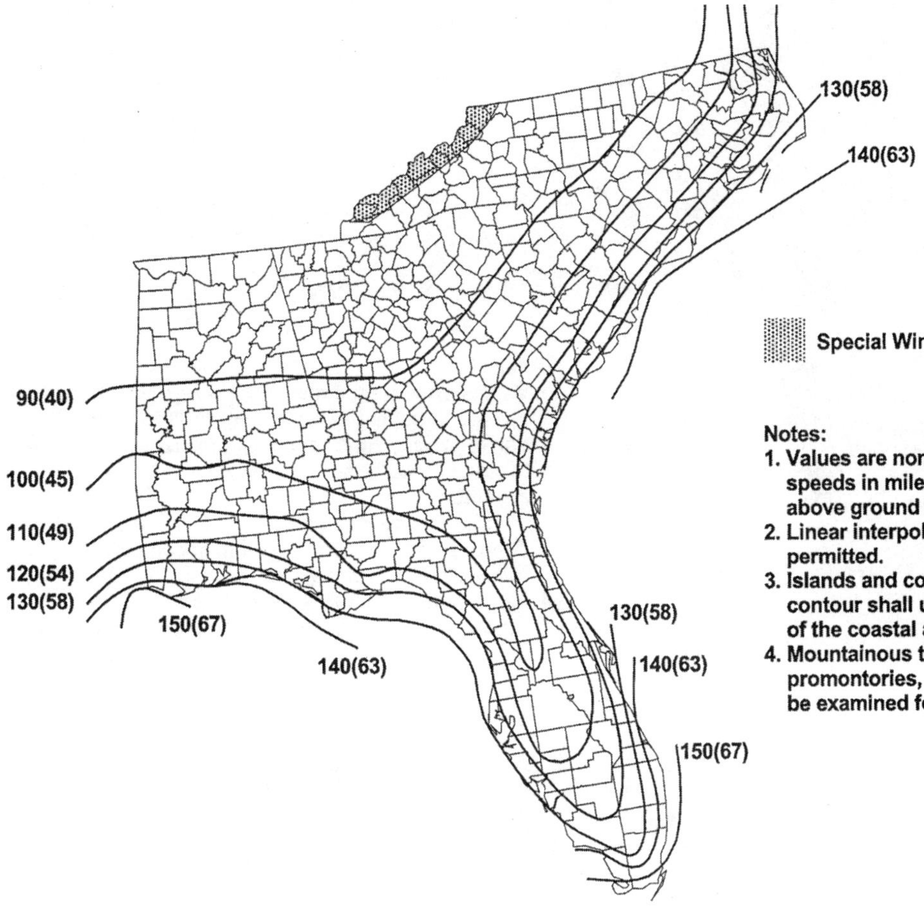

Special Wind Region

Notes:

1. Values are nominal design 3-second gust wind speeds in miles per hour (m/s) at 33 ft (10 m) above ground for Exposure C category.
2. Linear interpolation between wind contours is permitted.
3. Islands and coastal areas outside the last contour shall use the last wind speed contour of the coastal area.
4. Mountainous terrain, gorges, ocean promontories, and special wind regions shall be examined for unusual wind conditions.

FIGURE 1609–continued
BASIC WIND SPEED (3-SECOND GUST)
EASTERN GULF OF MEXICO AND SOUTHEASTERN U.S. HURRICANE COASTLINE

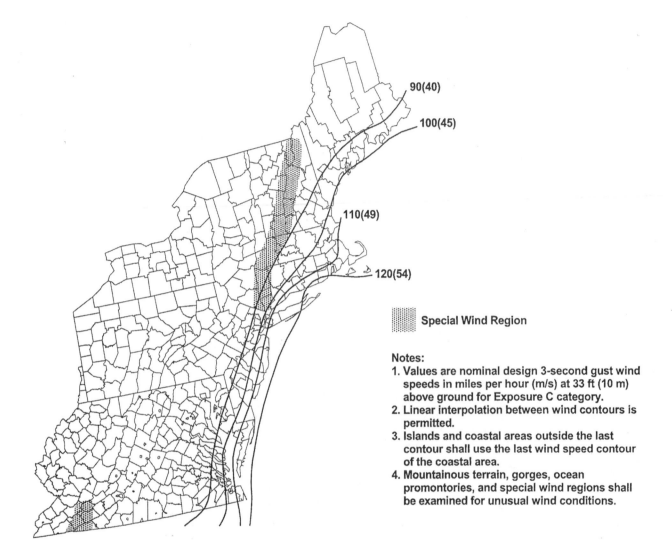

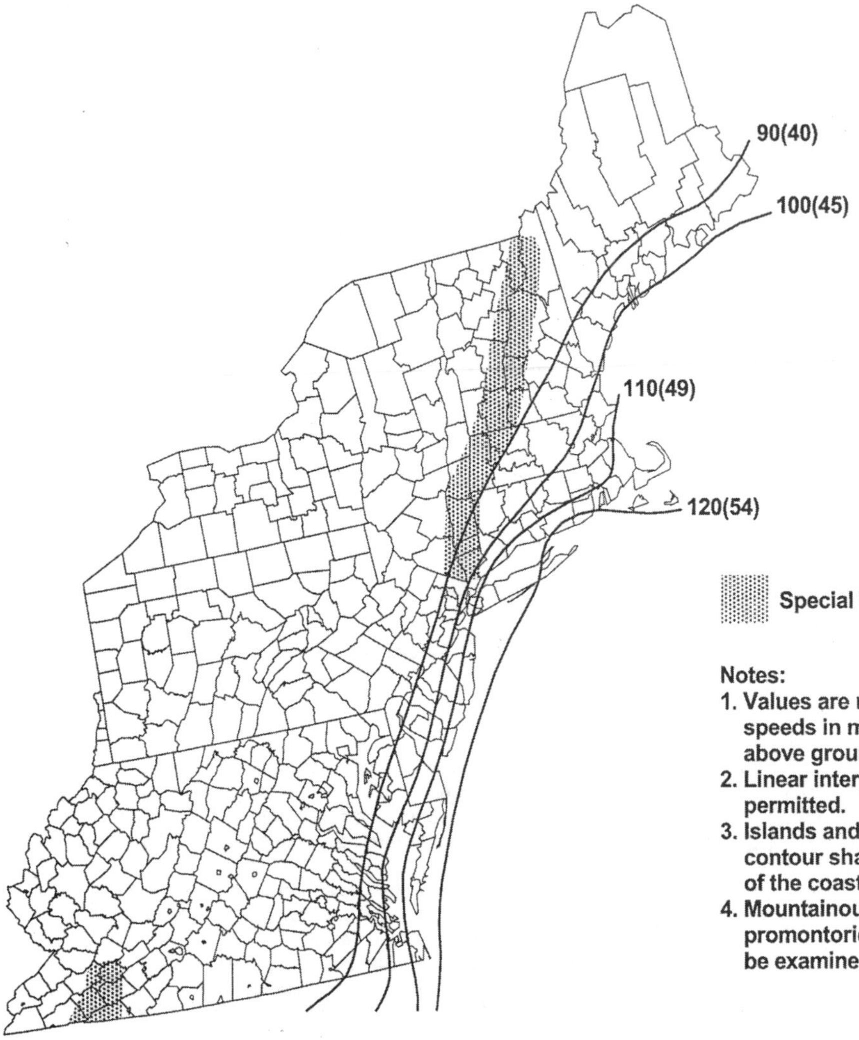

Special Wind Region

Notes:

1. Values are nominal design 3-second gust wind speeds in miles per hour (m/s) at 33 ft (10 m) above ground for Exposure C category.
2. Linear interpolation between wind contours is permitted.
3. Islands and coastal areas outside the last contour shall use the last wind speed contour of the coastal area.
4. Mountainous terrain, gorges, ocean promontories, and special wind regions shall be examined for unusual wind conditions.

FIGURE 1609–continued
BASIC WIND SPEED (3-SECOND GUST)
MID AND NORTHERN ATLANTIC HURRICANE COASTLINE

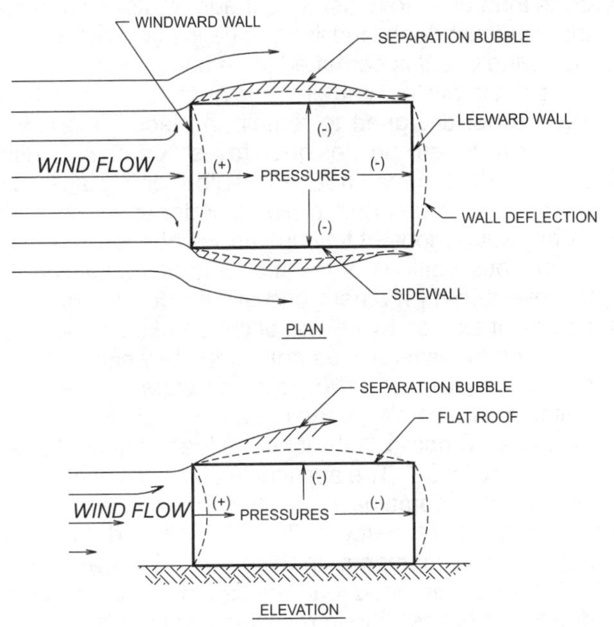

Figure 1609.1
WIND PRESSURES CAUSED BY EXTERNAL WIND FLOW

1609.1.1 Determination of wind loads. Wind loads on every building or structure shall be determined in accordance with Section 6 of ASCE 7. Wind shall be assumed to come from any horizontal direction and wind pressures shall be assumed to act normal to the surface considered.

Exceptions:

1. Wind loads determined by the provisions of Section 1609.6.
2. Subject to the limitations of Section 1609.1.1.1, the provisions of *SBCCI SSTD 10 Standard for Hurricane Resistant Residential Construction* shall be permitted for applicable Group R-2 and R-3 buildings.
3. Subject to the limitations of Section 1609.1.1.1, residential structures using the provisions of the *AF&PA Wood Frame Construction Manual for One- and Two-Family Dwellings*.
4. Designs using *NAAMM FP 1001 Guide Specification for Design of Metal Flagpoles*.
5. Designs using TIA/EIA-222 for antenna-supporting structures and antennas.

❖ The intent of Section 1609 is to require that buildings and structures be designed and constructed to resist the wind loads described in Section 6 of ASCE 7. The wind speeds in the ASCE 7 are also based on the 3-second-gust wind speed. In Section 6 of ASCE 7, there are three methods for determining wind loads on a building.

Method 1 is the simplified procedure and is the basis for Section 1609.6 of the code. Method 2 in ASCE 7 is the analytical procedure. Technically, there are two analytical procedures: rigid buildings of all heights and low-rise buildings [with mean roof heights less than or equal to 60 feet (18 288 mm)]. The simplified procedure in ASCE 7 and Section 1609.6 of the code is based on the low-rise buildings method. Method 3 is the wind tunnel procedure. Section 1609.1.1 provides four exceptions to using the provisions of ASCE 7 for the determination of wind loads.

Exception 1 makes reference to the simplified provisions for low-rise buildings in Section 1609.6. The simplified provisions apply to buildings that are 60 feet (18 288 mm) or less in height. See Section 1609.6 for other requirements as to the conditions where this design method can be used and for additional commentary on these provisions.

Exception 2 provides for the use of the SBCCI SSTD 10 for Group R-2 and R-3 buildings where they are located within Exposure A, B or C as defined in Section 1609.4 and not sited on the upper half of an isolated hill, escarpment or ridge with the characteristics described in Section 1609.1.1.1. SBCCI SSTD 10 has prescriptive construction requirements and required load capacity tables that replace the requirement for structural analysis, which is intended to provide improved design construction details to achieve greater structural performance for single- and multiple-family dwellings in a high-wind event. SBCCI SSTD 10 applies to one- and two-story residential buildings of conventional wood-frame, masonry and concrete wall construction. The standard provides information for buildings sited in three fastest-mile wind climates: 90, 100 and 110 mph. See Table 1609.3.1 for conversions between the 3-second-gust wind speed and the fastest-mile wind speed. See SCBBI SSTD 10 for other detailed application limitations.

Exception 3 provides for the use of the AF&PA *Wood Frame Construction Manual for One- and Two-Family Dwellings* where the building is sited within Exposure B or C as defined in Section 1609.4 and not on the upper half of an isolated hill, escarpment or ridge with the characteristics described in Section 1609.1.1.1. The AF&PA *Wood Frame Construction Manual* has prescriptive construction requirements and required load-resistance tables that replace the requirement for structural analysis. The tabulated engineered and prescriptive design provisions apply to one- and two-family wood-frame dwellings where the fastest-mile basic wind speed is between 90 and 120 mph. See Table 1609.3.1 for conversion between the 3-second-gust and fastest-mile wind speeds (see the AF&PA *Wood Frame Construction Manual*, Chapter 1, for other detailed application limitations).

Exceptions 4 and 5 simply refer to national standards dealing specifically with the design of flagpoles and telecommunication towers.

1609.1.1.1 Applicability. The provisions of SSTD 10 are applicable only to buildings located within Exposure, B or C as defined in Section 1609.4. The provisions of SSTD 10 and the *AF&PA Wood Frame construction Manual for One- and Two-Family Dwellings* shall not apply to buildings sited on the upper half of an isolated hill, ridge or escarpment meeting the following conditions:

1. The hill, ridge or escarpment is 60 feet (18 288 mm) or higher if located in Exposure B or 30 feet (9144 mm) or higher if located in Exposure C;
2. The maximum average slope of the hill exceeds 10 percent; and
3. The hill, ridge or escarpment is unobstructed upwind by other such topographic features for a distance from the high point of 50 times the height of the hill or 1 mile (1.61 km), whichever is greater.

❖ This section places limitations on the use of SBCCI SSTD 10 and the AF&PA *Wood Frame Construction Manual*. SBCCI SSTD 10 is based on the provisions of Section 1606.2 of the *Standard Building Code*, which is based on fastest-mile wind speed criteria. Neither of these standards takes into account the effects of isolated hills, ridges or escarpments as is required in ASCE 7. Additionally, these standards are limited to buildings sited in Exposure B or C.

1609.1.2 Minimum wind loads. The wind loads used in the design of the main wind-force-resisting system shall not be less than 10 psf (0.479 kN/m^2) multiplied by the area of the building or structure projected on a vertical plane normal to the wind direction. In the calculation of design wind loads for components and cladding for buildings, the algebraic sum of the pressures acting on opposite faces shall be taken into account. The design pressure for components and cladding of buildings shall not be less than 10 psf (0.479 kN/m^2) acting in either direction normal to the surface. The design force for open buildings and other structures shall not be less than 10 psf (0.479 kN/m^2) multiplied by the area A_f.

❖ The purpose of this section is to establish the minimum design wind pressure for the main wind-force-resisting system, components and cladding for buildings, open buildings and other structures. The floor on the main windforce-resisting system establishes a lower bound value on the total base shear on a building due to wind loading.

1609.1.3 Anchorage against overturning, uplift and sliding. Structural members and systems and components and cladding in a building or structure shall be anchored to resist wind-induced overturning, uplift and sliding and to provide continuous load paths for these forces to the foundation. Where a portion of the resistance to these forces is provided by dead load, the dead load, including the weight of soils and foundations, shall be taken as the minimum dead load likely to be in place during a design wind event. Where the alternate basic load combinations of Section 1605.3.2 are used, only two-thirds of the minimum dead load likely to be in place during a design wind event shall be used.

❖ This section requires that all elements of a building be anchored to resist wind-induced actions, such as overturning, uplift and sliding. Often, a considerable portion of this resistance is provided by the dead load of and within a building or structure, including the weight of foundations and any soil directly above them. This section requires the designer to give consideration to the dead load used to resist wind loads by stating that only the minimum dead load likely to be in place during a design wind event is permitted to be used. This, however, does not imply that certain parts or elements of a building are not designed to remain in place. This criteria simply requires the designer to use caution in using dead loads that may not be installed, or in place, such as may occur in certain projects under construction for significant periods of time in various phases.

Previous editions of the model codes specified that the overturning moment and sliding due to wind load could not exceed two-thirds of the dead load stabilizing moment; however, it was not completely clear whether this provision applied to uplift, all elements or to the building as a whole. In the code this limitation on dead load is accomplished through the load combinations of Section 1605.3.1. The applicable combination is $0.6D + W$. This load combination limits the dead load resisting wind loads to 60 percent (2/3 = 0.67; round down) and applies to all elements. In this form, it is clear that this safety factor on dead load applies to all actions where dead load is assisting in resisting wind loads.

Where the alternative basic load combinations of Section 1605.3.2 are used, it limits the dead load to two-thirds of the dead load likely to be in place during a design wind event. This accomplishes the desired safety factor on dead load as does the section above. Figure 1609.1.3 illustrates the effects of wind on the total structure. To keep a structure from sliding horizontally, the dead load must generate enough friction at the base of the structure (foundation) to overcome the horizontal base shear created by wind forces. Otherwise, adequate anchorage must be provided to resist the base shear.

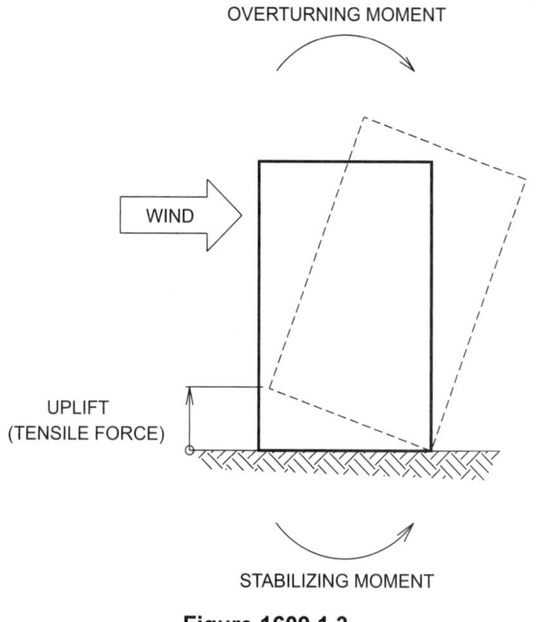

Figure 1609.1.3
OVERTURNING

1609.1.4 Protection of openings. In wind-borne debris regions, glazing that receives positive external pressure in the lower 60 feet (18 288 mm) in buildings shall be assumed to be openings unless such glazing is impact resistant or protected with an impact-resistant covering meeting the requirements of an approved impact-resisting standard or ASTM E 1996 and of ASTM E 1886 referenced therein as follows:

1. Glazed openings located within 30 feet (9144 mm) of grade shall meet the requirements of the Large Missile Test of ASTM E 1996.
2. Glazed openings located more than 30 feet (9144 mm) above grade shall meet the provisions of the Small Missile Test of ASTM E 1996.

Exceptions:

1. Wood structural panels with a minimum thickness of $^{7}/_{16}$ inch (11.1 mm) and maximum panel span of 8 feet (2438 mm) shall be permitted for opening protection in one- and two-story buildings. Panels shall be precut to cover the glazed openings with attachment hardware provided. Attachments shall be designed to resist the components and cladding loads determined in accordance with the provisions of Section 1609.6.1.2. Attachment in accordance with Table 1609.1.4 is permitted for buildings with a mean roof height of 33 feet (10 058 mm) or less where wind speeds do not exceed 130 mph (57.2 m/s).
2. Buildings in Category I as defined in Table 1604.5, including production greenhouses as defined in Section 1608.3.3.

❖ The purpose of this section is to address risks associated with wind-borne debris in high-wind areas. See the definitions of "Wind-borne debris region" and "Hurricane-prone regions" in Section 1609.2 for an explanation of where these provisions apply. Note that this section does not require that certain openings be protected from wind-borne debris. The section allows the glazing areas to be assumed as openings as an alternative to wind-borne debris protection.

This section requires glazing that is subjected to positive external pressure and located in the lower 60 feet (18 288 mm) of buildings in wind-borne debris regions to be assumed as openings unless the glazing is either protected from impact from wind-borne debris or impact resistant. During a hurricane, buildings are impacted from wind-borne debris due to violent high-velocity winds. This debris can and has impacted glazing, causing breakage and creating an opening within the building envelope. The presence of openings in the building envelope can have a significant effect on the magnitude of the total wind pressure required to be resisted by each structural element of a building. Depending on the location of these openings with respect to wind direction and the amount of background porosity, external and internal pressures may act in the same direction to produce higher forces on some walls and the roof.

An example of this is shown in Figure 1609.1.4. In this scenario, as the wind flows over the building, pressures are developed on the external surface as shown. Introduction of an opening in the windward wall causes the wind to rush into the building, exerting internal pressures (positive) against all interior surfaces. This type of opening has the net effect of producing potentially high internal pressures that will act in the same direction as the external pressures on the roof, side and leeward walls. Considering the high probability of wind-borne debris during a hurricane and the effect of an unintended opening in the building envelope, the code requires glazing in designated regions to be protected from wind-borne debris or to be considered openings.

If the designer chooses to consider the glazed areas as openings, then he or she must determine the opening classification of the building and include the higher internal pressures in the design of the building, as necessary. It is important to note again that providing wind-borne debris protection is an option, and designing a building as partially enclosed is permitted.

Wind-borne debris protection can be provided by installing impact-resistant glazing or an impact-resistant covering over the glazing. Where wind-borne protection is provided, the section specifies two types of tests to demonstrate adequate resistance: the large missile test (2 by 4) to simulate large debris within 30 feet (9144 mm) above grade and the small missile test to simulate smaller debris up to 60 feet (18 288 mm) above grade, both of which are common during very high winds. An example of small debris is gravel from the surrounding area that becomes airborne.

Impact-resistant coverings or glazing must meet the test requirements of an approved impact standard or ASTM E 1886 and ASTM E 1996. Other impact standards the building official may consider are SBCCI SSTD 12 and Miami-Dade County Protocols PA 201, 202 and 203. These standards specify similar-type testing with a large missile test (2 by 4), small missile test (2 gram balls) and cyclic pressure loading test. ASTM E 1886 and ASTM E 1996 work together with the standard test method (E 1886) and the test specification (E 1996), including scoping, technical requirements and pass/fail criteria.

Exception 1 permits the use of $^{7}/_{16}$-inch (11.1 mm) wood structural panels with maximum spans of 8 feet (2438 mm) as an impact-resistant covering for one- and two-story buildings. Panel attachments have to be designed to resist the component and cladding loads from Table 1609.6.2.1(2) or attached in accordance with Table 1609.1.4. This protective system has been tested and meets the requirements of SBCCI SSTD 12. The intent is that precut panel coverings and attachment hardware are provided on site.

Exception 2 exempts low-hazard buildings from this requirement for protecting openings against wind-borne debris.

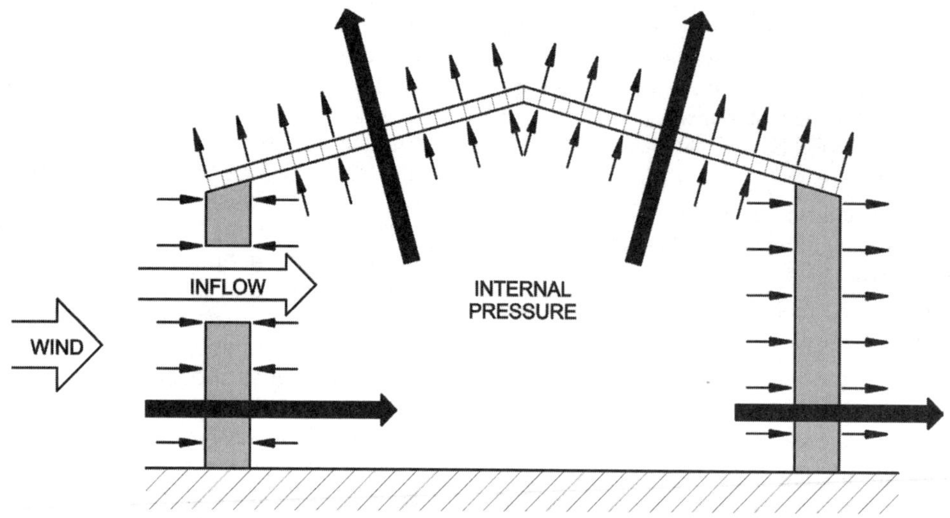

Figure 1609.1.4
EFFECTS OF OPENINGS IN THE BUILDING ENVELOPE

TABLE 1609.1.4
WIND-BORNE DEBRIS PROTECTION FASTENING SCHEDULE FOR WOOD STRUCTURAL PANELS[a,b,c]

FASTENER TYPE	FASTENER SPACING (inches)			
	Panel span ≤ 2 feet	2 feet < Panel span ≤ 4 feet	4 feet < Panel span ≤ 6 feet	6 feet < Panel span ≤ 8 feet
$2^1/_2$ No. 6 Wood screws	16	16	12	9
$2^1/_2$ No. 8 Wood screws	16	16	16	12

For SI: 1 inch = 25.4 mm, 1 foot = 304.8 mm, 1 pound = 4.4 N, 1 mile per hour = 0.44 m/s.

a. This table is based on a maximum wind speed (3-second gust) of 130 mph and mean roof height of 33 feet or less.

b. Fasteners shall be installed at opposing ends of the wood structural panel.

c. Where screws are attached to masonry or masonry/stucco, they shall be attached utilizing vibration-resistant anchors having a minimum withdrawal capacity of 490 pounds.

❖ This table provides the connections for the wood structural panel impact-resistant covering that is described in the exception to Section 1609.1.4. The table lists the spacing of the $2^1/_2$-inch-long (64 mm) No. 6 wood screws around the perimeter of the panel for the indicated panel spans. Note that Table 1609.1.4 is only applicable to buildings with a mean roof height of 33 feet (10 058 mm) or less and located where the basic wind speed is 130 mph or less.

1609.1.4.1 Building with openings. Where glazing is assumed to be an opening in accordance with Section 1609.1.4, the building shall be evaluated to determine if the openings are of sufficient area to constitute an open or partially enclosed building as defined in Section 1609.2. Open and partially enclosed buildings shall be designed in accordance with the applicable provisions of ASCE 7.

❖ The presence of openings requires an evaluation and classification of the building as either open, partially open or enclosed based on definitions given in Section 1609.2. Only enclosed buildings are within the scope of the simplified procedure of Section 1609.6. Otherwise the wind load analysis must be accomplished using ASCE 7.

1609.1.5 Wind and seismic detailing. Lateral-force-resisting systems shall meet seismic detailing requirements and limitations prescribed in this code, even when wind code prescribed load effects are greater than seismic load effects.

❖ This section is needed to specify that the ductility requirements of the code for seismic effects still apply even where the wind load effects are higher than the seismic load effects.

For example, consider the case where the earthquake load provisions apply in accordance with Section 1614.1 and the structural system is an ordinary concentrically braced steel frame where Section 2205 requires that the AISC seismic provisions are to be met. For this case, the ductility requirements in Section 14 of AISC would apply to the design of the steel frame, even if the wind load effects are higher than the seismic load effects.

1609.2 Definitions. The following words and terms shall, for the purposes of Section 1609.6, have the meanings shown herein.

❖ This section provides the definitions that apply to Section 1609.6, the simplified wind provisions for low-rise buildings that apply where the building is 60 feet (18 288 mm) or less in height. The additional wind-related definitions listed in Section 6.2 of ASCE 7 apply when using the standard for design in accordance with Section 1609.1.1.

BUILDINGS AND OTHER STRUCTURES, FLEXIBLE. Slender buildings and other structures that have a fundamental natural frequency less than 1 Hz.

❖ Flexible buildings are not intended to be designed according to the provisions of Section 1609.6, but according to the requirements in ASCE 7. Buildings are generally considered flexible when their height to least horizontal dimension exceeds four.

BUILDING, ENCLOSED. A building that does not comply with the requirements for open or partially enclosed buildings.

❖ This definition is needed since the simplified wind provisions in Section 1609.6 are limited to simple diaphragm buildings that by definition are enclosed buildings. Enclosed buildings are not prohibited from having openings, but those openings must be somewhat small or uniformly distributed. Buildings not meeting the definition of either an open or partially enclosed building are considered enclosed.

BUILDING, LOW-RISE. Enclosed or partially enclosed buildings that comply with the following conditions:

1. Mean roof height, *h*, less than or equal to 60 feet (18 288 mm).
2. Mean roof height, *h*, does not exceed least horizontal dimension.

❖ This definition is needed since the simplified wind provisions in Section 1609.6 are limited to low-rise buildings by this definition.

BUILDING, OPEN. A building having each wall at least 80 percent open. This condition is expressed for each wall by the equation:

$$A_o \geq 0.8 A_g \quad \textbf{(Equation 16-31)}$$

where:

A_o = Total area of openings in a wall that receives positive external pressure, in square feet (m^2).

A_g = The gross area of that wall in which A_o is identified, in square feet (m^2).

❖ This definition is needed since this term is used in the definition of an enclosed building. To be considered an open building, all walls have to be at least 80 percent open.

BUILDING, PARTIALLY ENCLOSED. A building that complies with both of the following conditions:

1. The total area of openings in a wall that receives positive external pressure exceeds the sum of the areas of openings in the balance of the building envelope (walls and roof) by more than 10 percent; and
2. The total area of openings in a wall that receives positive external pressure exceeds 4 square feet (0.37 m^2) or 1 percent of the area of that wall, whichever is smaller, and the percentage of openings in the balance of the building envelope does not exceed 20 percent.

These conditions are expressed by the following equations:

$$A_o > 1.10 A_{oi} \quad \textbf{(Equation 16-32)}$$

A_o > 4 square feet (0.37 m^2) or > $0.01A_g$, whichever is smaller, and $A_{oi}/A_{gi} \leq 0.20$ **(Equation 16-33)**

where:

A_o, A_g are as defined for an open building.

A_{oi} = The sum of the areas of openings in the building envelope (walls and roof) not including A_o, in square feet (m^2).

A_{gi} = The sum of the gross surface areas of the building envelope (walls and roof) not including A_g, in square feet (m^2).

❖ This definition is needed since this term is used in the definition of an enclosed building.

BUILDING, SIMPLE DIAPHRAGM. A building in which wind loads are transmitted through floor and roof diaphragms to the vertical lateral-force-resisting systems.

❖ This definition is needed since the applicability of the simplified wind load analysis (see Section 1609.6) to the design of the main windforce-resisting system is limited to simple diaphragm buildings. See Section 1609.6.1.1 for additional conditions on use of the simplified method.

COMPONENTS AND CLADDING. Elements of the building envelope that do not qualify as part of the main windforce-resisting system.

❖ This definition is needed to identify the types of building elements that are regulated by the provisions of Section 1609.6. Components and cladding generally receive wind loads directly and transfer these loads to the main windforce-resisting system. Examples of component and cladding elements are exterior wall panels; windows; doors and roofing; roof sheathing; roof rafters; wall girts and exterior wall studs. Note that components and cladding are intended to include some building elements that are sometimes considered as nonstructural items; thus, these elements are to be designed and constructed to meet the wind loading requirements of Section 1609.6 or ASCE 7. See the commentary to Section 1609.6.2.2 to determine when certain elements are considered components and cladding and part of the main wind-forceresisting system.

EFFECTIVE WIND AREA. The area used to determine GC_p. For component and cladding elements, the effective wind area in Tables 1609.6.2.1(2) and 1609.6.2.1(3) is the span length multiplied by an effective width that need not be less than one-third the span length. For cladding fasteners, the effective wind area shall not be greater than the area that is tributary to an individual fastener.

❖ This definition is needed for the application of component and cladding tables that are cited. Effective wind area is basically the tributary area of the building surface that delivers the force to the element considered; however, this area need not be less than one-third the square of the span of the element considered. Components such as studs and rafters are closely spaced and the tributary area of these elements is long and narrow. Using one-third of the square of the span better approximates the actual load distribution.

HURRICANE-PRONE REGIONS. Areas vulnerable to hurricanes defined as:

1. The U.S. Atlantic Ocean and Gulf of Mexico coasts where the basic wind speed is greater than 90 mph (39.6 m/s) and
2. Hawaii, Puerto Rico, Guam, Virgin Islands and American Samoa.

❖ This definition specifies the area wherein hurricane-force winds are expected to reach. These areas are also distinguished from nonhurricane wind speed regions by the inclusion of a hurricane importance factor implied in the wind speed contours of Figure 1609.

IMPORTANCE FACTOR, I_w. A factor that accounts for the degree of hazard to human life and damage to property.

❖ This definition is needed for the understanding of Table 1604.5 where the wind importance factor is listed. The importance factor is used to adjust the MRI. For normal buildings and structures, a MRI of 50 years is used. For essential or critical facilities, a MRI of 100 years is used. I_w is included in the wind pressure equation and has the effect of adjusting the wind speed up or down based on the nature of the occupancy.

MAIN WINDFORCE-RESISTING SYSTEM. An assemblage of structural elements assigned to provide support and stability for the overall structure. The system generally receives wind loading from more than one surface.

❖ This definition identifies the types of structural assemblies that are to comply with Section 1609.6.4. The main windforce-resisting system is the global structural system responsible for transferring wind loads from the building or structure to the ground, including moment-resisting frames, braced frames, diaphragms and shear walls. See the commentary to Section 1609.6.2.2 to determine when certain elements are considered components and cladding and part of the main windforce-resisting system.

MEAN ROOF HEIGHT. The average of the roof eave height and the height to the highest point on the roof surface, except that eave height shall be used for roof angle of less than or equal to 10 degrees (0.1745 rad).

❖ This definition is needed for understanding one of the limitations on the simplified wind load method in Section 1609.6.1. For the low-rise methodology in ASCE 7 and the simplified procedure of Section 1609.6, wind loads are based on the mean roof height of the building.

WIND-BORNE DEBRIS REGION. Areas within hurricane-prone regions within 1 mile (1.61 km) of the coastal mean high water line where the basic wind speed is 110 mph (48.4 m/s) or greater; or where the basic wind speed is 120 mph (52.8 m/s) or greater; or Hawaii.

❖ This definition identifies those areas that require consideration of the impact of wind-borne debris on the building envelope. See the commentary to Section 1609.1.4 for information on the effects of openings in the building envelope.

1609.3 Basic wind speed. The basic wind speed, in mph, for the determination of the wind loads shall be determined by Figure 1609 or by ASCE 7 Figure 6-1 when using the provisions of ASCE 7. Basic wind speed for the special wind regions indicated, near mountainous terrain, and near gorges, shall be in accordance with local jurisdiction requirements. Basic wind speeds determined by the local jurisdiction shall be in accordance with Section 6.5.4 of ASCE 7.

In nonhurricane-prone regions, when the basic wind speed is estimated from regional climatic data, the basic wind speed shall be not less than the wind speed associated with an annual probability of 0.02 (50-year mean recurrence interval), and the estimate shall be adjusted for equivalence to a 3-second gust wind speed at 33 feet (10 m) above ground in exposure Category C. The data analysis shall be performed in accordance with Section 6.5.4 of ASCE 7.

❖ This section establishes the basic wind speed that is to be used for design. The basic wind speed map identifies special wind regions where the speeds vary substantially within a very short distance due to topographic effects, such as mountains and valleys. This section specifies appropriate recurrence interval criteria to be used for estimating the basic wind speeds from regional climatic data in other than hurricane-prone regions. See Section 6.5.4 of ASCE 7 for issues to be addressed in the data analysis and the associated commentary for further details (also see commentary, Figure 1609).

1609.3.1 Wind speed conversion. When required, the 3-second gust wind velocities of Figure 1609 shall be converted to fastest-mile wind velocities using Table 1609.3.1.

❖ Some referenced standards contain criteria or applications that are based on the fastest-mile wind speed data. For example, SBCCI SSTD 10 specifies prescriptive procedures that use the fastest-mile wind speed designation versus the 3-second-gust wind speed that is the basis of Figure 1609. Table 1609.3.1 provides a means by which these standards can be used with new wind speeds until their requirements have been updated. The conversions were calculated using the gust factor curve of Figure C6-1 in the ASCE 7 commentary. It is important to note that the conversions provide equivalent wind speeds based on averaging times, and do not necessarily provide equivalent wind pressures.

TABLE 1609.3.1. See page 16-43.

❖ This table is used where any referenced standards are based on the fastest-mile wind speed. The table converts the 3-second-gust wind speed of a geographical location to the fastest-mile wind speed for use within the standards cited in Section 1609.1.1.

1609.4 Exposure category. For each wind direction considered, an exposure category that adequately reflects the characteristics of ground surface irregularities shall be determined for the site at which the building or structure is to be constructed. For a site located in the transition zone between categories, the category resulting in the largest wind forces shall apply. Account shall be taken of variations in ground surface roughness that arise from natural topography and vegetation as well as from constructed features. For any given wind direction, the exposure in which a specific building or other structure is sited shall be assessed as being one of the following categories. When applying the simplified wind load method of Section 1609.6, a single exposure category shall be used based upon the most restrictive for any given wind direction.

1. **Exposure A.** This exposure category is no longer used in ASCE 7.
2. **Exposure B.** Urban and suburban areas, wooded areas or other terrain with numerous closely spaced obstructions having the size of single-family dwellings or larger. Exposure B shall be assumed unless the site meets the definition of another type of exposure.
3. **Exposure C.** Open terrain with scattered obstructions, including surface undulations or other irregularities, having heights generally less than 30 feet (9144 mm) extending more than 1,500 feet (457.2 m) from the building site in any quadrant. This exposure shall also apply to any building located within Exposure B-type terrain where the building is directly adjacent to open areas of Exposure C-type terrain in any quadrant for a distance of more than 600 feet (182.9 m). This category includes flat open country, grasslands and shorelines in hurricane-prone regions.
4. **Exposure D.** Flat, unobstructed areas exposed to wind flowing over open water (excluding shorelines in hurricane-prone regions) for a distance of at least 1 mile (1.61 km). Shorelines in Exposure D include inland waterways, the Great Lakes and coastal areas of California, Oregon, Washington and Alaska. This exposure shall apply only to those buildings and other structures exposed to the wind coming from over the water. Exposure D extends inland from the shoreline a distance of 1,500 feet (460 m) or 10 times the height of the building or structure, whichever is greater.

❖ The concept of exposure categories provides a means to define the relative roughness of the boundary layer. The earth's surface exerts a horizontal drag force on wind due to ground obstructions that retard the flow of air close to the ground. The reduction in the flow of air is a function of height above ground and terrain roughness. Wind speeds increase with height above ground, and the relationship between height above ground and wind speed is exponential. The rate of increase in wind speeds with height is a function of the terrain features. The rougher the terrain (such as large city centers), the shallower the slope of the wind speed profile. The smoother the terrain (open water), the steeper the slope of the wind speed profile.

Exposure categories are used to define this roughness in the boundary layer. Exposure B is considered the roughest boundary layer condition. Exposure D is considered the smoothest boundary layer condition. Accordingly, calculated wind loads are less for Exposure B, which has more surface obstructions, as compared to Exposure D, with all other variables the same.

Exposure B is the most common type of exposure category in the country. A recent study by the National Association of Home Builders (NAHB) indicated that perhaps up to 80 percent of all buildings were located in Exposure B. Recognizing this, the code sets Exposure B as the default exposure category.

A significant philosophical change in exposure categories occurred with the inclusion of shorelines in hurricane-prone regions in the definition of Exposure C. Exposure D had been used for wind flowing over open water. New research determined that wave action at the water's surface in a hurricane, due to the intensity of the turbulence, produced substantial surface obstructions and friction to reduce the wind profile values that are more in line with Exposure C as opposed to Exposure D. Exposure D would still apply to inland waterways and shorelines that are not in the hurricane-prone regions, such as coastal California, Oregon, Washington and Alaska.

1609.5 Importance factor. Buildings and other structures shall be assigned a wind load importance factor, I_w, in accordance with Table 1604.5.

❖ The importance factor is used to adjust the MRI for a given building or structure. The value of the importance factors in Table 1604.5 is related directly to the classification of the building. The higher the importance factor, the higher the "importance" of the building from a safety standpoint. Design wind pressures are directly proportional to the importance factor (see Section 1609.6.2).

1609.6 Simplified wind load method.

TABLE 1609.3.1
EQUIVALENT BASIC WIND SPEEDS[a,b,c]

V_{3S}	85	90	100	105	110	120	125	130	140	145	150	160	170
V_{fm}	70	75	80	85	90	100	105	110	120	125	130	140	150

For SI: 1 mile per hour = 0.44 m/s.
a. Linear interpolation is permitted.
b. V_{3S} is the 3-second gust wind speed (mph).
c. V_{fm} is the fastest mile wind speed (mph).

1609.6.1 Scope. The procedures in Section 1609.6 shall be permitted to be used for determining and applying wind pressures in the design of enclosed buildings with flat, gabled and hipped roofs and having a mean roof height not exceeding the least horizontal dimension or 60 feet (18 288 mm), whichever is less, subject to the limitations of Sections 1609.6.1.1 and 1609.6.1.2. If a building qualifies only under Section 1609.6.1.2 for design of its components and cladding, then its main windforce-resisting system shall be designed in accordance with Section 1609.1.1.

Exception: The provisions of Section 1609.6 shall not apply to buildings sited on the upper half of an isolated hill or escarpment meeting all of the following conditions:

1. The hill or escarpment is 60 feet (18 288 mm) or higher if located in Exposure B or 30 feet (9144 mm) or higher if located in Exposure C.
2. The maximum average slope of the hill exceeds 10 percent.
3. The hill or escarpment is unobstructed upwind by other such topographic features for a distance from the high point of 50 times the height of the hill or 1 mile (1.61 km), whichever is less.

❖ This section provides the scoping criteria for the use of the simplified provisions for low-rise buildings. The building must meet the dimensional limitations specified in this section in order to use this procedure.

The pressures specified in Tables 1609.6.2.1(1), 1609.6.2.1(2) and 1609.6.2.1(3) do not consider the wind speed-up effects associated with buildings located on the upper half of an isolated hill, ridge or escarpment. This effect is accounted for in ASCE 7 by the topographic factor (K_{zt}), which for the simplified provisions in the code was taken as 1.0.

There are numerous benefits to having a simplified procedure for determining wind loads in the code. First and foremost, the simplified procedure increases the likelihood of the user getting the correct answer. Additionally, it is an easier solution for the occasional user.

The goal in developing the simplified method in the code was to devise a method that would align the complexity of the solution to the complexity of the problem, but not to penalize all designs; however, as this method has evolved, very few designs will be penalized as a result of using this simplified method. For a given simple diaphragm building, a user will get the same answer from the simplified method in the code as would be determined by performing the necessary calculations for the low-rise analytical method in ASCE 7.

The simplification concepts include calculation simplifications and building simplifications. Calculation simplifications are accomplished by reducing the number of variables (limit its use) and putting the equations into a simpler form. Building simplification is accomplished by determining the controlling load case for certain types of buildings.

The controlling calculations for determination of wind loads on low-rise buildings are velocity pressure at the mean roof height

$$Q_h = 0.00256 K_z K_{zt} K_d V^2 I_w$$

and design pressure

$P = q_h(GC_{pf} - Gc_{pi})$ Main windforce-resisting system (MWFRS)

$P = q_h(GC_p - Gc_{pi})$ Components and cladding

For the velocity pressure:

K_z = Velocity pressure exposure coefficient. Varies with height and exposure.

K_{zt} = Topographic factor. Limit the use of the simplified procedure to buildings with no topographic effects and K_{zt} = 1.

K_d = Directionality factor, 0.85 for buildings.

V = Basic wind speed, varies.

I_w = Importance factor, Table 1604.5

From this equation, K_{zt} and K_d are constant and I_w is a straight adjustment from Table 1604.5. Therefore, wind speeds can be tabulated for each map contour, for a given height and a given exposure, and a range for a single factor to be applied to account for different heights and exposures. Tables 1609.6.2.1(1), 1609.6.2.1(2) and 1609.6.2.1(3) provide design pressures for all wind speed contours shown on Figure 1609 for buildings with a mean roof height of 30 feet (9144 mm) and located in Exposure Category B. Table 1609.6.2.1(4) specifies the height and exposure adjustment coefficients for buildings of heights other than 30 feet (9144 mm) or located in an exposure category other than B.

For the design pressure:

GC_{pf} = External pressure coefficient for the MWFRS, varies depending on building geometry.

GC_p = External pressure coefficient for components and cladding, varies depending on the building geometry.

GC_{pi} = Internal pressure coefficients, ± 0.18 for enclosed buildings.

From this equation, GC_{pi} is constant since this procedure is limited to enclosed buildings, and GC_{pf} and GC_p vary. For the MWFRS, building simplification is accomplished by predetermining the controlling load case for each wind direction considered. Wind pressures are tabulated in Table 1609.6.2.1(1) for the transverse direction (perpendicular to the ridge) for three ranges of roof slopes, and for the longitudinal direction (parallel to the ridge) for all roof angles.

External pressure coefficients for components and cladding are dependent upon location in the building envelope and the effective wind area of the element considered. Table 1609.6.2.1(2) accounts for these variables by tabulating pressures for all applicable zones and for effective wind areas of 10, 20, 50 and 100 square feet (0.93, 1.86, 4.65 and 9.29 m^2). These zones account for the higher localized pressures that occur at discontinuities of the building.

1609.6.1.1 Main windforce-resisting systems. For the design of main windforce-resisting systems, the building must meet all of the following conditions:

1. The building is a simple diaphragm building as defined in Section 1609.2.
2. The building is not classified as a flexible building as defined in Section 1609.2.
3. The building does not have response characteristics making it subject to across wind loading, vortex shedding, instability due to galloping or flutter; and does not have a site location for which channeling effects or buffeting in the wake of upwind obstructions warrant special consideration.
4. The building structure has no expansion joints or separations.
5. The building is regular shaped and has an approximately symmetrical cross section in each direction with roof slopes not exceeding 45 degrees (0.78 rad.).

❖ The simplified wind load method is based on several simplifying assumptions. For proper application it is necessary that these assumptions be stated as conditions or limitations in the scope of this procedure. In addition to general limitations in Section 1609.6.1, this section lists conditions that must be met in order to use these wind pressures for designing the MWFRS. Note that there are different conditions to be met in order to apply these wind pressures to the design of components and cladding. As stated in Section 1609.6.1, based on these conditions it is possible that a building may qualify for use of these simplified wind loads for designing components and cladding, but not for designing the MWFRS.

1609.6.1.2 Components and cladding. For the design of components and cladding, the building must meet all of the following conditions:

1. The building does not have response characteristics making it subject to across wind loading, vortex shedding, instability due to galloping or flutter; and does not have a site location for which channeling effects or buffeting in the wake of upwind obstructions warrant special consideration.
2. The building is regular shaped with roof slopes not exceeding 45 degrees (0.78 rad.) for gable roofs, or 27 degrees (0.47 rad.) for hip roofs.

❖ For proper application of the simplified wind load method it is necessary that any simplifying assumptions be stated as conditions or limitations in the scope of this procedure. In addition to general limitations in Section 1609.6.1, this section lists conditions that must be met in order to use these wind pressures for designing components and cladding. Note that there are different conditions to be met in order to apply these wind pressures to the design of the MWFRS. As stated in Section 1609.6.1, based on these conditions it is possible that a building may qualify for the use of these simplified wind loads for designing components and cladding, but not for designing the MWFRS.

1609.6.2 Design procedure.

1. The basic wind speed, V, shall be determined in accordance with Section 1609.3. The wind shall be assumed to come from any horizontal direction.
2. An importance factor I_w shall be determined in accordance with Section 1609.5.
3. An exposure category shall be determined in accordance with Section 1609.4.
4. A height and exposure adjustment coefficient, λ, shall be determined from Table 1609.6.2.1(4).

❖ This section lists the criteria that are needed for determining wind pressures under the simplified wind load method. It also provides cross references to the appropriate section or table. Note that Section 1609.4 requires that in the simplified wind load method a single exposure category must be used based on the most restrictive exposure category determined for any wind direction.

1609.6.2.1 Main windforce-resisting system. Simplified design wind pressures, p_s, for the main windforce-resisting systems represent the net pressures (sum of internal and external) to be applied to the horizontal and vertical projections of building surfaces as shown in Figure 1609.6.2.1. For the horizontal pressures (Zones A, B, C, D), p_s is the combination of the windward and leeward net pressures. p_s shall be determined from Equation 16-34).

$$p_s = \lambda I_w p_{s30} \quad \textbf{(Equation 16-34)}$$

where:

λ = Adjustment factor for building height and exposure from Table 1609.6.2.1(4).

I_w = Importance factor as defined in Section 1609.5

p_{s30} = Simplified design wind pressure for Exposure B, at h = 30 feet (9144 mm), and for I_w = 1.0, from Table 1609.6.2.1(1).

❖ As described in the commentary to Section 1609.6.1, Table 1609.6.2.1(1) tabulates MWFRS pressures for buildings with a mean roof height of 30 feet (9144 mm) located in Exposure Category B. For other heights and exposures, the values simply have to be multiplied by the appropriate height and exposure adjustment factor, λ, from Table 1609.6.2.1(4). Additionally, the loads are required to be multiplied by the importance factor, I_w, from Table 1604.5 (see examples following Section 1609).

FIGURE 1609.6.2.1. See page 16-47.

❖ This code figure is to be used with Table 1609.6.2.1(1) for the application of MWFRS design wind load pressures. The purpose of this figure is to illustrate that the wind pressures in Table 1609.6.2.1(1) are applied to the horizontal and vertical projections of the building, and that the horizontal loads are applied from one side only (simplification for a simple diaphragm building).

Note 2 indicates that the building is to be rotated in 90-degree (1.57 rad) increments, applying the applicable loads to all four corners.

By limiting this method to simple diaphragm buildings, the equal and opposite internal pressures on the walls cancel out in the diaphragm and need not be considered [see commentary, Table 1609.2.1(1)]. This is not the case for the wind load on roof members. For roof slopes up to 25 degrees (.43 rad) it can be assumed that the uplift due to positive internal pressure is the controlling load case. At 25 degrees (.43 rad) the external windward roof pressure becomes positive. Note 4 points out that roof slopes greater than 25 degrees (.43 rad) require consideration of Load Cases 1 and 2, respectively representing positive and negative internal pressure load cases. Load Cases 1 and 2 are identified in the fourth column of Table 1609.2.1(1).

Where the horizontal wind pressure on the vertical projection of the roof (Zones B and D) is negative, it acts opposite to the positive wind pressure on the walls, reducing the total horizontal load for MWFRS design. Note 7 indicates that the horizontal wind load must not be less than the load determined with the pressure on the vertical projection of the roof (Zones B and D) set equal to zero.

TABLE 1609.6.2.1(1). See page 16-48.

❖ This table specifies the wind pressures for the design of the MWFRS for simple diaphragm buildings. See Section 1609.6.2.1 for additional factors to be applied to determine the main windforce pressures for design.

There are several important aspects of this table. First, the tabulated loads are to be applied to the vertical and horizontal projections of the building as shown in Figure 1609.6.2.1. Additionally, horizontal loads are to be applied only to one side of the building for each wind direction. The MWFRS of simple diaphragm buildings is not sensitive to the differences in the magnitudes of wind loads on the windward and leeward walls (or vertical surfaces). For simple diaphragm buildings, the wind load is collected into the horizontal diaphragm and distributed to the vertical lateral-force-resisting system; therefore, the horizontal loads shown in Table 1609.6.2.1(1) are the sum of the windward and leeward vertical surfaces exposed to wind.

TABLE 1609.6.2.1(2). See page 16-50

❖ This table specifies the design wind pressures for components and cladding building elements, such as doors, windows, siding and garage doors. See Figure 1609.6.2.2 for the application of the tabulated forces on the building, Also see Section 1609.6.2.2 for additional factors to be applied to determine the component and cladding wind pressures to be used for design.

TABLE 1609.6.2.1(3). See page 16-51.

❖ This table specifies the component and cladding wind pressures for roof overhangs. See Figure 1609.6.2.2 for the application of the tabulated forces on the building. Also see Section 1609.6.2.2 for additional factors to be applied to the values in this table to determine the roof component and cladding wind pressures to be used for design.

These pressures should not be confused with the roof overhang pressures specified in Table 1609.6.2.1(1). These pressures are used for the design of the individual members that comprise the roof overhang. The overhang values in Table 1609.6.2.1(1) account for the effects of roof overhangs on the MWFRS.

TABLE 1609.6.2.1(4). See page 16-51.

❖ Table 1609.6.2.1(4) provides the coefficients used to adjust the wind pressures from Tables 1609.6.2.1(1), 1609.6.2.1(2) and 1609.6.2.1(3) for height and exposure category. This factor accomplishes the same effect as the velocity exposure coefficient, K_z, in ASCE 7.

1609.6.2.1.1 Minimum pressures. The load effects of the design wind pressures from Section 1609.6.2.1 shall not be less than assuming the pressures , p_s, for Zones A, B, C and D all equal to +10 psf (0.48 kN/m^2), while assuming Zones E, F, G, and H all equal to 0 psf.

❖ This section essentially reiterates Section 1609.1.2 requirements for minimum wind loads required in the design of the MWFRS. It puts them in terms of the simplified wind load method by establishing the minimum pressure to be applied horizontally (Zones A through D) while assuming the vertical pressures (Zones E through H) are zero.

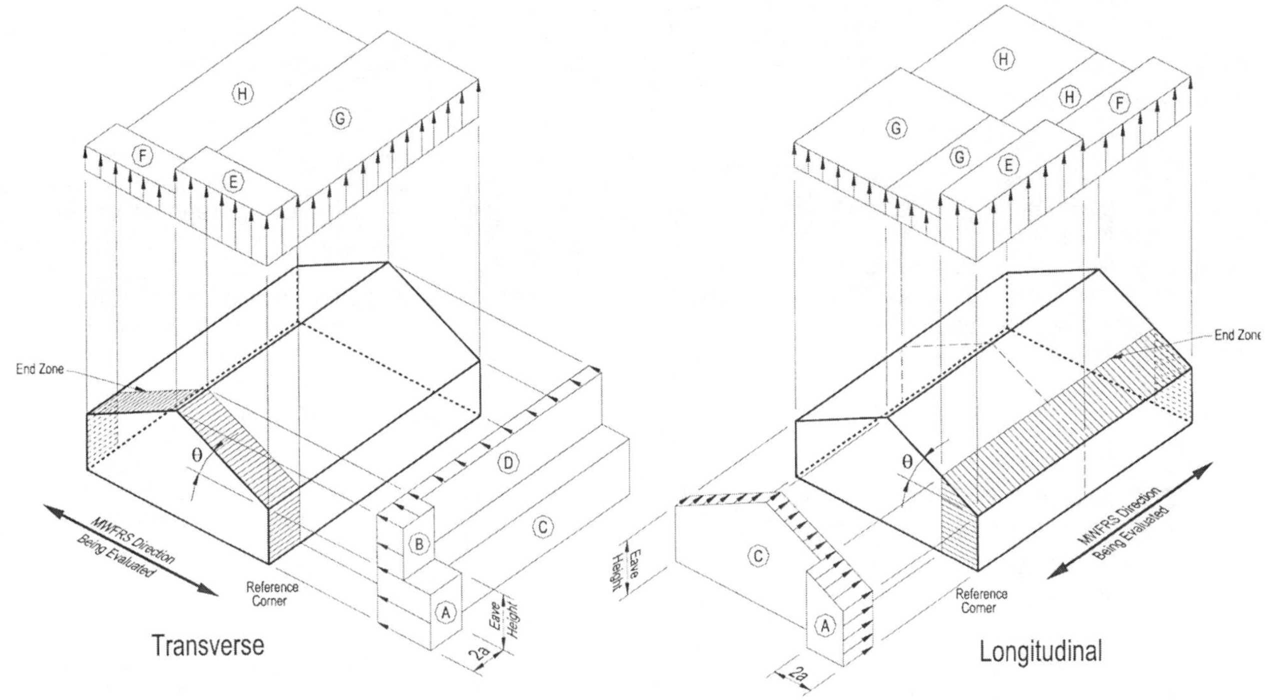

FIGURE 1609.6.2.1
MAIN WINDFORCE LOADING DIAGRAM

For SI: 1 foot = 304.8 mm, 1 degree = 0.0174 rad.

Notes:

1. Pressures are applied to the horizontal and vertical projections for Exposure B, at h = 30 feet, for I_w = 1.0. Adjust to other exposures and heights with adjustment factor λ.
2. The load patterns shown shall be applied to each corner of the building in turn as the reference corner.
3. For the design of the longitudinal MWFRS, use θ = 0°, and locate the Zone E/F, G/H boundary at the mid-length of the building.
4. Load Cases 1 and 2 must be checked for 25° < θ ≤ 45°. Load Case 2 at 25° is provided only for interpolation between 25° to 30°.
5. Plus and minus signs signify pressures acting toward and away from the projected surfaces, respectively.
6. For roof slopes other than those shown, linear interpolation is permitted.
7. The total horizontal load shall not be less than that determined by assuming p_S = 0 in Zones B and D.
8. The zone pressures represent the following:
   Horizontal pressure zones — Sum of the windward and leeward net (sum of internal and external) pressures on vertical projection of:
   A – End zone of wall C – Interior zone of wall
   B – End zone of roof D – Interior zone of roof
   Vertical pressure zones — Net (sum of internal and external) pressures on horizontal projection of:
   E – End zone of windward roof G – Interior zone of windward roof
   F – End zone of leeward roof H – Interior zone of leeward roof
9. Where Zone E or G falls on a roof overhang on the windward side of the building, use E_{OH} and G_{OH} for the pressure on the horizontal projection of the overhang. Overhangs on the leeward and side edges shall have the basic zone pressure applied.
10. Notation:
    a: 10 percent of least horizontal dimension or 0.4h, whichever is smaller, but not less than either 4 percent of least horizontal dimension or 3 feet.
    h: Mean roof height, in feet (meters), except that eave height shall be used for roof angles <10°.
    θ: Angle of plane of roof from horizontal, in degrees.

TABLE 1609.6.2.1(1)
SIMPLIFIED DESIGN WIND PRESSURE (MAIN WINDFORCE-RESISTING SYSTEM), p_{s30} (Exposure B at h = 30 feet with I_w = 1.0) (psf)

BASIC WIND SPEED (mph)	ROOF ANGLE (degrees)	ROOF RISE IN 12"	LOAD CASE	ZONES									
				Horizontal Pressures				Vertical Pressures				Overhangs	
				A	B	C	D	E	F	G	H	E_{OH}	G_{OH}
85	0 to 5°	Flat	1	11.5	-5.9	7.6	-3.5	-13.8	-7.8	-9.6	-6.1	-19.3	-15.1
	10°	2	1	12.9	-5.4	8.6	-3.1	-13.8	-8.4	-9.6	-6.5	-19.3	-15.1
	15°	3	1	14.4	-4.8	9.6	-2.7	-13.8	-9.0	-9.6	-6.9	-19.3	-15.1
	20°	4	1	15.9	-4.2	10.6	-2.3	-13.8	-9.6	-9.6	-7.3	-19.3	-15.1
	25°	6	1	14.4	2.3	10.4	2.4	-6.4	-8.7	-4.6	-7.0	-11.9	-10.1
			2	—	—	—	—	-2.4	-4.7	-0.7	-3.0	—	—
	30° to 45°	7 to 12	1	12.9	8.8	10.2	7.0	1.0	-7.8	0.3	-6.7	-4.5	-5.2
			2	12.9	8.8	10.2	7.0	5.0	-3.9	4.3	-2.8	-4.5	-5.2
90	0 to 5°	Flat	1	12.8	-6.7	8.5	-4.0	-15.4	-8.8	-10.7	-6.8	-21.6	-16.9
	10°	2	1	14.5	-6.0	9.6	-3.5	-15.4	-9.4	-10.7	-7.2	-21.6	-16.9
	15°	3	1	16.1	-5.4	10.7	-3.0	-15.4	-10.1	-10.7	-7.7	-21.6	-16.9
	20°	4	1	17.8	-4.7	11.9	-2.6	-15.4	-10.7	-10.7	-8.1	-21.6	-16.9
	25°	6	1	16.1	2.6	11.7	2.7	-7.2	-9.8	-5.2	-7.8	-13.3	-11.4
			2	—	—	—	—	-2.7	-5.3	-0.7	-3.4	—	—
	30° to 45°	7 to 12	1	14.4	9.9	11.5	7.9	1.1	-8.8	0.4	-7.5	-5.1	-5.8
			2	14.4	9.9	11.5	7.9	5.6	-4.3	4.8	-3.1	-5.1	-5.8
100	0 to 5°	Flat	1	15.9	-8.2	10.5	-4.9	-19.1	-10.8	-13.3	-8.4	-26.7	-20.9
	10°	2	1	17.9	-7.4	11.9	-4.3	-19.1	-11.6	-13.3	-8.9	-26.7	-20.9
	15°	3	1	19.9	-6.6	13.3	-3.8	-19.1	-12.4	-13.3	-9.5	-26.7	-20.9
	20°	4	1	22.0	-5.8	14.6	-3.2	-19.1	-13.3	-13.3	-10.1	-26.7	-20.9
	25°	6	1	19.9	3.2	14.4	3.3	-8.8	-12.0	-6.4	-9.7	-16.5	-14.0
			2	—	—	—	—	-3.4	-6.6	-0.9	-4.2	—	—
	30° to 45°	7 to 12	1	17.8	12.2	14.2	9.8	1.4	-10.8	0.5	-9.3	-6.3	-7.2
			2	17.8	12.2	14.2	9.8	6.9	-5.3	5.9	-3.8	-6.3	-7.2
110	0 to 5°	Flat	1	19.2	-10.0	12.7	-5.9	-23.1	-13.1	-16.0	-10.1	-32.3	-25.3
	10°	2	1	21.6	-9.0	14.4	-5.2	-23.1	-14.1	-16.0	-10.8	-32.3	-25.3
	15°	3	1	24.1	-8.0	16.0	-4.6	-23.1	-15.1	-16.0	-11.5	-32.3	-25.3
	20°	4	1	26.6	-7.0	17.7	-3.9	-23.1	-16.0	-16.0	-12.2	-32.3	-25.3
	25°	6	1	24.1	3.9	17.4	4.0	-10.7	-14.6	-7.7	-11.7	-19.9	-17.0
			2	—	—	—	—	-4.1	-7.9	-1.1	-5.1	—	—
	30° to 45°	7 to 12	1	21.6	14.8	17.2	11.8	1.7	-13.1	0.6	-11.3	-7.6	-8.7
			2	21.6	14.8	17.2	11.8	8.3	-6.5	7.2	-4.6	-7.6	-8.7
120	0 to 5°	Flat	1	22.8	-11.9	15.1	-7.0	-27.4	-15.6	-19.1	-12.1	-38.4	-30.1
	10°	2	1	25.8	-10.7	17.1	-6.2	-27.4	-16.8	-19.1	-12.9	-38.4	-30.1
	15°	3	1	28.7	-9.5	19.1	-5.4	-27.4	-17.9	-19.1	-13.7	-38.4	-30.1
	20°	4	1	31.6	-8.3	21.1	-4.6	-27.4	-19.1	-19.1	-14.5	-38.4	-30.1
	25°	6	1	28.6	4.6	20.7	4.7	-12.7	-17.3	-9.2	-13.9	-23.7	-20.2
			2	—	—	—	—	-4.8	-9.4	-1.3	-6.0	—	—
	30° to 45°	7 to 12	1	25.7	17.6	20.4	14.0	2.0	-15.6	0.7	-13.4	-9.0	-10.3
			2	25.7	17.6	20.4	14.0	9.9	-7.7	8.6	-5.5	-9.0	-10.3
130	0 to 5°	Flat	1	26.8	-13.9	17.8	-8.2	-32.2	-18.3	-22.4	-14.2	-45.1	-35.3
	10°	2	1	30.2	-12.5	20.1	-7.3	-32.2	-19.7	-22.4	-15.1	-45.1	-35.3
	15°	3	1	33.7	-11.2	22.4	-6.4	-32.2	-21.0	-22.4	-16.1	-45.1	-35.3
	20°	4	1	37.1	-9.8	24.7	-5.4	-32.2	-22.4	-22.4	-17.0	-45.1	-35.3
	25°	6	1	33.6	5.4	24.3	5.5	-14.9	-20.4	-10.8	-16.4	-27.8	-23.7
			2	—	—	—	—	-5.7	-11.1	-1.5	-7.1	—	—
	30° to 45°	7 to 12	1	30.1	20.6	24.0	16.5	2.3	-18.3	0.8	-15.7	-10.6	-12.1
			2	30.1	20.6	24.0	16.5	11.6	-9.0	10.0	-6.4	-10.6	-12.1

continued

TABLE 1609.6.2.1(1)-continued
SIMPLIFIED DESIGN WIND PRESSURE (MAIN WINDFORCE-RESISTING SYSTEM), p_{s30} (Exposure B at h = 30 feet with I_w = 1.0) (psf)

BASIC WIND SPEED (mph)	ROOF ANGLE (degrees)	ROOF RISE IN 12"	LOAD CASE	ZONES									
				Horizontal Pressures				Vertical Pressures				Overhangs	
				A	B	C	D	E	F	G	H	E_{OH}	G_{OH}
140	0 to 5°	Flat	1	31.1	-16.1	20.6	-9.6	-37.3	-21.2	-26.0	-16.4	-52.3	-40.9
	10°	2	1	35.1	-14.5	23.3	-8.5	-37.3	-22.8	-26.0	-17.5	-52.3	-40.9
	15°	3	1	39.0	-12.9	26.0	-7.4	-37.3	-24.4	-26.0	-18.6	-52.3	-40.9
	20°	4	1	43.0	-11.4	28.7	-6.3	-37.3	-26.0	-26.0	-19.7	-52.3	-40.9
	25°	6	1 2	39.0 —	6.3 —	28.2 —	6.4 —	-17.3 -6.6	-23.6 -12.8	-12.5 -1.8	-19.0 -8.2	-32.3 —	-27.5 —
	30° to 45°	7 to 12	1 2	35.0 35.0	23.9 23.9	27.8 27.8	19.1 19.1	2.7 13.4	-21.2 -10.5	0.9 11.7	-18.2 -7.5	-12.3 -12.3	-14.0 -14.0
150	0 to 5°	Flat	1	35.7	-18.5	23.7	-11.0	-42.9	-24.4	-29.8	-18.9	-60.0	-47.0
	10°	2	1	40.2	-16.7	26.8	-9.7	-42.9	-26.2	-29.8	-20.1	-60.0	-47.0
	15°	3	1	44.8	-14.9	29.8	-8.5	-42.9	-28.0	-29.8	-21.4	-60.0	-47.0
	20°	4	1	49.4	-13.0	32.9	-7.2	-42.9	-29.8	-29.8	-22.6	-60.0	-47.0
	25°	6	1 2	44.8 —	7.2 —	32.4 —	7.4 —	-19.9 -7.5	-27.1 -14.7	-14.4 -2.1	-21.8 -9.4	-37.0 —	-31.6 —
	30° to 45°	7 to 12	1 2	40.1 40.1	27.4 27.4	31.9 31.9	22.0 22.0	3.1 15.4	-24.4 -12.0	1.0 13.4	-20.9 -8.6	-14.1 -14.1	-16.1 -16.1
170	0 to 5°	Flat	1	45.8	-23.8	30.4	-14.1	-55.1	-31.3	-38.3	-24.2	-77.1	-60.4
	10°	2	1	51.7	-21.4	34.4	-12.5	-55.1	-33.6	-38.3	-25.8	-77.1	-60.4
	15°	3	1	57.6	-19.1	38.3	-10.9	-55.1	-36.0	-38.3	-27.5	-77.1	-60.4
	20°	4	1	63.4	-16.7	42.3	-9.3	-55.1	-38.3	-38.3	-29.1	-77.1	-60.4
	25°	6	1 2	57.5 —	9.3 —	41.6 —	9.5 —	-25.6 -9.7	-34.8 -18.9	-18.5 -2.6	-28.0 -12.1	-47.6 —	-40.5 —
	30° to 45°	7 to 12	1 2	51.5 51.5	35.2 35.2	41.0 41.0	28.2 28.2	4.0 19.8	-31.3 -15.4	1.3 17.2	-26.9 -11.0	-18.1 -18.1	-20.7 -20.7

For SI: 1 inch = 25.4 mm, 1 foot = 304.8 mm, 1 degree = 0.0174 rad, 1 mile per hour = 0.44 m/s, 1 pound per square foot = 47.9 N/m^2.

TABLE 1609.6.2.1(2)
NET DESIGN WIND PRESSURE (COMPONENT AND CLADDING), p_{net30} (Exposure B at h = 30 feet with I_w = 1.0) (psf)

	ZONE	EFFECTIVE WIND AREA	BASIC WIND SPEED V (mph—3-second gust)																	
			85		90		100		110		120		130		140		150		170	
Roof 0 to 7 degrees	1	10	5.3	-13.0	5.9	-14.6	7.3	-18.0	8.9	-21.8	10.5	-25.9	12.4	-30.4	14.3	-35.3	16.5	-40.5	21.1	-52.0
	1	20	5.0	-12.7	5.6	-14.2	6.9	-17.5	8.3	-21.2	9.9	-25.2	11.6	-29.6	13.4	-34.4	15.4	-39.4	19.8	-50.7
	1	50	4.5	-12.2	5.1	-13.7	6.3	-16.9	7.6	-20.5	9.0	-24.4	10.6	-28.6	12.3	-33.2	14.1	-38.1	18.1	-48.9
	1	100	4.2	-11.9	4.7	-13.3	5.8	-16.5	7.0	-19.9	8.3	-23.7	9.8	-27.8	11.4	-32.3	13.0	-37.0	16.7	-47.6
	2	10	5.3	-21.8	5.9	-24.4	7.3	-30.2	8.9	-36.5	10.5	-43.5	12.4	-51.0	14.3	-59.2	16.5	-67.9	21.1	-87.2
	2	20	5.0	-19.5	5.6	-21.8	6.9	-27.0	8.3	-32.6	9.9	-38.8	11.6	-45.6	13.4	-52.9	15.4	-60.7	19.8	-78.0
	2	50	4.5	-16.4	5.1	-18.4	6.3	-22.7	7.6	-27.5	9.0	-32.7	10.6	-38.4	12.3	-44.5	14.1	-51.1	18.1	-65.7
	2	100	4.2	-14.1	4.7	-15.8	5.8	-19.5	7.0	-23.6	8.3	-28.1	9.8	-33.0	11.4	-38.2	13.0	-43.9	16.7	-56.4
	3	10	5.3	-32.8	5.9	-36.8	7.3	-45.4	8.9	-55.0	10.5	-65.4	12.4	-76.8	14.3	-89.0	16.5	-102.2	21.1	-131.3
	3	20	5.0	-27.2	5.6	-30.5	6.9	-37.6	8.3	-45.5	9.9	-54.2	11.6	-63.6	13.4	-73.8	15.4	-84.7	19.8	-108.7
	3	50	4.5	-19.7	5.1	-22.1	6.3	-27.3	7.6	-33.1	9.0	-39.3	10.6	-46.2	12.3	-53.5	14.1	-61.5	18.1	-78.9
	3	100	4.2	-14.1	4.7	-15.8	5.8	-19.5	7.0	-23.6	8.3	-28.1	9.8	-33.0	11.4	-38.2	13.0	-43.9	16.7	-56.4
Roof > 7 to 27 degrees	1	10	7.5	-11.9	8.4	-13.3	10.4	-16.5	12.5	-19.9	14.9	-23.7	17.5	-27.8	20.3	-32.3	23.3	-37.0	30.0	-47.6
	1	20	6.8	-11.6	7.7	-13.0	9.4	-16.0	11.4	-19.4	13.6	-23.0	16.0	-27.0	18.5	-31.4	21.3	-36.0	27.3	-46.3
	1	50	6.0	-11.1	6.7	-12.5	8.2	-15.4	10.0	-18.6	11.9	-22.2	13.9	-26.0	16.1	-30.2	18.5	-34.6	23.8	-44.5
	1	100	5.3	-10.8	5.9	-12.1	7.3	-14.9	8.9	-18.1	10.5	-21.5	12.4	-25.2	14.3	-29.3	16.5	-33.6	21.1	-43.2
	2	10	7.5	-20.7	8.4	-23.2	10.4	-28.7	12.5	-34.7	14.9	-41.3	17.5	-48.4	20.3	-56.2	23.3	-64.5	30.0	-82.8
	2	20	6.8	-19.0	7.7	-21.4	9.4	-26.4	11.4	-31.9	13.6	-38.0	16.0	-44.6	18.5	-51.7	21.3	-59.3	27.3	-76.2
	2	50	6.0	-16.9	6.7	-18.9	8.2	-23.3	10.0	-28.2	11.9	-33.6	13.9	-39.4	16.1	-45.7	18.5	-52.5	23.8	-67.4
	2	100	5.3	-15.2	5.9	-17.0	7.3	-21.0	8.9	-25.5	10.5	-30.3	12.4	-35.6	14.3	-41.2	16.5	-47.3	21.1	-60.8
	3	10	7.5	-30.6	8.4	-34.3	10.4	-42.4	12.5	-51.3	14.9	-61.0	17.5	-71.6	20.3	-83.1	23.3	-95.4	30.0	-122.5
	3	20	6.8	-28.6	7.7	-32.1	9.4	-39.6	11.4	-47.9	13.6	-57.1	16.0	-67.0	18.5	-77.7	21.3	-89.2	27.3	-114.5
	3	50	6.0	-26.0	6.7	-29.1	8.2	-36.0	10.0	-43.5	11.9	-51.8	13.9	-60.8	16.1	-70.5	18.5	-81.0	23.8	-104.0
	3	100	5.3	-24.0	5.9	-26.9	7.3	-33.2	8.9	-40.2	10.5	-47.9	12.4	-56.2	14.3	-65.1	16.5	-74.8	21.1	-96.0
Roof > 27 to 45 degrees	1	10	11.9	-13.0	13.3	-14.6	16.5	-18.0	19.9	-21.8	23.7	-25.9	27.8	-30.4	32.3	-35.3	37.0	-40.5	47.6	-52.0
	1	20	11.6	-12.3	13.0	-13.8	16.0	-17.1	19.4	-20.7	23.0	-24.6	27.0	-28.9	31.4	-33.5	36.0	-38.4	46.3	-49.3
	1	50	11.1	-11.5	12.5	-12.8	15.4	-15.9	18.6	-19.2	22.2	-22.8	26.0	-26.8	30.2	-31.1	34.6	-35.7	44.5	-45.8
	1	100	10.8	-10.8	12.1	-12.1	14.9	-14.9	18.1	-18.1	21.5	-21.5	25.2	-25.2	29.3	-29.3	33.6	-33.6	43.2	-43.2
	2	10	11.9	-15.2	13.3	-17.0	16.5	-21.0	19.9	-25.5	23.7	-30.3	27.8	-35.6	32.3	-41.2	37.0	-47.3	47.6	-60.8
	2	20	11.6	-14.5	13.0	-16.3	16.0	-20.1	19.4	-24.3	23.0	-29.0	27.0	-34.0	31.4	-39.4	36.0	-45.3	46.3	-58.1
	2	50	11.1	-13.7	12.5	-15.3	15.4	-18.9	18.6	-22.9	22.2	-27.2	26.0	-32.0	30.2	-37.1	34.6	-42.5	44.5	-54.6
	2	100	10.8	-13.0	12.1	-14.6	14.9	-18.0	18.1	-21.8	21.5	-25.9	25.2	-30.4	29.3	-35.3	33.6	-40.5	43.2	-52.0
	3	10	11.9	-15.2	13.3	-17.0	16.5	-21.0	19.9	-25.5	23.7	-30.3	27.8	-35.6	32.3	-41.2	37.0	-47.3	47.6	-60.8
	3	20	11.6	-14.5	13.0	-16.3	16.0	-20.1	19.4	-24.3	23.0	-29.0	27.0	-34.0	31.4	-39.4	36.0	-45.3	46.3	-58.1
	3	50	11.1	-13.7	12.5	-15.3	15.4	-18.9	18.6	-22.9	22.2	-27.2	26.0	-32.0	30.2	-37.1	34.6	-42.5	44.5	-54.6
	3	100	10.8	-13.0	12.1	-14.6	14.9	-18.0	18.1	-21.8	21.5	-25.9	25.2	-30.4	29.3	-35.3	33.6	-40.5	43.2	-52.0
Wall	4	10	13.0	-14.1	14.6	-15.8	18.0	-19.5	21.8	-23.6	25.9	-28.1	30.4	-33.0	35.3	-38.2	40.5	-43.9	52.0	-56.4
	4	20	12.4	-13.5	13.9	-15.1	17.2	-18.7	20.8	-22.6	24.7	-26.9	29.0	-31.6	33.7	-36.7	38.7	-42.1	49.6	-54.1
	4	50	11.6	-12.7	13.0	-14.3	16.1	-17.6	19.5	-21.3	23.2	-25.4	27.2	-29.8	31.6	-34.6	36.2	-39.7	46.6	-51.0
	4	100	11.1	-12.2	12.4	-13.6	15.3	-16.8	18.5	-20.4	22.0	-24.2	25.9	-28.4	30.0	-33.0	34.4	-37.8	44.2	-48.6
	4	500	9.7	-10.8	10.9	-12.1	13.4	-14.9	16.2	-18.1	19.3	-21.5	22.7	-25.2	26.3	-29.3	30.2	-33.6	38.8	-43.2
	5	10	13.0	-17.4	14.6	-19.5	18.0	-24.1	21.8	-29.1	25.9	-34.7	30.4	-40.7	35.3	-47.2	40.5	-54.2	52.0	-69.6
	5	20	12.4	-16.2	13.9	-18.2	17.2	-22.5	20.8	-27.2	24.7	-32.4	29.0	-38.0	33.7	-44.0	38.7	-50.5	49.6	-64.9
	5	50	11.6	-14.7	13.0	-16.5	16.1	-20.3	19.5	-24.6	23.2	-29.3	27.2	-34.3	31.6	-39.8	36.2	-45.7	46.6	-58.7
	5	100	11.1	-13.5	12.4	-15.1	15.3	-18.7	18.5	-22.6	22.0	-26.9	25.9	-31.6	30.0	-36.7	34.4	-42.1	44.2	-54.1
	5	500	9.7	-10.8	10.9	-12.1	13.4	-14.9	16.2	-18.1	19.3	-21.5	22.7	-25.2	26.3	-29.3	30.2	-33.6	38.8	-43.2

For SI: 1 foot = 304.8 mm, 1 degree = 0.0174 rad, 1 mile per hour = 0.44 m/s, 1 pound per square foot = 47.9 N/m^2.

TABLE 1609.6.2.1(3)
ROOF OVERHANG NET DESIGN WIND PRESSURE (COMPONENT AND CLADDING), p_{net30} (Exposure B at h = 30 feet with I_w = 1.0) (psf)

	ZONE	EFFECTIVE WIND AREA (sq. ft.)	BASIC WIND SPEED V (mph—3-second gust)							
			90	100	110	120	130	140	150	170
Roof 0 to 7 degrees	2	10	-21.0	-25.9	-31.4	-37.3	-43.8	-50.8	-58.3	-74.9
	2	20	-20.6	-25.5	-30.8	-36.7	-43.0	-49.9	-57.3	-73.6
	2	50	-20.1	-24.9	-30.1	-35.8	-42.0	-48.7	-55.9	-71.8
	2	100	-19.8	-24.4	-29.5	-35.1	-41.2	-47.8	-54.9	-70.5
	3	10	-34.6	-42.7	-51.6	-61.5	-72.1	-83.7	-96.0	-123.4
	3	20	-27.1	-33.5	-40.5	-48.3	-56.6	-65.7	-75.4	-96.8
	3	50	-17.3	-21.4	-25.9	-30.8	-36.1	-41.9	-48.1	-61.8
	3	100	-10.0	-12.2	-14.8	-17.6	-20.6	-23.9	-27.4	-35.2
Roof > 7 to 27 degrees	2	10	-27.2	-33.5	-40.6	-48.3	-56.7	-65.7	-75.5	-96.9
	2	20	-27.2	-33.5	-40.6	-48.3	-56.7	-65.7	-75.5	-96.9
	2	50	-27.2	-33.5	-40.6	-48.3	-56.7	-65.7	-75.5	-96.9
	2	100	-27.2	-33.5	-40.6	-48.3	-56.7	-65.7	-75.5	-96.9
	3	10	-45.7	-56.4	-68.3	-81.2	-95.3	-110.6	-126.9	-163.0
	3	20	-41.2	-50.9	-61.6	-73.3	-86.0	-99.8	-114.5	-147.1
	3	50	-35.3	-43.6	-52.8	-62.8	-73.7	-85.5	-98.1	-126.1
	3	100	-30.9	-38.1	-46.1	-54.9	-64.4	-74.7	-85.8	-110.1
Roof > 27 to 45 degrees	2	10	-24.7	-30.5	-36.9	-43.9	-51.5	-59.8	-68.6	-88.1
	2	20	-24.0	-29.6	-35.8	-42.6	-50.0	-58.0	-66.5	-85.5
	2	50	-23.0	-28.4	-34.3	-40.8	-47.9	-55.6	-63.8	-82.0
	2	100	-22.2	-27.4	-33.2	-39.5	-46.4	-53.8	-61.7	-79.3
	3	10	-24.7	-30.5	-36.9	-43.9	-51.5	-59.8	-68.6	-88.1
	3	20	-24.0	-29.6	-35.8	-42.6	-50.0	-58.0	-66.5	-85.5
	3	50	-23.0	-28.4	-34.3	-40.8	-47.9	-55.5	-63.8	-82.2
	3	100	-22.2	-27.4	-33.2	-39.5	-46.4	-53.8	-61.7	-79.3

For SI: 1 foot = 304.8 mm, 1 degree = 0.0174 rad, 1 mile per hour = 0.45 m/s, 1 pound per square foot = 47.9 N/m^2.
Note: For effective areas between those given above, the load is permitted to be interpolated, otherwise use the load associated with the lower effective area.

TABLE 1609.6.2.1(4)
ADJUSTMENT FACTOR FOR BUILDING HEIGHT AND EXPOSURE, (λ)

MEAN ROOF HEIGHT (feet)	EXPOSURE		
	B	C	D
15	1.00	1.21	1.47
20	1.00	1.29	1.55
25	1.00	1.35	1.61
30	1.00	1.40	1.66
35	1.05	1.45	1.70
40	1.09	1.49	1.74
45	1.12	1.53	1.78
50	1.16	1.56	1.81
55	1.19	1.59	1.84
60	1.22	1.62	1.87

For SI: 1 foot = 304.8 mm.
a. All table values shall be adjusted for other exposures and heights by multiplying by the above coefficients.

1609.6.2.2 Components and cladding. Net design wind pressures, p_{net}, for the components and cladding of buildings represent the net pressures (sum of internal and external) to be applied normal to each building surface as shown in Figure 1609.6.2.2. The net design wind pressure, p_{net}, shall be determined from Equation 16-35:

$$p_{net} = \lambda I_w p_{net30} \quad \textbf{(Equation 16-35)}$$

where:

λ = Adjustment factor for building height and exposure from Table 1609.6.2.1(4).

I_w = Importance factor as defined in Section 1609.5.

p_{net30} = Net design wind pressure for Exposure B, at h = 30 feet (9144 mm), and for I_w = 1.0, from Tables 1609.6.2.1(2) and 1609.6.2.1(3).

❖ As described in the commentary to Section 1609.6.1, Tables 1609.6.2.1(2) and 1609.6.2.1(3) tabulate component and cladding pressures for buildings with a mean roof height of 30 feet (9144 mm) located in Exposure Category B. For other heights and exposures, the values simply have to be multiplied by the appropriate height and exposure adjustment factor from Table 1609.6.2.1(4). Additionally, the loads are required to be multiplied by the importance factor, I_w, from Table 1604.5 (see examples following Section 1609).

FIGURE 1609.6.2.2. See page 16-53.

❖ This figure is to be used with Tables 1609.6.2.1(2) and 1609.6.2.1(3) for the determination of component and cladding wind pressures. The figure identifies the location and dimensions of the various zones to which the component and cladding pressures must be applied. The zones are given in the second column of Tables 1609.6.2.1(2) and 1609.6.2.1(3).

1609.6.2.2.1 Minimum pressures. The positive design wind pressures, p_{net}, from Section 1609.6.2.2 shall not be less than +10 psf (0.48 kN/m^2), and the negative design wind pressures, p_{net}, from Section 1609.6.2.2 shall not be less than -10 psf (-0.48 kN/m^2).

❖ This section reiterates Section 1609.1.2 requirements for minimum wind loads in terms of the minimum positive and negative values for component and cladding design wind pressure, p_{net}.

1609.6.2.3 Load case. Members that act as both part of the main windforce-resisting system and as components and cladding shall be designed for each separate load case.

❖ Some elements of a building will function as part of the MWFRS and components and cladding including, but not limited to, roof panels, rafters and wall studs. These elements are required to be designed using loads that would occur by considering the element as part of the MWFRS, and separately checked (or designed) for the loads that would occur by considering the element as components and cladding. The element should be sized according to the more critical loading condition.

The maximum horizontal wall loads from Table 1609.6.2.1(1) are the maximum calculated horizontal loads for a windward or leeward wall from MWFRS loads. These loads are used when considering a wall element as part of the MWFRS to evaluate the effects of interaction with vertical loads from the construction above. The wall element should then be checked (or designed) for component and cladding loads based on its location in the building and its effective wind area.

1609.7 Roof systems.

1609.7.1 Roof deck. The roof deck shall be designed to withstand the wind pressures determined under either the provisions of Section 1609.6 for buildings with a mean roof height not exceeding 60 feet (18 288 mm) or Section 1609.1.1 for buildings of any height.

❖ This section specifies the wind load criteria for the roof deck. The roof deck is a structural component of the building and is required to resist the applicable wind pressures from Section 1609.6 or ASCE 7. This section is referenced by Section 1609.7.2 as the criteria for the wind design for roof coverings.

1609.7.2 Roof coverings. Roof coverings shall comply with Section 1609.7.1.

Exception: Rigid tile roof coverings that are air permeable and installed over a roof deck complying with Section 1609.7.1 are permitted to be designed in accordance with Section 1609.7.3.

❖ This section establishes the wind design criteria for roof coverings. The exception references the use of Section 1609.7.3 for air-permeable rigid tile roof coverings. If the roof deck is relatively impermeable wind pressures will act through it to the building frame system. The roof covering may or may not be subjected to the same wind pressures as the roof deck. If the roof covering is also relatively impermeable and fastened to the roof deck, the two components will react to and resist the same wind pressures. If the roof covering is not impermeable, the wind pressures will be able to develop on both the top of, and underneath, the roof covering. This “venting action” will negate some wind pressure on the roof covering.

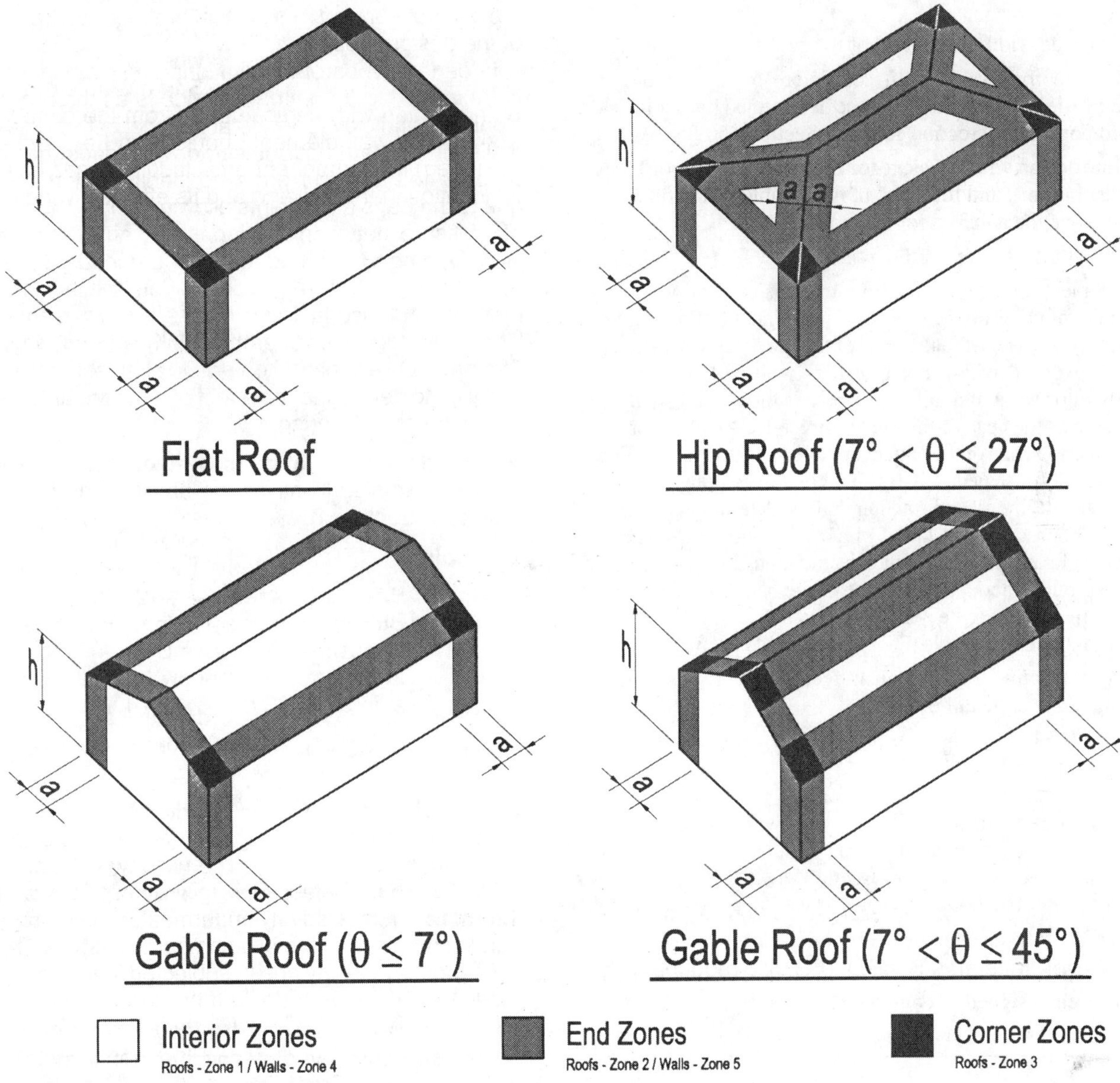

FIGURE 1609.6.2.2
COMPONENT AND CLADDING PRESSURE

For SI: 1 foot = 304.8 mm, 1 degree = 0.0174 rad.

Notes:

1. Pressures are applied normal to the surface for Exposure B, at h = 30 feet, for I_w = 1.0. Adjust to other exposures and heights with adjustment factor λ.
2. Plus and minus signs signify pressures acting toward and away from the surfaces, respectively.
3. For hip roofs with $\theta \leq 25°$, Zone 3 shall be treated as Zone 2.
4. For effective areas between those given, the value is permitted to be interpolated, otherwise use the value associated with the lower effective area.
5. Notation:
   a: 10 percent of least horizontal dimension or 0.4h, whichever is smaller, but not less than either 4 percent of least horizontal dimension or 3 feet.
   h: Mean roof height, in feet (meters), except that eave height shall be used for roof angles <10°.
   θ: Angle of plane of roof from horizontal, in degrees.

1609.7.3 Rigid tile. Wind loads on rigid tile roof coverings shall be determined in accordance with the following equation:

$$M_a = q_h C_L b L L_a [1.0 - Gc_p] \qquad \textbf{(Equation 16-36)}$$

For SI: $M_a = \dfrac{q_h C_L b L L_a [1.0 - Gc_p]}{1{,}000}$

where:

b = Exposed width, feet (mm) of the roof tile.

C_L = Lift coefficient. The lift coefficient for concrete and clay tile shall be 0.2 or shall be determined by test in accordance with Section 1715.2.

GC_p = Roof pressure coefficient for each applicable roof zone determined from Section 6 of ASCE 7. Roof coefficients shall not be adjusted for internal pressure.

L = Length, feet (mm) of the roof tile.

L_a = Moment arm, feet (mm) from the axis of rotation to the point of uplift on the roof tile. The point of uplift shall be taken at 0.76*L* from the head of the tile and the middle of the exposed width. For roof tiles with nails or screws (with or without a tail clip), the axis of rotation shall be taken as the head of the tile for direct deck application or as the top edge of the batten for battened applications. For roof tiles fastened only by a nail or screw along the side of the tile, the axis of rotation shall be determined by testing. For roof tiles installed with battens and fastened only by a clip near the tail of the tile, the moment arm shall be determined about the top edge of the batten with consideration given for the point of rotation of the tiles based on straight bond or broken bond and the tile profile.

M_a = Aerodynamic uplift moment, feet-pounds (N-mm) acting to raise the tail of the tile.

q_h = Wind velocity pressure, psf (kN/m^2) determined from Section 6.5.10 of ASCE 7.

Concrete and clay roof tiles complying with the following limitations shall be designed to withstand the aerodynamic uplift moment as determined by this section.

1. The roof tiles shall be either loose laid on battens, mechanically fastened, mortar set or adhesive set.
2. The roof tiles shall be installed on solid sheathing which has been designed as components and cladding.
3. An underlayment shall be installed in accordance with Chapter 15.
4. The tile shall be single lapped interlocking with a minimum head lap of not less than 2 inches (51 mm).
5. The length of the tile shall be between 1.0 and 1.75 feet (305 mm and 533 mm).
6. The exposed width of the tile shall be between 0.67 and 1.25 feet (204 mm and 381 mm).
7. The maximum thickness of the tail of the tile shall not exceed 1.3 inches (33 mm).
8. Roof tiles using mortar set or adhesive set systems shall have at least two-thirds of the tile's area free of mortar or adhesive contact.

❖ This section includes the wind design method for clay or concrete rigid tile roofs. The method consists of the calculation of the aerodynamic uplift moment from the wind that acts to raise the end of the tile. This section includes the characteristics and the type of installation of the concrete or clay roof tile that are required for the use of the design method.

In certain types of installations, the roof covering is not exposed to the same wind loads as the roof deck. Concrete and clay roof tiles are typical of this type of installation and are not subject to wind loads that would be obtained from current wind loading criteria. This is due to the gaps at tile joints allowing some equalization of pressure between the inner and outer face of the tiles, leading to reduced loads. A procedure has been developed through research for determining the uplift moment on loose-laid and mechanically fastened roof tiles when laid over sheathing with an underlayment. The procedure is based on practical measurements on real tiles to determine the effect of air being able to penetrate the roof covering.

Examples of calculation of wind pressures: The following examples illustrate the application of the wind criteria in Section 1609.

Example 1

Given: Basic wind speed, *V* = 120 mph
Building mean roof height, *H* = 45 feet
Exposure Category = D
20-square-foot window located in an edge strip of wall (Zone 5) I_w = 1.0

Find: Design component and cladding wind pressure

Obtain the component and cladding design wind pressures for a building with *H* = 30 feet and Exposure B, from Table 1609.6.2.1(2). The window is located in a wall, Zone 5, an effective wind area of 20 ft^2, and *V* = 120 mph. From Table 1609.6.2.1(2), the design pressures are:

p_{net30} = + 24.7 psf, -32.4 psf

These pressures have to be modified for mean roof height, exposure category and importance factor using Equation 16-35. From Table 1609.6.2.1(4), the height and exposure adjustment factor, λ, for *H* = 45 feet (13 716 mm) and Exposure D is 1.78.

Therefore, the design wind pressures are calculated as:

p_{net} = (+24.7 psf) x 1.78 x 1.0 = + 44 psf

p_{net} = (-32.4 psf) x 1.78 x 1.0 = -57.7 psf

Example 2

See Commentary Figure 1609.6(4) below.

Given: Simple diaphragm building
V = 130 mph
Exposure Category = C
Roof slope = 7:12
Building width, W = 48 feet
Building length, L = 50 feet
Wall height = 18 ft. I_w = 1.0

Find: Horizontal MWFRS wind loads for end zones.

First, the mean roof height, H, has to be determined. From the building geometry, H is calculated as follows:

$$H = 18 \text{ ft} + {}^{1}/_{2}(7/12)(48\text{ft}/^{2}) = 25 \text{ feet}$$

Roof slope = 7:12 . 30 degrees

Horizontal MWFRS wind loads for end zones, A and B in Figure 1609.2.1 are given in Table 1609.6.2.1(1) for H = 30 feet and Exposure B. For V = 130 and roof angle of 30 degrees, the applicable end zone horizontal loads are

Transverse direction (Zone A - wall)

p_{s30} = 30.1 psf

Transverse direction (Zone B - roof)

p_{s30} = 20.6 psf

Longitudinal direction (Zone A)

p_{s30} = 30.1 psf

These pressures have to be modified for mean roof height, exposure category and importance factor using Equation 16-34. From Table 1609.6.2.1(4), the height and exposure adjustment factor, λ, for H = 25 (7620 mm) feet and Exposure C is 1.35.

Therefore, the design horizontal wind loads are as follows:

p_s = (30.1 psf) x 1.35 x 1.0 = 40.6 psf (Zone A)

p_s = (20.6 psf) x 1.35 x 1.0 = 27.8 psf (Zone B)

These loads are to be applied to the MWFRS in accordance with Figure 1609.6(1) as shown in Commentary Figures 1609.6(5) and 1609.6(6).

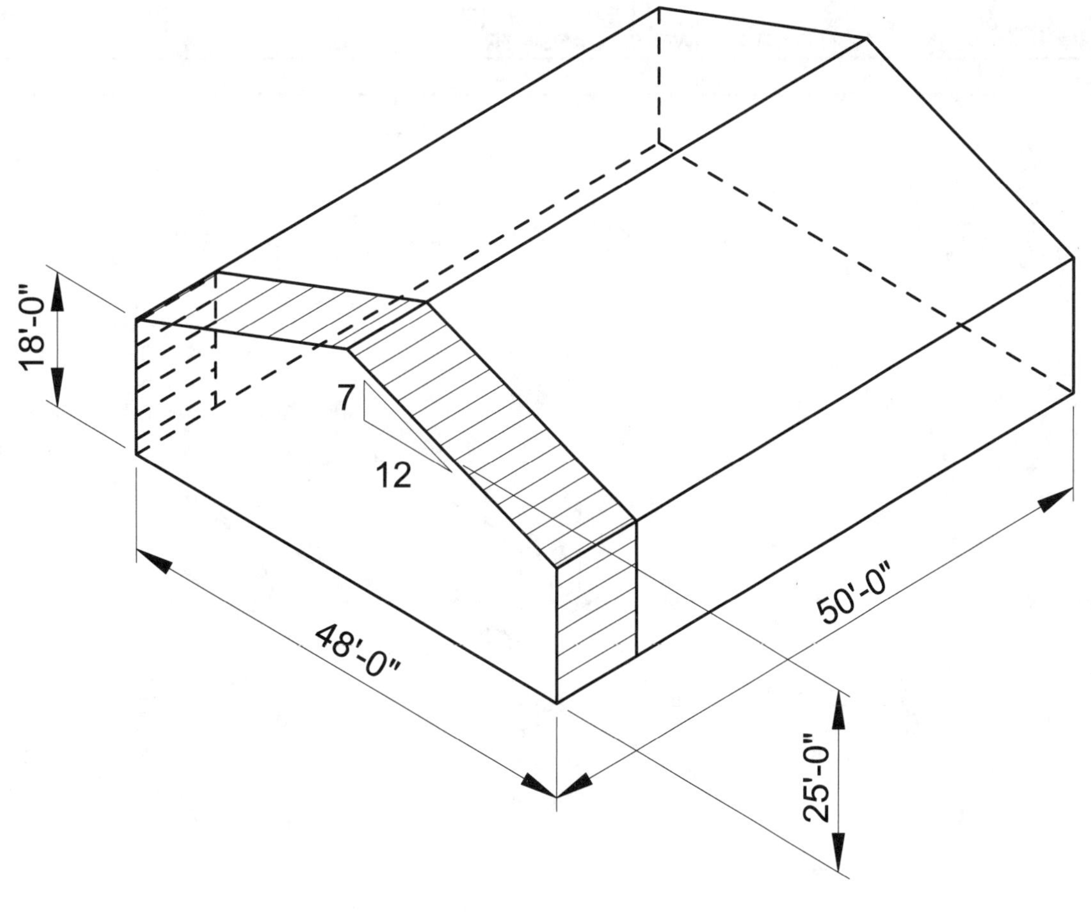

Figure 1609.6(4)
EXAMPLE 2 BUILDING

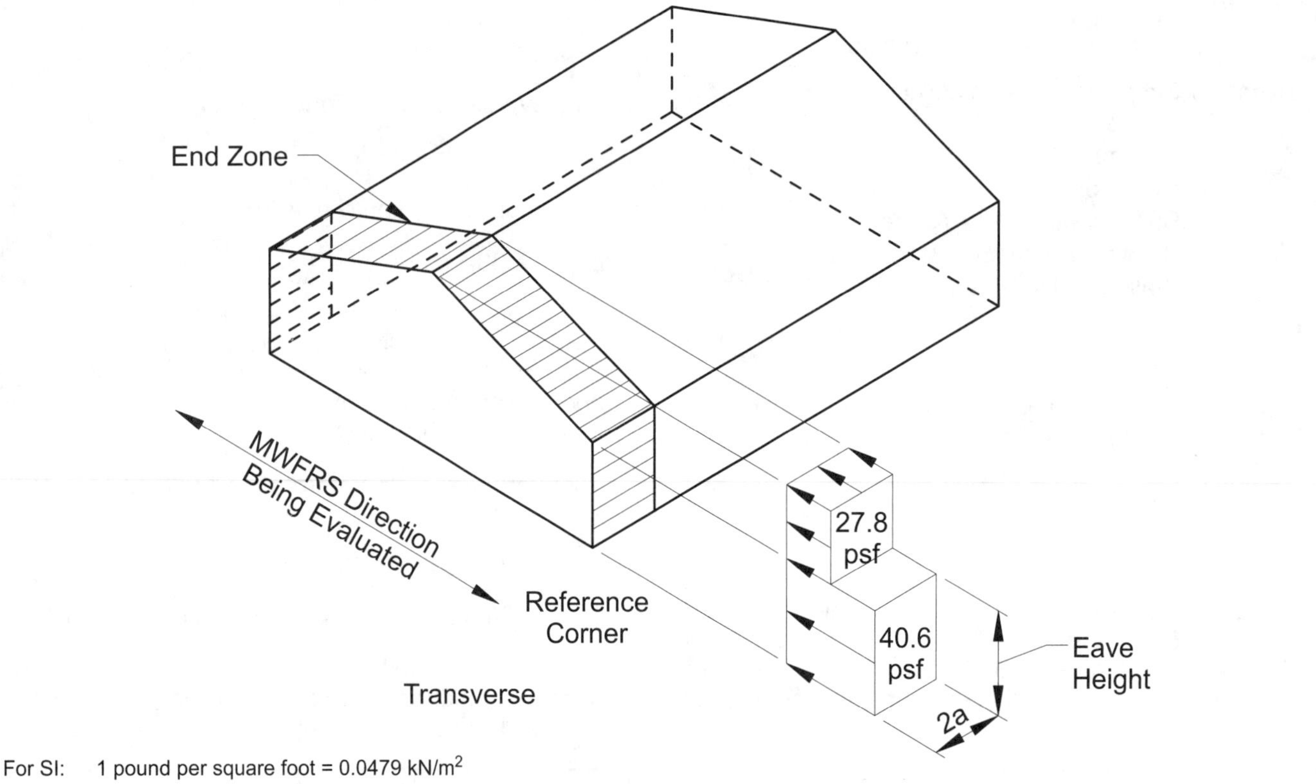

For SI: 1 pound per square foot = 0.0479 kN/m^2

Figure 1609.6(5)
HORIZONTAL MWFRS LOADS—TRANSVERSE DIRECTION

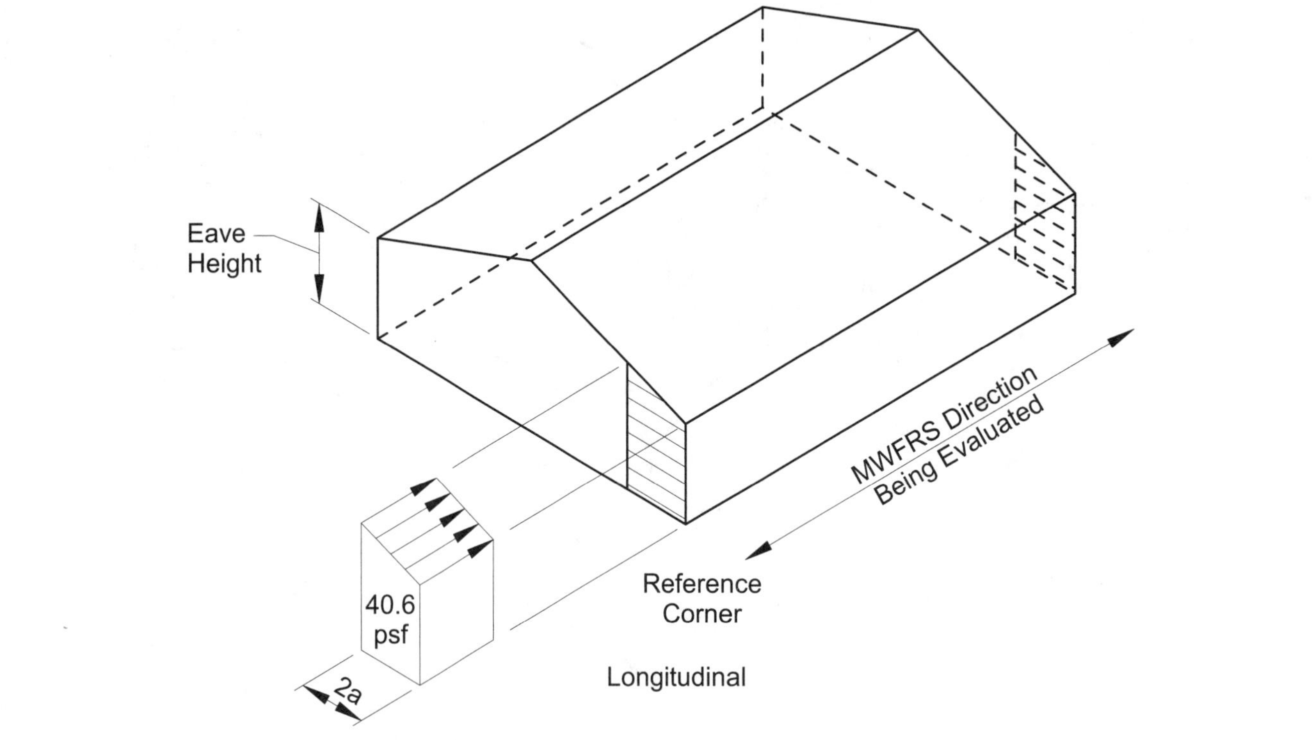

For SI: 1 pound per square foot = 0.0479 kN/m^2.

Figure 1609.6(6)
HORIZONTAL MWFRS LOADS—LONGITUDINAL DIRECTION

SECTION 1610
SOIL LATERAL LOAD

1610.1 General. Basement, foundation and retaining walls shall be designed to resist lateral soil loads. Soil loads specified in Table 1610.1 shall be used as the minimum design lateral soil loads unless specified otherwise in a soil investigation report approved by the building official. Basement walls and other walls in which horizontal movement is restricted at the top shall be designed for at-rest pressure. Retaining walls free to move and rotate at the top are permitted to be designed for active pressure. Design lateral pressure from surcharge loads shall be added to the lateral earth pressure load. Design lateral pressure shall be increased if soils with expansion potential are present at the site.

Exception: Basement walls extending not more than 8 feet (2438 mm) below grade and supporting flexible floor systems shall be permitted to be designed for active pressure.

❖ This section provides lateral loads for various soil types. This section requires that foundation and retaining walls be designed to be capable of resisting the lateral soil loads specified by Table 1610.1 where a specific soil investigation has not occurred.

Table 1610.1. See below.

❖ The table lists at-rest and active soil pressures for a number of different types of moist soils. The basis of the soil classification into the various types listed is ASTM D 2487. Soils identified by Note b in Table 1610.1 have unpredictable characteristics. These are called expansive soils. Because of their ability to absorb water, they shrink and swell to a higher degree than other soils. As expansive soils swell, they are capable of exerting large forces on soil-retaining structures; thus, these types of soils are not to be used as backfill.

SECTION 1611
RAIN LOADS

1611.1 Design rain loads. Each portion of a roof shall be designed to sustain the load of rainwater that will accumulate on it if the primary drainage system for that portion is blocked plus

TABLE 1610.1
SOIL LATERAL LOAD

DESCRIPTION OF BACKFILL MATERIAL[c]	UNIFIED SOIL CLASSIFICATION	DESIGN LATERAL SOIL LOAD[a] (pound per square foot per foot of depth)	
		Active pressure	At-rest pressure
Well-graded, clean gravels; gravel-sand mixes	GW	30	60
Poorly graded clean gravels; gravel-sand mixes	GP	30	60
Silty gravels, poorly graded gravel-sand mixes	GM	40	60
Clayey gravels, poorly graded gravel-and-clay mixes	GC	45	60
Well-graded, clean sands; gravelly sand mixes	SW	30	60
Poorly graded clean sands; sand-gravel mixes	SP	30	60
Silty sands, poorly graded sand-silt mixes	SM	45	60
Sand-silt clay mix with plastic fines	SM-SC	45	100
Clayey sands, poorly graded sand-clay mixes	SC	60	100
Inorganic silts and clayey silts	ML	45	100
Mixture of inorganic silt and clay	ML-CL	60	100
Inorganic clays of low to medium plasticity	CL	60	100
Organic silts and silt clays, low plasticity	OL	Note b	Note b
Inorganic clayey silts, elastic silts	MH	Note b	Note b
Inorganic clays of high plasticity	CH	Note b	Note b
Organic clays and silty clays	OH	Note b	Note b

For SI: 1 pound per square foot per foot of depth = 0.157 kPa/m, 1 foot = 304.8 mm.

a. Design lateral soil loads are given for moist conditions for the specified soils at their optimum densities. Actual field conditions shall govern. Submerged or saturated soil pressures shall include the weight of the buoyant soil plus the hydrostatic loads.

b. Unsuitable as backfill material.

c. The definition and classification of soil materials shall be in accordance with ASTM D 2487.

the uniform load caused by water that rises above the inlet of the secondary drainage system at its design flow.

$$R = 5.2\,(d_s + d_h) \qquad \textbf{(Equation 16-37)}$$

For SI: $R = 0.0098\,(d_s + d_h)$

where:

d_h = Additional depth of water on the undeflected roof above the inlet of secondary drainage system at its design flow (i.e., the hydraulic head), in inches (mm).

d_s = Depth of water on the undeflected roof up to the inlet of secondary drainage system when the primary drainage system is blocked (i.e., the static head), in inches (mm).

R = Rain load on the undeflected roof, in psf (kN/m^2). When the phrase "undeflected roof" is used, deflections from loads (including dead loads) shall not be considered when determining the amount of rain on the roof.

❖ While the intent in designing a roof drainage system is to prevent accumulation of water, it is not uncommon to find that roof drains have become clogged, roof pitches are inadequate to direct water to the drains, drains are located at levels higher than the low points of a roof and such other reasons that will cause roof ponding. Numerous roof failures have been caused by increased loads from ponding water and ice buildup. The roof is to be capable of resisting the maximum water depth, which occurs when all primary means of the roof drainage system are assumed to be blocked. The blockage is generally caused by debris at the entrance to the primary roof drains but can occur anywhere along the primary piping system, such as an under-slab collapse. Elevated independent secondary roof drains or overflow scuppers are commonly provided for roofs bounded by parapets in order to alleviate blockage of the primary roof drainage system. Note that primary drain lines are considered part of the primary roof drainage system; therefore, because of the possibility of the primary drain lines being blocked, the secondary drain lines are to be independent of the primary drain lines. This text is consistent with Section 1107 of the *International Plumbing Code*® (IPC®) and Section 8.3 of ASCE 7.

This section requires the utilization of two variables to determine rain load: the depth of the water on the undeflected roof, as measured from the low point elevation to the inlet elevation of the secondary drain, and the additional depth of water at the secondary drainage flow. The sum of these depths is the design depth for rainwater.

1611.2 Ponding instability. Ponding refers to the retention of water due solely to the deflection of relatively flat roofs. Roofs with a slope less than one-fourth unit vertical in 12 units horizontal (2-percent slope) shall be investigated by structural analysis to ensure that they possess adequate stiffness to preclude progressive deflection (i.e., instability) as rain falls on them or meltwater is created from snow on them. The larger of snow load or rain load shall be used in this analysis. The primary drainage system within an area subjected to ponding shall be considered to be blocked in this analysis.

❖ This section requires a check for ponding instability if the roof slope is less than $^1/_4$ inch per foot (1 in 48). Ponding instability is the continuing deflection under a rain or snow load that results in roof failure before the design snow load or the design drainage flow rate is reached. A ponding instability check is to be made assuming the primary roof drains are blocked. The determination of ponding instability is typically done by an iterative structural analysis where the incremental deflection is determined and the resulting increased ponding load from the deflection is added to the orginal ponding load.

1611.3 Controlled drainage. Roofs equipped with hardware to control the rate of drainage shall be equipped with a secondary drainage system at a higher elevation that limits accumulation of water on the roof above that elevation. Such roofs shall be designed to sustain the load of rainwater that will accumulate on them to the elevation of the secondary drainage system plus the uniform load caused by water that rises above the inlet of the secondary drainage system at its design flow determined from Section 1611.1. Such roofs shall also be checked for ponding instability in accordance with Section 1611.2.

❖ Controlled drainage is the limitation of the drainage flow rate to a low flow rate that is less than the rainfall rate such that the depth of the water builds up on the roof during a heavy rainfall. A secondary roof drain system is needed to limit the buildup of water to a specific depth for roof design. The depth of water on the roof is to include also the depth of the water above the inlet of the secondary drain when the design flow rate is reached.

SECTION 1612
FLOOD LOADS

1612.1 General. Within flood hazard areas as established in Section 1612.3, all new construction of buildings, structures and portions of buildings and structures, including substantial improvements and restoration of substantial damage to buildings and structures, shall be designed and constructed to resist the effects of flood hazards and flood loads.

❖ This section addresses additional requirements for all buildings and structures in designated flood hazard areas. These areas are commonly referred to as "floodplains" and are shown on a community's Flood Insurance Rate Map (FIRM), prepared by FEMA, or other adopted flood hazard map. Through the adoption of the code, communities meet a significant portion of the floodplain management regulation requirements necessary to participate in the NFIP. To participate in the NFIP, a jurisdiction must adopt, in addition to the flood-resistant requirements found in the code, Appendix G *Flood Resistant Construction* or a floodplain management ordinance that contains, at a minimum, the provisions contained in Appendix G. In either case, floodplain management requirements must be applied to all developments in designated flood hazard areas. Appendix G, Section 201.2, defines "Development" as "any man-made change to improved or unimproved real estate, including but not limited to buildings or other structures; temporary or permanent storage of materials; mining; dredging; filling; grading; paving; excavations; operations and other land-disturbing activities."

The NFIP was established to reduce flood losses to better indemnify individuals from flood losses and to reduce federal expenditures for disaster assistance. A community that has been determined to have flood hazard areas participates in the NFIP to protect health, safety and property and so that its citizens can purchase federally backed flood insurance. FEMA administers the NFIP, which includes monitoring community compliance with the floodplain management requirements of the NFIP.

Some communities choose to apply flood-resistant construction provisions to flood hazard areas that are more extensive than those shown on the community's FIRM. New buildings and structures, and substantial improvements to existing buildings and structures in flood hazard areas, are to be designed and constructed to resist flood forces to minimize damage. Flood forces include flotation, lateral (hydrostatic) pressures, moving water (hydrodynamic) pressures, wave impact and debris impact.

It is important to note that many states and communities have elected to regulate floodplain development to a higher standard than the minimum required to participate in the NFIP. Communities considering using the code and other *International Codes*® to meet the floodplain management requirements of the NFIP are advised to consult with their state NFIP coordinator or the appropriate FEMA regional office.

If located in designated flood hazard areas, buildings and structures that are damaged by any cause are to be examined to determine if the damage constitutes substantial damage, in which the cost of repairing/restoring the building or structure to its predamaged condition equals or exceeds 50 percent of its market value before the damage occurred. All substantial improvements and repairs of buildings and structures that are substantially damaged are to meet the flood-resistant provisions of the code.

To provide additional guidance on how communities may use the code and other *International Codes* to participate in the NFIP, FEMA, the International Code Council® (ICC®), the American Society of Civil Engineers (ASCE), the Association of State Floodplain Managers (ASFPM) and others have developed a publication, *Reducing Flood Losses Through the International Code Series: Meeting the Requirements of National Flood Insurance Program.*

1612.2 Definitions. The following words and terms shall, for the purposes of this section, have the meanings shown herein.

❖ The definitions in this section are needed to apply the flood load requirements of the code.

BASE FLOOD. The flood having a 1-percent chance of being equaled or exceeded in any given year.

❖ This term is used to define the land area along a body of water that is subject to flooding and within which flood-resistant design and construction requirements are applied. Typically, an authority such as FEMA or a state or local jurisdiction prepares flood hazard maps. In doing so, the base flood is derived by examining regional or local flood history to determine the 1-percent annual chance event, also referred to as the "100-year flood." The 1-percent annual chance event has been the standard used to regulate flood hazard areas for more than 30 years.

BASE FLOOD ELEVATION. The elevation of the base flood, including wave height, relative to the National Geodetic Vertical Datum (NGVD), North American Vertical Datum (NAVD) or other datum specified on the Flood Insurance Rate Map (FIRM).

❖ The base flood elevation is the height to which floodwaters will rise during passage or occurrence of the base flood. FEMA uses commonly accepted computer models that estimate hydrologic and hydraulic conditions to determine the annual chance (base) flood. Along rivers and streams, statistical methods and computer models may have been used to estimate runoff and to develop flood elevations. The models take into consideration watershed characteristics and the shape and nature of the floodplain, including natural ground contours and the presence of buildings, bridges and culverts. Along coastal areas, base flood elevations may be developed using models that take into account offshore bathymetry, historical storms and typical wind patterns. In many coastal areas, the base flood elevation includes wave heights.

BASEMENT. The portion of a building having its floor subgrade (below ground level) on all sides.

❖ This term, which is used only for buildings in flood hazard areas that are subject to the provisions of Section 1612, differs from the definition in Section 202. The general definition includes as basements those portions of buildings that are partially below grade. For the purpose of resisting flood damage, lack of drainage and a corresponding buildup of hydrostatic pressure on foundation walls are of concern when an area is below grade on all sides. In terms of federal flood insurance, buildings with basements (below grade on all sides) are subject to higher premium rates.

DESIGN FLOOD. The flood associated with the greater of the following two areas:

1. Area with a flood plain subject to a 1-percent or greater chance of flooding in any year; or
2. Area designated as a flood hazard area on a community's flood hazard map, or otherwise legally designated.

❖ The design flood is either the base flood (as defined above) or another flood determined based on other criteria. A state or local jurisdiction may choose to prepare and adopt flood hazard maps that show flood hazard areas that are not on maps prepared by FEMA. These may be areas that FEMA did not study or were studied with different criteria. For example, some communities elect to prepare flood hazard maps based on the assumption that the upland watershed is built out to existing zoning, often called ultimate development.

DESIGN FLOOD ELEVATION. The elevation of the "design flood," including wave height, relative to the datum specified on the community's legally designated flood hazard map. In areas designated as Zone AO, the design flood elevation shall be the elevation of the highest existing grade of the building's perimeter plus the depth number (in feet) specified on the flood hazard map. In areas designated as Zone AO where a depth number is not specified on the map, the depth number shall be taken as being equal to 2 feet (610 mm).

❖ The design flood elevation is the height to which floodwaters will rise during passage or occurrence of the design flood. The datum specified on the flood hazard map is important because it may differ from that used locally for other purposes.

Some flood hazard maps have areas denoted as AO zones, identifying areas subject to sheet flow flooding or ponding. Specific instructions for determining the design flood elevation of these zones are provided, because the maps specify a flood depth rather than a height above datum.

DRY FLOODPROOFING. A combination of design modifications that results in a building or structure, including the attendant utility and sanitary facilities, being water tight with walls substantially impermeable to the passage of water and with structural components having the capacity to resist loads as identified in ASCE 7.

❖ All residential buildings and most nonresidential buildings in flood hazard areas are elevated so that the lowest floors are at or above the design flood elevation. Under certain circumstances, nonresidential buildings may be designed and constructed with enclosed areas below the design flood elevation, provided they incorporate dry floodproofing modifications. In addition to keeping water out of enclosed areas, dry floodproofing modifications are to be designed to withstand hydrostatic and hydrodynamic loads.

EXISTING CONSTRUCTION. Any buildings and structures for which the "start of construction" commenced before the effective date of the community's first flood plain management code, ordinance or standard. "Existing construction" is also referred to as "existing structures."

❖ For the NFIP, buildings and structures are considered to be "existing" if construction started before the local jurisdiction adopted its first regulation governing development in flood hazard areas. This distinction is important because new construction, for which the start of construction is after the date of the first regulation, is to be in full compliance with the regulation. When activities are proposed for existing construction or structures, the costs of those activities are to be examined, and, if the proposed activities are a "substantial improvement" as defined below, then the entire existing structure is to be brought into compliance with the flood-resistant provisions of the code.

EXISTING STRUCTURE. See "Existing construction."

❖ See the commentary for the definition of "Existing construction."

FLOOD or FLOODING. A general and temporary condition of partial or complete inundation of normally dry land from:

1. The overflow of inland or tidal waters.
2. The unusual and rapid accumulation or runoff of surface waters from any source.

❖ The term "flood" is broadly defined because the condition may occur at all types and sizes of bodies of water. Along streams and riverine areas, flooding results from the accumulation of rainfall runoff that drains from upland watersheds. Along coasts and the shorelines of large lakes, flooding is caused by wind-driven surges and waves that push water onshore, often augmented by tidal influences.

FLOOD DAMAGE-RESISTANT MATERIALS. Any construction material capable of withstanding direct and prolonged contact with floodwaters without sustaining any damage that requires more than cosmetic repair.

❖ A building or structure located within a flood hazard area will have certain structural and nonstructural elements below the design flood elevation, even when it is elevated to be compliant with the requirements of Section 1612. To minimize damage and facilitate cleanup, the materials composing those elements are to be resistant to damage by floodwater. In flood hazard areas subject to high velocity wave action (commonly called "V Zones"), walls that are designed to break away during the design flood may be wetted during floods of lesser magnitude. When flood damage-resistant materials are used, the owners need only perform routine cleanup. In riverine and inland coastal flood hazard areas (commonly called "A Zones"), certain enclosures may be allowed if designed and constructed with openings that allow floodwater to enter and exit without causing structural damage. To minimize costs and facilitate cleanup, materials used for those enclosures are to resist flood damage. The National Evaluation Service (NES) has developed a protocol providing guidance for testing flood damage-resistant materials, entitled *Evaluation Plan for Determination of Flood Resistance of Building Elements* (April 2000).

FLOOD HAZARD AREA. The greater of the following two areas:

1. The area within a flood plain subject to a 1-percent or greater chance of flooding in any year.

2. The area designated as a flood hazard area on a community's flood hazard map, or otherwise legally designated.

❖ FEMA prepares FIRMs that delineate the land area that is subject to inundation by the 1-percent annual chance flood. Some states and local jurisdictions develop and adopt maps of flood hazard areas that are more extensive than the areas shown on FEMA's maps. For the purpose of the code, the flood hazard area within which the requirements are to be applied is the greater of the two delineated areas.

FLOOD HAZARD AREA SUBJECT TO HIGH VELOCITY WAVE ACTION. Area within the flood hazard area that is subject to high velocity wave action, and shown on a Flood Insurance Rate Map (FIRM) or other flood hazard map as Zone V, VO, VE or V1-30.

❖ Some coastal and lake shorelines experience flooding that includes wind-driven waves. Flood hazard areas that are anticipated to have high velocity wave action, which changes anticipated flood loads, are shown on FEMA's flood hazard area maps and commonly are referred to as "V Zones."

FLOOD INSURANCE RATE MAP (FIRM). An official map of a community on which the Federal Emergency Management Agency (FEMA) has delineated both the special flood hazard areas and the risk premium zones applicable to the community.

❖ The FIRM shows the flood hazard areas along bodies of water where some level of assessment or study by FEMA has determined that there is a risk of flooding. In addition to showing the areal extent of flood hazard, base flood elevations are shown where studies using detailed methods were conducted.

FLOOD INSURANCE STUDY. The official report provided by the Federal Emergency Management Agency containing the Flood Insurance Rate Map (FIRM), the Flood Boundary and Floodway Map (FBFM), the water surface elevation of the base flood and supporting technical data.

❖ A Flood Insurance Study is prepared for each local jurisdiction for which FEMA has evaluated flood hazards. The report summarizes the information that describes and supports the flood hazard area maps, including narratives of records of past floods, descriptions of the methodologies used and (if applicable) floodway data tables with additional information for waterways for which floodways have been designated.

FLOODWAY. The channel of the river, creek or other watercourse and the adjacent land areas that must be reserved in order to discharge the base flood without cumulatively increasing the water surface elevation more than a designated height.

❖ FEMA delineates floodways along most rivers and streams that are studied using detailed methods. The floodway is the area that is to be kept clear of encroachments, such as fill and buildings, in order to pass the base flood without increasing the water surface more than a designated height. The designated height is found by referencing the Floodway Data Table in the Flood Insurance Study. In general, velocities are higher and water depths are greater in floodways than in adjacent fringe areas.

LOWEST FLOOR. The floor of the lowest enclosed area, including basement, but excluding any unfinished or flood-resistant enclosure, usable solely for vehicle parking, building access or limited storage provided that such enclosure is not built so as to render the structure in violation of this section.

❖ The lowest floor is the most important reference point when designing and constructing a building or structure in a flood hazard area. The term is specifically defined to include basements, which are any areas that are below grade on all sides. For elevated buildings, compliance with the flood-resistant design and construction provisions of the code is determined by a survey that certifies that the lowest floor, including basement, is at or above the design flood elevation. It is important to note that certain enclosures may be built below otherwise elevated buildings provided they meet both the design criteria and use limitations (see Section 1612.5.1). If compliant with those limitations, then enclosures are not considered the lowest floor. This distinction also is important because the premium rate used to determine the cost of a federal flood insurance policy is very dependent on the elevation of the lowest floor.

SPECIAL FLOOD HAZARD AREA. The land area subject to flood hazards and shown on a Flood Insurance Rate Map or other flood hazard map as Zone A, AE, A1-30, A99, AR, AO, AH, V, VO, VE or V1-30.

❖ This term is used in the definition of "Flood Insurance Study" and in Section 1612.3, in which the flood hazard area is established. The special flood hazard area is the flood hazard area shown on maps prepared by FEMA.

START OF CONSTRUCTION. The date of permit issuance for new construction and substantial improvements to existing structures, provided the actual start of construction, repair, reconstruction, rehabilitation, addition, placement or other improvement is within 180 days after the date of issuance. The actual start of construction means the first placement of permanent construction of a building (including a manufactured home) on a site, such as the pouring of a slab or footings, installation of pilings or construction of columns.

Permanent construction does not include land preparation (such as clearing, excavation, grading or filling), the installation of streets or walkways, excavation for a basement, footings, piers or foundations, the erection of temporary forms or the installation of accessory buildings such as garages or sheds not occupied as dwelling units or not part of the main building. For a substantial improvement, the actual "start of construction" means the first alteration of any wall, ceiling, floor or other structural part of a building, whether or not that alteration affects the external dimensions of the building.

❖ This term is used in the definition of "Existing construction" and is applicable to improvements of existing buildings in order to capture all improvement work.

SUBSTANTIAL DAMAGE. Damage of any origin sustained by a structure whereby the cost of restoring the structure to its before-damaged condition would equal or exceed 50 percent of the market value of the structure before the damage occurred.

❖ This term is used in the definition of "Substantial improvement." Substantial damage is a special case of substantial improvement, and if the cost of restoring damage equals or exceeds 50 percent of the market value of the structure, then compliance of the existing building is required. It is notable that a substantial damage determination is to be made regardless of what causes damage. Buildings have sustained substantial damage due to flood, fire, wind, earthquake, deterioration and other causes.

SUBSTANTIAL IMPROVEMENT. Any repair, reconstruction, rehabilitation, addition or improvement of a building or structure, the cost of which equals or exceeds 50 percent of the market value of the structure before the improvement or repair is started. If the structure has sustained substantial damage, any repairs are considered substantial improvement regardless of the actual repair work performed. The term does not, however, include either:

1. Any project for improvement of a building required to correct existing health, sanitary or safety code violations identified by the building official and that are the minimum necessary to assure safe living conditions.
2. Any alteration of a historic structure provided that the alteration will not preclude the structure's continued designation as a historic structure.

❖ One of the long-range objectives of the NFIP is to reduce the exposure of older buildings that were built in flood hazard areas before local jurisdictions adopted flood hazard area maps and regulations. Section 105.3 directs the applicant to state the valuation of the proposed work as part of the information submitted to obtain a permit. To make a determination as to whether a proposed repair, reconstruction, rehabilitation, addition or improvement constitutes substantial improvement or damage, the cost of the proposed work is to be compared to the market value of the building or structure before the work is started. In order to determine market value, the building official may require the applicant to provide such information, as allowed under Section 105.3. For additional guidance, refer to FEMA 213, *Answers to Questions About Substantially Damaged Buildings* and FEMA 311, *Guidance on Estimating Substantial Damage Using the NFIP Residential Substantial Damage Estimator.*

1612.3 Establishment of flood hazard areas. To establish flood hazard areas, the governing body shall adopt a flood hazard map and supporting data. The flood hazard map shall include, at a minimum, areas of special flood hazard as identified by the Federal Emergency Management Agency in an engineering report entitled "The Flood Insurance Study for [INSERT NAME OF JURISDICTION]," dated [INSERT DATE OF ISSUANCE], as amended or revised with the accompanying Flood Insurance Rate Map (FIRM) and Flood Boundary and Floodway Map (FBFM) and related supporting data along with any revisions thereto. The adopted flood hazard map and supporting data are hereby adopted by reference and declared to be part of this section.

❖ Flood maps and studies are prepared by FEMA and are to be used as a community's official map, unless the community chooses to adopt a map that shows more extensive flood hazard areas. From time to time, FEMA's floodplain maps and studies may be revised and republished. When maps are revised and flood hazard areas are changed, FEMA involves the community and provides a formal opportunity to review the documents. Once the revisions are finalized, FEMA requires specific adoption of the new maps by the community. Many communities minimize having to adopt each revision by referencing the date of the original map and study and future revisions that may be issued. This is a method by which subsequent revisions to flood maps and studies may be adopted administratively without requiring legislative action on the part of the community.

1612.4 Design and construction. The design and construction of buildings and structures located in flood hazard areas, including flood hazard areas subject to high velocity wave action, shall be in accordance with ASCE 24.

❖ FEMA uses multiple designations for flood hazard areas shown on each FIRM, including AO, AH, A1-30, AE, A99, AR, AR/A1-30, AR/AE, AR/AO, AR/AH, AR/A, VO, or V1-30, VE and V. Along many open coasts and lake shores where wind-driven waves are predicted, the flood hazard area is commonly referred to as the "V Zone." For the purpose of the code and consistency with ASCE 7 and ASCE 24, these areas are called "flood hazard areas subject to high velocity wave action." Flood hazard areas that are inland of areas subject to high velocity wave action, and flood hazard areas along rivers and streams, are commonly referred to as "A Zones." Due to waves, the flood loads in areas subject to high velocity wave action differ from those in other flood hazard areas. ASCE 24 outlines in detail the specific requirements that are to be applied to buildings and structures in all flood hazard areas.

1612.5 Flood hazard documentation. The following documentation shall be prepared and sealed by a registered design professional and submitted to the building official:

1. For construction in flood hazard areas not subject to high-velocity wave action:
   - 1.1. The elevation of the lowest floor, including basement, as required by the lowest floor elevation inspection in Section 109.3.3.

1.2. For fully enclosed areas below the design flood elevation where provisions to allow for the automatic entry and exit of floodwaters do not meet the minimum requirements in Section 2.6.1.1, ASCE 24, construction documents shall include a statement that the design will provide for equalization of hydrostatic flood forces in accordance with Section 2.6.1.2, ASCE 24.

1.3. For dry floodproofed nonresidential buildings, construction documents shall include a statement that the dry floodproofing is designed in accordance with ASCE 24.

2. For construction in flood hazard areas subject to high-velocity wave action:

2.1. The elevation of the bottom of the lowest horizontal structural member as required by the lowest floor elevation inspection in Section 109.3.3.

2.2. Construction documents shall include a statement that the building is designed in accordance with ASCE 24, including that the pile or column foundation and building or structure to be attached thereto is designed to be anchored to resist flotation, collapse and lateral movement due to the effects of wind and flood loads acting simultaneously on all building components, and other load requirements of Chapter 16.

2.3. For breakaway walls designed to resist a nominal load of less than 10 psf (0.48 kN/m^2) or more than 20 psf (0.96 kN/m^2), construction documents shall include a statement that the breakaway wall is designed in accordance with ASCE 24.

❖ The NFIP requires that certain documentation be submitted in order to demonstrate compliance with provisions of the code that cannot be easily verified during a site inspection. The most common, described in Section 1612.5, Item 1.1, is the documentation of the lowest floor elevation. It provides evidence that the lowest floor is at or above the required minimum elevation. Elevation is one of the most important aspects of flood-resistant construction and is a significant factor used to determine flood insurance premium rates. FEMA Form 81-31, *Elevation Certificate,* which includes illustrations and instructions, is recommended [see Figure 1612.5.1(1)]. Building owners need elevation certificates to obtain flood insurance, and insurance agents use the certificates to compute the proper flood insurance premium rates.

The criteria for the minimum number and size of flood openings are set by the NFIP to allow free inflow and outflow of floodwaters under all types of flood conditions. The statement described in Section 1612.5, Item 1.2 is required in the construction documents if the flood openings do not meet the minimum number and size criteria set forth in ASCE 24. For further guidance, refer to FEMA FIA-TB #1, *Openings in Foundation Walls for Buildings Located in Special Flood Hazard Areas.*

The statement described in Section 1612.5, Item 1.3 is to be included in the construction documents for nonresidential buildings that are designed to be dry floodproofed. It is important to note that dry floodproofing is allowed only for nonresidential buildings and structures that are located in flood hazard areas not subject to high-velocity wave action. The registered design professional who seals the construction documents is indicating that, based upon development or review of the structural design, specifications and plans for construction, the design and methods of construction are in accordance with accepted standards of practice for meeting the following provisions: (a) the structure, together with attendant utilities and sanitary facilities, is water tight to the floodproofed design elevation indicated with walls that are substantially impermeable to the passage of water and (b) all structural components are capable of resisting hydrostatic and hydrodynamic flood forces, including the effects of bouyancy and anticipated debris impact forces. The use of FEMA Form 81-65, *Floodproofing Certification,* is recommended [see Figure 1612.5.1(2)]. This certificate is used by insurance agents to determine flood insurance premium rates for dry floodproofed nonresidential buildings. For further guidance, refer to FEMA FIA-TB #3, *Nonresidential Floodproofing—Requirements and Certification for Buildings Located in Special Flood Hazard Areas.*

Certain documentation must be submitted in order to demonstrate compliance with provisions of the code that cannot be verified readily during a site inspection. The most common, in Section 1612.5, Item 2.1, provides evidence that the bottom of the lowest horizontal members of buildings constructed in flood hazard areas subject to high-velocity wave action (V Zones) are elevated to or above the minimum required height.

Buildings located in flood hazard areas subject to high-velocity wave action and winds are expected to experience significant flood and wind loads simultaneously. FEMA and coastal communities report significant damage to buildings that are not built to current code. The statement described in Section 1612.5, Item 2.2 is included in the construction documents to indicate that the design meets the flood load provisions of ASCE 24 and other loads required by this chapter.

The documentation described in Section 1612.5, Item 2.3 is used only for specific situations in which properly elevated buildings in flood hazard areas subject to high-velocity wave action have enclosures beneath them, and then only if the walls of the enclosures are not designed to fail under a nominal load of between 10 psf (0.48 kN/m^2) and 20 psf (0.96 kN/m^2). Such enclosures are allowed only if their uses are restricted to building access, parking and limited storage. In these cases, the walls of the enclosures are to be designed to fail—or break away—under certain loads without transferring significant loads or causing damage to the building. Because breakaway walls will fail under flood conditions, building materials can become water-borne debris that may damage adjacent buildings. Refer to FEMA FIA-TB #5, *Free of Obstruction Requirements for Buildings Located in Coastal High Hazard Areas* and FEMA FIA-TB #9, *Design and Construction Guidance for Breakaway Walls Below Elevated Buildings.*

FEDERAL EMERGENCY MANAGEMENT AGENCY
NATIONAL FLOOD INSURANCE PROGRAM

ELEVATION CERTIFICATE

O.M.B. No. 3067-0077
Expires July 31, 2002

Important: Read the instructions on pages 1 - 7.

SECTION A - PROPERTY OWNER INFORMATION

For Insurance Company Use:
Policy Number
Company NAIC Number

BUILDING OWNER'S NAME

BUILDING STREET ADDRESS (Including Apt., Unit, Suite, and/or Bldg. No.) OR P.O. ROUTE AND BOX NO.

CITY STATE ZIP CODE

PROPERTY DESCRIPTION (Lot and Block Numbers, Tax Parcel Number, Legal Description, etc.)

BUILDING USE (e.g., Residential, Non-residential, Addition, Accessory, etc. Use Comments section if necessary.)

LATITUDE/LONGITUDE (OPTIONAL) (##° - ##' - ##.##" or ##.#####°)
HORIZONTAL DATUM: |__| NAD 1927 |__| NAD 1983
SOURCE: |__| GPS (Type):________ |__| USGS Quad Map |__| Other:________

SECTION B - FLOOD INSURANCE RATE MAP (FIRM) INFORMATION

B1. NFIP COMMUNITY NAME & COMMUNITY NUMBER	B2. COUNTY NAME	B3. STATE

B4. MAP AND PANEL NUMBER	B5. SUFFIX	B6. FIRM INDEX DATE	B7. FIRM PANEL EFFECTIVE/REVISED DATE	B8. FLOOD ZONE(S)	B9. BASE FLOOD ELEVATION(S) (Zone AO, use depth of flooding)

B10. Indicate the source of the Base Flood Elevation (BFE) data or base flood depth entered in B9.
|__| FIS Profile |__| FIRM |__| Community Determined |__| Other (Describe): ________
B11. Indicate the elevation datum used for the BFE in B9: |__| NGVD 1929 |__| NAVD 1988 |__| Other (Describe): ________
B12. Is the building located in a Coastal Barrier Resources System (CBRS) area or Otherwise Protected Area (OPA)? |__| Yes |__| No
Designation Date:________

SECTION C - BUILDING ELEVATION INFORMATION (SURVEY REQUIRED)

C1. Building elevations are based on: |__|Construction Drawings* |__|Building Under Construction* |__|Finished Construction
*A new Elevation Certificate will be required when construction of the building is complete.

C2. Building Diagram Number _____ (Select the building diagram most similar to the building for which this certificate is being completed - see pages 6 and 7. If no diagram accurately represents the building, provide a sketch or photograph.)

C3. Elevations – Zones A1-A30, AE, AH, A (with BFE), VE, V1-V30, V (with BFE), AR, AR/A, AR/AE, AR/A1-A30, AR/AH, AR/AO
Complete Items C3a-i below according to the building diagram specified in Item C2. State the datum used. If the datum is different from the datum used for the BFE in Section B, convert the datum to that used for the BFE. Show field measurements and datum conversion calculation. Use the space provided or the Comments area of Section D or Section G, as appropriate, to document the datum conversion.
Datum ________ Conversion/Comments ________
Elevation reference mark used________ Does the elevation reference mark used appear on the FIRM? |__| Yes |__| No

- ❑ a) Top of bottom floor (including basement or enclosure) ________ . ___ ft.(m)
- ❑ b) Top of next higher floor ________ . ___ ft.(m)
- ❑ c) Bottom of lowest horizontal structural member (V zones only) ________ . ___ ft.(m)
- ❑ d) Attached garage (top of slab) ________ . ___ ft.(m)
- ❑ e) Lowest elevation of machinery and/or equipment servicing the building ________ . ___ ft.(m)
- ❑ f) Lowest adjacent grade (LAG) ________ . ___ ft.(m)
- ❑ g) Highest adjacent grade (HAG) ________ . ___ ft.(m)
- ❑ h) No. of permanent openings (flood vents) within 1 ft. above adjacent grade ________
- ❑ i) Total area of all permanent openings (flood vents) in C3h ________ sq. in. (sq. cm)

License Number, Embossed Seal, Signature, and Date

SECTION D - SURVEYOR, ENGINEER, OR ARCHITECT CERTIFICATION

This certification is to be signed and sealed by a land surveyor, engineer, or architect authorized by law to certify elevation information.
I certify that the information in Sections A, B, and C on this certificate represents my best efforts to interpret the data available.
I understand that any false statement may be punishable by fine or imprisonment under 18 U.S. Code, Section 1001.

CERTIFIER'S NAME LICENSE NUMBER

TITLE COMPANY NAME

ADDRESS CITY STATE ZIP CODE

SIGNATURE DATE TELEPHONE

FEMA Form 81-31, AUG 99 SEE REVERSE SIDE FOR CONTINUATION REPLACES ALL PREVIOUS EDITIONS

Figure 1612.5.1(1)
ELEVATION CERTIFICATE

(continued)

IMPORTANT: In these spaces, copy the corresponding information from Section A.			For Insurance Company Use:
BUILDING STREET ADDRESS (Including Apt., Unit, Suite, and/or Bldg. No.) OR P.O. ROUTE AND BOX NO.			Policy Number
CITY	STATE	ZIP CODE	Company NAIC Number

SECTION D - SURVEYOR, ENGINEER, OR ARCHITECT CERTIFICATION (CONTINUED)

Copy both sides of this Elevation Certificate for (1) community official, (2) insurance agent/company, and (3) building owner.

COMMENTS

|__| Check here if attachments

SECTION E - BUILDING ELEVATION INFORMATION (SURVEY NOT REQUIRED) FOR ZONE AO AND ZONE A (WITHOUT BFE)

For Zone AO and Zone A (without BFE), complete Items E1 through E4. *If the Elevation Certificate is intended for use as supporting information for a LOMA or LOMR-F, Section C must be completed.*

E1. Building Diagram Number _____ (Select the building diagram most similar to the building for which this certificate is being completed – see pages 6 and 7. If no diagram accurately represents the building, provide a sketch or photograph.)

E2. The top of the bottom floor (including basement or enclosure) of the building is |__|__| ft.(m) |__|__|in.(cm) |__| above or |__| below (check one) the highest adjacent grade.

E3. For Building Diagrams 6-8 with openings (see page 7), the next higher floor or elevated floor (elevation b) of the building is |__|__| ft.(m) |__|__|in.(cm) above the highest adjacent grade.

E4. For Zone AO only: If no flood depth number is available, is the top of the bottom floor elevated in accordance with the community's floodplain management ordinance? |__| Yes |__| No |__| Unknown. The local official must certify this information in Section G.

SECTION F - PROPERTY OWNER (OR OWNER'S REPRESENTATIVE) CERTIFICATION

The property owner or owner's authorized representative who completes Sections A, B, and E for Zone A (without a FEMA-issued or community-issued BFE) or Zone AO must sign here.

PROPERTY OWNER'S OR OWNER'S AUTHORIZED REPRESENTATIVE'S NAME

ADDRESS CITY STATE ZIP CODE

SIGNATURE DATE TELEPHONE

COMMENTS

|__| Check here if attachments

SECTION G - COMMUNITY INFORMATION (OPTIONAL)

The local official who is authorized by law or ordinance to administer the community's floodplain management ordinance can complete Sections A, B, C (or E), and G of this Elevation Certificate. Complete the applicable item(s) and sign below.

G1. |__| The information in Section C was taken from other documentation that has been signed and embossed by a licensed surveyor, engineer, or architect who is authorized by state or local law to certify elevation information. (Indicate the source and date of the elevation data in the Comments area below.)

G2. |__| A community official completed Section E for a building located in Zone A (without a FEMA-issued or community-issued BFE) or Zone AO.

G3. |__| The following information (Items G4-G9) is provided for community floodplain management purposes.

G4. PERMIT NUMBER	G5. DATE PERMIT ISSUED	G6. DATE CERTIFICATE OF COMPLIANCE/OCCUPANCY ISSUED

G7. This permit has been issued for: |__| New Construction |__| Substantial Improvement

G8. Elevation of as-built lowest floor (including basement) of the building is: ___________ . ___ ft.(m) Datum: ___________

G9. BFE or (in Zone AO) depth of flooding at the building site is: ___________ . ___ ft.(m) Datum: ___________

LOCAL OFFICIAL'S NAME TITLE

COMMUNITY NAME TELEPHONE

SIGNATURE DATE

COMMENTS

|__| Check here if attachments

FEMA Form 81-31, AUG 99 REPLACES ALL PREVIOUS EDITIONS

Figure 1612.5.1(1)
ELEVATION CERTIFICATE

O.M.B. NO. 3067-0077
Expires July 31, 1999

FEDERAL EMERGENCY MANAGEMENT AGENCY
NATIONAL FLOOD INSURANCE PROGRAM

FLOODPROOFING CERTIFICATE

FOR NON-RESIDENTIAL STRUCTURES

The floodproofing of non-residential buildings may be permitted as an alternative to elevating to or above the Base Flood Elevation; however, a floodproofing design certification is required. This form is to be used for that certification. Floodproofing of a residential building does not alter a community's floodplain management elevation requirements or effect the insurance rating unless the community has been issued an exception by FEMA to allow floodproofed residential basements. The permitting of a floodproofed residential basement requires a separate certification specifying that the design complies with the local floodplain management ordinance.

	FOR INSURANCE COMPANY USE
BUILDING OWNER'S NAME	POLICY NUMBER
STREET ADDRESS (Including Apt., Unit, Suite and/or Bldg. Number) OR P.O. ROUTE AND BOX NUMBER	COMPANY NAIC NUMBER

OTHER DESCRIPTION (Lot and Block Numbers, etc.)

CITY STATE ZIP CODE

SECTION I FLOOD INSURANCE RATE MAP (FIRM) INFORMATION

Provide the following from the proper FIRM:

COMMUNITY NUMBER	PANEL NUMBER	SUFFIX	DATE OF FIRM INDEX	FIRM ZONE	BASE FLOOD ELEVATION (in AO Zones, use depth)

SECTION II FLOODPROOFING INFORMATION (By a Registered Professional Engineer or Architect)

Floodproofing Design Elevation Information:

Building is floodproofed to an elevation of ☐☐☐☐☐.☐ feet NGVD. (Elevation datum used must be the same as that on the FIRM.)

Height of floodproofing on the building above the lowest adjacent grade is ☐☐.☐ feet.

(NOTE: for insurance rating purposes, the building's floodproofed design elevation must be at least one foot above the Base Flood Elevation to receive rating credit. If the building is floodproofed only to the Base Flood Elevation, then the building's insurance rating will result in a higher premium.)

SECTION III CERTIFICATION (By a Registered Professional Engineer or Architect)

Non-Residential Floodproofed Construction Certification:

I certify that based upon development and/or review of structural design, specifications, and plans for construction that the design and methods of construction are in accordance with accepted standards of practice for meeting the following provisions:

The structure, together with attendant utilities and sanitary facilities, is watertight to the floodproofed design elevation indicated above, with walls that are substantially impermeable to the passage of water.

All structural components are capable of resisting hydrostatic and hydrodynamic flood forces, including the effects of buoyancy, and anticipated debris impact forces.

I certify that the information on this certificate represents my best efforts to interpret the data available. I understand that any false statement may be punishable by fine or imprisonment under 18 U.S. Code, Section 1001.

CERTIFIER'S NAME LICENSE NUMBER (or Affix Seal)

TITLE COMPANY NAME

ADDRESS CITY STATE ZIP

SIGNATURE DATE PHONE

Copies should be made of this Certificate for: 1) community official, 2) insurance agent/company, and 3) building owner.

FEMA Form 81-65, FEB. 97 REPLACES ALL PREVIOUS EDITIONS 000056 (2/97)

Figure 1612.5.1(2)
FLOODPROOFING CERTIFICATE

SECTION 1613
EARTHQUAKE LOADS DEFINITIONS

❖ The earthquake provisions begin in Section 1613 and continue through Section 1623. These seismic requirements are based, for the most part, on the NEHRP *Recommended Provisions for Seismic Regulations for New Buildings and Other Structures.* A new set of NEHRP Recommended Provisions has been prepared by the BSSC every three years since the first edition in 1985. FEMA has funded the BSSC project.

The code uses the NEHRP Recommended Provisions as the technical basis for seismic design requirements because of the nationwide input into the development of these design criteria. The NEHRP Recommended Provisions present up-to-date criteria for the design and construction of buildings subject to earthquake ground motions that are applicable anywhere in the nation. The requirements are intended to minimize the hazard to life for all buildings, increase the expected performance of higher occupancy buildings as compared to ordinary buildings and improve the capability of essential facilities to function during and after an earthquake. These minimum criteria are considered to be prudent and economically justified for the protection of life safety in buildings subject to earthquakes. The "design earthquake" ground motion levels specified may result in damage, both structural and nonstructural, from the high stresses that occur because of the dynamic nature of seismic events. For most structures, damage from the "design earthquake" ground motion would be repairable but might be so costly as to make it economically undesirable. For essential facilities, it is expected that damage from the "design earthquake" ground motion would not be so severe as to prevent continued occupancy and function of the facility. For ground motions greater than the design levels, the intent is that there be a low likelihood of structural collapse. Achieving the intended performance, however, depends on a number of factors, including the structural framing type, configuration, construction materials and as-built details of construction.

The major provisions include the following:

1. Design and construction requirements that are a function of seismic design categories rather than seismic "zones." The seismic design category classification includes consideration of the seismicity at the site, type of soil present at the site and the nature of the building occupancy.
2. Design is required for the effect of two-thirds of the maximum considered earthquake as defined in Section 1613, which has an approximate average return period of 2,500 years in most of the United States. Considering the margin of safety of 1.5 inherent in recent and current U.S. seismic design practice, which considers a design earthquake with an average return period of 475 years, this would prevent collapse of structures in the maximum considered earthquake.
3. Maximum considered earthquake ground motion maps that consist of contour lines of spectral response accelerations at periods of 0.2 second and 1 second on Site Class B soil.
4. A redundancy/reliability factor that is applied to increase the effect of the horizontal earthquake ground motion to compensate for the lack of structural redundancy in the lateral-force-resisting system.
5. A table containing a wide array of seismic-force-resisting systems that provides design coefficients and limitations.
6. A complete set of design requirements for mechanical, electrical and architectural building components and nonbuilding structures.
7. A complete set of design requirements for seismically isolated structures.

Note that the effect of vertical earthquake ground motion is taken into account when gravity and earthquake effects are combined.

For buildings three stories or less in height that meet a given set of criteria, a simplified analysis procedure is provided.

1613.1 Definitions. The following words and terms shall, for the purposes of this section, have the meanings shown herein.

❖ Definitions of terms that are associated with the content of Sections 1614 through 1623 are contained herein. These definitions can help the user to understand and apply the code requirements. It is important to emphasize that these definitions are applicable everywhere the terms are used in the code where the text relates to earthquake effects. The purpose of including these definitions in this section is to provide more convenient access to them (that is, the user does not need to refer back to Chapter 2).

ACTIVE FAULT/ACTIVE FAULT TRACE. A fault for which there is an average historic slip rate of 1 mm per year or more and geologic evidence of seismic activity within Holocene (past 11,000 years) times. Active fault traces are designated by the appropriate regulatory agency and/or registered design professional subject to identification by a geologic report.

❖ Sections 1615.2.3 and 1616.3.1 refer to active faults and active fault traces. This definition explains what qualifies as an active fault. Although faults have been in existence for eons, only faults that have moved within the past 11,000 years, which is "recent" geologically speaking, are considered active.

ATTACHMENTS, SEISMIC. Means by which components and their supports are secured or connected to the seismic-force-resisting system of the structure. Such attachments include anchor bolts, welded connections and mechanical fasteners.

❖ Section 1621 and the referenced ASCE 7 provisions include a number of terms that can become confusing if they are not clearly defined. The terms used include "attachment," "support," "fastener," "component seismic attachment" and "connection." The definition for "Seismic attachment" is very helpful in that it specifically lists the following examples of what attaches the equipment to the structure: anchor bolts, welded connections and mechanical fasteners. Examples of mechanical fasteners are nails, screws, bolts and metal connector plates.

BASE. The level at which the horizontal seismic ground motions are considered to be imparted to the structure.

❖ This definition is needed to determine the building height. The lateral seismic forces are transferred from the upper stories to the base of the building. The motion of the building's base and the earth are assumed to be the same during an earthquake.

BOUNDARY ELEMENTS. Chords and collectors at diaphragm and shear wall edges, interior openings, discontinuities and reentrant corners.

❖ Boundary elements are simply the items at the perimeter of horizontal diaphragms and vertical shear walls that are responsible for transferring forces from the diaphragms or the shear walls to other elements in the load path. It should be noted that this term is not used in the seismic provisions of Chapter 16 but rather in the structural materials chapters, such as Chapter 19.

BRITTLE. Systems, members, materials and connections that do not exhibit significant energy dissipation capacity in the inelastic range.

❖ This term is defined because certain seismic provisions use it in the context of what should be avoided. In terms of response to an earthquake, the goal is for a building and its components to "go along for the ride" without experiencing any permanent damage. Unfortunately, this isn't always possible, so the next best scenario is for some portions of the structure to permanently deform—that is, stretch, twist or bend—in response to ground motion. Those parts of the structure where permanent deformation takes place can be said to have entered the "inelastic range" and the stretching, twisting and bending is an "exhibition of significant energy dissipation." If a portion of the structure snaps instead of stretches, it is referred to as a "brittle failure."

COLLECTOR. A diaphragm or shear wall element parallel to the applied load that collects and transfers shear forces to the vertical-force-resisting elements or distributes forces within a diaphragm or shear wall.

❖ Collectors are a type of boundary element that is placed parallel to the applied load. They collect and transfer shear forces to vertical elements of the lateral-force-resisting system. Special requirements for collector elements can be found in Section 1620.

COMPONENT. A part or element of an architectural, electrical, mechanical or structural system.

❖ This term is used in Section 1620 to refer to parts of the structural system and in Section 1621 to refer to architectural elements such as walls, cantilevers, ceilings and cabinets as well as mechanical and electrical equipment and utilities.

Component, equipment. A mechanical or electrical component or element that is part of a mechanical and/or electrical system within or without a building system.

❖ This term is extensively referred to in the provisions referenced in Section 1621.

Component, flexible. Component, including its attachments, having a fundamental period greater than 0.06 second.

❖ The term "flexible component" is defined because it is referred to in the provisions referenced in Section 1621. A component is considered to be either flexible or rigid and is placed in the appropriate category depending on the period of the component, including its attachments. The period of a component can be defined as the time necessary to complete one cycle of back-and-forth motion. Something that is flexible will take longer to complete a cycle of back-and-forth motion than something that is rigid. In the case of components, 0.06 second is the cutoff period to qualify as a rigid component. Components with a period greater than 0.06 second are considered flexible. According to the NEHRP Recommended Provisions Commentary (see Section 6.3.3 of the NEHRP Provisions), mechanical equipment that may have fundamental periods up to approximately 0.125 second and would qualify as flexible include: vertical immersion and deep well pumps; belt-driven and vane axial fans; heaters; air handlers; chillers; boilers; heat exchangers; filters and evaporators. All electrical equipment components discussed in the NEHRP Recommended Provisions Commentary qualify as being flexible.

Component, rigid. Component, including its attachments, having a fundamental period less than or equal to 0.06 second.

❖ The term "rigid component" is defined because it is referred to in the provisions referenced in Section 1621. See the definition for "Component, flexible" for a description of the period and the basis for criteria to qualify as a rigid component. According to the NEHRP Recommended Provisions Commentary (see Section 6.3.3 of the NEHRP Provisions), the following types of mechanical equipment components have fundamental periods below 0.06 second and may be considered rigid: horizontal pumps, engine generators, motor generators, air compressors and motor-driven centrifugal blowers.

DESIGN EARTHQUAKE. The earthquake effects that buildings and structures are specifically proportioned to resist in Sections 1613 through 1622.

❖ A structure is required by the code to be designed for the ground motion corresponding to the design earthquake. The design earthquake spectral response acceleration is two-thirds of the maximum considered earthquake spectral response acceleration. The maximum considered earthquake spectral response includes the effect of the site soil or rock in addition to the earthquake map spectral acceleration values (see Sections 1615.1.2 and 1615.1.3 for specific requirements). The design earthquake is graphically represented by the design response spectrum pictured in Figure 1615.1.4.

DESIGNATED SEISMIC SYSTEM. Those architectural, electrical and mechanical systems and their components that require design in accordance with Section 1621 that have a component importance factor, I_p, greater than one.

❖ Designated seismic systems are those architectural, electrical and mechanical systems in the provisions referenced by Section 1621 that have an importance factor of 1.5. The importance factor, I_p, indicates that there is a higher level of importance placed on the system so that it will remain operational during and after an earthquake. This definition is important when applying the quality assurance provisions of Section 1705 that refer to designated seismic systems.

DISPLACEMENT.

❖ The following three displacement definitions are used in the provisions referenced by Section 1623 and are illustrated in Figure 1623.1. Seismically isolated structures have been "decoupled" from the motion of the surrounding ground, very much like shock absorbers on a car reduce the impact of bumps in the road. The isolators absorb the earthquake forces before they are transmitted to the structure and displacement takes place at the location of the isolators. Seismic isolation is a relatively new technique but is especially effective when it is necessary to protect the building's interior systems and contents (for example, for laboratories, hospitals and museums). It is used in both the construction of new buildings and in retrofitting buildings and has proven to be especially useful when historic buildings are being retrofitted to resist earthquake ground motions since it reduces the force levels transmitted to the building, which makes it less necessary to add structural components and thereby destroy historic materials and finishes.

Design displacement. The design earthquake lateral displacement, excluding additional displacement due to actual and accidental torsion, required for design of the isolation system.

❖ This is the displacement expected at the center of a base-isolated building in the design earthquake. It is used in the design of the isolation system. Because the displacement is at the center of the building, any displacement due to actual and accidental torsion is excluded.

Total design displacement. The design earthquake lateral displacement, including additional displacement due to actual and accidental torsion, required for design of the isolation system.

❖ This is the displacement expected at the corner of a base-isolated building in the design earthquake. It is used in the design of the isolation system. Because the displacement is at the corner of the building, any displacement due to actual and accidental torsion is included.

Total maximum displacement. The maximum considered earthquake lateral displacement, including additional displacement due to actual and accidental torsion, required for verification of the stability of the isolation system or elements thereof, design of building separations and vertical load testing of isolator unit prototype.

❖ This is the displacement expected at the corner of a base-isolated building in the maximum considered earthquake. It is used in the design of the isolator system. Because the displacement is at the corner of the building, any displacement due to actual and accidental torsion is included.

DISPLACEMENT RESTRAINT SYSTEM. A collection of structural elements that limits lateral displacement of seismically isolated structures due to the maximum considered earthquake.

❖ When a building is seismically isolated, the greatest portion of the response occurs across the base isolation bearings. Almost all of the response displacement occurs between the base of the structure and the ground. In the provisions referenced by Section 1623, the isolation system is permitted to include a displacement restraint that limits the lateral displacement due to the maximum considered earthquake.

EFFECTIVE DAMPING. The value of equivalent viscous damping corresponding to energy dissipated during cyclic response of the isolation system.

❖ This term and the definition is needed for the proper application of the provisions referenced by Section 1623.

EFFECTIVE STIFFNESS. The value of the lateral force in the isolation system, or an element thereof, divided by the corresponding lateral displacement.

❖ This is the equivalent linear stiffness for the isolation system that recognizes its actual force-displacement relationship is nonlinear. Obviously, the effective stiffness varies with the displacement at which it is determined. The same isolation system would have different effective stiffnesses under the total design displacement and the total maximum displacement.

HAZARDOUS CONTENTS. A material that is highly toxic or potentially explosive and in sufficient quantity to pose a significant life-safety threat to the general public if an uncontrolled release were to occur.

❖ This term clarifies what qualifies as being "hazardous." Additionally, it helps in determining the correct seismic use group and importance factor.

INVERTED PENDULUM-TYPE STRUCTURES. Structures that have a large portion of their mass concentrated near the top, and thus have essentially one degree of freedom in horizontal translation. The structures are usually T-shaped with a single column supporting the beams or framing at the top.

❖ This definition is needed to apply the system limitations on basic seismic-force-resisting systems in Section 1617.6. (e.g., see Item 7 in Table 1617.6.2). Inverted pendulum structures exhibit little redundancy or overstrength. Ineleastic behavior is concentrated at the base. Compared to other systems, inverted pendulum systems have less energy dissipation capacity.

ISOLATION INTERFACE. The boundary between the upper portion of the structure, which is isolated, and the lower portion of the structure, which moves rigidly with the ground.

❖ Used in the provisions referenced by Section 1623, the isolation interface can be assumed to pass through the midheight of elastomeric bearings or the sliding surface of sliding bearings (see Figure C 13.3.4 in Part 2 of the NEHRP Recommended Provisions Commentary or Figure C151-1 of the *SEAOC Blue Book Commentary*). The isolation interface need not be a horizontal plane, but could change elevation if isolator units were located at different levels.

ISOLATION SYSTEM. The collection of structural elements that includes individual isolator units, structural elements that transfer force between elements of the isolation system and connections to other structural elements.

❖ Used in the provisions referenced by Section 1623, the isolation system includes the isolator units, connections of isolator units to the structural system and all structural elements required for isolator stability. Isolator systems also may include supplemental damping devices.

ISOLATOR UNIT. A horizontally flexible and vertically stiff structural element of the isolation system that permits large lateral deformations under design seismic load. An isolator unit is permitted to be used either as part of or in addition to the weight-supporting system of the building.

❖ Used in the provisions referenced by Section 1623, isolator units include bearings that support the building weight and provide lateral flexibility. These bearings typically also provide damping and restraint against wind-induced displacement.

LOAD.

❖ Because seismic forces are a function of the gravity load, it is the only type of load defined in this section.

Gravity load (*W*). The total dead load and applicable portions of other loads as defined in Sections 1613 through 1622.

❖ The gravity load includes the prescribed dead load and applicable portions of the design and snow loads to be used in the calculation of the design seismic forces. See the definition of "*W*" in Section 1617.5.1 for further clarification of what specific gravity loads need to be considered.

MAXIMUM CONSIDERED EARTHQUAKE. The most severe earthquake effects considered by this code.

❖ The term "maximum considered earthquake" is used in the titles of Code Figures 1615(1) through (10) and reflects the maximum level of earthquake ground shaking that is reasonable for design structures to resist. The maximum considered earthquake concept results in an approximately uniform margin against collapse of structures designed for two-thirds of the maximum considered earthquake.

NONBUILDING STRUCTURE. A structure, other than a building, constructed of a type included in Section 1622.

❖ The requirements for nonbuilding structures that are self-supporting are contained in Section 1622.

OCCUPANCY IMPORTANCE FACTOR. A factor assigned to each structure according to its seismic use group as prescribed in Table 1604.5.

❖ The occupancy importance factor to be used for seismic design varies from 1.0 to 1.5, depending upon the importance associated with the use of a structure and the need for that use to continue immediately after an earthquake. Importance factors larger than 1 increase the earthquake force for which a building is designed (see also commentary, Section 1604.5 and Table 1604.5).

SEISMIC DESIGN CATEGORY. A classification assigned to a structure based on its seismic use group and the severity of the design earthquake ground motion at the site.

❖ Each building is assigned to a Seismic Design Category A, B, C, D, E or F depending upon the soil, the mapped spectral response accelerations at its site and the use of the building. Tables 1616.3(1) and 1616.3(2) are the last step in the process of determining the seismic design category of a building. See the example in Section 1614.1 for step-by-step instructions on how to determine the seismic design category.

SEISMIC-FORCE-RESISTING SYSTEM. The part of the structural system that has been considered in the design to provide the required resistance to the seismic forces prescribed herein.

❖ This definition is needed for the application of the seismic design provisions (see Table 1617.6.2 for a listing of seismic-force-resisting systems). The detailing sections referenced in the second column of Table 1617.6.2 often include the definition for the specific type of seismic-force-resisting system.

SEISMIC FORCES. The assumed forces prescribed herein, related to the response of the structure to earthquake motions, to be used in the design of the structure and its components.

❖ These are the forces to be used in the seismic design of structures and both structural and nonstructural components in accordance with Sections 1613 through 1623.

SEISMIC USE GROUP. A classification assigned to a building based on its use as defined in Section 1616.2.

❖ Seismic use group is to be determined by the general descriptions of use in Sections 1616.2.1 through 1616.2.3 and the specific use descriptions in Table 1604.5. The seismic use group is used in the determination of the seismic design category in Tables 1616.3(1) and 1616.3(2).

SHEAR WALL. A wall designed to resist lateral forces parallel to the plane of the wall.

❖ A shear wall carries primarily in-plane (rather then out-of-plane) lateral loads in addition to gravity loads. A shear wall relies on the in-plane properties (height and length) of the wall, coupled with the structural materials properties of the wall, to resist lateral loads. The lateral loads can be due to wind, earthquake or other external forces. A partition wall that is not designed to resist in-plane loads is not a shear wall. The term "shear wall" is used in the earthquake provisions to refer to specific types of seismic-force-resisting systems.

SHEAR WALL-FRAME INTERACTIVE SYSTEM. A structural system that uses combinations of shear walls and frames designed to resist lateral forces in proportion to their rigidities, considering interaction between shear walls and frames on all levels.

❖ A definition for this term is needed because it is a recognized type of seismic-force-resisting system in Section 1617.6.

SITE CLASS. A classification assigned to a site based on the types of soils present and their engineering properties as defined in Section 1615.1.5.

❖ This definition is necessary because the earthquake load on a structure is greatly affected by the soil at the site. There are six separate site classes, A through F, which are summarized in Table 1615.1.1.

SITE COEFFICIENTS. The values of, F_a, and, F_v, indicated in Tables 1615.1.2(1) and 1615.1.2(2), respectively.

❖ The site coefficients F_a and F_v are used when determining the design spectral response accelerations, S_{DS} and S_{D1}. They are coefficients that are functions of the soil (site class) and the seismicity at the site of the structure. The "*F*" stands for "factor," the subscript "*a*" stands for "acceleration" and the subscript "v" stands for "velocity."

STORY DRIFT RATIO. The story drift divided by the story height.

❖ This definition is needed to apply the requirements of Section 1617.3, which limit allowable story drift for building stability and control of damage to nonstructural building elements.

TORSIONAL FORCE DISTRIBUTION. The distribution of horizontal seismic forces through a rigid diaphragm when the center of mass of the structure at the level under consideration does not coincide with the center of rigidity (sometimes referred to as a "diaphragm rotation").

❖ This definition is helpful in the application of the provisions referenced by Section 1617.4. Once the design base shear is determined, it is distributed vertically to each level of the structure. Subsequently, the story shear at each level is distributed horizontally to the seismic-force-resisting elements. If the horizontal diaphragm is rigid, the manner in which the load is distributed is more complex than if the diaphragm is considered flexible. If the diaphragm at a particular floor level is rigid, the center of both mass and rigidity of the story supporting the diaphragm needs to be determined and, if they do not coincide, a torsional moment is created that needs to be translated into additional design forces for the seismic-force-resisting elements.

TOUGHNESS. The ability of a material to absorb energy without losing significant strength.

❖ The terms "toughness" and "ductility" often are used interchangeably. It is desirable for an assembly to exhibit toughness and ductility, as this allows the assembly to accommodate increased loading without sudden failure.

WIND-RESTRAINT SEISMIC SYSTEM. The collection of structural elements that provides restraint of the seismic-isolated structure for wind loads. The wind-restraint system may be either an integral part of isolator units or a separate device.

❖ This term is used in the provisions referenced by Section 1623. While lateral flexibility in a seismic isolation system is highly desirable for high seismic forces, it is clearly undesirable to have perceptible building movement under frequently occurring excitations, such as those caused by minor earthquakes or wind. Certain types of bearing provide the desired initial rigidity because of their high elastic stiffness. Some other isolation systems require a wind-restraint device for this purpose—typically a rigid component designed to fail under a given level of lateral load.

SECTION 1614
EARTHQUAKE LOADS—GENERAL

1614.1 Scope. Every structure, and portion thereof, shall as a minimum, be designed and constructed to resist the effects of earthquake motions and assigned a seismic design category as set forth in Section 1616.3. Structures determined to be in Seismic Design Category A need only comply with Section 1616.4.

Exceptions:

1. Structures designed in accordance with the provisions of Sections 9.1 through 9.6, 9.13 and 9.14 of ASCE 7 shall be permitted.
2. Detached one- and two-family dwellings as applicable in Section 101.2 in Seismic Design Categories A, B and C, or located where the mapped short-period spectral response acceleration, S_S, is less than 0.4 g, are exempt from the requirements of Sections 1613 through 1622.
3. The seismic-force-resisting system of wood frame buildings that conform to the provisions of Section 2308 are not required to be analyzed as specified in Section 1616.1.
4. Agricultural storage structures intended only for incidental human occupancy are exempt from the requirements of Sections 1613 through 1623.
5. Structures located where mapped short-period spectral response acceleration, S_S, determined in accordance with Section 1615.1, is less than or equal to 0.15g and where the mapped spectral response acceleration at 1-second period, S_1, determined in accordance with Section 1615.1, is less than or equal to 0.04g shall be categorized as Seismic Design Category A. Seismic Design Category A structures need only comply with Section 1616.4.
6. Structures located where the short-period design spectral response acceleration, S_{DS}, determined in accordance with Section 1615.1, is less than or equal to 0.167g and the design spectral response acceleration at 1-second period, S_{D1}, determined in accordance with Section 1615.1, is less than or equal to 0.067g, shall be categorized as Seismic Design Category A and need only comply with Section 1616.4.

❖ See Section 1613 for introductory commentary regarding earthquake provisions.

The earthquake design and construction requirements are intended to minimize hazard to life, increase the expected performance of buildings and nonbuilding structures when subjected to earthquake effects and improve the capability of facilities that are required for post-earthquake recovery to function during and after an earthquake. The code provisions are the minimum criteria considered to be prudent and economically justified for life safety. The loads in this section are to be used with the strength design or ASD methods cited in the structural material sections of the code.

One of the first steps in the design of a structure for earthquake forces is the determination of the seismic design category. The seismic design category is needed even to determine if a structure need only comply with nominal requirements for Seismic Design Category A (Section 1616.4) except where the mapped values S_s and S_1, are less than or equal to 0.15g and 0.04g, respectively, or the design spectral response accelerations, S_{DS} and S_{D1}, are less than or equal to 0.167g and 0.067g, respectively. The following example provides step-by-step instructions on how to determine a structure's seismic design category.

Determination of seismic design category

Step 1: Determine the mapped maximum considered earthquake spectral response acceleration at short periods, S_S, and at 1-second period, S_1, for the site location from Code Figures 1615(1) through 1615(10).

Step 2: Determine the (soil) site class in accordance with Table 1615.1.1.

Step 3: Determine the site coefficients F_a and F_v from Tables 1615.1.2(1) and 1615.1.2(2), respectively.

Step 4: Determine the 5-percent damped design spectral response acceleration at short periods, $S_{DS,}$ and at 1-second period, S_{DI}, as follows:

$$S_{DS} = (2/3)(F_a)(S_S)$$

$$S_{D1} = (2/3)(F_V)(S_1)$$

Step 5: Determine seismic use group in accordance with Table 1604.5 and Section 1616.2.

Step 6: Determine the seismic design category as prescribed by Tables 1616.3(1) and 1616.3(2). The highest of the seismic design categories from the two tables is the category assigned to the building, unless the exception to Section 1616.3 is applicable. For example, if the seismic design category from Table 1616.3(1) was D and from Table 1616.3(2) was C, then the building would be assigned a Seismic Design Category D.

There are six exceptions to seismic design requirements included in this section. The following discussion addresses each of the exceptions.

Exception 1 permits application of ASCE 7 seismic load provisions as an alternative to those in the code. Like the code, the ASCE 7 seismic load provisions are largely based on the NEHRP *Recommended Provisions for Seismic Regulations for New Buildings and Other Structures* (see commentary, Section 1613).

Exception 2 exempts detached one- and two-family dwellings from the requirements of Sections 1613 through 1623 under two conditions. The first applies to structures assigned to Seismic Design Category A, B or C (meaning that both S_{DS} < 0.5g *and* S_{D1} < 0.2g). The second is for structures having a value for mapped short-period spectral response acceleration, S_S, less than 0.4g. The latter condition is derived from the ASCE 7 seismic provisions. Since it is based solely on the mapped value for short periods, it may allow structures to qualify more directly for this exception than will the first condition. In other words, the value of S_S should be checked first. If the structure qualifies based on S_S, then it is not necessary to check the seismic design category.

Exception 3 basically exempts conventional light-frame wood construction from the seismic requirements in this chapter. It should be noted that the limitations for conventional light-frame wood construction are included in Section 2308.2. There is no limitation on the use of the structure, except that Seismic Design Category F structures do not qualify under the conventional light-frame wood construction provisions nor do irregular portions of Seismic Design Category D and E structures.

The conventional light-frame wood construction provisions are deemed to provide equivalent seismic resistance as compared to construction designed in accordance with the requirements of this chapter based on the history of such conventional construction.

In Exception 4, agricultural buildings are exempt because they present a minimal life safety hazard due to the low probability of human occupancy.

Exception 5 is triggered only by mapped spectral response accelerations and applies regardless of the soil type or building use. In order for the exception to apply, both the short-period and the long-period spectral accelerations must not be exceeded. These thresholds define areas that are considered to have extremely low seismic risk. Commentary Figure 1614.1(1) shows these areas that must only comply with the nominal requirements applicable for Seismic Design Category A structures. Note that Commentary Figures 1614.1(2) and (3) show locations where only one of the two criteria is satisfied.

Exception 6 is not as much an exception as it is a restatement of the parameters that define Seismic Design Category A structures in Tables 1616.3(1) and 1616.3(2).

[EB] 1614.1.1 Additions to existing buildings. An addition that is structurally independent from an existing structure shall be designed and constructed as required for a new structure in accordance with the seismic requirements for new structures. An addition that is not structurally independent from an existing structure shall be designed and constructed such that the entire structure conforms to the seismic-force resistance requirements for new structures unless the following conditions are satisfied:

1. The addition conforms with the requirements for new structures,
2. The addition does not increase the seismic forces in any structural element of the existing structure by more than 5 percent, unless the element has the capacity to resist the increased forces determined in accordance with Sections 1613 through 1622, and
3. Additions do not decrease the seismic resistance of any structural element of the existing structure by more than 5 percent cumulative since the original construction, unless the element has the capacity to resist the forces determined in accordance with Sections 1613 through 1622.

❖ In general, additions are governed by Section 3403. This section of the code gives more specific guidance with respect to the impact of an addition on the seismic resistance of an existing structure. The requirements of this section do not affect an existing structure where the addition is structurally independent. Only the independent addition is required to comply with the seismic requirements for new structures. Additions that are not structurally independent from the existing structure may increase the lateral load effect on the existing structure or the existing structure could impose additional lateral loads on the addition.

Because the addition and the existing structure resist earthquake effects as a single structure, designing the entire structure to resist the total lateral seismic forces is necessary. A small addition that is not structurally independent is permitted without alteration to the seismic-force-resisting system of the existing structure, provided the three conditions cited in the section are met. Condition 1 is reasonable since the addition represents new construction, and requiring compliance of the addition itself is not a hardship.

Condition 2 allows a 5-percent increase of the seismic forces in any structural element. It also allows the addition to increase the element's seismic forces by more than 5 percent as determined by the applicable provisions of Sections 1613 through 1622, provided that the existing structural element has the capacity to resist such increased forces.

Condition 3 permits a maximum cumulative decrease of 5 percent in the seismic resistance of any structural element in existing structures. It also allows the addition to decrease the element's seismic resistance by more than 5 percent, provided that the reduced capacity of the existing structural element is sufficient to resist the seismic forces prescribed for a new structure as determined by the applicable provisions of Sections 1613 through 1622.

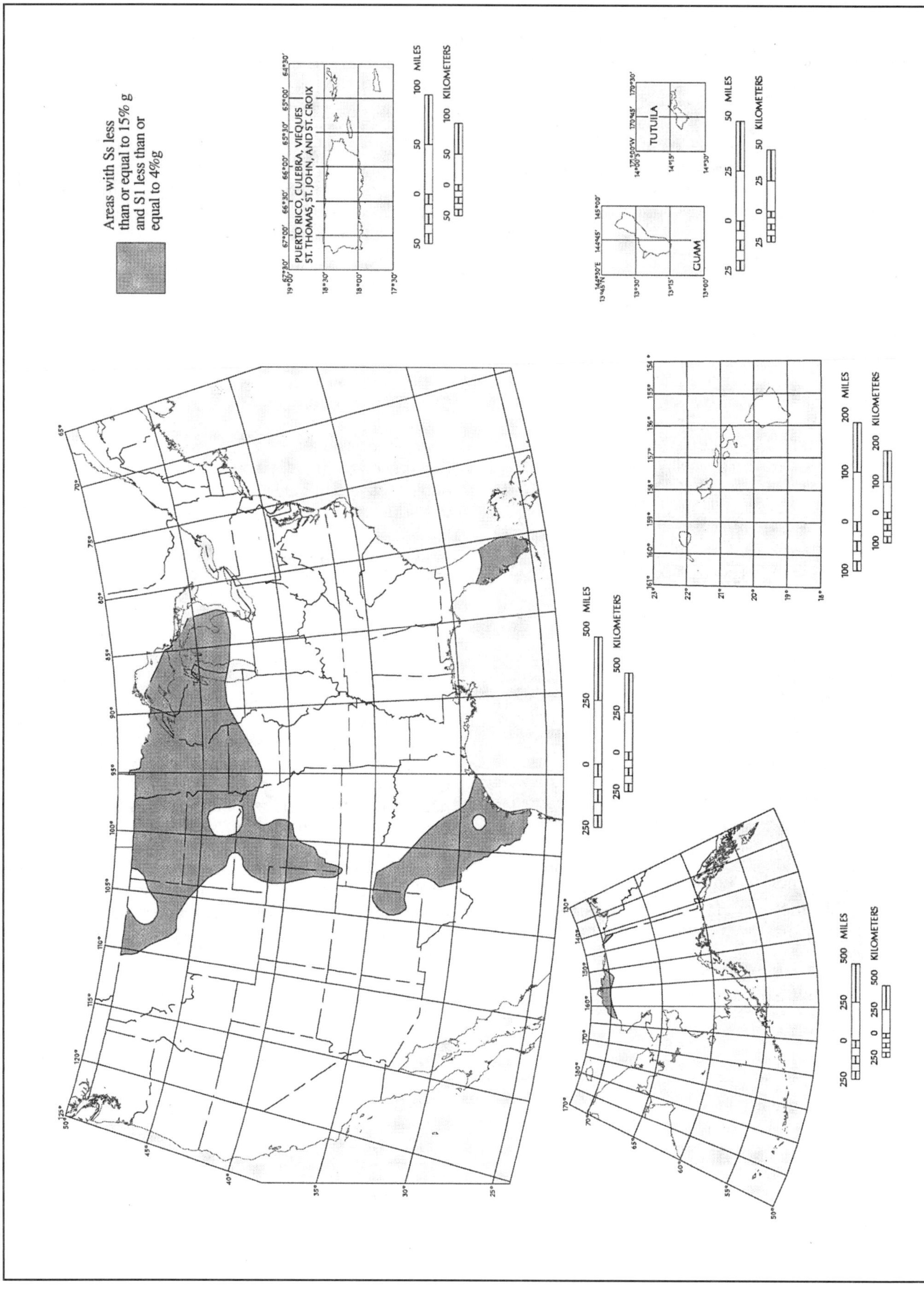

Figure 1614.1(1)
MAP AREAS WHERE EXCEPTION 5 IN SECTION 1614.1 APPLIES

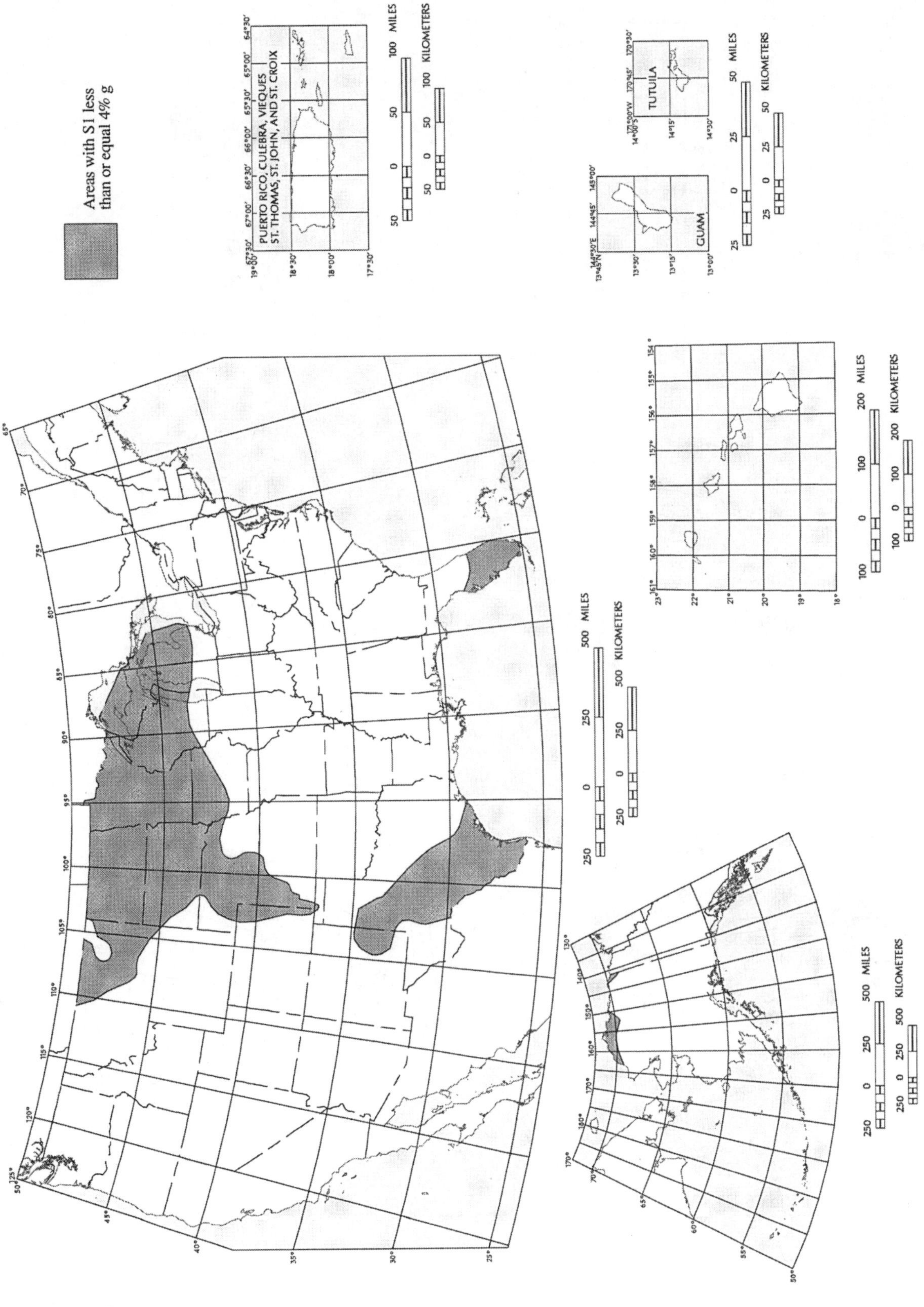

Figure 1614.1(2)
MAP AREAS WHERE S_1 IS LESS THAN OR EQUAL TO 4%g

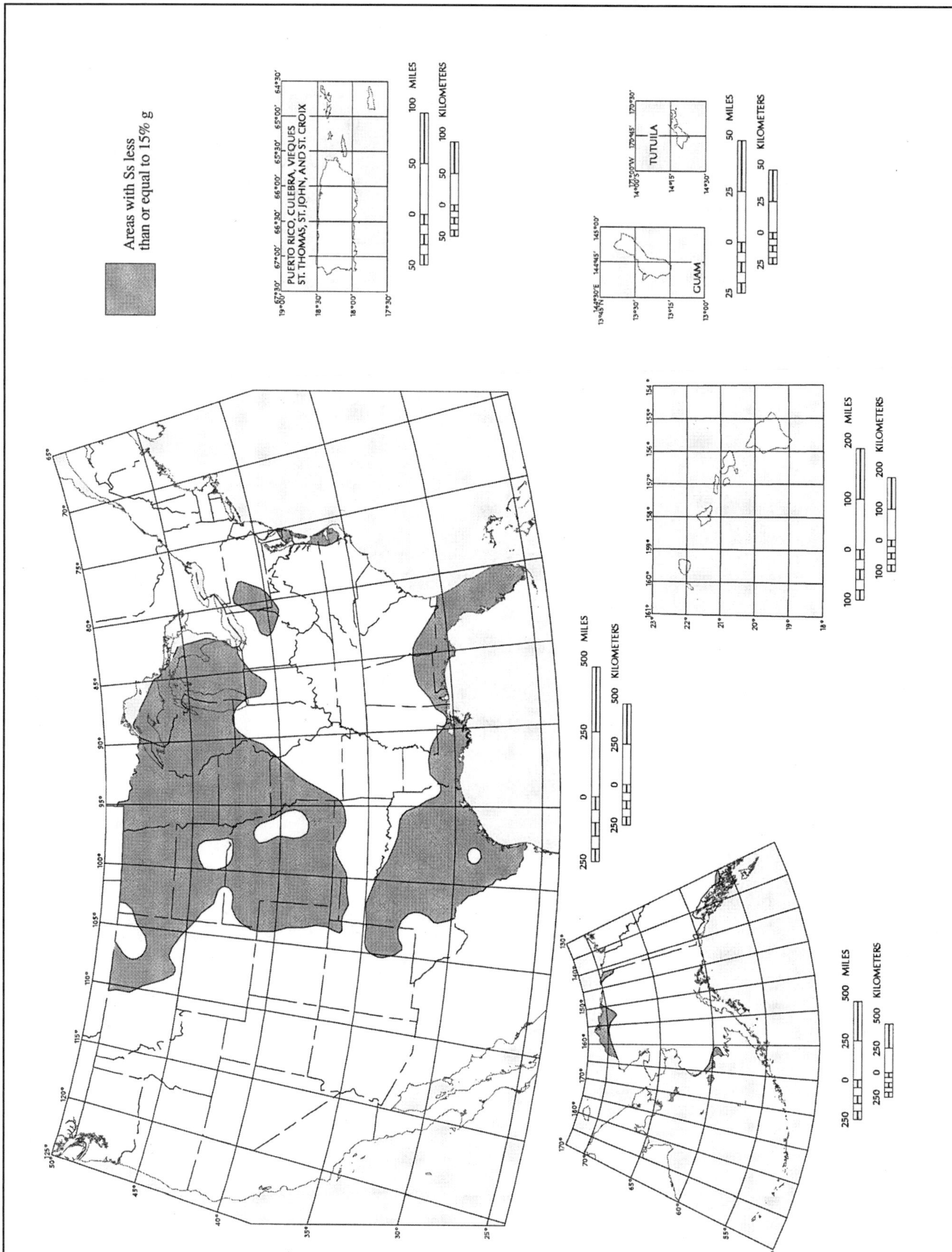

Figure 1614.1(3)
MAP AREAS WHERE S_s IS LESS THAN OR EQUAL TO 15%g

[EB] 1614.2 Change of occupancy. When a change of occupancy results in a structure being reclassified to a higher seismic use group, the structure shall conform to the seismic requirements for a new structure.

Exceptions:

1. Specific detailing provisions required for a new structure are not required to be met where it can be shown an equivalent level of performance and seismic safety contemplated for a new structure is obtained. Such analysis shall consider the regularity, overstrength, redundancy and ductility of the structure within the context of the specific detailing provided.
2. When a change of use results in a structure being reclassified from Seismic Use Group I to Seismic Use Group II and the structure is located in a seismic map area where $S_{DS} < 0.33$, compliance with this section is not required.

❖ In general, a change in occupancy is governed by Section 3406. This section of the code gives more specific guidance with respect to the impact the occupancy change has on the seismic use group of the structure. When a change of occupancy occurs that results in a higher seismic use group classification, the risk to safety is increased. Thus, an existing structure reclassified to a higher seismic use group because of an occupancy change is required to comply with current seismic provisions unless compliance with either of the exceptions can be demonstrated. Without Exception 1, it would be impossible in many instances to make an existing structure comply with the seismic requirements for a new structure. Exception 1 provides guidance to the building official and designer of areas that need to be investigated when compliance with the seismic requirements set forth in the code cannot be accomplished.

Exception 2 has its origins in ASCE 7. The purpose is to permit a change from Seismic Use Group I to Seismic Use Group II for structures subjected to low earthquake accelerations without meeting the requirements for a new structure. This would alleviate the 25-percent seismic force increase that would otherwise be required due to the importance factor, I_E, from Table 1604.5.

[EB] 1614.3 Alterations. Alterations are permitted to be made to any structure without requiring the structure to comply with Sections 1613 through 1623 provided the alterations conform to the requirements for a new structure. Alterations that increase the seismic force in any existing structural element by more than 5 percent or decrease the design strength of any existing structural element to resist seismic forces by more than 5 percent shall not be permitted unless the entire seismic-force-resisting system is determined to conform to Sections 1613 through 1623 for a new structure.

Exception: Alterations to existing structural elements or additions of new structural elements that are not required by Sections 1613 through 1623 and are initiated for the purpose of increasing the strength or stiffness of the seismic-force-resisting system of an existing structure need not be designed for forces conforming to Sections 1613 through 1623 provided that an engineering analysis is submitted indicating the following:

1. The design strength of existing structural elements required to resist seismic forces is not reduced.
2. The seismic force to required existing structural elements is not increased beyond their design strength.
3. New structural elements are detailed and connected to the existing structural elements as required by this chapter.
4. New or relocated nonstructural elements are detailed and connected to existing or new structural elements as required by this chapter.
5. The alterations do not create a structural irregularity as defined in Section 1616.5 or make an existing structural irregularity more severe.
6. The alterations do not result in the creation of an unsafe condition.

❖ The term "alteration" is defined in Section 202 as "any construction or renovation to an existing structure other than repair or addition." In general, alterations are governed by Section 3403. This section of the code gives more specific guidance with respect to the impact of an alteration on the seismic resistance of the structure. Alterations that increase the seismic load on existing structural elements by more than 5 percent or decrease the seismic resistance of existing structural elements by more than 5 percent require the entire seismic-force-resisting system to be in compliance with the seismic load provisions applicable to a new structure. Alterations that affect existing structural elements to a lesser extent are permitted without requiring the existing structure to comply with the provisions for new structures, as long as the alteration itself complies.

The exception addresses the issue of upgrading existing structures voluntarily for improved seismic performance. It does not apply to situations where other code sections trigger full compliance with the code. Otherwise, it allows an owner to initiate an improvement to the seismic-force-resisting system to the extent that it is viable to do so and provided the required engineering analysis is furnished.

1614.4 Quality assurance. A quality assurance plan shall be provided where required by Chapter 17.

❖ This section provides a cross reference to Chapter 17, as Section 1705 contains provisions for a quality assurance plan for earthquake-resistant construction.

1614.5 Seismic and wind. When the code-prescribed wind design produces greater effects, the wind design shall govern, but detailing requirements and limitations prescribed in this and referenced sections shall be followed.

❖ This section provides direction on what is required if the wind design loads govern over seismic effects. This is very often the case, particularly in areas of low seismicity. However, even where the wind design loads govern, considering the possibility of an earthquake occurring, the structure must be protected by the seismic detailing requirements to provide the necessary code-prescribed seismic ductility and resistance. For example, interior partitions, mechanical systems and ceilings are still to be designed for earthquake effects unless one of the exceptions of Section 1614.1 or Section 9.6.1 of ASCE 7 applies.

SECTION 1615
EARTHQUAKE LOADS—SITE GROUND MOTION

1615.1 General procedure for determining maximum considered earthquake and design spectral response accelerations. Ground motion accelerations, represented by response spectra and coefficients derived from these spectra, shall be determined in accordance with the general procedure of Section 1615.1, or the site-specific procedure of Section 1615.2. The site-specific procedure of Section 1615.2 shall be used for structures on sites classified as Site Class F, in accordance with Section 1615.1.1.

The mapped maximum considered earthquake spectral response acceleration at short periods (S_S) and at 1-second period (S_1) shall be determined from Figures 1615(1) through (10). Where a site is between contours, straight-line interpolation or the value of the higher contour shall be used.

The site class shall be determined in accordance with Section 1615.1.1. The maximum considered earthquake spectral response accelerations at short period and 1-second period adjusted for site class effects, S_{MS} and S_{M1}, shall be determined in accordance with Section 1615.1.2. The design spectral response accelerations at short period, S_{DS}, and at 1-second period, S_{D1}, shall be determined in accordance with Section 1615.1.3. The general response spectrum shall be determined in accordance with Section 1615.1.4.

❖ See Section 1613 for introductory commentary regarding earthquake provisions.

Section 1615.1 gives a general procedure for determining design spectral response accelerations that define the seismic design forces or the seismic design spectrum. Section 1615.2 gives a site-specific procedure for determining the same quantities. For structures founded on Site Class F (see Section 1615.1.1), the site-specific procedure must be used.

The procedure is outlined in Figure 1615.1. The mapped maximum considered earthquake spectral response accelerations at 0.2-second period (S_S) and 1-second period (S_1) for a particular site are to be determined from Code Figures 1615 (1) through (10). Where a site is between contours, as would usually be the case, straight-line interpolation or the value of the higher contour may be used.

The mapped maximum considered earthquake spectral response accelerations for a site may also be obtained using a design CD that has been prepared by the United States Geological Survey (USGS) in cooperation with the BSSC and FEMA.

The design CD is written in PDF format (Acrobat Reader 4.0 is included) to operate on a PC with a Windows 95, 97 or NT operating system. The data are interpolated for a specific latitude-longitude or zip code, which the user enters. Output for an entry uses the built-in database to interpolate for the specific site. Caution should be used when using a zip code. In regions of rapidly changing ground motion, the design parameters within a zip code may vary considerably from the value at the centroid of the zip code area.

The IBC/NEHRP maximum considered earthquake (MCE) output for a site are the two spectral values required for design. The user may request that site coefficients be included in the two spectral values. The user may also use the program to calculate an MCE response spectrum, with or without site coefficients. Site coefficients can be calculated and included in calculations by simply selecting the site class; the program then calculates the site coefficient.

FIGURE 1615(1). See page 16-82.

❖ This code figure gives the 5-percent damped maximum considered earthquake spectral response accelerations at 0.2-second period on Site Class B for the 48 contiguous United States.

FIGURE 1615(2). See page 16-84.

❖ This code figure gives the 5-percent damped maximum considered earthquake spectral response accelerations at 1-second period on Site Class B for the 48 contiguous United States.

FIGURE 1615(3). See page 16-86.

❖ The 5-percent damped maximum considered earthquake spectral response accelerations at 0.2-second period on Site Class B for California and Nevada are given on this code figure.

FIGURE 1615(4). See page 16-88.

❖ The 5-percent damped maximum considered earthquake spectral response accelerations at 1-second period on Site Class B for California and Nevada are given on this code figure.

FIGURE 1615(5). See page 16-90.

❖ The 5-percent damped maximum credible earthquake spectral response accelerations at 0.2-second period on Site Class B for the Intermountain Region (Idaho, Montana, western Wyoming, northwestern Utah, northeastern Nevada) are given on this code figure.

FIGURE 1615(6). See page 16-92.

❖ The 5-percent damped maximum credible earthquake spectral response accelerations at 1-second period on Site Class B for the Intermountain Region are given on this code figure.

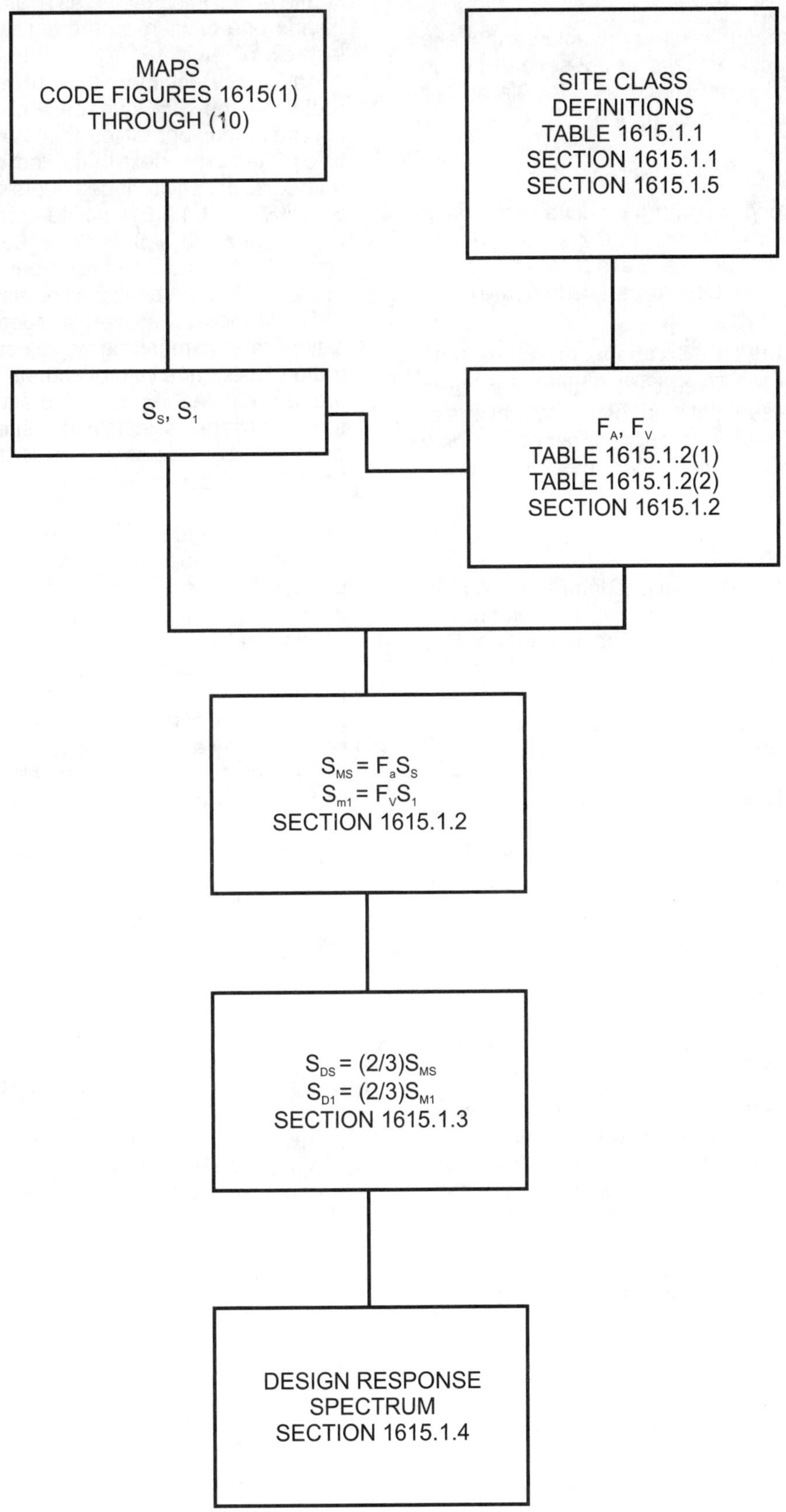

Figure 1615.1
GENERAL PROCEDURE FOR DETERMINING DESIGN RESPONSE SPECTRUM

FIGURE 1615(7). See page 16-94.

❖ The 5-percent damped maximum credible earthquake spectral response accelerations at 0.2-second period and at 1-second period on Site Class B for Alaska are given on this two-part code figure.

FIGURE 1615(8). See page 16-96.

❖ The 5-percent damped maximum credible earthquake spectral response accelerations at 0.2-second period and at 1-second period on Site Class B for the Hawaiian Islands are given on this two-part code figure.

FIGURE 1615(9). See page 16-98.

❖ This code figure gives the 5-percent damped maximum credible earthquake spectral response accelerations at 0.2-second period and at 1-second period on Site Class B for Puerto Rico, Culebra, Vieques, St. Thomas, St. John and St. Croix.

FIGURE 1615(10). See page 16-99.

❖ This code figure gives the 5-percent damped maximum credible earthquake spectral response accelerations at 0.2-second period and at 1-second period on Site Class B for Guam and Tutuilla.

1615.1.1 Site class definitions. The site shall be classified as one of the site classes defined in Table 1615.1.1. Where the soil shear wave velocity, $\bar{v}_s$, is not known, site class shall be determined, as permitted in Table 1615.1.1, from standard penetration resistance, $\bar{N}$, or from soil undrained shear strength, $\bar{s}_u$, calculated in accordance with Section 1615.1.5. Where site-specific data are not available to a depth of 100 feet (30 480 mm), appropriate soil properties are permitted to be estimated by the registered design professional preparing the soils report based on known geologic conditions.

When the soil properties are not known in sufficient detail to determine the site class, Site Class D shall be used unless the building official determines that Site Class E or F soil is likely to be present at the site.

❖ Each site is to be classified as one of six site classes (A through F), as defined in Table 1615.1.1, based on one of three soil properties measured over the top 100 feet (30 480 mm) of the site. If the top 100 feet (30 480 mm) are not homogeneous, as would typically be the case, Section 1615.1.5 indicates how to determine average properties. Site Class A is hard rock typically found in the eastern United States. Site Class B is softer rock, typical of the western parts of the country. Site Class C, D or E indicates progressively softer soils. From an earthquake-resistance perspective, rock is the best material for most structures to be founded on. Site Class F indicates soil so poor that site-specific geotechnical investigation and dynamic site response analysis are needed to determine appropriate site coefficients.

Site Class F is defined as any profile containing soils having one or more of the characteristics listed in the last row of Table 1615.1.1.

Any profile with more than 10 feet (30 480 mm) of soil (between the rock surface and the underside of the spread footing or mat foundation) having the characteristics listed in the next-to-the-last row of Table 1615.1.1, related to plasticity index (determined in accordance with ASTM D 4318), moisture content (determined in accordance with ASTM D 2216) and undrained shear strength (determined in accordance with ASTM D 2166 or D 2850), must be classified as Site Class E.

The three soil properties forming the basis of site classification are: shear wave velocity, standard penetration resistance or blow count (determined in accordance with ASTM D 1586) and undrained shear strength (determined in accordance with ASTM D 2166 or 2850). Site Class A and B designations must be based on shear wave velocity measurements or estimates.

Where soil property measurements to a depth of 100 feet (30 480 mm) are not feasible, the registered design professional preparing the soils report may estimate appropriate soil properties based on known geologic conditions.

When soil properties are not known in sufficient detail to determine the site class, the default site class is D, unless the building official determines that Site Class E or F soil may exist at the site.

Section 1615.1.5.1 provides step-by-step instructions on how to classify the soil at a site.

TABLE 1615.1.1. See page 16-81.

❖ This table is used to classify each site as one of six site classes based on one of three measured properties of soil: shear wave velocity, standard penetration resistance and undrained shear strength, except in the case of Site Class F. Plasticity index and moisture content are also used to classify a site as Site Class E or F.

1615.1.2 Site coefficients and adjusted maximum considered earthquake spectral response acceleration parameters. The maximum considered earthquake spectral response acceleration for short periods, S_{MS}, and at 1-second period, S_{M1}, adjusted for site class effects, shall be determined by Equations 16-38 and 16-39, respectively:

$$S_{MS} = F_a S_s \quad \textbf{(Equation 16-38)}$$

$$S_{M1} = F_v S_1 \quad \textbf{(Equation 16-39)}$$

where:

F_a = Site coefficient defined in Table 1615.1.2(1).

F_v = Site coefficient defined in Table 1615.1.2(2).

S_S = The mapped spectral accelerations for short periods as determined in Section 1615.1.

S_1 = The mapped spectral accelerations for a 1-second period as determined in Section 1615.1.

❖ Table 1615.1.2(1) defines an acceleration-related or short-period site coefficient, F_a, as a function of site class and the mapped spectral response acceleration at 0.2-second period, S_S. Table 1615.1.2(2) similarly defines a velocity-related or long-period site coefficient, F_V, as a function of site class and the mapped spectral response acceleration at 1-second period, S_1. The acceleration-related or short-period site coefficient F_a times S_s [see Figures 1615(1) through (10)] is S_{MS}, the 5-percent damped soil-modified maximum considered earthquake spectral response acceleration at short periods. The velocity-related or long-period site coefficient F_v times S_1 [see Figures 1615(1) through (10)] is S_{M1}, the 5-percent damped soil-modified maximum considered earthquake spectral response acceleration at 1-second period. Such modification by site coefficients is necessary because the mapped quantities are for Site Class B soils. Softer soils (Site Classes C through E) would typically amplify, and stiffer soils (Site Class A) would deamplify ground motion referenced to Site Class B.

As one would expect, both the short-period site coefficient, F_a, and the long-period site coefficient, F_v, are equal to unity for the benchmark Site Class B, irrespective of seismicity of the site. For Site Class A, both coefficients are smaller than unity, indicating reduction of benchmark Site Class B ground motion caused by the stiffer soils. For Site Classes C through E, both site coefficients, with the exception of F_a for Site Class E where $S_s \geq 1.00$, are larger than 1, indicating amplification of benchmark Site Class B ground motion on softer soils. For the same site class, each site coefficient is typically larger in areas of low seismicity than in areas of high seismicity. The basis of this lies in observations that low-magnitude subsurface rock motion is amplified to a larger extent by overlying softer soils than is high-magnitude rock motion. The site coefficients typically become larger for progressively softer soils. The only exception is provided by the short-period site coefficient in areas of high seismicity (S_s > 0.75), which remain unchanged or even decrease as the site class changes from D to E. The basis for this also lies in observations that very soft soils are not capable of amplifying the short-period components of subsurface rock motion; deamplification, in fact, takes place when the subsurface rock motion is high in magnitude.

TABLE 1615.1.2(1). See page 16-100.

❖ F_a, the acceleration-related or short-period site coefficient, is defined in this table as a function of site class and the seismicity at the site, in the form of the mapped spectral response acceleration at 0.2-second period, S_s [see Figures 1615(1) through (10)].

TABLE 1615.1.2(2). See page 16-100.

❖ F_v, the velocity-related or long-period site coefficient, is defined in this table as a function of site class and the seismicity at the site in the form of the mapped spectral response acceleration at 1-second period, S_1 [see Figures 1615(1) through (10)].

TABLE 1615.1.1
SITE CLASS DEFINITIONS

SITE CLASS	SOIL PROFILE NAME	AVERAGE PROPERTIES IN TOP 100 feet, AS PER SECTION 1615.1.5		
		Soil shear wave velocity, $\bar{v}_s$, (ft/s)	Standard penetration resistance, $\bar{N}$	Soil undrained shear strength, $\bar{s}_u$, (psf)
A	Hard rock	$\bar{v}_s$ > 5,000	N/A	N/A
B	Rock	2,500 < $\bar{v}_s$ ≤ 5,000	N/A	N/A
C	Very dense soil and soft rock	1,200 < $\bar{v}_s$ ≤ 2,500	$\bar{N}$ > 50	$\bar{s}_u$ ≥ 2,000
D	Stiff soil profile	600 ≤ $\bar{v}_s$ ≤ 1,200	15 ≤ $\bar{N}$ ≤ 50	1,000 ≤ $\bar{s}_u$ ≤ 2,000
E	Soft soil profile	$\bar{v}_s$ < 600	$\bar{N}$ < 15	$\bar{s}_u$ < 1,000
E	—	Any profile with more than 10 feet of soil having the following characteristics: 1. Plasticity index *PI* > 20, 2. Moisture content *w* ≥ 40%, and 3. Undrained shear strength $\bar{s}_u$ < 500 psf		
F	—	Any profile containing soils having one or more of the following characteristics: 1. Soils vulnerable to potential failure or collapse under seismic loading such as liquefiable soils, quick and highly sensitive clays, collapsible weakly cemented soils. 2. Peats and/or highly organic clays (*H* > 10 feet of peat and/or highly organic clay where *H* = thickness of soil) 3. Very high plasticity clays (*H* >25 feet with plasticity index *PI* >75) 4. Very thick soft/medium stiff clays (*H* >120 feet)		

For SI: 1 foot = 304.8 mm, 1 square foot = 0.0929 m^2, 1 pound per square foot = 0.0479 kPa. N/A = Not applicable

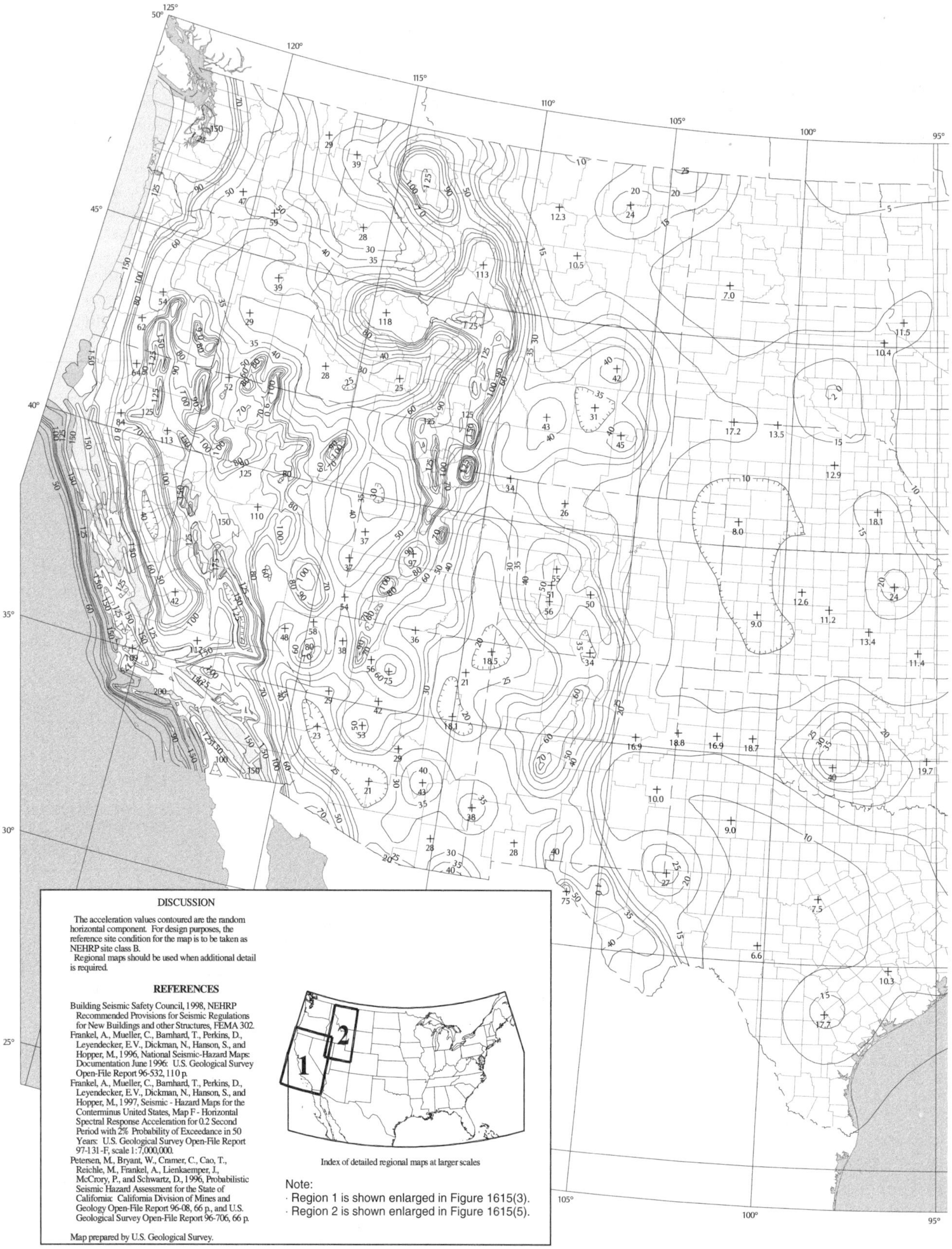

FIGURE 1615(1)
MAXIMUM CONSIDERED EARTHQUAKE GROUND MOTION FOR THE CONTERMINOUS UNITED STATES OF 0.2 SEC SPECTRAL RESPONSE ACCELERATION (5 PERCENT OF CRITICAL DAMPING), SITE CLASS B

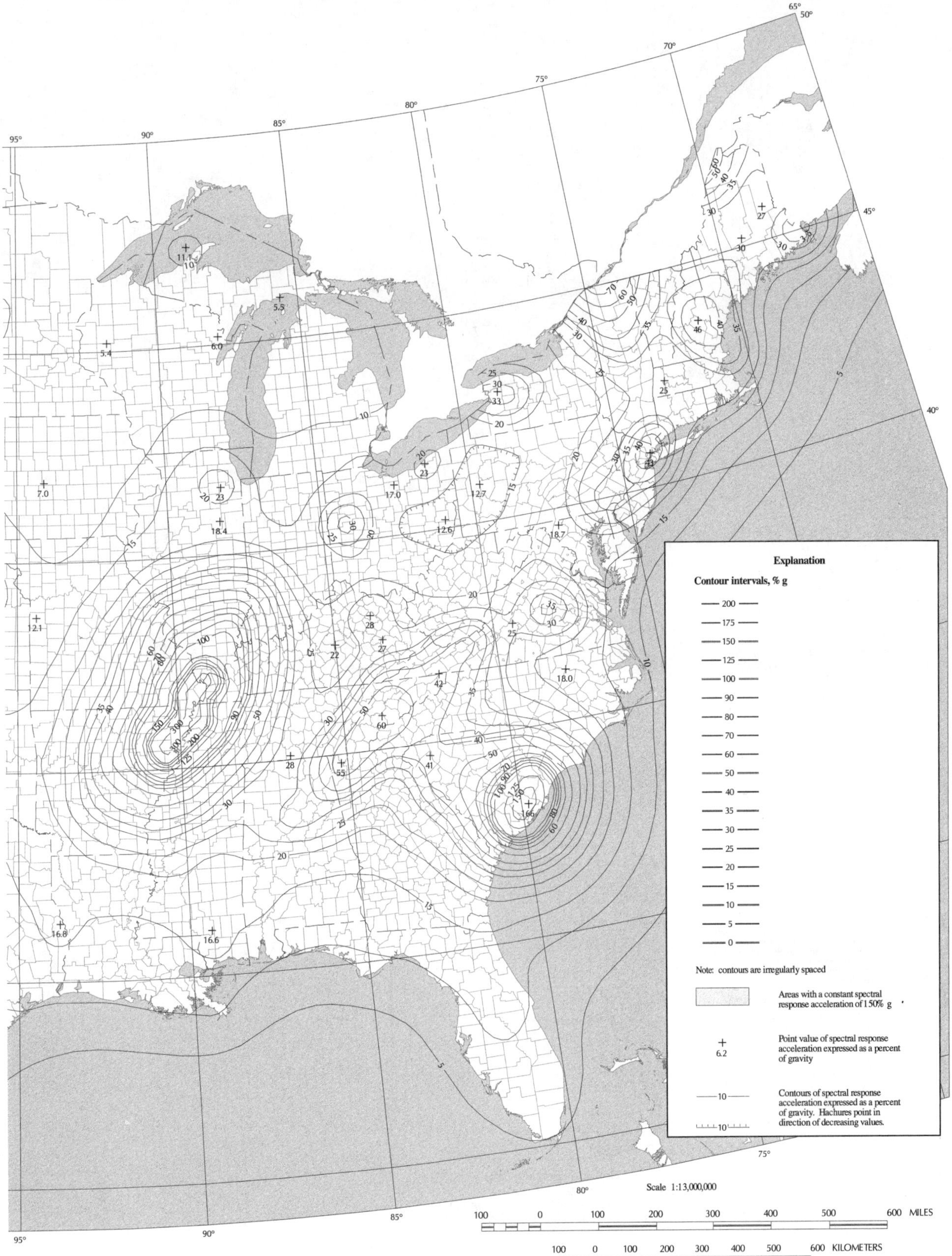

FIGURE 1615(1)—continued
MAXIMUM CONSIDERED EARTHQUAKE GROUND MOTION FOR THE CONTERMINOUS UNITED STATES OF 0.2 SEC SPECTRAL RESPONSE ACCELERATION (5 PERCENT OF CRITICAL DAMPING), SITE CLASS B

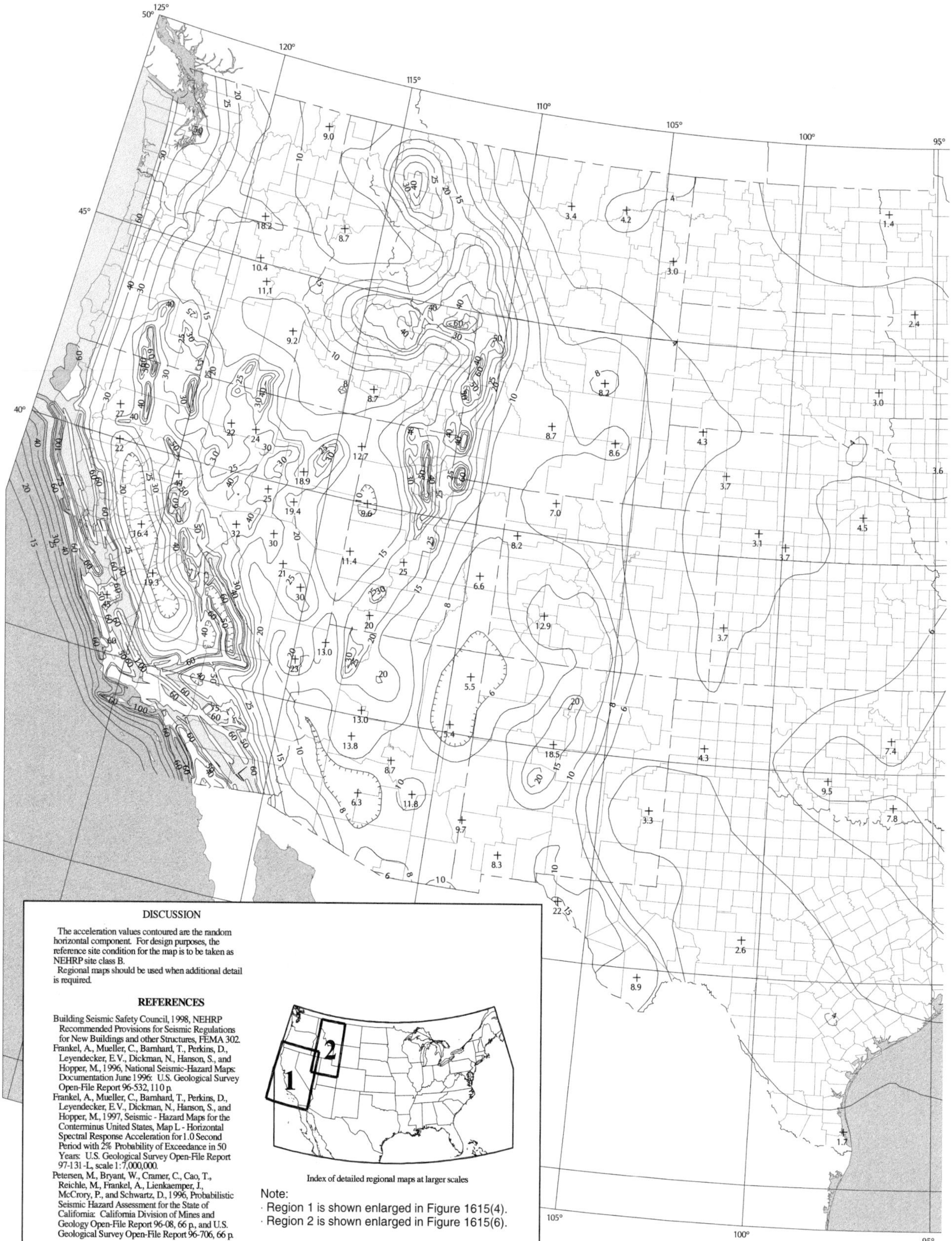

DISCUSSION

The acceleration values contoured are the random horizontal component. For design purposes, the reference site condition for the map is to be taken as NEHRP site class B.

Regional maps should be used when additional detail is required.

REFERENCES

Building Seismic Safety Council, 1998, NEHRP Recommended Provisions for Seismic Regulations for New Buildings and other Structures, FEMA 302.

Frankel, A., Mueller, C., Barnhard, T., Perkins, D., Leyendecker, E.V., Dickman, N., Hanson, S., and Hopper, M., 1996, National Seismic-Hazard Maps: Documentation June 1996: U.S. Geological Survey Open-File Report 96-532, 110 p.

Frankel, A., Mueller, C., Barnhard, T., Perkins, D., Leyendecker, E.V., Dickman, N., Hanson, S., and Hopper, M., 1997, Seismic - Hazard Maps for the Conterminus United States, Map L - Horizontal Spectral Response Acceleration for 1.0 Second Period with 2% Probability of Exceedance in 50 Years: U.S. Geological Survey Open-File Report 97-131-L, scale 1:7,000,000.

Petersen, M., Bryant, W., Cramer, C., Cao, T., Reichle, M., Frankel, A., Lienkaemper, J., McCrory, P., and Schwartz, D., 1996, Probabilistic Seismic Hazard Assessment for the State of California: California Division of Mines and Geology Open-File Report 96-08, 66 p., and U.S. Geological Survey Open-File Report 96-706, 66 p.

Map prepared by U.S. Geological Survey.

Note:
- Region 1 is shown enlarged in Figure 1615(4).
- Region 2 is shown enlarged in Figure 1615(6).

FIGURE 1615(2)
MAXIMUM CONSIDERED EARTHQUAKE GROUND MOTION FOR THE CONTERMINOUS UNITED STATES OF 1.0 SEC SPECTRAL RESPONSE ACCELERATION (5 PERCENT OF CRITICAL DAMPING), SITE CLASS B

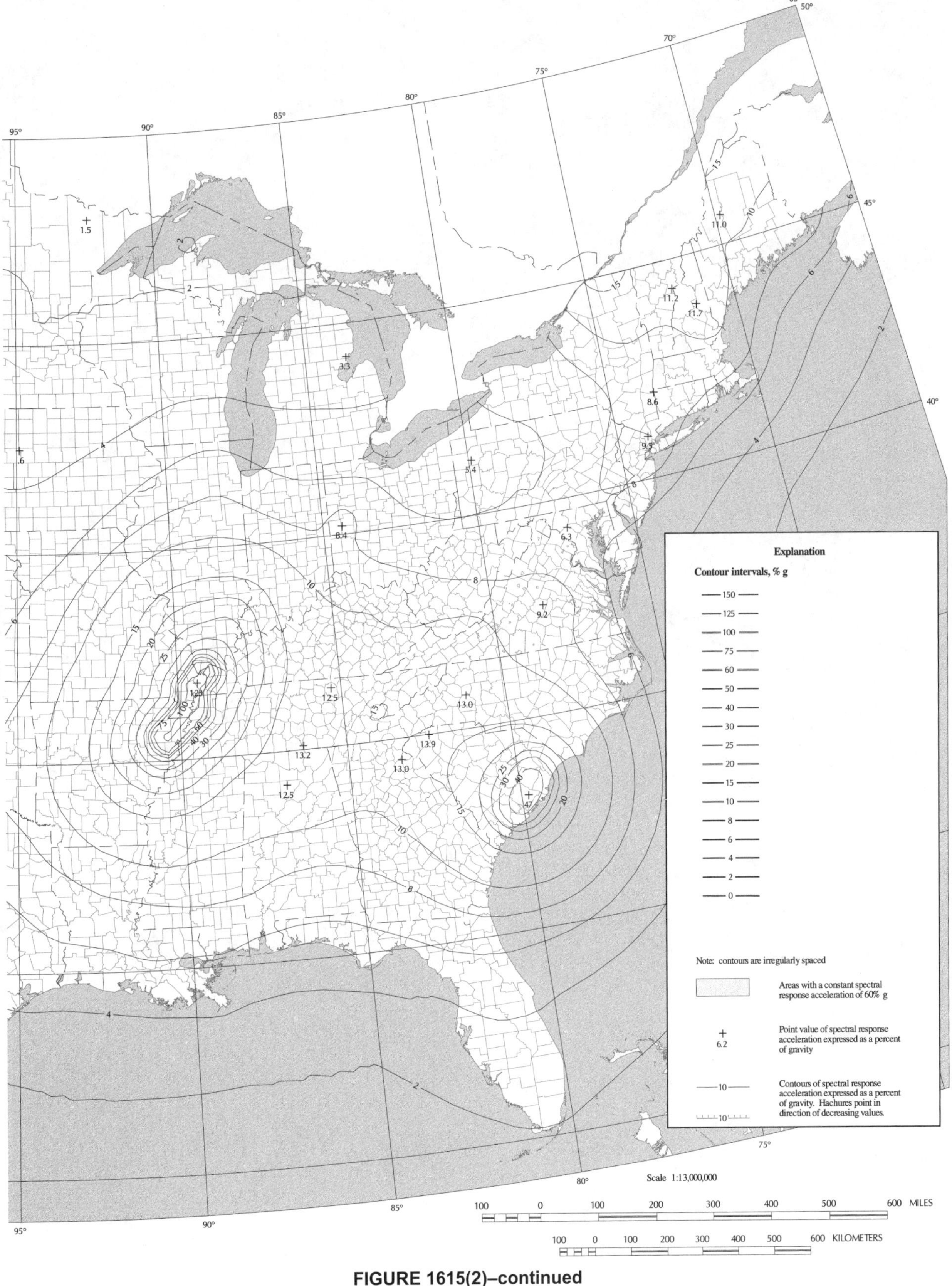

FIGURE 1615(2)–continued
MAXIMUM CONSIDERED EARTHQUAKE GROUND MOTION FOR THE CONTERMINOUS UNITED STATES OF 1.0 SEC SPECTRAL RESPONSE ACCELERATION (5 PERCENT OF CRITICAL DAMPING), SITE CLASS B

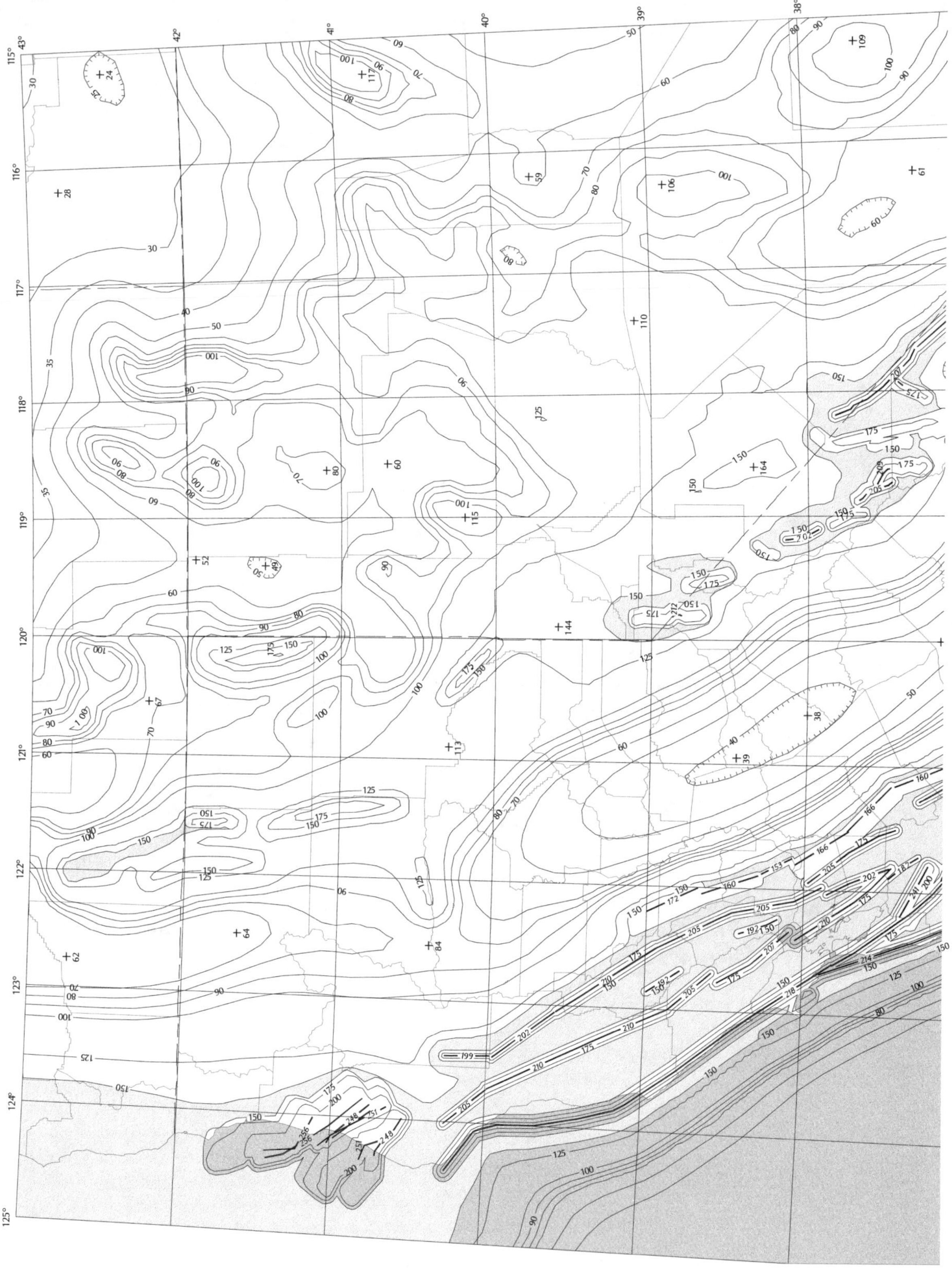

FIGURE 1615(3)
MAXIMUM CONSIDERED EARTHQUAKE GROUND MOTION FOR REGION 1 OF 0.2 SEC SPECTRAL RESPONSE ACCELERATION (5 PERCENT OF CRITICAL DAMPING), SITE CLASS B

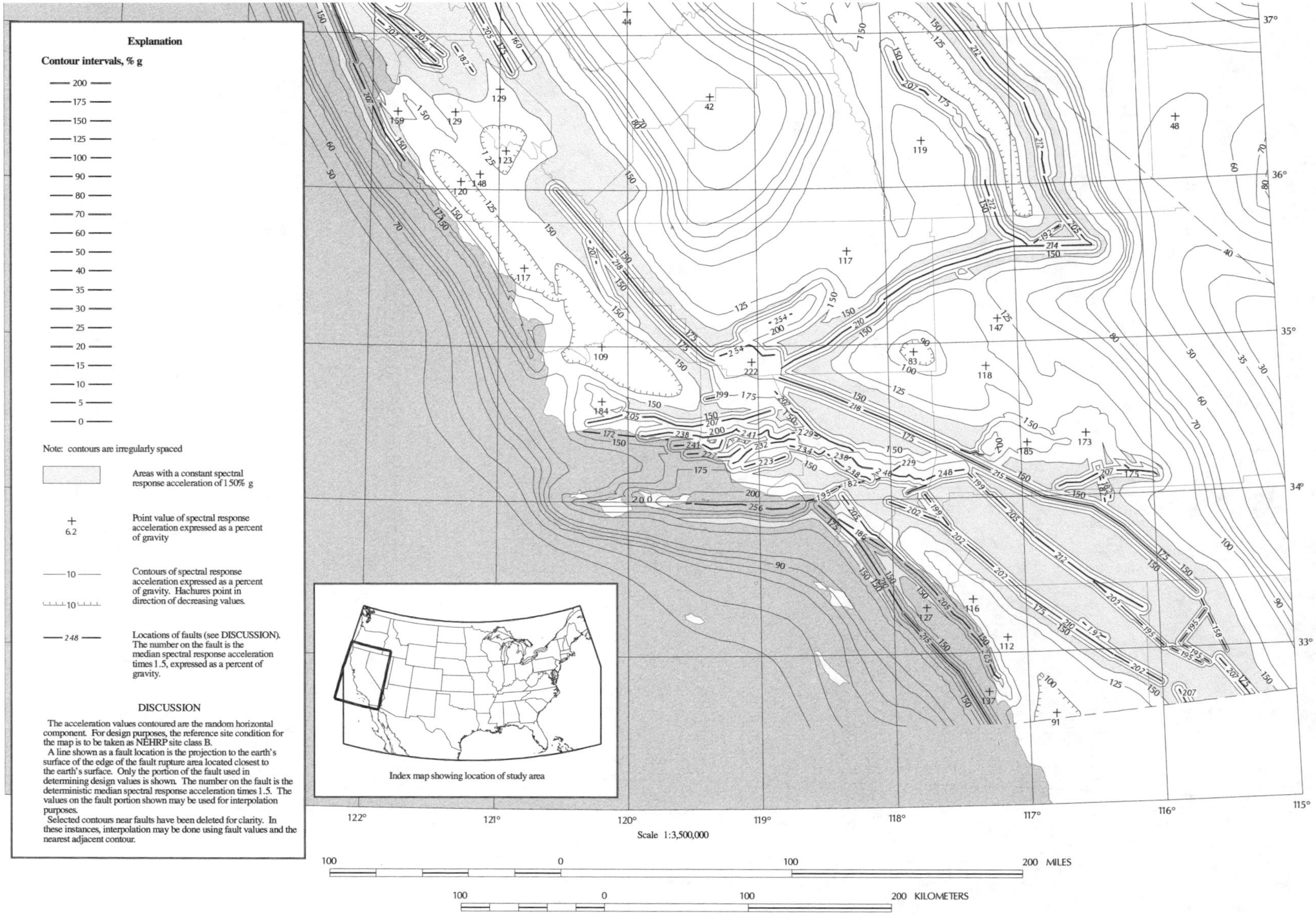

FIGURE 1615(3)–continued
MAXIMUM CONSIDERED EARTHQUAKE GROUND MOTION FOR REGION 1 OF 0.2 SEC SPECTRAL RESPONSE ACCELERATION (5 PERCENT OF CRITICAL DAMPING), SITE CLASS B

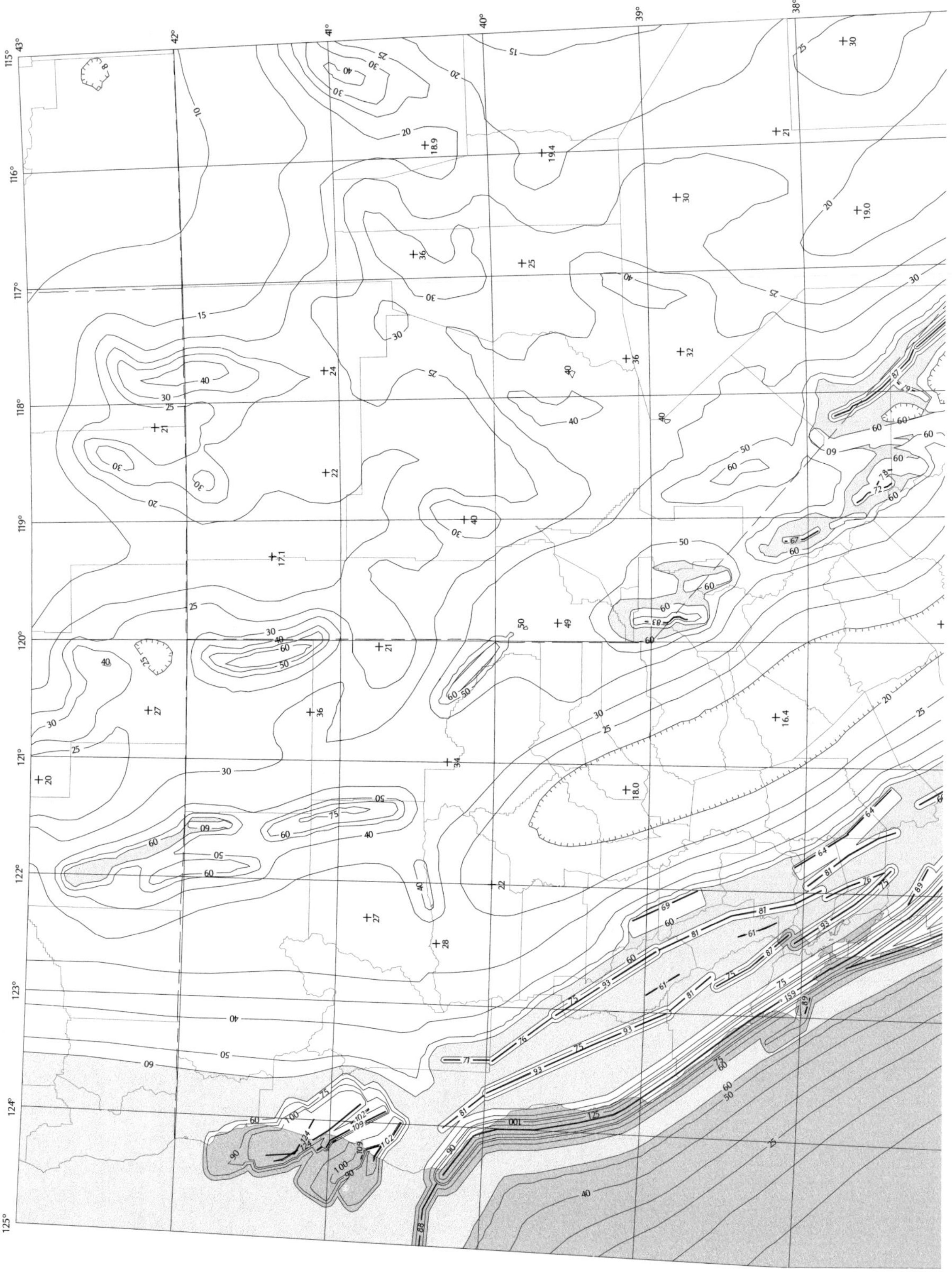

FIGURE 1615(4)
MAXIMUM CONSIDERED EARTHQUAKE GROUND MOTION FOR REGION 1 OF 1.0 SEC SPECTRAL RESPONSE ACCELERATION (5 PERCENT OF CRITICAL DAMPING), SITE CLASS B

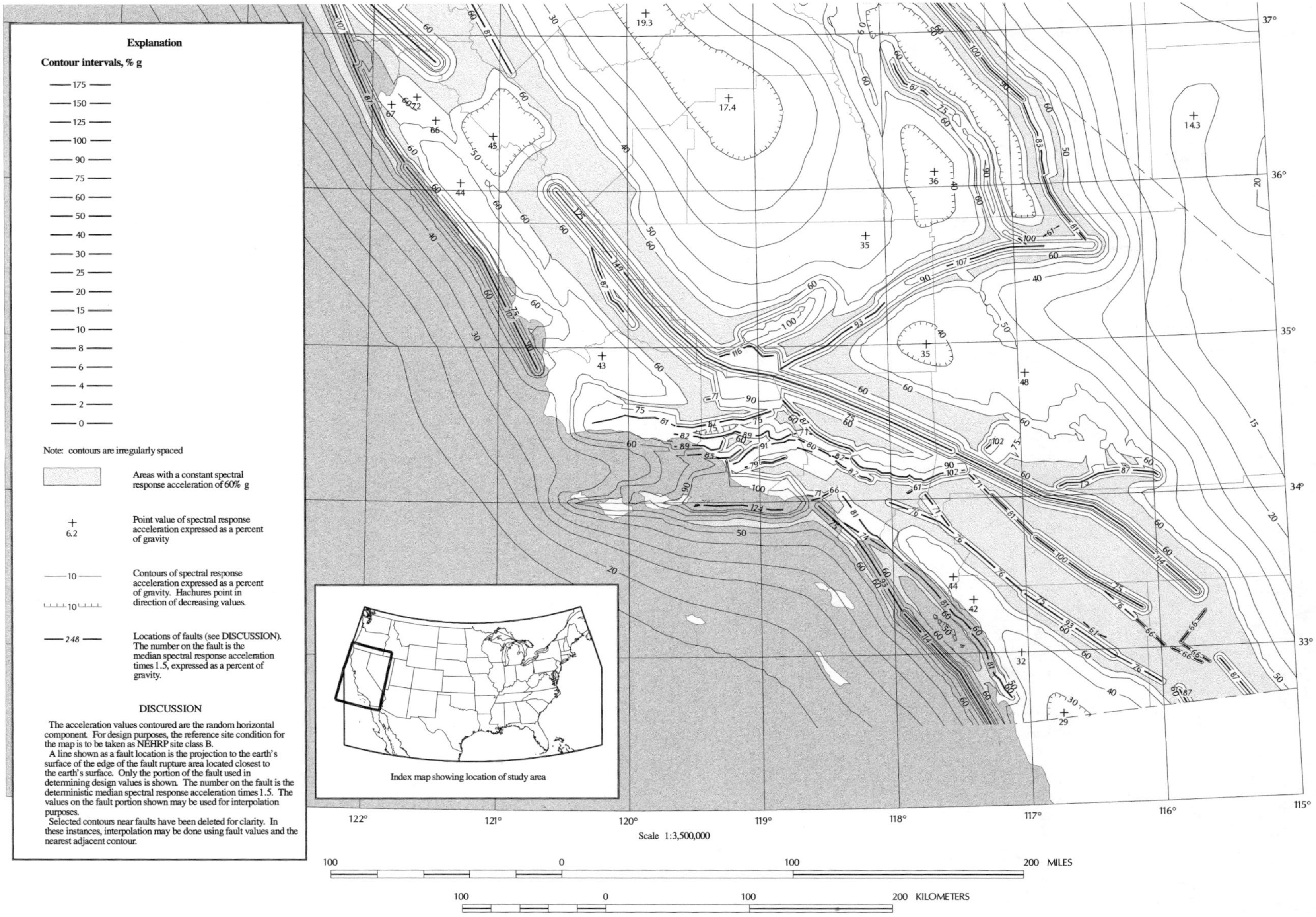

FIGURE 1615(4)—continued
MAXIMUM CONSIDERED EARTHQUAKE GROUND MOTION FOR REGION 1 OF 1.0 SEC SPECTRAL RESPONSE ACCELERATION (5 PERCENT OF CRITICAL DAMPING), SITE CLASS B

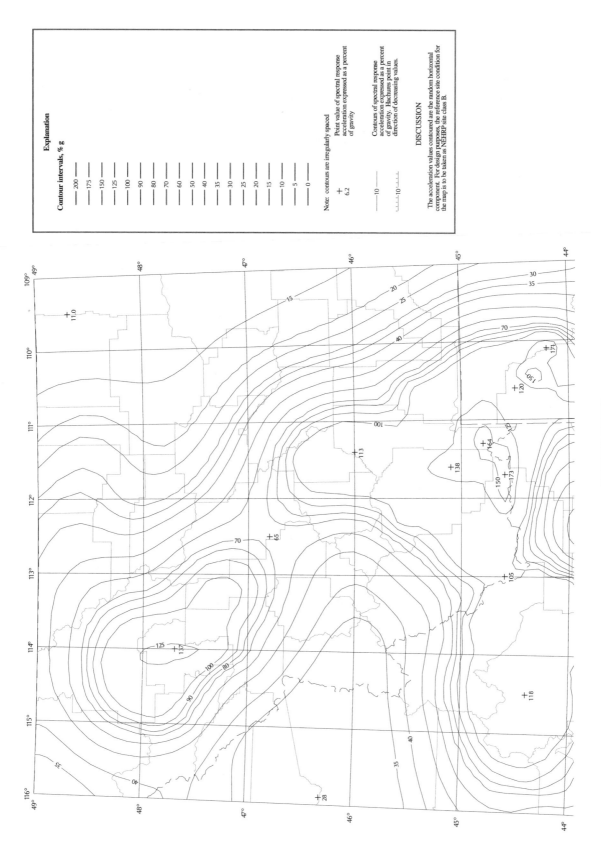

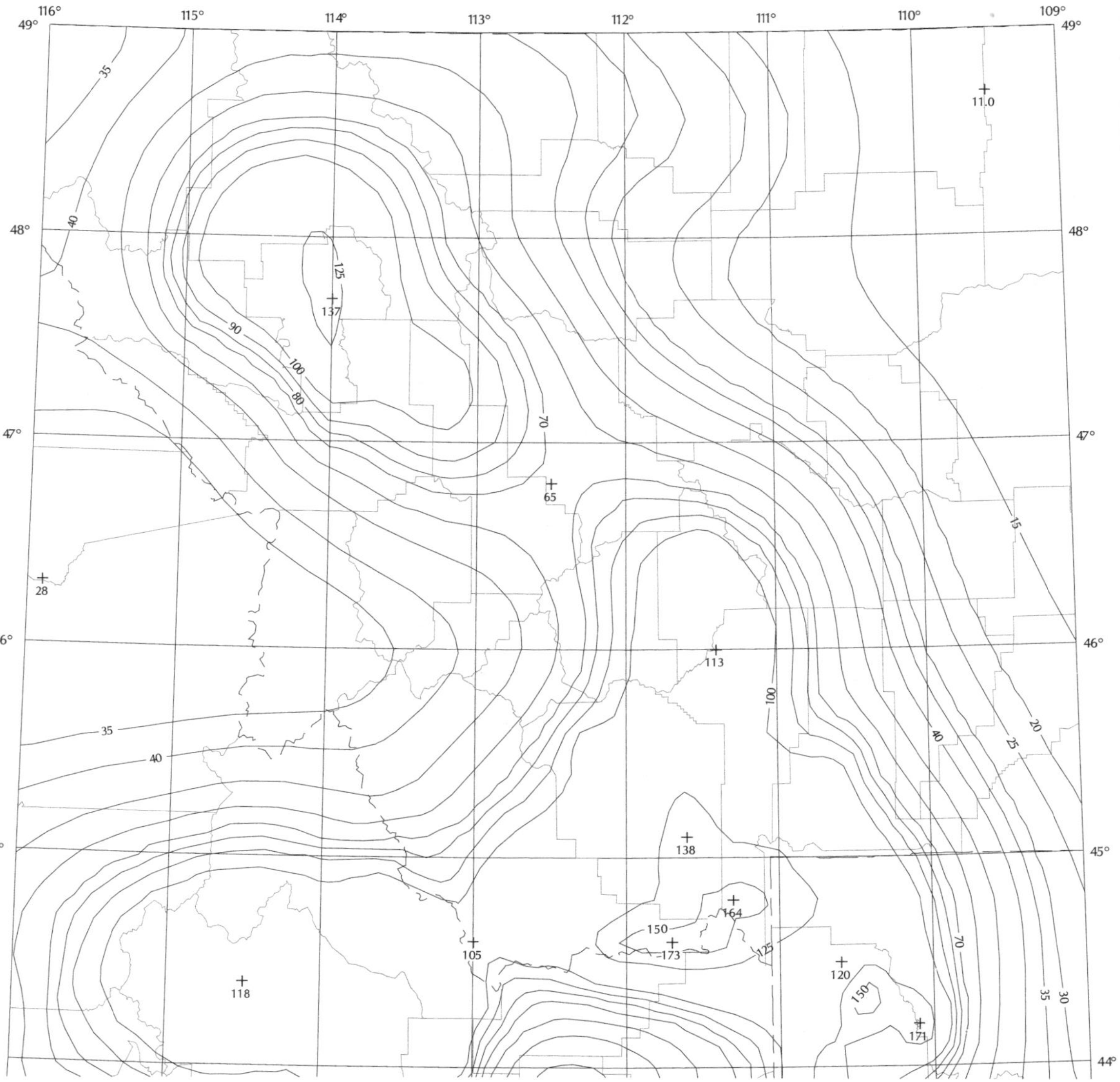

Explanation

Contour intervals, % g

200
175
150
125
100
90
80
70
60
50
40
35
30
25
20
15
10
5
0

Note: contours are irregularly spaced

+ 6.2	Point value of spectral response acceleration expressed as a percent of gravity
—10—	Contours of spectral response acceleration expressed as a percent of gravity. Hachures point in direction of decreasing values.
⊥⊥10⊥⊥	

DISCUSSION

The acceleration values contoured are the random horizontal component. For design purposes, the reference site condition for the map is to be taken as NEHRP site class B.

FIGURE 1615(5)
MAXIMUM CONSIDERED EARTHQUAKE GROUND MOTION FOR REGION 2 OF 0.2 SEC SPECTRAL RESPONSE ACCELERATION (5 PERCENT OF CRITICAL DAMPING), SITE CLASS B

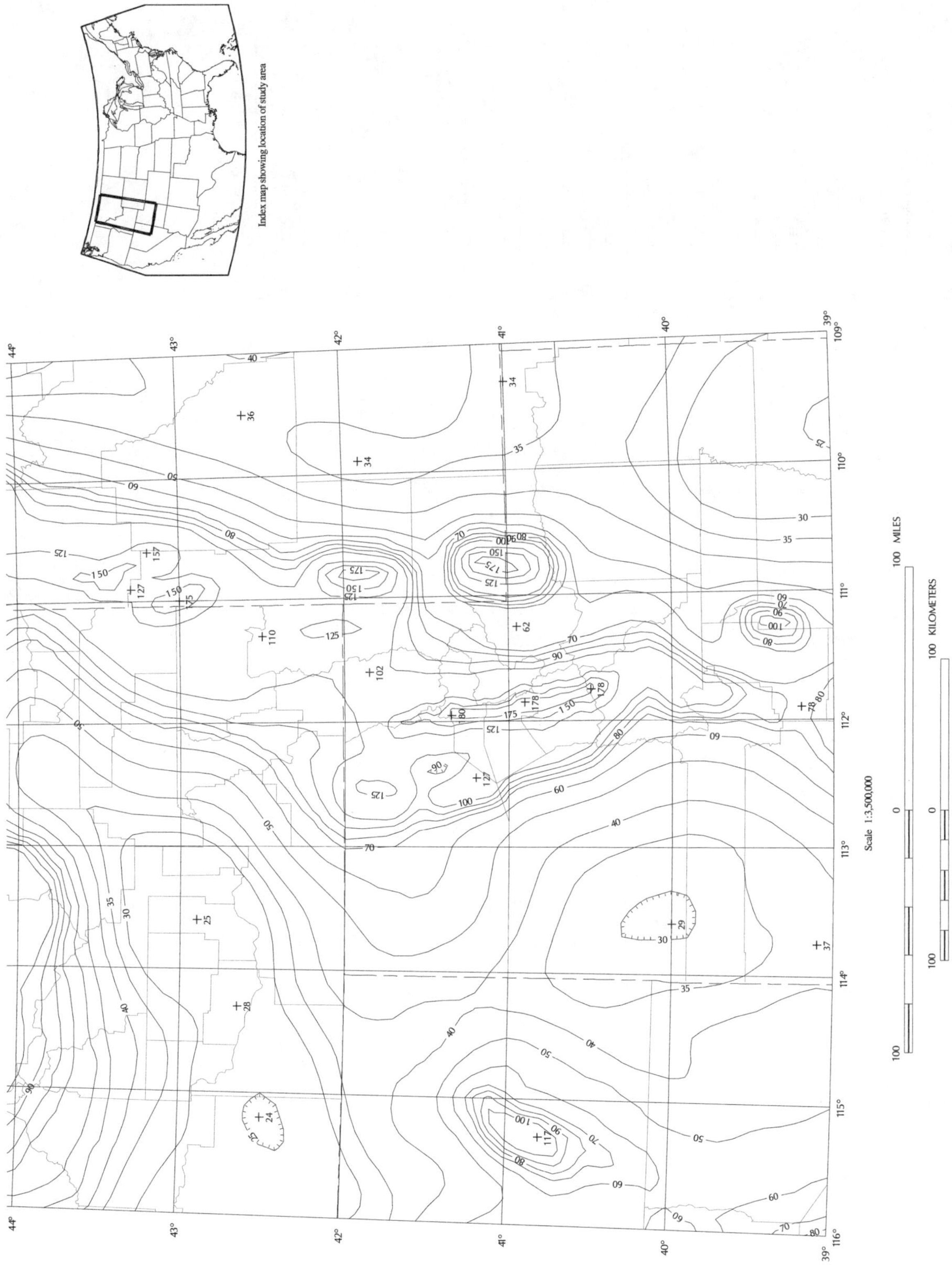

FIGURE 1615(5)—continued
MAXIMUM CONSIDERED EARTHQUAKE GROUND MOTION FOR REGION 2 OF 0.2 SEC SPECTRAL RESPONSE ACCELERATION (5 PERCENT OF CRITICAL DAMPING), SITE CLASS B

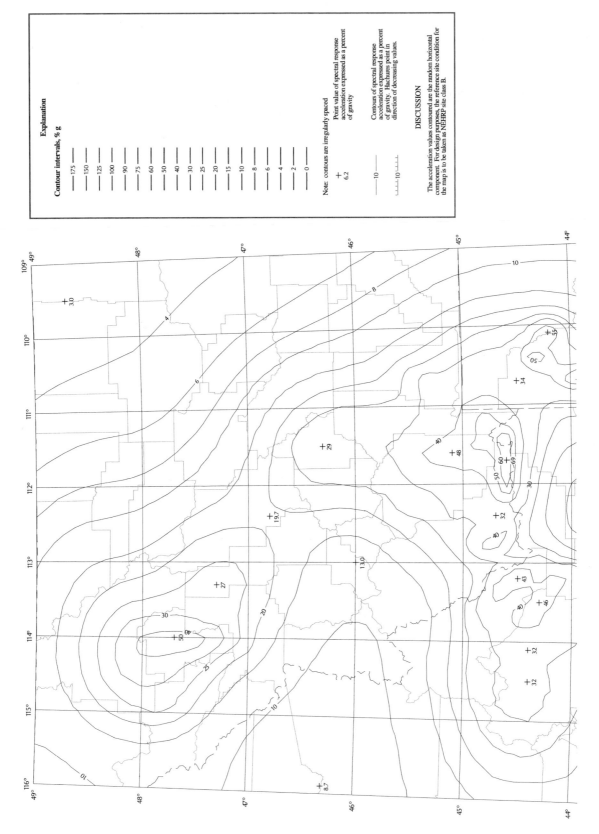

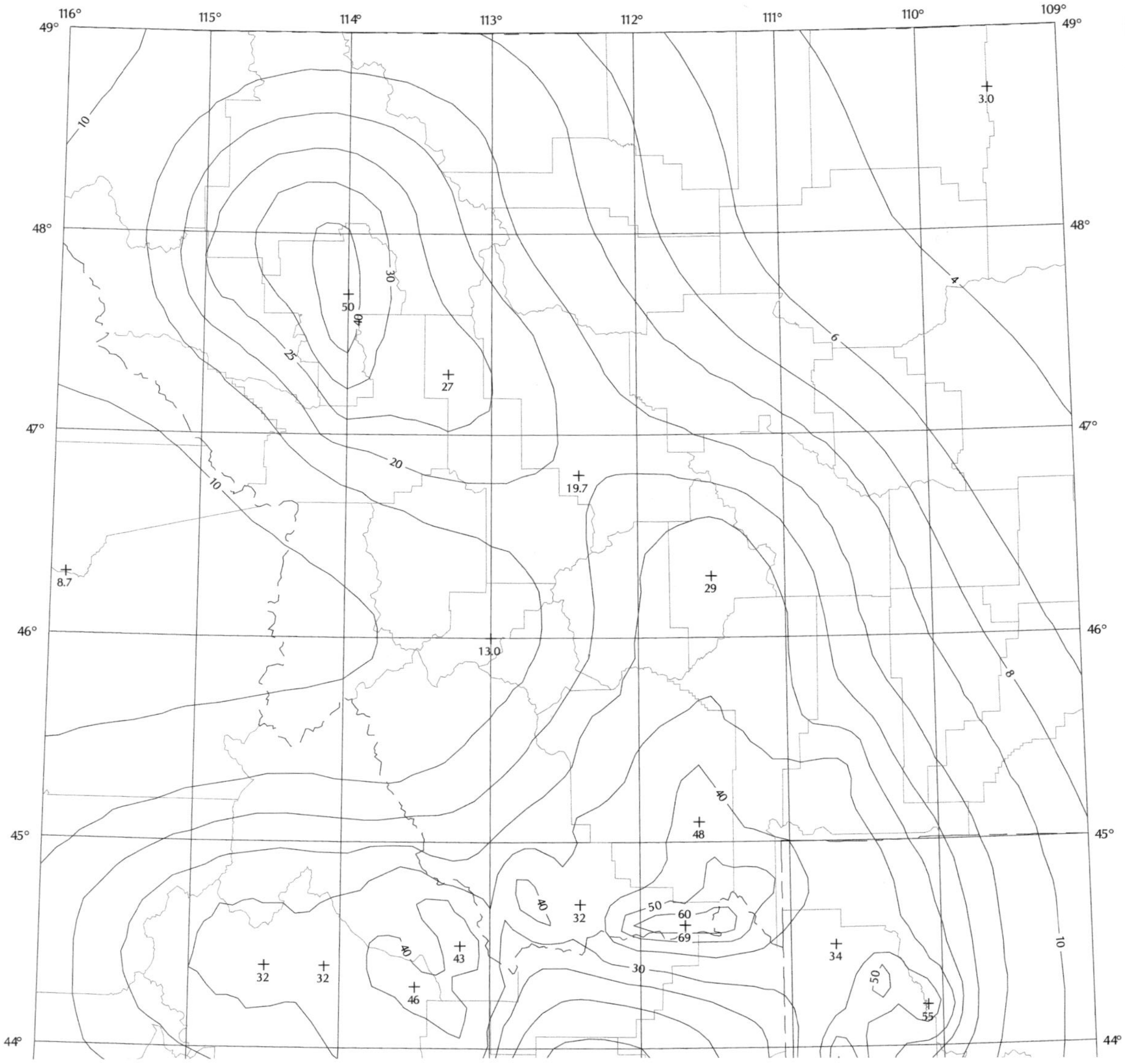

Explanation

Contour intervals, % g

— 175 —
— 150 —
— 125 —
— 100 —
— 90 —
— 75 —
— 60 —
— 50 —
— 40 —
— 30 —
— 25 —
— 20 —
— 15 —
— 10 —
— 8 —
— 6 —
— 4 —
— 2 —
— 0 —

Note: contours are irregularly spaced

+ 6.2 — Point value of spectral response acceleration expressed as a percent of gravity

—10— Contours of spectral response acceleration expressed as a percent of gravity. Hachures point in direction of decreasing values.

DISCUSSION

The acceleration values contoured are the random horizontal component. For design purposes, the reference site condition for the map is to be taken as NEHRP site class B.

FIGURE 1615(6)
MAXIMUM CONSIDERED EARTHQUAKE GROUND MOTION FOR REGION 2 OF 1.0 SEC SPECTRAL RESPONSE ACCELERATION (5 PERCENT OF CRITICAL DAMPING), SITE CLASS B

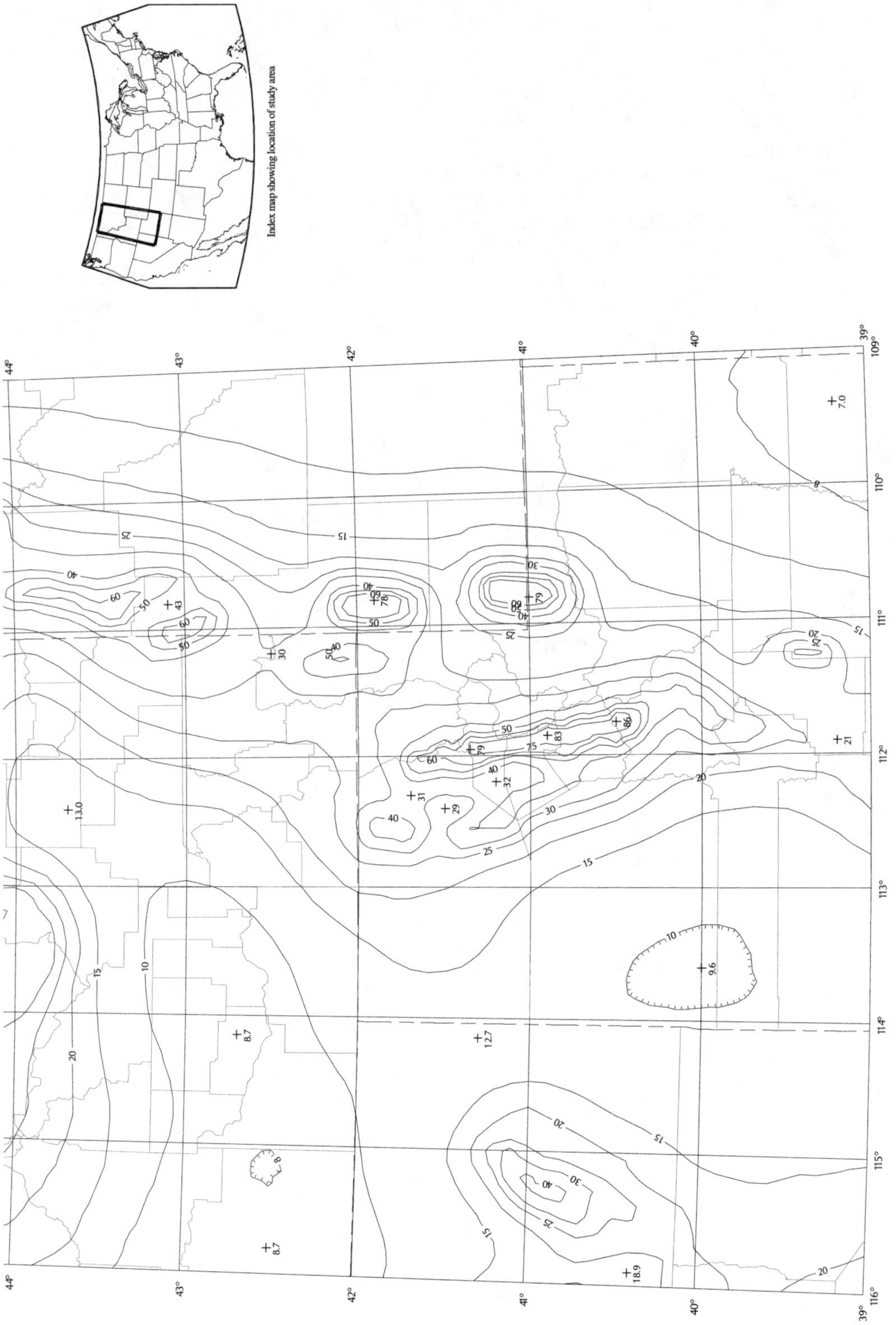

FIGURE 1615(6)—continued
MAXIMUM CONSIDERED EARTHQUAKE GROUND MOTION FOR REGION 2 OF 1.0 SEC SPECTRAL RESPONSE ACCELERATION (5 PERCENT OF CRITICAL DAMPING), SITE CLASS B

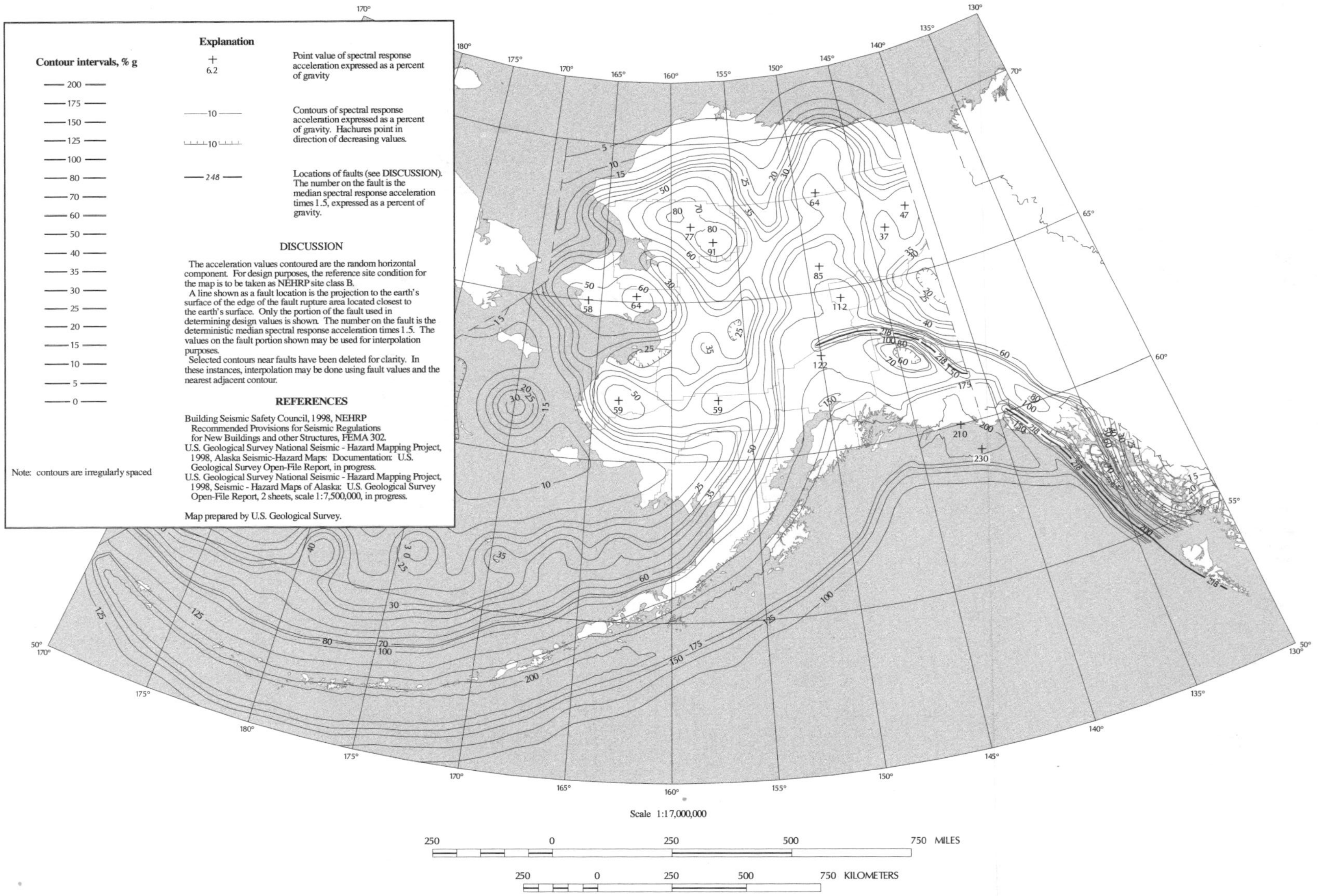

FIGURE 1615(7)
MAXIMUM CONSIDERED EARTHQUAKE GROUND MOTION FOR ALASKA OF 0.2 SEC SPECTRAL RESPONSE ACCELERATION (5 PERCENT OF CRITICAL DAMPING), SITE CLASS B

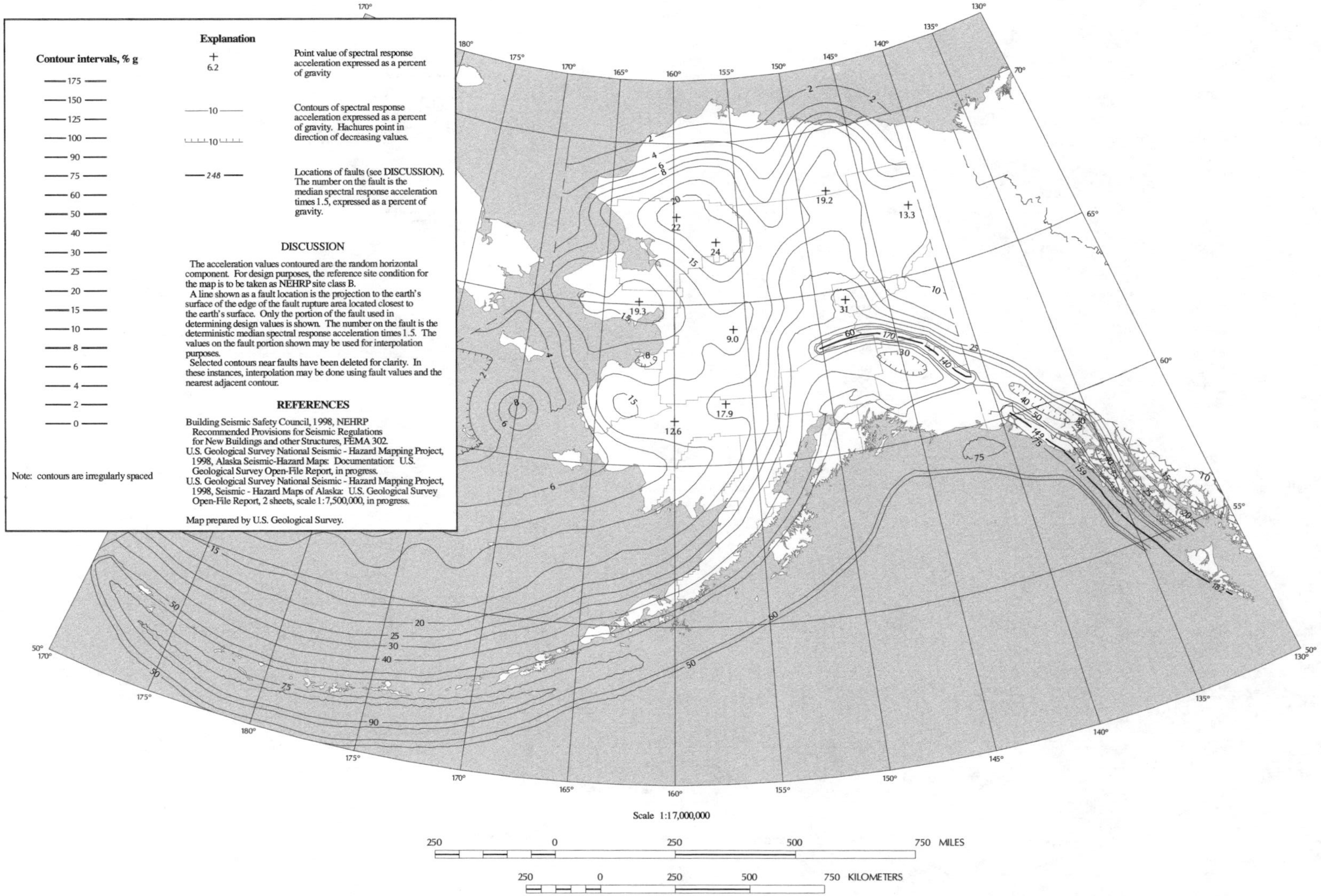

FIGURE 1615(7)–continued
MAXIMUM CONSIDERED EARTHQUAKE GROUND MOTION FOR ALASKA OF 1.0 SEC SPECTRAL RESPONSE ACCELERATION (5 PERCENT OF CRITICAL DAMPING), SITE CLASS B

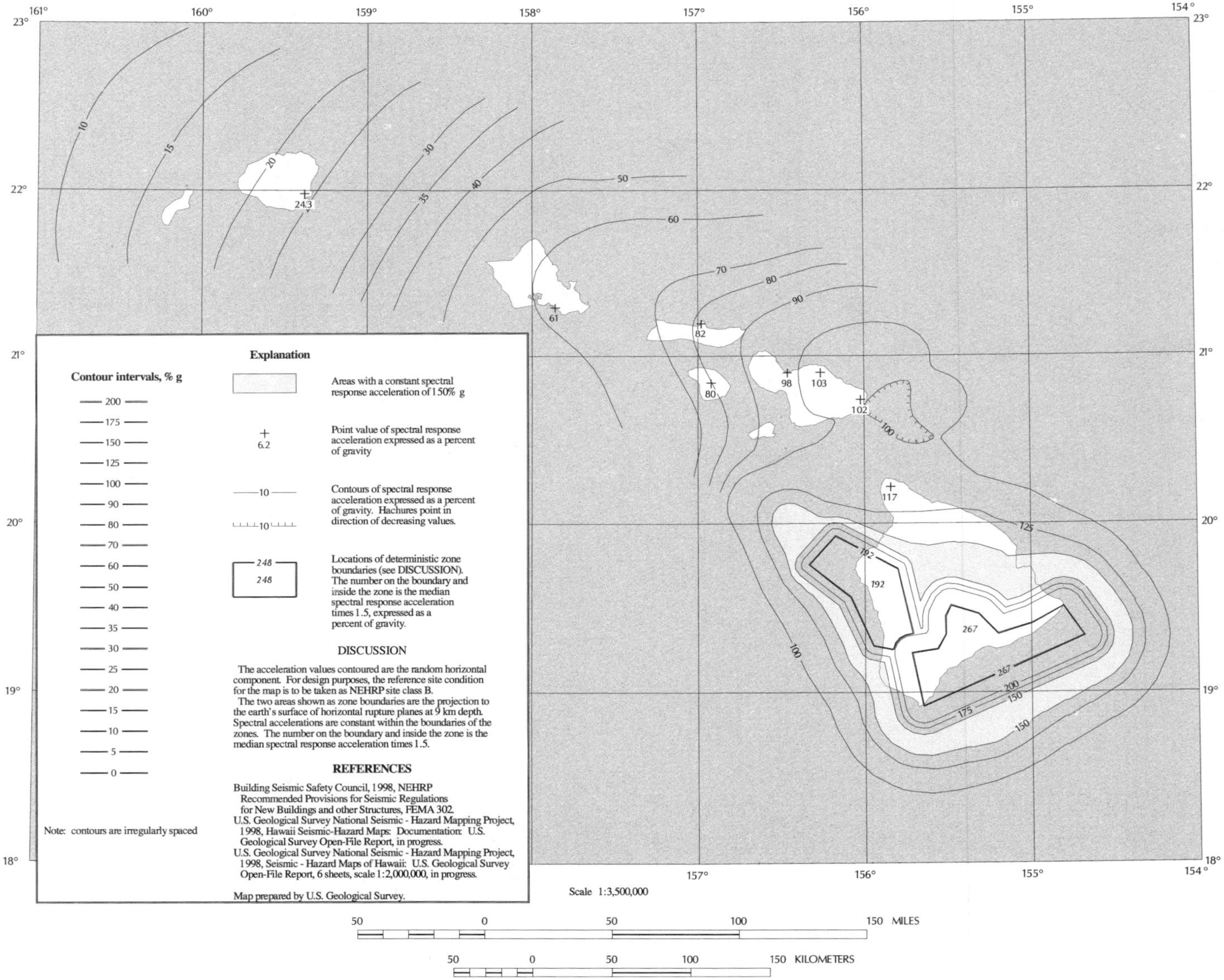

FIGURE 1615(8)
MAXIMUM CONSIDERED EARTHQUAKE GROUND MOTION FOR HAWAII OF 0.2 SEC SPECTRAL RESPONSE ACCELERATION (5 PERCENT OF CRITICAL DAMPING), SITE CLASS B

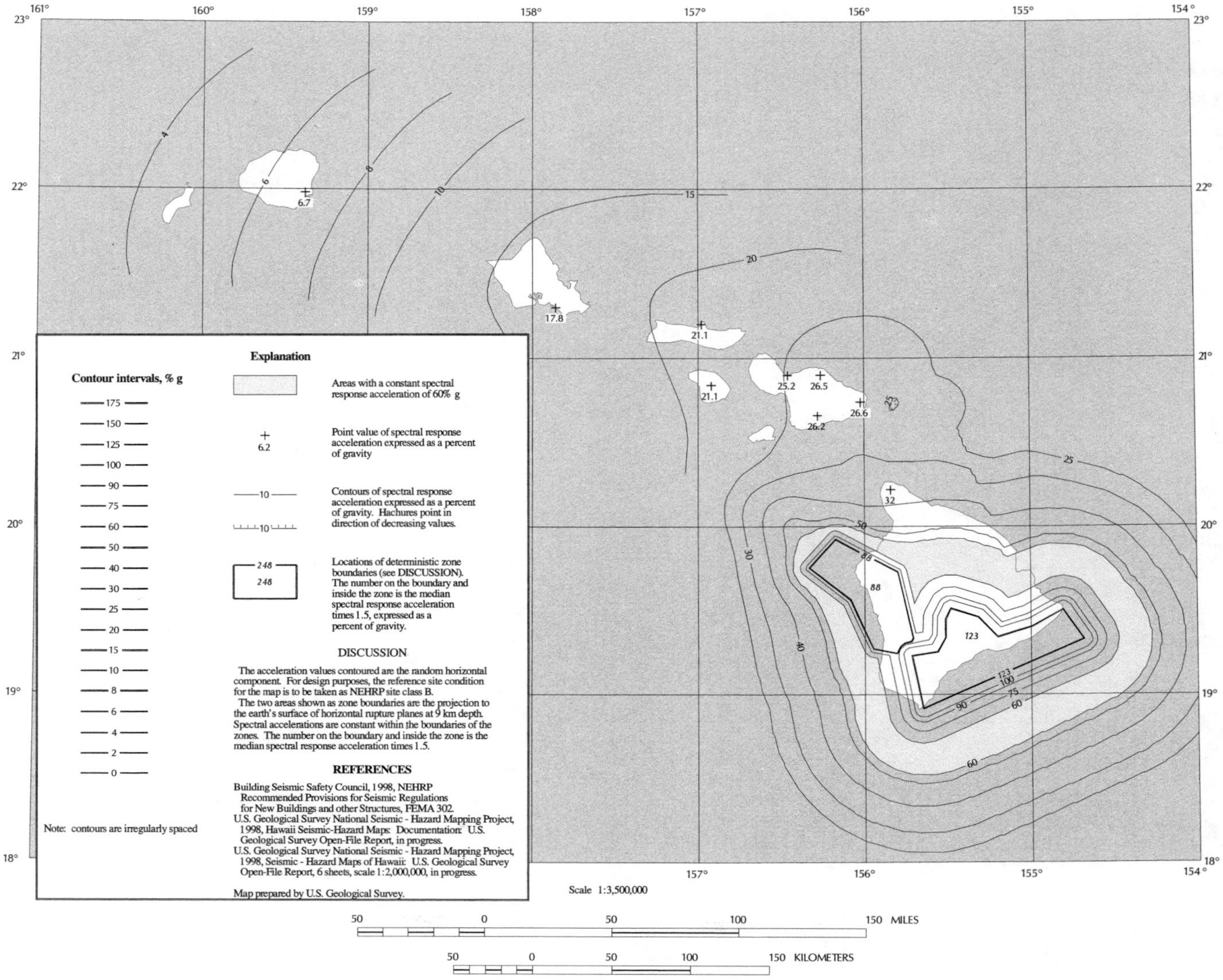

FIGURE 1615(8)–continued
MAXIMUM CONSIDERED EARTHQUAKE GROUND MOTION FOR HAWAII OF 1.0 SEC SPECTRAL RESPONSE ACCELERATION (5 PERCENT OF CRITICAL DAMPING), SITE CLASS B

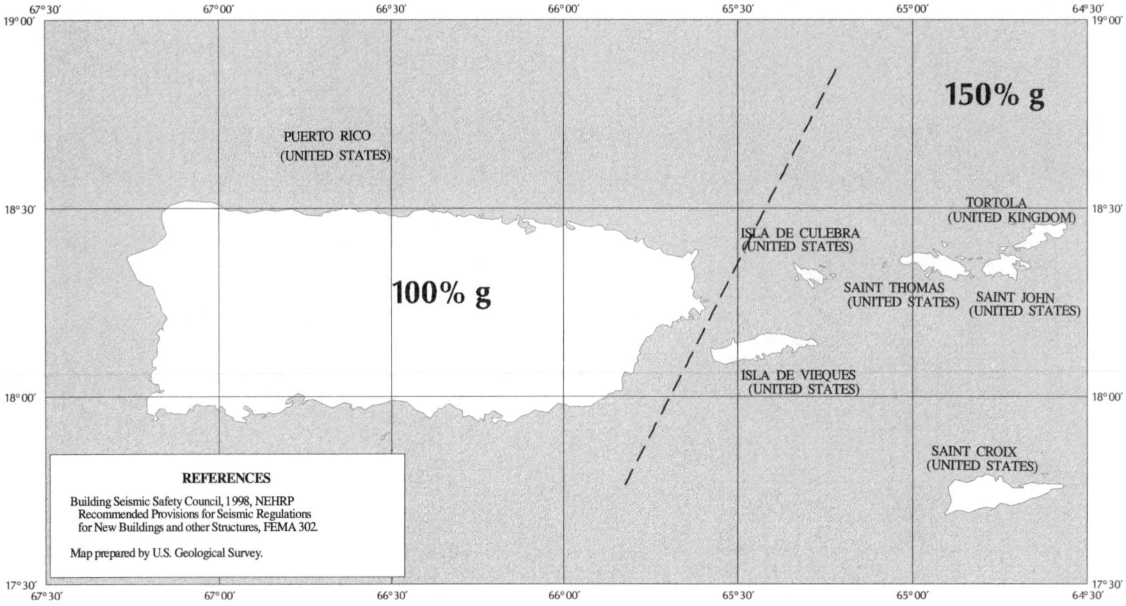

0.2 SEC SPECTRAL RESPONSE ACCELERATION (5% OF CRITICAL DAMPING)

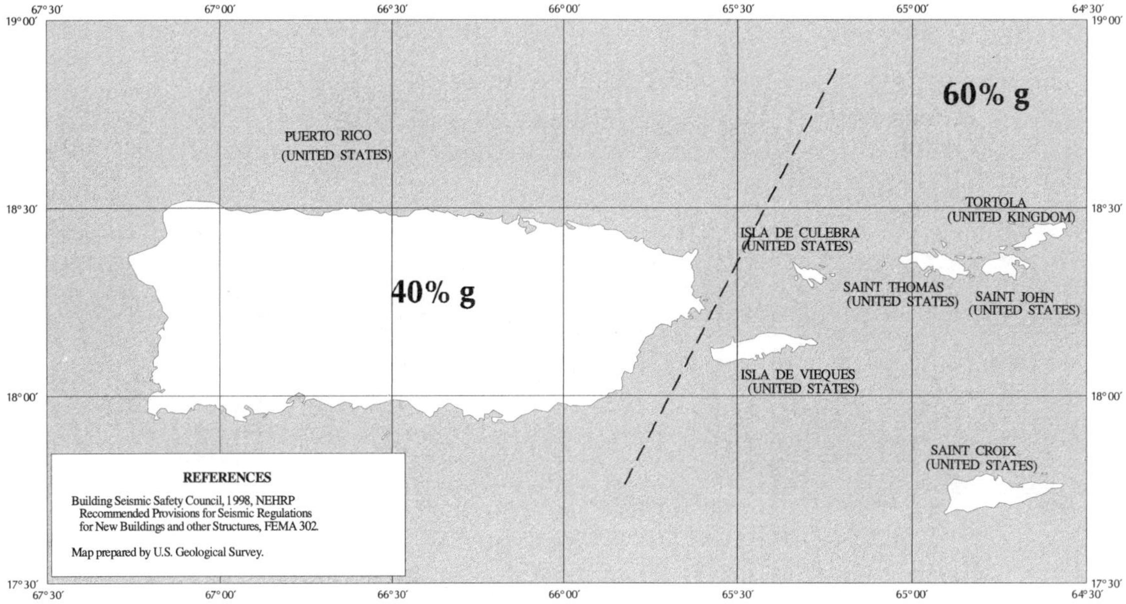

1.0 SEC SPECTRAL RESPONSE ACCELERATION (5% OF CRITICAL DAMPING)

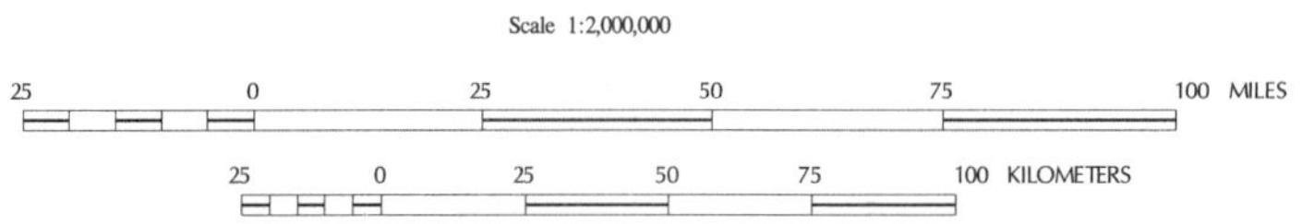

FIGURE 1615(9)
MAXIMUM CONSIDERED EARTHQUAKE GROUND MOTION FOR PUERTO RICO, CULEBRA, VIEQUES, ST. THOMAS, ST. JOHN, AND ST. CROIX OF 0.2 AND 1.0 SEC SPECTRAL RESPONSE ACCELERATION (5 PERCENT OF CRITICAL DAMPING), SITE CLASS B

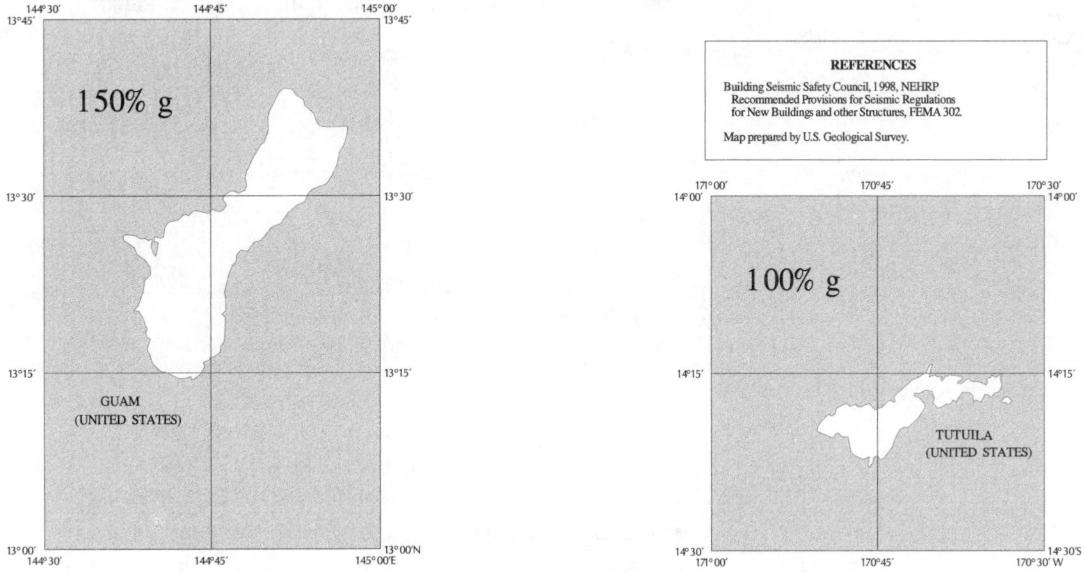

0.2 SEC SPECTRAL RESPONSE ACCELERATION (5% OF CRITICAL DAMPING)

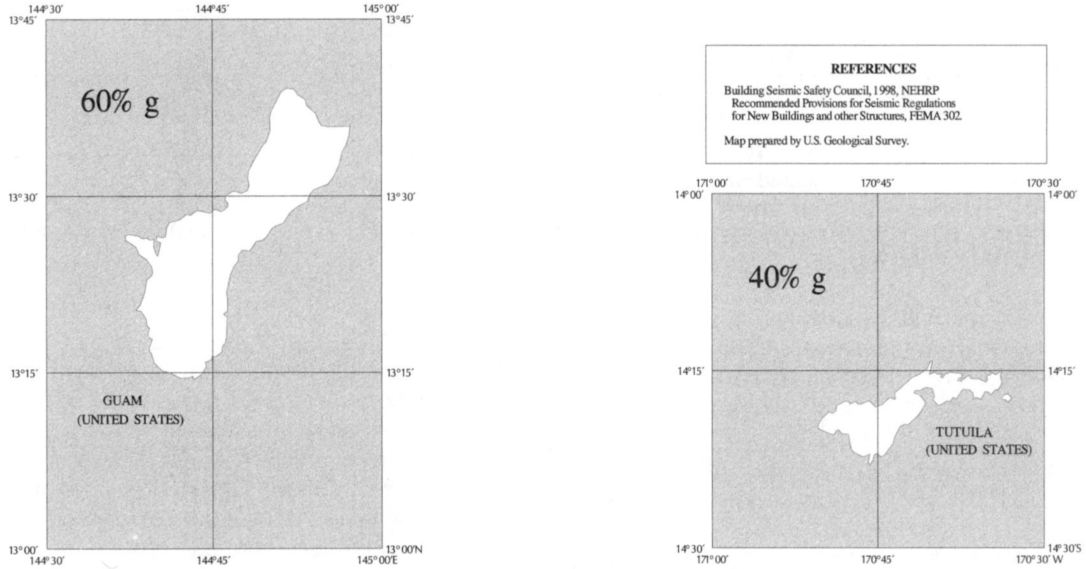

1.0 SEC SPECTRAL RESPONSE ACCELERATION (5% OF CRITICAL DAMPING)

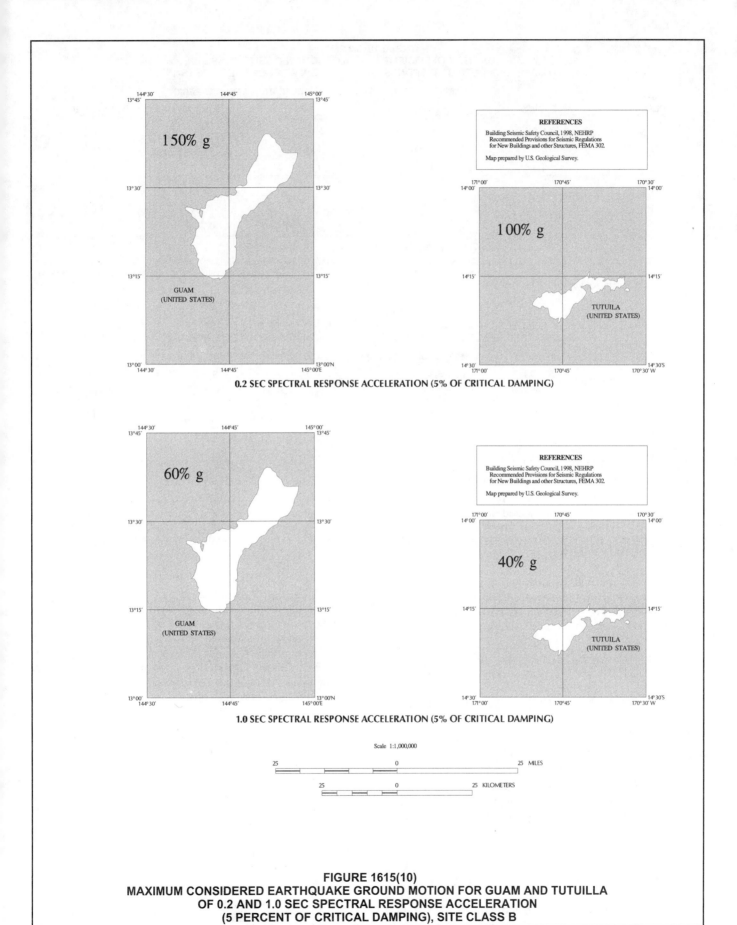

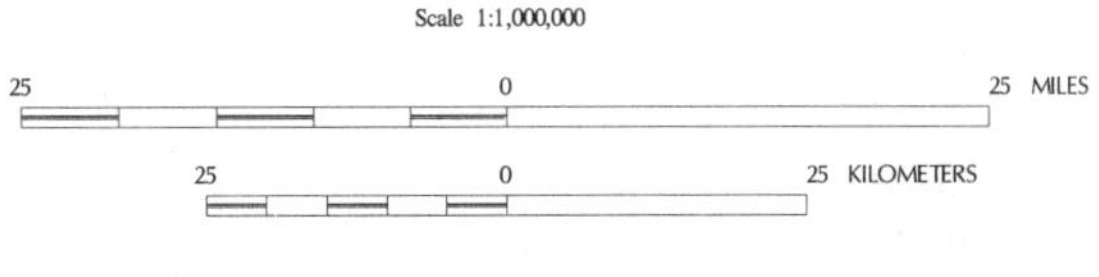

FIGURE 1615(10)
MAXIMUM CONSIDERED EARTHQUAKE GROUND MOTION FOR GUAM AND TUTUILLA OF 0.2 AND 1.0 SEC SPECTRAL RESPONSE ACCELERATION (5 PERCENT OF CRITICAL DAMPING), SITE CLASS B

TABLE 1615.1.2(1)
VALUES OF SITE COEFFICIENT F_a AS A FUNCTION OF SITE CLASS AND MAPPED SPECTRAL RESPONSE ACCELERATION AT SHORT PERIODS (S_S)[a]

SITE CLASS	MAPPED SPECTRAL RESPONSE ACCELERATION AT SHORT PERIODS				
	$S_S \le 0.25$	$S_S = 0.50$	$S_S = 0.75$	$S_S = 1.00$	$S_S \ge 1.25$
A	0.8	0.8	0.8	0.8	0.8
B	1.0	1.0	1.0	1.0	1.0
C	1.2	1.2	1.1	1.0	1.0
D	1.6	1.4	1.2	1.1	1.0
E	2.5	1.7	1.2	0.9	0.9
F	Note b	Note b	Note b	Note b	Note b

a. Use straight-line interpolation for intermediate values of mapped spectral response acceleration at short period, S_S.
b. Site-specific geotechnical investigation and dynamic site response analyses shall be performed to determine appropriate values, except that for structures with periods of vibration equal to or less than 0.5 second, values of F_a for liquefiable soils are permitted to be taken equal to the values for the site class determined without regard to liquefaction in Section 1615.1.5.1.

TABLE 1615.1.2(2)
VALUES OF SITE COEFFICIENT F_V AS A FUNCTION OF SITE CLASS AND MAPPED SPECTRAL RESPONSE ACCELERATION AT 1-SECOND PERIOD (S_1)[a]

SITE CLASS	MAPPED SPECTRAL RESPONSE ACCELERATION AT SHORT PERIODS				
	$S_1 \le 0.1$	$S_1 = 0.2$	$S_1 = 0.3$	$S_1 = 0.4$	$S_1 \ge 0.5$
A	0.8	0.8	0.8	0.8	0.8
B	1.0	1.0	1.0	1.0	1.0
C	1.7	1.6	1.5	1.4	1.3
D	2.4	2.0	1.8	1.6	1.5
E	3.5	3.2	2.8	2.4	2.4
F	Note b	Note b	Note b	Note b	Note b

a. Use straight-line interpolation for intermediate values of mapped spectral response acceleration at 1-second period, S_1.
b. Site-specific geotechnical investigation and dynamic site response analyses shall be performed to determine appropriate values, except that for structures with periods of vibration equal to or less than 0.5 second, values of F_v for liquefiable soils are permitted to be taken equal to the values for the site class determined without regard to liquefaction in Section 1615.1.5.1.

1615.1.3 Design spectral response acceleration parameters. Five-percent damped design spectral response acceleration at short periods, S_{DS}, and at 1-second period, S_{D1}, shall be determined from Equations 16-40 and 16-41, respectively:

$$S_{DS} = \frac{2}{3} S_{MS} \qquad \textbf{(Equation 16-40)}$$

$$S_{D1} = \frac{2}{3} S_{M1} \qquad \textbf{(Equation 16-41)}$$

where:

S_{MS} = The maximum considered earthquake spectral response accelerations for short period as determined in Section 1615.1.2.

S_{M1} = The maximum considered earthquake spectral response accelerations for 1-second period as determined in Section 1615.1.2.

❖ The design spectral response acceleration at 0.2-second period, S_{DS}, is two-thirds of the S_{MS} value calculated in accordance with Section 1615.1.2. The design spectral response acceleration at 1-second period, S_{D1}, is two-thirds of the S_{M1} value calculated in accordance with Section 1615.1.2. Two-thirds is the reciprocal of 1.5; thus, the design ground motion is 1/1.5 times the soil-modified maximum considered earthquake ground motion. This is in recognition of the inherent margin contained in the NEHRP Provisions that would make collapse unlikely under 1.5 times the design-level ground motion. The idea is to avoid collapse when a structure is subjected to the soil-modified maximum considered earthquake ground motion.

1615.1.4 General procedure response spectrum. The general design response spectrum curve shall be developed as indicated in Figure 1615.1.4 and as follows:

1. For periods less than or equal to T_O, the design spectral response acceleration, S_a, shall be determined by Equation 16-42.

2. For periods greater than or equal to T_O and less than or equal to T_S, the design spectral response acceleration, S_a, shall be taken equal to S_{DS}.

3. For periods greater than T_S, the design spectral response acceleration, S_a, shall be determined by Equation 16-43.

$$S_a = 0.6\frac{S_{DS}}{T_O}T + 0.4S_{DS}$$ **(Equation 16-42)**

$$S_a = \frac{S_{D1}}{T}$$ **(Equation 16-43)**

where:

S_{DS} = The design spectral response acceleration at short periods as determined in Section 1615.1.3.

S_{D1} = The design spectral response acceleration at 1-second period as determined in Section 1615.1.3.

T = Fundamental period (in seconds) of the structure (see Section 9.5.5.3 of ASCE 7).

T_O = $0.2\ S_{D1}/S_{DS}$

T_S = S_{D1}/S_{DS}

❖ This section provides a general method for obtaining a 5-percent damped response spectrum from the design spectral response acceleration parameters S_{DS} and S_{D1} at the site. The spectrum consists of a region of constant spectral response accelerations in the short-period range ["flat top" in Figure 1615.1.4(1)] and a region of constant spectral response velocities in the long-period range. The constant displacement domain of the response spectrum is not included in the generalized response spectrum because relatively few structures have a period long enough to fall into this range of very long periods.

The period $T_s = S_{D1}/S_{DS}$ divides the short-period range from the long-period range. Spectral response acceleration at any period in the short-period range ($T < T_s$) is given by $S_a = S_{DS}$.

The spectral response acceleration at any period in the long-period range ($T \geq T_s$) is given by Equation 16-43. The ramp building up to the flat top of the design spectrum in Figure 1615.1.4(1) is defined by specifying that the spectral response acceleration at zero period is equal to 40 percent of the spectral response acceleration corresponding to the flat top, S_{DS}, and that the period T_O at which the ramp ends is 20 percent of the period T_s [see Figure 1615.1.4(1)].

FIGURE 1615.1.4. See below.

❖ This code figure illustrates the design response spectrum established following the procedure of Section 1615.1.4.

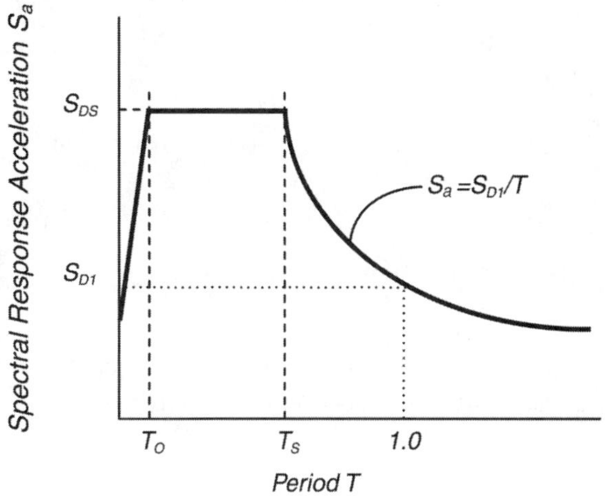

FIGURE 1615.1.4
DESIGN RESPONSE SPECTRUM

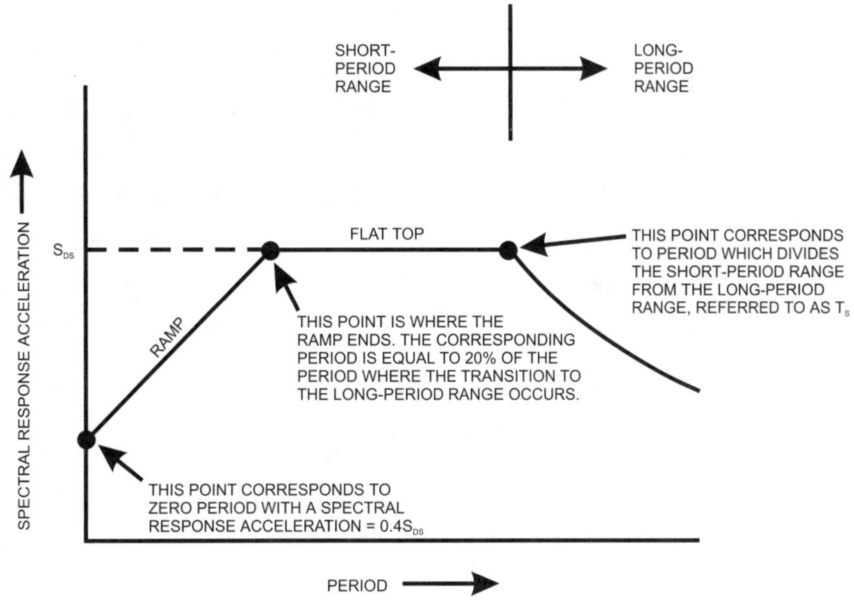

Figure 1615.1.4(1)
SCHEMATIC OF GENERAL PROCEDURE RESPONSE SPECTRUM

1615.1.5 Site classification for seismic design. Site classification for Site Class C, D or E shall be determined from Table 1615.1.5.

The notations presented below apply to the upper 100 feet (30 480 mm) of the site profile. Profiles containing distinctly different soil layers shall be subdivided into those layers designated by a number that ranges from 1 to n at the bottom where there is a total of n distinct layers in the upper 100 feet (30 480 mm). The symbol, i, then refers to any one of the layers between 1 and n.

where:

v_{si} = The shear wave velocity in feet per second (m/s).

d_i = The thickness of any layer between 0 and 100 feet (30 480 mm).

$$\bar{v}_s = \frac{\sum_{i=1}^{n} d_i}{\sum_{i=1}^{n} \frac{d_i}{v_{si}}} \qquad \textbf{(Equation 16-44)}$$

$$\sum_{i=1}^{n} d_i = 100 \text{ feet } (30\,480 \text{ mm})$$

N_i is the Standard Penetration Resistance (ASTM D 1586) not to exceed 100 blows/foot (mm) as directly measured in the field without corrections.

$$\bar{N} = \frac{\sum_{i=1}^{n} d_i}{\sum_{i=1}^{n} \frac{d_i}{N_i}} \qquad \textbf{(Equation 16-45)}$$

$$\bar{N}_{ch} = \frac{d_s}{\sum_{i=1}^{m} \frac{d_i}{N_i}} \qquad \textbf{(Equation 16-46)}$$

where:

$\sum_{i=1}^{m} d_i = d_s$ Use only d_i and N_i for cohesionless soils.

d_s = The total thickness of cohesionless soil layers in the top 100 feet (30 480 mm).

s_{ui} = The undrained shear strength in psf (kPa), not to exceed 5,000 psf (240 kPa), ASTM D 2166 or D 2850.

$$\bar{s}_u = \frac{d_c}{\sum_{i=1}^{k} \frac{d_i}{s_{ui}}} \qquad \textbf{(Equation 16-47)}$$

where:

$\sum_{i=1}^{k} d_i = d_c$

d_c = The total thickness $(100 - d_s)$ (For SI: $30\,480 - d_s$) of cohesive soil layers in the top 100 feet (30 480 mm).

PI = The plasticity index, ASTM D 4318.

w = The moisture content in percent, ASTM D 2216.

The shear wave velocity for rock, Site Class B, shall be either measured on site or estimated by a geotechnical engineer or engineering geologist/seismologist for competent rock with moderate fracturing and weathering. Softer and more highly fractured and weathered rock shall either be measured on site for shear wave velocity or classified as Site Class C.

The hard rock, Site Class A, category shall be supported by shear wave velocity measurements either on site or on profiles of the same rock type in the same formation with an equal or greater degree of weathering and fracturing. Where hard rock conditions are known to be continuous to a depth of 100 feet (30 480 mm), surficial shear wave velocity measurements are permitted to be extrapolated to assess $\bar{v}_s$.

The rock categories, Site Classes A and B, shall not be used if there is more than 10 feet (3048 mm) of soil between the rock surface and the bottom of the spread footing or mat foundation.

❖ Where different soil layers make up the upper 100 feet (30 480 mm) of soil profile at a site, this section provides formulas that average soil properties over the upper 100 feet (30 480 mm). There are specific considerations given for the determination of Site Class A or B using shear wave velocity. The rock categories, Site Class A or B, must not be used if there is more than 10 feet (3048 mm) of soil between the rock surface and the underside of the spread footing or mat foundation.

TABLE 1615.1.5. See below.

❖ This table repeats the criteria of Table 1615.1.1 that apply specifically to the assignment of Site Class C, D or E by either of the methods specified in Item 3 of Section 1615.1.5.1.

1615.1.5.1 Steps for classifying a site.

1. Check for the four categories of Site Class F requiring site-specific evaluation. If the site corresponds to any of these categories, classify the site as Site Class F and conduct a site-specific evaluation.

TABLE 1615.1.5
SITE CLASSIFICATION[a]

SITE CLASS	$\bar{v}_s$	$\bar{N}$ or $\bar{N}_{ch}$	$\bar{s}_u$
E	< 600 ft/s	< 15	< 1,000 psf
D	600 to 1,200 ft/s	15 to 50	1,000 to 2,000 psf
C	1,200 to 2,500 ft/s	> 50	> 2,000

For SI: 1 foot per second = 304.8 mm per second, 1 pound per square foot = 0.0479 kN/m^2.

a. If the $\bar{s}_u$ method is used and the $\bar{N}_{ch}$ and $\bar{s}_u$ criteria differ, select the category with the softer soils (for example, use Site Class E instead of D).

2. Check for the existence of a total thickness of soft clay > 10 feet (3048 mm) where a soft clay layer is defined by: $\bar{s}_u$ < 500 psf (25 kPa), $w \geq 40$ percent, and $PI > 20$. If these criteria are satisfied, classify the site as Site Class E.
3. Categorize the site using one of the following three methods with $\bar{v}_s$, $\bar{N}$, and $\bar{s}_u$ computed in all cases as specified.
   - 3.1. $\bar{v}_s$ for the top 100 feet (30 480 mm) ($\bar{v}_s$ method).
   - 3.2. $\bar{N}$ for the top 100 feet (30 480 mm) ($\bar{N}$ method).
   - 3.3. $\bar{N}_{ch}$ for cohesionless soil layers ($PI < 20$) in the top 100 feet (30 480 mm) and average, $\bar{s}_u$, for cohesive soil layers ($PI > 20$) in the top 100 feet (30 480 mm) ($\bar{s}_u$ method).

❖ This section outlines the specific steps leading to the classification of a site for other than Site Class A or B, since those are assigned solely on the basis of shear wave velocity as discussed in Section 1615.1.5.

The first step is to rule out the possibility that the soil meets the criteria of Table 1615.1.1 for Site Class F soil. The second step is to rule out the possibility that the soil meets the criteria of Table 1615.1.1 for Site Class E soil.

If none of the above conditions are satisfied, the third step is to classify the soil profile as either Site Class C, D or E on the basis of (a) the average shear wave velocity for the top 100 feet (30 480 mm), (b) the average standard penetration resistance for the top 100 feet (30 480 mm) or (c) the average standard penetration resistance of the cohesionless soil layers and the average undrained shear strength of the cohesive soil layers in the top 100 feet (30 480 mm). It should be noted that Table 1615.1.5 is actually an excerpt of Table 1615.1.1, except for the footnote added to Table 1615.1.5.

1615.2 Site-specific procedure for determining ground motion accelerations. A site-specific study shall account for the regional seismicity and geology; the expected recurrence rates and maximum magnitudes of events on known faults and source zones; the location of the site with respect to these; near source effects if any and the characteristics of subsurface site conditions.

❖ While the general procedure for determining design spectral response accelerations will be used in an overwhelming majority of cases, the code does mandate the use of a site-specific procedure for Site Classes E and F where the short-period mapped spectral response acceleration exceeds 1.25g, as would happen in the vicinity of known faults capable of generating large-magnitude earthquakes.

Five significant aspects must be accounted for in an investigation to determine site-specific ground motion, as detailed in this section.

Section 1615.2.1 defines a probabilistic site-specific maximum considered earthquake acceleration response spectrum. Section 1615.2.2 defines an acceleration response spectrum that represents a deterministic limit on maximum considered earthquake ground motion. Section 1615.2.3 defines a deterministic site-specific maximum considered earthquake acceleration response spectrum. As indicated in the exception to Section 1615.2.1, the site-specific maximum considered earthquake ground motion spectrum is required to be taken as the lesser of the probabilistic maximum considered earthquake ground motion spectrum of Section 1615.2.1 or the deterministic maximum considered earthquake ground motion spectrum of Section 1615.2.3, subject to a minimum of the deterministic limit ground motion spectrum of Section 1615.2.2.

1615.2.1 Probabilistic maximum considered earthquake. Where site-specific procedures are used as required or permitted by Section 1615, the maximum considered earthquake ground motion shall be taken as that motion represented by an acceleration response spectrum having a 2-percent probability of exceedance within a 50-year period. The maximum considered earthquake spectral response acceleration at any period, S_{aM}, shall be taken from the 2-percent probability of exceedance within a 50-year period spectrum.

> **Exception:** Where the spectral response ordinates at 0.2 second or 1 second for a 5-percent damped spectrum having a 2-percent probability of exceedance within a 50-year period exceed the corresponding ordinates of the deterministic limit of Section 1615.2.2, the maximum considered earthquake ground motion spectrum shall be taken as the lesser of the probabilistic maximum considered earthquake ground motion or the deterministic maximum considered earthquake ground motion spectrum of Section 1615.2.3, but shall not be taken as less than the deterministic limit ground motion of Section 1615.2.2.

❖ The probabilistic maximum considered earthquake acceleration response spectrum is defined as corresponding to a 2-percent probability of exceedance in 50 years (an approximate return period of 2,500 years). S_{aM}, the probabilistic maximum considered earthquake spectral response acceleration at any period, is to be taken from this spectrum. The generalized procedure for determining maximum considered earthquake spectral response accelerations of Section 1615.1 uses the same probability of exceedance over the same period of time.

1615.2.2 Deterministic limit on maximum considered earthquake ground motion. The deterministic limit for the maximum considered earthquake ground motion shall be the response spectrum determined in accordance with Figure 1615.2.2, where site coefficients, F_a and F_v, are determined in accordance with Section 1615.1.2, with the value of the mapped short-period spectral response acceleration, S_S, taken as 1.5g and the value of the mapped spectral response acceleration at 1 second, S_1, taken as 0.6g.

❖ The acceleration response spectrum of Code Figure 1615.2.2 represents this limit. The coefficients F_a and F_v are as given by Tables 1615.1.2(1) and 1615.1.2(2), respectively. In Code Figure 1615.2.2, the value of the mapped short-period spectral response acceleration, S_S, is taken as 1.5g and the value of the mapped spectral response acceleration at 1-second period, S_1, is taken as 0.6g.

FIGURE 1615.2.2 See below.

❖ The acceleration response spectrum of this figure represents a deterministic limit on the probabilistic maximum considered earthquake ground motion.

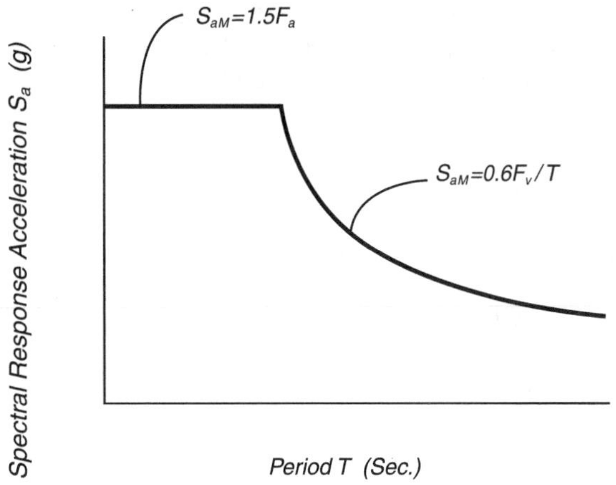

FIGURE 1615.2.2
DETERMINISTIC LIMIT ON MAXIMUM CONSIDERED EARTHQUAKE RESPONSE SPECTRUM

1615.2.3 Deterministic maximum considered earthquake ground motion. The deterministic maximum considered earthquake ground motion response spectrum shall be calculated as 150 percent of the median spectral response accelerations, S_{aM}, at all periods resulting from a characteristic earthquake on any known active fault within the region.

❖ The deterministic maximum considered earthquake acceleration response spectrum is defined by 150 percent of the median spectral response accelerations at all periods resulting from a characteristic earthquake on any known active fault within the region. The median values are increased by 50 percent to represent maximum considered earthquake ground motion values. The deterministic maximum considered earthquake spectral response acceleration, S_{aM}, is to be taken from this spectrum.

1615.2.4 Site-specific design ground motion. Where site-specific procedures are used to determine the maximum considered earthquake ground motion response spectrum, the design spectral response acceleration, S_a, at any period shall be determined from Equation 16-48:

$$S_a = \frac{2}{3} S_{aM} \qquad \textbf{(Equation 16-48)}$$

and shall be greater than or equal to 80 percent of the design spectral response acceleration, S_a, determined by the general response spectrum in Section 1615.1.4.

❖ The design spectral response acceleration, S_a, at any period is to be taken as two-thirds of the maximum considered earthquake response spectral acceleration, S_{aM}, at that period, but no smaller than 80 percent of the design spectral response acceleration, S_a, at the same period determined from the general response spectrum of Code Figure 1615.1.4. The two-thirds factor is in recognition of the margin of safety inherent in the NEHRP Provisions that makes collapse unlikely under 1.5 times the design-level ground motion.

1615.2.5 Design spectral response coefficients. Where the site-specific procedure is used to determine the design ground motion in accordance with Section 1615.2.4, the parameter S_{DS} shall be taken as the spectral acceleration, S_a, obtained from the site-specific spectra at a period of 0.2 second, except that it shall not be taken as less than 90 percent of the peak spectral acceleration, S_a, at any period. The parameter S_{D1} shall be taken as the greater of the spectral acceleration, S_a, at a period of 1 second or two times the spectral acceleration, S_a, at a period of 2 seconds. The parameters S_{MS} and S_{M1} shall be taken as 1.5 times S_{DS} and S_{D1}, respectively. The values so obtained shall not be taken as less than 80 percent of the values obtained from the general procedures of Section 1615.1.

❖ S_{DS}, the design spectral response acceleration coefficient at short periods, is to be taken as the value of S_a as defined in Section 1615.2.4, at a period of 0.2 second. S_{D1}, the design spectral response acceleration coefficient at a period of 1 second, is to be taken as the value of S_a as defined in Section 1615.2.4, at a period of 1 second. Neither value is to be taken as less than 80 percent of the corresponding value obtained from the general procedures of Section 1615.1.

SECTION 1616
EARTHQUAKE LOADS—CRITERIA SELECTION

1616.1 Structural design criteria. Each structure shall be assigned to a seismic design category in accordance with Section 1616.3. Seismic design categories are used in this code to determine permissible structural systems, limitations on height and irregularity, those components of the structure that must be designed for seismic resistance and the types of lateral force analysis that must be performed. Each structure shall be provided with complete lateral- and vertical-force-resisting systems capable of providing adequate strength, stiffness and energy dissipation capacity to withstand the design earthquake ground motions determined in accordance with Section 1615 within the prescribed deformation limits of Section 1617.3. The design ground motions shall be assumed to occur along any horizontal direction of a structure. A continuous load path, or paths, with adequate strength and stiffness to transfer forces induced by the design earthquake ground motions from the points of application to the final point of resistance shall be provided.

Allowable stress design is permitted to be used to evaluate sliding, overturning and soil bearing at the soil-structure interface regardless of the approach used in the design of the structure, provided load combinations of Section 1605.3 are utilized. When using allowable stress design for proportioning foundations, the value of 0.2 $S_{DS}D$ in Equations 16-50, 16-51, 16-52 and 16-53 or Equations 9.5.2.7-1, 9.5.2.7-2, 9.5.2.7.1-1 and 9.5.2.7.1-2 of ASCE 7 is permitted to be taken equal to zero. When the load combinations of Section 1605.3.2 are utilized, a one-third increase in soil allowable stresses is permitted for all load combinations that include *W* or *E*.

❖ See Section 1613 for introductory commentary regarding earthquake provisions.

Section 1616 provides the ground rules by which a structure needs to be designed depending on seismic design category assignment. The seismic design category is the trigger mechanism for a host of seismic code requirements, including the following:

1. Permissible seismic-force-resisting system.
2. Limitations on height and irregularity.
3. Components of the structure that must be designed for seismic forces.
4. The type of lateral force analysis that may be used.

An example in the commentary to Section 1614.1 provides a step-by-step procedure on how to determine the seismic design category of a structure.

Section 1616.1 simply states that a structure needs to be designed to adequately resist the seismic forces and gravity loads prescribed by the code. The section also states that a complete load path needs to be provided from the point of application of a load to the final point of resistance. Unfortunately, a common cause of failure in past earthquakes has been the lack of a continuous connection path or "load path" for the seismic forces from the point of origin to the foundation. It is stressed that deformations caused by design ground motions must be within prescribed deformation limits, which would require adequate structural stiffness. The need for adequate energy dissipation is also stressed because the code-prescribed design seismic forces anticipate inelastic structural response to the design ground motions.

The last paragraph addresses the use of allowable stress design for the evaluation of forces at the soil-structure interface. Typically, sizing of the foundation and checking for overturning and sliding are done by the ASD method. This paragraph provides clarification that although a structure is designed using the strength design method and corresponding load combinations, the soil-structure interface elements may be designed using the ASD method and corresponding load combinations.

1616.2 Seismic use groups and occupancy importance factors. Each structure shall be assigned a seismic use group and a corresponding occupancy importance factor (I_E) as indicated in Table 1604.5.

❖ This section requires each structure to be assigned a seismic use group and an importance factor (I_E). Although the seismic use group classification has no direct effect on the seismic design force, that force is increased for structures in higher use groups through importance factors larger than 1. An increase in the seismic use group classification (for example, from Group II to III) affects the seismic design category classification. The structural member and connection detailing requirements increase; as the seismic design category classification increases. Thus, a higher seismic use group classification indirectly increases the structural detailing requirements. The seismic use group classification also affects the seismic design requirements for architectural, mechanical and electrical components and systems in accordance with Section 1621.1.

The importance factor increases the required seismic design force. Figure 1616.2 tabulates the earthquake importance factor in accordance with the assigned seismic use group.

Structures assigned occupancy importance factors exceeding 1 must be designed for larger forces and are expected to sustain less damage than a structure with an occupancy importance factor of 1. This factor enables a structure's seismic performance capability to be somewhat controlled depending upon its use and its need to be functional after an earthquake.

SEISMIC USE GROUP	I_E
I	1.0
II	1.25
III	1.5

Figure 1616.2
OCCUPANCY IMPORTANCE FACTORS

1616.2.1 Seismic Use Group I. Seismic Use Group I structures are those not assigned to either Seismic Use Group II or III.

❖ Seismic Use Group I contains all occupancies other than those listed in Occupancy Categories I and II in Table 1604.5. Structures in Seismic Use Group I represent a lesser hazard to life because of fewer building occupants and smaller building size.

1616.2.2 Seismic Use Group II. Seismic Use Group II structures are those, the failure of which would result in a substantial public hazard due to occupancy or use as indicated by Table 1604.5, or as designated by the building official.

❖ Seismic Use Group II includes those occupancies that have large numbers of occupants because of the overall size of the building or number of stories; the character of the use, such as public assembly, schools or colleges or a height that exposes the occupants to a greater life safety hazard. This group also includes uses where the occupants are restrained or otherwise impaired, such as hospitals and jails.

1616.2.3 Seismic Use Group III. Seismic Use Group III structures are those having essential facilities that are required for postearthquake recovery and those containing substantial quantities of hazardous substances, as indicated in Table 1604.5, or as designated by the building official.

Where operational access to a Seismic Use Group III structure is required through an adjacent structure, the adjacent structure shall conform to the requirements for Seismic Use Group III structures. Where operational access is less than 10 feet (3048 mm) from an interior lot line or less than 10 feet (3048 mm) from another structure, access protection from potential falling debris shall be provided by the owner of the Seismic Use Group III structure.

❖ Table 1604.5 considers the relative seismic hazard because of the nature of the occupancy or the essential nature of a structure's function. Most of the nature-of-occupancy items in Table 1604.5 are self-explanatory, except those regarding power-generating stations and primary communication facilities. A power-generating station or other utility (such as a natural gas facility) is to be classified as Seismic Use Group III only if the facility serves an emergency backup function for a Seismic Use Group III structure such as a fire station, police station and the like. Otherwise, the power-generating station or utility should be classified as Seismic Use Group II.

Two of the bulleted items in the Seismic Use Group III description use the term "designated." This term is referring to designation by the building official that structures are required for post-earthquake response and recovery.

The second paragraph in Section 1616.2.3 provides for access to and occupancy of facilities in Seismic Use Group III after the design earthquake. These requirements are the result of lessons learned from previous earthquakes in which Seismic Use Group III facilities have been rendered out of service because of the failure of an adjacent structure or a structure's exterior components.

1616.2.4 Multiple occupancies. Where a structure is occupied for two or more occupancies not included in the same seismic use group, the structure shall be assigned the classification of the highest seismic use group corresponding to the various occupancies.

Where structures have two or more portions that are structurally separated in accordance with Section 1620, each portion shall be separately classified. Where a structurally separated portion of a structure provides required access to, required egress from or shares life safety components with another portion having a higher seismic use group, both portions shall be assigned the higher seismic use group.

❖ These requirements provide for the seismic resistance needed in new buildings or structures, including occupancies of more than one seismic use group. Unless multiple occupancies are structurally separated, the entire structure is to be designed in accordance with the requirements of the most critical seismic use group because it resists the earthquake effects as a unit.

1616.3 Determination of seismic design category. All structures shall be assigned to a seismic design category based on their seismic use group and the design spectral response acceleration coefficients, S_{DS} and S_{D1}, determined in accordance with Section 1615.1.3 or 1615.2.5. Each building and structure shall be assigned to the most severe seismic design category in accordance with Table 1616.3(1) or 1616.3(2), irrespective of the fundamental period of vibration of the structure, T.

Exception: The seismic design category is permitted to be determined from Table 1616.3(1) alone when all of the following apply:

1. The approximate fundamental period of the structure, T_a, in each of the two orthogonal directions determined in accordance with Section 9.5.5.3.2 of ASCE 7, is less than 0.8 T_s determined in accordance with Section 1615.1.4,
2. Equation 9.5.5.2.1-1 of ASCE 7 is used to determine the seismic response coefficient, C_s, and
3. The diaphragms are rigid as defined in Section 1602.

❖ The seismic design category classification system reflects the relative hazard from earthquake effects. The seismic design category considers the seismicity of the site in terms of the mapped spectral response accelerations, the site soil classification (A through F) and the nature of the structure's occupancy or essential post-earthquake function seismic use group (see the example in the commentary to Section 1614.1 for step-by-step instructions on how to determine the seismic design category). It is important to note that there are two code tables [1616.3(1) and 1616.3(2)], both of which need to be used to determine the seismic design category classification. The exception permits the seismic design category of structures meeting certain conditions to be based solely on the short-period response acceleration coefficient. This is similar to what is permitted under the *International Residential Code®* (IRC®)

Note that there are actually two conditions that allow the seismic design category to be determined based upon mapped spectral accelerations, making it unnecessary to go through the process described above. The first is Exception 5 under Section 1614.1. This exception identifies areas that are considered to have a low seismic hazard and they are directly placed in Seismic Design Category A. The second instance is the footnote to Table 1616.3(1) or 1616.3(2), which identifies areas that are close to a major active fault where $S_1 \geq 0.75g$ (see Figure 1616.3.1). These are areas of considerable seismic hazard, and the seismic design category is classified as E or F depending on the structure's seismic use group.

The seismic design category classification system is the key to the use and understanding of the seismic requirements because the analysis method, general design, structural material detailing and the structure's component and system design requirements are determined by the seismic design category. Some of the special inspection requirements in Section 1707 and structural observation requirements in Section 1709 are dependent on the seismic design category classification as well.

Generally, beginning with Seismic Design Category B, seismic requirements increase and are cumulative as the seismic design category becomes higher. For example, a structure that is classified as Seismic Design Category D must comply with all of the seismic requirements for Seismic Design Category B and C structures, in addition to those for Seismic Design Category D structures as shown in Figure 1616.3. Seismic Design Category A requirements are self-contained in Section 1616.4.

TABLE 1616.3(1)
SEISMIC DESIGN CATEGORY BASED ON SHORT-PERIOD RESPONSE ACCELERATIONS

VALUE OF S_{DS}	SEISMIC USE GROUP		
	I	II	III
$S_{DS} < 0.167g$	A	A	A
$0.167g \le S_{DS} < 0.33g$	B	B	C
$0.33g \le S_{DS} < 0.50g$	C	C	D
$0.50g \le S_{DS}$	D[a]	D[a]	D[a]

a. Seismic Use Group I and II structures located on sites with mapped maximum considered earthquake spectral response acceleration at 1-second period, S_1, equal to or greater than 0.75g, shall be assigned to Seismic Design Category E, and Seismic Use Group III structures located on such sites shall be assigned to Seismic Design Category F.

❖ This table defines "Seismic design category" as a function of seismic use group and the short-period (0.2 second) spectral response acceleration at the site of a structure. As the S_{DS} increases, the structure is assigned a higher seismic design category and the design requirements become more restrictive. Note a to the table reads the same as Note a to Table 1616.3(2) and is dependent upon the mapped 1-second spectral response acceleration. Any structure located where $S_1 \ge 0.75g$ will fall under either the Seismic Design Category E or F classification. Figure 1616.3.1 depicts U.S. locations where $S_1 \ge 0.75g$.

TABLE 1616.3(2)
SEISMIC DESIGN CATEGORY BASED ON 1-SECOND PERIOD RESPONSE ACCELERATION

VALUE OF S_{D1}	SEISMIC USE GROUP		
	I	II	III
$S_{D1} < 0.067g$	A	A	A
$0.067g \le S_{D1} < 0.133g$	B	B	C
$0.133g \le S_{D1} < 0.20g$	C	C	D
$0.20g \le S_{D1}$	D[a]	D[a]	D[a]

a. Seismic Use Group I and II structures located on sites with mapped maximum considered earthquake spectral response acceleration at 1-second period, S_1, equal to or greater than 0.75g, shall be assigned to Seismic Design Category E, and Seismic Use Group III structures located on such sites shall be assigned to Seismic Design Category F.

❖ This table defines "Seismic design category" as a function of seismic use group and the long-period (1-second) spectral response acceleration at the site of a structure. As the S_{D1} increases, the structure is assigned a higher seismic design category and the design requirements become more restrictive [see Table 1616.3(1) for information regarding Note a].

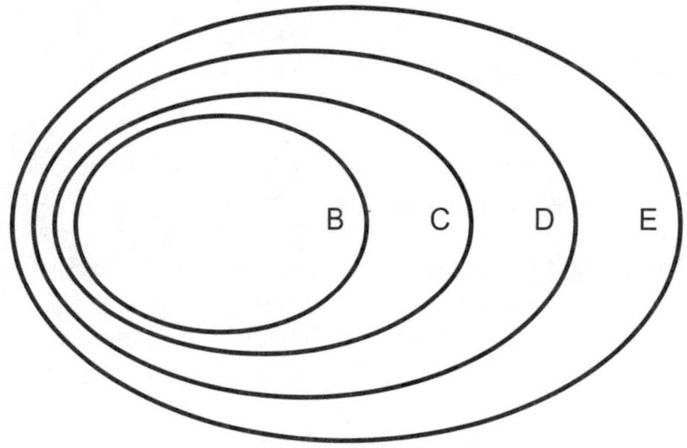

Figure 1616.3
CUMULATIVE NATURE OF PROVISIONS BY SEISMIC DESIGN CATEGORY

1616.3.1 Site limitation for Seismic Design Category E or F. A structure assigned to Seismic Design Category E or F shall not be sited over an identified active fault trace.

Exception: Detached Group R-3 as applicable in Section 101.2 of light-frame construction.

❖ This section precludes Seismic Design Category E or F structures from being damaged as a result of an identified active fault trace (see definition, Section 1613). Siting such critical structures over a fault could result in extremely large forces for which it may not be possible to reliably design to withstand.

As indicated in Section 1613, active fault traces are designated by the appropriate regulatory agency or registered design professional subject to identification by a geologic report. The site limitation does not apply to one- and two-family dwellings of wood or light-gage steel construction. See Figure 1616.3.1 for areas where structures are Seismic Design Category E or F.

1616.4 Design requirements for Seismic Design Category A. Structures assigned to Seismic Design Category A need only comply with the requirements of Sections 1616.4.1 through 1616.4.5.

❖ This section contains all of the seismic design requirements for Seismic Design Category A structures. This is the lowest category of seismic hazard and the requirements of this section are appropriate minimum requirements. Exception 5 to Section 1614.1 exempts Seismic Design Category A structures from all seismic design requirements except for those found in this section.

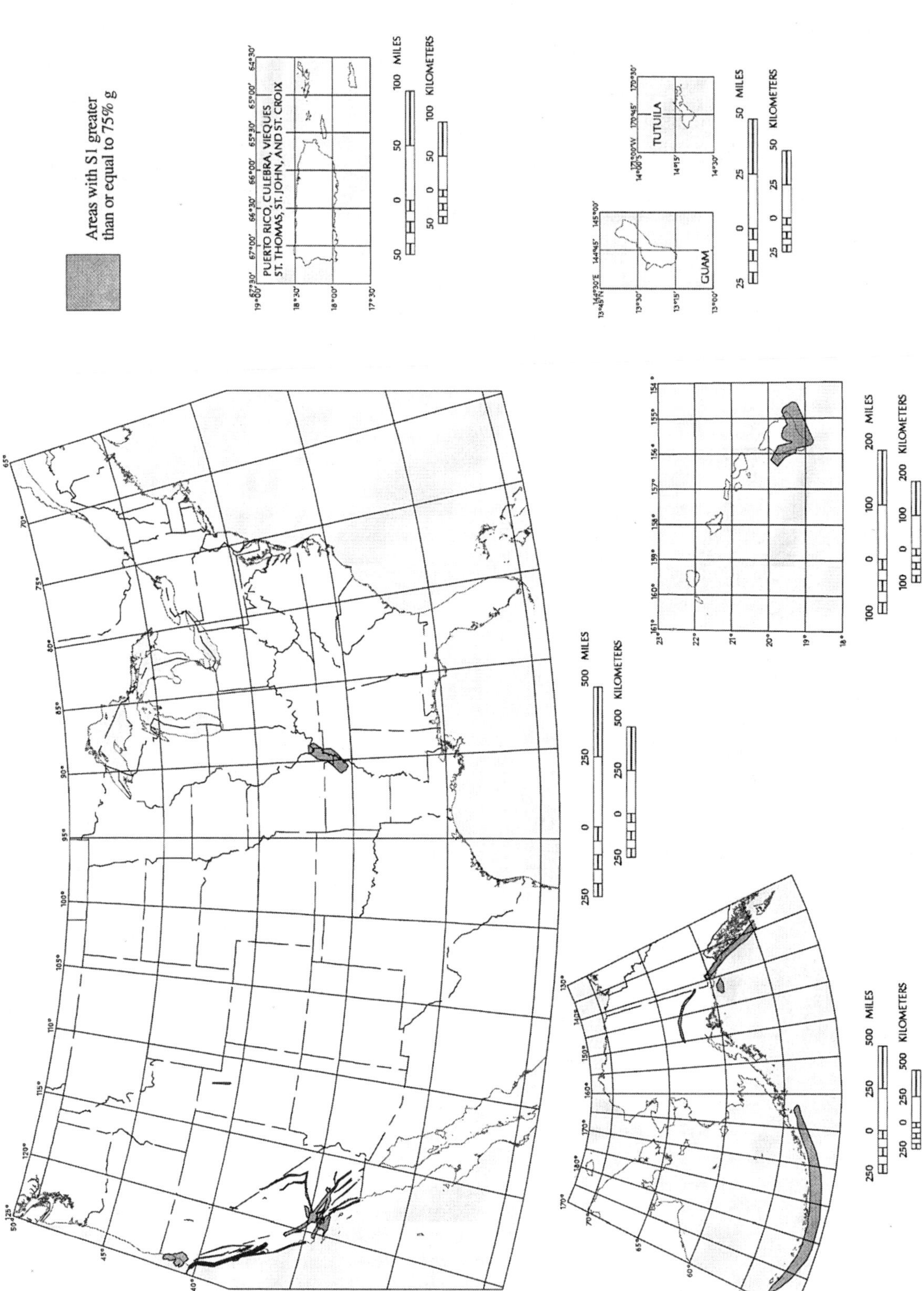

For SI: 1 mile = 1609 344 m.

Figure 1616.3.1
AREAS WHERE STRUCTURES ARE SEISMIC DESIGN CATEGORY E OR F

1616.4.1 Minimum lateral force. Structures shall be provided with a complete lateral-force-resisting system designed to resist the minimum lateral force, F_x, applied simultaneously at each floor level given by Equation 16-49:

$$F_x = 0.01\, w_x \qquad \textbf{(Equation 16-49)}$$

where:

F_x = The design lateral force applied at Level x.

w_x = The portion of the total gravity load of the structure, W, located or assigned to Level x.

W = The total dead load and other loads listed below:

1. In areas used for storage, a minimum of 25 percent of the reduced floor live load (floor live load in public garages and open parking structures need not be included).
2. Where an allowance for partition load is included in the floor load design, the actual partition weight or a minimum weight of 10 psf (0.479 kN/m^2) of floor area, whichever is greater.
3. Total operating weight of permanent equipment.
4. Twenty percent of flat roof snow load where flat roof snow load exceeds 30 psf (1.44 kN/m^2).

The direction of application of seismic forces used in design shall be that which will produce the most critical load effect in each component. The design seismic forces are permitted to be applied separately in each of two orthogonal directions and orthogonal effects are permitted to be neglected.

The effect of this lateral force shall be taken as E in the load combinations prescribed in Section 1605.2 for strength or load and resistance factor design methods, or Section 1605.3 for allowable stress design methods. Special seismic load combinations that include E_m need not be considered.

❖ Seismic Design Category A structures are located in areas of very low seismicity and, therefore, are required to be designed for a minimum lateral force equal to 1 percent of weight. This minimal design force is required basically so that a complete lateral-force-resisting system is provided for all structures. Typically, the design wind loads will govern for Seismic Design Category A structures.

Seismic Design Category A structures are not required to be designed considering orthogonal effects. The seismic forces need only be applied along each of the principal axes of the building. Figure 1616.4 illustrates what is meant by orthogonal effect, where the seismic forces affect the building in a direction other than along the principal axes, potentially creating higher stresses in structural members, such as columns that are common to orthogonal lateral-force-resisting systems.

1616.4.2 Connections. All parts of the structure between separation joints shall be interconnected, and the connections shall be capable of transmitting the seismic force, F_p, induced in the connection by the parts being connected. Any smaller portion of the structure shall be tied to the remainder of the structure for F_p equal to 0.05 times the weight of the smaller portion. A positive connection for resisting horizontal forces acting on the member shall be provided for each beam, girder or truss to its support. The connection shall have strength sufficient to resist 5 percent of the dead and live load vertical reaction applied horizontally.

❖ A common goal of the seismic detailing requirements is to tie a building together to act as a unit. In Seismic Design Category A structures, the elements that tie any smaller portion of a structure to the remainder of the structure need to be designed for a nominal 5 percent of the weight of the smaller portion being connected.

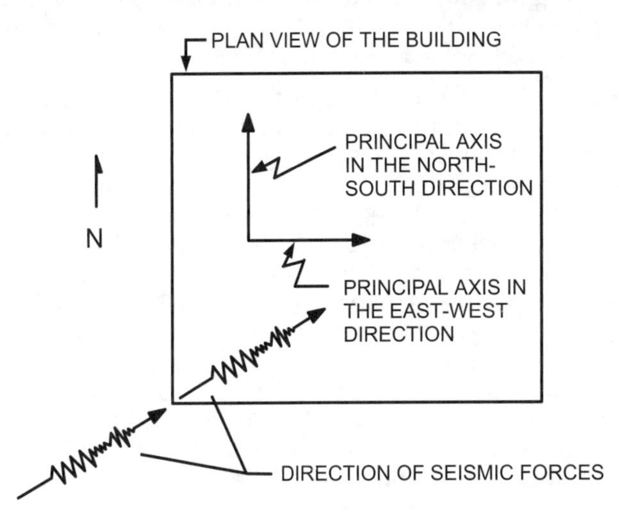

Figure 1616.4
ORTHOGONAL EFFECTS

1616.4.3 Anchorage of concrete or masonry walls. See Section 1604.8.2.

❖ Extensive damage can be caused by the pulling away of concrete or masonry walls from the roof. This may result in partial or even total collapse. Seismic Design Category A structures are required to have connections between masonry or concrete walls and structural elements providing lateral support to such walls designed in accordance with Section 1604.8.2.

1616.4.4 Conventional light-frame construction. Buildings constructed in compliance with Section 2308 are deemed to comply with Sections 1616.4.1, 1616.4.2 and 1616.4.3.

❖ Conventional light-frame wood construction is deemed to provide seismic resistance equivalent to that provided by the requirements set forth in Section 1616.4.

1616.4.5 Tank freeboard. Tanks in Seismic Use Group III according to Table 9.14.5.1.2 of ASCE 7 shall also comply with the freeboard requirements of Section 9.14.7.3.6.1.2 of ASCE 7.

❖ The referenced ASCE 7 provision is only applicable to tanks that are considered essential facilities or that contain highly toxic materials and are, therefore, classified as Seismic Use Group III (see commentary, Section 1616.2.3). The objective of providing freeboard is to accommodate sloshing of the material being held in the tank in order to limit damage to the tank itself. This must be considered for Seismic Design Category A tanks because the earthquake ground motion maps of the code do not account for the long-period accelerations that pose a problem for these tanks.

1616.5 Building configuration. Buildings shall be classified as regular or irregular based on the criteria in Section 9.5.2.3 of ASCE 7.

Exception: Buildings designed using the simplified analysis procedure in Section 1617.5 shall be classified in accordance with Section 1616.5.1.

❖ Building configuration provisions are necessary because past earthquakes have repeatedly shown that structures having irregular geometric configurations suffer greater damage than those having regular configurations. Once irregularities are introduced, compensating measures need to be taken to achieve the desired performance of the structure.

This section requires use of ASCE 7 provisions for classifying a building as regular or irregular, but an exception requires use of the code provisions for building configuration when the simplified analysis procedure of the code is used. Since the building configuration requirements of the code simplified procedure are comparable to those of ASCE 7, the following code commentary can provide insights into the ASCE 7 provisions as well.

1616.5.1 Building configuration (for use in the simplified analysis procedure of Section 1617.5). Buildings designed using the simplified analysis procedure in Section 1617.5 shall be classified as regular or irregular based on the criteria in this section. Such classification shall be based on the plan and vertical configuration. Buildings shall not exceed the limitations of Section 1616.6.1.

❖ This section on building configuration is only to be used when the simplified analysis procedure of the code is used. While these provisions were written for and previously applied to all methods of analysis, in order to qualify, the building must be within the limits established by Section 1616.6.1.

1616.5.1.1 Plan irregularity. Buildings having one or more of the features listed in Table 1616.5.1.1 shall be designated as having plan structural irregularity and shall comply with the requirements in the sections referenced in that table.

❖ Table 1616.5.1.1 describes each type of plan structural irregularity. Examples of regular structure configurations are shown in Figure 1616.5.1(1). Examples of buildings that have plan irregularities are shown in Figure 1616.5.1(2).

Table 1616.5.1.1 is very useful in that it not only provides a description of the various plan irregularities, but also indicates which seismic design categories are affected and which sections apply.

TABLE 1616.5.1.1. See page 16-111.

❖ Note that most of the additional requirements for irregular structures apply to those assigned to Seismic Design Categories C, D, E and F. The additional provisions and higher force requirements are designed to encourage, but not require, that structures with regular configurations be used.

1616.5.1.2 Vertical irregularity. Buildings having one or more of the features listed in Table 1616.5.1.2 shall be designated as having vertical irregularity and shall comply with the requirements in the sections referenced in that table.

Exceptions:

1. Structural irregularities of Type 1a, 1b or 2 in Table 1616.5.1.2 do not apply where no story drift ratio under design lateral load is greater than 130 percent of the story drift ratio of the next story above. Torsional effects need not be considered in the calculation of story drifts for the purpose of this determination. The story drift ratio relationship for the top two stories of the building is not required to be evaluated.
2. Irregularities of Types 1a, 1b and 2 of Table 1616.5.1.2 are not required to be considered for one-story buildings in any seismic design category or for two-story buildings in Seismic Design Category A, B, C or D.

❖ Table 1616.5.1.2 describes each type of vertical structural irregularity. Examples of irregular structure configurations are shown in Figure 1616.5.1.2.

Table 1616.5.1.2 is very useful in that it not only provides a description of the various vertical irregularities, but also indicates which seismic design categories are affected and which sections apply.

The two exceptions in Section 1616.5.1.2 apply to building structural arrangements that do not warrant application of the additional design requirements of Table 1616.5.1.2

TABLE 1616.5.1.2 See page 16-111.

❖ Note that most of the additional requirements for irregular structures apply to those assigned to Seismic Design Categories C, D, E and F. The additional requirements in this section encourage, but do not require, that structures with regular configurations be used.

TABLE 1616.5.1.1
PLAN STRUCTURAL IRREGULARITIES

IRREGULARITY TYPE AND DESCRIPTION		REFERENCE SECTION	SEISMIC DESIGN CATEGORY[a] APPLICATION
1a	Torsional Irregularity—to be considered when diaphragms are not flexible as determined in Section 1602.1.1 Torsional irregularity shall be considered to exist when the maximum story drift, computed including accidental torsion, at one end of the structure transverse to an axis is more than 1.2 times the average of the story drifts at the two ends of the structure.	9.5.5.5.2 of ASCE 7 1620.4.1 9.5.2.5.1 of ASCE 7 9.5.5.7.1 of ASCE 7	C, D, E and F D, E and F D, E and F C, D, E and F
1b	Extreme Torsional Irregularity—to be considered when diaphragms are not flexible as determined in Section 1602.1. Extreme torsional irregularity shall be considered to exist when the maximum story drift, computed and including accidental torsion, at one end of the structure transverse to an axis is more than 1.4 times the average of the story drifts at the two ends of the structure.	9.5.5.5.2 of ASCE 7 1620.4.1 1620.5.1 9.5.2.5.1 of ASCE 7 9.5.5.7.1 of ASCE 7	C, D, E and F D E and F D, E and F C, D, E and F
2	Reentrant Corners Plan configurations of a structure and its lateral-force-resisting system contain reentrant corners where both projections of the structure beyond a reentrant corner are greater than 15 percent of the plan dimension of the structure in the given direction.	1620.4.1	D, E and F
3	Diaphragm Discontinuity Diaphragms with abrupt discontinuities or variations in stiffness, including those having cutout or open areas greater than 50 percent of the gross enclosed diaphragm area, or changes in effective diaphragm stiffness of more than 50 percent from one story to the next.	1620.4.1	D, E and F
4	Out-of-Plane Offsets Discontinuities in a lateral-force-resistance path, such as out-of-plane offsets of the vertical elements.	1620.4.1 9.5.2.5.1 of ASCE 7 1620.2.9	D, E and F D, E and F B, C, D, E and F
5	Nonparallel Systems The vertical lateral-force-resisting elements are not parallel to or symmetric about the major orthogonal axes of the lateral-force-resisting system.	1620.3.2	C, D, E and F

a. Seismic design category is determined in accordance with Section 1616.

TABLE 1616.5.1.2
VERTICAL STRUCTURAL IRREGULARITIES

IRREGULARITY TYPE AND DESCRIPTION		REFERENCE SECTION	SEISMIC DESIGN CATEGORY[a] APPLICATION
1a	Stiffness Irregularity—Soft Story A soft story is one in which the lateral stiffness is less than 70 percent of that in the story above or less than 80 percent of the average stiffness of the three stories above.	9.5.2.5.1 of ASCE 7	D, E, and F
1b	Stiffness Irregularity—Extreme Soft Story An extreme soft story is one in which the lateral stiffness is less than 60 percent of that in the story above or less than 70 percent of the average stiffness of the three stories above.	1620.5.1 9.5.2.5.1 of ASCE 7	E and F D, E and F
2	Weight (Mass) Irregularity Mass irregularity shall be considered to exist where the effective mass of any story is more than 150 percent of the effective mass of an adjacent story. A roof that is lighter than the floor below need not be considered.	9.5.2.5.1 of ASCE 7	D, E and F
3	Vertical Geometric Irregularity Vertical geometric irregularity shall be considered to exist where the horizontal dimension of the lateral-force-resisting system in any story is more than 130 percent of that in an adjacent story.	9.5.2.5.1 of ASCE 7	D, E and F
4	In-plane Discontinuity in Vertical Lateral-Force-Resisting Elements An in-plane offset of the lateral-force-resisting elements greater than the length of those elements or a reduction in stiffness of the resisting element in the story below.	1620.4.1 9.5.2.5.1 of ASCE 7 1620.2.9	D, E and F D, E and F B, C, D, E and F
5	Discontinuity in Capacity—Weak Story A weak story is one in which the story lateral strength is less than 80 percent of that in the story above. The story strength is the total strength of seismic-resisting elements sharing the story shear for the direction under consideration.	1620.2.3 9.5.2.5.1 of ASCE 7 1620.5.1	B, C, D, E and F D, E and F E and F

a. Seismic design category is determined in accordance with Section 1616.

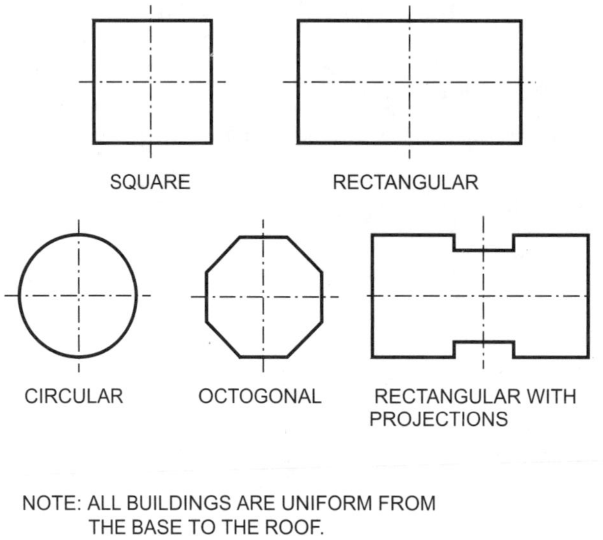

Figure 1616.5.1(1)
EXAMPLE OF REGULAR BUILDING SHAPES (PLAN VIEWS)

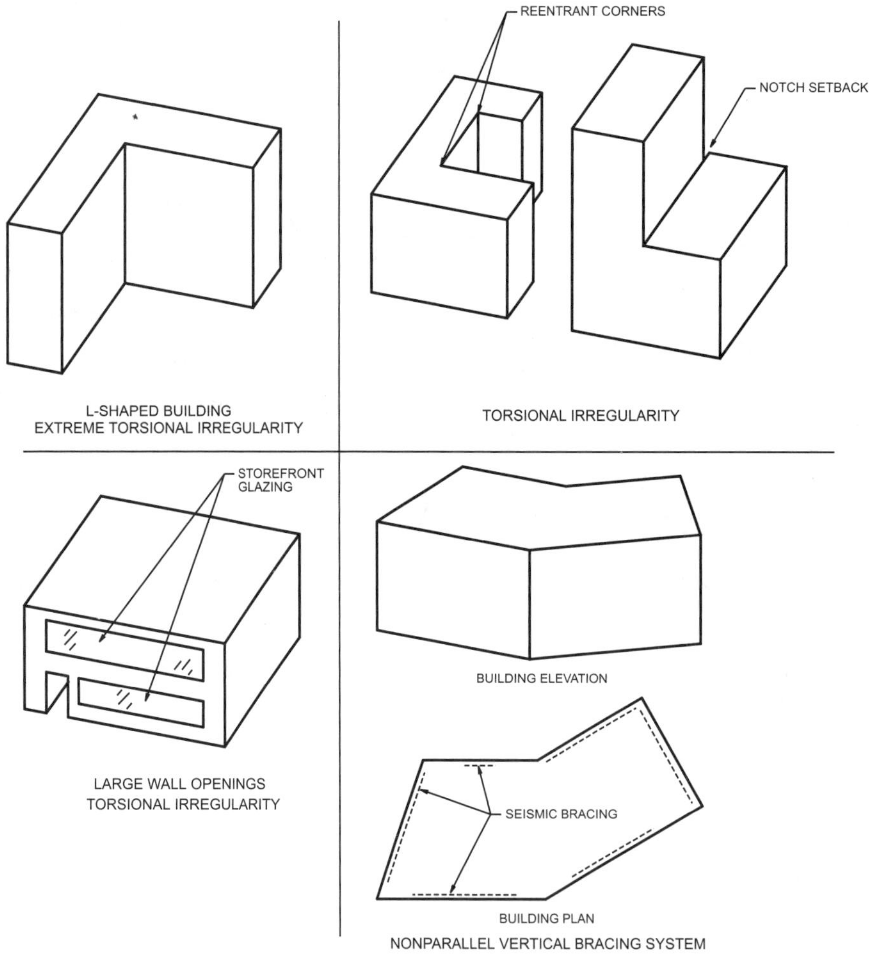

Figure 1616.5.1(2)
EXAMPLES OF BUILDINGS WITH PLAN IRREGULARITIES

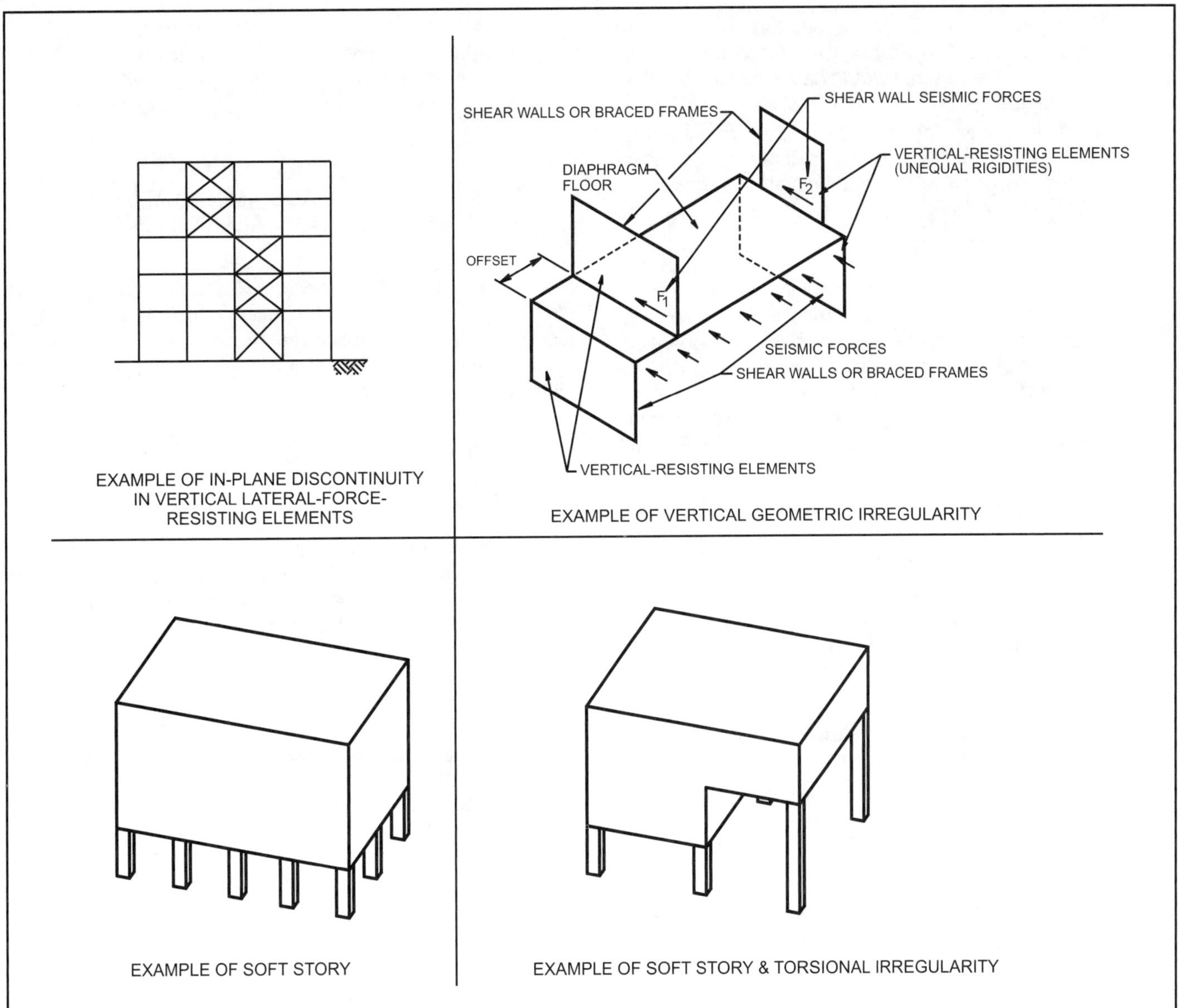

Figure 1616.5.1.2
EXAMPLES OF BUILDINGS WITH VERTICAL STRUCTURAL IRREGULARITIES

1616.6 Analysis procedures. A structural analysis conforming to one of the types permitted in Section 9.5.2.5.1 of ASCE 7 or to the simplified procedure in Section 1617.5 shall be made for all structures. The analysis shall form the basis for determining the seismic forces, E and E_m, to be applied in the load combinations of Section 1605 and shall form the basis for determining the design drift as required by Section 9.5.2.8 of ASCE 7 or Section 1617.3.

Exceptions:

1. Structures assigned to Seismic Design Category A.
2. Design drift need not be evaluated in accordance with Section 1617.3 when the simplified analysis method of Section 1617.5 is used.

❖ A structural analysis of the seismic-force-resisting system is required unless one or more of the exceptions in Section 1614 apply to the structure under consideration. As a rule, the greater the seismic hazard (as expressed by a structure's seismic design category) and the complexity of the building, the more rigorous the analysis that must be performed. The types of analysis that are permitted are listed in Table 9.5.2.5.1 of ASCE 7 based on seismic design category and the building's configuration. The following summarizes these requirements:

- Structures assigned to Seismic Design Category A and complying with Section 1616.4 require no further analysis due to the low seismic hazard in the geographical areas where such structures are located; however, any other analysis method permitted by the code could be performed at the discretion of the designer.
- Structures meeting the limits of Section 1616.6.1 are permitted to use the simplified analysis procedure of Section 1617.5 regardless of their seismic

design category. Because Section 1616.6.1 limits the applicability to Seismic Use Group I buildings, such structures would never be classified as Seismic Design Category F.

- The equivalent lateral force procedure (see Section 1617.4) is deemed appropriate for all structures assigned to Seismic Design Category B or C in view of the low-to-moderate seismicity of the building sites.
- Seismic Design Category D, E and F structures are permitted to use the equivalent lateral force procedure within certain limits. The approximations and assumptions on which the equivalent lateral force procedure is based are inappropriate for long-period structures as well as certain irregular structures. In such cases, more sophisticated analysis methods of Section 1618.1 are required.

The soil-structure interaction analysis of Section 1619 or the base isolation technique of Section 1623 might also be considered; both are optional.

1616.6.1 Simplified analysis. A simplified analysis, in accordance with Section 1617.5, shall be permitted to be used for any structure in Seismic Use Group I, subject to the following limitations, or a more rigorous analysis shall be made:

1. Buildings of light-framed construction not exceeding three stories in height, excluding basements.
2. Buildings of any construction other than light-framed construction, not exceeding two stories in height, excluding basements, with flexible diaphragms at every level as defined in Section 1602.

❖ Code users have been asking for a simplified approach to seismic design for years. This section has been provided in response to this request. The simplified procedure provides many time-saving benefits for the designer in that the design base shear is determined from one simple formula; the vertical distribution of the base shear is also very simple and is given by the same formula, except for the weight used in the calculation. The drift check is also simplified by allowing the assumption that the drift is equal to 1 percent of the story height, unless a more detailed analysis is performed.

However, this approach may only be used for a select subset of designed structures. Figure 1616.6.1 includes the qualifying criteria for the simplified analysis approach.

In other than light-frame construction, flexible diaphragms are required because the simplified analysis approach does not take into account torsional effects.

SECTION 1617 EARTHQUAKE LOADS—MINIMUM DESIGN LATERAL FORCE AND RELATED EFFECTS

1617.1 Seismic load effect *E* and E_m. The seismic load effect, *E*, for use in the basic load combinations of Sections 1605.2 and 1605.3 shall be determined from Section 9.5.2.7 of ASCE 7. The maximum seismic load effect, E_m, for use in the special seismic load combination of Section 1605.4 shall be the special seismic load determined from Section 9.5.2.7.1 of ASCE 7.

Exception: For structures designed using the simplified analysis procedure in Section 1617.5, the seismic load effects, *E* and E_m, shall be determined from Section 1617.1.1.

❖ See Section 1613 for introductory commentary regarding earthquake provisions. This section defines how the seismic load effects are determined for structures classified in Seismic Design Categories B through F in order to apply them to the appropriate load combinations of Section 1605. Load effect requirements for Seismic Design Category A structures are given in Section 1616.4.1.

The code requires the use of ASCE 7 for determining seismic load effects, but the exception requires structures designed using the simplified analysis procedure to determine load effects under Section 1617.1.1. Since the seismic load effects required for the code's simplified procedure are comparable to those of ASCE 7, the following code commentary can provide insights into the referenced ASCE 7 requirements as well.

1617.1.1 Seismic load effects, *E* and E_m (for use in the simplified analysis procedure of Section 1617.5). Seismic load effects, *E* and E_m, for use in the load combinations of Section 1605 for structures designed using the simplified analysis procedure in Section 1617.5 shall be determined as follows.

❖ *E* and E_m are defined by the formulas in the following sections. They are required input for the load combinations of Section 1605. These earthquake terms are quite complex and the following discussions will explain what they mean and how they are determined.

SEISMIC USE GROUP CRITERIA:

SEISMIC USE GROUP I

HEIGHT LIMITS:

Light-frame (wood or steel) construction	3 stories maximum above basement.
All other construction	2 stories maximum above basement, with flexible diaphragms at each level.

Figure 1616.6.1
QUALIFYING CRITERIA FOR SIMPLIFIED ANALYSIS APPROACH

1617.1.1.1 Seismic load effect, *E*. Where the effects of gravity and the seismic ground motion are additive, seismic load, *E*, for use in Equations 16-5, 16-10 and 16-17, shall be defined by Equation 16-50:

$$E = \rho Q_E + 0.2 S_{DS} D \qquad \textbf{(Equation 16-50)}$$

where:

D = The effect of dead load.

E = The combined effect of horizontal and vertical earthquake-induced forces.

ρ = A redundancy coefficient obtained in accordance with Section 1617.2.

Q_E = The effect of horizontal seismic forces.

S_{DS} = The design spectral response acceleration at short periods obtained from Section 1615.1.3 or 1615.2.5.

Where the effects of gravity and seismic ground motion counteract, the seismic load, *E*, for use in Equations 16-6, 16-12 and 16-18 shall be defined by Equation 16-51.

$$E = \rho Q_E - 0.2 S_{DS} D \qquad \textbf{(Equation 16-51)}$$

Design shall use the load combinations prescribed in Section 1605.2 for strength or load and resistance factor design methodologies, or Section 1605.3 for allowable stress design methods.

❖ An earthquake creates vibrational movement of the earth in all directions. Code writers have quantified this vibrational movement into the following formula:

$$E = \rho Q_E \pm 0.2 S_{DS} D$$

This formula can be broken down into two separate effects: one due to horizontal vibrational movement (ρQ_E) and the other due to vertical vibrational movement ($\pm 0.2 S_{DS} D$). All of the coefficients in this formula have been discussed earlier except for the reliability factor, which is discussed in Section 1617.2. In order to quantify *E* for the load combinations in Section 1605, one must complete the following steps:

Step 1: Determine axial force (P_{DL}), shear force (V_{DL}) and bending moment (M_{DL}) in element due to dead load.

Step 2: Determine axial force ($Q_{E(axial)}$), shear force ($Q_{E(shear)}$) and bending moment ($Q_{E(moment)}$) in element due to the design base shear, *V*, distributed as required in accordance with Section 1617.5.1.

Step 3: Create the table described in Figure 1617.1.1.1.

1617.1.1.2 Maximum seismic load effect, E_m. The maximum seismic load effect, E_m, shall be used in the special seismic load combinations in Section 1605.4.

Where the effects of the seismic ground motion and gravity loads are additive, seismic load, E_m, for use in Equation 16-19, shall be defined by Equation 16-52.

$$E_m = \Omega_0 Q_E + 0.2 S_{DS} D \qquad \textbf{(Equation 16-52)}$$

Where the effects of the seismic ground and gravity loads counteract, seismic load, E_m, for use in Equation 16-20, shall be defined by Equation 16-53.

$$E_m = \Omega_0 Q_E - 0.2 S_{DS} D \qquad \textbf{(Equation 16-53)}$$

where *E*, Q_E, S_{DS} are as defined above and Ω_0 is the system overstrength factor as given in Table 1617.6.2.

The term $\Omega_0 Q_E$ need not exceed the maximum force that can be transferred to the element by the other elements of the lateral-force-resisting system.

Where allowable stress design methodologies are used with the special load combinations of Section 1605.4, design strengths are permitted to be determined using an allowable stress increase of 1.7 and a resistance factor, ϕ, of 1.0. This increase shall not be combined with increases in allowable stresses or load combination reductions otherwise permitted by this code or the material reference standard except that combination with the duration of load increases in Chapter 23 is permitted.

❖ The maximum seismic load effect is required for the design of vulnerable elements critical to the stability of a structure. This maximum load effect generated in the structural and nonstructural components of a building can be much greater than those due to the design level force.

The overstrength factor, Ω_0, increases the design level effects to represent the actual forces that may be

	APPLIED LOAD = DEAD LOAD	$0.2S_{DS}D$	APPLIED LOAD = DESIGN BASE SHEAR, *V*	ρQ_E	*E*
Axial reaction	P_{DL}	$0.2S_{DS}P_{DL}$	$Q_{E(AXIAL)}$	$\rho Q_{E(AXIAL)}$	$\rho Q_{E(AXIAL)} + 0.2S_{DS}P_{DL}$ $\rho Q_{E(AXIAL)} - 0.2S_{DS}P_{DL}$
Shear reaction	V_{DL}	$0.2S_{DS}V_{DL}$	$Q_{E(SHEAR)}$	$\rho Q_{E(SHEAR)}$	$\rho Q_{E(SHEAR)} + 0.2S_{DS}V_{DL}$ $\rho Q_{E(SHEAR)} - 0.2S_{DS}V_{DL}$
Moment reaction	M_{DL}	$0.2S_{DS}M_{DL}$	$Q_{E(MOMENT)}$	$\rho Q_{E(MOMENT)}$	$\rho Q_{E(MOMENT)} + 0.2S_{DS}M_{DL}$ $\rho Q_{E(MOMENT)} - 0.2S_{DS}M_{DL}$

Figure 1617.1.1.1
SEISMIC LOAD EFFECT *E*

experienced by an element as a result of the design ground motion. The overstrength factor is set forth in Section 1617.6 and varies between 2 and 3.

Because E_m is a strength level force effect, adjustments need to be made if ASD is used. In cases where ASD is used, the allowable stresses may be increased by a factor of 1.7. No other stress increases or load reductions are permitted; however, the load duration factor for wood allowable stresses can still be used.

1617.2 Redundancy. The provisions given in Section 9.5.2.4 of ASCE 7 shall be used.

Exception: Structures designed using the simplified analysis procedure in Section 1617.5 shall use the redundancy provisions in Sections 1617.2.2.

❖ Just as regular structures have proven themselves to outperform irregular structures in earthquakes, structures with redundant lateral-force-resisting systems have performed better than those with little or no redundancy. What is meant by redundancy? Very simply, it is the concept of multiple load paths—more than one way for a load to be transmitted from its point of origin down to the soils underlying the foundations where all loads ultimately belong.

Compare two slabs: one supported by a single column and another supported by 20 different columns. If one of the columns supporting the second slab suffers significant damage to the point that it loses its vertical load-carrying capacity (for whatever reason), there is likely to be no detriment to the supported slab. The 19 undamaged columns will share in picking up the load that was being carried by the failed column. If, however, the single column supporting the first slab suffers similar significant damage, the supported slab will collapse to the ground. The second slab with multiple paths for the vertical loads to flow to the soils beneath the foundation has a redundant vertical-force-resisting system; the first slab with a single load path does not.

This section requires that the redundancy provisions of ASCE 7 be applied, and for structures analyzed under the simplified analysis procedure, the exception requires that Section 1617.2.2 be followed. Since the redundancy requirements of ASCE 7 are comparable to those of Section 1617.2.2, the accompanying commentary provides guidance that is applicable to the ASCE 7 provisions.

1617.2.1 ASCE 7, Sections 9.5.2.4.2 and 9.5.2.4.3. Modify Sections 9.5.2.4.2 and 9.5.2.4.3 as follows:

9.5.2.4.2 Seismic Design Category D: For structures in Seismic Design Category D, ρ shall be taken as the largest of the values of ρ_x calculated at each story "x" of the structure in accordance with Equation 9.5.2.4.2-1 as follows:

$$\rho_x = 2 - \frac{20}{r_{\max_x} \sqrt{A_x}}$$

where:

r_{max_x} = The ratio of the design story shear resisted by the single element carrying the most shear force in the story to the total story shear, for a given direction of loading. For braced frames, the value of r_{max_x} is equal to the lateral force component in the most heavily loaded brace element divided by the story shear. For moment frames, r_{max_x} shall be taken as the maximum of the sum of the shears in any two adjacent columns in the plane of a moment frame divided by the story shear. For columns common to two bays with moment-resisting connections on opposite sides at the level under consideration, 70 percent of the shear in that column is permitted to be used in the column shear summation. For shear walls, r_{max_x} shall be taken equal to shear in the most heavily loaded wall or wall pier multiplied by $10/l_w$ (the metric coefficient is $3.3/l_w$), divided by the story shear, where l_w is the wall or wall pier length in feet (m). The value of the ratio of $10/l_w$ need not be greater than 1.0 for buildings of light-framed construc--tion. For dual systems, r_{max_x} shall be taken as the maximum value defined above, considering all lateral-load-resisting elements in the story. The lateral loads shall be distributed to elements based on relative rigidities considering the interaction of the dual system. For dual systems, the value of ρ need not exceed 80 percent of the value calculated above.

A_x = The floor area in square feet of the diaphragm level immediately above the story.

Calculation of r_{max_x} need not consider the effects of accidental torsion and any dynamic amplification of torsion required by Section 9.5.5.5.2.

For a story with a flexible diaphragm immediately above, r_{max_x} shall be permitted to be calculated from an analysis that assumes rigid diaphragm behavior and ρ_x, need not exceed 1.25.

The value of ρ need not exceed 1.5, which is permitted to be used for any structure. The value of ρ shall not be taken as less than 1.0.

Exception: For structures with seismic-force-resisting systems in any direction comprised solely of special moment frames, the seismic-force-resisting system shall be configured such that the value of ρ calculated in accordance with this section does not exceed 1.25. The calculated value of ρ is permitted to exceed this limit when the design story drift, Δ, as determined in Section 9.5.5.7, does not exceed Δ_a/ρ for any story where Δ_a is the allowable story drift from Table 9.5.2.8.

The metric equivalent of Equation 9.5.2.4.2-1 is:

$$\rho_x = 2 - \frac{6.1}{r_{max_x} \sqrt{A_x}}$$

where: A_x is in square meters.

The value ρ shall be permitted to be taken equal to 1.0 in the following circumstances:

1. When calculating displacements for dynamic amplification of torsion in Section 9.5.5.5.2.
2. When calculating deflections, drifts and seismic shear forces related to Sections 9.5.5.7.1 and 9.5.5.7.2.
3. For design calculations required by Section 9.5.2.6, 9.6 or 9.14.

For structures with vertical combinations of seismic-force-resisting systems, the value of ρ shall be determined independently for each seismic-force-resisting system. The redundancy coefficient of the lower portion shall not be less than the following:

$$\rho_L = \frac{R_L \rho_u}{R_u}$$

where:

$\rho_L = \rho$ of lower portion.

$R_L = R$ of lower portion.

$\rho_u = \rho$ of upper portion.

$R_u = R$ of upper portion.

9.5.2.4.3 Seismic Design Categories E and F. For structures in Seismic Design Categories E and F, the value of ρ shall be calculated as indicated in Section 9.5.2.4.2, above.

Exception: For structures with lateral-force-resisting systems in any direction consisting solely of special moment frames, the lateral-force-resisting system shall be configured such that the value of ρ calculated in accordance with Section 9.5.2.4.2 does not exceed 1.1. The calculated value of ρ is permitted to exceed this limit when the design story drift, Δ, as determined in Section 9.5.5.7, does not exceed Δ_a/ρ for any story where Δ_a is the allowable story drift from Table 9.5.2.8.

❖ This section modifies the redundancy provisions of ASCE 7 that apply to all methods of analysis other than the simplified method of Section 1617.5 (also see commentary, Section 1617.2). These modifications to the ASCE 7 redundancy provisions make them virtually identical to those in Section 1617.2.2.

1617.2.2 Redundancy (for use in the simplified analysis procedure of Section 1617.5). A redundancy coefficient, ρ, shall be assigned to each structure designed using the simplified analysis procedure in Section 1617.5 in accordance with this section. Buildings shall not exceed the limitations of Section 1616.6.1.

❖ See the commentary to Section 1617.2 and the following subsections. This section is applicable only to those buildings analyzed using the simplified method.

1617.2.2.1 Seismic Design Category A, B or C. For structures assigned to Seismic Design Category A, B or C (see Section 1616), the value of the redundancy coefficient ρ is 1.0.

❖ The redundancy coefficient does not apply (meaning that it may be taken equal to 1) in Seismic Design Category A, B or C; seismic design forces are, therefore, unaffected by the redundancy of the lateral-force-resisting system.

1617.2.2.2 Seismic Design Category D, E or F. For structures in Seismic Design Category D, E or F (see Section 1616), the redundancy coefficient, ρ, shall be taken as the largest of the values of, ρ_i, calculated at each story "i" of the structure in accordance with Equation 16-54, as follows:

$$\rho_i = 2 - \frac{20}{r_{\max_i}\sqrt{A_i}}$$ **(Equation 16-54)**

For SI:

$$\rho_i = 2 - \frac{6.1}{r_{\max_i}\sqrt{A_i}}$$

where:

r_{max_i} = The ratio of the design story shear resisted by the most heavily loaded single element in the story to the total story shear, for a given direction of loading.

r_{max_i} = For braced frames, the value r_{max_i}, is equal to the horizontal force component in the most heavily loaded brace element divided by the story shear.

r_{max_i} = For moment frames, r_{max_i}, shall be taken as the maximum of the sum of the shears in any two adjacent columns in a moment frame divided by the story shear. For columns common to two bays with moment-resisting connections on opposite sides at the level under consideration, it is permitted to use 70 percent of the shear in that column in the column shear summation.

r_{max_i} = For shear walls, r_{max_i}, shall be taken as the maximum value of the product of the shear in the wall or wall pier and $10/l_w$ ($3.3/l_w$ for SI), divided by the story shear, where l_w is the length of the wall or wall pier in feet (m). In light-framed construction, the value of the ratio of $10/l_w$ need not be greater than 1.0.

r_{max_i} = For dual systems, r_{max_i}, shall be taken as the maximum value defined above, considering all lateral-load-resisting elements in the story. The lateral loads shall be distributed to elements based on relative rigidities considering the interaction of the dual system. For dual systems, the value of ρ need not exceed 80 percent of the value calculated above.

A_i = The floor area in square feet of the diaphragm level immediately above the story.

For a story with a flexible diaphragm immediately above, r_{max_i} shall be permitted to be calculated from an analysis that assumes rigid diaphragm behavior and ρ need not exceed 1.25.

The value, ρ, shall not be less than 1.0, and need not exceed 1.5.

Calculation of r_{max_i} need not consider the effects of accidental torsion and any dynamic amplification of torsion required by Section 9.5.5.5.2 of ASCE 7.

For structures with seismic-force-resisting systems in any direction comprised solely of special moment frames, the seismic-force-resisting system shall be configured such that the value of ρ calculated in accordance with this section does not exceed 1.25 for structures assigned to Seismic Design Category D, and does not exceed 1.1 for structures assigned to Seismic Design Category E or F.

> **Exception:** The calculated value of ρ is permitted to exceed these limits when the design story drift, Δ, as determined in Section 1617.5.4, does not exceed Δ_a/ρ for any story where Δ_a is the allowable story drift from Table 1617.3.1.

The value ρ shall be permitted to be taken equal to 1.0 in the following circumstances:

1. When calculating displacements for dynamic amplification of torsion in Section 9.5.5.5.2 of ASCE 7.
2. When calculating deflections, drifts and seismic shear forces related to Sections 9.5.5.7.1 and 9.5.5.7.2 of ASCE 7.
3. For design calculations required by Section 1620, 1621 or 1622.

For structures with vertical combinations of seismic-force-resisting systems, the value, ρ, shall be determined independently for each seismic-force-resisting system. The redundancy coefficient of the lower portion shall not be less than the following:

$$\rho_L = \frac{R_L \rho_u}{R_u} \qquad \textbf{(Equation 16-55)}$$

where:

ρ_L = ρ of lower portion.

R_L = R of lower portion.

ρ_u = ρ of upper portion.

R_u = R of upper portion.

❖ The value of the redundancy coefficient varies from 1.0 to 1.5 and is a factor that is used to amplify the effects of the design base shear when combined with the effects of gravity and other loads. The formula to determine the redundancy coefficient is rather complicated and has been broken down in Figure 1617.2.2.2 for clarification purposes.

If the lateral-force-resisting systems are different in the two orthogonal directions (e.g., north-south direction and east-west direction) of a building, the redundancy coefficient should be calculated separately in each direction, as is similarly done when calculating the period of the structure.

The code clarifies where the redundancy coefficient is permitted to be taken as equal to 1 such as for archi-

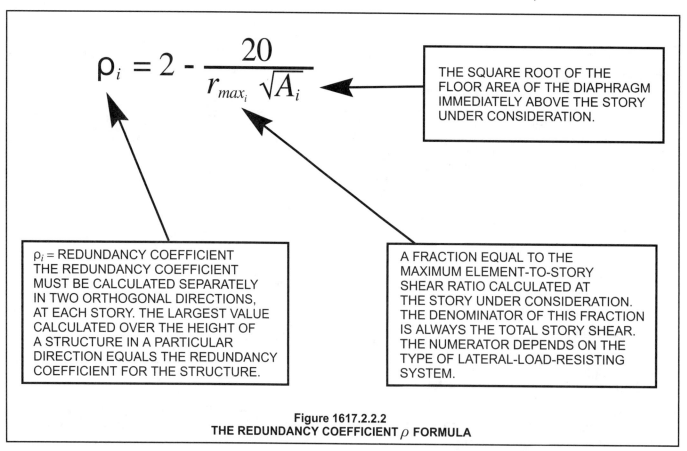

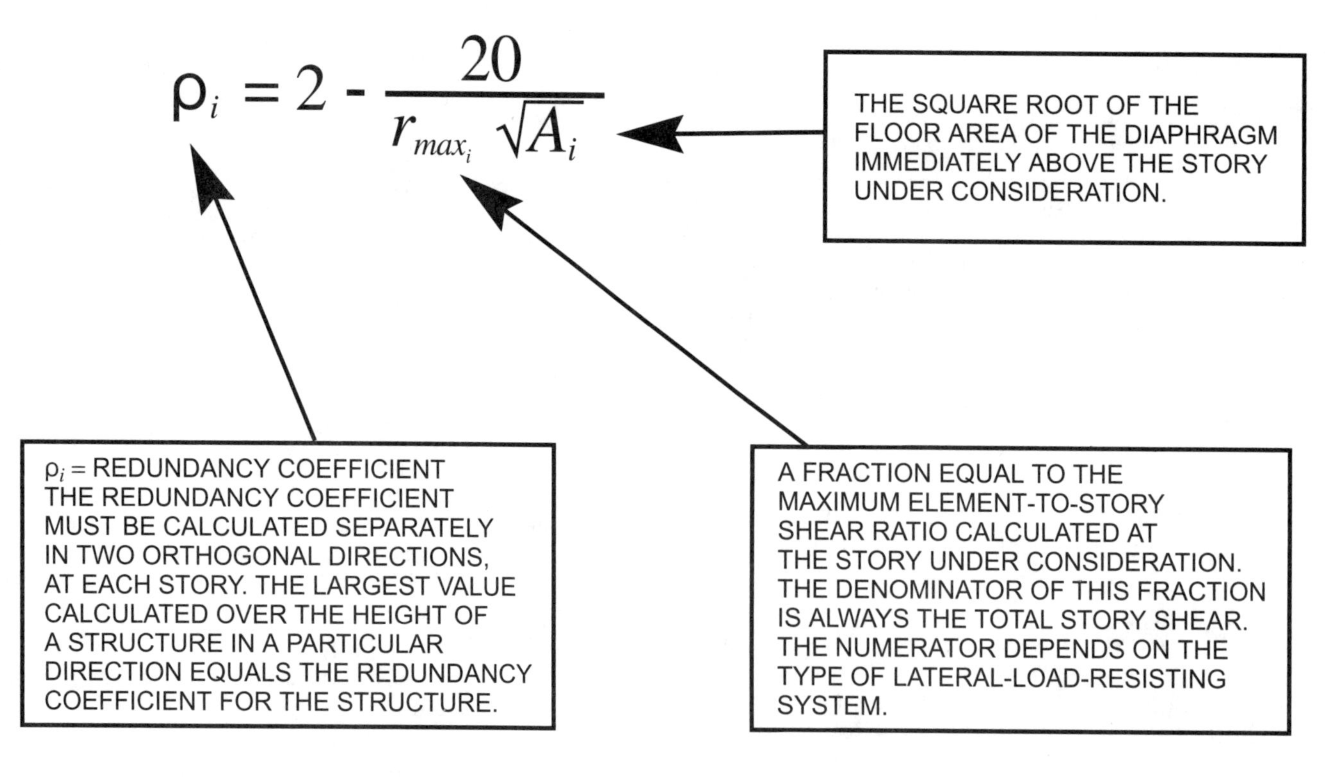

Figure 1617.2.2.2
THE REDUNDANCY COEFFICIENT ρ FORMULA

tectural, mechanical and electrical component design, when calculating drift, as well as for nonbuilding structures that are not similar to buildings.

The designer's goal should be to minimize the fraction r_{maxi}. In other words, no one element should be too heavily loaded or, as the saying goes, "don't put all of your eggs in one basket." If the designer sets $\rho_i \leq 1.0$, then one can solve for r_{maxi} as follows:

$$\rho_i \leq 1$$

$$1 \leq 2 - \frac{20}{(r_{max_i})(\sqrt{A_i})}$$

$$\frac{20}{(r_{max_i})(\sqrt{A_i})} \geq 1$$

$$r_{max_i} \leq \frac{20}{\sqrt{A_i}}$$

One quickly realizes that as the floor area increases, the value of r_{maxi} needs to decrease in order to achieve the goal of $\rho_i \leq 1.0$. This makes sense because as the floor area increases, there needs to be more lateral-force-resisting elements to share the load.

Even though a ρ value greater than 1.5 may be calculated, 1.5 is the maximum value that needs to be used in the code formulas, with two exceptions. The first is the case where special moment frames are used exclusively in one direction. In that case, the code places more restrictive limits on the calculated redundancy coefficient. These limits were originally imposed because special moment frame designs are usually drift controlled and, therefore, are not affected by a redundancy coefficient that modifies only the seismic load effect, *E*. In some cases, however, these limitations have the unintended effect of steering engineers away from what would otherwise be a highly desirable seismic-force-resisting system. The exception to these more restrictive limits allows "stiff" special moment frames that have drifts well within the limits of Section 1617.3. To qualify, the story drift must be limited to Δ_a/ρ, where Δ_a is the allowable story drift.

Another exception to the redundancy coefficient maximum value of 1.5 permits r_{max} to be calculated by assuming rigid diaphragm behavior for a story with a flexible diaphragm immediately above and limits ρ_i to 1.25; thus, for a structure with flexible diaphragms throughout, the maximum redundancy coefficient is effectively reduced from 1.5 to 1.25. Recognizing that the provisions for redundancy were developed with rigid diaphragm structures in mind, this attempts to lessen an unintended penalty on flexible diaphragm structures where additional lines of resistance are introduced.

Where different lateral-force-resisting systems are used over the height of a building, different ρ values may be used for the portions, subject to the limitation given in Section 1617.2.2.2. One example would be a multistory building constructed with concrete shear walls for the first story and steel-braced frames for the upper stories. The following minimum on ρ_L (redundancy coefficient for the lower portion) is dependent on both the redundancy coefficient ρ_u for the upper portion and the response modification factors R_L and R_u for the lower and upper portions, respectively (response modification factor is discussed in Section 1617.6).

$$\frac{\rho_L}{R_L} \leq \frac{\rho_U}{R_U}$$

Both factors ρ and R modify the seismic force effect for which structural members are designed. The design seismic force effect is proportional to the fraction ρ/R. In order to assure that this fraction is no greater for the upper portion than for the lower portion, which would create a weakening of lateral force resistance down the height of a building, a minimum is placed on ρ_L.

1617.3 Deflection and drift limits. The provisions given in Section 9.5.2.8 of ASCE 7 shall be used.

Exception: Structures designed using the simplified analysis procedure in Section 1617.5 shall meet the provisions in Section 1617.3.1.

❖ This section requires that the drift provisions of ASCE 7 be applied, and for structures analyzed under the simplified analysis procedure, the exception requires that Section 1617.3.1 be followed.

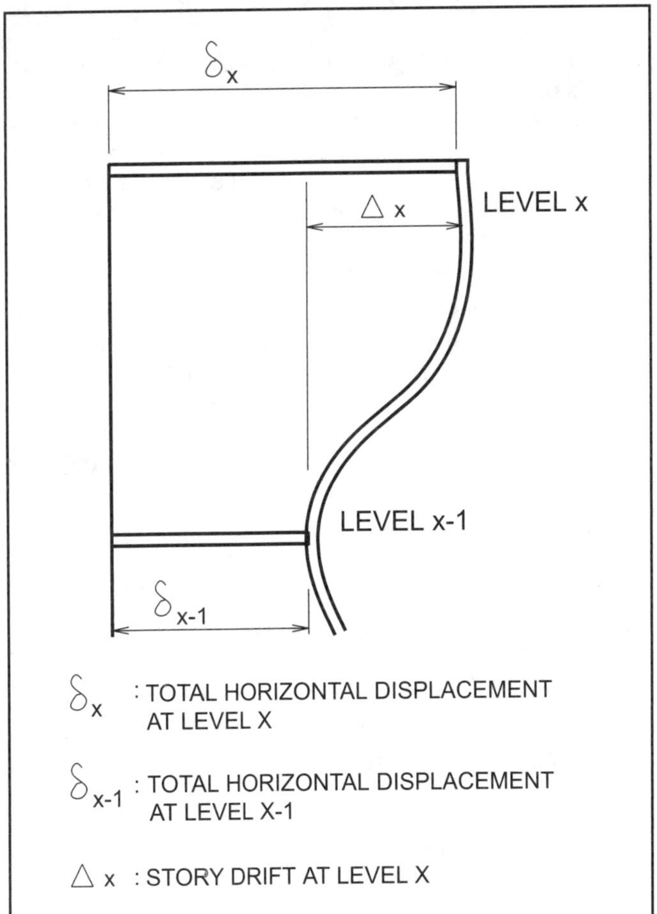

Figure 1617.3
EXAMPLE OF STORY DRIFT

1617.3.1 Deflection and drift limits (for use in the simplified analysis procedure of Section 1617.5). The design story drift, Δ, as determined in Section 1617.5.4, shall not exceed the allowable story drift, Δ_a, as obtained from Table 1617.3.1 for any story. All portions of the building shall be designed to act as an integral unit in resisting seismic forces unless separated structurally by a distance sufficient to avoid damaging contact under total deflection as determined in Section 1617.5.4. Buildings shall not exceed the limitations of Section 1616.6.1.

❖ Computation of story drift starts with the relative lateral displacement of a story caused by the design lateral forces and includes contributions of translational and torsional deflections.

The design story drift is the calculated elastic story drift that has been amplified in accordance with the requirements of Section 1617.5.3. An example of story drift is shown in Figure 1617.3. Limiting the drift of a story controls or reduces the potential for building damage, particularly as it concerns nonstructural items, such as glass, plaster and drywall materials, suspended ceilings and suspended fixtures and equipment. These can be sources of injury or even loss of life. Unreinforced masonry construction is particularly sensitive to potential damage from lateral displacement. The drift limitation is more stringent for structures where the interior partition walls have not been designed to accommodate story drift that is expected to occur during a major earthquake.

Experience has shown that buildings or separate parts of the same building, when not separated by adequate space, have often suffered severe damage caused by pounding of one building by the other during earthquakes. Thus, the total deflection of adjacent buildings is not to encroach on any interior lot line.

TABLE 1617.3.1. See below.

❖ Two factors having a major impact on structural member sizing are the story drift requirements of this table and the design base shear. Because of their lower stiffness, steel structures, especially moment-resisting frames, are more affected by the drift limitations than are stiffer reinforced concrete structures.

This table is organized such that buildings that can sustain greater drifts without damage are allowed greater drifts. More restrictive drift limits are precribed for higher seismic use groups. This table is the same as Table 5.2.8 of the 1997 NEHRP Recommended Provisions, and the commentary to that document contains further discussion.

Note that this table is for use under the simplified procedure that can only be used for Seismic Use Group I buildings, which is one of the limitations in Section 1616.6.1 that are referred to in Section 1617.3.1.

1617.4 Equivalent lateral force procedure for seismic design of buildings. The provisions given in Section 9.5.5 of ASCE 7 shall be used.

❖ This section requires the use of the equivalent lateral force procedure of ASCE 7. This procedure approximates the time-dependent dynamic inertia forces from an earthquake by equivalent static horizontal forces applied at the floor levels for simplicity. The "base" of structure is defined in Section 1613.

The equivalent static force method generally considers only the first or fundamental mode of vibration and the associated fundamental period of the structure. The static lateral forces are distributed over the height of the structure approximately according to the fundamental mode shape until the fundamental period exceeds 0.5 seconds.

The formula used for calculating the total lateral-seismic-force resistance to be built into a structure, more commonly referred to as the "design base shear" (*V*), is the most important and fundamental mathematical expression needed for the design of earthquake-resistant buildings. The design base shear is the sum of the lateral seismic design forces considered acting at the various floor levels of a building. For analytical purposes,

TABLE 1617.3.1
ALLOWABLE STORY DRIFT, Δ_a (inches)[a]

BUILDING	SEISMIC USE GROUP		
	I	II	III
Buildings, other than masonry shear wall or masonry wall frame buildings, four stories or less in height with interior walls, partitions, ceilings and exterior wall systems that have been designed to accommodate the story drifts	0.025 h_{sx}[b]	0.020 h_{sx}	0.015 h_{sx}
Masonry cantilever shear wall buildings[c]	0.010 h_{sx}	0.010 h_{sx}	0.010 h_{sx}
Other masonry shear wall buildings	0.007 h_{sx}	0.007 h_{sx}	0.007 h_{sx}
Masonry wall frame buildings	0.013 h_{sx}	0.013 h_{sx}	0.010 h_{sx}
All other buildings	0.020 h_{sx}	0.015 h_{sx}	0.010 h_{sx}

For SI: 1 inch = 25.4 mm.

a. There shall be no drift limit for single-story buildings with interior walls, partitions, ceilings and exterior wall systems that have been designed to accommodate the story drifts.

b. h_{sx} is the story height below Level *x*.

c. Buildings in which the basic structural system consists of masonry shear walls designed as vertical elements cantilevered from their base or foundation support which are so constructed that moment transfer between shear walls (coupling) is negligible.

lateral seismic design forces are typically assumed to act nonconcurrently in the direction of each of the principal axes of a structure.

1617.5 Simplified analysis procedure for seismic design of buildings. See Section 1616.6.1 for limitations on the use of this procedure. For purposes of this analytical procedure, a building is considered to be fixed at the base.

❖ Structures that qualify for use of the simplified analysis procedure are described in Section 1616.6. The simplified analysis procedure is advantageous in that the determination of the design base shear and its distribution along the height of a structure are simpler than in the equivalent lateral force procedure. Also, drift computation is considerably simpler.

The design story drift may be taken equal to 1 percent of story height, unless a more detailed analysis is performed. Note that the redundancy factor ρ may not be automatically taken equal to 1 when the simplified analysis procedure is used.

The simplified analysis is penalizing in that the structure is required to be designed for higher forces than if the equivalent lateral force procedure was used.

1617.5.1 Seismic base shear. The seismic base shear, *V*, in a given direction shall be determined in accordance with the following equation:

$$V = \frac{1.2 S_{DS}}{R} W \qquad \textbf{(Equation 16-56)}$$

where:

S_{DS} = The design elastic response acceleration at short period as determined in accordance with Section 1615.1.3.

R = The response modification factor from Table 1617.6.2.

W = The effective seismic weight of the structure, including the total dead load and other loads listed below:

1. In areas used for storage, a minimum of 25 percent of the reduced floor live load (floor live load in public garages and open parking structures need not be included).
2. Where an allowance for partition load is included in the floor load design, the actual partition weight or a minimum weight of 10 psf of floor area, whichever is greater (0.48 kN/m^2).
3. Total weight of permanent operating equipment.
4. 20 percent of flat roof snow load where flat snow load exceeds 30 psf (1.44 kN/m^2).

❖ The design base shear for the simplified analysis method is 20 percent greater than what is required by the equivalent lateral force method (see the commentary to Section 1617.4 for discussion on this method). An illustrative comparison of the base shear used in the two methods is provided in Figure 1617.5.1. Because the simplified analysis method cannot be used for structures greater than three stories in height, one can assume that the structure will be a short-period structure. Also, because the structure must be in Seismic Use Group I, the earthquake importance factor is equal to 1. Such a short-period structure would be required to be designed for $V = S_{DS}/R$, which is 20 percent less than the value given by Equation 16-56.

The effective seismic weight (*W*) used in the base shear equation is the total weight of both the structure and the contents that might reasonably be expected to be attached to the structure at the time the design earthquake occurs. For example, freshly fallen snow will not be firmly attached to the structure, unlike ice buildup that will be attached; therefore, only 20 percent of the flat roof snow load needs be included in the weight

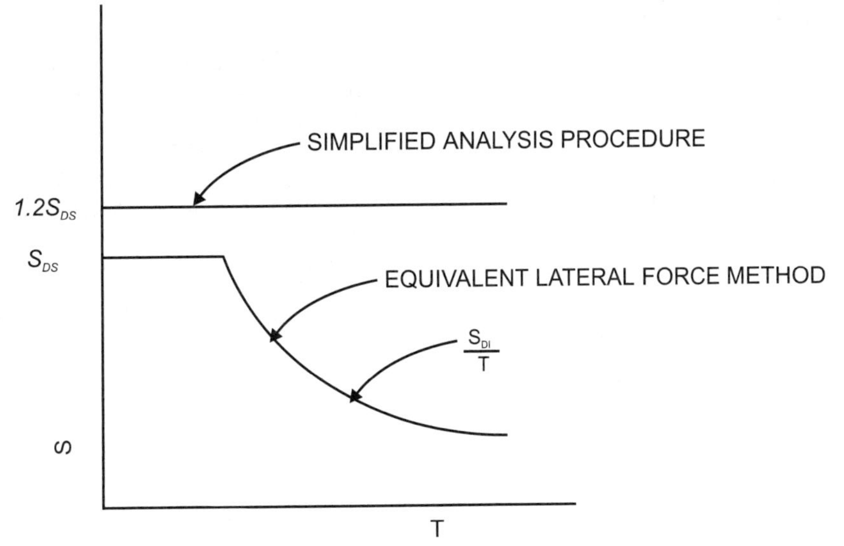

Figure 1617.5.1
SIMPLIFIED ANALYSIS METHOD VERSUS EQUIVALENT LATERAL FORCE METHOD

❖ when the snow load exceeds 30 psf (1.44 kN/m²). For the other variables used in determining the base shear, see the commentary to Section 1615.1.3 for S_{DS}, and Section 1617.6 for the response modification factor.

1617.5.2 Vertical distribution. The forces at each level shall be calculated using the following equation:

$$F_x = \frac{1.2 S_{DS}}{R} w_x \qquad \textbf{(Equation 16-57)}$$

where:

w_x = The portion of the effective seismic weight of the structure, *W*, at Level *x*.

❖ The base shear determined by the simplified analysis method is simply distributed to the various floor levels over the height of the building in proportion to the floor-level weight. If the weight at each floor level is the same, then the seismic force applied at each floor is the same.

1617.5.3 Horizontal distribution. Diaphragms constructed of untopped steel decking or wood structural panels or similar light-framed construction are permitted to be considered as flexible.

❖ This section permits diaphragms of light-frame construction to be considered flexible diaphragms, thus eliminating the need to demonstrate by calculations that these diaphragms meet the definition of "Flexible diaphragm" in Section 1602.1. Note that one of the limitations on the simplified procedure in Section 1616.6.1 states that, for other than light-frame construction, all horizontal diaphragms must be flexible.

1617.5.4 Design drift. For the purposes of Sections 1617.3.1 and 1620.4.6, the design story drift, Δ, shall be taken as 1 percent of the story height unless a more exact analysis is provided.

❖ Structures designed by the simplified analysis method are still subject to the maximum drift limits (see Section 1617.3) as well as the building separation requirement for Seismic Design Category D or E structures (see Section 1620.4.6); however, the drift may be assumed to simply be 1 percent of the story height.

1617.6 Seismic-force-resisting systems. The provisions given in Section 9.5.2.2 of ASCE 7 shall be used except as modified in Section 1617.6.1.

> **Exception:** For structures designed using the simplified analysis procedure in Section 1617.5, the provisions of Section 1617.6.2 shall be used.

❖ This section requires use of the ASCE 7 provisions for seismic-force-resisting systems for all analysis methods except the simplified procedure, in which case Section 1617.6.2 is applicable. Since the latter provisions are comparable to those of ASCE 7, the accompanying commentary can provide guidance on the corresponding ASCE 7 provisions as well.

1617.6.1 Modifications to ASCE 7, Section 9.5.2.2.

❖ In referencing ASCE 7 provisions, the code makes various modifications to these provisions. See the discussions below for explanations of the specific modifications.

1617.6.1.1 ASCE 7, Table 9.5.2.2. Modify Table 9.5.2.2 as follows:

1. Bearing wall systems: Ordinary reinforced masonry shear walls shall use a response modification coefficient of 2$^1/_2$. Light-framed walls sheathed with wood structural panels rated for shear resistance or steel sheets shall use a response modification coefficient of 6$^1/_2$. Table 1617.6.2 entries for ordinary plain prestressed masonry shear walls, intermediate prestressed masonry shear walls and special prestressed masonry shear walls shall apply.
2. Building frame systems: Ordinary reinforced masonry shear walls shall use a response modification coefficient of 3. Light-framed walls sheathed with wood structural panels rated for shear resistance or steel sheets shall use a response modification coefficient of 7. Table 1617.6.2 entries for ordinary plain prestressed masonry shear walls, intermediate prestressed masonry shear walls and special prestressed masonry shear walls shall apply.
3. Dual systems with intermediate moment frames capable of resisting at least 25 percent of prescribed seismic forces. Special steel concentrically braced frames shall use a deflection amplification factor of 4.
4. The table column titled Detailing Reference Section in Table 1617.6.2 shall apply.

❖ There are three distinct types of modifications that are made to the ASCE 7 design coefficient table. One type of modification is to the design factors for specific basic seismic-force-resisting systems that reflect the decisions made by the IBC Structural Committee. The second type is to include prestressed masonry shear wall system design values that are not included in ASCE 7 by referring to the corresponding entries in Table 1617.6.2. The third type of modification is a general reference to the detailing column of Table 1617.6.2, which provides useful cross references (not included in ASCE 7) to the detailing requirements in either the code or standards that are referenced by the code, such as ACI 318.

1617.6.1.2 ASCE 7, Section 9.5.2.2.2.1. Modify Section 9.5.2.2.2.1 by adding Exception 3 as follows:

3. The following two-stage static analysis procedure is permitted to be used for structures having a flexible upper portion supported on a rigid lower portion where both

portions of the structure considered separately can be classified as being regular, the average story stiffness of the lower portion is at least 10 times the average story stiffness of the upper portion and the period of the entire structure is not greater than 1.1 times the period of the upper portion considered as a separate structure fixed at the base:

3.1. The flexible upper portion shall be designed as a separate structure using the appropriate values of R and ρ.

3.2. The rigid lower portion shall be designed as a separate structure using the appropriate values of R and ρ. The reactions from the upper portion shall be those determined from the analysis of the upper portion amplified by the ratio of the R/ρ of the upper portion over R/ρ of the lower portion. This ratio shall not be less than 1.0.

❖ This modifies the provisions in ASCE 7 that are applicable to vertical combinations of framing systems. Those provisions are comparable to Section 1617.6.2.3.1, and the accompanying commentary is applicable to the ASCE 7 requirements.

The modification incorporates an exception permitting a so-called "two-stage analysis" for buildings meeting certain specific criteria. This provision has been part of the *Uniform Building Code*™ since the 1988 edition. It is beneficial in designing a combination consisting of a relatively light, flexible structure sitting atop a much more massive (and stiffer) lower structure. It often is applied to buildings that also are designed in accordance with the special provisions of Section 508.2.

1617.6.1.3 ASCE 7, Section 9.5.2.2.4.3. Modify Section 9.5.2.2.4.3 by changing exception to read as follows:

Exception: Reinforced concrete frame members not designed as part of the seismic-force-resisting system and slabs shall comply with Section 21.11 of Ref. 9.9-1.

❖ This modifies the deformation compatibility provisions in ASCE 7. Those provisions are comparable to Section 1617.6.2.4.3, and the accompanying commentary is applicable to the ASCE 7 requirements. The modification makes the exception applicable to slabs in addition to the concrete frame members that are not part of the seismic-force-resisting system.

1617.6.2 Seismic-force-resisting systems (for use in the Simplified analysis procedure of Section 1617.5). The basic lateral and vertical seismic-force-resisting systems shall conform to one of the types indicated in Table 1617.6.2 subject to the limitations on height indicated in the table based on seismic design category as determined in Section 1616. The appropriate response modification coefficient, R, system overstrength factor, Ω_0, and deflection amplification factor, C_d, indicated in Table 1617.6.2 shall be used in determining the base shear, element design forces and design story drift. For seismic-force-resisting systems not listed in Table 1617.6.2, analytical and test data shall be submitted that establish the dynamic characteristics and demonstrate the lateral-force resistance and energy dissipation capacity to be equivalent to the structural systems listed in Table 1617.6.2 for equivalent response modification coefficient, R, system overstrength coefficient, Ω_0, and deflection amplification factor, C_d, values. Buildings shall not exceed the limitations of Section 1616.6.1.

Exception: Structures assigned to Seismic Design Category A.

❖ This section quantifies the reduction in design earthquake forces by the response modification factor (R) because of the ductility of the seismic-resisting system. The purpose of the deflection amplification factor (C_d) is to increase the elastic deflection calculated under the code-prescribed seismic forces to an approximation of the actual displacement expected in the design earthquake. The purpose of the overstrength factor (Ω_0) is to increase the elastic member forces calculated under the code-prescribed seismic forces to reflect the actual forces the elements will experience during the design earthquake.

TABLE 1617.6.2. See page 16-124.

❖ The R-factor decreases the design base shear for those structural systems that have demonstrated good earthquake performance. Essentially, the coefficient R serves as an indication of the inherent ductility and damping of a structural system used to resist the seismic lateral forces (base shear). While it would be technically feasible to design buildings and other structures to remain fully elastic during their response to the seismic forces and displacements caused by the design-level ground shaking, it would be economically prohibitive, particularly in areas of moderate to strong earthquakes. Figure 1617.6 shows the schematic relationship of seismic base shear on a structure that is designed to remain in the elastic stress range versus the seismic base shear required by the code, which is based on inelastic response. The basic concept in the design of earthquake-resisting structures is to provide a ductile system that allows inelastic yielding to accommodate the seismic forces as long as such yielding does not weaken the gravity-load-carrying capacity of the structure. Some seismic-force-resisting systems have structure height limitations in Table 1617.6.2 because of a lack of reliable data on the behavior of high-rise structures having these seismic-force-resisting systems in areas with frequent major earthquakes. Some structural systems are not permitted in areas with frequent major earthquakes because of low ductility.

TABLE 1617.6.2
DESIGN COEFFICIENTS AND FACTORS FOR BASIC SEISMIC-FORCE-RESISTING SYSTEMS

BASIC SEISMIC-FORCE-RESISTING SYSTEM	DETAILING REFERENCE SECTION	RESPONSE MODIFICATION COEFFICIENT, R^a	SYSTEM OVERSTRENGTH FACTOR, Ω_o^g	DEFLECTION AMPLIFICATION FACTOR, C_d^b	SYSTEM LIMITATIONS AND BUILDING HEIGHT LIMITATIONS (FEET) BY SEISMIC DESIGN CATEGORY AS DETERMINED IN SECTION 1616.3[c]				
					A or B	C	D[d]	E[e]	F[e]
1. Bearing Wall Systems									
A. Ordinary steel braced frames in light-frame construction	2211	4	2	$3^1/_2$	NL	NL	65	65	65
B. Special reinforced concrete shear walls	1910.2.4	$5^1/_2$	$2^1/_2$	5	NL	NL	160	160	100
C. Ordinary reinforced concrete shear walls	1910.2.3	$4^1/_2$	$2^1/_2$	4	NL	NL	NP	NP	NP
D. Detailed plain concrete shear walls	1910.2.2	$2^1/_2$	$2^1/_2$	2	NL	NP	NP	NP	NP
E. Ordinary plain concrete shear walls	1910.2.1	$1^1/_2$	$2^1/_2$	$1^1/_2$	NL	NP	NP	NP	NP
F. Special reinforced masonry shear walls	1.13.2.2.5[o]	5	$2^1/_2$	$3^1/_2$	NL	NL	160	160	100
G. Intermediate reinforced masonry shear walls	1.13.2.2.4[o]	$3^1/_2$	$2^1/_2$	$2^1/_4$	NL	NL	NP	NP	NP
H. Ordinary reinforced masonry shear walls	1.13.2.2.3[o]	$2^1/_2$	$2^1/_2$	$1^3/_4$	NL	160	NP	NP	NP
I. Detailed plain masonry shear walls	1.13.2.2.2[o]	2	$2^1/_2$	$1^3/_4$	NL	NP	NP	NP	NP
J. Ordinary plain masonry shear walls	1.13.2.2.1[o]	$1^1/_2$	$2^1/_2$	$1^1/_4$	NL	NP	NP	NP	NP
K. Light frame walls with shear panels—wood structural panels/sheet steel panels	2306.4.1/ 2211	$6^1/_2$	3	4	NL	NL	65	65	65
L. Light framed walls with shear panels—all other materials	2306.4.5/ 2211	2	$2^1/_2$	2	NL	NL	35	NP	NP
M. Ordinary plain prestressed masonry shear walls	2106.1.1.1	$1^1/_2$	$2^1/_2$	$1^1/_4$	NL	NP	NP	NP	NP
N. Intermediate prestressed masonry shear walls	2106.1.1.2, 1.13.2.2.4[o]	$2^1/_2$	$2^1/_2$	$2^1/_2$	NL	35	NP	NP	NP
O. Special prestressed masonry shear walls	2106.1.1.3, 1.13.2.2.5[o]	$4^1/_2$	$2^1/_2$	$3^1/_2$	NL	35	35	35	35
2. Building Frame Systems									
A. Steel eccentrically braced frames, moment-resisting, connections at columns away from links	(15)[j]	8	2	4	NL	NL	160	160	100
B. Steel eccentrically braced frames, nonmoment resisting, connections at columns away from links	(15)[j]	7	2	4	NL	NL	160	160	100
C. Special steel concentrically braced frames	(13)[j]	6	2	5	NL	NL	160	160	100
D. Ordinary steel concentrically braced frames	(14)[j]	5	2	$4^1/_2$	NL	NL	35[n]	35[n]	NP[n]
E. Special reinforced concrete shear walls	1910.2.4	6	$2^1/_2$	5	NL	NL	160	160	100
F. Ordinary reinforced concrete shear walls	1910.2.3	5	$2^1/_2$	$4^1/_2$	NL	NL	NP	NP	NP
G. Detailed plain concrete shear walls	1910.2.2	3	$2^1/_2$	$2^1/_2$	NL	NP	NP	NP	NP

(continued)

TABLE 1617.6.2—continued
DESIGN COEFFICIENTS AND FACTORS FOR BASIC SEISMIC-FORCE-RESISTING SYSTEMS

BASIC SEISMIC-FORCE-RESISTING SYSTEM	DETAILING REFERENCE SECTION	RESPONSE MODIFICATION COEFFICIENT, R[a]	SYSTEM OVERSTRENGTH FACTOR, Ω_o[g]	DEFLECTION AMPLIFICATION FACTOR, C_d[b]	SYSTEM LIMITATIONS AND BUILDING HEIGHT LIMITATIONS (FEET) BY SEISMIC DESIGN CATEGORY[c] AS DETERMINED IN SECTION 1616.3[c]				
					A or B	C	D[d]	E[e]	F[e]
H. Ordinary plain concrete shear walls	1910.2.1	2	2 1/2	2	NL	NP	NP	NP	NP
I. Composite eccentrically braced frames	(14)[k]	8	2	4	NL	NL	160	160	100
J. Composite concentrically braced frames	(13)[k]	5	2	4 1/2	NL	NL	160	160	100
K. Ordinary composite braced frames	(12)[k]	3	2	3	NL	NL	NP	NP	NP
L. Composite steel plate shear walls	(17)[k]	6 1/2	2 1/2	5 1/2	NL	NL	160	160	100
M. Special composite reinforced concrete shear walls with steel elements	(16)[k]	6	2 1/2	5	NL	NL	160	160	100
N. Ordinary composite reinforced concrete shear walls with steel elements	(15)[k]	5	2 1/2	4 1/2	NL	NL	NP	NP	NP
O. Special reinforced masonry shear walls	1.13.2.2.5[o]	5 1/2	2 1/2	4	NL	NL	160	160	100
P. Intermediate reinforced masonry shear walls	1.13.2.2.4[o]	4	2 1/2	2 1/2	NL	NL	NP	NP	NP
Q. Ordinary reinforced masonry shear walls	1.13.2.2.3[o]	3	2 1/2	2 1/4	NL	160	NP	NP	NP
R. Detailed plain masonry shear walls	1.13.2.2.2[o]	2 1/2	2 1/2	2 1/4	NL	NP	NP	NP	NP
S. Ordinary plain masonry shear walls	1.13.2.2.1[o]	1 1/2	2 1/2	1 1/4	NL	NP	NP	NP	NP
T. Light frame walls with shear panels—wood structural panels/sheet steel panels	2306.4.1/ 2211	7	2 1/2	4 1/2	NL	NL	65	65	65
U. Light framed walls with shear panels—all other materials	2306.4.5/ 2211	2 1/2	2 1/2	2 1/2	NL	NL	35	NP	NP
V. Ordinary plain prestressed masonry shear walls	2106.1.1.1	1 1/2	2 1/2	1 1/4	NL	NP	NP	NP	NP
W. Intermediate prestressed masonry shear walls	2106.1.1.2, 1.13.2.2.4[o]	3	2 1/2	2 1/2	NL	35	NP	NP	NP
X. Special prestressed masonry shear walls	2106.1.1.3, 1.13.2.2.5[o]	4 1/2	2 1/2	4	NL	35	35	35	35
3. Moment-resisting Frame Systems									
A. Special steel moment frames	(9)[j]	8	3	5 1/2	NL	NL	NL	NL	NL
B. Special steel truss moment frames	(12)[j]	7	3	5 1/2	NL	NL	160	100	NP
C. Intermediate steel moment frames	(10)[j]	4 1/2	3	4	NL	NL	35[h]	NP[h,i]	NP[h,i]
D. Ordinary steel moment frames	(11)[j]	3 1/2	3	3	NL	NL	NP[h,i]	NP[h,i]	NP[h,i]
E. Special reinforced concrete moment frames	(21.1)[l]	8	3	5 1/2	NL	NL	NL	NL	NL

(continued)

TABLE 1617.6.2—continued
DESIGN COEFFICIENTS AND FACTORS FOR BASIC SEISMIC-FORCE-RESISTING SYSTEMS

BASIC SEISMIC-FORCE-RESISTING SYSTEM	DETAILING REFERENCE SECTION	RESPONSE MODIFICATION COEFFICIENT, R^a	SYSTEM OVER-STRENGTH FACTOR, Ω_o^g	DEFLECTION AMPLIFICATION FACTOR, C_d^b	SYSTEM LIMITATIONS AND BUILDING HEIGHT LIMITATIONS (FEET) BY SEISMIC DESIGN CATEGORY AS DETERMINED IN SECTION 1616.3[c]				
					A or B	C	D[d]	E[e]	F[e]
F. Intermediate reinforced concrete moment frames	(21.1)[l]	5	3	4 1/2	NL	NL	NP	NP	NP
G. Ordinary reinforced concrete moment frames	(21.1)[l]	3	3	2 1/2	NL	NP	NP	NP	NP
H. Special composite moment frames	(9)[k]	8	3	5 1/2	NL	NL	NL	NL	NL
I. Intermediate composite moment frames	(10)[k]	5	3	4 1/2	NL	NL	NP	NP	NP
J. Composite partially restrained moment frames	(8)[k]	6	3	5 1/2	160	160	100	NP	NP
K. Ordinary composite moment frames	(11)[k]	3	3	2 1/2	NL	NP	NP	NP	NP
L. Masonry wall frames	2106	5 1/2	3	5	NL	NL	160	160	100
4. Dual Systems with Special Moment Frames									
A. Steel eccentrically braced frames, moment-resisting connections, at columns away from links	(15)[j]	8	2 1/2	4	NL	NL	NL	NL	NL
B. Steel eccentrically braced frames, nonmoment-resisting connections, at columns away from links	(15)[j]	7	2 1/2	4	NL	NL	NL	NL	NL
C. Special steel concentrically braced frames	(13)[j]	8	2 1/2	6 1/2	NL	NL	NL	NL	NL
D. Special reinforced concrete shear walls	1910.2.4	8	2 1/2	6 1/2	NL	NL	NL	NL	NL
E. Ordinary reinforced concrete shear walls	1910.2.3	7	2 1/2	6	NL	NL	NP	NP	NP
F. Composite eccentrically braced frames	(14)[k]	8	2 1/2	4	NL	NL	NL	NL	NL
G. Composite concentrically braced frames	(13)[k]	6	2 1/2	5	NL	NL	NL	NL	NL
H. Composite steel plate shear walls	(17)[k]	8	2 1/2	6 1/2	NL	NL	NL	NL	NL
I. Special composite reinforced concrete shear walls with steel elements	(16)[k]	8	2 1/2	6 1/2	NL	NL	NL	NL	NL
J. Ordinary composite reinforced concrete shear walls with steel elements	(15)[k]	7	2 1/2	6	NL	NL	NP	NP	NP
K. Special reinforced masonry shear walls	1.13.2.2.5[o]	7	3	6 1/2	NL	NL	NL	NL	NL
L. Special reinforced masonry shear walls	1.13.2.2.4[o]	6 1/2	3	5 1/2	NL	NL	NP	NP	NP
5. Dual Systems with Intermediate Moment Frames[m]									
A. Special steel concentrically braced frames[f]	(13)[j]	4 1/2	2 1/2	4	NL	NL	35[h]	NP[h,i]	NP
B. Special reinforced concrete shear walls	1910.2.4	6	2 1/2	5	NL	NL	160	100	100
C. Ordinary reinforced concrete shear walls	1910.2.3	5 1/2	2 1/2	4 1/2	NL	NL	NP	NP	NP

(continued)

TABLE 1617.6.2–continued
DESIGN COEFFICIENTS AND FACTORS FOR BASIC SEISMIC-FORCE-RESISTING SYSTEMS

BASIC SEISMIC-FORCE-RESISTING SYSTEM	DETAILING REFERENCE SECTION	RESPONSE MODIFICATION COEFFICIENT, R^a	SYSTEM OVER-STRENGTH FACTOR, Ω_o^g	DEFLECTION AMPLIFICATION FACTOR, C_d^b	SYSTEM LIMITATIONS AND BUILDING HEIGHT LIMITATIONS (FEET) BY SEISMIC DESIGN CATEGORY AS DETERMINED IN SECTION 1616.3[c]				
					A or B	C	D[d]	E[e]	F[e]
D. Ordinary reinforced masonry shear walls	1.13.2.2.3[o]	3	3	$2^1/_2$	NL	160	NP	NP	NP
E. Intermediate reinforced masonry shear walls	1.13.2.2.4[o]	5	3	$4^1/_2$	NL	NL	NP	NP	NP
F. Composite concentrically braced frames	(13)[k]	5	$2^1/_2$	$4^1/_2$	NL	NL	160	100	NP
G. Ordinary composite braced frames	(12)[k]	4	$2^1/_2$	3	NL	NL	NP	NP	NP
H. Ordinary composite reinforced concrete shear walls with steel elements	(15)[k]	$5^1/_2$	$2^1/_2$	$4^1/_2$	NL	NL	NP	NP	NP
6. Shear Wall-frame Interactive System with Ordinary Reinforced Concrete Moment Frames and Ordinary Reinforced Concrete Shear Walls	21.1[l] 1910.2.3	$5^1/_2$	$2^1/_2$	5	NL	NP	NP	NP	NP
7. Inverted Pendulum Systems									
A. Cantilevered column systems		$2^1/_2$	2	$2^1/_2$	NL	NL	35	35	35
B. Special steel moment frames	(9)[j]	$2^1/_2$	2	$2^1/_2$	NL	NL	NL	NL	NL
C. Ordinary steel moment frames	(11)[j]	$1^1/_4$	2	$2^1/_2$	NL	NL	NP	NP	NP
D. Special reinforced concrete moment frames	21.1[l]	$2^1/_2$	2	$1^1/_4$	NL	NL	NL	NL	NL
8. Structural Steel Systems Not Specifically Detailed for Seismic Resistance	AISC—335 AISC—LRFD AISI AISC—HSS	3	3	3	NL	NL	NP	NP	NP

For SI: 1 foot = 304.8 mm, 1 pound per square foot = 0.0479 KN/m^2.

a. Response modification coefficient, *R*, for use throughout.
b. Deflection amplification factor, C_d.
c. NL = not limited and NP = not permitted.
d. See Section 1617.6.2.4.1 for a description of building systems limited to buildings with a height of 240 feet or less.
e. See Section 1617.6.2.4.1 for building systems limited to buildings with a height of 160 feet or less.
f. Ordinary moment frame is permitted to be used in lieu of intermediate moment frame in Seismic Design Categories B and C.
g. The tabulated value of the overstrength factor, Ω_o, may be reduced by subtracting $^1/_2$ for structures with flexible diaphragms but shall not be taken as less than 2.0 for any structure.
h. Steel ordinary moment frames and intermediate moment frames are permitted in single story buildings up to a height of 60 feet, when the moment joints of field connections are constructed of bolted end plates and the dead load of the roof does not exceed 15 pounds per square foot. The dead weight of the portion of walls more than 35 feet above the base shall not exceed 15 pounds per square foot.
i. Steel ordinary moment frames are permitted in buildings up to a height of 35 feet, where the dead load of the walls, floors and roof does not exceed 15 pounds per square foot.
j. AISC 341 Part I or Part III section number.
k. AISC 341 Part II section number.
l. ACI 318, Section number.
m. Steel intermediate moment resisting frames as part of a dual system are not permitted in Seismic Design Categories D, E, and F.
n. Steel ordinary concentrically braced frames are permitted in penthouse structures and in single-story buildings up to a height of 60 feet when the dead load of the roof does not exceed 15 pounds per square foot.
o. ACI 530/ASCE 5/TMS 402 section number.

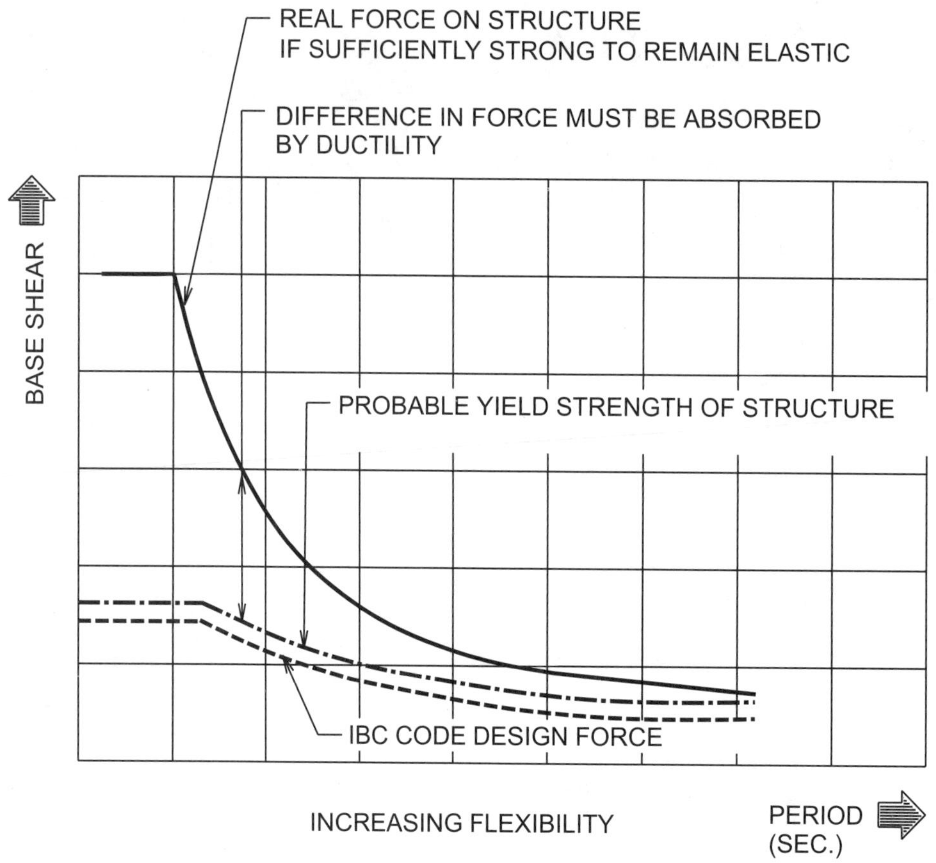

Figure 1617.6
BASE SHEAR VS. BUILDING PERIOD

1617.6.2.1 Dual systems. For a dual system, the moment frame shall be capable of resisting at least 25 percent of the design forces. The total seismic force resistance is to be provided by the combination of the moment frame and the shear walls or braced frames in proportion to their stiffness.

❖ Dual structural systems consist of moment-resisting frames in combination with shear walls or braced frames that act together to resist seismic lateral forces. The reason for the 25-percent requirement is so that the moment-resisting frame is designed and detailed for adequate performance, even if the shear walls or bracing systems can resist nearly the entire lateral force. In essence, the moment-resisting frame serves as a second line of defense and, thus, justifies a higher *R* value for the dual system than for a system consisting exclusively of shear walls. Care must be taken in calculations to distribute the design forces in accordance with the relative rigidities of the component systems and their structural elements.

1617.6.2.2 Combination along the same axis. For other than dual systems and shear wall-frame interactive systems, where a combination of different structural systems is utilized to resist lateral forces in the same direction, the value, *R*, used for design in that direction shall not be greater than the least value for any of the systems utilized in that same direction.

Exception: For light-framed, flexible diaphragm buildings, of Seismic Use Group I and two stories or less in height: Resisting elements are permitted to be designed using the least value of *R* for the different structural systems found on each independent line of resistance. The value of *R* used for design of diaphragms in such structures shall not be greater than the least value for any of the systems utilized in that same direction.

❖ If more than one type of lateral-force-resisting system is used in the same direction of loading, the *R* values of the different systems need to be compared, and the lowest value needs to be used in the determination of the base shear. This rule does not apply if the different structural systems together form and are designed as a dual system or a shear wall frame interactive system. There is another exception that addresses a very common condition in light-frame construction where one side of the structure has an open front and an ordinary moment-resisting frame of steel ($R = 3\frac{1}{2}$) is used to resist the lateral forces. The remainder of the structure uses wood structural panel shear walls to resist the lateral forces ($R = 6\frac{1}{2}$). Rather than penalize the entire structure by requiring use of an $R = 3\frac{1}{2}$, this exception allows each line of resistance to use the *R* of its respective lateral-load-resisting system. The flexible diaphragm, which distributes the lateral load among the lines of resistance, must be designed with the lowest *R* value of the structural systems used in the given direction.

1617.6.2.3 Combinations of framing systems. Where different seismic-force-resisting systems are used along the two orthogonal axes of the structure, the appropriate response modification coefficient, *R*, system overstrength factor, Ω_0, and deflection amplification factor, C_d, indicated in Table 1617.6.2 for each system shall be used.

❖ This section provides for structures that use different seismic-force-resisting systems in two orthogonal directions. For example, for a structure that has moment frames for the seismic-force-resisting system in the north-south direction and shear walls in the east-west direction, the appropriate *R*, C_d and Ω_0 values may be used in each direction.

If, however, one of the seismic-force-resisting systems has an *R* value less than 5 and the structure is assigned to Seismic Design Category D, E or F (note that buildings meeting the limitations of Section 1616.6.1 could not be classified as Seismic Design Category F), the lowest *R* value must be used for the entire structure in both directions (see Section 1617.6.2.3.1 and exceptions).

1617.6.2.3.1 Combination framing factor. The response modification coefficient, *R*, in the direction under consideration at any story shall not exceed the lowest response modification coefficient, *R*, for the seismic-force-resisting system in the same direction considered above that story, excluding penthouses. The system overstrength factor, Ω_0, in the direction under consideration at any story, shall not be less than the largest value of this factor for the seismic-force-resisting system in the same direction considered above that story. In structures assigned to Seismic Design Category D, E or F, if a system with a response modification coefficient, *R*, with a value less than five is used as part of the seismic-force-resisting system in any direction of the structure, the lowest such value shall be used for the entire structure.

Exceptions:

1. Detached one- and two-family dwellings constructed of light framing.
2. The response modification coefficient, *R*, and system overstrength factor, Ω_0, for supported structural systems with a weight equal to or less than 10 percent of the weight of the structure are permitted to be determined independent of the values of these parameters for the structure as a whole.
3. The following two-stage static analysis procedure is permitted to be used for structures having a flexible upper portion supported on a rigid lower portion where both portions of the structure considered separately can be classified as being regular, the average story stiffness of the lower portion is at least 10 times the average story stiffness of the upper portion and the period of the entire structure is not greater than 1.1 times the period of the upper portion considered as a separate structure fixed at the base:
   - 3.1. The flexible upper portion shall be designed as a separate structure using the appropriate values of *R* and ρ.
   - 3.2. The rigid lower portion shall be designed as a separate structure using the appropriate values of *R* and ρ. The reactions from the upper portion shall be those determined from the analysis of the upper portion amplified by the ratio of, R/ρ, of the upper portion over, R/ρ, of the lower portion. This ratio shall not be less than 1.0.

❖ For a given force or earthquake motion direction, the *R*-factor is not allowed to increase from the top to the bottom of the structure. For example, consider a building with ordinary reinforced masonry shear walls on the upper floors and an ordinary moment frame of steel in the first story. While this arrangement would be permitted by the code, the *R*-factor for the steel moment frame is limited to $2^1/_2$ (the *R*-factor for a bearing-wall system of ordinary reinforced masonry shear walls) instead of the $3^1/_2$ value in Table 1617.6.2. The design base shear is based on the less ductile upper-story structural system.

There are three exceptions to the two rules given in Section 1617.6.2.3. Detached one- and two-family dwellings of light-frame construction are exempt. Also exempt are supported structural systems with a weight less than or equal to 10 percent of the total weight of the structure. The exception is included to permit the use of such systems as a braced frame penthouse on a moment frame building without affecting the *R* value to be used in the design of the entire building, provided the penthouse does not represent a significant portion of the total building weight. A two-stage analysis procedure outlined in Items 3.1 and 3.2 of Exception 3 is allowed for structures meeting the specified conditions.

1617.6.2.3.2 Combination framing detailing requirements. The detailing requirements of Section 1620 required by the higher response modification coefficient, *R*, shall be used for structural components common to systems having different response modification coefficients.

❖ This section requires that for structural components common to systems having different *R* factors, the detailing requirements commensurate with the highest *R* value must be used.

1617.6.2.4 System limitations for Seismic Design Category D, E or F. In addition to the system limitation indicated in Table 1617.6.2, structures assigned to Seismic Design Category D, E or F shall be subject to the following.

❖ This section introduces four different subsections that are applicable to structures assigned to Seismic Design Categories D, E and F (note that buildings meeting the limitations of Section 1616.6.1 could not be classified as Seismic Design Category F).

1617.6.2.4.1 Limited building height. For buildings that have steel-braced frames or concrete cast-in-place shear walls, the height limits in Table 1617.6.2 for Seismic Design Category D or E are increased to 240 feet (73 152 mm) and for Seismic Design Category F to 160 feet (48 768 mm) provided that the buildings are configured such that the braced frames or shear walls arranged in any one plane conform to the following:

1. The braced frames or shear walls in any one plane shall resist no more than 50 percent of the total seismic forces in each direction, neglecting torsional effects.
2. The seismic force in the braced frames or shear walls in any one plane resulting from torsional effects shall not exceed 20 percent of the total seismic force in the braced frames or shear walls.

❖ This section relaxes the building height limitations of Table 1617.6.2 for certain structural systems that are expected to demonstrate good seismic performance due to the arrangement of the seismic-force-resisting elements specified. The simplified procedure has a height limit in stories, but not in feet. While it is not likely that buildings of this height could be economically designed under the simplified procedure, it is still possible to do so and comply with the code; thus, this height limitation must be listed as being applicable to the simplified procedure.

The height limitation difference from 240 feet (73 152 mm) (for Seismic Design Categories D and E) to 160 feet (48 768 mm) (for Seismic Design Category F) is necessary for the integrity of essential facilities located close to major faults after a major earthquake. The building must meet the stiffness and structural arrangement provisions of Section 1617.6.2.4.1 to qualify for the 160-foot (48 768 mm) height limitation. Otherwise, the height limitation must be in accordance with the more stringent requirements of Table 1617.6.2 for Seismic Design Category F.

1617.6.2.4.2 Interaction effects. Moment-resisting frames that are enclosed or adjoined by stiffer elements not considered to be part of the seismic-force-resisting system shall be designed so that the action or failure of those elements will not impair the vertical load and seismic-force-resisting capability of the frame. The design shall consider and provide for the effect of these rigid elements on the structural system at deformations corresponding to the design story drift, Δ, as determined in Section 1617.5.4. In addition, the effects of these elements shall be considered when determining whether a structure has one or more of the irregularities defined in Section 1616.5.1.

❖ The vertical-load-carrying and lateral-force-resisting capabilities of moment-resisting frames must not be compromised by the failure or action of stiffer elements not considered to be part of the lateral-force-resisting system, which enclose or adjoin such frames. This particular aspect of design is addressed in this section.

1617.6.2.4.3 Deformational compatibility. Every structural component not included in the seismic-force-resisting system in the direction under consideration shall be designed to be adequate for vertical load-carrying capacity and the induced moments and shears resulting from the design story drift, Δ, as determined in accordance with Section 1617.5.4. Where allowable stress design is used, Δ shall be computed without dividing the earthquake force by 1.4. The moments and shears induced in components that are not included in the seismic-force-resisting system in the direction under consideration shall be calculated including the stiffening effects of adjoining rigid structural and nonstructural elements.

Exception: Reinforced concrete frame members not designed as part of the seismic-force-resisting system and slabs shall comply with Section 21.11 of ACI 318.

❖ All structural members designated not to be part of the lateral-force-resisting system must maintain their capacity to carry full-factored gravity loads as they deform together with the lateral-force-resisting system through the design story drift, Δ. Reinforced concrete structural members that are not designed as part of the seismic-force-resisting system and slabs must satisfy the comprehensive requirements of Section 21.11 of ACI 318.

1617.6.2.4.4 Special moment frames. A special moment frame that is used but not required by Table 1617.6.2 is permitted to be discontinued and supported by a stiffer system with a lower response modification coefficient, R, provided the requirements of Sections 1620.2.3 and 1620.4.1 are met. Where a special moment frame is required by Table 1617.6.2, the frame shall be continuous to the foundation.

❖ This section specifies when special moment frames in the upper parts of a building can be supported on a stiffer system in the lower portions, and when such frames must be continuous all the way down to the foundation.

SECTION 1618
DYNAMIC ANALYSIS PROCEDURE FOR THE SEISMIC DESIGN OF BUILDINGS

1618.1 Dynamic analysis procedures. The following dynamic analysis procedures are permitted to be used in lieu of the equivalent lateral force procedure of Section 1617.4:

1. Modal Response Spectral Analysis.
2. Linear Time-history Analysis.
3. Nonlinear Time-history Analysis.

The dynamic analysis procedures listed above shall be performed in accordance with the requirements of Sections 9.5.6, 9.5.7 and 9.5.8, respectively, of ASCE 7.

❖ See Section 1613 for introductory commentary regarding earthquake provisions. This section requires use of the ASCE 7 provisions for dynamic analysis procedures.

Analysis procedures are addressed in Section 1616.6. Since it is a more rigorous analysis method, a dynamic analysis is always acceptable for design purposes under the code. Static procedures are allowed for structures assigned to the higher seismic design categories only under certain restrictions of regularity and height.

The code formally recognizes three dynamic analysis procedures: (1) modal analysis, (2) elastic time-history analysis and (3) inelastic time-history analysis. The reader should refer to the commentary to the *NEHRP Recommended Provisions for the Development of Seismic Regulations for New Buildings* and to the 1999 *SEAOC Blue Book Commentary* (Section 106.6) for in-depth discussion of the provisions of Section 1618.

SECTION 1619
EARTHQUAKE LOADS
SOIL-STRUCTURE INTERACTION EFFECTS

1619.1 Analysis procedure. If soil-structure interaction is considered in the determination of seismic design forces and corresponding displacements in the structure, the procedure given in Section 9.5.9 of ASCE 7 shall be used.

❖ See Section 1613 for introductory commentary regarding earthquake provisions. This section requires use of the ASCE 7 provisions for soil-structure interaction, if it is to be considered; however, use of soil-structure interaction is optional at the discretion of the engineer.

Soil-structure interaction analysis considers the dissipation of vibrational energy from the supporting soil. Additionally, the structure-rocking effect is considered because of the elastic nature of the supporting soil. This type of analysis is permitted because it is capable of predicting the seismic response of a structure more accurately than the analysis procedures in other sections of the code (see ASCE 7 for details of this analysis procedure).

SECTION 1620
EARTHQUAKE LOADS—DESIGN, DETAILING REQUIREMENTS AND STRUCTURAL COMPONENT LOAD EFFECTS

1620.1 Structural component design and detailing. The design and detailing of the components of the seismic-force-resisting system shall comply with the requirements of Section 9.5.2.6 of ASCE 7 in addition to the nonseismic requirements of this code except as modified in Sections 1620.1.1, 1620.1.2 and 1620.1.3.

> **Exception:** For structures designed using the simplified analysis procedure in Section 1617.5, the provisions of Sections 1620.2 through 1620.5 shall be used.

❖ See Section 1613 for introductory commentary regarding earthquake provisions. This section requires use of the ASCE 7 provisions for seismic detailing requirements. Note that the referenced provisions include requirements for Seismic Design Category A, most of which are comparable to the requirements in Section 1616.4. Wherever the ASCE 7 Seismic Design Category A requirements differ from those of Section 1616.4, a conflict exists. Where that is the case, Section 102.4 requires that the code provision governs; thus, Section 1616.4 would apply.

1620.1.1 ASCE 7, Section 9.5.2.6.2.5. Section 9.5.2.6.2.5 of ASCE 7 shall not apply.

❖ This modification specifically excludes Section 9.5.2.6.2.5 of ASCE 7, which requires that the design considers the potentially adverse effect that the failure of a single member of the seismic-force-resisting system may have on the stability of the structure. The reason for not making this provision applicable is that it is deemed to be unenforceable.

1620.1.2 ASCE 7, Section 9.5.2.6.2.11. Modify ASCE 7, Section 9.5.2.6.2.11, to read as follows:

9.5.2.6.2.11 Elements supporting discontinuous walls or frames. Columns, beams, trusses or slabs supporting discontinuous walls or frames of structures and the connections of the discontinuous element to the supporting member having plan irregularity Type 4 of Table 9.5.2.3.2 or vertical irregularity Type 4 of Table 9.5.2.3.3 shall have the design strength to resist the maximum axial force that can develop in accordance with the special seismic loads of Section 9.5.2.7.1.

Exceptions:

1. The quantity *E* in Section 9.5.2.7.1 need not exceed the maximum force that can be transmitted to the element by the lateral-force-resisting system at yield.
2. Concrete slabs supporting light-framed walls.

❖ This modified ASCE 7 provision corresponds to Section 1620.2.9, and the commentary to that section provides additional explanation of these requirements.

The first modification clarifies that the strength requirement is to be met not only by the members that support the discontinuous system, but to the connections as well. This addresses a potential weak link at the connection of the discontinuous system to the supporting member.

The second modification retains the two exceptions

that were included in the first edition of the code (see 2000 IBC Section 1620.1.9).

1620.1.3 ASCE 7, Section 9.5.2.6.3. Modify ASCE 7, Section 9.5.2.6.3, to read as follows:

9.5.2.6.3 Seismic Design Category C. Structures assigned to Category C shall conform to the requirements of Section 9.5.2.6.2 for Category B and to the requirements of this section. Structures that have plan structural irregularity Type 1a or 1b of Table 9.5.2.3.2 along both principal plan axes, or plan structural irregularity Type 5 of Table 9.5.2.3.2, shall be analyzed for seismic forces in compliance with Section 9.5.2.5.2.2. When the square root of the sum of the squares method of combining directional effects is used, each term computed shall be assigned the sign that will yield the most conservative result.

The orthogonal combination procedure of Section 9.5.2.5.2.2, Item a, shall be required for any column or wall that forms part of two or more intersecting seismic-force-resisting systems and is subjected to axial load due to seismic forces acting along either principal plan axis equaling or exceeding 20 percent of the axial load design strength of the column or wall.

❖ This modified ASCE 7 provision corresponds to Section 1620.3, and the commentary to that section provides additional explanation of these requirements.

The modification made to ASCE 7 retains provisions for direction of seismic load that were included in the first edition of the code (see 2000 IBC Section 1620.2.2). Section 1620.3.2 is comparable and the accompanying commentary is applicable with respect to this modification.

1620.2 Structural component design and detailing (for use in the simplified analysis procedure of Section 1617.5). The design and detailing of the components of the seismic-force-resisting system for structures designed using the simplified analysis procedure in Section 1617.5 shall comply with the requirements of Sections 1620.2 through 1620.5 in addition to the nonseismic requirements of this code. Buildings shall not exceed the limitations of Section 1616.6.1.

Exception: Structures assigned to Seismic Design Category A.

Structures assigned to Seismic Design Category B (see Section 1616) shall conform to Sections 1620.2.1 through 1620.2.10.

❖ Sections 1620.2 through 1620.5 contain the structural element strength and detailing requirements for seismic-force-resisting systems that apply to those structures that utilize the simplified analysis procedure of Section 1617.5. These provisions are necessary for the strength and inelastic deformability of a structure to withstand the time-varying ground motion during an earthquake.

Seismic Design Category A structures have no special material or structural system requirements other than those of Section 1616.4 because of their relatively low seismic risk.

Sections 1620.2.1 through 1620.2.10 contain special detailing requirements for Seismic Design Category B structures.

1620.2.1 Second-order load effects. Where θ exceeds 0.10 as determined in Section 9.5.5.7.2 in ASCE 7, second-order load effects shall be included in the evaluation of component and connection strengths.

❖ P-delta effects must be taken into account in the design of components and connections, whenever the secondary-to-primary moment ratio exceeds 10 percent. P-delta effects tend to be more important for structures assigned to lower, rather than higher, seismic design categories, because lower seismic design category structures are typically more flexible.

1620.2.2 Openings. Where openings occur in shear walls, diaphragms or other plate-type elements, reinforcement at the edges of the openings shall be designed to transfer the stresses into the structure. The edge reinforcement shall extend into the body of the wall or diaphragm a distance sufficient to develop the force in the reinforcement.

❖ Stresses along the edges of openings must be transferred to the main body of a structural element, so as to maintain the continuity of load path for seismic forces.

1620.2.3 Discontinuities in vertical system. Structures with a discontinuity in lateral capacity, vertical irregularity Type 5, as defined in Table 1616.5.1.2, shall not be over two stories or 30 feet (9144 mm) in height where the "weak" story has a calculated strength of less than 65 percent of the story above.

Exception: Where the "weak" story is capable of resisting a total seismic force equal to the overstrength factor, Ω_0, as given in Table 1617.6.2, multiplied by the design force prescribed in Section 1617.5, the height limitation does not apply.

❖ The height limitation of two stories or 30 feet (9144 mm) for "weak" story structures is based on the poor performance of such structures in the past. The exception provides an alternative lateral strength design method, which allows a modified weak story structure of greater height.

1620.2.4 Connections. All parts of the structure, except at separation joints, shall be interconnected and the connections shall be designed to resist the seismic force, F_p, induced by the parts being connected. Any smaller portion of the structure shall be tied to the remainder of the structure for the greater of:

$$F_p = 0.133\,S_{DS}\,w_p \qquad \textbf{(Equation 16-58)}$$

or

$$F_p = 0.05\,w_p \qquad \textbf{(Equation 16-59)}$$

where:

S_{DS} = The design, 5-percent damped, spectral response acceleration at short periods as defined in Section 1615.

w_p = The weight of the smaller portion.

A positive connection for resisting a horizontal force acting parallel to the member shall be provided for each beam, girder or truss to its support for a force not less than 5 percent of the dead plus live load reaction.

❖ This section requires all parts of a structure (except at separation joints) to be interconnected so that the building is tied together to act as a unit. As a minimum, any smaller portion of the building must be tied to the remainder of the building with elements that are capable of resisting the larger of (0.133 S_{DS}) or 0.05 times the weight of the smaller portion. A similar requirement for Seismic Design Catagory A structures can be found in Section 1616.4.2.

Additionally, it is required that a positive connection for resisting a horizontal force acting parallel to the member be provided for each beam, girder or truss, which cannot be less the 5 percent of the dead plus live load reaction. This same requirement is found for Seismic Design Category A structures in Section 1616.4.

1620.2.5 Diaphragms. Permissible deflection shall be that deflection up to which the diaphragm and any attached distributing or resisting element will maintain its structural integrity under design load conditions, such that the resisting element will continue to support design loads without danger to occupants of the structure.

Floor and roof diaphragms shall be designed to resist F_p as follows:

$$F_p = 0.2 I_E S_{DS} w_p + V_{px} \quad \textbf{(Equation 16-60)}$$

where:

F_p = The seismic force induced by the parts.

I_E = Occupancy importance factor (Table 1604.5).

S_{DS} = The short-period site design spectral response acceleration coefficient (Section 1615).

w_p = The weight of the diaphragm and other elements of the structure attached to the diaphragm.

V_{px} = The portion of the seismic shear force at the level of the diaphragm, required to be transferred to the components of the vertical seismic-force-resisting system because of the offsets or changes in stiffness of the vertical components above or below the diaphragm.

Diaphragms shall provide for both shear and bending stresses resulting from these forces. Diaphragms shall have ties or struts to distribute the wall anchorage forces into the diaphragm. Diaphragm connections shall be positive, mechanical or welded-type connections.

❖ This section contains the diaphragm design requirements necessary for lateral seismic force transfer for load path continuity. The connections between the diaphragm and supported or supporting elements are required to be positive, mechanical or welded-type connections because other connections, such as friction type, do not perform well under dynamic loads.

The floors and roof of a building are designed primarily to function as structural elements that support gravity loads and transfer these loads to the building foundation by means of columns or walls. Floors and roofs, however, are also designed as structural elements (diaphragms) that distribute horizontal (or in-plane) forces due to wind or earthquakes to vertical lateral-force-resisting systems, such as moment-resisting frames, braced frames and shear walls. These, in turn, transmit the forces to the foundation of the structure.

As a simple analogy, a diaphragm may be regarded as a horizontal beam composed of a web and flanges. The web, which is the shear-resisting element, is represented by the structural floor or roof deck. The flanges are the chords or boundary elements of the floor or roof deck that serve to resist the axial tension or compression resulting from flexural deformations. The horizontal loads are distributed (by beam action) to the diaphragm supports (the vertical-resisting system). One-directional diaphragm action is shown in Figure 1620.2.5(1).

Essentially, this section requires that the floor or roof diaphragm at every level of the structure be designed to span horizontally between vertical elements of the lateral-force-resisting system and to transfer the calculated lateral force to these elements. Where a diaphragm is required to transfer the shears from other vertical lateral-force-resisting elements above the diaphragm to vertical lateral-force-resisting elements below the diaphragm (such as in the case of an out-of-plane offset in a vertical lateral-force-resisting element along the height of a structure), the resulting forces are to be added to the lateral force. This condition is illustrated in Figure 1620.2.5(2).

In addition to designing the diaphragms for the load combinations set forth in Section 1605, this section requires that the diaphragm be designed for the load determined in accordance with Equation 16-60, which is independent of the *R*-factor.To reinforce a wood diaphragm, the code permits the use of subdiaphragms to transfer and distribute local forces to the primary diaphragm struts and the main diaphragm. Subdiaphragms usually constitute an integral portion of a wood diaphragm system. Figure 1620.2.5(3) shows typical subdiaphragm construction.

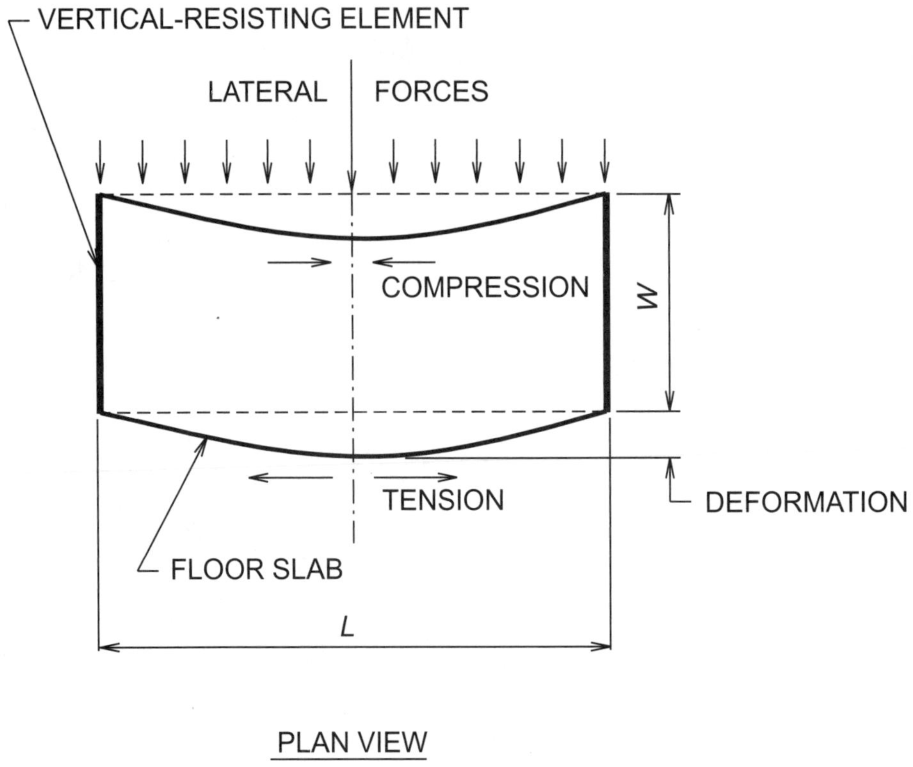

Figure 1620.2.5(1)
DIAPHRAGM ACTION

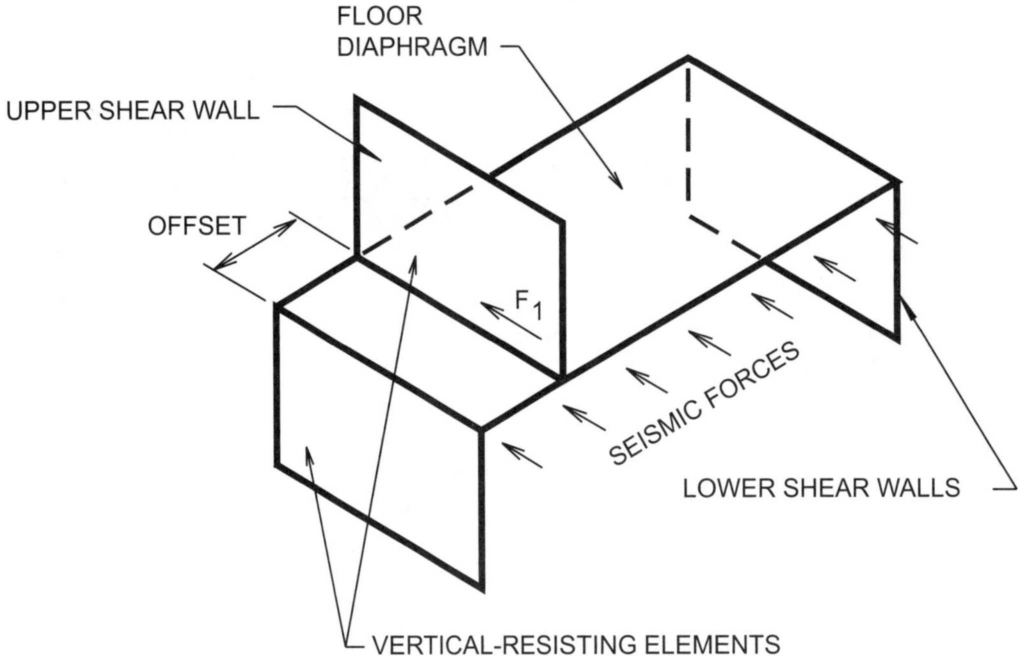

THE FLOOR DIAPHRAGM MUST TRANSFER THE UPPER SHEAR WALL FORCE F_1 TO THE LOWER SHEAR WALLS

Figure 1620.2.5(2)
EXAMPLE OF OFFSET ADDING TO DIAPHRAGM FORCE

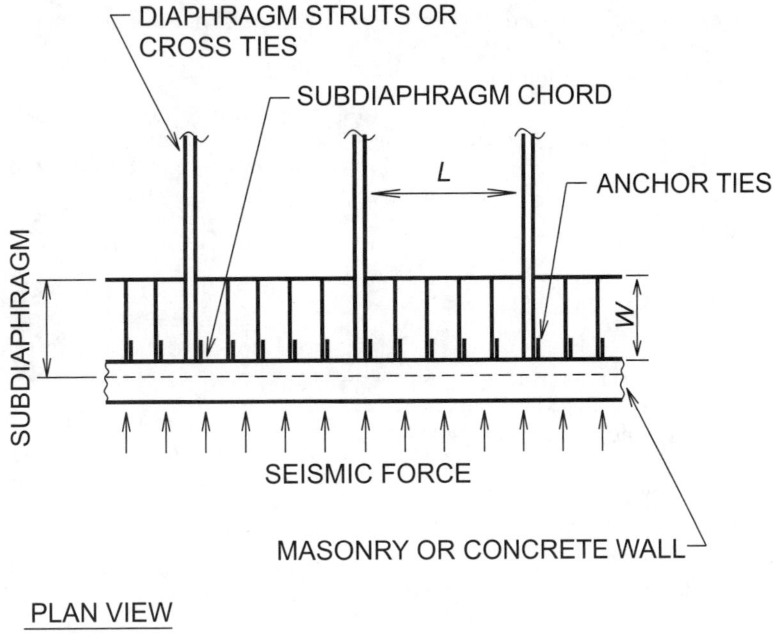

Figure 1620.2.5(3)
WOOD SUBDIAPHRAGM

1620.2.6 Collector elements. Collector elements shall be provided that are capable of transferring the seismic forces originating in other portions of the structure to the element providing the resistance to those forces. Collector elements, splices and their connections to resisting elements shall have the design strength to resist the special load combinations of Section 1605.4.

> **Exception:** In structures or portions thereof braced entirely by light-framed shear walls, collector elements, splices and connections to resisting elements need only have the strength to resist the load combinations of Section 1605.2 or 1605.3.

❖ Collectors are defined in Section 1613.1. Collectors typically transfer the seismic force from the diaphragm to the vertical lateral-force-resisting element. These structural elements must have the strength to transfer their seismic forces to the supporting structure for load path continuity. An example of a collector element is shown in Figure 1620.2.6. The design of the collector and its connections is critical so that seismic forces are transferred into the designated resisting elements. In order to guarantee a continuous load path, this section requires that collector elements, splices and their connections in other than light-frame construction be designed for the following seismic load combinations (see Section 1605.4):

$1.2D + f_{1L} + E_m$

$0.9D + E_m$

These load combinations are intended to reflect the actual maximum force the collector is likely to experience in the design earthquake (see Section 1605.4 for further discussion).

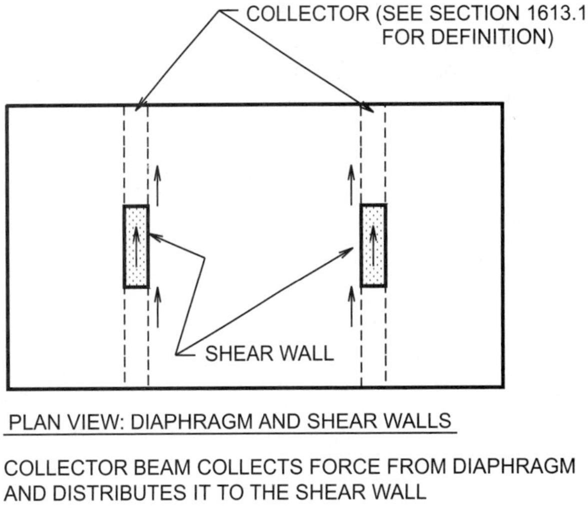

Figure 1620.2.6
COLLECTORS AND TIE BEAMS

1620.2.7 Bearing walls and shear walls. Bearing walls and shear walls and their anchorage shall be designed for an out-of-plane force, F_p, that is the greater of 10 percent of the weight of the wall, or the quantity given by Equation 16-61:

$$F_p = 0.40 I_E S_{DS} w_w \qquad \textbf{(Equation 16-61)}$$

where:

I_E = Occupancy importance factor (Table 1604.5).

S_{DS} = The short-period site design spectral response acceleration coefficient (Section 1615.1.3 or 1615.2.5).

w_w = The weight of the wall.

In addition, concrete and masonry walls shall be anchored to the roof and floors and members that provide lateral support for the wall or that are supported by the wall. The anchorage shall provide a direct connection between the wall and the supporting construction capable of resisting the greater of the force, F_p, as given by Equation 16-61 or (400 $S_{DS} I_E$) pounds per linear foot of wall. For SI: 5838 $S_{DS} I_E$ N/m. Walls shall be designed to resist bending between anchors where the anchor spacing exceeds 4 feet (1219 mm). Parapets shall conform to the requirements of Section 9.6.2.2 of ASCE 7.

❖ This section defines required out-of-plane design forces for load-bearing walls and specifies connections.

Heavy elements, such as concrete or masonry walls, may develop significant out-of-plane forces. Accordingly, when a diaphragm is designed to provide lateral support to either concrete or masonry walls, it must have continuous ties between diaphragm chords, as well as positive anchorage between the diaphragm and the walls in order to properly distribute the anchorage forces into the diaphragm. In the past, flexible diaphragms, such as those made of wood, have been inadequately tied to rigid walls of concrete or masonry and, as a result, suffered extensive damage in severe earthquakes.

The out-of-plane design force needs to be equal to the largest of the following:

1. $0.10w_w$
2. $0.40 I_E S_{DS} w_w$
3. 400 $S_{DS} I_E$ lbs/ft (5838 $S_{DS} I_E$ N/m)
4. 280 pounds per foot (4.10 kN/m) in accordance with Section 1604.8.2

The anchorage connector is typically spaced no more than 4 feet (1219 mm) on center so that the walls need not be designed to resist bending between anchors.

1620.2.8 Inverted pendulum-type structures. Supporting columns or piers of inverted pendulum-type structures shall be designed for the bending moment calculated at the base determined using the procedures given in Section 1617.4 and varying uniformly to a moment at the top equal to one-half the calculated bending moment at the base.

❖ Past frequent failures of the upper column section have resulted in requiring the design of the top of the column for 50 percent of the moment at the base. The commentary to the 1997 NEHRP Provisions describes the procedure to comply as follows:

> The bending moments due to the lateral force are first calculated for the base of the column...One-half of the calculated bending moment at the base is applied at the top and the moments along the column are varied from 1.5M at the base to 0.5M at the top.

The above explanation is a slight variance with the code language in Section 1620.2.8. Figure 1620.2.8 illustrates the procedure described in the commentary to the 1997 NEHRP Recommended Provisions.

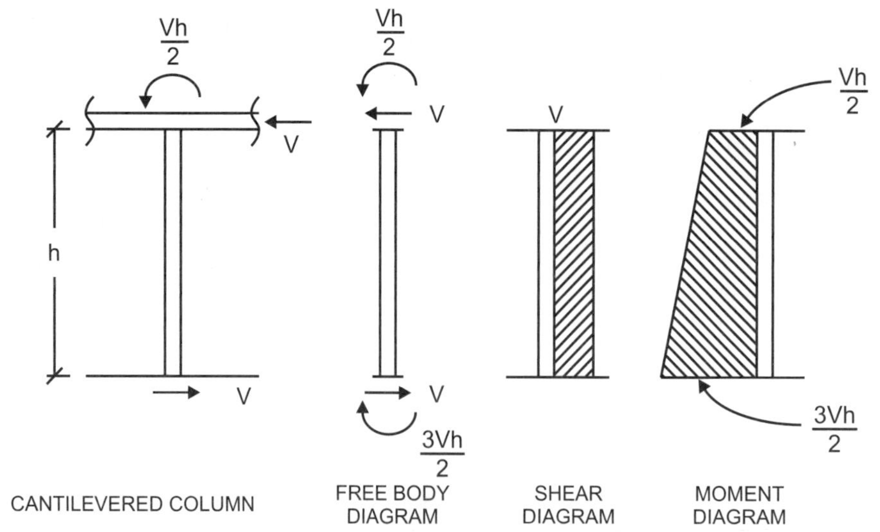

Figure 1620.2.8
ANALYSIS OF INVERTED PENDULUM-TYPE STRUCTURES

1620.2.9 Elements supporting discontinuous walls or frames. Columns or other elements subject to vertical reactions from discontinuous walls or frames of structures having plan irregularity Type 4 of Table 1616.5.1.1 or vertical irregularity Type 4 of Table 1616.5.1.2 shall have the design strength to resist special seismic load combinations of Section 1605.4. The connections from the discontinuous walls or frames to the supporting elements need not have the design strength to resist the special seismic load combinations of Section 1605.4.

Exceptions:

1. The quantity, E_m, in Section 1617.1.1.2 need not exceed the maximum force that can be transmitted to the element by the lateral-force-resisting system at yield.
2. Concrete slabs supporting light-framed walls.

❖ Figure 1620.2.9 illustrates situations where discontinuous shear walls occur. In these situations, it is critical that the elements supporting these shear walls are not damaged as a result of an earthquake, as this would jeopardize the vertical stability of the structure.

In order to guarantee that the supporting elements for the discontinuous walls or frames are adequate, this section requires that the supporting elements be designed for the following seismic load combinations (see Section 1605.4):

$1.2D + f_{1L} + E_m$

$0.9D + E_m$

These load combinations are intended to reflect the maximum loads the supporting elements are likely to experience in the design earthquake (see Section 1605.4 for further discussion).

1620.2.10 Direction of seismic load. The direction of application of seismic forces used in design shall be that which will produce the most critical load effect in each component. The requirement will be deemed satisfied if the design seismic forces are applied separately and independently in each of the two orthogonal directions.

❖ This section requires that seismic forces be applied in a direction that will produce the most critical load effect in each component. This requirement is deemed to be satisfied if the design seismic forces are applied separately and independently in each of the two orthogonal directions.

1620.3 Seismic Design Category C. Structures assigned to Seismic Design Category C (see Section 1616) shall conform to the requirements of Section 1620.2 for Seismic Design Category B and to Sections 1620.3.1 through 1620.3.2.

❖ Section 1620.3 contains special detailing requirements for Seismic Design Category C structures that are in addition to the requirements for Seismic Design Category B structures. In other words, a Seismic Design Category C structure needs to comply with the requirements in Sections 1620.2 and 1620.3.

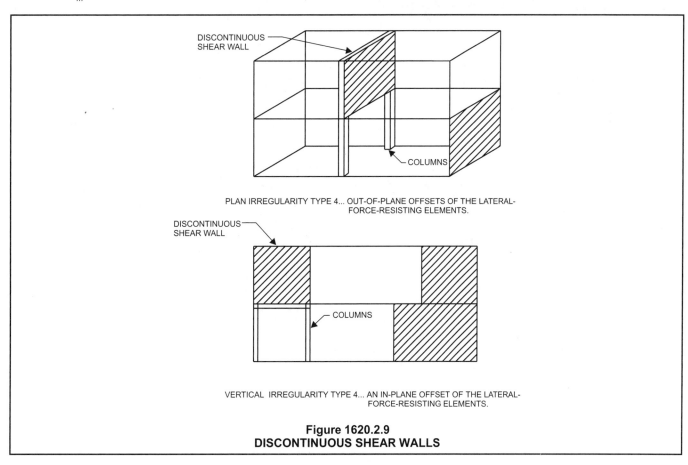

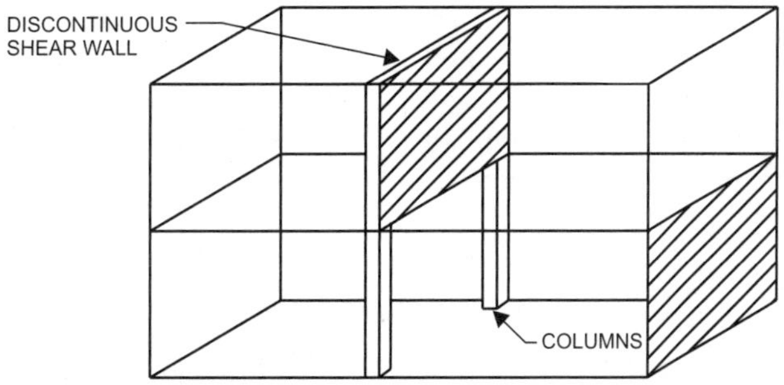

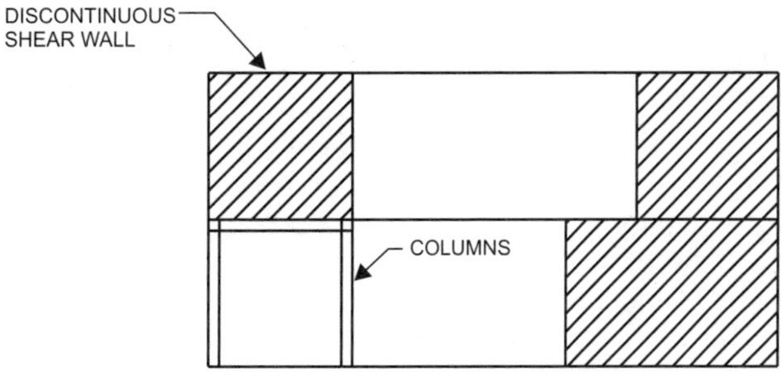

Figure 1620.2.9
DISCONTINUOUS SHEAR WALLS

1620.3.1 Anchorage of concrete or masonry walls. Concrete or masonry walls shall be anchored to floors and roofs and members that provide out-of-plane lateral support for the wall or that are supported by the wall. The anchorage shall provide a positive direct connection between the wall and floor or roof capable of resisting the horizontal forces specified in Equation 16-62 for structures with flexible diaphragms or in Section 9.6.1.3 of ASCE 7 (using a_p of 1.0 and R_p of 2.5) for structures with diaphragms that are not flexible.

$$F_p = 0.8\, S_{DS} I_E w_w \qquad \textbf{(Equation 16-62)}$$

where:

F_p = The design force in the individual anchors.

I_E = Occupancy importance factor in accordance with Section 1616.2.

S_{DS} = The design earthquake spectral response acceleration at short period in accordance with Section 1615.1.3.

w_w = The weight of the wall tributary to the anchor.

Diaphragms shall be provided with continuous ties or struts between diaphragm chords to distribute these anchorage forces into the diaphragms. Where added chords are used to form subdiaphragms, such chords shall transmit the anchorage forces to the main cross ties. The maximum length-to-width ratio of the structural subdiaphragm shall be $2^1/_2$ to 1. Connections and anchorages capable of resisting the prescribed forces shall be provided between the diaphragm and the attached components. Connections shall extend into the diaphragms a sufficient distance to develop the force transferred into the diaphragm.

The strength design forces for steel elements of the wall anchorage system shall be 1.4 times the force otherwise required by this section.

In wood diaphragms, the continuous ties shall be in addition to the diaphragm sheathing. Anchorage shall not be accomplished by use of toenails or nails subject to withdrawal, nor shall wood ledgers or framing be used in cross-grain bending or cross-grain tension. The diaphragm sheathing shall not be considered effective as providing the ties or struts required by this section.

In metal deck diaphragms, the metal deck shall not be used as the continuous ties required by this section in the direction perpendicular to the deck span.

Diaphragm-to-wall anchorage using embedded straps shall be attached to or hooked around the reinforcing steel or otherwise terminated so as to directly transfer force to the reinforcing steel.

❖ This provision is similar to Section 1620.2.7, except that more demands are placed on the anchorage connection. First, the force the connection needs to be designed for is increased to 0.8 $S_{DS}I_E w_w$ for structures with flexible diaphragms (see definition of "Diaphragm, flexible"). For structures with rigid diaphragms (see definition for "Diaphragm, rigid"), F_p is determined in accordance with Section 9.6.1.3 of ASCE 7, which includes a nonmandatory maximum and a mandatory minimum.

A restriction is placed on the maximum length-to-width ratio of the subdiaphragm [see Figure 1620.2.5(3)]. Such a maximum length-to-width ratio results in reasonable spacing between continuous ties.

This section also contains detailed requirements concerning:

1. Continuous ties between diaphragm chords.
2. Wood diaphragms.
3. Metal deck diaphragms.
4. Diaphragm-to-wall anchorage using embedded straps.

The anchorage to flexible diaphragm requirements have evolved over the years based largely on observation of performance of tilt-up concrete buildings in earthquakes.

1620.3.2 Direction of seismic load. For structures that have plan structural irregularity Type 1a or 1b of Table 1616.5.1.1 along both principal plan axes, or plan structural irregularity Type 5 in Table 1616.5.1.1, the critical direction requirement of Section 1620.2.10 shall be deemed satisfied if components and their foundations are designed for the following orthogonal combination of prescribed loads.

One hundred percent of the forces for one direction plus 30 percent of the forces for the perpendicular direction. The combination requiring the maximum component strength shall be used. Alternatively, the effects of the two orthogonal directions are permitted to be combined on a square root of the sum of the squares (SRSS) basis. When the SRSS method of combining directional effects is used, each term computed shall be assigned the sign that will result in the most conservative result.

The orthogonal combination procedure above shall be required for any column or wall that forms part of two or more intersecting seismic-force-resisting systems and is subjected to axial load due to seismic forces acting along either principal plan axis equaling or exceeding 20 percent of the axial load design strength of the column or wall.

❖ Earthquake forces act in all directions; however, it is not likely that the maximum effect will occur simultaneously along both principal axes of the building. Therefore, the code prescribes how to combine the earthquake effects acting along the principal axes of a building. Orthogonal effects may be significant for columns or other vertical members that participate in resisting earthquake forces in both principal directions of a building.

The orthogonal combinations specified in this section account for the seismic effect on columns and walls meeting the specified criteria as well as on a structure that has nonparallel vertical lateral-force-resisting systems (plan structural irregularity Type 5 in Table 1616.5.1.1), such as skewed braced frames, shear walls or moment frames. For an example of a structure that has nonparallel vertical bracing, see Figure

1616.5.1(2). An example of the required seismic load cases is as follows:

Seismic Load Case 1	100% East-West Seismic Force Effect + 30% North-South Seismic Force Effect
Seismic Load Case 2	30% East-West Seismic Force Effect + 100% North-South Seismic Force Effect

Alternatively, a method called the "square root of the sum of the squares" can be used where the force effects in the two orthogonal directions are squared, added and then the square root of the sum is taken.

1620.4 Seismic Design Category D. Structures assigned to Seismic Design Category D shall conform to the requirements of Section 1620.3 for Seismic Design Category C and to Sections 1620.4.1 through 1620.4.6.

❖ This section adds another layer of strength and detailing requirements for Seismic Design Category D structures.

1620.4.1 Plan or vertical irregularities. For buildings having a plan structural irregularity of Type 1a, 1b, 2, 3 or 4 in Table 1616.5.1.1 or a vertical structural irregularity of Type 4 in Table 1616.5.1.2, the design forces determined from Section 1617.5 shall be increased 25 percent for connections of diaphragms to vertical elements and to collectors, and for connections of collectors to the vertical elements.

Exception: When connection design forces are determined using the special seismic load combinations of Section 1605.4

❖ The increase in the connection design forces required in this section (connections of diaphragm to vertical elements and to collectors, and connections of collectors to vertical elements) accounts for the random vibratory motion of irregular structures. Historically, the connections of the seismic-force-resisting system have failed in irregular structures during a major earthquake; thus, the design forces for connections are to be increased by 25 percent.

1620.4.2 Vertical seismic forces. In addition to the applicable load combinations of Section 1605, horizontal cantilever and horizontal prestressed components shall be designed to resist a minimum net upward force of 0.2 times the dead load.

❖ This section provides the necessary uplift design requirements for certain structural members. Horizontal prestressed concrete components, such as double tees and hollow core floor members, may be overstressed by the uplift effect of an earthquake unless designed in accordance with this section.

1620.4.3 Diaphragms. Floor and roof diaphragms shall be designed to resist design seismic forces determined in accordance with Equation 16-63 as follows:

$$F_{px} = \frac{\sum_{i=x}^{n} F_i}{\sum_{i=x}^{n} w_i} w_{px} \qquad \textbf{(Equation 16-63)}$$

where:

F_i = The design force applied to Level i.

F_{px} = The diaphragm design force.

w_i = The weight tributary to Level i.

w_{px} = The weight tributary to the diaphragm at Level x.

The force determined from Equation 16-63 need not exceed $0.4S_{DS}I_E w_{px}$ but shall not be less than $0.2S_{DS}I_E w_{px}$ where S_{DS} is the design spectral response acceleration at short period determined in Section 1615.1.3 and I_E is the occupancy importance factor determined in Section 1616.2. When the diaphragm is required to transfer design seismic force from the vertical-resisting elements above the diaphragm to other vertical-resisting elements below the diaphragm due to offsets in the placement of the elements or to changes in relative lateral stiffness in the vertical elements, these forces shall be added to those determined from Equation 16-63 and to the upper and lower limits on that equation.

❖ For discussion on the mechanics of diaphragms, see Section 1620.2.5. In addition to designing the diaphragms for the load combinations set forth in Section 1605, this section requires that the diaphragm be designed for the load determined in accordance with Equation 16-63 with a maximum and minimum check. The diaphragm design force given by Equation 16-62 for Seismic Design Category C buildings is changed by Equation 16-63 for buildings assigned to Seismic Design Category D and higher. The V_{px} force of Equation 16-60 is to be added to the right side of Equation 16-63 when the circumstances are similar.

1620.4.4 Collector elements. Collector elements shall be provided that are capable of transferring the seismic forces originating in other portions of the structure to the element providing resistance to those forces.

Collector elements, splices and their connections to resisting elements shall resist the forces determined in accordance with Equation 16-63. In addition, collector elements, splices and their connections to resisting elements shall have the design strength to resist the earthquake loads as defined in the special load combinations of Section 1605.4.

Exception: In structures, or portions thereof, braced entirely by light-framed shear walls, collector elements, splices and their connections to resisting elements need only be designed to resist forces in accordance with Equation 16-63.

❖ See Section 1620.2.6 for discussion on collector element requirements. In addition, the collector and its connections are required to be designed for the diaphragm forces determined by Equation 16-63.

1620.4.5 Building separations. All structures shall be separated from adjoining structures. Separations shall allow for the displacement δ_M. Adjacent buildings on the same property shall be separated by at least δ_{MT} where

$$\delta_{MT} = \sqrt{(\delta_{M1})^2 + (\delta_{M2})^2} \qquad \textbf{(Equation 16-64)}$$

and δ_{M1} and δ_{M2} are the displacements of the adjacent buildings.

When a structure adjoins a property line not common to a public way, that structure shall also be set back from the property line by at least the displacement, δ_M, of that structure.

Exception: Smaller separations or property line setbacks shall be permitted when justified by rational analyses based on maximum expected ground motions.

❖ Buildings need to be able to respond to earthquake ground motion independently of each other. In order to prevent buildings from pounding against each other during an earthquake, they need to be adequately separated. The required separation distance is the statistical (square root of the sum of the squares) sum of the estimated design earthquake displacements (C_d times the elastically computed displacements under the specified seismic design forces) of the adjoining buildings.

1620.4.6 Anchorage of concrete or masonry walls to flexible diaphragms. In addition to the requirements of Section 1620.3.1, concrete and masonry walls shall be anchored to flexible diaphragms based on the following:

1. When elements of the wall anchorage system are not loaded concentrically or are not perpendicular to the wall, the system shall be designed to resist all components of the forces induced by the eccentricity.
2. When pilasters are present in the wall, the anchorage force at the pilasters shall be calculated considering the additional load transferred from the wall panels to the pilasters. The minimum anchorage at a floor or roof shall not be less than that specified in Item 1.

❖ This requirement is based on a similar provision that was added to the *Uniform Building Code* after the Northridge earthquake. Concrete or masonry walls are typically anchored to diaphragms with metal strap ties that are embedded in the wall and attached to wood roof framing using nails or bolts. Post-earthquake evaluation of buildings cited many examples of failed wall anchors that were compliant with the version of the code in effect at that time.

Item 1 states the obvious, with respect to accounting for eccentricities in connection design, but it is included because it appeared to be a contributing factor to anchorage failures.

Item 2 requires the anchorage force at pilasters to be calculated, considering that the wall panels between the pilasters are supported on all four sides. It is customary to assume that the wall is simply supported at the ground and roof. The presence of pilasters necessitates that their stiffening effect be considered; thus, anchorage forces at pilasters are specifically required to include the additional forces concentrated at that location.

1620.5 Seismic Design Category E or F. Structures assigned to Seismic Design Category E or F (Section 1616) shall conform to the requirements of Section 1620.4 for Seismic Design Category D and to Section 1620.5.1.

❖ This section sets forth additional requirements for structures assigned to Seismic Design Category E or F (note that buildings meeting the limitations of Section 1616.6.1 could not be classified as Seismic Design Category F).

1620.5.1 Plan or vertical irregularities. Structures having plan irregularity Type 1b of Table 1616.5.1.1 or vertical irregularities Type 1b or 5 of Table 1616.5.1.2 shall not be permitted.

❖ Structures with the following irregularities are not permitted in Seismic Design Categories E and F:

IRREGULARITY	TYPE
Plan Structural Irregularity	Extreme Torsional Irregularity
Vertical Structural Irregularity	Stiffness Irregularity (Extreme Soft Story) Discontinuity in Capacity (Weak Story)

SECTION 1621 ARCHITECTURAL, MECHANICAL AND ELECTRICAL COMPONENT SEISMIC DESIGN REQUIREMENTS

1621.1 Component design. Architectural, mechanical, electrical and nonstructural systems, components and elements permanently attached to structures, including supporting structures and attachments (hereinafter referred to as "components"), and nonbuilding structures that are supported by other structures, shall meet the requirements of Section 9.6 of ASCE 7 except as modified in Sections 1621.1.1, 1621.1.2 and 1621.1.3, excluding Section 9.6.3.11.2, of ASCE 7, as amended in this section.

❖ Section 1621 requires that the provisions of ASCE 7 be used for nonstructural components. Those provisions provide detailed requirements for architectural, electrical and mechanical components, which include specific component descriptions and force factors. The failure of nonstructural components during a major earthquake has been a common occurrence in the past. The failure of partition walls, ceilings, pipe supports, exterior veneers, lighting fixtures, etc., are potential life safety hazards. The requirements referenced in Section 1621 are the minimum criteria for life safety. As a function of weight, the component design force (F_p) is higher than a comparative force used for the seismic-force-resisting system for the following reasons:

1. The accelerations acting on elements higher up within a building are greater than at the ground level because of dynamic response of the structure to earthquake ground motion.
2. If an element itself is not rigid, then an amplified dynamic element response occurs.
3. Some elements lack the energy-absorbing properties of a ductile structure.

Nonstructural components are required to comply with the force and displacement provisions of Section 1621. Factors considered in determining to what extent nonstructural components need to comply with this section are as follows:

1. Seismicity at the location.
2. Height of a component above floor level and whether or not occupants would be injured should the component fall.
3. Whether the component stores, uses or transports hazardous substances that could endanger the occupants should the component fail.
4. Whether or not the component's function is required to maintain life safety. For example, failure of suspended ceilings and light fixtures may render a required exit unusable.
5. How the failure of one component may affect the functionality of another component.

1621.1.1 ASCE 7, Section 9.6.3.11.2: Section 9.6.3.11.2 of ASCE 7 shall not apply.

❖ The modification of ASCE 7 to exclude Section 9.6.3.11.2 eliminates the provision that indicates NFPA 13 installations are deemed to comply with the requirements of Section 1621. The current edition of NFPA 13 does not meet the seismic design requirements in general and, therefore, cannot be considered a complying standard due to force and fastener capacity concerns. While NFPA 13 is required in Section 903.3.1.1 for the installation of sprinklers, it is not included to comply with the seismic provisions; therefore, it is necessary to verify that such installations meet the force and displacement requirements that are applicable to piping systems and that anchorages are properly designed. It is likely that future editions of NFPA 13 will be more in line with the earthquake requirements of the code.

1621.1.2 ASCE 7, Section 9.6.2.8.1. Modify ASCE 7, Section 9.6.2.8.1, to read as follows:

9.6.2.8.1 General. Partitions that are tied to the ceiling and all partitions greater than 6 feet (1829 mm) in height shall be laterally braced to the building structure. Such bracing shall be independent of any ceiling splay bracing. Bracing shall be spaced to limit horizontal deflection at the partition head to be compatible with ceiling deflection requirements as determined in Section 9.6.2.6 for suspended ceilings and Section 9.6.2.6 for other systems.

> **Exception:** Partitions not taller than 9 feet (2743 mm) when the horizontal seismic load does not exceed 5 psf (0.240 KN/m^2) required in Section 1607.13.

❖ The modification to Section 9.6.2.8.1 of ASCE 7 retains an exception that was part of the first edition of the code (see 2000 IBC Section 1621.2.7). This provision requires partitions of all types of construction to be laterally braced if they are tied to the ceiling or are greater than 6 feet (1829 mm) in height. The exception excludes partitions up to 9 feet (2743 mm) in height if the horizontal seismic force, F_p, does not exceed the 5 psf (0.240 kN/m^2) horizontal live load required in Section 1607.13.

1621.1.3 ASCE 7, Section 9.6.3.13. Modify ASCE 7, Section 9.6.3.13, to read as follows:

9.6.3.13 Mechanical equipment, attachments and supports. Attachments and supports for mechanical equipment not covered in Sections 9.6.3.8 through 9.6.3.12 or Section 9.6.3.16 shall be designed to meet the force and displacement provisions of Section 9.6.1.3 and 9.6.1.4 and the additional provisions of this section. In addition to their attachments and supports, such mechanical equipment designated as having an I_p = 1.5, which contains hazardous or flammable materials in quantities that exceed the maximum allowable quantities for an open system listed in Section 307, shall, itself, be designed to meet the force and displacement provisions of Sections 9.6.1.3 and 9.6.1.4 and the additional provisions of this section. The seismic design of mechanical equipment, attachments and their supports shall include analysis of the following: the dynamic effects of the equipment, its contents and, when appropriate, its supports. The interaction between the equipment and the supporting structures, including other mechanical and electrical equipment, shall also be considered.

❖ The modification of Section 9.6.3.13 of ASCE 7 retains a portion of this same provision that was part of the first edition of the code (see 2000 IBC Section 1621.3.12). In the case of mechanical equipment with an I_p equaling 1.0, only the attachments and supports need to be designed for seismic forces and displacements. Where I_p equals 1.5, the ASCE 7 provision requires the entire system to be designed. This includes the mechanical equipment itself, in addition to its supports and attachments. The modification made by the code is that the mechanical equipment contains hazardous or flammable materials in quantities that exceed the maximum allowable quantities for an open system in addition to having a component importance factor of 1.5.

SECTION 1622
NONBUILDING STRUCTURES SEISMIC DESIGN REQUIREMENTS

1622.1 Nonbuilding structures. The requirements of Section 9.14 of ASCE 7 shall apply to nonbuilding structures except as modified by Sections 1622.1.1, 1622.1.2 and 1622.1.3.

❖ See Section 1613 for introductory commentary regarding earthquake provisions. This section requires that the provisions of ASCE 7 be used for nonbuilding structures.

There are many structures supported on the ground that are not buildings, mainly because they do not have any occupants. Some examples of nonbuilding structures are tanks, bins and towers. Industry standards are permitted to be used for the seismic design of nonbuilding structures when approved by the building official.

Different base shear formulas are used depending upon whether or not the nonbuilding structure is:

1. Supported by grade or another structure.
2. Rigid or flexible as determined by the nonbuilding structure period.
3. Similar or not similar to that of a building.

Because there is such a wide range of nonbuilding structure types, the ASCE 7 provisions that are referenced by Section 1622 can only provide seismic design requirements in a general sense; thus, specific design standards unique to a nonbuilding structure type will commonly be used. It is important to identify whether or not the specific design standard is based on ASD or strength design. If the standard is based on ASD, the seismic forces determined in accordance with Section 1622 need to be reduced by a factor of 1.4 to make them compatible with ASD.

1622.1.1 ASCE 7, Section 9.14.5.1. Modify Section 9.14.5.1, Item 9, to read as follows:

9.Where an approved national standard provides a basis for the earthquake-resistant design of a particular type of nonbuilding structure covered by Section 9.14, such a standard shall not be used unless the following limitations are met:

1. The seismic force shall not be taken as less than 80 percent of that given by the remainder of Section 9.14.5.1.
2. The seismic ground acceleration, and seismic coefficient, shall be in conformance with the requirements of Sections 9.4.1 and 9.4.1.2.5, respectively.
3. The values for total lateral force and total base overturning moment used in design shall not be less than 80 percent of the base shear value and overturning moment, each adjusted for the effects of soil structure interaction that is obtained by using this standard.

❖ This modification of Section 9.14.5.1 of ASCE 7 retains a provision that was part of the first edition of the code (see 2000 IBC Section 1622.2.5).

This section of ASCE 7 establishes the general design criteria for nonbuilding structures. The modification is to the limitations that are applicable where an approved national standard is used as the basis for earthquake-resistant design of a nonbuilding structure. In this case, the additional limitation in Item 1 requires that the force be no less that 80 percent of the force that is otherwise required by this section.

1622.1.2 ASCE 7, Section 9.14.7.2.1. Modify Section 9.14.7.2.1 to read as follows:

9.14.7.2.1 General. This section applies to all earth-retaining walls. The applied seismic forces shall be determined in accordance with Section 9.7.5.1 with a geotechnical analysis prepared by a registered design professional.

The seismic use group shall be determined by the proximity of the retaining wall to other nonbuilding structures or buildings. If failure of the retaining wall would affect an adjacent structure, the seismic use group shall not be less than that of the adjacent structure, as determined in Section 9.1.3. Earth-retaining walls are permitted to be designed for seismic loads as either yielding or nonyielding walls. Cantilevered reinforced concrete retaining walls shall be assumed to be yielding walls and shall be designed as simple flexural wall elements.

❖ This modification of Section 9.14.7.2.1 of ASCE 7 retains a provision that was part of the first edition of the code (see 2000 IBC Section 1622.4.2).

A registered design professional is responsible for determining the seismic forces and design methodology for earth-retaining structures. The retaining wall is probably the most common type of earth-retaining structure. Retaining walls that are free to rotate are "yielding walls" and are designed for active soil pressures. Walls that are not free to rotate, such as walls surrounding the basement of a building, are considered "nonyielding walls" and are typically designed for at-rest pressures.

It should be noted that for Seismic Design Categories D, E and F, Section 1802.2.7 requires the soils investigation report to include a determination of lateral pressures on basement and retaining walls due to earthquake motions.

1622.1.3 ASCE 7, Section 9.14.7.9. Add a new Section 9.14.7.9 to read as follows:

9.14.7.9 Buried structures. As used in this section, the term "buried structures" means subgrade structures such as tanks, tunnels and pipes. Buried structures that are designated as Seismic Use Group II or III, as determined in Section 9.1.3, or are of such a size or length as to warrant special seismic design as determined by the registered design professional, shall be identified in the geotechnical report. Buried structures shall be designed to resist seismic lateral forces determined from a substantiated analysis using standards approved by the building official. Flexible couplings shall be provided for buried structures where changes in the support system, configurations or soil condition occur.

❖ This modification of ASCE 7 retains a provision that was part of the first edition of the code (see 2000 IBC Section 1622.4.8).

Buried structures can suffer earthquake damage, and such structures identified in this section need to be designed by a registered design professional.

SECTION 1623
SEISMICALLY ISOLATED STRUCTURES

1623.1 Design requirements. Every seismically isolated structure and every portion thereof shall be designed and constructed in accordance with the requirements of Section 9.13 of ASCE 7, except as modified in Section 1623.1.1.

❖ This section requires the use of Section 9.13 of ASCE 7 which provides design as well as construction requirements for base-isolated structures and portions thereof. Also see Section 1613 for introductory commentary regarding earthquake design requirements.

A seismically isolated structure is designed to keep most of the horizontal movement of the ground from being transmitted to the building. This can lead to a significant reduction in floor accelerations and interstory drifts, two problematic aspects of conventional (fixed-base) construction. High floor accelerations are often experienced in stiff buildings, and large interstory drifts are experienced in flexible structures. These two occurrences make it difficult to limit the damage that may be sustained by the building components and contents, which at times may be more valuable than the building itself. Also, in certain types of buildings such as hospitals, protection of the components and contents may be vital to uninterrupted functioning of the facility during and following an earthquake.

In seismic retrofit applications where conventional techniques are impractical because either the historical character of a building must be preserved or disruptions of occupancy and use must be avoided, isolation provides a viable alternative.

In a base-isolated structure, the columns are typically supported on individual base isolation bearings that usually have relatively low horizontal stiffness and relatively high vertical stiffness. The bearings usually behave nonlinearly with high equivalent viscous damping. The relatively low horizontal stiffness increases the period of the base-isolated building. The high damping contributed by the bearings increases the damping of the base-isolated structure. The longer period and increased damping decrease the acceleration response of the isolated building to earthquake ground motion. As the period shifts to larger values, however, the earthquake response displacements increase. The increased damping has a moderating effect on this increase. Almost all response displacement occurs between the base of the structure and the ground. This is because, with base isolation, the largest portion of the response occurs across the base isolation bearings. It is not uncommon for utility lines and entranceways to have to accommodate 6 to 24 inches (152 to 610 mm) of relative displacement in the design earthquake ground motion in regions of high seismicity. The high vertical stiffness of the bearings minimizes rocking and amplification of vertical response. Four types of bearing commonly available include: laminated, high-damping rubber bearings; laminated rubber bearings with lead core; steel sliding bearings using dished surfaces and coil spring systems. Laminated rubber bearings (high damping and lead core) are used most commonly. They consist of layered sheets of alternating rubber and steel plates vulcanized together. Typically square or round in plan view, these bearings range in width from 14 to 45 inches (356 to 1143 mm) and some contain a lead core to provide additional damping. While lateral flexibility is most desirable under high levels of seismic excitation, it is clearly not desirable to have a structural system vibrate perceptibly under low levels of excitation, such as caused by wind and minor earthquakes; thus, a means of providing rigidity under low levels of excitation is a necessary element of a seismic isolation system.

The stability of the vertical-load-carrying elements of the isolated system must be verified under the total maximum displacement, which is defined in Section 1613. That section, in fact, defines three different displacements: design displacement, total design displacement and total maximum displacement, which are illustrated in Figure 1623.1, which is a reproduction of Figure C13.3.3 of the commentary to the NEHRP Recommended Provisions, Part 2. These definitions are identical to those in Section 9.2.1 of ASCE 7. It is important to understand the distinctions among the three displacements. Maximum displacement is caused by the maximum considered earthquake, while design displacement is caused by the design basis earthquake. The word "total" implies that torsional displacements are included; the absence of the word indicates that torsional displacements are excluded.

1623.1.1 ASCE 7, Section 9.13.6.2.3. Modify ASCE 7, Section 9.13.6.2.3, to read as follows:

9.13.6.2.3 Fire resistance. Fire-resistance ratings for the isolation system shall comply with Section 714.7 of the *International Building Code.*

❖ This section provides a cross reference to the code requirements for the fire-resistance rating of structural members. These detailed requirements supercede the corresponding provisions in ASCE 7. The intent is that, in the event of a fire, the isolation system must be capable of supporting the weight of the building, as required for other vertical-load-carrying elements of the structure in accordance with the type of construction requirements in Table 601.

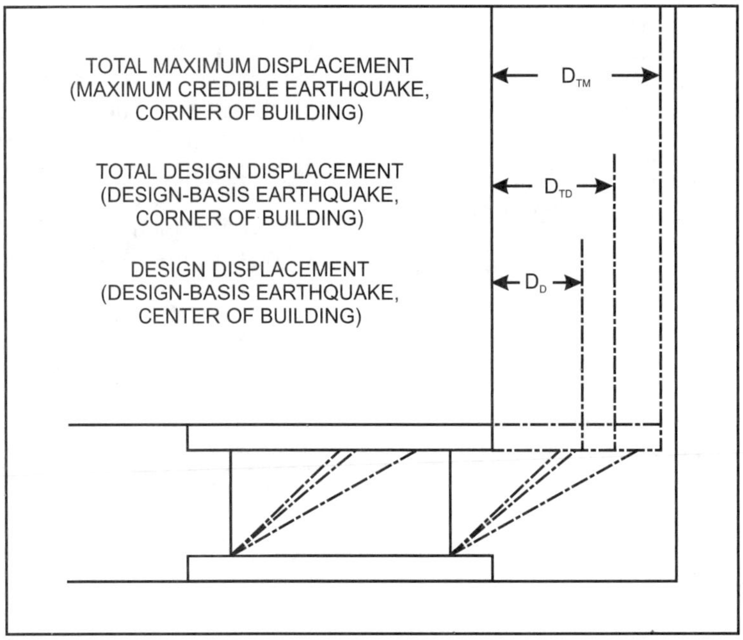

Figure 1623.1
DISPLACEMENT TERMINOLOGY

Bibliography

The following resource materials are referenced in this chapter or are relevant to the subject matter addressed in this chapter.

ADM 1-00, *Aluminum Design Manual: Part 1-A Aluminum Structures, Allowable Stress Design; and Part 1-BCAluminum Structures, Load and Resistance Factor Design of Buildings and Similar Type Structures.* Washington, DC: Aluminum Association, 2000.

ACI 318-02, *Building Code Requirements for Structural Concrete.* Farmington Hills, MI: American Concrete Institute, 2002.

ACI 530/ASCE 5/TMS 402-02, *Building Code Requirements for Masonry Structures*. Farmington Hills, MI: American Concrete Institute; New York: American Society of Civil Engineers; Boulder, CO: The Masonry Council, 2002.

AISC 341-02, *Seismic Provisions for Structural Steel Buildings*. Chicago: American Institute of Steel Construction, Inc., 2002.

AISC 355-89s1, *Specification for Structural Steel Buildings C Allowable Stress Design and Plastic Design, including Supplement No. 1, 2001*. Chicago: American Institute of Steel Construction, Inc., 1989 (2001).

AISC LRFD-99, *Load and Resistance Factor Design Specification for Structural Steel Buildings.* Chicago: American Institute of Steel Construction, Inc., 1999.

AISI NASPEC-01, *North American Specification for Design of Cold-Formed Steel Structural Members*. Washington, DC: American Iron and Steel Institute, 2001.

Algermissen, S.T. *An Introduction to the Seismicity of the United States*. Oakland, CA: Earthquake Engineering Research Institute, 1983.

Ambrose, James and Dimitry Vergun. *Design for Lateral Forces*. New York: John A. Wiley & Sons, 1987.

ASCE 3-91, *Standard Practice for the Construction and Inspection of Composite Slabs*. Reston, VA: American Society of Civil Engineers, 1991.

ASCE 7-02, *Minimum Design Loads for Buildings and Other Structures*. New York: American Society of Civil Engineers, 2002.

ASCE 7-88, *Minimum Design Loads for Buildings and Other Structures*. Reston, VA: American Society of Civil Engineers, 1988.

ASCE 7-95, *Minimum Design Loads for Buildings and Other Structures*. New York: American Society of Civil Engineers, 1995.

ASCE 8-90, *Standard Specification for Design of Cold-formed Stainless Steel Structural Members*. Reston, VA: American Society of Civil Engineers, 1990.

ASCE 24-98, *Flood Resistance Design and Construction Standard*. Reston, VA: American Society of Civil Engineers, 1998.

ASME A17.1-00, *Safety Code for Elevators and Escalators*. New York: American Society of Mechanical Engineers, 2000.

ASME B31.4-95, *Boilers and Pressure Vessels Code.* New York: American Society of Mechanical Engineers, 1995.

ASTM D 2487-00, *Standard Classification of Soils for Engineering Purposes Unified Soil Classification System.* West Conshohocken, PA: ASTM International, 2000.

Becker, Roy, Farzad Naeim and Edward J. Teal. *Seismic Design Practice for Steel Buildings.* Los Angeles: The Steel Committee of California, Revised 1988.

Bolt, Bruce A. *Earthquakes.* New York: W.H. Freeman and Company, 1988.

Chen, Wai-Fah and Charles Scawthorn. *Earthquake Engineering Handbook.* Boca Raton, FL: CRC Press LLC, 2003.

Choppra, Anil K. *Dynamics of Structures: A Primer.* Oakland, CA: Earthquake Engineering Research Institute, 1981.

Cook, N.J. *The Designer's Guide to Wind Loading of Building Structures, Part I: Building Research Establishment.* Cambridge: The University Press, 1985.

Diaphragms. Tacoma, WA: American Plywood Association, 1989.

Dowrick, David J. *Earthquake-Resistant Design for Engineers and Architects,* 2nd ed. New York: John Wiley & Sons, 1987.

Earthquake Design Requirements. Country Club Hills, IL: Building Officials and Code Administrators International, Inc., 1998.

Fanella, David A. and S. K. Ghosh, *Seismic and Wind Design of Concrete Buildings (2000 IBC, ASCE 7-98, ACI 318-99).* Falls Church, VA: International Code Council, 2003.

FEMA 55, *Coastal Construction Manual.* Washington, DC: Federal Emergency Management Agency, 1999.

FEMA 213, *Answers to Questions About Substantially Damaged Buildings.* Washington, DC: Federal Emergency Management Agency, 1991.

FEMA 259, *Engineering Principles and Practices for Retrofitting Flood-Prone Residential Buildings.* Washington, DC: Federal Emergency Management Agency, 1995.

FEMA 302, (NEHRP) *Recommended Provisions for Seismic Regulations for New Buildings and Other Structures, Part 1.* Washington, DC: Prepared by the Building Seismic Safety Council for the Federal Emergency Management Agency, 1998.

FEMA 303, (NEHRP) *Recommended Provisions for Seismic Regulations for New Buildings and Other Structures Commentary, Part 2 Commentary.* Washington, DC: Prepared by the Building Seismic Safety Council for the Federal Emergency Management Agency, 1998.

FEMA 311, *Guidance on Estimating Substantial Damage Using the NFIP Residential Substantial Damage Estimator.* Washington, DC: Federal Emergency Management Agency, 1998.

FEMA 348, *Protecting Building Utilities From Flood Damage: Principles and Practices for the Design and Construction of Flood-Resistant Building Utility Systems.* Washington, DC: Federal Emergency Management Agency, 1999.

FEMA 368, (NEHRP) *Recommended Provisions for Seismic Regulations for New Buildings and Other Structures, Part 1.* Washington, DC: Prepared by the Building Seismic Safety Council for the Federal Emergency Management Agency, 2001.

FEMA 369, (NEHRP) *Recommended Provisions for Seismic Regulations for New Buildings and Other Structures Commentary, Part 2 Commentary.* Washington, DC: Prepared by the Building Seismic Safety Council for the Federal Emergency Management Agency, 2001.

FEMA. (1996) *Floodproofing Certificate* (FEMA Form 81-65) [Online]. Available: http://www.fema.gov/nfip/ff81-65.pdf [1999, December 30].

FEMA. (1999) *Elevation Certificate* (FEMA Form 81-31). [Online]. Available: http://www.fema.gov/library/elvcert.pdf [1999, December 30].

FEMA FIA-TB #1: *Openings in Foundation Walls for Buildings Located in Special Flood Hazard Areas,* 1993.

FEMA FIA-TB #2: *Flood-Resistant Material Requirements for Buildings Located in Special Flood Hazard Areas,* 1993.

FEMA FIA-TB #3: *Nonresidential Floodproofing—Requirements and Certification for Buildings Located in Special Flood Hazard Areas,* 1993.

FEMA FIA-TB #4: *Elevator Installation for Buildings Located in Special Flood Hazard Areas.* 1993.

FEMA FIA-TB #5: *Free of Obstruction Requirements for Buildings Located in Coastal High Hazard Areas,* 1993.

FEMA FIA-TB #6: *Below-Grade Parking Requirements for Buildings Located in Special Flood Hazard Areas,* 1993.

FEMA FIA-TB #7: *Wet Floodproofing Requirements for Structures Located in Special Flood Hazard Areas,* 1993.

FEMA FIA-TB #8: *Corrosion Protection for Metal Connectors In Coastal Areas for Structures Located in Special Flood Hazard Areas,* 1996.

FEMA FIA-TB #9: *Design and Construction Guidance for Breakaway Walls Below Elevated Buildings.*

FEMA. (various dates). *NFIP Technical Bulletin Series.* Washington, DC: National Flood Insurance Program. [Online]. Available: http://www.fema.gov/mit/techbul.htm.

Fuller, Myron. *The New Madrid Earthquake*. Marion, IL: U.S. Department of the Interior/Geological Survey, 1912 (reprinted 1988).

Ghosh, S. K. *Seismic Design Using Structural Dynamics (2000 IBC)*. Falls Church, VA: International Code Council, 2003.

Glen, Berg V. *Seismic Design Codes and Procedures*. Oakland, CA: Earthquake Engineering Research Institute, 1983.

Green, Norman. *Earthquake-Resistant Building Design and Construction*. New York: Elsevier Science Publishing Company, 1987.

Guide for the Design and Construction of Mill Buildings (AISE Technical Report #13). Washington, DC: Association of Iron and Steel Engineers, August 1, 1979.

Hall, W.J. and N.M. Newmark. *Earthquake Spectra and Design*. Oakland, CA: Earthquake Engineering Research Institute, 1982.

IPC-03, *International Plumbing Code*. Falls Church, VA: International Code Council, 2003.

IRC-03, International Residential Code. Falls Church, VA: International Code Council, 2003.

Isyumor, N. and T. Tschanz. *Building Motion in Wind*. New York: American Society of Civil Engineers, 1986.

Liu, Henry. *Wind Engineering: A Handbook for Structural Engineers*. Englewood Cliffs, NJ: Prentice Hall Inc., 1991.

Low Rise Building Systems Manual. Cleveland, OH: Metal Building Manufacturers Association, 1986.

Lubbock, Richard Marshall, Kishor C. Mehta and Dale Perry. *Guide to the Use of the Wind Load Provisions of ASCE 7-88 (formerly ANSI A58.1)*. New York: American Society of Civil Engineers, 1991.

Mehta, Kishor C. and Joseph E. Minor. "Wind Damage Observations and Implications," *Journal of the Structural Division*. New York: American Society of Civil Engineers, Vol. 105, November 1979.

NAAMM 1001-90, *Guide Specification for Design of Metal Flag Poles* Chicago, IL: National Association of Architectural Metal Manufacturers, 1990.

Naeim, Farzad. *The Seismic Design Handbook*. Norwell, MA: Kluwer Academic Publishers, 2001.

NCMA - TEK No. 109A, *Concrete Masonry Design for Seismic Forces*. Herndon, VA: National Concrete Masonry Association, Rev. 1989.

NFPA 13-99, *Installation of Sprinkler Systems*. Qunicy, MA: National Fire Protection Association, 1999.

Pakiser, Louis C. *Earthquakes*. Washington, DC: U.S. Department of the Interior/Geological Survey, 1988.

Proceedings of the 20th Joint Meeting of the U.S.-Japan Cooperative Program in Natural Resources, Wind and Seismic Effects. National Institute of Standards and Technology, NIST SP 760, January 1989.

Proceedings of the First International Conference on Snow Engineering. Santa Barbara, CA: U.S. Army Corps of Engineers-Cold Regions Research & Engineering Laboratory, Special Report 89-6, July 1988.

Recommendations for Direct-Hung Acoustical Tile and Lay-in Panel Ceilings, Seismic Zones 0 - 2. St. Charles, IL: Ceiling and Interior Systems Construction Association, 1991.

Recommendations for Direct-Hung Acoustical Tile and Lay-in Panel Ceilings, Seismic Zones 3 - 4. St Charles, IL: Ceiling and Interior Systems Construction Association, 1991.

Recommended Lateral Force Requirements and Commentary (SEAOC Blue Book). Sacramento, CA: Structural Engineers Association of California, 1999.

Recommended Lateral Force Requirements and Tentative Commentary. San Francisco: Structural Engineers Association of California, 1988.

Reducing Earthquake Hazards: Lessons Learned from Earthquakes. Publication No. 86-02. Oakland, CA: Earthquake Engineering Research Institute, 1986.

Reducing Flood Losses Through the International Code Series: Meeting the Requirements of the National Flood Insurance Program. Falls Church, VA: International Code Council, May 2000.

RM1-97, *Specifications for the Design, Testing, and Utilization of Industrial Steel Storage Racks.* Charlotte, NC: Rack Manufacturers Institute, 1997.

SBCCI SSTD 10-99, *Standard for Hurricane Resistant Residential Construction.* Birmingham, AL: Southern Building Code Congress International, 1999.

SJI-94, *Standard Specifications, Load Tables and Weight Tables for Steel Joists and Joist Girders*. Myrtle Beach, SC: Steel Joist Institute, 1994.

SMACNA-HVAC-95, *HVAC Duct Construction Standards, Metal and Flexible.* Chantilly, VA: Sheet Metal & Air Conditioning Contractors' National Association, Inc., 1995.

SMACNA-Seismic-98, *Seismic Restraint Manual Guidelines for Mechanical Systems, 1991, including Appendix B, 1998.* Chantilly, VA: Sheet Metal & Air Conditioning Contractors' National Association, Inc., 1998.

Specifications for Electric Overhead Traveling Cranes (CMAA Specification #70). New York: Crane Manufacturers Association of America, Inc., 1988.

Structural Engineering Loads. Design Manual 2.02. Naval Facilities Engineering Command, September 1986.

Taly, Narendra. *Loads and Loads Paths in Buildings Principles of Structural Design*. Falls Church, VA: International Code Council, 2003.

The Loma Prieta Earthquake of October 17, 1989. Washington, DC: U.S. Department of the Interior/Geological Survey, 1989.

The Severity of an Earthquake. Washington, DC: U.S Department of the Interior/Geological Survey, 1985.

UBC-97, *Uniform Building Code*. Whittier, CA: International Conference of Building Officials, 1997.

WFCM-01, *Wood Frame Construction Manual for One- and Two-Family Dwellings*. Washington, DC: American Forest and Paper Association, 2001.

Williams, Alan. *Seismic and Wind Forces Structural Design Examples*. Falls Church, VA: International Code Council, 2003.

Wind Loading and Wind-Induced Structural Response. Committee on Wind Effects. New York: American Society of Civil Engineers, 1987.

Chapter 17: Structural Tests and Special Inspections

General Comments

In this chapter, the code sets minimum quality standards for the acceptance of materials used in building construction. It also establishes requirements for special inspections, quality assurance and structural observations and load testing.

Section 1701 contains the scope statement and the general statement for new and used materials.

The definitions of terms primarily related to this chapter are in Section 1702.

Section 1703 addresses the approval process.

Special inspections that are required are specified in Section 1704.

Section 1705 includes the quality assurance requirements based on the seismic design category of the structure.

Quality assurance requirements based on wind speed and wind exposure classification are in Section 1706.

Section 1707 contains the special inspection requirements based on the seismic design category of the structure.

Structural testing for seismic resistance is addressed in Section 1708.

Section 1709 establishes when structural observation by a registered design professional is required for high-seismic and high-wind areas.

The general requirements for determining the design strengths of materials are in Section 1710.

Section 1711 provides for an alternative test procedure in the absence of approved standards.

Provisions for a test load are addressed in Section 1712.

Section 1713 includes requirements for field load testing of a structure.

Preconstruction load testing of materials and methods of construction that are not capable of being designed by an approved analysis is covered by Section 1714.

Section 1715 includes specific material and test standards.

Chapter 17 provides information regarding the evaluation, inspection and approval process for any material or system proposed for use as a component of a structure. These are general requirements that expand on the requirements of Chapter 1 relating to the roles and responsibilities of the building official regarding approval of building components. Additionally, the chapter includes general requirements relating to the roles and responsibilities of the owner, contractor, special inspectors and architects or engineers.

Purpose

This chapter provides procedures and criteria for: testing materials or assemblies, labeling materials, systems and assemblies and special inspections of structural assemblies.

SECTION 1701 GENERAL

1701.1 Scope. The provisions of this chapter shall govern the quality, workmanship and requirements for materials covered. Materials of construction and tests shall conform to the applicable standards listed in this code.

❖ This chapter gives provisions for quality, workmanship, testing and labeling of all materials used in the construction of buildings and structures. In general, all construction materials and tests must conform to the applicable standards listed in the code. This chapter provides requirements for materials and tests when there are no applicable standards; specific tests and standards are referenced in other chapters of the code. Additionally, this chapter provides basic requirements for labeling construction materials and assemblies, and for special inspections of structural systems and components.

1701.2 New materials. New building materials, equipment, appliances, systems or methods of construction not provided for in this code, and any material of questioned suitability proposed for use in the construction of a building or structure, shall be subjected to the tests prescribed in this chapter and in the approved rules to determine character, quality and limitations of use.

❖ Testing is required to be performed on materials that are not specifically provided for in the code. For example, the manufacturer of a sandwich panel consisting of aluminum skins and a foam plastic core wishes to use this panel as an exterior weather covering. The material does not conform to any of the standards referenced in Chapter 14, so an appropriate test protocol must be developed. The same provision for acceptance of alternative materials is already given in Section 104.11. That section provides a strong, definitive statement for performance requirements for alternative materials, requiring the proposed alternative to be equivalent to that prescribed by the code in quality, strength, effectiveness, durability and safety. Section 1701.2 simply reasserts that alternative materials (new materials) may be used, as long as the performance characteristics and quality can be established.

1701.3 Used materials. The use of second-hand materials that meet the minimum requirements of this code for new materials shall be permitted.

❖ Materials and assemblies may be reused, provided that they meet the requirements of the code for new materials (see Section 104.9.1 of the code regarding reuse of materials and equipment). Caution should be exercised in approving a used material for reuse. The applicable material standards must be consulted to determine if certain reuses are prohibited and to determine the characteristics of the used material that must be carefully checked before reuse is approved.

One example is a high-strength structural steel bolt. Reuse of the bolt is specifically prohibited by the AISC ASD standard. Even a piece of structural steel, such as a wide flange, would need to be carefully checked to determine that dimensional tolerances for a new piece of structural steel are met (see ASTM A 6 and A 36).

SECTION 1702
DEFINITIONS

1702.1 General. The following words and terms shall, for the purposes of this chapter and as used elsewhere in this code, have the meanings shown herein.

❖ This section contains definitions of terms that are associated with the subject matter of this chapter. It is important to emphasize that these terms are not exclusively related to this chapter, but are applicable everywhere the term is used in the code.

Definitions of terms can help in the understanding and application of the code requirements. The purpose for including these definitions within this chapter is to provide more convenient access to them without having to refer back to Chapter 2. For convenience, these terms are also listed in Chapter 2 with a cross reference to this section.

The use and application of all defined terms, including those defined herein, are set forth in Section 201.

APPROVED AGENCY. An established and recognized agency regularly engaged in conducting tests or furnishing inspection services, when such agency has been approved.

❖ In order to identify the basic criteria or to understand what agencies are being referred to in Section 1703, there is a need to define the term "approved agency." The word "approved" means "acceptable to the building official" (see the definition of "Approved" in Section 202). The basis for approval of an agency for a particular activity by the building official may include, but is not necessarily limited to, the capacity and capability of the agency to perform the work in accordance with Section 1704 and other applicable sections. This is typically done through review of the résumés and references of the agency and its personnel.

APPROVED FABRICATOR. An established and qualified person, firm or corporation approved by the building official pursuant to Chapter 17 of this code.

❖ An approved fabricator is one who has received approval to perform work without a code-required special inspection. The approval is based upon review of the fabricator's written procedural and quality-control manuals and periodic auditing of fabrication practices by an approved special inspection agency.

CERTIFICATE OF COMPLIANCE. A certificate stating that materials and products meet specified standards or that work was done in compliance with approved construction documents.

❖ A certificate of compliance is a document issued by a supplier of materials and products that certifies they meet the specified requirements.

FABRICATED ITEM. Structural, load bearing or lateral load-resisting assemblies consisting of materials assembled prior to installation in a building or structure, or subjected to operations such as heat treatment, thermal cutting, cold working or reforming after manufacture and prior to installation in a building or structure. Materials produced in accordance with standard specifications referenced by this code, such as rolled structural steel shapes, steel-reinforcing bars, masonry units and plywood sheets, shall not be considered "fabricated items."

❖ The term "fabricated items" can easily be misinterpreted to encompass a number of items for which the code does not intend special inspections; therefore, the term is defined to clarify the intent of the code (see Section 1704).

INSPECTION CERTIFICATE. An identification applied on a product by an approved agency containing the name of the manufacturer, the function and performance characteristics, and the name and identification of an approved agency that indicates that the product or material has been inspected and evaluated by an approved agency (see Section 1703.5 and "Label," "Manufacturer's Designation" and "Mark").

❖ An inspection certificate is an identification applied to a product indicating that the individual product has been inspected by a third-party agency. The identification should include three items:

1. The name of the manufacturer;
2. The function and performance characteristics of the product; and
3. The name of the approved agency completing the inspection and evaluation for the product.

Note that the requirements for a "label" (see Section 1702 for the definition of "Label") differ from those of an inspection certificate. A label is issued by a third-party inspection agency that performs periodic inspections, while an inspection certificate is issued for the product at the time of inspection. The issuance of inspection certificates is an ongoing procedure.

LABEL. An identification applied on a product by the manufacturer that contains the name of the manufacturer, the function and performance characteristics of the product or material, and the name and identification of an approved agency and that indicates that the representative sample of the product or material has been tested and evaluated by an approved agency (see Section 1703.5 and "Inspection Certificate," "Manufacturer's Designation" and "Mark").

❖ A label is used to delineate materials and assemblies that are required to bear the identification of the manufacturer, as well as a third-party quality control agency. The quality control agency allows the use of its label based on periodic audits and inspections of the manufacturer's plant. The code contains specific requirements for labeling (see commentary, Section 1703.5).

MANUFACTURER'S DESIGNATION. An identification applied on a product by the manufacturer indicating that a product or material complies with a specified standard or set of rules (see also "Inspection Certificate," "Label" and "Mark").

❖ This represents terminology for a manufacturer's self-certification that a product complies with a given standard (see commentary, Section 1703.4).

MARK. An identification applied on a product by the manufacturer indicating the name of the manufacturer and the function of a product or material (see also "Inspection Certificate," "Label" and "Manufacturer's Designation").

❖ A mark represents the manufacturer's identification placed on a product, stating who made the product and describing its function. There is, however, no certification of compliance to any particular standard and no third-party quality control (see commentary, Section 1703.4).

SPECIAL INSPECTION. Inspection as herein required of the materials, installation, fabrication, erection or placement of components and connections requiring special expertise to ensure compliance with approved construction documents and referenced standards (see Section 1704).

❖ This category of inspection is intended to include those material connections or installations that require a special level of knowledge and attention. For example, special inspections are required for the installation of high-strength bolts, welded connections, concrete reinforcement, prestressed concrete, fabrication of laminated wood structural elements and pile installations to comply with the contract documents and the standards under which they are assembled.

SPECIAL INSPECTION, CONTINUOUS. The full-time observation of work requiring special inspection by an approved special inspector who is present in the area where the work is being performed.

❖ Continuous special inspection requires full-time monitoring, by a special inspector, of the work designated by the code or the quality assurance plan as requiring continuous inspection.

SPECIAL INSPECTION, PERIODIC. The part-time or intermittent observation of work requiring special inspection by an approved special inspector who is present in the area where the work has been or is being performed and at the completion of the work.

❖ Periodic special inspection requires part-time monitoring, by a special inspector, of the work designated by the code or the quality assurance plan as requiring periodic inspection.

SPRAYED FIRE-RESISTANT MATERIALS. Cementitious or fibrous materials that are spray-applied to provide fire-resistant protection of the substrates.

❖ The cementitious or fibrous material is pneumatically projected onto a surface such that the density, thickness and cohesion/adhesion of the material will provide fire resistance to the surface.

STRUCTURAL OBSERVATION. The visual observation of the structural system by a registered design professional for general conformance to the approved construction documents at significant construction stages and at completion of the structural system. Structural observation does not include or waive the responsibility for the inspection required by Section 109, Section 1704 or other sections of this code.

❖ The registered design professional in responsible charge is required to visit the site to visually determine general conformance to the approved construction documents. Structural observation is in addition to and does not substitute for any required inspections by Section 109, 1704 or other sections of the code.

SECTION 1703
APPROVALS

1703.1 Approved agency. An approved agency shall provide all information as necessary for the building official to determine that the agency meets the applicable requirements.

❖ This section specifies the information that an approved agency must provide to the building official to enable him or her to determine if the agency meets the applicable requirements.

1703.1.1 Independent. An approved agency shall be objective and competent. The agency shall also disclose possible conflicts of interest so that objectivity can be confirmed.

❖ As part of the basis for a building official's approval of a particular inspection agency, the agency must demonstrate its objectivity and competence. The judgement of objectivity is linked to the financial and fiduciary independence of the agency. The competence of the agency is judged by its experience and organization, and the experience of its personnel.

For example, suppose that ACME Agency is the inspection agency employed by Builder's, Inc. for factory-built fireplaces. During investigation of the agency, it is discovered that ACME and Builder's are subsidiaries of the same parent company, Conglomerate, Inc. The inspection agency and the manufacturer clearly have a relationship that is undesirable from the standpoint of independence.

1703.1.2 Equipment. An approved agency shall have adequate equipment to perform required tests. The equipment shall be periodically calibrated.

❖ As part of judging the ability of a testing or inspection agency, the building official should determine that the agency has the proper equipment to perform the required tests or inspections.

1703.1.3 Personnel. An approved agency shall employ experienced personnel educated in conducting, supervising and evaluating tests and/or inspections.

❖ The competence of an inspection or testing agency is also based on the experience and background of its personnel. For example, if 10 engineering graduates form an agency, the building official should question whether or not this newly formed agency is sufficiently experienced to perform the tests.

If the services being provided by the inspection agency or test agency come within the purview of the professional registration laws of the state in which the building is being constructed, the building official must request evidence that the personnel are qualified to perform the work in accordance with this professional registration law as well.

1703.2 Written approval. Any material, appliance, equipment, system or method of construction meeting the requirements of this code shall be approved in writing after satisfactory completion of the required tests and submission of required test reports.

❖ In order to have a documented record of the approval and the basis for it, including any conditions or limitations, materials and systems must be approved in writing by the building official. The code also requires the approval to be granted within a reasonable period of time, after all documentation has been satisfactorily developed and submitted, so as to avoid any unnecessary delay in completion of construction.

1703.3 Approved record. For any material, appliance, equipment, system or method of construction that has been approved, a record of such approval, including the conditions and limitations of the approval, shall be kept on file in the building official's office and shall be open to public inspection at appropriate times.

❖ Written approvals must be kept on file by the building official, and be available and open to the public. This provides reasonable access to the records on approvals of materials and systems should there be any subsequent investigation or further evaluation.

1703.4 Performance. Specific information consisting of test reports conducted by an approved testing agency in accordance with standards referenced in Chapter 35, or other such information as necessary, shall be provided for the building official to determine that the material meets the applicable code requirements.

❖ When conformance to the code is predicated on the performance and quality of materials, the building official must require the submittal of testing reports from an approved agency. In the absence of such reports, the building official must accept specific information and details that prove compliance with the intent of the applicable code requirements.

1703.4.1 Research and investigation. Sufficient technical data shall be submitted to the building official to substantiate the proposed use of any material or assembly. If it is determined that the evidence submitted is satisfactory proof of performance for the use intended, the building official shall approve the use of the material or assembly subject to the requirements of this code. The cost offsets, reports and investigations required under these provisions shall be paid by the permit applicant.

❖ This section is usually used in conjunction with Section 104.11 when analysis of any construction material, such as new and innovative materials, is required to determine code compliance. The analysis is based entirely upon technical data. All costs of testing and investigations must be paid by the applicant.

1703.4.2 Research reports. Supporting data, where necessary to assist in the approval of materials or assemblies not specifically provided for in this code, shall consist of valid research reports from approved sources.

❖ Evaluation reports prepared by approved agencies, such as those published by organizations affiliated with model code groups, may be accepted as part of the information needed by the building official to form the basis for approval. Such reports can supplement the building department resources by eliminating the need for the building official to conduct a detailed analysis on each new product, material or system. It is critically important that such material be truly objective and credible and not consist merely of manufacturer's brochures or similar proprietary information. It is also important to note that when the building official is utilizing research reports in evaluating compliance with the code, such as those issued by organizations affiliated with model code groups, he or she is not mandated to approve these research reports just because the code is the legally adopted building code in the jurisdiction. These reports are not code text; they are advisory only and intended for technical reference.

1703.5 Labeling. Where materials or assemblies are required by this code to be labeled, such materials and assemblies shall be labeled by an approved agency in accordance with Section 1703. Products and materials required to be labeled shall be labeled in accordance with the procedures set forth in Sections 1703.5.1 through 1703.5.3.

❖ This section provides requirements for third-party inspection of a manufacturer of a material or assembly when the code says that the material or assembly must be labeled. The materials or assemblies required to be labeled are given in other chapters of the code and the *International Mechanical Code*® (IMC®), *International Fire Code*® (IFC®) and *International Plumbing Code*® (IPC®). Labeling provides a readily available source of information that is useful for field inspection of installed products. The label identifies the product or material and provides other information that can be investigated further if there is any question as to its suitability for the specific installation.

Some examples are gas appliances, fire doors, prefabricated construction (when the building official does not inspect it), electrical appliances, glass, factory-built fireplaces, plywood and other wood members when used structurally, lumber and foam plastics.

1703.5.1 Testing. An approved agency shall test a representative sample of the product or material being labeled to the relevant standard or standards. The approved agency shall maintain a record of the tests performed. The record shall provide sufficient detail to verify compliance with the test standard

❖ As a basis for the allowed use of an agency's label, the agency is required to perform testing on the material or product in accordance with the standard referenced by the code. For example, Section 903.1 of the IMC requires that factory-built fireplaces be tested in accordance with the referenced standard UL 127. This section states that factory-built fireplaces are required to be listed and labeled by an approved agency.

1703.5.2 Inspection and identification. The approved agency shall periodically perform an inspection, which shall be in-plant if necessary, of the product or material that is to be labeled. The inspection shall verify that the labeled product or material is representative of the product or material tested.

❖ The approved agency whose label is to be applied to a product must perform periodic inspections. The primary objective of these inspections is to determine that the manufacturer is, indeed, making the same product that was tested. For example, using the factory-built fireplace discussed in the commentary to Section 1703.5.1, if the fire chamber wall in the test was $^{3}/_{8}$-inch-thick (9.5 mm) steel, the inspection agency must check to see that this thickness is being used. If the manufacturer has decided to use $^{1}/_{4}$-inch (6.4 mm) steel, then the inspection agency would be forced to withdraw the use of its label and listing.

1703.5.3 Label information. The label shall contain the manufacturer's or distributor's identification, model number, serial number, or definitive information describing the product or material's performance characteristics and approved agency's identification.

❖ This section states what information is required on a label (see Figure 1703.5.3). The purpose of this is so that there will be sufficient information for the inspector to determine that the installed product is the same as that which was approved during plan review.

Figure 1703.5.3
TYPICAL LABEL INFORMATION

1703.6 Heretofore approved materials. The use of any material already fabricated or of any construction already erected, which conformed to requirements or approvals heretofore in effect, shall be permitted to continue, if not detrimental to life, health or safety to the public.

❖ If a material or system had been approved before the code took effect, it can continue to be used as long as it can be shown that the material or system is not detrimental to the health or safety of the building occupants or the public. In other words, the code is not retroactive.

1703.7 Evaluation and follow-up inspection services. Where structural components or other items regulated by this code are not visible for inspection after completion of a prefabricated assembly, the permit applicant shall submit a report of each prefabricated assembly. The report shall indicate the complete details of the assembly, including a description of the assembly and the assembly's components, the basis upon which the assembly is being evaluated, test results and similar information, and other data as necessary for the building official to determine conformance to this code. Such a report shall be approved by the building official.

❖ As an alternative to physical inspection by the building official in the plant or location where prefabricated components are manufactured, such as modular homes, trusses, etc., the building official has the option of accepting an evaluation report from an approved agency detailing such inspections.

1703.7.1 Follow-up inspection. The permit applicant shall provide for special inspections of fabricated items in accordance with Section 1704.2.

❖ The owner is required to provide special inspections of fabricated assemblies at the fabrication plant in accordance with Section 1704.2.

1703.7.2 Test and inspection records. Copies of necessary test and inspection records shall be filed with the building official.

❖ All testing and inspection records related to a fabricated assembly must be filed with the building official so as to maintain a complete and legal record of the assembly and erection of the building.

SECTION 1704 SPECIAL INSPECTIONS

1704.1 General. Where application is made for construction as described in this section, the owner or the registered design professional in responsible charge acting as the owner's agent shall employ one or more special inspectors to provide inspections during construction on the types of work listed under Section 1704. The special inspector shall be a qualified person who shall demonstrate competence, to the satisfaction of the building official, for inspection of the particular type of construction or operation requiring special inspection. These inspections are in addition to the inspections specified in Section 109.

Exceptions:

1. Special inspections are not required for work of a minor nature or as warranted by conditions in the jurisdiction as approved by the building official.
2. Special inspections are not required for building components unless the design involves the practice of professional engineering or architecture as defined by applicable state statutes and regulations governing the professional registration and certification of engineers or architects.
3. Unless otherwise required by the building official, special inspections are not required for occupancies in Group R-3 as applicable in Section 101.2 and occupancies in Group U that are accessory to a residential occupancy including, but not limited to, those listed in Section 312.1.

❖ The permit applicant is responsible for hiring the special inspector and must incur all associated costs. According to Section 105.1, the permit applicant may be the owner or authorized agent in connection with the project (see Section 105.1 for further details).

Exceptions to the requirement for special inspections are minor work and work not required to be designed or sealed by a registered design professional, as regulated by the jurisdiction in which the project is located. Occupancies in Group R-3 or U that are accessory to an R-3 occupancy are typically not required to be designed by a registered design professional; however, this is not true in all cases, with Group R-3 and accessory Group U occupancies being specifically excluded.

It should be noted that Exception 1 does not mean that the inspections listed are not required. It only means that they are not required to be made by a special inspector. Additionally, Exception 1 refers to "conditions in the jurisdiction" as a possible exception. The primary "condition" envisioned is one in which the jurisdiction has the resources and skills to perform the inspection tasks, instead of a special inspector. This exception should not be interpreted as one that can be invoked by the permit applicant. A local jurisdiction should not be obligated to invoke this exception. The purpose of this exception is merely to allow jurisdictions to continue doing inspections if they so desire.

Exception 2 eliminates the special inspection requirement for projects where a design professional is not required. The type of projects that do not require a design professional varies from state to state.

1704.1.1 Building permit requirement. The permit applicant shall submit a statement of special inspections prepared by the registered design professional in responsible charge in accordance with Section 106.1 as a condition for permit issuance. This statement shall include a complete list of materials and work requiring special inspections by this section, the inspections to be performed and a list of the individuals, approved agencies or firms intended to be retained for conducting such inspections.

❖ The applicant must submit for approval a detailed outline of the special inspection program, including the building plans and specifications, before issuance of the building permit. This section places the burden of identifying which materials, components and work require special inspections on the permit applicant. This detailed outline, or statement of special inspections, is required to be prepared by the registered design professional responsible for the building or structure. This is because the special inspections statement relates directly to the construction and design documents, which are the responsibility of the registered design professional.

This section also details the areas to be addressed in the statement. A complete list of materials and work requiring special inspections, the types of inspections and inspection agencies or firms must be provided to the building official. The qualifications and credentials of such individuals, agencies or firms should be submitted for review by the building official.

1704.1.2 Report requirement. Special inspectors shall keep records of inspections. The special inspector shall furnish inspection reports to the building official, and to the registered design professional in responsible charge. Reports shall indicate that work inspected was done in conformance to approved construction documents. Discrepancies shall be brought to the immediate attention of the contractor for correction. If the discrepancies are not corrected, the discrepancies shall be brought to the attention of the building official and to the registered design professional in responsible charge prior to the completion of that phase of the work. A final report documenting required special inspections and correction of any discrepancies noted in the inspections shall be submitted at a point in time agreed upon by the permit applicant and the building official prior to the start of work.

❖ Records of each inspection must be submitted to the building official so as to compile a complete legal record of the project. These records must include all inspections made, violations and discrepancies. Before a certificate of occupancy is issued, a final report must be submitted indicating that all special inspections have been made and all discrepancies resolved or removed in order to show compliance with the applicable code requirements. It is the responsibility of the special inspector to document and submit inspection records to the building official and to the registered design professional in responsible charge of the project.

1704.2 Inspection of fabricators. Where fabrication of structural load-bearing members and assemblies is being performed on the premises of a fabricator's shop, special inspection of the fabricated items shall be required by this section and as required elsewhere in this code.

❖ Inspection of in-plant fabrications and the requirements for special in-plant inspections are addressed herein. This section should be used in conjunction with Section 1703.7 relating to evaluation and follow-up inspections.

1704.2.1 Fabrication and implementation procedures. The special inspector shall verify that the fabricator maintains detailed fabrication and quality control procedures that provide a basis for inspection control of the workmanship and the fabricator's ability to conform to approved construction documents and referenced standards. The special inspector shall review the procedures for completeness and adequacy relative to the code requirements for the fabricator's scope of work.

Exception: Special inspections as required by Section 1704.2 shall not be required where the fabricator is approved in accordance with Section 1704.2.2.

❖ The special inspector is required to verify not only that the fabricator complies with the design details and in-house quality control procedures at the plant, but also its ability to conform to the approved drawings, standards and specifications. An example of this would be an inspection of proper placement and rolling of truss-plate connectors at a wood-truss manufacturing plant. Improper procedures could result in the connectors "popping out" or "peeling back" after the truss is concealed and loaded, thus causing structural failure. Special inspections are not required if an approved independent agency conducts in-house inspections.

1704.2.2 Fabricator approval. Special inspections required by this code are not required where the work is done on the premises of a fabricator registered and approved to perform such work without special inspection. Approval shall be based upon review of the fabricator's written procedural and quality control manuals and periodic auditing of fabrication practices by an approved special inspection agency. At completion of fabrication, the approved fabricator shall submit a certificate of compliance to the building official stating that the work was performed in accordance with the approved construction documents.

❖ Special inspections are not required by the code where the work is done on the premises of an approved fabricator. If the fabricator meets the qualifications described in the code and is approved by the building official, then its quality control procedures are expected to be such that a special inspector is not required for work accomplished in the fabricator's shop.

1704.3 Steel construction. The special inspections for steel elements of buildings and structures shall be as required by Section 1704.3 and Table 1704.3. Where required, special inspection of steel shall also comply with Section 1715.

Exceptions:

1. Special inspection of the steel fabrication process shall not be required where the fabricator does not perform any welding, thermal cutting or heating operation of any kind as part of the fabrication process. In such cases, the fabricator shall be required to submit a detailed procedure for material control that demonstrates the fabricator's ability to maintain suitable records and procedures such that, at any time during the fabrication process, the material specification, grade and mill test reports for the main stress-carrying elements are capable of being determined.
2. The special inspector need not be continuously present during welding of the following items, provided the materials, welding procedures and qualifications of welders are verified prior to the start of the work; periodic inspections are made of the work in progress; and a visual inspection of all welds is made prior to completion or prior to shipment of shop welding.
   - 2.1. Single pass fillet welds not exceeding $^5/_{16}$ inch (7.9 mm) in size.
   - 2.2. Floor and roof deck welding.
   - 2.3. Welded studs when used for structural diaphragm.
   - 2.4. Welded sheet steel for cold-formed steel framing members such as studs and joists.
   - 2.5. Welding of stairs and railing systems.

❖ The requirements to be followed by the special inspector for the erection and fabrication of structural steel elements of building construction are described in this section. This section also refers to Table 1704.3 for special inspection of steel construction.

An exemption is allowed if the fabrication plant does not utilize any facilities or methods that may alter the physical characteristics or properties of the steel members or components, such as welding, thermal cutting or heating operations. The fabricator would, in any case, need to provide evidence that procedures are used that verify that the proper material specification and grade for the main stress-carrying elements are supplied in accordance with the job specifications and shop drawings.

TABLE 1704.3. See page 17-9.

❖ Table 1704.3 lists the types of materials and inspections, the verification required and the referenced standards to be used in evaluating conformance to and code compliance with main stress-carrying elements of steel construction.

1704.3.1 Welding. Welding inspection shall be in compliance with AWS D1.1. The basis for welding inspector qualification shall be AWS D1.1.

❖ The referenced standard for this section is AWS D1.1 from the American Welding Society (AWS). This standard covers 10 areas of welding: general provisions; design of welded connections; workmanship; technique; qualification; inspection; strengthening and repair of existing structures; design of new buildings; and the design of new bridges and new tubular structures. The tables in AWS D1.1 are intended only to provide prequalified joint geometry, such as root opening, angles and clearances, which will permit a qualified welder to deposit sound weld metal.

Weld joints other than those specified by AWS may be qualified, provided that they are tested and qualified in accordance with AWS D1.1.

All prequalified or qualified welds should have a written welding procedure specification prepared by the fabricator, following the detail and outline suggested by AWS D1.1. This specification should be available to the engineer, inspector and building official.

Weld inspectors are also required to be qualified in accordance with AWS D1.1, which references AWS QC1.

1704.3.2 Details. The special inspector shall perform an inspection of the steel frame to verify compliance with the details shown on the approved construction documents, such as bracing, stiffening, member locations and proper application of joint details at each connection.

❖ The special inspector is required to perform an inspection of the entire steel frame to verify compliance with the applicable code requirements and the approved engineering drawings.

1704.3.3 High-strength bolts. Installation of high-strength bolts shall be periodically inspected in accordance with AISC specifications.

❖ Installation and periodic inspection of high-strength (ASTM A 325 or ASTM A 490 or equivalent) bolts must be in accordance with AISC ASD or AISC LRFD.

1704.3.3.1 General. While the work is in progress, the special inspector shall determine that the requirements for bolts, nuts, washers, and paint; bolted parts; and installation and tightening in such standards are met. For bolts requiring pretensioning, the special inspector shall observe the pre-installation testing and calibration procedures when such procedures are required by the installation method or by project plans or specification; determine that all plies of connected materials have been drawn together and properly snugged; and monitor the installation of bolts to verify that the selected procedure for installation is properly used to tighten bolts. For joints required to be tightened only to the snug tight condition, the special inspector need only verify that the connected materials have been drawn together and properly snugged.

❖ Inspection procedures that provide the greatest ensurance that bolts are properly installed and tensioned are provided first by observing the calibration testing of the fasteners using the selected installation procedure, and then by monitoring the work in progress to ensure that the procedure that provided the specified tension is routinely adhered to. When such a program is followed, no further evidence of proper bolt tension is required.

1704.3.3.2 Periodic monitoring. Monitoring of bolt installation for pretensioning is permitted to be performed on a periodic basis when using the turn-of-nut method with matchmarking techniques, the direct tension indicator method, or the alternate design fastener (twist-off bolt) method. Joints designated as snug tight need be inspected only on a periodic basis.

❖ Periodic monitoring by a special inspector is permitted for high-strength bolts installed and tightened by the turn-of-nut method with matchmarking techniques, the direct tension indicator method, the alternate design fastener method or joints designated as snug tight.

1704.3.3.3 Continuous monitoring. Monitoring of bolt installation for pretensioning using the calibrated wrench method or the turn-of-nut method without matchmarking shall be performed on a continuous basis.

❖ Continuous monitoring by a special inspector is required for high-strength bolts installed and tightened by the calibrated wrench method or the turn-of-nut method without matchmarking.

TABLE 1704.3
REQUIRED VERIFICATION AND INSPECTION OF STEEL CONSTRUCTION

VERIFICATION AND INSPECTION	CONTINUOUS	PERIODIC	REFERENCED STANDARD[a]	IBC REFERENCE
1. Material verification of high-strength bolts, nuts and washers:				
a. Identification markings to conform to ASTM standards specified in the approved construction documents.	—	X	Applicable ASTM material specifications; AISC ASD, Section A3.4; AISC LRFD, Section A3.3	—
b. Manufacturer's certificate of compliance required.	—	X	—	—
2. Inspection of high-strength bolting:				
a. Bearing-type connections.	—	X	AISC LRFD Section M2.5	1704.3.3
b. Slip-critical connections.	X	X		
3. Material verification of structural steel:				
a. Identification markings to conform to ASTM standards specified in the approved construction documents.	—	—	ASTM A 6 or ASTM A 568	1708.4
b. Manufacturers' certified mill test reports.	—	—	ASTM A 6 or ASTM A 568	
4. Material verification of weld filler materials:				—
a. Identification markings to conform to AWS specification in the approved construction documents.	—	—	AISC, ASD, Section A3.6; AISC LRFD, Section A3.5	—
b. Manufacturer's certificate of compliance required.	—	—	—	—
5. Inspection of welding: a. Structural steel:	—	—		
1) Complete and partial penetration groove welds.	X	—	AWS D1.1	1704.3.1
2) Multi-pass fillet welds.	X	—		
3) Single-pass fillet welds > $^5/_{16}''$	X	—		
4) Single-pass fillet welds < $^5/_{16}''$	—	X		
5) Floor and deck welds.	—	X	AWS D1.3	—
b. Reinforcing steel:	—	—	AWS D1.4 ACI 318: 3.5.2	1903.5.2
1) Verification of weldability of reinforcing steel other than ASTM A 706.	—	X		
2) Reinforcing steel-resisting flexural and axial forces in intermediate and special moment frames, and boundary elements of special reinforced concrete shear walls and shear reinforcement.	X	—		
3) Shear reinforcement.	X	—		
4) Other reinforcing steel.	—	X		
6. Inspection of steel frame joint details for compliance with approved construction documents:		X	—	1704.3.2
a. Details such as bracing and stiffening.	—	—		
b. Member locations.	—	—		
c. Application of joint details at each connection.	—			

For SI: 1 inch = 25.4 mm.

a. Where applicable, see also Section 1707.1, Special inspection for seismic resistance.

1704.4 Concrete construction. The special inspections and verifications for concrete construction shall be as required by this section and Table 1704.4.

Exception: Special inspections shall not be required for:

1. Isolated spread concrete footings of buildings three stories or less in height that are fully supported on earth or rock.
2. Continuous concrete footings supporting walls of buildings three stories or less in height that are fully supported on earth or rock where:
   - 2.1. The footings support walls of light frame construction;
   - 2.2. The footings are designed in accordance with Table 1805.4.2; or
   - 2.3. The structural design is based on a f'_c no greater than 2,500 pounds per square inch (psi) (17.2 Mpa).
3. Nonstructural concrete slabs supported directly on the ground, including prestressed slabs on grade, where the effective prestress in the concrete is less than 150 psi (1.03 Mpa).
4. Concrete foundation walls constructed in accordance with Table 1805.5(1), 1805.5(2), 1805.5(3) or 1805.5(4).
5. Concrete patios, driveways and sidewalks, on grade.

❖ This section establishes criteria for special inspections of elements of buildings and structures of concrete construction. Exceptions to the requirements of this section address concrete components that have little or no load-carrying requirements, such as nonstructural slabs on grade, driveways, patios, etc., or footings and foundations that require no reinforcement and carry relatively low loads.

TABLE 1704.4
REQUIRED VERIFICATION AND INSPECTION OF CONCRETE CONSTRUCTION

VERIFICATION AND INSPECTION	CONTINUOUS	PERIODIC	REFERENCED STANDARD[a]	IBC REFERENCE
1. Inspection of reinforcing steel, including prestressing tendons, and placement.	—	X	ACI 318: 3.5, 7.1-7.7	1903.5, 1907.1, 1907.7, 1914.4
2. Inspection of reinforcing steel welding in accordance with Table 1704.3, Item 5B.	—	—	AWS D1.4 ACI 318: 3.5.2	1903.5.2
3. Inspect bolts to be installed in concrete prior to and during placement of concrete where allowable loads have been increased.	X	—	—	1912.5
4. Verifying use of required design mix.	—	X	ACI 318: Ch. 4, 5.2-5.4	1904, 1905.2-1905.4, 1914.2, 1914.3
5. At the time fresh concrete is sampled to fabricate specimens for strength tests, perform slump and air content tests, and determine the temperature of the concrete.	X	—	ASTM C 172 ASTM C 31 ACI 318: 5.6, 5.8	1905.6, 1914.10
6. Inspection of concrete and shotcrete placement for proper application techniques.	X	—	ACI 318: 5.9, 5.10	1905.9, 1905.10, 1914.6, 1914.7, 1914.8
7. Inspection for maintenance of specified curing temperature and techniques.	—	X	ACI 318: 5.11-5.13	1905.11, 1905.13, 1914.9
8. Inspection of prestressed concrete: a. Application of prestressing forces. b. Grouting of bonded prestressing tendons in the seismic-force-resisting system.	 X X	—	 ACI 318: 18.20 ACI 318: 18.18.4	—
9. Erection of precast concrete members.	—	X	ACI 318: Ch. 16	—
10. Verification of in-situ concrete strength, prior to stressing of tendons in posttensioned concrete and prior to removal of shores and forms from beams and structural slabs.	—	X	ACI 318: 6.2	1906.2

For SI: 1 inch = 25.4 mm.

a. Where applicable, see also Section 1707.1, Special inspection for seismic resistance.

TABLE 1704.4. See page 17-10.

❖ Required verifications and inspections during concrete construction operations are listed in Table 1704.4. This table lists the types of inspections required and the referenced standards for the placing, curing, prestressing and erection of concrete construction.

1704.4.1 Materials. In the absence of sufficient data or documentation providing evidence of conformance to quality standards for materials in Chapter 3 of ACI 318, the building official shall require testing of materials in accordance with the appropriate standards and criteria for the material in Chapter 3 of ACI 318. Weldability of reinforcement, except that which conforms to ASTM A 706, shall be determined in accordance with the requirements of Section 1903.5.2.

❖ Concrete materials, such as cement, aggregates, admixtures and water, must comply with the standards of Chapter 3 of ACI 318, which regulates materials and addresses specific standards. In the absence of sufficient data or documentation, the building official must require testing in accordance with the standards listed in Chapter 3 of ACI 318.

ASTM A 706 is the standard for reinforcing steel that is weldable, meaning that the chemical composition and manufacturing processes are such that the material is well suited for an acceptable quality of weld. Section 1903.5.2 states that any standard other than ASTM A 706 used for reinforcement material would need to be supplemented for weldability requirements. The intent of this provision is that, where welding of reinforcing steel is required, the steel specified and delivered must be checked for weldability.

1704.5 Masonry construction. Masonry construction shall be inspected and evaluated in accordance with the requirements of this section, depending on the classification of the building or structure or nature of occupancy, as defined by this code (see Table 1604.5 and Section 1617.2).

Exception: Special inspections shall not be required for:

1. Empirically designed masonry, glass unit masonry or masonry veneer designed by Section 2109, 2110 or ACI 530/ASCE 5/TMS 402, Chapters 5, 6 or 7, when they are part of nonessential buildings (see Table 1604.5 and Section 1617.2).
2. Masonry foundation walls constructed in accordance with Table 1805.5(1), 1805.5(2), 1805.5(3) or 1805.5(4).

❖ This section establishes whether special inspection is required, and the necessary level, for masonry construction. This is determined by the design method (empirical versus engineered) and whether or not a structure is classified as an essential facility. These requirements are summarized in Table 1704.5. An essential facility would be an Occupancy Category IV in Table 1604.5 and/or Seismic Use Group III in accordance with Section 1616.2.3. Additionally, masonry foundation walls complying with the prescriptive tables in Section 1805.5 are exempt from special inspection.

TABLE 1704.5
SPECIAL INSPECTION OF MASONRY CONSTRUCTION

	TYPE OF FACILITY			
	Nonessential		Essential	
Design method	Empirical	Engineered	Empirical	Engineered
Special inspection	Exempt	Level 1	Level 1	Level 2

1704.5.1 Empirically designed masonry, glass unit masonry and masonry veneer in essential facilities. The minimum inspection program for masonry designed by Chapter 14, Section 2109 or 2110, or by Chapter 5, 6, or 7 of ACI 530/ASCE 5/TMS 402, in essential facilities listed in Table 1604.5 and Section 1616.2, shall comply with Table 1704.5.1.

❖ This section defines the minimum level of special inspection required for empirically designed masonry, glass unit masonry and masonry veneer in essential facilities (see Table 1704.5.1).

TABLE 1704.5.1. See page 17-12.

❖ The minimum required special inspections and verifications during masonry construction are listed in Table 1704.5.1. This table also lists the required criteria from the code, ACI 530 and ACI 530.1. This table applies to empirically designed masonry in essential facilities and to engineered design in nonessential facilities.

1704.5.2 Engineered masonry in nonessential facilities. The minimum special inspection program for masonry designed by Section 2106, 2107 or 2108, or by chapters other than Chapters 5, 6, or 7 of ACI 530/ASCE 5/TMS 402, in nonessential facilities (see Table 1604.5 and Section 1617.2) shall comply with Table 1704.5.1.

❖ Engineered masonry in nonessential facilities requires special inspection. The minimum special inspections required are listed in Table 1704.5.1.

1704.5.3 Engineered masonry in essential facilities. The minimum special inspection program for masonry designed by Section 2106, 2107 or 2108, or by chapters other than Chapters 5, 6 or 7 of ACI 530/ASCE5/TMS 402, in essential facilities (see Table 1604.5 and Section 1616.2) shall comply with Table 1704.5.3.

❖ Engineered masonry in essential facilities requires special inspection. The minimum special inspections required are listed in Table 1704.5.3.

TABLE 1704.5.3. See page 17-13.

❖ The minimum required special inspections and verifications during masonry construction of engineered design masonry in essential facilities are listed in Table 1704.5.3. This table also lists the required criteria from the code, ACI 530 and ACI 530.1.

TABLE 1704.5.1
LEVEL 1 SPECIAL INSPECTION

INSPECTION TASK	FREQUENCY OF INSPECTION		REFERENCE FOR CRITERIA		
	Continuous during task listed	Periodically during task listed	IBC section	ACI 530/ ASCE 5/TMS 402[a]	ACI 530.1/ ASCE 6/TMS 602[a]
1. As masonry construction begins, the following shall be verified to ensure compliance:					
a. Proportions of site-prepared mortar.	—	X	—	—	Art. 2.6A
b. Construction of mortar joints.		X			Art. 3.3B
c. Location of reinforcement and connectors.		X			Art. 3.4, 3.6A
d. Prestressing technique.	—	X	—	—	Art. 3.6B
e. Grade and size of prestressing tendons and anchorages.	—	X	—	—	Art. 2.4B, 2.4H
2. The inspection program shall verify:					
a. Size and location of structural elements.	—	X	—	—	Art. 3.3G
b. Type, size and location of anchors, including other details of anchorage of masonry to structural members, frames or other construction.	—	X	—	Sec. 1.2.2(e), 2.1.4, 3.1.6	—
c. Specified size, grade and type of reinforcement.	—	X	—	Sec. 1.12	Art. 2.4, 3.4
d. Welding of reinforcing bars.	X	—	—	Sec. 2.1.10.6.2 3.2.3.4(b)	—
e. Protection of masonry during cold weather (temperature below 40°F) or hot weather (temperature above 90°F).	—	X	Sec. 2104.3, 2104.4	—	Art. 1.8C, 1.8D
f. Application and measurement of prestressing force.	—	X	—	—	Art. 3.6B
3. Prior to grouting, the following shall be verified to ensure compliance:					
a. Grout space is clean.	—	X	—	—	Art. 3.2D
b. Placement of reinforcement and connectors and prestressing tendons and anchorages.		X		Sec. 1.12	Art. 3.4
c. Proportions of site-prepared grout and prestressing grout for bonded tendons.		X		—	Art. 2.6B
d. Construction of mortar joints.		X		—	Art. 3.3B
4. Grout placement shall be verified to ensure compliance with code and construction document provisions.	X	—	—	—	Art 3.5
a. Grouting of prestressing bonded tendons.	X	—	—	—	Art. 3.6C
5. Preparation of any required grout specimens, mortar specimens and/or prisms shall be observed.	X	—	Sec. 2105.2.2, 2105.3	—	Art. 1.4
6. Compliance with required inspection provisions of the construction documents and the approved submittals shall be verified.	—	X	—	—	Art. 1.5

For SI: °C = (°F – 32)/1.8.

a. The specific standards referenced are those listed in Chapter 35.

1704.6 Wood construction. Special inspections of the fabrication process of prefabricated wood structural elements and assemblies shall be in accordance with Section 1704.2. Special inspections of site-built assemblies shall be in accordance with Section 1704.1.

❖ The fabrication process of wood structural elements and assemblies (such as wood trusses) that is being performed on the premises of a fabricator's shop must receive special inspection in accordance with Section 1704.2.

1704.6.1 Fabrication of high-load diaphragms. High-load diaphragms using values from Table 2306.3.2 shall be installed with special inspections as indicated in Section 1704.1. The special inspector shall inspect the wood structural panel sheathing to ascertain whether it is of the grade and thickness shown on the approved building plans. Additionally, the special inspector must verify the nominal size of framing members at adjoining panel edges, the nail or staple diameter and length, the number of fastener lines and that spacing between fasteners in each line and at edge margins agrees with the approved building plans.

❖ This section requires special inspection of specific portions of diaphragms that are designed in accordance with Table 2306.3.2. By their very nature, these "high-load" diaphragms are likely to carry more signifi-

TABLE 1704.5.3
LEVEL 2 SPECIAL INSPECTION

INSPECTION TASK	FREQUENCY OF INSPECTION		REFERENCE FOR CRITERIA		
	Continuous during task listed	Periodically during task listed	IBC section	ACI 530/ ASCE 5/ TMS 402[a]	ACI 530.1/ ASCE 6/ TMS 602[a]
1. From the beginning of masonry construction, the following shall be verified to ensure compliance:					
a. Proportions of site-mixed mortar, grout and prestressing grout for bonded tendons.	—	X	—	—	Art. 2.6A
b. Placement of masonry units and construction of mortar joints.	—	X	—	—	Art. 3.3B
c. Placement of reinforcement, connectors and prestressing tendons and anchorges.	—	X	—	Sec. 1.12	Art. 3.4, 3.6 A
d. Grout space prior to grouting.	X	—	—	—	Art. 3.2D
e. Placement of grout.	X	—	—	—	Art. 3.5
f. Placement of prestressing grout.	X	—	—	—	Art. 3.6C
2. The inspection program shall verify:					
a. Size and location of structural elements.	—	X	—	—	Art. 3.3G
b. Type, size and location of anchors, including other details of anchorage of masonry to structural members, frames or other construction.	X	—	—	Sec. 1.2.2(e), 2.1.4, 3.1.6	—
c. Specified size, grade and type of reinforcement.		X	—	Sec. 1.12	Art. 2.4, 3.4
d. Welding of reinforcement.	X	—	—	Sec.2.1.10.6.2, 3.2.3.4(b)	—
e. Protection of masonry during cold weather and (temperature below 40°F) or hot weather (temperature above 90°F).	—	X	Sec. 2104.3, 2104.4	—	Art. 1.8C, 1.8D
f. Application and measurement of prestressing force.	X	—	—	—	Art. 3.6B
3. Preparation of any required grout specimens, mortar specimens and/or prisms shall be observed.	X	—	Sec. 2105.2.2, 2105.3	—	Art. 1.4
4. Compliance with required inspection provisions of the construction documents and the approved submittals shall be verified.	—	X	—	—	Art 1.5

For SI: °C = (°F – 32)/1.8.
a. The specific standards referenced are those listed in Chapter 35.

cant wind or seismic loads, making it necessary to perform special inspection.

1704.7 Soils. The special inspections for existing site soil conditions, fill placement and load-bearing requirements shall follow Sections 1704.7.1 through 1704.7.3. The approved soils report, required by Section 1802.2, shall be used to determine compliance.

Exception: Special inspections not required during placement of fill less than 12 inches (305 mm) deep.

❖ The increasing use of prepared fill as a load-bearing soil strata has established the need to make specific mention of this construction in the special inspection requirements. The load-bearing capacity of the supporting soil has a significant impact on the structural integrity of any building. The amount of compaction and the methods vary depending on the particular design. Use of proper compaction, lift and density, however, is critical to achieving the desired bearing capacity.

1704.7.1 Site preparation. Prior to placement of the prepared fill, the special inspector shall determine that the site has been prepared in accordance with the approved soils report.

❖ The first important step for prepared fill is to verify that the site preparation meets specified requirements, including proper excavation depth, removal of all deleterious material and any other special requirements that the soils engineer deems necessary for the design.

1704.7.2 During fill placement. During placement and compaction of the fill material, the special inspector shall determine that the material being used and the maximum lift thickness comply with the approved report, as specified in Section 1803.4.

❖ The fill placement operation is clearly an item that falls into the special inspection category. Without observing and documenting that the proper material is used and the specified compaction techniques and lifts are employed, the load-bearing capacity may not be consistent with the design requirements.

1704.7.3 Evaluation of in-place density. The special inspector shall determine, at the approved frequency, that the in-place dry density of the compacted fill complies with the approved report.

❖ A major factor in the design of the fill is the in-place density. This evaluation is needed so that the compaction methods result in adequate soil-bearing capacity.

1704.8 Pile foundations. A special inspector shall be present when pile foundations are being installed and during tests. The special inspector shall make and submit to the building official records of the installation of each pile and results of load tests. Records shall include the cutoff and tip elevation of each pile relative to a permanent reference.

❖ Special inspections of pile foundations must conform to the criteria and standards of Section 1807.2, which establishes inspection procedures and record submittal information.

1704.9 Pier foundations. Special inspection is required for pier foundations for buildings assigned to Seismic Design Category C, D, E or F in accordance with Section 1616.3.

❖ For structures located in Seismic Design Category C, D, E or F, the pier foundations will typically be of reinforced concrete that is subject to special inspections in accordance with Section 1704.4. It should be noted, however, that the exceptions in Section 1704.4 do not apply to pier foundations in Seismic Design Category C, D, E or F.

1704.10 Wall panels and veneers. Special inspection is required for exterior and interior architectural wall panels and the anchoring of veneers for buildings assigned to Seismic Design Category E or F in accordance with Section 1616.3. Special inspection of such masonry veneer shall be in accordance with Section 1704.5.

❖ A building frame is often adequately designed to withstand an earthquake, but the veneer connections may not adequately perform in the event of an earthquake. Falling bricks, glass or other materials can cause serious injury or damage. Buildings in Seismic Design Category E or F are in the most severe earthquake category.

1704.11 Sprayed fire-resistant materials. Special inspections for sprayed fire-resistant materials applied to structural elements and decks shall be in accordance with Sections 1704.11.1 through 1704.11.5. Special inspections shall be based on the fire-resistance design as designated in the approved construction documents.

❖ To ensure that a sprayed fire-resistant material, or SFRM as it is commonly referred to, performs as intended by the manufacturer, certain conditions are required to be met during its application. The conditions include: temperature at time of application; substrate conditions; protection provided; and thickness and density of material. These conditions must be checked and verified at the time of, or prior to, the application of the SFRM. This section stipulates that special inspections are required in accordance with Sections 1704.11.1 through 1704.11.5.

1704.11.1 Structural member surface conditions. The surfaces shall be prepared in accordance with the approved fire-resistance design and the approved manufacturer's written instructions. The prepared surface of structural members to be sprayed shall be inspected before the application of the sprayed fire-resistant material.

❖ The integrity of an SFRM system depends, foremost, on the conditions of the surface of the steel member to which it is to be applied. The system must be fully adhered to the surface for proper performance, in accordance with design values.

1704.11.2 Application. The substrate shall have a minimum ambient temperature before and after application as specified in the approved manufacturer's written instructions. The area for application shall be ventilated during and after application as required by the approved manufacturer's written instructions.

❖ During application of SFRMs, and immediately thereafter during cure of the material, several items must be controlled, including ambient temperature during the application and temperature of the substrate and the sprayed fire-resistant materials. Temperature control is important to determine that the necessary chemical reactions needed to make a particular material bond to the steel surfaces and hold together, do, in fact, happen. The minimum or maximum temperatures necessary for proper bond and cure depend on the specific type of material. The scope of the special inspection also includes the proper ventilation for curing.

1704.11.3 Thickness. The average thickness of the sprayed fire-resistant materials applied to structural elements shall not be less than the thickness required by the approved fire-resistance design. Individual measured thickness, which exceeds the thickness specified in a design by $^1/_4$ inch (6.4 mm) or more shall be recorded as the thickness specified in the design plus $^1/_4$ inch (6.4 mm). For design thicknesses 1 inch (25 mm) or greater, the minimum allowable individual thickness shall be the design thickness minus $^1/_4$ inch (6.4 mm). For design thicknesses less than 1 inch (25 mm), the minimum allowable individual thickness shall be the design thickness minus 25 percent. Thickness shall be determined in accordance with ASTM E 605. Samples of the sprayed fire-resistant materials shall be selected in accordance with Sections 1704.11.3.1 and 1704.11.3.2.

❖ For the system to provide the required design fire-resistance rating, it must be applied at the appropriate thickness. The required sampling provided for in Sections 1704.11.3.1 and 1704.11.3.2 is based on ASTM E 605. This standard also provides testing methods that are commonly used by the industry.

1704.11.3.1 Floor, roof and wall assemblies. The thickness of the sprayed fire-resistant material applied to floor, roof and wall assemblies shall be determined in accordance with ASTM E 605, taking the average of not less than four measurements for each 1,000 square feet (93 m^2) of the sprayed area on each floor or part thereof.

❖ Sampling of a SFRM for membrane components (floors, roofs or walls) is based on the square footage of the components. The size of the sample and the amount of area per sample are taken from ASTM E 605.

1704.11.3.2 Structural framing members. The thickness of the sprayed fire-resistant material applied to structural members shall be determined in accordance with ASTM E 605. Thickness testing shall be performed on not less than 25 percent of the structural members on each floor.

❖ Sampling of a SFRM for structural elements is based on the number of structural elements. Again, sample size and number of elements represented by one sample are taken from ASTM E 605.

1704.11.4 Density. The density of the sprayed fire-resistant material shall not be less than the density specified in the approved fire-resistant design. Density of the sprayed fire-resistant material shall be determined in accordance with ASTM E 605.

❖ The density of a SFRM will have an impact on the fire-resistance rating of the system. Therefore, it is important that the density of the material be measured to verify that the product is as designed. The sampling requirements given are the same as those for thickness measurements in Sections 1704.11.3.1 and 1704.11.3.2. The required method of determining density is provided in ASTM E 605.

1704.11.5 Bond strength. The cohesive/adhesive bond strength of the cured sprayed fire-resistant material applied to structural elements shall not be less than 150 pounds per square foot (psf) (7.18 kN/m^2). The cohesive/adhesive bond strength shall be determined in accordance with the field test specified in ASTM E 736 by testing in-place samples of the sprayed fire-resistant material selected in accordance with Sections 1704.11.5.1 and 1704.11.5.2.

❖ Bond strength of the material is an essential variable in the design equation for an appropriate fire-resistance rating of a SFRM. A minimum cohesive/adhesive bond strength of 150 pounds per square foot (psf) (7.18 kN/m^2) is required in this section based on the American Institute of Architects (AIA) Master Specification and the recommendations of the General Services Administration (GSA) for durability and serviceability of the material.

1704.11.5.1 Floor, roof and wall assemblies. The test samples for determining the cohesive/adhesive bond strength of the sprayed fire-resistant materials shall be selected from each floor, roof and wall assembly at the rate of not less than one sample for every 10,000 square feet (929 m^2) or part thereof of the sprayed area in each story.

❖ The sampling rate for bond in this section is less than the samples required for thickness and density, since the thickness and density samples minimize the need for bond strength sampling.

1704.11.5.2 Structural framing members. The test samples for determining the cohesive/adhesive bond strength of the sprayed fire-resistant materials shall be selected from beams, girders, joists, trusses and columns at the rate of not less than one sample for each type of structural framing member for each 10,000 square feet (929 m^2) of floor area or part thereof in each story.

❖ The bond strength sampling rate in this section is the same as that indicated in Section 1704.11.5.1 for floor, roof and wall assemblies.

1704.12 Exterior insulation and finish systems (EIFS). Special inspections shall be required for all EIFS applications.

Exceptions:

1. Special inspections shall not be required for EIFS applications installed over a water-resistive barrier with a means of draining moisture to the exterior.
2. Special inspections shall not be required for EIFS applications installed over masonry or concrete walls.

❖ Special inspections are required for all EIFS installations except for the two exceptions in this section. The first exception recognizes that EIFS, which are installed over a water-resistive barrier and incorporate flashings at penetrations and terminations and a means of drainage to the exterior, afford a built-in redundancy to water penetration that makes the need for special inspections less critical.

The second exception recognizes that concrete and masonry substrates are relatively durable and the exposure to moisture in wall conditions does not necessarily have a detrimental effect on these materials.

1704.13 Special cases. Special inspections shall be required for proposed work that is, in the opinion of the building official, unusual in its nature, such as, but not limited to, the following examples:

1. Construction materials and systems that are alternatives to materials and systems prescribed by this code.
2. Unusual design applications of materials described in this code.
3. Materials and systems required to be installed in accordance with additional manufacturer's instructions that prescribe requirements not contained in this code or in standards referenced by this code.

❖ This section requires special inspections for proposed work that is unique and not specifically addressed in the code or in standards referenced by the code. For example, a designer chooses to utilize a new type of wood-laminated beam system in lieu of standard steel beam construction. Because the laminated beam is a new product, the inspector must rely on load tables, connection details and bearing length charts from the manufacturer, provided that the system is previously approved for installation by the building official.

1704.14 Special inspection for smoke control. Smoke control systems shall be tested by a special inspector.

❖ This section requires that all smoke control systems be tested by special inspection.

1704.14.1 Testing scope. The test scope shall be as follows:

1. During erection of ductwork and prior to concealment for the purposes of leakage testing and recording of device location.
2. Prior to occupancy and after sufficient completion for the purposes of pressure difference testing, flow measurements, and detection and control verification.

❖ Testing of the smoke control system must be performed in at least two stages: the first testing is during erection of ductwork and prior to concealment, and the second testing is before the certificate of occupancy is issued. Both stages require testing by a special inspection agency to show compliance with the requirement of this section.

1704.14.2 Qualifications. Special inspection agencies for smoke control shall have expertise in fire-protection engineering, mechanical engineering and certification as air balancers.

❖ An approved special inspection agency of smoke control systems must have expertise and certification in fire protection engineering, mechanical engineering or air balancing.

SECTION 1705
QUALITY ASSURANCE FOR SEISMIC RESISTANCE

1705.1 Scope. A quality assurance plan for seismic requirements shall be provided in accordance with Section 1705.2 for the following:

1. The seismic-force-resisting systems in structures assigned to Seismic Design Category C, D, E or F, in accordance with Section 1616.
2. Designated seismic systems in structures assigned to Seismic Design Category D, E or F.
3. The following additional systems in structures assigned to Seismic Design Category C:
   - 3.1. Heating, ventilating and air-conditioning (HVAC) ductwork containing hazardous materials and anchorage of such ductwork.
   - 3.2. Piping systems and mechanical units containing flammable, combustible or highly toxic materials.
   - 3.3. Anchorage of electrical equipment used for emergency or standby power systems.
4. The following additional systems in structures assigned to Seismic Design Category D:
   - 4.1. Systems required for Seismic Design Category C.
   - 4.2. Exterior wall panels and their anchorage.
   - 4.3. Suspended ceiling systems and their anchorage.
   - 4.4. Access floors and their anchorage.
   - 4.5. Steel storage racks and their anchorage, where the factor, *Ip*, determined in Section 9.6.1.5 of ASCE 7, is equal to 1.5.
5. The following additional systems in structures assigned to Seismic Design Category E or F:
   - 5.1. Systems required for Seismic Design Categories C and D.

5.2. Electrical equipment.

Exceptions:

1. A quality assurance plan is not required for structures designed and constructed in accordance with the conventional construction provisions of Section 2308.
2. A quality assurance plan is not required for structures designed and constructed in accordance with the following:
   - 2.1. The structure is constructed of light wood framing or light framed cold-formed steel; the design spectral response acceleration at short periods, S_{DS}, as determined in Section 1615.1, does not exceed 0.5g, and the height of the structure does not exceed 35 feet (10 668 mm) above grade plane; or
   - 2.2. The structure is constructed using a reinforced masonry structural system or reinforced concrete structural system; the design spectral response acceleration at short periods, S_{DS}, as determined in Section 1615.1, does not exceed 0.5g, and the height of the structure does not exceed 25 feet (7620 mm) above grade plane; or
   - 2.3. The structure is a detached one- or two-family dwelling not exceeding two stories in height; and
     - 2.3.1. The structure is classified as Seismic Use Group I, as determined in Section 1616.2; and
     - 2.3.2. The structure does not have any of the following plan or vertical irregularities as defined in Section 1616.5:
       - a. Torsional irregularity.
       - b. Nonparallel systems.
       - c. Stiffness irregularity–extreme soft story and soft story.
       - d. Discontinuity in capacity—weak story.

❖ This section provides minimum requirements for quality assurance for seismic-force-resisting systems and designated seismic systems. These requirements supplement the testing and inspection requirements contained in the referenced standards given in other sections of the code. The seismic provisions of the code rely heavily on the concept of quality control for good construction.

The exceptions to the preparation of a quality assurance plan are intended for conventional light-frame wood construction and structures constructed of light wood framing and light gauge cold-formed steel framing with a height no greater than 35 feet (10 668 mm) above grade plane that are located in areas of low seismic risk (S_{DS} does not exceed 0.50g) and that satisfy all of the criteria indicated, or structures constructed of reinforced masonry not more than 25 feet (7620 mm) above grade plane that are located in areas of low seismic risk (S_{DS} does not exceed 0.50g), and that satisfy all of the criteria indicated.

Exception 2.3 applies to detached one- or two-family dwellings not exceeding two stories in height that are included in Seismic Use Group I. The exception is also limited to those structures that do not have any of the following irregularities: torsional irregularity; extreme torsional irregularity; nonparallel systems; stiffness irregularity (soft story); stiffness irregularity (extreme soft story) or discontinuity in capacity (weak story). Any structure that does not satisfy all of the criteria included in the exception or is otherwise exempted by the code is required to have a quality assurance plan prepared by a registered design professional. It is important to emphasize that this exception is for the preparation of a quality assurance plan, not for the design of the structure in accordance with the requirements of the code.

1705.2 Quality assurance plan preparation. The design of each designated seismic system shall include a quality assurance plan prepared by a registered design professional. The quality assurance plan shall identify the following:

1. The designated seismic systems and seismic-force-resisting systems that are subject to quality assurance in accordance with Section 1705.1.
2. The special inspections and testing to be provided as required by Sections 1704 and 1708 and other applicable sections of this code, including the applicable reference standards referred to by this code.
3. The type and frequency of testing required.
4. The type and frequency of special inspections required.
5. The required frequency and distribution of testing and special inspection reports.
6. The structural observations to be performed.
7. The required frequency and distribution of structural observation reports.

❖ The quality assurance plan must be prepared by the registered design professional responsible for the design of each designated seismic system that is subject to quality assurance, whether it be architectural, electrical, mechanical or structural in nature. The quality assurance plan may be a very simple listing of those elements of each system that have been designated as being important enough to receive special inspections or testing. The extent and duration of the inspections must be set forth in the quality assurance plan, as well as the specific tests and the frequency of testing that is required.

1705.3 Contractor responsibility. Each contractor responsible for the construction of a seismic-force-resisting system, designated seismic system, or component listed in the quality assurance plan shall submit a written contractor's statement of responsibility to the building official and to the owner prior to the commencement of work on the system or component. The contractor's statement of responsibility shall contain the following:

1. Acknowledgment of awareness of the special requirements contained in the quality assurance plan.
2. Acknowledgment that control will be exercised to obtain conformance with the construction documents approved by the building official.
3. Procedures for exercising control within the contractor's organization, the method and frequency of reporting, and the distribution of the reports.
4. Identification and qualifications of the person(s) exercising such control and their position(s) in the organization.

❖ The authority having jurisdiction must approve the quality assurance plan and obtain from each contractor a written statement that he or she understands the requirements of the quality assurance plan and will exercise the necessary control to obtain conformance. The exact methods of control are the responsibility of the individual contractors, subject to approval by the authority having jurisdiction. However, special inspections of the work are required in specific situations to provide the authority having jurisdiction reasonable assurance that there is compliance with the approved construction documents.

The extent of the qualifications of contractors and subcontractors can vary considerably, which is why the extent of quality control can also vary considerably. The quality assurance plan, therefore, is an opportunity to identify those areas of special concern that must be addressed during the construction process. Those areas include but are not limited to: types of testing; frequency of testing; types of inspections; frequency of inspections and the extent of the structural observations to be performed.

SECTION 1706 QUALITY ASSURANCE FOR WIND REQUIREMENTS

1706.1 Scope. A quality assurance plan shall be provided in accordance with Section 1706.1.1.

❖ This section provides minimum requirements for quality assurance of windforce-resisting systems and designated wind systems. These requirements supplement the testing and inspection requirements contained in the referenced standards given in other sections of the code. The wind provisions of the code are written to rely heavily on the concept of quality controls for good construction.

1706.1.1 Where required. A quality assurance plan for wind requirements shall be provided for all structures constructed in the following areas:

1. In wind exposure Categories A and B, where the 3-second-gust basic wind speed is 120 miles per hour (mph) (52.8 m/sec) or greater.
2. In wind exposure Categories C and D, where the 3-second-gust basic wind speed is 110 mph (49 m/sec) or greater.

Exception: A quality assurance plan is not required for structures designed and constructed in accordance with the *International Residential Code* or the conventional construction provisions of Section 2308 of this code, provided that all of the applicable items listed in Section 1706.1.2 are inspected during construction by a qualified person approved by the building official.

❖ A quality assurance plan for wind requirements is only required for high-wind areas. The exception to the preparation of a quality assurance plan is intended for conventional light-frame wood structures designed and constructed in accordance with the *International Residential Code*® (IRC®), provided that all of the applicable items listed in Section 1706.1.2 are inspected.

1706.1.2 Detailed requirements. Where required by Section 1706.1.1, a quality assurance plan shall be provided for the following:

1. Roof cladding and roof framing connections.
2. Wall connections to roof and floor diaphragms and framing.
3. Roof and floor diaphragm systems, including collectors, drag struts, and boundary elements.
4. Vertical windforce-resisting systems, including braced frames, moment frames and shear walls.
5. Windforce-resisting system connections to the foundation.
6. Fabrication and installation of components and assemblies required to meet the impact resistance requirements of Section 1609.1.4.

Exception: Fabrication of manufactured components and assemblies that have a label indicating compliance with the wind-load and impact-resistance requirements of this code.

❖ This section designates the wind systems that the quality assurance plan must address when one is required. Note that these items will require inspection during construction for those structures listed in the exception to Section 1706.1.1, even though a quality assurance plan is not required. The exception exempts manufactured components and assemblies that are labeled as complying with the required wind load and impact resistance.

1706.2 Quality assurance plan preparation. The design of each main windforce-resisting system and each wind-resisting component shall include a quality assurance plan prepared by a registered design professional.

Exception: For construction that is not required to be designed by a registered design professional, the quality assurance plan may be prepared by a qualified person approved by the building official.

The quality assurance plan shall identify the following:

1. The main windforce-resisting systems and wind-resisting components that are subject to quality assurance in accordance with Section 1706.1.
2. The special inspections and testing to be provided as required by Section 1704 and other applicable sections of this code, including the applicable referenced standards referred to by this code.
3. The type and frequency of testing required.
4. The type and frequency of special inspections required.
5. The required frequency and distribution of testing and special inspection reports.
6. The structural observations to be performed.
7. The required frequency and distribution of structural observation reports.

❖ The quality assurance plan must be prepared by a registered design professional. It may be a very simple listing of elements of each system designated as important enough to receive special inspections or testing. The extent and duration of the inspections must be set forth in the quality assurance plan, as well as the specific tests and frequency of testing required. The exception allows the quality assurance plan to be prepared by a qualified person other than a registered design professional if the construction is not required to be designed by a registered design professional.

1706.3 Contractor responsibility. Each contractor responsible for the construction of a main windforce-resisting system or a wind-resisting component listed in the quality assurance plan shall submit a written contractor's statement of responsibility to the building official and to the owner prior to the commencement of work on the system or component. The contractor's statement of responsibility shall contain the following:

1. Acknowledgment of awareness of the special requirements contained in the quality assurance plan;
2. Acknowledgment that control will be exercised to obtain conformance with the construction documents approved by the building official;
3. Procedures for exercising control within the contractor's organization, the method and frequency of reporting, and the distribution of the reports; and
4. Identification and qualifications of the person(s) exercising such control and their position(s) in the organization.

❖ The authority having jurisdiction must approve the quality assurance plan and must obtain from each contractor a written statement that he or she understands the requirements of the quality assurance plan and will exercise the necessary control to obtain conformance. The exact methods of control are the responsibility of individual contractors, subject to approval by the authority having jurisdiction. However, special inspections of the work are required in specific situations to provide the authority having jurisdiction reasonable assurance that there is compliance with the approved construction documents.

The extent of the qualifications of the contractor and subcontractors can vary considerably. The quality assurance plan, therefore, is an opportunity to identify those areas of special concern that must be addressed during the construction process, including, but not limited to, types of testing, frequency of testing, types of inspections, frequency of inspections and the extent of the structural observations to be performed.

SECTION 1707
SPECIAL INSPECTIONS FOR SEISMIC RESISTANCE

1707.1 Special inspections for seismic resistance. Special inspection as specified in this section is required for the following, where required in Section 1704.1. Special inspections itemized in Sections 1707.2 through 1707.8 are required for the following:

1. The seismic-force-resisting systems in structures assigned to Seismic Design Category C, D, E or F, as determined in Section 1616.
2. Designated seismic systems in structures assigned to Seismic Design Category D, E or F.
3. Architectural, mechanical and electrical components in structures assigned to Seismic Design Category C, D, E or F that are required in Sections 1707.6 and 1707.7.

❖ The requirements in this section supplement the inspection requirements contained in the referenced standards in other sections of the code. The seismic provisions of the code are written to rely heavily on the concept of special inspections for good construction.

1707.2 Structural steel. Continuous special inspection for structural welding in accordance with AISC 341.

Exceptions:

1. Single-pass fillet welds not exceeding $^5/_{16}$ inch (7.9 mm) in size.
2. Floor and roof deck welding.

❖ Continuous special inspection is required for all structural steel welding except single-pass fillet welds not exceeding $^5/_{16}$ inch (7.9 mm) and floor and roof deck welding.

1707.3 Structural wood. Continuous special inspection during field gluing operations of elements of the seismic-force-resisting system. Periodic special inspections for nailing, bolting, anchoring and other fastening of components within the seismic-force-resisting system, including drag struts, braces and hold-downs.

Exception: Fastening of wood sheathing used for wood shear walls, shear panels and diaphragms where the fastener spacing is more than 4 inches (102 mm) on center (o.c.).

❖ Continuous special inspection of field gluing operations of structural wood is required, while only periodic special inspection of fastenings of components within the seismic-force-resisting system is required. The exception waives periodic special inspection of wood sheathing fastening in shear walls and diaphragms that resist relatively low shear forces. Rather than being based on the component's design shear, fastener spacing is considered a more practical reason for special inspection. All things being equal, as the fastener spacing decreases, not only is the component's design load higher, but also the potential for splitting framing members increases. The latter concern is mainly associated with nailing.

1707.4 Cold-formed steel framing. Periodic special inspections during welding operations of elements of the seismic-force-resisting system. Periodic special inspections for screw attachment, bolting, anchoring and other fastening of components within the seismic-force-resisting system, including struts, braces, and hold-downs.

❖ Periodic special inspection is required for cold-formed steel framing and its fastening.

1707.5 Storage racks and access floors. Periodic special inspection during the anchorage of access floors and storage racks 8 feet (2438 mm) or greater in height in structures assigned to Seismic Design Category D, E or F.

❖ The anchorage of storage racks and access floors in structures assigned to Seismic Design Category D, E or F requires periodic special inspection.

1707.6 Architectural components. Periodic special inspection during the erection and fastening of exterior cladding, interior and exterior nonbearing walls and interior and exterior veneer in structures assigned to Seismic Design Category D, E or F.

Exceptions:

1. Special inspection is not required for architectural components in structures 30 feet (9144 mm) or less in height.
2. Special inspection is not required for cladding and veneer weighing 5 pounds psf (24.5 N/m^2) or less.
3. Special inspection is not required for interior nonbearing walls weighing 15 psf (73.5 N/m^2) or less.

❖ It is anticipated that the minimum requirements for architectural components will be complied with when the special inspector is satisfied that the method of anchoring or fastening and the number, spacing and types of fasteners actually used conform with the approved construction documents for the component installed. It is noted that such special inspection requirements are only for those components in Seismic Design Category D, E or F.

1707.7 Mechanical and electrical components. Periodic special inspection during the anchorage of electrical equipment for emergency or standby power systems in structures assigned to Seismic Design Category C, D, E or F. Periodic special inspection during the installation of anchorage of other electrical equipment in structures assigned to Seismic Design Category E or F. Periodic special inspection during installation of piping systems intended to carry flammable, combustible, or highly toxic contents and their associated mechanical units in structures assigned to Seismic Design Category C, D, E or F. Periodic special inspection during the installation of HVAC ductwork that will contain hazardous materials in structures assigned to Seismic Design Category C, D, E or F.

❖ It is anticipated that the minimum requirements for mechanical and electrical components will be complied with when the special inspector is satisfied that the method of anchoring or fastening and the number, spacing and types of fasteners actually used conform with the approved construction documents for the components installed. It is noted that such special inspection requirements are for selected electrical, lighting, piping and ductwork components in Seismic Design Category C, D, E or F.

1707.7.1 Component inspection. Special inspection is required for the installation of the following components where the component has a Component Importance Factor of 1.0 or 1.5 in accordance with Section 9.6.1.5 of ASCE 7. Evidence of the quality control program shall be permanently identified on each piece of equipment by a label.

1. Equipment using combustible energy sources.
2. Electrical motors, transformers, switchgear unit substations and motor control centers.
3. Reciprocating and rotating-type machinery.
4. Piping distribution systems 3 inches (76 mm) and larger.
5. Tanks, heat exchangers and pressure vessels.

❖ This section delineates the mechanical and electrical components that are required to have special inspection.

1707.7.2 Component and attachment testing. The component manufacturer shall test or analyze the component and the component mounting system or anchorage for the design forces in Chapter 16 for those components having a Component Importance Factor of 1.0 or 1.5 in accordance with Chapter 16. The manufacturer shall submit a certificate of compliance for review and acceptance by the registered design professional responsible for the design, and for approval by the building official. The basis of certification shall be by test on a shaking table, by three-dimensional shock tests, by an analytical method using dynamic characteristics and forces from Chapter 16 or by more rigorous analysis. The special inspector shall inspect the component and verify that the label, anchorage or mounting conforms to the certificate of compliance.

❖ This section requires that components having a Component Importance Factor of 1.0 or 1.5 must demonstrate seismic performance by testing or analytical methods. A certificate of compliance for review and acceptance is also required. The registered design professional responsible for the design then must review and accept the certificate of compliance, and the building official must approve it. Verification is required by the special inspector that the component label, anchorage or mounting conforms to the certificate of compliance.

1707.7.3 Component manufacturer certification. Each manufacturer of equipment to be placed in a building assigned to Seismic Design Categories E and F, in accordance with Chapter 16, where the equipment has a Component Importance Factor of 1.0 or 1.5 in accordance with Chapter 16, shall maintain an approved quality control program. Evidence of the quality control program shall be permanently identified on each piece of equipment by a label.

❖ The manufacturer of equipment that has a Component Importance Factor of 1.0 or 1.5 and is to be used in Seismic Design Categories E and F must maintain an approved quality control program. Each piece of equipment must be labeled as permanent evidence of the quality control program.

1707.8 Seismic isolation system. Provide periodic special inspection during the fabrication and installation of isolator units and energy dissipation devices if used as part of the seismic isolation system.

❖ Seismic isolation units and energy dissipation devices must receive periodic special inspection during fabrication and installation.

SECTION 1708
STRUCTURAL TESTING FOR SEISMIC RESISTANCE

1708.1 Masonry. Testing and verification of masonry materials and assemblies prior to construction shall comply with the requirements of this section, depending on the classification of building or structure or nature of occupancy, as defined in this code (see Table 1604.5 or Section 1616.2).

❖ This section establishes the level of quality assurance required for masonry construction. This is determined by the design method (empirical versus engineered) and whether or not a structure is classified as an essential facility. These requirements are summarized in Table 1708.1. An essential facility would be an Occupancy Category IV in Table 1604.5 and/or Seismic Use Group III in accordance with Section 1616.2.3. The quality assurance levels identified as Level 1 and Level 2 coincide with the level of special inspection required by Section 1704.5.

TABLE 1708.1
QUALITY ASSURANCE OF MASONRY CONSTRUCTION

	TYPE OF FACILITY			
	Nonessential		Essential	
Design method	Empirical	Engineered	Empirical	Engineered
Quality assurance	a	Level 1	Level 1	Level 2

a. Certificate of compliance (see Section 1708.1.1).

1708.1.1 Empirically designed masonry and glass unit masonry in nonessential facilities. For masonry designed by Section 2109 or 2110, or by Chapter 5 or 7 of ACI 530/ASCE 5/TMS 402, in nonessential facilities (see Table 1604.5 or Section 1616.2), certificates of compliance used in masonry construction shall be verified prior to construction.

❖ The minimum level of testing and verification required for empirically designed masonry and glass unit masonry in nonessential facilities is a certificate of compliance indicating that the masonry materials used comply with the code and contract documents.

1708.1.2 Empirically designed masonry and glass unit masonry in essential facilities. The minimum testing and verification prior to construction for masonry designed by Section 2109 or 2110, or by Chapter 5 or 7 of ACI 530/ASCE 5/TMS 402, in essential facilities (see Table 1604.5 or Section 1616.2) shall comply with the requirements of Table 1708.1.2, Level 1 Quality Assurance.

❖ The minimum level of testing and verification required for empirically designed masonry and glass unit masonry in essential facilities is a certificate of compliance indicating that the masonry materials used comply with the code and contract documents plus verification of f'_m prior to construction (see Table 1708.1.2, Level 1 Quality Assurance).

TABLE 1708.1.2
LEVEL 1 QUALITY ASSURANCE

MINIMUM TESTS AND SUBMITTALS
Certificates of compliance used in masonry construction.
Verification of f'_m prior to construction, except where specifically exempted by this code.

❖ See the commentary to Section 1708.1.4 for discussion.

1708.1.3 Engineered masonry in nonessential facilities. The minimum testing and verification prior to construction for masonry designed by Section 2107 or 2108, or by chapters other than Chapter 5, 6 or 7 of ACI 530/ASCE 5/TMS 402, in nonessential facilities (see Table 1604.5 or Section 1616.2) shall comply with Table 1708.1.2.

❖ Engineered masonry in nonessential facilities requires the same testing and verification as empirically designed in essential facilities (see Table 1708.1.2, Level 1 Quality Assurance).

1708.1.4 Engineered masonry in essential facilities. The minimum testing and verification prior to construction for masonry designed by Section 2107 or 2108, or by chapters other than Chapter 5, 6 or 7 of ACI 530/ASCE 5/TMS 402, in essential facilities (see Table 1604.5 or Section 1616.2) shall comply with Table 1708.1.4, Level 2 Quality Assurance.

❖ Engineered masonry in essential facilities requires testing and verification prior to construction. The minimum tests and submittals are listed in Table 1708.1.4. The requirements are the same as for engineered masonry in nonessential facilities, with the added requirements that f'_m must be verified every 5,000 square feet (465 m^2) during construction and the proportions of materials in mortar and grout must be verified as delivered to the site (see Table 1708.1.4, Level 2 Quality Assurance).

TABLE 1708.1.4
LEVEL 2 QUALITY ASSURANCE

MINIMUM TESTS AND SUBMITTALS
Certificates of compliance used in masonry construction.
Verification of f'_m prior to construction and every 5000 square feet during construction.
Verification of proportions of materials in mortar and grout as delivered to the site.

For SI: 1 square foot = 0.0929 m^2.

❖ See the commentary to Section 1708.1.4 for discussion.

1708.2 Testing for seismic resistance. The tests specified in Sections 1708.3 through 1708.6 are required for the following:

1. The seismic-force-resisting systems in structures assigned to Seismic Design Category C, D, E or F, as determined in Section 1616.
2. Designated seismic systems in structures assigned to Seismic Design Category D, E or F.
3. Architectural, mechanical and electrical components in structures assigned to Seismic Design Category C, D, E or F that are required in Section 1708.5.

❖ This section specifies when material seismic resistance tests for seismic-force-resisting systems and designated seismic systems are required. These requirements supplement the test requirements contained in the referenced standards given in other sections of the code. The seismic provisions of the code are written to rely heavily on the concept of material tests such that good quality material is used for seismic-resistant construction.

1708.3 Reinforcing and prestressing steel. Certified mill test reports shall be provided for each shipment of reinforcing steel used to resist flexural, shear and axial forces in reinforced concrete intermediate frames, special moment frames and boundary elements of special reinforced concrete or reinforced masonry shear walls. Where ASTM A 615 reinforcing steel is used to resist earthquake-induced flexural and axial forces in special moment frames and in wall boundary elements of shear walls in structures assigned to Seismic Design Category D, E or F, as determined in Section 1616, the testing requirements of ACI 318 shall be met. Where ASTM A 615 reinforcing steel is to be welded, chemical tests shall be performed to determine weldability in accordance with Section 1903.5.2.

❖ Certified material test reports are required for rebar. When ASTM A 615 is used in special moment frames and shear walls in structures in Seismic Design Category D, E or F, the testing requirements of ACI 318 must be used. Where ASTM A 615 rebar is to be welded, a chemical analysis must be provided to determine weldability of the steel in accordance with AWS D1.4.

1708.4 Structural steel. The testing contained in the quality assurance plan shall be as required by AISC 341 and the additional requirements herein. The acceptance criteria for nondestructive testing shall be as required in AWS D1.1 as specified by the registered design professional.

Base metal thicker than 1.5 inches (38 mm), where subject to through-thickness weld shrinkage strains, shall be ultrasonically tested for discontinuities behind and adjacent to such welds after joint completion. Any material discontinuities shall be accepted or rejected on the basis of ASTM A 435 or ASTM A 898 (Level 1 Criteria) and criteria as established by the registered design professional(s) in responsible charge and the construction documents.

❖ Structural steel must be tested as required by AISC 341. Any nondestructive testing must use the acceptance criteria in AWS D1.1 and as specified by the registered design professional. Base metal thicker than 1.5 inches (38 mm) is subject to laminations and discontinuities during the steel-making process. Any base metal thicker than 1.5 inches (38 mm) that will be subjected to through-thickness weld shrinkage strains must be ultrasonically tested after welding is completed. The acceptance criteria for ultrasonic testing are required to be on the basis of ASTM A 435 or A 898 (Level 1 criteria) and as established by the registered design professional.

1708.5 Mechanical and electrical equipment. Each manufacturer of designated seismic system components shall test or analyze the component and its mounting system or anchorage and shall submit a certificate of compliance for review and acceptance by the registered design professional in responsible charge of the design of the designated seismic system and for approval by the building official. The evidence of compliance shall be by actual test on a shake table, by three-dimensional shock tests, by an analytical method using dynamic characteristics and forces, by the use of experience data (i.e., historical data demonstrating acceptable seismic performance), or by more rigorous analysis providing for equivalent safety. The special inspector shall examine the designated seismic system and shall determine whether the anchorages and label conform with the evidence of compliance.

❖ Mechanical and electrical components having a Component Importance Factor of 1.0 or 1.5 must demonstrate seismic performance by testing or analytical methods. A certificate of compliance for review and acceptance is also required. The registered design professional responsible must then review and accept the certificate of compliance, and the building official must approve it. Verification is required by the special inspector that the component label, anchorage or mounting conforms to the certificate of compliance.

1708.6 Seismically isolated structures. For required system tests, see Section 9.13.9 of ASCE 7.

❖ The referenced section of ASCE 7 contains detailed provisions for isolation system testing. This testing provides effective stiffness and effective damping values to be used in the design of a seismically isolated structure. A minimum of two full-size specimens must be tested for each proposed type and size of isolator unit. These prototypes are not to be used in the construction, unless approved by both the registered design professional and the building official.

SECTION 1709 STRUCTURAL OBSERVATIONS

1709.1 Structural observations. Structural observations shall be provided for those structures included in Seismic Design Category D, E or F, as determined in Section 1616, where one or more of the following conditions exist:

1. The structure is included in Seismic Use Group II or III.
2. The height of the structure is greater than 75 feet (22 860 mm) above the base.
3. The structure is in Seismic Design Category E and Seismic Use Group I and greater than two stories in height.
4. When so designated by the registered design professional in responsible charge of the design.
5. When such observation is specifically required by the building official.

Structural observations shall also be provided for those structures sited where the basic wind speed exceeds 110 miles per hour (3 second gust) determined from Figure 1609, where one or more of the following conditions exist:

1. The structure is included in Category III or IV according to Table 1604.5.
2. The height of the structure is greater than 75 feet (22 860 mm).
3. When so designated by the registered design professional in responsible charge of the design,
4. When such observation is specifically required by the building official.

The owner shall employ a registered design professional to perform structural observation as defined in Section 1702. Deficiencies shall be reported in writing to the owner and the building official. At the conclusion of the work included in the permit, the structural observer shall submit to the building official a written statement that the site visits have been made and identify any reported deficiencies which, to the best of the structural observer's knowledge, have not been resolved.

❖ This section requires that a registered engineer or architect be employed by the owner to provide on-site visits to observe compliance with the structural drawings. Structural observations are required under certain conditions in high-seismic and high-wind locations (see Section 1702 for the definition of "Structural observation").

The intent of requiring structural observations by a registered design professional for these structures is to verify that the seismic-force-resisting systems, the designated seismic systems and the windforce-resisting systems are constructed in general conformance with the construction documents.

SECTION 1710 DESIGN STRENGTHS OF MATERIALS

1710.1 Conformance to standards. The design strengths and permissible stresses of any structural material that are identified by a manufacturer's designation as to manufacture and grade by mill tests, or the strength and stress grade is otherwise confirmed to the satisfaction of the building official, shall conform to the specifications and methods of design of accepted engineering practice or the approved rules in the absence of applicable standards.

❖ Structural materials must conform to applicable design standards, approved rules and accepted methods of engineering practice. Conformance to these provisions and to the manufacturer's designations provides the building official with the information needed to verify that the materials will perform their intended function satisfactorily.

1710.2 New materials. For materials that are not specifically provided for in this code, the design strengths and permissible stresses shall be established by tests as provided for in Section 1711.

❖ Materials that are not explicitly covered by the code are allowed when subjected to the appropriate testing demonstrating adequate performance (see Section 1701.2).

SECTION 1711 ALTERNATIVE TEST PROCEDURE

1711.1 General. In the absence of approved rules or other approved standards, the building official shall make, or cause to be made, the necessary tests and investigations; or the building official shall accept duly authenticated reports from approved agencies in respect to the quality and manner of use of new materials or assemblies as provided for in Section 104.11. The cost of all tests and other investigations required under the provisions of this code shall be borne by the permit applicant.

❖ Test reports from approved agencies may be used as a basis for approval of materials that are not within the purview of any approved rules (i.e., "new materials" as mentioned in Section 1701.2). This section directly references Section 104.11. It is within the power of the building official to accept reports from an approved agency. In determining the approval, the building official should check that the agency is an independent third-party agency with no financial or fiduciary affiliations with the applicant or material supplier. The capability and competency of the agency must also be examined. It should be noted that this section assigns responsibility for costs of the testing to the applicant.

SECTION 1712 TEST SAFE LOAD

1712.1 Where required. Where proposed construction is not capable of being designed by approved engineering analysis, or where proposed construction design method does not comply with the applicable material design standard, the system of construction or the structural unit and the connections shall be subjected to the tests prescribed in Section 1714. The building official shall accept certified reports of such tests conducted by an approved testing agency, provided that such tests meet the requirements of this code and approved procedures.

❖ Testing to determine safe load is required when a structural component cannot be designed in accordance with approved engineering practices or where the construction design method does not fully comply with the respective material design standard listed in Chapter 35. If either of these situations exist, the structural components are required to be subjected to the prescriptive tests listed in Section 1714, which address loading and deflection criteria.

An example of a structural component that cannot be designed by approved engineering practice is a composite concrete and steel slab in which the shear connector is some type of new configuration called a "widget." The horizontal shear that can be developed is unknown, therefore, a complete analysis cannot be performed. This section also restates the building official's option of accepting data from an approved testing agency, as previously stated in Section 1703.4.

SECTION 1713 IN-SITU LOAD TESTS

1713.1 General. Whenever there is a reasonable doubt as to the stability or load-bearing capacity of a completed building, structure or portion thereof for the expected loads, an engineering assessment shall be required. The engineering assessment shall involve either a structural analysis or an in-situ load test, or both. The structural analysis shall be based on actual material properties and other as-built conditions that affect stability or load-bearing capacity, and shall be conducted in accordance with the applicable design standard. If the structural assessment determines that the load-bearing capacity is less than that required by the code, load tests shall be conducted in accordance with Section 1713.2. If the building, structure or portion thereof is found to have inadequate stability or load-bearing capacity for the expected loads, modifications to ensure structural adequacy or the removal of the inadequate construction shall be required.

❖ The intent of this section is to utilize an engineering analysis to verify the adequacy of the structure, if possible. The load test requirement should only be done if an engineering analysis does not verify structural adequacy. Load tests are last resort options, and the building official should document his or her reasons for any load testing requirement.

An example of the structural analysis executed would be an analysis by a third-party engineering firm acceptable to both the building official and the owner. The structural integrity may be questioned for items such as visible signs of excessive settlement or lateral deflection, such as cracks in concrete foundation walls or excessive vibration when the assembly is loaded.

A load test procedure must simulate the actual load conditions to which the structure is subjected during normal use (see Section 1713.3 for details).

1713.2 Test standards. Structural components and assemblies shall be tested in accordance with the appropriate material standards listed in Chapter 35. In the absence of a standard that contains an applicable load test procedure, the test procedure shall be developed by a registered design professional and approved. The test procedure shall simulate loads and conditions of application that the completed structure or portion thereof will be subjected to in normal use.

❖ When load test procedures for materials are given by the applicable referenced material standard, the test procedure outlined in that specific standard must be adhered to without variation. If a referenced standard lacks a load test procedure or a material or assembly does not have a specific referenced standard, then such a test must be developed by a registered design professional and approved by the building official. The test procedure must be representative of and simulate the actual loading conditions that the completed structure or portion thereof will be subjected to during normal use.

1713.3 In-situ load tests. In-situ load tests shall be conducted in accordance with Section 1713.3.1 or 1713.3.2 and shall be supervised by a registered design professional. The test shall simulate the applicable loading conditions specified in Chapter 16 as necessary to address the concerns regarding structural stability of the building, structure or portion thereof.

❖ The criteria for in-situ load tests are set forth for two categories: procedures specified, which are regulated by Section 1713.3.1, and procedures not specified, which are regulated by Section 1713.3.2. This section further requires that the test be performed under the supervision of a registered design professional and that it simulate the actual loads and conditions of the completed structure or portion thereof.

1713.3.1 Load test procedure specified. Where a standard listed in Chapter 35 contains an applicable load test procedure and acceptance criteria, the test procedure and acceptance criteria in the standard shall apply. In the absence of specific load factors or acceptance criteria, the load factors and acceptance criteria in Section 1713.3.2 shall apply.

❖ The load test must be in accordance with the applicable referenced standard. Section 1713.3.2 must only be utilized in the absence of either a specific standard or specific load factors and acceptance criteria from applicable referenced standards.

1713.3.2 Load test procedure not specified. In the absence of applicable load test procedures contained within a standard referenced by this code or acceptance criteria for a specific material or method of construction, such existing structure shall be subjected to a test procedure developed by a registered design professional that simulates applicable loading and deformation conditions. For components that are not a part of the seismic load-resisting system, the test load shall be equal to two times the unfactored design loads. The test load shall be left in place for a period of 24 hours. The structure shall be considered to have successfully met the test requirements where the following criteria are satisfied:

1. Under the design load, the deflection shall not exceed the limitations specified in Section 1604.3.
2. Within 24 hours after removal of the test load, the structure shall have recovered not less than 75 percent of the maximum deflection.
3. During and immediately after the test, the structure shall not show evidence of failure.

❖ If the applicable standards do not specify load factor or testing criteria acceptance methods, then the testing criteria listed in this section must be followed. Note that the design load includes design live load and all dead loads that are not yet in place, such as the dead load from tenant walls in a speculative office building.

SECTION 1714
PRECONSTRUCTION LOAD TESTS

1714.1 General. In evaluating the physical properties of materials and methods of construction that are not capable of being designed by approved engineering analysis or that do not comply with applicable material design standards listed in Chapter 35, the structural adequacy shall be predetermined based on the load test criteria established in this section.

❖ This section establishes requirements for load testing structural assemblies that are either incapable of being designed or those that, for one reason or another, do not comply with the applicable material design standards. This section does not govern the load testing of existing buildings, which is governed by Section 1713.

The different categories of preconstruction load tests are addressed herein. Specified load test procedures are regulated by Section 1714.2. Load test procedures that are not specified are regulated by Section 1714.3. Wall and partition assemblies are regulated by Section 1714.4. Exterior window and door assemblies are regulated by Section 1714.5, and test specimens are regulated by Section 1714.6.

1714.2 Load test procedures specified. Where specific load test procedures, load factors and acceptance criteria are included in the applicable design standards listed in Chapter 35, such test procedures, load factors and acceptance criteria shall apply. In the absence of specific test procedures, load factors or acceptance criteria, the corresponding provisions in Section 1714.3 shall apply.

❖ This section has priority over Section 1714.3, provided that load factors and acceptance criteria are established in the applicable design standards.

1714.3 Load test procedures not specified. Where load test procedures are not specified in the applicable design standards listed in Chapter 35, the load-bearing and deformation capacity of structural components and assemblies shall be determined on the basis of a test procedure developed by a registered design professional that simulates applicable loading and deformation conditions. For components and assemblies that are not a part of the seismic load-resisting system, the test shall be as specified in Section 1714.3.1. Load tests shall simulate the applicable loading conditions specified in Chapter 16.

❖ In the absence of load factors and acceptance criteria in the applicable design standards and in accordance with Section 1714.2, this section is to be used by the building official to determine if conformance to the applicable code requirements has been achieved. Additionally, this section provides the design professional and the building official with specific loading and pass/fail criteria.

1714.3.1 Test procedure. The test assembly shall be subjected to an increasing superimposed load equal to not less than two times the superimposed design load. The test load shall be left in place for a period of 24 hours. The tested assembly shall be considered to have successfully met the test requirements if the assembly recovers not less than 75 percent of the maximum deflection within 24 hours after the removal of the test load. The test assembly shall then be reloaded and subjected to an increasing superimposed load until either structural failure occurs or the superimposed load is equal to two and one-half times the load at which the deflection limitations specified in Section 1714.3.2 were reached, or the load is equal to two and one-half times the superimposed design load. In the case of structural components and assemblies for which deflection limitations are not specified in Section 1714.3.2, the test specimen shall be subjected to an increasing superimposed load until structural failure occurs or the load is equal to two and one-half times the desired superimposed design load. The allowable superimposed design load shall be taken as the lesser of:

1. The load at the deflection limitation given in Section 1714.3.2.

2. The failure load divided by 2.5.
3. The maximum load applied divided by 2.5.

❖ Load test criteria relating to superimposed design loads are established herein. These requirements are a compilation of commonly accepted engineering practices to test adequately against structural failure.

In the case of structural components and assemblies for which maximum deflection limitations are not addressed in Section 1714.3.2, the test assemblies must be subjected to increasing superimposed loads until failure occurs or the load is equal to two and one-half times the superimposed design load, whichever occurs first.

1714.3.2 Deflection. The deflection of structural members under the design load shall not exceed the limitations in Section 1604.3.

❖ Acceptance criteria for deflection of structural systems when subjected to the allowable design load are used to demonstrate adequate structural performance and are addressed in Section 1604.3.

1714.4 Wall and partition assemblies. Load-bearing wall and partition assemblies shall sustain the test load both with and without window framing. The test load shall include all design load components. Wall and partition assemblies shall be tested both with and without door and window framing.

❖ Load-bearing wall and partition assemblies must sustain loads with and without window framing. It is not appropriate to assume that a wall will sustain loads better if window framing is involved in the test. Each individual design must be evaluated separately based on the construction of that assembly. All design loads, such as vertical and lateral forces, must be included in the test.

1714.5 Exterior window and door assemblies. The design pressure rating of exterior windows and doors in buildings shall be determined in accordance with Section 1714.5.1 or 1714.5.2.

Exception: Structural wind load design pressures for window units smaller than the size tested in accordance with Section 1714.5.1 or 1714.5.2 shall be permitted to be higher than the design value of the tested unit provided such higher pressures are determined by accepted engineering analysis. All components of the small unit shall be the same as the tested unit. Where such calculated design pressures are used, they shall be validated by an additional test of the window unit having the highest allowable design pressure.

❖ This section allows two methods of load test for exterior window and door assemblies. The first method, provided for in Section 1714.5.1, allows products to be tested and labeled as conforming to AAMA/NWWDA/101/I.S.2. The second method allows products to be tested in accordance with ASTM E 330 and the glazing must comply with Section 2403. The exception allows window units smaller than the size tested to have higher design pressures, provided the higher pressures are determined by accepted engineering analysis, all components of the smaller unit are the same as the tested unit and an additional test of the smaller unit having the highest calculated design pressure is performed in accordance with Section 1714.5.1 or 1714.5.2.

1714.5.1 Aluminum, vinyl and wood exterior windows and glass doors. Aluminum, vinyl and wood exterior windows and glass doors shall be labeled as conforming to AAMA/NWWDA 101/I.S.2 or 101/I.S.2/NAFS. The label shall state the name of the manufacturer, the approved labeling agency and the product designation as specified in AAMA/NWWDA 101/I.S.2 or 101/I.S.2/NAFS. Products tested and labeled as conforming to AAMA/NWWDA 101/I.S.2 or 101/I.S.2/NAFS shall not be subject to the requirements of Sections 2403.2 and 2403.3.

❖ This section requires exterior windows and doors complying with AAMA/NWWDA/101/I.S.2 or 101/I.S.2/NAFS to be labeled as such. Products so tested and labeled must not be required to meet Sections 2403.2 and 2403.3. By requiring the product to be labeled, the building official does not have to interpret test results and determine load-carrying capacities, or accept the manufacturer's interpretation of tests.

1714.5.2 Exterior windows and door assemblies not provided for in Section 1714.5.1. Exterior window and door assemblies shall be tested in accordance with ASTM E 330. Exterior window and door assemblies containing glass shall comply with Section 2403. The design pressure for testing shall be calculated in accordance with Chapter 16. Each assembly shall be tested for 10 seconds at a load equal to 1.5 times the design pressure.

❖ This section allows an alternate to the provisions of Section 1714.5.1.

This procedure is to be used to verify the integrity of door and window assemblies as a whole, and its results do not supersede, but rather complement, the requirements of Chapter 24. Glass thickness must be determined in accordance with the provisions of Chapter 24. The assemblies must comply with Section 2403, and the design pressure for testing is determined from Chapter 16. Each assembly is required to be tested for 10 seconds at a load equal to 1.5 times the design pressure. When testing a product line that has a variety of sizes, the most critical size can usually be tested and the results used to qualify other similar products within its family. In general, the larger size and the most heavily loaded of each particular design, type, construction or configuration should be tested. The code does not specify how many specimens are to be tested. ASTM E 330 states that if only one sample is tested, it should be selected by the specifying authority.

1714.6 Test specimens. Test specimens and construction shall be representative of the materials, workmanship and details normally used in practice. The properties of the materials used to construct the test assembly shall be determined on the basis of tests on samples taken from the load assembly or on representative samples of the materials used to construct the load test assembly. Required tests shall be conducted or witnessed by an approved agency.

❖ The test specimen must resemble and simulate as much as possible the design being tested using materials and workmanship that could be expected in the actual construction or fabrication the test itself must be witnessed or conducted by an agency acceptable to and approved by the building official.

SECTION 1715 MATERIAL AND TEST STANDARDS

1715.1 Test standards for joist hangers and connectors.

❖ This section prescribes the test standard and criteria to be used for joist hangers and connectors. This criteria is meant to be used in establishing the load capacity of joist hangers and connectors used in wood construction for which there is no calculated procedure recognized by the code.

1715.1.1 Test standards for joist hangers. The vertical load-bearing capacity, torsional moment capacity, and deflection characteristics of joist hangers shall be determined in accordance with ASTM D 1761, using lumber having a specific gravity of 0.49 or greater, but not greater than 0.55, as determined in accordance with AFPA NDS for the joist and hangers.

❖ The specified ASTM standard test method provides a procedure for evaluating the vertical load-carrying capacity, torsional moment capacity and deflection characteristics of joist hangers and similar devices used to connect wood joists to headers of wood or other materials. The lumber used for the test specimen must have a specific gravity equal to or greater than 0.49, but not greater than 0.55.

1715.1.2 Vertical load capacity for joist hangers. The vertical load capacity for the joist hanger shall be determined by testing three joist hanger assemblies as specified in ASTM D 1761. If the ultimate vertical load for any one of the tests varies more than 20 percent from the average ultimate vertical load, at least three additional tests shall be conducted. The allowable vertical load for a normal duration of loading of the joist hanger shall be the lowest value determined from the following:

1. The lowest ultimate vertical load from any test divided by three (where three tests are conducted and each ultimate vertical load does not vary more than 20 percent from the average ultimate vertical load).
2. The average ultimate vertical load for all tests divided by six (where six or more tests are conducted).
3. The vertical load at which the vertical movement of the joist with respect to the header is 0.125 inch (3.2 mm) in any test.
4. The allowable design load for nails or other fasteners utilized to secure the joist hanger to the wood members.
5. The allowable design load for the wood members forming the connection.

❖ The method prescribed establishes the allowable load for normal duration, as defined by the American Forest & Paper Association (AF&PA) *National Design Specification (NDS) for Wood Construction*. Additionally, allowable stresses cannot exceed those allowed by the code. For example, published allowable loads cannot contain nail loads higher than those allowed by NDS, nor can tension in steel strapping exceed that allowed by the steel design standards noted in Chapter 22.

For loads of other than normal duration, the stresses or loads may be increased or must be decreased as noted by the appropriate design standard, but in no case can the load exceed that which will produce $^1/_8$-inch (3.2 mm) deflection.

EXAMPLE:

GIVEN: A manufacturer's test results for a particular joist hanger are as follows:

Test 1 Ultimate load = 1,000 lbs with the $^1/_8$-inch deflection occurring at 400 lbs.

Test 2 Ultimate load = 1,100 lbs with the $^1/_8$-inch deflection occurring at 350 lbs.

Test 3 Ultimate load = 900 lbs with the $^1/_8$-inch deflection occurring 375 lbs.

The manufacturer submitted structural calculations indicating that the allowable design load of the nails is 250 pounds (114 kg), and the allowable shear load in the wood joists framing to the hangers is 280 pounds (127 kg). Joist hanger geometry does not allow for any meaningful calculation of stresses in the steel sections of the hanger.

FIND: The allowable load for the joist hanger.

SOLUTION:

Average ultimate load = (1,000 + 1,100 + 900) ÷ 3.0 = 1,000 lbs

Since 20 percent of 1,000 is 200, the test scatter is within the allowable range of plus and minus 20 percent of the average ultimate load; therefore, three tests are sufficient to establish allowable load.

Thus, the allowable load is 250 pounds (114 kg) based on the lesser of:

Lowest ultimate load ÷ 3.0 = 300 lbs (137 kg)

$^1/_8$-inch deflection in any test = 350 lbs (159 kg)

Allowable nail load = 250 lbs (114 kg)

Allowable joist shear = 280 lbs (127 kg)

In this case, the allowable vertical load for a normal duration is limited by the calculated allowable design load of the fastener [250 lbs (113 kg)] in accordance with Item 4 of Section 1715.1.2. This value is permitted to be modified by a duration of loading factor; however, the modified value cannot exceed the lowest test value as determined in accordance with Items 1, 2 and 3 of Section 1715.1.2.

1715.1.3 Torsional moment capacity for joist hangers. The torsional moment capacity for the joist hanger shall be determined by testing at least three joist hanger assemblies as specified in ASTM D 1761. The allowable torsional moment for normal duration of loading of the joist hanger shall be the average torsional moment at which the lateral movement of the top or bottom of the joist with respect to the original position of the joist is 0.125 inch (3.2 mm).

❖ The allowable torsional moment capacity for a joist hanger is determined by testing in accordance with ASTM D 1761 with the limitation that rotational deflection of the top or bottom of the joist with respect to the header must not exceed 0.125 inch (3.2 mm). This allowable torsional moment is not to be modified by duration of loading factors.

1715.1.4 Design value modifications for joist hangers. Allowable design values for joist hangers that are determined by Item 4 or 5 in Section 1715.1.2 shall be permitted to be modified by the appropriate duration of loading factors as specified in AFPA NDS but shall not exceed the direct loads as determined by Item 1, 2 or 3 in Section 1715.1.2. Allowable design values determined by Item 1, 2 or 3 in Sections 1715.1.2 and 2305.1 shall not be modified by duration of loading factors.

❖ The calculated allowable design values, as determined by Item 4 or 5 of Section 1715.1.2 and modified by duration of loading factors, must not exceed the lowest test value as determined by Item 1, 2 or 3 of Section 1715.1.2 or the value from Section 2305.1.

1715.2 Concrete and clay roof tiles.

❖ This section prescribes the test standards and criteria to be used to determine the overturning resistance and wind characteristics of concrete and clay roof tiles.

1715.2.1 Overturning resistance. Concrete and clay roof tiles shall be tested to determine their resistance to overturning due to wind in accordance with SBCCI SSTD 11 and Chapter 15.

❖ Section 1715.2.1 requires concrete and clay tiles to be tested to determine their overturning resistance in accordance with SBCCI SSTD 11. SSTD 11 prescribes methods for determining the allowable overturning moment for mechanically fastened, adhesive-set and mortar-set tiles. A test procedure is also prescribed for determining the allowable uplift loads on hip/ridge tiles.

1715.2.2 Wind tunnel testing. When roof tiles do not satisfy the limitations in Chapter 16 for rigid tile, a wind tunnel test shall be used to determine the wind characteristic of the concrete or clay tile roof covering in accordance with SBCCI SSTD 11 and Chapter 15.

❖ The wind tunnel test procedures in SSTD 11 must be used if the roof tiles do not meet the limitations of Chapter 16 for rigid tile.

Bibliography

The following resource materials are referenced in this chapter or are relevant to the subject matter addressed in this chapter.

AAMA/NAFS-1, *Voluntary Performance Specification for Window, Skylights and Glass Doors.* Schamburg, IL: American Architectural Manufacturers Association, 1997.

AAMA/NWWDA 101/I.S.2-97, *Voluntary Specifications for Aluminum Vinyl (PVC) and Wood Windows and Glass Doors.* Schaumburg, IL: American Architectural Manufacturers Association, 1997.

ACI 318-02, *Building Code Requirements for Structural Concrete.* Farmington Hills, MI: American Concrete Institute, 2002.

ACI 318-02, *Building Code Requirements for Structural Concrete*, Farmington Hills, MI: American Concrete Institute, 2002.

ACI 530/ASCE 5/TMS 402-02, *Building Code Requirements for Masonry Structures.* Farmington Hills, MI: American Concrete Institute; New York: American Society of Civil Engineers; Boulder, CO: The Masonry Council, 2002.

ACI 530.1/ASCE 6/TMS 602-02, *Specifications for Masonry Structures.* Farmington Hills, MI: American Concrete Institute; New York: American Society of Civil Engineers; Boulder, CO: The Masonry Council, 2002.

AF&PA NDS-01, *National Design Specification for Wood Construction.* Washington, DC: American Forest & Paper Association, 2001.

AISC ASD-89, *Specification for Structural Steel Buildings — Allowable Stress Design, Plastic Design.* Chicago: American Institute of Steel Construction, 1989.

AISC LRFD-99, *Load and Resistance Factor Design Specification for Structural Steel Buildings.* Chicago: American Institute of Steel Construction, 1999.

AISC Seismic (2002), *Seismic Provisions for Structural Steel Buildings.* Chicago: American Institute of Steel Construction, 2002.

ASCE 7-02, *Minimum Design Loads for Buildings and Other Structures.* New York: American Society of Civil Engineers, 2002.

ASTM A 6-01b, *Specification for General Requirements for Rolled Structural Bars, Plates, Shapes, and Sheet Piling.* West Conshohocken, PA: ASTM International, 2001.

ASTM A 36/A 36M-00, *Specification for Carbon Structural Steel.* West Conshohocken, PA: ASTM International, 2000.

ASTM A 325-91, *Specification for High-Strength Bolts for Structural Steel Joints.* West Conshohocken, PA: ASTM International, 1991.

ASTM A 435-90 (2001), *Specification for Straight Beam Ultrasound Examination of Steel Plates*. West Conshohocken, PA: ASTM International, 2001.

ASTM A 490-91, *Specification for Heat-Treated, Steel Structural Bolts, 150 ksi (1035 MPa) Tensile Strength*. West Conshohocken, PA: ASTM International, 1991.

ASTM A 568-01, *Specification for Steel Sheet, Carbon and High-Strength, Low-Alloy, Hot-Rolled and Cold-Rolled, General Requirements For*. West Conshohocken, PA: ASTM International, 2001.

ASTM A 615/A 615M-00, *Specification for Deformed and Plain Billet-Steel Bars for Concrete Reinforcement.* West Conshohocken, PA: ASTM International, 2000.

ASTM A 706-00, *Specification for Low-Alloy Steel Deformed and Plain Bars for Concrete Reinforcement.* West Conshohocken, PA: ASTM International, 2000.

ASTM A 898-91(2001), *Straight Beam Ultrasound Examination of Rolled Steel Structural Shapes*. West Conshohocken, PA: ASTM International, 2001.

ASTM C 31/C31M-98, *Standard Practice for Making and Curing Concrete Test Specimens in the Field.* West Conshohocken, PA: ASTM International, 1998.

ASTM C 172-99, *Standard Practice for Sampling Freshly Mixed Concrete*. West Conshohocken, PA: ASTM International, 1999.

ASTM D 1761-88 (2000), *Standard Test Method for Mechanical Fasteners in Wood*. West Conshohocken, PA: ASTM International, 2000.

ASTM E 330-97, *Standard Test Methods for Structural Performance of Exterior Windows, Curtain Walls, and Doors by Uniform Static Air Pressure Difference.* West Conshohocken, PA: ASTM International, 1997.

ASTM E 605-00, *Test Methods for Thickness and Density of Sprayed Fire-Resistive Material Applied to Structural Members*. West Conshohocken, PA: ASTM International, 2000.

AWS D1.1-00, *Structural Welding Code–Steel.* Miami, FL: American Welding Society, 2000.

AWS D1.3-98, *Structural Welding Code–Sheet Steel.* Miami, FL: American Welding Society, 1998.

AWS D1.4-98, *Structural Welding Code–Reinforced Steel.* Miami, FL: American Welding Society, 1998.

AWS QC1-88, *Specification for Qualification and Certificate of Welding Inspectors*. Miami, FL: American Welding Society, 1988.

FEMA 368, *NEHRP Recommended Provisions for Seismic Regulations for New Buildings and Other Structures*. Washington, DC: Federal Emergency Management Agency, 2001.

Henry, John R., "Special Inspection, Structural Observation and Quality Assurance Under the 2000 IBC," *Structural Engineer*, April 2002, pp 24-28.

IFC-2003, *International Fire Code*. Falls Church, VA: International Code Council, 2003.

IMC-2003, *International Mechanical Code*. Falls Church, VA: International Code Council, 2003.

IPC-2003, *International Plumbing Code*. Falls Church, VA: International Code Council, 2003.

IRC-2003, *International Residential Code*. Falls Church, VA: International Code Council, 2003.

Model Program for Special Inspection. Falls Church, VA: International Codes Council, 2002.

RCSC-85, *Specification for Structural Joints Using A325 or A490 Bolts—with 1988 Revisions.* Chicago: Research Council on Structural Connections (c/o American Institute of Steel Construction, Inc.), 1988.

SBCCI SSTD 11-97, *Standard for Determining Wind Resistance of Concrete or Clay Roof Tiles*. Birmingham, AL: Southern Building Code Congress International, 1997.

UL 127-99, *Factory-Built Fireplaces*. Northbrook, IL: Underwriters Laboratories Inc., 1999.

Chapter 18: Soils and Foundations

General Comments

Chapter 18 contains provisions regulating the design and construction of foundations for buildings and other structures and involves geotechnical and structural considerations in the selection and installation of adequate supports for the loads transferred from the structure above.

Section 1801 gives the general scope or purpose of the provisions in Chapter 18.

Section 1802 provides the requirements for foundation and soils investigations that are to be conducted at the site prior to final design and construction.

Section 1803 includes excavation, grading and backfill provisions.

Section 1804 establishes the allowable load-bearing values for soils where site soil tests do not verify that higher soil values are appropriate.

Section 1805 contains the regulations for spread footings and foundations.

Section 1806 provides requirements for retaining walls.

Section 1807 provides specifications for the dampproofing and waterproofing of floor slabs and below-grade walls.

Section 1808 gives the requirements for pier and pile foundations.

Section 1809 provides the specifications for driven pile foundations.

Section 1810 includes the provisions for cast-in-place concrete pile foundations.

Section 1811 contains the regulations for composite piles.

Section 1812 provides the pier foundation specifications that are required by the code.

The proper design and construction of a foundation system is critical to the satisfactory performance of the entire building structure that it supports.

Foundation problems are not uncommon and vary greatly. They may be of a simple or complex nature and may be manageable or without practical remedy. The uncertainties of foundation construction make it extremely difficult to seek and handle every potential for failure within the text of the code. The provisions of the code are meant to set forth and regulate the minimum standards and conventional practices needed for the design and construction of foundation systems so as to provide adequate safety to life and property. Due care must be exercised in the planning and design of foundation systems based on obtaining sufficient soils information, the use of accepted engineering procedures, experience and good technical judgement.

Essentially, there are two parts to the foundation system: the substructure and the soil. The substructure consists of structural components that serve as the medium through which the building loads are transmitted to the supporting earth (soil or rock). The substructure components may consist of shallow foundations, such as basement walls, grade walls, beams, isolated spread footings or combinations of these components. Shallow foundations may also involve the use of mat or raft foundations. As may be required, the substructure can consist of deep foundations involving the use of piles, drilled shafts or piers, caissons or other similar deep foundations. The second part of the foundation system involves the use of soil (including rock) as a structural material to carry the load of the building or any other load transmitted through the substructure.

The substructure and the soil are interdependent elements of the foundation system and must be understood and dealt with as a composite engineering consideration. Indeed, the selection of the kind of substructure to be used, whether it employs any of the different types of shallow foundations commonly used or adopts the use of a deep foundation, is a direct function of the nature of the soil encountered at the project site. Chapter 18 broadly outlines the conventional systems of foundation construction. Although it does not specifically include special or patented systems, it does not preclude their use because most such construction will fall into the general categories of foundation systems prescribed in the provisions and, thus, meet the intent of the code.

In determining the load-bearing capacity and other values of the soil mass, the code provisions address such considerations in terms of "undisturbed" soil. Special provisions are included where prepared fill is to be utilized for foundation support.

Purpose

The provisions of this chapter set forth the minimum requirements for the design, construction and resistance to water intrusion (for areas not subject to scour or water pressure by wind and wave action) of foundation systems for buildings and other structures.

SECTION 1801 GENERAL

1801.1 Scope. The provisions of this chapter shall apply to building and foundation systems in those areas not subject to scour or water pressure by wind and wave action. Buildings and foundations subject to such scour or water pressure loads shall be designed in accordance with Chapter 16.

❖ The provisions contained in this chapter for foundation design and construction do not apply to structures located in flood hazard areas where potential scour or water pressure from wind and wave action exist. Special design considerations are required for foundations subject to scour and water pressure that this chapter does not address. Section 1612 addresses flood loads and the establishment of flood hazard areas. Section 1612.4 requires buildings and structures located in a flood hazard area to be designed in accordance with ASCE 24. This standard provides extensive guidance to be used where scour by wind and wave action is expected to significantly impact the foundation of a building or structure. Once the expected loss of supporting soil has been determined, an effective foundation system can be designed and constructed. Failure to consider scour in areas subject to scour during a design flood event may lead to significant building damage or failure. For additional guidance, refer to FEMA 55, *Coastal Construction Manual.*

1801.2 Design. Allowable bearing pressures, allowable stresses and design formulas provided in this chapter shall be used with the allowable stress design load combinations specified in Section 1605.3. The quality and design of materials used structurally in excavations, footings and foundations shall conform to the requirements specified in Chapters 16, 19, 21, 22 and 23 of this code. Excavations and fills shall also comply with Chapter 33.

❖ Design requirements in Chapter 18 are based on an allowable stress design (ASD) approach. Allowable stresses and service loads should not be used with the load combinations for strength design, and vice versa. This section clarifies the applicable load combinations from Chapter 16 that are to be used for the design of foundations.

1801.2.1 Foundation design for seismic overturning. Where the foundation is proportioned using the strength design load combinations of Section 1605.2, the seismic overturning moment need not exceed 75 percent of the value computed from Section 9.5.5.6 of ASCE 7 for the equivalent lateral force method, or Section 1618 for the modal analysis method.

❖ When strength design is used to size foundations, the seismic overturning moment for the equivalent lateral force method or the modal analysis method can be reduced by 25 percent. This reduction in overturning moment accounts for the fact that any short-duration upward displacement of the foundation will have no or minimal permanent consequences. Prior to strength-based seismic code forces, foundations were traditionally proportioned for overturning moments based on working stress design loads. The 75-percent reduction is essentially equal to dividing the overturning moment by a factor of 1.4, thereby reducing it to a working stress level. This maintains parity for the sizing of foundations due to overturning moment regardless of whether strength design load combinations or ASD load combinations are used.

SECTION 1802 FOUNDATION AND SOILS INVESTIGATIONS

1802.1 General. Foundation and soils investigations shall be conducted in conformance with Sections 1802.2 through 1802.6. Where required by the building official, the classification and investigation of the soil shall be made by a registered design professional.

❖ Sections 1802.2 through 1802.6 specify the conditions for which foundation and soil investigations are required. Although this section states that the classification and investigation of soils is to be done by a registered design professional only when required by the building official, consultation with geotechnical engineers or individuals with significant experience in soil and foundation analysis is advised.

1802.2 Where required. The owner or applicant shall submit a foundation and soils investigation to the building official where required in Sections 1802.2.1 through 1802.2.7.

> **Exception:** The building official need not require a foundation or soils investigation where satisfactory data from adjacent areas is available that demonstrates an investigation is not necessary for any of the conditions in Sections 1802.2.1 through 1802.2.6.

❖ Soils investigations to determine subsurface conditions should be made prior to the design and construction of new buildings and other structures. Such investigations should also be conducted when additions or alterations to existing facilities are contemplated and are of such scope as would significantly increase or change the distribution of foundation loads.

There are two main objectives for conducting a soils investigation. The first is of a confirmatory nature. Its purpose is to obtain information already known from adjacent structures, such as soil boring records, field test results, laboratory test data and analyses and any other knowledge useful in the design of the foundation system. The second objective is of an exploratory nature. It is warranted where soils information does not exist or is insufficient or unsatisfactory for use in the design of the foundation system.

Regardless of the objective of the soils investigation, the information generally required includes one or more (or all) of the following items for determining subsurface conditions:

1. The depth, thickness and composition of each soil stratum;
2. For rock, the characteristics of the rock stratum (or strata), including the thickness of the rock to a reasonable depth;
3. The depth of ground water below the site surface and
4. The engineering properties of the soil and rock strata that are pertinent for the proper design and performance of the foundation system.

For shallow foundations, the soils investigation should yield sufficient information to establish the character and load-bearing capacity of the soil (or rock) at depths that will receive the foundations.

Foundation problems are not uncommon and may vary greatly, ranging from very simple and manageable problems to very complex situations that may be either manageable or without practical remedy. The field of soil mechanics and foundation engineering is diverse and complicated, and since it is not an exact science, its application requires specialized knowledge and judgmental decisions based on experience. Where subsurface conditions are found or suspected to be of a critical nature, the building official is encouraged to seek the professional advice of experienced foundation engineers.

As indicated in the exception, where geotechnical data from adjacent areas are well known, the building official can accept the use of local engineering practices for the design of foundations.

1802.2.1 Questionable soil. Where the safe-sustaining power of the soil is in doubt, or where a load-bearing value superior to that specified in this code is claimed, the building official shall require that the necessary investigation be made. Such investigation shall comply with the provisions of Sections 1802.4 through 1802.6.

❖ Whenever the safe-sustaining power of a soil is in doubt, or where a load-bearing value superior to that specified is claimed, the building official may order field load tests to be conducted. The investigation is to comply with the applicable provisions of Sections 1802.4 through 1802.6.

One such method of investigation includes construction test pits for field load-bearing tests; however, since field load-bearing tests are very expensive, they are seldom used. Sufficient information is usually available from soil borings taken at the site.

Test pits are usually required to be at least 4 square feet (0.37 m^2) in area and be excavated down to the elevation of the proposed bearing surface. The typical apparatus for making such tests involves placing the test loads on a platform supported on a post through which the applied loads are transferred to a bearing plate of a specified size and, in turn, to the soil below. A typical setup for field load tests is shown in Figure 1802.2.1.

It is important that the load (weights on platform) is applied such that all of it will be transmitted to the soil as a static load without impact, fluctuation or eccentricity.

The load should be applied incrementally, and continuous records of all settlements should be kept. Measurements are usually made by settlement recording devices, such as dial gauges, capable of measuring the settlement of the test bearing plate to an accuracy of at least 0.01 inch (0.25 mm).

The test is continued until either the maximum test load is reached or the ratio of load increment to settlement increment reaches a minimum, steady magnitude sustained for a period of 48 hours. After the load is released, the elastic rebound of the soil is also measured for a period of time.

Load test results are normally presented in a load settlement diagram in which the applied test load measured in tons per square foot is plotted in relation to the settlement readings recorded in fractions of an inch. The bearing capacity of the soil can be computed from the test results.

There are some drawbacks to the use of field load tests for determining soil-bearing capacity. Test results can be misleading if the soil under the footing is not uniform for the full depth of load influence, which is equal to about twice the width of the footing. Also, since a load test is conducted for a short duration, settlements that occur due to the consolidation of the soil over a very long time cannot be predicted.

Again, field load-bearing tests are very expensive and should be avoided when possible. This section permits the use of other methods for determining the safe

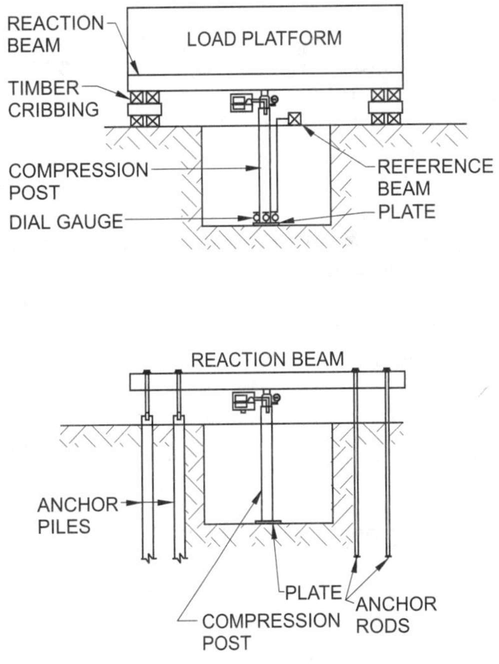

Figure 1802.2.1
TYPICAL SETUP FOR CONDUCTING STATIC LOAD TESTS

bearing capacity of soils. Standard laboratory tests can usually produce sufficient proof of satisfactory bearing capacity and settlement information. However, there are conditions when standard laboratory tests may not produce reliable results, such as when clay materials contain a pattern of cracks or when stiff clays may have suffered differential movement or expansion.

1802.2.2 Expansive soils. In areas likely to have expansive soil, the building official shall require soil tests to determine where such soils do exist.

❖ Expansive soils, often referred to as "swelling soils," contain montmorillonite minerals and have the characteristics of absorbing water and swelling, or shrinking and cracking when drying. Significant volume changes can cause serious damage to buildings and other structures as well as to pavements and sidewalks. Swelling soils are found throughout the nation, but are more prevalent in regions with dry or moderately arid climates.

1802.2.3 Ground-water table. A subsurface soil investigation shall be performed to determine whether the existing ground-water table is above or within 5 feet (1524 mm) below the elevation of the lowest floor level where such floor is located below the finished ground level adjacent to the foundation.

Exception: A subsurface soil investigation shall not be required where waterproofing is provided in accordance with Section 1807.

❖ There are several reasons for conducting a subsurface investigation to determine the level of ground water at a construction site. If the ground-water table is above subsurface slabs (i.e., basement floors), then walls and floors need to be designed to resist hydrostatic pressures. Foundation walls and basement slabs may need to be dampproofed or waterproofed, depending on the location of the ground-water table. A subsurface investigation will also determine the type of drainage system needed as a permanent installation, whether there will be any major water problems that could affect the excavation operations and construction of the foundation system, and if it is necessary to provide a temporary drainage system of a type and size that will control ground-water seepage.

Ground-water levels can vary significantly over a year's time, as well as from year to year. While it would be ideal to make ground-water table observations that encompass a full annual cycle, the reality is that such an undertaking would not normally facilitate a design/construction program and would be impractical. Ground-water observations must be made in shorter time intervals; however, this situation poses some real problems. For example, measurements of ground-water levels in bore holes taken 24 hours after completion of the soil borings can provide an acceptable indication of the water level in permeable soils such as sand, gravel or sand/gravel mixtures. Fine-grained soils of low permeability, such as clays, require the use of observation tubes (piezometers) and, depending on the specific properties of the soil, a time period of 10 weeks or longer to obtain acceptable readings.

Water levels established by either of the two methods described above are sufficient indication of the water conditions at the time of measurement, but do not necessarily represent the highest possible ground-water levels that can occur. For design purposes, the water levels established by field observations may need to be adjusted with the climatological and hydrological records of the region in order to establish the high and low points.

As indicated in the exception, a subsurface investigation is not required where floors, walls, joints and penetrations are waterproofed as required in Section 1807.3.

1802.2.4 Pile and pier foundations. Pile and pier foundations shall be designed and installed on the basis of a foundation investigation and report as specified in Sections 1802.4 through 1802.6 and Section 1808.2.1.

❖ A foundation investigation is required when pile and pier foundations are used. Such investigations are needed to define as accurately as possible the subsurface conditions of soil and rock materials, establish the soil and rock profiles across the construction site and locate the ground-water table. Sometimes, it may also be necessary to determine specific soil properties, such as shear strength, relative density, compressibility and other such technical data required for analyzing subsurface conditions. Foundation investigations may also be used to render such valuable data as information on existing construction at the site or on neighboring properties (including boring and test records), the type and condition of the existing structures, their age, the type of foundations used and performance over the years. Knowledge of existing deleterious substances in the soils that could affect the durability (as well as the performance) of the piles, data on geologic conditions at the site (including such information as the existence of mines, earth cavities, underground streams or other adverse water conditions) and a history of any seismic activity and hurricanes are also important.

The types of information described above are usually obtained by means of soil and rock borings; laboratory and field tests and engineering analyses. Such information is used for determining design loads; types and lengths of piles; driving criteria and selection of equipment and probable durability of pile materials in relation to subsurface conditions.

1802.2.5 Rock strata. Where subsurface explorations at the project site indicate variations or doubtful characteristics in the structure of the rock upon which foundations are to be constructed, a sufficient number of borings shall be made to a depth of not less than 10 feet (3048 mm) below the level of the foundations to provide assurance of the soundness of the foundation bed and its load-bearing capacity.

❖ Rock may be found at levels near or at the earth's surface, upon which shallow foundations can be supported or will range downward to very low levels, upon which piles and other types of deep foundations can bear.

Most intact rock will have compressive strengths that far exceed the requirements for foundation support. It is most common, however, to find cracks, joints and other defects in rock formations that will increase the compressibility of the material. Depending on the nature and extent of the defects, settlement may become the governing factor in determining allowable load-bearing capacity rather than rock strength.

Where the condition of the rock is in doubt, borings must be made at least 10 feet (305 mm) into the rock stratum below the bottom of the footings to verify the soundness of the material and to determine its load-bearing capacity.

1802.2.6 Seismic Design Category C. Where a structure is determined to be in Seismic Design Category C in accordance with Section 1616, an investigation shall be conducted, and shall include an evaluation of the following potential hazards resulting from earthquake motions: slope instability, liquefaction and surface rupture due to faulting or lateral spreading.

❖ The potential for liquefaction, surface rupture or slope instability at a building site is greater in areas of moderate and high seismicity than in areas of lower seismicity. Also, the consequences of damage resulting from such hazards are more severe for buildings in higher seismic use groups (such as essential facilities). This section thus requires a soils investigation report for sites of buildings assigned to Seismic Design Category C and higher, including an evaluation of specific hazards resulting from earthquake motions. The purpose of this section is to reduce the hazard of large ground movement and the damaging effect on the structure that might result if any of the conditions cited in the code text should occur.

Liquefaction of saturated granular soils has been a major source of building damage during past earthquakes. For example, many structures in Niigata, Japan suffered major damage as a consequence of liquefaction during its 1964 earthquake. Loss of bearing strength, differential settlement and differential horizontal displacement due to lateral spread were the direct causes of damage. Many structures have been similarly damaged by differential ground displacements during U.S. earthquakes, such as the San Fernando Valley Juvenile Hall during the 1971 San Fernando, California earthquake and the Marine Sciences Laboratory at Moss Landing, California during the 1989 Loma Prieta event.

For more information regarding evaluation of slope instability, liquefaction and surface rupture due to faulting or lateral spreading, see Section 7.4 of the 1997 NEHRP Provisions Commentary.

1802.2.7 Seismic Design Category D, E or F. Where the structure is determined to be in Seismic Design Category D, E or F, in accordance with Section 1616, the soils investigation requirements for Seismic Design Category C, given in Section 1802.2.6, shall be met, in addition to the following. The investigation shall include:

1. A determination of lateral pressures on basement and retaining walls due to earthquake motions.

2. An assessment of potential consequences of any liquefaction and soil strength loss, including estimation of differential settlement, lateral movement or reduction in foundation soil-bearing capacity, and shall address mitigation measures. Such measures shall be given consideration in the design of the structure and can include, but are not limited to, ground stabilization, selection of appropriate foundation type and depths, selection of appropriate structural systems to accommodate anticipated displacements or any combination of these measures. The potential for liquefaction and soil strength loss shall be evaluated for site peak ground acceleration magnitudes and source characteristics consistent with the design earthquake ground motions. Peak ground acceleration shall be determined from a site-specific study taking into account soil amplification effects, as specified in Section 1615.2.

Exception: A site-specific study need not be performed provided that peak ground acceleration equal to $S_{DS}/2.5$ is used, where S_{DS} is determined in accordance with Section 1615.2.1.

❖ This section includes additional requirements for the soil investigation report for sites of structures assigned to Seismic Design Category D and higher. The investigation must determine lateral pressures on basement and retaining walls due to earthquake motions. Earthquake motions create increased lateral soil pressure on walls below the ground surface, especially in soft soils in areas of high seismicity. The requirements of this section are necessary to make sure that the dynamic soil pressures are included in the design of basement and retaining walls. Section 7.5.1 of the 1997 NEHRP Provisions Commentary includes a discussion about how earth-retaining structures have been designed for dynamic loads.

Additionally, a thorough assessment of potential consequences of any liquefaction and soil strength loss needs to be made and considered in the design of the structure. See the commentary to Section 1802.2.6 for a discussion of earthquake damage due to liquefaction. Design to prevent damage due to liquefaction consists of three parts: evaluation of liquefaction hazard, evaluation of potential ground displacement and mitigating the hazard by: designing to resist ground displacement, reducing the potential for liquefaction or choosing an alternative site with less hazard.

The assessment is required to be made for site peak ground acceleration based on a site-specific study unless a peak ground acceleration equal to $S_{DS}/2.5$ is used. The peak ground acceleration needs to be known by the geotechnical engineer in order to evaluate the potential for liquefaction and soil strength loss. The exception allows an estimate of $S_{DS}/2.5$ to be used to represent the peak ground acceleration.

1802.3 Soil classification. Where required, soils shall be classified in accordance with Section 1802.3.1 or 1802.3.2.

❖ This section provides parameters for classification of soils in accordance with the code.

1802.3.1 General. For the purposes of this chapter, the definition and classification of soil materials for use in Table 1804.2 shall be in accordance with ASTM D 2487.

❖ Where required, soils are to be classified in accordance with ASTM D 2487. This standard provides a system for which to classify soils for engineering purposes based on laboratory determination of particle size characteristics, liquid limit and plasticity index. The classification system identifies three major soil divisions—course-grained, fine-grained and highly organic—which are further broken down into 15 basic soil groups. ASTM D 2487 is the ASTM version of the Unified Soil Classification System.

1802.3.2 Expansive soils. Soils meeting all four of the following provisions shall be considered expansive, except that tests to show compliance with Items 1, 2 and 3 shall not be required if the test prescribed in Item 4 is conducted:

1. Plasticity index (PI) of 15 or greater, determined in accordance with ASTM D 4318.
2. More than 10 percent of the soil particles pass a No. 200 sieve (75 μm), determined in accordance with ASTM D 422.
3. More than 10 percent of the soil particles are less than 5 micrometers in size, determined in accordance with ASTM D 422.
4. Expansion index greater than 20, determined in accordance with ASTM D 4829.

❖ Expansive soils, often referred to as "swelling soils," contain montmorillonite minerals that shrink or swell with changes in moisture content. Expansive soils tend to swell when absorbing water or shrink and crack when drying. Significant volume changes can cause serious damage to buildings and other structures as well as to pavements and sidewalks. Swelling soils are found throughout the nation, but are more prevalent in regions with dry or moderately arid climates.

There is a general relationship between the plasticity index (PI) of a soil as determined by the ASTM D 4318 standard test method and the potential for expansion, as shown in Figure 1802.3.2.

This section defines "Expansive soil" as any plastic material with a PI of 15 or greater with more than 10 percent of the soil particles passing a No. 200 sieve and less than 5 micrometers in size. As an alternate, tests in accordance with ASTM D 4829 can be used to determine if a soil is expansive. The expansion index is a measure of the swelling potential of the soil.

The amount and depth of potential swelling that can occur in a clay material are, to some extent, functions of the cyclical moisture content in the soil. In dryer climates where the moisture content in the soil near the ground surface is low because of evaporation, there is a greater potential for extensive swelling than the same soil in wetter climates where the variations of moisture content are not as severe. Volume changes in highly expansive soils range between 7 and 10 percent, but experience has shown that occasionally, under abnormal conditions, they can reach as high as 25 percent.

SWELLING POTENTIAL	PLASTICITY INDEX
Low	0-15
Medium	0-35
High	20-55
Very high	35 and above

Figure 1802.3.2
SWELLING POTENTIAL OF SOILS AND PLASTICITY INDEX

Source: R.B. Peck, W.E. Hanson and T.H. Thornburn. *Foundation Engineering,* 2nd ed. (New York: John Wiley & Sons, 1974).

1802.4 Investigation. Soil classification shall be based on observation and any necessary tests of the materials disclosed by borings, test pits or other subsurface exploration made in appropriate locations. Additional studies shall be made as necessary to evaluate slope stability, soil strength, position and adequacy of load-bearing soils, the effect of moisture variation on soil-bearing capacity, compressibility, liquefaction and expansiveness.

❖ When soils are required to be classified, the classification must be based on observations and tests such as borings or test pits. In addition to the situations specified that require a soils investigation and classification, the evaluation of slope stability, soil strength, position and adequacy of load-bearing soils, moisture effects, compressibility and liquefaction are required to be performed when deemed necessary by the building official or registered design professional.

1802.4.1 Exploratory boring. The scope of the soil investigation including the number and types of borings or soundings, the equipment used to drill and sample, the in-situ testing equipment and the laboratory testing program shall be determined by a registered design professional.

❖ Whenever the allowable load on a soil is in doubt and a field investigation is necessary, exploratory borings are to be made to determine the load-bearing value of the soil. The soil investigation is to be performed by a registered design professional, which in most cases would be a geotechnical engineer. Exploratory borings and their associated tests should be conducted in each area of relatively dissimilar subsoil conditions.

1802.5 Soil boring and sampling. The soil boring and sampling procedure and apparatus shall be in accordance with generally accepted engineering practice. The registered design professional shall have a fully qualified representative on the site during all boring and sampling operations.

❖ A qualified representative is required on site to verify that the information obtained from the investigation will be adequate, valid and acceptable to the building official.

1802.6 Reports. The soil classification and design load-bearing capacity shall be shown on the construction document. Where required by the building official, a written report of the investigation shall be submitted that includes, but need not be limited to, the following information:

1. A plot showing the location of test borings and/or excavations.
2. A complete record of the soil samples.
3. A record of the soil profile.
4. Elevation of the water table, if encountered.
5. Recommendations for foundation type and design criteria, including but not limited to: bearing capacity of natural or compacted soil; provisions to mitigate the effects of expansive soils; mitigation of the effects of liquefaction, differential settlement and varying soil strength; and the effects of adjacent loads.
6. Expected total and differential settlement.
7. Pile and pier foundation information in accordance with Section 1808.2.2.
8. Special design and construction provisions for footings or foundations founded on expansive soils, as necessary.
9. Compacted fill material properties and testing in accordance with Section 1803.5.

❖ When soils are required to be classified, their classification and load-bearing capacity must be shown on the construction documents. If a written report is required by the building official, it is required to include at a minimum the nine items listed in this section. These items will establish a retrievable and verifiable record of the soil conditions if problems are encountered in the future. These items also provide the minimum necessary information for compliance with the code and an adequate foundation system.

SECTION 1803
EXCAVATION, GRADING AND FILL

1803.1 Excavations near footings or foundations. Excavations for any purpose shall not remove lateral support from any footing or foundation without first underpinning or protecting the footing or foundation against settlement or lateral translation.

❖ The purpose of this section is to provide soil stability of adjacent foundations when excavations are made. Due to their lack of shear strength, cohesionless soils, such as sand, will slide to the bottom of an excavation until a certain slope of the sides is reached. This slope is known as the angle of repose of natural slope and is independent of the depth of the excavation.

Cohesive (fine-grained) soils, such as clay, behave much differently compared to granular materials. For example, unsupported vertical cuts of 20 feet (6096 mm) or more can be made in stiff plastic clay materials. This is due to a firm bond between the particles of the cohesive soil. But the strength of this bond (cohesiveness) will vary based on the conditions of the soil, such as density, water content, plasticity and sensitivity (loss of shear strength upon disturbance).

In cohesive soils, when a certain critical depth of excavation is reached, the sides of the cut will fail and the soil mass will fall to the bottom. Unlike granular materials, such as sand, the steepest slope at which a cohesive soil will stand decreases as the depth of the excavation increases. Technically, in cohesive soils, the resistance against sliding is a function of the shearing resistance of the material and its corresponding angle of internal friction (frictional resistance between particles).

The use of the angle of internal friction (and other factors) to calculate slope stability is applicable not only to cohesive soils, but also to granular materials. For example, when the slope angle of an excavation in a bed of sand exceeds the angle of internal friction of the material, the sand will slide down the slope; therefore, the steepest slope that sand can attain is equal to the angle of internal friction. The angle of repose (previously discussed) will be approximately the same value as the angle of internal friction only when the sand is in a dry and loose condition (such as in a stockpile) or is fully immersed in water.

For simplicity, what we have been dealing with in this part of the commentary is slope stability as it relates to homogeneous soils such as sand and clay. In nature, however, soils often occur as mixtures or layers (strata) of different materials, making the determination of slope stability a highly technical and complex subject. Normally for shallow excavations, determination of safe slopes is a matter of applying local experience. In cases of deep cuts, however, slope stability is best determined through tests and analytical methods performed by professionals experienced in foundation engineering.

Some texts dealing with soil mechanics contain tables that indicate the angles of slope that can be expected for various soil materials commonly found throughout the country. While such tables provide useful information, they should be used only as a guide. They should not be used for design purposes, nor employed when a critical subsurface condition exists or when the type of soil (established by borings) does not closely fit those described in the tables.

1803.2 Placement of backfill. The excavation outside the foundation shall be backfilled with soil that is free of organic material, construction debris, cobbles and boulders or a controlled low-strength material (CLSM). The backfill shall be placed in lifts and compacted, in a manner that does not damage the foundation or the waterproofing or dampproofing material.

Exception: Controlled low-strength material need not be compacted.

❖ Section 1803.2 requires that soils used for backfilling foundation excavations must be free of organic material, construction debris or large rocks. The type of soil used for backfill purposes becomes an important consideration in the design of foundation walls. For example, clean sand, gravel or a mixture of these two granular materials is considered the best kind of backfill to use because each is free draining and generally has frost-free properties. On the other hand, fine-grained soils, such as clays, tend to accumulate moisture and are susceptible to swelling and shrinking as well as frost action. Such backfill materials, particularly at times when shrinkage cracks occur, can become loaded with rainwater, thus subjecting foundation walls and basement floors to hydrostatic pressures and possible structural damage.

Backfilling and related work should be performed in such a way as to prevent the movement of the earth of adjoining properties or the subsequent caving in of backfilled areas. Backfilling should not be done until retaining walls, foundation walls or other construction against which backfill is to be placed is in suitable condition to resist lateral pressures. This section requires that backfill be free of organic materials, construction debris, cobbles and boulders.

In addition to carefully selecting the backfill material, the soil should be placed in lifts, usually 9 inches (229 mm) or less, and compacted to prevent significant subsidence due to consolidation under its own weight. While compaction is done by hand-operated tampers or other portable compaction equipment, care should be taken to prevent any possible damage to waterproofing or dampproofing installations and to avoid overcompaction of backfill since it may cause excessive earth pressure against foundation walls. As an alternative to compacted backfill the code also permits controlled low-strength material (CLSM) (see commentary, Section 1803.6).

1803.3 Site grading. The ground immediately adjacent to the foundation shall be sloped away from the building at a slope of not less than one unit vertical in 20 units horizontal (5-percent slope) for a minimum distance of 10 feet (3048 mm) measured perpendicular to the face of the wall or an approved alternate method of diverting water away from the foundation shall be used.

Exception: Where climatic or soil conditions warrant, the slope of the ground away from the building foundation is permitted to be reduced to not less than one unit vertical in 48 units horizontal (2-percent slope).

The procedure used to establish the final ground level adjacent to the foundation shall account for additional settlement of the backfill.

❖ This section requires that the ground immediately adjacent to the foundation be sloped away from the building at a rate of not less than 1 unit vertical in 20 units horizontal for a distance of at least 10 feet (3048 mm). Having finished grade slope away from the building facilitates water drainage and reduces the potential for water standing under and around the building. The exception permits the slope to be reduced to a rate of 1 unit vertical in 48 units horizontal where climatic or soil conditions warrant. This exception would be applicable in arid areas or sites that are surrounded by free-draining soils, such as sand.

1803.4 Grading and fill in floodways. In floodways shown on the flood hazard map established in Section 1612.3, grading and/or fill shall not be approved unless it has been demonstrated through hydrologic and hydraulic analyses performed by a registered design professional in accordance with standard engineering practice that the proposed grading or fill, or both, will not result in any increase in flood levels during the occurrence of the design flood.

❖ As the definition in Section 1612.2 indicates, a floodway is that portion of a flood hazard area that is reserved for the discharge of the design flood event. The National Flood Insurance Program (NFIP) requires that the impact of development or encroachment on the floodway should be considered. This section permits grading or fill in a floodway only if it is demonstrated that it will not adversely affect surrounding areas by increasing the design flood elevation.

1803.5 Compacted fill material. Where footings will bear on compacted fill material, the compacted fill shall comply with the provisions of an approved report, which shall contain the following:

1. Specifications for the preparation of the site prior to placement of compacted fill material.
2. Specifications for material to be used as compacted fill.
3. Test method to be used to determine the maximum dry density and optimum moisture content of the material to be used as compacted fill.
4. Maximum allowable thickness of each lift of compacted fill material.
5. Field test method for determining the in-place dry density of the compacted fill.
6. Minimum acceptable in-place dry density expressed as a percentage of the maximum dry density determined in accordance with Item 3.
7. Number and frequency of field tests required to determine compliance with Item 6.

Exception: Compacted fill material less than 12 inches (305 mm) in depth need not comply with an approved report, provided it has been compacted to a minimum of 90 percent Modified Proctor in accordance with ASTM D 1557. The compaction shall be verified by a qualified inspector approved by the building official.

❖ Where prepared fill is to be utilized for foundation support, the soils report required by Section 1803.5 is to contain detailed information for approval of the fill. The information on the prepared fill in the soils report is necessary to permit a reasonable prediction as to the load-bearing capacity of the fill material. Additionally, when prepared fill is to be utilized where special inspections are required, the fill operation itself must be performed under the scrutiny of a special inspection. For example, if one (or more) of the exceptions in Section 1704.1 applies, special inspection of the prepared fill operation is not required.

1803.6 Controlled low-strength material (CLSM). Where footings will bear on controlled low-strength material (CLSM), the CLSM shall comply with the provisions of an approved report, which shall contain the following:

1. Specifications for the preparation of the site prior to placement of the CLSM.
2. Specifications for the CLSM.
3. Laboratory or field test method(s) to be used to determine the compressive strength or bearing capacity of the CLSM.
4. Test methods for determining the acceptance of the CLSM in the field.
5. Number and frequency of field tests required to determine compliance with Item 4.

❖ As an alternative to compacted fill in accordance with Section 1803.5, this section permits the use of CLSM for the support of footings. CLSM must be placed in accordance with an approved report that includes requirements for the material strength and field verification.

In Chapter 2, CLSM is defined as "self-compacting cementitious materials." This class of material is commonly referred to by a variety of other names, which include flowable fill, controlled density fill, unshrinkable fill and soil-cement slurry. Guidance on the use of these materials can be found in ACI 229R. Additional documents that may be useful references for sampling and testing these materials include the following ASTM International standards:

- ASTM D 4832, *Standard Test Method for Preparation and Testing of Controlled Low-Strength Material (CLSM) Test Cylinders*.
- ASTM D 5971, *Standard Practice for Sampling Freshly Mixed Controlled Low-Strength Material*.
- ASTM D 6023, *Standard Test Method for Unit Weight, Yield, Cement Content, and Air Content (Gravimetric) of Controlled Low-Strength Material (CLSM)*.
- ASTM D 6024, *Standard Test Method for Ball Drop on Controlled Low-Strength Material (CLSM) to Determine Suitability for Load Application*.
- ASTM D 6103, *Standard Test Method for Flow Consistency of Controlled Low-Strength Material (CLSM)*.

SECTION 1804
ALLOWABLE LOAD-BEARING VALUES OF SOILS

1804.1 Design. The presumptive load-bearing values provided in Table 1804.2 shall be used with the allowable stress design load combinations specified in Section 1605.3.

❖ This section clarifies the use of the presumptive load-bearing values relating to foundations and footing design. Foundation design is based on the ASD approach. Since Chapter 16 includes provisions, including load combinations, for strength design [load and resistance factor design (LRFD)], it is necessary to clarify that these values are to be used with the ASD load combinations.

1804.2 Presumptive load-bearing values. The maximum allowable foundation pressure, lateral pressure or lateral sliding resistance values for supporting soils at or near the surface shall not exceed the values specified in Table 1804.2 unless data to substantiate the use of a higher value are submitted and approved.

Presumptive load-bearing values shall apply to materials with similar physical characteristics and dispositions.

Mud, organic silt, organic clays, peat or unprepared fill shall not be assumed to have a presumptive load-bearing capacity unless data to substantiate the use of such a value are submitted.

Exception: A presumptive load-bearing capacity is permitted to be used where the building official deems the load-bearing capacity of mud, organic silt or unprepared fill is adequate for the support of lightweight and temporary structures.

❖ Where the load-bearing capacity of the soil has not been determined by borings, as specified in Section 1802, the presumptive load-bearing values listed in Table 1804.2 are intended to apply in the design of shallow foundation systems.

While fill material (unconsolidated), mud, muck, peat, organic silt, soft clay and other unprepared fill materials are considered to have no presumptive load-bearing value, soil tests may show that they do have some limited load-bearing capacity and, based on this type of evidence, the building official may approve the construction of lightweight structures upon such soils.

The presumption is that the building official possesses sufficient technical knowledge on the character and behavior of subsurface materials to render a valid judgement on the adequacy of the soil to support satisfactorily the lightweight or temporary structure, or he or she has sought and gained specific advice through consultation with professionals who are competent in the field of foundation engineering. It would be an unwise practice to authorize construction on exceptionally weak soils without the benefit of technical knowledge to make judgmental decisions.

TABLE 1804.2. See page 18-10.

❖ The values in Table 1804.2 are intended to be lower-bound allowable pressures to be used where the bearing value is not determined by borings as specified in Section 1802. The classes of soil and rock listed in Table 1804.2 are those materials most commonly found at construction sites around the country. The allowable foundation pressures expressed in terms of pounds per square foot (psf) for each class of subsurface materials listed in Table 1804.2 are based on long experience with the behavior of these materials under loaded conditions.

Should local soil conditions be significantly different than those listed in Table 1804.2, then with the approval of the building official, local experience can be employed in the design of foundations, particularly where actual load-bearing values are less than the allowable unit pressures given in the table.

Whether the allowable load-bearing values of Table 1804.2 are used or local conditions on soil capacity prevail, many precautions must be taken in the design of foundation systems that are not regulatory functions of the code, but rather are the professional considerations and design applications of those who are engaged in foundation engineering. The building official needs to pay particular attention to proper selection of the allowable load-bearing capacity of the soil because it is the source of many foundation failures. The problem arises where the load-bearing material directly under a foundation overlies a stratum (or strata) of weaker material having a smaller allowable load-bearing capacity. The selection of the load-bearing value to be used in the design of the foundation should take into account the load distribution at the weaker stratum (or strata) so that the pressure on the soil does not exceed its allowable load-bearing capacity. In this respect, it is important to have a soil profile showing the classes of material at the construction site or at properties in the near vicinity of or adjacent to the area of construction. Such information can be a part of the records or other data required by Section 1802.6 or the results of a soil investigation.

With this information, the building official can consult with the registered design professional in charge of the foundation design to ascertain that due care was exercised in adopting proper soil load-bearing values.

1804.3 Lateral sliding resistance. The resistance of structural walls to lateral sliding shall be calculated by combining the values derived from the lateral bearing and the lateral sliding resistance shown in Table 1804.2 unless data to substantiate the use of higher values are submitted for approval.

For clay, sandy clay, silty clay and clayey silt, in no case shall the lateral sliding resistance exceed one-half the dead load.

❖ The lateral sliding resistance of structural walls is determined by combining the lateral bearing and lateral sliding values from Table 1804.2. The lateral sliding value from Table 1804.2 is determined by multiplying the applicable coefficient of friction by the dead load. For clay; sandy clay; silty clay and clayey silt; silt and sandy silt, the lateral sliding value is set at 130 psf (6.23 kPa). However, this section also stipulates that for clay, sandy clay, silty clay and clayey silt, the lateral sliding resistance cannot exceed one-half the dead load.

1804.3.1 Increases in allowable lateral sliding resistance. The resistance values derived from the table are permitted to be increased by the tabular value for each additional foot (305 mm) of depth to a maximum of 15 times the tabular value.

Isolated poles for uses such as flagpoles or signs and poles used to support buildings that are not adversely affected by a $^1/_2$-inch (12.7 mm) motion at the ground surface due to short-term lateral loads are permitted to be designed using lateral-bearing values equal to two times the tabular values.

❖ The lateral sliding resistance is calculated as the sum of the lateral bearing and lateral sliding values from Table 1804.2. The lateral bearing values are determined as the product of the tabular value and the depth below

TABLE 1804.2
ALLOWABLE FOUNDATION AND LATERAL PRESSURE

CLASS OF MATERIALS	ALLOWABLE FOUNDATION PRESSURE (psf)[d]	LATERAL BEARING (psf/f below natural grade)[d]	LATERAL SLIDING	
			Coefficient of friction[a]	Resistance (psf)[b]
1. Crystalline bedrock	12,000	1,200	0.70	—
2. Sedimentary and foliated rock	4,000	400	0.35	—
3. Sandy gravel and/or gravel (GW and GP)	3,000	200	0.35	—
4. Sand, silty sand, clayey sand, silty gravel and clayey gravel (SW, SP, SM, SC, GM and GC)	2,000	150	0.25	—
5. Clay, sandy clay, silty clay, clayey silt, silt and sandy silt (CL, ML, MH and CH)	1,500[c]	100	—	130

For SI: 1 pound per square foot = 0.0479 kPa, 1 pound per square foot per foot = 0.157 kPa/m.

a. Coefficient to be multiplied by the dead load.

b. Lateral sliding resistance value to be multiplied by the contact area, as limited by Section 1804.3.

c. Where the building official determines that in-place soils with an allowable bearing capacity of less than 1,500 psf are likely to be present at the site, the allowable bearing capacity shall be determined by a soils investigation.

d. An increase of one-third is permitted when using the alternate load combinations in Section 1605.3.2 that include wind or earthquake loads.

natural grade. This section essentially limits the depth to which the lateral bearing values are added to the lateral sliding values at 15 feet (4572 mm).

For isolated poles used as supports for flagpoles or signs, and poles used to support buildings that are not adversely affected by a $^1/_2$-inch (12.7 mm) motion at the ground surface due to short-term lateral loads, the lateral bearing values in Table 1804.2 are permitted to be multiplied by two.

SECTION 1805
FOOTINGS AND FOUNDATIONS

1805.1 General. Footings and foundations shall be designed and constructed in accordance with Sections 1805.1 through 1805.9. Footings and foundations shall be built on undisturbed soil, compacted fill material or CLSM. Compacted fill material shall be placed in accordance with Section 1803.5. CLSM shall be placed in accordance with Section 1803.6.

The top surface of footings shall be level. The bottom surface of footings is permitted to have a slope not exceeding one unit vertical in 10 units horizontal (10-percent slope). Footings shall be stepped where it is necessary to change the elevation of the top surface of the footing or where the surface of the ground slopes more than one unit vertical in 10 units horizontal (10-percent slope).

❖ Sections 1805.1 through 1805.9 address requirements related to proper design and installation of footings and foundations. It is important that footings and foundations be built on undisturbed soil of known bearing value or properly compacted fill with known bearing capacity. As an alternative to compacted fill the code permits the use of CLSM (see Section 1803.6 commentary). The tops and bottoms of footings are required to be essentially level, with a slope of 1 unit vertical in 10 units horizontal permitted for the bottom of footings. Where the slope of the surface of the ground exceeds 1:10, footings are required to be stepped. Although not specifically mentioned, crack propagation at the joints should be considered when determining the overlapping and vertical dimensions of the steps.

1805.2 Depth of footings. The minimum depth of footings below the undisturbed ground surface shall be 12 inches (305 mm). Where applicable, the depth of footings shall also conform to Sections 1805.2.1 through 1805.2.3.

❖ Footings are required to extend below the ground surface a minimum of 12 inches (305 mm). This is considered a minimum depth to protect the footing from movement of the soil caused by freezing and thawing in mild climate areas (see Section 1805.2.1 for general frost protection requirements).

1805.2.1 Frost protection. Except where otherwise protected from frost, foundation walls, piers and other permanent supports of buildings and structures shall be protected from frost by one or more of the following methods:

1. Extending below the frost line of the locality;
2. Constructing in accordance with ASCE-32; or
3. Erecting on solid rock.

Exception: Free-standing buildings meeting all of the following conditions shall not be required to be protected:

1. Classified in Importance Category I (see Table 1604.5);
2. Area of 400 square feet (37 m^2) or less; and
3. Eave height of 10 feet (3048 mm) or less.

Footings shall not bear on frozen soil unless such frozen condition is of a permanent character.

❖ Shallow foundations must be placed on soil strata with adequate load-bearing capacity and at depths to which freezing cannot penetrate. In winter, frost action can raise the ground level (frost heave), whereas in springtime, the same area will soften and settle back. If foundations are built in soil strata that can freeze, then the heave or vertical movement of the ground, which is rarely uniform, can cause serious damage to buildings and other structures. Frost heave can become particularly aggravated in clay soils. Well-drained soils, such as sand and gravel, are not as susceptible to extensive movements.

Except for structures classified as Category I in Table 1604.5 that are no more than 400 square feet (37 m^2) and having an eave height of no more than 10 feet (3048 mm), the foundation is to be protected from frost in accordance with this section. A common method of accomplishing this is by placing the footing bottom below the frost line. The "frost line" is defined as the lowest level below the ground surface to which a temperature of 32°F (0°C) extends. The factors determining the depth of the frost line are air temperature and the length of time it is below freezing [32°F (0°C)], as well as the ability of the soil to conduct heat and its level of thermal conductivity. Frost lines vary significantly throughout the country, ranging from 5 inches (127 mm) in the deep south to 100 inches (2540 mm) in the uppermost northern regions. The frost-free depth for shallow foundations is dependent on the frost line set for the particular locality of construction.

Another form of protection is the use of frost-protected shallow foundations (FPSF) in accordance with ASCE 32. This type of frost protection utilizes slab edge insulation to minimize heat loss at the slab edge. By retaining heat from the building in the ground, it has the effect of raising the frost line around the perimeter of the building.

Foundations are not to be placed on frozen soil because when the ground thaws, uneven settlement of the structure is apt to occur, thereby causing structural damage. This section does, however, permit footings to be constructed on permanently frozen soil. In permafrost areas, special precautions are necessary to prevent heat from the structure from thawing the soil beneath the foundation.

1805.2.2 Isolated footings. Footings on granular soil shall be so located that the line drawn between the lower edges of adjoining footings shall not have a slope steeper than 30 degrees (0.52 rad) with the horizontal, unless the material supporting the higher footing is braced or retained or otherwise laterally supported in an approved manner or a greater slope has been properly established by engineering analysis.

❖ The bottoms of adjacent footings bearing on granular soil are to be located so that a line drawn between their closest edges would not be steeper than 30 degrees (0.52 rad) from the horizontal. The exceptions are where the soil surrounding the higher footing is laterally braced or retained as approved by the building official or where engineering analysis shows that a greater slope can be tolerated. The purpose of this restriction is to provide conditions of safety against the possible influence of lateral and vertical soil pressures transmitted to the lower footing(s) by the loads of higher adjacent foundations (see Figure 1805.2.2).

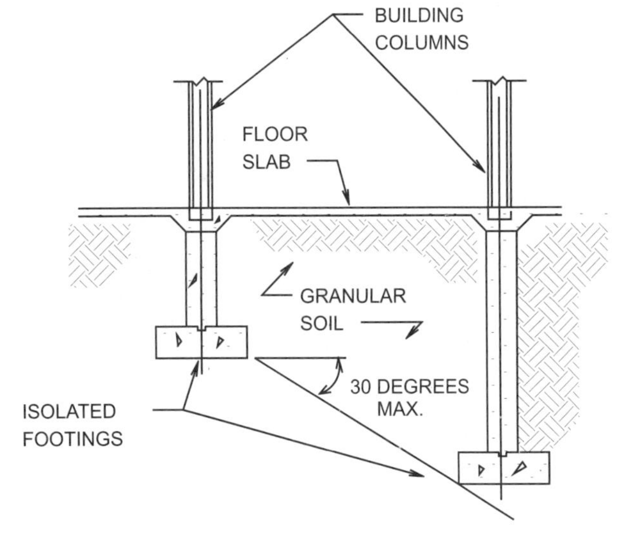

For SI: 1 degree = 0.01745 rad.

Figure 1805.2.2
ISOLATED FOOTINGS

1805.2.3 Shifting or moving soils. Where it is known that the shallow subsoils are of a shifting or moving character, footings shall be carried to a sufficient depth to ensure stability.

❖ See the commentary to Sections 1802.2.2 and 1802.3.2 for a discussion regarding expansive soils. Shallow foundations placed on or within a mass of shifting or moving soils must be designed with great rigidity and strength in order to adequately resist soil movement and avoid damage. This section requires footings to be carried to a sufficient depth for adequate stability. Adequate stability implies, among other things, the consideration of uplift and perpendicular forces exerted on the footings due to soil movement.

1805.3 Footings on or adjacent to slopes. The placement of buildings and structures on or adjacent to slopes steeper than one unit vertical in three units horizontal (33.3-percent slope) shall conform to Sections 1805.3.1 through 1805.3.5.

❖ The provisions of this section apply to buildings placed on or adjacent to slopes steeper than 1 unit vertical in 3 units horizontal.

1805.3.1 Building clearance from ascending slopes. In general, buildings below slopes shall be set a sufficient distance from the slope to provide protection from slope drainage, erosion and shallow failures. Except as provided for in Section 1805.3.5 and Figure 1805.3.1, the following criteria will be assumed to provide this protection. Where the existing slope is steeper than one unit vertical in one unit horizontal (100-percent slope), the toe of the slope shall be assumed to be at the intersection of a horizontal plane drawn from the top of the foundation and a plane drawn tangent to the slope at an angle of 45 degrees (0.79 rad) to the horizontal. Where a retaining wall is constructed at the toe of the slope, the height of the slope shall be measured from the top of the wall to the top of the slope.

❖ Commentary Figure 1805.3.1 depicts one of the criterion for locating buildings adjacent to the toe of an ascending slope. The setbacks required are intended to provide protection to the structure not only from shallow failures (sometimes referred to as "sloughing") but also from erosion and slope drainage. Furthermore, the space created by the setback provides access around the building and helps create a light and open-air environment. Where the existing slope is steeper than 1:1, the toe is assumed to be at a distance determined by the intersection of a horizontal line at the top of the foundation and a line drawn at a 45-degree (0.79 rad) angle to the horizontal line, and terminating at the top of the slope. This requirement is illustrated in Commentary Figure 1805.3.1.

1805.3.2 Footing setback from descending slope surface. Footings on or adjacent to slope surfaces shall be founded in firm material with an embedment and set back from the slope surface sufficient to provide vertical and lateral support for the footing without detrimental settlement. Except as provided for in Section 1805.3.5 and Figure 1805.3.1, the following setback is deemed adequate to meet the criteria. Where the slope is steeper than 1 unit vertical in 1 unit horizontal (100-percent slope), the required setback shall be measured from an imaginary plane 45 degrees (0.79 rad) to the horizontal, projected upward from the toe of the slope.

❖ The provisions of this section restrict the placement of footings adjacent to or on descending slopes so that vertical and lateral support for the footing are provided. The criteria for this condition is also shown in Code Figure 1805.3.1. When the existing slope is greater than 1:1, the setback is measured from an imaginary plane 45 degrees (0.79 rad) to the horizontal, projected from the toe of the slope. This requirement is illustrated in Commentary Figure 1805.3.2. For most conditions, the required setbacks will provide adequate lateral support for the foundations.

1805.3.3 Pools. The setback between pools regulated by this code and slopes shall be equal to one-half the building footing setback distance required by this section. That portion of the pool wall within a horizontal distance of 7 feet (2134 mm) from the top of the slope shall be capable of supporting the water in the pool without soil support.

❖ This section specifies the required setback distances for pools located near ascending or descending slopes. The minimum setback distance is established as one-half the required building setback from Section 1805.3.2. The pool wall that is within 7 feet (2134 mm) of the top slope is required to be self-supporting without support from the soil and is intended to provide additional safety measures should localized minor sliding and sloughing occur.

1805.3.4 Foundation elevation. On graded sites, the top of any exterior foundation shall extend above the elevation of the street gutter at point of discharge or the inlet of an approved drainage device a minimum of 12 inches (305 mm) plus 2 percent. Alternate elevations are permitted subject to the approval of the building official, provided it can be demonstrated that required drainage to the point of discharge and away from the structure is provided at all locations on the site.

❖ Figure 1805.3.4 depicts the requirements of this section regarding elevation for exterior foundations with respect to the street, gutter or point of inlet of a drainage device. The elevation of the street or gutter shown is that point at which drainage from the site reaches the street or gutter.

This requirement is intended to protect the building from water encroachment in case of heavy or unprecedented rain and may be modified on the approval of the building official if he or she finds that positive drainage

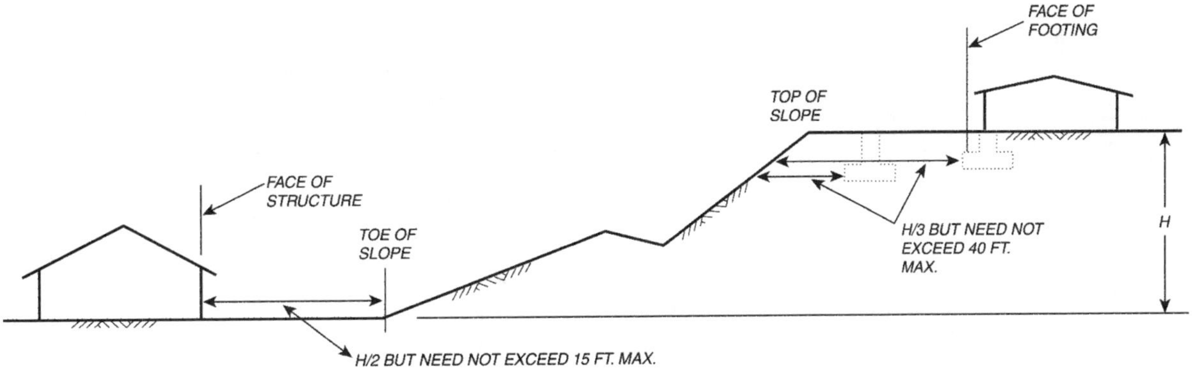

For SI: 1 inch = 25.4 mm.

FIGURE 1805.3.1
FOUNDATION CLEARANCES FROM SLOPES

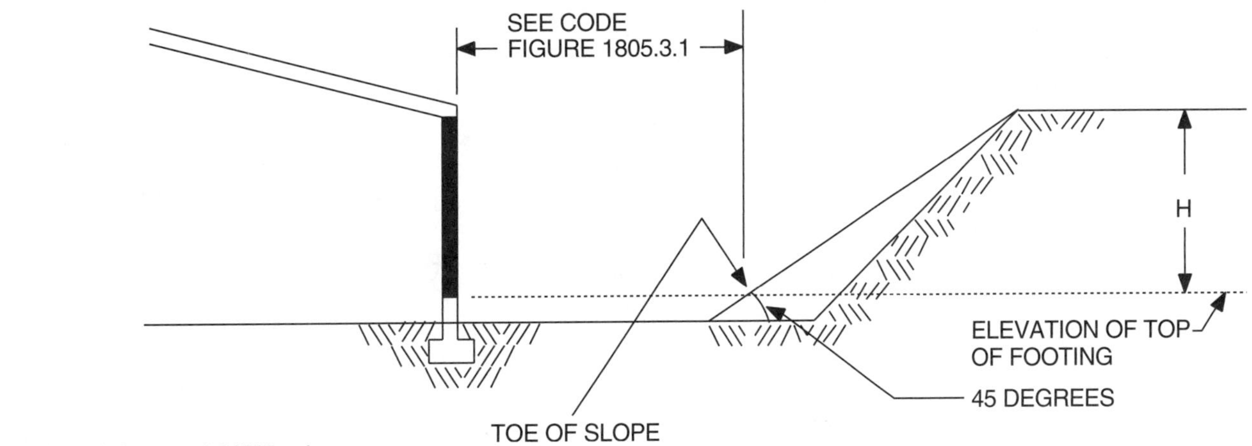

For SI: 1 degree = 0.01745 rad.

Figure 1805.3.1(2)
BUILDINGS ADJACENT TO ASCENDING SLOPE EXCEEDING 1 TO 1

slopes are provided to drain water away from the building and that the drainage pattern is not subject to temporary flooding due to landscaping or other impediments to drainage.

1805.3.5 Alternate setback and clearance. Alternate setbacks and clearances are permitted, subject to the approval of the building official. The building official is permitted to require an investigation and recommendation of a registered design professional to demonstrate that the intent of this section has been satisfied. Such an investigation shall include consideration of material, height of slope, slope gradient, load intensity and erosion characteristics of slope material.

❖ This section provides the building official the authority to approve alternative setbacks and clearances from slopes, provided he or she is satisfied that the intent of this section has been met. The building official has the authority to require a foundation investigation by a registered design professional to show that the intent of the code has been met. This item also specifies the parameters that must be considered by the registered design professional in the investigation.

1805.4 Footings. Footings shall be designed and constructed in accordance with Sections 1805.4.1 through 1805.4.6.

❖ Sections 1805.4.1 through 1805.4.6 pertain to the design and construction of concrete, masonry-unit, steel grillage and timber footings.

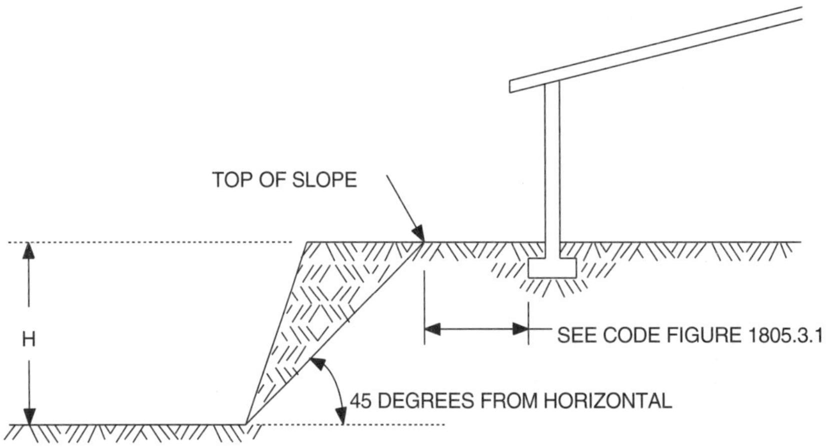

For SI: 1 degree = 0.01745 rad.

Figure 1805.3.2
BUILDINGS ADJACENT TO DESCENDING SLOPE EXCEEDING 1 TO 1

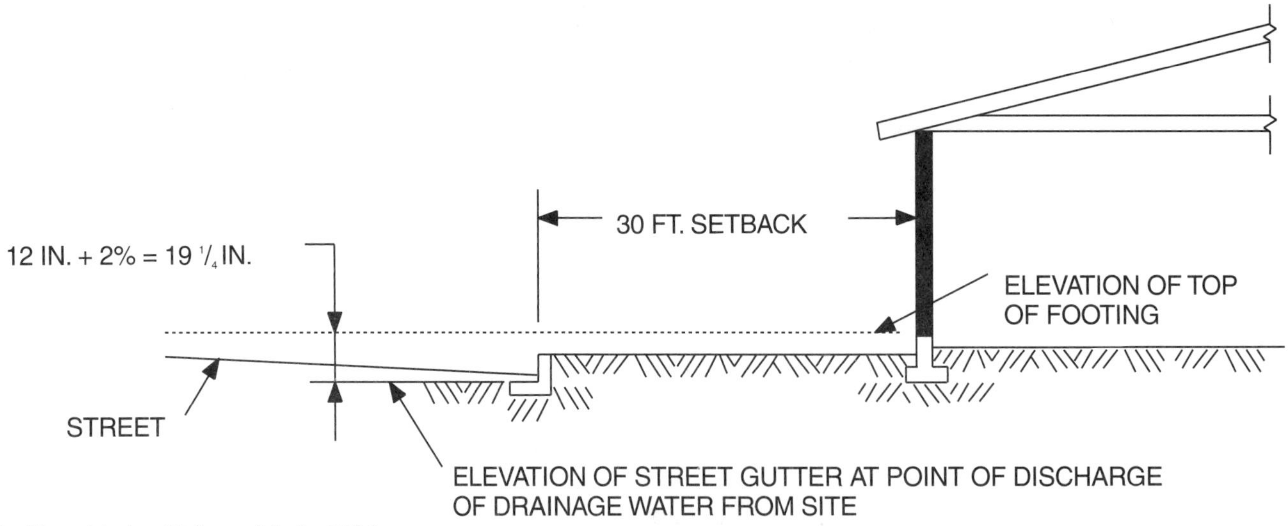

For SI: 1 inch = 25.4 mm, 1 foot = 304.8 mm.

Figure 1805.3.4
FOOTING ELEVATION ON GRADED SITES

1805.4.1 Design. Footings shall be so designed that the allowable bearing capacity of the soil is not exceeded, and that differential settlement is minimized. The minimum width of footings shall be 12 inches (305 mm).

Footings in areas with expansive soils shall be designed in accordance with the provisions of Section 1805.8.

❖ Regardless of the type of shallow foundation used, the allowable bearing capacity of soil must not be exceeded. There are two premises by which allowable soil-bearing pressures are established. The first premise requires that the safety factor against ultimate shear failure of the soil be adequate. The second premise requires that settlements under allowable bearing pressures not exceed tolerable values. In most cases, settlement governs the value established for allowable soil-bearing capacity. Bearing capacity is usually determined from a soils investigation and engineering analysis.

When the soils profile of a construction site is established by a sufficient number of test borings, and it indicates that a nonuniform soil condition exists where the strata of suitable bearing materials occurs at varying thicknesses or different depths, the foundation design must be adjusted to the subsurface condition to provide for the proper and safe performance of the foundation system.

Under such circumstances, it becomes necessary in the design of shallow foundations to determine the different depths at which isolated or continuous stepped footings need to be placed in order to obtain equal bearing pressures and avoid serious structural damage caused by differential (unequal) settlement of the different parts of the foundation system.

Another design method for obtaining equal bearing pressures and keeping the footings at a common elevation is to size the footings in accordance with the allowable bearing capacity of the soil at each location, thus producing a balanced design of the foundation system and preventing differential settlement.

Section 1805.8 stipulates required methods and design of foundations on expansive soils.

1805.4.1.1 Design loads. Footings shall be designed for the most unfavorable effects due to the combinations of loads specified in Section 1605.3. The dead load shall include the weight of foundations, footings and overlying fill. Reduced live loads, as specified in Section 1607.9, are permitted to be used in designing footings.

❖ Foundations are to be designed using those load combinations of full dead load (including overlying fill materials), floor or roof live loads, snow loads, wind or seismic forces and such other loads referenced in Section 1605.3 that will produce the most severe structural effects. The specified combinations allow for certain live load reductions in Section 1607.9, based on the low probabilities of simultaneous occurrences of the various possible load applications.

Foundation design is based on the ASD approach. Since Chapter 16 includes load combinations for strength design [load and resistance factor design (LRFD)], it is necessary to clarify that these values are to be used with the ASD load combinations in Section 1605.3.

1805.4.1.2 Vibratory loads. Where machinery operations or other vibrations are transmitted through the foundation, consideration shall be given in the footing design to prevent detrimental disturbances of the soil.

❖ Vibrations emanating from machinery operations and transmitted to the soil through the foundation may cause serious settlement to occur, particularly to foundations bearing on granular materials. While granular soils generally have a considerable volume of voids, the foundation pressure is usually carried and distributed by the bearing of grain on grain without detrimental deformations. Granular materials subjected to strong vibrations, however, may result in the particles readjusting and slipping into the void spaces. Essentially, the soil mass is consolidated and reduced in volume, causing vertical settlement.

Vibratory loads that cause a disturbance of the soil will flush water out of the material. In saturated granular soils, such as sand, the flushing action may cause a "quick" condition that allows sudden flow beneath the foundation, sometimes resulting in serious structural damage.

In soils that are rather impermeable, such as clays, vibration may also cause water to flush out, but it will take a very long time. Also, foundation settlement will occur over a prolonged period.

Care should be taken in the design of foundations to eliminate completely or minimize the transmission of vibratory loads to load-bearing soils.

1805.4.2 Concrete footings. The design, materials and construction of concrete footings shall comply with Sections 1805.4.2.1 through 1805.4.2.6 and the provisions of Chapter 19.

> **Exception:** Where a specific design is not provided, concrete footings supporting walls of light-frame construction are permitted to be designed in accordance with Table 1805.4.2.

❖ The design, construction and materials used for concrete footings are to comply with Chapter 19 and ACI 318, with the applicable requirements of Sections 1805.4.2.1 through 1805.4.2.6.

The exception applies to footings supporting walls of light-frame construction when a specific footing design is not provided. The purpose of Table 1805.4.2 is to specify footing sizes that can be used to safely support walls of light-frame construction. The table is based on anticipated loads on foundations due to wall, floor and roof systems.

TABLE 1805.4.2. See page 18-16.

❖ Table 1805.4.2 provides prescriptive designs for concrete and masonry-unit footings supporting walls of light-frame construction when a specific design is not provided. The first column of the table is specifically intended to apply to the number of floors supported by the footing, not the number of stories of a building. This ta-

TABLE 1805.4.2
FOOTINGS SUPPORTING WALLS OF LIGHT-FRAME CONSTRUCTION[a, b, c, d, e]

NUMBER OF FLOORS SUPPORTED BY THE FOOTING[f]	WIDTH OF FOOTING (inches)	THICKNESS OF FOOTING (inches)
1	12	6
2	15	6
3	18	8[g]

For SI: 1 inch = 25.4 mm, 1 foot = 304.8 mm.

a. Depth of footings shall be in accordance with Section 1805.2.

b. The ground under the floor is permitted to be excavated to the elevation of the top of the footing.

c. Interior-stud-bearing walls are permitted to be supported by isolated footings. The footing width and length shall be twice the width shown in this table, and footings shall be spaced not more than 6 feet on center.

d. See Section 1910 for additional requirements for footings of structures assigned to Seismic Design Category C, D, E or F.

e. For thickness of foundation walls, see Section 1805.5.

f. Footings are permitted to support a roof in addition to the stipulated number of floors. Footings supporting roof only shall be as required for supporting one floor.

g. Plain concrete footings for Group R-3 occupancies are permitted to be 6 inches thick.

ble provides for the minimum width and thickness of footings.

1805.4.2.1 Concrete strength. Concrete in footings shall have a specified compressive strength (f'_c) of not less than 2,500 pounds per square inch (psi) (17 237 kPa) at 28 days.

❖ This section of the code requires a minimum concrete compressive strength for footings of 2,500 psi (17 237 kPa) at 28 days. The code's intent is that concrete used as structural material for footing construction will be of sufficient strength and durability to satisfy safety requirements.

1805.4.2.2 Footing seismic ties. Where a structure is assigned to Seismic Design Category D, E or F in accordance with Section 1616, individual spread footings founded on soil defined in Section 1615.1.1 as Site Class E or F shall be interconnected by ties. Ties shall be capable of carrying, in tension or compression, a force equal to the product of the larger footing load times the seismic coefficient S_{DS} divided by 10 unless it is demonstrated that equivalent restraint is provided by reinforced concrete beams within slabs on grade or reinforced concrete slabs on grade.

❖ This section requires that spread footings on soft soil profiles be interconnected by ties when they support structures assigned to Seismic Design Category D and higher. The purpose of this section is to preclude excessive movement of one column or wall with respect to another (differential settlement). One of the prerequisites of adequate structural performance during an earthquake is that the foundation of the structure act as a unit. This is typically accomplished by tying together the pile caps, piers or piles with ties capable of carrying, in tension or compression, a force equal to 10 percent of the larger pile cap or column load multiplied by S_{DS}. S_{DS} is the design spectral response acceleration at short periods as determined in Section 1615.1.3. This can be accomplished through the use of concrete floor slabs or tie beams. The differential movement of the foundation should not exceed that included in the design of the seismic-force-resisting system.

1805.4.2.3 Plain concrete footings. The edge thickness of plain concrete footings supporting walls of other than light-frame construction shall not be less than 8 inches (203 mm) where placed on soil.

Exception: For plain concrete footings supporting Group R-3 occupancies, the edge thickness is permitted to be 6 inches (152 mm), provided that the footing does not extend beyond a distance greater than the thickness of the footing on either side of the supported wall.

❖ An isolated spread footing of square or rectangular shape is the most common type of shallow foundation used in building construction. Basically, the footing is a slab of concrete supporting a column or other concentrated load. The footing thickness or depth is generally a function of shear strength and is established by the more severe of two structural conditions. The thickness of a footing determined by one-way shear action may be compared to the shear in a beam where the vertical shear plane extends across the entire structural section. Thickness established by two-way shear action (punching shear) is based on the consideration that the column (or column pedestal) tends to punch through the footing and failure occurs as a fracture in the form of a truncated pyramid concentrated under the load (see Figure 1805.4.2.3). Foundation designs will normally result in concrete footing thicknesses that are greater than the minimum thicknesses prescribed in the code. The minimum thickness requirements for plain concrete footings are intended to provide adequate construction based on experience.

In compliance with the requirements of ACI 318, the edge thickness for plain concrete footings is not to be less than 8 inches (203 mm). In computing footing stresses (flexure, combined flexure, axial load and shear), however, the overall depth is to be 2 inches (51 mm) less than the actual thickness of the footing for all foundations on soil (see ACI 318 for design of plain concrete members) to allow for the irregularities of excavated earth surfaces and for possible contamination of the concrete by the soil in contact with the construction.

The exception permits the edge thickness of plain concrete footings to be reduced to 6 inches (152 mm) for occupancies in Group R-3 of light-frame construction where the footing does not extend beyond a distance greater than the thickness of the footing on either side of the supported wall. For lightweight construction with the dimensional limitations given, shear stresses in the concrete will not usually govern the design thicknesses of such foundations. Plain concrete is not to be used for footings on piles.

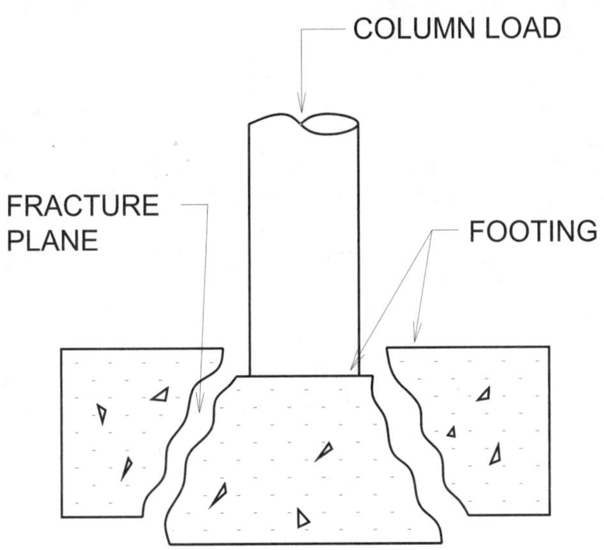

Figure 1805.4.2.3
PUNCHING SHEAR
(EXAGGERATED ILLUSTRATION)

1805.4.2.4 Placement of concrete. Concrete footings shall not be placed through water unless a tremie or other method approved by the building official is used. Where placed under or in the presence of water, the concrete shall be deposited by approved means to ensure minimum segregation of the mix and negligible turbulence of the water.

❖ Placing concrete under water should be avoided wherever possible. The risk of segregation of the concrete mixture is much greater when depositing under water as opposed to air, however, when concrete must be placed under water, it should be done by any one of several accepted methods used for such construction, including tremie. Due care must be exercised in the concreting operations so as to avoid or minimize segregation of the mix and turbulence of the water.

Generally, when concrete is to be placed under water, the mixture should be proportioned to provide a good plastic mix and high workability so that it will flow without segregation. The slump of the concrete should be 5 inches (127 mm) or greater. This desired consistency can be obtained by the use of rounded aggregates, a higher percentage of fines and entrained air. Cement content should be increased by 10 to 15 percent above the quantities required for similar mixtures placed in air to compensate for increases in water-cement ratios (see Chapter 19). In no case should the cement content be less than 600 pounds per cubic yard (355 kg/m^3) of concrete.

1805.4.2.5 Protection of concrete. Concrete footings shall be protected from freezing during depositing and for a period of not less than five days thereafter. Water shall not be allowed to flow through the deposited concrete.

❖ Concrete for footings should not be placed during rain, sleet or snow or in freezing weather unless adequate protection, as approved by the building official, is provided. Such protection, when required, is to be provided during the concreting operations and for a period of not less than five days thereafter. Rainwater or water from other sources must not be allowed to flow through freshly deposited concrete so as to increase the mixing water content or to damage the surface finish. For detailed information on materials, methods and procedures used in cold-weather operations, refer to ACI 306R, ACI 306.1 and Chapter 19.

1805.4.2.6 Forming of concrete. Concrete footings are permitted to be cast against the earth where, in the opinion of the building official, soil conditions do not require forming. Where forming is required, it shall be in accordance with Chapter 6 of ACI 318.

❖ Earth cuts are not normally used to form foundation construction, except for footings. Soil used to form the vertical sides of footings is to have sufficient stiffness to maintain the desired shape and dimensions before and during concreting operations. In the event that the soil is deemed to be unstable for such purpose, the building official is to require that formwork be built in accordance with the provisions of ACI 318.

1805.4.3 Masonry-unit footings. The design, materials and construction of masonry-unit footings shall comply with Sections 1805.4.3.1 and 1805.4.3.2, and the provisions of Chapter 21.

> **Exception:** Where a specific design is not provided, masonry-unit footings supporting walls of light-frame construction are permitted to be designed in accordance with Table 1805.4.2.

❖ Sections 1805.4.3.1 and 1805.4.3.2 provide general requirements for the construction of solid masonry footings. This type of foundation was widely used until the middle of the last century when it was replaced by steel or wood grillage and eventually by more economical plain and reinforced concrete foundations. At that time, masonry foundations were built of either stone cut to specific sizes or of rubble stone of random sizes bonded together with mortar. Masonry footings under columns were usually constructed as high piers in the shape of truncated pyramids. While stone footings may still exist in very old structures, they are extinct in modern construction. Today, although scarcely used, masonry footings may be constructed of hard-burned brick in cement mortar to support the walls of lightweight construction, such as for low residential buildings. Brick footings are usually set on a full bed of mortar spread upon the earth or a grout sill. Stepped footing courses are recommended to be built with the brick units laid on edge, these being capable of resisting a greater transverse stress than flat courses.

The exception applies to footings supporting walls of light-frame construction when a specific footing design is not provided. The purpose of Table 1805.4.2 is to specify footing sizes and depths that can be used to safely support walls of light-frame construction. The table is based on anticipated loads on foundations due to wall, floor and roof systems.

1805.4.3.1 Dimensions. Masonry-unit footings shall be laid in Type M or S mortar complying with Section 2103.7 and the depth shall not be less than twice the projection beyond the wall, pier or column. The width shall not be less than 8 inches (203 mm) wider than the wall supported thereon.

❖ Footings made of masonry units are to be laid in Type M or S mortar complying with the requirements of Section 2103.7. Type M is a high-strength mortar suitable for general use and, in particular, where maximum masonry compressive strength is required. It is also suitable for unreinforced masonry below grade and in construction that is in contact with earth. Type S is a general-use mortar where high lateral strength of masonry is required. It is specifically recommended for use in reinforced masonry.

Projections of footings beyond a wall, pier or column base are not to exceed dimensions by more than one-half the depth of the foundation. For example, a footing supporting a wall and consisting of three courses of brick with a depth of about 8 inches (203 mm) is not to project more than 4 inches (102 mm) ($^1/_2$ × 8 inches) beyond the base of the wall. This is to keep the shearing stresses in the footing within a safe limitation.

Masonry footings are not to be less than 8 inches (203 mm) wider than the thickness of the supported wall so as to provide adequate distribution of the wall load to the load-bearing soil.

1805.4.3.2 Offsets. The maximum offset of each course in brick foundation walls stepped up from the footings shall be $1^1/_2$ inches (38 mm) where laid in single courses, and 3 inches (76 mm) where laid in double courses.

❖ Foundation walls that entail stepping back successive courses of brick (called racking) are not to have horizontal offsets exceeding $1^1/_2$ inches (38 mm) from the face of the course below. If the steps are made in double course increments, then the offsets must not exceed 3 inches (76 mm).

Sometimes, wide footings are required so as not to exceed the safe load-bearing capacity of the supporting soil. Since footing projections beyond a wall are limited in dimension (see Section 1805.3.1), the wall must inevitably be stepped out to provide the proper base on the footing (see Figure 1805.4.3.2).

1805.4.4 Steel grillage footings. Grillage footings of structural steel shapes shall be separated with approved steel spacers and be entirely encased in concrete with at least 6 inches (152 mm) on the bottom and at least 4 inches (102 mm) at all other points. The spaces between the shapes shall be completely filled with concrete or cement grout.

❖ Steel grillage footings were extensively used during the latter part of the last century, but the development and

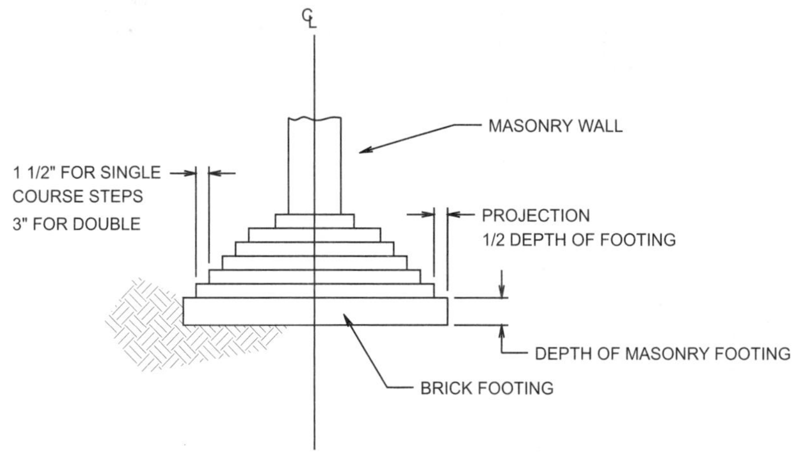

For SI: 1 inch = 25.4 mm.

Figure 1805.4.3.2
BRICK FOOTING OFFSET

use of reinforced concrete foundations have made this type of construction all but obsolete. They are, however, still used for underpinning purposes. There are many steel grillage footings in existence in old buildings.

A typical grillage footing consists of two or more tiers of steel beams (usually I sections) with each tier placed at right angles to the one below it. The beams in each tier are usually held together by a system of bolts and pipe spacers. The beams should be clean and unpainted and the whole system completely filled and encased in concrete with at least 6 inches (152 mm) of cover on the bottom and 4 inches (102 mm) at all other points. In lieu of concrete (other than the encasement), the spaces between the steel beams may be filled with cement grout.

1805.4.5 Timber footings. Timber footings are permitted for buildings of Type V construction and as otherwise approved by the building official. Such footings shall be treated in accordance with AWPA C2 or C3. Treated timbers are not required where placed entirely below permanent water level, or where used as capping for wood piles that project above the water level over submerged or marsh lands. The compressive stresses perpendicular to grain in untreated timber footings supported upon piles shall not exceed 70 percent of the allowable stresses for the species and grade of timber as specified in the AFPA NDS.

❖ The use of timber footings is allowed only in Type V construction, or as otherwise approved by the building official. Such footings are commonly built as grillages of heavy timbers, using large sections, such as railroad ties.

If the footings are to be constructed at depths above the water table, they are to be given a preservative treatment by pressure processes to protect the materials from decay, fungi and harmful insects. Pressure treatment is to be in accordance with the requirements of the C2 or C3 standard of the American-Wood Preservers' Association (AWPA). Except for use on cut surfaces, brush or spray applications of preservatives are not acceptable methods of treatment.

Preservative treatment by the pressure process within the limitations specified in AWPA C2 and C3 should not significantly affect the strength of the wood. Part of the process, however, involves the conditioning of timbers before treatment. Timbers conditioned by steaming or boiling under vacuum can suffer significant reductions in strength. This condition is recognized in the American Forest and Paper Association National Design Standard (AF&PA NDS) by the use of the untreated factor C_u.

The design values for treated round timber piles given in this specification are adjusted to compensate for strength reductions because of conditioning prior to treatment. It is recommended that the values given for the several species of wood be used in the design of treated timber footings. Design values given in other tables contained in the AF&PA referenced standard are for general structural purposes using untreated lumber. While the section of this specification for pressure-preservative treatment stipulates that the design values apply to wood products pressure impregnated by an approved process, it is not apparent that reductions have been made to allow for possible strength loss in the wood because of the treatment process.

Untreated timber may be used when the footings are completely embedded in soil below the water level. Long experience has shown that timber foundations permanently confined in water will stay sound and durable indefinitely. Wood submerged in fresh water cannot decay because the necessary air is excluded. It is not uncommon, however, to have changing ground-water levels because of changes in adjacent drainage systems and a variety of other subterranean conditions. Precautions should be taken so that untreated timber footings are placed at depths sufficiently below the ground-water level such that small drops in the water level will not expose the footings to air. Otherwise, decay will set in, causing settlements and possible structural failures.

The code also prescribes that compressive stresses perpendicular to grain in untreated timber footings supported upon piles are not to exceed 70 percent of the allowable stresses specified for the species and grade of timber given in the AF&PA referenced standard.

1805.4.6 Wood foundations. Wood foundation systems shall be designed and installed in accordance with AFPA Technical Report No. 7. Lumber and plywood shall be treated in accordance with AWPA C22 and shall be identified in accordance with Section 2303.1.8.1.

❖ This section covers the design and installation of a multicomponent wood foundation system complying with the specifications of AF&PA Technical Report No. 7. The construction is essentially a below-grade, load-bearing, wood frame system that serves as the enclosure for basements and crawl spaces and also as the structural foundation for the support of light frame structures.

This foundation system consists of several principal components:

1. Walls that consist of plywood panels attached to a wood stud framing system;
2. A composite footing consisting of a continuous wood plate set on a bed of granular materials, such as sand, gravel or crushed stone, which supports the foundation walls and transmits their loads to the bearing soil below;
3. Polyethylene film that serves as a vapor barrier and covers the exterior side of the plywood foundation walls from grade down to the footing plate;
4. Caulking compounds used for sealing joints in plywood walls as well as bonding agents for attaching the polyethylene film to the plywood and for sealing the film joints;
5. Metal fasteners made of silicon, bronze, copper or stainless steel or hot-dipped, zinc-coated steel nails or staples and

6. Pressure-treated plywood and lumber to protect the foundation material against decay, termites and other insects.

A typical cross section of a wood foundation basement wall is shown in Figure 1805.4.6.

All plywood and lumber used in the foundation system are required to receive a preservative treatment to protect the material against decay, termites and other insects in accordance with AWPA C22. This standard specifies that only Ammoniacal Copper Arsenate (ACA) or Chromated Copper Arsenate (CCA) water-borne preservatives are to be used in treating the softwoods typically used in the wood foundation system. Preservatives are to be applied by pressure processes conforming to the requirements of AWPA C1, C2 and C9. This section also requires the treated wood to be identified in accordance with Section 2303.1.8.1. See the commentary to that section for a discussion of identification requirements.

Ordinary metal fasteners in contact with preservatives used in the treatment of wood products will corrode over time and can cause serious structural problems. For this reason, AF&PA Technical Report No. 7 prescribes that only fasteners made of silicon bronze, copper or Type 304 or 316 stainless steel, as defined by the American Iron and Steel Institute (AISI), are permitted to be used in a wood foundation system. Hot-dipped, zinc-coated steel nails are permitted to be used in a wood foundation system when installed in accordance with the specific conditions contained in AF&PA Technical Report No. 7 for the surface treatment of nails and moisture protection of the foundation.

Certain cold-weather precautions should be taken in the installation of the wood foundation system. The composite footing consisting of a wood plate supported on a bed of stone or sand fill should not be placed on frozen ground. While the bottom of the wood plate footing would normally be placed below the frost line (i.e., basement construction), under certain drainage conditions, AF&PA Technical Report No. 7 permits the wood plate to be set above the frost line level. The building official should ascertain that such an alternative design satisfies the intent of the code. Most important is the use of proper sealants during very cold weather. All manufacturers of sealants and bonding agents impose temperature restrictions on the use of their products. Only sealants and bonding agents specifically produced for cold-weather conditions should be used.

Another cold-weather precaution relates to the use of sealants. All manufacturers of sealants and bonding agents place temperature limitations on their use. In

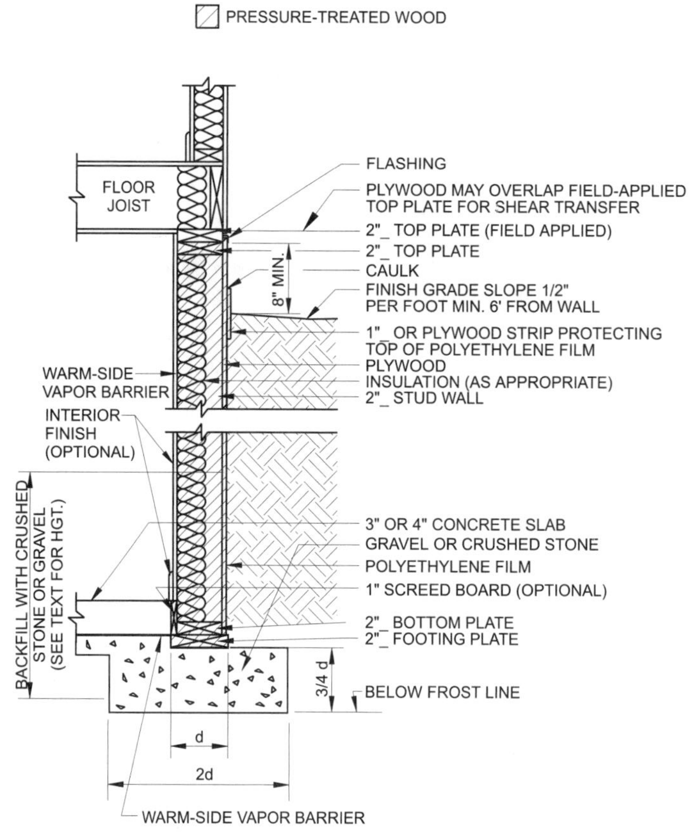

For SI: 1 inch = 25.4 mm, 1 foot = 304.8 mm.

Figure 1805.4.6
WOOD FOUNDATION

very cold weather, only those sealants that are specially produced for low-temperature applications should be used for the caulking of joints and for bonding purposes. This is an important consideration since sealants are a factor in verifying the water-tight integrity of the foundation installation.

The wood foundation system is an innovative design consisting of many parts and several kinds of materials. It can be expected to perform satisfactorily as a foundation for light construction only if the comprehensive provisions and recommendations of AF&PA Technical Report No. 7 are strictly followed.

1805.5 Foundation walls. Concrete and masonry foundation walls shall be designed in accordance with Chapter 19 or 21. Foundation walls that are laterally supported at the top and bottom and within the parameters of Tables 1805.5(1) through 1805.5(4) are permitted to be designed and constructed in accordance with Sections 1805.5.1 through 1805.5.5.

❖ This section covers foundation walls constructed of concrete or of masonry materials that are primarily intended for, but not necessarily limited to, basement construction in residential and light commercial buildings or other light structures.

Foundation walls are usually designed and constructed in accordance with the accepted engineering practices to carry the vertical loads from the structure above, resist wind and any lateral forces transmitted to the foundations and sustain earth pressures exerted against the walls, including any forces that may be imposed by frost action. For reinforced and plain concrete, the physical properties and design criteria are provided in Chapter 19. For masonry construction, the physical properties and design criteria are provided in Chapter 21.

This section also provides the option of using the design tables for plain masonry and concrete foundation walls, and the preengineered designs for reinforced masonry and concrete foundation walls. Foundation walls that are not laterally supported at the top and bottom are required to be designed by the specified standards, since the designs in the tables are only for walls that are laterally supported at the top and bottom.

1805.5.1 Foundation wall thickness. The minimum thickness of concrete and masonry foundation walls shall comply with Sections 1805.5.1.1 through 1805.5.1.3.

❖ Sections 1805.5.1.1 through 1805.5.1.2 prescribe the minimum thickness of foundation walls based on the thickness of the walls supported, the soil loads and heights for concrete or masonry foundation walls. The walls are to comply with the minimum thicknesses required for either section. Section 1805.5.1.3 prescribes the minimum thickness of foundation walls constructed of rough or random rubble stone.

1805.5.1.1 Thickness based on walls supported. The thickness of foundation walls shall not be less than the thickness of the wall supported, except that foundation walls of at least 8 inch (203 mm) nominal width are permitted to support brick-veneered frame walls and 10-inch-wide (254 mm) cavity walls provided the requirements of Section 1805.5.1.2 are met. Corbeling of masonry shall be in accordance with Section 2104.2. Where an 8-inch (203 mm) wall is corbeled, the top corbel shall be a full course of headers at least 6 inches (152 mm) in length, extending not higher than the bottom of the floor framing.

❖ The minimum thicknesses in this section are to facilitate the support of the wall above grade. The thickness requirements in this section are empirical and have been used successfully for many years. Where corbeling is desired to match the width of a masonry cavity wall above, see Section 2104.2 for construction requirements.

1805.5.1.2 Thickness based on soil loads, unbalanced backfill height and wall height. The thickness of foundation walls shall comply with the requirements of Table 1805.5(1) for plain masonry and plain concrete walls or Table 1805.5(2), 1805.5(3) or 1805.5(4) for reinforced concrete and masonry walls. When using the tables, masonry shall be laid in running bond and the mortar shall be Type M or S.

Unbalanced backfill height is the difference in height of the exterior and interior finish ground levels. Where an interior concrete slab is provided, the unbalanced backfill height shall be measured from the exterior finish ground level to the top of the interior concrete slab.

❖ The thickness of concrete and masonry walls are also required to meet the minimum requirements of Tables 1805.5.(1), 1805.5(2), 1805.5(3) and 1805.5(4).

As stated, unbalanced backfill height is the difference in height of the exterior and interior finished ground levels. The height of unbalanced backfill identifies the magnitude of the lateral soil load for the different classifications of soils presented in the tables. For an illustration of unbalanced backfill height, see Figure 1805.5.1.2.

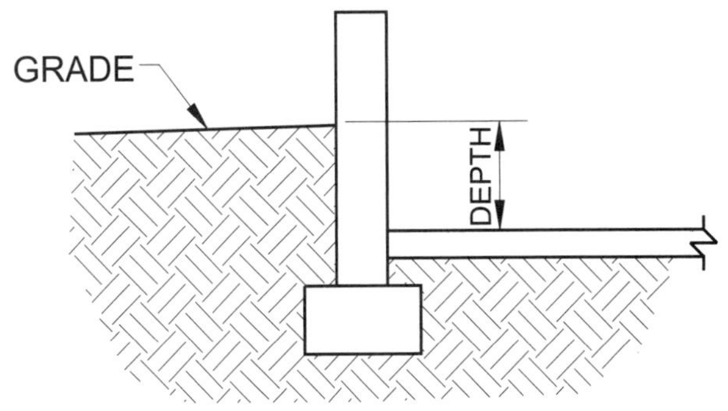

Figure 1805.5.1.2
UNBALANCED BACKFILL HEIGHT

1805.5.1.3 Rubble stone. Foundation walls of rough or random rubble stone shall not be less than 16 inches (406 mm) thick. Rubble stone shall not be used for foundations for structures in Seismic Design Category C, D, E or F.

❖ Foundation walls of rough or random rubble stone are limited to a thickness of 16 inches (406 mm). Because of the use of rough stones of irregular size and shape, a larger thickness is required (as compared to concrete or masonry walls) for adequate bonding of the stone and mortar to provide structural stability against soil pressures, as well as protection from water penetration.

1805.5.2 Foundation wall materials. Foundation walls constructed in accordance with Table 1805.5(1), 1805.5(2), 1805.5(3) or 1805.5(4) shall comply with the following:

1. Vertical reinforcement shall have a minimum yield strength of 60,000 psi (414 Mpa).
2. The specified location of the reinforcement shall equal or exceed the effective depth distance, *d*, noted in Tables 1805.5(2), 1805.5(3) and 1805.5(4) and shall be measured from the face of the soil side of the wall to the center of vertical reinforcement. The reinforcement shall be placed within the tolerances specified in ACI 530.1/ASCE 6/TMS 402, Article 3.4 B7 of the specified location.
3. Concrete shall have a specified compressive strength of not less than 2,500 psi (17.2 MPa) at 28 days.
4. Grout shall have a specified compressive strength of not less than 2,000 psi (13.8 MPa) at 28 days.
5. Hollow masonry units shall comply with ASTM C 90 and be installed with Type M or S mortar.

❖ This section specifies minimum requirements for materials and construction of foundation walls constructed in accordance with Tables 1805.5(1) through 1805.5(4). Item 2 provides specific placement requirements regarding the location of reinforcement in Tables 1805.5(2) through 1805.5(4). This dimension is necessary so that the wall is capable of developing the required bending capacity to resist the lateral soil loads. Additional guidance based on a structure's seismic design category is given in Section 1805.5.5.

TABLE 1805.5(1). See page 18-23.

❖ This table provides wall thickness designs for various wall heights and unbalanced backfill depths. The thicknesses in the table are proportional to the lateral soil loads for each classification of soils. The soil class designations are according to ASTM D 2487 (based on the Unified Soil Classification System that is commonly found in textbooks on soil mechanics). The lateral soil loads in the table are consistent with the soil loads in Table 1610.1. The description of the soils associated with each soil symbol (e.g., GW) is given in Table 1610.1.

The thicknesses in the table for plain masonry are based on research by the National Concrete Masonry Association (NCMA) in the report *Research Evaluation of the Flexural Strength of Concrete Masonry,* Project No. 93-172, December 7, 1993. The thickness for plain concrete masonry walls is based on ACI 318.

TABLE 1805.5(2). See page 18-24.

❖ The following commentary is applicable to Tables 1805.5(2) through 1805.5(4):

These tables apply to either concrete or masonry foundation walls. For general comments regarding this table, see the commentary to Table 1805.5(1). The thickness for the masonry walls in the table is based on the design requirements from ACI 530. The same wall thickness applies to either concrete or masonry foundation walls. The tables also specify the required size and spacing of the reinforcement.

Three different tables are shown for 8-, 10-, and 12-inch walls. Each table also specifies the required effective depth distance for each. For an explanation of the effective depth, see the commentary to Section 1805.5.2.

TABLE 1805.5(3). See page 18-24.

❖ See the commentary to Table 1805.5(2).

TABLE 1805.5(4). See page 18-25.

❖ See the commentary to Table 1805.5(2).

1805.5.3 Alternative foundation wall reinforcement. In lieu of the reinforcement provisions in Table 1805.5(2), 1805.5(3) or 1805.5(4), alternative reinforcing bar sizes and spacings having an equivalent cross-sectional area of reinforcement per linear foot (mm) of wall are permitted to be used, provided the spacing of reinforcement does not exceed 72 inches (1829 mm) and reinforcing bar sizes do not exceed No. 11.

❖ The purpose of this section is to provide an alternative for the reinforcement requirements in Tables 1805.5(2), 1805.5(3) and 1805.5(4) that would result in the same strength for lateral soil load resistance. For example, if No. 5 reinforcement bars were installed where the table indicates No. 4 at 48 inches (1219 mm) on center (o.c.), the spacing of the No. 5 bars for equivalent reinforcement area per foot of wall would be 72 inches (1829 mm) o.c. The increase in spacing would be allowed, since the cross-sectional area of a No. 5 bar is approximately 50 percent higher than a No. 4 reinforcement bar.

1805.5.4 Hollow masonry walls. At least 4 inches (102 mm) of solid masonry shall be provided at girder supports at the top of hollow masonry unit foundation walls.

❖ Foundation walls made of hollow load-bearing masonry units complying with the requirements of ASTM C 90 are to be reinforced with at least 4 inches (102 mm) of solid masonry at all locations where girders bear on masonry. This provides reinforcement against excessive stresses that may be induced in the wall through the support of heavy concentrated loads.

TABLE 1805.5(1)
PLAIN MASONRY AND PLAIN CONCRETE FOUNDATION WALLS[a, b, c]

PLAIN MASONRY				
		MINIMUM NOMINAL WALL THICKNESS (inches)		
		Soil classes and lateral soil load[a] (psf per foot below natural grade)		
WALL HEIGHT (feet)	HEIGHT OF UNBALANCED BACKFILL (feet)	GW, GP, SW and SP soils 30	GM, GC, SM, SM-SC and ML soils 45	SC, MH, ML-CL and Inorganic CL soils 60
7	4 (or less)	8	8	8
	5	8	10	10
	6	10	12	10 (solid[c])
	7	12	10 (solid[c])	10 (solid[c])
8	4 (or less)	8	8	8
	5	8	10	12
	6	10	12	12 (solid[c])
	7	12	12 (solid[c])	Note d
	8	10 (solid[c])	12 (solid[c])	Note d
9	4 (or less)	8	8	8
	5	8	10	12
	6	12	12	12 (solid[c])
	7	12 (solid[c])	12 (solid[c])	Note d
	8	12 (solid[c])	Note d	Note d
	9	Note d	Note d	Note d

PLAIN CONCRETE				
		MINIMUM NOMINAL WALL THICKNESS (inches)		
		Soil classes and lateral soil load[a] (psf per foot below natural grade)		
WALL HEIGHT (feet)	HEIGHT OF UNBALANCED BACKFILL (feet)	GW, GP, SW and SP soils 30	GM, GC, SM, SM-SC and ML soils 45	SC, MH, ML-CL and Inorganic CL soils 60
7	4 (or less)	$7^1/_2$	$7^1/_2$	$7^1/_2$
	5	$7^1/_2$	$7^1/_2$	$7^1/_2$
	6	$7^1/_2$	$7^1/_2$	8
	7	$7^1/_2$	8	10
8	4 (or less)	$7^1/_2$	$7^1/_2$	$7^1/_2$
	5	$7^1/_2$	$7^1/_2$	$7^1/_2$
	6	$7^1/_2$	$7^1/_2$	10
	7	$7^1/_2$	10	10
	8	10	10	12
9	4 (or less)	$7^1/_2$	$7^1/_2$	$7^1/_2$
	5	$7^1/_2$	$7^1/_2$	$7^1/_2$
	6	$7^1/_2$	$7^1/_2$	10
	7	$7^1/_2$	10	10
	8	10	10	12
	9	10	12	Note e

For SI: 1 inch = 25.4 mm, 1 foot = 304.8 mm, 1 pound per square foot per foot = 0.157 kPa/m.

a. For design lateral soil loads, see Section 1610. Soil classes are in accordance with the Unified Soil Classification System and design lateral soil loads are for moist soil conditions without hydrostatic pressure.

b. Provisions for this table are based on construction requirements specified in Section 1805.5.2.

c. Solid grouted hollow units or solid masonry units.

d. A design in compliance with Chapter 21 or reinforcement in accordance with Table 1805.5(2) is required.

e. A design in compliance with Chapter 19 is required.

TABLE 1805.5(2)
8-INCH CONCRETE AND MASONRY FOUNDATION WALLS WITH REINFORCING WHERE $d \geq 5$ INCHES[a, b, c]

WALL HEIGHT (feet)	HEIGHT OF UNBALANCED BACKFILL (feet)	VERTICAL REINFORCEMENT — Soil classes and lateral soil load[a] (psf per foot below natural grade)		
		GW, GP, SW and SP soils 30	GM, GC, SM, SM-SC and ML soils 45	SC, MH, ML-CL and Inorganic CL soils 60
7	4 (or less)	#4 at 48″ o.c.	#4 at 48″ o.c.	#4 at 48″ o.c.
	5	#4 at 48″ o.c.	#4 at 48″ o.c.	#4 at 40″ o.c.
	6	#4 at 48″ o.c.	#5 at 48″ o.c.	#5 at 40″ o.c.
	7	#4 at 40″ o.c.	#5 at 40″ o.c.	#6 at 48″ o.c.
8	4 (or less)	#4 at 48″ o.c.	#4 at 48″ o.c.	#4 at 48″ o.c.
	5	#4 at 48″ o.c.	#4 at 48″ o.c.	#4 at 40″ o.c.
	6	#4 at 48″ o.c.	#5 at 48″ o.c.	#5 at 40″ o.c.
	7	#5 at 48″ o.c.	#6 at 48″ o.c.	#6 at 40″ o.c.
	8	#5 at 40″ o.c.	#6 at 40″ o.c.	#7 at 40″ o.c.
9	4 (or less)	#4 at 48″ o.c.	#4 at 48″ o.c.	#4 at 48″ o.c.
	5	#4 at 48″ o.c.	#4 at 48″ o.c.	#5 at 48″ o.c.
	6	#4 at 48″ o.c.	#5 at 48″ o.c.	#6 at 48″ o.c.
	7	#5 at 48″ o.c.	#6 at 48″ o.c.	#7 at 48″ o.c.
	8	#5 at 40″ o.c.	#7 at 48″ o.c.	#8 at 48″ o.c.
	9	#6 at 40″ o.c.	#8 at 48″ o.c.	#8 at 32″ o.c.

For SI: 1 inch = 25.4 mm, 1 foot = 304.8 mm, 1 pound per square foot per foot = 0.157 kPa/m.

a. For design lateral soil loads, see Section 1610. Soil classes are in accordance with the Unified Soil Classification System and design lateral soil loads are for moist soil conditions without hydrostatic pressure.

b. Provisions for this table are based on construction requirements specified in Section 1805.5.2.

c. For alternative reinforcement, see Section 1805.5.3.

TABLE 1805.5(3)
10-INCH CONCRETE AND MASONRY FOUNDATION WALLS WITH REINFORCING WHERE $d \geq 6.75$ INCHES[a, b, c]

WALL HEIGHT (feet)	HEIGHT OF UNBALANCED BACKFILL (feet)	VERTICAL REINFORCEMENT — Soil classes and lateral soil load[a] (psf per foot below natural grade)		
		GW, GP, SW and SP soils 30	GM, GC, SM, SM-SC and ML soils 45	SC, MH, ML-CL and Inorganic CL soils 60
7	4 (or less)	#4 at 56″ o.c.	#4 at 56″ o.c.	#4 at 56″ o.c.
	5	#4 at 56″ o.c.	#4 at 56″ o.c.	#4 at 56″ o.c.
	6	#4 at 56″ o.c.	#4 at 48″ o.c.	#4 at 40″ o.c.
	7	#4 at 56″ o.c.	#5 at 56″ o.c.	#5 at 40″ o.c.
8	4 (or less)	#4 at 56″ o.c.	#4 at 56″ o.c.	#4 at 56″ o.c.
	5	#4 at 56″ o.c.	#4 at 56″ o.c.	#4 at 48″ o.c.
	6	#4 at 56″ o.c.	#4 at 48″ o.c.	#5 at 56″ o.c.
	7	#4 at 48″ o.c.	#4 at 32″ o.c.	#6 at 56″ o.c.
	8	#5 at 56″ o.c.	#5 at 40″ o.c.	#7 at 56″ o.c.
9	4 (or less)	#4 at 56″ o.c.	#4 at 56″ o.c.	#4 at 56″ o.c.
	5	#4 at 56″ o.c.	#4 at 56″ o.c.	#4 at 48″ o.c.
	6	#4 at 56″ o.c.	#4 at 40″ o.c.	#4 at 32″ o.c.
	7	#4 at 40″ o.c.	#5 at 48″ o.c.	#6 at 48″ o.c.
	8	#4 at 32″ o.c.	#6 at 48″ o.c.	#4 at 16″ o.c.
	9	#5 at 40″ o.c.	#6 at 40″ o.c.	#7 at 40″ o.c.

For SI: 1 inch = 25.4 mm, 1 foot = 304.8 mm, 1 pound per square foot per foot = 0.157 kPa/m.

a. For design lateral soil loads, see Section 1610. Soil classes are in accordance with the Unified Soil Classification System and design lateral soil loads are for moist soil conditions without hydrostatic pressure.

b. Provisions for this table are based on construction requirements specified in Section 1805.5.2.

c. For alternative reinforcement, see Section 1805.5.3.

TABLE 1805.5(4)
12-INCH CONCRETE AND MASONRY FOUNDATION WALLS WITH REINFORCING WHERE $d \geq 8.75$ INCHES[a, b, c]

WALL HEIGHT (feet)	HEIGHT OF UNBALANCED BACKFILL (feet)	VERTICAL REINFORCEMENT		
		Soil classes and lateral soil load[a] (psf per foot below natural grade)		
		GW, GP, SW and SP soils 30	GM, GC, SM, SM-SC and ML soils 45	SC, MH, ML-CL and Inorganic CL soils 60
7	4 (or less)	#4 at 72″ o.c.	#4 at 72″ o.c.	#4 at 72″ o.c.
	5	#4 at 72″ o.c.	#4 at 72″ o.c.	#4 at 72″ o.c.
	6	#4 at 72″ o.c.	#4 at 64″ o.c.	#4 at 48″ o.c.
	7	#4 at 72″ o.c.	#4 at 48″ o.c.	#5 at 56″ o.c.
8	4 (or less)	#4 at 72″ o.c.	#4 at 72″ o.c.	#4 at 72″ o.c.
	5	#4 at 72″ o.c.	#4 at 72″ o.c.	#4 at 72″ o.c.
	6	#4 at 72″ o.c.	#4 at 56″ o.c.	#5 at 72″ o.c.
	7	#4 at 64″ o.c.	#5 at 64″ o.c.	#4 at 32″ o.c.
	8	#4 at 48″ o.c.	#4 at 32″ o.c.	#5 at 40″ o.c.
9	4 (or less)	#4 at 72″ o.c.	#4 at 72″ o.c.	#4 at 72″ o.c.
	5	#4 at 72″ o.c.	#4 at 72″ o.c.	#4 at 64″ o.c.
	6	#4 at 72″ o.c.	#4 at 56″ o.c.	#5 at 64″ o.c.
	7	#4 at 56″ o.c.	#4 at 40″ o.c.	#6 at 64″ o.c.
	8	#4 at 64″ o.c.	#6 at 64″ o.c.	#6 at 48″ o.c.
	9	#5 at 56″ o.c.	#7 at 72″ o.c.	#6 at 40″ o.c.

For SI: 1 inch = 25.4 mm, 1 foot = 304.8 mm, 1 pound per square foot per foot = 0.157 kPa/m.

a. For design lateral soil loads, see Section 1610. Soil classes are in accordance with the Unified Soil Classification System and design lateral soil loads are for moist soil conditions without hydrostatic pressure.

b. Provisions for this table are based on construction requirements specified in Section 1805.5.2.

c. For alternative reinforcement, see Section 1805.5.3.

1805.5.5 Seismic requirements. Tables 1805.5(1) through 1805.5(4) shall be subject to the following limitations in Sections 1805.5.5.1 and 1805.5.5.2 based on the seismic design category assigned to the structure as defined in Section 1616.

❖ This section clarifies the application of the prescriptive foundation wall tables based on a structure's seismic design category.

1805.5.5.1 Seismic requirements for concrete foundation walls. Concrete foundation walls designed using Tables 1805.5(1) through 1805.5(4) shall be subject to the following limitations:

1. Seismic Design Categories A and B. No limitations, except provide not less than two No. 5 bars around window and door openings. Such bars shall extend at least 24 inches (610 mm) beyond the corners of the openings

2. Seismic Design Category C. Tables shall not be used except as allowed for plain concrete members in Section 1910.4.

3. Seismic Design Categories D, E and F. Tables shall not be used except as allowed for plain concrete members in ACI 318, Section 22.10.

❖ This section applies to concrete foundation walls. For structures classified as either Seismic Design Category A or B, the prescriptive tables apply along with the supplemental reinforcement specified at openings. For all other seismic design categories, use of these prescriptive tables is limited by the referenced code sections.

1805.5.5.2 Seismic requirements for masonry foundation walls. Masonry foundation walls designed using Tables 1805.5(1) through 1805.5(4) shall be subject to the following limitations:

1. Seismic Design Categories A and B. No additional seismic requirements.

2. Seismic Design Category C. A design using Tables 1805.5(1) through 1805.5(4) subject to the seismic requirements of Section 2106.4.

3. Seismic Design Category D. A design using Tables 1805.2(2) through 1805.5(4) subject to the seismic requirements of Section 2106.5.

4. Seismic Design Categories E and F. A design using Tables 1805.2(2) through 1805.5(4) subject to the seismic requirements of Section 2106.6.

❖ This section applies to masonry foundation walls. For structures classified as either Seismic Design Category A or B, the prescriptive tables apply. For all other seismic design categories, use of these prescriptive tables is limited by the referenced code sections. For structures classified as Seismic Design Category D or higher, only the tables that require reinforcement can be used.

1805.5.6 Foundation wall drainage. Foundation walls shall be designed to support the weight of the full hydrostatic pressure of undrained backfill unless a drainage system is installed in accordance with Sections 1807.4.2 and 1807.4.3.

❖ This section clarifies that if a foundation drainage system is not provided, the foundation wall has to be designed to support the weight of the full hydrostatic pressure of the undrained backfill. The lateral pressures exerted from undrained backfill will be significantly higher than those exerted from properly drained backfill.

1805.5.7 Pier and curtain wall foundations. Except in Seismic Design Categories D, E and F, pier and curtain wall foundations are permitted to be used to support light-frame construction not more than two stories in height, provided the following requirements are met:

1. All load-bearing walls shall be placed on continuous concrete footings bonded integrally with the exterior wall footings.
2. The minimum actual thickness of a load-bearing masonry wall shall not be less than 4 inches (102 mm) nominal or $3^5/_8$ inches (92 mm) actual thickness, and shall be bonded integrally with piers spaced 6 feet (1829 mm) on center (o.c.).
3. Piers shall be constructed in accordance with Chapter 21 and the following:
   - 3.1. The unsupported height of the masonry piers shall not exceed 10 times their least dimension.
   - 3.2. Where structural clay tile or hollow concrete masonry units are used for piers supporting beams and girders, the cellular spaces shall be filled solidly with concrete or Type M or S mortar.

     Exception: Unfilled hollow piers are permitted where the unsupported height of the pier is not more than four times its least dimension.
   - 3.3. Hollow piers shall be capped with 4 inches (102 mm) of solid masonry or concrete or the cavities of the top course shall be filled with concrete or grout.
4. The maximum height of a 4-inch (102 mm) load-bearing masonry foundation wall supporting wood frame walls and floors shall not be more than 4 feet (1219 mm) in height.
5. The unbalanced fill for 4-inch (102 mm) foundation walls shall not exceed 24 inches (610 mm) for solid masonry, nor 12 inches (305 mm) for hollow masonry.

❖ Pier and curtain wall foundation systems are now and have always been a very popular method of construction for many years, particularly throughout the eastern United States. The origin of the design is seen in many precolonial wood frame buildings. Pier and curtain wall foundations are only permitted to support structures of light-frame construction (wood or light-gage steel framing members) not more than two stories in height and assigned to Seismic Design Category A, B or C. Seismic detailing requirements for higher seismic design categories have not yet been developed. The term "pier" used in this application means pilaster rather than a type of footing. The term "curtain wall" refers to minimum 4-inch-thick (102 mm) masonry load-bearing walls. The provisions apply to simple wood frame buildings where the combined loads are minimal.

This type of foundation system is also addressed in Section R404.1.5.1 of the *International Residential Code*® (IRC®), which includes an accompanying figure illustrating in the code the code requirements. This IRC figure is reproduced as Figure 1805.5.7.

1805.6 Foundation plate or sill bolting. Wood foundation plates or sills shall be bolted or strapped to the foundation or foundation wall as provided in Chapter 23.

❖ Anchorage of the sill plate in Seismic Design Categories D, E and F is addressed in Section 2305.3.10. Sill plate anchorage for conventional light frame construction is addressed in Section 2308.6.

1805.7 Designs employing lateral bearing. Designs to resist both axial and lateral loads employing posts or poles as columns embedded in earth or embedded in concrete footings in the earth shall conform to the requirements of Sections 1805.7.1 through 1805.7.3.

❖ Section 1805.7 addresses design criteria for posts or poles to resist axial and lateral loads. While more accurate methods are available for determining the resistance of posts or poles to axial and lateral loads, the other methods are much more complex, and the results do not differ significantly from this method.

1805.7.1 Limitations. The design procedures outlined in this section are subject to the following limitations:

1. The frictional resistance for structural walls and slabs on silts and clays shall be limited to one-half of the normal force imposed on the soil by the weight of the footing or slab.
2. Posts embedded in earth shall not be used to provide lateral support for structural or nonstructural materials such as plaster, masonry or concrete unless bracing is provided that develops the limited deflection required.

Wood poles shall be treated in accordance with AWPA C2 or C4.

❖ The limitations imposed by this section are intended to address both structural stability and serviceability. The limitation of the frictional resistance for structural walls and slabs on silts and clays is consistent with Section 1804.3 and Table 1804.2, which also limit the sliding resistance to one-half the dead load. The limitations on the types of construction materials that utilize lateral support of poles are based on the brittle nature of the materials. In order to prevent excessive distortions, which would produce cracking of these brittle materials, this section limits the use of poles unless some type of rigid cross bracing is provided to limit the deflections to those that can be tolerated by the materials.

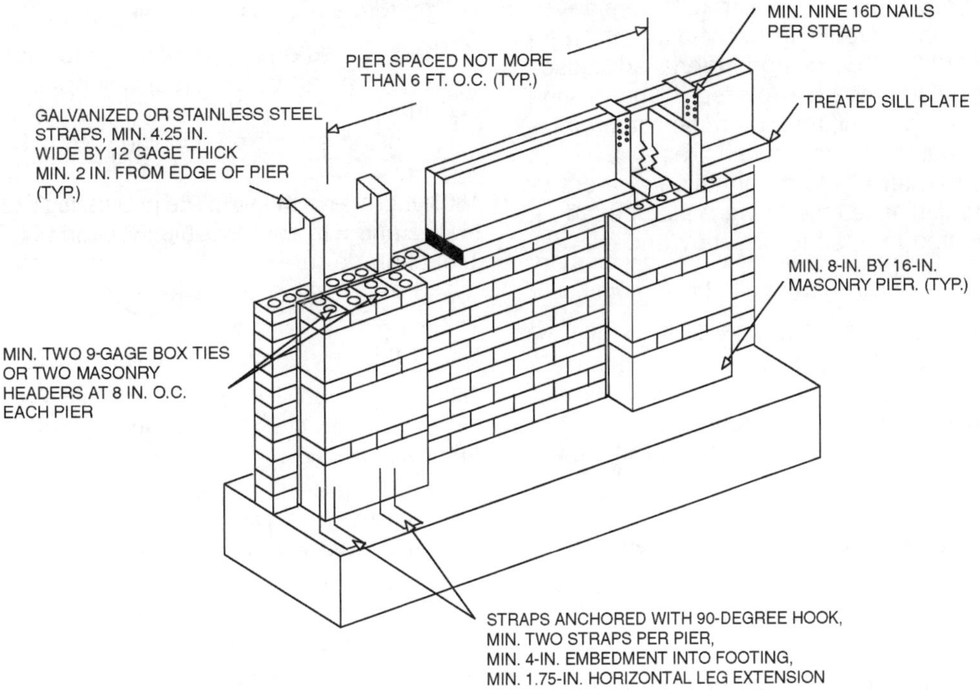

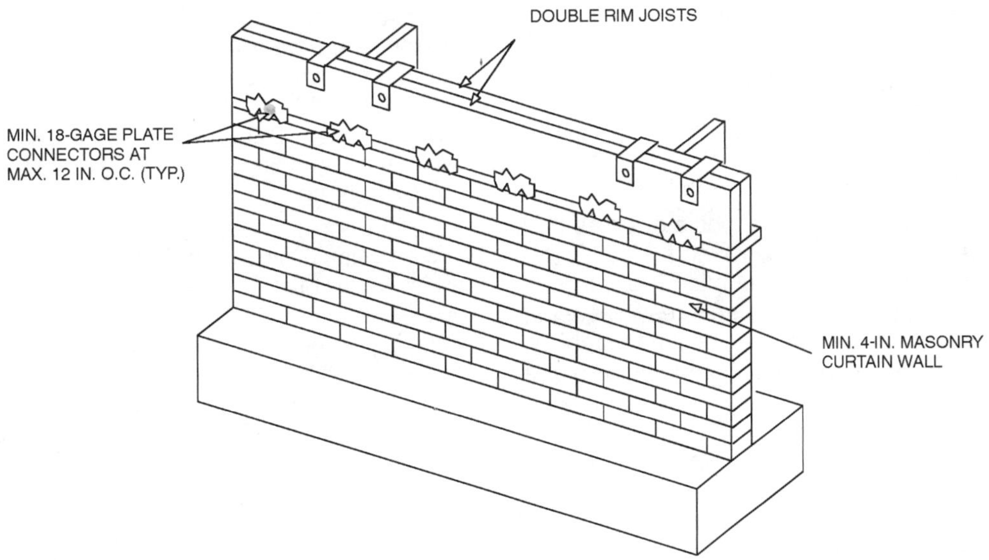

For SI: 1 inch = 25.4 mm, 1 foot = 304.8 mm, 1 degree = $^{0.79}/_{45}$ rad.

Figure 1805.5.7
FOUNDATION WALL CLAY MASONRY CURTAIN WALL WITH CONCRETE MASONRY PIERS

1805.7.2 Design criteria. The depth to resist lateral loads shall be determined by the design criteria established in Sections 1805.7.2.1 through 1805.7.2.3, or by other methods approved by the building official.

❖ The design criteria for the use of poles embedded in the ground or in concrete footings in the ground were developed for the Outdoor Advertising Association of America, Inc. in the 1940s. The design criteria addresses conditions where constraint is provided at the ground surface, such as a rigid floor, and where no constraint is provided. The original design criteria established a $^1/_2$-inch (12.7 mm) lateral pole deformation at the surface of the ground. These criteria were also based on field tests conducted in a range of sandy and gravelly soils and of silts and clays.

1805.7.2.1 Nonconstrained. The following formula shall be used in determining the depth of embedment required to resist lateral loads where no constraint is provided at the ground surface, such as rigid floor or rigid ground surface pavement, and where no lateral constraint is provided above the ground surface, such as a structural diaphragm.

$$d = 0.5A\{1 + [1 + (4.36h/A)]^{1/2}\}$$ **(Equation 18-1)**

where:

A = $2.34P/S_1\,b$.

b = Diameter of round post or footing or diagonal dimension of square post or footing, feet (m).

d = Depth of embedment in earth in feet (m) but not over 12 feet (3658 mm) for purpose of computing lateral pressure.

h = Distance in feet (m) from ground surface to point of application of "P."

P = Applied lateral force in pounds (kN).

S_1 = Allowable lateral soil-bearing pressure as set forth in Section 1804.3 based on a depth of one-third the depth of embedment in pounds per square foot (psf) (kPa).

❖ The required embedment depth for posts or poles that are not restrained at or above the ground surface is determined by Equation 18-1. This equation is applicable to posts or poles that have no lateral restraint at or above the ground surface. In terms of column stability, this is considered to be free at the end and pinned at the base.

1805.7.2.2 Constrained. The following formula shall be used to determine the depth of embedment required to resist lateral loads where constraint is provided at the ground surface, such as a rigid floor or pavement.

$$d^2 = 4.25(Ph/S_3\,b)$$ **(Equation 18-2)**

or alternatively

$$d^2 = 4.25\,(M_g/S_3\,b)$$ **(Equation 18-3)**

where:

M_g = moment in the post at grade, in foot-pounds (kN-m).

S_3 = Allowable lateral soil-bearing pressure as set forth in Section 1804.3 based on a depth equal to the depth of embedment in pounds per square foot (kPa).

❖ The required embedment depth for posts or poles that are restrained at the ground surface is determined by Equation 18-2 or 18-3. In terms of column stability, this is considered a fixed condition at the base.

1805.7.2.3 Vertical load. The resistance to vertical loads shall be determined by the allowable soil-bearing pressure set forth in Table 1804.2.

❖ See the commentary to Section 1804.2 and Table 1804.2.

1805.7.3 Backfill. The backfill in the annular space around columns not embedded in poured footings shall be by one of the following methods:

1. Backfill shall be of concrete with an ultimate strength of 2,000 psi (13.8 MPa) at 28 days. The hole shall not be less than 4 inches (102 mm) larger than the diameter of the column at its bottom or 4 inches (102 mm) larger than the diagonal dimension of a square or rectangular column.
2. Backfill shall be of clean sand. The sand shall be thoroughly compacted by tamping in layers not more than 8 inches (203 mm) in depth.
3. Backfill shall be of controlled low-strength material (CLSM).

❖ In order for the post or pole to meet the conditions and limitations of research for which the above criteria was established, backfill in the annular space around a column not embedded in a concrete footing must be either 2,000 psi (13.8 MPa) concrete, CLSM (see commentary, Section 1803.6) or clean sand thoroughly compacted by tamping in layers not more than 8 inches (203 mm) in depth.

1805.8 Design for expansive soils. Footings or foundations for buildings and structures founded on expansive soils shall be designed in accordance with Section 1805.8.1 or 1805.8.2.

Footing or foundation design need not comply with Section 1805.8.1 or 1805.8.2 where the soil is removed in accordance with Section 1805.8.3, nor where the building official approves stabilization of the soil in accordance with Section 1805.8.4.

❖ Expansive soils, often referred to as "swelling soils," contain montmorillonite minerals and have the characteristics of absorbing water and swelling or shrinking and cracking when drying. Significant volume changes can cause serious damage to buildings and other structures as well as to pavements and sidewalks. Swelling soils are found throughout the nation, but are more prevalent in regions with dry or moderately arid climates.

There is a general relationship between the PI of a soil as determined by the ASTM D 4318 standard test method and the potential for expansion. This relationship is shown in Figure 1802.3.2.

Section 1802.3.2 defines "Expansive soil" as any plastic material with a PI of 15 or greater with more than 10 percent of the soil particles passing a No. 200 sieve and less than 5 micrometers in size. As an alternative, tests in accordance with ASTM D 4829 can be used to determine if a soil is expansive. The expansion index is a measure of the swelling potential of the soil.

The amount and depth of potential swelling that can occur in a clay material are, to some extent, functions of the cyclical moisture content in the soil. In dryer climates where the moisture content in the soil near the ground surface is low because of evaporation, there is a greater potential for extensive swelling than the same soil in wetter climates where the variations of moisture content are not as severe.

Volume changes in highly expansive soils range between 7 and 10 percent, but experience has shown that occasionally, under abnormal conditions, they can reach as high as 25 percent.

1805.8.1 Foundations. Footings or foundations placed on or within the active zone of expansive soils shall be designed to resist differential volume changes and to prevent structural damage to the supported structure. Deflection and racking of the supported structure shall be limited to that which will not interfere with the usability and serviceability of the structure.

Foundations placed below where volume change occurs or below expansive soil shall comply with the following provisions:

1. Foundations extending into or penetrating expansive soils shall be designed to prevent uplift of the supported structure.
2. Foundations penetrating expansive soils shall be designed to resist forces exerted on the foundation due to soil volume changes or shall be isolated from the expansive soil.

❖ Shallow foundation systems placed on or within a mass of expansive soil must be designed with great rigidity and strength in order to adequately resist swelling pressures and avoid serious structural damage. Sometimes footings and piers, as well as foundation walls and grade beams, are isolated from the swelling soils by intervening fills or granular materials. While this type of insulation serves to cushion or diminish lateral pressures, it will not prevent structure heaving, except for grade beams constructed on collapsible forms.

The preferable foundation construction in expansive soils is the use of drilled piers with belled footings extending below the zone of swelling activity, or at least to a soil's stratum where the seasonal moisture content of the expansive soil will remain within a tolerable range. This type of construction is made possible in swelling soil's because the material usually consists of stiff clays that do not contain free water, providing excellent conditions for drilling holes in the ground. Piers with belled footings have been widely used for foundations in expansive soils, even in the construction of single-family dwellings.

The concrete shaft must be reinforced for its entire length because the swelling soil is apt to exert high uplift forces and subject the drilled pier construction to tensile stresses. This condition can be prevented or sufficiently mitigated by isolating the concrete from the swelling soil by surrounding the shaft with a vertical layer of granular soil or other suitable materials that possess little or no shearing strength (e.g., vermiculite).

1805.8.2 Slab-on-ground foundations. Slab-on-ground, mat or raft foundations on expansive soils shall be designed and constructed in accordance with WRI/CRSI *Design of Slab-on-Ground Foundations* or PTI *Design and Construction of Post-Tensioned Slabs-On-Ground*.

Exception: Slab-on-ground systems that have performed adequately in soil conditions similar to those encountered at the building site are permitted subject to the approval of the building official.

❖ It is not uncommon for concrete slabs on ground to be reinforced (prestressed) with post-tensioned strands so that the magnitude of tensile forces induced in the slab from uplift of the soil is reduced, thus avoiding cracking of the concrete at its top surface. It has also been a practice to cast the slab on cellular forms made of cardboard or other collapsible materials that will support wet concrete, but will yield when subjected to swelling pressures. Since there is no rebound of the form material after subsidence of the soil, floors must be designed as structural slabs. This section recognizes the methodologies in WRI/CRSI *Design of Slab-on-Ground Foundations* and PTI *Design and Construction of Post-Tensioned Slabs-on-Ground* for the design of slab-on-ground foundations located on expansive soils. The WRI/CRSI document provides design requirements for reinforced concrete slabs. The PTI document addresses prestressed (specifically post-tensioned) concrete.

1805.8.3 Removal of expansive soil. Where expansive soil is removed in lieu of designing footings or foundations in accordance with Section 1805.8.1 or 1805.8.2, the soil shall be removed to a depth sufficient to ensure a constant moisture content in the remaining soil. Fill material shall not contain expansive soils and shall comply with Section 1803.5 or 1803.6.

Exception: Expansive soil need not be removed to the depth of constant moisture, provided the confining pressure in the expansive soil created by the fill and supported structure exceeds the swell pressure.

❖ The best way to avoid the problems associated with expansive soils is to remove such soil from the construction site and, where necessary, replace it with suitable compacted fill material. Depending on site conditions, however, this method is not always economically feasible and is not widely used.

1805.8.4 Stabilization. Where the active zone of expansive soils is stabilized in lieu of designing footings or foundations in accordance with Section 1805.8.1 or 1805.8.2, the soil shall be stabilized by chemical, dewatering, presaturation or equivalent techniques.

❖ This section allows expansive soils to be stabilized by either chemical, presaturation or dewatering methods or by other equivalent techniques. These methods, however, have limited use.

Chemical stabilization may be effectively accomplished by the use of lime thoroughly mixed with the soil and compacted at approximately the optimum moisture content. The purpose of using lime is to reduce the plasticity of the soil, which will reduce the swelling potential (see commentary, Section 1805.8). Since this method requires a uniform mix of lime and soil, however, it is generally limited to compacted fills. There is a method of pressure injecting lime slurry into heavily fissured clay materials, but the method is not appropriate for all site conditions.

Presaturation by flooding the site is rarely a totally effective way to stabilize expansive soils, since it takes a very long time for the water to penetrate to any great depth. Dewatering of expansive soils, which consist generally of dense clays without free water, is not an effective way to control moisture content.

One good method of stabilization is to place sufficient compacted fill material on the site so that the downward pressures of the fill will, to the extent possible, balance the upward swelling pressures of the soil. The effectiveness of this balancing concept depends on the pressures developed by the expansive soil as a function of estimated volume change and the depth of compacted fill material necessary to counteract these pressures. The application of this stabilization method is usually economically feasible when the soil pressures are low, around 500 psf (23.9 kPa); however, pressures occasionally have reached as high as 20 tons per square foot (1915 kPa).

1805.9 Seismic requirements. See Section 1910 for additional requirements for footings and foundations of structures assigned to Seismic Design Category C, D, E or F.

For structures assigned to Seismic Design Category D, E or F, provisions of ACI 318, Sections 21.10.1 to 21.10.3, shall apply when not in conflict with the provisions of Section 1805. Concrete shall have a specified compressive strength of not less than 3,000 psi (20.68 MPa) at 28 days.

Exceptions:

1. Group R or U occupancies of light-framed construction and two stories or less in height are permitted to use concrete with a specified compressive strength of not less than 2,500 psi (17.2 MPa) at 28 days.
2. Detached one- and two-family dwellings of light-frame construction and two stories or less in height are not required to comply with the provisions of ACI 318, Sections 21.10.1 to 21.10.3.

❖ This section first provides an important reference to Section 1910 for additional requirements concerning footings and foundations of buildings assigned to Seismic Design Category C, D, E or F. As specified in Section 1910.4.3.2, all footings and foundations in these structures require reinforcement except for footings supporting walls in detached one- and two-family dwellings three stories or less in height and constructed with stud-bearing walls. The concern is that plain concrete footings may be susceptible to damage, which can be reduced or avoided if reinforcement is provided. Foundation damage is obviously very expensive to repair and is better avoided.

The second paragraph of this section refers to provisions in ACI 318 concerning the design and construction of foundations of buildings assigned to high seismic performance and design categories. Chapter 21 of ACI 318 is entitled "Special Provisions for Seismic Design," and Section 21.10 contains the requirements pertaining to foundations. This paragraph also requires a minimum concrete compressive strength for footings of 3,000 psi (20.68 MPa) at 28 days. Section 21.2.4.1 of ACI 318 prescribes such a limitation for reinforced concrete structural members that are part of the seismic-force-resisting system of a building. The code's intent is that concrete used as a structural material for footing construction will be of sufficient strength and durability. The exception permits the use of 2,500 psi (17.2 MPa) concrete in smaller residential-type structures. Additionally, relatively small, detached dwellings classified as Seismic Design Category D, E or F are exempted from the referenced ACI 318 provisions. These structures would only need to comply with requirements for Seismic Design Category C.

SECTION 1806 RETAINING WALLS

1806.1 General. Retaining walls shall be designed to ensure stability against overturning, sliding, excessive foundation pressure and water uplift. Retaining walls shall be designed for a safety factor of 1.5 against lateral sliding and overturning.

❖ This section provides design considerations for a retaining wall. The lateral pressure of the soil against the retaining wall is greatly influenced by soil moisture. Backfill is usually kept from being saturated for an extended length of time by placing drains near the base of the retaining wall to remove the water in the soil behind it.

SECTION 1807 DAMPPROOFING AND WATERPROOFING

1807.1 Where required. Walls or portions thereof that retain earth and enclose interior spaces and floors below grade shall be waterproofed and dampproofed in accordance with this section, with the exception of those spaces containing groups other than residential and institutional where such omission is not detrimental to the building or occupancy.

Ventilation for crawl spaces shall comply with Section 1203.4.

❖ Section 1807 covers the requirements for waterproofing and dampproofing those parts of substructure construction that need to be provided with moisture protection. Sections 1807.1 through 1807.4.3 identify the locations where moisture barriers are required and specify the materials to be used and the methods of application. The provisions also deal with subsurface water conditions, drainage systems and other protection requirements.

The term "waterproofing" is often used where dampproofing is really meant. Although both terms are intended to apply to the installation and use of moisture barriers, dampproofing does not furnish the same degree of moisture protection.

Dampproofing generally refers to the application of one or more coatings of a compound or other materials that are impervious to water, which are used to prevent the passage of water vapor through walls or other building components, and which restrict the flow of water under slight hydrostatic pressure. Waterproofing, on the other hand, refers to the application of coatings and sealing materials to walls or other building components to prevent moisture from penetrating in either a vapor or liquid form, even under conditions of significant hydrostatic pressure. Hydrostatic pressure is created by the presence of water under pressure. This pressure can occur when the ground-water table rises above the bottom of the foundation wall, or the soil next to the foundation wall becomes saturated with water caused by uncontrolled storm water runoff.

Section 1807.1 is an overall requirement that waterproofing and dampproofing applications are to be made to horizontal and vertical surfaces of below-ground spaces where the occupancy would normally be affected by the intrusion of water or moisture. Moisture or water in a floor below grade can cause damage to structural members such as columns, posts or load-bearing walls, as well as pose a health hazard by promoting the growth of bacteria or fungi and adversely affect any mechanical and electrical appliances that may be located at that level. It can also cause a great deal of damage to goods that may be located or stored in that lower level. These vertical and horizontal surfaces include foundation walls, retaining walls, underfloor spaces and floor slabs. Waterproofing and dampproofing are not required in locations other than residential and institutional occupancies where the omission of moisture barriers would not adversely affect the use of the spaces. An example of a location where waterproofing or dampproofing would not be required is in an open parking structure, as long as the structural components are individually protected against the effects of water. Waterproofing and dampproofing are not permitted to be omitted from residential and institutional occupancies where people may be sleeping or services are provided on the floor below grade. A person walking in a flooded basement may be in a very hazardous situation, particularly if the possibility of an electrical charge in the water exists caused by electrical service at that level.

Section 1807.1.1 addresses the type of problem faced when a portion of a story is above grade, while Section 1807.1.2 limits any infiltration of water into crawl spaces so as to protect this area from potential water damage and prevent ponding of water. Both of these sections reference other applicable sections of the code, as well as the exceptions.

1807.1.1 Story above grade. Where a basement is considered a story above grade and the finished ground level adjacent to the basement wall is below the basement floor elevation for 25 percent or more of the perimeter, the floor and walls shall be dampproofed in accordance with Section 1807.2 and a foundation drain shall be installed in accordance with Section 1807.4.2. The foundation drain shall be installed around the portion of the perimeter where the basement floor is below ground level. The provisions of Sections 1802.2.3, 1807.3 and 1807.4.1 shall not apply in this case.

❖ The provisions of this section, stated in another way, require that where a basement is deemed to be a story above grade (see definition, Section 202), the section of the basement floor that occurs below the exterior ground level and the walls that bound that part of the floor are to be dampproofed in accordance with the requirements of Section 1807.2.

The use of dampproofing, rather than waterproofing, is permitted here since hydrostatic pressure will not tend to develop against the walls if the basement is a story above grade and the ground level adjacent to the basement wall is below the basement floor elevation for no less than 50 percent of the basement perimeter.

Any water pressure that may occur against the walls below ground or under the basement floor would be relieved by the water drainage system required in this section. The drainage system would be installed at the base of the wall construction in accordance with Section 1807.4.2 for a minimum distance along those portions of the wall perimeter where the basement floor is below ground level.

Because of the relationship of grade to the basement floor and the inclusion of foundation drains, the potential for hydrostatic pressure buildup is not significant; Therefore, a ground-water table investigation, waterproofing and the basement floor gravel base course is not required.

The objective of Section 1807.4.1 is to prevent moisture migration in basement spaces. In story-above-grade construction that meets the requirements of this section, the basement floor would be only partly below ground level (sometimes a small part) and the need for moisture protection as required by Section 1807.4.1 would be unnecessary. Dampproofing of the floor slab would be required, however, in accordance with Section 1807.2.1.

1807.1.2 Under-floor space. The finished ground level of an under-floor space such as a crawl space shall not be located below the bottom of the footings. Where there is evidence that the ground-water table rises to within 6 inches (152 mm) of the ground level at the outside building perimeter, or that the surface water does not readily drain from the building site, the ground level of the under-floor space shall be as high as the outside finished ground level, unless an approved drainage system is provided. The provisions of Sections 1802.2.3, 1807.2, 1807.3 and 1807.4 shall not apply in this case.

❖ Essentially, the requirements of this section are designed to prevent any ponding of water in underfloor spaces, such as crawl spaces. Crawl spaces are particularly susceptible to ponding of water, since they are usually uninhabitable spaces that are observed very infrequently. Water can build up in these spaces and remain for an extended period of time without being noticed by the building occupants. This type of stagnant water under a building, which can harbor disease, mold and disease-carrying insects, such as mosquitoes, can result in a serious health concern. Water buildup in a crawl space can also damage the structural integrity of the building. Wood exposed to water will deteriorate and rot, while concrete and masonry exposed to water will deteriorate with a loss of strength.

Steel exposed to water or high humidity can eventually rust to the extent that effective structural capability is jeopardized. Water buildup in a crawl space can also damage any mechanical or electrical appliances, which may be located in the space, by causing corrosion of electrical parts or metal skins and deterioration to insulation used to protect heating elements.

Where it is known that the water table can rise to within 6 inches (152 mm) of the outside ground level, or where there is evidence that surface water cannot readily drain from the site, then the finished ground surface in underfloor spaces is to be set at an elevation equal to the outside ground level around the perimeter of the building unless an approved drainage system is provided. In order for the drainage system to be approved, it must be demonstrated to be adequate to prevent the infiltration of water into the underfloor space. This is done by determining the maximum possible flow of water near the foundation wall and footing and designing the drainage system to remove that flow of water as it occurs, without permitting the buildup of water at the foundation wall.

To prevent the ponding of water in the underfloor space from a rise in the ground-water table, or from storm water runoff, the finished ground level of an underfloor space is not to be located below the bottom of the foundation footings.

Dampproofing (see Section 1807.2) the foundation walls, waterproofing (see Section 1807.3) and providing subsoil drainage (see Section 1807.4) is not necessary if the ground level of the underfloor space is as high as the ground level at the outside of the building perimeter, as the foundation walls do not enclose an interior space below grade. Compliance with Sections 1807.2, 1807.3 and 1807.4 would still be required where the finished ground surface of the underfloor space is below the outside ground level.

1807.1.2.1 Flood hazard areas. For buildings and structures in flood hazard areas as established in Section 1612.3, the finished ground level of an under-floor space such as a crawl space shall be equal to or higher than the outside finished ground level.

Exception: Under-floor spaces of Group R-3 buildings that meet the requirements of FEMA/ FIA-TB-11.

❖ According to Section 1612.2, the definition of "Basement" for buildings and structures located in designated flood hazard areas is "the portion of a building having its floor subgrade (below grade) on all sides." This definition pertains to enclosed areas below elevated buildings and structures whether or not there is enough clearance for such areas to be occupied. An area without sufficient clearance to be occupied is often referred to as a "crawl space." Local officials may be called on to determine whether a crawl space is a basement under the flood-plain management requirements of the NFIP. Neither the use of the enclosed space nor the clearance height is the controlling factor as to whether an enclosed area below an elevated building is a basement under Section 1612.2. The controlling factor is whether the interior grade is below the exterior grade on all sides, even if the interior grade is established by the footing excavation that was not backfilled.

So that buildings and structures located in designated flood hazard areas receive favorable flood insurance premium rates, enclosed areas below elevated buildings or structures must meet the requirements of Section 1612.4, which references ASCE 24. In the event that a building or structure is not built in accordance with Section 1612.4, especially if it is determined to have a basement that is subgrade on all sides, flood insurance premium rates will be significantly higher.

In recognition of common construction practices, the exception permits crawl spaces to be as much as 2 feet (610 mm) below the lowest adjacent exterior grade in accordance with FEMA Technical Bulletin 11-01, *Crawlspace Construction for Buildings Located in Special Flood Hazard Areas*, provided the additional requirements of that document are met. Communities that choose to allow this practice must amend their flood-plain ordinance to include these additional requirements.

1807.1.3 Ground-water control. Where the ground-water table is lowered and maintained at an elevation not less than 6 inches (152 mm) below the bottom of the lowest floor, the floor and walls shall be dampproofed in accordance with Section 1807.2. The design of the system to lower the ground-water table shall be based on accepted principles of engineering that shall consider, but not necessarily be limited to, permeability of the soil, rate at which water enters the drainage system, rated capacity of pumps, head against which pumps are to operate and the rated capacity of the disposal area of the system.

❖ After completion of building construction, it is often necessary to maintain the water table at a level that is at least 6 inches (152 mm) below the bottom of the lowest floor in order to prevent the flow or seepage of water into the basement. Where the site consists of well-draining soil and the highest point of the water table occurs naturally at or lower than the required level stated above, there is no need to provide a site drainage system specifically designated to control the ground-water level. Where the soil characteristics and site topography are such that the water table can rise to a level that will produce a hydrostatic pressure against the basement structure, then a site drainage system may be installed to reduce the water level if there is sufficient land area to accomplish the purpose. When ground-water control in accordance with this section is provided, waterproofing in accordance with Section 1807.3 is not required.

There are many types of site drainage systems that can be employed to control ground-water levels. The most commonly used systems may involve the installation of drainage ditches or trenches filled with pervious materials, sump pits and discharge pumps, well point systems, drainage wells with deep-well pumps, sand-drain installations, etc. This section requires that all such systems be designed and constructed using accepted engineering principles and practices based upon considerations that include the permeability of the soil, amount and rate at which water enters the system, pump capacity, capacity of the disposal area and other such factors that are necessary for the complete design of an operable drainage system.

1807.2 Dampproofing required. Where hydrostatic pressure will not occur as determined by Section 1802.2.3, floors and walls for other than wood foundation systems shall be dampproofed in accordance with this section. Wood foundation systems shall be constructed in accordance with AFPA TR7.

❖ For a general definition of "Dampproofing," see the commentary to Section 1807. Where a ground-water table investigation made in accordance with the requirements of Section 1802.2.3 (see commentary) has established that the high water table will occur at such a level that the building substructure will not be subjected to hydrostatic pressure, then dampproofing in accordance with this section and a subsoil drain in accordance with Section 1807.4 are sufficient to control moisture in the floor below grade. Since the wall will not be subject to water under pressure, the more restrictive provisions of waterproofing, as outlined in Section 1807.3, are not required. Wood foundation systems specified in Section 1805.4.6 (see commentary) are to be dampproofed as required by AF&PA TR7.

1807.2.1 Floors. Dampproofing materials for floors shall be installed between the floor and the base course required by Section 1807.4.1, except where a separate floor is provided above a concrete slab.

Where installed beneath the slab, dampproofing shall consist of not less than 6-mil (0.006 inch; 0.152 mm) polyethylene with joints lapped not less than 6 inches (152 mm), or other approved methods or materials. Where permitted to be installed on top of the slab, dampproofing shall consist of mopped-on bitumen, not less than 4-mil (0.004 inch; 0.102 mm) polyethylene, or other approved methods or materials. Joints in the membrane shall be lapped and sealed in accordance with the manufacturer's installation instructions.

❖ Floors requiring dampproofing in accordance with Section 1807.2 are to employ materials as specified in Section 1807.2.1. The dampproofing materials must be placed between the floor construction and the supporting gravel or stone base as shown in Figure 1807.4.2. Even if a floor base in accordance with Section 1807.4.1 is not required, dampproofing is still to be placed under the slab unless otherwise specified in Section 1807.2.1. The installation is intended to provide a moisture barrier against the passage of water vapor or seepage into below-ground spaces.

The dampproofing material most commonly used for underslab installations consists of a polyethylene film no less than 6 mil [0.006 inch; (0.152 mm)] in thickness, which is applied over the gravel or stone base required in Section 1807.4.1. Care must be used in the installation of the material over the rough surface of the base and during the concreting operations so as not to puncture the polyethylene. Joints must be lapped at least 6 inches (152 mm). Other materials used in a similar way are made of neoprene or butyl rubber.

Dampproofing materials can also be applied on top of the base concrete slab if a separate floor is provided above the base slab, since dampproofing is provided to prevent moisture infiltration of the interior space, not the concrete slab.

Materials commonly used for dampproofing floors are listed in Figure 1807.2.2.

1807.2.2 Walls. Dampproofing materials for walls shall be installed on the exterior surface of the wall, and shall extend from the top of the footing to above ground level.

Dampproofing shall consist of a bituminous material, 3 pounds per square yard (16 N/m^2) of acrylic modified cement, $^1/_8$-inch (3.2 mm) coat of surface-bonding mortar complying with ASTM C 887, any of the materials permitted for waterproofing by Section 1807.3.2 or other approved methods or materials.

❖ Walls requiring dampproofing in accordance with Section 1807.2 are first to be prepared as required in Section 1807.2.2.1 and then coated with a bituminous material, cement or mortar as specified in Section 1807.2.2 or other approved materials and methods of application. Approved materials are those that will prevent moisture from penetrating the foundation wall when water is present but not under pressure.

Coatings are applied to cover prepared exterior wall surfaces extending from the top of the wall footings to slightly above ground level so that the entire wall that contacts the ground is protected. Surfaces are usually primed to provide a bond coat and then dampproofed with a protective coat of asphalt or tar pitch. Emulsion-type coatings may be applied directly on "green" concrete or unit masonry walls; however, because they are water-soluble materials, their use is not generally recommended. Installation should comply with the manufacturer's instructions.

Any of the materials specified in Section 1807.3.2 for waterproofing are also allowed to be used for dampproofing. Figure 1807.2.2 provides a list of bituminous materials that can be used, including the applicable standards that may be used as the basis of acceptance of such materials. Included in Figure 1807.2.2 is ASTM D 1668 for glass fabric that is treated with asphalt (Type I), coal-tar pitch (Type II) or organic resin (Type III).

Surface-bonding mortar complying with ASTM C 887 may be utilized. This specification covers the materials, properties and packaging of dry, combined materials for use as surface-bonding mortar with concrete masonry units that have not been prefaced, coated or painted. Since this specification does not address design or application, manufacturers' recommendations should be followed. This standard covers proportioning, physical requirements, sampling and testing. The minimum thickness of the coating is $^1/_8$ inch (3.2 mm).

Acrylic-modified cement coatings may be utilized at the rate of 3 pounds per square yard (16 N/m^2). These types of materials have been used and performed successfully as dampproofing materials for foundation walls. Surface-bonding mortar and acrylic-modified cement are limited in use to dampproofing. The ability of these two types of products to bridge nonstructural cracks, as required in Section 1807.3.2 for waterproofing materials, is not known. Therefore, their use is limited to dampproofing and they are not permitted to be used as waterproofing. Dampproofing may also include other materials and methods of installation acceptable to the building official.

MATERIAL	SPECIFICATION
Asphalt	ASTM D 449
Asphalt primer	ASTM D 41
Coal-tar	ASTM D 450
Concrete and masonry oil primer (for coal-tar applications only)	ASTM D 43
Treated glass fabric	ASTM D 1668

Figure 1807.2.2
MATERIALS FOR WATERPROOFING AND DAMPROOFING INSTALLATIONS

1807.2.2.1 Surface preparation of walls. Prior to application of dampproofing materials on concrete walls, holes and recesses resulting from the removal of form ties shall be sealed with a bituminous material or other approved methods or materials. Unit masonry walls shall be parged on the exterior surface below ground level with not less than $^3/_8$ inch (9.5 mm) of portland cement mortar. The parging shall be coved at the footing.

Exception: Parging of unit masonry walls is not required where a material is approved for direct application to the masonry.

❖ Before applying dampproofing materials, the concrete must be free of any holes or recesses that could affect the proper sealing of the wall surfaces. Air trapped beneath a dampproofing coating or membranes can cause blistering, while rocks and other sharp objects can puncture membranes. Irregular surfaces can also create uneven layering of coatings, which can result in vulnerable areas of dampproofing. Surface irregularities commonly associated with concrete wall construction can be sealed with bituminous materials or filled with portland cement grout or other approved methods.

Unit masonry walls are usually parged (plastered) with a $^1/_2$-inch-thick (12.7 mm) layer of portland cement and sand mix (1:2$^1/_2$ by volume) or with Type M mortar proportioned in accordance with the requirements of ASTM C 270, and applied in two $^1/_4$-inch-thick (6.4 mm) layers. In no case is parging to result in a final thickness of less than $^3/_8$ inch (9.5 mm). The parging is to be coved at the joint formed by the base of the wall and the top of the wall footing to prevent the accumulation of water at that location.

The moisture protection of unit masonry walls provided by the parging method may not be required where approved dampproofing materials, such as grout coatings, cement-based paints or bituminous coatings, can be applied directly to masonry surfaces.

1807.3 Waterproofing required. Where the ground-water investigation required by Section 1802.2.3 indicates that a hydrostatic pressure condition exists, and the design does not include a ground-water control system as described in Section 1807.1.3, walls and floors shall be waterproofed in accordance with this section.

❖ For a general definition of "Waterproofing," and the distinction between dampproofing and waterproofing, see the commentary to Section 1807.

The significance of waterproofing installations is that they are intended to provide moisture barriers against water seepage that may be forced into below-ground spaces by hydrostatic pressure.

Where a ground-water table investigation made in accordance with the requirements of Section 1802.2.3 (see commentary) has established that the high water table will occur at such a level that the building substructure will be subjected to hydrostatic pressure, and where the water table is not lowered by a water control system, as prescribed in the commentary to Section 1807.1.3, all floors and walls below ground level are to be waterproofed in accordance with Sections 1807.3.1, 1807.3.2 and 1807.3.3.

1807.3.1 Floors. Floors required to be waterproofed shall be of concrete, designed and constructed to withstand the hydrostatic pressures to which the floors will be subjected.

Waterproofing shall be accomplished by placing a membrane

of rubberized asphalt, butyl rubber, or not less than 6-mil (0.006 inch; 0.152 mm) polyvinyl chloride with joints lapped not less than 6 inches (152 mm) or other approved materials under the slab. Joints in the membrane shall be lapped and sealed in accordance with the manufacturer's installation instructions.

❖ Since floors that are required to be waterproofed are subjected to hydrostatic uplift pressures, such floors must, for all practical purposes, be made of concrete and designed and constructed to resist the maximum hydrostatic pressures possible. It is particularly important that the floor slab be properly designed, since severe cracking or movement of the concrete would allow water seepage into below-ground spaces because the ability of the waterproofing materials to bridge small cracks would be exceeded. Concrete floor construction is to comply with the applicable provisions of Chapter 19.

Below-ground floors subjected to hydrostatic uplift pressures are to be waterproofed with membrane materials placed as underslab or split-slab installations, including such materials as rubberized asphalt, butyl rubber and neoprene or with polyvinyl chloride (PVC) not less than 6 mil [0.006 inch; (0.152 mm)] in thickness, lapped at least 6 inches (152 mm). There are many proprietary membrane products available in the marketplace that are specifically made for waterproofing floors and walls (i.e., polyethylene sheets sandwiched between layers of asphalt), which may be used for that purpose when approved by the building official.

All membrane joints are to be lapped and sealed in accordance with the manufacturer's instructions to form a continuous, impermeable moisture barrier.

1807.3.2 Walls. Walls required to be waterproofed shall be of concrete or masonry and shall be designed and constructed to withstand the hydrostatic pressures and other lateral loads to which the walls will be subjected.

Waterproofing shall be applied from the bottom of the wall to not less than 12 inches (305 mm) above the maximum elevation of the ground-water table. The remainder of the wall shall be dampproofed in accordance with Section 1807.2.2. Waterproofing shall consist of two-ply hot-mopped felts, not less than 6-mil (0.006 inch; 0.152 mm) polyvinyl chloride, 40-mil (0.040 inch; 1.02 mm) polymer-modified asphalt, 6-mil (0.006 inch; 0.152 mm) polyethylene or other approved methods or materials capable of bridging nonstructural cracks. Joints in the membrane shall be lapped and sealed in accordance with the manufacturer's installation instructions.

❖ Walls that are required to be waterproofed in accordance with Section 1807.3 must first be prepared as required in Section 1807.3.2.1.

The walls must be designed to resist the hydrostatic pressure anticipated at the site, as well as any other lateral loads to which the wall will be subjected, such as soil pressures or seismic loads. As with floors required to be waterproofed, it is particularly important that walls required to be waterproofed be properly designed to resist all loads present, since cracking and other damage would allow water seepage into below-ground spaces. Water seepage can lead to deterioration of the foundation as wood rots, concrete and masonry erode and steel rusts. Such deterioration can cause structural failure of the foundation. More importantly, failure of the foundation wall can lead to structural failure of the building, since the foundation supports the building structure. Masonry or concrete construction must comply with the applicable provisions of Chapters 21 and 19, respectively.

Figures 1807.2.2 and 1807.3.2 list materials commonly used for the installation of moisture barriers in wall construction and the related standard that may be used as a basis for acceptance of such materials.

Asphalt and coal-tar products are not compatible and should not be used together.

Waterproofing installations are to extend from the bottom of the wall to a height no less than 12 inches (305 mm) above the maximum elevation of the ground-water table determined in accordance with the requirements of Section 1802.2.3 (see commentary). The remainder of the wall below ground level (if the height is small) may be waterproofed as a continuation of the installation or must be dampproofed in accordance with the requirements of Section 1807.2.2. If the ground-water table investigation is not conducted on the basis of the exception to Section 1802.2.3, then waterproofing should be provided from a point below the footing to above the ground level.

This section requires that waterproofing must consist of two-ply hot-mopped felts. The practice of the waterproofing industry is to select the number of plies of membrane material based on the hydrostatic head (height of water pressure against the wall. As a general practice, if the head of water is between 1 foot (305 mm) and 3 feet (914 mm), two plies of felt or fabric membrane are used; between 4 feet (1219 mm) and 10 feet (3048 mm), three-ply construction is needed and between 11 feet (3353 mm) and 25 feet (7620 mm), four-ply construction is necessary.

Waterproofing installations may also use polyvinyl chloride materials of no less than 6-mil [0.006 inch (0.152 mm)] thick, 40-mil [0.040 inch (1.02 mm)] polymer-modified asphalt or 6-mil [0.006 inch (0.152 mm)] polyethylene. These materials have been widely recognized for their effectiveness in bridging nonstructural cracks. Other approved materials and methods may be used provided that the same performance standards are met. All membrane joints must be lapped and sealed in accordance with the manufacturer's instructions.

MATERIAL	SPECIFICATION
Asphalt-saturated asbestos felt	ASTM D 250
Asphalt-saturated burlap fabric	ASTM D 1327
Asphalt-saturated cotton fabric	ASTM D 173
Asphalt-saturated organic felt	ASTM D 226
Coal-tar-saturated burlap fabric	ASTM D 1327
Coal-tar-saturated cotton fabric	ASTM D 173
Coal-tar-saturated organic felt	ASTM D 227

Figure 1807.3.2
MATERIALS FOR WATERPROOFING INSTALLATIONS

1807.3.2.1 Surface preparation of walls. Prior to the application of waterproofing materials on concrete or masonry walls, the walls shall be prepared in accordance with Section 1807.2.2.1.

❖ Before applying waterproofing materials to concrete or masonry walls, the surfaces must be prepared in accordance with the requirements of Section 1807.2.2.1, which requires the sealing of all holes and recesses. Surfaces to be waterproofed must also be free of any projections that might puncture or tear membrane materials that are applied over the surfaces.

1807.3.3 Joints and penetrations. Joints in walls and floors, joints between the wall and floor and penetrations of the wall and floor shall be made water-tight utilizing approved methods and materials.

❖ This section requires that all joints occurring in floors and walls and at locations where floors and walls meet, as well as all penetrations in floors and walls, be made water tight by approved methods. Sealing joints and penetrations is of primary importance to ensuring the effectiveness of the waterproofing. If the joints or penetrations are not sealed properly, they can develop leaks, which become a passageway for water to enter the building. Since the remainder of the foundation is wrapped in waterproofing, moisture can actually become trapped in the foundation walls or floor slab, and serious damage to these structural components can occur. Such methods may involve the use of construction keys (e.g., between the base of the wall and the top of the footing) or, if there is hydrostatic pressure, floor and wall joints may require the use of manufactured waterstops made of metal, rubber, plastic or mastic materials. Figures 1807.3.3(1) and 1807.3.3(2) illustrate examples of joint treatment and penetration treatment of waterproofing. Floor edges along walls and floor expansion joints may employ the use of any number of preformed expansion joint materials, such as asphalt, polyurethane, sponge rubber, self-expanding cork, cellular fibers bonded with bituminous materials, etc., which all comply with applicable ASTM or AASHTO standards or other federal specifications. A variety of sealants may be used together with the preformed joint materials. Gaskets made of neoprene and other materials are also available for use in concrete and masonry joints. The National Roofing Contractor's *NRCA Roofing and Waterproofing Manual* provides details for the reinforcement of membrane terminations, corners, intersections of slabs and walls, through-wall and slab penetrations and other locations.

Penetrations in walls and floors may be made water tight with grout or manufactured fill materials and sealants made for that purpose.

1807.4 Subsoil drainage system. Where a hydrostatic pressure condition does not exist, dampproofing shall be provided and a base shall be installed under the floor and a drain installed around the foundation perimeter. A subsoil drainage system designed and constructed in accordance with Section 1807.1.3 shall be deemed adequate for lowering the ground-water table.

❖ This section covers subsoil drainage systems in conjunction with dampproofing (see Section 1807.2) to protect below-ground spaces from water seepage. Such systems are not used where basements or other below-ground spaces are subject to hydrostatic pressure because they would not be effective in disposing of the amount of water anticipated if a hydrostatic pressure

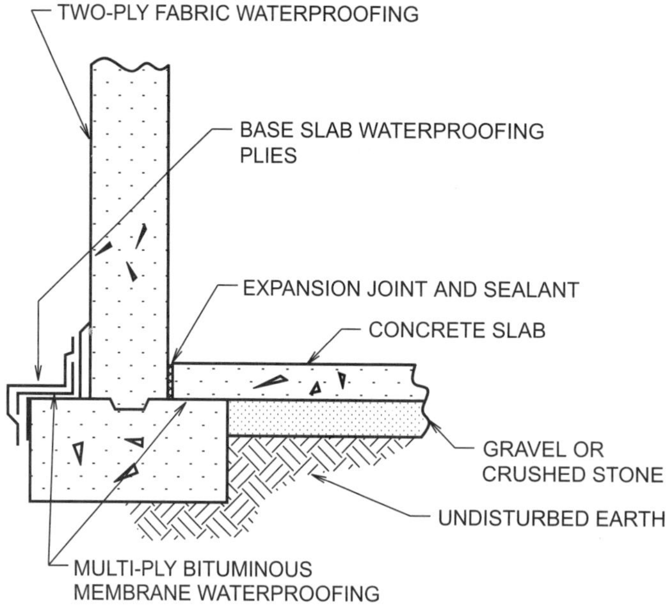

Figure 1807.3.3(1)
DETAIL—MEMBRANE PLACEMENT THROUGH A KEY JOINT

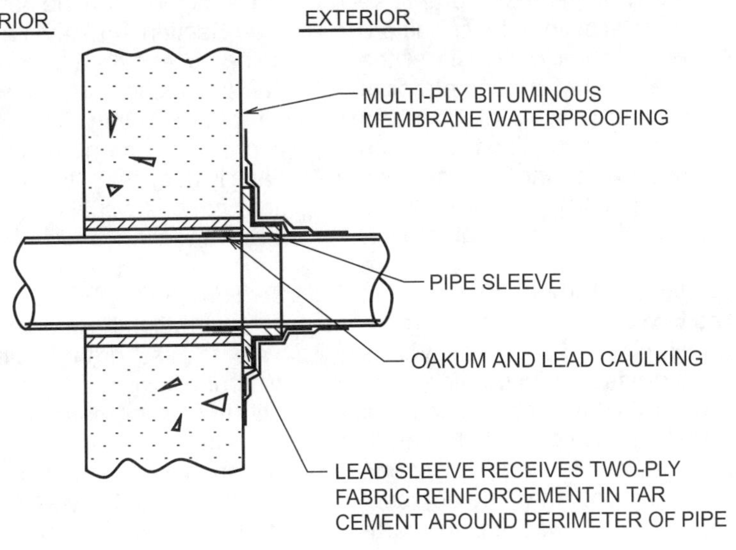

Figure 1807.3.3(2)
PIPE DETAIL—PROPER PLACEMENT OF WATERPROOFING ELEMENTS

condition exists. Ground-water tables may be reduced to acceptable levels by methods described in the commentary to Section 1807.1.3.

The details of subsoil drainage systems are covered in the requirements of Sections 1807.4.1 through 1807.4.3.

1807.4.1 Floor base course. Floors of basements, except as provided for in Section 1807.1.1, shall be placed over a floor base course not less than 4 inches (102 mm) in thickness that consists of gravel or crushed stone containing not more than 10 percent of material that passes through a No. 4 (4.75 mm) sieve.

> **Exception:** Where a site is located in well-drained gravel or sand/gravel mixture soils, a floor base course is not required.

❖ This section requires that basement floors, except for story-above-grade construction, must be placed on a gravel or stone base no less than 4 inches (102 mm) thick. Not more than 10 percent of the material is to pass a No. 4 sieve so as to provide a porous installation. Material that passes a No. 4 sieve would be fine silt or clay that would not permit the free movement of water through the floor base.

This requirement serves three purposes. The first is to provide an adjustment to the irregularities of a compacted subgrade so as to produce a level surface upon which to cast a concrete slab. The second is to provide a capillary break so that moisture from the soil below will not rise to the underside of the floor. Finally, where required, the porous base can act as a drainage system to expel underslab water by means of gravity or the use of a sump pump or other approved methods.

The exception allows for the omission of the floor base when the natural soils beneath the floor slab consist of well-draining granular materials such as sand, stone or mixtures of these materials. The exception is consistent with the requirements of Section 1807.4.3. Some caution, however, is justified in the use of this exception. If the granular soils contain an excessive percentage of fine materials, the porosity and the ability of the soil to provide a capillary break may be considerably diminished. The exception is only to be applied if the natural base is equivalent to the floor base otherwise required by this section.

1807.4.2 Foundation drain. A drain shall be placed around the perimeter of a foundation that consists of gravel or crushed stone containing not more than 10-percent material that passes through a No. 4 (4.75 mm) sieve. The drain shall extend a minimum of 12 inches (305 mm) beyond the outside edge of the footing. The thickness shall be such that the bottom of the drain is not higher than the bottom of the base under the floor, and that the top of the drain is not less than 6 inches (152 mm) above the top of the footing. The top of the drain shall be covered with an approved filter membrane material. Where a drain tile or perforated pipe is used, the invert of the pipe or tile shall not be higher than the floor elevation. The top of joints or the top of perforations shall be protected with an approved filter membrane material. The pipe or tile shall be placed on not less than 2 inches (51 mm) of gravel or crushed stone complying with Section 1807.4.1, and shall be covered with not less than 6 inches (152 mm) of the same material.

❖ This section describes in considerable detail the materials and features of construction required for the installation of foundation drain systems.

This type of drain system is suitable where the water table occurs at such elevation that there is no hydrostatic pressure exerted against the basement floor and walls and where the amount of seepage from the surrounding soil is so small that the water can be readily discharged by gravity or mechanical means into sewers or ditches. The objective is to combine the protection afforded by the dampproofing of walls and floors (see

Section 1807.2) and that given by perimeter drains so as to maintain below-ground spaces in a dry condition.

A foundation drain system generally consists of the installation of drain tiles made of clay or concrete or of drain pipes of corrugated metal or nonmetallic pipes surrounded by crushed stone or gravel. The foundation drain is set adjacent to the wall footing and extends around the perimeter of the building. Drain tiles are placed end to end with open joints to permit water to enter the system. Metallic and nonmetallic drains are made with perforations at the invert (bottom) section of the pipe and are installed with connected ends. Where drain tile or perforated drain pipe is used, the invert is not to be set higher than the basement floor line such that water conveyed by the drain does not seep into the filter material and create a hydrostatic pressure condition against the foundation wall and footing. The inverts should not be placed below the bottom of the adjacent wall footings so as to avoid carrying away fine soil particles that, in time, could undermine the footing and settlement of the foundation walls.

Tile joints or pipe perforations should be covered with an approved filter membrane material to prevent them from becoming clogged, preventing fine particles that may be contained in the surrounding soil from entering the system and being carried away by water. The filter material around the drain tiles or pipes (no to be confused with filter membrane material) is to consist of selected gravel and crushed stone containing no more than 10 percent of material that passes through a No. 4 sieve. The filter materials should be selected to prevent the movement of particles from the protected soil surrounding the drain installation into the drain. Filter material is to be placed in the excavation so that it will extend out from the edge of the wall footing a distance of at least 12 inches (305 mm), with the bottom of the fill being no higher than the bottom of the base under the floor (see Section 1807.4.1) and the top of the filter material being no less than 6 inches (152 mm) above the top of the wall footing so that water will not collect along the top of the footing. Requiring the bottom of the foundation drain to be no higher than the bottom of the floor base is necessary so that if the water table rises into the floor base, it will also be able to rise unobstructed into the foundation drain. The foundation drain will then drain the water away from the building, as required by Section 1807.4.3. The top of the filter fill material must be covered with an approved filter membrane to allow water to pass through to the perimeter drain tile or pipes without allowing, or at least greatly reducing, the possibility of fine soil materials entering the drainage system.

Drain tiles or pipes are to be installed in the filter bed and should be seated on at least 2 inches (51 mm) of filter material and covered with at least 6 inches (152 mm) of filter material to maintain good water flow into the drain tile or pipe. The requirements of this section are illustrated in Figure 1807.4.2.

1807.4.3 Drainage discharge. The floor base and foundation perimeter drain shall discharge by gravity or mechanical means into an approved drainage system that complies with the *International Plumbing Code.*

> **Exception:** Where a site is located in well-drained gravel or sand/gravel mixture soils, a dedicated drainage system is not required.

❖ This section references the *International Plumbing Code®* (IPC®) for the requirements of installing piping systems for the disposal of water from the floor base (see Section 1807.4.1) and the foundation drains (see Section 1807.4.2). Chapter 11 of the IPC deals with the piping materials, applicable standards and methods of

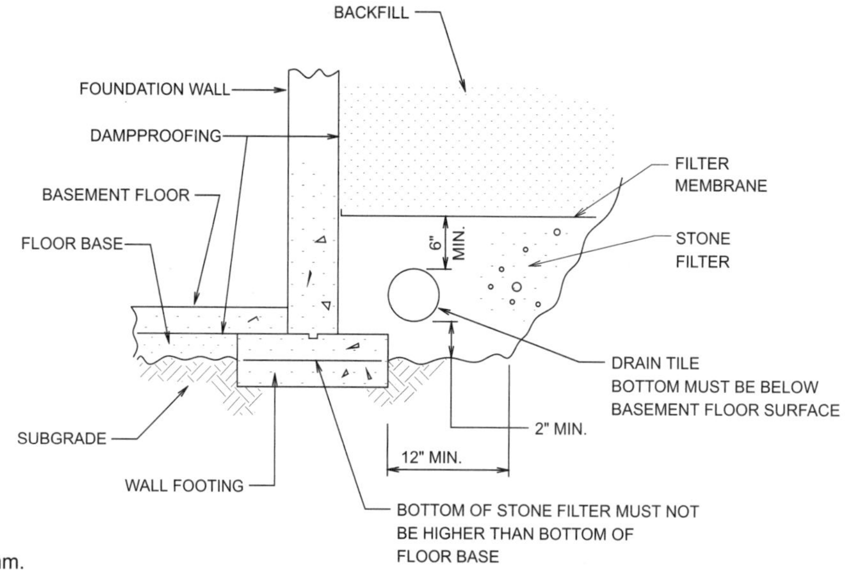

For SI: 1 inch = 25.4 mm.

Figure 1807.4.2
FOUNDATION DRAINAGE SYSTEM

installation of subsurface storm drains to facilitate water discharge either by gravity or mechanical means.

Where the soil at the site consists of well-drained granular materials (i.e., gravel or sand-gravel mixtures) to prevent the occurrence of hydrostatic pressure against the foundation walls and under the floor slab, the use of a dedicated drainage system as prescribed in the IPC is not required, since the site soils would permit natural drainage.

SECTION 1808 PIER AND PILE FOUNDATIONS

1808.1 Definitions. The following words and terms shall, for the purposes of this section, have the meanings shown herein.

❖ Section 1808 addresses the general requirements for the installation of various types of pier and pile foundations, and provides specific provisions related to allowable pier and pile loads, lateral support of piles, stability and other pertinent considerations required for the satisfactory construction of these deep foundation systems.

The following definitions are for the types of piers and piles addressed in Chapter 18.

FLEXURAL LENGTH. Flexural length is the length of the pile from the first point of zero lateral deflection to the underside of the pile cap or grade beam.

❖ This definition in used in connection with the special detailing requirements for the design of piles.

PIER FOUNDATIONS. Pier foundations consist of isolated masonry or cast-in-place concrete structural elements extending into firm materials. Piers are relatively short in comparison to their width, with lengths less than or equal to 12 times the least horizontal dimension of the pier. Piers derive their load-carrying capacity through skin friction, through end bearing, or a combination of both.

❖ See Section 1812 and the commentary for additional information on pier foundations.

Belled piers. Belled piers are cast-in-place concrete piers constructed with a base that is larger than the diameter of the remainder of the pier. The belled base is designed to increase the load-bearing area of the pier in end bearing.

❖ See Section 1812.6 and the commentary for additional information on belled piers.

PILE FOUNDATIONS. Pile foundations consist of concrete, wood or steel structural elements either driven into the ground or cast in place. Piles are relatively slender in comparison to their length, with lengths exceeding 12 times the least horizontal dimension. Piles derive their load-carrying capacity through skin friction, through end bearing, or a combination of both.

❖ See Sections 1809 through 1811 and the commentary for additional information on pile foundations.

Augered uncased piles. Augered uncased piles are constructed by depositing concrete into an uncased augered hole, either during or after the withdrawal of the auger.

❖ See Section 1810.3 and the commentary for additional information on drilled or augered uncased piles.

Caisson piles. Caisson piles are cast-in-place concrete piles extending into bedrock. The upper portion of a caisson pile consists of a cased pile that extends to the bedrock. The lower portion of the caisson pile consists of an uncased socket drilled into the bedrock.

❖ See Section 1810.7 and the commentary for additional information on caisson piles.

Concrete-filled steel pipe and tube piles. Concrete-filled steel pipe and tube piles are constructed by driving a steel pipe or tube section into the soil and filling the pipe or tube section with concrete. The steel pipe or tube section is left in place during and after the deposition of the concrete.

❖ See Section 1810.6 and the commentary for additional information on concrete-filled steel pipe and tube piles.

Driven uncased piles. Driven uncased piles are constructed by driving a steel shell into the soil to shore an unexcavated hole that is later filled with concrete. The steel casing is lifted out of the hole during the deposition of the concrete.

❖ See Section 1810.4 and the commentary for additional information on driven uncased piles.

Enlarged base piles. Enlarged base piles are cast-in-place concrete piles constructed with a base that is larger than the diameter of the remainder of the pile. The enlarged base is designed to increase the load-bearing area of the pile in end bearing.

❖ See Section 1810.2 and the commentary for additional information on enlarged base piles.

Steel-cased piles. Steel-cased piles are constructed by driving a steel shell into the soil to shore an unexcavated hole. The steel casing is left permanently in place and filled with concrete.

❖ See the commentary to Section 1810.5.

1808.2 Piers and piles—general requirements.

❖ This section provides general requirements applicable to the construction and installation of pier and pile foundations.

1808.2.1 Design. Piles are permitted to be designed in accordance with provisions for piers in Section 1808 and Sections 1812.3 through 1812.10 where either of the following conditions exists, subject to the approval of the building official:

1. Group R-3 and U occupancies not exceeding two stories of light-frame construction, or
2. Where the surrounding foundation materials furnish adequate lateral support for the pile.

❖ Subject to the approval of the building official, piles may be designed under the provisions for piers in either of two situations. One is where such piles are given enough lateral support by surrounding foundation materials to prevent or adequately limit flexural (bending) deformations. The second such allowance is for Group R-3 and U occupancies that are built of light-frame construction not exceeding two stories in height.

1808.2.2 General. Pier and pile foundations shall be designed and installed on the basis of a foundation investigation as defined in Section 1802, unless sufficient data upon which to base the design and installation is available.

The investigation and report provisions of Section 1802 shall be expanded to include, but not be limited to, the following:

1. Recommended pier or pile types and installed capacities.
2. Recommended center-to-center spacing of piers or piles.
3. Driving criteria.
4. Installation procedures.
5. Field inspection and reporting procedures (to include procedures for verification of the installed bearing capacity where required).
6. Pier or pile load test requirements.
7. Durability of pier or pile materials.
8. Designation of bearing stratum or strata.
9. Reductions for group action, where necessary.

❖ A foundations investigation is mandatory when pier or pile foundations are used. Such investigations are needed to define as accurately as possible the subsurface conditions of soil and rock materials, establish the soil and rock profiles across the construction site and locate the ground-water table. Sometimes it may also be necessary to determine specific soil properties, such as shear strength, relative density, compressibility and other such technical data required for analyzing subsurface conditions. Foundation investigations may also be used to render such valuable data as information on existing construction at the site or on neighboring properties (including boring and test records), the type and condition of the existing structures, their age, the type of foundations used and performance over the years. Knowledge of existing deleterious substances in the soils that could affect the durability (as well as the performance) of piers or piles, data on geologic conditions at the site (including such information as the existence of mines, earth cavities, underground streams or other adverse water conditions) and a history of any seismic activity, hurricanes, etc., are also important.

The types of information described above are usually obtained by means of soil and rock borings, laboratory and field tests, engineering analyses and research. Such information is used for determining design loads; type, spacing and lengths of piles; suitable bearing strata; driving criteria and selection of equipment and probable durability of pile materials in relation to subsurface conditions, etc.

1808.2.3 Special types of piles. The use of types of piles not specifically mentioned herein is permitted, subject to the approval of the building official, upon the submission of acceptable test data, calculations and other information relating to the structural properties and load capacity of such piles. The allowable stresses shall not in any case exceed the limitations specified herein.

❖ Piles are generally identified according to the materials used (concrete, steel or wood) or the methods of construction or installation. While the most commonly used types of piles are addressed in Sections 1808 through 1811, this section acknowledges that there are many variations of pile types used in the construction of deep foundations, including some special or proprietary types that are beyond the scope of the code.

However, while it is not the intent to preclude the use of such special or proprietary types of piles, it is necessary to ensure their structural integrity by submitting test data, calculations, information on structural properties and load capacity and installation procedures to the building official for approval.

1808.2.4 Pile caps. Pile caps shall be of reinforced concrete, and shall include all elements to which piles are connected, including grade beams and mats. The soil immediately below the pile cap shall not be considered as carrying any vertical load. The tops of piles shall be embedded not less than 3 inches (76 mm) into pile caps and the caps shall extend at least 4 inches (102 mm) beyond the edges of piles. The tops of piles shall be cut back to sound material before capping.

❖ Pile caps include all elements to which the piles are connected and are to be of reinforced concrete and designed in accordance with the requirements of ACI 318. For footings (pile caps) on piles, computations for moments and shears may be based on the assumption that the load reaction from any pile is concentrated at the pile center (see ACI 318 for loads and reactions of footings on piles).

The soil immediately under the pile cap is not considered to provide any support for vertical loads. For a detailed explanation of this requirement, see the commentary to Section 1808.2.9.3.

The heads of all piles are to be embedded no less than 3 inches (76 mm) into pile caps and the edges of the pile caps are to extend at least 4 inches (102 mm) beyond the closest sides of all piles. The degree of fixity between a pile head and the concrete cap depends on the method of connection required to satisfy design considerations.

1808.2.5 Stability. Piers or piles shall be braced to provide lateral stability in all directions. Three or more piles connected by a rigid cap shall be considered braced, provided that the piles are located in radial directions from the centroid of the group not less than 60 degrees (1 rad) apart. A two-pile group in a rigid cap shall be considered to be braced along the axis connecting the two piles. Methods used to brace piers or piles shall be subject to the approval of the building official.

Piles supporting walls shall be driven alternately in lines

spaced at least 1 foot (305 mm) apart and located symmetrically under the center of gravity of the wall load carried, unless effective measures are taken to provide for eccentricity and lateral forces, or the wall piles are adequately braced to provide for lateral stability. A single row of piles without lateral bracing is permitted for one- and two-family dwellings and lightweight construction not exceeding two stories or 35 feet (10 668 mm) in height, provided the centers of the piles are located within the width of the foundation wall.

❖ A group of piles designed to support a common load and, as may be required, to resist horizontal forces, must be braced or rigidly tied together to act as a single structural unit that will provide lateral stability in all directions. Piles connected by a rigid, reinforced concrete pile cap are deemed to be braced construction that serves the intent of this provision.

Three or more piles are generally used to support a building column load or other isolated concentrated load. In a three-pile group, lateral stability is assured by requiring that the piles are located such that they will not be less than 60 degrees (1.0 rad) apart as measured from the centroid of the group in a radial direction.

For stability of a pile group supporting a wall structure, the piles are braced by a continuous rigid footing and are alternately staggered in two lines at least 1 foot (305 mm) apart and symmetrically located on each side of the center of gravity of the wall. Other approved pile arrangements may be used to support walls, provided that the piles are adequately braced and the lateral stability of the foundation construction is ensured.

For lightweight construction, such as one- and two-family dwellings not exceeding two stories or 35 feet (10 668 mm) in height, a single row of piles located within the width of the wall is permitted. In this case, pile group stability is theoretically afforded only in the direction of the line of piles. However, in reality, cross walls, floor slabs and other structural components of building construction may contribute to the lateral stability of the foundation system.

1808.2.6 Structural integrity. Piers or piles shall be installed in such a manner and sequence as to prevent distortion or damage to piles being installed or already in place to the extent that such distortion or damage affects the structural integrity of the piles.

❖ The placement of piles is generally obtained by driving, vibrating, jacking, jetting, direct weight or a combination of these methods. Because of the generally harsh nature of pile installation operations, most piles are exposed to some degree of damage during placement. However, damage can be prevented or minimized by selecting the proper type of pile placement methods and techniques, as well as the right equipment to accomplish the work, all based on adequate knowledge of the soil conditions obtained from a foundation investigation (see Section 1808.2.1).

Piles must be placed in such a way as to insure their structural integrity and installed to such depths as determined by foundation investigation and engineering analysis to safely resist the loads that are to be imposed upon them. Due care must be exercised during pile placement operations to ensure the safety of adjacent piles or other structures leaving their strength and load capacity unimpaired.

Any pile damaged during installation to the extent that its structural integrity is affected must be satisfactorily repaired or rejected.

1808.2.7 Splices. Splices shall be constructed so as to provide and maintain true alignment and position of the component parts of the pier or pile during installation and subsequent thereto and shall be of adequate strength to transmit the vertical and lateral loads and moments occurring at the location of the splice during driving and under service loading. Splices shall develop not less than 50 percent of the least capacity of the pier or pile in bending. In addition, splices occurring in the upper 10 feet (3048 mm) of the embedded portion of the pier or pile shall be capable of resisting at allowable working stresses the moment and shear that would result from an assumed eccentricity of the pier or pile load of 3 inches (76 mm), or the pier or pile shall be braced in accordance with Section 1808.2.5 to other piers or piles that do not have splices in the upper 10 feet (3048 mm) of embedment.

❖ While it is physically and economically better to drive piles in one piece, site conditions sometimes necessitate that piles be driven in spliced sections. For example, when the soil or rock-bearing stratum is located so deep below the ground that the leads on the driving equipment will not receive full-length piles, it becomes necessary to install the piles sectionally or, where possible, to take up the extra length by setting the tip in a preexcavated hole (see commentary, Section 1808.2.13). When piles are installed in areas such as existing buildings with restricted headroom, they are also required to be placed in spliced sections. There are a number of other reasons for field splicing piles, such as restrictions on shipping lengths, the use of composite piles, etc.

This provision requires that splices be constructed so as to provide and maintain true alignment and position of the pile sections during installation. Furthermore, splices must be of sufficient strength to transmit safely the vertical and lateral loads on the piles, as well as to resist the bending stresses that may occur at splice locations during the driving operations and under long-term service loads. Splices are to develop at least 50 percent of the value of the pile in bending. Consideration should be given to the design of splices at locations where the piles may be subject to tension. Additionally, splices that occur in the upper 10 feet (3048 mm) of pile embedment are to be designed to resist the bending moments and shears at the allowable stress levels of the pile material, based on an assumed pile load eccentricity of 3 inches (76 mm), unless the pile is properly braced. Proper bracing of a spliced pile is deemed to exist if stability of the pile group is provided in accordance with the provisions of Section 1808.2.4, provided that the other piles in the group do not have splices in the upper 10 feet (3048 mm) of their embedded length.

There are different methods employed in splicing

piles based on the different materials used in pile construction. For example, timber piles are spliced by one of two commonly used methods. The first method uses a pipe sleeve with a length of about four to five times the diameter of the pile. The butting ends of the pile are sawn square for full contact of the two pile sections, and the spliced portions of the timber pile are trimmed smoothly around their periphery so as to fit tightly into the pipe sleeve. The other splicing method involves the use of steel straps and bolts. The butting ends of the pile sections are sawn square for full contact and proper alignment and the four sides are planed flat to receive the splicing straps. This type of splicing can resist some uplift forces.

Splicing of precast concrete piles usually occurs at the head portions of the piles where, after the piles are driven to their required depth, pile heads are cut off or spliced to the desired elevation for proper embedment in the concrete pile caps. Any portion of the pile that is cracked or shattered caused by the driving operations or cutting off of pile heads should be removed and spliced with fresh concrete. To cut off a precast concrete pile section, a deep groove is chiseled around the pile exposing the reinforcing bars, which are then cut off (by torch) to desired heights or extensions. The pile section above the groove is snapped off (by crane) and a new pile section is freshly cast to tie in with the precast pile.

Steel H-piles are spliced in the same manner as steel columns, normally by welding the sections together. Welded splices may be welded-plate or bar splices, butt-welded splices, special welded splice fittings or a combination of these. Spliced materials should be kept on the inner faces of the H-pile sections to avoid forcing a hole in the ground larger than the pile, causing at least a temporary loss in frictional value and lateral support that might result in excessive bending stresses.

Steel pipe piles may be spliced by butt welding, sometimes using straps to guide the sections and provide more strength to the welded joint. Another method is to use inside sleeves having a driving fit, with a flange extending between the pipe sections. By applying bituminous cement or compound on the outside of the ring before driving, a water-tight joint is obtained.

1808.2.8 Allowable pier or pile loads.

❖ This section provides the general requirements for the determination of allowable pile loads based on soil characteristics and capacity, as well as on installation conditions involving driveability; pier and pile stresses; overloads and other similar criteria.

1808.2.8.1 Determination of allowable loads. The allowable axial and lateral loads on piers or piles shall be determined by an approved formula, load tests or method of analysis.

❖ There are two general considerations for determining capacity as required for the design and installation of pier or pile foundations. The first consideration involves the determination of the underlying soil or rock characteristics. The second is the application of approved driving formulas, load tests or accepted methods of analysis to determine the pier or pile capacities required to resist the axial and lateral loads they will be subjected to, as well as to provide the basis for the proper selection of pile-driving equipment.

1808.2.8.2 Driving criteria. The allowable compressive load on any pile where determined by the application of an approved driving formula shall not exceed 40 tons (356 kN). For allowable loads above 40 tons (356 kN), the wave equation method of analysis shall be used to estimate pile driveability of both driving stresses and net displacement per blow at the ultimate load. Allowable loads shall be verified by load tests in accordance with Section 1808.2.8.3. The formula or wave equation load shall be determined for gravity-drop or power-actuated hammers and the hammer energy used shall be the maximum consistent with the size, strength and weight of the driven piles. The use of a follower is permitted only with the approval of the building official. The introduction of fresh hammer cushion or pile cushion material just prior to final penetration is not permitted.

❖ It has been accepted practice for many decades to predict the load capacity of a pile by its resistance to driving as determined by a pile-driving formula. The simple premise upon which a pile-driving formula is founded is that as the resistance of a pile to driving increases, the pile's capacity to support loads also increases. While several pile formulas have been developed over the years, none have been completely dependable.

The Engineering-News formula is the simplest and probably the most widely used in the United States. This calculation method, as well as other formulas in common use today, have generally shown poor correlations with load test results. However, the comparative differences between pile capacities as determined by driving formulas and the results of load tests are much smaller for soils consisting of free-draining, coarse-grained materials, such as sand and gravel, than for soils consisting of silt, clay or fine, dense sand.

The use of pile-driving formulas to determine pile capacities should generally be avoided, except on small jobs where the piles are to be driven in well-drained granular soils, and the cost of load testing cannot be justified.

1808.2.8.3 Load tests. Where design compressive loads per pier or pile are greater than those permitted by Section 1808.2.10, or where the design load for any pier or pile foundation is in doubt, control test piers or piles shall be tested in accordance with ASTM D 1143 or ASTM D 4945. At least one pier or pile shall be test loaded in each area of uniform subsoil conditions. Where required by the building official, additional piers or piles shall be load tested where necessary to establish the safe design capacity. The resulting allowable loads shall not be more than one-half of the ultimate load capacity of the test pier or pile as assessed by one of the published methods listed in Section 1808.2.8.3.1 with consideration for the test type, duration and subsoil. The ultimate load capacity shall be determined by a registered design professional, but shall be no greater than two times the test load that produces a settlement of 0.3 inches (7.6 mm). In subsequent installation of the balance of foundation piles, all piles shall be deemed to have a supporting capacity equal to the control pile where such piles are of the same type, size and relative length as the test pile; are installed using the same or comparable methods and equipment as the test pile; are installed in similar subsoil conditions as the test pile; and, for driven piles, where the rate of penetration (e.g., net displacement per blow) of such piles is equal to or less than that of the test pile through a comparable driving distance.

❖ The safest method for determining pile capacity is by a load test. A load test should be conducted wherever feasible and used where the pile capacity is intended to exceed 40 tons (356 kN) per pile (see Section 1808.2.8.2). Test piles are to be of the same type and size intended for use in the permanent foundation and installed with the same equipment, by the same procedure and in the same soils intended or specified for the work.

Load tests are to be conducted in accordance with the requirements of ASTM D 1143 or D 4945, which covers procedures for testing vertical or batter foundation piles, individually or in groups, to determine the ultimate pile load (pile capacity) and whether the pile or pile group is capable of supporting the load(s) without excessive or continuous settlement. Recognition, however, must be given to the fact that load-settlement characteristics and pile capacity determinations are based on data derived at the time and under conditions of the test. The long-term performance of a pile or group of piles supporting actual loads may produce behaviors that are different than those indicated by load test results. Judgement based on experience must be used to predict pile capacity and expected behavior.

The load-bearing capacity of all piles, except those seated on rock, does not reach the ultimate load until after a period of rest. The results of load tests cannot be deemed accurate or reliable unless there is an allowance for a period of adjustment. For piles driven in permeable soils, such as coarse-grained sand and gravel, the waiting period may be as little as two or three days. For test piles driven in silt, clay or fine sand, the waiting period may be 30 days or longer. The waiting period may be determined by testing (i.e., by redriving piles) or from previous experience.

This section requires that at least one pile be tested in each area of uniform subsoil conditions. The statement should not be misconstrued to mean that the tested area is to have only one uniform stratum of subsurface material, but rather that the soil profile, which may consist of several layers (strata) of different materials, must represent a substantially unchanging cross section in each area to be tested.

The allowable pile load to be used for design purposes is not to be more than one-half of the ultimate pile capacity, as determined in Section 1808.2.8.3.1, nor more than two times the load that produces a settlement of 0.3 inches (7.6 mm). The rate of penetration of permanent foundation piles must be equal to or less than that of the test pile(s).

All production piles should be of the same type, size and approximate length as the prototype test pile, as well as installed with the same or comparable equipment and methods. They should also be installed in soils similar to those for the test pile.

1808.2.8.3.1 Load test evaluation. It shall be permitted to evaluate pile load tests with any of the following methods:

1. Davisson Offset Limit.
2. Brinch-Hansen 90% Criterion.
3. Chin-Konder Extrapolation.
4. Other methods approved by the building official.

❖ This section lists generally accepted methods that can be used to determine the ultimate load capacity of test piles. Since no single method applies to all situations that are encountered, this listing provides the necessary latitude to select a method of load test evaluation that is appropriate for the type of pile being tested, the test procedure and the subsurface conditions.

1808.2.8.4 Allowable frictional resistance. The assumed frictional resistance developed by any pier or uncased cast-in-place pile shall not exceed one-sixth of the bearing value of the soil material at minimum depth as set forth in Table 1804.2, up to a maximum of 500 psf (24 kPa), unless a greater value is allowed by the building official after a soil investigation as specified in Section 1802 is submitted. Frictional resistance and bearing resistance shall not be assumed to act simultaneously unless recommended by a soil investigation as specified in Section 1802.

❖ Under certain circumstances, such as when the pile extends through cohesive soils, like clays, to a bearing stratum of compact sand and gravels, both skin friction and end bearing act together to support the pile. However, the nature of load sharing between the two and whether in fact both exist simultaneously can be determined only by a soils investigation. Thus, the code requires that in order to assume that both skin friction and end bearing act simultaneously, the assumption must be justified by means of a soils investigation.

1808.2.8.5 Uplift capacity. Where required by the design, the uplift capacity of a single pier or pile shall be determined by an approved method of analysis based on a minimum factor of safety of three or by load tests conducted in accordance with ASTM D 3689. The maximum allowable uplift load shall not exceed the ultimate load capacity as determined in Section 1808.2.8.3 divided by a factor of safety of two. For pile groups subjected to uplift, the allowable working uplift load for the group shall be the lesser of:

1. The proposed individual pile uplift working load times the number of piles in the group.
2. Two-thirds of the effective weight of the pile group and the soil contained within a block defined by the perimeter of the group and the length of the pile.

❖ Piles and piers subjected to uplift forces act in tension and are actually friction piles. The amount of tension that can be sustained by a pier or pile depends on the strength of the pier or pile material and the frictional or cohesive properties of the soil. Tensile resistance is not necessarily correlated with the bearing capacity of a pier or pile under compressive load. For example, the tensile resistance of a friction pile in clay will usually be about the same value as its bearing capacity because the skin friction developed in cohesive soils is very large. In comparison, a friction pile installed in granular materials (noncohesive), such as sand, will develop a tensile resistance that is considerably less than its bearing capacity.

Analytical methods can be used to determine the ultimate uplift resistance of a pier or pile, provided that the properties of the soil are well known. When the ultimate uplift resistance is established by analysis, a safety factor of three must be applied to determine the allowable uplift load of the pile.

The response of a pier or vertical or batter pile to an axially applied uplift force is best determined by an extraction test performed in accordance with the requirements of the ASTM D 3689 standard listed in Chapter 35 and in accordance with the provisions of this section.

Piers and piles must be well anchored into the pile cap by adequate connection devices in order to be effective in resisting uplift forces. In turn, the pile cap must also be designed to resist uplift stresses. Sometimes, it is necessary to give special consideration in the design of the pile itself to take the tensile stresses imposed by uplift conditions. For example, a cast-in-place or precast concrete pile must be designed so that the tensile reinforcement will extend the full length of the pile. Special consideration should also be given to the design of pile splices that are to act in tension.

1808.2.8.6 Load-bearing capacity. Piers, individual piles and groups of piles shall develop ultimate load capacities of at least twice the design working loads in the designated load-bearing layers. Analysis shall show that no soil layer underlying the designated load-bearing layers causes the load-bearing capacity safety factor to be less than two.

❖ The bearing capacity of a pier or pile, whether it is a single acting pile or part of a group, is determined as a pier- or pile-soil system. In this respect, pier and pile-bearing capacity is a function of either the strength properties of the pier or pile or the supporting strength of the soil. Obviously, the bearing capacity is controlled by the smaller value obtained in the two considerations.

In most cases, the supporting strength of the soil governs the bearing capacity of a pier or pile foundation. This section requires that the bearing capacity of piers or an individual pile or group of piles must not be more than one-half of the ultimate load capacities of the piers or piles as a function of the bearing capacity of the soil.

Sometimes, soils investigations show that weaker layers of soil underlie the intended bearing strata. To avoid damaging settlements, the weaker soils must have a safety factor of two or more as determined by analytical methods. Where the safety factor is less than two, piers or piles must either be driven to deeper bearing soils to obtain adequate and safe support or the design capacity of the piers or piles must be reduced, thus increasing the total number of piles in the foundation system.

1808.2.8.7 Bent piers or piles. The load-bearing capacity of piers or piles discovered to have a sharp or sweeping bend shall be determined by an approved method of analysis or by load testing a representative pier or pile.

❖ This section requires that piers or piles that are discovered to have sharp or sweeping bends, usually occurring because of obstructions encountered during driving operations, must be analyzed by an approved method or load tested by a representative pier or pile to determine their load-carrying capacity. Where acceptable, such piers or piles may be used at a reduced capacity; otherwise, they should be abandoned and replaced.

1808.2.8.8 Overloads on piers or piles. The maximum compressive load on any pier or pile due to mislocation shall not exceed 110 percent of the allowable design load.

❖ Because of subsurface obstructions or other reasons, it is sometimes necessary to offset piers or piles a small distance from their intended locations or they may be driven out of position. In such cases, the load distribution in a group of piles may be changed from the design requirements and cause some of the piles to be overloaded. This section requires that the maximum compressive load on any pier or pile caused by mislocation not exceed 110 percent of the allowable design load. Piers or piles exceeding this limitation must be extracted and installed in the proper location or other approved remedies applied, such as installing additional piers or piles to balance the group.

1808.2.9 Lateral support.

❖ This section provides specific requirements for lateral support of piers and piles and the allowable lateral load.

1808.2.9.1 General. Any soil other than fluid soil shall be deemed to afford sufficient lateral support to the pier or pile to prevent buckling and to permit the design of the pier or pile in accordance with accepted engineering practice and the applicable provisions of this code.

❖ Long experience with pile installations and performance under loaded conditions has shown that piles surrounded by earth, including even very soft and compressible clays, provide sufficient lateral restraint to prevent buckling.

1808.2.9.2 Unbraced piles. Piles standing unbraced in air, water or in fluid soils shall be designed as columns in accordance with the provisions of this code. Such piles driven into firm ground can be considered fixed and laterally supported at 5 feet (1524 mm) below the ground surface and in soft material at 10 feet (3048 mm) below the ground surface unless otherwise prescribed by the building official after a foundation investigation by an approved agency.

❖ Piles driven in fluid soils, such as saturated silts, or in water, or piles that extend up into open air are not laterally supported and may be subject to high bending stresses and subsequent buckling. Under such conditions, piles must be designed as columns in accordance with the applicable provisions of the code. This section designates a point of fixity [5 feet (1524 mm) below ground surface for firm soils and 10 feet (3048 mm) below ground surface for soft soils] without requiring a complete soils investigation. This provision essentially provides a starting point for analyzing the fixity of the pile.

1808.2.9.3 Allowable lateral load. Where required by the design, the lateral load capacity of a pier, a single pile or a pile group shall be determined by an approved method of analysis or by lateral load tests to at least twice the proposed design working load. The resulting allowable load shall not be more than one-half of that test load that produces a gross lateral movement of 1 inch (25 mm) at the ground surface.

❖ Because of wind loads, unbalanced building loads, earth pressures and the like, it is inevitable that piers, individual piles or groups of vertical piles supporting buildings or other structures will be subjected to lateral forces. The distribution of these lateral forces to piers or piles largely depends on how the loads are carried down through the structural framing system and transferred through the supporting foundation to the piers or piles. The amount of lateral load that can be taken by the pier or pile foundation is a function of the type of pier or pile used; the soil characteristics, particularly in the upper 10 feet (3048 mm) of the piers or piles; the embedment of the pier or pile head (fixity); the magnitude of the axial compressive load on the pier or pile; the nature of the lateral forces and the amount of horizontal pier or pile movement deemed acceptable.

The degree of fixity of the pier or pile head is an important design consideration under very high lateral loading unless some other method, such as the use of batter piles, is employed to resist lateral loads. The fixing of the pier or pile head against rotation reduces the lateral deflection. In general, pile butts are embedded 3 inches (76 mm) to 4 inches (102 mm) into the pile cap (see Section 1808.2.3) with no ties to the cap. These pile heads are neither fixed nor free, but somewhere in the middle. Such construction is satisfactory for most loading conditions.

The magnitude of friction developed between the surfaces of two structural elements in contact with each other is a function of the weight or loads applied. The larger the weight, the greater the frictional resistance developed. In the design of pile foundations, frictional resistance between the soil and the bottom of the pile caps (footings) should not be relied on to provide lateral restraint, since the vertical loads are transmitted through the piles to the supporting soil below and not to the ground immediately under the pile caps. Only the weights of the pile caps can supply some frictional resistance because such footings are constructed by placing fresh concrete on the soil, thus providing a positive contact. The weights of the pile caps in comparison to the magnitude of loads and lateral forces transmitted to the piles is nominal, however, and not too significant from a structural design standpoint. Also, in rare occurrences, soil has been known to settle under pile caps, leaving open spaces and thus eliminating the development of any frictional restraint.

Generally, about $^1/_4$-inch (6.4 mm) horizontal movement of a pier or pile is considered acceptable without tests. Piers or piles with their upper sections embedded in deep strata of very soft or soft clays and silts should not be relied on to resist lateral forces of more than 1,000 pounds (4.45 kN) per pile.

Where vertical piles are subjected to lateral forces exceeding acceptable limitations, the use of batter piles may be required. Lateral forces on many structures are also resisted by the embedded foundation walls and the sides of the pile caps.

The allowable lateral load capacity of a pier, single pile or group of piles is to be determined either by approved analytical methods or load tests. Load tests are to be conducted to produce lateral forces that are twice the proposed design load; however, in no case is the allowable pier or pile load to exceed one-half of the test load, which produces a gross lateral pier or pile movement of 1 inch (25 mm) as measured at the ground surface.

1808.2.10 Use of higher allowable pier or pile stresses. Allowable stresses greater than those specified for piers or for each pile type in Sections 1808 and 1809 are permitted where supporting data justifying such higher stresses is filed with the building official. Such substantiating data shall include:

1. A soils investigation in accordance with Section 1802.
2. Pier or pile load tests in accordance with Section 1808.2.8.3, regardless of the load supported by the pier or pile.

The design and installation of the pier or pile foundation shall be under the direct supervision of a registered design professional knowledgeable in the field of soil mechanics and pier or pile foundations who shall certify to the building official that the piers or piles as installed satisfy the design criteria.

❖ In other parts of this chapter, limitations are specified for the stress values used for design purposes. These allowable stresses are stated as a percentage of some limiting strength property of the pier or pile material. For example, allowable design stresses for piles made of steel are stated as a percentage of the yield strengths of the several grades of steel typically used for pile construction. For concrete, the allowable design stress is prescribed as a percentage of the 28-day specified compression strength. The allowable design stresses permitted for timber piles are based on the natural strengths of the different species of wood used for pile construction. The values have been developed and tabulated by the AF&PA and include reductions in pile strengths because of preservative treatment.

The allowable design stresses stipulated in the code for the different types of piles provide an adequate factor of safety against the dynamic forces of pile driving that may cause damage to the piles and prevent overstresses because of loading and subsoil conditions.

This section allows the use of higher allowable stresses when evidence supporting the values is submitted and approved by the building official. The data submitted to the building official should include analytical evaluations and findings from a foundation investigation as specified in Section 1808.2.1, and the results of load tests performed in accordance with the requirements of Section 1808.2.8.3. The technical data and the recommendation for the use of higher stress values must come from a registered engineer who is knowledgeable in soil mechanics and experienced in the design of pile foundations. This engineer is to supervise the pile design work and witness the installation of the pile foundation so as to certify to the building official that the construction satisfies the design criteria. In any case, the use of greater design stresses is not to result in permitting design loads that are larger than one-half of the test loads (see Section 1808.2.8.3).

1808.2.11 Piles in subsiding areas. Where piles are driven through subsiding fills or other subsiding strata and derive support from underlying firmer materials, consideration shall be given to the downward frictional forces that may be imposed on the piles by the subsiding upper strata.

Where the influence of subsiding fills is considered as imposing loads on the pile, the allowable stresses specified in this chapter are permitted to be increased where satisfactory substantiating data are submitted.

❖ Where compacted fill is placed over compressible soils, the underlying material will consolidate because of the added weight. The depth to suitable bearing material will, over a period of time, be shifted downward by the forces of the subsiding soil. Such forces caused by the weight of the fill are transmitted to the piles by skin friction and, in effect, serve as added loads on the piles. The magnitude of such loads must be determined by accepted analytical methods and be taken into account in the design of pile foundations.

1808.2.12 Settlement analysis. The settlement of piers, individual piles or groups of piles shall be estimated based on approved methods of analysis. The predicted settlement shall cause neither harmful distortion of, nor instability in, the structure, nor cause any stresses to exceed allowable values.

❖ A settlement analysis is performed to design a pier or pile foundation system that will maintain the stability and structural integrity of the supported building or structure. Essentially, every soil-bearing stratum must support the loads transferred from the pier or pile system without detrimental settlement. Foundation systems that suffer serious settlements, particularly differential settlements, can cause great structural damage to the supported structure as well as to the foundation itself.

The settlement analysis of an individual pier or pile is a complex procedure. In comparison, the analysis of a group of piles is even more complex because of the overlapping soil stresses caused by closely spaced piles. Analytical procedures vary with the type of piers or piles and especially with the type of soil.

Settlements are of two basic types: immediate settlements are those that occur as soon as the load is applied and usually take place within a period of less than 7 days; consolidation settlements are time dependent and take place over a long period of time. All cohesionless soils, such as granular materials consisting of sand, gravel or a mixture of both, which have a large coefficient of permeability (rapid draining properties), undergo immediate settlements. All fine-grained, saturated, cohesive soils, such as clays, undergo time-dependent consolidation settlements.

Settlement analysis would generally include cases involving piers and end-bearing piles on rock or hard soils as well as friction-type piles in both granular and cohesive soils. Load tests are often performed to provide data that will aid the settlement analysis.

1808.2.13 Preexcavation. The use of jetting, augering or other methods of preexcavation shall be subject to the approval of the building official. Where permitted, preexcavation shall be carried out in the same manner as used for piers or piles subject to load tests and in such a manner that will not impair the carrying capacity of the piers or piles already in place or damage adjacent structures. Pile tips shall be driven below the preexcavated depth until the required resistance or penetration is obtained.

❖ The use of preexcavation methods to facilitate the installation of piers or piles is often necessary for a number of reasons, including:

- To drive piles through upper layers of hard soil;
- To penetrate through subsurface obstructions;
- To eliminate or reduce the possibility of ground heave that could result in lifting adjacent piles already driven;
- To reduce ground pressures that could result in the lateral movement of adjacent piles or structures;
- To reduce the amount of driving necessary to seat the piles in the required bearing stratum;
- To reduce the possibility of damaging vibrations;
- To reduce the amount of noise associated with pile-driving operations and
- To accommodate the placement of piles that are longer than the leads of the driving equipment.

Methods commonly used for preexcavation are jetting and predrilling. The jetting method has been found to be more effective in granular soils, such as sand and fine gravel, than in cohesive materials, such as clay. However, jetting should not be done in granular soils containing very coarse gravel, cobbles and small boulders because such material cannot be removed by the jet stream and will result in a collection of stones at the bottom of the hole, making it very difficult, if not impossible, to drive a pile through the mass of stone material. The jetting operations must be controlled to avoid excessive losses of soil that could affect the stability of adjacent structures or the required bearing capacity of previously installed piles.

Jetting operations must be carefully controlled to avoid excessive loss of soil, which could affect the load-bearing capacity of piles already installed or the stability of adjacent structures.

Piles should be driven below the depth of the jetted hole until the required resistance or penetration is obtained. Before this preexcavation method is used, consideration should be given to the possibility that jetting, unless strictly controlled, can adversely affect load transfer, particularly as it involves the placement of nontapered piles.

In comparison to the jetting method described, a more controllable form of preexcavation is by predrilling or coring. This method greatly reduces the possibility of detrimental effects on adjacent piles or structures and can be performed as a dry operation or a wet rotary process. Dry drilling can be done by the use of a continuous-flight auger or a short-flight auger attached to the end of a drill stem or kelly bar. Wet drilling requires a hollow-stem, continuous-flight auger or a hollow drill stem employing the use of spade bits. When the wet rotary process of predrilling is used, bentonite slurry or plain water is circulated to keep the hole open. As in the case of jetting, piles should be driven with tips below the predrilled hole. This is necessary to prevent any voids or very loose or soft soils from occurring below the pile tip.

There are other methods used for preexcavation purposes, such as the dry tube method and spudding, but such procedures are seldom used. The methods to be employed for preexcavation are subject to the approval of the building official.

1808.2.14 Installation sequence. Piles shall be installed in such sequence as to avoid compacting the surrounding soil to the extent that other piles cannot be installed properly, and to prevent ground movements that are capable of damaging adjacent structures.

❖ As piles are driven sequentially within a group, progressive compaction of the surrounding soil occurs and there can be a buildup of unequal pressures around the piles, possibly causing some of them to be deflected off-line.

This situation is more apt to develop when piles are closely spaced. More importantly, soil compaction during driving operations can cause variations in pile lengths within a group to the extent that some of the piles may stop considerably short of reaching specified bearing material. Ground heave is another possible effect of soil compaction (see Section 1808.2.19).

To prevent or significantly reduce the problems associated with soil compaction, the driving sequence becomes an important consideration in the planning of pile installation work. For example, in placing a group of piles, if the outer piles within the group are driven first, the inner piles may be short of their required length because of soil compaction that builds up progressively toward the center of the cluster. Starting the pile-driving sequence at the edge of a group makes the piles progressively more difficult to drive, resulting in a one-sided bearing group. The generally accepted driving sequence is to work from the center of a cluster of piles outward. For pile arrangements consisting of rows of widely spaced piles (i.e., under a continuous concrete mat), driving can be done sequentially from one side to the other.

1808.2.15 Use of vibratory drivers. Vibratory drivers shall only be used to install piles where the pile load capacity is verified by load tests in accordance with Section 1808.2.8.3. The installation of production piles shall be controlled according to power consumption, rate of penetration or other approved means that ensure pile capacities equal or exceed those of the test piles.

❖ The use of vibratory drivers for the installation of piles is not applicable to all types of soil conditions. They are effective in granular soils with the use of nondisplacement piles, such as steel H-piles and pipe piles driven open ended. Vibratory drivers are also used for extracting piles or temporary casings employed in the construction of cast-in-place concrete piles.

Vibratory drivers, either low or high frequency, cause the pile to penetrate the soil by longitudinal vibrations. While this type of pile driver can produce remarkable results in the installation of nondisplacement piles under favorable soil conditions, the greatest difficulty is the lack of a reliable method of estimating the load-bearing capacity. After the pile has been installed with a vibratory driver, pile capacity can best be determined by using an impact-type hammer to set the pile in its final position.

An acceptable means of controlling pile capacity is by determining the power consumption in relation to the rate of penetration. Nonetheless, the use of a vibratory driver is only permitted where the pile load capacity is established by load tests in accordance with the requirements of Section 1808.2.8.3.

1808.2.16 Pile driveability. Pile cross sections shall be of sufficient size and strength to withstand driving stresses without damage to the pile, and to provide sufficient stiffness to transmit the required driving forces.

❖ Piles must be of size, strength and stiffness so as to be capable of resisting without damage:

- Crushing caused by impact forces during driving;
- Bending stresses during handling;
- Tension from uplift forces or from rebound during driving;
- Bending stresses caused by horizontal forces during driving and
- Bending stresses caused by pile curvatures.

Additionally, the pile must be capable of transmitting dynamic driving forces to mobilize the required ultimate pile capacity within the soil without severe elastic energy losses. Pile driveability depends on the pile stiffness, which is a function of pile length, cross-sectional area and modulus of elasticity (MOE). It should be noted that the yield strength does not affect stiffness; thus, caution should be observed in the use of high-yield-strength steels for higher loads on smaller cross sections requiring higher dynamic driving energy. A wave equation analysis would reflect pile stiffness or driveability.

The selection of pile types and dimensional requirements for driveability is a function of soil characteristics.

1808.2.17 Protection of pile materials. Where boring records or site conditions indicate possible deleterious action on pier or pile materials because of soil constituents, changing water levels or other factors, the pier or pile materials shall be adequately protected by materials, methods or processes approved by the building official. Protective materials shall be applied to the piles so as not to be rendered ineffective by driving. The effectiveness of such protective measures for the particular purpose shall have been thoroughly established by satisfactory service records or other evidence.

❖ Piles are often exposed to the deteriorating effects of biological, chemical and physical actions caused by a hostile underground environment that may exist at the time of their installation or that may later develop at the site. Under such conditions, pile materials must be properly protected to ensure their expected durability.

The problems associated with pile durability relate directly to the type of pile materials used. For example, concrete piles that are entirely embedded in undisturbed soil are deemed to be permanent installations. The level of the ground-water table is generally not a factor affecting the durability of concrete piles. Ground water that contains deleterious substances and readily flows through disturbed or granular soils, such as sand and gravel (regardless of the level of the water table), can have a deteriorating effect on concrete piles. Concrete piles embedded in fine grained, impervious soils, such as clay, generally are not adversely affected by ground water containing harmful substances. Concrete can also be affected by exposure to soils having a high sulfate content, unless Type II or V portland cement is used in making the concrete mixture.

Concrete piles installed in saltwater, such as for buildings or other structures in waterfront construction, are subject to chemical action from polluted waters, frost action on porous concrete, spalling and rusting of steel reinforcement. Spalling action may become particularly critical under tidal conditions where alternate wetting and drying of the concrete occurs in conjunction with cycles of freezing and thawing. Generally, concrete can be protected from damage by such adverse conditions with the use of special cements, dense concrete mixtures rich in cement content, adequate concrete cover over the reinforcement, air entrainment, suitable concrete admixtures or special surface coatings.

Piles made of steel materials and embedded entirely in undisturbed soil (regardless of its type) are not significantly affected by corrosion due to oxidation, mainly because undisturbed soil is so deficient in oxygen that progressive corrosion is repressed. However, steel may be subject to serious corrosion and structural deterioration where ground water contains deleterious substances from coal piles, alkali soils, active cinder fills, chemical waste from manufacturing operations, etc. Under such conditions, steel piles may be protected by encasement in concrete or by applying protective coatings such as coal-tar or other suitable materials. Steel piles installed in saltwater or exposed to a saltwater environment can corrode severely and should be protected by encasement in concrete or by the application of approved coatings. Piles that extend above ground level and are exposed to air should be painted in the same way as any type of structural steel construction to prevent rusting. Corrosion of load-bearing steel can also occur because of electrolytic action, and in such cases, cathodic protection may be required.

Timber piles totally embedded in the earth below the low point of ground-water level or entirely submerged in fresh water will last indefinitely without preservative treatment. However, timber piles that extend above the ground-water level or are exposed to air or saltwater are subject to decay as well as attacks by insects and marine borers. The piles may also be subjected to damage by the percolation of ground water heavily charged with alkali and acids. Under such conditions, timber piles must be pressure treated with preservatives in accordance with AWPA Standard C1 or C2 listed in Chapter 35.

Generally, at any site where piles are to be installed and where the soil is suspect or there is sufficient evidence of an adverse underground environment, a soils investigation should be conducted to determine the need and method to protect the piles against possible deterioration.

1808.2.18 Use of existing piers or piles. Piers or piles left in place where a structure has been demolished shall not be used for the support of new construction unless satisfactory evidence is submitted to the building official, which indicates that the piers or piles are sound and meet the requirements of this code. Such piers or piles shall be load tested or redriven to verify their capacities. The design load applied to such piers or piles shall be the lowest allowable load as determined by tests or redriving data.

❖ After the demolition of an existing building, piers or piles remaining in place cannot be reused to support a new structure unless sufficient and reliable information is provided to the building official showing that the new loads to be imposed on the existing pier or pile system will be adequately supported. This requirement is necessary because of the lack of adequate soil data and technical information on the pier or piling material used and the unavailability of pile-driving records made during the construction of the existing pier or pile foundation. The true condition of the piers or piles is not known, since they may have deteriorated over time, possibly reducing their load capacity. Pier or pile capacities may be determined by load tests, or the piles may be retracted and redriven to verify their load capacities.

1808.2.19 Heaved piles. Piles that have heaved during the driving of adjacent piles shall be redriven as necessary to develop the required capacity and penetration, or the capacity of the pile shall be verified by load tests in accordance with Section 1808.2.8.3.

❖ When piles are driven into cohesive soils, particularly saturated plastic clay materials, they often displace a volume of soil equal to that of the piles themselves. Such soil displacement usually occurs as ground heaves and may cause adjacent piles already driven to lift up and become unseated, causing a loss of pile load capacity. When this happens, heaved piles must be redriven to firm bearing in order to regain their required capacity. If heaved piles are not redriven, their capacities must be verified by means of load tests.

It should be noted that not all types of piles can be redriven. For example, heaved uncased cast-in-place concrete piles, or sectional piles whose splices cannot take tension, should be abandoned and replaced.

In redriving heaved piles, the same or comparable driving equipment should be employed as used in the original installation. However, there are exceptions to this rule. For example, in the installation of concrete-filled pipe piles, only the empty pipes were first driven. In redriving this type of pile, the pipes may now be filled with concrete resulting in much stiffer and heavier sections than were driven initially. In such cases, the driving technique must be adjusted to accommodate a considerably lesser driving resistance.

One method commonly used to prevent or reduce objectionable displacements caused by pile installations in soils subject to heaving is to preexcavate pile holes in accordance with the requirements of Section 1808.2.13.

1808.2.20 Identification. Pier or pile materials shall be identified for conformity to the specified grade with this identity maintained continuously from the point of manufacture to the point of installation or shall be tested by an approved agency to determine conformity to the specified grade. The approved agency shall furnish an affidavit of compliance to the building official.

❖ All pier or pile materials must be identified for conformity to construction specifications, providing information such as strength (species and grade for timber piles) and dimensions and other pertinent information as may be required. Such identification must be provided for all piers and piles, whether they are taken from manufacturer's stock or made for a particular project. Identification is to be maintained from the place of manufacture to the shipment, on-site handling, storage and installation of the piers or piles. Manufacturers, upon request, usually furnish certificates of compliance with construction specifications. In the absence of adequate data, piers and piles are to be tested to prove conformity to the specified grade.

In addition to mill certificates (steel piles), identification is made through plant manufacturing or inspection reports (precast concrete piers or piles and timber piles) and delivery tickets (concrete). Timber piles are stamped (branded) with information such as producer, species, treatment and length.

Identification is essential when high-yield-strength steel is specified. Frequently, pile cutoff lengths are reused and pile material may come from a jobber, a contractor's yard or a material supplier. In such cases, mill certificates are not available and the steel should be tested to see if it complies with the specifications.

1808.2.21 Pier or pile location plan. A plan showing the location and designation of piers or piles by an identification system shall be filed with the building official prior to installation of such piers or piles. Detailed records for piers or individual piles shall bear an identification corresponding to that shown on the plan.

❖ This section requires that a pier or pile location plan clearly showing the designation of all piers or piles in the foundation system be submitted to the building official prior to installation of piles. Preferably, such plans should be submitted before delivery of the piers or piles to the construction site and prior to staking the pier or pile locations.

The pier or pile location plan is not only an important tool for the installation of piers or piles, but also serves to communicate information between the owner, engineer, contractor, manufacturer, special inspector, building official and other interested persons. The special inspector (see commentary, Section 1808.2.22) must keep piling logs, records and reports based on this identification system. The use of the pier or pile plan is particularly important at sites where the variations in soil profiles are so extensive that it becomes necessary to manufacture piers or piles of different lengths to reach proper bearing levels. The building official should receive revised copies of the pier or pile location plan whenever field changes are made that add, delete or relocate piles.

1808.2.22 Special inspection. Special inspections in accordance with Sections 1704.8 and 1704.9 shall be provided for piles and piers, respectively.

❖ This section requires special inspections (see definition in Section 1702) in accordance with Sections 1704.8 and 1704.9 to ensure proper and safe installations. To be approved by the building official, an inspector should be experienced in pier or pile installation work. An inspector must be able to read and understand foundation plans and specifications, maintain accurate records, have a thorough understanding of the scope of the pile work to be performed, give concise and timely reports to the owner or engineer and submit in writing to the building official whatever field information is required.

The duties of the pier or pile inspector include the inspection and approval of pile-driving equipment, the inspection and acceptance of piles furnished for the work and the verification of all pile location stakes and spacings. Duties also include the observation and recording of required information relating to the installation of the pier or pile foundation, including keeping a log for each pile showing either the number of hammer blows for each foot of penetration and the final penetration in blows per inch or the driving record for only the last few feet of penetration. The driving log should also include information on the duration and cause of any delays, such as splicing time, equipment breakdown, changing cushions, etc. The record should also show the final tip and butt (cutoff) elevations, as well as the depths of preexcavation holes (see Section 1808.2.13). Finally, there should also be a record of any pile damage and repair work, pile extractions and replacements; observations on pile heaving and depth of redriving and information on pile alignment.

1808.2.23 Seismic design of piers or piles.

❖ This section includes additional requirements concerning the seismic design of piers and piles (as defined in Section 1808.1) supporting structures assigned to Seismic Design Category C and higher. For more information regarding piers and piles, consult the following publications, which are listed in the bibliography to this chapter: *Recommendations for Design, Manufacture, and Installation of Concrete Piles* (ACI 543R-74), *Design and Construction of Drilled Piers* (ACI 336.3R-93) (also *ACI Manual of Concrete Practice*, Part 4) and *Recommended Practice for Design, Manufacture and installation of Prestressed Concrete Piling, PCI Journal.*

1808.2.23.1 Seismic Design Category C. Where a structure is assigned to Seismic Design Category C in accordance with Section 1616, the following shall apply. Individual pile caps, piers or piles shall be interconnected by ties. Ties shall be capable of carrying, in tension and compression, a force equal to the product of the larger pile cap or column load times the seismic coefficient, S_{DS}, divided by 10 unless it can be demonstrated that equivalent restraint is provided by reinforced concrete beams within slabs on grade or reinforced concrete slabs on grade or confinement by competent rock, hard cohesive soils or very dense granular soils.

Exception: Piers supporting foundation walls, isolated interior posts detailed so the pier is not subject to lateral loads, lightly loaded exterior decks and patios, of Group R-3 and U occupancies not exceeding two stories of light-frame construction, are not subject to interconnection if it can be shown the soils are of adequate stiffness, subject to the approval of the building official.

❖ The purpose of this section is to preclude excessive movement of one pile or pile group with respect to another. The section is similar to Section 1805.4.2.2. One of the prerequisites of adequate structural performance during an earthquake is that the foundation of the structure must act as a unit. This is typically accomplished by tying together the pile caps, piers or piles with ties capable of carrying, both in tension and compression, a force equal to 10 percent of the larger pile cap or column load multiplied by S_{DS}, which is the design spectral response accelerations at short periods as determined in Section 1615.1.3. This can be accomplished through the use of concrete floor slabs or tie beams. Reliance upon lateral soil pressure is typically not a good idea because the motion is imparted from soil to structure (not inversely as is commonly assumed), and if the soil is soft enough to require piles, little reliance can be placed on soft-soil passive pressure to restrain relative pile displacement under dynamic conditions. If soils are shown to be of adequate stiffness, the building official may allow the use of the exception for the following piers in Group R-3 and U occupancies not exceeding two stories of light-frame construction:

1. Piers supporting foundation walls.
2. Piers supporting isolated interior posts detailed so the pier is not subject to lateral loads.
3. Piers supporting lightly loaded exterior decks and patios.

1808.2.23.1.1 Connection to pile cap. Concrete piles and concrete-filled steel pipe piles shall be connected to the pile cap by embedding the pile reinforcement or field-placed dowels anchored in the concrete pile in the pile cap for a distance equal to the development length. For deformed bars, the development length is the full development length for compression or tension, in the case of uplift, without reduction in length for excess area. Alternative measures for laterally confining concrete and maintaining toughness and ductile-like behavior at the top of the pile will be permitted provided the design is such that any hinging occurs in the confined region.

Ends of hoops, spirals and ties shall be terminated with seismic hooks, as defined in Section 21.1 of ACI 318, turned into the confined concrete core. The minimum transverse steel ratio for confinement shall not be less than one-half of that required for columns.

For resistance to uplift forces, anchorage of steel pipe (round HSS sections), concrete-filled steel pipe or H-piles to the pile cap shall be made by means other than concrete bond to the bare steel section.

Exception: Anchorage of concrete-filled steel pipe piles is permitted to be accomplished using deformed bars developed into the concrete portion of the pile.

Splices of pile segments shall develop the full strength of the pile, but the splice need not develop the nominal strength of the pile in tension, shear and bending when it has been designed to resist axial and shear forces and moments from the load combinations of Section 1605.4.

❖ This section prescribes special detailing requirements between the pile and pile cap, including required development of the longitudinal pile reinforcement in the cap and associated transverse reinforcement. This reinforcement is required to extend into the pile cap to tie the elements together and to assist in load transfer at the top of the pile to the pile cap. The connection must consist of embedment of the pile reinforcement in the pile cap for a distance equal to the development length as specified in ACI 318. Field-placed dowels anchored in the plastic concrete piles are acceptable. The development length to be provided is that for compression or, where uplift is indicated by analysis, tension without reduction in length for excess area. Where seismic confinement reinforcement at the top of the pile is required, alternative measures for laterally confining concrete and maintaining toughness and ductile-like behavior at the top of the pile are permitted, provided the design would force the hinge to occur in the confined region. Splices of pile segments are required to develop the nominal pile strength or splices can be designed for the special seismic load combinations of Section 1605.4.

1808.2.23.1.2 Design details. Pier or pile moments, shears and lateral deflections used for design shall be established considering the nonlinear interaction of the shaft and soil, as recommended by a registered design professional. Where the ratio of the depth of embedment of the pile-to-pile diameter or width is less than or equal to six, the pile may be assumed to be rigid.

Pile group effects from soil on lateral pile nominal strength shall be included where pile center-to-center spacing in the direction of lateral force is less than eight pile diameters. Pile group effects on vertical nominal strength shall be included where pile center-to-center spacing is less than three pile diameters. The pile uplift soil nominal strength shall be taken as the pile uplift strength as limited by the frictional force developed between the soil and the pile.

Where a minimum length for reinforcement or the extent of closely spaced confinement reinforcement is specified at the top of the pier or pile, provisions shall be made so that those specified lengths or extents are maintained after pier or pile cutoff.

❖ This section addresses miscellaneous issues unique to the seismic design of piers and piles. If the length is less than or equal to six times the least horizontal dimension of the pier, the pier can be assumed rigid and moments, shears and lateral deflections can be calculated accordingly. Where the length exceeds six times the least horizontal dimension, the nonlinear interaction of the pier or pile and soil effects are to be included in the design. The effect of abrupt changes in soil deposits, such as changes from soft to firm or loose to dense soils, should be included in the design.

This section also prescribes conditions under which group effects on the nominal pile strength, lateral as well as vertical, need to be included in the design.

The last paragraph of this section is intended to account for the condition where a pile encounters refusal at a shallower depth than intended and an unanticipated portion of the pile is cut off. It is imperative that the required reinforcement be provided at the top of the pile even when excess pile length is cut off.

1808.2.23.2 Seismic Design Category D, E or F. Where a structure is assigned to Seismic Design Category D, E or F in accordance with Section 1616, the requirements for Seismic Design Category C given in Section 1808.2.23.1 shall be met, in addition to the following. Provisions of ACI 318, Section 21.10.4, shall apply when not in conflict with the provisions of Sections 1808 through 1812. Concrete shall have a specified compressive strength of not less than 3,000 psi (20.68 MPa) at 28 days.

Exceptions:

1. Group R or U occupancies of light-framed construction and two stories or less in height are permitted to use concrete with a specified compressive strength of not less than 2,500 psi (17.2 MPa) at 28 days.
2. Detached one- and two-family dwellings of light-frame construction and two stories or less in height are not required to comply with the provisions of ACI 318, Section 21.10.4.
3. Section 21.10.4.4(a) of ACI 318 need not apply to concrete piles.

❖ This section includes additional requirements applicable to buildings assigned to Seismic Design Category D, E or F. The section is similar to Section 1805.9, except that Section 21.10.4 of ACI 318 is referenced.

Concrete piers and piles in buildings assigned to Seismic Design Category D, E or F are required to have a specified compressive strength of no less than 3,000 psi (20.68 MPa) at 28 days. This same requirement is also in Section 1805.9, which is applicable to foundations in buildings assigned to Seismic Design Category D, E or F.

An exception is provided for the use of 2,500 psi (17.2 MPa) concrete in smaller residential-type structures. In addition, small detached dwellings are exempt from the referenced ACI 318 provisions and merely need to meet the requirements for Seismic Design Category C. Exception 3 eliminates an ACI 318 requirement that is instead incorporated (with modification) in Section 1808.2.23.2.1.

1808.2.23.2.1 Design details for piers, piles and grade beams. Piers or piles shall be designed and constructed to withstand maximum imposed curvatures from earthquake ground motions and structure response. Curvatures shall include free-field soil strains modified for soil-pile-structure interaction coupled with pier or pile deformations induced by lateral pier or pile resistance to structure seismic forces. Concrete piers or piles on Site Class E or F sites, as determined in Section 1615.1.1, shall be designed and detailed in accordance with Sections 21.4.4.1, 21.4.4.2 and 21.4.4.3 of ACI 318 within seven pile diameters of the pile cap and the interfaces of soft to medium stiff clay or liquefiable strata. For precast prestressed concrete piles, detailing provisions as given in Sections 1809.2.3.2.1 and 1809.2.3.2.2 shall apply.

Grade beams shall be designed as beams in accordance with ACI 318, Chapter 21. When grade beams have the capacity to resist the forces from the load combinations in Section 1605.4, they need not conform to ACI 318, Chapter 21.

❖ The first paragraph of this section requires special consideration of flexural loads on piers or piles due to earthquake motions. The following discussion taken from the 1997 NEHRP Provisions Commentary, Section 7.5.3, provides justification for these requirements:

Special consideration is required in the design of concrete piles subject to significant bending during earthquake shaking. Bending can become crucial to pile design where portions of the foundation piles are supported in soils, such as loose granular materials or soft soils that are susceptible to large deformations or strength degradation. Severe pile bending problems can result from various combinations of soil conditions during strong ground shaking. For example:

1. Soil settlement at the pile-cap interface either from consolidation of soft soil prior to the earthquake or from soil compaction during the earthquake can create a free-standing short column adjacent to the pile cap.
2. Large deformations, or a reduction in strength or both, resulting from liquefaction of loose granular materials can cause bending or conditions of free-standing columns.
3. Large deformations in soft soils can cause varying degrees of pile bending. The degree of pile bending will depend upon thickness and strength of the soft soil layer(s) and the properties of the soft/stiff soil interface(s).

The designer needs to consider the variation in soil conditions and driven pile lengths in providing for pile ductility at potential high curvature interfaces. Interaction between the geotechnical and structural engineers is essential.

It is prudent to design piles to remain functional during and following earthquakes in view of the fact that it is difficult to repair foundation damage. The desired foundation performance can be accomplished by proper selection and detailing of the pile foundation system. Such design should accommodate bending from both reaction to the building's inertial loads and those induced by the motions of the soils themselves.

The second paragraph of this section allows grade beams to be designed to have the strength to resist the design forces given by the special seismic load combinations instead of designing the grade beams as beams in accordance with ACI 318, Chapter 21. The special seismic load combinations estimate the maximum forces that can realistically develop in the grade beam in an earthquake situation.

1808.2.23.2.2 Connection to pile cap. For piles required to resist uplift forces or provide rotational restraint, design of anchorage of piles into the pile cap shall be provided considering the combined effect of axial forces due to uplift and bending moments due to fixity to the pile cap. Anchorage shall develop a minimum of 25 percent of the strength of the pile in tension. Anchorage into the pile cap shall be capable of developing the following:

1. In the case of uplift, the lesser of the nominal tensile strength of the longitudinal reinforcement in a concrete pile, or the nominal tensile strength of a steel pile, or the pile uplift soil nominal strength factored by 1.3 or the axial tension force resulting from the load combinations of Section 1605.4.
2. In the case of rotational restraint, the lesser of the axial and shear forces, and moments resulting from the load combinations of Section 1605.4 or development of the full axial, bending and shear nominal strength of the pile.

❖ The requirements of this section address the need for conservatism in the design of connections between piles and pile caps. They are intended to allow energy dissipating mechanisms, such as rocking, to occur in the soil without failure of the pile.

1808.2.23.2.3 Flexural strength. Where the vertical lateral-force-resisting elements are columns, the grade beam or pile cap flexural strengths shall exceed the column flexural strength.

The connection between batter piles and grade beams or pile caps shall be designed to resist the nominal strength of the pile acting as a short column. Batter piles and their connection shall be capable of resisting forces and moments from the load combinations of Section 1605.4.

❖ This section requires that the grade beam or pile cap flexural strength exceed that of the supported column flexural strength if the column is a part of the lateral-force-resisting system.

Additional requirements are specified in this section for batter piles in order to limit earthquake damage to these systems. By their nature, batter pile systems have limited ductility and have performed poorly under strong ground motions. They are required to be designed using the special load combinations of Section 1605.4, which estimate the maximum force expected to be developed in these elements during a seismic event.

SECTION 1809
DRIVEN PILE FOUNDATIONS

1809.1 Timber piles. Timber piles shall be designed in accordance with the AFPA NDS.

❖ Round timber piles are best suited as friction-type piles. They are not recommended as piles to be driven through dense gravel, boulders or till, or for end-bearing piles on rock. While timber piles do not have the high allowable load capacities of structural steel or concrete piles, they are probably the most commonly used type of pile for deep foundation construction throughout the United States, mainly because of their availability, ease of handling and, sometimes, for economic reasons.

Timber piles are shaped from tree trunks and their lengths are dependent on the heights that the various species of trees used for making piles will grow. Piles are made as tapered sections because of the natural taper of tree trunks.

Round timber piles are usually made from Southern pine in lengths up to about 80 feet (24 384 mm) and from Pacific Coast Douglas fir in lengths up to about 125 feet (38 100 mm). Other species commonly used for piles are red oak and red pine. Timber piles that are 40 feet (12 192 mm) to 60 feet (18 288 mm) in length are common, but longer lengths cannot be obtained economically in all areas of the country.

Untreated timber piles that are embedded below the permanent ground-water level (fresh water only) may last indefinitely. Under conditions where piles are required to extend above ground-water level but still remain totally embedded in the ground, the untreated wood material may be subject to decay. In cases where untreated timber piles extend above the ground surface into the air, they are exposed to decay and insect attack. The durability of timber piles is best served by applying treatment with an approved preservative.

This section requires timber piles to be designed in accordance with the AF&PA's *National Design Specification for Wood Construction* (NDS). Section 6 of the NDS provides the appropriate design values, different species of timber piles and the applicable adjustment factors. While timber piles have to be designed in accordance with the NDS, they are usually only considered for design loads of 10 (89 kN) to 50 tons (445 kN).

One of the significant problems associated with timber pile installations is the possibility of damage caused by overdriving. Overdriving of wood piles may cause failure by bending, brooming at the tip, crushing or brooming at the butt end or splitting or breaking along the pile section (see Figure 1809.1). Another problem is the difficulties encountered when splicing timber piles to achieve greater lengths.

1809.1.1 Materials. Round timber piles shall conform to ASTM D 25. Sawn timber piles shall conform to DOC PS-20.

❖ The ASTM D 25 standard referenced in this section covers the physical characteristics of treated and untreated round timber piles. Essentially, the standard divides timber piles into two classifications: friction and end-bearing piles. The dimensional requirements for each classification are tabulated, giving the minimum circumference requirements for pile heads and the corresponding smaller tip circumference based on pile taper and lengths between 20 (6096 mm) and 120 feet (36 576 mm), measured in increments of 10 feet (3048 mm). The ASTM D 25 standard also includes requirements for the quality of wood, tolerances on pile straightness, twist of grain, knots, holes, scars, wood checks, shakes, splits and other necessary information. However, the requirements stated in the standard do not relate to the several species of wood used for making timber piles, nor to their relative strengths. A timber pile with typical dimensions from ASTM D 25 is shown in Figure 1809.1.1.

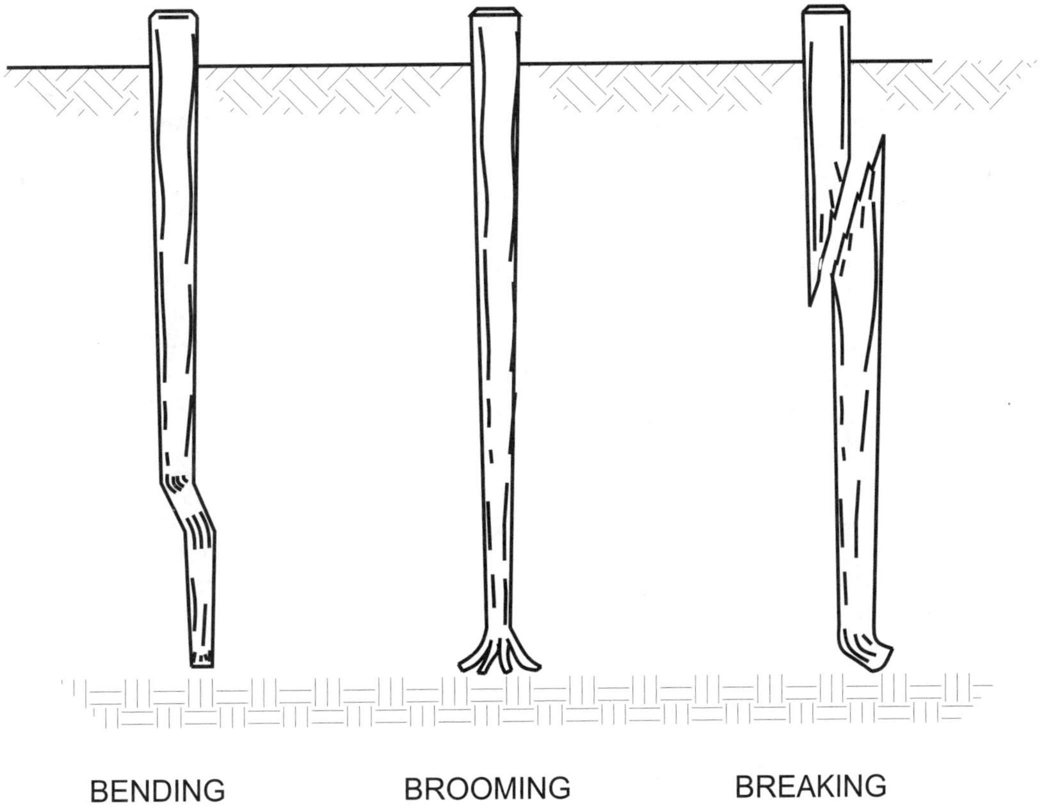

Figure 1809.1
TYPICAL EFFECTS OF OVERDRIVING TIMBER PILES

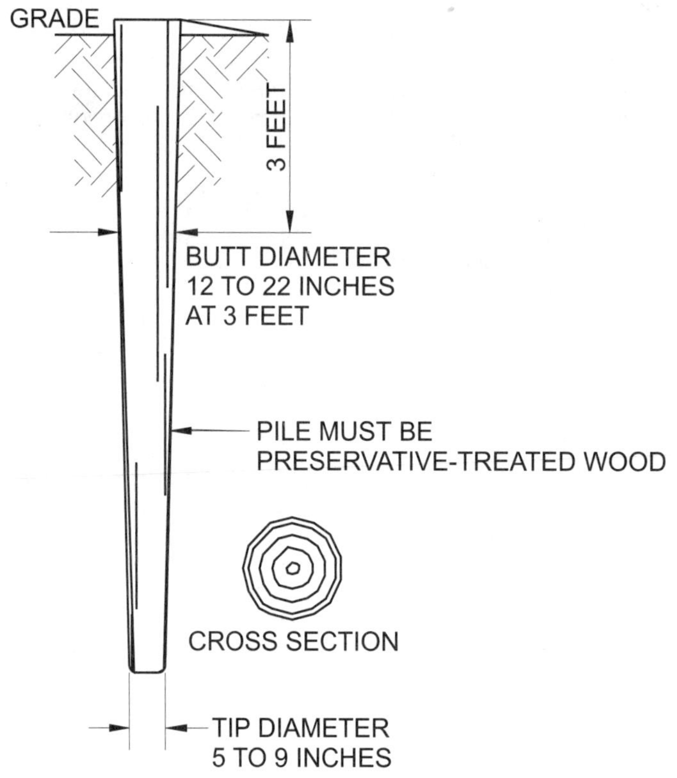

For SI: 1 inch = 25.4 mm, 1 foot = 304.8 mm.

Figure 1809.1.1
TIMBER PILES

1809.1.2 Preservative treatment. Timber piles used to support permanent structures shall be treated in accordance with this section unless it is established that the tops of the untreated timber piles will be below the lowest ground-water level assumed to exist during the life of the structure. Preservative and minimum final retention shall be in accordance with AWPA C3 for round timber piles and AWPA C24 for sawn timber piles. Preservative-treated timber piles shall be subject to a quality control program administered by an approved agency. Pile cutoffs shall be treated in accordance with AWPA M4.

❖ For a general discussion on the need for treating wood with preservatives, see the commentary to Sections 1805.4.5, 1808.2.17 and 1809.1. This section of the code specifically requires that the preservative treatment of timber piles by pressure processes conform to the applicable requirements of AWPA C3. It also covers all species of wood used for making piles. AWPA C24 is the standard to be used for the preservative treatment of sawn timber piles. The main difference in the respective standards is that round timber piles conform to ASTM D 25 and sawn timber piles comply with the grading rules of the species for strength.

This section also requires that, for timber piles subjected to saltwater exposure, the treatment of piles with water-borne preservatives and creosote complies with quality control procedures administered by an approved agency. Guides used in establishing these procedures are those contained in the American Wood Preservers' Bureau (AWPB) publications such as:

- MP1, *Quality Control and Inspection Procedures for Dual Treatment of Marine Piling Pressure Treated with Waterborne Preservatives and Creosote for Use in Marine Waters*;
- MP2, *Quality Control and Inspection Procedures for Marine Piling Pressure Treated with Creosote for Use in Marine Waters* or
- MP4, *Quality Control and Inspection Procedures for Marine Piling Pressure Treated with Waterborne Preservatives for Use in Marine Waters.*

The field treatment of cuts and injuries to timber piles, including the treatment at pile cutoffs, with preservatives made for applications of creosote and creosote mixtures or with water-borne preservatives is covered by AWPA M4.

1809.1.3 End-supported piles. Any sudden decrease in driving resistance of an end-supported timber pile shall be investigated with regard to the possibility of damage. If the sudden decrease in driving resistance cannot be correlated to load-bearing data, the pile shall be removed for inspection or rejected.

❖ A significant problem associated with the use of timber piles is the possibility of structural damage caused by overdriving. Overdriving usually occurs upon reaching rock or hard soil stratas and may result in pile bending, brooming at the tip, crushing or brooming at the head or splitting or breaking along the pile section (see commentary, Section 1809.1). To avoid such damage, if the penetration resistance suddenly increases, pile driving should be stopped immediately. If the penetration resistance suddenly decreases, serious damage to the pile could have occurred (including breaking), and the condition must be investigated. The pile can be withdrawn and inspected, and if required, abandoned and replaced.

1809.2 Precast concrete piles.

❖ This section sets forth minimum material and design requirements for precast nonprestressed (conventionally reinforced) and prestressed concrete piles.

1809.2.1 General. The materials, reinforcement and installation of precast concrete piles shall conform to Sections 1809.2.1.1 through 1809.2.1.4.

❖ Precast concrete piles are manufactured as nonprestressed (conventionally reinforced) or prestressed. Both types can be formed by bedcasting, spinning (centrifugal casting), vertical casting, slip forming or extrusion methods. They are usually made in square, octagonal or round shapes (see Figure 1809.2.1). Precast piles are manufactured as solid units or may be made with a hollow core. They can also be made with internal jet pipes or inspection ducts.

Precast piles are generally ordered in predetermined lengths based on the findings and analysis of soil exploratory work conducted at the project site.

This type of pile is a displacement pile that is normally installed with pile-driving equipment.

Precast concrete piles are to be designed in accordance with ACI 318. Nonprestressed concrete piles are usually considered for lengths of 40 feet (12 192 mm) to 50 feet (15 240 mm), while prestressed concrete piles are usually considered for lengths of 60 to 100 feet (18 288 to 30 480 mm). The general loading range for precast concrete piles is 40 to 400 tons (355 to 3558 kN).

Advantages of using precast concrete piles include their high load capacities and corrosion resistance. Disadvantages include their vulnerability to damage associated with handling, high breakage rates particularly when spliced, and the considerable displacement of soil.

1809.2.1.1 Design and manufacture. Piles shall be designed and manufactured in accordance with accepted engineering practice to resist all stresses induced by handling, driving and service loads.

❖ Precast piles must be properly designed to resist the stresses induced both by handling and driving operations, and later imposed by service loads. Care must be given during handling and installation to minimize or avoid possible damage to the piles, such as cracking, crushing or spalling.

This section requires the minimum lateral dimension to be 8 inches (203 mm) and corners of square piles to be chamfered. Chamfered corners consist of rounding off or smoothing the corners of a square pile. The triangular portions of the corners are typically weak spots and chamfering the corners reduces the probability of cracking or breakage of the pile at these points.

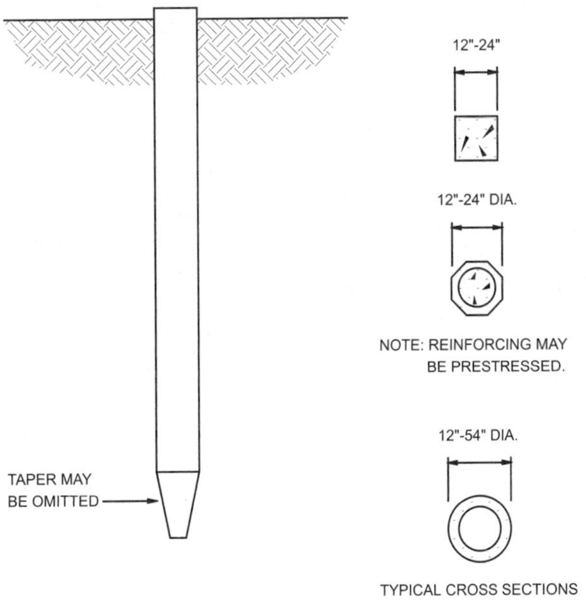

For SI: 1 inch = 25.4 mm.

Figure 1809.2.1
PRECAST NONPRESTRESSED AND PRESTRESSED CONCRETE PILES

1809.2.1.2 Minimum dimension. The minimum lateral dimension shall be 8 inches (203 mm). Corners of square piles shall be chamfered.

❖ This section provides the minimum dimension for precast concrete piles based on the size required to withstand the pile-driving operation. Chamfered corners are required to reduce concrete spalling during driving.

1809.2.1.3 Reinforcement. Longitudinal steel shall be arranged in a symmetrical pattern and be laterally tied with steel ties or wire spiral spaced not more than 4 inches (102 mm) apart, center to center, for a distance of 2 feet (610 mm) from the ends of the pile; and not more than 6 inches (152 mm) elsewhere except that at the ends of each pile, the first five ties or spirals shall be spaced 1 inch (25 mm) center to center. The gage of ties and spirals shall be as follows:

For piles having a diameter of 16 inches (406 mm) or less, wire shall not be smaller than 0.22 inch (5.6 mm) (No. 5 gage).

For piles having a diameter of more than 16 inches (406 mm) and less than 20 inches (508 mm), wire shall not be smaller than 0.238 inch (6 mm) (No. 4 gage).

For piles having a diameter of 20 inches (508 mm) and larger, wire shall not be smaller than $^1/_4$ inch (6.4 mm) round or 0.259 inch (6.6 mm) (No. 3 gage).

❖ For precast piles, the longitudinal steel must be set in a symmetrical arrangement in the pile. To resist high-impact stresses, lateral steel ties used for confining the longitudinal reinforcement are to be provided at each end of the pile for a distance of 2 feet (610 mm) or more and be closely spaced at 3 inches (76 mm) o.c., except that the first five ties from each end are to be spaced at 1-inch (25 mm) centers. Between these two closely tied ends, the longitudinal steel must be similarly tied at spacings not to exceed 6 inches (152 mm) o.c. so as to provide a reinforcing cage that will keep the pile from buckling or cracking during handling.

1809.2.1.4 Installation. Piles shall be handled and driven so as not to cause injury or overstressing, which affects durability or strength.

❖ This section requires that all precast piles be handled in such a manner and driven so as not to cause injury or overstressing, which affects pile durability and structural strength. Sometimes cracking or spalling occurs during the driving of both conventionally reinforced and prestressed concrete piles. Such damage can be classified into four basic types of failure:

1. The spalling of concrete at the butt end (head) of the pile caused by very high or irregular compressive stress concentrations may be the result of a number of causes, such as: insufficient cushioning at the head of the pile; the butt is not square or perpendicular to the longitudinal axis of the pile; the driving hammer and the pile are not aligned, causing eccentric impact forces; the reinforcing steel is not cut flush at the top of the pile, allowing the impact force of the hammer to be transmitted to the concrete through the projecting steel; insufficient spiral reinforcing at the head of the pile, resulting in spalling or splitting of the concrete; pile corners that spall on impact of the hammer because they were not chamfered (see Section 1809.2.1.2) and concrete fatigue caused by a large number of hammer blows at a very high stress level.
2. Spalling at the tip of the pile is usually caused by extremely high driving resistance, particularly when the pile point is seated on rock.
3. The breaking or transverse cracking of a pile is a complex occurrence. When a hammer strikes the cushion or head of a pile, a compressive stress is produced that travels as a wave along the length of the pile down to the tip. This wave action may be reflected from the point of the pile as either a compressive or tensile stress based on the soil resistance at the point, or the wave may simply pass into the soil. In cases where there is very little or no soil resistance, the compressive stress wave may be reflected back up the pile as a tensile stress and, depending on its magnitude, may cause damage along the pile. This phenomenon usually occurs in piles that are 50 feet (15 240 mm) or longer. When tip resistance is very low, such as in extremely soft soils, voids or in predrilled (or jetted) holes, the driving energy should be reduced to avoid high tensile stresses.
4. Spiral or transverse cracking of precast concrete piles may be caused by a combination of torsional stress (diagonal tension) and reflected tensile stress. Torsion is usually caused by excessive restraint of the pile in the leads, resulting in rotation of the pile during driving. Another reason is that the helmet or drive cap may fit too tightly on the pile, thus preventing the pile from rotating or adjusting slightly because of the soil action on the embedded section of the pile.

Good driving practices will generally eliminate the problems associated with the installation of precast concrete piles, as described. Good practice procedures are given in ACI 543R, which consists of recommendations for the design, manufacture and installation of concrete piles.

1809.2.2 Precast nonprestressed piles. Precast nonprestressed concrete piles shall conform to Sections 1809.2.2.1 through 1809.2.2.5.

❖ Precast nonprestressed (conventionally reinforced) concrete piles are manufactured from concrete and have reinforcement consisting of a steel reinforcing cage made up of several longitudinal bars or tie steel in the form of individual hoops or spirals.

1809.2.2.1 Materials. Concrete shall have a 28-day specified compressive strength (f'_c) of not less than 3,000 psi (20.68 Mpa).

❖ Concrete used for precast nonprestressed concrete piles is to have a minimum specified 28-day compressive stress (f'_c) of 3,000 psi (20.68 Mpa).

While ACI 318 specifically excludes concrete piles from its design provisions, the material, quality and mix requirements contained in this standard generally apply to the manufacture of precast concrete piles.

To provide for handling stresses, concrete and reinforcing steel are often designed to allow for an overstress up to 50 percent. Nonetheless, this section requires that the longitudinal reinforcement consist of not less than four bars and be at least 0.8 percent of the cross-sectional area of the pile, regardless that the calculated lifting stresses may sometimes indicate lesser requirements. Four bars are a practical minimum for symmetrical placement in square piles, which are the most commonly used type.

1809.2.2.2 Minimum reinforcement. The minimum amount of longitudinal reinforcement shall be 0.8 percent of the concrete section and shall consist of at least four bars.

❖ This section specifies the minimum amount of longitudinal reinforcement where seismic effects are minimal or low. See Sections 1809.2.2.2.1 and 1809.2.2.2.2 for reinforcement requirements for moderate and high seismic regions.

1809.2.2.2.1 Seismic reinforcement in Seismic Design Category C. Where a structure is assigned to Seismic Design Category C in accordance with Section 1616, the following shall apply. Longitudinal reinforcement with a minimum steel ratio of 0.01 shall be provided throughout the length of precast concrete piles. Within three pile diameters of the bottom of the pile cap, the longitudinal reinforcement shall be confined with closed ties or spirals of a minimum $^3/_8$ inch (9.5 mm) diameter. Ties or spirals shall be provided at a maximum spacing of eight times the diameter of the smallest longitudinal bar, not to exceed 6 inches (152 mm). Throughout the remainder of the pile, the closed ties or spirals shall have a maximum spacing of 16 times the smallest longitudinal-bar diameter, not to exceed 8 inches (203 mm).

❖ This section includes moderately ductile detailing requirements for precast nonprestressed piles in buildings assigned to Seismic Design Category C. The minimum longitudinal and transverse reinforcement requirements are so that piles will be able to accommodate seismically induced ground deformations that are expected to affect Seismic Design Category C buildings. The 1-percent minimum longitudinal reinforcement is a standard requirement for reinforced concrete columns. A 6-bar-diameter or 6-inch (152 mm) spacing of transverse reinforcement is a fairly common requirement to prevent buckling of longitudinal compression reinforcement. The transverse reinforcement spacing requirement of this section for the confinement region of the pile is somewhat less than that, allowing an 8-bar-diameter or 6-inch (152 mm) spacing. Outside of this region the spacing requirements for transverse reinforcing is less stringent.

1809.2.2.2.2 Seismic reinforcement in Seismic Design Category D, E or F. Where a structure is assigned to Seismic Design Category D, E or F in accordance with Section 1616, the requirements for Seismic Design Category C in Section 1809.2.2.2.1 shall apply except as modified by this section. Transverse confinement reinforcement consisting of closed ties or equivalent spirals shall be provided in accordance with Sections 21.4.4.1, 21.4.4.2 and 21.4.4.3 of ACI 318 within three pile diameters of the bottom of the pile cap. For other than Site Class E or F, or liquefiable sites and where spirals are used as the transverse reinforcement, it shall be permitted to use a volumetric ratio of spiral reinforcement of not less than one-half that required by Section 21.4.4.1(a) of ACI 318.

❖ This section requires that the provisions for Seismic Design Category C in Section 1809.2.2.1 must, as a minimum, be met. The confinement region of the pile requires the more stringent and demanding detailing provided in the referenced sections of ACI 318, which are the requirements for special moment frames. These increased transverse reinforcement requirements are intended to provide a high degree of ductility in the upper portion (confinement region) of the pile. Experience has shown that concrete piles tend to hinge or sustain damage immediately below the pile cap; therefore, tie spacing is reduced in this area to better confine the concrete.

1809.2.2.3 Allowable stresses. The allowable compressive stress in the concrete shall not exceed 33 percent of the 28-day specified compressive strength (f'_c) applied to the gross cross-sectional area of the pile. The allowable compressive stress in the reinforcing steel shall not exceed 40 percent of the yield strength of the steel (f_y) or a maximum of 30,000 psi (207 MPa). The allowable tensile stress in the reinforcing steel shall not exceed 50 percent of the yield strength of the steel (f_y) or a maximum of 24,000 psi (165 Mpa).

❖ This section specifies the allowable stresses in precast nonprestressed concrete piles. The allowable stresses are significantly less than the yield strength.

1809.2.2.4 Installation. A precast concrete pile shall not be driven before the concrete has attained a compressive strength of at least 75 percent of the 28-day specified compressive strength (f'_c), but not less than the strength sufficient to withstand handling and driving forces.

❖ This section requires that precast reinforced concrete piles must not be driven until the concrete has acquired at least three-quarters of its 28-day specified compressive strength. For example, if a pile is manufactured at the minimum compressive strength of 4,000 psi (27.6 MPa), it must not be driven until it has obtained a strength of at least 3,000 psi (20.68 MPa) ($^3/_4$ x 4,000) (27.6 MPa). In all cases, however, concrete strength must have developed to the point that it is sufficient to sustain the stresses imposed on the pile by handling and driving operations.

1809.2.2.5 Concrete cover. Reinforcement for piles that are not manufactured under plant conditions shall have a concrete cover of not less than 2 inches (51 mm).

Reinforcement for piles manufactured under plant control conditions shall have a concrete cover of not less than $1^1/_4$ inches (32 mm) for No. 5 bars and smaller, and not less than $1^1/_2$ inches (38 mm) for No. 6 through No. 11 bars except that longitudinal bars spaced less than $1^1/_2$ inches (38 mm) clear distance apart shall be considered bundled bars for which the minimum concrete cover shall be equal to that for the equivalent diameter of the bundled bars.

Reinforcement for piles exposed to seawater shall have a concrete cover of not less than 3 inches (76 mm).

❖ The minimum concrete cover is intended to provide protection to the reinforcement for piles exposed to the weather and against the effects of deleterious substances that may be present in the ground.This section permits the reinforcement of precast piles manufactured under plant control conditions to have reduced concrete cover requirements in comparison to piles produced under other precasting conditions. However, the term "manufactured under plant control conditions" should not be construed as meaning that piles must be produced in a plant. Precast piles made at the job site will also qualify under this section, provided that the control of form dimensions, placing of reinforcement, concrete quality and curing procedure are all equal to the level of control exercised in a plant.

1809.2.3 Precast prestressed piles. Precast prestressed concrete piles shall conform to the requirements of Sections 1809.2.3.1 through 1809.2.3.5.

❖ Precast prestressed concrete piles are either of the pretensioned or post-tensioned type. Pretensioned type of piles are normally cast in a plant in their full lengths, as predetermined by soils investigations and engineering analyses. The reinforcement in the concrete pile consists of tendons (stressing steel) that are tensioned before the concrete is placed. After casting and when the concrete has attained sufficient strength, the stretched tendons are released from their anchorage, thus placing the pile in continuous compression. In comparison, post-tensioned type of piles are made in a plant or on the job site, and the tendons are released after the concrete has hardened. In addition to the tendons, this type of pile also contains mild reinforcing steel to resist the handling stresses before the stressing steel is tensioned.

One of the primary advantages of prestressed over nonprestressed concrete piles is durability. Since the concrete is under continuous compression, hairline cracks are kept tightly closed and, thus, are more durable. Another advantage is that the tensile stresses that can develop in the concrete under certain driving conditions are less critical. Prestressed concrete piles are best suited for friction piles in sand, gravel and clays.

The purpose for prestressing concrete piles is explained in the commentary to Section 1809.2.3.4.

1809.2.3.1 Materials. Prestressing steel shall conform to ASTM A 416. Concrete shall have a 28-day specified compressive strength (f'_c) of not less than 5,000 psi (34.48 Mpa).

❖ Rods, strands or wires conforming to the requirements of ASTM A 416 are used as stressing steel in the manufacture of precast, prestressed concrete piles of both the pretensioned and the post-tensioned type. The most commonly used stressing steel is the seven wire, uncoated, stress relieved strand.

This section requires that concrete used in the manufacture of prestressed piles must have a 28-day specified compressive strength of no less than 5,000 psi (34.48 MPa). The use of higher strength concrete results in smaller volumetric changes with corresponding reductions in prestress losses and higher effective prestress values.

1809.2.3.2 Design. Precast prestressed piles shall be designed to resist stresses induced by handling and driving as well as by loads. The effective prestress in the pile shall not be less than 400 psi (2.76 MPa) for piles up to 30 feet (9144 mm) in length, 550 psi (3.79 MPa) for piles up to 50 feet (15 240 mm) in length and 700 psi (4.83 MPa) for piles greater than 50 feet (15 240 mm) in length.

Effective prestress shall be based on an assumed loss of 30,000 psi (207 MPa) in the prestressing steel. The tensile stress in the prestressing steel shall not exceed the values specified in ACI 318.

❖ For prestressed concrete piles, the effective prestress (the stress remaining in the pile after all losses have occurred—excluding the effect of superimposed loads and the weight of the pile) is not to be less than the specified values for the various lengths. Experience has shown that an effective prestress less than the minimum value prescribed herein is sometimes inadequate in preventing or controlling cracking of the concrete during the handling and installation operations. This section also requires the effective prestress to be based on an assumed loss of 30,000 psi (207 MPa) in the prestressing steel and the tensile stresses in the steel not to exceed the values set forth in ACI 318.

1809.2.3.2.1 Design in Seismic Design Category C. Where a structure is assigned to Seismic Design Category C in accordance with Section 1616, the following shall apply. The minimum volumetric ratio of spiral reinforcement shall not be less than 0.007 or the amount required by the following formula for the upper 20 feet (6096 mm) of the pile.

$$\rho_s = 0.12 f'_c / f_{yh} \qquad \textbf{(Equation 18-4)}$$

where:

f'_c = Specified compressive strength of concrete, psi (MPa)

f_{yh} = Yield strength of spiral reinforcement ≤ 85,000 psi (586 MPa).

ρ_s = Spiral reinforcement index (vol. spiral/vol. core).

At least one-half the volumetric ratio required by Equation 18-4 shall be provided below the upper 20 feet (6096 mm) of the pile.

The pile cap connection by means of dowels as indicated in Section 1808.2.23.1 is permitted. Pile cap connection by means of developing pile reinforcing strand is permitted provided that the pile reinforcing strand results in a ductile connection.

❖ This section gives requirements for precast prestressed piles supporting structures assigned to Seismic Design Category C. The minimum spiral reinforcement requirement that results in ductile prestressed concrete piles is based on PCI's "Recommended Practice for Design, Manufacture and Installation of Prestressed Concrete Piling." This document incorporates work done on piles in the United States and New Zealand.

These piles exhibit larger curvatures in the top 20 feet (6096 mm). This section requires (confinement) spiral reinforcement as determined by Equation 18-4. Based on ACI 318, this formula is deemed to provide for moderate ductility. In the lower portion of the pile, the required reinforcing is reduced by one-half.

1809.2.3.2.2 Design in Seismic Design Category D, E or F. Where a structure is assigned to Seismic Design Category D, E or F in accordance with Section 1616, the requirements for Seismic Design Category C in Section 1809.2.3.2.1 shall be met, in addition to the following:

1. Requirements in ACI 318, Chapter 21, need not apply, unless specifically referenced.
2. Where the total pile length in the soil is 35 feet (10 668 mm) or less, the lateral transverse reinforcement in the ductile region shall occur through the length of the pile. Where the pile length exceeds 35 feet (10 668 mm), the ductile pile region shall be taken as the greater of 35 feet (10 668 mm) or the distance from the underside of the pile cap to the point of zero curvature plus three times the least pile dimension.
3. In the ductile region, the center-to-center spacing of the spirals or hoop reinforcement shall not exceed one-fifth of the least pile dimension, six times the diameter of the longtitudinal strand, or 8 inches (203 mm), whichever is smaller.
4. Circular spiral reinforcement shall be spliced by lapping one full turn and bending the end of the spiral to a 90-degree hook or by use of a mechanical or welded splice complying with Sec. 12.14.3 of ACI 318.
5. Where the transverse reinforcement consists of circular spirals, the volumetric ratio of spiral transverse reinforcement in the ductile region shall comply with the following:

$$\rho_s = 0.25(f'_c/f_{yh})(A_g/A_{ch} - 1.0)[0.5 + 1.4P/(f'_c A_g)]$$

(Equation 18-5)

but not less than:

$$\rho_s = 0.12(f'_c/f_{yh})[0.5 + 1.4P/(f'_c A_g)]$$

(Equation 18-6)

and need not exceed:

$$\rho_s = 0.021$$

(Equation 18-7)

where:

A_g = Pile cross-sectional area, square inches (mm²).

A_{ch} = Core area defined by spiral outside diameter, square inches (mm²).

f'_c = Specified compressive strength of concrete, psi (MPa)

f_{yh} = Yield strength of spiral reinforcement ≤ 85,000 psi (586 MPa).

P = Axial load on pile, pounds (kN), as determined from Equations 16-5 and 16-6.

ρ_s = Volumetric ratio (vol. spiral/ vol. core).

This required amount of spiral reinforcement is permitted to be obtained by providing an inner and outer spiral.

6. When transverse reinforcement consists of rectangular hoops and cross ties, the total cross-sectional area of lateral transverse reinforcement in the ductile region with spacings, and perpendicular to dimension, h_c, shall conform to:

$$A_{sh} = 0.3sh_c(f'_c/f_{yh})(A_g/A_{ch} - 1.0)[0.5 + 1.4P/(f'_c A_g)]$$

(Equation 18-8)

but not less than:

$$A_{sh} = 0.12sh_c(f'_c/f_{yh})[0.5 + 1.4P/(f'_c A_g)]$$

(Equation 18-9)

where:

f_{yh} = ≤ 70,000 psi (483 MPa).

h_c = Cross-sectional dimension of pile core measured center to center of hoop reinforcement, inch (mm).

s = Spacing of transverse reinforcement measured along length of pile, inch (mm).

A_{sh} = Cross-sectional area of tranverse reinforcement, square inches (mm²)

f'_c = Specified compressive strength of concrete, psi (MPa)

The hoops and cross ties shall be equivalent to deformed bars not less than No. 3 in size. Rectangular hoop ends shall terminate at a corner with seismic hooks.

Outside of the length of the pile requiring transverse confinement reinforcing, the spiral or hoop reinforcing with a volumetric ratio not less than one-half of that required for transverse confinement reinforcing shall be provided.

❖ The provisions for Seismic Design Category C must be followed in addition to these requirements. This section clarifies that Chapter 21 of ACI 318 does not generally apply to precast prestressed concrete piles unless it is explicitly referenced. This chapter was never intended for such piles.

The provisions found in this section are based on PCI's "Recommended Practice for Design, Manufacture and Installation of Prestressed Concrete Piling." This document incorporates work done on piles in the United States and New Zealand.

The maximum volumetric ratio set forth in Equation 18-7 is based on pile testing that has shown the 0.021 maximum is sufficient for the smaller square precast prestressed concrete piles to perform in a ductile manner.

1809.2.3.3 Allowable stresses. The maximum allowable design compressive stress, f'_c, in concrete shall be determined as follows:

$$f_c = 0.33 f'_c - 0.27 f_{pc} \qquad \textbf{(Equation 18-10)}$$

where:

f'_c = The 28-day specified compressive strength of the concrete.

f_{pc} = The effective prestress stress on the gross section.

❖ The maximum allowable design compressive stress (f_c) in concrete is to be determined by the following formula:

$f_c = 0.33\ f'_c - 0.27 f_{pc}$

The term "0.33 f_c" is the same as the allowable compressive stress used for conventionally reinforced concrete piles (see Section 1809.2.2.3). Since prestressing places the concrete in compression, the concrete compressive stress thus induced and designated by the term "0.27f_{pc}" must be subtracted from the gross allowable stress in the pile (0.33f'_c), with the remainder (f'_c) representing the net amount of compressive stress value available for the service loads on the pile. The notation f_{pc} is the effective prestress value on the gross area of the pile section, which is the stress remaining in the pile after all losses have occurred, excluding the effect of the service loads and the weight of the piles.

The initial prestressing force can be measured with some accuracy during and immediately after the tensioning operations. The final prestress varies with time, because of losses caused by the elastic shortening of concrete, creep, the shrinkage of concrete, relaxation of the steel stress and frictional losses caused by curvature in the stressing tendons. Such losses in prestress are included in the notation 0.27 f_{pc} in the formula (equation) first stated above. Furthermore, the allowable design load (P_a) can be expressed by the following equation:

$P_a = (0.33\ f'_c - 0.27\ f_{pc})\ A_c = f_c A_c$

where A_c is the area of the concrete section, including the prestressing steel.

1809.2.3.4 Installation. A prestressed pile shall not be driven before the concrete has attained a compressive strength of at least 75 percent of the 28-day specified compressive strength (f'_c), but not less than the strength sufficient to withstand handling and driving forces.

❖ The requirements for the installation of prestressed concrete piles are the same as the installation provisions for conventionally reinforced precast concrete piles (see commentary, Section 1809.2.2.4).

The reason for prestressing concrete piles is to place the concrete under continuous compression so that any hairline cracks that may develop will be kept tightly closed and to prevent possible injury to the pile from tensile stresses that may occur during installation operations. The handling and driving of prestressed piles do not require the same degree of care as needed to install conventionally reinforced concrete piles. As a general rule, prestressed piles are more durable than precast reinforced piles.

1809.2.3.5 Concrete cover. Prestressing steel and pile reinforcement shall have a concrete cover of not less than $1^1/_4$ inches (32 mm) for square piles of 12 inches (305 mm) or smaller size and $1^1/_2$ inches (38 mm) for larger piles, except that for piles exposed to seawater, the minimum protective concrete cover shall not be less than $2^1/_2$ inches (64 mm).

❖ The minimum concrete cover is intended to provide protection to the reinforcement for piles exposed to the weather and against the effects of deleterious substances that may be present in the ground. The concrete cover requirement for piles exposed to seawater is increased to $2^1/_2$ inches (64 mm) because of the highly corrosive effects of salt.

1809.3 Structural steel piles. Structural steel piles shall conform to the requirements of Sections 1809.3.1 through 1809.3.5.

❖ The piles in this category are made of load-bearing steel materials that can develop high load capacities, and include H-piles and open-ended steel pipe piles. In addition to stating the applicable material specifications for the manufacturing of steel piles, this section also prescribes the allowable design stresses and dimensional requirements (see Figure 1809.3).

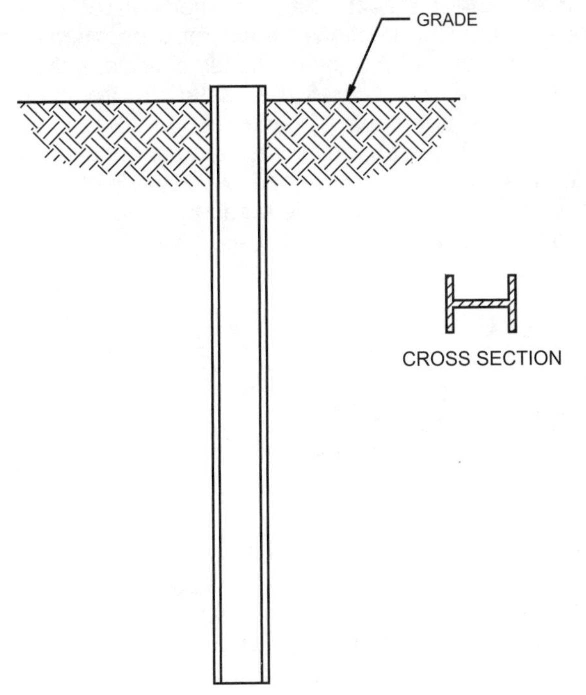

Figure 1809.3
STEEL H-PILE SECTIONS

1809.3.1 Materials. Structural steel piles, steel pipe and fully welded steel piles fabricated from plates shall conform to ASTM A 36, ASTM A 252, ASTM A 283, ASTM A 572, ASTM A 588 or ASTM A 913.

❖ The materials used in the manufacture of steel piles must comply with the requirements of ASTM A 36, A 252, A 283, A 572, A 588 or A 913.

While H-piles are ordinarily made with steel materials conforming to ASTM A 36 or ASTM A 572 requirements, the employment of other standards is not excluded, provided it can be shown that such steels meet the applicable chemical and mechanical properties of one of the listed specifications to establish the suitability of the material for pile use.

Welded or seamless pipe piles must conform to ASTM A 252 requirements.

1809.3.2 Allowable stresses. The allowable axial stresses shall not exceed 35 percent of the minimum specified yield strength (F_y).

Exception: Where justified in accordance with Section 1808.2.10, the allowable axial stress is permitted to be increased above $0.35F_y$, but shall not exceed $0.5F_y$.

❖ In the elastic design of steel foundation piles, the compressive stresses are limited to a percentage of the steel yield strength (F_y). For structural steel piles, the maximum allowable design stresses are not to exceed 0.35 F_y (35 percent of the yield strength). Allowable design stresses may, however, be increased to 0.50 F_y on the basis that justification for such an increase is provided by both approved analytical methods and load tests conducted in accordance with the requirements of Section 1808.2.10.

H-piles consist of rolled wide-flange steel H-beams. They are manufactured in standard sizes with nominal beam depths in the range of 8 inches (203 mm) to 14 inches (355 mm).

H-piles are usually employed as deep end-bearing piles because of their high load-carrying capacity and, with few exceptions, they are nondisplacement piles that can readily penetrate soil strata to reach rock or other suitable point-supporting subsurface materials. Ideally, steel H-piles are driven to hard or medium-hard rock. Many load tests have shown that, for H-piles driven to refusal on rock, the stresses at failure closely equate to the yield strength of the material. Under such conditions, consideration should be given to designing H-pile foundations at a high load capacity per pile based on design stress levels up to 0.50 F_y and using pile-tip reinforcement.

High load-carrying capacity can also be obtained by H-piles driven to practical refusal in load-bearing strata of soft rock, hardpan overlying a rock layer or in very dense granular soils. Unlike piles seated on hard or medium-hard rock, however, the ability of these strata to support pile loads up to the yield point of the steel may be questionable. A thorough analytical study combined with an extensive load test program is required before considering using design stresses up to 0.50 F_y.

Experience indicates that corrosion is not a practical problem for steel piles driven in natural soil, due primarily to the absence of oxygen in the soil, however, in fill ground at or above the ground-water table, moderate corrosion may occur and protection may be needed. Commonly used protection methods include coatings applied before driving.

H-piles can also be used as friction piles. Because of the many variables involved in the design of this type of pile, as compared to piles driven to end-bearing on rock, however, it is not possible to produce conclusive information on maximum loads and stresses. It is to be recognized that the soil, rather than the structural strength of the H-pile, will be the controlling factor in the design. Additionally, the design of friction piles is directly influenced by such considerations as group action, driveability and settlement limitations, as opposed to piles driven to end-bearing rock.

As required by this section, an increase in the allowable design stresses from 0.35 F_y up to 0.50 F_y, representing a maximum step-up of 43 percent in the design load, needs to be justified by approved analytical methods as well as by the results of load testing, particularly where it involves steel pile installations in subsurface load-bearing materials other than hard or medium-hard rock. When possible, building officials should seek the recommendations of registered engineers who are experienced in soil mechanics and in the design of deep foundations before authorizing increases in allowable compressive stresses.

Steel pipe piles are made of seamless, welded or spiral-welded pipes and are most frequently filled with concrete. Pipe piles conforming to ASTM A 252 are used as both friction piles and end-bearing installations. Piles may be driven open ended or closed ended. Open-ended piles are generally specified where the soil investigation indicates that rock load-bearing strata are found at levels reasonably close to the ground surface, especially where the loads to be supported are great. The earth forced into the interior core of the pipe pile by driving operations is cleaned out by a blowing technique or auguring, depending on the type of soil encountered. The pile is then filled with concrete.

The use of closed-ended piles is indicated when soil investigations show the absence of rock formations or when rock is found at excessive depths. Under such conditions, closed-ended piles are driven as friction piles.

The yield strengths *(F_y)* of welded or seamless pipe piles specified in ASTM A 252 and upon which the allowable compressive stresses are based are as follows: For Grade 1 steel, the yield strength is 30,000 psi (207 MPa); for Grade 2, 35,000 psi (241 MPa) and for Grade 3, 45,000 psi (310 MPa). Grade 1 steel is seldom used for pipe pile fabrication.

1809.3.3 Dimensions of H-piles. Sections of H-piles shall comply with the following:

1. The flange projections shall not exceed 14 times the minimum thickness of metal in either the flange or the web and the flange widths shall not be less than 80 percent of the depth of the section.
2. The nominal depth in the direction of the web shall not be less than 8 inches (203 mm).
3. Flanges and web shall have a minimum nominal thickness of $^3/_8$ inch (9.5 mm).

❖ Dimensional requirements for manufacturing structural steel H-piles are provided in this section. H-piles are proportioned to withstand the large impact stresses imposed on piles during hard driving. The thicknesses of the flanges and the web of a rolled-steel H-pile section are made equal in order to avoid damage that could occur during hard driving if the piles were proportioned with a mixture of thick and thinner parts. Flange widths are proportioned in relation to the depth of the pile section to provide rigidity in the weak axis.

It would seem unnecessary to repeat the dimensioning requirements for H-piles in the code, since they were originally created by the steel industry and have for decades been the industry standard used in the manufacture of hot-rolled steel shapes. The main purpose is to provide the dimensional basis for the fabrication of similar pile products made principally of welded steel plates and other necessary steel parts.

While it is the general preference and practice to use rolled-steel piles, occasionally it becomes necessary to fabricate special pile sections because of: time problems imposed by mill scheduling and delivery, a special need for heavier H-pile sections than are customarily available, an immediate need for replacement piles or for any other reason. The dimensional requirements contained in this section regulate the fabrication of such special piles and provide the building official with a basis for approval. Fabricated pile materials are to comply with the requirements of Section 1809.3.1.

This section requires that the flange projection not exceed 14 times the minimum thickness of metal in either the flange or the web. The measurement of the flange projection is shown in Figure 1809.3 and can be calculated by the indicated formula:

$$P = \frac{W - t_w}{2} \text{ or } \frac{W - t_f}{2}$$

where W is the flange width, t_w is the web thickness, t_f is the flange thickness and P is the flange projection.

For example, the dimensions of an HP 14 × 73 pile section are as follows:

The width of flange, W = 14.585 inches.
The thickness of the flange, t_f = 0.505 inch.
The thickness of the web, t_w = .505 inch.
Therefore, the actual flange projection is:

$$P = \frac{14.585 - 0.505}{2} = 7.04 \text{ inches (179 mm)}$$

The maximum allowable flange projection is 14 × t_f or 14 × t_w (whichever is smaller).

Check: 14 × 0.505 = 7.07 inches (179.5 mm), which is greater than the actual flange projection of 7.04 inches (178.8 mm); therefore, an HP 14 73 pile is acceptable.

This section also specifies that flange widths are not to be less than 80 percent of the depth of the pile section. In actuality, practically all H-piles are manufactured so that their flanges are slightly greater in width (by fractions of an inch) than the depths of the sections. In fact, piles are squared so that their flange widths are about equal to their depth.

Flange and web thickness is not to be less than $^3/_8$ inch (9.5 mm). In H-piles, the flange thicknesses are equal to the web thickness.

1809.3.4 Dimensions of steel pipe piles. Steel pipe piles driven open ended shall have a nominal outside diameter of not less than 8 inches (203 mm). The pipe shall have a minimum of 0.34 square inch (219 mm^2) of steel in cross section to resist each 1,000 foot-pounds (1356 N×m) of pile hammer energy or the equivalent strength for steels having a yield strength greater than 35,000 psi (241 MPa). Where pipe wall thickness less than 0.188 inch (4.8 mm) is driven open ended, a suitable cutting shoe shall be provided.

❖ Because of their uniform cross section from butt to tip, steel pipe piles provide unvarying resistance to bending and lateral forces applied in any direction.

A pipe to be driven open ended must have an outside diameter of no less than 8 inches (204 mm). Smaller diameter pipes could allow soil to bridge the sides and cause plugging at their tips.

SECTION 1810
CAST-IN-PLACE CONCRETE PILE FOUNDATIONS

1810.1 General. The materials, reinforcement and installation of cast-in-place concrete piles shall conform to Sections 1810.1.1 through 1810.1.3.

❖ Section 1810 addresses materials, design and installation of cast-in-place concrete piles. Cast-in-place concrete piles are installed by placing concrete into preformed holes in the ground extending to required load-bearing depths. Since this type of pile is not subjected to driving stresses, only the stresses imposed by the service loads are considered in the design. The predetermination of pile lengths is not as important as it is with manufactured piles since adjustments can readily be made in the field to accommodate subsurface conditions. Reinforcement, if required, is installed prior to or during concrete placement.

Cast-in-place concrete piles basically consist of two types: uncased and cased. The design and construction of the various types of uncased piles are covered in Sections 1810.2, 1810.3 and 1810.4. The cased type of pile is covered in Section 1810.5.

Section 1810.6 addresses concrete-filled steel pipe and tube piles. Section 1810.7 addresses caisson piles.

Concrete materials, the proportioning and quality of concrete mixtures and the reinforcement used for cast-in-place piles are to comply with Sections 1810.1.1 and 1810.1.2, and specifically with the applicable provisions of Chapter 19. General installation requirements applicable to all types of cast-in-place concrete pile construction covered by Section 1810 are to comply with Section 1810.1.3.

1810.1.1 Materials. Concrete shall have a 28-day specified compressive strength (f'_c) of not less than 2,500 psi (17.24 MPa). Where concrete is placed through a funnel hopper at the top of the pile, the concrete mix shall be designed and proportioned so as to produce a cohesive workable mix having a slump of not less than 4 inches (102 mm) and not more than 6 inches (152 mm). Where concrete is to be pumped, the mix design including slump shall be adjusted to produce a pumpable concrete.

❖ Concrete used for cast-in-place piles is to have a minimum 28-day compressive strength of 2,500 psi (17.24 MPa). Higher concrete compressive strengths are usually specified to increase pile load capacity as well as for economic reasons.

This section requires that concrete be designed and proportioned to produce a cohesive workable mix with a consistency (slump) ranging between 4 inches (102 mm) and 6 inches (152 mm). This provision is within the bounds required for the proper placement of conventional structural concrete for cased cast-in-place piles. The slump requirements stated here, however, should not be considered so absolute that, with the use of engineering judgement, the consistency of the concrete cannot be adjusted to satisfy prevailing conditions of construction. For example, concrete placement conducted under difficult conditions—such as for piles containing heavy reinforcement, for very long cased piles, for pile shells driven on steep batters or for other demanding situations—may require the employment of special concrete mixes using reduced quantities of coarse aggregates with corresponding increases in sand and cement content. Under such conditions, slumps of 4 inches (102 mm) with a tolerance of plus 2 inches (51 mm) or minus 1 inch (25 mm) are in order.

The consistency of concrete required for uncased cast-in-place concrete piles may be altogether different than the slumps specified and used for cased construction. For example, concrete placed in drilled holes should have a slump of at least 6 inches (152 mm) so that the concrete flows properly and full bond is attained with the reinforcement.

Furthermore, high slump concrete is used so that the complete volume of the hole is filled, including natural soil crevices and pockets and surface irregularities caused by the drilling operations. High slumps are also used where the concrete must displace drilling slurry.

Concrete slumps as high as 8 inches (203 mm) are sometimes needed where temporary casings are used to facilitate pile construction and are subsequently extracted as the concrete is placed.

Concrete is usually placed through a funnel hopper at the top of the pile opening, but other approved methods, such as pumping or tremie, may also be used.

1810.1.2 Reinforcement. Except for steel dowels embedded 5 feet (1524 mm) or less in the pile and as provided in Section 1810.3.4, reinforcement where required shall be assembled and tied together and shall be placed in the pile as a unit before the reinforced portion of the pile is filled with concrete except in augered uncased cast-in-place piles. Tied reinforcement in augered uncased cast-in-place piles shall be placed after piles are concreted, while the concrete is still in a semifluid state.

❖ The typical procedure for constructing cast-in-place concrete piles is to tie the head of the pile with the pile cap, inserting dowels in the freshly placed concrete pile to obtain an embedment of about 5 feet (1524 mm). Main reinforcement consisting of a cage of longitudinal deformed reinforcing bars tied together with individual steel hoops or a spiral must be placed and securely held in position in the pile opening (cased or uncased) prior to the placement of concrete. This method, however, is not applicable to auger-injected concrete piles (see commentary, Section 1810.3.4).

Reinforcement required for resisting tensile stresses imposed by uplift forces may consist of a single bar or other structural steel unit or a cluster of bars placed at the center for the full length of the pile.

1810.1.2.1 Reinforcement in Seismic Design Category C. Where a structure is assigned to Seismic Design Category C in accordance with Section 1616, the following shall apply. A minimum longitudinal reinforcement ratio of 0.0025 shall be provided for uncased cast-in-place concrete drilled or augered piles, piers or caissons in the top one-third of the pile length, a minimum length of 10 feet (3048 mm) below the ground or that required by analysis, whichever length is greatest. The minimum reinforcement ratio, but no less than that ratio required by rational analysis, shall be continued throughout the flexural length of the pile. There shall be a minimum of four longitudinal bars with closed ties (or equivalent spirals) of a minimum $^3/_8$ inch (9 mm) diameter provided at 16-longitudinal-bar diameter maximum spacing. Transverse confinement reinforcing with a maximum spacing of 6 inches (152 mm) or 8-longitudinal-bar diameters, whichever is less, shall be provided within a distance equal to three times the least pile dimension of the bottom of the pile cap.

❖ Longitudinal and transverse reinforcement requirements prescribed by this section result in moderate ductility in buildings assigned to Seismic Design Category C to withstand seismically induced ground deformations that they can encounter. Separate transverse reinforcement is specified for the portions of the pile within and outside of the potential plastic hinge zone.

The purpose of this section is to include pile bending, which is a result of ground horizontal movement during an earthquake, in the structural design. The reinforcement in the pile, required to resist the tension caused by pile bending, increases the ductility of the foundation such that bending or shear failure is precluded.

1810.1.2.2 Reinforcement in Seismic Design Category D, E or F. Where a structure is assigned to Seismic Design Category D, E or F in accordance with Section 1616, the requirements for Seismic Design Category C given above shall be met, in addition to the following. A minimum longitudinal reinforcement ratio of 0.005 shall be provided for uncased cast-in-place drilled or augered concrete piles, piers or caissons in the top one-half of the pile length, a minimum length of 10 feet (3048 mm) below ground or throughout the flexural length of the pile, whichever length is greatest. The flexural length shall be taken as the length of the pile to a point where the concrete section cracking moment strength multiplied by 0.4 exceeds the required moment strength at that point. There shall be a minimum of four longitudinal bars with transverse confinement reinforcing provided in the pile in accordance with Sections 21.4.4.1, 21.4.4.2 and 21.4.4.3 of ACI 318 within three times the least pile dimension of the bottom of the pile cap. It shall be permitted to use a transverse spiral reinforcing ratio of not less than one-half of that required in Section 21.4.4.1(a) of ACI 318 for other than Class E, F or liquefiable sites. Tie spacing throughout the remainder of the concrete section shall not exceed 12-longitudinal-bar diameters, one-half the least dimension of the section, nor 12 inches (305 mm). Ties shall be a minimum of No. 3 bars for piles with a least dimension up to 20 inches (508 mm), and No. 4 bars for larger piles.

❖ The requirements of Section 1810.1.2.1 apply in addition to the requirements of this section. This section is similar in intent, with increased minimum reinforcement requirements for ductility during earthquake ground motion that may be experienced by buildings assigned to Seismic Design Category D and higher. Separate transverse reinforcement requirements are given for portions of a pile within and beyond the potential plastic hinge zone. The specified confinement reinforcing is similar to ACI 318 requirements for special moment frames, except that a reduction is permitted for competent soils in recognition of the confinement attributed to those soils. Note that flexural length is redefined for purposes of this section.

1810.1.3 Concrete placement. Concrete shall be placed in such a manner as to ensure the exclusion of any foreign matter and to secure a full-sized shaft. Concrete shall not be placed through water except where a tremie or other approved method is used. When depositing concrete from the top of the pile, the concrete shall not be chuted directly into the pile but shall be poured in a rapid and continuous operation through a funnel hopper centered at the top of the pile.

❖ The pile opening must be free of all foreign matter prior to placing concrete. If there is excessive water in the hole [about 4 inches (102 mm) is permitted], the concrete may be placed by tremie or pumping methods. Concrete is usually deposited from the top through a funnel hopper, which should have a discharge opening that is sized smaller than the smallest pile cross section in the foundation system. Vibrators should not be used in the concreting operations of cast-in-place, uncased concrete piles because they may damage the earth walls and cause contamination of the concrete. Improper vibration may also cause excessive bleeding.

Other installation problems associated with cast-in-place concrete piles are discussed in the commentary to Section 1810.3.3.

1810.2 Enlarged base piles. Enlarged base piles shall conform to the requirements of Sections 1810.2.1 through 1810.2.5.

❖ Enlarged base cast-in-place concrete piles are of two types. The first type is the compacted concrete base pile, which consists of a bulb-shaped footing formed after driving a steel casing to its final depth. This type of pile is also known as a pressure-injected footing. The second type of enlarged base pile consists of a truncated cone or pyramid-shaped precast concrete tip larger than the casing diameter, which is driven into the soil together with the steel casing. For more comprehensive descriptions of the two types of enlarged base piles, see the commentary to Section 1810.2.3.

Enlarged base piles are a displacement type of piles that are installed only in granular soils, since the voids between the particles of soil allow densification (compression) of the material surrounding the pile without creating excessive pressures that could damage adjacent piles by causing lateral movements or heaving.

Enlarged base piles generally function as end-bearing piles, deriving their bearing capacity from the strength of the granular materials upon which they are seated. See Figure 1810.2 for illustrations of typical enlarged base cast-in-place concrete piles.

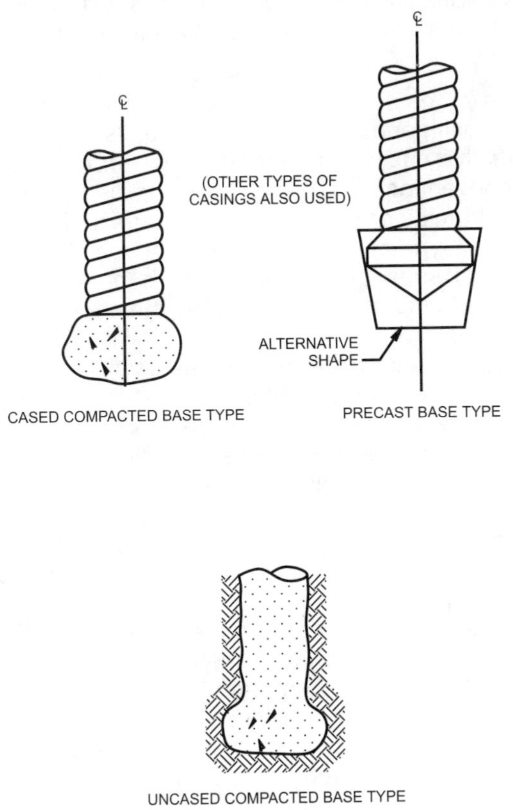

Figure 1810.2
ENLARGED BASE PILES—UNCASED OR CASED SHAFTS

1810.2.1 Materials. The maximum size for coarse aggregate for concrete shall be $^3/_4$ inch (19.1 mm). Concrete to be compacted shall have a zero slump.

❖ Concrete materials used for enlarged base piles are to comply with the applicable requirements of Chapter 19. Coarse aggregate materials used in the concrete mix are not to exceed $^3/_4$ inch (19.1 mm) in size. Concrete used for the bulb type of pile described in Section 1810.2 is to have a zero slump in order to provide a stiff mix capable of being compacted by a heavy drop weight.

1810.2.2 Allowable stresses. The maximum allowable design compressive stress for concrete not placed in a permanent steel casing shall be 25 percent of the 28-day specified compressive strength (f'_c). Where the concrete is placed in a permanent steel casing, the maximum allowable concrete stress shall be 33 percent of the 28-day specified compressive strength (f'_c).

❖ For uncased cast-in-place concrete piles formed by driving temporary steel casings, the allowable design stress in the concrete is not to exceed 0.25 f'_c (25 percent of the 28-day specified compressive strength). Pile capacity based on concrete strength alone (without consideration for soil capacity) must be calculated on the basis of the internal cross-sectional area of the pile casing (see commentary, Section 1810.4.1).

For cased piles where a steel casing is left permanently in place, the allowable design stress in the concrete is not to exceed 0.33 f'_c (33 percent of the 28-day specified compressive strength).

1810.2.3 Installation. Enlarged bases formed either by compacting concrete or driving a precast base shall be formed in or driven into granular soils. Piles shall be constructed in the same manner as successful prototype test piles driven for the project. Pile shafts extending through peat or other organic soil shall be encased in a permanent steel casing. Where a cased shaft is used, the shaft shall be adequately reinforced to resist column action or the annular space around the pile shaft shall be filled sufficiently to reestablish lateral support by the soil. Where pile heave occurs, the pile shall be replaced unless it is demonstrated that the pile is undamaged and capable of carrying twice its design load.

❖ The pressure-injected footing type of pile is normally installed by first driving a steel casing into the ground. This is done by means of a drop weight (hammer) operating inside the casing and impacting against a concrete or gravel plug confined in the tip of the steel casing. Next, after the required depth has been reached, the plug is driven out of the casing and an enlarged base is formed in the granular bearing soil by progressively adding and driving out small batches of zero-slump concrete by the operation of the drop weight. Caged reinforcement, when required, is inserted in the casing and joined to the footing by placing and compacting (by drop weight) a small amount of concrete. In the final step of the installation procedure, the uncased pile shaft is formed by placing zero-slump concrete in small lifts and compacting with the drop weight as the casing is withdrawn. Where caged reinforcement is used, the drop weight operates inside the cage. An illustration of the pressure-injected footing method of installation is shown in Figure 1810.2.3.

The shaft of this type of pile may also be constructed by placing high-slump concrete into the core as the drive casing is withdrawn.

A permanent cased shaft pile may be formed by inserting a steel shell in the drive casing after forming the footing, as described above, then filling the steel shell with conventional concrete and removing the drive casing.

The second type of enlarged base pile generally described in the introductory remarks of Section 1810.4 is installed with an enlarged precast concrete base at the end of a mandrel-driven steel casing. A major problem associated with this type of pile installation is that in driving the permanent casing into the ground, the precast base creates a hole that is larger than the diameter of the pile shaft, leaving an open space around the pile and thus losing the lateral support usually provided by the surrounding soil. The same problem also applies to the pressure-injected footing type of pile where a permanent shell is used to form the pile shaft (as described in the previous paragraph), creating an open space around the shaft after the drive casing is withdrawn. In such cases, the annular space around the pile shaft must be filled to provide the necessary lateral support.

The customary practice is to fill the annular space by pumping grout or washing in granular material.

The spacings of enlarged base piles are normally greater than most other types of conventional piles in order to avoid base interferences and the close overlapping of soil-bearing areas that would serve to reduce pile capacity.

1810.2.4 Load-bearing capacity. Pile load-bearing capacity shall be verified by load tests in accordance with Section 1808.2.8.3.

❖ Due to irregularities in pile shapes, specifically as they relate to pressure injected footing types of piles, the bearing capacity must be ascertained by means of load tests performed in accordance with the requirements of Section 1808.2.8.3.

1810.2.5 Concrete cover. The minimum concrete cover shall be $2^1/_2$ inches (64 mm) for uncased shafts and 1 inch (25 mm) for cased shafts.

❖ In lined shafts, the clearances required for the installation of reinforcement in the pile casings can be closely controlled with spacers; therefore, this section permits the minimum concrete cover dimension to be as small as 1 inch (25 mm). Due to surface irregularities in the walls of unlined piles, however, greater dimensional tolerance is needed to install the reinforcement; therefore, the minimum cover requirement is set at $2^1/_2$ inches (63 mm).

1810.3 Drilled or augered uncased piles. Drilled or augered uncased piles shall conform to Sections 1810.3.1 through 1810.3.5.

❖ Holes for the drilled type of piles are formed by machine drilling with augers or bucket-type drills (with or without protective casings) and then filling them with concrete by conventional methods, which include funnel hoppers, tremies or pumping. In comparison, holes for the augered type of piles are formed by a hollow-stem auger, and subsequently, concrete is pressure injected into the holes through the hollow stem as the auger is withdrawn.

The difference in allowable design stresses between the drilled and augured type of cast-in-place concrete pile is based on the problems associated with the installation of these piles. Because the difficulties of installing augured piles are much greater in comparison to the drilled type of piles, the code requires a larger factor of safety. This is reflected in the lower value of the allowable design stress for augered piles (see Section 1810.3.1).

There are many problems associated with drilled hole piles that are discussed in the commentary to Section 1810.3.3; however, there are several advantages to using this type of pile, which include economy complete nondisplacement, minimal driving vibration, high skin friction, good contact on rock for end bearing and no splicing is required.

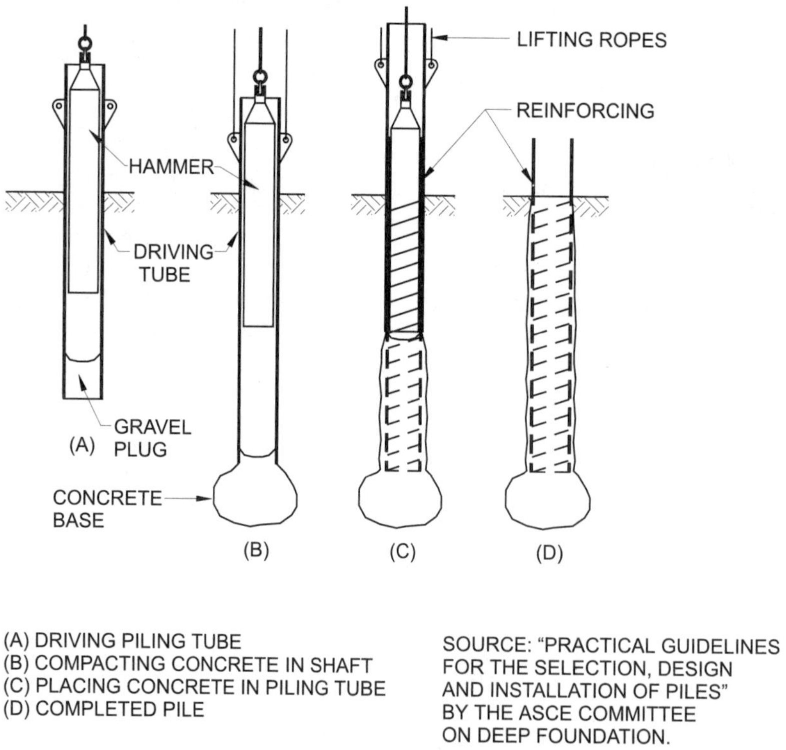

Figure 1810.2.3
PRESSURE-INJECTED FOOTING

1810.3.1 Allowable stresses. The allowable design stress in the concrete of drilled uncased piles shall not exceed 33 percent of the 28-day specified compressive strength (f'_c). The allowable design stress in the concrete of augered cast-in-place piles shall not exceed 25 percent of the 28-day specified compressive strength (f'_c). The allowable compressive stress of reinforcement shall not exceed 34 percent of the yield strength of the steel or 25,500 psi (175.8 Mpa).

❖ The allowable design stresses for drilled uncased piles are not to exceed 0.33 f'_c (33 percent of the 28-day specified concrete compressive strength). These permissible design stresses relate to cast-in-place concrete piles in which holes are formed by machine drilling with augers or bucket-type drills, with or without the use of temporary protective casings. Concrete is placed by conventional methods, which include funnel hoppers, tremies or pumping.

The allowable design stresses for augured cast-in-place piles are not to exceed 0.25 f'_c (25 percent of the 28-day specified concrete compressive strength). For this type of installation, pile holes are formed and the concrete is injected through the hollow-stem auger as the auger is withdrawn.

The problems associated with the installation of augered piles as compared to drilled piles are greater; therefore, a larger factor of safety is needed. This is reflected in the lower design stresses allowed for augered piles.

1810.3.2 Dimensions. The pile length shall not exceed 30 times the average diameter. The minimum diameter shall be 12 inches (305 mm).

Exception: The length of the pile is permitted to exceed 30 times the diameter, provided that the design and installation of the pile foundation are under the direct supervision of a registered design professional knowledgeable in the field of soil mechanics and pile foundations. The registered design professional shall certify to the building official that the piles were installed in compliance with the approved construction documents.

❖ The dimensional relationship between the diameter and length of a pile has been established for many decades and is based on the premise that a pile under axial load acts as a column and is subject to buckling. However, it has also been established through technological advancements, research and experience that piles completely contained in soils, even in soft materials, do not behave as free-standing columns and the danger of pile buckling is extremely remote. Not withstanding this kind of evidence, the code limitations placed on dimensional requirements based on diameter-to-length ratios have become accepted practices.

1810.3.3 Installation. Where pile shafts are formed through unstable soils and concrete is placed in an open-drilled hole, a steel liner shall be inserted in the hole prior to placing the concrete. Where the steel liner is withdrawn during concreting, the level of concrete shall be maintained above the bottom of the liner at a sufficient height to offset any hydrostatic or lateral soil pressure.

Where concrete is placed by pumping through a hollow-stem auger, the auger shall be permitted to rotate in a clockwise direction during withdrawal. The auger shall be withdrawn in a continuous manner in increments of about 12 inches (305 mm) each. Concreting pumping pressures shall be measured and maintained high enough at all times to offset hydrostatic and lateral earth pressures. Concrete volumes shall be measured to ensure that the volume of concrete placed in each pile is equal to or greater than the theoretical volume of the hole created by the auger. Where the installation process of any pile is interrupted or a loss of concreting pressure occurs, the pile shall be redrilled to 5 feet (1524 mm) below the elevation of the tip of the auger when the installation was interrupted or concrete pressure was lost and reformed. Augered cast-in-place piles shall not be installed within six pile diameters center to center of a pile filled with concrete less than 12 hours old, unless approved by the building official. If the concrete level in any completed pile drops during installation of an adjacent pile, the pile shall be replaced.

❖ There are many problems associated with drilled hole piles. Most of these problems are related to soil conditions, including soil or rock debris accumulating at the base of the pile or occurring in the pile shaft; reductions in the shaft cross section caused by the necking of soil walls because of soft materials or earth pressures; discontinuities in the pile shaft; hollows on the surface of the shaft and other problems linked to drilling operations. Such problems are usually addressed by the use of proper installation techniques.

Whenever unstable soils are encountered in drilling operations, such as loose granular soils, organic soils, very soft silts or clays and water-bearing subsurface materials, a temporary steel liner is to be placed in the hole to prevent the collapse of the earth walls or sloughing off of the soil during concrete placement.

In placing concrete in temporarily lined holes, the top of the concrete should be kept well above the bottom edge of the steel liner as it is withdrawn in order to offset any hydrostatic or lateral soil pressures. Stiff (low slump) concrete should not be used, so as to avoid the potential problem of concrete arching in the liner tube and causing discontinuities or voids to occur in the pile shaft as the liner is withdrawn. For a discussion on concrete consistency (slump values), see the commentary to Section 1810.1.1.

While many of the problems associated with auger-placed concrete piles are similar to those described above for drilled hole piles, other problems may also be introduced because of the particular installation technique. The pile installation procedure using a hollow stem continuous-flight auger, both for drilling and concrete placement purposes, does not permit the use of temporary steel liners. Damage to this type of pile, because of improper installation procedures, can result in the incomplete filling of the pile hole, concrete discontinuities along the pile shaft, reductions in the cross-sectional area of the shaft (necking) and soil inclusions.

Also, the loss of side support of the drilled hole or vertical displacement caused by ground pressures or soil movement may cause serious damage to the pile shaft.

In drilling the pile hole, the hollow-stem auger should be rotated and advanced continuously until the required tip elevation is reached. At that point in the drilling operation, rotation of the auger should be stopped to avoid removing excess soils that could result in damaging adjacent piles and to keep the auger flights full of soil as a means of retaining the hole walls. Concrete is then pumped through the hollow stem, filling the hole from the bottom up as the auger is withdrawn.

Concrete should be pumped under continuous pressure, and the rate of withdrawal of the auger should be carefully controlled for a continuous and full-sized shaft. The auger should be withdrawn in a smooth, continuous operation to prevent the surrounding soil from squeezing into the hole (necking) and possibly contaminating the concrete.

Discontinuities in the concrete shaft can develop if the auger is improperly withdrawn.

Concrete pumping pressures at the auger outlet should be greater than any hydrostatic or lateral pressures occurring in the hole. Excessively high pumping pressures, however, should be avoided in placing concrete surrounded by soft soils because it could cause upward or lateral movement of adjacent piles.

The volume of concrete placed in the hole is to be measured and should normally exceed the theoretical volume of the augered hole by approximately 10 percent. If the amount of the concrete pumped into the hole is considerably more than the quantity indicated above, however, then the cause should be investigated. It could mean, for instance, that concrete (grout) is being pumped into soft soil strata, solution cavities, underground pipelines or other underground structures.

If, for any reason, the pile-concreting operation is interrupted, a pressure drop occurs in pumping the concrete, the auger is improperly handled and is raised too fast or for any other cause that could damage the pile or cause a reduction in the required size of the pile section, then the hollow-stem auger must be redrilled to the required tip elevation and the pile reformed from the bottom up.

Unless approved by the building official, no new pile holes are to be drilled or injected with concrete if they are located within a center-to-center distance of six pile diameters from other adjacent piles containing fresh concrete that have not been allowed to set for a period of 12 hours or more. This is to reduce the possibility of damage to adjacent piles.

If the concrete surface of a completed pile is observed to drop below its cast elevation during the drilling of adjacent piles, then the completed pile should be rejected and replaced. Such an occurrence could indicate some deformation along the pile shaft that could affect its structural integrity.

1810.3.4 Reinforcement. For piles installed with a hollow-stem auger, where full-length longitudinal steel reinforcement is placed without lateral ties, the reinforcement shall be placed through ducts in the auger prior to filling the pile with concrete. All pile reinforcement shall have a concrete cover of not less than $2^1/_2$ inches (64 mm).

> **Exception:** Where physical constraints do not allow the placement of the longitudinal reinforcement prior to filling the pile with concrete or where partial-length longitudinal reinforcement is placed without lateral ties, the reinforcement is allowed to be placed after the piles are completely concreted but while concrete is still in a semifluid state.

❖ Unlike drilled cast-in-place concrete piles, caged reinforcement for auger-placed piles cannot be installed prior to filling the pile hole with concrete because the hollow-stem auger must be positioned in the hole at all times during drilling and concreting operations. To facilitate this, the cage must be pushed through the concrete in fluid form after the auger has been withdrawn. The problem associated with the placement of caged reinforcement in this way, particularly where it involves long cages, is that it cannot be determined whether the assembly has been positioned appropriately within the filled hole such that the reinforcement will have a minimum $2^1/_2$-inch (63 mm) concrete cover at all places (see Section 1810.1.2 for a description of caged reinforcement).

As specified in this section, pile reinforcement may consist of individual longitudinal reinforcing bars without lateral ties or a spiral when placed through special ducts built into the auger prior to filling the hole with concrete.

The exception recognizes that it is not possible to always place full-length longitudinal steel reinforcement in the hollow stem of the auger prior to placing the concrete. In low headroom piles, the connector or splice in the reinforcing is often too large to fit in the hollow stem. Many specified single-center bars are too large to fit in the hollow stem and allow placement of concrete containing standard aggregates. Additionally, many large-center bars have a 90- (1.57 rad) or 180-degree (3.14 rad) hook, which prevents the bars from going into the hollow stem. This exception permits the placement of the longitudinal steel into the pile after placing concrete is complete.

1810.3.5 Reinforcement in Seismic Design Category C, D, E or F. Where a structure is assigned to Seismic Design Category C, D, E or F in accordance with Section 1616, the corresponding requirements of Sections 1810.1.2.1 and 1810.1.2.2 shall be met.

❖ This section provides a convenient cross reference to Sections 1810.1.2.1 and 1810.1.2.2 for seismic reinforcement detailing requirements for drilled or augered uncased piles. Augered uncased piles are defined in Section 1808.1.

1810.4 Driven uncased piles. Driven uncased piles shall conform to Sections 1810.4.1 through 1810.4.4.

❖ A driven uncased pile is a displacement type of cast-in-place concrete pile formed by driving a temporary steel pile casing into the ground, filling the opening with concrete and then extracting the casing. Pile casings are driven either with a closed end using a detachable tip made of steel or concrete that remains permanently in the ground upon withdrawal of the casing or by using a mandrel that seals off the tip of the casing, keeping soil from entering the core during the driving operation.

A reinforcing cage can be installed in the casing before placing concrete. Sometimes the top part of the pile is formed with a short section of corrugated steel shell that is left permanently in place to protect the pile from contamination by loose soil or entry of surface water when the driven casing is withdrawn and the level of concrete subsides to fill in the spaces previously occupied by the casing. The space created at the top of the pile caused by the drop in the concrete level is subsequently filled with concrete up to the pile cutoff elevation.

Sections 1810.4.1 through 1810.4.4 provide detailed requirements for the design and construction of driven uncased piles.

1810.4.1 Allowable stresses. The allowable design stress in the concrete shall not exceed 25 percent of the 28-day specified compressive strength (f'_c) applied to a cross-sectional area not greater than the inside area of the drive casing or mandrel.

❖ The allowable stress in the concrete is not to exceed 0.25 f'_c (25 percent of the 28-day specified concrete compressive strength). Pile capacity based on concrete strength is to be calculated on the basis of a cross-sectional area no greater than the inside area of the drive casing or the area of the mandrel. The reason for this requirement is that the exact or average area of the pile hole cannot be established with any certainty because of the dimensional irregularities of the pile hole after removal of the casing.

1810.4.2 Dimensions. The pile length shall not exceed 30 times the average diameter. The minimum diameter shall be 12 inches (305 mm).

> **Exception:** The length of the pile is permitted to exceed 30 times the diameter, provided that the design and installation of the pile foundation is under the direct supervision of a registered design professional knowledgeable in the field of soil mechanics and pile foundations. The registered design professional shall certify to the building official that the piles were installed in compliance with the approved design.

❖ See the commentary to Section 1810.3.2 for discussion of pile length to diameter limits.

1810.4.3 Installation. Piles shall not be driven within six pile diameters center to center in granular soils or within one-half the pile length in cohesive soils of a pile filled with concrete less than 48 hours old unless approved by the building official. If the concrete surface in any completed pile rises or drops, the pile shall be replaced. Piles shall not be installed in soils that could cause pile heave.

❖ Pile casings driven in granular soils are to be spaced at least six pile diameters center to center. Pile casings driven in cohesive soils are not to be spaced less than one-half of the pile length when the concrete (in adjacent piles) is less than 48 hours old. This latter provision should not be construed to mean that piles in clay materials cannot be centered at spacings closer than 15 times their length (one-half by 30 times the diameter; see Section 1810.4.2). Indeed, they can be installed at the same spacings as piles in granular materials. Piles driven in cohesive materials, however, can cause significant soil displacement that can induce high lateral pressures on adjacent piles. Unless the concrete in these adjacent piles has had sufficient time to set (48 hours or more) and acquire some of its ultimate strength, considerable pile damage can be caused by the earth pressures. Because of the voids between the particles in granular soils, the effects of soil displacement are not as serious as for cohesive soils, such as clay.

The 48-hour concrete time-set requirement as stated in this section should not be confused with the 12-hour requirement referred to in Section 1810.3.3 for drilled or augered piles. While both types are cast-in-place concrete piles, unlike the driven uncased piles specified in Section 1810.4, drilled or augered piles are not of the displacement type, which essentially compacts the soil and may cause great lateral pressures and earth movement. The soil in the drilled method of installation is removed rather than displaced, and the subsurface influence on adjacent piles is far less than on the driven pile, particularly in cohesive soils.

Driven uncased piles are exposed to various types of potential damage. Besides the possibility of concrete contamination by the surrounding soil and surface water at the top of the pile, as described in Section 1810.4, the pile cross section may be subjected to squeezing or necking from lateral soil pressures and intrusions of displaced soils or other obstructions. Also, the concrete may be damaged by loss of support caused by removal of adjacent pile casings from the soil surrounding the pile. Driven uncased piles are not recommended under conditions where significant ground heave could occur or where highly unstable soils exist.

Subsurface investigations and load testing may be required to establish pile capacity, since there can be no correlation between the driving resistance of the pile casing and pile capacity. This is because the casing is eventually removed.

1810.4.4 Concrete cover. Pile reinforcement shall have a concrete cover of not less than $2^1/_2$ inches (64 mm), measured from the inside face of the drive casing or mandrel.

❖ For driven uncased piles, pile reinforcement, where required, is to be caged and installed so that the concrete cover over the reinforcement is no less than $2^1/_2$ inches (64 mm), measured from the inside face of the pile casing or inward from the periphery of the mandrel.

1810.5 Steel-cased piles. Steel-cased piles shall comply with the requirements of Sections 1810.5.1 through 1810.5.4.

❖ Steel-cased piles are the most widely used type of cast-in-place concrete piles. Essentially, they consist of mandrel-driven, light-gage steel shells or thin-walled pipes that are left permanently in place, reinforced as required by the design and filled with concrete. The shell is either a constant section for the full length of the pile or a step-tapered shape. A typical illustration is shown in Figure 1810.5.

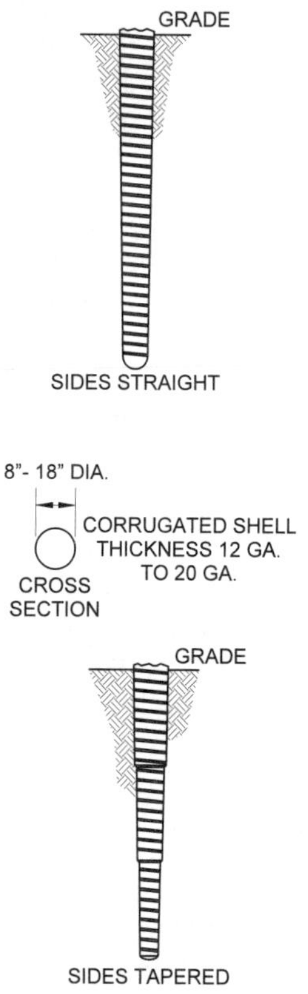

For SI: 1 inch = 25.4 mm.

Figure 1810.5
STEEL-CASED PILES

1810.5.1 Materials. Pile shells or casings shall be of steel and shall be sufficiently strong to resist collapse and sufficiently water tight to exclude any foreign materials during the placing of concrete. Steel shells shall have a sealed tip with a diameter of not less than 8 inches (203 mm).

❖ Steel pipe shells or casings driven with a mandrel can be torn or otherwise damaged because of underground obstructions, such as rock crevices. If this happens, the water tightness and structural integrity of the pile may be affected. Sometimes the buildup of ground pressures during driving may cause pile shells to squeeze or even collapse after the mandrel has been withdrawn. This section requires that steel-cased piles have shells of adequate thickness to resist serious damage or collapse and to maintain sufficient water tightness so as to prevent any foreign materials from entering the pile opening during the placing of concrete.

Pile tips are to have a closed end and be no less than 8 inches (203 mm) in diameter.

1810.5.2 Allowable stresses. The allowable design compressive stress in the concrete shall not exceed 33 percent of the 28-day specified compressive strength (f'_c). The allowable concrete compressive stress shall be 0.40 (f'_c) for that portion of the pile meeting the conditions specified in Sections 1810.5.2.1 through 1810.5.2.4.

❖ For cased cast-in-place concrete piles formed by driving permanent steel casings, the allowable design compressive stress in the concrete is not to exceed 0.33 f'_c (33 percent of the 28-day specified compressive strength). Except for the minimum pile tip diameter of 8 inches (203 mm), the pile section designed for a maximum allowable stress of 0.33 f'_c may be of any diameter, limited only by the standard sizes available from the manufacturers of steel pile casings.

The allowable concrete compressive stress may be increased to 0.40 f'_c, provided that the piles comply with the limiting provisions contained in Sections 1810.5.2.1 through 1810.5.2.4. The basis for this increase in allowable concrete stress is the added strength given to the concrete by the confining action of the steel shell. The general formula for increased allowable stress caused by confinement is:

$$f_c = 0.33f'_c\left(\frac{1+7.5tf_y}{Df'_c}\right)$$

where:

F_c = Allowable concrete stress.

f'_c = Specified 28-day concrete strength.

T = Thickness of steel shell.

F_y = Yield strength of steel.

D = Diameter of steel shell.

This formula is from the Portland Cement Association's (PCA) *Report on Allowable Stresses in Concrete Piles.*

When values for the various terms as set forth in Sections 1810.5.2.1 through 1810.5.2.4 are inserted in the given formula, the resulting allowable stress is 0.40 f'_c. Higher allowable stresses would result if, for example, the shell thickness was increased or the shell diameter was decreased. This increased allowable stress caused by confinement applies only to nonaxial load-bearing steel where the stress in the steel is taken in hoop tension instead of axial compression.

1810.5.2.1 Shell thickness. The thickness of the steel shell shall not be less than manufacturer's standard gage No. 14 gage (0.068 inch) (1.75 mm) minimum.

❖ Steel pile shells are to be No. 14 gage (U.S. standard) or thicker, but are not to be considered in the design of the pile to carry a portion of the pile load. The equivalent thickness for No.14-gage material is approximately 0.068 inches (1.7 mm).

1810.5.2.2 Shell type. The shell shall be seamless or provided with seams of strength equal to the basic material and be of a configuration that will provide confinement to the cast-in-place concrete.

❖ The shell for this type of pile must be seamless or have spirally welded seams and be of the strength and configuration required to provide structural confinement of the concrete fill. "Confinement" is the technical qualification that permits the use of increased allowable compressive stresses. Simply stated, the pile shell restrains the concrete in directions perpendicular to the applied stresses (see commentary, Section 1810.5.2).

1810.5.2.3 Strength. The ratio of steel yield strength (f_y) to 28-day specified compressive strength (f'_c) shall not be less than six.

❖ This section requires that the ratio of the yield strength (f_y) of the steel used in pile casings to the design compressive strength of concrete (f'_c) is not to be less than six. The yield strength of the steel used for pile casings of the type specified in this section is normally 30,000 psi (207 MPa) or greater. For example, in selecting a casing with a yield strength (f_y) of 30,000 psi (207 MPa) and a concrete compressive strength of 3,000 psi (21 MPa), the resulting ratio (f_y / f'_c) would be 10, which is greater than the minimum required ratio of six; therefore, the strengths of the pile materials are acceptable. For comparison, use the same steel casing and a specified concrete compressive strength (f'_c) of 5,000 psi (34 MPa). The resulting ratio would be exactly six. It can readily be seen that for concrete strengths greater than 5,000 psi (34 MPa), the type of steel used for the casing material would need yield strengths greater than 30,000 psi (207 MPa). For example, in order to meet the minimum ratio of six as required by this section, if the 28-day concrete compressive strength (f'_c) was specified at 6,000 psi (41 MPa), the material to be used for the casing would require a steel yield strength of at least 36,000 psi (248 MPa) conforming to ASTM A 36.

1810.5.2.4 Diameter. The nominal pile diameter shall not be greater than 16 inches (406 mm).

❖ The nominal pile diameter is not to exceed 16 inches (406 mm) in order to qualify for the "confined concrete" provision (see Section 1810.5.2) that allows for an increase in the allowable design compressive stress.

1810.5.3 Installation. Steel shells shall be mandrel driven their full length in contact with the surrounding soil.

The steel shells shall be driven in such order and with such spacing as to ensure against distortion of or injury to piles already in place. A pile shall not be driven within four and one-half average pile diameters of a pile filled with concrete less than 24 hours old unless approved by the building official. Concrete shall not be placed in steel shells within heave range of driving.

❖ For a general description of steel-cased cast-in-place concrete piles, refer to the commentary to Section 1810.5.

One advantage to this type of pile is that after the driving and removal of the mandrel before the concrete is placed, the steel casing can be inspected internally along its full length; therefore, any damage that has resulted from the pile-driving operation can be readily discovered and corrected.

Concrete placement is to comply with the applicable provisions of Section 1810.1.3.

Steel casings are not to be driven closer than four and one-half pile diameters of any adjacent piles filled with concrete until the concrete in the shells has cured for at least 24 hours and achieved an early strength. The basic reason for requiring four and one-half pile diameters between driving and concreting is a general reluctance to expose fresh concrete to vibrations as it sets. Several independent tests have shown, however, that there is no detrimental effect caused by vibrations on setting concrete. In some cases, there was a strength gain. The code requirement of four and one-half pile diameters is reasonable. It is often impractical to place concrete in pile shells right next to piles being driven. Pile casings can be driven earlier than the minimum 24-hour period when approved by the building official.

Driven light-gage steel pile shells that are left open can also be damaged by earth pressures when driving other piles in the close vicinity. Under such circumstances, pile shells are sometimes protected by inserting dummy mandrels.

Pile shells that are in place, adjacent to and within heave range of a pile being driven are normally left open if the soil condition is such that it can cause pile heave. This condition is particularly critical in cohesive soils, such as clay. Heaved piles must be redriven before filling with concrete. If a concreted pile has heaved, however, it can be safely redriven if proper techniques and a suitably designed pile cushion are used.

1810.5.4 Reinforcement. Reinforcement shall not be placed within 1 inch (25 mm) of the steel shell. Reinforcing shall be required for unsupported pile lengths or where the pile is designed to resist uplift or unbalanced lateral loads.

❖ Reinforcing steel is required in steel-cased, cast-in-place concrete piles for those lengths of pile that are laterally unsupported (such as piles projecting above the ground surface) and piles that must be designed for uplift forces (tension loads) or unbalanced lateral loads. Such reinforcement is to be designed to resist the bending and tensile stresses induced on the pile unit.

Steel-reinforcing cages must be installed at least 1

inch (25 mm) clear of the inside surface of the pile casing to allow sufficient space for the proper bonding of the steel reinforcing bars and the fill concrete.

1810.5.4.1 Seismic reinforcement. Where a structure is assigned to Seismic Design Category C, D, E or F in accordance with Section 1616, the reinforcement requirements for drilled or augered uncased piles in Section 1810.3.5 shall be met.

Exception: A spiral-welded metal casing of a thickness not less than manufacturer's standard gage No. 14 gage (0.068 inch) is permitted to provide concrete confinement in lieu of the closed ties or equivalent spirals required in an uncased concrete pile. Where used as such, the metal casing shall be protected against possible deleterious action due to soil constituents, changing water levels or other factors indicated by boring records of site conditions.

❖ The purpose of this section is to include pile bending, which is a result of ground horizontal movement during an earthquake, in the structural design. The reinforcement in the pile, required to resist tension caused by pile bending, increases the ductility of the foundation such that bending or shear failure is precluded. The shear strength and confining ability of spiral-welded metal casing eliminates the need for special pile ties, which is the basis for the exception in this section.

1810.6 Concrete-filled steel pipe and tube piles. Concrete-filled steel pipe and tube piles shall conform to the requirements of Sections 1810.6.1 through 1810.6.5.

❖ This section covers concrete-filled steel piles consisting of either seamless or welded pipe or closed-ended tubular steel piles with either straight or tapered sections. These piles may be installed as friction or end-bearing units. If driven open ended, the pipe is cleaned out (earth core removed) before being filled with concrete. The steel pipe may be driven with an internal mandrel (see Figure 1810.6).

1810.6.1 Materials. Steel pipe and tube sections used for piles shall conform to ASTM A 252 or ASTM A 283. Concrete shall conform to Section 1810.1.1. The maximum coarse aggregate size shall be $^{3}/_{4}$ inch (19.1 mm).

❖ Seamless or welded-steel pipe piles must conform to the requirements of ASTM A 252. Steel tube piles are to be made of materials conforming to the requirements of ASTM A 283, or such other approved specifications suitable for use in pile fabrication.

The concrete used for filling the piles is to consist of $^{3}/_{4}$-inch (19.1 mm) maximum-sized coarse aggregates and conform to the requirements of Section 1810.1.1. Concrete should not contain any admixtures (e.g., calcium chloride) that would be corrosive to steel.

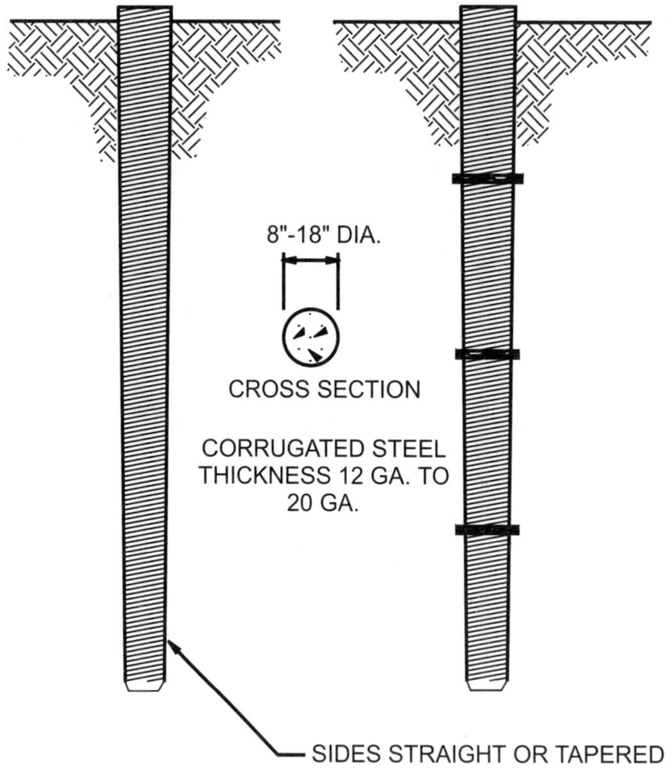

For SI: 1 inch = 25.4 mm.

Figure 1810.6
CONCRETE-FILLED STEEL PIPE PILE

1810.6.2 Allowable stresses. The allowable design compressive stress in the concrete shall not exceed 33 percent of the 28-day specified compressive strength (f'_c). The allowable design compressive stress in the steel shall not exceed 35 percent of the minimum specified yield strength of the steel (F_y), provided F_y shall not be assumed greater than 36,000 psi (248 MPa) for computational purposes.

Exception: Where justified in accordance with Section 1808.2.10, the allowable stresses are permitted to be increased to 0.50 F_y.

❖ The allowable design compressive stress in concrete must not exceed 0.33 f'_c. The allowable design compressive stress in steel must not exceed 0.35 f_y, except that the allowable design compressive stress may be increased up to 50 percent of the minimum specified yield strength of steel where substantiated by Section 1808.2.10.

The allowable stress provisions for both concrete and steel materials used for this type of pile are adequately covered by this section and are self explanatory. For concrete-filled pipe piles, the full design load is shared by both steel and concrete; however, in driving an empty pipe pile, the full dynamic stresses resulting from the required ultimate capacity (twice design load) must be taken only by steel. Increasing allowable stresses to 0.50 f_y could lead to pile deformations during driving.

In comparison, for mandrel-driven piles, the full design load is taken by concrete fill because the pile walls are so thin [$^1/_{10}$-inch (2.5 mm) minimum or 12 gage] that the strength of the steel shells is totally disregarded, with one exception. In accordance with the recommendations of a report on the design, manufacture and installation of concrete piles (ACI 543R), plain or fluted casings of the minimum thickness stated in Section 1810.6.3 may be designed as a load-carrying pile component, provided the cross-sectional area of the casing is at least 3 percent of the gross pile area.

Steel pipe piles are made of welded or seamless pipe in three grades of steel having yield strengths of 30,000, 35,000 and 45,000 psi in accordance with ASTM A 252. They are used as either end-bearing or friction-type piles and may be driven with open or closed ends. Open-ended piles are generally used where rock bearing materials are found at levels reasonably close to the surface, and especially where very large loads are to be supported. In driving an open-ended pipe pile, the earth is forced into the core of the pipe and must be cleaned out either by a blowing technique or augering, depending on the type of soils encountered. As may be specified, the pipe is then filled with concrete.

When soils investigations show the absence of rock formations, or where rock occurs at excessive depths and it is not practical or economically feasible to reach such depths, closed-end pipe piles are driven as friction piles.

1810.6.3 Minimum dimensions. Piles shall have a nominal outside diameter of not less than 8 inches (203 mm) and a minimum wall thickness in accordance with Section 1809.3.4. For mandrel-driven pipe piles, the minimum wall thickness shall be $^1/_{10}$ inch (2.5 mm).

❖ This section requires that seamless or welded-steel pipe piles and tube piles have a nominal outside diameter of no less than 8 inches (203 mm). The 8-inch (203 mm) minimum diameter is necessary in order to inspect the inside of the pile to observe any damage that may have occurred during the driving process. Additionally, the minimum diameter is necessary to provide stiffness for driving purposes. The nominal wall thicknesses of steel pipe piles are to be in accordance with Section 1809.3.4.

1810.6.4 Reinforcement. Reinforcement steel shall conform to Section 1810.1.2. Reinforcement shall not be placed within 1 inch (25 mm) of the steel casing.

❖ Steel reinforcement placed in concrete-filled cores of steel pipe or tube piles to resist the bending stresses induced by lateral forces acting upon the piles or by other structural loads is to comply with the requirements of Section 1810.1.2.

Internal pile reinforcement, where required, consists of a cage made of several longitudinal deformed reinforcing bars tied together with lateral steel in the form of individual hoops or a spiral. Metal reinforcement is to comply with the applicable ASTM specifications referenced in ACI 318.

Preassembled cages should be provided with spacer bars to facilitate accurate placement and allow for a minimum clearance of 1 inch (25 mm) between the reinforcement and the pile casing. Longitudinal bars may be discontinued along the pile when no longer needed because of load transfer into the soil; no more than two bars should be stopped off at any one point along the pile. This provision is especially critical in thin-walled piles where the casings do not carry any of the axial or bending loads.

1810.6.4.1 Seismic reinforcement. Where a structure is assigned to Seismic Design Category C, D, E or F in accordance with Section 1616, the following shall apply. Minimum reinforcement no less than 0.01 times the cross-sectional area of the pile concrete shall be provided in the top of the pile with a length equal to two times the required cap embedment anchorage into the pile cap, but not less than the tension development length of the reinforcement. The wall thickness of the steel pipe shall not be less than $^3/_{16}$ inch (5 mm).

❖ This section prescribes seismic reinforcement requirements for concrete-filled steel pipe and tube piles in structures assigned to Seismic Design Category C, D, E or F. These types of piles are defined in Section 1808.1. The minimum wall thickness is based on testing and practice that has resulted in successful installations in moderate to high seismic regions.

1810.6.5 Placing concrete. The placement of concrete shall conform to Section 1810.1.3.

❖ The placement of concrete in the core of a pipe pile or thin-walled tube pile must be performed in accordance with the requirements of Section 1810.1.3 to verify that the required design strength of the concrete is fully developed throughout the cross section and length of the pile.

Overdriving steel pipe, thin-walled tube piles or encountering damaging subsurface obstructions can result in bent piles, torn pile walls and distorted or ruptured pile tips. Such damage can allow the entry of soils or water into the pile cores. Pile walls may also collapse because of the buildup of excessive soil pressures during driving operations, especially in the case of thin-walled piles after removal of the mandrel.

Piles should be inspected internally to determine their condition after driving. If serious damage is discovered that would affect the structural performance of the pile, appropriate measures should be taken to correct the condition before the placement of concrete.

1810.7 Caisson piles. Caisson piles shall conform to the requirements of Sections 1810.7.1 through 1810.7.6.

❖ The caisson pile, commonly known as a "drilled-in caisson," is installed as a special type of high-load-capacity pile. Its construction features are described in Section 1810.7.1. The design and allowable stress provisions, as well as the material, steel core and installation requirements, are contained in Sections 1810.7.2 through 1810.7.6.

Some advantages of using caisson piles include the control of construction during installation, nondisplacement, the ability to clean out the shaft and drive the pile further, its high load capacities and ease in splicing. The main disadvantage of caisson piles is their significant cost.

An illustration of a typical caisson pile is shown in Figure 1810.7.

1810.7.1 Construction. Caisson piles shall consist of a shaft section of concrete-filled pipe extending to bedrock with an uncased socket drilled into the bedrock and filled with concrete. The caisson pile shall have a full-length structural steel core or a stub core installed in the rock socket and extending into the pipe portion a distance equal to the socket depth.

❖ The caisson pile is a high-load capacity, end-bearing type of pile. Essentially, it is constructed as a cased, cast-in-place concrete pile formed by driving a thick-walled, open-ended steel pipe down to suitable rock material, cleaning out the soil materials within the pipe, drilling a socket into the rock, inserting a structural steel core into the pipe and then filling the pipe and drilled socket with concrete. The core material used in the pipe casing is of hot rolled structural shapes such as wide flange or I-beam sections. Steel rails are also used. Core material is installed to extend from the bottom of the rock socket up to the head of the pile or part of the way up the casing, depending on design requirements.

1810.7.2 Materials. Pipe and steel cores shall conform to the material requirements in Section 1809.3. Pipes shall have a minium wall thickness of $^3/_8$ inch (9.5 mm) and shall be fitted with a suitable steel-driving shoe welded to the bottom of the pipe. Concrete shall have a 28-day specified compressive strength (f'_c) of not less than 4,000 psi (27.58 MPa). The concrete mix shall be designed and proportioned so as to produce a cohesive workable mix with a slump of 4 inches to 6 inches (102 mm to 152 mm).

❖ The pipe used in the construction of caisson piles must have a minimum thickness of $^3/_8$ inch (9.5 mm) and comply with the requirements of the ASTM A 252 standard specification for welded and seamless pipe piles. While this specification tabulates pipe pile sections up to 20 inches (508 mm) in outside diameter, larger sections may be ordered from the manufacturers of such products. Caisson piles as large as 36 inches (914 mm) in diameter have been used.

Caisson pipes must be fitted with a hardened steel drive shoe to facilitate hard driving to the rock and to sufficiently seat the tip of the pipe into the rock and form a seal that will prevent water and foreign matter from entering the pile hole during the concrete filling operations.

The materials used for the concrete fill must be proportioned to provide a workable mix that will have a minimum 28-day specified compressive strength (f'_c) of 4,000 psi (27.6 MPa) and a consistency (slump) of 4 inches (102 mm) to 6 inches (152 mm). Concrete is to comply with the applicable provisions of Chapter 19.

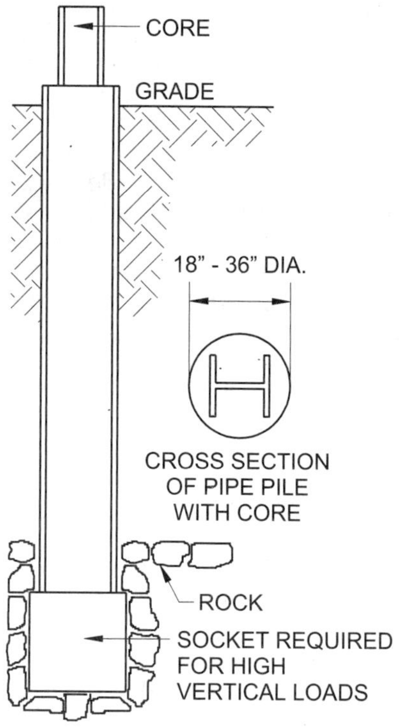

For SI: 1 inch - 25.4 mm.

Figure 1810.7
CAISSON PILES

1810.7.3 Design. The depth of the rock socket shall be sufficient to develop the full load-bearing capacity of the caisson pile with a minimum safety factor of two, but the depth shall not be less than the outside diameter of the pipe. The design of the rock socket is permitted to be predicated on the sum of the allowable load-bearing pressure on the bottom of the socket plus bond along the sides of the socket. The minimum outside diameter of the caisson pile shall be 18 inches (457 mm), and the diameter of the rock socket shall be approximately equal to the inside diameter of the pile.

❖ The principal structural feature of the caisson type of pile is the rock socket, which is filled with concrete and designed to take the full load of the pile by end bearing and the frictional resistance offered by the rough walls of the socket.

The relationship between the depth of the socket and the bearing capacity of the rock cannot be determined with any great accuracy because of the natural joints, bedding planes and fissures generally found in rock formations. Experience has shown, however, that the frictional resistance in hard rock is often sufficient to carry the load. In the case of soft rock, the shear strength of the socket walls should be determined by testing rock samples.

1810.7.4 Structural core. The gross cross-sectional area of the structural steel core shall not exceed 25 percent of the gross area of the caisson. The minimum clearance between the structural core and the pipe shall be 2 inches (51 mm). Where cores are to be spliced, the ends shall be milled or ground to provide full contact and shall be full-depth welded.

❖ Core material made of structural steel or rails must comply with ASTM A 36. Other steels used as permanent load-bearing structural cores in concrete piles are referenced in ACI 543R, which deals with the design, manufacture and installation of concrete piles. The purpose of the steel core is to bond with the concrete fill to act as a composite section and provide additional strength and load-bearing capacity.

1810.7.5 Allowable stresses. The allowable design compressive stresses shall not exceed the following: concrete, $0.33\,f'_c$; steel pipe, $0.35\,F_y$ and structural steel core, $0.50\,F_y$.

❖ Load-carrying capacity is determined as a composite of the strengths of the three components of the caisson piles: the driven steel pipe, the concrete and the structural core. As in other types of cased cast-in-place concrete piles, the concrete stress is not to exceed $0.33\,f'_c$ (33 percent of the 28-day specified concrete compressive strength). Steel pipe stresses are not to exceed $0.35\,f_y$ (35 percent of the steel yield strength) and the structural steel core material stresses are not to exceed $0.50\,f_y$.

1810.7.6 Installation. The rock socket and pile shall be thoroughly cleaned of foreign materials before filling with concrete. Steel cores shall be bedded in cement grout at the base of the rock socket. Concrete shall not be placed through water except where a tremie or other approved method is used.

❖ A brief description of the installation sequence of caisson piles is given in the commentary to Section 1810.7.1. One of the problems in the construction of caisson piles is that an amount of water will seep into the bottom of the pile opening through the rock joints or fissures in the socket. Sometimes water may seep into the opening at the end of the pipe casing because of incomplete seating of the pipe into the rock material. If the water in the hole cannot be controlled by ejection or other methods rendering the placement of concrete under reasonably dry conditions, the concrete may be placed by a tremie or other methods when approved by the building official.

SECTION 1811
COMPOSITE PILES

1811.1 General. Composite piles shall conform to the requirements of Sections 1811.2 through 1811.5.

❖ Composite piles are made of two or more sections of different materials or of pile types that are spliced together to form a single pile unit. These piles are usually used when significant pile lengths are needed. However, due to economic considerations and problems with splicing, composite piles are seldom used today.

An illustration of typical pile combinations is shown in Figure 1811.1.

Sections 1811.2 through 1811.5 set forth the minimum design, construction and installation requirements for composite piles.

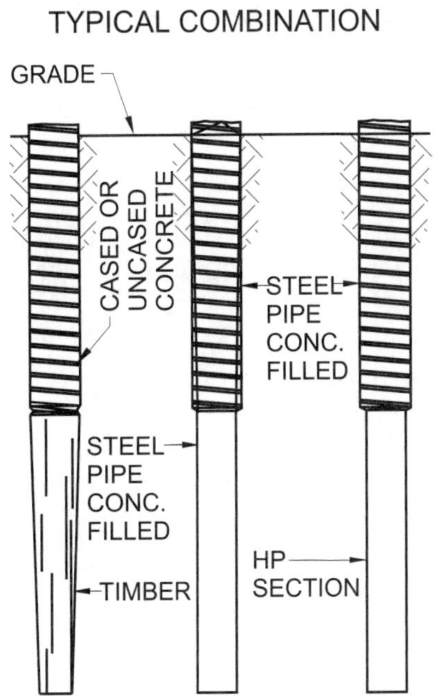

Figure 1811.1
COMPOSITE PILES

1811.2 Design. Composite piles consisting of two or more approved pile types shall be designed to meet the conditions of installation.

❖ Like any other type of pile, the design of composite piles should consider: the installation requirements involving the handling and storage of piling materials; the selection of pile-driving equipment, with special regard for the physical properties of the weaker section, and the application of procedures that would prevent possible installation problems and prevent any adverse effect on the structural integrity of the pile during driving and its long-term performance. The driveability of a pile consisting of two or more different materials or pile types should be given special consideration to produce a pile design that will have sufficient stiffness and rigidity at the splice points to transmit the driving energy adequately, as well as withstand the driving forces.

1811.3 Limitation of load. The maximum allowable load shall be limited by the capacity of the weakest section incorporated in the pile.

❖ The strength of a composite pile consisting of two or more sections of different materials or pile types is governed by the weaker section of the spliced components. This section requires that the maximum load-carrying capacity of a composite pile be established by the physical properties of the weakest section. This limitation, while necessary from an engineering standpoint, provides for the inefficient use of the stronger pile sections. It is an inherent disadvantage in the use of composite piles.

1811.4 Splices. Splices between concrete and steel or wood sections shall be designed to prevent separation both before and after the concrete portion has set, and to ensure the alignment and transmission of the total pile load. Splices shall be designed to resist uplift caused by upheaval during driving of adjacent piles, and shall develop the full compressive strength and not less than 50 percent of the tension and bending strength of the weaker section.

❖ The strength of a composite pile is governed not only by the weaker section, but also by the strength and details of the splice that joins and holds the pile sections together.

There are several problems associated with the splicing of sections comprising a composite pile. For example, if the length of a composite pile is of such dimension that it will fit in the leaders of the driving equipment, then the full length of the spliced pile can be driven as a continuous operation. However, if the pile is too long to fit in the leaders as a single unit, the pile must be installed in sections, and the driving would have to be interrupted in order to make the splice in the leaders.

It is most important that splicing devices be made in such a way that their installation will be simple and quick. When pile driving must be stopped to make a splice in the leaders, the time expended should not be of a duration that would allow the soil to set up (soil freeze) so that continued driving would produce excessive stresses in the pile.

Sometimes when the predetermined length of a composite pile is too long to fit in the leaders, but not overly long, the difference in length between the pile and the leaders may be made up by inserting the lower end of the spliced unit in a preexcavated hole, thus allowing for continuous driving of the pile (see Section 1808.2.13).

Another problem related to splicing occurs in trying to accurately align the pile sections. When the lower section of a composite pile is driven nearly full length into the ground, it is very difficult to determine its true direction, particularly if the pile has drifted off line because of soil pressures, subsurface obstructions, improper driving or for any other reason. When a pile section is connected to a section already driven in the ground, the upper section should be installed in the same direction as the longitudinal axis of the lower section, even though the direction may not be vertical. It is better to maintain a misdirection rather than to try to correct a situation and create a bend at the pile joint.

The selection of the splicing method to be used for assembling a composite pile should be based on the driving and service load stresses that must be resisted by the pile. Splices should be designed to prevent separation of the pile sections during construction and thereafter. Special consideration must be given to the design of splices to resist uplift forces, whether the piles are subject to heaving or specifically designed as tension piles. Engineering practice is to design the splice strong enough to develop about one and one-half to two times the design uplift force. Pile splices must also be designed to resist compressive, bending and shear stresses imposed by construction and service loads.

1811.5 Seismic reinforcement. Where a structure is assigned to Seismic Design Category C, D, E or F in accordance with Section 1616, the following shall apply. Where concrete and steel are used as part of the pile assembly, the concrete reinforcement shall comply with that given in Sections 1810.1.2.1 and 1810.1.2.2 or the steel section shall comply with Section 1809.3.5 or 1810.6.4.1.

❖ This section includes seismic requirements for composite piles, which are piles constructed as composites of two or more of the pile types defined in Section 1808.1. As such, these piles must comply with the applicable seismic requirements for the different types of piles used as indicated by the specific sections referenced.

SECTION 1812
PIER FOUNDATIONS

1812.1 General. Isolated and multiple piers used as foundations shall conform to the requirements of Sections 1812.2 through 1812.10, as well as the applicable provisions of Section 1808.2.

❖ Section 1812 addresses general requirements for the construction of isolated pier foundations (drilled or excavated shafts with or without belled bottoms) using plain or reinforced concrete or piers encased by steel shafts.

Piers can be used for both shallow and deep foundation construction, but are principally employed as deep foundations. There are basically two methods for constructing piers. The first method involves hand excavation of pits or holes that must be lined to prevent caving of the earth sides. The openings are subsequently filled with concrete. The "Chicago Caisson" method is an example of this type of construction. A circular hole is excavated at least $3^1/_2$ feet (1067 mm) in diameter to allow for working space; a larger dimension may be necessary as required by the design. On the way down, the excavation is lined with vertical boards (lagging) that are held in place by horizontal steel rings. The bottom of the shaft may be belled out to increase the bearing area. The hole is then filled with concrete, leaving the lagging and steel rings in place as part of the permanent foundation. It was not uncommon to carry foundations of this type down 50 or 60 feet (is 240 or 18 288 mm) below ground. Hand-excavated sheeted pits or cased holes are seldom used today, except for shallow foundations or for underpinning purposes.

The second and most common method used for constructing piers consists of drilling a hole in the earth (by machine), belling out the bottom where necessary to obtain the required bearing area and placing concrete in the hole. Pier holes may be cased or left without such protection based on the stability of the soil. Piers may contain reinforcement or be constructed without reinforcement, depending on design requirements. Drilled piers are also known by other names, such as drilled shafts or drilled caissons.

Piers, unlike piles, are not normally used in groups, but are usually constructed as isolated foundations. Piers are generally designed as end-bearing units to carry loads greater than the capacity of single piles.

Sections 1812.2 through 1812.10 provide the minimum design, construction and installation requirements for piers. This section also references Section 1808.2, which addresses general requirements for piers and piles, including the required foundation investigation.

1812.2 Lateral dimensions and height. The minimum dimension of isolated piers used as foundations shall be 2 feet (610 mm), and the height shall not exceed 12 times the least horizontal dimension.

❖ Pier foundations are not to have widths (or diameters) less than 2 feet (610 mm) and should be proportioned so that the height (length) of the pier is no more than 12 times the least lateral dimension. This is an empirical requirement to prevent effects similar to those experienced in slender columns where, because of the flexibility of slender compression members, excessive bending stresses that cause structural failure, including buckling, may be induced.

1812.3 Materials. Concrete shall have a 28-day specified compressive strength (f'_c) of not less than 2,500 psi (17.24 MPa). Where concrete is placed through a funnel hopper at the top of the pier, the concrete mix shall be designed and proportioned so as to produce a cohesive workable mix having a slump of not less than 4 inches (102 mm) and not more than 6 inches (152 mm). Where concrete is to be pumped, the mix design including slump shall be adjusted to produce a pumpable concrete.

❖ Concrete used for piers is to have a minimum 28-day compressive strength of 2,500 psi (17.24 MPa). Higher concrete compressive strengths are usually specified to increase pier load capacity as well as for economic reasons.

This section requires that concrete be designed and proportioned to produce a cohesive workable mix with a consistency (slump) of no less than 4 inches (102 mm) and 6 inches (152 mm). The slump requirements stated here, however, should not be considered so absolute that, with the use of engineering judgement, the consistency of the concrete cannot be adjusted to satisfy prevailing conditions of construction. For example, concrete placement conducted under difficult conditions—such as for piles containing heavy reinforcement, very long cased piles, pile shells driven on steep batters or other demanding situations—may require the employment of special concrete mixes using reduced quantities of coarse aggregates with corresponding increases in sand and cement content. Under such conditions, slumps of 4 inches (102 mm) with a tolerance of plus 2 inches (51 mm) or minus 1 inch (25 mm) are in order.

High slump concrete is used so that the complete volume of the hole is filled, including natural soil crevices and pockets and surface irregularities caused by the drilling operations. High slumps are also used where the concrete must displace drilling slurry.

Concrete is usually placed through a funnel hopper at the top of the pile opening, but other approved methods such as pumping or tremie may also be used.

1812.4 Reinforcement. Except for steel dowels embedded 5 feet (1524 mm) or less in the pier, reinforcement where required shall be assembled and tied together and shall be placed in the pier hole as a unit before the reinforced portion of the pier is filled with concrete.

Exception: Reinforcement is permitted to be wet set and the $2^1/_2$- inch (64 mm) concrete cover requirement be reduced to 2 inches (51 mm) for Group R-3 and U occupancies not exceeding two stories of light-frame construction, provided the construction method can be demonstrated to the satisfaction of the building official.

Reinforcement shall conform to the requirements of Sections 1810.1.2.1 and 1810.1.2.2.

Exceptions:

1. Isolated piers supporting posts of Group R-3 and U occupancies not exceeding two stories of light-frame construction are permitted to be reinforced as required by rational analysis but not less than a minimum of one No. 4 bar, without ties or spirals, when detailed so the pier is not subject to lateral loads and the soil is determined to be of adequate stiffness.

2. Isolated piers supporting posts and bracing from decks and patios appurtenant to Group R-3 and U occupancies not exceeding two stories of light-frame construction are permitted to be reinforced as required by rational analysis but not less than one No. 4 bar, without ties or spirals, when the lateral load, E, to the top of the pier does not exceed 200 pounds (890 N) and the soil is determined to be of adequate stiffness.
3. Piers supporting the concrete foundation wall of Group R-3 and U occupancies not exceeding two stories of light-frame construction are permitted to be reinforced as required by rational analysis but not less than two No. 4 bars, without ties or spirals, when it can be shown the concrete pier will not rupture when designed for the maximum seismic load, E_m, and the soil is determined to be of adequate stiffness.
4. Closed ties or spirals where required by Section 1810.1.2.2 are permitted to be limited to the top 3 feet (914 mm) of the piers 10 feet (3048 mm) or less in depth supporting Group R-3 and U occupancies of Seismic Design Category D, not exceeding two stories of light-frame construction.

❖ This section prescribes the necessary seismic reinforcement for piers. Piers are defined in Section 1808.1 and are relatively short in comparison to their width. Piers are required to comply with the same seismic reinforcement requirements as cast-in-place concrete piles in Sections 1810.1.2.1 and 1810.1.2.2. However, there are quite a number of exceptions applicable to Group R-3 and U occupancies. Often piers are used in residential applications because of expansive soils and high ground-water tables. The requirement for transverse reinforcement is excessive for the interior piers of wood frame residential structures that only support vertical loads imposed by wood posts and for lightly loaded piers. Exception 1 exempts such interior piers from the transverse reinforcement requirements of this section. Exceptions 2 through 4 allow lesser reinforcement than what is required by this section for residential structures given certain conditions.

1812.5 Concrete placement. Concrete shall be placed in such a manner as to ensure the exclusion of any foreign matter and to secure a full-sized shaft. Concrete shall not be placed through water except where a tremie or other approved method is used. When depositing concrete from the top of the pier, the concrete shall not be chuted directly into the pier but shall be poured in a rapid and continuous operation through a funnel hopper centered at the top of the pier.

❖ The drilled hole must be free of all foreign matter prior to placing concrete. If there is excessive water in the hole [about 4 inches (102 mm) is permitted], the concrete may be placed by tremie or pumping methods. Concrete is usually deposited from the top through a funnel hopper that should have a discharge opening that is sized smaller than the smallest pier cross section in the foundation system. Vibrators should not be used in the concreting operations because they may damage the earth walls and cause contamination of the concrete. Improper vibration may also cause excessive bleeding.

1812.6 Belled bottoms. Where pier foundations are belled at the bottom, the edge thickness of the bell shall not be less than that required for the edge of footings. Where the sides of the bell slope at an angle less than 60 degrees (1 rad) from the horizontal, the effects of vertical shear shall be considered.

❖ Piers constructed as deep foundations are sometimes belled out at their bottoms to increase bearing areas. Belled bottoms can be made by either hand excavation or a mechanical method involving underreaming with a belling bucket, as is commonly used in the construction of drilled piers.

The soil should be sufficiently cohesive so the roof of the bell will not collapse during excavation, cleanout operations and the placement of fresh concrete. Where the character of the soil is such that attempts at belling out pier bottoms might not produce acceptable results, it would be preferable to carry the piers down as straight shafts into soil strata with better load-bearing values, allowing the loads to be carried by the smaller pier bottoms or to such depths that would permit the soils to carry the loads by side friction. The building official should ascertain the suitability of the soil for bell construction through soil investigations and reports and by the recommendations of the architect or engineer.

Section 1812.6 requires bells to have vertical edges at their bottoms that are equal to the thickness requirements for concrete footings. The purpose of this specification is to prevent shear breaks in angled edges caused by soil pressures. Bell slopes (sides) are not to have sides that are less than 60 degrees (1 rad) from the horizontal, unless the effects of vertical shear are considered in the design (see Figure 1812.6).

1812.7 Masonry. Where the unsupported height of foundation piers exceeds six times the least dimension, the allowable working stress on piers of unit masonry shall be reduced in accordance with ACI 530/ASCE 5/TMS 402.

❖ ACI 530 contains provisions for reducing the allowable working stress on masonry columns based on the height to radius of gyration of the column.

1812.8 Concrete. Where adequate lateral support is not provided, and the unsupported height to least lateral dimension does not exceed three, piers of plain concrete shall be designed and constructed as pilasters in accordance with ACI 318. Where the unsupported height to least lateral dimension exceeds three, piers shall be constructed of reinforced concrete, and shall conform to the requirements for columns in ACI 318.

Exception: Where adequate lateral support is furnished by the surrounding materials as defined in Section 1808.2.9, piers are permitted to be constructed of plain or reinforced concrete. The requirements of ACI 318 for bearing on concrete shall apply.

❖ The slenderness ratio (unsupported height to least lateral dimension) is limited to three for piers constructed of plain concrete. When the slenderness ratio exceeds three, the pier is required to be constructed of reinforced concrete in accordance with ACI 318. Where the surrounding soil provides lateral support for the pier, this section permits the pier to be of plain or reinforced concrete. The bearing on concrete requirements of ACI 318 will govern.

1812.9 Steel shell. Where concrete piers are entirely encased with a circular steel shell, and the area of the shell steel is considered reinforcing steel, the steel shall be protected under the conditions specified in Section 1808.2.17. Horizontal joints in the shell shall be spliced to comply with Section 1808.2.7.

❖ The use of permanent steel shells as reinforcement for concrete and as may be required to facilitate excavation conditions is generally associated with drilled pier construction. A drilled concrete pier consisting of a steel shell is analogous to a pipe pile. The engineering significance is the same for both types of construction, except that pipe piles are driven in groups of two or more for the support of a building column while drilled piers serving the same purpose and of larger diameters are designed as single, isolated foundations. Shells may be installed during or after excavation operations, depending on the stability of the soils.

Steel materials must be protected from corrosive soils and shell joints must be properly spliced, all in accordance with the applicable provisions of Section 1808.2.7.

1812.10 Dewatering. Where piers are carried to depths below water level, the piers shall be constructed by a method that will provide accurate preparation and inspection of the bottom, and the depositing or construction of sound concrete or other masonry in the dry.

❖ Where pier foundations are carried to depths below water level, the construction operations should be conducted under dry conditions. Operating in dry conditions becomes particularly important where it involves inspections of drilled excavations to ascertain the accuracy of shaft alignment and dimensions; when necessary, to verify the load-bearing capacity of the supporting soil or rock stratum; to verify the removal of loose material from pier bottoms; to inspect bell-outs for possible failures and to ascertain visually that the opening will remain stable during concreting operations.

Where water inflow is expected or encountered in deep excavations and dewatering becomes necessary, it can be accomplished by any of a number of commonly used methods, ranging from the simple practice of pumping out excavations to the more complex systems employed by lowering water tables below foundation bottoms.

The choice of the dewatering method depends mainly on the site conditions, including the soil characteristics and soil profile, the level of the water table and the type and depth of the foundation construction. Serious water problems occur in granular soils rather than in cohesive clay materials.

Pier foundations are best used in soils that will permit construction in dry conditions for the reasons stated above. When the findings of soil investigations indicate a high probability of encountering serious water problems and it is apparent that the control of water inflow will be most difficult and, perhaps, economically imprac-

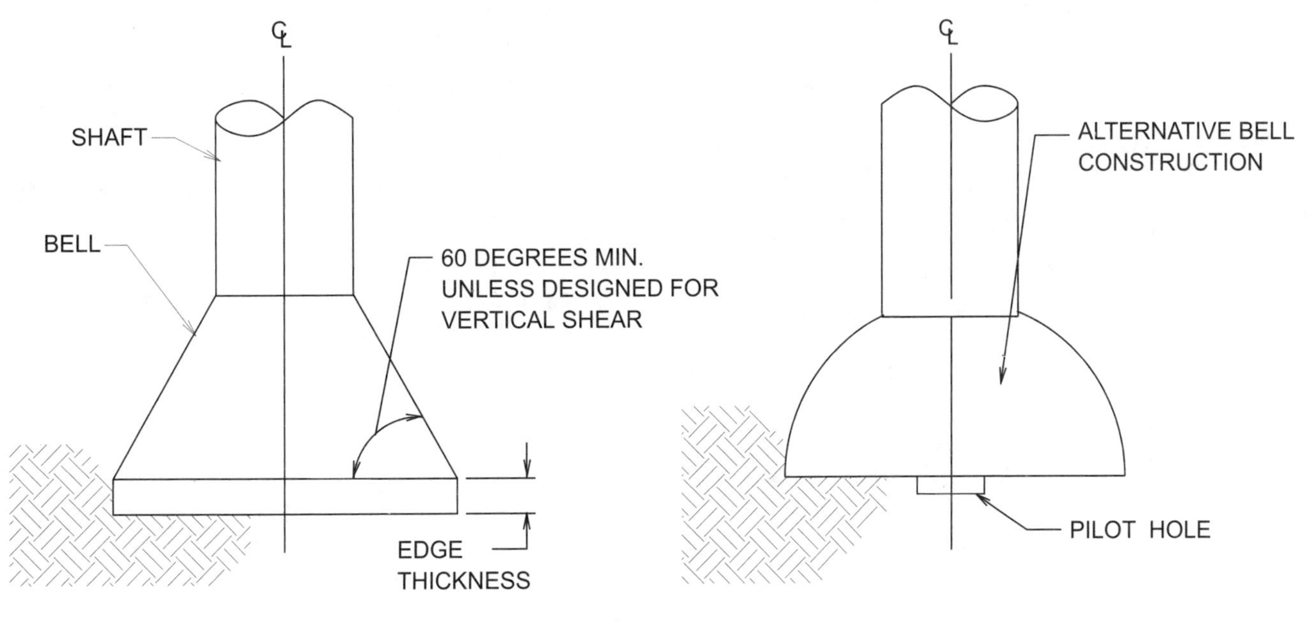

Figure 1812.6
BELLED BOTTOMS OF PIER FOUNDATIONS

tical, it is preferable to select another type of foundation construction (i.e., driven piles or other suitable foundations).

One of the most common methods used for stopping the inflow of water from pockets or strata of cohesionless soils (granular materials) encountered during the drilling of deep pier holes is to introduce a slurry consisting of soil, bentonite and water. The fluid is mixed by rotating, lifting and lowering the drilling auger. When the walls of the pier opening at the locations of water inflow have been sealed by the fluid mixture, a pipe casing is inserted to protect the hole and the slurry is then bailed out, allowing the drilling to proceed under dry conditions.

Bibliography

The following resource materials are referenced in this chapter or are relevant to the subject matter addressed in this chapter.

ACI 229R-99, *Controlled Low-Strength Materials.* Farmington Hills, MI: American Concrete Institute, 1999.

ACI 306R-88, *Cold Weather Concreting.* Farmington Hills, MI: American Concrete Institute, 1988.

ACI 306.1-90, *Standard Specification for Cold Weather Concrete.* Farmington Hills, MI: American Concrete Institute, 1990.

ACI 318-02, *Building Code Requirements for Structural Concrete.* Farmington Hills, MI: American Concrete Institute, 2002.

ACI 530/ASCE 5/TMS 402-02, *Building Code Requirements for Masonry Structures.* Farmington Hills, MI: American Concrete Institute; Reston, VA: American Society of Civil Engineers; Boulder, CO: The Masonry Society, 2002.

ACI 530.1/ASCE 6/TMS 602-02, *Specification for Masonry Structures.* Farmington Hills, MI: American Concrete Institute; Reston, VA: American Society of Civil Engineers; Boulder, CO: The Masonry Society, 2002.

ACI 543R-74 (1980), *Recommendations for Design, Manufacture and Installation of Concrete Piles.* Farmington Hills, MI: American Concrete Institute, 1980.

AF&PA NDS-01, *National Design Specification for Wood Construction—with 1997 Supplement; Design Values for Wood Construction.* Washington, DC: American Forest and Paper Association, 2001.

AF&PA Technical Report 7-87, *Basic Requirements for Permanent Wood Foundation Systems.* Washington, DC: American Forest and Paper Association, 1987.

ASCE 24-98, *Flood Resistant Design and Construction Standard.* Reston, VA: American Society of Civil Engineers, 1998.

ASCE 32-01, *Design of Frost Protected Shallow Foundations.* Reston, VA: American Society of Civil Engineers, 2001.

ASTM A 36M-00, *Specification for Carbon Structural Steel.* West Conshohocken, PA: ASTM International, 2000.

ASTM A 252-E01, *Specification for Welded and Seamless Steel Pipe Piles.* West Conshohocken, PA: ASTM International, 2001.

ASTM A 283/A 283M-00, *Specification for Low and Intermediate Tensile Strength Carbon Steel Plates.* West Conshohocken, PA: ASTM International, 2000.

ASTM A 416/416M-99, *Specification for Steel Strand, Uncoated Seven-Wire for Prestressed Concrete.* West Conshohocken, PA: ASTM International, 1999.

ASTM A 572/A 572M-00, *Specification for High-Strength Low-Alloy Columbian-Vanadium Structural Steel.* West Conshohocken, PA: ASTM International, 2001.

ASTM A 588/A 588M-01, *Specification for High-Strength Low-Alloy Structural Steel with 50 ksi (345 Mpa) Minimum Yield Point to 4 inches Thick.* West Conshohocken, PA: ASTM International, 2001.

ASTM A 913/A 913M-01, *Specification for High-Strength Low-Alloy Steel Shapes of Structural Quality, Produced by Quenching and Self-Tempering Process (QST).* West Conshohocken, PA: ASTM International, 2001.

ASTM C 90-01A, *Specification for Load-bearing Concrete Masonry Units.* West Conshohocken, PA: ASTM International, 2001.

ASTM C 270-01A, *Specification for Mortar for Unit Masonry.* West Conshohocken, PA: ASTM International, 2001.

ASTM C 887-01, *Specification for Packaged, Dry, Combined Materials for Surface Bonding Mortar.* West Conshohocken, PA: ASTM International, 2001.

ASTM D 25-99E01, *Specification for Round Timber Piles.* West Conshohocken, PA: ASTM International, 1999.

ASTM D 422-98, *Standard Test Method for Particle Size Analysis of Soils.* West Conshohocken, PA: ASTM International, 1998.

ASTM D 1143-81 (1994)e1, *Standard Test Method for Piles Under Static Axial Compressive Load.* West Conshohocken, PA: ASTM International, 1994.

ASTM D 1557-00, *Standard Test Method for Laboratory Compaction Characteristics of Soil Using Modified Effort.* West Conshohocken, PA: ASTM International, 2000.

ASTM D 1668-97a, *Standard Specification for Glass Fabrics (Woven and Treated) for Roofing and Waterproofing.* West Conshohocken, PA: ASTM International, 1997.

ASTM D 2487-00, *Standard Practice for Classification of Soils for Engineering Purposes.* West Conshohocken, PA: ASTM International, 2000.

ASTM D 3689-90 (1995), *Specification of Individual Piles Under Static Axial Tensile Load.* West Conshohocken, PA: ASTM International, 1995.

ASTM D 4318-00, *Standard Test Method for Liquid Limit, Plastic Limit, and Plasticity Index of Soils.* West Conshohocken, PA: ASTM International, 2000.

ASTM D 4829-95, *Test Method for Expansion Index of Soils.* West Conshohocken, PA: ASTM International, 1995.

ASTM D 4832-02, *Standard Test Method for Preparation and Testing of Controlled Low Strength Material (CLSM) Test Cylinders.* West Conshohocken, PA: ASTM International, 2002.

ASTM D 4945-00, *Test Method for High-Strain Dynamic Testing of Piles.* West Conshohocken, PA: ASTM International, 2000.

ASTM D 5971-01, *Standard Practice for Sampling Freshly Mixed Controlled Low-Strength Material.* West Conshohocken, PA: ASTM International, 2001.

ASTM D 6023-02, *Standard Test Method for Unit Weight, Yield, Cement Content, and Air Content (Gravimetric) of Controlled Low-Strength Material (CLSM).* West Conshohocken, PA: ASTM International, 2002.

ASTM D 6024-02, *Standard Test Method for Ball Drop on Controlled Low-Strength Material (CLSM) to Determine Suitability for Load Application.* West Conshohocken, PA: ASTM International, 2002.

ASTM D 6103-97, *Standard Test Method for Flow Consistency of Controlled Low Strength Material (CLSM).* West Conshohocken, PA: ASTM International, 1997.

AWPA C1-00, *All Timber Products — Preservative Treatment by Pressure Process.* Grandbury, TX: American Wood-Preservers' Association, 2000.

AWPA C2-01, *Lumber, Timber, Bridge Ties and Mine Ties — Preservative Treatment by Pressure Process.* Grandbury, TX: American Wood-Preservers' Association, 2001.

AWPA C3-99, Piles — *Preservative Treatment by Pressure Processes.* Grandbury, TX: American Wood-Preservers' Association, 1999

AWPA C4-99, Poles — *Preservative Treatment by Pressure Processes.* Grandbury, TX: American Wood-Preservers' Association, 1999.

AWPA C9-00, *Plywood — Preservative Treatment by Pressure Processes.* Grandbury, TX: American Wood-Preservers' Association, 2000.

AWPA C22-96, *Lumber and Plywood for Permanent Wood Foundations — Preservative Treatment by Pressure Processes.* Grandbury, TX: American Wood-Preservers' Association, 1996.

AWPA C24-96, *Sawn Timber Piles Used for Residential Commercial Buildings.* Grandbury, TX: American Wood-Preservers' Association, 1996.

AWPA M4-01, *Standard for the Care of Preservative-Treated Wood Products.* Grandbury, TX: American Wood-Preservers' Association, 2001.

AWPB MP1, *Quality Control and Inspection Procedures for Dual Treatment of Marine Piling Pressure Treated with Waterborne Preservatives and Creosote for Use in Marine Waters.* American Wood Preservers Bureau.

AWPB MP2, *Quality Control and Inspection Procedures for Dual Treatment of Marine Piling Pressure Treated with Waterborne Preservatives and Creosote for Use in Marine Waters.* American Wood Preservers Bureau.

AWPB MP4, *Quality Control and Inspection Procedures for Dual Treatment of Marine Piling Pressure Treated with Waterbrone Preservatives and Creosote for Use in Marine Waters.* American Wood Preservers Bureau.

Design and Construction of Drilled Piers (ACI 336.3R-93), American Concrete Institute, Farmington Hills, MI: 1993. Also *ACI Manual of Concrete Practice*, Part 4.

DOC PS-20-99, *American Softwood Lumber Standard.* Gaithersburg, MD: U.S. Department of Commerce; National Institute of Standards and Technology, 1999.

FEMA 55, *Coastal Construction Manual*. Washington, DC: Federal Emergency Management Agency, 1999.

FEMA TB11-01, *Crawlspace Construction for Buildings Located in Special Flood Hazard Areas.* Washington, DC: Federal Emergency Management Agency, 2001.

Hanson, W.E., Peck, R.B. and T.H. Thornburn. *Foundation Engineering.* 2nd ed. New York, NY: John Wiley & Sons, 1974

IPC-03, *International Plumbing Code.* Falls Church, VA: International Code Council, 2003.

IRC-03, *International Residential Code*. Falls Church, VA: International Code Council, 2003.

NCMA report, *Research Evaluation of the Flexural Strength of Concrete Masonry, Project No. 93-172, Dec. 7, 1993.* Herndon VA: National Concrete Masonry Association, 1993.

NRCA Roofing and Waterproofing Manual, 3rd ed. Chicago, IL: National Roofing Contractors Association, 1989.

PTI-1996, *Design and Construction of Post-Tensioned Slabs-on-Ground, 2nd ed.* Phoenix, AZ: Post-Tensioning Institute, 1996.

"Recommended Practice for Design, Manufacture and installation of Prestressed Concrete Piling." *PCI Journal*, V. 38, No. 2, March-April 1993, pp. 14-41.

WRI/CRSI-96, *Design of Slab-on-Ground Foundations.* Leesburg, VA: Wire Reinforcing Institute, 1996.

Chapter 19: Concrete

General Comments

The words "cement" and "concrete" are used in several sections in this chapter. At the outset of this commentary, the distinction between cement and concrete should be made. Often, one may hear of references to cement floors, cement sidewalks and driveways, or of cement walls in describing walls made of cast-in-place or precast concrete or of concrete masonry, and the like. Cement and concrete are not interchangeable words. Cement is only one of four basic ingredients (cement, water, sand and stone) that make up the concrete material used in construction. In the end, there are only concrete structures and concrete building components. Structures of cement alone simply do not exist.

While concrete is the most versatile of the structural materials, it is also one of the most variable and complex materials used in building construction. Concrete is used in a number of different ways. For building substructures, it is used in spread footings, mats, piles and foundation walls. Above-ground concrete is used for building walls and enclosures, for floors and roof construction, and for structural framework. Concrete also has a number of ancillary uses in building facilities: the construction of sidewalks, patios and driveways; for sewers and basins; for equipment bases; for fire protection purposes; as an insulating material; and so on. Concrete can be cast into any desired shape that can be structurally formed in the field or in a precasting plant with a variety of architectural finishes and surface configurations to satisfy aesthetic requirements. Mixtures can be varied to achieve strengths [2,500 to 20,000 pounds per square inch (psi)] that meet the structural needs of the design. Concrete mixes can also be made to achieve good workability of the fresh material and to provide durability of the hardened product exposed to adverse environmental conditions, as well as to the conditions of wear imposed by the use of the facilities.

Because of the variable properties of the material and the numerous design and construction options available in the uses of concrete, due care and control throughout the entire construction process is necessary. This process includes: selection of materials and design of the mixture; mixing and transport of the concrete; proper construction and use of formwork; setting of the reinforcement and, in prestressed construction, the tensioning of the tendons and finally, the placement, finishing, curing and testing of the concrete material.

Purpose

The provisions of this chapter are intended to set the minimum accepted practices that apply to the design and construction of buildings or structural components using concrete.

Italics are used for text within Sections 1902 through 1908 of this code to indicate provisions that differ from ACI 318.

SECTION 1901
GENERAL

1901.1 Scope. The provisions of this chapter shall govern the materials, quality control, design and construction of concrete used in structures.

❖ Section 1901 provides information on the application and organization of the concrete provisions of Chapter 19. For additional background information on the provisions of this chapter, particularly as they relate to the inspection of concrete construction, consult PCA-EB115, *Concrete Inspection Handbook,* listed in the bibliography. Section 1901.1 provides a definitive statement on the important aspects of concrete construction covered by Chapter 19. Proper attention to these areas will improve the chances of a quality concrete structure that is code compliant.

1901.2 Plain and reinforced concrete. Structural concrete shall be designed and constructed in accordance with the requirements of this chapter and ACI 318 as amended in Section 1908 of this code. Except for the provisions of Sections 1904 and 1911, the design and construction of slabs on grade shall not be governed by this chapter unless they transmit vertical loads or lateral forces from other parts of the structure to the soil.

❖ The structural concrete provisions of the code comply strictly with the provisions of ACI 318/318R, except for the modifications contained in Section 1908. Slabs on grade are another important exception. Slabs on grade typically only convey the gravity dead load of the slab and the gravity live load superimposed on the slab directly to the soil; thus, slabs are not considered to be "structural concrete" that must comply with the design provisions of ACI 318. Slabs on grade, however, must comply with the durability provisions of Section 1904 and the minimum thickness and vapor barrier provisions of Section 1911.

The last sentence of Section 1901.2 clarifies that slabs on grade used to transmit "vertical loads", "lateral forces" or both from other parts of the structure to the soil must comply with all the applicable portions of Chapter 19 and ACI 318. In many cases, slabs on grade are used to transmit lateral forces due to wind or seismic into the ground. A common example is a typical preengineered metal building in which the shear at the bottom of the columns is resisted by placing "hairpins" around the column anchor bolts and extending the hairpins into the slab on grade. Although this may appear to be a new requirement, codes have for many years required that structures be designed so that they do not slide or overturn.

1901.3 Source and applicability. The contents of Sections 1902 through 1907 of this chapter are patterned after, and in general conformity with, the provisions for structural concrete in ACI 318. Where sections within Chapters 2 through 7 of ACI 318 are referenced in other chapters and appendices of ACI 318, the provisions of Sections 1902 through 1907 of this code shall apply.

❖ Section 1901.3 references ACI 318; thus, all of the requirements in ACI 318 become an enforceable part of the code. Sections 1902 through 1907 of the code closely parallel Chapters 2 through 7 of ACI 318; however, only those construction provisions that are needed by code enforcement personnel in the field have been transcribed (reprinted) in the code. For example, Section 1903.6 on admixtures provides abbreviated provisions when compared with Section 3.6 of ACI 318, even though the more complete provisions of ACI 318 must be followed. Architects, engineers, building officials and others who are responsible for the design and construction of concrete must use ACI 318 as modified by this chapter to obtain the full requirements of the code.

The 1999 edition of ACI 318 includes requirements for all structural concrete. The term "structural concrete" refers to all concrete used for structural purposes, including both plain and reinforced. By definition, "Plain concrete" is either unreinforced or contains less reinforcement than would be required for "reinforced concrete" by ACI 318. The definition of "Reinforced concrete" includes conventionally reinforced and prestressed concrete. ACI 318 is equally applicable to cast-in-place concrete and precast concrete. Additionally it has provisions for composite concrete and steel compression members utilizing structural steel shapes and steel pipe or tubing.

The general purpose of ACI 318 is to provide a minimum acceptable standard for use in governing concrete material properties and the design and construction of concrete structural elements. While engineering knowledge, experience and sound judgement must prevail, nonetheless, structural elements of concrete must comply with ACI 318 unless preconstruction load tests are conducted in accordance with Section 1714 of the code.

ACI 318 generally does not govern special concrete structures, such as arches, tanks, reservoirs, bins and silos, blast-resistant structures and chimneys; however, portions of it may be applicable (e.g., durability requirements).

ACI 318 does not govern the design and construction of composite concrete slabs on steel deck, as this type of construction is regulated by ASCE 3 (see Section 2209.2). However, ASCE 3 references the applicable provisions of ACI 318 (i.e., Chapters 2 through 7 of ACI 318, which are partially transcribed in Sections 1902 through 1907 of this chapter). The design and construction of concrete slabs cast on noncomposite, stay-in-place steel deck are regulated by this chapter and ACI 318.

ACI 318 does not govern portions of concrete piles, caissons and drilled piers embedded in soil capable of providing lateral support, except in buildings assigned to Seismic Design Category D, E or F (see Section 1808.2.23). In general, portions of concrete pier and pile foundations embedded in soil capable of providing lateral support to the foundation element must comply with Sections 1808 through 1812. Portions of concrete piles, caissons and drilled piers in soil not capable of providing lateral support, projecting above the ground surface or in water must comply with this chapter and ACI 318 (see Section 1808.2.9.2).

Generally, foundation walls and shallow footings must comply with Chapter 19 and ACI 318; however, the code user should be aware of requirements located in Section 1805 that may prevail over those in this chapter.

ACI 318 cites numerous material and test standards of the ASTM International (ASTM) that are applicable to concrete construction. These ASTM standards are declared to be part of the ACI document and, therefore, also become enforceable under the code. A few of the standards referenced in ACI 318 are repeated in the code under Section 1903 and other sections of Chapter 19.

Chapter 1 of ACI 318 covers general requirements that may be in conflict with the administration and enforcement sections of the code, particularly those provisions that relate to drawings and inspections. In cases of conflict, the provisions in the code will govern in accordance with Section 102.4.

1901.4 Construction documents. The construction documents for structural concrete construction shall include:

1. The specified compressive strength of concrete at the stated ages or stages of construction for which each concrete element is designed.
2. The specified strength or grade of reinforcement.
3. The size and location of structural elements, reinforcement, and anchors.
4. Provision for dimensional changes resulting from creep, shrinkage and temperature.
5. The magnitude and location of prestressing forces.
6. Anchorage length of reinforcement and location and length of lap splices.
7. Type and location of mechanical and welded splices of reinforcement.

8. Details and location of contraction or isolation joints specified for plain concrete.
9. Minimum concrete compressive strength at time of posttensioning.
10. Stressing sequence for posttensioning tendons.
11. For structures assigned to Seismic Design Category D, E or F, a statement if slab on grade is designed as a structural diaphragm (see Section 21.10.3.4 of ACI 318).

❖ This section provides a list of important information that must be included in construction documents. These items, which are specifically related to concrete construction, are in addition to the structural design information required by Section 1603, and additional requirements of Section 106. The list is not all-inclusive and may be expanded at the discretion of the building official.

1901.5 Special inspection. The special inspection of concrete elements of buildings and structures and concreting operations shall be as required by Chapter 17.

❖ This section provides a cross reference to the code's special inspection provisions for concrete found in Section 1704, which specifies that most structural concrete, with the exception of some foundations, needs to have a special inspection. ACI 318 does not contain provisions for the special inspection of concrete, except that Section 1.3.5 requires continuous inspection of the placement of reinforcement and concrete in special moment frames that resist seismic forces in regions of high seismic risk, which correspond to Seismic Design Category D, E or F of the code.

SECTION 1902 DEFINITIONS

1902.1 General. The following words and terms shall, for the purposes of this chapter and as used elsewhere in this code, have the meanings shown herein.

❖ Section 1902 contains definitions of terms that are associated with the subject matter of this chapter. It is important to emphasize that these terms are not exclusively related to this chapter but are applicable everywhere the term is used in the code. The code has transcribed only those definitions for concrete construction that are used in Chapter 19. Additional definitions and related commentary for Chapter 19 are contained in ACI 318, Chapter 2.

Definitions of terms can help in the understanding and application of code requirements. The purpose for including these definitions within this chapter is to provide more convenient access to them without having to refer back to Chapter 2 of the code. Also, for convenience, these terms are listed in Chapter 2 with a cross reference to this section. The use and application of all defined terms, including those defined herein, are set forth in Section 201.

ADMIXTURE. Material other than water, aggregate or hydraulic cement, used as an ingredient of concrete and added to concrete before or during its mixing to modify its properties.

❖ This definition identifies the types of materials regulated by the requirements of Section 1903.6. Admixtures are incorporated into the concrete mixture to modify some aspect of the concrete's performance. For example, admixtures can retard the setting of concrete in hot weather or accelerate the setting and strength gain in cold-weather conditions, enhance the workability of the concrete and improve the concrete's resistance to freezing and thawing, and to deicer chemicals.

AGGREGATE. Granular material, such as sand, gravel, crushed stone and iron blast-furnace slag, used with a cementing medium to form a hydraulic cement concrete or mortar.

❖ Concrete and mortar are composed of portland cement or any other hydraulic cement, aggregate and water. Concrete generally contains fine and coarse aggregate, while mortar only contains fine aggregate. This definition lists the types of material that may be used as fine and coarse aggregate as regulated by Section 1903.3.

AGGREGATE, LIGHTWEIGHT. Aggregate with a dry, loose weight of 70 pounds per cubic foot (pcf) (1120 kg/m^3) or less.

❖ Lightweight aggregates have densities significantly lower than normal-weight aggregates, ranging from 35 to 70 pounds per cubic foot (pcf) (560 to 1120 kg/m^3), as compared to 75 to 110 pcf (1200 to 1760 kg/m^3) for normal-weight aggregates.

CEMENTITIOUS MATERIALS. Materials as specified in Section 1903 that have cementing value when used in concrete either by themselves, such as portland cement, blended hydraulic cements and expansive cement, or such materials in combination with fly ash, other raw or calcined natural pozzolans, silica fume, and/or ground granulated blast-furnace slag.

❖ This definition identifies the types of materials regulated by the requirements of Section 1904.1. Note that cementitious materials include materials in addition to portland cement. Cementitious materials, when combined with water, form the paste that binds the aggregate together to form the hardened concrete.

COLUMN. A member with a ratio of height-to-least-lateral dimension exceeding three, used primarily to support axial compressive load.

❖ This definition is necessary to distinguish between structural elements that must be designed as a column or wall versus those that may be designed as a pedestal (i.e., height-to-least-lateral-dimension ratio of three or less). The column definition does not require that the element be vertical. By definition, columns may also resemble a wall, as long as their primary purpose is to support axial compressive loads and they meet the prescribed ratio of height-to-least-lateral dimension (see the definition of "Pedestal").

CONCRETE. A mixture of portland cement or any other hydraulic cement, fine aggregate, coarse aggregate and water, with or without admixtures.

❖ This definition provides insight as to the necessary materials that must be included in concrete. The definition is necessary to distinguish between concrete and admixtures used in concrete and for the application of the requirements of Chapter 19 (see the definition of "Admixture").

CONCRETE, SPECIFIED COMPRESSIVE STRENGTH OF, (f'_c). The compressive strength of concrete used in design and evaluated in accordance with the provisions of Section 1905, expressed in pounds per square inch (psi) (MPa). Whenever the quantity f'_c is under a radical sign, the square root of the numerical value only is intended, and the result has units of psi (Mpa).

❖ Compressive strength is one of the primary physical properties of concrete that is a measurement of maximum resistance of a concrete specimen to axial loading. It also provides an index of its durability to resist anticipated exposure conditions. Specified compressive strength, (f'_c), is generally expressed in pounds per square inch (psi) (MPa) at an age of 28 days and is selected by the architect or engineer to meet loading and exposure conditions.

CONTRACTION JOINT. Formed, sawed or tooled groove in a concrete structure to create a weakened plane and regulate the location of cracking resulting from the dimensional change of different parts of the structure.

❖ Contraction joints (also called "control joints") provide predetermined locations for cracks to form to relieve tensile stresses due to restraint of movement in the plane of a slab or wall caused by drying or thermal shrinkage or both. Contraction joints should be constructed to permit transfer of loads perpendicular to the plane of the slab or wall across the joint.

DEFORMED REINFORCEMENT. Deformed reinforcing bars, bar mats, deformed wire, welded plain wire fabric and welded deformed wire fabric conforming to ACI 318, Section 3.5.3.

❖ The code requires reinforcement used in concrete to be of steel and meet one of several ASTM standards in accordance with ACI 318, Section 3.5.3. Each standard establishes one or more grades for classification of the bars and wires. In addition, the standards establish minimum requirements for: spacing and height of deformations, nominal weight, diameter, cross-sectional area and perimeter and minimum tensile and yield strengths, among other requirements. It should be noted that plain welded wire fabric is considered to be deformed reinforcement, since the welded wire intersections serve to "bond" the reinforcement to the concrete in the same way as deformations on a deformed bar.

DUCT. A conduit (plain or corrugated) to accommodate prestressing steel for post-tensioned installation.

❖ This term is used by the prestressing industry to refer to the sleeves used for installing prestressing steel in post-tensioned elements. Requirements for post-tensioning ducts can be found in ACI 318, Section 18.17.

EFFECTIVE DEPTH OF SECTION (*d*). The distance measured from extreme compression fiber to the centroid of tension reinforcement.

❖ Steel reinforcement used in flexural members, such as beams and slabs, must be located where it will be effective in resisting flexural tension stresses under loading conditions. Construction documents generally do not specify this location as the effective depth of section, (*d*). The effective depth, *d*, is usually the distance from the side of the member furthermost from the steel. For example, in a slab, *d* for the bottom steel is the distance from the top of the slab to the centerline of the bar. In the case of a slab, typically *d* is determined by subtracting the bottom cover plus one-half the bar diameter from the slab thickness. Tolerances for depth, *d*, are provided in Section 1907.5.2.

ISOLATION JOINT. A separation between adjoining parts of a concrete structure, usually a vertical plane, at a designed location such as to interfere least with performance of the structure, yet to allow relative movement in three directions and avoid formation of cracks elsewhere in the concrete and through which all or part of the bonded reinforcement is interrupted.

❖ Isolation joints (often mistakenly called "expansion joints") permit both horizontal and vertical differential movement at adjoining parts of a structure. Isolation joints are usually necessary around the perimeter of a slab on grade where it abuts a wall, and around columns penetrating a slab on grade to prevent the concrete slab from bonding with the previously placed concrete.

PEDESTAL. An upright compression member with a ratio of unsupported height-to-average-least-lateral dimension of three or less.

❖ By definition, a pedestal is a short column that is required to be upright. It should be noted that a column has a ratio of height-to-least-lateral dimension exceeding three, whereas a pedestal has a ratio of height-to-least-lateral dimension of three or less. Pedestals are permitted to be constructed of plain concrete by Section 22.8 of ACI 318 and their use is common in residential construction. See Section 1910 for limitations on the use of plain concrete pedestals in buildings assigned to Seismic Design Category C, D, E or F.

PLAIN CONCRETE. Structural concrete with no reinforcement or with less reinforcement than the minimum amount specified for reinforced concrete.

❖ Plain concrete is generally used for residential and light commercial buildings, mainly in the construction of walls, footings and pedestals. Where plain concrete is used in structural members, it is called "structural plain concrete." Code provisions for structural plain concrete are contained in Section 1909. The use of plain concrete in structures assigned to Seismic Design Category C, D, E or F is either limited or prohibited (see Section 1910 of the code and Section 22.10 of ACI 318).

PLAIN REINFORCEMENT. Reinforcement that does not conform to the definition of "Deformed reinforcement" (see ACI 318, Section 3.5.4).

❖ Plain reinforcement does not have deformations or "ridges" to qualify as deformed reinforcement according to ACI 318. An example of plain reinforcement is smooth wire.

POSTTENSIONING. Method of prestressing in which prestressing steel is tensioned after concrete has hardened.

❖ This term refers to prestressing that is accomplished by tensioning prestressing steel after the concrete has hardened (see the definition for "Prestressed concrete").

PRECAST CONCRETE. A structural concrete element cast elsewhere than its final position in the structure.

❖ Precast concrete members may be cast on site and subsequently lifted into place (e.g., tilt-up wall construction) or cast off site, usually in a plant, and transported to the construction site for erection. Precast concrete manufactured under plant control conditions has closer tolerances than cast-in-place concrete construction. This results in the need for lesser minimum concrete cover for reinforcement versus that needed for cast-in-place concrete (see ACI 318, Sections 7.7.1 and 7.7.2). Code provisions for precast concrete are contained in ACI 318, Chapter 16.

PRESTRESSED CONCRETE. Structural concrete in which internal stresses have been introduced to reduce potential tensile stresses in concrete resulting from loads.

❖ Internal stresses are created in prestressed concrete by tendons consisting of wires, strands or bars that are loaded in tension. The tensile force in the tendon is transferred to the concrete as a compressive force, thereby increasing the load-carrying capacity of the concrete.

There are two prestressing methods. In the "pretensioning" method, tendons are pretensioned prior to the placement of concrete. While the tendons are in an elongated state, and after the concrete has reached a predetermined strength, the tendons are cut and the prestress force is transferred to the concrete through the bond between the steel and concrete. Pretensioned concrete members are normally cast off site; therefore, they are referred to as precast, prestressed concrete members.

In the "posttensioning" method, the tendons are contained in conduit or sleeves (normally a plastic sheathing) and placed in the forms in the desired profile prior to placement of the concrete. The sheathing prevents the concrete from bonding to the tendons. After the concrete has achieved sufficient strength, the tendons are tensioned by a jacking machine and anchored in place with special fittings generally located at the ends of the concrete member. Tendons that are inserted into hollow conduits are normally grouted in their conduits after tensioning to create "bonded prestressing tendons." Tendons that are encased or wrapped in a plastic sheathing cannot be grouted, and are called "unbonded prestressed tendons." Posttensioned concrete members are generally cast in place the same as conventional nonprestressed concrete.

Code provisions for prestressed concrete are contained in ACI 318, Chapter 18.

PRESTRESSING STEEL. High-strength steel element such as wire, bar or strand, or a bundle of such elements, used to impart prestress forces to concrete.

❖ In prestressed concrete, this term refers to the steel that is used to apply the prestressing force to the concrete and consists of wires, cables, bars, rods or strands that are loaded in tension. Section 3.5.5 of ACI 318 specifies the materials that can be used.

PRETENSIONING. Method of prestressing in which prestressing steel is tensioned before concrete is placed.

❖ This term refers to prestressing that is accomplished by tensioning prestressing steel before the concrete has hardened (see the definition of "Prestressed concrete").

REINFORCED CONCRETE. Structural concrete reinforced with no less than the minimum amounts of prestressing steel or nonprestressed reinforcement specified in ACI 318, Chapters 1 through 21 and ACI 318 Appendices A through C.

❖ Reinforced concrete is structural concrete in which steel reinforcement or prestressing steel is placed within the concrete member so the two materials act together to resist forces. Reinforcement may be provided to resist tensile, compressive or shear stresses. Welded wire fabric is typically placed in concrete slabs on grade to distribute cracks caused by drying, shrinkage or temperature changes or both, and normally does not constitute structural reinforcement.

REINFORCEMENT. Material that conforms to Section 1903.5, excluding prestressing steel unless specifically included.

❖ Reinforcement is the term used to describe the steel bars or wire fabric placed in concrete. The reinforcement may be used to resist flexural tension or compressive stresses in beams, slabs, walls and columns; axial tension and compressive stresses in columns and walls; shear stresses in beams and columns and torsional stresses in beams. In addition, ties or spirals are used in members with compression reinforcement to laterally restrain the longitudinal reinforcement. Plain bars (e.g., those used as dowels for shear transfer at contraction joints), anchor bolts and other metal inserts are generally not considered to be reinforcement under the provisions of the code. Prestressing tendons are excluded from the definition because they are not to be considered as reinforcement unless in a tensioned condition (see the definition of "Tendon").

RESHORES. Shores placed snugly under a concrete slab or other structural member after the original forms and shores have been removed from a larger area, thus requiring the new slab or structural member to deflect and support its own weight and existing construction loads applied prior to the installation of the reshores.

❖ Reshoring is similar to shoring; however, reshores are placed after the shoring has been removed and formwork has been stripped and before the new structural members are allowed to carry additional construction loads. The reshores transfer construction loads to other structural members (typically floors) and minimize deflection of the "green" concrete member. Code provisions for reshores are contained in Section 1906.2.

SHORES. Vertical or inclined support members designed to carry the weight of the formwork, concrete and construction loads above.

❖ Some structures require support for the formwork and "green" concrete other than the actual forms themselves. The code uses the term "shores" for this temporary supporting structure. Code provisions for shores are contained in Section 1906.2.

SPIRAL REINFORCEMENT. Continuously wound reinforcement in the form of a cylindrical helix.

❖ All longitudinal bars in compression members, such as columns, must be enclosed with lateral reinforcement. Spiral reinforcement is commonly used for lateral reinforcement in circular columns because it provides a cost efficient and structurally effective substitute for traditional ties. Code provisions for spiral reinforcement are contained in ACI 318, Section 7.10.4.

STIRRUP. Reinforcement used to resist shear and torsion stresses in a structural member; typically bars, wires or welded wire fabric (plain or deformed) either single leg or bent into L, U or rectangular shapes and located perpendicular, or at an angle to, longitudinal reinforcement. (The term "stirrups" is usually applied to lateral reinforcement in flexural members and the term "ties" to those in compression members.)

❖ Stirrups are used to resist diagonal tension that develops in a beam as a result of shear in excess of that which the concrete is permitted to resist. Stirrups are typically composed of smaller reinforcing bars bent in the shape of a U and tied vertically around the sides and bottom of the main longitudinal reinforcement in a beam. Stirrups are generally spaced closer together at the ends of beams where shear loads are greater. Occasionally one may encounter the use of inclined stirrups or bent-up bottom bars used as shear reinforcement. It is critical that inclined stirrups slope up toward the nearer support if they are to be effective in resisting shear. Code provisions for stirrups are contained in ACI 318, Sections 11.5 and 12.13.

STRUCTURAL CONCRETE. Concrete used for structural purposes, including plain and reinforced concrete.

❖ The majority of Chapter 19 is consistent with ACI 318/318R. The definition of "Structural concrete" clarifies that the concrete design and construction provisions of the code apply to both plain and reinforced concrete (see the definitions of "Plain concrete" and "Reinforced concrete").

TENDON. In pretensioning applications, the tendon is the prestressing steel. In posttensioned applications, the tendon is a complete assembly consisting of anchorages, prestressing steel and sheathing with coating for unbonded applications or ducts with grout for bonded applications.

❖ In prestressed concrete that is pretensioned, this term refers to the prestressing steel, while in post-tensioned applications it is a collective term for the complete assembly. Tendons may not be used as reinforcement in an unstressed physical state (see the definitions of "Reinforcement" and "Prestressed concrete").

SECTION 1903
SPECIFICATIONS FOR TESTS AND MATERIALS

1903.1 General. Materials used to produce concrete and testing thereof shall comply with the applicable standards listed in ACI 318 and this section. *Tests of concrete and the materials used in concrete shall be in accordance with ACI 318, Section 3.8. Where required, special inspections and tests shall be in accordance with Chapter 17.*

❖ Section 1903 regulates the materials commonly used for both reinforced and plain concrete construction and provides the basis for the selection and acceptance of such materials. The requirements conform to the provisions of ACI 318.

Section 1903.1 sets the basis for the code provisions covering the selection and acceptance of materials used in plain and reinforced concrete construction, including the admixtures in concrete. It clarifies that the code user must use ACI 318 for tests for concrete and its ingredients.

1903.2 Cement. Cement used to produce concrete shall comply with ACI 318, Section 3.2.

❖ The words "cement" and "concrete" are used in several sections of Chapter 19. It is important to understand the distinction between cement and concrete. Often, one may hear of references to cement floors, or cement sidewalks and driveways, or of cement walls in describing walls made of cast-in-place or precast concrete or of concrete masonry. Cement and concrete are not interchangeable words. Cement is only one of three basic ingredients (cement, water and aggregate) that make up the concrete material used in construction. In the end use, there are only concrete structures and concrete building components. Structures of cement alone simply do not exist. The code permits the use of cements that meet ASTM C 150, ASTM C 595, ASTM C 845 and ASTM C 1157. General information on each of these cements is provided below.

Portland Cement—ASTM C 150:

ASTM C 150 specifies the chemical composition, physical properties and test methods for producing eight types of portland cement as follows:

1. Type I portland cement is for general use where the special properties of other types of cement are not required. While Type I is the most widely used type of portland cement produced, it should not be used in concrete structures that will be subject to sulfate attack, such as from soil or ground water, or in large concrete placements where the temperature rise from the heat of hydration produced by the chemical reaction of cement and water can become an objectionable condition in the construction process.
2. Type IA portland cement is an air-entraining cement for general use, as is Type I portland cement, but it is used specifically where the benefits of air entrainment are desired. The purpose of air entrainment is discussed in other parts of this commentary.
3. Type II portland cement is for general use when moderate sulfate resistance or moderate heat of hydration is desired.
4. Type IIA portland cement is for general use, as is Type II portland cement, but is used where the benefits of air entrainment are also desired.
5. Type III portland cement is a high-early-strength cement especially used to accelerate the hardening of wet concrete to facilitate construction scheduling.
6. Type IIIA portland cement has the same properties and is used for the same purposes as Type III portland cement, but is also used where the benefits of air entrainment are desired.
7. Type IV portland cement is used especially where a low heat of hydration is required. It is used principally in mass concrete construction such as dams, bridge abutments, heavy foundations, etc. It is seldom used for building construction.
8. Type V portland cement is for special use where high sulfate resistance is desired, and is principally used in areas where the soils or ground waters have high concentrations of sulfate.

Blended Hydraulic Cement—ASTM C 595:

ASTM C 595 specifies five types of blended cements as follows:

1. Portland blast-furnace slag cement, Type IS, is for use in general concrete construction. Portland cement (ASTM C 150) may be modified by blending fine-granulated, blast-furnace slag to provide moderate sulfate resistance, Type IS (MS); air entrainment, Type IS (A); moderate heat of hydration, Type IS (MH) or any combination by adding the indicated suffixes.
2. Portland-pozzolan cement, Type IP, is for general use and will perform similar to Type I portland cement except that it can have slightly lower 28-day concrete compressive strength and can be less resistant to deicer scaling. The cement can be modified to provide moderate sulfate resistance, Type IP (MS); air entrainment, Type IP (A); moderate heat of hydration, Type IP (MH) or any combination by adding the indicated suffixes.
3. Slag cement, Type S, is used in combination with portland cement in making concrete or in combination with hydrated lime in making masonry mortar. It can be modified to provide air entrainment, Type S (A). Section 3.2.1 of ACI 318 prohibits the use of slag cement in structural concrete.
4. Pozzolan-modified portland cement, Type I (PM), is used in general concrete construction and can be modified to provide moderate sulfate resistance, Type I (PM)(MS); air entrainment, Type I (PM)(A); or moderate heat of hydration, Type I (PM)(MH), respectively.
5. Slag-modified portland cement, Type I (SM), is used for general concrete construction and may be modified to provide moderate sulfate resistance, air entrainment or moderate heat of hydration by adding the suffixes (MS), (A) or (MH), respectively.

Expansive Hydraulic Cements—ASTM C 845:

Expansive hydraulic cements expand during the early hardening period after the concrete has set. The cements are essentially composed of hydraulic calcium aluminates and calcium sulfates. These cements are used on projects where shrinkage-compensating concrete is specified. When restrained, the expansion of shrinkage-compensating concrete induces compressive stresses that can offset tensile stresses due to drying shrinkage in the concrete member. Typically, this type of concrete is used for floor slabs and pavements to minimize or eliminate drying-shrinkage cracking. The cement is classified as Type E-1 and will include the letter-designating suffix (K), (M) or (S), depending on the chemical composition of the cement.

Performance Blended Hydraulic Cement—ASTM C 1157:

ASTM C 1157 differs from ASTM C 150 and ASTM C 595 in that it does not establish the chemical composition of the different types of cements; however, individual constituents used to manufacture ASTM C 1157 cements must comply with the requirements specified in the standard. The standard also provides for several optional requirements, including one for cement with low reactivity to alkali-reactive aggregates.

1903.3 Aggregates. Aggregates used in concrete shall comply with ACI 318, Section 3.3.

❖ Section 3.3 of ACI 318 requires that concrete aggregates conform to ASTM C 33 or ASTM C 330. All fine and coarse aggregates used in producing normal-weight concrete [135 to 160 psf (6.5 to 7.6 kNm^2)] must comply with the provisions of ASTM C 33. All aggregates used in making lightweight structural concrete [85 to 115 psf (4 to 5.5 kNm^2)] must comply with the requirements of ASTM C 330. Both ASTM C 33 and ASTM C 330 specify physical property requirements and provide for sampling and testing of aggregate materials.

Fine aggregates used in normal-weight concrete consist of natural sand, manufactured sand or a combination of both materials. Coarse aggregates consist of either gravel, crushed gravel, crushed stone, air-cooled blast-furnace slag, crushed hydraulic-cement concrete or combinations of any of these materials. Aggregates used for making lightweight structural concrete are expanded shale, clay, slate and slag. Lightweight aggregates, such as pumice, pearlite and vermiculite, are generally used in making insulating concrete, but their use is prohibited in structural concrete.

The performance characteristics of concrete are greatly affected by the selection and proportioning of aggregates in the concrete mixture. Aggregates not only affect the workability of fresh concrete, but in the hardened state they are a factor in providing adequate resistance to abrasion; freezing and thawing, drying shrinkage and fire exposure. They also affect the strength of concrete.

Harmful substances may be present in aggregates in such quantities that the performance of the concrete could be seriously affected. ASTM C 33 and ASTM C 330 set content limitations for such deleterious materials.

Sometimes the practicalities of construction necessitate deviations from normal practices. Aggregates that conform to ASTM material standards are not always economically or readily available, and local, nonconforming aggregates may be successfully used in concrete. In such cases, ACI 318 permits evidence of a long history of satisfactory performance or special tests conducted to show adequate strength and durability of the concrete to serve as the basis for acceptance of nonconforming aggregate materials.

1903.4 Water. Water used in mixing concrete shall be clean and free from injurious amounts of oils, acids, alkalis, salts, organic materials or other substances that are deleterious to concrete or steel reinforcement and shall comply with ACI 318, Section 3.4.

❖ The water used in concrete mixtures is required to be clear and clean. It must not contain injurious amounts of oils, acids, alkalis, salts, organic materials or other substances that are harmful to concrete or concrete reinforcement.

Potable water obtained from municipal water supply systems is normally suitable for use as mixing water in concrete. However, this does not mean that nonpotable water cannot be used in making concrete. ACI 318 permits the use of nonpotable water under specific conditions of acceptance. Water that appears discolored or has an unusual or repulsive taste or odor must undergo an analysis using applicable ASTM test methods. Evidence of its acceptability can be provided by checking service records of concrete made with such water, or from other information that demonstrates the water will not be detrimental to the concrete. In the absence of such evidence, water suspected of being of questionable quality can be subject to the acceptance criteria for concrete strength and time set contained in the ASTM C 94.

Environmental concerns have caused many ready-mixed producers to reuse washout water (i.e., water used to reclaim aggregate or clean out returning trucks) as mix water. This is permitted by ASTM C 94, provided the water meets the criteria outlined in the standard. In addition, the standard permits the concrete purchaser to require optional testing to determine chloride ion, sulfate, alkali and total solids in the water. This is particularly important where chloride ion content of the concrete must be limited for corrosion protection, where aggregate susceptible to alkali reactivity will be used, or both (see Section 1904.4 or ACI 318, Section 4.4.1).

1903.5 Steel reinforcement. Reinforcement and welding of reinforcement to be placed in concrete construction shall conform to the requirements of this section.

❖ This section specifies the types and shapes of steel that are permitted to be used as reinforcement in structural concrete and the provisions under which reinforcement may be welded.

1903.5.1 Reinforcement type. Reinforcement shall be deformed reinforcement, except plain reinforcement is permitted for spirals or prestressing steel, and reinforcement consisting of structural steel, steel pipe or steel tubing is permitted where specified in ACI 318. Reinforcement shall comply with ACI 318, Section 3.5.

❖ Steel reinforcement conforming to the requirements of ACI 318 must be of the deformed type (with lugs or protrusions), except that plain steel reinforcement can be used for spirals and prestressing tendons. Under certain conditions, reinforcement can also consist of structural steel, steel pipe or steel tubing.

ACI 318, Section 3.5, requires that reinforcement for concrete must comply with one of the following ASTM material standards:

- Deformed reinforcement;
- Deformed and plain billet-steel bars, ASTM A 615;
- Deformed and plain low-alloy steel bars, ASTM A 706. (This special reinforcement is used primarily in structures assigned to Seismic Design Category D, E or F and is not readily available in all parts of the country.); or
- Rail-steel and axle-steel deformed bars, ASTM A996.

Bar mats for concrete reinforcement must conform to ASTM A 184 and be fabricated from reinforcing bars that conform to either ASTM A 615, ASTM A 616, ASTM A 617 or ASTM A 706.

Deformed wire must conform to ASTM A 496, except that the wire must not be smaller than Size D-4; and welded deformed wire fabric for concrete reinforcement must conform to ASTM A 497. Welded intersections must not be spaced farther apart than 16 inches (406 mm) in the direction of the principal reinforcement, except when used as stirrups.

Welded plain wire fabric for concrete reinforcement must conform to ASTM A 185. The code considers welded plain fabric to be deformed reinforcement.

Reinforcing bars may be zinc coated (galvanized) or epoxy coated to provide corrosion resistance.

Galvanized bars must conform to ASTM A 767. Epoxy-coated bars must conform to ASTM A 775.

Steel wire and welded wire fabric may be epoxy coated in accordance with ASTM A 884 to provide corrosion resistance.

ASTM A 996 is a replacement for ASTM A 616 and ASTM A 617. Unlike its predecessors, ASTM A 996 does not provide for plain bars. Although manufacturers will be producing bars in accordance with the new standard, there are no additional changes that will affect those involved in the design, construction and inspection of reinforced concrete structures.

Plain reinforcement:
Plain bars for spiral reinforcement must conform to either ASTM A 615 or ASTM A 706. Smooth wire for spiral reinforcement must conform to ASTM A 82.

Prestressing steel:
Steel used for prestressed concrete must conform to one of the following ASTM standards:

1. Wire, including low-relaxation wire, must be uncoated, stress-relieved steel wire, ASTM A 421;
2. Strand, including low-relaxation wire, must be uncoated, seven-wire, stress-relieved steel, ASTM A 416;
3. High-strength steel bars, ASTM A 722; or
4. Wire, strands and bars that are not specifically listed in ASTM A 421, ASTM A 416 or ASTM A 722 can be used, provided they conform to the minimum requirements of these specifications and do not have properties that make the materials less satisfactory than intended by these standard specifications.

Structural steel:
Structural steel used with reinforcing bars in composite compression members must conform to one of the following ASTM standards:

1. Carbon steel shapes of structural quality, ASTM A 36;
2. High-strength, low-alloy structural steel shapes, ASTM A 242;
3. High-strength, low-alloy columbium-vanadium steels of structural quality, ASTM A 572; or
4. High-strength, low-alloy structural steel with 50,000 psi (344 750 kPa) minimum yield point to 4 inches (102 mm) thick, ASTM A 588.

Steel pipe or tubing:
Steel pipe or tubing for composite compression members consisting of a steel-encased concrete core must comply with one of the following ASTM standards:

1. Grade B seamless and welded black and hot-dipped galvanized steel pipe, ASTM A 53;
2. Cold-formed welded and seamless carbon steel structural tubing of round, square, rectangular or special shape, ASTM A 500; or
3. Hot-formed welded and seamless carbon steel structural tubing of round, square, rectangular or special shape, ASTM A 501.

See Section 1916 for concrete-filled pipe columns.

1903.5.2 Welding. Welding of reinforcing bars shall conform to AWS D1.4. Type and location of welded splices and other required welding of reinforcing bars shall be indicated on the design drawings or in the project specifications. The ASTM reinforcing bar specifications, except for ASTM A 706, shall be supplemented to require a report of material properties necessary to conform to the requirements in AWS D1.4.

❖ The most important considerations before welding reinforcing bars are the weldability of the steel and the compatibility of the welding process. The American Welding Society (AWS) specifies the proper welding practices in the D1.4 standard.

The ASTM standards for concrete reinforcement permitted in the code each contain the material properties of the specific types of metal. The steel chemistry set forth in any of these standards must be restricted to a given range to provide compatibility with the welding requirements of AWS D1.4. Therefore, to assure the weldability of the steel and conformance with AWS requirements, this section stipulates that the ASTM reinforcing bar specifications must be supplemented with a report addressing the material properties of the particular type of steel to be used in the welding process. However, the steel produced using ASTM A 706 is exempt from the requirement for a supplemental report because it was purposely developed for welding and already has a restricted chemistry.

Where there is a need to weld existing reinforcing bars in making alterations to existing buildings, and mill reports of the steel are not available, AWS D1.4 requires that a chemical analysis be made to determine the composition of the steel. As a practical alternative, an assumption can be made that the carbon equivalent is above 0.75 percent for the reinforcement, and the steel can be welded after a preheat temperature of 500°F (260°C) is applied. While this procedure solves the welding problem, consideration must be given to the possible effects of heat damage to the concrete and the stress levels in the reinforcing steel.

1903.6 Admixtures. Admixtures to be used in concrete shall be subject to prior approval by the registered design professional and shall comply with ACI 318, Section 3.6.

❖ By definition, an admixture is a material other than water, aggregates or hydraulic cement, used as an ingredient in concrete (or mortar) and added to the concrete immediately before or during mixing. There are four reasons for using admixtures:

1. To achieve certain desired properties in concrete more readily and effectively than can be obtained by other means;
2. To insure the quality of concrete during the mixing, transporting, placing and curing operations under adverse weather conditions;
3. To overcome certain adverse conditions during concreting operations; and
4. To reduce the cost of concrete construction.

While each of the several different types of admixtures used in concrete is intended to have a beneficial effect on certain properties of concrete, at the same time, the admixture can have an adverse effect on other properties. Therefore, to make sure the overall quality of concrete intended by the design engineer and the code is met, the use of admixtures must have prior approval of the engineer.

The effectiveness of an admixture in concrete is dependent on such factors as the type and the amount of cement; the shape, gradation and proportions of the aggregate used in the concrete mixture; the water content and the mixing time, slump and temperatures of the surrounding air and concrete. To insure proper proportioning of the concrete in accordance with the requirements of Section 1905.2, trial mixtures should be made with the admixture and the job materials at temperatures and humidities anticipated at the job site. Trial mixtures will establish the compatibility of the admixture with the other admixtures that may be used in the concrete, as well as determine the effects of the admixture on the properties of fresh and hardened concrete. The amount of admixture used in concrete should not exceed the amount recommended by the manufacturer or the optimum amount determined by laboratory tests.

Because concentrations of chloride ions can cause severe corrosive effects on metallic materials, admixtures containing chloride as part of the formulation, including the use of calcium chloride as an antifreeze admixture during cold weather, are not allowed in prestressed concrete, concrete with aluminum embedment or concrete that is in contact with permanent installations of galvanized formwork.

Admixtures used in concrete must comply with the applicable ASTM material specifications referenced in ACI 318. There are several classes of admixtures covered by the ASTM specifications. Each class of admixture, as well as each type of admixture within a class, serve different purposes and affect the properties of concrete in different ways.

Air-Entraining Admixtures—ASTM C 260:

Air-entraining admixtures incorporate microscopic air bubbles in the concrete mixture. Air entrainment serves three main purposes:

1. It improves the resistance of moist concrete to the effects of freezing and thawing cycles;
2. It improves the resistance of concrete to the actions of deicing chemicals; and
3. It improves the workability of fresh concrete.

Entrained air can be introduced in concrete by means of air-entraining cements specified in Section 1903.2, or as a separate ingredient (i.e., an admixture) in the concrete mixture (ASTM C 260) introduced before or during the mixture operations. If desired, both methods can be used together, but difficulties may arise in controlling the amount of air entrainment.

Water-Reducing, Retarding and Accelerating Admixtures—ASTM C 494 and ASTM C 1017:

The ASTM C 494 material specification for chemical admixtures gives the physical requirements for each of seven types of admixture—Types A through G listed below—and prescribes the applicable test methods and other information necessary for the acceptance of admixture materials:

Type A—Water-reducing admixtures
Type B—Retarding admixtures
Type C—Accelerating admixtures
Type D—Water-reducing and retarding admixtures
Type E—Water-reducing and accelerating admixtures
Type F—Water-reducing and high-range admixtures
Type G—Water-reducing, high-range and retarding admixtures.

Type A water-reducing (and set-controlling) admixtures can be used in several ways in proportioning concrete mixtures to achieve certain desired results:

1. The cement content can be maintained while reducing the amount of water, thus lowering the water-cementitious materials (w/cm) ratio and increasing the strength of concrete as well as its durability.

2. The w/cm ratio can be maintained by reducing both the water and cement contents of the concrete resulting in a cost savings (less cement used); and

3. The slump of the concrete can be increased without increasing the w/cm ratio, thus maintaining the same strength level while improving the workability of the concrete.

The purpose of Type B chemical admixtures is to retard the rate of the setting of concrete. Such admixtures are normally used in hot weather when an accelerated rate of hardening of concrete occurs, making placing and finishing operations very difficult. On the negative side, the use of retarding admixtures will cause reductions in concrete strength at very early ages. More important, the effects of retarding admixtures on other properties of concrete, particularly shrinkage, are unpredictable.

Type C accelerating admixtures are used mainly in cold-weather operations to shorten the setting time of concrete and accelerate the strength development at early ages. The problem is that calcium chloride is the prevalent ingredient used in producing these admixtures. Calcium chloride can cause an increase in drying shrinkage, reinforcement corrosion, discoloration of concrete (darkens concrete), loss of strength at later ages and possibly scaling. This type of admixture is prohibited in prestressed concrete construction.

Type D chemical admixtures perform as a combination of the properties of Types A and B. They are used to reduce the quantity of mixing water required to produce concrete of a given consistency (slump) and to retard the setting time of the concrete.

Type E chemical admixtures perform as a combination of the properties of Type A and C admixtures. They serve to reduce the quantity of mixing water required to produce concrete of a given consistency (slump) and accelerate the setting time and early strength development of the concrete.

Type F chemical admixtures are water-reducing admixtures similar to Type A, except that they can reduce the quantity of mixing water by 12 percent or more to produce concrete of a given consistency. Up to 30-percent water reduction can be achieved with these admixtures.

Type G chemical admixtures are high-range, water-reducing and retarding admixtures similar in function to Types F and B, except they can reduce the quantity of mixing water by 12 percent or more to produce concrete of a given consistency and retard the setting time of the concrete.

ASTM C 1017 regulates admixtures that are similar to a high-range water reducer, but are used in a different manner. These admixtures are typically referred to as "superplasticizers." Instead of being used to reduce the amount of mix water and maintain the same consistency (i.e., as a water reducer), the superplasticizer is used with the same amount of mix water that would be necessary to produce relatively low-slump concrete [e.g., 3$^1/_2$ inches (89 mm)]. The superplasticizer causes an increase in the slump, which makes the concrete flowable. Two types of admixtures are provided for under ASTM C 1017: Type I, plasticizing; and Type II, plasticizing and retarding. The requirements of ASTM C 1017 are similar to those of ASTM C 494 (Types F and G), however, in order to comply with ASTM C 1017, the admixture must increase the slump at least 3 $^1/_2$ inches (89 mm) compared to the control specimen.

Pozzolans—ASTM C 618:

Admixtures made of raw or natural pozzolans and fly ash are covered by ASTM C 618 material specifications. While these mineral admixtures will benefit as well as adversely affect the properties of concrete, they are widely used as partial replacements for portland cement and often result in more economical concrete construction.

By definition, a pozzolan is a siliceous or aluminosiliceous material that in itself possesses little or no cementitious value, but will, in finely divided form and in the presence of water, chemically react with the calcium hydroxide released by the hydration of portland cement to form compounds possessing cementitious properties. Pozzolanic materials used for making admixtures come from a number of natural sources, such as diatomaceous earth, opaline cherts, clays, shales, volcanic tuffs and pumicites. The ASTM C 618 standard designates these materials as Class N pozzolans.

Fly ash is the most widely used of all the mineral admixtures in concrete. Essentially, it is a finely divided residue (a powder) that results from the combustion of pulverized coal. Large quantities of fly ash are produced in coal-burning electric power-generating plants. Fly ash is primarily silicate glass containing silica, aluminum, iron and calcium. It also contains small amounts of magnesium, sulfur, sodium, potassium and carbon. ASTM C 618 designates these materials as Class F and Class C fly ashes, depending on the percentages of calcium and carbon contents. Each class of fly ash affects the properties of concrete in different ways. It is a complex subject, and for more information on fly ash and other mineral admixtures, refer to the Portland Cement Association publications cited in the bibliography. Aside from economic reasons, pozzolan and fly ash admixtures are used in concrete for any one or more of the following purposes: increase sulfate resistance, reduce expansions in concrete due to alkali-silica reactions, reduce the permeability of concrete, decrease heat generation or improve the workability and finishing qualities of the concrete.

The properties of pozzolans and fly ash may vary

widely. Fly ash, in particular, can have varying amounts of carbon, silica, sulfur, alkalis and other ingredients that can adversely affect the strength, air content and durability of concrete. Because it is a residue and not a controlled product, the properties of fly ash can vary widely, not only between productions coming from different electric power-generating plants, but also between batches coming from the same plant.

Slag—ASTM C 989:

Ground-granulated blast-furnace slag complying with the requirements of ASTM C 989 provides characteristics similar to those exhibited by pozzolan admixtures. Ground-granulated blast-furnace slag and fly ash are often used in combination as admixtures in portland cement (ASTM C 150) concrete. The practice has been growing in recent years due to energy conservation considerations, but mainly because of savings in the cost of concrete realized when fly ash or slag or both fly ash and slag are used to partially replace cement.

Silica Fume—ASTM C 1240:

Silica fume is a byproduct of the manufacture of elemental silicon or ferro-silicon alloys in electric arc furnaces. When used in conjunction with a high-range water-reducing admixture, it is possible to produce concrete with compressive strengths of 20,000 psi (138 MPa) or higher. It is also used to achieve a very dense cement paste matrix to reduce the permeability of concrete, thus providing better corrosion protection to reinforcing steel in concrete subjected to deicing chemicals (e.g., parking garages). Tests have shown that silica fume greatly improves the sulfate resistance of concrete. For this reason, the code requires that concrete to be subjected to very severe sulfate exposure be made with Type V ASTM C 150 cement plus a pollozan, such as silica fume, that has shown through tests or service records to improve sulfate resistance (see Table 1904.3).

1903.7 Storage of materials. The storage of materials for use in concrete shall comply with the provisions of Sections 1903.7.1 and 1903.7.2.

❖ This section regulates the storage of concrete ingredients and prohibits use of such materials if they have suffered contamination or deterioration.

1903.7.1 Manner of storage. Cementitious materials and aggregates shall be stored in such a manner as to prevent deterioration or intrusion of foreign matter.

❖ Cement stored in contact with damp air or moisture sets more slowly and has less strength than cement that is maintained in a dry condition. Bulk cement can be stored in weather-tight containers, such as steel bins or concrete silos.

Bags of cement are typically stored in warehouses or sheds where the relative humidity is kept as low as possible. Bags of cement must not be stored on damp floors or on earth (job-site conditions), but should rest on pallets and be stacked close together to reduce air circulation. Bags should be covered with a waterproof covering to prevent deterioration when exposed to weather.

Aggregates are normally stockpiled on the ground in areas that have been stripped of all vegetation and debris and have been leveled to facilitate handling operations. When two or more types or sizes of aggregates are involved, the stockpiles must be so arranged as to avoid crowding or overlapping with other materials or other sizes of like aggregates. In removing material from a stockpile, a layer of aggregate of sufficient thickness must be left on the ground to keep the handling equipment, such as a front-end loader or clamshell, from picking up earth along with the aggregate. Stockpiles typically are built up in centric uniform layers to prevent coarse aggregates from rolling down the sides and segregating the material.

1903.7.2 Unacceptable material. Any material that has deteriorated or has been contaminated shall not be used for concrete.

❖ Ordinarily, cement does not remain in storage long, but it can be stored for long periods without deterioration. At the time of use, however, cement is required to be free flowing and relatively free of lumps. If the lumps do not readily break up or there is other reason to doubt the quality of the cement, the material can be tested for strength or loss on ignition in accordance with the ASTM standard under which it was manufactured before it is used in concrete.

If there is reason to doubt that any aggregate taken from a stockpile is not clean or appears to be contaminated, any of the applicable standard ASTM tests that would detect impurities should be made before the material is used for making concrete.

1903.8 Glass fiber reinforced concrete. *Glass fiber reinforced concrete (GFRC) and the materials used in such concrete shall be in accordance with the PCI MNL 128 standard.*

❖ This section references a document that addresses the quality of materials used in glass fiber reinforced concrete panels. Glass fiber reinforced concrete is utilized in the prescriptive fire-resistance ratings for wall assemblies [see Table 720.1(2), Item 15].

SECTION 1904
DURABILITY REQUIREMENTS

1904.1 Water-cementitious materials ratio. The water-cementitious materials ratios specified in Tables 1904.2.2 and 1904.3 shall be calculated using the weight of cement meeting ASTM C 150, ASTM C 595, ASTM C 845 or ASTM C 1157, plus the weight of fly ash and other pozzolans meeting ASTM C 618, slag meeting ASTM C 989 and silica fume meeting ASTM C 1240, if any, except that where concrete is exposed to deicing chemicals, Section 1904.2.3 further limits the amount of fly ash, pozzolans, silica fume, slag or the combination of these materials.

❖ The provisions of Section 1904 specify the proportioning of concrete that is necessary to obtain the required strengths and to provide the properties of concrete that will resist the effects of adverse environmental conditions. This includes concrete: that will be subjected to freeze-thaw cycles; in contact with deicer chemicals or exposed to the deteriorating effects of sulfates, corrosive actions of chloride ions on steel reinforcement or other harmful exposures.

Section 1904.1 clarifies that in addition to including traditional cements in the weight of cement used to calculate the water-cementitious materials ratio, it is permissible to include the weights of fly ash or other pozzolan, slag and silica fume. The user is also alerted to the fact that Table 1904.2.3 limits the amount of fly ash or other pozzolan, slag and silica fume in concrete to be exposed to deicer chemicals.

1904.2 Freezing and thawing exposures. Concrete that will be exposed to freezing and thawing or deicing chemicals shall comply with Sections 1904.2.1 through 1904.2.3.

❖ This section requires that concrete be proportioned such that the required strength, and other properties of hardened concrete needed to resist the effects of freezing and thawing, will be achieved.

1904.2.1 Air entrainment. Normal-weight and lightweight concrete exposed to freezing and thawing or deicing chemicals shall be air entrained with air content indicated in Table 1904.2.1. Tolerance on air content as delivered shall be ±1.5 percent. For specified compressive strength (f'_c) greater than 5,000 psi (34.47 MPa), reduction of air content indicated in Table 1904.2.1 by 1.0 percent is permitted.

❖ Freezing and thawing cycles can be one of the most destructive weathering factors for concrete when it is wet. For nonair-entrained concrete, deterioration is caused by the freezing of water in the cement matrix, in the aggregate or both. Studies have documented that concrete provided with proper air entrainment is highly resistant to this deterioration. Table 1904.2.1 specifies the required air content based on the aggregate size and exposure condition. An exposure condition would be considered severe if the concrete is located in a cold climate and could be continuously in contact with moisture prior to freezing, or exposed to deicing chemicals. A parking structure is an example of a severe exposure because vehicles may track in deicing chemicals from the streets. The exposure condition would be considered moderate if concrete is in a cold climate and in contact with moisture prior to freezing only occasionally, and no deicing chemicals are used. Exterior walls, beams, girders and slabs not in direct contact with the ground are examples of moderate exposure.

Normal-weight and lightweight concrete subjected to freezing and thawing cycles or deicer chemicals must be air entrained in compliance with the air content requirements set forth in Table 1904.2.1.

Concrete with compressive strengths in excess of 5,000 psi (34.47 MPa) has characteristics that improve resistance to freezing and thawing (e.g., lower water-cementitious materials ratio, lower porosity), and therefore are permitted to have 1 percent less air content than required by Table 1904.2.1. For example, the required total air content for a 5,500 psi (37.9 MPa) mix that will be made with $^3/_8$-inch (9.5 mm) maximum size aggregate and subjected to a "severe exposure" is $7^1/_2$ - 1 = $6^1/_2$ percent.

TABLE 1904.2.1
TOTAL AIR CONTENT FOR FROST-RESISTANT CONCRETE

NOMINAL MAXIMUM AGGREGATE SIZE[a] (inches)	AIR CONTENT (percent)	
	Severe exposure[b]	Moderate exposure[b]
$^3/_8$	$7^1/_2$	6
$^1/_2$	7	$5^1/_2$
$^3/_4$	6	5
1	6	$4^1/_2$
$1^1/_2$	$5^1/_2$	$4^1/_2$
2[c]	5	4
3[c]	$4^1/_2$	$3^1/_2$

For SI: 1 inch = 25.4 mm.

a. See ASTM C 33 for tolerance on oversize for various nominal maximum size designations.

b. The severe and moderate exposures referenced in this table are not based on the weathering regions shown in Figure 1904.2.2. For the purposes of this section, severe and moderate exposures shall be defined as follows:

1. Severe exposure occurs where concrete will be in almost continuous contact with moisture prior to freezing, or where deicing salts are used. Examples are pavements, bridge decks, sidewalks, parking garages and water tanks.
2. Moderate exposure occurs where concrete will be only occasionally exposed to moisture prior to freezing, and where deicing salts are not used. Examples are certain exterior walls, beams, girders and slabs not in direct contact with soil.

c. These air contents apply to total mix, as for the preceding aggregate sizes. When testing these concretes, however, aggregate larger than $1^1/_2$ inches is removed by hand picking or sieving and air content is determined on the minus $1^1/_2$-inch fraction of the mix (tolerance on air content as delivered applies to this value). Air content of total mix is computed from value determined on the minus 1½-inch fraction.

❖ Air is introduced in concrete by the use of air-entraining cements, air-entraining admixtures or a combination of both.

The required air content in Table 1904.2.1, expressed as a percentage of the volume of concrete, is based on the size of coarse aggregates used in the mix. For smaller aggregate sizes, the percentage is increased to account for the increase in mortar volume that results when aggregate size decreases.

Note b of the table defines both "moderate" and "severe" exposures and gives examples of the types of structures most affected. Examples of severe exposure are pavements, bridge decks, sidewalks, parking garages and water tanks. Examples of moderate exposure are certain exterior walls, beams, girders and slabs not in direct contact with soil.

The code requires that all concrete exposed to freezing and thawing conditions or to deicer chemicals must be air entrained. While air entrainment is mandatory, to do otherwise would result in exposed concrete that is afforded less protection than is necessary to provide durability in cold-weather climates. For comprehensive information on both air content and selecting proportions for normal-weight concrete, refer to ACI 211.1.

1904.2.2 Concrete properties. Concrete that will be subject to the exposures given in Table 1904.2.2(1) shall conform to the corresponding maximum water-cementitious materials ratios and minimum specified concrete compressive strength requirements of that table. In addition, concrete that will be exposed to deicing chemicals shall conform to the limitations of Section 1904.2.3.

> ***Exception:*** *For occupancies and appurtenances thereto in Group R occupancies that are in buildings less than four stories in height, normal-weight aggregate concrete that is subject to weathering (freezing and thawing), as determined from Figure 1904.2.2, or deicer chemicals shall comply with the requirements of Table 1904.2.2(2).*

❖ Concrete subject to the exposures listed in Table 1904.2.2(1) must comply with the limitations on water-cementitious materials ratio, or specified compressive strength, or both. The intent of the upper limit on the water-cementitious materials ratio and lower limit on specified compressive strength is to achieve dense, impermeable or water-tight concrete. Note that the minimum specified compressive strength applies to both normal-weight and lightweight aggregate concrete, while the maximum water-cementitious materials ratio applies only to normal-weight concrete. The variability of moisture absorption in lightweight aggregate makes calculation of the water-cementitious materials ratio uncertain.

In the exception, in climatic areas where concrete construction is exposed to freeze-thaw cycles or deicer chemicals, concrete must be of the quality necessary to resist the harmful effects of such weather conditions. In all buildings that are subject to special exposure conditions, except those of Group R occupancies that are less than four stories in height, the water-cementitious materials ratio and specified compressive strength of the concrete is required to be designed to conform to the requirements of Table 1904.2.2(1). The specified compressive strengths of concrete in low-rise Group R occupancies (less than four stories) need only comply with Table 1904.2.2(2), which mandates a minimum specified compressive strength as a function of the concrete element and exposure.

TABLE 1904.2.2(1). See page 19-15.

❖ Table 1904.2.2(1) provides water-cementitious materials (w/cm) ratios for normal-weight aggregate concrete in conjunction with various exposure conditions. The values for w/cm ratios are the maximum permissible without regard for the relative strength of concrete. A control on the w/cm ratio is provided in addition to the minimum specified compressive strength values (f'_c) for normal-weight concrete, because the w/cm ratio is a more reliable indication of durability and watertightness of concrete. The table only gives a minimum specified compressive strength value for proportioning concrete mixtures using lightweight aggregates because the amount of water absorbed by lightweight aggregates is difficult to determine, making the calculation of w/cm ratios unreliable.

TABLE 1904.2.2(2). See page 19-15.

❖ This table gives the minimum specified compressive strength values (f'_c) required for concrete used in various components for construction of Group R occupancy buildings that are less than four stories, and which are subject to either freezing and thawing while wet, or exposure to deicer chemicals, such as walls, slabs and foundations. The main purpose of this table is to assure adequate performance of the concrete by requiring a minimum compressive strength (f'_c) based on the classifications of weathering determined from Code Figure 1904.2.2.

FIGURE 1904.2.2. See page 19-15.

❖ Figure 1904.2.2 shows the geographic regions within the U.S. mainland, which are classified as having negligible, moderate and severe weather exposures. The sole purpose of the map is to locate those weathering areas in the country where the requirements for minimum specified compressive strength of concrete of Table 1904.2.2(2) apply. Note that the boundary lines defining the weather regions are only approximate. Therefore, building officials in communities located near a boundary line must establish applicable exposure classifications (severe, moderate or negligible) based on their knowledge of weather conditions in the area and the relative performance of concrete. Figure 1904.2.2 is not intended to be used to define "moderate exposure" and "severe exposure" as used in Table 1904.2.1.

1904.2.3 Deicing chemicals. For concrete exposed to deicing chemicals, the maximum weight of fly ash, other pozzolans, silica fume or slag that is included in the concrete shall not exceed the percentages of the total weight of cementitious materials given in Table 1904.2.3.

❖ The purpose of this section is to limit the quantity of certain materials in a concrete mixture to assure adequate durability of the concrete that will be exposed to deicing chemicals. Although research has shown that the use of fly ash, slag and silica fume produces a finer pore structure in concrete resulting in a lower permeability and improved durability, their effectiveness is limited by the relative amounts of these materials and portland or blended cement. The limitations of this section do not apply if the concrete will not be exposed to deicing chemicals.

TABLE 1904.2.3. See page 19-16.

❖ See the commentary to Section 1904.2.3.

TABLE 1904.2.2(1)
REQUIREMENTS FOR SPECIAL EXPOSURE CONDITIONS

EXPOSURE CONDITION	MAXIMUM WATER-CEMENTITIOUS MATERIALS RATIO, BY WEIGHT, NORMAL-WEIGHT AGGREGATE CONCRETE	MINIMUM f'_c, NORMAL-WEIGHT AND LIGHTWEIGHT AGGREGATE CONCRETE (psi)
Concrete intended to have low permeability when exposed to water	0.50	4,000
Concrete exposed to freezing and thawing in a moist condition or to deicing chemicals	0.45	4,500
For corrosion protection of reinforcement in concrete exposed to chlorides from deicing chemicals, salt, saltwater, brackish water, seawater or spray from these sources	0.40	5,000

For SI: 1 pound per square inch = 0.00689 MPa.

TABLE 1904.2.2(2)
MINIMUM SPECIFIED COMPRESSIVE STRENGTH (f'_c)

TYPE OR LOCATION OF CONCRETE CONSTRUCTION	MINIMUM SPECIFIED COMPRESSIVE STRENGTH (f'_c at 28 days, psi)		
	Negligible exposure	Moderate exposure	Severe exposure
Basement walls[c] and foundations not exposed to the weather	2,500	2,500	2,500[a]
Basement slabs and interior slabs on grade, except garage floor slabs	2,500	2,500	2,500[a]
Basement walls[c], foundation walls, exterior walls and other vertical concrete surfaces exposed to the weather	2,500	3,000[b]	3,000[b]
Driveways, curbs, walks, patios, porches, carport slabs, steps and other flatwork exposed to the weather, and garage floor slabs	2,500	3,000[b]	3,500[b]

For SI: 1 pound per square inch = 0.00689 MPa.

a. Concrete in these locations that can be subjected to freezing and thawing during construction shall be of air-entrained concrete in accordance with Table 1904.2.1.

b. Concrete shall be air entrained in accordance with Table 1904.2.1.

c. Structural plain concrete basement walls are exempt from the requirements for special exposure conditions of Section 1904.2.2 (see Section 1909.1.1).

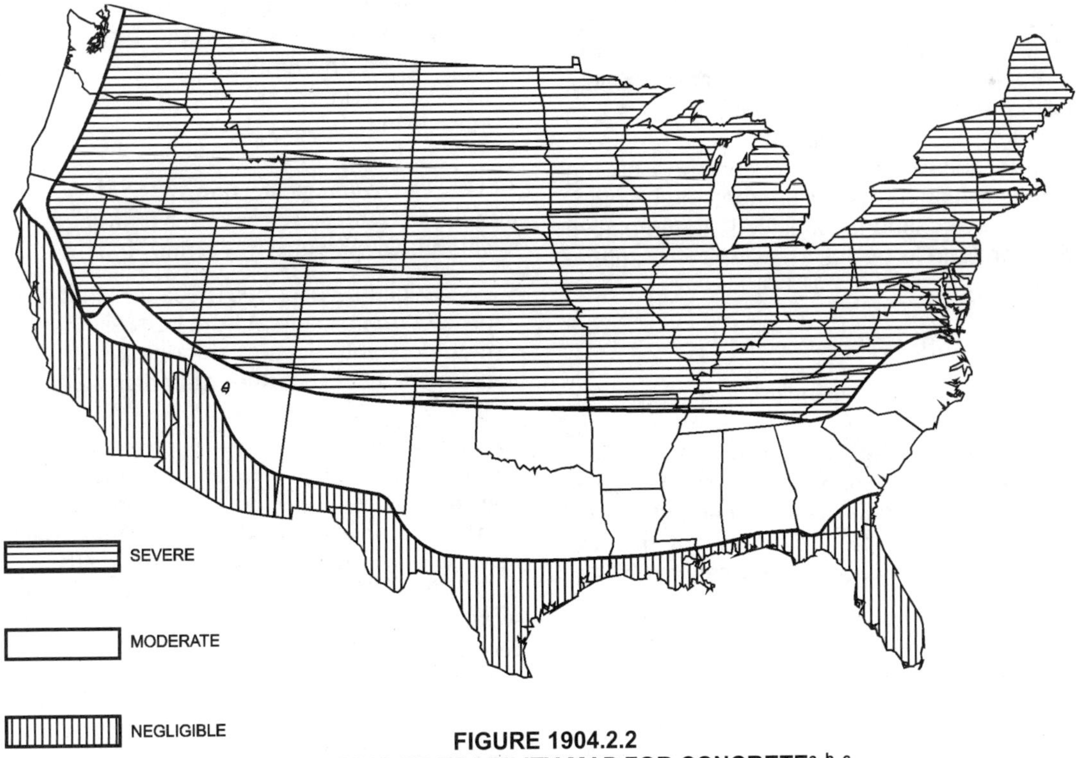

FIGURE 1904.2.2
WEATHERING PROBABILITY MAP FOR CONCRETE[a, b, c]

TABLE 1904.2.3
REQUIREMENTS FOR CONCRETE EXPOSED TO DEICING CHEMICALS

CEMENTITIOUS MATERIALS	MAXIMUM PERCENT OF TOTAL CEMENTITIOUS MATERIALS BY WEIGHT[a, b]
Fly ash or other pozzolans conforming to ASTM C 618	25
Slag conforming to ASTM C 989	50
Silica fume conforming to ASTM C 1240	10
Total of fly ash or other pozzolans, slag and silica fume	50[c]
Total of fly ash or other pozzolans and silica fume	35[c]

a. The total cementitious material also includes ASTM C 150, ASTM C 595, ASTM C 845 and ASTM C 1157 cement.
b. The maximum percentages shall include:
1. Fly ash or other pozzolans present in Type IP or I (PM) blended cement, ASTM C 595, or ASTM C 1157.
2. Slag used in the manufacture of an IS or I (SM) blended cement, ASTM C 595, or ASTM C 1157.
3. Silica fume, ASTM C 1240, present in a blended cement.
c. Fly ash or other pozzolans and silica fume shall constitute no more than 25 and 10 percent, respectively, of the total weight of the cementitious materials.

1904.3 Sulfate exposures. Where concrete will be exposed to sulfate-containing solutions, it shall comply with the provisions of Sections 1904.3.1 and 1904.3.2.

❖ This section regulates the proportioning and types of cement permitted in concrete that will be exposed to sulfate solutions.

1904.3.1 Concrete quality. Concrete to be exposed to sulfate-containing solutions or soils shall conform to the requirements of Table 1904.3 or shall be concrete made with a cement that provides sulfate resistance and that has a maximum water-cementitious materials ratio and minimum compressive strength from Table 1904.3.

❖ Many naturally occurring soils contain sufficient amounts of sulfates to cause deterioration of concrete that does not possess certain physical and chemical qualities. Referred to as alkaline soils, they contain concentrations of sodium and magnesium sulfates (salts). Water or solutions coming in contact with concrete can also contain sulfates (e.g., seawater and sewage). Concrete made with sulfate-resisting cement and having limitations on its maximum water-cementitious materials ratio (w/cm), minimum specified compressive strength (f'_c) or both provides resistance to sulfate attack.

TABLE 1904.3. See page 19-17.

❖ While the title of Table 1904.3 is misleading, it gives the requirements for concrete exposed to sulfate-containing soils and solutions based on four exposure levels: negligible, moderate, severe and very severe. The table specifies the types of sulfate-resisting cements to be used in concrete mixtures, the maximum water-cementitious materials ratio for normal-weight aggregate concrete and the minimum specified compressive strengths for normal-weight aggregate and lightweight aggregate concrete—all relative to the four exposure conditions stated above.

In addition to the proper selection of cement, water-cementitious materials ratio, or strength of concrete as specified in Table 1904.3, or both, other requirements for producing sulfate-resisting concrete include: adequate air entrainment, low slump, sufficient consolidation of the concrete mixture, adequate cover of the reinforcement and proper curing.

As indicated by Note b, seawater is considered to be a moderate exposure.

1904.3.2 Calcium chloride. Calcium chloride as an admixture shall not be used in concrete to be exposed to severe or very severe sulfate-containing solutions as defined in Table 1904.3.

❖ This section prohibits the use of calcium chloride as an admixture in concrete that is to be exposed to severe and very severe sulfate-containing soil or solutions, as defined in Table 1904.3. The principal consideration in producing cements for sulfate resistance is to limit the tricalcium aluminate (C3A) content in accordance with the requirements of ASTM C 150. The use of calcium chloride as an admixture, because of its calcium content, would adversely affect the performance of cement in resisting sulfate attack.

1904.4 Corrosion protection of reinforcement. Reinforcement in concrete shall be protected from corrosion and exposure to chlorides as provided by Sections 1904.4.1 and 1904.4.2.

❖ This section prescribes measures that must be taken to protect reinforcement from corrosion where exposed to high concentrations of chloride ion (Cl-) present in calcium chloride and other materials having chloride ion.

1904.4.1 General. For corrosion protection of reinforcement in concrete, the maximum water-soluble chloride ion concentrations in hardened concrete at ages from 28 to 42 days contributed from the ingredients including water, aggregates, cementitious materials and admixtures shall not exceed the limits of Table 1904.4.1. When testing is performed to determine water-soluble chloride ion content, test procedures shall conform to ASTM C 1218.

❖ The maximum concentration of water-soluble chloride ion permitted in hardened concrete from all sources is given in Table 1904.4.1. Chloride in concrete could be in a water-soluble state or be chemically combined with other ingredients. The soluble chlorides induce corrosion, while chlorides in combined form have little or no corrosive effect. Sampling and testing for water-soluble chloride ions in concrete must comply with the requirements of ASTM C 1218.

Additional information on the effects of chlorides on the corrosion of reinforcing steel is given in the ACI 201.2R publication, *Guide to Durable Concrete,* and ACI 222R, *Corrosion of Metals in Concrete*.

TABLE 1904.4.1. See below.

❖ For corrosion protection, Table 1904.4.1 limits the maximum soluble chloride ion concentration in hardened concrete that is contributed by the ingredients (i.e., mix water, aggregates, cementitious materials and admixtures). Tests for water-soluble chloride ion concentrations can be performed on hardened concrete in accordance with ASTM C 1218. An initial evaluation of the chloride ion content is possible by testing the individual concrete ingredients and comparing the summation, based on the concrete proportions, to values in Table 1904.4.1. If the total chloride ion content exceeds the table values, it is advisable to test samples of the hardened concrete. The total value for the ingredients can be higher than that permitted by Table 1904.4.1, because some of the chloride ions present in the individual ingredients are insoluble or will have reacted with the cement during hydration, and therefore will not be available in hardened concrete.

1904.4.2 Exposure to chlorides. Where concrete with reinforcement will be exposed to chlorides from deicing chemicals, salt, saltwater, brackish water, seawater or spray from these sources, the requirements of Table 1904.2.2(1) for water-cementitious materials ratio and concrete strength, and the minimum concrete cover requirements of Section 1907.7, shall be satisfied. See ACI 318, Section 18.16, for corrosion protection of unbonded tendons.

❖ To provide the properties of concrete that will resist the harmful effects of exposure to chlorides from deicing chemicals, salts, seawater or brackish water, concrete mixtures containing normal-weight aggregates must con-

TABLE 1904.3
REQUIREMENTS FOR CONCRETE EXPOSED TO SULFATE-CONTAINING SOLUTIONS

SULFATE EXPOSURE	WATER SOLUBLE SULFATE (SO_4) IN SOIL, PERCENT BY WEIGHT	SULFATE (SO_4) IN WATER (ppm)	CEMENT TYPE			MAXIMUM WATER-CEMENTITIOUS MATERIALS RATIO, BY WEIGHT, NORMAL-WEIGHT AGGREGATE CONCRETE[a]	MINIMUM f'_c, NORMAL-WEIGHT AND LIGHTWEIGHT AGGREGATE CONCRETE (psi)[a]
			ASTM C 150	ASTM C 595	*ASTM C 1157*		
Negligible	0.00 - 0.10	0 - 150	—	—	—	—	—
Moderate[b]	0.10 - 0.20	150 - 1,500	II	II, IP (MS), IS (MS), P (MS), I (PM)(MS), I (SM)(MS)	*MS*	0.50	4,000
Severe	0.20 - 2.00	1,500 - 10,000	V	—	*HS*	0.45	4,500
Very severe	Over 2.00	Over 10,000	V plus pozzolan[c]	—	*HS plus pozzolan*[d]	0.45	4,500

For SI: 1 pound per square inch = 0.00689 MPa.

a. A lower water-cementitious materials ratio or higher strength may be required for low permeability or for protection against corrosion of embedded items or freezing and thawing (see Table 1904.2.2).

b. Seawater.

c. Pozzolan that has been determined by test or service record to improve sulfate resistance when used in concrete containing Type V cement.

d. Pozzolan that has been determined by test or service record to improve sulfate resistance when used in concrete containing Type HS blended cement.

TABLE 1904.4.1
MAXIMUM CHLORIDE ION CONTENT FOR CORROSION PROTECTION OF REINFORCEMENT

TYPE OF MEMBER	MAXIMUM WATER SOLUBLE CHLORIDE ION (Cl) IN CONCRETE, PERCENT BY WEIGHT OF CEMENT
Prestressed concrete	0.06
Reinforced concrete exposed to chloride in service	0.15
Reinforced concrete that will be dry or protected from moisture in service	1.00
Other reinforced concrete construction	0.30

form to the maximum water-cementitious materials ratios and the minimum specified compressive strengths (f'_c) indicated in Table 1904.2.2(1). Concrete mixtures containing lightweight aggregates must conform to the minimum specified compressive strength (f'_c) requirements given in the same table.

It should be noted that the requirements of this section apply to reinforced concrete and to plain concrete containing steel reinforcement. Minimum concrete cover requirements for the protection of reinforcement are stated in Section 1907.7. Additionally, Section 1907.7.5 requires that in corrosive environments or other severe exposure conditions, minimum cover requirements must be suitably increased (see Section 1907.7.5).

SECTION 1905 CONCRETE QUALITY, MIXING AND PLACING

1905.1 General. The required strength and durability of concrete shall be determined by compliance with the proportioning, testing, mixing and placing provisions of Sections 1905.1.1 through 1905.13.

❖ Section 1905 specifies the methods for proportioning concrete mixtures to obtain the proper workability of fresh concrete and to secure properties of strength, durability, watertightness and abrasion resistance in hardened concrete as necessary to meet performance requirements. The section also regulates the mixing; handling and conveying; depositing and curing of concrete to insure the intended performance of the finished product.

Section 1905.1 covers the methods used in proportioning mixtures to provide adequate workability of fresh concrete and to render those qualities of strength and durability in hardened concrete that are necessary to meet performance requirements. This section includes methods for testing concrete and criteria for its evaluation and acceptance based on strength requirements, as well as requirements for mixing, handling, transporting, placing and curing of concrete for field quality.

1905.1.1 Strength. Concrete shall be proportioned to provide an average compressive strength as prescribed in Section 1905.3, and shall satisfy the durability criteria of Section 1904. Concrete shall be produced to minimize the frequency of strengths below f'_c as prescribed in Section 1905.6.3.3. *For concrete designed and constructed in accordance with this chapter, f'_c shall not be less than 2,500 psi (17.22 MPa).* No maximum specified compressive strength shall apply unless restricted by a specific provision of this code or ACI 318.

❖ Concrete mix designs are proportioned for strength based on probabilistic concepts so that adequate strength will be developed in the concrete. This section requires that the average strength of concrete produced must exceed the larger of (1) the value of f'_c specified by the registered design professional for the structural design requirements and (2) the minimum strength required for any special exposure condition prescribed in Section 1904. Concrete proportioned to the code requirements may produce strength tests that fall below the specified compressive strength, f'_c. This section introduces this concept by noting that it is the code's intent to "minimize frequency of strength below f'_c." The acceptability of concrete based on the measurement of strength is provided for in Section 1905.6. If a concrete strength test falls below f'_c, the acceptability of this lower-strength concrete is provided for in Section 1905.6.3.3.

This section also prescribes that all concrete designed and constructed in accordance with Chapter 19 must have a minimum specified compressive strength, f'_c, of 2,500 psi (17.22 MPa). This limit has been in model building codes for many years.

1905.1.2 Cylinder tests. Requirements for f'_c shall be based on tests of cylinders made and tested as prescribed in Section 1905.6.3.

❖ This section prescribes that f'_c must be verified by making and testing cylinders in accordance with Section 1905.6.3. Typical test specimens for evaluating compressive strength of concrete are cylinders 6 inches (152 mm) in diameter by 12 inches (305 mm) high (see commentary, Section 1905.6.3).

1905.1.3 Basis of f'_c. Unless otherwise specified, f'_c shall be based on 28-day tests. If other than 28 days, test age for f'_c shall be as indicated in construction documents.

❖ Most laboratory tests to determine compressive strength levels are based on 28-day cured specimens. If the specified compressive strength of concrete used in the structural design is other than a 28-day age, the specific age must be stated in the construction documents and the concrete must be tested accordingly. For example, certain conditions may require a concrete structure to receive its service load at an age earlier than 28 days. Under such a circumstance, the compressive strength of the concrete must be specified and obtained in a time frame that will accommodate the loading requirement. Another example is when early concrete strength is provided to facilitate a faster schedule for removal of forms and shores. In many such situations, obtaining earlier strength in concrete is desirable. On some projects, the architect or engineer may require that the specified compressive strength of concrete, f'_c, be based on a period of time longer than 28 days. A typical example is concrete in columns in the lower stories of a high-rise building. Since these columns will not be subjected to their full design load for several months, a longer period of time can be allowed since the concrete continues to gain strength after the first 28 days. In these cases, 56-day and 84-day strengths are commonly specified.

1905.1.4 Lightweight aggregate concrete. Where design criteria in ACI 318, Sections 9.5.2.3, 11.2 and 12.2.4, provide for use of a splitting tensile strength value of concrete (f_{ct}), laboratory tests shall be made in accordance with ASTM C 330 to establish the value of f_{ct} corresponding to the specified value of f'_c.

❖ Sections 9.5.2.3 (modulus of rupture), 11.2 (concrete shear strength) and 12.2.4 (development of reinforcement) of ACI 318 require that design criteria be modified for lightweight aggregate concrete based on a relationship between splitting tensile strength, f_{ct}, and specified compressive strength, f'_c. Testing for splitting tensile strength must be in compliance with ASTM C 330.

1905.1.5 Field acceptance. Splitting tensile strength tests shall not be used as a basis for field acceptance of concrete.

❖ This section prohibits the use of the splitting tensile strength test (ASTM C 330) as a substitute for the more rigorous testing and acceptance criteria required in Section 1905.6.

1905.2 Selection of concrete proportions. Concrete proportions shall be determined in accordance with the provisions of Sections 1905.2.1 through 1905.2.3.

❖ This section provides the basic criteria that must be met when proportioning concrete mixes (i.e., preparing a concrete mix design). The objective in designing concrete mixtures is to determine the most economical combination of available ingredients to produce concrete that will satisfy the performance requirements under particular conditions of use.

1905.2.1 General. Proportions of materials for concrete shall be established to provide:

1. Workability and consistency to permit concrete to be worked readily into forms and around reinforcement under the conditions of placement to be employed, without segregation or excessive bleeding.
2. Resistance to special exposures as required by Section 1904.
3. Conformance with the strength test requirements of Section 1905.6.

❖ Materials must be properly proportioned to produce concrete mixtures that meet minimum workability (and consistency), special exposure and strength requirements. Workability is the measure of the ease of placing, consolidating and finishing freshly mixed concrete. Consistency is the measure of its ability to flow. The code requires that concrete must be readily workable into forms and around reinforcement and to have a plastic consistency that will prevent segregation and excessive bleeding. Segregation occurs when the aggregates in freshly mixed concrete are not held in suspension, which produces a mixture that is not homogeneous. Bleeding is the migration of water to the top surface of freshly placed concrete. Excessive bleeding increases the water-cementitious materials ratio near the top surface of the concrete, creating a potential for poor durability of the top surface.

Resistance to special exposure conditions provides minimum safeguards to protect against deterioration of concrete and reinforcement due to freeze-thaw exposure and exposure to chemicals found in concrete and the surrounding environment (see commentary, Section 1904).

The requirement that proportioning provide "conformance with the strength test requirements of Section 1905.6" means that after the mixture proportions have been selected and the job has started, concrete delivered to the job must be evaluated and found acceptable in accordance with the provisions of Section 1905.6. This clarifies that the ultimate goal of concrete proportioning is to provide quality concrete throughout the term of construction. Testing and evaluation of concrete during the construction process is necessary to insure that the selected mixture proportions continue to perform should materials and job-related conditions vary during the construction process.

1905.2.2 Different materials. Where different materials are to be used for different portions of proposed work, each combination shall be evaluated.

❖ When different materials like aggregate or cement are to be used in different portions of the concrete work, a separate evaluation of the concrete mixture for conformance to the code and job specifications must be performed.

1905.2.3 Basis of proportions. Concrete proportions shall be established in accordance with Section 1905.3 or Section 1905.4, and shall comply with the applicable requirements of Section 1904.

❖ The preferred methods for selecting concrete mixture proportions are those using field experience or laboratory trial mixtures in accordance with Section 1905.3. Estimation of the water-cementitious materials ratio in accordance with Section 1905.4 is permitted when no prior field experience or trial mixture data is available, and only when approved by the engineer or architect as Section 5.4 of ACI 318 indicates.

1905.3 Proportioning on the basis of field experience and/or trial mixtures. Concrete proportioning determined on the basis of field experience and/or trial mixtures shall be done in accordance with ACI 318, Section 5.3.

❖ In proportioning concrete on the basis of field experience, a concrete production facility (e.g., a ready-mixed concrete supplier) that has a suitable record of at least 30 consecutive strength tests, or two groups of 30 consecutive tests of similar materials and conditions that will compare closely with the work to be performed, may use the standard deviation methods of ACI 318, Section 5.3.1, to calculate the required average compressive strength. The concrete thus produced must meet a compressive strength that does not deviate more than 1,000 psi (6.9 MPa) from the required specified compressive strength of concrete, (f'_c).

Experience has shown that there is enough variation in concrete mixtures, even if they are made of the same materials and proportions, to render variations in test results. This happens even when test specimens are made from the same batch of concrete. Therefore, in using field experience as the basis for proportioning new concrete, it is necessary to account for such differences by measuring how a group of test results varies from the average. This is done by applying ACI 318 calculation methods for determining standard deviations.

The standard deviation of strength tests of a concrete mixture with at least 30 consecutive tests can be determined as follows:

$$s = \left[\frac{\Sigma(x_i - x)^2}{n-1} \right]^{1/2}$$

where:

s = Standard deviation, psi.

X_i = Individual strength tests as defined in Section 1905.6.2.4.

x = Average number of consecutive strength test results.

n = Number of consecutive strength tests.

If two records are used to obtain at least 30 tests, the standard deviation used should be the statistical average of the values calculated from each test record in accordance with the following formula:

$$s = \left[\frac{(n_1-1)(s_1)^2 + (n_2-1)(s_2)^2}{(n_1+n_2-2)} \right]^{1/2}$$

where:

s = Statistical average standard deviation where two test records are used to estimate the standard deviation.

s_1, s_2 = Standard deviations calculated from two test records, 1 and 2, respectively.

n_1, n_2 = Number of tests in each test record, respectively.

Example:

Calculations for determining the standard deviation are as follows:

In this example, a record of 30 consecutive strength tests is available. Each test represents the average of two cylinders tested at 28-day strengths. The specified compressive strength (f'_c) required for the design is 4,000 psi (27.6 MPa) and therefore, from experience, the average compressive strength (f'_{cr}) would be in the vicinity of 4,500 psi (31.0 MPa) more or less. The record of 30 tests to be used in the analysis would have to be at about that level of strength.

Step 1: Calculate x, which is the average strength of the entire group of 30 tests. Add all the individual test strengths (x_i), and divide the total by the number of tests (n). For simplicity, each of the individual tests is not shown here, and only the totals are used in this example.

Assume the strength value for the entire group of 30 tests adds up to 132,510 psi (913.5 MPa).

$$\text{Therefore, } x = \frac{132,510}{30} = 4,417$$

Step 2: Determine the value of the term $(x_i - x)^2$. This value needs to be calculated for each of the 30 tests: x was calculated in Step 1 above. x_i represents the value of each individual test.

Therefore, as an example:

Test 1 = $(4,760 - 4,417)^2 = 343^2$ 117,649 psi
Test 2 = $(4,300 - 4,417)^2 = -117^2$ 13,689 psi
Test 3 = $(4,840 - 4,417)^2 = 423^2$ 178,929 psi

and so on until all 30 expressions are determined.

Step 3: Add up all the numbers obtained in Step 2 above. The total figure represents the expression, $\Sigma (x_i - x)^2$. For this example, assume the total adds up to 5,241,720 psi.

Step 4: Determine the standard deviation.

$$s = \left[\frac{5,241,720}{30-1} \right]^{1/2} = 425 \text{ psi}$$

In practice, the calculations illustrated above are normally done by a laboratory experienced in the design of concrete mixtures or by a capable concrete supplier. Engineers rely on these sources of information. Today, computers perform the calculations, and considerably more than a group of 30 consecutive strength tests are used to determine a standard deviation.

When available records consist of less than the required number of 30 consecutive tests [as specified in ACI 318, Section 5.3.1.1(c)], but are between 15 and 29 tests, the standard deviation must be increased by the applicable modification factor listed in ACI 318, Table 5.3.1.2. This will result in obtaining a more conservative required average strength as a protection against the possibility that the use of a smaller number of sample tests may cause an underestimation of the standard deviation.

When the record of strength tests is insufficient (less than 15 tests), and the procedures given in ACI 318, Section 5.3.1.1 cannot be applied, the required strength must exceed the specified design strength by an amount that ranges from 1,000 to 1,400 psi (6.9 to 7.9 MPa) in accordance with ACI 318, Table 5.3.2.2, and documented in accordance with ACI 318, Section 5.3.3.

1905.4 Proportioning without field experience or trial mixtures. Concrete proportioning determined without field experience or trial mixtures shall be done in accordance with ACI 318, Section 5.4.

❖ The cost of determining trial mixture data may not be justifiable on projects involving small quantities of concrete. Also, on occasion, the concrete supply to a project may be interrupted and there is not sufficient time to perform tests and evaluation on a new concrete supply. This provision permits concrete mixtures with no prior experience or trial mixture data, as required in Section 1905.3, to be based on other experience or information, subject to the approval of the engineer or architect. The concrete produced must have an average compressive strength at least 1,200 psi (8.3 MPa) greater than the specified compressive strength, f'_c, and meet minimum durability prescribed in Section 1904. In addition, the provisions should not be used where the specified compressive strength, f'_c, of the concrete exceeds 4,000 psi (27.6 MPa).

1905.5 Average strength reduction. As data become available during construction, it is permissible to reduce the amount by which the average compressive strength (f'_c) is required to exceed the specified value of f'_c in accordance with ACI 318, Section 5.5.

❖ During a construction project, if the strength test results of the concrete are running appreciably higher than the required average strength established for the work in accordance with Section 1905.3, the engineer may order a reduction in the required strength. Such adjustment must be based on average test results, as required by ACI 318, Section 5.5, and the procedures for proportioning concrete on the basis of field experience as specified in ACI 318, Sections 5.3.1 and 5.3.2. The reduced value is still subject to the special exposure requirements of Sections 1904.2 through 1904.4. Normally, strength reductions are warranted on construction projects that use large quantities of concrete.

1905.6 Evaluation and acceptance of concrete. The criteria for evaluation and acceptance of concrete shall be as specified in Sections 1905.6.2 through 1905.6.5.5.

❖ The specified frequency of testing and application of minimum acceptance criteria for concrete used in the construction process is covered in this section.

1905.6.1 Qualified technicians. Concrete shall be tested in accordance with the requirements in Sections 1905.6.2 through 1905.6.5. Qualified field testing technicians shall perform tests on fresh concrete at the job site, prepare specimens required for curing under field conditions, prepare specimens required for testing in the laboratory and record the temperature of the fresh concrete when preparing specimens for strength tests. Qualified laboratory technicians shall perform all required laboratory tests.

❖ This section contains a new provision in ACI 318 that requires technicians performing code-related tests on fresh concrete in the field, including fabricating strength test specimens (concrete cylinders) and code-related tests on concrete in the laboratory, to be qualified to perform such tests. Technicians desiring to perform field tests can demonstrate their qualifications to the satisfaction of the building official by becoming certified by ACI as a "Concrete Field Test Technician – Grade I," or by meeting the requirements of ASTM C 1077, or a comparable program. Similarly, laboratory technicians can demonstrate their competence by becoming certified by ACI as a "Concrete Laboratory Testing Technician," "Concrete Strength Testing Technician," or by meeting the requirements of ASTM C 1077, or a comparable program.

1905.6.2 Frequency of testing. The frequency of conducting strength tests of concrete shall be as specified in Sections 1905.6.2.1 through 1905.6.2.4.

❖ Frequency of testing is a significant factor in the effectiveness of quality control of concrete. To assure that test results will be representative of the concrete in the structure, samples of fresh concrete to determine its compressive strength must strictly follow the provisions of this section.

1905.6.2.1 Minimum frequency. Samples for strength tests of each class of concrete placed each day shall be taken not less than once a day, nor less than once for each 150 cubic yards (115 m^3) of concrete, nor less than once for each 5,000 square feet (465 m^2) of surface area for slabs or walls.

❖ This section prescribes three minimum sampling criteria for strength tests of each class of concrete (i.e., each mix design) based on the amount of concrete placed on a given day. (An example problem using the criteria in this section, as well as Sections 1905.6.2.2, 1905.6.2.3 and 1905.6.2.4, is provided below.) For test samples to provide a true measure of concrete strength, they must be taken on a random basis to provide a fair representation of the quality of concrete used in the construction. To be representative, the choice of both times and batches of concrete to be sampled must be made on the basis of chance alone. Samples should not be selected on the basis of the appearance of the concrete, convenient times or for any other reason that would cause the test results to lose their statistical validity. Procedures for random selection of batches are described in ASTM D 3665. Procedures for sampling freshly mixed concrete are provided in ASTM C 172.

1905.6.2.2 Minimum number. On a given project, if the total volume of concrete is such that the frequency of testing required by Section 1905.6.2.1 would provide less than five strength tests for a given class of concrete, tests shall be made from at least five randomly selected batches or from each batch if fewer than five batches are used.

❖ The representative nature of strength test data becomes less reliable when there are fewer than five tests. This criterion is intended to provide the largest possible sample size for projects that use small volumes of concrete.

1905.6.2.3 Small volume. When the total volume of a given class of concrete is less than 50 cubic yards (38 m^3), strength tests are not required when evidence of satisfactory strength is submitted to and approved by the building official.

❖ On a given project, if the quantity of a specific class (i.e., mix design) of concrete totals less than 50 cubic yards (38 m^3), strength tests may be waived with the approval of the building official if evidence is provided to show that proper concrete strength can be obtained. Such evidence may include certification from the concrete supplier that the material meets strength requirements based on field experience of previous projects or by trial batches in accordance with ACI 318 (see commentary, Section 1905.3).

1905.6.2.4 Strength test. A strength test shall be the average of the strengths of two cylinders made from the same sample of concrete and tested at 28 days or at the test age designated for the determination of f'_c.

❖ The statistical approach used in ACI 318 to proportion concrete mixtures for strength is based on strength test results that are the average of two cylinders. Likewise, the statistical approach is based on 28-day cylinder breaks, or other test age as specified in the construction documents. Using criteria other than that prescribed in this section will result in strength data that is not representative of concrete in the structure.

Example:

The quantity of concrete required to cast a 4-inch-thick (102 mm) slab for a floor, 3,000 square feet (279 m^2) in area, would be about 37 cubic yards (28 m^3). This represents the total volume of concrete involved in the project. About four truckloads of ready-mixed concrete using 10 cubic yard (7.6 m^3) capacity mixers (or agitators) are needed. How many strength tests are required?

According to the provisions of Section 1905.6.2.1, only one strength test is required because less than 150 cubic yards (115 m^3) of concrete and less than 5,000 square feet (465 m^2) of floor are to be cast in one day. Section 1905.6.1.2 modifies this requirement by specifying that when less than five strength tests for the total volume of a particular class of concrete are involved on a given project, at least one must be made for each batch of concrete. In this example, each truckload of concrete is a "batch" and, therefore, at least four strength tests are required. Section 1905.6.1.4 prescribes that a strength test be the average strengths of two cylinders made from the same sample (batch) of concrete. This means that a minimum of eight cylinders must be tested for the 37 cubic yards (28 m^3) of concrete used in this project.

However, in applying the provisions of Section 1905.6.1.3, the building official may waive the need for strength tests when the quantity of concrete to be cast is less than 50 cubic yards (38 m^3), but only if, in his or her judgement, there is sufficient evidence that the concrete will meet strength requirements.

1905.6.3 Laboratory-cured specimens. Laboratory-cured specimens shall comply with the provisions of Sections 1905.6.3.1 through 1905.6.3.4.

❖ This section provides criteria for determining the acceptability of concrete strength. As soon as test results become known during the course of the work, they can be immediately evaluated and the concrete accepted if it meets specified strength requirements. However, should concrete fail to meet the strength requirements in the acceptance criteria, changes should be made to the concrete to increase the average strength of subsequent test results.

1905.6.3.1 Sampling. Samples for strength tests shall be taken in accordance with ASTM C 172.

❖ ASTM C 172 provides a procedure for obtaining a representative sample of fresh concrete as delivered to the project site. The sample of concrete is then used to make strength test specimens (cylinders) that are tested to determine compliance with the quality control requirements of the code. The timing and manner of taking samples of freshly mixed concrete are critical to the production of quality concrete. Unless the sample is truly representative, the test results may be misleading.

1905.6.3.2 Cylinders. Cylinders for strength tests shall be molded and laboratory cured in accordance with ASTM C 31 and tested in accordance with ASTM C 39.

❖ After the sample is taken, specimens for strength testing must be prepared in accordance with ASTM C 31, which provides a standardized procedure for making test cylinders under field conditions, as well as for proper curing, protection and transporting to the laboratory for testing. In the laboratory the cylinders are tested in accordance with ASTM C 39 to determine the compressive strength of concrete. ASTM C 31 requires that whenever cylinders are fabricated for strength tests, the slump, air content and temperature of the sample of concrete must be determined. It should be pointed out that the requirement for determining the air content applies to all concrete, including nonair-entrained concrete.

1905.6.3.3 Acceptance of results. The strength level of an individual class of concrete shall be considered satisfactory if both of the following requirements are met:

1. Every arithmetic average of any three consecutive strength tests equals or exceeds f'_c.
2. No individual strength test (average of two cylinders) falls below f'_c by more than 500 psi (3.45 MPa) when f'_c is 5,000 psi (34.45 MPa) or less, or by more than $0.10f'_c$ when f'_c is more than 5,000 psi.

❖ Concrete strength is deemed satisfactory as long as the average of any three consecutive strength tests (average of two cylinders) equals or exceeds the specified compressive strength (f'_c), and no individual strength test (average of two cylinders) falls below the specified compressive strength of concrete, f'_c, by the amount specified.

1905.6.3.4 Correction. If either of the requirements of Section 1905.6.3.3 is not met, steps shall be taken to increase the average of subsequent strength test results. The requirements of Section 1905.6.5 shall be observed if the requirement of Section 1905.6.3.3, Item 2, is not met.

❖ When concrete fails to satisfy the strength requirements in Section 1905.6.3.3, changes to the concrete mix must be made to increase the average of subsequent strength test results. The changes that must be made to improve the concrete mixture will depend on particular circumstances and job specifications, but could include changes to the content of cement, water and air. Modifications that can be made to the concrete mixture to increase strength include changes in cement content, slump and air entrainment. Changes that only involve modifying mix proportions do not require resubmittal or approval of the concrete mixture. However, if the modifications involve changes to the ingredients of concrete, such as the use of a different type of cement, materials obtained from different sources or the addition of admixtures, evidence of an acceptable improvement in the concrete strength must be presented.

1905.6.4 Field-cured specimens. Field-cured specimens shall comply with the provisions of Sections 1905.6.4.1 through 1905.6.4.4.

❖ This section provides requirements for making, protecting and curing concrete cylinders in the field to determine if protection and curing of concrete in the structure is adequate in unusual weather conditions. They are also required to determine when forms may be removed (see Section 1906.2.2.1).

1905.6.4.1 When required. Where required by the building official, the results of strength tests of cylinders cured under field conditions shall be provided.

❖ Normally, the requirement for field-cured cylinders is found in the construction specifications. Although field-cured cylinders are not required by the code to verify adequacy of curing and protection, this section authorizes the building official to require them when necessary.

The building official should consider several things when deciding whether or not field-cured cylinders are necessary. One reason for testing field-cured cylinders is to determine if the concrete in the structure is being properly cured. Typically, when concrete is kept in a moist condition at a temperature above 50°F (10°C) for seven days (three days for high-early-strength concrete), it will undergo proper curing as required. Another reason is to confirm the concrete strength prior to removal of forms and shores. Section 1906.2.2.1, Item 3, requires confirmation of concrete strength by the testing of field-cured cylinders or other approved methods. Deciding whether or not to require field-cured cylinders will usually be based on one or more of the following:

1. Prior experience with the contractor;
2. Prior test results of field-cured cylinders;
3. Temperature conditions during construction;
4. Cold-weather protection method(s) for the concrete;
5. Protection method(s) to minimize moisture loss from the concrete; or
6. Method(s) for supplying additional moisture to the concrete.

1905.6.4.2 Curing. Field-cured cylinders shall be cured under field conditions in accordance with ASTM C 31.

❖ ASTM C 31 provides procedures not only for making and curing concrete test specimens that are to be field cured, but also for laboratory curing specimens as prescribed in Section 1905.6.3. Field-cured cylinders are stripped from their molds after a minimum initial curing and are stored in or on the structure as near as possible to the representative concrete. All surfaces of the cylinder are protected from the elements in as near as possible the same way as the formed work, thus providing the cylinders with the same temperatures and moisture environments as the structural work.

1905.6.4.3 Sampling. Field-cured test cylinders shall be molded at the same time and from the same samples as laboratory-cured test cylinders.

❖ This section stipulates that test cylinders to be field cured must be molded from the same sample of concrete as laboratory-cured test cylinders, thus providing companion cylinders for comparison testing as required in the following section.

1905.6.4.4 Correction. Procedures for protecting and curing concrete shall be improved when the strength of field-cured cylinders at the test age designated for determination of f'_c is less than 85 percent of that of companion laboratory-cured cylinders. The 85-percent limitation shall not apply if the field-cured strength exceeds f'_c by more than 500 psi (3.45 Mpa).

❖ The purpose of field-cured specimens is to simulate the curing of concrete under field conditions and to compare the test results of the field cylinders with those of laboratory-cured specimens made and cured under ideal moisture conditions. The comparison is made between the actual strengths of the field-cured cylinders and the laboratory-cured specimens, and not between the test results of the field specimens and the specified compressive strength of concrete, f'_c. Experience has shown that field-cured specimens generally do not test less than 85 percent of the standard laboratory moist cured cylinders. However, in cases when the field specimens fail to meet the 85-percent criterion specified in Section 1905.6.4.4, the test results are still acceptable if the strengths exceed the specified compressive strength by more than 500 psi (3.45 MPa).

1905.6.5 Low-strength test results. The investigation of low-strength test results shall be in accordance with the provisions of Sections 1905.6.5.1 through 1905.6.5.5.

❖ This section prescribes the procedures that must be applied for structural safety when test results fail to meet the acceptance criteria. The building official must apply judgement as to the true significance of low-strength test results and whether they indicate need for concern. For example, strength tests failing to meet the acceptance criteria of the following section can be expected to occur in approximately one in every 100 tests, even though concrete strength and uniformity may be satisfactory. Allowance should be made for such statistically expected variations in deciding whether the strength level being produced is adequate.

1905.6.5.1 Precaution. If any strength test (see Section 1905.6.2.4) of laboratory-cured cylinders falls below the specified value of f'_c by more than the values given in Section 1905.6.3.3, Item 2, or if tests of field-cured cylinders indicate deficiencies in protection and curing (see Section 1905.6.4.4), steps shall be taken to assure that the load-carrying capacity of the structure is not jeopardized.

❖ When test results for laboratory or field-cured cylinders indicate a strength value lower than permitted in this section, an assessment of the significance of the difference must be made. It is imperative that the investigation follow a logical sequence of possible cause and effect. First, test reports should be reviewed and results analyzed before any action is taken. Of special interest in the early investigation should be the procedures used in making, handling and testing the cylinders, since almost all deficiencies in failure to follow prescribed procedures will lower cylinder strength. Examples include: extra days in the field; curing over 80°F (26.7 °C); frozen cylinders; impact during transporting to lab; delay in moist curing at the lab; improper caps and insufficient care in breaking cylinders. Second, engineering computations should be made to determine if the design of the concrete in question, including the effects of reinforcement, provides sufficient safety factors to support the intended service loads at the lower strength indicated by the test results. Nondestructive tests, such as the Schimdt rebound hammer (ASTM C 805), the Windor probe (ASTM C 803), or an ultrasonic pulse velocity test may be used to determine whether or not a portion of the structure actually contains low-strength concrete. In extreme cases, compressive strength tests of cores, as provided for in the following section, may be necessary. For in-place concrete investigations, it is essential to know where in the structure the questionable concrete was deposited. This information should be part of the data recorded at the time the test cylinders were molded. This information is both important and necessary when determining the location to core drill, should the need arise.

1905.6.5.2 Core tests. If the likelihood of low-strength concrete is confirmed and calculations indicate that load-carrying capacity is significantly reduced, tests of cores drilled from the area in question in accordance with ASTM C 42 are permitted. In such cases, three cores shall be taken for each strength test that falls below the values given in Section 1905.6.3.3, Item 2.

❖ If a preliminary investigation determines that core tests should be required, core drilling in the area in question should be performed according to procedures outlined in ASTM C 42. The testing of cores requires great care in the operation itself and in the interpretation of the test results; thus, it is critical that ASTM C 42 is followed in detail.

1905.6.5.3 Condition of cores. Cores shall be prepared for transport and storage by wiping drilling water from their surfaces and placing the cores in water-tight bags or containers immediately after drilling. Cores shall be tested no earlier than 48 hours and not later than seven days after coring unless approved by the registered design professional.

❖ This section provides conditioning requirements for core samples that are intended to produce strength tests that are indicative of the in-situ concrete strength for the area in question.

1905.6.5.4 Test results. Concrete in an area represented by core tests shall be considered structurally adequate if the average of three cores is equal to at least 85 percent of f'_c and if no single core is less than 75 percent of f'_c. Additional testing of cores extracted from locations represented by erratic core strength results is permitted.

❖ In evaluating core test results, the fact that core strengths may not equal the specified compressive strength of concrete, f'_c., should not be a cause for concern, provided they are within the limits specified. It is realistic to expect the compressive strength of cores to be less than f'_c. because the size of the core differs from that of a standard concrete test cylinder, and procedures for obtaining and curing the cores are different. Therefore, the concrete can be considered acceptable from the standpoint of strength if test results from the cores are at least 85 percent of f'_c. for the average of three cores and 75 percent of f'_c. for a single core. Cores must be obtained and tested in accordance with ASTM C 42. Cores for compressive strength testing must have a length-to-diameter ratio between 1.0 and 2.10. For ratios less than 1.94, the compressive strength is multiplied by a correction factor (less than one) given in ASTM C 42 to correlate the strength of the core to that of a standard concrete test cylinder that has a length-to-diameter ratio of 2. Because of the less than delicate nature of the coring and sawing process, this section also permits erratic core and cube strengths to be discarded from the sample under consideration.

1905.6.5.5 Strength evaluation. If the criteria of Section 1905.6.5.4 are not met and the structural adequacy remains in doubt, the building official is permitted to order a strength evaluation in accordance with ACI 318, Chapter 20, for the questionable portion of the structure, or take other appropriate action.

❖ As a last resort, load tests may be required to check the adequacy of structural members that are seriously in doubt. Generally, such tests are suited only for flexural members, such as floors or beams, but they may sometimes apply to other members. Load testing is a highly specialized procedure that must be performed and interpreted by an engineer fully qualified in the proper techniques. Another "appropriate action" may be approving the structural element for a reduced live load.

1905.7 Preparation of equipment and place of deposit. Preparation before concrete placement shall include the following:

1. Equipment for mixing and transporting concrete shall be clean.
2. Debris and ice shall be removed from spaces to be occupied by concrete.
3. Forms shall be properly coated.
4. Masonry filler units that will be in contact with concrete shall be well drenched.
5. Reinforcement shall be thoroughly clean of ice or other deleterious coatings.
6. Water shall be removed from the place of deposit before concrete is placed unless a tremie is to be used or unless otherwise permitted by the building official.
7. Laitance and other unsound material shall be removed before additional concrete is placed against hardened concrete.

❖ This section specifies the preparatory work on formwork, reinforcement and concrete equipment that is necessary prior to the placement of concrete. The section addresses the need for using clean equipment and for cleaning forms and reinforcement before placing concrete. All debris—including sawdust, nails, wood pieces and other litter—that usually collects inside the forms must be removed. The surface condition of reinforcement must comply with Section 1907.4. All standing water, snow or ice must also be removed from the forms.

1905.8 Mixing. Mixing of concrete shall be performed in accordance with Sections 1905.8.1 through 1905.8.3.

❖ This section provides basic requirements for concrete mixing.

1905.8.1 General. Concrete shall be mixed until there is a uniform distribution of materials and shall be discharged completely before the mixer is recharged.

❖ Concrete is to be mixed thoroughly until it is uniform in appearance and all ingredients are evenly distributed. If concrete has been adequately mixed, samples taken from different portions of a batch will have essentially the same unit weight, air content, slump and coarse aggregate content.

1905.8.2 Ready-mixed concrete. Ready-mixed concrete shall be mixed and delivered in accordance with the requirements of ASTM C 94 or ASTM C 685.

❖ Ready-mixed concrete is commonly mixed in a central plant and transported to the construction site in agitating or nonagitating trucks, mixed entirely in transit or after reaching the job or mixed partially in a central plant and completed in transit or at the job site. ASTM C 94 and ASTM C 685 cover methods of mixing and transporting concrete.

1905.8.3 Job-mixed concrete. Job-mixed concrete shall comply with ACI 318, Section 5.8.3.

❖ Concrete is sometimes mixed at the construction site in a stationary mixer. Job-mixed concrete must conform to the requirements of ACI 318, Section 5.8.3, which provides general provisions for materials handling, batching and the mixing of concrete, and includes requirements for job records.

1905.9 Conveying. The method and equipment for conveying concrete to the place of deposit shall comply with Sections 1905.9.1 and 1905.9.2.

❖ The handling and transporting of concrete needs to be carefully controlled in each step of the operation to maintain uniformity of the materials. This section requires the equipment for handling and transporting concrete to be capable of supplying concrete to the place of deposit continuously and reliably under all conditions and methods of placement, including by crane buckets; wheelbarrows or power buggies; chutes; belt conveyors or pumps or by any other acceptable means.

1905.9.1 Method of conveyance. Concrete shall be conveyed from the mixer to the place of final deposit by methods that will prevent separation or loss of materials.

❖ To maintain uniformity and an even distribution of ingredients in the concrete mixture, it is important to handle and transport the material in such a way as to avoid segregation of the coarse aggregates from the mortar or the water from the other ingredients comprising the mixture.

1905.9.2 Conveying equipment. The conveying equipment shall be capable of providing a supply of concrete at the site of placement without separation of ingredients and without interruptions sufficient to permit the loss of plasticity between successive increments.

❖ This section requires that equipment used for handling and transporting concrete be capable of supplying concrete to the place of deposit continuously and reliably under all conditions. Equipment that may be used to convey concrete includes pumps, belt conveyors, pneumatic systems, wheelbarrows, power buggies, crane buckets and tremies.

1905.10 Depositing. The depositing of concrete shall comply with the provisions of Sections 1905.10.1 through 1905.10.8.

❖ This section addresses the placement of fresh concrete in its final position as well as concerns such as placement timing, concrete that has partially hardened, continuous concreting, retempering, depositing in walls, construction joints and consolidation.

1905.10.1 Segregation. Concrete shall be deposited as nearly as practicable to its final position to avoid segregation due to rehandling or flowing.

❖ Segregation is the separation of coarse aggregate from mortar. Concrete should be deposited at or near its final position in the structure. There is a tendency for aggregates to segregate when concrete is flowed laterally into place or deposited in locations that necessitate rehandling.

1905.10.2 Placement timing. Concreting operations shall be carried on at such a rate that the concrete is at all times plastic and flows readily into spaces between reinforcement.

❖ At all times during the placement operations, concrete is required to remain in a plastic state to allow it to flow readily into the spaces between reinforcement. Placing should be carried on at such a rate that the concrete, which is being integrated with fresh concrete, is still plastic. Concrete is deemed to be in a "plastic" or semifluid state when it is capable of being easily molded by hand. A very wet or soupy concrete mixture is not within the definition of "Plastic." A plastic mix keeps all grains of sand and pieces of coarse aggregate encased and held in place. The parts of such mixtures are not apt to segregate while being transported, placed or worked in the forms.

1905.10.3 Unacceptable concrete. Concrete that has partially hardened or been contaminated by foreign materials shall not be deposited in the structure.

❖ Concrete that has become partially hardened (i.e., advanced beyond an initial set) prior to placement will have significantly less strength than concrete that has been placed while in a plastic state and allowed to gain strength under appropriate curing conditions. Concrete that has been contaminated by foreign materials, such as dirt or sawdust, will have lower strength, be less durable and have other less desirable properties depending on the amount of contaminants.

1905.10.4 Retempering. Retempered concrete or concrete that has been remixed after initial set shall not be used unless approved by the registered design professional.

❖ Concrete may be remixed after first achieving its initial set without losing appreciable strength, however, it should be done infrequently and with caution. Because of the potential for lower strength, it must be approved by the registered design professional.

1905.10.5 Continuous operation. After concreting has started, it shall be carried on as a continuous operation until placing of a panel or section, as defined by its boundaries or predetermined joints, is completed, except as permitted or prohibited by Section 1906.4.

❖ Concrete should be deposited continuously or in layers of such thickness that it will not be deposited on concrete that has hardened sufficiently to cause flow lines, seams or planes of weakness within the section. If a section cannot be placed continuously, preplanned construction joints (in accordance with Section 1906.4) should be provided as called for on the drawings or as may otherwise be permitted.

1905.10.6 Placement in vertical lifts. The top surfaces of vertically formed lifts shall be generally level.

❖ This section requires that the top surface of concrete being placed in a vertical member, such as walls, be kept as level as practical. The top surface of a concrete lift provides a potential location for seams or planes of weakness. The structural consequence of these potential planes of weakness is less if they are approximately level or horizontal.

1905.10.7 Construction joints. When construction joints are required, they shall be made in accordance with Section 1906.4.

❖ A construction joint occurs where freshly mixed concrete is placed in contact with existing hardened concrete. Special details and procedures are necessary to enable construction joints to provide structural integrity and other desirable characteristics, as outlined in Section 1906.4.

1905.10.8 Consolidation. Concrete shall be thoroughly consolidated by suitable means during placement and shall be thoroughly worked around reinforcement and embedded fixtures and into corners of the forms.

❖ Consolidation is the process of compacting fresh concrete to mold it within forms and around embedded items and reinforcement to eliminate stone pockets and entrapped air. Consolidation is accomplished by hand rodding or mechanical methods, such as vibration. Consolidation around reinforcement is particularly important to develop bond strength between the concrete and the reinforcement.

1905.11 Curing. The curing of concrete shall be in accordance with Sections 1905.11.1 through 1905.11.3.

❖ This section identifies the scope of the sections regarding concrete curing. Proper curing is needed in order to achieve the required field strength of the concrete and to provide a durable finished product.

1905.11.1 Regular. Concrete (other than high early strength) shall be maintained above 50°F (10°C) and in a moist condition for at least the first seven days after placement, except when cured in accordance with Section 1905.11.3.

❖ Immediately after placement and finishing, concrete must be protected and maintained with minimal moisture loss at a relatively constant temperature for the period necessary for the hydration of cement and the hardening of concrete so that the desired properties of strength and durability may develop. Concrete (other than high early strength) that is maintained in a moist condition and at a temperature above 50°F (10°C) for seven days can be expected to obtain its specified compressive strength. While the amount of mixing water in the concrete mixture is normally sufficient for curing purposes, excessive water loss by evaporation should be prevented by properly applying curing materials and methods. Evaporation may reduce the amount of water below what is necessary for the development of the desired properties of concrete.

1905.11.2 High early strength. High-early-strength concrete shall be maintained above 50°F (10°C) and in a moist condition for at least the first three days, except when cured in accordance with Section 1905.11.3.

❖ Concrete made with high-early-strength cement develops strength much earlier than concrete made with normal (Type I) or moderate sulfate-resistance (Type II) cements. Earlier development of strength reduces the length of time (from seven to three days) that it is necessary to provide moist, temperate curing conditions.

1905.11.3 Accelerated curing. Accelerated curing of concrete shall comply with ACI 318, Section 5.11.3.

❖ The principal reason for using accelerated methods for curing concrete is to speed up strength development and reduce the time of curing. Accelerated curing is commonly used in the manufacturing of precast structural concrete elements. ACI 318, Section 5.11.3, prescribes acceptable methods, testing and acceptance criteria for accelerated curing. Curing by these special methods can produce compressive strengths that are lower than concrete moist cured by conventional methods. As a consequence, concrete mixtures should be proportioned to meet the specified compressive strength requirements under such curing conditions and be provided with supplementary strength testing.

1905.12 Cold weather requirements. Concrete that is to be placed during freezing or near-freezing weather shall comply with the following:

1. Adequate equipment shall be provided for heating concrete materials and protecting concrete during freezing or near-freezing weather.
2. Concrete materials and reinforcement, forms, fillers and ground with which concrete is to come in contact shall be free from frost.
3. Frozen materials or materials containing ice shall not be used.

❖ Concrete placed in cold weather can adversely affect the strength and durability of the hardened material unless proper protection is provided to prevent the fresh concrete from freezing. This section describes the preparatory work necessary before concrete is placed in cold weather. Cold weather is defined as a period when, for more than three consecutive days, the mean daily temperature drops below 40°F (4.4°C). When temperatures are above 50°F (10°C) during more than half of any 24-hour period, the concrete should not be regarded as coming under the cold weather requirements of this section. For more information, refer to ACI 306R, which provides requirements for protection, temperatures during and after concrete placement, curing methods and other detailed information needed for cold weather concreting.

1905.13 Hot weather requirements. During hot weather, proper attention shall be given to ingredients, production methods, handling, placing, protection and curing to prevent excessive concrete temperatures or water evaporation that could impair the required strength or serviceability of the member or structure.

❖ Hot weather cannot be simply defined in terms of time/temperature limits as is used to define cold weather (see commentary, Section 1905.12) because of the influence of other environmental factors. Hot weather, as it affects concrete, can best be described as any combination of high temperature, low relative humidity and wind velocity that tends to adversely affect the properties of fresh or hardened concrete.

Hot weather can create difficulties in fresh concrete such as:

- Increased water demand;
- Accelerated slump loss;
- Increased rate of setting and the ensuing problems of placing and finishing concrete;
- Increased tendency for plastic cracking and
- Control of air content of air-entrained concrete.

Although adding water to concrete during hot weather would seem to be the appropriate solution, it can have adverse effects on the properties of hardened concrete, such as:

- Decreased strength;
- Decreased durability and watertightness;
- Nonuniform surface appearance and
- Increased tendency for drying shrinkage.

Detailed information on the materials, methods and procedures necessary to alleviate the difficulties described above is contained in ACI 305R.

SECTION 1906 FORMWORK, EMBEDDED PIPES AND CONSTRUCTION JOINTS

1906.1 Formwork. The design, fabrication and erection of forms shall comply with Sections 1906.1.1 through 1906.1.6.

❖ These provisions give the minimum requirements for the design, construction and materials for formwork to provide installations that are adequate and safe to receive the concrete work as shown on the plans and specified in the construction documents.

1906.1.1 General. Forms shall result in a final structure that conforms to shapes, lines and dimensions of the members as required by the construction documents.

❖ This section provides minimum performance requirements for formwork with respect to shape, alignment and dimension of concrete members. Formwork is very seldom constructed to exactly the dimensions shown on the construction documents, nor is it perfectly straight, level and plumb. Fortunately, it seldom needs to be perfect. Tolerances for alignment and dimensions are normally found in the construction documents. In the absence of such information, allowable tolerances can be found in ACI 117 and ACI 347. The recommended practices covered in this document include preparation of construction documents, design criteria for vertical and horizontal forces and lateral pressures, design considerations (including capacities of formwork accessories), preparation of formwork design drawings, construction and use of forms (including safety considerations), materials for formwork and formwork for special structures and special methods of construction.

While the information contained in ACI 347 is pertinent to the work of engineers, architects and building officials, it is important to understand that the recommended practices are based on the premise that the layout, design and construction of formwork are the responsibility of the contractor.

1906.1.2 Strength. Forms shall be substantial and sufficiently tight to prevent leakage of mortar.

❖ Forms must have adequate strength and be tightly fitted to prevent gaps from developing at joints and seams that would allow the loss of cement paste (mortar). Loss of mortar can cause honeycombing and rock pockets that diminish concrete appearance and may adversely affect strength and durability.

1906.1.3 Bracing. Forms shall be properly braced or tied together to maintain position and shape.

❖ Formwork should be straight, free of warping and adequately braced and tied together to resist concrete pressure without bulging or failure.

1906.1.4 Placement. Forms and their supports shall be designed so as not to damage previously placed structures.

❖ Shoring and formwork supports must be designed so they do not impose undue loading or are not anchored in a manner that will cause damage to previously placed concrete.

1906.1.5 Design. Design of formwork shall comply with ACI 318, Section 6.1.5.

❖ Although forms have a temporary function, they are engineered structures that must be designed to withstand the loads and vibrations encountered during concrete placement. ACI 318, Section 6.1.5, requires that formwork be designed to consider rate and method of concrete placement, construction loads (i.e., vertical, horizontal and impact) and special forming requirements (e.g., shells, folded plates, domes and architectural elements).

1906.1.6 Forms for prestressed concrete. Forms for prestressed concrete members shall be designed and constructed to permit movement of the member without damage during application of the prestressing force.

❖ When prestressing forces are induced in post-tensioned concrete members, the members tend to shorten and, in the case of beams and slabs, also tend to "lift." The shortening of the elements may damage forming if it is not designed and constructed to accommodate this movement. Because prestressing forces generally induce compression in portions of beams and slabs that are resisting flexural tension stresses when loaded (i.e., the bottom portion of the mid-span region of a member), the member tends to "lift" when the prestressing forces are applied to the concrete. This tendency to lift may temporarily remove all gravity loads from the shores that have been supporting the formwork and element prior to application of the prestressing. If the shores are not mechanically attached to the formwork and braced laterally, they may partially or completely fall out, leaving the post-tensioned member with no support for construction loads that may be applied later.

1906.2 Removal of forms, shores and reshores. The removal of forms and shores and the installation of reshores shall comply with Sections 1906.2.1 through 1906.2.2.3.

❖ This section provides general requirements for the removal of forms and the installation of reshores that are necessary to provide for worker safety and prevent damage to the concrete during the construction process.

1906.2.1 Removal of forms. Forms shall be removed in such a manner so as not to impair safety and serviceability of the structure. Concrete to be exposed by form removal shall have sufficient strength not to be damaged by the removal operation.

❖ Forms must not be removed until the concrete has developed sufficient tensile strength to resist cracking, spalling or, worse yet, structural failure. Usually, the side (vertical) forms of beams, girders, columns and walls can be removed after 12 hours. Before the removal of formwork supporting the bottom (horizontal) portions of structural elements, such as slabs, beams and girders, additional evaluation is required to determine the appropriate time for removal to assure that the integrity of the structure is maintained.

1906.2.2 Removal of shores and reshores. The provisions of Sections 1906.2.2.1 through 1906.2.2.3 shall apply to slabs and beams, except where cast on the ground.

❖ This section provides general requirements for the removal of forms and shores and installation of reshores for structural slabs and beams (i.e., those not supported on the ground) to insure a safe structure during construction and prevent damage to the concrete during the construction process.

1906.2.2.1 Removal schedule. Before starting construction, the contractor shall develop a procedure and schedule for removal of shores and installation of reshores and for calculating the loads transferred to the structure during the process.

1. The structural analysis and concrete strength data used in planning and implementing form removal and shoring shall be furnished by the contractor to the building official when so requested.
2. No construction loads shall be supported on, nor any shoring removed from, any part of the structure under construction except when that portion of the structure in combination with the remaining forming and shoring system has sufficient strength to support safely its weight and the loads placed thereon.
3. Sufficient strength shall be demonstrated by structural analysis considering the proposed loads, the strength of the forming and shoring system and concrete strength data. Concrete strength data shall be based on tests of field-cured cylinders or, when approved by the building official, on other procedures to evaluate concrete strength.

❖ This section requires the contractor to develop a procedure and schedule for the removal of forms and placement of reshores supporting slabs and beams (other than those supported by the ground) before construction is started. The time for removal of the slab and beam forms must be determined by structural analysis, giving consideration to the effect of construction loads and the actual strength of the concrete. Before forms are stripped, concrete must acquire sufficient strength to avoid collapse or excessive deflection or distortions that would cause permanent structural damage. The strength of concrete must be determined prior to form removal by testing field-cured cylinders in accordance with ASTM C 39 or by other procedures approved by the building official such as:

- Compressive strength tests, ASTM C 873 [These tests are limited to slabs where the depth of concrete is 5 to 12 inches (127 to 305 mm).];
- Penetration resistance of hardened concrete, ASTM C 803;
- Pullout strength tests, ASTM C 900 and
- Maturity factor determinations in accordance with ASTM C 1074.

The strength analysis and concrete data used to develop the procedures and time schedule must be furnished to the building official, if requested.

For safety purposes, special loading conditions, such as unsymmetrical placement of concrete, impact loads, uplift and concentrated loads must be considered in the design of formwork, including the loads transmitted from the floors above in multistory construction.

1906.2.2.2 Construction loads. No construction loads exceeding the combination of superimposed dead load plus specified live load shall be supported on any unshored portion of the structure under construction, unless analysis indicates adequate strength to support such additional loads.

❖ Construction loads include combinations of dead loads, such as the weights of formwork and fresh concrete, and live loads, such as workmen, equipment, stored materials, runways and other loads imposed by the concreting operations. These loads are often greater than the design live load and must be supported by shores unless structural analysis and test data show that shores may be omitted.

1906.2.2.3 Prestressed members. Form supports for prestressed concrete members shall not be removed until sufficient prestressing has been applied to enable prestressed members to carry their dead load and anticipated construction loads.

❖ Before the dead load and any superimposed construction loads can be safely supported by an unshored, posttensioned, prestressed member, the concrete must attain the compressive strength established by the engineer, and prestressing must be applied to the concrete by tensioning the tendons to the value specified in the construction documents. If shores are removed from such a member prior to the concrete being prestressed or from one in which inadequate prestress has been applied to the concrete, the member may be damaged or experience structural failure.

1906.3 Conduits and pipes embedded in concrete. Conduits, pipes and sleeves of any material not harmful to concrete and within the limitations of ACI 318, Section 6.3, are permitted to be embedded in concrete with approval of the registered design professional.

❖ Conduits, pipes and sleeves may be embedded in concrete, provided that the properties of the material or the size and location of embedments will not damage the concrete. In the design of structural concrete, embedments must not be considered as structurally replacing any of the displaced concrete, except in the case of concrete under compression in accordance with ACI 318, Section 6.3.6.

The use of bare aluminum in structural concrete is prohibited because the material reacts with concrete, and in the presence of chloride ions, may produce an electrolytic reaction with steel reinforcement causing corrosion of the steel that may result in cracking or spalling of the concrete or both. Further, in conjunction with the use of aluminum materials in concrete, particularly aluminum conduits, stray electric currents may accelerate the electrolytic reaction. Therefore, any aluminum embedments in concrete must be suitably coated or covered to prevent direct contact with the concrete.

Care must be taken in the design and construction of slabs, walls or beams to insure that any conduits, pipes or sleeves passing through these structural elements will not be located in places that would create planes of weakness or distress and compromise the strength of the concrete construction. The locations of conduits, pipes and sleeves passing through structural elements should be shown and properly located on the project plans. In the absence of such information, the building official should seek the approval of the design engineer before allowing installation.

Conduits and pipes are permitted to be embedded within a column, provided that such installations do not displace more than 4 percent of the column cross-sectional area on which the strength is calculated or that is required for fire protection purposes. This allowance is based on the premise that a column is a compression member, and that a nominal displacement of concrete with steel will not seriously affect structural integrity.

ACI 318, Section 6.3.5, gives empirical rules for acceptance of conduits and pipes embedded in slabs, walls and beams (other than those merely passing through) that apply unless the structural engineer specifically approves drawings showing the size and location of the conduit and pipe.

1906.4 Construction joints. Construction joints shall comply with the provisions of Sections 1906.4.1 through 1906.4.6.

❖ There are three basic types of joints used in concrete construction: construction joints, contraction or control joints and isolation joints. This section deals only with construction joints, which are mainly used in structural floor and roof systems. Construction joints are stopping places in the process of concrete placement. A true construction joint should bond new concrete to existing concrete and permit no movement. This condition of construction is essential in structural concrete. See the commentary to Section 1909.3 for information on contraction and isolation joints.

1906.4.1 Surface cleaning. The surface of concrete construction joints shall be cleaned and laitance removed.

❖ The surface of previously placed concrete must be clean and free of laitance before concrete placement is continued. Laitance is the light gray or nearly white substance that may form during the finishing operation. The laitance layer has little strength, and if not removed will reduce the bond between hardened and fresh concrete.

1906.4.2 Joint treatment. Immediately before new concrete is placed, construction joints shall be wetted and standing water removed.

❖ If the hardened concrete joint is not lightly wetted prior to placement of fresh concrete, its dry surface may absorb water from the fresh concrete that is needed for proper strength development, thus creating a weakened plane. Overwetting, especially when it creates standing water, will increase the water-cementitious materials ratio of the concrete at the joint, which can also reduce localized strength and bonding potential.

1906.4.3 Location for force transfer. Construction joints shall be so made and located as not to impair the strength of the structure. Provision shall be made for the transfer of shear and other forces through construction joints (see ACI 318, Section 11.7.9).

❖ The location of construction joints is an important structural consideration that must be specified by the designer either in the construction documents or on a case-by-case basis. Where shear due to gravity is not significant, which typically occurs in the middle of the span of flexural members such as slabs and beams, a vertical joint (without shear key, diagonal dowels or other shear transfer method) may be acceptable. ACI 318, Section 11.7.9, requires that the construction joint be clean and free of laitance to ensure proper bond between the hardened joint and the fresh concrete. Joint surfaces (where the designer has assumed a coefficient of friction for shear transfer) must be roughened to an amplitude of approximately $^1/_4$ inch (6.4 mm).

1906.4.4 Location in slabs, beams and girders. Construction joints in floors shall be located within the middle third of spans of slabs, beams and girders. Joints in girders shall be offset a minimum distance of two times the width of intersecting beams.

❖ As a general rule, construction joints should be located where shear due to gravity loads is at a minimum. In slabs, beams and girders, this location is within the middle third of the span, preferably nearer the center where the shear stress approaches zero. Where a beam intersects a girder, the joint in the girder must be offset a distance equal to twice the width of the beam.

1906.4.5 Vertical support. Beams, girders or slabs supported by columns or walls shall not be cast or erected until concrete in the vertical support members is no longer plastic.

❖ Where horizontal members such as beams, girders and slabs are placed monolithically with their supporting columns or walls, cracking may develop at the interface of the horizontal members and the vertical supporting member due to settlement and bleeding of the plastic concrete in the supporting member. To lessen the potential for such cracking, concrete in walls and columns supporting slabs, beams or girders must be allowed to harden before additional concrete is placed.

1906.4.6 Monolithic placement. Beams, girders, haunches, drop panels and capitals shall be placed monolithically as part of a slab system, unless otherwise shown in the design drawings or specifications.

❖ When beams, girders, brackets, column capitals, haunches and drop panels are not placed monolithically with the floor or roof slab, there is potential for development of a weakened plane that will reduce the flexural capacity of the composite member. By definition, monolithic concrete contains no joints other than construction joints.

SECTION 1907 DETAILS OF REINFORCEMENT

1907.1 Hooks. Standard hooks on reinforcing bars used in concrete construction shall comply with ACI 318, Section 7.1.

❖ The ends of most reinforcing bars, except at lap splices, must be bent to form a hook in order to provide better anchorage to the concrete. ACI 318 requires that hooks be bent 90 degrees (1.6 rad) ("L" shape), 135 degrees (2.4 rad) or 180 degrees (3.1 rad) depending on their function and loading condition.

1907.2 Minimum bend diameters. Minimum reinforcement bend diameters utilized in concrete construction shall comply with ACI 318, Section 7.2.

❖ If reinforcement is bent too sharply, the outside surface of the bend may develop cracks that will reduce anchorage strength and increase the potential for corrosion. To prevent this from occurring, the code restricts the minimum inside diameter of reinforcing bar bends based on the bar size; requiring larger minimum-diameter bends for larger bar sizes.

1907.3 Bending. The bending of reinforcement shall comply with Sections 1907.3.1 and 1907.3.2.

❖ This section identifies the scope of the code requirements regarding bending of reinforcement. Proper bending is needed so that the reinforcement will not have a reduction in tensile strength.

1907.3.1 Cold bending. Reinforcement shall be bent cold, unless otherwise permitted by the registered design professional.

❖ Normally, bends in reinforcement complying with the minimum diameters specified in Section 7.2 of ACI 318 can be accomplished with the reinforcement at ambient temperatures. In some unusual construction conditions, normal-bend diameters may not be practical and bends with smaller diameters than specified in Section 7.2 of ACI 318 may be necessary. In order to accomplish such bends, the reinforcement may need to be heated prior to bending. This section permits this practice provided it is approved by the registered design professional.

1907.3.2 Embedded reinforcement. Reinforcement partially embedded in concrete shall not be field bent, except as shown on the construction documents or permitted by the registered design professional.

❖ Reinforcement that is partially embedded in the concrete must be bent in the field on occasion. Under such circumstances, the registered design professional must decide whether the bars are to be bent cold or heated. Heating is usually ordered for large-sized bars that cannot be easily bent cold. Reinforcing bars can be bent successfully if they are first preheated to between 1,100 and 1,200°F (593 and 649°C) and then gently bent in as gradual an arc as possible. The decision to heat partially embedded reinforcement is normally based on the proximity of the bar bend to the adjacent concrete. Heating, where necessary, must be controlled at all times to prevent damage to the bars that would adversely affect structural strength. In the case of partially embedded reinforcement, heating must also be done in a manner to avoid spalling or otherwise causing damage to the concrete.

1907.4 Surface conditions of reinforcement. The surface conditions of reinforcement shall comply with the provisions of Sections 1907.4.1 through 1907.4.3.

❖ This section identifies the code requirements regarding surface conditions of reinforcement. Surface conditions are important for proper bond strength to the concrete.

1907.4.1 Coatings. At the time concrete is placed, reinforcement shall be free from mud, oil or other nonmetallic coatings that decrease bond. Epoxy coatings of steel reinforcement in accordance with ACI 318, Sections 3.5.3.7 and 3.5.3.8, are permitted.

❖ At the time of concrete placement, all metal reinforcement must be free from mud, oil or other deleterious materials that may adversely affect or reduce the bond between the reinforcement and the concrete. Bonding of the two materials is necessary in order for the concrete to take full advantage of the tensile strength provided by the reinforcement.

1907.4.2 Rust or mill scale. Except for prestressing steel, steel reinforcement with rust, mill scale or a combination of both, shall be considered satisfactory, provided the minimum dimensions, including height of deformations and weight of a hand-wire-brushed test specimen, comply with applicable ASTM specifications (see Section 1903.5).

❖ Research has shown that the amount of rust and mill scale that normally remains on reinforcement after handling actually improves the bond between concrete and the reinforcement. Rust becomes a concern when it has progressed to a state where there is a measurable loss in height of deformations and weight of a specimen subjected to a hand-wire-brushed test.

1907.4.3 Prestressing steel. Prestressing steel shall be clean and free of oil, dirt, scale, pitting and excessive rust. A light coating of rust is permitted.

❖ Excessive rust, pitting or both must not be permitted to develop on the surface of prestressing steel because these highly stressed elements are more susceptible to corrosion. Oil and dirt must also be removed to maintain a proper bond between the steel and the concrete.

1907.5 Placing reinforcement. The placement of concrete reinforcement shall comply with the provisions of Sections 1907.5.1 through 1907.5.4.

❖ This section identifies the scope of the code requirements regarding reinforcement placement.

1907.5.1 Support. Reinforcement, including tendons, and posttensioning ducts shall be accurately placed and adequately supported before concrete is placed, and shall be secured against displacement within tolerances permitted in Section 1907.5.2. Where approved by the registered design professional, embedded items (such as dowels or inserts) that either protrude from precast concrete members or remain exposed for inspection are permitted to be embedded while the concrete is in a plastic state, provided the following conditions are met:

1. Embedded items are not required to be hooked or tied to reinforcement within the concrete.
2. Embedded items are maintained in the correct position while the concrete remains plastic.
3. The concrete is properly consolidated around the embedded item.

❖ Reinforcement, tendons and posttensioning ducts must be accurately positioned, supported and secured to prevent displacement by construction loads or the placing of concrete beyond the tolerances permitted by Section 1907.5.2. This section recognizes the practical necessity of placing some items in precast elements after the concrete is in place and relies on the registered design professional to determine proper installation.

1907.5.2 Tolerances. Unless otherwise specified by the registered design professional, reinforcement, including tendons, and posttensioning ducts shall be placed within the tolerances specified in Sections 1907.5.2.1 and 1907.5.2.2.

❖ Reinforcement, including bars and welded wire fabric, tendons and posttensioning ducts must be placed and held in position as shown on the project drawings or as otherwise specified by the engineer. The strength of a concrete member can be adversely affected by the improper positioning of reinforcement that exceeds the tolerances permitted by Sections 1907.5.2.1 and 1907.5.2.2.

1907.5.2.1 Depth and cover. Tolerance for depth, *d*, and minimum concrete cover in flexural members, walls and compression members shall be as shown in Table 1907.5.2.1, except that tolerance for the clear distance to formed soffits shall be minus $^1/_4$ inch (6.4 mm) and tolerance for cover shall not exceed minus one-third the minimum concrete cover required in the design drawings or specifications.

❖ This section provides tolerances that apply simultaneously to concrete cover and member effective depth (*d*). Cover is measured from the outside of the reinforcement to the face of the concrete. "Effective depth" or (*d*) is defined as the distance from the extreme compression fiber to the centroid of tension reinforcement. The effective depth (*d*) is a very important dimension, since deviation in this dimension, especially for members of lesser depth, can have an adverse effect on the strength of the member. The permitted tolerances take this into account, with a smaller permitted variation [i.e., $^3/_8$ inch (9.5 mm)] for smaller members; for example, members where *d* is 8 inches (203 mm) or less]. Note that a plus (+) tolerance increases the dimension and a minus (-) tolerance decreases the dimension. Where only a minus (-) tolerance is indicated for cover, there is no limit in the other direction; however, normally when cover is increased beyond that specified, effective depth (*d*) is reduced. The last sentence of the text allows a reduced cover dimension for formed soffits, because they provide additional exposure protection. Measurement of effective depth (*d*) and cover is illustrated in Figure 1907.5.2.1.

TABLE 1907.5.2.1
TOLERANCES

DEPTH (*d*) (inches)	TOLERANCE ON *d* (inch)	TOLERANCE ON MINIMUM CONCRETE COVER (inch)
$d > 8$	$\pm ^3/_8$	$- ^3/_8$
$d > 8$	$\pm ^1/_2$	$- ^1/_2$

For SI: 1 inch = 25.4 mm.

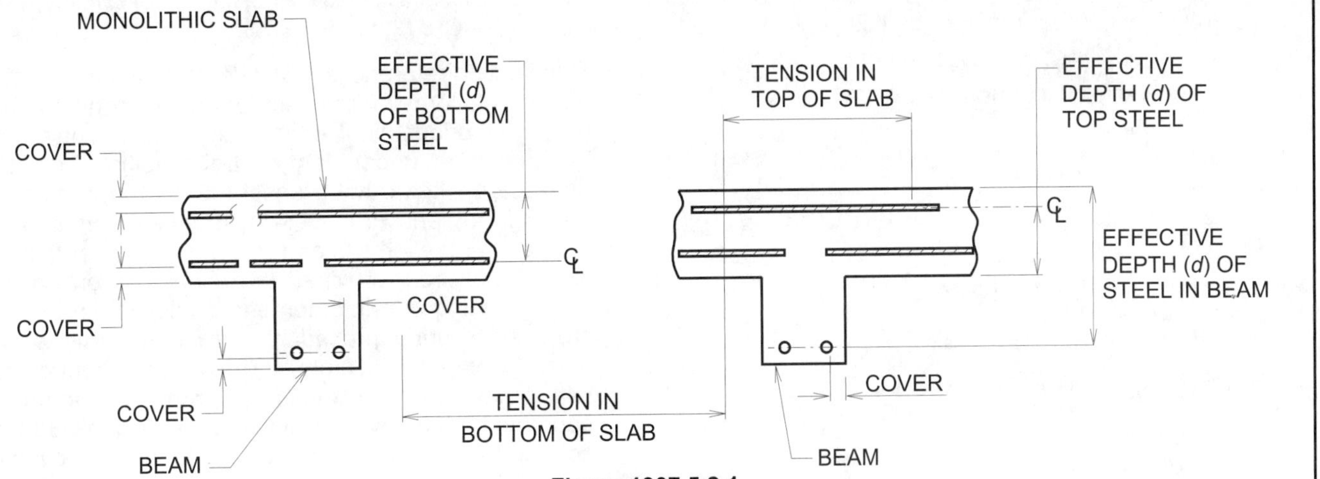

Figure 1907.5.2.1
TYPICAL SLAB & BEAM CONSTRUCTION
(NOT TO SCALE)

1907.5.2.2 Bends and ends. Tolerance for longitudinal location of bends and ends of reinforcement shall be ± 2 inches (± 51 mm) except the tolerance shall be ± $^1/_2$ inch (± 12.7 mm) at the discontinuous ends of brackets and corbels, and ± 1 inch (25 mm) at the discontinuous ends of other members. The tolerance for minimum concrete cover of Section 1907.5.2.1 shall also apply at discontinuous ends of members.

❖ The locations of bends and ends of reinforcement as required by the construction documents are important to the structural integrity of concrete members. In addition to the code-prescribed rebar placing tolerances, job specifications may contain more stringent reinforcement tolerances that must be followed during construction.

1907.5.3 Welded wire fabric. Welded wire fabric with wire size not greater than W5 or D5 used in slabs not exceeding 10 feet (3048 mm) in span is permitted to be curved from a point near the top of the slab over the support to a point near the bottom of the slab at midspan, provided such reinforcement is either continuous over, or securely anchored at support.

❖ This section permits the draping of welded wire fabric in short-span slabs [10 feet (3048 mm) or less] where the fabric consists of smaller sizes of wire that are flexible enough to curve readily over the supports. "A point near the top of the slab" and "a point near the bottom of the slab" should be interpreted as distances from the top and bottom of the slab that provide the cover required by Section 1907.7. Provisions for anchorage requirements at supports are found in Sections 12.11 and 12.12 of ACI 318.

1907.5.4 Welding. Welding of crossing bars shall not be permitted for assembly of reinforcement unless authorized by the registered design professional.

❖ Unless otherwise approved by the engineer, this section prohibits tack welding of crossing bars. Metallurgical notches are often created as a result of tack welding at the welded point in which the material is quickly heated at a high temperature and then allowed to cool rapidly. Welding crossbars can seriously weaken the bar at the point of welding.

1907.6 Spacing limits for reinforcement. The clear distance between reinforcing bars, bundled bars, tendons and ducts shall comply with ACI 318, Section 7.6.

❖ The code requires that reinforcement be separated by a minimum distance to allow concrete to flow readily around it and to prevent too close a spacing between bars on a line that can cause shear and shrinkage cracking.

1907.7 Concrete protection for reinforcement. The minimum concrete cover for reinforcement shall comply with Sections 1907.7.1 through 1907.7.7.

❖ A minimum thickness of concrete cover must be provided to protect reinforcement against weather and corrosive environments and may be required to protect against exposure to fire. In some extremely corrosive atmospheres, such as in the presence of chemical fumes, or in the vicinity of saltwater or where there are severe moisture and temperature changes, the amount of specified cover may need to be increased. Galvanized or epoxy-coated reinforcement is also used to provide corrosion protection.

1907.7.1 Cast-in-place concrete (nonprestressed). Minimum concrete cover shall be provided for reinforcement in nonprestressed, cast-in-place concrete construction in accordance with Table 1907.7.1, but shall not be less than required by Sections 1907.7.5 and 1907.7.7.

❖ This section references Table 1907.7.1 for the minimum concrete cover generally required for cast-in-place concrete. Other requirements such as fire protection or exposure to corrosive environments could necessitate cover in excess of that required by this section.

TABLE 1907.7.1
MINIMUM CONCRETE COVER

CONCRETE EXPOSURE	MINIMUM COVER (inches)
1. Concrete cast against and permanently exposed to earth	3
2. Concrete exposed to earth or weather	
No. 6 through No. 18 bar	2
No. 5 bar, W31 or D31 wire, and smaller	$1^1/_2$
3. Concrete not exposed to weather or in contact with ground	
Slabs, walls, joists:	
No. 14 and No. 18 bars	$1^1/_2$
No. 11 bar and smaller	$^3/_4$
Beams, columns:	
Primary reinforcement, ties, stirrups, spirals	$1^1/_2$
Shells, folded plate members:	
No. 6 bar and larger	$^3/_4$
No. 5 bar, W31 or D31 wire, and smaller	$^1/_2$

For SI: 1 inch = 25.4 mm.

❖ Concrete cover is measured from the concrete surface to the outermost surface of the steel to which the cover requirement applies. The values contained in the table above provide minimum concrete cover requirements for reinforcement in cast-in-place concrete exposed to most environments. Section 1904.4.2 requires that if concrete with reinforcement will be exposed to chlorides from deicing chemicals, salts, saltwater, brackish water, seawater or spray from these sources, it must comply with Table 1904.2.2(1) and the minimum concrete cover requirements of Section 1907.7. The intent of the reference to Section 1907.7 is that the provisions of Section 1907.7.5 apply (see commentary, Section 1907.7.5). For concrete elements required to achieve a fire-resistance rating, see the commentary to Section 1907.7.7.

1907.7.2 Cast-in-place concrete (prestressed). The minimum concrete cover for prestressed and nonprestressed reinforcement, ducts and end fittings in cast-in-place prestressed concrete shall comply with ACI 318, Section 7.7.2.

❖ This section references ACI 318 for minimum cover requirements for reinforcement in prestressed concrete. The minimum concrete cover values equally apply to prestressed and nonprestressed reinforcement, ducts and end fittings. See the commentaries to Section 1907.7.5 for prestressed concrete exposed to corrosive environments and Section 1907.7.7 for prestressed concrete elements required to be fire-resistance rated.

1907.7.3 Precast concrete (manufactured under plant control conditions). The minimum concrete cover for prestressed and nonprestressed reinforcement, ducts and end fittings in precast concrete manufactured under plant control conditions shall comply with ACI 318, Section 7.7.3.

❖ This section references ACI 318 for minimum cover requirements for reinforcement for precast concrete manufactured under plant control conditions. The phrase "manufactured under plant control conditions" does not necessarily mean that precast members must be produced in a plant. Precast structural elements can also be made at the job site and may qualify under this section of the code, provided the control of form dimensions, placing of reinforcement, quality control of concrete and curing procedures are equal to practices normally expected in a plant. These tighter controls permit lesser cover thickness requirements for precast concrete versus cast-in-place construction. See the commentaries to Section 1907.7.5 for precast concrete exposed to corrosive environments and Section 1907.7.7 for precast concrete elements required to be fire-resistance rated.

1907.7.4 Bundled bars. The minimum concrete cover for bundled bars shall comply with ACI 318, Section 7.7.4.

❖ In construction that requires heavy concentrations of reinforcement, bundles of standard bar sizes can save space and reduce congestion, thus allowing for proper placement and consolidation of concrete. Bundling of bars is permitted, but only if such bundles are in conformance with ACI 318, Section 7.6.6. The minimum cover for bundled bars must be equal to the equivalent diameter of the bundle, but need to be greater than 2 inches (51 mm), except for concrete cast against and permanently exposed to earth, where the minimum cover is 3 inches (76 mm). Equivalent diameter is determined from the total cross-sectional area of the bundled bars.

1907.7.5 Corrosive environments. In corrosive environments or other severe exposure conditions, prestressed and nonprestressed reinforcement shall be provided with additional protection in accordance with ACI 318, Section 7.7.5.

❖ Where severe corrosive environments are encountered, the amount of concrete protection must be increased, in accordance with Section 7.7.5 of ACI 318. Section 1904.2.2 cites several examples of concrete that will be exposed to external sources of chloride in service. The amount of increase for the cover over reinforcing steel is generally based on the judgement of the registered design professional. The commentary to this section of ACI 318 recommends (but does not require) a minimum cover for reinforcement of 2 inches (51 mm) for walls and slabs and $2^1/_2$ inches (64 mm) for other structural members. These two dimensions can be reduced to $1^1/_2$ and and 2 inches (38.1 and 64 mm), respectively, for precast concrete manufactured under plant control conditions. The provisions also require that the denseness and nonporosity of the protecting concrete also be considered. Section 1904.4.2 and Table

1904.2.2(1) mandate a lower water-cementitious materials ratio, a minimum specified compressive strength, or both for concrete exposed to these environments, thus insuring a more impermeable (i.e., denser) cement paste matrix. Even higher densities can be achieved with the use of some admixtures, such as silica fume. As an alternative to providing increased cover and denser cement paste, other protection may be provided, such as the use of galvanized or epoxy-coated reinforcement, or cathodically protecting the reinforcement.

1907.7.6 Future extensions. Exposed reinforcement, inserts and plates intended for bonding with future extensions shall be protected from corrosion.

❖ The exposed surfaces of reinforcement, inserts and plates that will be used for extensions of a structure must be protected to prevent corrosion that will cause them to fail the criteria of Section 1907.4 and otherwise reduce their ability to bond with future concrete.

1907.7.7 Fire protection. When this code requires a thickness of cover for fire protection greater than the minimum concrete cover specified in Section 1907.7, such greater thickness shall be used.

❖ The cover requirements in Section 1907.7.1 for reinforcement in cast-in-place concrete construction will generally equal or exceed the requirements for fire protection purposes. There are, however, a few exceptions for higher fire-resistance ratings, particularly those for members having 3- and 4-hour fire endurance ratings. Three of these exceptions should be noted:

1. The reinforcement in unrestrained, reinforced concrete floor and roof slabs requiring a fire-resistance rating of 3 hours must have a $1^{1}/_{4}$-inch (32 mm) minimum concrete cover [see Table 721.2.3(1)].
2. The main reinforcement in unrestrained beams less than 8 inches (203 mm) in width and requiring a fire-resistance rating of 3 hours may require a cover of more than $1^{1}/_{2}$ inches (38.1 mm) [see Table 721.2.3(2)].
3. The main reinforcement in columns having a required fire-resistance rating of 2 hours or greater may require a cover of 2 inches (51 mm). This should be achieved with No. 4 ties (see Section 721.2.4.2).

Monolithic cast-in-place concrete is usually considered restrained construction in accordance with Appendix X3 of ASTM E 119 and, therefore, the exceptions listed above may not totally apply or may apply on a limited basis under certain conditions of construction. An unrestrained condition is one in which the load-carrying element is free to expand and rotate at the supports. As an example, this may apply to precast concrete elements. For a more comprehensive guide to determining these conditions, refer to Appendix X3 of ASTM E 119.

For additional information on fire resistance of concrete, see Sections 721.1 and 721.2.

1907.8 Special reinforcement details for columns. Offset bent longitudinal bars in columns and load transfer in structural steel cores of composite compression members shall comply with the provisions of ACI 318, Section 7.8.

❖ The provisions of ACI 318, Section 7.8, only apply to special reinforcement details in columns where they change in dimension and columns of composite construction that contain structural steel cores. Offset bends in column longitudinal reinforcement are permitted only where column faces are offset less than 3 inches (76 mm).

1907.9 Connections. Connections between concrete framing members shall comply with the provisions of ACI 318, Section 7.9.

❖ The provisions of ACI 318, Section 7.9, require that at connections of principal framing elements (such as beam-to-column connections), ties, spirals, stirrups or other forms of confinement must be provided for splices of continuing reinforcement and anchorage of reinforcement that terminates in the connection. This is necessary so the joint will maintain its flexural capacity without deteriorating under repeated loadings.

1907.10 Lateral reinforcement for compression members. Lateral reinforcement for concrete compression members shall comply with the provisions of ACI 318, Section 7.10.

❖ Section 7.10 of ACI 318 provides minimum design and detailing requirements that apply to lateral reinforcement used in compression members such as columns. Lateral reinforcement may consist of ties or spirals or both that encircle and confine the longitudinal reinforcement. In some cases, the lateral reinforcement may be called upon to provide resistance to shear or torsion or both, in which case the ties are referred to as "stirrups."

1907.11 Lateral reinforcement for flexural members. Lateral reinforcement for compression reinforcement in concrete flexural members shall comply with the provisions of ACI 318, Section 7.11.

❖ Section 7.11 of ACI 318 provides minimum design and detailing requirements that apply to lateral reinforcement used to confine compression reinforcement in flexural members such as beams and girders. In some cases, lateral reinforcement may be called upon to provide resistance to shear or torsion or both, in which case the ties are referred to as "stirrups."

1907.12 Shrinkage and temperature reinforcement. Reinforcement for shrinkage and temperature stresses in concrete members shall comply with the provisions of ACI 318, Section 7.12.

❖ Section 7.12 of ACI 318 requires that a minimum amount of reinforcement be placed in one-way structural slabs to minimize cracking caused by concrete shrinkage and large temperature fluctuations and to tie the structure together. One-way structural slabs have flexural reinforcement extending in one direction only. Shrinkage and temperature reinforcement is placed at a right angle to the principal reinforcement. This requirement does not apply to soil-supported slabs.

1907.13 Requirements for structural integrity. The detailing of reinforcement and connections between concrete members shall comply with the provisions of ACI 318, Section 7.13, to improve structural integrity.

❖ The intent of ACI 318, Section 7.13, is to improve structural redundancy and ductility and in general to offer the user a prescriptive method of complying with the intent of ASCE 7, Section 1.4. The provisions require minimum reinforcement detailing so that concrete members of a structure are effectively tied together to improve overall structural integrity should it suffer severe local damage from abnormal loads (i.e., explosions, vehicle impact, and high wind events such as tornadoes). It is not the intent that a structure be designed to resist either general collapse caused by gross misuse or severe abnormal loads acting on a large portion of it.

SECTION 1908 MODIFICATIONS TO ACI 318

1908.1 General. The text of ACI 318 shall be modified as indicated in Sections 1908.1.1 through 1908.1.9.

❖ Section 1908 contains modifications to ACI 318, the standard adopted by reference in Section 1901.2. Most of the modifications are related to the seismic design of concrete structures. Text that differs from or is not found in ACI 318 is printed in italics to identify the changes.

1908.1.1 ACI 318, Section 21.1. Modify existing definitions and add the following definitions to ACI 318, Section 21.1.

DESIGN DISPLACEMENT. Total lateral displacement expected for the design-basis earthquake, *as specified by Section 9.5.5.7 of ASCE 7 or 1617.5.4 of the International Building Code.*

❖ This is defined to be the same as δ_x (or interstory drift based on δ_x) as given in Section 9.5.5.7 of ASCE 7. When the simplified analysis procedure of Section 1617.5 is used, design displacement may be based on an interstory drift of 1 percent of the story height (see Section 1617.5.4).

STORY DRIFT RATIO. *The design displacement over a story divided by the story height.*

❖ This definition is used in the modification to ACI 318 as shown in Section 1908.1.6.

WALL PIER. *A wall segment with a horizontal length-to-thickness ratio of at least 2.5, but not exceeding six, whose clear height is at least two times its horizontal length.*

❖ See Figure 1908.1.1.

1908.1.2 ACI 318, Section 21.2.1. Modify Sections 21.2.1.2, 21.2.1.3 and 21.2.1.4 to read as follows:

21.2.1.2 For structures assigned to Seismic Design Category A or B, provisions of Chapters 1 through 18 and 22 shall apply except as modified by the provisions of this chapter. Where the seismic design loads are computed using provisions for intermediate or special concrete systems, the requirements of Chapter 21 for intermediate or special systems, as applicable, shall be satisfied.

21.2.1.3 For structures assigned to Seismic Design Category C, intermediate or special moment frames, *or ordinary or special reinforced concrete structural walls* shall be used to resist *seis-*

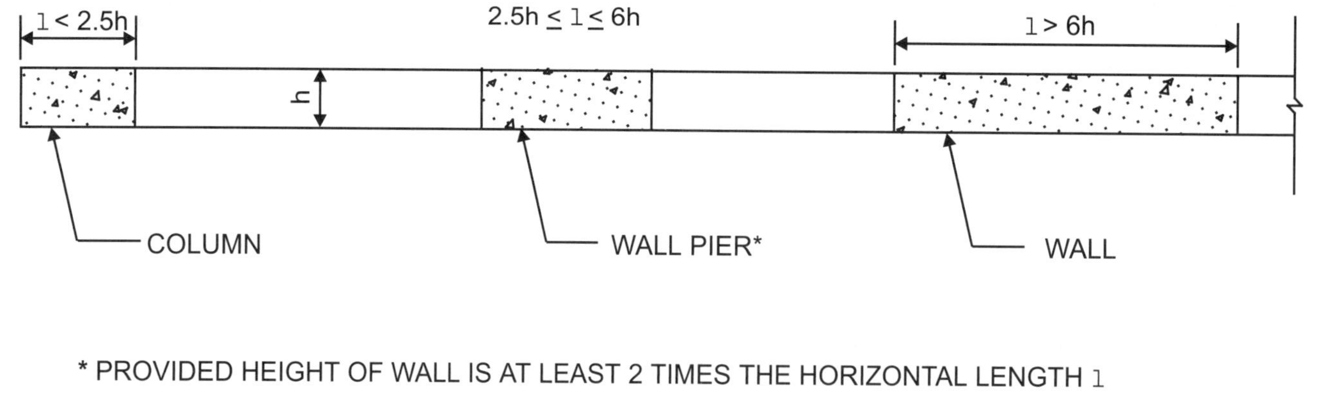

Figure 1908.1.1
PLAN OF WALL WITH OPENINGS

mic forces induced by earthquake motions. Where the design seismic loads are computed using provisions for special concrete systems, the requirements of Chapter 21 for special systems, as applicable, shall be satisfied.

21.2.1.4 For structures assigned to Seismic Design Category D, E or F, special moment frames, *special reinforced concrete structural walls,* diaphragms and trusses *and foundations* complying with Sections 21.2 through 21.10 shall be used to resist forces induced by earthquake motions. Frame members not proportioned to resist earthquake forces shall comply with Section 21.11.

❖ This modification provides a clarification that is necessary for correct application of the earthquake design provisions.

1908.1.3 ACI 318, Section 21.2.5. Modify ACI 318, Section 21.2.5, by renumbering as Section 21.2.5.1 and adding new Sections 21.2.5.2, 21.2.5.3 and 21.2.5.4 to read as follows:

21.2.5 Reinforcement in members resisting earthquake-induced forces.

21.2.5.1 Except as permitted in Sections 21.2.5.2 through 21.2.5.4, reinforcement resisting earthquake-induced flexural and axial forces in frame members and in structural wall boundary elements shall comply with ASTM A 706. ASTM 615, Grades 40 and 60 reinforcement, shall be permitted in these members if (a) the actual yield strength based on mill tests does not exceed the specified yield strength by more than 18,000 psi (retests shall not exceed this value by more than an additional 3,000 psi), and (b) the ratio of the actual ultimate tensile strength to the actual tensile yield strength is not less than 1.25.

21.2.5.2 Prestressing steel shall be permitted in flexural members of frames, provided the average prestress, f_{pc}, calculated for an area equal to the member's shortest cross-sectional dimension multiplied by the perpendicular dimension shall be the lesser of 700 psi (4.83 MPa) or $f'_c/6$ at locations of nonlinear action where prestressing steel is used in members of frames.

21.2.5.3 Unless the seismic-force-resisting frame is qualified for use through structural testing as required by the ACI T1.1, for members in which prestressing steel is used together with mild reinforcement to resist earthquake-induced forces, prestressing steel shall not provide more than one-quarter of the strength for either positive or negative moments at the nonlinear action location and shall be anchored at the exterior face of the joint or beyond.

21.2.5.4 Anchorages for tendons must be demonstrated to perform satisfactorily for seismic loadings. Anchorage assemblies shall withstand, without failure, a minimum of 50 cycles of loading ranging between 40 and 85 percent of the minimum specified tensile strength of the prestressing steel.

❖ This modification renumbers ACI 318, Section 21.2.5, as 21.2.5.1, and adds an important exception in new Sections 21.2.5.2 through 21.2.5.4. All of these requirements apply to structures assigned to Seismic Design Category D, E or F.

21.2.5.1: ACI 318, Section 21.2.5 requires reinforcement resisting earthquake-induced flexural and axial forces in frame members and in shear wall boundary elements to comply with ASTM A 706. ASTM 615 Grade 40 and 60 reinforcement is permitted in these members subject to two supplementary requirements that are included in ACI 318 (IBC modified Section 21.2.5.1). The use of prestressing steel to resist earthquake-induced flexural and axial forces in frame members and shear wall boundary elements would thus have been precluded, but for the added exception. The exception allows such usage, subject to the conditions given in Sections 21.2.5.2 through 21.2.5.4.

21.2.5.2: This restriction originated from research carried out by Ishizuka and Hawkins at the University of Washington, Seattle in the mid-1980s. f_{pc} is the average compressive stress in concrete due to effective prestress force only (after allowance for all prestress losses). Based on the noted research, the 1994 NEHRP Provisions for the first time allowed the use of prestressing tendons in conjunction with deformed reinforcing bars in frames resisting earthquake forces, provided f_{pc} would be kept below the greater of 350 psi and $f'_c/12$. When the 1997 Uniform Building Code adopted the same provision, the limit on f_{pc} was changed to the lesser of 350 psi (2.4 mPa) or $f'_c/12$. In the 1997 NEHRP Provisions, the limit was increased to the lesser of 700 psi (4.8 mPa) or $f'_c/6$, based on newer research by Stanton, Stone and Cheok; this has been incorporated into the code.

21.2.5.3: The added restrictions of this section are based directly on research by Ishizuka and Hawkins discussed above. Also, qualifying a moment frame system by structural testing that complies with the criteria of ACI T1.1 is another option. This referenced acceptance criteria is meant to fulfill the requirement in Section 21.2.1.5 of ACI 318 that allows the use of reinforced concrete structures not satisfying Chapter 21 requirements if testing demonstrates performance at least equal to that of a comparable monolithic reindorced concrete structure.

21.2.5.4: The requirement to demonstrate the satisfactory seismic performance of tendon anchorages was part of the 1994 NEHRP Provisions; however, it was not adopted into the 1997 UBC. The provision was retained in the 1997 NEHRP Provisions and is now part of the code.

1908.1.4 ACI 318, Section 21.7. Modify ACI 318, Section 21.7, by adding a new Section 21.7.10 to read as follows:

21.7.10 Wall piers and wall segments.

21.7.10.1 Wall piers not designed as a part of a special moment frame shall have transverse reinforcement designed to satisfy the requirements in Section 21.7.10.2.

Exceptions:

1. *Wall piers that satisfy Section 21.11.*

2. *Wall piers along a wall line within a story where other shear wall segments provide lateral support to the wall piers, and such segments have a total stiffness of at least six times the sum of the stiffness of all the wall piers.*

21.7.10.2 Transverse reinforcement shall be designed to resist the shear forces determined from Sections 21.3.4.2 and 21.4.5.1. Where the axial compressive force, including earthquake effects, is less than $A_g f'_c/20$, transverse reinforcement in wall piers is permitted to have standard hooks at each end in lieu of hoops. Spacing of transverse reinforcement shall not exceed 6 inches (152 mm). Transverse reinforcement shall be extended beyond the pier clear height for at least the development length of the largest longitudinal reinforcement in the wall pier.

21.7.10.3 Wall segments with a horizontal length-to-thickness ratio less than 2.5 shall be designed as columns.

❖ This modification adds the new Section 21.7.10 to ACI 318, Section 21.7, which applies to structures assigned to Seismic Design Category D, E or F.

21.7.10: This added section is titled "Wall piers and Wall segments." Wall piers are defined in Section 1908.1.1.

21.7.10.1: Wall piers either may be designed as part of a special moment frame or must conform to the requirements of Section 21.7.10.2. Two important exceptions to this requirement are provided. A wall pier satisfying the requirement of deformation compatibility with the lateral-force-resisting system need not be detailed by Section 21.7.10.2. Wall piers laterally supported by much stiffer shear walls along the same line within a story also need not be detailed by Section 21.7.10.2.

21.7.10.2: These detailing requirements were added to the 1988 UBC out of concern that thin column-like elements between openings in shear walls were being designed without proper transverse reinforcement. A wall pier is required to be designed for the maximum shear force that can develop in the pier, computed in accordance with ACI 318, Section 21.3.4.1, if $P_u \leq 0.1A_g f'_c$ or in accordance with ACI 318, Section 21.3.5.1, if $P_u > 0.1A_g f'_c$. The transverse reinforcement is required to be spaced at no more than 6 inches (152 mm), and is required to be extended beyond the clear pier height by at least one development length of the largest longitudinal bar. The transverse reinforcement may have a standard hook at each end, as long as $P_u < A_g f'_c/20$.

21.7.10.3: This section clarifies that a segment of wall with a horizontal length-to-thickness ratio of less than $2^1/_2$ must be designed as a column.

1908.1.5 ACI 318, Section 21.10.1.1. Modify ACI 318, Section 21.10.1.1, to read as follows:

21.10.1.1 Foundations resisting earthquake-induced forces or transferring earthquake-induced forces between a structure and the ground shall comply with the requirements of Section 21.10 and other applicable provisions of ACI 318 *unless modified by Chapter 18 of the International Building Code.*

❖ This ACI 318 modification points out that there are foundation requirements in the code that, in accordance with Section 102.4, would take precedence over those of ACI 318.

1908.1.6 ACI 318, Section 21.11. Modify ACI Sections 21.11.1 and 21.11.2.2 and add Sections 21.11.5 through 21.11.7 as follows:

21.11.1 Frame members assumed not to contribute to lateral resistance shall be detailed according to Section 21.11.2 or 21.11.3 depending on the magnitude of moments induced in those members when subjected to the design displacement. If effects of design displacements are not explicitly checked, it shall be permitted to apply the requirements of Section 21.11.3. *Slab-column connections shall comply with Sections 21.11.5 through 21.11.7. Conformance to Section 21.11 satisfies the deformation compatibility requirements of Section 9.5.2.2.4.3 of ASCE 7.*

21.11.2.2 Members with factored gravity axial forces exceeding ($A_g f'_c/10$) shall satisfy Sections 21.4.3, 21.4.4.1(c), 21.4.4.3 and 21.4.5. The maximum longitudinal spacing of ties shall be, s_o, for the full column height. The spacing, s_o, shall not be more than six diameters of the smallest longitudinal bar enclosed or 6 inches (152 mm), whichever is smaller. *Lap splices of longitudinal reinforcement in such members need not satisfy Section 21.4.3.2 in structures where the seismic- force-resisting system does not include special moment frames.*

21.11.5 Reinforcement to resist punching shear shall be provided in accordance with Sections 21.11.5.1 and 21.11.5.2 at slab column connections where story drift ratio exceeds [0.035 – 0.05 ($V_u/\phi V_c$)] except that Sections 21.11.4.1 and 21.11.4.2 need not be satisfied where $V_u/\phi V_c$ is less than 0.2 or where the story drift ratio is less than 0.005. V_u equals the factored punching shear from gravity load excluding shear stress from unbalanced moment. V_u is calculated for the load combination 1.2D + 1.0L + 0.2S. The load factor on L is permitted to be reduced to 0.5 in accordance with Section 9.2.1(a). In no case shall shear reinforcement be less than that required in Section 11.12 for loads without consideration of seismic effects.

21.11.5.1 — The slab shear reinforcement shall provide V_s not less than $3.5\sqrt{f'_c}$.

21.11.5.2 — Slab shear reinforcement shall extend not less than five times the slab thickness from the face of column.

21.11.6 — Bottom bars or wires within the column strip shall conform to Section 13.3.8.5 except that splices shall be Class B.

21.11.7 — Within the effective slab width defined in Section 13.5.3.2, the ratio of nonprestressed bottom reinforcement to gross concrete area shall not be less than 0.004. Where bottom reinforcement is not required to be continuous, such reinforcement shall extend a minimum of five times the slab thickness plus one development length beyond the face of the column or terminated at the slab edge with a standard hook.

❖ This section introduces amendments to Section 21.11 requirements of ACI 318 for frame members not proportioned to resist forces induced by earthquake motions in structures assigned to Seismic Design Category D, E or F.

21.11.1: The modification to this section of ACI 318 adds the reference to the added provisions for slab-to-column connections. The intention of these provisions is to prevent punching shear failures of flat slabs in the vicinity of columns. Additionally, a clarification is added stating that compliance with this section satisfies the deformation compatibility requirements of Section 1617.6.

21.11.2.2: While ACI 318 requires such members with factored gravity axial forces exceeding $A_g f'_c/10$ to satisfy ACI 318, Section 21.4.3, the code modification excludes the requirement of ACI 318, Section 21.4.3.2, for lap splices to be in the center half of the column. The amendment regarding the location of lap splices is sensible because gravity columns are required to sustain only their gravity load-carrying capacity under the design earthquake displacements, whereas columns forming part of the seismic-force-resisting system are required to sustain full gravity as well as lateral loads under the same displacements. While a requirement to keep lap splices away from regions of potential plastic hinging near the ends of the column may be justified in the latter case, it appears to be unjustified and unnecessary for a gravity column.

21.11.5: This section establishes the conditions under which shear reinforcement is required in slabs, based on gravity load shear stress and the story drift. It also determines the amount of shear reinforcement that is necessary and the location of that added reinforcement.

21.11.6: This section increases the lap splice requirement for bottom bars within the column strip from Class A to Class B. This recognizes that these critical slab bottom bars may yield as a result of earthquake deformations.

21.11.7: This section provides a minimum amount of bottom reinforcement in the region near the column where earthquake deformations may result in yielding of the steel reinforcing. It also requires the extension of bottom bars in posttensioned slabs in which continuous bottom reinforcement is not normally required.

1908.1.7 ACI 318, Section 21.13.2. Modify ACI 318, Section 21.13.2, to read as follows:

21.13.2 In connections between wall panels, or between wall panels and the foundation, *yielding shall be restricted to reinforcement.*

❖ This modification to ACI 318 affects connections between intermediate precast concrete wall panels by deleting the term "steel elements." In doing so, the code limits yielding in those connections to the reinforcing steel.

SECTION 1909 STRUCTURAL PLAIN CONCRETE

1909.1 Scope. The design and construction of structural plain concrete, both cast-in-place and precast, shall comply with the minimum requirements of Section 1909 and ACI 318, Chapter 22.

❖ Section 1909 covers the minimum requirements for structural members regulated by the code and constructed of plain concrete. By definition, plain concrete contains no reinforcement, or has less reinforcement than necessary to meet the requirements for reinforced concrete in ACI 318. In addition to the provisions of Section 1909, there are other provisions in Chapter 22 of ACI 318 that must be met for plain concrete used for structural purposes.

1909.1.1 Special structures. For special structures, such as arches, underground utility structures, gravity walls and shielding walls, the provisions of this section shall govern where applicable.

❖ Special structures regulated by the code, such as an underground utility vault with plain concrete walls, plain concrete retaining walls designed to resist overturning by gravity, plain concrete walls constructed for shielding purposes and plain concrete walls used in conjunction with outdoor handball courts are also required to meet the applicable requirements of this section.

1909.2 Limitations. The use of structural plain concrete shall be limited to:

1. Members that are continuously supported by soil, such as walls and footings, or by other structural members capable of providing continuous vertical support.
2. Members for which arch action provides compression under all conditions of loading.
3. Walls and pedestals.

The use of structural plain concrete columns and structural plain concrete footings on piles is not permitted. See Section 1910 for additional limitations on the use of structural plain concrete.

❖ Since the structural integrity of plain concrete members depends solely on the properties of the concrete, their use is limited to members that are generally under compression. In addition, such members must be able to tolerate random cracking. For these reasons, such members are limited to uses where (1) they are continuously supported by soil, such as footings, or by other structural members capable of providing continuous vertical support; (2) arch action provides compression under all conditions of loading or (3) as walls and pedestals. Examples of structural members permitted to be of plain concrete include footings, foundation and basement walls, above-grade walls, arches over openings in a concrete wall in lieu of a reinforced lintel and pedestals. Pedestals are frequently used in crawl spaces or under decks to support floor framing, and when used as such they are usually referred to as piers. A "pedestal" must be proportioned so the height does not exceed three times the least lateral dimension. If the height-to-least-lateral dimension exceeds three, the member is defined as a "column" and must be designed as a reinforced member according to ACI 318.

Although pile caps (i.e., footings supported on piles) are generally cast on the soil, they are designed to be supported by the piles, much like a structural floor or roof slab is supported by beams that are in turn supported by columns. For this reason, footings supported by piles are not permitted to be of plain concrete.

It is generally desirable that structural plain concrete members be under compression at all times; however, this is not always possible, especially in the case of walls of plain concrete. When a wall is subjected to lateral loads from wind or soil, or both, a portion of the wall cross section may be under net tension. The tension is a result of lateral loads causing flexural tension exceeding the compressive stress from the axial load on the wall, if any.

A common type of construction in areas of the country where expansive soils are present is drilled pier foundations supporting reinforced concrete grade beams, which in turn support concrete basement walls. Under this scenario, the use of a structural plain concrete basement wall is permitted provided the wall is proportioned to resist the axial and lateral loads to which it will be subjected.

Section 1910 generally restricts the use of structural plain concrete members to buildings of Seismic Design Category A or B. In some cases, plain concrete members are permitted in higher seismic design categories if some reinforcement is provided to give the members some ductility and hold the concrete together should cracking occur during a seismic event. The presence of the nominal amount of reinforcement specified by Section 1910 does not change the classification of a member from plain to reinforced concrete.

1909.3 Joints. Contraction or isolation joints shall be provided to divide structural plain concrete members into flexurally discontinuous elements in accordance with ACI 318, Section 22.3.

❖ Generally, considerably more water is added to a concrete mix than is necessary to react with the cement. The additional water that is added for workability must eventually evaporate and will continue to do so until the moisture content of the hardened concrete reaches equilibrium with the surrounding environment. As the concrete dries, it tries to shrink. If the member, such as a slab or basement wall, is restrained so it cannot shrink, tensile stresses develop in the concrete. If the tensile strength of the concrete is exceeded, the slab or wall will crack. Temperature changes and creep have the same effect as moisture changes.

Section 22.3 provides two alternatives for addressing shrinkage of concrete. In the first case, the concrete element can be isolated from other elements that may restrain it by using isolation joint material, typically referred to as "expansion joint material." This is not always feasible; therefore, a second alternative allows the element to be subdivided into flexurally discontinuous smaller elements with contraction joints, often called "control joints." A contraction joint is a weakened cross section of the concrete member where tensile stresses in the restrained concrete will be concentrated as it tries to shrink. Because the concrete is intentionally weakened at the contraction joint, the crack will occur at that location. The contraction joint can be formed while the concrete is still plastic, such as with a groover, or in the case of a slab, the hardened concrete can be sawed before it has dried sufficiently enough to crack. In the case of a wall, the joint is typically created by installing strips on the inside surfaces of the wall forms. To be effective, the joint should be at least one-quarter the slab or wall thickness. This means that for walls, each strip located on opposing form surfaces should be at least one-eighth the wall thickness. To be effective, joints need to be located where the tensile stress will be greatest; typically at offsets in slabs and at openings in walls.

1909.4 Design. Structural plain concrete walls, footings and pedestals shall be designed for adequate strength in accordance with ACI 318, Sections 22.4 through 22.8.

Exception: For Group R-3 as applicable in Section 101.2 occupancies and buildings of other occupancies less than two stories in height of light-frame construction, the required edge thickness of ACI 318 is permitted to be reduced to 6 inches (152 mm), provided that the footing does not extend more than 4 inches (102 mm) on either side of the supported wall.

❖ Sections 22.4 through 22.8 of ACI 318 contain design procedures for elements permitted to be of structural plain concrete, including walls, footings and pedestals. The procedures do not allow credit to be given for any reinforcement that may be present in the member. This means that if the required strength of a member exceeds its design strength when calculated without benefit of any reinforcement present, it must be designed as reinforced concrete.

ACI 318 does not permit columns of plain concrete; however, plain concrete pedestals are permitted, provided the ratio of unsupported height-to-average-least-lateral dimension does not exceed three. For example, if the unsupported height of a 10-inch (254 mm) square pier is 28 inches (711 mm), the ratio is 2.8 [28 ÷ 10 = 2.8 (711 ÷ 254 = 2.8)]; therefore, the pier qualifies as a "pedestal."

For the exception, Section 22.7.4 of ACI 318 requires the thickness of a structural plain concrete footing to be a minimum of 8 inches (203 mm). This exception permits the thickness of the footing to be reduced to 6 inches (152 mm). The reduced footing thickness only applies to dwellings and for buildings of all other occupancies one story in height that support light-frame construction, like wood or cold-formed steel. The other limitation is that the footing cannot extend more than 4 inches (102 mm) beyond the face of the supported wall. The 4-inch (102 mm) limitation is to reduce the likelihood that the footing will experience a flexural failure at the junction of the footing and the wall supported above.

1909.5 Precast members. The design, fabrication, transportation and erection of precast, structural plain concrete elements shall be in accordance with ACI 318, Section 22.9.

❖ ACI 318 requires that the design of precast members must include all loading conditions that the member will be subjected to from fabrication to in-place use in the completed structure. The load considerations must include form removal, storage, transportation and erection at the job site. The end use of precast members is subject to the same limitations as previously cited in the commentary to Section 1909.2.

1909.6 Walls. In addition to the requirements of this section, structural plain concrete walls shall comply with the applicable requirements of ACI 318, Chapter 22.

❖ Section 22.6 of ACI 318 has two design procedures for structural plain concrete walls. For walls where the resultant of all factored loads (axial and lateral) falls in the middle one-third of the wall, an empirical procedure that considers only the axial load is permitted. Where the resultant of all factored loads falls outside the middle one-third of the wall, a rational procedure specified in ACI 318 that considers both axial and lateral loads must be used. Regardless of which one of the two design approaches is used, the provisions of Section 1909.6.1 through 1909.6.3 apply. These provisions only apply to walls that are laterally supported at the top and bottom with elements such as slabs on ground and structural floors/roofs. Walls designed in accordance with these provisions are intended for use in low-rise buildings assigned to Seismic Design Category A or B (see Section 1910). Chapter 32 of PCA-EB070 contains numerous aids to facilitate the design of structural plain concrete walls and footings.

1909.6.1 Basement walls. The thickness of exterior basement walls and foundation walls shall be not less than $7^1/_2$ inches (191 mm). Structural plain concrete exterior basement walls shall be exempt from the requirements for special exposure conditions of Section 1904.2.2.

❖ This section requires exterior basement and foundation walls to be a minimum of $7^1/_2$ inches (191 mm) in thickness. Though the code states exterior walls, the intent of the section is for all basement and foundation walls retaining earth to meet the $7^1/_2$ inch (191 mm) minimum dimension. Such a situation may occur in a building with a partial basement. The provisions also exempt basement walls from the special exposure conditions described in Section 1904.2.2.

1909.6.2 Other walls. Except as provided for in Section 1909.6.1, the thickness of bearing walls shall be not less than $^1/_{24}$ the unsupported height or length, whichever is shorter, but not less than $5^1/_2$ inches (140 mm).

❖ This section applies to interior and exterior bearing walls above grade, and to interior bearing walls in a basement. It requires the thickness to be a minimum of $5^1/_2$ inches (140 mm), but in no case less than $^1/_{24}$ of the unsupported height or length of the wall, whichever is less. For example, a concrete wall with a height of 8 feet (2438 mm) is required to be a minimum of $5^1/_2$ inches (140 mm), since $^1/_{24}$ of the height of the wall is less than $5^1/_2$ inches [8 ft. x 12 in./ft. x $^1/_{24}$ = 4 inches (2438 mm x $^1/_{24}$ = 102 mm)]. For a 12-foot-high (3658 mm) wall, the minimum thickness required is 6 inches [12 ft. x 12 in./ft. x $^1/_{24}$ = 6 in. (3658 mm x $^1/_{24}$ = 152 mm)].

1909.6.3 Openings in walls. Not less than two No. 5 bars shall be provided around window and door openings. Such bars shall extend at least 24 inches (610 mm) beyond the corners of openings.

❖ All openings for windows and doors in structural plain concrete walls must be bound by two No. 5 (16) bars as follows. For openings not extending to the bottom of the wall, the bars must be placed on all four sides of the opening. For openings extending to the bottom of the wall, the bars are only necessary along the sides and top. To provide continuity at the corners of the openings, the bars must extend 24 inches (610 mm) past the corners. These extensions should occur at all four corners of an opening not extending to the bottom of the wall, and at the two top corners for other openings.

SECTION 1910
SEISMIC DESIGN PROVISIONS

1910.1 General. The design and construction of concrete components that resist seismic forces shall conform to the requirements of this section and to ACI 318 except as modified by Section 1908.

❖ The design forces and their effects calculated for the design earthquake in accordance with Sections 1616 and 1617 are coupled with the type of seismic-force-resisting system used to resist these forces. Since the forces and their effects used in the design assume a specific structural response, it is necessary that the structure be designed, detailed and constructed to respond accordingly. The provisions of this section are intended to provide a cross reference between the code and ACI 318, so the appropriate detailing prescribed in Chapter 21 of ACI 318 is properly implemented in the design. Since structures of Seismic Design Category A are exempt from all the seismic design provisions except those in Section 1616.4, the requirements in this section are applicable only to buildings assigned to Seismic Design Category B, C, D, E or F.

The design and detailing requirements for plain and reinforced concrete are in ACI 318. Section 1910.1 coordinates requirements of the code with those of ACI 318. The coordination is necessary because of differences in terminology between the two documents. In addition, the section modifies and adds to ACI 318 provisions to assure structural response consistent with the design earthquake forces and effects calculated in accordance with Chapter 16. These revisions and modifications are found in both Section 1908 and this section.

1910.2 Classification of shear walls. Structural concrete shear walls that resist seismic forces shall be classified in accordance with Sections 1910.2.1 through 1910.2.4.

❖ Section 1617.6 recognizes four types of concrete shear walls that may be permitted as part of the basic seismic-force-resisting system of a structure, depending upon its seismic design category and height. Requirements for these four types of walls are given in Sections 1910.2.1 through 1910.2.4. Once the response modification coefficient, R, of a specific structural system utilizing shear walls is used to calculate the base shear and other design quantities dependent upon it, the corresponding deflection amplification factor, C_d, system overstrength factor, Ω_0, and detailing requirements for the selected system must be used. For example, in a structure assigned to Seismic Design Category C, if the seismic-force-resisting system is to utilize concrete shear walls, Section 1617.6 permits either ordinary or special reinforced concrete shear walls. If it is decided to use special reinforced concrete shear walls because the response modification coefficient for this structural system is higher, resulting in a smaller base shear, the detailing requirements of Sections 21.2 and 21.7 of ACI 318 for special reinforced concrete shear walls must be used (see Section 1910.2.4).

1910.2.1 Ordinary plain concrete shear walls. Ordinary plain concrete shear walls are walls conforming to the requirements of Chapter 22 of ACI 318.

❖ Ordinary plain concrete shear walls must comply with the requirements of Chapter 22 of ACI 318. No reinforcement is required in these walls, other than around openings in accordance with Section 1909.6.3 (i.e., Section 22.6.6.5 of ACI 318). Ordinary plain concrete shear walls may only be used in structures assigned to Seismic Design Category A or B, and in those structures exempt from seismic design by Exceptions 2, 4, 5 and 6 of Section 1614.1.

1910.2.2 Detailed plain concrete shear walls. Detailed plain concrete shear walls are walls conforming to the requirements for ordinary plain concrete shear walls and shall have reinforcement as follows: Vertical reinforcement of at least 0.20 square inch (129 mm^2) in cross-sectional area shall be provided continuously from support to support at each corner, at each side of each opening and at the ends of walls. The continuous vertical bar required beside an opening is permitted to substitute for one of the two No. 5 bars required by Section 22.6.6.5 of ACI 318. Horizontal reinforcement at least 0.20 square inch (129 mm^2) in cross-sectional area shall be provided:

1. Continuously at structurally connected roof and floor levels and at the top of walls;
2. At the bottom of load-bearing walls or in the top of foundations where doweled to the wall; and
3. At a maximum spacing of 120 inches (3048 mm).

Reinforcement at the top and bottom of openings, where used in determining the maximum spacing specified in Item 3 above, shall be continuous in the wall.

❖ Detailed plain concrete shear walls must conform to the requirements for ordinary plain concrete shear walls (Section 1910.2.1), and have some reinforcement to provide limited ductility and to keep the concrete from disintegrating should cracking occur before or during a seismic event. Detailed plain concrete shear walls may only be used in structures assigned to Seismic Design Category A or B, and those exempt from seismic design by Exceptions 2, 4, 5 and 6 of Section 1614.1. The area of vertical reinforcement of 0.20 square inch (129 mm^2) corresponds to the area of a No. 4 (13) reinforcing bar. The provision requiring vertical reinforcement at the "ends of walls" is intended to also require a bar on each side of a contraction (i.e., control) joint. If the provisions of Item 1 require a closer spacing of horizontal reinforcement than Item 3 [i.e., 120 inches or 10 feet (3048 mm)], then the closer spacing applies. Item 1 requires that a bar be located at the top of a wall. Where a roof is connected to the wall by a ledger and the wall extends above the roof, the bar required to be located at a structurally connected roof can also serve as the bar at the top of the wall if it is within 12 inches (305 mm) of the top.

1910.2.3 Ordinary reinforced concrete shear walls. Ordinary reinforced concrete shear walls are walls conforming to the requirements of ACI 318 for ordinary reinforced concrete structural walls.

❖ Ordinary reinforced concrete shear walls must conform to the requirements for ordinary reinforced concrete structural walls as defined in Section 21.1 of ACI 318. Such walls must comply with the applicable provisions of Chapters 1 through 18 of ACI 318. The difference in terminology between the code and ACI 318 (i.e., "ordinary reinforced concrete shear walls" versus "ordinary reinforced concrete structural walls") is intentional. ACI Committee 318 maintains that since tall slender walls are flexure governed in their behavior as well as failure mode, it is inappropriate to apply the term "shear walls" to them; hence the term "structural walls." Ordinary reinforced concrete shear walls may only be used in structures assigned to Seismic Design Category A, B or C, and in those structures exempt from seismic design by Exceptions 2, 4, 5 and 6 of Section 1614.1.

1910.2.4 Special reinforced concrete shear walls. Special reinforced concrete shear walls are walls conforming to the requirements of ACI 318 for special reinforced concrete structural walls or special precast structural walls.

❖ Special reinforced concrete shear walls must conform to the requirements for special reinforced concrete structural walls or special precast structural walls as defined in Section 21.1 of ACI 318. Special reinforced concrete structural walls must comply with the applicable provisions of ACI 318, Chapters 1 through 18, and Sections 21.2 and 21.7. The difference in terminology between the code and ACI 318 (i.e., "special reinforced concrete shear walls" versus "special reinforced concrete structural walls") is intentional. ACI Committee 318 maintains that since tall slender walls are flexure-governed in their behavior as well as failure mode, it is inappropriate to apply the term "shear walls" to them; hence the term "structural walls." Special reinforced concrete shear walls are required in structures assigned to Seismic Design Category D, E or F within the height limits of Section 1617.6, and may be used in all other structures. Similarly, special precast structural walls are intended for use where the seismic hazard is above average. These walls must comply with ACI 318 Chapters 1 through 18 as well as Sections 21.2 and 21.18.

1910.3 Seismic Design Category B. Structures assigned to Seismic Design Category B, as determined in Section 1616, shall conform to the requirements for Seismic Design Category A and to the additional requirements for Seismic Design Category B of this section.

❖ Structures assigned to Seismic Design Category B are at increased seismic risk compared to those assigned to Category A (see the modification to Section 21.2.1.2 of ACI 318 given in Section 1908.1.2). The requirements of this section reflect the additional demands placed on these structures and are over and above those found in ACI 318.

1910.3.1 Ordinary moment frames. In flexural members of ordinary moment frames forming part of the seismic-force-resisting system, at least two main flexural reinforcing bars shall be provided continuously top and bottom throughout the beams, through or developed within exterior columns or boundary elements.

Columns of ordinary moment frames having a clear height-to-maximum-plan-dimension ratio of five or less shall be designed for shear in accordance with Section 21.12.3 of ACI 318.

❖ Ordinary moment frames of reinforced concrete are required by Section 21.1 of ACI 318 to comply with the applicable provisions of Chapters 1 through 18 of that document. Thus no special detailing is required by ACI 318 to provide increased energy dissipation capacity and ductility. In editions of the *NEHRP Recommended Provisions for Seismic Regulations for New Buildings* prior to 1994, the response modification coefficient, *R*, for ordinary moment frames of concrete was two. After presentation of considerable analytical and experimental findings justifying a higher value of three, it was decided to allow the higher value provided additional continuity reinforcement was also required to increase the ductility of the frame. The additional continuity reinforcement must be coordinated with that required by ACI 318, Section 7.13.

Investigation of damage in many earthquakes including the 1991 Northridge, California, earthquake has shown that failures of short stiff columns is a cause for concern. This is not surprising since the shear force generated in a column is inversely proportional to the clear height of the column. In an attempt to preclude shear failure that might take away from the degree of inelastic deformability anticipated by an *R*-value of 3, short stiff columns of ordinary moment frames must be detailed as required for intermediate moment frames. Therefore, columns of ordinary moment frames having a clear height to maximum plan dimension ratio of five or less are required to be designed for shear in accordance with Section 21.12.3 of ACI 318.

1910.4 Seismic Design Category C. Structures assigned to Seismic Design Category C, as determined in Section 1616, shall conform to the requirements for Seismic Design Category B and to the additional requirements for Seismic Design Category C of this section.

❖ Structures assigned to Seismic Design Category C are at a greater seismic risk compared to those assigned to Categories A or B. The requirements of this section reflect the additional demands placed on these structures and are over and above those found in ACI 318, or address subject matter not found in that standard (see the modification of Section 21.2.1.3 of ACI 318 made in Section 1908.1.2).

1910.4.1 Seismic-force-resisting systems. Moment frames used to resist seismic forces shall be intermediate moment frames or special moment frames. Shear walls used to resist seismic forces shall be ordinary reinforced concrete shear walls or special reinforced concrete shear walls. Ordinary reinforced concrete shear walls constructed of precast concrete elements shall comply with the additional requirements of Section 21.13 of ACI 318 for intermediate precast concrete structural walls, as modified by Section 1908.1.7.

❖ Section 1617.6 requires concrete moment frames in structures assigned to Seismic Design Category C to be either intermediate moment frames or special moment frames. Ordinary moment frames are not permitted. Intermediate moment frames are required by ACI 318, Section 21.1, to comply with the applicable provisions of Chapters 1 through 18, plus Sections 21.2.2.3 and 21.12 of that document. Special moment frames are required by ACI 318, Section 21.1, to comply with the applicable provisions of Chapters 1 through 18, plus Sections 21.2 through 21.6 of that document. If concrete shear walls are used as the seismic-force-resisting system, they must either be ordinary or special reinforced concrete shear walls complying with Section 1910.2.3 or 1910.2.4, respectively. Ordinary or detailed plain concrete shear walls are not permitted (see Sections 1910.2.1 and 1910.2.2). Ordinary reinforced concrete shear wall systems utilizing precast elements are permitted. Section 21.13 of ACI 318 has provisions for the connections between panels as well as between panels and the foundation that must be complied with. Also refer to Section 1908.1.7 for the modification the code makes to this section.

1910.4.2 Discontinuous members. Columns supporting reactions from discontinuous stiff members, such as walls, shall be designed for the special load combinations in Section 1605.4 and shall be provided with transverse reinforcement at the spacing, s_o, as defined in Section 21.12.5.2 of ACI 318 over their full height beneath the level at which the discontinuity occurs. This transverse reinforcement shall be extended above and below the column as required in Section 21.4.4.5 of ACI 318.

❖ Because of a common desire on the part of building owners for an open ground story (for stores, ballrooms in hotels and so forth), shear walls are often discontinued above that story. The columns supporting such shear walls are particularly vulnerable to damage, as has been observed in many earthquakes. Thus, there are two safeguards concerning these elements built into the code—one related to strength and the other related to inelastic deformability or detailing. The strength of columns supporting discontinued stiff members must be augmented by designing them using the special load combinations of Section 1605.4, with E_m defined in Section 1617.1. These combinations apply a multiplier of Ω_0 on the effects of the horizontal seismic forces, Q_E, whereas the regular strength design load combinations of Section 1605.2 apply a multiplier of Ω. While Ω varies between 1 and 1.5, Ω_0 varies between 2 and 3. Additionally, these columns must have special transverse reinforcement (conforming to ACI 318, Sections 21.4.4.1 through 21.4.4.3) not only within the regions of potential plastic hinging at the ends, but also throughout their height. The transverse reinforcement must continue into the discontinued and any supporting wall for one development length of the largest longitudinal column bar. If the lower end of a column supporting a discontinued stiff member is supported on a footing or mat, the special transverse reinforcement must continue at least 12 inches (305 mm) into the foundation.

1910.4.3 Plain concrete. Structural plain concrete members in structures assigned to Seismic Design Category C shall conform to ACI 318 and with Sections 1910.4.3.1 through 1910.4.3.3.

❖ Section 22.2.1.3 of ACI 318 implicitly prohibits the use of structural plain concrete shear walls in structures "in regions of moderate seismic risk or for structures assigned to intermediate seismic performance or design categories." This category of structure under ACI 318 is deemed to be the same as structures assigned to Seismic Design Category C by the code. ACI 318 does not explicitly address whether or not other structural elements of plain concrete can be used in structures assigned to Seismic Design Category C. Therefore, this section provides criteria whereby such members can be used in limited cases.

1910.4.3.1 Walls. Structural plain concrete walls are not permitted in structures assigned to Seismic Design Category C.

Exception: Structural plain concrete basement, foundation or other walls below the base are permitted in detached one- and two-family dwellings constructed with stud-bearing walls. Such walls shall have reinforcement in accordance with Section 22.6.6.5 of ACI 318.

❖ Structural plain concrete walls are prohibited in structures assigned to Seismic Design Category C due to their inability to respond inelastically, which is considered to be essential if they are to perform satisfactorily during the design earthquake. It should be pointed out, however, that such walls are permitted in structures assigned to Seismic Design Category C if exempted from the seismic provisions of the code by Exception 2 to Section 1614.1. The exception to Section 1910.4.3.1 to permit structural plain concrete basement or foundation walls in certain one- and two-family dwellings is no longer needed. It is a carry-over from the final draft of the code in which not all Seismic Design Category C detached one- and two-family dwellings were exempt from the seismic design requirements as they are under this edition of the code (see Section 1614.1, Exception 2).

1910.4.3.2 Footings. Isolated footings of plain concrete supporting pedestals or columns are permitted provided the projection of the footing beyond the face of the supported member does not exceed the footing thickness.

Exception: In detached one- and two-family dwellings three stories or less in height, the projection of the footing beyond the face of the supported member is permitted to exceed the footing thickness.

Plain concrete footings supporting walls shall be provided with not less than two continuous longitudinal reinforcing bars. Bars shall not be smaller than No. 4 and shall have a total area of not less than 0.002 times the gross cross-sectional area of the footing. For footings which exceed 8 inches (203 mm) in thickness, a minimum of one bar shall be provided at the top and bottom of the footing. For foundation systems consisting of a plain concrete footing and a plain concrete stemwall, a minimum of one bar shall be provided at the top of the stemwall and at the bottom of the footing. Continuity of reinforcement shall be provided at corners and intersections.

Exceptions:

1. In detached one- and two-family dwellings three stories or less in height and constructed with stud-bearing walls, plain concrete footings supporting walls are permitted without longitudinal reinforcement.

2. Where a slab-on-ground is cast monolithically with the footing, one No. 5 bar is permitted to be located at either the top or bottom of the footing.

❖ Plain concrete footings supporting columns or pedestals are permitted if the projection of the footing beyond the face of the column or pedestal does not exceed the footing thickness. The limitation on the projection was judged to be prudent because of the empirical manner in which the vertical component of earthquake ground motion is considered (see Section 1617.1). With the limitation on the projection, a flexural or shear failure in the footing is virtually impossible. The exception to Section 1910.4.3.2 to waive the projection limitation in certain one- and two-family dwellings is no longer needed. It is a carry-over from the final draft of the code in which not all Seismic Design Category C detached one- and two-family dwellings were exempt from the seismic design requirements as they are under this edition of the code (see Section 1614.1, Exception 2).

Wall footings of structural plain concrete are permitted in all structures assigned to Seismic Design Category C. Some longitudinal reinforcement is specified in order to keep the footing together should it crack before or during an earthquake. The reinforcement will also allow the footing to act as a "reinforced" concrete member should the earthquake-induced ground motion cause the supported structure to be subjected to differential vertical movement. For footings 8 inches or less (203 mm) in thickness, the bars will typically be located with 3 inches (76 mm) cover to the bottom as required by Item 1 of Section 1907.7.1. For footings thicker than 8 inches (203 mm), one bar will be located with respect to the bottom as noted previously, and the other with $1^1/_2$ or 2 inches (38 or 51 mm) of cover from the top of the footing as required by Item 2 of Section 1907.7.1. Typically, two No. 4 (13) [total of 0.40 square inches (258 mm^2)] or two No. 5 (16) [total of 0.62 square inches (400 mm^2)] bars will suffice, with the former being adequate for footings up to 200 square inches (0.129 m^2) in cross-sectional area [0.40 $in.^2/.002 = 200\ in.^2$ (258 $mm^2/.002 \times 10^6 = 0.129\ m^2$)], and the latter being suitable for footings up to 310 square inches (0.200 m^2) in cross-sectional area [0.62 $in.^2/.002 = 310\ in.^2$ (400 $mm^2/.002 \times 10^6 = 0.200\ m^2$)]. Where one bar will be placed in the bottom of the footing and the other will be placed in the top of the stem wall, the concrete surface of the top of the footing should be intentionally roughened so that a good bond develops between the two concrete placements. Exception 1, which permits structural plain concrete footings without longitudinal reinforcement in certain one- and two-family dwellings, is no longer needed. It is a carry-over from the final draft of the code in which not all Seismic Design Category C detached one- and two-family dwellings were exempt from the seismic design requirements as they are under this edition of the code (see Section 1614.1, Exception 2). Where a slab-on-ground is cast monolithically with the footing, typically referred to as a "turned-down slab," Exception 2 allows one No. 5 (16) bar [versus two No. 4 (13) bars] to be located in either the top or bottom of the footing. In this case, the top of the footing is the top of the slab.

1910.4.3.3 Pedestals. Plain concrete pedestals shall not be used to resist lateral seismic forces.

❖ In Section 1902.1, "Pedestal" is defined as "an upright compression member with a ratio of unsupported height-to-least-lateral dimension of three or less." Pedestals or piers, as they are often called, are commonly used in dwellings with crawl spaces. If the ratio of height-to-least-lateral dimension exceeds three or the pedestal is used to resist lateral forces due to earthquake ground motion, the member must be designed as a column in accordance with ACI 318 and reinforced.

1910.5 Seismic Design Category D, E or F. Structures assigned to Seismic Design Category D, E or F, as determined in Section 1616, shall conform to the requirements for Seismic Design Category C and to the additional requirements of this section.

❖ Structures assigned to Seismic Design Category D, E or F are at increased seismic risk compared to those assigned to Category A, B or C. The provisions of this section are intended to call attention to the detailing requirements of ACI 318. Also see the modification in Section 1908.1.2 made to Section 21.2.1.4 of ACI 318.

1910.5.1 Seismic-force-resisting systems. Moment frames used to resist seismic forces shall be special moment frames. Shear walls used to resist seismic forces shall be special reinforced concrete shear walls.

❖ Section 1617.6 requires concrete moment frames in structures assigned to Seismic Design Category D, E or F to be special moment frames. Ordinary and intermediate moment frames are not permitted. Special moment frames are required by ACI 318, Section 21.1, to comply with the applicable provisions of Chapters 1 through 18, plus Sections 21.2 through 21.5 of that document. If concrete shear walls are used as part of the seismic-force-resisting system and the structure is within the height limits of Section 1617.6, they must be special reinforced concrete shear walls complying with Section 1910.2.4. Ordinary and detailed plain concrete shear walls and ordinary reinforced concrete shear walls are not permitted (see Sections 1910.2.1, 1910.2.2, and 1910.2.3).

1910.5.2 Frame members not proportioned to resist forces induced by earthquake motions. Frame components assumed not to contribute to lateral force resistance shall conform to ACI 318, Section 21.11, as modified by Section 1908.1.6 of this chapter.

❖ In U.S. design practice, it is common to designate certain structural members as being outside of the lateral-force-resisting system of a building. Members that are not part of the designated lateral-force-resisting system are not required to meet all the detailing requirements of members that are relied upon to resist lateral forces. They must, however, be able to sustain their full gravity-load-carrying capacity under the displacements the lateral-force-resisting system will undergo in the design-basis earthquake by conforming to ACI 318, Section 21.11. The above displacements are called the "design displacements" in ACI 318. The definition is developed particularly for the code in Section 1908.1.1.

When the induced moments and shears under design displacements, combined with factored gravity moments and shears (1.05D + 1.28L or 0.9D, whichever is critical), do not exceed the design moment and shear strength, respectively, of a gravity frame member, detailing may be by ACI 318, Section 21.11.2. When the induced moments or shears under design displacements, combined with factored gravity moments and shears, do exceed the design moment or shear strength, respectively, of a gravity frame member, detailing must be by ACI 318, Section 21.11.3. Section 21.11.2 prescribes detailing requirements intended to provide a system capable of sustaining gravity loads under moderate excursions into the inelastic range. Section 21.11.3 prescribes detailing requirements intended to provide a system capable of sustaining gravity loads under larger displacements.

SECTION 1911
MINIMUM SLAB PROVISIONS

1911.1 General. The thickness of concrete floor slabs supported directly on the ground shall not be less than $3^1/_2$ inches (89 mm). A 6-mil (0.006 inch; 152 mm) polyethylene vapor retarder with joints lapped not less than 6 inches (152 mm) shall be placed between the base course or subgrade and the concrete floor slab, or other approved equivalent methods or materials shall be used to retard vapor transmission through the floor slab.

Exception: A vapor retarder is not required:

1. For detached structures accessory to occupancies in Group R-3 as applicable in Section 101.2, such as garages, utility buildings or other unheated facilities.
2. For unheated storage rooms having an area of less than 70 square feet (6.5 m^2) and carports attached to occupancies in Group R-3 as applicable in Section 101.2.
3. For buildings of other occupancies where migration of moisture through the slab from below will not be detrimental to the intended occupancy of the building.
4. For driveways, walks, patios and other flatwork which will not be enclosed at a later date.
5. Where approved based on local site conditions.

❖ Section 1911 requires that slabs on grade be a minimum thickness so superimposed loads are transmitted to the subgrade without causing structural distress of the slab. Additionally, slabs must be protected on the underside with material that will retard the transmission of water vapor through the floor slab into the enclosed space so moisture levels are not excessive.

The coarse aggregate most commonly used in concrete slabs supported directly on the ground is allowed up to 5 percent (by weight) of the stone larger than 1 inch (25 mm) in size, with all coarse aggregates in this grading classification passing a $1^1/_2$-inch (38 mm) sieve. Section 3.3.2 of ACI 318 requires that the nominal maximum size of coarse aggregate be no larger than one-third the depth of the slab, thus making it necessary that slabs be a minimum thickness of $3^1/_2$ inches (89 mm) to accommodate coarse aggregate stones between 1 and $1^1/_2$ inches (25 and 38 mm) in size. Also, as a matter of experience, a $3^1/_2$-inch (89 mm) minimum slab thickness is needed to support typical concentrated loads without damage to the concrete, particularly at locations where small voids or pockets in the subgrade may develop because of improper compaction during construction or from other causes.

An approved 6-mil vapor retarder (see Section 202 for definition) placed between the concrete slab and the supporting ground is required to prevent the migration of water vapor through the slab into the space above. The five exceptions acknowledge conditions where the migration of limited moisture through the slab will not adversely affect the occupancy of the structure. Exception 3 is intended to give relief to occupancies in which additional moisture will not be detrimental, such as indoor swimming pools, industrial processes generating large amounts of moisture and certain storage facilities. Exception 5 is intended to give relief based on environmental or geological conditions, such as locations either with arid climates where the relative humidity is typically low or with well-drained soils and low water tables.Vapor retarders, while preventing moisture from migrating through the slab from below, can adversely affect the construction and performance of concrete slabs on grade. Finishing operations may be delayed since it will normally take longer for the bleed water to evaporate from the surface. Additionally, the tendency for concrete to develop both plastic and drying shrinkage cracking may be exacerbated. Another problem that occurs is the tendency for the slab to curl or warp upward at the edges. Curling occurs because the bottom of the slab remains moist longer than the top portion due to the vapor retarder; therefore, the top shortens more than the bottom, causing curling. If none of the exceptions apply and the vapor retarder must be provided, one solution to the above problems is to place a 2- or 3-inch (51- or 76 mm) layer of sand on top of the vapor retarder prior to concrete placement. The sand will absorb some of the excess water from the concrete and allow earlier finishing. This will also partially or wholly mitigate some of the other adverse effects. If a layer of sand is used, it must be compacted by wetting the day before the concrete is to be placed. However, if the sand is to function effectively as a "blotter," it must be free of drainable water at the time of concrete placement.

SECTION 1912 ANCHORAGE TO CONCRETE— ALLOWABLE STRESS DESIGN

1912.1 Scope. The provisions of this section shall govern the allowable stress design of headed bolts and headed stud anchors cast in normal-weight concrete for purposes of transmitting structural loads from one connected element to the other. These provisions do not apply to anchors installed in hardened concrete or where load combinations include earthquake loads or effects. The bearing area of headed anchors shall be not less than one and one-half times the shank area. Where strength design is used, or where load combinations include earthquake loads or effects, the design strength of anchors shall be determined in accordance with Section 1913. Bolts shall conform to ASTM A 307 or an approved equivalent.

❖ Section 1912 contains requirements for design of anchorage to concrete by the ASD method, also called "working stress design." Provisions for strength design of anchorage to concrete are given in Section 1913. This section must not be used for load combinations that include seismic forces or their effects. In those cases, the provisions of Section 1913 are to be used.

The code provides two methods of design for anchors cast in concrete. Section 1912 provides for limited use of the ASD method and Section 1913 provides an expanded approach for the strength design method. The ASD provisions of this section are adapted from the 1997 edition of the UBC. They are based on limited test data of headed bolts cast in normal-weight concrete and subjected to static loading. Therefore, its application may not be used for hooked (J- or L-) bolts, lightweight concrete, post-installed anchors (i.e., anchors installed in hardened concrete) or when load combinations include earthquake loads.

1912.2 Allowable service load. The allowable service load for headed anchors in shear or tension shall be as indicated in Table 1912.2. Where anchors are subject to combined shear and tension, the following relationship shall be satisfied:

$$(P_s / P_t)^{5/3} + (V_s / V_t)^{5/3} \leq 1 \qquad \textbf{(Equation 19-1)}$$

where:

P_s = Applied tension service load, pounds (newtons).

P_t = Allowable tension service load from Table 1912.2, pounds (newtons).

V_s = Applied shear service load, pounds (newtons).

V_t = Allowable shear service load from Table 1912.2, pounds (newtons).

❖ Table 1912.2 gives allowable service loads for tension or shear, but not at the same time. In real-world applications, both forces are frequently acting at the same time. This section provides a formula for determining the maximum values that may be assigned to bolts that are subjected to combinations of shear and tension loading. The term "allowable service load" signifies that unfactored shear and tension loads are to be used. They may include live, dead, wind and other loading conditions, except for earthquake loads.

TABLE 1912.2. See below.

❖ Table 1912.2 provides shear and tension loads that may be assigned to headed bolts or stud anchors that are cast in normal-weight concrete. Note that the table as-

TABLE 1912.2
ALLOWABLE SERVICE LOAD ON EMBEDDED BOLTS (pounds)

BOLT DIAMETER (inches)	MINIMUM EMBEDMENT (inches)	EDGE DISTANCE (inches)	SPACING (inches)	MINIMUM CONCRETE STRENGTH (psi)					
				f'_c = 2,500		f'_c = 3,000		f'_c = 4,000	
				Tension	Shear	Tension	Shear	Tension	Shear
1/4	2 1/2	1 1/2	3	200	500	200	500	200	500
3/8	3	2 1/4	4 1/2	500	1,100	500	1,100	500	1,100
1/2	4	3	6	950	1,250	950	1,250	950	1,250
	4	5	5	1,450	1,600	1,500	1,650	1,550	1,750
5/8	4 1/2	3 3/4	7 1/2	1,500	2,750	1,500	2,750	1,500	2,750
	4 1/2	6 1/4	7 1/2	2,125	2,950	2,200	3,000	2,400	3,050
3/4	5	4 1/2	9	2,250	3,250	2,250	3,560	2,250	3,560
	5	7 1/2	9	2,825	4,275	2,950	4,300	3,200	4,400
7/8	6	5 1/4	10 1/2	2,550	3,700	2,550	4,050	2,550	4,050
1	7	6	12	3,050	4,125	3,250	4,500	3,650	5,300
1 1/8	8	6 3/4	13 1/2	3,400	4,750	3,400	4,750	3,400	4,750
1 1/4	9	7 1/2	15	4,000	5,800	4,000	5,800	4,000	5,800

For SI: 1 inch = 25.4 mm, 1 pound per square inch = 0.00689 MPa, 1 pound = 4.45 N.

signs values for either shear or tension. Both values cannot be assigned to the same bolt at the same time. The table values are based on a particular combination of bolt size [from $^1/_4$ to $1^1/_4$ inches (6.4 to 32 mm)] and concrete strength [i.e., 2,500, 3,000 and 4,000 psi (17, 21 and 28 MPa)]. When using the table values, minimum bolt embedments, edge distances and spacing between bolts must be observed. Edge distances are measured from the face of the bolt to the sides and ends of concrete members. Decreases up to 50 percent in edge distance or spacing or both are provided for in Section 1912.3.

1912.3 Required edge distance and spacing. The allowable service loads in tension and shear specified in Table 1912.2 are for the edge distance and spacing specified. The edge distance and spacing are permitted to be reduced to 50 percent of the values specified with an equal reduction in allowable service load. Where edge distance and spacing are reduced less than 50 percent, the allowable service load shall be determined by linear interpolation.

❖ This section expands the applicability of Table 1912.2 when bolt locations cannot meet the required minimum edge distances or spacing or both specified by the table. Tabulated minimum edge distances and spacings may be reduced up to 50 percent for a corresponding 50-percent reduction of service loads. For reductions in edge distance or spacing or both less than 50 percent, corresponding service loads may be determined by interpolation. For example, if $^5/_8$-inch (15.9 mm) bolts are embedded $4^1/_2$ inches (114 mm) in 2,500 psi (17 MPa) concrete, the table assigns shear and tension values of 2,750 and 1,500 pounds (12,100 and 6,600 N), respectively, if at least $3^3/_4$-inch (95 mm) edge distance and $7^1/_2$-inch (191 mm) spacing between bolts are provided. If only a 25-percent reduction in edge distance or spacing or both is desired, the corresponding shear and tension values (determined by linear interpolation) are 2,063 and 1,125 pounds (9,077 and 4,950 N), respectively.

1912.4 Increase in allowable load. Increase of the values in Table 1912.2 by one-third is permitted where the provisions of Section 1605.3.2 permit an increase in allowable stress for wind loading.

❖ This section follows a time-honored principle of permitting a one-third increase in allowable stress for wind. The rationale for this increase is that design wind loads occur infrequently and their maximum effects last only a few seconds. Therefore, an overstress of one-third for a very short time at infrequent intervals is considered acceptable.

1912.5 Increase for special inspection. Where special inspection is provided for the installation of anchors, a 100-percent increase in the allowable tension values of Table 1912.2 is permitted. No increase in shear value is permitted.

❖ The tension and shear values in Table 1912.2 are based on load test data that has been reduced by a safety factor. The code acknowledges that special inspection improves construction practices sufficiently to allow a 100-percent increase in the tension values provided in the table, thus allowing a lower safety factor when special inspection is provided.

The shear values of the table are not permitted to be increased when special inspection is provided, because the table values reflect a safety factor that is not overly conservative. The provisions of this section were taken from the footnotes of Table 19D (Allowable Service Load on Embedded Bolts) of the 1997 *Uniform Building Code* (UBC). The footnotes first appeared in the 1976 UBC and were accompanied by a 50-percent increase in the table's shear values, but no increase in the table's tensile values. The footnotes also enabled the tensile values to be increased 50 percent, but only when special inspection is provided.

SECTION 1913 ANCHORAGE TO CONCRETE— STRENGTH DESIGN

1913.1 Scope. The provisions of this section shall govern the strength design of anchors installed in concrete for purposes of transmitting structural loads from one connected element to the other. Headed bolts, headed studs and hooked (J- or L-) bolts cast in concrete and expansion anchors and undercut anchors installed in hardened concrete shall be designed in accordance with Appendix D of ACI 318, provided they are within the scope of Appendix D.

> **Exception:** Where the basic concrete breakout strength in tension of a single anchor, N_b, is determined in accordance with Equation (D-7), the concrete breakout strength requirements of Section D.4.2.2 shall be considered satisfied by the design procedures of Sections D.5.2 and D.6.2 for anchors exceeding 2 inches (51 mm) in diameter or 25 inches (635 mm) tensile embedment depth.

The strength design of anchors that are not within the scope of Appendix D of ACI 318, and as amended above, shall be in accordance with an approved procedure.

❖ Section 1913 is restricted in scope to anchors that transmit structural loads from attachments into concrete members and vice versa under typical in-service conditions. Other standards, such as those applicable to construction and/or material handling, for instance, can require more stringent safety levels and must be followed where they apply.

The first edition of the code included provisions for strength design of anchorage to concrete in Section 1913. At that time, the provisions that were being developed for inclusion in ACI 318 were not yet available in a standard that could be referenced by the code. For that reason, the strength design requirements for concrete anchorage that were under development were incorporated into Section 1913 with some minor changes. At that time, one of the limitations was that the strength design for concrete anchorage was only applicable to cast-in-place anchors, such as the headed studs and headed bolts shown in Figure 1913.1.

With the publication of the 2002 edition of ACI 318, requirements for the strength design of concrete anchorage were included as Appendix D; thus, the bulk of the code provisions have been replaced with a reference to Appendix D of ACI 318, since it is a more current version of the same requirements. Based on the content of these newer requirements, the scope of this section also includes post-installed anchors, such as expansion anchors and undercut anchors, provided they are evaluated and categorized in accordance with ACI 355.2. Design of anchors that are not within the scope of this section must be in accordance with an approved procedure. Examples of anchors that fall into the latter category are specialty inserts, through-bolts, adhesive anchors, grouted anchors and powder-actuated fasteners.

Section D.2.4 of the referenced ACI 318 provisions excludes load applications producing high-cycle fatigue or extremely short duration impact (such as blast or shock wave). This statement does not exclude seismic loads, and, in fact, Section 1912 for design of anchorage to concrete by the ASD method requires the use of strength design for load combinations that include seismic forces or their effects.

Section D.4.2.2 of ACI 318 allows determination of the concrete breakout strength in both tension and shear to be determined according to the referenced sections that are based on the concrete capacity design (CCD) method. Consistent with the range of test data that was available, this section restricts the use of the option to compute concrete breakout strength of the anchorage based on an anchor's diameter or embedment length. Instead it requires testing of anchors exceeding 2 inches (51 mm) in diameter or with a depth of embedment greater than 25 inches (635 mm) in order to determine concrete breakout strength. The exception is an important modification to this anchor size and embedment limit that is based on subsequent analysis of test data. It indicates that if Equation D-7 is used to establish the concrete breakout strength of a single anchor, then these limits on size and embedment depth do not apply.

SECTION 1914
SHOTCRETE

1914.1 General. Shotcrete is mortar or concrete that is pneumatically projected at high velocity onto a surface. Except as specified in this section, shotcrete shall conform to the requirements of this chapter for plain or reinforced concrete.

❖ Section 1914 regulates the materials and test procedures for shotcrete construction. Shotcrete is the commonly accepted and generic name for pneumatically projected mortar or concrete. Terms such as gunite, sprayed concrete, spraycrete, pneumatically applied mortar and concrete are often used to refer to shotcrete. ACI 506R, the guide for shotcreting, defines "Shotcrete" as mortar or concrete conveyed through a hose and pneumatically projected at high velocity onto a surface. The force of the jet impacting on the shotcrete surface compacts the material, and when applied properly, the shotcrete surface will support itself without sagging or sloughing, even for vertical and overhead applications. Shotcrete must conform to the requirements for structural concrete specified in this chapter and ACI 318 unless modified by Section 1914.

1914.2 Proportions and materials. Shotcrete proportions shall be selected that allow suitable placement procedures using the delivery equipment selected and shall result in finished in-place hardened shotcrete meeting the strength requirements of this code.

❖ Although the compressive strength of shotcrete must be verified prior to beginning work, its resulting strength and acceptability relies more on the suitability of placement procedures and the delivery equipment. Prior to construction, strength is established by the submittal of

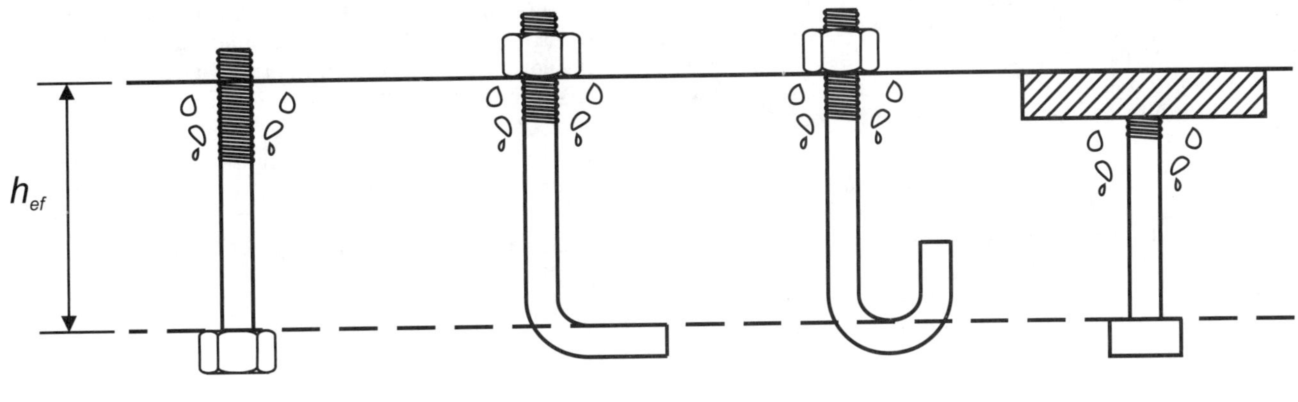

Figure 1913.1
TYPES OF CAST-IN-PLACE ANCHORS

a mix design that is supported by test results. Mix designs from previous jobs can be used. Product data sheets for any admixtures, and graduations for the sand and aggregate, need to be included in the submittal. Verification of in-place hardened shotcrete strength is determined by testing in accordance with Section 1914.10.

1914.3 Aggregate. Coarse aggregate, if used, shall not exceed $^3/_4$ inch (19.1 mm).

❖ For construction sections that are several inches thick and where adequate gunning equipment is available, the use of shotcrete mixtures containing coarse aggregates may be advantageous. Coarse aggregate must not exceed $^3/_4$ inch (19.1 mm) in size and the gradation limitations for a combination of fine (sand) and coarse aggregates should be in accordance with the requirements of ACI 506.2. Normal-weight coarse aggregate for concrete must comply with the requirements of ASTM C 33. Lightweight coarse aggregate for concrete must comply with the requirements of ASTM C 330 (see commentary, Section 1903.3).

1914.4 Reinforcement. Reinforcement used in shotcrete construction shall comply with the provisions of Sections 1914.4.1 through 1914.4.4.

❖ The best results will be obtained when the reinforcing steel is designed and placed to cause the least interference with the high-velocity placement of shotcrete. When proper techniques are used, shotcrete can be successfully placed through two or three layers of reinforcing steel. The size, spacing and splicing of reinforcement is specified in Sections 1914.4.1, 1914.4.2 and 1914.4.3, respectively.

1914.4.1 Size. The maximum size of reinforcement shall be No. 5 bars unless it is demonstrated by preconstruction tests that adequate encasement of larger bars will be achieved.

❖ Depending on the thickness and nature of the work, reinforcement will consist of either deformed bars or welded wire fabric. Bar sizes are limited to a No. 5 (16) bar. Larger bar sizes can be used if it can be shown that adequate bonding with and encasement in the concrete can be accomplished. Preconstruction tests may be required by the building official to provide verification that proper bonding can be achieved (see Section 1914.5).

1914.4.2 Clearance. When No. 5 or smaller bars are used, there shall be a minimum clearance between parallel reinforcement bars of $2^1/_2$ inches (64 mm). When bars larger than No. 5 are permitted, there shall be a minimum clearance between parallel bars equal to six diameters of the bars used. When two curtains of steel are provided, the curtain nearer the nozzle shall have a minimum spacing equal to 12 bar diameters and the remaining curtain shall have a minimum spacing of six bar diameters.

Exception: Subject to the approval of the building official, required clearances shall be reduced where it is demonstrated by preconstruction tests that adequate encasement of the bars used in the design will be achieved.

❖ Experience has shown that a $2^1/_2$-inch (64 mm) space between No. 5 (16) bars, and a six-bar diameter space between larger bars, is sufficient clearance for placement of shotcrete where there is one layer of reinforcement. Experience has also shown that shotcrete application can be effective when the spacing between bars in the closer of two layers (nearer to nozzle) of steel is 12 bar diameters, and in the remaining layer the bars have a spacing of six bar diameters. The "exception" clarifies that lesser clearance dimensions are permitted if they can be verified by preconstruction tests.

For welded wire fabric, ACI 506R recommends a minimum spacing of 2 inches (51 mm) between wires in both directions. The minimum clearance between the reinforcement and the form or other back-up surface may vary depending on whether a mortar mix (fine aggregate only) or concrete (fine and coarse aggregate) is used for shotcrete and whether conventional reinforcing steel or welded wire fabric is used. This spacing should be specified or shown on the construction documents. ACI 506R recommends that no less than a 2-inch (51 mm) clearance should be used for No. 5 (16) reinforcing bars. The minimum cover for reinforcement must comply with the requirements of Section 1907.7.

1914.4.3 Splices. Lap splices of reinforcing bars shall utilize the noncontact lap splice method with a minimum clearance of 2 inches (51 mm) between bars. The use of contact lap splices necessary for support of the reinforcing is permitted when approved by the building official, based on satisfactory preconstruction tests that show that adequate encasement of the bars will be achieved, and provided that the splice is oriented so that a plane through the center of the spliced bars is perpendicular to the surface of the shotcrete.

❖ Bars and welded wire fabric must be lapped in such a manner so as not to create weak sections in the shotcrete. Lapped reinforcing bars should not be tied together, but rather be separated by at least 2 inches (51 mm) wherever possible. The building official should only permit contact lap splices when it can be shown by preconstruction tests that the reinforcement will be completely encased. Welded wire fabric must be lapped 1 $^1/_2$ squares in all directions. Continuous chair support should not be placed parallel and directly underneath bars because chairs may prevent full shotcrete encasement or cause sand pockets. Individual chairs should also to be offset to prevent a line of weakness in the shotcrete.

1914.4.4 Spirally tied columns. Shotcrete shall not be applied to spirally tied columns.

❖ Typically plain reinforcement (i.e., no surface deformations) is used for spiral reinforcement in columns. Shotcrete is not permitted to be applied to spirally tied columns because of the lack of these deformations. Additionally, ACI 318 requires that the clear spacing between spirals must not exceed 3 inches (76 mm). These conditions make it difficult to achieve adequate bond between the concrete and the reinforcement.

1914.5 Preconstruction tests. When required by the building official, a test panel shall be shot, cured, cored or sawn, examined and tested prior to commencement of the project. The sample panel shall be representative of the project and simulate job conditions as closely as possible. The panel thickness and reinforcing shall reproduce the thickest and most congested area specified in the structural design. It shall be shot at the same angle, using the same nozzleman and with the same concrete mix design that will be used on the project. The equipment used in preconstruction testing shall be the same equipment used in the work requiring such testing, unless substitute equipment is approved by the building official.

❖ When required by the building official, preconstruction test specimens must be made for each type of construction application (flat, vertical or overhead) using the materials, mix proportions, equipment and the nozzle operator proposed for the project. Test panels should be at least 30 inches (762 mm) square, and must have the same reinforcement and thickness [but not less than 3 inches (76 mm)] as specified and shown on the construction documents for permanent construction. Drilled cores should be taken from the panels and tested in accordance with ASTM C 42 procedures for compliance with strength requirements. Preconstruction testing may not be necessary if it can be shown to the satisfaction of the building official that mix, materials, equipment and construction personnel have produced satisfactory results on similar work.

1914.6 Rebound. Any rebound or accumulated loose aggregate shall be removed from the surfaces to be covered prior to placing the initial or any succeeding layers of shotcrete. Rebound shall not be used as aggregate.

❖ Rebound is loose aggregate and cement paste that bounces off application surfaces after colliding with formwork, reinforcement or the shotcrete surface itself. Factors that affect the amount of rebound include the position of the work (flat, vertical or overhead applications), air pressure, cement content, water content, maximum size and grading of aggregates, amount of reinforcement and thickness of the layer. Rebound must not be reused as aggregate or in any way worked back into the construction. If rebound does not fall clear of the work, it must be removed.

1914.7 Joints. Except where permitted herein, unfinished work shall not be allowed to stand for more than 30 minutes unless edges are sloped to a thin edge. For structural elements that will be under compression and for construction joints shown on the approved construction documents, square joints are permitted. Before placing additional material adjacent to previously applied work, sloping and square edges shall be cleaned and wetted.

❖ Construction joints are generally tapered to a thin edge over a width of approximately 12 inches (305 mm). Square construction joints (nontapered) should be avoided wherever possible in shotcrete construction because they form a trap for rebound. Where the joint will be subjected to compressive stresses, however, square joints are frequently used and, in such cases, care must be taken to avoid or remove trapped rebound at the joint. All joints, tapered or square, must be thoroughly cleaned of loose material and wetted prior to the application of additional shotcrete.

1914.8 Damage. In-place shotcrete that exhibits sags, sloughs, segregation, honeycombing, sand pockets or other obvious defects shall be removed and replaced. Shotcrete above sags and sloughs shall be removed and replaced while still plastic.

❖ After placement, any shotcrete that lacks uniformity or exhibits segregation, honeycombing or delamination, or that contains dry patches, slugs, voids or sand pockets (porous areas low in cement content), or that sags or sloughs should be removed and replaced. As a practical matter, small holes created by cores taken for testing purposes cannot be filled solidly by shooting mortar at high velocity. The patching mixture to be applied by hand should be made of the same materials and approximately of the same proportions as used for the shotcrete, except that coarse aggregate, if used, should be omitted. Chapter 9 of ACI 301 provides additional recommendations for repairing shotcrete, including filling of holes created where cores were obtained for testing.

1914.9 Curing. During the curing periods specified herein, shotcrete shall be maintained above 40°F (4°C) and in moist condition.

❖ Like cast-in-place concrete, shotcrete must be cured in a moist condition to achieve its potential strength and durability. Good curing practice for shotcrete construction generally requires that surfaces be kept continuously moist for a period of at least 7 days. During curing, the air that is in contact with shotcrete surfaces must be maintained at temperatures above freezing, with the temperature of the shotcrete maintained above 40°F (4°C). For recommendations on winter protection, see the commentary to Section 1905.12 or refer to ACI 306R.

1914.9.1 Initial curing. Shotcrete shall be kept continuously moist for 24 hours after shotcreting is complete or shall be sealed with an approved curing compound.

❖ Immediately after finishing operations are complete, shotcrete must be kept continuously moist for a period of 24 hours by one of the following methods:

1. Ponding or continuous sprinkling;
2. Use of absorptive mat or fabric, sand or other protective covering kept continuously moist;

3. Continuous steam not exceeding 150°F (66°C) or vapor mist bath; or

4. Membrane-forming curing compounds and application conforming to ASTM C 309.

1914.9.2 Final curing. Final curing shall continue for seven days after shotcreting, or for three days if high- early-strength cement is used, or until the specified strength is obtained. Final curing shall consist of the initial curing process or the shotcrete shall be covered with an approved moisture-retaining cover.

❖ Immediately following the initial curing specified in Section 1914.9.1, additional curing must be provided using one of the following materials or methods:

1. Continue the method used in the initial curing process (see commentary, Section 1914.9.1);
2. Use sheet materials conforming to ASTM C 171 or
3. Use other moisture-retaining covers acceptable to the registered design professional and approved by the building official.

Curing must be continued for at least the first seven days after shotcreting, or for the first three days if high-early-strength cement is used (Type III or IIIA portland cement), or until the specified strength is obtained. During the curing period, shotcrete must be maintained at a temperature above 40°F (4°C) and in a moist condition.

1914.9.3 Natural curing. Natural curing shall not be used in lieu of that specified in this section unless the relative humidity remains at or above 85 percent, and is authorized by the registered design professional and approved by the building official.

❖ Natural curing is exposure to the atmosphere without any cover or water spray. If the atmospheric conditions surrounding the shotcrete construction are satisfactory, such as when the relative humidity is at or above 85 percent, the registered design professional, with the approval of the building official, may authorize natural curing. Where natural curing is used, the shotcrete must be maintained above 40°F (4°C) for the period specified in Section 1914.9.2.

1914.10 Strength tests. Strength tests for shotcrete shall be made by an approved agency on specimens that are representative of the work and which have been water soaked for at least 24 hours prior to testing. When the maximum-size aggregate is larger than $^3/_8$ inch (9.5 mm), specimens shall consist of not less than three 3-inch-diameter (76 mm) cores or 3-inch (76 mm) cubes. When the maximum-size aggregate is $^3/_8$ inch (9.5 mm) or smaller, specimens shall consist of not less than 2-inch-diameter (51 mm) cores or 2-inch (51 mm) cubes.

❖ This section provides basic strength test criteria for shotcrete. Experience has shown that aggregate size has a direct bearing on the accuracy of compressive tests. Therefore, core and cube specimens containing aggregates that are larger than $^3/_8$ inch (9.5 mm) must be a minimum of 3 inches (76 mm) in diameter (or cross section). Test specimens are required to be soaked in water for at least 24 hours prior to testing to maintain a common environmental condition for all specimens and to insure uniformity between test results. Cores should be obtained and tested in accordance with ASTM C 42.

1914.10.1 Sampling. Specimens shall be taken from the in-place work or from test panels, and shall be taken at least once each shift, but not less than one for each 50 cubic yards (38.2 m^3) of shotcrete.

❖ Frequency of testing is a significant factor in the effectiveness of quality control of concrete. To ensure that test results will be representative of the concrete in the structure, specimens must be taken at the minimum frequency required by this section.

1914.10.2 Panel criteria. When the maximum-size aggregate is larger than $^3/_8$ inch (9.5 mm), the test panels shall have minimum dimensions of 18 inches by 18 inches (457 mm by 457 mm). When the maximum size aggregate is $^3/_8$ inch (9.5 mm) or smaller, the test panels shall have minimum dimensions of 12 inches by 12 inches (305 mm by 305 mm). Panels shall be shot in the same position as the work, during the course of the work and by the nozzlemen doing the work. The conditions under which the panels are cured shall be the same as the work.

❖ When test panels are used, they are required to meet the criteria in this section. Experience has shown that aggregate size has a direct bearing on the ability to produce test panels that accurately portray the condition of in-place shotcrete. Evidence suggests that test panels containing aggregates larger than $^3/_8$ inch (9.5 mm) must be a minimum of 18 inches (457 mm) square, and that test panels with aggregate $^3/_8$ inch (9.5 mm) or smaller must be a minimum 12 inches (305 mm) square. The ability of the nozzle operator is a significant factor in achieving quality shotcrete. Therefore, it is critical that the nozzle operator doing the work also prepares the test panels. Since the position of the work (flat, vertical or overhead) and the method of curing have an impact on the quality of work, they must also be accounted for when making test panels.

1914.10.3 Acceptance criteria. The average compressive strength of three cores from the in-place work or a single test panel shall equal or exceed $0.85f'_c$ with no single core less than $0.75f'_c$. The average compressive strength of three cubes taken from the in-place work or a single test panel shall equal or exceed f'_c with no individual cube less than $0.88f'_c$. To check accuracy, locations represented by erratic core or cube strengths shall be retested.

❖ In evaluating core test results, the fact that core strengths may not equal the specified compressive strength of concrete, f'_c, should not be a cause for concern, provided they are within the limits specified. It is realistic to expect the compressive strength of cores to be less than f'_c because the size of the core differs from that of a standard concrete test cylinder, and procedures for obtaining and curing the cores are different. Therefore, the shotcrete can be considered acceptable from the standpoint of strength if test results from the cores are at least 85 percent of f'_c for the average of three cores and 75 percent of f'_c for a single core. Cores should be obtained and tested in accordance with ASTM C 42. Cores for compressive strength testing must have a length-to-diameter ratio between 1.0 and 2.10. For ratios less than 1.94, the compressive strength is multiplied by a correction factor (less than 1) given in ASTM C 42 to correlate the strength of the core to that of a standard concrete test cylinder that has a length-to-diameter ratio of 2. Experience has shown that test results from cubes can be held to a higher standard, thus requiring the average of three cubes to equal or exceed f'_c with no individual cube less than 0.88 f'_c. Since the height-to-least-lateral dimension of a cube is 1, ACI 506R indicates that the compressive strength of a cube can be correlated to that of a standard test cylinder by multiplying the test results by 0.85. Either correlated or uncorrelated results for cubes can be used to determine if the specified acceptance criteria are met. Because of the less than delicate nature of the coring and sawing process, this section also permits erratic core and cube strengths to be discarded from the sample under consideration.

SECTION 1915
REINFORCED GYPSUM CONCRETE

1915.1 General. Reinforced gypsum concrete shall comply with the requirements of ASTM C 317 and ASTM C 956.

❖ The provisions of Section 1915 apply to poured-in-place reinforced gypsum concrete that is used to construct structural roof decks (when placed over permanent formboards). Gypsum concrete is a manufactured product consisting of calcined gypsum to which the producer may add aggregates complying with ASTM C 35, wood chips or wood shavings. Only water is added at the job site.

Similar proprietary products manufactured under various trade names are intended to be used in nonstructural applications such as floor toppings, where increased sound resistance is desired. These products can also be used as floor levers, and this application is prevalent in rehabilitation of existing structures. This section is not intended to govern the use of these materials where used in nonstructural applications.

Gypsum concrete (the product) used in reinforced gypsum concrete construction must conform to the requirements of ASTM C 317. This standard provides for two classes (i.e., A and B) of gypsum concrete based on minimum compressive strength. Class A must achieve a minimum compressive strength of 500 psi (3.5 MPa) and have a density not exceeding 60 lb/ft^3 (960 kg/m^3). Class B must achieve a compressive strength of 1,000 psi (6.9 MPa) with no restriction on density. Before the development of newer insulating materials suitable for installation on roof decks, reinforced gypsum concrete roof decks were used where energy efficiency was a concern. To increase the energy efficiency of the material, an aggregate complying with ASTM C 35, such as perlite or vermiculite, is used to reduce the density of the gypsum concrete, thus increasing its *R*-value.

The design and installation of reinforced gypsum concrete roof decks must comply with the requirements of ASTM C 956. Gypsum concrete roof decks must be reinforced with zinc-coated (galvanized) welded or woven wire mesh or fabric and must be designed to support anticipated loads without exceeding the allowable stresses specified in Appendix Section X.2 of ASTM C 956. ASTM C 956 indicates that "methods of design [of reinforced gypsum concrete] shall follow established principles of mechanics and principles of design for reinforced concrete in accordance with ACI 318." Requirements of Section 7.12 of ACI 318 for shrinkage and temperature reinforcement in reinforced concrete do not apply to reinforced gypsum concrete.

1915.2 Minimum thickness. The minimum thickness of reinforced gypsum concrete shall be 2 inches (51 mm) except the minimum required thickness shall be reduced to $1^1/_2$ inches (38 mm), provided the following conditions are satisfied:

1. The overall thickness, including the formboard, is not less than 2 inches (51 mm).
2. The clear span of the gypsum concrete between supports does not exceed 33 inches (838 mm).
3. Diaphragm action is not required.
4. The design live load does not exceed 40 pounds per square foot (psf) (1915 Pa).

❖ This section provides an exception to the ASTM C 956 minimum 2-inch (51 mm) thickness requirement for reinforced gypsum concrete roof decks. The design live load mentioned in Item 4 should be interpreted to mean the larger of the design roof live load and the design snow load.

SECTION 1916
CONCRETE-FILLED PIPE COLUMNS

1916.1 General. Concrete-filled pipe columns shall be manufactured from standard, extra-strong or double-extra-strong steel pipe or tubing that is filled with concrete so placed and manipulated as to secure maximum density and to ensure complete filling of the pipe without voids.

❖ Section 1916 regulates the design and construction of concrete-filled steel pipe or tube columns that rely on the concrete and steel acting compositely to resist loads.

Although the code requires concrete-filled pipe columns to be standard, extra-strong or double extra-strong steel pipe, this language is out of date since it only refers to pipe manufactured in accordance with ASTM A 53. Both ACI 318 and AISC LRFD require that steel tube or pipe used in composite concrete-filled pipe columns comply with ASTM A 53 Grade B, ASTM A 500 or ASTM A 501. In addition, both ACI 318 and AISC LRFD provide limitations on minimum steel wall thickness and either of these referenced standards should be consulted for up-to-date design information.

During placement, the concrete must be thoroughly consolidated with a mechanical vibrator to develop a cohesive mass that fills all voids in the steel pipe or tube. This assures a complete bonding of the concrete to the steel shell, and any reinforcing steel and other embedded items that may be present in the pipe or tube.

1916.2 Design. The safe supporting capacity of concrete-filled pipe columns shall be computed in accordance with the approved rules or as determined by a test.

❖ The load-carrying capacity of concrete-filled pipe columns should be calculated in accordance with the provisions contained in ACI 318 or AISC LRFD for composite compression members. In lieu of using analytical procedures, the load-carrying capacity may be determined by preconstruction load tests in accordance with Section 1714.

1916.3 Connections. Caps, base plates and connections shall be of approved types and shall be positively attached to the shell and anchored to the concrete core. Welding of brackets without mechanical anchorage shall be prohibited. Where the pipe is slotted to accommodate webs of brackets or other connections, the integrity of the shell shall be restored by welding to ensure hooping action of the composite section.

❖ In general terms, this section prescribes conditions for making structural connections to concrete-filled pipe columns. For the column to act compositely, all loads must be transferred to the concrete core, not just to the steel shell, as would be the case with a bracket welded to the exterior surface of the shell. Since increased structural capacity is achieved with concrete fill, the integrity of the shell must be preserved so it can function to laterally confine the concrete core. If a connection requires welding to the steel shell, it must be done prior to filling the core with concrete unless it can be shown that the concrete will not be damaged from the heat of welding.

1916.4 Reinforcement. To increase the safe load-supporting capacity of concrete-filled pipe columns, the steel reinforcement shall be in the form of rods, structural shapes or pipe embedded in the concrete core with sufficient clearance to ensure the composite action of the section, but not nearer than 1 inch (25 mm) to the exterior steel shell. Structural shapes used as reinforcement shall be milled to ensure bearing on cap and base plates.

❖ This section does not require steel reinforcement in the concrete core; however, if it is added it must comply with the requirements of this section and the applicable provisions of ACI 318 or AISC LRFD. Since the concrete fill and the steel must serve as a composite section to support the loads on the column, construction must be such that reinforcement placed in the core of the pipe has sufficient clearance to allow the concrete to flow around it for proper encasement and bonding.

1916.5 Fire-resistance-rating protection. Pipe columns shall be of such size or so protected as to develop the required fire-resistance ratings specified in Table 601. Where an outer steel shell is used to enclose the fire-resistant covering, the shell shall not be included in the calculations for strength of the column section. The minimum diameter of pipe columns shall be 4 inches (102 mm) except that in structures of Type V construction not exceeding three stories or 40 feet (12 192 mm) in height, pipe columns used in the basement and as secondary steel members shall have a minimum diameter of 3 inches (76 mm).

❖ Composite pipe columns required to be fire-resistant rated may be designed to achieve the required rating with the steel left unprotected on the exterior. This is accomplished by allowing the concrete to act as a heat sink, and Kodur and MacKinnon have developed a detailed methodology for providing the required fire resistance in this manner.

Some prefabricated steel pipe columns consist of an inner pipe filled with concrete and an outer steel shell encasing conventional concrete or a proprietary insulating concrete that enables the column assembly to achieve a fire-resistance rating. Where such a column is required to have a fire-resistance rating, the structural calculations for the load-carrying capacity of the column must not include the outer steel shell unless it is protected with materials that provide a fire-resistance rating as required in Table 601.

Since the concrete in these columns is completely encased, if it is heated sufficiently by a fire, free moisture in the concrete will be turned into steam. If provisions are not made to relieve the resulting pressure, the pipe or tube may burst causing premature failure of the column. To prevent this, Kodur and MacKinnon recommend that a 1-inch-diameter (25 mm) vent hole be provided near the top and bottom of the column. If the column is continuous through more than one story of the structure, pairs of holes as described previously should be provided in each story. Listed columns may have other venting arrangements. Care must be taken to ensure that the holes are not inadvertently or intentionally plugged during the construction process.

1916.6 Approvals. Details of column connections and splices shall be shop fabricated by approved methods and shall be approved only after tests in accordance with the approved rules. Shop-fabricated concrete-filled pipe columns shall be inspected by the building official or by an approved representative of the manufacturer at the plant.

❖ Concrete-filled pipe columns, including their connection details and splices, which are shop fabricated as preengineered sections, are to be inspected at the plant by the building official or an approved manufacturer's representative. Approvals of such manufactured products are to be based on accepted fabrication practices and tests.

Bibliography

The following resource materials are referenced in this chapter or are relevant to the subject matter addressed in this chapter.

ACI 116R-00, *Cement and Concrete Terminology*. Farmington Hills, MI: American Concrete Institute, 1990.

ACI 117-90/117R-90, *Standard Specifications for Tolerances for Concrete Construction and Materials and Commentary*. Farmington Hills, MI: American Concrete Institute, 1990.

ACI 201.2R-92 (Reapproved 1997), *Guide to Durable Concrete*. Farmington Hills, MI: American Concrete Institute, 1992.

ACI 211.1-91 (Reapproved 1997), *Standard Practice for Selecting Proportions for Normal, Heavyweight, and Mass Concrete*. Farmington Hills, MI: American Concrete Institute, 1991.

ACI 211.2-98, *Standard Practice for Selecting Proportions for Structural Lightweight Concrete*. Farmington Hills, MI: American Concrete Institute, 1998.

ACI 212.3R-91, *Chemical Admixtures for Concrete*. Farmington Hills, MI: American Concrete Institute, 1991.

ACI 222.1-96, *Provisional Standard Test Method for Water-Soluble Chloride Available for Corrosion of Embedded Steel in Mortar and Concrete Using the Soxhlet Extractor*. Farmington Hills, MI: American Concrete Institute, 1996.

ACI 301-99, *Specifications for Structural Concrete*. Farmington Hills, MI: American Concrete Institute, 1999.

ACI 304R-00, *Guide for Measuring, Mixing, Transporting and Placing Concrete*. Farmington Hills, MI: American Concrete Institute, 2000.

ACI 305R-99, *Report on Hot Weather Concreting*. Farmington Hills, MI: American Concrete Institute, 1999.

ACI 306R-88 (Reapproved 1997), *Report on Cold Weather Concreting*. Farmington Hills, MI: American Concrete Institute, 1988.

ACI 308-92, *Standard Practice for Curing Concrete*. Farmington Hills, MI: American Concrete Institute, 1992.

ACI 309R-96, *Guide for Consolidation of Concrete*. Farmington Hills, MI: American Concrete Institute, 1996.

ACI 318-02/318R-02, *Building Code Requirements for Structural Concrete and Commentary*. Farmington Hills, MI: American Concrete Institute, 2002.

ACI 347R-94 (Reapproved 1999), *Guide to Formwork for Concrete*. Farmington Hills, MI: American Concrete Institute, 1994.

ACI 349-97/349R-97, *Code Requirements for Nuclear Safety Related Concrete Structures and Commentary*. Farmington Hills, MI: American Concrete Institute, 1997.

ACI 355.2-01, *Evaluating the Performance of Post-Installed Mechanical Anchors in Concrete*. Farmington Hills, MI: American Concrete Institute, 2001.

ACI 355.2R-01, *Commentary on Evaluating the Performance of Post-Installed Mechanical Anchors in Concrete*. Farmington Hills, MI: American Concrete Institute, 2001.

ACI 423.6-01/423.6R-01, *Specification for Unbonded Single-Strand Tendons and Commentary*. Farmington Hills, MI: American Concrete Institute, 2001.

ACI 506R-90 (Reapproved 1995), *Guide to Shotcrete*. Farmington Hills, MI: American Concrete Institute, 1990.

ACI 506.2-95, *Specification for Shotcrete*. Farmington Hills, MI: American Concrete Institute, 1995.

ACI T1.1-01, *Acceptance Criteria for Moment Frames Based on Structural Testing*. Farmington Hills, MI: American Concrete Institute, 2001.

ACI T1.1R-01, *Commentary on Acceptance Criteria for Moment Frames Based on Structural Testing*. Farmington Hills, MI: American Concrete Institute, 2001.

AISC LRFD (1999), *Load and Resistance Factor Design for Structural Steel Buildings*. Chicago: American Institute of Steel Construction, 1999.

ANSI A58.1-1982, *Minimum Design Loads for Buildings and Other Structures* (Also, ANSI A58.1-1972). New York: American National Standards Institute, 1972, 1982.

ASCE 3-91, *Standard for the Structural Design of Composite Slabs*. New York: American Society of Civil Engineers, 1992.

ASCE 7-02, *Minimum Design Loads for Buildings and Other Structures* (also ASCE 7-88, ASCE 7-93, ASCE 7-95). New York: American Society of Civil Engineers, 1990, 1993, 1995; ASCE 7-02, ASCE: Reston, VA, 2002.

ASME B1.1-89, *Unified Inch Screw Threads (UN and UNR Thread Form)*. Fairfield, NJ: American Society of Mechanical Engineers, 1989.

ASME B18.2.1-96, *Square and Hex Bolts and Screws, Inch Series*. Fairfield, NJ: American Society of Mechanical Engineers, 1996.

ASME B18.2.6-96, *Fasteners for Use in Structural Applications*. Fairfield, NJ: American Society of Mechanical Engineers, 1996.

ASTM A 36/A 36M-00, *Standard Specification for Carbon Structural Steel*. West Conshohocken, PA: ASTM International, 2000.

ASTM A 53-97, *Standard Specification for Pipe, Steel, Black and Hot-Dipped, Zinc-Coated Welded and Seamless*. West Conshohocken, PA: ASTM International, 1997.

ASTM A 82-01, *Standard Specification for Steel Wire, Plain, for Concrete Reinforcement*. West Conshohocken, PA: ASTM International, 2001.

ASTM A 184-96, *Standard Specification for Fabricated Deformed Steel Bar Mats for Concrete Reinforcement*. West Conshohocken, PA: ASTM International, 1996.

ASTM A 185-01, *Standard Specification for Steel Welded Wire Fabric, Plain, for Concrete Reinforcement*. West Conshohocken, PA: ASTM International, 2001.

ASTM A 242-93a/A 242M-93a, *Standard Specification for High-Strength Low-Alloy Structural Steel*. West Conshohocken, PA: ASTM International, 1993.

ASTM A 416-99, *Standard Specification for Steel Strand, Uncoated Seven-Wire for Prestressed Concrete*. West Conshohocken, PA: ASTM International, 1999.

ASTM A 421/A 421M-98, *Standard Specification for Uncoated Stress-Relieved Steel Wire for Prestressed Concrete*. West Conshohocken, PA: ASTM International, 1998.

ASTM A 496-01, *Standard Specification for Steel Wire, Deformed, for Concrete Reinforcement*. West Conshohocken, PA: ASTM International, 2001.

ASTM A 497-97, *Standard Specification for Steel Welded Wire Fabric, Deformed, for Concrete Reinforcement*. West Conshohocken, PA: ASTM International, 1997.

ASTM A 500-96, *Standard Specification for Cold-Formed Welded and Seamless Carbon Steel Structural Tubing in Rounds and Shapes*. West Conshohocken, PA: ASTM International, 1996.

ASTM A 501-96, *Standard Specification for Hot-Formed Welded and Seamless Carbon Steel Structural Tubing*. West Conshohocken, PA: ASTM International, 1996.

ASTM A 572/A 572M-01, *Standard Specification for High-Strength Low-Alloy Columbium-Vanadium Steels of Structural Quality*. West Conshohocken, PA: ASTM International, 2001.

ASTM A 588/A 588M-01, *Standard Specification for High-Strength Low-Alloy Structural Steel with 50 ksi [345 MPa] Minimum Yield Point to 4 inches [100 mm] Thick*. West Conshohocken, PA: ASTM International, 2001.

ASTM A 615-00, *Standard Specification for Deformed and Plain Billet-Steel Bars for Concrete Reinforcement*. West Conshohocken, PA: ASTM International, 2000.

ASTM A 706/A 706M-00, *Standard Specification for Low-Alloy Steel Deformed and Plain Bars for Concrete Reinforcement*. West Conshohocken, PA: ASTM International, 2000.

ASTM A 722/A 722M-98, *Standard Specification for Uncoated High-Strength Steel Bar for Prestressing Concrete*. West Conshohocken, PA: ASTM International, 1998.

ASTM A 767/A 767M-00b, *Standard Specification for Zinc-Coated (Galvanized) Steel Bars for Concrete Reinforcement*. West Conshohocken, PA: ASTM International, 2000.

ASTM A 775-01, *Standard Specification for Epoxy-Coated Reinforcing Steel Bars*. West Conshohocken, PA: ASTM International, 2001.

ASTM A 884-99, *Standard Specification for Epoxy-Coated Steel Wire and Welded Wire Fabric for Reinforcement*. West Conshohocken, PA: ASTM International, 1999.

ASTM A 996/A 996M-00, *Specification for Rail-Steel and Axle-Steel Deformed Bars for Concrete Reinforcement*. West Conshohocken, PA: ASTM International, 2000.

ASTM C 31/C 31M-98, *Standard Practice for Making and Curing Concrete Test Specimens in the Field*. West Conshohocken, PA: ASTM International, 1998.

ASTM C 33-99ae1, *Standard Specification for Concrete Aggregates*. West Conshohocken, PA: ASTM International, 1999.

ASTM C 35-95 (2001), *Standard Specification for Inorganic Aggregates for Use in Gypsum Plaster*. West Conshohocken, PA: ASTM International, 1995.

ASTM C 39-99ae1, *Standard Test Method for Compressive Strength of Cylindrical Concrete Specimens*. West Conshohocken, PA: ASTM International, 1999.

ASTM C 42/C 42M-99, *Standard Test Method for Obtaining and Testing Drilled Cores and Sawed Beams of Concrete*. West Conshohocken, PA: ASTM International, 1999.

ASTM C 94/C 94M-00, *Standard Specification for Ready-Mixed Concrete*. West Conshohocken, PA: ASTM International, 2000.

ASTM C 142-78(1990), *Standard Test Method for Clay Lumps and Friable Particles in Aggregates*. West Conshohocken, PA: ASTM International, 1990.

ASTM C 150-01, *Standard Specification for Portland Cement*. West Conshohocken, PA: ASTM International, 2001.

ASTM C 171-92, *Standard Specification for Sheet Materials for Curing Concrete*. West Conshohocken, PA: ASTM International, 1992.

ASTM C 172-99, *Standard Practice For Sampling Freshly Mixed Concrete*. West Conshohocken, PA: ASTM International, 1999.

ASTM C 231-97, *Standard Test Method for Air Content of Freshly Mixed Concrete by the Pressure Method*. West Conshohocken, PA: ASTM International, 1997.

ASTM C 260-95, *Standard Specification for Air-Entraining Admixtures for Concrete*. West Conshohocken, PA: ASTM International, 1995.

ASTM C 309-93, *Standard Specification for Liquid Membrane-Forming Compounds for Curing Concrete*. West Conshohocken, PA: ASTM International, 1993.

ASTM C 317-93a, *Standard Specification for Gypsum Concrete*. West Conshohocken, PA: ASTM International, 1993.

ASTM C 330-99, *Standard Specification for Lightweight Aggregates for Structural Concrete*. West Conshohocken, PA: ASTM International, 1999.

ASTM C 494-92, *Standard Specification for Chemical Admixtures for Concrete*. West Conshohocken, PA: ASTM International, 1992.

ASTM C 595-00, *Standard Specification for Blended Hydraulic Cements*. West Conshohocken, PA: ASTM International, 2000

ASTM C 618-99, *Standard Specification for Fly Ash and Raw or Calcined Natural Pozzolan for Use as a Mineral Admixture in Portland Cement Concrete*. West Conshohocken, PA: ASTM International, 1999.

ASTM C 685/C 685M-98a, *Standard Specification for Concrete Made by Volumetric Batching and Continuous Mixing*. West Conshohocken, PA: ASTM International, 1998.

ASTM C 803-90, *Standard Test Method for Penetration Resistance of Hardened Concrete*. West Conshohocken, PA: ASTM International, 1990.

ASTM C 805-94, *Standard Test Method for Rebound Number of Hardened Concrete*. West Conshohocken, PA: ASTM International, 1994.

ASTM C 845-96, *Standard Specification for Expansive Hydraulic Cement*. West Conshohocken, PA: ASTM International, 1996.

ASTM C 873-94, *Standard Test Method for Compressive Strength of Concrete Cylinders Cast in Place in Cylindrical Molds*. West Conshohocken, PA: ASTM International, 1994.

ASTM C 900-87(1993), *Standard Test Method of Pullout Strength of Hardened Concrete*. West Conshohocken, PA: ASTM International, 1993.

ASTM C 956-97, *Standard Specification for Installation of Cast-in-Place Reinforced Gypsum Concrete*. West Conshohocken, PA: ASTM International, 1997.

ASTM C 989-99, *Standard Specification for Ground Granulated Blast-Furnace Slag for Use in Concrete and Mortars*. West Conshohocken, PA: ASTM International, 1999.

ASTM C 1017-92, *Standard Specification for Chemical Admixtures for Use in Producing Flowing Concrete*. West Conshohocken, PA: ASTM International, 1992.

ASTM C 1074-93, *Standard Practice of Estimating Concrete Strength by the Maturity Method*. West Conshohocken, PA: ASTM International, 1993.

ASTM C 1077-96, *Standard Practice for Laboratories Testing Concrete and Concrete Aggregates for Use in Construction and Criteria for Laboratory Evaluation*. West Conshohocken, PA: ASTM International, 1996.

ASTM C 1157M-00, *Standard Performance Specification for Blended Hydraulic Cement*. West Conshohocken, PA: ASTM International, 2000.

ASTM C 1218/C 1218M-99, *Standard Test Method for Water-Soluble Chloride in Mortar and Concrete*. West Conshohocken, PA: ASTM International, 1999.

ASTM C 1240-00E1, *Standard Specification for Silica Fume for Use as a Mineral Admixture in Hydraulic-Cement Concrete, Mortar, and Grout*. West Conshohocken, PA: ASTM International, 2000.

ASTM D 3665-93, *Standard Practice for Random Sampling of Construction Materials*. West Conshohocken, PA: ASTM International, 1993.

ASTM E 119-00, *Standard Test Methods for Fire Tests of Building Construction Materials*. West Conshohocken, PA: ASTM International, 2000.

ATC 3-06, *Tentative Provisions for the Development of Seismic Regulations for Buildings*. Applied Technology Council. NBS Special Publication 510, NSF Publication 78-8. Washington, DC: U.S. Government Printing Office, 1978.

AWS D1.1-00, *Structural Welding Code — Steel*. Miami: American Welding Society, 2000.

AWS D1.4-98, *Structural Welding Code — Reinforcing Steel*. Miami: American Welding Society, 1998.

BOCA National Building Code/1999. Country Club Hills, Il: Building Officials and Code Administrators International, Inc., 1999.

Cook, R. A. and R. E. Klingner. "Behavior of Ductile Multiple-Anchor Steel-to-Concrete Connections with Surface-Mounted Baseplates." *Anchors in Concrete: Design and Behavior.* American Concrete Institute Special Publication SP-130, February 1992, pp. 61–122.

Cook, R. A. and R. E. Klingner. "Ductile Multiple-Anchor Steel-to-Concrete Connections." *Journal of Structural Engineering*. ASCE, Vol. 118, No. 6, June 1992, pp. 1645-1665.

Design of Fasteners in Concrete. Comite Euro-International du Beton (CEB). London: Thomas Telford Services Ltd., Jan. 1997.

DeVries, Richard A. *Anchorage of Headed Reinforcement in Concrete*. Ph.D. Dissertation. The University of Texas at Austin, December 1996.

Eligehausen, R., W. Fuchs, and B. Mayer. "Load Bearing Behavior of Anchor Fastenings in Tension." Betonwerk

+ Fertigteiltechnik, December 1987, pp. 826–832, and January 1988, pp. 29–35.

Eligehausen, R. and W. Fuchs. "Load Bearing Behavior of Anchor Fastenings under Shear, Combined Tension and Shear or Flexural Loadings." Betonwerk + Fertigteiltechnik. pp. 48-56, February 1988.

Eligehausen, R. and T. Balogh. "Behavior of Fasteners Loaded in Tension in Cracked Reinforced Concrete." *ACI Structural Journal*, Vol. 92, No. 3, May-June 1995, pp. 365-379.

Farrow, C. B. and R. E. Klingner. "Tensile Capacity of Anchors with Partial or Overlapping Failure Surfaces: Evaluation of Existing Formulas on an LRFD Basis." *ACI Structural Journal*, Vol. 92, No. 6, November-December 1995, pp. 698-710.

Fastenings to Concrete and Masonry Structures, State of the Art Report. Comite Euro-International du Beton, (CEB), Bulletin No. 216. London: Thomas Telford Services Ltd., 1994.

Fuchs, W., R. Eligehausen, and J. Breen. "Concrete Capacity Design (CCD) Approach for Fastening to Concrete." *ACI Structural Journal*, Vol. 92, No. 1, January-February 1995, pp. 73-93. Discussion—*ACI Structural Journal*, Vol. 92, No. 6, November-December 1995, pp. 787-802.

Furche, J. and R. Eligehausen. "Lateral Blow-out Failure of Headed Studs Near a Free Edge." *Anchors in Concrete: Design and Behavior*. American Concrete Institute Special Publication SP-130, February 1992, pp. 235-252.

Ghosh, S.K. "Design of Reinforced Concrete Buildings Under the 1997 UBC." *Building Standards*. International Conference of Building Officials, May-June 1998, pp. 20-24.

Ghosh, S.K. "Needed Adjustments in 1997 UBC." *Proceedings*, 1998 Convention, Structural Engineers Association of California, October 7-10, 1998, Reno-Sparks, NV, pp. T9.1-T9.15.

Ghosh, S.K., S.D. Nakaki, and K. Krishnan. "Precast Structures in Regions of High Seismicity: 1997 UBC Design Provisions." *PCI Journal*, Vol. 42, No. 6, November-December 1997, pp. 76-93.

Ishizuka, T. and N.M. Hawkins. "Effect of Bond Deterioration on the Seismic Response of Reinforced and Partially Prestressed Concrete and Ductile Moment Resistant Frames." Report SM87-2. Seattle: Department of Civil Engineering, University of Washington, 1987.

Klingner, R., J. Mendonca, and J. Malik. "Effect of Reinforcing Details on the Shear Resistance of Anchor Bolts under Reversed Cyclic Loading." *Journal of the American Concrete Institute*, Vol. 79, No. 1, 1982, pp. 3-12.

Kodur, V. and D. MacKinnon. "Design of Concrete-Filled Hollow Structural Steel Columns for Fire Endurance." *AISC Engineering Journal*.

Kuhn, D. and F. Shaikh. "Slip-Pullout Strength of Hooked Anchors." Research Report. University of Wisconsin—Milwaukee, 1996.

Lotze, D. and R.E. Klingner. "Behavior of Multiple-Anchor Attachments to Concrete from the Perspective of Plastic Theory." Report PMFSEL 96-4. Ferguson Structural Engineering Laboratory, the University of Texas at Austin, March 1997.

Lutz, L. "Discussion to Concrete Capacity Design (CCD) Approach for Fastening to Concrete." *ACI Structural Journal*, November/December 1995, pp. 791-792 and author's closure, pp. 798-799.

Manual of Standard Practice, 26th edition. Schaumburg, IL: Concrete Reinforcing Steel Institute, 1997.

Matlock, A. H., J. Yamazaki, and B. T. Kattula. "Comparative Study of Prestressed Concrete Beams, with and without Bond." *ACI Journal, Proceedings*, Vol. 68, No. 2, February 1971, pp. 116-125.

Muspratt, M. A. "Behavior of a Prestressed Concrete Waffle Slab with Unbonded Tendons." *ACI Journal, Proceedings*, Vol. 66, No. 12, December 1969, pp. 1001-1004.

NEHRP-85, *NEHRP (National Earthquake Hazards Reduction Program) Recommended Provisions for the Development of Seismic Regulations for New Buildings*. Washington, DC: Building Seismic Safety Council for the Federal Emergency Management Agency, 1985.

NEHRP-88, *NEHRP (National Earthquake Hazards Reduction Program) Recommended Provisions for the Development of Seismic Regulations for New Buildings*. Washington, DC: Building Seismic Safety Council for the Federal Emergency Management Agency, 1988.

NEHRP-91, *NEHRP (National Earthquake Hazards Reduction Program) Recommended Provisions for the Development of Seismic Regulations for New Buildings*. Washington, DC: Building Seismic Safety Council for the Federal Emergency Management Agency, 1991.

NEHRP-94, *NEHRP (National Earthquake Hazards Reduction Program) Recommended Provisions for the Development of Seismic Regulations for New Buildings*. Washington, DC: Building Seismic Safety Council for the Federal Emergency Management Agency, 1994.

NEHRP-97, *NEHRP (National Earthquake Hazards Reduction Program) Recommended Provisions for the Development of Seismic Regulations for New Buildings and Other Structures*. Washington, DC: Building Seismic Safety Council for the Federal Emergency Management Agency, 1997.

NEHRP-00, *NEHRP (National Earthquake Hazards Reduction Program) Recommended Provisions for the Development of Seismic Regulations for New Buildings and Other Structures*. Washington, DC: Building Seismic Safety Council for the Federal Emergency Management Agency, 2001.

Odello, R. J. and B. M. Mehta. “Behavior of a Continuous Prestressed Concrete Slab with Drop Panels.” Report. Division of Structural Engineering and Structural Mechanics, University of California, Berkeley, 1967.

PCA-EB001, *Design and Control of Concrete Mixtures*, 13th edition. Skokie, IL: Portland Cement Association, 1988.

PCA-EB070, *Notes on ACI 318-99 Building Code Requirements for Structural Concrete — with Design Applications*. Skokie, IL: Portland Cement Association, 1999.

PCA-EB080, *Strength Design of Anchorage to Concrete*. Skokie, IL: Portland Cement Association, 2002.

PCA-EB115, *Concrete Inspection Handbook*. Skokie, IL: Portland Cement Association, 2000.

PCA-IS521, *Strength Design Load Combinations for Concrete Elements*. Skokie, IL: Portland Cement Association, 1998.

PCA-SP040, *Fly Ash in Cement and Concrete*. Skokie, IL: Portland Cement Association, 1987.

PCI Design Handbook, 4th edition. Chicago: Precast/Prestressed Concrete Institute, 1992.

Placing Reinforcing Bars, 7th edition. Schaumburg, IL: Concrete Reinforcing Steel Institute, 1997.

Primavera, E.J ., J. P. Pinelli, and E. H. Kalajian. “Tensile Behavior of Cast-in-Place and Undercut Anchors in High-Strength Concrete.” *ACI Structural Journal*, Vol. 94, No. 5, September-October 1997, pp. 583-594.

Shaikh, A. F. and W. Yi. “In-Place Strength of Welded Studs.” *PCI Journal*, Vol. 30 (2), March-April 1985.

Standard Building Code, 1999 ed. Birmingham, AL: Southern Building Code Congress, International, 1999.

Stanton, J., W. C. Stone, and G. S. Cheok. “A Hybrid Reinforced Precast Frame for Seismic Regions.”*PCI Journal*, Vol. 42, No. 2, March-April 1997, pp. 20-32.

Uniform Building Code, 1997 ed. Whittier, CA: International Conference of Building Officials, 1997.

Wong, T. L. “Stud Groups Loaded in Shear.” M.S. Thesis, Oklahoma State University, 1988.

Zhang, Y. “Dynamic Behavior of Multiple Anchor Connections in Cracked Concrete.” Ph.D. Dissertation, the University of Texas at Austin, August 1997.

Chapter 20: Aluminum

General Comments

Chapter 20 contains standards for the use of aluminum in building construction. Only provisions for structural applications of these materials are included.

This chapter does not seek to establish standards for aluminum specialty products, such as storefront framing, architectural hardware, etc. The use of aluminum in HVAC and plumbing systems is addressed in the *International Mechanical Code*® (IMC®) and the *International Plumbing Code*® (IPC®), respectively. This chapter applies to the structural requirements.

Purpose

Aluminum has certain physical properties, structural characteristics and nonstructural characteristics that set it apart as a building material. By utilizing the standards set forth in this chapter, a proper application of this material can be obtained.

SECTION 2001
GENERAL

2001.1 Scope. This chapter shall govern the quality, design, fabrication and erection of aluminum.

❖ The scope for Chapter 20 is provided in this section.

SECTION 2002
MATERIALS

2002.1 General. Aluminum used for structural purposes in buildings and structures shall comply with AA ASM 35 and Parts 1-A and 1-B of the Aluminum Design Manual. The nominal loads shall be the minimum design loads required by Chapter 16.

❖ The referenced standards to be applied in the utilization of aluminum in buildings and other structures are established. Parts 1-A and 1-B of the Aluminum Association's (AA) *Aluminum Design Manual* apply to the design of aluminum building-type structural load-carrying members and elements according to Allowable Stress Design (ASD) and Load and Resistance Factor Design (LRFD) criteria, respectively. ASM 35 sets criteria for the use and installation of flashings, sheet roofing and similar applications.

Bibliography

The following resource materials are referenced in this chapter or are relevant to the subject matter addressed in this chapter.

AA ASM 35-80, *Specification for Aluminum Sheet Metal Work in Building Construction*. Washington, DC: The Aluminum Association, 1980.

AA ADM1-00, *Aluminum Design Manual*; "Part 1-A: Specifications for Aluminum Structures, Allowable Stress Design" and "Part 1-B: Specifications for Aluminum Structures, Load and Resistance Factor Design of Buildings and Similar Type Structures." Washington, DC: The Aluminum Association, 2000.

IMC-2003, *International Mechanical Code*. Falls Church, VA: International Code Council, 2003.

IPC-2003, *International Plumbing Code*. Falls Church, VA: International Code Council, 2003.

Chapter 21: Masonry

General Comments

Masonry construction has been used for at least 10,000 years in a variety of structures—homes, private and public buildings and historical monuments. The masonry of ancient times involved two major materials: brick manufactured from sun-dried mud or burned clay and shale; and natural stone.

The first masonry structures were unreinforced and intended to support mainly gravity loads. The weight of these structures stabilized them against lateral loads from wind and earthquakes.

Masonry construction has progressed through several stages of development. Fired clay brick became the principal building material in the United States during the middle 1800s. Concrete masonry was introduced to construction during the early 1900s and, along with clay masonry, expanded in use to all types of structures.

Historically, "rules of thumb" (now termed "empirical design") were the only available methods of masonry design. Only in recent times have masonry structures been engineered using structural calculations. In the last 45 years, the introduction of engineered reinforced masonry has resulted in structures that are stronger and more stable against lateral loads, such as wind and seismic.

Masonry consists of a variety of materials. Raw materials are made into masonry units of different sizes and shapes, each having specific physical and mechanical properties. Both the raw materials and the method of manufacture affect masonry unit properties.

The word "masonry" is a general term that applies to construction using hand-placed units of clay, concrete, structural clay tile, glass block, natural stones and the like. One or more types of masonry units are bonded together with mortar, metal ties, reinforcement and accessories to form walls and other structural elements.

Proper masonry construction depends on correct design, materials, handling, installation and workmanship.

With a fundamental understanding of the functions and properties of the materials that comprise masonry construction and with proper design and construction, quality masonry structures are not difficult to obtain.

During the pioneer era of U.S. history, the fireplace was the central focus of residential cooking and heating. Today, the fireplace is essentially a decorative feature of residential construction. For energy conservation, existing fireplaces are sometimes converted and new fireplaces are designed to provide supplemental heat.

Of the many types of fireplaces, the most common are single face. Multifaced fireplaces, such as a corner fireplace with two adjacent open sides, fireplaces with two opposite faces open (common exposure to two rooms) or fireplaces with three or all faces open also occur, but are less common.

While the provisions of this chapter are for single-faced fireplaces, almost all types of masonry fireplaces include the same basic construction features: the base assembly, which consists of a foundation and hearth support; the firebox assembly, which consists of a fireplace opening, a hearth, a firebox or combustion chamber and the throat and the smoke chamber, which supports the chimney liner.

Masonry fireplaces are made primarily of clay brick or natural stones, but also of concrete masonry or cast-in-place concrete. Chimneys for medium- and high-heat appliances require special attention for fire safety.

Purpose

Chapter 21 provides comprehensive and practical requirements for masonry construction, based on the latest state of technical knowledge. The provisions of Chapter 21 require minimum accepted practices and the use of standards for the design and construction of masonry structures and elements of structures. The provisions address: material specifications and test methods; types of wall construction; criteria for engineered design (by working stress and strength design methods); criteria for empirical design; required details of construction and other aspects of masonry, including execution of construction. The provisions are intended to result in safe and durable masonry. The provisions of Chapter 21 are also intended to prescribe minimum accepted practices for the design and construction of glass unit masonry, masonry fireplaces, masonry heaters and masonry chimneys.

SECTION 2101 GENERAL

2101.1 Scope. This chapter shall govern the materials, design, construction and quality of masonry.

❖ Section 2101 prescribes general requirements for masonry designed in accordance with Chapter 21 of the code. It identifies masonry design methods and the conditions required for the use of each method. The methods are intended as a practical means for safety under a variety of potential service conditions.

Minimum requirements for construction documents and fireplace drawings are also included in Section 2101.

Chapter 21 contains the minimum code requirements for acceptance of masonry design and construction by the building official. Compliance with these requirements is intended to result in masonry construction with the minimum required structural adequacy and durability. Requirements more stringent than these are appropriate where mandated by sound engineering and judgement. Less restrictive requirements, however, are not permitted.

2101.2 Design methods. Masonry shall comply with the provisions of one of the following design methods in this chapter as well as the requirements of Sections 2101 through 2104. Masonry designed by the working stress design provisions of Section 2101.2.1, the strength design provisions of Section 2101.2.2 or the prestressed masonry provisions of Section 2101.2.3 shall comply with Section 2105.

❖ This section requires masonry to comply with one of six design methods and the requirements contained in Sections 2101 through 2104 for construction documents, materials and construction.

The six design methods listed in Sections 2101.2.1 through 2101.2.6 can be categorized into two general design approaches for masonry. The first approach, engineered design, encompasses working stress, prestressed masonry and strength design. Use of these design methods necessitates a quality assurance program in accordance with Section 2105. The second approach, prescriptive design, includes the empirical design method, provisions for glass unit masonry and provisions for masonry veneer. Prescriptive design is permitted only under limited conditions as noted in Section 2109.1.1.

When the design professional chooses engineered design, the prescriptive masonry requirements of this chapter do not apply. For example, Section 2109 does not apply to engineered masonry.

Other provisions of the code also apply to masonry. For example, fire-resistant construction using masonry is required to comply with Chapter 7. Design loads and related requirements, including seismic forces and detailing, are required to comply with Chapter 16. Masonry foundations are required to comply with the provisions of Chapter 18. Special inspections of masonry construction are required in Chapter 17. Masonry veneer is addressed in Chapter 14.

2101.2.1 Working stress design. Masonry designed by the working stress design method shall comply with the provisions of Sections 2106 and 2107.

❖ This section requires that masonry designed by the working stress design method meets both the working stress design requirements in Section 2107 and the seismic design requirements in Section 2106. Section 2107 requires working stress design to comply with Chapters 1 and 2 of ACI 530/ASCE 5/TMS 402 with minor exceptions. Additional information on these procedures is given in the commentaries to Section 2107 and ACI 530/ASCE 5/TMS 402.

ACI 530/ASCE 5/TMS 402 and ACI 530.1/ASCE 6/TMS 602 are referenced throughout Chapter 21. A description of these standards is warranted here. Both are joint publications of the American Concrete Institute (ACI), the Structural Engineering Institute of the American Society of Civil Engineers (ASCE) and The Masonry Society (TMS) and are produced through a joint committee of those societies, called the Masonry Standards Joint Committee (MSJC). These standards are typically referred to as the MSJC Code and Specification to reflect their joint authorship and sponsorship of the committee that oversees their development. The standards are developed through an ANSI-regulated consensus process and reflect the current state of technical knowledge on masonry design and construction.

The MSJC Code (ACI 530/ASCE 5/TMS 402) contains minimum requirements for masonry elements of structures. Topics include: construction documents; quality assurance; materials; analysis and design; strength and serviceability; flexural and axial stresses; shear; reinforcement; walls; columns; pilasters; beams and lintels and empirical design.

The engineered method in ACI 530/ASCE 5/TMS 402 is a working stress design method, which assumes linearly elastic material behavior and properties and uses working loads (see Chapter 16). The strength design method is also specified in the standard.

The MSJC Specification (ACI 530.1/ASCE 6/TMS 602) sets minimum acceptable levels of construction. It includes minimum requirements for composition; preparation and placement of materials; quality assurance for materials and masonry; execution of masonry construction; inspection and verification of quality. ACI 530.1/ASCE 6/TMS 602 contains both mandatory and optional requirements. The mandatory requirements are enforceable code requirements; the optional requirements may be invoked by the design professional. The specification is meant to be modified for use with the particular project under design.

2101.2.2 Strength design. Masonry designed by the strength design method shall comply with the provisions of Sections 2106 and 2108.

❖ Masonry is required to meet the strength design provisions referenced in Section 2108 and the seismic design requirements in Section 2106.

2101.2.3 Prestressed masonry. Prestressed masonry shall be designed in accordance with Chapters 1 and 4 of ACI 530/ASCE 5/TMS 402 and Section 2106. Special inspection during construction shall be provided as set forth in Section 1704.5.

❖ Prestressed masonry must comply with the applicable chapters of the ACI referenced standard, *Building Code Requirements for Masonry Structures*. Additional requirements for prestressed masonry shear walls used to resist earthquake loads are found in Section 2106.

2101.2.4 Empirical design. Masonry designed by the empirical design method shall comply with the provisions of Sections 2106 and 2109 or Chapter 5 of ACI 530/ASCE 5/TMS 402.

❖ This section permits the empirical design of masonry either by the provisions of Sections 2106 and 2109, or Chapter 5 of ACI 530/ASCE 5/TMS 402. This is because nearly all of the requirements in Section 2109 are based on the requirements in Chapter 5 of ACI 530/ASCE 5/TMS 402. Additional information on these provisions is given in the commentaries to Section 2109 and ACI 530/ASCE 5/TMS 402.

2101.2.5 Glass masonry. Glass masonry shall comply with the provisions of Section 2110 or with the requirements of Chapter 7 of ACI 530/ASCE 5/TMS 402.

❖ Glass masonry must comply with either the provisions of Section 2110 or Chapter 7 of ACI 530/ASCE 5/TMS 402. The provisions in Section 2110 are based on the requirements in Chapter 7 of ACI 530/ASCE 5/TMS 402. Additional information on these provisions is given in the commentaries to Section 2109 and ACI 530/ASCE 5/TMS 402.

2101.2.6 Masonry veneer. Masonry veneer shall comply with the provisions of Chapter 14.

❖ This section requires masonry veneer to comply with the provisions of Chapter 14; specifically, Sections 1405.5 for anchored masonry veneer and 1405.9 for adhered masonry veneer. These sections reference the provisions in Chapter 6 of ACI 530/ASCE 5/TMS 402. Additional information on these provisions is given in the commentaries to Chapter 14 and ACI 530/ASCE 5/TMS 402.

2101.3 Construction documents. The construction documents shall show all of the items required by this code including the following:

1. Specified size, grade, type and location of reinforcement, anchors and wall ties.
2. Reinforcing bars to be welded and welding procedure.
3. Size and location of structural elements.
4. Provisions for dimensional changes resulting from elastic deformation, creep, shrinkage, temperature and moisture.

❖ Construction requirements must be clearly identified in the contract documents so that the structure is properly constructed using appropriate materials and methods. This section requires that, as a minimum, critical items required by the code and by the particular design be shown in the construction documents. The list is a minimum and should not be considered all-inclusive by the design professional. Both the design professional and the building official are permitted to require additional items as needed for a particular structure.

2101.3.1 Fireplace drawings. The construction documents shall describe in sufficient detail the location, size and construction of masonry fireplaces. The thickness and characteristics of materials and the clearances from walls, partitions and ceilings shall be clearly indicated.

❖ This section requires the submission of construction documents for all fireplaces so that compliance with appropriate code sections can be properly determined during plan review. The type of information and its format for plan review are established in this section. Construction documents are required showing relationships of components, as well as details related to the specific characteristics of the materials and techniques to be used when erecting the fireplace and chimney system. Such details are to include the type of brick or stone; refractory brick; concrete masonry; mortar requirements; wall thicknesses; clearances; dimensions of openings and dimensions of the firebox and the hearth extension.

SECTION 2102
DEFINITIONS AND NOTATIONS

2102.1 General. The following words and terms shall, for the purposes of this chapter and as used elsewhere in this code, have the meanings shown herein.

❖ This section contains definitions of terms associated with the subject matter of this chapter. Definitions of terms can help in the understanding and application of code requirements. Definitions are included within this chapter to provide convenient access to them without having to refer back to Chapter 2.

ADOBE CONSTRUCTION. Construction in which the exterior load-bearing and nonload-bearing walls and partitions are of unfired clay masonry units, and floors, roofs and interior framing are wholly or partly of wood or other approved materials.

Adobe, stabilized. Unfired clay masonry units to which admixtures, such as emulsified asphalt, are added during the manufacturing process to limit the units' water absorption so as to increase their durability.

Adobe, unstabilized. Unfired clay masonry units that do not meet the definition of "Adobe, stabilized."

❖ Adobe masonry was popular in the southwest United States due to the availability of soil for units, the frequent exposure to intense sunlight to dry the units, the thermal mass provided by the completed adobe structure and the low cost of this form of construction. This form of construction has relatively low strength, a lack of formalized design procedures and labor-intensive manufacture of units and construction of the building, and accordingly, has not been used as much in recent years.

The two types of adobe masonry, stabilized and unstabilized, are briefly described below. Prescriptive design requirements for adobe masonry are contained in Section 2109.8.

Adobe, stabilized. Admixtures are added to the soil to produce more durable units.

Adobe, unstabilized. Unstabilized adobe does not contain stabilizers in the soil and is, therefore, not as durable as stabilized adobe.

ANCHOR. Metal rod, wire or strap that secures masonry to its structural support.

❖ Anchors are fasteners connecting two components. Figure 2102.1(1) shows examples of anchor bolts that can be used in masonry and Figure 2102.1(3) shows some uses of anchor bolts to connect wood floors and masonry walls.

In this chapter, anchors are required where masonry walls meet intersecting walls, floors, roofs or the foundation below. Requirements for strength and durability of metal anchors are given in Section 2103.11.

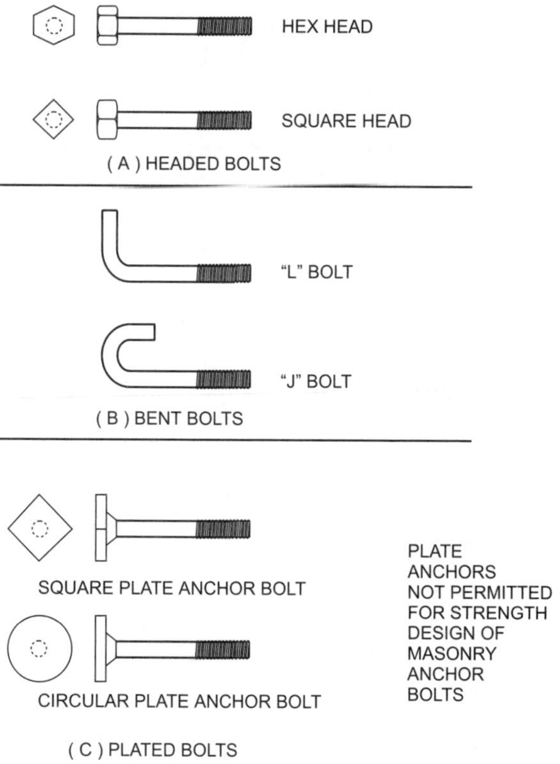

Figure 2102.1(1)
ANCHOR BOLTS

ARCHITECTURAL TERRA COTTA. Plain or ornamental hard-burned modified clay units, larger in size than brick, with glazed or unglazed ceramic finish.

❖ Architectural terra cotta refers to fired clay units with architectural shape and fired glazed coating. While rarely used today in new construction, repairs to terra cotta in existing building construction are not uncommon. These clay masonry units are usually produced for custom-made, anchored, ornamental veneers.

AREA.

Bedded. The area of the surface of a masonry unit that is in contact with mortar in the plane of the joint.

Gross cross-sectional. The area delineated by the out-to-out specified dimensions of masonry in the plane under consideration.

Net cross-sectional. The area of masonry units, grout and mortar crossed by the plane under consideration based on out-to-out specified dimensions.

❖ **Area.** Different areas are used in different calculations throughout this chapter. It is important to use the correct area, as each different area may give dramatically different results that may not be appropriate.

Bedded. The bedded area is simply the area of the unit's surface on which mortar is placed and through which stresses are transferred to the adjacent work.

Gross cross-sectional. The gross cross-sectional area of the masonry is the specified masonry width (thickness) multiplied by the specified length, as illustrated in Figure 2102.1(2). While subtraction of core areas of the masonry unit is not required, subtraction of the space between wythes is required in noncomposite walls. Empirical compressive stress design is based on the gross cross-sectional area of the masonry.

Net cross-sectional. The net cross-sectional area encompasses the area of units, grout and mortar contained within the plane under consideration. For ungrouted masonry, this area is sometimes equal to the bedded area, or more often to the minimum specified area of the face shells. For grouted masonry, this also includes that area of cores, cells or spaces filled with grout.

BED JOINT. The horizontal layer of mortar on which a masonry unit is laid.

❖ This is a horizontal mortar joint [see Figure 2102.1(4)] that separates a course of masonry units from the ones above and supports the weight of the masonry. Unlike the head or collar joint, it is easily closed when the masonry unit is placed. For a masonry unit in the typical (stretcher) orientation, the bed joint faces are the top and bottom, while the bed surface of the masonry unit is the underside. A special type of bed joint is the base-course joint or starting joint placed over foundations. See Section 2104.1.2 for requirements for thicknesses, placement and permitted tolerances for bed joints.

BOND BEAM. A horizontal grouted element within masonry in which reinforcement is embedded.

❖ Bond beams permit horizontal reinforcement to be placed in masonry. For hollow masonry unit walls, special units can either be manufactured or the webs and end shells can be reduced by saw cutting to allow horizontal reinforcing bars to be placed in the wall.

BOND REINFORCING. The adhesion between steel reinforcement and mortar or grout.

❖ This term describes the adhesion between reinforcing steel and grout or mortar that transfers stresses between those elements.

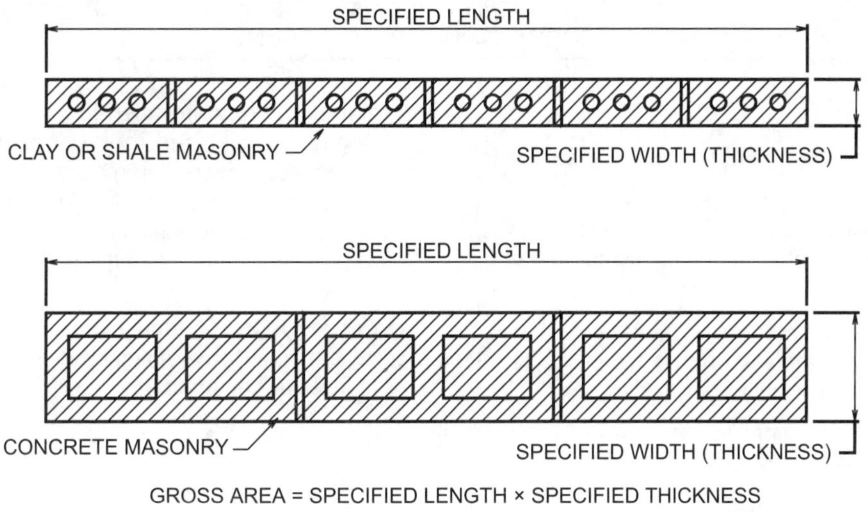

Figure 2102.1(2)
MASONRY/FLOOR ANCHOR BOLT CONNECTOR

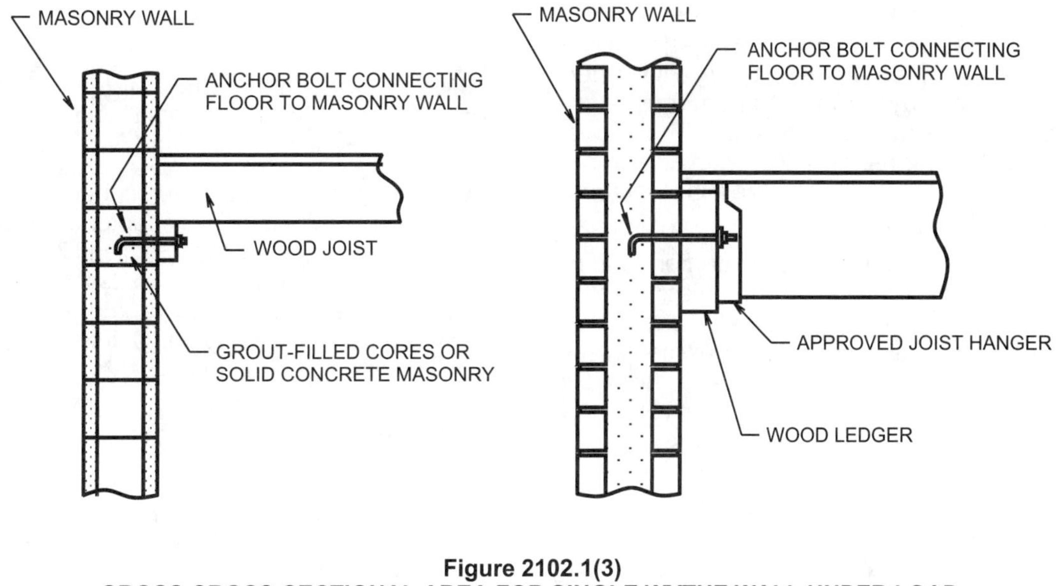

Figure 2102.1(3)
GROSS CROSS-SECTIONAL AREA FOR SINGLE WYTHE WALL UNDER LOAD

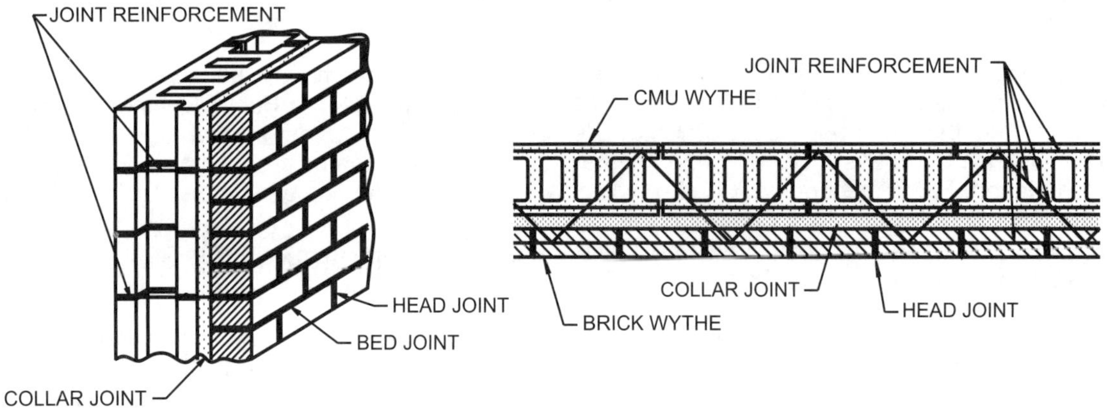

Figure 2102.1(4)
MASONRY CONSTRUCTION TERMS

BRICK.

Calcium silicate (sand lime brick). A masonry unit made of sand and lime.

Clay or shale. A masonry unit made of clay or shale, usually formed into a rectangular prism while in the plastic state and burned or fired in a kiln.

Concrete. A masonry unit having the approximate shape of a rectangular prism and composed of inert aggregate particles embedded in a hardened cementitious matrix.

❖ Brick is composed of masonry units that are generally prismatic (rectangular) in shape.

Calcium silicate brick (sand lime brick). This solid brick unit is made principally from high-silica sand and lime.

Clay or shale. These masonry units are manufactured from surface clay, shale or fire clay. Different manufacturing processes and physical properties are associated with each material. Surface clays are found in sedimentary layers near the surface. Shales are clays subjected to geologic pressure, resulting in a solid state similar to slate. Fire clays are mined from deeper layers, resulting in more uniform properties. These units are formed into the desired shape by extrusion, molding or pressing. They are then fired in a kiln to increase their strength and durability.

Concrete. Concrete brick units are made from a zero-slump mix of portland cement and possibly other cementitious materials, aggregates, water and admixtures. These units are solid or have a shallow depression called a frog. Slump brick, for example, is a decorative concrete brick with bulged sides resulting from the consistency of the mix and the manufacturing process.

BUTTRESS. A projecting part of a masonry wall built integrally therewith to provide lateral stability.

❖ These elements serve one or more purposes and are sometimes called "pilasters." As secondary structural members, they are used to limit the unbraced horizontal length of walls (and their corresponding length-to-thickness ratios) by providing horizontally spaced points of lateral support. As lateral-load-resisting beam elements, or as vertical-load-resisting beam columns built integrally with the wall, they are primary structural members.

CAST STONE. A building stone manufactured from portland cement concrete precast and used as a trim, veneer or facing on or in buildings or structures.

❖ Cast stone is a simulated stone precast from portland cement concrete. This material is typically used for veneer, but can also be used in other applications.

CELL. A void space having a gross cross-sectional area greater than $1^1/_2$ square inches (967 mm^2).

❖ This term defines a large intentional void within a masonry unit. Grout and reinforcing steel are often placed in cells to form reinforced masonry.

CHIMNEY. A primarily vertical enclosure containing one or more passageways for conveying flue gases to the outside atmosphere.

❖ A chimney is a primarily vertical enclosure containing one or more flues. This chapter regulates masonry chimneys and fireplaces in Sections 2111 through 2113. Chimneys differ from metal vents in the materials from which they are constructed and the type of appliance they are designed to serve. Chimneys can vent much hotter flue gases than metal vents.

CHIMNEY TYPES.

High-heat appliance type. An approved chimney for removing the products of combustion from fuel-burning, high-heat appliances producing combustion gases in excess of 2,000°F (1093°C) measured at the appliance flue outlet (see Section 2113.11.3).

Low-heat appliance type. An approved chimney for removing the products of combustion from fuel-burning, low-heat appliances producing combustion gases not in excess of 1,000°F (538°C) under normal operating conditions, but capable of producing combustion gases of 1,400°F (760°C) during intermittent forces firing for periods up to 1 hour. Temperatures shall be measured at the appliance flue outlet.

Masonry type. A field-constructed chimney of solid masonry units or stones.

Medium-heat appliance type. An approved chimney for removing the products of combustion from fuel-burning, medium-heat appliances producing combustion gases not exceeding 2,000°F (1093°C) measured at the appliance flue outlet (see Section 2113.11.2).

❖ Provisions for several types of chimneys are contained in Chapter 21, as described below.

High-heat appliance type. High-heat chimneys are used in industrial applications, such as incinerators, kilns and blast furnaces. Section 2113.11.3 contains requirements for the construction and installation of chimneys for high-heat appliances.

Low-heat appliance type. Most domestic fuel-burning appliances are low-heat appliances. Low-heat appliances include solid-fuel-burning appliances, such as room heaters and wood stoves. Section 2113 contains requirements for the construction and installation of chimneys for low-heat appliances.

Masonry type. Masonry chimneys can have one or more flues and are field constructed of masonry units, stone, concrete and fired-clay materials. Masonry chimneys can stand alone or be part of a masonry fireplace. Section 2113 contains requirements for the construction and installation of masonry chimneys.

Most masonry chimneys require a chimney liner, resistant to heat and the corrosive action of the products of combustion. Chimney liners are generally made of fired-clay tile, refractory brick, poured-in-place refractory materials or stainless steel.

Medium-heat appliance type. Some examples of medium-heat appliances are annealing furnaces, galvanizing furnaces, pulp dryers and charcoal furnaces. Section 2113.11.2 contains requirements for the construction and installation of chimneys for medium-heat appliances.

CLEANOUT. An opening to the bottom of a grout space of sufficient size and spacing to allow the removal of debris.

❖ These openings allow debris to be removed from a space to be grouted. The code references ACI 530.1/ASCE 6/TMS 602 for minimum construction requirements for masonry, including the minimum size and maximum spacing of cleanouts for grouted masonry.

COLLAR JOINT. Vertical longitudinal joint between wythes of masonry or between masonry and backup construction that is permitted to be filled with mortar or grout.

❖ A collar joint is a filled space between masonry wythes [see Figure 2102.1(4)]. Care is necessary for proper construction of collar joints, especially where solid filling is required.

COLUMN, MASONRY. An isolated vertical member whose horizontal dimension measured at right angles to its thickness does not exceed three times its thickness and whose height is at least four times its thickness.

❖ Masonry columns typically resist moment and axial compression and sometimes axial tension from uplift. Masonry elements falling within the dimensional limits for columns must be designed and detailed accordingly, with minimum column ties and minimum vertical reinforcement.

The requirements for masonry columns vary. Some members meeting the geometric requirements for columns do not have significant structural demands placed on them. In light of this, Section 2107.2.2 exempts some column-type elements from these column detailing requirements.

COMPOSITE ACTION. Transfer of stress between components of a member designed so that in resisting loads, the combined components act together as a single member.

❖ This definition is needed in order to fully explain what is meant by composite masonry.

COMPOSITE MASONRY. Multiwythe masonry members acting with composite action.

❖ When masonry wythes are connected or bonded together so that stresses can be transferred adequately between them, the masonry is considered as composite masonry. Section 2.1.3.2 of ACI 530/ASCE 5/TMS 402 provides minimum bonding requirements for masonry to be considered as composite.

COMPRESSIVE STRENGTH OF MASONRY. Maximum compressive force resisted per unit of net cross-sectional area of masonry, determined by the testing of masonry prisms or a function of individual masonry units, mortar and grout.

❖ The specified compressive strength of masonry, f'_m, is used for the engineered design of masonry (working stress design of Section 2107 and strength design of Section 2108). The average compressive strength of masonry, determined by the prism test method or the unit strength method (Section 2105.2), must equal or exceed the specified compressive strength.

CONNECTOR. A mechanical device for securing two or more pieces, parts or members together, including anchors, wall ties and fasteners.

❖ A few types of steel connectors are illustrated in Figures 2102.1(2) and 2102.1(4). Masonry connectors attach intersecting components and also act as bonding elements. The specified size, grade, type and location of connectors are required on construction documents, in accordance with Section 2101.3.

COVER. Distance between surface of reinforcing bar and edge of member.

❖ An adequate thickness of masonry materials is needed between reinforcing steel and the surface of the masonry for two important reasons. The first reason is for proper transfer of stresses between the reinforcing steel and the masonry. This cover, often referred to as the "structural cover," is noted by the term *K* in Equation 21-2. The second reason is so that the reinforcement is protected from corrosion or degradation. Accordingly, larger cover is required in the MSJC standards for masonry with more severe exposure.

DIAPHRAGM. A roof or floor system designed to transmit lateral forces to shear walls or other lateral-load-resisting elements.

❖ A diaphragm is a planar horizontal structural element (for example, a floor or roof) designed to transmit horizontal forces to vertical resisting elements (for example, shear walls or frames). Diaphragms are essential elements in the lateral-load-resisting system of a structure. Diaphragms are considered as rigid or flexible in their own planes. Flexible diaphragms deflect more than rigid diaphragms under imposed loads.

DIMENSIONS.

Actual. The measured dimension of a masonry unit or element.

Nominal. A dimension equal to a specified dimension plus an allowance for the joints with which the units are to be laid. Thickness is given first, followed by height and then length.

Specified. The dimensions specified for the manufacture or construction of masonry, masonry units, joints or any other component of a structure.

❖ Different dimensions are used to designate sizes of masonry units and masonry elements. The terms below denote the common meanings of various types of dimensions used in the chapter.

Actual. The actual dimensions of a masonry unit are its measured dimensions. Actual dimensions should equal the specified dimensions within the construction or manufacturing tolerances.

Nominal. The nominal dimensions of a masonry unit are the specified dimensions, plus the specified thickness of one mortar joint. Nominal dimensions are used for architectural layout of masonry structures. Figure 2102.1(5) shows nominal dimensions for a specific concrete masonry unit.

Specified. The specified dimensions are prescribed in the construction documents. Actual dimensions should equal the specified dimensions, within the construction or manufacturing tolerances. Figure 2102.1(5) shows specified and nominal dimensions for a concrete masonry unit.

EFFECTIVE HEIGHT. For braced members, the effective height is the clear height between lateral supports and is used for calculating the slenderness ratio. The effective height for unbraced members is calculated in accordance with engineering mechanics.

❖ Effective height is a theoretical distance used to predict the buckling load (compressive capacity as governed by

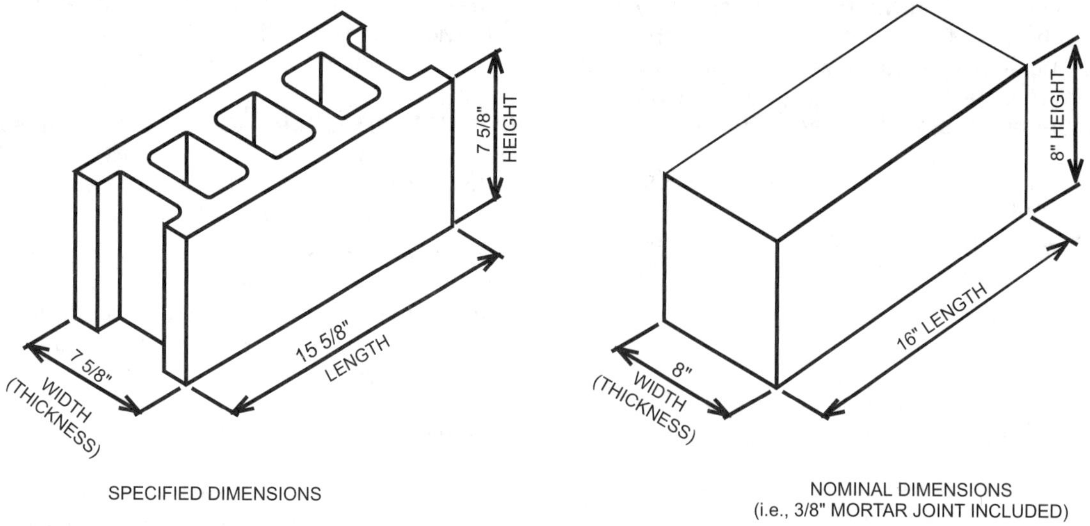

For SI: 1 inch = 25.4 mm.

Figure 2102.1(5)
SPECIFIED AND NOMINAL DIMENSIONS FOR NOMINAL 8 × 8 × 16 CONCRETE MASONRY UNITS

stability) of a wall or column. For a braced condition (no sidesway), effective height is conservatively required to be assumed as the clear height between points of lateral support. For an unbraced condition (sidesway permitted), the effective height is greater than the clear height and must be calculated.

FIREPLACE. A hearth and fire chamber or similar prepared place in which a fire may be made and which is built in conjunction with a chimney.

❖ Requirements for masonry fireplaces are contained in Section 2111.

FIREPLACE THROAT. The opening between the top of the firebox and the smoke chamber.

❖ This definition is necessary for proper understanding of the code criteria for location and minimum cross-sectional area. This criterion is based on many years of successful performance and is needed to provide proper construction requirements (see Section 2111.7).

GROUTED MASONRY.

Grouted hollow-unit masonry. That form of grouted masonry construction in which certain designated cells of hollow units are continuously filled with grout.

Grouted multiwythe masonry. That form of grouted masonry construction in which the space between the wythes is solidly or periodically filled with grout.

❖ Masonry with grout, either in the cells of hollow units or in the collar joint, is considered grouted masonry. Grouted masonry has a greater surface area to resist loads and a better transfer of stresses to reinforcing steel.

Grouted hollow-unit masonry. Hollow masonry units are often reinforced and grouted to provide stronger elements. Table 7 of ACI 530.1/ASCE 6/TMS 602 provides requirements on fine or coarse grout based on the dimensions of the cell to be grouted.

Grouted multiwythe masonry. The space between wythes of multiwythe masonry can be grouted to provide stronger elements. Table 7 of ACI 530.1/ASCE 6/TMS 602 provides requirements on fine or coarse grout, based on the dimensions of the space to be grouted.

HEAD JOINT. Vertical mortar joint placed between masonry units within the wythe at the time the masonry units are laid.

❖ Vertically oriented joints between masonry units are head joints [see Figure 2102.1(4)].

HEADER (Bonder). A masonry unit that connects two or more adjacent wythes of masonry.

❖ Masonry bond between adjacent masonry wythes, and masonry bond anchorage between intersecting masonry walls, are occasionally accomplished with connecting units called "headers" or "bonders." The units may be visible on the outside of either wythe, or may not be visible on one or more wythes. If not visible, they are referred to as "blind headers." Headers must have a minimum embedment in each wythe.

HEIGHT, WALLS. The vertical distance from the foundation wall or other immediate support of such wall to the top of the wall.

❖ This term means the actual height, measured from the bottom to the top of the wall, for free-standing cantilever walls, or as the vertical distance between points of lateral support for walls spanning between floor or roof levels.

MASONRY. A built-up construction or combination of building units or materials of clay, shale, concrete, glass, gypsum, stone or other approved units bonded together with or without mortar or grout or other accepted method of joining.

Ashlar masonry. Masonry composed of various sized rectangular units having sawed, dressed or squared bed surfaces, properly bonded and laid in mortar.

Coursed ashlar. Ashlar masonry laid in courses of stone of equal height for each course, although different courses shall be permitted to be of varying height.

Glass unit masonry. Nonload-bearing masonry composed of glass units bonded by mortar.

Plain masonry. Masonry in which the tensile resistance of the masonry is taken into consideration and the effects of stresses in reinforcement are neglected.

Random ashlar. Ashlar masonry laid in courses of stone set without continuous joints and laid up without drawn patterns. When composed of material cut into modular heights, discontinuous but aligned horizontal joints are discernible.

Reinforced masonry. Masonry construction in which reinforcement acting in conjunction with the masonry is used to resist forces.

Solid masonry. Masonry consisting of solid masonry units laid contiguously with the joints between the units filled with mortar.

❖ The materials (other than gypsum) and elements constructed as stated in this definition are considered masonry construction and are regulated by Chapter 21. This term identifies the building elements of plain (unreinforced) masonry, reinforced masonry, grouted masonry, glass unit masonry and masonry veneer.

Ashlar masonry. Units for ashlar masonry construction are rectangular in shape but variable in size.

Coursed ashlar. In coursed ashlar masonry, all units in one course are the same height, although different courses may have different heights.

Glass unit masonry. Glass unit masonry is required to be designed in accordance with Sections 2101.2.4 and 2110.

Plain masonry. Plain masonry has historically been referred to as "unreinforced masonry." Since such masonry may actually contain some reinforcement, however, the term "unreinforced" has fallen out of favor. When reinforcement is contained in plain masonry, its contribution to the strength of the system is required to be ignored. The bond between the masonry units and mortar is critical in the performance of plain masonry.

Random ashlar. Random ashlar has discontinuous bed joints because units have different height.

Reinforced masonry. Reinforced masonry contains reinforcement (currently limited to steel reinforcement) and is designed considering the tensile strength of that reinforcement. Not all masonry containing reinforcement is considered reinforced masonry. Some plain masonry contains reinforcement (usually to reduce the size of any cracks that may form), but the contribution of that reinforcement is required to be neglected.

Solid masonry. This term describes single- or multi-wythe walls composed of solid masonry units, including the thickness of the collar joint if it is filled with mortar or grout.

MASONRY UNIT. Brick, tile, stone, glass block or concrete block conforming to the requirements specified in Section 2103.

Clay. A building unit larger in size than a brick, composed of burned clay, shale, fired clay or mixtures thereof.

Concrete. A building unit or block larger in size than 12 inches by 4 inches by 4 inches (305 mm by 102 mm by 102 mm) made of cement and suitable aggregates.

Hollow. A masonry unit whose net cross-sectional area in any plane parallel to the load-bearing surface is less than 75 percent of its gross cross-sectional area measured in the same plane.

Solid. A masonry unit whose net cross-sectional area in every plane parallel to the load-bearing surface is 75 percent or more of its gross cross-sectional area measured in the same plane.

❖ Masonry units are natural stone units or manufactured units of fired clay, shale, cementitious materials or glass.

Clay. Clay masonry units are manufactured from fired clay or shale (also see the definition of "Brick, clay or shale").

Concrete. Concrete masonry units are manufactured from a zero-slump mixture of portland cement (and possibly other cementitious materials), aggregates, water and sometimes admixtures.

Hollow. Hollow masonry units are those having a specified net cross-sectional area less than 75 percent of their corresponding gross cross-sectional area. Where the specified net cross-sectional area is equal to or greater than 75 percent of the gross cross-sectional area, the unit is considered to be solid.

Solid. Solid masonry units have a specified net cross-sectional area 75 percent or greater of their corresponding gross cross-sectional area. Where the specified net cross-sectional area is less than 75 percent of the gross cross-sectional area, the unit is considered to be hollow.

MEAN DAILY TEMPERATURE. The average daily temperature of temperature extremes predicted by a local weather bureau for the next 24 hours.

❖ This is the predicted average daily temperature to confirm when cold-weather and hot-weather construction techniques are required to be followed.

MORTAR. A plastic mixture of approved cementitious materials, fine aggregates and water used to bond masonry or other structural units.

❖ Mortar is the material that bonds units and accessories together and compensates for dimensional variations of the units. Both the plastic and hardened properties of mortar are important for strong, durable, water-tight construction. Material requirements and referenced standards for several permitted mortar types are given in Section 2103.7.

MORTAR, SURFACE-BONDING. A mixture to bond concrete masonry units that contains hydraulic cement, glass fiber reinforcement with or without inorganic fillers or organic modifiers and water.

❖ This mortar is a packaged, dry, combined material permitted for use in the surface bonding of concrete masonry units that have not been prefaced, coated or painted. Masonry units are stacked without mortar joints and surface-bonding mortar is then applied to both sides of the wall surface, creating a structural element.

PLASTIC HINGE. The zone in a structural member in which the yield moment is anticipated to be exceeded under loading combinations that include earthquakes.

❖ The portion of a member where the yield moment is expected to be exceeded under seismic loads is considered a plastic hinge zone. Location and detailing of plastic hinge zones are a critical part of strength design for seismic loads.

PRESTRESSED MASONRY. Masonry in which internal stresses have been introduced to counteract potential tensile stresses in masonry resulting from applied loads.

❖ This definition provides an understanding of the term "prestressed masonry," which is used to define particular types of shear wall systems recognized under the code.

PRISM. An assemblage of masonry units and mortar with or without grout used as a test specimen for determining properties of the masonry.

❖ Compliance with the specified compressive strength of masonry, f'_m, can be verified by prism tests. The prism configuration and construction methods for such tests are prescribed in ASTM C 1314.

RUBBLE MASONRY. Masonry composed of roughly shaped stones.

Coursed rubble. Masonry composed of roughly shaped stones fitting approximately on level beds and well bonded.

Random rubble. Masonry composed of roughly shaped stones laid without regularity of coursing but well bonded and fitted together to form well-divided joints.

Rough or ordinary rubble. Masonry composed of unsquared field stones laid without regularity of coursing but well bonded.

❖ Rubble consists of pieces of stone that are irregular in shape and size. Rubble is often laid to form walls, foundations and paving.

Coursed rubble. Coursed rubble consists of roughly shaped stones that are laid with continuous bed joints.

Random rubble. Random rubble has approximately level beds, but discontinuous bed joints because of the varying heights of the individual units.

Rough or ordinary rubble. This type of rubble is laid without regular coursing.

RUNNING BOND. The placement of masonry units such that head joints in successive courses are horizontally offset at least one-quarter the unit length.

❖ Figure 2102.1(6) illustrates the required overlap for running bonds. The minimum overlap is necessary to provide strength between units when masonry spans horizontally. Reinforcement or reinforced bond beams are required in Section 2109.6.5.2 for masonry in other than running bond. Masonry not laid in running bond is often referred to as "stackbonded."

SHEAR WALL.

Detailed plain masonry shear wall. A masonry shear wall designed to resist lateral forces neglecting stresses in reinforcement, and designed in accordance with Section 2106.1.1.

Intermediate prestressed masonry shear wall. A prestressed masonry shear wall designed to resist lateral forces considering stresses in reinforcement, and designed in accordance with Section 2106.1.1.2.

Intermediate reinforced masonry shear wall. A masonry shear wall designed to resist lateral forces considering stresses in reinforcement, and designed in accordance with Section 2106.1.1.

Ordinary plain masonry shear wall. A masonry shear wall designed to resist lateral forces neglecting stresses in reinforcement, and designed in accordance with Section 2106.1.1.

Ordinary plain prestressed masonry shear wall. A prestressed masonry shear wall designed to resist lateral forces considering stresses in reinforcement, and designed in accordance with Section 2106.1.1.1.

Ordinary reinforced masonry shear wall. A masonry shear wall designed to resist lateral forces considering stresses in reinforcement, and designed in accordance with Section 2106.1.1.

Special prestressed masonry shear wall. A prestressed masonry shear wall designed to resist lateral forces considering stresses in reinforcement and designed in accordance with

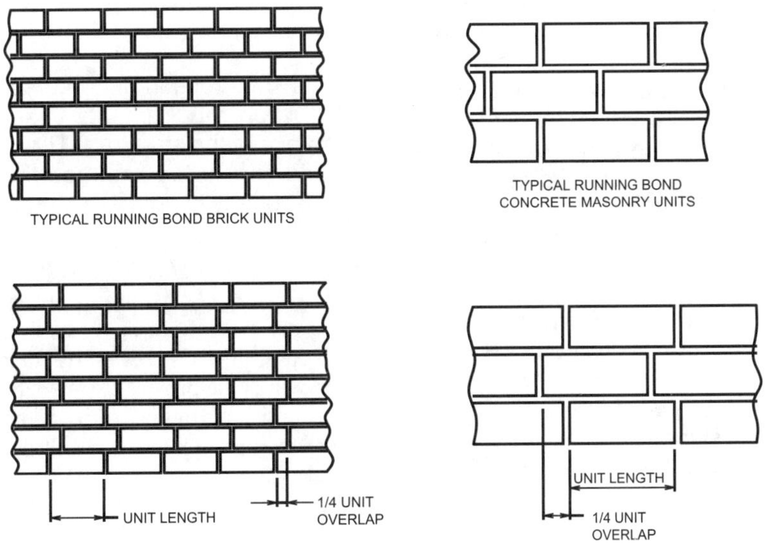

MASONRY THAT IS NOT OVERLAPPED A MINIMUM OF ONE-QUARTER OF THE UNIT LENGTH IS CONSIDERED TO BE LAID IN STACK BOND.

Figure 2102.1(6)
RUNNING BOND MASONRY

Section 2106.1.1.3 except that only grouted, laterally restrained tendons are used.

Special reinforced masonry shear wall. A masonry shear wall designed to resist lateral forces considering stresses in reinforcement, and designed in accordance with Section 2106.1.1.

❖ Shear walls are vertical diaphragms resisting vertical and in-plane lateral loads. They are part of the lateral-force-resisting system and basic seismic-force-resisting system. The various types of shear wall systems defined are necessary to correctly characterize their expected performance level in resisting earthquake forces (also see the definition of "Basic seismic-force-resisting system" in Chapter 16). Shear walls must be adequately connected to floor and roof diaphragms and to the foundation, so that loads can be transferred effectively between these elements.

Detailed plain masonry shear wall. Such shear walls are designed as plain (unreinforced) masonry, but contain a minimum amount of reinforcement in the horizontal and vertical directions. Because of this reinforcement, these walls have more favorable seismic design parameters, including higher response modification factors, *R*, than ordinary plain masonry shear walls.

Intermediate prestressed masonry shear wall. This definition distinguishes intermediate prestressed masonry shear walls from the other types of prestressed masonry shear walls that are recognized by the code in order to classify the lateral-force-resisting system in determining earthquake loads.

Intermediate reinforced masonry shear wall. These shear walls are designed as reinforced masonry and also must contain a minimum amount of prescriptive reinforcement. Because they contain reinforcement, their seismic performance will be better than that of plain masonry shear walls in seismic events and they are accordingly permitted in areas of moderate as well as low seismic risk. These walls have more favorable seismic design parameters, including higher response modification factors, *R*, than plain masonry shear walls and ordinary reinforced masonry shear walls.

Ordinary plain masonry shear wall. Such shear walls meet only minimum requirements, without minimum amounts of horizontal and vertical reinforcement. Thus, they may be used only in areas of low seismic risk. Plain masonry walls are designed as unreinforced masonry (by the noted sections), although they may in fact contain reinforcement.

Ordinary plain prestressed shear wall. This definition distinguishes ordinary plain prestressed masonry shear walls from the other types of prestressed masonry shear walls that are recognized by the code in order to classify the lateral-force-resisting system in determining earthquake loads.

Ordinary reinforced masonry shear wall. These shear walls are designed as reinforced masonry. Because they contain reinforcement, their seismic performance is expected to be better than that of plain masonry shear walls and they are accordingly permitted in areas of moderate as well as low seismic risk. These walls have more favorable seismic design parameters, including higher response modification factors, *R*, than plain masonry shear walls. When used in areas of moderate seismic risk (Seismic Design Category C), however, minimum reinforcement is required as noted in Section 2106.4.

Special prestressed masonry shear wall. This definition distinguishes special prestressed masonry shear walls from the other types of prestressed masonry shear walls that are recognized by the code in order to classify the lateral-force-resisting system in determining earthquake loads.

Special reinforced masonry shear wall. These shear walls are designed as reinforced masonry and must also meet prescriptive reinforcement limits and material limitations. Because of these requirements, they are permitted to be used in all seismic risk areas. These walls have the most favorable seismic design parameters, including the highest response modification factors, *R*, of any of the masonry shear wall types.

SHELL. The outer portion of a hollow masonry unit as placed in masonry.

❖ The shells of a masonry unit are defined by how the unit is used in construction. They are the portions of a hollow masonry unit exposed on the faces of elements and may include face shells and end webs.

SPECIFIED. Required by construction documents.

❖ The construction documents contain material and construction requirements essential to the proper performance of the structure. These requirements are considered minimums and material or construction that does not comply is not permitted by the code.

SPECIFIED COMPRESSIVE STRENGTH OF MASONRY, f'_m. Minimum compressive strength, expressed as force per unit of net cross-sectional area, required of the masonry used in construction by the construction documents, and upon which the project design is based. Whenever the quantity f'_m is under the radical sign, the square root of numerical value only is intended and the result has units of pounds per square inch (psi) (Mpa).

❖ Engineered design of structural masonry is based on the specified compressive strength of the masonry, f'_m.

This strength is required to be shown on the contract documents. Strength of the constructed masonry, determined by the unit strength method or prism strength method, is required to equal or exceed the specified compressive strength of masonry.

STACK BOND. The placement of masonry units in a bond pattern is such that head joints in successive courses are vertically aligned. For the purpose of this code, requirements for stack bond shall apply to masonry laid in other than running bond.

❖ Figure 2102.1(6) illustrates the required overlap for running bonds. If this required overlap is not provided, the wall is considered to be laid in stack bond. Reinforcement, or reinforced bond beams, are required by Section 2109.6.5.2 for stack bond masonry.

STONE MASONRY. Masonry composed of field, quarried or cast stone units bonded by mortar.

Ashlar stone masonry. Stone masonry composed of rectangular units having sawed, dressed or squared bed surfaces and bonded by mortar.

Rubble stone masonry. Stone masonry composed of irregular-shaped units bonded by mortar.

❖ Stone masonry is comprised of natural marble, limestone, granite, sandstone and slate for building purposes. Ashlar stone is further distinguished as coursed or random. Rubble stone masonry is further distinguished as coursed, random or rough.

Ashlar stone masonry. Units for ashlar masonry construction are rectangular in shape but variable in size.

Rubble stone masonry. Unlike ashlar stone masonry, which has rectangular units, rubble stone masonry units are irregular in shape. Rubble stone masonry is further distinguished as coursed, random or rough.

STRENGTH.

Design strength. Nominal strength multiplied by a strength reduction factor.

Nominal strength. Strength of a member or cross section calculated in accordance with these provisions before application of any strength-reduction factors.

Required strength. Strength of a member or cross section required to resist factored loads.

❖ The term "strength" is used in both general and specific senses. In the general sense, the strength of a member is its capacity to resist internal forces and moments. In the specific sense, strength is further categorized by type. The tensile strength of a member, for instance, refers to how much tensile force the member can support.

In the context of strength design, the force resulting from factored design actions is referred to as the required strength. An approximation to the "minimum expected" strength of the member is referred to as the "nominal strength" (see the commentary on nominal strength). This nominal strength is then multiplied by a strength reduction factor to account for material, design and construction variabilities to determine the design strength. The design strength must equal or exceed the required strength.

In the context of working stress design, the force resulting from unfactored design loads is referred to as the "applied (actual) force." The anticipated strength would then be reduced by appropriate safety factors to either an allowable working stress or an allowable working strength (for example, for anchor bolts). This working stress or strength is required to equal or exceed the applied (actual) stress or force.

"Design strength," "Nominal strength" and "Required strength" are defined in more detail immediately below. Working stresses and strengths are defined in ACI 530/ACI 5/TMS 402.

Strength can also refer to the load at which a test specimen fails (for example, prism strength).

Design strength. The strength design procedures of Section 2108 use the term "design strength" to indicate a realistic capacity of a member considering material, design and construction variabilities. The design strength is obtained by multiplying the nominal strength by a strength reduction factor. The design strength must equal or exceed the required strength (see separate definition in Section 2102.1).

Nominal strength. The strength design procedures of Section 2108 use the term "nominal strength" to refer to the capacity of a masonry member, determined based on the assumptions contained in Section 2108. It is sometimes referred to as the "expected" strength of the member; however, this is a misnomer. The nominal strength would equal the expected strength if: the masonry member were constructed of materials complying exactly with the minimum material requirements, design equations were perfect and construction tolerances were zero. Since material strengths commonly exceed minimum requirements, expected strength is often much higher than nominal strength. For instance, Grade 60 reinforcement often has a yield strength of around 66,000 psi (455 MPa), even though the minimum specified yield strength requirement is 60,000 psi (414 MPa). Therefore, the nominal strength is not the expected strength of the member, although it can be grossly classified as the minimum expected strength. Expected strength design is not included in Chapter 21 and is currently beyond the scope of the code. To determine the design strength, the nominal strength is multiplied by a strength reduction factor to account for material, design and construction variability. The design strength is required to equal or exceed the required strength.

Required strength. In the strength design of masonry (see Section 2108), the required strength is that which corresponds to the factored design loads on the structure. The design strength (nominal strength times the appropriate strength reduction factor) must equal or exceed the required strength.

TIE, LATERAL. Loop of reinforcing bar or wire enclosing longitudinal reinforcement.

❖ Lateral ties enclose longitudinal reinforcement. They are typically used in columns to support the compression reinforcement and masonry core so that these can support extreme loads even with some degradation of the masonry. Reinforcement can be assumed to be effective in carrying compressive forces only when supported by lateral ties.

Lateral ties also resist shear loads. Ties must form a closed rectangle or loop to completely surround the longitudinal reinforcement. Stirrups, in contrast, can be open.

TIE, WALL. A connector that connects wythes of masonry walls together.

❖ Ties are used to connect adjacent wythes and are subject to requirements for strength, durability and installation. Ties are adjustable or nonadjustable. A typical nonadjustable "Z" wire tie is shown in Figure 2109.6.3(1).

TILE. A ceramic surface unit, usually relatively thin in relation to facial area, made from clay or a mixture of clay or other ceramic materials, called the body of the tile, having either a "glazed" or "unglazed" face and fired above red heat in the course of manufacture to a temperature sufficiently high enough to produce specific physical properties and characteristics.

❖ Ceramic tile units are manufactured from nonmetallic materials and fired at high temperatures to obtain specific properties. Tile is considered a thin, nonstructural finish that must be supported by a strong, stiff, dimensionally stable backing.

TILE, STRUCTURAL CLAY. A hollow masonry unit composed of burned clay, shale, fire clay or mixture thereof, and having parallel cells.

❖ These clay masonry units are produced as end tiles and side tiles and differ from clay brick by having required cells with thinner webs between them.

WALL. A vertical element with a horizontal length-to-thickness ratio greater than three, used to enclose space.

Cavity wall. A wall built of masonry units or of concrete, or a combination of these materials, arranged to provide an airspace within the wall, and in which the inner and outer parts of the wall are tied together with metal ties.

Composite wall. A wall built of a combination of two or more masonry units bonded together, one forming the backup and the other forming the facing elements.

Dry-stacked, surface-bonded walls. A wall built of concrete masonry units where the units are stacked dry, without mortar on the bed or head joints, and where both sides of the wall are coated with a surface-bonding mortar.

Masonry-bonded hollow wall. A wall built of masonry units so arranged as to provide an airspace within the wall, and in which the facing and backing of the wall are bonded together with masonry units.

Parapet wall. The part of any wall entirely above the roof line.

❖ Masonry walls typically enclose space. They are generally required to be designed and installed for weather resistance, durability and adequate structural strength. The given dimensional requirements differentiate walls from columns.

Cavity wall. Cavity walls are made up of solid or hollow masonry units separated by a continuous airspace or cavity. This continuous airspace adds insulating value and acts as a barrier to moisture when detailed with flashing and weep holes. In many cavity walls, thermal insulation is placed between the wythes to further enhance thermal efficiency.

Composite wall. A composite wall is a multiwythe wall with wythes that act together to resist loads. The distinction of having wythes with different mechanical properties is important in engineering design. Walls constructed of different materials must be evaluated for lateral- and vertical-load-bearing performance and for differential movement between the wythes.

Dry-stacked, surface-bonded walls. Although this type of wall is dry stacked, a leveling course must be set in a full bed of mortar. Dry-stacked walls are also required to be placed in a running bond pattern (see commentary to the definition of "Mortar, surface-bonding").

Masonry-bonded hollow wall. Hollow walls are similar to cavity walls in that they are made up of solid or hollow units separated by an airspace. Unlike a cavity wall, however, the wythes are bonded together by masonry units, which causes the wythes to act together under load.

Parapet wall. These portions of masonry walls project above the roof. A parapet wall is exposed to weather on both sides and is laterally unsupported at the top. Parapets often have copings.

WEB. An interior solid portion of a hollow masonry unit as placed in masonry.

❖ The webs of hollow units are provided to support and strengthen the face shells. Web heights are permitted to be reduced so that horizontal reinforcement can be placed in the element.

WYTHE. Each continuous, vertical section of a wall, one masonry unit in thickness.

❖ Sometimes referred to as a "leaf" or "tier," each wythe is one thickness of a masonry unit.

NOTATIONS.

A_n = Net cross-sectional area of masonry, square inches (mm^2).

b = Effective width of rectangular member or width of flange for T and I sections, inches (mm).

d_b = Diameter of reinforcement, inches (mm).

f_r = Modulus of rupture, psi (MPa).

f_y = Specified yield stress of the reinforcement or the anchor bolt, psi (MPa).

f'_m = Specified compressive strength of masonry at age of 28 days, psi (MPa).

K = The lesser of the masonry cover, clear spacing between adjacent reinforcement, or five times d_b, inches (mm).

L_s = Distance between supports, inches (mm).

L_w = Length of wall, inches (mm).

l_d = Required development length of reinforcement, inches (mm).

l_{de} = Embedment length of reinforcement, inches (mm).

P_w = Weight of wall tributary to section under consideration, pounds (N).

t = Specified wall thickness dimension or the least lateral dimension of a column, inches (mm).

V_n = Nominal shear strength, pounds (N).

V_u = Required shear strength due to factored loads, pounds (N).

W = Wind load, or related internal moments in forces.

γ = Reinforcement size factor.

ρ_n = Ratio of distributed shear reinforcement on plane perpendicular to plane of A_{mv}.

ρ_{max} = Maximum reinforcement ratio.

ϕ = Strength reduction factor.

❖ Explanations of notations used in Chapter 21 listed above clarify the differences among these terms and the appropriate units, if any, that apply to them.

SECTION 2103 MASONRY CONSTRUCTION MATERIALS

2103.1 Concrete masonry units. Concrete masonry units shall conform to the following standards: ASTM C 55 for concrete brick; ASTM C 73 for calcium silicate face brick; ASTM C 90 for load-bearing concrete masonry units or ASTM C 744 for prefaced concrete and calcium silicate masonry units.

❖ Proper selection of materials is essential to produce masonry with adequate strength and durability. This section sets forth prescriptive and performance-based requirements (referenced standards) for masonry materials. Test procedures and criteria for establishing and verifying quality are included. Concrete masonry refers to solid and hollow concrete units, including concrete brick, concrete block, split-face block, slump block and other special units. This section requires conformance to ASTM standards for each specific type of concrete masonry unit. The standards include requirements for materials; manufacture; physical properties; moisture content; strength; absorption; minimum dimensions and permissible variations; inspection; testing and rejection.

Concrete masonry units are selected based on the desired use and appearance. Units are typically specified by weight, type and strength.

Hollow load-bearing concrete masonry units are manufactured in accordance with the requirements of ASTM C 90 using portland cement, water and mineral aggregates. Other suitable materials, such as approved admixtures, are permitted in accordance with ASTM C 90.

Hollow load-bearing concrete masonry units have three weight classifications: normal-weight units of 125 pounds per cubic foot (pcf) (2000 kg/m^3) or more; medium-weight units of between 105 pcf (1680 kg/m^3) and 125 pcf (2000 kg/m^3) and lightweight units of less than 105 pcf (1680 kg/m^3). Lightweight aggregates generally are expanded blast-furnace slag or similar suitable materials. Normal-weight units are made from sand, gravel, crushed stone or air-cooled, blast-furnace-slag aggregates.

Hollow load-bearing concrete masonry units are classified into two types. Type I are moisture-controlled units complying with the moisture-content requirements of ASTM C 90; Type II are nonmoisture-controlled units.

Concrete brick is required to comply with ASTM C 55. Concrete building brick and other solid concrete veneer and facing units are typically smaller than concrete masonry units conforming to ASTM C 90. They are made from portland cement, water and mineral aggregates, with or without inclusion of other approved materials.

Concrete brick is manufactured as lightweight, medium-weight and normal-weight units as described above for hollow concrete masonry units. Concrete brick is manufactured in two grades: Grade N and Grade S. Grade N units are used as architectural veneer and facing units in exterior walls, where high strength and resistance to moisture penetration and freeze-thaw cycling are required. Grade S units are used when moderate strength and resistance to freeze-thaw action and moisture penetration are required.

These units are classified into two types for each of the grades defined. Type I are moisture-controlled units complying with the moisture-content requirements of ASTM C 55. Type II are nonmoisture-controlled units.

Calcium silicate face brick is a solid masonry unit complying with ASTM C 73. These units are manufactured principally from silica sand, hydrated lime and water.

Calcium silicate face brick is manufactured in two grades: Grade SW and Grade MW. Grade SW is required where exposure to moisture in the presence of freezing temperatures is anticipated. Grade MW is permitted where the anticipated exposure has freezing temperatures, but without water saturation.

Grade SW units have a minimum permitted compressive strength on the gross area of 4,500 pounds per square inch (psi) (31.0 MPa) (average of three units), but not less than 3,500 psi (24.1MPa) for any individual unit. Grade MW units have a minimum permitted compressive strength on the gross area of 2,500 psi (17.2 MPa) (average of three units), but not less than 2,000 psi (13.8 MPa) for any individual unit.

ASTM C 744 is referenced for the manufacture of prefaced concrete and calcium silicate masonry units, commonly referred to as "glazed" concrete masonry units. The specified exposed surfaces of these units are covered during their manufacture with resin, resin and inert filler or cement and inert filler to produce a smooth resinous tile-like facing.

Facing requirements of that standard address resistance to chemicals; failure of adhesion of the facing material; abrasion surface-burning characteristics, color and color change; soiling and cleansability. The standard also covers dimensional tolerances, including face dimensions and distortions.

2103.2 Clay or shale masonry units. Clay or shale masonry units shall conform to the following standards: ASTM C 34 for structural clay load-bearing wall tile; ASTM C 56 for structural clay nonload-bearing wall tile; ASTM C 62 for building brick (solid masonry units made from clay or shale); ASTM C 1088 for solid units of thin veneer brick; ASTM C 126 for ceramic-glazed structural clay facing tile, facing brick and solid masonry units; ASTM C 212 for structural clay facing tile; ASTM C 216 for facing brick (solid masonry units made from clay or shale) and ASTM C 652 for hollow brick (hollow masonry units made from clay or shale).

Exception: Structural clay tile for nonstructural use in fireproofing of structural members and in wall furring shall not be required to meet the compressive strength specifications. The fire-resistance rating shall be determined in accordance with ASTM E 119 and shall comply with the requirements of Table 602.

❖ Section 2103.2 requires conformance with ASTM standards for masonry units manufactured from clay or shale. The various standards also include requirements for materials; manufacture; physical properties; minimum dimensions and permissible variations; inspections and testing.

Clay or shale masonry units are manufactured from clay and shale, which are compounds of silica or alumina. Shale is simply a hardened clay. The raw materials are formed into the desired shape by extrusion and cutting, molding or pressing while in the plastic state. The units are then fired in a kiln. The raw materials and the manufacturing process influence the physical properties of the manufactured unit.

Clay or shale masonry units are selected for their intended use from a variety of shapes, sizes and strengths. Units are specified based on grades and type.

Solid face brick units are required to conform to ASTM C 216. This standard covers clay brick intended to be used in masonry and supplying structural or facing components, or both, to the structure. These units are available in a variety of sizes, textures, colors and shapes.

ASTM C 216 contains requirements for two grades of durability: Grade SW and Grade MW. Grade SW brick is intended for use where high and uniform resistance to damage caused by freeze-thaw cycling is desired and where the brick may be subjected to such cycling while saturated with water. Grade MW brick is intended for use where moderate resistance to cyclic freeze-thaw damage is permissible or where the brick may be damp but not saturated with water when such cycling occurs. ASTM C 216 further classifies face brick into three types of appearance: Types FBS, FBX and FBA. Type FBS (face brick standard) units are permitted for general use in masonry. Type FBX (face brick select) units are also for general use, but have a higher degree of precision and lower permissible variation in size than Type FBS units. Type FBA (face brick architectural) units are also for general use, but are intentionally manufactured to produce characteristic architectural effects resulting from nonuniformity in size and texture of the units.

Solid units of building brick are required to conform to ASTM C 62. This specification covers brick intended for use in both structural and nonstructural masonry where external appearance is not critical. This brick was formerly called "common brick," and the standard does not contain appearance requirements.

ASTM C 62 contains requirements for three grades of durability: Grade SW, Grade MW and Grade NW. Grade SW brick is intended for use where high and uniform resistance to freeze-thaw damage is desired and where the brick may be subjected to such cycling while saturated with water. Grade MW brick is intended for use where moderate resistance to cyclic freeze-thaw damage is permissible or where the brick may be damp but not saturated with water when such cycling occurs. Grade NW is used where little resistance to cyclic freeze-thaw damage is required and is acceptable for applications protected from water absorption and freezing.

Hollow brick is required to conform to ASTM C 652, which regulates hollow building and hollow facing brick. Hollow brick differs from structural clay tile in that it has more stringent physical property requirements, such as thicker shell and web dimensions and higher minimum compressive strengths.

Hollow brick has two grades of durability: Grade SW and Grade MW. Grade SW is used where a high and uniform degree of resistance to freeze-thaw degradation and disintegration by weathering is desired and when the brick may be saturated while frozen. Grade MW is intended for use where a moderate and somewhat nonuniform degree of resistance to freeze-thaw degradation is permissible.

ASTM C 652 also classifies hollow brick into four types of appearance: HBS, HBX, HBA and HBB. Type HBS (hollow brick standard) units are permitted for general use. Type HBX (hollow brick select) units are also for general use, but have a higher degree of dimensional precision than Type HBS units. Type HBA (hollow brick architectural) units are also for general use, but are intentionally manufactured to produce characteristic architectural effects resulting from nonuniformity in size and texture of the units. Type HBB (hollow building brick) units are for general use in masonry where a particular color, texture, finish, uniformity or limits on cracks, warpage or other imperfections detracting from their appearance are not a consideration.

ASTM C 1088 regulates thin brick used in adhered veneer having a maximum actual thickness of $1^3/_4$ inches (44 mm). Thin veneer brick units are specified in three types of appearance and two grades of durability.

Exterior-grade units are for exposure to weather;

interior-grade units are not. Type TBS (standard) units are permitted for general use. Type TBX (select) units are also for general use, but have a higher degree of precision and lower permissible variation in size than Type TBS units. Type TBA (architectural) units are also for general use, but are intentionally manufactured to produce characteristic effects in variations of size, color and texture.

ASTM C 34 is referenced for the manufacture of structural clay load-bearing wall tile. This type of tile has two grades. Grade LBX is suitable for general use in masonry construction and can be used in masonry exposed to weathering, provided that the units meet durability requirements for Grade SW of ASTM C 216 for solid units of face brick. This tile grade is also suitable for the direct application of stucco.

Grade LB is suitable for general use in masonry not exposed to freeze-thaw action, or for exposed masonry where protected with a minimum 3-inch (76 mm) facing of stone, brick or other masonry materials.

Fire-resistant tile intended for use in load-bearing masonry is required to conform to ASTM C 34. Tile intended for use in fireproofing structural members is required to be of such sizes and shapes as to cover completely the exposed surfaces of the members.

ASTM C 212, the referenced standard regulating structural clay facing tile, covers two types of structural clay load-bearing facing tile: Types FTX and FTS.

Type FTX clay facing tile is a smooth-face tile suitable for use in exposed exterior and interior masonry walls and partitions where low absorption, easy cleaning and resistance to staining are required. The physical characteristics of this tile require a high degree of mechanical perfection, narrow color range and minimum variation in face dimensions.

Type FTS clay facing tile is a smooth- or rough-textured face tile suitable for general use in exposed exterior and interior masonry walls and partitions where moderate absorption, variation in face dimensions, minor defects in surface finish and moderate color variations are permissible.

There are two classes of tile for the types stated above: standard (for general use) and special duty (having superior resistance to impact and moisture transmission and supporting greater lateral and vertical loads).

ASTM C 56 is the standard regulating the manufacture of nonload-bearing tile, which is made from the same materials as other types of tile previously mentioned and is used for partitions, fireproofing and furring.

ASTM C 126 prescribes requirements for ceramic-glazed structural clay load-bearing facing tile and brick and other solid masonry units made from clay, shale, fire clay or combinations thereof. The standard specifies two grades and two types of ceramic-glazed masonry: Grade S is used with comparatively narrow mortar joints, while Grade SS is used where variations in face dimensions are very small; Type I units are used where only one finished face is to be exposed, while Type II units are used where both finished faces are to be exposed.

The finish of glazed units is important. Requirements and test methods are prescribed for imperviousness, opacity, resistance to fading and crazing, hardness, abrasion resistance and flame spread smoke density. Compressive strength requirements, dimensional tolerances and permissible distortions are also prescribed by the standard.

The exception to this section allows structural clay tile that does not meet the compressive strength specifications to be used as nonstructural fireproofing for structural members. The fire-resistance rating must be determined in accordance with ASTM E 119.

2103.3 Stone masonry units. Stone masonry units shall conform to the following standards: ASTM C 503 for marble building stone (exterior): ASTM C 568 for limestone building stone; ASTM C 615 for granite building stone; ASTM C 616 for sandstone building stone or ASTM C 629 for slate building stone.

❖ Many types of natural stone, including marble, granite, slate, limestone and sandstone, are used in building construction. This section covers natural stones that are sawed, cut, split or otherwise shaped for masonry purposes. Natural stones for building purposes are specified in various grades, textures and finishes and are required to have physical properties appropriate to their intended use.

Where applicable to specific kinds of natural stone, the finished units are required to be sound and free from spalls, cracks, open seams, pits and other defects that would impair structural strength and durability and, when applicable, fire resistance.

This section requires conformance to the following ASTM standards for natural stone masonry: ASTM C 503 for marble building stone (exterior); ASTM C 568 for limestone building stone; ASTM C 615 for granite building stone; ASTM C 616 for sandstone building stone or ASTM C 629 for slate building stone. The standards contain requirements for absorption, density (except for slate), compressive strength (except for slate), modulus of rupture, abrasion resistance (except for granite) and acid resistance (slate only).

2103.4 Ceramic tile. Ceramic tile shall be as defined in, and shall conform to the requirements of, ANSI A137.1.

❖ Ceramic tile is made from clay, possibly mixed with other ceramic materials. Metallic oxides may be included for glaze coloring. Ceramic tile products are available in a broad range of sizes, appearances, characteristics and function.

ANSI A137.1 is the recognized industry standard for the manufacture, testing and labeling of ceramic tile. According to this standard, tile should be shipped in sealed cartons with the grade of contents indicated by grade seals with a distinctive coloring: blue for standard grade (units as perfect and free from defects as is possible in the manufacturing process) and yellow for "seconds" (units having slight defects, but free from structural defects and cracks).

ANSI A137.1 groups ceramic tile into four major

types: glazed wall tile, mosaic tile, quarry tile and paving tile.

2103.5 Glass unit masonry. Hollow glass units shall be partially evacuated and have a minimum average glass face thickness of $^3/_{16}$ inch (4.8 mm). Solid glass-block units shall be provided when required. The surfaces of units intended to be in contact with mortar shall be treated with a polyvinyl butyral coating or latex-based paint. Reclaimed units shall not be used.

❖ Consensus national standards have not been written to establish minimum material properties or test methods for glass blocks. Reliance must be placed on the manufacturer's specifications and glass block is required to meet the minimum specified dimensions of those specifications. Glass block has generally performed well in service.

2103.6 Second-hand units. Second-hand masonry units shall not be reused unless they conform to the requirements of new units. The units shall be of whole, sound materials and free from cracks and other defects that will interfere with proper laying or use. Old mortar shall be cleaned from the unit before reuse.

❖ This section allows for the use of salvaged brick and other second-hand masonry units, provided that their quality and condition meet the requirements for new masonry units. Second-hand units must be whole, of sound material, clean and free from defects that would interfere with proper laying or use.

Most second-hand masonry units come from the demolition of old buildings. Masonry units manufactured in the past do not generally compare with the quality of masonry made by modern manufacturing methods under controlled conditions. Therefore, designers should expect salvaged masonry units to have lower strength and durability than new units.

Generally, the difference between walls laid up with new masonry units and second-hand units of the same type is the adhesion of the mortar to the masonry surfaces. When new masonry units are laid in fresh mortar, water and fine cementitious particles are absorbed into the masonry, thereby improving bond strength. In contrast, pores in the bed faces of second-hand masonry units, regardless of cleaning, are filled with particles of cement, lime and deleterious substances that impede adequate absorption, thereby adversely affecting bond between the mortar and the masonry.

2103.7 Mortar. Mortar for use in masonry construction shall conform to ASTM C 270 and shall conform to the proportion specifications of Table 2103.7(1) or the property specifications of Table 2103.7(2). Type S or N mortar shall be used for glass unit masonry. The amount of water used in mortar for glass unit masonry shall be adjusted to account for the lack of absorption. Retempering of mortar for glass unit masonry shall not be permitted after initial set. Unused mortar shall be discarded within $2^1/_2$ hours after initial mixing except that unused mortar for glass unit masonry shall be discarded within $1^1/_2$ hours after initial mixing.

❖ Masonry mortar bonds masonry units to form an integral structure. Mortar provides a tight and weather-resistant seal between units; bonds with steel joint reinforcement, ties and other metal accessories and compensates for dimensional variations in masonry units. It can also serve aesthetic purposes through contrasts of color, texture and shadow lines created by different types of tooled joints.

This section of the code establishes ASTM C 270 as the standard regulating mortar to be used in masonry construction. The standard covers mortars for use in the construction of nonreinforced and reinforced unit masonry structures. It specifies types of mortar and two alternative specifications—the proportion specification and the property specification.

Type M mortar has the highest compressive and tensile bond strength. It is suitable for general use, particularly where maximum masonry compressive strength is required and in construction that is in contact with earth.

Type S mortar is a general-purpose mortar with high compressive and tensile bond strengths. It is often used in reinforced masonry and in unreinforced masonry that requires high strength to resist out-of-plane lateral loads.

Type N mortar is a general-purpose mortar with medium compressive and tensile bond strengths. It is used where high vertical and lateral loads are not expected. Type O is a low-strength mortar suitable for use in nonload-bearing construction and where the masonry will not be subject to severe weathering or freeze-thaw cycling.

The materials used in mortar are required to conform to the following standard specifications referenced in ASTM C 270, ASTM C 5, ASTM C 91, ASTM C 144, ASTM C 150, ASTM C 207, ASTM C 595 and ASTM C 1329.

In addition to type, mortar for masonry is further classified according to its primary cementitious materials. In portland cement-lime mortar, the cementitious materials are portland cement and hydrated mason's lime. In masonry-cement mortar and mortar-cement mortar, the principal cementitious material is portland cement, which is contained in the masonry cement and the mortar cement.

Regardless of a masonry mortar's principal cementitious constituents, each ingredient—cement, lime (when used), sand and water—contributes to the mortar's overall performance. Portland cement, mortar cement, masonry cement or blended hydraulic cement make the hardened mortar strong and durable. Lime contributes to the workability and water retention of fresh mortar and to the flexibility, compressive strength and tensile bond strength of hardened mortar. Sand serves as a filler material, increases strength and reduces shrinkage. Water determines the consistency of fresh mortar and also hydrates the cementitious materials, resulting in hardening of the mortar.

Tables 2103.7(1) and 2103.7(2) are based on Tables 1 and 2 of ASTM C 270. These tables provide requirements for the production of mortar by the proportion method (by volume) and by the property method (by compressive strength and other properties), respectively. Whether the proportion or property specification governs depends on the contract documents. When neither proportion nor property specification is prescribed, the proportion specification governs, unless data are presented to and accepted by the specifier to show that mortar meets the requirements of the property specification.

ASTM C 270 also covers materials and testing requirements for water retention, compressive strength and air content.

The types of mortar recommended for use in various types of masonry construction are shown in the appendix to ASTM C 270. The performance of masonry is influenced by mortar workability, water retentivity, bond strength, compressive strength and long-term deformability (creep). Each mortar type has corresponding use limitations. For example, allowable compressive stresses permitted in Section 2109.3.2 for empirical design are greater for Types M and S mortar than for Type N mortar.

Glass-block units are required to be laid in Type S or N mortar. Admixtures containing set accelerators or antifreeze compounds are not permitted in mortars for glass-block masonry.

TABLE 2103.7(1). See page 21-20.

❖ This table prescribes proportions for each mortar type when mortar is specified by proportion. The table covers portland cement-lime mortars, masonry-cement mortars and mortar-cement mortars. Materials are measured by volume. The required volume of mason's sand used is based on the combined volume of all cementitious materials (including lime, if applicable).

TABLE 2103.7(2). See page 21-20.

❖ This table prescribes minimum required physical properties of plastic and hardened mortar for each type when mortar is specified by property. The table covers portland cement-lime mortars, masonry-cement mortars and mortar-cement mortars. The specified properties are compressive strength, water retention and air content. Values are for mortars prepared in a laboratory in accordance with ASTM C 270. These values do not directly relate to specimens of field mortar tested in accordance with ASTM C 780.

2103.8 Surface-bonding mortar. Surface-bonding mortar shall comply with ASTM C 887. Surface bonding of concrete masonry units shall comply with ASTM C 946.

❖ Specifications for materials to be used in premixed surface-bonding mortar and for the properties of the mortar are contained in ASTM C 887. Requirements for masonry units and other materials used in constructing dry-stacked, surface-bonded masonry for walls and for the construction itself are contained in ASTM C 946.

In addition to its primary function of bonding masonry units, surface-bonding mortar can provide resistance to penetration by wind-driven rain.

2103.9 Mortars for ceramic wall and floor tile. Portland cement mortars for installing ceramic wall and floor tile shall comply with ANSI A108.1A and ANSI A108.1B and be of the compositions indicated in Table 2103.9.

❖ This section pertains to cement mortars and organic adhesives used for setting ceramic wall and floor tiles. Each mortar type has certain qualities that make it suitable for installing tile over different kinds of backing materials or under a certain set of conditions.

Cement mortars are "thick-bed" mortars applied in thicknesses of $^3/_4$ to $1^1/_4$ inches (19.1 to 32 mm) on floors and $^3/_4$ to 1 inch (19.1 to 25 mm) on walls to achieve the specified slopes and flatness in the finished tile work. Portland cement mortars are suitable for setting ceramic tile in most installations. They can be applied over properly prepared backings of clay or concrete masonry; concrete; wood frame; rough wood floors and plywood floors; foam insulation board; gypsum wallboard and portland cement or gypsum plaster. Cement mortars can be reinforced with metal lath or wire mesh; in such cases, however, additional mortar thickness may be required. Cement mortars have good structural strength and are not affected by prolonged contact with water.

Complete material and installation specifications are contained in ANSI A108.1A and A108.1B, which include required specifications for: installation of wire lath and scratch coats; mortar mixes, bond coat mixes and mortar application; installation methods on floors, walls and countertops; grouting of tile and general requirements for tile installations.

TABLE 2103.9
CERAMIC TILE MORTAR COMPOSITIONS

LOCATION	MORTAR	COMPOSITION
Walls	Scratchcoat	1 cement; $^1/_5$ hydrated lime; 4 dry or 5 damp sand
	Setting bed and leveling coat	1 cement; $^1/_2$ hydrated lime; 5 damp sand to 1 cement 1 hydrated lime, 7 damp sand
Floors	Setting bed	1 cement; $^1/_{10}$ hydrated lime; 5 dry or 6 damp sand; or 1 cement; 5 dry or 6 damp sand
Ceilings	Scratchcoat and sand bed	1 cement; $^1/_2$ hydrated lime; $2^1/_2$ dry sand or 3 damp sand

❖ Portland cement-lime mortars are a mixture of portland cement (ASTM C 150), hydrated lime (ASTM C 207) and damp mason's sand (ASTM C 144). Proportions of these ingredients are given in Table 2103.9, dealing with the portland cement-lime mortars most commonly used for setting ceramic tile on wall, floor and ceiling construction and measured in parts by volume. Complete material and installation specifications are contained in ANSI A108.1.

TABLE 2103.7(1) MORTAR PROPORTIONS

MORTAR	TYPE	PROPORTIONS BY VOLUME (cementitious materials): Portland cement[a] or blended cement[b]	Masonry cement[c] M	Masonry cement[c] S	Masonry cement[c] N	Mortar cement[d] M	Mortar cement[d] S	Mortar cement[d] N	HYDRATED LIME[e] OR LIME PUTTY	AGGREGATE MEASURED IN A DAMP, LOOSE CONDITION
Cement-lime	M	1	—	—	—	—	—	—	$^1/_4$	Not less than $2^1/_4$ and not more than 3 times the sum of the separate volumes of cementitious materials
	S	1	—	—	—	—	—	—	over $^1/_4$ to $^1/_2$	
	N	1	—	—	—	—	—	—	over $^1/_2$ to $1^1/_4$	
	O	1	—	—	—	—	—	—	over $1^1/_4$ to $2^1/_2$	
Mortar cement	M	1	—	—	—	—	—	1	—	
	M	—	—	—	—	1	—	—	—	
	S	$^1/_2$	—	—	—	—	—	1	—	
	S	—	—	—	—	—	1	—	—	
	N	—	—	—	—	—	—	1	—	
	O	—	—	—	—	—	—	1	—	
Masonry cement	M	1	—	—	1	—	—	—	—	
	M	—	1	—	—	—	—	—	—	
	S	$^1/_2$	—	—	1	—	—	—	—	
	S	—	—	1	—	—	—	—	—	
	N	—	—	—	1	—	—	—	—	
	O	—	—	—	1	—	—	—	—	

a. Portland cement conforming to the requirements of ASTM C 150.
b. Blended cement conforming to the requirements of ASTM C 595.
c. Masonry cement conforming to the requirements of ASTM C 91.
d. Mortar cement conforming to the requirements of ASTM C 1329.
e. Hydrated lime conforming to the requirements of ASTM C 207.

TABLE 2103.7(2) MORTAR PROPERTIES[a]

MORTAR	TYPE	AVERAGE COMPRESSIVE[b] STRENGTH AT 28 DAYS minimum (psi)	WATER RETENTION minimum (%)	AIR CONTENT maximum (%)
Cement-lime	M	2,500	75	12
	S	1,800	75	12
	N	750	75	14[c]
	O	350	75	14[c]
Mortar cement	M	2,500	75	12
	S	1,800	75	12
	N	750	75	14[c]
	O	350	75	14[c]
Masonry cement	M	2,500	75	18
	S	1,800	75	18
	N	750	75	20[d]
	O	350	75	20[d]

For SI: 1 inch = 25.4 mm, 1 pound per square inch = 6.895 kPa.
a. This aggregate ratio (measured in damp, loose condition) shall not be less than $2^1/_4$ and not more than 3 times the sum of the separate volumes of cementitious materials.
b. Average of three 2-inch cubes of laboratory prepared mortar, in accordance with ASTM C 270.
c. When structural reinforcement is incorporated in cement-lime or mortar cement mortars, the maximum air content shall not exceed 12 percent.
d. When structural reinforcement is incorporated in masonry cement mortar, the maximum air content shall not exceed 18 percent.

2103.9.1 Dry-set portland cement mortars. Premixed prepared portland cement mortars, which require only the addition of water and are used in the installation of ceramic tile, shall comply with ANSI A118.1. The shear bond strength for tile set in such mortar shall be as required in accordance with ANSI A118.1. Tile set in dry-set portland cement mortar shall be installed in accordance with ANSI A108.5.

❖ Dry-set portland cement mortar for ceramic tile is required to comply with ANSI A118.1, which describes test methods and minimum requirements for dry-set mortar, including sampling, free water content, setting characteristics, shrinkage, shear strength and staining.

Dry-set mortar is a mixture of portland cement, sand and perhaps water-retention admixtures. Dry-set mortars are suitable for use over properly prepared backings of clay or concrete masonry; concrete; cut-cell expanded polystyrene or rigid closed-cell urethane insulation board; gypsum wallboard; lean portland cement mortar and hardened wall and floor setting beds.

Installation must comply with ANSI A108.5. The "thin-bed" mortars are applied in a single layer as thin as $^3/_{32}$ inch (2.4 mm), but usually in thicknesses of $^1/_8$ to $^1/_4$ inch (3.2 to 6.4 mm). The use of dry-set mortar for leveling work is limited to a maximum thickness of $^1/_4$ inch (6.4 mm).

This material has excellent water and impact resistance. It is also water-cleanable, nonflammable, good for exterior use and requires no presoaking of the tile. Dry-set mortar is intended for use with glazed wall tile, ceramic mosaics, pavers and quarry tile.

Shear bond strength (of tile set in mortar) is required to be tested in accordance with standards applicable to the mortar used. Tile set with dry-set mortar must be installed in accordance with ANSI A108.5.

2103.9.2 Electrically conductive dry-set mortars. Premixed prepared portland cement mortars, which require only the addition of water and comply with ANSI A118.2, shall be used in the installation of electrically conductive ceramic tile. Tile set in electrically conductive dry-set mortar shall be installed in accordance with ANSI A108.7.

❖ Mortars used in the installation of electrically conductive ceramic tile are required to conform to ANSI A108.7 for premixed prepared mortars and to ANSI A108.7 for conductive dry-set mortars. ANSI A108.7 is the correct reference for tile set in conductive dry-set mortar.

2103.9.3 Latex-modified portland cement mortar. Latex-modified portland cement thin-set mortars in which latex is added to dry-set mortar as a replacement for all or part of the gauging water that are used for the installation of ceramic tile shall comply with ANSI A118.4. Tile set in latex-modified portland cement shall be installed in accordance with ANSI A108.5.

❖ Latex-modified portland cement mortar is a mixture of portland cement, sand and special latex admixtures. It is applied as a thin-bed material, like dry-set portland cement mortar (see commentary, Section 2103.9.1). Since the latex used in these mortars varies among manufacturers, instructions for mixing and use must be followed carefully. Applicable material and installation standards are ANSI A108.5 and A118.4.

2103.9.4 Epoxy mortar. Ceramic tile set and grouted with chemical-resistant epoxy shall comply with ANSI A118.3. Tile set and grouted with epoxy shall be installed in accordance with ANSI A108.6.

❖ Epoxy mortar is a two-part system consisting of epoxy resin and hardener. It is applied as a layer of $^1/_{16}$ to $^1/_8$ inch (1.6 to 3.2 mm) in thickness and is suitable for use on properly prepared floors of concrete, wood, plywood, steel plate or ceramic tile. It is particularly suitable where chemical resistance or high bond strength is required. Deflection control is critical for the successful use of this material, and floor systems should not deflect more than $^1/_{360}$ of their span. Epoxy mortar is recommended for setting ceramic mosaics, quarry tile and paver tile. Applicable material and installation standards are ANSI A108.6 and A118.3.

2103.9.5 Furan mortar and grout. Chemical-resistant furan mortar and grout that are used to install ceramic tile shall comply with ANSI A118.5. Tile set and grouted with furan shall be installed in accordance with ANSI A108.8.

❖ Furan mortar is a two-part system consisting of a furan resin and a hardener. It is suitable where chemical resistance is critical. It is used primarily on floors in laboratories and industrial plants. Acceptable subfloors include concrete, steel plate and ceramic tile. Furan grout is intended for quarry tile and pavers, mainly in industrial areas requiring maximum chemical resistance.

ANSI A118.5 is the standard regulating both furan mortar and grout. Installation must comply with ANSI A108.8.

2103.9.6 Modified epoxy-emulsion mortar and grout. Modified epoxy-emulsion mortar and grout that are used to install ceramic tile shall comply with ANSI A118.8. Tile set and grouted with modified epoxy-emulsion mortar and grout shall be installed in accordance with ANSI A108.9.

❖ Modified epoxy-emulsion mortar and grout used to install ceramic tile are required to comply with ANSI A118.8 and to be installed in accordance with ANSI A108.9.

ANSI A118.8 describes the test methods and the minimum requirements for modified epoxy-emulsion mortar and grout. The chemical and solvent resistance of these mortars and grouts tends to exceed those of organic adhesives and equal those of latex-modified portland-cement mortars. They are not, however, designed to meet the requirements of ANSI A108.6 or A118.3.

These types of mortars and grouts are three-part systems that include emulsified epoxy resins and hardeners, preblended portland cement and silica sand. They are used as a bond-coat setting mortar or grout. They can be cleaned from wall and floor surfaces using a wet sponge prior to initial set.

ANSI A118.8 regulates water absorption, flexural

strength, thermal expansion, linear shrinkage, tensile strength and compressive strength.

2103.9.7 Organic adhesives. Water-resistant organic adhesives used for the installation of ceramic tile shall comply with ANSI A136.1. The shear bond strength after water immersion shall not be less than 40 psi (275 kPa) for Type I adhesive, and not less than 20 psi (138 kPa) for Type II adhesive, when tested in accordance with ANSI A136.1. Tile set in organic adhesives shall be installed in accordance with ANSI A108.4.

❖ Organic adhesives are prepared materials that cure or set by evaporation and that are ready to use without adding liquid or powder. Adhesives are suitable for installing tiles on prepared wall and floor surfaces, including: brick and concrete masonry; concrete; gypsum wallboard; portland cement or gypsum plaster and wood-flooring systems.

Organic adhesives (mastics) are applied as a thin layer approximately $^1/_{16}$ inch (1.6 mm) thick. An underlayment is used to level and true surfaces. Organic adhesive does not permit the soaking of tiles and is not suitable for exterior use. Bond strength varies greatly among the numerous brands of organic adhesives available for use in construction. Adhesives must meet minimum bond-strength requirements of this section of ANSI A136.1 for Type I and II adhesive. The installation of tile with organic adhesives is required to conform to ANSI A108.4.

2103.9.8 Portland cement grouts. Portland cement grouts used for the installation of ceramic tile shall comply with ANSI A118.6. Portland cement grouts for tile work shall be installed in accordance with ANSI A108.10.

❖ Portland cement grouts are the most commonly used grouts for tile walls. The mixture of portland cement and other ingredients is water resistant and uniform in color. The water in the grout is essential to promote a good bond and to develop full grout strength. This type of grout is required to comply with ANSI A118.6 and to be installed in accordance with ANSI A108.10.

2103.10 Grout. Grout shall conform to Table 2103.10 or to ASTM C 476. When grout conforms to ASTM C 476, the grout shall be specified by proportion requirements or property requirements.

❖ Grout intended for use in the construction of engineered and empirically designed masonry structures is required to comply with ASTM C 476, which regulates materials, measurement, mixing and storage of materials.

Two types of grout are used in masonry: fine and coarse. Fine grout is made of cement, sand and water, with optional small quantities of lime. Coarse grout includes the same ingredients, plus pea gravel or a larger $^3/_4$-inch (19.1 mm) coarse aggregate.

Whether fine or coarse grout is used depends on the size of the grout space to be filled. Coarse grout is permitted to be used in cavities that are 2 inches (51 mm) or more in width and in the cells of hollow units that are 4 inches (102 mm) or more in both directions. Smaller spaces require the use of fine grout.

The materials used in masonry grout are required to be listed in ASTM C 476 and must conform to the following standard specifications: ASTM C 5; ASTM C 150; ASTM C 207; ASTM C 404 and ASTM C 595.

Grout may be specified either by proportion or property. When the proportion method is specified, Table 2103.10 provides required proportions of grout materials by volume. Information on property requirements is given in ASTM C 476.

TABLE 2103.10
GROUT PROPORTIONS BY VOLUME FOR MASONRY CONSTRUCTION

TYPE	PARTS BY VOLUME OF PORTLAND CEMENT OR BLENDED CEMENT	PARTS BY VOLUME OF HYDRATED LIME OR LIME PUTTY	AGGREGATE, MEASURED IN A DAMP, LOOSE CONDITION	
			Fine	Coarse
Fine grout	1	0-$^1/_{10}$	$2^1/_4$-3 times the sum of the volumes of the cementitious materials	—
Coarse grout	1	0-$^1/_{10}$	$2^1/_4$-3 times the sum of the volumes of the cementitious materials	1-2 times the sum of the volumes of the cementitious materials

❖ This table, which is based on ASTM C 476, prescribes volume proportions of grout for masonry. Information on fine and coarse grout is given in the commentary to ACI 530.1/ASCE 6/TMS 602. Additionally, ACI 530.1/ASCE 6/TMS 602 provides specific requirements as to where fine and coarse grout are permitted to be used.

2103.11 Metal reinforcement and accessories. Metal reinforcement and accessories shall conform to Sections 2103.11.1 through 2103.11.7.

❖ This section contains standards and material requirements for anchors, joint reinforcement, wire accessories, ties, wire fabric and for required corrosion protection of these items.

Several referenced standards originally applied to steel reinforcement for concrete, but are suitable and required for masonry construction as well.

2103.11.1 Deformed reinforcing bars. Deformed reinforcing bars shall conform to one of the following standards: ASTM A 615 for deformed and plain billet-steel bars for concrete reinforcement; ASTM A 706 for low-alloy steel deformed bars for concrete reinforcement; ASTM A 767 for zinc-coated reinforcing steel bars; ASTM A 775 for epoxy-coated reinforcing steel bars and ASTM A 996 for rail steel and axle steel deformed bars for concrete reinforcement.

❖ ASTM A 615 regulates the manufacture of plain and deformed reinforcement, Grades 40 and 60, for new or recycled steel.

ASTM A 706 regulates the manufacture of plain and

deformed reinforcement, Grade 60 only, from mold-cast or strand-cast steel. This steel is intended for welding, in accordance with the procedures of AWS D1.4. It is more ductile than other steel and is especially suitable for structures in zones of high seismicity (see commentary, Section 2108.3).

Where masonry is expected to be exposed to especially corrosive conditions, galvanized reinforcement meeting ASTM A 767 or epoxy-coated reinforcement meeting ASTM A 775 can be used.

ASTM A 996 regulates rail steel and axle steel deformed bars.

Deformed reinforcement is recommended for nearly every application in reinforced masonry and is required in most instances.

2103.11.2 Joint reinforcement. Joint reinforcement shall comply with ASTM A 951. The maximum spacing of crosswires in ladder-type joint reinforcement and of point of connection of cross wires to longitudinal wires of truss-type reinforcement shall be 16 inches (400 mm).

❖ ASTM A 951 contains material requirements, mechanical properties and tolerances for joint reinforcement. Longitudinal wires are required to be deformed in accordance with this standard to provide a mechanical bond with the surrounding mortar. Cross wires are plain.

ASTM A 82 regulates the manufacture of cold-drawn, plain steel wire for joint reinforcement, wire anchors and ties. Joint reinforcement can be epoxy coated to provide added corrosion protection (see commentary, Section 2103.11.6).

2103.11.3 Deformed reinforcing wire. Deformed reinforcing wire shall conform to ASTM A 496.

❖ ASTM A 496 regulates the manufacture of cold-worked, deformed wire having yield stresses of 70 ksi (483 MPa) (welded wire fabric) and 75 ksi (517 MPa). This wire is used as reinforcement and in the manufacture of welded, deformed-wire fabric. It is not commonly used in masonry.

2103.11.4 Wire fabric. Wire fabric shall conform to ASTM A 185 for plain steel-welded wire fabric for concrete reinforcement or ASTM A 496 for welded deformed steel wire fabric for concrete reinforcement.

❖ ASTM A 185 regulates the manufacture of plain welded-wire fabric, using ASTM A 82 steel. ASTM A 497 regulates the manufacture of welded deformed wire fabric using steels conforming to ASTM A 82 and A 496, alone or in combination. Wire fabric is seldom used in masonry.

2103.11.5 Anchors, ties and accessories. Anchors, ties and accessories shall conform to the following standards: ASTM A 36 for structural steel; ASTM A 82 for plain steel wire for concrete reinforcement; ASTM A 185 for plain steel-welded wire fabric for concrete reinforcement; ASTM A 167, Type 304, for stainless and heat-resisting chromium-nickel steel plate, sheet and strip and ASTM A 366 for cold-rolled carbon steel sheet, commercial quality.

❖ Anchors, ties and accessories are manufactured in a variety of ways with a number of materials. Applicable ASTM specifications prescribe minimum requirements for these materials.

ASTM A 36 regulates the manufacture of hot-rolled steel with a minimum specified yield stress of 36 ksi (248 MPa), for general structural purposes. This material is suitable for welding.

ASTM A 82 regulates the manufacture of cold-drawn, plain-steel wire for joint reinforcement, wire anchors and ties.

ASTM A 185 regulates the manufacture of plain-steel welded-wire fabric, using ASTM A 82 steel, or ASTM A 167 Type 304, corrosion-resistant steel.

ASTM A 366 regulates the manufacture of sheet steel used for metal anchors and ties. The specification covers cold-rolled carbon steel sheet of commercial quality, in coils or cut lengths. This material is intended for exposed or unexposed parts made by bending, moderate drawing, forming or welding. This standard also specifies mechanical properties and tests relative to bending, hardness and moderate deformability and sets requirements for chemical composition. See Figure 2109.6.3(1) for typical masonry accessories.

2103.11.6 Prestressing tendons. Prestressing tendons shall conform to one of the following standards:

a. Wire . . . . ASTM A 421

b. Low-relaxation wire . . ASTM A 421

c. Strand . . . . ASTM A 416

d. Low-relaxation strand. . ASTM A 416

e. Bar . . . . ASTM A 722

Exceptions:

1. Wire, strands and bars not specifically listed in ASTM A 421, ASTM A 416 or ASTM A 722 are permitted, provided they conform to the minimum requirements in ASTM A 421, ASTM A 416, or ASTM A 722 and are approved by the architect/engineer.
2. Bars and wires of less than 150 kips per square inch (ksi) (1034 MPa) tensile strength and conforming to ASTM A 82, ASTM A 510, ASTM A 615, ASTM A 616, ASTM A 996 or ASTM A 706/A 706 M are permitted to be used as prestressed tendons provided that:
   2.1. The stress relaxation properties have been assessed by tests according to ASTM E 328 for the maximum permissible stress in the tendon.
   2.2. Other nonstress-related requirements of ACI 530/ASCE 5/TMS 402, Chapter 4, addressing prestressing tendons are met.

❖ This section specifies the materials for components of the prestressing system. They are similar to those used in prestressed concrete.

2103.11.7 Corrosion protection. Corrosion protection for prestressing tendons, prestressing anchorages, couplers and end block shall comply with the requirements of ACI 530.1/ASCE 6/TMS 602, Article 2.4G. Corrosion protection for carbon steel accessories used in exterior wall construction or interior walls exposed to a mean relative humidity exceeding 75 percent shall comply with either Section 2103.11.7.1 or 2103.11.7.2. Corrosion protection for carbon steel accessories used in interior walls exposed to a mean relative humidity equal to or less than 75 percent shall comply with either Section 2103.11.7.1, 2103.11.7.2 or 2103.11.7.3.

❖ Joint reinforcement, anchors, wall ties and accessories are required to be galvanized to protect the steel from deterioration and corrosion, unless the steel is inherently corrosion resistant, such as ASTM A 167 Type 304 stainless steel. The application and thickness of the galvanizing depends on the intended location or severity of exposure. Carbon steel accessories located in exterior masonry wall construction are required to be either hot-dipped galvanized in accordance with Section 2103.11.7.1 or epoxy coated in accordance with Section 2103.11.7.2. Carbon steel accessories for use in interior-wall construction are permitted to comply with less-restrictive mill galvanized requirements as noted in Section 2103.11.7.3.

2103.11.7.1 Hot-dipped galvanized. Apply a hot-dipped galvanized coating after fabrication as follows:

1. For joint reinforcement, wall ties, anchors and inserts, apply a minimum coating of 1.5 ounces per square foot (psf) (458 g/m^2) complying with the requirements of ASTM A 153, Class B.
2. For sheet metal ties and sheet metal anchors, comply with the requirements of ASTM A 153, Class B.
3. For steel plates and bars, comply with the requirements of either ASTM A 123 or ASTM A 153, Class B.

❖ Carbon steel accessories located in exterior masonry wall construction are required to meet the protection requirements of ASTM A 153, which regulates zinc-coated iron and steel hardware. The minimum required coating of 1.5 ounces per square foot of surface (458 g/m^2) is derived from Table 1 of ASTM A 153 for rolled, pressed and forged articles under $^3/_{16}$ inch (4.8 mm) in thickness and over 15 inches (381 mm) in length.

The finish and appearance of zinc-coated articles are specified in ASTM A 153. Zinc-coated articles are required to be free from uncoated areas, blisters, flux deposits, black spots and other inclusions that would interfere with the intended use. The coating is required to be smooth and uniform in thickness.

2103.11.7.2 Epoxy coatings. Carbon steel accessories shall be epoxy coated as follows:

1. For joint reinforcement, comply with the requirements of ASTM A 884 Class B, Type 2 – 18 mils (457μm).
2. For wire ties and anchors, comply with the requirements of ASTM A 899 Class C —20 mils (508μm).
3. For sheet metal ties and anchors, provide a minimum thickness of 20 mils (508μm) or in accordance with the manufacturer's specification.

❖ As an alternative to hot-dipped galvanizing, carbon steel accessories in exterior walls can be epoxy coated. This section specifies the appropriate coating type and minimum coating thickness applicable to each type of accessory.

2103.11.7.3 Mill galvanized. Apply a mill galvanized coating as follows:

1. For joint reinforcement, wall ties, anchors and inserts, apply a minimum coating of 0.1 ounce psf (31g/m^2) complying with the requirements of ASTM A 641.
2. For sheet metal ties and sheet metal anchors, apply a minimum coating complying with Coating Designation G-60 according to the requirements of ASTM A 653.
3. For anchor bolts, steel plates or bars not exposed to the earth, weather or a mean relative humidity exceeding 75 percent, a coating is not required.

❖ Carbon steel accessories for use in interior-wall construction are permitted to comply with less-restrictive material specifications (ASTM A 641), as noted.

2103.11.8 Tests. Where unidentified reinforcement is approved for use, not less than three tension and three bending tests shall be made on representative specimens of the reinforcement from each shipment and grade of reinforcing steel proposed for use in the work.

❖ Reinforcing bars are to be rolled with raised symbols or letters impressed on the metal identifying the manufacturing mill. When required by the building official, the grade of material is to be identified by a satisfactory mill test. When the manufacturing mill is not identified but the reinforcement is approved for use under ordinary material procedures, at least three tension and three bending tests are required on representative specimens of the reinforcement from each shipment and grade of reinforcing steel proposed for use. The tension and bending tests are required to be performed by an approved testing agency.

SECTION 2104
CONSTRUCTION

2104.1 Masonry construction. Masonry construction shall comply with the requirements of Sections 2104.1.1 through 2104.5 and with ACI 530.1/ASCE 6/TMS 602.

❖ This section establishes the requirements, based on accepted practice and referenced standards, regulating materials and construction methods used in engineered and empirically designed masonry construction. Engineered masonry construction is further regulated by Sections 2101.2.1 and 2101.2.2.

2104.1.1 Tolerances. Masonry, except masonry veneer, shall be constructed within the tolerances specified in ACI 530.1/ASCE 6/TMS 602.

❖ The MSJC Specification, ACI 530.1/ASCE 6/TMS 602, prescribes tolerances for placement of reinforcement and masonry units. Specific tolerances listed include cross sections of elements; thicknesses of mortar joints; widths of grout spaces; variation from level; variation from plumb; trueness to a line; alignment of columns and walls; location of elements and placement of reinforcement and accessories. These required tolerances are intended to inhibit corrosion; provide placement locations reflecting those assumed in design; provide clearance for mortar and grout and maintain compatible lateral deflections of parallel wythes. Tolerances are relatively strict since masonry is a structural material and is also exposed to weather. Aesthetics are not a factor in these requirements.

2104.1.2 Placing mortar and units. Placement of mortar and units shall comply with Sections 2104.1.2.1 through 2104.1.2.5.

❖ Workmanship is of primary importance in providing strong and durable masonry construction. This is especially true for the placement of mortar and masonry unit. Water penetration can often be traced to bond breaks at the mortar-unit interface. Additionally, masonry strength relies on the mortar bedded areas required in this section and ACI 530.1/ASCE 6/TMS 602.

2104.1.2.1 Bed and head joints. Unless otherwise required or indicated on the construction documents, head and bed joints shall be $^3/_8$ inch (9.5 mm) thick, except that the thickness of the bed joint of the starting course placed over foundations shall not be less than $^1/_4$ inch (6.4 mm) and not more than $^3/_4$ inch (19.1 mm).

❖ The required thickness for bed joints has a permitted tolerance of plus or minus $^1/_8$ inch (3 mm). This thickness is intended to provide bonding of reinforcement and ties and to allow for the dimensional tolerances of the masonry units. The greater permitted thickness of the starting mortar course is intended to accommodate variations in the elevation of the top surface of the foundation.

2104.1.2.1.1 Open-end units. Open-end units with beveled ends shall be fully grouted. Head joints of open-end units with beveled ends need not be mortared. The beveled ends shall form a grout key that permits grouts within $^5/_8$ inch (15.9 mm) of the face of the unit. The units shall be tightly butted to prevent leakage of the grout.

❖ Open-end units can be placed around vertical reinforcing steel, rather than having to be lifted up and over it, as closed-end units must be. Such units are manufactured with one or no end shells.

The special type of open-end unit described in this section is manufactured with beveled ends of the faceshells, so that when the unit is grouted, grout fills the head joint seam (see Figure 2104.1.2.1.1). Because of this, head joints of such construction need not be filled with mortar. Open-end units without this bevel must be laid with mortared head joints.

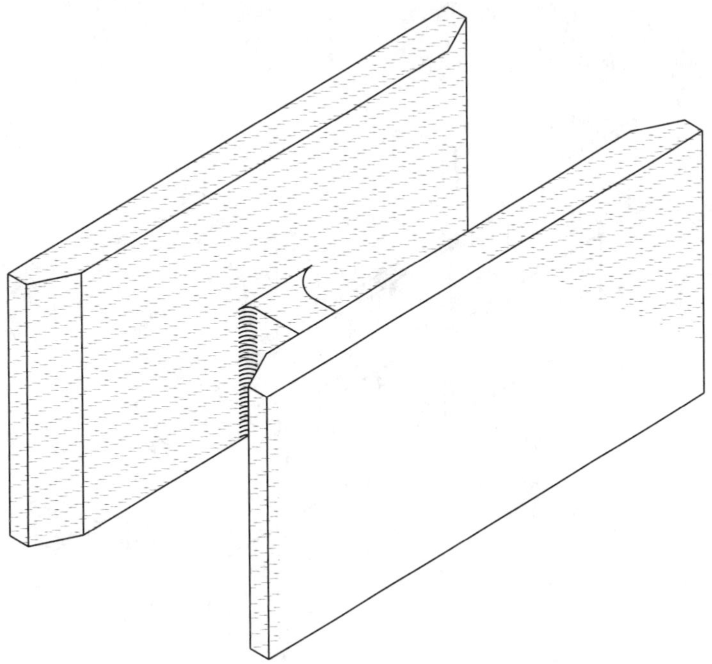

Figure 2104.1.2.1.1
MORTARLESS HEAD JOINT UNIT

2104.1.2.2 Hollow units. Hollow units shall be placed such that face shells of bed joints are fully mortared. Webs shall be fully mortared in all courses of piers, columns, pilasters, in the starting course on foundations where adjacent cells or cavities are to be grouted, and where otherwise required. Head joints shall be mortared a minimum distance from each face equal to the face shell thickness of the unit.

❖ Figure 2104.1.2.2 shows the typical placement of mortar for hollow-unit masonry walls. Except for the initial bed joint at the starter course, the cross webs of hollow units in walls are not usually mortared except at the cross webs of cells that are to be grouted in partially grouted walls. Cross webs of hollow units in piers, columns and pilasters, however, are required to be mortared.

2104.1.2.3 Solid units. Unless otherwise required or indicated on the construction documents, solid units shall be placed in fully mortared bed and head joints. The ends of the units shall be completely buttered. Head joints shall not be filled by slushing with mortar. Head joints shall be constructed by shoving mortar tight against the adjoining unit. Bed joints shall not be furrowed deep enough to produce voids.

❖ Solid units result in fully mortared bed joints since the 25-percent core area is effectively crossed with mortar. Height-to-thickness ratios and calculated stresses are based on the fully mortared bed joint.

Filling head joints by slushing from above is not permitted, since this could result in voids and lead to water penetration. The head-joint mortar may be placed on the end of the masonry unit prior to placing the unit in the wall. The furrow typically formed during bed joint placement should not be too deep since mortar voids would result (see Figure 2104.1.2.3).

2104.1.2.4 Glass unit masonry. Glass units shall be placed so head and bed joints are filled solidly. Mortar shall not be furrowed.

Unless otherwise required, head and bed joints of glass unit masonry shall be $^1/_4$ inch (6.4 mm) thick, except that vertical joint thickness of radial panels shall not be less than $^1/_8$ inch (3.2 mm). The bed joint thickness tolerance shall be minus $^1/_{16}$ inch (1.6 mm) and plus $^1/_8$ inch (3.2 mm). The head joint thickness tolerance shall be plus or minus $^1/_8$ inch (3.2 mm).

❖ Glass unit masonry is required to be placed with full head and bed joints. Glass units are manufactured to be modular with a joint thickness of $^1/_4$ inch (6.4 mm), smaller than the typical $^3/_8$ inch (9.6 mm) for concrete masonry and the $^3/_8$ inch (9.6 mm) or $^1/_2$ inch (12.8 mm) joint thickness for clay brick masonry. Designers should recognize this difference and use correct nominal dimensions when laying out glass unit masonry.

Tolerances for glass unit masonry are much tighter than for other types of masonry because glass unit masonry units are manufactured to very tight tolerances and are more dimensionally stable than other types of masonry.

Head joints on radial panels of glass unit masonry are permitted to be as thin as $^1/_8$ inch (3.2 mm), because placing rectangular units in a curved wall requires wider joints on one face of the wall and narrower joints on the other face.

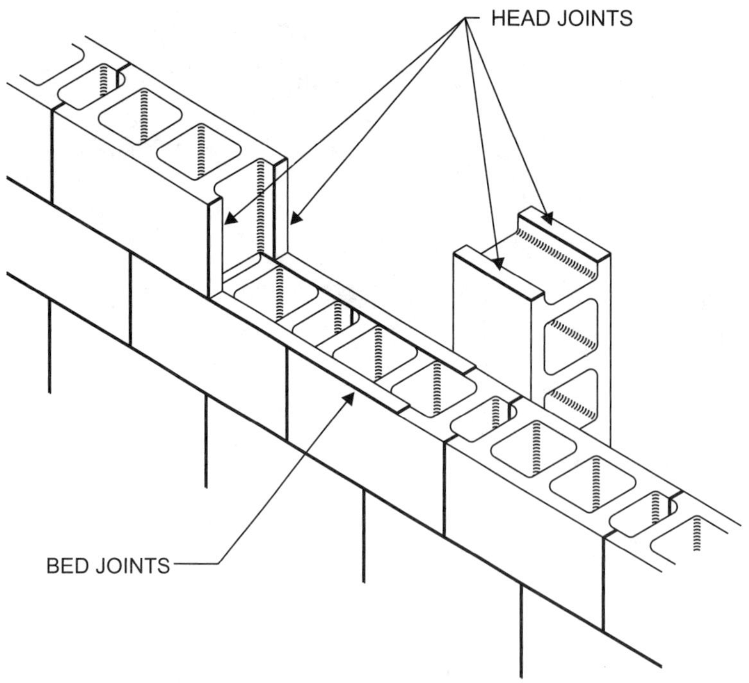

Figure 2104.1.2.2
MORTAR PLACEMENT ON HOLLOW UNITS IN WALLS

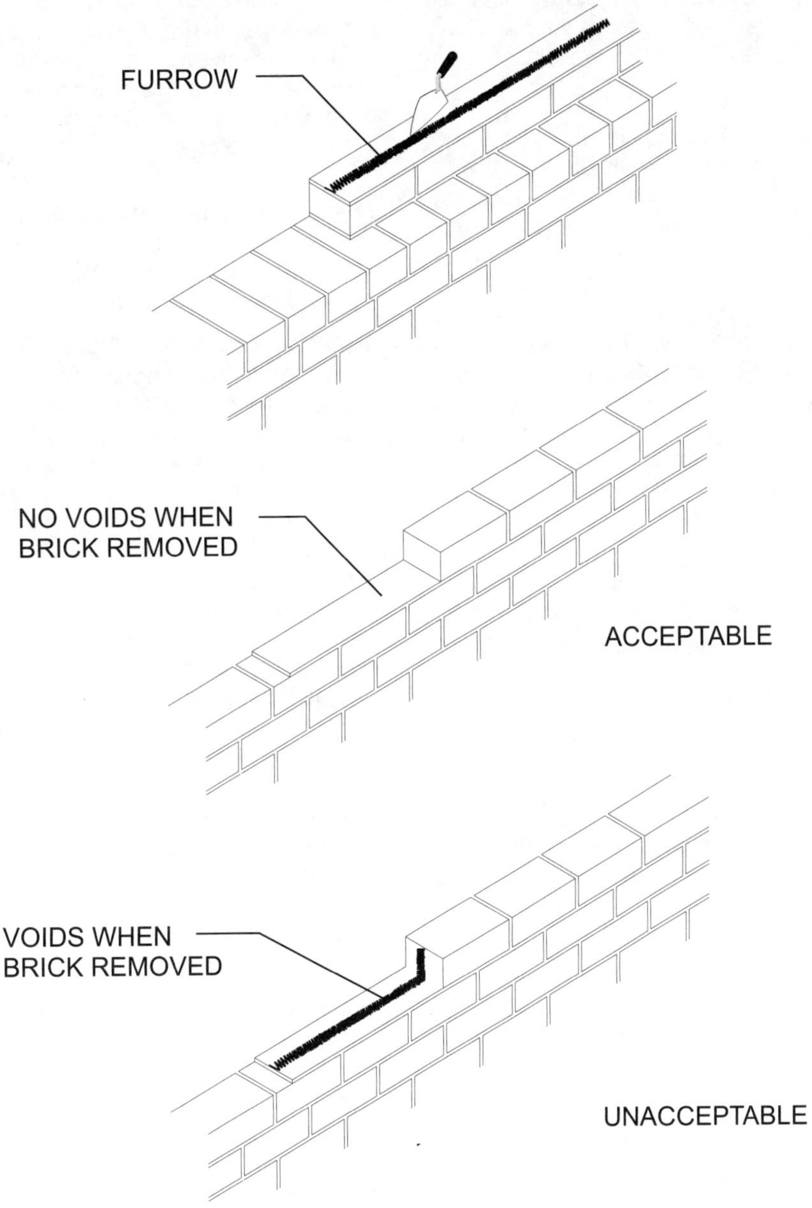

Figure 2104.1.2.3
FURROWING

2104.1.2.5 All units. Units shall be placed while the mortar is soft and plastic. Any unit disturbed to the extent that the initial bond is broken after initial positioning shall be removed and relaid in fresh mortar.

❖ Mortar should not be spread too far ahead of units, as it will stiffen and lose plasticity, especially in hot weather. Mortar that has stiffened should not be used. ASTM C 270 requires that mortar be used within 2½ hours of initial mixing.

A test for mortar workmanship includes removal of a unit. Bonded mortar will stick to the removed unit as well as the remaining masonry; stiff, unbonded mortar will not. Broken joints should be remortared.

2104.1.3 Installation of wall ties. The ends of wall ties shall be embedded in mortar joints. Wall tie ends shall engage outer face shells of hollow units by at least $^1/_2$ inch (12.7 mm). Wire wall ties shall be embedded at least $1^1/_2$ inches (38 mm) into the mortar bed of solid masonry units or solid-grouted hollow units. Wall ties shall not be bent after being embedded in grout or mortar.

❖ Installation of wall ties requires adequate embedment in mortar, adequate strength and durability of ties. Wall ties are not permitted to be bent after being embedded in grout or mortar, since movement of the ties will reduce the effectiveness of the embedment and subsequent bonding.

2104.1.4 Chases and recesses. Chases and recesses shall be constructed as masonry units are laid. Masonry directly above chases or recesses wider than 12 inches (305 mm) shall be supported on lintels.

❖ Chases and recesses are designed so that they do not reduce the required strength of the walls or affect the required fire-resistance rating of the wall. Any required chases must be formed during construction of the masonry wall and not by cutting out a section of the finished wall, which could create a weak point and jeopardize the wall strength.

This empirical limitation requires lintels for chases and recesses wider than 12 inches (305 mm). Although arching action reduces lintel moments, the potential of weaknesses created by chases and recesses makes them subject to engineered design. The supporting lintel is permitted to be of masonry, concrete or steel and must be engineered as required in Section 2104.1.5.

2104.1.5 Lintels. The design for lintels shall be in accordance with the masonry design provisions of either Section 2107 or 2108. Minimum length of end support shall be 4 inches (102 mm).

❖ Masonry lintels are required to be engineered as load-bearing beam elements by either the working stress design method of Section 2107 or the strength design method of Section 2108. This requirement is consistent with the empirical provisions, which do not include horizontally spanning elements such as beams. Bearing on masonry is required to be a minimum of 4 inches (102 mm) to adequately distribute stresses.

2104.1.6 Support on wood. Masonry shall not be supported on wood girders or other forms of wood construction except as permitted in Section 2304.12.

❖ Masonry is not permitted to be supported on wood construction, except as permitted in Section 2304.12, because of both the danger of collapse in case of fire and the serviceability considerations noted below.

Masonry is brittle and relatively weak in tension. It can crack when subjected to deformations. Wood is flexible and usually exhibits elastic deformation and creep (additional deflection from long-term loads). Because of this, masonry supported on wood tends to crack. Therefore, masonry is not permitted to be supported by wood members, unless such masonry meets the exceptions in Section 2304.12.

Glass-block panels are typically limited in size and are often used in windows as decoration. For this limited use, support on wood is permitted. The deflection limitation in Section 2110.4.2 is consistent with the requirements for masonry. In calculating the probable deflection of the supporting wood member, consideration of creep is recommended.

2104.1.7 Masonry protection. The top of unfinished masonry work shall be covered to protect the masonry from the weather.

❖ Water can accumulate within masonry walls and it evaporates relatively slowly. Moisture migrates to the surface, dissolves salts in the masonry and deposits them on the surface as a white powder called "efflorescence." To prevent this, the tops of walls exposed to weather are required to be covered.

2104.1.8 Weep holes. Weep holes provided in the outside wythe of masonry walls shall be at a maximum spacing of 33 inches (838 mm) on center (o.c.). Weep holes shall not be less than $^{3}/_{16}$ inch (4.8 mm) in diameter.

❖ Masonry is not water tight. Water can penetrate into and through it. Weep holes are to be provided in all masonry wall construction to help water leave the wall. Without weep holes, moisture would remain within the masonry construction and evaporation would be its only means of exit.

Many methods for providing weep holes are available, including open head joints; $^{3}/_{8}$-inch-diameter (9.6 mm) holes in joints and wicking cord laid in the joint and behind the brick.

The 33-inch (838 mm) maximum spacing of weep holes, while apparently odd since it is not modular for typical masonry, was selected because weep holes are often spaced at 32 inches (813 mm) on center, plus or minus typical construction tolerances.

2104.2 Corbeled masonry. The maximum corbeled projection beyond the face of the wall shall not be more than one-half of the wall thickness nor one-half the wythe thickness for hollow walls. The maximum projection of one unit shall neither exceed one-half the height of the unit nor one-third the thickness at right angles to the wall.

❖ This section covers the limitations and the construction features of corbeled masonry.

A corbel is formed by projecting courses of masonry with the first or lowest course projecting out from the face of the wall and each successive course projecting out from the supporting course below. While corbels may be constructed using various load-bearing masonry units, they are most commonly applied to clay masonry and are used for aesthetic purposes and to support offset or thickened walls above them.

The permitted extent of corbeling is illustrated in Figure 2104.2. The total horizontal projection of a corbel from the face of the wall is limited to no more than one-half the wall thickness of a solid wall or more than one-half the thickness of the wythe in a cavity wall. Furthermore, the projection of a single course is not to exceed one-half of the unit height, nor one-third of the unit bed depth, whichever is less.

These limitations establish a maximum angle of corbelled brick masonry, measured from the plane of the wall, of about $26^{1}/_{2}$ degrees (0.46 rad). A smaller angle can be achieved by decreasing the brick projections to less than the maximum allowed by this section.

Corbels produce and often transfer eccentric loads, which must be considered in the design of masonry walls.

Corbels constructed in walls of hollow units or in hollow masonry walls are required to be of solid masonry. This requirement can be satisfied by using solid units, grouting the cores of hollow units or grouting open spaces in hollow walls.

2104.2.1 Molded cornices. Unless structural support and anchorage are provided to resist the overturning moment, the center of gravity of projecting masonry or molded cornices shall lie within the middle one-third of the supporting wall. Terra cotta and metal cornices shall be provided with a structural frame of approved noncombustible material anchored in an approved manner.

❖ Figure 2104.2.1 illustrates this requirement. Unless cornices or other projections are specifically designed to be supported and anchored to the wall or other adequate structural supports, such as columns, beams, spandrels or floor slabs, the center of gravity of the projecting element is required to fall within the middle third (kern) of the supporting wall.

Cornices are usually ornamental and may or may not support other loads. They can be made of cast-in-place or precast concrete; cast or natural stone; terra cotta; brick or even metal.

Metal and terra-cotta cornices are commonly furnished with a structural frame of noncombustible material that can be anchored to the masonry wall or to other building supports as approved by the building official.

2104.3 Cold weather construction. The cold weather construction provisions of ACI 530.1/ASCE 6/TMS 602, Article 1.8 C, or the following procedures shall be implemented when either the ambient temperature falls below 40°F (4°C) or the temperature of masonry units is below 40°F (4°C).

❖ When masonry construction is conducted in temperatures below 40°F (4°C), both the masonry components and the structure are required to be protected in accordance with this section. One of the main goals of these provisions is to prevent fresh mortar from freezing.

During cold-weather construction, especially in temperatures below freezing, the curing and subsequent performance of masonry are influenced by the temperature and properties of the mortar and the masonry units, as well as the severity of the exposure (temperature and wind).

2104.3.1 Preparation.

1. Temperatures of masonry units shall not be less than 20°F (-7°C) when laid in the masonry. Masonry units containing frozen moisture, visible ice or snow on their surface shall not be laid.

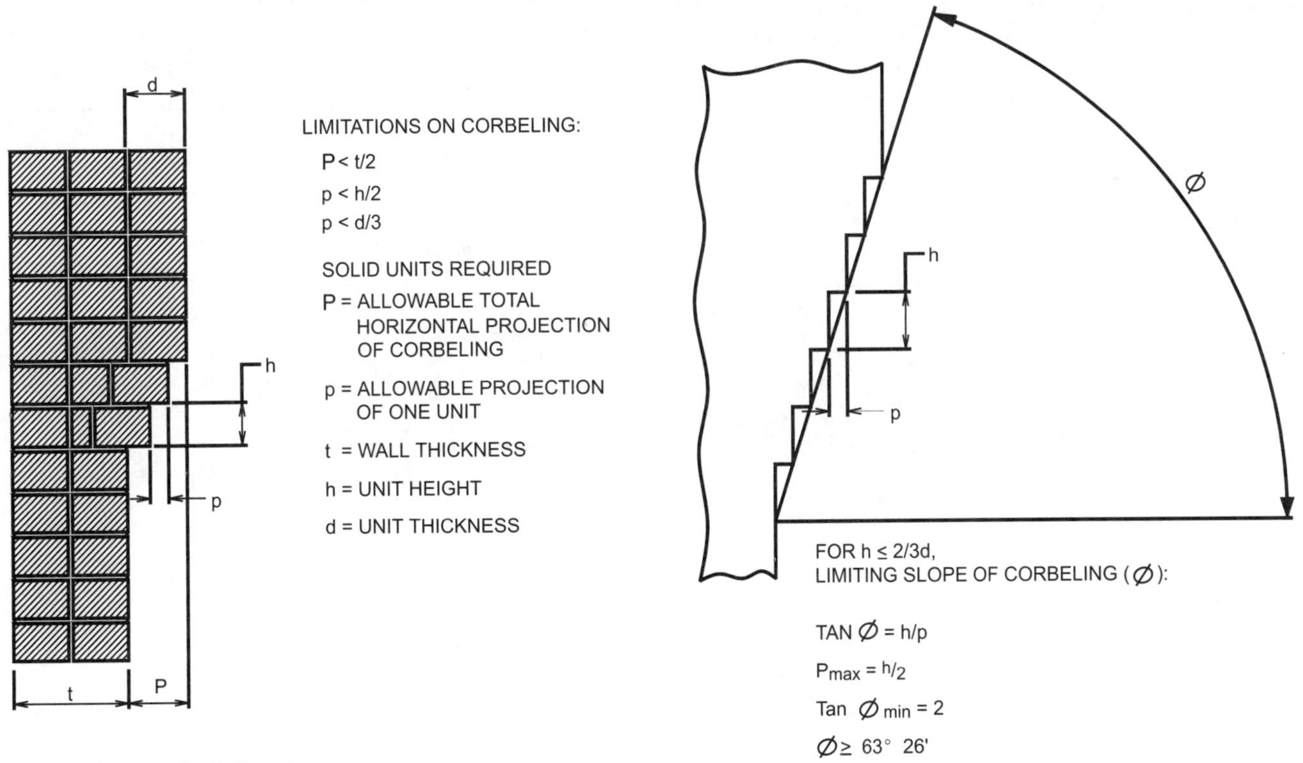

For SI: 1 degree = 0.01745 rad.

Figure 2104.2
CORBELED MASONRY

2. Visible ice and snow shall be removed from the top surface of existing foundations and masonry to receive new construction. These surfaces shall be heated to above freezing, using methods that do not result in damage.

❖ Under cold-weather conditions masonry construction should not proceed if these requirements are not met. Masonry units to be used during cold-weather construction should be dry. Masonry must not be laid on snow or ice-covered surfaces, because bond cannot be developed between the units and the mortar bed under those conditions. Additionally, the units can move when the mortar thaws.

2104.3.2 Construction. The following requirements shall apply to work in progress and shall be based on ambient temperature.

❖ This section sets minimum requirements to be met as masonry work proceeds based on the ambient temperature. Mortars (or grout) mixed at low temperatures using cold (not frozen) materials have plastic properties that are significantly different from those mixed at temperatures above 40°F (4°C). If the water content of the mortar is more than 6 to 8 percent when it freezes, expansion and possible damage could occur. If the water content of the mortar is less than 6 to 8 percent, the expected mortar expansion is less severe.

Antifreeze admixtures should not be permitted as a means of lowering the freezing point of mortar. Simply keeping the mortar from freezing does not result in proper hydration. Some commercially available mortar admixtures that claim to have antifreeze qualities may be improperly described or labeled because they do not significantly lower the freezing point of the mortar, but rather serve as accelerators of cement hydration. Calcium chloride is the most commonly used accelerator and is the main ingredient in many commercially available mortar admixtures. Because chlorides enhance the corrosion of embedded steel, however, admixtures containing chlorides should not be used in masonry containing metal ties, joint reinforcement or other metal accessories in contact with mortar.

Mortar mixed at a low temperature requires longer curing times and gains strength more slowly than mortar mixed at higher temperatures. To counteract the slow strength development of mortar at low temperatures, mixing water is required to be heated. In some instances, the sand used in the mortar is also heated. To accelerate early strength development, Type III (high-early-strength portland cement) can be substituted for Type I cement.

Cold-weather conditions do not require any changes to the mortar or grout mix proportions of cement, lime and sand (and coarse aggregate for grout).

2104.3.2.1 Construction requirements for temperatures between 40°F (4°C) and 32°F (0°F). The following construction requirements shall be met when the ambient temperature is between 40°F (4°C) and 32°F (0°C):

1. Glass unit masonry shall not be laid.

2. Water and aggregates used in mortar and grout shall not be heated above 140°F (60°C).

3. Mortar sand or mixing water shall be heated to produce mortar temperatures between 40°F (4°C) and 120°F

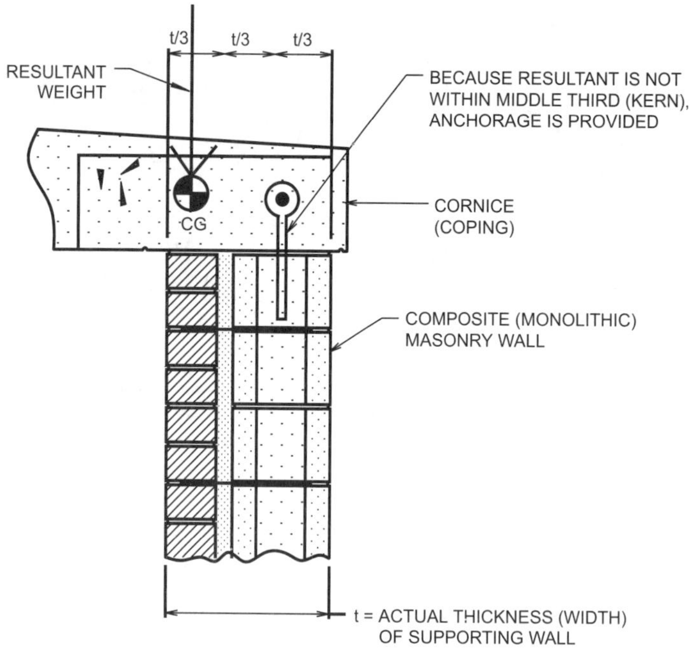

Figure 2104.2.1
MOLDED CORNICE

(49°C) at the time of mixing. When water and aggregates for grout are below 32°F(0°C), they shall be heated.

❖ See the commentary to Section 2104.3.2.

2104.3.2.2 Construction requirements for temperatures between 32°F (0°C) and 25°F (-4°C). The requirements of Section 2104.3.2.1 and the following construction requirements shall be met when the ambient temperature is between 32°F (0°C) and 25°F (-4°C):

1. The mortar temperature shall be maintained above freezing until used in masonry.
2. Aggregates and mixing water for grout shall be heated to produce grout temperature between 70°F (21°C) and 120°F (49°C) at the time of mixing. Grout temperature shall maintained above 70°F (21°C) at the time of grout placement.

❖ See the commentary to Section 2104.3.2.

2104.3.2.3 Construction requirements for temperatures between 25°F (-4°C) and 20°F (-7°C). The requirements of Sections 2104.3.2.1 and 2104.3.2.2 and the following construction requirements shall be met when the ambient temperature is between 25°F (-4°C) and 20°F (-7°C):

1. Masonry surfaces under construction shall be heated to 40°F (4°C).
2. Wind breaks or enclosures shall be provided when the wind velocity exceeds 15 miles per hour (mph) (24 km/h).
3. Prior to grouting, masonry shall be heated to a minimum of 40°F (4°C).

❖ Low temperatures do not significantly affect the performance characteristics of masonry units. The absorption (suction) characteristics of cold, wet and frozen masonry units are decreased, however, which may affect mortar bond. Preheating masonry units prevents the sudden cooling of warm mortar in contact with cold units. A heated unit will absorb more water from the mortar because of the absorptive characteristics of a cooling body.

In selecting masonry units for cold-weather construction, consideration should be given to the absorption rates (see ASTM C 67). Units with initial rates of absorption of 30 grams per 30 square inches (19 355 mm^2) per minute are preferable for cold-weather construction because they greatly reduce the possibility of disruptive expansion of the mortar due to freezing. Units with an initial rate of absorption of 5 to 6 grams per 30 square inches (19 355 mm^2) per minute or less may not absorb sufficient water to prevent disruptive expansion if the mortar freezes (also see commentary, Section 2104.3.2).

2104.3.2.4. Construction requirements for temperatures below 20°F (-7°C). The requirements of Sections 2104.3.2.1, 2104.3.2.2 and 2104.3.2.3 and the following construction requirement shall be met when the ambient temperature is below 20°F (-7°C): Enclosures and auxiliary heat shall be provided to maintain air temperature within the enclosure to above 32°F (0°C).

❖ In addition to all of the above requirements, ambient temperatures below 20°F (-7°C) require the masonry construction to be enclosed and the use of auxiliary heat (also see commentary, Section 2104.3.2).

2104.3.3 Protection. The requirements of this section and Sections 2104.3.3.1 through 2104.3.3.4 apply after the masonry is placed and shall be based on anticipated minimum daily temperature for grouted masonry and anticipated mean daily temperature for ungrouted masonry.

❖ This section sets minimum requirements to be met after masonry work is completed based on the anticipated temperature. Lower expected temperatures require more extensive means of protection.

2104.3.3.1 Glass unit masonry. The temperature of glass unit masonry shall be maintained above 40°F (4°C) for 48 hours after construction.

❖ See the commentary to Sections 2104.3 and 2104.3.3.

2104.3.3.2 Protection requirements for temperatures between 40°F (4°C) and 25°F (-4°C). When the temperature is between 40°F (4°C) and 25°F (-4°C), newly constructed masonry shall be covered with a weather-resistive membrane for 24 hours after being completed.

❖ See the commentary to Sections 2104.3 and 2104.3.3.

2104.3.3.3 Protection requirements for temperatures between 25°F (-4°C) and 20°F (-7°C). When the temperature is between 25°F (-4°C) and 20°F (-7°C), newly constructed masonry shall be completely covered with weather-resistive insulating blankets, or equal protection, for 24 hours after being completed. The time period shall be extended to 48 hours for grouted masonry, unless the only cement in the grout is Type III portland cement.

❖ See the commentary to Sections 2104.3 and 2104.3.3.

2104.3.3.4 Protection requirements for temperatures below 20°F (-7°C). When the temperature is below 20°F (-7°C), newly constructed masonry shall be maintained at a temperature above 32°F (0°C) for at least 24 hours after being completed by using heated enclosures, electric heating blankets, infrared lamps or other acceptable methods. The time period shall be extended to 48 hours for grouted masonry, unless the only cement in the grout is Type III portland cement.

❖ When temperatures are less than 20°F (-7°C), temporary heating is required to protect the masonry (also see commentary, Sections 2104.3 and 2104.3.3).

2104.4 Hot weather construction. The hot weather construction provisions of ACI 530.1/ASCE 6/TMS 602, Article 1.8 D, or the following procedures shall be implemented when the temperature or the temperature and wind-velocity limits of this section are exceeded.

❖ Temperature, solar radiation, humidity and wind influence the rate of absorption of masonry units and the rate of mortar set. During hot-weather conditions, special precautions must be taken for adequate strength gain of the masonry.

2104.4.1 Preparation. The following requirements shall be met prior to conducting masonry work.

❖ This section imposes minimum required precautions before masonry work proceeds under hot-weather conditions. The purpose is to maintain mortar temperatures less than that which causes a flash set.

2104.4.1.1 Temperature. When the ambient temperature exceeds 100°F (38°C), or exceeds 90°F (32°C) with a wind velocity greater than 8 mph (13 km/h):

1. Necessary conditions and equipment shall be provided to produce mortar having a temperature below 120°F (49°C).
2. Sand piles shall be maintained in a damp, loose condition.

❖ When hot-weather conditions exceed those specified, procedures are required to keep mortar and sand in an acceptable condition.

2104.4.1.2 Special conditions. When the ambient temperature exceeds 115°F (46°C), or 105°F (40°C) with a wind velocity greater than 8 mph (13 km/h), the requirements of Section 2104.4.1.1 shall be implemented, and materials and mixing equipment shall be shaded from direct sunlight.

❖ When hot-weather conditions are extreme, materials and mixing equipment are required to be shaded, in addition to the requirements in Section 2104.4.1.1. Unshaded equipment becomes so hot that mortar can experience flash set.

2104.4.2 Construction. The following requirements shall be met while masonry work is in progress.

❖ This section imposes minimum precautions to be taken as masonry work proceeds under hot-weather conditions. Their purpose is to keep mortar temperatures lower than that which causes flash set. Ice is prohibited because of the potential for weak mortar and grout.

2104.4.2.1 Temperature. When the ambient temperature exceeds 100°F (38°C), or exceeds 90°F (32°C) with a wind velocity greater than 8 mph (13 km/h):

1. The temperature of mortar and grout shall be maintained below 120°F (49°C).
2. Mixers, mortar transport containers and mortar boards shall be flushed with cool water before they come into contact with mortar ingredients or mortar.
3. Mortar consistency shall be maintained by retempering with cool water.
4. Mortar shall be used within 2 hours of initial mixing.

❖ When hot-weather conditions exceed the limitations of this section, procedures are required to keep mortar and grout cool. These procedures include flushing mortar board, containers and other surfaces in contact with mortar and grout with water and retempering.

2104.4.2.2 Special conditions. When the ambient temperature exceeds 115°F (46°C), or exceeds 105°F (40°C) with a wind velocity greater than 8 mph (13 km/h), the requirements of Section 2104.4.2.1 shall be implemented and cool mixing water shall be used for mortar and grout. The use of ice shall be permitted in the mixing water prior to use. Ice shall not be permitted in the mixing water when added to the other mortar or grout materials.

❖ In extreme hot-weather conditions, ice can be added to the mixing water for mortar and grout to maintain their temperatures at acceptable levels. Ice cannot, however, be present in the mixing water when it is added to the dry materials in the mortar or grout.

2104.4.3 Protection. When the mean daily temperature exceeds 100°F (38°C), or exceeds 90°F (32°C) with a wind velocity greater than 8 mph (13 km/h), newly constructed masonry shall be fog sprayed until damp at least three times a day until the masonry is three days old.

❖ When hot-weather conditions are extended (as defined by the mean daily temperature), masonry is required to be fogsprayed to help hydrate it. Excessive wetting of the wall, however, can cause adverse effects such as efflorescence and moisture expansion in concrete masonry.

2104.5 Wetting of brick. Brick (clay or shale) at the time of laying shall require wetting if the unit's initial rate of water absorption exceeds 30 grams per 30 square inches (19 355 mm^2) per minute or 0.035 ounce per square inch (1 g/645 mm^2) per minute, as determined by ASTM C 67.

❖ Clay brick laid on fresh mortar generally absorbs water from the mortar along with fine particles of cementitious materials, which help to bond the mortar with the brick. If the brick absorbs too much water, the mortar may have insufficient water for hydration. It is therefore necessary to determine the initial rate of absorption (IRA) of the brick.

Mortar bonds best with brick whose IRA is between 5 and 30 grams per 30 square inches (19 355 mm^2) of surface per minute. Where the water absorption rate exceeds 30 grams per 30 square inches (19 355 mm^2) per minute (as determined by ASTM C 67), brick is required to be wetted before placement, but should be surface dry when laid.

SECTION 2105
QUALITY ASSURANCE

2105.1 General. A quality assurance program shall be used to ensure that the constructed masonry is in compliance with the construction documents.

❖ This section requires that a quality assurance program be used so that the masonry is constructed according to the contract documents. In addition, inspection and testing complying with Chapter 17 also requires verification of the masonry compressive strength.

The quality assurance program shall comply with the inspection and testing requirements of Chapter 17.

2105.2 Acceptance relative to strength requirements.

❖ The quality assurance provisions in this section emphasize verification of masonry compressive strengths. This is accomplished by comparing conservatively estimated strengths (based on unit strength and mortar type) or tested prism strengths to the specified compressive strength of the masonry, f'_m, and, when required, by mortar, grout or both to see that they are in compliance.

These quality assurance methods are for general consistency of the constructed masonry. Masonry is relatively strong in compression and thus rarely fails in that manner, but rather in flexural tension. Compression tests are required, however, because they are simple ways of assessing quality.

As implied in the above paragraph, two methods are prescribed in Section 2105.2 to judge acceptance relative to the compressive strength of the masonry assemblage—the unit strength method and the prism test method. These are described in Section 2105.2 and in this commentary. When the strength of constructed masonry is questioned, testing of prisms that have been saw cut from the masonry is permitted in accordance with Section 2105.3.

2105.2.1 Compliance with f'_m. Compressive strength of masonry shall be considered satisfactory if the compressive strength of each masonry wythe and grouted collar joint equals or exceeds the value of f'_m.

❖ Design of structural masonry is based on the specified compressive strength of the masonry, f'_m. This strength is required to be shown on the contract documents, since structural design is based on it. Strength of the constructed masonry determined by the unit strength method or the prism strength method is required to equal or exceed the specified compressive strength of the masonry, f'_m.

In a multiwythe wall designed as a composite wall, the compressive strength of masonry for each wythe or grouted collar joint must equal or exceed f'_m.

2105.2.2 Determination of compressive strength. The compressive strength for each wythe shall be determined by the unit strength method or by the prism test method as specified herein.

❖ There are two means of determining the compressive strength of masonry: the unit strength method and the prism test method. The first eliminates the expense of prism tests, but is more conservative.

2105.2.2.1 Unit strength method.

❖ This section describes a prescriptive procedure to estimate the expected compressive strength of the masonry based on the compressive strength of masonry units and the mortar type. This so-called unit strength method was generated using prism test data. Mortar joint thickness is limited because it influences the compressive strength of masonry.

2105.2.2.1.1 Clay masonry. The compressive strength of masonry shall be determined based on the strength of the units and the type of mortar specified using Table 2105.2.2.1.1, provided:

1. Units conform to ASTM C 62, ASTM C 216 or ASTM C 652 and are sampled and tested in accordance with ASTM C 67.
2. Thickness of bed joints does not exceed $^5/_8$ inch (15.9 mm).
3. For grouted masonry, the grout meets one of the following requirements:
   3.1. Grout conforms to ASTM C 476.
   3.2. Minimum grout compressive strength equals f'_m but not less than 2,000 psi (13.79 MPa). The compressive strength of grout shall be determined in accordance with ASTM C 1019.

❖ This section prescribes the conditions under which the unit strength method can be used for clay masonry. These conditions include requirements for clay masonry units, maximum mortar joint thickness and grout.

TABLE 2105.2.2.1.1
COMPRESSIVE STRENGTH OF CLAY MASONRY

NET AREA COMPRESSIVE STRENGTH OF CLAY MASONRY UNITS (psi)		NET AREA COMPRESSIVE STRENGTH OF MASONRY (psi)
Type M or S mortar	Type N mortar	
1,700	2,100	1,000
3,350	4,150	1,500
4,950	6,200	2,000
6,600	8,250	2,500
8,250	10,300	3,000
9,900	—	3,500
13,200	—	4,000

For SI: 1 pound per square inch = 0.00689 Mpa.

❖ Table 2105.2.2.1.1 lists the compressive strength of masonry in terms of the strength of the clay masonry unit and the mortar type. This table is based on the research results cited in the commentary to ACI 530.1/ASCE 6/TMS 602. A similar table has been used successfully in both ACI 530.1/ASCE 6/TMS 602 and the *Uniform Building Code*™ (UBC) since 1988.

The designer can use this table to estimate a specified compressive strength of masonry to use in design, based on the expected strength of the clay masonry units and the specified mortar type. The contractor can use the table to find what clay unit masonry strength and mortar type are needed to comply with the specified strength of the masonry, f'_m, given in the contract documents. The column entitled "Net Area Compressive Strength of Masonry" must equal or exceed the specified strength of the masonry, f'_m.

2105.2.2.1.2 Concrete masonry. The compressive strength of masonry shall be determined based on the strength of the unit and type of mortar specified using Table 2105.2.2.1.2, provided:

1. Units conform to ASTM C 55 or ASTM C 90 and are sampled and tested in accordance with ASTM C 140.
2. Thickness of bed joints does not exceed $^5/_8$ inch (15.9 mm).
3. For grouted masonry, the grout meets one of the following requirements:
   3.1. Grout conforms to ASTM C 476.
   3.2. Minimum grout compressive strength equals f'_m but not less than 2,000 psi (13.79 MPa). The compressive strength of grout shall be determined in accordance with ASTM C 1019.

❖ This section prescribes the conditions under which the unit strength method can be used for concrete masonry. These conditions include requirements for the concrete masonry units, maximum mortar joint thickness and grout. Concrete masonry units must be tested in accordance with ASTM C 140.

TABLE 2105.2.2.1.2
COMPRESSIVE STRENGTH OF CONCRETE MASONRY

NET AREA COMPRESSIVE STRENGTH OF CONCRETE MASONRY UNITS (psi)		NET AREA COMPRESSIVE STRENGTH OF MASONRY (psi)[a]
Type M or S mortar	Type N mortar	
1,250	1,300	1,000
1,900	2,150	1,500
2,800	3,050	2,000
3,750	4,050	2,500
4,800	5,250	3,000

For SI: 1 inch = 25.4 mm, 1 pound per square inch = 0.00689 MPa.
a. For units less than 4 inches in height, 85 percent of the values listed.

❖ Table 2105.2.2.1.2 lists the compressive strength of masonry in terms of the strengths of the concrete masonry units and the mortar type. This table is based on the research cited in the commentary to ACI 530.1/ASCE 6/TMS 602. A similar table has been used successfully in both ACI 530.1/ASCE 6/TMS 602 and the UBC since 1988.

The designer can use this table to estimate a specified compressive strength of masonry for use in design, based on the expected strength of the concrete masonry units and the specified mortar type. The contractor can use the table to find what concrete masonry strength and mortar type are needed to comply with the specified strength of the masonry, f'_m, given in the contract documents. The column entitled "Net Area Compressive Strength of Masonry" must equal or exceed the specified strength of the masonry, f'_m.

2105.2.2.2 Prism test method.

❖ The prism test method is used when required in the project specifications or when the restrictions of Section 2105.2.2.1 do not apply. Prisms are required to be constructed in accordance with ASTM C 1314 using the same materials and workmanship as in the structure.

ASTM C 1314 replaced ASTM E 447 for field-constructed prism specimens, which was referenced in editions of the specification prior to 1999. The use of ASTM C 1314 is intended to address many of the concerns over the difficulty and imprecision of ASTM E 447 for large prisms.

2105.2.2.2.1 General. The compressive strength of masonry shall be determined by the prism test method:

1. Where specified in the construction documents.
2. Where masonry does not meet the requirements for application of the unit strength method in Section 2105.2.2.1.

❖ Prism tests are required whenever specified and whenever the masonry does not meet the restrictions for the unit strength method.

2105.2.2.2.2 Number of prisms per test. A prism test shall consist of three prisms constructed and tested in accordance with ASTM C 1314.

❖ Whenever prism testing is specified or used, three prism specimens must be constructed and tested in accordance with ASTM C 1314.

2105.3 Testing prisms from constructed masonry. When approved by the building official, acceptance of masonry that does not meet the requirements of Section 2105.2.2.1 or 2105.2.2.2 shall be permitted to be based on tests of prisms cut from the masonry construction in accordance with Sections 2105.3.1, 2105.3.2 and 2105.3.3.

❖ While uncommon, there are times when the strength of masonry determined by the unit strength method or prism test method may be questioned or may be lower than the specified strength. Because low strengths could result from inappropriate testing procedures or unintentional damage to the test specimens, prisms may be saw cut from the completed masonry wall and tested. This section prescribes procedures for such tests.

Such testing is difficult, requires at least 28 days and requires replacement of the affected wall area. Therefore, every effort should be taken so that strengths determined by the unit strength method or the prism test method are adequate.

2105.3.1 Prism sampling and removal. A set of three masonry prisms that are at least 28 days old shall be saw cut from the masonry for each 5,000 square feet (465 m^2) of the wall area that is in question but not less than one set of three masonry prisms for the project. The length, width and height dimensions of the prisms shall comply with the requirements of ASTM C 1314. Transporting, preparation and testing of prisms shall be in accordance with ASTM C 1314.

❖ Removal of prisms from a constructed wall requires care so that the prism is not damaged and that damage to the wall is minimal. Prisms must be representative of the wall, yet not contain reinforcing steel, which would bias the results. As with a prism test of newly constructed masonry, a prism test from existing masonry requires three prism specimens.

2105.3.2 Compressive strength calculations. The compressive strength of prisms shall be the value calculated in accordance ASTM C 1314, except that the net cross-sectional area of the prism shall be based on the net mortar bedded area.

❖ Compressive strength calculations from saw-cut specimens must be based on the net mortar bedded area, which must be determined before the prism is tested. The testing agency must determine this area accurately.

2105.3.3 Compliance. Compliance with the requirement for the specified compressive strength of masonry, f'_m, shall be considered satisfied provided the modified compressive strength equals or exceeds the specified f'_m. Additional testing of specimens cut from locations in question shall be permitted.

❖ Strengths determined from saw-cut prisms must equal or exceed the specified strength of masonry, f'_m.

SECTION 2106 SEISMIC DESIGN

2106.1 Seismic design requirements for masonry. Masonry structures and components shall comply with the requirements in Section 1.13.2.2 of ACI 530/ASCE 5/TMS 402 and Section 1.13.3, 1.13.4, 1.13.5, 1.13.6 or 1.13.7 of ACI 530/ASCE 5/TMS 402 depending on the structure's seismic design category as determined in Section 1616.3. All masonry walls, unless isolated on three edges from in-plane motion of the basic structural systems, shall be considered to be part of the seismic-force-resisting system. In addition, the following requirements shall be met.

❖ Section 2106 contains minimum requirements for masonry structures based upon their seismic design category. This section requires the use of MSJC Code seismic design criteria. Requirements established for various seismic risk categories are cumulative from lower to higher categories. These prescriptive and design-oriented provisions have been established to improve the performance of masonry during seismic events by providing additional structural strength, ductility and stability against the dynamic effects of earthquakes. As seismic demand increases, the provisions require more positive connection between structural elements, increased ductility and greater material reliability.

Chapter 16 contains state-of-the-art criteria for seismic design, including provisions applicable to masonry. The requirements of Chapter 16 for seismic resistance (for example, design forces and masonry detailing) remain applicable and are discussed in the commentary to that chapter. Compliance with Chapter 21 is not a substitute for compliance with the seismic provisions of Chapter 16. More information on seismic design is contained in the commentaries to Chapter 16 and the NEHRP Provisions.

To comply with the provisions in Section 2106, the seismic design category must be determined for the building or structure under consideration. Refer to the commentary to Chapter 16 for information on determining the seismic design category and other seismic parameters. Based on the seismic design category, the designer needs to meet the minimum requirements of Section 2106.2, 2106.3, 2106.4, 2106.5 or 2106.6 as well as the referenced MSJC Code sections.

2106.1.1 Basic seismic-force-resisting system. Buildings relying on masonry shear walls as part of the basic seismic-force-resisting system shall comply with Section 1.13.2.2 of ACI 530/ASCE 5/TMS 402 or with Section 2106.1.1.1, 2106.1.1.2 or 2106.1.1.3.

❖ A basic seismic-force-resisting system must be defined for all buildings. Most masonry buildings use shear walls to serve as the basic seismic-force-resisting system, although other systems are sometimes used (such as concrete or steel frames with masonry infill). Such shear walls must be designed by the engineered methods in Section 2107 or 2108, unless the structure is assigned to Seismic Design Category A, in which case the empirical provisions of Section 2109 may be used.

There are three types of masonry shear wall systems that are recognized by the IBC, but are not specifically listed in the MSJC Code. They are ordinary plain prestressed shear walls (Section 2106.1.1.1), intermediate prestressed masonry shear walls (Section 2106.1.1.2) and special prestressed masonry shear walls (Section 2106.1.1.3). The shear wall systems recognized by the MSJC Code are discussed below.

Ordinary plain masonry shear walls (see Section 1.13.2.2.1 of ACI 530/ASCE 5/TMS 402) meet minimum requirements only and thus may be used only in areas of low seismic risk. Plain masonry walls are designed as unreinforced masonry (by the noted section), although they may in fact contain reinforcement.

Ordinary reinforced masonry shear walls (see Section 1.13.2.2.3 of ACI 530/ASCE 5/TMS 402) are required to meet minimum requirements for reinforced masonry as noted in the referenced section. Because they contain reinforcement, their performance is expected to be better than that of plain masonry shear walls and they are accordingly permitted in areas of both low and moderate seismic risk. Additionally, these walls have more favorable seismic design parameters, including higher response modification factors, *R*, than plain masonry shear walls. When assigned to moderate seismic risk areas (Seismic Design Category C), however, minimum reinforcement is required as noted in Section 2106.4.

Detailed plain masonry shear walls (see Section 1.13.2.2.2 of ACI 530/ASCE 5/TMS 402) are designed as unreinforced masonry in accordance with the section noted, but contain minimum reinforcement in the horizontal and vertical directions. Because of this reinforcement, these walls have more favorable seismic design parameters, including higher response modification factors, *R*, than ordinary plain masonry shear walls.

Intermediate reinforced masonry shear walls (see Section 1.13.2.2.4 of ACI 530/ASCE 5/TMS 402) are designed as reinforced masonry as noted in the referenced section and are also required to contain a minimum amount of prescriptive reinforcement. Because they contain reinforcement, their seismic performance is better than that of plain masonry shear walls and they are accordingly permitted in both areas of low and moderate seismic risk. Additionally, these walls have more favorable seismic design parameters, including higher response modification factors, *R*, than plain masonry shear walls and ordinary reinforced masonry shear walls.

Special reinforced masonry shear walls (see Section 1.13.2.2.5 of ACI 530/ASCE 5/TMS 402) are designed as reinforced masonry as noted in the referenced section and are also required to meet restrictive reinforcement and material requirements. Because of these reinforcement and material requirements, they are permitted to be used in all seismic risk areas. Additionally, these walls have the most favorable seismic design parameters, including the highest response modification factors, *R*, of any of the masonry shear wall types.

2106.1.1.1 Ordinary plain prestressed masonry shear walls. Ordinary plain prestressed masonry shear walls shall comply with the requirements of Chapter 4 of ACI 530/ASCE 5/TMS 402.

❖ This type of shear wall is recognized as a basic seismic-force-resisting system under the code and it must comply with the limitations for these systems in Section 1617.6. The only other stipulation is that it comply with the prestressed masonry requirements of the MSJC Code.

2106.1.1.2 Intermediate prestressed masonry shear walls. Intermediate prestressed masonry shear walls shall comply with the requirements of Section 1.13.2.2.4 of ACI 530/ASCE 5/TMS 402 and shall be designed by Chapter 4, Section 4.5.3.3, of ACI 530/ASCE 5/TMS 402 for flexural strength and by Section 3.2.4.1.2 of ACI 530/ASCE 5/TMS 402 for shear strength. Sections 1.13.2.2.5(a), 3.2.3.5 and 3.2.4.3.2(c) of ACI 530/ASCE 5/TMS 402 shall be applicable for reinforcement. Flexural elements subjected to load reversals shall be symmetrically reinforced. The nominal moment strength at any section along a member shall not be less than one-fourth the maximum moment strength. The cross-sectional area of bonded tendons shall be considered to contribute to the minimum reinforcement in Section 1.13.2.2.4 of ACI 530/ASCE 5/TMS 402. Tendons shall be located in cells that are grouted the full height of the wall.

❖ This type of shear wall is recognized as a basic seismic-force-resisting system under the code and it must comply with the limitations for these systems in Section 1617.6. Besides requiring that it comply with the prestressed masonry provisions of the MSJC Code, additional requirements must be met that are consistent with the results of research and testing carried out in New Zealand. These additional requirements intend that the walls develop their flexural capacity prior to shear failure. Tendons are limited to cells that are fully grouted, since testing has not substantiated that laterally unrestrained tendons are satisfactory for moderate seismic hazards.

2106.1.1.3 Special prestressed masonry shear walls. Special prestressed masonry shear walls shall comply with the requirements of Section 1.13.2.2.5 of ACI 530/ASCE 5/TMS 402 and shall be designed by Chapter 4, Section 4.5.3.3, of ACI 530/ASCE 5/TMS 402 for flexural strength and by Section 3.2.4.1.2 of ACI 530/ASCE 5/TMS 402 for shear strength. Sections 1.13.2.2.5(a), 3.2.3.5 and 3.2.4.3.2(c) of ACI 530/ASCE 5/TMS 402 shall be applicable for reinforcement. Flexural elements subjected to load reversals shall be symmetrically reinforced. The nominal moment strength at any section along a member shall not be less than one-fourth the maximum moment strength. The cross-sectional area of bonded tendons shall be considered to contribute to the minimum reinforcement in Section 1.13.2.2.5 of ACI 530/ASCE 5/TMS 402. Special prestressed masonry shear walls shall also comply with the requirements of Section 3.2.3.5 of ACI 530/ASCE 5/TMS 402.

❖ This type of shear wall is recognized as a basic seismic-force-resisting system under the code and it must comply with the limitations for these systems in Section 1617.6. Besides requiring that it comply with the prestressed masonry provisions of the MSJC Code, additional limitations apply that are based on the testing used to substantiate these systems. These additional requirements intend that the walls develop their flexural capacity prior to shear failure.

2106.1.1.3.1 Prestressing tendons. Prestressing tendons shall consist of bars conforming to ASTM A 722.

❖ Test specimens for prestressed masonry shear wall systems used only high-strength bar tendons, indicating that is an appropriate restriction for these systems that are intended for exposure to high seismic hazards.

2106.1.1.3.2 Grouting. All cells of the masonry wall shall be grouted.

❖ Testing of prestressed masonry shear wall systems indicates that exposure to high seismic hazard is only permissible if all cells of the hollow concrete masonry unit (CMU) wall are grouted solid. This is because partially grouted walls have demonstrated somewhat brittle behavior.

2106.2 Anchorage of masonry walls. Masonry walls shall be anchored to the roof and floors that provide lateral support for the wall in accordance with Section 1604.8.2.

❖ This provision applies to all masonry structures, regardless of seismic design category. Because masonry walls typically depend on lateral support from floors and roofs, they are required to be anchored directly to those elements. Section 1604.8.2 prohibits reliance on friction (from dead load) alone to hold walls, floors and roofs together and prescribes minimum forces for design of anchorage.

2106.3 Seismic Design Category B. Structures assigned to Seismic Design Category B shall conform to the requirements of Section 1.13.4 of ACI 530/ASCE 5/TMS 402 and to the additional requirements of this section.

❖ Requirements in Seismic Design Category (SDC) B are slightly more restrictive than in SDC A. Since the requirements are cumulative with each successive SDC, masonry assigned to SDC B must meet the requirements for SDC A (Section 1.13.3 of the MSJC Code) as well as Section 1.13.4 of the MSJC Code and the requirements in Section 2106.3. Therefore, besides those requirements in Section 2106.2, masonry shear walls must also be one of the types discussed in the commentary to Section 2106.1.1 and be rationally designed in this SDC and above.

2106.3.1 Masonry walls not part of the lateral-force-resisting system. Masonry partition walls, masonry screen walls and other masonry elements that are not designed to resist vertical or lateral loads, other than those induced by their own mass, shall be isolated from the structure so that the vertical and lateral forces are not imparted to these elements. Isolation joints and connectors between these elements and the structure shall be designed to accommodate the design story drift.

❖ So that seismic loads are not inadvertently transferred into elements such as masonry partition and screen walls, they are required to be isolated from the seismic-force-resisting system. However, these elements need out-of-plane support. Appropriate connectors are available and should be used. This is, in effect, a modification to the seismic requirements of the MSJC Code, since in that standard this is a requirement for SDC C structures and higher.

2106.4 Additional requirements for structures in Seismic Design Category C. Structures assigned to Seismic Design Category C shall conform to the requirements of Section 1.13.5 of ACI 530/ASCE 5/TMS 402 and the additional requirements of this section.

❖ In addition to the requirements of SDC, minimum levels of reinforcement and detailing are required to enhance ductility and load-transfer capability in masonry structures assigned to SDC C. The minimum provisions for improved performance of masonry construction in SDC C must be met, regardless of the method of design.

2106.4.1 Design of discontinuous members that are part of the lateral-force-resisting system. Columns and pilasters that are part of the lateral-force-resisting system and that support reactions from discontinuous stiff members such as walls shall be provided with transverse reinforcement spaced at no more than one-fourth of the least nominal dimension of the column or pilaster. The minimum transverse reinforcement ratio shall be 0.0015. Beams supporting reactions from discontinuous walls or frames shall be provided with transverse reinforcement spaced at no more than one-half of the nominal depth of the beam. The minimum transverse reinforcement ratio shall be 0.0015.

❖ The requirements in this section are intended to reduce the chance of local failure or collapse in elements supporting discontinuous portions of the lateral-force-resisting system by increasing the strength and toughness of those elements. Elements used to redistribute or transfer the effects of seismic overturning are susceptible to local inelastic response that can significantly impair the ability of the lateral-force-resisting system to achieve the required overall ductility. The provisions in this section are intended to increase the ability of elements to resist inelastic deformations. This allows inelastic deformations to be distributed throughout the buildings, as is implied by the large response modification factors used in design.

2106.5 Additional requirements for structures in Seismic Design Category D. Structures assigned to Seismic Design Category D shall conform to the requirements of Section 2106.4, Section 1.13.6 of ACI 530/ASCE 5/TMS 402 and the additional requirements of this section.

❖ Requirements in this section parallel most of the requirements in the MSJC Code.

2106.5.1 Loads for shear walls designed by the working stress design method. When calculating in-plane shear or diagonal tension stresses by the working stress design method, shear walls that resist seismic forces shall be designed to resist 1.5 times the seismic forces required by Chapter 16. The 1.5 multiplier need not be applied to the overturning moment.

❖ This provision is based on a similar provision from the UBC. It requires that the in-plane shear stresses due to seismic loading be increased by 50 percent for design purposes and is intended to provide adequate strength and ductility in shear walls of structures in seismically active areas.

2106.5.2 Shear wall shear strength. For a shear wall whose nominal shear strength exceeds the shear corresponding to development of its nominal flexural strength, two shear regions exist.

For all cross sections within a region defined by the base of the shear wall and a plane at a distance L_w above the base of the shear wall, the nominal shear strength shall be determined by Equation 21-1.

$$V_n = A_n \rho_n f_y \quad \textbf{(Equation 21-1)}$$

The required shear strength for this region shall be calculated at a distance $L_w/2$ above the base of the shear wall, but not to exceed one-half story height.

For the other region, the nominal shear strength of the shear wall shall be determined from Section 2108.

❖ The intent of this provision is to provide a ductile flexural limit state. The plastic hinge region is considered to extend vertically from the base of the wall to a distance equal to the plan length of the wall. In this region, the shear strength of the wall is based on the transverse reinforcement only. Above the plastic hinge, the shear strength of the wall is based on both the masonry and the transverse reinforcement.

2106.6 Additional requirements for structures in Seismic Design Category E or F. Structures assigned to Seismic Design Category E or F shall conform to the requirements of Section 2106.5 and Section 1.13.7 of ACI 530/ASCE 5/TMS 402.

❖ Additional restrictions are imposed on buildings assigned to the highest seismic risk categories.

SECTION 2107
WORKING STRESS DESIGN

2107.1 General. The design of masonry structures using working stress design shall comply with Section 2106 and the requirements of Chapters 1 and 2, except Section 2.1.2.1 and 2.1.3.3 of ACI 530/ASCE 5/TMS 402. The text of ACI 530/ASCE 5/TMS 402 shall be modified as follows.

❖ Section 2107 adopts the working stress design method of Chapters 1 and 2 of the MSJC Code (ACI 530/ASCE 5/TMS 402) with modifications that the IBC Structural Committee felt were needed. This method of engineered masonry design has been used successfully for years and remains the dominant method of designing masonry structures.

Refer to the commentary to Chapters 1 and 2 of the MSJC Code for additional information on the working stress design method for masonry. This section also requires conformance to the seismic design provisions of Section 2106. The sections of the MSJC Code specifically not adopted exclude redundant or conflicting provisions such as the MSJC load combinations, which are only intended to apply where no such load combinations are included in the building code.

The pseudo-strength design provision of ACI 530/ASCE 5/TMS 402, Section 2.1.3.3, is also excluded. This method allows masonry to be designed using strength-based seismic loads and working stress provisions. Since Chapter 16 contains service-level load combinations, however, pseudo-strength design is not needed and is therefore not permitted.

2107.2 Modifications to ACI 530/ASCE 5/TMS 402.

❖ The IBC Structural Subcommittee considered and adopted several modifications to the MSJC Code regarding special inspection, column requirements, splice requirements for reinforcement (both lap splices and mechanical or welded splices) and maximum bar size. These modifications supersede the MSJC Code provisions when working stress design is used.

2107.2.1 ACI 530/ASCE 5/TMS 402, Chapter 2. Special inspection during construction shall be provided as set forth in Section 1704.5.

❖ The inspection provisions for masonry in Chapter 17 were based on the inspection requirements in the MSJC Code and Specification, with several key modifications. Masonry designed by working stress procedures must be inspected in accordance with the IBC inspection requirements.

2107.2.2 ACI 530/ASCE 5/TMS 402, Section 2.1.6. Masonry columns used only to support light-frame roofs of carports, porches, sheds or similar structures with a maximum area of 450 square feet (41.8 m^2) assigned to Seismic Design Category A, B or C are permitted to be designed and constructed as follows:

1. Concrete masonry materials shall be in accordance with Section 2103.1. Clay or shale masonry units shall be in accordance with Section 2103.2.
2. The nominal cross-sectional dimension of columns shall not be less than 8 inches (203 mm).
3. Columns shall be reinforced with not less than one No. 4 bar centered in each cell of the column.
4. Columns shall be grouted solid.
5. Columns shall not exceed 12 feet (3658 mm) in height.
6. Roofs shall be anchored to the columns. Such anchorage shall be capable of resisting the design loads specified in Chapter 16.
7. Where such columns are required to resist uplift loads, the columns shall be anchored to their footings with two No. 4 bars extending a minimum of 24 inches (610 mm) into the columns and bent horizontally a minimum of 15 inches (381 mm) in opposite directions into the footings. One of these bars is permitted to be the reinforcing bar specified in Item 3 above. The total weight of a column and its footing shall not be less than 1.5 times the design uplift load.

❖ Section 2107.2.2 exempts lightly loaded columns (such as those that support carport roofs, which experience primarily axial tension and flexure in high-wind events) from the prescriptive requirements of Section 2.1.6 of the MSJC Code. All other columns need to comply with Section 2.1.6 of the MSJC Code. This provision is similar to one in the *Standard Building Code*© (SBC©)and is intended to relax an aspect of the MSJC Code, which defines columns by geometry rather than function. According to that document, masonry members of a certain geometry, even though they act primarily in flexure, are classified as columns and thus must meet minimum reinforcement requirements for columns.

2107.2.3 ACI 530/ASCE 5/TMS 402, Section 2.1.10.6.1.1, lap splices. The minimum length of lap splices for reinforcing bars in tension or compression, l_{ld}, shall be calculated by Equation 21-2, but shall not be less than 15 inches (380 mm).

$$l_{ld} = \frac{0.16 d_b^2 f_y \gamma}{K\sqrt{f'_m}}$$ **(Equation 21-2)**

For SI: $l_{ld} = \dfrac{1.95 d_b^2 f_y \gamma}{K\sqrt{f'_m}}$

where:

d_b = Diameter of reinforcement, inches (mm).

f_y = Specified yield stress of the reinforcement or the anchor bolt, psi (MPa).

f'_m = Specified compressive strength of masonry at age of 28 days, psi (MPa).

l_{ld} = Minimum lap splice length, inches (mm).

K = The lesser of the masonry cover, clear spacing between adjacent reinforcement or five times d_b, inches (mm).

γ = 1.0 for No. 3 through No. 5 reinforcing bars. 1.4 for No. 6 and No. 7 reinforcing bars. 1.5 for No. 8 through No. 9 reinforcing bars.

❖ This modification brings consistency to the requirements for splice lengths for reinforcement according to working stress and strength design. The IBC Structural Committee accepted this modification based on broad support from the masonry industry and engineers, since it was noted that existing splice-length requirements are overly conservative for small bar sizes and unconservative for large ones. Note that Section 2107.2.5 prohibits lap spicing of bars that are larger than No. 9.

Traditional requirements for development lengths and splices of reinforcing bars in masonry have long been questioned because of the disparity between required development lengths (and lap lengths) typically used for masonry and those used in reinforced concrete. This provision is based on requirements that have been in the UBC for several years, but also incorporates updates based on extensive research that supports the position that development lengths and lap splice lengths traditionally used were often overly conservative for small bars and unconservative for large ones. Equation 21-2 is based on that research and indicates that the required lap length and development length depend on bar diameter, the cover depth of masonry over the reinforcing steel, the strength of the masonry and the strength of the reinforcing steel.

2107.2.4 ACI 530/ASCE 5/TMS 402, maximum bar size. The bar diameter shall not exceed one-eighth of the nominal wall thickness and shall not exceed one-quarter of the least dimension of the cell, course or collar joint in which it is placed.

❖ Allowing large bars in the masonry walls can result in an overreinforced section that increases the risk of a brittle failure. This section specifies limits on the diameter of reinforcement that are not in the MSJC Code in order to prevent potential problems with overreinforcement and congestion of reinforcement. The requirements are based on tests of splices and successful performance in construction.

2107.2.5 ACI 530/ASCE 5/TMS 402, splices for large bars. Reinforcing bars larger than No. 9 in size shall be spliced using mechanical connectors in accordance with ACI 530/ASCE 5/TMS 402, Section 2.1.10.6.3.

❖ Research has showed that effectively lap splicing large reinforcing bars in masonry is difficult and impractical because of the excessive lap lengths required. This section adds a provision not in the MSJC Code requiring mechanical splices for all bars larger than No. 9 (32 mm) in diameter. Such large bars are rarely required in masonry and do not perform as well as a larger number of smaller bars.

2107.2.6 ACI 530/ASCE 5/TMS 402, Maximum reinforcement percentage. Special reinforced masonry shear walls having a shear span ratio, M/Vd, equal to or greater than 1.0 and having an axial load, P greater than $0.05 f'_m A_n$ which are subjected to in-plane forces, shall have a maximum reinforcement ratio, ρ_{max}, not greater than that computed as follows:

$$\rho_{\max} = \frac{n f'_m}{2 f_y \left(n + \dfrac{f_y}{f'_m} \right)}$$ **(Equation 21-3)**

❖ The allowable stress design (ASD) provisions of the MSJC Code have no limit on the maximum reinforcement ratio. This section places such a limit on the maximum reinforcement ratio in special reinforced masonry shear walls. It is necessary to account for the relatively high potential for inelastic response in such systems.

This requirement is based on the recommendations of a blue-ribbon panel appointed by the Masonry Alliance for Codes and Standards (MACS). It was the panels' conclusion that these requirements should only apply to the design of shear walls for in-plane forces that have significant axial load ($P > .05 f'_m A_n$) and that are controlled by flexure (i.e., $M/Vd \geq 1.0$).

SECTION 2108
STRENGTH DESIGN OF MASONRY

2108.1 General. The design of masonry structures using strength design shall comply with Section 2106 and the requirements of Chapters 1 and 3 of ACI 530/ASCE 5/TMS 402.

The minimum nominal thickness for hollow clay masonry

in accordance with Section 3.2.5.5 of ACI 530/ASCE 5/TMS 402 shall be 4 inches (102 mm).

❖ The first edition of the IBC contained extensive strength design requirements, since a consensus standard for the strength design of masonry did not yet exist. Since strength design procedures were incorporated in the 2002 edition of the MSJC Code (ACI 530/ASCE 5/TMS 402), this section now requires that strength design be in accordance with Chapters 1 and 3 of the MSJC Code with some modifications to specific sections. For instance, the minimum nominal thickness of 6 inches (152 mm) for hollow day units is reduced to 4 inches (102 mm). This section also invokes the minimum seismic requirements of Section 2106 for all masonry designed by this method. Additionally, masonry designed by this method must be inspected during construction in accordance with the special inspection provisions of Section 1704.5.

2108.2 ACI 530/ASCE 5/TMS 402, Section 3.2.2(g). Modify Section 3.2.2(g) as follows:

3.2.2(g). The relationship between masonry compressive stress and masonry strain shall be assumed to be defined by the following:

Masonry stress of 0.80 f'_m shall be assumed uniformly distributed over an equivalent compression zone bounded by edges of the cross section and a straight line located parallel to the neutral axis at a distance, $a = 0.80\ c$, from the fiber of maximum compressive strain. The distance, c, from the fiber of maximum strain to the neutral axis shall be measured perpendicular to that axis. For out-of-plane bending, the width of the equivalent stress block shall not be taken greater than six times the nominal thickness of the masonry wall or the spacing between reinforcement, whichever is less. For in-plane bending of flanged walls, the effective flange width shall not exceed six times the thickness of the flange.

❖ This section modifies the design assumptions of the MSJC Code. These principles have traditionally been used for reinforced masonry members designed by the strength method. The values for the maximum usable strain are based on extensive research of masonry materials and are different from the values used in the 1997 UBC. Concerns have been raised regarding the implied precision of the values. However, the reported values for the maximum usable strain accurately represent those observed during testing.

While tension may still develop in the masonry of a reinforced element, it is not considered effective in resisting design loads. However, the tensile resistance of masonry is considered implicitly in computing the stiffness of reinforced masonry. If it were not, the effective moment of inertia would always be just the cracked transformed moment of inertia.

The modification made by the IBC provides direction on the stress block width for out-of-plane bending as well as for flanged walls. These limitations are consistent with the NEHRP Provisions and are based on usual practice under previous codes.

2108.3 ACI 530/ASCE 5/TMS 402, Section 3.2.3.4. Modify Section 3.2.3.4 (b) and (c) as follows:

3.2.3.4 (b). A welded splice shall have the bars butted and welded to develop at least 125 percent of the yield strength, f_y, of the bar in tension or compression, as required. Welded splices shall be of ASTM A 706 steel reinforcement. Welded splices shall not be permitted in plastic hinge zones of intermediate or special reinforced walls or special moment frames of masonry.

3.2.3.4 (c). Mechanical splices shall be classified as Type 1 or 2 according to Section 21.2.6.1 of ACI 318. Type 1 mechanical splices shall not be used within a plastic hinge zone or within a beam-column joint of intermediate or special reinforced masonry shear walls or special moment frames. Type 2 mechanical splices are permitted in any location within a member.

❖ This section modifies the strength design splice requirements for consistency with the 2000 edition of the NEHRP Provisions. Splices for reinforcement can be achieved by lapping the reinforcement, welding the reinforcement or mechanical splicing.

Two modifications are made to welded splice requirements [Item (b)]. Splices in reinforcing steel used in the lateral-force-resisting system subjected to high seismic strains must be able to develop the strength of the steel in order to achieve the required performance. To be successfully welded, the chemistry of the steel must be controlled to limit carbon content as well as other elements, such as sulfur and phosphorus. Since the chemistry of reinforcing steel conforming to ASTM A 615, for example, is not controlled and is likely to be unknown, a weld that develops the strength of the steel is not guaranteed. ASTM A 706 steel, on the other hand, has controlled chemistry and can be reliably welded; therefore, if splices are to be accomplished by welding, the use of ASTM A 706 reinforcing steel is mandatory.

Welded splices are required to be able to develop at least 125 percent of the yield strength of the spliced rebars; however, reinforcing steel conforming to ASTM A 706 or for that matter ASTM A 615 or A 996 can have an actual yield strength greater than 125 percent of the minimum required yield strength. This means a code-conforming welded splice may conceivably fail before the spliced bars have yielded, thereby limiting the inelastic deformability of that structural member. The use of welded splices is, therefore, prohibited at locations of potential plastic hinging of members in structural systems that are expected to undergo significant inelastic response in resisting forces due to earthquakes.

The modification to mechanical splices in Item (c) requires that the splice be classified in accordance with ACI 318 as Class 1 or Class 2. This is due to the fact that reinforcing steel is predominantly produced from remelted steel scrap, making it difficult to control the strength. The resulting products tend to have a strength considerably higher than the specified yield strength. This is similar to the situation that has occurred in structural steel where the actual yield strength can be much greater than the specified yield strength. Since there is no upper limit on the yield strength (except for ASTM A 706) and only a minimum required yield strength, most reinforcing steel will have a higher yield point than that specified.

Testing by the California Department of Transportation (CALTRANS) indicates the overstrength can be as much as 60 percent over the specified strength. Cyclic tests of splices meeting only the 125-percent criterion (i.e. Class 1) show that, in many cases, they cannot survive several excursions in the post yield range as imposed by cyclic testing. Splices in reinforcing steel used in the lateral-force-resisting system in plastic hinge zones and beam-column joints are subjected to high seismic strains, and they must be able to develop the strength of the steel in order to achieve the required performance. Hence, the requirement for a Type 2 splice that develops the specified tensile strength of the bar.

2108.4 ACI 530/ASCE 5/TMS 402, Section 3.2.3.5.1. Add the following text to Section 3.2.3.5.1:

For special prestressed masonry shear walls, strain in all prestressing steel shall be computed to be compatible with a strain in the extreme tension reinforcement equal to five times the strain associated with the reinforcement yield stress, f_y. The calculation of the maximum reinforcement shall consider forces in the prestressing steel that correspond to these calculated strains.

❖ This MSJC Code section limits the percentages of flexural reinforcement in order to provide ductile behavior. Overreinforced flexural members can fail in a brittle mode by the crushing of the masonry. Such failures are sudden and catastrophic and, therefore, must be avoided.

Section 2106.1.1.3, covering special prestressed masonry shear walls, requires compliance with this code section, which includes a modification of the MSJC Code section specifically for those shear wall types. This modification clarifies how the reinforcing limitation is to be met for this structural system.

SECTION 2109 EMPIRICAL DESIGN OF MASONRY

2109.1 General. Empirically designed masonry shall conform to this chapter or Chapter 5 of ACI 530/ASCE 5/TMS 402.

❖ This section permits empirical design of masonry by either the provisions of Section 2109 or Chapter 5 of ACI 530/ASCE 5/TMS 402. This is because nearly all of the requirements in Section 2109 are based on the requirements in Chapter 5 of ACI 530/ASCE 5/TMS 402 with minor modifications. Additional information on these provisions can be found in the commentary to ACI 530/ASCE 5/TMS 402.

Empirical provisions are design rules developed by experience rather than engineering analysis. The empirical rules in these provisions are based on records dating back as far as 1889 in *A Treatise on Masonry Construction*, by Ira Baker. The most recent publication providing the basis for these empirical provisions is ANSI A41.1.

This empirical design method is based on several premises: gravity loads are reasonably centered on bearing walls; effects of reinforcement are neglected; walls are laid in running bond and buildings have limited height, seismic risk and wind loading. The requirements of and limitations regarding the use of empirical design reflect these assumptions.

2109.1.1 Limitations. Empirical masonry design shall not be utilized for any of the following conditions:

1. The design or construction of masonry in buildings assigned to Seismic Design Category D, E or F as specified in Section 1616, and the design of the seismic-force-resisting system for buildings assigned to Seismic Design Category B or C.
2. The design or construction of masonry structures located in areas where the basic wind speed exceeds 110 mph (177 km/hr).
3. Buildings more than 35 feet (10 668 mm) in height which have masonry wall lateral-force-resisting systems.

In buildings that exceed one or more of the above limitations, masonry shall be designed in accordance with the engineered design provisions of Section 2107 or 2108, or the foundation wall provisions of Section 1805.5.

❖ Empirical design is permitted for structures having limited seismic risk, wind loading and height. These limitations are justified, since buildings that were representative of the historically based empirical provisions are uncommon today. For example, buildings of the past were smaller, had more interior masonry walls and typically had different floor construction than modern buildings.

Where any one of the three stated limitations exists, the masonry structure is not permitted to be empirically designed. Engineered design in accordance with Section 2107 or 2108 is required in such instances. Foundation walls complying with Section 1805.5 are also acceptable.

1. The empirical provisions of Section 2109 and the required referenced standards, plus the seismic loading and detailing requirements of Chapter 16, are adequate for the level of risk associated with the seismic-force-resisting systems of buildings located in Seismic Design Category A. Where masonry is used for purposes other than the seismic-force-resisting system, the empirical design method may be used for buildings assigned to Seismic Design Category A, B or C. Engineered design is required for buildings in higher seismic design categories.

2. This requirement applies individually to the lateral-load-resisting system and to building elements not effectively participating in the lateral-load-resisting system. For example, empirical design of the lateral-load-resisting system is permitted only where the basic wind speed does not exceed 110 mph (145 km/hr). Otherwise, engineered design is required. These are similar to, but not the same as, the wind load restrictions

in Chapter 5 of the MSJC Code (ACI 530/ASCE 5/TMS 402).

3. The limitation is justified since today's buildings are taller than those on which these empirical provisions are based. Empirical design is permitted for elements not effectively participating in the lateral-load-resisting system. Engineered design is required where the lateral-load-resisting system of a building has substantial height.

2109.2 Lateral stability.

❖ The lateral stability requirements of Section 2109.2 are required for buildings using empirical design for lateral load resistance. The requirements of this section do not apply when the engineered masonry design method of Section 2101.1.1 is used. This section contains requirements for empirical design of lateral-load-resisting systems composed of diaphragms and shear walls.

Requirements include minimum lengths of masonry shear walls in both principal plan directions of the building and maximum span-to-width (depth) ratios of floor and roof diaphragms. Requirements for roofs and dry-stacked, surface-bonded walls are also part of this section.

Lateral load resistance is a basic requirement for structural design. The distribution of loads within the lateral-load-resisting system is a function of the relative rigidities of diaphragms and shear walls. The lateral-load-resisting system transfers lateral wind and seismic forces to the foundation in the form of base shear and overturning moment and maintains stability of the structure under gravity loads.

2109.2.1 Shear walls. Where the structure depends upon masonry walls for lateral stability, shear walls shall be provided parallel to the direction of the lateral forces resisted.

❖ The shear wall requirements in this section apply where empirically designed masonry is used for lateral load resistance. Shear walls are required in both principal plan directions of the structure, parallel to the lateral loads required in Chapter 16. Load-bearing walls serve as shear walls.

2109.2.1.1 Shear wall thickness. Minimum nominal thickness of masonry shear walls shall be 8 inches (203 mm).

Exception: Shear walls of one-story buildings are permitted to be a minimum nominal thickness of 6 inches (152 mm).

❖ An empirical minimum nominal shear wall thickness of 8 inches (203 mm) is required to transfer lateral loads to the foundation. This minimum is based on experience. Shear walls that are also load-bearing walls are required to comply with the compressive stress requirements of Section 2106 and the lateral support requirements of Section 2107. Those requirements may govern.

The exception addresses single-story buildings that have limited lateral loads and, therefore, limited base shears and overturning moments. This justifies the permitted minimum nominal thickness of 6 inches (152 mm), as listed in the exception to this section.

2109.2.1.2 Cumulative length of shear walls. In each direction in which shear walls are required for lateral stability, shear walls shall be positioned in two separate planes. The minimum cumulative length of shear walls provided shall be 0.4 times the long dimension of the building. Cumulative length of shear walls shall not include openings or any element whose length is less than one-half its height.

❖ Figure 2109.2.1.2 diagrams the lengths in each direction of the building, parallel to the lateral loads. The term "cumulative" refers to the sum of the lengths of shear wall segments in a single direction. The required length of shear wall in each direction is 0.4 times the long dimension of the building.

Walls above and below openings such as windows and doors are not considered to be shear wall segments. Neither are portions of a wall that have a height of two or more times its length.

2109.2.1.3 Maximum diaphragm ratio. Masonry shear walls shall be spaced so that the length-to-width ratio of each diaphragm transferring lateral forces to the shear walls does not exceed the values given in Table 2109.2.1.3.

❖ For empirical design, this section limits length- (or span)-to-width ratios of floor and roof diaphragms between shear walls, based on the inherent rigidity of the diaphragm construction. This effectively distributes shear walls throughout the structure at adequate intervals for the diaphragm span.

The assumed deflected shape of a diaphragm is illustrated in Figure 2109.2.1.3(1). Flexible diaphragms deflect more than rigid ones, possibly resulting in detachment of components. A rigid diaphragm deflects little and transfers loads to shear walls in proportion to their relative rigidities (an important engineered masonry consideration). Symmetrical distribution of shear walls in the building plan is desirable, reducing torsional rotation of the building under lateral loads.

TABLE 2109.2.1.3
DIAPHRAGM LENGTH-TO-WIDTH RATIOS

FLOOR OR ROOF DIAPHRAGM CONSTRUCTION	MAXIMUM LENGTH-TO-WIDTH RATIO OF DIAPHRAGM PANEL
Cast-in-place concrete	5:1
Precast concrete	4:1
Metal deck with concrete fill	3:1
Metal deck with no fill	2:1
Wood	2:1

❖ Table 2109.2.1.3 sets the maximum span-to-width ratio of diaphragms constructed of the listed materials. See Figure 2109.2.1.3(2) for an illustration of the terms "diaphragm span" (length) and "width" (depth). Concrete diaphragms are permitted to have high span- (length-) to-width ratios because of their rigidity, whereas wood diaphragms are required to have low span- (length-) to-width ratios because of their flexibility.

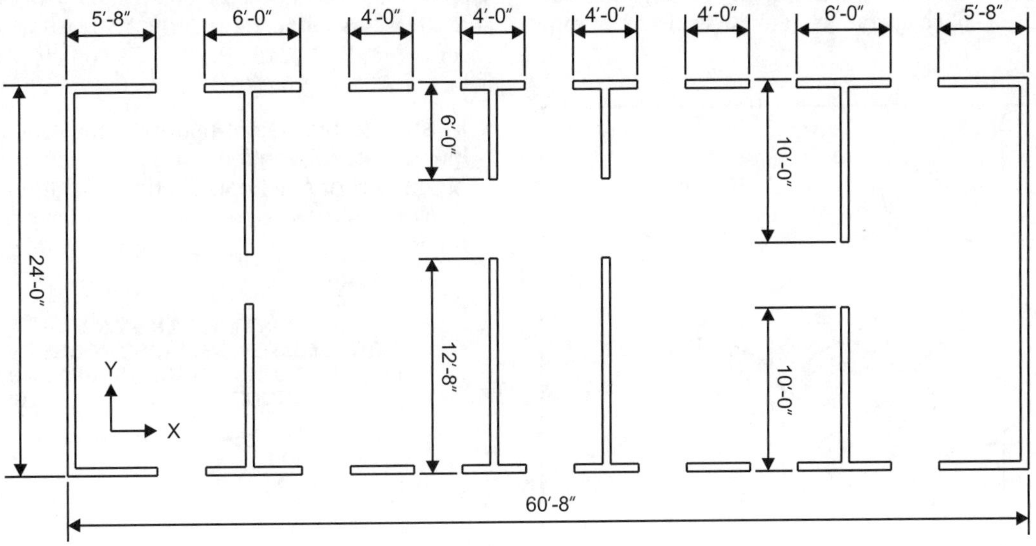

MINIMUM CUMULATIVE SHEAR WALL LENGTH = 0.4 x LONG DIMENSION
MINIMUM L = 0.4(60.7′) = 24.3′
X - DIRECTION: L = 2(5.7 + 6.0 + 4.0 + 4.0 + 4.0 + 4.0 + 6.0 + 5.7) = 78.8 ft. > 24.3 ft. OK
Y - DIRECTION: L = 2(24.0 + 10.0 + 10.0 + 12.7 + 6.0) = 125.4 ft. > 24.3 ft. OK

For SI: 1 foot = 304.8 mm.

Figure 2109.2.1.2
CUMULATIVE LENGTH OF SHEAR WALL REQUIREMENTS FOR EMPIRICALLY DESIGNED MASONRY

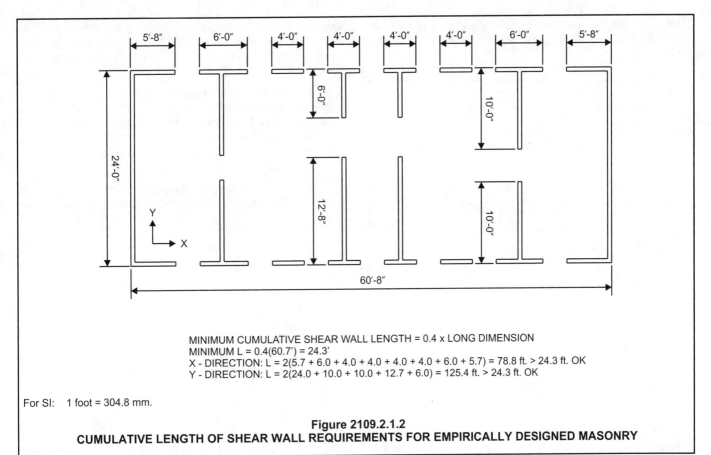

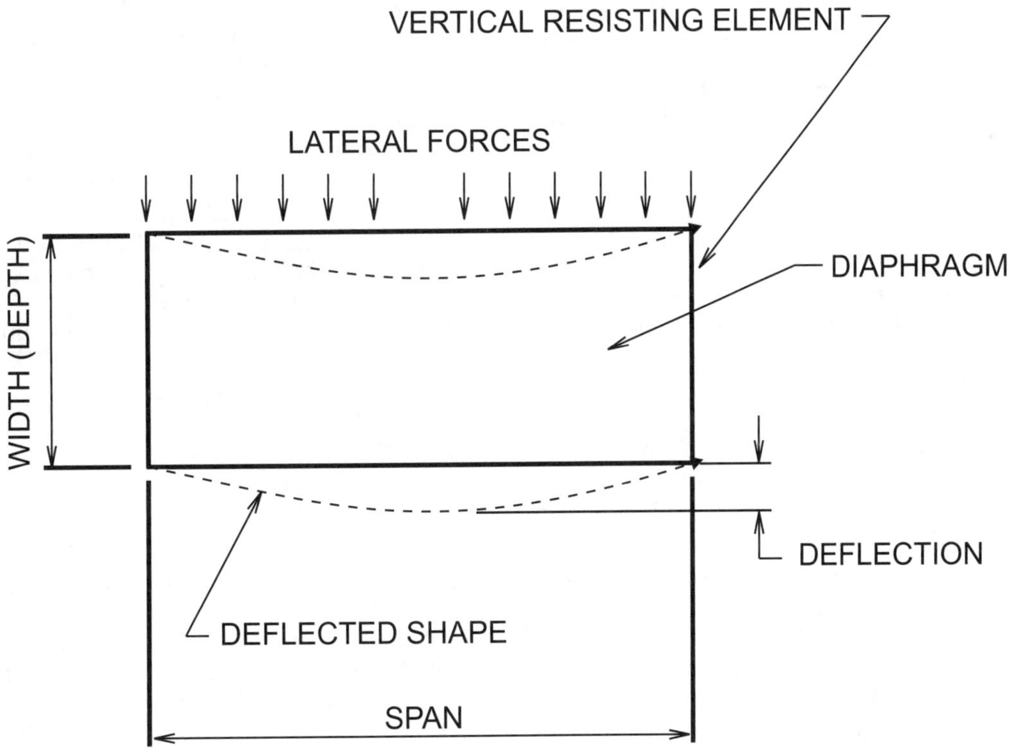

Figure 2109.2.1.3(1)
DIAPHRAGM ACTION (PLAN VIEW)

2109.2.2 Roofs. The roof construction shall be designed so as not to impart out-of-plane lateral thrust to the walls under roof gravity load.

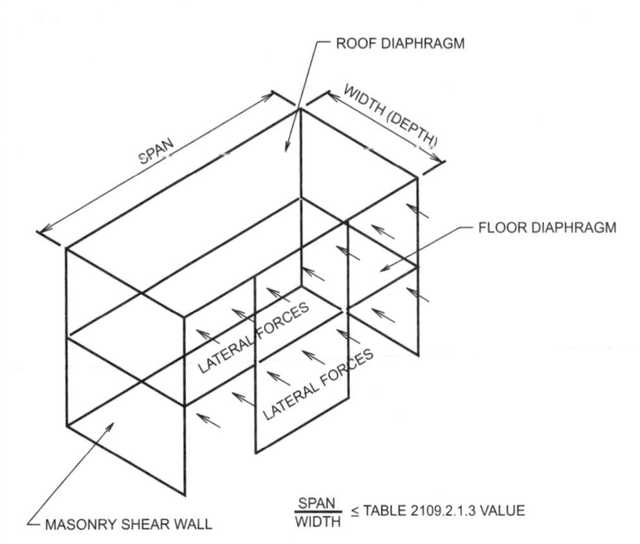

Figure 2109.2.1.3(2)
EMPIRICAL MAXIMUM DIAPHRAGM RATIO

❖ Roofs are not permitted to rely on masonry walls to resist thrust perpendicular to the wall. The low tensile capacity of masonry and the lack of engineered design for these loads in the empirical provisions result in this requirement. Connections that apply thrust perpendicular to the wall are not permitted.

As another example, a wood frame cathedral ceiling does not, by design, have ceiling joists to resist outward thrust. If the ridge beam is not designed for vertical support and limited deflection, opposing sloped rafters usually transfer thrust to the outside walls.

Consideration of horizontal thrust is also necessary for truss connections, especially scissor trusses. In a truss, the bottom chord is usually in tension. Chord elongation can impart significant lateral thrust perpendicular to masonry bearing walls. In general, the lower the roof slope, the greater the lateral thrust.

2109.2.3 Surface-bonded walls. Dry-stacked, surface-bonded concrete masonry walls shall comply with the requirements of this code for masonry wall construction, except where otherwise noted in this section.

❖ Dry-stacked, surface-bonded masonry walls consist of courses of concrete masonry units without mortar joints assembled to form unreinforced walls. Both sides of the walls are coated with a $^1/_{16}$- to $^1/_8$- inch-thick layer (1.6 to 3.2 mm) of cementitious mortar reinforced with glass fibers capable of increasing the tensile strength of the masonry and unifying the construction.

2109.2.3.1 Strength. Dry-stacked, surface-bonded concrete masonry walls shall be of adequate strength and proportions to support all superimposed loads without exceeding the allowable stresses listed in Table 2109.2.3.1. Allowable stresses not specified in Table 2109.2.3.1 shall comply with the requirements of ACI 530/ASCE 5/TMS 402.

❖ In surface-bonded masonry construction, strength is lower than conventional masonry construction because of the lack of solid contact between the units.

Where stresses are not specified in Table 2109.2.3.1, engineered design in accordance with Section 2107 or 2108 is required.

TABLE 2109.2.3.1
ALLOWABLE STRESS GROSS CROSS-SECTIONAL AREA FOR DRY-STACKED, SURFACE-BONDED CONCRETE MASONRY WALLS

DESCRIPTION	MAXIMUM ALLOWABLE STRESS (psi)
Compression standard block	45
Shear	10
Flexural tension	
Vertical span	18
Horizontal span	30

For SI: 1 pound per square inch = 0.006895 Mpa.

❖ The values for allowable stresses based on the gross cross-sectional area of masonry units are given in this table. The gross cross-sectional area is the actual area of a section perpendicular to the direction of the load, without subtraction of the core areas of hollow masonry units.

The flexural strength of surface-bonded walls is about the same as conventional masonry with mortar joints. In the vertical direction, where walls are supported top and bottom, Table 2109.2.3.1 allows a maximum flexural tensile stress of 18 psi (0.12 MPa) based on the gross area. In the horizontal direction, where walls span laterally between supports, a maximum flexural stress of 30 psi (0.21 MPa) is permitted based on the gross area when units are dry stacked in running bond.

The shear strength of surface-bonded walls is less than that of conventional, mortar-jointed walls. This table allows a shear strength of 10 psi (0.069 MPa) based on the gross area.

2109.2.3.2 Construction. Construction of dry-stacked, surface-bonded masonry walls, including stacking and leveling of units, mixing and application of mortar and curing and protection shall comply with ASTM C 946.

❖ The construction of dry-stacked, surface-bonded walls must conform to the requirements of ASTM C 946.

It is not practical to construct the horizontal surface of a wall footing or foundation level enough to receive the base (first) course of masonry without additional leveling. Therefore, it is customary to lay the base course of masonry on a mortar bed so that the remainder of the dry-stacked units will be erected level. As the units are erected, their ends should be butted together as tightly as possible.

If the bearing surfaces of the concrete units are not ground smooth and flat, shims may be required between the units to erect the wall plumb and level. Such shims should be of metal, mortar, wood or plastic.

Because dry-stacked walls have no mortar joints, it is not possible to use horizontal steel joint reinforcement to reduce the size of cracks associated with temperature and moisture movements. Reinforced bond beams and sufficient control joints in surface-bonded masonry can be used to reduce the widths of such cracks.

The joints of dry-stacked units are tight, with no space for connectors to be embedded in the wall. The face shells or cross webs of such concrete masonry units therefore, should be notched or depressed to accommodate ties and anchors that must be embedded in grout.

Packaged dry-bonding mortar should be mixed with water at the job site in accordance with ASTM C 946 or the manufacturer's recommendations, including curing and protection procedures after application of the material. Surface-bonding mortars are usually applied by hand troweling to thicknesses between $^1/_{16}$ and $^1/_8$ inch (1.6 to 3.2 mm). While they may also be sprayed on, this is usually followed by hand or mechanical troweling to obtain the desired finish.

2109.3 Compressive stress requirements.

❖ This section applies to empirically designed masonry and, as with the other empirical design requirements of Section 2109, not to masonry designed by the engineered approaches of Sections 2107 and 2108.

Vertical dead and live loads, as required in Chapter 16, encompass a wide range of possibilities, as does the conceivable configuration of supported floor spans. Consequently, specifying empirical minimum sizes that would account for all vertical loading and span conditions would be impractical. Section 2109.3 contains an empirical compressive stress design procedure for single- and multiwythe masonry addressing actual required loading and resulting in minimum areas of masonry to resist vertical loads.The design is based on an average compressive stress on the gross cross-sectional area, using specified instead of nominal dimensions.

The result is a way of sizing and proportioning masonry without complete engineering analysis and design. The required areas are intended to be conservative with respect to engineered design and, along with the other empirical design provisions, to adequately address buildings and elements permitted to be empirically designed by the provisions of Section 2109.

2109.3.1 Calculations. Compressive stresses in masonry due to vertical dead plus live loads, excluding wind or seismic loads, shall be determined in accordance with Section 2109.3.2.1. Dead and live loads shall be in accordance with Chapter 16, with live load reductions as permitted in Section 1607.9.

❖ This section identifies the vertical design loads that must be used in calculating average compressive stresses. The required dead plus live loads are intended to include all loads except wind and seismic loads. Design roof loads given in Chapter 16 are required, including applicable design snow loads.

Effects of wind and seismic forces on empirically designed masonry structures are addressed by the lateral stability requirements of Section 2109.2 and by the seismic, wind and building height limitations of Section 2109.1.1.

2109.3.2 Allowable compressive stresses. The compressive stresses in masonry shall not exceed the values given in Table 2109.3.2. Stress shall be calculated based on specified rather than nominal dimensions.

❖ The maximum permitted compressive stresses on the gross cross-sectional area are given in Table 2109.3.2. Calculation of gross-area compressive stresses is discussed in the commentary to Section 2109.3.2.1. Section 2109.3.2 requires the use of specified dimensions (see also the commentary for the definition of "Dimensions" in Section 2102.1).

TABLE 2109.3.2. See page 21-46.

❖ The compressive strength of the unit, as well as the mortar type used, limit the allowable compressive stress on the gross area.

2109.3.2.1 Calculated compressive stresses. Calculated compressive stresses for single wythe walls and for multiwythe composite masonry walls shall be determined by dividing the design load by the gross cross-sectional area of the member. The area of openings, chases or recesses in walls shall not be included in the gross cross-sectional area of the wall.

❖ Using the vertical design load required by Section 2109.3.1, the compressive stress on the gross cross-sectional area of the masonry must be calculated.

Gross cross-sectional area is illustrated in Figure 2102.1(3) for two single-wythe wall examples. Specified (not nominal) dimensions are used and mortared head joints are included. Cores are not required to be subtracted. Other openings in the wall, including chases and recesses, are required to be subtracted.

The calculated compressive stress is the total required design load divided by the gross cross-sectional area, as illustrated in Figure 2109.3.2.1(1) for a single-wythe wall. The dead weight of the masonry units is part of the total required design load. The calculated compressive stresses are not to exceed the allowable values listed in Table 2109.3.2.

The allowable compressive stresses for masonry directly under concentrated loads (bearing stresses) are recommended in the commentary to ACI 530/ASCE 5/TMS 402: 125 percent of the Table 2109.3.2 value if the load acts on the full wall thickness; or 150 percent of the Table 2109.3.2 value if the load acts on concentrically placed bearing plates greater than one-half, but less than the full supporting area. Concentrated loads on load-bearing walls transmitted from beams, girders or other structural elements normally bear on units of solid masonry or on hollow masonry units with grout-filled cores at least 4 inches (102 mm) high.

Bearing plates are often used to distribute concentrated loads and to prevent damage to the bearing areas of the supporting masonry. Masonry bond beams can also be used for this purpose.

Concentrated loads must be distributed over an area whose length cannot exceed the center-to-center distance between loads, nor one-half of the wall height [see Figure 2109.3.2.1(2)]. Loads are not to be distributed across continuous vertical joints, such as expansion or control joints in masonry walls, or across head joints in stack bond construction. For large concentrated loads, masonry pilasters may be required.

2109.3.2.2 Multiwythe walls. The allowable stress shall be as given in Table 2109.3.2 for the weakest combination of the units used in each wythe.

❖ Multiwythe masonry walls are often constructed with units having different mechanical properties. An example of a multiwythe wall is a composite 8-inch (204 mm) nominal thickness wall using Type S mortar, consisting of one wythe of 4-inch-wide (102 mm) common brick and one wythe of 4-inch-wide (102 mm) lightweight concrete block, bonded together with required joint reinforcement and having completely filled collar joints.

In this example, if the compressive strength of the concrete masonry unit is less than that of the common brick, the allowable compressive stress for both wythes is to be based on the strength of the concrete masonry

TABLE 2109.3.2
ALLOWABLE COMPRESSIVE STRESSES FOR EMPIRICAL DESIGN OF MASONRY

CONSTRUCTION; COMPRESSIVE STRENGTH OF UNIT GROSS AREA (psi)	ALLOWABLE COMPRESSIVE STRESSES[a] GROSS CROSS-SECTIONAL AREA (psi)	
	Type M or S mortar	Type N mortar
Solid masonry of brick and other solid units of clay or shale; sand-lime or concrete brick:		
8,000 or greater	350	300
4,500	225	200
2,500	160	140
1,500	115	100
Grouted masonry, of clay or shale; sand-lime or concrete:		
4,500 or greater	225	200
2,500	160	140
1,500	115	100
Solid masonry of solid concrete masonry units:		
3,000 or greater	225	200
2,000	160	140
1,200	115	100
Masonry of hollow load-bearing units:		
2,000 or greater	140	120
1,500	115	100
1,000	75	70
700	60	55
Hollow walls (noncomposite masonry bonded)[b]		
Solid units:		
2,500 or greater	160	140
1,500	115	100
Hollow units	75	70
Stone ashlar masonry:		
Granite	720	640
Limestone or marble	450	400
Sandstone or cast stone	360	320
Rubble stone masonry		
Coursed, rough or random	120	100

For SI: 1 pound per square inch = 0.006895 MPa.

a. Linear interpolation for determining allowable stresses for masonry units having compressive strengths which are intermediate between those given in the table is permitted.

b. Where floor and roof loads are carried upon one wythe, the gross cross-sectional area is that of the wythe under load; if both wythes are loaded, the gross cross-sectional area is that of the wall minus the area of the cavity between the wythes. Walls bonded with metal ties shall be considered as noncomposite walls unless collar joints are filled with mortar or grout.

unit.

For loading on only one wythe of a composite (grouted collar joint), multiwythe masonry wall, the wythes and the collar joint are to be considered part of the gross cross-sectional area. For loading on only one wythe of a noncomposite (ungrouted collar joint) multiwythe masonry wall, the gross cross-sectional area is limited to the area of the wythe under load and the allowable compressive stresses in Table 2109.3.2 are based on the units and mortar comprising the wythe under load.

The use of the term "composite" refers to multicomponent masonry members that act as a unit. Note b of Table 2109.3.2 considers a multiwythe masonry wall to act compositely (monolithically) if it is bonded with metal ties (or joint reinforcement) and has completely filled collar joints.

Openings in the wall, including chases and recesses, are required to be subtracted from the gross wall area, as stated in Section 2106.2.1.

2109.4 Lateral support.

❖ Section 2109.4 contains empirical design provisions for the spacing of lateral support locations of masonry walls. Requirements include maximum ratios of wall length or wall height to wall thickness. The requirements of this section do not apply when the engineered masonry design method of Section 2101.1.1 is selected.

Compression elements such as columns and walls may have axial capacities limited by buckling, based on slenderness effects (which depends on their stiffness and unsupported length). Limits are also included for nonbearing walls to account for resistance to out-of-plane loads.

The design professional is responsible for indicating in the construction documents the method of lateral support of masonry walls.

2109.4.1 Intervals. Masonry walls shall be laterally supported in either the horizontal or vertical direction at intervals not exceeding those given in Table 2109.4.1.

❖ The requirements of this section prescribe maximum length or height-to-thickness ratios for locations of lateral support (see Figure 2109.4.1). Providing lateral support at the resulting maximum heights increases the buckling capacity to an acceptable level. Spacing of lateral support must be designed either vertically or horizontally, but not in both directions.

TABLE 2109.4.1
WALL LATERAL SUPPORT REQUIREMENTS

CONSTRUCTION	MAXIMUM WALL LENGTH TO THICKNESS OR WALL HEIGHT TO THICKNESS
Bearing walls	
Solid units or fully grouted	20
All others	18
Nonbearing walls	
Exterior	18
Interior	36

❖ This table shows maximum length- or height-to-thickness ratios between locations of lateral support. Lateral support is required to be provided either vertically or horizontally, but not in both directions.

Figure 2109.4.1 illustrates use of the table for an em-

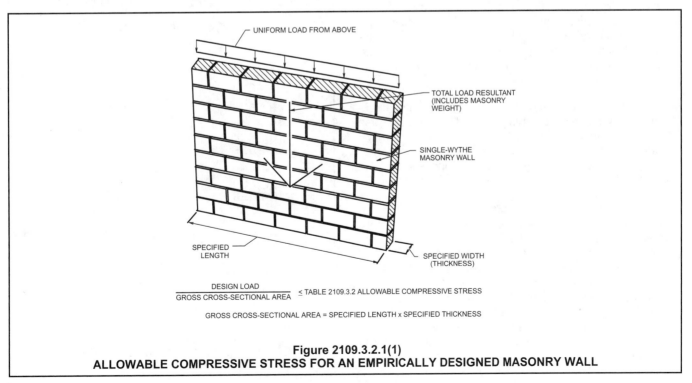

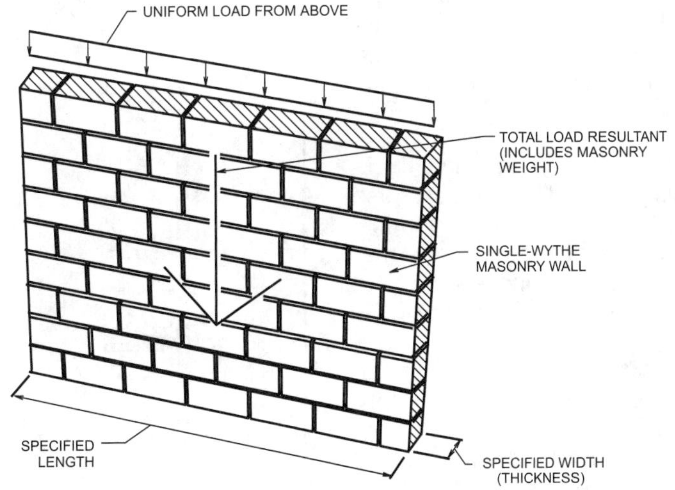

$$\frac{\text{DESIGN LOAD}}{\text{GROSS CROSS-SECTIONAL AREA}} \leq \text{TABLE 2109.3.2 ALLOWABLE COMPRESSIVE STRESS}$$

GROSS CROSS-SECTIONAL AREA = SPECIFIED LENGTH x SPECIFIED THICKNESS

Figure 2109.3.2.1(1)
ALLOWABLE COMPRESSIVE STRESS FOR AN EMPIRICALLY DESIGNED MASONRY WALL

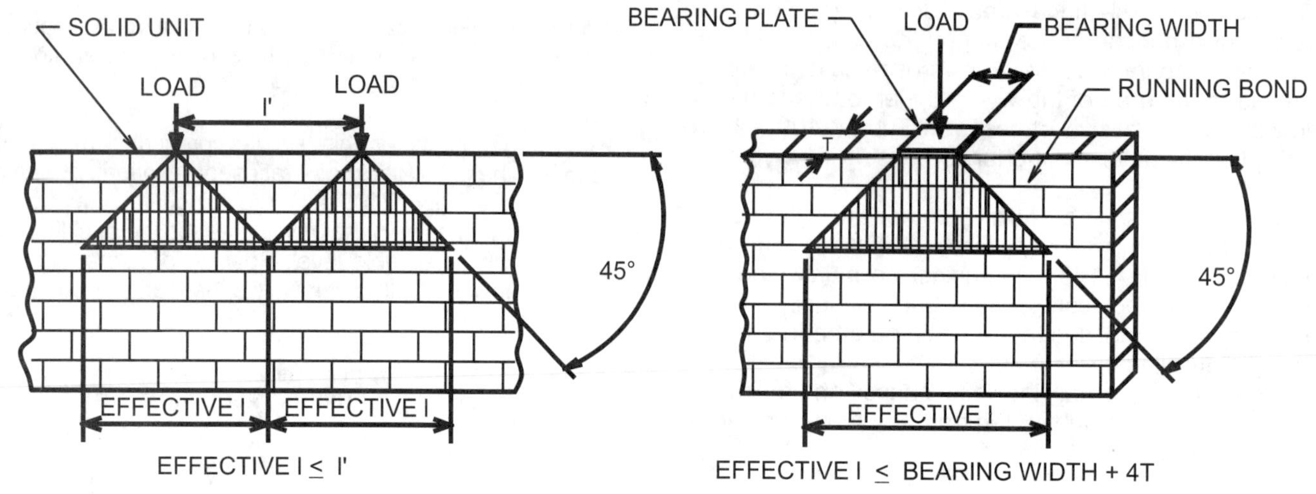

For SI: 1 degree = .01745 rad.

Figure 2109.3.2.1(2)
ASSUMED DISTRIBUTION OF COMPRESSIVE STRESS UNDER CONCENTRATED LOAD

pirically designed masonry building. As shown, the height is measured between points of lateral support.

2109.4.2 Thickness. Except for cavity walls and cantilever walls, the thickness of a wall shall be its nominal thickness measured perpendicular to the face of the wall. For cavity walls, the thickness shall be determined as the sum of the nominal thicknesses of the individual wythes. For cantilever walls, except for parapets, the ratio of height-to-nominal thickness shall not exceed six for solid masonry or four for hollow masonry. For parapets, see Section 2109.5.5.

❖ The nominal thickness or the sum of the nominal thicknesses is used in determining the wall thickness for use in Table 2109.4.1. Figure 2102.1(5) illustrates nominal versus specified thickness of a particular concrete masonry unit. Nominal dimensions are discussed in the commentary to Section 2102.1 (see the definition of "Dimensions").

The use of the nominal thickness is permitted for calculating thickness ratios for empirical design. In contrast, the engineered design method uses the rational Euler buckling equation and the accuracy of specified dimensions is appropriately required. Table 2109.4.1 is intended, however, to produce conservative results with respect to the engineered method in most cases.

For cavity walls, the adjacent wythes do not act together because of the absence of a shear connection. Thus, the wall thickness consists of only the sum of the wythes.

This section does not include retaining walls, which

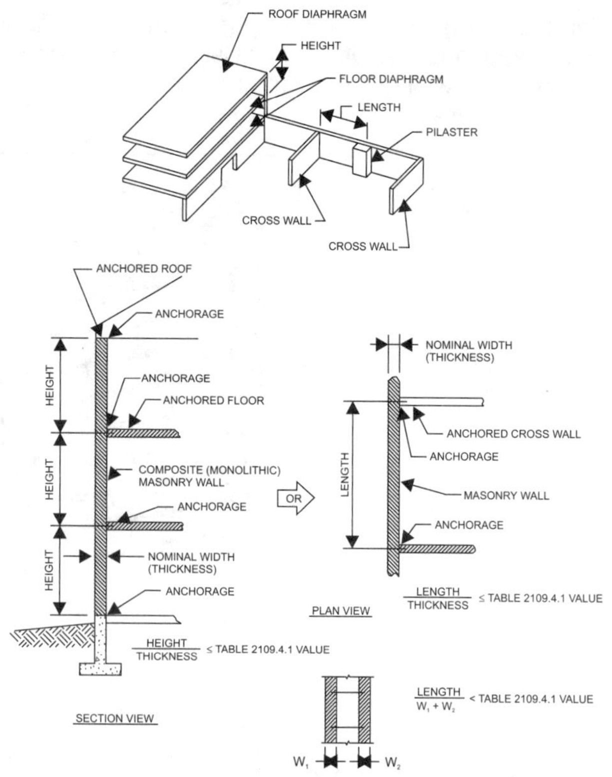

Figure 2109.4.1
LATERAL SUPPORT REQUIREMENTS FOR EMPIRICALLY DESIGNED MASONRY WALLS

must comply with the requirements of Chapter 18. For definitions of "Solid" and "Hollow" masonry, refer to Section 2102.1.

Free-standing cantilever walls have a lower allowable height-to-thickness ratio because of the lack of lateral support at the free (unsupported) end. Parapets are subject to the more restrictive requirements of Section 2109.5.5 because of their location and exterior exposure.

2109.4.3 Support elements. Lateral support shall be provided by cross walls, pilasters, buttresses or structural frame members when the limiting distance is taken horizontally, or by floors, roofs acting as diaphragms or structural frame members when the limiting distance is taken vertically.

❖ Lateral support points complying with the requirements of this section are achieved by required anchorage complying with Section 2109.7.

The code requires only one direction of span for lateral support, either vertical or horizontal. The lateral support of masonry walls can be achieved by floors or roofs when the limiting distance is measured vertically, or by columns, buttresses (pilasters) or cross walls when the limiting distance is measured horizontally.

A masonry pier is an isolated column designed to support vertical loads. A pilaster is a masonry column integrally bonded to a wall.

A buttress is a pilaster whose outside edge (face running in the same direction as the wall) slants toward the wall and whose horizontal cross section increases from top to bottom. Buttresses are used mainly for out-of-plane lateral support of high walls. Both pilasters and buttresses may project out from one or both faces of the wall.

For empirical design, solid bearing walls are permitted to span 20 times their thickness between supports, while hollow walls or walls of hollow masonry units can span 18 times their thickness. For example, a 16-inch (406 mm) hollow block wall or wall constructed with hollow units can have an unsupported height or length of 24 feet (7315 mm) [18 times 1.33 feet (405 mm)].

2109.5 Thickness of masonry. Minimum thickness requirements shall be based on nominal dimensions of masonry.

❖ Section 2109.5 provides requirements for minimum nominal thicknesses of empirically designed walls of masonry, including rubble stone (see the definition of "Dimensions"). Changes in thickness and parapet walls are also included in this section.

2109.5.1 Thickness of walls. The thickness of masonry walls shall conform to the requirements of Section 2109.5.

❖ Section 2109.5.1 contains provisions for minimum nominal thicknesses of empirically designed masonry walls. The requirements of this section do not apply when engineered masonry design is used.

The MSJC has concluded that the minimum thickness ratios listed in Section 2109.4.1 (which are derived from ANSI A41.1) are not always conservative when compared to the results achieved from a working stress analysis.

2109.5.2 Minimum thickness. The minimum thickness of masonry bearing walls more than one story high shall be 8 inches (203 mm). Bearing walls of one-story buildings shall not be less than 6 inches (152 mm) thick.

❖ The minimum thicknesses required in this section are nominal thicknesses (in accordance with the definition in Section 2102.1) and apply to bearing walls. Like the measurement of thickness in Section 2109.4, the space between cavity walls and multiwythe, noncomposite walls is to be excluded when determining wall thickness.

2109.5.3 Rubble stone walls. The minimum thickness of rough or random or coursed rubble stone walls shall be 16 inches (406 mm).

❖ Rubble stone walls are composed of stone masonry having irregularly shaped units bonded by mortar. The greater thickness required for these walls is justified by this irregularity. Accordingly, Table 2109.3.2 permits relatively low compressive stresses for this material. Nominal thickness in this case is the average thickness. This section is not applicable to ashlar masonry (rectangular units).

2109.5.4 Change in thickness. Where walls of masonry of hollow units or masonry bonded hollow walls are decreased in thickness, a course or courses of solid masonry shall be interposed between the wall below and the thinner wall above, or special units or construction shall be used to transmit the loads from face shells or wythes above to those below.

❖ Where hollow walls are decreased in thickness, one or more courses of solid masonry are to be placed between the thicker wall below and the thinner wall above. Alternatively, special construction can be introduced to transmit the load from the wall above to the supporting wall below. For walls constructed with concrete masonry units, a bond beam as thick as the lower wall may be placed between the two wall sections.

2109.5.5 Parapet walls.

❖ Parapet walls are cantilever walls located above the roof line that are typically exposed to weather on both faces; therefore, they may require special consideration.

2109.5.5.1 Minimum thickness. Unreinforced parapet walls shall be at least 8 inches (203 mm) thick, and their height shall not exceed three times their thickness.

❖ Because of their exposure conditions and the hazard associated with unreinforced parapets, a minimum thickness of 8 inches (203 mm) and a maximum height-to-thickness ratio of 3:1 is required. These requirements are minimums and thicker parapets or smaller height-to-thickness ratios may be necessary.

2109.5.5.2 Additional provisions. Additional provisions for parapet walls are contained in Sections 1503.2 and 1503.3.

❖ See the commentary to Sections 1504.2, 1504.3 and 1504.4 regarding the application of these provisions to parapets.

2109.5.6 Foundation walls. Foundation walls shall comply with the requirements of Sections 2109.5.6.1 and 2109.5.6.2.

❖ Masonry foundation walls must comply with Section 1805.5 if the requirements of this section cannot be met.

2109.5.6.1 Minimum thickness. Minimum thickness for foundation walls shall comply with the requirements of Table 2109.5.6.1. The provisions of Table 2109.5.6.1 are only applicable where the following conditions are met:

1. The foundation wall does not exceed 8 feet (2438 mm) in height between lateral supports,
2. The terrain surrounding foundation walls is graded to drain surface water away from foundation walls,
3. Backfill is drained to remove ground water away from foundation walls,
4. Lateral support is provided at the top of foundation walls prior to backfilling,
5. The length of foundation walls between perpendicular masonry walls or pilasters is a maximum of three times the basement wall height,
6. The backfill is granular and soil conditions in the area are nonexpansive, and
7. Masonry is laid in running bond using Type M or S mortar.

❖ This section provides empirical criteria for foundation walls that are based on similar requirements in the MSJC Code. It is necessary to satisfy the seven listed conditions in order to use this approach.

TABLE 2109.5.6.1
FOUNDATION WALL CONSTRUCTION

WALL CONSTRUCTION	NOMINAL WALL THICKNESS (inches)	MAXIMUM DEPTH OF UNBALANCED BACKFILL (feet)
Hollow unit masonry	8 10 12	5 6 7
Solid unit masonry	8 10 12	5 7 7
Fully grouted masonry	8 10 12	7 8 8

For SI: 1 inch = 25.4 mm, 1 foot = 304.8 mm.

❖ The minimum foundation wall thickness can be established from this table based on the depth of fill that is supported, as well as the proposed wall construction.

2109.5.6.2 Design requirements. Where the requirements of Section 2109.5.6.1 are not met, foundation walls shall be designed in accordance with Section 1805.5.

❖ If the conditions of Section 2109.5.6.1 cannot be met, the more general requirements for foundation walls in Chapter 18 must be complied with.

2109.6 Bond.

❖ Section 2109.6 contains provisions for bonding adjacent wythes of empirically designed multiwythe masonry walls. The requirements of this section do not apply to engineered masonry design methods of Section 2107 or 2108.

2109.6.1 General. The facing and backing of multiwythe masonry walls shall be bonded in accordance with Section 2109.6.2, 2109.6.3 or 2109.6.4.

❖ This section establishes the requirements and methods of bonding together the facing and backing of adjacent wythes of multiwythe masonry walls. The use of masonry units, nonadjustable metal ties, adjustable metal ties and metal joint reinforcement as bonding elements is provided for in this section.

Bonding increases structural integrity, including load transfer between adjacent wythes and across head and collar joints. The requirements include both transverse (through-wall) and longitudinal (in-wall) bonding. Longitudinal bonding is provided for in Section 2109.5, with requirements for running bond and stack bond.

Adjacent wythes of multiwythe masonry walls are usually brick-to-brick, brick-to-block or block-to-block.

Collar joints are usually $^3/_8$ to 4 inches (9.5 to 102 mm) thick.

Adjacent wythes of multi-wythe masonry walls are considered composite (monolithic) when connected by metal ties or joint reinforcement and by a collar joint solidly filled with mortar or grout. Accordingly, Note b of Table 2109.3.2 requires completely filled collar joints if adjacent wythes bonded with metal ties are to be considered composite (monolithic) for empirical compressive stress design.

2109.6.2 Bonding with masonry headers.

❖ Masonry headers (bonders) are permitted to be used to connect adjacent wythes, in accordance with this section. Differential thermal movement, especially at exterior walls, can crack masonry bonders. Metal ties or joint reinforcement are more ductile and are recommended at these locations. Walls having the specified masonry bonders are considered composite (monolithic). Header requirements are based on successful past performance.

2109.6.2.1 Solid units. Where the facing and backing (adjacent wythes) of solid masonry construction are bonded by means of masonry headers, no less than 4 percent of the wall surface of each face shall be composed of headers extending not less than 3 inches (76 mm) into the backing. The distance between adjacent full-length headers shall not exceed 24 inches (610 mm) either vertically or horizontally. In walls in which a single header does not extend through the wall, headers from the opposite sides shall overlap at least 3 inches (76 mm), or headers from opposite sides shall be covered with another header course overlapping the header below at least 3 inches (76 mm).

❖ Solid masonry units are defined in Section 2102.1. The 4-percent requirement applies to each surface of the wall and 24-inch (610 mm) spacing is required in both horizontal and vertical directions. Spacing is measured between the nearest surfaces of masonry units. The 3-inch (76 mm) overlap is a projected measurement, since headers are spaced a maximum of 24 inches (610 mm) vertically.

Bonding requirements for solid units are less restrictive than for hollow units because of the greater cross-sectional area crossing the collar joint.

2109.6.2.2 Hollow units. Where two or more hollow units are used to make up the thickness of a wall, the stretcher courses shall be bonded at vertical intervals not exceeding 34 inches (864 mm) by lapping at least 3 inches (76 mm) over the unit below, or by lapping at vertical intervals not exceeding 17 inches (432 mm) with units that are at least 50 percent greater in thickness than the units below.

❖ Hollow masonry units are defined in Section 2102.1. An entire course of headers is required at the stated vertical spacing. Spacing is measured between the nearest surfaces of masonry units. The halved spacing of 17 inches (432 mm) is permitted for units having twice the thickness, because of the greater cross-sectional area crossing the collar joint. Bonding requirements for hollow units are more restrictive than for solid units because of the smaller cross-sectional area crossing the collar joint.

2109.6.2.3 Masonry bonded hollow walls. In masonry bonded hollow walls, the facing and backing shall be bonded so that not less than 4 percent of the wall surface of each face is composed of masonry bonded units extending not less than 3 inches (76 mm) into the backing. The distance between adjacent bonders shall not exceed 24 inches (610 mm) either vertically or horizontally.

❖ This section prescribes procedures for bonding hollow walls using masonry units.

2109.6.3 Bonding with wall ties or joint reinforcement.

❖ Figure 2109.6.3(1) illustrates typical masonry accessories. Shown are ladder-type and truss-type joint reinforcement and a "Z" wire wall tie. The spacing requirements given in this section are illustrated in Figure 2109.6.3(2). Strength and durability requirements for metal ties and joint reinforcement are given in Section 2103.11. Wall ties are required to be placed in accordance with Section 2104.1.3.

Metal wall ties or joint reinforcement provide a more ductile connection between adjacent wythes than do masonry headers. The size, spacing and number of ties required in this section are based on experience.

Adjacent wythes bonded with metal ties or joint reinforcement are considered composite (monolithic) only if the collar joint between them is completely filled with mortar or grout.

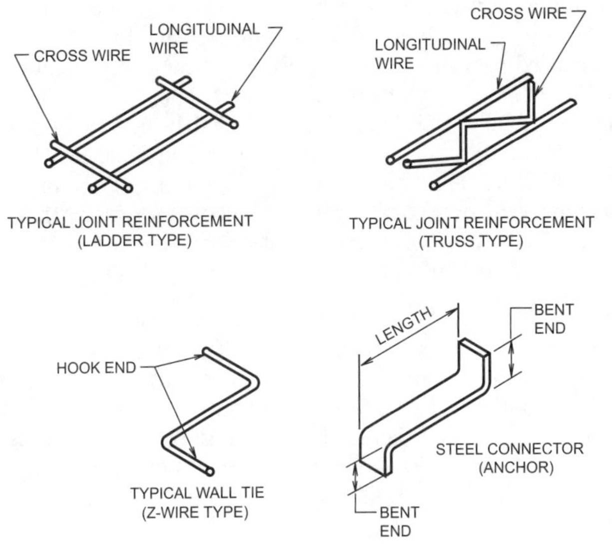

Figure 2109.6.3(1)
TYPICAL MASONRY ACCESSORIES

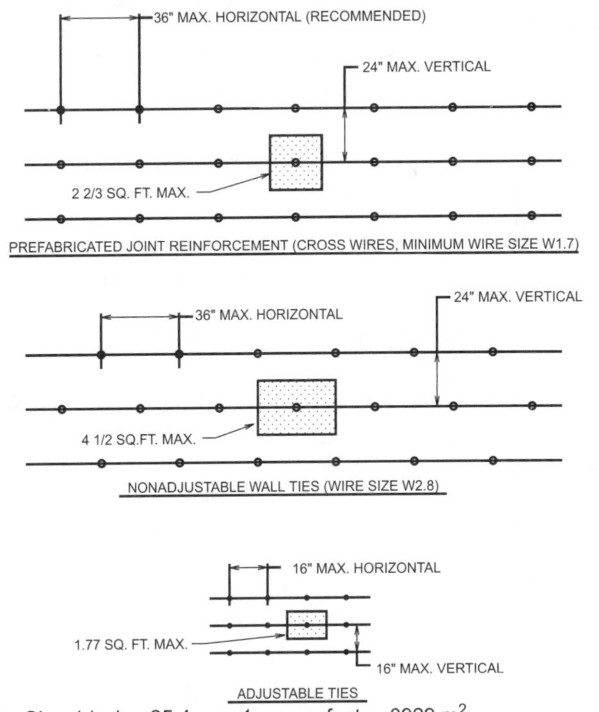

For SI: 1 inch = 25.4 mm, 1 square foot = .0929 m^2.

Figure 2109.6.3(2)
EMPIRICAL REQUIREMENTS FOR BONDING MULTIWYTHE WALLS WITH TIES OR JOINT REINFORCEMENT

2109.6.3.1 Bonding with wall ties. Except as required by Section 2109.6.3.1.1, where the facing and backing (adjacent wythes) of masonry walls are bonded with wire size W2.8 (MW18) wall ties or metal wire of equivalent stiffness embedded in the horizontal mortar joints, there shall be at least one metal tie for each $4^{1}/_{2}$ square feet (0.42 m^2) of wall area. The maximum vertical distance between ties shall not exceed 24 inches (610 mm), and the maximum horizontal distance shall not exceed 36 inches (914 mm). Rods or ties bent to rectangular shape shall be used with hollow masonry units laid with the cells vertical. In other walls, the ends of ties shall be bent to 90-degree (1.57 rad) angles to provide hooks no less than 2 inches (51 mm) long. Wall ties shall be without drips. Additional bonding ties shall be provided at all openings, spaced not more than 36 inches (914 mm) apart around the perimeter and within 12 inches (305 mm) of the opening.

❖ Requirements for spacing of nonadjustable wall ties are shown in Figure 2109.6.3(2). The minimum wire size is W2.8. Where wythes of hollow units are bonded and cores are required to be vertical, rectangular-shaped ties are required. "Z" ties, illustrated in Figure 2109.6.3(1), are used in other applications. Drips on wall ties can adversely affect the tie strength and are, therefore, prohibited.

2109.6.3.1.1 Bonding with adjustable wall ties. Where the facing and backing (adjacent wythes) of masonry are bonded with adjustable wall ties, there shall be at least one tie for each 1.77 square feet (0.164 m^2) of wall area. Neither the vertical nor horizontal spacing of the adjustable wall ties shall exceed 16 inches (406 mm). The maximum vertical offset of bed joints from one wythe to the other shall be $1^{1}/_{4}$ inches (32 mm). The maximum clearance between connecting parts of the ties shall be $^{1}/_{16}$ inch (1.6 mm). When pintle legs are used, ties shall have at least two wire size W2.8 (MW18) legs.

❖ Spacing requirements for bonding with adjustable ties are shown in Figure 2109.6.3(2) and are more restrictive because of the lower stiffness of those ties. A single tie of the pintle- and eye-type is required to have two legs. The maximum clearance between the eye and pintle is $^{1}/_{16}$ inch (1.6 mm). Figures 2109.6.3.1.1 (1), (2) and (3) show requirements for adjustable anchors.

2109.6.3.2 Bonding with prefabricated joint reinforcement. Where the facing and backing (adjacent wythes) of masonry are bonded with prefabricated joint reinforcement, there shall be at least one cross wire serving as a tie for each $2^{2}/_{3}$ square feet (0.25 m^2) of wall area. The vertical spacing of the joint reinforcing shall not exceed 24 inches (610 mm). Cross wires on prefabricated joint reinforcement shall not be less than W1.7 (MW11) and shall be without drips. The longitudinal wires shall be embedded in the mortar.

❖ Truss and ladder-type joint reinforcement is illustrated in Figures 2102.1(4), 2109.6.3(1), 2109.6.3.1.1(1) and 2109.6.3.1.1(2). Ladder-type joint reinforcement better accommodates differential movement between wythes. Cross wires are required to have a minimum wire size of W1.7 for the transfer of stresses between adjacent wythes across the collar joint. Cross wires are permitted to be plain, while longitudinal wires are required to be deformed for increased bond strength.

2109.6.4 Bonding with natural or cast stone.

❖ Because of the irregularity of mortar joints permitted with coursed, random or rough stone masonry, as well as the need for peripheral mortar area in through-wall bonding units, metal ties do not adequately bond and are not required in this type of construction.

2109.6.4.1 Ashlar masonry. In ashlar masonry, bonder units, uniformly distributed, shall be provided to the extent of not less than 10 percent of the wall area. Such bonder units shall extend not less than 4 inches (102 mm) into the backing wall.

❖ Ashlar is often classified as either coursed or random. Bonding requirements apply equally to both types. The 10-percent requirement applies to one wall surface only. The "uniform distribution" requirement is intended to allow for slight deviations necessary for the nonmodular units typically encountered.

2109.6.4.2 Rubble stone masonry. Rubble stone masonry 24 inches (610 mm) or less in thickness shall have bonder units with a maximum spacing of 36 inches (914 mm) vertically and 36 inches (914 mm) horizontally, and if the masonry is of greater thickness than 24 inches (610 mm), shall have one bonder unit for each 6 square feet (0.56 m^2) of wall surface on both sides.

❖ Rubble stone masonry is required to be through bonded using stones placed at a maximum spacing of 3 feet (914 mm) vertically and horizontally. Where the wall is more than 24 inches (610 mm) thick, at least one bond unit is required for each 6 square feet (0.56 m^2) of wall surface on both sides.

2109.6.5 Masonry bonding pattern.

❖ Masonry may be constructed with a variety of bond patterns to provide a variety of decorative appearances. Because some of these bond patterns are stronger than others, two broad categories have been established to describe them. Section 2109.5.1 applies to masonry laid in running bond. Section 2109.6.5.2 applies to masonry laid in "other than running bond," which in that section is referred to as "stack bond," although that term can also be used to describe a very specific bond pattern in which the head joints are aligned continuously from course to course, rather than being offset. See the definitions of "Running bond" and "Stack bond" in Section 2102.1 and related commentary.

2109.6.5.1 Masonry laid in running bond. Each wythe of masonry shall be laid in running bond, head joints in successive courses shall be offset by not less than one-fourth the unit length or the masonry walls shall be reinforced longitudinally as required in Section 2109.6.5.2.

❖ The requirement for running bond is illustrated in Figure 2102.1(6). The minimum overlap is intended to provide structural continuity across head joints.

2109.6.5.2 Masonry laid in stack bond. Where unit masonry is laid with less head joint offset than in Section 2109.6.5.1, the minimum area of horizontal reinforcement placed in mortar bed joints or in bond beams spaced not more than 48 inches (1219 mm) apart, shall be 0.0003 times the vertical cross-sectional area of the wall.

❖ Where the overlap required for running bond, as illustrated in Figure 2102.1(6), is not provided, additional structural continuity must be provided across head joints in accordance with this section. Each mortar joint or bond beam is required to have the minimum stated reinforcement to provide distributed reinforcement in the wall.

2109.7 Anchorage.

❖ This section contains provisions for anchorage of empirically designed masonry elements at locations of lateral support, including intersecting walls, floors, roofs and adjoining structural framing. The requirements of this section do not apply to the engineered masonry design methods of Sections 2107 and 2108.

2109.7.1 General. Masonry elements shall be anchored in accordance with Sections 2109.7.2 through 2109.7.4.

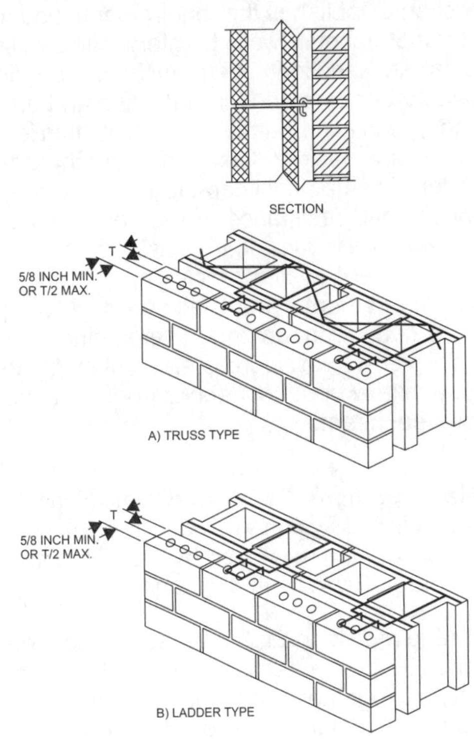

For SI: 1 inch = 25.4 mm.

Figure 2109.6.3.1.1(2)
ADJUSTABLE ASSEMBLY DETAILS

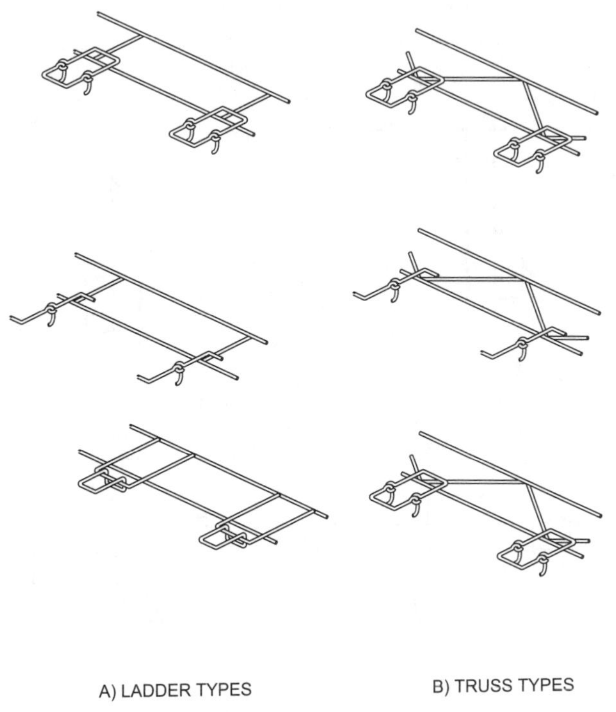

Figure 2109.6.3.1.1(1)
ADJUSTABLE JOINT REINFORCEMENT ASSEMBLIES

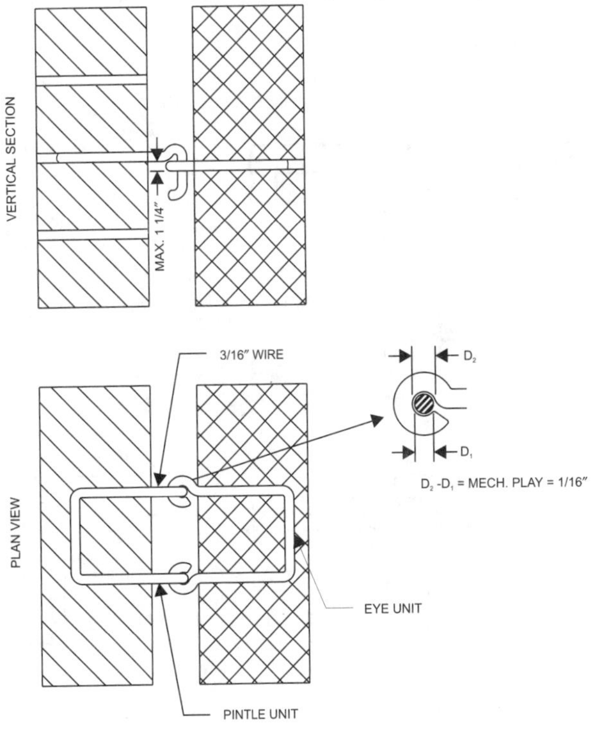

For SI: 1 inch = 25.4 mm.

Figure 2109.6.3.1.1(3)
ADJUSTABLE WALL TIES

❖ This section establishes the requirements and methods of anchoring masonry walls to intersecting walls, floors and roofs. This section covers anchoring requirements for masonry bonders and for metal ties and anchors.

Anchorage in accordance with this section is required for empirically designed masonry. Anchorage provisions address empirical lateral load resistance, but not the vertical load resistance of the connection, floor or roof system. The vertical load resistance of the connection, floor or roof system, including permitted spans, bearing requirements and bracing, must comply with applicable code requirements. The empirical provisions in this chapter do not address elements with horizontal spans, such as wood, floor decks, roof systems or concrete or steel beams.

2109.7.2 Intersecting walls. Masonry walls depending upon one another for lateral support shall be anchored or bonded at locations where they meet or intersect by one of the methods indicated in Sections 2109.7.2.1 through 2109.7.2.5.

❖ Where required by the lateral bracing requirements in Section 2109.4, intersecting walls and partitions must be anchored together by any of the methods described in the sections that follow (see Figure 2109.7.2). These requirements, however, do not prohibit other methods for connecting intersecting walls.

For example, long walls, by design, often require control joints or expansion joints at cross walls. The construction details for these walls differ considerably from walls that are required to be anchored or bonded monolithically.

Intersecting walls, such as tees and corners, may be tied together using interlocking masonry units or heavy metal ties placed in the bed joints.

2109.7.2.1 Bonding pattern. Fifty percent of the units at the intersection shall be laid in an overlapping masonry bonding pattern, with alternate units having a bearing of not less than 3 inches (76 mm) on the unit below.

❖ Masonry bonding requires that at least 50 percent of the masonry units cross at the juncture of the walls as bonding units. The units must be laid with a 3-inch (76 mm) minimum bearing. Figure 2109.7.2 shows typical brick walls bonded at a tee intersection and a corner. While these coursings are common, others are possible.

2109.7.2.2 Steel connectors. Walls shall be anchored by steel connectors having a minimum section of $^1/_4$ inch (6.4 mm) by $1^1/_2$ inches (38 mm), with ends bent up at least 2 inches (51 mm) or with cross pins to form anchorage. Such anchors shall be at least 24 inches (610 mm) long and the maximum spacing shall be 48 inches (1219 mm).

❖ Rigid steel connectors must be used at vertical intervals of 4 feet (1219 mm) or less. For this type of anchorage, the most commonly used fastener is a steel strip with cross-sectional dimensions not less than $^1/_4$ by $1^1/_2$ inches (6.4 by 38 mm) in lengths of 2 feet (610 mm) or more and with 2-inch (51 mm) hooked ends [see Figure 2109.6.3(1)]. Figure 2109.7.2 shows a tee and corner connection made with steel connectors.

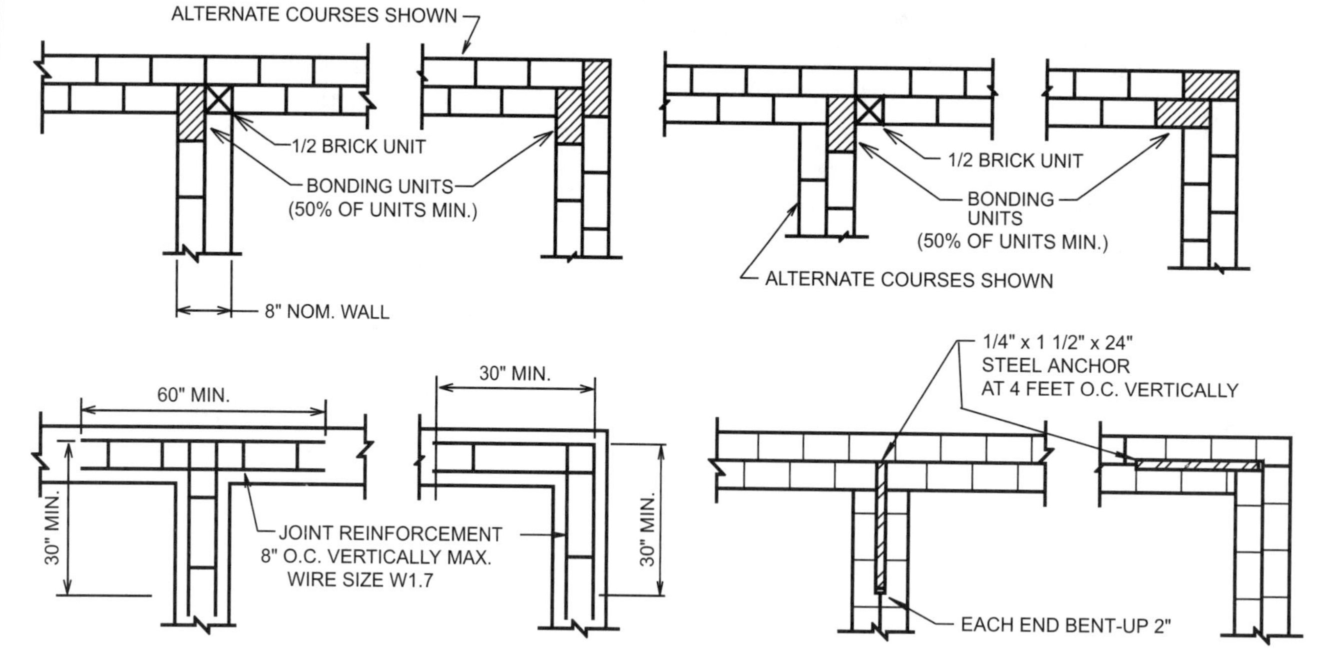

For SI: 1 inch = 25.4 mm.

Figure 2109.7.2
ANCHORAGE OF INTERSECTING WALLS

2109.7.2.3 Joint reinforcement. Walls shall be anchored by joint reinforcement spaced at a maximum distance of 8 inches (203 mm). Longitudinal wires of such reinforcement shall be at least wire size W1.7 (MW 11) and shall extend at least 30 inches (762 mm) in each direction at the intersection.

❖ Figure 2109.7.2 shows a tee and corner connection made with joint reinforcement. Prefabricated joint reinforcement made for that purpose is commonly used. Joint reinforcement is spaced not more than 8 inches (204 mm) vertically.

The ladder-type joint reinforcement shown is made with $^{3}/_{16}$-inch-diameter (4.8 mm) deformed side bars cross connected with a minimum wire size of W1.7.

2109.7.2.4 Interior nonload-bearing walls. Interior nonload-bearing walls shall be anchored at their intersection, at vertical intervals of not more than 16 inches (406 mm) with joint reinforcement or $^{1}/_{4}$-inch (6.4 mm) mesh galvanized hardware cloth.

❖ Where interior nonload-bearing masonry walls or partitions intersect, they may be connected by bonders described in Section 2109.7.2.1 or, if constructed separately, by joint reinforcement or $^{1}/_{4}$-inch (6.4 mm) galvanized steel mesh spaced vertically at a maximum interval of 16 inches (406 mm).

2109.7.2.5 Ties, joint reinforcement or anchors. Other metal ties, joint reinforcement or anchors, if used, shall be spaced to provide equivalent area of anchorage to that required by this section.

❖ Truss-type, prefabricated metal ties and steel strip anchors bent at the ends can be hooked into vertical mortar joints or embedded in grout-filled cores of concrete block units. Other methods of reinforcement can be used if their cross-sectional area and distribution are equivalent.

2109.7.3 Floor and roof anchorage. Floor and roof diaphragms providing lateral support to masonry shall comply with the live loads in Section 1607.3 and shall be connected to the masonry in accordance with Sections 2109.7.3.1 through 2109.7.3.3.

❖ Where the lateral support locations required in Section 2109.4 are provided by intersecting floors or roofs, the minimum anchorage requirements of this section apply. The specified anchorage is intended to transfer shear between a floor or roof diaphragm and the wall, as well as provide locations of lateral support.

This section does not address vertical load resistance of floors or roofs. The connections required by the loads transferred from floors and roofs must be in accordance with applicable requirements for those elements, including required design loads in Chapter 16. The minimum requirements stated here do not address floor and roof loads to be resisted by the connection.

2109.7.3.1 Wood floor joists. Wood floor joists bearing on masonry walls shall be anchored to the wall at intervals not to exceed 72 inches (1829 mm) by metal strap anchors. Joists parallel to the wall shall be anchored with metal straps spaced not more than 72 inches (1829 mm) o.c. extending over or under and secured to at least three joists. Blocking shall be provided between joists at each strap anchor.

❖ Anchorage with metal straps is specified by this section. Blocking is required for stability against rotation of joists parallel to the wall. Strength and material requirements for metal straps are specified in Section 2103.11.

2109.7.3.2 Steel floor joists. Steel floor joists bearing on masonry walls shall be anchored to the wall with $^{3}/_{8}$-inch (9.5 mm) round bars, or their equivalent, spaced not more than 72 inches (1829 mm) o.c. Where joists are parallel to the wall, anchors shall be located at joist bridging.

❖ The strength and material requirements for $^{3}/_{8}$-inch (9.5 mm) round bars are specified in Section 2103.11. The requirements for steel joist construction, including materials, design, load-bearing capacity and cross bridging, are in Chapter 22.

2109.7.3.3 Roof diaphragms. Roof diaphragms shall be anchored to masonry walls with $^{1}/_{2}$-inch-diameter (12.7 mm) bolts, 72 inches (1829 mm) o.c. or their equivalent. Bolts shall extend and be embedded at least 15 inches (381 mm) into the masonry, or be hooked or welded to not less than 0.20 square inch (129 mm^2) of bond beam reinforcement placed not less than 6 inches (152 mm) from the top of the wall.

❖ Roofs must be anchored to bond beams and to the tops of other walls. The material requirements for the bolts are given in Section 2103.11.5.

2109.7.4 Walls adjoining structural framing. Where walls are dependent upon the structural frame for lateral support, they shall be anchored to the structural members with metal anchors or otherwise keyed to the structural members. Metal anchors shall consist of $^{1}/_{2}$-inch (12.7 mm) bolts spaced at 48 inches (1219 mm) o.c. embedded 4 inches (102 mm) into the masonry, or their equivalent area.

❖ Selection of connectors for anchoring masonry to steel beams and columns is based on the loads that must be resisted. Lateral restraint perpendicular to the wall is required as a location of lateral support. Diaphragm transfer of shear forces is also required parallel to the wall, if such forces occur. Alternative methods of reinforcement can be used if the cross-sectional area and spacing are equivalent.

2109.8 Adobe construction. Adobe construction shall comply with this section and shall be subject to the requirements of this code for Type V construction.

❖ Adobe masonry was popular in the southwest United States due to the availability of soil for units, the limited rainfall and low humidity to dry the units, the thermal mass provided by the completed adobe structure and the low cost of this form of construction. Requirements for adobe construction are based on previous requirements in the SBC and the UBC. They are a combination of empirical provisions and rudimentary engineering. Since there are no ASTM standards for adobe materials, test methods have been included in the code. Design is based on gross cross-sectional dimensions.

Requirements for unstabilized adobe are contained in Section 2109.8.1. Requirements for stabilized adobe are contained in Section 2109.8.2. Requirements in Sections 2109.8.3 and 2109.8.4 apply to both unstabilized and stabilized adobe. This is one of the few sources for such design information.

2109.8.1 Unstabilized adobe.

❖ Unstabilized adobe does not contain stabilizers and is generally not as durable or dimensionally stable as stabilized adobe.

2109.8.1.1 Compressive strength. Adobe units shall have an average compressive strength of 300 psi (2068 kPa) when tested in accordance with ASTM C 67. Five samples shall be tested and no individual unit is permitted to have a compressive strength of less than 250 psi (1724 kPa).

❖ Average compressive strength, based on five specimens tested in accordance with ASTM C 67, must be at least 300 psi (2068 kPa).

2109.8.1.2 Modulus of rupture. Adobe units shall have an average modulus of rupture of 50 psi (345 kPa) when tested in accordance with the following procedure. Five samples shall be tested and no individual unit shall have a modulus of rupture of less than 35 psi (241 kPa).

❖ Average modulus of rupture, based on five specimens tested in accordance with Sections 2109.8.1.2.1 through 2109.8.1.2.4, must be at least 50 psi (345 kPa).

2109.8.1.2.1 Support conditions. A cured unit shall be simply supported by 2-inch-diameter (51 mm) cylindrical supports located 2 inches (51 mm) in from each end and extending the full width of the unit.

❖ These required support conditions are typical for modulus of rupture tests.

2109.8.1.2.2 Loading conditions. A 2-inch-diameter (51 mm) cylinder shall be placed at midspan parallel to the supports.

❖ Loading through a hydraulic cylinder at midspan is common for these tests.

2109.8.1.2.3 Testing procedure. A vertical load shall be applied to the cylinder at the rate of 500 pounds per minute (37 N/s) until failure occurs.

❖ The required application of vertical load is easily controlled in testing laboratories.

2109.8.1.2.4 Modulus of rupture determination. The modulus of rupture shall be determined by the equation:

$$f_r = 3WL_s/2bt^2 \qquad \textbf{(Equation 21-4)}$$

where, for the purposes of this section only:

b = Width of the test specimen measured parallel to the loading cylinder, inches (mm).

f_r = Modulus of rupture, psi (MPa).

L_s = Distance between supports, inches (mm).

t = Thickness of the test specimen measured parallel to the direction of load, inches (mm).

W = The applied load at failure, pounds (N).

❖ Equation 21-4 is based on simple engineering mechanics and is valid for all rectangular specimens tested in this fashion.

2109.8.1.3 Moisture content requirements. Adobe units shall have a moisture content not exceeding 4 percent by weight.

❖ This section limits the moisture content of unstabilized adobe units to acceptable levels.

2109.8.1.4 Shrinkage cracks. Adobe units shall not contain more than three shrinkage cracks and any single shrinkage crack shall not exceed 3 inches (76 mm) in length or $^1/_8$ inch (3.2 mm) in width.

❖ As adobe units dry, they shrink and can crack. This section places limits on those potential cracks to keep the masonry structurally sound and reasonably water resistant.

2109.8.2 Stabilized adobe.

❖ This type of adobe is manufactured with stabilizers to increase its durability and decrease its water absorption.

2109.8.2.1 Material requirements. Stabilized adobe shall comply with the material requirements of unstabilized adobe in addition to Sections 2109.8.2.1.1 and 2109.8.2.1.2.

❖ Stabilized adobe must comply with the few material requirements for unstabilized adobe in Section 2109.8.1. Stabilized units must also comply with soil compatibility and absorption requirements in Sections 2109.2.1.1 and 2109.2.1.2.

2109.8.2.1.1 Soil requirements. Soil used for stabilized adobe units shall be chemically compatible with the stabilizing material.

❖ The soil and stabilizing materials must be chemically compatible, so that the stabilized units will be durable.

2109.8.2.1.2 Absorption requirements. A 4-inch (102 mm) cube, cut from a stabilized adobe unit dried to a constant weight in a ventilated oven at 212°F to 239°F (100°C to 115°C), shall not absorb more than $2^1/_2$- percent moisture by weight when placed upon a constantly water-saturated, porous surface for seven days. A minimum of five specimens shall be tested and each specimen shall be cut from a separate unit.

❖ This section prescribes a test method to verify that stabilized adobe units meet absorption limits.

2109.8.3 Working stress. The allowable compressive stress based on gross cross-sectional area of adobe shall not exceed 30 psi (207 kPa).

❖ This section prescribes the allowable compressive stress of adobe based on its gross cross-sectional area.

2109.8.3.1 Bolts. Bolt values shall not exceed those set forth in Table 2109.8.3.1.

❖ This section requires the capacity of bolts to be based on Table 2109.8.3.1. Specific types of bolts are not identified, but headed, bent-bar and plate anchors should all be acceptable.

**TABLE 2109.8.3.1
ALLOWABLE SHEAR ON BOLTS IN ADOBE MASONRY**

DIAMETER OF BOLTS (inches)	MINIMUM EMBEDMENT (inches)	SHEAR (pounds)
$^1/_2$	—	—
$^5/_8$	12	200
$^3/_4$	15	300
$^7/_8$	18	400
1	21	500
$1^1/_8$	24	600

For SI: 1 inch = 25.4 mm, 1 pound = 4.448 N.

❖ The allowable shear values in this table are based on the capacity of the adobe masonry. The capacity of the anchor-bolt steel is much higher, so the lower strength of the adobe controls.

2109.8.4 Construction.

❖ This section contains general construction requirements for height restrictions, mortar restrictions, mortar joint construction and water-resistance requirements for parapet walls and also specific requirements for wall thickness; foundations; isolated piers and columns; tie beams; exterior finish and lintels.

2109.8.4.1 General.

❖ This section contains general construction requirements for height restrictions, mortar restrictions, mortar joint construction and water-resistance requirements for parapet walls.

2109.8.4.1.1 Height restrictions. Adobe construction shall be limited to buildings not exceeding one story, except that two-story construction is allowed when designed by a registered design professional.

❖ Because of the low strength of adobe masonry, it is limited to use in single-story buildings, unless a registered design professional is hired, in which case two-story buildings are permitted.

2109.8.4.1.2 Mortar restrictions. Mortar for stabilized adobe units shall comply with Chapter 21 or adobe soil. Adobe soil used as mortar shall comply with material requirements for stabilized adobe. Mortar for unstabilized adobe shall be portland cement mortar.

❖ A variety of mortars are acceptable for stabilized adobe as noted; however, portland cement-lime mortars are required for unstabilized adobe. Selection of a relatively weak mortar that is compatible with the units is appropriate.

2109.8.4.1.3 Mortar joints. Adobe units shall be laid with full head and bed joints and in full running bond.

❖ Full mortar joints, the same as for other solid units, are required for adobe construction. Units are required to be laid in running bond.

2109.8.4.1.4 Parapet walls. Parapet walls constructed of adobe units shall be waterproofed.

❖ Waterproofing parapets reduce moisture infiltration into the adobe.

2109.8.4.2 Wall thickness. The minimum thickness of exterior walls in one-story buildings shall be 10 inches (254 mm). The walls shall be laterally supported at intervals not exceeding 24 feet (7315 mm). The minimum thickness of interior load-bearing walls shall be 8 inches (203 mm). In no case shall the unsupported height of any wall constructed of adobe units exceed 10 times the thickness of such wall.

❖ Because of the low strength of adobe masonry walls, thicker walls with more closely spaced supports are required.

2109.8.4.3 Foundations.

❖ This section prescribes foundation requirements for adobe masonry.

2109.8.4.3.1 Foundation support. Walls and partitions constructed of adobe units shall be supported by foundations or footings that extend not less than 6 inches (152 mm) above adjacent ground surfaces and are constructed of solid masonry (excluding adobe) or concrete. Footings and foundations shall comply with Chapter 18.

❖ So that adobe masonry is properly supported, solid masonry or concrete foundations are required.

2109.8.4.3.2 Lower course requirements. Stabilized adobe units shall be used in adobe walls for the first 4 inches (102 mm) above the finished first-floor elevation.

❖ Because of their greater durability, stabilized adobe units are required at the base of adobe walls. Conventional masonry units can also be used to satisfy this requirement.

2109.8.4.4 Isolated piers or columns. Adobe units shall not be used for isolated piers or columns in a load-bearing capacity. Walls less than 24 inches (610 mm) in length shall be considered isolated piers or columns.

❖ Adobe units are not strong enough to carry significant loads and are, therefore, not permitted to be used as isolated piers or columns.

2109.8.4.5 Tie beams. Exterior walls and interior load-bearing walls constructed of adobe units shall have a continuous tie beam at the level of the floor or roof bearing and meeting the following requirements.

❖ To distribute loads more evenly into the adobe, tie beams are required at the floor or roof levels. Tie beams can be constructed of concrete or wood as described in Sections 2109.8.4.5.1 and 2108.8.4.5.2, respectively.

2109.8.4.5.1 Concrete tie beams. Concrete tie beams shall be a minimum depth of 6 inches (152 mm) and a minimum width of 10 inches (254 mm). Concrete tie beams shall be continuously reinforced with a minimum of two No. 4 reinforcing bars. The ultimate compressive strength of concrete shall be at least 2,500 psi (17.2 MPa) at 28 days.

❖ This section provides requirements for concrete tie beams to be cast above adobe masonry walls to distribute loads from floors and roofs.

2109.8.4.5.2 Wood tie beams. Wood tie beams shall be solid or built up of lumber having a minimum nominal thickness of 1 inch (25 mm), and shall have a minimum depth of 6 inches (152 mm) and a minimum width of 10 inches (254 mm). Joints in wood tie beams shall be spliced a minimum of 6 inches (152 mm). No splices shall be allowed within 12 inches (305 mm) of an opening. Wood used in tie beams shall be approved naturally decay-resistant or pressure-treated wood.

❖ This section provides requirements for wood tie beams to be constructed above adobe masonry walls to distribute loads from floors and roofs.

2109.8.4.6 Exterior finish. Exterior walls constructed of unstabilized adobe units shall have their exterior surface covered with a minimum of two coats of portland cement plaster having a minimum thickness of $^3/_4$ inch (19.1 mm) and conforming to ANSI A42.2. Lathing shall comply with ANSI A42.3. Fasteners shall be spaced at 16 inches (406 mm) o.c. maximum. Exposed wood surfaces shall be treated with an approved wood preservative or other protective coating prior to lath application.

❖ Unstabilized adobe must be coated with plaster to increase its durability.

2109.8.4.7 Lintels. Lintels shall be considered structural members and shall be designed in accordance with the applicable provisions of Chapter 16.

❖ Lintels over door and window openings are required to be structurally designed to carry imposed loads and to distribute those loads into the supporting adobe.

SECTION 2110
GLASS UNIT MASONRY

2110.1 Scope. This section covers the empirical requirements for nonload-bearing glass unit masonry elements in exterior or interior walls.

❖ Section 2110 contains provisions for glass unit masonry walls, which are nearly identical to the glass unit masonry provisions in Chapter 7 of ACI 530/ASCE 5/TMS 402. Because those provisions are essentially the same, IBC Section 2101.2.4 permits glass unit masonry to comply with the provisions of Chapter 7 of ACI 530/ASCE 5/TMS 402 or of Section 2110.

Glass unit masonry panels are permitted to be used in interior or exterior walls, provided that they are nonload bearing and comply with the requirements of Section 2110, which are partly empirical and partly based on tests.

2110.1.1 Limitations. Solid or hollow approved glass block shall not be used in fire walls, party walls, fire barriers or fire partitions, or for load-bearing construction. Such blocks shall be erected with mortar and reinforcement in metal channel-type frames, structural frames, masonry or concrete recesses, embedded panel anchors as provided for both exterior and interior walls or other approved joint materials. Wood strip framing shall not be used in walls required to have a fire-resistance rating by other provisions of this code.

Exceptions:

1. Glass-block assemblies having a fire protection rating of not less than $^3/_4$ hour shall be permitted as opening protectives in accordance with Section 715 in fire barriers and fire partitions that have a required fire-resistance rating of 1 hour or less and do not enclose exit stairways or exit passageways.
2. Glass-block assemblies as permitted in Section 404.5, Exception 2.

❖ Structural glass blocks are not permitted in fire walls, party walls, fire barrier walls or fire partitions, with two exceptions. Exception 1 permits glass blocks that have been tested and classified for a $^3/_4$-hour fire protection rating in openings to be used in fire barrier walls or fire partitions with a required fire-resistance rating of 1 hour or less. Since 1-hour fire barrier walls can be utilized to enclose interior exit stairways and exit ramps (see Section 1019.1), as well as exit passageways (see Section 1020.3), the exception does not apply to those locations. This is consistent with Sections 1019.1.1 and 1020.4, which limit openings in these exit components to those that are necessary for egress purposes. Exception 2 permits glass blocks to be installed in accordance with the requirements in Section 404.5, Exception 2 for the enclosure of atriums. Because Section 404.5 requires a 1- hour fire barrier wall to enclose an atrium, this exception is redundant since Exception 1 already permits this.

2110.2 Units. Hollow or solid glass-block units shall be standard or thin units.

❖ This section contains minimum requirements for glass masonry units, since a corresponding ASTM standard does not exist. Units are permitted to be either hollow or solid and are required to meet requirements for either standard or thin units.

Glass units are usually factory coated at their edges. Uncoated glass-block units can be field coated by following the manufacturer's instructions.

2110.2.1 Standard units. The specified thickness of standard units shall be $3^7/_8$ inches (98 mm).

❖ This section requires a specified thickness for standard glass masonry units of $3\ ^7/_8$ inches (98 mm).

2110.2.2 Thin units. The specified thickness of thin units shall be $3^1/_8$ inches (79 mm) for hollow units or 3 inches (76 mm) for solid units.

❖ Thicknesses for thin glass unit masonry are given in this section.

2110.3 Panel size.

❖ This section provides limits on the size of exterior standard-unit and thin-unit panels, interior panels, solid glass-block panels and curved glass-block panels.

The glass-block panels must be restrained laterally and be capable of resisting horizontal forces. Panels exceeding these size limits require intermediate structural supports so that loads can be adequately resisted.

2110.3.1 Exterior standard-unit panels. The maximum area of each individual exterior standard-unit panel shall be 144 square feet (13.4 m^2) when the design wind pressure is 20 psf (958 N/m^2). The maximum panel dimension between structural supports shall be 25 feet (7620 mm) in width or 20 feet (6096 mm) in height. The panel areas are permitted to be adjusted in accordance with Figure 2110.3.1 for other wind pressures.

❖ Single panels of glass block are limited to a maximum length of 25 feet (7620 mm) and a maximum height of 20 feet (6096 mm) between structural supports. When subjected to a wind pressure of 20 psf (958 N/m^2), any single panel cannot exceed 144 square feet (13 m^2) in area. This maximum panel permissible area can be adjusted, however, for other wind pressures by the use of Code Figure 2110.3.1.

FIGURE 2110.3.1. See below.

❖ The wind load resistance curve represents capacities for a variety of panel conditions. The 144-square-feet (13 m^2) area limit is based on a safety factor of 2.7 when the design wind pressure is 20 psf (958 N/m^2).

2110.3.2 Exterior thin-unit panels. The maximum area of each individual exterior thin-unit panel shall be 85 square feet (7.9 m^2). The maximum dimension between structural supports shall be 15 feet (4572 mm) in width or 10 feet (3048 mm) in height. Thin units shall not be used in applications where the design wind pressure exceeds 20 psf (958 N/m^2).

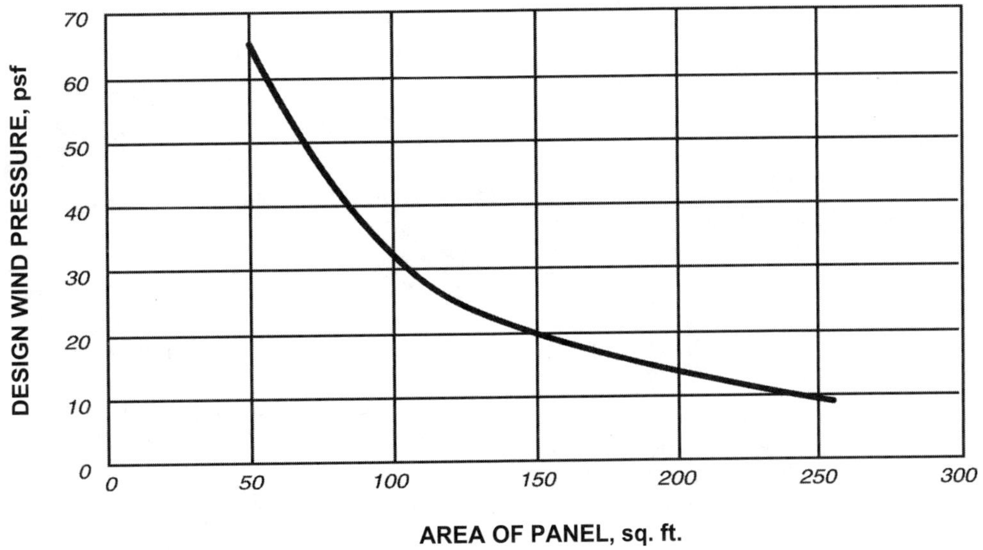

For SI: 1 square foot = 0.0929 m^2, 1 pound per square foot = 47.9 N/m^2.

FIGURE 2110.3.1
GLASS MASONRY DESIGN WIND LOAD RESISTANCE

❖ The limitations on the use of exterior panels with thin-unit glass block are more restrictive than those in Section 2110.3.1 for exterior panels with standard units, since thin units are not as strong as standard ones.

2110.3.3 Interior panels. The maximum area of each individual standard-unit panel shall be 250 square feet (23.2 m^2). The maximum area of each thin-unit panel shall be 150 square feet (13.9 m^2). The maximum dimension between structural supports shall be 25 feet (7620 mm) in width or 20 feet (6096 mm) in height.

❖ Interior panels can be larger than exterior ones since wind pressures are expected to be lower.

2110.3.4 Solid units. The maximum area of solid glass-block wall panels in both exterior and interior walls shall not be more than 100 square feet (9.3 m^2).

❖ Panels constructed of solid glass block must also meet the requirements of this section.

2110.3.5 Curved panels. The width of curved panels shall conform to the requirements of Sections 2110.3.1, 2110.3.2 and 2110.3.3, except additional structural supports shall be provided at locations where a curved section joins a straight section, and at inflection points in multicurved walls.

❖ Curved panels of glass block must meet the appropriate requirements of Sections 2110.3.1 through 2110.3.3, plus the requirements for additional supports at critical locations as defined in this section.

2110.4 Support.

❖ This section requires that glass-unit masonry panels be laterally supported by panel anchors, channel-type restraints or a combination of both. Channel-type restraints can be made of concrete, masonry, metal, wood or other materials, provided that adequate lateral support is achieved.

2110.4.1 Isolation. Glass unit masonry panels shall be isolated so that in-plane loads are not imparted to the panel.

❖ Isolation joints are needed at the top and sides of glass-unit masonry panels so that in-plane forces are not transferred to the panels.

2110.4.2 Vertical. Maximum total deflection of structural members supporting glass unit masonry shall not exceed $l/_{600}$.

❖ The sizes of structural members supporting glass-block panels must be determined by structural analysis in order to avoid excessive deflections that could damage the glass-block construction. Deflections of supporting members are limited to a maximum of $l/600$, where l is the span of the supporting member.

2110.4.3 Lateral. Glass unit masonry panels more than one unit wide or one unit high shall be laterally supported along their tops and sides. Lateral support shall be provided by panel anchors along the top and sides spaced not more than 16 inches (406 mm) o.c. or by channel-type restraints. Glass unit masonry panels shall be recessed at least 1 inch (25 mm) within channels and chases. Channel-type restraints shall be oversized to accommodate expansion material in the opening and packing and sealant between the framing restraints and the glass unit masonry perimeter units. Lateral supports for glass unit masonry panels shall be designed to resist applied loads, or a minimum of 200 pounds per lineal feet (plf) (2919 N/m) of panel, whichever is greater.

Exceptions:

1. Lateral support at the top of glass unit masonry panels that are no more than one unit wide shall not be required.
2. Lateral support at the sides of glass unit masonry panels that are no more than one unit high shall not be required.

❖ Glass-block panels in exterior masonry walls or in openings of structural framing systems (curtain walls) are required to be restrained laterally to resist both external and internal pressures caused by wind and horizontal forces from earthquakes. Lateral support can be provided by channel-type or panel anchors.

Adhering to the dimensional limitations imposed on glass-block panels and complying with the requirements for construction will generally produce glass-block elements that adequately resist normal wind conditions. In regions with very high winds or high seismic risk, however, glass-block panels and their anchorage to supporting structural elements should be checked for adequacy.

The exceptions in this section recognize that loads and expected movement of small glass-unit panels are small enough to permit installation without isolation joints.

2110.4.3.1 Single unit panels. Single unit glass unit masonry panels shall conform to the requirements of Section 2110.4.3, except lateral support shall not be provided by panel anchors.

❖ Single-unit panels generally have the same requirements as multiple-unit panels described in Section 2110.4.3.

2110.5 Expansion joints. Glass unit masonry panels shall be provided with expansion joints along the top and sides at all structural supports. Expansion joints shall have sufficient thickness to accommodate displacements of the supporting structure, but shall not be less than $^3/_8$ inch (9.5 mm) in thickness. Expansion joints shall be entirely free of mortar or other debris and shall be filled with resilient material. The sills of glass-block panels shall be coated with approved water-based asphaltic emulsion, or other elastic waterproofing material, prior to laying the first mortar course.

❖ Sills supporting glass-block panel on a mortar bed are required to be first made water resistant with a heavy coat of asphalt emulsion or other approved water-resistant material. When the emulsion on the sill has dried, the full mortar bed can be placed, followed by the lowest course of glass block.

Before any glass-block units are placed, however, the head and jamb areas over the full height and width of the glass-block panel should be provided with expansion strips [3/8 inches (9.5 mm) thick] made for this purpose.

Glass-block units should be placed in successive courses on mortar beds containing panel reinforcement as required by the project documents, either directly or through reference to the manufacturer's instructions. Panel reinforcement should not be placed across expansion joints. Joints should be tooled smooth and concave before the mortar sets and joints around the perimeter of the panel should be raked out sufficiently to receive filler and caulking materials.

After the mortar sets, expansion joints are typically closed with a sealant.

2110.6 Mortar. Mortar for glass unit masonry shall comply with Section 2103.7.

❖ See the commentary to Section 2103.7. Glass-unit masonry panels are to be laid in Type N or S mortar.

2110.7 Reinforcement. Glass unit masonry panels shall have horizontal joint reinforcement spaced not more than 16 inches (406 mm) on center, located in the mortar bed joint, and extending the entire length of the panel but not across expansion joints. Longitudinal wires shall be lapped a minimum of 6 inches (152 mm) at splices. Joint reinforcement shall be placed in the bed joint immediately below and above openings in the panel. The reinforcement shall have not less than two parallel longitudinal wires of size W1.7 (MW11), and have welded cross wires of size W1.7 (MW11).

❖ Panel reinforcement made especially for glass block is typically of the ladder type, formed of two W1.7 galvanized wires spaced 2 inches (51 mm) apart with W1.7 galvanized cross wires welded at 16-inch (406 mm) intervals. Where placed continuously, the sections are required to lap at least 6 inches (152 mm). Such reinforcement must not extend across expansion joints because it would compromise their effectiveness.

SECTION 2111
MASONRY FIREPLACES

2111.1 Definition. A masonry fireplace is a fireplace constructed of concrete or masonry. Masonry fireplaces shall be constructed in accordance with this section, Table 2111.1 and Figure 2111.1.

❖ The provisions of this section apply to the design and installation of concrete and masonry fireplaces, which for simplicity are referred to as "masonry fireplaces."

TABLE 2111.1. See page 21-62.

❖ This table summarizes many of the requirements for fireplaces contained in Section 2111. Commentary to the listed section references should be reviewed for information on each item in the table.

This table does not address all the requirements of Section 2111. Each code section must be checked to see that a fireplace is in compliance.

FIGURE 2111.1. See page 21-63.

❖ This figure graphically shows many of the requirements for fireplaces contained in Section 2111. This figure does not address all the requirements of Section 2111.

2111.2 Footings and foundations. Footings for masonry fireplaces and their chimneys shall be constructed of concrete or solid masonry at least 12 inches (305 mm) thick and shall extend at least 6 inches (153 mm) beyond the face of the fireplace or foundation wall on all sides. Footings shall be founded on natural undisturbed earth or engineered fill below frost depth. In areas not subjected to freezing, footings shall be at least 12 inches (305 mm) below finished grade.

❖ Masonry fireplaces and chimneys must be supported on adequate footings due to their weight and the forces imposed on them by wind, earthquakes and other effects. This section prescribes minimum footing requirements that are typically adequate for standard fireplaces and chimneys. Extremely large, tall or heavy fireplaces and chimneys, however, may need more substantial foundations (see Item T in Figure 2111.1).

2111.2.1 Ash dump cleanout. Cleanout openings, located within foundation walls below fireboxes, when provided, shall be equipped with ferrous metal or masonry doors and frames constructed to remain tightly closed, except when in use. Cleanouts shall be accessible and located so that ash removal will not create a hazard to combustible materials.

❖ Noncombustible, tightly sealed cleanout doors are required to reduce the danger of fire spread through the cleanout openings. Cleanout openings are required to be easily accessible to allow ash to be readily removed.

2111.3 Seismic reinforcing. Masonry or concrete fireplaces shall be constructed, anchored, supported and reinforced as required in this chapter. In Seismic Design Category D, masonry and concrete fireplaces shall be reinforced and anchored as detailed in Sections 2111.3.1, 2111.3.2, 2111.4 and 2111.4.1 for chimneys serving fireplaces. In Seismic Design Category A, B or C, reinforcement and seismic anchorage is not required. In Seismic Design Category E or F, masonry and concrete chimneys shall be reinforced in accordance with the requirements of Sections 2101 through 2109.

❖ Unreinforced fireplaces and chimneys subjected to strong ground motion have sustained damage in past earthquakes. The requirements in this section provide minimum reinforcement in an effort to increase structural integrity during such events. More substantial reinforcement may, however, be required in areas of high seismicity or for atypical fireplaces and chimneys.

2111.3.1 Vertical reinforcing. For fireplaces with chimneys up to 40 inches (1016 mm) wide, four No. 4 continuous vertical bars, anchored in the foundation, shall be placed in the concrete, between wythes of solid masonry or within the cells of hollow unit masonry and grouted in accordance with Section 2103.10. For fireplaces with chimneys greater than 40 inches (1016 mm) wide, two additional No. 4 vertical bars shall be provided for

TABLE 2111.1
SUMMARY OF REQUIREMENTS FOR MASONRY FIREPLACES AND CHIMNEYS[a]

ITEM	LETTER	REQUIREMENTS	SECTION
Hearth and hearth extension thickness	A	4-inch minimum thickness for hearth, 2-inch minimum thickness for hearth extension.	2111.9
Hearth extension (each side of opening)	B	8 inches for fireplace opening less than 6 square feet. 12 inches for fireplace opening greater than or equal to 6 square feet.	2111.10
Hearth extension (front of opening)	C	16 inches for fireplace opening less than 6 square feet. 20 inches for fireplace opening greater than or equal to 6 square feet.	2111.10
Firebox dimensions	—	20-inch minimum firebox depth. 12-inch minimum firebox depth for Rumford fireplaces.	2111.6
Hearth and hearth extension reinforcing	D	Reinforced to carry its own weight and all imposed loads.	2111.9
Thickness of wall of firebox	E	10 inches solid masonry or 8 inches where firebrick lining is used.	2111.5
Distance from top of opening to throat	F	8 inches minimum.	2111.7 2111.7.1
Smoke chamber wall thickness dimensions	G	6 inches lined; 8 inches unlined. Not taller than opening width; walls not inclined more than 45 degrees from vertical for prefabricated smoke chamber linings or 30 degrees from vertical for corbeled masonry.	2111.8
Chimney vertical reinforcing	H	Four No. 4 full-length bars for chimney up to 40 inches wide. Add two No. 4 bars for each additional 40 inches or fraction of width, or for each additional flue.	2111.3.1, 2113.3.1
Chimney horizontal reinforcing	J	$^{1}/_{4}$-inch ties at each 18 inches, and two ties at each bend in vertical steel.	2111.3.2, 2113.3.2
Fireplace lintel	L	Noncombustible material with 4-inch bearing length of each side of opening.	2111.7
Chimney walls with flue lining	M	4-inch-thick solid masonry with $^{5}/_{8}$-inch fireclay liner or equivalent. $^{1}/_{2}$-inch grout or airspace between fireclay liner and wall.	2113.11.1
Effective flue area (based on area of fireplace opening and chimney)	P	See Section 2113.16.	2113.16
Clearances From chimney From fireplace From combustible trim or materials Above roof	R	 2 inches interior, 1 inch exterior or 12 inches from lining. 2 inches back or sides or 12 inches from lining. 6 inches from opening 3 feet above roof penetration, 2 feet above part of structure within 10 feet.	 2113.19 2111.11 2111.12 2113.9
Anchorage strap Number required Embedment into chimney Fasten to Number of bolts	S	$^{3}/_{16}$ inch by 1 inch Two 12 inches hooked around outer bar with 6-inch extension. 4 joists Two $^{1}/_{2}$-inch diameter.	2111.4 2113.4.1
Footing Thickness Width	T	 12-inch minimum. 6 inches each side of fireplace wall.	2111.2

For SI: 1 inch = 25.4 mm, 1 foot = 304.8 mm, 1 square foot = 0.0929 m^2, 1 degree = 0.017 rad.

a. This table provides a summary of major requirements for the construction of masonry chimneys and fireplaces. Letter references are to Figure 2111.1, which shows examples of typical construction. This table does not cover all requirements, nor does it cover all aspects of the indicated requirements. For the actual mandatory requirements of the code, see the indicated section of text.

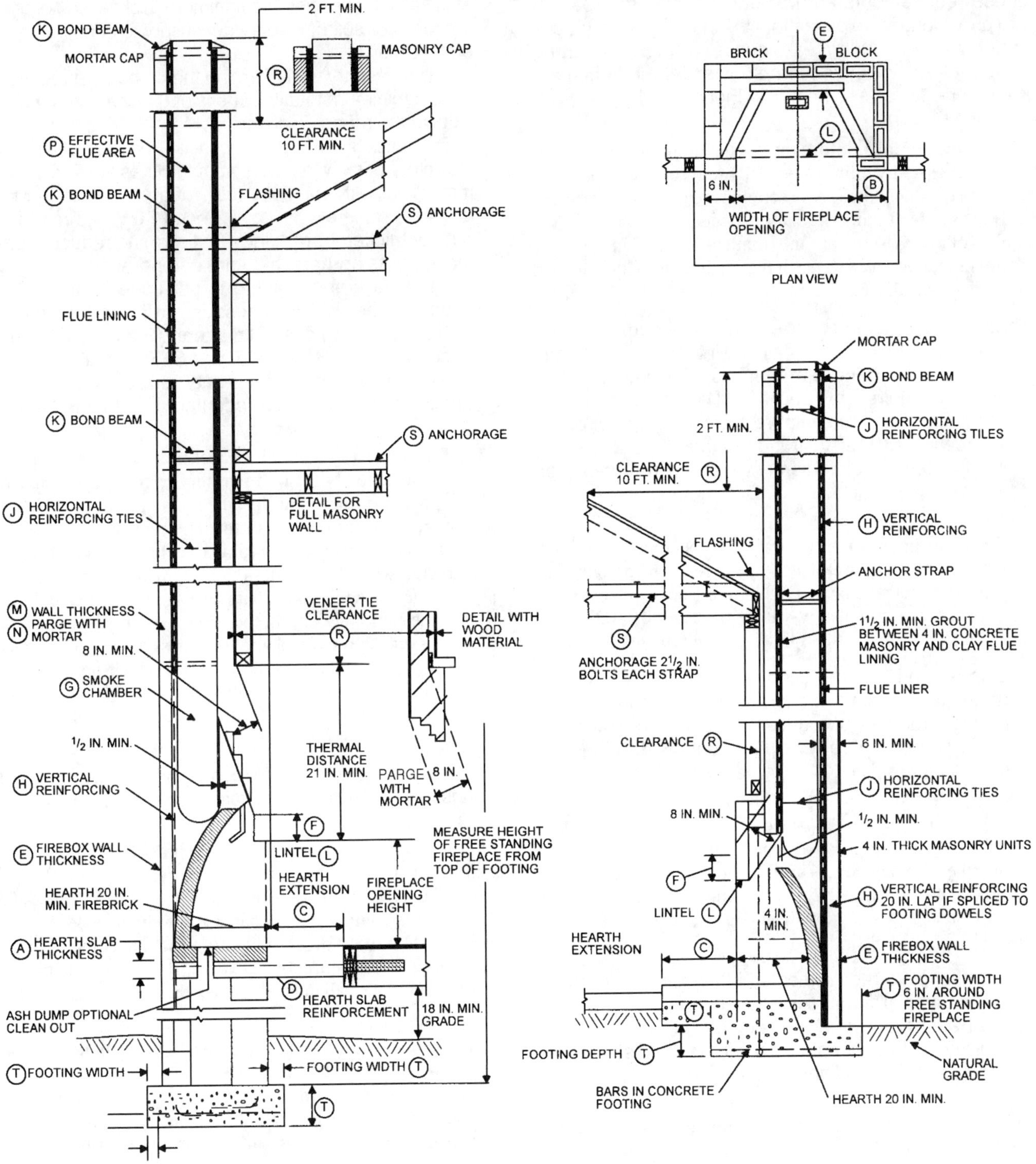

BRICK FIREBOX AND CHIMNEY—
SECTIONAL SIDE VIEW ON WOOD FLOOR

BRICK FIREBOX AND BLOCK CHIMNEY—
SECTIONAL SIDE VIEW ON CONCRETE SLAB

For SI: 1 inch = 25.4 mm, 1 foot = 304.8 mm.

FIGURE 2111.1
FIREPLACE AND CHIMNEY DETAILS

each additional 40 inches (1016 mm) in width or fraction thereof.

❖ These requirements are traditional minimum prescriptive provisions to help maintain the integrity of fireplaces and chimneys during earthquakes. To resist strong earthquakes, however, more substantial reinforcement may be required (see Item H in Figure 2111.1).

2111.3.2 Horizontal reinforcing. Vertical reinforcement shall be placed enclosed within $^1/_4$-inch (6.4 mm) ties or other reinforcing of equivalent net cross-sectional area, spaced not to exceed 18 inches (457 mm) on center in concrete; or placed in the bed joints of unit masonry at a minimum of every 18 inches (457 mm) of vertical height. Two such ties shall be provided at each bend in the vertical bars.

❖ These requirements are traditional minimum prescriptive provisions to help maintain the integrity of fireplaces and chimneys during earthquakes. The vertical reinforcement required by Section 2111.3.1 must be enclosed within the horizontal reinforcement required by this section. To resist strong earthquakes, however, more substantial reinforcement may be required (see Item J in Figure 2111.1).

2111.4 Seismic anchorage. Masonry and concrete chimneys in Seismic Design Category D shall be anchored at each floor, ceiling or roof line more than 6 feet (1829 mm) above grade, except where constructed completely within the exterior walls. Anchorage shall conform to the following requirements.

❖ Fireplaces and chimneys can fail by overturning during earthquakes. Seismic anchorage to floors and roof diaphragms is required to reduce the cantilevered height of the chimney and thereby help prevent overturning (see Item S in Figure 2111.1).

2111.4.1 Anchorage. Two $^3/_{16}$-inch by 1-inch (4.8 mm by 25.4 mm) straps shall be embedded a minimum of 12 inches (305 mm) into the chimney. Straps shall be hooked around the outer bars and extend 6 inches (152 mm) beyond the bend. Each strap shall be fastened to a minimum of four floor joists with two $^1/_2$-inch (12.7 mm) bolts.

❖ The prescriptive requirements in this section are traditional for typical fireplaces and chimneys. More substantial anchorage may be required in areas of high seismicity, for large fireplaces or where the distance between floor and roof diaphragms is large.

2111.5 Firebox walls. Masonry fireboxes shall be constructed of solid masonry units, hollow masonry units grouted solid, stone or concrete. When a lining of firebrick at least 2 inches (51 mm) in thickness or other approved lining is provided, the minimum thickness of back and sidewalls shall each be 8 inches (203 mm) of solid masonry, including the lining. The width of joints between firebricks shall not be greater than $^1/_4$ inch (6.4 mm). When no lining is provided, the total minimum thickness of back and sidewalls shall be 10 inches (254 mm) of solid masonry. Firebrick shall conform to ASTM C 27 or ASTM C 1261 and shall be laid with medium-duty refractory mortar conforming to ASTM C 199.

❖ This section specifies the minimum thicknesses of refractory brick or solid masonry necessary to contain the generated heat.

Solid masonry walls forming the firebox are required to have a minimum total thickness of 8 inches (204 mm), including the refractory lining.

The refractory lining is to consist of a low-duty, fire-clay refractory brick with a minimum thickness of 2 inches (52 mm), laid with medium-duty refractory mortar. Mortar joints are generally $^1/_{16}$ to $^3/_{16}$ inch (1.6 to 4.8 mm) thick, but not thicker than $^1/_4$ inch (6.4 mm), to reduce thermal movements and prevent joint deterioration.

Where a firebrick lining is not used in firebox construction, the wall thickness is not to be less than 10 inches (254 mm) of solid masonry. Firebrick is required to conform to ASTM C 27 or ASTM C 1261. Firebrick must be laid with medium-duty refractory mortar conforming to ASTM C 199 (see Item E in Figure 2111.1).

2111.5.1 Steel fireplace units. Steel fireplace units are permitted to be installed with solid masonry to form a masonry fireplace provided they are installed according to either the requirements of their listing or the requirements of this section. Steel fireplace units incorporating a steel firebox lining shall be constructed with steel not less than $^1/_4$ inch (6.4 mm) in thickness, and an air-circulating chamber which is ducted to the interior of the building. The firebox lining shall be encased with solid masonry to provide a total thickness at the back and sides of not less than 8 inches (203 mm), of which not less than 4 inches (102 mm) shall be of solid masonry or concrete. Circulating air ducts employed with steel fireplace units shall be constructed of metal or masonry.

❖ This section provides minimum requirements for steel linings used in masonry fireplaces to improve heat flow.

2111.6 Firebox dimensions. The firebox of a concrete or masonry fireplace shall have a minimum depth of 20 inches (508 mm). The throat shall not be less than 8 inches (203 mm) above the fireplace opening. The throat opening shall not be less than 4 inches (102 mm) in depth. The cross-sectional area of the passageway above the firebox, including the throat, damper and smoke chamber, shall not be less than the cross-sectional area of the flue.

> **Exception:** Rumford fireplaces shall be permitted provided that the depth of the fireplace is at least 12 inches (305 mm) and at least one-third of the width of the fireplace opening, and the throat is at least 12 inches (305 mm) above the lintel, and at least $^1/_{20}$ the cross-sectional area of the fireplace opening.

❖ The proper functioning of the fireplace depends on the size of the face opening and the chimney dimensions, which in turn are related to the room size [see Figure 2111.6(1)]. This section specifies a minimum depth of 20 inches (508 mm) for the combustion chamber because that depth influences the draft requirement. The dimensions of the firebox (depth, opening size and shape) are usually based on two considerations: aesthetics and the need to prevent the room from overheating. Suggested dimensions for single-opening fireboxes are given in technical publications of the Brick Institute of America (BIA) and the National Concrete Masonry Association (NCMA).

This section also provides additional criteria for the throat's location and minimum cross-sectional area. Those criteria are based on many years of construction of successfully functioning fireplaces.

The exception permits the use of Rumford fireplaces and the dimensions associated with this design style. Rumford fireplaces are tall, shallow fireplaces that can radiate a large amount of heat into a room.

The code reference to the depth of the fireplace is interpreted as the depth of the firebox [see Figure 2111.6(2)]. The throat is required to be made at least 12 inches (305 mm) above the lintel and at least one-twentieth of the cross-sectional area of the fireplace opening. Smoke chambers and flues for Rumford fireplaces should be sized and built like other masonry fireplaces. While those who build Rumford fireplaces do not totally agree about how they work, many books and guides address their construction.

2111.7 Lintel and throat. Masonry over a fireplace opening shall be supported by a lintel of noncombustible material. The minimum required bearing length on each end of the fireplace opening shall be 4 inches (102 mm). The fireplace throat or damper shall be located a minimum of 8 inches (203 mm) above the top of the fireplace opening.

❖ Permanent support for the masonry above the fireplace opening is provided by noncombustible lintels of steel, masonry or concrete (see Item L of Figure 2111.1). Combustible lintels (for example, those made from wood) are not appropriate due to the risk of fire damage and the probable collapse of the masonry above the opening.

The minimum bearing requirement of 4 inches (102 mm) is empirical, based on typical masonry fireplace openings. Lintels that support more than the typical weight of masonry above the fireplace opening may require a larger bearing area.

The throat of a fireplace is the slot-like opening above the firebox, through which flames, smoke and hot combustion gases pass into the smoke chamber. The throat is as wide as the combustion chamber and is required to be located at least 8 inches (204 mm) above the lintel for conventional fireplaces. The back wall of the combustion chamber extends up to the throat, which is provided with a metal damper.

2111.7.1 Damper. Masonry fireplaces shall be equipped with a ferrous metal damper located at least 8 inches (203 mm) above

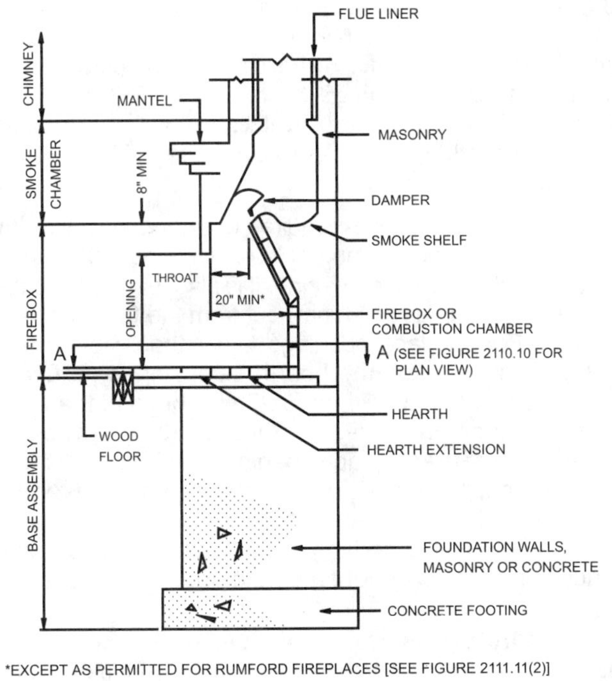

For SI: 1 inch = 25.4 mm.

Figure 2111.6(1)
SECTION THROUGH FIREPLACE

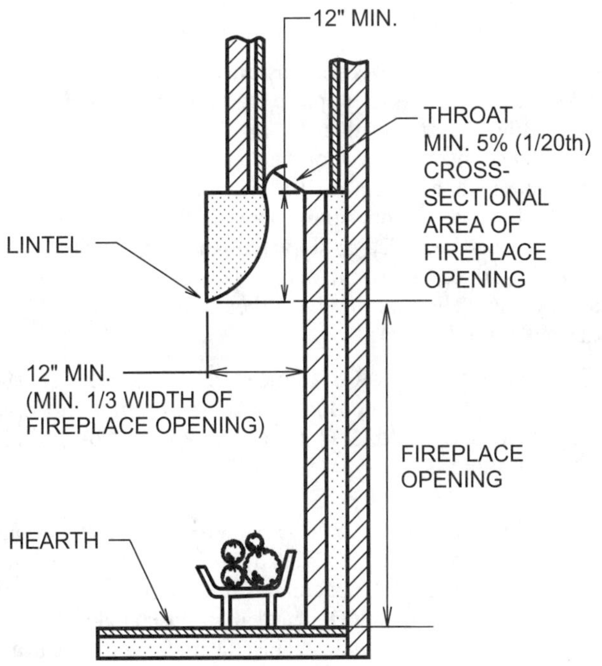

For SI: 1 inch = 25.4 mm.

Figure 2111.6(2)
RUMFORD FIREPLACE

the top of the fireplace opening. Dampers shall be installed in the fireplace or at the top of the flue venting the fireplace, and shall be operable from the room containing the fireplace. Damper controls shall be permitted to be located in the fireplace.

❖ A damper is used to close the chimney flue when the fireplace is not in use. This section provides guidance on its location and construction.

2111.8 Smoke chamber walls. Smoke chamber walls shall be constructed of solid masonry units, hollow masonry units grouted solid, stone or concrete. Corbeling of masonry units shall not leave unit cores exposed to the inside of the smoke chamber. The inside surface of corbeled masonry shall be parged smooth. Where no lining is provided, the total minimum thickness of front, back and sidewalls shall be 8 inches (203 mm) of solid masonry. When a lining of firebrick at least 2 inches (51 mm) thick, or a lining of vitrified clay at least $^5/_8$ inch (15.9 mm) thick, is provided, the total minimum thickness of front, back and sidewalls shall be 6 inches (152 mm) of solid masonry, including the lining. Firebrick shall conform to ASTM C 27 or ASTM C 1261 and shall be laid with refractory mortar conforming to ASTM C 199.

❖ The minimum wall thickness specified for the throat and smoke chamber is required to provide support for the flue construction, as well as adequate thermal insulation for the adjacent combustible construction.

In conventional fireplace construction, the smoke chamber is a tapering section whose vertical dimension is measured from the damper or throat to the bottom of the chimney flue. The actual height is a function of the fireplace opening.

2111.8.1 Smoke chamber dimensions. The inside height of the smoke chamber from the fireplace throat to the beginning of the flue shall not be greater than the inside width of the fireplace opening. The inside surface of the smoke chamber shall not be inclined more than 45 degrees (0.76 rad) from vertical when prefabricated smoke chamber linings are used or when the smoke chamber walls are rolled or sloped rather than corbeled. When the inside surface of the smoke chamber is formed by corbeled masonry, the walls shall not be corbeled more than 30 degrees (0.52 rad) from vertical.

❖ This section specifies the smoke chamber configuration that is needed for the proper function of the masonry chimney. Also see Table 2111.1, Item G, for smoke chamber requirements.

2111.9 Hearth and hearth extension. Masonry fireplace hearths and hearth extensions shall be constructed of concrete or masonry, supported by noncombustible materials, and reinforced to carry their own weight and all imposed loads. No combustible material shall remain against the underside of hearths or hearth extensions after construction.

❖ The fireplace hearth consists of two parts. The hearth, commonly called the "inner hearth," is the floor of the combustion chamber and is obviously constructed of noncombustible material. The outer hearth, commonly known as the "hearth extension," projects beyond the face of the fireplace into the room and also consists of noncombustible materials, such as brick or concrete masonry, concrete, floor tile or stone (see Figure 2111.1, Items A and D). The hearth extension may continue out from the inner hearth at the same level or be stepped down to a lower level. Minimum dimensions for the hearth extension are prescribed in Section 2111.10 and are discussed in the commentary to that section.

Combustible forms and centers could ignite from exposure to heat from the adjacent fireplace and from burning embers that escape the firebox; these and other similar concealed, combustible components must be removed.

2111.9.1 Hearth thickness. The minimum thickness of fireplace hearths shall be 4 inches (102 mm).

❖ The required minimum thickness of 4 inches (102 mm) is an empirical requirement that has historically been successful.

2111.9.2 Hearth extension thickness. The minimum thickness of hearth extensions shall be 2 inches (51 mm).

> **Exception:** When the bottom of the firebox opening is raised at least 8 inches (203 mm) above the top of the hearth extension, a hearth extension of not less than $^3/_8$-inch-thick (9.5 mm) brick, concrete, stone, tile or other approved noncombustible material is permitted.

❖ These requirements are empirical and have historically been successful.

2111.10 Hearth extension dimensions. Hearth extensions shall extend at least 16 inches (406 mm) in front of, and at least 8 inches (203 mm) beyond, each side of the fireplace opening. Where the fireplace opening is 6 square feet (0.557 m^2) or larger, the hearth extension shall extend at least 20 inches (508 mm) in front of, and at least 12 inches (305 mm) beyond, each side of the fireplace opening.

❖ The hearth extension is required to extend the full width of the fireplace opening and at least 8 inches (203 mm) beyond each side of the opening. It is also required to extend at least 16 inches (406 mm) out from the face of the fireplace. For fireplace openings larger than 6 square feet (0.557 m^2) in area, the hearth extension is required to extend at least 20 inches (508 mm) beyond the face of the fireplace and at least 12 inches (305 mm) beyond each side of the fireplace opening (see Figure 2111.10).

The hearth extension is intended to serve as a fire-protective separation between the firebox and adjacent combustible flooring or furnishings and to prevent accidental spills of hot embers or logs from the fire.

2111.11 Fireplace clearance. Any portion of a masonry fireplace located in the interior of a building or within the exterior wall of a building shall have a clearance to combustibles of not less than 2 inches (51 mm) from the front faces and sides of masonry fireplaces and not less than 4 inches (102 mm) from the back faces of masonry fireplaces. The airspace shall not be filled, except to provide fireblocking in accordance with Section 2111.13.

Exceptions:

1. Masonry fireplaces listed and labeled for use in contact with combustibles in accordance with UL 127, and installed in accordance with the manufacturer's installation instructions, are permitted to have combustible material in contact with their exterior surfaces.
2. When masonry fireplaces are constructed as part of masonry or concrete walls, combustible materials shall not be in contact with the masonry or concrete walls less than 12 inches (306 mm) from the inside surface of the nearest firebox lining.
3. Exposed combustible trim and the edges of sheathing materials, such as wood siding, flooring and drywall, are permitted to abut the masonry fireplace sidewalls and hearth extension, in accordance with Figure 2111.11, provided such combustible trim or sheathing is a minimum of 12 inches (306 mm) from the inside surface of the nearest firebox lining.
4. Exposed combustible mantels or trim is permitted to be placed directly on the masonry fireplace front surrounding the fireplace opening provided such combustible materials shall not be placed within 6 inches (153 mm) of a fireplace opening. Combustible material within 12 inches (306 mm) of the fireplace opening shall not project more than 1/8 inch (3.2 mm) for each 1-inch (25 mm) distance from such opening.

❖ Combustible materials, such as framing studs and joists, must not be installed closer than 2 inches (51 mm) to the exterior surface of fireplace walls because of the fire hazard to materials in this location. Heat transmitted through fireplace walls can ignite combustible structural materials in contact with the walls. For this reason, a minimum required clearance has been established from the fireplace to combustibles.

❖ This figure clarifies Exception 3 to the clearance requirement for masonry fireplaces. The edge abutment of combustible sheathing materials where there is an adequate thickness of masonry is a long-standing practice that is considered safe, provided the minimum clearance to the firebox lining is maintained.

2111.12 Mantel and trim. Woodwork or other combustible materials shall not be placed within 6 inches (152 mm) of a fireplace opening. Combustible material within 12 inches (305 mm) of the fireplace opening shall not project more than 1/8 inch (3.2 mm) for each 1-inch (25 mm) distance from such opening.

❖ Combustible materials attached to the fireplace face, such as wood trim and mantels, are not permitted to be installed closer than 6 inches (152 mm) from the fireplace opening. Materials located above the opening and that project excessively create a severe fire hazard, however, and are required to have a minimum clearance of 12 inches (305 mm) from the fireplace opening.

2111.13 Fireplace fireblocking. All spaces between fireplaces and floors and ceilings through which fireplaces pass shall be fireblocked with noncombustible material securely fastened in place. The fireblocking of spaces between wood joists, beams or headers shall be to a depth of 1 inch (25 mm) and shall only be placed on strips of metal or metal lath laid across the spaces between combustible material and the chimney.

❖ Fireblocking is required to prevent the travel of flames, smoke or hot gases to other areas of the building through gaps between the chimney and the floor or ceiling assemblies. The 1-inch (25 mm) depth requirement is intended to be both a minimum and a maximum.

2111.14 Exterior air. Factory-built or masonry fireplaces covered in this section shall be equipped with an exterior air supply to ensure proper fuel combustion unless the room is mechani-

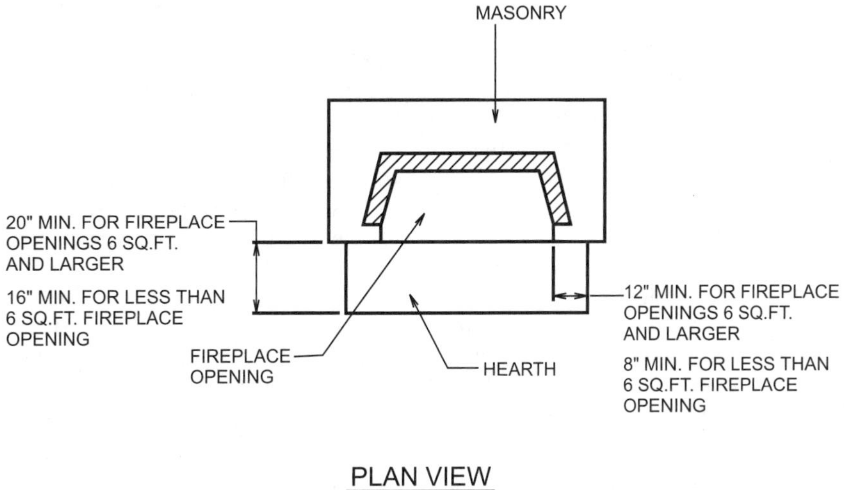

For SI: 1 inch = 25.4 mm, 1 square foot = .0929 m^2.

Figure 2111.10
HEARTH EXTENSION

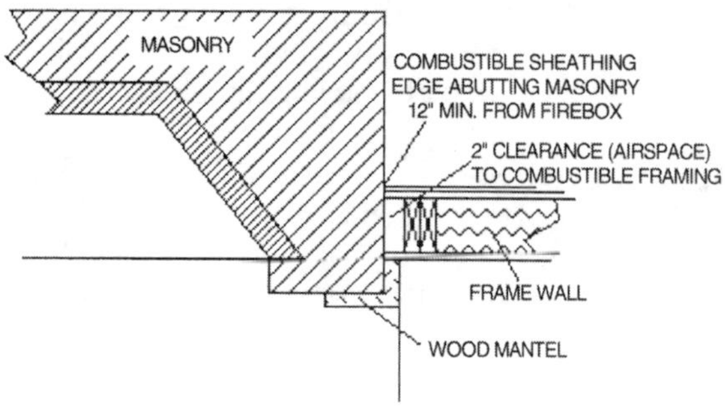

For SI: 1 inch = 25.4 mm

FIGURE 2111.11
ILLUSTRATION OF EXCEPTION TO FIREPLACE CLEARANCE PROVISION

cally ventilated and controlled so that the indoor pressure is neutral or positive.

❖ Adequate airflow is needed to provide exterior oxygen for the fire and to maintain draft through the chimney so that toxic combustion gases can be exhausted. Adequate airflow is especially important in modern, tightly sealed buildings.

2111.14.1 Factory-built fireplaces. Exterior combustion air ducts for factory-built fireplaces shall be listed components of the fireplace, and installed according to the fireplace manufacturer's instructions.

❖ Factory-built fireplaces are required to include exterior combustion air ducts. To function properly, these units need to be installed according to the manufacturer's recommendations.

2111.14.2 Masonry fireplaces. Listed combustion air ducts for masonry fireplaces shall be installed according to the terms of their listing and manufacturer's instructions.

❖ For proper functioning and airflow, air ducts for masonry fireplaces must be installed according to the manufacturer's recommendations (see also commentary, Section 2111.14).

2111.14.3 Exterior air intake. The exterior air intake shall be capable of providing all combustion air from the exterior of the dwelling. The exterior air intake shall not be located within the garage, attic, basement or crawl space of the dwelling nor shall the air intake be located at an elevation higher than the firebox. The exterior air intake shall be covered with a corrosion-resistant screen of $^1/_4$-inch (6.4 mm) mesh.

❖ Air intakes are required to provide air from outside of the building. The air intakes are not permitted to be placed in garages, basements, crawl spaces or other areas where gases from the exhaust of automobiles, furnaces and other sources could be brought into the building.

Air intakes should be covered with screens to prevent pests from entering the building when the fireplace is not functioning.

2111.14.4 Clearance. Unlisted combustion air ducts shall be installed with a minimum 1-inch (25 mm) clearance to combustibles for all parts of the duct within 5 feet (1524 mm) of the duct outlet.

❖ The 1-inch (25 mm) clearance is required to reduce the risk of fire through the air intakes.

2111.14.5 Passageway. The combustion air passageway shall be a minimum of 6 square inches (3870 mm^2) and not more than 55 square inches (0.035 m^2), except that combustion air systems for listed fireplaces or for fireplaces tested for emissions shall be constructed according to the fireplace manufacturer's instructions.

❖ These minimum requirements are intended to provide adequate airflow through the fireplace.

2111.14.6 Outlet. The exterior air outlet is permitted to be located in the back or sides of the firebox chamber or within 24 inches (610 mm) of the firebox opening on or near the floor. The outlet shall be closable and designed to prevent burning material from dropping into concealed combustible spaces.

❖ The requirements on the location of the outlet in the firebox chamber are needed for adequate airflow to the fire while avoiding the direct flow of air into the adjacent room. Such openings must not become clogged with ash, which would restrict airflow. Even more important, burning material must not drop into concealed combustible spaces due to the risk of fire damage to the building. Outlets must therefore be both closable and adequately designed to prevent this.

SECTION 2112
MASONRY HEATERS

2112.1 Definition. A masonry heater is a heating appliance constructed of concrete or solid masonry, hereinafter referred to as "masonry," having a mass of at least 1,760 pounds (800 kg), excluding the chimney and foundation, which is designed to absorb and store heat from a solid fuel fire built in the firebox by routing the exhaust gases through internal heat exchange channels in which the flow path downstream of the firebox includes at least one 180-degree (3.14 rad) change in flow direction before entering the chimney, and that delivers heat by radiation from the masonry surface of the heater that shall not exceed 230°F (110°C) except within 8 inches (203 mm) surrounding the fuel loading door(s).

❖ Masonry heaters are appliances designed to absorb and store heat from a relatively small fire and to radiate that heat into the building interior. They are thermally more efficient than traditional fireplaces because of their design. Interior passageways through the heater allow hot exhaust gases from the fire to transfer heat into the masonry, which then radiates into the building.

2112.2 Installation. Masonry heaters shall be listed or installed in accordance with ASTM E 1602.

❖ASTM E 1602 provides guidelines for installing masonry heaters.

2112.3 Seismic reinforcing. Seismic reinforcing shall not be required within the body of a masonry heater whose height is equal to or less than 2.5 times its body width and where the masonry chimney serving the heater is not supported by the body of the heater. Where the masonry chimney shares a common wall with the facing of the masonry heater, the chimney portion of the structure shall be reinforced in accordance with Sections 2113 and 2113.4.

❖Because of the large bulk and squat geometry of these heaters, seismic reinforcement is not typically required. Flexural tensile stresses, which typically cause damage to unreinforced masonry, rarely occur. Where chimneys extend above these heaters, however, seismic reinforcement is required by Section 2113. See the commentary to that section for additional information on this requirement.

2112.4 Masonry heater clearance. Wood or other combustible framing shall not be placed within 4 inches (102 mm) of the outside surface of a masonry heater, provided the wall thickness of the firebox is not less than 8 inches (203 mm) and the wall thickness of the heat exchange channels is not less than 5 inches (127 mm). A clearance of at least 8 inches (203 mm) shall be provided between the gas-tight capping slab of the heater and a combustible ceiling. The required space between the heater and combustible material shall be fully vented to permit the free flow of air around all heater surfaces.

❖Heat conducted through masonry heater walls can ignite combustible structural materials in contact with these walls. For this reason, a minimum required clearance to combustibles from masonry heaters has been established. Because masonry heaters typically generate more heat for a longer period of time than traditional fireplaces, greater clearances to combustible materials are needed to reduce the risk of fire.

SECTION 2113
MASONRY CHIMNEYS

2113.1 General. A masonry chimney is a chimney constructed of concrete or masonry, hereinafter referred to as "masonry." Masonry chimneys shall be constructed, anchored, supported and reinforced as required in this chapter.

❖A masonry chimney is a field-constructed assembly that can consist of masonry units, grout, reinforced concrete, rubble stone, fire-clay liners and mortars. A masonry chimney is permitted to serve residential (low-heat), medium- and high-heat appliances. This section outlines the general code requirements regarding construction details for all masonry chimneys, including those serving masonry fireplaces regulated by Section 2111.

2113.2 Footings and foundations. Foundations for masonry chimneys shall be constructed of concrete or solid masonry at least 12 inches (305 mm) thick and shall extend at least 6 inches (152 mm) beyond the face of the foundation or support wall on all sides. Footings shall be founded on natural undisturbed earth or engineered fill below frost depth. In areas not subjected to freezing, footings shall be at least 12 inches (305 mm) below finished grade.

❖Masonry fireplaces and chimneys must be supported on adequate foundations due to their weight and the forces imposed on them by wind, earthquakes and other effects. This section prescribes minimum foundation requirements that are typically adequate for standard chimneys.

Extremely large, tall or heavy chimneys, however, may need more substantial foundations. Also, a chimney foundation probably will support a larger load than the adjacent building foundations. For this reason, chimney footings and foundations are often separated from the building foundation. A chimney foundation monolithic with the building foundation is permitted, provided the soil pressure and anticipated settlement are approximately uniform.

2113.3 Seismic reinforcing. Masonry or concrete chimneys shall be constructed, anchored, supported and reinforced as required in this chapter. In Seismic Design Category D, masonry and concrete chimneys shall be reinforced and anchored as detailed in Sections 2113.3.1, 2113.3.2 and 2113.4. In Seismic Design Category A, B or C, reinforcement and seismic anchorage is not required. In Seismic Design Category E or F, masonry and concrete chimneys shall be reinforced in accordance with the requirements of Sections 2101 through 2108.

❖Unreinforced fireplaces and chimneys subjected to strong ground motion have been severely damaged in past earthquakes. The requirements in this section provide minimum reinforcement in an effort to keep these structures together during such events. More substantial reinforcement, however, may be required in areas of high seismicity or for atypical chimneys.

2113.3.1 Vertical reinforcing. For chimneys up to 40 inches (1016 mm) wide, four No. 4 continuous vertical bars anchored in the foundation shall be placed in the concrete, between wythes of solid masonry or within the cells of hollow unit masonry and grouted in accordance with Section 2103.10. Grout shall be prevented from bonding with the flue liner so that the flue liner is free to move with thermal expansion. For chimneys greater than 40 inches (1016 mm) wide, two additional No. 4 vertical bars shall be provided for each additional 40 inches (1016 mm) in width or fraction thereof.

❖These requirements are traditional minimum prescriptive provisions to help maintain the structural integrity of fireplaces and chimneys during earthquakes. More reinforcement may be required in areas of high seismicity or for atypical chimneys.

2113.3.2 Horizontal reinforcing. Vertical reinforcement shall be placed enclosed within $^1/_4$-inch (6.4 mm) ties, or other reinforcing of equivalent net cross-sectional area, spaced not to exceed 18 inches (457 mm) o.c. in concrete, or placed in the bed joints of unit masonry, at a minimum of every 18 inches (457 mm) of vertical height. Two such ties shall be provided at each bend in the vertical bars.

❖ These requirements are traditional minimum prescriptive provisions intended to help maintain the integrity of fireplaces and chimneys during earthquakes. The vertical reinforcement required by Section 2113.3.1 must be enclosed within the horizontal reinforcement required by this section. More reinforcement may be required in areas of high seismicity or for atypical chimneys.

2113.4 Seismic anchorage. Masonry and concrete chimneys and foundations in Seismic Design Category D shall be anchored at each floor, ceiling or roof line more than 6 feet (1829 mm) above grade, except where constructed completely within the exterior walls. Anchorage shall conform to the following requirements.

❖ Fireplaces and chimneys must be connected to floor and roof diaphragms to prevent overturning during earthquakes. Chimneys must be anchored at the ceiling line of roof or ceiling assemblies and at floor levels below the roof. Such anchorage is of lesser importance where the floor assembly is 6 feet (1829 mm) or less above grade.

2113.4.1 Anchorage. Two $^3/_{16}$-inch by 1-inch (4.8 mm by 25 mm) straps shall be embedded a minimum of 12 inches (305 mm) into the chimney. Straps shall be hooked around the outer bars and extend 6 inches (152 mm) beyond the bend. Each strap shall be fastened to a minimum of four floor joists with two $^1/_2$-inch (12.7 mm) bolts.

❖ The prescriptive requirements in this section are traditional for typical fireplaces and chimneys. More substantial anchorage may be required in areas of high seismicity, for large fireplaces or where the distance between floor and roof diaphragms is large.

2113.5 Corbeling. Masonry chimneys shall not be corbeled more than half of the chimney's wall thickness from a wall or foundation, nor shall a chimney be corbeled from a wall or foundation that is less than 12 inches (305 mm) in thickness unless it projects equally on each side of the wall, except that on the second story of a two-story dwelling, corbeling of chimneys on the exterior of the enclosing walls is permitted to equal the wall thickness. The projection of a single course shall not exceed one-half the unit height or one-third of the unit bed depth, whichever is less.

❖ A corbel is formed by projecting courses of masonry with the first or lowest course projecting out from the face of the wall and each successive course projecting out from the supporting course below. The angle of a corbel is limited so that the structural capacities of the masonry are not exceeded.

Figure 2113.5 illustrates the limitations of corbeling in an eccentrically loaded masonry chimney. The total corbeled projection cannot exceed 6 inches (152 mm). Where corbeling projects from the wall on one side only, the minimum allowable thickness of the wall is 12 inches (305 mm). The maximum allowable horizontal projection of an individual course of brick cannot exceed one-half the height and one-third the thickness of the masonry unit.

A chimney that projects equally on each side of the wall loads the wall concentrically. The requirement for a minimum wall thickness of 12 inches (305 mm) does not apply to this chimney configuration.

This section contains an exception to the limitations on corbel projection. In single-family homes, traditional chimneys originated in the second story and commonly served heating appliances in bedrooms, studies and sewing rooms. The exception allows corbeling to project from the exterior masonry wall a distance equivalent to the masonry wall thickness. This exception, however, does not affect the limitation on minimum wall thickness stated in the second sentence of this section. For example, a chimney in the second story of a two-story building cannot be corbeled at all if the wall is less than 12 inches (305 mm) thick and the corbeling is to project from one side only. If the corbeling is to project equally from both sides of the wall, the minimum wall thickness does not apply.

2113.6 Changes in dimension. The chimney wall or chimney flue lining shall not change in size or shape within 6 inches (152 mm) above or below where the chimney passes through floor components, ceiling components or roof components.

❖ At changes in shape or direction, masonry chimneys can have thinner walls, making them susceptible to leakage of water from the outside and to hot spots and leakage of combustion gases from the inside. This is a fire hazard. The intent of this provision is to prohibit changes in shape and size near combustible construction. This section prohibits any change in dimension or direction of a masonry chimney from 6 inches (152 mm) below the combustible construction to 6 inches (152 mm) above it (see Figure 2113.6).

2113.7 Offsets. Where a masonry chimney is constructed with a fireclay flue liner surrounded by one wythe of masonry, the maximum offset shall be such that the centerline of the flue above the offset does not extend beyond the center of the chimney wall below the offset. Where the chimney offset is supported by masonry below the offset in an approved manner, the maximum offset limitations shall not apply. Each individual corbeled masonry course of the offset shall not exceed the projection limitations specified in Section 2113.5.

❖ An offset requires two changes in direction and causes two vertical sections of a chimney to be offset (misaligned) from each other. The intent of this section is to provide an upper portion of an offset that is structurally stable with respect to the lower portion.

The offset limitation does not apply when the inclined portion of the chimney and the portion above the offset are supported in an approved manner by underlying masonry construction.

2113.8 Additional load. Chimneys shall not support loads other than their own weight unless they are designed and constructed to support the additional load. Masonry chimneys are permitted to be constructed as part of the masonry walls or concrete walls of the building.

❖ Because the requirements for chimneys in Section 2113 are based on past performance, chimneys should not be used to support other loads unless they are specifically designed to do so.

2113.9 Termination. Chimneys shall extend at least 2 feet (610 mm) higher than any portion of the building within 10 feet (3048 mm), but shall not be less than 3 feet (914 mm) above the highest point where the chimney passes through the roof.

❖ Chimneys must be terminated well above adjacent portions of the building so that flue gases are exhausted away from combustible materials and for proper airflow through the chimney (see Figure 2113.9).

2113.9.1 Spark arrestors. Where a spark arrestor is installed on a masonry chimney, the spark arrestor shall meet all of the following requirements:

1. The net free area of the arrestor shall not be less than four times the net free area of the outlet of the chimney flue it serves.
2. The arrestor screen shall have heat and corrosion resistance equivalent to 19-gage galvanized steel or 24-gage stainless steel.
3. Openings shall not permit the passage of spheres having a diameter greater than $^1/_2$ inch (13 mm) nor block the pas-

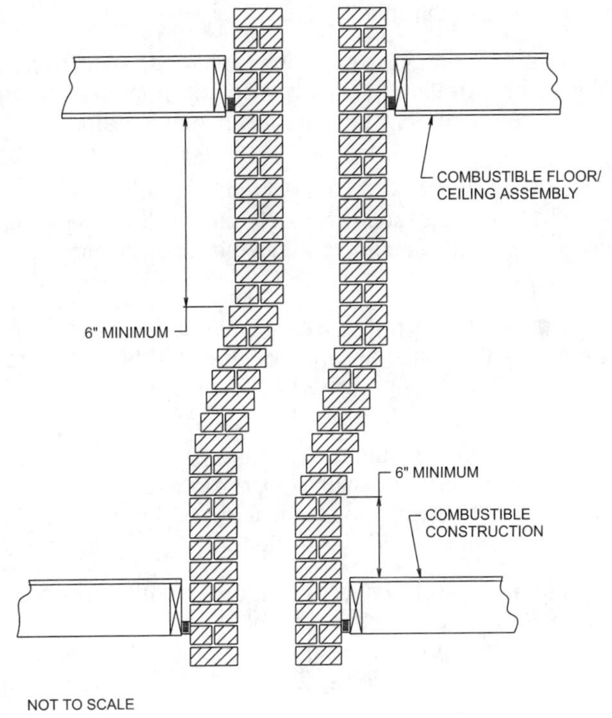

For SI: 1 inch = 25.4 mm.

Figure 2113.6
MASONRY CHIMNEY CHANGE IN SHAPE (OFFSET)

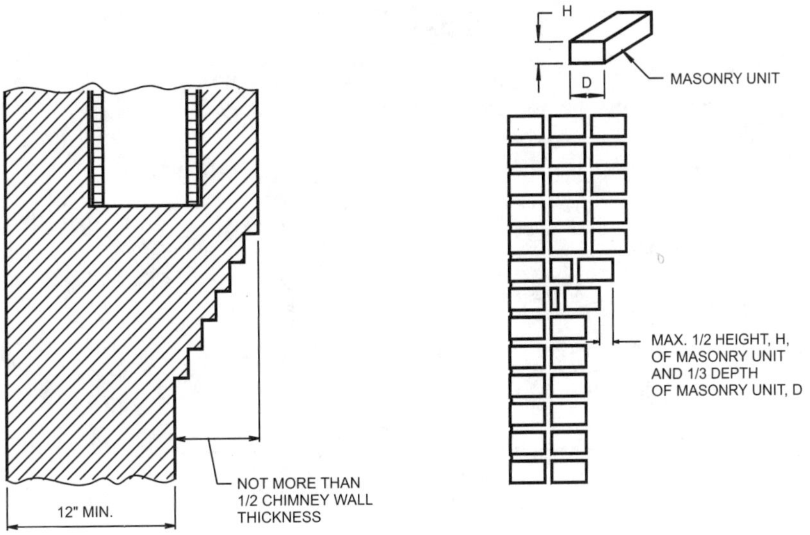

For SI: 1 inch = 25.4 mm.

Figure 2113.5
CORBELING LIMITATIONS

sage of spheres having a diameter less than $^3/_8$ inch (11 mm).

4. The spark arrestor shall be accessible for cleaning and the screen or chimney cap shall be removable to allow for cleaning of the chimney flue.

❖ This section provides specifications for spark arrestors, if they are provided. Their use is not mandated by the code, but owners or builders often install them.

2113.10 Wall thickness. Masonry chimney walls shall be constructed of concrete, solid masonry units or hollow masonry units grouted solid with not less than 4 inches (102 mm) nominal thickness.

❖ The minimum chimney wall thickness is necessary to achieve a thermal mass that will predictably control heat transmission through the walls of the chimney.

2113.11 Flue lining (material). Masonry chimneys shall be lined. The lining material shall be appropriate for the type of appliance connected, according to the terms of the appliance listing and the manufacturer's instructions.

❖ The liner forms the flue passageway and is the actual conduit for all products of combustion. The flue lining is required to withstand exposure to high temperatures and corrosive chemicals. It protects the masonry construction of the chimney walls and allows the chimney to be gas tight.

2113.11.1 Residential-type appliances (general). Flue lining systems shall comply with one of the following:

1. Clay flue lining complying with the requirements of ASTM C 315, or equivalent.
2. Listed chimney lining systems complying with UL 1777.
3. Factory-built chimneys or chimney units listed for installation within masonry chimneys.
4. Other approved materials that will resist corrosion, erosion, softening or cracking from flue gases and condensate at temperatures up to 1,800°F (982°C).

❖ This section requires that the lining for residential-type appliances comply with ASTM C 315 or other approved equivalent and be capable of resisting degradation from flue gas. Chimney liner systems that are tested and labeled by an approved agency in accordance with UL 1777 are also permitted.

2113.11.1.1 Flue linings for specific appliances. Flue linings other than those covered in Section 2113.11.1 intended for use with specific appliances shall comply with Sections 2113.11.1.2 through 2113.11.1.4 and Sections 2113.11.2 and 2113.11.3.

❖ This section identifies flue lining materials to be used for specific appliances.

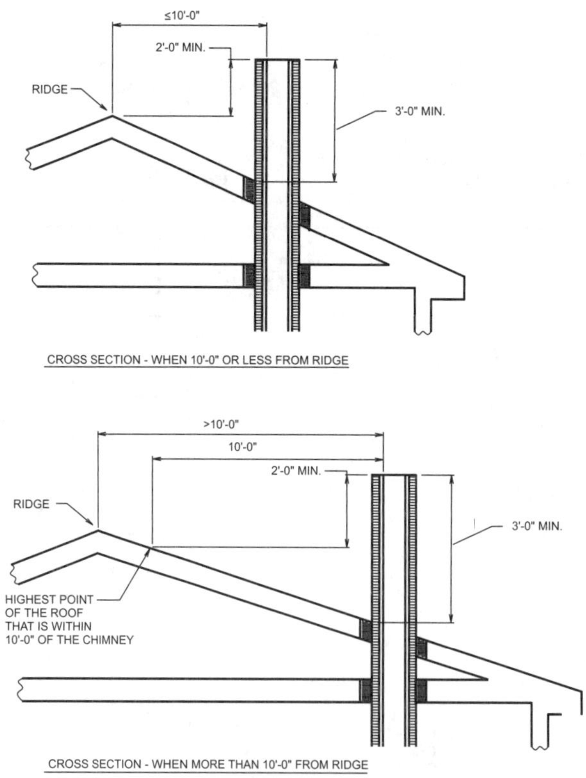

For SI: 1 foot = 304.8 mm.

Figure 2113.9
MINIMUM TERMINATION OF CHIMNEYS

2113.11.1.2 Gas appliances. Flue lining systems for gas appliances shall be in accordance with the *International Fuel Gas Code*.

❖ The *International Fuel Gas Code*® (IFGC®) must be used to determine appropriate flue-lining systems for gas appliances.

2113.11.1.3 Pellet fuel-burning appliances. Flue lining and vent systems for use in masonry chimneys with pellet fuel-burning appliances shall be limited to flue lining systems complying with Section 2113.11.1 and pellet vents listed for installation within masonry chimneys (see Section 2113.11.1.5 for marking).

❖ Flue-lining and vent systems in masonry chimneys of pellet fuel-burning appliances can either conform to Section 2113.11.1 or be an approved listed system.

2113.11.1.4 Oil-fired appliances approved for use with L-vent. Flue lining and vent systems for use in masonry chimneys with oil-fired appliances approved for use with Type L vent shall be limited to flue lining systems complying with Section 2113.11.1 and listed chimney liners complying with UL 641 (see Section 2113.11.1.5 for marking).

❖ Flue lining and vent systems in masonry chimneys of oil-fired appliances with Type L vents can either conform to Section 2113.11.1 or be an approved listed system (UL 641).

2113.11.1.5 Notice of usage. When a flue is relined with a material not complying with Section 2113.11.1, the chimney shall be plainly and permanently identified by a label attached to a wall, ceiling or other conspicuous location adjacent to where the connector enters the chimney. The label shall include the following message or equivalent language: "This chimney is for use only with (type or category of appliance) that burns (type of fuel). Do not connect other types of appliances."

❖ Clearly displayed information on the appropriate use and fuel for appliances is required to protect against the use of improper types of fuel that could cause fire.

2113.11.2 Concrete and masonry chimneys for medium-heat appliances.

❖ This section establishes requirements for masonry chimneys serving medium-heat appliances, including the chimney materials, lining, termination height and proper clearances to combustibles.

2113.11.2.1 General. Concrete and masonry chimneys for medium-heat appliances shall comply with Sections 2113.1 through 2113.5.

❖ Chimneys serving a medium-heat appliance must comply with the general chimney requirements in Sections 2113.1 through 2113.5. These include minimum requirements for footings and foundations; seismic reinforcement; anchorage and corbeling.

2113.11.2.2 Construction. Chimneys for medium-heat appliances shall be constructed of solid masonry units or of concrete with walls a minimum of 8 inches (203 mm) thick, or with stone masonry a minimum of 12 inches (305 mm) thick.

❖ Masonry chimneys for medium-heat appliances must be constructed of concrete in accordance with Chapter 19, or of solid masonry units in accordance with this chapter. Chimneys constructed of solid masonry units or reinforced concrete are required to have a minimum wall thickness of 8 inches (203 mm). The minimum thickness requirement is necessary to achieve a thermal mass that will predictably control heat transmission through the walls of the chimney. Chimneys constructed of rubble stone masonry (irregular or roughly shaped stones) are required to have a wall thickness of no less than 12 inches (305 mm). The rate of heat transmission through stone walls is not predictable because of the varying thickness of the individual stones and the nonuniform mortar joints between the stones. For these reasons, stone wall chimneys are required to be thicker to provide a reasonable margin of safety.

2113.11.2.3 Lining. Concrete and masonry chimneys shall be lined with an approved medium-duty refractory brick a minimum of $4^1/_2$ inches (114 mm) thick laid on the $4^1/_2$-inch bed (114 mm) in an approved medium-duty refractory mortar. The lining shall start 2 feet (610 mm) or more below the lowest chimney connector entrance. Chimneys terminating 25 feet (7620 mm) or less above a chimney connector entrance shall be lined to the top.

❖ A medium-heat appliance produces flue gas temperatures up to 2,000°F (1093°C). The chimney lining reduces heat transmission to the walls of the chimney and contains flue gases within a continuous duct until they are away from the building. This section requires at least a medium-duty refractory brick. A $4^1/_2$-inch (114 mm) medium-duty refractory brick lining tested and classified in accordance with ASTM C 27 meets this criterion. Each course of the refractory brick liner is required to be installed on a full bed of refractory mortar to form a structurally stable, continuous, gas-tight liner.

Figure 2113.11.2.3 depicts a masonry chimney serving a medium-heat appliance. The liner extends from 2 feet (610 mm) below the lowest inlet to the top of the chimney, which is located less than 25 feet (7620 mm) above the highest inlet.

A chimney liner is required in all portions of the chimney that are exposed to flue gases. The liner is required to extend below the lowest inlet to provide protection to the masonry from flue gases, which can deteriorate the masonry and the mortar joints.

2113.11.2.4 Multiple passageway. Concrete and masonry chimneys containing more than one passageway shall have the liners separated by a minimum 4-inch-thick (102 mm) concrete or solid masonry wall.

❖ When a chimney requires multiple passageways, a 4-inch (102 mm) partition of solid masonry is required between them to act as a barrier between passageways and to enhance structural integrity.

2113.11.2.5 Termination height. Concrete and masonry chimneys for medium-heat appliances shall extend a minimum of 10 feet (3048 mm) higher than any portion of any building within 25 feet (7620 mm).

❖ Chimneys for medium-heat appliances are required to extend at least 10 feet (3048 mm) above the highest portion of the building within 25 feet (7620 mm) horizontally. This is intended to provide an acceptable height to carry away the flue gases safely and to provide adequate clearance to the roof and surrounding structures to allow any burning embers to extinguish before landing.

2113.11.2.6 Clearance. A minimum clearance of 4 inches (102 mm) shall be provided between the exterior surfaces of a concrete or masonry chimney for medium-heat appliances and combustible material.

❖ A 4-inch (102 mm) minimum airspace clearance to combustibles is required for a medium-heat masonry chimney. This large clearance is needed because of the higher temperatures produced by medium-heat appliances.

2113.11.3 Concrete and masonry chimneys for high-heat appliances.

❖ This section establishes the requirements for concrete and masonry chimneys serving high-heat appliances, including chimney materials, lining, termination height and proper clearances to combustibles.

2113.11.3.1 General. Concrete and masonry chimneys for high-heat appliances shall comply with Sections 2113.1 through 2113.5.

❖ Chimneys serving high-heat appliances are required to comply with the general chimney requirements in Sections 2113.1 through 2113.5, including minimum provisions for footings and foundations; seismic reinforcement; anchorage and corbeling.

2113.11.3.2 Construction. Chimneys for high-heat appliances shall be constructed with double walls of solid masonry units or of concrete, each wall to be a minimum of 8 inches (203 mm) thick with a minimum airspace of 2 inches (51 mm) between the walls.

❖ The flue gases produced by a high-heat appliance can have temperatures above 2,000°F (1,093°C). To prevent fire, such chimneys must be enclosed by two wythes of masonry with an airspace between them.

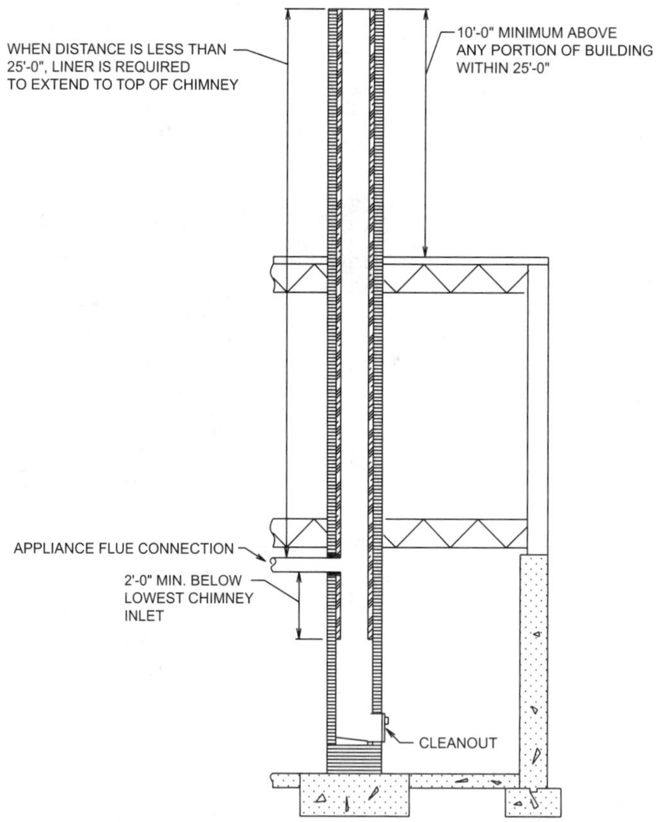

For SI: 1 foot = 304.8 mm.

Figure 2113.11.2.3
MASONRY CHIMNEY FOR MEDIUM-HEAT APPLIANCE

Thus, the chimney liner is required to be protected by two solid masonry unit walls or reinforced concrete walls, each a minimum of 8 inches (204 mm) thick with an airspace between them of at least 2 inches (51 mm). The airspace is intended to provide thermal insulation and to allow for thermal expansion of the chimney components.

2113.11.3.3 Lining. The inside of the interior wall shall be lined with an approved high-duty refractory brick, a minimum of $4^1/_2$ inches (114 mm) thick laid on the $4^1/_2$-inch bed (114 mm) in an approved high-duty refractory mortar. The lining shall start at the base of the chimney and extend continuously to the top.

❖ High-heat appliances can produce flue gases with temperatures exceeding 2,000°F (1,093°C). The chimney lining reduces heat transmission to the walls of the chimney and contains flue gases within a continuous passageway until they are outside the building. This section permits only a high-duty refractory brick having a minimum thickness of $4^1/_2$ inches (114 mm) to be used in a masonry chimney serving a high-heat appliance. An approved high-duty refractory lining usually consists of brick tested and classified in accordance with ASTM C 27. To provide a structurally stable and continuous gas-tight liner, each course of the refractory brick liner must be laid on a full bed of refractory mortar approved for high-heat appliances.

2113.11.3.4 Termination height. Concrete and masonry chimneys for high-heat appliances shall extend a minimum of 20 feet (6096 mm) higher than any portion of any building within 50 feet (15 240 mm).

❖ Chimneys for high-heat appliances must terminate at least 20 feet (6096 mm) above the highest portion of the building within 50 feet (15 240 mm) horizontally. This allows flue gases to be safely carried away and provides enough clearance to allow any burning embers to extinguish before landing on combustibles.

2113.11.3.5 Clearance. Concrete and masonry chimneys for high-heat appliances shall have approved clearance from buildings and structures to prevent overheating combustible materials, permit inspection and maintenance operations on the chimney and prevent danger of burns to persons.

❖ The extremely high-temperature flue gas produced by a high-heat appliance is a major fire safety concern and requiring sufficient clearance from buildings protects nearby combustible material, permits inspection and maintenance of the chimney and minimizes danger to persons.

2113.12 Flue lining (installation). Flue liners shall be installed in accordance with ASTM C 1283 and extend from a point not less than 8 inches (203 mm) below the lowest inlet or, in the case of fireplaces, from the top of the smoke chamber, to a point above the enclosing walls. The lining shall be carried up vertically, with a maximum slope no greater than 30 degrees (0.52 rad) from the vertical.

Fireclay flue liners shall be laid in medium-duty refractory mortar conforming to ASTM C 199, with tight mortar joints left smooth on the inside and installed to maintain an airspace or insulation not to exceed the thickness of the flue liner separating the flue liners from the interior face of the chimney masonry walls. Flue lining shall be supported on all sides. Only enough mortar shall be placed to make the joint and hold the liners in position.

❖ The liner forms the flue passageway and is the actual conduit of all products of combustion. It must withstand exposure to high temperatures and corrosive chemicals from the flue gases. The chimney lining protects the masonry construction of the chimney walls and allows the chimney to be constructed gas tight.

Installation must comply with ASTM C 1283. The liner is required to extend below the lowest inlet to provide protection to the masonry from flue gases, which can deteriorate the masonry and mortar joints.

Refractory mortar must comply with ASTM C 199.

2113.13 Additional requirements.

❖ This section contains requirements for listed materials to be used as flue linings and for spaces surrounding chimney lining systems.

2113.13.1 Listed materials. Listed materials used as flue linings shall be installed in accordance with the terms of their listings and the manufacturer's instructions.

❖ Listed materials for flue linings must be installed in accordance with the manufacturer's instructions. Such instructions are not listed in the code since they vary with each system.

2113.13.2 Space around lining. The space surrounding a chimney lining system or vent installed within a masonry chimney shall not be used to vent any other appliance.

Exception: This shall not prevent the installation of a separate flue lining in accordance with the manufacturer's instructions.

❖ The space surrounding a chimney lining system provides a thermal buffer between the flue and the surrounding masonry. Using it for another purpose is prohibited.

2113.14 Multiple flues. When two or more flues are located in the same chimney, masonry wythes shall be built between adjacent flue linings. The masonry wythes shall be at least 4 inches (102 mm) thick and bonded into the walls of the chimney.

Exception: When venting only one appliance, two flues are permitted to adjoin each other in the same chimney with only the flue lining separation between them. The joints of the adjacent flue linings shall be staggered at least 4 inches (102 mm).

❖ A single masonry chimney can contain any number of flue-gas passageways. When more than two flues are contained in the same chimney, the flue liners are required to be subdivided by masonry wythes into groups of not more than two (see Figure 2113.14). The purpose of the masonry wythes is to unify the chimney structurally and to isolate pairs of flues serving dissimilar appliances.

2113.15 Flue area (appliance). Chimney flues shall not be smaller in area than the area of the connector from the appliance. Chimney flues connected to more than one appliance shall not be less than the area of the largest connector plus 50 percent of the areas of additional chimney connectors.

Exceptions:

1. Chimney flues serving oil-fired appliances sized in accordance with NFPA 31.
2. Chimney flues serving gas-fired appliances sized in accordance with the *International Fuel Gas Code*.

❖ Flues are sized to match the opening at the top of the appliance so that appliances function properly. Smaller flues are not permitted since they may constrict the passage of gases out from the appliance. The *International Residential Code*® (IRC®) provides guidelines on the sizing of these flues.

2113.16 Flue area (masonry fireplace). Flue sizing for chimneys serving fireplaces shall be in accordance with Section 2113.16.1 or 2113.16.2.

❖ This section provides requirements for the net cross-sectional area of the flue and throat between the firebox and the smoke chamber. Airflow through the fireplace is affected by the dimensions of the firebox opening, the shape and cross-sectional area of the flue and the height of the chimney (see Code Figure 2113.16). For proper fireplace operation, the required cross-sectional area is about one-tenth that of the fireplace opening. This ratio may vary somewhat with the height of the chimney and the configuration (round or rectangular) of the chimney flue.

Section 2113.16.1 prescribes the minimum flue area based on the fireplace opening alone, while Section 2113.16.2 prescribes it based on the height of the fireplace, the fireplace opening area and flue type.

FIGURE 2113.16. See page 21-77.

❖ This figure prescribes minimum flue sizes as a function of chimney height and the fireplace opening area. For example, for a 20-foot-high (6096 mm) chimney and a fireplace opening area of 2,000 square inches (1 290 320 mm^2), the minimum cross-sectional area is 205 $^1/_2$ square inches (132 548 mm^2) for a round flue or 241 $^1/_2$ square inches (155 767 mm^2) for a square or rectangular flue. Using Table 2113.16 (1), that minimum required cross-sectional area can be met by a circular flue with a diameter of 18 inches (457 mm). Using Table 2113.16(2), the minimum cross-sectional area would require a square flue with an inside dimension of 19 $^1/_2$ inches (495 mm).

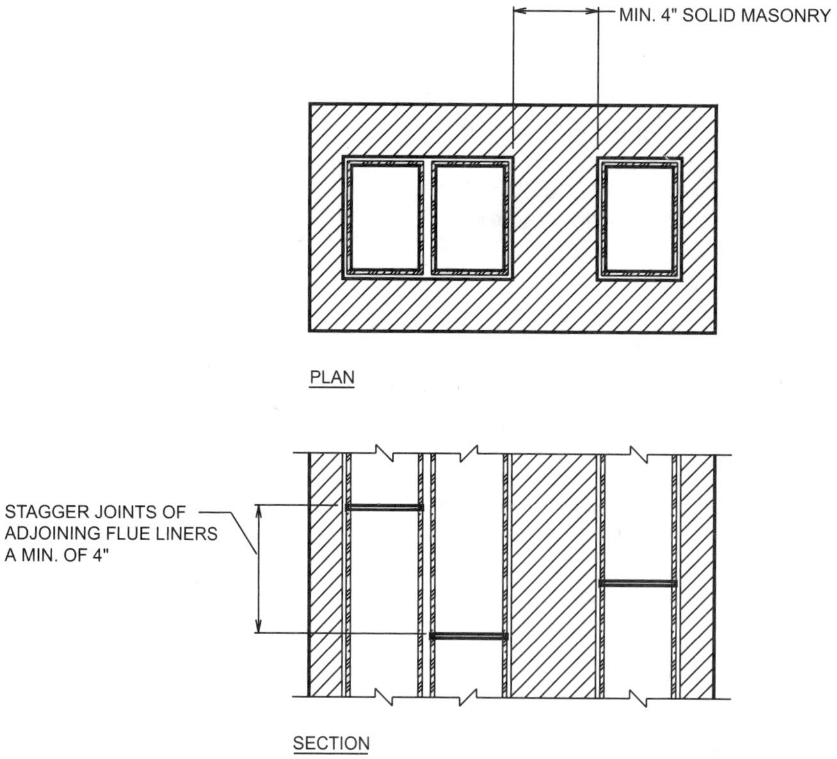

For SI: 1 inch = 25.4 mm.

Figure 2113.14
LOW-HEAT CHIMNEY WITH MORE THAN TWO FLUES

2113.16.1 Minimum area. Round chimney flues shall have a minimum net cross-sectional area of at least $^1/_{12}$ of the fireplace opening. Square chimney flues shall have a minimum net cross-sectional area of at least $^1/_{10}$ of the fireplace opening. Rectangular chimney flues with an aspect ratio less than 2 to 1 shall have a minimum net cross-sectional area of at least $^1/_{10}$ of the fireplace opening. Rectangular chimney flues with an aspect ratio of 2 to 1 or more shall have a minimum net cross-sectional area of at least $^1/_8$ of the fireplace opening.

❖ This section prescribes the minimum flue area based on the fireplace opening and the shape of the flue. For a given size fireplace, circular chimney flues are permitted to have a smaller area than rectangular flues. Wide, narrow flues require a larger area for adequate airflow.

2113.16.2 Determination of minimum area. The minimum net cross-sectional area of the flue shall be determined in accordance with Figure 2113.16. A flue size providing at least the equivalent net cross-sectional area shall be used. Cross-sectional areas of clay flue linings are as provided in Tables 2113.16(1) and 2113.16(2) or as provided by the manufacturer or as measured in the field. The height of the chimney shall be measured from the firebox floor to the top of the chimney flue.

❖ This section provides an alternate method to determine flue size based on the height of the chimney, the fireplace opening area and the flue type [see commentary, Tables 2113.16(1) and 2113.16(2) and Code Figure 2113.16].

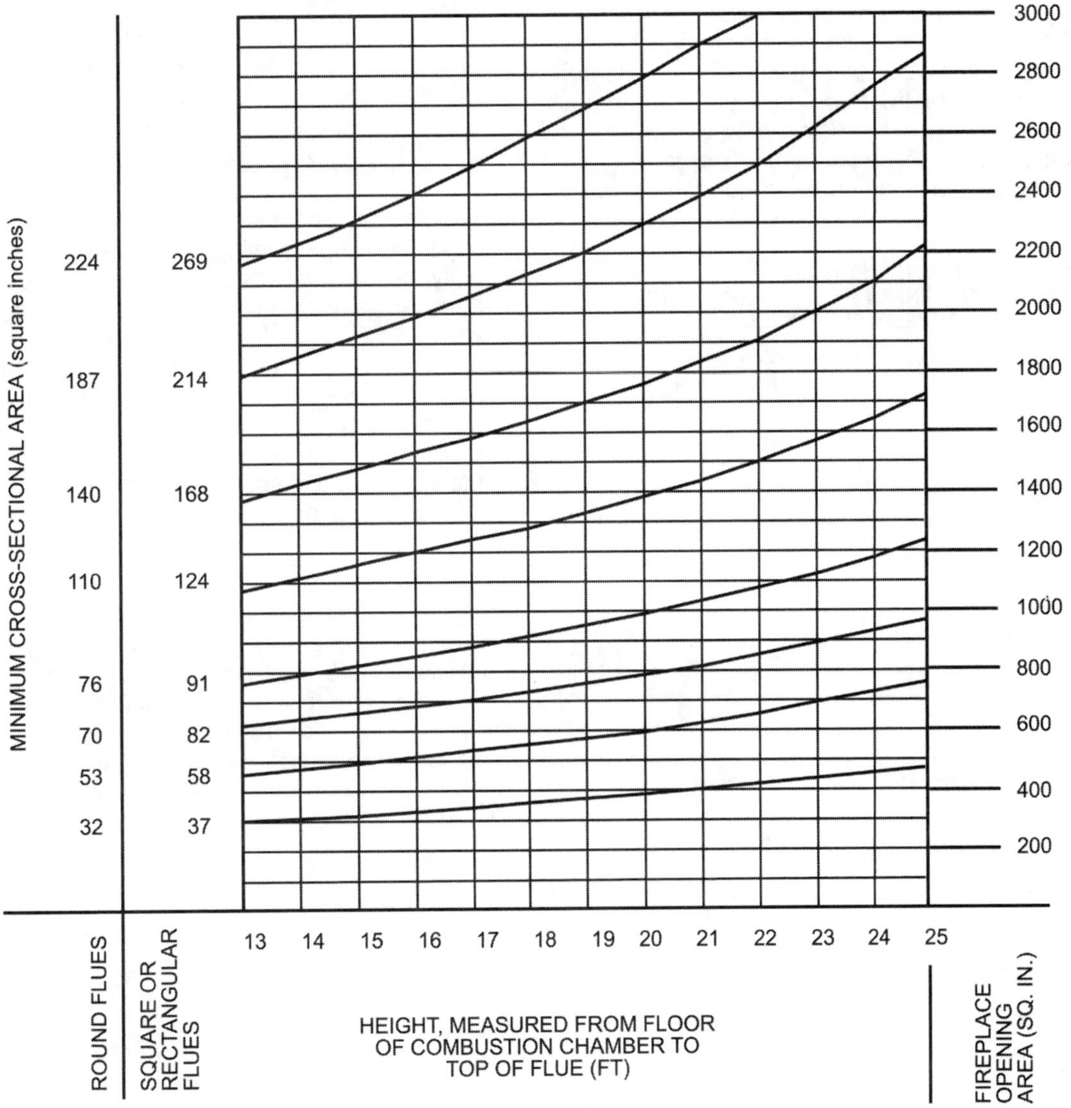

For SI: 1 inch = 25.4 mm, 1 square inch = 645 mm^2.

FIGURE 2113.16
FLUE SIZES FOR MASONRY CHIMNEYS

TABLE 2113.16(1)
NET CROSS-SECTIONAL AREA OF ROUND FLUE SIZES[a]

FLUE SIZE, INSIDE DIAMETER (inches)	CROSS-SECTIONAL AREA (square inches)
6	28
7	38
8	50
10	78
$10\,^3/_4$	90
12	113
15	176
18	254

For SI: 1 inch = 25.4 mm, 1 square inch = 645.16 mm^2.
a. Flue sizes are based on ASTM C 315.

❖ This table gives the areas of circular flues of standard sizes so that users of the code can easily determine a flue size complying with Code Figure 2113.16. The flue areas shown in this table are determined by the equation $A = \pi \times (d/2)^2$, where A is the cross-sectional area, d is the inside diameter of the flue and π is approximated as 3.14. Alternately, a minimum diameter can be calculated as $d = [(A/\pi)^{½}]/4$.

TABLE 2113.16(2)
NET CROSS-SECTIONAL AREA OF SQUARE AND RECTANGULAR FLUE SIZES[a]

FLUE SIZE, INSIDE DIMENSION (inches)	CROSS-SECTIONAL AREA (square inches)
$4^1/_2 \times 13$	34
$7^1/_2 \times 7^1/_2$	37
$8^1/_2 \times 8^1/_2$	47
$7^1/_2 \times 11^1/_2$	58
$8^1/_2 \times 13$	74
$7^1/_2 \times 15^1/_2$	82
$11^1/_2 \times 11^1/_2$	91
$8^1/_2 \times 17^1/_2$	101
13×13	122
$11^1/_2 \times 15^1/_2$	124
$13 \times 17^1/_2$	165
$15^1/_2 \times 15^1/_2$	168
$15^1/_2 \times 19^1/_2$	214
$17^1/_2 \times 17^1/_2$	226
$19^1/_2 \times 19^1/_2$	269
20×20	286

For SI: 1 inch = 25.4 mm, 1 square inch = 645.16 mm^2.
a. Flue sizes are based on ASTM C 315.

❖ This table gives the area of square and rectangular flues of standard sizes so that code users can easily determine a flue size complying with Code Figure 2113.16. Because the flue corners are rounded, the flue areas shown in this table are not determined by simply multiplying the inside dimensions shown. The areas are accordingly slightly smaller than a true rectangle of the dimensions shown in the table.

2113.17 Inlet. Inlets to masonry chimneys shall enter from the side. Inlets shall have a thimble of fireclay, rigid refractory material or metal that will prevent the connector from pulling out of the inlet or from extending beyond the wall of the liner.

❖ The inlet must be noncombustible and strong enough not to be pulled out.

2113.18 Masonry chimney cleanout openings. Cleanout openings shall be provided within 6 inches (152 mm) of the base of each flue within every masonry chimney. The upper edge of the cleanout shall be located at least 6 inches (152 mm) below the lowest chimney inlet opening. The height of the opening shall be at least 6 inches (152 mm). The cleanout shall be provided with a noncombustible cover.

Exception: Chimney flues serving masonry fireplaces, where cleaning is possible through the fireplace opening.

❖ This section requires a cleanout to be installed in a chimney to facilitate cleaning and inspection. A fireplace inherently provides access to its chimney through the firebox, throat and smoke chamber. The cleanout cover and opening frame are to be of an approved noncombustible material, such as cast iron or precast concrete and must be arranged to remain tightly closed. The requirement for placing the cleanout at least 6 inches (152 mm) below the lowest connection to the chimney is intended to minimize the possibility of combustion products exiting the chimney through the cleanout.

2113.19 Chimney clearances. Any portion of a masonry chimney located in the interior of the building or within the exterior wall of the building shall have a minimum airspace clearance to combustibles of 2 inches (51 mm). Chimneys located entirely outside the exterior walls of the building, including chimneys that pass through the soffit or cornice, shall have a minimum airspace clearance of 1 inch (25 mm). The airspace shall not be filled, except to provide fireblocking in accordance with Section 2113.20.

Exceptions:

1. Masonry chimneys equipped with a chimney lining system listed and labeled for use in chimneys in contact with combustibles in accordance with UL 1777, and installed in accordance with the manufacturer's instructions, are permitted to have combustible material in contact with their exterior surfaces.
2. Where masonry chimneys are constructed as part of masonry or concrete walls, combustible materials shall not be in contact with the masonry or concrete wall less than 12 inches (305 mm) from the inside surface of the nearest flue lining.

3. Exposed combustible trim and the edges of sheathing materials, such as wood siding, are permitted to abut the masonry chimney sidewalls, in accordance with Figure 2113.19, provided such combustible trim or sheathing is a minimum of 12 inches (305 mm) from the inside surface of the nearest flue lining. Combustible material and trim shall not overlap the corners of the chimney by more than 1 inch (25 mm).

❖ Clearance between the external surfaces of the masonry chimney and all combustible materials must be maintained. The intent of this section is to require a 2-inch (51 mm) minimum airspace clearance between combustibles and surfaces of all chimneys located in the interior of a building or within an exterior wall assembly. A 1-inch (25 mm) airspace clearance is allowed only where the chimney is located entirely outside of the building.

If any portion of a chimney is located in an exterior wall, that chimney must be considered as an interior chimney and must have a 2-inch (51 mm) minimum airspace clearance. The 1-inch (25 mm) clearance is allowed because the exterior surface of an outdoor chimney can dissipate heat. An outdoor chimney is exposed to outdoor ambient temperatures, allowing it to operate with cooler surface temperatures. Like all airspace clearances, the 1-inch (25 mm) airspace clearance is not permitted to be filled with any material except the noncombustible material necessary for fireblocking. The required clearance for exterior chimneys applies to all combustible materials, including sheathing, siding, insulation, framing and trim.

The exception recognizes a variety of chimney liners that are tested and labeled in accordance with UL 1777.

UL 1777 covers metallic and nonmetallic chimney liners intended for field installation into new and existing masonry chimneys used for natural draft venting of gas, oil and solid-fuel-burning appliances having maximum continuous flue-gas temperatures not exceeding 1,000°F (538°C). Some lining systems are labeled for reduced clearance applications and could allow the construction or rehabilitation of chimneys in contact with combustibles, without compromising the safety normally provided by the code-prescribed airspace clearances.

Chimney liner systems are typically metal or poured-in-place concrete and incorporate insulation to retard the transfer of heat to the surrounding masonry walls of the chimney.

FIGURE 2113.19. See below.

❖ This figure clarifies Exception 3 to the clearance requirements for masonry chimneys. The edge abutment of combustible sheathing or trim where there is an adequate thickness of masonry is a long-standing practice that is considered safe, provided the minimum clearance to the flue lining is maintained.

2113.20 Chimney fireblocking. All spaces between chimneys and floors and ceilings through which chimneys pass shall be fireblocked with noncombustible material securely fastened in place. The fireblocking of spaces between wood joists, beams or headers shall be to a depth of 1 inch (25 mm) and shall only be placed on strips of metal or metal lath laid across the spaces between combustible material and the chimney.

❖ Fireblocking is required to prevent the travel of flames, smoke and hot gases to other areas of the building through the gaps between the chimney and the floor or ceiling assemblies. The 1-inch (25 mm) depth requirement is intended to be both the minimum and the maximum [see Figures 2113.20(1) and (2)].

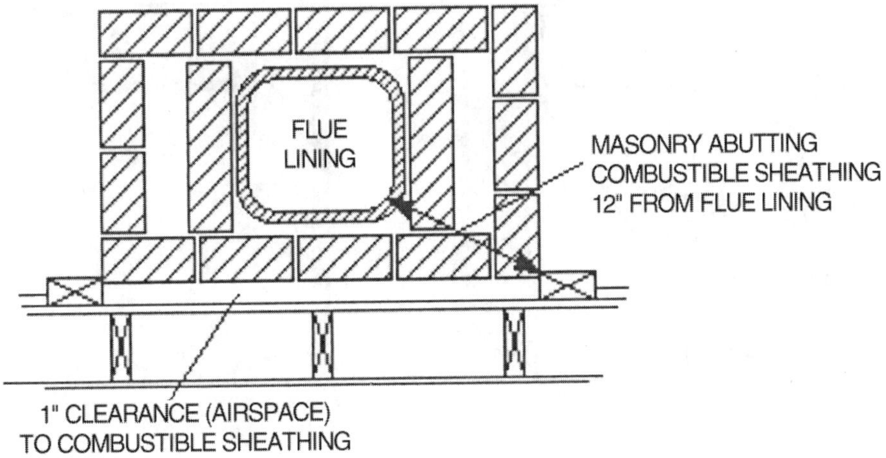

For SI: 1 inch = 25.4 mm

**FIGURE 2113.19
ILLUSTRATION OF EXCEPTION TO
CHIMNEY CLEARANCE PROVISION**

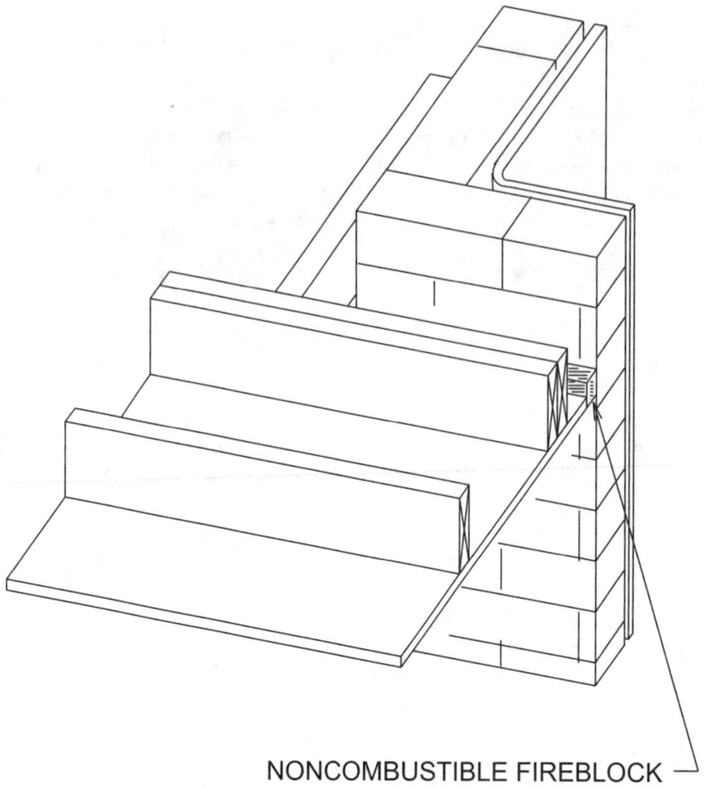

Figure 2113.20(1)
FIREBLOCKING

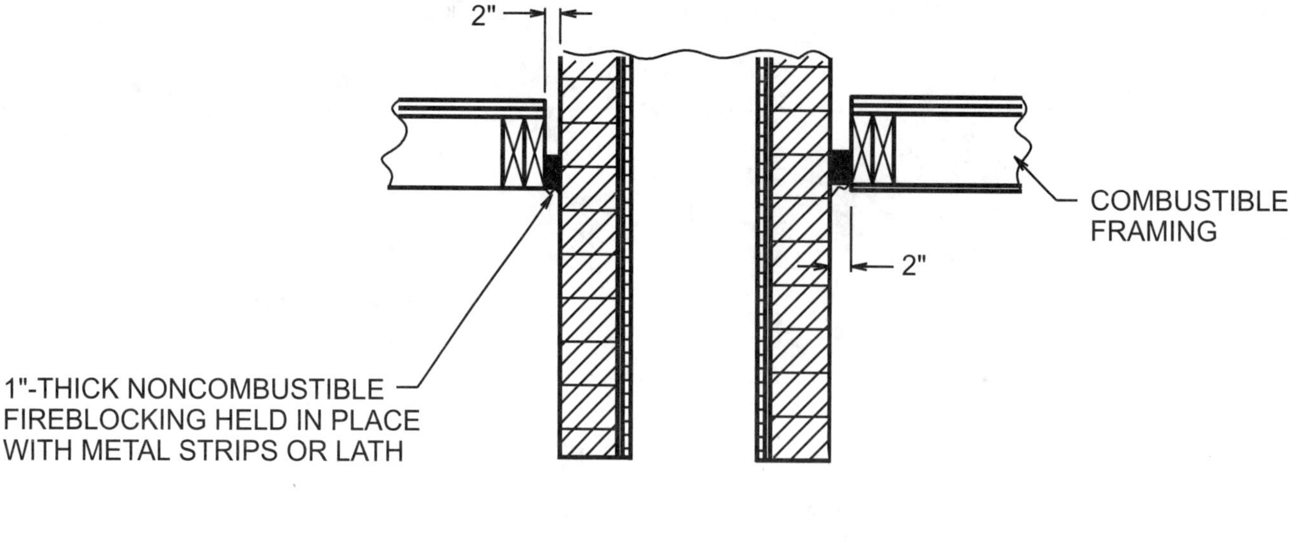

For SI: 1 inch = 25.4 mm.

Figure 2113.20(2)
FIREBLOCKING (SECTION)

Bibliography

The following resource materials are referenced in this chapter or are relevant to the subject matter addressed in this chapter.

ACI 530-02/ASCE 5-02/TMS 402-02, *Building Code Requirements for Masonry Structures* (1999 MSJC Code). Farmington Hills, MI: American Concrete Institute; Reston, VA: Structural Engineering Institute of the American Society of Civil Engineers; Boulder, CO: The Masonry Society, 2002.

ACI 530.1-02/ASCE 6-02/TMS 602-02, *Specifications for Masonry Structures*. Farmington Hills, MI: American Concrete Institute; Reston, VA: Structural Engineering Institute of the American Society of Civil Engineers; Boulder, CO: The Masonry Society, 2002.

ANSI A41.1-53, *American Standard Building Code Requirements for Masonry*. New York: American National Standards Institute, 1953.

ANSI A108.1A-99, *Glazed Wall Tile, Ceramic Mosaic Tile, Quarry Tile and Paver Tile Installed with Portland Cement Mortar*. New York: American National Standards Institute, 1999.

ANSI A108.1B-99, *Glazed Wall Tile, Ceramic Mosaic Tile, Quarry Tile and Paver Tile Installed with Portland Cement Mortar*. New York: American National Standards Institute, 1999.

ANSI A108.4-99, *Ceramic Tile Installed with Organic Adhesives or Water Cleanable Tile Setting Epoxy Adhesive*. New York: American National Standards Institute, 1999.

ANSI A108.5-99, *Ceramic Tile Installed with Dry-Set Portland Cement Mortar or Latex-Portland Cement Mortar*. New York: American National Standards Institute, 1999.

ANSI A108.6-99, *Ceramic Tile Installed with Chemical Resistant, Water Cleanable Tile Setting and Grouting Epoxy*. New York: American National Standards Institute, 1999.

ANSI A108.7-92, *Specifications for Electrically Conductive Ceramic Tile Installed with Conductive Dry-Set Portland Cement Mortar*. New York: American National Standards Institute, 1992.

ANSI A108.8-99, *Ceramic Tile Installed with Chemical Resistant Furan Mortar and Grout*. New York: American National Standards Institute, 1999.

ANSI A108.9-99, *Ceramic Tile Installed with Modified Epoxy Emulsion Mortar/Grout*. New York: American National Standards Institute, 1999.

ANSI A108.10-99, *Installation of Grout in Tilework*. New York: American National Standards Institute, 1999.

ANSI A118.1-99, *Dry-Set Portland Cement Mortar*. New York: American National Standards Institute, 1999.

ANSI A118.2-99, *Conductive Dry-Set Portland Cement Mortar*. New York: American National Standards Institute, 1999.

ANSI A118.3-99, *Chemical Resistant Water Cleanable Tile Setting and Grouting Epoxy and Water Cleanable Tile Setting Epoxy Adhesive*. New York: American National Standards Institute, 1999.

ANSI A118.4-99, *Latex-Portland Cement Mortar*. New York: American National Standards Institute, 1999.

ANSI A118.5-99, *Chemical Resistant Furan*. New York: American National Standards Institute, 1999.

ANSI A118.6-99, *Ceramic Tile Grouts*. New York: American National Standards Institute, 1999.

ANSI A118.8-99, *Modified Epoxy Emulsion Mortar/Grout*. New York: American National Standards Institute, 1999.

ANSI A136.1-99, *Organic Adhesives for Installation of Ceramic Tile*. New York: American National Standards Institute, 1999.

ANSI A137.1-88, *Ceramic Tile*. New York: American National Standards Institute, 1988.

ASCE 7-02, *Minimum Design Loads for Buildings and Other Structures*. Reston, VA: American Society of Civil Engineers, 2002.

ASTM A 36/A 36 M-00, *Specification for Structural Steel*. West Conshohocken, PA: ASTM International, 2000.

ASTM A 82-01, *Specification for Steel Wire, Plain, for Concrete Reinforcement*. West Conshohocken, PA: ASTM International, 2001.

ASTM A 153-01a, *Specification for Zinc Coating (Hot Dip) on Iron and Steel Hardware*. West Conshohocken, PA: ASTM International, 2001.

ASTM A 167-99, *Specification for Stainless and Heat-Resisting Chromium-Nickel Steel Plate, Sheet and Strip*. West Conshohocken, PA: ASTM International, 1999.

ASTM A 185-01, *Specification for Steel Welded Wire Fabric, Plain, for Concrete Reinforcement*. West Conshohocken, PA: ASTM International, 2001.

ASTM A 496-01, *Specification for Steel Wire, Deformed, for Concrete Reinforcement*. West Conshohocken, PA: ASTM International, 2001.

ASTM A 497-97, *Specification for Welded Deformed Steel Wire Fabric, for Concrete Reinforcement*. West Conshohocken, PA: ASTM International, 1997.

ASTM A 615-00, *Specification for Deformed and Plain Billet-Steel Bars for Concrete Reinforcement*. West Conshohocken, PA: ASTM International, 2000.

ASTM A 641-98, *Specification for Zinc-Coated (Galvanized) Carbon Steel Wire*. West Conshohocken, PA: ASTM International, 1998.

ASTM A 706/A706M-00, *Specification for Low-Alloy Steel Deformed Bars for Concrete Reinforcement*. West Conshohocken, PA: ASTM International, 2000.

ASTM A 767/A 767M-00, *Specification for Zinc-Coated (Galvanized) Bars for Concrete Reinforcement*. West Conshohocken, PA: ASTM International, 2000.

ASTM A 775/A 775M-01, *Specification for Epoxy-Coated Reinforcing Steel Bars*. West Conshohocken, PA: ASTM International, 2001.

ASTM A 951-00, *Specification for Masonry Joint Reinforcement*. West Conshohocken, PA: ASTM International, 2000.

ASTM C 5-97, *Specification for Quicklime for Structural Purposes*. West Conshohocken, PA: ASTM International, 1997.

ASTM C 27-98, *Standard Classification of Fireclay and High-Alumina Refractory Brick*. West Conshohocken, PA: ASTM International, 1998.

ASTM C 34-96 (2001), *Specification for Structural Clay Load-Bearing Wall Tile*. West Conshohocken, PA: ASTM International, 2001.

ASTM C 55-01A, *Specification for Concrete Building Brick*. West Conshohocken, PA: ASTM International, 2001.

ASTM C 56-96 (2001), *Specification for Structural Clay Non-Load-Bearing Tile*. West Conshohocken, PA: ASTM International, 2001.

ASTM C 62-01, *Specification for Building Brick (Solid Masonry Units Made from Clay or Shale)*. West Conshohocken, PA: ASTM International, 2001.

ASTM C 67-02, *Test Methods of Sampling and Testing Brick and Structural Clay Tile*. West Conshohocken, PA: ASTM International, 2002.

ASTM C 73-99a, *Specification for Calcium Silicate Face Brick (Sand Lime Brick)*. West Conshohocken, PA: ASTM International, 1999.

ASTM C 90-01A, *Specification for Load-Bearing Concrete Masonry Units*. West Conshohocken, PA: ASTM International, 2001.

ASTM C 91-01, *Standard Specification for Masonry Cement*. West Conshohocken, PA: ASTM International, 2001.

ASTM C 97-83, *Standard Test Methods for Absorption and Bulk Specific Gravity of Natural Stone*. West Conshohocken, PA: ASTM International, 1983.

ASTM C 99-87, *Standard Test Method for Modulus of Rupture of Natural Building Stone*. West Conshohocken, PA: ASTM International, 1987.

ASTM C 120-85, *Standard Test Method for Flexural Testing of Slate (Modulus of Rupture, Modulus of Elasticity)*. West Conshohocken, PA: ASTM International, 1985.

ASTM C 121-85, *Standard Test Method for Water Absorption of Slate*. West Conshohocken, PA: ASTM International, 1985.

ASTM C 126-99, *Specification for Ceramic Glazed Structural Clay Facing Tile, Facing Brick and Solid Masonry Units*. West Conshohocken, PA: ASTM International, 1999.

ASTM C 140-01 ae1, *Methods of Sampling and Testing Concrete Masonry Units*. West Conshohocken, PA: ASTM International, 2001.

ASTM C 144-89, *Standard Specification for Aggregate for Masonry Mortar*. West Conshohocken, PA: ASTM International, 1989.

ASTM C 150-01, *Specification for Portland Cement*. West Conshohocken, PA: ASTM International, 2001.

ASTM C 170-87, *Standard Test Method for Compressive Strength of Natural Building Stone*. West Conshohocken, PA: ASTM International, 1987.

ASTM C 199-84 (2000), *Standard Test Method for Pier Test for Refractory Mortars*. West Conshohocken, PA: ASTM International, 2000.

ASTM C 207-97, *Standard Specification for Hydrated Lime for Masonry Purposes*. West Conshohocken, PA: ASTM International, 1997.

ASTM C 212-00, *Specification for Structural Clay Facing Tile*. West Conshohocken, PA: ASTM International, 2000.

ASTM C 216-01A, *Specification for Facing Brick (Solid Masonry Units Made from Clay or Shale)*. West Conshohocken, PA: ASTM International, 2001.

ASTM C 217-01A, *Standard Test Method for Weather Resistance of Natural Slate*. West Conshohocken, PA: ASTM International, 2001.

ASTM C 270-01A, *Specification for Mortar for Unit Masonry*. West Conshohocken, PA: ASTM International, 2001.

ASTM C 305-82, *Standard Method for Mechanical Mixing of Hydraulic Cement Pastes and Mortars for Plastic Consistency*. West Conshohocken, PA: ASTM International, 1982.

ASTM C 315-00, *Specification for Clay Flue Linings*. West Conshohocken, PA: ASTM International, 2000.

ASTM C 331-01, *Standard Specification for Lightweight Aggregates for Concrete Masonry Units*. West Conshohocken, PA: ASTM International, 2001.

ASTM C 404-87, *Standard Specification for Aggregates for Masonry Grout*. West Conshohocken, PA: ASTM International, 1987.

ASTM C 476-01, *Specification for Grout for Masonry*. West Conshohocken, PA: ASTM International, 2001.

ASTM C 503-99eo1, *Specification For Marble Dimension Stone (Exterior)*. West Conshohocken, PA: ASTM International, 1999.

ASTM C 568-99, *Specification for Limestone Dimension Stone*. West Conshohocken, PA: ASTM International, 1999.

ASTM C 595-00, *Specification for Blended Hydraulic Cements*. West Conshohocken, PA: ASTM International, 2000.

ASTM C 615-99, *Specification for Granite Dimension Stone*. West Conshohocken, PA: ASTM International, 1999.

ASTM C 616-99, *Specification for Quartz-Based Dimension Stone*. West Conshohocken, PA: ASTM International, 1999.

ASTM C 629-99, *Specification for Slate Dimension Stone*. West Conshohocken, PA: ASTM International, 1999.

ASTM C 652-01A, *Specification for Hollow Brick (Hollow Masonry Units Made from Clay or Shale)*. West Conshohocken, PA: ASTM International, 2001.

ASTM C 744-99, *Specification for Prefaced Concrete and Calcium Silicate Masonry Units*. West Conshohocken, PA: ASTM International, 1999.

ASTM C 887-01, *Specification for Packaged, Dry, Combined Materials for Surface Bonding Mortar*. West Conshohocken, PA: ASTM International, 2001.

ASTM C 946-91 (2001), *Practice for Construction of Dry-Stacked, Surface-Bonded Walls*. West Conshohocken, PA: ASTM International, 2001.

ASTM C 1019-00B (1993), *Test Method for Sampling and Testing Grout*. West Conshohocken, PA: ASTM International, 2000.

ASTM C 1088-01A, *Specification for Thin Veneer Brick Units Made from Clay or Shale*. West Conshohocken, PA: ASTM International, 2001.

ASTM C 1261-98, *Firebox Brick for Residential Fireplaces*. West Conshohocken, PA: ASTM International, 1998.

ASTM C 1283-00, *Standard Practice for Installing Clay Flue Liners*. West Conshohocken, PA: ASTM International, 2000.

ASTM C 1314-02, *Standard Method for Constructing and Testing Masonry Prisms Used to Determine Compliance with Specified Compressive Strength of Masonry*. West Conshohocken, PA: ASTM International, 2002.

ASTM C 1327-97, *Specification for Mortar Cement*. West Conshohocken, PA: ASTM International, 1997.

ASTM E 84-01, *Test Method for Surface Burning Characteristics of Building Materials*. West Conshohocken, PA: ASTM International, 2001.

ASTM E 119-00, *Test Methods for Fire Tests of Building Construction and Materials*. West Conshohocken, PA: ASTM International, 2000.

ASTM E 447-92b, *Tests Methods for Compressive Strength of Masonry Prisms*. West Conshohocken, PA: ASTM International, 1992.

ASTM E 1602-01, *Standard Guide for Construction of Solid Fuel Burning Masonry Heaters*. West Conshohocken, PA: ASTM International, 2001.

AWS D 1.4-98, *Structural Welding Code—Reinforced Steel*. Miami, FL: American Welding Society, 1998.

Baker, Ira. *A Treatise on Masonry Construction*. Chicago: University of Illinois, 1889.

Commentary on Building Code Requirements for Masonry Structures (ACI 530-02/ASCE 5-02/TMS 402-02). Farmington Hills, MI: American Concrete Institute; Reston, VA: Structural Engineering Institute of the American Society of Civil Engineers; Boulder, CO: The Masonry Society, 2002.

Commentary on Specifications for Masonry Structures (ACI 530.1-02/ASCE 6-02/TMS 602-02). Farmington Hills, MI: American Concrete Institute; Reston, VA: Structural Engineering Institute of the American Society of Civil Engineers; Boulder, CO: The Masonry Society, 2002.

Compressive Strength Testing of Masonry Mortar, A TMS Monograph. Boulder, CO: The Masonry Society, 1996.

IFC-03, *International Fire Code*. Falls Church, VA: International Code Council, 2003.

IFGC-03, *International Fuel Gas Code*. Falls Church, VA: International Code Council, 2003.

IRC-03, *International Residential Code*. Falls Church, VA: International Code Council, 2003.

NEHRP Recommended Provisions for Seismic Regulations for New Buildings and Other Structures. Washington, DC: Building Seismic Safety Council, 1997.

Standard Building Code©. Birmingham, AL: Southern Building Code Congress International, Inc., 1997.

UL 641-95. *Type L Low-Temperature Venting Systems*. Northbrook, IL: Underwriters Laboratories Inc., 1995.

UL 1777-98, *Chimney Liners*. Northbrook, IL: Underwriters Laboratories Inc., 1998.

Uniform Building Code™. Whittier, CA: International Conference of Building Officials, 1997.

Chapter 22: Steel

General Comments

Chapter 22 contains provisions governing the materials, design, construction and quality of steel structural members.

Section 2201 contains the general requirements.

Section 2202 includes the definitions of terms related to structural steel construction.

Section 2203 includes provisions for the identification and protection of steel for structures.

Section 2204 addresses requirements for welded and bolted connections.

Section 2205 references the design standards for structural steel construction.

Section 2206 addresses steel joists.

Section 2207 specifies the design standard for steel cable structures.

Section 2208 references the specification for the design and installation of steel storage racks.

Section 2209 specifies the appropriate referenced standard for cold-formed steel.

Section 2210 references standards for light-framed construction utilizing cold-formed steel.

Section 2211 provides detailed provisions for light-frame cold-formed steel shear walls.

Steel is a noncombustible material commonly associated with Type 1 and 2 construction; however, it is permitted to be used and commonly found in all types of construction. There are two main families of steel members: the first is hot-rolled structural shapes and members that are made up of a combination of rolled shapes and plates, and the second is composed of sections that are cold formed from steel sheets, strips, plates or flat bars in roll-forming machines, or by bending in brake press operations.

The code requires that materials used in the design of structural steel members conform to designated national standards. Chapter 22 is involved largely with the design and use of steel materials using the specifications and standards of the American Institute of Steel Construction (AISC), the American Iron and Steel Institute (AISI), the Steel Joist Institute (SJI) and the American Society of Civil Engineers (ASCE).

Purpose

The purpose of Chapter 22 is to provide the requirements necessary for the design and construction of structural steel, cold-formed steel, steel joists, steel cable structures, steel storage racks and composite construction. This chapter specifies the appropriate design and construction standards for these types of structures. It also provides a road map of the applicable technical requirements for steel structures. Detailed provisions are included for the wind and seismic requirements for light-frame cold-formed steel walls.

SECTION 2201 GENERAL

2201.1 Scope. The provisions of this chapter govern the quality, design, fabrication and erection of steel used structurally in buildings or structures.

❖ Chapter 22 is formatted in a very similar fashion to the existing model codes, since the code essentially gives the user a road map on the use of design standards appropriate for various types of steel construction.

It is important to note that the scope of the various sections that adopt design specifications is often different, since each section is intended to reflect the scope of the adopted standard. The following text will discuss the background of each section.

SECTION 2202 DEFINITIONS AND NOMENCLATURE

2202.1 Definitions. The following words and terms shall, for the purposes of this chapter and as used elsewhere in this code, have the meaning shown herein.

❖ This section contains definitions of terms associated with the subject matter of this chapter. Definitions of terms can help in the understanding and application of code requirements. Definitions are included within this chapter to provide convenient access to them without having to refer back to Chapter 2.

ADJUSTED SHEAR RESISTANCE. In Type II shear walls, the unadjusted shear resistance multiplied by the shear resistance adjustment factors of Table 2211.3.

❖ Type II shear walls require the application of empirically determined shear resistance adjustment factors as described in Section 2211.3.2, Item 4. The term "adjusted shear resistance" is used to refer to the resulting shear resistance.

STEEL CONSTRUCTION, COLD-FORMED. That type of construction made up entirely or in part of steel structural members cold formed to shape from sheet or strip steel such as roof deck, floor and wall panels, studs, floor joists, roof joists and other structural elements.

❖ This definition is necessary to distinguish structural members of cold-formed steel from structural steel.

STEEL JOIST. Any steel structural member of a building or structure made of hot-rolled or cold-formed solid or open-web sections, or riveted or welded bars, strip or sheet steel members, or slotted and expanded, or otherwise deformed rolled sections.

❖ This definition is necessary for the application of Section 2206.

STEEL MEMBER, STRUCTURAL. Any steel structural member of a building or structure consisting of a rolled steel structural shape other than cold-formed steel, or steel joist members.

❖ This definition is necessary to distinguish structural steel members from cold-formed steel members used for structural purposes.

TYPE I SHEAR WALL. A wall designed to resist in-plane lateral forces that is fully sheathed and provided with hold-down anchors at each end of the wall segment. Type I walls are permitted to have openings where detailing for force transfer around the openings is provided (see Figure 2202.1).

❖ This term is used to describe traditionally designed shear walls, commonly referred to as "segmented shear walls," consisting of fully sheathed wall sections, each of which are restrained against overturning by using hold-downs. This design approach is covered in Section 2211.2, and the top portion of Figure 2202.1 schematically depicts this type of shear wall.

TYPE II SHEAR WALL. A wall designed to resist in-plane lateral forces that is sheathed with wood structural panel or sheet steel that contains openings, that have not been specifically designed and detailed for force transfer around wall openings. Hold-down anchors for Type II shear walls are only required at the ends of the wall (see Figure 2202.1).

❖ This term is used to describe shear walls designed using an alternative to traditional shear wall design methodology, commonly referred to as "perforated shear walls." This method consists of prescriptive provisions along with empirical adjustments to the tabulated nominal shear wall values. Shear walls designed by this method require hold-downs only at the ends. This design method is covered in Section 2211.3, and the bottom portion of Figure 2202.1 schematically depicts this type of shear wall.

TYPE II SHEAR WALL SEGMENT. A section of shear wall with full-height sheathing and which meets the aspect ratio limits of Section 2211.3.2(3).

❖ This term refers to a fully sheathed portion of a Type II shear wall that also meets the height-to-width, or aspect ratio, limit. Only these portions of the Type II shear wall are considered in determining the percentage of full-height sheathing as described in Section 2211.3.2, Item 1.

UNADJUSTED SHEAR RESISTANCE. In Type II walls, the unadjusted shear resistance is based on the design shear and the limitations of Section 2211.3.1.

❖ Type II shear walls designed in accordance with Section 2211.3 use design shear values determined as described in Section 2211.3.2, Item 3, and subject to the limitations of Section 2211.3.1. The term "unadjusted shear resistance" is used to refer to the resulting design shear value.

FIGURE 2202.1.. See page 22-3

❖ This figure schematically depicts the two types of light-framed shear walls that are covered in Section 2211. Also see the definitions of "Type I shear wall" and "Type II shear wall."

2202.2 Nomenclature. The following symbols shall, for the purposes of this chapter and as used elsewhere in this code, have the meanings shown herein.

ϕ = Resistance factor (see Section 2211.2.1).

Ω = Factor of safety (see Section 2211.2.1).

Ω_o = System overstrength factor (see Section 1617.6).

C_o = Shear resistance adjustment factor from Table 2211.3.

ΣL_i = Sum of widths of Type II shear wall segments, feet (mm/1,000).

C = Compression chord uplift force, lbs (kN).

V = Shear force in Type II shear wall, lbs (kN).

h = The height of a shear wall measured as:

1. The maximum clear height from top of foundation to bottom of diaphragm framing above or,
2. The maximum clear height from top of a diaphragm to bottom of diaphragm framing above.

v = Unit shear force, plf (kN/m).

w = The width of a shear wall or wall pier in the direction of application of force measured as the sheathed dimension of the shear wall.

❖ This section includes the nomenclature that is needed for the application of the shear wall provisions of Section 2211.

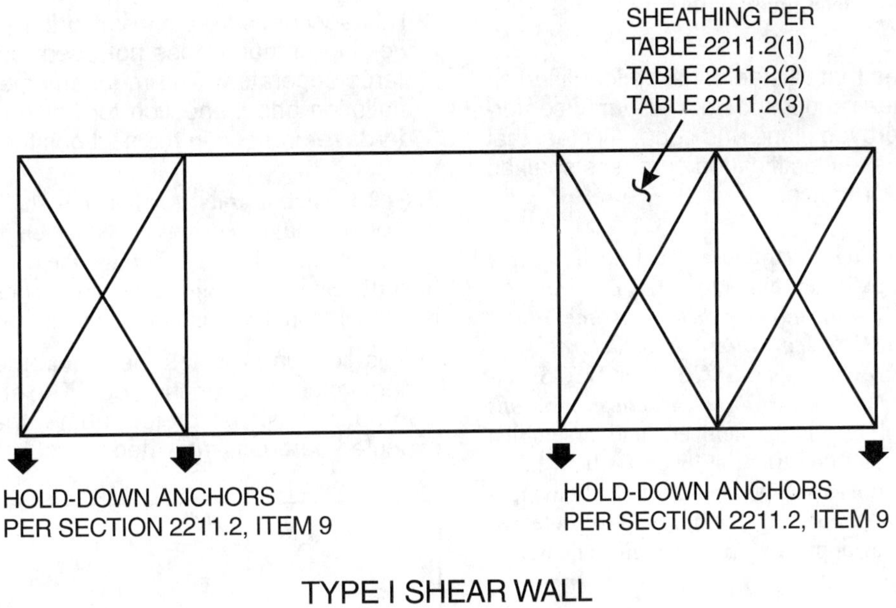

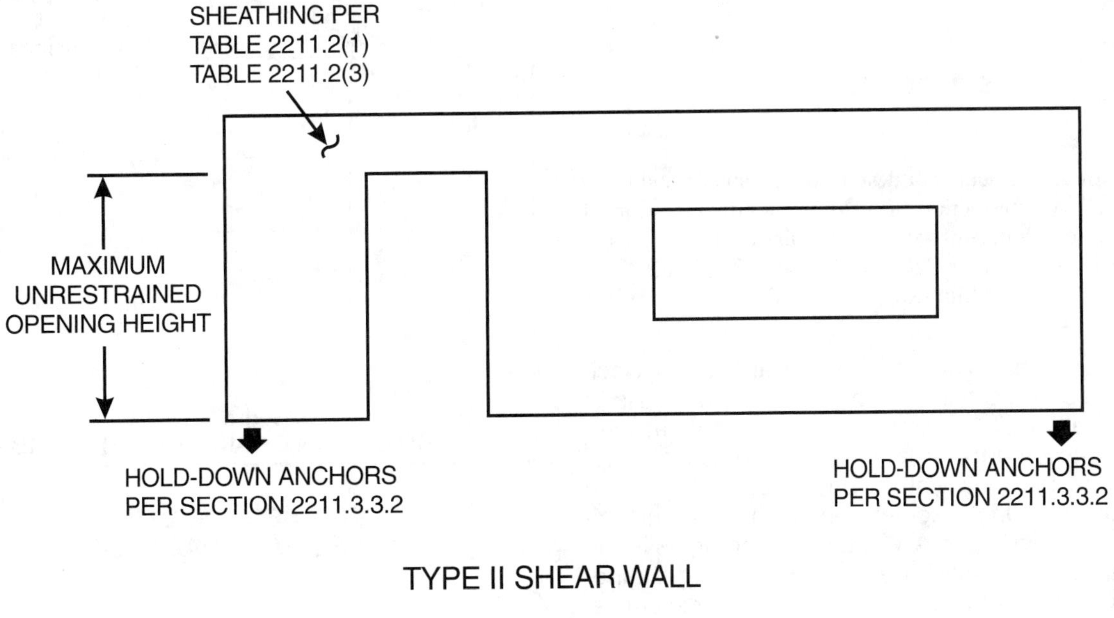

FIGURE 2202.1
TYPE I AND TYPE II SHEAR WALLS

SECTION 2203 IDENTIFICATION AND PROTECTION OF STEEL FOR STRUCTURAL PURPOSES

2203.1 Identification. Steel furnished for structural load-carrying purposes shall be properly identified for conformity to the ordered grade in accordance with the specified ASTM standard or other specification and the provisions of this chapter. Steel that is not readily identifiable as to grade from marking and test records shall be tested to determine conformity to such standards.

❖ This section requires that all steel must be identified so that both the engineer and inspector can verify conformance with the approved plans and code. All steel that is not marked for these identification purposes shall be tested to verify conformance.

2203.2 Protection. Painting of structural steel shall comply with the requirements contained in either the *AISC Load and Resistance Factor Design Specification for Structural Steel Buildings* (AISC-LRFD), *AISC Specification for Structural Steel Buildings—Allowable Stress Design* (AISC 335) or *AISC Specification for the Design of Steel Hollow Structural Sections* (AISC-HSS). Individual structural members and assembled panels of cold-formed steel construction, except where fabricated of approved corrosion-resistant steel or of steel having a corrosion resistant or other approved coating, shall be protected against corrosion with an approved coat of paint, enamel or other approved protection.

❖ This section alerts the user that protection of structural steel members is required, and must be in accordance with the appropriate AISC specifications. Also, cold-formed steel structural members need to be of corrosion-resistant steel, or have a corrosion-resistant metallic coating, paint or other approved coating.

SECTION 2204 CONNECTIONS

2204.1 Welding. The details of design, workmanship and technique for welding, inspection of welding and qualification of welding operators shall conform to the requirements of the specifications listed in Sections 2205, 2206, 2207, 2209 and 2210. Special inspection of welding shall be provided where required by Section 1704.

❖ This section does not adopt the American Welding Society (AWS) specifications for the design and details of welding; rather, it requires that welding be accomplished in accordance with the requirements of the appropriate design specification. This is done because the referenced standards sometimes adopt different editions of the appropriate AWS specification or, as in the case of steel joists, they have their own welding requirements contained within the SJI standards. For example, since the AISC allowable stress design (ASD) and load and resistance factor design (LRFD) methods were published several years apart, these two specifications adopt and make separate modifications to different editions of AWS D1.1.

2204.2 Bolting. The design, installation and inspection of bolts shall be in accordance with the requirements of the specifications listed in Sections 2205, 2206, 2209 and 2210. Special inspection of the installation of high-strength bolts shall be provided where required by Section 1704.

❖ This section is very similar to the previous section on welding in that it does not adopt bolt installation standards separately. The requirements for the design, installation and inspection for bolting are found in standards referenced in the sections that are noted.

2204.2.1 Anchor rods. Anchor rods shall be set accurately to the pattern and dimensions called for on the plans. The protrusion of the threaded ends through the connected material shall be sufficient to fully engage the threads of the nuts, but shall not be greater than the length of the threads on the bolts.

❖ This section provides the requirements for the setting and projection of anchor rods. The end of the anchor rod threads must not project above the face of the connected part as is indicated in Figure 2204.2.1.

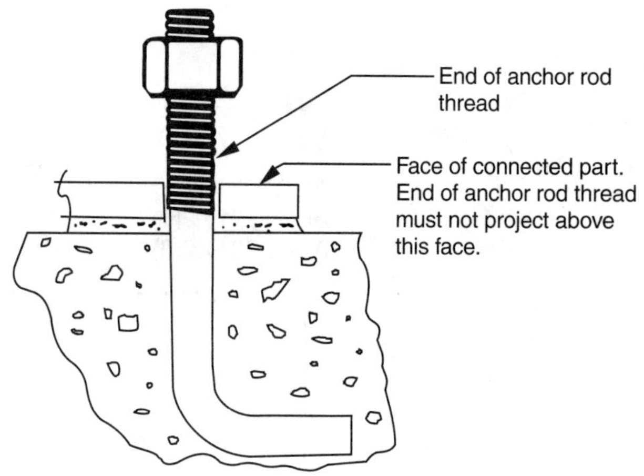

Figure 2204.2.1
PROJECTION OF ANCHOR ROD THREADS

SECTION 2205 STRUCTURAL STEEL

2205.1 General. The design, fabrication and erection of structural steel for buildings and structures shall be in accordance with either the AISC-LRFD, AISC 335 or AISC-HSS. Where required, the seismic design of steel structures shall be in accordance with the additional provisions of Section 2205.2.

❖ This section requires that structural steel (see the definition in Section 2202) be designed in accordance with one of three specifications published by the AISC. The code cites referenced standards by short titles in the body of the code, with the complete title and appropriate date cited in Chapter 35. The three standards permitted for structural steel members are:

AISC 335	*Specification for Structural Steel Buildings—Allowable Stress Design and Plastic Design, including Supplement No.1, 2001*
AISC-LRFD	*Load and Resistance Factor Design Specification for Structural Steel Buildings*
AISC-HSS	*Specification for the Design of Steel Hollow Structural Sections*

The latter is a new standard specifically developed for the design of round, square and rectangular steel tubular structural sections. Both AISC-ASD and AISC-LRFD have been adopted by the codes for many years.

2205.2 Seismic requirements for steel structures. The design of structural steel structures to resist seismic forces shall be in accordance with the provisions of Section 2205.2.1 or 2205.2.2 for the appropriate seismic design category.

❖ The code adopts state-of-the-art seismic requirements based on AISC 341, *Seismic Provisions for Structural Steel Buildings*. It incorporates significant research results from the SAC Research Program, which began as a result of the damage that was caused by the Northridge earthquake to steel moment frames. As a result of this FEMA-funded research project, it is expected that the AISC 341 provisions will be undergoing annual changes for the next several years (see the commentary to AISC 341 for a detailed discussion).

AISC 341 is divided into three parts. Parts I and III are the requirements for structural steel, and Part II is applicable to composite construction. Part III also contains requirements that allow the use of ASD.

2205.2.1 Seismic Design Category A, B or C. Structural steel structures assigned to Seismic Design Category A, B or C, in accordance with Section 1616, shall be of any construction permitted in Section 2205. An *R* factor as set forth in Section 1617.6 for the appropriate steel system is permitted where the structure is designed and detailed in accordance with the provisions of AISC 341, Parts I and III. Systems not detailed in accordance with the above shall use the *R* factor in Section 1617.6 designated for "steel systems not detailed for seismic."

❖ For structures classified as Seismic Design Category A, B or C, a designer is permitted to use a response modification coefficient (*R*) of 3, and design the structure without the detailing required in AISC 341. If the designer wishes to use a response modification coefficient (*R*) greater than 3, the AISC 341 provisions are mandatory.

2205.2.2 Seismic Design Category D, E or F. Structural steel structures assigned to Seismic Design Category D, E or F shall be designed and detailed in accordance with AISC 341, Part I or III.

❖ In Seismic Design Category D, E and F, the provisions of AISC 341, Part I or Part IIII , are mandatory.

2205.3 Seismic requirements for composite construction. The design, construction and quality of composite steel and concrete components that resist seismic forces shall conform to the requirements of the AISC LRFD and ACI 318. An *R* factor as set forth in Section 1617.6 for the appropriate composite steel and concrete system is permitted where the structure is designed and detailed in accordance with the provisions of AISC 341, Part II. In Seismic Design Category B or above, the design of such systems shall conform to the requirements of AISC 341, Part II.

❖ The code contains the first concise technical provisions for combining structural steel and concrete in the lateral-force-resisting system. The use of composite construction has been growing in low-risk seismic areas, but has been hampered in high-risk seismic areas due to the lack of technical design requirements. Since these are new requirements, the designer must submit substantiating evidence to the building official for approval. Where the design relies on composite elements that are expected to undergo inelastic demands, the evidence must include cyclic tests in accordance with the test protocol and acceptance criteria established by AISC 341.

Composite structures classified as Seismic Design Category A are permitted to be designed in accordance with ACI and AISC-LRFD. Structures classified as Seismic Design Category B or above are required to be designed in accordance with Part II of ASIC 341.

2205.3.1 Seismic Design Categories D, E and F. Composite structures are permitted in Seismic Design Categories D , E and F, subject to the limitations in Section 1617.6, where substantiating evidence is provided to demonstrate that the proposed system will perform as intended by AISC 341, Part II. The substantiating evidence shall be subject to building official approval. Where composite elements or connections are required to sustain inelastic deformations, the substantiating evidence shall be based on cyclic testing.

❖ In Seismic Design Category D, E and F, composite structures are permitted when the designer presents test data or other evidence to the building official that the connections have the necessary rotational capacity to resist the expected deformations. In addition to this data, the use of Part II of AISC 341 is required.

SECTION 2206
STEEL JOISTS

2206.1 General. The design, manufacturing and use of open web steel joists and joist girders shall be in accordance with one of the following Steel Joist Institute specifications:

1. *Standard Specifications for Open Web Steel Joists, K Series.*
2. *Standard Specifications for Longspan Steel Joists, LH Series and Deep Longspan Steel Joists, DLH Series.*
3. *Standard Specifications for Joist Girders.*

Where required, the seismic design of buildings shall be in accordance with the additional provisions of Section 2205.2 or 2211.

❖ Steel joists are required to be designed and manufactured in accordance with the specifications published by the SJI.

SECTION 2207
STEEL CABLE STRUCTURES

2207.1 General. The design, fabrication and erection including related connections, and protective coatings of steel cables for buildings shall be in accordance with ASCE 19.

❖ The code adopts ASCE 19 for the design of steel cable structures. This standard addresses the use of carbon steel or stainless steel cables formed from coated or uncoated helically twisted wire strand, wire rope or a parallel wire strand.

2207.2 Seismic requirements for steel cable. The design strength of steel cables shall be determined by the provisions of ASCE 19 except as modified by these provisions.

1. A load factor of 1.1 shall be applied to the prestress force included in T_3 and T_4 as defined in Section 3.12.
2. In Section 3.2.1, Item (c) shall be replaced with "1.5 T_3" and Item (d) shall be replaced with "1.5 T_4."

❖ This section provides the user with appropriate modifications for structural cables that are intended for use in seismic applications. These modifications were first developed by the Building Seismic Safety Council for incorporation into the first edition of the National Earthquake Hazards Reduction Program (NEHRP) Provisions.

SECTION 2208
STEEL STORAGE RACKS

2208.1 Storage racks. The design, testing and utilization of industrial steel storage racks shall be in accordance with the *RMI Specification for the Design, Testing and Utilization of Industrial Steel Storage Racks*. Racks in the scope of this specification include industrial pallet racks, movable shelf racks and stacker racks, and does not apply to other types of racks, such as drive-in and drive-through racks, cantilever racks, portable racks or rack buildings. Where required, the seismic design of storage racks shall be in accordance with the provisions of Section 9.6.2.9 of ASCE 7.

❖ This section requires that the design of steel storage racks be in accordance with the Rack Manufactures Institute (RMI) standard. The types of racks that are included in the scope of that standard are noted. Other types of racks, such as drive-in or drive-through racks or cantilever racks, portable racks or rack buildings, that are beyond the scope of the RMI standard, should be designed in accordance with applicable referenced specifications such as AISC 335, AISC-LRFD or the *North American Specification for the Design of Cold-Formed Steel Structural Members* (AISI-NASPEC). Where the racks are located in structures classified as Seismic Design Category C or higher, they also must be designed in accordance with the requirements of Chapter 16. According to the referenced ASCE 7 provision, racks supported at or below the base of the structure must comply with the requirements for nonbuilding structures, while racks supported above the base of the structure must comply with the requirements for architectural components.

SECTION 2209
COLD-FORMED STEEL

2209.1 General. The design of cold-formed carbon and low-alloy steel structural members shall be in accordance with the *North American Specification for the Design of Cold-Formed Steel Structural Members* (AISI-NASPEC). The design of cold-formed stainless-steel structural members shall be in accordance with ASCE 8. Cold-formed steel light-framed construction shall comply with Section 2210.

❖ This section adopts AISI-NASPEC, which has been developed by the AISI in cooperation with the Canadian Standards Association (CSA) and Camara Nacional de la Industria del Hierro y del Acero (CANACERO). It supercedes the standards previously published individually by AISI and CSA and is intended for use throughout North America. The latter is accomplished by including three appendix chapters with the country-specific provisions applicable in either the United States, Canada or Mexico. Like its predecessor, it incorporates LRFD and ASD into a single document.

Cold-formed stainless steel is required to comply with ASCE 8. This standard applies to structural members cold formed from annealed and cold-rolled stainless steels. ASCE 8 includes both the LRFD and ASD methods.

2209.2 Composite slabs on steel decks. Composite slabs of concrete and steel deck shall be designed and constructed in accordance with ASCE 3.

❖ This section adopts ASCE 3, which covers the design and testing of both normal and lightweight concrete placed over a cold-formed steel deck. The steel deck performs the dual function of acting as a form for the concrete during construction and as positive reinforcement for the slab during service.

SECTION 2210
COLD-FORMED STEEL
LIGHT-FRAMED CONSTRUCTION

2210.1 General. The design, installation and construction of cold-formed carbon or low-alloy steel, structural and nonstructural steel framing, shall be in accordance with the *Standard for Cold-Formed Steel Framing—General Provisions, American Iron and Steel Institute* (AISI-General) *and* AISI-NASPEC.

❖ Section 2210 applies to light-framed construction utilizing cold-formed steel. Light-framed construction is characterized by vertical (i.e., walls) and horizontal (i.e., floors and roofs) structural elements formed by a system of repetitive members (see Chapter 2 definition). This section references AISI-NASPEC in addition to AISI-General. As its name implies, AISI-General covers general requirements for the design and installation of cold-formed steel framing members in walls, floors and roof assemblies. For specific design information, AISI-General refers the reader to AISI *Specification for the Design of Cold-Formed Steel Structural Members.* Since this section also references the successor to that standard, it is necessary to use the corresponding requirement of AISI-NASPEC under the code.

2210.2 Headers. The design and installation of cold-formed steel box and back-to-back headers, and double L-headers used in single-span conditions for load- carrying purposes shall be in accordance with the *Standard for Cold-Formed Steel Framing—Header Design*, American Iron and Steel Institute (AISI-Header), subject to the limitations therein.

❖ The code references AISI *Standard for Cold-Formed Steel Framing-Header Design*, which has design and installation requirements for headers. Figure 2210.2 schematically depicts the header types covered by the standard. Headers consist of C sections (either "back-to-back" or "box") as well as double L headers.

2210.3 Trusses. The design, quality assurance, installation and testing of cold-formed steel trusses shall be in accordance with the *Standard for Cold-Formed Steel Framing–Trusses*, American Iron and Steel Institute (AISI-Truss), subject to the limitations therein.

❖ The code references AISI *Standard for Cold-Formed Steel Framing–Truss Design*. This document covers truss design, quality criteria, installation and bracing as well as test methods.

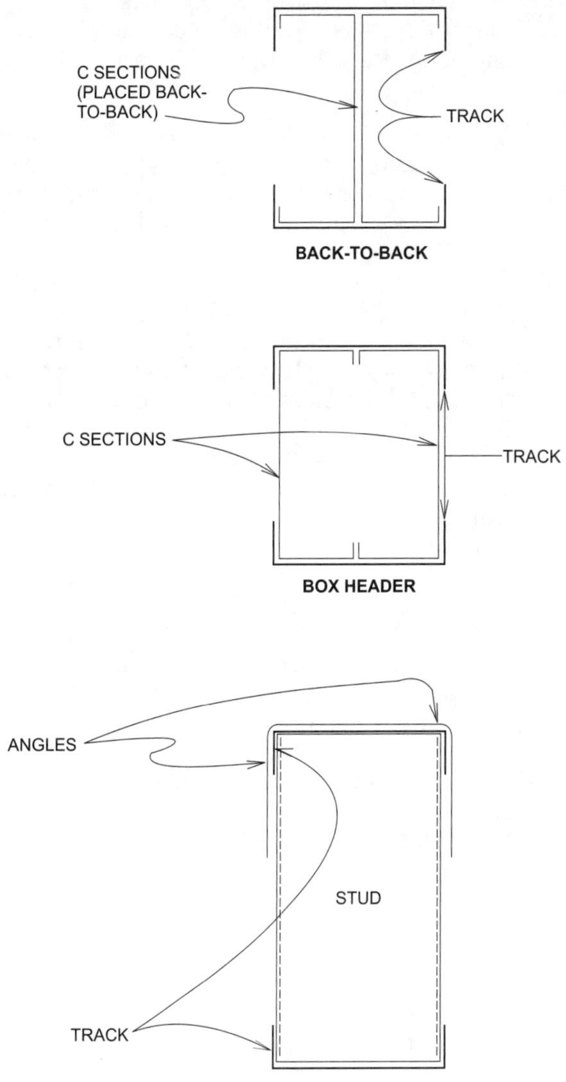

Figure 2210.2
HEADER TYPES

SECTION 2211
COLD-FORMED STEEL
LIGHT-FRAMED SHEAR WALLS

2211.1 General. In addition to the requirements of Section 2210, the design of cold-formed steel light-framed shear walls, to resist wind and seismic loads shall be in accordance with the requirements of Section 2211.2 for Type I (segmented) shear walls or Section 2211.3 for Type II (perforated) shear walls.

Light-framed structures assigned to Seismic Design Categories A, B or C, in accordance with Section 1616, shall be of any construction permitted in Section 2210. An *R* factor as set forth in Section 1617.6 for the appropriate steel system is permitted where the lateral design of the structure is in accordance with the provisions of Section 2211.4. Systems not detailed in accordance with Section 2211.4 shall use the *R* factor in Section 1617.6 designated for "steel systems not detailed for seismic."

In Seismic Design Categories D, E and F, the lateral design of light-framed structures shall also comply with the requirements in Section 2211.4

❖ This section regulates light-framed shear walls constructed with cold-formed steel framing. Two distinct approaches to the design of shear walls are included: Type I and Type II.

Type I shear walls are commonly referred to as "segmented shear walls" and consist of sheathed wall sections, each of which are restrained against overturning by using hold-downs. This design approach is covered in Section 2211.2. Unless specifically designed for force transfer around openings, no contribution from the sheathing above or below an opening is allowed.

Type II shear walls are an alternative to segmented shear wall design and are commonly referred to as "perforated shear walls." This method is addressed in Section 2211.3 and is comprised of specific prescriptive requirements along with empirical adjustments to the tabulated shear wall values. The adjustment factor varies with the maximum opening height ratio for a particular shear wall and the percentage of the wall length that has full-height sheathing. A Type II shear wall requires hold-downs only at the ends, but its capacity will typically be less than that of a comparable Type I shear wall with fully restrained segments.

For structures classified as Seismic Design Category A, B or C, a designer is permitted to use a response modification coefficient (*R*) of 3 for design of the structure without the detailing required by Section 2211.4; however, if the designer prefers to use a response modification coefficient (*R*) greater than 3 in accordance with Section 1617.6, the requirements of Section 2211.4 must be used. The detail requirements of Section 2211.4 are mandatory for any structure classified as Seismic Design Category D, E or F.

2211.2 Type I shear walls. The design of Type I shear walls, of cold-formed steel light-framed construction, to resist wind and seismic loads, shall be in accordance with the requirements of this section.

1. The nominal shear value for Type I shear walls, as shown in Table 2211.2(1) for wind loads, Table 2211.2(2) for wind or seismic loads or Table 2211.2(3) for seismic loads, is permitted to establish allowable shear values or design shear values.
2. Boundary members, chords, collectors and connections thereto shall be proportioned to transmit the induced forces.
3. As an alternative to the values in Tables 2211.2(1), 2211.2(2) and 2211.2(3), shear values are permitted to be calculated by the principles of mechanics by using approved fastener values and shear values appropriate for the sheathing material attached.
4. Type I shear walls sheathed with wood structural or sheet steel panels are permitted to have window openings, between hold-down anchors at each end of a wall segment, where details are provided to account for force transfer around openings.
5. The aspect ratio limitations of Section 2211.2.2, Item 5, shall apply to the entire Type I segment and to each wall pier at the side of each opening.
6. The height of the wall pier (*h*) shall be defined as the clear height of the pier at the side of an opening.
7. The width of a pier (*w*) shall be defined as the sheathed width of the pier.
8. The width of wall piers shall not be less than 24 inches (102 mm).
9. Hold-down anchors shall be provided at each end of a Type I shear wall capable of resisting the design forces.

❖ These provisions are based upon monotonic and cyclic tests performed under the direction of Dr. R. Serrette at Santa Clara University in 1996 and 1997, which were sponsored by AISI. The provisions in the code are essentially the same as those in prior codes but have been updated to reflect the work done by Dr. Serrette in 1997 to include data on narrow plywood, OSB shear walls and the inclusion of shear values for sheet steel. The assemblies listed in Tables 2211.2(1) through 2211.2(3) are based upon tested assemblies. Limitations listed herein are intended to reflect the assemblies as tested. Designers wishing to extrapolate this data should use engineering judgement with caution.

TABLE 2211.2(1). See page 22-9.

❖ This table provides the nominal shear resistance to wind loads for use with the indicated sheathing materials. It is important to note that unlike Chapter 23 for wood, where the tabulated shear values are allowable values, the shear values in Chapter 22 are nominal values (sometimes referred to as "ultimate") and are not to be used directly in design. These values must be divided by a factor of safety (*W*) for ASD or multiplied by a strength reduction factor (*f*) for LRFD, in accordance with Section 2211.2.1. Presenting these as nominal values is consistent with the format of AISI-NASPEC and provides the user with a seamless transition between ASD and LRFD.

TABLE 2211.2(2). See page 22-9.

❖ This table provides the nominal shear resistance values for a shear wall with gypsum board facing on both sides. See the commentary to Table 2211.2(1) for further discussion regarding the application of this table.

TABLE 2211.2(3). See page 22-10.

❖ This table provides the nominal shear resistance to seismic forces for a shear wall with the indicated sheathing materials. See the commentary to Table 2211.2(1) for further discussion regarding the application of this table.

TABLE 2211.2(1)
NOMINAL SHEAR VALUES FOR WIND FORCES IN POUNDS PER FOOT FOR SHEAR WALLS FRAMED WITH COLD-FORMED STEEL STUDS[a]

ASSEMBLY DESCRIPTION	MAXIMUM HEIGHT/LENGTH RATIO *h/w*	FASTENER SPACING AT PANEL EDGES[b] (inches)				MAXIMUM FRAMING SPACING (inches o.c.)
		6	4	3	2	
$^{15}/_{32}$-inch structural 1 sheathing (4-ply) plywood one side	2:1	1,065[c]	—	—	—	24
$^{7}/_{16}$-inch rated sheathing (OSB), one side	2:1	910[c]	1,410	1,735	1,910	24
$^{7}/_{16}$-inch rated sheathing (OSB), one side, oriented perpendicular to framing	2:1	1,020[c]	—	—	—	24
$^{7}/_{16}$-inch rated sheathing (OSB), one side	4:1[d]	—	1,025	1,425	1,825	24
0.018-inch steel sheet, one side	2:1	485	—	—	—	24
0.027-inch steel sheet, one side	4:1[d]	—	1,000	—	—	24

For SI: 1 inch = 25.4 mm, 1 pound per foot = 14.5939 N/m.

a. Nominal shear values shall be multiplied by the resistance factor (ϕ) to determine design strength or divided by the safety factor (Ω) to determine allowable shear values as set forth in Section 2211.2.1.

b. Screws shall be attached to intermediate supports at 12 inches on center unless otherwise shown.

c. Where fully blocked gypsum board is applied to the opposite side of this assembly, in accordance with Table 2211.2(2) with screw spacing at 7 inches o.c. edge and 7 inches o.c. field, these nominal values are permitted to be increased by 30 percent.

d. Where aspect ratio (*h/w*) is greater than 2:1, the design shear shall be reduced as required by Section 2211.2.2, Item 5.

TABLE 2211.2(2)
NOMINAL SHEAR VALUES FOR WIND AND SEISMIC FORCES IN POUNDS PER FOOT FOR SHEAR WALLS FRAMED WITH COLD-FORMED STEEL STUDS AND FACED WITH GYPSUM BOARD[a,b]

WALL CONSTRUCTION	MAXIMUM HEIGHT/LENGTH RATIO *h/w*	ORIENTATION	SCREW SPACING (inches)		NOMINAL SHEAR VALUE (plf)
			Edge	Field	
$^{1}/_{2}$-inch gypsum board on both sides of wall; Studs maximum 24 inches o.c.	2:1	Gypsum board applied perpendicular to framing with strap blocking behind the horizontal joint and with solid blocking between the first two end studs	7	7	585
			4	4	850

For SI: 1 inch = 25.4 mm, 1 pound per foot = 14.5939 N/m.

a. Nominal shear values shall be multiplied by the resistance factor (ϕ) to determine design strength or divided by the safety factor (Ω) to determine allowable shear values as set forth in Section 2211.2.1.

b. Walls resisting seismic loads shall be subject to the limitations in Section 1617.6.

2211.2.1 Design shear determination. Where allowable stress design (ASD) is used, the allowable shear value shall be determined by dividing the nominal shear value, shown in Tables 2211.2(1), 2211.2(2) and 2211.2(3), by a factor of safety (Ω) of 2.5.

Where load and resistance factor design (LRFD) is used, the design shear value shall be determined by multiplying the nominal shear value, shown in Tables 2211.2(1), 2211.2(2) and 2211.2(3), by a resistance factor (ϕ) of 0.55.

❖ See the commentary to Table 2211.2(1).

2211.2.2 Limitations for systems. The lateral-resistant systems listed in Tables 2211.2(1), 2211.2(2) and 2211.2(3) shall conform to the following requirements:

1. Studs shall be a minimum $1^{5}/_{8}$ inches (41.3 mm) by $3^{1}/_{2}$ inches (89 mm) with a $^{3}/_{8}$-inch (9.5 mm) return lip. As a minimum, studs shall be doubled (back to back) at shear wall ends.

2. Track shall be a minimum $1^{1}/_{4}$ inches (31.8 mm) by $3^{1}/_{2}$ inches (89 mm).

3. Both studs and track shall have a minimum uncoated base metal thickness of 33 mils (0.84 mm) and shall be of the following grades of structural quality steel: ASTM A 653 SS Grade 33, ASTM A 792 SS Grade 33 or ASTM A 875 SS Grade 33.

4. Fasteners along the edges in shear panels shall be placed not less than $^{3}/_{8}$ inch (9.5 mm) in from panel edges.

5. The height-to-width shear wall aspect ratio (*h/w*) of wall systems shall not exceed the values in Tables 2211.2(1), 2211.2(2) and 2211.2(3). Where the limiting ratio of *h/w* is greater than 2:1, the shear values shall be multiplied by 2*w/h*.

6. Panel thicknesses shown are minimums. Panels less than 12 inches (305 mm) wide shall not be used. All panel edges shall be fully blocked.
7. Where horizontal strap blocking is used to provide edge blocking, it shall be a minimum $1^1/_2$ inches (38 mm) wide and of the same material and equal or greater thickness as the track and studs.
8. The design shear values for shear panels with different nominal shear values applied to the same side of a wall are not cumulative except as permitted in Tables 2211.2(1), 2211.2(2) and 2211.2(3). For walls with material applied to both faces of the same wall, the design shear value of material of the same capacity is cumulative. Where the material nominal shear values are not equal, the design shear value shall be either two times the design shear value of the material with the smaller values or shall be taken as the value of the stronger side, whichever is greater. Summing shear values of dissimilar material applied to opposite faces or to the same wall line is not allowed unless permitted by Table 2211.2(1).

❖ This section provides limitations that apply to all Type I shear walls used for wind or seismic resistance. The limitations herein are those that apply to all wall assemblies in Tables 2211.2(1) through 2211.2(3). This includes the minimum size and thickness of the studs and track used in shear wall systems. Larger and thicker studs and track are permitted for wind applications, but the thickness of framing members in seismic applications has additional limitations in Section 2211.4.5.

Generally, the maximum height-to-width aspect ratio permitted by Tables 2211.2(1) through 2211.2(3) is 2:1. Item 5 permits the aspect ratio of certain wall assemblies to exceed this limit, provided the reduction factor shown is applied to the tabulated nominal shear value. In Table 2211.2(1) for wind loads, Note d identifies two such assemblies. In Table 2211.2(3) for seismic loads, Note c identifies three such assemblies and further restricts this increased aspect ratio to Seismic Design Category A, B or C structures. The basis for allowing this increase stems from testing that demonstrated steel stud wall systems with an aspect ratio of 4:1 can, in fact, develop the strengths given in the shear wall tables. Since the stiffness of these shear walls is reduced, design level deflections increase accordingly; thus, applying the reduction factor to the nominal shear wall values is a means of controlling deflections.

The assemblies listed in Tables 2211.2(1) and 2211.2(3) all indicate sheathing is applied to one side only. Item 8 provides a rationale for determining shear values of double-sided shear walls. Testing indicates that this approach is conservative. It permits the summing of shear values of assemblies with different nominal (tabulated) shear values under certain circumstances.

2211.2.2.1 Sheet steel sheathing. Steel sheets, attached to cold-formed steel framing, are permitted to resist horizontal forces produced by wind or seismic loads.

1. Steel sheets shall have a minimum base metal thickness as shown in Table 2211.2(1) or 2211.2(3), and shall be of the following grades of structural quality steel: ASTM A653

TABLE 2211.2(3)
NOMINAL SHEAR VALUES FOR SEISMIC FORCES IN POUNDS PER FOOT FOR SHEAR WALLS FRAMED WITH COLD-FORMED STEEL STUDS[a]

ASSEMBLY DESCRIPTION	MAXIMUM HEIGHT/LENGTH RATIO *h/w*	FASTENER SPACING AT PANEL EDGES[b] (inches)				MAXIMUM FRAMING SPACING (inches o.c.)
		6	4	3	2	
$^{15}/_{32}$-inch Structural 1 Sheathing (4-ply) plywood one side	2:1[c]	780	990	1,465	1,625	24
$^{15}/_{32}$-inch Structural 1 Sheathing (4-ply) plywood one side; end studs 0.043 inch minimum thickness	2:1	—	—	1,775	2,190	24
$^{15}/_{32}$-inch Structural 1 Sheathing (4-ply) plywood one side; all studs and track 0.043 inch minimum thickness	2:1	890	1,330	1,775	2,190	24
$^{7}/_{16}$-inch OSB one side	2:1[c]	700	915	1,275	1,625	24
$^{7}/_{16}$-inch OSB one side end studs, 0.043 inch minimum thickness	2:1	—	—	1,520	2,060	24
0.018-inch minimum thickness steel sheet one side	2:1	390	—	—	—	24
0.027-inch minimum thickness steel sheet one side	2:1[c]	—	1,000	1,085	1,170	24

For SI: 1 inch = 25.4 mm, 1 pound per foot = 14.5939 N/m.

a. Nominal shear values shall be multiplied by the resistance factor (ϕ) to determine design strength or divided by the safety factor (Ω) to determine allowable shear values as set forth in Section 2211.2.1.

b. Screws shall be attached to intermediate supports at 12 inches o.c. unless otherwise shown.

c. In Seismic Design Category A, B and C the aspect ratio (h/w) is permitted to be 4:1 where the design shear is reduced as required by Section 2211.2.2, Item 5.

SS Grade 33, ASTM A792 SS Grade 33 or ASTM A 875 SS Grade 33.

2. Nominal shear values, used to establish the allowable shear value or design shear value, are given in Tables 2211.2(1) for wind loads and 2211.2(3) for seismic loads.
3. Steel sheets are permitted to be applied either parallel or perpendicular to framing. All edges of steel sheets shall be attached to framing members, strap blocking or shall be overlapped and attached to each other with screw spacing as required for edges.
4. Screws used to attach steel sheets shall be a minimum No. 8 modified truss head.

❖ Sheathing of steel sheets is allowed for both wind [see Table 2211.2(1)] and seismic [see Table 2211.2(3)] applications. This section clarifies how the tables are to be applied for steel sheet attached to the cold-formed steel wall framing. The orientation of the steel sheets does not affect the nominal values in the tables. This section specifies the screws that are to be used, and the spacing is to be as shown in Tables 2211.2(1) and 2211.2(3).

2211.2.2.2 Wood structural panel sheathing. Cold-formed steel framed wall systems, sheathed with wood structural panels, are permitted to resist horizontal forces produced by wind or seismic loads subject to the following:

1. Nominal shear values, used to establish the allowable shear value or design shear value, are given in Tables 2211.2(1), for wind loads, and 2211.2(3), for seismic loads.
2. Wood structural panels shall comply with DOC PS 1 or PS 2 and shall be manufactured using exterior glue.
3. Wood structural panels shall be attached to steel framing with flat-head self-drilling tapping screws with a minimum head diameter of 0.292 inch (8 mm).
4. Where $^{7}/_{16}$-inch oriented strand board (OSB) is specified, $^{15}/_{32}$-inch structural 1 sheathing (plywood) is permitted.
5. Structural panels are permitted to be applied either parallel or perpendicular to framing.
6. Increases of the nominal loads shown in Tables 2211.2(1) and 2211.2(3) shall not be permitted for duration of load as permitted in Chapter 23.

❖ This section provides detailed design and construction information for shear walls having wood structural panel sheathing applied to the cold-formed steel wall framing. The general limitations for wood-based sheathing products are given here. As noted, $^{15}/_{32}$-inch (12 mm) plywood may be substituted for $^{7}/_{16}$-inch (11 mm) OSB. It is important to note that the shear values for wood sheathing are nominal loads (ultimate) and are not permitted to be increased for duration of load. These values apply whether the wood panels are installed vertically or horizontally. This section also specifies the screws to be used in order to achieve the shear values with the spacing as indicated in the applicable table.

2211.2.2.3 Gypsum board panel sheathing. Cold-formed steel framed wall systems, sheathed with gypsum board, are permitted to resist horizontal forces produced by wind or seismic loads subject to the following:

1. Nominal shear values, used to establish the allowable shear value or design shear value, are given in Table 2211.2(2).
2. The shear values listed in Table 2211.2(2) shall not be cumulative with the shear values of other materials applied to the same wall unless otherwise permitted herein.
3. The nominal shear values shown are for gypsum board that is applied to both sides of the wall.
4. Where gypsum board is only applied to one side of the wall, the nominal shear values shall be taken as one-half of the value shown.
5. Where gypsum board is applied perpendicular to studs, end joints of adjacent courses of gypsum board sheets shall not occur over the same stud.
6. Screws used to attach gypsum board shall be a minimum No. 6 in accordance with ASTM C 954.
7. Walls resisting seismic loads shall be subject to the limitations in Section 1617.6.

❖ This section specifies how Table 2211.2(2) is to be used for gypsum board applied to cold-formed steel wall framing. The gypsum board panel sheathing assemblies tested were double sided; thus, the values in the table apply where the gypsum sheathing is placed on both sides of the wall. When applied only to one side of a shear wall, the nominal shear value is taken as one-half the tabulated values. The table requires that the gypsum board is placed with the long dimension as horizontal (perpendicular to the studs). The vertical joints in adjacent courses of gypsum board are to be staggered on different wall studs. Note that gypsum board can be utilized to increase the shear values of certain assemblies utilizing wood structural panel sheathing in accordance with Note c in Table 2211.2(1). Otherwise, the values are not cumulative with other materials.

2211.3 Type II shear walls. Type II shear walls sheathed with wood structural panels or sheet steel are permitted to resist wind and seismic loads when designed in accordance with this section. Type II walls shall meet the requirements for Type I walls except as revised by this section.

❖ The Type II shear wall provisions apply to wood structural panels and sheet steel sheathing. This section is modeled after the perforated shear wall requirements for wood-framed shear walls in the 2000 NEHRP Provisions. Testing has validated the applicability of this methodology to shear walls framed with cold-formed steel members. It is important to note that all the requirements and limitations for Type I shear walls are applicable to Type II shear walls unless this section modifies the requirement.

TABLE 2211.3
SHEAR RESISTANCE ADJUSTMENT FACTOR—C_o

WALL HEIGHT (*h*)	MAXIMUM OPENING HEIGHT RATIO[a] AND HEIGHT				
	h/3	h/2	2h/3	5h/6	h
8′0″	2′8″	4′0″	5′4″	6′8″	8′0″
10′0″	3′4″	5′0″	6′8″	8′4″	10′0″
Percent full-height sheathing[b]	Shear Resistance Adjustment Factor				
10%	1.00	0.69	0.53	0.43	0.36
20%	1.00	0.71	0.56	0.45	0.38
30%	1.00	0.74	0.59	0.49	0.42
40%	1.00	0.77	0.63	0.53	0.45
50%	1.00	0.80	0.67	0.57	0.50
60%	1.00	0.83	0.71	0.63	0.56
70%	1.00	0.87	0.77	0.69	0.63
80%	1.00	0.91	0.83	0.77	0.71
90%	1.00	0.95	0.91	0.87	0.83
100%	1.00	1.00	1.00	1.00	1.00

a. See Section 2211.3.2(2).
b. See Section 2211.3.2(1).

2211.3.1 Limitations. The following limitations shall apply to the use of Type II shear walls:

1. A Type II shear wall segment, meeting the minimum aspect ratio (*h/w*) of Section 2211.3.2, Item 3, shall be located at each end of a Type II shear wall. Openings shall be permitted to occur beyond the ends of the Type II shear wall; however, the width of such openings shall not be included in the width of the perforated shear wall.
2. In Seismic Design Categories B, C, D, E and F, the nominal shear values shall be based upon edge screw spacing not less than 4 inches o.c.
3. A Type II shear wall shall not have out-of-plane (horizontal) offsets. Where out-of-plane offsets occur, portions of the wall on each side of the offset shall be considered as separate perforated shear walls.
4. Collectors for shear transfer shall be provided through the full length of the Type II shear wall.
5. A Type II shear wall shall have uniform top of wall and bottom of wall elevations. Type II shear walls not having uniform elevations shall be designed by other methods.
6. Type II shear wall height, *h*, shall not exceed 20 feet (6096 mm).

❖ Item 2 restricts the nominal shear wall values to those that are based on screws spaced 4 inches (103 mm) on center or more. This rules out high-demand shear wall assemblies and is consistent with the 2000 NEHRP approach to wood-framed perforated shear walls.

2211.3.2 Type II shear wall resistance. The Type II shear wall resistance shall be equal to the adjusted shear resistance multiplied by the sum of the widths (ΣL_i) of the Type II shear wall segments and shall be calculated in accordance with the following:

1. The percent of full-height sheathing shall be calculated as the sum of widths (ΣL_i) of Type II shear wall segments divided by the total width of the Type II shear wall including openings.
2. The maximum opening height ratio shall be calculated by dividing the maximum opening clear height by the shear wall height, *h*.
3. The unadjusted shear resistance shall be the design shear values calculated in accordance with Section 2211.2.1 based upon the values in Tables 2211.2(1) and 2211.2(3). The aspect ratio of all Type II shear wall segments used in calculations shall not exceed 2:1.

   Exception: Where permitted by Tables 2211.2.1(1) and 2211.2(3), the aspect ratio (*h/w*) of Type II wall segments greater than 2:1, but in no case greater than 4:1, is permitted to be included in the calculation of the unadjusted shear resistance for the wall, provided the values are multiplied by 2*w/h*.

4. The adjusted shear resistance shall be calculated by multiplying the unadjusted shear resistance by the shear resistance adjustment factors of Table 2211.3. For intermediate percentages of full-height sheathing, the values are permitted to be determined by interpolation.

❖ This section describes the determination of Type II shear wall resistance for a given configuration. First, the unadjusted shear resistance is determined for seismic and wind loading from the applicable table as described in Item 3. For the wall under consideration determine the percentage of full-height sheathing (see Item 1) and the maximum opening height ratio (see Item 2). Based on these two parameters, determine the shear resistance adjustment factor from Table 2211.3 and then calculate the adjusted shear resistance according to item 4. The product of the adjusted shear resistance and the total widths of Type II shear wall segment in the wall under consideration is the shear resistance.

2211.3.3 Anchorage and load path. Design of Type II shear wall anchorage and load path shall conform to the requirements of this section, or shall be calculated using principles of mechanics.

❖ This section prescribes minimum design forces for anchorage and compression chords in Type II shear walls. The intent is to conservatively address the nonuniform shear distribution occurring in Type II shear walls. Alternative designs based on principles of mechanics are permitted.

2211.3.3.1 Anchorage for in-plane shear. The unit shear force ,*v*, transmitted into the top and out of the base of the Type II shear wall full-height sheathing segments, and into collectors (drag struts) connecting shear wall segments, shall be calculated in accordance with the following:

$$v = \frac{V}{C_o \Sigma L_i} \qquad \textbf{(Equation 22-1)}$$

where:

v = Unit shear force, plf (kN/m).

V = Shear force in Type II shear wall, lbs (kN).

C_o = Shear resistance adjustment factor from Table 2211.3.

ΣL_i = Sum of widths of Type II shear wall segments, feet (mm/1,000).

❖ This section prescribes the unit shear force, *v*, to be applied to the top and bottom of each Type II shear wall segment. It represents the shear when full overturning restraint is provided, making it a conservative approximation of the nonuniform shear distribution. Designing for this unit shear, as well as the uniform uplift force of Section 2211.3.3.3, both of which are based on the unadjusted shear capacity provides a sheathing-to-bottom plate attachment such that the capacity of the shear wall will not be governed by this connection.

2211.3.3.2 Uplift anchorage at Type II shear wall ends. Anchorage for uplift forces due to overturning shall be provided at each end of the Type II shear wall. Where seismic loads govern, the uplift anchorage shall be determined in accordance with the requirements of Section 2211.4.3.

❖ The uplift force due to overturning should be computed using Equation 22-2 for the compression chord force.

2211.3.3.3. Uplift anchorage between Type II shear wall ends. In addition to the requirements of Section 2211.3.3.1, Type II shear wall bottom plates at full-height sheathing shall be anchored for a uniform uplift force, *t*, equal to the unit shear force, *v*, determined in Section 2211.3.3.1.

❖ The unit uplift force, *t*, is equal to the unit shear force, *v* (see commentary, Section 2211.3.3.1).

2211.3.3.4. Compression chords. Vertical elements at each end of each Type II shear wall segment shall be designed for a compression force, *C*, from each story calculated in accordance with the following:

$$C = \frac{Vh}{C_o \Sigma L_i} \qquad \textbf{(Equation 22-2)}$$

where:

C = Compression chord uplift force, lbs (kN).

V = Shear force in Type II shear wall, lbs (kN).

h = Shear wall height feet, (mm/1,000).

C_o = Shear resistance adjustment factor from Table 2211.3.

ΣL_i = Sum of widths of Type II shear wall segments, feet (mm/1,000).

❖ The vertical members on either side of all Type II shear wall segments must be designed for the required compression chord force.

2211.3.3.5. Load path. A load path to the foundation shall be provided for the uplift shear and compression forces as determined from Sections 2211.3.3.1 through 2211.3.3.4, inclusive. Elements resisting shear wall forces contributed by multiple stories shall be designed for the sum of forces contributed by each story.

❖ A load path for the forces prescribed in this section is necessary. Consideration must be given to the shear wall forces from a story, or stories, above in multistory buildings.

2211.4 Seismic Design Categories D, E and F.

2211.4.1 General. In addition to the requirements of Sections 2211.2 and 2211.3, light-framed cold-formed steel wall systems, that resist seismic loads, in buildings assigned to Seismic Design Category D, E or F, shall comply with the requirements of this section.

❖ In addition to the limitations in Sections 2211.2 and 2211.3, structures classified as Seismic Design Category D and above are subject to the limitations in this section.

2211.4.2 Connections. Connections for diagonal bracing members, top chord splices, boundary members and collectors shall be designed to develop the lesser of the nominal tensile strength of the member or the design seismic force multiplied by the seismic overstrength factor, Ω_o, from Section 1617.6. The pull-out resistance of screws shall not be used to resist design seismic forces.

❖ Connections for critical members need to be designed to develop the strength of the connected members or they can be designed for the seismic forces that may be expected to be delivered by applying the system overstrength factor. Since screws exhibit a brittle failure mode in pullout, they are not permitted to resist seismic forces in direct tension.

2211.4.3 Anchorage of braced wall segments. Studs or other vertical boundary members at the ends of wall segments, that resist seismic loads, braced with either sheathing or diagonal braces, shall be anchored such that the bottom track is not required to resist uplift by bending of the track web. Both flanges of the studs shall be braced to prevent lateral torsional buckling. Studs or other vertical boundary members and anchorage thereto shall have the nominal strength to resist design seismic force multiplied by the seismic overstrength factor, Ω_o, from Section 1617.6.

❖ Since the bottom web of the bottom track in steel framing has limited bending resistance, hold-down anchors are required to be attached to the wall to avoid bending of the track web. This is typically achieved by attaching the hold-down anchor directly to the end studs in a braced bay. Additionally, these end studs need to have the capacity (at ultimate strength) to assure that the forces delivered by the sheathing or strapping do not exceed its capacity in tension or compression.

2211.4.4 Sheet steel sheathing. Where steel sheathing provides lateral resistance, the design and construction of such walls shall be in accordance with the additional requirements of this section. Perimeter members at openings shall be provided and shall be detailed to distribute the shearing stresses. Wall studs and track shall have a minimum uncoated base metal thickness of 33 mils (0.84 mm) and shall not have an uncoated base metal thickness greater than 48 mils (1.10 mm). The nominal shear value for light-framed wall systems for buildings in Seismic Design Category D, E or F shall be based upon values from Table 2211.2(3).

❖ The code limits the framing members in these systems to thicknesses of 33 mils (0.84 mm) and 43 mils (1.10 mm) (commonly known as 20 and 18 gage). Research has shown that when attaching sheathing to 54 mils (1.4 mm) studs (16 gage) with No. 8 screws, the maximum load is governed by the shear failure of the screws, which should be avoided. If a designer wishes to develop shear values for sheathed walls for thicker studs, using No. 10 screws prevents this screw failure mode. It is expected that future code changes will provide clarification. For the time being, the designer should use sound engineering judgement.

2211.4.5 Wood structural panel sheathing. Where wood structural panels provide lateral resistance, the design and construction of such walls shall be in accordance with the additional requirements of this section. Perimeter members at openings shall be provided and shall be detailed to distribute the shearing stresses. Wood sheathing shall not be used to splice these members. Wall studs and track shall have a minimum uncoated base metal thickness of 33 mils (0.84 mm) and shall not have an uncoated base metal thickness greater than 48 mils (1.10 mm). The nominal shear value for light-framed wall systems for buildings in Seismic Design Category D, E or F shall be based upon values from Table 2211.2(3).

❖ See the commentary to Section 2211.4.4.

2211.4.6 Diagonal bracing. Where diagonal bracing is provided for lateral resistance, provisions shall be made for pretensioning or other methods of installing tension-only bracing shall be used to guard against loose diagonal straps. The *l/r* of the brace is permitted to exceed 200.

❖ Where diagonal straps are used, the contractor needs to take precautions to ensure that the straps are not in a slack condition. When straps are installed with too much slack, their effectiveness is greatly reduced.

2211.4.7 Gypsum board panel sheathing. Gypsum board panel sheathing is permitted to resist seismic loads, subject to the limitations in Table 2211.2(2) and Section 1617.6.

❖ This section reiterates that Table 2211.2(2) is the table applicable to gypsum board sheathing and cross references the basic seismic force limitations found in Section 1617.6.

BIBLIOGRAPHY

The following resource material is referenced in this chapter or is relevant to the subject matter addressed in this chapter.

AISI-96, *Specification for the Design of Cold-Formed Steel Structural Members, including 1999 Supplement*. Washington, DC: American Iron and Steel Institute, 1996.

AISC 335-89, *Specification for Structural Steel Buildings—Allowable Stress Design and Plastic Design, including Supplement No. 1*, dated 2001. Chicago, IL: American Institute of Steel Construction, 1989.

AISC 341-02, *Seismic Provisions for Structural Steel Buildings*. Chicago, IL: American Institute of Steel Construction, 2002.

AISC HSS-00, *Specification for the Design of Steel Hollow Structural Sections*. Chicago, IL: American Institute of Steel Construction, 2000.

AISC LRFD-99, *Load and Resistance Factor Design Specification for Structural Steel Buildings*. Chicago, IL: American Institute of Steel Construction, 1999.

AISI *General-01, Standard for Cold-Formed Steel Framing-General Provisions*. Washington, DC: American Iron and Steel Institute, 2001.

AISI *Header-01, Standard for Cold-Formed Steel Framing-Header Design*. Washington, DC: American Iron and Steel Institute, 2001.

AISI NASPEC-01, *North American Specification for the Design of Cold-Formed Steel Structural Members*. Washington, DC: American Iron and Steel Institute, 2001.

AISI Truss-01, *Standard for Cold-Formed Steel Framing-Truss Design*. Washington, DC: American Iron and Steel Institute, 2001.

ASCE 3-91, *Standard for the Construction and Inspection of Composite Slabs*. Reston, VA: American Society of Civil Engineers, 1991.

ASCE 7-02, *Minimum Design Loads for Buildings and Other Structures*. New York: American Society of Civil Engineers, 2002.

ASCE 8-90, *Standard Specification for the Design of Cold-Formed Stainless Steel Structural Members*. Reston, VA: American Society of Civil Engineers, 1990.

ASCE 19-96, *Structural Applications of Steel Cables for Buildings*. Reston, VA: American Society of Civil Engineers, 1996.

AWS D1.1-00, *Structural Welding Code—Steel*. Miami, FL: American Welding Society, 2000.

FEMA 368, *NEHRP Recommended Provisions for Seismic Regulations for New Buildings and Other Structures*. Washington, DC: Federal Emergency Management Agency, 2001.

RMI-97, *Specification for the Design, Testing and Utilization of Industrial Steel Storage Racks*. Charlotte, NC: Rack Manufacturers Institute, 1997.

Chapter 23: Wood

General Comments

This chapter contains information required to design and construct buildings or structures that include wood or wood-based structural elements, and is organized around the application of three design methodologies: allowable stress design (ASD), load and resistance factor design (LRFD) and conventional construction. Included are references to design and manufacturing standards for various wood and wood-based products; general construction requirements; design criteria for lateral-force-resisting systems and specific requirements for the application of the three design methods (ASD, LRFD and conventional construction). Chapter 23 includes elements of all three previous regional model codes—the *BOCA® National Building Code* (BNBC), the *Standard Building Code* (SBC) and the *Uniform Building Code™* (UBC™). It most closely follows the format of the UBC.

Acceptable standards for the manufacture of wood or wood-based products include provisions for sizes, grades (labels), quality control and certification programs, or similar methods of identification. Specific requirements and tables have been developed using referenced design methods to provide a minimum level of safety. This chapter also contains requirements both for the use of products in conjunction with wood and wood-based structural elements, and for prevention of decay.

In general, only Type 3, 4, and 5 buildings may be constructed of wood. Accordingly, Chapter 23 is referenced when the combination of the occupancy (determined in Chapter 3) and the height and area of the building or structure (determined in Chapter 5) indicate that the construction (specified in Chapter 6) can be Type 3, 4 or 5. Another basis for referencing Chapter 23 is when wood elements are used in Type 1 or 2 structures as permitted in Section 603. This chapter gives information on the application of fire-retardant-treated wood (FRTW), interior wood elements and trim in these structures. All structural criteria for application of referenced standards and procedures included in Chapter 23 are based on the loading requirements of Chapter 16 or on historical performance.

Chapter 23 is not a textbook on construction. It is assumed that the reader has both the training and experience needed to understand the principles and practices of wood design and construction. Without such understanding, some sections may be misunderstood and misapplied. This commentary should help to promote better understanding of the structure and application of the methods specified in Chapter 23.

Section 2301 identifies three methods of design, and compliance with one or more is required.

Section 2302 contains the most commonly used terms to describe wood and wood-based products. Some engineering terms are also included.

Section 2303 provides reference to manufacturing standards, necessary specification criteria and use and application provisions.

Section 2304 contains provisions for the proper design and construction of all wood structures and the use of all wood products. This section also includes the fastening schedule.

Section 2305 contains general design requirements for lateral-force-resisting systems. Whether the structure is engineered using ASD or LRFD, the provisions of this section apply.

Section 2306 contains provisions for the design of structures using ASD. Historically, all of the industry publications have been developed for ASD. More recently, LRFD has been introduced, making it necessary to distinguish clearly which provisions are appropriate for ASD or LRFD. The provisions of this section are not appropriate for LRFD.

Section 2307 is simply a reference to the consensus standard, which governs the design of structures using the LRFD methodology.

Although fairly limited in application, Section 2308 contains the provisions for conventional construction that may be used to construct certain wood-frame structures. Limitations on the use of this section are provided in Section 2308.2 for a quick determination.

Purpose

This chapter provides minimum guidance for the design of buildings and structures that use wood and wood-based products in their framing and fabrication. Alternative methods and materials can be used where justified by engineering analysis and testing. In all cases, the provisions of Section 2304 apply to all elements of wood-frame construction.

SECTION 2301 GENERAL

2301.1 Scope. The provisions of this chapter shall govern the materials, design, construction and quality of wood members and their fasteners.

❖ Section 2301 includes specifications for use of and standards for production of wood and wood-based products such as boards, dimensional lumber and engineered wood products, such as I-joists, glued-laminated timber, structural panels, trusses, particleboard, fiberboard and hardboard. Also included are criteria and specifications for the use of other materials such as connectors used in conjunction with wood or wood-based products. Other chapters of the code also affect the use of wood materials in buildings and should be referenced prior to making final decisions on the use of any product.

The scope of this chapter is established in Section 2301.1, and broadly encompasses wood products and the limitations placed on them and their various applications within the code.

2301.2 General design requirements. The design of structural elements or systems, constructed partially or wholly of wood or wood-based products, shall be based on one of the following methods.

❖ This chapter includes three methods of designing with wood or wood-based products. This section limits designs to one of these three methods unless an alternate method has been proven to be acceptable as permitted in Section 104.11. It is not uncommon for only one element of a structure to require engineered design. This section recognizes "partial" design.

2301.2.1 Allowable stress design (ASD). Design using allowable stress design methods shall resist the applicable load combinations of Chapter 16 in accordance with the provisions of Sections 2304, 2305 and 2306.

❖ ASD is the traditional method of engineering wood structures. This section prescribes the use of the load combinations specified in Chapter 16. Additionally, it requires the use of the general requirements in Section 2304, lateral resistance requirements in Section 2305 and ASD requirements in Section 2306.

2301.2.2 Load and resistance factor design (LRFD). Design using load and resistance factor design (LRFD) methods shall resist the applicable load combinations of Chapter 16 in accordance with the provisions of Sections 2304, 2305 and 2307.

❖ LRFD is a relatively new method of engineering wood structures. This section also requires the use of the load combinations in Chapter 16, general requirements in Section 2304 and lateral resistance requirements in Section 2305. The requirements for design using the LRFD method are found in Section 2307.

2301.2.3 Conventional light-frame wood construction. The design and construction of conventional light-frame wood construction shall be in accordance with the provisions of Sections 2304 and 2308.

> **Exception:** Buildings designed in accordance with the provisions of the *AF&PA Wood Frame Construction Manual for One- and Two-Family Dwellings* shall be deemed to meet the requirements of the provisions of Section 2308.

❖ The term "conventional construction" in this case means the design and construction of buildings using typical configurations and methods, which do not require the calculation of loads or analysis by a design professional. The descriptive and prescriptive provisions of the conventional light-frame construction sections are based on commonly accepted engineering practice and experience. The use of these provisions is limited to buildings of relatively small volume that do not incorporate unusual configurations, elements or loadings. Conventional light-frame wood construction is not really new, but may appear to be unique to some code users. This section references the general criteria in Section 2304, which is applicable to all wood construction. Section 2308 contains prescriptive limits and methods applicable to wood construction that has not been designed using the methodologies of Section 2301.2.1 or 2301.2.2. The American Forest and Paper Association (AF&PA) *Wood Frame Construction Manual For One- and Two-Family Dwellings* is also permitted as an alternative to the conventional construction provisions of the code.

2301.3 Nominal sizes. For the purposes of this chapter, where dimensions of lumber are specified, they shall be deemed to be nominal dimensions unless specifically designated as actual dimensions (see Section 2304.2).

❖ The use of nominal sizes for lumber is part of the traditional nomenclature for grading and identification of manufactured pieces. "Nominal" simply refers to the short-hand term such as "2 by 4" when the actual piece of lumber has a real dimension of 1 $^1/_2$ inches by 3 $^1/_2$ inches (38 mm by 89 mm)—not 2 inches by 4 inches (51 mm by 102 mm). Section 2304.2, however, prescribes that in determining the required size for design purposes, computations must be based on the actual size, rather than the nominal size, of the lumber.

SECTION 2302 DEFINITIONS

2302.1 Definitions. The following words and terms shall, for the purposes of this chapter, have the meanings shown herein.

❖ Definitions are included in Chapter 23 for terms common to wood products that are not found elsewhere in the code. It is important to emphasize that these terms are not exclusively related to this chapter, but are applicable everywhere the term is used in the code.

Definitions of terms can help in understanding the application of the code requirements. The purpose for including these definitions within this chapter is to provide more convenient access to them without having to refer back to Chapter 2. For convenience, these terms are also listed in Chapter 2 with a cross reference to this section. The use and application of all defined terms, including those defined herein, are set forth in Section 202.

ACCREDITATION BODY. An approved, third-party organization that is independent of the grading and inspection agencies, and the lumber mills, and that initially accredits and subsequently monitors, on a continuing basis, the competency and performance of a grading or inspection agency related to carrying out specific tasks.

❖ The process of determining the grade of lumber not only includes the actual method of applying the grade stamp to the product, but also the certification of the grading agency, and its methods of quality control, its rules and the work of its agents. For example, an independent third-party quality control program meeting the accreditation requirements of the American Lumber Standards Committee, Inc. (ALSC) or an equivalent process is required for all softwood lumber that is to be graded for use in construction in the United States.

ADJUSTED SHEAR RESISTANCE. The unadjusted shear resistance multiplied by the shear resistance adjustment factors of Table 2305.3.7.2.

❖ Perforated shear walls require the application of empirically determined shear resistance adjustment factors as described in Section 2305.3.7.2.2, Item 3. The term "adjusted shear resistance" is used to refer to the resulting shear resistance.

BRACED WALL LINE. A series of braced wall panels in a single story that meets the requirements of Section 2308.3 or 2308.12.4.

❖ A braced wall line consists of conventionally framed shear walls that provide lateral force resistance for buildings constructed in accordance with Section 2308.

BRACED WALL PANEL. A section of wall braced in accordance with Section 2308.9.3 or 2308.12.4.

❖ A braced wall panel is a segment of a braced wall line. Braced wall panels are discussed in Section 2308.9.3.

COLLECTOR. A horizontal diaphragm element parallel and in line with the applied force that collects and transfers diaphragm shear forces to the vertical elements of the lateral-force-resisting system and/or distributes forces within the diaphragm.

❖ The collector is a portion of the diaphragm that transfers the load from the point of application into the adjoining elements, usually the braced wall lines.

CONVENTIONAL LIGHT-FRAME WOOD CONSTRUCTION. A type of construction whose primary structural elements are formed by a system of repetitive wood-framing members. See Section 2308 for conventional light-frame wood construction provisions.

❖ Conventional light-frame wood construction is one of the three methods for designing wood buildings or structures that is recognized in this chapter. Section 2308 provides limits for use of prescriptive provisions and specifications for construction when this method is chosen.

CRIPPLE WALL. A framed stud wall extending from the top of the foundation to the underside of floor framing for the lowest occupied floor level.

❖ Cripple walls are built on the top of footings or foundation walls. They can typically be found along the top of stepped foundation walls where the grade adjoining the structure changes height. Cripple walls must be properly braced to resist lateral forces. They are often treated the same as a first-story wall. Provisions for the bracing of cripple walls are in Section 2308.9.4.

DIAPHRAGM, UNBLOCKED. A diaphragm that has edge nailing at supporting members only. Blocking between supporting structural members at panel edges is not included. Diaphragm panels are field nailed to supporting members.

❖ Unblocked diaphragms can resist low and moderate shear forces. Diaphragm sheathing may be applied with the long dimension of the sheathing either perpendicular or parallel to the main framing members. When the edge of the sheathing is not supported by the main framing member, the diaphragm is considered to be unblocked. Various tables in this chapter provide different values for structural sheathing when it is blocked and when it is not blocked.

DRAG STRUT. See "Collector."

❖ A drag strut is a type of collector for lateral loads when the diaphragm is not continuous. This can be found when there are openings in a floor, or the floor is not all on one level.

FIBERBOARD. A fibrous, homogeneous panel made from lignocellulosic fibers (usually wood or cane) and having a density of less than 31 pounds per cubic foot (pcf) (497 kg/m^3) but more than 10 pcf (160 kg/m^3).

❖ Fiberboard is used primarily as an insulating board and for decorative purposes, but may also be used as wall or roof sheathing. The cellulosic components of fiberboard are broken down to individual fibers and molded to create the bond between the fibers. Other ingredients may be added during processing to provide or improve certain properties such as strength and water resistance, in addition to surface finishes for decorative products. Fiberboard is used in all locations where panels are desirable, including wall sheathing, insulation of walls and roofs, roof decking, doors and interior finish.

GLUED BUILT-UP MEMBER. A structural element, the section of which is composed of built-up lumber, wood structural panels or wood structural panels in combination with lumber, all parts bonded together with structural adhesives.

❖ Typically these built-up beam and column sections consist of one or more webs with glued-lumber flanges and stiffeners. They are custom-designed structural elements that require the application of ASD or LRFD standards.

GRADE (LUMBER). The classification of lumber in regard to strength and utility in accordance with American Softwood Lumber Standard DOC PS 20 and the grading rules of an approved lumber rules-writing agency.

❖ The grade identifies how capable a particular piece of lumber is to resist applied loads. The mark is applied to solid sawn pieces of wood and includes the species, grade and whether it was finished (surfaced) green or dry. The species and grade designation on the grade mark, when used in conjunction with the design value in the *National Design Specifications (NDS®) Supplement* or the LRFD Manual, provides all the information necessary to determine what load the piece of lumber is capable of holding. The ALSC provides facsimile sheets for agencies accredited to grade lumber. Although rare, fake grade stamps can be found in the marketplace.

HARDBOARD. A fibrous-felted, homogeneous panel made from lignocellulosic fibers consolidated under heat and pressure in a hot press to a density not less than 31 pcf (497 kg/m^3).

❖ Hardboard is used for various interior applications, as well as siding applications. Other ingredients may be added during processing to provide or improve properties such as strength, water resistance and general utility.

NAILING, BOUNDARY. A special nailing pattern required by design at the boundaries of diaphragms.

❖ When designing a diaphragm or a shear wall, the sheathing for those elements must be attached to provide the design resistance to the applied load. Shear capacity tables for wood structural panels specify the nail size and spacing to be used along the edge and in the field of each panel. Boundary nailing is the required nailing pattern for panels located along the edges of diaphragms where the stresses are typically highest.

NAILING, EDGE. A special nailing pattern required by design at the edges of each panel within the assembly of a diaphragm or shear wall.

❖ When designing a diaphragm or shear panel, the sheathing for those elements must be attached in a fashion that will provide design resistance to the applied load. Shear capacity tables for wood structural panels used as sheathing specify nail size and spacing. For example, Table 2306.3.1 requires 6 penny (6d) nails to penetrate 1 $^1/_4$ inches (32 mm) when a $^5/_{16}$-inch (7.9 mm) panel is used on 2-inch-wide (51 mm) framing. To use a shear capacity of 250 pounds per linear feet (plf) (3648 N/m), the nail spacing at boundary conditions where the diaphragm is blocked is recommended to be 4 inches (102 mm) on center (o.c.).

NAILING, FIELD. Nailing required between the sheathing panels and framing members at locations other than boundary nailing and edge nailing.

❖ When designing a diaphragm or shear panel, the sheathing for those elements must be attached in a fashion that will provide design resistance to the applied load. Shear capacity tables for wood structural panels used as sheathing specify nail size and spacing. For example, Note b in Table 2306.3.1 indicates that the maximum spacing for nails along intermediate framing members must be 12 inches (305 mm) o.c.

NATURALLY DURABLE WOOD. The heartwood of the following species with the exception that an occasional piece with corner sapwood is permitted if 90 percent or more of the width of each side on which it occurs is heartwood.

❖ Because of their natural ability to resist deterioration, the harder portions of some species of wood are considered to be naturally durable. The code specifies that "occasional" sapwood is permitted if heartwood constitutes 90 percent of each side.

Decay resistant. Redwood, cedar, black locust and black walnut.

❖ Redwood, cedar, black locust and black walnut lumber are known to resist deterioration due to the action of microbes that enter the wood fibers. The code defines these species of lumber as being decay resistant.

Termite resistant. Redwood and Eastern red cedar.

❖ Redwood and eastern red cedar are considered to be resistant to infestation by termites and are thus listed as naturally durable. The Formosan Termite, however, is capable of destroying all naturally durable species of wood.

NOMINAL SIZE (LUMBER). The commercial size designation of width and depth, in standard sawn lumber and glued-laminated lumber grades; somewhat larger than the standard net size of dressed lumber, in accordance with DOC PS 20 for sawn lumber and with the *National Design Specification for Wood Construction* (NDS) for glued-laminated lumber.

❖ Unless specifically required by the code, all dimensions listed in the code for wood member sizes are nominal, not actual. DOC PS 20-99 specifies the required minimum dimension of lumber for each stated nominal size. The process used to smooth and finish the surface of the lumber and to dry the wood removes a certain thickness of wood on each side of the piece; therefore, a 2 by 4 is approximately 1 $^1/_2$ inches by 3 $^1/_2$ inches (38 mm by 89 mm). The actual dimension of the lumber, otherwise known as dressed size, is limited to ensure consistency in lumber sizes, which is essential to the satisfactory construction of a wood-frame building.

PARTICLEBOARD. A generic term for a panel primarily composed of cellulosic materials (usually wood), generally in the form of discrete pieces or particles, as distinguished from fibers. The cellulosic material is combined with synthetic resin or other suitable bonding system by a process in which the interparticle bond is created by the bonding system under heat and pressure.

❖ Particleboard is one of the family of wood-based products that can be used as wood panels. The definition describes the characteristics unique to particleboard as compared to other types of wood-based composite panels. Particleboard intended for use as wood structural panels is subject to the requirements of Section 2303.1.7. Particleboard intended for use in shear panels is regulated by the requirements of Section 2306.4.3.

PERFORATED SHEAR WALL. A wood structural panel sheathed wall with openings, that has not been specifically designed and detailed for force transfer around openings.

❖ This term is used to describe shear walls designed using the provisions of Section 2305.3.7.2 as an alternative to segmented shear wall design methodology (see commentary, Section 2305.3.7.2).

PERFORATED SHEAR WALL SEGMENT. A section of shear wall with full-height sheathing that meets the aspect ratio limits of Section 2305.3.3.

❖ This term refers to a fully sheathed portion of a perforated shear wall that also meets the aspect ratio limit. Only these portions of the perforated shear wall are considered in determining the percentage of full-height sheathing as described in Section 2305.3.7.2.2, Item 1.

PRESERVATIVE-TREATED WOOD. Wood (including plywood) pressure treated with preservatives in accordance with Section 2303.1.8.

❖ Wood that is exposed to high levels of moisture or heat is susceptible to decay both from fungus and other organisms and insect attack. The damage caused by decay or insects can jeopardize the performance of wood members so as to reduce the performance below that required by the code. Section 2304.11 identifies the locations where the use of preservative-treated wood is required. It is important to note that preservative treatment requires a pressure-treatment process. Painting, coating or other surface treatment does not produce preservative-treated wood that will perform as required. The treatment process uses pressure to achieve the depth of penetration of preservative into the wood that is needed to verify that the wood will be resistant to decay and insects over time. Surface treatments may be washed away by rain or ground water, or may chip or peel. The American Wood Preservers' Association (AWPA) is the consensus standards-writing organization for treated wood.

REFERENCE RESISTANCE (*D*). The resistance (force or moment as appropriate) of a member or connection computed at the reference end use conditions.

❖ The LRFD method for the design of diaphragms utilizes *D* and *D′* as the reference and adjusted diaphragm shear resistance per unit length. Table 2.6-1 in AF&PA/ASCE 16 identifies all the adjustments to be used for LRFD based on the material and service conditions.

SHEAR WALL. A wall designed to resist lateral forces parallel to the plane of a wall.

❖ Shear walls are the transfer element for the lateral forces that are applied to a floor or roof diaphragm. They function structurally to provide a path for these loads to the foundation. Typically in wood framing these walls are framed with studs and have a structural wood panel attached to the outside using specific nailing patterns to provide the necessary stiffness. They also may have a tie-down attached at both ends of the wall to prevent overturning of the wall.

STRUCTURAL GLUED-LAMINATED TIMBER. Any member comprising an assembly of laminations of lumber in which the grain of all laminations is approximately parallel longitudinally, in which the laminations are bonded with adhesives.

❖ Structural wood members can be created by laminating smaller pieces of lumber together in a specific way using glues to create high-performing structural members. Typically, these 2-inch-wide (51 mm) members are arranged in a specific order so that their strength characteristics are used most efficiently. Although these members are made of smaller pieces of lumber, they act together to create sizes that are included in the requirements for heavy timber framing members. Section 2303.1.3 specifies the method of manufacturing these members, and the means of identifying their performance capabilities.

SUBDIAPHRAGM. A portion of a larger wood diaphragm designed to anchor and transfer local forces to primary diaphragm struts and the main diaphragm.

❖ Subdiaphragms can be used by the designer to distribute large diaphragm forces into walls. They are especially common in buildings that incorporate concrete and masonry walls, and are subject to seismic forces.

TIE-DOWN (HOLD-DOWN). A device used to resist uplift of the chords of shear walls.

❖ A shear wall collects lateral forces, which are typically applied at the top of and parallel to the wall. This load will rack the wall and attempt to turn it over. If the strength of the sheathing and framing is correct for the load and appropriately attached, the sheathing will resist the tendency to rack the wall. To prevent the wall from overturning, it must be adequately anchored to the structure below. Anchor bolts are not typically designed to resist these loads. Hold-downs are often necessary at the ends of the shear wall designed to resist high lateral loads due to wind or seismic events. Section 2305.3.6 specifies the conditions where hold-downs are required.

TREATED WOOD. Wood impregnated under pressure with compounds that reduce its susceptibility to flame spread or to deterioration caused by fungi, insects or marine borers.

❖ Wood is treated to reduce its ability to propagate flame or resist damage caused by fungus or insects. Several AWPA standards are prescribed in Section 2303.1.8. These include water-borne as well as several oil-borne standards for treatment, testing and quality control of various wood and wood-based products. The ALSC also accredits third-party inspection agencies in a manner similar to their grading rules certification, for the quality control of preservative-treated wood products. Facsimiles of accredited agencies' quality marks for treatment are available from ALSC.

UNADJUSTED SHEAR RESISTANCE. The allowable shear set forth in Table 2306.4.1 where the aspect ratio of any perforated shear wall segment used in calculation of perforated shear wall resistance does not exceed 2:1. Where the aspect ratio of any perforated shear wall segment used in calculation of perforated shear wall resistance is greater than 2:1, but not exceeding 3.5:1, the unadjusted shear resistance shall be the allowable shear set forth in Table 2306.4.1, multiplied by $2w/h$.

❖ Perforated shear walls designed in accordance with Section 2305.3.7.2 use design shear values determined in accordance with this definition. The term "unadjusted shear resistance" is used to refer to the resulting design shear value.

WOOD SHEAR PANEL. A wood floor, roof or wall component sheathed to act as a shear wall or diaphragm.

❖ Wood shear panels are horizontal (diaphragms) or vertical (wood shear walls) panels, which transmit forces in the plane of the sheathing caused by lateral loads transmitted from one component of the structure to another. Wood shear panels usually consist of wood framing (studs or joists) either singly or doubly sheathed. The code gives design provisions for wood shear panels that are sheathed in 4-foot by 8-foot (1219 mm by 2438 mm) wood structural panels.

WOOD STRUCTURAL PANEL. A panel manufactured from veneers, or wood strands or wafers, or a combination of veneer and wood strands or wafers, bonded together with waterproof synthetic resins or other suitable bonding systems. Examples of wood structural panels are:

❖ The code defines wood structural panels in terms of the materials commonly used in their manufacture. Wood structural panels intended for structural use must comply with DOC PS1 or PS2 in accordance with Section 2303.1.4. Both of these documents specify the required structural performance characteristics of wood structural panels, and detail the requirements for the third-party label each panel is to bear.

Composite panels. A structural panel that is made of layers of veneer and wood-based material;

❖ Composite panels are structural panels that are made up of a combination of wood veneers and wood-based materials. The wood veneer usually forms the two outer layers and may also be used in the core of the panel. These layers of veneer and wood flakes or strands, called "furnish," may be cross laminated.

Oriented strand board (OSB). A wood structural panel that is a mat-formed product composed of thin rectangular wood strands or wafers arranged in oriented layers; or

❖ Oriented strand board (OSB) is one of three types of wood structural panels. OSB panels are fabricated out of multiple layers made up of wood flakes or strands called "furnish." Like plywood, these layers of furnish are oriented at 90 degrees (1.57 rad) to each other. This gives OSB panels properties very similar to those of plywood.

Plywood. A wood structural panel comprised of plies of wood veneer arranged in cross-aligned layers.

❖ Plywood is one of three types of wood structural panels. It is manufactured by gluing together three or more cross-laminated layers of wood. The code does not differentiate between OSB and plywood, since manufacturing is done according to either the PS1 standard for industrial plywood or the PS2 standard for OSB and performance-rated plywood.

SECTION 2303
MINIMUM STANDARDS AND QUALITY

2303.1 General. Structural lumber, end-jointed lumber, prefabricated I-joists, structural glued-laminated timber, wood structural panels, fiberboard sheathing (when used structurally), hardboard siding (when used structurally), particleboard, preservative-treated wood, fire-retardant-treated wood, hardwood, plywood, trusses and joist hangers shall conform to the applicable provisions of this section.

❖ When designed and built in accordance with the standards listed in Section 2303, a building or structure is deemed to comply with the code. These standards contain most of the information needed to adequately design a structure. It is necessary for the designer to have a working knowledge of engineering principles and experience with construction to properly interpret the recommendations and meet the provisions of other applicable sections of the code.

Section 2303.1 lists the various materials that have production and quality control standards. The use of these standards is fundamental for manufacturers in producing products and maintaining quality control procedures. Designers, owners and building officials must understand these standards and the methods prescribed in them to be able to identify products that have been produced in accordance with their criteria. Without the knowledge that the products meet the standard, there is little assurance that a safe and efficient building or structure will be constructed.

2303.1.1 Lumber. Lumber used for load-supporting purposes, including end-jointed or edge-glued lumber, machine stress-rated or machine evaluated lumber, shall be identified by the grade mark of a lumber grading or inspection agency that has been approved by an accreditation body that complies with DOC PS 20 or equivalent. Grading practices and identification shall comply with rules published by an agency approved in accordance with the procedures of DOC PS 20 or equivalent procedures. In lieu of a grade mark on the material, a certificate of inspection as to species and grade issued by a lumber-grading or inspection agency meeting the requirements of this section is permitted to be accepted for precut, remanufactured or rough-sawn lumber, and for sizes larger than 3 inches (76 mm) nominal thickness.

Approved end-jointed lumber is permitted to be used interchangeably with solid-sawn members of the same species and grade.

❖ All lumber used to support loads in a building or structure is required to be properly identified. Every species and grade of lumber has a unique inherent strength value. These values are further modified in sawn timber by the presence of growth characteristics that vary from piece to piece, such as knots, slope of grain, checks, etc. Without adequate identification, it would be impossible to verify the material used in the field. The required grade mark must identify the species or species grouping, grade and moisture content at the time of surfacing, the grading agency and the mill name or grader's number. Figure 2303.1.1 illustrates typical grade mark labels.

Certification is an acceptable alternative to a grade mark from both United States and Canadian grading agencies. Grading agencies are certified by the ALSC.

Design values are published by lumber grade rules-writing agencies for both individual and grouped species. A grouped species is lumber that is cut and marketed in lots containing two or more species, such as Spruce-Pine-Fir. These species grow together in large areas. It is more economical to market the lumber as a species group than attempt segregation. The assigned strength values include those applicable to the weaker species in the group.

The code also allows certain types of structural lumber to have a certificate of inspection instead of a grade mark. A certificate of inspection is acceptable for precut, remanufactured, or rough-sawn lumber and for sizes larger than 3 inches (76 mm) nominal in thickness. It is industry practice to place only one label (grade mark) on a piece of lumber, which may be removed on precut and remanufactured lumber. Each piece of lumber is graded after it has been cut to a standard size. The grade of the piece is determined based on its size, number and location of strength-reducing characteristics. Therefore, one log may produce lumber of two or more different grades. It is also industry practice not to label lumber having a nominal thickness larger than 3 inches (76 mm), or rough-sawn material where the label may be illegible. A certificate of inspection from an approved agency is acceptable instead of the label for these types of lumber. The certificate should be filed with the permanent records of the building or structure.

If defects exceeding those permitted for the grade allegedly installed are visible, then a grader would be able to determine that the wood is definitely not of a suitable grade. To determine if the wood in question is definitely of a suitable grade, the grader must inspect all four faces of the piece. This cannot happen once the lumber is installed in the building, as other components of the building will be covering up some of the faces of the pieces.

End-joined or edge-glued lumber is acceptable when identified by an appropriate grade mark. The NDS permits the use of such lumber in Section 4.1.6 of the standard.

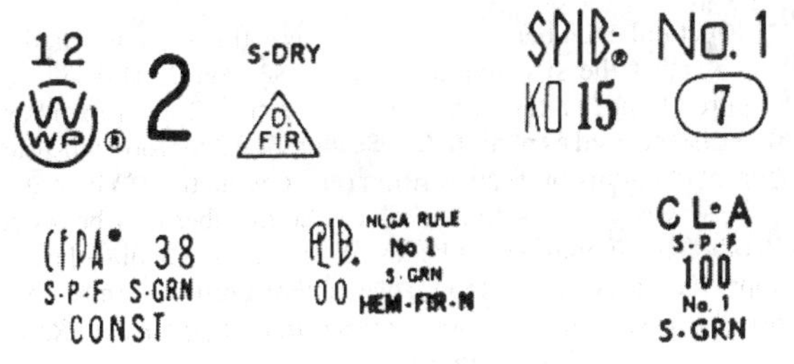

Figure 2303.1.1
TYPICAL LABELS
(Grade Marks)
American Lumber Standards Committee, U.S.D.C.

2303.1.2 Prefabricated wood I-joists. Structural capacities and design provisions for prefabricated wood I-joists shall be established and monitored in accordance with ASTM D 5055.

❖ This section specifies that the shear, moment and stiffness capacities of prefabricated wood I-joists be established and monitored by ASTM D 5055. This standard also specifies that application details such as bearing length and web openings are to be considered in determining structural capacity. Wood I-joists are structural members typically used in floor and roof construction manufactured out of sawn or structural composite lumber flanges and structural panel webs, bonded together with exterior adhesives forming an "I" cross section. The standard requires I-joist manufacturers to employ an independent inspection agency to monitor the procedures for quality assurance. Finally, the standard specifies that proper installation instructions accompany the product to the job site. The instructions are required to include weather protection, handling requirements and, where required, web reinforcement, connection details, lateral support, bearing details, web hole-cutting limitations and any special situation.

2303.1.3 Structural glued-laminated timber. Glued-laminated timbers shall be manufactured and identified as required in AITC A190.1 and ASTM D 3737.

❖ Glued-laminated timbers are required by this section to be manufactured following ANSI/AITC 90.1 and ASTM D 3737. Knowing the standard these products must meet makes it easier to determine that the product found in the field will meet the design requirements.

2303.1.4 Wood structural panels. Wood structural panels, when used structurally (including those used for siding, roof and wall sheathing, subflooring, diaphragms and built-up members), shall conform to the requirements for their type in DOC PS 1 or PS 2. Each panel or member shall be identified for grade and glue type by the trademarks of an approved testing and grading agency. Wood structural panel components shall be designed and fabricated in accordance with the applicable standards listed in Section 2306.1 and identified by the trademarks of an approved testing and inspection agency indicating conformance with the applicable standard. In addition, wood structural panels when permanently exposed in outdoor applications shall be of exterior type, except that wood structural panel roof sheathing exposed to the outdoors on the underside is permitted to be interior type bonded with exterior glue, Exposure 1.

❖ Wood structural panels is the inclusive term for plywood, OSB and other composite panels of wood-based materials. Identification of grade for plywood includes N, A, B, C Plugged, C and D (from no knots or patches to large knots and knotholes). The identification of glue type is by means of the exposure durability classifications: Exterior, Exposure 1, Exposure 2 and Interior.

2303.1.5 Fiberboard. Fiberboard for its various uses shall conform to ANSI/AHA A194.1 or ASTM C 208. Fiberboard sheathing, when used structurally, shall be so identified by an approved agency as conforming to ANSI/AHA A194.1 or ASTM C 208.

❖ All fiberboard must meet the requirements of AHA A194.1 or ASTM C 208, as well as being verminproof, resistant to rot-producing fungi and water repellent. These standards give physical requirements for construction grades of fiberboard, including sheathing grade and roof-insulating grade. The sheathing grade of fiberboard is further broken down into regular and intermediate densities.

2303.1.5.1 Jointing. To ensure tight-fitting assemblies, edges shall be manufactured with square, shiplapped, beveled, tongue-and-groove or U-shaped joints.

❖ Tight-fitting joints in the fiberboard are required for all applications, including insulation, siding and wall sheathing.

2303.1.5.2 Roof insulation. Where used as roof insulation in all types of construction, fiberboard shall be protected with an approved roof covering.

❖ Fiberboard is not intended for prolonged exposure to sunlight, wind, rain or snow. Where fiberboard is used as roof insulation, it must be protected with an approved roof covering to prevent water saturation and subsequent delamination and to avoid decay and destruction of the glue bond by moisture.

2303.1.5.3 Wall insulation. Where installed and fireblocked to comply with Chapter 7, fiberboards are permitted as wall insulation in all types of construction. In fire walls and fire barriers, unless treated to comply with Section 803.1 for Class A materials, the boards shall be cemented directly to the concrete, masonry or other noncombustible base and shall be protected with an approved noncombustible veneer anchored to the base without intervening airspaces.

❖ Fiberboard is permitted without any fire-resistance treatment in the walls of all types of construction (see Section 603.1). When used in fire walls and fire barrier walls, fiberboard must be either treated to comply with Class A flame spread or adhered directly to a noncombustible base and protected by a tight-fitting, noncombustible veneer that is fastened through the fiberboard to the base. This will prevent the fiberboard from contributing to the spread of fire.

2303.1.5.3.1 Protection. Fiberboard wall insulation applied on the exterior of foundation walls shall be protected below ground level with a bituminous coating.

❖ Fiberboard insulation applied to the exterior side of foundation walls is required to be protected from the weather to improve its service life and maintain its performance characteristics. Of particular concern is foundation insulation that is in close proximity to grade and runs the risk of being damaged by a lawn mower, rocks or soil that is kicked up against it, water from a garden hose or rainwater splash back, etc. Protection is required for all fiberboard insulation on the exterior face of foundation walls.

2303.1.5.4 Insulating roof deck. Where used as roof decking in open beam construction, fiberboard insulation roof deck shall have a nominal thickness of not less than 1 inch (25 mm).

❖ Fiberboard roof panels are used on open-beam-type roofs for insulation and structural purposes. A minimum thickness of 1 inch (25 mm) is specified to reduce the possibility of early burnthrough in a fire.

2303.1.6 Hardboard. Hardboard siding used structurally shall be identified by an approved agency conforming to AHA A135.6. Hardboard underlayment shall meet the strength requirements of $^7/_{32}$-inch (5.6 mm) or $^1/_4$-inch (6.4 mm) service class hardboard planed or sanded on one side to a uniform thickness of not less than 0.200 inch (5.1 mm). Prefinished hardboard paneling shall meet the requirements of AHA A135.5. Other basic hardboard products shall meet the requirements of AHA

A135.4. Hardboard products shall be installed in accordance with manufacturer's recommendations.

❖ Hardboard siding that is to be used structurally must be manufactured in accordance with AHA A135.6 and marked to indicate conformance with the standard, whether primed or unprimed, and to identify the producer and the type, either lap or panel. Hardboard products are produced primarily from inter-felted ligno-cellulosic fibers. There are five classes based on strength values. Underlayments are limited to $^7/_{32}$-inch (5.6 mm) or $^1/_4$-inch (6.4 mm) service class.

Prefinished hardboard is required to be manufactured to AHA A135.5 standards, and must be marked to indicate the standard and to identify the producer, flame spread index, finish class, type of gloss and type of substrate, or must be accompanied by written certification of the same information.

2303.1.7 Particleboard. Particleboard shall conform to ANSI A208.1. Particleboard shall be identified by the grade mark or certificate of inspection issued by an approved agency. Particleboard shall not be utilized for applications other than indicated in this section unless the particleboard complies with the provisions of Section 2306.4.3.

❖ Sponsored by the Composite Panel Association (CPA), ANSI A208.1 is the basic specification for the manufacture of particleboard, which establishes a system of grade marks for the boards' grade, density and strength.

Particleboard used in construction is medium density, and is designated by "M" as the first digit. The second digit in the designation is related to grade. The designations range from 1 to 3, with higher designations being the strongest. Grade M-S refers to medium density, "special grade" particleboard. This grade was added to ANSI A208.1 after the M-1, M-2 and M-3 grades had been established. Grade M-S falls between M-1 and M-2 in physical properties.

An optional third digit of the grade designation indicates that the particleboard has a special characteristic. The grades of particleboard specified in Table 2306.4.3, M-S and M-2 "Exterior Glue," are manufactured with exterior glue to increase their durability characteristics.

While ANSI A208.1 has provisions for M-3-grade particleboard, panels that meet the requirements of this higher grade are more commonly evaluated and used as wood structural panels, in accordance with DOC PS-2. Therefore, Grade M-3 material is not addressed in Table 2306.4.3.

2303.1.7.1 Floor underlayment. Particleboard floor underlayment shall conform to Type PBU of ANSI A208.1. Type PBU underlayment shall not be less than $^1/_4$-inch (6.4 mm) thick and shall be installed in accordance with the instructions of the Composite Panel Association.

❖ Although similar to medium density, Grade 1 particleboard—particleboard intended for use as floor underlayment—is designated "PBU" and has stricter limits on levels of formaldehyde emission permitted than those placed on Grade "M" particleboard. Particleboard intended for use as floor underlayment is not commonly manufactured with exterior glue, which could emit higher levels of formaldehyde than that permitted by ANSI A208.1 for "PBU" grade floor underlayment.

Particleboard underlayment is often applied over a structural subfloor to provide a smooth surface for resilient-finish or textile floor coverings. The minimum $^1/_4$-inch (6.4 mm) thickness is applicable over panel-type subflooring. Particleboard underlayment installed over board or deck subflooring that has multiple joints should have a thickness of $^3/_8$ inch (9.5 mm). Joints in the underlayment should not be over joints in the subflooring.

All particleboard underlayment with thicknesses of $^1/_4$ through $^3/_4$ inch (6.4 through 19.1 mm) should be attached with a minimum of 6d annular threaded nails spaced 6 inches (152 mm) o.c. on the edges and 10 inches (254 mm) o.c. for intermediate supports.

2303.1.8 Preservative-treated wood. Lumber, timber, plywood, piles and poles supporting permanent structures required by Section 2304.11 to be preservative treated shall conform to the requirements of the applicable AWPA Standard C1, C2, C3, C4, C9, C14, C15, C16, C22, C23, C24, C28, C31, C33 and M4, for the species, product, preservative and end use. Preservatives shall conform to AWPA P1/P13, P2, P5, P8 and P9. Lumber and plywood used in wood foundation systems shall conform to Chapter 18.

❖ Wood is able to absorb chemicals because of its cellular characteristics. Preservative treatment procedures that will repel termites and destroy decay-causing fungus utilize this capability. The process includes placing wood in a large cylinder and applying a vacuum to remove as much air as possible from the wood. The chemical solution is then introduced and pressure is applied to force the solution into the wood. The pressure is maintained until the desired absorption is obtained. A final vacuum is often applied to remove as much excess preservative as possible.

There are several variations of this basic process, all of which must provide the required retention of preservatives. AWPA P1/P13, P2, P5, P8 and P9 prescribe the requirements for the preservatives that are used, while AWPA C1, C2, C9 and C22 specify the minimum results of the treatment process, including the depth of penetration of the chemical into the wood, and the amount of chemical retained by the wood once the process is completed. Additional requirements for wood used in foundation systems are given in Section 1805.4.6.

2303.1.8.1 Identification. Wood required by Section 2304.11 to be preservative treated shall bear the quality mark of an inspection agency that maintains continuing supervision, testing and inspection over the quality of the reservative-treated wood. Inspection agencies for preservative-treated wood shall be listed by an accreditation body that complies with the requirements of the American Lumber Standards Treated Wood Program, or equivalent. The quality mark shall be on a stamp or label affixed to the preservative-treated wood, and shall include the following information:

1. Identification of treating manufacturer.
2. Type of preservative used.
3. Minimum preservative retention (pcf).
4. End use for which the product is treated.
5. AWPA standard to which the product was treated.
6. Identity of the accredited inspection agency.

❖ Quality marks are necessary to determine that preservative-treated wood conforms to applicable standards. The identifying mark must be by an approved inspection agency that has continuous follow-up services. Additionally, the inspection agency must be listed and certified as being competent by an approved organization. The American Lumber Standards Corporation (ALSC) provides certification of treating agencies. Facsimiles of their quality marks are available by contacting them. The required quality mark is not a substitute for a grade mark. When wood or wood-based materials are being used structurally, both the quality mark and grade mark must be displayed on the piece.

2303.1.8.2 Moisture content. Where preservative-treated wood is used in enclosed locations where drying in service cannot readily occur, such wood shall be at a moisture content of 19 percent or less before being covered with insulation, interior wall finish, floor covering or other materials.

❖ Water-borne preservatives are subject to leaching unless properly dried and protected. The requirement for reducing the moisture content to 19 percent or less is intended to aid in preventing such leaching. Also, all structural members are presumed to have a moisture content of 19 percent or less.

Preservative treatment does not require any adjustment of design values. Some species have a high surface tension, which means that it is difficult to penetrate the surface of the lumber. These species are often incised to break the surface tension and allow the chemicals to penetrate the member. Incising requires a reduction in the design values for the member.

2303.1.9 Structural composite lumber. Structural capacities for structural composite lumber shall be established and monitored in accordance with ASTM D 5456.

❖ The purpose of this section is to specify the appropriate standard for structural composite lumber. Included within the standard are criteria for laminated veneer lumber and parallel strand lumber. The ASTM standard includes requirements for testing, criteria for determining allowable stresses, requirements for independent inspection and quality assurance procedures.

2303.2 Fire-retardant-treated wood. Fire-retardant-treated wood is any wood product which, when impregnated with chemicals by a pressure process or other means during manufacture, shall have, when tested in accordance with ASTM E 84, a listed flame spread index of 25 or less and show no evidence of significant progressive combustion when the test is continued for an additional 20-minute period. In addition, the flame front shall not progress more than 10.5 feet (3200 mm) beyond the centerline of the burners at any time during the test.

❖ FRTW is plywood and lumber that has been pressure impregnated with chemicals to improve its flame spread characteristics beyond that of untreated wood. The effectiveness of the pressure-impregnated fire-retardant treatment is determined by subjecting the material to tests conducted in accordance with ASTM E 84, with the modification that the test is extended to 30 minutes rather than 15 minutes. Using this procedure, a flame spread index is established during the standard 10-minute test period. The test is continued for an additional 20 minutes. During this added time period, there must not be any significant flame spread. At no time must the flame spread more than 10 $^{1}/_{2}$ feet (3200 mm) past the centerline of the burners.

The result of impregnating wood with fire-retardant chemicals is a chemical reaction at certain temperature ranges. This reaction reduces the release of certain intermediate products which contribute to the flaming of wood, and also results in the formation of a greater percentage of charcoal and water. Some chemicals are also effective in reducing the oxidation rate for charcoal residue. Fire-retardant chemicals also reduce the heat release rate of fire-retardant-treated wood when burning over a wide range of temperatures. This section gives provisions for the treatment and use of fire-retardant-treated wood.

2303.2.1 Labeling. Fire-retardant-treated lumber and wood structural panels shall be labeled. The label shall contain the following items:

1. The identification mark of an approved agency in accordance with Section 1703.5.
2. Identification of the treating manufacturer.
3. The name of the fire-retardant treatment.
4. The species of wood treated.
5. Flame spread and smoke-developed index.
6. Method of drying after treatment.
7. Conformance with appropriate standards in accordance with Sections 2303.2.2 through 2303.2.5.
8. For fire-retardant-treated wood exposed to weather, damp or wet locations, include the words "No increase in the listed classification when subjected to the Standard Rain Test" (ASTM D 2898).

❖ For continued quality, each piece of FRTW must be identified by an approved agency having a reinspection service. The identification must show the performance rating of the material, including the 30-minute ASTM E 84 test results determined in Section 2303.2, and design adjustment values determined in Section 2303.2.2. The third-party agency that provides the FRTW label is also required to state on the label that the FRTW complies with the requirements of Section 2303.2, and that design adjustment values have been determined for the FRTW in compliance with the provisions of Section 2303.2.2.

The FRTW label must be distinct from the grading label to avoid confusion between the two. The grading label provides information about the properties of wood before it is fire-retardant treated. The FRTW label provides properties of the wood after FRTW treatment. It is imperative that the FRTW label be presented in such a manner that it complements the grading label, and does not create confusion over which label takes precedence.

2303.2.2 Strength adjustments. Design values for untreated lumber and wood structural panels, as specified in Section 2303.1, shall be adjusted for fire-retardant-treated wood. Adjustments to design values shall be based on an approved method of investigation that takes into consideration the effects of the anticipated temperature and humidity to which the fire-retardant-treated wood will be subjected, the type of treatment and redrying procedures.

❖ Experience has shown that certain factors can affect the physical properties of FRTW. Among these factors are the pressure treatment and redrying processes used, and the extremes of temperature and humidity that the FRTW will be subjected to once installed. The design values for all FRTW must be adjusted for the effects of the treatment and environmental conditions, such as high temperature and humidity in attic installations. This section requires the determination of these design adjustment values, based on an investigation procedure that includes subjecting the FRTW to similar temperatures and humidities, and has been approved by the building official. The FRTW tested must be identical to that which is produced. Items to be considered by the building official reviewing the test procedure include species and grade of the untreated wood, and conditioning of wood, such as drying before the fire-retardant treatment process. A fire-retardant wood treater may choose to have its treatment process evaluated by model code evaluation services.

The FRTW is required by Section 2303.2.1 to be labeled with the design adjustment values. These can take the form of factors that are multiplied by the original design values of the untreated wood to determine its allowable stresses, or new allowable stresses that have already been factored down in consideration of the FRTW treatment.

2303.2.2.1 Wood structural panels. The effect of treatment and the method of redrying after treatment, and exposure to high temperatures and high humidities on the flexure properties of fire-retardant-treated softwood plywood shall be determined in accordance with ASTM D 5516. The test data developed by ASTM D 5516 shall be used to develop adjustment factors, maximum loads and spans, or both, for untreated plywood design values in accordance with ASTM D 6305. Each manufacturer shall publish the allowable maximum loads and spans for service as floor and roof sheathing for its treatment.

❖ This section references the test standard developed to evaluate the flexural properties of fire-retardant-treated plywood that is exposed to high temperatures. Note that while the section title refers to wood structural panels, the referenced standard is limited to softwood plywood. Therefore judgement is required in determining the effects of elevated temperature and humidity on other types of wood structural panels.

2303.2.2.2 Lumber. For each species of wood treated, the effect of the treatment and the method of redrying after treatment and exposure to high temperatures and high humidities on the allowable design properties of fire-retardant-treated lumber shall be determined in accordance with ASTM D 5664. The test data developed by ASTM D 5664 shall be used to develop modification factors for use at or near room temperature and at elevated temperatures and humidity in accordance with an approved method of investigation. Each manufacturer shall publish the modification factors for service at temperatures of not less than 80°F (26.7°C) and for roof framing. The roof framing modification factors shall take into consideration the climatological location.

❖ This section references the test standard developed to determine the necessary adjustments to design values for lumber that has been fire-retardant treated and includes the effects of elevated temperatures.

2303.2.3 Exposure to weather, damp or wet locations. Where fire-retardant-treated wood is exposed to weather, or damp or wet locations, it shall be identified as "Exterior" to indicate there is no increase in the listed flame spread index as defined in Section 2303.2 when subjected to ASTM D 2898.

❖ Some fire-retardant treatments are soluble when exposed to the weather or used under high-humidity conditions. Note that the humidity threshold established for interior applications in Section 2303.2.4 is 92 percent. Therefore, FRTW used in an interior location that will exceed this threshold should comply with this section. When a FRTW product is to be exposed to any of the conditions noted in this section, it must be further tested in accordance with ASTM D 2898. Testing requires the material to meet the performance criteria listed in Section 2303.2. The material is then subjected to the ASTM weathering test and retested after drying. There must not be any significant differences in the performance recorded before and after the weathering test.

2303.2.4 Interior applications. Interior fire-retardant-treated wood shall have moisture content of not over 28 percent when tested in accordance with ASTM D 3201 procedures at 92-percent relative humidity. Interior fire-retardant-treated wood shall be tested in accordance with Section 2303.2.2.1 or 2303.2.2.2. Interior fire-retardant-treated wood designated as

Type A shall be tested in accordance with the provisions of this section.

❖ The environment in which the FRTW is used can affect its performance. In order to make sure that the performance will be adequate in a humid interior condition, testing in accordance with ASTM D 3201 as well as that specified in Section 2303.2.2.1 or 2303.2.2.2 is required for all interior wood. Requiring all interior wood to be tested for the effects of high temperature and humidity reduces the chance of premature failure due to improper use of FRTW.

2303.2.5 Moisture content. Fire-retardant-treated wood shall be dried to a moisture content of 19 percent or less for lumber and 15 percent or less for wood structural panels before use. For wood kiln dried after treatment (KDAT), the kiln temperatures shall not exceed those used in kiln drying the lumber and plywood submitted for the tests described in Section 2303.2.2.1 for plywood and 2303.2.2.2 for lumber.

❖ These moisture content thresholds are necessary to prevent leaching of the fire retardant from the wood. Section 2303.2.2 requires that the strength adjustments consider the redrying procedure and this section clarifies that the drying temperatures for wood kiln dried after treatment must be consistent with those on which the adjustment factors were based.

2303.2.6 Type I and II construction applications. See Section 603.1 for limitations on the use of fire-retardant-treated wood in buildings of Type I or II construction.

❖ Although Type I and II construction typically would only allow noncombustible materials, Section 603.1 allows various parts of a building to be combustible. Specifically, there are 22 items listed that are exceptions to the strict limits on Types I and II. There are three conditions listed in that section that will allow FRTW to be used: in nonbearing partitions where their fire resistance is two hours or less; in nonbearing exteriors that do not require a fire-resistance rating and in roof construction of buildings not over two stories high.

2303.3 Hardwood plywood. Hardwood and decorative plywood shall be manufactured and identified as required in HPVA HP-1.

❖ Hardwood plywood, like any other product that undergoes a change in the process of being configured for use in construction, has standard criteria for the manufacturing process and the quality of the materials that are to be used. ANSI/HPVA HP-1 is the standard for hardwood plywood, and is one of the family of standards for this type of product.

2303.4 Trusses. Metal-plate-connected wood trusses shall be manufactured as required by TPI 1. Each manufacturer of trusses using metal plate connectors shall retain an approved agency to make unscheduled inspections of truss manufacturing and delivery operations. The inspection shall cover all phases of truss operations, including lumber storage, handling, cutting fixtures, presses or rollers, manufacturing, bundling and banding.

❖ Metal-plate-connected wood trusses have been a popular product for construction of wood-frame buildings for many years. There are significant benefits to building with these engineered systems of wood and metal connectors; however, there are also criteria associated with the manufacturing, transportation and placement of trusses that are critical to their productive use in a structure. ANSI/TPI 1 is the industry-developed standard for managing this process.

A manufacturer of trusses is also required by this section to have an agency review its operations to determine compliance with this standard. This type of evaluation by a qualified person is necessary for the quality of the product and its ability to perform as designed.

2303.4.1 Truss design drawings. Truss construction documents shall be prepared by a registered design professional and shall be provided to the building official and approved prior to installation. These construction documents shall include, at a minimum, the information specified below. Truss shop drawings shall be provided with the shipment of trusses delivered to the job site.

1. Slope or depth, span and spacing;
2. Location of joints;
3. Required bearing widths;
4. Design loads as applicable;
5. Top chord live load (including snow loads);
6. Top chord dead load;
7. Bottom chord live load;
8. Bottom chord dead load;
9. Concentrated loads and their points of application;
10. Controlling wind and earthquake loads;
11. Adjustments to lumber and metal connector plate design value for conditions of use;
12. Each reaction force and direction;
13. Metal connector plate type, size, thickness or gage, and the dimensioned location of each metal connector plate except where symmetrically located relative to the joint interface;
14. Lumber size, species and grade for each member;
15. Connection requirements for:
    - 15.1. Truss to truss girder;
    - 15.2. Truss ply to ply; and
    - 15.3. Field splices.
16. Calculated deflection ratio or maximum deflection for live and total load;

17. Maximum axial compression forces in the truss members to design the size, connections and anchorage of the permanent continuous lateral bracing. Forces shall be shown on the truss construction documents or on supplemental documents; and
18. Required permanent truss member bracing location.

❖ Truss design drawings are an important subject for most designers and contractors. The code requires information be provided to document what the truss is supposed to do and how it is supposed to do it. Designers and contractors must depend on the fabricator in most instances to design the truss. Contracts for the fabricator do not typically track with the general contractor, making the availability of the truss drawings for review sometimes problematic.

The majority of the items on the list are readily available from any designer of trusses. The code specifies the information needed based on the difficulty in determining the appropriateness of the trusses in the field. Since documentation is included with the submitted plans, the inspector has an easy time verifying that the trusses meet the designer's intent.

2303.5 Test standard for joist hangers and connectors. For the required test standards for joist hangers and connectors, see Section 1715.1.

❖ Tests for joist hangers and connections are included in the criteria for special inspections for materials and test standards in Section 1715. They reference ASTM D 1761 and establish the criteria for use of lumber with a specific gravity of 0.49 or larger, but not more than 0.55 in accordance with the NDS. Additionally, Section 1715.2 adds limits on the vertical load capacity for joist hangers by requiring that there be consistency of the tested capacity within 20 percent of the average ultimate load. If this cannot be achieved, additional testing is required.

2303.6 Nails and staples. Nails and staples shall conform to requirements of ASTM F 1667. Nails used for framing and sheathing connections shall have minimum average bending yield strengths as follows: 80 kips per square inch (ksi) (551 MPa) for shank diameters larger than 0.177 inch (4.50 mm) but not larger than 0.254 inch (6.45 mm), 90 ksi (620 MPa) for shank diameters larger than 0.142 inch (3.61 mm) but not larger than 0.177 inch (4.50 mm) and 100 ksi (689 MPa) for shank diameters of 0.142 inch (3.61 mm) or less.

❖ The code specifies the minimum criteria for nails and staples that is necessary to resist the design loads. The normal connections are included in the nailing schedule in Table 2304.9.1.

2303.7 Shrinkage. Consideration shall be given in design to the possible effect of cross-grain dimensional changes considered vertically which may occur in lumber fabricated in a green condition.

❖ Lumber described as in a "green condition" has a moisture content higher than that used to define a dry condition under the applicable grading rules, which is reflected on the grade mark (see commentary, Figure 2303.1.1). Because it will shrink more than dry lumber, use of lumber that is fabricated in a green condition requires consideration of the effects of cross-grain shrinkage. The extent of cross-grain shrinkage is a function of the lumber's moisture at the time of construction and the amount of drying that occurs after construction.

SECTION 2304
GENERAL CONSTRUCTION REQUIREMENTS

2304.1 General. The provisions of this section apply to design methods specified in Section 2301.2.

❖ The general criteria for construction with wood as the structural framing material are contained in Section 2304.

Section 2301.2 lists all the methods of designing a wood structural member. These include the ASD, LRFD and conventional light-frame wood construction as prescribed in Sections 2304 and 2308. When a design is based on any of these methods, it must still conform to the requirements of Section 2304.

2304.2 Size of structural members. Computations to determine the required sizes of members shall be based on the net dimensions (actual sizes) and not nominal sizes.

❖ ASD and LRFD designs assume that actual member sizes [i.e., $1^1/_2$ inches by $3^1/_2$ inches (38 mm by 89 mm] are being used, rather than nominal sizes (i.e., 2 by 4). Conventional construction design (specified in Section 2308) does not require computations to determine the sizes of members, since by its very nature all the computations have been completed. However, there are limitations to the applicable buildings and occupancies for which conventional construction design could be used. This provision is an attempt to make sure that if design principles are applied to a specific element, the designer uses design values consistent with the actual load-carrying capacity of the element, as reflected in the referenced design standards.

2304.3 Wall framing. The framing of exterior and interior walls shall be in accordance with the provisions specified in Section 2308 unless a specific design is furnished.

❖ The conventional construction criteria in Section 2308 provide great detail and specifications for the construction of walls. Under the majority of circumstances, the methods and procedures for framing a wall in accordance with Section 2308.9 will provide the necessary resistance to all loads for a typical wall. The exception may be construction conditions that fall outside the limitations stated in Section 2308.

2304.3.1 Bottom plates. Studs shall have full bearing on a 2-inch-thick (actual $1^1/_2$-inch, 38 mm) or larger plate or sill having a width at least equal to the width of the studs.

❖ In order to insure that the loads imposed on a stud are fully transferred into the supporting structural members below, it is required that the studs bear on a transfer member that is required to be the minimum thickness of a nominal 2-inch (51 mm) framing member. The plate is required to be at least the same width as the stud so that the full capacity of the stud is being realized. If the plate were narrower than the stud and the entire area of the stud were required to transfer the bearing load without crushing the stud's fibers, there would likely be a local failure of the wall. If the wall were designed so that the bearing on the bottom plate was not a critical issue (i.e., an interior nonbearing wall), the design would have to demonstrate that the plate can be smaller than the studs to compensate for the irregular construction.

2304.3.2 Framing over openings. Headers, double joists, trusses or other approved assemblies that are of adequate size to transfer loads to the vertical members shall be provided over window and door openings in load-bearing walls and partitions.

❖ Walls depend upon the studs to be continuous and carry the imposed loads from the top plate to the bottom plate and eventually to the foundation. When there is an opening in the wall, this continuity is interrupted. It is necessary to transfer the loads from above the opening to other structural supports. These could be adjoining studs, ganged studs, or cripple studs that support the member framing over the opening, or even columns installed as part of the wall.

An important structural consideration is the connection between the member spanning the opening and the support. Whether that connection is a simple direct support, a hanger or some other type of connection, it is necessary to know its capacity.

2304.3.3. Shrinkage. Wood walls and bearing partitions shall not support more than two floors and a roof unless an analysis satisfactory to the building official shows that shrinkage of the wood framing will not have adverse effects on the structure or any plumbing, electrical or mechanical systems, or other equipment installed therein due to excessive shrinkage or differential movements caused by shrinkage. The analysis shall also show that the roof drainage system and the foregoing systems or equipment will not be adversely affected or, as an alternate, such systems shall be designed to accommodate the differential shrinkage or movements.

❖ The adverse effects of shrinkage are greatest in multistory buildings. This section states conditions under which shrinkage must be considered in the design of wall framing. If shrinkage is not considered in multistory wood-frame buildings it can adversely affect utilities that are not able to accommodate the resulting movement of the structure. This section provides performance criteria that apply to wood-bearing walls supporting more than two floors and a roof.

Another example of the adverse effects of shrinkage is the impact it can have on shear wall performance. Shrinkage has been recognized as a factor in loose hold-down nuts in earthquake-damaged buildings. In this situation the hold-down cannot resist tension until the slack is taken up, resulting in excessive movement of the shear wall under lateral loading. Since in most cases the shear wall sheathing conceals the hold-downs, this problem often goes undetected.

2304.4 Floor and roof framing. The framing of wood-joisted floors and wood framed roofs shall be in accordance with the provisions specified in Section 2308 unless a specific design is furnished.

❖ Similar to the criteria for wall construction, floors may be designed using the methods prescribed in Section 2308. Among the problems associated with the use of the tables for conventional construction are the limitations on allowable loads. Span Tables 2308.8(1) and (2); 2308.10.2(1) and (2); and 2308.10.3(1), (2), (3), (4), (5) and (6) each describe a different loading condition for floors, roof and rafters. The maximum floor load in Table 2308.8, is for a 40 pound per square foot (psf) (1.9 kN/m^2) live load. Table 2308.10.2, covers only dead loads of 5 psf (2.4 kN/m^2) for ceiling rafters, and Table 2308.10.3 only provides for up to a 50 psf (2.4 kN/m^2) ground snow load. The range of required loading is much larger, as demonstrated in Table 1607.1. The only structures for which a 40 psf (2 kN/m^2) live load on the floor is allowed are catwalks, fire escapes on single-family dwellings, private rooms and wards in a hospital, cell blocks, school classrooms, and all areas of residential occupancies other than public areas. All other occupancies require a higher load capacity, in which case, these tables do not apply.

2304.5 Framing around flues and chimneys. Combustible framing shall be a minimum of 2 inches (51 mm), but shall not be less than the distance specified in Sections 2111 and 2113 and the *International Mechanical Code*, from flues, chimneys and fireplaces, and 6 inches (152 mm) away from flue openings.

❖ This provision is duplicated in the *International Mechanical Code*® (IMC®), but must be included in both places in order to alert the framer, installer and inspector of the requirement. When exposed to heat for a prolonged period of time the characteristics of wood may change, which may cause the temperature at which it will ignite to be lowered. By providing separation, the wood will not be directly exposed because of the airspace between it and the heat source.

2304.6 Wall sheathing. Except as provided for in Section 1405 for weatherboarding or where stucco construction that complies with Section 2510 is installed, enclosed buildings shall be sheathed with one of the materials of the nominal thickness specified in Table 2304.6 or any other approved material of equivalent strength or durability.

❖ Wall sheathing can be constructed of many materials. The fundamental reason for having sheathing is to stabilize the wall-framing members [typically 2-inch-thick (51 mm) pieces of lumber] when the wall is loaded. Studs nailed to the top and bottom plate are stable to a degree, but will collapse with significant force placed on them unless there is some type of stabilizing membrane or strap. Sheathing, whether it is a heavy structural membrane found on exterior bearing walls or a light one that would be used for nonbearing interior partitions, provides necessary stability for the walls. Some sheathing may be strictly for the purpose of insulating the building, or in addition to providing lateral bracing, may provide a finish for the wall. Sheathing is not the only means of bracing a wall.

This section establishes that only the sheathing types listed in Table 2304.6, which includes board and various panel products (structural panels, boards, gypsum and reinforced cement mortar) are to be used for walls. There is an allowance that other materials may be used if it is shown they are capable of providing equivalent strength and durability. Strapping or "let-in" braces can perform adequately in some situations where they are designed as part of the wall system.

2304.6.1 Wood structural panel sheathing. Where wood structural panel sheathing is used as the exposed finish on the exterior of outside walls, it shall have an exterior exposure durability classification. Where wood structural panel sheathing is used on the exterior of outside walls but not as the exposed finish, it shall be of a type manufactured with exterior glue (Exposure 1 or Exterior). Where wood structural panel sheathing is used elsewhere, it shall be of a type manufactured with intermediate or exterior glue.

❖ Wood structural panels are defined by this chapter as being plywood, OSB and composite panels. Each of these types of panel products has identical performance characteristics when manufactured according to the PS 2 standard. This section requires equivalent exposure durability performance. It requires the classification to be Exposure 1 or Exterior when the material is likely to be exposed to the weather.

TABLE 2304.6. See below.

❖ Table 2304.6 reinforces Section 2304.6, which requires all sheathing to meet minimum limitations. Because of loading conditions or the design of the assembly to which the wall sheathing is attached, the sheathing may have to be thicker. The table is simple to use: determine the type and thickness of the sheathing, and the maximum wall stud spacing that must be used to install the material is provided.

2304.6.2 Interior paneling. Softwood wood structural panels used for interior paneling shall conform with the provisions of Chapter 8 and shall be installed in accordance with Table 2304.9.1. Panels shall comply with DOC PS 1 or PS 2. Prefinished hardboard paneling shall meet the requirements of AHA A135.5, *Prefinished Hardboard Paneling*. Hardwood plywood shall conform to HPVA HP-1, *The American National Standard for Hardwood and Decorative Plywood*.

❖ Interior paneling, similar to exterior paneling, must meet minimum standards of acceptance. DOC PS 1-95 and PS 2-95 standards establish the general standards for all softwood veneer panel products. ANSI/AHA A135.5 and ANSI/HPVA HP-1 standards address the criteria for hardwood veneered plywoods.

2304.7 Floor and roof sheathing.

❖ Floor and roof sheathing perform several functions. They are the primary base for both the walking surface for the floor and the roofing materials. Floor sheathing may be the only structural element transferring live loads to the floor joists. Most floor covering is nonstructural. The sheathing may also serve as a subfloor or a substrate for another finish, such as hardwood.

TABLE 2304.6
MINIMUM THICKNESS OF WALL SHEATHING

SHEATHING TYPE	MINIMUM THICKNESS	MAXIMUM WALL STUD SPACING
Wood boards	$^5/_8$ inch	24 inches on center
Fiberboard	$^1/_2$ inch	16 inches on center
Wood structural panel	In accordance with Tables 2308.9.3(2) and 2308.9.3(3)	—
M-S "Exterior Glue" and M-2 "Exterior Glue" Particleboard	In accordance with Tables 2306.4.3 and 2308.9.3(5)	—
Gypsum sheathing	$^1/_2$ inch	16 inches on center
Gypsum wallboard	$^1/_2$ inch	24 inches on center
Reinforced cement mortar	1 inch	24 inches on center

For SI: 1 inch = 25.4 mm.

Roof sheathing is similar in that it is typically covered by the roofing material. Some roof covering can be of structural value, but sheathing typically provides the only means by which loads from wind and snow are transferred to the structural framing members.

Sheathing is a critical element in the transfer mechanism of lateral forces.

2304.7.1 Structural floor sheathing. Structural floor sheathing shall be designed in accordance with the general provisions of this code and the special provisions in this section.

Floor sheathing conforming to the provisions of Table 2304.7(1), 2304.7(2), 2304.7(3) or 2304.7(4) shall be deemed to meet the requirements of this section.

❖ The tables referenced are used for both floor and roof applications (see commentary, Section 2304.7.2).

2304.7.2 Structural roof sheathing. Structural roof sheathing shall be designed in accordance with the general provisions of this code and the special provisions in this section.

Roof sheathing conforming to the provisions of Table 2304.7(1), 2304.7(2), 2304.7(3) or 2304.7(5) shall be deemed to meet the requirements of this section. Wood structural panel roof sheathing shall be bonded by exterior glue.

❖ There are four tables referenced in this section that describe the various spans for structural sheathing typically produced by mills in North America. Using the tables is straightforward: simply enter the table with the span between structural framing members, and depending on the direction of the framing, determine the minimum thickness required by the code. It may be necessary to install thicker material if the loads require additional resistance.

Two additional pieces of information that may be needed are the grade and span rating of the material being used. Lumber and structural sheathing both are required to be marked with grade stamps indicating the capabilities of the lumber or panels, based on the testing and conditions of the material.

TABLE 2304.7(1). See below.

❖ Table 2304.7(1) applies to "surfaced dry" or "surfaced unseasoned" lumber. These terms refer to the condition of the lumber at the time it was surfaced, or finished. This is only critical when determining the actual dimension of the lumber. A piece that is surfaced dry is narrower than a piece that is surfaced unseasoned (often referred to as surfaced green). The grade stamp will indicate whether it is surfaced dry or surfaced green.

TABLE 2304.7(2). See below.

❖ The grading rules are listed and are an important part of the criteria used in determining the adequacy of lumber to perform the necessary work. In this table, the grading rules are listed, but there are agencies not listed that use these rules and will be marking the lumber. The NDS lists all agencies that write grading rules, as well as those that use the rules to grade the lumber. A full up-to-date listing of all approved agencies can be obtained from the ALSC.

TABLE 2304.7(1)
ALLOWABLE SPANS FOR LUMBER FLOOR AND ROOF SHEATHING[a,b]

SPAN (inches)	MINIMUM NET THICKNESS (inches) OF LUMBER PLACED			
	Perpendicular to supports		Diagonally to supports	
	Surfaced dry[c]	Surfaced unseasoned	Surfaced dry[c]	Surfaced unseasoned
Floors				
24	$^{3}/_{4}$	$^{25}/_{32}$	$^{3}/_{4}$	$^{25}/_{32}$
16	$^{5}/_{8}$	$^{11}/_{16}$	$^{5}/_{8}$	$^{11}/_{16}$
Roofs				
24	$^{5}/_{8}$	$^{11}/_{16}$	$^{3}/_{4}$	$^{25}/_{32}$

For SI: 1 inch = 25.4 mm.

a. Installation details shall conform to Sections 2304.6.1 and 2304.6.2 for floor and roof sheathing, respectively.

b. Floor or roof sheathing conforming with this table shall be deemed to meet the design criteria of Section 2304.6.

c. Maximum 19-percent moisture content.

TABLE 2304.7(2)
SHEATHING LUMBER, MINIMUM GRADE REQUIREMENTS: BOARD GRADE

SOLID FLOOR OR ROOF SHEATHING	SPACED ROOF SHEATHING	GRADING RULES
Utility	Standard	NLGA, WCLIB, WWPA
4 common or utility	3 common or standard	NLGA, WCLIB, WWPA, NSLB or NELMA
No. 3	No. 2	SPIB
Merchantable	Construction common	RIS

TABLE 2304.7(3). See below.

❖ Structural wood panels are also graded, but the type of grade stamp used identifies the thickness and the span capabilities of the piece. The span rating (first column of the table) will appear on the grade stamp of the panel. The first number on the span rating indicates the allowable spacing of support members (rafters or trusses) in a roof application. The second number indicates the allowable spacing of floor joists or trusses in a floor application. For instance, if the span rating is 24/16, the panel may be used on rafters spaced 24 inches (610 mm) o.c. and on floor joists spaced 16 inches (406 mm) o.c. Span ratings were developed for easy identification and correct application of panels in the field.

For the span rating to be applicable, other conditions of the table for loading and use must be met. For instance, for a 24/16 to be used on roof rafters spaced 16 inches (406 mm) o.c. the snow load for the roof cannot exceed 40 psf (1915 Pa) (listed in the sixth column of the table). The allowable spacing of support members is also dependent on the type of edge support provided for the panels (see Note f). The table is applicable only when the strength axis of the panel is perpendicular to the supports—in other words, when the long edge of the panel is perpendicular to the joists or rafters.

TABLE 2304.7(3)
ALLOWABLE SPANS AND LOADS FOR WOOD STRUCTURAL PANEL SHEATHING AND SINGLE-FLOOR GRADES CONTINUOUS OVER TWO OR MORE SPANS WITH STRENGTH AXIS PERPENDICULAR TO SUPPORTS[a,b]

SHEATHING GRADES		ROOF[c]				FLOOR[d]
Panel span rating roof/floor span	Panel thickness (inches)	Maximum span (inches)		Load[e] (psf)		Maximum span (inches)
		With edge support[f]	Without edge support	Total load	Live load	
12/0	5/16	12	12	40	30	0
16/0	5/16, 3/8	16	16	40	30	0
20/0	5/16, 3/8	20	20	40	30	0
24/0	3/8, 7/16, 1/2	24	20[g]	40	30	0
24/16	7/16, 1/2	24	24	50	40	16
32/16	15/32, 1/2, 5/8	32	28	40	30	16[h]
40/20	19/32, 5/8, 3/4, 7/8	40	32	40	30	20[h,i]
48/24	23/32, 3/4, 7/8	48	36	45	35	24
54/32	7/8, 1	54	40	45	35	32
60/32	7/8, 1 1/8	60	48	45	35	32
SINGLE FLOOR GRADES		**ROOF[c]**				**FLOOR[d]**
Panel span rating	Panel thickness (inches)	Maximum span (inches)		Load[e] (psf)		Maximum span (inches)
		With edge support[f]	Without edge support	Total load	Live load	
16 o.c.	1/2, 19/32, 5/8	24	24	50	40	16[h]
20 o.c.	19/32, 5/8, 3/4	32	32	40	30	20[h,i]
24 o.c.	23/32, 3/4	48	36	35	25	24
32 o.c.	7/8, 1	48	40	50	40	32
48 o.c.	1 3/32, 1 1/8	60	48	50	40	48

For SI: 1 inch = 25.4 mm, 1 pound per square foot = 0.0479 kN/m^2.

a. Applies to panels 24 inches or wider.
b. Floor and roof sheathing conforming with this table shall be deemed to meet the design criteria of Section 2304.7.
c. Uniform load deflection limitations $^1/_{180}$ of span under live load plus dead load, $^1/_{240}$ under live load only.
d. Panel edges shall have approved tongue-and-groove joints or shall be supported with blocking unless $^1/_4$-inch minimum thickness underlayment or 1 $^1/_2$ inches of approved cellular or lightweight concrete is placed over the subfloor, or finish floor is $^3/_4$-inch wood strip. Allowable uniform load based on deflection of $^1/_{360}$ of span is 100 pounds per square foot except the span rating of 48 inches on center is based on a total load of 65 pounds per square foot.
e. Allowable load at maximum span.
f. Tongue-and-groove edges, panel edge clips (one midway between each support, except two equally spaced between supports 48 inches on center), lumber blocking or other. Only lumber blocking shall satisfy blocked diaphragm requirements.
g. For $^1/_2$-inch panel, maximum span shall be 24 inches.
h. Span is permitted to be 24 inches on center where $^3/_4$-inch wood strip flooring is installed at right angles to joist.
i. Span is permitted to be 24 inches on center for floors where 1$^1/_2$ inches of cellular or lightweight concrete is applied over the panels.

TABLE 2304.7(4). See below.

❖ Often a combination of a wood subfloor and a finish floor is used and there is additional capacity for that composite floor action. This table provides the additional capacity associated with that installation. It is necessary to go to DOC PS 1 to determine what species groups are represented by the numbers in the table.

TABLE 2304.7(5). See below.

❖ The table provides criteria for panels that may be used on roofs with significant snow loads. The allowable spacing of rafters or trusses and the allowable load are determined by the grade and thickness of the panel.

2304.8 Mechanically laminated floors and decks.

TABLE 2304.7(4)
ALLOWABLE SPAN FOR WOOD STRUCTURAL PANEL COMBINATION SUBFLOOR-UNDERLAYMENT (SINGLE FLOOR)[a,b]
(Panels Continuous Over Two or More Spans and Strength Axis Perpendicular to Supports)

IDENTIFICATION	MAXIMUM SPACING OF JOISTS (inches)				
	16	20	24	32	48
Species group[c]	Thickness (inches)				
1	$^1/_2$	$^5/_8$	$^3/_4$	—	—
2, 3	$^5/_8$	$^3/_4$	$^7/_8$	—	—
4	$^3/_4$	$^7/_8$	1	—	—
Single floor span rating[d]	16 o.c.	20 o.c.	24 o.c.	32 o.c.	48 o.c.

For SI: 1 inch = 25.4 mm, 1 pound per square foot = 0.0479 kN/m^2.

a. Spans limited to value shown because of possible effects of concentrated loads. Allowable uniform loads based on deflection of $^1/_{360}$ of span is 100 pounds per square foot except allowable total uniform load for 1 $^1/_8$-inch wood structural panels over joists spaced 48 inches on center is 65 pounds per square foot. Panel edges shall have approved tongue-and-groove joints or shall be supported with blocking, unless $^1/_4$-inch minimum thickness underlayment or 1 $^1/_2$ inches of approved cellular or lightweight concrete is placed over the subfloor, or finish floor is $^3/_4$-inch wood strip.

b. Floor panels conforming with this table shall be deemed to meet the design criteria of Section 2304.7.

c. Applicable to all grades of sanded exterior-type plywood. See DOC PS 1 for plywood species groups.

d. Applicable to Underlayment grade, C-C (Plugged) plywood, and Single Floor grade wood structural panels.

TABLE 2304.7(5)
ALLOWABLE LOAD (PSF) FOR WOOD STRUCTURAL PANEL ROOF SHEATHING CONTINUOUS OVER TWO OR MORE SPANS AND STRENGTH AXIS PARALLEL TO SUPPORTS
(Plywood Structural Panels Are Five-Ply, Five-Layer Unless Otherwise Noted)[a,b]

PANEL GRADE	THICKNESS (inch)	MAXIMUM SPAN (inches)	LOAD AT MAXIMUM SPAN (psf)	
			Live	Total
Structural I sheathing	$^7/_{16}$	24	20	30
	$^{15}/_{32}$	24	35[c]	45[c]
	$^1/_2$	24	40[c]	50[c]
	$^{19}/_{32}$, $^5/_8$	24	70	80
	$^{23}/_{32}$, $^3/_4$	24	90	100
Sheathing, other grades covered in DOC PS 1 or DOC PS 2	$^7/_{16}$	16	40	50
	$^{15}/_{32}$	24	20	25
	$^1/_2$	24	25	30
	$^{19}/_{32}$	24	40[c]	50[c]
	$^5/_8$	24	45[c]	55[c]
	$^{23}/_{32}$, $^3/_4$	24	60[c]	65[c]

For SI: 1 inch = 25.4 mm, 1 pound per square foot = 0.0479 kN/m^2.

a. Roof sheathing conforming with this table shall be deemed to meet the design criteria of Section 2304.7.

b. Uniform load deflection limitations $^1/_{180}$ of span under live load plus dead load, $^1/_{240}$ under live load only. Edges shall be blocked with lumber or other approved type of edge supports.

c. For composite and four-ply plywood structural panel, load shall be reduced by 15 pounds per square foot.

2304.8.1 General. A laminated lumber floor or deck built up of wood members set on edge, when meeting the following requirements, is permitted to be designed as a solid floor or roof deck of the same thickness, and continuous spans are permitted to be designed on the basis of the full cross section using the simple span moment coefficient.

Nail lengths shall not be less than two and one-half times the net thickness of each lamination. Where deck supports are 4 feet (1219 mm) on center (o.c.) or less, side nails shall be spaced not more than 30 inches (762 mm) o.c. alternately near top and bottom edges, and staggered one-third of the spacing in adjacent laminations. Where supports are spaced more than 4 feet (1219 mm) o.c., side nails shall be spaced not more than 18 inches (457 mm) o.c. alternately near top and bottom edges, and staggered one-third of the spacing in adjacent laminations. Two side nails shall be used at each end of butt-jointed pieces.

Laminations shall be toenailed to supports with 20d or larger common nails. Where the supports are 4 feet (1219 mm) o.c. or less, alternate laminations shall be toenailed to alternate supports; where supports are spaced more than 4 feet (1219 mm) o.c., alternate laminations shall be toenailed to every support. A single-span deck shall have all laminations full length. A continuous deck of two spans shall not have more than every fourth lamination spliced within quarter points adjoining supports. Joints shall be closely butted over supports or staggered across the deck but within the adjoining quarter spans. No lamination shall be spliced more than twice in any span.

❖ One method of floor construction is the use of individual wood members set on edge, connected with nails and mechanically laminated to produce a monolithic surface.

This section gives the fastening requirements for floors that are constructed of individual wood members set on edge and mechanically laminated to produce a continuous surface. For the floor to act as a unit, fastening is critical, and therefore it is prescribed in this section. If the floor is constructed in accordance with these fastening requirements, then for purposes of design it can be considered a solid wood floor, acting as one element.

2304.9 Connections and fasteners.

2304.9.1 Fastener requirements. Connections for wood members shall be designed in accordance with the appropriate methodology in Section 2301.2. The number and size of nails connecting wood members shall not be less than that set forth in Table 2304.9.1.

❖ Section 2301.2 gives alternative design methodologies for wood construction: the ASD method (the applicable standard in accordance with Section 2306 is AF&PA's *National Design Specification for Wood Construction*), the LRFD method (the applicable standard is ASCE 16, in accordance with Section 2307), and the conventional light-frame methodologies which are embodied in Sections 2304 and 2308 of the code. The ASD and LRFD methods contain design provisions for fasteners, and the specification of fastening based on design calculations is appropriate when those methods are used. In addition, the prescriptive requirements of Table 2304.9.1 must be met. If conventional provisions are used and there is no design being conducted by means of the ASD or LRFD methodologies, then the prescriptive fastening of Table 2304.9.1 is all that is required. However, when the ASD or LRFD methods are used, fastening must comply with the associated design standards as well as the prescriptive fastenings specified in Table 2304.9.1.

TABLE 2304.9.1. See page 23-20.

❖ Familiarity with the terms and configurations of conventional frame construction is necessary to apply this table. For an understanding of terms such as "sole plate" or "rim joist," consult a basic framing text. See Figure 2304.9.1 for an illustration of certain nail types.

This table provides the minimum fastening required for all lightweight wood-frame constructions. Buildings designed using either the ASD or LRFD methodologies because of greater loadings must have fasteners that comply with the criteria of this table. It would be common for designed structures to have greater fastening requirements than prescribed in Table 2304.9.1.

2304.9.2 Sheathing fasteners. Sheathing nails or other approved sheathing connectors shall be driven so that their head or crown is flush with the surface of the sheathing.

❖ This requirement is a matter of workmanship. Protruding nails do not provide the intended connecting capacity and could be hazardous. Likewise, nails overdriven into structural sheathing may not perform as expected.

2304.9.3 Joist hangers and framing anchors. Connections depending on joist hangers or framing anchors, ties and other mechanical fastenings not otherwise covered are permitted where approved. The vertical load-bearing capacity, torsional moment capacity and deflection characteristics of joist hangers shall be determined in accordance with Section 1715.1.

❖ This section provides for the use of joist and framing anchors instead of conventional nailing or the use of ledger strips. Joist and framing anchors are engineered components and must be used in accordance with the manufacturer's instructions and design specifications. The capacity for each hanger must be specified by the manufacturer and is determined in accordance with the test methods outlined in Section 1715.1. Typically the dimensions of the framing elements connected, the number and size of nails used for installation and the loading conditions are all specified by the manufacturer. Approval by the building official should depend on verification that the anchors have been tested and are being used in accordance with the manufacturer's specifications.

TABLE 2304.9.1
FASTENING SCHEDULE

CONNECTION	FASTENING[a,m]	LOCATION
1. Joist to sill or girder	3 - 8d common 3 - 3″ × 0.131″ nails 3 - 3″ 14 gage staples	toenail
2. Bridging to joist	2 - 8d common 2 - 3″ × 0.131″ nails 2 - 3″14 gage staples	toenail each end
3. 1″ × 6″ subfloor or less to each joist	2 - 8d common	face nail
4. Wider than 1″ × 6″ subfloor to each joist	3 - 8d common	face nail
5. 2″ subfloor to joist or girder	2 - 16d common	blind and face nail
6. Sole plate to joist or blocking	16d at 16″ o.c. 3″ × 0.131″ nails at 8″ o.c. 3″ 14 gage staples at 12″ o.c.	typical face nail
Sole plate to joist or blocking at braced wall panel	3 - 16d at 16″ 4 - 3″ × 0.131″ nails at 16″ 4 - 3″ 14 gage staples per 16″	braced wall panels
7. Top plate to stud	2 - 16d common 3 - 3″ × 0.131″ nails 3 - 3″ 14 gage staples	end nail
8. Stud to sole plate	4 - 8d common 4 - 3″ × 0.131″ nails 3 - 3″ 14 gage staples	toenail
	2 - 16d common 3 - 3″ × 0.131″ nails 3 - 3″ 14 gage staples	end nail
9. Double studs	16d at 24″ o.c. 3″ × 0.131″ nail at 8″ o.c. 3″ 14 gage staple at 8″ o.c.	face nail
10. Double top plates	16d at 16″ o.c. 3″ × 0.131″ nail at 12″ o.c. 3″ 14 gage staple at 12″ o.c.	typical face nail
Double top plates	8-16d common 12 - 3″ × 0.131″ nails 12 - 3″ 14 gage staples	lap splice
11. Blocking between joists or rafters to top plate	3 - 8d common 3 - 3″ × 0.131″ nails 3 - 3″ 14 gage staples	toenail
12. Rim joist to top plate	8d at 6″ (152 mm) o.c. 3″ × 0.131″ nail at 6″ o.c. 3″ 14 gage staple at 6″ o.c.	toenail
13. Top plates, laps and intersections	2 - 16d common 3 - 3″ × 0.131″ nails 3 - 3″ 14 gage staples	face nail
14. Continuous header, two pieces	16d common	16″ o.c. along edge
15. Ceiling joists to plate	3 - 8d common 5 - 3″ × 0.131″ nails 5 - 3″ 14 gage staples	toenail
16. Continuous header to stud	4 - 8d common	toenail

(continued)

TABLE 2304.9.1—continued
FASTENING SCHEDULE

CONNECTION	FASTENING[a,m]	LOCATION
17. Ceiling joists, laps over partitions (see Section 2308.10.4.1, Table 2308.10.4.1)	3 - 16d common minimum, Table 2308.10.4.1 4 - 3″ × 0.131″ nails 4 - 3″ 14 gage staples	face nail
18. Ceiling joists to parallel rafters (see Section 2308.10.4.1, Table 2308.10.4.1)	3 - 16d common minimum, Table 2308.10.4.1 4 - 3″ × 0.131″ nails 4 - 3″ 14 gage staples	face nail
19. Rafter to plate (see Section 2308.10.1, Table 2308.10.1)	3 - 8d common 3 - 3″ × 0.131″ nails 3 - 3″ 14 gage staples	toenail
20. 1″ diagonal brace to each stud and plate	2 - 8d common 2 - 3″ × 0.131″ nails 2 - 3″ 14 gage staples	face nail
21. 1″ × 8″ sheathing to each bearing wall	2 - 8d common	face nail
22. Wider than 1″× 8″ sheathing to each bearing	3 - 8d common	face nail
23. Built-up corner studs	16d common 3″ × 0.131″ nails 3″ 14 gage staples	24″ o.c. 16″ o.c. 16″ o.c.
24. Built-up girder and beams	20d common 32″ o.c. 3″ × 0.131″ nail at 24″ o.c. 3″ 14 gage staple at 24″ o.c.	face nail at top and bottom staggered on opposite sides
	2 - 20d common 3 - 3″ × 0.131″ nails 3 - 3″ 14 gage staples	face nail at ends and at each splice
25. 2″ planks	16d common	at each bearing
26. Collar tie to rafter	3 - 10d common 4 - 3″ × 0.131″ nails 4 - 3″ 14 gage staples	face nail
27. Jack rafter to hip	3 - 10d common 4 - 3″ × 0.131″nails 4 - 3″ 14 gage staples	toenail
	2 - 16d common 3 - 3″ × 0.131″ nails 3 - 3″ 14 gage staples	face nail
28. Roof rafter to 2-by ridge beam	2 - 16d common 3 - 3″ × 0.131″ nails 3 - 3″ 14 gage staples	toenail
	2 - 16d common 3 - 3″ × 0.131″ nails 3 - 3″ 14 gage staples	face nail
29. Joist to band joist	3 - 16d common 5 - 3″ × 0.131″ nails 5 - 3″ 14 gage staples	face nail

(continued)

TABLE 2304.9.1—continued
FASTENING SCHEDULE

CONNECTION	FASTENING[a,m]		LOCATION
30. Ledger strip	3 - 16d common 4 - 3″ × 0.131″ nails 4 - 3″ 14 gage staples		face nail
31. Wood structural panels and particleboard:[b] Subfloor, roof and wall sheathing (to framing):	1/2″ and less	6d[c,l] 2 3/8″ × 0.113″ nail[n] 1 3/4″ 16 gage[o]	
	19/32″ to 3/4″	8d[d] or 6d[e] 2 3/8″ × 0.113″ nail[p] 2″ 16 gage[p]	
	7/8″ to 1″	8d[c]	
	1 1/8″ to 1 1/4″	10d[d] or 8d[e]	
Single Floor (combination subfloor-underlayment to framing):	3/4″ and less	6d[e]	
	7/8″ to 1″	8d[e]	
	1 1/8″ to 1 1/4″	10d[d] or 8d[e]	
32. Panel siding (to framing)	1/2″ or less	6d[f]	
	5/8″	8d[f]	
33. Fiberboard sheathing:[g]	1/2″	No. 11 gage roofing nail[h] 6d common nail No. 16 gage staple[i]	
	25/32″	No. 11 gage roofing nail[h] 8d common nail No. 16 gage staple[i]	
34. Interior paneling	1/4″	4d[j]	
	3/8″	6d[k]	

For SI: 1 inch = 25.4 mm.

a. Common or box nails are permitted to be used except where otherwise stated.

b. Nails spaced at 6 inches on center at edges, 12 inches at intermediate supports except 6 inches at supports where spans are 48 inches or more. For nailing of wood structural panel and particleboard diaphragms and shear walls, refer to Section 2305. Nails for wall sheathing are permitted to be common, box or casing.

c. Common or deformed shank.

d. Common.

e. Deformed shank.

f. Corrosion-resistant siding or casing nail.

g. Fasteners spaced 3 inches on center at exterior edges and 6 inches on center at intermediate supports.

h. Corrosion-resistant roofing nails with 7/16-inch-diameter head and 1 1/2-inch length for 1/2-inch sheathing and 1 3/4-inch length for 25/32-inch sheathing.

i. Corrosion-resistant staples with nominal 7/16-inch crown and 1 1/8-inch length for 1/2-inch sheathing and 1 1/2-inch length for 25/32-inch sheathing. Panel supports at 16 inches (20 inches if strength axis in the long direction of the panel, unless otherwise marked).

j. Casing or finish nails spaced 6 inches on panel edges, 12 inches at intermediate supports.

k. Panel supports at 24 inches. Casing or finish nails spaced 6 inches on panel edges, 12 inches at intermediate supports.

l. For roof sheathing applications, 8d nails are the minimum required for wood structural panels.

m. Staples shall have a minimum crown width of 7/16 inch.

n. For roof sheathing applications, fasteners spaced 4 inches on center at edges, 8 inches at intermediate supports.

o. Fasteners spaced 4 inches on center at edges, 8 inches at intermediate supports for subfloor and wall sheathing and 3 inches on center at edges, 6 inches at intermediate supports for roof sheathing.

p. Fasteners spaced 4 inches on center at edges, 8 inches at intermediate supports.

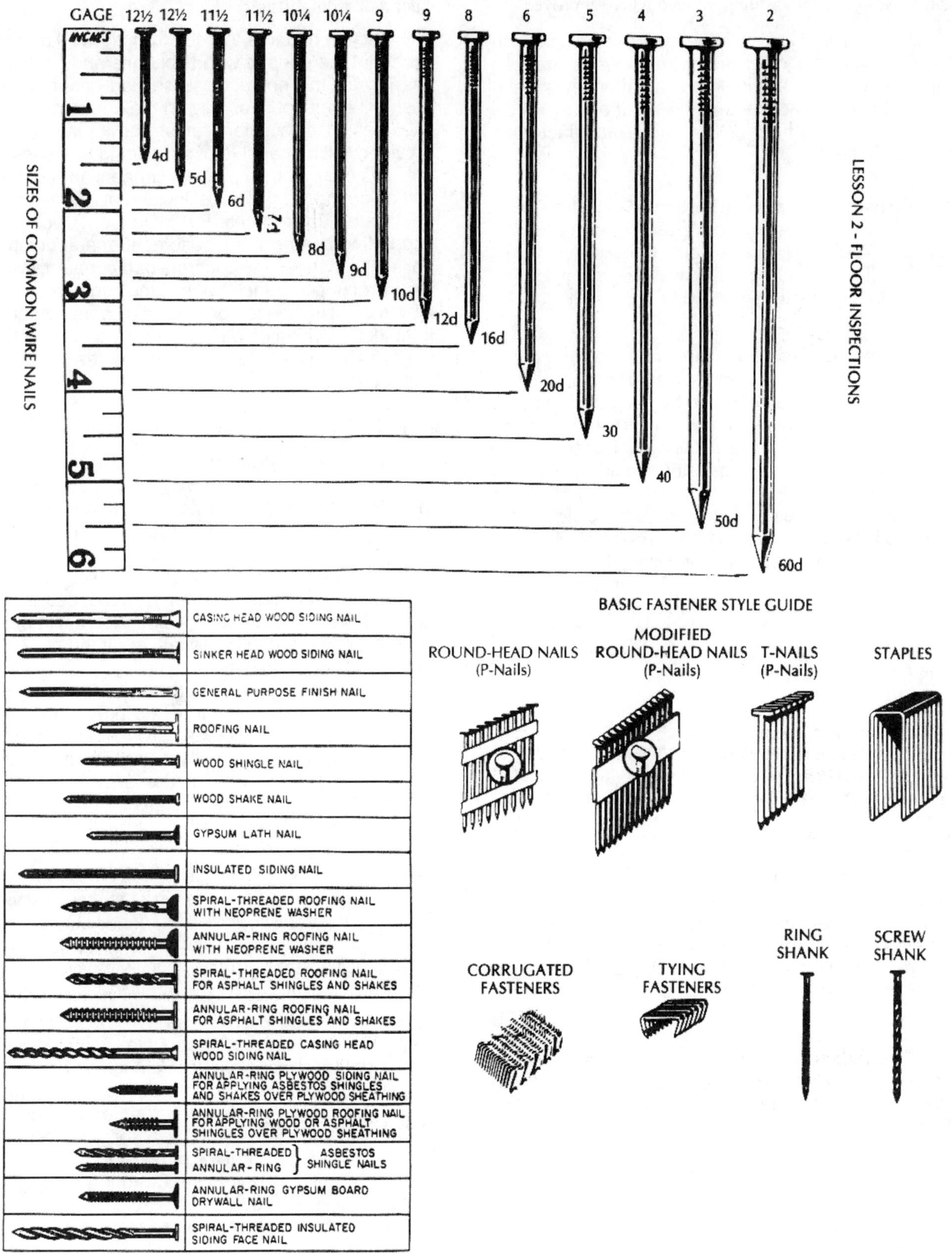

Figure 2304.9.1
STANDARD NAILS AND STAPLES
Independent Nail Packing Co., Bridgewater, MA

2304.9.4 Other fasteners. Clips, staples, glues and other approved methods of fastening are permitted where approved.

❖ In addition to joist hangers (see commentary, Section 2304.9.3), there are other fastening methods that can be approved. Approval should be based on testing evidence by the manufacturer and the fact that they will be used in accordance with the manufacturer's recommendations.

2304.9.5 Fasteners in preservative-treated and fire-retardant-treated wood. Fasteners for preservative-treated and fire-retardant-treated wood shall be of hot-dipped zinc-coated galvanized steel, stainless steel, silicon bronze or copper. Fastenings for wood foundations shall be as required in AF&PA Technical Report No. 7.

❖ Corrosion-resistant fasteners are required for treated wood, since treated wood is associated with conditions where wood and fasteners could be subjected to moisture. This applies to FRTW as well as pressure-preservative treated wood, since FRTW is also commonly used in applications where moisture could occur, such as the roofs of buildings.

If treated wood is used for purposes of termite resistance and in locations where moisture accumulation is not likely, there is no practical necessity for the use of corrosion-resistant fasteners. AF&PA Technical Report No. 7 requires specific types of corrosion-resistant nails for locations below grade and above grade in wood foundations, based on the increased exposure to moisture in below-grade applications.

2304.9.6 Load path. Where wall framing members are not continuous from foundation sill to roof, the members shall be secured to ensure a continuous load path. Where required, sheet metal clamps, ties or clips shall be formed of galvanized steel or other approved corrosion-resistant material not less than 0.040 inch (1.01 mm) nominal thickness.

❖ In conventional construction, the fastening required by Table 2304.9.1 provides for a continuous connection between framing elements as required by this section. If the loads on the building exceed the parameters established in Section 2308.2, the building must be designed in accordance with the ASD or LRFD design methodologies and the connections must be designed in accordance with those standards as well to ensure a continuous load path in accordance with this section. The designed connections may incorporate engineered tie or strap components with capacities specified by the manufacturer. This section specifies the minimum thickness of sheet metal for fabricated straps or ties.

2304.9.7 Framing requirements. Wood columns and posts shall be framed to provide full end bearing. Alternatively, column-and-post end connections shall be designed to resist the full compressive loads, neglecting end-bearing capacity. Column-and-post end connections shall be fastened to resist lateral and net induced uplift forces.

❖ This section requires that the framing workmanship be such that the ends of wood columns and posts be fully bearing for the entire cross section of the column or post. If it does not bear fully on the entire cross section, then design calculations must show that the bearing provided will be capable of supporting the loads. It also requires there to be a positive connection (not merely a friction connection) to resist incidental lateral forces and anticipated uplift forces. For instance, a wood column supporting beams in the basement of a structure could not simply bear on the concrete pad beneath it without a means of positive anchorage to the concrete, to prevent it from being displaced on impact by a person or object. If the combination of loads on the structure could result in uplift forces, then a connection to resist uplift would also be required.

2304.10 Heavy timber construction.

❖ Heavy timber construction originated in New England to serve the needs of the growing textile industry. As the industry modernized, the need for larger open space, unobstructed by columns, gradually reduced the demand for this type of construction.

Today, due primarily to its aesthetic value, heavy timber construction is used in many other occupancies. It is commonly used for assembly and mercantile buildings, such as schools and churches, auditoriums, gymnasiums and supermarkets.

The provisions of this section must be used with the minimum dimension requirements in Section 602.4. For a building to be classified as heavy timber, all structural elements must comply with the minimum dimensions and detailing provisions of this section. A structural analysis must be performed to ensure that the minimum dimensions are adequate.

2304.10.1 Columns. Columns shall be continuous or superimposed throughout all stories by means of reinforced concrete or metal caps with brackets, or shall be connected by properly designed steel or iron caps, with pintles and base plates, or by timber splice plates affixed to the columns by metal connectors housed within the contact faces, or by other approved methods.

❖ The AF&PA publication, *Heavy Timber Construction Details—Wood Construction Data 5*, provides details for the proper connection of columns and beams. Traditionally, heavy timber structures were designed and constructed using a prescriptive approach, much like today's conventional construction for light-frame buildings. The number and size of bolts, lag screws or connectors should be determined through analysis of the loads to be supported.

Pintles provide a method to fasten the butt ends of beams or columns. A pintle acts like a short column, connecting the ends of the structural elements.

2304.10.1.1 Column connections. Girders and beams shall be closely fitted around columns and adjoining ends shall be cross tied to each other, or intertied by caps or ties, to transfer horizontal loads across joints. Wood bolsters shall not be placed on tops of columns unless the columns support roof loads only.

❖ Where columns and girders or beams intersect, care must be taken for load transfer. Proper detailing at the intersection of columns and girders will limit the effects of shrinkage. Elements that are not continuous must be properly tied to transfer design forces. Today's engineered wood products are available in multiple depths and widths, and in lengths up to 60 feet (18 288 mm), so continuous span structural elements are readily available. Bolster blocks are used to reduce the compression perpendicular to grain stresses where roof beams bear on columns. The shrinkage of the bolster block must be considered in the roof design. Roof beams must have a positive connection to the column to resist uplift.

2304.10.2 Floor framing. Approved wall plate boxes or hangers shall be provided where wood beams, girders or trusses rest on masonry or concrete walls. Where intermediate beams are used to support a floor, they shall rest on top of girders, or shall be supported by ledgers or blocks securely fastened to the sides of the girders, or they shall be supported by an approved metal hanger into which the ends of the beams shall be closely fitted.

❖ Wood members must not bear directly on masonry or concrete. Typically, a steel bearing plate or custom steel hanger is used to support heavy timber elements. Bearing plates provide an even distribution of load on the wall. Wall plate boxes should be designed for positive connection to the masonry. The plate should also prevent lateral movement of the beam, but allow the beam to fall away if it is burnt through. Wood bearing directly on masonry or concrete may absorb moisture, resulting in the loss of bearing capacity. Where direct bearing on top of the girders is not possible or desirable, a metal hanger designed to transfer all induced loads is permitted. An additional advantage to using a hanger is less shrinkage. Green beams should be installed so their top edge is slightly above the top edge of the girder.

2304.10.3 Roof framing. Every roof girder and at least every alternate roof beam shall be anchored to its supporting member; and every monitor and every sawtooth construction shall be anchored to the main roof construction. Such anchors shall consist of steel or iron bolts of sufficient strength to resist vertical uplift of the roof.

❖ Anchorage of roof beams and girders should be in accordance with the engineered design. This section provides minimum acceptable anchorage practices. Sawtooth roof construction is used to provide large amounts of daylight into the floor immediately beneath the roof. The roof line looks much like the cutting edge of a saw, with regularly spaced peaks and valleys. The vertical or near vertical line connecting the peaks and valleys is filled with glazing. This unique framing pattern must be anchored to the main roof supports.

2304.10.4 Floor decks. Floor decks and covering shall not extend closer than $^1/_2$ inch (12.7 mm) to walls. Such $^1/_2$-inch (12.7 mm) spaces shall be covered by a molding fastened to the wall either above or below the floor and arranged such that the molding will not obstruct the expansion or contraction movements of the floor. Corbeling of masonry walls under floors is permitted in place of such molding.

❖ Flooring will expand and contract as the relative humidity in the building changes. To accommodate expansion, a $^1/_2$-inch (12.7 mm) gap is required at the wall. The gap will also prevent moisture in the walls from migrating into the floor boards. The fire integrity of the floor assembly must be maintained, so either molding or corbeling of the wall below is required to prevent the passage of smoke and hot gases.

2304.10.5 Roof decks. Where supported by a wall, roof decks shall be anchored to walls to resist uplift forces determined in accordance with Chapter 16. Such anchors shall consist of steel or iron bolts of sufficient strength to resist vertical uplift of the roof.

❖ The roof deck must be anchored frequently enough to resist uplift forces. Anchorage must be in accordance with the engineered design.

2304.11 Protection against decay and termites.

2304.11.1 General. Where required by this section, protection from decay and termites shall be provided by the use of naturally durable or preservative-treated wood.

❖ Conditions favorable for decay and fungus attack are discussed in Sections 2304.11.2 and 2304.11.4, along with preservative treatments. This section sets forth specific locations of concern and minimum construction practices to prevent such an attack. Termite protection is also defined.

2304.11.2 Wood used above ground. Wood installed above ground in the locations specified in Sections 2304.11.2.1 through 2304.11.2.6 shall be naturally durable wood or preservative-treated wood that uses water-borne preservatives, and shall be treated in accordance with AWPA C2 or C9 or applicable AWPA standards for above-ground use.

❖ The American Wood Preservers' Association (AWPA) is the principal standards-writing body for the wood preserving industry in the United States. AWPA standards cover all aspects of treated wood from the treating process to appropriate treatment retention levels for specific applications. Wood products used in locations identified in this section are particularly susceptible to deterioration. Deterioration can result from continuous or intermittent contact with moisture or termites.

Decay is caused by fungi, which are low forms of plant life that feed on wood. For fungi to attack wood in service, the following conditions must be present: (1) temperature in the range of 35° to 100° F (1° to 38° C), (2) adequate supply of oxygen and (3) wood moisture content in excess of 20 percent.

2304.11.2.1 Joists, girders and subfloor. Where wood joists or the bottom of a wood structural floor without joists are closer than 18 inches (457 mm), or wood girders are closer than 12 inches (305 mm) to the exposed ground in crawl spaces or unexcavated areas located within the perimeter of the building foundation, the floor assembly (including posts, girders, joists and subfloor) shall be of naturally durable or preservative-treated wood.

❖ The separations specified in this section are recommended for all buildings and are considered necessary to: (1) maintain wood elements in permanent structures at safe moisture contents for decay protection, (2) provide a termite barrier and (3) facilitate periodic inspection. When it is not possible or practical to comply with the clearance specified, the use of naturally durable or pressure preservative-treated wood is recommended. However, when termites are known to exist, space must be provided for periodic inspection.

2304.11.2.2 Framing. Wood framing members, including wood sheathing, which rest on exterior foundation walls and are less than 8 inches (203 mm) from exposed earth shall be of naturally durable or preservative-treated wood. Wood framing members and furring strips attached directly to the interior of exterior masonry or concrete walls below grade shall be of approved naturally durable or preservative-treated wood.

❖ Traditionally, termite shields were used to obstruct the construction of tunnels from the soil, up the face of the foundation and into the frame. Eight inches (203 mm) of exposed foundation wall provide adequate surface area for a visual inspection for termite tubes. Wood attached directly to masonry or concrete will draw moisture from these materials, and may reach a moisture content greater than 20 percent, depending on the relative humidity of the basement.

2304.11.2.3 Sleepers and sills. Sleepers and sills on a concrete or masonry slab that is in direct contact with earth shall be of naturally durable or preservative-treated wood.

❖ Wood in contact with either the earth or material that retains moisture must be naturally durable or preservative treated. The moisture content of wood will change based on the relative humidity of the air and the moisture of products with which it is in contact. Interior slabs that are protected from the earth by an impervious moisture barrier should not be considered in direct contact with the earth for purposes of this requirement.

2304.11.2.4 Girder ends. The ends of wood girders entering exterior masonry or concrete walls shall be provided with a $^1/_2$-inch (12.7 mm) air space on top, sides and end, unless naturally durable or preservative-treated wood is used.

❖ Clearance around the ends of girders prevents moisture from migrating from the concrete into the wood. The $^1/_2$-inch (12.7 mm) space allows for air circulation and periodic inspection for termites.

2304.11.2.5 Wood siding. Clearance between wood siding and earth on the exterior of a building shall not be less than 6 inches (152 mm) except where siding, sheathing and wall framing are of naturally durable or preservative-treated wood.

❖ Exterior wood siding should extend at least 1 inch (25 mm) below the top of the foundation to provide a drip line protecting the sill from rainwater. The siding is typically held off the foundation with a starter strip attached to the sheathing that ends at the bottom of the plate. The 1-inch (25 mm) overhang causes the rainwater to drip off the bottom edge of the siding, rather than being absorbed by the underlying sheathing.

2304.11.2.6 Posts or columns. Posts or columns supporting permanent structures and supported by a concrete or masonry slab or footing that is in direct contact with the earth shall be of naturally durable or preservative-treated wood.

Exceptions:

1. Posts or columns that are either exposed to the weather or located in basements or cellars, supported by concrete piers or metal pedestals projected at least 1 inch (25 mm) above the slab or deck and 6 inches (152 mm) above exposed earth, and are separated therefrom by an impervious moisture barrier.
2. Posts or columns in enclosed crawl spaces or unexcavated areas located within the periphery of the building, supported by a concrete pier or metal pedestal at a height greater than 8 inches (203 mm) from exposed ground, and are separated therefrom by an impervious moisture barrier.

❖ A permanent structure is not defined in the code. It is a term used by the AWPA in Standard C-15 to describe appropriate retention levels of pressure-treated wood. Permanent structures are those suitable for human occupancy and are placed on a footing.

When the post or column is directly embedded in the earth or in concrete in direct contact with the earth, it must be naturally durable or treated in accordance with AWPA standards. The exceptions permit the use of common framing lumber when the post or column is isolated from the earth or concrete. Generally, structural elements in basements or cellars are easier to inspect, which explains the lower clearances. Posts or columns in crawl spaces are required to have a greater separation distance to safeguard against termites between inspections. In both exceptions, an impervious moisture barrier must be provided to insure the post or column does not absorb water from the earth or concrete pier.

2304.11.3 Laminated timbers. The portions of glued-laminated timbers that form the structural supports of a building or other structure and are exposed to weather and not properly protected by a roof, eave or similar covering shall be pressure treated with preservative, or be manufactured from naturally durable or preservative-treated wood.

❖ It is common practice to design large glued-laminated arches that "spring" or are connected to foundations near ground level. Exterior walls are built inside the span of these arches, leaving the initial few feet of the wood arches exposed to the weather. Experience has shown that covering the tops of these arches with metal or other water seals is not sufficient to prevent decay. Therefore, arches and other exposed wood members not protected by roofs or similar covers must be laminated of naturally durable or pressure-treated wood.

2304.11.4 Wood in contact with the ground or fresh water. Wood in contact with the ground (exposed earth) that supports permanent structures shall be of naturally durable (species for both decay and termite resistance) or preservative-treated wood using water-borne preservatives and shall be treated in accordance with AWPA C2, C9 or other applicable AWPA standard for soil or fresh water contact, where used in the locations specified in Sections 2304.11.4.1 and 2304.11.4.2.

Exception: Untreated wood is permitted where such wood is continuously and entirely below the ground-water level or submerged in fresh water.

❖ Wood that has been completely submerged below the ground-water table or in fresh water has been known to last in good condition for thousands of years. This method of preservation is the only exception to requiring preservative treatment of all wood in contact with or embedded in the ground.

2304.11.4.1 Posts or columns. Posts and columns supporting permanent structures that are embedded in concrete in direct contact with the earth or embedded in concrete exposed to the weather, or in direct contact with the earth, shall be of preservative-treated wood.

❖ This requirement is similar to Section 2304.11.2.6, but emphasizes that posts and columns encased or embedded in concrete must be of preservative-treated wood. The condition is generally more severe than wood placed directly on concrete or embedded in the earth. The embedded portion of the column will most likely be at a moisture content in excess of 20 percent. At some point very near where the concrete ends, the conditions will be ideal for decay—having both a high moisture content and the presence of air. Unless treated wood is used, decay will occur very near where the encasing begins.

2304.11.4.2 Wood structural members. Wood structural members that support moisture-permeable floors or roofs that are exposed to the weather, such as concrete or masonry slabs, shall be of naturally durable or preservative-treated wood unless separated from such floors or roofs by an impervious moisture barrier.

❖ The framing condition described is uncommon, but the requirement for naturally durable or preservative-treated wood is appropriate. Concrete or masonry can contain sufficient moisture to raise the moisture content in supporting wood structural elements above 20 percent. Therefore, an impervious moisture barrier or naturally durable or preservative-treated wood is required.

2304.11.5 Supporting member for permanent appurtenances. Naturally durable or preservative-treated wood shall be utilized for those portions of wood members that form the structural supports of buildings, balconies, porches or similar permanent building appurtenances where such members are exposed to the weather without adequate protection from a roof, eave, overhang or other covering to prevent moisture or water accumulation on the surface or at joints between members.

Exception: When a building is located in a geographical region where experience has demonstrated that climatic conditions preclude the need to use durable materials where the structure is exposed to the weather.

❖ Decks and porches exposed to rain and snow are subject to deterioration. In desert areas where rainfall is slight, there may not be enough moisture to cause any deterioration; therefore, the exception is provided.

2304.11.6 Termite protection. In geographical areas where the hazard of termite damage is known to be very heavy, the floor framing shall be of naturally durable or preservative-treated wood, or provided with approved methods of termite protection.

❖ Most termites found in the United States are classified as subterranean termites, which indicates their need to live in the ground. For many years, metal termite shields were considered the best method of preventing infestation. Properly installed, such shields are still a good defense. Today, soil poisoning is considered the more efficient method of controlling termites; however, the chemicals used are highly toxic.

Additionally, all possible locations of infestation must be poisoned. Therefore, the work should be carefully monitored and performed by professionals who are competent and understand the referenced document. Under no circumstances should soil poisoning be attempted by those unfamiliar with toxic products and their application.

2304.11.7 Wood used in retaining walls and cribs. Wood installed in retaining or crib walls shall be of preservative-treated wood treated in accordance with AWPA C2 or C9 for soil and fresh water contact.

❖ Where the structure is regulated by the code, wood used to retain soil is to be preservative treated. Presumably, where the failure of the retaining wall or crib may threaten life safety or result in property damage, the durability provided by preservative-treated wood is necessary.

2304.11.8 Attic ventilation. For attic ventilation, see Section 1203.2.

❖ Adequate ventilation is necessary to ensure that moisture-laden air can readily escape from the attic space (see Section 1203.2 for further discussion).

2304.11.9 Under-floor ventilation (crawl space). For under-floor ventilation (crawl space), see Section 1203.3.

❖ Enclosed spaces under floors of buildings are referred to as crawl spaces when not designed and constructed as basements. The earth is usually exposed or covered with a vapor retarder. Regardless, there is always the potential for a large amount of moisture vapor from the ground being present. This moisture must be removed to prevent wetting and drying cycles that can cause decay and fungus attack in wood members. Experience has shown that crawl spaces constructed in conformance with this section and ventilated to the exterior as required by Section 1203.3 do not develop decay problems.

Where a high ground-water table exists or moisture is abnormally high, the crawl space must be adequately drained in accordance with Section 1807.1.2. It should be recognized that the moisture problem may be exaggerated in colder months. Therefore, operable vents should be used with caution. Once insulation in such spaces becomes wet, it seldom dries out, thus negating its performance.

2304.12 Wood supporting masonry or concrete. Wood members shall not be used to permanently support the dead load of any masonry or concrete.

Exceptions:

1. Masonry or concrete nonstructural floor or roof surfacing not more than 4 inches (102 mm) thick is permitted to be supported by wood members.
2. Any structure is permitted to rest upon wood piles constructed in accordance with the requirements of Chapter 18.
3. Veneer of brick, concrete or stone applied as specified in Section 1405.5 having an installed weight of 40 pounds per square foot (psf) (1.9 kN/m^2) or less is permitted to be supported by an approved treated wood foundation when the maximum height of veneer does not exceed 30 feet (9144 mm) above the foundation. Such veneer used as an interior wall finish is permitted to be supported on wood floor construction. The wood floor construction shall be designed to support the additional weight of the veneer plus any other loads and to limit the deflection and shrinkage to $^1/_{600}$ of the span of the supporting members.
4. Glass unit masonry having an installed weight of 20 psf (0.96 kN/m^2) or less is permitted to be installed in accordance with the provisions of Section 2110. The wood construction supporting the glass unit masonry shall be designed for dead and live loads to limit deflection and shrinkage to $^1/_{600}$ of the span of the supporting members.

❖ It is common for wood structural elements to support masonry and concrete construction. When properly designed, taking into consideration long-term deflection, wood can adequately support these products. The limitations on wood supporting masonry and concrete are recommended by that industry. Masonry and concrete are brittle materials, which do not tolerate movement. Wood is hygroscopic, meaning it changes dimensions as a result of absorbing or releasing (drying) water. When improperly detailed, dimensional change in wood can result in damage to masonry and concrete.

The exceptions provide some relief. Flooring and roof coverings meeting specific thickness criteria may be supported by wood. The wood must be designed to carry the weight of the material. A strict interpretation would prohibit wood piles from supporting buildings containing masonry or concrete components; an exception permits this. Likewise, brick veneers are permitted to rest on foundation walls constructed using the permanent wood foundation system. Lastly, glass block can rest on wood floors, provided consideration is given to deflection and shrinkage.

SECTION 2305
GENERAL DESIGN REQUIREMENTS FOR LATERAL-FORCE-RESISTING SYSTEMS

2305.1 General. Structures using wood shear walls and diaphragms to resist wind, seismic and other lateral loads shall be designed and constructed in accordance with the provisions of this section.

❖ General design requirements for lateral-force-resisting systems are described in this section and are applicable to engineered structures. Elements discussed in this section are designed using either ASD or LRFD methodologies.

When elements within structures constructed according to Section 2308 need to be engineered, provisions of Section 2305 can be applied without engineering the entire structure. The extent of design to be provided must be determined by the design professional and building official. However, the minimum acceptable extent is often taken to be force transfer into the element, design of the element and force transfer out of the element. When more than one braced wall line or diaphragm in any area of conventional structures requires engineering analysis, the assumptions used to develop Section 2308 may have changed, and a complete analysis may be appropriate for the entire lateral-force-resisting system. For example, the absence of a ceiling diaphragm may create a nonconventional configuration.

2305.1.1 Shear resistance based on principles of mechanics. Shear resistance of diaphragms and shear walls are permitted to be calculated by principles of mechanics using values of fastener strength and sheathing shear resistance.

❖ In lieu of using the tabular values for shear capacity, such as those in Table 2306.3.1, this section permits calculation of shear wall and diaphragm resistance by principles of engineering mechanics. Methodologies are described in the technical literature that allow for an analytical determination of shear capacity. Design values for wood members and fastener strength in wood members can be determined in accordance with the *National Design Specification (NDS) for Wood Construction* and AF&PA/ASCE 16, *Load and Resistance Factor Design (LRFD) for Engineered Wood Construction.*

2305.1.2 Framing. Boundary elements shall be provided to transmit tension and compression forces. Perimeter members at openings shall be provided and shall be detailed to distribute the shearing stresses. Diaphragm and shear wall sheathing shall not be used to splice boundary elements. Diaphragm chords and collectors shall be placed in, or tangent to, the plane of the diaphragm framing unless it can be demonstrated that the moments, shears and deformations, considering eccentricities resulting from other configurations can be tolerated without exceeding the adjusted resistance and drift limits.

❖ The transfer of forces into and out of diaphragms and shear walls is necessary in the analysis of load path continuity. Boundary elements must be sized and connected to the diaphragm to ensure force transfer. This section provides basic framing requirements for boundary elements in shear walls and diaphragms. Good construction practice for boundary elements utilizes framing members in the plane or tangent to the plane of the diaphragm. Where splices occur, transfer of forces is usually through addition of framing members or metal connectors—not the diaphragm or shear wall sheathing.

2305.1.2.1 Framing members. Framing members shall be at least 2 inch (51 mm) nominal width. In general, adjoining panel edges shall bear and be attached to the framing members and butt along their centerlines. Nails shall be placed not less than $^3/_8$ inch (9.5 mm) from the panel edge, not more than 12 inches (305 mm) apart along intermediate supports, and 6 inches (152 mm) along panel edge bearings, and shall be firmly driven into the framing members.

❖ The prescriptive requirements in this section for the use of dimension lumber and minimum panel edge distance for nailing are intended to provide an adequate connection between sheathing and framing in shear wall and diaphragm construction. Edge spacing is intended to prevent splitting of the wood structural panel. The fastener spacing is consistent with the minimum spacing provided in Table 2304.9.1.

2305.1.3 Openings in shear panels. Openings in shear panels that materially affect their strength shall be fully detailed on the plans, and shall have their edges adequately reinforced to transfer all shearing stresses.

❖ Openings occur in shear walls and diaphragms for stairs, windows, doors, shafts and other purposes. The transfer of shear forces around openings must be effectively addressed in the design to assure adequate capacity exists around the perimeter of the opening to transfer shear forces.

2305.1.4 Shear panel connections. Positive connections and anchorages, capable of resisting the design forces, shall be provided between the shear panel and the attached components. In Seismic Design Category D, E or F, toenails shall not be used to transfer lateral forces in excess of 150 pounds per foot (2189 N/m) from diaphragms to shear walls, drag struts (collectors) or other elements, or from shear walls to other elements.

❖ This section states the requirement for connections and anchorage capable of resisting design forces with a limitation on the use of toe-nail connections for shear transfer in Seismic Design Category D, E or F. Where the transfer of lateral forces from seismic loads is in excess of 150 pounds per foot (2188 N/m), other connection methods should be utilized.

2305.1.5 Wood members resisting horizontal seismic forces contributed by masonry and concrete. Wood shear walls, diaphragms, horizontal trusses and other members shall not be used to resist horizontal seismic forces contributed by masonry or concrete construction in structures over one story in height.

Exceptions:

1. Wood floor and roof members are permitted to be used in horizontal trusses and diaphragms to resist horizontal seismic forces contributed by masonry or concrete construction (including those due to masonry veneer, fireplaces and chimneys) provided such forces do not result in torsional force distribution through the truss or diaphragm.
2. Wood structural panel sheathed shear walls are permitted to be used to provide resistance to seismic forces contributed by masonry or concrete construction in two-story structures of masonry or concrete construction, provided the following requirements are met:
   - 2.1. Story-to-story wall heights shall not exceed 12 feet (3658 mm).
   - 2.2. Diaphragms shall not be designed to transmit lateral forces by rotation. Diaphragms shall not cantilever past the outermost supporting shear wall.
   - 2.3. Combined deflections of diaphragms and shear walls shall not permit story drift of supported masonry or concrete walls to exceed the limit of Section 1617.3.
   - 2.4. Wood structural panel sheathing in diaphragms shall have unsupported edges blocked. Wood structural panel sheathing for both stories of shear walls shall have unsupported edges blocked and, for the lower story, shall have a minimum thickness of $^{15}/_{32}$ inch (11.9 mm).

2.5. There shall be no out-of-plane horizontal offsets between the first and second stories of wood structural panel shear walls.

❖ The use of wood diaphragms with masonry or concrete walls is common practice. Often referred to as "tilt-up" construction, buildings of this type can be very large in floor area. Another common construction type addressed in this section is glass front stores. Buildings constructed of three masonry or concrete walls with the fourth wall being all glass often have wood-frame roofs. The compatibility of wood and other less ductile construction materials is regulated in this section. Many of the provisions of this section are subjective and based on field observations following major seismic events.

2305.2 Design of wood diaphragms.

2305.2.1 General. Wood diaphragms are permitted to be used to resist horizontal forces provided the deflection in the plane of the diaphragm, as determined by calculations, tests or analogies drawn therefrom, does not exceed the permissible deflection of attached distributing or resisting elements. Connections shall extend into the diaphragm a sufficient distance to develop the force transferred into the diaphragm.

❖ General requirements for wood diaphragms include consideration of diaphragm strength and deflection. The transfer of forces into and out of the diaphragm must be included in the design. General requirements for diaphragms can be found in AF&PA/ASCE 16, supplements to the LRFD and ASD Manuals, and APA Research Report 138—*Plywood Diaphragms*.

2305.2.2 Deflection. Permissible deflection shall be that deflection up to which the diaphragm and any attached distributing or resisting element will maintain its structural integrity under design load conditions, such that the resisting element will continue to support design loads without danger to occupants of the structure. Calculations for diaphragm deflection shall account for the usual bending and shear components as well as any other factors, such as nail deformation, which will contribute to deflection.

The deflection (Δ) of a blocked wood structural panel diaphragm uniformly nailed throughout is permitted to be calculated by using the following formula. If not uniformly nailed, the constant 0.188 (For SI: 1/1627) in the third term must be modified accordingly.

$$\Delta = \frac{5vL^3}{8EAb} + \frac{vL}{4Gt} + 0.188Le_n + \frac{\Sigma(\Delta_c X)}{2b} \qquad \textbf{(Equation 23-1)}$$

For SI: $$\Delta = \frac{0.052L^3}{EAb} + \frac{vL}{4Gt} + \frac{Le_n}{1627} + \frac{\Sigma(\Delta_c X)}{2b}$$

where:

A = Area of chord cross section, in square inches (mm²).

b = Diaphragm width, in feet (mm).

E = Elastic modulus of chords, in pounds per square inch (N/mm²).

e_n = Nail deformation, in inches (mm).

G = Modulus of rigidity of wood structural panel, in pounds per square inch (N/mm²).

L = Diaphragm length, in feet (mm).

t = Effective thickness of wood structural panel for shear, in inches (mm).

v = Maximum shear due to design loads in the direction under consideration, in pounds per linear foot (plf) (N/mm).

Δ = The calculated deflection, in inches (mm).

$\sum(\Delta_c X)$ = Sum of individual chord-splice values on both sides of the diaphragm, each multiplied by its distance to the nearest support.

❖ Equation 23-1 provides the estimate of the deflection of a blocked, uniformly nailed wood structural panel diaphragm. Total deflection represents the sum of four components contributing to the overall deflection: bending deflection, shear deflection, nail slip and chord splice slip. Guidance on the selection of various equation inputs such as nail deformation, effective panel thickness and sum of individual chord-splice values can be obtained from AF&PA/ASCE 16 and supplements to the AF&PA ASD Manual.

2305.2.3 Diaphragm aspect ratios. Size and shape of diaphragms shall be limited as set forth in Table 2305.2.3.

❖ This section simply references the table for maximum diaphragm ratios. Maximum diaphragm ratios are based on their performance during high wind and seismic events.

TABLE 2305.2.3
MAXIMUM DIAPHRAGM DIMENSION RATIOS HORIZONTAL AND SLOPED DIAPHRAGM

TYPE	MAXIMUM LENGTH - WIDTH RATIO
Wood structural panel, nailed all edges	4:1
Wood structural panel, blocking omitted at intermediate joints	3:1
Diagonal sheathing, single	3:1
Diagonal sheathing, double	4:1

❖ This table provides maximum aspect ratios for floor and roof diaphragms using wood structural panel or diagonal board sheathing. Additional requirements for open-front structures and cantilevered diaphragms are described in Section 2305.2.5 and Figures 2305.2.5(1) and (2).

2305.2.4 Construction. Shear panels shall be constructed of wood structural panels, manufactured with exterior glue, not less than 4 feet by 8 feet (1219 mm by 2438 mm), except at boundaries and changes in framing. Boundary elements shall be connected at corners. Wood structural panel thickness for horizontal diaphragms shall not be less than set forth in Tables 2304.7(3) and 2304.7(5) for corresponding joist spacing and loads, except that $^1/_4$ inch (6.4 mm) is permitted to be used where perpendicular loads permit. Sheet-type sheathing shall be arranged so that the width of a sheet in a shear wall shall not be less than 2 feet (610 mm).

❖ Basic prescriptive requirements for diaphragm construction specify the use of 4-foot by 8-foot (1219 mm by 2438 mm) sheets except that a sheet not less than 2 feet (610 mm) wide may be used to fit the dimensions of the particular diaphragm. In addition to the panel thicknesses included in Tables 2304.7(3) and 2304.7(5), $^1/_4$-inch-thick (6.4 mm) panels may be used where the design for loads perpendicular to the panel indicates sufficient strength and stiffness. This may occur for light loads or in situations where framing members are closely spaced.

2305.2.4.1 Seismic Design Category F. Structures assigned to Seismic Design Category F shall conform to the requirements in Section 1620.5 or Section 9.5.2.6.5 of ASCE 7, and to the additional requirements of this section.

Wood structural panel sheathing used for diaphragms and shear walls that are part of the seismic-force-resisting system shall be applied directly to the framing members.

> **Exception:** Wood structural panel sheathing in a diaphragm is permitted to be fastened over solid lumber planking or laminated decking provided the panel joints and lumber planking or laminated decking joints do not coincide.

❖ As an additional precaution in structures assigned to Seismic Design Category F, requirements are provided on attachment of wood structural panel sheathing to framing members, solid lumber planking or laminated decking.

2305.2.5 Rigid diaphragms. Design of structures with rigid diaphragms shall conform to the structure configuration requirements of Section 9.5.2.3 of ASCE 7 and the horizontal shear distribution requirements of Section 9.5.5.5 of ASCE 7.

Open front structures with rigid wood diaphragms resulting in torsional force distribution are permitted provided the length, *l*, of the diaphragm normal to the open side does not exceed 25 feet (7620 mm), the diaphragm sheathing conforms to Section 2305.2.4, and the *l/w* ratio [as shown in Figure 2305.2.5(1)] is less than 1.0 for one-story structures or 0.67 for structures over one story in height.

> **Exception:** Where calculations show that diaphragm deflections can be tolerated, the length, *l*, normal to the open end is permitted to be increased to a *l/w* ratio not greater than 1.5 where sheathed in compliance with Section 2305.2.4 or to 1.0 where sheathed in compliance with Section 2306.3.4 or 2306.3.5.

Rigid wood diaphragms are permitted to cantilever past the outermost supporting shear wall (or other vertical resisting element) a length, *l*, of not more than 25 feet (7620 mm) or two-thirds of the diaphragm width, *w*, whichever is the smaller. Figure 2305.2.5(2) illustrates the dimensions of *l* and *w* for a cantilevered diaphragm.

Structures with rigid wood diaphragms having a torsional irregularity in accordance with Table 1616.5.1.1, Item 1, shall meet the following requirements: The *l/w* ratio shall not exceed 1.0 for one-story structures or 0.67 for structures over one story in height, where *l* is the dimension parallel to the load direction for which the irregularity exists.

> **Exception:** Where calculations demonstrate that the diaphragm deflections can be tolerated, the width is permitted to be increased and the *l/w* ratio is permitted to be increased to 1.5 where sheathed in compliance with Section 2305.2.4 or 1.0 where sheathed in compliance with Section 2306.3.4 or 2306.3.5.

❖ For diaphragms defined as rigid, rotational or torsional behavior is expected and results in redistribution of shear to the vertical-force-resisting elements. Additional prescriptive limits are placed on rigid diaphragms based on the type of irregularity and diaphragm construction. Further discussion of these limits is available in Chapter 12 of the NEHRP Provisions Commentary.

FIGURES 2305.2.5(1) and 2305.2.5(2). See page 23-32.

❖ The source of Figures 2305.2.5(1) and (2) is Chapter 12 of the NEHRP Provisions. The figures provide the meaning of the *l* and *W* terms that are specified in Section 2305.2.5.

2305.3 Design of wood shear walls.

2305.3.1 General. Wood shear walls are permitted to resist horizontal forces in vertical distributing or resisting elements, provided the deflection in the plane of the shear wall, as determined by calculations, tests or analogies drawn therefrom, does not exceed the more restrictive of the permissible deflection of attached distributing or resisting elements or the drift limits of Section 1617.3. Shear wall sheathing other than wood structural panels shall not be permitted in Seismic Design Category E or F (see Section 1617.6).

❖ General requirements for wood shear walls include consideration of shear wall strength and deflection. General requirements for wood shear walls can be found in AF&PA/ASCE 16, Supplements to the LRFD and ASD Manuals, and APA Research Report 154—*Wood Structural Panel Shear Wall.*

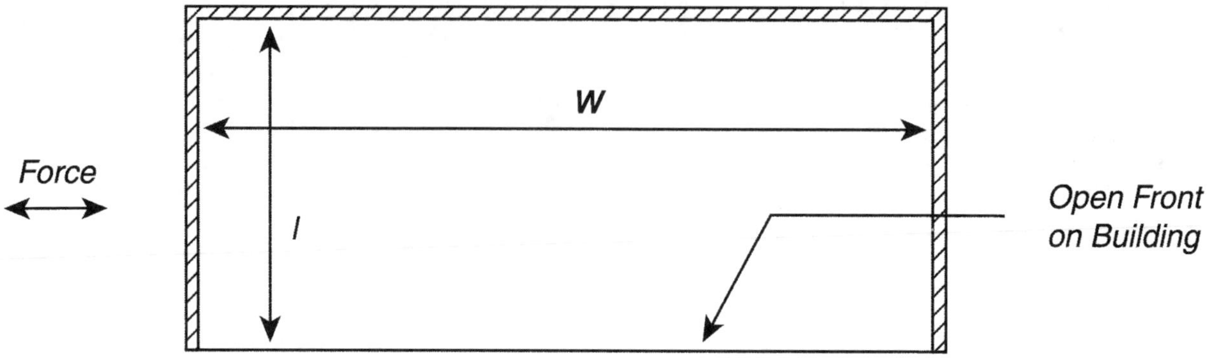

FIGURE 2305.2.5(1)
DIAPHRAGM LENGTH AND WIDTH FOR PLAN VIEW OF OPEN-FRONT BUILDING

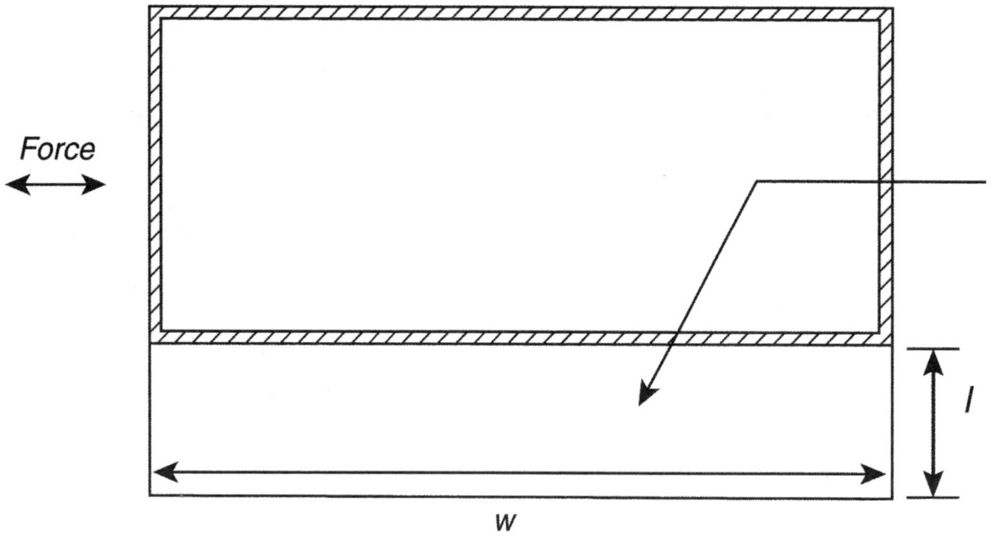

FIGURE 2305.2.5(2)
DIAPHRAGM LENGTH AND WIDTH FOR PLAN VIEW OF CANTILEVERED DIAPHRAGM

2305.3.2 Deflection. Permissible deflection shall be that deflection up to which the shear wall and any attached distributing or resisting element will maintain its structural integrity under design load conditions, i.e., continue to support design loads without danger to occupants of the structure.

The deflection (Δ) of a blocked wood structural panel shear wall uniformly fastened throughout is permitted to be calculated by the use of the following formula:

$$\Delta = \frac{8vh^3}{EAb} + \frac{vh}{Gt} + 0.75he_n + d_a \qquad \textbf{(Equation 23-2)}$$

$$\text{For SI: } \Delta = \frac{vh^3}{3EAb} + \frac{vh}{Gt} + \frac{he_n}{406.7} + d_a$$

where:

A = Area of boundary element cross section in square inches (mm²) (vertical member at shear wall bound ary).

b = Wall width, in feet (mm).

d_a = Deflection due to anchorage details (rotation and slip at tie-down bolts).

E = Elastic modulus of boundary element (vertical member at shear wall boundary), in pounds per square inch (N/mm²).

e_n = Deformation of mechanically fastened connections, in inches (mm²).

G = Modulus of rigidity of wood structural panel, in pounds per square inch (N/mm²).

h = Wall height, in feet (mm).

t = Effective thickness of wood structural panel for shear, in inches (mm).

v = Maximum shear due to design loads at the top of the wall, in pounds per linear foot (N/mm).

Δ = The calculated deflection, in inches (mm).

❖ The equation in this section provides an estimate of the deflection of a blocked wood structural panel shear wall. Total deflection represents the sum of four components contributing to overall deflection: bending deflection, shear deflection, fastener slip and anchorage deformation. Guidance on the selection of various equation inputs can be obtained from AF&PA/ASCE 16 and Supplements to the AF&PA ASD Manual.

2305.3.3 Shear wall aspect ratios. Size and shape of shear walls and shear wall segments within shear walls containing openings shall be limited as set forth in Table 2305.3.3.

❖ See the commentary to Table 2305.3.3.

TABLE 2305.3.3
MAXIMUM SHEAR WALL ASPECT RATIOS

TYPE	MAXIMUM HEIGHT-WIDTH RATIO
Wood structural panels or particleboard, nailed edges	For other than seismic: 3¹/₂:1 For seismic: 2:1[a]
Diagonal sheathing, single	2:1
Fiberboard	1¹/₂:1
Gypsum board, gypsum lath, cement plaster	1¹/₂:1[b]

a. For design to resist seismic forces, shear wall height-width ratios greater than 2:1, but not exceeding 3¹/₂:1, are permitted provided the allowable shear values in Table 2306.4.1 are multiplied by $2w/h$.

b. Ratio shown is for unblocked construction. Aspect ratio is permitted to be 2:1 where the wall is installed as blocked construction in accordance with Section 2306.4.5.1.2.

❖ Table 2305.3.3 provides maximum shear wall aspect ratios for wood-frame shear walls sheathed with wood structural panel, particleboard, fiberboard, diagonal sheathing, gypsum board, gypsum lath or cement plaster.

For shear walls with wood structural panel or particleboard sheathing that resist seismic loads, the maximum aspect ratio is 2:1 unless the tabulated allowable shear values are adjusted in accordance with Note a. This approach for addressing larger aspect ratios is based on the *Recommended Provisions for Seismic Regulations for New Buildings and Other Structures* (NEHRP 2000). Based on review of test data of larger aspect ratio shear walls, this adjustment addresses the concern of the reduced stiffness associated with these larger aspect ratios. Also note that the resistance to loading other than sesimic permits the use of a larger aspect ratio.

2305.3.4 Shear wall height definition. The height of a shear wall shall be defined as:

1. The maximum clear height from top of foundation to bottom of diaphragm framing above; or
2. The maximum clear height from top of diaphragm to bottom of diaphragm framing above [see Figure 2305.3.4(a)].

❖ The shear wall height definition is needed for the evaluation of the maximum aspect ratios in Table 2305.3.3. This height definition applies to shear walls that have not been designed for force transfer around openings in accordance with Section 2305.7.1.

FIGURE 2305.3.4. See page 23-34.

❖ Figure 2305.3.4 is provided to illustrate the determination of the height and width of a shear wall.

2305.3.5 Shear wall width definition. The width of a shear wall shall be defined as the sheathed dimension of the shear wall in the direction of application of force [see Figure 2305.3.4(a)].

❖ Shear wall width, as defined in this section, is provided for the purpose of calculating maximum height-to-width ratios for shear walls. When used to calculate shear wall aspect ratio, it is intended that a standard 4-foot by 8-foot (1219 mm by 2438 mm) panel used in standard 8-foot (2438 mm) wall construction should qualify as meeting an aspect ratio of 2:1.

2305.3.5.1 Shear wall segment width definition. The width of full-height sheathing adjacent to unrestrained openings in a shear wall.

❖ Shear wall segments refer to those portions of a shear wall on either side of an opening. Only portions of the wall with full-height sheathing can be considered as contributing to the width of a segment and in satisfying the aspect ratio limits of Section 2305.3.3.

2305.3.6 Overturning restraint. Where the dead load stabilizing moment in accordance with Chapter 16 allowable stress design load combinations is not sufficient to prevent uplift due to overturning moments on the wall, an anchoring device shall be provided. Anchoring devices shall maintain a continuous load path to the foundation.

❖ This section provides a general requirement to maintain a continuous load path to the foundation for uplift due to wall overturning forces. The location or frequency of this restraint depends on the design method used. The segmented shear wall design approach requires that each segment be evaluated individually for overturning. The perforated shear wall method requires that anchorage be provided at each end of the shear wall. Typically, these tie-downs, or hold-downs, consist of metal straps or other hardware attached to the shear wall end posts.

2305.3.7 Shear walls with openings. The provisions of this section shall apply to the design of shear walls with openings. Where framing and connections around the openings are designed for force transfer around the openings, the provisions of

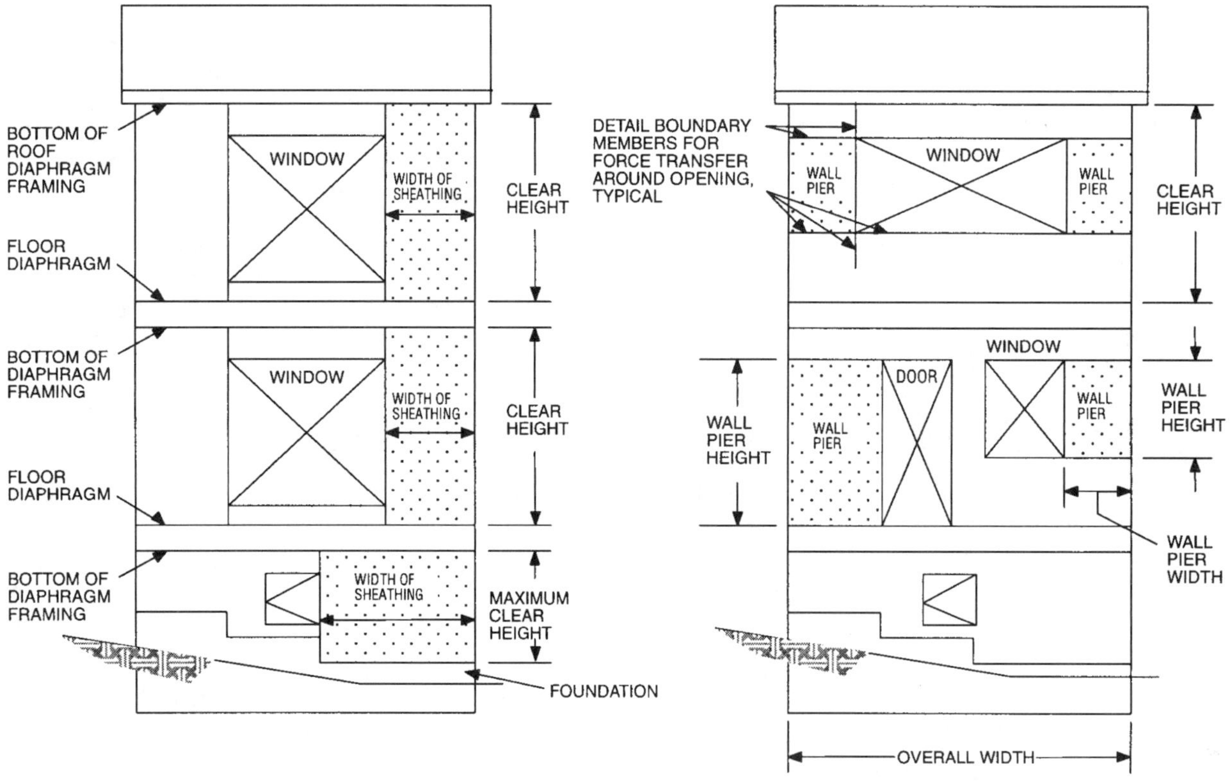

(a) HEIGHT-TO-WIDTH RATIO

(b) HEIGHT-TO-WIDTH RATIO WITH DESIGN FOR FORCE TRANSFER AROUND OPENINGS

FIGURE 2305.3.4
GENERAL DEFINITION OF SHEAR WALL HEIGHT, WIDTH AND HEIGHT-TO-WIDTH RATIO

Section 2305.3.7.1 shall apply. Where framing and connections around the openings are not designed for force transfer around the openings, the provisions of Section 2305.3.7.2 shall apply.

❖ This section suggests that there are only two options for designing shear walls with openings. If one considers the commonly used segmented shear wall design approach, there are, in fact, three methods that could be used.

When framing and connections around the openings are specifically designed and detailed for force transfer around the opening, the provisions of Section 2305.3.7.1 apply. This method is an alternative to segmented shear wall design wherein the shear wall segments are as shown in Figure 2305.3.4(a). In Section 2305.3.7.1, design and detailing around openings are based on rigid body behavior and utilize the full shear resistance developed in the wood structural panel shear wall. The benefit of this additional design and detailing is that the height-to-width of wall piers can be determined as indicated in Figure 2305.3.4(b), potentially allowing a narrower width pier to comply with the shear wall aspect ratio limits of Section 2305.3.3.

As an option to either the segmented shear wall design method or design for force transfer around openings, the perforated shear wall provisions of Section 2305.3.7.2 can be utilized. In the perforated shear wall method, the wall is not assumed to act as a rigid body and shear capacity is reduced, accounting for the reduced strength and stiffness of unrestrained shear wall segments within the perforated shear wall.

2305.3.7.1 Force transfer around openings. Where shear walls with openings are designed for force transfer around the openings, the limitations of Table 2305.3.3 shall apply to the overall shear wall including openings and to each wall pier at the side of an opening. The height of a wall pier shall be defined as the clear height of the pier at the side of an opening. The width of a wall pier shall be defined as the sheathed width of the pier at the side of an opening. Design for force transfer shall be based on a rational analysis. Detailing of boundary elements around the opening shall be provided in accordance with the provisions of this section [see Figure 2305.3.4(b)].

❖ Shear walls designed using provisions for force transfer around openings utilize aspect ratio limits defined in Table 2305.3.3; however, definitions of wall pier height and width are used to calculate the aspect ratio. Wall pier width and height are shown in Figure 2305.3.4(b). In this approach for design of shear walls with openings, design force transfer around the openings involves developing a system of piers and coupling beams within the shear wall. Load paths for the shear and flexure developed in the piers and coupling beams generally require blocking and strapping extending from each corner of the opening to some distance beyond. This approach is often utilized in highly loaded walls when limited amounts of sheathing are present around openings and often results in shear wall detailing involving many metal tension straps around the openings.

2305.3.7.2 Perforated shear walls. The provisions of Section 2305.3.7.2 shall be permitted to be used for the design of perforated shear walls.

❖ Shear walls meeting the provisions of this section are commonly referred to as perforated shear walls. Aspect ratio limits of Section 2305.3.3 apply to shear wall segments within the perforated shear wall as described by the definition of "Perforated shear wall segment" in Section 2302.

This method of shear wall design utilizes empirically determined reductions (see Table 2305.3.7.2) of wood structural panel shear wall capacities that are based on the maximum opening height as well as the percentage of a particular shear wall that qualifies as a perforated shear wall segment(s). This method recognizes the strength and stiffness of sheathed areas above and below openings without the need to specifically design and detail for force transfer as required under Section 2305.3.7.1. It is not expected that sheathed wall areas above and below openings behave as coupling beams acting end to end, but rather that they provide local restraint at their ends. As a consequence, significantly reduced capacities are attributed to interior perforated shear wall segments with limited overturning restraint. Further background on the development of provisions for perforated shear walls and example problems are provided in the 2000 NEHRP Commentary.

TABLE 2305.3.7.2. See page 23-36.

❖ See the commentary to Section 2305.3.7.2.

2305.3.7.2.1 Limitations. The following limitations shall apply to the use of Section 2305.3.7.2:

1. A perforated shear wall segment shall be located at each end of a perforated shear wall. Openings shall be permitted to occur beyond the ends of the perforated shear wall; however, the width of such openings shall not be included in the width of the perforated shear wall.
2. The allowable shear set forth in Table 2306.4.1 shall not exceed 490 plf (7150 N/m).
3. Where out-of-plane offsets occur, portions of the wall on each side of the offset shall be considered as separate perforated shear walls.
4. Collectors for shear transfer shall be provided through the full length of the perforated shear wall.
5. A perforated shear wall shall have uniform top of wall and bottom of wall elevations. Perforated shear walls not having uniform elevations shall be designed by other methods.
6. Perforated shear wall height, *h*, shall not exceed 20 feet (6096 mm).

❖ This section limits the configuration of perforated shear walls. Item 2 limits the allowable shear values of the shear wall assemblies in order to rule out high-demand shear wall assemblies that are not considered suitable for this method.

TABLE 2305.3.7.2
SHEAR RESISTANCE ADJUSTMENT FACTOR, C_o

WALL HEIGHT, H	MAXIMUM OPENING HEIGHT[a]				
	H/3	H/2	2H/3	5H/6	H
8′ wall	2′-8″	4′-0″	5′-4″	6′-8″	8′-0″
10′ wall	3′-4″	5′-0″	6′-8″	8′-4″	10′-0″
Percent full-height sheathing[b]	**Shear resistance adjustment factor**				
10%	1.00	0.69	0.53	0.43	0.36
20%	1.00	0.71	0.56	0.45	0.38
30%	1.00	0.74	0.59	0.49	0.42
40%	1.00	0.77	0.63	0.53	0.45
50%	1.00	0.80	0.67	0.57	0.50
60%	1.00	0.83	0.71	0.63	0.56
70%	1.00	0.87	0.77	0.69	0.63
80%	1.00	0.91	0.83	0.77	0.71
90%	1.00	0.95	0.91	0.87	0.83
100%	1.00	1.00	1.00	1.00	1.00

For SI: 1 inch = 25.4 mm, 1 foot = 304.8 mm.

a. See Section 2305.3.7.2.2, Item 2.

b. See Section 2305.3.7.2.2, Item 1.

2305.3.7.2.2 Perforated shear wall resistance. The resistance of a perforated shear wall shall be calculated in accordance with the following:

1. The percent of full-height sheathing shall be calculated as the sum of the widths of perforated shear wall segments divided by the total width of the perforated shear wall including openings.
2. The maximum opening height shall be taken as the maximum opening clear height. Where areas above and below an opening remain unsheathed, the height of opening shall be defined as the height of the wall.
3. The adjusted shear resistance shall be calculated by multiplying the unadjusted shear resistance by the shear resistance adjustment factors of Table 2305.3.7.2. For intermediate percentages of full-height sheathing, the values in Table 2305.3.7.2 are permitted to be interpolated.
4. The perforated shear wall resistance shall be equal to the adjusted shear resistance times the sum of the widths of the perforated shear wall segments.

❖ This section describes the determination of perforated shear wall resistance. First, however, it is necessary to determine the unadjusted shear resistance as described in the definition in Section 2302. For the wall under consideration, determine the percentage of full-height sheathing (see Item 1) and the maximum opening height ratio (see Item 2). Based on these two parameters, determine the shear resistance adjustment factor from Table 2305.3.7.2 and then calculate the adjusted shear resistance according to Item 3. The product of the adjusted shear resistance and the total widths of perforated shear wall segments in the wall under consideration is the shear resistance of that shear wall.

2305.3.7.2.3 Anchorage and load path. Design of perforated shear wall anchorage and load path shall conform to the requirements of Sections 2305.3.7.2.4 through 2305.3.7.2.8, or shall be calculated using principles of mechanics. Except as modified by these sections, wall framing, sheathing, sheathing attachment and fastener schedules shall conform to the requirements of Section 2305.2.4 and Table 2306.4.1.

❖ This section prescribes minimum design forces for uplift anchorage and compression chords in perforated shear walls. These requirements conservatively address the nonuniform shear distribution occurring in perforated shear walls. Alternative designs based on principles of mechanics are permitted.

2305.3.7.2.4 Uplift anchorage at perforated shear wall ends. Anchorage for uplift forces due to overturning shall be provided at each end of the perforated shear wall. The uplift anchorage shall conform to the requirements of Section 2305.3.6 except that for each story the minimum tension chord uplift force, T, shall be calculated in accordance with the following:

$$T = \frac{Vh}{C_o \Sigma L_i} \qquad \text{(Equation 23-3)}$$

where:

T = Tension chord uplift force, pounds (N).

V = Shear force in perforated shear wall, pounds (N).

h = Shear wall height, feet (mm).

C_o = Shear resistance adjustment factor from Table 2305.3.7.2.

ΣL_i = Sum of widths of perforated shear wall segments, feet (mm).

❖ Anchorage against uplift is required only at the ends of the perforated shear wall as opposed to providing full restraint of each segment as is the case when designing using a segmented shear wall approach. The uplift force due to overturning should be computed using Equation 23-3.

2305.3.7.2.5 Anchorage for in-plane shear. The unit shear force, *v*, transmitted into the top of a perforated shear wall, out of the base of the perforated shear wall at full-height sheathing and into collectors (drag struts) connecting shear wall segments, shall be calculated in accordance with the following:

$$v = \frac{V}{C_o \Sigma L_i} \qquad \textbf{(Equation 23-4)}$$

where:

v = Unit shear force, pounds per lineal feet (N/m).

V = Shear force in perforated shear wall, pounds (N).

C_o = Shear resistance adjustment factor from Table 2305.3.7.2.

ΣL_i = Sum of widths of perforated shear wall segments, feet (mm).

❖ This section prescribes the unit shear force, *v*, to be applied to the top and bottom of each perforated shear wall segment. Designing for a unit shear, as well as the uniform uplift force of Section 2305.3.7.2.6, that are based on the unadjusted shear capacity provides a sheathing-to-bottom plate attachment such that the capacity of the shear wall will not be governed by this connection.

2305.3.7.2.6 Uplift anchorage between perforated shear wall ends. In addition to the requirements of Section 2305.3.7.2.4, perforated shear wall bottom plates at full-height sheathing shall be anchored for a uniform uplift force, *t*, equal to the unit shear force, *v*, determined in Section 2305.3.7.2.5.

❖ The unit uplift force, *t*, is taken equal to the unit shear force, *v* (see commentary, Section 2305.3.7.2.5).

2305.3.7.2.7 Compression chords. Each end of each perforated shear wall segment shall be designed for a compression chord force, *C*, equal to the tension chord uplift force, *T*, calculated in Section 2305.3.7.2.4.

❖ The vertical members on either side of all perforated shear wall segments must be designed for the required compression chord force.

2305.3.7.2.8 Load path. A load path to the foundation shall be provided for each uplift force, *T* and *t*, for each shear force, *V* and *v*, and for each compression chord force, *C*. Elements resisting shear wall forces contributed by multiple stories shall be designed for the sum of forces contributed by each story.

❖ A load path for the forces prescribed in this section is necessary. Consideration must be given to shear wall forces from a story, or stories, above in multistory buildings.

2305.3.7.2.9 Deflection of shear walls with openings. The controlling deflection of a blocked shear wall with openings uniformly nailed throughout shall be taken as the maximum individual deflection of the shear wall segments calculated in accordance with Section 2305.3.2, divided by the appropriate shear resistance adjustment factors of Table 2305.3.7.2.

❖ The provision for calculation of deflection of a perforated shear wall is based on the observation from testing that the reduction in strength is comparable to the reduction in stiffness.

2305.3.8 Summing shear capacities. The shear values for shear panels of different capacities applied to the same side of the wall are not cumulative except as allowed in Table 2306.4.1.

The shear values for material of the same type and capacity applied to both faces of the same wall are cumulative. Where the material capacities are not equal, the allowable shear shall be either two times the smaller shear capacity or the capacity of the stronger side, whichever is greater.

Summing shear capacities of dissimilar materials applied to opposite faces or to the same wall line is not allowed.

Exception: For wind design, the allowable shear capacity of shear wall segments sheathed with a combination of wood structural panels and gypsum wallboard on opposite faces, fiberboard structural sheathing and gypsum wallboard on opposite faces or hardboard panel siding and gypsum wallboard on opposite faces shall equal the sum of the sheathing capacities of each face separately.

❖ This section provides general requirements for summing shear capacities of sheathing materials applied to either one side of a wall or both sides of the same wall. The capacity of a wall that is sheathed on opposite sides, with the same material type but different capacities, is taken as the greater of either the capacity of the stronger side or twice the shear value of the lesser capacity.

Generally, summing shear capacities of different sheathing materials is not permitted. However, the exception reflects the results of testing conducted to confirm that the summing of shear resistance of different sheathing materials is justifiable for certain combinations of materials. Test results for wood structural panels in combination with gypsum wallboard were provided in APA Report 157.

2305.3.9 Adhesives. Adhesive attachment of shear wall sheathing is not permitted as a substitute for mechanical fasteners, and shall not be used in shear wall strength calculations alone, or in combination with mechanical fasteners in Seismic Design Category D, E or F.

❖ The restriction against adhesive attachment of shear wall sheathing in high seismic areas addresses a concern that the increased stiffness associated with use of adhesives will attract more load to the shear wall than anticipated and could cause brittle failure of anchoring devices.

2305.3.10 Sill plate size and anchorage in Seismic Design Category D, E or F. Two-inch (51 mm) nominal wood sill plates for shear walls shall include steel plate washers, a minimum of $^3/_{16}$ inch by 2 inches by 2 inches (4.76 mm by 51 mm by 51 mm) in size, between the sill plate and nut. Sill plates resisting a design load greater than 490 plf (LRFD) (7154 N/m) or 350 plf (ASD) (5110 N/m) shall not be less than a 3-inch (76 mm) nominal member. Where a single 3-inch (76 mm) nominal sill plate is used, 2-20d box end nails shall be substituted for 2-16d common end nails found in Line 8 of Table 2304.9.1.

> **Exception:** In shear walls where the design load is less than 840 plf (LRFD) (12 264 N/m) or 600 plf (ASD) (8760 N/m), the sill plate is permitted to be a 2-inch (51 mm) nominal member if the sill plate is anchored by two times the number of bolts required by design and $^3/_{16}$ inch by 2 inch by 2 inch (4.76 mm by 51 mm by 51 mm) plate washers are used.

❖ This provision addresses a concern over bottom plate damage observed in highly loaded sill plates in Seismic Design Category D, E or F. A minimum $^3/_{16}$-inch by 2-inch by 2-inch (4.8 mm by 51 mm by 51 mm) plate washer is required between the sill plate and nut when attaching the sill plate to the foundation. It is expected that this plate washer will reduce cross-grain bending forces in the sill plate. In addition, a 3-inch (76 mm) sill plate is required when the design load is greater than 350 plf (5110 N/m) (ASD) as an added measure of strength against cross-grain bending forces. Where the single 3-inch (76 mm) sill plate is used, longer end nails should be substituted to provide sufficient penetration into the stud.

The exception permitting 2-inch (51 mm) plates enables the use of 2-inch (51 mm) sill plates for design loads up to 600 plf (8760 N/m) (ASD) provided that twice the number of anchor bolts required by design are used to anchor the sill plate. While this requirement may not be practical for new construction, it is expected that this option would be used to upgrade existing structures with 2-inch (51 mm) sill plates.

SECTION 2306
ALLOWABLE STRESS DESIGN

2306.1 Allowable stress design. The structural analysis and construction of wood elements in structures using allowable design methods shall be in accordance with the following applicable standards:

❖ This section contains a number of specifications intended to provide guidance to nonprofessional as well as professional users of the code. These specifications, based on accepted engineering practices and experiences, are considered to be the minimum acceptable methods for constructing wood elements in structures. When designed and built in accordance with the standards listed in this section, a building or structure is deemed to comply with the code. These standards contain most of the information needed to adequately design a structure in accordance with the ASD method. It is necessary for the designer to have a working knowledge of engineering principles and experience to properly interpret these recommendations and meet the provisions of other applicable sections of the code.

The most common and applicable practices are summarized in the standards listed in Section 2306.1. These standards contain references to additional technical publications that address special problems and design criteria for wood structures. The users of these standards should have training and expertise in engineering and construction in order to properly interpret the standards' requirements. These standards employ ASD—the traditional format for presenting safety checking equations in structural design standards. The basic design equation for ASD requires that the specified product allowable design value meet or exceed the actual (applied) stress or other effect imposed by the specified loads. In ASD, the allowable design values are set very low and the nominal load magnitudes are set at once-in-a-lifetime levels. This combination produces designs that maintain high safety levels yet remain economically feasible.

While preventing structural collapse remains the primary purpose of structural design, the designer must also consider how the design will perform from the perspective of serviceability, durability and fire safety. The ASD approach is to provide adequate resistance to certain limit states for serviceability. Serviceability limit states are those that restrict the normal use and occupancy of the structure, such as excessive deflection and vibration. The code defines these limits as a ratio of the member span, for example, *L*/360 computed under live load or *L*/240 under total load for floors.

American Forest & Paper Association.
NDS National Design Specification for Wood Construction

❖ The AF&PA *National Design Specification (NDS) for Wood Construction* is promulgated and distributed by the American Forest & Paper Association (AF&PA), listed in Chapter 35. The NDS® was first adopted in 1944 and has been updated periodically to reflect new knowledge under the auspices of AF&PA and its predecessor organizations, the National Lumber Manufacturers Association (NLMA) and the National Forest Products Association (NFPA). A supplement to the NDS collates all of the recommended working stress values published by the approved grade-rules writing agencies. The NDS is included in AF&PA's *Allowable Stress Design Manual for Wood Construction* (ASD Manual). The ASD Manual provides guidance for selection of most wood-based structural products used in the construction of wood buildings.

AF&PA also publishes *Wood Construction Data (WCD) No. 5, Heavy Timber Construction Details*. It is written in code language for easy reference. WCD No. 5 contains detailed specifications for sizes of members required and other provisions, which collectively provide a definition of heavy timber construction. Recommendations are also included for the fabrication and erection of this unique construction. It should be used in conjunction with the NDS and the American Institute of Timber Construction (AITC) specifications when the members are glued laminated.

The *Supplement to the National Design Specification (NDS®)* is a collation of the allowable working stress values for sawn lumber as published by seven grade-rules writing agencies: National Lumber Grades Authority (Canada), Northern Softwood Lumber Bureau, Northeastern Lumber Manufacturers Association, Redwood Inspection Bureau, Southern Pine Inspection Bureau, West Coast Lumber Inspection Bureau and Western Wood Products Association.

These allowable working stress values, often called design values, have been approved by the Board of Review of the ALSC with advice from the U.S. Forest Products Laboratory of the U.S. Department of Agriculture. They have been certified for conformance with the U.S. Department of Commerce Voluntary Standard PS 20. Regional grading agencies formulate and publish grading rules, and ALSC approves these rules as conforming with PS 20.

PS 20 requires that design values for visually graded lumber be developed in accordance with appropriate ASTM standards or other technically sound criteria. The design values reflected in the NDS Supplement are derived from ASTM D 1990, which is utilized to derive strength values from full-sized samples of lumber.

Design values for visually graded timbers, decking and some species and grades of dimension lumber are based on the provisions of ASTM D 245. The methods in ASTM D 245 involve adjusting the strength properties of small clear specimens of wood for the effects of knots, slope of grain, splits, checks, size, duration of load, moisture content and other influencing factors, to obtain design values applicable to normal conditions of service.

The design values given in the supplement are for normal loading conditions and under dry service conditions in a covered building. If other conditions of loadings or wet end-use conditions exist, these values must be adjusted in accordance with the recommendations of the NDS.

American Institute of Timber Construction.

AITC 104	Typical Construction Details
AITC 110	Standard Appearance Grades for Structural Glued Laminated Timber
AITC 112	Standard for Tongue-and-Groove Heavy Timber Roof Decking
AITC 113	Standard for Dimensions of Structural Glued Laminated Timber
AITC 117	Standard Specifications for Structural Glued Laminated Timber of Softwood Species
AITC 119	Structural Standard Specifications for Glued Laminated Timber of Hardwood Species
AITC A190.1	Structural Glued Laminated Timber
AITC 200	Inspection Manual
AITC 500	Determination of Design Values for Structural Glued Laminated Timber

❖ The AITC standards listed are primarily intended for glued-laminated structural members, but also include recommendations for sawn timber members. These standards supplement the provisions of the NDS and WCD No. 5 with no conflicting provisions. All the standards should be used collectively.

AITC Standard A190.1 is intended to provide nationally recognized requirements for the production, inspection, testing and certification of structural glued-laminated timber. Section 6 of the standard provides detailed requirements for in-plant quality control and the qualification of third-party inspection and testing agencies. These specifications include provisions for required testing during and after fabrication. Section 7 outlines the requirements for marking, which must be distinctive and must identify the edition of AITC A190.1 used, the inspection and testing agency and the laminating plant.

The marking on other than custom members meeting specific job specifications must also include information on: species or species group of lumber, design values used (if not standard), intended use of member (simple beam, tension, or compression member, etc.), type of adhesive (dry-wet-use), appearance grade (industrial-architectural-premium), treatment (if any) and proofloading (if applicable).

Truss Plate Institute, Inc.

TPI 1 National Design Standard for Metal Plate Connected Wood Truss Construction

❖ TPI 1 is the standard for the design, fabrication, quality control and erection of light-frame metal-plate-connected wood trusses. The standard was prepared through a joint effort by the American National Standards Institute (ANSI) and the Truss Plate Institute (TPI). This standard should be used in conjunction with other standards for wood construction.

American Society of Agricultural Engineers.

ASAE EP 484.2	Diaphragm Design of Metal-Clad, Post-Frame Rectangular Buildings
ASAE EP 486.1	Shallow Post Foundation Design
ASAE 559	Design Requirements and Bending Properties for Mechanically Laminated Columns

❖ ASAE develops standards that are needed for the construction of agricultural and utility buildings.

APA—The Engineered Wood Association.

Plywood Design Specification

Plywood Design Specification Supplement 1 - Design & Fabrication of Plywood Curved Panels.

Plywood Design Specification Supplement 2 - Design & Fabrication of Glued Plywood-Lumber beams.

Plywood Design Specification Supplement 3 - Design & Fabrication of Plywood Stressed-Skin Panels.

Plywood Design Specification Supplement 4 - Design & Fabrication of Plywood Sandwich Panels.

Plywood Design Specification Supplement 5 - Design & Fabrication of All-Plywood Beams.

EWS T300	Glulam Connection Details
EWS S560	Field Notching and Drilling of Glued Laminated Timber Beams
EWS S475	Glued Laminated Beam Design Tables
EWS X450	Glulam in Residential Construction
EWS X440	Product and Application Guide: Glulam
EWS R540	Builders Tips: Proper Storage and Handling of Glulam Beams

❖ APA recommendations and design standards are considered to provide the best criteria for the design and erection of structures incorporating plywood and other structural-use panel products.

2306.1.1 Joists and rafters. The design of rafter spans is permitted to be in accordance with the *AF&PA Span Tables for Joists and Rafters*.

❖ Spans for joists and rafters may be designed in accordance with the AF&PA *Span Tables for Joists and Rafters* or in accordance with approved engineering standards. When sizes of floor joists are other than standard sizes conforming to U.S. Department of Commerce PS 20, they must conform to the NDS. When applicable, other span tables may be used, such as those from the *International Residential Code*® (IRC®), Southern Forest Products Association, Western Wood Products Association and Canadian Wood Council. The spans given in these tables are the same as those published in AF&PA *Span Tables for Joists and Rafters*. The AF&PA span tables were developed to provide uniformity in use of wood joists and rafters for light wood-frame structures where the live loads do not exceed those for which the tables were designed. The tables for floor joists were designed with the live load, dead load and deflection criteria shown in Chapter 16.

2306.1.2 Plank and beam flooring. The design of plank and beam flooring is permitted to be in accordance with the *AF&PA Wood Construction Data No. 4*.

❖ The plank-and-beam method for framing floors and roofs has been used in buildings for many years. The adaptation of this system to residential construction has raised technical issues concerning details of application. AF&PA's *Wood Construction Data No. 4, Plank and Beam Framing for Residential Buildings*, contains information pertaining to design principles, advantages and limitations, construction details and structural requirements for the plank-and-beam method of framing.

2306.1.3 Treated wood stress adjustments. The allowable unit stresses for preservative-treated wood need no adjustment for treatment, but are subject to other adjustments.

The allowable unit stresses for fire-retardant-treated wood, including fastener values, shall be developed from an approved method of investigation that considers the effects of anticipated temperature and humidity to which the fire-retardant-treated wood will be subjected, the type of treatment and the redrying process. Other adjustments are applicable except that the impact load duration shall not apply.

❖ The allowable stresses for FRTW products must be determined by testing that takes into account anticipated temperatures and humidity levels that will be encountered during service. The allowable stresses for untreated products are not applicable to fire-retardant-treated products, because the treatment chemicals can reduce the strength of materials. This is particularly true for products such as roof sheathing which is subject to conditions of high temperature and humidity.

Before approving the use of FRTW products, the building official should require the manufacturer of each subject product to submit its recommended allowable stresses and provide documentation on how those values were determined. This documentation should be required for each chemical that is to be used, because performance will vary between formulations. Some manufacturers maintain evaluation reports which can be used by the building official in reviewing these products.

Additionally, the building official should enforce any restriction on the use and location of any such treated product recommended by the manufacturer or justified based on the nature of the testing performed. For example, when these products are proposed for use in attics, the building official should make sure that they have been tested under conditions that will insure suitable performance under conditions of high temperature and humidity typically found in attics.

The building official should also be alerted to the fact that the ventilation requirement in the code takes on added importance when such products are used in attics. Ventilation requirements should be strictly enforced, and the building official should consult the manufacturer to determine if any additional ventilation above the code requirements might be necessary. Ventilation is needed to control temperature and humidity levels to which the products will be subjected. A lack of adequate ventilation can change the service conditions to which the products are subjected and, as a result, reduce the strength of the products below the values published by the manufacturer.

2306.2 Wind provisions for walls.

2306.2.1 Wall stud bending stress increase. The NDS fiber stress in bending (F_b) design values for wood studs resisting wind shall be increased by the factors in Table 2306.2.1, in lieu of the 1.15 repetitive member factor, to take into consideration the load sharing and composite actions provided by the wood structural panels as defined in Section 2302.1, where the studs are designed for bending in accordance with Section 1609.6 spaced no more than 16 inches (406 mm) o.c, covered on the inside with a minimum of 1/2-inch (12.7 mm) gypsum board fastened in accordance with Table 2306.4.5, and sheathed on the exterior with a minimum of 3/8-inch (9.5 mm) wood structural panel sheathing that is attached to the studs using a minimum of 8d common nails spaced a maximum of 6 inches o.c. (152 mm) at panel edges and 12 inches o.c. (305 mm) in the field of the panels.

❖ See the commentary to Table 2306.2.1.

TABLE 2306.2.1
WALL STUD BENDING STRESS INCREASE FACTORS

STUD SIZE	SYSTEM FACTOR
2 × 4	1.5
2 × 6	1.4
2 × 8	1.3
2 × 10	1.2
2 × 12	1.15

❖ Design values for structural lumber are published in the supplement to the *National Design Specification (NDS) for Wood Construction*. The tabulated design values are for normal load duration under the moisture conditions specified. Bending design values, F_b, for dimension lumber may be multiplied by the repetitive member factor, C_r = 1.15, when such members are used as joists, truss chords, rafters, studs, planks, decking or for similar members that are in contact or spaced not more than 24 inches (610 mm) o.c., are not less than three in number and are joined by floor, roof or other load-distributing elements adequate to support the design load.

For wall systems composed of 2-inch by 4-inch (51 mm by 102 mm) to 2-inch by 10-inch (51 mm by 254 mm) studs faced on one side a minimum of 3/8-inch (9.5 mm) wood structural panel and with a minimum of 1/2-inch (12.7 mm) gypsum board on the other side, the repetitive member factor may be increased to the values in Table 2306.2.1 as a result of load sharing and composite actions of the wall framing system in a high-wind condition.

2306.3 Wood diaphragms.

❖ A diaphragm is a relatively thin structural element, e.g., floor or roof assembly, usually rectangular in plan, capable of resisting shear parallel to its edges. The three major elements of a diaphragm are the web (wood structural panels and joists/rafters), the chords (edge members) and the fasteners (such as nails). The web is usually composed of several panels. The chords are also almost always one piece. Therefore, fasteners and connection chord splices must be considered as major factors in the performance of a structural diaphragm.

Most floors, roofs and walls can function as a diaphragm. Full diaphragm action can be obtained by detailing the sheathing fasteners and framing connections. Diaphragm construction is used extensively to resist horizontal forces of wind and earthquakes. Diaphragms may be either vertical (often referred to as shear walls) or horizontal. In special designs, they may be oriented in any direction. A diaphragm acts in the same manner as an I-beam. The plywood panel skin acts as the web and the edge members act as the flanges, which are called chords. However, due to the relatively greater depth of a diaphragm, the reactions to stress are somewhat different. The diaphragm design assumes that stresses are uniformly distributed across the panel and the web does not contribute to the resistance of tension and compression stresses as it does in a shallow beam.

A series of diaphragms may be tied together to develop a very rigid structure. Their ability to resist large shear stresses has led to their use in innovative construction methods such as folded plate roof elements, geodesic domes and space frames.

Diaphragms must be securely connected to supporting members and to foundations that are capable of resisting all design forces. Anchorage for floors and roof generally presents no special problems. It is only required that the stresses be transferred to the chords, assuming they are adequately supported. Shear walls, however, require special attention to their anchorage to the foundation. The foundation and the connections thereto must be designed to resist the uplift, compression and shear forces applied.

Openings in diaphragms, such as skylights, doors and windows, disrupt the uniform distribution of shear stresses across the diaphragm. Provisions must be made for continuity and compensation for this disruption. For instance, reinforcement may be necessary at windows.

2306.3.1 Shear capacities modifications. The allowable shear capacities in Table 2306.3.1 for horizontal wood structural panel diaphragms shall be increased 40 percent for wind design.

❖ See the commentary to Table 2306.3.1.

TABLE 2306.3.1. See page 23-42.

❖ The permitted 40-percent increase is based on reducing the load factor of 2.8, used to develop shear capacities for structural sheathing, to 2.0. The rationale for this increase is that it is no longer prudent to retain the previous minimum safety factor of 2.8 on wood structural sheathing materials. The judgement is that a factor of 2.0 is adequate—resulting in the 1.4 factor relative to today's allowable values.

TABLE 2306.3.1
RECOMMENDED SHEAR (POUNDS PER FOOT) FOR WOOD STRUCTURAL PANEL DIAPHRAGMS WITH FRAMING OF DOUGLAS-FIR-LARCH, OR SOUTHERN PINE[a] FOR WIND OR SEISMIC LOADING

PANEL GRADE	COMMON NAIL SIZE OR STAPLE[f] LENGTH AND GAGE	MINIMUM FASTENER PENETRATION IN FRAMING (inches)	MINIMUM NOMINAL PANEL THICKNESS (inch)	MINIMUM NOMINAL WIDTH OF FRAMING MEMBER (inches)	BLOCKED DIAPHRAGMS				UNBLOCKED DIAPHRAGMS	
					Fastener spacing (inches) at diaphragm boundaries (all cases) at continuous panel edges parallel to load (Cases 3, 4), and at all panel edges (Cases 5 and 6)[b]				Fasteners spaced 6" max. At supported edges[b]	
					6	4	$2\,^1/_2$[c]	2[c]		
					Fastener spacing (inches) at other panel edges (Cases 1, 2, 3 and 4)[b]					
					6	6	4	3	Case 1 (No unblocked edges or continuous joints parallel to load)	All other configurations (Cases 2, 3, 4, 5 and 6)
Structural I Grades	6d[e]	$1\,^1/_4$	$^5/_{16}$	2	185	250	375	420	165	125
				3	210	280	420	475	185	140
	$1\,^1/_2$ 16 Gage	1		2	155	205	310	350	135	105
				3	175	230	345	390	155	115
	8d	$1\,^3/_8$	$^3/_8$	2	270	360	530	600	240	180
				3	300	400	600	675	265	200
	$1\,^1/_2$ 16 Gage	1		2	175	235	350	400	155	115
				3	200	265	395	450	175	130
	10d[d]	$1\,^1/_2$	$^{15}/_{32}$	2	320	425	640	730	285	215
				3	360	480	720	820	320	240
	$1\,^1/_2$ 16 Gage	1		2	175	235	350	400	155	120
				3	200	265	395	450	175	130
Sheathing, single floor and other grades covered in DOC PS 1 and PS 2	6d[e]	$1\,^1/_4$	$^5/_{16}$	2	170	225	335	380	150	110
				3	190	250	380	430	170	125
	$1\,^1/_2$ 16 Gage	1		2	140	185	275	315	125	90
				3	155	205	310	350	140	105
	6d[e]	$1\,^1/_4$	$^3/_8$	2	185	250	375	420	165	125
				3	210	280	420	475	185	140
	8d	$1\,^3/_8$		2	240	320	480	545	215	160
				3	270	360	540	610	240	180

(continued)

TABLE 2306.3.1—continued
RECOMMENDED SHEAR (POUNDS PER FOOT) FOR WOOD STRUCTURAL PANEL DIAPHRAGMS WITH FRAMING OF DOUGLAS-FIR-LARCH, OR SOUTHERN PINE[a] FOR WIND OR SEISMIC LOADING

PANEL GRADE	COMMON NAIL SIZE OR STAPLE[f] LENGTH AND GAGE	MINIMUM FASTENER PENETRATION IN FRAMING (inches)	MINIMUM NOMINAL PANEL THICKNESS (inch)	MINIMUM NOMINAL WIDTH OF FRAMING MEMBER (inches)	BLOCKED DIAPHRAGMS: Fastener spacing (inches) at diaphragm boundaries (all cases) at continuous panel edges parallel to load (Cases 3, 4), and at all panel edges (Cases 5 and 6)[b]				UNBLOCKED DIAPHRAGMS: Fasteners spaced 6" max. at supported edges[b]	
					6	4	2 1/2[c]	2[c]		
					Fastener spacing (inches) at other panel edges (Cases 1, 2, 3 and 4)[b]					
					6	6	4	3	Case 1 (No unblocked edges or continuous joints parallel to load)	All other configurations (Cases 2, 3, 4, 5 and 6)
Sheathing, single floor and other grades covered in DOC PS 1 and PS 2 (continued)	1 1/2 16 Gage	1	3/8	2	160	210	315	360	140	105
				3	180	235	355	400	160	120
	8d	1 3/8	7/16	2	255	340	505	575	230	170
				3	285	380	570	645	255	190
	1 1/2 16 Gage	1		2	165	225	335	380	150	110
				3	190	250	375	425	165	125
	8d	1 3/8	15/32	2	270	360	530	600	240	180
				3	300	400	600	675	265	200
	10d[d]	1 1/2		2	290	385	575	655	255	190
				3	325	430	650	735	290	215
	1 1/2 16 Gage	1		2	160	210	315	360	140	105
				3	180	235	355	405	160	120
	10d[d]	1 1/2	19/32	2	320	425	640	730	285	215
				3	360	480	720	820	320	240
	1 3/4 16 Gage	1		2	175	235	350	400	155	115
				3	200	265	395	450	175	130

(continued)

TABLE 2306.3.1—continued
RECOMMENDED SHEAR (POUNDS PER FOOT) FOR WOOD STRUCTURAL PANEL DIAPHRAGMS WITH FRAMING OF DOUGLAS-FIR-LARCH, OR SOUTHERN PINE[a] FOR WIND OR SEISMIC LOADING

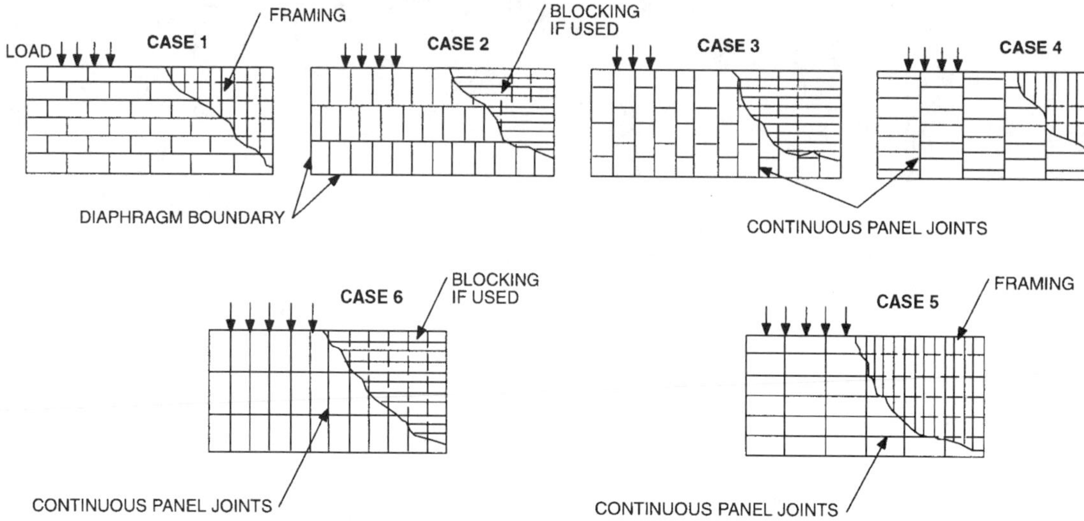

For SI: 1 inch = 25.4 mm, 1 pound per foot = 14.5939 N/m.

a. For framing of other species: (1) Find specific gravity for species of lumber in AFPA National Design Specification. (2) For staples find shear value from table above for Structural I panels (regardless of actual grade) and multiply value by 0.82 for species with specific gravity of 0.42 or greater, or 0.65 for all other species. (3) For nails find shear value from table above for nail size for actual grade and multiply value by the following adjustment factor: Specific Gravity Adjustment Factor = [1-(0.5 - SG)], where SG = Specific Gravity of the framing lumber. This adjustment factor shall not be greater than 1.

b. Space fasteners maximum 12 inches o.c. along intermediate framing members (6 inches o.c. where supports are spaced 48 inches o.c.).

c. Framing at adjoining panel edges shall be 3 inches nominal or wider, and nails shall be staggered where nails are spaced 2 inches o.c. or 2 $^1/_2$ inches o.c.

d. Framing at adjoining panel edges shall be 3 inches nominal or wider, and nails shall be staggered where both of the following conditions are met: (1) 10d nails having penetration into framing of more than 1$^1/_2$ inches and (2) nails are spaced 3 inches o.c. or less.

e. 8d is recommended minimum for roofs due to negative pressures of high winds.

f. Staples shall have a minimum crown width of $^7/_{16}$ inch.

2306.3.2 Wood structural panel diaphragms. Structural panel diaphragms with wood structural panels are permitted to be used to resist horizontal forces not exceeding those set forth in Table 2306.3.1 or 2306.3.2 or calculated by principles of mechanics without limitations by using values for fastener strength in the NDS structural design properties for wood structural panels based on DOC PS-1 and DOC PS-2 or plywood design properties given in the APA Plywood Design Specification.

❖ Wood structural panel diaphragms may be designed using the information in Table 2306.3.1 or 2306.3.2. Note that the panel thickness and joist spacings must also comply with the requirements for sheathing of floors and roofs. Wood structural panel diaphragms may also be designed using accepted engineering principles. AF&PA *National Design Specification (NDS) for Wood Construction*® provides information on shear resistance of mechanical fasteners and chord design. Working stresses for wood structural panels and other materials are contained in other references in the code.

The design of horizontal diaphragms to resist shear stresses depends on the direction of continuous panel joints relative to the direction of load and not on the direction of the long dimension of the panel or framing. Six cases of panel orientation are illustrated in Table 2306.3.1. The nailing schedule and the resultant shear resistance will be affected by the panel layout (relative to load) selected. A diaphragm usually must resist loads in both transverse and longitudinal directions. Each direction must be considered separately with a different case of panel orientation; thus, the allowable load will be applicable to each direction.

Blocked diaphragms have all panel edges supported by and fastened to framing members. Blocking is used at the edges of panels which are not over joists to provide for connecting units and to assist in the transfer of shear forces to the adjacent panel. Unblocked diaphragms may be used provided the applicable nailing schedule in Table 2304.9.1 is used and the shear stresses do not exceed those allowed for unblocked diaphragms in Table 2306.3.1. Unblocked diaphragms are often controlled by the buckling of the panel (skin). With the same nail spacing, allowable design loads for blocked diaphragms are 1 $^1/_2$ to 2 times the design loads for unblocked diaphragms. In addition, the maximum loads for which blocked diaphragms may be designed are many times greater. The forces that must be resisted usually serve as the basis for choosing blocked or unblocked construction.

TABLE 2306.3.2. See page 23-45.

❖ This table contains wood structural panel diaphragm assemblies that have higher capacities compared to those of Table 2306.3.1. These can be useful for larger flat-roofed structures subjected to high wind or seismic loads. The capacities reflected in this table are based upon calculations using the European yield method and were also confirmed by testing.

2306.3.3 Diagonally sheathed lumber diaphragms. Diagonally sheathed lumber diaphragms shall be nailed in accordance with Table 2306.3.3.

❖ See the commentary to Table 2306.3.3.

TABLE 2306.3.3. See page 23-46.

❖ This table presents a prescriptive schedule for nailing of nominal 1-inch (25 mm) and 2-inch (51 mm) lumber sheathing placed diagonally to the structural framing. The nail schedule is intended to be used in accordance with the requirements of Section 2306.3.4 for single diagonally sheathed lumber diaphragms and Section 2306.3.5 for double diagonally sheathed lumber diaphragms.

TABLE 2306.3.2
ALLOWABLE SHEAR IN POUNDS PER FOOT FOR HORIZONTAL BLOCKED DIAPHRAGMS UTILIZING MULTIPLE ROWS OF FASTENERS (HIGH LOAD DIAPHRAGMS) WITH FRAMING OF DOUGLAS FIR, LARCH OR SOUTHERN PINE[a] FOR WIND OR SEISMIC LOADING[b]

PANEL GRADE[c]	FASTENER AND SIZE	MINIMUM FASTENING PENETRATION IN FRAMING (inches)	MINIMUM NOMINAL PANEL THICKNESS (inch)	MINIMUM NOMINAL WIDTH OF FRAMING MEMBER[e] (inches)	LINES OF FASTENERS	BLOCKED DIAPHRAGMS					
						Cases 1 and 2[d]					
						Fastener Spacing Per Line at Boundaries (inches)					
						4		$2^1/_2$		2	
						Fastener Spacing Per Line at Other Panel Edges (inches)					
						6	4	4	3	3	2
Structural I grades	10d common nails	$1^1/_2$	$^{15}/_{32}$	3	2	605	815	875	1,150	—	—
				4	2	700	915	1,005	1,290	—	—
				4	3	875	1,220	1,285	1,395	—	—
			$^{19}/_{32}$	3	2	670	880	965	1,255	—	—
				4	2	780	990	1,110	1,440	—	—
				4	3	965	1,320	1,405	1,790	—	—
			$^{23}/_{32}$	3	2	730	955	1,050	1,365	—	—
				4	2	855	1,070	1,210	1,565	—	—
				4	3	1,050	1,430	1,525	1,800	—	—
	14 gage staples	2	$^{15}/_{32}$	3	2	600	600	860	960	1,060	1,200
				4	3	860	900	1,160	1,295	1,295	1,400
			$^{19}/_{32}$	3	2	600	600	875	960	1,075	1,200
				4	3	875	900	1,175	1,440	1,475	1,795
Sheathing single floor and other grades covered in DOC PS 1 and PS 2	10d common nails	$1^1/_2$	$^{15}/_{32}$	3	2	525	725	765	1,010	—	—
				4	2	605	815	875	1,105	—	—
				4	3	765	1,085	1,130	1,195	—	—
			$^{19}/_{32}$	3	2	650	860	935	1,225	—	—
				4	2	755	965	1,080	1,370	—	—
				4	3	935	1,290	1,365	1,485	—	—
			$^{23}/_{32}$	3	2	710	935	1,020	1,335	—	—
				4	2	825	1,050	1,175	1,445	—	—
				4	3	1,020	1,400	1,480	1,565	—	—
	14 gage staples	2	$^{15}/_{32}$	3	2	540	540	735	865	915	1,080
				4	3	735	810	1,005	1,105	1,105	1,195
			$^{19}/_{32}$	3	2	600	600	865	960	1,065	1,200
				4	3	865	900	1,130	1,430	1,370	1,485
			$^{23}/_{32}$	4	3	865	900	1,130	1,490	1,430	1,545

For SI: 1 inch = 25.4 mm, 1 pound per foot = 14.5939 N/m.

a. For framing of the other species: (1) Find specific gravity for species of framing lumber in AFPA National Design Specification, (2) Find shear value from table above for nail size of actual grade, and (3) Multiply value by the following adjustment factor = [1 - (0.5 - *SG*)], where *SG* = Specific gravity of the framing lumber. This adjustment factor shall not be greater than 1.

b. Fastening along intermediate framing members: Space nails 12 inches on center, except 6 inches on center for spans greater than 32 inches.

c. Panels conforming to PS 1 or PS 2.

d. This table gives shear values for Cases 1 and 2 as shown in Table 2306.3.1. The values shown are applicable to Cases 3, 4, 5 and 6 as shown in Table 2306.3.1, providing fasteners at all continuous panel edges are spaced in accordance with the boundary fastener spacing.

e. The minimum depth of framing members shall be 3 inches.

TABLE 2306.3.3
DIAGONALLY SHEATHED LUMBER DIAPHRAGM NAILING SCHEDULE

SHEATHING NOMINAL DIMENSION	NAILING TO INTERMEDIATE AND END-BEARING STUDS		NAILING AT THE SHEAR PANEL BOUNDARIES	
	Type, size and number of nails per board			
	Common nails	Box nails	Common nails	Box nails
1 × 6	2 - 8d	3 - 8d	3 - 8d	5 - 8d
1 × 8	3 - 8d	4 - 8d	4 - 8d	6 - 8d
2 × 6	2 - 16d	3 - 16d	3 - 16d	5 - 16d
2 × 8	3 - 16d	4 - 16d	4 - 16d	6 - 16d

2306.3.4 Single diagonally sheathed lumber diaphragms. Single diagonally sheathed lumber diaphragms shall be constructed of minimum 1-inch (25 mm) thick nominal sheathing boards laid at an angle of approximately 45 degrees (0.78 rad) to the supports. The shear capacity for single diagonally sheathed lumber diaphragms of southern pine or Douglas fir-larch shall not exceed 300 plf (4378 N/m) of width. The shear capacities shall be adjusted by reduction factors of 0.82 for framing members of species with a specific gravity equal to or greater than 0.42 but less than 0.49 and 0.65 for species with a specific gravity of less than 0.42, as contained in the NDS.

❖ The reduction factors are based on common wire nail shear design values for three classifications of lumber as follows:

A. Specific gravity, G, ≥ 0.49 (for example, Southern pine and Douglas fir-larch);

B. 0.42 ≤ G < 0.49 (for example, spruce-pine-fir and hem-fir), and

C. 0.42 > G (for example, aspen, balsam fir and cedars).

The specific gravity of lumber species can be found in Table 12A of AF&PA *National Design Specification (NDS) for Wood Construction.*

2306.3.4.1 End joints. End joints in adjacent boards shall be separated by at least one stud or joist space and there shall be at least two boards between joints on the same support.

❖ End joints of individual boards are required to be staggered in location as described in this section to maintain the shear capacity specified in Section 2306.3.4.2.

2306.3.4.2 Single diagonally sheathed lumber diaphragms. Single diagonally sheathed lumber diaphragms made up of 2-inch (51 mm) nominal diagonal lumber sheathing fastened with 16d nails shall be designed with the same shear capacities as shear panels using 1-inch (25 mm) boards fastened with 8d nails, provided there are not splices in adjacent boards on the same support and the supports are not less than 4 inch (102 mm) nominal depth or 3 inch (76 mm) nominal thickness.

❖ The shear capacity referenced in this section is conservative, since tests of wood shear panels with 2-inch (51 mm) nominal lumber sheathing are not readily available.

2306.3.5 Double diagonally sheathed lumber diaphragms. Double diagonally sheathed lumber diaphragms shall be constructed of two layers of diagonal sheathing boards at 90 degrees (1.57 rad) to each other on the same face of the supporting members. Each chord shall be considered as a beam with uniform load per foot equal to 50 percent of the unit shear due to diaphragm action. The load shall be assumed as acting normal to the chord in the plan of the diaphragm in either direction. The span of the chord or portion thereof shall be the distance between framing members of the diaphragm, such as the joists, studs and blocking that serve to transfer the assumed load to the sheathing. The shear capacity of double diagonally sheathed diaphragms of Southern pine or Douglas fir-larch shall not exceed 600 plf (8756 kN/m) of width. The shear capacity shall be adjusted by reduction factors of 0.82 for framing members of species with a specific gravity equal to or greater than 0.42 but less than 0.49 and 0.65 for species with a specific gravity of less than 0.42, as contained in the NDS. Nailing of diagonally sheathed lumber diaphragms shall be in accordance with Table 2306.3.3.

❖ This section provides the shear capacity for various configurations of double diagonal board-sheathed shear panels. This type of construction is seldom used because of the labor costs of individual board installation. Double diagonal construction has a higher shear capacity than the single diagonal construction described in Section 2306.3.4. For discussion on the reduction factor, see the commentary to Section 2306.3.4.

2306.3.6 Gypsum board diaphragm ceilings. Gypsum board diaphragm ceilings shall be in accordance with Section 2508.5.

❖ See the commentary to Section 2508.5.

2306.4 Shear walls. Panel sheathing joints in shear walls shall occur over studs or blocking. Adjacent panel sheathing joints shall occur over and be nailed to common framing members (see Section 2305.3.1 for limitations on shear wall bracing materials).

❖ A shear wall is a vertical diaphragm designed to resist lateral forces parallel to the plane of the wall. Therefore, much of the commentary in Section 2306.3 on diaphragms is also applicable to this section.

Detailing is critical in the design of shear walls so that forces are properly distributed to top plates, ledgers, bond beams or any other continuous element at the perimeter to the diaphragm. Therefore, it is important that splices be designed to transmit tension or compression forces occurring at the location of a splice. Panel sheathing joints must occur over studs or blocking and common framing members.

2306.4.1 Wood structural panel shear walls. The allowable shear capacities for wood structural panel shear walls shall be in accordance with Table 2306.4.1. These capacities are permitted to be increased 40 percent for wind design. Shear walls are permitted to be calculated by principles of mechanics without limitations by using values for nail strength given in the NDS and wood structural panel design properties given in the APA/PDS.

❖ See the commentary to Table 2306.4.1.

TABLE 2306.4.1. See page 23-48.

❖ The allowable shear capacities for wood structural panel shear walls in Table 2306.4.1 were developed on engineering principles and monotonic testing.

The 40-percent increase in shear capacities is based on engineering principles. Full-scale monotonic testing was conducted to validate the model. The increase takes into effect the short time duration of loading associated with a wind event.

Shear walls may also be designed without limitations by a rational method using the values for nail strength in the AF&PA NDS or by using wood structural panel design properties in APA/PDS. For shear walls utilizing wood framing, energy dissipation is almost entirely due to nail bending. Since fasteners other than nails and staples have not been extensively tested under cyclic load applications, the rational method is limited to nail strength values. When screws or adhesives have been tested in assemblies subjected to cyclic loading, they have tended to have a brittle mode of failure.

2306.4.2 Lumber sheathed shear walls. Single and double diagonally sheathed lumber diaphragms are permitted using the construction and allowable load provisions of Sections 2306.3.4 and 2306.3.5.

❖ Although wood structural panel sheathed shear walls are most prevalent in use today, diagonally sheathed lumber shear walls are still used for new construction in some regions and are encountered in the alteration of existing structures. The shear resistance values are applicable to buildings exposed to seismic or wind forces.

2306.4.3 Particleboard shear walls. The design shear capacity of particleboard shear walls shall be in accordance with Table 2306.4.3. Shear panels shall be constructed with particleboard sheets not less than 4 feet by 8 feet (1219 mm by 2438 mm), except at boundaries and changes in framing. Particleboard panels shall be designed to resist shear only, and chords, collector members and boundary elements shall be connected at all corners. Panel edges shall be backed with 2-inch (51 mm) nominal or wider framing. Sheets are permitted to be installed either horizontally or vertically. For $^{3}/_{8}$-inch (9.5 mm) particleboard sheets installed with the long dimension parallel to the studs spaced 24 inches (610 mm) o.c, nails shall be spaced at 6 inches (152 mm) o.c. along intermediate framing members. For all other conditions, nails of the same size shall be spaced at 12 inches (305 mm) o.c. along intermediate framing members. Particleboard panels less than 12 inches (305 mm) wide shall be blocked. Particleboard shall not be used to resist seismic forces in structures in Seismic Design Category D, E or F.

❖ See the commentary to Table 2306.4.3.

TABLE 2306.4.3. See page 23-49.

❖ Limitations of use in shear walls are greater for particleboard than for wood structural panels and lumber. The design shear capacity of particleboard shear walls is limited to the values listed in Table 2306.4.3, and particleboard shear walls may only be used to resist shear forces. Further, this section specifies that special care must be taken so that all panel edges are properly attached to and fully blocked with 2-inch (51 mm) nominal framing for transfer of shear forces to boundary elements.

2306.4.4 Fiberboard shear walls. The design shear capacity of fiberboard shear walls shall be in accordance with Table 2308.9.3(4). The fiberboard sheathing shall be applied vertically or horizontally to wood studs not less than 2 inch (51 mm) nominal thickness spaced 16 inches (406 mm) o.c. Blocking not less than 2 inch (51 mm) nominal in thickness shall be provided at horizontal joints. Fiberboard shall not be used to resist seismic forces in structures in Seismic Design Category D, E or F.

❖ For structural fiberboard to be used in the construction of shear walls, the long dimension of the panel must be parallel to the studs. A 2:1 aspect ratio is established. The shear capacity of fiberboard shear walls is limited to the values listed in Table 2308.9.3(4). Further, this section specifies that special care is to be taken so that all panel edges are properly attached to and fully blocked with 2-inch (51 mm) nominal framing for transfer of shear forces to boundary elements.

2306.4.5 Shear walls sheathed with other materials. Shear capacities for walls sheathed with lath and plaster, and gypsum board shall be in accordance with Table 2306.4.5. Shear walls sheathed with lath, plaster and gypsum board shall be constructed in accordance with Chapter 25 and Section 2306.4.5.1. Walls resisting seismic loads shall be subject to the limitations in Section 1617.6.

❖ See the commentary to Table 2306.4.5.

TABLE 2306.4.1
ALLOWABLE SHEAR (POUNDS PER FOOT) FOR WOOD STRUCTURAL PANEL SHEAR WALLS WITH FRAMING OF DOUGLAS-FIR-LARCH, OR SOUTHERN PINE[a] FOR WIND OR SEISMIC LOADING[b, h, i, j]

PANEL GRADE	MINIMUM NOMINAL PANEL THICKNESS (inch)	MINIMUM FASTENER PENETRATION IN FRAMING (inches)	PANELS APPLIED DIRECT TO FRAMING					PANELS APPLIED OVER 1/2″ OR 5/8″ GYPSUM SHEATHING				
			NAIL (common or galvanized box) or staple size[k]	Fastener spacing at panel edges (inches)				NAIL (common or galvanized box) or staple size[k]	Fastener spacing at panel edges (inches)			
				6	4	3	2[e]		6	4	3	2[e]
Structural I Sheathing	5/16	1 1/4	6d	200	300	390	510	8d	200	300	390	510
		1	1 1/2 16 Gage	165	245	325	415	2 16 Gage	125	185	245	315
	3/8	1 3/8	8d	230[d]	360[d]	460[d]	610[d]	10d	280	430	550[f]	730
		1	1 1/2 16 Gage	155	235	315	400	2 16 Gage	155	235	310	400
	7/16	1 3/8	8d	255[d]	395[d]	505[d]	670[d]	10d	280	430	550[f]	730
		1	1 1/2 16 Gage	170	260	345	440	2 16 Gage	155	235	310	400
	15/32	1 3/8	8d	280	430	550	730	10d	280	430	550[f]	730
		1	1 1/2 16 Gage	185	280	375	475	2 16 Gage	155	235	300	400
		1 1/2	10d	340	510	665[f]	870	10d	—	—	—	—
Sheathing, plywood siding[g] except Group 5 Species	5/16 or 1/4[c]	1 1/4	6d	180	270	350	450	8d	180	270	350	450
		1	1 1/2 16 Gage	145	220	295	375	2 16 Gage	110	165	220	285
	3/8	1 1/4	6d	200	300	390	510	8d	200	300	390	510
		1 3/8	8d	220[d]	320[d]	410[d]	530[d]	10d	260	380	490[f]	640
		1	1 1/2 16 Gage	140	210	280	360	2 16 Gage	140	210	280	360
	7/16	1 3/8	8d	240[d]	350[d]	450[d]	585[d]	10d	260	380	490[f]	640
		1	1 1/2 16 Gage	155	230	310	395	2 16 Gage	140	210	280	360
	15/32	1 3/8	8d	260	380	490	640	10d	260	380	490[f]	640
		1 1/2	10d	310	460	600[f]	770	—	—	—	—	—
		1	1 1/2 16 Gage	170	255	335	430	2 16 Gage	140	210	280	360
	19/32	1 1/2	10d	340	510	665[f]	870	—	—	—	—	—
		1	1 3/4 16 Gage	185	280	375	475	—	—	—	—	—
			Nail Size (galvanized casing)					Nail Size (galvanized casing)				
	5/16[c]	1 1/4	6d	140	210	275	360	8d	140	210	275	360
	3/8	1 3/8	8d	160	240	310	410	10d	160	240	310[f]	410

For SI: 1 inch = 25.4 mm, 1 pound per foot = 14.5939 N/m.

a. For framing of other species: (1) Find specific gravity for species of lumber in AF&PA National Design Specification. (2) For staples find shear value from table above for Structural I panels (regardless of actual grade) and multiply value by 0.82 for species with specific gravity of 0.42 or greater, or 0.65 for all other species. (3) For nails find shear value from table above for nail size for actual grade and multiply value by the following adjustment factor: Specific Gravity Adjustment Factor = [1-(0.5 - SG)], where SG = Specific Gravity of the framing lumber. This adjustment factor shall not be greater than 1.

b. Panel edges backed with 2-inch nominal or wider framing. Install panels either horizontally or vertically. Space fasteners maximum 6 inches on center along intermediate framing members for 3/8-inch and 7/16-inch panels installed on studs spaced 24 inches on center. For other conditions and panel thickness, space fasteners maximum 12 inches on center on intermediate supports.

c. 3/8-inch panel thickness or siding with a span rating of 16 inches on center is the minimum recommended where applied direct to framing as exterior siding.

d. Shears are permitted to be increased to values shown for 15/32-inch sheathing with same nailing provided (a) studs are spaced a maximum of 16 inches on center, or (b) if panels are applied with long dimension across studs.

e. Framing at adjoining panel edges shall be 3 inches nominal or wider, and nails shall be staggered where nails are spaced 2 inches on center.

f. Framing at adjoining panel edges shall be 3 inches nominal or wider, and nails shall be staggered where both of the following conditions are met: (1) 10d nails having penetration into framing of more than 1 1/2 inches and (2) nails are spaced 3 inches on center.

g. Values apply to all-veneer plywood. Thickness at point of fastening on panel edges governs shear values.

h. Where panels are applied on both faces of a wall and nail spacing is less than 6 inches o.c. on either side, panel joints shall be offset to fall on different framing members. Or framing shall be 3 inch nominal or thicker and nails on each side shall be staggered.

i. In Seismic Design Category D, E or F, where shear design values exceed 490 pounds per lineal foot (LRFD) or 350 pounds per lineal foot (ASD) all framing members receiving edge nailing from abutting panels shall not be less than a single 3-inch nominal member. Plywood joint and sill plate nailing shall be staggered in all cases. See Section 2305.3.10 for sill plate size and anchorage requirements.

j. Galvanized nails shall be hot dipped or tumbled.

k. Staples shall have a minimum crown width of 7/16 inch.

TABLE 2306.4.3
ALLOWABLE SHEAR FOR PARTICLEBOARD SHEAR WALL SHEATHING

PANEL GRADE	MINIMUM NOMINAL PANEL THICKNESS (inch)	MINIMUM NAIL PENETRATION IN FRAMING (inches)	PANELS APPLIED DIRECT TO FRAMING				
			Nail size (common or galvanized box)	Allowable shear (pounds per foot) nail spacing at panel edges (inches)[a]			
				6	4	3	2
M-S "Exterior Glue" and M-2 "Exterior Glue"	$^3/_8$	$1\ ^1/_2$	6d	120	180	230	300
	$^3/_8$	$1\ ^1/_2$	8d	130	190	240	315
	$^1/_2$			140	210	270	350
	$^1/_2$	$1\ ^5/_8$	10d	185	275	360	460
	$^5/_8$			200	305	395	520

For SI: 1 inch = 25.4 mm, 1 pound per foot = 14.5939 N/m.
a. Values are not permitted in Seismic Design Category D, E or F.

TABLE 2306.4.5. See page 23-50.

❖ This table recognizes the shear capacities of gypsum board sheathing, expanded metal or woven wire lath and portland cement plaster in shear walls. Because of the brittleness of these materials, their use to resist seismic forces is limited for structures classified as Seismic Design Category D and prohibited in structures classified as Seismic Design Category E or F (see Section 1617.6). In wind design, the shear capacities of these materials are often added to the shear capacity of the wood structural panel or wood sheathing to determine the total shear capacity of the shear wall to resist wind forces.

2306.4.5.1 Application of gypsum board or lath and plaster to wood framing.

❖ This section presents general prescriptive details on the application of gypsum board and lath and plaster to wood framing. It is deemed that compliance with these will develop a structurally sound and serviceable wall surface.

2306.4.5.1.1 Joint staggering. End joints of adjacent courses of gypsum board shall not occur over the same stud.

❖ Staggering the joints is common practice, and serves to add to the overall shear capacity of the wall.

2306.4.5.1.2 Blocking. Where required in Table 2306.4.5, wood blocking having the same cross-sectional dimensions as the studs shall be provided at joints that are perpendicular to the studs.

❖ Full-sized blocking is required at unsupported edges. This increases the overall shear capacity by allowing direct transfer of shear forces from the gypsum board to the framing elements.

2306.4.5.1.3 Nailing. Studs, top and bottom plates and blocking shall be nailed in accordance with Table 2304.9.1.

❖ Table 2304.9.1 is the fastener schedule for all wood connections in conventional framing. This section simply emphasizes that the proper connections are essential to the wall's performance in resisting shear.

2306.4.5.1.4 Fasteners. The size and spacing of nails shall be set forth in Table 2306.4.5. Nails shall be spaced not less than $^3/_8$ inch (9.5 mm) from edges and ends of gypsum boards or sides of studs, blocking and top and bottom plates.

❖ The minimum edge spacing of nails is necessary so that they develop their full load-resisting capacity without splitting the material at the edge.

2306.4.5.1.5 Gypsum lath. Gypsum lath shall be applied perpendicular to the studs. Maximum allowable shear values shall be as set forth in Table 2306.4.5.

❖ The values in Table 2306.4.5 assume a perpendicular application of the gypsum lath.

2306.4.5.1.6 Gypsum sheathing. Four-foot-wide (1219 mm) pieces of gypsum sheathing shall be applied parallel or perpendicular to studs. Two-foot-wide (610 mm) pieces of gypsum sheathing shall be applied perpendicular to the studs. Maximum allowable shear values shall be as set forth in Table 2306.4.5.

❖ For gypsum sheathing, the values in Table 2306.4.5 assume application in either parallel or perpendicular orientation, except for sheets 2 feet (610 mm) wide in the first row of the gypsum sheathing portion of the table.

2306.4.5.1.7 Other gypsum boards. Gypsum board shall be applied parallel or perpendicular to studs. Maximum allowable shear values shall be as set forth in Table 2306.4.5.

❖ For typical gypsum board, the values in Table 2306.4.5 assume application in either parallel or perpendicular orientation.

TABLE 2306.4.5
ALLOWABLE SHEAR FOR WIND OR SEISMIC FORCES FOR SHEAR WALLS OF LATH AND PLASTER OR GYPSUM BOARD WOOD FRAMED WALL ASSEMBLIES

TYPE OF MATERIAL	THICKNESS OF MATERIAL	WALL CONSTRUCTION	FASTENER SPACING[b] MAXIMUM (inches)	SHEAR VALUE[a,e] (plf)	MINIMUM FASTENER SIZE[c,d,j,k]
1. Expanded metal or woven wire lath and portland cement plaster	7/8″	Unblocked	6	180	No. 11 gage 1 1/2″ long, 7/16″ head 16 Ga. Galv. Staple, 7/8″ legs
2. Gypsum lath, plain or perforated	3/8″ lath and 1/2″ plaster	Unblocked	5	100	No. 13 gage, 1 1/8″ long, 19/64″ head, plasterboard nail 16 Ga. Galv. Staple, 1 1/8″ long 0.120″ Nail, min. 3/8″ head, 1 1/4″ long
3. Gypsum sheathing	1/2″ × 2′ × 8′	Unblocked	4	75	No. 11 gage, 1 3/4″ long, 7/16″ head, diamond-point, galvanized 16 Ga. Galv. Staple, 1 3/4″ long
	1/2″ × 4′	Blocked[f] Unblocked	4 7	175 100	
	5/8″ × 4′	Blocked	4″ edge/ 7″ field	200	6d galvanized 0.120″ Nail, min. 3/8″ head, 1 3/4″ long
4. Gypsum board, gypsum veneer base, or water-resistant gypsum backing board	1/2″	Unblocked[f]	7	75	5d cooler or wallboard 0.120″ Nail, min. 3/8″ head, 1 1/2″ long 16 Gage Staple, 1 1/2″ long
		Unblocked[f]	4	110	
		Unblocked	7	100	
		Unblocked	4	125	
		Blocked[g]	7	125	
		Blocked[g]	4	150	
		Unblocked	8/12[h]	60	No. 6-1 1/4″ screws[i]
		Blocked[g]	4/16[h]	160	
		Blocked[g]	4/12[h]	155	
		Blocked[f, g]	8/12[h]	70	
		Blocked[g]	6/12[h]	90	
	5/8″	Unblocked[f]	7	115	6d cooler or wallboard 0.120″ Nail, min. 3/8″ head, 1 3/4″ long 16 Gage Staple, 1 1/2″ legs, 1 5/8″ long
			4	145	
		Blocked[g]	7	145	
			4	175	
		Blocked[g] Two-ply	Base ply: 9 Face ply: 7	250	Base ply—6d cooler or wallboard 1 3/4″ × 0.120″ Nail, min. 3/8″ head 1 5/8″ 16 Ga. Galv. Staple Face ply—8d cooler or wallboard 0.120″ Nail, min. 3/8″ head, 2 3/8″ long 15 Ga. Galv. Staple, 2 1/4″ long
		Unblocked	8/12[h]	70	No. 6-1 1/4″ screws[i]
		Blocked[g]	8/12[h]	90	

For SI: 1 inch = 25.4 mm, 1 foot = 304.8 mm, 1 pound per foot = 14.5939 N/m.

a. These shear walls shall not be used to resist loads imposed by masonry or concrete construction (see Section 2305.1.5). Values shown are for short-term loading due to wind or seismic loading in Seismic Design Categories A, B and C. Walls resisting seismic loads shall be subject to the limitations in Section 1617.6. Values shown shall be reduced 25 percent for normal loading.

b. Applies to nailing at studs, top and bottom plates and blocking.

c. Alternate nails are permitted to be used if their dimensions are not less than the specified dimensions. Drywall screws are permitted to be substituted for the 5d, 6d (cooler) nails listed above. 1 1/4 inches Type S or W, No. 6 for 6d (cooler) nails.

d. For properties of cooler nails, see ASTM C 514.

e. Except as noted, shear values are based on a maximum framing spacing of 16 inches on center.

f. Maximum framing spacing of 24 inches on center.

g. All edges are blocked, and edge nailing is provided at all supports and all panel edges.

h. First number denotes fastener spacing at the edges; second number denotes fastener spacing in the field.

i. Screws are Type W or S.

j. Staples shall have a minimum crown width of 7/16 inch, measured outside the legs.

k. Staples for the attachment of gypsum lath and woven-wire lath shall have a minimum crown width of 3/4 inch, measured outside the legs.

SECTION 2307
LOAD AND RESISTANCE FACTOR DESIGN

2307.1 Load and resistance factor design (LRFD). The structural analysis and construction of wood elements and structures using load and resistance factor design (LRFD) methods shall be in accordance with ASCE 16.

❖ The design of wood structures has historically been governed by the general design provisions and recommended practices of the *National Design Specification (NDS) for Wood Construction*, an allowable stress design specification. ASCE 16 (AF&PA's) *Load and Resistance Factor Design (*LRFD*) Standard for Engineered Wood Construction* was created to provide alternate design provisions based on reliability theory and a uniform practice in the design of engineered wood structures. ASCE 16 is included in AF&PA's *Load and Resistance Factor Design (*LRFD*) Manual of Engineered Wood Construction*, a multipart design manual for engineers. Theoretical reliability-based analysis has been used for many years in the electronics and aerospace industries with great success. The extension of theoretical reliability concepts to building applications has proven to be somewhat more difficult, but still achievable. LRFD has evolved to become the preferred format for converting structural design standards to a so-called limit states approach. The LRFD standard is intended for use in conjunction with competent engineering design, accurate fabrication and adequate supervision of construction.

The basic design equation for LRFD requires that the specified product strength or resistance meet or exceed the stress or other effect imposed by the specified loads. In allowable stress design (ASD), the permissible stress levels are set very low and the load magnitudes are set at once-in-a-lifetime levels. This combination produces designs that maintain high safety levels yet remain economically feasible. In LRFD, the basic design equation follows a similar format, in which factored resistance must be greater than or equal to the factored load effects.

ASCE 16 and its supporting ASTM standards adopt the reliability refinements embodied in the load factors of ASCE 7, *Minimum Design Loads for Buildings and Other Structures*. The procedures in ASTM D 5457 are consistent with other standards in their use of reliability concepts in the background of calculations.

SECTION 2308
CONVENTIONAL LIGHT-FRAME CONSTRUCTION

2308.1 General. The requirements of this section are intended for conventional light-frame construction. Other methods are permitted to be used provided a satisfactory design is submitted showing compliance with other provisions of this code. Interior nonload-bearing partitions, ceilings and curtain walls of conventional light-frame construction are not subject to the limitations of this section. Alternatively, compliance with the following standard shall be permitted subject to the limitations therein and the limitations of this code: *American Forest and Paper Association (AF&PA) Wood Frame Construction Manual for One- and Two-Family Dwellings* (WCFM).

❖ The provisions of Section 2308 recognize that light-frame wood construction, as that term is discussed in Section 2308.1, is deemed to comply with the engineering requirements of the code. Conventional construction provisions are prescriptive regulations that result in construction that will successfully resist gravity and lateral loads, within the limits of Section 2308.2, to which the structure can reasonably be expected to be exposed. Certain provisions of conventional construction rely upon historic performance, rather than principles of engineering.

These specifications are generally only applicable to light wood-frame building construction having closely spaced framing, using studs up to 2 inches by 6 inches (51 mm by 152 mm) in size and joists and rafters up to 2 inches by 12 inches (51 mm by 305 mm) in size. With a few exceptions, a "conventional" structure is framed in the "platform" style. Repetitive, closely spaced framing is not specifically defined by the code, however, it is usually assumed to include framing that does not exceed a spacing of 24 inches (610 mm) o.c. This is the greatest spacing commonly found in these types of structures, the greatest spacing covered by AF&PA *Span Tables for Joists and Rafters*, and the maximum spacing for which referenced standards allow increases in bending stresses for repetitive member use.

For structures within the scope of conventional construction, however, other methods of construction are permitted, provided that they are designed to comply with either ASD methodology or LRFD methodology, as discussed in the commentary to Section 2301.2.

Some interior nonbearing applications are permitted to use the provisions of Section 2308, even though the remainder of the building does not fall within the limitations of Section 2308.2. These buildings would be of light wood frame construction as well as other types of construction in which wood framing is permitted under other portions of the code.

2308.2 Limitations. Buildings are permitted to be constructed in accordance with the provisions of conventional light-frame construction, subject to the following limitations, and to further limitations of Sections 2308.11 and 2308.12.

❖ It is acknowledged that conventional construction provisions concerning framing members and sheathing that carry gravity loads are adequate. For resistance to lateral loads (wind and seismic), however, experience has shown that additional requirements—or limits on the application of conventional construction provisions—are needed.

1. Buildings shall be limited to a maximum of three stories above grade. For the purposes of this section, for buildings

in Seismic Design Category D or E as determined in Section 1616, cripple stud walls shall be considered to be a story.

Exception: Solid blocked cripple walls not exceeding 14 inches (356 mm) in height need not be considered a story.

❖ The requirements of conventional construction are based on anticipated loads, both gravity and lateral. Buildings greater than three stories high load the lower stories to a level higher than addressed by these provisions.

2. Bearing wall floor-to-floor heights shall not exceed 10 feet (3048 mm).

❖ Studs in bearing walls have a tendency to bow, or move laterally, along the plane of the wall. Although this lateral movement can be restrained by design, the prescriptive provisions of Section 2308 do not attempt to address the problem for studs taller than 10 feet (3048 mm), which is assumed as the maximum stable height using the typical grades and species (type) of wood in general use in conventional construction.

3. Loads as determined in Chapter 16 shall not exceed the following:

   3.1. Average dead loads shall not exceed 15 psf (718 N/m^2) for roofs and exterior walls, floors and partitions.

   3.2. Live loads shall not exceed 40 psf (1916 N/m^2) for floors.

   3.3. Ground snow loads shall not exceed 50 psf (2395 N/m^2).

❖ As the requirements of Section 2308 are based on anticipated loads, this item limits both dead and live loads in the structure. The prescriptive provisions of conventional construction are based on relatively light dead loads and do not address support of masonry or concrete other than veneer. Additionally, the span tables for joists are based on live loads of residential uses, and other elements that support floor loads assume residential loading.

4. Wind speeds shall not exceed 100 miles per hour (mph) (44 m/s) (3-second gust).

Exception: Wind speeds shall not exceed 110 mph (48.4 m/s) 3-second gust for buildings in Exposure Category A or B.

❖ While some elements of conventional construction are acceptable for use in buildings in high-wind areas (as will be seen in the documents referenced in Section 2308.2.1), the general framing and sheathing requirements are not. Buildings in Exposure Categories A and B, as addressed in Section 1609.4, are somewhat sheltered by other buildings as well as by trees and other obstructions. For that reason, the exception recognizes that conventional construction will provide adequate lateral resistance in buildings in those areas.

5. Roof trusses and rafters shall not span more than 40 feet (12 192 mm) between points of vertical support.

❖ In buildings with roof framing spans in excess of 40 feet (12 192 mm), the horizontal thrust of that framing on the top plate on which it rests is greater than can be resisted by the ceiling joist and rafter connections required in Section 2308.10.4.1. Note that the limitation is on the span of the truss or rafter and not on the width of the building. The building width could exceed 40 feet (12 192 mm) as long as the actual span of the roof framing is no more than 40 feet (12 192 mm).

6. The use of the provisions for conventional light-frame construction in this section shall not be permitted for buildings in Seismic Design Category B, C, D, E or F for Seismic Use Group III, as determined in Section 1616.

❖ Seismic Use Group III structures classified in Seismic Design Categories B through F require an engineered design. These are considered essential facilities and the design seismic forces in such structures are greater than can be resisted by conventional construction.

7. Conventional light-frame construction is limited in irregular structures in Seismic Design Category D or E, as specified in Section 2308.12.6.

❖ Irregular portions of buildings (see commentary, Section 2308.12.6) are not permitted to use conventional construction in the higher seismic design categories.

2308.2.1 Basic wind speed greater than 100 mph (3-second gust). Where the basic wind speed exceeds 100 mph (3-second gust), the provisions of either the *AF&PA Wood Frame Construction Manual for One- and Two-Family Dwellings* (WFCM), or the SBCCI *Standard for Hurricane-Resistant Residential Construction* (SSTD 10), are permitted to be used.

❖ Where the basic wind speed exceeds the limitation in Section 2308.2, Item 4, the provisions of the section cannot be used in constructing the building. Engineering design, however, is not always required. The engineered yet prescriptive provisions of AF&PA's *Wood Frame Construction Manual for One- and Two-Family Dwellings* (WFCM) and SBCCI SSTD-10 may be used on buildings that are within the scope of those documents.

2308.2.2 Buildings in Seismic Design Category B, C, D or E. Buildings of conventional light-frame construction in Seismic Design Category B or C, as determined in Section 1616, shall comply with the additional requirements in Section 2308.11.

Exceptions:

1. Detached one- and two-family dwellings as applicable in Section 101.2 in Seismic Design Category B.
2. Detached one- and two-family dwellings as applicable in Section 101.2 in Seismic Design Category C where masonry veneer is limited to the first two stories above grade.

Buildings of conventional light-frame construction in Seismic Design Category D or E, as determined in Section 1616, shall comply with the additional requirements in Section 2308.12.

❖ In recognition of additional lateral forces anticipated in Seismic Design Category B or C and those in D or E, additional limits are placed on the application of conventional construction (see commentary, Sections 2308.11 and 2308.12).

2308.3 Braced wall lines. Buildings shall be provided with exterior and interior braced wall lines as described in Section 2308.9.3 and installed in accordance with Sections 2308.3.1 through 2308.3.4.

❖ Exterior walls of buildings that do not require a seismic analysis in accordance with Section 1614.1 and are located in regions where hurricane winds are not anticipated (see Section 2308.2) are required to be braced in accordance with this section. In addition, if the spacing between the braced exterior walls exceeds 35 feet (10 668 mm) (see Section 2308.3.1), interior braced wall lines are also required.

The requirements for wall bracing vary, depending upon the level of seismic activity for the region in which the building under consideration is located and whether the story in question is supporting other stories. The bracing requirements increase for higher seismic design categories and for walls that support the weight of additional stories above. Various types of structural sheathing materials are permitted, although not every material is permitted in every instance.

Nailing requirements are critical to the performance of wall bracing. To insure that the bracing material is effective in resisting racking of the wall, it must be installed in the manner prescribed.

2308.3.1 Spacing. Spacing of braced wall lines shall not exceed 35 feet (10 668 mm) o.c. in both the longitudinal and transverse directions in each story.

❖ While model building codes in the past have been quite specific as to the type of bracing materials to be used and the amount of bracing required in any wall, no limits on the number or maximum separation between braced walls have been established. However, by mandating the maximum spacing of braced wall lines and thereby limiting the lateral forces acting on these vertical elements, this spacing requirement provides a lateral-force-resisting system that will be less prone to overstressing (see Figure 2308.9.3).

2308.3.2 Braced wall panel connections. Forces shall be transferred from the roofs and floors to braced wall panels and from the braced wall panels in upper stories to the braced wall panels in the story below by the following:

1. Braced wall panel top and bottom plates shall be fastened to joists, rafters or full-depth blocking. Braced wall panels shall be extended and fastened to roof framing at intervals not to exceed 50 feet (15 240 mm) between parallel braced wall lines.

   Exception: Where roof trusses are used, lateral forces shall be transferred from the roof diaphragm to the braced wall by blocking of the ends of the trusses or by other approved methods.

2. Bottom plate fastening to joist or blocking below shall be with not less than 3-16d nails at 16 inches (406 mm) o.c.
3. Blocking shall be nailed to the top plate below with not less than 3-8d toenails per block.
4. Joists parallel to the top plates shall be nailed to the top plate with not less than 8d toenails at 6 inches (152 mm) o.c.

In addition, top plate laps shall be nailed with not less than 8-16d face nails on each side of each break in the top plate.

❖ The connection described in this section serves to effectively tie the structure together, so that forces can be adequately transferred from walls and roofs to the braced wall lines. In this regard, the connections are critical.

In regard to Item 1, when the maximum dimension of a roof exceeds 50 feet (15 240 mm) in any direction without being fastened to a braced wall line, the force resulting from the mass of the roof during an earthquake would exceed the resistance capability of the braced walls. Since the spacing of the walls cannot exceed 35 feet (10 668 mm) in accordance with Section 2308.3.1, if the roof is fastened to all braced wall lines then this requirement will be met.

2308.3.3 Sill anchorage. Where foundations are required by Section 2308.3.4, braced wall line sills shall be anchored to concrete or masonry foundations. Such anchorage shall conform to the requirements of Section 2308.6 except that such anchors shall be spaced at not more than 4 feet (1219 mm) o.c. for structures over two stories in height. The anchors shall be distributed along the length of the braced wall line. Other anchorage devices having equivalent capacity are permitted.

❖ The more restrictive spacing is required since a braced wall is designed to resist greater loads than a nonbraced wall (see commentary, Section 2308.6).

2308.3.3.1 Anchorage to all-wood foundations. Where all-wood foundations are used, the force transfer from the braced wall lines shall be determined based on calculation and shall have a capacity greater than or equal to the connections required by Section 2308.3.3.

❖ There are no prescriptive provisions currently in the code for anchorage of braced wall lines to wood foundations. Therefore, transfer of the forces from the wall line to the foundation must be provided. Connections must be designed to meet the capacity of the bolts described in Section 2308.6.

2308.3.4 Braced wall line support. Braced wall lines shall be supported by continuous foundations.

Exception: For structures with a maximum plan dimension not over 50 feet (15 240 mm), continuous foundations are required at exterior walls only.

❖ An exception allows horizontal dimensions of up to 50 feet (15 240 mm) between continuous foundations. This is intended to permit residential cripple walls to be braced at exterior walls only, within the limitations described.

Connections between horizontal and vertical-resisting elements and transferring of forces to the foundations are prescribed in this section. Some suggested details are available in Chapter 12 of the NEHRP Provisions Commentary.

2308.4 Design of portions. Where a building of otherwise conventional construction contains nonconventional structural elements, those elements shall be designed to resist the forces specified in Chapter 16. The extent of such design need only demonstrate compliance of nonconventional elements with other applicable provisions of this code, and shall be compatible with the performance of the conventional framed system.

❖ Although the code envisions the use of conventional construction provisions without the need to formally design elements of the building, there may be certain portions of the building's structural system that do not comply with the requirements of Section 2308. In such an instance, that structural component must be designed to comply with the requirements of Chapter 16 (see the commentary to Section 2305.1 for additional information).

2308.5 Connections and fasteners. Connections and fasteners used in conventional construction shall comply with the requirements of Section 2304.9.

❖ As discussed in the commentary to Section 2304.9, the section on fasteners and connectors applies to all types of wood construction addressed in Chapter 23, and the number of fasteners stipulated in Table 2304.9.1 is the minimum to be used. This section emphasizes that requirement and also requires compliance with other portions of the section that address specific fastener applications.

2308.6 Foundation plates or sills. Foundations and footings shall be as specified in Chapter 18. Foundation plates or sills resting on concrete or masonry foundations shall comply with Section 2304.3.1. Foundation plates or sills shall be bolted or anchored to the foundation with not less than $^1/_2$-inch-diameter (12.7 mm) steel bolts or approved anchors. Bolts shall be embedded at least 7 inches (178 mm) into concrete or masonry, and spaced not more than 6 feet (1829 mm) apart. There shall be a minimum of two bolts or anchor straps per piece with one bolt or anchor strap located not more than 12 inches (305 mm) or less than 4 inches (102 mm) from each end of each piece. A properly sized nut and washer shall be tightened on each bolt to the plate.

❖ This section prescribes the size and spacing of foundation bolts or anchors for wood structures that are permitted to use conventional construction. Note that regulations pertaining to the foundation or footing are found in Chapter 18. Also note that types of anchors other than bolts and anchor straps are permitted when approved.

2308.7 Girders. Girders for single-story construction or girders supporting loads from a single floor shall not be less than 4 inches by 6 inches (102 mm by 152 mm) for spans 6 feet (1829 mm) or less, provided that girders are spaced not more than 8 feet (2438 mm) o.c. Spans for built-up 2-inch (51 mm) girders shall be in accordance with Table 2308.9.5 or 2308.9.6. Other girders shall be designed to support the loads specified in this code. Girder end joints shall occur over supports.

Where a girder is spliced over a support, an adequate tie shall be provided. The ends of beams or girders supported on masonry or concrete shall not have less than 3 inches (76 mm) of bearing.

❖ Girders constructed of a single piece of wood must meet the size requirement of this section. Those that are built up—constructed of two or more pieces of 2-inch (51 mm) (nominal) wide lumber—must comply with the specified tables. Other girders are permitted if properly designed. Where girders are long enough to require the use of more than a single piece of wood, the joints between the pieces must be supported and cannot occur in mid-span. If several members comprise a beam, they should be adequately tied together to provide continuity and to insure that the members act compositely.

The minimum bearing length for girders supported by concrete or masonry walls is 3 inches (76 mm). This is based on the anticipated loads for the construction typical in this chapter, the allowable compressive stresses perpendicular to the grain for beam sizes and grades typical of the requirements in this chapter, and the consideration of shear failure of the masonry.

2308.8 Floor joists. Spans for floor joists shall be in accordance with Table 2308.8(1) or 2308.8(2). For other grades and or species, refer to the *AF&PA Span Tables for Joists and Rafters.*

❖ See the commentary to Tables 2308.8(1) and (2).

TABLES 2308.8(1) and (2). See pages 23-55 - 23-58.

❖ The spans shown in Tables 2308.8(1) and (2) were derived from the AF&PA *Span Tables for Joists and Rafters*. For simplicity, spans of only the most common species, or types, and grades of wood are shown. When other species or grades of wood are used, the AF&PA tables should be consulted to determine the proper span. Additionally, span tables published by the Southern Forest Products Association, the Western Wood Products Association and the Canadian Wood Council can also be consulted. The spans published in those tables were derived in the same manner as those published in the AF&PA tables.

Spans are based on species of wood, grade, live load, dead load and deflection criteria (see Section 2304.4 for further discussion).

TABLE 2308.8(1)
FLOOR JOIST SPANS FOR COMMON LUMBER SPECIES
(Residential Sleeping Areas, Live Load = 30 psf, L/Δ = 360)

JOIST SPACING (inches)	SPECIES AND GRADE		DEAD LOAD = 10 psf				DEAD LOAD = 20 psf			
			2x6	2x8	2x10	2x12	2x6	2x8	2x10	2x12
			Maximum floor joist spans							
			(ft. - in.)	(ft. - in.)	(ft. - in.)	(ft. - in.)	(ft. - in.)	(ft. - in.)	(ft. - in.)	(ft. - in.)
12	Douglas Fir-Larch	SS	12-6	16-6	21-0	25-7	12-6	16-6	21-0	25-7
	Douglas Fir-Larch	#1	12-0	15-10	20-3	24-8	12-0	15-7	19-0	22-0
	Douglas Fir-Larch	#2	11-10	15-7	19-10	23-0	11-6	14-7	17-9	20-7
	Douglas Fir-Larch	#3	9-8	12-4	15-0	17-5	8-8	11-0	13-5	15-7
	Hem-Fir	SS	11-10	15-7	19-10	24-2	11-10	15-7	19-10	24-2
	Hem-Fir	#1	11-7	15-3	19-5	23-7	11-7	15-2	18-6	21-6
	Hem-Fir	#2	11-0	14-6	18-6	22-6	11-0	14-4	17-6	20-4
	Hem-Fir	#3	9-8	12-4	15-0	17-5	8-8	11-0	13-5	15-7
	Southern Pine	SS	12-3	16-2	20-8	25-1	12-3	16-2	20-8	25-1
	Southern Pine	#1	12-0	15-10	20-3	24-8	12-0	15-10	20-3	24-8
	Southern Pine	#2	11-10	15-7	19-10	24-2	11-10	15-7	18-7	21-9
	Southern Pine	#3	10-5	13-3	15-8	18-8	9-4	11-11	14-0	16-8
	Spruce-Pine-Fir	SS	11-7	15-3	19-5	23-7	11-7	15-3	19-5	23-7
	Spruce-Pine-Fir	#1	11-3	14-11	19-0	23-0	11-3	14-7	17-9	20-7
	Spruce-Pine-Fir	#2	11-3	14-11	19-0	23-0	11-3	14-7	17-9	20-7
	Spruce-Pine-Fir	#3	9-8	12-4	15-0	17-5	8-8	11-0	13-5	15-7
16	Douglas Fir-Larch	SS	11-4	15-0	19-1	23-3	11-4	15-0	19-1	23-0
	Douglas Fir-Larch	#1	10-11	14-5	18-5	21-4	10-8	13-6	16-5	19-1
	Douglas Fir-Larch	#2	10-9	14-1	17-2	19-11	9-11	12-7	15-5	17-10
	Douglas Fir-Larch	#3	8-5	10-8	13-0	15-1	7-6	9-6	11-8	13-6
	Hem-Fir	SS	10-9	14-2	18-0	21-11	10-9	14-2	18-0	21-11
	Hem-Fir	#1	10-6	13-10	17-8	20-9	10-4	13-1	16-0	18-7
	Hem-Fir	#2	10-0	13-2	16-10	19-8	9-10	12-5	15-2	17-7
	Hem-Fir	#3	8-5	10-8	13-0	15-1	7-6	9-6	11-8	13-6
	Southern Pine	SS	11-2	14-8	18-9	22-10	11-2	14-8	18-9	22-10
	Southern Pine	#1	10-11	14-5	18-5	22-5	10-11	14-5	17-11	21-4
	Southern Pine	#2	10-9	14-2	18-0	21-1	10-5	13-6	16-1	18-10
	Southern Pine	#3	9-0	11-6	13-7	16-2	8-1	10-3	12-2	14-6
	Spruce-Pine-Fir	SS	10-6	13-10	17-8	21-6	10-6	13-10	17-8	21-4
	Spruce-Pine-Fir	#1	10-3	13-6	17-2	19-11	9-11	12-7	15-5	17-10
	Spruce-Pine-Fir	#2	10-3	13-6	17-2	19-11	9-11	12-7	15-5	17-10
	Spruce-Pine-Fir	#3	8-5	10-8	13-0	15-1	7-6	9-6	11-8	13-6

(continued)

TABLE 2308.8(1)—continued
FLOOR JOIST SPANS FOR COMMON LUMBER SPECIES
(Residential Sleeping Areas, Live Load = 30 psf, L/Δ = 360)

JOIST SPACING (inches)	SPECIES AND GRADE		DEAD LOAD = 10 psf				DEAD LOAD = 20 psf			
			2x6	2x8	2x10	2x12	2x6	2x8	2x10	2x12
			Maximum floor joist spans							
			(ft. - in.)	(ft. - in.)	(ft. - in.)	(ft. - in.)	(ft. - in.)	(ft. - in.)	(ft. - in.)	(ft. - in.)
19.2	Douglas Fir-Larch	SS	10-8	14-1	18-0	21-10	10-8	14-1	18-0	21-0
	Douglas Fir-Larch	#1	10-4	13-7	16-9	19-6	9-8	12-4	15-0	17-5
	Douglas Fir-Larch	#2	10-1	12-10	15-8	18-3	9-1	11-6	14-1	16-3
	Douglas Fir-Larch	#3	7-8	9-9	11-10	13-9	6-10	8-8	10-7	12-4
	Hem-Fir	SS	10-1	13-4	17-0	20-8	10-1	13-4	17-0	20-7
	Hem-Fir	#1	9-10	13-0	16-4	19-0	9-6	12-0	14-8	17-0
	Hem-Fir	#2	9-5	12-5	15-6	17-1	8-11	11-4	13-10	16-1
	Hem-Fir	#3	7-8	9-9	11-10	13-9	6-10	8-8	10-7	12-4
	Southern Pine	SS	10-6	13-10	17-8	21-6	10-6	13-10	17-8	21-6
	Southern Pine	#1	10-4	13-7	17-4	21-1	10-4	13-7	16-4	19-6
	Southern Pine	#2	10-1	13-4	16-5	19-3	9-6	12-4	14-8	17-2
	Southern Pine	#3	8-3	10-6	12-5	14-9	7-4	9-5	11-1	13-2
	Spruce-Pine-Fir	SS	9-10	13-0	16-7	20-2	9-10	13-0	16-7	19-6
	Spruce-Pine-Fir	#1	9-8	12-9	15-8	18-3	9-1	11-6	14-1	16-3
	Spruce-Pine-Fir	#2	9-8	12-9	15-8	18-3	9-1	11-6	14-1	16-3
	Spruce-Pine-Fir	#3	7-8	9-9	11-10	13-9	6-10	8-8	10-7	12-4
24	Douglas Fir-Larch	SS	9-11	13-1	16-8	20-3	9-11	13-1	16-2	18-9
	Douglas Fir-Larch	#1	9-7	12-4	15-0	17-5	8-8	11-0	13-5	15-7
	Douglas Fir-Larch	#2	9-1	11-6	14-1	16-3	8-1	10-3	12-7	14-7
	Douglas Fir-Larch	#3	6-10	8-8	10-7	12-4	6-2	7-9	9-6	11-0
	Hem-Fir	SS	9-4	12-4	15-9	19-2	9-4	12-4	15-9	18-5
	Hem-Fir	#1	9-2	12-0	14-8	17-0	8-6	10-9	13-1	15-2
	Hem-Fir	#2	8-9	11-4	13-10	16-1	8-0	10-2	12-5	14-4
	Hem-Fir	#3	6-10	8-8	10-7	12-4	6-2	7-9	9-6	11-0
	Southern Pine	SS	9-9	12-10	16-5	19-11	9-9	12-10	16-5	19-11
	Southern Pine	#1	9-7	12-7	16-1	19-6	9-7	12-4	14-7	17-5
	Southern Pine	#2	9-4	12-4	14-8	17-2	8-6	11-0	13-1	15-5
	Southern Pine	#3	7-4	9-5	11-1	13-2	6-7	8-5	9-11	11-10
	Spruce-Pine-Fir	SS	9-2	12-1	15-5	18-9	9-2	12-1	15-0	17-5
	Spruce-Pine-Fir	#1	8-11	11-6	14-1	16-3	8-1	10-3	12-7	14-7
	Spruce-Pine-Fir	#2	8-11	11-6	14-1	16-3	8-1	10-3	12-7	14-7
	Spruce-Pine-Fir	#3	6-10	8-8	10-7	12-4	6-2	7-9	9-6	11-0

Check sources for availability of lumber in lengths greater than 20 feet.

For SI: 1 inch = 25.4 mm, 1 foot = 304.8 mm, 1 pound per square foot = 47.8 N/m^2.

TABLE 2308.8(2)
FLOOR JOIST SPANS FOR COMMON LUMBER SPECIES
(Residential Living Areas, Live Load = 40 psf, L/Δ = 360)

JOIST SPACING (inches)	SPECIES AND GRADE		DEAD LOAD = 10 psf				DEAD LOAD = 20 psf			
			2x6	2x8	2x10	2x12	2x6	2x8	2x10	2x12
			Maximum floor joist spans							
			(ft. - in.)	(ft. - in.)	(ft. - in.)	(ft. - in.)	(ft. - in.)	(ft. - in.)	(ft. - in.)	(ft. - in.)
12	Douglas Fir-Larch	SS	11-4	15-0	19-1	23-3	11-4	15-0	19-1	23-3
	Douglas Fir-Larch	#1	10-11	14-5	18-5	22-0	10-11	14-2	17-4	20-1
	Douglas Fir-Larch	#2	10-9	14-2	17-9	20-7	10-6	13-3	16-3	18-10
	Douglas Fir-Larch	#3	8-8	11-0	13-5	15-7	7-11	10-0	12-3	14-3
	Hem-Fir	SS	10-9	14-2	18-0	21-11	10-9	14-2	18-0	21-11
	Hem-Fir	#1	10-6	13-10	17-8	21-6	10-6	13-10	16-11	19-7
	Hem-Fir	#2	10-0	13-2	16-10	20-4	10-0	13-1	16-0	18-6
	Hem-Fir	#3	8-8	11-0	13-5	15-7	7-11	10-0	12-3	14-3
	Southern Pine	SS	11-2	14-8	18-9	22-10	11-2	14-8	18-9	22-10
	Southern Pine	#1	10-11	14-5	18-5	22-5	10-11	14-5	18-5	22-5
	Southern Pine	#2	10-9	14-2	18-0	21-9	10-9	14-2	16-11	19-10
	Southern Pine	#3	9-4	11-11	14-0	16-8	8-6	10-10	12-10	15-3
	Spruce-Pine-Fir	SS	10-6	13-10	17-8	21-6	10-6	13-10	17-8	21-6
	Spruce-Pine-Fir	#1	10-3	13-6	17-3	20-7	10-3	13-3	16-3	18-10
	Spruce-Pine-Fir	#2	10-3	13-6	17-3	20-7	10-3	13-3	16-3	18-10
	Spruce-Pine-Fir	#3	8-8	11-0	13-5	15-7	7-11	10-0	12-3	14-3
16	Douglas Fir-Larch	SS	10-4	13-7	17-4	21-1	10-4	13-7	17-4	21-0
	Douglas Fir-Larch	#1	9-11	13-1	16-5	19-1	9-8	12-4	15-0	17-5
	Douglas Fir-Larch	#2	9-9	12-7	15-5	17-10	9-1	11-6	14-1	16-3
	Douglas Fir-Larch	#3	7-6	9-6	11-8	13-6	6-10	8-8	10-7	12-4
	Hem-Fir	SS	9-9	12-10	16-5	19-11	9-9	12-10	16-5	19-11
	Hem-Fir	#1	9-6	12-7	16-0	18-7	9-6	12-0	14-8	17-0
	Hem-Fir	#2	9-1	12-0	15-2	17-7	8-11	11-4	13-10	16-1
	Hem-Fir	#3	7-6	9-6	11-8	13-6	6-10	8-8	10-7	12-4
	Southern Pine	SS	10-2	13-4	17-0	20-9	10-2	13-4	17-0	20-9
	Southern Pine	#1	9-11	13-1	16-9	20-4	9-11	13-1	16-4	19-6
	Southern Pine	#2	9-9	12-10	16-1	18-10	9-6	12-4	14-8	17-2
	Southern Pine	#3	8-1	10-3	12-2	14-6	7-4	9-5	11-1	13-2
	Spruce-Pine-Fir	SS	9-6	12-7	16-0	19-6	9-6	12-7	16-0	19-6
	Spruce-Pine-Fir	#1	9-4	12-3	15-5	17-10	9-1	11-6	14-1	16-3
	Spruce-Pine-Fir	#2	9-4	12-3	15-5	17-10	9-1	11-6	14-1	16-3
	Spruce-Pine-Fir	#3	7-6	9-6	11-8	13-6	6-10	8-8	10-7	12-4

(continued)

TABLE 2308.8(2)—continued
FLOOR JOIST SPANS FOR COMMON LUMBER SPECIES
(Residential Living Areas, Live Load = 40 psf, L/Δ = 360)

JOIST SPACING (inches)	SPECIES AND GRADE		DEAD LOAD = 10 psf				DEAD LOAD = 20 psf			
			2x6	2x8	2x10	2x12	2x6	2x8	2x10	2x12
			Maximum floor joist spans							
			(ft. - in.)	(ft. - in.)	(ft. - in.)	(ft. - in.)	(ft. - in.)	(ft. - in.)	(ft. - in.)	(ft. - in.)
19.2	Douglas Fir-Larch	SS	9-8	12-10	16-4	19-10	9-8	12-10	16-4	19-2
	Douglas Fir-Larch	#1	9-4	12-4	15-0	17-5	8-10	11-3	13-8	15-11
	Douglas Fir-Larch	#2	9-1	11-6	14-1	16-3	8-3	10-6	12-10	14-10
	Douglas Fir-Larch	#3	6-10	8-8	10-7	12-4	6-3	7-11	9-8	11-3
	Hem-Fir	SS	9-2	12-1	15-5	18-9	9-2	12-1	15-5	18-9
	Hem-Fir	#1	9-0	11-10	14-8	17-0	8-8	10-11	13-4	15-6
	Hem-Fir	#2	8-7	11-3	13-10	16-1	8-2	10-4	12-8	14-8
	Hem-Fir	#3	6-10	8-8	10-7	12-4	6-3	7-11	9-8	11-3
	Southern Pine	SS	9-6	12-7	16-0	19-6	9-6	12-7	16-0	19-6
	Southern Pine	#1	9-4	12-4	15-9	19-2	9-4	12-4	14-11	17-9
	Southern Pine	#2	9-2	12-1	14-8	17-2	8-8	11-3	13-5	15-8
	Southern Pine	#3	7-4	9-5	11-1	13-2	6-9	8-7	10-1	12-1
	Spruce-Pine-Fir	SS	9-0	11-10	15-1	18-4	9-0	11-10	15-1	17-9
	Spruce-Pine-Fir	#1	8-9	11-6	14-1	16-3	8-3	10-6	12-10	14-10
	Spruce-Pine-Fir	#2	8-9	11-6	14-1	16-3	8-3	10-6	12-10	14-10
	Spruce-Pine-Fir	#3	6-10	8-8	10-7	12-4	6-3	7-11	9-8	11-3
24	Douglas Fir-Larch	SS	9-0	11-11	15-2	18-5	9-0	11-11	14-9	17-1
	Douglas Fir-Larch	#1	8-8	11-0	13-5	15-7	7-11	10-0	12-3	14-3
	Douglas Fir-Larch	#2	8-1	10-3	12-7	14-7	7-5	9-5	11-6	13-4
	Douglas Fir-Larch	#3	6-2	7-9	9-6	11-0	5-7	7-1	8-8	10-1
	Hem-Fir	SS	8-6	11-3	14-4	17-5	8-6	11-3	14-4	16-10[a]
	Hem-Fir	#1	8-4	10-9	13-1	15-2	7-9	9-9	11-11	13-10
	Hem-Fir	#2	7-11	10-2	12-5	14-4	7-4	9-3	11-4	13-1
	Hem-Fir	#3	6-2	7-9	9-6	11-0	5-7	7-1	8-8	10-1
	Southern Pine	SS	8-10	11-8	14-11	18-1	8-10	11-8	14-11	18-1
	Southern Pine	#1	8-8	11-5	14-7	17-5	8-8	11-3	13-4	15-11
	Southern Pine	#2	8-6	11-0	13-1	15-5	7-9	10-0	12-0	14-0
	Southern Pine	#3	6-7	8-5	9-11	11-10	6-0	7-8	9-1	10-9
	Spruce-Pine-Fir	SS	8-4	11-0	14-0	17-0	8-4	11-0	13-8	15-11
	Spruce-Pine-Fir	#1	8-1	10-3	12-7	14-7	7-5	9-5	11-6	13-4
	Spruce-Pine-Fir	#2	8-1	10-3	12-7	14-7	7-5	9-5	11-6	13-4
	Spruce-Pine-Fir	#3	6-2	7-9	9-6	11-0	5-7	7-1	8-8	10-1

Check sources for availability of lumber in lengths greater than 20 feet.

For SI: 1 inch = 25.4 mm, 1 foot = 304.8 mm, 1 pound per square foot = 47.8 N/m^2.

a. End bearing length shall be increased to 2 inches.

2308.8.1 Bearing. Except where supported on a 1-inch by 4-inch (25.4 mm by 102 mm) ribbon strip and nailed to the adjoining stud, the ends of each joist shall not have less than $1^1/_2$ inches (38 mm) of bearing on wood or metal, or less than 3 inches (76 mm) on masonry.

❖ This section is needed so that floor joists have adequate bearing surface on their supports. A minimum bearing distance is stipulated for bearing on wood or steel. The 3 inches (76 mm) on masonry, which should also apply to bearing on concrete, is to minimize the potential for shear failure of the masonry or concrete. As an exception to the minimum $1^1/_2$-inch (38 mm) bearing surface on wood, joists are permitted to bear on a ribbon strip nailed to the narrow face of the studs, as long as the ends of the joists are also nailed to the adjoining studs.

2308.8.2 Framing details. Joists shall be supported laterally at the ends and at each support by solid blocking except where the ends of the joists are nailed to a header, band or rim joist or to an adjoining stud or by other means. Solid blocking shall not be less than 2 inches (51mm) in thickness and the full depth of the joist. Notches on the ends of joists shall not exceed one-fourth the joist depth. Holes bored in joists shall not be within 2 inches (51 mm) of the top or bottom of the joist, and the diameter of any such hole shall not exceed one-third the depth of the joist. Notches in the top or bottom of joists shall not exceed one-sixth the depth and shall not be located in the middle third of the span.

Joist framing from opposite sides of a beam, girder or partition shall be lapped at least 3 inches (76 mm) or the opposing joists shall be tied together in an approved manner.

Joists framing into the side of a wood girder shall be supported by framing anchors or on ledger strips not less than 2 inches by 2 inches (51 mm by 51 mm).

❖ The general requirement of this section is that joists be laterally supported by solid blocking at each end and at each support. The blocking is intended to prevent lateral torsional buckling of the joist. The blocking is permitted to be omitted where the ends of the joists are restrained either by nailing or by the use of some other means, such as a joist hanger, to resist buckling.

Wood is comprised of relatively small elongated cells oriented parallel to each other and bound together physically and chemically by lignin. The cells are continuous from end to end of a piece of lumber. When a beam or joist is notched or cut, the cell strands are interrupted. In notched beams, the effective depth is reduced equal to the depth of the notch (see Figure 2308.8.2).

Some designs and installation practices require that limited notching and cutting occur. Notching should be avoided when possible. Holes bored in beams and joists create the same problems as notches. When necessary, the holes should be located in areas with the least stress concentration, generally along the neutral axis of the joist. Beams subject to high horizontal shear stress (short span/heavy load) should never be cut. Wood members having a thickness of over 4 inches (102 mm) should never be notched, except at the ends of the members.

Joists must be adequately tied together to provide continuity and prevent horizontal separation of the building and must be adequately supported when not bearing on the top of girders.

2308.8.2.1 Engineered wood products. Cuts, notches and holes bored in trusses, laminated veneer lumber, glue-laminated members or I-joists are not permitted unless the effects of such penetrations are specifically considered in the design of the member.

❖ The notching and boring allowances that are permitted for solid sawn joist framing cannot be applied to engineered wood products. Because these types of products are commonly incorporated into conventional construction, this section points out that they must be treated differently. However, it does permit notching and boring that is considered in the design of the member.

2308.8.3 Framing around openings. Trimmer and header joists shall be doubled, or of lumber of equivalent cross section, where the span of the header exceeds 4 feet (1219 mm). The ends of header joists more than 6 feet (1829 mm) long shall be supported by framing anchors or joist hangers unless bearing on a beam, partition or wall. Tail joists over 12 feet (3658 mm) long shall be supported at the header by framing anchors or on ledger strips not less than 2 inches by 2 inches (51 mm by 51 mm).

❖ Figure 2308.8.3 illustrates the criteria for framing around openings.

2308.8.4 Supporting bearing partitions. Bearing partitions parallel to joists shall be supported on beams, girders, doubled joists, walls or other bearing partitions. Bearing partitions perpendicular to joists shall not be offset from supporting girders, walls or partitions more than the joist depth unless such joists are of sufficient size to carry the additional load.

❖ The floor joist spans provided in Tables 2308.8(1) and (2) assume uniform loading conditions and will support only limited concentrated loads. Unless designed by engineering analysis, bearing partitions must be supported by other bearing partitions, beams or girders. This may be accomplished one way by continuing a bearing partition through a joist space to the lower partition, installing headers between joists that rest on the lower partition, and placing headers on top of adequately supported flooring.

A major problem that frequently occurs is the orientation of heavy loads, such as bathtubs, parallel with the floor joists. Additional joists should be installed to support these loads (see Figure 2308.8.4).

Bearing partitions oriented perpendicular to joists cannot be offset from supporting girders, walls or partitions more than the depth of the joists, unless the joists are of sufficient size to carry the additional load (see Figure 2308.8.4).

Although not specifically required, it is typically advantageous to double up floor joists supporting nonload-bearing partitions oriented parallel to the joists or to provide solid blocking between the joists to transfer the wall load to the supporting joists (see Figure 2308.8.4).

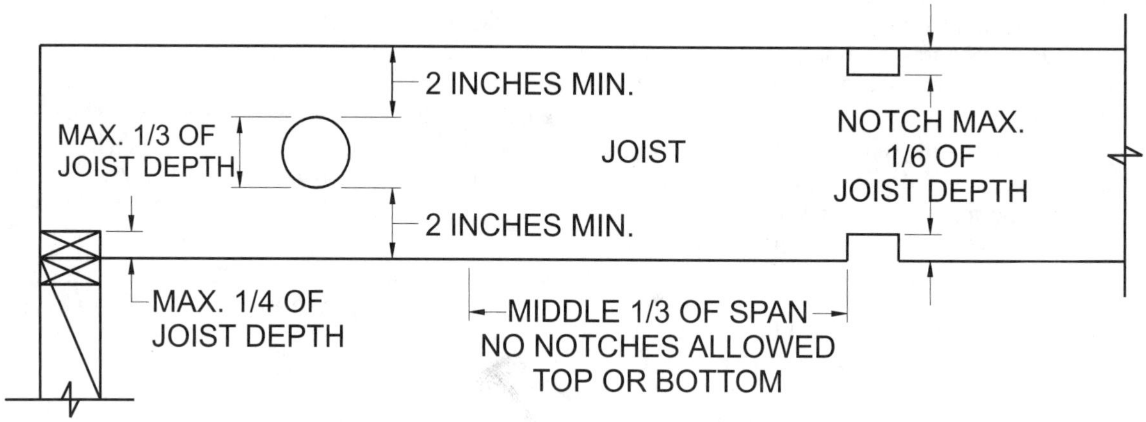

For SI: 1 inch = 25.4 mm.

Figure 2308.8.2
CUTTING, NOTCHING AND BORING JOISTS

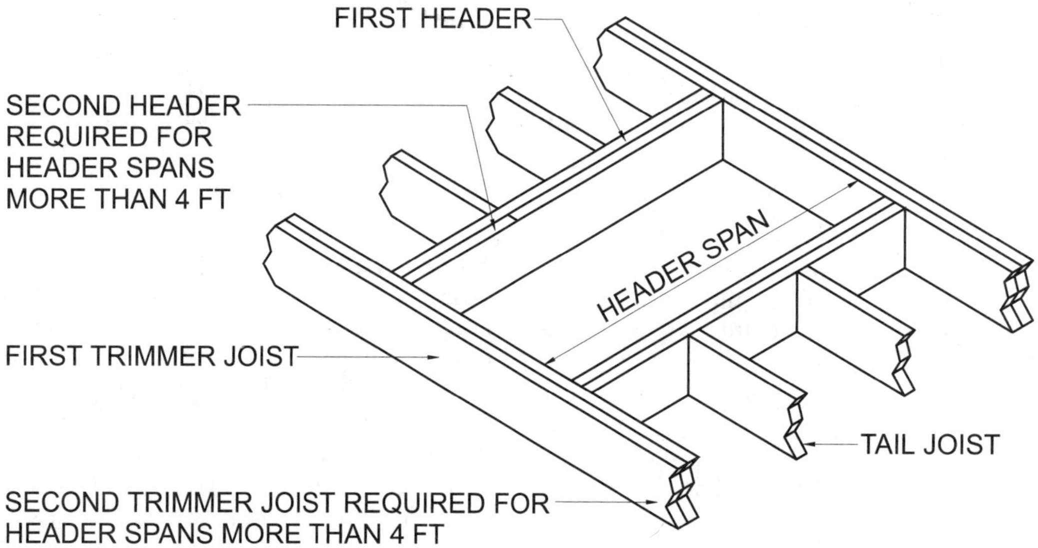

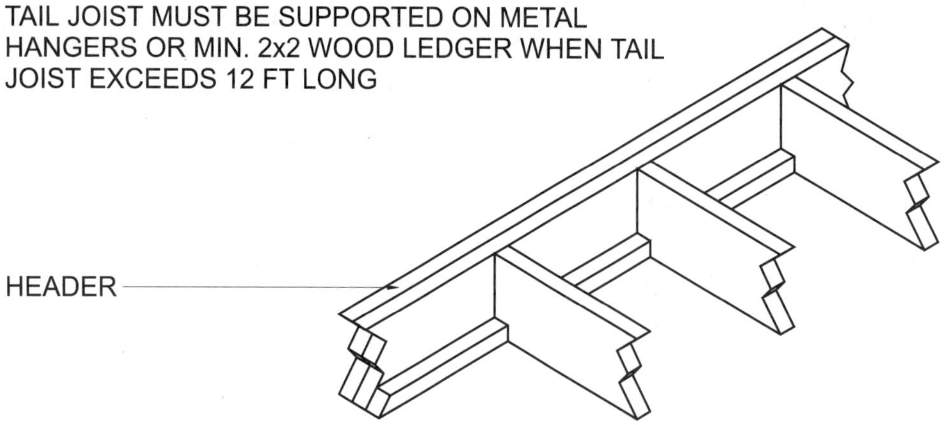

For SI: 1 foot = 304.8 mm.

Figure 2308.8.3
FRAMING AROUND OPENING

2308.8.5 Lateral support. Floor, attic and roof framing with a nominal depth-to-thickness ratio greater than or equal to 5:1 shall have one edge held in line for the entire span. Where the nominal depth-to-thickness ratio of the framing member exceeds 6:1, there shall be one line of bridging for each 8 feet (2438 mm) of span, unless both edges of the member are held in line. The bridging shall consist of not less than 1-inch by 3-inch (25 mm by 76 mm) lumber, double nailed at each end, of equivalent metal bracing of equal rigidity, full-depth solid blocking or other approved means. A line of bridging shall also be required at supports where equivalent lateral support is not otherwise provided.

❖ When the depth-to-thickness ratio of joists and rafters exceeds 5:1, as would be the case in members larger than 2 inches by 10 inches (51 mm by 254 mm), the lateral support required by Section 2308.8.2 is not sufficient to prevent lateral buckling between supports. Additional resistance is required. Sheathing, subflooring, decking and similar materials attached to each joist or rafter are considered to provide edge restraint.

These requirements are cumulative. The support required by Section 2308.8.2 applies to all joists. Additionally, members greater than 2 inches by 10 inches (51 mm by 254 mm) must have one edge held in line, and members greater than 2 inches by 12 inches (51 mm by 305 mm) must have one edge held in line as well as a line of bridging at each 8 feet (2438 mm) of span (which may be omitted if both edges are held in line).

2308.8.6 Structural floor sheathing. Structural floor sheathing shall comply with the provisions of Section 2304.7.1.

❖ Subflooring is considered to be a structural element, whereas underlayment is related to serviceability. Both must comply with the appropriate tables for maximum span referenced in Section 2304.7.1.

2308.8.7 Under-floor ventilation. For under-floor ventilation, see Section 1203.3.

❖ Enclosed spaces under floors of buildings are referred to as crawl spaces when not designed and constructed as basements. The earth is usually exposed or covered with a vapor retarder. Regardless, there is always the potential for a large amount of moisture vapor from the

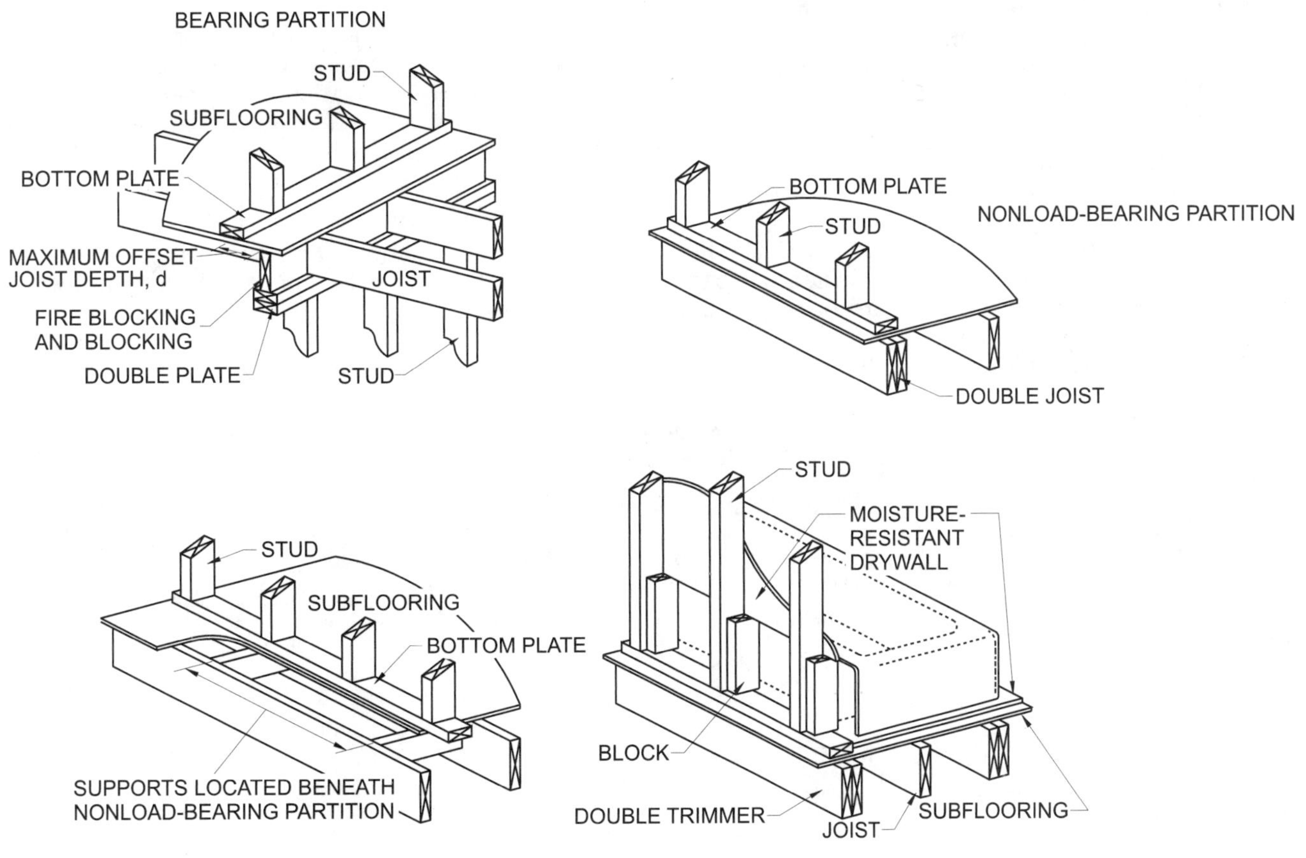

SOURCE: AF&PA *Wood Frame Construction Manual for One- and Two-Family Dwellings, SBC High Wind Edition*, 1995.

Figure 2308.8.4
JOISTS SUPPORTING PARTITIONS AND HEAVY LOADS

ground. This moisture must be removed to prevent wetting and drying cycles that can cause decay and fungus attack in wood members. Experience has shown that crawl spaces constructed in conformance with Section 2308 and ventilated to the exterior as required by Section 1203.3 do not develop decay problems. Access to crawl spaces is required as specified in Section 1209.1.

It should be recognized that moisture problems may be exaggerated in colder months; therefore, operable vents should be used with caution. Once insulation in such spaces becomes wet, it seldom dries out, thus negating its performance.

2308.9 Wall framing.

2308.9.1 Size, height and spacing. The size, height and spacing of studs shall be in accordance with Table 2308.9.1 except that utility-grade studs shall not be spaced more than 16 inches (406 mm) o.c., or support more than a roof and ceiling, or exceed 8 feet (2438 mm) in height for exterior walls and load-bearing walls or 10 feet (3048 mm) for interior nonload-bearing walls.

❖ The specifications in these sections are empirical interpretations of accepted engineering design practices and traditional designs for light wood-frame buildings. When stud heights exceed 10 feet (3048 mm) or the structure is outside the scope of applicability of conventional construction requirements, studs must be designed in accordance with accepted engineering practice.

TABLE 2308.9.1. See below.

❖ Wood studs in conventional construction are limited to the maximum height and spacing as indicated in this table. The height and spacing limits are based on the dead and live loads associated with conventional construction that are specified in Section 2308.2.

2308.9.2 Framing details. Studs shall be placed with their wide dimension perpendicular to the wall. Not less than three studs shall be installed at each corner of an exterior wall.

Exception: At corners, two studs are permitted, provided wood spacers or backup cleats of $^3/_8$-inch-thick (9.5 mm) wood structural panel, $^3/_8$-inch (9.5 mm) Type M "Exterior Glue" particleboard, 1-inch-thick (25 mm) lumber or other approved devices that will serve as an adequate backing for the attachment of facing materials are used. Where fire-resistance ratings or shear values are involved, wood spacers, backup cleats or other devices shall not be used unless specifically approved for such use.

❖ Wood-frame exterior walls support all vertical imposed loads and also must act as braced walls to resist lateral loads, such as from wind and earthquakes. Adequately secured and sufficient corner posts are required to assist in this function, in addition to providing a degree of continuity in the exterior walls. Traditionally, three studs were installed. Experience has shown that two studs will adequately perform this function, provided other means are employed to support any interior finish that may be installed. Note that these other means may not be acceptable if they affect the fire resistance or shear resisting capacity of the walls.

2308.9.2.1 Top plates. Bearing and exterior wall studs shall be capped with double top plates installed to provide overlapping at corners and at intersections with other partitions. End joints in double top plates shall be offset at least 48 inches (1219 mm), and shall be nailed with not less than eight 16d face nails on each side of the joint. Plates shall be a nominal 2 inches (51 mm) in depth and have a width at least equal to the width of the studs.

Exception: A single top plate is permitted, provided the plate is adequately tied at joints, corners and intersecting walls by at least the equivalent of 3-inch by 6-inch (76 mm by 152 mm) by 0.036-inch-thick (0.914 mm) galvanized steel that is nailed to each wall or segment of wall by six 8d nails or

TABLE 2308.9.1
SIZE, HEIGHT AND SPACING OF WOOD STUDS

STUD SIZE (inches)	BEARING WALLS				NONBEARING WALLS	
	Laterally unsupported stud height[a] (feet)	Supporting roof and ceiling only	Supporting one floor, roof and ceiling	Supporting two floors, roof and ceiling	Laterally unsupported stud height[a] (feet)	Spacing (inches)
		Spacing (inches)				
2 × 3[b]	—	—	—	—	10	16
2 × 4	10	24	16	—	14	24
3 × 4	10	24	24	16	14	24
2 × 5	10	24	24	—	16	24
2 × 6	10	24	24	16	20	24

For SI: 1 inch = 25.4 mm, 1 foot = 304.8 mm.

a. Listed heights are distances between points of lateral support placed perpendicular to the plane of the wall. Increases in unsupported height are permitted where justified by an analysis.

b. Shall not be used in exterior walls.

equivalent, provided the rafters, joists or trusses are centered over the studs with a tolerance of no more than 1 inch (25 mm).

❖ Double top plates serve three major functions:

1. They overlap at corners and interior wall intersections, thus tying the building together;
2. They serve as beams to support joists and rafters that are not located directly over the studs (see Section 2308.9.2.2); and
3. They serve as chords for floor and roof diaphragms (see Section 2305.2).

The tolerance of 1 inch (25 mm) with which the rafters or joists are to be centered over the studs is required to prevent overstress or excessive deflection in the single top plate. The steel strap provides a minimum level of continuity.

2308.9.2.2 Top plates for studs spaced at 24 inches (610 mm). Where bearing studs are spaced at 24-inch (610 mm) intervals and top plates are less than two 2-inch by 6-inch (51 mm by 152 mm) or two 3-inch by 4-inch (76 mm by 102 mm) members and where the floor joists, floor trusses or roof trusses that they support are spaced at more than 16-inch (406 mm) intervals, such joists or trusses shall bear within 5 inches (127 mm) of the studs beneath or a third plate shall be installed.

❖ Ideally, joists, rafters, or trusses should be located directly over bearing studs. The empirical provisions of this section allow joists or trusses to bear within 5 inches (127 mm) of the stud below when the joists or trusses are spaced more than 16 inches (406 mm) o.c., and the studs are spaced 24 inches (610 mm) o.c. This provision assumes there are double top (bearing) plates that have all joints located over the studs. Additional support is required if the offset is more than 5 inches (127 mm).

2308.9.2.3 Nonbearing walls and partitions. In nonbearing walls and partitions, studs shall be spaced not more than 28 inches (711 mm) o.c. and are permitted to be set with the long dimension parallel to the wall. Interior nonbearing partitions shall be capped with no less than a single top plate installed to provide overlapping at corners and at intersections with other walls and partitions. The plate shall be continuously tied at joints by solid blocking at least 16 inches (406 mm) in length and equal in size to the plate or by $^{1}/_{2}$-inch by $1^{1}/_{2}$-inch (12.7 mm by 38 mm) metal ties with spliced sections fastened with two 16d nails on each side of the joint.

❖ Nonbearing partitions are not intended to add to the rigidity or structural performance of the building. They are, however, expected to stay in place and have the ability to support finished membranes applied. The code limits stud spacing to 28 inches (711 mm) o.c., but allows the stud to be oriented with the wide dimension parallel to the wall length. Partitions must be able to support the lateral loads to which they are subjected, but not less than 5 psf (0.240 kN/m^2) lateral load, in accordance with Section 1607.13.

Support of the ceiling membranes next to nonbearing partitions requires special consideration. Partitions may be held in place by two floor or ceiling joists or spacers. If a spacer is used, additional support for a ceiling membrane may be obtained by attaching a 1-inch (25 mm) board to the top of the partition. Nonbearing partitions should never be attached rigidly to trussed roofs or floors. Many instances have been cited where such connections have caused partitions to be raised above floor level.

2308.9.2.4 Plates or sills. Studs shall have full bearing on a plate or sill not less than 2 inches (51 mm) in thickness having a width not less than that of the wall studs.

❖ The single bottom plate serves to anchor the wall to the floor. Studs are attached to the sill by end nailing or toe nailing.

2308.9.3 Bracing. Braced wall lines shall consist of braced wall panels that meet the requirements for location, type and amount of bracing as shown in Figure 2308.9.3, specified in Table 2308.9.3(1), and are in line or offset from each other by not more than 4 feet (1219 mm). Braced wall panels shall start not more than 8 feet (2438 mm) from each end of a braced wall line. A designed collector shall be provided if the bracing begins more than 12.5 feet (3810 mm) from an end of a braced wall line. Braced wall panels shall be clearly indicated on the plans. Construction of braced wall panels shall be by one of the following methods:

1. Nominal 1-inch by 4-inch (25 mm by 102 mm) continuous diagonal braces let into top and bottom plates and intervening studs, placed at an angle not more than 60 degrees (1.0 rad) or less than 45 degrees (0.79 rad) from the horizontal and attached to the framing in conformance with Table 2304.9.1.
2. Wood boards of $^{5}/_{8}$-inch (15.9 mm) net minimum thickness applied diagonally on studs spaced not over 24 inches (610 mm) o.c.
3. Wood structural panel sheathing with a thickness not less than $^{5}/_{16}$ inch (7.9 mm) for a 16-inch (406 mm) stud spacing and not less than $^{3}/_{8}$ inch (9.5 mm) for a 24-inch (610 mm) stud spacing in accordance with Tables 2308.9.3(2) and 2308.9.3(3).
4. Fiberboard sheathing panels not less than $^{1}/_{2}$ inch (12.7 mm) thick applied vertically or horizontally on studs spaced not over 16 inches (406 mm) o.c. where installed with fasteners in accordance with Section 2306.4.4 and Table 2308.9.3(4).
5. Gypsum board [sheathing $^{1}/_{2}$ inch (12.7 mm) thick by 4 feet (1219 mm) wide wallboard or veneer base] on studs spaced not over 24 inches (610 mm) o.c. and nailed at 7 inches (178 mm) o.c. with nails as required by Table 2306.4.5.
6. Particleboard wall sheathing panels where installed in accordance with Table 2308.9.3(5).
7. Portland cement plaster on studs spaced 16 inches (406 mm) o.c. installed in accordance with Section 2510.
8. Hardboard panel siding where installed in accordance with Section 2303.1.6 and Table 2308.9.3(6).

For cripple wall bracing, see Section 2308.9.4.1. For Methods 2, 3, 4, 6, 7 and 8, each panel must be at least 48 inches (1219 mm) in length, covering three stud spaces where studs are spaced 16 inches (406 mm) apart and covering two stud spaces where studs are spaced 24 inches (610 mm) apart.

For Method 5, each panel must be at least 96 inches (2438 mm) in length where applied to one face of a panel and 48 inches (1219 mm) where applied to both faces.

All vertical joints of panel sheathing shall occur over studs and adjacent panel joints shall be nailed to common framing members. Horizontal joints shall occur over blocking or other framing equal in size to the studding except where waived by the installation requirements for the specific sheathing materials.

Sole plates shall be nailed to the floor framing and top plates shall be connected to the framing above in accordance with Section 2308.3.2. Where joists are perpendicular to braced wall lines above, blocking shall be provided under and in line with the braced wall panels.

❖ This section continues the provisions that were started in Section 2308.3 for bracing of walls to resist lateral forces. Braced wall panels are portions of walls, as required by Table 2308.9.3(1), composed of the bracing materials discussed below. Portions of braced wall lines are permitted to be offset from each other, and braced wall panels are not required to extend to the end of the braced wall line (see Figure 2308.9.3).

TABLE 2308.9.3(1). See page 23-66.

❖ See the commentary to Section 2308.9.3.

TABLES 2308.9.3(2) through 2308.9.3(6). See page 23-66.

❖ Tables 2308.9.3(2) through 2308.9.3(6) are for the materials that are prescriptively permitted to be used to form the braced wall panels in braced wall lines. The location and minimum acceptable length of braced wall panels are specified in Table 2308.9.3(1). Detailed requirements for various materials—grades, thickness, nailing schedule, stud spacing and similar requirements—are also contained in the tables referenced for each material. Also see the commentary to Section 2303 concerning the minimum standards and quality of these materials.

2308.9.3.1 Alternative bracing. Any bracing required by Section 2308.9.3 is permitted to be replaced by the following:

1. In one-story buildings, each panel shall have a length of not less than 2 feet 8 inches (813 mm) and a height of not more than 10 feet (3048 mm). Each panel shall be sheathed on one face with $^3/_8$-inch-minimum-thickness (9.5 mm) wood structural panel sheathing nailed with 8d common or galvanized box nails in accordance with Table 2304.9.1 and blocked at wood structural panel edges. Two anchor bolts installed in accordance with Section 2308.6 shall be provided in each panel. Anchor bolts shall be placed at each panel outside quarter points. Each panel end stud shall have a tie-down device fastened to the foundation, capable of providing an approved uplift capacity of not less than 1,800 pounds (8006 N). The tie-down device shall be installed in accordance with the manufacturer's recommendations. The panels shall be supported directly on a foundation or on floor framing supported directly on a foundation that is continuous across the entire length of the braced wall line. This foundation shall be reinforced with not less than one No. 4 bar top and bottom.

   Where the continuous foundation is required to have a depth greater than 12 inches (305 mm), a minimum 12-inch by 12-inch (305 mm by 305 mm) continuous footing or turned down slab edge is permitted at door openings in the braced wall line. This continuous footing or turned down slab edge shall be reinforced with not less than one No. 4 bar top and bottom. This reinforcement shall be lapped 15 inches (381 mm) with the reinforcement required in the continuous foundation located directly under the braced wall line.

2. In the first story of two-story buildings, each wall panel shall be braced in accordance with Section 2308.9.3.1, Item 1, except that the wood structural panel sheathing shall be provided on both faces, three anchor bolts shall be placed at one-quarter points, and tie-down device uplift capacity shall not be less than 3,000 pounds (13 344 N).

❖ This provision allows a minimum 32-inch (813 mm) wall bracing length in lieu of the 48-inch (1219 mm) length required elsewhere. This bracing detail allows more flexibility in constructing braced wall panels adjacent to garage doors and other similar openings. This detail was developed by APA—Engineered Wood Association and is supported in APA Research Report No. 156.

For two-story construction, Item 2 allows alternative bracing to be used only in the first story, provided it is installed as prescribed by Item 1 and, in addition, the sheathing is installed on both faces of the segment, three anchor bolts are installed at panel quarter points and the tie-downs have a capacity of 3,000 pounds (13 344 N)

FIGURE 2308.9.3.1. See page 23-68.

2308.9.4 Cripple walls. Foundation cripple walls shall be framed of studs not less in size than the studding above with a minimum length of 14 inches (356 mm), or shall be framed of solid blocking. Where exceeding 4 feet (1219 mm) in height, such walls shall be framed of studs having the size required for an additional story.

❖ Cripple walls (sometimes referred to as foundation stud walls or knee walls) are stud walls usually less than 8 feet (2438 mm) in height that rest on the foundation plate and support the first immediate floor above. The minimum stud length of 14 inches (356 mm) is based on the length necessary to properly fasten the studs to the foundation wall plate and the double plate above. Where the studs are less than 14 inches (356 mm) in length, they should be installed with wall plates and with the solid blocking tightly fit between each stud. This blocking performs two purposes: it provides a level, uniform bearing surface for the support of the floor above and transmits lateral forces from the floor to the foundation.

SEISMIC DESIGN CATEGORY	MAXIMUM WALL SPACING (feet)	REQUIRED BRACING LENGTH, b
A, B, and C	35′-0″	Table 2308.9.3(1) and Section 2308.9.3
D and E	25′-0″	Table 2308.12.4

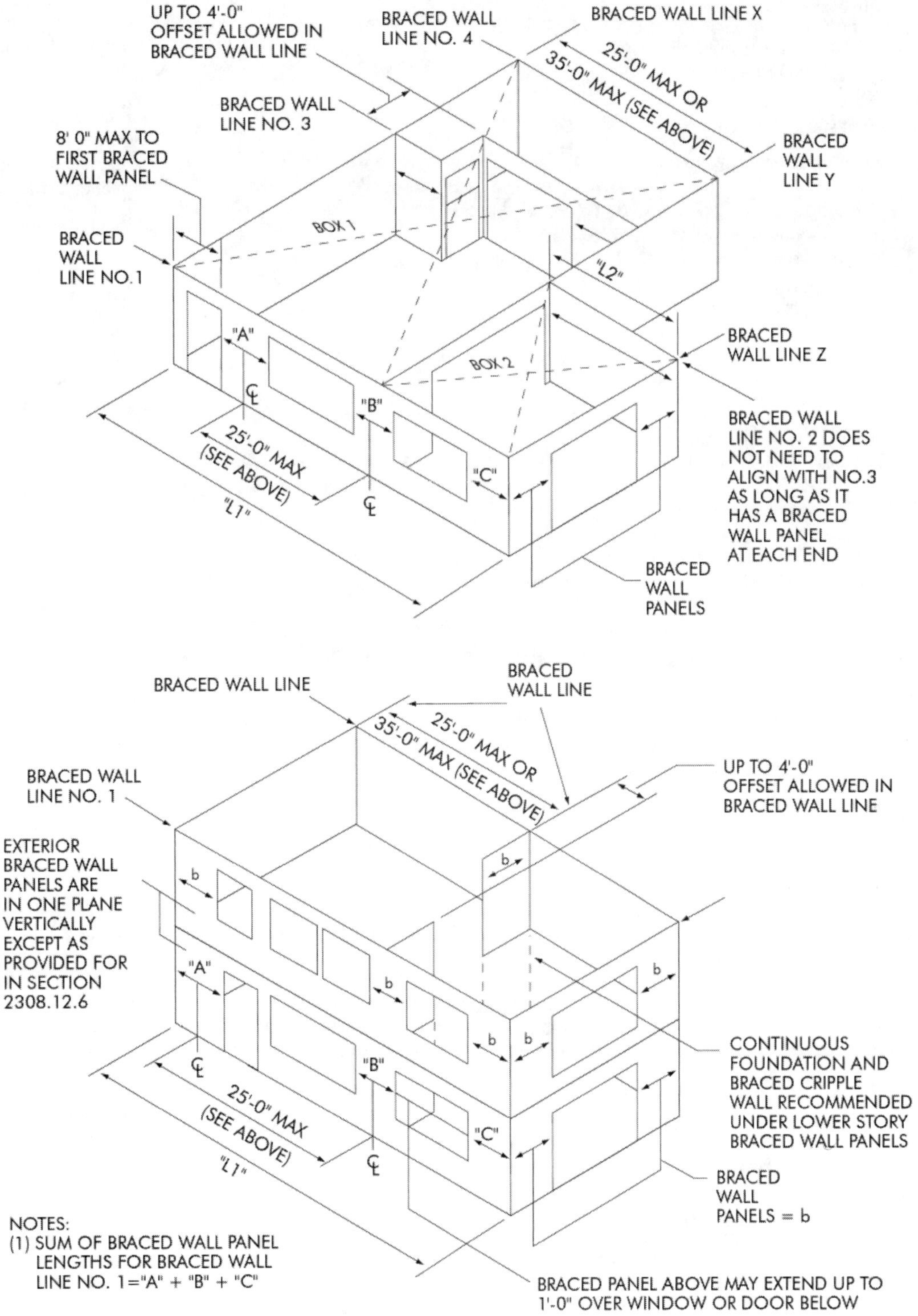

For SI: 1 foot = 304.8 mm.

FIGURE 2308.9.3
BASIC COMPONENTS OF THE LATERAL BRACING SYSTEM

TABLE 2308.9.3(1)
BRACED WALL PANELS[a]

SEISMIC DESIGN CATEGORY	CONDITION	CONSTRUCTION METHODS[b,c] 1	2	3	4	5	6	7	8	BRACED PANEL LOCATION AND LENGTH[d]
A and B	One story, top of two or three story	X	X	X	X	X	X	X	X	Each end and not more than 25 feet on center
	First story of two story or second story of three story	X	X	X	X	X	X	X	X	
	First story of three story	—	X	X	X	X[e]	X	X	X	
C	One story, top of two or three story	—	X	X	X	X	X	X	X	Each end and not more than 25 feet on center
	First story of two story or second story of three story	—	X	X	X	X[e]	X	X	X	Each end and not more than 25 feet on center but total length shall not be less than 25% of building length[f]
	First story of three story	—	X	X	X	X[e]	X	X	X	Each end and not more than 25 feet on center but total length shall not be less than 40% of building length[f]

For SI: 1 inch = 25.4 mm, 1 foot = 304.8 mm.

a. This table specifies minimum requirements for braced panels that form interior or exterior braced wall lines.

b. See Section 2308.9.3 for full description.

c. See Section 2308.9.3.1 for alternative braced panel requirement.

d. Building length is the dimension parallel to the braced wall length.

e. Gypsum wallboard applied to framing supports that are spaced at 16 inches on center.

f. The required lengths shall be doubled for gypsum board applied to only one face of a braced wall panel.

TABLE 2308.9.3(2)
EXPOSED PLYWOOD PANEL SIDING

MINIMUM THICKNESS[a] (inch)	MINIMUM NUMBER OF PLIES	STUD SPACING (inches) Plywood siding applied directly to studs or over sheathing
$^3/_8$	3	16[b]
$^1/_2$	4	24

For SI: 1 inch = 25.4 mm.

a. Thickness of grooved panels is measured at bottom of grooves.

b. Spans are permitted to be 24 inches if plywood siding applied with face grain perpendicular to studs or over one of the following: (1) 1-inch board sheathing, (2) $^7/_{16}$ -inch wood structural panel sheathing or (3) $^3/_8$ –inch wood structural panel sheathing with strength axis (which is the long direction of the panel unless otherwise marked) of sheathing perpendicular to studs.

TABLE 2308.9.3(3)
WOOD STRUCTURAL PANEL WALL SHEATHING[b]
(Not Exposed to the Weather, Strength Axis Parallel or Perpendicular to Studs Except as Indicated Below)

MINIMUM THICKNESS (inch)	PANEL SPAN RATING	STUD SPACING (inches) Siding nailed to studs	Nailable sheathing: Sheathing parallel to studs	Nailable sheathing: Sheathing perpendicular to studs
$^5/_{16}$	12/0, 16/0, 20/0 Wall–16″ o.c.	16	—	16
$^3/_8$, $^{15}/_{32}$, $^1/_2$	16/0, 20/0, 24/0, 32/16 Wall–24″ o.c.	24	16	24
$^7/_{16}$, $^{15}/_{32}$, $^1/_2$	24/0, 24/16, 32/16 Wall–24″ o.c.	24	24[a]	24

For SI: 1 inch = 25.4 mm.

a. Plywood shall consist of four or more plies.

b. Blocking of horizontal joints shall not be required except as specified in Sections 2306.4 and 2308.12.4.

TABLE 2308.9.3(4)
ALLOWABLE SHEAR VALUES (plf) FOR WIND OR SEISMIC LOADING ON VERTICAL DIAPHRAGMS OF FIBERBOARD SHEATHING BOARD CONSTRUCTION FOR TYPE V CONSTRUCTION ONLY[a, b, c, d, e, f, g, h]

THICKNESS AND GRADE	FASTENER SIZE	SHEAR VALUE (pounds per linear foot) 3-INCH NAIL SPACING AROUND PERIMETER AND 6-INCH AT INTERMEDIATE POINTS
$^1/_2$″ Structural	No. 11 gage galvanized roofing nail $1^1/_2$″ long, $^7/_{16}$″ head	125[g]
$^{25}/_{32}$″ Structural	No. 11 gage galvanized roofing nail $1^3/_4$″ long, $^7/_{16}$″ head	175[g]

For SI: 1 inch = 25.4 mm, 1 pound per foot = 14.5939 N/m.

a. Fiberboard sheathing diaphragms shall not be used to brace concrete or masonry walls.
b. Panel edges shall be backed with 2 inch or wider framing of Douglas fir-larch or Southern pine.
c. Fiberboard sheathing on one side only.
d. Fiberboard panels are installed with their long dimension parallel or perpendicular to studs.
e. Fasteners shall be spaced 6 inches on center along intermediate framing members.
f. For framing of other species: (1) Find specific gravity for species of lumber in AF&PA National Design Specification, and (2) Multiply the shear value from the above table by 0.82 for species with specific gravity of 0.42 or greater, or 0.65 for all other species.
g. The same values can be applied when staples are used as described in Table 2304.9.1.
h. Values are not permitted in Seismic Design Category D, E or F.

TABLE 2308.9.3(5)
ALLOWABLE SPANS FOR PARTICLEBOARD WALL SHEATHING
(Not Exposed to the Weather, Long Dimension of the Panel Parallel or Perpendicular to Studs)

GRADE	THICKNESS (inch)	STUD SPACING (inches)	
		Siding nailed to studs	Sheathing under coverings specified in Section 2308.9.3 parallel or perpendicular to studs
M-S "Exterior Glue" and M-2"Exterior Glue"	$^3/_8$	16	—
	$^1/_2$	16	16

For SI: 1 inch = 25.4 mm.

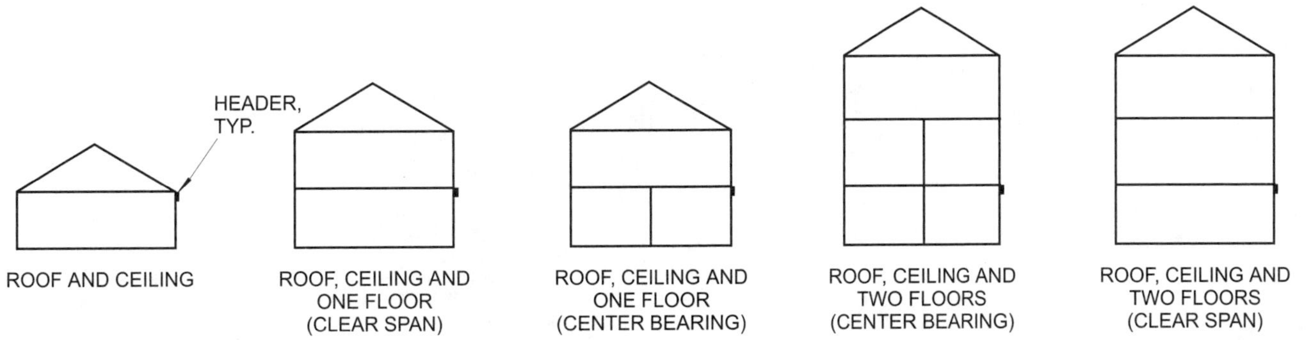

Figure 2308.9.5
HEADER SUPPORT CONDITIONS FOR EXTERIOR LOAD BEARING WALLS (Source: AF&PA WOOD-FRAME CONSTRUCTION MANUAL)

TABLE 2308.9.3(6)
HARDBOARD SIDING

SIDING	MINIMUM NOMINAL THICKNESS (inch)	2 × 4 FRAMING MAXIMUM SPACING	NAIL SIZE[a,b,d]	NAIL SPACING General	NAIL SPACING Bracing panels[c]
1. Lap siding					
Direct to studs	$^3/_8$	16″ o.c.	8d	16″ o.c.	Not applicable
Over sheathing	$^3/_8$	16″ o.c.	10d	16″ o.c.	Not applicable
2. Square edge panel siding					
Direct to studs	$^3/_8$	24″ o.c.	6d	6″ o.c. edges; 12″ o.c. at intermediate supports	4″ o.c. edges; 8″ o.c. at intermediate supports
Over sheathing	$^3/_8$	24″ o.c.	8d	6″ o.c. edges; 12″ o.c. at intermediate supports	4″ o.c. edges; 8″ o.c. at intermediate supports
3. Shiplap edge panel siding					
Direct to studs	$^3/_8$	16″ o.c.	6d	6″ o.c. edges; 12″ o.c. at intermediate supports	4″ o.c. edges; 8″ o.c. at intermediate supports
Over sheathing	$^3/_8$	16″ o.c.	8d	6″ o.c. edges; 12″ o.c. At intermediate supports	4″ o.c. edges; 8″ o.c. at intermediate supports

For SI: 1 inch = 25.4 mm.

a. Nails shall be corrosion resistant.

b. Minimum acceptable nail dimensions:

	Panel Siding (inch)	Lap Siding (inch)
Shank diameter	0.092	0.099
Head diameter	0.225	0.240

c. Where used to comply with Section 2308.9.3.

d. Nail length must accommodate the sheathing and penetrate framing $1^1/_2$ inches.

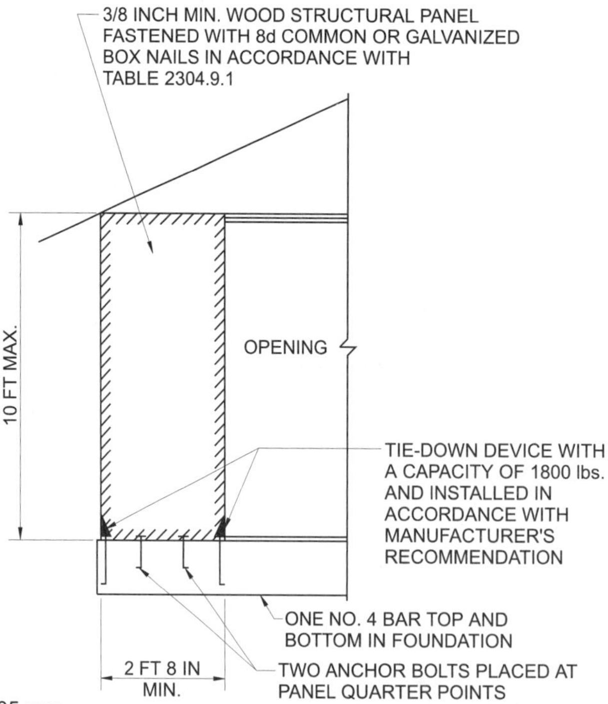

For SI: 1 inch = 25.4 mm, 1 foot = 305 mm, 1 pound = .004448 kN.

Figure 2308.9.3.1
ALTERNATIVE BRACED WALL PANELS—ONE-STORY

2308.9.4.1 Bracing. For the purposes of this section, cripple walls having a stud height exceeding 14 inches (356 mm) shall be considered a story and shall be braced in accordance with Table 2308.9.3(1) for Seismic Design Category A, B or C. See Section 2308.12.4 for Seismic Design Category D or E.

❖ When the cripple wall stud height is more than 14 inches (356 mm), the wall is to be considered as a first-story wall and the bracing requirements of Table 2308.9.3(1) will apply for buildings in lower seismic design categories. In Seismic Design Category D or E, the increased bracing requirements of Table 2308.12.4 will apply.

2308.9.4.2 Nailing of bracing. Spacing of edge nailing for required wall bracing shall not exceed 6 inches (152 mm) o.c. along the foundation plate and the top plate of the cripple wall. Nail size, nail spacing for field nailing and more restrictive boundary nailing requirements shall be as required elsewhere in the code for the specific bracing material used.

❖ As is the case with wall bracing in general, nailing of the specific structural sheathing material is critical for the racking resistance of the cripple wall.

2308.9.5 Openings in exterior walls.

TABLE 2308.9.5. See page 23-70.

❖ The header support conditions utilized in this table are shown in Figure 2308.9.5 (see commentary, Section 2308.9.5.1).

2308.9.5.1 Headers. Headers shall be provided over each opening in exterior-bearing walls. The spans in Table 2308.9.5 are permitted to be used for one- and two-family dwellings. Headers for other buildings shall be designed in accordance with Section 2301.2.1 or 2301.2.2. Headers shall be of two pieces of nominal 2-inch (51 mm) framing lumber set on edge as permitted by Table 2308.9.5 and nailed together in accordance with Table 2304.9.1 or of solid lumber of equivalent size.

❖ Headers are required to transfer loads that are received from the wall and floor/roofs above. Header spans for exterior bearing walls of one- and two-family dwellings may be sized and supported in accordance with Table 2308.9.5. Headers for other structures must be designed.

Table 2308.9.5, as well as Table 2308.9.6, provides an empirical procedure for determining header sizes in exterior walls. The tables were developed using the following design load criteria:

Roof LL	=	20 psf
Roof DL	=	10 psf
Ceiling LL	=	5 psf
Ceiling DL	=	5 psf
Floor LL	=	40 psf
Floor DL	=	10 psf
Wall DL	=	11 psf

The spans were developed by using the given loads and performing moment and deflection calculations to determine the maximum span length for the specific size headers shown in the tables. The span length calculations also included a load combination reduction factor. Current ASD design loads do not address load combination reduction factors. The load combination reduction factors were developed in accordance with LRFD load combinations, the smaller of $1.6L + 0.5S$ or $1.6S + 0.5L$, where S is the snow load and L is the live load. The appropriate factor is then divided by 1.6 to adjust from LRFD to ASD.

The header span in the tables is determined based on the ground snow load in accordance with Figure 1608.2, the building width, the supporting conditions and the size of the header to be used. The header sizes must comprise the required lumber sizes shown in the tables, or may be of solid lumber of equivalent size.

2308.9.5.2 Header support. Wall studs shall support the ends of the header in accordance with Table 2308.9.5. Each end of a lintel or header shall have a length of bearing of not less than $1^1/_2$ inches (38 mm) for the full width of the lintel.

❖ The header studs (jack studs) on which the headers rest should be continuous from the header to the bottom plate (or sill plate). Cutting the header stud to support a sill is not allowed. Headers should be adequately nailed together and to the wall studs. This critical nailing is specified in Table 2304.9.1 [see Figures 2308.9.5.2(1) and (2)].

2308.9.6 Openings in interior bearing partitions. Headers shall be provided over each opening in interior bearing partitions as required in Section 2308.9.5. The spans in Table 2308.9.6 are permitted to be used for one- and two-family dwellings. Wall studs shall support the ends of the header in accordance with Table 2308.9.5 or 2308.9.6 as appropriate.

❖ As is required for exterior bearing walls, loads from the framing above must be transferred around openings in interior bearing walls.

TABLE 2308.9.6. See page 23-73.

❖ The header support conditions that are used in this table are illustrated in Figure 2308.9.6 (see commentary, Section 2308.9.5.1).

2308.9.7 Openings in interior nonbearing partitions. Openings in nonbearing partitions are permitted to be framed with single studs and headers. Each end of a lintel or header shall have a length of bearing of not less than $1^1/_2$ inches (38 mm) for the full width of the lintel.

❖ Nonbearing partitions are not intended to add to the rigidity or structural performance of the building. They are, however, expected to stay in place and have the ability to support finished membranes applied.

TABLE 2308.9.5
HEADER AND GIRDER SPANS[a] FOR EXTERIOR BEARING WALLS
(Maximum Spans for Douglas Fir-Larch, Hem-Fir, Southern Pine and Spruce-Pine-Fir[b] and Required Number of Jack Studs)

HEADERS SUPPORTING	SIZE	GROUND SNOW LOAD (psf)[e]											
		30						50					
		Building width[c] (feet)											
		20		28		36		20		28		36	
		Span	NJ[d]	Span	NJ[d]	Span	NJ[d]	Span	NJ[d]	Span	NJ[d]	Span	NJ[d]
Roof & Ceiling	2-2 × 4	3-6	1	3-2	1	2-10	1	3-2	1	2-9	1	2-6	1
	2-2 × 6	5-5	1	4-8	1	4-2	1	4-8	1	4-1	1	3-8	2
	2-2 × 8	6-10	1	5-11	2	5-4	2	5-11	2	5-2	2	4-7	2
	2-2 × 10	8-5	2	7-3	2	6-6	2	7-3	2	6-3	2	5-7	2
	2-2 × 12	9-9	2	8-5	2	7-6	2	8-5	2	7-3	2	6-6	2
	3-2 × 8	8-4	1	7-5	1	6-8	1	7-5	1	6-5	2	5-9	2
	3-2 × 10	10-6	1	9-1	2	8-2	2	9-1	2	7-10	2	7-0	2
	3-2 × 12	12-2	2	10-7	2	9-5	2	10-7	2	9-2	2	8-2	2
	4-2 × 8	9-2	1	8-4	1	7-8	1	8-4	1	7-5	1	6-8	1
	4-2 × 10	11-8	1	10-6	1	9-5	2	10-6	1	9-1	2	8-2	2
	4-2 × 12	14-1	1	12-2	2	10-11	2	12-2	2	10-7	2	9-5	2
Roof Ceiling & 1 Center-Bearing Floor	2-2 × 4	3-1	1	2-9	1	2-5	1	2-9	1	2-5	1	2-2	1
	2-2 × 6	4-6	1	4-0	1	3-7	2	4-1	1	3-7	2	3-3	2
	2-2 × 8	5-9	2	5-0	2	4-6	2	5-2	2	4-6	2	4-1	2
	2-2 × 10	7-0	2	6-2	2	5-6	2	6-4	2	5-6	2	5-0	2
	2-2 × 12	8-1	2	7-1	2	6-5	2	7-4	2	6-5	2	5-9	3
	3-2 × 8	7-2	1	6-3	2	5-8	2	6-5	2	5-8	2	5-1	2
	3-2 × 10	8-9	2	7-8	2	6-11	2	7-11	2	6-11	2	6-3	2
	3-2 × 12	10-2	2	8-11	2	8-0	2	9-2	2	8-0	2	7-3	2
	4-2 × 8	8-1	1	7-3	1	6-7	1	7-5	1	6-6	1	5-11	2
	4-2 × 10	10-1	1	8-10	2	8-0	2	9-1	2	8-0	2	7-2	2
	4-2 × 12	11-9	2	10-3	2	9-3	2	10-7	2	9-3	2	8-4	2
Roof Ceiling & 1 Clear Span Floor	2-2 × 4	2-8	1	2-4	1	2-1	1	2-7	1	2-3	1	2-0	1
	2-2 × 6	3-11	1	3-5	2	3-0	2	3-10	2	3-4	2	3-0	2
	2-2 × 8	5-0	2	4-4	2	3-10	2	4-10	2	4-2	2	3-9	2
	2-2 × 10	6-1	2	5-3	2	4-8	2	5-11	2	5-1	2	4-7	3
	2-2 × 12	7-1	2	6-1	3	5-5	3	6-10	2	5-11	3	5-4	3
	3-2 × 8	6-3	2	5-5	2	4-10	2	6-1	2	5-3	2	4-8	2
	3-2 × 10	7-7	2	6-7	2	5-11	2	7-5	2	6-5	2	5-9	2
	3-2 × 12	8-10	2	7-8	2	6-10	2	8-7	2	7-5	2	6-8	2
	4-2 × 8	7-2	1	6-3	2	5-7	2	7-0	1	6-1	2	5-5	2
	4-2 × 10	8-9	2	7-7	2	6-10	2	8-7	2	7-5	2	6-7	2
	4-2 × 12	10-2	2	8-10	2	7-11	2	9-11	2	8-7	2	7-8	2

TABLE 2308.9.5—continued
HEADER AND GIRDER SPANS[a] FOR EXTERIOR BEARING WALLS
(Maximum Spans for Douglas Fir-Larch, Hem-Fir, Southern Pine and Spruce-Pine-Fir[b] and Required Number of Jack Studs)

HEADERS SUPPORTING	SIZE	GROUND SNOW LOAD (psf)[e]											
		30						50					
		Building width[c] (feet)											
		20		28		36		20		28		36	
		Span	NJ[d]	Span	NJ[d]	Span	NJ[d]	Span	NJ[d]	Span	NJ[d]	Span	NJ[d]
Roof Ceiling & 2 Center-Bearing Floors	2-2 × 4	2-7	1	2-3	1	2-0	1	2-6	1	2-2	1	1-11	1
	2-2 × 6	3-9	2	3-3	2	2-11	2	3-8	2	3-2	2	2-10	2
	2-2 × 8	4-9	2	4-2	2	3-9	2	4-7	2	4-0	2	3-8	2
	2-2 × 10	5-9	2	5-1	2	4-7	3	5-8	2	4-11	2	4-5	3
	2-2 × 12	6-8	2	5-10	3	5-3	3	6-6	2	5-9	3	5-2	3
	3-2 × 8	5-11	2	5-2	2	4-8	2	5-9	2	5-1	2	4-7	2
	3-2 × 10	7-3	2	6-4	2	5-8	2	7-1	2	6-2	2	5-7	2
	3-2 × 12	8-5	2	7-4	2	6-7	2	8-2	2	7-2	2	6-5	3
	4-2 × 8	6-10	1	6-0	2	5-5	2	6-8	1	5-10	2	5-3	2
	4-2 × 10	8-4	2	7-4	2	6-7	2	8-2	2	7-2	2	6-5	2
	4-2 × 12	9-8	2	8-6	2	7-8	2	9-5	2	8-3	2	7-5	2
Roof, Ceiling & 2 Clear Span Floors	2-2 × 4	2-1	1	1-8	1	1-6	2	2-0	1	1-8	1	1-5	2
	2-2 × 6	3-1	2	2-8	2	2-4	2	3-0	2	2-7	2	2-3	2
	2-2 × 8	3-10	2	3-4	2	3-0	3	3-10	2	3-4	2	2-11	3
	2-2 × 10	4-9	2	4-1	3	3-8	3	4-8	2	4-0	3	3-7	3
	2-2 × 12	5-6	3	4-9	3	4-3	3	5-5	3	4-8	3	4-2	3
	3-2 × 8	4-10	2	4-2	2	3-9	2	4-9	2	4-1	2	3-8	2
	3-2 × 10	5-11	2	5-1	2	4-7	3	5-10	2	5-0	2	4-6	3
	3-2 × 12	6-10	2	5-11	3	5-4	3	6-9	2	5-10	3	5-3	3
	4-2 × 8	5-7	2	4-10	2	4-4	2	5-6	2	4-9	2	4-3	2
	4-2 × 10	6-10	2	5-11	2	5-3	2	6-9	2	5-10	2	5-2	2
	4-2 × 12	7-11	2	6-10	2	6-2	3	7-9	2	6-9	2	6-0	3

For SI: 1 inch = 25.4 mm, 1 foot = 304.8 mm, 1 pound per square foot = 47.8 N/m^2.

a. Spans are given in feet and inches (ft-in).

b. Tabulated values are for No. 2 grade lumber.

c. Building width is measured perpendicular to the ridge. For widths between those shown, spans are permitted to be interpolated.

d. NJ - Number of jack studs required to support each end. Where the number of required jack studs equals one, the header is permitted to be supported by an approved framing anchor attached to the full-height wall stud and to the header.

e. Use 30 pounds per square foot ground snow load for cases in which ground snow load is less than 30 pounds per square foot and the roof live load is equal to or less than 20 pounds per square foot.

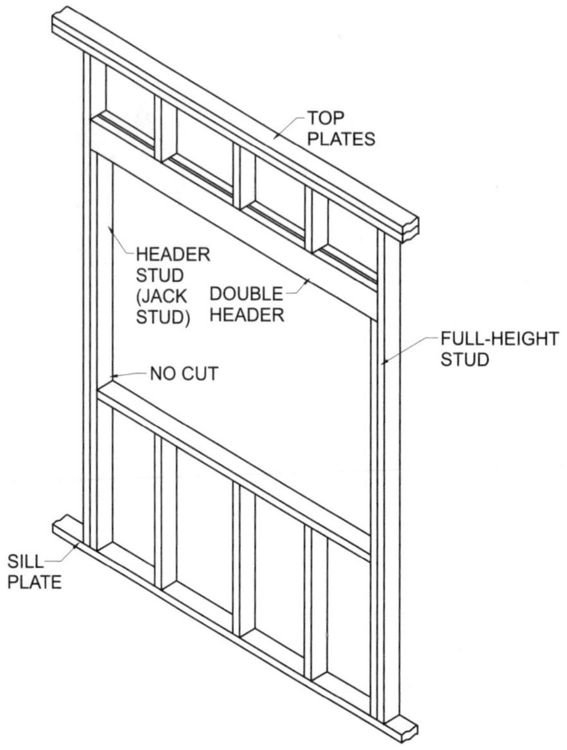

Figure 2308.9.5.2(1)
STUDS AND HEADERS AROUND WALL OPENINGS

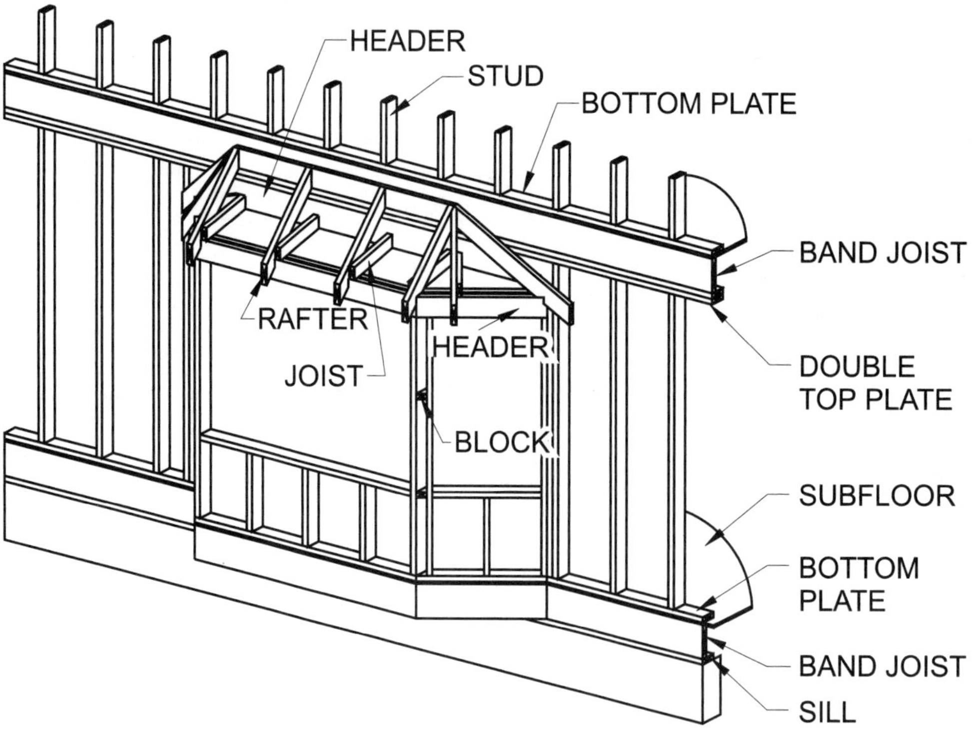

Figure 2308.9.5.2(2)
STUDS AND HEADERS AROUND WALL OPENINGS—BAY WINDOWS
(Source: AF&PA *Wood Frame Construction Manual for One- and Two-Family Dwellings, SBC High Wind Edition*, 1995.)

TABLE 2308.9.6
HEADER AND GIRDER SPANS[a] FOR INTERIOR BEARING WALLS
(Maximum Spans for Douglas Fir-Larch, Hem-Fir, Southern Pine and Spruce-Pine-Fir[b] and Required Number of Jack Studs)

HEADERS AND GIRDERS SUPPORTING	SIZE	BUILDING WIDTH[c] (feet)					
		20		28		36	
		Span	NJ[d]	Span	NJ[d]	Span	NJ[d]
One Floor Only	2-2 × 4	3-1	1	2-8	1	2-5	1
	2-2 × 6	4-6	1	3-11	1	3-6	1
	2-2 × 8	5-9	1	5-0	2	4-5	2
	2-2 × 10	7-0	2	6-1	2	5-5	2
	2-2 × 12	8-1	2	7-0	2	6-3	2
	3-2 × 8	7-2	1	6-3	1	5-7	2
	3-2 × 10	8-9	1	7-7	2	6-9	2
	3-2 × 12	10-2	2	8-10	2	7-10	2
	4-2 × 8	9-0	1	7-8	1	6-9	1
	4-2 × 10	10-1	1	8-9	1	7-10	2
	4-2 × 12	11-9	1	10-2	2	9-1	2
Two Floors	2-2 × 4	2-2	1	1-10	1	1-7	1
	2-2 × 6	3-2	2	2-9	2	2-5	2
	2-2 × 8	4-1	2	3-6	2	3-2	2
	2-2 × 10	4-11	2	4-3	2	3-10	3
	2-2 × 12	5-9	2	5-0	3	4-5	3
	3-2 × 8	5-1	2	4-5	2	3-11	2
	3-2 × 10	6-2	2	5-4	2	4-10	2
	3-2 × 12	7-2	2	6-3	2	5-7	3
	4-2 × 8	6-1	1	5-3	2	4-8	2
	4-2 × 10	7-2	2	6-2	2	5-6	2
	4-2 × 12	8-4	2	7-2	2	6-5	2

For SI: 1 inch = 25.4 mm, 1 foot = 304.8 mm.

a. Spans are given in feet and inches (ft-in).
b. Tabulated values are for No. 2 grade lumber.
c. Building width is measured perpendicular to the ridge. For widths between those shown, spans are permitted to be interpolated.
d. NJ - Number of jack studs required to support each end. Where the number of required jack studs equals one, the headers are permitted to be supported by an approved framing anchor attached to the full-height wall stud and to the header.

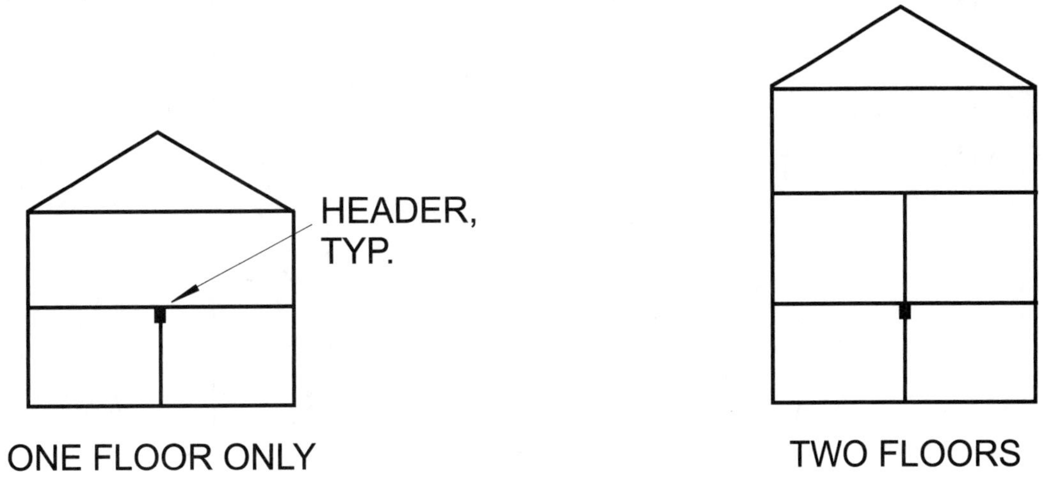

Figure 2308.9.6
HEADER SUPPORT CONDITIONS FOR INTERIOR LOAD BEARING WALLS
(Source: AF&PA WOOD-FRAME CONSTRUCTION MANUAL)

2308.9.8 Pipes in walls. Stud partitions containing plumbing, heating or other pipes shall be so framed and the joists underneath so spaced as to give proper clearance for the piping. Where a partition containing such piping runs parallel to the floor joists, the joists underneath such partitions shall be doubled and spaced to permit the passage of such pipes and shall be bridged. Where plumbing, heating or other pipes are placed in or partly in a partition, necessitating the cutting of the soles or plates, a metal tie not less than 0.058 inch (1.47 mm) (16 galvanized gage) and $1^1/_2$ inches (38 mm) wide shall be fastened to each plate across and to each side of the opening with not less than six 16d nails.

❖ This section requires that installation of piping in partitions and through floors does not compromise the framing. Rather than cutting joists or leaving partitions improperly supported, planning is necessary to provide clearance for pipes that extend through floors and adequate support for partitions that contain piping.

2308.9.9 Bridging. Unless covered by interior or exterior wall coverings or sheathing meeting the minimum requirements of this code, stud partitions or walls with studs having a height-to-least-thickness ratio exceeding 50 shall have bridging not less than 2 inches (51 mm) in thickness and of the same width as the studs fitted snugly and nailed thereto to provide adequate lateral support. Bridging shall be placed in every stud cavity and at a frequency such that no stud so braced shall have a height-to-least-thickness ratio exceeding 50 with the height of the stud measured between horizontal framing and bridging or between bridging, whichever is greater.

❖ Where stud partitions do not have adequate sheathing to restrain the studs laterally in their weaker, or smaller, dimension, stud walls or partitions of such a height that the height-to-least-thickness ratio exceeds 50 are required to have bridging (in this case, solid blocking) cut in between the studs with a minimum thickness of 2 inches (51 mm) and the same width as the studs. Location will vary with the height of the wall, with the blocking installed to reduce the height-to-least-width ratio to less than 50.

2308.9.10 Cutting and notching. In exterior walls and bearing partitions, any wood stud is permitted to be cut or notched to a depth not exceeding 25 percent of its width. Cutting or notching of studs to a depth not greater than 40 percent of the width of the stud is permitted in nonbearing partitions supporting no loads other than the weight of the partition.

❖ Studs should not be cut, notched or bored when possible to maintain the cross-sectional bearing area. When the damage exceeds that specified in Section 2308.9.10, the studs should be doubled or otherwise reinforced to provide the required strength (see Figure 2308.9.10).

2308.9.11 Bored holes. A hole not greater in diameter than 40 percent of the stud width is permitted to be bored in any wood stud. Bored holes not greater than 60 percent of the width of the

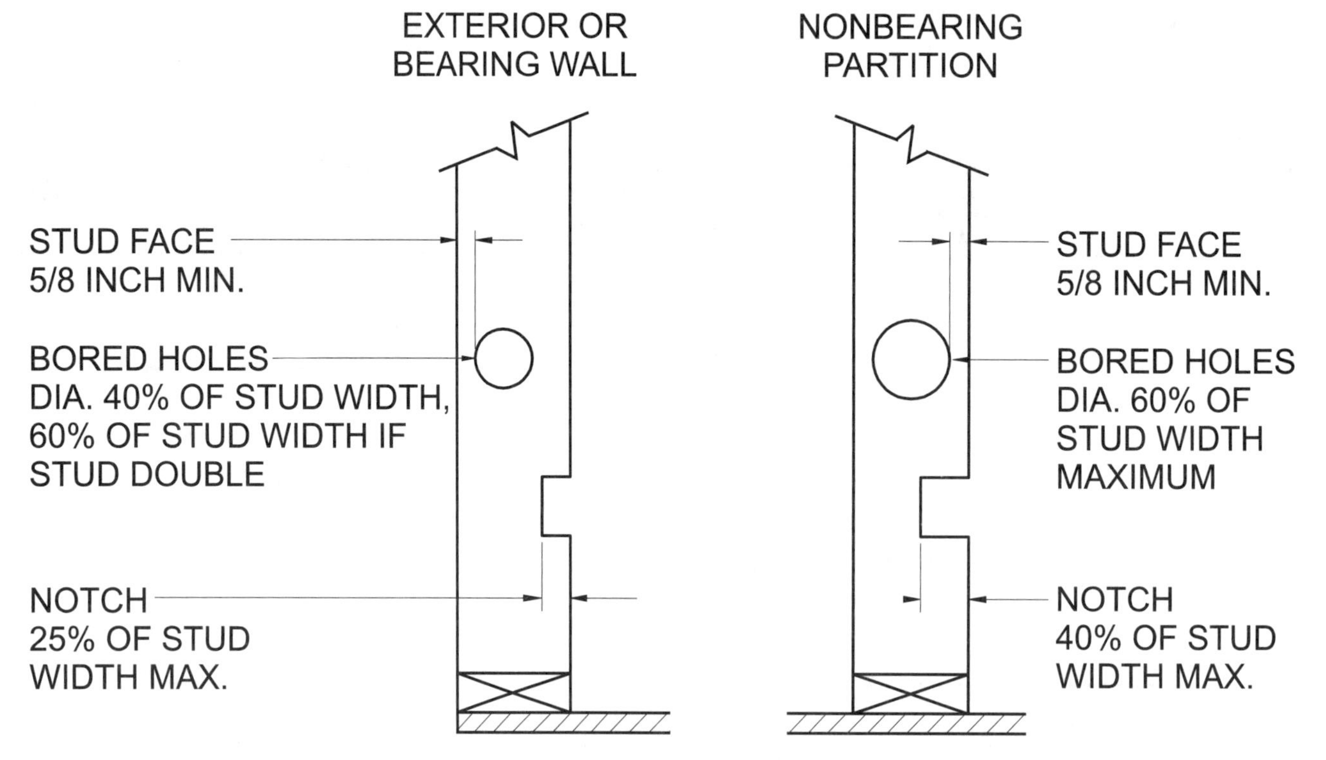

For SI: 1 inch = 25.4 mm.

Figure 2308.9.10
CUTTING, NOTCHING, AND BORED HOLES

stud are permitted in nonbearing partitions or in any wall where each bored stud is doubled, provided not more than two such successive doubled studs are so bored.

In no case shall the edge of the bored hole be nearer than $^5/_8$ inch (15.9 mm) to the edge of the stud.

Bored holes shall not be located at the same section of stud as a cut or notch.

❖ Due to the redundancy of this type of construction, limited notching and hole boring are allowed (see commentary, Section 2308.9.10).

2308.10 Roof and ceiling framing. The framing details required in this section apply to roofs having a minimum slope of three units vertical in 12 units horizontal (25-percent slope) or greater. Where the roof slope is less than three units vertical in 12 units horizontal (25-percent slope), members supporting rafters and ceiling joists such as ridge board, hips and valleys shall be designed as beams.

❖ The code intends that the framing details for roofs in Section 2308.10 apply only to roofs having a minimum slope of 3 units vertical in 12 units horizontal. For roofs flatter than 3:12, the side thrust becomes so large as to exceed the capabilities of conventional construction. In that situation, members that support rafters must be designed as beams and must be supported by exterior or interior bearing walls.

2308.10.1 Wind uplift. Roof assemblies shall have rafter and truss ties to the wall below. Resultant uplift loads shall be transferred to the foundation using a continuous load path. The rafter or truss to wall connection shall comply with Tables 2304.9.1 and 2308.10.1.

❖ The subject of resisting wind uplift on roofs in areas within the limitations of conventional construction that are not considered to be high-wind areas was not addressed in previous building codes. However, in developing the provisions of Section 2308, it was believed that there was justification for addressing this situation since significant uplift occurs in areas outside of high-wind areas.

This section requires compliance with both Table 2304.9.1, which is the fastening schedule, and Table 2308.10.1, which specifies the minimum uplift resistance to be provided between the roof framing and the wall below. Although the toe-nail connections required by Table 2304.9.1 could conceivably provide enough uplift resistance to satisfy Table 2308.10.1 for short roof spans in the lower wind speed areas, in almost all instances, additional uplift resistance is required in the form of approved connectors. When uplift connectors are required by the table, one is required on every rafter or truss to the stud below, assuming the roof framing is spaced 24 inches (610 mm) o.c. in accordance with Note b.

Additionally, the other notes to Table 2308.10.1 may modify the tabular requirements. The text of this section requires that in addition to tying the roof framing to the wall below, a load path must then be established to the foundation. Note f modifies the load requirements for the connections that establish the load path.

TABLE 2308.10.1. See page 23-76.

❖ See the commentary to Section 2308.10.1.

2308.10.2 Ceiling joist spans. Allowable spans for ceiling joists shall be in accordance with Table 2308.10.2(1) or 2308.10.2(2). For other grades and species, refer to the *AF&PA Span Tables for Joists and Rafters*.

❖ Allowable spans for ceiling joists are to be in accordance with Table 2308.10.2(1) or 2308.10.2(2). For other grades and species, refer to the AF&PA *Span Tables for Joists and Rafters* (see commentary, Section 2308.10.3).

TABLES 2308.10.2(1) and (2). See pages 23-76 to 23-80.

❖ For Tables 2308.10.2(1) and (2), see the commentary to Sections 2308.10.3 and 2304.4.

TABLES 2308.10.3(1) through (6). See page 23-81 to 23-92.

❖ For Tables 2308.10.3(1) through (6), see the commentary to Sections 2308.10.3 and 2304.4.

2308.10.3 Rafter spans. Allowable spans for rafters shall be in accordance with Table 2308.10.3(1), 2308.10.3(2), 2308.10.3(3), 2308.10.3(4), 2308.10.3(5) or 2308.10.3(6). For other grades and species, refer to the *AF&PA Span Tables for Joists and Rafters*.

❖ The spans shown in the referenced tables were derived from the AF&PA *Span Tables for Joists and Rafters*. For simplicity, spans of only the most common species, or types, and grades of wood are shown. When other species or grades of wood are used, the AF&PA tables should be consulted to determine the proper span. Additionally, span tables published by the Southern Forest Products Association, the Western Wood Products Association and the Canadian Wood Council can be consulted. The spans in those tables were derived in the same manner as those published in the AF&PA tables.

Spans are based on species of wood, grade, live load, dead load and deflection criteria.

2308.10.4 Ceiling joist and rafter framing. Rafters shall be framed directly opposite each other at the ridge. There shall be a ridge board at least 1-inch (25 mm) nominal thickness at ridges and not less in depth than the cut end of the rafter. At valleys and hips, there shall be a single valley or hip rafter not less than 2-inch (51 mm) nominal thickness and not less in depth than the cut end of the rafter.

❖ Traditional practice is to provide both a ridge board between opposite rafters as a nailing base and a full bearing for the rafter. Rafters must be placed directly opposite each other, and the ridge board must have a depth equal to or greater than the cut end of the rafter. The rafter ends must be flush against the ridge board to avoid excessive horizontal shear in the rafters (see Figure 2308.10.4).

TABLE 2308.10.1
REQUIRED RATING OF APPROVED UPLIFT CONNECTORS (pounds)[a,b,c,e,f,g,h]

BASIC WIND SPEED (3-second gust)	ROOF SPAN (feet)							OVERHANGS (pounds/feet)[d]
	12	20	24	28	32	36	40	
85	-72	-120	-145	-169	-193	-217	-241	-38.55
90	-91	-151	-181	-212	-242	-272	-302	-43.22
100	-131	-281	-262	-305	-349	-393	-436	-53.36
110	-175	-292	-351	-409	-467	-526	-584	-64.56

For SI: 1 inch = 25.4 mm, 1 foot = 304.8 mm, 1 mile per hour = 1.61 km/hr, 1 pound = 0.454 Kg, 1 pound/foot = 14.5939 N/m.

a. The uplift connection requirements are based on a 30-foot mean roof height located in Exposure B. For Exposure C or D and for other mean roof heights, multiply the above loads by the adjustment coefficients in Table 1609.6.2.1(4).

b. The uplift connection requirements are based on the framing being spaced 24 inches on center. Multiply by 0.67 for framing spaced 16 inches on center and multiply by 0.5 for framing spaced 12 inches on center.

c. The uplift connection requirements include an allowance for 10 pounds of dead load.

d. The uplift connection requirements do not account for the effects of overhangs. The magnitude of the above loads shall be increased by adding the overhang loads found in the table. The overhang loads are also based on framing spaced 24 inches on center. The overhang loads given shall be multiplied by the overhang projection and added to the roof uplift value in the table.

e. The uplift connection requirements are based upon wind loading on end zones as defined in Section 1609.6.3. Connection loads for connections located a distance of 20 percent of the least horizontal dimension of the building from the corner of the building are permitted to be reduced by multiplying the table connection value by 0.7 and multiplying the overhang load by 0.8.

f. For wall-to-wall and wall-to-foundation connections, the capacity of the uplift connector is permitted to be reduced by 100 pounds for each full wall above. (For example, if a 500-pound rated connector is used on the roof framing, a 400-pound rated connector is permitted at the next floor level down.)

g. Interpolation is permitted for intermediate values of basic wind speeds and roof spans.

h. The rated capacity of approved tie-down devices is permitted to include up to a 60-percent increase for wind effects where allowed by material specifications.

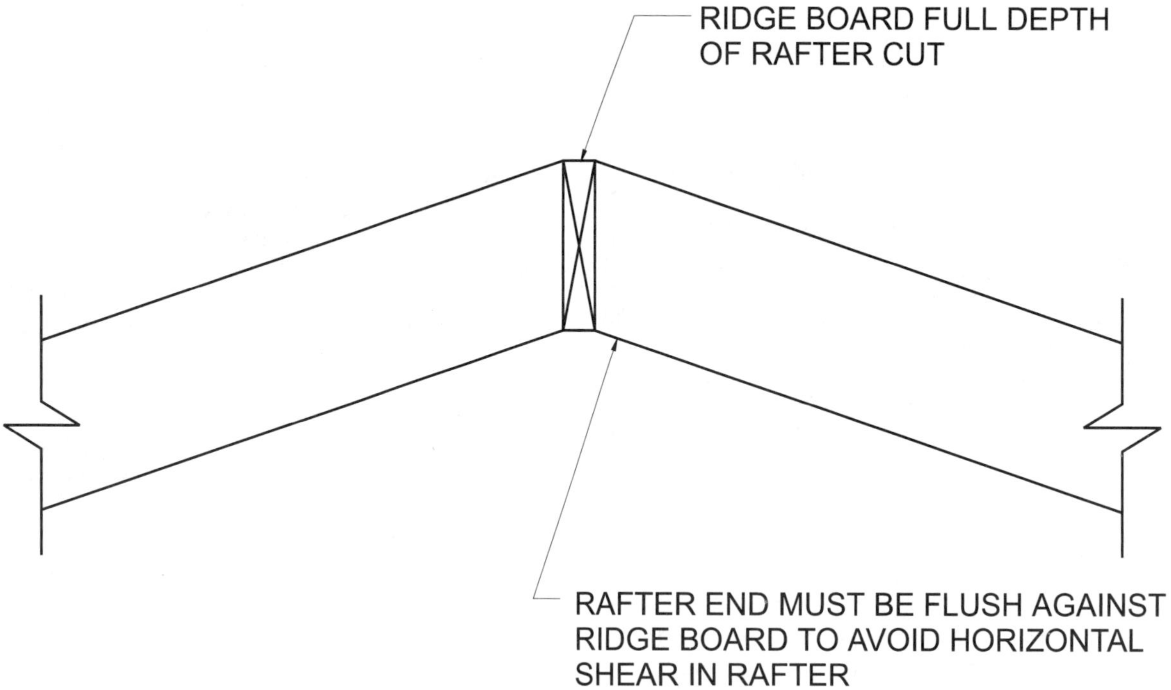
RIDGE BOARD FULL DEPTH OF RAFTER CUT
RAFTER END MUST BE FLUSH AGAINST RIDGE BOARD TO AVOID HORIZONTAL SHEAR IN RAFTER

Figure 2308.10.4
RAFTER END CUT AT RIDGE

TABLE 2308.10.2(1)
CEILING JOIST SPANS FOR COMMON LUMBER SPECIES
(Uninhabitable Attics Without Storage, Live Load = 10 pounds psf, L/Δ = 240)

CEILING JOIST SPACING (inches)	SPECIES AND GRADE		DEAD LOAD = 5 pounds per square foot			
			2 × 4	2 × 6	2 × 8	2 × 10
			Maximum ceiling joist spans			
			(ft. - in.)	(ft. - in.)	(ft. - in.)	(ft. - in.)
12	Douglas Fir-Larch	SS	13-2	20-8	Note a	Note a
	Douglas Fir-Larch	#1	12-8	19-11	Note a	Note a
	Douglas Fir-Larch	#2	12-5	19-6	25-8	Note a
	Douglas Fir-Larch	#3	10-10	15-10	20-1	24-6
	Hem-Fir	SS	12-5	19-6	25-8	Note a
	Hem-Fir	#1	12-2	19-1	25-2	Note a
	Hem-Fir	#2	11-7	18-2	24-0	Note a
	Hem-Fir	#3	10-10	15-10	20-1	24-6
	Southern Pine	SS	12-11	20-3	Note a	Note a
	Southern Pine	#1	12-8	19-11	Note a	Note a
	Southern Pine	#2	12-5	19-6	25-8	Note a
	Southern Pine	#3	11-6	17-0	21-8	25-7
	Spruce-Pine-Fir	SS	12-2	19-1	25-2	Note a
	Spruce-Pine-Fir	#1	11-10	18-8	24-7	Note a
	Spruce-Pine-Fir	#2	11-10	18-8	24-7	Note a
	Spruce-Pine-Fir	#3	10-10	15-10	20-1	24-6
16	Douglas Fir-Larch	SS	11-11	18-9	24-8	Note a
	Douglas Fir-Larch	#1	11-6	18-1	23-10	Note a
	Douglas Fir-Larch	#2	11-3	17-8	23-0	Note a
	Douglas Fir-Larch	#3	9-5	13-9	17-5	21-3
	Hem-Fir	SS	11-3	17-8	23-4	Note a
	Hem-Fir	#1	11-0	17-4	22-10	Note a
	Hem-Fir	#2	10-6	16-6	21-9	Note a
	Hem-Fir	#3	9-5	13-9	17-5	21-3
	Southern Pine	SS	11-9	18-5	24-3	Note a
	Southern Pine	#1	11-6	18-1	23-1	Note a
	Southern Pine	#2	11-3	17-8	23-4	Note a
	Southern Pine	#3	10-0	14-9	18-9	22-2
	Spruce-Pine-Fir	SS	11-0	17-4	22-10	Note a
	Spruce-Pine-Fir	#1	10-9	16-11	22-4	Note a
	Spruce-Pine-Fir	#2	10-9	16-11	22-4	Note a
	Spruce-Pine-Fir	#3	9-5	13-9	17-5	21-3

(continued)

TABLE 2308.10.2(1)—continued
CEILING JOIST SPANS FOR COMMON LUMBER SPECIES
(Uninhabitable Attics Without Storage, Live Load = 10 pounds psf, L/Δ = 240)

CEILING JOIST SPACING (inches)	SPECIES AND GRADE		DEAD LOAD = 5 pounds per square foot			
			2 × 4	2 × 6	2 × 8	2 × 10
			Maximum ceiling joist spans			
			(ft. - in.)	(ft. - in.)	(ft. - in.)	(ft. - in.)
19.2	Douglas Fir-Larch	SS	11-3	17-8	23-3	Note a
	Douglas Fir-Larch	#1	10-10	17-0	22-5	Note a
	Douglas Fir-Larch	#2	10-7	16-7	21-0	25-8
	Douglas Fir-Larch	#3	8-7	12-6	15-10	19-5
	Hem-Fir	SS	10-7	16-8	21-11	Note a
	Hem-Fir	#1	10-4	16-4	21-6	Note a
	Hem-Fir	#2	9-11	15-7	20-6	25-3
	Hem-Fir	#3	8-7	12-6	15-10	19-5
	Southern Pine	SS	11-0	17-4	22-10	Note a
	Southern Pine	#1	10-10	17-0	22-5	Note a
	Southern Pine	#2	10-7	16-8	21-11	Note a
	Southern Pine	#3	9-1	13-6	17-2	20-3
	Spruce-Pine-Fir	SS	10-4	16-4	21-6	Note a
	Spruce-Pine-Fir	#1	10-2	15-11	21-0	25-8
	Spruce-Pine-Fir	#2	10-2	15-11	21-0	25-8
	Spruce-Pine-Fir	#3	8-7	12-6	15-10	19-5
24	Douglas Fir-Larch	SS	10-5	16-4	21-7	Note a
	Douglas Fir-Larch	#1	10-0	15-9	20-1	24-6
	Douglas Fir-Larch	#2	9-10	14-10	18-9	22-11
	Douglas Fir-Larch	#3	7-8	11-2	14-2	17-4
	Hem-Fir	SS	9-10	15-6	20-5	Note a
	Hem-Fir	#1	9-8	15-2	19-7	23-11
	Hem-Fir	#2	9-2	14-5	18-6	22-7
	Hem-Fir	#3	7-8	11-2	14-2	17-4
	Southern Pine	SS	10-3	16-1	21-2	Note a
	Southern Pine	#1	10-0	15-9	20-10	Note a
	Southern Pine	#2	9-10	15-6	20-1	23-11
	Southern Pine	#3	8-2	12-0	15-4	18-1
	Spruce-Pine-Fir	SS	9-8	15-2	19-11	25-5
	Spruce-Pine-Fir	#1	9-5	14-9	18-9	22-11
	Spruce-Pine-Fir	#2	9-5	14-9	18-9	22-11
	Spruce-Pine-Fir	#3	7-8	11-2	14-2	17-4

For SI: 1 inch = 25.4 mm, 1 foot = 304.8 mm, 1 pound per square foot = 47.8 N/m^2.

a. Span exceeds 26 feet in length. Check sources for availability of lumber in lengths greater than 20 feet.

TABLE 2308.10.2(2)
CEILING JOIST SPANS FOR COMMON LUMBER SPECIES
(Uninhabitable Attics With Limited Storage, Live Load = 20 pounds per square foot, $L/\Delta = 240$)

CEILING JOIST SPACING (inches)	SPECIES AND GRADE		DEAD LOAD = 10 pounds per square foot			
			2 × 4	2 × 6	2 × 8	2 × 10
			Maximum ceiling joist spans			
			(ft. - in.)	(ft. - in.)	(ft. - in.)	(ft. - in.)
12	Douglas Fir-Larch	SS	10-5	16-4	21-7	Note a
	Douglas Fir-Larch	#1	10-0	15-9	20-1	24-6
	Douglas Fir-Larch	#2	9-10	14-10	18-9	22-11
	Douglas Fir-Larch	#3	7-8	11-2	14-2	17-4
	Hem-Fir	SS	9-10	15-6	20-5	Note a
	Hem-Fir	#1	9-8	15-2	19-7	23-11
	Hem-Fir	#2	9-2	14-5	18-6	22-7
	Hem-Fir	#3	7-8	11-2	14-2	17-4
	Southern Pine	SS	10-3	16-1	21-2	Note a
	Southern Pine	#1	10-0	15-9	20-10	Note a
	Southern Pine	#2	9-10	15-6	20-1	23-11
	Southern Pine	#3	8-2	12-0	15-4	18-1
	Spruce-Pine-Fir	SS	9-8	15-2	19-11	25-5
	Spruce-Pine-Fir	#1	9-5	14-9	18-9	22-11
	Spruce-Pine-Fir	#2	9-5	14-9	18-9	22-11
	Spruce-Pine-Fir	#3	7-8	11-2	14-2	17-4
16	Douglas Fir-Larch	SS	9-6	14-11	19-7	25-0
	Douglas Fir-Larch	#1	9-1	13-9	17-5	21-3
	Douglas Fir-Larch	#2	8-9	12-10	16-3	19-10
	Douglas Fir-Larch	#3	6-8	9-8	12-4	15-0
	Hem-Fir	SS	8-11	14-1	18-6	23-8
	Hem-Fir	#1	8-9	13-5	16-10	20-8
	Hem-Fir	#2	8-4	12-8	16-0	19-7
	Hem-Fir	#3	6-8	9-8	12-4	15-0
	Southern Pine	SS	9-4	14-7	19-3	24-7
	Southern Pine	#1	9-1	14-4	18-11	23-1
	Southern Pine	#2	8-11	13-6	17-5	20-9
	Southern Pine	#3	7-1	10-5	13-3	15-8
	Spruce-Pine-Fir	SS	8-9	13-9	18-1	23-1
	Spruce-Pine-Fir	#1	8-7	12-10	16-3	19-10
	Spruce-Pine-Fir	#2	8-7	12-10	16-3	19-10
	Spruce-Pine-Fir	#3	6-8	9-8	12-4	15-0

(continued)

TABLE 2308.10.2(2)—continued
CEILING JOIST SPANS FOR COMMON LUMBER SPECIES
(Uninhabitable Attics With Limited Storage, Live Load = 20 pounds per square foot, L/Δ = 240)

CEILING JOIST SPACING (inches)	SPECIES AND GRADE		DEAD LOAD = 10 pounds per square foot			
			2 × 4	2 × 6	2 × 8	2 × 10
			Maximum ceiling joist spans			
			(ft. - in.)	(ft. - in.)	(ft. - in.)	(ft. - in.)
19.2	Douglas Fir-Larch	SS	8-11	14-0	18-5	23-4
	Douglas Fir-Larch	#1	8-7	12-6	15-10	19-5
	Douglas Fir-Larch	#2	8-0	11-9	14-10	18-2
	Douglas Fir-Larch	#3	6-1	8-10	11-3	13-8
	Hem-Fir	SS	8-5	13-3	17-5	22-3
	Hem-Fir	#1	8-3	12-3	15-6	18-11
	Hem-Fir	#2	7-10	11-7	14-8	17-10
	Hem-Fir	#3	6-1	8-10	11-3	13-8
	Southern Pine	SS	8-9	13-9	18-1	23-1
	Southern Pine	#1	8-7	13-6	17-9	21-1
	Southern Pine	#2	8-5	12-3	15-10	18-11
	Southern Pine	#3	6-5	9-6	12-1	14-4
	Spruce-Pine-Fir	SS	8-3	12-11	17-1	21-8
	Spruce-Pine-Fir	#1	8-0	11-9	14-10	18-2
	Spruce-Pine-Fir	#2	8-0	11-9	14-10	18-2
	Spruce-Pine-Fir	#3	6-1	8-10	11-3	13-8
24	Douglas Fir-Larch	SS	8-3	13-0	17-1	20-11
	Douglas Fir-Larch	#1	7-8	11-2	14-2	17-4
	Douglas Fir-Larch	#2	7-2	10-6	13-3	16-3
	Douglas Fir-Larch	#3	5-5	7-11	10-0	12-3
	Hem-Fir	SS	7-10	12-3	16-2	20-6
	Hem-Fir	#1	7-6	10-11	13-10	16-11
	Hem-Fir	#2	7-1	10-4	13-1	16-0
	Hem-Fir	#3	5-5	7-11	10-0	12-3
	Southern Pine	SS	8-1	12-9	16-10	21-6
	Southern Pine	#1	8-0	12-6	15-10	18-10
	Southern Pine	#2	7-8	11-0	14-2	16-11
	Southern Pine	#3	5-9	8-6	10-10	12-10
	Spruce-Pine-Fir	SS	7-8	12-0	15-10	19-5
	Spruce-Pine-Fir	#1	7-2	10-6	13-3	16-3
	Spruce-Pine-Fir	#2	7-2	10-6	13-3	16-3
	Spruce-Pine-Fir	#3	5-5	7-11	10-0	12-3

For SI: 1 inch = 25.4 mm, 1 foot = 304.8 mm, 1 pound per square foot = 47.8 N/m^2.

a. Span exceeds 26 feet in length. Check sources for availability of lumber in lengths greater than 20 feet.

TABLE 2308.10.3(1)
RAFTER SPANS FOR COMMON LUMBER SPECIES
(Roof Live Load = 20 pounds per square foot, Ceiling Not Attached to Rafters, L/Δ = 180)

RAFTER SPACING (inches)	SPECIES AND GRADE		DEAD LOAD = 10 pounds per square foot					DEAD LOAD = 20 pounds per square foot				
			2 × 4	2 × 6	2 × 8	2 × 10	2 × 12	2 × 4	2 × 6	2 × 8	2 × 10	2 × 12
			Maximum rafter spans									
			(ft. - in.)	(ft. - in.)	(ft. - in.)	(ft. - in.)	(ft. - in.)	(ft. - in.)	(ft. - in.)	(ft. - in.)	(ft. - in.)	(ft. - in.)
12	Douglas Fir-Larch	SS	11-6	18-0	23-9	Note a	Note a	11-6	18-0	23-5	Note a	Note a
	Douglas Fir-Larch	#1	11-1	17-4	22-5	Note a	Note a	10-6	15-4	19-5	23-9	Note a
	Douglas Fir-Larch	#2	10-10	16-7	21-0	25-8	Note a	9-10	14-4	18-2	22-3	25-9
	Douglas Fir-Larch	#3	8-7	12-6	15-10	19-5	22-6	7-5	10-10	13-9	16-9	19-6
	Hem-Fir	SS	10-10	17-0	22-5	Note a	Note a	10-10	17-0	22-5	Note a	Note a
	Hem-Fir	#1	10-7	16-8	21-10	Note a	Note a	10-3	14-11	18-11	23-2	Note a
	Hem-Fir	#2	10-1	15-11	20-8	25-3	Note a	9-8	14-2	17-11	21-11	25-5
	Hem-Fir	#3	8-7	12-6	15-10	19-5	22-6	7-5	10-10	13-9	16-9	19-6
	Southern Pine	SS	11-3	17-8	23-4	Note a	Note a	11-3	17-8	23-4	Note a	Note a
	Southern Pine	#1	11-1	17-4	22-11	Note a	Note a	11-1	17-3	21-9	25-10	Note a
	Southern Pine	#2	10-10	17-0	22-5	Note a	Note a	10-6	15-1	19-5	23-2	Note a
	Southern Pine	#3	9-1	13-6	17-2	20-3	24-1	7-11	11-8	14-10	17-6	20-11
	Spruce-Pine-Fir	SS	10-7	16-8	21-11	Note a	Note a	10-7	16-8	21-9	Note a	Note a
	Spruce-Pine-Fir	#1	10-4	16-3	21-0	25-8	Note a	9-10	14-4	18-2	22-3	25-9
	Spruce-Pine-Fir	#2	10-4	16-3	21-0	25-8	Note a	9-10	14-4	18-2	22-3	25-9
	Spruce-Pine-Fir	#3	8-7	12-6	15-10	19-5	22-6	7-5	10-10	13-9	16-9	19-6
16	Douglas Fir-Larch	SS	10-5	16-4	21-7	Note a	Note a	10-5	16-0	20-3	24-9	Note a
	Douglas Fir-Larch	#1	10-0	15-4	19-5	23-9	Note a	9-1	13-3	16-10	20-7	23-10
	Douglas Fir-Larch	#2	9-10	14-4	18-2	22-3	25-9	8-6	12-5	15-9	19-3	22-4
	Douglas Fir-Larch	#3	7-5	10-10	13-9	16-9	19-6	6-5	9-5	11-11	14-6	16-10
	Hem-Fir	SS	9-10	15-6	20-5	Note a	Note a	9-10	15-6	19-11	24-4	Note a
	Hem-Fir	#1	9-8	14-11	18-11	23-2	Note a	8-10	12-11	16-5	20-0	23-3
	Hem-Fir	#2	9-2	14-2	17-11	21-11	25-5	8-5	12-3	15-6	18-11	22-0
	Hem-Fir	#3	7-5	10-10	13-9	16-9	19-6	6-5	9-5	11-11	14-6	16-10
	Southern Pine	SS	10-3	16-1	21-2	Note a	Note a	10-3	16-1	21-2	Note a	Note a
	Southern Pine	#1	10-0	15-9	20-10	25-10	Note a	10-0	15-0	18-10	22-4	Note a
	Southern Pine	#2	9-10	15-1	19-5	23-2	Note a	9-1	13-0	16-10	20-1	23-7
	Southern Pine	#3	7-11	11-8	14-10	17-6	20-11	6-10	10-1	12-10	15-2	18-1
	Spruce-Pine-Fir	SS	9-8	15-2	19-11	25-5	Note a	9-8	14-10	18-10	23-0	Note a
	Spruce-Pine-Fir	#1	9-5	14-4	18-2	22-3	25-9	8-6	12-5	15-9	19-3	22-4
	Spruce-Pine-Fir	#2	9-5	14-4	18-2	22-3	25-9	8-6	12-5	15-9	19-3	22-4
	Spruce-Pine-Fir	#3	7-5	10-10	13-9	16-9	19-6	6-5	9-5	11-11	14-6	16-10

(continued)

TABLE 2308.10.3(1)—continued
RAFTER SPANS FOR COMMON LUMBER SPECIES
(Roof Live Load = 20 pounds per square foot, Ceiling Not Attached to Rafters, L/Δ = 180)

RAFTER SPACING (inches)	SPECIES AND GRADE		DEAD LOAD = 10 pounds per square foot					DEAD LOAD = 20 pounds per square foot				
			2 × 4	2 × 6	2 × 8	2 × 10	2 × 12	2 × 4	2 × 6	2 × 8	2 × 10	2 × 12
			Maximum rafter spans									
			(ft. - in.)	(ft. - in.)	(ft. - in.)	(ft. - in.)	(ft. - in.)	(ft. - in.)	(ft. - in.)	(ft. - in.)	(ft. - in.)	(ft. - in.)
19.2	Douglas Fir-Larch	SS	9-10	15-5	20-4	25-11	Note a	9-10	14-7	18-6	22-7	Note a
	Douglas Fir-Larch	#1	9-5	14-0	17-9	21-8	25-2	8-4	12-2	15-4	18-9	21-9
	Douglas Fir-Larch	#2	8-11	13-1	16-7	20-3	23-6	7-9	11-4	14-4	17-7	20-4
	Douglas Fir-Larch	#3	6-9	9-11	12-7	15-4	17-9	5-10	8-7	10-10	13-3	15-5
	Hem-Fir	SS	9-3	14-7	19-2	24-6	Note a	9-3	14-4	18-2	22-3	25-9
	Hem-Fir	#1	9-1	13-8	17-4	21-1	24-6	8-1	11-10	15-0	18-4	21-3
	Hem-Fir	#2	8-8	12-11	16-4	20-0	23-2	7-8	11-2	14-2	17-4	20-1
	Hem-Fir	#3	6-9	9-11	12-7	15-4	17-9	5-10	8-7	10-10	13-3	15-5
	Southern Pine	SS	9-8	15-2	19-11	25-5	Note a	9-8	15-2	19-11	25-5	Note a
	Southern Pine	#1	9-5	14-10	19-7	23-7	Note a	9-3	13-8	17-2	20-5	24-4
	Southern Pine	#2	9-3	13-9	17-9	21-2	24-10	8-4	11-11	15-4	18-4	21-6
	Southern Pine	#3	7-3	10-8	13-7	16-0	19-1	6-3	9-3	11-9	13-10	16-6
	Spruce-Pine-Fir	SS	9-1	14-3	18-9	23-11	Note a	9-1	13-7	17-2	21-0	24-4
	Spruce-Pine-Fir	#1	8-10	13-1	16-7	20-3	23-6	7-9	11-4	14-4	17-7	20-4
	Spruce-Pine-Fir	#2	8-10	13-1	16-7	20-3	23-6	7-9	11-4	14-4	17-7	20-4
	Spruce-Pine-Fir	#3	6-9	9-11	12-7	15-4	17-9	5-10	8-7	10-10	13-3	15-5
24	Douglas Fir-Larch	SS	9-1	14-4	18-10	23-4	Note b	8-11	13-1	16-7	20-3	23-5
	Douglas Fir-Larch	#1	8-7	12-6	15-10	19-5	22-6	7-5	10-10	13-9	16-9	19-6
	Douglas Fir-Larch	#2	8-0	11-9	14-10	18-2	21-0	6-11	10-2	12-10	15-8	18-3
	Douglas Fir-Larch	#3	6-1	8-10	11-3	13-8	15-11	5-3	7-8	9-9	11-10	13-9
	Hem-Fir	SS	8-7	13-6	17-10	22-9	Note b	8-7	12-10	16-3	19-10	23-0
	Hem-Fir	#1	8-4	12-3	15-6	18-11	21-11	7-3	10-7	13-5	16-4	19-0
	Hem-Fir	#2	7-11	11-7	14-8	17-10	20-9	6-10	10-0	12-8	15-6	17-11
	Hem-Fir	#3	6-1	8-10	11-3	13-8	15-11	5-3	7-8	9-9	11-10	13-9
	Southern Pine	SS	8-11	14-1	18-6	23-8	Note b	8-11	14-1	18-6	22-11	Note a
	Southern Pine	#1	8-9	13-9	17-9	21-1	25-2	8-3	12-3	15-4	18-3	21-9
	Southern Pine	#2	8-7	12-3	15-10	18-11	22-2	7-5	10-8	13-9	16-5	19-3
	Southern Pine	#3	6-5	9-6	12-1	14-4	17-1	5-7	8-3	10-6	12-5	14-9
	Spruce-Pine-Fir	SS	8-5	13-3	17-5	21-8	25-2	8-4	12-2	15-4	18-9	21-9
	Spruce-Pine-Fir	#1	8-0	11-9	14-10	18-2	21-0	6-11	10-2	12-10	15-8	18-3
	Spruce-Pine-Fir	#2	8-0	11-9	14-10	18-2	21-0	6-11	10-2	12-10	15-8	18-3
	Spruce-Pine-Fir	#3	6-1	8-10	11-3	13-8	15-11	5-3	7-8	9-9	11-10	13-9

For SI: 1 inch = 25.4 mm, 1 foot = 304.8 mm, 1 pound per square foot = 47.9 N/m^2.

a. Span exceeds 26 feet in length. Check sources for availability of lumber in lengths greater than 20 feet.

TABLE 2308.10.3(2)
RAFTER SPANS FOR COMMON LUMBER SPECIES
(Roof Live Load = 20 pounds per square foot, Ceiling Not Attached to Rafters, $L/\Delta = 240$)

RAFTER SPACING (inches)	SPECIES AND GRADE		DEAD LOAD = 10 pounds per square foot					DEAD LOAD = 20 pounds per square foot				
			2 × 4	2 × 6	2 × 8	2 × 10	2 × 12	2 × 4	2 × 6	2 × 8	2 × 10	2 × 12
			Maximum rafter spans									
			(ft. - in.)	(ft. - in.)	(ft. - in.)	(ft. - in.)	(ft. - in.)	(ft. - in.)	(ft. - in.)	(ft. - in.)	(ft. - in.)	(ft. - in.)
12	Douglas Fir-Larch	SS	10-5	16-4	21-7	Note a	Note a	10-5	16-4	21-7	Note a	Note a
	Douglas Fir-Larch	#1	10-0	15-9	20-10	Note a	Note a	10-0	15-4	19-5	23-9	Note a
	Douglas Fir-Larch	#2	9-10	15-6	20-5	25-8	Note a	9-10	14-4	18-2	22-3	25-9
	Douglas Fir-Larch	#3	8-7	12-6	15-10	19-5	22-6	7-5	10-10	13-9	16-9	19-6
	Hem-Fir	SS	9-10	15-6	20-5	Note a	Note a	9-10	15-6	20-5	Note a	Note a
	Hem-Fir	#1	9-8	15-2	19-11	25-5	Note a	9-8	14-11	18-11	23-2	Note a
	Hem-Fir	#2	9-2	14-5	19-0	24-3	Note a	9-2	14-2	17-11	21-11	25-5
	Hem-Fir	#3	8-7	12-6	15-10	19-5	22-6	7-5	10-10	13-9	16-9	19-6
	Southern Pine	SS	10-3	16-1	21-2	Note a	Note a	10-3	16-1	21-2	Note a	Note a
	Southern Pine	#1	10-0	15-9	20-10	Note a	Note a	10-0	15-9	20-10	25-10	Note a
	Southern Pine	#2	9-10	15-6	20-5	Note a	Note a	9-10	15-1	19-5	23-2	Note a
	Southern Pine	#3	9-1	13-6	17-2	20-3	24-1	7-11	11-8	14-10	17-6	20-11
	Spruce-Pine-Fir	SS	9-8	15-2	19-11	25-5	Note a	9-8	15-2	19-11	25-5	Note a
	Spruce-Pine-Fir	#1	9-5	14-9	19-6	24-10	Note a	9-5	14-4	18-2	22-3	25-9
	Spruce-Pine-Fir	#2	9-5	14-9	19-6	24-10	Note a	9-5	14-4	18-2	22-3	25-9
	Spruce-Pine-Fir	#3	8-7	12-6	15-10	19-5	22-6	7-5	10-10	13-9	16-9	19-6
16	Douglas Fir-Larch	SS	9-6	14-11	19-7	25-0	Note a	9-6	14-11	19-7	24-9	Note a
	Douglas Fir-Larch	#1	9-1	14-4	18-11	23-9	Note a	9-1	13-3	16-10	20-7	23-10
	Douglas Fir-Larch	#2	8-11	14-1	18-2	22-3	25-9	8-6	12-5	15-9	19-3	22-4
	Douglas Fir-Larch	#3	7-5	10-10	13-9	16-9	19-6	6-5	9-5	11-11	14-6	16-10
	Hem-Fir	SS	8-11	14-1	18-6	23-8	Note a	8-11	14-1	18-6	23-8	Note a
	Hem-Fir	#1	8-9	13-9	18-1	23-1	Note a	8-9	12-11	16-5	20-0	23-3
	Hem-Fir	#2	8-4	13-1	17-3	21-11	25-5	8-4	12-3	15-6	18-11	22-0
	Hem-Fir	#3	7-5	10-10	13-9	16-9	19-6	6-5	9-5	11-11	14-6	16-10
	Southern Pine	SS	9-4	14-7	19-3	24-7	Note a	9-4	14-7	19-3	24-7	Note a
	Southern Pine	#1	9-1	14-4	18-11	24-1	Note a	9-1	14-4	18-10	22-4	Note a
	Southern Pine	#2	8-11	14-1	18-6	23-2	Note a	8-11	13-0	16-10	20-1	23-7
	Southern Pine	#3	7-11	11-8	14-10	17-6	20-11	6-10	10-1	12-10	15-2	18-1
	Spruce-Pine-Fir	SS	8-9	13-9	18-1	23-1	Note a	8-9	13-9	18-1	23-0	Note a
	Spruce-Pine-Fir	#1	8-7	13-5	17-9	22-3	25-9	8-6	12-5	15-9	19-3	22-4
	Spruce-Pine-Fir	#2	8-7	13-5	17-9	22-3	25-9	8-6	12-5	15-9	19-3	22-4
	Spruce-Pine-Fir	#3	7-5	10-10	13-9	16-9	19-6	6-5	9-5	11-11	14-6	16-10

(continued)

TABLE 2308.10.3(2)—continued
RAFTER SPANS FOR COMMON LUMBER SPECIES
(Roof Live Load = 20 pounds per square foot, Ceiling Not Attached to Rafters, L/Δ = 240)

RAFTER SPACING (inches)	SPECIES AND GRADE		DEAD LOAD = 10 pounds per square foot					DEAD LOAD = 20 pounds per square foot				
			2 × 4	2 × 6	2 × 8	2 × 10	2 × 12	2 × 4	2 × 6	2 × 8	2 × 10	2 × 12
			Maximum rafter spans									
			(ft. - in.)	(ft. - in.)	(ft. - in.)	(ft. - in.)	(ft. - in.)	(ft. - in.)	(ft. - in.)	(ft. - in.)	(ft. - in.)	(ft. - in.)
19.2	Douglas Fir-Larch	SS	8-11	14-0	18-5	23-7	Note a	8-11	14-0	18-5	22-7	Note a
	Douglas Fir-Larch	#1	8-7	13-6	17-9	21-8	25-2	8-4	12-2	15-4	18-9	21-9
	Douglas Fir-Larch	#2	8-5	13-1	16-7	20-3	23-6	7-9	11-4	14-4	17-7	20-4
	Douglas Fir-Larch	#3	6-9	9-11	12-7	15-4	17-9	5-10	8-7	10-10	13-3	15-5
	Hem-Fir	SS	8-5	13-3	17-5	22-3	Note a	8-5	13-3	17-5	22-3	25-9
	Hem-Fir	#1	8-3	12-11	17-1	21-1	24-6	8-1	11-10	15-0	18-4	21-3
	Hem-Fir	#2	7-10	12-4	16-3	20-0	23-2	7-8	11-2	14-2	17-4	20-1
	Hem-Fir	#3	6-9	9-11	12-7	15-4	17-9	5-10	8-7	10-10	13-3	15-5
	Southern Pine	SS	8-9	13-9	18-1	23-1	Note a	8-9	13-9	18-1	23-1	Note a
	Southern Pine	#1	8-7	13-6	17-9	22-8	Note a	8-7	13-6	17-2	20-5	24-4
	Southern Pine	#2	8-5	13-3	17-5	21-2	24-10	8-4	11-11	15-4	18-4	21-6
	Southern Pine	#3	7-3	10-8	13-7	16-0	19-1	6-3	9-3	11-9	13-10	16-6
	Spruce-Pine-Fir	SS	8-3	12-11	17-1	21-9	Note a	8-3	12-11	17-1	21-0	24-4
	Spruce-Pine-Fir	#1	8-1	12-8	16-7	20-3	23-6	7-9	11-4	14-4	17-7	20-4
	Spruce-Pine-Fir	#2	8-1	12-8	16-7	20-3	23-6	7-9	11-4	14-4	17-7	20-4
	Spruce-Pine-Fir	#3	6-9	9-11	12-7	15-4	17-9	5-10	8-7	10-10	13-3	15-5
24	Douglas Fir-Larch	SS	8-3	13-0	17-2	21-10	Note a	8-3	13-0	16-7	20-3	23-5
	Douglas Fir-Larch	#1	8-0	12-6	15-10	19-5	22-6	7-5	10-10	13-9	16-9	19-6
	Douglas Fir-Larch	#2	7-10	11-9	14-10	18-2	21-0	6-11	10-2	12-10	15-8	18-3
	Douglas Fir-Larch	#3	6-1	8-10	11-3	13-8	15-11	5-3	7-8	9-9	11-10	13-9
	Hem-Fir	SS	7-10	12-3	16-2	20-8	25-1	7-10	12-3	16-2	19-10	23-0
	Hem-Fir	#1	7-8	12-0	15-6	18-11	21-11	7-3	10-7	13-5	16-4	19-0
	Hem-Fir	#2	7-3	11-5	14-8	17-10	20-9	6-10	10-0	12-8	15-6	17-11
	Hem-Fir	#3	6-1	8-10	11-3	13-8	15-11	5-3	7-8	9-9	11-10	13-9
	Southern Pine	SS	8-1	12-9	16-10	21-6	Note a	8-1	12-9	16-10	21-6	Note a
	Southern Pine	#1	8-0	12-6	16-6	21-1	25-2	8-0	12-3	15-4	18-3	21-9
	Southern Pine	#2	7-10	12-3	15-10	18-11	22-2	7-5	10-8	13-9	16-5	19-3
	Southern Pine	#3	6-5	9-6	12-1	14-4	17-1	5-7	8-3	10-6	12-5	14-9
	Spruce-Pine-Fir	SS	7-8	12-0	15-10	20-2	24-7	7-8	12-0	15-4	18-9	21-9
	Spruce-Pine-Fir	#1	7-6	11-9	14-10	18-2	21-0	6-11	10-2	12-10	15-8	18-3
	Spruce-Pine-Fir	#2	7-6	11-9	14-10	18-2	21-0	6-11	10-2	12-10	15-8	18-3
	Spruce-Pine-Fir	#3	6-1	8-10	11-3	13-8	15-11	5-3	7-8	9-9	11-10	13-9

For SI: 1 inch = 25.4 mm, 1 foot = 304.8 mm, 1 pound per square foot = 47.9 N/m^2.

a. Span exceeds 26 feet in length. Check sources for availability of lumber in lengths greater than 20 feet.

TABLE 2308.10.3(3)
RAFTER SPANS FOR COMMON LUMBER SPECIES
(Ground Snow Load = 30 pounds per square foot, Ceiling Not Attached to Rafters, $L/\Delta = 180$)

RAFTER SPACING (inches)	SPECIES AND GRADE		DEAD LOAD = 10 pounds per square foot					DEAD LOAD = 20 pounds per square foot				
			2 × 4	2 × 6	2 × 8	2 × 10	2 × 12	2 × 4	2 × 6	2 × 8	2 × 10	2 × 12
			Maximum rafter spans									
			(ft. - in.)	(ft. - in.)	(ft. - in.)	(ft. - in.)	(ft. - in.)	(ft. - in.)	(ft. - in.)	(ft. - in.)	(ft. - in.)	(ft. - in.)
12	Douglas Fir-Larch	SS	10-0	15-9	20-9	Note a	Note a	10-0	15-9	20-1	24-6	Note a
	Douglas Fir-Larch	#1	9-8	14-9	18-8	22-9	Note a	9-0	13-2	16-8	20-4	23-7
	Douglas Fir-Larch	#2	9-5	13-9	17-5	21-4	24-8	8-5	12-4	15-7	19-1	22-1
	Douglas Fir-Larch	#3	7-1	10-5	13-2	16-1	18-8	6-4	9-4	11-9	14-5	16-8
	Hem-Fir	SS	9-6	14-10	19-7	25-0	Note a	9-6	14-10	19-7	24-1	Note a
	Hem-Fir	#1	9-3	14-4	18-2	22-2	25-9	8-9	12-10	16-3	19-10	23-0
	Hem-Fir	#2	8-10	13-7	17-2	21-0	24-4	8-4	12-2	15-4	18-9	21-9
	Hem-Fir	#3	7-1	10-5	13-2	16-1	18-8	6-4	9-4	11-9	14-5	16-8
	Southern Pine	SS	9-10	15-6	20-5	Note a	Note a	9-10	15-6	20-5	Note a	Note a
	Southern Pine	#1	9-8	15-2	20-0	24-9	Note a	9-8	14-10	18-8	22-2	Note a
	Southern Pine	#2	9-6	14-5	18-8	22-3	Note a	9-0	12-11	16-8	19-11	23-4
	Southern Pine	#3	7-7	11-2	14-3	16-10	20-0	6-9	10-0	12-9	15-1	17-11
	Spruce-Pine-Fir	SS	9-3	14-7	19-2	24-6	Note a	9-3	14-7	18-8	22-9	Note a
	Spruce-Pine-Fir	#1	9-1	13-9	17-5	21-4	24-8	8-5	12-4	15-7	19-1	22-1
	Spruce-Pine-Fir	#2	9-1	13-9	17-5	21-4	24-8	8-5	12-4	15-7	19-1	22-1
	Spruce-Pine-Fir	#3	7-1	10-5	13-2	16-1	18-8	6-4	9-4	11-9	14-5	16-8
16	Douglas Fir-Larch	SS	9-1	14-4	18-10	23-9	Note a	9-1	13-9	17-5	21-3	24-8
	Douglas Fir-Larch	#1	8-9	12-9	16-2	19-9	22-10	7-10	11-5	14-5	17-8	20-5
	Douglas Fir-Larch	#2	8-2	11-11	15-1	18-5	21-5	7-3	10-8	13-6	16-6	19-2
	Douglas Fir-Larch	#3	6-2	9-0	11-5	13-11	16-2	5-6	8-1	10-3	12-6	14-6
	Hem-Fir	SS	8-7	13-6	17-10	22-9	Note a	8-7	13-6	17-1	20-10	24-2
	Hem-Fir	#1	8-5	12-5	15-9	19-3	22-3	7-7	11-1	14-1	17-2	19-11
	Hem-Fir	#2	8-0	11-9	14-11	18-2	21-1	7-2	10-6	13-4	16-3	18-10
	Hem-Fir	#3	6-2	9-0	11-5	13-11	16-2	5-6	8-1	10-3	12-6	14-6
	Southern Pine	SS	8-11	14-1	18-6	23-8	Note a	8-11	14-1	18-6	23-8	Note a
	Southern Pine	#1	8-9	13-9	18-1	21-5	25-7	8-8	12-10	16-2	19-2	22-10
	Southern Pine	#2	8-7	12-6	16-2	19-3	22-7	7-10	11-2	14-5	17-3	20-2
	Southern Pine	#3	6-7	9-8	12-4	14-7	17-4	5-10	8-8	11-0	13-0	15-6
	Spruce-Pine-Fir	SS	8-5	13-3	17-5	22-1	25-7	8-5	12-9	16-2	19-9	22-10
	Spruce-Pine-Fir	#1	8-2	11-11	15-1	18-5	21-5	7-3	10-8	13-6	16-6	19-2
	Spruce-Pine-Fir	#2	8-2	11-11	15-1	18-5	21-5	7-3	10-8	13-6	16-6	19-2
	Spruce-Pine-Fir	#3	6-2	9-0	11-5	13-11	16-2	5-6	8-1	10-3	12-6	14-6

(continued)

TABLE 2308.10.3(3)—continued
RAFTER SPANS FOR COMMON LUMBER SPECIES
(Ground Snow Load = 30 pounds per square foot, Ceiling Not Attached to Rafters, L/Δ = 180)

RAFTER SPACING (inches)	SPECIES AND GRADE		DEAD LOAD = 10 pounds per square foot					DEAD LOAD = 20 pounds per square foot				
			2 × 4	2 × 6	2 × 8	2 × 10	2 × 12	2 × 4	2 × 6	2 × 8	2 × 10	2 × 12
			Maximum rafter spans									
			(ft. - in.)	(ft. - in.)	(ft. - in.)	(ft. - in.)	(ft. - in.)	(ft. - in.)	(ft. - in.)	(ft. - in.)	(ft. - in.)	(ft. - in.)
19.2	Douglas Fir-Larch	SS	8-7	13-6	17-9	21-8	25-2	8-7	12-6	15-10	19-5	22-6
	Douglas Fir-Larch	#1	7-11	11-8	14-9	18-0	20-11	7-1	10-5	13-2	16-1	18-8
	Douglas Fir-Larch	#2	7-5	10-11	13-9	16-10	19-6	6-8	9-9	12-4	15-1	17-6
	Douglas Fir-Larch	#3	5-7	8-3	10-5	12-9	14-9	5-0	7-4	9-4	11-5	13-2
	Hem-Fir	SS	8-1	12-9	16-9	21-4	24-8	8-1	12-4	15-7	19-1	22-1
	Hem-Fir	#1	7-9	11-4	14-4	17-7	20-4	6-11	10-2	12-10	15-8	18-2
	Hem-Fir	#2	7-4	10-9	13-7	16-7	19-3	6-7	9-7	12-2	14-10	17-3
	Hem-Fir	#3	5-7	8-3	10-5	12-9	14-9	5-0	7-4	9-4	11-5	13-2
	Southern Pine	SS	8-5	13-3	17-5	22-3	Note a	8-5	13-3	17-5	22-0	25-9
	Southern Pine	#1	8-3	13-0	16-6	19-7	23-4	7-11	11-9	14-9	17-6	20-11
	Southern Pine	#2	7-11	11-5	14-9	17-7	20-7	7-1	10-2	13-2	15-9	18-5
	Southern Pine	#3	6-0	8-10	11-3	13-4	15-10	5-4	7-11	10-1	11-11	14-2
	Spruce-Pine-Fir	SS	7-11	12-5	16-5	20-2	23-4	7-11	11-8	14-9	18-0	20-11
	Spruce-Pine-Fir	#1	7-5	10-11	13-9	16-10	19-6	6-8	9-9	12-4	15-1	17-6
	Spruce-Pine-Fir	#2	7-5	10-11	13-9	16-10	19-6	6-8	9-9	12-4	15-1	17-6
	Spruce-Pine-Fir	#3	5-7	8-3	10-5	12-9	14-9	5-0	7-4	9-4	11-5	13-2
24	Douglas Fir-Larch	SS	7-11	12-6	15-10	19-5	22-6	7-8	11-3	14-2	17-4	20-1
	Douglas Fir-Larch	#1	7-1	10-5	13-2	16-1	18-8	6-4	9-4	11-9	14-5	16-8
	Douglas Fir-Larch	#2	6-8	9-9	12-4	15-1	17-6	5-11	8-8	11-0	13-6	15-7
	Douglas Fir-Larch	#3	5-0	7-4	9-4	11-5	13-2	4-6	6-7	8-4	10-2	11-10
	Hem-Fir	SS	7-6	11-10	15-7	19-1	22-1	7-6	11-0	13-11	17-0	19-9
	Hem-Fir	#1	6-11	10-2	12-10	15-8	18-2	6-2	9-1	11-6	14-0	16-3
	Hem-Fir	#2	6-7	9-7	12-2	14-10	17-3	5-10	8-7	10-10	13-3	15-5
	Hem-Fir	#3	5-0	7-4	9-4	11-5	13-2	4-6	6-7	8-4	10-2	11-10
	Southern Pine	SS	7-10	12-3	16-2	20-8	25-1	7-10	12-3	16-2	19-8	23-0
	Southern Pine	#1	7-8	11-9	14-9	17-6	20-11	7-1	10-6	13-2	15-8	18-8
	Southern Pine	#2	7-1	10-2	13-2	15-9	18-5	6-4	9-2	11-9	14-1	16-6
	Southern Pine	#3	5-4	7-11	10-1	11-11	14-2	4-9	7-1	9-0	10-8	12-8
	Spruce-Pine-Fir	SS	7-4	11-7	14-9	18-0	20-11	7-1	10-5	13-2	16-1	18-8
	Spruce-Pine-Fir	#1	6-8	9-9	12-4	15-1	17-6	5-11	8-8	11-0	13-6	15-7
	Spruce-Pine-Fir	#2	6-8	9-9	12-4	15-1	17-6	5-11	8-8	11-0	13-6	15-7
	Spruce-Pine-Fir	#3	5-0	7-4	9-4	11-5	13-2	4-6	6-7	8-4	10-2	11-10

For SI: 1 inch = 25.4 mm, 1 foot = 304.8 mm, 1 pound per square foot = 47.9 N/m^2.

a. Span exceeds 26 feet in length. Check sources for availability of lumber in lengths greater than 20 feet.

TABLE 2308.10.3(4)
RAFTER SPANS FOR COMMON LUMBER SPECIES
(Ground Snow Load = 50 pounds per square foot, Ceiling Not Attached to Rafters, L/Δ = 180)

RAFTER SPACING (inches)	SPECIES AND GRADE		DEAD LOAD = 10 pounds per square foot					DEAD LOAD = 20 pounds per square foot				
			2 × 4	2 × 6	2 × 8	2 × 10	2 × 12	2 × 4	2 × 6	2 × 8	2 × 10	2 × 12
			Maximum rafter spans									
			(ft. - in.)	(ft. - in.)	(ft. - in.)	(ft. - in.)	(ft. - in.)	(ft. - in.)	(ft. - in.)	(ft. - in.)	(ft. - in.)	(ft. - in.)
12	Douglas Fir-Larch	SS	8-5	13-3	17-6	22-4	26-0	8-5	13-3	17-0	20-9	24-10
	Douglas Fir-Larch	#1	8-2	12-0	15-3	18-7	21-7	7-7	11-2	14-1	17-3	20-0
	Douglas Fir-Larch	#2	7-8	11-3	14-3	17-5	20-2	7-1	10-5	13-2	16-1	18-8
	Douglas Fir-Larch	#3	5-10	8-6	10-9	13-2	15-3	5-5	7-10	10-0	12-2	14-1
	Hem-Fir	SS	8-0	12-6	16-6	21-1	25-6	8-0	12-6	16-6	20-4	23-7
	Hem-Fir	#1	7-10	11-9	14-10	18-1	21-0	7-5	10-10	13-9	16-9	19-5
	Hem-Fir	#2	7-5	11-1	14-0	17-2	19-11	7-0	10-3	13-0	15-10	18-5
	Hem-Fir	#3	5-10	8-6	10-9	13-2	15-3	5-5	7-10	10-0	12-2	14-1
	Southern Pine	SS	8-4	13-0	17-2	21-11	Note a	8-4	13-0	17-2	21-11	Note a
	Southern Pine	#1	8-2	12-10	16-10	20-3	24-1	8-2	12-6	15-9	18-9	22-4
	Southern Pine	#2	8-0	11-9	15-3	18-2	21-3	7-7	10-11	14-1	16-10	19-9
	Southern Pine	#3	6-2	9-2	11-8	13-9	16-4	5-9	8-5	10-9	12-9	15-2
	Spruce-Pine-Fir	SS	7-10	12-3	16-2	20-8	24-1	7-10	12-3	15-9	19-3	22-4
	Spruce-Pine-Fir	#1	7-8	11-3	14-3	17-5	20-2	7-1	10-5	13-2	16-1	18-8
	Spruce-Pine-Fir	#2	7-8	11-3	14-3	17-5	20-2	7-1	10-5	13-2	16-1	18-8
	Spruce-Pine-Fir	#3	5-10	8-6	10-9	13-2	15-3	5-5	7-10	10-0	12-2	14-1
16	Douglas Fir-Larch	SS	7-8	12-1	15-10	19-5	22-6	7-8	11-7	14-8	17-11	20-10
	Douglas Fir-Larch	#1	7-1	10-5	13-2	16-1	18-8	6-7	9-8	12-2	14-11	17-3
	Douglas Fir-Larch	#2	6-8	9-9	12-4	15-1	17-6	6-2	9-0	11-5	13-11	16-2
	Douglas Fir-Larch	#3	5-0	7-4	9-4	11-5	13-2	4-8	6-10	8-8	10-6	12-3
	Hem-Fir	SS	7-3	11-5	15-0	19-1	22-1	7-3	11-5	14-5	17-8	20-5
	Hem-Fir	#1	6-11	10-2	12-10	15-8	18-2	6-5	9-5	11-11	14-6	16-10
	Hem-Fir	#2	6-7	9-7	12-2	14-10	17-3	6-1	8-11	11-3	13-9	15-11
	Hem-Fir	#3	5-0	7-4	9-4	11-5	13-2	4-8	6-10	8-8	10-6	12-3
	Southern Pine	SS	7-6	11-10	15-7	19-11	24-3	7-6	11-10	15-7	19-11	23-10
	Southern Pine	#1	7-5	11-7	14-9	17-6	20-11	7-4	10-10	13-8	16-2	19-4
	Southern Pine	#2	7-1	10-2	13-2	15-9	18-5	6-7	9-5	12-2	14-7	17-1
	Southern Pine	#3	5-4	7-11	10-1	11-11	14-2	4-11	7-4	9-4	11-0	13-1
	Spruce-Pine-Fir	SS	7-1	11-2	14-8	18-0	20-11	7-1	10-9	13-8	16-8	19-4
	Spruce-Pine-Fir	#1	6-8	9-9	12-4	15-1	17-6	6-2	9-0	11-5	13-11	16-2
	Spruce-Pine-Fir	#2	6-8	9-9	12-4	15-1	17-6	6-2	9-0	11-5	13-11	16-2
	Spruce-Pine-Fir	#3	5-0	7-4	9-4	11-5	13-2	4-8	6-10	8-8	10-6	12-3

(continued)

TABLE 2308.10.3(4)—continued RAFTER SPANS FOR COMMON LUMBER SPECIES (Ground Snow Load = 50 pounds per square foot, Ceiling Not Attached to Rafters, $L/\Delta = 180$)

RAFTER SPACING (inches)	SPECIES AND GRADE		DEAD LOAD = 10 pounds per square foot					DEAD LOAD = 20 pounds per square foot				
			2 × 4	2 × 6	2 × 8	2 × 10	2 × 12	2 × 4	2 × 6	2 × 8	2 × 10	2 × 12
			Maximum rafter spans									
			(ft. - in.)	(ft. - in.)	(ft. - in.)	(ft. - in.)	(ft. - in.)	(ft. - in.)	(ft. - in.)	(ft. - in.)	(ft. - in.)	(ft. - in.)
19.2	Douglas Fir-Larch	SS	7-3	11-4	14-6	17-8	20-6	7-3	10-7	13-5	16-5	19-0
	Douglas Fir-Larch	#1	6-6	9-6	12-0	14-8	17-1	6-0	8-10	11-2	13-7	15-9
	Douglas Fir-Larch	#2	6-1	8-11	11-3	13-9	15-11	5-7	8-3	10-5	12-9	14-9
	Douglas Fir-Larch	#3	4-7	6-9	8-6	10-5	12-1	4-3	6-3	7-11	9-7	11-2
	Hem-Fir	SS	6-10	10-9	14-2	17-5	20-2	6-10	10-5	13-2	16-1	18-8
	Hem-Fir	#1	6-4	9-3	11-9	14-4	16-7	5-10	8-7	10-10	13-3	15-5
	Hem-Fir	#2	6-0	8-9	11-1	13-7	15-9	5-7	8-1	10-3	12-7	14-7
	Hem-Fir	#3	4-7	6-9	8-6	10-5	12-1	4-3	6-3	7-11	9-7	11-2
	Southern Pine	SS	7-1	11-2	14-8	18-9	22-10	7-1	11-2	14-8	18-7	21-9
	Southern Pine	#1	7-0	10-8	13-5	16-0	19-1	6-8	9-11	12-5	14-10	17-8
	Southern Pine	#2	6-6	9-4	12-0	14-4	16-10	6-0	8-8	11-2	13-4	15-7
	Southern Pine	#3	4-11	7-3	9-2	10-10	12-11	4-6	6-8	8-6	10-1	12-0
	Spruce-Pine-Fir	SS	6-8	10-6	13-5	16-5	19-1	6-8	9-10	12-5	15-3	17-8
	Spruce-Pine-Fir	#1	6-1	8-11	11-3	13-9	15-11	5-7	8-3	10-5	12-9	14-9
	Spruce-Pine-Fir	#2	6-1	8-11	11-3	13-9	15-11	5-7	8-3	10-5	12-9	14-9
	Spruce-Pine-Fir	#3	4-7	6-9	8-6	10-5	12-1	4-3	6-3	7-11	9-7	11-2
24	Douglas Fir-Larch	SS	6-8	10-3	13-0	15-10	18-4	6-6	9-6	12-0	14-8	17-0
	Douglas Fir-Larch	#1	5-10	8-6	10-9	13-2	15-3	5-5	7-10	10-0	12-2	14-1
	Douglas Fir-Larch	#2	5-5	7-11	10-1	12-4	14-3	5-0	7-4	9-4	11-5	13-2
	Douglas Fir-Larch	#3	4-1	6-0	7-7	9-4	10-9	3-10	5-7	7-1	8-7	10-0
	Hem-Fir	SS	6-4	9-11	12-9	15-7	18-0	6-4	9-4	11-9	14-5	16-8
	Hem-Fir	#1	5-8	8-3	10-6	12-10	14-10	5-3	7-8	9-9	11-10	13-9
	Hem-Fir	#2	5-4	7-10	9-11	12-1	14-1	4-11	7-3	9-2	11-3	13-0
	Hem-Fir	#3	4-1	6-0	7-7	9-4	10-9	3-10	5-7	7-1	8-7	10-0
	Southern Pine	SS	6-7	10-4	13-8	17-5	21-0	6-7	10-4	13-8	16-7	19-5
	Southern Pine	#1	6-5	9-7	12-0	14-4	17-1	6-0	8-10	11-2	13-3	15-9
	Southern Pine	#2	5-10	8-4	10-9	12-10	15-1	5-5	7-9	10-0	11-11	13-11
	Southern Pine	#3	4-4	6-5	8-3	9-9	11-7	4-1	6-0	7-7	9-0	10-8
	Spruce-Pine-Fir	SS	6-2	9-6	12-0	14-8	17-1	6-0	8-10	11-2	13-7	15-9
	Spruce-Pine-Fir	#1	5-5	7-11	10-1	12-4	14-3	5-0	7-4	9-4	11-5	13-2
	Spruce-Pine-Fir	#2	5-5	7-11	10-1	12-4	14-3	5-0	7-4	9-4	11-5	13-2
	Spruce-Pine-Fir	#3	4-1	6-0	7-7	9-4	10-9	3-10	5-7	7-1	8-7	10-0

For SI: 1 inch = 25.4 mm, 1 foot = 304.8 mm, 1 pound per square foot = 47.9 N/m^2.

a. Span exceeds 26 feet in length. Check sources for availability of lumber in lengths greater than 20 feet.

TABLE 2308.10.3(5)
RAFTER SPANS FOR COMMON LUMBER SPECIES
(Ground Snow Load = 30 pounds per square foot, Ceiling Attached to Rafters, L/Δ = 240)

RAFTER SPACING (inches)	SPECIES AND GRADE		DEAD LOAD = 10 pounds per square foot					DEAD LOAD = 20 pounds per square foot				
			2 × 4	2 × 6	2 × 8	2 × 10	2 × 12	2 × 4	2 × 6	2 × 8	2 × 10	2 × 12
			Maximum rafter spans									
			(ft. - in.)	(ft. - in.)	(ft. - in.)	(ft. - in.)	(ft. - in.)	(ft. - in.)	(ft. - in.)	(ft. - in.)	(ft. - in.)	(ft. - in.)
12	Douglas Fir-Larch	SS	9-1	14-4	18-10	24-1	Note a	9-1	14-4	18-10	24-1	Note a
	Douglas Fir-Larch	#1	8-9	13-9	18-2	22-9	Note a	8-9	13-2	16-8	20-4	23-7
	Douglas Fir-Larch	#2	8-7	13-6	17-5	21-4	24-8	8-5	12-4	15-7	19-1	22-1
	Douglas Fir-Larch	#3	7-1	10-5	13-2	16-1	18-8	6-4	9-4	11-9	14-5	16-8
	Hem-Fir	SS	8-7	13-6	17-10	22-9	Note a	8-7	13-6	17-10	22-9	Note a
	Hem-Fir	#1	8-5	13-3	17-5	22-2	25-9	8-5	12-10	16-3	19-10	23-0
	Hem-Fir	#2	8-0	12-7	16-7	21-0	24-4	8-0	12-2	15-4	18-9	21-9
	Hem-Fir	#3	7-1	10-5	13-2	16-1	18-8	6-4	9-4	11-9	14-5	16-8
	Southern Pine	SS	8-11	14-1	18-6	23-8	Note a	8-11	14-1	18-6	23-8	Note a
	Southern Pine	#1	8-9	13-9	18-2	23-2	Note a	8-9	13-9	18-2	22-2	Note a
	Southern Pine	#2	8-7	13-6	17-10	22-3	Note a	8-7	12-11	16-8	19-11	23-4
	Southern Pine	#3	7-7	11-2	14-3	16-10	20-0	6-9	10-0	12-9	15-1	17-11
	Spruce-Pine-Fir	SS	8-5	13-3	17-5	22-3	Note a	8-5	13-3	17-5	22-3	Note a
	Spruce-Pine-Fir	#1	8-3	12-11	17-0	21-4	24-8	8-3	12-4	15-7	19-1	22-1
	Spruce-Pine-Fir	#2	8-3	12-11	17-0	21-4	24-8	8-3	12-4	15-7	19-1	22-1
	Spruce-Pine-Fir	#3	7-1	10-5	13-2	16-1	18-8	6-4	9-4	11-9	14-5	16-8
16	Douglas Fir-Larch	SS	8-3	13-0	17-2	21-10	Note a	8-3	13-0	17-2	21-3	24-8
	Douglas Fir-Larch	#1	8-0	12-6	16-2	19-9	22-10	7-10	11-5	14-5	17-8	20-5
	Douglas Fir-Larch	#2	7-10	11-11	15-1	18-5	21-5	7-3	10-8	13-6	16-6	19-2
	Douglas Fir-Larch	#3	6-2	9-0	11-5	13-11	16-2	5-6	8-1	10-3	12-6	14-6
	Hem-Fir	SS	7-10	12-3	16-2	20-8	25-1	7-10	12-3	16-2	20-8	24-2
	Hem-Fir	#1	7-8	12-0	15-9	19-3	22-3	7-7	11-1	14-1	17-2	19-11
	Hem-Fir	#2	7-3	11-5	14-11	18-2	21-1	7-2	10-6	13-4	16-3	18-10
	Hem-Fir	#3	6-2	9-0	11-5	13-11	16-2	5-6	8-1	10-3	12-6	14-6
	Southern Pine	SS	8-1	12-9	16-10	21-6	Note a	8-1	12-9	16-10	21-6	Note a
	Southern Pine	#1	8-0	12-6	16-6	21-1	25-7	8-0	12-6	16-2	19-2	22-10
	Southern Pine	#2	7-10	12-3	16-2	19-3	22-7	7-10	11-2	14-5	17-3	20-2
	Southern Pine	#3	6-7	9-8	12-4	14-7	17-4	5-10	8-8	11-0	13-0	15-6
	Spruce-Pine-Fir	SS	7-8	12-0	15-10	20-2	24-7	7-8	12-0	15-10	19-9	22-10
	Spruce-Pine-Fir	#1	7-6	11-9	15-1	18-5	21-5	7-3	10-8	13-6	16-6	19-2
	Spruce-Pine-Fir	#2	7-6	11-9	15-1	18-5	21-5	7-3	10-8	13-6	16-6	19-2
	Spruce-Pine-Fir	#3	6-2	9-0	11-5	13-11	16-2	5-6	8-1	10-3	12-6	14-6

(continued)

TABLE 2308.10.3(5)—continued
RAFTER SPANS FOR COMMON LUMBER SPECIES
(Ground Snow Load = 30 pounds per square foot, Ceiling Attached to Rafters, L/Δ = 240)

RAFTER SPACING (inches)	SPECIES AND GRADE		DEAD LOAD = 10 pounds per square foot					DEAD LOAD = 20 pounds per square foot				
			2 × 4	2 × 6	2 × 8	2 × 10	2 × 12	2 × 4	2 × 6	2 × 8	2 × 10	2 × 12
			Maximum rafter spans									
			(ft. - in.)	(ft. - in.)	(ft. - in.)	(ft. - in.)	(ft. - in.)	(ft. - in.)	(ft. - in.)	(ft. - in.)	(ft. - in.)	(ft. - in.)
19.2	Douglas Fir-Larch	SS	7-9	12-3	16-1	20-7	25-0	7-9	12-3	15-10	19-5	22-6
	Douglas Fir-Larch	#1	7-6	11-8	14-9	18-0	20-11	7-1	10-5	13-2	16-1	18-8
	Douglas Fir-Larch	#2	7-4	10-11	13-9	16-10	19-6	6-8	9-9	12-4	15-1	17-6
	Douglas Fir-Larch	#3	5-7	8-3	10-5	12-9	14-9	5-0	7-4	9-4	11-5	13-2
	Hem-Fir	SS	7-4	11-7	15-3	19-5	23-7	7-4	11-7	15-3	19-1	22-1
	Hem-Fir	#1	7-2	11-4	14-4	17-7	20-4	6-11	10-2	12-10	15-8	18-2
	Hem-Fir	#2	6-10	10-9	13-7	16-7	19-3	6-7	9-7	12-2	14-10	17-3
	Hem-Fir	#3	5-7	8-3	10-5	12-9	14-9	5-0	7-4	9-4	11-5	13-2
	Southern Pine	SS	7-8	12-0	15-10	20-2	24-7	7-8	12-0	15-10	20-2	24-7
	Southern Pine	#1	7-6	11-9	15-6	19-7	23-4	7-6	11-9	14-9	17-6	20-11
	Southern Pine	#2	7-4	11-5	14-9	17-7	20-7	7-1	10-2	13-2	15-9	18-5
	Southern Pine	#3	6-0	8-10	11-3	13-4	15-10	5-4	7-11	10-1	11-11	14-2
	Spruce-Pine-Fir	SS	7-2	11-4	14-11	19-0	23-1	7-2	11-4	14-9	18-0	20-11
	Spruce-Pine-Fir	#1	7-0	10-11	13-9	16-10	19-6	6-8	9-9	12-4	15-1	17-6
	Spruce-Pine-Fir	#2	7-0	10-11	13-9	16-10	19-6	6-8	9-9	12-4	15-1	17-6
	Spruce-Pine-Fir	#3	5-7	8-3	10-5	12-9	14-9	5-0	7-4	9-4	11-5	13-2
24	Douglas Fir-Larch	SS	7-3	11-4	15-0	19-1	22-6	7-3	11-3	14-2	17-4	20-1
	Douglas Fir-Larch	#1	7-0	10-5	13-2	16-1	18-8	6-4	9-4	11-9	14-5	16-8
	Douglas Fir-Larch	#2	6-8	9-9	12-4	15-1	17-6	5-11	8-8	11-0	13-6	15-7
	Douglas Fir-Larch	#3	5-0	7-4	9-4	11-5	13-2	4-6	6-7	8-4	10-2	11-10
	Hem-Fir	SS	6-10	10-9	14-2	18-0	21-11	6-10	10-9	13-11	17-0	19-9
	Hem-Fir	#1	6-8	10-2	12-10	15-8	18-2	6-2	9-1	11-6	14-0	16-3
	Hem-Fir	#2	6-4	9-7	12-2	14-10	17-3	5-10	8-7	10-10	13-3	15-5
	Hem-Fir	#3	5-0	7-4	9-4	11-5	13-2	4-6	6-7	8-4	10-2	11-10
	Southern Pine	SS	7-1	11-2	14-8	18-9	22-10	7-1	11-2	14-8	18-9	22-10
	Southern Pine	#1	7-0	10-11	14-5	17-6	20-11	7-0	10-6	13-2	15-8	18-8
	Southern Pine	#2	6-10	10-2	13-2	15-9	18-5	6-4	9-2	11-9	14-1	16-6
	Southern Pine	#3	5-4	7-11	10-1	11-11	14-2	4-9	7-1	9-0	10-8	12-8
	Spruce-Pine-Fir	SS	6-8	10-6	13-10	17-8	20-11	6-8	10-5	13-2	16-1	18-8
	Spruce-Pine-Fir	#1	6-6	9-9	12-4	15-1	17-6	5-11	8-8	11-0	13-6	15-7
	Spruce-Pine-Fir	#2	6-6	9-9	12-4	15-1	17-6	5-11	8-8	11-0	13-6	15-7
	Spruce-Pine-Fir	#3	5-0	7-4	9-4	11-5	13-2	4-6	6-7	8-4	10-2	11-10

For SI: 1 inch = 25.4 mm, 1 foot = 304.8 mm, 1 pound per square foot = 47.9 N/m^2.

a. Span exceeds 26 feet in length. Check sources for availability of lumber in lengths greater than 20 feet.

TABLE 2308.10.3(6)
RAFTER SPANS FOR COMMON LUMBER SPECIES
(Ground Snow Load = 50 pounds per square foot, Ceiling Attached to Rafters, L/Δ = 240)

RAFTER SPACING (inches)	SPECIES AND GRADE		DEAD LOAD = 10 pounds per square foot					DEAD LOAD = 20 pounds per square foot				
			2 × 4	2 × 6	2 × 8	2 × 10	2 × 12	2 × 4	2 × 6	2 × 8	2 × 10	2 × 12
			Maximum rafter spans									
			(ft. - in.)	(ft. - in.)	(ft. - in.)	(ft. - in.)	(ft. - in.)	(ft. - in.)	(ft. - in.)	(ft. - in.)	(ft. - in.)	(ft. - in.)
12	Douglas Fir-Larch	SS	7-8	12-1	15-11	20-3	24-8	7-8	12-1	15-11	20-3	24-0
	Douglas Fir-Larch	#1	7-5	11-7	15-3	18-7	21-7	7-5	11-2	14-1	17-3	20-0
	Douglas Fir-Larch	#2	7-3	11-3	14-3	17-5	20-2	7-1	10-5	13-2	16-1	18-8
	Douglas Fir-Larch	#3	5-10	8-6	10-9	13-2	15-3	5-5	7-10	10-0	12-2	14-1
	Hem-Fir	SS	7-3	11-5	15-0	19-2	23-4	7-3	11-5	15-0	19-2	23-4
	Hem-Fir	#1	7-1	11-2	14-8	18-1	21-0	7-1	10-10	13-9	16-9	19-5
	Hem-Fir	#2	6-9	10-8	14-0	17-2	19-11	6-9	10-3	13-0	15-10	18-5
	Hem-Fir	#3	5-10	8-6	10-9	13-2	15-3	5-5	7-10	10-0	12-2	14-1
	Southern Pine	SS	7-6	11-0	15-7	19-11	24-3	7-6	11-10	15-7	19-11	24-3
	Southern Pine	#1	7-5	11-7	15-4	19-7	23-9	7-5	11-7	15-4	18-9	22-4
	Southern Pine	#2	7-3	11-5	15-0	18-2	21-3	7-3	10-11	14-1	16-10	19-9
	Southern Pine	#3	6-2	9-2	11-8	13-9	16-4	5-9	8-5	10-9	12-9	15-2
	Spruce-Pine-Fir	SS	7-1	11-2	14-8	18-9	22-10	7-1	11-2	14-8	18-9	22-4
	Spruce-Pine-Fir	#1	6-11	10-11	14-3	17-5	20-2	6-11	10-5	13-2	16-1	18-8
	Spruce-Pine-Fir	#2	6-11	10-11	14-3	17-5	20-2	6-11	10-5	13-2	16-1	18-8
	Spruce-Pine-Fir	#3	5-10	8-6	10-9	13-2	15-3	5-5	7-10	10-0	12-2	14-1
16	Douglas Fir-Larch	SS	7-0	11-0	14-5	18-5	22-5	7-0	11-0	14-5	17-11	20-10
	Douglas Fir-Larch	#1	6-9	10-5	13-2	16-1	18-8	6-7	9-8	12-2	14-11	17-3
	Douglas Fir-Larch	#2	6-7	9-9	12-4	15-1	17-6	6-2	9-0	11-5	13-11	16-2
	Douglas Fir-Larch	#3	5-0	7-4	9-4	11-5	13-2	4-8	6-10	8-8	10-6	12-3
	Hem-Fir	SS	6-7	10-4	13-8	17-5	21-2	6-7	10-4	13-8	17-5	20-5
	Hem-Fir	#1	6-5	10-2	12-10	15-8	18-2	6-5	9-5	11-11	14-6	16-10
	Hem-Fir	#2	6-2	9-7	12-2	14-10	17-3	6-1	8-11	11-3	13-9	15-11
	Hem-Fir	#3	5-0	7-4	9-4	11-5	13-2	4-8	6-10	8-8	10-6	12-3
	Southern Pine	SS	6-10	10-9	14-2	18-1	22-0	6-10	10-9	14-2	18-1	22-0
	Southern Pine	#1	6-9	10-7	13-11	17-6	20-11	6-9	10-7	13-8	16-2	19-4
	Southern Pine	#2	6-7	10-2	13-2	15-9	18-5	6-7	9-5	12-2	14-7	17-1
	Southern Pine	#3	5-4	7-11	10-1	11-11	14-2	4-11	7-4	9-4	11-0	13-1
	Spruce-Pine-Fir	SS	6-5	10-2	13-4	17-0	20-9	6-5	10-2	13-4	16-8	19-4
	Spruce-Pine-Fir	#1	6-4	9-9	12-4	15-1	17-6	6-2	9-0	11-5	13-11	16-2
	Spruce-Pine-Fir	#2	6-4	9-9	12-4	15-1	17-6	6-2	9-0	11-5	13-11	16-2
	Spruce-Pine-Fir	#3	5-0	7-4	9-4	11-5	13-2	4-8	6-10	8-8	10-6	12-3

(continued)

TABLE 2308.10.3(6)—continued
RAFTER SPANS FOR COMMON LUMBER SPECIES
(Ground Snow Load = 50 pounds per square foot, Ceiling Attached to Rafters, $L/\Delta = 240$)

RAFTER SPACING (inches)	SPECIES AND GRADE		DEAD LOAD = 10 pounds per square foot					DEAD LOAD = 20 pounds per square foot				
			2 × 4	2 × 6	2 × 8	2 × 10	2 × 12	2 × 4	2 × 6	2 × 8	2 × 10	2 × 12
			Maximum rafter spans									
			(ft. - in.)	(ft. - in.)	(ft. - in.)	(ft. - in.)	(ft. - in.)	(ft. - in.)	(ft. - in.)	(ft. - in.)	(ft. - in.)	(ft. - in.)
19.2	Douglas Fir-Larch	SS	6-7	10-4	13-7	17-4	20-6	6-7	10-4	13-5	16-5	19-0
	Douglas Fir-Larch	#1	6-4	9-6	12-0	14-8	17-1	6-0	8-10	11-2	13-7	15-9
	Douglas Fir-Larch	#2	6-1	8-11	11-3	13-9	15-11	5-7	8-3	10-5	12-9	14-9
	Douglas Fir-Larch	#3	4-7	6-9	8-6	10-5	12-1	4-3	6-3	7-11	9-7	11-2
	Hem-Fir	SS	6-2	9-9	12-10	16-5	19-11	6-2	9-9	12-10	16-1	18-8
	Hem-Fir	#1	6-1	9-3	11-9	14-4	16-7	5-10	8-7	10-10	13-3	15-5
	Hem-Fir	#2	5-9	8-9	11-1	13-7	15-9	5-7	8-1	10-3	12-7	14-7
	Hem-Fir	#3	4-7	6-9	8-6	10-5	12-1	4-3	6-3	7-11	9-7	11-2
	Southern Pine	SS	6-5	10-2	13-4	17-0	20-9	6-5	10-2	13-4	17-0	20-9
	Southern Pine	#1	6-4	9-11	13-1	16-0	19-1	6-4	9-11	12-5	14-10	17-8
	Southern Pine	#2	6-2	9-4	12-0	14-4	16-10	6-0	8-8	11-2	13-4	15-7
	Southern Pine	#3	4-11	7-3	9-2	10-10	12-11	4-6	6-8	8-6	10-1	12-0
	Spruce-Pine-Fir	SS	6-1	9-6	12-7	16-0	19-1	6-1	9-6	12-5	15-3	17-8
	Spruce-Pine-Fir	#1	5-11	8-11	11-3	13-9	15-11	5-7	8-3	10-5	12-9	14-9
	Spruce-Pine-Fir	#2	5-11	8-11	11-3	13-9	15-11	5-7	8-3	10-5	12-9	14-9
	Spruce-Pine-Fir	#3	4-7	6-9	8-6	10-5	12-1	4-3	6-3	7-11	9-7	11-2
24	Douglas Fir-Larch	SS	6-1	9-7	12-7	15-10	18-4	6-1	9-6	12-0	14-8	17-0
	Douglas Fir-Larch	#1	5-10	8-6	10-9	13-2	15-3	5-5	7-10	10-0	12-2	14-1
	Douglas Fir-Larch	#2	5-5	7-11	10-1	12-4	14-3	5-0	7-4	9-4	11-5	13-2
	Douglas Fir-Larch	#3	4-1	6-0	7-7	9-4	10-9	3-10	5-7	7-1	8-7	10-0
	Hem-Fir	SS	5-9	9-1	11-11	15-12	18-0	5-9	9-1	11-9	14-5	16-8
	Hem-Fir	#1	5-8	8-3	10-6	12-10	14-10	5-3	7-8	9-9	11-10	13-9
	Hem-Fir	#2	5-4	7-10	9-11	12-1	14-1	4-11	7-3	9-2	11-3	13-0
	Hem-Fir	#3	4-1	6-0	7-7	9-4	10-9	3-10	5-7	7-1	8-7	10-0
	Southern Pine	SS	6-0	9-5	12-5	15-10	19-3	6-0	9-5	12-5	15-10	19-3
	Southern Pine	#1	5-10	9-3	12-0	14-4	17-1	5-10	8-10	11-2	13-3	15-9
	Southern Pine	#2	5-9	8-4	10-9	12-10	15-1	5-5	7-9	10-0	11-11	13-11
	Southern Pine	#3	4-4	6-5	8-3	9-9	11-7	4-1	6-0	7-7	9-0	10-8
	Spruce-Pine-Fir	SS	5-8	8-10	11-8	14-8	17-1	5-8	8-10	11-2	13-7	15-9
	Spruce-Pine-Fir	#1	5-5	7-11	10-1	12-4	14-3	5-0	7-4	9-4	11-5	13-2
	Spruce-Pine-Fir	#2	5-5	7-11	10-1	12-4	14-3	5-0	7-4	9-4	11-5	13-2
	Spruce-Pine-Fir	#3	4-1	6-0	7-7	9-4	10-9	3-10	5-7	7-1	8-7	10-0

For SI: 1 inch = 25.4 mm, 1 foot = 304.8 mm, 1 pound per square foot = 47.9 N/m^2.

2308.10.4.1 Ceiling joist and rafter connections. Ceiling joists and rafters shall be nailed to each other and the assembly shall be nailed to the top wall plate in accordance with Tables 2304.9.1 and 2308.10.1. Ceiling joists shall be continuous or securely joined where they meet over interior partitions and fastened to adjacent rafters in accordance with Tables 2308.10.4.1 and 2304.9.1 to provide a continuous rafter tie across the building where such joists are parallel to the rafters. Ceiling joists shall have a bearing surface of not less than $1^1/_2$ inches (38 mm) on the top plate at each end.

Where ceiling joists are not parallel to rafters, an equivalent rafter tie shall be installed in a manner to provide a continuous tie across the building, at a spacing of not more than 4 feet (1219 mm) o.c. The connections shall be in accordance with Tables 2308.10.4.1 and 2304.9.1, or connections of equivalent capacities shall be provided. Where ceiling joists or rafter ties are not provided at the top of the rafter support walls, the ridge formed by these rafters shall also be supported by a girder conforming to Section 2308.4.

Rafter ties shall be spaced not more than 4 feet (1219 mm) o.c. Rafter tie connections shall be based on the equivalent rafter spacing in Table 2308.10.4.1. Where rafter ties are spaced at 32 inches (813 mm) o.c., the number of 16d common nails shall be two times the number specified for rafters spaced 16 inches (406 mm) o.c., with a minimum of 4-16d common nails where no snow loads are indicated. Where rafter ties are spaced at 48 inches (1219 mm) o.c., the number of 16d common nails shall be two times the number specified for rafters spaced 24 inches (610 mm) o.c., with a minimum of 6-16d common nails where no snow loads are indicated. Rafter/ceiling joist connections and rafter/tie connections shall be of sufficient size and number to prevent splitting from nailing.

❖ Sloped rafters will exert lateral loads on the walls of buildings. Therefore, unless the ridge is otherwise supported, as in post and beam framing, the building must provide horizontal ties to prevent spreading of walls (see Figure 2308.10.4.1). When ceiling joists are parallel to the rafters and adequately connected to the top plate and rafters, the continuous tie requirements of this section are met.

Ceiling joists are sometimes raised above the wall support of the rafters to provide additional ceiling height. Only when such design is accompanied by engineering analysis or tests should this be allowed.

Structural problems are often found in houses having cathedral ceilings. Since no joists are present to tie the walls together across the building, properly designed beams and posts must be used to support the rafters to satisfy code requirements.

Wind blowing across the roof develops negative pressures on the rafters. Rafter ties are intended to prevent separation of the rafters at the ridge under such conditions. Rafter ties must never be considered as a means of reducing the span of rafters.

TABLE 2308.10.4.1. See page 23-94.

❖ See the commentary to Section 2308.10.4.1.

2308.10.4.2 Notches and holes. Notching at the ends of rafters or ceiling joists shall not exceed one-fourth the depth. Notches in the top or bottom of the rafter or ceiling joist shall not exceed one-sixth the depth and shall not be located in the middle one-third of the span, except that a notch not exceeding one-third of the depth is permitted in the top of the rafter or ceiling joist not further from the face of the support than the depth of the member.

Holes bored in rafters or ceiling joists shall not be within 2 inches (51 mm) of the top and bottom and their diameter shall not exceed one-third the depth of the member.

❖ Wood is comprised of relatively small elongated cells oriented parallel to each other and bound together physically and chemically by lignin. The cells are continuous from end to end of a piece of lumber. When a rafter is notched or cut, the cell strands are interrupted. In notched rafters, the effective depth is reduced equal to the depth of the notch (see Figure 2308.8.2).

Some designs and installation practices require that limited notching and cutting occur. Notching should be avoided when possible. Holes bored in rafters create the same problems as notches. When necessary, the holes should be located in areas with the least stress concentration, generally along the neutral axis of the rafter. Beams subject to high horizontal shear stress (short span/heavy load) should not be cut. Wood members having a thickness of over 4 inches (102 mm) should not be notched, except at the ends of the members.

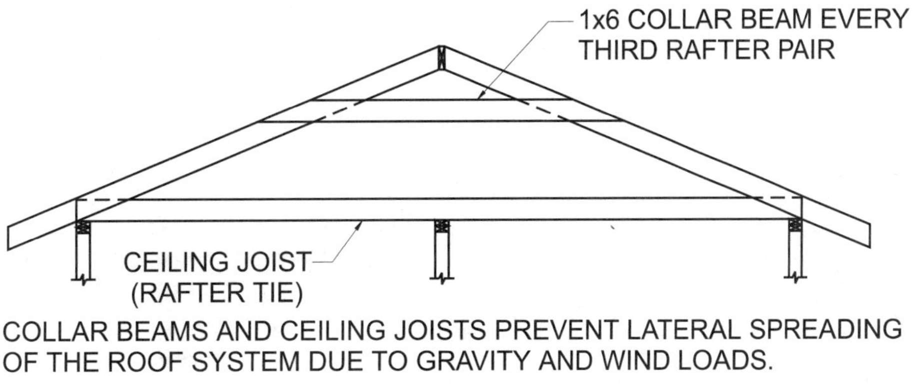
1x6 COLLAR BEAM EVERY
THIRD RAFTER PAIR
CEILING JOIST
(RAFTER TIE)
COLLAR BEAMS AND CEILING JOISTS PREVENT LATERAL SPREADING
OF THE ROOF SYSTEM DUE TO GRAVITY AND WIND LOADS.

Figure 2308.10.4.1
RAFTER TIES

TABLE 2308.10.4.1
RAFTER TIE CONNECTIONS[g]

RAFTER SLOPE	TIE SPACING (inches)	NO SNOW LOAD				GROUND SNOW LOAD (pound per square foot)							
						30 pounds per square foot				50 pounds per square foot			
						Roof span (feet)							
		12	20	28	36	12	20	28	36	12	20	28	36
		Required number of 16d common nails[a,b] per connection[c,d,e,f]											
3:12	12	4	6	8	10	4	6	8	11	5	8	12	15
	16	5	7	10	13	5	8	11	14	6	11	15	20
	24	7	11	15	19	7	11	16	21	9	16	23	30
	32	10	14	19	25	10	16	22	28	12	27	30	40
	48	14	21	29	37	14	32	36	42	18	32	46	60
4:12	12	3	4	5	6	3	5	6	8	4	6	9	11
	16	3	5	7	8	4	6	8	11	5	8	12	15
	24	4	7	10	12	5	9	12	16	7	12	17	22
	32	6	9	13	16	8	12	16	22	10	16	24	30
	48	8	14	19	24	10	18	24	32	14	24	34	44
5:12	12	3	3	4	5	3	4	5	7	3	5	7	9
	16	3	4	5	7	3	5	7	9	4	7	9	12
	24	4	6	8	10	4	7	10	13	6	10	14	18
	32	5	8	10	13	6	10	14	18	8	14	18	24
	48	7	11	15	20	8	14	20	26	12	20	28	36
7:12	12	3	3	3	4	3	3	4	5	3	4	5	7
	16	3	3	4	5	3	4	5	6	3	5	7	9
	24	3	4	6	7	3	5	7	9	4	7	10	13
	32	4	6	8	10	4	8	10	12	6	10	14	18
	48	5	8	11	14	6	10	14	18	9	14	20	26
9:12	12	3	3	3	3	3	3	3	4	3	3	4	5
	16	3	3	3	4	3	3	4	5	3	4	5	7
	24	3	3	5	6	3	4	6	7	3	6	8	10
	32	3	4	6	8	4	6	8	10	5	8	10	14
	48	4	6	9	11	5	8	12	14	7	12	16	20
12:12	12	3	3	3	3	3	3	3	3	3	3	3	4
	16	3	3	3	3	3	3	3	4	3	3	4	5
	24	3	3	3	4	3	3	4	6	3	4	6	8
	32	3	3	4	5	3	5	6	8	4	6	8	10
	48	3	4	6	7	4	7	8	12	6	8	12	16

For SI: 1 inch = 25.4 mm, 1 foot = 304.8 mm, 1 pound per square foot = 47.8 N/m^2.

a. 40d box or 16d sinker box nails are permitted to be substituted for 16d common nails.

b. Nailing requirements are permitted to be reduced 25 percent if nails are clinched.

c. Rafter tie heel joint connections are not required where the ridge is supported by a load-bearing wall, header or ridge beam.

d. When intermediate support of the rafter is provided by vertical struts or purlins to a load-bearing wall, the tabulated heel joint connection requirements are permitted to be reduced proportionally to the reduction in span.

e. Equivalent nailing patterns are required for ceiling joist to ceiling joist lap splices.

f. Connected members shall be of sufficient size to prevent splitting due to nailing.

g. For snow loads less than 30 pounds per square foot, the required number of nails is permitted to be reduced by multiplying by the ratio of actual snow load plus 10 divided by 40, but not less than the number required for no snow load.

2308.10.4.3 Framing around openings. Trimmer and header rafters shall be doubled, or of lumber of equivalent cross section, where the span of the header exceeds 4 feet (1219 mm). The ends of header rafters more than 6 feet (1829 mm) long shall be supported by framing anchors or rafter hangers unless bearing on a beam, partition or wall.

❖ See Figure 2308.8.3 for an illustration of the criteria for framing around openings.

2308.10.5 Purlins. Purlins to support roof loads are permitted to be installed to reduce the span of rafters within allowable limits and shall be supported by struts to bearing walls. The maximum span of 2-inch by 4-inch (51 mm by 102 mm) purlins shall be 4 feet (1219 mm). The maximum span of the 2-inch by 6-inch (51 mm by 152 mm) purlin shall be 6 feet (1829 mm), but in no case shall the purlin be smaller than the supported rafter. Struts shall not be smaller than 2-inch by 4-inch (51 mm by 102 mm) members. The unbraced length of struts shall not exceed 8 feet (2438 mm) and the minimum slope of the struts shall not be less than 45 degrees (0.79 rad) from the horizontal.

❖ This section permits the use of purlins and struts to reduce the span of rafters. Where the roof slope is less than 3 units vertical in 12 units horizontal, the code requires that members supporting rafters such as ridge boards, for example, be designed as beams. However, struts can be installed from a partition up to the ridge board to reduce the span (see Figure 2308.10.5).

2308.10.6 Blocking. Roof rafters and ceiling joists shall be supported laterally to prevent rotation and lateral displacement in accordance with the provisions of Section 2308.8.5.

❖ In the unusual situation in which the ends of rafters are not provided with built-in lateral support, the ends must be braced specifically to resist the tendency to twist or lay over.

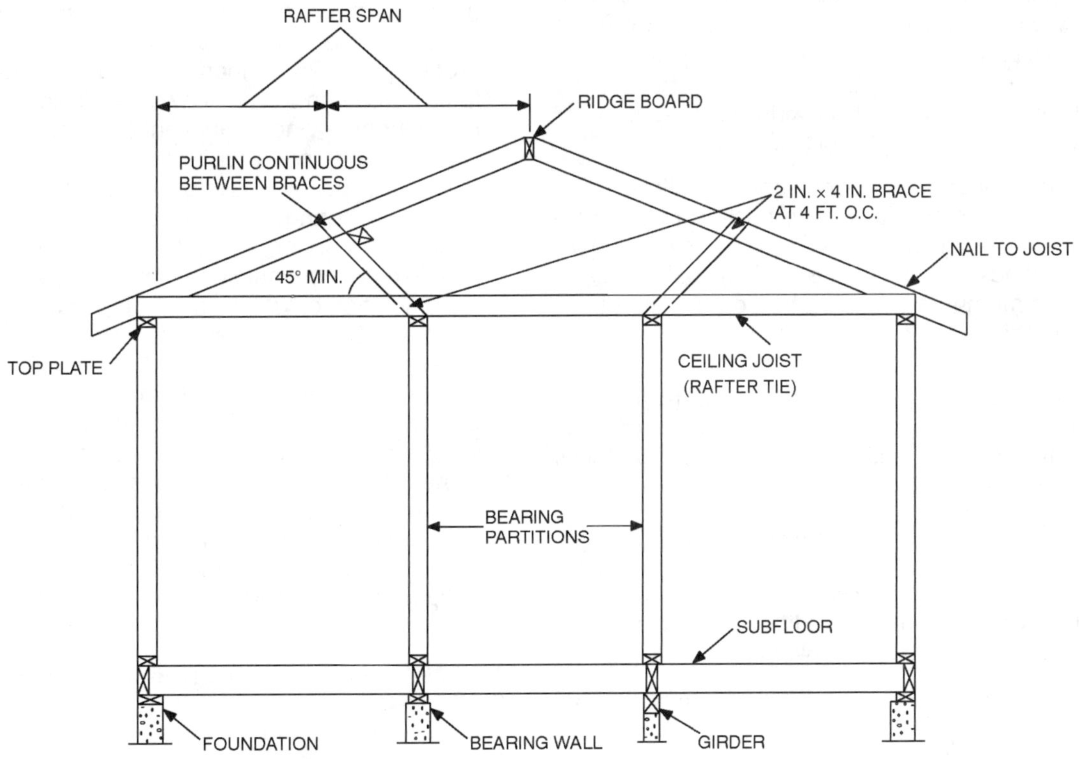

For SI: 1 inch = 25.4 mm, 1 foot = 304.8 mm.

Figure 2308.10.5
BRACED RAFTER CONSTRUCTION

2308.10.7 Wood trusses.

2308.10.7.1 Design. Wood trusses shall be designed in accordance with the requirements of Chapter 23 and accepted engineering practice. Members are permitted to be joined by nails, glue, bolts, timber connectors, metal connector plates or other approved framing devices.

❖ Trusses are to be designed in accordance with Chapter 16 and are to comply with Section 2303.4 (see the commentary on that section). While trusses used in most light frame construction have members connected with metal connector plates, this section recognizes that other methods exist to connect truss members.

2308.10.7.2 Bracing. The bracing of wood trusses shall comply with their appropriate engineered design.

❖ To prevent collapse during construction, and until permanent bracing is installed, trusses should be adequately braced. When final bracing is in place, trusses should be positioned as vertical as possible; tilted trusses will not perform as designed.

2308.10.7.3 Alterations to trusses. Truss members and components shall not be cut, notched, drilled, spliced or otherwise altered in any way without written concurrence and approval of a registered design professional. Alterations resulting in the addition of loads to any member (e.g., HVAC equipment, water heater) shall not be permitted without verification that the truss is capable of supporting such additional loading.

❖ Trusses are engineered products, and any alteration to specific elements may prevent the truss from performing as intended. Additionally, trusses are designed to resist specific loads—changing the loading on the truss or on any of its members should not be done without approval of a registered design professional.

2308.10.8 Roof sheathing. Roof sheathing shall be in accordance with Tables 2304.7(3) and 2304.7(5) for wood structural panels, and Tables 2304.7(1) and 2304.7(2) for lumber and shall comply with Section 2304.7.2.

❖ See the commentary to Section 2304.7.2.

2308.10.8.1 Joints. Joints in lumber sheathing shall occur over supports unless approved end-matched lumber is used, in which case each piece shall bear on at least two supports.

❖ When board sheathing is used, the ends of the boards must be located over the joists or rafters. Such support is not required when end-matched tongue-and-groove lumber is used or when the ends are otherwise prevented from moving relative to each other.

2308.10.9 Roof planking. Planking shall be designed in accordance with the general provisions of this code.

In lieu of such design, 2-inch (51 mm) tongue-and- groove planking is permitted in accordance with Table 2308.10.9. Joints in such planking are permitted to be randomly spaced, provided the system is applied to not less than three continuous spans, planks are center matched and end matched or splined, each plank bears on at least one support, and joints are separated by at least 24 inches (610 mm) in adjacent pieces.

❖ Lumber planking on roofs must conform to the requirements of Section 2304.7.2, and Table 2304.7(1) for typical lumber roof sheathing. This section provides an alternative using 2-inch (51 mm) tongue-and-groove planking.

TABLE 2308.10.9. See page 23-97.

❖ Table 2308.10.9 gives allowable span (spacing of roof supports) based on the live load and deflection limits (from the code) and the design values for the planking used.

2308.10.10 Attic ventilation. For attic ventilation, see Section 1202.2.

❖ Large amounts of water vapor migrate by movement of air-carrying water or by diffusion through the building envelope materials because of a vapor pressure difference. The sources of water vapor include cooking, laundering, bathing, and breathing and perspiration of persons. These can account for an average daily production of 25 pounds (11.3 kg) of water vapor for a family of four in a typical dwelling. This average can be much higher where appliances such as humidifiers, washers and dryers are used.

As the vapor moves into the attic, it may reach its dew point, thus condensing on wood roof components. This wetting and drying action will cause rotting and decay. To avoid this, the attic must be ventilated to prevent the accumulation of water on building components. The installation of a vapor retarder acts to prevent the passage of moisture to the attic. An effective vapor retarder allows a decrease in ventilation. Vapor retarders are ineffective when openings in the barrier are allowed where moisture can be carried by air into the attic. This is the reason exhaust fans must terminate outdoors and not in the attic. Care should be exercised so that attic vent openings remain unobstructed. Attic ventilation is also required for the removal of excess heat. Access to attic spaces is required by Section 1208.2.

2308.11 Additional requirements for conventional construction in Seismic Design Category B or C. Structures of conventional light-frame construction in Seismic Design Category B or C, as determined in Section 1616, shall comply with Sections 2308.11.1 through 2308.11.3, in addition to the provisions of Sections 2308.1 through 2308.10.

❖ See Sections 1615 and 1616 for the procedure to determine the seismic design category. This section references these sections where construction limits apply due to seismic considerations. The limits on conventional construction are based on the NEHRP Provisions.

TABLE 2308.10.9
ALLOWABLE SPANS FOR 2-INCH TONGUE-AND-GROOVE DECKING

SPAN[a] (feet)	LIVE LOAD (pound per square foot)	DEFLECTION LIMIT	BENDING STRESS (f) (pound per square inch)	MODULUS OF ELASTICITY (E) (pound per square inch)
Roofs				
4	20	1/240 1/360	160	170,000 256,000
	30	1/240 1/360	210	256,000 384,000
	40	1/240 1/360	270	340,000 512,000
4.5	20	1/240 1/360	200	242,000 305,000
	30	1/240 1/360	270	363,000 405,000
	40	1/240 1/360	350	484,000 725,000
5.0	20	1/240 1/360	250	332,000 500,000
	30	1/240 1/360	330	495,000 742,000
	40	1/240 1/360	420	660,000 1,000,000
5.5	20	1/240 1/360	300	442,000 660,000
	30	1/240 1/360	400	662,000 998,000
	40	1/240 1/360	500	884,000 1,330,000
6.0	20	1/240 1/360	360	575,000 862,000
	30	1/240 1/360	480	862,000 1,295,000
	40	1/240 1/360	600	1,150,000 1,730,000
6.5	20	1/240 1/360	420	595,000 892,000
	30	1/240 1/360	560	892,000 1,340,000
	40	1/240 1/360	700	1,190,000 1,730,000
7.0	20	1/240 1/360	490	910,000 1,360,000
	30	1/240 1/360	650	1,370,000 2,000,000
	40	1/240 1/360	810	1,820,000 2,725,000

(continued)

TABLE 2308.10.9–continued
ALLOWABLE SPANS FOR 2-INCH TONGUE-AND-GROOVE DECKING

SPAN[a] (feet)	LIVE LOAD (pound per square foot)	DEFLECTION LIMIT	BENDING STRESS (*f*) (pound per square inch)	MODULUS OF ELASTICITY (*E*) (pound per square inch)
Roofs				
7.5	20	1/240 1/360	560	1,125,000 1,685,000
	30	1/240 1/360	750	1,685,000 2,530,000
	40	1/240 1/360	930	2,250,000 3,380,000
8.0	20	1/240 1/360	640	1,360,000 2,040,000
	30	1/240 1/360	850	2,040,000 3,060,000
Floors				
4 4.5 5.0	40	1/360	840 950 1,060	1,000,000 1,300,000 1,600,000

For SI: 1 inch = 25.4 mm, 1 foot = 304.8 mm, 1 pound per square foot = 0.0479 kN/m^2, 1 pound per square inch = 0.00689 N/mm^2.

a. Spans are based on simple beam action with 10 pounds per square foot dead load and provisions for a 300-pound concentrated load on a 12-inch width of decking. Random layup is permitted in accordance with the provisions of Section 2308.10.9. Lumber thickness is $1^1/_2$ inches nominal.

2308.11.1 Number of stories. Structures of conventional light-frame construction shall not exceed two stories in height in Seismic Design Category C.

Exception: Detached one- and two-family dwellings are permitted to be three stories in height in Seismic Design Category C.

❖ This section limits conventional construction to two stories in height in Seismic Design Category (SDC) C.

When conventional construction provisions are used for one- and two-family dwellings in SDC C, the allowable number of stories is increased to three.

2308.11.2 Concrete or masonry. Concrete or masonry walls, or masonry veneer shall not extend above the basement.

Exceptions:

1. Masonry veneer is permitted to be used in the first two stories above grade or the first three stories above grade where the lowest story has concrete or masonry walls in Seismic Design Category B, provided that structural use panel wall bracing is used, and the length of bracing provided is 1.5 times the required length as determined in Table 2308.9.3(1).
2. Masonry veneer is permitted to be used in the first story above grade or the first two stories above grade where the lowest story has concrete or masonry walls in Seismic Design Category B or C.
3. Masonry veneer is permitted to be used in the first two stories above grade in Seismic Design Categories B and C provided the following criteria are met:
   3.1. Type of brace per Section 2308.9.3 shall be Method 3 and the allowable shear capacity in accordance with Table 2306.4.1 shall be a minimum of 350 plf (5108 N/m) (ASD).
   3.2. The bracing of the top story shall be located at each end and at least every 25 feet (7620 mm) o.c. but not less than 40 percent of the braced wall line. The bracing of the first story shall be located at each end and at least every 25 feet (7620 mm) o.c. but not less than 35 percent of the braced wall line.
   3.3. Hold-down connectors shall be provided at the ends of braced walls for the second floor to first floor wall assembly with an allowable design of 2,000 pounds (907.0 kg). Hold-down connectors shall be provided at the ends of each wall segment of the braced walls for the first floor to foundation with an allowable design of 3,900 pounds (1768 kg). In all cases, the hold-down connector force shall be transferred to the foundation.
   3.4. Cripple walls shall not be permitted.

❖ Because of their weight, concrete walls, masonry walls and masonry veneer can impose lateral loads from seismic events that are beyond the loading considered acceptable for conventional light-frame construction. For this reason, the code prohibits the use of concrete or masonry walls above the basement. The use of masonry veneer above grade is permitted in accordance with the exceptions. Exception 2 allows one story of masonry veneer above grade (i.e., the basement level) and two stories above grade if the lower story walls are constructed of concrete or masonry, which are designed for seismic effects. Exception 1 allows masonry veneer either two or three stories above the basement of structures classified as Seismic Design Category B under certain conditions that include increasing the amount of wall bracing. It would not be necessary to meet these conditions for masonry veneer on the first story only, since the second exception permits this. Exception 3 allows up to two stories of masonry veneer above grade, provided the added requirements for wall bracing and hold-downs are complied with.

2308.11.3 Framing and connection details. Framing and connection details shall conform to Sections 2308.11.3.1 through 2308.11.3.3.

❖ Sections 2308.11.3.1 through 2308.11.3.3 contain specific connection requirements for Seismic Design Category B or C.

2308.11.3.1 Anchorage. Braced wall lines shall be anchored in accordance with Section 2308.6 at foundations.

❖ Braced wall lines must be anchored to the foundation so that they will not overturn because of a moment created by lateral loads. Anchorage for overturning will nearly always be more critical for the individual shear panel than for the entire wall.

2308.11.3.2 Stepped footings. Where the height of a required braced wall panel extending from foundation to floor above varies more than 4 feet (1219 mm), the following construction shall be used:

1. Where the bottom of the footing is stepped and the lowest floor framing rests directly on a sill bolted to the footings, the sill shall be anchored as required in Section 2308.3.3.
2. Where the lowest floor framing rests directly on a sill bolted to a footing not less than 8 feet (2438 mm) in length along a line of bracing, the line shall be considered to be braced. The double plate of the cripple stud wall beyond the segment of footing extending to the lowest framed floor shall be spliced to the sill plate with metal ties, one on each side of the sill and plate. The metal ties shall not be less than 0.058 inch [1.47 mm (16 galvanized gage)] by 1.5 inches (38 mm) wide by 48 inches (1219 mm) with eight 16d common nails on each side of the splice location (see Figure 2308.11.3.2). The metal tie shall have a minimum yield of 33,000 pounds per square inch (psi) (227 Mpa).
3. Where cripple walls occur between the top of the footing and the lowest floor framing, the bracing requirements for a story shall apply.

❖ The primary objectives of footing design are to provide a level surface for construction of the foundation wall, to provide adequate transfer and distribution of the building loads to the underlying soil, to provide adequate strength to prevent differential settlement of the building and to provide adequate anchorage or mass to resist potential uplift and overturning forces from lateral loads placed on the structure. The most common footing type in residential construction is a continuous concrete spread footing. These concrete footings are recognized in prescriptive footing size tables for most typical conditions. In contrast, special conditions such as stepped footings, which might be required in steeply sloped sites, require special consideration.

FIGURE 2308.11.3.2. See page 23-100.

❖ This figure illustrates the code requirements for stepped footing details (see Section 2308.11.3.2 for further discussion).

2308.11.3.3 Openings in horizontal diaphragms. Openings in horizontal diaphragms with a dimension perpendicular to the joist that is greater than 4 feet (1.2 m) shall be constructed in accordance with the following:

1. Blocking shall be provided beyond headers.
2. Metal ties not less than 0.058 inch [1.47 mm (16 galvanized gage)] by 1.5 inches (38 mm) wide with eight 16d common nails on each side of the header-joist intersection shall be provided (see Figure 2308.11.3.3). The metal ties shall have a minimum yield of 33,000 psi (227 Mpa).

❖ Horizontal diaphragms are floor and roof assemblies that are usually clad with structural wood sheathing panels such as plywood or OSB. Though more complicated and difficult to visualize, lateral forces that are applied to a building from wind or seismic events follow a load path that distributes and transfers shear and overturning forces from the lateral loads. When openings are punched into the diaphragm, it creates the problem of how to transfer the lateral loads around the opening. Another concern is the stiffness of the diaphragm. These provisions are a prescriptive solution for openings not greater than 4 feet (1219 mm) in dimension and provide a general means for a load path in these specific cases in lieu of an engineered design.

FIGURE 2308.11.3.3. See page 23-100.

❖ See the commentary to Section 2308.11.3.3.

2308.12 Additional requirements for conventional construction in Seismic Design Category D or E. Structures of conventional light-frame construction in Seismic Design Category D or E, as determined in Section 1616, shall conform to Sections 2308.12.1 through 2308.12.9, in addition to the requirements for Seismic Design Category B or C in Section 2308.11.

❖ Conventional construction is a type of construction that has been used successfully for many years and relies on standard practice as governed by prescriptive building code requirements. For conditions where the lateral loading on the building is greater, such as in higher seismic areas of the country with Seismic Design Categories D and E, additional prescriptive methods or limitations are considered necessary. The basis for the requirements for Seismic Design Category D or E is the NEHRP Provisions.

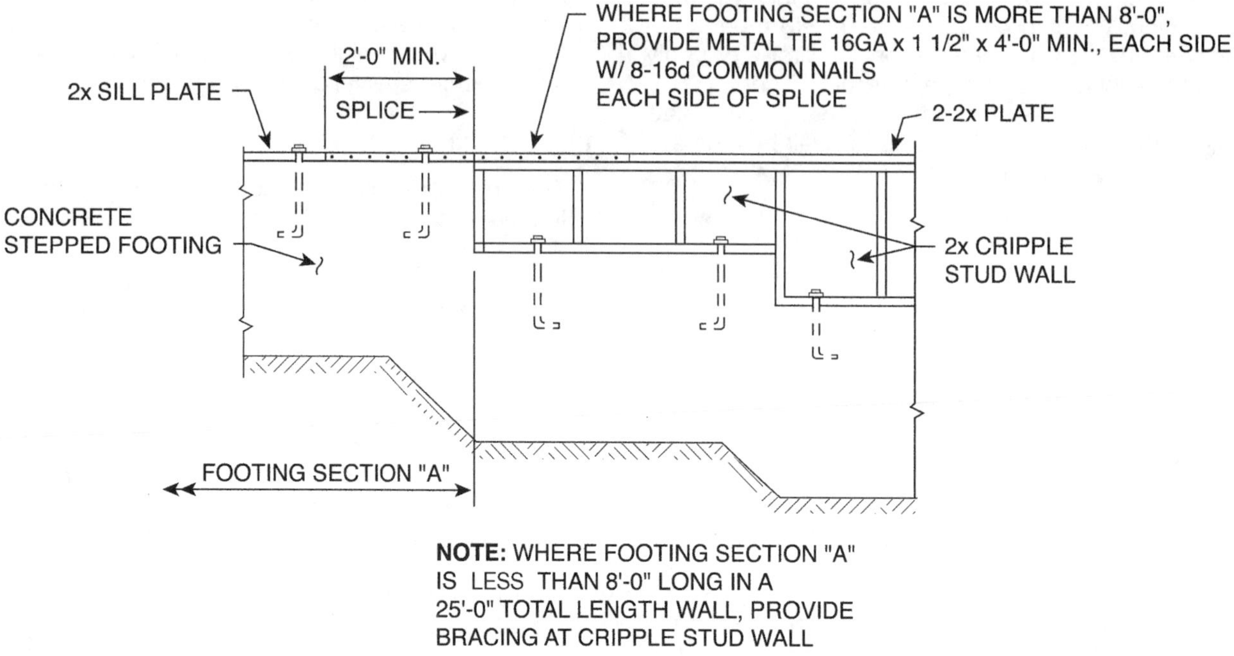

For SI: 1 inch = 25.4 mm, 1 foot = 304.8 mm.

FIGURE 2308.11.3.2
STEPPED FOOTING CONNECTION DETAILS

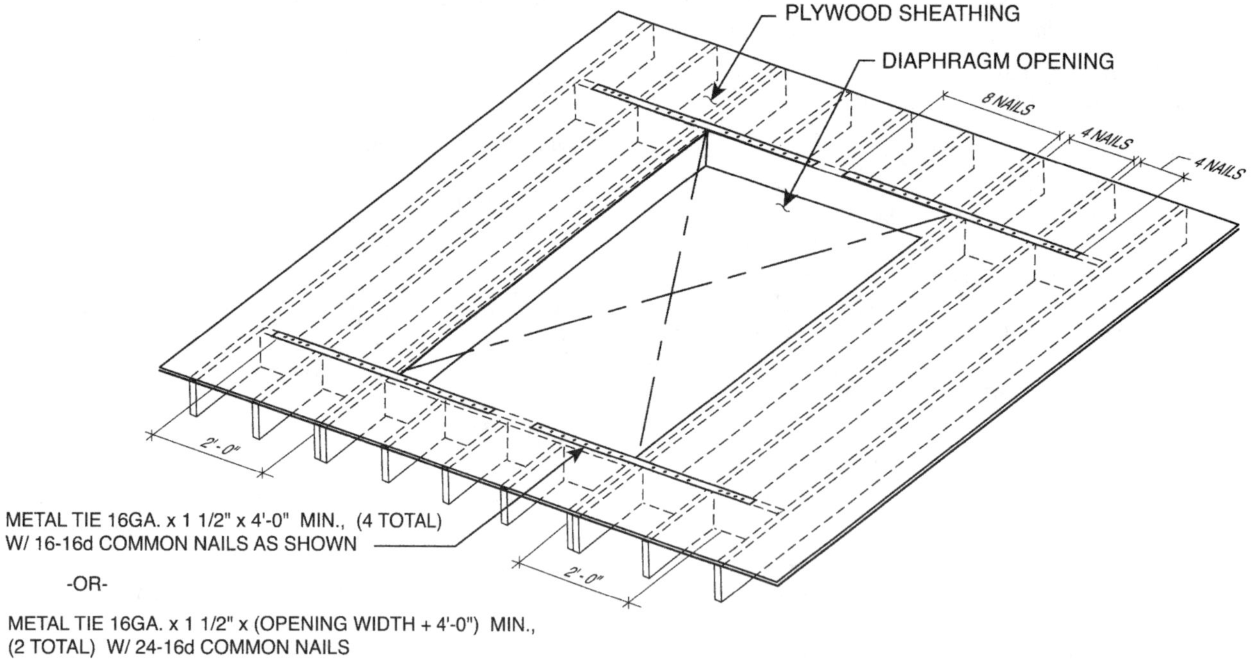

For SI: 1 inch = 25.4 mm, 1 foot = 304.8 mm.

FIGURE 2308.11.3.3
OPENINGS IN HORIZONTAL DIAPHRAGMS

2308.12.1 Number of stories. Structures of conventional light-frame construction shall not exceed one story in height in Seismic Design Category D or E.

Exception: Detached one- and two-family dwellings are permitted to be two stories high in Seismic Design Category D or E.

❖ This section limits the building height where conventional construction is permitted. Essentially the one-story limit on buildings other than detached one- and two-family dwellings requires those buildings to be engineered design.

2308.12.2 Concrete or masonry. Concrete or masonry walls, or masonry veneer shall not extend above the basement.

Exception: Masonry veneer is permitted to be used in the first story above grade in Seismic Design Category D provided the following criteria are met:

1. Type of brace in accordance with Section 2308.9.3 shall be Method 3 and the allowable shear capacity in accordance with Table 2306.4.1 shall be a minimum of 350 plf (5108 N/m) (ASD).
2. The bracing of the first story shall be located at each end and at least every 25 feet (7620 mm) o.c. but not less than 45 percent of the braced wall line.
3. Hold-down connectors shall be provided at the ends of braced walls for the first floor to foundation with an allowable design of 2,100 pounds (1768 kg).
4. Cripple walls shall not be permitted.

❖ Because of their weight, concrete walls, masonry walls and masonry veneer can impose lateral loads from seismic events in excess of the loading considered acceptable for conventional light-frame construction. For this reason, the code prohibits the use of concrete or masonry walls above the basement. The use of masonry veneer one story above grade is permitted for structures classified as Seismic Design Category D in accordance with the exception, provided the requirements for wall bracing and hold-downs are met.

2308.12.3 Braced wall line spacing. Spacing between interior and exterior braced wall lines shall not exceed 25 feet (7620 mm).

❖ In order to maintain the existing bracing element capacities that have been used in conventional construction for years, braced walls have been introduced. They limit the lateral forces supported by the elements by limiting the wind and seismic tributary area and the building area. This concept introduced mandating the maximum spacing between lines of bracing elements and limiting offsets within a line. A series of braced wall panels comprise a braced wall line. A two-dimensional grid of interior and exterior braced wall lines breaks up the building into a series of boxes as illustrated in Figure 2308.9.3. The size of the box is either 25 or 35 feet (7620 or 10 668 mm), depending on seismic design category and wind speed. The 25-foot (7620 mm) dimension is related to the maximum framing spans, which are included in other sections of this chapter.

2308.12.4 Braced wall line sheathing. Braced wall lines shall be braced by one of the types of sheathing prescribed by Table 2308.12.4 as shown in Figure 2308.9.3. The sum of lengths of braced wall panels at each braced wall line shall conform to Table 2308.12.4. Braced wall panels shall be distributed along the length of the braced wall line and start at not more than 8 feet (2438 mm) from each end of the braced wall line. A designed collector shall be provided where the bracing begins more than 8 feet (2438 mm) from each end of a braced wall line. Panel sheathing joints shall occur over studs or blocking. Sheathing shall be fastened to studs and top and bottom plates and at panel edges occurring over blocking. Wall framing to which sheathing used for bracing is applied shall be nominal 2 inch wide (actual $1^1/_2$ inch, 38 mm) or larger members.

Cripple walls having a stud height exceeding 14 inches (356 mm) shall be considered a story for the purpose of this section and shall be braced as required for braced wall lines in accordance with Table 2308.12.4. Where interior braced wall lines occur without a continuous foundation below, the length of parallel exterior cripple wall bracing shall be one and one-half times the lengths required by Table 2308.12.4. Where the cripple wall sheathing type used is Type S-W, and this additional length of bracing cannot be provided, the capacity of Type S-W sheathing shall be increased by reducing the spacing of fasteners along the perimeter of each piece of sheathing to 4 inches (102 mm) o.c.

❖ Braced wall lines are required to be braced in accordance with one of the methods specified in Table 2308.12.4, as illustrated in Figure 2308.9.3. All vertical joints of the panel are required to occur over studs. Horizontal joints are required to occur over blocking equal in size to the studding unless waived by the installation requirements for the specific sheathing materials. A collector or drag strut, which is usually a system of members in light frame construction, collects and transfers loads by tension or compression to the shear-resisting segments of a wall line. In a typical home, special design of chord members may involve some modest detailing of splices at the diaphragm boundary. Compressive forces are rarely a concern when at least a double top plate is used as a collector, particularly when the collector is braced against lateral buckling by attachment to other construction. Therefore, it is typical practice to design the collector and any splices in the collector to resist a tension force as calculated by general engineering procedures.

TABLE 2308.12.4. See page 23-102.

❖ See the commentary to Section 2308.12.4.

2308.12.5 Attachment of sheathing. Fastening of braced wall panel sheathing shall not be less than that prescribed in Table 2308.12.4 or 2304.9.1. Wall sheathing shall not be attached to framing members by adhesives.

❖ The nailing required by Table 2308.12.4 is specified in Note d. The connections in Table 2304.9.1 are the minimum allowed for conventional construction.

TABLE 2308.12.4
WALL BRACING IN SEISMIC DESIGN CATEGORIES D AND E
(Minimum Length of Wall Bracing per each 25 Linear Feet of Braced Wall Line[a])

STORY LOCATION	SHEATHING TYPE[b]	$0.50 \le S_{DS} < 0.75$	$0.75 \le S_{DS} \le 1.00$	$1.00 < S_{DS}$
Top or only story	G-P[d]	14 feet 8 inches	18 feet 8 inches[c]	25 feet 0 inches[c]
	S-W	8 feet 0 inches	9 feet 4 inches[c]	12 feet 0 inches[c]
Story below top story	G-P[d]	NP	NP	NP
	S-W	13 feet 4 inches[c]	17 feet 4 inches[c]	21 feet 4 inches[c]
Bottom story of three stories	G-P[d]	Conventional construction not permitted; conformance with Section 2301.2.1 or 2301.2.2 is required.		
	S-W			

For SI: 1 inch = 25.4 mm, 1 foot = 304.8 mm.

a. Minimum length of panel bracing of one face of wall for S-W sheathing or both faces of wall for G-P sheathing; *h/w* ratio shall not exceed 2:1. For S-W panel bracing of the same material on two faces of the wall, the minimum length is permitted to be one-half the tabulated value but the *h/w* ratio shall not exceed 2:1 and design for uplift is required.

b. G-P = gypsum board, fiberboard, particleboard, lath and plaster, or gypsum sheathing boards; S-W = wood structural panels and diagonal wood sheathing. NP = not permitted.

c. Applies to one- and two-family detached dwellings only.

d. Nailing as specified below shall occur at all panel edges at studs, at top and bottom plates, and, where occurring, at blocking:
For $^1/_2$-inch gypsum board, 5d (0.113 inch diameter) cooler nails at 7 inches on center;
For $^5/_8$-inch gypsum board, No. 11 gage (0.120 inch diameter) at 7 inches on center;
For gypsum sheathing board, $1^3/_4$ inches long by $^7/_{16}$-inch head, diamond point galvanized nails at 4 inches on center;
For gypsum lath, No. 13 gage (0.092 inch) by $1^1/_8$ inches long, $^{19}/_{64}$-inch head, plasterboard at 5 inches on center;
For portland cement plaster, No. 11 gage (0.120 inch) by $1^1/_2$ inches long, $^7/_{16}$- inch head at 6 inches on center;
For fiberboard and particleboard, No. 11 gage (0.120 inch) by $1^1/_2$ inches long, $^7/_{16}$-inch head, galvanized nails at 3 inches on center.

2308.12.6 Irregular structures. Conventional light-frame construction shall not be used in irregular portions of structures in Seismic Design Category D or E. Such irregular portions of structures shall be designed to resist the forces specified in Chapter 16 to the extent such irregular features affect the performance of the conventional framing system. A portion of a structure shall be considered to be irregular where one or more of the conditions described in Items 1 through 6 below are present.

❖ Irregular structures are those that require engineered design in Seismic Design Category D or E because of their unusual shape or discontinuities in the lateral-force-resisting system. These conditions produce torsional response or result in forces considered too high to be addressed prescriptively. When a structure falls within the description of irregular, it is required that either the entire structure or the nonconventional portions be engineered. The design professional is left to judge the extent of the portion to be designed. This often involves design of the nonconforming element, force transfer into the element and a load path from the element to the foundation. A nonconforming portion will sometimes have enough of an impact on the behavior of a structure to warrant that the entire lateral-force-resisting system receive an engineered design.

1. Where exterior braced wall panels are not in one plane vertically from the foundation to the uppermost story in which they are required, the structure shall be considered to be irregular [see Figure 2308.12.6(1)].

   Exception: Floors with cantilevers or setbacks not exceeding four times the nominal depth of the floor joists [see Figure 2308.12.6(2)] are permitted to support braced wall panels provided:

   1. Floor joists are 2 inches by 10 inches (51 mm by 254 mm) or larger and spaced not more than 16 inches (406 mm) o.c.
   2. The ratio of the back span to the cantilever is at least 2:1.
   3. Floor joists at ends of braced wall panels are doubled.
   4. A continuous rim joist is connected to the ends of cantilevered joists. The rim joist is permitted to be spliced using a metal tie not less than 0.058 inch (1.47 mm) (16 galvanized gage) and $1^1/_2$ inches (38 mm) wide fastened with six 16d common nails on each side. The metal tie shall have a minimum yield of 33,000 psi (227 Mpa).
   5. Joists at setbacks or the end of cantilevered joists shall not carry gravity loads from more than a single story having uniform wall and roof loads, nor carry the reactions from headers having a span of 8 feet (2438 mm) or more.

❖ This limit applies when braced wall panels are offset out of plane from floor to floor. In-plane offsets are discussed in another item. Ideally, braced wall panels would always stack above each other from floor to floor with the length stepping down at upper floors as less length of bracing is required.

Because cantilevers and setbacks are very often incorporated into structures, the exception offers rules by which limited cantilevers and setbacks can be considered conventional.

2. Where a section of floor or roof is not laterally supported by braced wall lines on all edges, the structure shall be considered to be irregular [see Figure 2308.12.6(3)].

Exception: Portions of roofs or floors that do not support braced wall panels above are permitted to extend up to 6 feet (1829 mm) beyond a braced wall line [see Figure 2308.12.6(4)].

❖ This limitation applies to open-front structures or portions of structures. The conventional construction bracing concept is based on using braced wall lines to divide a structure into a series of boxes of limited dimension, with the seismic force to each box being limited by the size. The intent is that each box be supported by braced wall lines on all four sides, limiting the amount of torsion that can occur. The exception, which permits portions of roofs or floors to extend past the braced wall line, is intended to permit construction such as porch roofs and bay windows. Walls with no lateral resistance are allowed in areas where braced walls are prohibited.

3. Where the end of a required braced wall panel extends more than 1 foot (305 mm) over an opening in the wall below, the structure shall be considered to be irregular. This requirement is applicable to braced wall panels offset in plane and to braced wall panels offset out of plane as permitted by the exception to Item 1 above in this section [see Figure 2308.12.6(5)].

Exception: Braced wall panels are permitted to extend over an opening not more than 8 feet (2438 mm) in width where the header is a 4-inch by 12-inch (102 mm by 305 mm) or larger member.

❖ This limitation applies when braced wall panels are offset in plane. Ends of braced wall panels supported on window or door headers can be calculated to transfer large vertical reactions to headers that may not be of adequate size to resist these reactions. The exception permits a 1-foot (305 mm) extension of the braced wall panel over a 4-inch by 12-inch [actual $3^1/_2$-inch by $11^1/_2$-inch (89 mm by 286 mm)] header on the basis that the vertical reaction is within a 45-degree (0.78 rad) line of the header support and therefore will not result in critical shear or flexure. All other header conditions require an engineered design. Walls with no lateral resistance are allowed in areas where braced walls are prohibited.

4. Where portions of a floor level are vertically offset such that the framing members on either side of the offset cannot be lapped or tied together in an approved manner, the structure shall be considered to be irregular [see Figure 2308.12.6(6)].

Exception: Framing supported directly by foundations need not be lapped or tied directly together.

❖ This limitation results from observation of damage that is somewhat unique to split-level wood-frame construction. If floors on either side of an offset move in opposite directions due to earthquake or wind loading, the short bearing wall in the middle becomes unstable and vertical support for the upper joists can be lost, resulting in a collapse. If the vertical offset is limited to a dimension equal to or less than the joist depth, then a simple strap tie directly connecting joists on different levels can be provided, and the irregularity eliminated.

5. Where braced wall lines are not perpendicular to each other, the structure shall be considered to be irregular [see Figure 2308.12.6(7)].

❖ This limitation applies to nonperpendicular braced wall lines. When braced wall lines are not perpendicular to each other, further evaluation is needed to determine force distributions and required bracing.

6. Where openings in floor and roof diaphragms having a maximum dimension greater than 50 percent of the distance between lines of bracing or an area greater than 25 percent of the area between orthogonal pairs of braced wall lines are present, the structure shall be considered to be irregular [see Figure 2308.12.6(8)].

❖ This limitation attempts to place a practical limit on openings in floors and roofs. Because stair openings are essential to residential construction and have long been used without any report of life safety hazards resulting, these are felt to be acceptable conventional construction.

FIGURES 2308.12.6(1) through (8). See page 23-104 - 23-107.

❖ For Figures 2308.12.6(1) through (8), see the commentary to Section 2308.12.6.

2308.12.7 Exit facilities. Exterior exit balconies, stairs and similar exit facilities shall be positively anchored to the primary structure at not over 8 feet (2438 mm) o.c. or shall be designed for lateral forces. Such attachment shall not be accomplished by use of toenails or nails subject to withdrawal.

❖ Because of their importance, the exterior means of egress components shown must be anchored to the structure in buildings classified as Seismic Design Category D or E to provide resistance to lateral forces.

2308.12.8 Steel plate washers. Steel plate washers shall be placed between the foundation sill plate and the nut. Such washers shall be a minimum of $^3/_{16}$ inch by 2 inches by 2 inches (4.76 mm by 51 mm by 51mm) in size.

❖ Walls subjected to the lateral loads anticipated in Seismic Design Category D or E, particularly tall walls, will tend to try to tip over. To resist this overturning, the code increases the bearing area on the sill plate at each bolt by stipulating a minimum-size plate washer that spreads the load over a wider area.

2308.12.9 Anchorage in Seismic Design Category E. Steel bolts with a minimum nominal diameter of $^5/_8$ inch (15.9 mm) shall be used in Seismic Design Category E.

❖ Lateral seismic loads on buildings cause the bottom of the walls to slide, and the tendency of sill plate bolts is to rip through the wood. Dowel-type connectors (bolts, nails, screws and pins) rely on metal-to-wood bearing for the transfer of lateral loads. One of the factors that determines dowel bearing strength is the diameter of the bolt. By increasing the minimum diameter of the bolts from ½ inch (12.7 mm) as specified in Section 2308.6, the bearing surface area of the bolt is increased to provide a connection to resist the increased seismic forces of Seismic Design Category E.

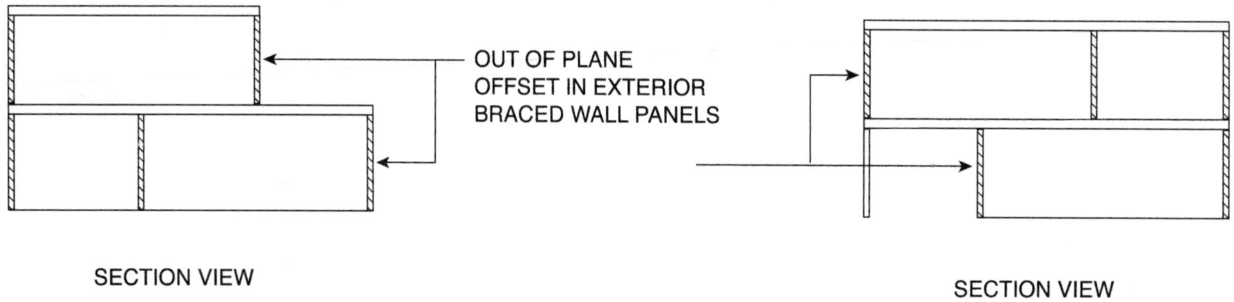

For SI: 1 foot = 304.8 mm.

FIGURE 2308.12.6(1)
BRACED WALL PANELS OUT OF PLANE

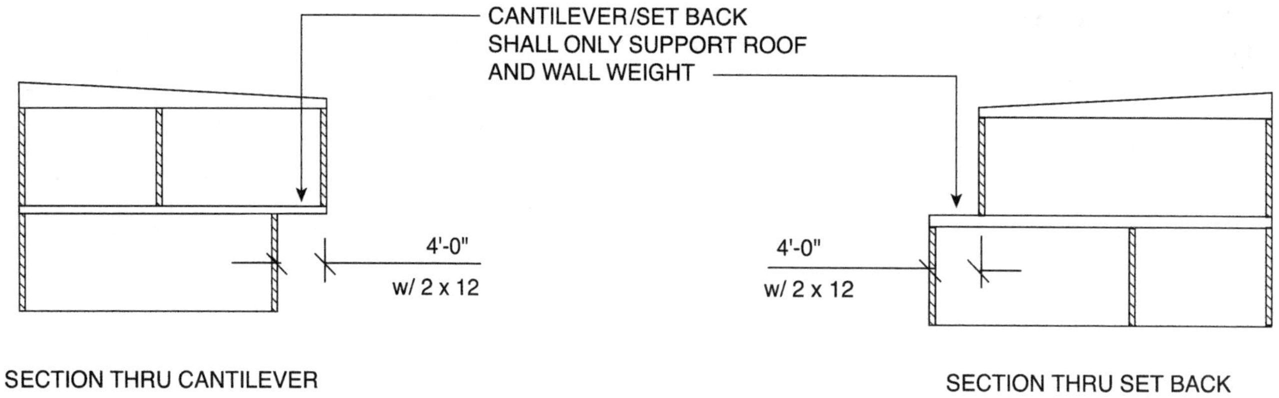

For SI: 1 foot = 304.8 mm.

FIGURE 2308.12.6(2)
BRACED WALL PANELS SUPPORTED BY CANTILEVER OR SET BACK

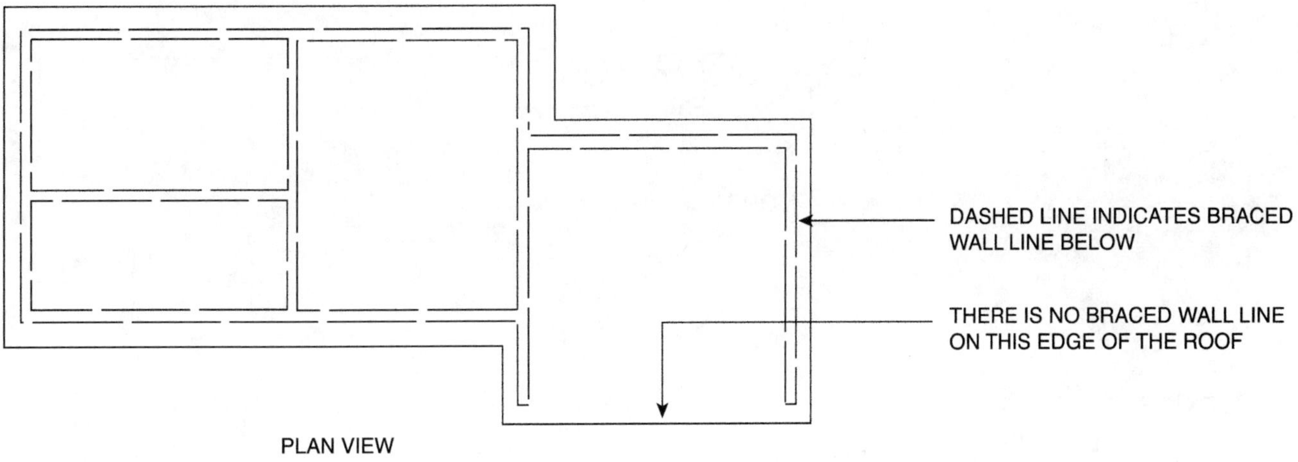
DASHED LINE INDICATES BRACED WALL LINE BELOW
THERE IS NO BRACED WALL LINE ON THIS EDGE OF THE ROOF
PLAN VIEW

FIGURE 2308.12.6(3)
FLOOR OR ROOF NOT SUPPORTED ON ALL EDGES

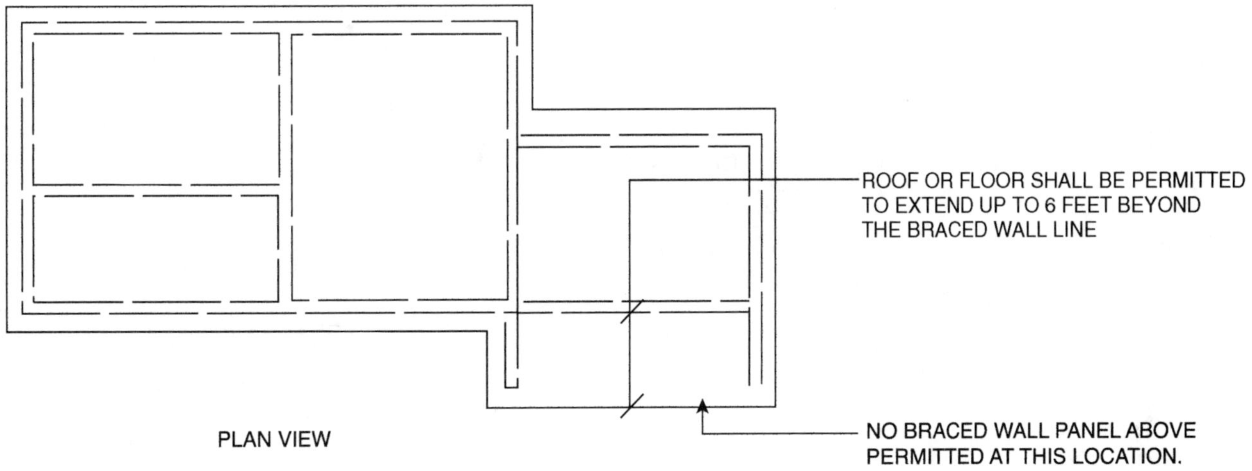
ROOF OR FLOOR SHALL BE PERMITTED TO EXTEND UP TO 6 FEET BEYOND THE BRACED WALL LINE
NO BRACED WALL PANEL ABOVE PERMITTED AT THIS LOCATION.
PLAN VIEW

For SI: 1 foot = 304.8 mm.

FIGURE 2308.12.6(4)
ROOF OR FLOOR EXTENSION BEYOND BRACED WALL LINE

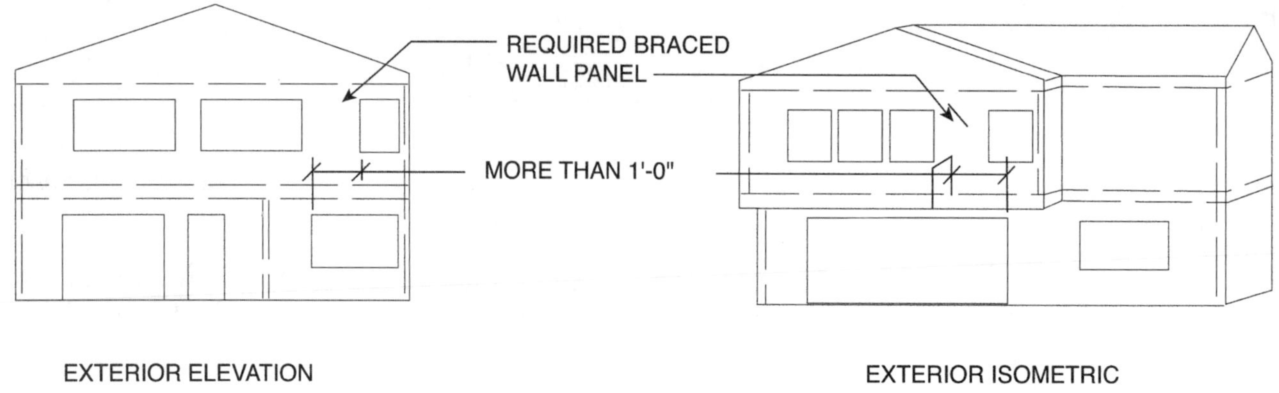

For SI: 1 foot = 304.8 mm.

**FIGURE 2308.12.6(5)
BRACED WALL PANEL EXTENSION OVER OPENING**

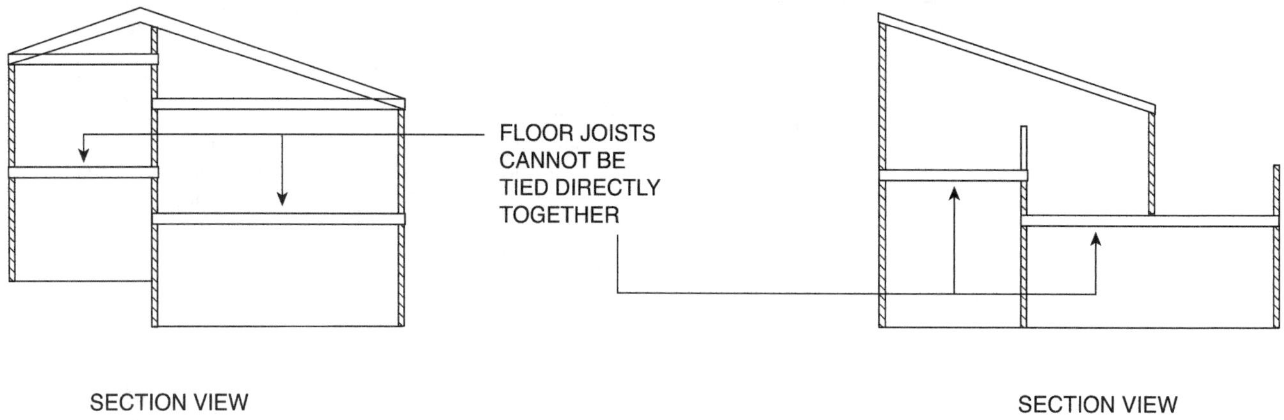

**FIGURE 2308.12.6(6)
PORTIONS OF FLOOR LEVEL OFFSET VERTICALLY**

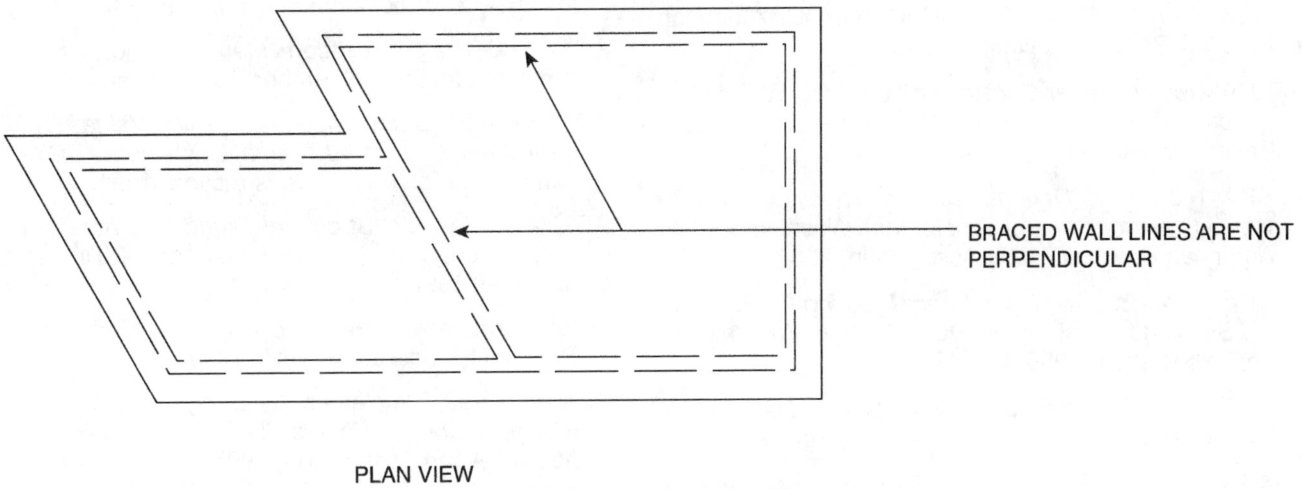

FIGURE 2308.12.6(7)
BRACED WALL LINES NOT PERPENDICULAR

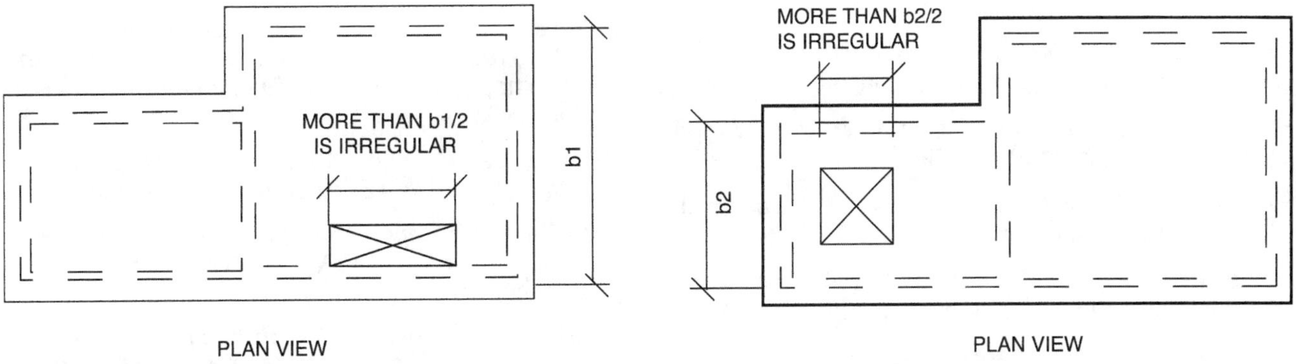

FIGURE 2308.12.6(8)
OPENING LIMITATIONS FOR FLOOR AND ROOF DIAPHRAGMS

Bibliography

The following resource materials are referenced in this chapter or are relevant to the subject matter addressed in this chapter.

AF&PA, *Allowable Stress Design Manual for Wood Construction (ASD Manual).* Washington, DC: American Forest & Paper Association.

AF&PA, *Heavy Timber Construction Details—Wood Construction Data 5.* Washington, DC: American Forest & Paper Association.

AF&PA NDS—01, *National Design Specification® for Wood Construction and Supplement.* Washington, DC: American Forest & Paper Association, 2001.

AF&PA No.4—89, *Plank and Beam Framing for Residential Buildings.* Washington, DC: American Forest & Paper Association, 1989.

AF&PA, *Span Tables for Joists and Rafters.* Washington, DC: American Forest & Paper Association, 1993.

AF&PA Technical Report 7—87, *Basic Requirements for Permanent Wood Foundation System.* Washington, DC: American Forest & Paper Association, 1987.

AF&PA, *Wood Frame Construction Manual for One- and Two-Family Dwellings (WFCM).* Washington, D.C. American Forest and Paper Association, 2001.

AF&PA/ASCE 16— *Supplements to the LRFD and ASD Manuals.* Reston, VA: American Society of Civil Engineers,1995.

AF&PA/ASCE 16—95, *Standard for Load and Resistance Factor Design (LRFD) for Engineered Wood Construction.* Reston, VA: American Society of Civil Engineers,1995.

AHA A135.4—95, *Basic Hardboard.* Palatine, IL: American Hardwood Association, 1995.

AHA A135.5—95, *Prefinished Hardboard Paneling.* Palatine, IL: American Hardwood Association, 1995.

AHA A135.6—98, *Hardboard Siding.* Palatine, IL: American Hardwood Association, 1998.

AHA A194.1—85, *Cellulosic Fiber Board.* Palatine, IL: American Hardwood Association, 1985.

AITC 104—84, *Typical Construction Details.* Englewood, CO: American Institute of Timber Construction, 1984.

AITC 110—01, *Standard Appearance Grades for Structural Glued Laminated Timber.* Englewood, CO: American Institute of Timber Construction, 2001.

AITC 112—93, *Standard for Tongue-and-groove Heavy Timber Roof Decking.* Englewood, CO: American Institute of Timber Construction, 1993.

AITC 113—93, *Standard for Dimensions of Structural Glued Laminated Timber.* Englewood, CO: American Institute of Timber Construction, 1993.

AITC 117—01, *Structural Glued-Laminated Timber.* Englewood, CO: American Institute of Timber Construction, 2001.

AITC 119—96, *Standard Specifications for Hardwood Glued-Laminated Timber.* Englewood, CO: American Institute of Timber Construction, 1996.

AITC 200—92, *Inspection Manual.* Englewood, CO: American Institute of Timber Construction, 1992.

AITC 500—91, *Determination of Design Values for Structural Glued Laminated Timber.* Englewood, CO: American Institute of Timber Construction, 1991.

AITC A 190.1—92, *Structural Glued-Laminated Timber.* Englewood, CO: American Institute of Timber Construction, 1992.

ANSI A 208.1—99, *Particleboard.* New York: American National Standards Institute, 1999.

APA PDS—90, Supplement 1—*Design and Fabrication of Plywood Curved Panels.* Tacoma, WA: APA—Engineered Wood Association, 1995.

APA PDS—90, Supplement 3—*Design and Fabrication of Plywood Stressed-Skin Panels.* Tacoma, WA: APA—Engineered Wood Association, 1996.

APA PDS—90, Supplement 4—*Design and Fabrication of Plywood Sandwich Panels.* Tacoma, WA: APA—Engineered Wood Association, 1993.

APA PDS—92, Supplement 2—*Design and Fabrication of Plywood-Lumber Beams.* Tacoma, WA: APA—Engineered Wood Association, 1998.

APA PDS—95, Supplement 5—*Design and Fabrication of All-Plywood Beams.* Tacoma, WA: APA—Engineered Wood Association, 1995.

APA PDS—97, *Plywood Design Specification.* Tacoma, WA: APA—Engineered Wood Association, 1997.

APA Report 157—95, *Wood Structural Panel Shear Walls with Gypsum Wallboard and Window/Door Openings.* Tacoma, WA: APA—The Engineered Wood Association, 1995.

APA Research Report 138, *Plywood Diaphragms.* Tacoma, WA: APA—Engineered Wood Association.

APA Research Report 154, *Wood Structural Panel Shear Wall.* Tacoma, WA: APA—Engineered Wood Association.

ASCE 7—02, *Minimum Design Loads for Buildings and Other Structures.* Reston, VA: American Society of Civil Engineers, 2002.

ASTM C 208-95, *Specification for Celluosic Fiber Insulating Board.* West Conshohocken, PA: ASTM International, 1995.

ASTM D 2898-94(1999), *Test Methods for Accelerated Weathering of Fire-Retardant-Treated Wood for Fire Testing.* West Conshohocken, PA: ASTM International, 1994.

ASTM D 3201-94(1998), *Test Method for Hygroscopic Properties of Fire-Retardant-Treated Wood and Wood-base Products*. West Conshohocken, PA: ASTM International, 1994.

ASTM D 3737-01B, *Practice for Establishing Allowable Properties for Structural Glued Laminated Timber (Glulam)*. West Conshohocken, PA: ASTM International, 2001.

ASTM D 5055-00, *Specification for Establishing and Monitoring Structural Capacities of Prefabricated Wood I-Joists*. West Conshohocken, PA: ASTM International, 2000.

AWPA C1—00, *All Timber Products-Preservative Treatment by Pressure Processes.* Grandbury, TX: American Wood-Preservers' Association, 2000.

AWPA C2—01, *Lumber, Timber, Bridge Ties and Mine Ties-Preservatives Treatment by Pressure Processes.* Grandbury, TX: American Wood-Preservers' Association, 2001.

AWPA C9—00, *Plywood-Preservative Treatment by Pressure Processes.* Grandbury, TX: American Wood-Preservers' Association, 2000.

AWPA C15—00, *Wood for Commercial-Residential Construction Preservative Treatment.* Grandbury, TX: American Wood-Preservers' Association, 2000.

AWPA C22—96, *Lumber and Plywood for Permanent Wood Foundations-Preservative Treatment by Pressure Processes.* Grandbury, TX: American Wood-Preservers' Association, 1996.

AWPA P1/13—01, *Standard for Coal Tar Creosote for Land and Fresh Water and Marine (Costal Water) Use.* Grandbury, TX: American Wood-Preservers' Association, 2001.

AWPA P2—01, *Standard for Creosote Solutions.* Grandbury, TX: American Wood-Preservers' Association, 2001.

AWPA P5—01, *Standard for Waterborne Preservatives.* Grandbury, TX: American Wood-Preservers' Association, 2001.

AWPA P8—01, *Standard for Oil-borne Preservatives.* Grandbury, TX: American Wood-Preservers' Association, 2001.

AWPA P9—01, *Standard for Solvents and Formulations for Organic Preservative Systems.* Grandbury, TX: American Wood-Preservers' Association, 2001.

DOC PS 1—95, *Construction and Industrial Plywood.* Gaithersburg, MD: U.S. Department of Commerce, 1995.

DOC PS 2—92, *Performance Standard for Wood-Based Structural-Use Panels.* Gaithersburg, MD: U.S. Department of Commerce, 1992.

DOC PS 20—99, *American Softwood Lumber Standard.* Gaithersburg, MD: U.S. Department of Commerce, 1999.

EWS R540—96, *Builders Tips: Proper Storage and Handling of Glulam Beams.* Tacoma, WA: APA—Engineered Wood Association, 1996.

EWS S475—99, *Glued Laminated Beam Design Tables.* Tacoma, WA: APA—Engineered Wood Association, 1999.

EWS S560—99, *Field Notching and Drilling of Glued Laminated Timber Beams.* Tacoma, WA: APA—Engineered Wood Association, 1999.

EWS T300—99, *Glulam Connection Details.* Tacoma, WA: APA—Engineered Wood Association, 1999.

EWS X440—00, *Product and Application Guide: Gluman.* Tacoma, WA: APA—Engineered Wood Association, 2000.

EWS X450—97, *Glulam in Residential Construction.* Tacoma, WA: APA—Engineered Wood Association, 1997.

Henry, John R. "A Better Way to Think: LRFD for Engineered Wood Construction." *Structural Engineer*, February 2003, pp 20-25.

HPVA HP-1—00, *The American National Standard for Hardwood and Decorative Plywood.* Reston, VA: Hardwood Plywood Veneer Association, 2000.

NEHRP 2000, *Recommended Provisions for New Buildings and Other Structures Commentary.* Washington, DC: National Earthquake Hazards Reduction Program (FEMA Document #303), 2000.

NEHRP 2000, *Recommended Provisions for Seismic Regulations for New Buildings and Other Structures.* Washington, DC: National Earthquake Hazards Reduction Program (FEMA Document #302), 2000.

SBCCI SSTD 10—99, *Standard for Hurricane Resistant Construction.* Birmingham, AL: Southern Building Code Congress International, 1999.

TPI 1—02, *National Design Standard for Metal Plate Connected Wood Trusses Construction.* New York: American National Standards Institute, 2002.

Wood Handbook, Wood as an Engineering Material, 1999 Washington, DC: USDA Forest Products Laboratory, U.S. Government Printing Office, 1999.

Chapter 24: Glass and Glazing

General Comments

Chapter 24 includes design provisions and quality requirements for glazing consisting of glass and, to a limited extent, light-transmitting plastics.

Section 2401 contains the general contents and scope of this chapter.

Section 2402 provides meanings for words and phrases necessary for a complete understanding of the requirements of this chapter.

Section 2403 establishes the identification requirements for safety and nonsafety glazing, glass support and framing systems, glass thicknesses and requirements for glass at interior locations. This section also addresses special provisions for louvered and jalousie windows.

Section 2404 provides the design requirements and limitations for vertical glass exposed to wind loads, and sloped glass exposed to wind, snow and dead loads. Design factors for the more common types of glass are included in the design tables.

Section 2405 defines the allowable glazing types and related requirements for sloped glazing and skylights.

Section 2406 provides a detailed listing of the applications where safety glazing is required. Test standards category classifications for safety glazing are also included in this section.

Section 2407 considers glass used as either structural balustrades or in-fill panels in handrail and guardrail systems.

Section 2408 clarifies the special requirements for glass areas exposed to extraordinary impact from people. These are applicable to walls and doors in certain athletic facilities.

Section 2409 provides standards, materials and design criteria for glass installed in walking surfaces, including areas such as floors and sidewalks.

As is true with other chapters in the code that deal with building materials and components, this chapter evolved from the considerations of safety, public welfare and the integrity of a building's construction. There is an ever-growing demand from building owners and architects to incorporate larger quantities of glass in buildings. This in turn increases the concerns for such things as safety and energy conservation. The glass industry is being called on to meet these demands. Model building codes are changing almost daily to stay abreast of the ever-changing technology being employed in the glass industry.

Recognizing the risks should sloped glass fail, the model code groups and the glass and glazing industry have developed changes that only allow safety glazing materials over occupiable areas. Complying materials include laminated, fully tempered and wired glasses and acrylic and polycarbonate plastics. Except in residential occupancies, fully tempered glass is required to have a screen below the glass.

In addition to the concern for overhead glazing, a need developed to codify and address glazing materials in locations that will be exposed to human impact including, but not necessarily limited to: sliding patio doors; side-swinging storm and primary use doors; tub and shower enclosures; and doors and certain fixed and operable glass panels. Building officials and the glass industry joined together to develop and enact safety glazing requirements to fill this need.

As a result, safety glazing laws have been adopted in many states. Based on the recommendation of the Consumer Product Safety Commission (CPSC), a federal law was enacted in 1977 that mandated safety glazing in all areas that might reasonably be exposed to impact from people. The requirements of this federal law were similar to the laws adopted by these states.

This chapter also addresses the design and selection of glass to withstand wind loads, snow loads, dead loads and other imposed loads.

To assist in evaluating the requirements for glass in buildings, properties of the various types are listed below:

- **Annealed glass**—has been subjected to the annealing process, which consists of controlled cooling in a cooling or annealing oven. Types of annealed glass include float, plate and sheet.
  - **Float glass**—is produced by floating a continuous ribbon of molten glass on molten tin. As the glass flows downstream, the temperature of the glass and tin is reduced. When the glass ribbon solidifies, it is lifted from the tin onto rollers and slowly cooled in a controlled cooling oven or annealing lehr. There is no further treatment of the glass. Most clear, gray, bronze and green-tinted glass for architectural use is presently produced by the float process.
  - **Plate glass**—is produced by grinding and polishing rough glass blanks. The glass blanks are formed by drawing semimolten glass through textured rollers, which are then fed through an annealing lehr prior to grinding and polishing. It should be noted that there is limited domestic production of plate glass material.
  - **Sheet glass**—is manufactured by continuously drawing molten glass from a bath, horizontally or vertically over rollers, through an annealing lehr. Because of the manufacturing process, sheet glass has an inherent waviness characteristic, and is, therefore, not typically used in applications where good optical quality is desired, such

as architectural glazing, reflective glazing or show windows. The most common use of sheet glass is in residential applications.

- **Tempered or fully tempered glass**—is annealed glass that has been heated and quenched in a controlled operation to provide a high level of surface compression. For most types of loading conditions, it has approximately four times the strength of regular annealed glass. Tempered glass breaks into small, relatively harmless fragments, and is used as safety glazing material. After tempering, this glass cannot be cut, drilled or otherwise altered without breaking.
- **Heat-strengthened glass**—is produced in much the same way as tempered glass, except the level of surface compression is less. Heat-strengthened glass has approximately twice the strength of regular annealed glass and breaks into relatively large pieces. It is not a safety glazing material.
- **Laminated glass**—is a sandwich of two or more plies of glass bonded together with resilient plastic interlayers, usually polyvinyl butyral. The glass plies may be regular, tempered or heat-strengthened glass. The plastic interlayer thickness is generally 0.030 inch (0.76 mm), but may range up to 0.090 inch (2.3 mm) for some applications. When this glass breaks, the fragments are held together by the plastic interlayer.
- **Wired glass**—is a monolithic glass with wire mesh embedded in approximately the center of the glass thickness. Coupled with a suitable framing system, some wire glass is fire resistant. For low levels of impact, the wire in the glass will retain the broken fragments. However, the use of wire glass as safety glazing in specific hazardous locations is limited to use in fire-resistance-rated assemblies in other than Group E occupancies, except where the wire glass passes the impact test requirements of CPSC 16 CFR, Part 1201.

  See Section 2406.1.2 for impact test requirements. The surface of wire glass may be smooth or textured. Wire glass cannot be tempered or heat strengthened.
- **Insulating glass**—is a factory-assembled glazing unit consisting of two or more panes of glass separated by airspaces. The edges of the finished glazing unit, including the airspaces, are hermetically sealed. The entrapped air is removed, creating a vacuum, which is necessary to prevent condensation within the sealed airspaces.

  The airspaces between the glass can be filled with specialized gas to add to the energy efficiency of the glazing unit. The panes of glass may be any type, except for patterned glasses with deep patterns.
- **Patterned glass**—has a texture or pattern embossed on one or both surfaces during the manufacturing process. Some glass with shallow patterns can be tempered and heat strengthened.
- **Spandrel glass**—is manufactured from either float glass or plate glass that has an opaque coating of fired-on ceramic glaze on the interior surface (or room side). This glass is typically heat strengthened or tempered to resist the thermal stresses (movement) caused by the high absorption of solar radiation by the opaque coating.

Purpose

The primary purpose of this chapter is to provide information to establish the adequacy of glazing from the standpoint of life safety and performance. The sections on glass in handrails and guards, glazing in athletic facilities and safety glazing dictate a performance standard that will minimize risks to a level consistent with other provisions of the code.

SECTION 2401 GENERAL

2401.1 Scope. The provisions of this chapter shall govern the materials, design, construction and quality of glass, light-transmitting ceramic and light-transmitting plastic panels for exterior and interior use in both vertical and sloped applications in buildings and structures.

❖ This chapter establishes regulations for glass and glazing used in buildings and structures that, when installed, are subjected to wind loads, snow loads and dead loads. Engineering and design requirements in the form of tables, formulas and design loads are contained in this chapter. Additional structural requirements are found in Chapter 16.

A second concern of this chapter is glass and glazing used in areas where likely to be broken by human contact. Requirements for identification of safety glass, human impact loads, identification of hazardous locations and the types of glass that are allowed in railings and walls are included in this chapter.

This chapter details certain prescriptive and performance requirements for glazing used in buildings, including thickness limitations, design of glass supports and allowable sizes based on wind, snow and dead loads. The limitations on glass used in sloped glazing and areas exposed to human impact are also included. The information contained in this chapter on light-transmitting plastics is limited. For additional information on plastic materials, referenced standards and other code requirements for plastics, refer to Chapter 26.

2401.2 Glazing replacement. The installation of replacement glass shall be as required for new installations.

❖ Any glazing installed in existing construction is required to meet all the requirements and standards of the code. This includes installing new glass in an existing window, door or other type of opening, even where the glass being replaced did not comply with the standards of the code.

Relocating any type of manufactured unit containing glass, such as a door or window, from one location in an existing building or structure to another location in the same building or structure, or even a different building or structure, is to be considered "new glazing," even though the unit containing the glass is existing.

SECTION 2402 DEFINITIONS

2402.1 Definitions. The following words and terms shall, for the purposes of this chapter and as used elsewhere in this code, have the meanings shown herein.

❖ No attempt has been made in the code or this chapter to define commonly used words or terms that are used in accordance with their established and accepted dictionary meanings, except where a word or term has been loosely used and it is necessary to define its meaning to avoid misunderstanding. For compatibility, when one of the *International Codes*® is referenced to establish the meaning of a word, term, phrase or abbreviation, the referenced code must bear the same publication date as the *International Building Code*® (IBC®) to which it is applied.

Definitions of terms can help in the understanding and application of code requirements. The purpose for including these definitions in this chapter is to provide more convenient access without having to refer back to Chapter 2. For convenience, these terms are also listed in Chapter 2 with a cross reference to this section. The use and application of all defined terms, including those defined herein, are set forth in Chapter 2.

DALLE GLASS. A decorative composite glazing material made of individual pieces of glass that are embedded in a cast matrix of concrete or epoxy.

❖ Dalle glass is a form of "stained" or "decorative" glass, and is more commonly known as "faceted" glass. Dalle glass is an outdated term and is rarely used. It can be prefabricated in panels that are set into framed openings, or it can be constructed in place, similar to the way a mason would construct a single wythe masonry wall. The actual glazing can be heavy pieces of flat or textured glass, irregular-shaped chunks of glass, glass shapes or translucent minerals cut to form geometric patterns or artistic scenes. Knowing that dalle glass or faceted glass is a type of decorative glass is necessary because decorative glass is not required to be safety glazing in hazardous locations as required in Section 2406.3 (see Section 2406.3.1, Exception 2, for specific exceptions).

DECORATIVE GLASS. A carved, leaded or Dalle glass or glazing material whose purpose is decorative or artistic, not functional; whose coloring, texture or other design qualities or components cannot be removed without destroying the glazing material and whose surface, or assembly into which it is incorporated, is divided into segments.

❖ This definition is needed to understand and properly apply Section 2406.3.1, Exception 2. Decorative glass is not to be included in the requirements for specific hazardous locations, and thus is not required to be safety glazing.

The code uses the terms "glass" and "glazing" interchangeably as if there is no difference between the two. This can, and quite often does, cause confusion when talking about "safety glazing" or "safety glass."

Glass is a hard, brittle, inorganic material that is manufactured from a mixture of silica, flux and a stabilizer to form a transparent, translucent or opaque material. While in its molten (semiliquid) form, it can be blown, drawn, rolled, pressed or cast into a variety of shapes. Most of the glass referenced in this code is in sheet form. Glass, safety or not, as well as other materials can be used as glazing.

Glazing is to furnish or fit with glass, such as to install

glass in an opening in a door. Glass installed in an exterior wall would be properly identified as a glazed opening, or more commonly, a window. An enclosure around a shower can be constructed of many materials, but it is considered to be a glazed enclosure if the enclosing material is glass. This code does permit the limited use of some plastics as glazing materials.

SECTION 2403 GENERAL REQUIREMENTS FOR GLASS

2403.1 Identification. Each pane shall bear the manufacturer's label designating the type and thickness of the glass or glazing material. The identification shall not be omitted unless approved and an affidavit is furnished by the glazing contractor certifying that each light is glazed in accordance with approved construction documents that comply with the provisions of this chapter. Safety glazing shall be identified in accordance with Section 2406.2.

Each pane of tempered glass, except tempered spandrel glass, shall be permanently identified by the manufacturer. The identification label shall be acid etched, sand blasted, ceramic fired, embossed or shall be of a type that once applied cannot be removed without being destroyed.

Tempered spandrel glass shall be provided with a removable paper marking by the manufacturer.

❖ Basic marking for identifying type and thickness of glass, as well as requirements and precautionary statements regarding the glass support system, are prescribed in this section. Detailed performance requirements are addressed in other sections.

Glass is a product that is manufactured and fabricated as assemblies at locations other than the building site. Identification of glass and glazing with a label bearing the type and thickness is needed by the building inspector to perform a visual inspection (see Figure 2403.1). The inspection is needed to verify that the glazing is installed in the proper location in the building in compliance with this chapter for hazardous locations and structural loads. The code requires that each light, which is an individual pane of glass, be labeled. Only tempered glass is required by the code to be permanently labeled. Laminated, annealed, float and spandrel, unlike tempered glass, can be cut and fabricated without fracturing. Laminated glass is stockpiled in standard-size sheets by manufacturers. It consists of two or more sheets of glass held together by an intervening layer of plastic material. The stock sheets are labeled by the manufacturer in one corner, which is cut off in the fabrication process. A permanent label is therefore impractical, and the building official relies on an affidavit from the installer of the glazing.

The glazing contractor is responsible for the final installation of glass and glazing in the building, and must submit, with the approval of the building official, an affidavit as an alternate to labeling. The affidavit must certify that each light is glazed under the approved plans and specifications. The affidavit relieves the building inspector of the responsibility for inspecting the glass and glazing for type and thickness. The affidavit does not relieve the plans examiner from the responsibility of reviewing the plans and specifications for compliance with the code. An affidavit may not be submitted for tempered glass that must be permanently identified in compliance with the requirements of this section.

Only tempered glass is required by the code to be permanently identified; the identification is to be visible after installation. Tempered glass, both heat strengthened and fully tempered, is produced by reheating and rapidly cooling annealed glass. Heat-strengthened and fully tempered glass have increased mechanical strength and resistance to thermal stresses. Fully tem-

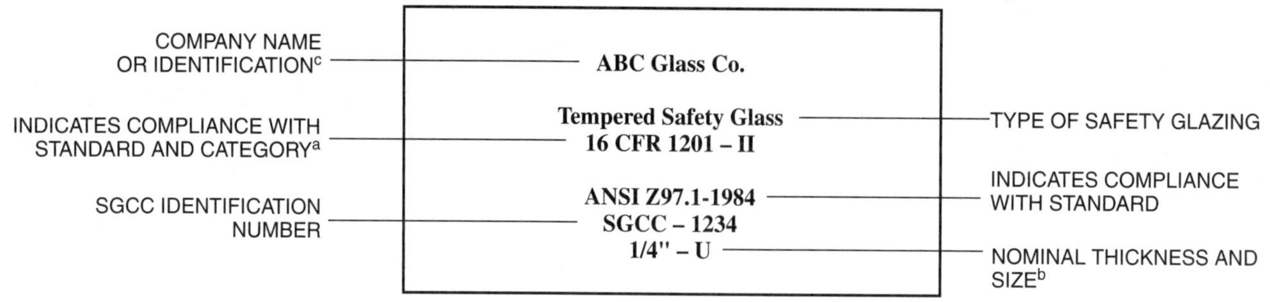

For SI: 1 inch = 25.4 mm, 1 square foot = 0.0929 m^2.

a. "I" after "16 CFR 1201" indicates sizes 9 square feet and less are covered; "II" indicates all sizes are covered.
b. "U" indicates all sizes are covered; "L" indicates only sizes listed by the Safety Glazing Certification Council (SGCC) are covered.
c. Numbers or letters for date coding and plant identification may also be included.

Figure 2403.1
EXAMPLE IDENTIFYING MARK

pered glass is twice as strong as heat-strengthened glass. Fully tempered glass will pulverize into innumerable small, cube-like fragments when broken at any point. Heat-strengthened glass breaks similarly to annealed glass. Heat-strengthened and fully tempered glass cannot be altered after fabrication. Tempered glass cannot be cut, drilled, ground or polished without fracture. The manufacturer of tempered glass must furnish the exact size and shape.

Permanent identification of spandrel glass is not required by the code, since a permanent label may alter the performance of the glass and cause breakage. Spandrel glass is an opaque glass that is heat strengthened through the process of fire fusing an opaque ceramic color to the interior surface of the glass. Spandrel glass is used as exterior windows and curtain walls to conceal internal construction and to provide solar control and energy conservation. An exposed permanent label on tinted and reflective spandrel glass could change the coefficient of expansion of the glass. The label reacts differently to the heating and cooling effects of the atmosphere than the coated glass surface, which can cause increased incidence of glass breakage. The code, therefore, permits a removable paper label for identification of spandrel glass.

2403.2 Glass supports. Where one or more sides of any pane of glass are not firmly supported, or are subjected to unusual load conditions, detailed construction documents, detailed shop drawings and analysis or test data assuring safe performance for the specific installation shall be prepared by a registered design professional.

❖ Most glass in buildings is firmly supported on all four edges. Tables 2404.1 and 2404.2 and Figures 2404(1) through 2404(12) apply to this condition. With many contemporary designs, glass is installed with support at the top and bottom edges only and, in some cases, with support on three edges. The free edges usually abut adjacent glass panels. When one or more edges of the glass are unsupported, the resistance to wind load is reduced. In this case, or in other cases where glass is subjected to unusual loading conditions, the code requires special engineering analysis and detailed construction documents to demonstrate compliance.

2403.3 Framing. To be considered firmly supported, the framing members for each individual pane of glass shall be designed so the deflection of the edge of the glass perpendicular to the glass pane shall not exceed $^1/_{175}$ of the glass edge length or $^3/_4$ inch (19.1 mm), whichever is less, when subjected to the larger of the positive or negative load where loads are combined as specified in Section 1605.

❖ Framing for glazing systems includes, but is not necessarily limited to: aluminum storefront framing; hollow metal framing; job-fabricated custom wood framing; factory-manufactured window units constructed of wood, metal, vinyl or a combination of these materials, most of which are industry standards acknowledged and approved by the code. Regardless of the framing system, it is to comply with the structural requirements of Chapter 16.

The charts, tables and equations provided throughout this chapter are for calculating glass size based on rectangular shapes with all edges firmly supported by any of a variety of the above-listed framing systems. It is important to note that "all edges" of each piece of glass must be supported within the parameters described in this section and elsewhere in the code. If not, detailed engineering analysis must be done as indicated in Section 2403.2.

2403.4 Interior glazed areas. Where interior glazing is installed adjacent to a walking surface, the differential deflection of two adjacent unsupported edges shall not be greater than the thickness of the panels when a force of 50 pounds per linear foot (plf) (730 N/m) is applied horizontally to one panel at any point up to 42 inches (1067 mm) above the walking surface.

❖ This provision addresses the installation of glass panels in interior locations adjacent to walking surfaces, such as storefronts within covered mall buildings. Unlike exterior, interior installations, do not require the placement of mullions between adjacent panels to provide weather resistance. Without intermediate mullions, individual glass panels are free to move independently of each other. When only one panel is subjected to horizontal pressure, the deflection of the panel may cause gaps between adjacent panels not subjected to the same horizontal pressure. These gaps pose a potential pinching hazard to anyone who may contact the glass panels and cause the differential deflection, especially children. This hazard is greater when the glass panels are not mounted in the same plane and the meeting edges form an angle.

The design criteria provided in this section limit the use of glass panels to those that will not deflect more than the thickness of the glass panel itself, when the specified force is applied at a height up to 42 inches (1067 mm) above the walking surface. This limitation protects against the occurrence of gaps between adjacent panels that would otherwise create a pinching hazard. The 42-inch (1067 mm) dimension represents the height range in which a person's body will likely contact the glass. The load must be assumed to occur at the point anywhere within the 42-inch (1067 mm) range that will cause the greatest differential deflection between the glass panels.

2403.5 Louvered windows or jalousies. Float, wired and patterned glass in louvered windows and jalousies shall be no thinner than nominal $^3/_{16}$ inch (4.8 mm) and no longer than 48 inches (1219 mm). Exposed glass edges shall be smooth.

Wired glass with wire exposed on longitudinal edges shall not be used in louvered windows or jalousies.

Where other glass types are used, the design shall be submitted to the building official for approval.

❖ Louvered and jalousie windows consist of a series of overlapping horizontal glass louvers that pivot simultaneously in the window frame. During opening, the bottom edge of each louver swings toward the exterior and the top edge swings toward the interior. The glass panels are supported on two vertical sides with the horizontal edges of the glass exposed.

For safety reasons, the exposed edges of the louvered panels are required to be smooth; thus, wire glass panels, if used, must have no exposed projecting wire. To provide structural strength, the panels are limited to a minimum thickness of $^3/_{16}$ inch (4.8 mm) and a maximum length of 48 inches (1219 mm).

Louvered and jalousie windows are exempt from safety glazing requirements in all applications, including those where a flat pane of glass is otherwise required to be safety glass. This exemption is based on records that show the injuries associated with this use of glass are primarily from persons impacting the glass edge with no cutting or piercing injuries resulting from glass breakage. Safety glass would not have an effect on this type of injury. There are also practical production reasons associated with fabricating safety glazing for the relatively long, thin slats.

SECTION 2404 WIND, SNOW, SEISMIC AND DEAD LOADS ON GLASS

2404.1 Vertical glass. Glass sloped 15 degrees (0.26 rad) or less from vertical in windows, curtain and window walls, doors and other exterior applications shall be designed to resist the wind loads in Section 1609 for components and cladding. Glass in glazed curtain walls, glazed storefronts and glazed partitions shall meet the seismic requirements of ASCE 7, Section 9.6.2.10. Glazing firmly supported on all four edges is permitted to be designed by the following provisions. Where the glass is not firmly supported on all four edges, analysis or test data ensuring safe performance for the specific installation shall be prepared by a registered design professional.

The design of vertical glazing shall be based on the following equation:

$$F_{gw} \leq F_{ga} \qquad \textbf{(Equation 24-1)}$$

where:

F_{gw} is the wind load on the glass computed in accordance with Section 1609 and F_{ga} is the maximum allowable load on the glass computed by the following formula:

$$F_{ga} = c_1 F_{ge} \qquad \textbf{(Equation 24-2)}$$

where:

F_{ge} = Maximum allowable equivalent load, pounds per square foot (psf) (kN/m^2) determined from Figures 2404(1) through 2404(12) for the applicable glass dimensions and thickness.

c_1 = Factor determined from Table 2404.1 based on glass type.

❖ This section is based on the guidelines and requirements established by ASTM E 1300. The standard presents the procedure for determining glass requirements for wind, snow and dead loads. It also considers the influence of width-to-height ratios, effects of weathering and exposure and statistical influence of glass size.

The procedures for determining compliance of the more common types of glass used in vertical applications are explained in this section. For a given glass type, thickness, plate length and plate width, determine the allowable equivalent load from the appropriate Figures 2404(1) through 2404(12). Multiply that load by the appropriate factor in Table 2404.1 based on the type of glass being used. The result is the allowable uniform wind load for that glass type, thickness and size. The allowable wind load must be at least the required wind loads determined from Section 1609 for components and cladding. If the glass allowable wind load is less than the required wind load from Section 1609, thicker glass is needed.

In buildings that are classified as Seismic Design Category D, E or F, glazed partitions, glazed storefronts and glazed curtain walls must be capable of withstanding the relative seismic displacements between the points of attachment to the supporting structure(s). The objective of the referenced ASCE 7 provision is to limit the hazard of falling glass during and after an earthquake.

FIGURES 2404(1) through 2404(12). See pages 24-9 through 24-14.

❖ Figures 2404(1) through 2404(12) provide allowable design loads, F_{ge}, in pounds per square foot (psf), for a range of thicknesses and glass plate dimensions. The allowable design loads from the graphs are used in the formulas in Sections 2404.1 and 2404.2. The curved lines on each graph correspond to the indicated allowed design load. Where the design load lines are dashed, the glass deflection is greater than $^3/_4$ inch (19.1 mm) when subjected to the indicated allowable design load. The code does not limit the deflection to $^3/_4$ inch (19.1 mm). The $^3/_4$ inch (19.1 mm) deflection on the graphs is for information only. The graphs are a plot of loads that were applied for a 60-second duration.

Use the point determined by the glass plate length and width to determine the allowable design load, F_{ge}.

TABLE 2404.1. See page 24-7.

❖ The factors in Table 2401.1 adjust the allowable equivalent loads from the appropriate Figures 2404(1) through 2404(12) based on the specific type of glass that is being used. For example, fully tempered, single-thickness glass plate has an allowable uniform load four times that determined from the appropriate code Figures 2404(1) through 2404(12), because the multiplying factor in Table 2404.1 for single glass, fully tempered is 4.0.

Table 2401.1 is to be used for all vertical glazing and for sloped glazing 15 degrees (0.26 rad) or less from vertical when the wind load is dominant. The wind load is dominant where load combination 1 (Equation 24-3) or 2 (Equation 24-4) results in the highest required design load in Section 2404.2 for sloped glass installations.

TABLE 2404.1
c_1 FACTORS FOR VERTICAL AND SLOPED GLASS[a]
[For use with Figures 2404(1) through 2404(12)]

GLASS TYPE	FACTOR
Single Glass	
Regular (annealed)	1.0
Heat strengthened	2.0
Fully tempered	4.0
Wired	0.50
Patterned[c]	1.0
Sandblasted[d]	0.50
Laminated—regular plies[e]	0.7/0.90[f]
Laminated—heat-strengthened plies[e]	1.5/1.8[f]
Laminated—fully tempered plies[e]	3.0/3.6[f]
Insulating Glass[b]	
Regular (annealed)	1.8
Heat strengthened	3.6
Fully tempered	7.2
Laminated—regular plies[e]	1.4/1.6[f]
Laminated—heat-strengthened plies[e]	2.7/3.2[f]
Laminated—fully tempered plies[e]	5.4/6.5[f]

a. Either Table 2404.1 or 2404.2 shall be appropriate for sloped glass depending on whether the snow or wind load is dominant (see Section 2404.2). For glass types (vertical or sloped) not included in the tables, refer to ASTM E 1300 for guidance.
b. Values apply for insulating glass with identical panes.
c. The value for patterned glass is based on the thinnest part of the pattern; interpolation between graphs is permitted.
d. The value for sandblasted glass is for moderate levels of sandblasting.
e. Values for laminated glass are based on the total thickness of the glass and apply for glass with two equal glass ply thicknesses.
f. The lower value applies if, for any laminated glass pane, either the ratio of the long to short dimension is greater than 2.0 or the lesser dimension divided by the thickness of the pane is 150 or less; the higher value applies in all other cases.

2404.2 Sloped glass. Glass sloped more than 15 degrees (0.26 rad) from vertical in skylights, sunrooms, sloped roofs and other exterior applications shall be designed to resist the most critical of the following combinations of loads.

$$F_g = W_o - D \qquad \textbf{(Equation 24-3)}$$

$$F_g = W_i + D + 0.5\,S \qquad \textbf{(Equation 24-4)}$$

$$F_g = 0.5\,W_i + D + S \qquad \textbf{(Equation 24-5)}$$

where:

D = Glass dead load (psf)

For glass sloped 30 degrees (0.52 rad) or less from horizontal,

$D = 13\,t_g$ (For SI: $0.0245\,t_g$)

For glass sloped more than 30 degrees (0.52 rad) from horizontal,

$D = 13\,t_g \cos\theta$ (For SI: $0.0245\,t_g \cos\theta$).

F_g = Total load, psf (kN/m^2) on glass.

S = Snow load, psf (kN/m^2) as determined in Section 1608.

t_g = Total glass thickness, inches (mm) of glass panes and plies.

W_i = Inward wind force, psf (kN/m^2) as calculated in Section 1609.

W_o = Outward wind force, psf (kN/m^2) as calculated in Section 1609.

θ = Angle of slope from horizontal.

Exception: Unit skylights shall be designed in accordance with Section 2405.5.

The design of sloped glazing shall be based on the following equation:

$$F_g \leq F_{ga} \qquad \textbf{(Equation 24-6)}$$

where F_g is the maximum load on the glass determined from Equations 24-3 through 24-5, and F_{ga} is the maximum allowable load on the glass.

If F_g is determined by Equation 24-3 or 24-4 above, F_{ga} shall be computed as for vertical glazing in Section 2404.1. If F_g is determined by Equation 24-5 above, F_{ga} shall be computed by the following equation:

$$F_{ga} = c_2\,F_{ge} \qquad \textbf{(Equation 24-7)}$$

where:

F_{ge} = Maximum allowable equivalent load (psf) determined from Figures 2404(1) through 2404(12) for the applicable glass dimensions and thickness.

c_2 = Factor determined from Table 2404.2 based on glass type.

❖ This section provides the procedure for determining the maximum allowable equivalent load that may be supported by a sloped glazing system. The maximum allowable equivalent load must be greater than the required load determined from the highest value of the three load combinations listed (Equations 24-3, 24-4 and 24-5).The required design loads are first determined from the three load combinations listed as Equations 24-3, 24-4 and 24-5. The applicable wind loads from the components and cladding requirements of Section 1609 and the snow loads from Section 1608 are used to determine the required design loads from the three load combinations.

After the required design loads are determined, the maximum allowable equivalent load is determined. First, from applicable Figures 2404(1) through 2404(12) for a selected thickness, determine the allowable equivalent load. If the highest required load is from load combination 1 (Equation 24-3) or 2 (Equation 24-4), multiply the value from Figures 2404(1) through 2404(12) by the appropriate factor in Table 2404.1. If the highest required load is from load combination 3 (Equation 24-5), multiply the value from Figures 2404(1) through 2404(12) by the appropriate factor in Table 2404.2.

If the maximum allowable equivalent load is not greater than that required from the load combinations, choose a greater thickness and recalculate the maximum allowable equivalent load.

The ability of glass panes to resist loads varies with the duration of the load. Glass will withstand greater short-term loads (wind loads) than it will long-term loads (snow loads). The percentage difference in load resistance related to the duration of the load varies among different types of glass (regular annealed glass, heat-strengthened glass and tempered glass). Adjustments in long-term loads to convert them into equivalent short-term loads vary with glass type. This is the basis for the different values in Table 2404.2, related to snow load, compared to those values in Table 2404.1, related to wind load. Thus, the values in Table 2404.2 are less than those in Table 2404.1 for the same type of glass.

The exception refers to design provisions that are specific to unit skylights (i.e., single panel, factory-assembled units).

TABLE 2404.2
c_2 FACTORS FOR SLOPED GLASS[a]
[For use with Figures 2404(1) through 2404(12)]

GLASS TYPE	FACTOR
Single Glass	
Regular (annealed)	0.6
Heat strengthened	1.6
Fully tempered	3.6
Wired	0.3
Patterned[c]	0.6
Laminated — regular plies[d]	0.3/0.45[e]
Laminated — heat-strengthened plies[d]	0.8/1.2[e]
Laminated — fully tempered plies[d]	1.8/2.7[e]
Insulating Glass[b]	
Regular (annealed)	1.1
Heat strengthened	2.9
Fully tempered	6.5
Laminated — regular plies[d]	0.54/0.81[e]
Laminated — heat-strengthened plies[d]	1.4/2.2[e]
Laminated — fully tempered plies[d]	3.3/4.9[e]

a. Either Table 2404.1 or 2404.2 shall be appropriate for sloped glass depending on whether the snow or wind load is dominant (see Section 2404.2). For glass types (vertical or sloped) not included in the tables, refer to ASTM E 1300 for guidance.
b. Values apply for insulating glass with identical panes.
c. The value for patterned glass is based on the thinnest part of the pattern; interpolation between graphs is permitted.
d. Values for laminated glass are based on the total thickness of the glass and apply for glass with two equal glass ply thicknesses.
e. The lower value applies where, for any laminated glass pane, either the ratio of the long to short dimension is greater than 2.0 or the lesser dimension divided by the thickness of the pane is 150 or less. The higher value applies in all other cases.

❖ Table 2404.2 should be used only with the sloped glass design procedure in Section 2404.2 when load combination 3 (Equation 24-5) results in the highest required design load. The factors in this table are based on the snow load being dominant in the load combinations in Section 2404.2. In sloped applications, asymmetrical insulating glass is often used. A common construction is single heat-strengthened glass outboard, and laminated regular or heat-strengthened glass inboard. The asymmetrical types of insulating glass are not included in Table 2404.2 because of the large number of possible combinations. Additional information is available in *Glass Design for Sloped Glazing*, published by the American Architectural Manufacturers Association (AAMA).

SECTION 2405
SLOPED GLAZING AND SKYLIGHTS

2405.1 Scope. This section applies to the installation of glass and other transparent, translucent or opaque glazing material installed at a slope more than 15 degrees (0.26 rad) from the vertical plane, including glazing materials in skylights, roofs and sloped walls.

❖ This section covers all applications of glazing and light-transmitting plastics installed on a slope more than 15 degrees (0.26 rad) from vertical. The provisions for determining the combination of wind, snow and dead loads that sloped glazing must resist are found in Section 2404. For additional information on the design of sloped glazing and skylights, refer to *Glass Design for Sloped Glazing*, available from the AAMA.

Sloped glazing is defined as any glazing system or material installed on a slope more than 15 degrees (0.26 rad) from vertical. Where glass is installed within 15 degrees (0.26 rad) of vertical, the gravity force (dead load) of the glass normal to its surface is very small and does not contribute materially to the total load. From an engineering standpoint, glass within 15 degrees (0.26 rad) of vertical will perform essentially the same as glass installed vertically.

Sloped glazing, by definition, is not intended to specifically exclude any type of glass or glazing system from the requirements of Chapter 24. Opaque glass, such as spandrel glass, as well as transparent and translucent glass, both opaque and light-transmitting plastics, decorative glass and glazing and any other glazing-type material contained in the chapter are subject to the design and structural requirements of this chapter, Chapter 16 and related chapters elsewhere in code.

Providing occupants with a safe environment is tantamount in the design and construction of buildings and structures. Sound engineering principles should always be employed and building code requirements strictly adhered to when dealing with issues that will affect the life, safety and welfare of people using or occupying buildings or structures. This is especially true when using building materials such as glass, which can cause severe injury to people when it is not used in a safe and proven manner.

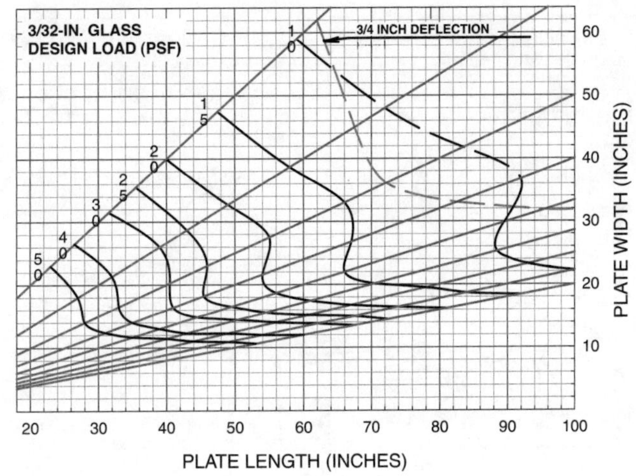

For SI: 1 inch = 25.4 mm, 1 pound per square foot = 0.0479 kPa.

FIGURE 2404(1)[a, b, c, d, e, f]
MAXIMUM ALLOWABLE LOAD FOR VERTICAL AND SLOPED RECTANGULAR GLASS SUPPORTED ON ALL EDGES

NOTES:

a. In each graph, the vertical axis is the lesser dimension; the horizontal axis is the greater dimension.
b. The diagonal number on each graph shows the equivalent design load in psf.
c. The dashed lines indicate glass that has deflection in excess of $^3/_4$ inch.
d. Interpolation between lines is permitted. Extrapolation is not allowed.
e. For laminated glass, the applicable glass thickness is the total glass thickness.
f. For insulating glass panes, the applicable glass thickness is the thickness of one pane.

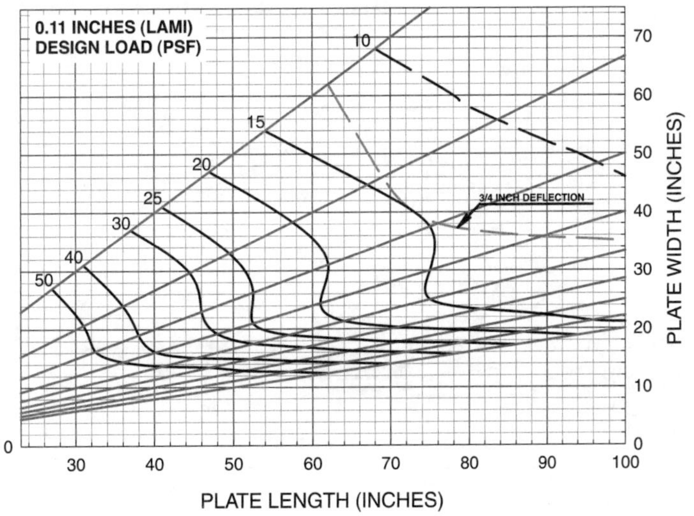

For SI: 1 inch = 25.4 mm, 1 pound per square foot = 0.0479 kPa.

FIGURE 2404(2)[a, b, c, d, e, f]
MAXIMUM ALLOWABLE LOAD FOR VERTICAL AND SLOPED RECTANGULAR GLASS SUPPORTED ON ALL EDGES

NOTES:

a. In each graph, the vertical axis is the lesser dimension; the horizontal axis is the greater dimension.
b. The diagonal number on each graph shows the equivalent design load in psf.
c. The dashed lines indicate glass that has deflection in excess of $^3/_4$ inch.
d. Interpolation between lines is permitted. Extrapolation is not allowed.
e. For laminated glass, the applicable glass thickness is the total glass thickness.
f. For insulating glass panes, the applicable glass thickness is the thickness of one pane.

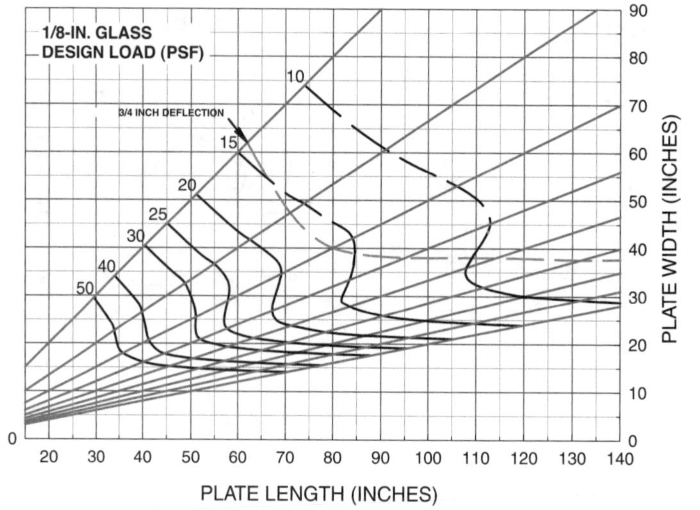

For SI: 1 inch = 25.4 mm, 1 pound per square foot = 0.0479 kPa.

FIGURE 2404(3)[a, b, c, d, e, f]
MAXIMUM ALLOWABLE LOAD FOR VERTICAL AND SLOPED RECTANGULAR GLASS SUPPORTED ON ALL EDGES

NOTES:

a. In each graph, the vertical axis is the lesser dimension; the horizontal axis is the greater dimension.
b. The diagonal number on each graph shows the equivalent design load in psf.
c. The dashed lines indicate glass that has deflection in excess of $^3/_4$ inch.
d. Interpolation between lines is permitted. Extrapolation is not allowed.
e. For laminated glass, the applicable glass thickness is the total glass thickness.
f. For insulating glass panes, the applicable glass thickness is the thickness of one pane.

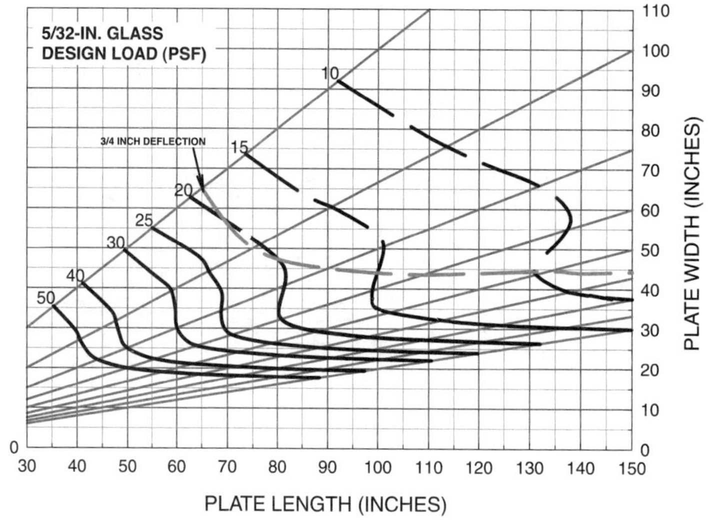

For SI: 1 inch = 25.4 mm, 1 pound per square foot = 0.0479 kPa.

FIGURE 2404(4)[a, b, c, d, e, f]
MAXIMUM ALLOWABLE LOAD FOR VERTICAL AND SLOPED RECTANGULAR GLASS SUPPORTED ON ALL EDGES

NOTES:

a. In each graph, the vertical axis is the lesser dimension; the horizontal axis is the greater dimension.
b. The diagonal number on each graph shows the equivalent design load in psf.
c. The dashed lines indicate glass that has deflection in excess of $^3/_4$ inch.
d. Interpolation between lines is permitted. Extrapolation is not allowed.
e. For laminated glass, the applicable glass thickness is the total glass thickness.
f. For insulating glass panes, the applicable glass thickness is the thickness of one pane.

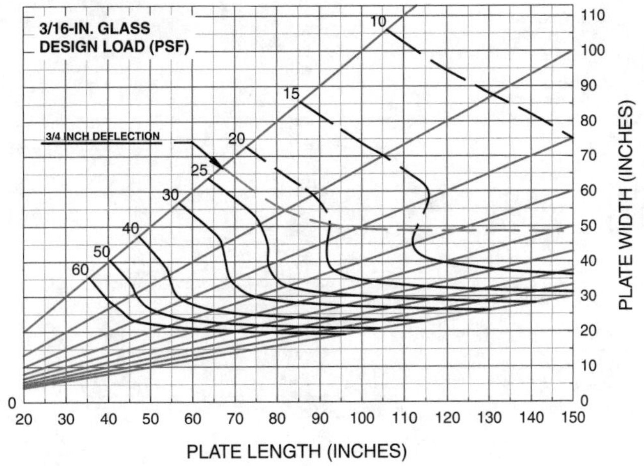

For SI: 1 inch = 25.4 mm, 1 pound per square foot = 0.0479 kPa.

FIGURE 2404(5)[a, b, c, d, e, f]
MAXIMUM ALLOWABLE LOAD FOR VERTICAL AND SLOPED RECTANGULAR GLASS SUPPORTED ON ALL EDGES

NOTES:

a. In each graph, the vertical axis is the lesser dimension; the horizontal axis is the greater dimension.
b. The diagonal number on each graph shows the equivalent design load in psf.
c. The dashed lines indicate glass that has deflection in excess of $^3/_4$ inch.
d. Interpolation between lines is permitted. Extrapolation is not allowed.
e. For laminated glass, the applicable glass thickness is the total glass thickness.
f. For insulating glass panes, the applicable glass thickness is the thickness of one pane.

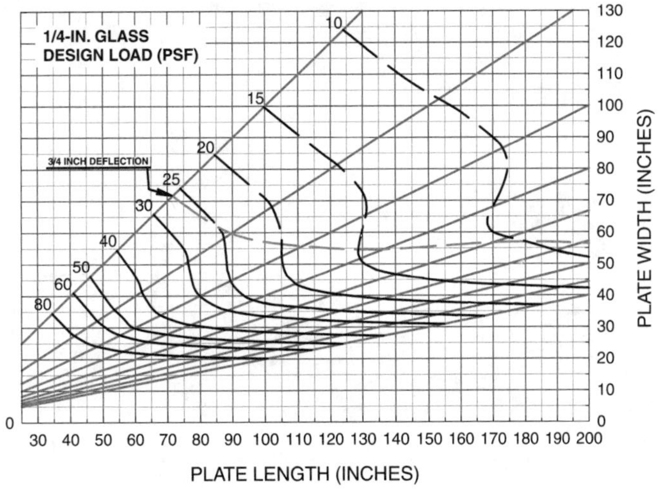

For SI: 1 inch = 25.4 mm, 1 pound per square foot = 0.0479 kPa.

FIGURE 2404(6)[a, b, c, d, e, f]
MAXIMUM ALLOWABLE LOAD FOR VERTICAL AND SLOPED RECTANGULAR GLASS SUPPORTED ON ALL EDGES

NOTES:

a. In each graph, the vertical axis is the lesser dimension; the horizontal axis is the greater dimension.
b. The diagonal number on each graph shows the equivalent design load in psf.
c. The dashed lines indicate glass that has deflection in excess of $^3/_4$ inch.
d. Interpolation between lines is permitted. Extrapolation is not allowed.
e. For laminated glass, the applicable glass thickness is the total glass thickness.
f. For insulating glass panes, the applicable glass thickness is the thickness of one pane.

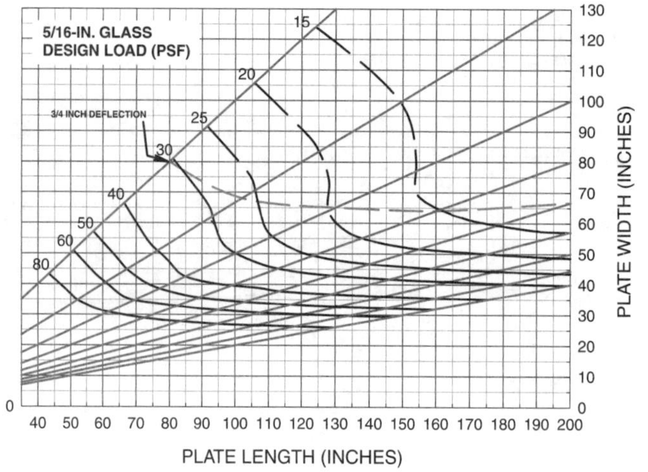

For SI: 1 inch = 25.4 mm, 1 pound per square foot = 0.0479 kPa.

FIGURE 2404(7)[a, b, c, d, e, f]
MAXIMUM ALLOWABLE LOAD FOR VERTICAL AND SLOPED RECTANGULAR GLASS SUPPORTED ON ALL EDGES

NOTES:

a. In each graph, the vertical axis is the lesser dimension; the horizontal axis is the greater dimension.
b. The diagonal number on each graph shows the equivalent design load in psf.
c. The dashed lines indicate glass that has deflection in excess of $^3/_4$ inch.
d. Interpolation between lines is permitted. Extrapolation is not allowed.
e. For laminated glass, the applicable glass thickness is the total glass thickness.
f. For insulating glass panes, the applicable glass thickness is the thickness of one pane.

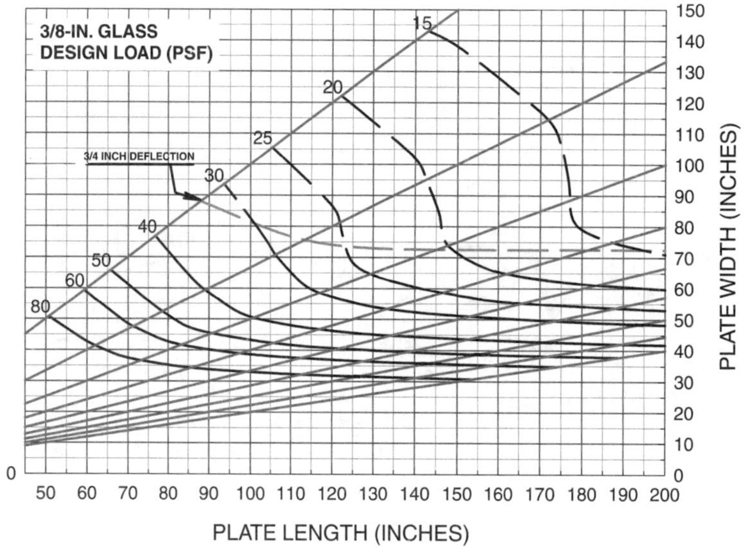

For SI: 1 inch = 25.4 mm, 1 pound per square foot = 0.0479 kPa.

FIGURE 2404(8)[a, b, c, d, e, f]
MAXIMUM ALLOWABLE LOAD FOR VERTICAL AND SLOPED RECTANGULAR GLASS SUPPORTED ON ALL EDGES

NOTES:

a. In each graph, the vertical axis is the lesser dimension; the horizontal axis is the greater dimension.
b. The diagonal number on each graph shows the equivalent design load in psf.
c. The dashed lines indicate glass that has deflection in excess of $^3/_4$ inch.
d. Interpolation between lines is permitted. Extrapolation is not allowed.
e. For laminated glass, the applicable glass thickness is the total glass thickness.
f. For insulating glass panes, the applicable glass thickness is the thickness of one pane.

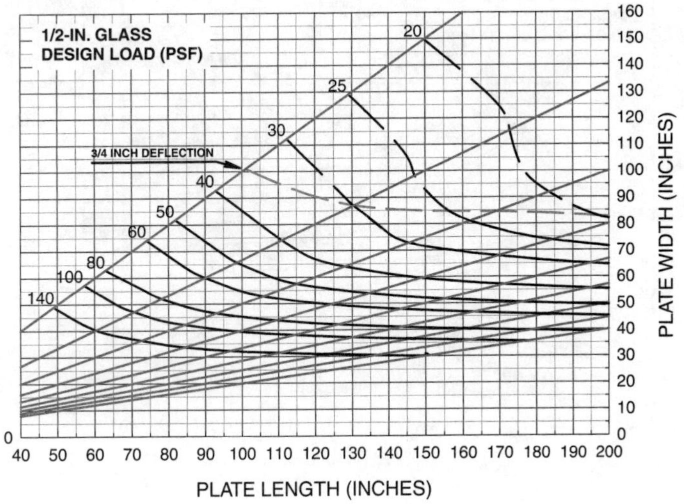

For SI: 1 inch = 25.4 mm, 1 pound per square foot = 0.0479 kPa.

FIGURE 2404(9)[a, b, c, d, e, f]
MAXIMUM ALLOWABLE LOAD FOR VERTICAL AND SLOPED RECTANGULAR GLASS SUPPORTED ON ALL EDGES

NOTES:

a. In each graph, the vertical axis is the lesser dimension; the horizontal axis is the greater dimension.
b. The diagonal number on each graph shows the equivalent design load in psf.
c. The dashed lines indicate glass that has deflection in excess of $^3/_4$ inch.
d. Interpolation between lines is permitted. Extrapolation is not allowed.
e. For laminated glass, the applicable glass thickness is the total glass thickness.
f. For insulating glass panes, the applicable glass thickness is the thickness of one pane.

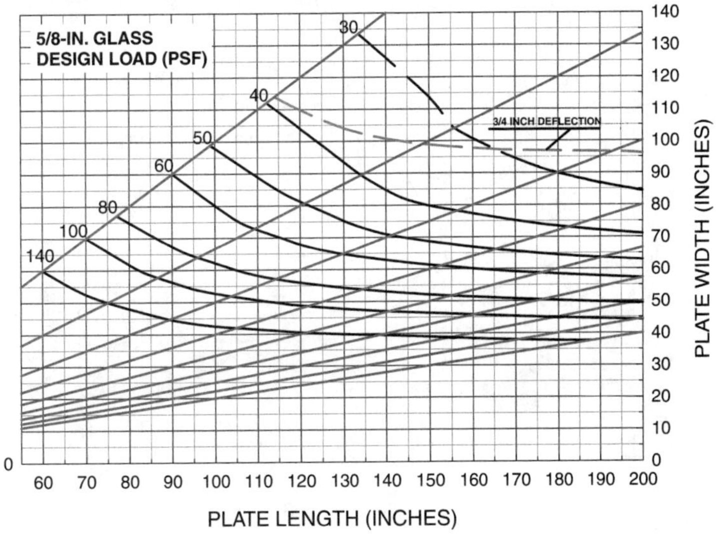

For SI: 1 inch = 25.4 mm, 1 pound per square foot = 0.0479 kPa.

FIGURE 2404(10)[a, b, c, d, e, f]
MAXIMUM ALLOWABLE LOAD FOR VERTICAL AND SLOPED RECTANGULAR GLASS SUPPORTED ON ALL EDGES

NOTES:

a. In each graph, the vertical axis is the lesser dimension; the horizontal axis is the greater dimension.
b. The diagonal number on each graph shows the equivalent design load in psf.
c. The dashed lines indicate glass that has deflection in excess of $^3/_4$ inch.
d. Interpolation between lines is permitted. Extrapolation is not allowed.
e. For laminated glass, the applicable glass thickness is the total glass thickness.
f. For insulating glass panes, the applicable glass thickness is the thickness of one pane.

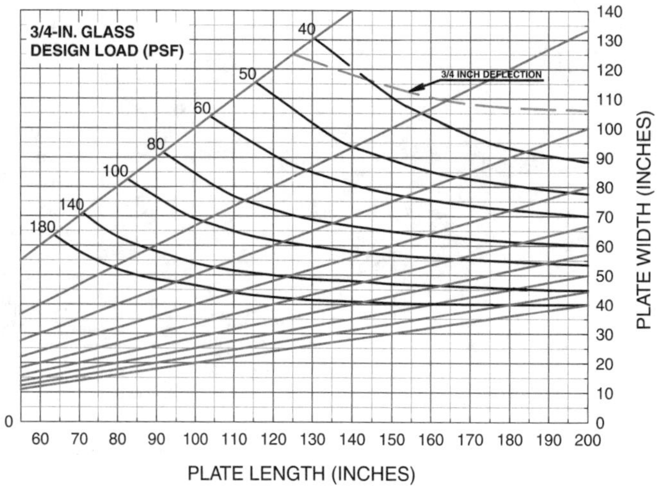

For SI: 1 inch = 25.4 mm, 1 pound per square foot = 0.0479 kPa.

FIGURE 2404(11)[a, b, c, d, e, f]
MAXIMUM ALLOWABLE LOAD FOR VERTICAL AND SLOPED RECTANGULAR GLASS SUPPORTED ON ALL EDGES

NOTES:

a. In each graph, the vertical axis is the lesser dimension; the horizontal axis is the greater dimension.
b. The diagonal number on each graph shows the equivalent design load in psf.
c. The dashed lines indicate glass that has deflection in excess of $^3/_4$ inch.
d. Interpolation between lines is permitted. Extrapolation is not allowed.
e. For laminated glass, the applicable glass thickness is the total glass thickness.
f. For insulating glass panes, the applicable glass thickness is the thickness of one pane.

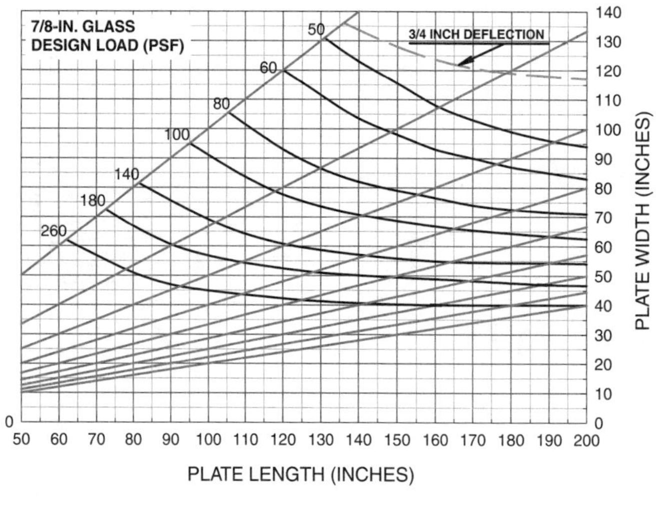

For SI: 1 inch = 25.4 mm, 1 pound per square foot = 0.0479 kPa.

FIGURE 2404(12)[a, b, c, d, e, f]
MAXIMUM ALLOWABLE LOAD FOR VERTICAL AND SLOPED RECTANGULAR GLASS SUPPORTED ON ALL EDGES

NOTES:

a. In each graph, the vertical axis is the lesser dimension; the horizontal axis is the greater dimension.
b. The diagonal number on each graph shows the equivalent design load in psf.
c. The dashed lines indicate glass that has deflection in excess of $^3/_4$ inch.
d. Interpolation between lines is permitted. Extrapolation is not allowed.
e. For laminated glass, the applicable glass thickness is the total glass thickness.
f. For insulating glass panes, the applicable glass thickness is the thickness of one pane.

2405.2 Allowable glazing materials and limitations. Sloped glazing shall be any of the following materials, subject to the listed limitations.

1. For monolithic glazing systems, the glazing material of the single light or layer shall be laminated glass with a minimum 30-mil (0.76 mm) polyvinyl butyral (or equivalent) interlayer, wired glass, light-transmitting plastic materials meeting the requirements of Section 2607, heat-strengthened glass or fully tempered glass.

2. For multiple-layer glazing systems, each light or layer shall consist of any of the glazing materials specified in Item 1 above.

Annealed glass is permitted to be used as specified within Exceptions 2 and 3 of Section 2405.3.

For additional requirements for plastic skylights, see Section 2610. Glass-block construction shall conform to the requirements of Section 2101.2.5.

❖ All configurations of glazing and plastic are covered in this section. This includes single (monolithic) glass and plastic, insulating glass units consisting of any number of glass panes (but usually two), double-walled plastic skylights and any combination of glass or plastic panels separately framed and inserted into a single opening (such as a prime window and a storm window).

Laminated glass is required to have a minimum 30-mil-thick (0.76 mm) plastic interlayer. Approved plastic materials are those that comply with the provisions of the code for strength, durability, sanitation and fire performance required for installation. Acrylics (e.g., Plexiglas) and polycarbonates (e.g., Lexan) are examples of plastics that are commonly found in sloped glazing installations.

2405.3 Screening. Where used in monolithic glazing systems, heat-strengthened glass and fully tempered glass shall have screens installed below the glazing material. The screens and their fastenings shall: (1) be capable of supporting twice the weight of the glazing; (2) be firmly and substantially fastened to the framing members and (3) be installed within 4 inches (102 mm) of the glass. The screens shall be constructed of a noncombustible material not thinner than No. 12 B&S gage (0.0808 inch) with mesh not larger than 1 inch by 1 inch (25 mm by 25 mm). In a corrosive atmosphere, structurally equivalent noncorrosive screen materials shall be used. Heat-strengthened glass, fully tempered glass and wired glass, when used in multiple-layer glazing systems as the bottom glass layer over the walking surface, shall be equipped with screening that conforms to the requirements for monolithic glazing systems.

Exception: In monolithic and multiple-layer sloped glazing systems, the following applies:

1. Fully tempered glass installed without protective screens where glazed between intervening floors at a slope of 30 degrees (0.52 rad) or less from the vertical plane shall have the highest point of the glass 10 feet (3048 mm) or less above the walking surface.

2. Screens are not required below any glazing material, including annealed glass, where the walking surface below the glazing material is permanently protected from the risk of falling glass or the area below the glazing material is not a walking surface.

3. Any glazing material, including annealed glass, is permitted to be installed without screens in the sloped glazing systems of commercial or detached noncombustible greenhouses used exclusively for growing plants and not open to the public, provided that the height of the greenhouse at the ridge does not exceed 30 feet (9144 mm) above grade.

4. Screens shall not be required within individual dwelling units in Groups R-2, R-3 and R-4 as applicable in Section 101.2 where fully tempered glass is used as single glazing or as both panes in an insulating glass unit, and the following conditions are met:

   4.1. Each pane of the glass is 16 square feet (1.5 m^2) or less in area.

   4.2. The highest point of the glass is 12 feet (3658 mm) or less above any walking surface or other accessible area.

   4.3. The glass thickness is $^3/_{16}$ inch (4.8 mm) or less.

5. Screens shall not be required for laminated glass with a 15-mil (0.38 mm) polyvinyl butyral (or equivalent) interlayer used within individual dwelling units in Groups R-2, R-3 and R-4 as applicable in Section 101.2 within the following limits:

   5.1. Each pane of glass is 16 square feet (1.5 m^2) or less in area.

   5.2. The highest point of the glass is 12 feet (3658 mm) or less above a walking surface or other accessible area.

❖ In sloped applications, certain glass types have a tendency to fall from the opening when broken. As a result, a retaining net or screening is required in specific locations. Regular (annealed) glass may not be used in sloped glazing installations, except as defined in Section 2405.3. Heat-strengthened and fully tempered glazing are two types of glass that may fall from an opening when broken by wind or snow loads. For these glasses, a retaining screen is required on the interior side of the glazing unit. This would apply to single glazing or when the glass is the inboard pane in an insulating glass unit. Other exceptions for fully tempered glass are listed in Section 2405.3.

Wire glass, when used as the inboard pane in an insulating glass unit, has a tendency to crack from thermal stresses. This is due to weakening at the edges where the wire protrudes. Where fire-resistance-rated glazing material is needed, wire glass is mandatory. In this case, because of the increased likelihood of breakage, a retaining screen is also required.

There are both performance and prescriptive requirements for the retaining screen. Any screen that meets the performance requirements is acceptable.

Five exceptions to the requirements for a retaining

screen are listed. For the purposes of this section, patterned glass that is annealed is to be considered regular (annealed) glass.

Exception 1: Fully tempered glass does not require a protective screen when the glass is sloped 30 degrees (0.52 rad) or less from vertical, and the top edge is less than 10 feet (3048 mm) above a walking surface. The sloped walls in many commercial buildings are within these limitations. The glass has been satisfactory with no known risks. When broken, fully tempered glass will break into small fragmented pieces which are unlikely to cause harm upon impact [see Figure 2405.3(1)].

Exception 2: In cases where the area below the glass is either permanently protected or inaccessible, falling glass does not present a risk. When these conditions are met, screens are not required below the sloped glazing [see Figure 2405.3(2)].

Exception 3: Glass in greenhouses does not pose a serious risk to the general public. The failure of a sloped glazing system in a private greenhouse is not considered a high risk, as the majority of floor space is taken up by plants and the building is not occupied on a regular basis by a large number of people. Where a greenhouse is not attached to any other structure, has restricted access and meets the specified height limitation, screens are not required below the sloped glazing.

Exception 4: Sloped glazing installations using fully tempered glass are not required to have screens installed below the glass when the glazing units or skylights are installed within individual dwelling units of Group R-2, R-3 and R-4 occupancies, and the glazing complies with the size, thickness and location requirements specified in Items 4.1, 4.2 and 4.3. This exception is applicable only to glazing installations that are located within a dwelling unit; it does not apply to any public or common areas. It is also not applicable to areas in buildings containing two-family and multiple family dwellings that are common to all occupants (e.g., entry foyers, main corridors, community storage and laundry rooms).

This exception is based on a survey conducted by the glass and glazing industry. The information produced from the survey indicated an 0.081-percent incidence of breakage among fully tempered sun spaces and skylights. More importantly, the survey found no incidences of injury were reported when the glazing was fully tempered glass installed within the limitations specified in the exception.

Exception 5: This exception is permitted because of the protection afforded by the polyvinyl butyral interlayer located within the laminated glass panel. The interlayer serves to keep the broken pieces of glass in place so that little or no glass falls when broken by impact. This exception applies only to individual dwelling units within R-2, R-3 and R-4 occupancies when sized and located as prescribed in Items 5.1 and 5.2.

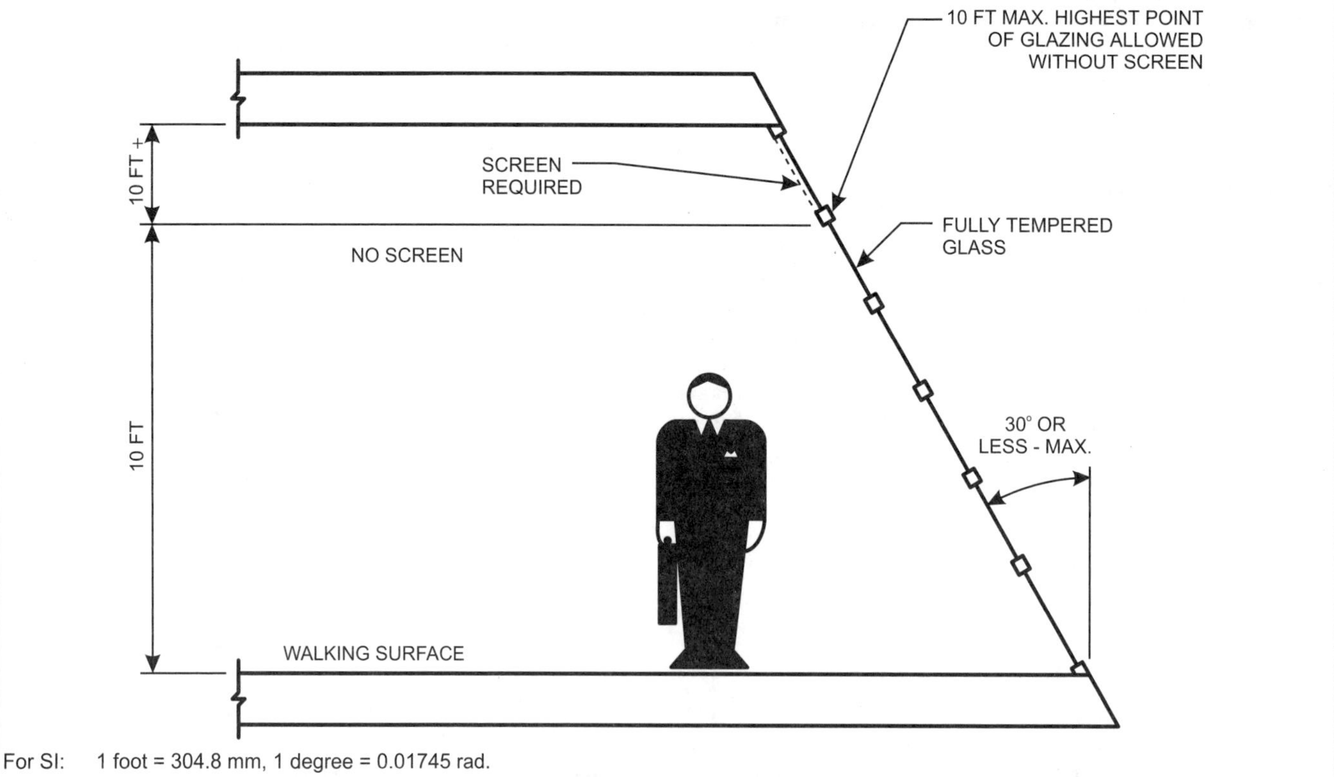

For SI: 1 foot = 304.8 mm, 1 degree = 0.01745 rad.

Figure 2405.3(1)
FULLY TEMPERED GLAZING—30° MAXIMUM SLOPE

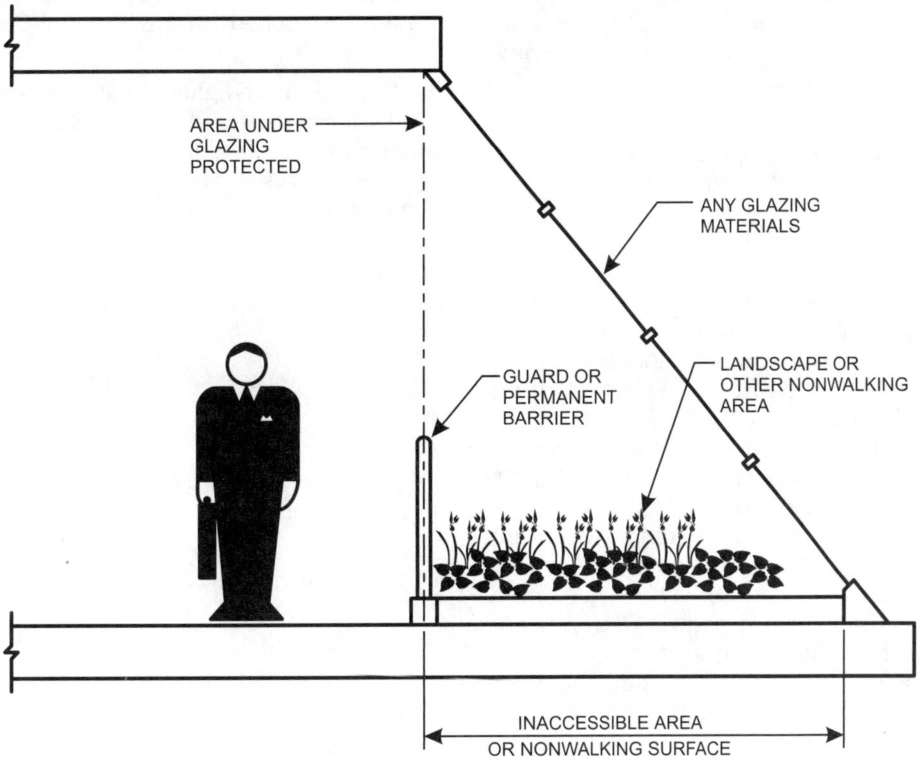

Figure 2405.3(2)
INACCESSIBLE AREA UNDER SLOPED GLAZING

2405.4 Framing. In Type 1 and 2 construction, sloped glazing and skylight frames shall be constructed of noncombustible materials. In structures where acid fumes deleterious to metal are incidental to the use of the buildings, approved pressure-treated wood or other approved noncorrosive materials are permitted to be used for sash and frames. Framing supporting sloped glazing and skylights shall be designed to resist the tributary roof loads in Chapter 16. Skylights set at an angle of less than 45 degrees (0.79 rad) from the horizontal plane shall be mounted at least 4 inches (102 mm) above the plane of the roof on a curb constructed as required for the frame. Skylights shall not be installed in the plane of the roof where the roof pitch is less than 45 degrees (0.79 rad) from the horizontal.

> **Exception:** Installation of a skylight without a curb shall be permitted on roofs with a minimum slope of 14 degrees (three units vertical in 12 units horizontal) in Group R-3 occupancies as applicable in Section 101.2. All unit skylights installed in a roof with a pitch flatter than 14 degrees (0.25 rad) shall be mounted at least 4 inches (102 mm) above the plane of the roof on a curb constructed as required for the frame unless otherwise specified in the manufacturer's installation instructions.

❖ The requirement for noncombustible (metal) framing for glazing systems in Type 1 and 2 construction must be consistent with the requirements for the remainder of the structure. In cases where acidic fumes harmful to metal framing may occur, pressure-treated wood or noncorrosive material may be used.

Regardless of the type of construction, when installed on surfaces having a slope of less than 45 degrees (0.79 rad) from the horizontal, all sloped glazing and skylights are required to be mounted on curbs at least 4 inches (102 mm) high. The raised curb will provide additional protection to the skylight from burnt or burning debris in the event of a fire. Burnt or burning pieces of wood are known technically as "brands." When airborne, they are known as "flying brands." For slopes 45 degrees (0.79 rad) and greater, the protection provided by the curbs is not necessary. In these cases, the glass may be installed flush with the surrounding roof.

The exception for Group R-3 occupancies is based on field experience and general practice for one- and two-family dwellings. For the 3:12 slope typical for Group R-3, neither flying brands nor water runoff has been a problem.

2405.5 Unit skylights. Unit skylights shall be tested and labeled as complying with 101/I.S.2/NAFS *Voluntary Performance Specification for Windows, Skylights and Glass*. The label shall state the name of the manufacturer, the approved labeling agency, the product designation and the performance grade rating as specified in 101/I.S.2/NAFS. If the product manufacturer has chosen to have the performance grade of the skylight rated separately for positive and negative design pressure, then the label shall state both performance grade ratings as specified in 101/I.S.2/NAFS and the skylight shall comply with Section 2405.5.2. If the skylight is not rated separately for positive and negative pressure, then the performance grade rating shown on the label shall be the performance grade rating determined in accordance with 101/I.S.2/NAFS for both positive and negative design pressure, and the skylight shall conform to Section 2405.5.1.

❖ This section establishes design provisions for unit skylights (single-panel, factory-assembled units) that are consistent with those provided for sloped glazing in Section 2402, but permit the unit skylights to be designed separately for maximum positive and negative pressure. The most critical load on a skylight is determined by the climate in which it is installed. In a colder climate with heavier snow loads and moderate design wind speeds, the positive load on a skylight from the combined snow and dead loads will be more critical than the negative load from wind uplift. The opposite will be the case in warmer, coastal climates with higher design wind speeds, and little or no snow load.

This method of rating skylights is addressed in the AAMA (American Architectural Manufacturers Association) referenced standard 101/I.S.2/NAFS *Voluntary Performance Specification for Windows, Skylights and Glass Doors*. The separate rating system for positive and negative pressure on skylights allows the manufacturer to design and fabricate products that are best suited for the climate in which they will be used. 101/I.S.2/NAFS establishes the performance requirements for skylights based on the desired performance grade rating, which includes minimum requirements for resistance to air leakage, water infiltration and design load pressures. The resulting performance grade rating states the design load pressure used to rate the product, but also includes consideration of these additional performance characteristics. For skylights certified for only one performance grade, the rating is based on the minimum requirements met for both positive and negative design pressure. Skylights certified for two performance grades are rated separately for positive and negative design pressure.

2405.5.1 Unit skylights rated for the same performance grade for both positive and negative design pressure. The design of unit skylights shall based on the following equation:

$F_g \leq PG$ **(Equation 24-8)**

where:

F_g is the maximum load on the skylight determined from Equations 24-3 through 24-5 in Section 2404.2.

PG is the performance grade rating of the skylight.

❖ For skylights that are not rated separately for positive and negative design pressure, this section requires that the minimum performance grade rating of the skylight must not be exceeded by any of the load combinations given in Section 2404.2.

2405.5.2 Unit skylights rated for separate performance grades for positive and negative design pressure. The design of unit skylights rated for performance grade for both positive and negative design pressures shall be based on the following equations:

$F_{gi} \leq PG_{Pos}$ **(Equation 24-9)**

$F_{go} \leq PG_{Neg}$ **(Equation 24-10)**

where:

PG_{Pos} is the performance grade rating of the skylight under positive design pressure,

PG_{Neg} is the performance grade rating of the skylight under negative design pressure, and

F_{gi} and F_{go} are determined in accordance with the following:

If $W_O \geq D$, where W_o is the outward wind force, psf (kN/m^2) as calculated in Section 1609 and D is the dead weight of the glazing, psf (kN/m^2) as determined in Section 2404.2 for glass, or by the weight of the plastic, psf (kN/m^2) for plastic glazing.

F_{gi} is the maximum load on the skylight determined from Equations 24-4 and 24-5 in Section 2404.2,

F_{go} is the maximum load on the skylight determined from Equation 24-3.

If $W_O < D$, where W_o is the outward wind force, psf (kN/m^2) as calculated in Section 1609 and D is the dead weight of the glazing, psf (kN/m^2) as determined in Section 2404.2 for glass, or by the weight of the plastic for plastic glazing.

F_{gi} is the maximum load on the skylight determined from Equations 24-3 through 24-5 in Section 2404.2,

$F_{go} = 0$.

❖ This section establishes requirements for skylights that are rated separately for positive and negative design pressure. The performance grade rating for positive design pressure is not to be exceeded by load combinations that include dead load, snow and wind acting towards the face of the skylight, and the performance grade rating for negative design pressure is not to be exceeded by the load combination that considers wind acting away from the face of the skylight.

SECTION 2406
SAFETY GLAZING

2406.1 Human impact loads. Individual glazed areas, including glass mirrors, in hazardous locations as defined in Section 2406.3 shall comply with Sections 2406.1.1 through 2406.1.5.

❖ With the increased use of sliding patio doors, glass tub and shower enclosures and similar applications where large pieces of glass are used, cutting and piercing injuries from contact with broken glass have become a safety concern. In recognition of a lack of code provisions controlling these risks, codes included limited requirements that regulate the use of glass. It was recognized that safety glazing materials must be required for glazed areas that would reasonably be exposed to human impact. Some questions arose as to the proper definition of these areas (Section 2406.3 identifies these "hazardous locations"). Section 2406 provides comprehensive regulations for the use of safety glazing materials and installation parameters.

Safety glazing requirements apply to both replacement glass and new construction. Replacement glass includes windows, doors and other assemblies containing glass that are moved, partially or intact, from one location in a building to another location in the same building, or to another building on the same site or a different site.

While glass mirrors are included within the scope of glazing, mirrors with a continuous backing support are not required to be safety glazed according to Exception 7 in Section 2406.3.1.

2406.1.1 CPSC 16 CFR, Part 1201. Except as provided in Sections 2406.1.2 through 2406.1.5, all glazing shall pass the test requirements of CPSC 16 CFR 1201, listed in Chapter 35. Glazing shall comply with the CPSC 16 CFR, Part 1201 criteria, for Category I or II as indicated in Table 2406.1.

❖ In 1977, the CPSC, in cooperation with building officials and the glass industry, drafted and approved CPSC 16 CFR, Part 1201. This is the standard recognized in the code. Laminated glass with an 0.030-inch (0.76 mm) interlayer readily meets CPSC test requirements. The test in CPSC 16 CFR, Part 1201 eliminated wire glass and laminated glass that is more than 9 square feet (0.84 m^2) in area having an 0.015-inch (0.38 mm) interlayer, as they could not meet the CPSC tests for approved safety glazing materials. Sections 2406.1.2 through 2406.1.5 are essentially exceptions to CPSC 16 CFR, Part 1201.

2406.1.2 Wired glass. In other than Group E, wired glass installed in fire doors, fire windows and view panels in fire-resistant walls shall be permitted to comply with ANSI Z97.1.

❖ This section reduces the impact test requirement from the CPSC tests to the ANSI Z97.1 impact tests for fire-resistance-rated installations. Generally, wired glass is not permitted for nonrated openings at human impact locations, since wired glass is required to meet only the less-stringent test requirements of ANSI Z97.1. Wired glass is required to meet the CPSC 16 CFR requirements for installations subject to human impact loads in Group E occupancies and in all occupancies where the opening is not required to have a fire-resistance rating. If the wired glass does not meet the CPSC test criteria, other types of glazing that meet the CPSC criteria must be used.

The ANSI Z97.1 test, which has been developed over many years, consists of an impact from a free-swinging punching bag filled with lead shot. The glass is impacted from different drop heights, depending on the predicted kinetic energy at impact for various glass applications. In the test, the glass must break in a manner that will materially reduce the risk of cutting and piercing injuries, or not break at all. This standard has been periodically updated, but remains basically unchanged.

There are other fire-resistance-rated glazing materials that meet both the CPSC impact tests and fire-resistance requirements. These are subject to compliance with Chapter 7 requirements for fire doors and fire windows.

2406.1.3 Plastic glazing. Plastic glazing shall meet the weathering requirements of ANSI Z97.1.

❖ This section has requirements for plastic safety glazing that have been eliminated from CPSC 16 CFR, Part 1201, necessitating the inclusion of this exception for plastic glazing. Requirements for plastic glazing are in Chapter 26.

2406.1.4 Glass block. Glass-block walls shall comply with Section 2101.2.5.

❖ The testing requirements for glass-block walls are described in Chapter 21, which eliminates the CPSC test requirement, and references the glass-block wall section of the code. Glass-block walls and panels are also not required to meet the test requirements of CPSC 16 CFR; however, that does not mean there are no safety requirements placed on the installation of glass block. Structural limitations to prevent failure caused by excessive loading, size limitations of panels, location and extent of panels in fire-resistance-rated assemblies and other safety issues are defined elsewhere in the code.

2406.1.5 Louvered windows and jalousies. Louvered windows and jalousies shall comply with Section 2403.5.

❖ Louvered and jalousie windows are discussed in detail in Section 2403.5.

TABLE 2406.1. See page 24-20.

❖ The classifications in this table are according to the CPSC 16 CFR, Part 1201 regulation. The tests in the federal regulation vary with the category classification. Category I glass is impacted from a drop height of 18 inches (457 mm) and Category II is impacted from a drop height of 48 inches (1219 mm).

TABLE 2406.1
MINIMUM CATEGORY CLASSIFICATION OF GLAZING

EXPOSED SURFACE AREA OF ONE SIDE OF ONE LITE	GLAZING IN STORM OR COMBINATION DOORS (Category class)	GLAZING IN DOORS (Category class)	GLAZED PANELS REGULATED BY ITEM 7 OF SECTION 2406.3 (Category class)	GLAZED PANELS REGULATED BY ITEM 6 OF SECTION 2406.3 (Category class)	DOORS AND ENCLOSURES REGULATED BY ITEM 5 OF SECTION 2406.3 (Category class)	SLIDING GLASS DOORS PATIO TYPE (Category class)
9 square feet or less	I	I	No requirement	I	II	II
More than 9 square feet	II	II	II	II	II	II

For SI: 1 square foot = $0.0929m^2$.

2406.2 Identification of safety glazing. Except as indicated in Section 2406.2.1, each pane of safety glazing installed in hazardous locations shall be identified by a label specifying the labeler, whether the manufacturer or installer, and the safety glazing standard with which it complies, as well as the information specified in Section 2403.1. The label shall be acid etched, sand blasted, ceramic fired or an embossed mark, or shall be of a type that once applied cannot be removed without being destroyed.

Exceptions:

1. For other than tempered glass, labels are not required, provided the building official approves the use of a certificate, affidavit or other evidence confirming compliance with this code.
2. Tempered spandrel glass is permitted to be identified by the manufacturer with a removable paper label.

❖ This section requires that each individual piece of safety glazing must be permanently marked with the manufacturer's or installer's designation. The designation is a label of identification applied directly to the glass by the manufacturer or installer stating that the product or material complies with the specified code standards or requirements as defined in Section 2406.1.

Many manufacturers of safety glazing materials have their products certified by the Safety Glazing Certification Council (SGCC) to meet either ANSI Z97.1 or CPSC 16 CFR, Part 1201. These provide the basis for the marking of safety glazing materials.

2406.2.1 Multilight assemblies. Multilight glazed assemblies having individual lights not exceeding 1 square foot (0.09 square meter) in exposed area shall have at least one light in the assembly marked as indicated in Section 2406.2. Other lights in the assembly shall be marked "CPSC 16 CFR 1201" or "ANSI Z97.1," as appropriate.

❖ In the case of glass material, many times the glass is purchased by the fabricator/installer in large sheets from the manufacturer, and cut into many pieces to fit various openings. This is particularly true with laminated or plastic glazing material. In this instance, it is impracticable to apply the designation label in such a manner that each cut piece will have a manufacturer's permanent designation. Where this occurs, the code permits multilight glazed assemblies to have only one light marked with the required information.

2406.3 Hazardous locations. The following shall be considered specific hazardous locations requiring safety glazing materials:

1. Glazing in swinging doors except jalousies (see Section 2406.3.1).
2. Glazing in fixed and sliding panels of sliding door assemblies and panels in sliding and bifold closet door assemblies.
3. Glazing in storm doors.
4. Glazing in unframed swinging doors.
5. Glazing in doors and enclosures for hot tubs, whirlpools, saunas, steam rooms, bathtubs and showers. Glazing in any portion of a building wall enclosing these compartments where the bottom exposed edge of the glazing is less than 60 inches (1524 mm) above a standing surface.
6. Glazing in an individual fixed or operable panel adjacent to a door where the nearest exposed edge of the glazing is within a 24-inch (610 mm) arc of either vertical edge of the door in a closed position and where the bottom exposed edge of the glazing is less than 60 inches (1524 mm) above the walking surface.

    Exceptions:

    1. Panels where there is an intervening wall or other permanent barrier between the door and glazing.
    2. Where access through the door is to a closet or storage area 3 feet (914 mm) or less in depth. Glazing in this application shall comply with Section 2406.3, Item 7.
    3. Glazing in walls perpendicular to the plane of the door in a closed position, other than the wall towards which the door swings when opened, in one- and two-family dwellings or within dwelling units in Group R-2.

7. Glazing in an individual fixed or operable panel, other than in those locations described in preceding Items 5 and 6, which meets all of the following conditions:

    7.1. Exposed area of an individual pane greater than 9 square feet (0.84 m^2);

    7.2. Exposed bottom edge less than 18 inches (457 mm) above the floor;

    7.3. Exposed top edge greater than 36 inches (914 mm) above the floor; and

7.4. One or more walking surface(s) within 36 inches (914 mm) horizontally of the plane of the glazing.

Exception: Safety glazing for Item 7 is not required for the following installations:

1. A protective bar $1^1/_2$ inches (38 mm) or more in height, capable of withstanding a horizontal load of 50 pounds plf (730 N/m) without contacting the glass, is installed on the accessible sides of the glazing 34 inches to 38 inches (864 mm to 965 mm) above the floor.
2. The outboard pane in insulating glass units or multiple glazing where the bottom exposed edge of the glass is 25 feet (7620 mm) or more above any grade, roof, walking surface or other horizontal or sloped (within 45 degrees of horizontal) (0.78 rad) surface adjacent to the glass exterior.

8. Glazing in guards and railings, including structural baluster panels and nonstructural in-fill panels, regardless of area or height above a walking surface.
9. Glazing in walls and fences enclosing indoor and outdoor swimming pools, hot tubs and spas where all of the following conditions are present:

9.1. The bottom edge of the glazing on the pool or spa side is less than 60 inches (1524 mm) above a walking surface on the pool or spa side of the glazing; and

9.2. The glazing is within 60 inches (1524 mm) horizontally of the water's edge of a swimming pool or spa.

10. Glazing adjacent to stairways, landings and ramps within 36 inches (914 mm) horizontally of a walking surface; when the exposed surface of the glass is less than 60 inches (1524 mm) above the plane of the adjacent walking surface.
11. Glazing adjacent to stairways within 60 inches (1524 mm) horizontally of the bottom tread of a stairway in any direction when the exposed surface of the glass is less than 60 inches (1524 mm) above the nose of the tread.

Exception: Safety glazing for Item 10 or 11 is not required for the following installations where:

1. The side of a stairway, landing or ramp which has a guardrail or handrail, including balusters or in-fill panels, complying with the provisions of Sections 1012 and 1607.7; and
2. The plane of the glass is greater than 18 inches (457 mm) from the railing.

❖ The provisions of this section apply to all occupancies and building types, except as specifically excluded in this chapter. This section identifies 11 locations that are considered by the code to be hazardous and require the use of safety glass for glazed openings or glazed partitioning. Several figures are included to help illustrate how and where safety glazing must be used.

The first four items included in this section identify various types of doors and door assemblies. Four separate types of doors are presented independently to help avoid any confusion as to what types of doors are required to have safety glass. Collectively, Items 1 through 4 can be summarized by saying that any door containing glazing must be glazed with safety glass or other safety glazing material recognized by the code for that intended purpose. Items 1 through 4 are not restricted to egress or exit doors [see Figure 2406.3(1)].

Jalousie assemblies in doors, as described in Section 2403.5, are not required to have safety glazing. There are other limited exceptions discussed in Item 6 that apply to doors and fixed panels.

Because of the increased possibility of injury caused by falling in the areas identified in Section 2406.3, Item 5, the provisions for safety glazing are intentionally stringent. Falling by itself can cause enough injury without adding cutting from broken glass to the list of possible injuries. Glass wall partitions and glass doors that are not part of the primary building wall construction, interior or exterior, used to enclose these areas, must be constructed from safety glass or other safety glazing material allowed by the code.

Glazed openings in primary building walls, interior or exterior, located within the enclosed compartment areas identified in Section 2406.3, Item 5, where the bottom edge of the glass is within 60 inches (1524 mm) of the standing or walking surface, must be glazed with safety glass or other safety glazing material allowed by the code [see Figure 2406.3(2)]. Glazed openings (i.e., windows) located outside the identified enclosures are not required to be glazed with safety glass, unless where required by other sections in this chapter or elsewhere in the code.

The purpose for including the area in Item 6 as a hazardous location is to provide protection in cases where a person may slip or mistake the glass panel adjacent to a door for a passageway and walk into the glass, or where a person may push against the sidelight with one hand for support while opening the door with the other hand. There are reported accidents where a person's hand has slipped from the doorknob, impacted and broken the glass adjacent to the door, thereby causing injury. This item is applicable to glass adjacent to both exterior and interior doors used for passage for all occupancies and types of buildings.

It is not necessary for an entire piece of glass in a glazed wall or opening to be within the 24-inch (610 mm) arc to require it to be safety glass. If any portion of an individual piece of glass is within the arc, that piece of glass must be safety glass [see Figure 2406.3(3)].

There are three exceptions to Item 6 that eliminate the need for safety glazing adjacent or in close proximity to glass doors. These exceptions are included because they eliminate or greatly reduce the possibility of human contact with fixed glazing.

Exception 1: An intervening wall provides a permanent barrier that prevents people from having physical contact with any glass that is beyond the intervening wall, but is still within the 24-inch (610 mm) arc. This barrier eliminates many of the safety issues addressed in Item 6. This exception may be applied to an interior or exterior wall/glazing condition [See Figure 2406.3(4)]. The code does not specify a minimum height requirement for the intervening wall; therefore, where the wall does not extend to the full height of the rooms or spaces where the door/wall is located, the 60-inch (1524 mm) requirement must be applied. If, for example, the top of the wall is 60 inches (1524 mm) above the floor, any individual piece of glass that extends below the top of the wall must be safety glass, or the top of the wall must be extended to provide a barrier for any piece of nonsafety glazing within 60 inches (1524 mm) of the floor. The same must be applied to the 24-inch (610 mm) arc in the plan dimension. Any glazing within the arc must be safety glazing, or the wall must be extended to include glazing not protected by the wall [see Figure 2406.3(5)].

Exception 2: This exception addresses closets and other storage areas that are not considered "walk-ins." The assumption is that at such locations a building occupant would not need to pass through the door opening to access the storage area, thus reducing the risk of impact with any adjacent glazing.

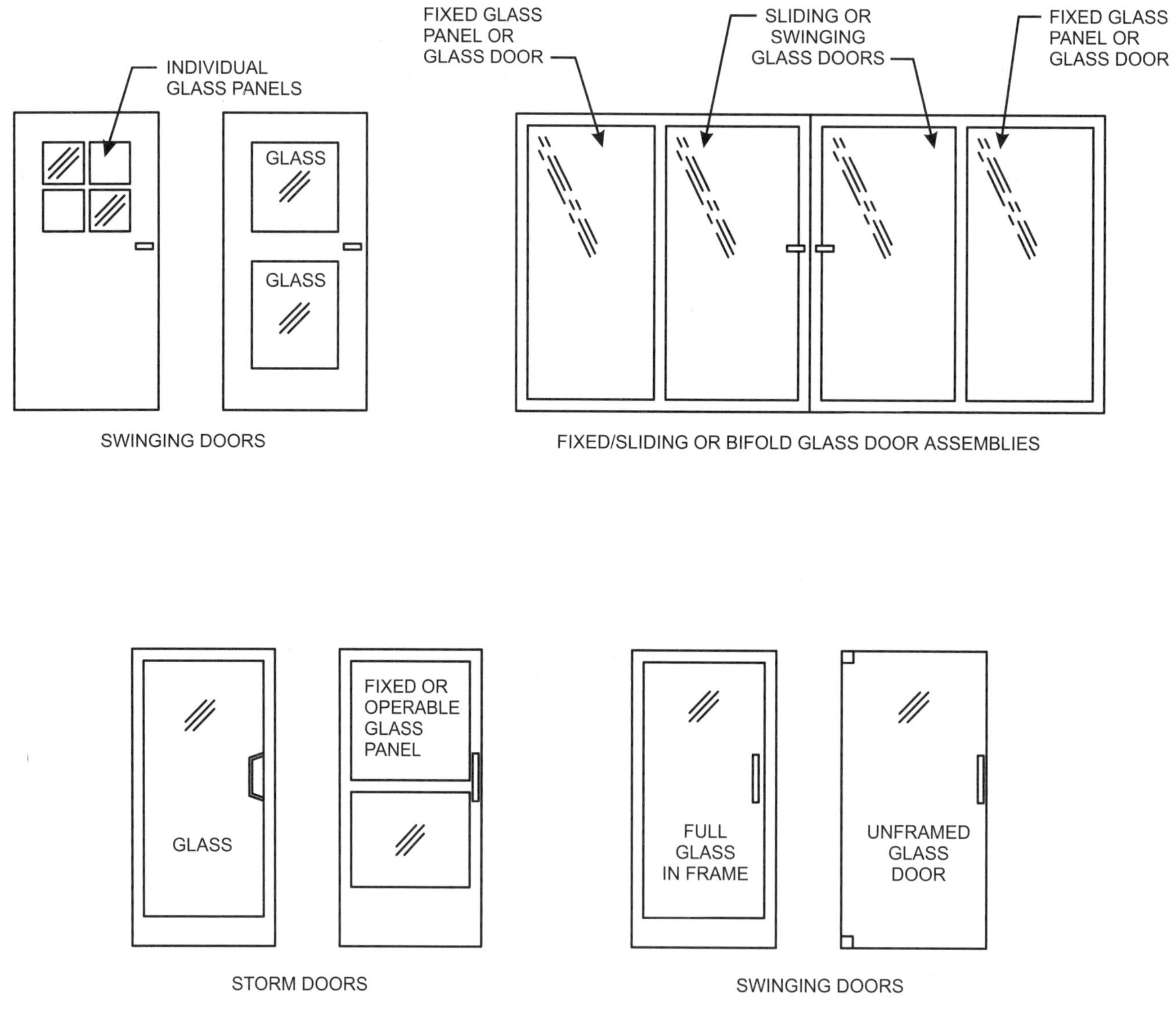

Figure 2406.3(1)
EXAMPLES OF DOORS REQUIRING SAFETY GLAZING

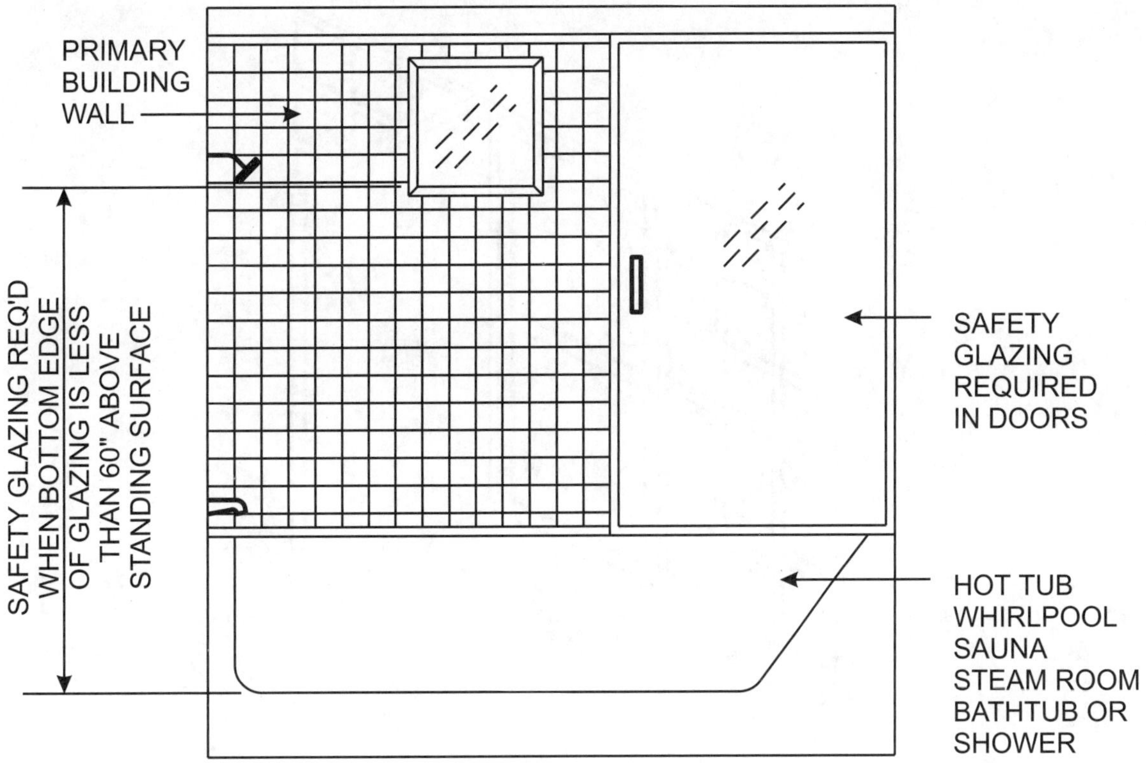

For SI: 1 inch = 25.4 mm.

Figure 2406.3(2)
SAFETY GLAZING IN DOORS AT WET AREAS

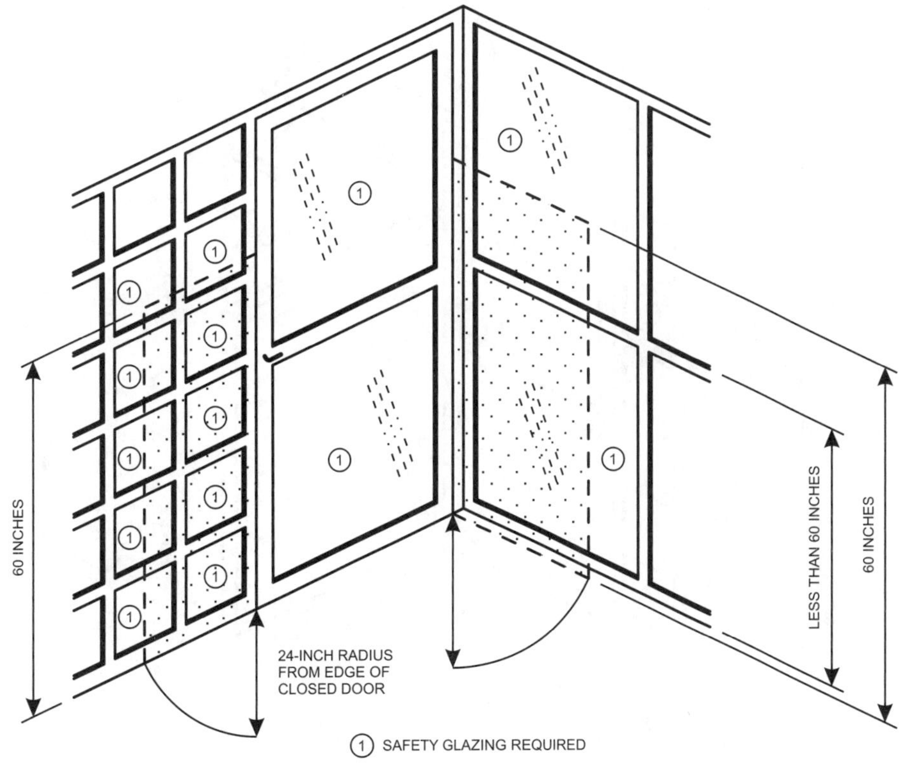

For SI: 1 inch = 25.4 mm.

Figure 2406.3(3)
SAFETY GLAZING IN CLOSE PROXIMITY TO DOORS

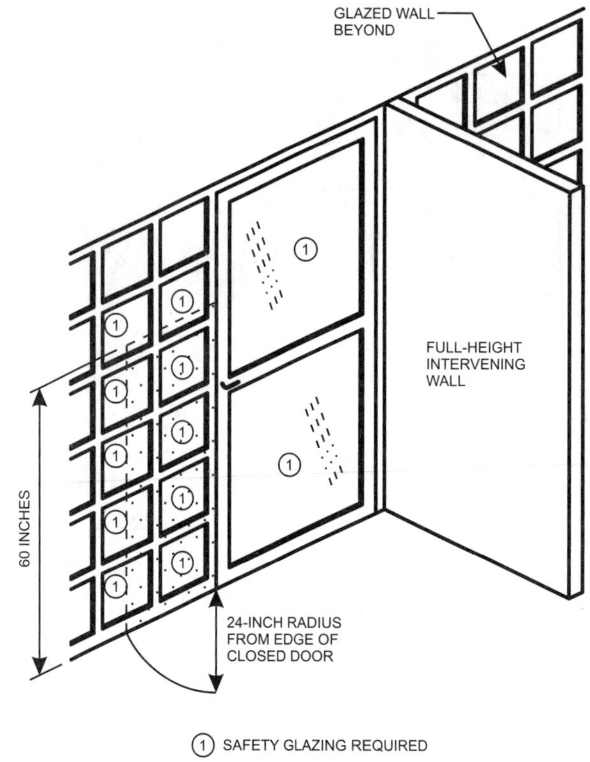

For SI: 1 inch = 25.4 mm.

Figure 2406.3(4)
SAFETY GLAZING AT FULL-HEIGHT INTERVENING WALL

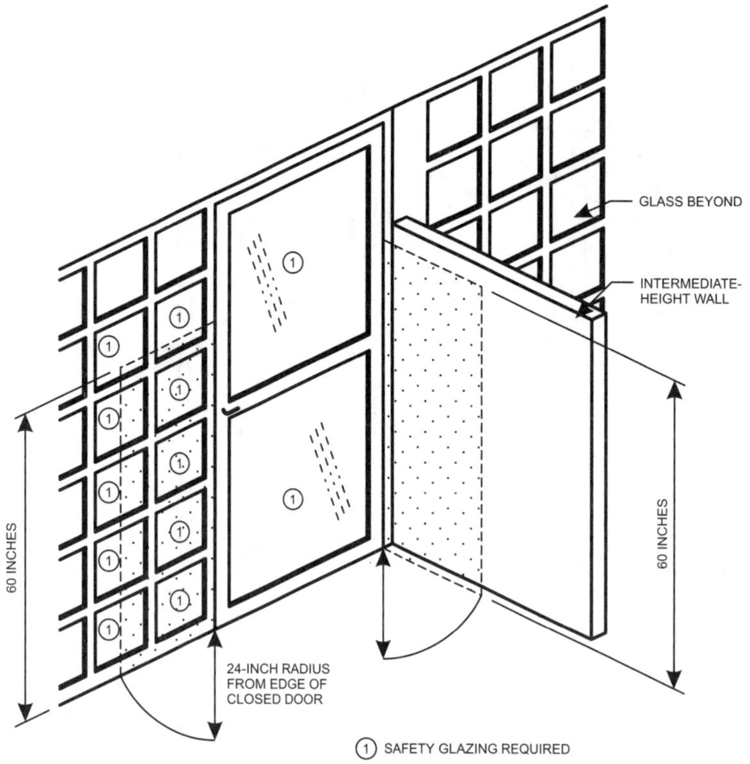

For SI: 1 inch = 25.4 mm.

Figure 2406.3(5)
SAFETY GLAZING AT INTERMEDIATE-HEIGHT INTERVENING WALL

Exception 3: Because the occupants of one- and two-family dwellings are presumed to be familiar with their environment, and walls that are perpendicular to the plane of a door are, by definition, parallel to the direction of travel of an occupant using the door, the risk of impact with glazing is considered to be lower. This decreased risk is recognized by Exception 3. Since there is a chance that an opening door could push a person into the wall towards which the door opens, the exception does not apply, and safety glazing would be required at that location.

Individual dwelling units within Group R-2 occupancies are allowed the same exemption; however, it does not extend to the use of glazing in close proximity to doors in public use areas.

The reason for including the type of glazed opening in Item 7 as a hazardous location is to provide protection where the glazed opening could be mistaken for a passageway, or a clear opening that someone might be able to walk through, fall into or otherwise be accidentally forced into. In the case of a child, the opening would not need to be very large to provide the necessary setting to encourage an accident. A glazed opening sized and located within the parameters described in Item 7 and not protected with safety glazing would most certainly present an unsafe condition [see Figure 2406.3(6)].

The requirements in Item 7 are modeled after the criteria established in CPSC 16 CFR, Part 1201, which requires the use of safety glazing where all of the conditions listed in Items 7.1 through 7.4 occur.

In Item 7.1, the 9 square feet (0.84 m^2) is meant to include an individual piece of glass. It is not intended to limit the number of individual pieces that can be placed next to each other within a single glazed opening. It just requires that every individual piece exceeding the size specified be safety glass.

In Item 7.2, when the bottom edge of a glazed opening is less than 18 inches (457 mm) above a floor or walking surface, it does not present enough of a visual barrier to provide ample warning to prevent someone from mistaking the glazed opening for a passageway or clear opening. People have a natural tendency to look forward while walking rather than looking down; therefore, a more substantial visual barrier is needed at the floor level.

Where Item 7.2 addresses the sill height of glazed openings with no regard to how high the head of the opening may be, it assumes the opening is high enough for an adult to walk through. Item 7.3 addresses the head height of the glazed opening with the assumption that the sill is at the floor level. Item 7.3 is aimed at the same basic set of safety concerns. It is merely directing the focus at a different group of people, that being children. Just as adults can mistake a tall glazed opening for a passageway or clear opening, a small child can do the same for a glazed opening with a low head height. Other examples of potential accidents from impacting glass at this type of opening might include: people in wheelchairs, shopping carts, baby strollers (with or

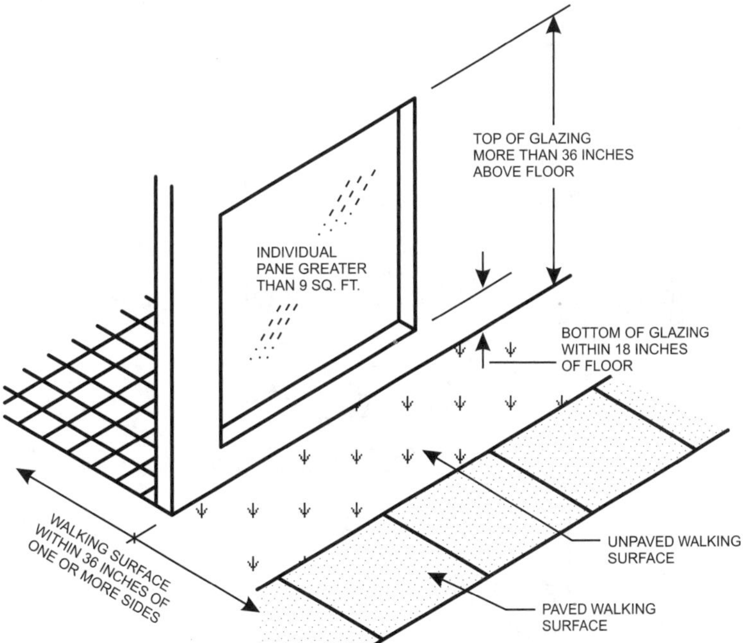

For SI: 1 inch = 25.4 mm, 1 square foot = 0.0929 m^2.

Figure 2406.3(6)
FIXED OR OPERABLE GLAZED OPENING

without baby inside), children on tricycles or other small riding toys. The accident potential for this type of glazed opening is high; therefore, safety glass or other approved safety glazing material is demanded accordingly.

The focus of Item 7.4 for requiring safety glass is no different than it has been throughout Item 7, which is preventing unsafe glazed openings that people can mistake for passageways or clear openings. In Items 7.1 through 7.3, it states that where a walking surface is adjacent to the wall where the glazed opening is located, human contact with the glass in the opening is inevitable. Item 7.4 states that if there is not a walking surface within 36 inches (914 mm) horizontally, in conjunction with the other requirements, the risks described throughout Item 7 are minimal, and safety glass is not needed. It is only when walking surfaces are within 36 inches (914 mm) that human contact becomes a safety concern that must be addressed.

Exceptions 1 and 2 to Item 7 are allowed because the concerns of Item 7 are resolved or they do not exist. When these exceptions are met, there is no need for the use of safety glass or other approved safety glazing.

The protective bar described in Exception 1 will provide a visual as well as a physical barrier, both of which will prevent someone from attempting to walk through a glazed opening [see Figure 2406.3(7)].

The outboard pane of insulating glass as described in Exception 2 is well outside the area that any person could possibly come into contact with and certainly not mistake for a passageway or clear opening. It should be noted that this exception applies only to the outboard pane. The criteria for requiring the inboard pane to be safety glass are prescribed elsewhere in this chapter [see Figure 2406.3(8)].

Item 8 states that regardless of any provision for safety glazing or exemption from requiring safety glazing at any location or application contained in this section or elsewhere in the code—where glass or other approved glazing material is used in guards or railings, either wholly or partly, structural or nonstructural, regardless of size, location, type or any other circumstances not specifically mentioned herein—it must be safety glass or other approved safety glazing material. If the type of glazing material is not available as an approved safety glazing material, it may not be used in guards or railings. There are no exceptions to Item 8.

The reason for requiring safety glazing at walls or fencing used for enclosing swimming pools, hot tubs and spas, as stated in Item 9, is the high amount of activity coupled with the wet floor and walking surface conditions that are always present in these types of facilities or spaces. The conditions presented in Item 9 are focused on glazing in openings as viewed from the pool or spa side of the wall, fence or enclosure. Glazed openings in pool or spa walls viewed from the side of the wall away from the pool or spa are discussed elsewhere in this section. As a point of clarity, a wall, fence or enclosure is any wall that encloses a pool or spa area. It can be a wall constructed specifically to enclose an outdoor pool or spa, an existing or new exterior

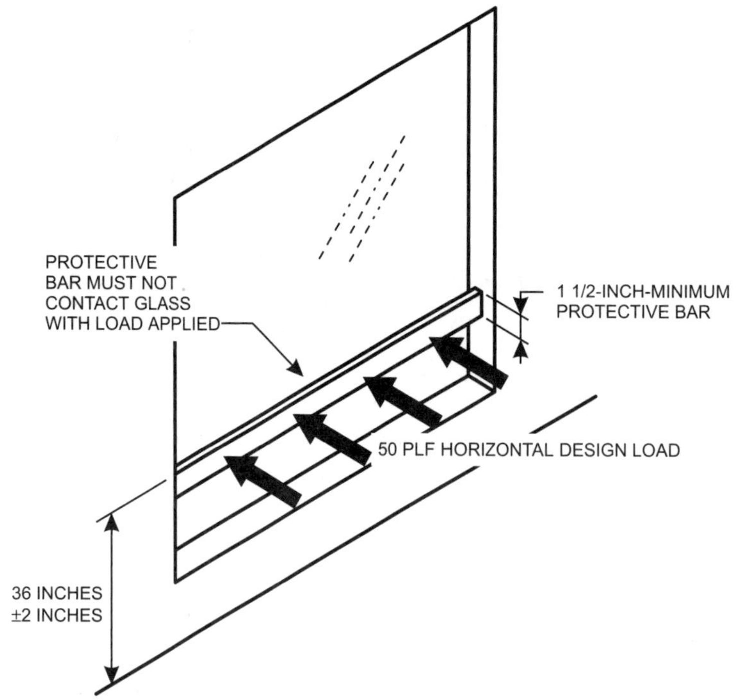

For SI: 1 inch = 25.4 mm, 1 lb/ft = 14.5 N/m.

Figure 2406.3(7)
PROTECTIVE BAR AT GLAZED OPENINGS

building wall, a residential garage wall, a dressing cabana or a pool equipment shed. It can be an interior or exterior wall. It applies to any wall that can support a glazed opening.

Item 9.1 has the requirement for safety glass when the sill of the glazed opening is within 60 inches (1524 mm) of a walking surface that is similar to Item 5 because of the similarity in environmental and usage conditions. In both areas people are exposed to wet floor or walking surfaces in high-activity facilities. Compare the 18-inch (457 mm) bottom edge height in Condition 7.2 of Item 7 to the 60 inches (1524 mm) required here and in Item 5. The 18-inch (457 mm) standard is primarily a visual barrier whereas the 60-inch (1524 mm) standard is a human impact consideration [see Figure 2406.3(9)].

All the things that have been discussed in Exceptions 9 and Item 9.1 can be applied to Item 9.2 as well. The standard here is for any glazing within 60 inches (1524 mm) horizontally when measured from the edge of the swimming pool or spa to the water.

All of the elements in both of these conditions must be present or safety glazing is not required. Glazing in doors is addressed elsewhere in this and other sections.

Stairways and ramps present users with a greater risk for injury caused by falling than a flat surface. Not only is the risk of falling greater when using a stair, but the injuries are generally more severe. Unlike falling on a flat surface where the floor will, for the most part, break a person's fall, there is nothing to stop someone from continuing to fall until he or she reaches the bottom of the stair. The increased risks inherent in stairways, as well as attempting to be consistent with other chapters in the code that mandate more restrictive requirements when addressing safety issues involving stairs and ramps, account for the more restrictive requirements for glazing in and around stairways and ramps.

According to Item 10, this includes any glazing within 36 inches (914 mm) horizontally of any walking surface when the exposed surface of that glazing is within 60 inches (1524 mm) of the walking surface. The walking surface in question would be part of a stair or ramp itself, including top, bottom and intermediate landings. It does not include adjacent floors or other walking surfaces.

In Item 11, the concern is with any glass that may be located within 60 inches (1524 mm) from the bottom tread in a run of stairs and within 60 inches (1524 mm) vertically of the walking surface of a stair. The code does not distinguish between a bottom tread at the primary floor level or at an intermediate landing. The last tread in a run of steps is the bottom tread. The 60-inch (1524 mm) dimension is from any point on the bottom tread, horizontally in any direction to any surface of any glazing within that range.

There is an exception to hazardous location conditions of Item 10 or 11 that relieves the requirement for safety glazing because they effectively eliminate the risk of cutting or piercing injury caused by broken glass upon human contact. Both conditions of the exception to items 10 and 11 must be present to eliminate the need for safety glazing as described in Item 10 or 11.

Guards and railings that comply with the means of egress requirements prescribed in Chapter 10, the structural requirements of Chapter 16 and are located where directed in Item 1 of the exception will effectively serve to prevent someone from crashing into glazed walls or glazed openings in walls located in close vertical proximity to stairs or ramps. Once the guards or railings required in Item 1 are in place, no glazing may be closer than 18 inches (457 mm) to guards or railings when measured horizontally from the guard or railing.

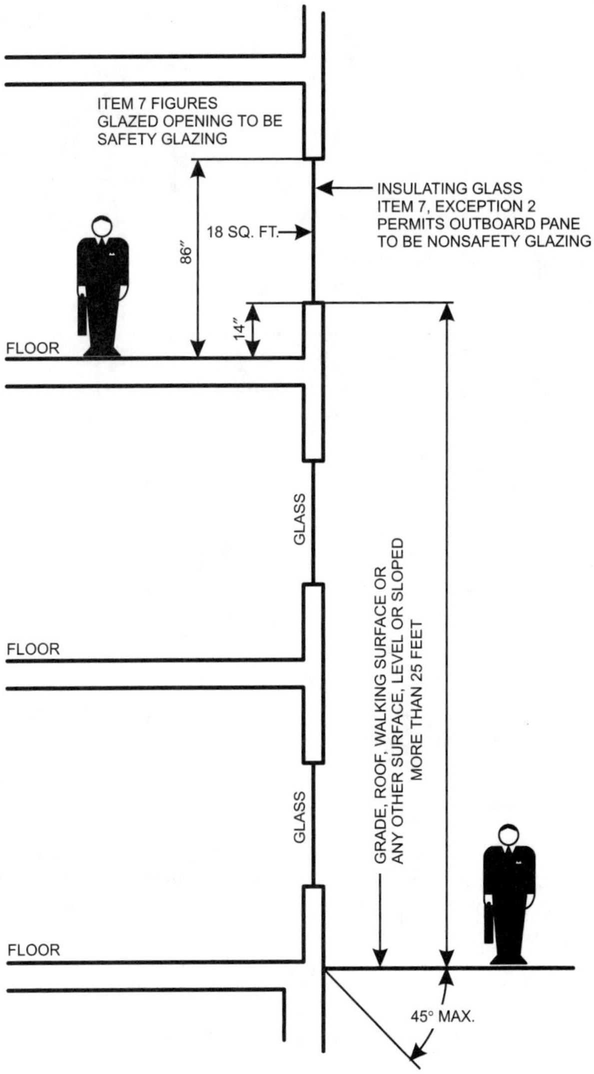

For SI: 1 inch = 25.4 mm, 1 foot = 304.8 mm, 1 square foot = 0.0929 m^2, 1 degree = 0.01745 rad.

Figure 2406.3(8)
HIGH GLAZED OPENINGS

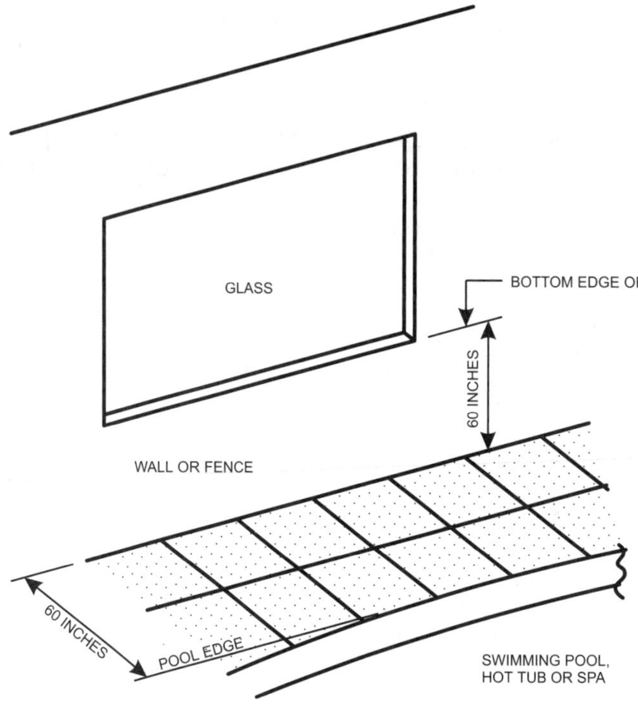

For SI: 1 inch = 25.4 mm.

Figure 2406.3(9)
GLAZING IN POOL ENCLOSURE

2406.3.1 Exceptions. The following products, materials and uses shall not be considered specific hazardous locations:

1. Openings in doors through which a 3-inch (76 mm) sphere is unable to pass.
2. Decorative glass in Section 2406.3, Item 1, 6 or 7.
3. Glazing materials used as curved glazed panels in revolving doors.
4. Commercial refrigerated cabinet glazed doors.
5. Glass-block panels complying with Section 2101.2.5.
6. Louvered windows and jalousies complying with the requirements of Section 2403.5.
7. Mirrors and other glass panels mounted or hung on a surface that provides a continuous backing support.

❖ There are seven specific glazing applications that are not classified as hazardous locations and are not required to be glazed with safety glass or other approved safety glazing material.

1. Glass view panels 3 inches (76 mm) or less in diameter installed in doors are not likely to create a safety hazard if the glass is broken.
2. Decorative glass and decorative glass assemblies are not required to be safety glass for most applications (see Item 8 of Section 2406.3 regarding guards and railings). Stained glass assemblies in doors or walls are readily visible and easily identified, and are not likely to be mistaken for an open passageway. As a rule, stained glass assemblies are made from small pieces of glass that would cause little or no damage if broken.
3. The reason for not classifying revolving doors, which are generally made from curved glass panels, as a hazardous location is that the two primary safety concerns are either not present or pose minimal or no unsafe condition in the case of human impact.

   Several factors must be considered when classifying revolving doors. The first thing to be considered is that they cannot be used as part of the required means of egress. Another thing to consider is that revolving doors are not fixed glass panels that can be mistaken for a passageway or clear opening. In addition, the individual pieces of glass to make the large door panels, by necessity, are required to be very thick. The extra thickness of glass, coupled with the added strength of the curved form and the movable panel, make it highly unlikely that injury could result from glass breakage caused by human impact.
4. This exception is based on CPSC 16 CFR, Part 1201. Technically, glass in doors of refrigerator cabinets should comply with the standard when the door is open. The intent, however, was not to cover doors a person would not ordinarily use for egress. Refrigerated cabinets in food markets would be in this category and are, therefore, ex-

empt. There are also no records of injuries from glass in this application.

5. Glass block, also referred to as "glass unit masonry," is not classified as glass or glazing. It is considered a type of masonry and is subject to complying with Chapter 21 (see Sections 2101.2.4 and 2110).

6. Jalousie and louvered windows are not classified as hazardous locations, and are not required to be glazed with safety glass or other approved safety glazing material. The reasons why these types of glazed openings are exempt are addressed in Section 2403.5.

7. This exception recognizes the potential for cutting or piercing injuries to be minimized when the glass or glazing material is completely supported by a solid, continuous backing material, as is often the case for mounted mirrors or other glass panels. When glazing is fully supported with a solid backing material, it is still subject to breakage by human impact; however, a person's limb or other body part would be protected because the backing material would prevent it from being in the path of falling glass.

2406.4 Fire department access panels. Fire department glass access panels shall be of tempered glass. For insulating glass units, all panes shall be tempered glass.

❖ Because of the need to provide safety for fire-fighting access, this section requires that all glass be safety glazing. When fire-fighting operations include breaking through openings, all glass will obviously be broken, which will subject fire fighters to the hazards of broken glass from all panes.

SECTION 2407
GLASS IN HANDRAILS AND GUARDS

2407.1 Materials. Glass used as structural balustrade panels in railings shall be constructed of either single fully tempered glass, laminated fully tempered glass or laminated heat-strengthened glass. Glazing in railing in-fill panels shall be of an approved safety glazing material that conforms to the provisions of Section 2406.1.1. For all glazing types, the minimum nominal thickness shall be $^1/_4$ inch (6.4 mm). Fully tempered glass and laminated glass shall comply with Category II of CPSC 16 CFR 1201, listed in Chapter 35.

❖ This section provides requirements for glass applications in balusters and in-fill panels for handrails and guardrails.

Glass in handrail and guardrail applications is often exposed to impact from people, carried objects and other items. Early code requirements were vague regarding the appropriate types of glass required for these installations.

Glass used as structural balustrade panels must resist the design loads applied to the railing with an adequate factor of safety (see Figure 2407.1). On a practical basis, the only glass types that are structurally adequate are single and laminated tempered glass and laminated heat-strengthened glass. Other glass types and all safety plastics would need to be excessively thick to resist the design loads on the railing.

Baluster panels are almost always installed in a hard-setting expanding cement. This is to provide a rigid, secure support for the glass. Large compressive stresses are exerted on the glass at the bottom edge. Only the glass types listed in the previous paragraph will routinely resist these forces without breakage. This is another reason why only single and laminated tempered glass and laminated heat-strengthened glass are allowed.

In-fill panels that do not support the railing have less stringent requirements. All safety glazing materials are allowed. A minimum thickness of $^1/_4$ inch (6.4 mm) is listed to verify reasonable penetration and breakage resistance. All types must meet the requirements of CPSC 16 CFR, Part 1201.

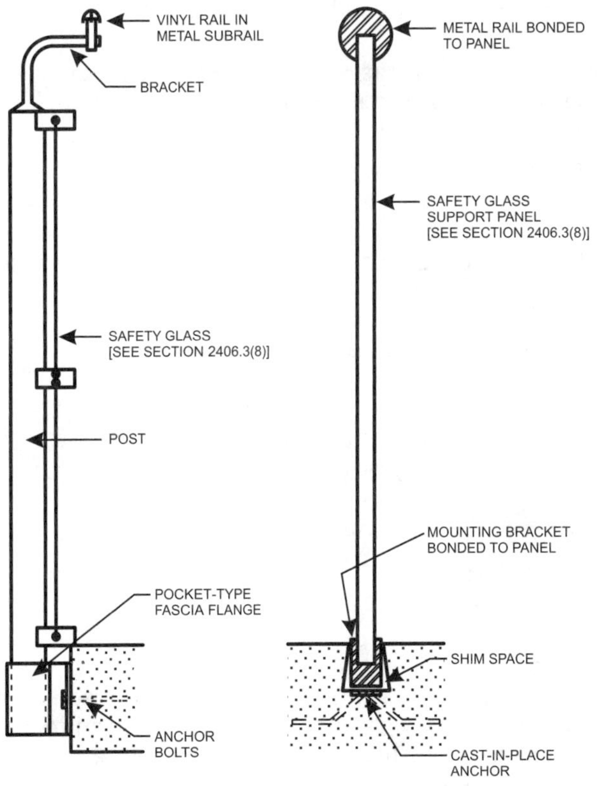

Figure 2407.1
GLASS GUARDS

2407.1.1 Loads. The panels and their support system shall be designed to withstand the loads specified in Section 1607.7. A safety factor of four shall be used.

❖ This section requires that railing systems using glass balusters be designed based on a safety factor of four. Nominally identical panes of glass inherently have a wide variation in strength. The safety factor of four is used in the design to minimize the likelihood that breakage will occur below the design loads. It is not intended that an in-place glass railing system be tested for or capable of withstanding four times the design load.

2407.1.2 Support. Each handrail or guard section shall be supported by a minimum of three glass balusters or shall be otherwise supported to remain in place should one baluster panel fail. Glass balusters shall not be installed without an attached handrail or guard.

❖ Glass balustrade panels have been used for many decades without a significant history of problems. Glass of any type may break, however, if exposed to severe impact or other unusual or unexpected loads. A railing system should be designed so that the failure of a single glass panel does not cause the collapse of the railing. As a result, the code does not allow glass balusters to be used without a railing, and it requires that railings be fastened to a minimum of three glass balustrade panels or otherwise supported if one panel fails.

2407.1.3 Parking garages. Glazing materials shall not be installed in railings in parking garages except for pedestrian areas not exposed to impact from vehicles.

❖ This section is included to remove any ambiguity. The large impact loads from vehicles cannot be resisted by glass balustrades. As a result, they are not allowed under any circumstances where the railing might reasonably be impacted by vehicles. Glass and plastic in-fill panels are also not allowed because of the high likelihood of failure if impacted by vehicles.

SECTION 2408
GLAZING IN ATHLETIC FACILITIES

2408.1 General. Glazing in athletic facilities and similar uses subject to impact loads, which forms whole or partial wall sections or which is used as a door or part of a door, shall comply with this section.

❖ In most applications of safety glazing materials, human impact into glass is an unexpected and infrequent occurrence. Safety glazing is required to minimize injury associated with the unintentional impact and subsequent breakage of fixed glass or glazing. Glass in walls or doors of enclosures for athletic activity is exposed to frequent and intentional impact. For proper performance, the glass not only must break in a relatively safe manner, but also withstand breakage under normal use. A further requirement is that the glass be sufficiently rigid to provide a resilient playing surface. This section includes the performance requirements for glass in this special application.

The code requirement for safety glazing applies to any athletic facility as well as similar areas where the glass is subject to frequent human impact in the opinion of the building official.

2408.2 Racquetball and squash courts.

❖ The primary focus of this section is the walls and doors of racquetball, handball, squash and volleyball courts. In these applications, the glass is intentionally and frequently impacted in the normal course of its use.

2408.2.1 Testing. Test methods and loads for individual glazed areas in racquetball and squash courts subject to impact loads shall conform to those of CPSC 16 CFR, Part 1201, listed in Chapter 35, with impacts being applied at a height of 59 inches (1499 mm) above the playing surface to an actual or simulated glass wall installation with fixtures, fittings and methods of assembly identical to those used in practice.

Glass walls shall comply with the following conditions:

1. A glass wall in a racquetball or squash court, or similar use subject to impact loads, shall remain intact following a test impact.
2. The deflection of such walls shall not be greater than $1^1/_2$ inches (38 mm) at the point of impact for a drop height of 48 inches (1219 mm).

Glass doors shall comply with the following conditions:

1. Glass doors shall remain intact following a test impact at the prescribed height in the center of the door.
2. The relative deflection between the edge of a glass door and the adjacent wall shall not exceed the thickness of the wall plus $^1/_2$ inch (12.7 mm) for a drop height of 48 inches (1219 mm).

❖ The acceptance testing for the glass walls is similar in some respects to that included in CPSC 16 CFR, Part 1201. There are some major variations in the impact testing and end-point criteria. There are two major variations in the testing: the glass is impacted at a height of 59 inches (1499 mm) and the actual (or simulated) construction, complete with fixtures and attachments, is tested. The impact level of 59 inches (1499 mm) is the average shoulder height for an adult male as listed in handbooks for architectural design.

The end-point criteria require that the glass assembly not fail upon impact and the maximum deflection not exceed $1^1/_2$ inches (38 mm). Although the latter is not related to safety, it is included to provide a reasonable playing surface. Both are unique and are not ordinary requirements for safety glazing materials.

Glass doors in the applications cited have been particularly hazardous in some designs. A player may impact and deflect the access door, while the adjacent wall remains in its original position. Occasionally, a player's fingers have become trapped between the door and the wall when the force on the door is released. This section of the code is directed at reducing this hazard.

Though not stated, the intent is that the impact for the glass panels be at the midpoint of the horizontal dimension.

2408.3 Gymnasiums and basketball courts. Glazing in multipurpose gymnasiums, basketball courts and similar athletic facilities subject to human impact loads shall comply with Category II of CPSC 16 CFR 1201, listed in Chapter 35.

❖ Unlike racquetball and squash courts, the intentional impacting of gymnasium and basketball court enclosures is not anticipated. Nevertheless, due to the nature of their use, these facilities have a higher incidence of unintentional glazing impacts. This section requires any glazing that is subject to human impact loads in gymnasiums, basketball courts and similar athletic facilities to be Category II safety glazing (see commentary, Table 2406.1).

SECTION 2409
GLASS IN FLOORS AND SIDEWALKS

2409.1 General. Glass installed in the walking surface of floors, landings, stairwells and similar locations shall comply with Sections 2409.2 through 2409.4.

❖ This section addresses the use of glass in floors, sidewalks and other walking surfaces. It includes the design loads, material specifications and applicable design formulas.

Sections 2409.2 through 2409.4 are applicable wherever glass is installed in a walking surface.

2409.2 Design load. The design for glass used in floors, landings, stair treads and similar locations shall be determined as indicated in Section 2409.4 based on the load that produces the greater stresses from the following:

1. The uniformly distributed unit load (F_u) from Section 1605;
2. The concentrated load (F_c) from Table 1607.1; or
3. The actual load (F_a) produced by the intended use.

The dead load (D) for glass in psf (kN/m^2) shall be taken as the total thickness of the glass plies in inches by 13 (For SI: glass plies in mm by 0.0245). Load reductions allowed by Section 1607.9 are not permitted.

❖ This section establishes the design load criteria that are to be met. The text requires the highest of the three loads specified to be the design basis. The full, unreduced live loads are to be used in the design.

2409.3 Laminated glass. Laminated glass having a minimum of two plies shall be used. The glass shall be capable of supporting the total design load, as indicated in Section 2409.4, with any one ply broken.

❖ Laminated glass is required because it is capable of supporting loads even if one of the glass plies is broken. The text provides for an additional safety factor by specifying that the glass be able to support the entire design load with one of the glass plies broken. As a result, the glass is able to sustain a significant impact, even one that would break one ply of glass without someone falling through it.

2409.4 Design formula. Glass in floors and sidewalks shall be designed to resist the most critical of the following combinations of loads:

$$F_g = 2\,F_u + D \qquad \textbf{(Equation 24-11)}$$

$$F_g = (8F_c/A) + D \qquad \textbf{(Equation 24-12)}$$

$$F_g = F_a + D \qquad \textbf{(Equation 24-13)}$$

where:

A = Area of rectangular glass, ft^2 (m^2).

D = Glass dead load (psf) = 13 t_g (for SI: 0.0245 t_g, kN/m^2).

t_g = Total glass thickness, inches (mm).

F_a = Actual intended use load, psf (kN/m^2).

F_c = Concentrated load, pounds (kN).

F_g = Total load, psf (kN/m^2) on glass.

F_u = Uniformly distributed load, psf (kN/m^2).

The design of the glazing shall be based on

$$F_g \le F_{ga} \qquad \textbf{(Equation 24-14)}$$

where F_g is the maximum load on the glass determined from the load combinations above, and F_{ga} is the maximum allowable load on the glass, computed by the following formula:

$$F_{ga} = 0.67\ c_2\ F_{ge} \qquad \textbf{(Equation 24-15)}$$

where:

F_{ge} = Maximum allowable equivalent load, psf (kN/m^2), determined from Figures 2404(1) through 2404(12) for the applicable glass dimensions and thickness; and

c_2 = Factor determined from Table 2404.2 based on glass type.

The factor, c_2, for laminated glass found in Table 2404.2 shall apply to two-ply laminates only. The value of F_a shall be doubled for dynamic applications.

❖ The first load combination (Equation 24-11) is for the uniform load, the second load combination (Equation 24-12) is for the concentrated load and the third load combination (Equation 24-13) applies to the actual load.

The maximum load from these load combinations must not exceed the glass load-carrying resistance. The 0.67 factor in each of the three equations reduces the probability of breakage to one per 1,000 at the design load.

BIBLIOGRAPHY

The following resource materials are referenced in this chapter or are relevant to the subject matter addressed in this chapter.

AAMA NAFS 1-02, *Voluntary Specification for Windows, Skylights and Glass Doors*. Schaumburg, IL: American Architectural Manufacturers Association, 2002

ANSI Z97.1-84 (R1994), *Safety Performance Specifications and Methods of Test for Safety Glazing Materials Used in Buildings*. New York: American National Standards Institute, 1984.

ASCE 7-02, *Minimum Design Loads for Buildings and Other Structures*. New York: American Society of Civil Engineers, 2002.

ASTM C 1036-91, *Specification for Flat Glass*. West Conshohocken, PA: ASTM International, 1991.

ASTM C 1048-92, *Heat-Treated Flat Glass: Kind HS, Kind FT Coated and Uncoated Glass*. West Conshohocken, PA: ASTM International, 1992.

ASTM E 1300-00, *Standard Practice for Determining Load Resistance of Glass in Buildings*. West Conshohocken, PA: ASTM International, 2000.

CPSC 16 CFR, Part 1201-00, *Safety Standard for Architectural Glazing*. Washington, DC: Consumer Product Safety Commission, 1998.

Glass Design for Sloped Glazing. Des Plaines, IL: American Architectural Manufacturers Association, 1987.

Chapter 25: Gypsum Board And Plaster

General Comments

Chapter 25 contains the provisions and referenced standards that regulate the design, construction and quality of gypsum board and plaster. These represent the most common interior and exterior finish materials in the building industry, and the provisions dealing with them are contained within this single chapter. This chapter deals primarily with quality-control-related issues with regards to material specifications and installation requirements. This chapter addresses weather ratings, and also provides cross references to the requirements for shear walls constructed of either wood stud or light-gage steel stud framing. In addition, provisions for constructing horizontal gypsum board ceiling diaphragms are included.

Where materials described in this chapter are used or required for fire-resistant construction, the code requires that they also comply with the provisions of Chapter 7.

Section 2501 addresses the scope of this chapter.

Section 2502 provides the meaning of words and phrases that are used frequently throughout this chapter. Additional definitions are listed at the beginning of each chapter and in Chapter 2.

Section 2503 explains the responsibility of the contractor to have work inspected by the building official.

Section 2504 defines the construction and application requirements for gypsum board and plaster for use on walls and ceilings.

Section 2505 allows for the use of lightweight shear walls that are sheathed with gypsum board or lath and plaster.

Section 2506 outlines the quality standards for gypsum board and related materials and accessories.

Section 2507 outlines the quality standards for lath, plaster and related materials and accessories.

Section 2508 includes material and construction standards for the use of gypsum board and gypsum plaster materials with reference to fire-resistance-rated assemblies. It also provides requirements for constructing gypsum board ceiling diaphragms using wood joists.

Section 2509 establishes the requirements for the use of gypsum board in wet areas.

Section 2510 describes the material standards and construction requirements for cement plaster and related accessories.

Section 2511 defines the material standards, construction requirements and limitations for plaster and related accessories when applied to an interior surface.

Section 2512 defines the material standards, construction requirements and limitations for plaster and related accessories when applied to exterior and weather-exposed surfaces.

Section 2513 addresses the use of applied aggregate finishes to interior and exterior plaster.

Gypsum board and plaster products represent a line of building products that are, for the most part, manufactured under the control of industry standards. The building official or inspector only needs to verify that the appropriate product has been used and properly installed for the intended use and location. Several key points of installation have been addressed in this chapter where these practices have been known to create problems or when not addressed by other gypsum or plaster product standards.

Purpose

Chapter 25 provides the minimum design requirements through referenced and industry standards for the use of gypsum board, gypsum plaster and cement plaster materials in construction.

SECTION 2501 GENERAL

2501.1 Scope.

❖ Chapter 25 includes the requirements that govern the materials, design, construction, quality and application for both interior and exterior gypsum board and plaster products. Although these products are most commonly used as interior or exterior finished wall and ceiling covering the surface of buildings, proper design and application are necessary to provide weather resistiveness and required fire protection for both structural and nonstructural building components, as well as structural requirements for gypsum board and plaster products and assemblies.

Provisions for the materials, quality, construction and labeling of all gypsum board and plaster materials used in the construction of wall and ceiling coverings are included, along with their method of fastening and, in the case of plaster, the permitted materials for lath, plaster and aggregate in all buildings and structures. In general, all gypsum board and plaster materials are required to conform to the applicable standards as referenced in this chapter.

2501.1.1 General. Provisions of this chapter shall govern the materials, design, construction and quality of gypsum board, lath, gypsum plaster and cement plaster.

❖ Although plaster has many uses in construction, including ornamental and decorative work, its use in the code is regulated purely as a wall and ceiling covering material. The code regulates the installation of wall and ceiling covering materials, as well as the quality standards for the materials themselves. Chapter 25 regulates gypsum board, lath and plaster.

2501.1.2 Performance. Lathing, plastering and gypsum board construction shall be done in the manner and with the materials specified in this chapter, and when required for fire protection, shall also comply with the provisions of Chapter 7.

❖ The extensive use of gypsum board and plaster throughout the construction industry for many years has led to the development of a comprehensive index of tested and proven performance standards. This chapter is directed at identifying the performance standards required to insure a quality of construction that will provide a safe, habitable environment for the life of the building or structure. The standards contained and referenced in this chapter and elsewhere in the code establish the minimum performance standards and requirements for all methods and materials contained in this chapter.

Gypsum plaster, portland cement plaster and gypsum board products and materials have inherent fire-resistant qualities that provide outstanding passive fire protection of building components and assemblies. When these materials are used in the construction of fire-resistive assemblies, the details of construction are to comply with the requirements established by the official reports for the fire tests conducted by recognized testing laboratories in accordance with Chapter 7 and the referenced standards listed in Chapter 35.

2501.1.3 Other materials. Other approved wall or ceiling coverings shall be permitted to be installed in accordance with the recommendations of the manufacturer and the conditions of approval.

❖ Gypsum board and plaster are currently the primary building materials being used for wall and ceiling covering, but there are many other materials in use today for covering interior and exterior walls, ceilings and soffits. This chapter makes no attempt to address the use of any other building materials such as wood, concrete, concrete masonry, brick, metal or vinyl, nor is it intended to limit the use of these or any other materials for similar installations. The material standards and construction regulations for other approved building materials are addressed elsewhere in the code.

SECTION 2502 DEFINITIONS

2502.1 Definitions. The following words and terms shall, for the purposes of this chapter and as used elsewhere in this code, have the meanings shown herein.

❖ Definitions of terms can help in the understanding and application of the code requirements. The purpose for including these definitions within this chapter is to provide more convenient access to them without having to refer back to Chapter 2.

For convenience, these terms are also listed in Chapter 2 with a cross reference to this section. Terms that are italicized provide a visual identification throughout the code text that a definition exists for that term.

The use and application of all defined terms, including those defined herein, are set forth in Section 201.

CEMENT PLASTER. A mixture of portland or blended cement, portland cement or blended cement and hydrated lime, masonry cement or plastic cement and aggregate and other approved materials as specified in this code.

❖ Cement plaster (often referred to as "portland cement plaster" or "stucco") is a cementitious-based plaster material with excellent water-resistant properties. It is the only type of plaster that is permitted by the code to be used as an exterior wall covering and as a base coat and finish coat. Cement plaster can also be used as an interior finish wall covering, but is required by the code to be used in all interior wet areas such as toilet rooms, showers, saunas, steam rooms, indoor swimming pools or any other area that will be exposed to excessive amounts of moisture or humidity for prolonged periods of time.

EXTERIOR SURFACES. Weather-exposed surfaces.

❖ Exterior surfaces are the vertical and horizontal surfaces located on the outside of a building or structure and are subject to weather conditions such as wind, changes in temperature, humidity, moisture and dampness. The commentary on "Weather-exposed surfaces" included in this section allows for specific exceptions to areas on the exterior of a building or structure that will not be exposed to direct contact with water, such as rain, sleet and frozen or melting snow.

GYPSUM BOARD. Gypsum wallboard, gypsum sheathing, gypsum base for gypsum veneer plaster, exterior gypsum soffit board, predecorated gypsum board or water-resistant gypsum backing board complying with the standards listed in Tables 2506.2, 2507.2 and Chapter 35.

❖ Gypsum board is the most commonly used material for interior wall covering. Gypsum board is also used for exterior sheathing, plaster lath and ceiling covering. Because it is installed in sheet form, it is less labor intensive and generally considered more cost effective than other wall and ceiling materials, such as plaster. Gypsum board requires a minimal amount of finishing and will readily accept paint, wallpaper, vinyl fabric, special textured paint and similar surface finish materials.

Gypsum board will be subject to severe failure when placed in direct contact with water or continuous moisture. For this reason, the code does not allow gypsum board to be used in wet areas unless it is provided with a finish material impervious to moisture.

GYPSUM PLASTER. A mixture of calcined gypsum or calcined gypsum and lime and aggregate and other approved materials as specified in this code.

❖ Gypsum plaster is available in several forms and is designed for a variety of applications, including those where fire protection of building components, construction of fire-resistance-rated assemblies within the building or control of sound transmission is needed. Gypsum plaster can be applied using a two- or three-coat method, depending upon the backing or lathing system. Gypsum plaster provides a hard, smooth finish surface that will readily receive most decorative finishes, such as paint, vinyl wall fabric, wallpaper and textured paint. Various aggregates can be added to the plaster composition to achieve different surface textures, from a smooth troweled to a heavily textured finish.

However, the code does not permit the use of gypsum plaster at exterior locations or at wet areas on the interior of the building.

GYPSUM VENEER PLASTER. Gypsum plaster applied to an approved base in one or more coats normally not exceeding $^1/_4$ inch (6.4 mm) in total thickness.

❖ Gypsum veneer plaster is specifically designed for use as a one-coat plaster veneer finish surface. Gypsum veneer plaster must be applied over a solid base such as gypsum board lath, gypsum base for veneer plaster, concrete or concrete unit masonry. Although gypsum veneer plaster is designed to be applied in one coat, it can be applied in two coats, as long as the maximum thickness does not exceed $^1/_4$ inch (6.4 mm).

Gypsum veneer plaster provides a hard, smooth finish surface, much like three-coat gypsum plaster, that will readily receive a variety of decorative finishes. Like gypsum plaster, the code does not permit the use of gypsum veneer plaster at exterior locations or at wet areas on the interior of the building.

INTERIOR SURFACES. Surfaces other than weather-exposed surfaces.

❖ Interior surfaces are all surfaces, exposed and unexposed to view, on the inside of a building or structure that are not subject to damage or failure of their intended purpose because of the effects of uncontrollable weather conditions, such as rain, sleet, snow and wind. Interior surfaces are generally exposed to mechanically controlled climatic conditions through air conditioning, heating and humidity control. Because of these controls, interior surfaces require far less restrictions than exterior surfaces; therefore, the quantities and types of materials and finishes permitted for use on interior surfaces are greatly increased.

WEATHER-EXPOSED SURFACES. Surfaces of walls, ceilings, floors, roofs, soffits and similar surfaces exposed to the weather except the following:

1. Ceilings and roof soffits enclosed by walls, fascia, bulkheads or beams that extend a minimum of 12 inches (305 mm) below such ceiling or roof soffits.
2. Walls or portions of walls beneath an unenclosed roof area, where located a horizontal distance from an open exterior opening equal to at least twice the height of the opening.
3. Ceiling and roof soffits located a minimum horizontal distance of 10 feet (3048 mm) from the outer edges of the ceiling or roof soffits.

❖ Outside surfaces of a building or structure are generally exposed to elements of the weather; therefore, it is necessary to take special precautions to protect them from damage. The most common way to protect exterior surfaces from damage caused by rain, sleet, wind, ice and snow or water from nonweather and other sources is by the use of building materials that are designed, manufactured and installed to resist or withstand the extremes produced by weather. There are many building materials other than those identified in this chapter that provide excellent protection from weather damage, and are permitted by the *International Codes*® for use on weather-exposed surfaces. Requirements for other approved materials, such as, but not limited to, brick, concrete, precast concrete panels, concrete unit masonry, wood, vinyl, aluminum, glass and steel, are specified in other chapters of the code.

There are isolated surfaces located on the exterior of a building or structure that are not directly exposed to or in contact with the damaging effects of the weather. These areas are identified in Items 1, 2 and 3 above. Although these surfaces are located on the exterior of the building or structure, they are provided with a permissible amount of protection and are not directly exposed to the weather with regards to moisture. This must be taken into consideration when wall and ceiling covering materials are being selected. It is not required to use waterproof materials at these locations, but materials approved by the code and the product manufacturer for exterior applications must be used.

WIRE BACKING. Horizontal strands of tautened wire attached to surfaces of vertical supports which, when covered with the building paper, provide a backing for cement plaster.

❖ When metal lath attached to exterior wall studs is used as backing for a plaster base coat, additional reinforcing is necessary to stiffen the wall framing system and minimize unwanted movement to the wall itself. Plaster is a fairly rigid material and excessive movement will cause it to crack. Eighteen-gage metal wire attached horizontally to the face of the wall studs and covered with building paper will provide the additional stiffness required to minimize cracking in the finished surface (see Section 2510.5 for additional commentary on the requirements for wire backing).

SECTION 2503 INSPECTION

2503.1 Inspection. Lath and gypsum board shall be inspected in accordance with Section 109.3.5.

❖ This section establishes guidelines and requirements for the inspection of all components of gypsum board and plaster products, assemblies and installations as outlined in the code and as required or requested by the building official.

The building official is allowed to inspect the installation of all plaster lathing system components prior to being concealed from view by the application of a plaster base coat or finished coats. It is essential not only that the lath be the proper type for the plaster system being used, but also that the components of the system are installed properly.

Plaster wall and ceiling assemblies as well as gypsum board wall and ceiling assemblies are quite often, by design, part of a building's passive fire protection system, as they provide not only the fire-resistant construction necessary for the proper separation of defined areas within a structure, but also fire-protected components of the structure. The building official/inspector is to be given the opportunity by the permit holder to verify and be assured that the assemblies will serve the purpose for which they are designed. This can only be done by allowing the building official to visually inspect the components of the various systems while they are still accessible to allow for visual and physical inspection.

SECTION 2504 VERTICAL AND HORIZONTAL ASSEMBLIES

2504.1 Scope. The following requirements shall be met where construction involves gypsum board, lath and plaster in vertical and horizontal assemblies.

❖ Section 2504, along with the referenced standards and structural limitations and requirements addressed elsewhere, define the minimum requirements for the structural members supporting the weight of gypsum board and plaster where either or both are used in the construction of walls and ceilings.

Vertical assemblies (walls and partitions) must be designed in accordance with the engineering chapters of the code for proper structural support of gypsum board or plaster wall covering. If the wall or partition construction and assembly are not designed to resist the loads prescribed in the code, the wall covering materials will perform inadequately. The general design provisions of Chapter 16 and the deflection limitations are particularly important in this regard.

The code requirements for the design and framing of horizontal assemblies (ceilings) are intended to achieve the same results as described for vertical assemblies. Gypsum board and plaster ceilings are generally supported by a suspended metal grid framing system of metal channels or T-bars suspended by steel wires tied to the structure above. The ceiling covering may also be fastened directly to the structural framing member or furring which is fastened to the structural frame. All ceiling support systems must comply with the structural load requirements of Chapter 16.

2504.1.1 Wood framing. Wood supports for lath or gypsum board, as well as wood stripping or furring, shall not be less than 2 inches (51 mm) nominal thickness in the least dimension.

Exception: The minimum nominal dimension of wood furring strips installed over solid backing shall not be less than 1 inch by 2 inches (25 mm by 51 mm).

❖ The minimum dimensions required by this section for wood supports for gypsum board or lath and plaster are intended to provide adequate support and reduce surface distortion due to warping of wood supports.

Wood furring strips attached to a solid backing may be used to support wall or ceiling covering. The furring must be installed with the wide face against the solid backing. This will provide a nominal 2-inch-wide (51 mm) nailing surface equal to the nailing surface of a standard wood stud, joist or rafter.

2504.1.2 Studless partitions. The minimum thickness of vertically erected studless solid plaster partitions of $^{3}/_{8}$-inch (9.5 mm) and $^{3}/_{4}$-inch (19.1 mm) rib metal lath or $^{1}/_{2}$-inch-thick (12.7 mm) long-length gypsum lath and gypsum board partitions shall be 2 inches (51 mm).

❖ As the name implies, a studless partition is constructed of gypsum board lath or ribbed metal lath with plaster applied to each face, without the stud frame wall that typically provides support for the wall covering [see Figures 2504.1.2(1) and 2504.1.2(2)]. The lath and plaster form the structural elements of the partition that must resist the lateral loads applied to the wall. Studless partitions may only be used as nonload-bearing partitions; however, they must comply with the deflection and general design requirements specified in Chapter 16.

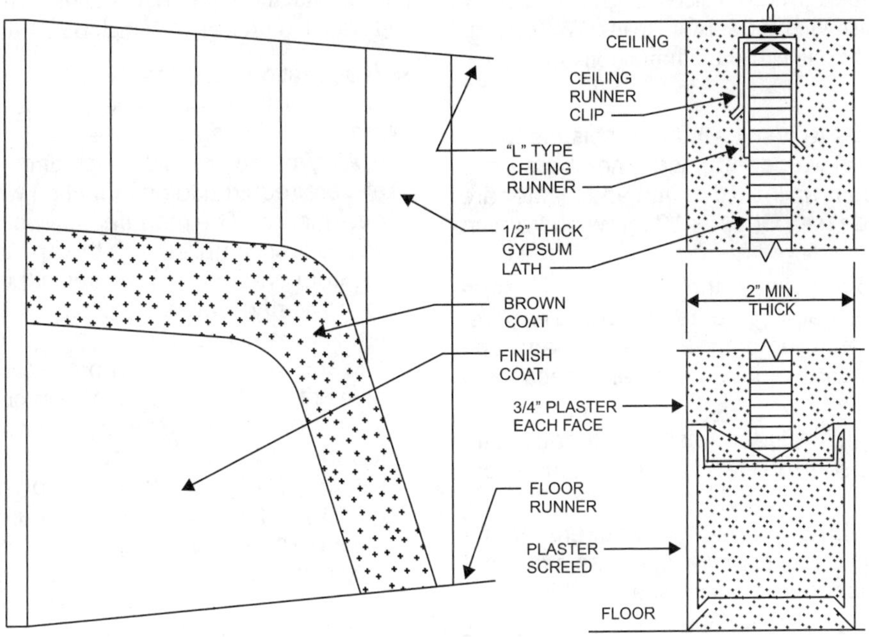

For SI: 1 inch = 25 mm.

Figure 2504.1.2(1)
STUDLESS PLASTER PARTITION WITH GYPSUM BOARD LATH

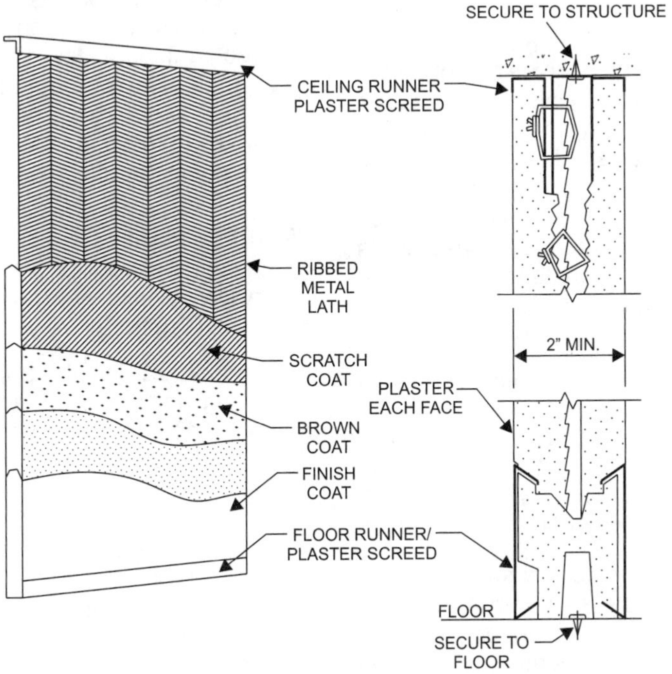

For SI: 1 inch = 25 mm.

Figure 2504.1.2(2)
STUDLESS PLASTER PARTITION WITH RIBBED METAL LATH

SECTION 2505 SHEAR WALL CONSTRUCTION

2505.1 Resistance to shear (wood framing). Wood-framed shear walls sheathed with gypsum board, lath and plaster shall be designed and constructed in accordance with Section 2306.4 and are permitted to resist wind and seismic loads. Walls resisting seismic loads shall be subject to the limitations in Section 1617.6.

❖ Shear walls designed to resist wind and seismic loads and constructed of wood framing members and sheathed with gypsum wallboard or lath and plaster are permitted by the code (see Chapter 23 for wood framing requirements).

- Section 2306.4 defines the structural requirements for wood framing members, nailing requirements, diaphragm bracing, allowable loads and all other structural requirements and necessary incidentals.
- Section 2306.4.5 includes seismic application limitations for gypsum, plaster and wood frame shear walls.
- Section 2306.4.5.1 outlines the requirements for the application of gypsum wallboard as a component of a wood frame shear wall assembly.
- Table 2306.4.5 includes the design guidelines and construction requirements necessary for use of wood frame shear walls using gypsum board or lath and plaster as the finished wall surface.

The cross reference is to the limitations for basic seismic force resisting systems found in Section 1617.6.

Care must be taken when selecting gypsum board or plaster materials if they are being used as part of an exterior wall that will be exposed to the weather. Gypsum board, gypsum board lath, gypsum plaster and gypsum veneer plaster are not water resistant and are intended for interior applications only. Specific gypsum board materials covered with an appropriate exposed finish material or metal lath with cement plaster base coats and finish coat (stucco) are appropriate for exterior applications exposed to the weather.

2505.2 Resistance to shear (steel framing). Cold-formed steel framed shear walls sheathed with gypsum board and constructed in accordance with the materials and provisions of Sections 2211.1, 2211.2, 2211.2.1 and 2211.2.2.3 are permitted to resist wind and seismic loads. Walls resisting seismic loads shall be subject to the limitations in Section 1617.6.

❖ Shear walls designed to resist wind and seismic loads and constructed of light-gage steel stud framing members and sheathed with gypsum board are permitted by the code. Structural design guidelines and construction requirements for these shear walls are provided in Chapter 22. The cross reference is to the limitations for basic seismic force resisting systems found in Section 1617.6.

SECTION 2506 GYPSUM BOARD MATERIALS

2506.1 General. Gypsum board materials and accessories shall be identified by the manufacturer's designation to indicate compliance with the appropriate standards referenced in this section and stored to protect such materials from the weather.

❖ This section contains the appropriate referenced standards for the proper use and quality of gypsum board materials and accessories.

All gypsum products, including gypsum board materials, are required to be identified with the manufacturer's designation. The manufacturer is required to certify that its product complies with the applicable standard for the gypsum product. This enables the installer and inspector to verify that the product is being used for its intended installation.

Generally, all gypsum products and accessories are delivered to the job site in their original packages, containers or bundles. These products will be identified with the manufacturer's name, trademark and any applicable standard designations. For example, for gypsum board products, the thickness and type are shown on the end-bundling tapes or the board panel. The premixed paste products are identified on their container, and bag products are appropriately identified on the face of the bag. With regard to weather protection, all gypsum products must be kept dry because of the deleterious effect of moisture. If properly stored above ground and fully protected from the weather and exposure to direct sunlight, gypsum products and accessories may be stored outside; however, the storage of gypsum board is not recommended in exterior locations. Exposure to the elements may result in water stain, light discoloration, mildew, loss of structural strength or sagging, all of which could hinder the performance of the product structurally and in fire performance, as well as the ability to be finished in the intended fashion. Gypsum board should be stored flat—never on edge or end.

2506.2 Standards. Gypsum board materials shall conform to the appropriate standards listed in Table 2506.2 and Chapter 35 and, where required for fire protection, shall conform to the provisions of Chapter 7.

❖ Table 2506.2 lists the appropriate American Society for Testing and Materials (ASTM) standards for gypsum board materials and accessories. The most common building material used in Table 2506.2 is gypsum wallboard (also referred to as "gypsum board," "gyp. board" or "sheetrock"). ASTM C 36 provides the design specifications for gypsum board used for walls and ceilings. Gypsum board consists of a noncombustible core, which is primarily gypsum, and is surfaced with paper firmly bonded to the core. The paper surfacing material is less than 0.125-inches (3.18 mm) thick and has a flame spread index less than 25 when tested in accordance with ASTM E 84. This requirement is in accordance with fire-resistant requirements of Chapter 7 for composite materials classified as noncombustible.

Gypsum materials that are required to have a fire-resistance rating in accordance with the code must also comply with the applicable provisions of Chapter 7. In addition, Chapter 7 establishes and defines the requirements for the fire-resistance ratings of building assemblies and structural components. Such assemblies are to be constructed in accordance with the test procedures as prescribed in the appropriate referenced standards listed in Chapter 35.

TABLE 2506.2
GYPSUM BOARD MATERIALS AND ACCESSORIES

MATERIAL	STANDARD
Accessories for gypsum board	ASTM C 1047
Gypsum sheathing	ASTM C 79
Gypsum wallboard	ASTM C 36
Joint reinforcing tape and compound	ASTM C 474; C 475
Nails for gypsum boards	ASTM C 514, F 547, F 1667
Steel screws	ASTM C 954; C 1002
Steel studs, nonload bearing	ASTM C 645
Steel studs, load bearing	ASTM C 955
Water-resistant gypsum backing board	ASTM C 630
Exterior soffit board	ASTM C 931
Fiber-reinforced gypsum panels	ASTM C 1278
Gypsum backing board and gypsum shaftliner board	ASTM C 442
Gypsum ceiling board	ASTM C 1395
Standard specification for gypsum board	ASTM C 1396
Predecorated gypsum board	ASTM C 960
Adhesives for fastening gypsum wallboard	ASTM C 557
Testing gypsum and gypsum products	ASTM C 22; C 472; C 473
Glass mat gypsum substrate	ASTM C 1177
Glass mat gypsum backing panel	ASTM C 1178

❖ Table 2506.2 includes a listing of the materials and accessories approved by the code for use in gypsum board construction, along with the appropriate ASTM standard for each. The installation of each of these materials must conform to all the requirements of the referenced standard. It is not the intent of the code, however, by way of this table, to limit or restrict alternative materials or accessories in the use of gypsum board and its associated incidentals as a building material acknowledged and approved by the code. Any and all materials not specifically mentioned, described or referenced, however, are subject to the requirements of the code, as well as the entire family of *International Codes.*

The gypsum industry has established a phase-in program for the use of the new "umbrella" standard ASTM C 1396. Manufacturers have begun to dual-label products using both the current standard as well as ASTM C 1396. Over time, the individual gypsum product standards will be phased out.

2506.2.1 Other materials. Metal suspension systems for acoustical and lay-in panel ceilings shall conform with ASTM C 635 listed in Chapter 35 and Section 9.6.2.6 of ASCE 7 for installation in high seismic areas.

❖ Gypsum board is often attached to a light-gage metal ceiling framing system constructed from hat channels and C channels installed perpendicular to each other, forming a grid. This grid is suspended from the floor or roof structure overhead by steel wires. The structural requirements for this suspended ceiling support system are outlined in Chapter 16.

Once the suspended ceiling support grid is installed as required by Chapter 16, the gypsum board ceiling is attached to the grid the same as any other horizontal assembly described in this chapter. Ceilings constructed using suspended hat channels and C channels must comply with the guidelines of ASTM C 754.

SECTION 2507
LATHING AND PLASTERING

2507.1 General. Lathing and plastering materials and accessories shall be marked by the manufacturer's designation to indicate compliance with the appropriate standards referenced in this section and stored in such a manner to protect them from the weather.

❖ These basic requirements apply to lathing and plastering materials, products and accessories referenced in this chapter. This section requires that all lath and plaster materials be identified with the manufacturer's designation and indicate compliance with the appropriate standards. This enables the installer and the inspector to verify that the product is being used for its intended installation. All accessories must also be marked with a reference to the appropriate standard. The manufacturer is required to certify that its product complies with the applicable standard for lathing and plastering material. Lath and plaster products and accessories are delivered to the job site in their original packages, containers or bundles. These products must be identified with the manufacturer's name, trademark and any applicable standard designations. With regard to weather protection, plaster products must be kept dry because of the deleterious effect of moisture. If properly stored above ground, fully protected from the weather and exposure to direct sunlight and away from condensation and damp surfaces, lath and plaster products and accessories may be stored outside; however, it is not recommended.

2507.2 Standards. Lathing and plastering materials shall conform to the standards listed in Table 2507.2 and Chapter 35 and, where required for fire protection, shall also conform to the provisions of Chapter 7.

❖ The appropriate referenced standards that regulate all lath and plastering materials are contained in this section. It also addresses the proper marking of materials, protection from weather and installation requirements. Additionally, provisions of Chapter 7 are referenced for building assemblies utilizing lath and plastering materials that require a fire-resistance rating; therefore, the assemblies must comply with the provisions of Chapter 7.

TABLE 2507.2
LATH, PLASTERING MATERIALS AND ACCESSORIES

MATERIAL	STANDARD
Accessories for gypsum veneer base	ASTM C 1047
Exterior plaster bonding compounds	ASTM C 932
Gypsum base for veneer plasters	ASTM C 588
Gypsum casting and molding plaster	ASTM C 59
Gypsum Keene's cement	ASTM C 61
Gypsum lath	ASTM C 37
Gypsum plaster	ASTM C 28
Gypsum veneer plaster	ASTM C 587
Interior bonding compounds, gypsum	ASTM C 631
Lime plasters	ASTM C 5; C 206
Masonry cement	ASTM C 91
Metal lath	ASTM C 847
Plaster aggregates	
Sand	ASTM C 35; C 897
Perlite	ASTM C 35
Vermiculite	ASTM C 35
Plastic cement	ASTM C 1328
Blended cement	ASTM C 595
Portland cement	ASTM C 150
Steel studs and track	ASTM C 645; C 955
Steel screws	ASTM C 1002; C 954
Welded wire lath	ASTM C 933
Woven wire plaster base	ASTM C 1032

❖ Table 2507.2 includes a listing of the materials and accessories approved by the code for use in plaster construction, along with the appropriate ASTM standard for each. The installation of each of these materials must conform to all the requirements of the referenced standard. It is not the intent of the code to limit or restrict alternative materials or accessories in the use of plaster and its associated incidentals as a building material acknowledged and approved by the code. Any and all materials not specifically mentioned, described or referenced herein are subject to the requirements of the code, as well as the entire family of *International Codes*.

SECTION 2508
GYPSUM CONSTRUCTION

2508.1 General. Gypsum board and gypsum plaster construction shall be of the materials listed in Tables 2506.2 and 2507.2. These materials shall be assembled and installed in compliance with the appropriate standards listed in Tables 2508.1 and 2511.1, and Chapter 35.

❖ This section contains the referenced standards and requirements for the proper installation of gypsum construction.

This section references Table 2506.2 for the appropriate materials required in gypsum board and gypsum plaster construction, as well as the corresponding material standard. Table 2508.1 addresses the appropriate installation standards required for use in gypsum construction. Gypsum board in fire-resistance-rated assemblies is to be installed in accordance with the fastening schedule utilized in the tested assembly. The type and spacing of the fasteners are reflected in the fire test reports of various assemblies, including the summary of fire tests contained in various resource documents that compile summaries of fire test results, such as Underwriters Laboratories' *Fire Resistance Directory*, Warnock Hersey's *Certification Listing* or the Gypsum Association's *Fire Resistance Design Manual* D (GA-600). Refer to Chapter 7 and its commentary for additional information related to fire-resistance-rated assemblies.

TABLE 2508.1
INSTALLATION OF GYPSUM CONSTRUCTION

MATERIAL	STANDARD
Gypsum sheathing	ASTM C 1280
Gypsum veneer base	ASTM C 844
Gypsum board	GA-216; ASTM C 840
Interior lathing and furring	ASTM C 841
Steel framing for gypsum boards	ASTM C 754; C 1007

❖ Table 2508.1 includes the installation requirements from the code for the use of gypsum and gypsum-related construction products, along with the appropriate ASTM standard for each. The installation of each of these materials must conform to all the requirements of the referenced standard. There is no intent to limit or restrict alternative methods of construction acknowledged and approved by the code. Any and all methods not specifically mentioned, described or referenced herein are subject to the requirements of the code, as well as the entire family of *International Codes.*

2508.2 Limitations. Gypsum wallboard or gypsum plaster shall not be used in any exterior surface where such gypsum construction will be exposed directly to the weather. Gypsum wallboard shall not be used where there will be direct exposure to water or continuous high humidity conditions. Gypsum sheathing shall be installed on exterior surfaces in accordance with ASTM C 1280.

❖ Because of the detrimental effect of moisture on gypsum products, gypsum-based materials and construction must be protected from the weather. See Section 2502 for the definition of "Weather-exposed surfaces," which includes some limited conditions where gypsum board is permitted to be used on the "outside" of a building or structure.

2508.2.1 Weather protection. Gypsum wallboard, gypsum lath or gypsum plaster shall not be installed until weather protection for the installation is provided.

❖ This section relates to the necessary environmental conditions for proper plaster application and hydration. Temperature limitations are needed to provide a uniform heating condition for a minimum of one week prior to the start of plastering, and are the same as the requirements stated in ASTM C 842. This minimizes the possibility of plaster cracking because of movements or stresses caused by thermal changes.

2508.3 Single-ply application. Edges and ends of gypsum board shall occur on the framing members, except those edges and ends that are perpendicular to the framing members. Edges and ends of gypsum board shall be in moderate contact except in concealed spaces where fire-resistance-rated construction, shear resistance or diaphragm action is not required.

❖ The application of gypsum board is specified in this chapter for nonfire-resistant construction or construction where diaphragm (shear wall) action is not required. Chapter 7 and fire test reports will establish the means of fastening and supporting the ends and edges of gypsum board for nonfire-resistant assemblies. In the case of stud walls required for diaphragm action, Chapters 22 and 23 establish structural requirements such as sizes and spacing fasteners, allowable loads and the conditions of end and edge supports.

Gypsum board is the generic name for a family of sheet products consisting of a noncombustible core, primarily composed of gypsum, with a paper surfacing on each face. The installation and finishing of gypsum board materials must comply with the referenced standards in Table 2508.1. *Application and Finishing of Gypsum Board*, referred to as "GA 216," published by the Gypsum Association, is the most commonly recommended specification for the installation and finishing of gypsum board.

- **Framing:** As with any gypsum board or gypsum sheet product, it is critical to inspect the framing on which the gypsum material is to be attached. Finishing of gypsum board will be made extremely difficult when installed over unlevel, unplumb, out of square or structurally inadequate framing. All of these concerns must be satisfied before the installation of any gypsum board product can begin.

  Gypsum board can be attached to a variety of framing members, including but not necessarily limited to, wood or metal studs; joists and rafters; suspended metal frame ceiling systems and wood or metal furring on concrete or masonry walls. Framing systems that are not installed square, level and plumb will create stress points where the gypsum board is attached to the framing. Stress points can cause the gypsum board to crack, shear or delaminate. This can become a significant issue when fire-resistance-rated assemblies are compromised.
- **Fastening:** The type of fastening devices and methods for attaching gypsum board to framing are dependent upon the type of framing system being used and the type of assembly being constructed. Screws, nails and construction adhesives are approved types of fasteners. Fasteners are to be selected and installed as required by the specifications referenced in GA 216, except as required elsewhere in the code for fire-resistance-rated assemblies, shear wall construction or where more restrictive requirements are provided by the referenced standards listed in Chapter 35.
- **Finishing:** Gypsum must be finished with materials and accessories designed and approved for the application for which they are being used. Joint tape, joint compound and all accessories are to be approved and installed as recommended by the manufacturer and as required by the referenced standards.
- **Multiple-ply application:** This procedure for gypsum board is generally intended to increase the fire-resistance rating of an assembly, the shear-resisting strength of a shear wall assembly or the sound transmission class for an assembly. The construction and technical requirements for these and similar types of specialized assemblies are covered elsewhere in the *International Codes.* The requirements for gypsum board itself are as defined in this chapter, except where more restrictive requirements are mandated for a specialized assembly.

2508.3.1 Floating angles. Fasteners at the top and bottom plates of vertical assemblies, or the edges and ends of horizontal assemblies perpendicular to supports, and at the wall line are permitted to be omitted except on shear resisting elements or fire-resistance-rated assemblies. Fasteners shall be applied in such a manner as not to fracture the face paper with the fastener head.

❖ Floating angles are the edges of gypsum board that are not directly fastened to a framing member where a change in the plane of the gypsum board occurs. Gypsum board used as part of a shear wall assembly or as a component part of any fire-resistance-rated assembly must have all edges fastened to a framing member.

It is given that gypsum board used as part of a vertical assembly (wall) is fastened to the wall studs along the edges and in the field of the gypsum board as defined in GA 216. Adequate support for the weight of the gypsum board itself will be provided, therefore eliminating the need for additional fastening at the top and bottom plates of the wall framing. Fastening is not to be omitted at any inside or outside corner of a wall assembly.

Fasteners may be omitted at the edges of horizontal assemblies (ceiling and soffits) when the edge of the gypsum board is perpendicular to the framing member. As with wall assemblies, it is given that fasteners located along the edges and in the field of the gypsum board in accordance with GA 216 will provide adequate support for the weight of the gypsum board itself. Edges of horizontal gypsum board assemblies that are parallel to framing members must be fastened to the framing member except where the parallel edge is at the intersection of the wall and ceiling. Typically, the gypsum board will be installed on the ceiling before it is installed on the wall. The edge of the gypsum board on the wall will be butted against the underside edge of the ceiling board providing additional edge support for the ceiling.

This section permits the elimination of unnecessary fastening of gypsum board in specific locations and conditions. It is not the intent of this section to compromise the integrity of the gypsum board support or fastening system. Gypsum board must be fastened to prevent sagging, warping, waviness or any visual defects.

2508.4 Joint treatment. Gypsum board fire-resistance-rated assemblies shall have joints and fasteners treated.

Exception: Joint and fastener treatment need not be provided where any of the following conditions occur:

1. Where the gypsum board is to receive a decorative finish such as wood paneling, battens, acoustical finishes or any similar application that would be equivalent to joint treatment.
2. On single-layer systems where joints occur over wood framing members.
3. Square edge or tongue-and-groove edge gypsum board (V-edge), gypsum backing board or gypsum sheathing.
4. On multilayer systems where the joints of adjacent layers are offset from one to another.
5. Assemblies tested without joint treatment.

❖ This section requires joint and fastener treatment for all gypsum board fire-resistance-rated assemblies. The code also identifies and permits exceptions where the gypsum board is to receive a decorative finish or any similar application that is considered to be equivalent to the joint or fastener treatment.

These exceptions indicate that joint treatment is not required for assemblies tested without it:

1. Where the finish layer of gypsum board is to receive a decorative covering or similar type of applied material that will cover exposed joints, finishing of joints is not required; however, the applied material must provide the same fire-resistance rating of the assembly that it is being applied to.
2. Joint treatment is not required in single-layer applications where joints occur directly over wood framing members. In effect, the additional joint treatment does not materially increase the fire rating of the assembly, and many partitions have been tested and passed the fire test without the added protection of joint and fastener treatment as part of the test design.
3. Gypsum board with factory-designed edges that will prevent the penetration of smoke or fire into a rated assembly does not have to be finished. The factory edges listed will minimize gaps at joints and maintain the integrity of the assembly.
4. Multilayer assemblies with offset joints will eliminate any gaps in the system; therefore, fire smoke and harmful gases cannot penetrate the fire-resistance-rated assembly. This makes finishing of joints unnecessary.
5. Where approved fire tests indicate that joint and fastener treatment is included as a part of the tested assembly, the code requires that joint and fastener treatment be applied in the same manner and extent as the approved test assembly. Where the approved test assembly does not specifically state whether joint or fastener treatment is part of the test assembly, the code requires that joint and fastener treatment is to be applied to any fire-resistance-rated assembly. Assemblies tested without joint treatment need not be finished.

As indicated elsewhere in this section, with regard to gypsum board application, joint and fastener treatment is primarily applied for aesthetic reasons, and to provide a satisfactory finish surface for the application of decorative finishes, such as paint, wallpaper, vinyl wall fabric, textured paint or a number of other finishes used for the appearance of a rated or nonrated assembly.

2508.5 Horizontal gypsum board diaphragm ceilings. Gypsum board shall be permitted to be used on wood joists to create a horizontal diaphragm ceiling in accordance with Table 2508.5.

❖ Generally a gypsum board ceiling does not serve as a load-carrying structural element for other than its own weight. This section provides installation requirements [refer to Figure 2508.5(1)] that allow the use of a horizontal gypsum board ceiling diaphragm. This permits a wood frame ceiling to carry limited lateral loads, such as a portion of the component earthquake force (see Section 1621) on a partition as shown in Figure 2508.5(2).

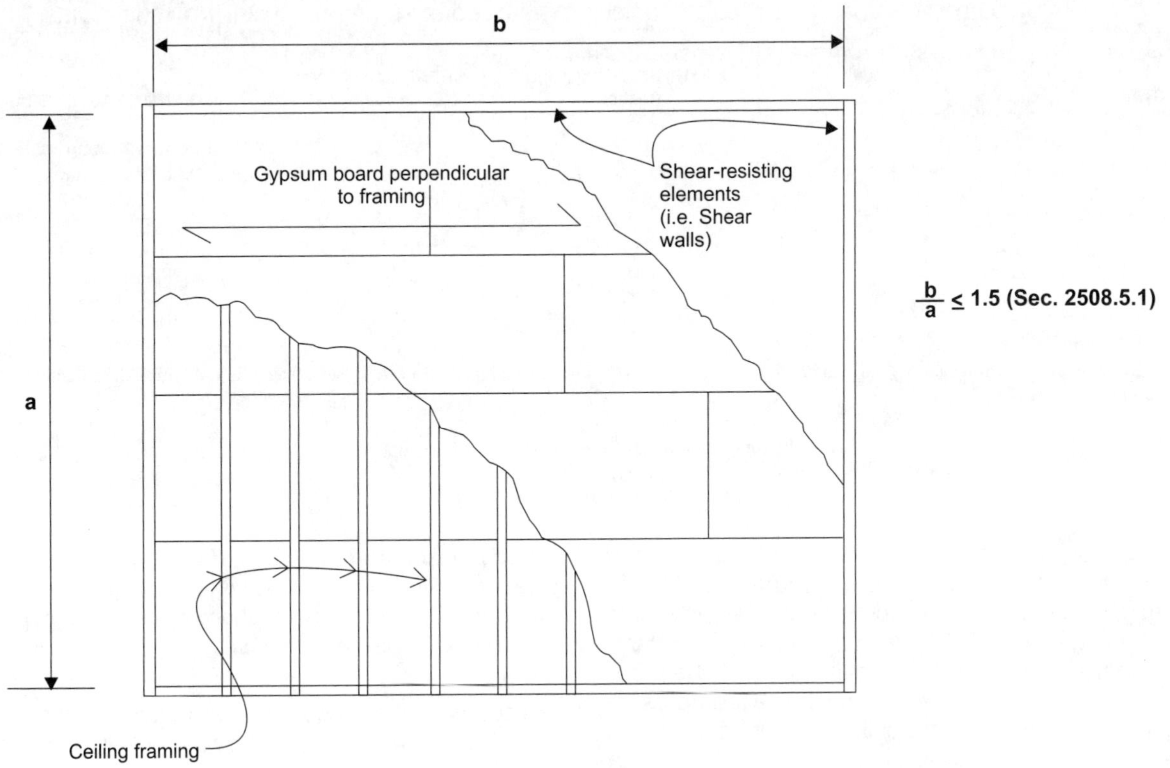

Figure 2508.5(1)
HORIZONTAL GYPSUM BOARD CEILING DIAPHRAGM CONSTRUCTION

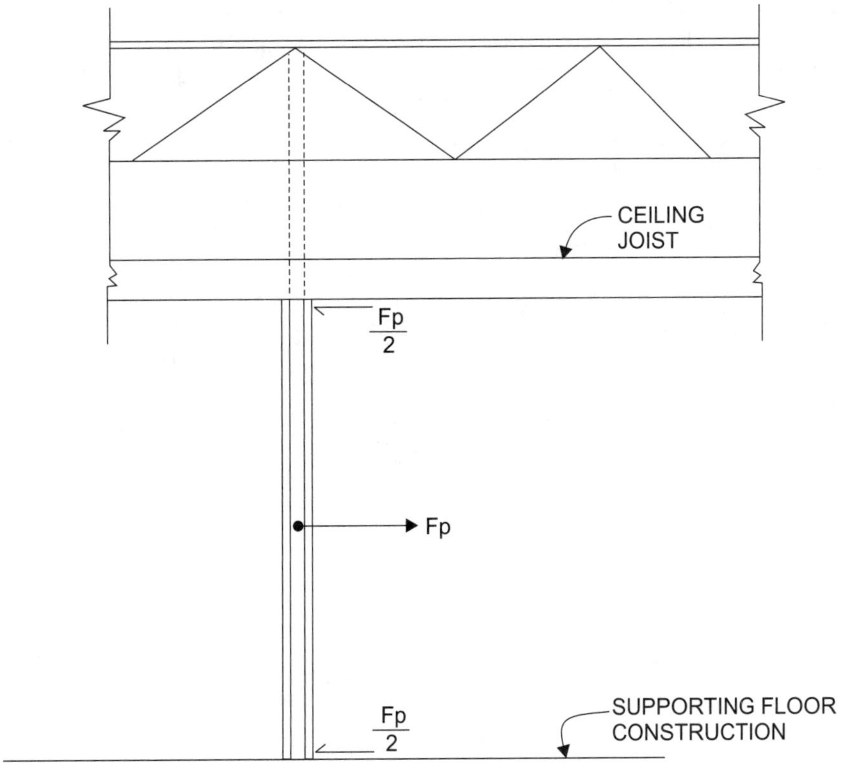

Figure 2508.5(2)
PARTITION SUPPORTED LATERALLY BY GYPSUM BOARD CEILING DIAPHRAGM

TABLE 2508.5
SHEAR CAPACITY FOR HORIZONTAL WOOD-FRAMED GYPSUM BOARD DIAPHRAGM CEILING ASSEMBLIES

MATERIAL	THICKNESS OF MATERIAL (MINIMUM) (inches)	SPACING OF FRAMING MEMBERS (MAXIMUM) (inches)	SHEAR VALUE[a,b] (plf of ceiling)	MIMIMUM FASTENER SIZE
Gypsum board	$^1/_2$	16 o.c.	90	5d cooler or wallboard nail; $1^5/_8$-inch long; 0.086-inch shank; $^{15}/_{64}$ -inch head[c]
Gypsum board	$^1/_2$	24 o.c.	70	5d cooler or wallboard nail; $1^5/_8$-inch long; 0.086-inch shank; $^{15}/_{64}$ -inch head[c]

For SI:1 inch = 25.4 mm.

a. Values are not cumulative with other horizontal diaphragm values and are for short-term loading due to wind or seismic loading. Values shall be reduced 25 percent for normal loading.

b. Values shall be reduced 50 percent in Seismic Categories D, E and F.

c. $1^1/_4$-inch, No. 6 Type S or W screws are permitted to be substituted for the listed nails.

❖ The table provides the shear capacity that is to be used in designing a horizontal gypsum board ceiling diaphragm. Since the material thickness and fastener size do not vary, the capacity is strictly a function of the framing member spacing. Further qualifications on the tabulated shear capacities that are stated in the footnotes include: ceiling diaphragm capacities are not cumulative with other diaphragm capacities; the capacities given are only for "short-term" loads due to wind and earthquakes; in structures classified as seismic design category D, E or F the capacities must be reduced by 50 percent.

2508.5.1 Diaphragm proportions. The maximum allowable diaphragm proportions shall be $1^1/_2$:1 between shear resisting elements. Rotation or cantilever conditions shall not be permitted.

❖ Limiting the diaphragm proportions and prohibiting cantilevers and rotation controls the deflection and distortion of the ceiling diaphragm [see Figure 2508.5(1)].

2508.5.2 Installation. Gypsum board used in a horizontal diaphragm ceiling shall be installed perpendicular to ceiling framing members. End joints of adjacent courses of gypsum board shall not occur on the same joist.

❖ The installation requirements of this section are necessary to obtain the shear capacities indicated in Table 2508.5.

2508.5.3 Blocking of perimeter edges. All perimeter edges shall be blocked using a wood member not less than 2-inch by 6-inch (51 mm by 159 mm) nominal dimension. Blocking material shall be installed flat over the top plate of the wall to provide a nailing surface not less than 2 inches (51 mm) in width for the attachment of the gypsum board.

❖ This section states the minimum required edge blocking that is illustrated in Figure 2508.5.3.

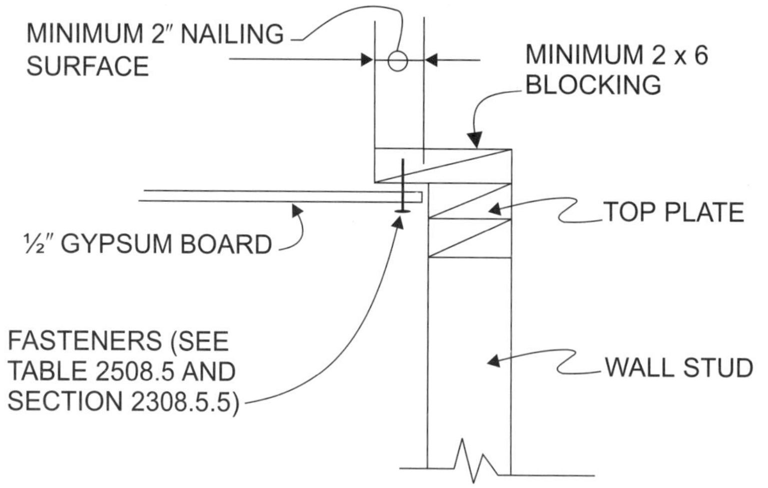

For SI: 1 inch = 25.4 mm.

Figure 2508.5.3
PERIMETER BLOCKING

2508.5.4 Fasteners. Fasteners used for the attachment of gypsum board to a horizontal diaphragm ceiling shall be as defined in Table 2508.5. Fasteners shall be spaced not more than 7 inches (178 mm) on center (o.c.) at all supports, including perimeter blocking, and not more than $^3/_8$ inch (9.5 mm) from the edges and ends of the gypsum board.

❖ This establishes maximum fastener spacing and minimum edge distances.

2508.5.5 Lateral force restrictions. Gypsum board shall not be used in diaphragm ceilings to resist lateral forces imposed by masonry or concrete construction.

❖ This section prohibits the use of a gypsum board ceiling diaphragm in resisting lateral loads due to masonry or concrete.

SECTION 2509 GYPSUM BOARD IN SHOWERS AND WATER CLOSETS

2509.1 Wet areas. Showers and public toilet walls shall conform to Sections 1210.2 and 1210.3.

❖ Special caution must be taken when gypsum board is used in applications where it will be exposed to moisture, because gypsum is not naturally resistant to water, humidity or other forms of moisture. Failure of gypsum board can be caused by direct or indirect exposure to all forms of moisture. Moisture-resistant gypsum board, which is faced with moisture-resistant paper, must be used in high moisture areas, and additional protection such as paint must be applied to prevent gypsum board from being exposed to moisture. Moisture-resistant gypsum board is known and referred to as “green board,” because of the distinctive green moisture-resistant paper that is applied to it, which also makes it readily identifiable as moisture resistant.

Wet areas at interior locations such as those identified in Sections 1209.2 and 1209.3 must be protected from moisture by a nonabsorbent wall or ceiling material. Cement plaster (gypsum plaster is not permitted in wet areas) and tile or other like finish materials will provide moisture protection for the wall surface. An additional waterproofing membrane must be provided where the wall is constructed of wood framing (see related commentary to Section 2511.5.1).

2509.2 Base for tile. When gypsum board is used as a base for tile or wall panels for tubs, shower or water closet compartment walls, water-resistant gypsum backing board shall be used as a substrate. Regular gypsum wallboard is permitted under tile or wall panels in other wall and ceiling areas when installed in accordance with GA-216 or ASTM C 840.

❖ The code requires that moisture-resistant gypsum board (green board) is to be used as a base for ceramic and other hard tiles or wall panels for tubs, showers or water-closet compartment walls. It must be used in these locations because of its moisture-resistant qualities. All joints in gypsum board that will be covered by tile or wall panels must be taped and finished even though they will be unexposed to view when the installation of the wall-covering material is complete. This is necessary to prevent moisture from migrating through the unfinished joints and causing damage to the gypsum board and the wall framing system. Corrosion-resistant fasteners must be used in wet areas.

Water-resistant gypsum board is the minimum requirement of the code for tile substrate. Certain tile manufacturers may require a cementitious tile backer board as a substrate for their product as a condition of its warranty. The manufacturer’s requirement for a tile backer board, which may be in excess of the material requirements of the code, is specified by the designer and coordinated/verified by the supplier, installer, designer and manufacturer.

2509.3 Limitations. Water-resistant gypsum backing board shall not be used in the following locations:

1. Over a vapor retarder in shower or bathtub compartments.
2. Where there will be direct exposure to water or in areas subject to continuous high humidity.
3. On ceilings where frame spacing exceeds 12 inches (305 mm) o.c. for $^1/_2$-inch-thick (12.7 mm) water-resistant gypsum backing board and more than 16 inches (406 mm) o.c. for $^5/_8$-inch-thick (15.9 mm) water-resistant gypsum backing board.

❖ Although there are many gypsum board sheet products that are manufactured and approved for use in wet areas or areas exposed to moisture or humidity, there are still some extreme conditions where even water-resistant gypsum board will not provide the level of moisture protection necessary:

1. Gypsum board installed on the walls and ceilings at shower and bathtub areas must be finished to prevent moisture from penetrating the wall or ceiling finish and contacting the gypsum board. The finish applied to the exposed face of the gypsum board, in effect, creates a water-resistant barrier that not only stops water from getting to the gypsum board, but also prevents the release of moisture from within the wall or the gypsum board itself. For this reason, gypsum board must not be installed over the outboard side of any vapor barrier or retarder. This will create a waterproof membrane on both faces of the gypsum board, causing moisture to be trapped in the gypsum board that will ultimately cause it to decompose and fail.
2. Water-resistant gypsum board is not to be used in areas that will be subject to direct exposure to water or continuous exposure to high humidity at locations such as saunas, steam rooms, gang showers or indoor pools. Gypsum board products, including the water-resistant type, are not intended for these extreme conditions and will provide unsatisfactory performance. Nongypsum wall and ceiling materials such as concrete ma-

sonry, ceramic tile on cement backer board, cement plaster (stucco) or other materials designed and recommended for high moisture exposure must be used in these locations.

3. The additional weight, due to moisture, that is commonly applied as live loads to ceilings in high moisture or wet areas must be considered when selecting the thickness of the gypsum board and the spacing of the structural support members necessary to prevent sagging of the ceiling. The more restrictive frame spacing and gypsum board thicknesses required in this section are established to allow for the additional weight. Where other structural requirements in the code are less restrictive for the conditions contained in this section, these requirements apply.

SECTION 2510 LATHING AND FURRING FOR CEMENT PLASTER (STUCCO)

2510.1 General. Exterior and interior cement plaster and lathing shall be done with the appropriate materials listed in Table 2507.2 and Chapter 35.

❖ This section contains the appropriate referenced standards for materials, proper application and protection of interior and exterior lath and furring for the installation of portland cement plaster (stucco).

Table 2507.2 lists the appropriate ASTM standards for portland cement, metal lath and plastering materials. These referenced standards are the minimum material specifications required by the code to meet the requirements of this section, as well as related requirements elsewhere in the code for the proper application and procedures for lath and furring for cement plaster.

2510.2 Weather protection. Materials shall be stored in such a manner as to protect such materials from the weather.

❖ Like gypsum products, portland cement stucco lathing and plastering materials must be stored to prevent the deleterious effect of the weather.

2510.3 Installation. Installation of these materials shall be in compliance with ASTM C 926 and ASTM C 1063.

❖ Table 2511.1 refers to ASTM C 1063 as the standard specification for interior and exterior lathing and furring for portland cement-based plastering, as specified in ASTM C 926. Where a fire-resistance rating is required for plastered assemblies and construction, details of construction must be in accordance with reports of fire tests of assemblies that have met the requirements of the fire-resistance rating imposed. Where a specific degree of sound control is required for plastered assemblies and construction, details of construction must be in accordance with official reports of tests conducted in recognized testing laboratories in accordance with the applicable requirements of ASTM E 90.

Table 2511.1 also refers to ASTM C 1063 as the standard specification for the installation of portland cement-based plaster. ASTM C 1063 provides the minimum requirements for the application of portland cement-based plaster for exterior (stucco) and interior work, along with the tables necessary for proportioning various plaster mixes and thicknesses (refer to Section 2512 for additional requirements for cement plaster).

2510.4 Corrosion resistance. Metal lath and lath attachments shall be of corrosion-resistant material.

❖ Because of the amount of water that is present in plaster when it is mixed and applied to a surface, all metal components and accessories, such as metal lath, screws, screeds, grounds, metal runners at edges of the plaster, expansion and control joint molding or any other metal component that will be in contact with the plaster, must be protected from corrosion and rusting from prolonged contact with wet plaster. Corrosion of metal components will cause failure and create voids in the plaster that will allow moisture to get into the wall and cause damage to the structural and nonstructural elements of the building. For these reasons, the code requires metal components be made of corrosion-resistant materials. Table 2507.2 provides a listing of the required ASTM standards for lath and related accessories.

2510.5 Backing. Backing or a lath shall provide sufficient rigidity to permit plaster applications.

❖ Proper application of exterior plaster requires that the base coat is thoroughly pushed through the lath to provide the required mechanical bond. For this reason, it is required by the code that the lath and its backing provide sufficient rigidity to permit the proper application of the plaster. A common method of providing a rigid backing for the lath for exterior plaster over open wood or metal stud framing is to fasten 18-gage wire backing (see Figure 2510.5) to the studs at 6 inches (153 mm) on center horizontally. It is critical to stretch the wire tightly across the face of the studs to insure the desired rigidity for the backing. The wire backing must be attached to each stud to prevent slippage or movement of the wire.

2510.5.1 Support of lath. Where lath on vertical surfaces extends between rafters or other similar projecting members, solid backing shall be installed to provide support for lath and attachments.

❖ It is important for the lath to be supported continuously along all edges. Continuous edge support is necessary to prevent excessive movement of the finished plaster, which will cause unwanted movement cracking. This is especially critical on exterior vertical surfaces of cement plaster because cracking provides an entry point for water into the building. At locations where the wall plane is interrupted at regular intervals, such as eaves and soffits where rafter ends project beyond the face of the wall, the code requires that the lath be supported by solid backing at all edges. In addition to supporting the edges of the lath, the code requires that all plaster trim, edging, screeds, grounds and other accessories be rigidly supported with solid backing at all edges.

The damage that will be caused to finished plaster on an exterior vertical surface due to unsupported edges is much more obvious than what would be expected on an interior surface; however, failure to interior wall surfaces is of no less concern. For this reason, the code makes no distinction between interior and exterior surfaces. Solid backing is required for lath at all locations, both interior and exterior.

2510.5.2 Use of gypsum backing board.

❖ The code does not permit the use of gypsum board or gypsum lath as a backing for cement plaster, except as noted in Section 2510.5.2.1.

2510.5.2.1 Use of gypsum board as a backing board. Gypsum lath or gypsum wallboard shall not be used as a backing for cement plaster.

Exception: Gypsum lath or gypsum wallboard is permitted, with a weather-resistant barrier, as a backing for self-furred metal lath or self-furred wire fabric lath and cement plaster where either of the following conditions occur:

1. On horizontal supports of ceilings or roof soffits.
2. On interior walls.

❖ Because of its excellent water-resistant properties, cement plaster is the only type permitted by the code for use as a finished exterior surface exposed to the weather, or at an interior location subject to excessive moisture. Gypsum has none of the water-resistant properties possessed by cement-based plasters or cementitious backer boards, which are designed specifically for use in high moisture areas. Due to the inability of gypsum-based materials to resist moisture, the code does not permit the use of gypsum board or gypsum lath as an interior or exterior backing in direct contact with cement plaster. Gypsum board lath subjected to moisture will decompose and cause failure to the entire plaster system, including the outer or finished coats.Exceptions to this restriction are permitted when gypsum is protected from contact with cement plaster as allowed in specific locations described below.

Gypsum lath and gypsum wallboard are permitted to be used as a backing for cement plaster when a weather-resistant barrier, such as Grade D building paper, is installed at the back of the lath or to the face of the gypsum board to prevent the cement plaster from contacting the gypsum backing. The building paper must be securely attached to the face of the gypsum board before installing the self-furring metal lath or self-furring wire fabric to create a weather-resistive barrier. Items 1 and 2 include the extent to which exceptions may be applied.

2510.5.2.2 Use of gypsum sheathing backing. Gypsum sheathing is permitted as a backing for metal or wire fabric lath and cement plaster on walls. A weather-resistant barrier shall be provided in accordance with Section 2510.6.

❖ Gypsum board sheathing is permitted to be used as a backer for cement plaster when installed with a

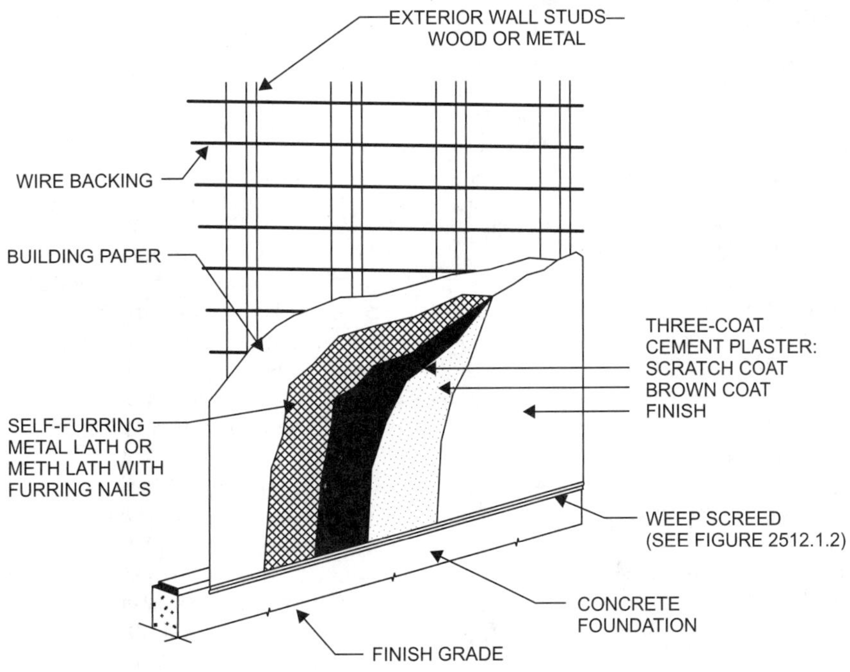

Figure 2510.5
THREE-COAT PLASTER ON UNSHEATHED EXTERIOR STUD WALL

weather-resistant barrier as described in Section 2510.5.2.1 and as required in Section 2510.6.

2510.5.3 Backing not required. Wire backing is not required under expanded metal lath or paperbacked wire fabric lath.

❖ Because metal lath or paper-backed wire fabric lath has reinforcing ribs to provide necessary rigidity, the code requires no additional backing for these materials.

2510.6 Weather-resistant barriers. Weather-resistant barriers shall be installed as required in Section 1404.2 and, where applied over wood-based sheathing, shall include a weather-resistant vapor-permeable barrier with a performance at least equivalent to two layers of Grade D paper.

❖ The code requires that a weather-resistive barrier be installed behind exterior plaster for the reasons provided in Section 1404.2. The code also requires that when the barrier is applied over wood-based sheathing such as plywood, the barrier is to be two layers of Grade D building paper. This requirement is based on the observed problems where one layer of typical Type 15-pound (6.8 kg) felt is applied over wood sheathing. The wood sheathing eventually exhibits dry rot due to the penetration of moisture. Cracking is then created in the plaster due to movement of the sheathing caused by alternate expansion and contraction. Field experience has shown that where two layers of building paper are used, the penetration of moisture is considerably decreased, as is the cracking of the plaster due to movement of the sheathing caused by the wet and dry cycles. Grade D building paper is specified because it has the proper water vapor permeability to prevent entrapment of moisture between the paper and the sheathing.

2510.7 Preparation of masonry and concrete. Surfaces shall be clean, free from efflorescence, sufficiently damp and rough for proper bond. If the surface is insufficiently rough, approved bonding agents or a portland cement dash bond coat mixed in proportions of not more than two parts volume of sand to one part volume of portland cement or plastic cement shall be applied. The dash bond coat shall be left undisturbed and shall be moist cured not less than 24 hours.

❖ The masonry surface must be properly prepared to receive the cement plaster base coat to insure proper bond between the masonry base and the plaster. Masonry surfaces must be clean, damp, free of dust and efflorescence and reasonably flat, without excessive protrusions or indentations. These irregularities will affect the evenness and uniform thickness of the finished plaster. Mortar joints should be struck flush to help minimize any surface irregularities.

Masonry bases for plaster have varying degrees of suction (water absorption rate) but generally provide a good adhesive as well as mechanical bond. Most masonry bases have enough suction so that special bonding agents will not be required to achieve the adequate bond between the plaster and the base. Where denser and harder masonry materials that have very low suction characteristics are used, however, preparation of the surface, similar to that of concrete, may be necessary. Such preparation can include mechanical scoring or the use of chemical bonding agents.

Concrete generally has a denser and tighter finished surface than masonry. Concrete does not have the same suction properties as masonry; therefore, the bond between concrete and plaster must be completely mechanical. Like masonry, the concrete surface must be clean, rough, damp and free of efflorescence and form-releasing agents. If the surface of the concrete is not adequately rough to provide the required mechanical bond, it may be prepared as follows:

- If preparation of the surface is to begin before the concrete is completely set, the surface can be scarified with a metal tool or by using a stiff, heavy-duty wire brush.
- If preparation of the concrete surface must be done after the concrete has hardened, the surface may be bush hammered or scarified with chisels or tools designed for that purpose.
- Another common method, which is specified in this section, is to apply a rich mixture of portland cement, sand and water as a dash coat. This application, in which the material is "dashed" onto the concrete surface with a course brush, may only be used where portland cement plaster is going to be applied to the concrete surface.
- Application of a liquid bonding agent.

SECTION 2511
INTERIOR PLASTER

2511.1 General. Plastering gypsum plaster or cement plaster shall not be less than three coats where applied over metal lath or wire fabric lath and not less than two coats where applied over other bases permitted by this chapter.

Exception: Gypsum veneer plaster and cement plaster specifically designed and approved for one-coat applications.

❖ This section is intended to define the standards and guidelines for the materials and installation of interior plaster. The requirements of this section include gypsum plaster, gypsum veneer plaster and portland cement plaster. This section also addresses plaster applied to nonlath surfaces and the environmental conditions for which interior plaster is stored and installed.

It is the consensus of the industry that multicoat work is necessary for control of plaster thickness and density, particularly where plaster application is done by hand because most of the materials used for plaster compact under hand application as a result of pressure applied to the trowel. Experience has proven that this change in density is more controllable and will be more uniform when the plaster is applied in thin, successive layers. For this reason, the code requires three-coat plastering over metal lath or wire fabric lath and two-coat work applied over other plaster bases [see Figures 2511.1(1) and 2511.1(2)]. Reducing the requirements for plaster bases other than metal or wire lath depends on the rigidity of the plaster base itself. More rigid plaster bases are not as susceptible to variations in thickness and flatness of the surface. It may be considered that the first coat in three-coat work on a flexible base, such as wire lath, is used to stiffen that base to provide the rigidity necessary to attain uniform thickness and surface flatness.

Gypsum veneer plaster as specified in ASTM C 843 is specifically designed to be applied as a one-coat plaster, and is approved by the code for that application. Any one-coat system must be applied over a solid base as approved by the code, the referenced standard or the product manufacturer.

TABLE 2511.1
INSTALLATION OF PLASTER CONSTRUCTION

MATERIAL	STANDARD
Gypsum plaster	ASTM C 842
Gypsum veneer plaster	ASTM C 843
Interior lathing and furring (gypsum plaster)	ASTM C 841
Lathing and furring (cement plaster)	ASTM C 1063
Portland cement plaster	ASTM C 926
Steel framing	ASTM C 754; C 1007

❖ Table 2511.1 provides the ASTM installation standards required by the code for plaster construction. This table also includes installation standards for related construction materials. Chapter 35 contains the complete title and appropriate issue date for the standard referenced.

2511.1.1 Installation. Installation of lathing and plaster materials shall conform with Table 2511.1 and Section 2507.

❖ This section refers to Section 2508 for the appropriate standards to regulate the proper installation of gypsum lath and related gypsum board materials and accessories. The expected performance of a product or material is dependent on proper installation. Table 2511.1 lists the appropriate ASTM standards for plaster construction.

2511.2 Limitations. Plaster shall not be applied directly to fiber insulation board. Cement plaster shall not be applied directly to gypsum lath or gypsum plaster except as specified in Sections 2510.5.1 and 2510.5.2.

❖ Fiber insulation board does not have the qualities necessary to provide an acceptable base for the application of plaster. It absorbs excessive moisture from the plaster mix that creates problems of workmanship, and it does not have the structural properties to provide the stability and rigidity required for a proper functioning plaster base. In colder and damper climates, fiberboard insulation retains the moisture absorbed from the plaster for a relatively longer period of time than other, more dense bases. This lack of density and excessive retention of moisture will cause the premature failure of the plaster. For these reasons, the code prohibits the use of fiber insulation as a plaster base.

Portland cement plaster will not bond properly to gyp-

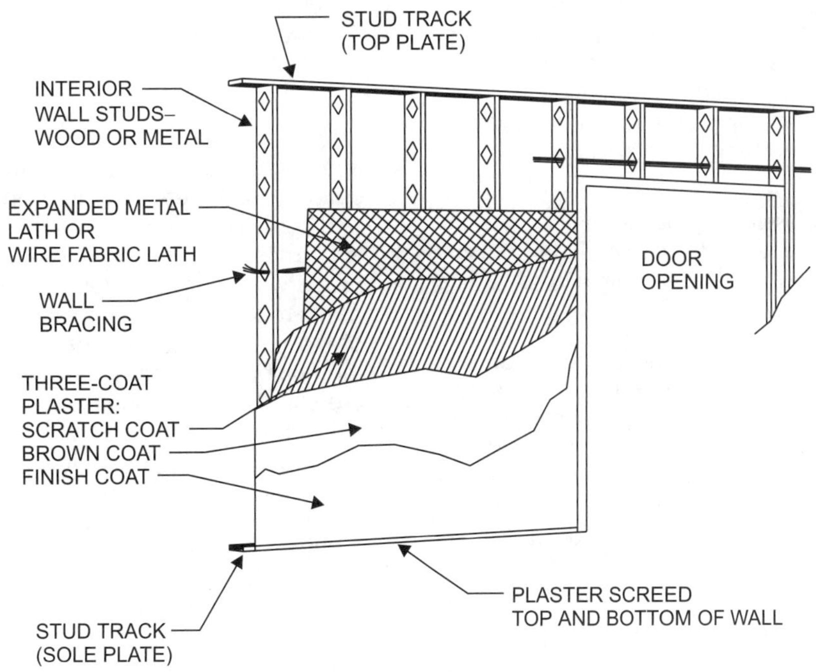

Figure 2511.1(1)
THREE-COAT PLASTER ON INTERIOR STUD WALL

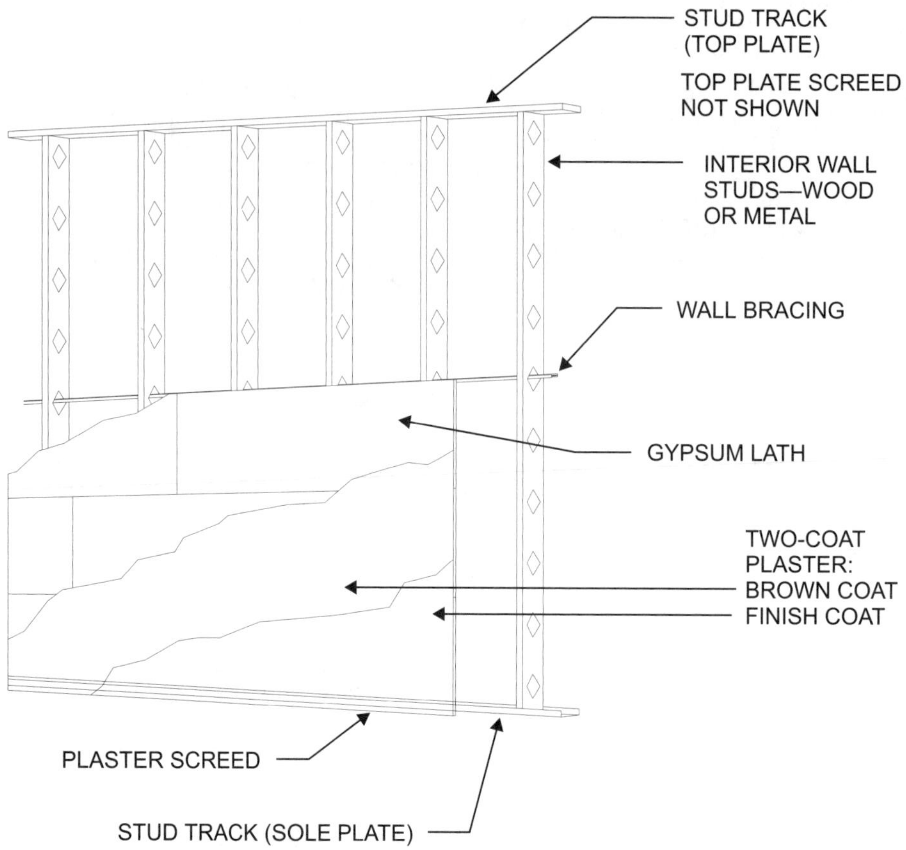

Figure 2511.1(2)
TWO-COAT PLASTER ON INTERIOR STUD WALL

sum plaster bases; therefore, the code prohibits the use of portland cement plaster on a gypsum plaster base. The code does, however, permit the use of exterior plaster to be applied on horizontal surfaces, such as ceilings and soffits, over gypsum lath and gypsum board when used as a backing for metal lath.

2511.3 Grounds. Where installed, grounds shall ensure the minimum thickness of plaster as set forth in ASTM C 842 and ASTM C 926. Plaster thickness shall be measured from the face of lath and other bases.

❖ Plaster grounds are utilized to establish and maintain the required thickness of plaster. Grounds are usually wood or metal strips of a known thickness attached to the plaster base. The intent is that plaster grounds are used as a guide for the straightedge in determining the thickness of the plaster. Door or window frames are used as plaster grounds. Dollops of plaster of the desired thickness may also be used as grounds.

2511.4 Interior masonry or concrete. Condition of surfaces shall be as specified in Section 2510.7. Approved specially prepared gypsum plaster designed for application to concrete surfaces or approved acoustical plaster is permitted. The total thickness of base coat plaster applied to concrete ceilings shall be as set forth in ASTM C 842 or ASTM C 926. Should ceiling surfaces require more than the maximum thickness permitted in ASTM C 842 or ASTM C 926, metal lath or wire fabric lath shall be installed on such surfaces before plastering.

❖ Special gypsum plaster and acoustical plaster that is specifically designed and manufactured for direct application to concrete masonry unit walls, concrete walls and ceilings without a gypsum board or metal lath base is permitted by the code. Special care must be taken to properly prepare the masonry or concrete surfaces that are to receive a plaster finish to insure both proper bond of the plaster and the desired or required levelness of the finished surface. Preparation of masonry and concrete surfaces is discussed in Section 2510.7.

Where a concrete ceiling is to receive a plaster finish, the code requires a total thickness of base coat plaster as required by the ASTM standards referenced in this section. For applications of plaster on concrete ceiling surfaces, a thinner coat is required than for other applications. This is due primarily to the difficulty of achieving an adequate bond between the plaster and the concrete surface. Because of the horizontal application, the weight of the plaster base coat must be kept to a minimum. If it becomes necessary to increase the thickness of the plaster beyond the maximum allowed, due for example to the unevenness of the concrete surface or for some similar reason, metal lath or wire-fabric lath must be attached to the concrete ceiling surface to provide anchorage for the weight of the additional plaster.

2511.5 Wet areas. Showers and public toilet walls shall conform to Sections 1210.2 and 1210.3. When wood frame walls and partitions are covered on the interior with cement plaster or tile of similar material and are subject to water splash, the framing shall be protected with an approved moisture barrier.

❖ Wet areas at interior locations such as those identified in Sections 1210.2 and 1210.3 must be protected from moisture by nonabsorbent wall or ceiling material. Cement plaster (gypsum plaster is not permitted in wet areas) and tile or other like materials will provide moisture protection for the wall surface. An additional waterproofing membrane must be provided where the wall is constructed of wood framing.

SECTION 2512
EXTERIOR PLASTER

2512.1 General. Plastering with cement plaster shall not be less than three coats where applied over metal lath or wire fabric lath and not less than two coats where applied over masonry, concrete or gypsum board backing as specified in Section 2510.5. If the plaster surface is to be completely covered by veneer or other facing material, or is completely concealed by another wall, plaster application need be only two coats, provided the total thickness is as set forth in ASTM C 926.

❖ This section is intended to define the standards and guidelines for the type of materials and installation of exterior plaster. The requirements of this section include the limitations of gypsum plaster and gypsum veneer plaster. This section will also address plaster applied to solid backings, alternative application methods, curing and plaster additives. Guidelines and requirements for preventing water infiltration into the building are discussed in this section.

Portland cement plaster (stucco) is the only type approved by the code for the use of exterior plaster finish. Exterior portland cement plasters are required by the code to be applied in no less than three coats when applied over expanded metal or wire-fabric lath for the same reasons discussed for interior plaster (see commentary, Section 2511.1 and Figure 2512.1). When portland cement plaster is applied over other approved plaster bases, the code permits the application of a two-coat plaster system. The code also allows plaster work that is going to be completely concealed to be applied as a two-coat system, provided the total thickness meets the requirements of ASTM C 926, because the finish coat (second coat) of plaster will be the surface that will receive the exterior finish, such as paint, and it is critical for the finish coat to provide a clean, smooth and visually acceptable surface. Where the plaster surface is to be completely concealed, it is not necessary to provide a finish coat.

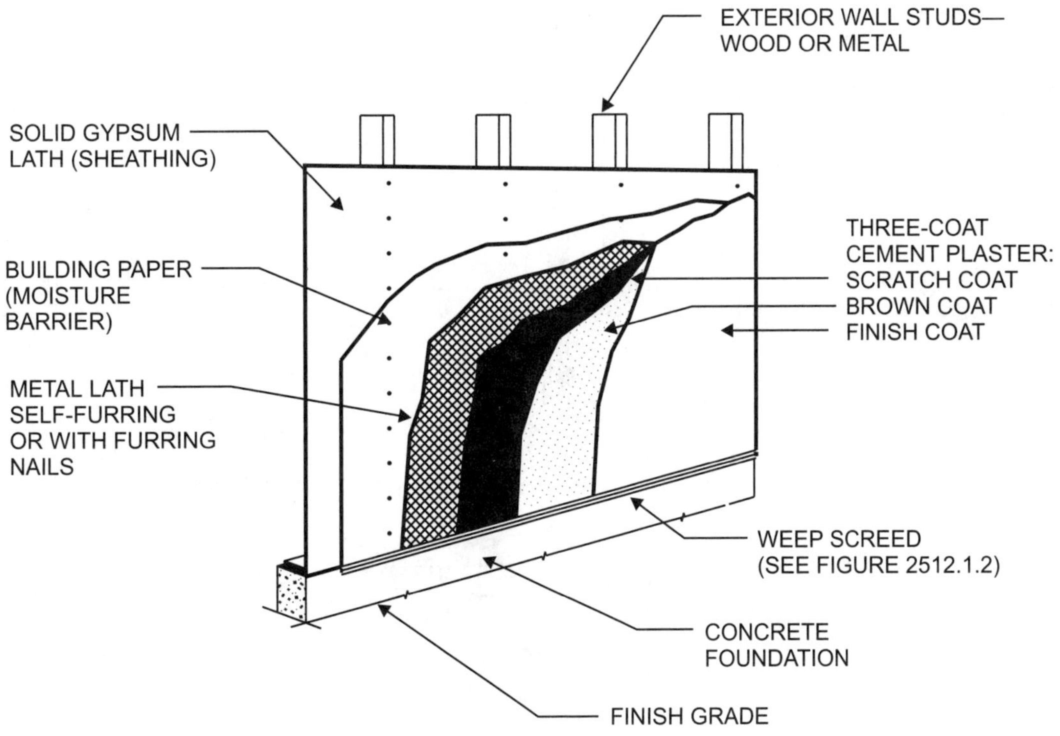

Figure 2512.1
THREE-COAT PLASTER ON EXTERIOR STUD WALL

2512.1.1 On-grade floor slab. On wood framed or steel stud construction with an on-grade concrete floor slab system, exterior plaster shall be applied in such a manner as to cover, but not to extend below, the lath and paper. The application of lath, paper and flashing or drip screeds shall comply with ASTM C 1063.

❖ The code requires that the exterior plaster be installed to completely cover, but not extend below, the lath and paper (moisture barrier membrane) on wood or metal stud exterior wall construction supported by a concrete slab-on-grade floor. This requirement, combined with the requirement for a continuous weep screed as described in Section 2512.1.2, are intended to prevent the entrapment and subsequent channeling of free moisture to the interior of the building.

2512.1.2 Weep screeds. A minimum 0.019-inch (0.48 mm) (No. 26 galvanized sheet gage), corrosion-resistant weep screed with a minimum vertical attachment flange of $3^1/_2$ inches (89 mm) shall be provided at or below the foundation plate line on exterior stud walls in accordance with ASTM C 926. The weep screed shall be placed a minimum of 4 inches (102 mm) above the earth or 2 inches (51 mm) above paved areas and be of a type that will allow trapped water to drain to the exterior of the building. The weather-resistant barrier shall lap the attachment flange. The exterior lath shall cover and terminate on the attachment flange of the weep screed.

❖ Water and moisture can penetrate an exterior plaster wall for a variety of reasons and in a number of ways. It is expected that some moisture will penetrate the plaster in an exterior wall; therefore, the design of the wall should include a weep screed, which will provide a way to release the moisture (see Figure 2512.1.2). Once water or moisture penetrates the plaster, it will migrate down the exterior wall face of the weather-resistive barrier until it reaches the sill plate or mud sill. At this point, the water will seek a way out of the wall. If the exterior plaster system is not detailed and constructed with provisions to allow the moisture to escape to the exterior, it will find its own way out. This exit will almost certainly be through the interior of the wall and cause leaking, and damage to the interior of the building. For this reason, the code requires a continuous weep screed at the bottom of exterior walls to permit the moisture to escape to the exterior of the building.

2512.2 Plasticity agents. Only approved plasticity agents and approved amounts thereof shall be added to portland cement. When plastic cement or masonry cement is used, no additional lime or plasticizers shall be added. Hydrated lime or the equivalent amount of lime putty used as a plasticizer is permitted to be added to cement plaster or cement and lime plaster in an amount not to exceed that set forth in ASTM C 926.

❖ Admixtures such as plasticizers should not be added to portland cement unless approved by the building official. Some admixtures can create harmful effects that more than offset the desired improvement in plasticity. It is preferable that plasticizers be added during the man-

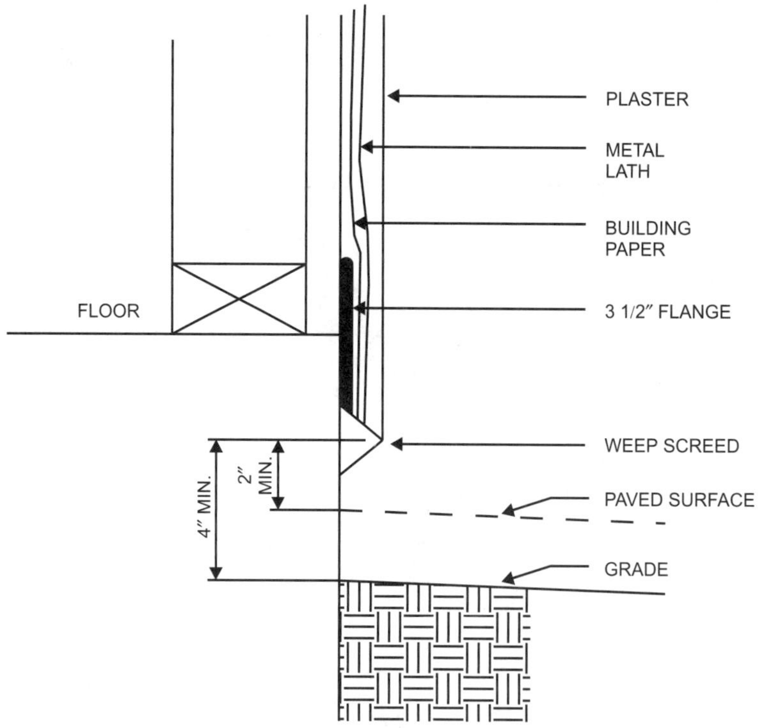

For SI: 1 inch = 25.4 mm.

Figure 2512.1.2
WEEP SCREED FOR EXTERIOR PLASTER ON STUD WALL AT SLAB ON GRADE

ufacturing of the cement for product uniformity and proper proportions. When plastic cement is used, the code does not allow any further additions of plasticizers because the amount added during the manufacturing process is adequate and is the maximum amount permitted. Hydrated lime and lime putty are time-tested plasticizers used with portland cement plaster, and their use is permitted by the code in the amounts specified in ASTM C 926 as referenced herein.

2512.3 Limitations. Gypsum plaster shall not be used on exterior surfaces.

❖ Gypsum veneer plaster, gypsum plaster or any other similar gypsum-based plaster materials are not permitted for use as an exterior finish material. Gypsum-based plaster is subject to deterioration and ultimate failure when it is exposed to exterior weather conditions such as excessive heat, humidity and moisture, which are natural and unavoidable elements.

2512.4 Cement plaster. Plaster coats shall be protected from freezing for a period of not less than 24 hours after set has occurred. Plaster shall be applied when the ambient temperature is higher than 40°F (4°C), unless provisions are made to keep cement plaster work above 40°F (4°C) during application and 48 hours thereafter.

❖ Portland cement plaster contains a high percentage of water when it is mixed and applied, and before the process of curing is complete. For this reason, it is adversely affected by cold weather and freezing in much the same way that portland cement mortar and portland cement concrete are affected. The application of portland cement plaster to a frozen base or to a base covered with ice or frost will not only weaken the bond of the plaster to its base, but also cause the plaster itself to freeze.

When portland cement plaster is applied in freezing weather or to a frozen base, or is mixed with frozen ingredients, it loses a high proportion of its strength and, therefore, does not comply with the requirements of the code. Plaster materials must be stored to protect them from adverse weather conditions and should only be installed when weather conditions are within the requirements of the code and will remain so for the duration of the curing process.

2512.5 Second-coat application. The second coat shall be brought out to proper thickness, rodded and floated sufficiently rough to provide adequate bond for the finish coat. The second coat shall have no variation greater than $^{1}/_{4}$ inch (6.4 mm) in any direction under a 5-foot (1524 mm) straight edge.

❖ In plaster work, a base coat is any coat beneath the finish coat. This is true whether the plaster is a two-coat or three-coat application. In a three-coat application, the first coat is usually referred to as the "scratch" coat. It is usually applied over a flexible base, such as expanded metal or wire fabric lath, and is intended to stiffen the base and provide a mechanical bond to the base. As the name implies, the first coat is scratched with a scarifying tool to provide a series of horizontal ridges or scratches that are intended to provide mechanical keys for the application of the second coat (or brown coat). The brown coat usually constitutes the major bulk of the plaster and, consequently, materially affects the membrane strength. As a result, proportioning and workability are critical, and the mix should have high plasticity for proper application. The term "brown coat" is utilized by the trade to differentiate the color of the second coat from the finish coat. The finish coat is usually much lighter in color than the second coat.

2512.6 Curing and interval. First and second coats of cement plaster shall be applied and moist cured as set forth in ASTM C 926 and Table 2512.6.

❖ Sufficient time between coats is to be allowed to permit each coat to cure or develop enough rigidity to resist cracking or other physical damage when the next coat is applied. The proper amount of moisture must be provided to the plaster mix to allow for the proper curing time for each coat. The most effective procedure for curing and the time required between each coat will depend on the climate and job conditions. Moist or fog curing will permit and control the continuous hydration of cementitious materials.

TABLE 2512.6
CEMENT PLASTERS[a]

COAT	MINIMUM PERIOD MOIST CURING	MINIMUM INTERVAL BETWEEN COATS
First	48 hours[a]	48 hours[b]
Second	48 hours	7 days[c]
Finish	—	Note c

a. The first two coats shall be as required for the first coats of exterior plaster, except that the moist-curing time period between the first and second coats shall not be less than 24 hours. Moist curing shall not be required where job and weather conditions are favorable to the retention of moisture in the cement plaster for the required time period.

b. Twenty-four-hour minimum interval between coats of interior cement plaster. For alternate method of application, see Section 2512.8.

c. Finish coat plaster is permitted to be applied to interior portland cement base coats after a 48-hour period.

❖ The timing between coats will vary with climate conditions and the type of plaster base being used. Table 2512.6 specifies the minimum curing times between coats; however, temperature and relative humidity will extend or reduce the curing time required between consecutive coats. Cold or wet weather will increase the required curing time, while hot or dry weather will shorten it. Moderate changes in temperature and humidity can be controlled by providing additional heating during cold and damp weather, and by reducing the loss of moisture by prewetting during hot or dry weather.

All the conditions mentioned above must be taken into consideration when selecting the most appropriate method of curing that will provide the best results for the application being considered as well as the job conditions present during the curing process. Any of the three curing methods listed below, or any combination of the three, are permitted by the code:

1. Moist curing is accomplished by applying a fine fog spray of water as often as required to retain moisture and control the rate of curing. Moisture is generally applied two times a day—once in the morning and again in the late afternoon or evening—although job conditions are more of a factor than a predetermined time schedule. Care must be taken to avoid erosion damage to portland cement-based plaster surfaces due to excessive amounts of water running down vertical surfaces. Except in extreme or severe drying conditions, the wetting of the finish coat should be avoided.
2. Plastic film, when taped or weighted down around the perimeter of the plastered work area, will provide a vapor barrier to retain the moisture between the membrane and the plaster. Care must be taken when placing the plastic film around the work area. If the film is placed too soon before the plaster is allowed to stiffen, the film can come into contact with the wet plaster and damage the surface texture. If too much time is allowed before placing the film, an excessive amount of moisture will have escaped causing premature curing, which can cause cracking to the finished plaster.
3. Canvas, cloth or sheet material barriers can be erected to deflect sunlight and wind, both of which will affect the rate of evaporation. If the humidity is very low, this method alone may not provide adequate protection.

2512.7 Application to solid backings. Where applied over gypsum backing as specified in Section 2510.5 or directly to unit masonry surfaces, the second coat is permitted to be applied as soon as the first coat has attained sufficient hardness.

❖ When plaster is applied directly to a solid backing, such as gypsum board lath, concrete or concrete masonry, the minimum time intervals listed in Table 2512.6 are not required between the first (scratch) coat and the second (brown) coat. The brown coat may be applied as soon as the scratch coat has reached sufficient hardness to support the weight of the additional plaster without causing damage to the scratch coat.

2512.8 Alternate method of application. The second coat is permitted to be applied as soon as the first coat has attained sufficiently rigidity to receive the second coat.

❖ While the majority of experts on the application of plaster generally agree that the first two coats of portland cement plaster should be cured as required by Section 2512.8.2, there is an increasingly large number who feel that the brown coat can be applied without requiring moist curing of the scratch coat, but at least allowing the scratch coat to harden to a condition of sufficient rigidity to accept the second coat. This alternate method of application is also considered by its proponents to be stronger than when the scratch coat has been cured as required by the code. This is due to a manufacturer's claim that a superior bond is produced between the two coats of plaster when the second coat is applied to a set, but uncured, coat.

2512.8.1 Admixtures. When using this method of application, calcium aluminate cement up to 15 percent of the weight of the portland cement is permitted to be added to the mix.

❖ In order to develop a quicker set, the code permits the addition of calcium aluminate cement up to 15 percent of the weight of the portland cement. These high-alumina cements are characterized by a faster set time than portland. Because high-alumina cements do not possess the same high strength and qualities as portland cement in plaster mixes, the code limits their proportion of the total mix.

2512.8.2 Curing. Curing of the first coat is permitted to be omitted and the second coat shall be cured as set forth in ASTM C 926 and Table 2512.6.

❖ When applied as specified in ASTM C 926, the complete curing of the first (scratch) coat may be omitted. The second (brown) coat may be applied as soon as the scratch coat achieves sufficient rigidity; however, the second coat must be fully cured as indicated in Table 2512.6 prior to the application of the finished coat (for additional commentary on curing, see Section 2512.6).

2512.9 Finish coats. Cement plaster finish coats shall be applied over base coats that have been in place for the time periods set forth in ASTM C 926. The third or finish coat shall be applied with sufficient material and pressure to bond and to cover the brown coat and shall be of sufficient thickness to conceal the brown coat.

❖ The base coats in plaster work provide the strength for the plaster membrane but generally do not provide a proper texture for a finished surface; therefore, a thin, veneer-like coat of plaster or stucco is applied to the base coats as a finish coat. The finish coat can be applied as a textured, ornamental or decorative finish, or it can be applied as a smooth, flat surface appropriate to receive paint, wallpaper, applied textured finishes or a variety of architectural finishes. Finish plaster coats must be allowed to properly cure before applying any adhered or wet finishes.

SECTION 2513
EXPOSED AGGREGATE PLASTER

2513.1 General. Exposed natural or integrally colored aggregate is permitted to be partially embedded in a natural or colored bedding coat of cement plaster or gypsum plaster, subject to the provisions of this section.

❖ This section is provided in the code to establish the requirements for the use of exposed aggregate plaster, both portland cement plaster and gypsum plaster. Exposed aggregate plaster is approved by the code for both interior and exterior applications, although it is more commonly used with portland cement plaster as an exterior wall finish surface. The plaster systems, accessories, materials, assemblies, installation procedures and installation applications are essentially the same as other plastering systems with the exception of the bedding coat that is prepared to receive the exposed aggregate material.

Exposed aggregate plaster is a plaster finish, which as the name implies, has small chips of exposed aggregate embedded in its finish surface. Exposed aggregate finish is used most commonly on exterior cement plaster walls, although its use is not limited to exterior applications or cement plaster. Exposed aggregate finishes may also be applied to gypsum plaster, but like all other forms of gypsum plaster, it is limited to interior and nonwet areas. The purpose of using an exposed aggregate plaster is to provide a variety of textured and colored surfaces.

2513.2 Aggregate. The aggregate shall be applied manually or mechanically and shall consist of marble chips, pebbles or similar durable, moderately hard (three or more on the Mohs hardness scale), nonreactive materials.

❖ Two of the more common exposed aggregate finishes are a washed pebble finish and a marble chip finish. There are other aggregates available, but the material requirements and application procedures are the same as those contained in this section. The mention of these two aggregates is not meant to exclude other materials from being acceptable.

The code requires that the materials used for the aggregate be nonreactive and moderately hard (three or harder on the Mohs hardness scale for minerals). The mineral that is used as the identifier for the hardness of three is calcite, which in its common form is represented by limestone and marble, to name only two. By requiring at least this hardness, the intent of the code is to approximate ordinary sand aggregate plaster in durability.

The aggregate is applied by rubbing or "dashing" it into or against the bedding coat and then using a trowel or other suitable tool to embed the aggregate firmly into the bedding coat. The aggregate often requires a light tamping to insure the proper bedding.

2513.3 Bedding coat proportions. The bedding coat for interior or exterior surfaces shall be composed of one-part portland cement, one-part Type S lime and a maximum of three parts of graded white or natural sand by volume. The bedding coat for interior surfaces shall be composed of 100 pounds (45.4 kg) of neat gypsum plaster and a maximum of 200 pounds (90.8 kg) of graded white sand. A factory-prepared bedding coat for interior or exterior use is permitted. The bedding coat for exterior surfaces shall have a minimum compressive strength of 1,000 pounds per square inch (psi) (6895 kPa).

❖ The plaster mix proportions for the exposed aggregate bedding coat are to be as defined in this section.

For the same reason the code establishes the hardness requirements for the aggregate used in exposed aggregate plaster, it also establishes durability requirements for the bedding coat. The minimum compressive strength of the exterior bedding coat is to be 1,000 pounds per square inch (psi) (6895 kPa).

Type S lime is a special hydrated lime for masonry purposes, and is required in the exterior bedding coat to increase high early plasticity and water retention.

For the increased plasticity required of the bedding coat on an interior application, the code requires the use of a relatively richer gypsum plaster.

2513.4 Application. The bedding coat is permitted to be applied directly over the first (scratch) coat of plaster, provided the ultimate overall thickness is a minimum of $^7/_8$ inch (22 mm), including lath. Over concrete or masonry surfaces, the overall thickness shall be a minimum of $^1/_2$ inch (12.7 mm).

❖ The requirements for the scratch coat used as part of the exposed aggregate plaster are the same as those for other scratch coats described and specified elsewhere in the code. The bedding coat may be used as the second coat, provided the total plaster thicknesses required in this section are maintained through the entire system.

2513.5 Bases. Exposed aggregate plaster is permitted to be applied over concrete, masonry, cement plaster base coats or gypsum plaster base coats installed in accordance with Section 2511 or 2512.

❖ The requirements for the base coat used as part of the exposed aggregate plaster are the same as those for other base coats described and specified elsewhere in the code.

2513.6 Preparation of masonry and concrete. Masonry and concrete surfaces shall be prepared in accordance with the provisions of Section 2510.7.

❖ See the commentary for Section 2510.7 for the preparation of masonry and concrete.

2513.7 Curing of base coats. Cement plaster base coats shall be cured in accordance with ASTM C 926. Cement plaster bedding coats shall retain sufficient moisture for hydration (hardening) for 24 hours minimum or, where necessary, shall be kept damp for 24 hours by light water spraying.

❖ The curing requirements for cement plaster base coat are required to be the same as that shown in Table 2512.6 and as described in the related commentary, except that moist curing is only necessary for 24 hours in lieu of the 48 hours shown in Table 2512.6.

BIBLIOGRAPHY

The following resource materials are referenced in this chapter or are relevant to the subject matter addressed in this chapter.

1995 Certification Listing. Ontario, Canada: Warnock Hersey, 1995.

ASTM C 5-97, *Specification for Quicklime for Structural Purposes.* West Conshohocken, PA: ASTM International, 1997.

ASTM C 22/C 22M-00, *Specification for Gypsum.* West Conshohocken, PA: ASTM International, 2000.

ASTM C 28/C 28M-00, *Specification for Gypsum Plasters.* West Conshohocken, PA: ASTM International, 2000.

ASTM C 35-95(2001), *Specification for Inorganic Aggregates for Use in Gypsum Plaster.* West Conshohocken, PA: ASTM International, 1995.

ASTM C 36/C 36M-01, *Specification for Gypsum Wallboard.* West Conshohocken, PA: ASTM International, 2001.

ASTM C 37/C 37M-01, *Specification for Gypsum Lath.* West Conshohocken, PA: ASTM International, 2001.

ASTM C 59/C 59M-00, *Specification for Gypsum Casting and Molding Plaster.* West Conshohocken, PA: ASTM International, 2000.

ASTM C 61/C 61M-00, *Specification for Gypsum Keene's Cement.* West Conshohocken, PA: ASTM International, 2000.

ASTM C 79-01, *Specification for Treated Core and Nontreated Core Gypsum Sheathing Board.* West Conshohocken, PA: ASTM International, 2001.

ASTM C 91-01, *Specification for Masonry Cement.* West Conshohocken, PA: ASTM International, 2001.

ASTM C 150-01, *Specification for Portland Cement.* West Conshohocken, PA: ASTM International, 2001.

ASTM C 206-84(1997), *Specification for Finishing Hydrated Lime.* West Conshohocken, PA: ASTM International, 1997.

ASTM C 442/C 442M-01, *Specification for Gypsum Backing Board Coreboard and Gypsum Shaftliner Board.* West Conshohocken, PA: ASTM International, 2001.

ASTM C 472-99, *Specification for Standard Test Methods for Physical Testing of Gypsum, Gypsum Plasters and Gypsum Concrete.* West Conshohocken, PA: ASTM International, 1999.

ASTM C 473-00, *Specification for Standard Test Methods for Physical Testing of Gypsum Panel Products.* West Conshohocken, PA: ASTM International, 2000.

ASTM C 474-01, *Test Methods for Joint Treatment Materials for Gypsum Board Construction.* West Conshohocken, PA: ASTM International, 2001.

ASTM C 475-01, *Specification for Joint Compound and Joint Tape for Finishing Gypsum Board.* West Conshohocken, PA: ASTM International, 2001.

ASTM C 514-01, *Specification for Nails for the Application of Gypsum Board.* West Conshohocken, PA: ASTM International, 2001.

ASTM C 557-99, *Specification for Adhesives for Fastening Gypsum Wallboard to Wood Framing.* West Conshohocken, PA: ASTM International, 1999.

ASTM C 587-97, *Specification for Gypsum Veneer Plaster.* West Conshohocken, PA: ASTM International, 1997.

ASTM C 588/C 588M-01, *Specification for Gypsum Base for Veneer Plasters.* West Conshohocken, PA: ASTM International, 2001.

ASTM C 595-01, *Specification for Blended Hydraulic Cements.* West Conshohocken, PA: ASTM International, 2001.

ASTM C 630/C630M-01, *Specification for Water-Resistant Gypsum Backing Board.* West Conshohocken, PA: ASTM International, 2001.

ASTM C 631-00, *Specification for Bonding Compounds for Interior Gypsum Plastering.* West Conshohocken, PA: ASTM International, 2000.

ASTM C 635-00, *Specification for the Manufacture, Performance, and Testing of Metal Suspension Systems for Acoustical Tile and Lay-in Panel Ceilings.* West Conshohocken, PA: ASTM International, 2000.

ASTM C 645-00, *Specification for Nonstructural Steel Framing Members.* West Conshohocken, PA: ASTM International, 2000.

ASTM C 754-00, *Specification for Installation of Steel Framing Members to Receive Screw-Attached Gypsum Panel Products.* West Conshohocken, PA: ASTM International, 2000.

ASTM C 840-01, *Specification for Application and Finishing of Gypsum Board.* West Conshohocken, PA: ASTM International, 2001.

ASTM C 841-99, *Specification for Installation of Interior Lathing and Furring.* West Conshohocken, PA: ASTM International, 1999.

ASTM C 842-99, *Specification for Application of Interior Gypsum Plaster.* West Conshohocken, PA: ASTM International, 1999.

ASTM C 843-99, *Specification for Application of Gypsum Veneer Plaster.* West Conshohocken, PA: ASTM International, 1999.

ASTM C 844-99, *Specification for Application of Gypsum Base to Receive Gypsum Veneer Plaster.* West Conshohocken, PA: ASTM International, 1999.

ASTM C 847-00, *Specification for Metal Lath.* West Conshohocken, PA: ASTM International, 2000.

ASTM C 897-00, *Standard Practice for Vapor Attack on Refractories for Furnace Superstructures.* West Conshohocken, PA: ASTM International, 2000.

ASTM C 926-98a, *Specification for Application of Portland Cement-Based Plaster.* West Conshohocken, PA: ASTM International, 1998.

ASTM C 931/C 931M-01, *Specification for Exterior Gypsum Soffit Board.* West Conshohocken, PA: ASTM International, 2001.

ASTM C 932-98a, *Specification for Surface-Applied Bonding Agents for Exterior Plastering.* West Conshohocken, PA: ASTM International, 1998.

ASTM C 933-96a, *Specification for Welded Wire Lath.* West Conshohocken, PA: ASTM International, 1996.

ASTM C 954-00, *Specification for Steel Drill Screws for the Application of Gypsum Panel Products or Metal Plaster Bases to Steel Studs From 0.033 in. (0.84 mm) to 0.112 in. (2.84 mm) in Thickness.* West Conshohocken, PA: ASTM International, 2000.

ASTM C 955-01, *Specification for Load-Bearing (Transverse and Axial) Steel Studs, Runners (Track), and Bracing or Bridging for Screw Application of Gypsum Panel Products and Metal Plaster Bases.* West Conshohocken, PA: ASTM International, 2001.

ASTM C 960-97, *Specification for Predecorated Gypsum Board.* West Conshohocken, PA: ASTM International, 1997.

ASTM C 1002-01, *Specification for Steel Drill Screws for the Application of Gypsum Panel Products or Metal Plaster Bases.* West Conshohocken, PA: ASTM International, 2001.

ASTM C 1007-00, *Specification for Installation of Load-Bearing (Transverse and Axial) Steel Studs and Related Accessories.* West Conshohocken, PA: ASTM International, 2000.

ASTM C 1032-96, *Specification for Woven Wire Plaster Base.* West Conshohocken, PA: ASTM International, 1996.

ASTM C 1047-99, *Specification for Accessories for Gypsum Wallboard and Gypsum Veneer Base.* West Conshohocken, PA: ASTM International, 1999.

ASTM C 1063-99, *Specification for Installation of Lathing and Furring to Receive Interior and Exterior Portland Cement-Based Plaster.* West Conshohocken, PA: ASTM International, 1999.

ASTM C 1177/C 1177M-01, *Specification for Glass Mat Gypsum Substrate for Use as Sheathing.* West Conshohocken, PA: ASTM International, 2001.

ASTM C 1178/C 1178M-01, *Specification for Glass Mat Water-Resistant Gypsum Backing Panel.* West Conshohocken, PA: ASTM International, 2001.

ASTM C 1280-99, *Specification for Application of Gypsum Sheathing.* West Conshohocken, PA: ASTM International, 1999.

ASTM C 1328-00, *Specification for Plastic (Stucco) Cement.* West Conshohocken, PA: ASTM International, 2000.

ASTM C 1395/C 1395M-01, *Specification for Gypsum Ceiling Board.* West Conshohocken, PA: ASTM International, 2001.

ASTM C 1396-99, *Standard Specification for Gypsum Board.* West Conshohocken, PA: ASTM International, 1999.

ASTM E 84-01, *Test Method for Surface Burning Characteristics of Building Materials.* West Conshohocken, PA: ASTM International, 2001.

ASTM E 90-99, *Test Method for Laboratory Measurement of Airborne Sound Transmission Loss of Building Partitions and Elements.* West Conshohocken, PA: ASTM International, 1999.

ASTM E 119-00, *Test Methods for Fire Tests of Building Construction and Materials.* West Conshohocken, PA: ASTM International, 2000.

ASTM F 547-77(95), *Terminology of Nails for Use with Wood and Wood-Base Materials.* West Conshohocken, PA: ASTM International, 1995.

ASTM F 1667-01A, *Specification for Driven Fasteners: Nails, Spikes, and Staples.* West Conshohocken, PA: ASTM International, 2001.

GA 216-00, *Application and Finishing of Gypsum Board.* Washington, DC: Gypsum Association, 2000.

GA 230-91, *Vapor Retarders Over Water-Resistant Gypsum Backing Board.* Washington, DC: Gypsum Association, 1991.

GA 600-00, *Fire Resistance Design Manual.* Washington, DC: Gypsum Association, 2000.

Guidelines for Determining the Fireresistance of Building Elements. Country Club Hills, IL: Building Officials and Code Administrators International, Inc., 1993.

UL-96, *Fire Resistance Directory.* Northbrook, IL: Underwriters Laboratories Inc., 1996.

Chapter 26: Plastic

General Comments

Plastics are a very large group of synthetic materials whose structures are based on the chemistry of carbon. Plastics are also called polymers because they are made of extremely long chains of carbon atoms. Since plastics can be readily molded, extruded, cast into various shapes and forms or drawn into filaments and fibers, they are increasingly being used in the construction of buildings and structures. While progress in polymer technology makes it increasingly difficult to make general statements about these materials, the following properties are characteristic of many plastics:

- Low strength (in comparison to steel and other metals commonly used in construction);
- Low stiffness (modulus of elasticity) as compared to metals (except for reinforced plastics);
- Tendency to creep (increase in length under tensile stress);
- Low hardness (except for formaldehyde plastics);
- Low density (varies, plastics can be expanded or unexpanded);
- Brittleness at low temperatures and loss of strength and hardness at moderately elevated temperatures;
- Flammability (although many plastics do not burn);
- Outstanding electrical characteristics, such as electrical resistance and
- Degradation of some plastics, due to environmental agents, such as ultraviolet radiation (although most plastics are highly resistant to chemical attack).

When plastics and foam plastics were first proposed to be included in the code, consideration was given to several factors.

The following is the evolution of plastics being introduced into the model codes, ultimately leading to inclusion in the code.

It was recognized that most buildings classified as noncombustible construction contained combustible elements. These elements were not primary structural members but included materials such as: asphalt and pitch (in roof coverings); wood (in flooring systems, as wall and ceiling interior finish and trim and as furring and lath in walls); paper products (as slip sheets and vapor retarders) and cork (as insulation).

There was a sensitivity to the burning characteristics of all plastic materials based on the history of fires involving plastics in warehouses and in building applications not covered by any of the model codes. It was agreed, therefore, that plastic materials should be required to meet minimum fire performance properties and each application would have appropriate requirements related to characteristics such as rate of burning, surface flammability or smoke development.

This position is somewhat different from the regulation of traditional materials, such as wood, paper and asphalt, which frequently are not required to be separately fire tested when used within an assembly.

Plastic materials to be left exposed on the interior or exterior of buildings would have to meet the existing requirements for other finish materials as well as be limited in area and subject to separation requirements between adjoining areas. The primary concern was that plastic materials would be used in applications in which glass had traditionally been used, such as skylights and light-diffusing ceilings, and since little experience was available at the time, a conservative approach should be taken.

Foam plastics used as thermal insulation would be separated from the interior of buildings by a thermal barrier equivalent to $^1/_2$-inch-thick (12.7 mm) gypsum wallboard. This position was supported by industry because most of the applications at that time included plaster or wallboard finishes of $^1/_2$ inch (12.7 mm) or greater thickness, and the fire record for such applications was good.

Specific examples of commonly used acceptable applications would be included in the code, such as the use of foam plastics within wood frame and masonry walls.

Finally, the code would include a statement that "diversified tests" would be necessary for approval of any application or material not meeting the prescriptive requirements of the code. The reason for including diversified tests was because it was clear that it would not be possible or even desirable to include all applications specifically in the code. Diversified tests were intended to mean tests related to the end use and usually were full-scale assemblies exposed to fire conditions that were representative of actual fires that might be expected to occur.

Purpose

This chapter establishes minimum requirements for all light-transmitting plastics and foam plastics in all applications regulated by the code. Some plastics exhibit rapid flame spread and heavy smoke density characteristics when exposed to fire. Additionally, exposure to the heat generated by a fire can cause some plastics to deform and undergo a change in their structural stability. The requirements and limitations of this chapter are necessary to control the use of these combustible products such that they do not compromise the safety of building occupants.

SECTION 2601
GENERAL

2601.1 Scope. These provisions shall govern the materials, design, application, construction and installation of foam plastic, foam plastic insulation, plastic veneer, interior plastic finish and trim and light-transmitting plastics. See Chapter 14 for requirements for exterior wall finish and trim.

❖ Chapter 26 governs foam plastics and light-transmitting plastic materials regardless of their use. Requirements for foam-plastic insulation are found in Section 2603. Requirements for interior finish and trim are found in Section 2604. Requirements for plastic veneer are found in Section 2605. Requirements for light-transmitting plastics are found in Section 2606. Requirements for light-transmitting plastic wall panels are found in Section 2607. Requirements for light-transmitting plastic glazing are found in Section 2608. Requirements for light-transmitting plastic roof panels are found in Section 2609. Requirements for light-transmitting plastic skylight glazing are found in Section 2610. Requirements for light-transmitting plastic interior signs are found in Section 2611.

SECTION 2602
DEFINITIONS

2602.1 General. The following words and terms shall, for the purposes of this chapter and as used elsewhere in this code, have the meanings shown herein.

❖ Definitions of terms can help in the understanding and application of the code requirements. The purpose for including these definitions within this chapter is to provide more convenient access to them without having to refer back to Chapter 2. For convenience, these terms are also listed in Chapter 2 with a cross reference to this section. The use and application of all defined terms, including those defined herein, are set forth in Section 202.

Plastics include any of the large group of synthetic or natural organic materials that can be shaped, formed and molded at high temperatures and are capable of maintaining their form at low temperatures. Plastics can be divided into two basic classes: thermoplastic and thermosetting. Plastics can also be broken down into Class CC1 or CC2 (see Section 2606). Plastics are combustible materials and some can give off toxic gases when exposed to fire. Plastics can be synthetically designed to exhibit desirable strength, rigidity and flammability characteristics.

FOAM PLASTIC INSULATION. A plastic that is intentionally expanded by the use of a foaming agent to produce a reduced-density plastic containing voids consisting of open or closed cells distributed throughout the plastic for thermal insulating or acoustical purposes and that has a density less than 20 pounds per cubic foot (pcf) (320 kg/m^3).

❖ Foam-plastic insulation is plastic that is imbedded with gas.

LIGHT-DIFFUSING SYSTEM. Construction consisting in whole or in part of lenses, panels, grids or baffles made with light-transmitting plastics positioned below independently mounted electrical light sources, skylights or light-transmitting plastic roof panels. Lenses, panels, grids and baffles that are part of an electrical fixture shall not be considered as a light-diffusing system.

❖ Heat-resistant plastics, such as polystyrene and urea, are frequently used for light diffusion.

LIGHT-TRANSMITTING PLASTIC ROOF PANELS. Structural plastic panels other than skylights that are fastened to structural members, or panels or sheathing and that are used as light-transmitting media in the plane of the roof.

❖ Plastic roof panels are primarily used in areas where light transmission rather than visual clarity is the goal (i.e., factory and industrial applications). Even though they are used as light-transmitting media, they must support the loads and forces encountered, just like roof surfaces. Typical applications include factories, warehouses and greenhouses where glass breakage is a practical concern.

LIGHT-TRANSMITTING PLASTIC WALL PANELS. Plastic materials that are fastened to structural members, or to structural panels or sheathing, and that are used as light-transmitting media in exterior walls.

❖ This term is only applicable to plastic wall panels allowing the transmission of light. Even though they are used as light-transmitting media, they must support the loads and forces encountered, just like other wall surfaces. Foam-core insulated panels are regulated by Section 2603 and are not considered plastic wall panels.

PLASTIC, APPROVED. Any thermoplastic, thermosetting or reinforced thermosetting plastic material that conforms to combustibility classifications specified in the section applicable to the application and plastic type.

❖ This term applies to plastic that has properties that conform to the code requirements for its intended use.

PLASTIC GLAZING. Plastic materials that are glazed or set in frame or sash and not held by mechanical fasteners that pass through the glazing material.

❖ In lieu of glass, plastic is permitted to be used as a glazing material. Plastic glazing is required to meet the same load requirements as glass.

REINFORCED PLASTIC, GLASS FIBER. Plastic reinforced with glass fiber having not less than 20 percent of glass fibers by weight.

❖ Most commonly, translucent polyester (plastic) panels are reinforced with glass fibers to achieve a stronger, stiffer and more impact-resistant panel.

THERMOPLASTIC MATERIAL. A plastic material that is capable of being repeatedly softened by increase of temperature and hardened by decrease of temperature.

❖ Thermoplastic materials become plastic and deform easily at higher temperatures. Indeed, many of these materials have limited construction value at high temperatures. Examples of thermoplastics are cellulose acetate, nylon, polycarbonates, polyethylene (three different densities), methyl methacrylate (plexiglass), polypropylene, polybutylene, polystyrene, acrylonitrile-butadiene-styrene (ABS), polytetrafluro ethylene (Teflon®), polyvinyl chloride (PVC), chlorinated polyvinyl chloride (CPVC) and polyethylene terephthalate (polyester). Note there are both thermosetting and thermoplastic polyesters. These materials have high thermal expansion coefficients and, while they may burn slowly in a fire, they decompose rapidly to yield a heavy black smoke and acrid, sometimes very toxic, fumes depending on the specific material and burning conditions.

THERMOSETTING MATERIAL. A plastic material that is capable of being changed into a substantially nonreformable product when cured.

❖ This term is the opposite of a thermoplastic material. Common thermosetting plastic includes epoxy, melamine, urea, polyester and polyurethane.

SECTION 2603
FOAM PLASTIC INSULATION

2603.1 General. The provisions of this section shall govern the requirements and uses of foam plastic insulation in buildings and structures.

❖ Section 2603 contains all of the provisions for foam-plastic insulation products, such as rigid insulation board. Foam plastic is a general term given to insulating products that have been manufactured by injecting gas into the raw plastic product. Extruded polystyrene, expanded polystyrene (also called EPS or beadboard), isocyanurate, open-cell isocyanurate, phenolic foam and polyurethane are among the many types of foam insulation products subject to the requirements found in this section. Section 2603.3 specifies the maximum permissible surface-burning and smoke-developed properties of materials, requires product labeling and defines two methods of obtaining compliance: through the prescriptive installation instructions of Sections 2603.4 through 2603.7 or through an alternative approval based on the testing requirements of Section 2603.8.

2603.2 Labeling and identification. Packages and containers of foam plastic insulation and foam plastic insulation components delivered to the job site shall bear the label of an approved agency showing the manufacturer's name, the product listing, product identification and information sufficient to determine that the end use will comply with the code requirements.

❖ All foam plastics or packages of foam plastics delivered to the construction site must be labeled. Also, labels are required on containers (usually two components in 55-gallon drums) of ingredients delivered for the production of foam plastic at the construction site. The label must include identification of the approved agency and detailed product identification or information describing the performance characteristics of the product. It was intended that labeling printed on board stock or on packaging be acceptable. It was also intended that the label information include the name of the manufacturer or distributor, the type of foam plastic, the performance characteristics required and the name of the approved testing agency.

2603.3 Surface-burning characteristics. Unless otherwise indicated in this section, foam plastic insulation and foam plastic cores of manufactured assemblies shall have a flame spread index of not more than 75 and a smoke-developed index of not more than 450 where tested in the maximum thickness intended for use in accordance with ASTM E 84. Loose fill-type foam plastic insulation shall be tested as board stock for the flame spread index and smoke-developed index.

Exceptions:

1. Smoke-developed index for interior trim as provided for in Section 2604.2.
2. In cold storage buildings, ice plants, food plants, food processing rooms and similar areas, foam plastic insulation where tested in a thickness of 4 inches (102 mm) shall be permitted in a thickness up to 10 inches (254 mm) where the building is equipped throughout with an automatic fire sprinkler system in accordance with Section 903.3.1.1. The approved automatic sprinkler system shall be provided in both the room and that part of the building in which the room is located.
3. Foam plastic insulation that is a part of a Class A, B or C roof-covering assembly provided the assembly with the foam plastic insulation satisfactorily passes FM 4450 or UL 1256. The smoke-developed index shall not be limited for roof applications.
4. Foam plastic insulation greater than 4 inches (102 mm) in thickness shall have a maximum flame spread index of 75 and a smoke-developed index of 450 where tested at a minimum thickness of 4 inches (102 mm), provided the end use is approved in accordance with Section 2603.8 using the thickness and density intended for use.

5. Flame spread and smoke-developed indexes for foam plastic interior signs in covered mall buildings provided the signs comply with Section 402.14.

❖ All foam plastic materials are required to be tested in accordance with ASTM E 84, unless specifically exempted or modified by one of the listed exceptions. When testing in accordance with ASTM E 84 is required, the materials must be tested in the maximum thickness to be used. While there was no known correlation between thickness and burning characteristics, it was generally concluded that thicker materials would likely burn more rapidly and produce more smoke if involved in a building fire. Subsequent testing has shown that thicker materials do produce more smoke when involved in building fires; however, the assumption about more rapid burning has not been confirmed. The maximum values for flame spread and smoke-developed ratings are 75 and 450, respectively, unless specifically indicated otherwise. In selecting the appropriate limitation for flame spread value, consideration was given to the existing maximum limitations for interior finish, which were 25, 75, 200 and 500. It was agreed that it would be less confusing to use one of the existing values rather than create a new category. The maximum flame spread value of 75 was chosen on the basis that it was lower than untreated wood (which usually was 100 to 165). The maximum smoke-developed rating of 450 was selected because, at the time, the code permitted interior finish materials that gave off "smoke no more dense than that given off by untreated wood," and some interior finish materials were known to have smoke-developed ratings of 450. Section 803.1 stipulates a maximum smoke-developed rating of 450 for interior finish. In selecting the maximum flame spread and smoke values, it was believed that a conservative approach was being taken by requiring an insulation material to meet the same requirements as interior finish, even though the insulation was intended to be covered with an interior finish material.

The requirements for surface-burning characteristics of foam plastic apply to foam plastics used as cores of manufactured assemblies. The intent is that, while the finished assemblies are not required to be tested for surface-burning characteristics, the foam-plastic core is not exempt from the general requirement; therefore, foam plastic is regulated in factory-manufactured assemblies the same as it is in field-fabricated applications.

Five exceptions to the flame spread and smoke-developed rating requirements of 75 and 450, respectively, are permitted.

Exception 1 refers to specific code sections that modify the requirements established by Section 2603.3.

Because cold storage construction frequently uses insulation in greater thickness, Exception 2 is given to the requirement for the testing at thickness of intended use. In such a case, other protection, specifically automatic fire suppression, is required.

Exception 3 states that when used as roof insulation, foam plastic is not required to meet the flame spread rating of 75 when the roof deck assembly has been tested and performance is acceptable in accordance with FM 4450 or UL 1256. Smoke-developed ratings are not limited for any foam plastic used as roof insulation whether the assembly has been tested or not, as smoke generation is less of a concern on the exterior of a building.

Exception 4 is provided to accommodate occupancies, such as cold-storage facilities, which require thicknesses of foam-plastic insulation in excess of 4 inches (102 mm) up to a maximum of 10 inches (254 mm). Four inches (102 mm) is the maximum practical thickness that can be tested in the ASTM E 84 furnace. As such, an approval basis for greater thicknesses is necessary. The foam plastic is to be tested in the thickness to be installed according to Section 2603.8. The consequence of not providing for this exception would be to prohibit any foam-plastic installation over 4 inches (102 mm) in thickness. This would be unrealistic and unnecessary.

Flame spread index requirements were already established when the code began regulating foam plastics. It was known at that time that foam plastics were rapidly becoming the replacement for cork as insulation for commercial freezers and coolers. Many of the buildings insulated before 1975 did not have thermal barriers over the insulation. Additionally, many of the freezer and cooler buildings built before 1975 were not equipped with automatic sprinkler systems.

Exception 5 exempts plastic interior signs in covered mall buildings when they comply with Section 402.14. Section 402.14 requires compliance with UL 1975, which measures the rate of heat release of burning materials used in the manufacture of such signs. Section 402.14 sets the pass/fail criteria at a maximum heat release rate of 150 kW. This test meets the intent of what is trying to be achieved by Section 2603.3. This test is a significantly different measure of burning characteristics of materials, since it relates to the rate of heat release rather than flame spread; however, while the two tests are completely different and cannot be calibrated to each other, the rate-of-heat-release test does deal with the spirit and intent of Section 2603.3, which is to limit the spread of fire.

2603.4 Thermal barrier. Except as provided for in Sections 2603.4.1 and 2603.8, foam plastic shall be separated from the interior of a building by an approved thermal barrier of 0.5-inch (12.7 mm) gypsum wallboard or equivalent thermal barrier material that will limit the average temperature rise of the unexposed surface to not more than 250°F (120°C) after 15 minutes of fire exposure, complying with the standard time-temperature curve of ASTM E 119. The thermal barrier shall be installed in such a manner that it will remain in place for 15 minutes based on FM 4880, UL 1040, NFPA 286 or UL 1715. Combustible concealed spaces shall comply with Section 717.

❖ The use of an approved thermal barrier to separate foam plastics from the interior of a building is required in this section. Section 2603.4.1 describes circumstances where the thermal barrier is not required. An approved thermal barrier is defined as $^1/_2$-inch (12.7 mm) gypsum wallboard or the equivalent. This section sets forth the test methods and criteria by which alternative thermal barriers are to be qualified to limit the average temperature rise of the unexposed face to 250°F (121°C) for 15 minutes of fire exposure while complying with the time-temperature conditions of ASTM E 119.

Before 1975, experience had shown that foam plastics covered with plaster or $^1/_2$-inch (12.7 mm) gypsum wallboard had performed satisfactorily in building fires. For this reason, $^1/_2$-inch (12.7 mm) gypsum wallboard was included in the code as a minimum requirement. It was recognized that specifying a single material would not be desirable in a performance code; therefore, the words "or equivalent" were added. The thermal barrier test was selected as the appropriate method to determine equivalent performance. Although regular $^1/_2$-inch (12.7 mm) gypsum wallboard happened to have a 15-minute rating in the prescribed thermal barrier test, there was no intent that the 15-minute requirement for the thermal barrier would insure 15 minutes of escape time in an actual building fire. This is related to the fact that the time-temperature conditions of ASTM E 119 may not reflect actual fire conditions. Gypsum wallboard is still the most commonly used thermal barrier, but equivalent materials are permitted and have been used when shown to be equivalent.

This section also states that the thermal barrier must be installed in such a way that it remains in place for 15 minutes when exposed to fire conditions and that combustible concealed spaces comply with Section 717.

2603.4.1 Thermal barrier not required. The thermal barrier specified in Section 2603.4 is not required under the conditions set forth in Sections 2603.4.1.1 through 2603.4.1.13.

❖ The conditions described in Sections 2603.4.1.1 through 2603.4.1.13 do not warrant the requirement for the thermal barrier described in Section 2603.4. The prescriptive installation methods described in these sections must be strictly followed in order to serve as an alternative to the thermal barrier. Furthermore, installations of foam plastics in accordance with these sections need not meet the alternative approval testing requirements of Section 2603.8.

2603.4.1.1 Masonry or concrete construction. In a masonry or concrete wall, floor or roof system where the foam plastic insulation is covered on each face by a minimum of 1 inch (25 mm) thickness of masonry or concrete.

❖ No thermal barrier is required when 1 inch (25 mm) or more of masonry or concrete are placed between the foam plastic and the interior of the building. The intent is to accept 1 inch (25 mm) of masonry or concrete as equal to (or better than) $^1/_2$-inch (12.7 mm) gypsum wallboard. This condition can arise when foam plastics are installed either within a wall or on one side of a wall. Some common examples are when foam plastics are installed:

- In the cavity of a hollow masonry wall;
- As the core of a concrete-faced panel;
- On the exterior face of a masonry wall and covered with an exterior finish or
- Within the cores of hollow masonry units.

2603.4.1.2 Cooler and freezer walls. Foam plastic installed in a maximum thickness of 10 inches (254 mm) in cooler and freezer walls shall:

1. Have a flame spread index of 25 or less and a smoke-developed index of not more than 450, where tested in a minimum 4 inch (102 mm) thickness.
2. Have flash ignition and self-ignition temperatures of not less than 600°F and 800°F (316°C and 427°C), respectively.
3. Have a covering of not less than 0.032-inch (0.8 mm) aluminum or corrosion-resistant steel having a base metal thickness not less than 0.0160 inch (0.4 mm) at any point.
4. Be protected by an automatic sprinkler system. Where the cooler or freezer is within a building, both the cooler or freezer and that part of the building in which it is located shall be sprinklered.

❖ Cooler and freezer walls insulated with 10 inches (254 mm) or less of foam plastic do not require a thermal barrier when complying with all four criteria in Section 2603.4.1.2.

2603.4.1.3 Walk-in coolers. In nonsprinklered buildings, foam plastic having a thickness that does not exceed 4 inches (102 mm) and a maximum flame spread index of 75 is permitted in walk-in coolers or freezer units where the aggregate floor area does not exceed 400 square feet (37 m^2) and the foam plastic is covered by a metal facing not less than 0.032-inch-thick (0.81 mm) aluminum or corrosion-resistant steel having a minimum base metal thickness of 0.016 inch (0.41 mm). A thickness of up to 10 inches (254 mm) is permitted where protected by a thermal barrier.

❖ The thermal barrier may be omitted when foam plastics up to 4 inches (102 mm) thick with specified metal facings are used in coolers or freezers no greater than 400 square feet (37 m^2) in floor area. The required metal facing is intended to act as a barrier against ignition of the foam plastic. It is not intended to serve as a thermal barrier. This section recognizes the relatively small size of walk-in coolers and the accompanying lower hazard based on the small number of human occupants and the short travel distance to the cooler door; therefore, less-conservative requirements are permitted. Foam plastic up to 10 inches thick (254 mm) is permitted when protected by a thermal barrier.

2603.4.1.4 Exterior walls–one-story buildings. For one-story buildings, foam plastic having a flame spread index of 25 or less, and a smoke-developed index of not more than 450, shall be permitted without thermal barriers in or on exterior walls in a thickness not more than 4 inches (102 mm) where the foam plastic is covered by a thickness of not less than 0.032-inch-thick (0.81 mm) aluminum or corrosion-resistant steel having a base metal thickness of 0.0160 inch (0.41 mm) and the building is equipped throughout with an automatic sprinkler system in accordance with Section 903.3.1.1.

❖ Foam plastics may be used without a thermal barrier in one-story buildings provided that:

- The material is not more than 4 inches (102 mm) thick;
- It has a flame spread index not greater than 25;
- It has a smoke-developed index of not more than 450;
- It is covered with specified metal facings; and
- The building is equipped throughout with an automatic sprinkler system.

This provision is intended to permit the use of metal-faced panels, primarily in storage buildings other than cold-storage construction. Cold-storage construction was also covered in this section (see also Sections 2603.4.1.2 and 2603.4.1.3, and Exception 2 of Section 2603.3). The limitations are based on the results of large-scale tests performed by independent laboratories and sponsored by the plastics industry.

2603.4.1.5 Roofing. Foam plastic insulation under a roof assembly or roof covering that is installed in accordance with the code and the manufacturer's instructions shall be separated from the interior of the building by wood structural panel sheathing not less than 0.47 inch (11.9 mm) in thickness bonded with exterior glue, with edges supported by blocking, tongue-and-groove joints or other approved type of edge support, or an equivalent material. A thermal barrier is not required for foam plastic insulation that is a part of a Class A, B or C roof-covering assembly, provided the assembly with the foam plastic insulation satisfactorily passes FM 4450 or UL 1256.

❖ The thermal barrier is not required over plywood roof sheathing that is at least $^{15}/_{32}$ inch thick (12 mm) and is manufactured utilizing exterior-grade glue. Also, the plywood must be properly installed to provide adequate edge support. Other materials equivalent to the plywood described may be used; however, this applies only where plywood roof decks are acceptable in the type of construction of the building under consideration.

Also, when the assembly complies with FM 4450 or UL 1256, a thermal barrier need not to be installed. The intent is to recognize that roof assemblies tested to the criteria of the referenced standards have adequately demonstrated a resistance to fire from the underside of the roof deck; therefore, no additional testing is necessary, nor is an additional thermal barrier needed.

2603.4.1.6 Attics and crawl spaces. Within an attic or crawl space where entry is made only for service of utilities, foam plastic insulation shall be protected against ignition by 1.5-inch-thick (38 mm) mineral fiber insulation; 0.25-inch-thick (6.4 mm) wood structural panel, particleboard or hardboard; 0.375-inch (9.5 mm) gypsum wallboard, corrosion-resistant steel having a base metal thickness of 0.016 inch (0.4 mm) or other approved material installed in such a manner that the foam plastic insulation is not exposed. The protective covering shall be consistent with the requirements for the type of construction.

❖ Foam plastics installed in attics and crawl spaces are required to be protected from ignition by a covering that need not qualify as a thermal barrier. The protective covering must be consistent with the type of construction, which means that combustible coverings are permitted only where combustible materials are otherwise allowed. The intent is to recognize such spaces as separate from the areas of buildings normally occupied by people and, therefore, not subjected to the numerous fires caused by people. Also, it recognizes that requiring a thermal barrier over foam plastics while permitting unlimited use of other combustible materials would be inconsistent and of little or no benefit for fire safety.

2603.4.1.7 Doors not required to have a fire protection rating. Where pivoted or side-hinged doors are permitted without a fire protection rating, foam plastic insulation, having a flame spread index of 75 or less and a smoke-developed index of not more than 450, shall be permitted as a core material where the door facing is of metal having a minimum thickness of 0.032-inch (0.8 mm) aluminum or steel having a base metal thickness of not less than 0.016 inch (0.4 mm) at any point.

❖ Foam plastic may be used as a core material for doors without using a thermal barrier, provided that a specified metal facing is used and the door is not required to have a fire protection rating. The intent is that doors required to have a fire protection rating must be tested with the foam plastic cores in place in order to qualify.

This section recognizes the limited area of doors compared to the walls in which they are located and regulates foam plastics similar to other materials used in doors.

2603.4.1.8 Exterior doors in buildings of Group R-2 or R-3. In occupancies classified as Group R-2 or R-3 as applicable in Section 101.2, foam-filled exterior entrance doors to individual dwelling units that do not require a fire-resistance rating shall be faced with wood or other approved materials.

❖ Wood facings on exterior entrance doors of Group R-2 and R-3 buildings are permitted when the doors are not required to have a fire-resistance rating. Doors complying with this section are not required to comply with Section 2603.4.1.7.

2603.4.1.9 Garage doors. Where garage doors are permitted without a fire-resistance rating and foam plastic is used as a core material, the door facing shall be metal having a minimum thickness of 0.032-inch (0.8 mm) aluminum or 0.010-inch (0.25 mm) steel or the facing shall be minimum 0.125-inch-thick (3.2 mm) wood. Garage doors having facings other than those described above shall be tested in accordance with, and meet the acceptance criteria of DASMA 107.

Exception: Garage doors using foam plastic insulation complying with Section 2603.3 in detached and attached garages associated with one- and two-family dwellings need not be provided with a thermal barrier.

❖ The prescriptive requirements of this section regulate garage doors with foam-plastic cores used in nonfire-resistance-rated applications. The requirements in Section 2603.4.1.9, which governs overhead sectional, coiling or vertical lift-type garage doors, are slightly different from the requirements of Sections 2603.4.1.7 and 2603.4.1.8, which govern means of egress doors, such as side-swinging doors. These "garage doors" are not limited to such doors used in conjunction with a private or public garage occupancy but refer to doors for vehicles in any occupancies.

The minimum door-facing material and thickness requirements for garage doors with foam-plastic cores stated in this section were determined based on the results of a fire testing program sponsored by the National Association of Garage Door Manufacturers (NAGDM). This program exposed a variety of commercially manufactured garage doors to a room corner fire test. The results of the testing indicated that garage doors having facings in the minimum thicknesses stated did not spread fire to the edge of the specimen and did not cause flashover in the test room.

This section exempts exterior garage doors used in garages attached to one- and two-family dwellings and detached garages (Group R-3) from the thermal barrier requirements of Section 2603.4. This exemption is similar to that of Section 2603.4.1.8 for exterior egress doors in Group R-2 and R-3 occupancies.

2603.4.1.10 Siding backer board. Foam plastic insulation of not more than 2,000 British thermal units per square feet (Btu/sq. ft.) (22.7 MJ/m^2) as determined by NFPA 259 shall be permitted as a siding backer board with a maximum thickness of 0.5 inch (12.7 mm), provided it is separated from the interior of the building by not less than 2 inches (51 mm) of mineral fiber insulation or equivalent or where applied as insulation with re-siding over existing wall construction.

❖ Foam plastics of 2,000 Btu/sq. ft. (22.7 MJ/m^2) used as siding backer board must not be greater than $^1/_2$ inch thick (12.7 mm), and a thermal barrier is not required when mineral fiber insulation of at least 2 inches (51 mm) thick is installed. Conversely, if a thickness greater than $^1/_2$-inch (12.7 mm) siding backer board is used, the thermal barrier must be installed regardless of the thickness of the mineral fiber insulation (or equivalent). The intent of this section is twofold. First, this section permits continued use of a product that was widely used at the time this portion of the code was developed. In this case, the combustibility of the finished wall was recognized to be controlled by plywood or other common sheathing materials.

Second, the use of foam plastics is permitted in residing existing buildings. In such applications, the foam plastic is used primarily as a leveling board. Adding the foam plastic over existing siding is assumed not to have a substantive effect on the overall fire safety of the building in question. This is because fires typically occur more often on the inside than on the outside of buildings. Also, the code requires fire-resistance ratings from the exterior side of walls only when the fire separation distance is 5 feet (1524 mm) or less (see Section 704.5).

2603.4.1.11 Interior trim. Foam plastic used as interior trim in accordance with Section 2604 shall be permitted without a thermal barrier.

❖ This section specifies that a thermal barrier is not needed for interior trim that meets the limitations of Section 2604 and related subsections since the hazard is limited.

2603.4.1.12 Interior signs. Foam plastic used for interior signs in covered mall buildings in accordance with Section 402.14 shall be permitted without a thermal barrier.

❖ This section specifies that a thermal barrier is not needed for interior signs that meet the limits stated in Section 402.14, since the hazard from the interior signs is appropriately addressed by test standard UL1975 (see commentary, Section 402.14).

2603.4.1.13 Type V construction. Foam plastic spray applied to a sill plate and header of Type V construction is subject to all of the following:

1. The maximum thickness of the foam plastic shall be $3^1/_4$ inches (82.6 mm).
2. The density of the foam plastic shall be in the range of 1.5 to 2.0 pcf (24 to 32 kg/m^3).
3. The foam plastic shall have a flame spread index of 25 or less and an accompanying smoke-developed index of 450 or less when tested in accordance with ASTM E84.

❖ This section does not require a thermal barrier where foam plastic is sprayed upon sill plates and headers in Type V construction. This is only allowed if specific criteria related to thickness, density, flame spread and smoke-development criteria are met. This particular allowance was based upon testing in a room corner fire test that compared the performance of an all-wood floor system to that of a wood floor system with foam plastic sprayed on sill plates and headers.

2603.5 Exterior walls of buildings of any height. Exterior walls of buildings of Type I, II, III or IV construction of any height shall comply with Sections 2603.5.1 through 2603.5.7. Exterior walls of cold storage buildings required to be constructed of noncombustible materials, where the building is more than one story in height, shall also comply with the provisions of Sections 2603.5.1 through 2603.5.7. Exterior walls of buildings of Type V construction shall comply with Sections 2603.2, 2603.3 and 2603.4.

❖ All foam plastics used on exterior walls of all types of buildings, except wood frame, are to be installed in accordance with Sections 2603.5.1 through 2603.5.7. Installations on one-story, noncombustible walls of cold storage buildings are required to also comply with the provisions of Sections 2603.5.1 through 2603.5.7.

Foam plastics used on exterior walls of Type V buildings are required to comply with Sections 2603.2, 2603.3 and 2603.4. The intent is to regulate the use of an insulating envelope over the exterior of a structure when the envelope provides no structural support other than the transfer of wind loads. It is recognized that some envelopes will be constructed in place by installing a rigid foam plastic and covering it with an exterior finish while others would be installed as prefabricated panels complete with exterior finish.

2603.5.1 Fire-resistance-rated walls. Where the wall is required to have a fire-resistance rating, data based on tests conducted in accordance with ASTM E 119 shall be provided to substantiate that the fire-resistance rating is maintained.

❖ Foam plastics are permitted in walls that are required to have fire-resistance ratings. Such assemblies must be fire tested with the foam plastic in place.

2603.5.2 Thermal barrier. Any foam plastic insulation shall be separated from the building interior by a thermal barrier meeting the provisions of Section 2603.4, unless special approval is obtained on the basis of Section 2603.8.

Exception: One-story buildings complying with Section 2603.4.1.4.

❖ A thermal barrier is required, unless alternative approval is obtained in accordance with Section 2603.8. The intent is to make it clear that the thermal barrier requirement of Section 2603.4 is applicable. The exception provides for the reduced requirements of Section 2603.4.1.4 for one-story buildings.

2603.5.3 Potential heat. The potential heat of foam plastic insulation in any portion of the wall or panel shall not exceed the potential heat expressed in Btu per square feet (mJ/m^2) of the foam plastic insulation contained in the wall assembly tested in accordance with Section 2603.5.5. The potential heat of the foam plastic insulation shall be determined by tests conducted in accordance with NFPA 259 and the results shall be expressed in Btu per square feet (mJ/m^2).

Exception: One-story buildings complying with Section 2603.4.1.4.

❖ This section limits the combustible content of exterior walls based on the potential heat of the foam-plastic insulation that performs satisfactorily in the wall or panel assembly tested in accordance with Section 2603.5.5.

2603.5.4 Flame spread and smoke-developed indexes. Foam plastic insulation, exterior coatings and facings shall be tested separately in the thickness intended for use, but not to exceed 4 inches (102 mm), and shall each have a flame spread index of 25 or less and a smoke-developed index of 450 or less as determined in accordance with ASTM E 84.

Exception: Prefabricated or factory-manufactured panels having minimum 0.020-inch (0.51 mm) aluminum facings and a total thickness of 0.25 inch (6.4 mm) or less are permitted to be tested as an assembly where the foam plastic core is not exposed in the course of construction.

❖ Each component, including the foam plastic core, must be tested and have a maximum flame spread of 25 and a maximum smoke-developed index of 450 when tested in accordance with ASTM E 84. The material should be tested with the same thickness for use, except that ASTM E 84 is limited to a 4-inch (102 mm) thickness for testing. Section 2603.5.5 allows full-scale testing in lieu of the requirements in this section. Full-scale testing can address the actual thickness. Materials should not be restricted by the testing limitations of ASTM E 84 when they perform well at a higher thickness in the full scale test. The exception applies to prefabricated panels that can be installed without exposing the foam plastic.

2603.5.5 Test standard. The wall assembly shall be tested in accordance with and comply with the acceptance criteria of NFPA 285.

Exception: One-story buildings complying with Section 2603.4.1.4.

❖ This section is the only provision dealing with propagation of fire due to exposure from an exterior source. Other combustibles are not allowed as components of noncombustible building exterior walls.

2603.5.6 Label required. The edge or face of each piece of foam plastic insulation shall bear the label of an approved agency. The label shall contain the manufacturer's or distributor's identification, model number, serial number or definitive information describing the product or materials' performance characteristics and approved agency's identification.

❖ Each piece of foam plastic must be labeled. This requirement is somewhat different than the general marking requirement of Section 2603.2, which only requires a manufacturer's identifying mark on either the product or the packaging. The reason for this is that properties other than fire performance of the foam plastic, such as bond strength to the exterior finish, are of significance in this type of application. Additionally, the intent is to provide a means for specific product identification not only during construction but also after construction is complete by the removal of a sample product from the finished structure should a question arise.

2603.5.7 Ignition. Exterior walls shall not exhibit sustained flaming where tested in accordance with NFPA 268. Where a material is intended to be installed in more than one thickness, tests of the minimum and maximum thickness intended for use shall be performed.

Exception: Assemblies protected on the outside with one of the following:

1. A thermal barrier complying with Section 2603.4.
2. A minimum 1 inch (25 mm) thickness of concrete or masonry.

3. Glass-fiber-reinforced concrete panels of a minimum thickness of 0.375 inch (9.5 mm).
4. Metal-faced panels having minimum 0.019- inch-thick (0.48 mm) aluminum or 0.016-inch- thick (0.41 mm) corrosion-resistant steel outer facings.
5. A minimum 0.875 inch (22.2 mm) thickness of stucco complying with Section 2510.

❖ The foam plastic is not to support continued flaming when tested in accordance with NFPA 268. Where the foam plastic is to be used in more than one thickness, the test is to be performed on minimum and maximum thickness. If the assembly is protected on the outside with one of the specified materials, the test is not required.

2603.6 Roofing. Foam plastic insulation meeting the requirements of Sections 2603.2, 2603.3 and 2603.4 shall be permitted as part of a roof-covering assembly, provided the assembly with the foam plastic insulation is a Class A, B or C roofing assembly where tested in accordance with ASTM E 108 or UL 790.

❖ Foam-plastic insulation that meets the labeling and identification, surface-burning characteristics and thermal barrier criteria is permitted as part of a Class A, B or C roof-covering assembly that has been tested in accordance with ASTM E 108 or UL 790.

2603.7 Plenums. Foam plastic insulation shall not be used as interior wall or ceiling finish in plenums except as permitted in Section 2604 or when protected by a thermal barrier in accordance with Section 2603.4.

❖ This section clarifies how foam plastics are to be applied in plenums as opposed to their use as duct and pipe insulation. Foam plastic found in a plenum must meet one of the following options:

1. Must have a thermal barrier,
2. Passed an appropriate large scale test (Section 2603.8) or
3. Meets the requirements of Section 2604 for trim and finish (i.e., is dense enough, meets area and thickness limitations and complies with flame spread limitations).

2603.8 Special approval. Foam plastic shall not be required to comply with the requirements of Sections 2603.4 through 2603.7, where specifically approved based on large-scale tests such as, but not limited to, FM 4880, UL 1040, NFPA 286 or UL 1715. Such testing shall be related to the actual end-use configuration and be performed on the finished manufactured foam plastic assembly in the maximum thickness intended for use. Foam plastics that are used as interior finish on the basis of special tests shall also conform to the flame spread requirements of Chapter 8. Assemblies tested shall include seams, joints and other typical details used in the installation of the assembly and shall be tested in the manner intended for use.

❖ Foam plastic does not have to comply with the installation and use requirements of Sections 2603.4 through 2603.7 when alternative approval is obtained in accordance with this section. This section lists examples of specific tests, such as: FM 4880, UL 1040, NFPA 286 or UL 1715. The intent is to require testing based on the proposed end-use configuration of the foam-plastic assembly and on an exposing fire that is appropriate in size and location for the proposed application.

These tests are to be performed on full-scale assemblies. The tested assemblies must include typical seams, joints and other details that will occur in the finished installation. Thorough testing provides an accurate depiction of the in-place fire performance of assemblies and systems using foam plastics. Materials would still need to additionaly pass the flame spread requirements of Chapter 8. Again, Section 2603.8 is focused on the end-use configuration; therefore, the foam plastic will not likely be exposed to demonstrate compliance with Chapter 8. Chapter 8 also allows the use of NFPA 286 as a test method and provides specific pass/fail criteria.

SECTION 2604
INTERIOR FINISH AND TRIM

2604.1 General. Plastic materials installed as interior finish or trim shall comply with Chapter 8. Foam plastics shall only be installed as interior finish where approved in accordance with the special provisions of Section 2603.8. Foam plastics that are used as interior finish shall also meet the flame spread index requirements for interior finish in accordance with Chapter 8. Foam plastics installed as interior trim shall comply with Section 2604.2.

❖ Plastic materials used as interior finish that are neither light transmitting nor foam are required to comply with Chapter 8. Foam plastic used as interior finish is to be in accordance with the special approval provisions of Section 2603.8 and meet the flame spread index requirements of Chapter 8 for interior finish. Foam plastics installed as interior trim are required to comply with Section 2604.2.

[F] 2604.2 Interior trim. Foam plastic used as interior trim shall comply with Sections 2604.2.1 through 2604.2.4.

❖ This section identifies the applicable subsections that apply to the interior trim requirements.

[F] 2604.2.1 Density. The minimum density of the interior trim shall be 20 pcf (320 kg/m^3).

❖ In order to qualify as interior trim, the density of materials must be at least 20 pounds (320 kg/m^3) per cubic foot (pcf). The intent was to separate those materials used for trim from those intended for use as insulation. The 20-pcf (320 kg/m^3) value was selected because it was approximately midway between the densities commonly used for insulations and for trim when this provision was added to the code. Most foam-plastic insulation was, and still is, in the range of 1 to 2$^1/_2$ pcf (16.02 to 40 kg/m^3), with very few materials over 5 pcf (81 kg/m^3).

Foam-plastic trim was manufactured in a density range of 30 to 45 pcf (481 to 721 kg/m^3) when the density provision was originally incorporated in the code.

[F] 2604.2.2 Thickness. The maximum thickness of the interior trim shall be 0.5 inch (12.7 mm) and the maximum width shall be 8 inches (204 mm).

❖ Even though other trim materials are not limited in dimension, the maximum thickness and width of foam-plastic trim is limited to $^1/_2$ inch (12.7 mm) and 8 inches (203 mm), respectively. These dimensions were selected because they were typical of the maximums being produced at the time this provision was included in the code.

[F] 2604.2.3 Area limitation. The interior trim shall not constitute more than 10 percent of the aggregate wall and ceiling area of any room or space.

❖ Trim cannot constitute more than 10 percent of the aggregate area of the walls and ceiling of a room. This limitation is simply a restatement of the general requirement for all combustible trim, which appears in Section 805.1.2.

[F] 2604.2.4 Flame spread. The flame spread index shall not exceed 75 where tested in accordance with ASTM E 84. The smoke-developed index shall not be limited.

❖ The flame spread must not be greater than 75 when tested in accordance with ASTM E 84. The value of 75 was selected to be consistent with the requirement for foam-plastic insulation, even though other materials used as trim are permitted to have flame spread indexes of up to 200 in many locations. The smoke-developed index is not regulated.

SECTION 2605
PLASTIC VENEER

2605.1 Interior use. Where used within a building, plastic veneer shall comply with the interior finish requirements of Chapter 8.

❖ Plastic veneer may be used inside a building if it meets the finish requirements of Chapter 8.

2605.2 Exterior use. Exterior plastic veneer shall be permitted to be installed on the exterior walls of buildings of any type of construction in accordance with all of the following requirements:

1. Plastic veneer shall comply with Section 2606.4.
2. Plastic veneer shall not be attached to any exterior wall to a height greater than 50 feet (15 240 mm) above grade.
3. Sections of plastic veneer shall not exceed 300 square feet (27.9 m^2) in area and shall be separated by a minimum of 4 feet (1219 mm) vertically.

Exception: The area and separation requirements and the smoke-density limitation are not applicable to plastic veneer applied to buildings constructed of Type VB construction, provided the walls are not required to have a fire-resistance rating.

❖ Exterior plastic veneer must be an approved Class CC1 or CC2 plastic material in accordance with the specifications in Section 2606.4 and meet all three of the requirements in this section for height, area and separation. The requirements for rigid vinyl siding are listed in Section 1404.9 and ASTM D 3679.

SECTION 2606
LIGHT-TRANSMITTING PLASTICS

2606.1 General. The provisions of this section and Sections 2607 through 2611 shall govern the quality and methods of application of light-transmitting plastics for use as light-transmitting materials in buildings and structures. Foam plastics shall comply with Section 2603. Light-transmitting plastic materials that meet the other code requirements for walls and roofs shall be permitted to be used in accordance with the other applicable chapters of the code.

❖ Section 2606 contains the provisions for light-transmitting plastics. It specifies both the requirements for an approved light-transmitting plastic and the fire properties of materials. It also requires adequate structural properties of materials and assemblies and specific installation methods for various applications.

2606.2 Approval for use. Sufficient technical data shall be submitted to substantiate the proposed use of any light-transmitting material, as approved by the building official and subject to the requirements of this section.

❖ Before the material can be used, sufficient data is to be submitted to the building official for approval.

2606.3 Identification. Each unit or package of light-transmitting plastic shall be identified with a mark or decal satisfactory to the building official, which includes identification as to the material classification.

❖ The material or its packaging must be marked CC1 or CC2 plastic.

2606.4 Specifications. Light-transmitting plastics, including thermoplastic, thermosetting or reinforced thermosetting plastic material, shall have a self-ignition temperature of 650°F (343°C) or greater where tested in accordance with ASTM D 1929; a smoke-developed index not greater than 450 where tested in the manner intended for use in accordance with ASTM E 84, or not greater than 75 where tested in the thickness intended for use in accordance with ASTM D 2843 and shall conform to one of the following combustibility classifications:

Class CC1: Plastic materials that have a burning extent of 1 inch (25 mm) or less where tested at a nominal thickness of 0.060 inch (1.5 mm), or in the thickness intended for use, in accordance with ASTM D 635,

Class CC2: Plastic materials that have a burning rate of 2.5 inches per minute (1.06 mm/s) or less where tested at a nominal thickness of 0.060 inch (1.5 mm), or in the thickness intended for use, in accordance with ASTM D 635.

❖ An approved material may be either thermoplastic, thermosetting or reinforced thermosetting. Approved plastics require a minimum self-ignition temperature of 650°F (343°C) when tested in accordance with ASTM D 1929. This temperature excludes the use of easily ignited plastic materials, such as cellulose nitrate, which ignites at a temperature of about 300°F (149°C). For comparative purposes, untreated wood and paper ignite at about 500°F (260°C).

In order for a plastic to be approved, its smoke-density rating is limited and must be evaluated by one of two test methods prescribed (smoke-developed index limited to 450 by ASTM E 84 or 75 by ASTM D 2843). For comparative purposes, these smoke-developed values are the maximum that can be expected from some wood products tested under similar conditions; however, there is no intent to indicate a correlation between the results from the two test methods.

Approved light-transmitting plastics are to be tested in accordance with ASTM D 635 and be of combustibility Class CC1 or CC2. Class CC1 plastic materials burn at a slower rate than Class CC2 plastics. Class CC1 generally consists of polycarbonate materials whereas Class CC2 consists of acrylics.

Thermoplastic materials are those plastics that will melt when heated and are formed into their final shape utilizing this property. Most light-transmitting plastic products are flat or formed sheets.

Thermosetting materials are those plastics that cure with heat and may generate heat during curing but will not melt by heating after curing. They are formed into their final product shape before or during curing, except for possible machining.

2606.5 Structural requirements. Light-transmitting plastic materials in their assembly shall be of adequate strength and durability to withstand the loads indicated in Chapter 16. Technical data shall be submitted to establish stresses, maximum unsupported spans and such other information for the various thicknesses and forms used as deemed necessary by the building official.

❖ Light-transmitting plastic applications must meet the necessary structural load requirements. For example, glazing assemblies must be able to withstand the imposed wind loads, and skylight assemblies must be able to support expected snow loads. It is the intent of this section that durability includes weatherability for those applications subjected to outdoor exposure. Technical information to support the structural integrity is to be submitted to the building official as he or she deems necessary.

2606.6 Fastening. Fastening shall be adequate to withstand the loads in Chapter 16. Proper allowance shall be made for expansion and contraction of light-transmitting plastic materials in accordance with accepted data on the coefficient of expansion of the material and other material in conjunction with which it is employed.

❖ The design loads for the assembly, members and fasteners are specified in Chapter 16. Adequate provision must be made for movement caused by the differential expansion expected from the different types of materials connected. This is especially important with plastics since they usually have coefficients of expansion that are greater than the materials to which they are connected.

2606.7 Light-diffusing systems. Unless the building is equipped throughout with an automatic sprinkler system in accordance with Section 903.3.1.1, light-diffusing systems shall not be installed in the following occupancies and locations:

1. Group A with an occupant load of 1,000 or more.
2. Theaters with a stage and proscenium opening and an occupant load of 700 or more.
3. Group I-2.
4. Group I-3.
5. Exit stairways and exit passageways.

❖ Light-diffusing systems are not permitted in the four occupancies above, nor in exits in any group, unless the building is protected by an automatic sprinkler system in accordance with NFPA 13. The intent is to limit the use of combustible materials in occupancies without sprinkler protection. These occupancies are generally recognized as being potentially more hazardous for the building occupants in the event of fire. The specific restriction for nonsprinklered buildings of Groups I-2 and I-3 is based on the physical restraint or incapacitation of the occupants.

2606.7.1 Support. Light-transmitting plastic diffusers shall be supported directly or indirectly from ceiling or roof construction by use of noncombustible hangers. Hangers shall be at least No. 12 steel-wire gage (0.106 inch) galvanized wire or equivalent.

❖ This section specifies that all hangers supporting the plastic diffusers must be of steel wire at least 0.106 inch (2.7 mm) in diameter or equivalent. It is important that noncombustible supports be used because the original test data were generated on these types of assemblies.

2606.7.2 Installation. Light-transmitting plastic diffusers shall comply with Chapter 8 unless the light-transmitting plastic diffusers will fall from the mountings before igniting, at an ambient temperature of at least 200°F (93°C) below the ignition temperature of the panels. The panels shall remain in place at an ambient room temperature of 175°F (79°C) for a period of not less than 15 minutes.

❖ Diffusers must also comply with Chapter 8 unless the plastic panels will fall from their mountings at an ambient temperature of at least 200°F (93°C) below their ignition temperature. The intent is to regulate panels that remain in place the same as interior finish and to eliminate surface flame spread requirements for materials that will not burn while remaining in place.

The code requires that all panels must remain in place for at least 15 minutes at an ambient room temperature of 175°F (79°C). The intent of this requirement is to avoid a panic situation that could arise from premature dropping of ceiling panels.

2606.7.3 Size limitations. Individual panels or units shall not exceed 10 feet (3048 mm) in length nor 30 square feet (2.79 m^2) in area.

❖ This section states the maximum length [10 feet (3048 mm)] and area [30 square feet (2.8 m^2)] of any single panel. The intent is to limit the size of panels expected to fall from place if the ambient room temperature at the ceiling were to reach 175°F (79°C), in order to avoid possible obstructions and injury caused by the fallen panels.

2606.7.4 Fire suppression system. In buildings that are equipped throughout with an automatic sprinkler system in accordance with Section 903.3.1.1, plastic light-diffusing systems shall be protected both above and below unless the sprinkler system has been specifically approved for installation only above the light-diffusing system. Areas of light-diffusing systems that are protected in accordance with this section shall not be limited.

❖ When buildings are required to have approved automatic sprinkler systems throughout, sprinkler protection must be provided both above and below the diffusing system. A sprinkler system is permitted to be installed above the diffusing system only when the assembly is designed for this use and specifically approved by the building official. The intent is that when specific approval is granted, it is based on appropriate test results related to end use. When sprinkler protection is provided, the area of the light-diffusing system is not subject to the area limitations of Section 2606.7.3.

2606.7.5 Electrical lighting fixtures. Light-transmitting plastic panels and light-diffuser panels that are installed in approved electrical lighting fixtures shall comply with the requirements of Chapter 8 unless the light-transmitting plastic panels conform to the requirements of Section 2606.7.2. The area of approved light-transmitting plastic materials that are used in required exits or corridors shall not exceed 30 percent of the aggregate area of the ceiling in which such panels are installed, unless the building is equipped throughout with an automatic sprinkler system in accordance with Section 903.3.1.1.

❖ Plastic light-transmitting and diffuser panels in electrical fixtures are required to comply with Chapter 8 or the installation requirements of Section 2606.7.2. Additionally, the aggregate ceiling area of such panels is limited to 30 percent in required exits and corridors, unless the building is equipped throughout with an automatic sprinkler system in accordance with NFPA 13. The intent of this section is to regulate fixtures the same as other ceiling finish panels. It also restricts the use of combustible materials in required exits and corridors while giving recognition to the benefit of automatic sprinkler systems.

2606.8 Partitions. Light-transmitting plastics used in or as partitions shall comply with the requirements of Chapters 6 and 8.

❖ When light-transmitting plastics are used in or as partitions, they must comply with the requirements of Chapters 6 and 8. The intent is to regulate light-transmitting plastics with the same requirements as other materials used for this application.

2606.9 Bathroom accessories. Light-transmitting plastics shall be permitted as glazing in shower stalls, shower doors, bathtub enclosures and similar accessory units. Safety glazing shall be provided in accordance with Chapter 24.

❖ Approved plastics are permitted as glazing in a variety of uses in bathrooms. When an approved Class CC1 or CC2 plastic material is used for glazing shower stalls, shower doors, bathtub enclosures or other bathroom accessories, it must meet the human impact requirements of ANSI Z97.1 under Section 2406.1.3

2606.10 Awnings, patio covers and similar structures. Awnings constructed of light-transmitting plastics shall be constructed in accordance with provisions specified in Section 3105 and Chapter 32 for projections and appendages. Patio covers constructed of light-transmitting plastics shall comply with Section 2606. Light-transmitting plastics used in canopies at motor fuel-dispensing facilities shall comply with Section 2606 except as modified by Section 406.5.2.

❖ Light-transmitting plastics used as awnings and similar structures must meet the general performance requirements of other appropriate sections of the code.

2606.11 Greenhouses. Light-transmitting plastics shall be permitted in lieu of plain glass in greenhouses.

❖ Greenhouses may be constructed of any type of approved plastic material that meets the requirements of Section 2606.

2606.12 Solar collectors. Light-transmitting plastic covers on solar collectors having noncombustible sides and bottoms shall be permitted on buildings not over three stories in height or

9,000 square feet (836.1 m^2) in total floor area, provided the light-transmitting plastic cover does not exceed 33.33 percent of the roof area for CC1 materials or 25 percent of the roof area for CC2 materials.

Exception: Light-transmitting plastic covers having a thickness of 0.010 inch (0.3 mm) or less or shall be permitted to be of any plastic material provided the area of the solar collectors does not exceed 33.33 percent of the roof area.

❖ Approved light-transmitting plastics can be used for solar collector covers that meet the requirements of Section 2606.12. Any plastic material may be used as solar collector covers provided that the cover thickness does not exceed 0.010 inches (0.254 mm) and the collector does not exceed 33.33 percent of the roof area.

SECTION 2607
LIGHT-TRANSMITTING PLASTIC WALL PANELS

2607.1 General. Light-transmitting plastics shall not be used as wall panels in exterior walls in occupancies in Groups A-l, A-2, H, I-2 and I-3. In other groups, light-transmitting plastics shall be permitted to be used as wall panels in exterior walls, provided that the walls are not required to have a fire-resistance rating and the installation conforms to the requirements of this section. Such panels shall be erected and anchored on a foundation, waterproofed or otherwise protected from moisture absorption and sealed with a coat of mastic or other approved waterproof coating. Light-transmitting plastic wall panels shall also comply with Section 2606.

❖ Section 2607 establishes requirements for the use of plastic panels as exterior walls. Plastic wall panels are permitted to be used in most occupancies except those with dense occupant loads or where the occupants are confined or otherwise impaired. The size and quantity of plastic wall panels are also regulated by this section. These panels must also comply with the applicable provisions of Section 2606, which governs all light-transmitting plastics.

Plastic wall panels are permitted to be used in occupancies classified as Groups A-3, A-4, A-5, B, E, F-1, F-2, I-1, M, R-1, R-2, R-3, R-4, S-1, S-2 and U. They may not be used in occupancies where the generated products of combustion would significantly impair the egress of the impaired, confined or crowded occupants. These types of combustible wall panels are not permitted where the exterior walls are required to be fire-resistance rated.

As with any exterior covering, plastic wall panels must be adequately fastened in a permanent manner. To reduce the chance of water infiltration into the plastic wall panels, they must be erected and anchored on a base that is protected from water absorption. Additionally, the plastic wall panel itself must be sealed with a waterproof coating. These types of combustible wall panels are not permitted where the exterior walls are required to be fire-resistance rated.

2607.2 Installation. Exterior wall panels installed as provided for herein shall not alter the type of construction classification of the building.

❖ Plastic is a combustible material. The use of light-transmitting plastic wall panels is permitted, however, without downgrading the type of construction. This exception is also clarified in Section 603.1, Item 12; therefore, the use of light-transmitting plastic wall panels is permitted in all types of construction. These types of combustible wall panels, however, are not permitted where the exterior walls are required to be fire-resistance rated (see Section 2607.1).

2607.3 Height limitation. Light-transmitting plastics shall not be installed more than 75 feet (22 860 mm) above grade plane, except as allowed by Section 2607.5.

❖ Plastic wall panels are not permitted to be installed greater than 75 feet (22 860 mm) above grade plane in an unsprinklered building. The 75-foot (22 860 mm) threshold is the highest elevation that can be reached by hose lines on the ground. In structures sprinklered throughout, however, Section 2607.5 permits the maximum height to be unlimited. Be aware that Table 503 also places height limitations on structures. In some cases, the limitations of this table are more restrictive than those of Sections 2607.3 and 2607.5. At all times, the most restrictive requirement prevails.

2607.4 Area limitation and separation. The maximum area of a single wall panel and minimum vertical and horizontal separation requirements for exterior light-transmitting plastic wall panels shall be as provided for in Table 2607.4. The maximum percentage of wall area of any story in light-transmitting plastic wall panels shall not exceed that indicated in Table 2607.4 or the percentage of unprotected openings permitted by Section 704.8, whichever is smaller.

Exceptions:

1. In structures provided with approved flame barriers extending 30 inches (760 mm) beyond the exterior wall in the plane of the floor, a vertical separation is not required at the floor except that provided by the vertical thickness of the flame barrier projection.
2. Veneers of approved weather-resistant light-transmitting plastics used as exterior siding in buildings of Type V construction in compliance with Section 1406.
3. The area of light-transmitting plastic wall panels in exterior walls of greenhouses shall be exempt from the area limitations of Table 2607.4 but shall be limited as required for unprotected openings in accordance with Section 704.8.

❖ Refer to Table 2607.4 for area limitations of single wall panels and the minimum separation requirements between adjacent panels. The maximum percentage of wall area faced with plastic panels is limited by Table 2607.4, as well as by Section 704.8. At all times, the most restrictive requirement prevails.

Vertical separation of panels (see Table 2607.4) is not

required where continuous architectural projections create at least a 30-inch (762 mm) barrier (called a flame barrier) between adjacent panels. Under most conditions, this flame barrier will delay or prevent vertical fire spread from one story to the next. This horizontal barrier is seen as protection equivalent to the vertical separation typically required by Table 2607.4 (see Figure 2607.4).

Veneers of approved exterior light-transmitting plastic siding installed in compliance with Section 1406 on Type V construction are exempt from the area limitations and separation requirements.

Greenhouse walls can have an unlimited area of approved light-transmitting wall panels, unless limited by the unprotected opening requirements of Section 704.8.

TABLE 2607.4. See page 26-15.

❖ This table establishes size and separation limitations for exterior plastic wall panels. The requirements of this table are not applicable to plastic exterior siding meeting the exception to Section 2607.4. The requirements are determined by the type of plastic material (Class CC1 or CC2) and the fire separation distance. Plastic wall panels are only permitted where a minimum fire separation distance of 6 feet (1829 mm) is provided. The vertical separation requirements can be modified by Section 2607.4 (see Note b of Table 2607.4). The maximum area limitations of a single panel and the maximum percentage area of an exterior wall can be doubled when the structure is completely sprinklered in accordance with NFPA 13 (see Section 2607.5).

2607.5 Automatic sprinkler system. Where the building is equipped throughout with an automatic sprinkler system in accordance with Section 903.3.1.1, the maximum percentage area of exterior wall in any story in light-transmitting plastic wall panels and the maximum square footage of a single area given in Table 2607.4 shall be increased 100 percent, but the area of light-transmitting plastic wall panels shall not exceed 50 percent of the wall area in any story, or the area permitted by Section 704.8 for unprotected openings, whichever is smaller. These installations shall be exempt from height limitations.

❖ The extra protection provided by a complete NFPA 13 automatic fire sprinkler system enables the limitations placed on light-transmitting plastic wall panels to be reduced. In order for these modifications to apply, the structure must be sprinklered throughout. When this protection is achieved, the height to which plastic wall panels can be installed is not limited. Furthermore, the single panel size and exterior wall area limitations can be doubled. At no time, however, can the area of wall panels per story exceed 50 percent of the wall area of that story. The percentage limitation of Section 704.8 for unprotected openings is also applicable.

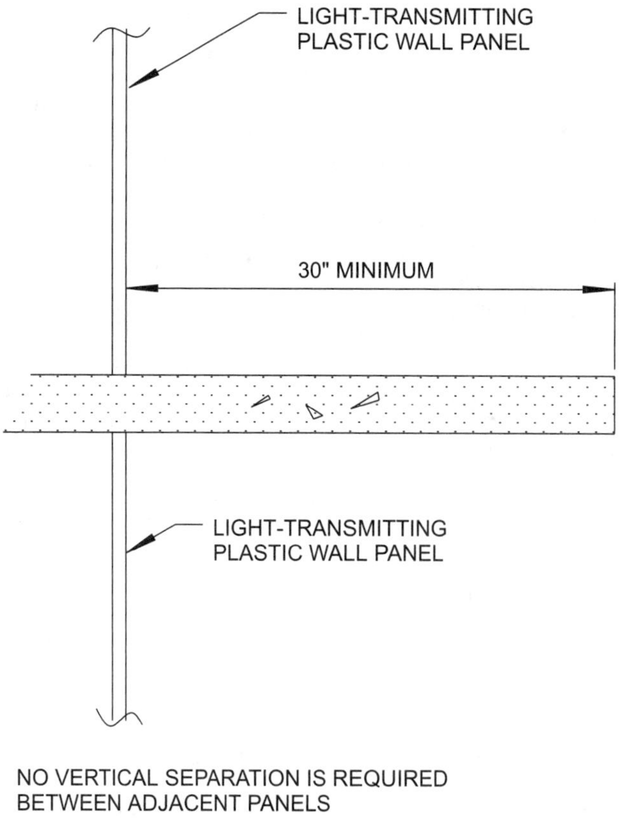

For SI: 1 inch = 25.4 mm.

Figure 2607.4
FLAME BARRIER

TABLE 2607.4
AREA LIMITATION AND SEPARATION REQUIREMENTS FOR LIGHT-TRANSMITTING PLASTIC WALL PANELS[a]

FIRE SEPARATION DISTANCE (feet)	CLASS OF PLASTIC	MAXIMUM PERCENTAGE AREA OF EXTERIOR WALL IN PLASTIC WALL PANELS	MAXIMUM SINGLE AREA OF PLASTIC WALL PANELS (square feet)	MINIMUM SEPARATION OF PLASTIC WALL PANELS (feet)	
				Vertical	Horizontal
Less than 6	—	Not Permitted	Not Permitted	—	—
6 or more but less than 11	CC1	10	50	8	4
	CC2	Not Permitted	Not Permitted	—	—
11 or more but less than or equal to 30	CC1	25	90	6	4
	CC2	15	70	8	4
Over 30	CC1	50	Not Limited	3[b]	0
	CC2	50	100	6[b]	3

For SI: 1 foot = 304.8 mm, 1 square foot = 0.0929 m^2.

a. For combinations of plastic glazing and plastic wall panel areas permitted, see Section 2607.6.

b. For reductions in vertical separation allowed, see Section 2607.4.

2607.6 Combinations of glazing and wall panels. Combinations of light-transmitting plastic glazing and light-transmitting plastic wall panels shall be subject to the area, height and percentage limitations and the separation requirements applicable to the class of light-transmitting plastic as prescribed for light-transmitting plastic wall panel installations.

❖ Plastic glazing and plastic wall panels display equivalent fire characteristics; therefore, where plastic glazing (regulated by Section 2608) and plastic wall panels are both used, the aggregate area of plastics must be in compliance with the limitations and requirements of Table 2607.4.

SECTION 2608 LIGHT-TRANSMITTING PLASTIC GLAZING

2608.1 Buildings of Type VB construction. Openings in the exterior walls of buildings of Type VB construction, where not required to be protected by Section 704, shall be permitted to be glazed or equipped with light-transmitting plastic. Light-transmitting plastic glazing shall also comply with Section 2606.

❖ This section restricts the use of plastics to openings that are not required to have a fire protection rating. The requirement for walls, and their openings, to be protected is dependent on their fire separation distance (see Section 704).

2608.2 Buildings of other types of construction. Openings in the exterior walls of buildings of types of construction other than Type VB, where not required to be protected by Section 704, shall be permitted to be glazed or equipped with light-transmitting plastic in accordance with Section 2606 and all of the following:

1. The aggregate area of light-transmitting plastic glazing shall not exceed 25 percent of the area of any wall face of the story in which it is installed. The area of a single pane of glazing installed above the first story above grade plane shall not exceed 16 square feet (1.5 m^2) and the vertical dimension of a single pane shall not exceed 4 feet (1219 mm).

   Exception: Where an automatic sprinkler system is provided throughout in accordance with Section 903.3.1.1, the area of allowable glazing shall be increased to a maximum of 50 percent of the wall face of the story in which it is installed with no limit on the maximum dimension or area of a single pane of glazing.

2. Approved flame barriers extending 30 inches (762 mm) beyond the exterior wall in the plane of the floor, or vertical panels not less than 4 feet (1219 mm) in height, shall be installed between glazed units located in adjacent stories.

   Exception: Buildings equipped throughout with an automatic sprinkler system in accordance with Section 903.3.1.1.

3. Light-transmitting plastics shall not be installed more than 75 feet (22 860 mm) above grade level.

   Exception: Buildings equipped throughout with an automatic sprinkler system in accordance with Section 903.3.1.1.

❖ In addition to restricting the use of plastic glazing to exterior wall openings that are not required to have a fire protection rating, this section specifies the maximum area within a wall that openings can occupy. An opening that has plastic as the glazing material is essentially treated the same as an unprotected glass opening.

To limit the spread of fire, the allowable area of plastic glazing is limited to the lesser of two conditions. The first condition restricts the area of plastic glazing to 25 percent of the wall area of the story. The second condition restricts the area of plastics to that of the unprotected openings permitted in Section 704.

The exception doubles the amount of wall area that is allowed to be glazed with plastic when an approved automatic sprinkler system is installed. The allowable areas are increased to 50 percent of the wall area of the story with no limit on the maximum dimension or area of a single pane of glazing. Consistent with other provisions to limit fire spread, this section places restrictions on the size and vertical separation of plastic glazing. The 4-foot (1219 mm) vertical separation is widely accepted as adequate for spandrels and other intervening construction between floors.

Consistent with other provisions to limit fire spread, this section places restrictions on the size and vertical separation of plastic glazing. The 4-foot (1219 mm) vertical separation is widely accepted as adequate for spandrels and other intervening construction between floors.

The 75-foot (22 860 mm) height limitation is generally accepted as the maximum height attainable by fire-fighting apparatus.

SECTION 2609 LIGHT-TRANSMITTING PLASTIC ROOF PANELS

2609.1 General. Light-transmitting plastic roof panels shall comply with this section and Section 2606. Light-transmitting plastic roof panels shall not be installed in Groups H, I-2 and I-3. In all other groups, light-transmitting plastic roof panels shall comply with any one of the following conditions:

1. The building is equipped throughout with an automatic sprinkler system in accordance with Section 903.3.1.1.
2. The roof construction is not required to have a fire-resistance rating by Table 601.
3. The roof panels meet the requirements for roof coverings in accordance with Chapter 15.

❖ This section establishes general requirements for light-transmitting plastic roof panels. These requirements are similar in nature to the other light-transmitting plastic provisions of this chapter. The requirements are intended to control the use of combustible construction material (i.e., light-transmitting plastics). These provisions specifically detail where roof panels may be used and any accompanying conditions that are required. Occupancies in Group H, I-2 or I-3 may not use light-transmitting plastic roof panels. Buildings or structures containing all other groups may use the roof panels, but at least one of the three conditions listed must be present. The first condition requires the structure under consideration to be equipped with an automatic sprinkler system installed in accordance with NFPA 13. The second condition allows a structure that is not suppressed to contain plastic roof panels if Table 601 of the code does not require any fire-resistance ratings for the roof construction, including beams, trusses, framing, arches and roof decks (based on the type of construction and the distance between the floor and lowest roof member). The final condition that allows plastic roof panels in certain structures is that panels meet the requirements for roof coverings in Chapter 15.

2609.2 Separation. Individual roof panels shall be separated from each other by a distance of not less than 4 feet (1219 mm) measured in a horizontal plane.

Exceptions:

1. The separation between roof panels is not required in a building equipped throughout with an automatic sprinkler system in accordance with Section 903.3.1.1.
2. The separation between roof panels is not required in low-hazard occupancy buildings complying with the conditions of Section 2609.4, Exception 2 or 3.

❖ The separation requirements for plastic roof panels are established herein. Individual light-transmitting plastic roof panels are limited in area by the provisions of Section 2609.4. This section provides the separation distances necessary between the individual roof panels to isolate the fire hazard associated with this combustible material.

The 4-foot (1219 mm) minimum separation distance required between each individual roof panel must be measured in a horizontal plane (see Figure 2609.2).

There are two exceptions in the code pertaining to the separation requirements. Any structure that is equipped with an automatic sprinkler system installed in accordance with NFPA 13, or low-hazard occupancy buildings complying with Exception 2 or 3 of Section 2609.4, are exempt from the separation requirements.

2609.3 Location. Where exterior wall openings are required to be protected by Section 704.8, a roof panel shall not be installed within 6 feet (1829 mm) of such exterior wall.

❖ This section provides requirements as to where light-transmitting plastic roof panels may be located in relation to the exterior walls of buildings or structures. Again, limitations are included as to where these panels of combustible construction can be located in relation to adjacent fire hazards. These provisions apply only when they are determined to be applicable from the requirements for fire-resistance-rated exterior walls and their exterior opening protectives found in Section 704.8. When certain openings in the exterior walls are required to be protected, these location requirements must be followed.

When an exterior wall contains required exterior opening protectives, the code text requires plastic roof panels to be installed at least 6 feet (1829 mm) from the wall. There is no exception to this requirement for structures equipped throughout with an approved automatic sprinkler system (see Figure 2609.3).

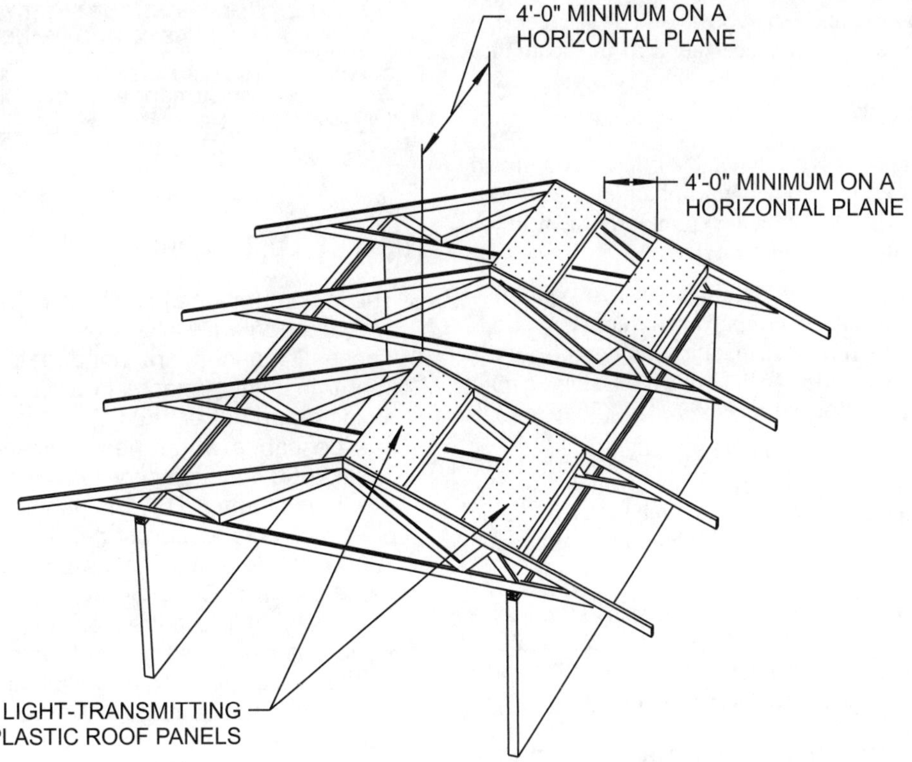

For SI: 1 inch = 25.4 mm, 1 foot = 304.8 mm.

Figure 2609.2
PLASTIC ROOF PANEL SEPARATION FROM OTHER PANELS

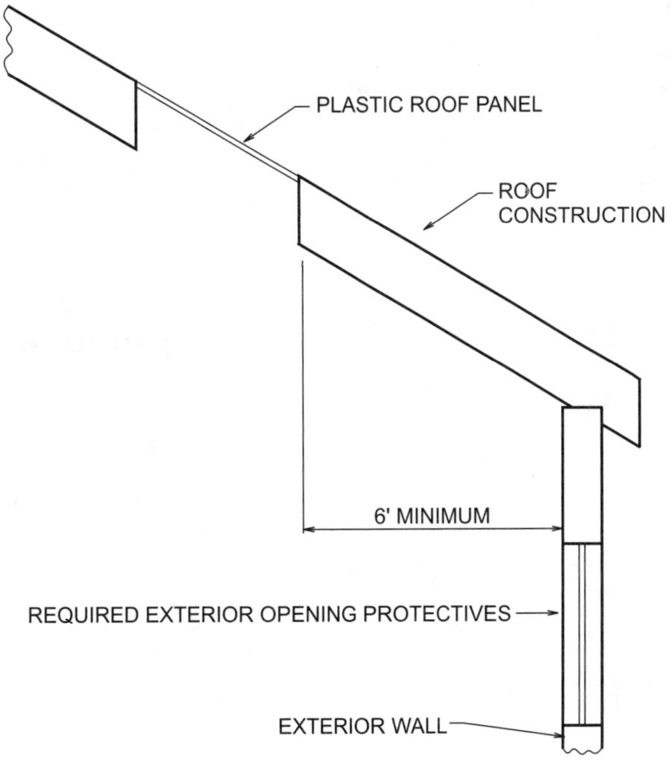

For SI: 1 foot = 304.8 mm.

Figure 2609.3
PLASTIC ROOF PANEL SEPARATION FROM EXTERIOR WALL

2609.4 Area limitations. Roof panels shall be limited in area and the aggregate area of panels shall be limited by a percentage of the floor area of the room or space sheltered in accordance with Table 2609.4.

Exceptions:

1. The area limitations of Table 2609.4 shall be permitted to be increased by 100 percent in buildings equipped throughout with an automatic sprinkler system in accordance with Section 903.3.1.1.
2. Low-hazard occupancy buildings, such as swimming pool shelters, shall be exempt from the area limitations of Table 2609.4, provided that the buildings do not exceed 5,000 square feet (465 m^2) in area and have a minimum fire separation distance of 10 feet (3048 mm).
3. Greenhouses that are occupied for growing plants on a production or research basis, without public access, shall be exempt from the area limitations of Table 2609.4 provided they have a minimum fire separation distance of 4 feet (1220 mm).
4. Roof coverings over terraces and patios in occupancies in Group R-3 as applicable in Section 101.2 shall be exempt from the area limitations of Table 2609.4 and shall be permitted with light-transmitting plastics.

❖ This section provides certain area limitations for plastic roof panels. Panels of combustible plastic construction with its high fire-hazard properties must be limited in terms of the area of the structure. These requirements not only list the area limitations for individual panels but also limit the aggregate or sum total of all roof panels located in the structure's roof. The aggregate area limitations are based on a floor area percentage of the room or space enclosed by the roof.

The area limitations, which are based on the class of the plastic roof panels, are detailed in Table 2609.4.

There are four exceptions to these area limitations. When certain additional protection is provided or the panels are located in structures of limited hazard, the area limitations can be altered. In some cases, these exceptions increase the limitations or, in other cases, exempt certain structures from the limitations. Exception 1 increases the area limitations by 100 percent for those structures equipped throughout with an automatic sprinkler system installed in accordance with NFPA 13. Exception 2 allows certain low-hazard buildings to be exempt from the roof panel limitations. These buildings must not exceed 5,000 square feet (465 m^2) in area and they must have at least 10 feet (3048 mm) of fire separation distance in order for the exception to apply. Exception 3 is for greenhouses that have at least a 4-foot (1219 mm) fire separation distance. Exception 4 allows occupancies in Group R-3 that have roofed-over terraces and patios to be exempt from the area limitations for roof panels. Such roof panels must be constructed with approved plastic meeting the requirements of Section 2603.

TABLE 2609.4
AREA LIMITATIONS FOR LIGHT-TRANSMITTING PLASTIC ROOF PANELS

CLASS OF PLASTIC	MAXIMUM AREA OF INDIVIDUAL ROOF PANELS (square feet)	MAXIMUM AGGREGATE AREA OF ROOF PANELS (percent of floor area)
CC1	300	30
CC2	100	25

For SI: 1 square foot = 0.0929 m^2.

❖ The maximum area limitations for both individual panels and aggregate areas are based on the class of plastic. These limitations are needed to restrict the use of light-transmitting plastic in roofs and roof coverings of buildings and structures. Section 2609.4 provides text that describes when these limitations are to be used. Table 2609.4 contains three different headings and only two lines of entry. The first heading is labeled "Class of plastic" and is to be used in conjunction with Section 2606.4 for determining whether the plastic used in roof panels is either Class CC1 (e.g., polycarbonate) or CC2 (e.g., acrylic). Once the class of plastic of the roof panels in question has been determined, the table is to be used to determine the area limitations for such panels.

The second heading in the table is identified as "Maximum area of individual roof panels (square feet)." These values provide the maximum area limitations for each individual roof panel based on the class of plastic used to manufacture the panel. These limitations are given in units of square feet.

The last heading, labeled "Maximum aggregate area of roof panels (percent of floor area)," provides the aggregate amount of area that can be occupied by plastic panels located in a roof. This limitation is based on a percentage of the floor area of the room or space that is sheltered by the roof in question. The floor area in units of square feet is multiplied by the appropriate percentage for the class of plastic used. This product provides the maximum area for all plastic panels used in the roof.

SECTION 2610
LIGHT-TRANSMITTING PLASTIC SKYLIGHT GLAZING

2610.1 Light-transmitting plastic glazing of skylight assemblies. Skylight assemblies glazed with light-transmitting plastic shall conform to the provisions of this section and Section 2606. Unit skylights glazed with light-transmitting plastic shall also comply with Section 2405.5.

Exception: Skylights in which the light-transmitting plastic conforms to the required roof-covering class in accordance with Section 1505.

❖ This section covers horizontal and sloped applications of light-transmitting plastic skylights. This section contains mounting, size and spacing limitations for these combustible assemblies. Plastic skylights sloped more than 15 degrees (.26 rad) from vertical are also required to comply with Section 2405.

Acrylic and polycarbonate plastics are commonly used glazing materials in skylight assemblies. The use of combustible skylights is permitted subject to the requirements of Section 2606. The reference to Section 2405.5 is intended to highlight the specific structural considerations for unit skylights, regardless of construction materials.

The exception eliminates any added requirements for plastic skylights that meet the roofing classification for the surrounding roof covering as described in Chapter 15.

2610.2 Mounting. The light-transmitting plastic shall be mounted above the plane of the roof on a curb constructed in accordance with the requirements for the type of construction classification, but at least 4 inches (102 mm) above the plane of the roof. Edges of light-transmitting plastic skylights or domes shall be protected by metal or other approved noncombustible material, or the light-transmitting plastic dome or skylight shall be shown to be able to resist ignition where exposed at the edge to a flame from a Class B brand as described in ASTM E 108 or UL 790.

Exceptions:

1. Curbs shall not be required for skylights used on roofs having a minimum slope of three units vertical in 12 units horizontal (25-percent slope) in occupancies in Group R-3 as applicable in Section 101.2 and on buildings with a nonclassified roof covering.
2. The metal or noncombustible edge material is not required where nonclassified roof coverings are permitted.

❖ The mounting requirements for plastic skylights are similar to those established in Section 2405.4 for glass skylights. The 4-inch (102 mm) curb requirement protects the skylight from flying brands, provides a wash for water running down the slope and provides a noticeable barrier to caution a person on the roof from stepping on the skylight; however, the skylight must be designed for the structural requirements of Section 2606.5. Since the edge of the plastic is more susceptible to ignition than the surface, noncombustible edge protection is required to reduce the chances of edge ignition. This requirement is only applicable when the plastic is of a type and design that does not resist the flying brand exposure as described in ASTM E 108 or UL 790.

Exception 1 is based on the general field experience with plastic skylights used in one- and two-family dwellings with slopes equal to or greater than three units vertical in 12 units horizontal (3:12). Additionally, skylights in unclassified roofs are not required to be mounted on curbs. Such a skylight need not provide fire performance beyond that required for the surrounding roof covering.

The basis for Exception 2 is similar to that for Exception 1. There is no need to require edge protection for the plastic when the adjacent area is an unclassified roof permitted by Section 1505.5.

2610.3 Slope. Flat or corrugated light-transmitting plastic skylights shall slope at least four units vertical in 12 units horizontal (4:12). Dome-shaped skylights shall rise above the mounting flange a minimum distance equal to 10 percent of the maximum span of the dome but not less than 3 inches (76 mm).

Exception: Skylights that pass the Class B Burning Brand Test specified in ASTM E 108 or UL 790.

❖ As are most of the restrictions for plastic skylights, the minimum slope requirements are based on limiting the risks involved with combustible roof components. Both the slope for flat and corrugated plastic skylights and the rise for domed skylights are selected to shed flying brands. Consistent with other sections, the slope requirements are not applicable when the skylight complies with the specified requirements of ASTM E 108 or UL 790.

2610.4 Maximum area of skylights. Each skylight shall have a maximum area within the curb of 100 square feet (9.30 m^2).

Exception: The area limitation shall not apply where the building is equipped throughout with an automatic sprinkler system in accordance with Section 903.3.1.1 or the building is equipped with smoke and heat vents in accordance with Section 910.

❖ The area limitation for unsprinklered buildings minimizes the fire risks for plastic skylights. The 100-square-foot (9 m^2) limitation does not apply to buildings provided throughout with an NFPA 13 automatic sprinkler system or equipped with smoke and heat vents in accordance with Section 910. This exception is largely academic, since contemporary plastic skylights do not normally exceed this area limitation.

2610.5 Aggregate area of skylights. The aggregate area of skylights shall not exceed $33^1/_3$ percent of the floor area of the room or space sheltered by the roof in which such skylights are installed where Class CC1 materials are utilized, and 25 percent where Class CC2 materials are utilized.

Exception: The aggregate area limitations of light-transmitting plastic skylights shall be increased 100 percent beyond the limitations set forth in this section where the building is equipped throughout with an automatic sprinkler system in accordance with Section 903.3.1.1 or the building is equipped with smoke and heat vents in accordance with Section 910.

❖ This section includes limitations directed to minimize the risks of fire spread. The $33\,^1/_3$- and 25-percent limitations based on floor area for Class CC1 (polycarbonate) and CC2 (acrylic) materials, respectively, are judged to inhibit sufficiently the spread of fire. An exception allows the percentage to be increased to 66 $^{2}/_{3}$ percent (Class CC1) and 50 percent (Class CC2) when an approved automatic sprinkler system is installed throughout the building or is equipped with smoke and heat vents in accordance with Section 910. To comply with the intent of this provision, the skylights should be distributed over the entire area of the roof and not concentrated in one area.

2610.6 Separation. Skylights shall be separated from each other by a distance of not less than 4 feet (1219 mm) measured in a horizontal plane.

Exceptions:

1. Buildings equipped throughout with an automatic sprinkler system in accordance with Section 903.3.1.1.
2. In Group R-3 as applicable in Section 101.2, multiple skylights located above the same room or space with a combined area not exceeding the limits set forth in Section 2610.4.

❖ As with the other items cited, the minimum 4-foot (1219 mm) spacing minimizes the risk of fire spread for buildings without automatic sprinklers (also see Figure 2609.2). The separation is not necessary when complete fire suppression is provided in accordance with NFPA 13 or in Group R-3, as applicable in Section 2610.4, where the skylights are located above the same room or space with a combined area not to exceed the limits of Section 2610.4.

2610.7 Location. Where exterior wall openings are required to be protected in accordance with Section 704, a skylight shall not be installed within 6 feet (1829 mm) of such exterior wall.

❖ The minimum 6-foot (1829 mm) spacing to a fire-resistance-rated wall is to minimize the risk of fire spread (also see Section 2609.3).

2610.8 Combinations of roof panels and skylights. Combinations of light-transmitting plastic roof panels and skylights shall be subject to the area and percentage limitations and separation requirements applicable to roof panel installations.

❖ This is a precautionary statement to clarify that skylights and roof panels must be considered together when establishing size and separation limitations. When combinations of plastic skylights and roof panels occur, the area limitations of Section 2609.4 govern.

SECTION 2611
LIGHT-TRANSMITTING PLASTIC INTERIOR SIGNS

2611.1 General. Light-transmitting plastic interior wall signs shall be limited as specified in Sections 2611.2 through 2611.4. Light-transmitting plastic interior wall signs in covered mall buildings shall comply with Section 402.14. Light-transmitting plastic interior signs shall also comply with Section 2606.

❖ This section contains the scoping statement that requires all light-transmitting plastic materials used for interior wall signage applications to comply with the requirements of Sections 2611.2, 2611.3 and 2611.4. These requirements include the maximum allowable aggregate area of signs, the maximum individual area of signs and encasement requirements. The only exception to these requirements is light-transmitting plastics that are used for interior wall signage in covered mall buildings, since they are regulated by Section 402.14.

These provisions help clarify the requirements for plastic signs in all occupancies other than covered mall buildings and exterior applications. One of the most common uses of light-transmitting plastic in interior signage applications is as the facing of internally illuminated or nonilluminated wall signs (sometimes referred to as "box" or "can" signs). Typical interior applications include tenant signage, building directories, promotional or advertising sign boxes and directional signage (other than exit signage).

2611.2 Aggregate area. The sign shall not exceed 20 percent of the wall area.

❖ The provisions of this section limit the maximum amount of wall area that may be covered by all plastic signs located on that wall, whereas Section 2611.3 governs the maximum permitted size of any individual sign. This limitation on the amount of wall area that can be covered by plastic signs is intended to minimize the potential for the spread of fire along a wall surface through involvement of the plastic signage.

2611.3 Maximum area. The sign shall not exceed 24 square feet (2.23 m^2).

❖ When applying the provisions of Sections 2611.2 and 2611.3, the most restrictive of each section must be applied. For example, if a single sign with the maximum area of 24 square feet (2 m^2) exceeds 20 percent of the wall area on which it is located, the area of the sign must be reduced to a size where the 20-percent limitation is not exceeded. Figures 2611.3(1) and 2611.3(2) give two examples of these provisions.

2611.4 Encasement. Edges and backs of the sign shall be fully encased in metal.

❖ The provisions of this section are intended to minimize the amount of surface that could be exposed to direct impingement by fire, and thus reduce the potential for ignition of the plastic.

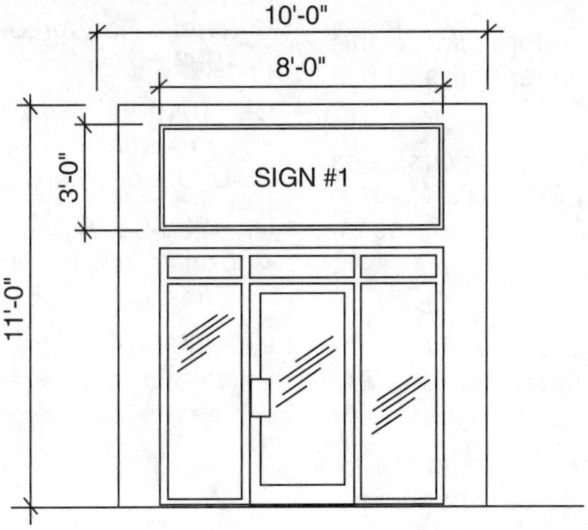

AREA OF SIGN #1 = 24 SQ. FT. (MAX. AREA PER SECTION 2611.3)
AGGREGATE AREA = 24 SQ. FT.

AREA OF WALL = 110 SQ. FT.
20% OF 110 SQ. FT. = 22 SQ. FT. (MAX. AREA PER SECTION 2611.2)
24 SQ. FT. $\leq$ 22 SQ. FT. ∴ AGGREGATE AREA OF SIGNS EXCEEDS AREA PERMITTED BY SECTION 2611.2

For SI: 1 inch = 25.4 mm, 1 foot = 304.8 mm, 1 square foot = 0.0929 m^2.

Figure 2611.3(1)
MAXIMUM AREA OF LIGHT-TRANSMITTING PLASTIC SIGNS—EXAMPLE 1

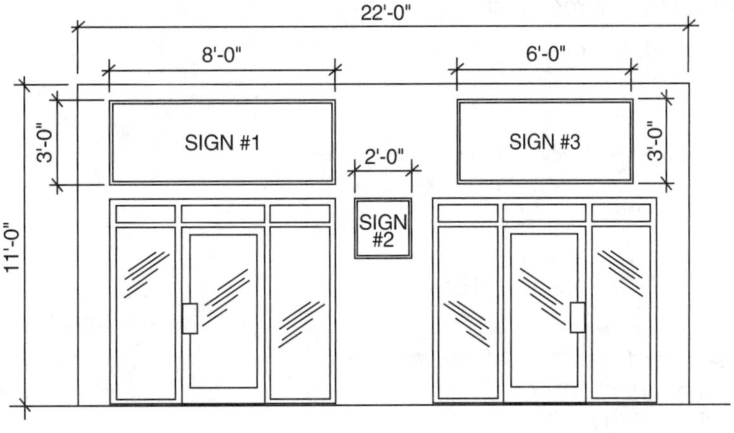

AREA OF SIGN #1 = 24 SQ. FT. (MAX. AREA PER SECTION 2611.3)
AREA OF SIGN #2 = 4 SQ. FT.
AREA OF SIGN #3 = 18 SQ. FT.
AGGREGATE AREA = 46 SQ. FT.

AREA OF WALL = 242 SQ. FT.
20% OF 242 SQ. FT. = 48.4 SQ. FT. (MAX. AREA PER SECTION 2611.2)
46 SQ. FT. $\leq$ 48.4 SQ. FT. ∴AGGREGATE AREA OF SIGNS COMPLIES WITH SECTION 2611.2

For SI: 1 inch = 25.4 mm, 1 foot = 304.8 mm, 1 square foot = 0.0929 m^2.

Figure 2611.3(2)
MAXIMUM AREA OF LIGHT-TRANSMITTING PLASTIC SIGNS—EXAMPLE 2

Bibliography

The following resource materials are referenced in this chapter or are relevant to the subject matter addressed in this chapter.

ANSI Z97.1-84, *Safety Performance Specifications and Methods of Test for Safety Glazing Materials Used in Buildings*. New York: American National Standards Institute, 1984.

ASTM D 635-00, *Standard Test Method for Rate of Burning and/or Extent and Time of Burning of Plastics in a Horizontal Position*. West Conshohocken, PA: ASTM International, 2000.

ASTM D 1929-96 (2001)E01, *Standard Test Method for Ignition Properties of Plastics*. West Conshohocken, PA: ASTM International, 2001.

ASTM D 2843-99, *Standard Test Method for Density of Smoke from the Burning or Decomposition of Plastics*. West Conshohocken, PA: ASTM International, 1999.

ASTM D 3679-01c, *Specification for Rigid Poly [Vinyl Chloride (PVC) Siding]*. West Conshohocken, PA: ASTM International, 2001.

ASTM E 75-76(1990)^{e1}, *Standard Test Methods for Chemical Analysis of Copper-Nickel and Copper-Nickel-Zinc Alloys*. West Conshohocken, PA: ASTM International, 1990.

ASTM E 84-01, *Standard Test Method for Surface-Burning Characteristics of Building Materials*. West Conshohocken, PA: ASTM International, 2001.

ASTM E 108-00, *Standard Test Method for Fire Tests of Roof Coverings*. West Conshohocken, PA: ASTM International, 2000.

ASTM E 119-00, *Standard Test Methods for Fire Tests of Building Construction and Materials*. West Conshohocken, PA: ASTM International, 2000.

CPSC 16 CFR, Part 1201-77, *Safety Standard for Architectural Glazing*. Washington, DC: Consumer Product Safety Commission, 1977.

FM 4450 (1989), *Approval Standard for Class 1 Insulated Steel Deck Roofs*—with Supplement (July 1992). Norwood, MA: Factory Mutual Standards Laboratories Department, 1989.

FM 4880 (2001), *Approval Standard for Class 1: a) Insulated Wall or Wall and Roof/Ceiling Panels, b) Plastic Interior Finish Materials, c) Plastic Exterior Building Panels, d) Wall/Ceiling Coating Systems and e) Interior or Exterior Finish Systems.* Norwood, MA: Factory Mutual Standards Laboratories Department, 1994.

NFPA 13-99, *Installation of Sprinkler Systems*. Quincy, MA: National Fire Protection Association, 2001.

NFPA 268-96, *Standard Test Method for Determining Ignitability of Exterior Wall Assemblies Using a Radiant Heat Energy Source.* Quincy, MA: National Fire Protection Association, 1999.

UL 790-98, *Tests of Fire Resistance of Roof Covering Materials.* Northbrook, IL: Underwriters Laboratories Inc., 1998.

UL 1040-96, *Fire Test of Insulated Wall Construction—with Revisions through April 2001.* Northbrook, IL: Underwriters Laboratories Inc., 2001.

UL 1256-98, *Standard for Safety Fire Test of Roof Deck Constructions*. Northbrook, IL: Underwriters Laboratories Inc., 1998.

UL 1715-97, *Fire Test of Interior Finish Material.* Northbrook, IL: Underwriters Laboratories Inc., 1997.

Chapter 27:
Electrical

General Comments

This chapter references the ICC *Electrical Code*® (ICC EC™) and contains specific provisions for emergency and standby power. The ICC EC regulates the design, construction, installation, operation, maintenance and use of electrical systems and equipment. The ICC EC references NFPA 70 in addition to providing administrative and technical provisions. Except for the scope section, this chapter is devoted to establishing where emergency and standby power systems are required.

Purpose

Since electrical systems and components are an integral part of almost all structures, it is necessary for the code to address them to protect life, limb, health and property. In addition to general lighting and power needs, structures depend on electricity for operation of many of the life safety systems required by the code, including fire alarm, smoke control, exhaust, fire suppression, communication and fire command systems. Electricity is also depended upon for elements of egress systems, including elevators, exit signage, egress path illumination, power doors, stair pressurization and smokeproof enclosures.

SECTION 2701
GENERAL

2701.1 Scope. This chapter governs the electrical components, equipment and systems used in buildings and structures covered by this code. Electrical components, equipment and systems shall be designed and constructed in accordance with the provisions of the ICC *Electrical Code*.

❖ Chapter 27 regulates the electrical portions of buildings and structures that fall under the scope of the code. Regulation is accomplished by reference to the ICC EC which, in turn, references NFPA 70.

[F] SECTION 2702
EMERGENCY AND STANDBY POWER SYSTEMS

2702.1 Installation. Emergency and standby power systems shall be installed in accordance with the ICC *Electrical Code*, NFPA 110 and NFPA 111.

❖ This section is consistent and coordinated with Section 1202.6 of the ICC EC. Emergency power systems are intended to provide electrical power for life safety systems, such as egress illumination, emergency communications, fire pumps, high-rise building elevators and processes involving the handling and use of hazardous materials. In other words, emergency power is required where the loss of normal power would endanger occupants. Such systems are covered in Article 700 of NFPA 70 and one of their key features is the required response time of 10 seconds or less. The time between loss of normal power and the provision of emergency power must be kept very short to prevent putting occupants at risk. This is especially important during an emergency event such as a building fire, but is important at all times to prevent occupant panic, which could happen if a crowded building is suddenly plunged into darkness.

Standby power systems are covered in Article 701 of NFPA 70 and are intended to provide electrical power for loads not as critical as those requiring emergency power. Standby power loads include smoke control systems; certain elevators; certain hazardous material operations; smokeproof enclosure systems; illumination; heating, ventilating and air-conditioning (HVAC) systems; refrigeration and sewage pumps. Standby power systems must provide power within 60 seconds of failure of primary power.

Sources of power for emergency power systems (NFPA 70, Section 701-11) include storage batteries, generators, uninterruptible power supplies and separate services. Sources of power for standby systems include those allowed for emergency systems plus a source that is taken from a point of connection ahead of the normal service disconnecting means.

NFPA 110 addresses the performance criteria and "nuts and bolts" of emergency and standby power systems and classifies them into types, classes and levels relative to maximum response time, minimum required operation time and life safety importance factor, respectively. NFPA 111 addresses stored emergency power supply systems and is similar in coverage to NFPA 110. Stored energy systems typically rely on batteries that store chemical energy.

2702.1.1 Stationary generators. Emergency and standby power generators shall be listed in accordance with UL 2200.

❖ UL 2200 was developed at the request of building officials to provide a standard to evaluate the safety and reliability of stationery engine generators. UL 2200 establishes a basis for this evaluation.

2702.2 Where required. Emergency and standby power systems shall be provided where required by Sections 2702.2.1 through 2702.2.19.

❖ Fires can cause the loss of utility power as a result of either damage to equipment and wiring or fire-fighter action to shut off power to eliminate sources of ignition and danger to personnel.

Sections 2702.2.1 through 2702.2.19 dictate where emergency and standby power systems are required based upon the nature of the electrical loads. Note that emergency and standby power systems have different characteristics, and subsequent sections will require one or the other (see commentary, Section 2702.1).

2702.2.1 Group A occupancies. Emergency power shall be provided for voice communication systems in Group A occupancies in accordance with Section 907.2.1.2.

❖ Emergency voice/alarm systems are designed to allow the fast, efficient and orderly egress of large numbers of people and are, therefore, considered a life safety system worthy of being provided with emergency power (see commentary, Section 907.2.1.2).

2702.2.2 Smoke control systems. Standby power shall be provided for smoke control systems in accordance with Section 909.11.

❖ Smoke control systems are intended to maintain a tenable environment in a building to allow the occupants ample time to evacuate or relocate to protected areas. As such, smoke control systems are life safety systems and must be dependable (see commentary, Section 909.11).

2702.2.3 Exit signs. Emergency power shall be provided for exit signs in accordance with Section 1011.5.3.

❖ Emergency power is warranted for exit signage illumination since guiding occupants to the exits is certainly a life safety function (see commentary, Section 1003.2.10.5).

2702.2.4 Means of egress illumination. Emergency power shall be provided for means of egress illumination in accordance with Section 1006.3.

❖ The path of travel to all exits must be illuminated to guide occupants and allow for safe egress; therefore, emergency power for illumination is warranted (see commentary, Section 1003.2.11.2).

2702.2.5 Accessible means of egress elevators. Standby power shall be provided for elevators that are part of an accessible means of egress in accordance with Section 1007.4.

❖ Elevators can be a component of an accessible means of egress in accordance with Section 1003.2.13.3 and must, therefore, be dependable at all times. Without backup power, an elevator could be a dead end for someone with physical disabilities who is trying to egress a building (see commentary, Section 1003.2.13.3).

2702.2.6 Horizontal sliding doors. Standby power shall be provided for horizontal sliding doors in accordance with Section 1008.1.3.3.

❖ Power-operated doors could be an obstruction to egress if the primary power supply fails; therefore, standby power is required to maintain door operation (see commentary, Section 1003.3.1.3.3).

2702.2.7 Semiconductor fabrication facilities. Emergency power shall be provided for semiconductor fabrication facilities in accordance with Section 415.9.10.

❖ Where hazardous materials are utilized, such as in hazardous production materials (HPM) facilities, many systems are depended upon to protect the occupants from exposure to hazardous materials, including exhaust/ventilation systems, gas cabinet exhaust systems, gas detection systems, alarm systems and suppression systems. Loss of power would endanger the occupants; thus, emergency power is essential for these occupancies (see commentary, Sections 415.9.10 and 415.9.10.1).

2702.2.8 Membrane structures. Standby power shall be provided for auxiliary inflation systems in accordance with Section 3102.8.2. Emergency power shall be provided for exit signs in temporary tents and membrane structures in accordance with the *International Fire Code*.

❖ Air-supported and air-inflated structures would collapse on the occupants if the inflation systems fail. Section 3102.8.1.1 requires redundant inflation equipment that would be as worthless as the primary system in the event of power failure; therefore, standby power is required. Emergency power is required for all exit signs in temporary tents and membrane structures (see commentary, Section 3102.8.2).

2702.2.9 Hazardous materials. Emergency or standby power shall be provided in occupancies with hazardous materials in accordance with Section 414.5.4.

❖ Where hazardous materials and processes are housed, occupant safety could be dependent upon one or more ventilation, treatment, temperature control, alarm and detection systems. Thus, emergency or standby power is required, depending upon the load (see commentary, Section 414.5.4).

2702.2.10 Highly toxic and toxic materials. Emergency power shall be provided for occupancies with highly toxic or toxic materials in accordance with the *International Fire Code.*

❖ See the commentary to Section 2702.2.9 and Chapter 27 of the *International Fire Code*® (IFC®) for additional information.

2702.2.11 Organic peroxides. Standby power shall be provided for occupancies with silane gas in accordance with the *International Fire Code.*

❖ See the commentary to Section 2702.2.9 and Chapter 39 of the IFC for additional information.

2702.2.12 Pyrophoric materials. Emergency power shall be provided for occupancies with silane gas in accordance with the *International Fire Code.*

❖ See the commentary to Section 2702.2.9 and Chapter 41 of the IFC for additional information.

2702.2.13 Covered mall buildings. Standby power shall be provided for voice/alarm communication systems in covered mall buildings in accordance with Section 402.12.

❖ See the commentary to Section 907.2.20 and Section 402.12 for additional information.

2702.2.14 High-rise buildings. Emergency and standby power shall be provided in high-rise buildings in accordance with Sections 403.10 and 403.11.

❖ Occupants of high-rise buildings are at greater risk due to longer egress travel times, lack of fire-fighter access and the danger of vertical spread of fire and upward smoke migration. In accordance with this chapter and Sections 403.10 and 403.11, some loads in a high-rise building will require standby power and some will require emergency power (see commentary, Sections 403.10 and 403.11).

2702.2.15 Underground buildings. Emergency and standby power shall be provided in underground buildings in accordance with Sections 405.9 and 405.10.

❖ In the event of power failure, occupants could be underground without light, ventilation and the numerous required life safety systems. These structures are analogous to inverted high-rise buildings. Section 405.9 requires standby power for specified loads and Section 405.10 requires emergency power for specified loads (see commentary, Sections 405.9 and 405.10).

2702.2.16 Group I-3 occupancies. Emergency power shall be provided for doors in Group I-3 occupancies in accordance with Section 408.4.2.

❖ In an emergency, occupants in detention and correctional facilities are at the mercy of door-locking mechanisms and those who control such locks; thus, emergency power is warranted (see commentary, Section 408.4.2).

2702.2.17 Airport traffic control towers. Standby power shall be provided in airport traffic control towers in accordance with Section 412.1.5.

❖ This requirement is similar in intent to that for standby power in high-rise buildings, but due to the limited number of occupants and other protective features required by Section 412, only standby power is required (see commentary, Section 412.1.5).

2702.2.18 Elevators. Standby power for elevators shall be provided as set forth in Section 3003.1.

❖ Section 3003.1 provides the "how to" related text for elevator standby power systems, whereas other sections, including Sections 2702.2.5 and 2702.2.14, dictate where such power must be provided (see commentary, Section 3003.1).

2702.2.19 Smokeproof enclosures. Standby power shall be provided for smokeproof enclosures as required by Section 909.20

❖ To protect egress elements from the invasion of smoke, mechanical ventilation/pressurization systems must be dependable and thus provided with standby power (see commentary, Sections 909.20.6.2 and 1005.3.2.5).

2702.3 Maintenance. Emergency and standby power systems shall be maintained and tested in accordance with the *International Fire Code.*

❖ Emergency and standby power systems must be maintained, serviced and tested to assure dependability (see commentary, Section 604 of the IFC).

Bibliography

The following resource materials are referenced in this chapter or are relevant to the subject matter addressed in this chapter.

ICC EC-2000, *ICC Electrical Code*. Falls Church, VA: International Code Council, 2000.

NFPA 70-99, *National Electrical Code*. Quincy, MA: National Fire Protection Association, 1999.

NFPA 110-99, *Emergency and Standby Power Systems*. Quincy, MA: National Fire Protection Association, 1999.

NFPA 111-96, *Stored Electrical Energy Emergency and Standby Power Systems*. Quincy, MA: National Fire Protection Association, 1996.

UL 2200, *Stationery Engine Generator Assemblies*. Northbrook, IL: Underwriters Laboratories, 2002.

Chapter 28: Mechanical Systems

General Comments

Chapter 28 provides for the approval, installation, construction, inspection, operation and maintenance of all mechanical appliances, equipment and systems by direct reference to the *International Mechanical Code®* (IMC®) and the *International Fuel Gas Code®* (IFGC®). All mechanical equipment and appliances, other than gas-fired equipment and appliances, must comply with the provisions of the IMC. Fuel gas piping, gas-fired appliances and gas-fired appliance venting must comply with the IFGC.

Purpose

The purpose of Chapter 28 is to verify that mechanical equipment, appliances and mechanical systems are regulated with respect to design, construction, installation, inspection and maintenance. This chapter references the IMC and IFGC. This eliminates the need to duplicate the text of those codes and utilizes the performance and specification criteria of those codes as the basis for regulation of the construction, inspection and maintenance of all mechanical and fuel gas equipment, appliances and systems.

While this chapter only consists of references to other chapters and codes, it is necessary since almost all newly constructed buildings and structures contain mechanical equipment, appliances and systems. Without this chapter, a necessary link between the code and the IMC and IFGC would not exist.

SECTION 2801 GENERAL

2801.1 Scope. Mechanical appliances, equipment and systems shall be constructed, installed and maintained in accordance with the *International Mechanical Code* and the *International Fuel Gas Code*. Masonry chimneys, fireplaces and barbecues shall comply with the *International Mechanical Code* and Chapter 21 of this code.

❖ The requirements for all mechanical systems are governed by this chapter. This chapter does not contain any requirements for mechanical equipment, appliances and systems; instead it references the IMC and IFGC. These two codes along with nine others, including the *International Building Code®* (IBC®), make up the family of the 2003 editions of the *International Codes®*.

In the process of producing the code, including actual mechanical equipment, appliances and system provisions were considered unnecessary since such provisions would duplicate and possibly conflict with those of the IMC or IFGC. Since most newly constructed buildings contain mechanical equipment, appliances and systems, this chapter is necessary to reference the IMC and IFGC and tie them together with the code.

The IMC and IFGC work together to regulate all mechanical systems. Therefore, whether designing the mechanical system for a building or enforcing the provisions of the code that deal with mechanical systems, both the IMC and IFGC will be required.

The IMC and IFGC are complete codes in themselves, and set forth necessary administrative and enforcement requirements, as well as the requirements for installation, construction, inspection and maintenance of mechanical equipment, appliances and systems. The IMC regulates all aspects of mechanical systems, including: air distribution systems and ductwork; boilers and hydronic systems; exhaust systems; fuel-oil piping systems; combustion air requirements; chimneys and vents; ventilation air requirements; refrigeration systems; solar-powered systems and specific appliances and equipment. The IFGC regulates all aspects of fuel-gas piping systems, fuel gas utilization equipment and related accessories, including: combustion air requirements for gas appliances; chimneys and vents for gas appliances; fuel gas piping and specific gas appliances and gas equipment.

A key component of both the IMC and IFGC is Chapter 3, which requires all mechanical equipment and appliances to be listed and labeled to the standard or standards relevant to that equipment or appliance (see Figure 2801.1).

Another important element of the IMC and IFGC is a set of unique definitions found in Chapter 2 of both codes. These definitions supersede any others with respect to mechanical and gas systems. Especially noteworthy is the definition of "Noncombustible material" in the IMC, which does not include composite materials as provided for in Section 703.4.2 of the code. In the context of the IMC, such composite materials are considered to be combustible.

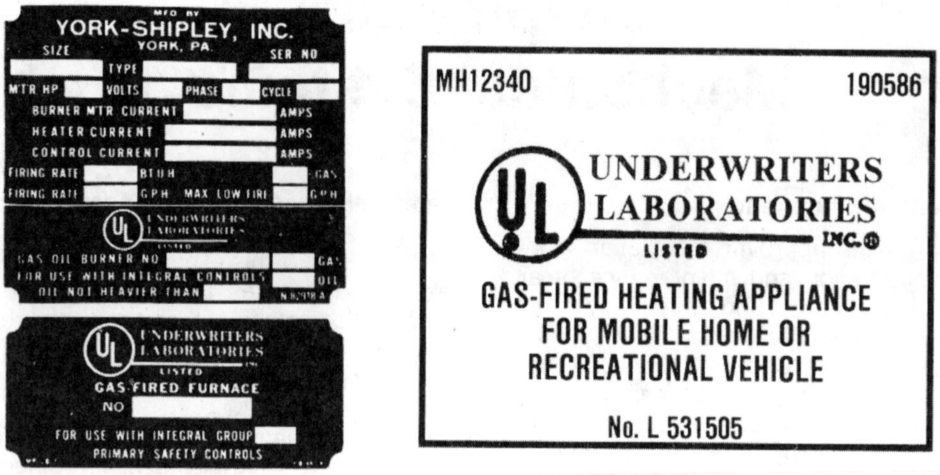

Figure 2801.1
EXAMPLE LABELS

Bibliography

The following resource material is referenced in this chapter or is relevant to the subject matter addressed in this chapter.

IFGC-2003, *International Fuel Gas Code*. Falls Church, VA: International Code Council, 2003.

IMC-2003, *International Mechanical Code*. Falls Church, VA: International Code Council, 2003.

Chapter 29: Plumbing Systems

General Comments

Determining the number of fixtures that are necessary for a given building has been a problem plaguing the plumbing industry for years.

There have been various methods used to evaluate the need for plumbing fixtures.

For office buildings, studies have indicated that design guidelines based on occupancy times, arrival rates, duration and patterns of fixture use provide the number of required plumbing fixtures for a desired level of service. The Building Technology Research Laboratory at the Stevens Institute of Technology conducted a study based on the "queueing theory" for determining the number of plumbing fixtures for a desired level of service. The theory is based on waiting times during peak periods of use, fixture utilization and the probability of finding a vacant fixture. This type of method provides answers for designing service systems. For residential-type buildings and health care facilities, plumbing fixtures are based on the minimum need, resulting in at least one water closet and one lavatory for each dwelling unit, guestroom or hospital room.

Studies completed by the U.S. military have been used for dormitories and prisons, which determine the number of fixtures based on a simultaneous need in a regimented society. This assumes that everyone rises at approximately the same time and has a limited amount of time to shower, use the water closet and lavatory.

The National Restaurant Association conducted a study based on the difference in use for a restaurant and a nightclub. It should be noted that the study did not take into account today's fast-food style restaurant, nor did it allow for restaurants located along heavily traveled routes, such as highway rest stops.

Fixture requirements for factory and industrial uses are based on the same requirements as for storage facilities. This method establishes a more realistic minimum requirement for factory occupancies. The reasonableness in the number of plumbing fixtures was established through a limited study of factory projects in Henrico County, Virginia.

The fixture needs for the remaining occupancies were determined based on empirical data, experience and tradition. There are no exact definitive studies providing values that are supportable; the values have been periodically modified based on general observations. Various studies are currently being undertaken to develop a rational method for determining the minimum number of necessary plumbing fixtures.

Purpose

The purpose of Chapter 29 is to provide a building with the necessary number of plumbing fixtures of a specific quality. The fixtures must be properly installed to be both accessible and usable by the individuals occupying the building. The quality and design of every fixture must be in accordance with the *International Plumbing Code®* (IPC®).

SECTION 2901 GENERAL

2901.1 Scope. The provisions of this chapter and the *International Plumbing Code* shall govern the erection, installation, alteration, repairs, relocation, replacement, addition to, use or maintenance of plumbing equipment and systems. Plumbing systems and equipment shall be constructed, installed and maintained in accordance with the *International Plumbing Code*. Private sewage disposal systems shall conform to the *International Private Sewage Disposal Code*.

❖ This section contains the scoping requirements for Chapter 29. Compliance with this chapter will result in a building or structure containing adequate plumbing fixtures for the sanitary, hygienic, cleaning, washing and food preparation needs of the occupants. It is important to note that the "P" in brackets before the main section number indicates that the entire content of Section 2902 is maintained by the International Plumbing Code Development Committee. This means that the entire content of Section 2902 can be found in Section 403 of the IPC. These provisions have been located in the code for the convenience of the architect, who is often responsible for the layout and number of plumbing fixtures when creating the floor plan of a building.

[P] SECTION 2902 MINIMUM PLUMBING FACILITIES

2902.1 Minimum number of fixtures. Plumbing fixtures shall be provided for the type of occupancy and in the minimum number shown in Table 2902.1 Types of occupancies not shown in Table 2902.1 shall be considered individually by the building official. The number of occupants shall be determined by this code. Occupancy classification shall be determined in accordance with Chapter 3.

❖ Table 2902.1 establishes the minimum number of plumbing fixtures required for each building (note that this is the same as Table 403.1 in the IPC). The occupant load used for calculating the number of fixtures required is the same occupant load used for determining means of egress. Methods for calculating occupant loads are found in Section 1004.1. By using the means of egress occupant loads, there is consistency in occupant load calculations for the application of the number of plumbing fixtures and the means of egress provisions. The means of egress occupant loads of a building do not always reflect typical day-to-day occupant loads; however, the table takes this into account by modifying the values for determining the number of fixtures.

TABLE 2902.1. See page 29-3

❖ Table 2902.1 provides simple straight-line ratios for determining the minimum number of plumbing fixtures. To aid in the use of the table, the type of building category has been listed by occupancy classification along with a brief description. The occupancy identification parallels the classification listed in Chapter 3.

The user simply identifies the occupancy of Table 2902.1 and applies the ratio to determine the amount of water closets, lavatories, bathtubs, showers (when applicable), drinking fountains and sinks needed for the structure.

Section 419.2 of the IPC permits a urinal to be substituted for a water closet for a maximum of 67 percent of the required number of water closets. For example, if five water closets are required in a men's room, the design professional may choose to install two water closets and three urinals, or three water closets and two urinals. If four urinals are installed, however, two water closets are also required.

In most of the building occupancy classifications, only one service sink is required for the entire building except in hospitals and dwelling units. The one service sink per building must be available from all portions of the building. For example, a service sink installed in an office building should not be located in an area that is not accessible from all tenant spaces.

Determining Number of Fixtures

Once the occupancy of the building is selected, determining the minimum required number of fixtures is simple and straightforward. The common question asked in applying the table is whether the occupant load is divided regarding the number of each sex and then the table is applied, or whether the table is applied to the entire occupant load and then the required number of fixtures is divided between the sexes. The rule of thumb (except for assembly occupancies required to be provided with different fixture ratios for females and males) is that the fixtures are split, not the number of people. Where a building occupant load is calculated as being 60-percent female and 40-percent male, it is important to remember that the fixtures are divided, not the number of people.

The division between male and female water closets is an attempt to resolve what is known as "potty parity" in assembly facilities. The table reflects a more equitable distribution in the number of plumbing fixtures provided at large places of assembly. The increase in the number of water closets required in all assembly occupancies is based on the following criteria:

1. Women sometimes take a longer period of time to use the facilities, thereby decreasing turnaround time;
2. Women outnumber men in the general population and this profile is represented in large places of assembly and
3. Women sometimes require more frequent use of plumbing facilities, thereby creating longer waiting lines.

In the case of assembly occupancies, the total design occupant load obtained from the code is first divided into a 50:50 ratio as required by Section 2902.3, unless statistical data approved by the building official indicate a different distribution of the sexes. The total number of water closets required for males is then calculated by multiplying the male occupant load by the corresponding water closet ratio found in Table 2902.1. The total number of female water closets is determined using the female occupant load and corresponding ratio. Such an application will result in a larger number of water closets required for females (even though the sex distribution is still considered as 50:50). The requirements for male/female water closets are the same in nightclub and restaurant occupancies because a larger number of fixtures is already required for these types of establishments.

When applying ratios listed in the table, any fraction of the quantity of people requires an additional fixture. For example, when the table requires one fixture per 25 people and the building occupant load is 76, a total of four fixtures is required.

The total number of required fixtures may result in an odd number. Because the fixtures must be equally distributed between the sexes, an additional fixture is required. For example, if an office building requires a total of three water closets, there must be two in the women's room and two in the men's room.

This is not the case with a mixed-use building, however. For a mixed-use building where a fraction of a number occurs in one of the occupancies, the number is rounded up.

TABLE 2902.1
MINIMUM NUMBER OF REQUIRED PLUMBING FIXTURES[a]

No.	CLASSIFICATION	OCCUPANCY	DESCRIPTION	WATER CLOSETS (SEE SECTION 419.2 OF THE *INTERNATIONAL PLUMBING CODE* FOR URINALS) MALE	WATER CLOSETS FEMALE	LAVATORIES MALE	LAVATORIES FEMALE	BATHUBS OR SHOWERS	DRINKING FOUNTAINS (SEE SECTION 410.1 OF THE *INTERNATIONAL PLUMBING CODE)*	OTHER
1	Assembly (see Sections 2902.2, 2902.5 and 2902.6)	A-1	Theaters usually with fixed seats and other buildings for the performing arts and motion pictures	1 per 125	1 per 65	1 per 200		—	1 per 500	1 service sink
		A-2	Nightclubs, bars, taverns, dance halls and buildings for similar purposes	1 per 40	1 per 40	1 per 75		—	1 per 500	1 service sink
			Restaurants, banquet halls and food courts	1 per 75	1 per 75	1 per 200		—	1 per 500	1 service sink
		A-3	Auditoriums without permanent seating, art galleries, exhibition halls, museums, lecture halls, libraries, arcades and gymnasiums	1 per 125	1 per 65	1 per 200		—	1 per 500	1 service sink
			Passenger terminals and transportation facilities	1 per 500	1 per 500	1 per 750		—	1 per 1,000	1 service sink
		A-3	Places of worship and other religious services. Churches without assembly halls	1 per 150	1 per 75	1 per 200		—	1 per 1,000	1 service sink
		A-4	Coliseums, arenas, skating rinks, pools and tennis courts for indoor sporting events and activities	1 per 75 for the first 1,500 and 1 per 120 for the remainder exceeding 1,500	1 per 40 for the first 1,500 and 1 per 60 for the remainder exceeding 1,500	1 per 200	1 per 150	—	1 per 1,000	1 service sink
		A-5	Stadiums, amusement parks, bleachers and grandstands for outdoor sporting events and activities	1 per 75 for the first 1,500 and 1 per 120 for the remainder exceeding 1,500	1 per 40 for the first 1,500 and 1 per 60 for the remainder exceeding 1,500	1 per 200	1 per 150	—	1 per 1,000	1 service sink
2	Business (see Sections 2902.2, 2902.4, 2902.4.1 and 2902.6)	B	Buildings for the transaction of business, professional services, other services involving merchandise, office buildings, banks, light industrial and similar uses	1 per 25 for the first 50 and 1 per 50 for the remainder exceeding 50		1 per 40 for the first 50 and 1 per 80 for the remainder exceeding 50		—	1 per 100	1 service sink

(continued)

TABLE 2902.1—continued
MINIMUM NUMBER OF REQUIRED PLUMBING FIXTURES[a]

No.	CLASSIFICATION	OCCUPANCY	DESCRIPTION	WATER CLOSETS (SEE SECTION 419.2 OF THE *INTERNATIONAL PLUMBING CODE* FOR URINALS) MALE	WATER CLOSETS FEMALE	LAVATORIES MALE	LAVATORIES FEMALE	BATHUBS OR SHOWERS	DRINKING FOUNTAINS (SEE SECTION 410.1 OF THE *INTERNATIONAL PLUMBING CODE)*	OTHER
3	Educational	E	Educational facilities	1 per 50		1 per 50		—	1 per 100	1 service sink
4	Factory and industrial	F-1 and F-2	Structures in which occupants are engaged in work fabricating, assembly or processing of products or materials	1 per 100		1 per 100		See Section 411 of the *International Plumbing Code*	1 per 400	1 service sink
5	Institutional	I-1	Residential care	1 per 10		1 per 10		1 per 8	1 per 100	1 service sink
		I-2	Hospitals, ambulatory nursing home patients[b]	1 per per room[c]		1 per per room[c]		1 per 15	1 per 100	1 service sink
			Employees, other than residential care[b]	1 per 25		1 per 35		—	1 per 100	—
			Visitors, other than residential care	1 per 75		1 per 100		—	1 per 500	—
		I-3	Prisons[b]	1 per cell		1 per cell		1 per 15	1 per 100	1 service sink
		I-3	Reformatories, detention centers and correctional centers[b]	1 per 15		1 per 15		1 per 15	1 per 100	1 service sink
		I-4	Adult day care and child care[b]	1 per 15		1 per 15		1 per 15[d]	1 per 100	1 service sink
6	Mercantile (see Section 2902.2, 2902.5 and 2902.6)	M	Retail stores, service stations, shops, salesrooms, markets and shopping centers	1 per 500		1 per 750		—	1 per 1,000	1 service sink
7	Residential	R-1	Hotels, motels, boarding houses (transient)	1 per guestroom		1 per guestroom		1 per guestroom	—	1 service sink
		R-2	Dormitories, fraternities, sororities and boarding house (not transient)	1 per 10		1 per 10		1 per 8	1 per 100	1 service sink
		R-2	Apartment house	1 per dwelling unit		1 per dwelling unit		1 per dwelling unit	—	1 kitchen sink per dwelling unit; 1 automatic clothes washer connection per 20 dwelling units[e]

(continued)

TABLE 2902.1—continued
MINIMUM NUMBER OF REQUIRED PLUMBING FIXTURES[a]

No.	CLASSIFICATION	OCCUPANCY	DESCRIPTION	WATER CLOSETS (SEE SECTION 419.2 OF THE *INTERNATIONAL PLUMBING CODE* FOR URINALS)		LAVATORIES		BATHUBS OR SHOWERS	DRINKING FOUNTAINS (SEE SECTION 410.1 OF THE *INTERNATIONAL PLUMBING CODE)*	OTHER
				MALE	FEMALE	MALE	FEMALE			
7	Residential	R-3	One- and two-family dwellings	1 per dwelling unit		1 per dwelling unit		1 per dwelling unit	—	1 kitchen sink per dwelling unit; 1 automatic clothes washer connection per 20 dwelling units[e]
		R-4	Residential care/assisted living facilities	1 per 10		1 per 10		1 per 8	1 per 100	1 service sink
8	Storage (see Sections 2902.2, 2902.4 and 2902.4.1)	S-1 S-2	Structures for the storage of goods, warehouses, storehouses and freight depots, low and moderate hazard	1 per 100		1 per 100		See Section 411 of the *International Plumbing Code*	1 per 1,000	1 service sink

a. The fixtures shown are based on one fixture being the minimum required for the number of persons indicated or any fraction of the number of persons indicated. The number of occupants shall be determined by this code.

b. Toilet facilities for employees shall be separate from facilities for inmates or patients.

c. A single-occupant toilet room with one water closet and one lavatory serving not more than two adjacent patient rooms shall be permitted where such room is provided with direct access from each patient room and with provisions for privacy.

d. For day nurseries, a maximum of one bathtub shall be required.

e. For attached one- and two-family dwellings, one automatic clothes washer connection shall be required per 20 dwelling units.

Example 1: A mixed-use building with a Group B (Business) occupant load of 538, Group A-3 (Libraries, halls, museums) occupant load of 115 and Group S-1 (Storage) occupant load of 82.

What is the minimum number of fixtures required for each sex?

The categories to apply are Use Group B, Use Group A-3 and Use Group S-1.

Refer to the table for Use Group B (business):

Number of water closets:

538 occupants:

$$\frac{\text{1 fixture}}{\text{25 occupants for the first 50 and 1 per 50 for the remainder exceeding 50}} = \frac{50}{25}$$

= 2 water closets

$$\text{488 occupants} \times \frac{\text{1 fixture}}{\text{50 occupants}} = \frac{488}{50}$$

= 9.76 water closets = 10

Total = 12 water closets

Divide between the sexes:

12 water closets × 50%

= 6 water closets/women
= 6 water closets/men

Number of lavatories:

538 occupants:

$$\frac{\text{1 fixture}}{\text{40 occupants for the first 50 and 1 per 80 for the remainder exceeding 50}} = \frac{50}{40}$$

= 1 lavatory

$$\text{488 occupants} \times \frac{\text{1 fixture}}{\text{80 occupants}} = \frac{488}{80}$$

6.1 lavatories = 7

Total = 8 lavatories

Divide between the sexes:

8 lavatories × 50%

= 4 lavatories/women
= 4 lavatories/men

Number of drinking fountains:

$$538 \text{ occupants} \times \frac{1 \text{ fixture}}{100 \text{ occupants}} = \frac{538}{100}$$

$$= 5.4 \text{ drinking fountains} = 6$$

Refer to the table for libraries, halls and museums:
Number of water closets:
Divide between the sexes:

115 occupants × 50%

= 57.5 women
= 57.5 men

$$57.5 \text{ male occupants} \times \frac{1 \text{ fixture}}{125 \text{ occupants}} = \frac{57.5}{125}$$

$$= .46 \text{ water closet} = 1$$

$$57.5 \text{ female occupants} \times \frac{1 \text{ fixture}}{65 \text{ occupants}} = \frac{57.5}{65}$$

$$= .88 \text{ water closet} = 1$$

Number of lavatories:

$$15 \text{ occupants} \times \frac{1 \text{ fixture}}{200 \text{ occupants}} = \frac{115}{200}$$

$$= .58 \text{ lavatories} = 1$$

Divide between the sexes:

1 lavatory × 50 percent

= .5 lavatory/women = 1
= .5 lavatory/men = 1

Number of drinking fountains:

$$115 \text{ occupants} \times \frac{1 \text{ fixture}}{500 \text{ occupants}} = \frac{115}{500}$$

$$= .23 \text{ drinking fountains} = 1$$

Refer to the table for storage:
Number of water closets:

$$82 \text{ occupants} \times \frac{1 \text{ fixture}}{100 \text{ occupants}} = \frac{82}{100}$$

$$= .82 \text{ water closet} = 1$$

Divide between the sexes:

1 water closet × 50%

= .5 water closet/women = 1
= .5 water closet/men = 1

The design professional elects to install five urinals in the men's room as permitted by Section 419.2. The table would change accordingly.

EXAMPLE 1
TOTAL NUMBER OF REQUIRED FIXTURES

FIXTURE	WOMEN'S	MEN'S
Water closet	8	3
Urinal	—	5
Lavatory	6	6
Drinking fountain	8	
Service sink	1	

EXAMPLE 1
TOTAL NUMBER OF REQUIRED FIXTURES
(Mixed-Use Building)

OCCUPANCY	OCCUPANT LOAD	WATER CLOSETS				LAVATORIES			DRINKING FOUNTAINS	SERVICE SINK
		RATIO	MEN	RATIO	WOMEN	RATIO	MEN	WOMEN		
Business	538	1 per 25 for the first 50 and 1 per 50 for the remainder exceeding 50	6	1 per 25 for the first 50 and 1 per 50 for the remainder exceeding 50	6	1 per 40 for the first 50 and 1 per 80 for the remainder exceeding 50	4	4	6	-
Libraries, halls, museums, etc.	115	1/125	1	1/65	1	1/200	1	1	1	-
Storage	82	1/100	1	1/100	1	1/100	1	1	1	-
Total Required		8		8		6		6	8	1

Number of lavatories:

$$82 \text{ occupants} \times \frac{1 \text{ fixture}}{100 \text{ occupants}} = \frac{82}{100}$$

$$= .82 \text{ lavatory} = 1$$

Divide between the sexes:

1 lavatory × 50 percent

= .5 water closet/women = 1
= .5 water closet/men = 1

Number of drinking fountains:

$$82 \text{ occupants} \times \frac{1 \text{ fixture}}{1{,}000 \text{ occupants}} = \frac{82}{1{,}000}$$

$$= .08 \text{ drinking fountains} = 1$$

Number of service sinks: = 1

Example 2 : A stadium has an occupant load of 5,200 seats. What is the minimum number of fixtures required for each sex?

Refer to the table for Use Group A-5 (stadiums):

Divide occupant load between the sexes:

5,200 occupants × 50 percent

= 2,600/female occupants
= 2,600/male occupants

Number of water closets:

$$2{,}600 \text{ female occupants} \times \frac{1 \text{ fixture}}{40 \text{ occupants for the first } 1{,}500 \text{ and } 1 \text{ per } 60 \text{ for the remainder exceeding } 1{,}500} = \frac{1{,}500}{40}$$

$$= 37.5 \text{ water closets} = 38$$

$$1{,}100 \text{ female occupants} \times \frac{1 \text{ fixture}}{60 \text{ occupants}} = \frac{1{,}100}{60}$$

$$= 18.3 \text{ water closets} = 19$$

$$= 57 \text{ water closets}$$

$$2{,}600 \text{ male occupants} \times \frac{1 \text{ fixture}}{75 \text{ occupants for the first } 1{,}500 \text{ and } 1 \text{ per } 120 \text{ for the remainder exceeding } 1{,}500} = \frac{1{,}500}{40}$$

$$= 20 \text{ water closets}$$

$$1{,}100 \text{ male occupants} \times \frac{1 \text{ fixture}}{120 \text{ occupants}} = \frac{1{,}100}{120}$$

$$= 9.1 \text{ water closets} = 10$$

$$= 30 \text{ water closets}$$

Number of lavatories:

Divide between the sexes:

5,200 occupants x 50%

= 2,600 men
= 2,600 women

$$2{,}600 \text{ male occupants} \times \frac{1 \text{ fixture}}{200 \text{ occupants}} = \frac{2{,}600}{200}$$

$$= 13 \text{ lavatories}$$

$$2{,}600 \text{ male occupants} \times \frac{1 \text{ fixture}}{150 \text{ occupants}} = \frac{2{,}600}{150}$$

$$= 17.3 \text{ or } 18 \text{ lavatories}$$

Number of drinking fountains:

$$5{,}200 \text{ occupants} \times \frac{1 \text{ fixture}}{1{,}000 \text{ occupants}} = \frac{5{,}200}{1{,}000}$$

$$= 5.2 \text{ or } 6 \text{ drinking fountains}$$

Number of service sinks = 1

EXAMPLE 2
TOTAL NUMBER OF REQUIRED FIXTURES

FIXTURE	WOMEN'S	MEN'S
Water closet	57	30
Lavatory	18	13
Drinking fountain	6	
Service sink	1	

The design professional is permitted to substitute up to 20 water closets with urinals. In this example, the designer chooses to install 20 urinals in the men's room. The resulting number of fixtures would change accordingly.

EXAMPLE 2
TOTAL NUMBER OF REQUIRED FIXTURES

FIXTURE	WOMEN'S	MEN'S
Water closet	57	10
Urinal	—	20
Lavatory	18	13
Drinking fountain	6	
Service sink	1	

In buildings with large areas or multiple stories, the distribution of plumbing fixtures must satisfy the requirements of the occupants. For example, a 20-story high-rise office building cannot have all of the required plumbing fixtures located on the 10th floor. The path of travel to the plumbing fixtures must not exceed a distance of 500 feet (152 400 mm) nor require travel beyond the adjacent story above or below. This travel distance limitation, both total and vertical distance, can result in fixtures not being located on every floor of a multistory building. When determining the number of fixtures required in a multistory building, the entire building occupant load is used. Fixtures are not required based on the occupant load of each floor. The distribution of fixtures, however, must satisfy the demand for each floor. The percentage of the number of fixtures is to be consistent with the percentage of the occupant load they serve.

In determining the minimum required number of plumbing fixtures in a child care facility or elementary school, a design professional might provide child-size fixtures based on the building's use in an attempt to accommodate children; however, this is subject to the approval of the building official. These facilities must be located so that they can be accessed by the public. It is not the intent of the code to prevent toilet facilities from being located in individual classrooms, but these facilities should be in excess of the fixtures required by Table 2902.1.

2902.1.1 Unisex toilet and bath fixtures. Fixtures located within unisex toilet bathing rooms complying with Section 404 of the *International Plumbing Code* are permitted to be included in determining the minimum required number of fixtures for assembly and mercantile occupancies.

❖ This section permits fixtures located in unisex toilet and bathing rooms to count towards the minimum required number for assembly and mercantile occupancies. Such fixtures can reduce the number required in either male or female toilet rooms, but not both (see Sections 1109.2.1 through 1109.2.1.4).

2902.2 Separate facilities. Where plumbing fixtures are required, separate facilities shall be provided for each sex.

Exceptions:

1. Separate facilities shall not be required for private facilities.
2. Separate employee facilities shall not be required in occupancies in which 15 or fewer people are employed.
3. Separate facilities shall not be required in structures or tenant spaces with a total occupant load, including both employees and customers, of 15 or less.
4. Separate facilities shall not be required in mercantile occupancies in which the maximum occupant load is 50 or less.

❖ Separate facilities must be provided for each sex, meaning a separate women's room and men's room. If the required number of fixtures based on Table 2902.1 is one water closet and one lavatory, two of each fixture must be provided. One group of fixtures must be installed in the women's room and one in the men's room. Single-occupant toilet rooms for use by both sexes are permitted for:

1. Private residential uses, such as within dwelling units and hotel guestrooms;
2. Employees in buildings in which 15 or fewer people are employed;
3. Small establishments in which the occupant load, including customers and employees, is 15 or fewer and
4. Mercantile occupancies in which the maximum occupant load is 50 or fewer.

Where fixtures are located in a mercantile occupancy, the number of employees is determined by the building owner. In other facilities, such as an office building, the number of employees is determined by the occupant load requirements in the code.

2902.3 Number of occupants of each sex. The required water closets, lavatories and showers or bathtubs shall be distributed equally between the sexes based on the percentage of each sex anticipated in the occupant load. The occupant load shall be composed of 50 percent of each sex, unless statistical data approved by the building official indicate a different distribution of the sexes.

❖ This section requires an equal number of plumbing fixtures for each sex, except as required by the ratios in Table 2902.1 for high-density assembly occupancies. Although this section attempts to provide equality between the sexes for other occupancies, it only regulates the minimum required number of fixtures. A design professional may provide additional plumbing fixtures to one of the sexes, such as a greater number of urinals in the men's room.

Another exception to providing an equal division of required fixtures between the sexes is a building owner or designer verifying that the use of the building involves an unequal distribution of the sexes. An example of this type of building use is a women's health club. It would be inappropriate to provide an equal number of men's plumbing fixtures in this building.

2902.4 Location of employee toilet facilities in occupancies other than assembly or mercantile. Access to toilet facilities in occupancies other than mercantile and assembly shall be from within the employees' working area. Employee facilities shall be either separate facilities or combined employee and public facilities.

> **Exception:** Facilities that are required for employees in storage structures or kiosks, and are located in adjacent structures under the same ownership, lease or control, shall be a maximum travel distance of 500 feet (152 m) from the employees' working area.

❖ The requirements for where employee facilities must be located have been divided into two categories: other than assembly or mercantile occupancies and assembly or mercantile occupancies. Employee facilities in occupancies other than assembly or mercantile must be located within the employee's work area.

An exception is allowed for small structures, including toll booths, photo processing booths, kiosks and parking lot booths. Such buildings are not required to have plumbing fixtures within the building; however, plumbing facilities must be available within 500 feet (152 400 mm) of these structures.

2902.4.1 Travel distance. The required toilet facilities in occupancies other than assembly or mercantile shall be located not more than one story above or below the employees' working area and the path or travel to such facilities shall not exceed a distance of 500 feet (152 m).

> **Exception:** The location and maximum travel distances to required employee toilet facilities in factory and industrial occupancies are permitted to exceed that required in Section 2902.4.1, provided the location and maximum travel distance are approved by the building official.

❖ What is considered to be an employee's working area is subject to interpretation; however, the code clearly provides two conditions that must be met (see Section 2904.4.1). The required toilet facilities must be located within a travel distance of 500 feet (152 400 mm) and the employee must not be required to travel beyond the next adjacent story above or below his or her working area. Where public facilities are provided in the employee's working area, such facilities can serve as the required employee facilities. Otherwise, separate employee facilities are required.

An exception provides for the unique nature of factory and industrial occupancies where the designer is better equipped to determine the location of employee toilet facilities.

2902.5 Location of employee toilet facilities in mercantile and assembly occupancies. Employees shall be provided with toilet facilities in building and tenant spaces utilized as restaurants, nightclubs, places of public assembly and mercantile occupancies. The employee facilities shall be either separate facilities or combined employee and public facilities. The required toilet facilities shall be located not more than one story above or below the employees' work area and the path of travel to such facilities, in other than covered malls, shall not exceed a distance of 500 feet (152 m). The path of travel to required facilities in covered malls shall not exceed a distance of 300 feet (91 m).

> **Exception:** Employee toilet facilities shall not be required in tenant spaces where the travel distance from the main entrance of the tenant spaces to a central toilet area does not exceed 300 feet (91 m) and such central toilet facilities are located not more than one story above or below the tenant space.

❖ Every mercantile or assembly tenant space or building is required to have employee facilities located within it. Where public facilities are provided within the assembly or mercantile area (a restaurant or large department store), the location requirements for the employee facility can be met with the public facilities. Otherwise, separate employee facilities must be provided within the tenant space.

The intent of this section is to prevent employees from having to leave the tenant space or building to utilize toilet facilities. Employees are more susceptible to crime if they must leave their workplace to use the toilet facilities. For example, a small clothing store in a covered mall may have a single employee and requiring him or her to secure the store and travel into public areas to access a toilet would be a security risk. The required facilities in buildings other than covered malls must be located within a travel distance of 500 feet (152 400 mm) and not more than one story above or below the employee's work area. Such facilities for covered malls must be located within a travel distance of 300 feet (91 440 mm).

An exception is provided for employees in tenant spaces where the travel distance from the main entrance to a central toilet area does not exceed 300 feet (91 440 mm) and such facilities are located no more than one story above or below the tenant space.

2902.6 Public facilities. Customers, patrons and visitors shall be provided with public toilet facilities in structures and tenant spaces intended for public utilization. Public toilet facilities shall be located not more than one story above or below the space required to be provided with public toilet facilities and the path of travel to such facilities shall not exceed a distance of 500 feet (152 m).

❖ Public facilities are required in restaurants, nightclubs, places of public assembly and business occupancies open to the public; however, such facilities may be centrally located in a covered mall building. It is inappropriate to locate them in a storage area, behind the kitchen or in other areas not open and available to the public. A travel distance limitation of 500 feet (152 400 mm) is mandated along with the location of not more than one story above or below the space required to have public toilet facilities.

2902.6.1 Covered malls. In covered mall buildings, the path of travel to required toilet facilities shall not exceed a distance of 300 feet (91 m). Facilities shall be installed in each individual store or in a central toilet area located in accordance with this section. The maximum travel distance to the central toilet facilities in covered mall buildings shall be measured from the main entrance of any store or tenant space.

❖ In covered malls, the path of travel to required toilet facilities must not exceed a distance of 300 feet (91 440 mm) and is measured from the main entrance of any store or tenant space. This addresses the needs of the elderly, persons with disabilities and children. Additionally, covered malls are frequently very congested and occupants are unfamiliar with their surroundings. The minimum number of required plumbing facilities is based on total square footage of the mall, including tenant spaces, and must be installed in each individual store or in a central toilet area. This section does not prohibit the installation of separate toilet facilities in individual tenant spaces (see commentary, Section 2902.6); however, such facilities are not deductible from the total common facilities required if the facilities in the tenant space are intended for employees only (see commentary, Section 2902.5).

2902.6.2 Pay facilities. Where pay facilities are installed, such facilities shall be in excess of the required minimum facilities. Required facilities shall be free of charge.

❖ Pay facilities have been included in some public areas to prevent vagrants from loitering in the public bathrooms. This section does not prevent pay facilities from being installed, but they must be in excess of those required in Table 2902.1.

2902.6.3 Signage. A legible sign designating the sex shall be provided in a readily visible location near the entrance to each toilet facility. Signs for accessible toilet facilities shall comply with ICC A117.1.

❖ Public facilities must be designated by a legible sign for each sex that is located near the entrance to such facilities. The code is silent on the specifics of the type of designation and is dependent on the approval of the building official.

BIBLIOGRAPHY

The following resource materials are referenced in this chapter or are relevant to the subject matter addressed in this chapter.

ASPE-95, *A Critique of Queueing Theory Approaches to Plumbing Design.* Westlake, CA: American Society of Plumbing Engineers, 1995.

ICC A117.1-98, *Accessible and Usable Buildings and Facilities.* Falls Church, VA: International Code Council, 1998.

IPC-2003, *International Plumbing Code*. Falls Church, VA: International Code Council, 2003.

Chapter 30: Elevators And Conveying Systems

General Comments

Chapter 30 contains the provisions that regulate vertical and horizontal transportation and material-handling systems installed in buildings. The installation must comply with the requirements in this chapter and the standards referenced herein.

Section 3001.1 contains the scope of the chapter.

Section 3001.2 identifies the standards to which elevators and conveying systems must comply.

Section 3001.3 contains requirements for accessible elevators.

Sections 3002.1 through 3002.7 contain the requirements for elevator hoistway enclosures.

Sections 3003.1 through 3003.2 include requirements for emergency operation of elevators.

Sections 3004.1 through 3004.5 identify requirements for hoistway venting.

Sections 3005.1 through 3005.4 address requirements for conveying systems and personnel and material hoists.

Sections 3006.1 through 3006.6 address machine room requirements.

Purpose

The purpose of this chapter is to regulate the installation, testing, inspection, maintenance, alteration and repair of vertical and horizontal transportation and material-handling systems installed in buildings. Compliance with the requirements in this chapter provides for life safety and promotes public welfare. The chapter is also intended to be used as a minimum safety standard by architects, engineers, insurance companies, manufacturers and contractors and as standard safety practice for owners and managers of buildings where equipment covered by this chapter is installed and used.

SECTION 3001 GENERAL

3001.1 Scope. This chapter governs the design, construction, installation, alteration and repair of elevators and conveying systems and their components.

❖ This section indicates that the requirements in this chapter are applicable to the design, construction, installation, alteration and repair of any elevator, conveying system or component thereof. In addition to requirements for design, construction and installation, requirements for alteration, repair, testing and inspections of elevators and conveying systems are located in the standards referenced in this chapter, and are, therefore, not included in the code.

3001.2 Referenced standards. Except as otherwise provided for in this code, the design, construction, installation, alteration, repair and maintenance of elevators and conveying systems and their components shall conform to ASME A17.1, ASME A90.1, ASME B20.1, ALI ALCTV, and ASCE 24 for construction in flood hazard areas established in Section 1612.3.

❖ The enforceability of a standard is established in this section, and applies wherever the provisions of this chapter do not otherwise indicate a requirement. Elevators installed in buildings and structures located in designated flood hazard areas may be subject to additional flood-resistant design and construction requirements. The application of these additional requirements depends on whether the elevator or components of the elevator equipment are to be located below the design flood elevation. Those not familiar with the subtle differences that make a standard applicable to a specific equipment design should review both the standard's scope and definitions. In general, the applications of the referenced standards are as follows:

- ASME A17.1 applies to transportation equipment that is permanently installed to transport people and freight.
- ASME A90.1 applies specifically to belt-type manlifts.
- ASME B20.1 applies to equipment that is permanently installed to transport freight. People are prohibited from riding this equipment.
- ALI ALCTV applies specifically to automotive lifts, such as those used in automated parking structures.
- ASCE 24 applies to the construction of elevators and conveying systems located in flood hazard areas, as established by application of Section 1612.3 of the code. Section 1612.3 requires flood hazard areas to be established by the governing body through the adoption of a flood hazard map, which locates the flood hazard areas within a given jurisdiction. Chapter 8 of ASCE 24-98 and FEMA FIA-TB #4, *Elevator Installation for Build-*

ings Located in Special Flood Hazard Areas provides extensive guidance on the proper method of installing elevators in buildings and structures located in designated flood hazard areas. In addition to raising elevator service equipment above the design flood elevation, a concern is determining that the emergency operation requirements of Section 3003 do not require the elevator cab to descend automatically to a level below the design flood elevation during conditions of flooding. The safety of any occupants may be jeopardized and the elevator cab may be damaged if it descends below the design flood elevation during flooding.

Referenced standards have a long history of providing minimum safety requirements for the equipment covered by their scope. As an example, ASME A17.1 was first published in 1921 and 14 subsequent editions have been published.

Interpretations of the referenced standards can usually be obtained from their publisher. The committee responsible for developing and maintaining the standard will respond to all interpretation requests. The ASME A17.1, A90.1 and B20.1 standards committees have also published interpretations and include copies of all recent interpretations to purchasers of these documents. The publisher of ASME A17.1 also has available a handbook that explains and augments the requirements. While every effort has been made to eliminate conflicts between the code and the referenced standards, such conflicts may occasionally occur. Where differences occur, the requirements of the code take precedence (see Section 102.4).

3001.3 Accessibility. Passenger elevators required to be accessible by Chapter 11 shall conform to ICC A117.1.

❖ Section 1109.6 requires elevators that form part of an accessible route to be accessible as required by ICC A117.1. ICC A117.1 contains requirements for elevator location and access; operation and leveling; door operation; door size; door protective and reopening devices; door delay (passenger service time) from hall calls and car calls; car inside dimensions, controls and position indicators and signals; telephone or intercom systems; floor coverings; minimum illumination; hall buttons; hall lanterns; door jamb markings and clearance between sills.

This standard provides acceptable and appropriate service for all users, and makes multistory buildings accessible to, and usable by, people with physical disabilities.

3001.4 Change in use. A change in use of an elevator from freight to passenger, passenger to freight, or from one freight class to another freight class shall comply with Part XII of ASME A17.1.

❖ This section requires enforcement of Part XII of ASME A17.1 whenever the elevator changes in use or freight class. The application of these requirements results in an elevator that will operate safely and comply with requirements that are unique to, and necessary for, the new use or freight class.

SECTION 3002
HOISTWAY ENCLOSURES

3002.1 Hoistway enclosure protection. Elevator, dumbwaiter and other hoistway enclosures shall have a fire-resistance rating not less than that specified in Chapter 6 and shall be constructed in accordance with Chapter 7.

❖ To address the migration of smoke and fire between floor levels in a building, this section requires the elevator hoistway to be protected (fire-resistance rated). Reference is made to Chapters 6 and 7 of the code for the required fire resistance and construction of the hoistway enclosure. Chapter 6 contains fire-resistance requirements for building elements, including bearing and nonload-bearing walls. Chapter 7 also contains fire-resistance rating requirements, as well as construction requirements for shafts, which is typically the method used to satisfy the requirements for a hoistway enclosure. However, other methods of providing protection to the floor opening that the elevator passes through can be utilized. For example, an elevator located totally within an atrium would not need its own enclosure because the floor openings are protected by virtue of the atrium requirements (smoke control, fire suppression system, etc.).

3002.1.1 Opening protectives. Openings in hoistway enclosures shall be protected as required in Chapter 7.

❖ To address the migration of smoke and fire between floor levels in a building, this section requires elevator hoistway openings to be protected (fire-resistance rated). Reference is made to Chapter 7 for protection of openings in hoistway enclosures. Chapter 7 contains minimum hourly rating requirements for openings based on the fire-resistance rating required for the enclosure. For example, if a hoistway enclosure was required to be 2-hour fire-resistance rated, a $1^1/_2$-hour fire door assembly would be required, in accordance with Table 715.3.

3002.1.2 Hardware. Hardware on opening protectives shall be of an approved type installed as tested, except that approved interlocks, mechanical locks and electric contacts, door and gate electric contacts and door-operating mechanisms shall be exempt from the fire test requirements.

❖ In order to verify that hardware used with elevator doors will not affect their fire-resistance rating, door hardware not specifically exempt by this section is required to be part of the door assembly during the fire test. The devices that are exempt from the fire test have minimum fire and physical requirements contained in ASME A17.1. Passenger elevator door hardware consists of a header, track hangers, pendant bolts, floor sill with guides, sill support plates, sill brackets, retaining angles and closer assemblies. Each component may bear a separate label, or the assembly may have one label to cover all components. The label should clearly identify the assembly or component to which it applies. Typically, labels are viewed from inside the hoistway.

3002.2 Number of elevator cars in a hoistway. Where four or more elevator cars serve all or the same portion of a building, the elevators shall be located in at least two separate hoistways. Not more than four elevator cars shall be located in any single hoistway enclosure.

❖ The maximum number of elevator cars in a single hoistway is four. This limits the potential of a single hoistway fire incident causing the removal of all elevators from service, particularly in larger structures, such as high-rise buildings.

3002.3 Emergency signs. An approved pictorial sign of a standardized design shall be posted adjacent to each elevator call station on all floors instructing occupants to use the exit stairways and not to use the elevators in case of fire. The sign shall read: IN FIRE EMERGENCY, DO NOT USE ELEVATOR. USE EXIT STAIRS. The emergency sign shall not be required for elevators that are part of an accessible means of egress complying with Section 1007.4.

❖ Elevators are unsafe during a fire because:

- Persons may push a corridor button and have to wait for an elevator that may never respond. Valuable time in which to leave the building safely is lost.
- Elevators cannot start until the car and hoistway doors are closed. A panic could lead to overcrowding of an elevator and blockage of the doors, thus preventing closing.
- Power failure during a fire can happen at any time and thus lead to entrapment.
- Elevators respond to car and corridor calls. One of these calls may be at the fire floor.

Fatal delivery of the elevator to the fire floor can be caused by:

- An elevator passenger pressing the car button for the fire floor.
- One or both of the corridor call buttons being pushed on the fire floor.
- Heat melting or deforming the corridor push button or its wiring at the fire floor.
- Normal functioning of the elevator, such as high-call or low-call reversal, occurring at the fire floor.

In a fire emergency, an occupant will normally try to exit the building by the route he or she entered (elevators). The pictorial sign will indicate to the occupants not to use the elevators, but rather to use the stairs to exit the building.

The exception to this requirement recognizes that elevators can be an integral part of the means of egress required for persons with physical disabilities. In such cases, the sign required by this section would not be appropriate since it would be communicating a contradictory message. Section 1007.4 dictates specific protection methods to permit the use of elevators as an element of an accessible means of egress. See Figure 3002.3 for an example of an emergency pictorial sign.

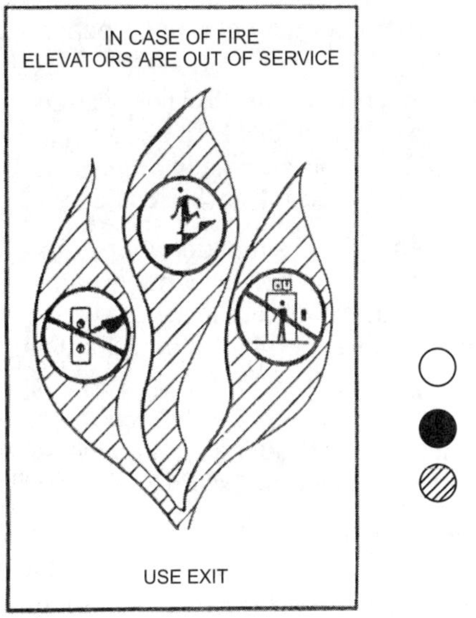

Figure 3002.3
ELEVATOR CORRIDOR CALL STATION PICTOGRAPH

3002.4 Elevator car to accommodate ambulance stretcher. In buildings four stories in height or more, at least one elevator shall be provided for fire department emergency access to all floors. Such elevator car shall be of such a size and arrangement to accommodate a 24-inch by 76-inch (610 mm by 1930 mm) ambulance stretcher in the horizontal, open position and shall be identified by the international symbol for emergency medical services (star of life). The symbol shall not be less than 3 inches (76 mm) high and shall be placed inside on both sides of the hoistway door frame.

❖ To aid fire-fighting rescue efforts in tall buildings (over three stories above grade), one of the provided elevators must be sized to accommodate a typical ambulance stretcher and emergency personnel. Additionally, to aid the emergency personnel in identifying which elevator is sized to accommodate a stretcher, the international symbol (star of life) is required to be located on each side of the hoistway door frame.

3002.5 Emergency doors. Where an elevator is installed in a single blind hoistway or on the outside of a building, there shall be installed in the blind portion of the hoistway or blank face of the building, an emergency door in accordance with ASME A17.1.

❖ To allow emergency personnel reasonable access to a blind elevator hoistway, emergency doors are required to be provided. A blind hoistway is a hoistway without openings. For example, an express elevator that opens at the lobby and serves only floors 10 through 20 is a blind hoistway from floors 2 through 9. Therefore, emergency doors are required in accordance with ASME A17.1, which contains requirements for opening sizes, opening access, door operation and door location.

3002.6 Prohibited doors. Doors, other than hoistway doors and the elevator car door, shall be prohibited at the point of access to an elevator car unless such doors are readily openable from the car side without a key, tool, special knowledge or effort.

❖ This section addresses the situation where there is a door installed in front of the hoistway door, perhaps for security purposes. If that door was locked, a building occupant could become trapped in the area between the set of doors.

3002.7 Common enclosure with stairway. Elevators shall not be in a common shaft enclosure with a stairway.

❖ To avoid the potential of a fire hazard spreading from the elevator or elevator hoistway and affecting an exit stair, separate protective enclosures are required for each.

[F] SECTION 3003 EMERGENCY OPERATIONS

3003.1 Standby power. In buildings and structures where standby power is required or furnished to operate an elevator, the operation shall be in accordance with Sections 3003.1.1 through 3003.1.4.

❖ Elevators are the most common and convenient means of providing access in multistory buildings. As such, elevators represent a prime candidate for accessible means of egress from such buildings, especially in light of the difficulties involved in carrying a person in a wheelchair up or down a stairway. For this reason, Chapter 10 requires elevators, which are used to provide an accessible means of egress, to be provided with standby power in accordance with this section. The primary consideration for elevators as an accessible means of egress is that the elevator most likely will be available and protected during a fire event. This section addresses the standby power and operation requirements for such an elevator. The emergency power requirement establishes a higher degree of reliability that the elevator will be available and usable by reducing the likelihood of power loss caused by fire or other conditions.

3003.1.1 Manual transfer. Standby power shall be manually transferable to all elevators in each bank.

❖ To allow emergency personnel to provide standby power to any elevator(s) at any time, this section requires a manual means of transferring standby power to all elevators in each bank. This manual transfer system could assist emergency personnel in evacuating building occupants.

3003.1.2 One elevator. Where only one elevator is installed, the elevator shall automatically transfer to standby power within 60 seconds after failure of normal power.

❖ In the case where a building is equipped only with one elevator, a standby power system must automatically restore full operation of the elevator within 60 seconds after failure of the primary power supply.

3003.1.3 Two or more elevators. Where two or more elevators are controlled by a common operating system, all elevators shall automatically transfer to standby power within 60 seconds after failure of normal power where the standby power source is of sufficient capacity to operate all elevators at the same time. Where the standby power source is not of sufficient capacity to operate all elevators at the same time, all elevators shall transfer to standby power in sequence, return to the designated landing and disconnect from the standby power source. After all elevators have been returned to the designated level, at least one elevator shall remain operable from the standby power source.

❖ In the case where a building is equipped with two or more elevators, this section provides allowable methods of standby power operation. When the provided standby power system is sized to accommodate all of the elevators, then all are required to automatically switch to standby power 60 seconds after failure of the primary power supply. There are provisions for a standby power system to be sized smaller and provide standby power to the elevators in sequence, rather than all at once. However, after all elevators have been returned to their designated landings by standby power, at least one must remain operable through standby power.

3003.1.4 Venting. Where standby power is connected to elevators, the machine room ventilation or air conditioning shall be connected to the standby power source.

❖ To allow elevator equipment in the machine room to operate normally, the venting system serving the machine room must be connected to the standby power system that serves the elevators.

3003.2 Fire-fighters' emergency operation. Elevators shall be provided with Phase I emergency recall operation and Phase II emergency in-car operation in accordance with ASME A17.1.

❖ A fire-fighter service is required for all elevators. There are two objectives of fire-fighter service required by ASME A17.1:

- Phase I emergency recall operation is the operation of an elevator wherein it is automatically or manually recalled to a specific landing and removed from normal service because of activation of the fire-fighter service; and

- Phase II emergency in-car operation is the operation of an elevator by fire fighters where such operation is under their control.

Elevators need to have recall capability to reduce the possibility of trapping passengers during a fire. Elevators need to have Phase II emergency in-car operation capability to aid fire fighters in the evacuation of persons with physical disabilities. An elevator needs to provide a useable means of egress for the disabled in a fire emergency.

SECTION 3004 HOISTWAY VENTING

3004.1 Vents required. Hoistways of elevators and dumbwaiters penetrating more than three stories shall be provided with a means for venting smoke and hot gases to the outer air in case of fire.

Exceptions:

1. In occupancies of other than Groups R-1, R-2, I-1, I-2 and similar occupancies with overnight sleeping quarters, venting of hoistways is not required where the building is equipped throughout with an approved automatic sprinkler system installed in accordance with Section 903.3.1.1 or 903.3.1.2.
2. Sidewalk elevator hoistways are not required to be vented.

❖ Ventilation of hoistways is required to prevent the accumulation and spread of hot smoke and gases from a fire to the upper stories of a building. The majority of deaths in fires are a result of smoke. For example, in the fire at the MGM Grand Hotel in Las Vegas, 70 of the 84 deaths occurred on the upper floors where smoke concentration was the greatest—even though the fire was at the first floor. This could be attributed to stack effect, which is a phenomenon that exists in high-rise buildings. Stack effect describes the movement of air inside and outside a building. During a fire, the presence of stack effect generally results in the movement of smoke and combustion products from lower levels to upper levels through shafts in the building. Stack effect can be reversed in an air-conditioned building when the outside temperature is relatively high (see Figure 3004.1).

Under the conditions listed in Exception 1, sprinklers are permitted as an alternative to the vents in all buildings except those of the groups noted. The top of sidewalk elevator hoistways is open to the outside; thus, no venting is required as indicated in Exception 2.

3004.2 Location of vents. Vents shall be located below the floor or floors at the top of the hoistway, and shall open either directly to the outer air or through noncombustible ducts to the outer air. Noncombustible ducts shall be permitted to pass through the elevator machine room provided that portions of the ducts located outside the hoistway or machine room are enclosed by construction having not less than the fire protection rating required for the hoistway. Holes in the machine room floors for the passage of ropes, cables or other moving elevator equipment shall be limited so as not to provide greater than 2 inches (51 mm) of clearance on all sides.

❖ The vent location is intended to aid in the exhaust of smoke to the exterior of the building, and to limit migration of smoke into the machine room. At one time, hoistways were vented into the machine room, a grate in the machine room floor vented to the exterior of the building. Allowing smoke and heat into the machine room may cause equipment malfunction. Therefore, vents must be located in the hoistway directly below the ceiling, and if ventilation through the machine room is advantageous by design, it must be provided through the use of noncombustible ductwork. Further, when noncombustible ductwork is utilized for the ventilation system, any portion of the ductwork that is outside of the hoistway or the machine room must be enclosed by fire-resistance-rated

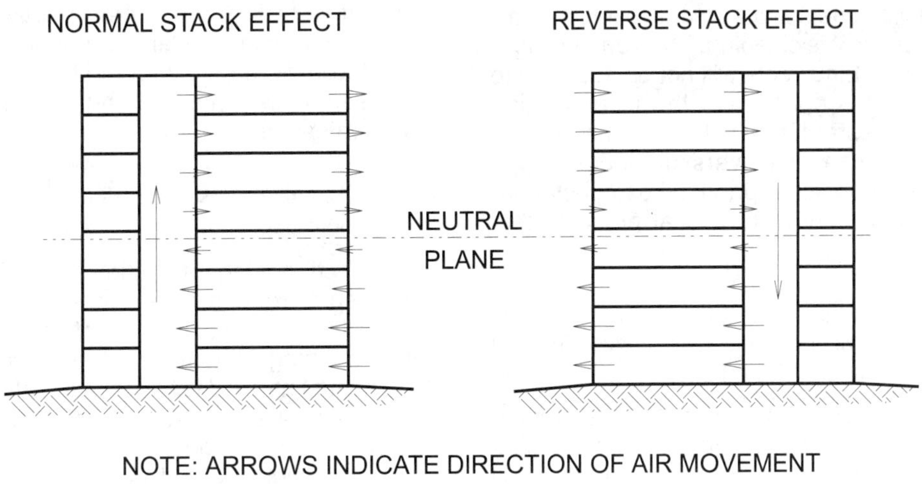

Figure 3004.1
STACK EFFECT

construction equivalent to that required for the hoistway. The clearance around elevator ropes and other moving equipment is limited in size in order to limit the passage of smoke into the machine room.

3004.3 Area of vents. Except as provided for in Section 3004.3.1, the area of the vents shall not be less than $3^1/_2$ percent of the area of the hoistway nor less than 3 square feet (0.28 m^2) for each elevator car, and not less than $3^1/_2$ percent nor less than 0.5 square foot (0.047 m^2) for each dumbwaiter car in the hoistway, whichever is greater. Of the total required vent area, not less than one-third shall be of the permanently open type unless all vents activate upon detection of smoke from any of the elevator lobby smoke detectors.

❖ Vent sizes are calculated such that there will not be a significant accumulation of smoke at the top of the hoistway. The vents may all be closed, provided that they open automatically upon activation of any of the elevator lobby smoke detectors. This will provide a degree of energy conservation while maintaining the ability to exhaust smoke.

3004.3.1 Reduced vent area. Where mechanical ventilation conforming to the *International Mechanical Code* is provided, a reduction in the required vent area is allowed provided that all of the following conditions are met:

1. The occupancy is not in Group R-1, R-2, I-1 or I-2 or of a similar occupancy with overnight sleeping quarters.
2. The vents required by Section 3004.2 do not have outside exposure.
3. The hoistway does not extend to the top of the building.
4. The hoistway and machine room exhaust fan is automatically reactivated by thermostatic means.
5. Equivalent venting of the hoistway is accomplished.

❖ If a mechanical means of ventilation is provided to the hoistway, then a reduction in the vent area is permitted. This option is not applicable to buildings with sleeping quarters, because mechanical systems are subject to malfunctioning, which could allow smoke to infiltrate areas in which occupants are sleeping. The vent arrangement is to be such that the vents will not be exposed to the elements, and thus not subject to malfunctioning due to weather exposure. Usually, this can be accomplished through the use of louvers or hoods. Engineering calculations should be provided as substantiation of venting that is equivalent to that prescribed in Section 3004.3.

3004.4 Closed vents. Closed portions of the required vent area shall consist of windows or duct openings glazed with annealed glass not more than 0.125 inch (3.2 mm) thick.

❖ Where the vents are permitted to be closed by Section 3004.3, the closure is to be by annealed glass. Annealed glass not more than $^1/_8$ inch (3.2 mm) thick will break from heat or when subject to water pressure from a fire hose, thus allowing the hoistway to vent as intended.

3004.5 Plumbing and mechanical systems. Plumbing and mechanical systems shall not be located in an elevator shaft.

> **Exception:** Floor drains, sumps and sump pumps shall be permitted at the base of the shaft provided they are indirectly connected to the plumbing system.

❖ If a pipe or duct that conveys gases, vapors or liquids, and passes through a hoistway were to fail, operation of the equipment could be affected in such a manner as to make it unsafe. Water could cause the malfunctioning of hoistway door interlocks, thereby allowing an elevator to leave a floor with the hoistway door open. A wet brake would also cause an extremely hazardous condition. The exception allows for a floor drain, sump or sump pump to be installed at the base of the shaft. However, the discharge is required to indirectly connect to the plumbing system, and not be connected directly to any drain. The intent of this requirement is to eliminate the possibility of sewer gases entering into the hoistway, which could cause an explosion hazard.

SECTION 3005 CONVEYING SYSTEMS

3005.1 General. Escalators, moving walks, conveyors, personnel hoists and material hoists shall comply with the provisions of this section.

❖ Conveying systems include escalators, moving walks, conveyors and personnel and material hoists. Section 3005 contains requirements specific to these types of conveying systems.

3005.2 Escalators and moving walks. Escalators and moving walks shall be constructed of approved noncombustible and fire-retardant materials. This requirement shall not apply to electrical equipment, wiring, wheels, handrails and the use of $^1/_{28}$-inch (0.9 mm) wood veneers on balustrades backed up with noncombustible materials.

❖ The intent of this requirement is to limit the fire fuel loading. The presence of large amounts of combustible material would create a large fuel load, and result in an increased migration of fire and smoke from floor to floor, or from space to space.

3005.2.1 Enclosure. Escalator floor openings shall be enclosed except where Exception 2 of Section 707.2 is satisfied.

❖ This section refers to Section 707.2, Exception 2, for the enclosure requirements of escalators. Exception 2 gives two methods of protection that would allow the floor opening without a shaft enclosure.

The first method is to provide draft curtains and sprinklers. The area of the floor opening is also limited to twice the area of the escalator. This is necessary since the protection method is not adequate for every size opening, especially when the floor area within the perimeter of the opening is used for stock storage, displays, etc. (the atrium provisions of Section 404 are more appropriate for those use conditions). The use of this method is acceptable for connecting an unlimited number of stories in buildings and occupancies of Groups B and M. In buildings other than Groups B and M, the alternative can only be utilized to connect four stories. This provision acknowledges the current uses of this concept but also limits its use in buildings with higher fuel load and more hazardous occupancy characteristics, such as sleeping occupancies (Groups I and R).

The sprinkler draft curtain method is detailed in NFPA 13. The method consists of surrounding the escalator opening, in an otherwise fully sprinklered building, with an 18-inch-deep (457 mm) draft curtain located on the underside of the floor to which the escalator or moving walk ascends. This serves to delay the heat, smoke and combustion gases (developed in the early stages of a fire) on that floor from entering into the escalator or moving walk well. A row of closely spaced automatic sprinklers located outside of the draftstop also surrounds the opening. When activated by heat, the sprinklers provide a water curtain. A typical installation is shown in Figure 3005.2.1(1). In combination with the sprinkler system in the building, this system delays fire spread to allow time for evacuation.

The second method is to provide power-operated automatic fire shutters at all floor openings. The rolling shutter completely closes off the opening between floors in the event of a fire. When the shutter is actuated, power to the escalator or moving walk must be automatically disconnected. To provide protection against entrapment of an occupant, the shutter is required to have a sensitive leading edge, such that it stops closing when it comes in contact with an obstacle. Once the obstacle is removed, the shutter continues to close.

3005.2.2 Escalators. Where provided in below-grade transportation stations, escalators shall have a clear width of 32 inches (815 mm) minimum.

> **Exception:** The clear width is not required in existing facilities undergoing alterations.

❖ A minimum 32-inch (864 mm) width for escalators servicing below-grade transportation systems, such as subways, is based on minimum accessibility requirements.

3005.3 Conveyors. Conveyors and conveying systems shall comply with ASME B20.1.

❖ ASME B20.1 applies to the design, construction, installation, maintenance, inspection and operation of conveyors and conveying systems in relation to hazards. The conveyors may be of bulk material, package or unit-handling types where the installation is designed for permanent, temporary or portable operation.This standard must apply, along with the requirements noted below, to all conveyor installations.

This standard specifically excludes any conveyor designed for, installed for or used primarily for the movement of human beings. It does, however, apply to certain conveying devices that have incorporated platforms or control rooms specifically designed for authorized operating personnel within their supporting structure.

This standard does not apply to equipment such as industrial trucks, tractors and trailers, tiering machines (except pallet load tierers), cranes, hoists, power shovels, power scoops, bucket drag lines, trenchers, elevators, manlifts, moving walks, escalators, highway or rail vehicles, cableways, tramways or pneumatic conveyors.

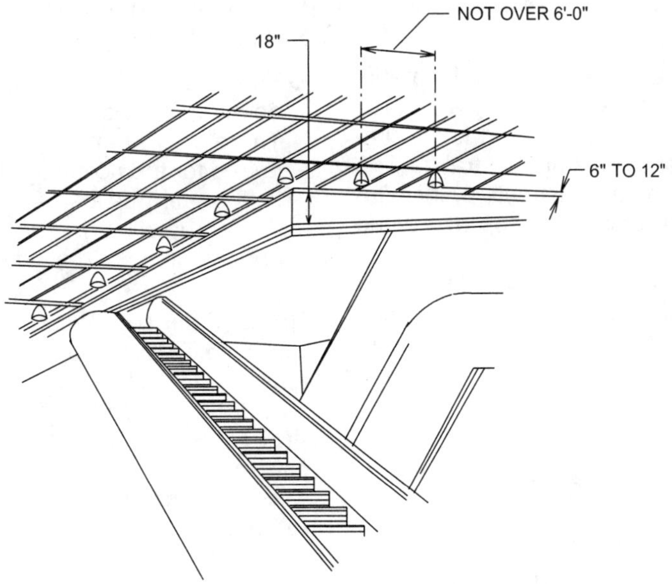

For SI: 1 inch = 25.4 mm, 1 foot = 304.8 mm.

Figure 3005.2.1(1)
DRAFT AND SPRINKLER WATER CURTAIN

3005.3.1 Enclosure. Conveyors and related equipment connecting successive floors or levels shall be enclosed with fire barrier walls and approved opening protectives complying with the requirements of Section 3002 and Chapter 7.

❖ Protection against the migration of smoke and fire must be maintained whenever a conveyor penetrates a floor/ceiling assembly, and thus connects floor levels. This section references Section 3002 and Chapter 7 for the requirements for fire barrier walls and opening protectives.

3005.3.2 Conveyor safeties. Power-operated conveyors, belts and other material-moving devices shall be equipped with automatic limit switches which will shut off the power in an emergency and automatically stop all operation of the device.

❖ Conveyor controls must be arranged such that in an emergency, the conveyor will stop all operation. All safety devices, including wiring of electrical safety devices, are required to be arranged to operate in a "fail-safe" manner; that is, if power failure or failure of the device itself occurs, a hazardous condition must not result.

3005.4 Personnel and material hoists. Personnel and material hoists shall be designed utilizing an approved method that accounts for the conditions imposed during the intended operation of the hoist device. The design shall include, but is not limited to, anticipated loads, structural stability, impact, vibration, stresses and seismic restraint. The design shall account for the construction, installation, operation and inspection of the hoist tower, car, machinery and control equipment, guide members and hoisting mechanism. Additionally, the design of personnel hoists shall include provisions for field testing and maintenance which will demonstrate that the hoist device functions in accordance with the design. Field tests shall be conducted upon the completion of an installation or following a major alteration of a personnel hoist.

❖ In lieu of a referenced standard, personnel and material hoists are to be designed and installed using a method that includes structural stability, anticipated loads, impact vibration, stresses and seismic restraint. The building official should examine construction, installation, operation of the hoist tower, etc., for approval of the design.

Personnel hoist design must also include field tests upon completion of installation or major alteration and maintenance, which demonstrate that hoist device functions are in accordance with the design.

SECTION 3006
MACHINE ROOMS

3006.1 Access. An approved means of access shall be provided to elevator machine rooms and overhead machinery spaces.

❖ A means of access to the elevator machine room is required by this section, and must be specifically approved. Things to consider in providing for this access are the maintenance and replacement of the equipment therein, and the necessary inspections to be performed.

3006.2 Venting. Elevator machine rooms that contain solid-state equipment for elevator operation shall be provided with an independent ventilation or air-conditioning system to protect against the overheating of the electrical equipment. The system shall be capable of maintaining temperatures within the range established for the elevator equipment.

❖ Elevator machine rooms may contain equipment that is temperature sensitive. If this equipment is not kept within its safe-temperature range, the possibility exists for a malfunction, which could trap an occupant or occupants within the elevator. Additionally, if this malfunction occurred during an emergency condition, it could impede emergency personnel in their efforts to perform fire-fighting or rescue duties.

3006.3 Pressurization. The elevator machine room serving a pressurized elevator hoistway shall be pressurized upon activation of a heat or smoke detector located in the elevator machine room.

❖ To prevent the migration of smoke and heat into a machine room from a pressurized hoistway enclosure, the machine room is required to be pressurized as well. However, the machine room only needs to be pressurized when heat or smoke is present. The pressurization is required to occur upon activation of a smoke or heat detector located in the machine room.

3006.4 Machine rooms and machinery spaces. Elevator machine rooms and machinery spaces shall be enclosed with construction having a fire-resistance rating not less than the required rating of the hoistway enclosure served by the machinery. Openings shall be protected with assemblies having a fire-resistance rating not less than that required for the hoistway enclosure doors.

❖ This section acknowledges that there will inherently be openings between the hoistway and the machine room. Therefore, to maintain the fire-resistance rating required for the hoistway, the machine room must be protected with fire barrier walls and opening protectives. This protection is required to be both vertical and horizontal.

3006.5 Shunt trip. Where elevator hoistways or elevator machine rooms containing elevator control equipment are protected with automatic sprinklers, a means installed in accordance with NFPA 72, Section 3-8.15, Elevator Shutdown, shall be provided to disconnect automatically the main line power supply to the affected elevator prior to the application of water. This means shall not be self-resetting. The activation of sprinklers outside the hoistway or machine room shall not disconnect the main line power supply.

❖ This section acknowledges the hazards of a sprinkler system, mainly electrical malfunction, to an operating elevator and elevator equipment contained in the machine room. Therefore, the main power supply line to the affected elevator(s) is required to automatically disconnect, and not be reset, before the suppression system is activated.

3006.6 Plumbing systems. Plumbing systems shall not be located in elevator equipment rooms.

❖ If a pipe or duct that conveys gases, vapors or liquids and passes through a machine room or machinery space were to fail, the operation of the equipment could be affected in such a manner as to make it unsafe. Water could cause the malfunctioning of hoistway door interlocks, thereby allowing an elevator to leave a floor with the hoistway door open.

Bibliography

The following resource materials are referenced in this chapter or are relevant to the subject matter addressed in this chapter.

ALI ALCTV-98, *Standard for Automobile Lifts—Safety Requirements for the Construction, Testing and Validation.* Indialantic, FL: Automotive Lift Institute, 1998.

ASME A17.1-00, *Safety Code for Elevators and Escalators—with A17.1a-97 and A17.1b-98 Addenda.* New York: American Society of Mechanical Engineers,2000.

ASME A90.1-97, *Safety Standard for Belt Manlifts*. New York: American Society of Mechanical Engineers, 1997.

ASME B20.1-00, *Safety Standard for Conveyors and Related Equipment*. New York: American Society of Mechanical Engineers,2000.

FEMA FIA-TB #4, *Elevator Installation for Buildings Located in Special Flood Hazard Areas.*

ICC A117.1-98, *Accessible and Usable Buildings and Facilities*. Falls Church, VA: International Code Council, 1998.

NFIP Technical Bulletin Series. Washington, DC: National Flood Insurance Program. [Online]. Available: *http://www.fema.gov/MIT/techbul.htm*

NFPA 13-99, *Installation of Sprinkler Systems*. Quincy, MA: National Fire Protection Association, 1999.

NFPA 72-99, *National Fire Alarm Code*. Quincy, MA: National Fire Protection Association, 1999.

Chapter 31: Special Construction

General Comments

Chapter 31 contains provisions that govern the construction and protection of structures with unique characteristics not addressed by other sections of the code and that require special consideration. This chapter includes requirements for outdoor signs; membrane structures; temporary structures; pedestrian walkways; radio and television towers; radio and television antennas; canopies and awnings; marquees and swimming pool enclosures.

Section 3101 addresses the scope of the chapter.

Section 3102 regulates membrane structures.

Section 3103 regulates structures that are erected for a period of less than 180 days.

Section 3104 regulates pedestrian walkways between buildings.

Section 3105 regulates the construction of awnings and canopies.

Section 3106 regulates the construction of marquees.

Section 3107 addresses the design, construction and maintenance of projecting signs.

Section 3108 regulates the construction of towers and antennas that are used for radio and television transmission.

Section 3109 regulates swimming pool enclosures.

The provisions of Chapter 31 are to be considered as additional requirements to those contained elsewhere in the code. These provisions may alter the general requirements found in other locations, but are only applicable to the specific element regulated therein. For example, the provisions for temporary construction regulate how the structure is to be built, but do not address means of egress. As such, the provisions of Chapter 10 are applicable.

The regulations contained in this chapter for outdoor signs were prompted by improper design, erection, placement, location and maintenance. Improperly designed and installed signs can pose a hazard to people and property. Improperly wired electrical signs that have not been maintained can cause fires.

The criteria in this chapter are straightforward and, in turn, lend themselves to straightforward enforcement.

Purpose

The purpose of Chapter 31 is to locate in one place the provisions that regulate construction or protection requirements for structures having unique characteristics, such as structures that are unusually tall and not used for human occupancy, structures occupied for short periods of time or recreational structures (swimming pools).

This chapter also provides the owner, designer and installer of outdoor signs with the minimum safe practices presented in performance language whenever possible, thus allowing the widest possible application.

SECTION 3101 GENERAL

3101.1 Scope. The provisions of this chapter shall govern special building construction including membrane structures, temporary structures, pedestrian walkways and tunnels, awnings and canopies, marquees, signs, and towers and antennas.

❖ This section makes it clear that the requirements of this chapter are to be used in conjunction with the general requirements of the code. When a specific provision is contained in Chapter 31, it will take precedence over the general requirement.

SECTION 3102 MEMBRANE STRUCTURES

3102.1 General. The provisions of this section shall apply to air-supported, air-inflated, membrane-covered cable and membrane-covered frame structures, collectively known as membrane structures, erected for a period of 180 days or longer. Those erected for a shorter period of time shall comply with the *International Fire Code*. Membrane structures covering water storage facilities, water clarifiers, water treatment plants, sewage treatment plants, greenhouses and similar facilities not used for human occupancy, are required to meet only the requirements of Sections 3102.3.1 and 3102.7.

❖ This section applies to membrane structures that are erected for a period of 180 days or longer. Some membrane structures are of the inflated type, where the air pressure is within a series of tubes, which then form the structure and keep it upright. In other structures, the interior atmosphere is pressurized and cables may be used to restrain the fabric membrane. Other types include cable-supported membrane structures in which the roof element is supported by cables and air pressure may or may not be provided. These are commonly used in sports stadiums. Membrane structures may also be framework-type structures in which the membrane is stretched over the frame to form the enclosing walls and the roof. Finally, membrane structures may consist of a membrane tensioned to provide weather protection without enclosing walls.

The history of regulating membrane structures dates back to 1944, when efforts were first made to control circus tent flammability after a circus fire in Hartford, Connecticut claimed 168 lives. Early regulations required that the tent fabric be flame resistant, that adequate exiting from both the grandstands and the tent be provided and that sufficient clear space around the tent be provided to control potential ignition sources and to permit rapid emergency exiting.

Large permanent membrane structures began to see significant use in the United States in the early 1960s. Prior to that time, the only membrane structures in common usage were air-inflated and air-supported covers for tennis courts, swimming pools, etc. Because of fire prevention code requirements at that time, such structures were often restricted to temporary use, which was defined as less than 90 days.

The reference to 180 days in Section 3102.1 is intended to establish membrane structures complying with this section as being outside the scope and provisions for temporary structures.

This section includes provisions for construction types, height and area limitations and structural support. Otherwise, the other applicable sections of the code apply.

So that a membrane structure will continue to provide its intended function, it must undergo continuous maintenance. Fire safety of such structures and the premises is regulated by the *International Fire Code®* (IFC®).

This section applies to all membrane structures, regardless of the supporting mechanisms or structure. Membrane structures erected for a period of less than 180 days are required to comply with the provisions of the IFC. Membrane structures that are not intended for human occupancy need only comply with the provisions of Section 3102.3.1 or 3102.7.

3102.2 Definitions. The following words and terms shall, for the purposes of this section and as used elsewhere in this code, have the meanings shown herein:

❖ Definitions of terms that are associated with the content of this section are contained herein. These definitions can help in the understanding and application of the code requirements. It is important to emphasize that these terms are not exclusively related to this section, but are applicable everywhere the term is used in the code. The purpose for including these definitions within this section is to provide more convenient access to them without having to refer back to Chapter 2.

For convenience, these terms are also listed in Chapter 2 with a cross reference to this section.

The use and application of all defined terms, including those defined herein, are set forth in Section 201.

AIR-INFLATED STRUCTURE. A building where the shape of the structure is maintained by air pressurization of cells or tubes to form a barrel vault over the usable area. Occupants of such a structure do not occupy the pressurized area used to support the structure.

❖ This type of membrane structure is characterized by multiple layers of membrane arranged such that small pockets or cells are formed. These pockets or cells are pressurized with air and form the structural capability of the membrane structure. Note that the occupants of the structure are not subjected to the pressurized areas, as they are in air-supported membrane structures.

AIR-SUPPORTED STRUCTURE. A building wherein the shape of the structure is attained by air pressure and occupants of the structure are within the elevated pressure area. Air-supported structures are of two basic types:

❖ An air-supported structure identifies those membrane structures that are completely pressurized for the purposes of supporting the membrane covering. Most "domed" sports arenas use air pressure within the structure to support the membrane covering. The membrane covering can consist of one layer or multiple layers; thus, air-supported structures are classified as either "single skin" or "double skin."

Double skin. Similar to a single skin, but with an attached liner that is separated from the outer skin and provides an airspace which serves for insulation, acoustic, aesthetic or similar purposes.

❖ A double-skin, air-supported structure contains multiple layers of membrane sheathing. The membranes are usually separated a distance to allow for pressurized air or other materials to be inserted between the plies. The pressurized air or other materials usually serve to increase the insulating and acoustical properties.

Single skin. Where there is only the single outer skin and the air pressure is directly against that skin.

❖ A single-skin, air-supported structure consists of just one membrane covering that is directly supported by the interior pressurized air. No other membranes are provided for insulating or acoustical purposes. If the membrane covering consists of several laminated plies, such an arrangement is still considered a single-skin, air-supported structure.

CABLE-RESTRAINED, AIR-SUPPORTED STRUCTURE. A structure in which the uplift is resisted by cables or webbings which are anchored to either foundations or dead men. Reinforcing cable or webbing is attached by various methods to the membrane or is an integral part of the membrane. This is not a cable-supported structure.

❖ This definition establishes a variation of the air-supported membrane structure. A single-skin or double-skin membrane is still pressurized by interior air, but does not have the strength or tearing resistance to support the air pressure load. A "fishnet" system of cable wires is placed over the exterior surface of the membrane or is integrally woven into the membrane material. The "fishnet" system is then tied to exterior walls of the building or other structural supports to resist the pressurized air loads.

MEMBRANE-COVERED CABLE STRUCTURE. A nonpressurized structure in which a mast and cable system provides support and tension to the membrane weather barrier and the membrane imparts stability to the structure.

❖ This definition identifies a structure in which the membrane fabric is draped over a cable framework. These structures do not involve any interior air pressurization. The cable framework system is supported by a mast or other tower system. The cables are usually pretensioned to provide structural support for the rest of the exterior structure.

MEMBRANE-COVERED FRAME STRUCTURE. A nonpressurized building wherein the structure is composed of a rigid framework to support a tensioned membrane which provides the weather barrier.

❖ This type of membrane structure is the same as the membrane-covered cable structure, except that a trussed framework constructed of structural shapes provides the support instead of pretensioned cables. Again, no air pressurization is required or provided.

NONCOMBUSTIBLE MEMBRANE STRUCTURE. A membrane structure in which the membrane and all component parts of the structure are noncombustible.

❖ Any membrane structure that is constructed of all noncombustible materials is classified by this definition. To qualify as noncombustible, the materials, including the membrane itself, must be tested in accordance with ASTM E 136 and must satisfy the criteria in Section 703.4.

3102.3 Type of construction. Noncombustible membrane structures shall be classified as Type IIB construction. Noncombustible frame or cable-supported structures covered by an approved membrane in accordance with Section 3102.3.1 shall be classified as Type IIB construction. Heavy timber frame-supported structures covered by an approved membrane in accordance with Section 3102.3.1 shall be classified as Type IV construction. Other membrane structures shall be classified as Type V construction.

Exception: Plastic less than 30 feet (9144 mm) above any floor used in greenhouses, where occupancy by the general public is not authorized, and for aquaculture pond covers, is not required to be flame resistant.

❖ Because of the limited fire-resistance capabilities of the membrane, membrane structures are usually classified as Type IIB, IV or V. These are defined as follows:

Type IIB—Membrane structures in which the membrane is noncombustible as defined by Section 703.4; or membrane-covered structures in which the frame or cable is noncombustible and the membrane is either noncombustible or flame resistant.

Type IV—Membrane-covered structures in which the frame is made from heavy timber and the membrane is noncombustible or flame resistant.

Type V—Membrane structures in which the membrane is flame resistant and combustible; or membrane-covered structures in which the frame or cable is combustible and the membrane is noncombustible or flame resistant.

In accordance with Section 602.2, a membrane-covered structure is to be considered as Type I or II construction, provided that the frame or cable complies with the provisions for Type I or II construction and the membrane structure is noncombustible, used exclusively as a roof and located more than 20 feet (6096 mm) above any floor, balcony or gallery floor level. This is similar to the provisions in Table 601 that do not require fire-resistance-rated roof construction under similar conditions.

The exception indicates that in recent years it is common practice to build greenhouses with higher ridge lines. Since this increase in height, there has been no record of related property loss.

3102.3.1 Membrane and interior liner material. Membranes and interior liners shall be either noncombustible as set forth in Section 703.4, or flame resistant as determined in accordance with NFPA 701 and the manufacturer's test protocol.

Exception: Plastic less than 20 mil (500 mm) in thickness used in greenhouses, where occupancy by the general public is not authorized, and for aquaculture pond covers, is not required to be flame resistant.

❖ The membrane is a thin, flexible, impervious material capable of being supported by an air pressure of $1^1/_2$ inches (38.1 mm) of water column. Membranes may be noncombustible. Except for limited applications of plastic greenhouses and pond covers, combustible membranes are required to be flame resistant.

Noncombustible membranes and liner materials must meet the criteria of Section 703.4, which references ASTM E 136.

Materials that are not noncombustible are required to be flame resistant. Flame resistance is determined by testing the membrane in accordance with NFPA 701. Although typically not applicable to membrane structures, it should be noted that recent tests have indicated a problem with combining two fabrics that individually passed the test procedure, but in combination would not be acceptable; therefore, if multiple layers are used, testing should demonstrate acceptable performance for the material combination to be utilized.

Both greenhouses that are not open to the public and aquaculture pond covers may be of plastic that is less than 20 mil (500 mm) in thickness and need not be flame resistant. This provision for greenhouses is based on the limited number of occupants.

3102.4 Allowable floor areas. The area of a membrane structure shall not exceed the limitations set forth in Table 503, except as provided in Section 506.

❖ The allowable area per floor of a membrane structure is determined using Sections 503 and 506, based on the construction types established in Section 3102.3.

3102.5 Maximum height. Membrane structures shall not exceed one story nor shall such structures exceed the height limitations in feet set forth in Table 503.

Exception: Noncombustible membrane structures serving as roofs only.

❖ Membrane structures are most commonly used to enclose large, open areas without intermediate structural supports, such as columns. For this reason, as well as the engineering difficulties associated with multistory buildings, membrane structures are limited to one story and the maximum height permitted in Table 503. The height is not limited, however, if a noncombustible membrane is used as a roof membrane only.

3102.6 Mixed construction. Membrane structures shall be permitted to be utilized as specified in this section as a portion of buildings of other types of construction. Height and area limits shall be as specified for the type of construction and occupancy of the building.

❖ The supporting structure of the membrane must be equal to or better than that of the building to which it is attached. Where air-supported or air-inflated structures are attached as an element of a building, their area limitations are based on the building to which they are attached. For example, an air-supported membrane building for large assembly without a working stage would be limited to 24,000 square feet (2230 m^2). When used as the roof of a Type AV building, such as a football stadium, the area is unlimited in accordance with the construction type of the building it covers.

3102.6.1 Noncombustible membrane. A noncombustible membrane shall be permitted for use as the roof or as a skylight of any building or atrium of a building of any type of construction provided it is at least 20 feet (6096 mm) above any floor, balcony or gallery.

❖ See Figure 3102.6.1 for an example of the application of this section.

3102.6.1.1 Flame-resistant membrane. A flame-resistant membrane shall be permitted to be used as the roof or as a skylight on buildings of Type IIB, III, IV and V construction provided it is at least 20 feet (6096 mm) above any floor, balcony or gallery.

❖ See Figure 3102.6.1 for an example of the application of this section.

3102.7 Engineering design. The structure shall be designed and constructed to sustain dead loads; loads due to tension or in-

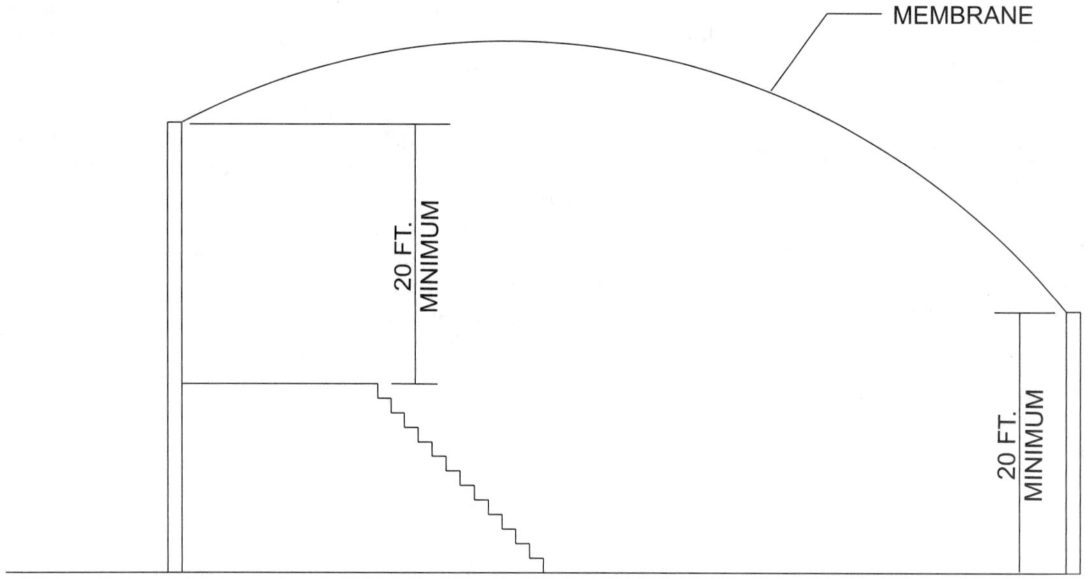

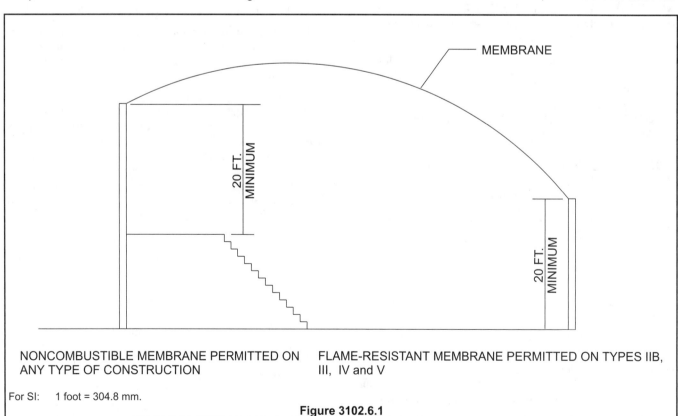

NONCOMBUSTIBLE MEMBRANE PERMITTED ON ANY TYPE OF CONSTRUCTION

FLAME-RESISTANT MEMBRANE PERMITTED ON TYPES IIB, III, IV and V

For SI: 1 foot = 304.8 mm.

Figure 3102.6.1
NONCOMBUSTIBLE AND FLAME-RESISTANT MEMBRANE CONSTRUCTION

flation; live loads including wind, snow or flood and seismic loads and in accordance with Chapter 16.

❖ The structure must be designed and constructed to sustain all dead loads, loads caused by tensioning or inflation and live loads, including wind, snow, flood and seismic. Chapter 16 provides criteria for design loads, except those associated with tensioning or inflation.

3102.8 Inflation systems. Air-supported and air-inflated structures shall be provided with primary and auxiliary inflation systems to meet the minimum requirements of Sections 3102.8.1 through 3102.8.3.

❖ While the primary inflation system is essential to the integrity of air-supported and air-inflated buildings, a secondary or auxiliary inflation system is required for all such membrane structures to prevent deflation and to allow the occupants to egress. A back-up structural support system is also required for air-supported and air-inflated structures with an occupant load of more than 50 or that cover a swimming pool because of life safety considerations.

3102.8.1 Equipment requirements. This inflation system shall consist of one or more blowers and shall include provisions for automatic control to maintain the required inflation pressures. The system shall be so designed as to prevent overpressurization of the system.

❖ The structural integrity of air-supported and air-inflated structures depends on a reliable inflation system.

3102.8.1.1 Auxiliary inflation system. In addition to the primary inflation system, in buildings exceeding 1,500 square feet (140 m^2) in area, an auxiliary inflation system shall be provided with sufficient capacity to maintain the inflation of the structure in case of primary system failure. The auxiliary inflation system shall operate automatically when there is a loss of internal pressure and when the primary blower system becomes inoperative.

❖ As a safety precaution, such structures that exceed 1,500 square feet (139 m^2) must also be provided with an auxiliary inflation system. The auxiliary system is required to operate automatically upon loss of internal pressure or if the primary system becomes inoperative. In accordance with Section 3102.8.2, when an auxiliary system is required, standby power is also to be provided for at least 4 hours.

3102.8.1.2 Blower equipment. Blower equipment shall meet the following requirements:

1. Blowers shall be powered by continuous-rated motors at the maximum power required for any flow condition as required by the structural design.
2. Blowers shall be provided with inlet screens, belt guards and other protective devices as required by the building official to provide protection from injury.
3. Blowers shall be housed within a weather-protecting structure.
4. Blowers shall be equipped with backdraft check dampers to minimize air loss when inoperative.
5. Blower inlets shall be located to provide protection from air contamination. The location of inlets shall be approved.

❖ Safety features of the blower system for the purposes of reliability, prevention of overpressurization and personal safety include:

- Blowers with motors rated for continuous operation at maximum power;
- Inlet screens, belt guards and other protective devices;
- Weather-protected housing for blowers;
- Backdraft check dampers to minimize air loss and
- Inlets located to prevent air contamination.

The *International Mechanical Code®* (IMC®) requires makeup air inlets to be at least 10 feet (3048 mm) from any potential source of contamination.

3102.8.2 Standby power. Wherever an auxiliary inflation system is required, an approved standby power-generating system shall be provided. The system shall be equipped with a suitable means for automatically starting the generator set upon failure of the normal electrical service and for automatic transfer and operation of all of the required electrical functions at full power within 60 seconds of such service failure. Standby power shall be capable of operating independently for a minimum of 4 hours.

❖ Air-supported and air-inflated structures exceeding 1,500 square feet (139 m^2) are to be provided with standby power for the auxiliary inflation system. The standby power system must comply with the provisions of Section 2702 and be capable of operating for a minimum of 4 hours.

3102.8.3 Support provisions. A system capable of supporting the membrane in the event of deflation shall be provided for in air-supported and air-inflated structures having an occupant load of more than 50 or where covering a swimming pool regardless of occupant load. The support system shall be capable of maintaining membrane structures used as a roof for Type I construction not less than 20 feet (6096 mm) above floor or seating areas. The support system shall be capable of maintaining other membranes at least 7 feet (2134 mm) above the floor, seating area or surface of the water.

❖ To allow safe occupant egress and access for fire fighting, air-supported and air-inflated structures that have an occupant load of more than 50 or cover a swimming pool are required to have a standby support system. Regardless of the occupant load, swimming pools are included because of the additional time necessary to egress when in a pool and because of the hazard that would exist if a membrane collapsed on a pool in which a person was swimming.

The support system is to be capable of maintaining all membrane structures at least 7 feet (2134 mm) above the floor, seating area or surface water, whichever is highest, to permit upright egress travel by the occupants, as well as to aid occupants who may be unable to egress. When the membrane structure is used as a roof on a Type I or II building, the support system must be capable of maintaining the membrane structure at least 20 feet (6096 mm) above the floor or seating area so that the contribution of the roof structure to the fire is minimized.

SECTION 3103
TEMPORARY STRUCTURES

3103.1 General. The provisions of this section shall apply to structures erected for a period of less than 180 days. Tents and other membrane structures erected for a period of less than 180 days shall comply with the *International Fire Code.* Those erected for a longer period of time shall comply with applicable sections of this code.

Exception: Provisions of the *International Fire Code* shall apply to tents and membrane structures erected for a period of less than 180 days.

❖ The purpose of Section 3103 is to provide public safety requirements in relation to fire, storm, collapse and crowd behavior in temporary structures. Temporary use is defined as a period of less than 180 days. Section 1024 is applicable to the means of egress and the use of bleachers, grandstands and folding or telescopic seating within the temporary structure. Temporary structures may also include temporary wiring and other fire hazards, which are different from those expected in permanent structures. These items are to be identified in the construction documents.

Tents have been a major fire protection concern since the disastrous circus fire in Hartford, Connecticut on July 6, 1944. This fire, which demonstrated the need for flame-retardant materials, resulted in 168 fatalities.

Any structure that is erected for a period of 180 days or more must comply with all other applicable sections of the code. It should be noted that "temporary" refers to the structure being erected, not the use of the structure; therefore, if a structure is erected for 365 days a year, but is used on a seasonal basis for less than 180 days, this section does not apply. Such structures are considered permanent.

3103.1.1 Permit required. Temporary structures that cover an area in excess of 120 square feet (11.16 m^2), including connecting areas or spaces with a common means of egress or entrance which are used or intended to be used for the gathering together of 10 or more persons, shall not be erected, operated or maintained for any purpose without obtaining a permit from the building official.

❖ Permits are required for temporary structures that are greater than 120 square feet (11 m^2) and are used for the gathering of 10 or more people. A permit from the building official is not required when the temporary structure is used for recreational camping or by less than 10 persons.

3103.2 Construction documents. A permit application and construction documents shall be submitted for each installation of a temporary structure. The construction documents shall include a site plan indicating the location of the temporary structure and information delineating the means of egress and the occupant load.

❖ Before the installation of any temporary structure, a permit application and construction documents are required to be submitted.

Construction documents, as defined in Chapter 2, include written, graphic and pictorial documents. These documents are to clearly provide pertinent information about the project, including the location, seating capacity, construction and all mechanical and electrical equipment, as a minimum.

3103.3 Location. Temporary structures shall be located in accordance with the requirements of Table 602 based on the fire-resistance rating of the exterior walls for the proposed type of construction.

❖ With respect to exposure hazards, temporary structures are no different from permanent structures; therefore, the fire-resistance rating of the temporary structure's exterior walls will depend upon the occupancy and the fire separation distance.

3103.4 Means of egress. Temporary structures shall conform to the means of egress requirements of Chapter 10 and shall have a maximum exit access travel distance of 100 feet (30 480 mm).

❖ The means of egress in temporary structures is to comply with the requirements of Chapter 10, except the maximum exit access travel distance is limited to 100 feet (30 480 mm). The reduced travel distance applies to all temporary uses and is based on concerns associated with the construction weaknesses, which may accelerate fire spread or rapidly decrease the stability of the structure during a fire.

SECTION 3104
PEDESTRIAN WALKWAYS AND TUNNELS

3104.1 General. This section shall apply to connections between buildings such as pedestrian walkways or tunnels, located at, above or below grade level, that are used as a means of travel by persons. The pedestrian walkway shall not contribute to the building area or the number of stories or height of connected buildings.

❖ Pedestrian walkways are normally seen as bridges connecting adjacent buildings. This section addresses the practical application of pedestrian walkways and tunnels. This section deals with walkways between buildings, including grade-level connections between buildings or corridors around the perimeter of one building. The area of the walkway itself is not to be considered a part of any building it serves, nor is it considered to add to the number of stories or height of any building it serves, provided that the walkway and its construction comply with this section.

3104.2 Separate structures. Connected buildings shall be considered to be separate structures.

Exceptions:

1. Buildings on the same lot in accordance with Section 503.1.3.
2. For purposes of calculating the number of Type B units required by Chapter 11, structurally connected buildings and buildings with multiple wings shall be considered one structure.

❖ Exception 1 to this section states that each building to be served by a pedestrian walkway is to be considered as a separate building except where, in accordance with Section 503.1.3, the owner chooses to have separate buildings on the same lot considered as one building. Exception 1 also applies to a connected building that may be of unlimited area as addressed in Section 507. Exception 2 establishes that, for purposes of evaluation of a building complex for required number of Type B dwelling units, buildings connected by pedestrian walkways must be evaluated as a single building. This provision was installed to coordinate with *Fair Housing Accessibility Guidelines* (FHAG). The requirement is based upon recommendations in the *Accessibility Analysis of Model Codes, International Building Code Analysis, Report for Public Comment,* dated October 20, 1999 by the U.S. Department of Housing and Urban Development (HUD). Based upon International Code Council® (ICC®) compliance with these recommendations, HUD has determined that compliance with the 2003 edition of the code is "safe harbor" for responsible parties in the design of a building for accessibility (see commentary, Section 1101).

3104.3 Construction. The pedestrian walkway shall be of noncombustible construction.

Exception: Combustible construction shall be permitted where connected buildings are of combustible construction.

❖ The walkway must be of noncombustible construction, unless all of the buildings it connects are of combustible construction.

It is intended that the construction materials of the walkway be consistent with the type of construction classification of the buildings connected by the walkway. Also note that this section does not require structural members of the walkway to have a fire-resistance rating, even if the connected buildings are of a protected construction type.

3104.4 Contents. Only materials and decorations approved by the building official shall be located in the pedestrian walkway.

❖ This restriction on the use of the floor area is similar to the restriction on the use of the floor of an atrium, except there is no relaxation of requirements if an automatic sprinkler system is installed. The reference to low-hazard uses implies that the space created by the walkway is not to be used for any purpose that would incorporate significant amounts of combustible materials. In general, the same parameters that would apply to exit access corridors and atriums can be applied here.

3104.5 Fire barriers between pedestrian walkways and buildings. Walkways shall be separated from the interior of the building by fire barrier walls with a fire-resistance rating of not less than 2 hours. This protection shall extend vertically from a point 10 feet (3048 mm) above the walkway roof surface or the connected building roof line, whichever is lower, down to a point 10 feet (3048 mm) below the walkway and horizontally 10 feet (3048 mm) from each side of the pedestrian walkway. Openings within the 10-foot (3048 mm) horizontal extension of the protected walls beyond the walkway shall be equipped with devices providing a $^{3}/_{4}$-hour fire protection rating in accordance with Section 715.

Exception: The walls separating the pedestrian walkway from a connected building are not required to have a fire-resistance rating by this section where any of the following conditions exist:

1. The distance between the connected buildings is more than 10 feet (3048 mm), the pedestrian walkway and connected buildings are equipped throughout with an automatic sprinkler system in accordance with NFPA 13 and the wall is constructed of a tempered, wired or laminated glass wall and doors subject to the following:
   - 1.1. The glass shall be protected by an automatic sprinkler system in accordance with NFPA 13 and the sprinkler system shall completely wet the entire surface of interior sides of the glass wall when actuated.
   - 1.2. The glass shall be in a gasketed frame and installed in such a manner that the framing system will deflect without breaking (loading) the glass before the sprinkler operates.
   - 1.3. Obstructions shall not be installed between the sprinkler heads and the glass.
2. The distance between the connected buildings is more than 10 feet (3048 mm), and both sidewalls of the pedestrian walkway are at least 50 percent open with the open area uniformly distributed to prevent the accumulation of smoke and toxic gases.
3. Buildings are on the same lot, in accordance with Section 503.1.3.
4. Where exterior walls of connected buildings are required by Section 704 to have a fire-resistance rating greater than 2 hours, the walkway shall be equipped throughout with an automatic sprinkler system installed in accordance with NFPA 13.

The previous exceptions shall apply to pedestrian walkways having a maximum height above grade of three stories or 40 feet

(12 192 mm), or five stories or 55 feet (16 764 mm) where sprinklered.

❖ It is this requirement for fire-resistance-rated separation that allows buildings and walkways to be considered as separate buildings. The extent of this fire-resistance-rated separation is shown in Figure 3104.5. If the wall of the connected building would not normally extend 10 feet (3048 mm) above the walkway roof, there is no requirement in this section to extend it.

The portion of the exterior wall that separates the walkway from the building is required to have a fire-resistance rating unless one of the exceptions of this section is met.

This fire-resistance rating of the wall separating the walkway from a connected building is not required when one of the following is met:

1. If the distance between the connected buildings is more than 10 feet (3048 mm) and the walkway and all connected buildings are fully sprinklered, then any glass in the wall between the walkway and a connected building must be protected as described, even though the wall need not be rated.
2. The distance between the connected buildings is more than 10 feet (3048 mm) and both side walls of the pedestrian walkway are open for at least 50 percent of each wall area, with the open area uniformly distributed to prevent the accumulation of smoke and toxic gases. This concept is synonymous with that used for open parking structures where free movement of smoke is achieved through the openness of the structure.
3. The buildings are designed under the provisions of Section 503.1.3 for two buildings on one lot. This exception is based on the concept that the exterior walls are considered to be interior walls for fire-resistance purposes.
4. If the walkway is fully sprinklered and if more than a 2-hour fire-resistance rating is required by Section 704, this exception can be taken and no fire-resistance rating would be required at the end wall of the walkway.

These four exceptions apply to the walkway rather than the buildings that the walkway connects.

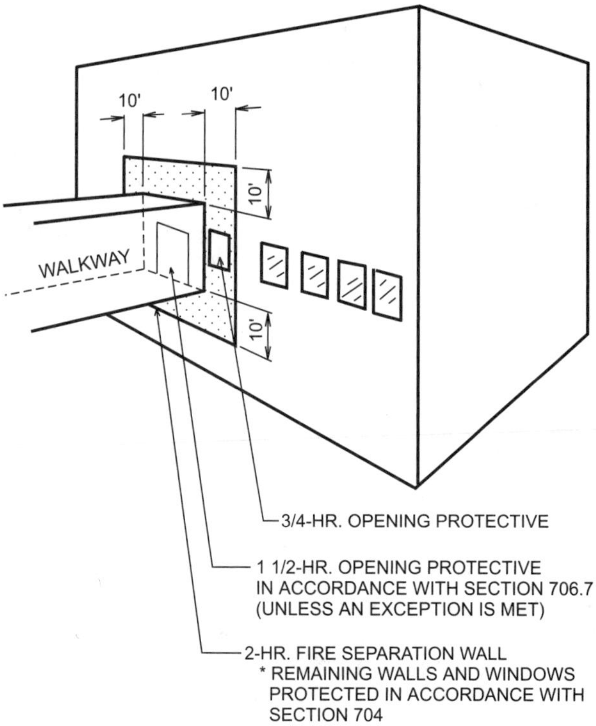

For SI: 1 foot = 304.8 mm.

Figure 3104.5
PEDESTRIAN WALKWAY-BUILDING INTERFACE

3104.6 Public way. Pedestrian walkways over a public way shall also comply with Chapter 32.

❖ The acceptance of a pedestrian walkway over a public street by the use of a bridge or any other means is required to comply with Chapter 32.

3104.7 Egress. Access shall be provided at all times to a pedestrian walkway that serves as a required exit.

❖ Pedestrian walkways may be used as exits. If they are, their doors must be equipped with hardware that will allow them to function as exits at all times.

3104.8 Width. The unobstructed width of pedestrian walkways shall not be less than 36 inches (914 mm). The total width shall not exceed 30 feet (9144 mm).

❖ A maximum width has been set to prevent entire floors from being extended to adjacent buildings, and thus from being called walkways. This section is intended to encompass bridge-type structures, not full-floor extensions.

The minimum width is also subject to other code requirements. For example, if the walkway provides a means of egress component, it would be sized in accordance with Section 1003.2.3.

3104.9 Exit access travel. The length of exit access travel shall not exceed 200 feet (60 960 mm).

Exceptions:

1. Exit access travel distance on a pedestrian walkway equipped throughout with an automatic sprinkler system in accordance with NFPA 13 shall not exceed 250 feet (76 200 mm).
2. Exit access travel distance on a pedestrian walkway constructed with both sides at least 50 percent open shall not exceed 300 feet (91 440 mm).

3. Exit access travel distance on a pedestrian walkway constructed with both sides at least 50 percent open, and equipped throughout with an automatic sprinkler system in accordance with NFPA 13, shall not exceed 400 feet (122 m).

❖ The maximum length of exit access travel distance within the pedestrian walkway has been set as a means of controlling the size. A 200-foot (60 960 mm) maximum length is consistent with the maximum length of exit access travel distance allowed in unsprinklered buildings for most occupancy classifications (see Table 1004.2.4). This section gives three means by which the exit access travel distance within the pedestrian walkway can be increased. Exception 1 allows the exit access travel distance to be increased to 250 feet (76 200 mm) if the walkway is equipped with an automatic sprinkler system. This increase in travel distance correlates with that allowed in Table 1004.2.4. Exception 2 allows the exit access travel distance to be increased to 300 feet (91 440 mm) if both sides of the walkway have openings to the exterior for at least 50 percent of the wall area, with the open area uniformly distributed to prevent the accumulation of smoke and toxic gases. This increase in travel distance correlates to that allowed in Table 1004.2.4 for an open parking structure (Group S-2). Exception 3 expands on Exception 2 by including the use of an automatic sprinkler system. The exit access travel distance is then increased to 400 feet (122 m^2). This increase in travel distance correlates to the allowances in Table 1004.2.4.

3104.10 Tunneled walkway. Separation between the tunneled walkway and the building to which it is connected shall not be less than 2-hour fire-resistant construction and openings therein shall be protected in accordance with Table 715.3.

❖ Tunnels present a greater hazard than other walkways. The concern is greater than for enclosed walkways because a tunnel's underground location makes it less accessible to fire-fighting efforts and it relies on a mechanical ventilation system. To prevent the spread of fire, 2-hour fire-resistant construction with $1^1/_2$-hour opening protectives are required at the tunnel and building interface (see Figure 3104.10).

3104.11 Ventilation. Smoke and heat vents shall be provided for enclosed walkways and tunneled walkways as required for Group F-1 occupancies in accordance with Section 910.

❖ Although NFPA 204 is written as a guide for smoke and heat venting of large undivided floor areas, the basic principles can be applied to an enclosed walkway or tunnel. By providing draft curtains to divide the ceiling and venting the areas defined by those draft curtains, the spread of smoke and heat can be checked. For example, in Figure 3104.10, Buildings A and B are connected by a tunnel. An uncontrolled fire in Building A is first arrested from spreading to the tunnel by the fire-re-

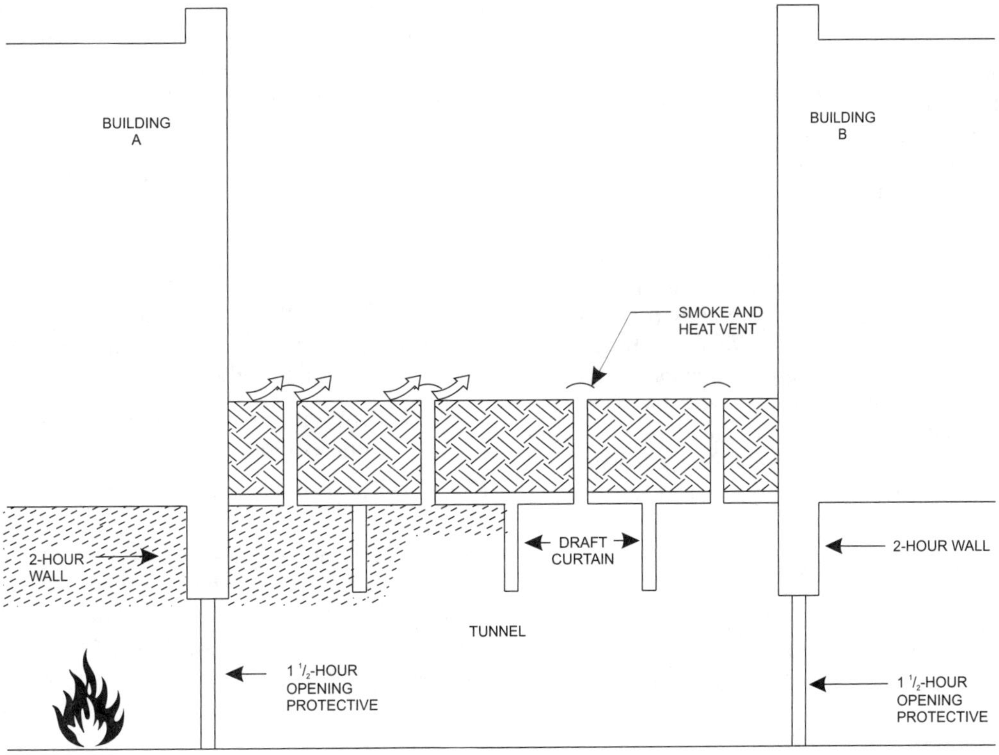

Figure 3104.10
TUNNELED WALKWAY

sistant wall with opening protectives. Eventually, a failure occurs and smoke and heat begin to enter the tunnel. As the nearest draft curtain area begins to fill with smoke and heat, a heat sensor or smoke detector actuates the vent, releasing the smoke and heat. If the vent cannot accommodate the volume, smoke will begin to spill over into the next draft curtain area, which will activate the next vent. There is still a fire-resistant barrier at the other end of the tunnel to protect Building B, so it is unlikely that the fire will spread from building to building through such an enclosed walkway or tunnel.

SECTION 3105 AWNINGS AND CANOPIES

3105.1 General. Awnings or canopies shall comply with the requirements of this section and other applicable sections of this code.

❖ This section addresses the variety of special structures that are defined as canopies and awnings. The intent of this section is that if a canopy or awning is used, it should be soundly designed so as not to present a hazard to its users or the public.

3105.2 Definition. The following term shall, for the purposes of this section and as used elsewhere in this code, have the meaning shown herein.

RETRACTABLE AWNING. A retractable awning is a cover with a frame that retracts against a building or other structure to which it is entirely supported.

❖ The key to this definition is that it provides for an element of building construction that is increasing in popularity, especially in residential construction. The need was to properly identify that this section of the code is applicable to these awnings.

3105.3 Design and construction. Awnings and canopies shall be designed and constructed to withstand wind or other lateral loads and live loads as required by Chapter 16 with due allowance for shape, open construction and similar features that relieve the pressures or loads. Structural members shall be protected to prevent deterioration. Awnings shall have frames of noncombustible material, fire-retardant-treated wood, wood of Type IV size, or 1-hour construction with combustible or noncombustible covers and shall be either fixed, retractable, folding or collapsible.

❖ Any of the structures addressed in Section 3105 are subject to wind and live load considerations under the provisions of Chapter 16. When Chapter 16 is applied, care should be taken to consider the shape of the awning or canopy in question with regard to snow buildup and wind effects.

Structural members of an awning or canopy should be of materials that will resist rust or decay. Chapter 23 should be referenced for wood members that are required to comply with the decay-resistance provision. Steel or aluminum structural members are to be protected to resist rust.

3105.4 Canopy materials. Canopies shall be constructed of a rigid framework with an approved covering, that is flame resistant in accordance with NFPA 701 or has a flame spread index not greater than 25 when tested in accordance with ASTM E 84.

❖ Canopies, as regulated by this section, are awnings that have ground support. This section specifically states the types of materials that are to be used and how a canopy is to be constructed. It limits the frame construction to metal and the coverings to only those that are deemed flame resistant when tested in accordance with NFPA 701, or those having a flame spread of 25 or less when tested in accordance with ASTM E 84.

SECTION 3106 MARQUEES

3106.1 General. Marquees shall comply with this section and other applicable sections of this code.

❖ The intent of this section is that if a marquee is used, it should be soundly designed so as not to present a hazard to its users or the public.

3106.2 Thickness. The maximum height or thickness of a marquee measured vertically from its lowest to its highest point shall not exceed 3 feet (914 mm) where the marquee projects more than two-thirds of the distance from the property line to the curb line, and shall not exceed 9 feet (2743 mm) where the marquee is less than two-thirds of the distance from the property line to the curb line.

❖ The restrictions placed on the size, projection and clearances (see Figure 3106.2) for marquees are intended to:

1. Prevent interference with the free movement of pedestrians.
2. Prevent interference with trucks and other tall vehicles using the public street.
3. Prevent interference with fire-fighting operations at a building.
4. Prevent interference with utilities.

3106.3 Roof construction. Where the roof or any part thereof is a skylight, the skylight shall comply with the requirements of Chapter 24. Every roof and skylight of a marquee shall be sloped to downspouts that shall conduct any drainage from the marquee in such a manner so as not to spill over the sidewalk.

❖ The purpose of this section is to require proper drainage to prevent buildup on the roof due to a marquee or skylight.

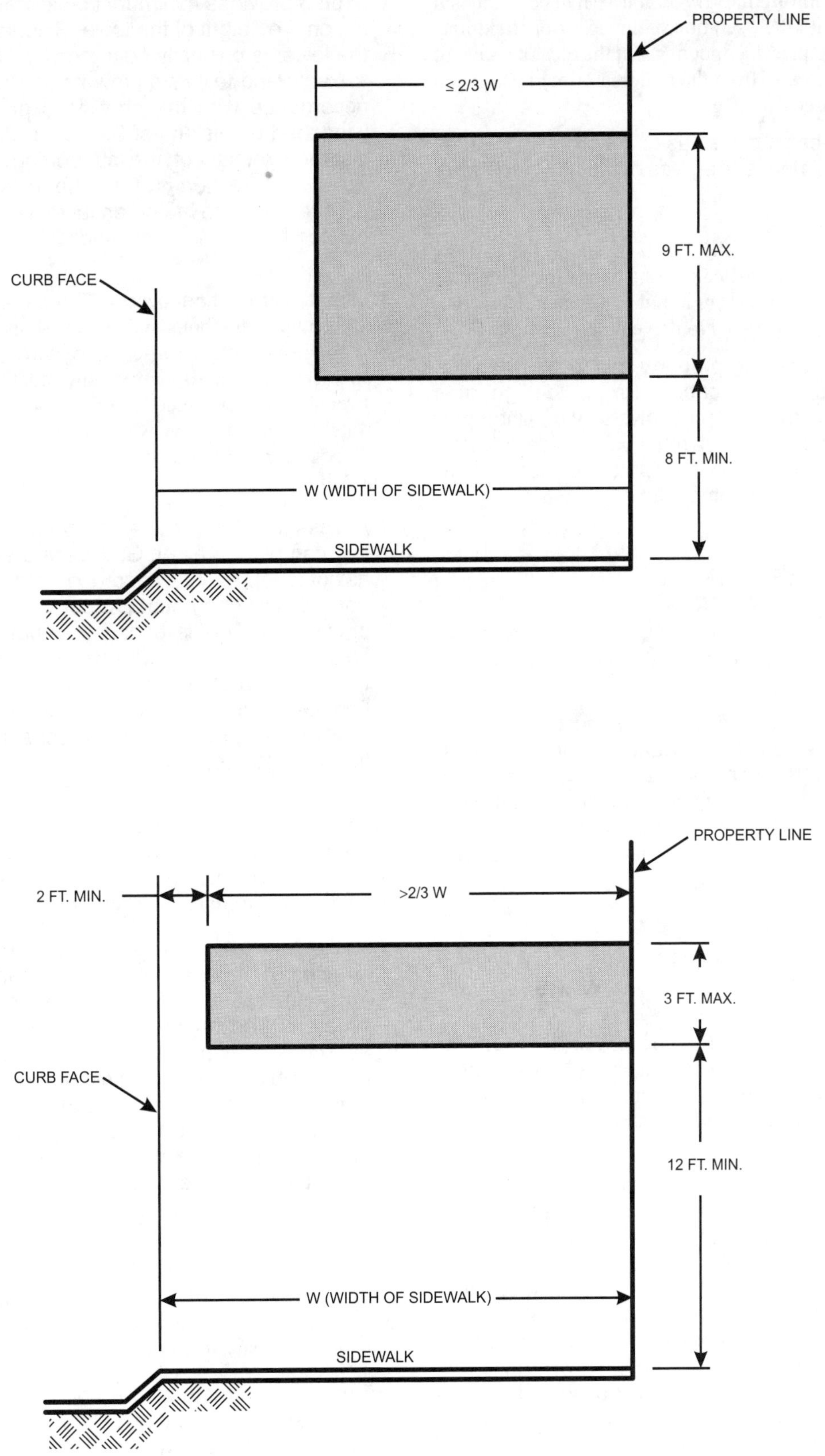

For SI: 1 foot = 304.8 mm.

Figure 3106.2
PERMISSIBLE PROJECTION, HEIGHT AND CLEARANCES FOR MARQUEES

3106.4 Location prohibited. Every marquee shall be so located as not to interfere with the operation of any exterior standpipe, and such that the marquee does not obstruct the clear passage of stairways or exit discharge from the building or the installation or maintenance of street lighting.

❖ A marquee is to be placed so as to not prohibit any type of fire protection, means of egress or maintenance work to the building or streetlights.

3106.5 Construction. A marquee shall be supported entirely from the building and constructed of noncombustible materials. Marquees shall be designed as required in Chapter 16. Structural members shall be protected to prevent deterioration.

❖ Damage to both life and property can occur from improperly constructed marquees. Proper design of a marquee involves a structural analysis by qualified persons. Wind and seismic loads must be considered in the design process, as outlined in Chapter 16. Additionally, all material must be noncombustible.

SECTION 3107 SIGNS

3107.1 General. Signs shall be designed, constructed and maintained in accordance with this code.

❖ The design of outdoor signs is to be in accordance with the code. If the design is inadequate, no amount of expert construction, repair or maintenance will result in a safe sign. As can be determined from the code, the enforcement of this section is confined to that which is displayed in any manner out of doors for recognized advertising purposes.

SECTION 3108 RADIO AND TELEVISION TOWERS

3108.1 General. Subject to the provisions of Chapter 16 and the requirements of Chapter 15 governing the fire-resistance ratings of buildings for the support of roof structures, radio and television towers shall be designed and constructed as herein provided.

❖ Radio and television towers are special structures in that they are not normally occupied and are open structures containing little, if any, fuel load. As such, this section provides specific provisions for radio and television towers, designed to maintain the structural integrity of the tower and to minimize its potential to present a hazard (such as serving as a potential fire source because of inadequate grounding).

Since the towers are not normally occupied and do not typically contain a fuel load, the structure is exempt from other provisions of the code, except for the wind load provisions of Section 3108.4 and the combustibility restrictions of the supporting structures as required by Section 1509 (see Section 3108.1). The height and area limitations of Chapter 5 do not apply; instead, Section 3108.3 provides minimum construction criteria depending on the height of the tower. The structural integrity of the tower is basically governed by Section 3108.4.

Radio and television towers are to be constructed in accordance with the provisions of Section 3108, the wind load provisions of Section 3108.4 and the fire-resistance ratings of the supporting structure, as governed by Section 1509. The provisions of Section 3108.4 minimize the potential for collapse of a tower as caused by forces associated with wind loads.

3108.2 Location and access. Towers shall be located and equipped with step bolts and ladders so as to provide ready access for inspection purposes. Guy wires or other accessories shall not cross or encroach upon any street or other public space, or over above-ground electric utility lines, or encroach upon any privately owned property without written consent of the owner of the encroached-upon property, space or above-ground electric utility lines.

❖ Access must be provided to the tower so that its condition can be surveyed. Guy wires are to be arranged so as not to represent a hazard to the public, and therefore, not cross or encroach a street, public space or electric power line. This is because of both the damage that could occur if a guy wire were suddenly released and the need to prohibit the wire from becoming an obstruction. Guy wires are not to encroach on any other property without previous written consent by the owner of the affected property.

3108.3 Construction. Towers shall be constructed of approved corrosion-resistant noncombustible material. The minimum type of construction of isolated radio towers not more than 100 feet (30 480 mm) in height shall be Type IIB.

❖ Radio and television towers are required to be constructed of material that can withstand exposure to the environment and, therefore, be corrosion resistant. Whereas Section 1509 addresses the minimum allowable construction of towers that are roof structures, Section 3108.3 provides minimum construction requirements for all towers. Most radio and television towers are constructed of noncombustible materials. Towers that are isolated and are neither roof structures nor more than 100 feet (30 480 mm) in height may be of Type IIB construction.

3108.4 Loads. Towers shall be designed to resist wind loads in accordance with TIA/EIA-222. Consideration shall be given to conditions involving wind load on ice-covered sections in localities subject to sustained freezing temperatures.

❖ To maintain structural integrity, radio and television towers must be designed to resist wind loads in accordance with EIA/TIA 222, which provides minimum load requirements for the design of buildings and other structures that are subject to code requirements. In areas subject to freezing conditions, consideration must be given to the effects of wind loads on ice-covered towers.

3108.4.1 Dead load. Towers shall be designed for the dead load plus the ice load in regions where ice formation occurs.

❖ To minimize the potential for collapse, a radio and television tower is to be designed to withstand the dead load of the structural elements of the tower, as well as the loads associated with a buildup of ice in areas subject to ice conditions.

3108.4.2 Wind load. Adequate foundations and anchorage shall be provided to resist two times the calculated wind load.

❖ To include a factor of safety, the foundation and anchorage must be designed and constructed to withstand twice the calculated wind uplift so as to minimize the potential for overturning.

3108.5 Grounding. Towers shall be permanently and effectively grounded.

❖ To minimize the potential for a fire resulting from lightning, radio and television towers are to be adequately grounded so that the electric charge is carried to the ground without affecting adjacent structures.

SECTION 3109 SWIMMING POOL ENCLOSURES AND SAFETY DEVICES

3109.1 General. Swimming pools shall comply with the requirements of this section and other applicable sections of this code.

❖ The provisions of this section apply to both public and private swimming pools. They are intended to provide for the use of swimming pools as well as to limit or delay unauthorized access to swimming pools by small children, particularly those five years old and younger. All but small pools not equipped with water-recirculating systems or those that involve structural materials must comply with the provisions of this section. Requirements are included for pool enclosures and barriers.

Small pools not equipped with water-recirculating systems or those that do not involve structural materials need not comply with the provisions of this section.

3109.2 Definition. The following word and term shall, for the purposes of this section and as used elsewhere in this code, have the meaning shown herein.

❖ The definition of terms that are associated with the content of this section is contained herein. This definition can help in the understanding and application of the code requirements. It is important to emphasize that this term is not exclusively related to this section but is applicable everywhere the term is used throughout the code. The purpose for including this definition within this chapter is to provide more convenient access to it without having to refer back to Chapter 2.

SWIMMING POOLS. Any structure intended for swimming, recreational bathing or wading that contains water over 24 inches (610 mm) deep. This includes in-ground, above-ground and on-ground pools; hot tubs; spas and fixed-in-place wading pools.

❖ The small above-ground pools frequently found in residential yards are exempt, provided that they do not exceed 2 feet (610 mm) in depth.

3109.3 Public swimming pools. Public swimming pools shall be completely enclosed by a fence at least 4 feet (1290 mm) in height or a screen enclosure. Openings in the fence shall not permit the passage of a 4-inch-diameter (102 mm) sphere. The fence or screen enclosure shall be equipped with self-closing and self-latching gates.

❖ Whether the pool is privately owned and operated is irrelevant to whether or not it is a public swimming pool. As long as the pool is not associated with an occupancy in Group R-3, it is considered a public swimming pool. Examples of public swimming pools include a pool in an apartment complex located in a common use area, a pool at a health club and a municipal swimming pool.

A public pool requires a barrier no lower than 4 feet (1219 mm). This barrier/wall limits or delays unauthorized access to the pool by small children, particularly those five years old or younger. It is important to note that this barrier is to enclose the swimming pool completely. Any construction or natural element that does not surround the pool will allow access at some point; hence, any gate, door or other access component must be self-closing and self-latching.

3109.4 Residential swimming pools. Residential swimming pools shall comply with Sections 3109.4.1 through 3109.4.3.

Exception: A swimming pool with a power safety cover or a spa with a safety cover complying with ASTM F 1346.

❖ Although the occupants and guests of a Group R-3 residence have access to the swimming pool regardless of its location, it is important to distinguish what elements separate the pool from the occupants. Pools that are totally within the structure are included in this section. Of critical concern is the easy access afforded to children by an indoor pool. For this reason, Sections 3109.4.1 through 3109.4.3 must be closely followed.

3109.4.1 Barrier height and clearances. The top of the barrier shall be at least 48 inches (1219 mm) above grade measured on the side of the barrier that faces away from the swimming pool. The maximum vertical clearance between grade and the bottom of the barrier shall be 2 inches (51 mm) measured on the side of the barrier that faces away from the swimming pool. Where the top of the pool structure is above grade, the barrier is authorized to be at ground level or mounted on top of the pool structure, the maximum vertical clearance between the top of the pool structure and the bottom of the barrier shall be 4 inches (102 mm).

❖ The barrier height requirement of 48 inches (1219 mm) above the ground is based on reports that documented the ability of children under the age of five to climb over barriers that are less than 48 inches (1219 mm) in height. The basis for the 4-inch (102 mm) criterion for an opening between the barrier and the top of pool frame is the same as for guardrail construction (see Figure 3109.4.1).

3109.4.1.1 Openings. Openings in the barrier shall not allow passage of a 4-inch-diameter (102 mm) sphere.

❖ The basis for the 4-inch (102 mm) criterion is the same as for guardrail construction (see Figure 3109.4.1.1). It is based on studies of the body measurements of children 13 to 18 months old.

3109.4.1.2 Solid barrier surfaces. Solid barriers which do not have openings shall not contain indentations or protrusions except for normal construction tolerances and tooled masonry joints.

❖ This provision is intended to reduce the potential for gaining a foothold and climbing the barrier.

3109.4.1.3 Closely spaced horizontal members. Where the barrier is composed of horizontal and vertical members and the distance between the tops of the horizontal members is less than 45 inches (1143 mm), the horizontal members shall be located on the swimming pool side of the fence. Spacing between vertical members shall not exceed 1.75 inches (44 mm) in width. Where there are decorative cutouts within vertical members, spacing within the cutouts shall not exceed 1.75 inches (44 mm) in width.

❖ This more stringent $1^3/_4$-inch (44 mm) provision for spacing between vertical members applies when the spacing between horizontal members is less than 45 inches (1143 mm). It acknowledges the potential for a child to gain both a handhold and a foothold on closely spaced horizontal members, and is intended to reduce that potential by limiting the space between the vertical members on the same barrier. If the horizontal members are spaced less than 45 inches (1143 mm), they must also be located on the swimming pool side of the fence (see Figure 3109.4.1.3) so that they are not available to be used to climb the barriers.

3109.4.1.4 Widely spaced horizontal members. Where the barrier is composed of horizontal and vertical members and the distance between the tops of the horizontal members is 45 inches (1143 mm) or more, spacing between vertical members shall not exceed 4 inches (102 mm). Where there are decorative cutouts within vertical members, spacing within the cutouts shall not exceed 1.75 inches (44 mm) in width.

❖ This requirement is the counterpart to Section 3109.4.1.3 in that it permits the opening in the barrier to be 4 inches (102 mm), provided the vertical spacing of the horizontal members equals or exceeds 45 inches (1143 mm) (see Figure 3104.10). It limits openings in the barrier to a 4-inch (102 mm) diameter. The spacing of horizontal members 45 inches (1143 mm) apart precludes them from being used by small children to climb the barrier.

3109.4.1.5 Chain link dimensions. Maximum mesh size for chain link fences shall be a 2.25 inch square (57 mm square) unless the fence is provided with slats fastened at the top or the bot-

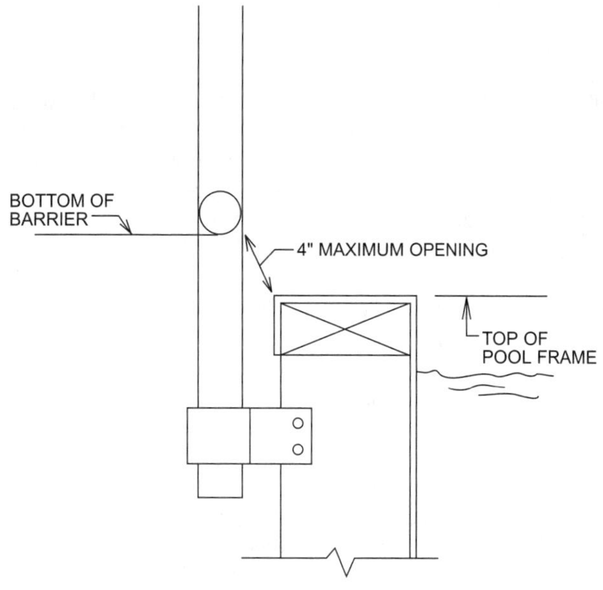

For SI: 1 inch = 25.4 mm.

Figure 3109.4.1
OPENING LIMITATIONS

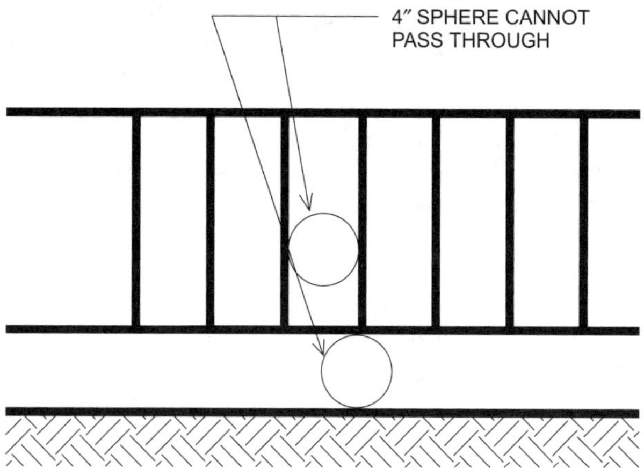

For SI: 1 inch = 25.4 mm.

Figure 3109.4.1.1
BARRIER OPENINGS

tom which reduce the openings to no more than 1.75 inches (44 mm).

❖ The $2^1/_4$-inch (57 mm) dimension is intended to reduce the potential to gain a foothold (see Figure 3109.4.1.6). The opening is permitted to be increased above this $2^1/_4$-inch (57 mm) dimension if decorative slats are used that reduce the size of the opening to $1^3/_4$ inches (44 mm), since the potential to obtain a foothold or handhold is reduced when such slats are installed.

3109.4.1.6 Diagonal members. Where the barrier is composed of diagonal members, the maximum opening formed by the diagonal members shall be no more than 1.75 inches (44 mm).

❖ A slightly bigger opening is permitted for barriers composed of diagonal members other than chain-link fences, on the basis that such barriers would be more difficult to gain a foothold and handhold on than a chain-link fence. The $1^3/_4$-inch (44 mm) dimension is consistent with Sections 3109.4.1.3, 3109.4.1.4 and 3109.4.1.5.

3109.4.1.7 Gates. Access gates shall comply with the requirements of Sections 3109.4.1.1 through 3109.4.1.6 and shall be equipped to accommodate a locking device. Pedestrian access

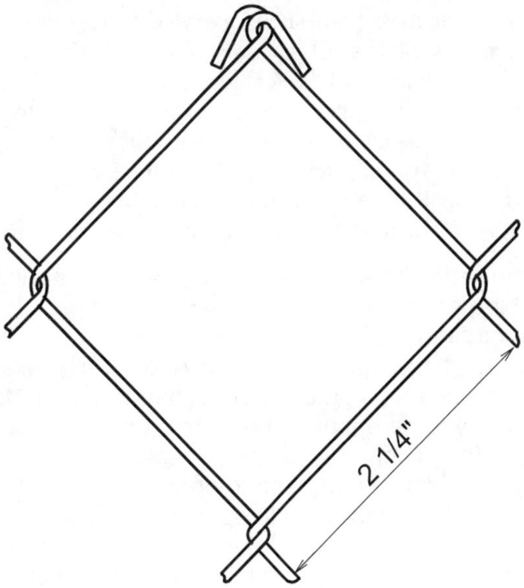

For SI: 1 inch = 25.4 mm.

Figure 3109.4.1.6
CHAIN-LINK FENCE MESH FOR PRIVATE SWIMMING POOLS

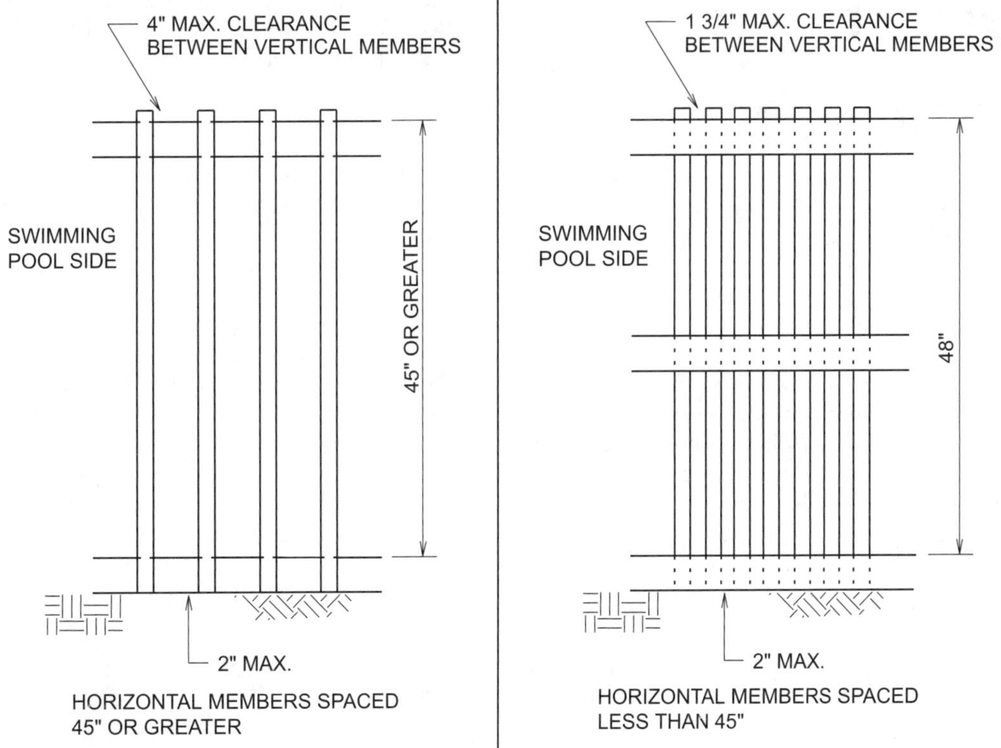

For SI: 1 inch = 25.4 mm.

Figure 3109.4.1.3
PRIVATE SWIMMING POOL BARRIER CONSTRUCTION

gates shall open outward away from the pool and shall be self-closing and have a self-latching device. Gates other than pedestrian access gates shall have a self-latching device. Where the release mechanism of the self-latching device is located less than 54 inches (1372 mm) from the bottom of the gate, the release mechanism shall be located on the pool side of the gate at least 3 inches (76 mm) below the top of the gate, and the gate and barrier shall have no opening greater than 0.5 inch (12.7 mm) within 18 inches (457 mm) of the release mechanism.

❖ A gate represents the same potential hazard relative to climbing as do other portions of the barrier and, therefore, must be constructed in accordance with Sections 3109.4.1.1 through 3109.4.1.6. Additionally, since the gate represents a potential breach of the barrier due to the ability to be opened, the code provides prescriptive details for its construction and operation. A self-closing pedestrian gate is required to open away from the pool because if the latch fails to operate, a child pushing on the gate will not gain immediate access to the pool. Pushing on the gate may also engage the latch. Large, nonpedestrian gates are not required to be self-closing due to the prohibitive cost coupled with the fact that these gates are typically operated by persons other than small children. The 54-inch (1372 mm) latch height requirement is intended to limit the potential for small children to reach and activate it. If located less than 54 inches (1372 mm), the code's prescriptive location requirements are intended to preclude the latch from being activated by small children who are not on the pool side of the gate.

3109.4.1.8 Dwelling wall as a barrier. Where a wall of a dwelling serves as part of the barrier, one of the following shall apply:

1. Doors with direct access to the pool through that wall shall be equipped with an alarm which produces an audible warning when the door and its screen are opened. The alarm shall sound continuously for a minimum of 30 seconds immediately after the door is opened and be capable of being heard throughout the house during normal household activities. The alarm shall automatically reset under all conditions. The alarm shall be equipped with a manual means to temporarily deactivate the alarm for a single opening. Such deactivation shall last no more than 15 seconds. The deactivation switch shall be located at least 54 inches (1372 mm) above the threshold of the door.
2. The pool shall be equipped with a power safety cover which complies with ASTM F 1346.
3. Other means of protection, such as self-closing doors with self-latching devices, which are approved by the administrative authority, shall be accepted so long as the degree of protection afforded is not less than the protection afforded by Section 3109.4.1.8, Item 1 or 2.

❖ Many residential settings with backyard pools utilize the dwelling as a portion of the barrier required around the pool, i.e., the fence bounding the property terminates at the dwelling. This limits access to the pool by unsupervised children around the perimeter of the fence but there is still a potential for children to access the pool from within the dwelling. Indeed, almost half the children involved in drowning or near-drowning accidents gained access to the pool from the dwelling.

The provisions of this section are intended to restrict such access by small children and are applicable to all doors in walls that form a portion of the barrier required around private swimming pools, both outdoor and indoor.

Protection of such door openings to pool areas can be achieved in any one of the methods described in Items 1 through 3. The alarm is configured to allow adults accessing the house to open the door, enter the house and deactivate the system to prevent a "false alarm." The touchpad permitted to deactivate the system is required to be mounted 54 inches (1372 mm) above the floor, which is presumed to be beyond the reach of small children.

An audible alarm specified in Item 1 on the doors leading to the pool area is intended to provide a warning to a supervising adult that the pool area has been entered.

Item 2 permits doors to pool areas to be protected by devices that render the door self-closing and self-latching. The release mechanism for the latching device is required to be located a minimum of 54 inches (1372 mm) above the floor, which is anticipated to be beyond the reach of small children. In addition to self-closing and self-latching devices, doors protected by the method specified in Item 2 are required to open away from the pool area. This is so that if the door failed to latch, a child outside the pool pushing against the door would cause it to close and not swing to an open position.

Item 3 does not require protection of the door itself; but limits access to the pool by means of a power safety cover. The performance criteria specified when this option is selected is to ensure that the power safety cover is an adequate and reliable barrier to the pool.

3109.4.1.9 Pool structure as barrier. Where an aboveground pool structure is used as a barrier or where the barrier is mounted on top of the pool structure, and the means of access is a ladder or steps, then the ladder or steps either shall be capable of being secured, locked or removed to prevent access, or the ladder or steps shall be surrounded by a barrier which meets the requirements of Sections 3109.4.1.1 through 3109.4.1.8. When the ladder or steps are secured, locked or removed, any opening created shall not allow the passage of a 4-inch-diameter (102 mm) sphere.

❖ The code permits the wall of the pool itself to serve as the barrier to the pool, provided that the wall extends at least 48 inches (1219 mm) above the finished ground level around the perimeter of the pool.

3109.4.2 Indoor swimming pools. Walls surrounding indoor swimming pools shall not be required to comply with Section 3109.4.1.8.

❖ Indoor pools are required to have the alarms, etc., required by Section 3108.4.1.8. The pool in an indoor environment is generally under closer scrutiny; therefore, it is less apt to be an area where small children would be allowed to stray into without being noticed.

3109.4.3 Prohibited locations. Barriers shall be located so as to prohibit permanent structures, equipment or similar objects from being used to climb the barriers.

❖ This section is especially important when the wall of the pool itself is used as the barrier (see Section 3109.4.1.9). Pumps and other equipment located adjacent to the pool wall (barrier) present a hazard in that they may provide a means by which small children could climb over the pool wall and access the pool area.

3109.5 Entrapment avoidance. Where the suction inlet system, such as an automatic cleaning system, is a vacuum cleaner system which has a single suction inlet, or multiple suction inlets which can be isolated by valves, each suction inlet shall protect against user entrapment by an approved antivortex cover, a 12-inch by 12-inch (304 mm by 304 mm) or larger grate, or other approved means.

In addition, all pools and spas shall be equipped with an alternative backup system which shall provide vacuum relief should grate covers be missing. Alternative vacuum relief devices shall include one of the following:

1. Approved vacuum release system.
2. Approved vent piping.
3. Other approved devices or means.

❖ Vacuum devices for suction inlet systems in pool water circulation are a safety hazard. Body or hair entrapment can cause drowning and evisceration; therefore, it is important to provide protection against possible entrapment at the pool entrances to suction inlets, as well as vacuum relief for the vacuum system.

Bibliography

The following resource materials are referenced in this chapter or are relevant to the subject matter addressed in this chapter.

24 CFR, *Fair Housing Accessibility Guidelines* (FHAG). Washington, DC: Department of Housing and Urban Development, 1991.

ASTM E 84-01, *Test Method for Surface Burning Characteristics of Building Materials*. West Conshohocken, PA: ASTM International, 2001.

ASTM E 136-99, *Test Method for Behavior of Materials in a Vertical Tube Furnace at 750° C*. West Conshohocken, PA: ASTM International, 1999.

EIA/TIA 222-F-96, *Structural Standards for Steel Antenna Towers and Antenna Supporting Structures*. Arlington, VA: Electronics Industries Association, 1996.

IFC-2003, *International Fire Code*. Falls Church, VA: International Code Council, 2003.

IMC-2003, *International Mechanical Code*. Falls Church, VA: International Code Council, 2003.

NFPA 701-99, *Methods of Fire Tests for Flame-Resistant Textiles and Films*. Quincy, MA: National Fire Protection Association, 1999.

NFPA 204 M-98, *Guide for Smoke and Heat Venting*. Quincy, MA: National Fire Protection Association, 1998.

NSPI 1-91 *Standard for Public Swimming Pools*. Alexandria, VA: National Spa and Pool Institute, 1991.

NSPI 4-92 *Standard for Above-Ground/Onground Residential Swimming Pools*. Alexandria, VA: National Spa and Pool Institute, 1992.

NSPI 5-94, *Standards for Residential Swimming Pools*. Alexandria, VA: National Spa and Pool Institute, 1994.

Present, Paula. *Child Drowning Study, A Report on the Epidemiology of Drownings in Residential Pools to Children Under Age 5*. Consumer Product Safety Commission, September 1987.

Chapter 32: Encroachments Into The Public Right-Of-Way

General Comments

Chapter 32 contains provisions that regulate a variety of structure encroachments and projections.

Section 3202 addresses encroachments on lot lines and street lot lines below grade, above grade, below 8 feet (2438 mm) in height and 8 feet (2438 mm) or more above grade.

Chapter 32 establishes the basis for the determination and application of many other code requirements, such as the minimum type of construction. Incorrect determination of the minimum type of construction will result in misapplication of the code.

The requirements of this chapter affect the decisions to be made when applying the rest of the code. Encroachments and projections have to be taken into consideration in addition to means of egress, fire resistance, fire protection or even structural and material considerations.

Purpose

The purpose of Chapter 32 is to regulate projections and encroachments of structures.

SECTION 3201 GENERAL

3201.1 Scope. The provisions of this chapter shall govern the encroachment of structures into the public right-of-way.

❖ Interior lot lines are ordinarily established by property ownership. The extent and location of the street lot line (public right-of-way) is determined by zoning or planning statutes.

3201.2 Measurement. The projection of any structure or appendage shall be the distance measured horizontally from the lot line to the outermost point of the projection.

❖ This section defines how the extent of encroachment is determined. The extent of encroachment of any structure or appendage is to be measured from the lot line to the outermost (furthest) point of the projection.

3201.3 Other laws. The provisions of this chapter shall not be construed to permit the violation of other laws or ordinances regulating the use and occupancy of public property.

❖ The projections permitted in this chapter are also subject to other limitations identified in applicable laws and ordinances. Other applicable laws, for example, are those established by the authorities in regulating right-of-way for highways, streets or public areas adjacent to the structure in question.

The street lot line occurs on each side of a structure that faces or fronts on a street. Many property descriptions have the centerline of the public way as the property line. In these cases, the lot line is defined not by the centerline of the public way, but by the line that identifies where the public way exists. The public right-of-way or other setback lines are usually established by zoning laws. For example, local zoning typically requires that residential uses have setbacks ranging from 25 to 60 feet (7620 to 18 288 mm), whereas business uses may be permitted to build up to the street lot line. While the walls of the structure may not project beyond this street lot line, certain projections are permitted as stated in this chapter.

The building lines for other parts of the structure are also usually established by local zoning ordinances in the form of required side and rear yards. This chapter and most zoning ordinances limit the projections into side setback areas. Projections beyond interior lot lines (i.e., lot lines that are not street or public way lot lines) are not permitted in this chapter. Once established, setback lines form the reference point for application of this chapter, especially when the street right-of-way line and setback are close or coincide.

Figure 3201.3 shows the building and street lot lines as established for a multiple-family occupancy.

3201.4 Drainage. Drainage water collected from a roof, awning, canopy or marquee, and condensate from mechanical equipment shall not flow over a public walking surface.

❖ A great number of street projections are canopies or mansard roofs. They have the potential to create a hazardous condition due to improper drainage on the walkways they cover. It is very important that all street projections have proper drainage systems.

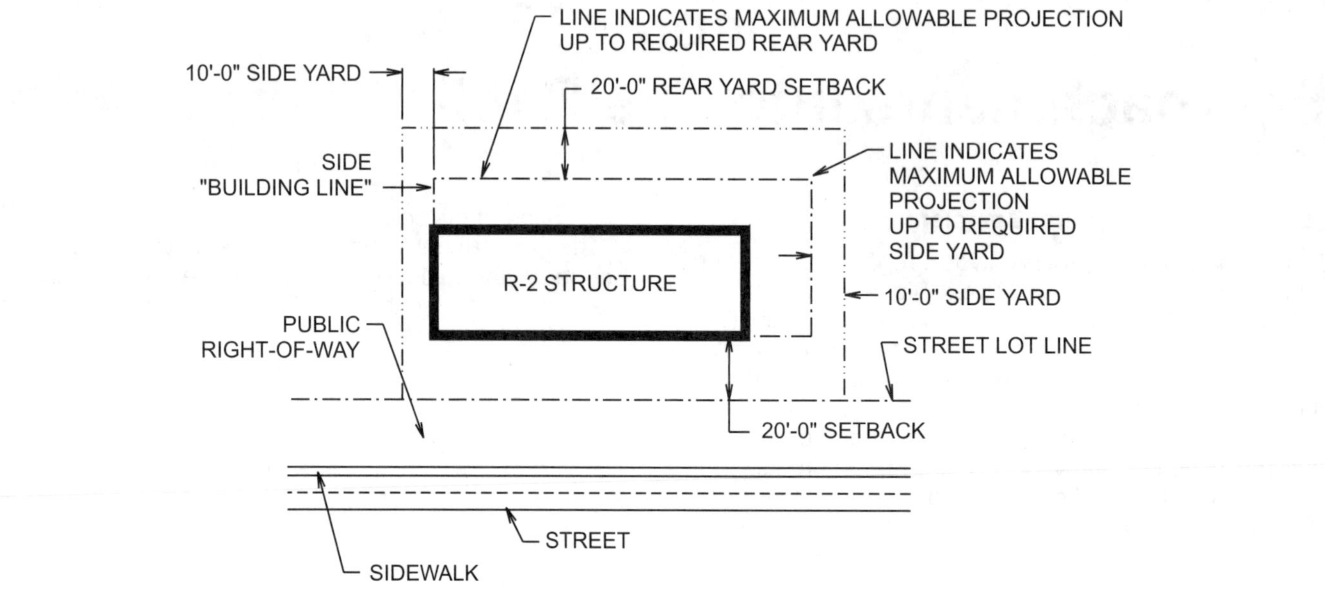

For SI: 1 inch = 25.4 mm, 1 foot = 304.8 mm.

Figure 3201.3
LIMITATIONS ON STREET PROJECTIONS

SECTION 3202
ENCROACHMENTS

3202.1 Encroachments below grade. Encroachments below grade shall comply with Sections 3202.1.1 through 3202.1.3.

❖ Allowable projections of below-grade structure components are described in Sections 3202.1.1 through 3202.1.3.

3202.1.1 Structural support. A part of a building erected below grade that is necessary for structural support of the building or structure shall not project beyond the lot lines, except that the footings of street walls or their supports which are located at least 8 feet (2438 mm) below grade shall not project more than 12 inches (305 mm) beyond the street lot line.

❖ One of the permissible street lot line encroachments (footings) is defined in this section. Where a structure is built immediately adjacent to the street lot line, the footing is permitted to project across that line (see Figure 3202.1.1). Projection of footings across side (interior) lot lines and onto adjacent property is not permitted.

3202.1.2 Vaults and other enclosed spaces. The construction and utilization of vaults and other enclosed space below grade shall be subject to the terms and conditions of the authority or legislative body having jurisdiction.

❖ Below-grade vaults and other enclosed spaces that open at the sidewalk level are usually covered with solid structural caps (e.g., concrete slabs) and are permitted to project beyond the building line or street lot line. However, since these structures occur in the public right-of-way, they are subject to local zoning laws and approval of the authorities having jurisdiction.

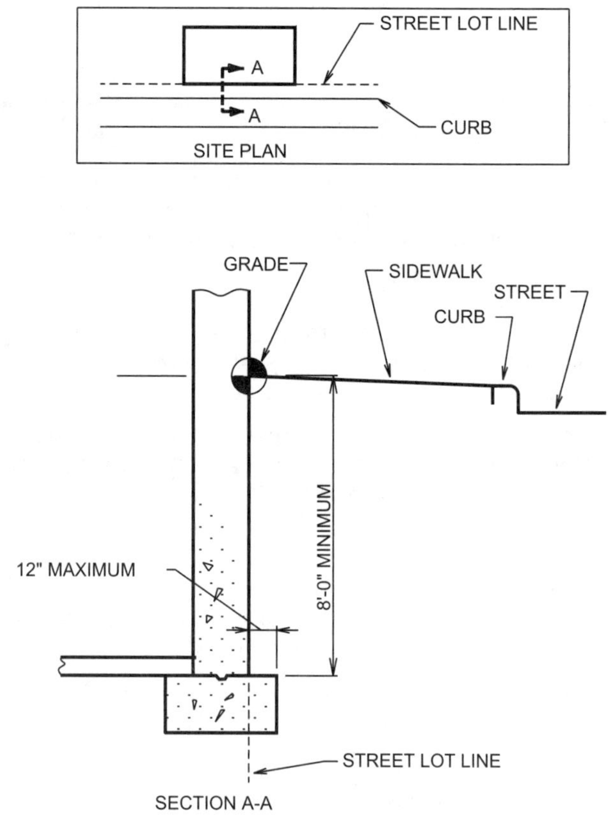

For SI: 1 inch = 25.4 mm, 1 foot = 304.8 mm.

Figure 3202.1.1
ENCROACHMENT BELOW GRADE

3202.1.3 Areaways. Areaways shall be protected by grates, guards or other approved means.

❖ Areaways, which provide natural light or ventilation to below-grade spaces, are required to be protected by grates, guards or other approved means of protection (see Figure 3202.1.3).

3202.2 Encroachments above grade and below 8 feet in height. Encroachments into the public right-of-way above grade and below 8 feet (2438 mm) in height shall be prohibited except as provided for in Sections 3202.2.1 through 3202.2.3. Doors and windows shall not open or project into the public right-of-way.

❖ This section prohibits all projections into the public right-of-way that are above grade and below 8 feet (2438 mm), except those specifically allowed by Sections 3202.2.1 through 3202.2.3 and 3202.2.4 (i.e., steps, architectural features such as columns or pilasters, awnings, temporary vestibules and storm enclosures). The maximum permitted encroachments and minimum clearances, as well as additional protection requirements, for the specified allowed components are detailed in their respective sections.

3202.2.1 Steps. Steps shall not project more than 12 inches (305 mm) and shall be guarded by approved devices not less than 3 feet (914 mm) high, or shall be located between columns or pilasters.

❖ Steps are permitted to project a maximum of 12 inches (305 mm) into the public right-of-way by this section. However, since steps may pose a tripping hazard, an approved device (guard), ornamental column or pilaster must be provided at a minimum height of 3 feet (914 mm) to minimize this hazard. Figure 3202.2.1 illustrates compliance with this provision.

3202.2.2 Architectural features. Columns or pilasters, including bases and moldings shall not project more than 12 inches (305 mm). Belt courses, lintels, sills, architraves, pediments and similar architectural features shall not project more than 4 inches (102 mm).

❖ This section addresses the variety of architectural appurtenances that are part of a structure's exterior wall or roof. Consistent with the projection limitations given for steps in Section 3202.2.1, the maximum encroachment into the public way is 12 inches (305 mm) for columns and pilasters (see Figure 3202.2.1). Other noted architectural features are limited to maximum encroachments of 4 inches (102 mm).

3202.2.3 Awnings. The vertical clearance from the public right-of-way to the lowest part of any awning, including valances, shall be 7 feet (2134 mm) minimum.

❖ Awnings are permitted to project into the public right-of-way. However, the clearance to any part of an awning, including its valance, is required to be a minimum of 7 feet (2134 mm) above grade (see Figure 3202.2.3).

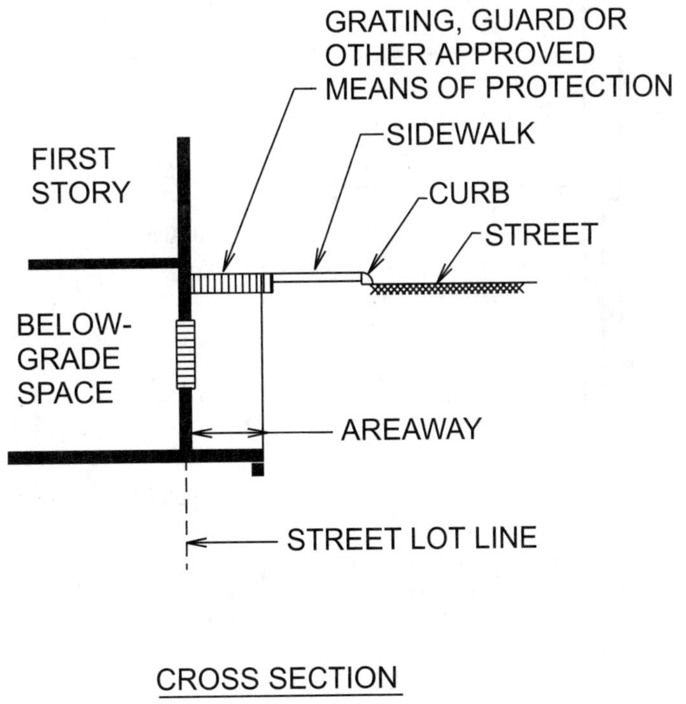

Figure 3202.1.3
AREAWAYS

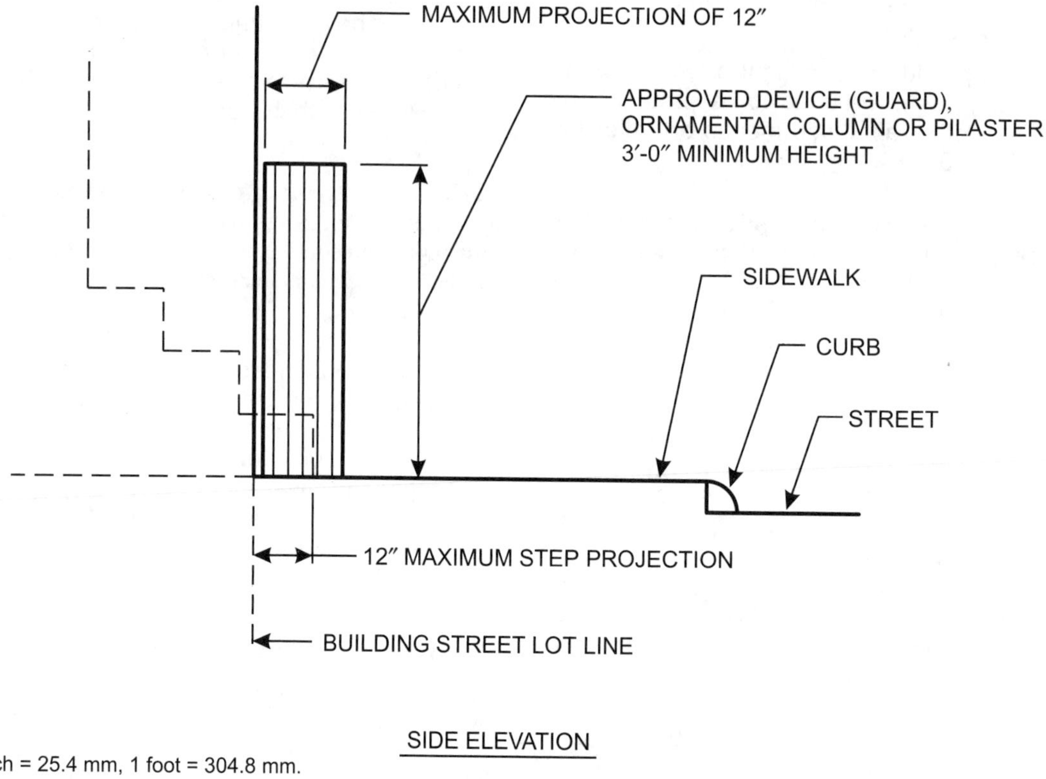

For SI: 1 inch = 25.4 mm, 1 foot = 304.8 mm.

Figure 3202.2.1
STEPS

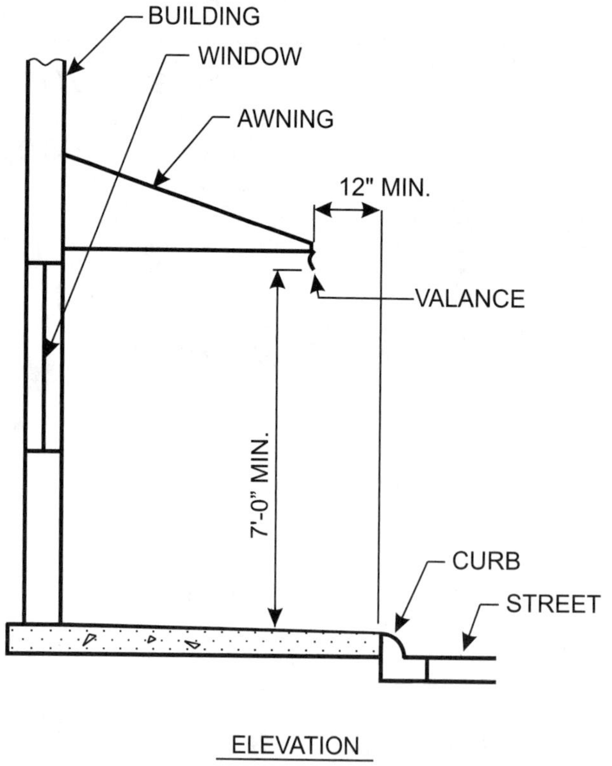

For SI: 1 inch = 25.4 mm, 1 foot = 304.8 mm.

Figure 3202.2.3
AWNINGS

3202.3 Encroachments 8 feet or more above grade. Encroachments 8 feet (2438 mm) or more above grade shall comply with Sections 3202.3.1 through 3202.3.4.

❖ This section prohibits all projections into the public right-of-way that are 8 feet (2438 mm) or more above grade, except those specifically allowed by Sections 3202.3.1 through 3202.3.4. The maximum permitted encroachments and minimum clearances, as well as additional approval requirements, for the specified allowed components are detailed in their respective sections.

3202.3.1 Awnings, canopies, marquees and signs. Awnings, canopies, marquees and signs shall be constructed so as to support applicable loads as specified in Chapter 16. Awnings, canopies, marquees and signs with less than 15 feet (4572 mm) clearance above the sidewalk shall not extend into or occupy more than two-thirds the width of the sidewalk measured from the building. Stanchions or columns that support awnings, canopies, marquees and signs shall be located not less than 2 feet (610 mm) in from the curb line.

❖ This section contains three distinct requirements. First, awnings, canopies, marquees and signs must be designed in accordance with the provisions of Chapter 16, with regard to the live loads that may be imposed on such a structure.

Second, the allowable projection of awnings, canopies, marquees and signs with less than 15 feet (4572 mm) of clearance above the sidewalk is limited to two-thirds of the width of the sidewalk measured from the building.

Finally, the location of vertical structural components used for supporting awnings, canopies, marquees and signs (stanchions or columns) is also regulated. A minimum of 2 feet (610 mm) of clearance between the curb line and the vertical structural support for awnings, canopies, marquees and signs is required (see Figure 3202.3.1).

3202.3.2 Windows, balconies, architectural features and mechanical equipment. Where the vertical clearance above grade to projecting windows, balconies, architectural features or mechanical equipment is more than 8 feet (2438 mm), 1 inch (25 mm) of encroachment is permitted for each additional 1 inch (25 mm) of clearance above 8 feet (2438 mm), but the maximum encroachment shall be 4 feet (1219 mm).

❖ Projections into the public right-of-way for appurtenances such as windows, balconies, architectural features or mechanical equipment more than 8 feet (2438 mm) above grade are allowed by this section. Such projections are limited to 1 inch (25 mm) of encroachment for each additional inch of clearance above the minimum 8-foot (2438 mm) clearance required. However, the maximum allowable projection is limited to 4 feet (1219 mm) (see Figure 3202.3.2).

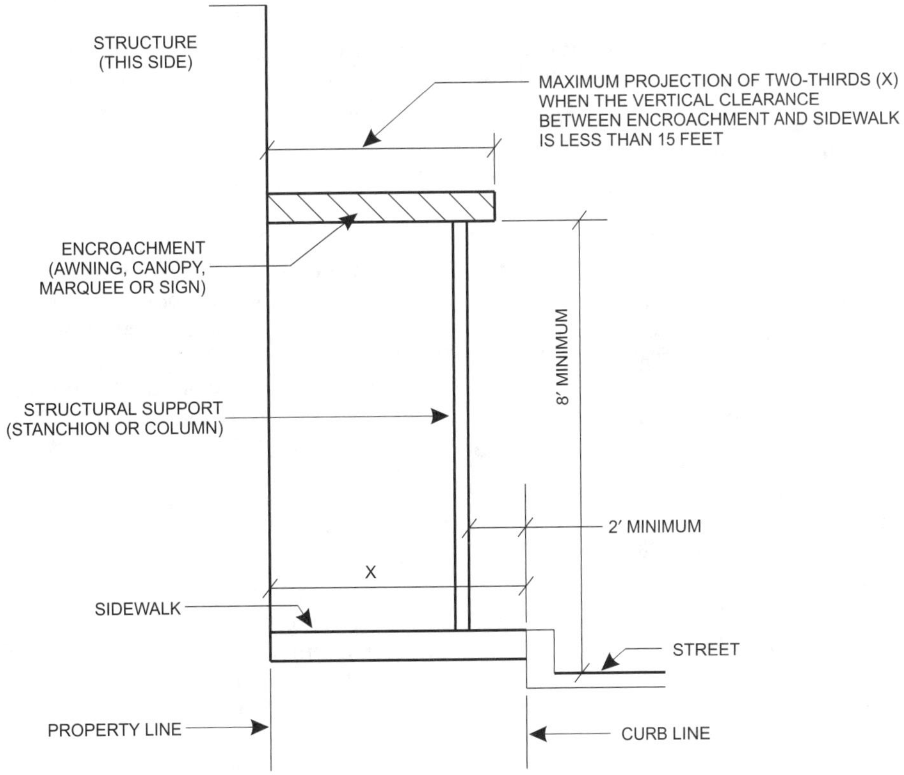

For SI: 1 foot = 304.8 mm.

Figure 3202.3.1
AWNINGS, CANOPIES, MARQUEES AND SIGNS 8 FEET OR MORE ABOVE GRADE

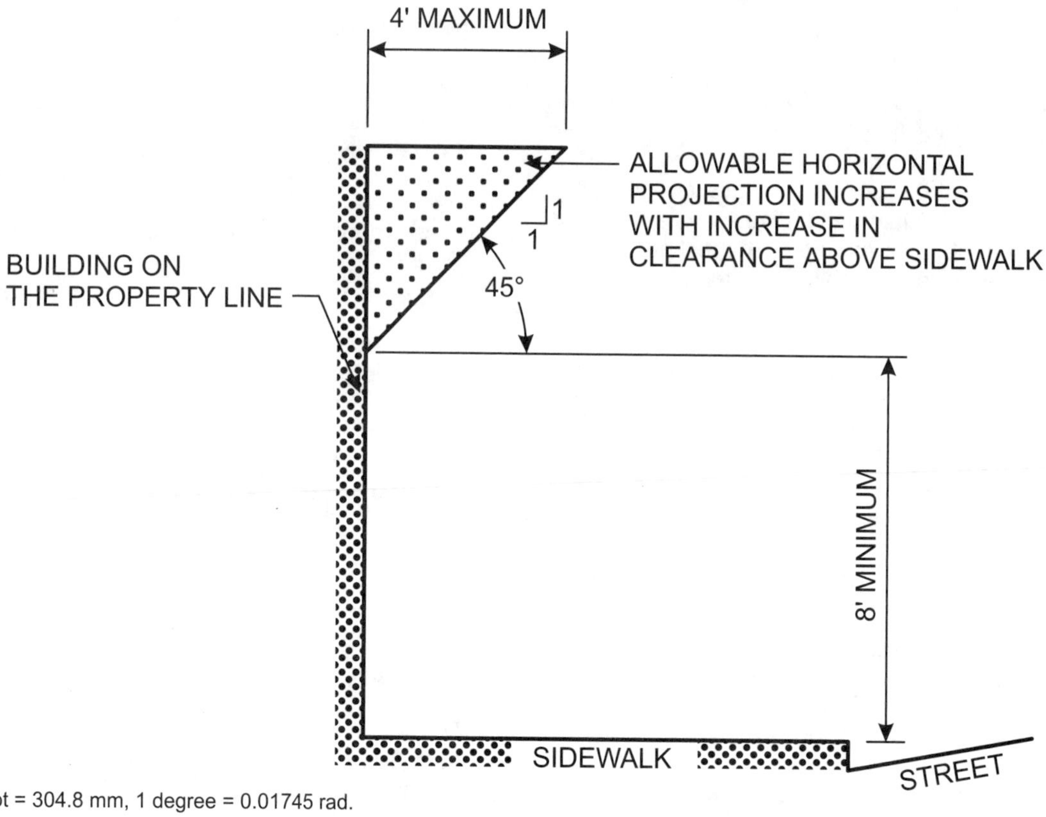

For SI: 1 foot = 304.8 mm, 1 degree = 0.01745 rad.

Figure 3202.3.2
WIDOWS, BALCONIES, ARCHITECTURAL FEATURES AND MECHANICAL EQUIPMENT

3202.3.3 Encroachments 15 feet or more above grade. Encroachments 15 feet (4572 mm) or more above grade shall not be limited.

❖ Because encroachments into the public right-of-way that are 15 feet (4572 mm) or more above grade do not interfere with or impede pedestrian or vehicular traffic, they are not regulated by the provisions of this chapter.

3202.3.4 Pedestrian walkways. The installation of a pedestrian walkway over a public right-of-way shall be subject to the approval of local authority having jurisdiction. The vertical clearance from the public right-of-way to the lowest part of a pedestrian walkway shall be 15 feet (4572 mm) minimum.

❖ As in Section 3202.3.3, projections of pedestrian walkways that have a minimum clearance of 15 feet (4572 mm) above grade do not interfere or impede pedestrian or vehicular traffic and thus are not prohibited by this section. The installation of pedestrian walkways over a public right-of-way is, however, required to be approved by the local authority having jurisdiction.

3202.4 Temporary encroachments. Where allowed by the local authority having jurisdiction, vestibules and storm enclosures shall not be erected for a period of time exceeding 7 months in any one year and shall not encroach more than 3 feet (914 mm) nor more than one-fourth of the width of the sidewalk beyond the street lot line. Temporary entrance awnings shall be erected with a minimum clearance of 7 feet (2134 mm) to the lowest portion of the hood or awning where supported on removable steel or other approved noncombustible support

❖ This section addresses a special situation in which temporary projections across the building line or street lot line are erected for climatic reasons (i.e., during the winter months some building owners may erect vestibule enclosures or canopies over entrances). The purpose of this section is to allow these energy-saving procedures, yet place limitations on such projections.

Specifically, enclosures temporarily appended to an entranceway are permitted to project not more than 3 feet (914 mm) beyond the building line or street lot line, or up to one-fourth the width of the sidewalk. Whichever dimension results in the lesser projection defines the maximum projection.

Additionally, when temporary awnings are erected, clearance between the sidewalk and the lowest part of the awning must be at least 7 feet (2134 mm), which is consistent with the requirements for permanent awnings in Section 3202.2.3. It is further required that the frame for the hood or awning is to be supported on removable steel or other approved noncombustible material. This last provision of this section is necessitated by unforeseen changes in street lot lines, for example, that would result in the encroachment becoming an obstruction in the public right-of-way.

Chapter 33: Safeguards During Construction

General Comments

The building construction process involves a number of known and unanticipated hazards. Chapter 33 establishes specific regulations in order to minimize the risk to the public and adjacent property. Some construction failures have resulted during the initial stages of grading, excavation and demolition. During these early stages, poorly designed and installed sheeting and shoring have resulted in ditch and embankment cave-ins. Also, inadequate underpinning of adjoining existing structures or careless removal of existing structures has produced construction failures.

The most critical period in building construction related to the physical safety of those on the job site is during the ongoing process when all building components have not yet been completed. Compounding this incomplete state is the use of dangerous construction methods, materials and equipment.

The importance of reasonable precautions is evidenced by the federal Occupational Safety and Health Act and related state and local regulations.

Purpose

The purpose of Chapter 33 is to cite safety requirements during construction or demolition of buildings and structures. These requirements are intended to protect the public from injury and adjoining property from damage.

SECTION 3301 GENERAL

3301.1 Scope. The provisions of this chapter shall govern safety during construction and the protection of adjacent public and private properties.

❖ The fundamental rationale behind this section is to establish that reasonable safety precautions are to be provided during the construction or demolition process to protect the public from injury and adjoining property from damage.

3301.2 Storage and placement. Construction equipment and materials shall be stored and placed so as not to endanger the public, the workers or adjoining property for the duration of the construction project.

❖ This section requires that construction materials and equipment must be located and protected pursuant to the governing provisions of this chapter so that the public and adjoining property are safeguarded at all times during construction or demolition processes.

SECTION 3302 CONSTRUCTION SAFEGUARDS

3302.1 Remodeling and additions. Required exits, existing structural elements, fire protection devices and sanitary safeguards shall be maintained at all times during remodeling, alterations, repairs or additions to any building or structure.

Exceptions:

1. When such required elements or devices are being remodeled, altered or repaired, adequate substitute provisions shall be made.
2. When the existing building is not occupied.

❖ Demolition and construction operations must not create a hazard for the occupants of a building during an alteration or addition. As such, the existing fire protection, means of egress elements and safety systems must remain in place and functional. Two options, however, are provided in this section thst permit the fire protection, means of egress elements and safety systems to be changed, modified or removed. The first option provides an alternative method that must be approved by the building official. The second option is simply to have an unoccupied building, which means that residents of a dwelling unit or employees of a business are not to be present during the construction or demolition process.

3302.2 Manner of removal. Waste materials shall be removed in a manner which prevents injury or damage to persons, adjoining properties and public rights-of-way.

❖ Safe and sanitary procedures for the removal of building construction and demolition waste must be provided. The method of waste removal must be controlled such that debris will not pose a hazard, eyesore or nuisance to the public and neighboring properties. Examples of acceptable practices include: evidence that a professional disposal service will haul away the debris; limiting areas used for storing and handling demolished materials; enclosing storage areas so that only authorized personnel can gain access; establishing routes that waste removal vehicles are permitted to use; covering or tarping debris to prevent flying objects; providing fully enclosed chutes to control falling objects and dust and scheduling waste removal when adjoining property and the public will be least exposed to unusual and possibly dangerous situations caused by fumes, noise, dust and unfamiliar events involved during a demolition process. Note that the accepted practice of waste removal is subject to the approval of the building official.

SECTION 3303 DEMOLITION

3303.1 Construction documents. Construction documents and a schedule for demolition must be submitted when required by the building official. Where such information is required, no work shall be done until such construction documents or schedule, or both, are approved.

❖ Some processes, methods and materials used during demolition can cause damage to adjoining property and increase the risk of injury to the general public. As such, this section indicates that the building official can request: verification that structural stability of existing buildings on the same lot or adjacent properties is maintained and not compromised; to know the materials and equipment necessary to perform the demolition involved; the timetable in which certain demolition activities will be performed; details to establish if the general public and adjoining property are exposed to any unfamiliar or greater hazard than normal; special inspection reports made by qualified firms, agencies or individuals; qualifications of the laborers involved in the operation, handling and removal of demolition equipment and materials; verification that any federal, state or local statue is followed, etc. Note that the building official must grant approval for any additional information requested before demolition can be started.

3303.2 Pedestrian protection. The work of demolishing any building shall not be commenced until pedestrian protection is in place as required by this chapter.

❖ Demolition must not be started until all of the necessary precautions are taken to protect the general public as indicated throughout this chapter.

3303.3 Means of egress. A party wall balcony or horizontal exit shall not be destroyed unless and until a substitute means of egress has been provided and approved.

❖ The exits provided through a party wall must be maintained operational and usable during demolition; however, based upon the extent of work, an alternative means of egress through a party wall is only permitted when approved by the authority having jurisdiction.

3303.4 Vacant lot. Where a structure has been demolished or removed, the vacant lot shall be filled and maintained to the existing grade or in accordance with the ordinances of the jurisdiction having authority.

❖ A demolition site must be restored if additional building construction or demolition operations are not scheduled to take place. The site must be filled and graded to the level of the surrounding site or maintained in accordance with local or state statues, which may set forth other or additional grading requirements, such as provisions for elevations, drainage and flood control.

The time frame imposed or stipulated to abate an imminent hazard created by a vacant lot is subject to the building official. In addition, footings, foundations, basement walls and floors must be removed when the building official determines that a threat is posed to human life or the public welfare or if any portion of the foundation system prevents proper grading from being done.

3303.5 Water accumulation. Provision shall be made to prevent the accumulation of water or damage to any foundations on the premises or the adjoining property.

❖ A vacant lot must be graded in accordance with Section 3303.4 in such a way that water is prevented from ponding and causing damage to structures on the premises or adjacent properties, particularly foundation systems and building elements in contact with the ground. Footings, foundations, basement walls or floors must be removed if this drainage is prevented.

3303.6 Utility connections. Service utility connections shall be discontinued and capped in accordance with the approved rules and the requirements of the authority having jurisdiction.

❖ The procedures for disconnecting and abandoning utility connections in a safe and satisfactory manner must be in accordance with local, state and federal statues. Utility providers, such as electrical, telephone, water, gas or sewer, often recommend or require notification and have certain guidelines to follow.

A related aspect is the precaution to investigate for any on-site underground obstructions, such as utility service lines and buried oil and gasoline or septic tanks. If the temporary use of some existing service, such as electricity or water, is requested during the demolition project, then temporary permits must be obtained with the appropriate stipulations, including any rerouting or protection requirements from the appropriate authority.

SECTION 3304 SITE WORK

3304.1 Excavation and fill. Excavation and fill for buildings and structures shall be constructed or protected so as not to endanger life or property. Stumps and roots shall be removed from the soil to a depth of at least 12 inches (305 mm) below the surface of the ground in the area to be occupied by the building. Wood forms which have been used in placing concrete, if within the ground or between foundation sills and the ground, shall be removed before a building is occupied or used for any purpose. Before completion, loose or casual wood shall be removed from direct contact with the ground under the building.

❖ A site that is to be excavated or filled must be kept out of contact with dead and decaying foliage. As such, the base of trees, roots and limbs that protrude above the ground as well as 12 inches (305 mm) into the ground must be taken out or separated from the earth where the building or structure will be erected. Additionally, any wood forms used for the concrete must be removed along with any wood product near the building that may decay or rot.

3304.1.1 Slope limits. Slopes for permanent fill shall not be steeper than one unit vertical in two units horizontal (50-percent slope). Cut slopes for permanent excavations shall not be steeper than one unit vertical in two units horizontal (50-percent slope). Deviation from the foregoing limitations for cut slopes shall be permitted only upon the presentation of a soil investigation report acceptable to the building official.

❖ The final grading around a foundation system must slope away from the building or structure, such that the ratio does not exceed 2:1. A trench or hole dug for a foundation system is required to have a stable embankment. As such, the embankment must not have a slope that exceeds a 2:1 ratio; however, due to the existing soil conditions or particular location, it may be impractical to achieve the required slope. This section, therefore, indicates that an engineering analysis can be submitted with recommendations for alternative methodology and safeguards, subject to the approval of the building official.

3304.1.2 Surcharge. No fill or other surcharge loads shall be placed adjacent to any building or structure unless such building or structure is capable of withstanding the additional loads caused by the fill or surcharge. Existing footings or foundations which can be affected by any excavation shall be underpinned adequately or otherwise protected against settlement and shall be protected against later movement.

❖ Gradient stability must be maintained for existing buildings or structures during the excavation, construction or demolition processes. As such, earth, construction materials, debris and even construction equipment must not impose a load that will cause the ground to give way and collapse or damage the foundation system of adjacent buildings or structures; therefore, components such as engineered bracing must be provided for protection if adjacent buildings or structures are to be exposed to loads that exceed those of the existing design.

3304.1.3 Footings on adjacent slopes. For footings on adjacent slopes, see Chapter 18.

❖ Any footings located on adjacent slopes must be handled pursuant to Chapter 18.

3304.1.4 Fill supporting foundations. Fill to be used to support the foundations of any building or structure shall comply with Section 1803.5. Special inspections of compacted fill shall be in accordance with Section 1704.7.

❖ Fill that will be installed to support a building or structure must be compacted and verified in accordance with Section 1803.4 and special inspection regulations stated in Section 1704.7.

SECTION 3305 SANITARY

3305.1 Facilities required. Sanitary facilities shall be provided during construction, remodeling or demolition activities in accordance with the *International Plumbing Code.*

❖ Construction employees must have plumbing facilities available during the construction or demolition process of a building. The facilities must conform to the requirements set forth in the *International Plumbing Code*® (IPC®).

SECTION 3306 PROTECTION OF PEDESTRIANS

3306.1 Protection required. Pedestrians shall be protected during construction, remodeling and demolition activities as required by this chapter and Table 3306.1. Signs shall be provided to direct pedestrian traffic.

❖ Safeguards are required to be in place during construction or demolition operations in accordance with Table 3306.1 and this chapter. In addition, since construction operations alter the familiar setting and path of travel, it is necessary to provide some form of visible directional sign to lead the public toward safety and away from potential hazards.

This table establishes the type of protection required based upon the location to overhead hazards and distance in relation to ground hazards. Once the type of protection is determined from the table, the applicable code section, such as Section 3306.4 for construction railings, Section 3306.5 for barriers and Section 3306.7 for covered walkways, must be provided as regulated.

TABLE 3306.1
PROTECTION OF PEDESTRIANS

HEIGHT OF CONSTRUCTION	DISTANCE FROM CONSTRUCTION TO LOT LINE	TYPE OF PROTECTION REQUIRED
8 feet or less	Less than 5 feet	Construction railings
	5 feet or more	None
More than 8 feet	Less than 5 feet	Barrier and covered walkway
	5 feet or more, but not more than one-fourth the height of construction	Barrier and covered walkway
	5 feet or more, but between one-fourth and one-half the height of construction	Barrier
	5 feet or more, but exceeding one-half the height of construction	None

For SI: 1 foot = 304.8 mm.

3306.2 Walkways. A walkway shall be provided for pedestrian travel in front of every construction and demolition site unless the authority having jurisdiction authorizes the sidewalk to be fenced or closed. Walkways shall be of sufficient width to accommodate the pedestrian traffic, but in no case shall they be less than 4 feet (1219 mm) in width. Walkways shall be provided with a durable walking surface. Walkways shall be accessible in accordance with Chapter 11 and shall be designed to support all imposed loads and in no case shall the design live load be less than 150 pounds per square foot (psf) (7.2 kN/m^2).

❖ Construction operations must not narrow or impede the normal flow of pedestrian traffic along a walkway by the placement of a fence or other enclosure. Note that the authority having jurisdiction must approve any construction operations that will cause the narrowing or impedance of a walkway, such as the sidewalk. If a walkway is narrowed or enclosed, the construction of another or wider walkway is required for all pedestrians. The walkway must be able to handle the normal anticipated flow of pedestrian traffic and must not be less than the minimum 4-foot (1219 mm) width. In addition, the walkway must be a stable surface that is capable of supporting all imposed loads. The minimum design load of the walkway must not be less than 150 pounds per square feet (psf) (7182 Pa).

3306.3 Directional barricades. Pedestrian traffic shall be protected by a directional barricade where the walkway extends into the street. The directional barricade shall be of sufficient size and construction to direct vehicular traffic away from the pedestrian path.

❖ Similar to the local, federal and state guidelines that govern provisions to protect employees while doing road work, this section establishes that a barrier must be erected. The barrier must be capable of redirecting and regulating the flow of vehicular traffic where constructed or projected into a street. An example of how to regulate traffic is to provide the necessary warnings, notice of caution and instructions to the operators of motor vehicles with signs and colors that are common to the local, federal and state guidelines. The intent of this visual barrier is to keep pedestrians out of vehicular traffic and prevent vehicles from encroaching on or using the pedestrian walkway as a roadway.

3306.4 Construction railings. Construction railings shall be at least 42 inches (1067 mm) in height and shall be sufficient to direct pedestrians around construction areas.

❖ A barrier consisting of a horizontal rail and supports must be constructed with a minimum height of 42 inches (1067 mm). In addition, this barrier must be capable of controlling the flow of pedestrian traffic by routing travel away from the hazards associated with a construction area.

3306.5 Barriers. Barriers shall be a minimum of 8 feet (2438 mm) in height and shall be placed on the side of the walkway nearest the construction. Barriers shall extend the entire length of the construction site. Openings in such barriers shall be protected by doors which are normally kept closed.

❖ When a barrier is required by Table 3306.1, it must be constructed to impede, separate and obstruct passage of pedestrians onto a construction site. The structure must be a minimum 8 feet (2438 mm) in height and located continuously along the walkway on the side where construction activities are being done. Openings, such as doors or gates, are permitted as long as the only time they are open is during use by authorized personnel. As such, when the doors or gates are not being used, they must remain closed.

3306.6 Barrier design. Barriers shall be designed to resist loads required in Chapter 16 unless constructed as follows:

1. Barriers shall be provided with 2-inch by 4-inch (51 mm by 102 mm) top and bottom plates.
2. The barrier material shall be a minimum of $^3/_4$-inch (19.1 mm) boards or $^1/_4$-inch (6.4 mm) wood structural use panels.
3. Wood structural use panels shall be bonded with an adhesive identical to that for exterior wood structural use panels.
4. Wood structural use panels $^1/_4$ inch (6.4 mm) or $^5/_{16}$ inch (23.8 mm) in thickness shall have studs spaced not more than 2 feet (610 mm) on center (o.c.).

5. Wood structural use panels $^3/_8$ inch (9.5 mm) or $^1/_2$ inch (12.7 mm) in thickness shall have studs spaced not more than 4 feet (1219 mm) o.c., provided a 2-inch by 4-inch (51 mm by 102 mm) stiffener is placed horizontally at midheight where the stud spacing exceeds 2 feet (610 mm) o.c.
6. Wood structural use panels $^5/_8$ inch (15.9 mm) or thicker shall not span over 8 feet (2438 mm).

❖ This section establishes two methods that can be used to design barriers. The first method establishes that the barrier can be designed using the same materials and assembly instructions as indicated in Items 1 through 6. The second method provides for a barrier that will resist the loads imposed on it in order to be part of a designed assembly. The loads, which need to be resisted, are the same as those established in Chapter 16.

3306.7 Covered walkways. Covered walkways shall have a minimum clear height of 8 feet (2438 mm) as measured from the floor surface to the canopy overhead. Adequate lighting shall be provided at all times. Covered walkways shall be designed to support all imposed loads. In no case shall the design live load be less than 150 psf (7.2 kN/m^2) for the entire structure.

Exception: Roofs and supporting structures of covered walkways for new, light-frame construction not exceeding two stories in height are permitted to be designed for a live load of 75 psf (3.6kN/m^2) or the loads imposed on them, whichever is greater. In lieu of such designs, the roof and supporting structure of a covered walkway are permitted to be constructed as follows:

1. Footings shall be continuous 2-inch by 6-inch (51 mm by 152 mm) members.
2. Posts not less than 4 inches by 6 inches (102 mm by 152 mm) shall be provided on both sides of the roof and spaced not more than 12 feet (3658 mm) o.c.
3. Stringers not less than 4 inches by 12 inches (102 mm by 305 mm) shall be placed on edge upon the posts.
4. Joists resting on the stringers shall be at least 2 inches by 8 inches (51 mm by 203 mm) and shall be spaced not more than 2 feet (610 mm) o.c.
5. The deck shall be planks at least 2 inches (51 mm) thick or wood structural panels with an exterior exposure durability classification at least $^{23}/_{32}$ inch (18.3 mm) thick nailed to the joists.
6. Each post shall be knee braced to joists and stringers by 2-inch by 4-inch (51 mm by 102 mm) minimum members 4 feet (1219 mm) long.
7. A 2-inch by 4-inch (51 mm by 102 mm) minimum curb shall be set on edge along the outside edge of the deck.

❖ Walkways that are provided with a covering that extends out over the pedestrian path of travel, pursuant to Table 3306.1, must be at least 8 feet (2438 mm) in height. The measurement must be taken from the top of the walking surface vertically upward to the underside of the roof covering. The walkway must maintain a satisfactory level of light to the space that will be equal to or better than that provided prior to the installation of the protective covering. The covered walkway must also be structurally capable of resisting the design loads established in Chapter 16, but not less than 150 psf (7182 Pa) for the live load.

Note that this section provides two alternative design options. One option permits the covered walkway to resist the design loads established in Chapter 16, but not less than the minimum 75 psf (3591 Pa) live load if nearby construction does not exceed two stories in height and the construction materials are lightweight, such as light-frame construction. The second option is to construct the covered walkway in accordance with the seven criteria set forth in this section.

3306.8 Repair, maintenance and removal. Pedestrian protection required by this chapter shall be maintained in place and kept in good order for the entire length of time pedestrians may be endangered. The owner or the owner's agent, upon the completion of the construction activity, shall immediately remove walkways, debris and other obstructions and leave such public property in as good a condition as it was before such work was commenced.

❖ Any safeguards required by this chapter must be kept in good functional condition for the entire duration of the construction or demolition activity so that the public will not be placed in harm's way. Once a building or structure is occupiable and the site is properly graded, the protection must be removed. As such, public property that was affected by the construction activity must be restored to or left in the condition that existed prior to the work.

3306.9 Adjacent to excavations. Every excavation on a site located 5 feet (1524 mm) or less from the street lot line shall be enclosed with a barrier not less than 6 feet (1829 mm) high. Where located more than 5 feet (1524 mm) from the street lot line, a barrier shall be erected when required by the building official. Barriers shall be of adequate strength to resist wind pressure as specified in Chapter 16.

❖ This section establishes that whenever excavation is to take place less than 5 feet (1524 mm) from the edge of a roadway, a barrier must be erected with a minimum height of 6 feet (1829 mm). If the excavation is located greater than 5 feet (1524 mm) from the edge of the roadway, the building official must evaluate the level of hazard for the public and the necessary precautions to take. As such, the building official is the authority to order the construction of a structural barrier. Any barrier erected must maintain other provisions of the code such that it is capable of handling and resisting design wind loads denoted in Chapter 16.

SECTION 3307
PROTECTION OF ADJOINING PROPERTY

3307.1 Protection required. Adjoining public and private property shall be protected from damage during construction, remodeling and demolition work. Protection must be provided for footings, foundations, party walls, chimneys, skylights and roofs. Provisions shall be made to control water runoff and erosion during construction or demolition activities. The person making or causing an excavation to be made shall provide written notice to the owners of adjoining buildings advising them that the excavation is to be made and that the adjoining buildings should be protected. Said notification shall be delivered not less than 10 days prior to the scheduled starting date of the excavation.

❖ This section emphasizes the need to protect all existing public and private property bordering the proposed construction or demolition operations. The term "property" only alludes to existing buildings. As such, any building element or system must be provided with a safeguard that will limit the damage that could be caused from the processes involved to the equipment and materials used. Additionally, soil erosion and land disbursement control resulting from the construction or demolition operations must be provided to prevent spillage and spread of disturbed soil debris. The site must be graded in accordance with Sections 3303.4 and 3303.5 for demolition and must be maintained in a similar manner while there is construction taking place. The owner or owner's agent has the responsibility to provide a written notice 10 days in advance for any demolition or construction activities that may warrant bordering lots to be protected from damage.

SECTION 3308
TEMPORARY USE OF STREETS, ALLEYS AND PUBLIC PROPERTY

3308.1 Storage and handling of materials. The temporary use of streets or public property for the storage or handling of materials or of equipment required for construction or demolition, and the protection provided to the public shall comply with the provisions of the authority having jurisdiction and this chapter.

❖ In addition to jurisdictional regulations, this chapter establishes procedures for the storage of construction materials and equipment to protect access to public safety equipment, utilities and public transportation facilities.

3308.1.1 Obstructions. Construction materials and equipment shall not be placed or stored so as to obstruct access to fire hydrants, standpipes, fire or police alarm boxes, catch basins or manholes, nor shall such material or equipment be located within 20 feet (6096 mm) of a street intersection, or placed so as to obstruct normal observations of traffic signals or to hinder the use of public transit loading platforms.

❖ This section indicates that precautions for material and equipment storage and placement must be provided so as not to block or obstruct access to fire hydrants; standpipes (including fire department siamese connections for sprinklers and standpipes); fire or police alarm boxes; utility boxes and meters; catch basins or manholes or any other vital facility whose function contributes to the health, safety and welfare of the public. Also, storage must not be placed within 20 feet (6096 mm) of a street intersection if it obstructs the normal observation of traffic signals or hinders the use of any mass-transit loading platforms, such as sidewalk bus stops, taxi waiting areas, etc.

3308.2 Utility fixtures. Building materials, fences, sheds or any obstruction of any kind shall not be placed so as to obstruct free approach to any fire hydrant, fire department connection, utility pole, manhole, fire alarm box or catch basin, or so as to interfere with the passage of water in the gutter. Protection against damage shall be provided to such utility fixtures during the progress of the work, but sight of them shall not be obstructed.

❖ Utility fixtures, such as those used by electrical, telephone, water, gas or sewer companies, as well as fire protection devices must be hidden from view and not be blocked from access and use. When construction operations are near a utility fixture, precautions must be provided so as not to cause damage. The utility companies and the authority having jurisdiction, including the fire department, may have specific requirements and guidelines to follow to limit the possibility of damage and obstruction.

SECTION 3309
FIRE EXTINGUISHERS

[F] 3309.1 Where required. All structures under construction, alteration or demolition shall be provided with not less than one approved portable fire extinguisher in accordance with Section 906 and sized for not less than ordinary hazard as follows:

1. At each stairway on all floor levels where combustible materials have accumulated.
2. In every storage and construction shed.
3. Additional portable fire extinguishers shall be provided where special hazards exist, such as the storage and use of flammable and combustible liquids.

❖ A means to provide fire protection during the construction and demolition process is required. As such, this section indicates provisions for portable fire extinguishers. In addition to provisions of this section, the regulations of Section 906 and the *International Fire Code*® (IFC®) must be maintained.

3309.2 Fire hazards. The provisions of this code and the *International Fire Code* shall be strictly observed to safeguard against all fire hazards attendant upon construction operations.

❖ Methods, procedures and construction materials each contribute, in some way, to creating a fire hazard; therefore, in addition to the provisions of the code, the IFC must be used to regulate the proper means of safety that must be provided for a building that is being demolished, altered or constructed.

SECTION 3310 EXITS

3310.1 Stairways required. Where a building has been constructed to a height greater than 50 feet (15 240 mm) or four stories, or where an existing building exceeding 50 feet (15 240 mm) in height is altered, at least one temporary lighted stairway shall be provided unless one or more of the permanent stairways are erected as the construction progresses.

❖ Temporary stairways must be constructed in accordance with Section 1009, including riser, tread, guard and handrail requirements.

3310.2 Maintenance of exits. Required means of egress shall be maintained at all times during construction, demolition, remodeling or alterations and additions to any building.

Exception: Approved temporary means of egress systems and facilities.

❖ Access to all existing required exits is required to remain unobstructed and usable by any occupants within the existing building while construction or demolition work is being done. Note that depending upon the extent of work, an exit may become obstructed; therefore, an alternative exit must be provided if the blocked exit is a required exit.

[F] SECTION 3311 STANDPIPES

3311.1 Where required. Buildings four stories or more in height shall be provided with not less than one standpipe for use during construction. Such standpipes shall be installed where the progress of construction is not more than 40 feet (12 192 mm) in height above the lowest level of fire department access. Such standpipe shall be provided with fire department hose connections at accessible locations adjacent to usable stairs. Such standpipes shall be extended as construction progresses to within one floor of the highest point of construction having secured decking or flooring.

❖ The scope of this section is to provide fire safety procedures during the construction operation in accordance with the code and the IFC. Standpipes that are required by Section 905 to be a permanent part of the building must be installed and remain functional as construction or demolition progresses. Functional standpipes are required so that fire-fighting capability is available at all times within a reasonable proximity to all potential fire locations. Standpipes must be in place when a building or structure under construction exceeds 40 feet (12 192 mm) in height and thereafter as it progresses to its completed height. During demolition or construction, the standpipe must be operational no lower than one floor below the highest point of construction.

3311.2 Buildings being demolished. Where a building is being demolished and a standpipe exists within such a building, such standpipe shall be maintained in an operable condition so as to be available for use by the fire department. Such standpipe shall be demolished with the building but shall not be demolished more than one floor below the floor being demolished.

❖ Standpipes in buildings under demolition must remain in service during the demolition process so that fire-fighting capability is maintained. Similar to buildings under construction, the standpipe is to remain in operation up to one floor level below the floor being demolished. The purpose of this section is to establish the requirement for fire safety procedures during the demolition process in accordance with the code and the IFC.

3311.3 Detailed requirements. Standpipes shall be installed in accordance with the provisions of Chapter 9.

Exception: Standpipes shall be either temporary or permanent in nature, and with or without a water supply, provided that such standpipes conform to the requirements of Section 905 as to capacity, outlets and materials.

❖ During the construction or demolition process, standpipes complying with Section 3311.1 must be provided in a temporary or permanent location so that fire fighters will have a sufficient means to supply water to their connection. These standpipes must either be connected to a permanent water supply source or have the capability to be connected to one in accordance with Section 3311.4. All standpipes must comply with Section 905 and must be installed and provided in a building pursuant to the regulations of Chapter 9.

3311.4 Water supply. Water supply for fire protection, either temporary or permanent, shall be made available as soon as combustible material accumulates.

❖ This section is to provide a fire safety provision in accordance with the code and the IFC. As such, when a standpipe system is present during the construction process pursuant to Section 3311.1, a water source must be readily available to provide fire-fighting capabilities at all times.

[F] SECTION 3312 AUTOMATIC SPRINKLER SYSTEM

3312.1 Completion before occupancy. In buildings where an automatic sprinkler system is required by this code, it shall be unlawful to occupy any portion of a building or structure until the automatic sprinkler system installation has been tested and approved, except as provided in Section 110.3.

❖ A certificate of occupancy must not be issued by the building official if a required automatic sprinkler system is not approved; however, a temporary occupancy may be granted at the discretion of the building official.

3312.2 Operation of valves. Operation of sprinkler control valves shall be permitted only by properly authorized personnel and shall be accompanied by notification of duly designated parties. When the sprinkler protection is being regularly turned off and on to facilitate connection of newly completed segments, the sprinkler control valves shall be checked at the end of each work period to ascertain that protection is in service.

❖ The scope of this section is to provide fire safety procedures during construction operations in accordance with the code and the IFC. A sprinkler system must remain operable unless work is being done on it. In such a case, the water must only be shut off by authorized personnel coupled with the notification of the proper authorities so that a form of check and balance is achieved to make certain the water is turned back on.

Bibliography

The following resource materials are referenced in this chapter or are relevant to the subject matter addressed in this chapter.

DOL 29 CFR; Part 1910-74, *Occupational Safety and Health Standards*. Washington, DC: U.S. Department of Labor-Occupational Safety and Health Administration, 1974.

IFC-2003, *International Fire Code*. Falls Church, VA: International Code Council, 2003.

IPC-2003, *International Plumbing Code.* Falls Church, VA: International Code Council, 2003.

Chapter 34
Existing Structures

General Comments

Chapter 34 contains provisions for the alteration, repair, addition and change of occupancy of existing buildings and structures. Chapter 34 includes or refers to the code requirements for existing buildings and structures, exclusive of the administrative provisions presented in Chapter 1 (Section 102.6 regulates unaltered existing buildings). The format of this chapter readily identifies the alternative methods of code compliance for existing buildings and structures.

Section 3401 establishes the provisions for maintenance, repairs and compliance with other technical codes.

Section 3402 provides definitions of terms that are primarily associated with Chapter 34.

Section 3403 states general requirements and also includes provisions and referenced sections that regulate existing structural loads, nonstructural elements and alteration or replacement of existing stairways.

Section 3404 contains the requirements for fire escapes.

Section 3405 contains provisions for the installation of new glass and glass replacement.

Section 3406 describes the conditions that constitute a change of occupancy.

Section 3407 addresses the requirements for historic buildings and structures.

Section 3408 addresses the requirements for structures that are moved into or within a jurisdiction.

Section 3409 contains provisions for accessible routes, as well as accessible elements, within existing buildings and structures.

Section 3410 contains provisions for an alternative method of evaluation of additions, alterations or changes of occupancy in existing buildings and structures based on a numerical scoring system involving various safety issues and the degree of code compliance for each issue. This section contains the provisions to be included in the evaluation and the conditions on which that evaluation is based.

There are 18 safety parameters of every existing building that must be evaluated, including: building height; building area; compartmentation; tenant and dwelling unit separations; corridor walls; vertical openings; heating, ventilating and air-conditioning (HVAC) systems; automatic fire detection; fire alarm systems; smoke control; means of egress capacity and number; dead-end pockets and corridors; maximum travel distance to an exit; elevator controls; means of egress emergency lighting; mixed occupancies; sprinklers and incidental use areas.

The procedure for evaluating the 18 safety parameters in a qualitative and quantitative matrix has evolved from several different sources. A particularly strong proponent for this methodology has been the National Institute of Building Sciences (NIBS). NIBS has published eight rehabilitation guidelines for the U.S. Department of Housing and Urban Development (HUD).

Section 3410 also contains the necessary database information, the form for the evaluation process and the mandatory safety scores by which each occupancy must comply. The procedure for determining whether a building meets the pass/fail criteria and which mandatory score is needed for mixed occupancies is also included in this section.

Purpose

A large number of existing buildings and structures do not comply with the current building code requirements for new construction. Although many of these buildings are potentially salvageable, rehabilitation is often cost prohibitive because they may not be able to comply with all the requirements for new construction. At the same time, it is necessary to regulate construction in existing buildings that undergo additions, alterations, renovations, extensive repairs or change of occupancy. Such activity represents an opportunity to ensure that new construction complies with the current building codes and that existing conditions are maintained, at a minimum, to their current level of compliance or are improved as required. To accomplish this objective, and to make the rehabilitation process easier, this chapter allows for a controlled departure from full compliance with the technical codes, without compromising the minimum standards for fire prevention and life safety features of the rehabilitated building.

[EB] SECTION 3401
GENERAL

3401.1 Scope. The provisions of this chapter shall control the alteration, repair, addition and change of occupancy of existing structures.

Exception: Existing bleachers, grandstands and folding and telescopic seating shall comply with ICC 300-02.

❖ This section states the scope of this chapter and references alternative methods of code compliance for alteration, repair, addition and change of occupancy of existing structures. This section also defines the responsibilities for maintenance, repairs, compliance with other codes and periodic testing.

3401.2 Maintenance. Buildings and structures, and parts thereof, shall be maintained in a safe and sanitary condition. Devices or safeguards which are required by this code shall be maintained in conformance with the code edition under which installed. The owner or the owner's designated agent shall be responsible for the maintenance of buildings and structures. To determine compliance with this subsection, the building official shall have the authority to require a building or structure to be reinspected. The requirements of this chapter shall not provide the basis for removal or abrogation of fire protection and safety systems and devices in existing structures.

❖ This section establishes the owner's responsibility to keep the building maintained in accordance with the code and the other referenced *International Codes*®.

The building official has the authority to rule on the performance of maintenance work when public health or safety is affected. The building official also has the authority to require a building to be maintained in compliance with the health and safety provisions required by the code.

Fire protection and safety systems in existing structures are to remain in place and be maintained. The removal of a previously required existing safety system is not permitted, even if the structure would meet the minimum requirements of this chapter without that system. For example, an existing sprinkler system in part of a building must remain functional even if the system was not required to meet the compliance alternative provisions of Section 3410.

3401.3 Compliance with other codes. Alterations, repairs, additions and changes of occupancy to existing structures shall comply with the provisions for alterations, repairs, additions and changes of occupancy in the *International Fire Code, International Fuel Gas Code, International Plumbing Code, International Property Maintenance Code, International Private Sewage Disposal Code, International Mechanical Code, International Residential Code* and ICC *Electrical Code.*

❖ This section clarifies the relationship between this chapter and the *International Fuel Gas Code*® (IFGC®), *International Mechanical Code*® (IMC®), *International Plumbing Code*® (IPC®) and the *International Residential Code*® (IRC®). When alterations and repairs are made to existing mechanical and plumbing systems, the provisions of the *International Codes* for alterations and repairs must be followed. Those codes indicate the extent to which existing systems must comply with the stated requirements. Where portions of existing building systems, such as plumbing, mechanical and electrical systems, are not being altered or repaired, those systems may continue to exist without being upgraded as long as they are not hazardous or unsafe to the building occupants.

[EB] SECTION 3402
DEFINITIONS

3402.1 Definitions. The following term shall, for the purposes of this chapter and as used elsewhere in the code, have the following meaning:

❖ Definitions of terms can be helpful in the understanding and application of the code requirements. The purpose for including this definition within this chapter is to provide more convenient access to it without having to refer back to Chapter 2. For convenience, this term is also in Chapter 2 with a cross reference to this section.

TECHNICALLY INFEASIBLE. An alteration of a building or a facility that has little likelihood of being accomplished because the existing structural conditions require the removal or alteration of a load-bearing member that is an essential part of the structural frame, or because other existing physical or site constraints prohibit modification or addition of elements, spaces or features which are in full and strict compliance with the minimum requirements for new construction and which are necessary to provide accessibility.

❖ This term is defined in order to provide a basis for the application of accessibility provisions to existing buildings (see Section 3409). Bringing any given existing site or building that is altered into full compliance with all accessibility requirements applicable to new construction may require extraordinary effort because of existing physical characteristics. The code utilizes the concept of technical infeasibility to provide a basis for exceptions from strict compliance with the provisions for new construction in an existing building.

[EB] SECTION 3403
ADDITIONS, ALTERATIONS OR REPAIRS

3403.1 Existing buildings or structures. Additions or alterations to any building or structure shall conform with the requirements of the code for new construction. Additions or alterations shall not be made to an existing building or structure which will cause the existing building or structure to be in violation of any provisions of this code. An existing building plus additions shall comply with the height and area provisions of Chapter 5. Portions of the structure not altered and not affected by the alteration are not required to comply with the code requirements for a new structure.

Exception: For buildings and structures in flood hazard areas established in Section 1612.3, any additions, alterations or repairs that constitute substantial improvement of the existing structure, as defined in Section 1612.2, shall comply with the flood design requirements for new construction and all aspects of the existing structure shall be brought into compliance with the requirements for new construction for flood design.

❖ The purpose of this section is to establish the guidelines for additions or alterations to existing buildings and structures.

An addition is an increase in the area to an existing building. When a new building is erected immediately adjacent to an existing building and they are separated by a fire wall, it is considered a separate building, not an addition to the existing structure. The new building must be designed to comply with the technical provisions of Chapters 1 through 33; not with the provisions of this chapter. The existing building must be evaluated considering the elimination of the adjacent open space now occupied by the new building.

An existing structure that is of a type of construction that does not comply with the height and area limitations of Table 503 may not be added to, unless the type of construction is upgraded or the allowable building height and area are increased through the addition of a sprinkler system throughout both the existing building and the addition.

A building with a proposed addition is to be evaluated based on the type of construction of the existing building or the addition, whichever is the lower type. When reviewing for compliance with Table 503, see Section 602.1.1.

Figure 3403.1 shows an example of this situation. If a sprinkler system is not planned for both the existing building and the addition, the proposed addition does not meet the requirements of Table 503. The area limitation in Table 503 for the existing construction Type VB, Group F-1 is 8,500 square feet (789.7 m^2) and the total proposed area is 11,000 square feet (1022 m^2). The area evaluation is based on Type VB construction, even though the proposed addition is Type VA construction, because the allowable area in Table 503 for Type VB construction is less than the allowable area for Type VA construction.

One way to satisfy the requirements of Table 503 is to upgrade the type of construction of the existing building to Type VA, which has an allowable area of 14,000 square feet (130 m^2). Another solution is to install an automatic sprinkler system that complies with NFPA 13 throughout the building. The revised allowable area of the building would be:

8,500 square feet + (3 × 8,500 square feet)
= 34,000 square feet (3,158.7 m^2)

in accordance with the sprinkler area increase, which is more than the proposed building area of 11,000 square feet (1022 m^2).

Alterations include renovations, which implies that something is changed in the structure. For example, the removal, rearrangement or replacement of partition walls in an office building is an alteration because, in part, of possible impact on the means of egress, fire resistance or other life safety features of the building. Conversely, the replacement of damaged trim pieces on a door frame is considered a repair, not an alteration.

Alterations are to conform to the requirements for a new structure. For example, consider the corridor of an office building that is to be extended 18 feet (5486 mm), as shown in Figure 3403.2. The existing office building has no sprinkler system (and none is proposed), so the corridor extension walls and doors are to be fire-resistance rated in accordance with Section 1016. This is applicable even if, for some reason, the walls and doors of the existing corridor system are not fire-resistance-rated construction.

This section also indicates that unaltered portions of a structure are not required to comply with code provisions for new construction. All requirements applicable to existing, unaltered areas are contained in the code (see Sections 102.6 and 115) or the *International Property Maintenance Code*® (IPMC®) and the *International Fire Code*® (IFC®).

Assume that the existing corridor walls in the example office building shown in Figure 3403.2 are not fire-resistance-rated construction.

Should this condition be deemed "unsafe" in accordance with Sections 102.6 and 115? The code gives the building official the authority to make these judgements based on the specific conditions unique to each existing building's characteristics. The IFC and the IPMC do not require fire-resistance-rated construction of existing corridor walls, nor do they require existing doors to be fire rated; however, they do require the doors to be self-closing. The corridor doors, therefore, must be brought into compliance with this requirement. The lack of a fire-resistance rating on an existing corridor is not, in itself, an example of an unsafe condition as defined in Section 115.1.

3403.2 Structural. Additions or alterations to an existing structure shall not increase the force in any structural element by more than 5 percent, unless the increased forces on the element are still in compliance with the code for new structures, nor shall the strength of any structural element be decreased to less than that required by this code for new structures. Where repairs are made to structural elements of an existing building, and uncovered structural elements are found to be unsound or otherwise structurally deficient, such elements shall be made to conform to the requirements for new structures.

❖ Wherever an addition to an existing building is made, the affected members of the original structure must be assessed to determine their ability to resist the increased forces. Because an addition typically adds new loads to the existing building, the architect or engineer responsible for the design of the project must analyze the original construction to determine whether any of the structural components, including foundations, require reinforcement. Where an existing building was initially designed to support the loads of future additions, the building official should seek verification to establish that the proposed addition will result in the affected building complying with the current code requirements. All new construction is also required to comply with current code requirements.

Structural element forces may be increased by no more than 5 percent, but only if the structural element stresses do not exceed those given by the material standards referenced by the code at the time of construction. This restriction does not apply if the structural element meets the current material standards and is permitted by the code as if it were new.

When remodeling or repairing an existing building, structural elements, including any uncovered structural elements, that are found to be deficient must be reinforced or replaced. If the structural integrity has been affected, repairs are to be made in compliance with current code provisions.

3403.2.1 Existing live load. Where an existing structure heretofore is altered or repaired, the minimum design loads for the structure shall be the loads applicable at the time of erection, provided that public safety is not endangered thereby.

❖ Where an existing building is undergoing alterations or repairs requiring changes to the structural loading, the applicable minimum design loads are those required by the code at the time of the original construction. Should the original design loads (typically live loads) and stresses be substantially different from the current requirements, or other pertinent design criteria are added to the code that would lead the building official to believe that the public would be endangered, he or she has the authority to order investigation and correction of the deficiencies based on the load requirements and other applicable provisions of the current code.

3403.2.2 Live load reduction. If the approved live load is less than required by Section 1607, the areas designed for the reduced live load shall be posted in with the approved load. Placards shall be of an approved design.

❖ It is not uncommon for the design live load requirements to change from the time a building is originally designed to when it undergoes a renovation or change of occupancy. The live loads used in the original design may have been adequate for the building's initial use and may have been in compliance with all the code requirements that were in effect at that time. Many years and many code changes can change the status of the structural design parameters and code requirements, which does not mean the existing structural system is inadequate and cannot be used when the building is renovated or has a change of occupancy. It just means that when the live loads that were used for the design of the existing building are lower than the those required by current standards, the design live loads used for the original design must be posted.

3403.3 Nonstructural. Nonstructural alterations or repairs to an existing building or structure are permitted to be made of the same materials of which the building or structure is constructed,

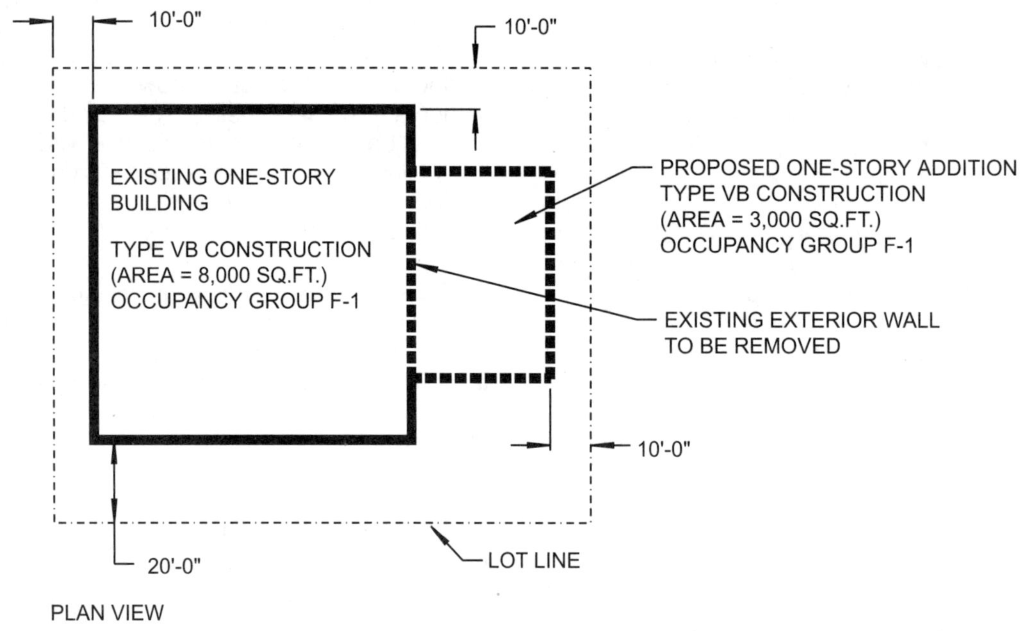

For SI: 1 inch = 25.4 mm, 1 foot = 304.8 mm, 1 square foot =0.0929 m^2.

Figure 3403.1
EXISTING BUILDING WITH ADDITION

provided that they do not adversely affect any structural member or the fire-resistance rating of any part of the building or structure.

❖ Where an existing building is altered or repaired, this section permits nonstructural materials that are removed to be replaced with the same type of materials that were used in the existing building. This provision is limited to the replacement use of nonstructural materials only, and only when the code does not specifically require otherwise. This section permits an existing wood glazing frame to be replaced with a new wood frame, but Section 3405.1 requires the replacement glazing to comply with the provisions for new construction.

3403.4 Stairways. An alteration or the replacement of an existing stairway in an existing structure shall not be required to comply with the requirements of a new stairway as outlined in Section 1009 where the existing space and construction will not allow a reduction in pitch or slope.

❖ An existing stair that is rebuilt in place is not required to comply with the provisions of Section 1009.3 for tread and riser dimensions. The rebuilt stair is permitted to have the same tread and risers as the existing stair. The term "rebuilt" in this section is intended to mean extensively repaired or removed and reconstructed in the same location.

[EB] SECTION 3404 FIRE ESCAPES

3404.1 Where permitted. Fire escapes shall be permitted only as provided for in Sections 3404.1.1 through 3404.1.4.

❖ Sections 3404.1.1 through 3404.5 address the current requirements for the use of exterior fire escapes. Their use and features as defined in this section should not be confused with the requirements for exterior exit stairways as defined in Section 1022.

The use of exterior fire escapes as a means of egress was popular in building designs of the past. They were used for economy of construction and to "save" usable space in the buildings they served. Being located outside the building, they were also considered safer than the unenclosed interior exit stairways of the time. Over the years, exterior fire escapes lost their appeal for many reasons:

- Fire escapes were never an integral part of the building's design; they were a necessary appendage hung from or attached to the building wall.
- Fire escapes were most commonly constructed of cast-iron or steel, both of which required a high degree of maintenance to protect the corrodible materials from the effects of weather.
- Because fire escapes are "open structures," they are subject to icing during the winter months. This makes them dangerous to use, and in extreme cases they can become completely unusable.

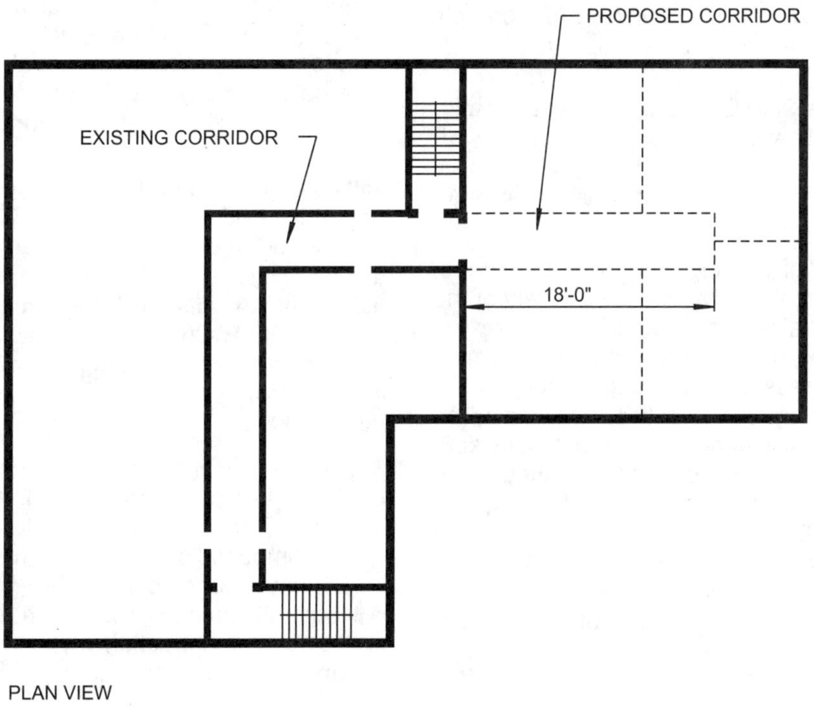

For SI: 1 foot = 304.8 mm.

Figure 3403.2
PROPOSED CORRIDOR EXTENSION

- People with a fear of heights can find exterior fire escapes difficult and sometimes impossible to use.
- Fire escapes were an unpleasant sight and an unwanted element in the architectural design.

The use of exterior fire escapes is all but obsolete except for existing buildings with a clear deficiency in the means of egress that cannot be reasonably rectified in other ways.

This section indicates that fire escapes are permitted only on existing buildings. New fire escapes may be installed on existing buildings only where exterior stairs cannot be used. Section 3404.2 addresses the location of fire escapes and Section 3404.3 addresses the construction of fire escapes.

3404.1.1 New buildings. Fire escapes shall not constitute any part of the required means of egress in new buildings.

❖ Because of the inherent hazards and lack of dependability, open exterior fire escapes are not permitted as part of the required means of egress in new construction.

3404.1.2 Existing fire escapes. Existing fire escapes shall be continued to be accepted as a component in the means of egress in existing buildings only.

❖ Exterior fire escapes in existing buildings are acceptable as a component of the required means of egress because, in most cases, it would be physically impractical and economically prohibitive to retrofit older buildings with interior or exterior exit stairways that comply with all the code requirements for new construction.

3404.1.3 New fire escapes. New fire escapes for existing buildings shall be permitted only where exterior stairs cannot be utilized due to lot lines limiting stair size or due to the sidewalks, alleys or roads at grade level. New fire escapes shall not incorporate ladders or access by windows.

❖ Continuing corrosion of existing exterior fire escapes, as well as other factors that may affect the safety of the structures, often make replacement of these fire escapes necessary. The building official may determine that the means of egress in an existing building is unsafe, requiring some corrective action, which may be the installation of a new exterior fire escape. This section permits new exterior fire escapes for existing buildings where exterior stairs conforming to the requirements of Section 1022 cannot be used because of either lot lines that limit stair size or encroachments on sidewalks, alleys or roads at grade level.

Because it was the accepted practice of the time, access to exterior fire escapes was quite often through windows located in corridors or individual offices. This practice is not permitted when replacement exterior fire escapes are used. The building must be altered to provide proper exit access to the fire escape.

The code does not permit the use of ladders as a component in the means of egress; therefore, ladders cannot be used as a component part of an exterior fire escape, nor can a ladder be used to access an exterior fire escape.

3404.1.4 Limitations. Fire escapes shall comply with this section and shall not constitute more than 50 percent of the required number of exits nor more than 50 percent of the required exit capacity.

❖ For reasons of overall life safety, exterior fire escapes may be used as a component of the means of egress only when the total number used does not exceed one-half the required number of exits and more than one-half the total required exit capacity.

3404.2 Location. Where located on the front of the building and where projecting beyond the building line, the lowest landing shall not be less than 7 feet (2134 mm) or more than 12 feet (3658 mm) above grade, and shall be equipped with a counterbalanced stairway to the street. In alleyways and thoroughfares less than 30 feet (9144 mm) wide, the clearance under the lowest landing shall not be less than 12 feet (3658 mm).

❖ In the past, exterior fire escapes were allowed to project beyond the property line and extend over sidewalks, alleyways and roads. Where such conditions are still accepted by the code (see Sections 3404.1.2 and 3404.1.3), the clearance to the lowest landing from grade level must be at least 7 feet (2134 mm), and not more than 12 feet (3658 mm) for exterior fire escapes located in front of the building. In alleyways and thoroughfares less than 30 feet (9144 mm) wide, the clearance under the lowest landing must be at least 12 feet (3658 mm). To facilitate these clearance requirements, exterior fire escapes are usually equipped with a counterbalanced stairway that retracts to a horizontal position when the fire escape is not in use.

3404.3 Construction. The fire escape shall be designed to support a live load of 100 pounds per square foot (4788 Pa) and shall be constructed of steel or other approved noncombustible materials. Fire escapes constructed of wood not less than nominal 2 inches (51 mm) thick are permitted on buildings of Type 5 construction. Walkways and railings located over or supported by combustible roofs in buildings of Type 3 and 4 construction are permitted to be of wood not less than nominal 2 inches (51 mm) thick.

❖ Traditionally and typically, exterior fire escapes are constructed of cast iron or steel, although the code does permit exterior fire escapes to be constructed of other noncombustible materials. The use of nominal 2-inch (51 mm) wood is permitted in Type V construction. Nominal 2-inch (51 mm) wood may also be used in Type III and IV construction when the exterior fire escape is supported by wood construction permitted by Table 601.

3404.4 Dimensions. Stairs shall be at least 22 inches (559 mm) wide with risers not more than, and treads not less than, 8 inches (203 mm) and landings at the foot of stairs not less than 40 inches (1016 mm) wide by 36 inches (914 mm) long, located not more than 8 inches (203 mm) below the door.

❖ The normal dimensions for interior and exterior exit stairways (see Section 1009) used in the means of egress do not apply to exterior fire escapes. The dimensional standard for exterior fire escapes has been in place for many years, thus many exterior fire escapes in use were designed to comply with this standard. Many of these fire escapes are installed in places where it would be difficult, and often impossible, to retrofit a new fire escape if a new set of dimensional standards were employed. For this reason it is judged appropriate to maintain the current standard. The minimum tread, riser, stair width and landing size dimensions for fire escape construction remain unchanged from the past.

3404.5 Opening protectives. Doors and windows along the fire escape shall be protected with 3/4-hour opening protectives.

❖ Safe exit access and safe exits are required for occupants using an exterior fire escape as a means of egress. Door and window openings in exterior walls adjacent to and along the path of travel of the exterior fire escape must be protected from the interior of the building with not less than $^{3}/_{4}$-hour protective. This is similar to the requirements for protection of openings on exterior stairs as described in Section 1020.4.

Where more restrictive opening protectives are required by other sections elsewhere in the code, the more restrictive protection must be used. The $^{3}/_{4}$-hour requirement for the exterior fire escape is the minimum when no other considerations are present (see Tables 601, 602, 704.8 and 714.2).

[EB] SECTION 3405
GLASS REPLACEMENT

3405.1 Conformance. The installation or replacement of glass shall be as required for new installations.

❖ The technical provisions for glass and glazing are detailed in Chapter 24. All technical requirements of the code for glass and glazing that apply to new construction also apply to all additions or alterations to existing buildings. There are no exceptions to this requirement. When glazing in an existing building is replaced or relocated within the same building, it must comply with current standards. Section 2401.2 contains specific requirements for replacement glazing.

[EB] SECTION 3406
CHANGE OF OCCUPANCY

3406.1 Conformance. No change shall be made in the use or occupancy of any building that would place the building in a different division of the same group of occupancy or in a different group of occupancies, unless such building is made to comply with the requirements of this code for such division or group of occupancy. Subject to the approval of the building official, the use or occupancy of existing buildings shall be permitted to be changed and the building is allowed to be occupied for purposes in other groups without conforming to all the requirements of this code for those groups, provided the new or proposed use is less hazardous, based on life and fire risk, than the existing use.

❖ A change in occupancy in an existing structure may change the level of inherent hazards that the code was initially intended to address.

Regardless of whether the change is to an occupancy considered to be more or less hazardous, this section applies the provisions of the code for new construction to an existing structure having a new occupancy. This is done so that the applicable code requirements adequately address the specific hazards of the new occupancy. For example, a change from an existing mercantile occupancy to a business occupancy renders all Group B provisions applicable to all portions of the structure where the occupancy has changed.

This section is one of the most frequently used provisions in the code for application to existing structures, since the occupancy in a building or structure is subject to change during the life of the building.

3406.2 Certificate of occupancy. A certificate of occupancy shall be issued where it has been determined that the requirements for the new occupancy classification have been met.

❖ An existing building that has been classified into a new occupancy group must receive a certificate of occupancy before tenancy. The code requirements for one occupancy group are not always the same as those for the new occupancy group. The new occupancy must be inspected to verify that all the applicable code requirements have been meet.

3406.3 Stairways. Existing stairways in an existing structure shall not be required to comply with the requirements of a new stairway as outlined in Section 1009 where the existing space and construction will not allow a reduction in pitch or slope.

❖ A stairway in an existing building does not have to be modified to comply with the dimensional provisions in Section 1009.3 for new stair construction. This provision is for the design of a particular stair and is aimed at addressing that stair only. This exception is not intended to eliminate any of the other technical provisions that might apply to an occupancy group, such as means of egress, accessibility, exit capacity, etc.

For example, when a building is altered or there is a change of tenant and the building is subject to a change in occupancy, the existing stair is permitted to remain as it was prior to the alteration or change of tenant. The means of egress capacity, however, must comply with the requirements for the new occupancy. The capacity may not be lessened to match the existing stair if it is not sufficient to provide the capacity required by the new occupancy. An additional stair or another approved means of egress must be provided to comply with the means of egress requirements for the new occupancy as indicated in Chapter 10.

[EB] SECTION 3407 HISTORIC BUILDINGS

3407.1 Historic buildings. The provisions of this code relating to the construction, repair, alteration, addition, restoration and movement of structures, and change of occupancy shall not be mandatory for historic buildings where such buildings are judged by the building official to not constitute a distinct life safety hazard.

❖ This section provides an exception from code requirements when the building in question has historic value. The most important criterion for application of this section is that the building must be essentially accredited as being of historic significance by a qualified party or agency. Usually this is done by a state or local authority after careful review of the historical value of the building. Most, if not all, states have such authorities, as do many local jurisdictions. The agencies with such authority can be located at the state or local government level or through the local chapter of the American Institute of Architects (AIA). Other considerations include the structural condition of the building (i.e., is the building structurally sound), its proposed use, its impact on life safety and how the intent of the code, if not the letter, will be achieved.

3407.2 Flood hazard areas. Within flood hazard areas established in accordance with Section 1612.3, where the work proposed constitutes substantial improvement as defined in Section 1612.2, the building shall be brought into conformance with Section 1612.

Exception: Historic buildings that are:

a. Listed or preliminarily determined to be eligible for listing in the National Register of Historic Places; or

b. Determined by the Secretary of the U.S. Department of Interior as contributing to the historical significance of a registered historic district or a district preliminarily determined to qualify as an historic district; or

c. Designated as historic under a state or local historic preservation program that is approved by the Department of Interior.

❖ This section is simply a reminder to the user of the code that, when substantial improvements are made to historic structures, there could be some ramifications regarding flood hazard that need to be considered, notwithstanding the exception stated in Section 3407.1. In other words, when there are substantial improvements to a building in a flood hazard area, as defined in Section 1612.2, the building official is not given the same discretion, as given in Section 3407.1, to waive code requirements for the alteration or restoration of a historic building.

[EB] SECTION 3408 MOVED STRUCTURES

3408.1 Conformance. Structures moved into or within the jurisdiction shall comply with the provisions of this code for new structures.

❖ Moved structures generally are required to comply with the provisions applicable to new construction. The moved structure may comply with the alternative provisions of Section 3410 instead of the code requirements for new structures, which may be particularly useful if the moved structure is older than the effective date of the adoption of building codes within the jurisdiction. The fire separation distance of the moved structure must comply with requirements for new structures even if the compliance alternative provisions of Section 3410 are used to meet the code requirements.

[EB] SECTION 3409 ACCESSIBILITY FOR EXISTING BUILDINGS

3409.1 Scope. The provisions of Sections 3409.1 through 3409.8 apply to maintenance, change of occupancy, additions and alterations to existing buildings, including those identified as historic buildings.

Exception: Type B dwelling or sleeping units required by Section 1107 are not required to be provided in existing buildings and facilities.

❖ The purpose of Section 3409 is to establish minimum criteria for accessibility when dealing with existing buildings and facilities. The history and efforts involved are similar to that discussed in Chapter 11 (see commentary, Chapter 11, "General Comments"). When a facility or element is altered, it must meet new code requirements. For example, if a door and frame are removed and replaced, the door must meet the requirements for width, height, maneuvering clearances and hardware. If just the doorknob is being removed, it must be replaced with lever hardware.

If the area undergoing alteration does not contain a primary function (see Section 3409.6), there are no additional requirements; however, if the area contains a primary function, there are additional criteria that may require work not in the original scope to achieve accessibility. This additional criteria is to provide an accessible route to the altered area, including any toilets and drinking fountains that serve it. Requirements for an accessible route might specify that the door previously discussed be removed and replaced because it did not have adequate width or maneuvering clearances.

The principle behind this approach to upgrading existing buildings is that they will become more accessible over time. A valid time to work towards that goal is when a structure is being altered. Special considerations are offered because of the difficulty involved in dealing with existing facilities that may not have been built with accessibility for physically disabled persons in mind. For example, when a historically registered home is being made into a museum, if changing the front door to allow for wheelchair access would alter the historical significance, alternatives are offered in Section 3409.8. Another example is the alternatives offered in Section 3409.7 if it is technically infeasible to provide full accessibility in an existing building. Please note that the term "technically infeasible" refers to either movement of a major structural element or other physical constraints. For example, a ramp to provide entrance or exit from a particular door may not be possible because of property lines or setback constraints.

Appendix E addresses accessibility in existing buildings for items that are not typically enforceable through the normal building code enforcement approach.

Section 3409.1 addresses accessibility in existing buildings that are being renovated or altered. The exception for Type B dwelling and sleeping units in existing buildings being altered or undergoing a change of occupancy is for consistency with the Fair Housing Amendments Act (FHAA). Additions that contained dwelling and sleeping units required to be Type B in new construction would be required to meet Type B dwelling unit requirements. Accessible and Type A dwelling and sleeping units are required in existing institutional and residential buildings undergoing additions or alterations (see Section 3409.7.7).

3409.2 Maintenance of facilities. A building, facility or element that is constructed or altered to be accessible shall be maintained accessible during occupancy.

❖ Continued compliance with the accessibility requirements of the code is dependent on maintenance of such facilities throughout the life of the building. For example, drinking fountains that are required to be accessible are of little value if they malfunction through deterioration or failure of any of the working parts. In other cases, inoperable elevators, locked accessible doors and obstructed accessible routes must be maintained such that they are readily usable by individuals with disabilities.

3409.3 Change of occupancy. Existing buildings, or portions thereof, that undergo a change of group or occupancy shall have all of the following accessible features:

1. At least one accessible building entrance.
2. At least one accessible route from an accessible building entrance to primary function areas.
3. Signage complying with Section 1110.
4. Accessible parking, where parking is being provided.
5. At least one accessible passenger loading zone, when loading zones are provided.
6. At least one accessible route connecting accessible parking and accessible passenger loading zones to an accessible entrance.

Where it is technically infeasible to comply with the new construction standards for any of these requirements for a change of group or occupancy, the above items shall conform to the requirements to the maximum extent technically feasible. Change of group or occupancy that incorporates any alterations or additions shall comply with this section and Sections 3409.4, 3409.5, 3409.6 and 3409.7.

❖ This section establishes that when an existing building or a portion thereof undergoes a change of occupancy, full compliance with accessibility requirements is expected and reasonable for the portion undergoing the change of occupancy, except where technical infeasibility can be demonstrated. If full compliance is technically infeasible, the element must be made accessible to the fullest extent that is feasible. This is consistent with the general approach that has always been taken relative to other matters regulated by the code. Changes of occupancy have always been viewed as a reasonable opportunity to secure a higher level of code compliance than is necessarily deemed acceptable for alterations.

Six items are required in a facility undergoing a change of occupancy. The intent is to provide the bare minimum to get people from a point of arrival into the building and to the area of primary function for the new use. This is not based on any specific provisions of the *Americans with Disabilities Act Accessibility Guidelines* (ADAAG), but parallels the intent of the requirements for removal of barriers. If the area undergoing a change of occupancy contains a primary function, the accessible route provisions in Section 3409.6 are also applicable (see commentary, Section 3409.6).

3409.4 Additions. Provisions for new construction shall apply to additions. An addition that affects the accessibility to, or contains an area of primary function, shall comply with the requirements in Section 3409.6 for accessible routes.

❖ Additions must comply with new construction. An addition, however, is also an alteration to an existing building; therefore, accessible route provisions for existing buildings are applicable (see commentary, Section 3409.6). For example, a new dining area is added in a restaurant. All accessible elements within the parameter of the addition must be constructed accessible. If the route to the addition or the bathrooms is in the existing building, it must be evaluated to see if it needs to be altered.

3409.5 Alterations. A building, facility or element that is altered shall comply with the applicable provisions in Chapter 11 and ICC A117.1, unless technically infeasible. Where compliance with this section is technically infeasible, the alteration shall provide access to the maximum extent technically feasible.

Exceptions:

1. The altered element or space is not required to be on an accessible route, unless required by Section 3409.6.

2. Accessible means of egress required by Chapter 10 are not required to be provided in existing buildings and facilities.

❖ The code approaches application of accessibility provisions to a facility that is altered by broadly requiring full conformance to new construction, meaning full accessibility is expected. Exceptions are then provided to indicate the conditions under which less than full accessibility is permitted.

The circumstance under which full compliance with accessibility provisions is not required is when it is deemed to be technically infeasible (see the commentary on the definition of "Technically infeasible" in Section 3402.1). This is considered reasonable since, if not provided for, plans for alterations may be otherwise abandoned by the building owner. The opportunity to upgrade and increase the current level of accessibility in an existing building would then be lost. This concern is also embodied in the requirement that an altered element or space is expected to be made accessible to the extent to which it is technically feasible to do so. In this manner, the code accomplishes the greatest degree of accessibility while recognizing the justifiable difficulties that may be involved in providing full accessibility.

The availability and usability of accessible elements are critically dependent on the presence of an accessible route leading to the accessible elements. The requirement for an accessible route represents one of the potential difficulties in an existing building that does not currently have adequate accessible routes. For example, in a multistory building, the accessible route to upper floors will most often be provided in the form of an elevator. It may be technically infeasible to provide an elevator in an existing building that does not currently contain an elevator. For this reason, Exception 1 indicates that an altered element or space is not required to be on an accessible route, except to the extent required in Section 3409.6.

Exception 2 indicates that accessible means of egress are not required as a result of undertaking alterations to existing buildings. Strict compliance with Section 1007 is often technically infeasible. Note that this is not an exception for accessible entrance requirements.

3409.5.1 Extent of application. An alteration of an existing element, space or area of a building or facility shall not impose a requirement for greater accessibility than that which would be required for new construction.

Alterations shall not reduce or have the effect of reducing accessibility of a building, portion of a building or facility.

❖ The purpose of this section is to clarify where alterations and scoping for alterations requirements apply. The requirements in Sections 3409.6 and 3409.7 do not impose a higher level of accessibility than the level required in new construction. At the same time, alterations cannot result in a lesser degree of accessibility than existed before the alterations were undertaken.

3409.6 Alterations affecting an area containing a primary function. Where an alteration affects the accessibility to, or contains an area of primary function, the route to the primary function area shall be accessible. The accessible route to the primary function area shall include toilet facilities or drinking fountains serving the area of primary function.

Exceptions:

1. The costs of providing the accessible route are not required to exceed 20 percent of the costs of the alterations affecting the area of primary function.
2. This provision does not apply to alterations limited solely to windows, hardware, operating controls, electrical outlets and signs.
3. This provision does not apply to alterations limited solely to mechanical systems, electrical systems, installation or alteration of fire protection systems and abatement of hazardous materials.
4. This provision does not apply to alterations undertaken for the primary purpose of increasing the accessibility of an existing building, facility or element.

❖ An area containing a primary function is one in which a major activity for which the building or facility is intended is carried out. For example, the lobby of a hotel in which the registration and check-out desk is located would be a primary function area. Other examples would be the dining area of a restaurant, the meeting rooms or exhibition halls in a conference center, virtually all office and work areas in a business building and retail display areas in a mercantile occupancy. The key concept is that a primary function area is one that contains a major activity of the facility. Areas that contain activities not related to the main purpose of the facility would not be considered a primary function area. For example, a mechanical equipment room, storage closet, toilet facilities, corridors, lounges and locker rooms would not be considered primary function areas. With this background, it is clear that areas containing a primary function are clearly more critical in terms of the purpose for which people enter and use the facility; therefore, this section reflects that when such areas are altered or added, it is important to require that an accessible route to the primary function area be provided. When an accessible route to a primary function area is required by this section, an accessible route to such facilities, including any restrooms and drinking fountains serving the primary function area, must also be made accessible, even though such facilities and areas may not by themselves be considered primary function areas.

There are conditions under which it may not be reasonable to enforce strictly this requirement for an accessible route to an altered or added primary function area. Exception 1 approaches this by utilizing the cost of the alterations or addition as a basis for determining if providing an accessible route is reasonable. The requirement for a complete accessible route does not apply when the cost of providing it exceeds 20 percent of the cost of the alterations or addition to the primary function area. These costs are intended to be based on the actual costs of the planned alterations or addition to the primary function area before consideration of the cost of providing an accessible route. For example, if the planned alterations will cost $100,000, not including the cost of an accessible route to a primary function area, this exception would apply if the additional cost of providing the accessible route would exceed $20,000.

It is not the intent to exempt all requirements for accessibility when the total cost for providing the accessible route exceeds the 20-percent threshold. Improvements to the accessible route are required to the extent that costs do not exceed 20 percent of the cost to the planned alteration or addition. It is not required that the full 20 percent be spent. If the accessible route (including accessible bathrooms) is already provided, no additional expenditure is required. Note that there is not a priority list given for where money should be spent on improving the accessible route. The logical progression is access to the site, accessible exterior routes to accessible entrances, access throughout the facility, access to services within the facility, toilet and bathing rooms and, finally, drinking fountains. Evaluation on how and where the money available should best be spent must be made on a case-by-case basis. For example, if an accessible route is not available to an upper level, and the cost of an elevator is more than 20 percent of the cost of the renovation, then other alternatives could be investigated, such as a platform lift or limited access elevator, or adding the elevator pit and shaft at this time, with elevator equipment added later. If all such items are in excess of the 20-percent limit, perhaps the money available could be spent towards making the toilet rooms accessible. The idea is that existing buildings would become fully accessible over time.

Exceptions 2 and 3 identify certain alterations that are not intended to trigger the requirement for providing an accessible route to a primary function area. Alterations limited to such elements as windows, hardware, operating controls, electrical outlets, signage, mechanical, electrical and fire protection systems, including alterations for the purpose of abating a hazardous materials circumstance, do not affect the usability of a primary function area in the same manner as alterations that affect the floor plan or the configuration, location or size of rooms or spaces. It is therefore considered unreasonable to require the installation of an accessible route when the scope of alterations is limited to that reflected in these exceptions.

Note that the costs for these items are not "backed out" of the total cost for the alteration before applying Exception 1. Exceptions 2 and 3 are alterations limited to the specific items referenced.

Exception 4 is intended to avoid penalizing a building owner who is undertaking alterations or additions for the purpose of increasing accessibility. It is appropriate to encourage owners to make such alterations without requiring them to do more work simply because they chose to increase the accessibility of the space. This could otherwise have the opposite effect of discouraging such alterations to avoid the expense of undertaking more work and expense than was originally planned. For example, federal law [Americans with Disabilities Act (ADA)] requires that owners of existing buildings remove certain existing barriers to accessibility. Removal of such barriers may require a permit from the building official. It would be unreasonable to have such activity trigger the mandatory requirement for further alterations to accomplish accessibility beyond the original planned work. In principle, the code takes the view that some extent of greater accessibility is positive progress and should be encouraged, not penalized.

3409.7 Scoping for alterations. The provisions of Sections 3409.7.1 through 3409.7.11 shall apply to alterations to existing buildings and facilities.

❖ The specific provisions of this section are intended to reflect conditions under which less than full accessibility, as would be required in new construction, is permitted in altered areas. As previously discussed, Section 3409.5 requires altered areas to comply with the full range of accessibility-related provisions of the code for new construction. This section reflects a reasonable set of conditions under which a different level of accessibility can be provided. Sections 3409.7.1 through 3409.7.12 are part of the code's coordination effort with ICC A117.1 and the recommendations for the ADAAG Review Federal Advisory Committee.

3409.7.1 Entrances. Accessible entrances shall be provided in accordance with Section 1105.

> **Exception:** Where an alteration includes alterations to an entrance, and the building or facility has an accessible entrance, the altered entrance is not required to be accessible, unless required by Section 3409.6. Signs complying with Section 1110 shall be provided.

❖ This provision is contained here to point to the accessibility provisions of Chapter 11 for entrances (see commentary, Section 1105.).

3409.7.2 Elevators. Altered elements of existing elevators shall comply with ASME A17.1 and ICC A117.1. Such elements shall also be altered in elevators programmed to respond to the same hall call control as the altered elevator.

❖ Requirements for new construction state that all elevators on an accessible route must be fully accessible in accordance with ICC A117.1. If a passenger elevator is altered, the altered element must be accessible in accordance with the requirements for existing elevators in ICC A117.1, Section 407.5. If the altered elevator is part of a bank of elevators, the same element must be made accessible in every elevator that is part of that bank. The purpose of this requirement is to have consistency among elevators in a bank so that disabled people are not required to wait for a specific elevator when the general population can take the first available elevator.

3409.7.3 Platform lifts. Platform (wheelchair) lifts complying with ICC A117.1 and installed in accordance with ASME A18.1 shall be permitted as a component of an accessible route.

❖ This section provides for the use of platform (wheelchair) lifts in existing buildings. In order to create an accessible route where there are changes in floor levels, the provisions for new construction would most often require the installation of an elevator or ramp. Platform lifts are allowed in new construction for limited conditions (see Section 1109.7). If the space in an existing building precludes the installation of an elevator or ramp, a platform lift may be the only practical solution. Given the choice between no accessibility or accessibility by a platform lift, accessibility is preferable. Note that in accordance with Section 1007.5, platform lifts are also permitted for an accessible means of egress; however, accessible means of egress are not required in existing buildings undergoing alterations in accordance with Section 3409.5, Exception 2.

3409.7.4 Stairs and escalators in existing buildings. In alterations where an escalator or stair is added where none existed previously, an accessible route shall be provided in accordance with Sections 1104.4 and 1104.5.

❖ If a stair or escalator is added as part of an alteration in a location where one did not previously exist, the alteration must also include an accessible route between the same two levels. If an accessible route is already available between the two levels, or if the stair or escalator is replacing an existing stair or escalator, this requirement is not applicable. In conjunction with Section 3409.5.1, if the requirement for the accessible route would be in excess of what is required for new construction, such as an accessible route to an area that was exempted by Section 1103.2, 1104.4, 1107 or 1108, this requirement is not applicable. The intent is that if a route is provided between accessible levels for a nondisabled person to use, it is reasonable to also expect an accessible route.

3409.7.5 Ramps. Where steeper slopes than allowed by Section [illegible]2 are necessitated by space limitations, the slope of ramps [illegible] providing access to existing buildings or facilities shall [illegible]ly with Table 3409.7.5.

❖ [illegible]his section recognizes the circumstances where, due to existing site or configuration constraints, a ramp with a slope of one unit vertical in 12 units horizontal (1:12) may not be feasible. A steeper slope is allowed where the elevation change does not exceed 6 inches (152 mm). The remainder of ramp requirements, such as width, landings, etc., are set forth in Section 1010.

TABLE 3409.7.5
RAMPS

SLOPE	MAXIMUM RISE
Steeper than 1:10 but not steeper than 1:8	3 inches
Steeper than 1:12 but not steeper than 1:10	6 inches

For SI: 1 inch = 25.4 mm.

❖ In existing buildings, ramps that rise 3 inches (76 mm) or less may have a slope as steep as one unit vertical in eight units horizontal (1:8). In existing buildings, ramps that rise 6 inches (152 mm) or less may have a slope as steep as one unit vertical in 10 units horizontal (1:10). If it is possible to provide a lesser slope, it is desirable to do so. These steeper slopes should only be utilized when the one unit vertical in 12 units horizontal (1:12) slope is not possible.

3409.7.6 Performance areas. Where it is technically infeasible to alter performance areas to be on an accessible route, at least one of each type of performance area shall be made accessible.

❖ This section recognizes that, because of the existing arrangement and location of performing areas (e.g., stages, platforms, orchestra pits, etc.), it may be infeasible to alter all performing areas to be on an accessible route. In such cases, it is reasonable to require that a minimum of one of each type of performing area be made accessible.

3409.7.7 Dwelling or sleeping units. Where I-1, I-2 , I-3, R-1, R-2 or R-4 dwelling or sleeping units are being altered or added, the requirements of Section 1107 for Accessible or Type A units and Chapter 9 for accessible alarms apply only to the quantity of spaces being altered or added.

❖ This section sets forth the rate for providing Accessible and Type A dwelling or sleeping units in Groups I-1, I-2, I-3, R-1, R-2 and R-4 when such facilities are altered. Assuming that Accessible or Type A units are not already provided, the number of Accessible or Type A units to be incorporated into each alteration is based on the number being altered. For example, if a nursing home was being altered a portion at a time, 50 percent of the units being altered each time would be required to be wheelchair accessible. It is not the intent that all units being altered are required to be Accessible until 50 percent of the units in the entire facility are accessible. The total number of Accessible units in the facility is not required to exceed that required for new construction, as indicated in Section 3409.5.1. It is unreasonable to require a greater level of accessibility in an existing building than is required in new construction.

This section also references visible and audible alarm requirements in Chapter 9. If the alarm system is part of the alteration, the alarms must comply with Section 907. Sleeping accommodations in Groups I-1 and R-1 are required to have visible alarms in accordance with Table 907.9.1.3. Section 907.9.1.4 also contains requirements for alarms within Group R-2 units. If the alarm system is not part of the alteration, it is not the intent of this section to require the fire alarm system to be upgraded (see Section 3403.1).

3409.7.8 Jury boxes and witness stands. In alterations, accessible wheelchair spaces are not required to be located within the defined area of raised jury boxes or witness stands and shall be permitted to be located outside these spaces where the ramp or lift access restricts or projects into the means of egress.

❖ This exception for jury boxes and witness stands is consistent with ADAAG Section 232. The intent is that if ramp access to a jury box or witness stand would have the ramp limiting or blocking the means of egress for the general population in the space, alternative locations for potential jurors or witnesses are viable.

3409.7.9 Toilet rooms. Where it is technically infeasible to alter existing toilet and bathing facilities to be accessible, an accessible unisex toilet or bathing facility is permitted. The unisex facility shall be located on the same floor and in the same area as the existing facilities.

❖ This section deals with circumstances in which it is technically infeasible to alter existing toilet facilities to be accessible. In new construction, both the men's and women's facilities would be required to be accessible. An alternative solution when it is technically infeasible to alter the existing toilet rooms would be the creation of a single unisex toilet or bathing room containing accessible facilities. If this alternative is selected, the room must be located on the same floor and in the same area as the existing toilet or bathroom. This is the best alternative to fully complying, separate men's and women's facilities. One might argue that it is technically infeasible to accomplish either of these alternatives, since the alternative to altering existing facilities involves creation of an additional toilet or bathroom that was not otherwise contemplated. This would not be a persuasive argument since there is likely to be space available somewhere in the facility to commit for use as a toilet or bathroom. In any case, the intent is that some form of accessible toilet room or bathing facility is necessary and must be provided. Signage must be provided at the inaccessible toilet rooms in accordance with Sections 1110.1 and 1110.2 to notify disabled persons when a facility is not accessible and direct them to the nearest accessible facilities. It should be noted that this alternative is not offered as a choice between making the existing separate-sex toilet rooms accessible or providing a unisex accessible toilet room. The existing separate-sex toilet rooms must be altered when it is technically feasible. Consideration of a unisex facility is only available when alteration of the existing toilet rooms is technically infeasible (see the definition for "Technically infeasible" in Section 3402.1).

3409.7.10 Dressing, fitting and locker rooms. Where it is technically infeasible to provide accessible dressing, fitting or locker rooms at the same location as similar types of rooms, one accessible room on the same level shall be provided. Where separate-sex facilities are provided, accessible rooms for each sex shall be provided. Separate-sex facilities are not required where only unisex rooms are provided.

❖ This section takes a similar approach for dressing rooms as provided for in Section 3409.7.9 for toilet rooms and bathing facilities. If it is technically infeasible to alter existing dressing rooms to be accessible, then space elsewhere on the level must be committed to providing no less than one accessible dressing room. In this case, if the existing dressing rooms provide separate rooms for each sex, then no less than one accessible dressing room for each sex must be provided.

3409.7.11 Check-out aisles. Where check-out aisles are altered, at least one of each check-out aisle serving each function shall be made accessible until the number of accessible check-out aisles complies with Section 1109.12.2.

❖ This section deals with accessible check-out aisles. As check-out aisles are altered, at least one serving each function is required to be made accessible until the total number of accessible check-out aisles for the function meets Section 1109.12.2. In facilities with 5,000 square feet (465 m^2) or less of selling space, only one check-out aisle is required to be made accessible (see Section 1109.12.2).

3409.7.12 Thresholds. The maximum height of thresholds at doorways shall be $^3/_4$ inch (19.1 mm). Such thresholds shall have beveled edges on each side.

❖ Thresholds at doorways may be $^3/_4$ inch (19.1 mm) maximum in existing buildings. In new construction, a typical threshold is $^1/_2$ inch (12.7 mm) maximum in accordance with Section 1008.1.6. This section recognizes that such things as differences in floor materials may create changes in elevation greater than that allowed in new construction. Edges of thresholds must be beveled to allow passage of a wheelchair.

3409.8 Historic buildings. These provisions shall apply to buildings and facilities designated as historic structures that undergo alterations or a change of occupancy, unless technically infeasible. Where compliance with the requirements for accessible routes, ramps, entrances or toilet facilities would threaten or destroy the historic significance of the building or facility, as determined by the authority having jurisdiction, the alternative requirements of Sections 3409.8.1 through 3409.8.5 for that element shall be permitted.

❖ For this section to be applicable, the building must be registered as historic. Historic buildings are treated much the same as provided for in Sections 3409.3 and 3409.6, in that a historic building that is altered or has undergone a change of occupancy is expected to comply with accessibility requirements, unless technical infeasibility can be demonstrated; however, this section also goes on to acknowledge that the historic character of a building may be adversely affected by strict compliance with accessibility provisions. For example, compliance with door width requirements may necessitate the removal of an existing set of doors that is critical to the historic character of the building. This section is intended to exempt such conditions in order to maintain the historic character of the building. Because limited extent of accessibility is desired in all facilities, Sections 3409.8.1 through 3409.8.5 allow for alternatives.

In an effort to coordinate with ADAAG requirements, items that cannot be enforced through the typical code enforcement process, but are associated with historical buildings, have been included in Appendix E.

3409.8.1 Site arrival points. At least one accessible route from a site arrival point to an accessible entrance shall be provided.

❖ Full compliance would require an accessible route from all site arrival points. If this requirement would adversely affect the historical significance of the building, the alternative available is to provide an accessible route from one site arrival point to an accessible entrance.

3409.8.2 Multilevel buildings and facilities. An accessible route from an accessible entrance to public spaces on the level of the accessible entrance shall be provided.

❖ Full compliance might require an accessible route to levels above or below, as well as throughout, the entrance level. If this requirement would adversely affect the historical significance of the building, the alternative is to provide an accessible route from the accessible entrance to all spaces open to the public on the entrance level. If elevators are provided, but are not accessible, signage in accordance with Section 1110.2 is required.

3409.8.3 Entrances. At least one main entrance shall be accessible.

Exceptions:

1. If a main entrance cannot be made accessible, an accessible nonpublic entrance that is unlocked while the building is occupied shall be provided; or
2. If a main entrance cannot be made accessible, a locked accessible entrance with a notification system or remote monitoring shall be provided.

Signs complying with Section 1110 shall be provided at the primary entrance and the accessible entrance.

❖ Full compliance would require 50 percent of the entrances to be accessible. If this requirement would adversely affect the historical significance of the building, only one main entrance is required to be made accessible. If a main entrance cannot be made accessible, then an employee or service entrance may serve as the accessible entrance, provided that it remains unlocked when the building is open. Alternatively, a locked entrance, where monitoring or a notification system is available, could be provided. Signage must be provided at inaccessible entrances in accordance with Sections 1110.1 and 1110.2.

3409.8.4 Toilet and bathing facilities. Where toilet rooms are provided, at least one accessible toilet room complying with Section 1109.2.1 shall be provided.

❖ Full compliance would require an accessible toilet/bathing facility at each location where toilet/bathing facilities are provided. If altering the existing facilities to be accessible would adversely affect the historical significance of the building, only one accessible toilet/bathing facility is required. Signage must be provided at inaccessible toilet rooms in accordance with Section 1110.2.

3409.8.5 Ramps. The slope of a ramp run of 24 inches (610 mm) maximum shall not be steeper than one unit vertical in eight units horizontal (12-percent slope).

❖ Full compliance would allow a maximum slope for ramps of one unit vertical in 12 units horizontal (1:12). If providing a fully compliant ramp would adversely affect the historical significance of the building, for an elevation change of 3 inches (76 mm) or less, a slope of one unit vertical in eight units horizontal (1:8) maximum is permitted.

[EB] SECTION 3410 COMPLIANCE ALTERNATIVES

3410.1 Compliance. The provisions of this section are intended to maintain or increase the current degree of public safety, health and general welfare in existing buildings while permitting repair, alteration, addition and change of occupancy without requiring full compliance with Chapters 2 through 33, or Sections 3401.3, and 3403 through 3407, except where compliance with other provisions of this code is specifically required in this section.

❖ Neither the designer nor the building official can physically inspect and evaluate every aspect of an existing building or structure, since many of its features may be concealed within the construction. It is therefore necessary to emphasize those items that can be evaluated. There are 18 critically important elements that can be quantified and evaluated to determine the level of safety for an existing building.

This type of analysis provides the designer and the building official with a rational basis for establishing the safety of an existing building or structure without having physical access to every part of the building, documentation of the original design or the construction history of a building.

3410.2 Applicability. Structures existing prior to [DATE TO BE INSERTED BY THE JURISDICTION. NOTE: IT IS RECOMMENDED THAT THIS DATE COINCIDE WITH THE EFFECTIVE DATE OF BUILDING CODES WITHIN THE JURISDICTION], in which there is work involving additions, alterations or changes of occupancy shall be made to conform to the requirements of this section or the provisions of Sections 3403 through 3407. The provisions in Sections 3410.2.1 through 3410.2.5 shall apply to existing occupancies that will continue to be, or are proposed to be, in Groups

A, B, E, F, M, R, S and U. These provisions shall not apply to buildings with occupancies in Group H or I.

❖ The adopting jurisdiction is to insert the desired date as indicated in this section. Section 3410 applies only to structures existing prior to the established date. The date that construction was first regulated through a comprehensive building code in the jurisdiction is recommended because buildings predating any building regulation are often not equipped with the types of systems and features that modern codes require. Newer buildings are more likely to be in compliance with current codes. These older buildings are assumed to face more difficulty in achieving a minimum level of life safety and are more likely to need the greater flexibility provisions of Section 3410.

The occupancies that qualify for the provisions of Section 3410 are listed in Section 3410.2.

Section 3410 does not apply to Groups H (high hazard) and I (institutional).

3410.2.1 Change in occupancy. Where an existing building is changed to a new occupancy classification and this section is applicable, the provisions of this section for the new occupancy shall be used to determine compliance with this code.

❖ When a building undergoes a change of occupancy classification and Section 3410 is applied, the evaluation method in Section 3410 must be applied to the new occupancy for determining whether the existing building meets the compliance alternative in the code.

3410.2.2 Partial change in occupancy. Where a portion of the building is changed to a new occupancy classification, and that portion is separated from the remainder of the building with fire barrier wall assemblies having a fire-resistance rating as required by Table 302.3.2 for the separate occupancies, or with approved compliance alternatives, the portion changed shall be made to conform to the provisions of this section.

Where a portion of the building is changed to a new occupancy classification, and that portion is not separated from the remainder of the building with fire separation assemblies having a fire-resistance rating as required by Table 302.3.2 for the separate occupancies, or with approved compliance alternatives, the provisions of this section which apply to each occupancy shall apply to the entire building. Where there are conflicting provisions, those requirements which secure the greater public safety shall apply to the entire building or structure.

❖ Where a portion of the building is changed to a new occupancy classification and is designed in accordance with Section 3410, the following options may be employed, dependent upon the fire separation of the portion of the building from the remainder of the existing building.

For a full fire separation, the new occupancy portion must be evaluated with the existing or proposed building design to be in full compliance with the provisions of Section 3410. The remainder of the existing building must also be evaluated in accordance with Section 3410. The mandatory safety scores for the new occupancy portion of the building and the existing occupancy are obtained from those listed in Table 3410.8 and are incorporated in the building's final evaluation score (see Table 3410.9).

For a nonfire-resistance-rated assembly of a new occupancy portion of an existing building, the provisions of Section 3410 for each occupancy must apply to the entire building. The requirements offering the greater public safety are applied to the entire building. Any proposed or existing building attributes and modifications must be reviewed in light of these requirements. See the examples described in Sections 3410.6.16 and 3410.9.1 for mixed occupancies.

3410.2.3 Additions. Additions to existing buildings shall comply with the requirements of this code for new construction. The combined height and area of the existing building and the new addition shall not exceed the height and area allowed by Chapter 5. Where a fire wall that complies with Section 705 is provided between the addition and the existing building, the addition shall be considered a separate building.

❖ The requirements in this section are applied in the same way as those in Section 3410.1. This section is included in Section 3410 so that provisions for additions are included for both optional methods of code compliance indicated in Section 3401.1. See the commentary to Section 3401.1 for a discussion of these options.

The evaluation method in Section 3410 may be used for an existing building that has been modified by an addition, provided that it meets the applicability requirements of Section 3410.2; however, the addition must comply with all the requirements of the code for new construction.

3410.2.4 Alterations and repairs. An existing building or portion thereof, which does not comply with the requirements of this code for new construction, shall not be altered or repaired in such a manner that results in the building being less safe or sanitary than such building is currently. If, in the alteration or repair, the current level of safety or sanitation is to be reduced, the portion altered or repaired shall conform to the requirements of Chapters 2 through 12 and Chapters 14 through 33.

❖ An existing building that is altered or repaired may be designed and evaluated in accordance with Section 3410, provided it meets the applicability provisions of Section 3410.2.

When an existing building is altered or repaired, materials or methods consistent with the original construction must be used. The alteration or repair must not cause the building to be less safe or sanitary than when it was originally constructed. Should the alteration or repair cause a reduction in safety or sanitation, such reductions must be corrected to meet the requirements of Chapters 2 through 12 and Chapters 14 through 33.

3410.2.5 Accessibility requirements. All portions of the buildings proposed for change of occupancy shall conform to the accessibility provisions of Chapter 11.

❖ Any building or part of a building that has a change of occupancy must meet the requirements for accessibility in Chapter 11; however, Section 3409 is an acceptable alternative since Section 3409 applies to all existing buildings that have alterations, repairs, additions or a change of occupancy.

3410.3 Acceptance. For repairs, alterations, additions and changes of occupancy to existing buildings that are evaluated in accordance with this section, compliance with this section shall be accepted by the building official.

❖ When an owner or designer of an existing building decides to apply Section 3410 and complies with all the provisions of the section, including the applicability requirements in Section 3410.2, the building official must accept for review the proposed work or change of occupancy. This mandatory acceptance is tempered by one exception, as stated in Section 3410.3.1.

3410.3.1 Hazards. Where the building official determines that an unsafe condition exists, as provided for in Section 115, such unsafe condition shall be abated in accordance with Section 115.

❖ When the building official finds an unsafe condition in the building that is not being corrected by the proposed work, he or she must order the abatement or correction of the unsafe condition or hazard just as would be ordered in an existing building that is not being renovated, as stipulated by Section 115. This section sets forth a required comprehensive performance objective of abating any condition that is unsafe, insanitary, illegal or of improper occupancy, egress deficient, fire hazardous, poorly maintained or otherwise dangerous to human life or the public welfare. Guidelines for abatement are provided in the code and in Chapter 1 of both the IPMC (see Sections 108 and 110) and the IFC.

3410.3.2 Compliance with other codes. Buildings that are evaluated in accordance with this section shall comply with the *International Fire Code* and *International Property Maintenance Code.*

❖ This section requires an existing building that is subjected to the evaluation scoring process of Section 3410.6 to also comply with the IFC and IPMC. Those codes provide minimum requirements for health and safety that all existing buildings are expected to meet, regardless of whether there are any changes being made to the building or occupancy. Regardless of an existing building's final safety scores, the requirements of these referenced codes must be followed so occupants are safeguarded from hazards. This provision is similar to the requirements of Section 3401.2.

3410.4 Investigation and evaluation. For proposed work covered by this section, the building owner shall cause the existing building to be investigated and evaluated in accordance with the provisions of this section.

❖ This section and the subsequent sections address what must be done by the owner or designer who elects to employ Section 3410 for a proposed rehabilitation program and the corresponding action by the building official to assess the program objectively for approval or denial. The following actions must be taken:

- Fully investigate the building using both on-site inspections and research of all available building construction documents;
- Evaluate the building for conformance to Sections 3410.5 through 3410.9.1;
- Analyze the structure and
- Submit to the building official documented results of the investigation and evaluation plus any proposed compliance alternatives.

The building official must then determine whether the existing building, proposed work or change in occupancy complies with Section 3409.

Thus, the election and implementation of Section 3410 by the designer or owner is part of a code compliance option that may be used for existing buildings that the building official has the responsibility to review and assess. A proper and satisfactorily prepared submission by the owner or designer offering qualitative and quantitative data requires review by the building official for approval. If the submission is rejected, the building official must specifically cite any deficiencies and violations.

The owner is to have the existing building and any proposed work investigated and evaluated for compliance with the 18 parameters specified in Sections 3410.6.1 through 3410.6.18.

3410.4.1 Structural analysis. The owner shall have a structural analysis of the existing building made to determine adequacy of structural systems for the proposed alteration, addition or change of occupancy. The existing building shall be capable of supporting the minimum load requirements of Chapter 16.

❖ The owner is required to perform a complete structural analysis of the building to ensure that it can support the required loads. This requires that all interior loads meet the minimum load requirements of Chapter 16. Existing and altered buildings must be shown to support the expected loading. Any existing exterior member not affected by either interior loading or additional exterior loading need not be evaluated, provided that the structural member has, over a period of time, proven its ability to withstand the forces that normally create stress. Loads imposed on existing structural members by alterations, additions or a change of occupancy must be shown to sustain the requirements of Chapter 16, as stated in Sections 3403.2 through 3403.2.2. This structural analysis provides the owner and the building official with reasonable assurance that the building is structurally safe.

3410.4.2 Submittal. The results of the investigation and evaluation as required in Section 3410.4, along with proposed compliance alternatives, shall be submitted to the building official.

❖ The results of the investigation, including structural analysis and evaluation, must be submitted to the building official. If alternative methods, materials or equivalency concepts are proposed, these must also be submitted to the building official for review and approval.

3410.4.3 Determination of compliance. The building official shall determine whether the existing building, with the proposed addition, alteration or change of occupancy, complies with the provisions of this section in accordance with the evaluation process in Sections 3410.5 through 3410.9.

❖ When the results of the investigation and evaluation are submitted to the building official, he or she must determine whether the proposed work conforms to the provisions of Section 3410 and whether the evaluation was performed in accordance with Sections 3410.5 through 3410.6.18.

3410.5 Evaluation. The evaluation shall be comprised of three categories: fire safety, means of egress and general safety, as defined in Sections 3410.5.1 through 3410.5.3.

❖ This section and the subsequent sections address three general areas of safety to be evaluated: fire safety (FS), means of egress (ME) and general safety (GS). Section 3410.6 and the subsequent sections address 18 safety parameters that reflect on those areas. Each of the 18 safety parameters indicated in Sections 3410.6.1 through 3410.6.18 must be carefully reviewed and assigned a numerical value that signifies the degree of safety influence on the three overall general safety categories.

The 18 safety parameters that are given assigned values are:

3410.6.1 Building height
3410.6.2 Building area
3410.6.3 Compartmentation
3410.6.4 Tenant and dwelling unit separations
3410.6.5 Corridor walls
3410.6.6 Vertical openings
3410.6.7 HVAC systems
3410.6.8 Automatic fire detection
3410.6.9 Fire alarm systems
3410.6.10 Smoke control
3410.6.11 Means of egress capacity and number
3410.6.12 Dead ends
3410.6.13 Maximum exit access travel distance
3410.6.14 Elevator control
3410.6.15 Means of egress emergency lighting
3410.6.16 Mixed occupancies
3410.6.17 Automatic sprinklers
3410.6.18 Incidental use

These 18 safety parameters have been determined to be the most critical factors related to the minimum degree of life safety and property protection needed in an existing building.

When mixed occupancies in an existing building are not separated by fire-resistance-rated assemblies or fire walls meeting the most restrictive fire rating of the different occupancies, the entire evaluation must be based on the occupancy with the most restrictive requirements. The evaluation process considers the score for the various occupancies and applies the lowest score to the entire building.

When the mixed occupancies are properly separated in compliance with Section 302.3.2, they are to be evaluated separately and the score for each occupancy will apply to each portion based on its use. Both height and area formulas are computed for each occupancy. See the commentary to Sections 3410.2.2 and 3410.6.16 for a further discussion of application for mixed occupancies.

If there are four different occupancies in the building, the values must be computed for each of the four occupancies. Each occupancy is required to meet only its own mandatory building score.

When mixed occupancies are separated by fire walls complying with the requirements of Section 705, separate buildings are created and must be evaluated separately for all 18 safety parameters. The assigning of numerical values to each of the 18 safety parameters establishes a measurable quantity of what each of the parameters contributes to the overall safety of the building. Some evaluated parameters have a negative influence; others have a positive one. In total, the parameters may or may not result in an acceptable building score. The evaluation will determine whether the existing building has enough positive factors to overcome the negative parameters, or will indicate the negative factors that must be upgraded by alternative modifications.

The evaluation is divided into three categories or areas, which are defined in Sections 3410.5.1 through 3410.5.3.

3410.5.1 Fire safety. Included within the fire safety category are the structural fire resistance, automatic fire detection, fire alarm and fire suppression system features of the facility.

❖ A partial list of the items used to evaluate fire safety in a building is given in this section.

3410.5.2 Means of egress. Included within the means of egress category are the configuration, characteristics and support features for means of egress in the facility.

❖ The means of egress features that are evaluated by Section 3410 fall into the general areas of configuration, characteristics and support features. The specific features include travel distance, dead ends, emergency lighting and exit capacity and number.

3410.5.3 General safety. Included within the general safety category are the fire safety parameters and the means of egress parameters.

❖ This category includes every item that is used in either the fire safety or means of egress evaluation.

3410.6 Evaluation process. The evaluation process specified herein shall be followed in its entirety to evaluate existing buildings. Table 3410.7 shall be utilized for tabulating the results of the evaluation. References to other sections of this code indicate that compliance with those sections is required in order to gain credit in the evaluation herein outlined. In applying this section to a building with mixed occupancies, where the separation between the mixed occupancies does not qualify for any category indicated in Section 3410.6.16, the score for each occupancy shall be determined and the lower score determined for each section of the evaluation process shall apply to the entire building.

Where the separation between the mixed occupancies qualifies for any category indicated in Section 3410.6.16, the score for each occupancy shall apply to each portion of the building based on the occupancy of the space.

❖ This section is the key to understanding the entire evaluation process. The first sentence of this section clearly states that every one of the 18 safety parameters indicated in Sections 3410.6.1 through 3410.6.18 must be evaluated and nothing may be omitted. After the 18 safety parameters have been evaluated and assigned a numerical value, the values are entered in Table 3410.7. The values must be tabulated to obtain the building score for each of the three general categories.

Some of the 18 safety parameters listed require mandatory compliance with other sections of the code and establish the foundation for determining a proper evaluation of those parameters, regardless of whether the result is positive or negative. This involves a coordination of mandatory basic requirements with the respective existing building conditions to arrive at the numerical evaluations prescribed in Sections 3410.6.1 through 3410.6.18. Section 3410.6.16 also addresses how mixed occupancies are handled in the evaluation. To apply this section, it is necessary to understand Section 302. See the commentary to Sections 302, 3410.2.2, 3410.6.2.1 and 3410.6.16 for a further discussion of the application of the evaluation procedure for mixed occupancies.

3410.6.1 Building height. The value for building height shall be the lesser value determined by the formula in Section 3410.6.1.1. Chapter 5 shall be used to determine the allowable height of the building, including allowable increases due to automatic sprinklers as provided for in Section 504.2. Subtract the actual building height from the allowable and divide by 12 $^1/_2$ feet. Enter the height value and its sign (positive or negative) in Table 3410.7 under Safety Parameter 3410.6.1, Building Height, for fire safety, means of egress and general safety. The maximum score for a building shall be 10.

❖ As a starting point for the actual evaluation process, this section and Section 3410.6.1.1 define in detail how to perform the building height evaluation. The exact values are to be computed and compared against the mandatory safety scores. The values are not to be rounded. For calculation purposes, two decimal places would be appropriate. It is not necessary to build in any inaccuracy that occurs through the rounding process.

The maximum score for a building is now 10. Regardless of the building's allowable height as compared to its actual height, the intent of the code is to limit all buildings to a maximum score of 10.

3410.6.1.1 Height formula. The following formulas shall be used in computing the building height value.

$$\text{Height value, feet} = \frac{(AH)-(EBH)}{12.5} \times CF$$

$$\text{Height value, stories} = (AS - EBS) \times CF$$

(Equation 34-1)

where:

AH = Allowable height in feet from Table 503.

EBH = Existing building height in feet.

AS = Allowable height in stories from Table 503.

EBS = Existing building height in stories.

CF = 1 if (AH) - (EBH) is positive.

CF = Construction-type factor shown in Table 3409.6.6(2) if (AH) - (EBH) is negative.

Note. Where mixed occupancies are separated and individually evaluated as indicated in Section 3410.6, the values AH, AS, EBH and EBS shall be based on the height of the fire area of the occupancy being evaluated.

❖ Two height formulas for calculating the score to be entered in Table 3410.7 are given. The formulas use the allowable height in both feet and stories from Table 503 and the height of the existing building. The denominator in the formula, 12.5, represents an average story height in feet.

The actual story height and the overall existing building height are to be directly compared to Table 503. Table 503 serves as a datum level that allows for establishing a numerical height value for the existing building by comparing its actual height and its type of construction as represented by a construction factor (*CF*).

If the existing building actual height is less than or equal to the allowable height of Table 503, then the *CF* value is 1 (no negative factor is assigned). If the actual height exceeds the Table 503 allowable height, the building is not in compliance with Table 503 and represents a safety deficiency. A deficiency rating is assigned that is dependent on the type of construction. The construction factors to establish this negative value are given in Table 3410.6.6(2).

When a building is not in compliance with Table 503, it is considered less safe than a building that does comply with Table 503. As a result, deficiency points are assessed. Additional safeguards must be provided to compensate for this condition. This is the primary reason that a different construction factor must be used for buildings not in compliance. The construction factor is the equivalent deficiency that must be overcome by providing additional protection in other areas that are to be evaluated.

Example 1:

A six-story building of Type IB construction is 60 feet (18 288 mm) tall. The building is not sprinklered. It contains a Group B business occupancy.

The allowable height from Table 503 is 11 stories, 160 feet (48,800 mm).

AH = 160 feet

AS = 11 stories

EBH = 60 feet

EBS = 6 stories

CF = 1

Height value in feet $= \dfrac{160 - 60}{12.5} \times CF$

$= 8 \times 1$

$= 8$

Height value in stories $= (11 - 6) \times CF$

$= 5 \times 1$

$= 5$

The building height value will be 5, because it is the lesser of the two height values.

Example 2:

A seven-story building of Type IIIB construction is 80 feet (24 400 mm) tall. The building is sprinklered. It contains a Group M mercantile occupancy.

The allowable height from Table 503 is four stories, 55 feet (16 775 mm).

AH = 55 + 20 (for sprinklers) = 75 feet

AS = 4 + 1 (for sprinklers) = 5 stories

EBS = 7 stories

EBH = 80 feet

CF = 3.5 from Table 3410.6.6(2)

Height value in feet $= \dfrac{75 - 80}{12.5} \times CF$

$= 0.4 \times 3.5$

$= -1.4$

Height value in stories $= (5 - 7) \times CF$

$= -2 \times 3.5$

$= -7$

The building height value will be -7, because it is the lesser of the two height values.

The above values determine the entries for Table 3410.7 under Safety Parameter 3410.6.1 Building Height.

In Example 1, the height value of 5 is entered into the columns for fire safety (FS), means of egress (ME) and general safety (GS).

In Example 2, the height value of -7 is entered into the columns for fire safety (FS), means of egress (ME) and general safety (GS).

The assessed height value parameter is only one of the 18 safety parameters that need to be evaluated. Example 1 shows a positive contribution; however, this may not be enough to result in a sufficient overall building score. Similarly, the negative value in Example 2 may not in itself result in an overall insufficient building score.

3410.6.2 Building area. The value for building area shall be determined by the formula in Section 3410.6.2.2. Section 503 and the formula in Section 3410.6.2.1 shall be used to determine the allowable area of the building. This shall include any allowable increases due to open perimeter and automatic sprinklers as provided for in Section 506. Subtract the actual building area from the allowable area and divide by 1,200 square feet (112 m^2). Enter the area value and its sign (positive or negative) in Table 3410.7 under Safety Parameter 3410.6.2, Building Area, for fire safety, means of egress and general safety. In determining the area value, the maximum permitted positive value for area is 50 percent of the fire safety score as listed in Table 3410.8, Mandatory Safety Scores.

❖ In this section, the code user is shown how to calculate the building score for the building area.

This section also requires the maximum score of a building in this category. The maximum value is 50 percent of just the fire safety score listed in Table 3410.8 (see commentary, Section 3410.6.2.2). This maximum score is applicable and equal for the three building score categories—fire safety, means of egress and general safety. The positive score is limited to prevent this one parameter from providing enough points to unjustifiably overcome too many deficiencies in other parameters.

3410.6.2.1 Allowable area formula. The following formula shall be used in computing allowable area:

$$AA = \frac{(SP + OP + 100) \times (\text{area, Table 503})}{100}$$

(Equation 34-2)

where:

AA = Allowable area.

SP = Percent increase for sprinklers (Section 506.3).

OP = Percent increase for open perimeter (Section 506.2).

❖ The formula used to calculate the allowable area of the building is a direct correlation of the area increases and reductions allowed for new buildings in Section 506.

The use of these increases and reductions is addressed in Chapter 5.

3410.6.2.2 Area formula. The following formula shall be used in computing the area value. Determine the area value for each occupancy fire area on a floor-by-floor basis. For each occupancy, choose the minimum area value of the set of values obtained for the particular occupancy.

$$\text{Area value } i = \frac{\text{Allowable area}_i}{1{,}200 \text{ square feet}}\left[1-\left(\frac{\text{Actual area}_i}{\text{Allowable area}_i}+\ldots+\frac{\text{Actual area}_n}{\text{Allowable area}_n}\right)\right]$$

(Equation 34-3)

where:

i = Value for an individual separated occupancy on a floor.

n = Number of separated occupancies on a floor.

❖ The area formula provides a numerical value for the actual building area to be entered in Table 3410.7 under Safety Parameter 3410.6.2, Building Area. If the area of the existing building is less than the allowable area, it is considered to be safer. Credit for area is given. If the area of the existing building is larger than the allowable area, it represents a condition that is judged to be less safe. The building receives a negative score. This can be overcome by providing additional safeguards.

To use the formula for existing buildings that contain mixed occupancies, an area value must be determined for each story of a building. If a building contains just one occupancy or just one occupancy on a particular floor separated from other floors, the formula simply reduces to the allowable area minus the actual area divided by the constant 1,200. The formula is also applicable to a story containing several separated occupancies. In such a situation, the formula requires each actual area to be divided by its respective allowable area. The resulting fractions are added together, subtracted from the constant of 1 and multiplied by the ratio of the allowable area of the particular use divided by the constant 1,200. This method of determining area values for several occupancies that are separated on a story is directly comparable to the unity formula in Section 302.3.2.

Example 1: Figure 3410.6.2.2(1) illustrates a Group B building of Type IIA construction. The building is five stories in height, unsprinklered.

The overall building area is 200 feet (60 960 mm) by 200 feet (60 960 mm).

$$\text{Area value} = \frac{75{,}000}{1{,}200}\left[1-\left(\frac{40{,}000}{75{,}000}\right)\right]$$

$$= 62.5\,[1 - 0.533]$$

$$= 29.2$$

Allowable area = 75,000 square feet
Actual area = 40,000 square feet

In Table 3410.7, enter the following values under Safety Parameter 3410.6.2, Building Area:

Fire safety (FS) = 12

Means of egress (ME) = 12

General safety (GS) = 12

These values are equal to 50 percent of the value in Table 3410.8 for the mandatory fire safety score (MFS) of 24 for a Group B occupancy. This is the maximum value credit permitted for the building area parameter of Section 3410.6.2, even though the area value has been computed as 29.2 in Section 3410.6.2.2. The purpose for limiting the positive score is to prevent this one parameter from providing enough points to unjustifiably overcome too many deficiencies in other parameters.

The positive values entered in Table 3410.7 under Safety Parameter 3410.6.2, Building Area (FS = 12, ME = 12 and GS = 12), represents the evaluation for only one of the 18 safety parameters that must be assessed to arriving at the overall building score evaluation of Table 3410.7. As previously described for the building height parameter values under line 3410.6.1 of Table 3410.7, a single parameter of the total 18 safety parameters to be assessed does not, in itself, determine the ultimate acceptable building score.

Example 2: Figure 3410.6.2.2(2) illustrates a separated mixed occupancy building (Groups B, M and S-1) of Type IIIB construction. The building is two stories in height, fully sprinklered.

The overall building area is 100 feet (30 480 mm) by 200 feet (60 960 mm).

Actual area for Group M on first story = 20,000 square feet.

Actual area for Group M on second story = 4,400 square feet.

Actual area for Group B on second story = 10,000 square feet.

Actual area for Group S-1 on second story = 5,600 square feet.

SP = 200% (from Section 506.3)

OP = 16.7% (from Section 506.2)

Tabular areas from Table 503:

Group B = 19,000 square feet

Group M = 12,500 square feet

Group S-1 = 17,500 square feet

Allowable Areas:

$$AA = \frac{(200 + 16.7 + 100) \times \text{Tabular Area}}{100}$$

AA for Group B = 60,173 square feet

AA for Group M = 39,587 square feet

AA for Group S-1 = 55,422 square feet

Area Values:

For Group M, 1st story

$$\text{Area value} = \frac{39{,}587}{1{,}200}\left[1-\left(\frac{20{,}000}{39{,}587}\right)\right]$$

$$= 33[1 - 0.505]$$

$$= 16.3$$

For Group B, 2nd story

$$\text{Area value} = \frac{60{,}173}{1{,}200}\left[1-\left(\frac{10{,}000}{60{,}173}+\frac{4{,}400}{39{,}587}+\frac{5{,}600}{55{,}422}\right)\right]$$

$$= 50[1 - 0.505]$$

$$= 50[1 - 0.378]$$

Area value = 31

For Group M, 2nd story

$$\text{Area value} = \frac{39{,}587}{1{,}200}[10.378]$$

$$= 32.9[1 - 0.378]$$

Area value = 20.4

For Group S-1, 2nd story

$$\text{Area value} = \frac{55{,}422}{1{,}200}[1-0.378]$$

$$= 46[1 - 0.378]$$

Area value = 28.6

Since these mixed occupancies are being separated so that one of the categories indicated in Section 3410.6.16 is applicable, a separate score must be computed for each occupancy.

In this example, the area value for the Group B occupancy is calculated to be 31, but the maximum area value permitted is 50 percent of the mandatory fire safety score listed in Table 3410.8. For a Group B occupancy, the mandatory fire safety score is 24; therefore, the maximum positive value that can be entered into Table 3410.7 is 12.

When an occupancy is located in more than one story, a separate area value must be calculated for each story. For input into Table 3410.7, the area value to be used is the lesser of all the individual area values for that occupancy group, but not greater than 50 percent of the mandatory fire safety score. In Figure

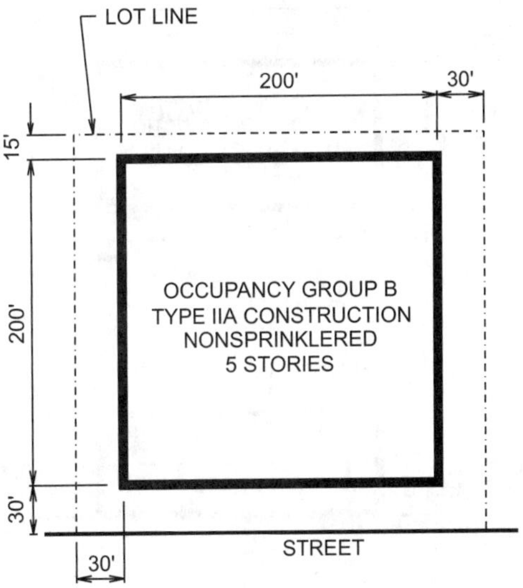

ACTUAL AREA = 200' × 200' = 40,000 SQ.FT.

SP = 0 (FROM SECTION 506.3)
OP = 100% (FROM SECTION 506.2)
ALLOWABLE AREA IN TABLE 503 = 37,500 SQ.FT.

$$AA = \frac{(0 + 100 + 100) \times 37{,}500}{100} = \frac{200 \times 37{,}500}{100} = 75{,}000 \text{ SQ.FT.}$$

For SI: 1 foot = 304.8 mm,
1 square foot = .0929 m^2.

Figure 3410.6.2.2(1)
ALLOWABLE AREA—SINGLE OCCUPANCY BUILDING

3410.6.2.2(2), there is an area classified as a Group M occupancy on both the first and second stories. The area value for the Group M occupancy on the first story is 16.3, and the area value for the second story is 20.4. Fifty percent of the mandatory fire safety score for Group M is 9.5. The lowest value for all the Group M occupant areas in the building is 9.5, which is 50 percent of the mandatory fire safety score.

The area value for the Group S-1 occupancy portion of the building is 28.6. This value, just as the values for the other occupancies in this example, exceeds the 50 percent maximum permitted by the mandatory fire safety score. The maximum for Group S-1 occupancy is 7.5, 50 percent of 15.

3410.6.3 Compartmentation. Evaluate the compartments created by fire barrier walls which comply with Sections 3410.6.3.1 and 3410.6.3.2 and which are exclusive of the wall elements considered under Sections 3410.6.4 and 3410.6.5. Conforming compartments shall be figured as the net area and do not include shafts, chases, stairways, walls or columns. Using Table 3410.6.3, determine the appropriate compartmentation value (*CV*) and enter that value into Table 3410.7 under Safety Parameter 3410.6.3, Compartmentation, for fire safety, means of egress and general safety.

❖ This section establishes and evaluates the compartments contained within an existing building by the effectiveness of the enclosing fire separation assemblies of both walls and floor/ceiling assemblies. Larger compartments are considered to be a greater safety risk than smaller compartments because the larger areas can become involved in a single fire incident affecting a greater portion of the building at one time.

Fire separation assemblies must comply with Sections 3410.6.3.1 and 3410.6.3.2. Fire separation assemblies are exclusive of the other separations or enclosures relating to Sections 3410.6.4 and 3410.6.5.

The evaluation of the compartments contained within an existing building is a linear function allowing interpolation between the various categories. This approach allows the compartmentation value to increase or decrease consistent with the actual changes in compartment sizes. Such an adjustment removes the previously built-in bias against smaller-sized buildings. Higher compartmentation values are assigned to buildings with smaller compartments.

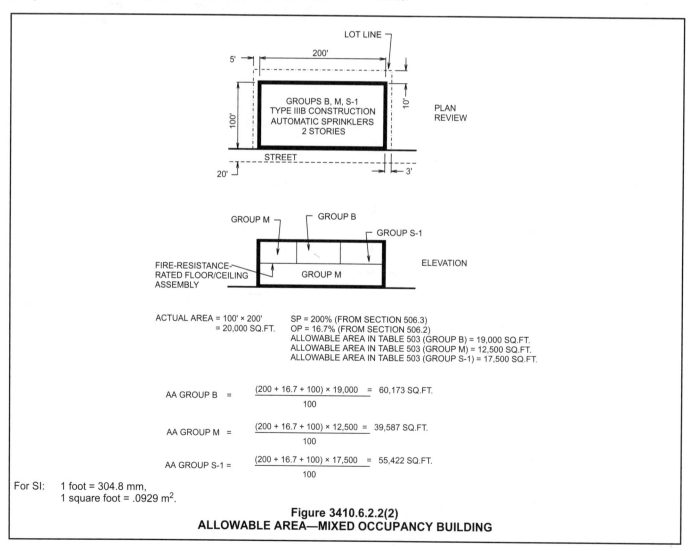

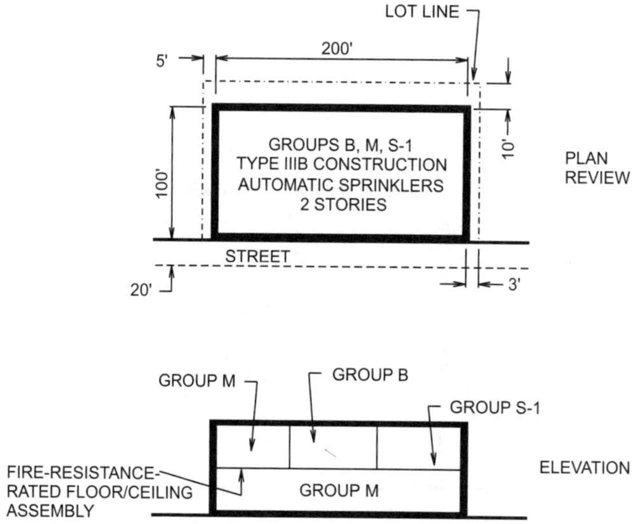

ACTUAL AREA = 100' × 200'
= 20,000 SQ.FT.

SP = 200% (FROM SECTION 506.3)
OP = 16.7% (FROM SECTION 506.2)
ALLOWABLE AREA IN TABLE 503 (GROUP B) = 19,000 SQ.FT.
ALLOWABLE AREA IN TABLE 503 (GROUP M) = 12,500 SQ.FT.
ALLOWABLE AREA IN TABLE 503 (GROUP S-1) = 17,500 SQ.FT.

$$\text{AA GROUP B} = \frac{(200 + 16.7 + 100) \times 19{,}000}{100} = 60{,}173 \text{ SQ.FT.}$$

$$\text{AA GROUP M} = \frac{(200 + 16.7 + 100) \times 12{,}500}{100} = 39{,}587 \text{ SQ.FT.}$$

$$\text{AA GROUP S-1} = \frac{(200 + 16.7 + 100) \times 17{,}500}{100} = 55{,}422 \text{ SQ.FT.}$$

For SI: 1 foot = 304.8 mm,
1 square foot = .0929 m^2.

Figure 3410.6.2.2(2)
ALLOWABLE AREA—MIXED OCCUPANCY BUILDING

TABLE 3410.6.3
COMPARTMENTATION VALUES

OCCUPANCY	CATEGORIES[a]				
	a Compartment size equal to or greater than 15,000 square feet	b Compartment size of 10,000 square feet	c Compartment size of 7,500 square feet	d Compartment size of 5,000 square feet	e Compartment size of 2,500 square feet
A-1, A-3	0	6	10	14	18
A-2	0	4	10	14	18
A-4, B, E, S-2	0	5	10	15	20
F, M, R, S-1	0	4	10	16	22

For SI: 1 square foot = 0.093 m^2.

a. For areas between categories, the compartmentation value shall be obtained by linear interpolation.

3410.6.3.1 Wall construction. A wall used to create separate compartments shall be a fire barrier conforming to Section 706 with a fire-resistance rating of not less than 2 hours. Where the building is not divided into more than one compartment, the compartment size shall be taken as the total floor area on all floors. Where there is more than one compartment within a story, each compartmented area on such story shall be provided with a horizontal exit conforming to Section 1021. The fire door serving as the horizontal exit between compartments shall be so installed, fitted and gasketed that such fire door will provide a substantial barrier to the passage of smoke.

❖ This section states that the walls determining the boundary of the compartment need to have fire barrier ratings of no less than 2 hours. These assemblies must be constructed in accordance with Section 706. For an existing building, this may need to be evaluated by both analyzing available plans and on-site investigations with a professional engineering determination of the required 2-hour fire-resistance rating. If the fire-resistance rating is less than 2 hours or cannot be reasonably determined, such walls should not be considered as compartmenting a fire area. The entire floor of an existing building must then be considered the compartment. If 2-hour-rated floor/ceiling assemblies are not present, the compartment size becomes the total area on all floors.

Openings and continuity requirements of Section 706 must be followed to maintain the integrity of the fire-resistance-rated wall assemblies and, thus, the compartments. Horizontal exits and their fire doors must comply with Section 1021.

The evaluation of an existing door condition requires an investigation of available engineering data and on-site inspections. Any uncertainty as to the performance of such doors may result in the openings being considered unprotected, expanding the compartment area or requiring an upgraded modification of the door to meet the current requirements of Section 1021.

3410.6.3.2 Floor/ceiling construction. A floor/ceiling assembly used to create compartments shall conform to Section 711 and shall have a fire-resistance rating of not less than 2 hours.

❖ The building features that provide the horizontal boundaries of the compartment need to provide effective fire-resistant integrity between floors. The existing floor/ceiling assemblies need to be rated for 2 hours and need to be tight against exterior walls. Penetrations in the floor/ceiling assemblies must be protected in accordance with Section 711 to maintain their fire-resistant integrity. The floor/ceiling assemblies must conform to all of the requirements of Section 711 to create the level of compartmentation required for this evaluation parameter.

3410.6.4 Tenant and dwelling unit separations. Evaluate the fire-resistance rating of floors and walls separating tenants, including dwelling units, and not evaluated under Sections 3410.6.3 and 3410.6.5. Under the categories and occupancies in Table 3410.6.4, determine the appropriate value and enter that value in Table 3410.7 under Safety Parameter 3410.6.4, Tenant and Dwelling Unit Separation, for fire safety, means of egress and general safety.

❖ This parameter is used to evaluate partitions in an existing building other than those used for the creation of compartments in Section 3410.3 or the enclosure of corridors in Section 3410.6.5. This section examines the level of separation between tenant spaces and dwelling units. The listed categories specifically reference Sections 706, 708 and 711 not only for fire-resistance ratings but also for continuity and opening purposes. Further credit is provided for existing buildings that have a 2-hour-rated separation between adjacent tenant spaces or dwelling units, which exceeds what is required for new construction.

TABLE 3410.6.4
SEPARATION VALUES

OCCUPANCY	CATEGORIES				
	a	b	c	d	e
A-1	0	0	0	0	1
A-2	-5	-3	0	1	3
R	-4	-2	0	2	4
A-3, A-4, B, E, F, M, S-1	-4	-3	0	2	4
S-2	-5	-2	0	2	4

❖ Table 3410.6.4 provides values for tenant space and dwelling unit separations that resist the spread of flames and smoke. The rationale for considering a nonfire-resistance-rated or incomplete separation as a safety liability is its unassumed role of impeding flame and smoke spread. Separations that compartmentalize the areas with significantly rated fire separation walls and floor/ceiling assemblies are considered the most effective. Buildings containing Group R occupancies have separation values based on the fact that dwelling unit separations are more critical than tenant separations in other occupancies.

This chapter places greater emphasis on compartmentation than on open floor areas when compared to the rest of the code (see commentary, Sections 3410 and 3410.6.4).

3410.6.4.1 Categories. The categories for tenant and dwelling unit separations are:

1. Category a — No fire partitions; incomplete fire partitions; no doors; doors not self-closing or automatic closing.
2. Category b — Fire partitions or floor assembly less than 1-hour fire-resistance rating or not constructed in accordance with Sections 708 or 711, respectively.
3. Category c — Fire partitions with 1 hour or greater fire-resistance rating constructed in accordance with Section 708 and floor assemblies with 1-hour but less than 2-hour fire-resistance rating constructed in accordance with Section 711, or with only one tenant within the fire area.
4. Category d — Fire barriers with 1-hour but less than 2-hour fire-resistance rating constructed in accordance with Section 706 and floor assemblies with 2-hour or greater fire-resistance rating constructed in accordance with Section 711.
5. Category e — Fire barriers and floor assemblies with 2-hour or greater fire-resistance rating and constructed in accordance with Sections 706 and 711, respectively.

❖ Tenant space and dwelling unit separations are categorized by the partitions being evaluated. The values of each category are listed in Table 3410.6.4 by occupancy classifications. Typical illustrations of the types of partitions are shown in Figures 3410.6.4.1(1) and 3410.6.4.1(2). The listed categories provide a graduated level of separation when compared to new construction requirements.

Category a addresses the situation where there is no separation or there are gaps in the separation provided between tenant spaces or dwelling units.

Category b accounts for those tenant spaces or dwelling units that are separated from one another with less than a 1-hour rating.

Category c represents what a newly constructed building would have for tenant space and dwelling unit separation. An existing building meeting this level of compliance gains no benefit or penalty and, therefore, the separation value is zero.

Category d is given additional credit because the existing building space or dwelling unit has tenant separation that is rated higher than what is required by the code for new construction. Additionally, the walls are required to

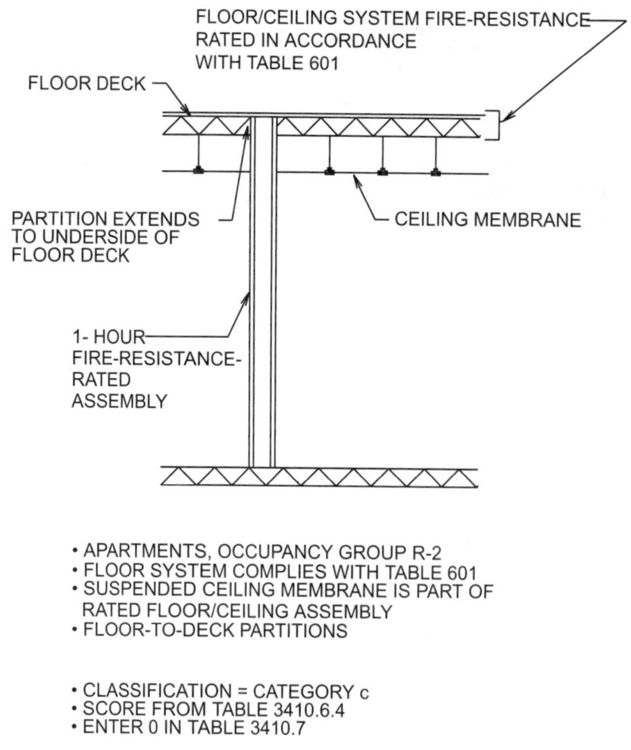

Figure 3410.6.4.1(1)
DWELLING UNIT SEPARATION

meet the fire barrier requirements of Section 706.

Category e is an added column that provides increased credit for existing buildings that have walls and floor/ceiling assemblies with fire-resistance ratings exceeding what is required for new buildings. This increased level of compartmentation in an existing building is credited accordingly with high separation values.

3410.6.5 Corridor walls. Evaluate the fire-resistance rating and degree of completeness of walls which create corridors serving the floor, and constructed in accordance with Section 1016. This evaluation shall not include the wall elements considered under Sections 3410.6.3 and 3410.6.4. Under the categories and groups in Table 3410.6.5, determine the appropriate value and enter that value into Table 3410.7 under Safety Parameter 3410.6.5, Corridor Walls, for fire safety, means of egress and general safety.

❖ Corridor walls are evaluated as fire partitions possessing an adequate fire-resistance rating and completeness to restrict fire and smoke migration into the corridor. Prescribed requirements are stated in Sections 708.4 and 1016. Existing corridor walls require investigation and analysis to determine equivalency to code requirements. Figures 3410.6.5(1) and 3410.6.5(2) illustrate various corridor wall values. Corridor walls contrast with the compartmentation and tenant dwelling unit separations of Sections 3410.6.3 and 3410.6.4 by requiring an appropriate fire-resistance rating and continuity (see commentary, Sections 708.4, 3410.6.3 and 3410.6.4).

The corridor wall evaluations do not include partitions required to establish a compartment (see Section 3410.6.3) or tenant and dwelling unit separations (see Section 3410.6.4).

TABLE 3410.6.5
CORRIDOR WALL VALUES

OCCUPANCY	CATEGORIES a	b	c[a]	d[a]
A-1	-10	-4	0	2
A-2	-30	-12	0	2
A-3, F, M, R, S-1	-7	-3	0	2
A-4, B, E, S-2	-5	-2	0	5

a. Corridors not providing at least one-half the travel distance for all occupants on a floor shall use Category b.

❖ Table 3410.6.5 assigns values to the different uses based on the relative degree of fire resistance and smoke resistance of corridor walls. Since corridors are enclosed (confined) spaces subject to the rapid buildup of smoke and heat, a degree of protection is necessary to minimize this hazard to the occupants egressing the building. The table reflects this emphasis through substantial negative scores in buildings without properly

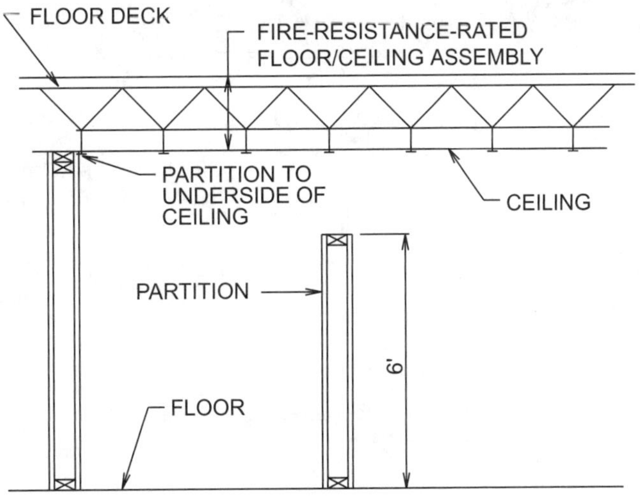

• FOOD COURT RESTAURANTS, OCCUPANCY GROUP A-3
• SOME PARTITIONS EXTEND TO CEILING OF FIRE-RESISTANCE-RATED FLOOR/CEILING ASSEMBLY
• NO CLOSERS ON DOORS BETWEEN ADJACENT TENANT SPACES
• SOME 6' PRIVACY PARTITIONS (PARTIAL PARTITIONS)

• CLASSIFICATION = CATEGORY a
• SCORE FROM TABLE 3410.6.4
• ENTER -4 IN TABLE 3410.7

For SI: 1 foot = 304.8 mm.

Figure 3410.6.4.1(2)
TENANT SEPARATION

enclosed corridors. Both Section 3410.6.5 and this table place an emphasis on corridor walls, which differs from that expressed elsewhere in the code. For example, Chapter 10 does not require corridors to be provided except in Group I-2 occupancies. The emphasis here is that when corridors are already present, they must provide basic protection. The table is an assessment of the relative risk represented by the corridor.

Note a to the table further controls the application of Categories c and d to existing buildings with certain means of egress arrangements. Although an existing building may have corridors with significant fire-resistance ratings and protected openings, little credit can be granted when the building occupants are protected for only a short period of time. Very short corridors in large floor plans or corridors located in just one tenant space of a floor plan provide a benefit to only a portion of the occupant load for a small portion of the overall exit access travel and, thus, do not accrue any positive points. Unless rated corridors are available for all occupants of that particular floor or they provide a protected path of travel for at least one-half of the occupants' overall travel length, negative corridor wall values are assigned. In existing buildings with very short or limited-use corridors, the code user is directed to use Category b, regardless of the corridors' fire-resistance ratings and opening protectives.

3410.6.5.1 Categories. The categories for corridor walls are:

1. Category a — No fire partitions; incomplete fire partitions; no doors; or doors not self-closing.
2. Category b — Less than 1-hour fire-resistance rating or not constructed in accordance with Section 708.4.
3. Category c — 1-hour to less than 2-hour fire-resistance rating, with doors conforming to Section 715 or without corridors as permitted by Section 1016.
4. Category d — 2-hour or greater fire-resistance rating, with doors conforming to Section 715.

❖ Corridor walls are categorized by the partitions being evaluated. The values of each category are listed in Table 3410.6.5 by occupancy classification.

3410.6.6 Vertical openings. Evaluate the fire-resistance rating of vertical exit enclosures, hoistways, escalator openings and other shaft enclosures within the building, and openings between two or more floors. Table 3410.6.6(1) contains the appropriate protection values. Multiply that value by the construction-type factor found in Table 3410.6.6(2). Enter the vertical opening value and its sign (positive or negative) in Table 3410.7 under Safety Parameter 3410.6.6, Vertical Openings, for fire

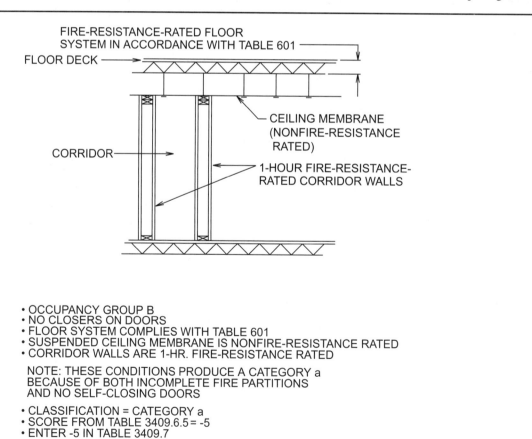

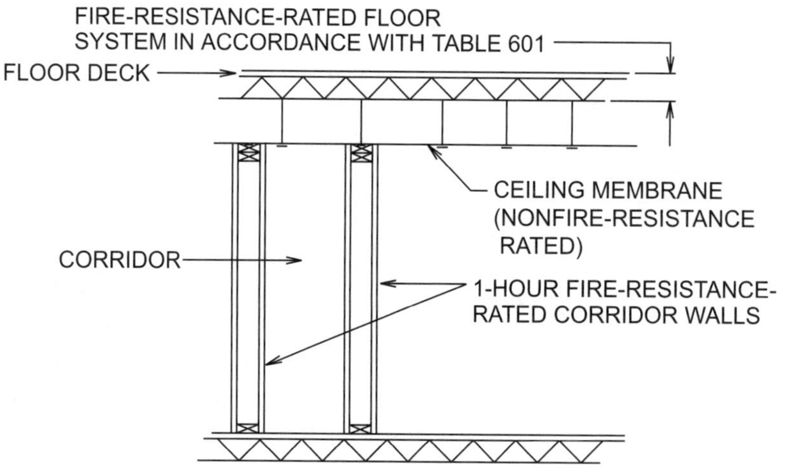

- OCCUPANCY GROUP B
- NO CLOSERS ON DOORS
- FLOOR SYSTEM COMPLIES WITH TABLE 601
- SUSPENDED CEILING MEMBRANE IS NONFIRE-RESISTANCE RATED
- CORRIDOR WALLS ARE 1-HR. FIRE-RESISTANCE RATED

NOTE: THESE CONDITIONS PRODUCE A CATEGORY a BECAUSE OF BOTH INCOMPLETE FIRE PARTITIONS AND NO SELF-CLOSING DOORS

- CLASSIFICATION = CATEGORY a
- SCORE FROM TABLE 3409.6.5 = -5
- ENTER -5 IN TABLE 3409.7

Figure 3410.6.5(1)
CORRIDOR WALL VALUES—CATEGORY a

safety, means of egress and general safety. If the structure is a one-story building, enter a value of 2. Unenclosed vertical openings that conform to the requirements of Section 707 shall not be considered in the evaluation of vertical openings.

❖ Vertical openings are used to evaluate the fire-resistance ratings of openings between floors of a building and between shaft enclosures, such as stairs, elevator hoistways and escalator openings. This section also gives the formula for determining the score to be entered in Table 3410.7. This section does not apply to any building element that conforms to the requirements of Section 707.2, which requires all floor openings connecting two or more stories to be protected by a shaft enclosure complying with Section 707; however, Section 707.2 includes a number of exceptions to these requirements, such as: Section 404 on atriums; a floor opening connecting no more than two floors that is not a means of egress and an escalator opening complying with Section 707.2 (also see commentary, Section 707).

3410.6.6.1 Vertical opening formula. The following formula shall be used in computing vertical opening value.

$$VO = PV \times CF \qquad \textbf{(Equation 34-4)}$$

VO = Vertical opening value.

PV = Protection value [Table 3410.6.6(1)]

CF = Construction type factor [Table 3410.6.6(2)]

❖ The vertical opening formula used in computing the values entered in Table 3410.7 is given in this section. See Section 3410.6.6 of the commentary for further discussion.

TABLE 3410.6.6(1)
VERTICAL OPENING PROTECTION VALUE

PROTECTION	VALUE
None (unprotected opening)	-2 times number floors connected
Less than 1 hour	-1 times number floors connected
1 to less than 2 hours	1
2 hours or more	2

❖ Table 3410.6.6(1) assigns relative protection values based on the fire protection ratings of the vertical openings in a building. The lower the fire protection, the greater the hazard to the rest of the building. The table reflects the varying levels of impact to an existing building based on the number of stories that are interconnected by unprotected openings. The greater the number of floors that are interconnected by unprotected openings, the greater the number of negative points assessed in the existing building's evaluation scores. The

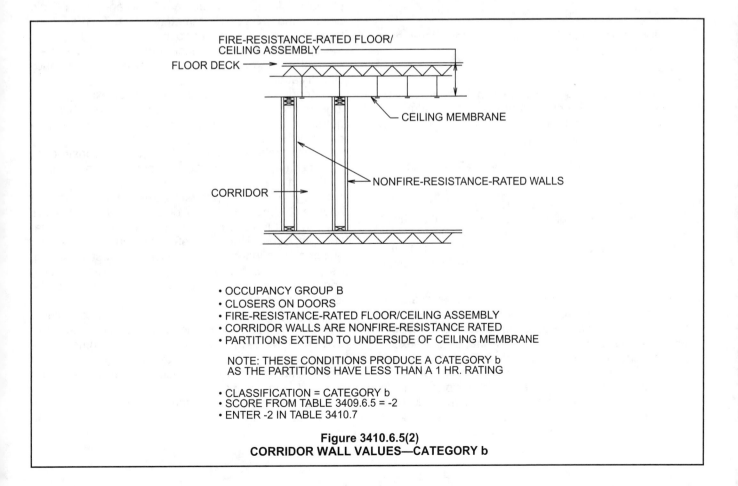

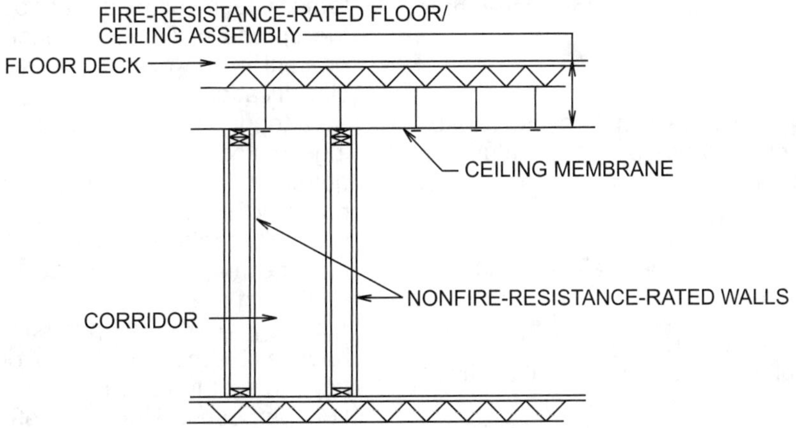

• OCCUPANCY GROUP B
• CLOSERS ON DOORS
• FIRE-RESISTANCE-RATED FLOOR/CEILING ASSEMBLY
• CORRIDOR WALLS ARE NONFIRE-RESISTANCE RATED
• PARTITIONS EXTEND TO UNDERSIDE OF CEILING MEMBRANE

NOTE: THESE CONDITIONS PRODUCE A CATEGORY b AS THE PARTITIONS HAVE LESS THAN A 1 HR. RATING

• CLASSIFICATION = CATEGORY b
• SCORE FROM TABLE 3409.6.5 = -2
• ENTER -2 IN TABLE 3410.7

Figure 3410.6.5(2)
CORRIDOR WALL VALUES—CATEGORY b

closer the building comes to meeting code compliance relative to rated shafts, the larger the positive protection value becomes. This table encourages and provides incentives to the existing building renovator to bring noncomplying situations into compliance with new construction requirements.

TABLE 3410.6.6(2)
CONSTRUCTION-TYPE FACTOR

	TYPE OF CONSTRUCTION								
	IA	IB	IIA	IIB	IIIA	IIIB	IV	VA	VB
FACTOR	1.2	1.5	2.2	3.5	2.5	3.5	2.3	3.3	7

❖ Relative values for each type of construction are assigned in Table 3410.6.6(2). These represent the relative degree of fire hazard of each type of construction when compared to other types of construction. Similar factors were considered in the original development of Table 503 for height and area limitations. When one building has two different opening circumstances that individually result in different values, the lower value must be used.

Example 1:

Assume a building of Type IIB construction, three stories in height. The building has a 2-hour fire-resistance-rated exhaust shaft with $1^1/_2$-hour fire dampers (complying with UL 555).

$VO = 2 \times 3.5 = 7.0$

Enter 7.0 in Table 3410.7 under Safety Parameter 3410.6.6, Vertical Openings.

Example 2:

Assume a building of Type IA construction, 12 stories in height. The building has a $^1/_2$-hour return-air shaft with rated dampers.

$VO = (-1 \times 12) \times 1.2 = -14.4$

Enter -14.4 in Table 3410.7 under Safety Parameter 3410.6.6, Vertical Openings.

3410.6.7 HVAC systems. Evaluate the ability of the HVAC system to resist the movement of smoke and fire beyond the point of origin. Under the categories in Section 3409.6.7.1, determine the appropriate value and enter that value into Table 3410.7 under Safety Parameter 3410.6.7, HVAC Systems, for fire safety, means of egress and general safety.

❖ The provisions of Chapter 34 are intended to evaluate the HVAC system's potential for resisting the movement and spread of fire, smoke or products of combustion. This section does not address HVAC systems that are used exclusively for smoke control in the building. The systems evaluated in this section are those that use either supply air, return air or exhaust air. For example, a typical building might have supply air ducts or shafts; return air ducts or shafts; toilet exhaust ducts or shafts and kitchen exhaust ducts or shafts, all of which are considered HVAC systems. All systems in the building are evaluated and the lowest score obtained by any of the systems is the score that must be assigned to the entire building.

These provisions include two other safety aspects that are also applicable for new construction: plenums and air movement in egress elements, such as exit access corridors and exit stairways. These factors can significantly affect the safety of the occupants of the existing building. In some cases, these safety aspects can be more important than just the number of stories connected by an HVAC system.

3410.6.7.1 Categories. The categories for HVAC systems are:

1. Category a — Plenums not in accordance with Section 602 of the *International Mechanical Code*. -10 points.
2. Category b — Air movement in egress elements not in accordance with Section 1016.4. -5 points.
3. Category c — Both categories a and b are applicable. -15 points.
4. Category d — Compliance of the HVAC system with Section 1016.4 and Section 602 of the *International Mechanical Code*. 0 points.
5. Category e — Systems serving one story; or a central boiler/chiller system without ductwork connecting two or more stories. 5 points.

❖ The five categories that must be used in the evaluation process are defined in this section, along with their applicable values. The values are to be entered in Table 3410.7 under Safety Parameter 3410.6.7, HVAC Systems. These values are not occupancy sensitive, since the spread of fire is dependent on the HVAC system present in the existing building, not on its occupancy classification.

Category a requires a value of -10 for existing buildings that contain plenums not in compliance with the requirements of Section 602 of the IMC. Locations of plenums within the existing buildings, the materials they are built of relative to the building's type of construction and the materials exposed to plenum air must all be evaluated.

Category b corresponds to existing buildings that have exit access corridors or exit stairways that are used to supply, return or exhaust air or for other ventilation purposes. That type of layout puts the existing building's occupants at greater risk and must, therefore, be penalized with a value of -5. An existing building complying with one of the exceptions of Section 1016.4 is not considered as Category b.

Category c is applicable when an existing building has both noncomplying plenums and corridors/stairways used for air movement. A value of -15 must be assigned to such a building because the movement of smoke and fire throughout would pose an even greater hazard to the occupants.

Category d represents the base value of zero. A newly constructed building is required to meet all the provisions of Section 1016.4 of the code and Section

602 of the IMC. An existing building that complies with these provisions is meeting this same minimum compliance level and, therefore, is neither penalized nor given benefit for that compliance.

Category e specifies a value of 5 points for HVAC systems that serve only one story of a building. The hazards of a fire spreading laterally through a story of a building via the HVAC system are minimal; therefore, the code assigns a positive value. A boiler/chiller system also does not lend itself to fire spread as long as there is no air movement in ducts.

Example 1:
Assume a Group R-2 apartment building, six stories in height. Each apartment has its own HVAC equipment located within the dwelling unit. Bathrooms are ventilated by fans connected to a central exhaust shaft; kitchen exhaust hoods connect to a central exhaust shaft serving all floors. Corridors have a direct outside supply-air system on each floor with no exhaust. The HVAC systems are classified as Category b because the corridors are being used as the make-up air source for the bathroom and kitchen exhausts. A score of -5 should be entered in Table 3410.7 under Safety Parameter 3410.6.7, HVAC Systems.

Example 2:
Assume a Group S-2 low-hazard storage occupancy is located in a one-story building. The building is preengineered steel Type IIB construction, and has a suspended gypsum board ceiling. Gypsum board has also been attached to the underside of the ceiling joists to serve as the upper membrane of a return-air plenum.

The plenum also contains plastic fire sprinkler piping for the automatic fire suppression system that is installed throughout the building.

The plenum is classified as noncombustible in accordance with Section 602 of the IMC. As long as the plastic fire sprinkler piping meets the optical density and flame spread limits, the plenum is in compliance with the code; therefore, the HVAC system will be classified as Category d. A value of zero should be entered in Table 3410.7 under Safety Parameter 3410.6.7, HVAC Systems.

3410.6.8 Automatic fire detection. Evaluate the smoke detection capability based on the location and operation of automatic fire detectors in accordance with Section 907 and the *International Mechanical Code*. Under the categories and occupancies in Table 3410.6.8, determine the appropriate value and enter that value into Table 3410.7 under Safety Parameter 3410.6.8, Automatic Fire Detection, for fire safety, means of egress and general safety.

❖ This section considers the use of smoke detectors in a building. To receive credit for the smoke detectors, they must be connected to audible alarms and installed in accordance with Section 907 of the code and Section 606 of the IMC.

TABLE 3410.6.8
AUTOMATIC FIRE DETECTION VALUES

OCCUPANCY	CATEGORIES a	b	c	d	e
A-1, A-3, F, M, R, S-1	-10	-5	0	2	6
A-2	-25	-5	0	5	9
A-4, B, E, S-2	-4	-2	0	4	8

❖ Table 3410.6.8 assigns values for each occupancy. The use of detectors increases the safety in a building by providing early warning to occupants. The lack of detectors and the associated early warning are considered less than optimum for safety in occupancies with high population densities and high combustible loads. Large deficiency points are accrued if adequate detection systems are not provided.

Example:
Assume a four-story Group R-1 motel with corridors and an elevator lobby on each story. Smoke detectors are installed throughout the corridors, closets, rooms and elevator lobbies. Only single-station detectors are installed in the guestrooms. There are smoke detectors in the HVAC return-air system and a fire alarm system is provided. An analysis of these building characteristics results in the selection of Category d. To be classified in Category e, guestrooms must have detectors connected to the building's emergency electrical system and annunciated by each room at a constantly attended location, such as the front desk. Additionally, the fire alarm system must be capable of being manually activated by the front desk when a smoke detector operates.

Category d value from Table 3410.6.8 = 2.

Enter 2 in Table 3410.7 under Safety Parameter 3410.6.8, Automatic Fire Detection.

3410.6.8.1 Categories. The categories for automatic fire detection are:

1. Category a — None.
2. Category b — Existing smoke detectors in HVAC systems and maintained in accordance with the *International Fire Code*.
3. Category c — Smoke detectors in HVAC systems. The detectors are installed in accordance with the requirements for new buildings in the *International Mechanical Code*.
4. Category d — Smoke detectors throughout all floor areas other than individual guestrooms, tenant spaces and dwelling units.
5. Category e — Smoke detectors installed throughout the fire area.

❖ The categories are based on the location and completeness of the smoke detectors. The categories represent a graduation in the levels of smoke detection that ranges from no detectors in Category a to full detection throughout all fire area spaces in Category e.

Category a is a facility that has no automatic smoke detection system.

Category b assumes that there are some smoke detectors in an existing building's HVAC system, but not to the extent required for new construction. If the limited HVAC system detectors are maintained in accordance with the IFC, the detection system qualifies as Category b.

Category c addresses existing buildings that have upgraded HVAC systems with duct detectors, as specified by the IMC. The detection values for Category c are zero, since this is the level of protection required for all new construction.

Category d is the classification for existing buildings that have full detection coverage throughout the public common spaces and secondary rooms and areas.

Category e has smoke detectors throughout fire areas in compliance with the provisions for new construction.

3410.6.9 Fire alarm systems. Evaluate the capability of the fire alarm system in accordance with Section 907. Under the categories and occupancies in Table 3410.6.9, determine the appropriate value and enter that value into Table 3410.7 under Safety Parameter 3410.6.9, Fire Alarm, for fire safety, means of egress and general safety.

❖ This section evaluates the capabilities of building fire alarm systems that are separate from the automatic fire detection system evaluated in Section 3410.6.8. A fire alarm system that is manually operated or activated by smoke detectors or sprinkler water-flow devices alerts the occupants to a fire situation. The fire alarm system will notify the occupants with visible or audible alarms so they may begin their egress and discharge from the building. Buildings of assembly, business, educational or residential occupancies can have large numbers of occupants in rooms with concentrated seating or people who are sleeping.

TABLE 3410.6.9
FIRE ALARM SYSTEM VALUES

OCCUPANCY	CATEGORIES			
	a	b[a]	c	d
A-1, A-2, A-3, A-4, B, E, R	-10	-5	0	5
F, M, S	0	5	10	15

a. For buildings equipped throughout with an automatic sprinkler system, add 2 points for activation by a sprinkler water flow device.

❖ Table 3410.6.9 gives values for each occupancy and type of fire alarm system provided. It reflects the idea that the presence of an alarm system in a building usually creates a safer condition for the occupants when compared to a building without a communication system.

Deficiency points are assigned to high population densities, such as Groups A-1, A-2, A-3, A-4 and E, which fit into Category a.

Section 3410.6.9 and this table have a different reasoning for alarm and communication systems than Section 907. The code, in Sections 907 and 908, does not require alarm systems in Groups A-1, A-2, A-3, M, F and S. This is based on the reasoning that alarm systems, especially manual alarms, are subject to false alarms. If the alarm is recognized, a panic situation could occur, unless the alarm system is a voice alarm or public address system.

The general concept is that it is safer to have no alarm in these occupancies than to have an alarm that sounds an unrecognizable signal or is subject to false alarming.

Example:

Assume a one-story assembly building has a complete manual fire alarm system, a voice alarm system, a public address system and a fire command station that does not contain status indicators and controls for the air-handling system. The fire command station does not have emergency power or lighting system controls. It also does not have a fire department communication panel. As a result, the building is classified in Category c.

Category c value from Table 3410.6.9 = 0.

Enter 0 in Table 3410.7 under Safety Parameter 3410.6.9, Fire Alarm System.

3410.6.9.1 Categories. The categories for fire alarm systems are:

1. Category a — None.
2. Category b — Fire alarm system with manual fire alarm boxes in accordance with Section 907.3 and alarm notification appliances in accordance with Section 907.9.
3. Category c — Fire alarm system in accordance with Section 907.
4. Category d — Category c plus a required emergency voice/alarm communications system and a fire command station that conforms to Section 403.8 and contains the emergency voice/alarm communications system controls, fire department communication system controls and any other controls specified in Section 911 where those systems are provided.

❖ These categories are defined by the fire alarm system that is provided within the existing building.

Category a means that there is no fire alarm system in the building or it does not conform to all the requirements of Section 907. This includes a system that does not have a secondary power supply in accordance with Section 907 or one where the zones of a floor exceed 20,000 square feet (1858.1 m^2) as noted in Section 907.8.

Category b applies when the manual fire alarm boxes comply with Section 907.3.1 and NFPA 72. The alarm-notification appliances, specifically the audible alarms, are in accordance with Section 907.9.2. Location, height and color of the manual fire alarm boxes must be specifically evaluated for compliance, along with the sound levels of the audible alarms when compared to the normal sound levels within the existing building.

Category c requires that the fire alarm system must comply with Section 907 and NFPA 72. This category indicates that a complete fire alarm system is present in the existing building and that the system complies with all the requirements for new construction.

Category d is applicable for a building that is provided with a fire command station for fire department operations that conforms to Section 403.8. Section 403.8 may be used when evaluating any building that complies with its performance requirements. This section specifies the fire command operations required, including:

- Voice/alarm communication system panels;
- Fire department communication panels;
- Fire detection and alarm system annunciator panels;
- An annunciator for visually indicating floor locations of elevators and whether they are operational;
- Status indicators and controls for air-handling systems;
- Controls for unlocking all stairway doors simultaneously;
- Sprinkler valve and water-flow detector display panels;
- Emergency and standby power and
- Status indicators and a telephone for fire department use with controlled access to the public telephone system.

3410.6.10 Smoke control. Evaluate the ability of a natural or mechanical venting, exhaust or pressurization system to control the movement of smoke from a fire. Under the categories and occupancies in Table 3410.6.10, determine the appropriate value and enter that value into Table 3410.7 under Safety Parameter 3410.6.10, Smoke Control, for means of egress and general safety.

❖ This section is used to evaluate characteristics that could limit smoke migration in the building, including operable windows, mechanical exhaust systems or pressurized stair or smokeproof enclosures.

TABLE 3410.6.10
SMOKE CONTROL VALUES

OCCUPANCY	CATEGORIES					
	a	b	c	d	e	f
A-1, A-2, A-3	0	1	2	3	6	6
A-4, E	0	0	0	1	3	5
B, M, R	0	2[a]	3[a]	3[a]	3[a]	4[a]
F, S	0	2[a]	2[a]	3[a]	3[a]	3[a]

a. This value shall be 0 if compliance with Category d or e in Section 3410.6.8.1 has not been obtained.

❖ Table 3410.6.10 assigns values for each occupancy and category of smoke control in the building. The table indicates the relative risk to the building occupants if adequate smoke control methods are not provided in the building.

In occupancies having a higher population density, zero points are indicated if no smoke control or limited smoke control is provided. The table also indicates that smoke control is extremely important in the exit stairs of a building. Note a of the table assigns zero credit points for Groups B, M, R, F and S unless the building complies with Category d or e in Section 3410.6.8.1, which are the automatic fire detection system requirements.

Example 1:
Assume a one-story Group A-4 church sanctuary. It has no operable windows and no smoke control system.

Category a from Table 3410.6.10 = 0.

Enter 0 in Table 3410.7 under Safety Parameter 3410.6.10, Smoke Control, for only means of egress and general safety (GS) parameters. No entry is to be made under the fire safety (FS) parameter.

Example 2:
Assume a three-story Group E high school. It has operable windows throughout the building. The stairways are interior without windows or a pressurization system. Category b from Table 3410.6.10 = 0.

Enter 0 in Table 3410.7 under Safety Parameter 3410.6.10, Smoke Control, for means of egress (ME) and general safety (GS) parameters.

Example 3:
Assume a two-story Group A-2 nightclub with a smoke control system that meets the requirements for Category e.

Category e from Table 3410.6.10 = 6.

Enter 6 in Table 3410.7 under Safety Parameter 3410.6.10, Smoke Control, for means of egress and general safety parameters.

Example 4:
Assume a six-story Group B office building with three stairways: one is a smokeproof enclosure conforming to Section 1019.1.8, one is pressurized in accordance with Section 909.20.5; and one has operable exterior windows. The building has smoke detectors throughout all floor spaces and fire areas.

Category f from Table 3410.6.10 = 4.

Enter 4 in Table 3410.7 under Safety Parameter 3410.6.10, Smoke Control, for means of egress (ME) and general safety (GS).

If detectors are omitted from some of the offices of the buildings, then a score of 0 must be entered in Table 3410.7. This is determined from Note a in Table 3410.6.10.

3410.6.10.1 Categories. The categories for smoke control are:

1. Category a — None.
2. Category b — The building is equipped throughout with an automatic sprinkler system. Openings are provided in exterior walls at the rate of 20 square feet (1.86 m^2) per 50 linear feet (15 240 mm) of exterior wall in each story and distributed around the building perimeter at intervals not exceeding 50 feet (15 240 mm). Such openings shall be readily openable from the inside without a key or separate

tool and shall be provided with ready access thereto. In lieu of operable openings, clearly and permanently marked tempered glass panels shall be used.

3. Category c — One enclosed exit stairway, with ready access thereto, from each occupied floor of the building. The stairway has operable exterior windows and the building has openings in accordance with Category b.
4. Category d — One smokeproof enclosure and the building has openings in accordance with Category b.
5. Category e — The building is equipped throughout with an automatic sprinkler system. Each fire area is provided with a mechanical air-handling system designed to accomplish smoke containment. Return and exhaust air shall be moved directly to the outside without recirculation to other fire areas of the building under fire conditions. The system shall exhaust not less than six air changes per hour from the fire area. Supply air by mechanical means to the fire area is not required. Containment of smoke shall be considered as confining smoke to the fire area involved without migration to other fire areas. Any other tested and approved design which will adequately accomplish smoke containment is permitted.
6. Category f — Each stairway shall be one of the following: a smokeproof enclosure in accordance with Section 1019.1.8; pressurized in accordance with Section 909.20.5; or shall have operable exterior windows.

❖ The six categories to be evaluated are compared to the occupancies to determine the score to be entered in Table 3410.7 under Safety Parameter 3410.6.10, Smoke Control.

Category a means there is no method of controlling smoke in the building.

Category b means the existing building is sprinklered and there are exterior windows that can be readily opened without the use of keys or tools. This category recognizes the fire safety benefits of the code's high-rise provisions, including automatic sprinkler systems. Operable panels or windows in the exterior walls must be provided at the rate of 20 square feet per 50 lineal feet (1.85 m^2 per 15 240 mm) of exterior wall in each story. The openings must be distributed around the perimeter of the building at intervals no more than 50 feet (15 240 mm) between windows. A story with very high ceilings may have a strip of windows located with latches or controls that are not easily reachable. These openings would not meet Category b standards, even if they are the correct sizes.

Category c requires at least one enclosed exit stairway to have operable windows that open to the exterior. Additionally, the building must have operable windows complying with the size and spacing requirements of Category b.

Category d requires a minimum of one smokeproof enclosure and operable windows in the building. These operable windows are the same as those addressed in Category b. By definition, a smokeproof enclosure refers to an enclosed interior exit stairway that conforms to Section 1009.3 as designated in Section 1019.1.8.

Section 1019.1.8 contains three methods for obtaining a satisfactory smokeproof enclosure: natural ventilation design (see Section 909.20.3), mechanical ventilation design (see Section 909.20.4) and stair pressurization design (see Section 909.20.5).

Additional requirements that must be considered when evaluating a smokeproof enclosure are: access (see Section 909.20.1), construction (see Section 909.20.2), ventilating equipment (see Section 909.20.6) and standby power (see Section 909.20.6.2).

Category e recognizes the use of a mechanical smoke control system designed in accordance with the provisions of the code. Each fire area within the building must have a mechanical smoke control system. Return and exhaust air from the system must be discharged directly to the exterior to achieve the necessary level of protection. A specific air change requirement is provided, independent of the fire area's volume or size.

Category f recognizes the merits of smokeproof enclosures, pressurized stairs and stairs with operable exterior windows. To be classified in this category, all stairs in the building must comply with the requirements of any one, or a combination of, the three types of stairs. It is possible for a single building to have a smokeproof enclosure, a pressurized stairway and a stairway with operable exterior windows.

Both the categories and the occupancy of the building are used in determining the score that will be entered in Table 3410.7. This score is determined from Table 3410.6.10. Note a in Table 3410.6.10 can have a significant impact on the scores. The note states that, even if the building has some level of smoke control, it is to receive no credit if it does not have an automatic fire detection system complying with Category d or e in Section 3410.6.8.1.

3410.6.11 Means of egress capacity and number. Evaluate the means of egress capacity and the number of exits available to the building occupants. In applying this section, the means of egress are required to conform to Sections 1003 through 1014 and 1016 through 1023 (except that the minimum width required by this section shall be determined solely by the width for the required capacity in accordance with Table 1005.1). The number of exits credited is the number that are available to each occupant of the area being evaluated. Existing fire escapes shall be accepted as a component in the means of egress when conforming to Section 3404. Under the categories and occupancies in Table 3410.6.11, determine the appropriate value and enter that value into Table 3410.7 under Safety Parameter 3410.6.11, Means of Egress Capacity, for means of egress and general safety.

❖ This section addresses the exit capacity and number of existing exits available to the building occupants. Before a building can be evaluated in this category, and Section 3410 in general, it must comply with the sections listed.

Exits are required to conform to the following sections before evaluation:

Sections 1013, 1014 and 1016. A mandatory requirement is not placed on the length of travel in accordance with Section 1015, because travel length does not directly impact exit capacity. Overall exit safety is influenced by travel length, however, and is evaluated in Section 3410.6.13.

Section 1004 establishes the minimum number of occupants the exit facilities must accommodate.

Section 1005 defines the capacity of the means of egress by identifying a minimum width of egress component per occupant. This is used to calculate the total capacity of the means of egress component.

Section 1018.1 establishes the minimum number of exits for the existing building occupant load.

Section 1023 requires exit stairs to discharge directly to the exterior or through recognized exit elements at the level of discharge.

Section 3404.1.2 permits the continued use of existing fire escapes in existing buildings only.

Evaluation of the means of egress capacity involves all egress components, including exit access, exits and exit discharge. This evaluation is correlated to the requirements for the means of egress for new construction.

TABLE 3410.6.11
MEANS OF EGRESS VALUES

OCCUPANCY	CATEGORIES				
	a[a]	b	c	d	e
A-1, A-2, A-3, A-4, E	-10	0	2	8	10
M	-3	0	1	2	4
B, F, S	-1	0	0	0	0
R	-3	0	0	0	0

a. The values indicated are for buildings six stories or less in height. For buildings over six stories in height, add an additional -10 points.

❖ Table 3410.6.11 assigns values for each occupancy and category from Section 3410.6.11.1. The table gives credit for providing additional exits and exit capacities beyond the minimum. These positive values indicate a greater degree of safety than is required under current code provisions. This additional evaluation contributes to the overall assessment of the building's safety and may offset other safety deficiencies [see Figures 3410.6.11(1) and 3410.6.11(2)]. Note a of the table provides a significant penalty for high-rise buildings that use fire escapes as part of the means of egress. In this case, if the building is seven stories or greater in height and uses fire escapes, -10 points must be added to the values already listed for Category a buildings.

3410.6.11.1 Categories. The categories for means of egress capacity and number of exits are:

1. Category a — Compliance with the minimum required means of egress capacity or number of exits is achieved through the use of a fire escape in accordance with Section 3403.
2. Category b — Capacity of the means of egress complies with Section 1004 and the number of exits complies with the minimum number required by Section 1018.
3. Category c — Capacity of the means of egress is equal to or exceeds 125 percent of the required means of egress capacity, the means of egress complies with the minimum required width dimensions specified in the code and the number of exits complies with the minimum number required by Section 1018.
4. Category d — The number of exits provided exceeds the number of exits required by Section 1018. Exits shall be located a distance apart from each other equal to not less than that specified in Section 1014.2.
5. Category e — The area being evaluated meets both Categories c and d.

❖ Five categories must be evaluated. These categories and the occupancy of the building will determine the score entered in Table 3410.7.

Category a is appropriate for buildings that comply with either the means of egress capacity or the number of exits, including the use of fire escapes. Although the code allows fire escapes to be used as an egress element in existing buildings, this is the least desirable option in any type of building. The code requires a building using fire escapes to use Category a and its corresponding negative points.

Category b is used for buildings that meet the minimum requirements of Sections 1004 for capacity of means of egress and 1018 for minimum number of exits and continuity..

Category c is used for buildings that meet or exceed the requirements for new construction. This category provides small positive points for existing buildings that meet all of the following requirements:

- The capacity of all the means of egress components is greater than 125 percent of the required capacity.
- All of the means of egress components comply with the minimum required widths [32 inch (813 mm) clear for doors, corridors that are 44 inches (1118 mm) wide and stairways].
- The minimum number of exits is provided based on the number of occupants on each floor level.

By providing oversized egress capacity and minimum egress width requirements, an existing building has added safety, which is rewarded with a small number of positive points.

Category d simply relates to buildings where more exits are provided than required by Section 1018, which contributes a positive factor to the building's exit capacity.

Before credit can be given to an existing building with a greater number of exits, the exits must be evaluated for their remoteness and independence from each other. All of the exits must be located at least one-half the length of the diagonal from each other before Category d can be used. If the building is equipped throughout with an automatic sprinkler system in accordance with NFPA 13 or 13R, the exits can be located one-fourth the length of the diagonal from each other.

Category e provides additional credit for existing buildings that have the characteristics of both Categories c and

d. Such a building, which has oversized capacities in its egress elements, minimum required clear widths for the egress components and additional exits that are all remotely located from one another, is considered to have the highest degree of safety; however, this is true for only some of the applicable occupancies. In Group B, F, S and R occupancies, any additional capacity or increased number of exits is not a significant factor. No positive points are awarded for this life safety issue for these occupancies.

3410.6.12 Dead ends. In spaces required to be served by more than one means of egress, evaluate the length of the exit access travel path in which the building occupants are confined to a single path of travel. Under the categories and occupancies in Table 3410.6.12, determine the appropriate value and enter that value into Table 3410.7 under Safety Parameter 3410.6.12, Dead Ends, for means of egress and general safety.

❖ This section is used to evaluate dead-end exit access conditions within the building. This section uses the terminology "confined to a single path of travel." Another way to illustrate the meaning of this is "a single direction of travel to reach an exit." A typical corridor is a single path in which a building occupant can travel in two directions to reach exits. A dead end may be a single path, but the key feature of a dead end is that only one direction is available to reach an exit. When building occupants have only one direction of travel available, a potentially hazardous condition is created because they may become trapped if the direction of travel is blocked by fire or smoke.

This section addresses only dead ends that are a component of exit access travel, which is "that portion of a means of egress that leads to an entrance to an exit." Although a room containing only one door may restrict exiting from the room to only one direction, it is not considered a dead-end condition as described in Section 1016.3. This section notes that the dead-end path of travel is limited to 20 feet (6096 mm) for exit access passageways and corridors that serve more than one exit; therefore, the single-door exit of a room enters an "exit access passageway or corridor," which is subject to a dead-end limitation of 20 feet (6096 mm). The travel distance within the room to the exit access is limited by Section 1015.1 and Table 1015.1. The total travel distance from the remote point within a room to the exit must include the room travel length plus exit access passageway or corridor length (see commentary, Section 3410.6.13). Section 3410.6.12 addresses the condition of dead-end passageways and corridors (see commentary, Section 1016.3).

TABLE 3410.6.12
DEAD-END VALUES

OCCUPANCY	CATEGORIES[a]		
	a	b	c
A-1, A-3, A-4, B, E, F, M, R, S	-2	0	2
A-2, E	-2	0	2

a. For dead-end distances between categories, the dead-end value shall be obtained by linear interpolation.

❖ The table reflects the relative degree of hazard associated with dead-end passageways and corridors. This is shown

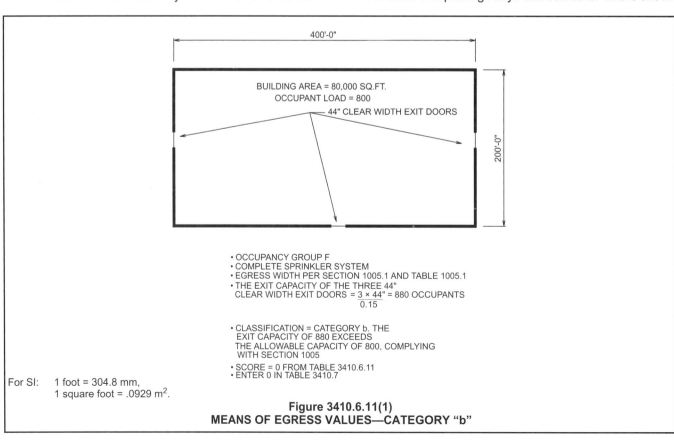

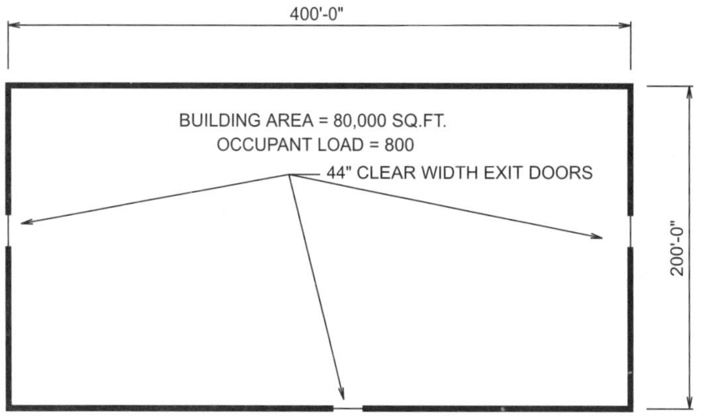

- OCCUPANCY GROUP F
- COMPLETE SPRINKLER SYSTEM
- EGRESS WIDTH PER SECTION 1005.1 AND TABLE 1005.1
- THE EXIT CAPACITY OF THE THREE 44" CLEAR WIDTH EXIT DOORS = $\frac{3 \times 44"}{0.15}$ = 880 OCCUPANTS
- CLASSIFICATION = CATEGORY b. THE EXIT CAPACITY OF 880 EXCEEDS THE ALLOWABLE CAPACITY OF 800, COMPLYING WITH SECTION 1005
- SCORE = 0 FROM TABLE 3410.6.11
- ENTER 0 IN TABLE 3410.7

For SI: 1 foot = 304.8 mm, 1 square foot = .0929 m^2.

Figure 3410.6.11(1)
MEANS OF EGRESS VALUES—CATEGORY "b"

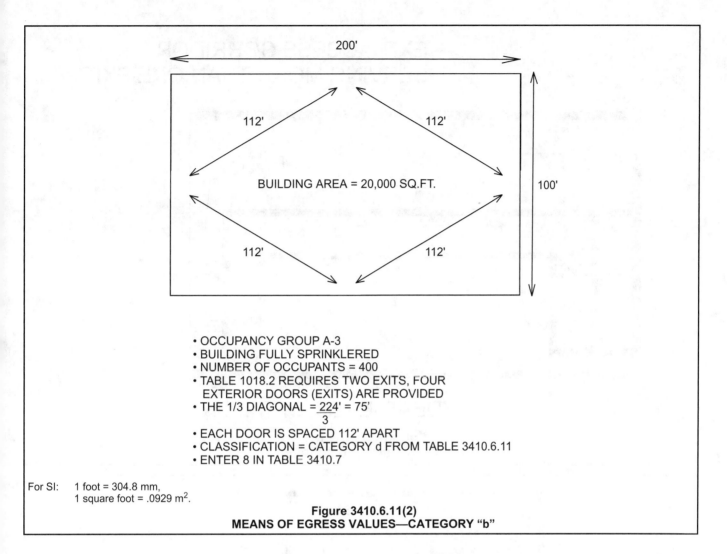

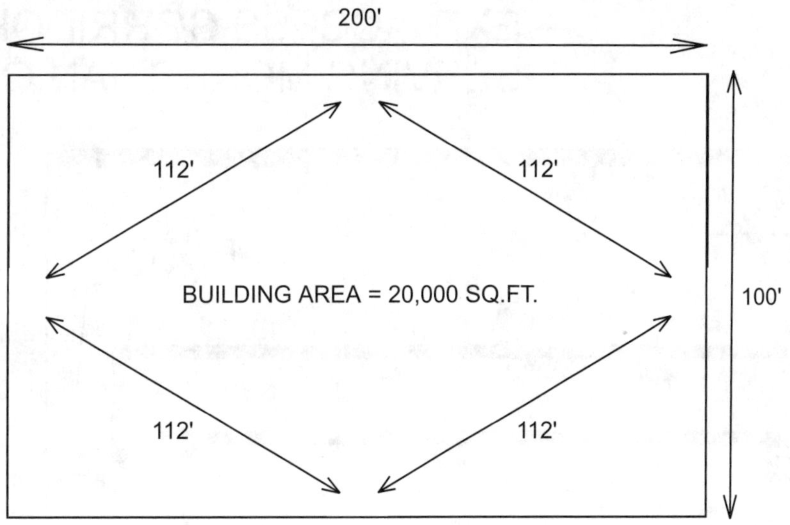

- OCCUPANCY GROUP A-3
- BUILDING FULLY SPRINKLERED
- NUMBER OF OCCUPANTS = 400
- TABLE 1018.2 REQUIRES TWO EXITS, FOUR EXTERIOR DOORS (EXITS) ARE PROVIDED
- THE 1/3 DIAGONAL = $\frac{224'}{3}$ = 75'
- EACH DOOR IS SPACED 112' APART
- CLASSIFICATION = CATEGORY d FROM TABLE 3410.6.11
- ENTER 8 IN TABLE 3410.7

For SI: 1 foot = 304.8 mm,
1 square foot = .0929 m^2.

Figure 3410.6.11(2)
MEANS OF EGRESS VALUES—CATEGORY "b"

by the deficiency points that apply where a dead-end exceeds 35 or 70 feet (10 668 or 21 336 mm) in an occupancy with a relatively high occupant load, such as an assembly occupancy. Note a allows for the interpolation for actual dead-end lengths between the distances specified in the categories. For example, if a building contains a corridor that has a dead-end length of 10 feet (3048 mm) at each end beyond the exits, the value of 1 is used; this is the midpoint between Categories b and c.

3410.6.12.1 Categories. The categories for dead ends are:

1. Category a — Dead end of 35 feet (10 670 mm) in nonsprinklered buildings or 70 feet (21 340 mm) in sprinklered buildings.
2. Category b — Dead end of 20 feet (6096 mm); or 50 feet (15 240 mm) in Group B in accordance with Section 1016.3 exception 2.
3. Category c — No dead ends; or ratio of length to width (l/w) is less than 2.5:1.

❖ This section defines the categories for dead-end exit access conditions.

Category a allows dead-end conditions of up to 35 feet (10 668 mm) for existing buildings that are not fully sprinklered, and up to 70 feet (21 336 mm) for existing buildings that are fully sprinklered. These distances correspond to the dead-end lengths listed in the IFC. Since these lengths far exceed the allowable lengths permitted for new construction, negative values are associated with this category.

Category b is the classification for buildings that comply with the requirements for new construction. This zero-based category provides no extra credit for complying with new construction requirements.

Category c represents conditions that exceed the requirements for new construction. If there are no dead-end corridors or the "corridor" is more of a "space," Category c can be used. In Category c the length-to-width ratio of 2.5 to define a space/corridor is the same as Exception 3 in Section 1016.3. Such a space allows the building user a more circular route and full view of the space; therefore, this does not constitute a dead-end situation.

These categories and occupancies of the building are used in Table 3410.6.12 to determine the score to be entered in Table 3410.7 under Safety Parameter 3410.6.12, Dead Ends [see Figures 3410.6.12.1(1), 3410.6.12.1(2) and 3410.6.12.1(3)].

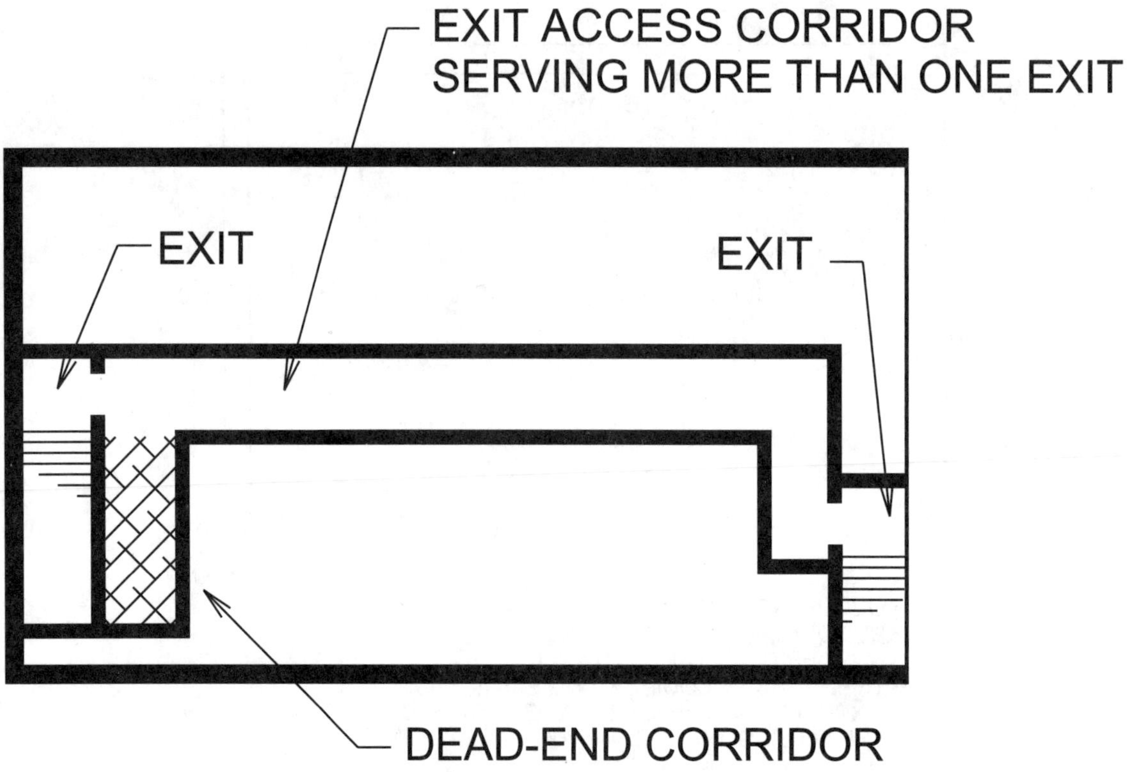

Figure 3410.6.12.1(1)
TYPICAL DEAD-END CORRIDOR

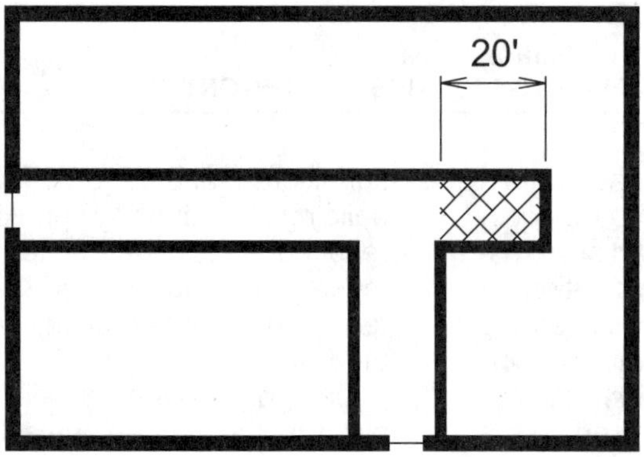

- OCCUPANCY GROUP B WITH MOVABLE PARTITIONS AND FURNITURE 6'-6" HIGH
- DEAD-END PASSAGEWAY IS 20'-0"
- CLASSIFICATION = CATEGORY b
- VALUE = 0 FROM TABLE 3410.6.12
- ENTER 0 IN TABLE 3410.7

For SI: 1 inch = 25.4 mm,
1 foot = 304.8 mm.

Figure 3410.6.12.1(2)
DEAD-END VALUES—CATEGORY b

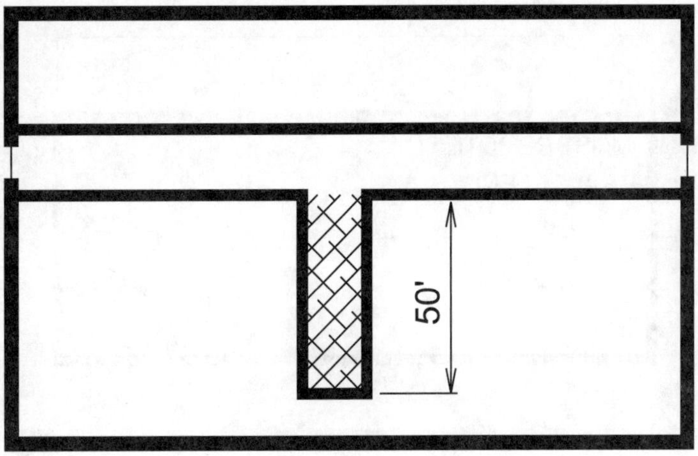

- OCCUPANCY GROUP R
- UNSPRINKLERED BUILDING
- DEAD-END CORRIDOR LENGTH IS 50'-0"
- CLASSIFICATION = CATEGORY a
- VALUE = -2 FROM TABLE 3410.6.12
- ENTER -2 IN TABLE 3410.7

For SI: 1 foot = 304.8 mm.

Figure 3410.6.12.1(3)
DEAD-END VALUES—CATEGORY a

3410.6.13 Maximum exit access travel distance. Evaluate the length of exit access travel to an approved exit. Determine the appropriate points in accordance with the following equation and enter that value into Table 3410.7 under Safety Parameter 3410.6.13, Maximum Exit Access Travel Distance, for means of egress and general safety. The maximum allowable exit access travel distance shall be determined in accordance with Section 1015.1.

$$\text{Points} = 20 \times \frac{\text{Maximum allowable travel distance} - \text{Maximum actual travel distance}}{\text{Max. allowable travel distance}}$$

❖ The length of exit access travel distance is evaluated in comparison to the travel distance allowed by Table 1015.1 for a particular occupancy. The exit access travel distance is measured from the most remote point in the building to the nearest exit. Total exit access travel length of Table 1015.1 includes travel distance within a room or space plus exit access corridor travel to the exit. Some modifications to the Table 1015.1 requirements are listed for certain occupancies and buildings as listed in the table's notes.

To determine the value to be assigned for maximum travel distances to an exit, the specified equation must be used. This equation allows a graduated scale to be used to evaluate compliance of an existing situation with new construction travel distance requirements. With the equation, a more definite evaluation can occur with a broader range of scores. Any existing building having overall travel distances less than those specified in Table 1015.1 will achieve a positive credit. Existing buildings with travel distances greater than those allowed for new construction will be assigned negative points based on the extent the travel distance exceeds the allowable limit.

For example, an existing Group B business building has a travel distance of 150 feet (45 720 mm). The distance is measured from the most remote corner of a partitioned office, down a corridor and to an exit door. Table 1015.1 permits an unsprinklered Group B building to have a travel distance of 200 feet (60 960 mm). The equation yields the following:

$$\text{Points} = 20 \times \left(\frac{200\text{ feet} - 155\text{ feet}}{200\text{ feet}} \right)$$

$$= 4.5$$

See Figures 3410.6.13(1), 3410.6.13(2) and 3410.6.13(3) for other examples.

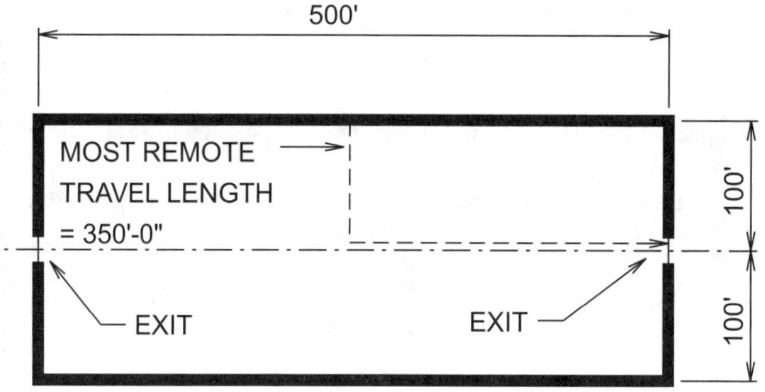

- COMBUSTIBLE STORAGE, OCCUPANCY GROUP S-1 WITH AUTOMATIC FIRE SUPPRESSION SYSTEM
- MOST REMOTE LENGTH OF EXIT ACCESS TRAVEL (SECTION 1015.1)
- TRAVEL DISTANCE IN BUILDING = 350'-0"
- TRAVEL DISTANCE LIMIT FROM TABLE 1015.1 = 250'-0"
- POINTS = $20 \times \frac{250 - 350}{250} = -8$

For SI: 1 foot = 304.8 mm.

Figure 3410.6.13(1)
EXIT ACCESS TRAVEL DISTANCE VALUES—OCCUPANCY GROUP S-1

- OCCUPANCY GROUP B WITHOUT FIRE SUPPRESSION SYSTEM
- MOST REMOTE TRAVEL LENGTH = 200'-0"
- TRAVEL DISTANCE IN BUILDING = 200'-0"
- TRAVEL DISTANCE LIMIT FROM TABLE 1015.1 = 200'-0"
- POINTS = $20 \times \frac{200 - 200}{200} = 0$

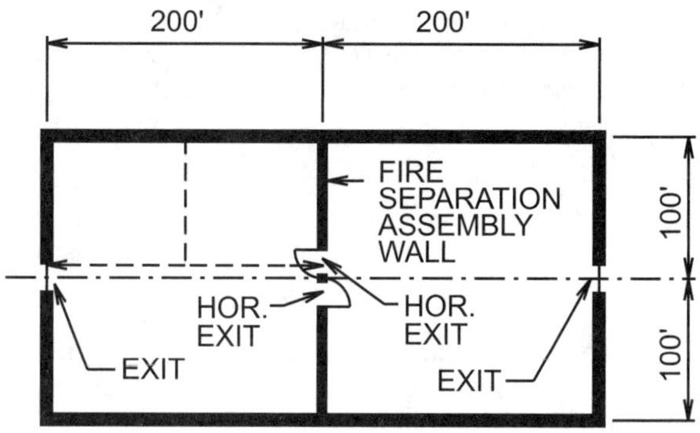

For SI: 1 foot = 304.8 mm.

Figure 3410.6.13(2)
EXIT ACCESS TRAVEL DISTANCE VALUES—OCCUPANCY GROUP B

- OCCUPANCY GROUP R
  WITH FIRE SUPPRESSION SYSTEM
- MOST REMOTE TRAVEL LENGTH = 100'-0"
- TRAVEL DISTANCE IN BUILDING = 100'-0"
- TRAVEL DISTANCE LIMIT FROM TABLE 1015.1 = 250'-0"

- POINTS = $20 \times \frac{250 - 100}{250} = +12$

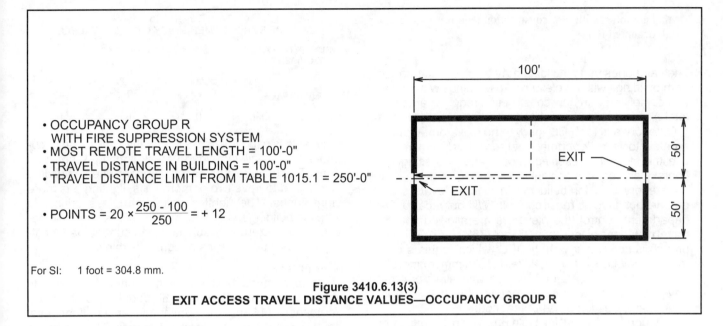

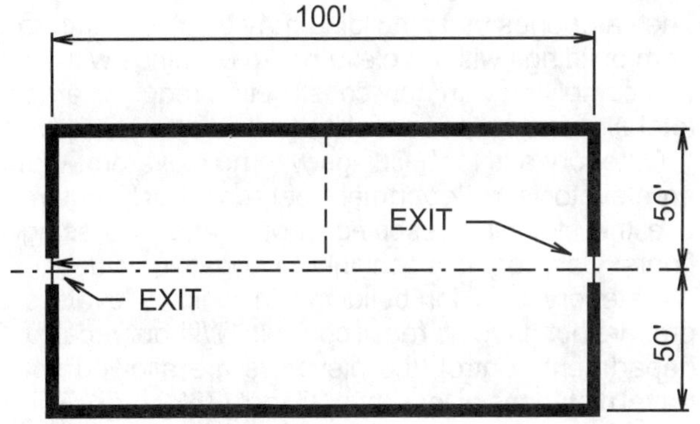

For SI: 1 foot = 304.8 mm.

Figure 3410.6.13(3)
EXIT ACCESS TRAVEL DISTANCE VALUES—OCCUPANCY GROUP R

3410.6.14 Elevator control. Evaluate the passenger elevator equipment and controls that are available to the fire department to reach all occupied floors. Elevator recall controls shall be provided in accordance with the *International Fire Code*. Under the categories and occupancies in Table 3410.6.14, determine the appropriate value and enter that value into Table 3410.7 under Safety Parameter 3410.6.14, Elevator Control, for fire safety, means of egress and general safety. The values shall be zero for a single-story building.

❖ The availability of elevators in the building and their incapability in an emergency are evaluated. This section addresses four different categories of elevators in existing buildings. This section does not address the requirements of Chapter 30. Access must be provided to all occupied floors by passenger elevators for the purposes of this evaluation. Freight elevators cannot be considered, since they may be in locations not readily accessible for fire department use. Elevator recall controls must comply with the IFC for either a Phase I or II category. One-story buildings, including those with mezzanines served by an elevator, are awarded zero value.

TABLE 3410.6.14
ELEVATOR CONTROL VALUES

ELEVATOR TRAVEL	CATEGORIES			
	a	b	c	d
Less than 25 feet of travel above or below the primary level of elevator access for emergency fire-fighting or rescue personnel	-2	0	0	+2
Travel of 25 feet or more above or below the primary level of elevator access for emergency fire-fighting or rescue personnel	-4	NP	0	+4

For SI: 1 foot = 304.8 mm.

❖ The table assigns values based on the types of controls the elevators have and the distance the elevators must travel to reach the floors they serve. The 25-foot (7620 mm) threshold of elevator travel is based on the IFC requirements and ASME A17.1. Elevator travel distance is based on the level where the elevator is accessed by the fire department. Usually, this is the ground floor or grade-level floor and corresponds to the level where fire command stations or fire alarm system annunciator panels are located. Any elevator that travels 25 feet (7620 mm) or more must always be provided with Phase I and II recall capabilities.

Example:

Assume a Group B business occupancy, three stories in height. The elevator lobbies are equipped with smoke detectors that recall the elevator to the main floor or to an alternate floor if the main floor detector is activated. The elevator travels more than 25 feet (7620 mm) above the main floor.

This building must be placed in Category c. Because the elevator travels more than 25 feet (7620 mm) above the main floor, it must be brought into compliance with the IFC, which requires Phase I or II recall. When the elevators are brought into compliance with the IFC, they are automatically placed in Category c. The value from Table 3410.6.14 is 0; therefore, 0 must be entered in Table 3410.7 under Safety Parameter 3410.6.14, Elevator Control.

3410.6.14.1 Categories. The categories for elevator controls are:

1. Category a — No elevator.
2. Category b — Any elevator without Phase I and II recall.
3. Category c — All elevators with Phase I and II recall as required by the *International Fire Code*.
4. Category d — All meet Category c; or Category b where permitted to be without recall; and at least one elevator

that complies with new construction requirements serves all occupied floors.

❖ The categories that a building may be placed into range from buildings with no elevators to buildings with elevators complying with new construction requirements and total elevator recall capability.

Category a is for buildings with no elevators. Without an elevator, fire department personnel are required to use the stairs for rescuing people and accessing fire floors. Category a is assigned a negative value.

Category b is for buildings in which elevators are present but have no recall controls. Without recall or fire department control, the elevators are allowed vertical travel distances of less than 25 feet (7620 mm). This approach is consistent with the IFC, which requires controls for elevators reaching 25 feet (7620 mm) or more.

Category c is used for buildings with elevators that have automatic recall when required by the IFC and, specifically, Section 607.1 of that code.

Category d is used for buildings when the elevator controls comply with either Category b or c and at least one of the elevators in the existing building serves all occupied floor levels. The controls must also comply with all the requirements for new construction. This category recognizes the benefits of having all elevators with Phase I and II controls, as well as having a newly constructed elevator with all the features to facilitate fire-fighting and rescue operations.

Based on the controls provided, or the lack of elevators, the appropriate category is determined. This category and the distance the elevator travels are then used along with Table 3410.6.14 to determine the value score the building will receive for this item. The value is then entered in Table 3410.7.

3410.6.15 Means of egress emergency lighting. Evaluate the presence of and reliability of means of egress emergency lighting. Under the categories and occupancies in Table 3410.6.15, determine the appropriate value and enter that value into Table 3410.7 under Safety Parameter 3410.6.15, Means of Egress Emergency Lighting, for means of egress and general safety.

❖ Lighting throughout the entire means of egress is evaluated in this section. Illumination of the means of egress is essential in a building during normal occupancy. During a fire condition, illumination becomes even more important, since visibility will more than likely be reduced by the buildup of smoke. Failure to provide adequate lighting in a fire condition has led to numerous deaths because occupants become disoriented and are unable to follow the means of egress to safety. It is appropriate to include egress lighting as one of the major building components that must be evaluated. The importance of this lighting is shown by the minimal values assigned by Table 3410.6.15 when egress lighting is not provided.

TABLE 3410.6.15
MEANS OF EGRESS EMERGENCY LIGHTING VALUES

NUMBER OF EXITS REQUIRED BY SECTION 1010	CATEGORIES		
	a	b	c
Two or more exits	NP	0	4
Minimum of one exit	0	1	1

❖ This table is used to assess the relative risk to the occupants of the building when egress lighting is not adequate.The table reinforces the idea that when a building is required to have two or more exits, lighting has more significance. The lighting values are less critical in smaller buildings where only one exit is required because the hazard of a substantial occupant load locating and using the means of egress is minimal.

Example:

A Group A church building with a number of classrooms that will accommodate up to 60 people has at least 1 footcandle (11 lux) of illumination at the floor level throughout the entire means of egress. The means of egress lighting in the sanctuary, corridors and stairs is provided with battery back-up power that is sized to provide lighting for 1 hour. The classroom lighting is wired to the main switch panel in the building and has no emergency power source.

The lighting must be placed in Category a because the classrooms are part of the means of egress (exit access) and the lack of emergency power does not comply with Section 1006.3. The classroom will accommodate up to 60 occupants; therefore, in accordance with Section 1014.1 and Table 1014.1, the rooms are required to have two means of egress. The lighting must be provided with an emergency power source in accordance with Section 1006.3. The value determined from Table 3410.6.15 is "NP." Emergency power must be added to the lighting in the classrooms. The resulting Category b has a value of 0. Enter 0 in Table 3410.7.

3410.6.15.1 Categories. The categories for means of egress emergency lighting are:

1. Category a — Means of egress lighting and exit signs not provided with emergency power in accordance with Section 2702.
2. Category b — Means of egress lighting and exit signs provided with emergency power in accordance with Section 2702.
3. Category c — Emergency power provided to means of egress lighting and exit signs which provides protection in the event of power failure to the site or building.

❖ There are three categories of egress lighting. These categories are consistent with the requirements established by the IFC .

Category a is used when there is no egress lighting in

any occupied room or space within the building or when the level of illumination is less than 1 footcandle (11 lux) at the floor level at any location in the means of egress, other than the exempted locations indicated in Section 1006 (see commentary, Section 1006). If means of egress lighting and exit signs are provided but are not connected to emergency power, Category a is also applicable. Without emergency power providing a 1-hour duration after primary power loss, the means of egress lighting and exit signs would not provide as reliable a level of safety to the building occupants.

Category b includes buildings where the means of egress lighting and exit signs comply with the requirements of Sections 1006 and 1011, respectively, including emergency power requirements. Section 1006.3 includes requirements for emergency power to maintain continuous illumination of the means of egress during a power failure. This category represents the zero-based criteria, which is the level required for new construction.

Category c is used when emergency power is provided for means of egress lighting and exit signs in excess of the minimum requirements for new construction. If the emergency power provides full protection to the site or building during power failure, Category c is applicable. Campus-type complexes or buildings that require extra security may have a power plant available to provide complete back-up power for indefinite periods of time. This will qualify for the positive points of this category.

These categories, along with the number of required exits, are used to determine the appropriate value from Table 3410.6.15. The value determined from Table 3410.6.15 is then entered in Table 3410.7.

3410.6.16 Mixed occupancies. Where a building has two or more occupancies that are not in the same occupancy classification, the separation between the mixed occupancies shall be evaluated in accordance with this section. Where there is no separation between the mixed occupancies or the separation between mixed occupancies does not qualify for any of the categories indicated in Section 3410.6.16.1, the building shall be evaluated as indicated in Section 3410.6 and the value for mixed occupancies shall be zero. Under the categories and occupancies in Table 3410.6.16, determine the appropriate value and enter that value into Table 3410.7 under Safety Parameter 3410.6.16, Mixed Occupancies, for fire safety and general safety. For buildings without mixed occupancies, the value shall be zero.

❖ These provisions are used to evaluate mixed occupancies within an existing building and whether the method of separating mixed occupancies conforms to the requirements of Section 302.3.2. If Section 302.3.2 is not completely understood, see the associated commentary before proceeding with the evaluation of the existing building in this section (see also commentary, Section 3410.6).

This section is applicable only to separated mixed occupancies. If a building is a single occupancy, the applicable value for this section is zero. If an existing mixed occupancy building has no fire-resistance-rated separation between the different uses, or the separation is fire-resistance rated for less than 1 hour, the applicable value is also zero. The building must also be evaluated in accordance with Section 3410.6. The zero-based category for this section is equivalent to full compliance with Section 302.3.2, which is the requirement for new construction.

TABLE 3410.6.16
MIXED OCCUPANCY VALUES[a]

OCCUPANCY	CATEGORIES		
	a	b	c
A-1, A-2, R	-10	0	10
A-3, A-4, B, E, F, M, S	-5	0	5

a. For fire-resistance ratings between categories, the value shall be obtained by linear interpolation.

❖ This table addresses the relative risk of a building in or close to compliance with the provisions for separated mixed occupancies. When mixed occupancies are not separated from each other, the risk from hazards is greater in high-density occupancies, such as Groups A-1 and A-2. This risk is also greater in residential occupancies, because occupants may be sleeping and not fully alert. For this reason, inadequate separation is given greater negative values. In buildings with lower occupant loads, and where the occupants are alert, the risks are relatively lower.

Note a permits linear interpolation of the corresponding values in each category. For example, an unsprinklered, Group B/Group F-1 mixed occupancy building is separated with a 2-hour fire-resistance-rated assembly. From Table 302.3.2, the required fire-resistance rating of the separation between a Group B and a Group F-1 occupancy for Category b is 3 hours. Since the rating that is provided is only 2 hours, an interpolation halfway between Category a and Category b can be used. The resulting points are 2.5.

Example 1:
Assume a three-story building. The first floor is a Group M mercantile occupancy and the upper two stories are Group R residential occupancies. The building is Type VB construction, 9,000 square feet (836.1 m^2) per floor. It is fully sprinklered and only 25 percent of the perimeter is accessible. Although Type VB construction is not required to be protected, the Group M occupancy on the first floor is separated from the Group R occupancy on the second floor with a 2-hour fire-resistance-rated floor/ceiling assembly. In this case, Exception 1 to Section 302.3.2 permits the required rating of 2 hours from Table 302.3.2 to be reduced to 1 hour because the building is equipped throughout with an automatic sprinkler system in accordance with Section 903. The provided 2-hour rating is twice that required; therefore, this building qualifies for Category c. A value of 10 points is assigned to the residential portion of the building, and a value of 5 points is assigned to the mercantile portion (see Section 3410.6 for commentary about mixed uses).

Example 2:
Assume a four-story building of Type IIB construction, fully sprinklered. It has 15,000 square feet (1393.5 m^2) per floor with 25-percent open perimeter. The building is to be a Group R-1 hotel, except for a portion of the first floor that will be used as a Group A-2 nightclub, occupying 1,000 square feet (92.9 m^2). Although the building is unprotected Type IIB construction, the floor/ceiling assemblies of the first and second floors are 3 hours and the fire barrier assembly at the first floor between the Group R-1 occupancy and the Group A-2 occupancy is also 3 hours. Considering these building characteristics, the following determinations are made:

Exception 1 to Section 302.3.2 is applicable in this example just as it was applicable in the previous example. Because of the presence of an automatic sprinkler system, the exception permits the rating of the separation required in Table 302.3.2 to be reduced by 1 hour. Although this building is required to have 1-hour separation, it has a 3-hour separation, which is more than double the minimum. This building is classified as Category c. A value of 10 points is awarded to both occupancies.

3410.6.16.1 Categories. The categories for mixed occupancies are:

1. Category a — Minimum 1-hour fire barriers between occupancies.
2. Category b — Fire barriers between occupancies in accordance with Section 302.3.2
3. Category c — Fire barriers between occupancies having a fire-resistance rating of not less than twice that required by Section 302.3.2.

❖ This section addresses three different conditions:

Category a is used when the separation of mixed occupancies does not conform to the fire-resistance ratings specified in Section 302.3.2. If the ratings between mixed occupancies are less than specified in Section 302.3.2, but at least a 1-hour rating is provided, Category a is applicable.

Category b is used when the separation of mixed occupancies conforms to Section 302.3.2 for fire-resistance ratings.

Category c is used when the fire-resistance ratings that separate occupancies in an existing building are no less than twice the requirements in Section 302.3.2. Category c gives bonus points or increased credits to existing buildings that have higher fire-resistance ratings.

Section 3410 does not require compliance with Section 302.3.2. Section 3410.6.16 evaluates whether the existing building's compliance with Section 302.3 provides relative safety within the building. If the building does comply with Section 302.3.2, Section 3410.6.16 acknowledges that there is no basis for assigning negative points, nor is there any basis for assigning positive points for safety; therefore, a neutral zero value is assigned. Whatever the value, it is entered in Table 3410.7 under Safety Parameter 3410.6.16, Mixed Occupancies.

3410.6.17 Automatic sprinklers. Evaluate the ability to suppress a fire based on the installation of an automatic sprinkler system in accordance with Section 903.3.1.1. "Required sprinklers" shall be based on the requirements of this code. Under the categories and occupancies in Table 3410.6.17, determine the appropriate value and enter that value into Table 3410.7 under Safety Parameter 3410.6.17, Automatic Sprinklers, for fire safety, means of egress divided by 2 and general safety.

❖ These provisions are used to determine the amount of credit that can be applied to the evaluation for the installation of an automatic sprinkler system in an existing building.

Sprinkler safety parameters appear to provide a double credit for automatic sprinkler systems. Basic credits are provided for: building height, Section 3410.6.1; building area, Section 3410.6.2 and maximum travel distance to an exit, Section 3410.6.13. Throughout the code additional credits are offered for the use of an automatic sprinkler system.

The evaluation of sprinklers in an existing building is based on whether an automatic sprinkler system is both required and installed. The criteria used to determine when an automatic sprinkler system is required are tied to the same requirements for new construction in Section 903. The thresholds listed in Sections 903.2 through 903.2.12 must be used to evaluate whether those characteristics and occupancies are present and whether a sprinkler system is needed.

The exceptions to Section 903; the requirements for stories and basements without openings of Section 903.2.10.1 and the suppression requirements for covered mall buildings, high-rise buildings, public garages and unlimited area buildings must also be evaluated. The determination of whether sprinklers are required in a building or a portion of a building must be done to correctly determine the category applicable to the existing building.

These guidelines for sprinklers allow for a more equitable evaluation of the contribution of that feature to overall building safety. This factor encourages the installation of automatic sprinkler systems in existing buildings by providing substantial negative and positive points.

The values for sprinklers are given in Table 3410.6.17. The appropriate credit values from Table 3410.6.17 are entered in Table 3410.7 under Safety Parameter 3410.6.17, Automatic Sprinklers, for both fire safety (FS) and general safety (GS), but only one-half the value is entered under the means of egress (ME).

The one-half credit for egress is allowed because some credits for sprinklers are incorporated into the parameters for means of egress capacity (see Section 3410.6.11), dead-ends (see Section 3410.6.12) and maximum travel distance to an exit (see Section 3410.6.13).

TABLE 3410.6.17
SPRINKLER SYSTEM VALUES

OCCUPANCY	CATEGORIES					
	a	b	c	d	e	f
A-1, A-3, F, M, R, S-1	-6	-3	0	2	4	6
A-2	-4	-2	0	1	2	4
A-4, B, E, S-2	-12	-6	0	3	6	12

❖ This table lists the credit values for the respective categories of Section 3410.6.17.1, based on the occupancy being evaluated in the existing building. The assembly occupancies containing large combustible fuel loads with large occupant loads are included with those occupancies containing large fuel loads. These occupancies represent buildings where experience has shown that adequate sprinkler systems save lives and where an on-site sprinkler system is necessary to supplement the local fire department capabilities.

Group A-2 buildings are contained in a separate line in the table for determining sprinkler system values. This occupancy with its densely packed high occupant loads, ill-defined seating and aisle arrangements and facilities services must be separately evaluated to adequately match its hazards with sprinkler system requirements. Churches and schools, which have high occupant/low-fuel loads, are combined with other low-fuel load occupancies in the last line item in the table. Category c is the zero-based category. The categories to the left of this column contain negative values that define buildings and occupancies that are required to be, but are not, sprinklered or are provided with inadequate sprinkler systems. The three categories to the right of this column contain positive values that represent buildings and occupancies with sprinkler systems that are deemed adequate or comply with current standards and requirements.

3410.6.17.1 Categories. The categories for automatic sprinkler system protection are:

1. Category a — Sprinklers are required throughout; sprinkler protection is not provided or the sprinkler system design is not adequate for the hazard protected in accordance with Section 903.
2. Category b — Sprinklers are required in a portion of the building; sprinkler protection is not provided or the sprinkler system design is not adequate for the hazard protected in accordance with Section 903.
3. Category c — Sprinklers are not required; none are provided.
4. Category d — Sprinklers are required in a portion of the building; sprinklers are provided in such portion; the system is one which complied with the code at the time of installation and is maintained and supervised in accordance with Section 903.
5. Category e — Sprinklers are required throughout; sprinklers are provided throughout in accordance with Chapter 9.
6. Category f — Sprinklers are not required throughout; sprinklers are provided throughout in accordance with Chapter 9.

❖ Six categories are defined in this section for evaluating the automatic sprinkler system in an existing building. These categories address all aspects, from existing buildings that are unsprinklered but are required to be sprinklered if new construction, to existing buildings that are sprinklered and are not required to be sprinklered if new construction. These categories reduce the impact of providing sprinkler protection required by Chapter 9 to be sprinklered in new construction. This greater range of points increases the flexibility in the use of this evaluation method.

Category a includes buildings or occupancies that exceed the requirements of Section 903 and are required to be sprinklered throughout the building. Category a buildings are provided with either no sprinkler system, or one that is inadequate and does not provide the required level of protection. To evaluate the existing sprinkler system, a trained fire protection engineer should evaluate the system's design against the applicable referenced standards. This is not the same type of consideration needed in Category d, where the existing system actually meets all of the requirements of any earlier edition of the applicable referenced standards. This category is considered the lowest acceptable level of compliance and is therefore associated with the largest negative values in Table 3410.6.17.

Category b is applicable when only a portion of the existing building is required by the provisions of Section 903 to be sprinklered. For example, a multistory building may have two of its stories qualify as windowless stories that require sprinklers, or a mercantile building may have one fire area exceeding 12,000 square feet (1114.8 m^2) and require sprinklers. In both cases the buildings are without an automatic sprinkler system, or the sprinkler systems that are in place are inadequate and do not comply with the technical provisions of the code. Since only portions of the buildings are required to be sprinklered, the negative sprinkler system values for this category in Table 3410.6.17 are exactly half of the Category a values.

Category c is the zero-based category for this safety parameter. The existing building is not required by Section 903 to be sprinklered and no sprinkler system is provided. Such buildings are neither penalized nor rewarded by Table 3410.6.17.

Category d is similar to Category b because only a portion of the existing building is required by Section 903 to be sprinklered. In this case there is a sprinkler system in place in that portion of the building and the system was designed at the time of its installation to comply with the requirements of an earlier edition of the applicable standards in Section 903. An example of this is a sprinkler system designed to comply with the hydraulic design criteria of the 1989 edition of NFPA 13 for an ordinary Group 2 hazard. As long as this existing sprinkler system is still properly maintained and supervised, the building qualifies for the small positive values listed in Table 3410.6.17.

Category e are existing buildings that are required by the provisions of Chapter 9 to be sprinklered throughout, and are protected throughout by a properly designed, installed and supervised sprinkler system. Although this category more closely represents the requirements for new construction, moderate positive values are awarded by Table 3410.6.17. The evaluation rewards existing buildings that are sprinklered with higher points.

Category f is the highest level of protection and is rewarded with maximum positive points in the accompanying table. Existing buildings, which are not required by Section 903 to be sprinklered but are voluntarily provided with a fully designed, installed and supervised system, provide an added level of protection that justifies substantial bonus points that may ultimately determine whether an existing building meets its mandatory safety scores.

3410.6.18 Incidental use. Evaluate the protection of incidental use areas in accordance with Section 302.1.1. Do not include those where this code requires suppression throughout the building including covered mall buildings, high-rise buildings, public garages and unlimited area buildings. Assign the lowest score from Table 3409.6.18 for the building or fire area being evaluated. If there are no specific occupancy areas in the building or fire area being evaluated, the value shall be zero.

❖ This section includes an evaluation system for the separation/protection requirements indicated for specific occupancy areas in an existing building. This evaluation is based on Table 302.1.1. If the building designer chooses to separate/protect the small rooms or areas in an existing building in accordance with this table, the building is classified according to its main use. Because some existing buildings may have been designed in this fashion, an evaluation procedure has been added to account for the level of protection provided. The lowest score must be assigned to the building or fire area for the specific occupancy areas. For example, an existing Group B building has six separate storage rooms. Each room is more than 100 square feet (9.3 m^2) in area. If five of the storage rooms are protected with limited area sprinkler systems and separated with smoke partitions, and the sixth room is not protected in the same manner, the lowest score for all the storage rooms is determined by the single unprotected storage room. If the building designer chooses to treat the specific occupancy areas as a separate occupancy, the provisions of Section 3410.9.1 are applicable and a zero is inserted in Table 3410.7 under Safety Parameter 3410.6.18, Incidental Use Area.

TABLE 3410.6.18
INCIDENTAL USE AREA VALUES[a]

PROTECTION REQUIRED BY TABLE 302.1.1	PROTECTION PROVIDED						
	None	1 Hour	AFSS	AFSS with SP	1 Hour and AFSS	2 Hours	2 Hours and AFSS
2 Hours and AFSS	-4	-3	-2	-2	-1	-2	0
2 Hours, or 1 Hour and AFSS	-3	-2	-1	-1	0	0	0
1 Hour and AFSS	-3	-2	-1	-1	0	-1	0
1 Hour	-1	0	-1	0	0	0	0
1 Hour, or AFSS with SP	-1	0	-1	0	0	0	0
AFSS with SP	-1	-1	-1	0	0	-1	0
1 Hour or AFSS	-1	0	0	0	0	0	0

a. AFSS = Automatic fire suppression system; SP = Smoke partitions (See Section 302.1.1.1).

NOTE: For Table 3409.7, see page 596.

❖ This table provides a matrix of characteristics arranged in a format of rows and columns for determining values for specific occupancy areas. The values are inserted into the building score sheet. The left-hand column of the table is arranged for the separation/protection specified by Table 302.1.1 for a particular occupancy room or area. Table 302.1.1 must first be consulted to determine the level of separation/protection required by the code for new construction. This same entry is then found in the left-hand column of Table 3410.6.18. The top row represents the actual level of protection that is provided in the existing building. The corresponding occupancy area value is read and then inserted into Table 3410.7 under Safety Parameter 3410.6.18, Incidental Use. Values of zero are assigned to all arrangements that represent compliance with the requirements for new construction. Negative values are assigned based on the degree of noncompliance with the requirements for new construction.

3410.7 Building score. After determining the appropriate data from Section 3410.6, enter those data in Table 3410.7 and total the building score.

❖ This section is the tally sheet for all of the 18 safety parameters evaluated in Sections 3410.6.1 through 3410.6.18, which determine the building's overall safety profile for fire safety (FS), means of egress (ME) and general safety (GS).

This section directs the data and values of the 18 safety parameters of Sections 3410.6.1 through 3410.6.18 to be entered in Table 3410.7 for totaling the building scores.

TABLE 3410.7
SUMMARY SHEET — BUILDING CODE

Existing occupancy ______________________ Proposed occupancy ______________________

Year building was constructed ______________________ Number of stories __________ Height in feet __________

Type of construction ______________________ Area per floor ______________________

Percentage of open perimeter ________% Percentage of height reduction ________%

Completely suppressed: Yes _______ No _______ Corridor wall rating ______________________

Compartmentation: Yes _______ No _______ Required door closers: Yes _______ No _______

Fire-resistance rating of vertical opening enclosures _

Type of HVAC system ______________________, serving number of floors _ _ _ _ _ _ _ _

Automatic fire detection: Yes _______ No _______, type and location ______________________

Fire alarm system: Yes _______ No _______, type ______________________

Smoke control: Yes _______ No _______, type ______________________

Adequate exit routes: Yes _______ No _______ Dead ends: __________ Yes _______ No _______

Maximum exit access travel distance ______________________ Elevator controls: Yes _______ No _______

Means of egress emergency lighting: Yes _______ No _______ Mixed occupancies: Yes _______ No _______

SAFETY PARAMETERS	FIRE SAFETY (FS)	MEANS OF EGRESS (ME)	GENERAL SAFETY (GS)
3410.6.1 Building Height 3410.6.2 Building Area 3410.6.3 Compartmentation			
3410.6.4 Tenant and Dwelling Unit Separations 3410.6.5 Corridor Walls 3410.6.6 Vertical Openings			
3410.6.7 HVAC Systems 3410.6.8 Automatic Fire Detection 3410.6.9 Fire Alarm System			
3410.6.10 Smoke control 3410.6.11 Means of Egress 3410.6.12 Dead ends	* * * * * * * * * * * *		
3410.6.13 Maximum Exit Access Travel Distance 3410.6.14 Elevator Control 3410.6.15 Means of Egress Emergency Lighting	* * * * * * * *		
3410.6.16 Mixed Occupancies 3410.6.17 Automatic Sprinklers 3410.6.18 Incidental Use		* * * * ÷ 2 =	
Building score — total value			

* * * *No applicable value to be inserted.

TABLE 3410.7. See above.

❖ Table 3410.7 is the summary sheet containing all of the relative attributes of the building. The summary sheet also contains a complete listing of the 18 safety parameters that have been evaluated. The upper portion of the summary sheet serves as a guide to the user to catalog and highlight existing building elements that relate to the 18 safety parameters and to the evaluation. The lower portion of the summary sheet is used to record the results of the 18 safety parameters that have been evaluated. These are added to produce the building score total values for fire safety (FS), means of egress (ME) and general safety (GS).

3410.8 Safety scores. The values in Table 3410.8 are the required mandatory safety scores for the evaluation process listed in Section 3410.6.

❖ This section lists the minimum scores for fire safety (FS), means of egress (ME) and general safety (GS) that must be obtained from the evaluation of the 18 safety parameters to be acceptable as a building meeting the code's objectives for public safety and health. This section summarizes the mandatory values of Table 3410.8 that must be met from the evaluation program.

TABLE 3410.8
MANDATORY SAFETY SCORES[a]

OCCUPANCY	FIRE SAFETY (MFS)	MEANS OF EGRESS (MME)	GENERAL SAFETY (MGS)
A-1	16	27	27
A-2	19	30	30
A-3	18	29	29
A-4, E	23	34	34
B	24	34	34
F	20	30	30
M	19	36	36
R	17	34	34
S-1	15	25	25
S-2	23	33	33

a. MFS = Mandatory Fire Safety;
MME = Mandatory Means of Egress;
MGS = Mandatory General Safety.

❖ The table lists the minimum mandatory safety scores for the evaluation of various occupancies for the three major mandatory safety scores of fire safety (MFS), means of egress (MME) and general safety (MGS). The mandatory safety values are based on the scores considered to be in compliance with the code for new construction. This is the zero-based concept. The scores have been determined as representing one level of compliance higher than the code's minimum requirements for new construction. The mandatory safety scores are consistent with the idea of establishing an equivalent level of safety, even though the existing building is evaluated only for the 18 safety parameters.

3410.9 Evaluation of building safety. The mandatory safety score in Table 3410.8 shall be subtracted from the building score in Table 3410.7 for each category. Where the final score for any category equals zero or more, the building is in compliance with the requirements of this section for that category. Where the final score for any category is less than zero, the building is not in compliance with the requirements of this section.

❖ The sections and table that follow are the final steps in the evaluation process. This section also discusses how mixed occupancies must to be treated during the final step of the evaluation process.

This section compares the three building scores from Table 3410.7 to the three mandatory safety scores from Table 3410.8. If the values in all three categories from Table 3410.7 exceed the corresponding mandatory safety scores in Table 3410.8, the building passes and is in compliance with the code. If the score in any one category is less than the mandatory safety score, the building is deemed to have failed and additional measures must be taken to bring the scores to a point that will at least equal the mandatory safety scores.

TABLE 3410.9. See below.

❖ Table 3410.9 shows in simple equations whether a building passes the evaluation by subtracting the mandatory safety score from Table 3410.7. This is done for each of the three general categories of evaluation: fire safety, means of egress and general safety. If the difference for each category is zero or greater, the existing building passes and is considered to comply with the code objectives for public safety.

Example:
A Group M mercantile occupancy receives evaluations for the 18 safety parameters that result in total building scores in Table 3410.7 as follows:

Fire safety (FS) = 23;

Means of egress (ME) = 41

General safety (GS) = 36

These scores are compared to the mandatory safety scores for a Group M occupancy from Table 3410.8.

Mandatory fire safety (MFS) = 19

Mandatory means of egress (MME) = 36

Mandatory general safety (MGS) = 36

TABLE 3410.9
EVALUATION FORMULAS[a]

FORMULA	T.3409.7			T.3409.8	SCORE	PASS	FAIL
FS-MFS ≥ 0	_____	(FS)	–	_____ (MFS) =	_____	_____	_____
ME-MME ≥ 0	_____	(ME)	–	_____ (MME) =	_____	_____	_____
GS-MGS ≥ 0	_____	(GS)	–	_____ (MGS) =	_____	_____	_____

a. FS = Fire Safety; MFS = Mandatory Fire Safety
ME = Means of Egress; MME = Mandatory Means of Egress
GS = General Safety; MGS = Mandatory General Safety

The mandatory score is subtracted from the building score.

Conclusion: The building is acceptable; all of the final safety scores are zero or greater.

Table 3410.9
EVALUATION FORMULAS[a]

Formula	Table 3410.7		Table 3410.8	Score	Pass	Fail
FS-MFS ≥ 0	23 (FS)	-	23 (MFS) =	0	X	___
ME-MME ≥ 0	41 (ME)	-	35 (MME)=	6	X	___
GS-MGS ≥ 0	36 (GS)	-	35 (MGS) =	1	X	___

Note a.
FS = Fire Safety
ME = Means of Egress
GS = General Safety
MFS = Mandatory Fire Safety
MME = Mandatory Means of Egress
MGS = Mandatory General Safety

Figure 3410.9
EXAMPLE OF EVALUATION FORMULAS

3410.9.1 Mixed occupancies. For mixed occupancies, the following provisions shall apply:

1. Where the separation between mixed occupancies does not qualify for any category indicated in Section 3410.6.16, the mandatory safety scores for the occupancy with the lowest general safety score in Table 3410.8 shall be utilized (see Section 3410.6.)
2. Where the separation between mixed occupancies qualifies for any category indicated in Section 3410.6.16, the mandatory safety scores for each occupancy shall be placed against the evaluation scores for the appropriate occupancy.

❖ This section explains how to determine whether a mixed occupancy building passes or fails the process of evaluation. It restates the information in Sections 3410.6 and 3410.6.16. The mixed occupancy evaluation is based on the requirements of Sections 302.3.1 and 302.3.2. The two procedures described are:

Procedure 1:
For unseparated occupancies in accordance with Section 302.3.1, or for occupancies that are separated with a fire-resistance rating of less than 1 hour, mandatory safety scores for the occupancy with the lowest general safety score of Table 3410.8 apply.

Procedure 2:
For separated occupancies in accordance with Section 302.3.2, or one of the categories listed in Section 3410.6.16, mandatory safety scores for each occupancy are compared to the evaluation scores for the appropriate occupancy. The total building score of Table 3410.7 is computed for each appropriate occupancy.

This section does not include a category or condition for a third option for mixed occupancies, which is the separation of multiple occupancies within a single building or structure with one or more fire walls. Likewise, this option is not addressed in Section 302.3.1. A fire wall creates separate and independent buildings; therefore, a separate evaluation must be done for each area within a building that is separated by fire walls.

Example:
A 40-foot (12 192 mm), three-story building has an open perimeter of 25 percent. It is Type IIB unprotected construction, unsprinklered. The building has 9,000 square feet (836.1 m^2) per floor. A Group M mercantile is on the first floor with Group B business occupancies on the second and third floors. Although the building is unprotected Type IIB construction, there is a 2-hour fire-resistance-rated floor/ceiling assembly separating the first and second floors. An analysis of this building will result in the following: "In accordance with Section 302.3.2 and Table 302.3.2, the mixed occupancies are separated by the required fire-rated assembly of 2 hours."

This building qualifies for Category b in Section 3410.6.16.1. A separate summary sheet for tabulating the building score in Table 3410.7 has to be completed for each of the two occupancies. The values of the building scores for fire safety, means of egress and general safety for the Group M mercantile occupancy and the Group B business occupancy must be calculated. The mercantile building scores are compared to the mandatory safety scores of 19, 36 and 36, and the business building scores are compared to the mandatory safety scores of 24, 34 and 34. If any one of the three building scores from either of the two occupancies does not equal or exceed the applicable safety score, the entire building fails the evaluation. For the building to pass, the safety parameters that are less than the mandatory safety score must be upgraded until a passing score is achieved.

BIBLIOGRAPHY

The following resource materials are referenced in this chapter or are relevant to the subject matter addressed in this chapter.

36 CFR Parts 1190 and 1191 Draft, *The Americans with Disabilities Act Accessibility Guidelines* (ADAAG), Washington, DC: Architectural and Transportation Barriers Compliance Board, April 2, 2002.

42 USC 3601-88, *Fair Housing Amendments Act* (FHAA). Washington, DC: United States Code, 1988.

"Accessibility and Egress for People with Physical Disabilities." CABO Board for the Coordination of the Model Codes Report, October 5, 1993.

ASME A17.1-00, *Safety Code for Elevators and Escalators*. New York: American Society of Mechanical Engineers, 2000.

ASME A17.3-96, *Safety Code for Existing Elevators and Escalators*. New York: American Society of Mechanical Engineers, 1996.

ASME 18.1-99, *Safety Standard for Platform Lifts and Stairway Chairlifts–with Addenda A18.1a-2001,* New York: American Society of Mechanical Engineers, 1999.

DOJ 28 CFR, Part 36-91, *Americans with Disabilities Act* (ADA). Washington, DC: Department of Justice, 1991.

DOJ 28 CFR, Part 36-91(Appendix A), *ADA Accessibility Guidelines for Buildings and Facilities* (ADAAG). Washington, DC: Department of Justice, 1991.

FED-STD-795-88, *Uniform Federal Accessibility Standards.* Washington, DC: General Services Administration; Department of Defense; Department of Housing and Urban Development; U.S. Postal Service, 1988.

"Final Report, Recommendations for a New ADAAG." ADAAG Review Federal Advisory Committee, September 30, 1996.

ICC A117.1-98, *Accessible and Usable Buildings and Facilities.* Falls Church, VA: International Code Council, 1998.

IMC-03, *International Mechanical Code.* Falls Church, VA: International Code Council, 2003.

IPMC-03, *International Property Maintenance Code.* Falls Church, VA: International Code Council, 2003.

NFPA 13-99, *Installation of Sprinkler Systems*. Quincy, MA: National Fire Protection Association, 1999.

NFPA 13D-99, *Installation of Sprinkler Systems in One- and Two-Family Dwellings and Manufactured Homes*. Quincy, MA: National Fire Protection Association, 1999.

NFPA 13R-99, *Installation of Sprinkler Systems in Residential Occupancies Up to and Including Four Stories in Height*. Quincy, MA: National Fire Protection Association, 1999.

NFPA 72-99, *National Fire Alarm Code*. Quincy, MA: National Fire Protection Association, 1999.

NFPA 101-00, *Alternative Approaches to Life Safety*. Quincy, MA: National Fire Protection Association, 2000.

Rehabilitation Guidelines/1980. Washington, DC: National Institute of Building Science for the U.S. Department of Housing and Urban Development, 1980.

1. Guideline for Setting and Adopting Standards for Building Rehabilitation;
2. Guideline for Approval of Building Rehabilitation;
3. Statutory Guideline for Building Rehabilitation;
4. Guideline for Managing Official Liability Associated with Building Rehabilitation;
5. Egress Guideline for Residential Rehabilitation;
6. Electrical Guideline for Residential Rehabilitation;
7. Plumbing DWV Guideline for Residential Rehabilitation; and
8. Guideline on Fire Ratings of Archaic Materials and Assemblies.

UL 555-96, *Fire Dampers*. Northbrook, IL: Underwriters Laboratories Inc., 1996.

Chapter 35: Referenced Standards

General Comments

Chapter 35 contains a comprehensive list of all standards that are referenced in the code. It is organized in a manner that makes it easy to locate specific document references.

This chapter lists the standards that are referenced in various sections of this document. The standards are listed herein by the promulgating agency of the standard, the standard identification, the date and title and the section or sections of this document that reference the standard. The application of the referenced standards shall be as specified in Section 102.4.

It is important to understand that not every document related to building design and construction is qualified to be a "referenced standard." The International Code Council® (ICC®) has adopted a criterion that standards referenced in the *International Codes®* and standards intended for adoption into the *International Codes* must meet in order to qualify as a referenced standard. The policy is summarized as follows:

- Code references: The scope and application of the standard must be clearly identified in the code text.
- Standard content: The standard must be written in mandatory language and appropriate for the subject covered. The standard shall not have the effect of requiring proprietary materials or prescribing a proprietary testing agency.
- Standard promulgation: The standard must be readily available and developed and maintained in a consensus process, such as ASTM or ANSI.

It should be noted that the ICC Code Development Procedures, of which the standards policy is a part, are updated periodically. A copy of the latest version can be obtained from the ICC offices.

Once a standard is incorporated into the code through the code development process, it becomes an enforceable part of the code. When the code is adopted by a jurisdiction, the standard also is part of that jurisdiction's adopted code. It is for this reason that the criteria were developed. Compliance with this policy provides that documents incorporated into the code are, among others, developed through the use of a consensus process, written in mandatory language and do not mandate the use of proprietary materials or agencies. The requirement for a standard to be developed through a consensus process is vital, as it means that the standard will be representative of the most current body of available knowledge on the subject as determined by a broad spectrum of interested or affected parties without dominance by any single interest group. A true consensus process has many attributes, including but not limited to:

- An open process that has formal (published) procedures that allow for the consideration of all viewpoints;
- A definitive review period that allows for the standard to be updated or revised;
- A process of notification to all interested parties and
- An appeals process.

Many available documents related to design, installation and construction, though useful, are not "standards" and are not appropriate for reference in the code. Often, these documents are developed or written with the intention of being used for regulatory purposes and are unsuitable for use as a regulation due to extensive use of recommendations, advisory comments and nonmandatory terms. Typical examples of such documents include installation instructions, guidelines and practices.

The objective of ICC's standards policy is to provide regulations that are clear, concise and enforceable— thus the requirement for standards to be written in mandatory language. This requirement is not intended to mean that a standard cannot contain informational or explanatory material that will aid the user of the standard in its application. When it is the desire of the standard's promulgating agency for such material to be included, however, the information must appear in a nonmandatory location, such as an annex or appendix, and be clearly identified as not being part of the standard.

Overall, standards referenced by the code must be authoritative, relevant, up to date and, most important, reasonable and enforceable. Standards that comply with ICC's standards policy fulfill these expectations.

Purpose

As a performance-oriented code, the code contains numerous references to documents that are used to regulate materials and methods of construction. The references to these documents within the code text consist of the promulgating agency's acronym and its publication designation (e.g., ASME A17.1) and a further indication that the document being referenced is the one that is listed in Chapter 35. Chapter 35 contains all of the information that is necessary to identify the specific referenced document. Included is the following information on a document's promulgating agency (see Figure 35):

- The promulgating agency (i.e., the agency's title);
- The promulgating agency's acronym and

- The promulgating agency's address.

For example, a reference to an ASME standard within the code indicates that the document is promulgated by the American Society of Mechanical Engineers (ASME), which is located in New York City. This chapter lists the standards agencies alphabetically for ease of identification.

This chapter also includes the following information on the referenced document itself (see Figure 35):

- The document's publication designation;
- The document's edition year;
- The document's title;
- Any addenda or revisions to the document that are applicable and
- Every section of the code in which the document is referenced.

For example, a reference to ASME A17.1 indicates that this document can be found in Chapter 35 under the heading ASME. The specific standards designation is A17.1. For convenience, these designations are listed in alphanumeric order. This chapter identifies that ASME A17.1 is titled *Safety Code for Elevators and Escalators,* the applicable edition (i.e., its year of publication) is 2000 and it is referenced in numerous sections of the code.

This chapter will also indicate when a document has been discontinued or replaced by its promulgating agency. When a document is replaced by a different one, a note will appear to tell the user the designation and title of the new document.

The key aspect of the manner in which standards are referenced by the code is that a specific edition of a specific standard is clearly identified. In this manner, the requirements necessary for compliance can be readily determined. The basis for code compliance is, therefore, established and available on an equal basis to the building official, contractor, designer and owner.

This chapter lists the standards that are referenced in various sections of this document. The standards are listed herein by the promulgating agency of the standard, the standard identification, the effective date and title and the section or sections of this document that reference the standard. The application of the referenced standards shall be as specified in Section 102.4.

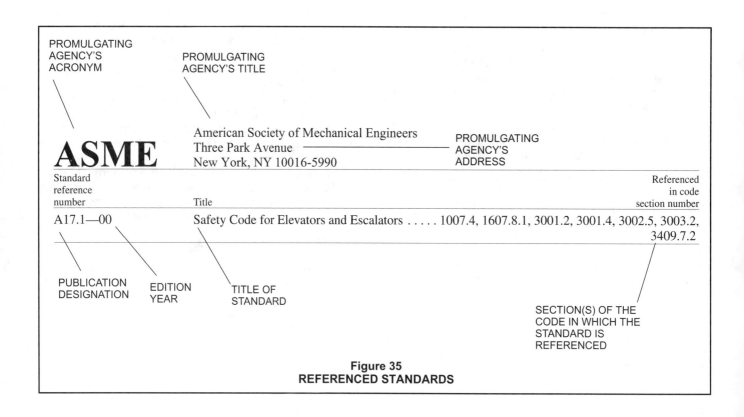

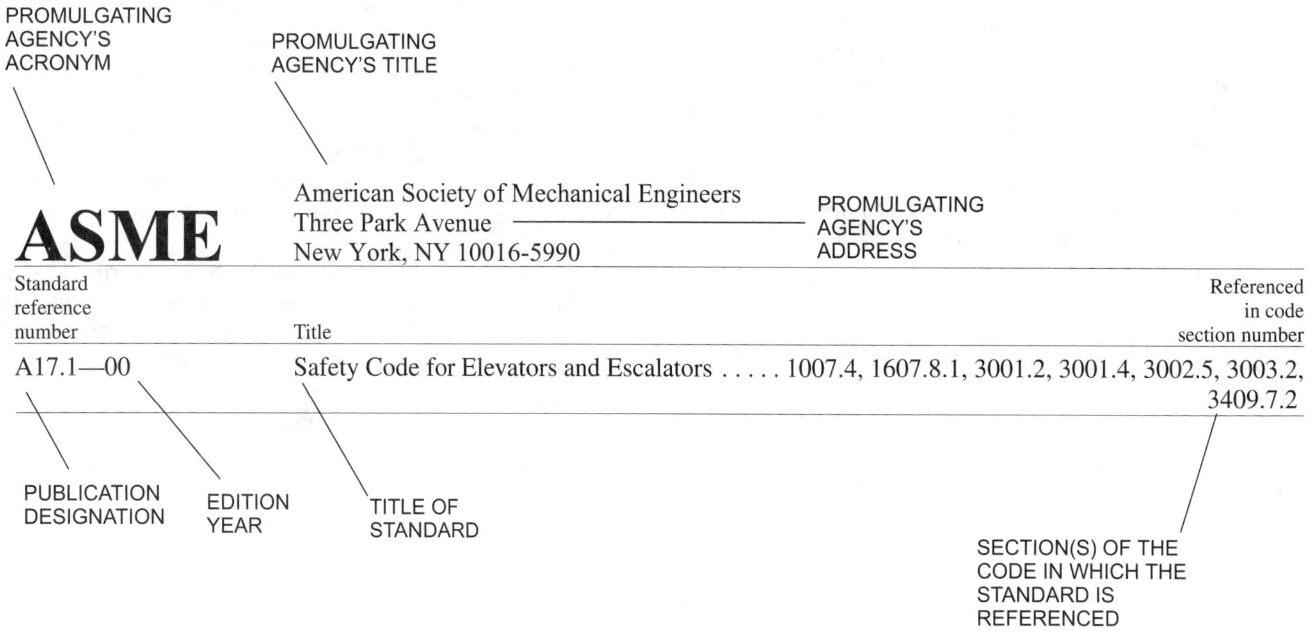

Figure 35
REFERENCED STANDARDS

This chapter lists the standards that are referenced in various sections of this document. The standards are listed herein by the promulgating agency of the standard, the standard identification, the effective date and title, and the section or sections of this document that reference the standard. The application of the referenced standards shall be as specified in Section 102.4.

AA

Aluminum Association
900 - 19th Street N.W., Suite 300
Washington, DC 20006

Standard reference number	Title	Referenced in code section number
ADM 1—00	Aluminum Design Manual: Part 1-A Aluminum Structures, Allowable Stress Design; and Part 1-B —Aluminum Structures, Load and Resistance Factor Design of Buildings and Similar Type Structures	1604.3.5, 2002.1
ASM 35—80	Aluminum Sheet Metal Work in Building Construction	2002.1

AAMA

American Architectural Manufacturers Association
1827 Waldon Office Square, Suite 104
Schaumburg, IL 60173

Standard reference number	Title	Referenced in code section number
1402—86	Standard Specifications for Aluminum Siding, Soffit and Fascia	1404.5.1
101/I.S.2—97	Voluntary Specifications for Aluminum, Vinyl (PVC) and Wood Windows and Glass Doors	1714.5.1
101/I.S.2/NAFS—02	Voluntary Performance Specification for Window, Skylights and Glass Doors	1714.5.1, 2405.5

ACI

American Concrete Institute
P.O. Box 9094
Farmington Hills, MI 48333-9094

Standard reference number	Title	Referenced in code section number
216.1—97	Standard Method for Determining Fire Resistance of Concrete and Masonry Construction Assemblies	Table 721.1(2), 721.1
318—02	Building Code Requirements for Structural Concrete	1604.3.2, Table 1617.6, 1617.6.2.4.3, Table 1704.3, 1704.4.1, Table 1704.4, 1708.3, 1805.4.2.6, 1805.9, 1808.2.23.1.1, 1808.2.23.2, 1808.2.23.2.1, 1808.2.23.2.2, 1809.2.3.2, 1809.2.3.2.2, 1810.1.2.2, 1812.8, 1901.2, 1901.3, 1901.4, 1902, 1903.1, 1903.2, 1903.3, 1903.4, 1903.5.1, 1903.6, 1904.4.2, 1905.1.4, 1905.3, 1905.4, 1905.5, 1905.6.5.5, 1905.8.3, 1905.11.3, 1906.1.5, 1906.3, 1906.4.3, 1907.1, 1907.2, 1907.4.1, 1907.6, 1907.7.2, 1907.7.3, 907.7.4, 1907.7.5, 1907.8, 1907.9, 1907.10, 1907.11, 1907.12, 1907.13, 1908 ,1908.1.1, 1908.1.2, 1908.1.3, 1908.1.4, 1908.1.5, 1908.1.6, 1908.1.7, 1908.1.8, 1908.1.9, 1909.1, 1909.3, 1909.4, 1909.5, 1909.6, 1910, 1910.2.1, 1910.2.2, 1910.2.3, 1910.2.4, 1910.3.1, 1910.4.1, 1910.4.2, 1910.4.3, 1910.4.3.1, 1910.5.2, 1913.1, 2108.3, 2205.3
530—02	Building Code Requirements for Masonry Structures	1405.5, 1405.5.3, 1405.9, 1604.3.4, 1704.5, 1704.5.1, Table 1704.5.1, 1704.5.2, Table 1703.3.1, 1708.1.1, 1708.1.2, 1708.1.3, 1805.5.2, 1812.7, 2101.2.3, 2101.2.4, 2101.2.5, 2103.11.6, 2106.1, 2106.1.1.1, 2106.1.1.2, 2106.1.1.3, 2106.3, 2106.4, 2106.5, 2106.6, 2107.1, 2107.2, 2107.2.1, 2107.2.2, 2107.2.4, 2107.2.5, 2107.2.6, 2108.1, 2108.2, 2108.4, 2109.1, 2109.2.3.1, 2109.2.3.2
530.1—02	Specifications for Masonry Structures	1405.5.1, 1405.9.1, Table1704.5.1, Table 1704.5.3, 1805.5.2, 2103.11.7, 2104.1, 2104.1.1, 2104.3,
TG/T1.1—01	Acceptance Criteria for Moment Frames Based on Testing	1908.1.3

AF&PA

American Forest & Paper Association
1111 19th St, NW Suite 800
Washington, DC 20036

Standard reference number	Title	Referenced in code section number
AF&PA/ASCE 16—95	Standard for Load and Resistance Factor Design (LRFD) for Engineered Wood Construction	2307.1
WCD No. 4—89	Plank and Beam Framing for Residential Buildings	2306.1.2
WFCM—01	Wood Frame Construction Manual for One-and Two-family Dwellings	2301.2.3, 2308.1, 2308.2.1
T.R. No. 7—87	Basic Requirements for Permanent Wood Foundation System	1805.4.6, 1807.2, 2304.9.5
NDS—01	National Design Specification (NDS) for Wood Construction with 2001Supplement	721.6.3.2, 1715.1.1, 1715.1.4, 1805.4.5, 1808.1, 2306.1, 2306.2.1, 2306.3.2, Table 2306.3.1, Table 2306.4.1, 2306.3.4, 2306.3.5, 2306.4.1, Table 2308.9.3(4)
AF&PA—93	Span Tables for Joists and Rafters	2306.1.1, 2308.8, 2308.10.2, 2308.10.3

AHA

American Hardwood Association
1210 West N.W. Highway
Palatine, IL 60067

Standard reference number	Title	Referenced in code section number
A135.4—95	Basic Hardboard	1404.3.1, 2303.1.6
A135.5—95	Prefinished Hardboard Paneling	2303.1.6, 2304.6.2
A135.6—98	Hardboard Siding	1404.3.2, 2303.1.6
A194.1—85	Cellulosic Fiber Board	2303.1.5

AISC

American Institute of Steel Construction
One East Wacker Drive, Suite 3100
Chicago, IL 60601-2001

Standard reference number	Title	Referenced in code section number
335—89s1	Specification for Structural Steel Buildings—Allowable Stress Design and Plastic Design, including Supplement No.1, 2001	1604.3.3, Table 1617.6.2, Table 1704.3, 2203.2, 2205.1
LRFD (1999)	Load and Resistance Factor Design Specification for Structural Steel Buildings	1604.3.3, Table 1617.6, Table 1704.3, 2203.2, 2205.1, 2205.3
HSS (2000)	Load and Resistance Factor Design Specification for Steel Hollow Structural Sections	1604.3.3, Table 1617.6, 2203.2, 2205.1
341—02	Seismic Provisions for Structural Steel Buildings	1602.1, Table 1617.6.2, 1707.2, 1708.4, 2205.2.1, 2205.2.2, 2205.3, 2205.3.1

AISI

American Iron and Steel Institute
1140 Connecticut Avenue
Suite 705
Washington, DC 20036

Standard reference number	Title	Referenced in code section number
NASPEC 2001	North American Specification for Design of Cold-Formed Steel Structural Members	1604.3.3, 2209.1
General	Standard for Cold-Formed Steel Framing-General Provisions, 2001	2210.1
Header	Standard for Cold-Formed Steel Framing-Header Design, 2001	2210.2
Truss	Standard for Cold-Formed Steel Framing-Truss Design, 2001	2210.3

AITC

American Institute of Timber Construction
Suite 140
7012 S. Revere Parkway
Englewood, CO 80112

Standard reference number	Title	Referenced in code section number
AITC A 190.1—92	Structural Glued Laminated Timber	2303.1.3, 2306.1
AITC Technical Note 7—96	Calculation of Fire Resistance of Glued Laminated Timbers	721.6.3.3
AITC 104—84	Typical Construction Details	2306.1
AITC 110—01	Standard Appearance Grades for Structural Glued Laminated Timber	2306.1
AITC 112—93	Standard for Tongue-and-Groove Heavy Timber Roof Decking	2306.1
AITC 113—01	Standard for Dimensions of Structural Glued Laminated Timber	2306.1
AITC 117—01	Standard Specifications for Structural Glued Laminated Timber of Softwood Species—Design Requirements—Standard Specifications for Structural Glued Laminated Timber of Softwood Species—Manufacturing Requirements	2306.1
AITC 119—96	Standard Specifications for Structural Glued Laminated Timber of Hardwood Species	2306.1
AITC 200—92	Inspection Manual	2306.1
AITC 500—91	Determination of Design Values for Structural Glued Laminated Timber	2306.1

ALI

Automotive Lift Institute
P.O. Box 33116
Indialantic, FL 32903-3116

Standard reference number	Title	Referenced in code section number
ALI ALCTV—98	Standard for Automotive Lifts—Safety Requirements for Construction, Testing and Validation (ANSI)	3001.2

ANSI

American National Standards Institute
25 West 43rd Street, Fourth Floor
New York, NY 10036

Standard reference number	Title	Referenced in code section number
A 13.1—96	Scheme for the Identification of Piping Systems	415.9.6.4
A 42.2—71	Portland Cement and Portland Cement Lime Plastering, Exterior (Stucco) and Interior	2109.8.4.6
A 42.3—71	Lathing and Furring for Portland Cement and Portland Cement Lime Plastering, Exterior Stucco and Interior	2109.8.4.6
A108.1A—99	Installation of Ceramic Tile in the Wet-set Method, with Portland Cement Mortar	2103.9
A108.1B—99	Installation of Ceramic Tile, Quarry Tile on a Cured Portland Cement Mortar Setting Bed with Dry-set or Latex-Portland Mortar	2103.9
A108.4—99	Installation of Ceramic Tile with Organic Adhesives or Water-cleanable Tile-Setting Epoxy Adhesive	2103.9.7
A108.5—99	Installation of Ceramic Tile with Dry-set Portland Cement Mortar or Latex Portland Cement Mortar	2103.9.1, 2103.9.2, 2103.9.3
A108.6—99	Installation of Ceramic Tile with Chemical-resistant, Water Cleanable Tile-setting-and-grouting Epoxy	2103.9.4
A108.7—92	Specification for Electrically Conductive Ceramic Tile Installed with Conductive Dry-set Portland Cement Mortar	2103.9.2
A108.8—99	Installation of Ceramic Tile with Chemical-resistant Furan Resin Mortar and Grout	2103.9.5
A108.9—99	Installation of Ceramic Tile with Modified Epoxy Emulsion Mortar/Grout	2103.9.6
A 108.10—99	Installation of Grout in Tilework	2103.9.8
A 118.1—99	American National Standard Specifications for Dry-set Portland Cement Mortar	2103.9.1
A 118.2—99	American National Standard Specifications for Conductive Dry-set Portland Cement Mortar	2103.9.2
A 118.3—99	American National Standard Specifications for Chemical-resistant, Water-cleanable Tile-setting and -Grouting Epoxy and Water Cleanable Tile-setting Epoxy Adhesive	2103.9.4
A 118.4—99	American National Standard Specifications for Latex-portland Cement Mortar	2103.9.3
A 118.5—99	American National Standard Specifications for Chemical Resistant Furan Mortar and Grouts for Tile Installation	2103.9.5
A 118.6—99	American National Standard Specifications for Cement Grouts for Tile Installation	2103.9.8
A 118.8—99	American National Standard Specifications for Modified Epoxy Emulsion Mortar/Grout	2103.9.6
A 136.1—99	American National Standard Specifications for Organic Adhesives for Installation of Ceramic Tile	2103.9.7
A 137.1—88	American National Standard Specifications for Ceramic Tile	2103.4
A 208.1—99	Particleboard	2303.1.7, 2303.1.7.1

ANSI—continued

B 31.3—99	Process Piping—Including Addendum	415.9.6.1
Z 97.1—84 (R1994)	Safety Glazing Materials Used in Buildings-safety Performance Specifications and Methods of Test (Reaffirmed 1994)	2406.1.3, 2406.1.2, 2407.1

APA

APA - Engineered Wood Association
P.O. Box 11700
Tacoma, WA 98411-0700

Standard reference number	Title	Referenced in code section number
APA PDS—97	Plywood Design Specification (revised 1998)	2306.1, Table 2306.3.1, 2306.3.2, 2306.4.1
APA PDS Supplement 1—90	Design and Fabrication of Plywood Curved Panels (revised 1995)	2306.1
APA PDS Supplement 2—92	Design and Fabrication of Plywood-lumber beams (revised 1998)	2306.1
APA PDS Supplement 3—90	Design and Fabrication of Plywood Stressed-skin Panels (revised 1996)	2306.1
APA PDS Supplement 4—90	Design and Fabrication of Plywood Sandwich Panels (revised 1993)	2306.1
APA PDS Supplement 5—95	Design and Fabrication of All-plywood Beams (revised 1995)	2306.1
EWS R540—96	Builders Tips: Proper Storage and Handling of Glulam Beams	2306.1
EWS S475—99	Glued Laminated Beam Design Tables	2306.1
EWS S560—99	Field Notching and Drilling of Glued Laminated Timber Beams	2306.1
EWS T300—99	Glulam Connection Details	2306.1
EWS X440—00	Product Guide-Glulam	2306.1
EWS X445—97	Glulam in Residential Construction —Southern Edition	2306.1
EWS X450—97	Glulam in Residential Construction —Western Edition	2306.1

ASAE

American Society of Agricultural Engineers
2950 Niles Road
St. Joseph, MI 49085-9659

Standard reference number	Title	Referenced in code section number
EP 484.2 (1998)	Diaphragm Design of Metal-Clad, Post-Frame Rectangular Buildings	2306.1
EP 486.1 (2000)	Shallow-Post Foundation Design	2306.1
EP 559 (1997)	Design Requirements and Bending Properties for Mechanically Laminated Columns	2306.1

ASCE/SEI

American Society of Civil Engineers
Structural Engineering Institute
1801 Alexander Bell Drive
Reston, VA 20191-4400

Standard reference number	Title	Referenced in code section number
3—91	Standard Practice for the Construction and Inspection of Composite Slabs	1604.3.3, 2209.2
5—02	Building Code Requirements for Masonry Structures	1405.5, 1405.5.3, 1405.9, 1604.3.4, 1704.5, 1704.5.1, Table 1704.5.1, 1704.5.2, Table 1703.3.1, 1708.1.1, 1708.1.2, 1708.1.3, 1805.5.2, 1812.7, 2101.2.3, 2101.2.4, 2101.2.5, 2103.11.6, 2106.1, 2106.1.1.1, 2106.1.1.2, 2106.1.1.3, 2106.3, 2106.4, 2106.5, 2106.6, 2107.1, 2107.2, 2107.2.1, 2107.2.2, 2107.2.4, 2107.2.5, 2107.2.6, 2108.1, 2108.2, 2108.4, 2109.1, 2109.2.3.1, 2109.2.3.2
6—02	Specifications for Masonry Structures	1405.5.1, 1405.9.1, Table1704.5.1, Table 1704.5.3, 1805.5.2, 2103.11.7, 2104.1, 2104.1.1, 2104.3,
7—02	Minimum Design Loads for Buildings and Other Structures	1605.1, 1605.2.2, 1605.3.1.2, 1605.3.2, 1608.1, 1608.3, 1608.3.4, 1608.3.5, 1608.4, 1608.5, 1608.6, 1608.7, 1608.8, 1608.9, 1609.1.1, 1609.1.4.1, 1609.3, Table 1609.3.1, 1609.7.3, 1612.2, 1614.1, 1616.1, 1616.3, 1616.4.5, 1616.5, Table 1616.5.1.1, Table 1616.5.1.2, 1616.6, 1617.1, 1617.2, 1617.2.1, 1617.2.2.2, 1617.3, 1617.4, 1617.6, 1617.6.1, 1617.6.1.1, 1618.1, 1619, 1620.1, 1620.1.1, 1620.1.2, 1620.1.3, 1620.2.1, 1620.2.7, 1620.3.1, 1621.1, 1621.1.1, 1621.1.2, 1621.1.3, 1622.1, 1622.1.1, 1622.1.2, 1622.1.3, 1623.1, 1623.1.1

ASCE/SEI—continued

8—90	Standard Specification for the Design of Cold-formed Stainless Steel Structural Members	1604.3.3, 2209.1
16—95	Standard for Load Resistance Factor Design (LRFD) for Engineered Wood Construction	2307.1
19—96	Structural Applications of Steel Cables for Buildings	2207.1, 2207.2
24—98	Flood Resistant Design and Construction	1203.3.2, 1612.4, 1612.5, 3001.2
29—99	Standard Calculation Methods for Structural Fire Protection	721.1
32—01	Design and Construction of Frost Protected Shallow Foundations	1805.2.1

ASME

American Society of Mechanical Engineers
Three Park Avenue
New York, NY 10016-5990

Standard reference number	Title	Referenced in code section number
A17.1—00	Safety Code for Elevators and Escalators	1007.4, 1607.8.1, 3001.2, 3001.4, 3002.5, 3003.2, 3409.7.2,
A18.1—99	Safety Standard for Platform Lifts and Stairway Chairlifts—with Addenda A18.1a-2001	1007.5, 1109.1, 3409.7.3
A90.1—97	Safety Standard for Belt Manlifts—with Addenda A90.1a-99	3001.2
B16.18—84 (R1994)	Cast Copper Alloy Solder Joint Pressure Fittings	909.13.1
B16.22—95	Wrought Copper and Copper Alloy Solder Joint Pressure Fittings—with Addenda B16.22a-98	909.13.1
B20.1—00	Safety Standard for Conveyors and Related Equipment	3001.2, 3005.3

ASTM

ASTM International
100 Barr Harbor Drive
West Conshohocken, PA 19428-2959

Standard reference number	Title	Referenced in code section number
A 6/A 6M— 01b	Specification for General Requirements for Rolled Steel, Structural Steel Bars, Plates, Shapes, and Sheet Piling	Table 1704.3
A 36/A 36M—00	Specification for Carbon Structural Steel	1809.3.1, 2103.11.5
A 82—01	Specification for Steel Wire, Plain, for Concrete Reinforcement	2103.11.5, 2103.11.6
A 123/A 123M—97e1	Specification for Zinc (Hot-Dip Galvanized) Coating on Iron and Steel Products	2103.11.7.1
A 153—01a	Specification for Zinc Coating (Hot-Dip) on Iron and Steel Hardware	2103.11.7.1
A 167—99	Specification for Stainless and Heat-Resisting Chromium-Nickel Steel Plate, Sheet and Strip	2103.11.5
A 185—01	Specification for Steel Welded Wire Reinforcement, Plain for Concrete	2103.11.4, 2103.11.5
A 252—E01	Specification for Welded and Seamless Steel Pipe Piles	1809.3.1, 1810.6.1
A 283/A 283M—00	Specification for Low and Intermediate Tensile Strength Carbon Steel Plates	1809.3.1, 1810.6.1
A 307—00	Specification for Carbon Steel Bolts and Studs, 60,000 psi Tensile Strength	1912.1
A 416—99	Specification for Steel Strand, Uncoated Seven-Wire for Prestressed Concrete	1809.2.3.1, 2103.11.6
A 421/A 421M—98	Specification for Uncoated Stress-Relieved Steel Wire for Prestressed Concrete	2103.11.6
A 435/A 435M—90 (2001)	Specification for Straight-Beam Ultrasonic Examination of Steel Plates	1708.4
A 496—01	Specification for Steel Wire, Deformed for Concrete Reinforcement	2103.11.3, 2103.11.4
A 510—00	Specification for General Requirements for Wire Rods and Coarse Round Wire, Carbon Steel	2103.11.6
A 568/A 568M—01	Specification for Steel, Sheet, Carbon, and High-Strength, Low-Alloy, Hot-Rolled and Cold-Rolled, General Requirements for	Table 1704.3
A 572/A 572M—01	Specification for High-Strength Low-Alloy Columbium-Vanadium Structural Steel	1809.3.1
A 588/A 588M—01	Specification for High-Strength Low-Alloy Structural Steel with 50 ksi (345 Mpa) Minimum Yield Point to 4 inches (100 mm) Thick	1809.3.1
A 615/A 615M—00	Specification for Deformed and Plain Billet-Steel Bars for Concrete Reinforcement	1708.3, 1908.1.8, 2103.11.1, 2103.11.6
A 641/A 641M—98	Specification for Zinc-coated (Galvanized) Carbon Steel Wire	2103.11.7.3
A 653/A 653M—01a	Specification for Steel Sheet, Zinc-coated Galvanized or Zinc-iron Alloy-Coated Galvannealed by the Hot-dip Process	Table 1507.4.3, 2211.2, 2211.2.2.1
A 706/A 706M—00	Specification for Low-Alloy Steel Deformed and Plain Bars for Concrete Reinforcement	1704.4.1, 1903.5.2, 1908.1.3
A 706/A 706M—01	Specification for Low-Alloy Steel Deformed and Plain Bars for Concrete Reinforcement	2103.11.1, 2103.11.6, 2108.3
A 722/A 722M—98	Specification for Uncoated High-Strength Steel Bar for Prestressing Concrete	2103.11.6, 2106.1.1.3.1
A 755/A 755M—01	Specification for Steel Sheet, Metallic-Coated by the Hot-Dip Process and Prepainted by the Coil-Coating Process for Exterior Exposed Building Products	Table 1507.4.3
A 767/A 767M—00b	Specification for Zinc-coated (Galvanized) Steel Bars for Concrete Reinforcement	2103.11.1
A 775/A 775M—01	Specification for Epoxy-Coated Steel Reinforcing Bars	2103.11.1

ASTM—continued

A 792/A 792M—01a	Specification for Steel Sheet, 55% Aluminum-Zinc Alloy-Coated by the Hot-Dip Process	Table 1507.4.3, 2211.2.2, 2211.2.2.1
A 875M—99a	Specification for Steel Sheet Zinc-54% Aluminum Alloy-Coated by the Hot Dip Process	2211.2.2, 2211.2.2.1
A 884—99	Specification for Epoxy-Coated Steel Wire and Welded Wire Fabric for Reinforcement	2103.11.7.2
A 898/A 898M—91 (2001)	Specification for Straight Beam Ultrasonic Examination of Rolled Steel Shapes	1708.4
A 899—91 (1999)	Specification for Steel Wire Epoxy-Coated	2103.11.7.2
A 913/A 913M—01	Specification for High-strength Low-Alloy Steel Shapes of Structural Quality, Produced by Quenching and Self-tempering Process (QST)	1809.3.1
A 951—00	Specification for Masonry Joint Reinforcement	2103.11.2
A 996/A 996M—00	Specification for Rail-Steel and Axle-Steel Deformed Bars Reinforcement for Concrete	2103.11.1, 2103.11.6
A1008—01A	Specification for Steel, Sheet, Cold-Rolled, Carbon, Structural, High-Strength Low-Alloy and High-Strength Low-Alloy with Improved Formability	2103.11.5
B 42—98	Specification for Seamless Copper Pipe, Standard Sizes	909.13.1
B 43—98	Specification for Seamless Red Brass Pipe, Standard Sizes	909.13.1
B 68M—99	Specification for Seamless Copper Tube, Bright Annealed (Metric)	909.13.1
B 88—99el	Specification for Seamless Copper Water Tube	909.13.1
B 101—01	Specification for Lead-Coated Copper Sheet and Strip for Building Construction	Table 1507.4.3
B 209—96	Specification for Aluminum and Aluminum Alloy Steel and Plate	Table 1507.4.3
B 251—97	Specification for General Requirements for Wrought Seamless Copper and Copper-alloy Tube	909.13.1
B 280—99el	Specification for Seamless Copper Tube for Air Conditioning and Refrigeration Field Service	909.13.1
B 633—98	Specification for Electrodeposited Coatings of Zinc on Iron and Steel	2211.2
C 5—97	Specification for Quicklime for Structural Purposes	Table 2507.2
C 22/C 22M—00	Specification for Gypsum	Table 2506.2
C 27—98	Specification for Standard Classification of Fireclay and High-Alumina Refractory Brick	2111.5, 2111.8
C 28/C 28M—00	Specification for Gypsum Plasters	Table 2507.2
C 31/31M—98	Practice for Making and Curing Concrete Test Specimens in the Field	Table 1704.4, 1905.6.3.2, 1905.6.4.2
C 33—99ae1	Specification for Concrete Aggregates	Table 1904.2.1
C 33—01a	Specification for Concrete Aggregates	721.3.1.4, 721.4.1.1.3
C 34—96 (2001)	Specification for Structural Clay Load-bearing Wall Tile	2103.2
C 35—95 (2001)	Specification for Inorganic Aggregates for Use in Gypsum Plaster	Table 2507.2
C 36/C 36M-01	Specification for Gypsum Wallboard	Figure 721.5.1(2), Figure 721.5.1(3), Table 721.5.1(2), Table 2506.2
C 37/C 37M-01	Specification for Gypsum Lath	Table 2507.2
C 39—99ae1	Test Method for Compressive Strength of Cylindrical Concrete Specimens	1905.6.3.2
C 42/C 42M—99	Test Method for Obtaining and Testing Drilled Cores and Sawed Beams of Concrete	1905.6.5.2
C 55—01A	Specification for Concrete Brick	Table 721.3.2, 2103.1, 2105.2.2.1.2
C 56—96 (2001)	Specification for Structural Clay Non-load Bearing Tile	2103.2
C 59/C 59M—00	Specification for Gypsum Casting and Molding Plaster	Table 2507.2
C 61/C 61M—00	Specification for Gypsum Keene's Cement	Table 2507.2
C 62—01	Specification for Building Brick (Solid Masonry Units Made from Clay or Shale)	2103.2, 2105.2.2.1.1
C 67—02	Test Methods of Sampling and Testing Brick and Structural Clay Tile	721.4.1.1.1, 1507.3.5, 2104.5, 2105.2.2.1.1, 2109.8.1.1
C 73—99a	Specification for Calcium Silicate Face Brick (Sand-lime Brick)	Table 721.3.2, 2103.1
C 79—01	Specification for Treated Core and Non-treated Core Gypsum Sheathing Board	Table 2506.2
C 90—01A	Specification for Loadbearing Concrete Masonry Units	Table 721.3.2, 1805.5.2, 2103.1, 2105.2.2.1.2
C 91—01	Specification for Masonry Cement	Table 2103.7(1), Table 2507.2
C 94/C 94M—00	Specification for Ready-Mixed Concrete	109.3.1, 1905.8.2
C 126—99	Specification for Ceramic Glazed Structural Clay Facing Tile, Facing Brick, and Solid Masonry Units	2103.2
C 140—01 ae1	Test Method Sampling and Testing Concrete Masonry Units and Related Units	721.3.1.2, 1507.3.5, 2105.2.2.1.2
C 150—99a	Specification for Portland Cement	1904.1, Table 1904.2.3
C 150—01	Specification for Portland Cement	Table 2103.7(1), Table 2507.2
C 172—99	Practice for Sampling Freshly Mixed Concrete	Table 1704.4, 1905.6.3.1
C 199—84 (2000)	Test Method for Pier Test for Refractory Mortars	2111.5, 2111.8, 2113.12
C 206—84 (1997)	Specification for Finishing Hydrated Lime	Table 2507.2
C 207—97	Specification for Hydrated Lime for Masonry Purposes	Table 2103.7(1)
C 208—95	Specification for Cellulosic Fiber Insulating Board	2303.1.5
C 212—00	Specification for Structural Clay Facing Tile	2103.2
C 216—01A	Specification for Facing Brick (Solid Masonry Units Made from Clay or Shale)	2103.2, 2105.2.2.1.1
C 270—01A	Specification for Mortar for Unit Masonry	2103.7, Table 2103.7(2)
C 315—00	Specification for Clay Flue Linings	2113.11.1, Table 2113.16(1), Table 2113.16(2)
C 317/C 317M—00	Specification for Gypsum Concrete	1915.1

ASTM—continued

C 330—99	Specification for Lightweight Aggregates for Structural Concrete	721.1.1, 1905.1.4
C 331—01	Specification for Lightweight Aggregates for Concrete Masonry Units	721.3.1.4, 721.4.1.1.3
C 406—00	Specification for Roofing Slate	1507.7.4
C 442/C 442M—01	Specification for Gypsum Backing Board and Coreboard and Gypsum Shaftliner Board	Table 2506.2
C 472— 99	Specification for Standard Test Methods for Physical Testing of Gypsum, Gypsum Plasters and Gypsum Concrete	Table 2506.2
C 473—00	Test Method for Physical Testing of Gypsum Panel Products	Table 2506.2
C 474—01	Test Methods for Joint Treatment Materials for Gypsum Board Construction	Table 2506.2
C 475—01	Specification for Joint Compound and Joint Tape for Finishing Gypsum Wallboard	Table 2506.2
C 476—01	Specification for Grout for Masonry	2103.10, 2105.2.2.1.1, 2105.2.2.1.2
C 503—99e01	Specification for Marble Dimension Stone (Exterior)	2103.3
C 514—01	Specification for Nails for the Application of Gypsum Board	Table 721.1(2), Table 721.1(3), Table 2306.4.5, Table 2506.2
C 516—E01	Specifications for Vermiculite Loose Fill Thermal Insulation	721.3.1.4, 721.4.1.1.3
C 547—00	Specification for Mineral Fiber Pipe Insulation	Table 721.1(2), Table 721.1(3)
C 549—81 (1995)	Specification for Perlite Loose Fill Insulation	721.3.1.4, 721.4.1.1.3
C 557—99	Specification for Adhesives for Fastening Gypsum Wallboard to Wood Framing	Table 2506.2
C 568—99	Specification for Limestone Dimension Stone	2103.3
C 587—97	Specification for Gypsum Veneer Plaster	Table 2507.2
C 588/C 588M—01	Specification for Gypsum Base for Veneer Plasters	Table 2507.2
C 595—00	Specification for Blended Hydraulic Cements	1904.1, Table 1904.2.3
C 595—01	Specification for Blended Hydraulic Cements	Table 2103.7(1), Table 2507.2
C 615—99	Specification for Granite Dimension Stone	2103.3
C 616—99	Specification for Quartz-based Dimension Stone	2103.3
C 618—99	Specification for Coal Fly Ash and Raw or Calcined Natural Pozzolan for Use as a Mineral Admixture in Concrete	1904.1, Table 1904.2.3
C 629—99	Specification for Slate Dimension Stone	2103.3
C 630/C 630M—01	Specification for Water-resistant Gypsum Backing Board	Table 2506.2
C 631—00	Specification for Bonding Compounds for Interior Gypsum Plastering	Table 2507.2
C 635—00	Specification for the Manufacturer, Performance, and Testing of Metal Suspension Systems for Acoustical Tile and Lay-in Panel Ceilings	803.9.1.1, 2506.2.1
C 636—96	Practice for Installation of Metal Ceiling Suspension Systems for Acoustical Tile and Lay-in Panels	803.9.1.1
C 645—00	Specification for Nonstructural Steel Framing Members	Table 2506.2, Table 2507.2
C 652—01A	Specification for Hollow Brick (Hollow Masonry Units Made from Clay or Shale)	2103.2, 2105.2.2.1.1
C 685/ C 685M—98a	Specification for Concrete Made by Volumetric Batching and Continuous Mixing	1905.8.2
C 744—99	Specification for Prefaced Concrete and Calcium Silicate Masonry Units	2103.1
C 754—00	Specification for Installation of Steel Framing Members to Receive Screw-attached Gypsum Panel Products	Table 2508.1, Table 2511.1
C 836—00	Specification for High-solids Content, Cold Liquid-applied Elastomeric Waterproofing Membrane for Use with Separate Wearing Course	1507.15.2
C 840—01	Specification for Application and Finishing of Gypsum Board	Table 2508.1, 2509.2
C 841—99	Specification for Installation of Interior Lathing and Furring	Table 2508.1, Table 2511.1
C 842—99	Specification for Application of Interior Gypsum Plaster	Table 2511.1, 2511.3, 2511.4
C 843—99	Specification for Application of Gypsum Veneer Plaster	Table 2511.1
C 844—99	Specification for Application of Gypsum Base to Receive Gypsum Veneer Plaster	Table 2508.1
C 845—96	Specification for Expansive Hydraulic Cement	1904.1, Table 1904.2.3
C 847—00	Specification for Metal Lath	Table 2507.2
C 887—(2001)	Specification for Packaged, Dry Combined Materials for Surface Bonding Mortar	1807.2.2, 2103.8
C 897—00	Specification for Aggregate for Job-Mixed Portland Cement-Based Plasters	Table 2507.2
C 926—98a	Specification for Application of Portland Cement Based-Plaster	2510.3, Table 2511.1, 2511.3, 2511.4, 2512.1, 2512.1.2 2512.2, 2512.6, 2512.8.2, 2513.7, 2512.9
C 931/C 931M—01	Specification for Exterior Gypsum Soffit Board	Table 2506.2
C 932—98a	Specification for Surface-applied Bonding Agents for Exterior Plastering	Table 2507.2
C 933—A(2001)	Specification for Welded Wire Lath	Table 2507.2
C 946—91 (2001)	Specification for Practice for Construction of Dry-stacked, Surface-bonded Walls	2103.8, 2109.2.3.2
C 954—00	Specification for Steel Drill Screws for the Application of Gypsum Panel Products or Metal Plaster Bases to Steel Studs from 0.033 inch (0.84 mm) to 0.112 inch (2.84 mm) in Thickness	2211.2.2.2, Table 2506.2, Table 2507.2
C 955—01	Specification for Load Bearing Transverse and Axial Steel Studs, RunnersTracks, and Bracing or Bridging, for Screw Application of Gypsum Panel Products and Metal Plaster Bases	Table 2506.2, Table 2507.2
C 956—97	Specification for Installation of Cast-in-place Reinforced Gypsum Concrete	1915.1

ASTM—continued

C 957—98	Specification for High-solids Content, Cold Liquid-applied Elastomeric Waterproofing Membrane with Integral Wearing Surface	1507.15.2
C 960—97	Specification for Predecorated Gypsum Board	Table 2506.2
C 989—99	Specification for Ground Granulated Blast-furnace Slag for Use in Concrete and Mortars	1904.1, Table 1904.2.3
C1002—01	Specification for Steel Self-Piercing Tapping Screws for the Application of Gypsum Panel Products or Metal Plaster Bases to Wood Studs or Steel Studs	Table 2506.2, Table 2507.2
C1007—00	Specification for Installation of Load Bearing (Transverse and Axial) Steel Studs and Related Accessories	Table 2508.1, Table 2511.1
C1019—00B	Test Method of Sampling and Testing Grout	2105.2.2.1.1, 2105.2.2.1.2
C1029—96	Specification for Spray-applied Rigid Cellular Polyurethane Thermal Insulation	1507.14.2
C1032—96	Specification for Woven Wire Plaster Base	Table 2507.2
C1047—99	Specification for Accessories for Gypsum Wallboard and Gypsum Veneer Base	Table 2506.2, Table 2507.2
C1063—99	Specification for Installation of Lathing and Furring to Receive Interior and Exterior Portland Cement Based Plaster	2510.3, Table 2511.1, 2512.1.1
C1088—01A	Specification for Thin Veneer Brick Units Made from Clay or Shale	2103.2
C1157—00	Performance Specification for Hydraulic Cement	1904.1, Table 1904.2.3
C1167—96	Specification for Clay Roof Tiles	1507.3.4, 1507.3.5
C1177/C1177M—01	Specification for Glass Mat Gypsum Substrate for Use as Sheathing	Table 2506.2
C1178/C1178M—01	Specification for Glass Mat Water-resistant Gypsum Backing Panel	Table 2506.2
C1186—99	Specification for Flat Nonasbestos Fiber Cement Sheets	1404.10
C1218/ C1218M—99	Test Method for Water-soluble Chloride in Mortar and Concrete	1904.4.1
C1240—00E1	Specification for Silica Fume for Use as a Mineral Admixture in Hydraulic-Cement Concrete, Mortar and Grout	1904.1, Table 1904.2.3
C1261—98	Specification for Firebox Brick for Residential Fireplaces	2111.5, 2111.8
C1278/C 1278M—01	Specification for Fiber-reinforced Gypsum Panels	Table 2506.2
C1280—99	Specification for Application of Gypsum Sheathing	Table 2508.1, 2508.1,Table 2508.2, 2508.2
C1283—00	Practice for Installing Clay Flue Liners	2113.12
C1314—02	Test Method for Compressive Strength of Masonry Prisms	2105.2.2.2.2, 2105.3.1, 2105.3.2
C1328—00	Specification for Plastic (Stucco Cement)	Table 2507.2
C1329—00	Specification for Mortar Cement	Table 2103.7(1)
C1395/1395M—01	Specification for Gypsum Ceiling Board	Table 2506.2
D 25—99E01	Specification for Round Timber Piles	1809.1.1
D 41—E01	Specification for Asphalt Primer Used in Roofing, Dampproofing and Waterproofing	Table 1507.10.2
D 43—00	Coal Tar Primer Used in Roofing, Dampproofing and Waterproofing	Table 1507.10.2
D 56—01	Test Method for Flash Point By Tag Closed Tester	307.2
D 86—01e01	Test Method for Distillation of Petroleum Products	307.2
D 93—00	Test Method for Flash Point By Pensky-Martens Closed Cup Tester	307.2
D 224—89 (1996)	Specification for Smooth-Surfaced Asphalt Roll Roofing (Organic Felt)	1507.2.9.2, 1507.6.4
D 225—01	Specification for Asphalt Shingles (Organic Felt) Surfaced with Mineral Granules	1507.2.5
D 226—97a	Specification for Asphalt-Saturated Organic Felt Used in Roofing and Waterproofing	1404.2, Table 1507.2, 1507.2.3, 1507.3.3, 1507.5.3, 1507.6.3, 1507.7.3, Table 1507.8, 1507.8.3, 1507.9.3, 1507.9.4, Table 1507.10.2
D 227—97a	Specification for Coal-tar-saturated Organic Felt Used in Roofing and Waterproofing	Table 1507.10.2
D 249—89 (1996)	Specification for Asphalt Roll Roofing (Organic Felt) Surfaced with Mineral Granules	1507.3.3, 1507.6.4
D 312—00	Specification for Asphalt Used in Roofing	Table 1507.10.2
D 371—89 (1996)	Specification for Asphalt Roll Roofing (Organic Felt) Surfaced with Mineral Granules: Wide-selvage	1507.6.4
D 422—98	Test Method for Particle-size Analysis of Soils	1802.3.2
D 450—00	Specification for Coal-tar Pitch Used in Roofing, Dampproofing and Waterproofing	Table 1507.10.2
D 635—98	Test Method for Rate of Burning and/or Extent and Time of Burning of Self-Supporting Plastics in a Horizontal Position	2606.4
D1143—81 (1994) E01	Test Method for Piles Under Static Axial Compressive Load	1808.2.8.3
D1227—00	Specification for Emulsified Asphalt Used as a Protective Coating for Roofing	Table 1507.10.2, 1507.15.2
D1557—00	Test Method for Laboratory Compaction Characteristics of Soil Using Modified Effort (56,000 ft-lb/ft^3 (2,700 KN m/m^3))	1803.5
D1586—99	Specification for Penetration Test and Split-barrel Sampling of Soils	1615.1.5
D1761—88(2000)	Test Method for Mechanical Fasteners in Wood	1715.1.1, 1715.1.2, 1715.1.3
D1863—93 (2000)	Specification for Mineral Aggregate Used on Built-up Roofs	Table 1507.10.2
D1929—96 (2001)E01	Test Method for Determining Ignition Properties of Plastics	402.14.4, 406.5.2, 1407.11.2.1, 2606.4
D1970—01	Specification for Self-Adhering Polymer Modified Bituminous Sheet Materials Used as Steep Roof Underlayment for Ice Dam Protection	1507.2.4, 1507.2.9.2
D2166—00	Test Method for Unconfined Compressive Strength of Cohesive Soil	1615.1.5

ASTM—continued

D2178—97a	Specification for Asphalt Glass Felt Used in Roofing and Waterproofing	Table 1507.10.2
D2216—98	Test Method for Laboratory Determination of Water(Moisture) Content of Soil and Rock by Mass	1615.1.5
D2487—00	Practice for Classification of Soils for Engineering Purposes (Unified Soil Classification System)	Table 1610.1, 1802.3.1
D2626—97b	Specification for Asphalt Saturated and Coated Organic Felt Base Sheet Used in Roofing	1507.3.3, Table 1507.10.2
D2822—91(97) el	Specification for Asphalt Roof Cement	Table 1507.10.2
D2823—90(97) el	Specification for Asphalt Roof Coatings	Table 1507.10.2
D2843—99	Test for Density of Smoke from the Burning or Decomposition of Plastics	2606.4
D2850—95(1999)	Test Method for Unconsolidated, Undrained Triaxial Compression Test on Cohesive Soils	1615.1.5
D2898—94 (1999)	Test Methods for Accelerated Weathering of Fire-Retardant-Treated Wood for Fire Testing	1505.1, 2303.2.1, 2303.2.3
D3019—e01(Supp)	Specification for Lap Cement Used with Asphalt Roll Roofing, Nonfibered, Asbestos Fibered, and Nonasbestos Fibered	Table 1507.10.2
D3161—99a	Test Method for a Wind Resistance of Asphalt Shingles (Fan Induced Method)	1507.2.7
D3201—94 (1998) el	Test Method for Hygroscopic Properties of Fire-Retardant Treated Wood and Wood-base Products	2303.2.4
D3278—96e01	Test Methods for Flash Point of Liquids by Small Scale Closed-Cup Apparatus	307.2
D3462—E01	Specification for Asphalt Shingles Made from Glass Felt and Surfaced with Mineral Granules	1507.2.5
D3468—99	Specification for Liquid-applied Neoprene and Chlorosulfonated Polyethylene Used in Roofing and Waterproofing	1507.15.2
D3679—01c(Supp)	Specification for Rigid Poly [Vinyl Chloride (PVC) Siding]	1404.9, 1405.13
D3689—90 (1995)	Method for Testing Individual Piles Under Static Axial Tensile Load	1808.2.8.5
D3737—01B	Practice for Establishing Allowable Properties for Structural Glued Laminated Timber (Glulam)	2303.1.3
D3746—85 (1996) el	Test Method for Impact Resistance of Bituminous Roofing Systems	1504.7
D3747—E01	Specification for Emulsified Asphalt Adhesive for Adhering Roof Insulation	Table 1507.10.2
D3909—97b	Specification for Asphalt Roll Roofing (Glass Felt) Surfaced with Mineral Granules	1507.6.4, Table 1507.10.2
D4022—00	Specification for Coal Tar Roof Cement, Asbestos Containing	Table 1507.10.2
D4272—99	Test Method for Total Energy Impact of Plastic Films by Dart Drop	1504.7
D4318—00	Test Methods for Liquid Limit, Plastic Limit, and Plasticity Index of Soils	1615.1.5, 1802.3.2
D4434—96	Specification for Poly (Vinyl Chloride) Sheet Roofing	1507.13.2
D4479—00	Specification for Asphalt Roof Coatings - Asbestos-Free	Table 1507.10.2
D4586—00	Specification for Asphalt Roof Cement, Asbestos Free	Table 1507.10.2
D4601—98	Specification for Asphalt-Coated Glass Fiber Base Sheet Used in Roofing	Table 1507.10.2
D4637—96	Specification for EPDM Sheet Used in Single-ply Roof Membrane	1507.12.2
D4829—95	Test Method for Expansion Index of Soils	1802.3.2
D4869—93	Specification for Asphalt-Saturated (Organic Felt) Underlayment Used in Steep Slope Roofing	Table1507.2, 1507.2.3
D4897—01	Specification for Asphalt-Coated Glass Fiber Venting Base Sheet Used in Roofing	Table 1507.10.2
D4990—97a	Specification for Coal Tar Glass Felt Used in Roofing and Waterproofing	Table 1507.10.2
D4945—00	Test Method for High-Strain Dynamic Testing of Piles P	1808.2.8.3
D5019—96	Specification for Reinforced Nonvulcanized Polymeric Sheet Used in Roofing Membrane	1507.12.2
D5055—00	Specification for Establishing and Monitoring Structural Capacities of Prefabricated Wood I-joists	2303.1.2
D5456—01AE01	Specification for Evaluation of Structural Composite Lumber Products	2303.1.9
D5516—99a	Test Method of Evaluating the Flexural Properties of Fire-Retardant Treated Softwood Plywood Exposed to the Elevated Temperatures	2303.2.2.1
D5643—00	Specification for Coal Tar Roof Cement, Asbestos-Free	Table 1507.10.2
D5664—01	Test Methods for Evaluating the Effects of Fire-Retardant Treatment and Elevated Temperatures on Strength Properties of Fire-Retardant Treated Lumber	2303.2.2.2
D5665—99a	Specification for Thermoplastic Fabrics Used in Cold-Applied Roofing and Waterproofing	Table 1507.10.2
D5726—98	Specification for Thermoplastic Fabrics Used in Hot-Applied Roofing and Waterproofing	Table 1507.10.2
D6083—97a	Specification for Liquid Applied Acrylic Coating Used in Roofing	Table 1507.10.2, 1507.15.2
D6162—00A	Specification for Styrene Butadiene Styrene (SBS) Modified Bituminous Sheet Materials Using a Combination of Polyester and Glass Fiber Reinforcements	1507.11.2
D6163—00 E01	Specification for Styrene Butadiene Styrene (SBS) Modified Bituminous Sheet Materials Using Glass Fiber Reinforcements	1507.11.2
D6164—00	Specification for Styrene Butadiene Styrene (SBS) Modified Bituminous Sheet Metal Materials Using Polyester Reinforcements	1507.11.2
D6222—98	Specification for Atactic Polypropylene (APP) Modified Bituminous Sheet Materials Using Polyester Reinforcements	1507.11.2
D6223—98	Specification for Atactic Polypropylene (APP) Modified Bituminous Sheet Materials Using a Combination of Polyester and Glass Fiber Reinforcements	1507.11.2
D6298—98	Specification for Fiberglass Reinforced Styrene-Butadiene-Styrene (SBS) Modified Bituminous Sheets with a Factory Applied Metal Surface	1507.11.2
D6305—99e1	Practice for Calculating Bending Strength Design Adjustment Factors for Fire-Retardant-Treated Plywood Roof Sheathing	2303.2.2.1

ASTM—continued

E 84—01	Test Methods for Surface Burning Characteristics of Building Materials	402.10, 402.14.4, 406.5.2, 410.3.5.3, 703.4.2, 719.1, 719.4, 802.1, 803.1, 803.5, 803.6.1, 803.6.2, 1407.10, 1407.10.1, 2303.2, 2603.3, 2603.4.1.13, 2603.5.4, 2604.2.4, 2606.4, 3105.3
E 90—99	Test Method for Laboratory Measurement of Airborne Sound Transmission Loss of Building Partitions and Elements	1207.2
E 96—00	Test Method for Water Vapor Transmission of Materials	1203.2
E 108—00	Test Methods for Fire Tests of Roof Coverings	1505.1, 2603.6, 2610.2, 2610.3
E 119—00	Test Methods for Fire Tests of Building Construction and Materials	410.3.5.2, 703.2, 703.2.1, 703.2.3, 703.3, 704.7, 704.9, 706.7, 711.3.2, 712.3.1, 712.4.1, 712.4.6, 713.1, 713.4, 714.7, 715.2, 716.5.2, 715.5.3.1, 715.6.2, 716.5.2, 716.5.3.1, 716.6.2, Table 721.1(1), 1407.10.2, 2103.2, 2603.3, 2603.4, 2603.4.1.13, 2603.5.4, 2604.2.4, 2606.4
E 136—99	Test Method for Behavior of Materials in a Vertical Tube Furnace at 750°C	703.4.1
E 328—86	Methods for Stress Relaxation for Materials and Structures	2103.11.6
E 330—97	Test Method for Structural Performance of Exterior Windows, Curtain Walls, and Doors by Uniform Static Air Pressure Difference	1714.5.2
E 331—00	Test Method for Water Penetration of Exterior Windows, Skylights, Doors, and Curtain Walls by Uniform Static Air Pressure Difference	1403.2
E 492—90 (1996)e1	Test Method for Laboratory Measurement of Impact Sound Transmission Through Floor-ceiling Assemblies Using the Tapping Machine	1207.3
E 605—00	Test Method for Thickness and Density of Sprayed Fire-resistive Material (SFRM) Applied to Structural Members	1704.11.3, 1704.11.3.1, 1704.11.3.2, 1704.11.4
E 681—01	Test Methods for Concentration Limits of Flammability of Chemical Vapors and Gases	307.2
E 736—00	Test Method for Cohesion/Adhesion of Sprayed Fire-resistive Materials Applied to Structural Members	1704.11.5
E 814—00	Test Method of Fire Tests of Through-penetration Firestops	702.1, 712.3.1.2, 712.4.1.2
E 970—00	Test Method for Critical Radiant Flux of Exposed Attic Floor Insulation Using a Radiant Heat Energy Source	719.3.1
E1300—00	Practice for Determining Load Resistance of Glass in Buildings	Table 2404.1, Table 2404.2
E1592—01	Test Method for Structural Performance of Sheet Metal Roof and Siding Systems by Uniform Static Air Pressure Difference	1504.3.2
E1602—01	Guide for Construction of Solid Fuel-Burning Masonry Heaters	2112.2
E1886—97	Test Method for Performance of Exterior Windows, Curtain Wall, Doors and Storm Shutters Impacted by Missiles and exposed to Cyclic Pressure Differentials	1609.1.4
E1966—00	Test Method for Fire-Resistant Joint Systems	702.1, 712.3
E1996—01	Specification for Performance of Exterior Windows, Glazed Curtain Walls, Doors and Storm Shutters Impacted by Windborne Debris in Hurricanes	1609.1.4
F 547—01	Terminology of Nails for Use with Wood and Wood-base Materials	Table 2506.2
F1346—91 (1996)	Performance Specification for Safety Covers and Labeling Requirements for All Covers for Swimming Pools, Spas and Hot Tubs	3104.9, 3109.4.1.8
F1667—01A	Specification for Driven Fasteners: Nails, Spikes, and Staples	Table 721.1(2), Table 721.1(3), 1507.2.6, 2303.6, Table 2506.2
G 152—00A	Practice for Operating Open Flame Carbon Arc Light Apparatus for Exposure of Nonmetallic Materials	1504.5
G 154—00A	Practice for Operating Fluorescent Light Apparatus for UV Exposure of Nonmetallic Materials	1504.5
G 155—00A	Practice for Operating Xenon Arc Light Apparatus for Exposure of Non-Metallic Materials	1504.5

AWPA

American Wood-Preservers' Association
P.O. Box 5690
Grandbury, TX 76049

Standard reference number	Title	Referenced in code section number
C1—00	All Timber Products—Preservative Treatment by Pressure Processes	1403.6, 1505.6, 2303.1.8
C2—01	Lumber, Timber, Bridge Ties and Mine Ties—Preservative Treatment by Pressure Processes	1403.6, Table 1507.9.5, 1805.4.5, 1805.7.1, 2303.1.8, 2304.11.2, 2304.11.4, 2304.11.7
C3—99	Piles—Preservative Treatment by Pressure Processes	1403.6, 1805.4.5, 1809.1.2, 2303.1.8
C4—99	Poles—Preservative Treatment by Pressure Processes	1403.6, 1805.7.1, 1808.1.2, 2303.1.8
C9—00	Plywood—Preservative Treatment by Pressure Processes	1403.6, 2303.1.8, 2304.11.2, 2304.11.4, 2304.11.7
C14—99	Wood for Highway Construction, Pressure Treatment by Pressure Process	2303.1.8
C15—00	Wood for Commercial-Residential Construction Preservative Treatment by Pressure Process	1403.6, 2303.1.8
C16—00	Wood Used on Farms, Pressure Treatment by Pressure Process	2303.1.8
C18—99	Standard for Pressure Treated Material in Marine Construction	1403.6
C22—96	Lumber and Plywood for Permanent Wood Foundations—Preservative Treatment by Pressure Processes	1403.6, 1805.4.6, 2303.1.8

AWPA—continued

C23—00	Round Poles and Posts Used in Building Construction—Preservative Treatment by Pressure Processes	1403.6, 2303.1.8
C24—96	Sawn Timber Piles Used to Support Residential and Commercial Structures	1403.6, 1809.1.2, 2303.1.8
C28—99	Standard for Preservative Treatment by Pressure Process of Structural Glued Laminated Members and Laminations before Gluing	1403.6, 2303.1.8
C31—00	Lumber Used Out of Contact with the Ground and Continuously Protected from Liquid Water—Treatment by Pressure Processes	2303.1.8
C33—00	Standard for Preservative Treatment of Structural Composite Lumber by Pressure Processes	2303.1.8
M4—01	Standard for the Care of Preservative-Treated Wood Products	1809.1.2, 2303.1.8
P1/13—01	Standard for Creosote Preservative	1403.6, 2303.1.8
P2—01	Standard for Creosote Solutions	1403.6, 2303.1.8
P3—01	Standard for Creosote-Petroleum Solution	1403.6
P5—01	Standard for Waterborne Preservatives	2303.1.8
P8—01	Standard for Oil-borne Preservatives	2303.1.8
P9—01	Standard for Solvents and Formulations for Organic Preservative Systems	2303.1.8

AWS

American Welding Society
550 N.W. LeJeune Road
Miami, FL 33126

Standard reference number	Title	Referenced in code section number
D1.1—00	Structural Welding Code—Steel	Table 1704.3, 1704.3.1, 1708.4
D1.3—98	Structural Welding Code—Sheet Steel	Table 1704.3
D1.4—98	Structural Welding Code—Reinforcing Steel	Table 1704.3, 1903.5.2

BHMA

Builders Hardware Manufacturers' Association
355 Lexington Avenue, 17th Floor
New York, NY 10017-6603

Standard reference number	Title	Referenced in code section number
A 156.10—99	American National Standard for Power Operated Pedestrian Doors	1008.1.3.2
A 156.19—97	American National Standard for Power Assist and Low Energy Operated Doors	1008.1.3.2

CGSB

Canadian General Standards Board
222 Queens Street
14th Floor, Suite 1402
Ottawa, Ontario, Canada KIA 1G6

Standard reference number	Title	Referenced in code section number
37-GP-52M (1984)	Roofing and Waterproofing Membrane, Sheet Applied, Elastomeric	1504.7, 1507.12.2
CAN/CGSB 37.54—95	Polyvinyl Chloride Roofing and Waterproofing Membrane	1507.13.2
37-GP-56M (1980)	Membrane, Modified, Bituminous, Prefabricated, and Reinforced for Roofing—with December 1985 Amendment	1507.11.2

CPSC

Consumer Product Safety Commission
4330 East West Highway
Bethesada, MD 20814-4408

Standard reference number	Title	Referenced in code section number
16 CFR Part 1201(1977)	Safety Standard for Architectural Glazing Material	2406.1.1, 2406.2.1, 2407.1, 2408.2.1, 2408.3
16 CFR Part 1209 (1979)	Interim Safety Standard for Cellulose Insulation	719.6
16 CFR Part 1404 (1979)	Cellulose Insulation	719.6
16 CFR Part 1500 (1991)	Hazardous Substances and Articles; Administration and Enforcement Regulations	307.2
16 CFR Part 1500.44 (2001)	Method for Determining Extremely Flammable and Flammable Solids	307.2
16 CFR Part 1507 (2001)	Fireworks Devices	307.2
16 CFR Part 1630 (2000)	Standard for the Surface Flammability of Carpets and Rugs	804.5.1

CSSB

Cedar Shake and Shingle Bureau
P.O. Box 1178
Sumas, WA 98295-1178

Standard reference number	Title	Referenced in code section number
CSSB—97	Grading and Packing Rules for Western Red Cedar Shakes and Western Red Shingles of the Cedar Shake and Shingle Bureau	Table 1507.8.4, Table 1507.9.5

DASMA

Door and Access Systems Manufacturers
Association International
1300 Summer Avenue
Cleveland, OH 44115-2851

Standard reference number	Title	Referenced in code section number
107—97	Room Fire Test Standard for Garage Doors Using Foam Plastic Insulation	2603.4.1.9

DOC

U.S. Department of Commerce
National Institute of Standards and Technology
100 Bureau Drive Stop 3460
Gaithersburg, MD 20899

Standard reference number	Title	Referenced in code section number
PS-1—95	Construction and Industrial Plywood	2211.2.2.2, 2303.1.4, 2304.6.2, Table 2304.7(4), 2306.3.2
PS-2—92	Performance Standard for Wood-based Structural-use Panels	1809.1.1, 2211.3.1, 2303.1.4, 2304.6.2, Table 2304.7(4), Table 2304.7(5), Table 2306.3.1, 2306.3.2
PS 20—99	American Softwood Lumber Standard	1809.1.1, 2302.1, 2303.1.1

DOL

U.S. Department of Labor
c/o Superintendent of Documents
U.S. Government Printing Office
Washington, DC 20402-9325

Standard reference number	Title	Referenced in code section number
29 CFR Part 1910.1000 (1974)	Air Contaminants	902.1

DOTn

U.S. Department of Transportation
c/o Superintendent of Documents
U.S. Government Printing Office
Washington, DC 20402-9325

Standard reference number	Title	Referenced in code section number
49 CFR Part 172 (1999)	Hazardous Materials Tables, Special Provisions, Hazardous Materials Communications, Emergency Response Information and Training Requirements	307.2
49 CFR Parts 173 (1999)	Specification of Transportation of Explosive and Other Dangerous Articles, UN 0335,UN 0336 Shipping Containers	307.2

FEMA

Federal Emergency Management Agency
Federal Center Plaza
500 C Street S.W.
Washington, DC 20472

Standard reference number	Title	Referenced in code section number
Pub 302	NEHRP Recommended Provisions for Seismic Regulations for New Buildings and Other Structures	Figure 1615(7), Figure 1615(8), Figure 1615(9), Figure 1615(10)
TB11-01	Crawlspace Construction for Buildings Located in Special Flood Hazard Areas	1807.1.2.1

FM

Factory Mutual
Standards Laboratories Department
1151 Boston-Providence Turnpike
Norwood, MA 02062

Standard reference number	Title	Referenced in code section number
4450 (1989)	Approval Standard for Class 1 Insulated Steel Deck Roofs—with Supplements thru 7/92	1504.3.1, 1508.1, 2603.3, 2603.4.1.5
4470 (1992)	Approval Standard for Class 1 Roof Covers	1504.3.1, 1504.7
4880 (2001)	American National Standard for Evaluating Insulated Wall or Wall and Roof/Ceiling Assemblies, Plastic Interior Finish Materials, Plastic Exterior Building Panels, Wall/Ceiling Coating Systems, Interior and Exterior Finish Systems	2603.4, 2603.8

GA

Gypsum Association
810 First Street N.E. #510
Washington, DC 20002-4268

Standard reference number	Title	Referenced in code section number
GA 216—00	Application and Finishing of Gypsum Board	Table 2508.1, 2509.2
GA 600—00	Fire-resistance Design Manual,16th Edition, April, 2000	Table 721.1(1), Table 721.1(2), Table 721.1(3)

HPVA

Hardwood Plywood Veneer Association
1825 Michael Faraday Drive
Reston, VA 20190-5350

Standard reference number	Title	Referenced in code section number
HP-1—2000	The American National Standard for Hardwood and Decorative Plywood	2303.3, 2304.6.2

ICC

International Code Council
5203 Leesburg Pike, Suite 600
Falls Church, VA 22041

Standard reference number	Title	Referenced in code section number
ICC/ANSI A117.1—98	Accessible and Usable Buildings and Facilities	406.2.2, 907.9.1.3,1007.6.5, 1010.1, 1010.6.5, 1010.9, 1011.3, 1101.2, 1102.1, 1103.2.13, 1106.6, 1107.2, 1109.2.2, 1109.3, 1109.4, 1109.8, 1109.15, 3001.3, 3409.5, 3409.7.2, 3409.7.3
ICC 300—02	ICC Standard on Bleachers, Folding and Telescopic Seating, and Grandstands	1024.1.1
ICC EC—03	ICC Electrical Code™	101.4.1, 107.3, 414.5.4, 414.9.2.8.1, 904.3.1, 907.5, 909.11, 909.12.1, 909.16.3, 1205.4.1, 1405.10.4, 2701.1, 2702.1, 3401.3
IEBC—03	International Existing Building Code™	101.2
IECC—03	International Energy Conservation Code®	101.4.7, 1202.3.2, 1301.1.1, 1403.2

ICC—continued

IFC—03	International Fire Code®	101.4.6, 102.6, 201.3, 307.9, Table 307.7(1), Table 307.7(2), 307.9, 404.2,406.5.1, 406.5.2, 406.6.1, 410.3.6, 411.1, 412.4.1, 413.1, 414.1.1, 414.1.2, 414.1.2.1, 414.2.4, Table 414.2.4, 414.3, 414.5, 414.5.1, Table 414.5.1, 414.5.2, 414.5.4, 414.5.5, 414.6, 415.1, 415.3, 415.3.1, Table 415.3.1, Table 415.3.2, 415.7, 415.7.1, 415.7.1.4, 415.7.2, 415.7.2.3, 415.7.2.5, 415.7.2.7, 415.7.2.8, 415.7.2.9, 415.7.3, 415.7.3.3.3, 415.7.3.5, 415.7.4, 415.8, 415.9.1, 415.9.2.7, 415.9.5.1, 415.9.7.2, 704.8.2, 706.1, 901.2, 901.3, 901.5, 901.6.2, 903.2.6.1, 903.2.11, Table 903.2.13, 903.5, 904.2.1, 905.1, 906.1, 907.2.5, 907.2.12.2, 907.2.14, 907.2.16, 907.19, 909.20, 910.2.3, Table 910.3, 1001.3, 1203.4.2, 1203.5, 2702.2.8, 2702.2.10, 2702.2.11, 2702.2.12, 2702.3, 3102.1, 3103.1, 3309.2, 3401.3, 3410.3.2, 3410.6.8.1, 3410.6.14, 3410.6.14.1
IFGC—03	International Fuel Gas Code®	101.4.2, 201.3, 415.7.3, 2113.11.1.2, 2113.15, 2801.1, 3401.3
IMC—03	International Mechanical Code®	101.4.3, 201.3, 307.9, 406.4.2, 406.6.3, 406.6.5, 409.3, 412.4.6, 414.1.2, 414.1.2.1, 414.1.2.2, 414.3, 415.7.1.4, 415.7.2, 415.7.2.8, 415.7.3, 415.7.4, 415.9.11.1, 416.3, 603.1, 707.2, 716.2.2, 716.5.4, 716.6.1, 716.6.2, 716.6.3, 717.5, 719.1, 903.2.12.1, 904.2.1, 904.11, 908.6, 909.1, 909.10.2, 1014.5, 1016.4.1, 1203.1, 1203.2.1, 1203.4.2, 1203.4.2.1, 1203.5, 1209.3, 2304.5, 3004.3.1, 3410.6.7.1, 3410.6.8
IPC—03	International Plumbing Code®	101.4.4, 201.3, 415.7.4, 717.5, 903.3.5, 1206.3.3, 1503.4, 1807.4.3, 2901.1, 2902.1.1, 3305.1, 3401.3,
IPMC—03	International Property Maintenance Code®	101.4.5, 102.6, 103.3, 3401.3, 3410.3.2.
IPSDC—03	International Private Sewage Disposal Code®	101.4.4, 2901.1, 3401.3
IRC—03	International Residential Code®	101.2, 308.3, 308.5 1706.1.1, 3401.3
IUWIC—03	International Urban—Wildland Interface Code™	Table 1505.1
SBCCI SSTD 10—99	Standard for Hurricane Resistant Residential Construction	1609.1.1, 2308.2.1
SBCCI SSTD 11—97	Test Standard for Determining Wind Resistance of Concrete or Clay Roof Tiles	1715.2.1, 1715.2.2
UBC Standard 18-2	Expansion Index Test	1802.3.2

NAAMM

National Association of Architectural
Metal Manufacturers
8 South Michigan Ave
Chicago, IL 60603

Standard reference number	Title	Referenced in code section number
FP 1001—90	Guide Specifications for Design of Metal Flag Poles	1609.1.1

NCMA

National Concrete Masonry Association
2302 Horse Pen Road
Herndon, VA 22071-3499

Standard reference number	Title	Referenced in code section number
TEK 5-8A (1996)	Details for Concrete Masonry Fire Walls	Table 719.1(2)

NFPA

National Fire Protection Association
1 Batterymarch Park
Quincy, MA 02269-9101

Standard reference number	Title	Referenced in code section number
11—98	Low Expansion Foam	904.7
11A—99	Medium- and High-expansion Foam Systems	904.7
12—00	Carbon Dioxide Extinguishing Systems	904.8, 904.11
12A—97	Halon 1301 Fire Extinguishing Systems	904.9
13—99	Installation of Sprinkler Systems	704.12, 707.2, 903.3.1.1, 903.3.2, 903.3.5.1.1, 904.11, 907.8, 1621.3.10.1, 3104.5, 3104.9
13D—99	Installation of Sprinkler Systems in One- and Two-family Dwellings and Manufactured Homes	903.1.2, 903.3.1.3, 903.3.5.1.1
13R—99	Installation of Sprinkler Systems in Residential Occupancies Up to and Including Four Stories in Height	903.1.2, 903.3.1.2, 903.3.5.1.1, 903.3.5.1.2, 903.4
14—00	Installation of Standpipe, Private Hydrants and Hose System	905.2, 905.3.4, 905.4.2, 905.8
16—99	Installation Foam-Water Sprinkler and Foam-Water Spray Systems	904.7, 904.11

NFPA—continued

17—98	Dry Chemical Extinguishing Systems	904.6, 904.11
17A—98	Wet Chemical Extinguishing Systems	904.5, 904.11
30—00	Flammable and Combustible Liquids Code	415.3
32—00	Drycleaning Plants	415.7.4
40—97	Storage and Handling of Cellulose Nitrate Motion Picture Film	409.1
61—99	Prevention of Fires and Dust Explosions in Agricultural and Food Product Facilities	415.7.1
72—99	National Fire Alarm Code	505.4, 901.6, 903.4.1, 904.3.5, 907.2,907.2.1, 907.2.1.1, 907.2.10, 907.2.10.4, 907.2.11.2, 907.2.11.3, 907.2.12.2.3, 907.2.12.3, 907.4, 907.5, 907.9.2, 907.10, 907.14, 907.16,907.17, 911.1, 3006.5
80—99	Fire Doors and Fire Windows	302.1.1.1, 715.3, 715.4.6.1, 715.4.4, 715.4.7.2, 715.5, 1008.1.3.3
85—01	Boiler and Combustion System Hazards Code (Note: NFPA 8503 has been incorporated into NFPA 85)	415.7.1
101—00	Life Safety Code	1024.6.2
110—99	Emergency and Standby Power Systems	2702.1
111—01	Stored Electrical Energy Emergency and Standby Power Systems	2702.1
120—99	Coal Preparation Plants	415.7.1
231C—98	Rack Storage of Materials	507.2
252—99	Standard Methods of Fire Tests of Door Assemblies	715.3.1, 715.3.2, 715.3.3, 715.3.4.1
253—00	Test for Critical Radiant Flux of Floor Covering Systems Using a Radiant Heat Energy Source	406.6.4, 804.2, 804.3
257—00	Standard for Fire Test for Window and Glass Block Assemblies	715.3.3, 715.4, 715.4.1, 715.4.2
259—98	Test Method for Potential Heat of Building Materials	2603.4.1.10, 2603.5.3
265—98	Standard Method of Fire Tests for Evaluating Room Fire Growth Contribution of Textile Wall Coverings	803.6.1, 803.6.1.1, 803.6.1.2
268—96	Standard Test Method for Determining Ignitibility of Exterior Wall Assemblies Using a Radiant Heat Energy Source	1406.2.1, 1406.2.1.1, 1406.2.1.2, 2603.5.7
285—98	Standard Method of Test for the Evaluation of Flammability Characteristics of Exterior Non-load-bearing Wall Assemblies Containing Combustible Components	1407.10.4, 2603.5.5
286—00	Standard Method of Fire Test for Evaluating Contribution of Wall and Ceiling Interior Finish to Room Fire Growth	402.14.4, 803.2, 803.2.1, 803.5, 2603.4, 2603.8
409—95	Standard on Aircraft Hangers	412.2.6, 412.4.5
418—01	Standard for Heliports	412.5.6
651—98	Machining and Finishing of Aluminum and the Production and Handling of Aluminum Powders	415.7.1
654—00	Prevention of Fire & Dust Explosions from the Manufacturing, Processing, and Handling of Combustible Particulate Solids	415.7.1
655—93	Prevention of Sulfur Fires and Explosions	415.7.1
664—98	Prevention of Fires Explosions in Wood Processing and Woodworking Facilities	415.7.1
701—99	Standard Methods of Fire Tests for Flame-Propagation of Textiles and Films	802.1, 805.1, 805.2, 3102.3.1, 3105.3,
704—96	Standard System for the Identification of the Hazards of Materials for Emergency Response	414.7.2, 415.2
1124—98	Manufacture, Transportation, and Storage of Fireworks and Pyrotechnic Articles	415.3.1
2001—00	Clean Agent Fire Extinguishing Systems	904.10

NIST

National Institute of Standards and Technology
U.S. Department of Commerce
100 Bureau Dr. – Stop 3460
Gaithersburg, MD 20899-3460

Standard reference number	Title	Referenced in code section number
BMS 71—41	Fire Tests of Wood and Metal-framed Partitions	721.7
TRBM-44—46	Fire-resistance and Sound-insulation Ratings for Walls, Partitions and Floors	721.7

PCI

Precast Prestressed Concrete Institute
175 W. Jackson Boulevard, Suite 1859
Chicago, IL 60604-9773

Standard reference number	Title	Referenced in code section number
MNL 124—89	Design for Fire Resistance of Precast Prestressed Concrete	721.2.3.1, Table 721.2.3(4)
MNL 128—01	Recommended Practice for Glass Fiber Reinforced Concrete Panels	1903.8

PTI

Post-Tensioning Institute
1717 W. Northern Avenue, Suite 114
Phoenix, AZ 85021

Standard reference number	Title	Referenced in code section number
PTI 1996	Design and Construction of Post-tensioned Slabs-on-ground, 2nd Edition	1805.8.2

RMA

Rubber Manufacturers Association
1400 K. Street, N.W. #900
Washington, DC 20005

Standard reference number	Title	Referenced in code section number
RP-1—90	Minimum Requirements for Non-reinforced Black EPDM Rubber Sheets	1507.12.2
RP-2—90	Minimum Requirements for Fabric-reinforced Black EPDM Rubber Sheets	1507.12.2
RP-3—85	Minimum Requirements for Fabric-reinforced Black Polychloroprene Rubber Sheets	1507.12.2

RMI

Rack Manufacturers Institute
8720 Red Oak Boulevard, Suite 201
Charlotte, NC 28217

Standard reference number	Title	Referenced in code section number
RMI (1997)	Design, Testing and Utilization of Industrial Steel Storage Racks	2208.1

SJI

Steel Joist Institute
3127 10th Avenue, North
Myrtle Beach, SC 29577-6760

Standard reference number	Title	Referenced in code section number
SJI—1994	Standard Specification for Joist Girders	1604.3.3, 2206
K-Series Specification—1994	Standard Specification for Open Web Steel Joists, K Series	2206
SJI—1994	Standard Specification for Longspan Steel Joists, LH Series and Deep Longspan Steel Joists, DLH Series	2206

SPRI

Single-Ply Roofing Institute
77 Rumford Ave.
Suite 3-B
Walthem, MA 02453

Standard reference number	Title	Referenced in code section number
ES-1—98	Wind Design Standard for Edge Systems Used with Low Slope Roofing Systems	1504.5
RP-4—88	Wind Design Guide for Ballasted Single-ply Roofing Systems	1504.4

TIA

Telecommunications Industry Association
2500 Wilson Boulevard
Arlington, VA 22201-3834

Standard reference number	Title	Referenced in code section number
TIA/EIA-222-F—96	Structural Standards for Steel Antenna Towers and Antenna Supporting Structures	1609.1.1, 3108.4

TMS

The Masonry Society
3970 Broadway, Unit 201-D
Boulder, CO 80304-1135

Standard reference number	Title	Referenced in code section number
0216—97	Standard Method for Determining Fire Resistance of Concrete and Masonry Construction Assemblies	Table 721.1(2), 721.1
402—02	Building Code Requirements for Masonry Structures	1405.5, 1405.5.3, 1405.9, 1604.3.4, 1704.5, 1704.5.1, Table 1704.5.1, 1704.5.2, Table 1703.3.1, 1708.1.1, 1708.1.2, 1708.1.3, 1805.5.2, 1812.7, 2101.2.3, 2101.2.4, 2101.2.5, 2103.11.6, 2106.1, 2106.1.1.1, 2106.1.1.2, 2106.1.1.3, 2106.3, 2106.4, 2106.5, 2106.6, 2107.1, 2107.2, 2107.2.1, 2107.2.2, 2107.2.4, 2107.2.5, 2107.2.6, 2108.1, 2108.2, 2108.4, 2109.1, 2109.2.3.1, 2109.2.3.2
602—02	Specification for Masonry Structures	1405.5.1, 1405.9.1, Table1704.5.1, Table 1704.5.3, 1805.5.2, 2103.11.7, 2104.1, 2104.1.1, 2104.3

TPI

Truss Plate Institute
583 D'Onofrio Drive, Suite 200
Madison, WI 53719

Standard reference number	Title	Referenced in code section number
TPI 1—2002	National Design Standards for Metal-Plate-Connected Wood Truss Construction	2303.4, 2306.1

UL

Underwriters Laboratories
333 Pfingsten Road
Northbrook, IL 60062-2096

Standard reference number	Title	Referenced in code section number
10A—98	Tin Clad Fire Doors—with Revisions through July, 1998	715.3
10B—97	Fire Tests of Door Assemblies	715.3.2
10C—98	Positive Pressure Fire Tests of Door Assemblies—with Revisions thru November, 2001	715.3.1, 715.3.3
14B—98	Sliding Hardware for Standard Horizontally Mounted Tin Clad Fire Doors—with Revisions through July, 2000	715.3
14C—96	Swinging Hardware for Standard Tin Clad Fire Doors Mounted Singly and in Pairs	715.3
103—98	Factory-Built Chimneys, for Residential Type and Building Heating Appliances—with Revisions through March 1999	717.2.5
127—99	Factory-Built Fireplaces—with Revisions through November,1999	717.2.5
268—96	Smoke Detectors for Fire Protective Signaling Systems—with Revisions through January, 1999	407.6, 907.2.6.1
300—96	Fire Testing of Fire Extinguishing Systems for Protection of Restaurant Cooking Areas —with Revisions through December, 1998	904.11
555—96	Fire Dampers—with Revisions through October, 2000	716.3
555C—96	Ceiling Dampers	716.3, 716.6.2
555S—99	Smoke Dampers—with Revisions through December, 1999	716.3, 716.3.1.1
580—94	Test for Uplift Resistance of Roof Assemblies—with Revisions through February, 1998	1504.3.1, 1504.3.2
641—95	Type L Low-Temperature Venting Systems—with Revisions through April, 1999	2113.11.1.4
790—97	Tests for Fire Resistance of Roof Covering Materials—with Revisions through July, 1998	1505.1, 2603.6, 2610.2, 2610.3
864—96	Control Units for Fire Protective Signaling Systems—with Revisions through March, 1999	909.12
1040—96	Fire Test of Insulated Wall Construction—with Revisions thru April, 2001	1407.10.3, 2603.4, 2603.8
1256—98	Fire Test of Roof Deck Construction—with Revisions through March, 2000	1508.1, 2603.3, 2603.4.1.5
1479—94	Fire Tests of Through-Penetration Firestops	712.3.1.2, 712.4.1.2
1715—97	Fire Test of Interior Finish Material	1407.10.2, 1407.10.3, 2603.4, 2603.8
1777—98	Chimney Liners—with Revisions through July, 1998	2113.11.1, 2113.19
1784—95	Air Leakage Tests of Door Assemblies	707.14.1, 710.5.2, 715.3.3, 715.3.5.1
1897—98	Uplift Tests for Roof Covering Systems—with Revisions through December, 1999	1504.3.1
1975—96	Fire Test of Foamed Plastics Used for Decorative Purposes	402.10, 402.14.5
2079—98	Tests for Fire Resistance of Building Joint Systems	702.1, 712.3
2200—98	Stationary Engine Generator Assemblies	2702.1.1

ULC

Underwriters Laboratories of Canada
7 Crouse Road
Scarborough, Ontario, Canada M1R3A9

Standard reference number	Title	Referenced in code section number
S102.2—M88	Standard Method of Test for Surface Burning Characteristics of Floor Coverings, and Miscellaneous Materials and Assemblies	719.4

USC

United States Code
c/o Superintendent of Documents
U.S. Government Printing Office
Washington, DC 20402-9325

Standard reference number	Title	Referenced in code section number
18 USC Part 1, Ch.40	Importation, Manufacture, Distribution and Storage of Explosive Materials	307.2

WDMA

Window and Door Manufacturers Association
1400 East Touhy Avenue #470
Des Plaines, IL 60018

Standard reference number	Title	Referenced in code section number
AAMA/NWWDA 101/I.S.2—97	Voluntary Specifications for Aluminum, Vinyl (PVC) and Wood Windows and Glass Doors	1714.5.1
AAMA/NWWDA 101/I.S.2/NAFS—02	Voluntary Performance Specification for Window, Skylights and Glass Doors	1714.5.1, 2405.5

WRI

Wire Reinforcement Institute, Inc.
203 Loudon Street, S.W.
2nd Floor, Suite 203C
Leesburg, VA 22075

Standard reference number	Title	Referenced in code section number
WRI/CRSI—96	Design of Slab-on-ground Foundations	1805.8.2

Appendix A: Employee Qualifications

The provisions contained in this appendix are not mandatory unless specifically referenced in the adopting ordinance.

General Comments

When adopted (see Section 101.2.1), this appendix provides jurisdictions with training, experience and certification requirements for those employees of the department of building safety responsible for enforcing the provisions of the building code, including the codes referenced in Section 101.4. The authorization to create these positions is found in Section 103 of the code. Note that jurisdictions may also be mandated by applicable state laws to employ only persons licensed by the state to perform certain duties, and may or may not be able to impose additional requirements.

Purpose

The purpose of this appendix is to provide optional criteria for qualifications for employees who enforce the code. A jurisdiction that wants to make this appendix a mandatory part of the code needs to specifically list this appendix in its adoption ordinance (see page v of the code for a sample ordinance for adoption).

SECTION A101 BUILDING OFFICIAL QUALIFICATIONS

A101.1 Building official. The building official shall have at least ten years experience or equivalent, as an architect, engineer, inspector, contractor, or superintendent of construction, or any combination of these, five years of which shall have been in supervisory experience. The building official should be certified as a building official through a recognized certification program. The building official shall be appointed or hired by the applicable governing authority.

❖ As the executive official in charge of administering the code enforcement program of the jurisdiction, the building official is required to possess significant experience in the construction industry, including at least five years as a supervisor. Although not mandated, it is recommended that the person hold certification as a building official, such as the Certified Building Official® (C.B.O.®) issued by the International Code Council® (ICC®). People with this certification have passed examinations that measure knowledge of legal and management principles as well as code technology. This position may have a different title, as described in the commentary to Section 103.1.

A101.2 Chief inspector. The building official can designate supervisors to administer the provisions of the *International Building*, *Mechanical*, and *Plumbing Codes*, *International Fuel Gas Code*, and the ICC *Electrical Code*. Each supervisor shall have at least ten years experience or equivalent, as an architect, engineer, inspector, contractor, or superintendent of construction, or any combination of these, five years of which shall have been in a supervisory capacity. They shall be certified through a recognized certification program for the appropriate trade.

❖ People who supervise inspectors and plans examiners reviewing construction for conformance to the building and referenced codes are expected to have considerable experience in their area of expertise, since they are responsible for overseeing the day-to-day operation of their divisions within the department. Large departments may have a separate division for each of the codes, while those of lesser size may group two or more together. Because of the responsibility placed on these persons, it is required that they have demonstrated knowledge of the codes that they administer by being certified in the appropriate category. The ICC provides certification programs that cover the range of categories involved.

A101.3 Inspector and plan examiner. The building official shall appoint or hire such number of officers, inspectors, assistants and other employees as shall be authorized by the jurisdiction. A person shall not be appointed or hired as inspector of construction or plan examiner who has not had at least five years experience as a contractor, engineer, architect, or as a superintendent, foreman, or competent mechanic in charge of construction. The inspector or plan examiner shall be certified, through a recognized certification program for the appropriate trade.

❖ Departments typically have inspectors and plans examiners who specialize in codes that regulate the various aspects of building construction. Additionally, there will be employees who handle clerical and various administrative duties. The building official will hire the number of employees authorized by the jurisdiction's elected officials. Newly hired inspectors and plans examiners are not necessarily expected to have had previous experience in code enforcement, but should be familiar with the construction industry, with at least five years experience in one or more of the fields listed. They also must possess certification in the job category in which they are employed. ICC provides certification programs that cover the range of categories involved.

A101.4 Termination of employment. Employees in the position of building official, chief inspector or inspector shall not be removed from office except for cause after full opportunity has been given to be heard on specific charges before such applicable governing authority.

❖ This section establishes that once employed, the building official and other technical employees of the department cannot be removed, except for cause, subject to a due process review of the specific charges.

SECTION A102
REFERENCED STANDARDS

IBC-2003	*International Building Code*—A101.2
IMC-2003	*International Mechanical Code*—A101.2
IPC-2003	*International Plumbing Code*—A101.2
IFGC-2003	*International Fuel Gas Code*—A101.2
ICC EC-2003	ICC *Electrical Code*—A101.2

Appendix B: Board Of Appeals

The provisions contained in this appendix are not mandatory unless specifically referenced in the adopting ordinance.

General Comments

When adopted (see Section 101.2.1), this appendix provides jurisdictions with detailed appeals board member qualifications and administrative procedures to supplement the basic requirements found in Section 112.

Purpose

The purpose of this appendix is to provide optional criteria for administrative procedures of the board of appeals and board member qualifications. A jurisdiction that wants to make this appendix a mandatory part of the code needs to specifically list this appendix in its adoption ordinance (see page v of the code for sample ordinance for adoption).

SECTION B101 GENERAL

B101.1 Application. The application for appeal shall be filed on a form obtained from the building official within 20 days after the notice was served.

❖ A party aggrieved by a decision of the building official, and desiring to appeal that decision, must file a written request for a hearing in a timely manner, in this case 20 days from the date of the decision. The building official's decision also needs to be in writing, as specified in Sections 105.3.1 for a permit application, 109.6 for an inspection and 113.2 for a violation, or in a letter written regarding a request for an approval, depending on the action being appealed. The request must be on the jurisdiction's standard appeals form, which requires the appellant to provide the information needed by the board members to make a decision. This is often supplemented by whatever additional information the appellant wants to provide to support his or her case.

B101.2 Membership of board. The board of appeals shall consist of persons appointed by the chief appointing authority as follows:

1. One for 5 years; one for 4 years; one for 3 years; one for 2 years; and one for 1 year.
2. Thereafter, each new member shall serve for 5 years or until a successor has been appointed.

The building official shall be an ex officio member of said board but shall have no vote on any matter before the board.

❖ The board of appeals is to consist of five members appointed by the "chief appointing authority" — typically, the mayor or city manager. One member must be appointed for five years, one for four, one for three, one for two and one for one year. This method of appointment allows for a smooth transition of board of appeals members, providing continuity of action over the years. The building official also sits as a member of the board, but cannot vote on matters regarding his or her own decisions.

B101.2.1 Alternate members. The chief appointing authority shall appoint two alternate members who shall be called by the board chairperson to hear appeals during the absence or disqualification of a member. Alternate members shall possess the qualifications required for board membership and shall be appointed for five years, or until a successor has been appointed.

❖ This section authorizes the chief appointing authority to appoint two alternate members who must be available if the principal members of board are absent or disqualified. Alternate members must possess the same qualifications as the principal members and are appointed for a term of five years, or until such time that a successor is appointed.

B101.2.2 Qualifications. The board of appeals shall consist of five individuals, one from each of the following professions or disciplines:

1. Registered design professional with architectural experience or a builder or superintendent of building construction with at least ten years' experience, five of which shall have been in responsible charge of work.
2. Registered design professional with structural engineering experience.
3. Registered design professional with mechanical and plumbing engineering experience or a mechanical contractor with at least ten years' experience, five of which shall have been in responsible charge of work.
4. Registered design professional with electrical engineering experience or an electrical contractor with at least ten years' experience, five of which shall have been in responsible charge of work.
5. Registered design professional with fire protection engineering experience or a fire protection contractor with at least ten years' experience, five of which shall have been in responsible charge of work.

❖ The board of appeals consists of five persons with the qualifications and experience indicated in this section. One must be a registered design professional (see Item 2) with structural experience. The others must be registered design professionals, construction superintendents or contractors with experience in the various areas of building construction. This is necessary because the board will be hearing appeals regarding the codes referenced in Section 101.4 as well as the building code. These requirements are important in that technical people rule on technical matters. The board of appeals is not the place for policy or political deliberations. It is intended that these matters be decided purely on their technical merits, with due regard for state-of-the-art construction technology.

B101.2.3 Rules and procedures. The board is authorized to establish policies and procedures necessary to carry out its duties.

❖ The board can set forth the detailed procedures by which it operates, supplementing those required by Section B101.3.2.

B101.2.4 Chairperson. The board shall annually select one of its members to serve as chairperson.

❖ It is customary to determine a chairperson annually so that a regular opportunity is available to evaluate and either reappoint the current chairperson or appoint a new one.

B101.2.5 Disqualification of member. A member shall not hear an appeal in which that member has a personal, professional or financial interest.

❖ All members must disqualify themselves regarding any appeal in which they have a personal, professional or financial interest.

B101.2.6 Secretary. The chief administrative officer shall designate a qualified clerk to serve as secretary to the board. The secretary shall file a detailed record of all proceedings in the office of the chief administrative officer.

❖ The chief administrative officer is to designate a qualified clerk, typically an employee in the department, to serve as secretary to the board. The secretary is required to keep detailed records of the proceedings. These may be needed for any future review of the board's decision.

B101.2.7 Compensation of members. Compensation of members shall be determined by law.

❖ Members of the board of appeals are not required to be compensated unless required by state or local law.

B101.3 Notice of meeting. The board shall meet upon notice from the chairperson, within ten days of the filing of an appeal or at stated periodic meetings.

❖ In order that an appellant's request be heard in a timely manner, the board must meet within 10 days of the filing of an appeal. In large jurisdictions, where there are likely to be more appeals, the board will often set a regular schedule of meeting dates, such as monthly, and the 10-day rule would not apply.

B101.3.1 Open hearing. All hearings before the board shall be open to the public. The appellant, the appellant's representative, the building official and any person whose interests are affected shall be given an opportunity to be heard.

❖ All hearings before the board must be open to the public. The appellant, the appellant's representative, the building official and any person whose interests are affected must be heard.

B101.3.2 Procedure. The board shall adopt and make available to the public through the secretary procedures under which a hearing will be conducted. The procedures shall not require compliance with strict rules of evidence, but shall mandate that only relevant information be received.

❖ The board is required to establish and make available to the public written procedures detailing how hearings are to be conducted. Additionally, this section provides that although strict rules of evidence are not applicable, the information presented must be deemed relevant so the hearings are not unnecessarily long.

B101.3.3 Postponed hearing. When five members are not present to hear an appeal, either the appellant or the appellant's representative shall have the right to request a postponement of the hearing.

❖ When all five members of the board are not present, either the appellant or the appellant's representative may request that the hearing be postponed. Note that this is not a requirement, and that the appellant can elect to have the case heard before the smaller board.

B101.4 Board decision. The board shall modify or reverse the decision of the building official by a concurring vote of two-thirds of its members.

❖ A concurring vote of two-thirds of the board is needed to modify or reverse the decision of the building official. This means that at least four members must vote to modify or reverse the building official, even if less than the full board is present in accordance with Section B101.3.3.

B101.4.1 Resolution. The decision of the board shall be by resolution. Certified copies shall be furnished to the appellant and to the building official.

❖ The building official is to be bound by the action of the board of appeals, unless it is the opinion of the building official that the board of appeals has acted improperly. In such cases, relief through the court having jurisdiction may be sought by the jurisdiction's counsel.

B101.4.2 Administration. The building official shall take immediate action in accordance with the decision of the board.

❖ To avoid any undue hindrance in the progress of construction, the building official is required to act without delay based on the board's decision. This action may be to enforce the decision or to seek judicial relief if the board's action can be demonstrated to be inappropriate.

Appendix C: Group U - Agricultural Buildings

The provisions contained in this appendix are not mandatory unless specifically referenced in the adopting ordinance.

General Comments

The provisions of Appendix C are supplemental to the remainder of the code's requirements for agricultural buildings and are only applicable when specifically adopted, as stated in Section 101.2.1. Appendix C contains provisions that may alter requirements found elsewhere in the code; however, the general requirements of the code still apply unless modified within this appendix. For example, the height and area limitations established in Chapter 5 apply to all Group U buildings and structures unless Appendix C is specifically adopted by the jurisdiction. In such a case, the allowable height and area limitations in Table C102.1 would supersede those in Table 503.

As with all of the appendices to the code, the provisions of Appendix C are not applicable unless the appendix is explicitly included in an adopting ordinance by the jurisdiction. Section 101.2.1 establishes the relationship of all appendices (see the commentary to that section for further discussion).

Section 312.1 of the code already contains a use and occupancy classification for buildings and structures considered as utility or miscellaneous in nature. Included in the Group U classification are agricultural buildings, barns, grain silos, greenhouses, livestock shelters, sheds and stables. Complete and detailed requirements are contained in the code for Group U buildings and structures; however, in most cases these provisions are more restrictive than a Group S-1 occupancy containing moderate-hazard contents that are likely to burn with moderate rapidity. Appendix C provides alternative provisions that can be adopted by a jurisdiction to realistically match safeguards with the actual hazards associated with buildings and structures that are minimally occupied by persons and pose a limited fire risk.

Purpose

Appendix C contains the requirements for protecting agricultural buildings. The provisions in this appendix reflect those occupancies and uses that deserve special consideration based on their limited occupant load. The code's foundation is the equivalent risk theory. Type of construction, fire-resistance ratings and means of egress requirements are balanced with the risks associated with agricultural buildings. This appendix contains general requirements, allowable heights and areas, mixed use provisions and exit requirements. The purpose of these provisions is to modify or supersede the basic code requirements for agricultural buildings. This approach allows the jurisdiction an alternative method that is more liberal and to address just the requirements for agricultural buildings that are included in Appendix C.

SECTION C101 GENERAL

C101.1 Scope. The provisions of this appendix shall apply exclusively to agricultural buildings. Such buildings shall be classified as Group U and shall include the following uses:

1. Livestock shelters or buildings, including shade structures and milking barns.
2. Poultry buildings or shelters.
3. Barns.
4. Storage of equipment and machinery used exclusively in agriculture.
5. Horticultural structures, including detached production greenhouses and crop protection shelters.
6. Sheds.
7. Grain silos.
8. Stables.

❖ This section of Appendix C contains the scope and applicability requirements for agricultural buildings. A specific list is provided to identify which agricultural buildings are intended to be controlled by the alternative requirements of the appendix. This list expands and clarifies agricultural buildings as stated in Section 312.1.

This section of the code contains an explicit list of those types of agricultural buildings that are eligible for the alternative requirements of Appendix C. Section 312.1 contains a similar list of utility and miscellaneous uses classified as Group U. Livestock shelters, barns, greenhouses, sheds, grain silos and stables are contained in both lists. Additionally, the term "livestock shelters" has been expanded to include weather protection structures and those used as milking barns. All of these buildings and structures are united in that they are used exclusively for agricultural purposes. As such, the occupant load for a Group U building is minimal. Although some of these structures may contain a considerable fuel load associated with their contents and uses, they would pose little fire hazards to persons or other occupiable buildings. The intent of the code is to match minimal requirements with these buildings based on the minimal fire and life hazards they pose.

SECTION C102
ALLOWABLE HEIGHT AND AREA

C102.1 General. Buildings classified as Group U agricultural building shall not exceed the area or height limits specified in Table C102.1.

❖ This section of Appendix C provides general building height and area limitations for Group U occupancies, just like Chapter 5 of the code. The height limitations, in terms of number of stories, and the area limitations from Table 503 for Group U buildings are in all cases the same or significantly less than those established for a Group S-1 structure. Those types of buildings are characterized by significant fuel loads that usually mandate the presence of an on-site automatic fire sprinkler system to abate the hazards to acceptable levels. A Group U building typically contains a lesser fire hazard and therefore, the allowable heights and areas should be greater than what is permitted for Group S-1 occupancies.

Revised entries for Table 503 are provided in this section for Group U buildings. Additionally, there are two modifications provided that supplement the requirements of Sections 507.1 and 507.3 for unlimited area buildings.

The provisions for governing the heights and areas of agricultural buildings are established in this section and are based on the Group U occupancy classification and the building's type of construction. Table C102.1 is referenced here as the primary determinant for the minimum type of construction required for Group U buildings.

TABLE C102.1. See below.

❖ This table is configured similar to Table 503. The nine types of construction are listed across the top of the table. Included are separate lines for the allowable area, allowable height in terms of the number of stories and allowable height in terms of feet. The values listed for height in terms of feet are the same values listed in Table 503. Only the allowable areas and heights in terms of number of stories have been increased. Although not specifically stated here, the rest of the provisions of Chapter 5 are still applicable for determining any modifications (see Sections 504 and 506).

C102.2 One-story unlimited area. The area of a one-story Group U agricultural building shall not be limited if the building is surrounded and adjoined by public ways or yards not less than 60 feet in width.

❖ Just as additional allowances are made for unlimited area buildings in Section 507, there are similar provisions for agricultural buildings. A single-story building without a basement level containing agricultural uses represents a low fire hazard due to the limited number of occupants that may be present. If such a building is further isolated with sufficient clear open space around it, the building area is not limited. No other structures may be located within the 60 feet (18 288 mm) of clear open space required around the building. The open space must be located on the same lot as the agricultural building to allow access for fire-fighting operations. The type of construction is not restricted with this option.

TABLE C102.1—BASIC ALLOWABLE AREA FOR A GROUP U, ONE STORY IN HEIGHT AND MAXIMUM HEIGHT OF SUCH OCCUPANCY

I		II		III and IV		V	
AA	BB	A	B	III A and IV	III B	A	B
ALLOWABLE AREA (square feet)[a]							
Unlimited	60,000	27,100	18,000	27,100	18,000	21,100	12,000
MAXIMUM HEIGHT IN STORIES							
Unlimited	12	4	2	4	2	3	2
MAXIMUM HEIGHT IN FEET							
Unlimited	160	65	55	65	55	50	40

For SI: 1 square foot = 0.0929 m^2.

a. See Section C102 for unlimited area under certain conditions.

C102.3 Two-story unlimited area. The area of a two-story Group U agricultural building shall not be limited if the building is surrounded and adjoined by public ways or yards not less than 60 feet (18 288 mm) in width and is provided with an approved automatic sprinkler system throughout in accordance with Section 903.3.1.1.

❖ This provision of the appendix, in effect, expands the allowance of Section 507.3. A two-story unlimited area agricultural building is permitted with the same open space requirement of Section C102.2 when it is equipped throughout with an automatic NFPA 13 fire sprinkler system. An agricultural building represents a lesser hazard than a Group B, F, M or S building. Therefore, a two-story fully suppressed agricultural building can also be of unlimited area as long as it is isolated. Once again, the type of construction is not restricted in these buildings.

SECTION C103 MIXED USES

C103.1 Mixed uses. Mixed uses shall be protected in accordance with Chapter 3.

❖ The appendix includes a reference for any mixed occupancies that may exist within the agricultural building; for example, a large milking barn with integral offices and lab areas used for the transaction of farm business and testing of milk products. Since these areas are occupiable spaces, they are not classified as part of the Group U portions. Most likely these areas are classified as Group B. As such, the building must be protected in accordance with Chapter 3. Another example would be a barn with sleeping accommodations for migrant farm workers. The sleeping rooms would be considered as guestrooms and would need to be classified as Group R.

Any of the options listed in Chapter 3 are available to the building designer to address the mixed uses and occupancies that may be present within an agricultural building. This would include accessory use areas in Section 302.2, nonseparated uses in Section 302.3.1 or separated uses in Section 302.3.2.

SECTION C104 EXITS

C104.1 Exit facilities. Exits shall be provided in accordance with Chapters 10 and 11.

Exceptions:

1. The maximum travel distance from any point in the building to an approved exit shall not exceed 300 feet (91 440 mm).
2. One exit is required for each 15,000 square feet (1393.5 m^2) of area or fraction thereof.

❖ The last section of this appendix provides a reminder to the permit applicant that the means of egress requirements of the code are still applicable to agricultural buildings. Although these buildings have a minimal number of occupants, provisions must still be made for their egress. This section includes two modifications to the Chapter 10 requirements that acknowledge the low occupant load and the limited hazards that the occupants are exposed to.

The means of egress for agricultural buildings must comply with the applicable provisions of Chapter 10. For example, Table 1004.1.2 specifies an occupant load for agricultural buildings based on a rate of one person for every 300 square feet (28 m^2) of gross floor area. Typically, Group U buildings are exempt from the requirements of Chapter 11 for accessibility to buildings by physically disabled persons. Section 1103.2.5 only mandates that access is required to paved work areas and those areas within a building or structure that are open to the general public.

The two exceptions modify or reinforce certain requirements of Chapter 10. Table 1015.1 limits the exit access travel distance for Group U buildings to 300 feet (91 440 mm). Exception 1 restates this 300-foot (91 440 mm) limit.

Exception 2 refines Table 1018.2 in determining the number of exits required for agricultural buildings. An exit is required for every 15,000 square feet (1,393 m^2) of floor area. This works out to the same 50 occupants specified in Table 1018.2; however, the single-story height limitation and 75-foot (22 860 mm) travel distance are deleted with Exception 2.

Appendix D: Fire Districts

The provisions contained in this appendix are not mandatory unless specifically referenced in the adopting ordinance.

General Comments

The provisions contained in this appendix are not mandatory unless specifically referenced in the adopting ordinance, as stated in Section 101.2.1.

Purpose

Fire districts, in one form or another, have been in the model codes for years. Over time, they have been revised, and in some cases deleted, as the codes have evolved to address the conflagration hazards that the fire district concept was intended to abate. However, there are jurisdictions that continue to utilize the provisions for fire districts. It is for this reason, in order to transition from the previously adopted model code to the *International Building Code*® (IBC®), that the provisions have been incorporated into the appendix.

SECTION D101 GENERAL

D101.1 Scope. The fire district shall include such territory or portion as outlined in an ordinance or law entitled "An Ordinance (Resolution) Creating and Establishing a Fire District." Wherever, in such ordinance creating and establishing a fire district, reference is made to the Fire District, it shall be construed to mean the fire district designated and referred to in this appendix.

❖ The establishment of fire districts is among the oldest recognized means of controlling the threat of urban conflagrations. (The term "conflagration" is used to describe a fire that involves three or more buildings and spreads across a natural or constructed barrier, such as a stream, street or other open space.) In fact, the use of fire district ordinances by local governments could be considered an early form of land use control similar to zoning as it is now practiced. Establishment of fire districts became popular at the beginning of the 20th century as a means of preventing the destruction of entire cities by fire. Massive urban conflagrations struck Chicago (1871), Baltimore (1872 and 1904) and San Francisco (1906), destroying all or most of the urbanized areas as they then existed just as these cities were experiencing perhaps their greatest development. Fire districts usually include areas, designated by ordinance or resolution by the local governing body, which are considered susceptible to mass fires involving multiple buildings or groups of buildings. While it may be true that fire district ordinances and building code adoption have significantly reduced the incidence of conflagrations in densely built or populated urbanized areas, continuing development of suburban and rural communities in areas where nearby wildlands are preserved or protected has spurred the increased incidence and severity of fires in urban/wildland interface and intermix areas, which often lie outside designated fire districts. This appendix is currently not designed to address this hazard. This appendix establishes regulatory guidelines for controlling construction, use and occupancy of buildings and conduct of dangerous operations or processes in urbanized areas that may be especially prone to the danger of fires spreading from one building to another. This appendix may be particularly useful in managing fire danger in areas built prior to the adoption and enforcement of a building code but, as stated, may not be suitable for managing the threat of urban/wildland interface and intermix area fire hazards. See the *International Urban-Wildland Interface Code*™ (IUWIC™) for guidance on managing fire hazards in urban/wildland interface and intermix areas.

D101.1.1 Mapping. The fire district complying with the provisions of Section D101.1 shall be shown on a map that shall be available to the public.

❖ The requirement to prepare a map facilitates communication to the public of the additional controls found in these guidelines. In effect, establishing a fire district is a form of land use regulation, and many communities include fire district controls in their zoning ordinance or establish fire districts through overlay zoning. Zoning maps are frequently prepared and used to communicate land use controls to the public and to illustrate the relationships between land uses, infrastructure and other features of the community.

D101.2 Establishment of area. For the purpose of this code, the fire district shall include that territory or area as described in Sections D101.2.1 through D101.2.3.

❖ The following sections describe the conditions typical of fire districts established by ordinances that reference this appendix. Note that these sections are not customarily intended to include an entire jurisdiction. In fact, this is contrary to the fundamental purpose of a fire district—protection from catastrophic fires that could destroy an entire town. By dividing all jurisdictions into at least two separate fire districts, a city presumably reduces its risk to destruction of just a portion of the city or town. In newer communities developed largely after the adoption of local land use regulations and building codes, fire districts provide little added protection against conflagrations.

If a community is so compact and densely developed that designation of separate areas is not possible or practical, an entire jurisdiction could be designated a fire district as a means of controlling hazardous occupancies and regulating combustible construction.

When adopted for use in areas developed prior to the adoption of a building code, fire districts may serve as an effective means of abating the hazard posed when no single building constitutes a public nuisance, but taken in aggregate, the hazard of groups of similarly noncode-conforming buildings is considered unreasonable. By preventing new construction and new occupancies from contributing to the nonconformity of existing hazards and isolating these nonconforming districts from new development, a city may gain a measure of security from a mass fire.

D101.2.1 Adjoining blocks. Two or more adjoining blocks, exclusive of intervening streets, where at least 50 percent of the ground area is built upon and more than 50 percent of the built-on area is devoted to hotels and motels of Group R-1; Group B occupancies; theaters, nightclubs, restaurants of Group A-1 and A-2 occupancies; garages, express and freight depots, warehouses and storage buildings used for the storage of finished products (not located with and forming a part of a manufactured or industrial plant); or Group S occupancy. Where the average height of a building is two and one-half stories or more, a block should be considered if the ground area built upon is at least 40 percent.

❖ The fire district should include areas densely developed with light and moderate fire hazard occupancies that have either significant fuel loads or significant transient occupant loads (see Figure D101.2.1). Nightclubs; theaters; restaurants; hotels and motels; certain types of Group B occupancies and warehouses not associated with industrial operations have been the sites of some of the nation's largest conflagrations. Intentionally set fires and delayed detection or notification of fire-fighting forces have contributed significantly to these losses. Occupancies and occupants typical of those listed in this section may be particularly prone to both of these factors.

Although Group F and M occupancies and Group S warehouses containing raw or waste materials are not listed, they are typically found in these areas as well. One- and two-family dwellings; multiple-family housing; schools or day care centers and institutional occupancies are not included in the description because they are typically found in less densely developed areas.

Local officials should recognize that redevelopment of industrial and warehouse buildings in existing fire districts for new residential and mercantile uses is increasingly common (see Section D103.2). These activities typically do not pose the same degree of mass fire danger in such buildings, despite the greater frequency of fires in residential occupancies. This may be due to reduced fuel loading and increased occupant activity, which aids in early detection and notification of a fire.

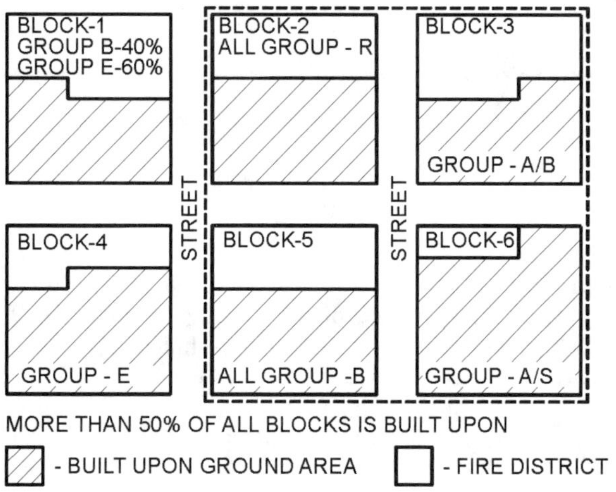

Figure D101.2.1

D101.2.2 Buffer zone. Where four contiguous blocks or more comprise a fire district, there shall be a buffer zone of 200 feet (60 960 mm) around the perimeter of such district. Streets, rights-of-way, and other open spaces not subject to building construction can be included in the 200-feet (60 960 mm) buffer zone.

❖ The buffer areas described by this section serve as fire or (preferably) fuel breaks by prohibiting the construction of buildings or structures that could become involved, allowing the fire to spread (see Figure D101.2.2). The green space or buffer zones may include the width of streets, streams or other dedicated rights-of-way that ensure segregation between adjacent fire district zones.

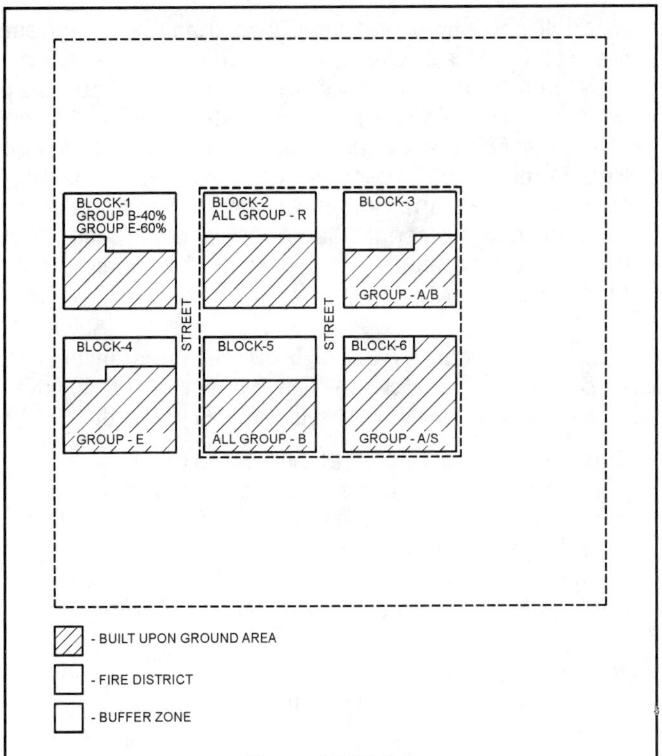

Figure D101.2.2

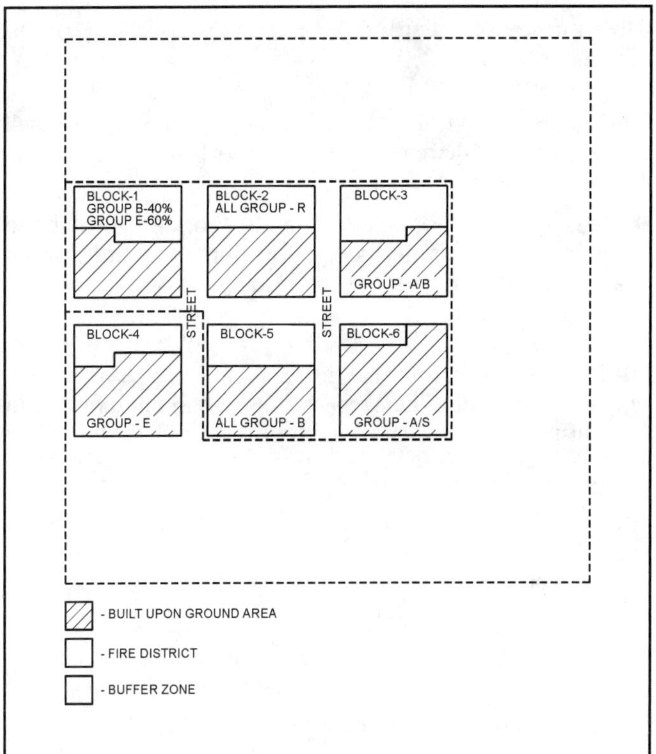

Figure D101.2.3

D101.2.3 Developed blocks. Where blocks adjacent to the fire district have developed to the extent that at least 25 percent of the ground area is built upon and 40 percent or more of the built-on area is devoted to the occupancies specified in Section D101.2.1, they can be considered for inclusion in the fire district, and can form all or a portion of the 200-foot (60 960 mm) buffer zone required in Section D101.2.2.

❖ Adjacent blocks may be added to the fire district as development of buffer spaces progresses (see Figure D101.2.3). However, care should be taken to ensure that the densities of buffer spaces do not increase to such an extent as to provide an avenue of fire spread between adjacent fire districts. If building densities remain between 25 and 50 percent of the land area, danger of fire spread between buildings in separate fire districts remains relatively low, as long as building separations required by Section D102.2 for the construction type are provided and maintained and combustible building features, such as roofing, sheathing, trim, canopies and signs, are similarly controlled.

SECTION D102
BUILDING RESTRICTIONS

D102.1 Types of construction permitted. Within the fire district every building hereafter erected shall be either Type I, II, III or IV, except as permitted in Section D104.

❖ The construction of Type V buildings is prohibited within a fire district. All other construction types are permitted, including unprotected Type IIB and IIIB structures with the further restrictions of Section D102.2 below.

D102.2 Other specific requirements.

D102.2.1 Exterior walls. Exterior walls of buildings located in the fire district shall comply with the requirements in Table 601 except as required in Section D102.2.6.

❖ Fire-resistant separations between buildings, either by distance or barriers, as described by Table 601, provide protection against fire spread by radiant and convective heat transfer and, to a limited extent, from small flying brands.

D102.2.2 Group H prohibited. Group H occupancies shall be prohibited from location within the fire district.

❖ Group H-1, H-2, H-3 and H-5 hazardous occupancies are prohibited within the fire district because of the fire hazard posed by these occupancies themselves. Group H-4 occupancies are prohibited due to the inherent danger of release and contamination posed by fire should occupancies containing highly toxic, toxic, radioactive or other health-hazard materials be exposed.

D102.2.3 Construction type. Every building shall be constructed as required based on the type of construction indicated in Chapter 6.

❖ This section reinforces the need for new buildings to be constructed as required based on height and area and type of construction considerations in the body of the code.

D102.2.4 Roof covering. Roof covering in the fire district shall conform to the requirements of ClassA or B roof coverings as defined in Section 1505.

❖ Class C and wood-shake shingle roofs are prohibited within the fire district due to the danger of fire spread and involvement through production of or exposure to burning brands.

D102.2.5 Structural fire rating. Walls, floors, roofs and their supporting structural members shall be a minimum of 1-hour fire-resistance-rated construction.

Exceptions:

1. Buildings of Type IV construction.
2. Buildings equipped throughout with an automatic sprinkler system in accordance with Section 903.3.1.1.
3. Automobile parking structures.
4. Buildings surrounded on all sides by a permanently open space of not less than 30 feet (9144 mm).
5. Partitions complying with Section 603.1(8).

❖ Unprotected Type IIB and IIIB construction is permitted within the fire district only for fully sprinklered buildings, automobile parking garages and buildings with 30 feet (9144 mm) of open space on all sides. All other buildings, except those of Type IV construction, must be of fire-resistance-rated construction.

D102.2.6 Exterior walls. Exterior load-bearing walls of Type II buildings shall have a fire-resistance rating of 2 hours or more where such walls are located within 30 feet (9144 mm) of a common property line or an assumed property line. Exterior nonload-bearing walls of Type II buildings located within 30 feet (9144 mm) of a common property line or an assumed property line shall have fireresistance ratings as required by Table 601, but not less than 1 hour. Exterior walls located more than 30 feet (9144 mm) from a common property line or an assumed property line shall comply with Table 601.

Exception: In the case of one-story buildings that are 2,000 square feet (186 m^2) or less in area, exterior walls located more than 15 feet (4572 mm) from a common property line or an assumed property line need only comply with Table 601.

❖ Although Type II buildings may be capable of resisting ignition because their components are generally noncombustible, their structural integrity could be threatened or contents ignited by radiant heat exposure from a fire in an adjacent building. Moreover, since Section 603.1, Item 1, permits the use of fire-retardant-treated wood for nonbearing partitions and roof systems of Type II buildings, these structures should not be assumed to be of noncombustible construction.

Small accessory buildings [less than 2,000 square feet (186 m^2)] are exempt from the exterior wall fire-resistance requirements of this section, provided they comply with Table 601 and are located at least 15 feet (4572 mm) from an adjacent or assumed property line. Bear in mind that, except as provided in Section D105, such structures may not be of Type V construction. Such buildings present little danger of contributing to a mass fire, provided they comply with the requirements of the code.

D102.2.7 Architectural trim. Architectural trim on buildings located in the fire district shall be constructed of approved noncombustible materials or fire-retardant-treated wood.

❖ The ignition of architectural trim on the exterior of buildings from radiant heat transfer may pilot the ignition of other building components or contents of the building, providing an avenue for further fire spread. Many types of architectural trim may also catch flying brands from a fire, sparking the ignition of the exposed building.

D102.2.8 Permanent canopies. Permanent canopies are permitted to extend over adjacent open spaces provided:

1. The canopy and its supports shall be of noncombustible material, fire-retardant-treated wood, Type IV construction, or of 1-hour fire resistance rated construction.

   Exception: Any textile covering for the canopy shall be flame resistant as determined by tests conducted in accordance with NFPA 701 after both accelerated water leaching and accelerating weathering.

2. Any canopy covering, other than textiles, shall have a flame spread index not greater than 25 when tested in accordance with ASTM E 84 in the form intended for use.
3. The canopy shall have at least one long side open.
4. The maximum horizontal width of the canopy shall not exceed 15 feet (4572 mm).
5. The fire resistance of exterior walls shall not be reduced.

❖ Although combustible exterior canopies may become involved in fire from radiant heat transfer, their relatively low mass and method of support and attachment to the building should minimize the risk that they will pilot the ignition of exterior walls or building contents. The requirements of this section restrict certain features of exterior canopies to ensure that their construction contributes little to fire growth or spread in the fire district.

D102.2.9 Roof structures. Structures, except aerial supports 12 feet (3658 mm)high or less, flag poles, water tanks and cooling towers, placed above the roof of any building within the fire district shall be of noncombustible material and shall be supported by construction of noncombustible material.

❖ Vertical structures with little combustible mass, or which will be cooled by their contents, are permitted, provided their failure in a fire does not pose a significant secondary hazard. Of course, rooftop water tanks supplying fire protection systems should not be of combustible construction because their failure may undermine the effectiveness of required fire protection systems.

D102.2.10 Plastic signs. The use of plastics complying with Section 2611 for signs is permitted provided the structure of the sign in which the plastic is mounted or installed is noncombustible.

❖ Free-standing and projecting signs are unlikely to contribute to fire spread between buildings, provided their structural supports are noncombustible and structural collapse of the sign supports will not directly expose a building or its contents. Although rooftop, marquee and wall-mounted signs may present a more direct hazard to buildings, Sections 2611.2 and 2611.3 limit the area of any plastic components and require all plastic materials to meet specified combustibility criteria (see Section 2606.4).

D102.2.11 Plastic veneer. Exterior plastic veneer is not permitted in the fire district.

❖ All plastic veneers, including vinyl siding, are prohibited in the fire district. Most plastics deform under high radiant heat conditions, some melt and drip and all will burn. Plastics represent a wide range of fire hazard conditions, generally all of which are unfavorable.

This prohibition is consistent with restrictions on other combustible exterior elements, such as roof and architectural trim. Although this appendix does not explicitly prohibit other combustible exterior veneers, Section D105.1, Item 11, excludes wood veneer from these provisions, and the use of such combustible cladding materials remains severely restricted.

SECTION D103 CHANGES TO BUILDINGS

D103.1 Existing buildings within the fire district. An existing building shall not hereafter be increased in height orarea unless it is of a type of construction permitted for new buildings within the fire district or is altered to comply with the requirements for such type of construction. Nor shall any existing building be hereafter extended on any side, nor square footage or floors added within the existing building unless such modifications are of a type of construction permitted for new buildings within the fire district.

❖ Bearing in mind that the principal purpose of Section D103 is to control the fire hazard posed by existing nonconforming construction by isolating areas where such buildings are located, it would be imprudent to allow any increase in the hazard within the fire district once it is established. Therefore, any new construction must comply with the requirements of the code, including limits on construction, area and height and fire protection (see Section D102.2.3).

D103.2 Other alterations. Nothing in Section D103.1 shall prohibit other alterations within the fire district provided there is no change of occupancy that is otherwise prohibited and provided the fire hazard is not increased by such alteration.

❖ All construction and any alteration of existing buildings must comply with the provisions of the code (see Section 101.2). Assuming all new work complies with the relevant requirements of the code and all applicable referenced standards, the hazard of any building in the fire district should not be increased.

Change of occupancy from one group to another is allowed within the fire district, provided that the new occupancy is permitted by the code for the type of construction, fire protection, means of egress and other requirements and is not a hazardous (Group H) occupancy (see Section D102.2.2).

D103.3 Moving buildings. Buildings shall not hereafter be moved into the fire district or to another lot in the fire district unless the building is of a type of construction permitted in the fire district.

❖ Relocated buildings, like new construction, must be of a type of construction permitted within the fire district. The type of construction will be determined by the area, height and occupancy of the building after its relocation. However, certain structural components and systems may require higher fire-resistance ratings than specified in Table 601, in accordance with Section D102.2.5.

SECTION D104 BUILDINGS LOCATED PARTIALLY IN THE FIRE DISTRICT

D104.1 General. Any building located partially in the fire district shall be of a type of construction required for the fire district, unless the major portion of such building lies outside of the fire district and no part is more than 10 feet (3048 mm) inside the boundaries of the fire district.

❖ Buildings that lie principally outside the fire district and project no more than 10 feet (3048 mm) within a fire district boundary are exempt from the fire district requirements and should be considered part of the buffer zone described in Sections D101.2.2 and D101.2.3.

In Figure D104(A), the building does not need to meet the provisions of Appendix D, since the majority of the building is outside and no more than 10 feet (3048 mm) of the building lies within the boundaries of the fire district.

In Figure D104(B), the building shall comply with the provisions of Appendix D, since more than 10 feet (3048 mm) of the building is within the boundaries of the fire district.

SECTION D105 EXCEPTIONS TO RESTRICTIONS IN FIRE DISTRICT

D105.1 General. The preceding provisions of this appendix shall not apply in the following instances:

1. Temporary buildings used in connection with duly authorized construction.
2. A private garage used exclusively as such, not more than one story in height, nor more than 650 square feet (60 m^2) in area, located on the same lot with a dwelling.
3. Fences not over 8 feet (2438 mm) high.
4. Coal tipples, material bins, and trestles constructed of Type IV construction.
5. Water tanks and cooling towers conforming to Sections 1509.3 and 1509.4.
6. Greenhouses less than 15 feet (4572 mm) high.
7. Porches on dwellings not over one story in height, and not over 10 feet (3048 mm) wide from the face of the building, provided such porch does not come within 5 feet (1524 mm) of any property line.
8. Sheds open on a long side not over 15 feet (4572 mm) high and 500 square feet (46 m^2) in area.
9. One- and two-family dwellings where of a type of construction not permitted in the fire district can be extended 25 percent of the floor area existing at the time of inclusion in the fire district by any type of construction permitted by this code.
10. Wood decks less than 600 square feet (56 m^2) where constructed of 2-inch (51 mm) nominal wood, pressure treated for exterior use.
11. Wood veneers on exterior walls conforming to Section 1405.4.
12. Exterior plastic veneer complying with Section 2605.2 where installed on exterior walls required to have a fire-resistance rating not less than 1 hour, provided the exterior plastic veneer does not exhibit sustained flaming as defined in NFPA 268.

❖ With the exception of Items 11 and 12, excluded structures are generally accessory to other buildings and pose little hazard of contributing to fire spreading from one building to another or between buildings in adjacent fire districts. Excluding these structures from the provisions of this appendix does not exempt any new structures of the types or uses listed from complying with the requirements of the code.

Water tanks for fire protection located within the fire district may not be governed by the requirements of this appendix.

Wood veneer applied to the exterior of structures must comply with the restrictions of Section 1405.4. The referenced provisions restrict the use of wood veneer to installation on noncombustible exterior walls of the fire resistance required by Table 601, and not exceeding two stories in height.

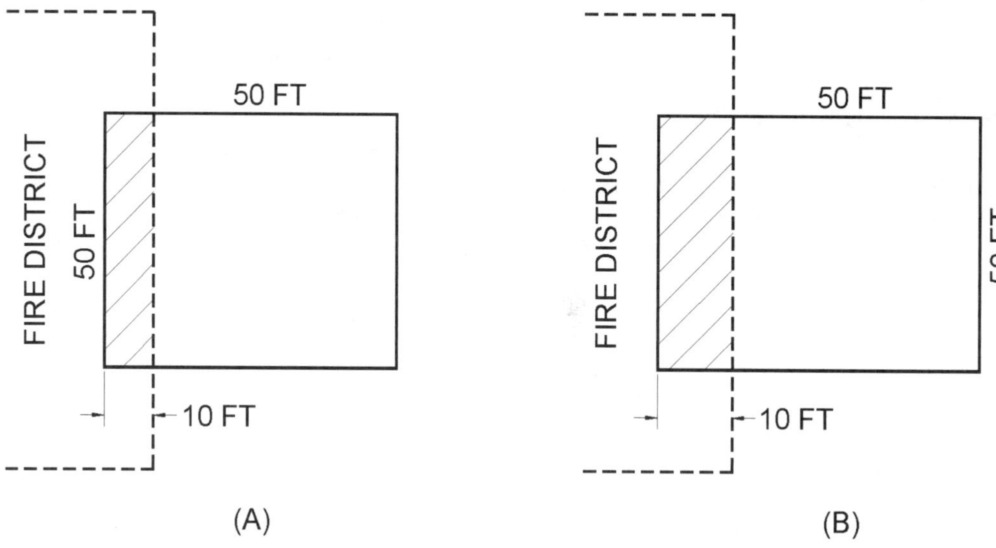

For SI: 1 foot = 304.8 mm.

Figure D104

SECTION D106
REFERENCED STANDARDS

ASTM E 84-01	Test Methods for Surface Burning Characteristics of Building Materials	D102.2.8
NFPA 268-96	Standard Test Method for Determining Ignitability of Exterior Wall Assemblies Using a Radiant Heat Energy Source	D105.1
NFPA 701-99	Methods of Fire Test for Flame-resistant Textiles and Films	D102.2.8

Appendix E: Supplementary Accessibility Requirements

The provisions contained in this appendix are not mandatory unless specifically referenced in the adopting ordinance.

General Comments

As stated in Section 101.2.1, the provisions of the appendices do not apply unless specifically adopted.

Chapter 11 sets forth requirements for accessibility to buildings and their associated sites and facilities for people with physical disabilities. Appendix E was added to address accessibility in construction for items that were not typically enforceable through the traditional building code enforcement process.

Section E101 contains the broad scope statement of the appendix and identifies the baseline criteria for accessibility as being in compliance with this appendix and ICC A117.1. ICC A117.1 is the consensus national standard that sets forth the details, dimensions and construction specifications for accessibility.

Section E102 contains definitions of terms that are associated with accessibility that are not found in Chapter 11.

Section E103 contains additional requirements for interior accessible routes. An accessible route is a key component of the built environment that provides a person with a disability access to spaces, elements, facilities and buildings.

Section E104 contains various accessibility requirements that are unique to specific occupancies and are applicable in addition to other general requirements of this chapter. Specific provisions unique to transient lodging and Group I-3 facilities are included.

Section E105 contains various requirements that are applicable to special equipment provided for public use, including requirements for water coolers; portable toilet and bathing rooms; laundry equipment; vending machines; automatic teller machines and fare machines and two-way communication systems.

Section E106 contains special criteria for public telephones, including TTY requirements.

Section E107 sets forth requirements for signage identifying permanent designations, such as room numbers in a hospital, directional and informational signs and signage specifically related to the transportation facilities identified in Sections E108, E109 and E110.

Section E108 lists criteria specific to bus stops and terminals.

Section E109 lists criteria specific to fixed transportation facilities and stations, such as train stations.

Section E110 lists criteria specific to airports.

Section E111 deals with historical facilities that are registered through the National Historic Preservation Act or a similar state or local law. If such a facility undergoes an alteration or change of occupancy, it must comply with accessibility regulations as specified in Section 3409.8. If the required alterations will adversely affect some historic aspect of the structure, alternatives are available. Before these alternatives can be utilized, the owner must go through a process to evaluate the requirements and implications of each. Section E111.3 describes the process if the building is registered through the National Historic Preservation Act. Section E111.4 describes the process if the building is registered through a state or local registration law. Section E111.5 provides alternatives for displays in historical buildings.

Section E112 enumerates the standards referenced in this appendix.

Purpose

Appendix E includes scoping requirements in the new *Americans with Disabilities Act Accessibility Guidelines* (ADAAG) that are not in Chapter 11 or mainstreamed throughout the code. Items in the appendix deal with topics not typically addressed in building codes. For example, items in Sections E102 through E107 are not part of the "built" environment, as they can be added, modified or removed without a building permit and without notifying the building official. Sections E108 through E110 deal specifically with transportation facilities. Section E111 specifies the process to follow regarding alterations or change of occupancy in historically registered buildings.

This appendix is being included in the code for the convenience of building owners and designers who are required to comply with new ADAAG, and for the benefit of jurisdictions that may wish to go beyond the traditional boundary of the code and formally adopt these additional ADAAG requirements. The requirements within the code indicate what is required and enforceable by the building official. The appendix includes items that are outside the scope of code enforcement, but must be complied with in addition to the building code requirements to satisfy ADAAG regulations.

SECTION E101
GENERAL

E101.1 Scope. The provisions of this appendix shall control the supplementary requirements for the design and construction of facilities for accessibility to physically disabled persons.

❖ This section establishes the scope of the appendix as providing for design and construction of facilities for accessibility to disabled persons other than items covered under Chapter 11. Items in the appendix are not enforceable under the typical building code review process because these items can be altered or removed without notification to the building official. Items in this appendix are included in new ADAAG.

E101.2 Design. Technical requirements for items herein shall comply with this code and ICC A117.1.

❖ This section establishes the primary and fundamental relationship of ICC A117.1 to the code. The code text is intended to "scope" or provide thresholds for application of required accessibility features. The referenced standard contains technical provisions indicating how compliance with the code is achieved. In short, this appendix specifies what, when and how many accessible features are required; the referenced standard indicates how to make these features accessible. Compliance with both the scoping and technical standard is necessary for accessibility.

SECTION E102
DEFINITIONS

E102.1 General. The following words and terms shall, for the purposes of this appendix, have the meanings shown herein.

❖ This section contains definitions of terms that are associated with the subject matter of this appendix.

Definitions of terms can help in the understanding and application of the code requirements. The purpose for including these definitions within this appendix is because appendices are not considered part of the code requirements unless specifically referenced in the adopting ordinance.

CLOSED-CIRCUIT TELEPHONE. A telephone with a dedicated line such as a house phone, courtesy phone or phone that must be used to gain entrance to a facility.

❖ Telephones described in this definition are utilized in hotels, apartments, airports, hospitals and similar facilities. Access to these systems is important to provide communication throughout a facility.

MAILBOXES. Receptacles for the receipt of documents, packages or other deliverable matter. Mailboxes include, but are not limited to, post office boxes and receptacles provided by commercial mail-receiving agencies, apartment houses and schools.

❖ The definition for "Mailboxes" clarifies that the scope of this provision is meant to address all types of facilities where a person could receive correspondence, packages or written notifications. The description includes post office boxes, mailboxes in the lobby of an apartment building or mailboxes in the lobby of a college dorm.

TRANSIENT LODGING. A building, facility or portion thereof, excluding inpatient medical care facilities and long-term care facilities, that contains one or more dwelling units or sleeping units. Examples of transient lodging include, but are not limited to, resorts, group homes, hotels, motels, dormitories, homeless shelters, halfway houses and social service lodging.

❖ Transient lodging facilities, such as hotels, typically provide short-term stays; however, this definition also includes facilities that provide longer term stays, such as dormitories. The 30-day or less stay, typically associated with transient lodging, is not applicable in this definition. The key is whether sleeping units or dwelling units are provided. The exclusions are for inpatient and long-term care facilities, such as hospitals and nursing homes. Additionally, this would not be applicable to permanent lodging facilities with dwelling units, such as apartments, condominiums or townhouses.

SECTION E103
ACCESSIBLE ROUTE

E103.1 Raised platforms. In banquet rooms or spaces where a head table or speaker's lectern is located on a raised platform, an accessible route shall be provided to the platform.

❖ An accessible route is required to a temporary raised platform utilized for a speaker or head table. An accessible route would not be required to a raised platform used for other purposes, such as a band. Access to permanently raised platforms or stages is addressed in Chapter 11.

SECTION E104
SPECIAL OCCUPANCIES

E104.1 General. Transient lodging facilities shall be provided with accessible features in accordance with Sections E104.2 and E104.3. Group I-3 occupancies shall be provided with accessible features in accordance with Sections E104.3 and E104.4.

❖ In the following sections there are bed and communication feature requirements for transient lodging facilities. Also included are communication and visitation booth requirements for jails (Group I-3).

E104.2 Accessible beds. In rooms or spaces having more than 25 beds, five percent of the beds shall have a clear floor space complying with ICC A117.1.

❖ If dormitory-type transient lodging facilities have 25 or more beds in a room, 5 percent of the beds must have adequate clearance to be accessible by a person using a wheelchair. If separate-sex facilities are provided, accessible beds must be provided for both sexes.

E104.2.1 Sleeping areas. A clear floor space complying with ICC A117.1 shall be provided on both sides of the accessible bed. The clear floor space shall be positioned for parallel approach to the side of the bed.

Exception: This requirement shall not apply where a single clear floor space complying with ICC A117.1 positioned for parallel approach is provided between two beds.

❖ To be considered accessible, the bed must be approachable from either side in order to accommodate wheelchair users who may prefer to transfer from one side to the other. The alternative offered is if a bed is provided on both sides of a single-approach location.

E104.3 Communication features. Communication features complying with ICC A117.1 shall be provided in accordance with Sections E104.3.1 through E104.3.4.

❖ Examples of communication features are telephones and doorbells.

E104.3.1 Transient lodging. In transient lodging facilities, sleeping units with accessible communication features shall be provided in accordance with Table E104.3.1. Units required to comply with Table E104.3.1 shall be dispersed among the various classes of units.

❖ In transient lodging facilities, an alternative notification system for persons with hearing impairments must be offered in the number of dwelling and sleeping units specified in Table E104.3.1 (see the definition for "Dwelling units" and "Sleeping units" in Chapter 2 and Section 310.2). The number of dwelling or sleeping units that are required to provide accessible communication features depends on the number of units provided. The number of accessible communication features is based on the number of units, not the number of beds. The rooms with accessible communication features are not required to be the same rooms that contain accessible beds.

E104.3.2 Group I-3. In Group I-3 occupancies at least 2 percent, but no fewer than one of the total number of general holding cells and general housing cells equipped with audible emergency alarm systems and permanently installed telephones within the cell, shall comply with Section E104.3.3.

❖ Jails are required to have a fire alarm system in accordance with Section 907.6.2. If cells also have permanently installed telephones, 2 percent of the telephones shall have a volume control and a place to plug in a TTY. The visual notification device for the phone cannot be connected to the visual alarms for the fire alarm system.

TABLE E104.3.1
DWELLING OR SLEEPING UNITS WITH ACCESSIBLE COMMUNICATION FEATURES

TOTAL NUMBER OF DWELLING OR SLEEPING UNITS PROVIDED	MINIMUM REQUIRED NUMBER OF DWELLING OR SLEEPING UNITS WITH ACCESSIBLE COMMUNICATION FEATURES
1	1
2 to 25	2
26 to 50	4
51 to 75	7
76 to 100	9
101 to 150	12
151 to 200	14
201 to 300	17
301 to 400	20
401 to 500	22
501 to 1,000	5% of total
1,001 and over	50 plus 3 for each 100 over 1,000

❖ This table specifies the number of dwelling and sleeping units in transient lodging facilities that are required to provide accessible communication features. In accordance the definition of "Sleeping unit," bedrooms in dwelling units are not considered sleeping units.

E104.3.3 Dwelling units and sleeping units. Where dwelling units and sleeping units are altered or added, the requirements of Section E104.3 shall apply only to the units being altered or added until the number of units with accessible communication features complies with the minimum number required for new construction.

❖ When transient lodging dwelling and sleeping units are altered or added, only these units are counted to determine the number of units that will require accessible communication features. Once the entire building meets the number required for new construction, no additional accessible communication features are required. This is consistent with the intent of Sections 3409.5.1 and 3409.7.7.

E104.3.4 Notification devices. Visual notification devices shall be provided to alert room occupants of incoming telephone calls and a door knock or bell. Notification devices shall not be connected to visual alarm signal appliances. Permanently installed telephones shall have volume controls and an electrical outlet complying with ICC A117.1 located within 48 inches (1219 mm) of the telephone to facilitate the use of a TTY.

❖ A type of visual notification, such as a flashing light, must be provided in some sleeping accommodations. The purpose of the visual notification is to alert a person with a hearing impairment to incoming phone calls or someone knocking at the door. The notification cannot be connected to any type of visual alarm system. If phones are provided within the dwelling unit or sleeping unit, they must be equipped with a volume control and be located within 48 inches (1219 mm) of an electrical outlet to accommodate a TTY.

E104.4 Partitions. Solid partitions or security glazing that separates visitors from detainees in Group I-3 occupancies shall provide a method to facilitate voice communication. Such methods are permitted to include, but are not limited to, grilles, slats, talk-through baffles, intercoms or telephone handset devices. The method of communication shall be accessible to individuals who use wheelchairs and individuals who have difficulty bending or stooping. Hand-operable communication devices, if provided, shall comply with Section E106.3.

❖ Section 1109.11 requires 5 percent of cubicles or counters in jail (Group I-3) visiting areas to be accessible in accordance with the built-in counter/work surface requirements. This section deals with the visiting situation where solid partitions of security glazing separates visitors from detainees. To facilitate communication, the system may include some type of handset, headphones or microphone. The system must consider in its design that either user may use a wheelchair or have difficulty bending or stooping. Any type of telephone or headphone system must have a volume control in accordance with Section 106.3.

SECTION E105 OTHER FEATURES AND FACILITIES

E105.1 Water coolers. Where water coolers are provided, at least 50 percent, but not less than one, of such units provided on each floor shall comply with ICC A117.1.

❖ Water coolers are sometimes used in place of drinking fountains. If water coolers are provided, they must have accessibility requirements similar to drinking fountains.

E105.2 Portable toilets and bathing rooms. Where multiple single-user portable toilet or bathing units are clustered at a single location, at least 5 percent, but not less than one toilet unit or bathing unit at each cluster, shall comply with ICC A117.1. Signs containing the International Symbol of Accessibility and complying with ICC A117.1 shall identify accessible portable toilets and bathing units.

Exception: Portable toilet units provided for use exclusively by construction personnel on a construction site.

❖ When portable toilets are provided for special events, at least 5 percent must be accessible. This is consistent with the requirements for sinks in Section 1109.3. Appropriate signage is also required. If portable toilets are provided for personnel on a construction site, they are not required to be accessible. This is consistent with Section 1103.2.6.

E105.3 Laundry equipment. Where provided in spaces required to be accessible, washing machines and clothes dryers shall comply with this section.

❖ When laundry facilities are provided that are open to the general public, or are associated with Accessible, Type A or B units, the washers and dryers must be accessible in accordance with Sections E105.3.1 and E105.3.2. This is consistent with Section 1107.3. The requirements are based on the washers or dryers provided in each space, not the total within the building. For example, a dorm with a single washer and dryer on each floor of a building should provide an accessible washer and dryer in each location.

E105.3.1 Washing machines. Where three or fewer washing machines are provided, at least one shall comply with ICC A117.1. Where more than three washing machines are provided, at least two shall comply with ICC A117.1.

❖ When three or fewer washing machines are provided in a laundry room, at least one must meet the requirements for accessibility in Section 611 in ICC A117.1. In larger facilities, at least two washers must be accessible. Requirements include criteria for clear floor space to access the unit; operable parts that meet height, graspability and force requirements and the location of doors to access the interior of the appliance.

E105.3.2 Clothes dryers. Where three or fewer clothes dryers are provided, at least one shall comply with ICC A117.1. Where more than three clothes dryers are provided, at least two shall comply with ICC A117.1.

❖ The same criteria for washers is expected for dryers (see commentary, Section E105.3.1).

E105.4 Depositories, vending machines, change machines and similar equipment. Where provided, at least one of each type of depository, vending machine, change machine and similar equipment shall comply with ICC A117.1.

Exception: Drive-up-only depositories are not required to comply with this section.

❖ All types of self-service machinery must be accessible. Examples include the book return at a library, a change machine at a laundry mat or a soda machine. A drive-up-only depository, such as a drive-up night deposit at a bank, is not required to be accessible. To be accessible the equipment must meet the provisions for clear floor space and have operable parts within reach ranges.

E105.5 Mailboxes. Where mailboxes are provided in an interior location, at least 5 percent, but not less than one, of each type shall comply with ICC A117.1. In residential and institutional facilities, where mailboxes are provided for each dwelling unit or sleeping unit, mailboxes complying with ICC A117.1 shall be provided for each unit required to be an Accessible unit.

❖ Whenever mailboxes are provided, such as in an office building lobby, at least 5 percent are required to be accessible. In residential and institutional facilities, if mailboxes are provided for each unit, the mailboxes for the Accessible units must have adequate clear floor space and meet operable parts provisions. Examples of such types of facilities would be dorms, hotels, assisted living facilities or nursing homes. The number of Accessible units depends on the type of facility (see Section 1107 for requirements).

E105.6 Automatic teller machines and fare machines. Where automatic teller machines or self-service fare vending, collection or adjustment machines are provided, at least one machine of each type at each location where such machines are provided shall be accessible. Where bins are provided for envelopes, wastepaper or other purposes, at least one of each type shall be accessible.

❖ When such machines are provided, such as at banks and in bus or train stations, at least one of each type of machine must be accessible. Specific criteria is provided in Section 707 in ICC A117.1.

E105.7 Two-way communication systems. Where two-way communication systems are provided to gain admittance to a building or facility or to restricted areas within a building or facility, the system shall comply with ICC A117.1.

❖ Apartments or condominiums with secured entrances often have some type of two-way communication system at the entry so the residents can screen persons entering the building. Specific criteria for systems connected with dwelling or sleeping units are provided in Sections 1004.6 and 1004.7 in ICC A117.1. Similar provisions would be applicable to two-way communication systems provided at other locations.

SECTION E106 TELEPHONES

E106.1 General. Where coin-operated public pay telephones, coinless public pay telephones, public closed-circuit telephones, courtesy phones or other types of public telephones are provided, accessible public telephones shall be provided in accordance with Sections E106.2 through E106.5 for each type of public telephone provided. For purposes of this section, a bank of telephones shall be considered two or more adjacent telephones.

❖ Many types of public phones are provided, such as pay telephones, courtesy phones throughout airports, house phones in hotels, entrance phones in apartment complexes, etc. Wherever such phones are provided, at least one of each type must comply with Sections E106.2 through E106.5. A bank of telephones is defined as two or more telephones in one location.

E106.2 Wheelchair-accessible telephones. Where public telephones are provided, wheelchair-accessible telephones complying with ICC A117.1 shall be provided in accordance with Table E106.2.

❖ For each type of public telephone provided, the number of phones specified in Table E106.2 must be accesible. Specific criteria is provided in Section 704.2 in ICC A117.1.

TABLE E106.2
WHEELCHAIR ACCESSIBLE TELEPHONES

NUMBER OF TELEPHONES PROVIDED ON A FLOOR OR LEVEL	MINIMUM REQUIRED NUMBER OF WHEELCHAIR-ACCESSIBLE TELEPHONES
1 or more single unit	1 per floor or level
1 bank	1 per floor or level
2 or more banks	1 per bank[a]

a. At least one telephone per floor shall provide a forward approach complying with ICC A117.1, except for exterior installations where dial-tone-first service is available.

❖ If single telephones are provided in several locations or one bank of telephones is provided on a floor level or over a site, then at least one of each type must be accessible. The accessible telephone can be accessed using a parallel approach or a forward approach. If two or more banks are provided on a floor level or over a site, at least one telephone at each location must be accessible.

E106.3 Volume controls. All public telephones provided shall have volume control complying with ICC A117.1.

❖ Every public telephone must have volume controls and be identified with appropriate signage. Specific criteria is found in Section 704.3 in ICC A117.1.

E106.4 TTYs. TTYs complying with ICC A117.1 shall be provided in accordance with Sections E106.4.1 through E106.4.9.

❖ TTYs are defined in ICC A117.1 as "machinery or equipment that employs interactive, graphic communications through the transmission of coded signals across the standard telephone network." The term "TTY" also refers to devices known as text telephones and TDDs. Text telephones are needed to enable a person with a hearing impairment to utilize a telephone to send and receive text messages. Sections E106.4.1 through E106.4.4 indicate when TTYs are required within a bank of telephones, on a floor, within a building or on a site. More than one section could apply. Sections E106.4.5 through E106.4.8 are special requirements to specific types of sites. Section E106.4.9 deals with appropriate signage.

E106.4.1 Bank requirement. Where four or more public pay telephones are provided at a bank of telephones, at least one public TTY shall be provided at that bank.

Exception: TTYs are not required at banks of telephones located within 200 feet (60 960 mm) of, and on the same floor as, a bank containing a public TTY.

❖ A TTY is required wherever four or more telephones are located together. If a bank is located within 200 feet (60 960 mm) of a TTY, that bank is not required to contain a TTY.

E106.4.2 Floor requirement. Where four or more public pay telephones are provided on a floor of a privately owned building, at least one public TTY shall be provided on that floor. Where at least one public pay telephone is provided on a floor of a publicly owned building, at least one public TTY shall be provided on that floor.

❖ This section is similar to Section E106.4.1; however, the number of phones on the floor is counted, instead of within a bank. The term "publicly owned building" is intended to apply to facilities such as government buildings, courthouses, libraries, schools, police stations, fire stations, park district buildings, etc.

E106.4.3 Building requirement. Where four or more public pay telephones are provided in a privately owned building, at least one public TTY shall be provided in the building. Where at least one public pay telephone is provided in a publicly owned building, at least one public TTY shall be provided in the building.

❖ This section is similar to Sections E106.4.1 and E106.4.2; however, the number of phones in a building is counted, instead of within a bank or on a floor. The term "publicly owned building" is intended to apply to facilities such as government buildings, courthouses, libraries, schools, police stations, fire stations, park district buildings, etc.

E106.4.4 Site requirement. Where four or more public pay telephones are provided on a site, at least one public TTY shall be provided on the site.

❖ This section is similar to Sections E106.4.1, E106.4.2 and E106.4.3; however, the number of phones on a site is counted.

E106.4.5 Rest stops, emergency road stops, and service plazas. Where a public pay telephone is provided at a public rest stop, emergency road stop or service plaza, at least one public TTY shall be provided.

❖ A TTY is required at these types of facilities regardless of the number of phones provided.

E106.4.6 Hospitals. Where a public pay telephone is provided in or adjacent to a hospital emergency room, hospital recovery room or hospital waiting room, at least one public TTY shall be provided at each such location.

❖ In hospitals, when phones are located for public use in an emergency room, recovery room or waiting room, a TTY is required at each location.

E106.4.7 Transportation facilities. Transportation facilities shall be provided with TTYs in accordance with Sections E109.2.5 and E110.2 in addition to the TTYs required by Sections E106.4.1 through E106.4.4.

❖ Transportation facilities, such as bus stations, train stations and airports, have specific requirements as stated in Sections E109.2.5 and E110.2. Additionally, they must comply with the requirements based on the number of telephones in a bank, on a floor, within a building or on a site.

E106.4.8 Detention and correctional facilities. In detention and correctional facilities, where a public pay telephone is provided in a secured area used only by detainees or inmates and security personnel, then at least one TTY shall be provided in at least one secured area.

❖ Specific criteria is provided for the detainee side of detention and correctional facilities. Phones available on the public side should comply with Sections E106.4.1 through E106.4.4.

E106.4.9 Signs. Public TTYs shall be identified by the International Symbol of TTY complying with ICC A117.1. Directional signs indicating the location of the nearest public TTY shall be provided at banks of public pay telephones not containing a public TTY. Additionally, where signs provide direction to public pay telephones, they shall also provide direction to public TTYs. Such signs shall comply with ICC A117.1 and shall include the International Symbol of TTY.

❖ Signage must be provided indicating the location of TTYs.

E106.5 Shelves for portable TTYs. Where a bank of telephones in the interior of a building consists of three or more public pay telephones, at least one public pay telephone at the bank shall be provided with a shelf and an electrical outlet in accordance with ICC A117.1.

Exceptions:

1. In secured areas of detention and correctional facilities, if shelves and outlets are prohibited for purposes of security or safety shelves and outlets for TTYs are not required to be provided.
2. The shelf and electrical outlet shall not be required at a bank of telephones with a TTY.

❖ If a bank of telephones has three or more public telephones, a shelf and electrical outlet must be provided to allow a person with a hearing impairment to utilize his or her own portable TTY. In a bank of four or more phones a public TTY is required (see Section E106.4.1); therefore, a shelf and outlet for a portable TTY are not required within the same bank. Due to security concerns, there may be situations in jails where the shelf and outlet for a portable TTY is not required.

SECTION E107 SIGNAGE

E107.1 Signs. Required accessible portable toilets and bathing facilities shall be identified by the International Symbol of Accessibility.

❖ Accessible portable toilets must be identified (see Section E105.2).

E107.2 Designations. Interior and exterior signs identifying permanent rooms and spaces shall be tactile. Where pictograms are provided as designations of interior rooms and spaces, the pictograms shall have tactile text descriptors. Signs required to provide tactile characters and pictograms shall comply with ICC A117.1.

Exceptions:

1. Exterior signs that are not located at the door to the space they serve are not required to comply.
2. Building directories, menus, seat and row designations in assembly areas, occupant names, building addresses and company names and logos are not required to comply.
3. Signs in parking facilities are not required to comply.
4. Temporary (seven days or less) signs are not required to comply.

❖ Where permanent signage is provided, such as room numbers in a hospital or hotel, tactile signage must also be provided. When pictographs are utilized, they must also have tactile text signage. For signage requirements, reference ICC A117.1, Section 703. Exceptions are provided for types of signs where it may not be effective or practical to provide tactile signage. Examples would be the remote sign at the street indicating the name or occupant of the building, the menu above the take-out counter, temporary signage or building directories.

E107.3 Directional and informational signs. Signs that provide direction to, or information about, permanent interior spaces of the site and facilities shall contain visual characters complying with ICC A117.1.

Exception: Building directories, personnel names, company or occupant names and logos, menus and temporary (seven days or less) signs are not required to comply with ICC A117.1.

❖ Directional information must also be tactile, with the exception of signage that may be changeable, such as building directories, personnel names, etc. For requirements for this signage, reference ICC A117.1, Section 703.

E107.4 Other signs. Signage indicating special accessibility provisions shall be provided as follows:

1. At bus stops and terminals, signage must be provided in accordance with Section E108.4.
2. At fixed facilities and stations, signage must be provided in accordance with Sections E109.2.2 through E109.2.2.3.
3. At airports, terminal information systems must be provided in accordance with Section E110.3.

❖ This section is a reference to additional signage requirements found in subsequent sections. This was done in an effort to include all signage requirements in one section.

SECTION E108 BUS STOPS

E108.1 General. Bus stops shall comply with Sections E108.2 through E108.5.

❖ Bus stops and terminals provide access to public transportation and are considered an essential service; therefore, access in these facilities is of special concern.

E108.2 Bus boarding and alighting areas. Bus boarding and alighting areas shall comply with Sections E108.2.1 through E108.2.4.

❖ "Kneeling" buses (those with a lift) deploy a ramp to allow access by persons with physical disabilities. When new bus stop pads are constructed to allow use of these buses, the requirements in Sections E108.2.1 through E108.2.4 are applicable.

E108.2.1 Surface. Bus boarding and alighting areas shall have a firm, stable surface.

❖ To allow proper deployment of the ramp and approach by a person in a wheelchair, a firm, stable surface must be used for the pad. Typically, this will be some form of concrete or asphalt.

E108.2.2 Dimensions. Bus boarding and alighting areas shall have a clear length of 96 inches (2440 mm) minimum, measured perpendicular to the curb or vehicle roadway edge, and a clear width of 60 inches (1525 mm) minimum, measured parallel to the vehicle roadway.

❖ To allow maneuvering for both the ramp and wheelchair user, a bus stop pad should be a minimum of 96 inches (2438 mm) deep and 60 inches (1524 mm) wide.

E108.2.3 Connection. Bus boarding and alighting areas shall be connected to streets, sidewalks or pedestrian paths by an accessible route complying with Section 104.

❖ All bus stop pads must have an accessible route leading from public access points, such as the sidewalk, to the bus stop pad. A pad allowing for kneeling buses is not beneficial if the pad itself is not accessible.

E108.2.4 Slope. Parallel to the roadway, the slope of the bus boarding and alighting area shall be the same as the roadway, to the maximum extent practicable. For water drainage, a maximum slope of 1:48 perpendicular to the roadway is allowed.

❖ The bus stop pad should be as flat as possible to allow a person in a wheelchair to wait at this location comfortably. A maximum slope of one unit vertical in 48 units horizontal (1:48) is allowed for drainage of rainwater to the curb and gutter drainage system provided along the edge of the road.

E108.3 Bus shelters. Where provided, new or replaced bus shelters shall provide a minimum clear floor or ground space complying with ICC A117.1, Section 305, entirely within the shelter. Such shelters shall be connected by an accessible route to the boarding area required by Section E108.2.

❖ A wheelchair parking space [30 inches by 48 inches (762 mm by 1219 mm)] must be provided within the bus shelter. If the shelter is small enough to meet the alcove requirements in ICC A117.1, Section 305.7, additional space is required for maneuvering. In addition, an accessible route is required from the bus shelter to the bus stop (see Section E108.2).

E108.4 Signs. New bus route identification signs shall have finish and contrast complying with ICC A117.1. Additionally, to the maximum extent practicable, new bus route identification signs shall provide visual characters complying with ICC A117.1.

Exception: Bus schedules, timetables and maps that are posted at the bus stop or bus bay are not required to meet this requirement.

❖ The bus routes must be identified with visual characters that have light/dark contrast and a nonglare surface so they are easily readable. As much as practicable, the signage must also meet the requirements in ICC A117.1, Section 703.4, for character form and spacing, line spacing and mounting height. There is an exception for schedules, timetables and maps due to the amount of information provided.

E108.5 Bus stop siting. Bus stop sites shall be chosen such that, to the maximum extent practicable, the areas where lifts or ramps are to be deployed comply with Sections E108.2 and E108.3.

❖ Whenever possible, bus stop sites should be chosen based on efforts to comply with the previous requirements for bus stop pads and bus shelters. Site constraints, such as a steeply inclined street, may not always make this practical.

SECTION E109 TRANSPORTATION FACILITIES AND STATIONS

E109.1 General. Fixed transportation facilities and stations shall comply with the applicable provisions of Sections E109.2 and E109.3.

❖ Fixed transportation facilities and stations, such as train or light rail stations, provide access to public transportation and are considered an essential service; therefore, access in these facilities is of special concern.

E109.2 New construction. New stations in rapid rail, light rail, commuter rail, intercity rail, high speed rail and other fixed guideway systems shall comply with Sections E109.2.1 through E109.2.8.

❖ All new construction involved with fixed guideway systems, including stations and terminals, is required to comply with this section. Criteria for existing buildings is listed in Section E109.3.

E109.2.1 Station entrances. Where different entrances to a station serve different transportation fixed routes or groups of fixed routes, at least one entrance serving each group or route shall comply with Section 1104 and ICC A117.1.

❖ A fixed guideway system, such as a commuter train system, may have separate platforms or entrances for a group of routes. When this is the case, at least one accessible route leading from public access points (e.g., public sidewalk, public parking, etc.) to the transit platform utilized for each group must be provided. When transit systems are underground, an accessible route must provide access to each subway station. As much as possible, the accessible entrance should be the same as the main entrance to the facility.

E109.2.2 Signs. Signage in fixed transportation facilities and stations shall comply with Sections E109.2.2.1 through E109.2.2.3.

❖ Specific requirements are listed for tactile, identification and informational signs.

E109.2.2.1 Tactile signs. Where signs are provided at entrances to stations identifying the station or the entrance, or both, at least one sign at each entrance shall be tactile. A minimum of one tactile sign identifying the specific station shall be provided on each platform or boarding area. Such signs shall be placed in uniform locations at entrances and on platforms or boarding areas within the transit system to the maximum extent practicable. Tactile signs shall comply with ICC A117.1.

Exceptions:

1. Where the station has no defined entrance but signs are provided, the tactile signs shall be placed in a central location.

2. Signs are not required to be tactile where audible signs are remotely transmitted to hand-held receivers, or are user or proximity actuated.

❖ When identification signs are provided at a facility entrance, a tactile sign for the visually impaired must also be provided. A tactile sign must also be provided at each platform or boarding area. The location of tactile signage should be consistent throughout the rail system. Tactile signage must also include braille and meet the requirements of ICC A117.1, Section 703. Tactile signage is not required when information is provided by alternative audible means.

E109.2.2.2 Identification signs. Stations covered by this section shall have identification signs containing visual characters complying with ICC A117.1. Signs shall be clearly visible and within the sightlines of a standing or sitting passenger from within the train on both sides when not obstructed by another train.

❖ When identification signs are provided so that persons on the trains can identify the station, the signs must meet the requirements of ICC A117.1, Section 703.

E109.2.2.3 Informational signs. Lists of stations, routes and destinations served by the station which are located on boarding areas, platforms or mezzanines shall provide visual characters complying with ICC A117.1 Signs covered by this provision shall, to the maximum extent practicable, be placed in uniform locations within the transit system.

❖ Signs indicating stations, routes and destinations served by the train system accessed from that platform must comply with ICC A117.1, Section 703. The location of such signage should be consistent throughout the system.

E109.2.3 Fare machines. Self-service fare vending, collection and adjustment machines shall comply with ICC A117.1, Section 707. Where self-service fare vending, collection or adjustment machines are provided for the use of the general public, at least one accessible machine of each type provided shall be provided at each accessible point of entry and exit.

❖ Self-service fare machines must be accessible in accordance with ICC A117.1, Section 707. If self-service machines are provided, at least one accessible machine must be provided along the accessible route between the entrance and the platform.

E109.2.4 Rail-to-platform height. Station platforms shall be positioned to coordinate with vehicles in accordance with the applicable provisions of 36 CFR, Part 1192. Low-level platforms shall be 8 inches (250 mm) minimum above top of rail.

Exception: Where vehicles are boarded from sidewalks or street level, low-level platforms shall be permitted to be less than 8 inches (250 mm).

❖ The section specifies the maximum space or gap between the platform and the train car. The front wheels of wheelchairs are small, so it is important that the maximum gap not be exceeded. In existing conditions, alternatives are permitted in accordance with the exceptions.

E109.2.5 TTYs. Where a public pay telephone is provided in a transit facility (as defined by the Department of Transportation) at least one public TTY complying with ICC A117.1, Section 704.4, shall be provided in the station. In addition, where one or more public pay telephones serve a particular entrance to a transportation facility, at least one TTY telephone complying with ICC A117.1, Section 704.4, shall be provided to serve that entrance.

❖ At least one public telephone in a transit facility is required to have a permanent TTY in accordance with ICC A117.1, Section 704.4. When a bank of four or more is provided at an entrance, at least one is required to have a permanent TTY in accordance with ICC A117.1, Section 704.4 (see also commentary, Section E106.4.7).

E109.2.6 Track crossings. Where a circulation path serving boarding platforms crosses tracks, an accessible route complying with ICC A117.1 shall be provided.

Exception: Openings for wheel flanges shall be permitted to be $2^1/_2$ inches (64 mm) maximum.

❖ If people utilizing the train station must cross tracks to access the platform, it is necessary to provide an accessible route that takes into account the changes in elevation and gaps associated with such a route. Front wheels of wheelchairs are small, and if a person in a wheelchair got stuck while crossing the tracks, the consequences could be severe. If it is not possible to meet the criteria specified, an alternative accessible route must be provided.

E109.2.7 Public address systems. Where public address systems convey audible information to the public, the same or equivalent information shall be provided in a visual format.

❖ If a public address system is utilized within a facility to provide information to the passengers, there must be an alternative means for providing the same information to persons with a hearing impairment.

E109.2.8 Clocks. Where clocks are provided for use by the general public, the clock face shall be uncluttered so that its elements are clearly visible. Hands, numerals and digits shall contrast with the background either light-on-dark or dark-on-light. Where clocks are mounted overhead, numerals and digits shall comply with ICC A117.1, Section 703.4.

❖ Clocks provided in transportation facilities must be readily distinguishable. When clocks are provided overhead, the numerals must comply with ICC A117.1, Section 703.4.

E109.3 Existing facilities: key stations. In rapid rail, light rail, commuter rail, intercity rail, high-speed rail and other fixed guideway systems, altered stations and intercity rail and key stations, as defined under criteria established by the Department of Transportation in Subpart C of 49 CFR Part 37, shall comply with Sections E109.3.1 through E109.3.3.

❖ Existing guideway system facilities are required to meet a lesser degree of accessibility specified in Sections E109.3.1 through E109.3.3.

E109.3.1 Accessible route. At least one accessible route from an accessible entrance to those areas necessary for use of the transportation system shall be provided. The accessible route shall include the features specified in Section E109.2, except that escalators shall comply with Section 3005.2.2. Where technical infeasibility in existing stations requires the accessible route to lead from the public way to a paid area of the transit system, an accessible fare collection machine complying with Section E109.2.3 shall be provided along such accessible route.

❖ At least one accessible route is required throughout passenger areas in an existing facility. Existing escalators in subway stations are not required to meet the criteria for 32-inch-minimum (813 mm) clear width in Section E109.2.9. If it is technically infeasible to provide an accessible route from the entrance to a ticket sales area, some type of fare-taking machinery may be provided along the accessible route to the train platform.

E109.3.2 Platform and vehicle floor coordination. Station platforms shall be positioned to coordinate with vehicles in accordance with the applicable provisions of 36 CFR Part 1192. Low-level platforms shall be 8 inches (250 mm) minimum above top of rail.

> **Exception:** Where vehicles are boarded from sidewalks or street level, low-level platforms shall be permitted to be less than 8 inches (250 mm).

❖ In existing facilities, the vertical separation between the platform and the cars is slightly less restrictive than in new construction. The gap between the platform and the cars is only measured at accessible cars. Retrofitted cars may have an even larger variance. When train cars cannot meet the vertical separation and gap requirements, alternatives such as lifts and ramps may be utilized.

E109.3.3 Direct connections. New direct connections to other facilities shall have an accessible route complying with Section 3409.6 from the point of connection to boarding platforms and transportation system elements used by the public. Any elements provided to facilitate future direct connections shall be on an accessible route connecting boarding platforms and transportation system elements used by the public.

❖ New buildings that provide a direct connection to the existing transit systems must also have an accessible route to the transit system. The exceptions in Section 3408.6 are applicable.

SECTION E110 AIRPORTS

E110.1 New construction. New construction of airports shall comply with Sections E110.2 through E110.4.

❖ Airports provide access to public transportation and are considered an essential service; therefore, access in these facilities is of special concern.

E110.2 TTYs. Where public pay telephones are provided, at least one TTY shall be provided in compliance with ICC A117.1, Section 704.4. Additionally, if four or more public pay telephones are located in a main terminal outside the security areas, a concourse within the security areas or a baggage claim area in a terminal, at least one public TTY complying with ICC A117.1, Section 704.4, shall also be provided in each such location.

❖ At least one public telephone in an airport facility is required to have a permanent TTY in accordance with ICC A117.1, Section 704.4. When a bank of four or more phones is provided at an entrance, at least one is required to have a permanent TTY in accordance with ICC A117.1, Section 704.4 (see also commentary, Section E106.4.7).

E110.3 Terminal information systems. Where terminal information systems convey audible information to the public, the same or equivalent information shall be provided in a visual format.

❖ If a public address system is utilized within a facility to provide information to passengers, there must be an alternative means for providing the same information to persons with a hearing impairment.

E110.4 Clocks. Where clocks are provided for use by the general public, the clock face shall be uncluttered so that its elements are clearly visible. Hands, numerals and digits shall contrast with their background either light-on-dark or dark-on-light. Where clocks are mounted overhead, numerals and digits shall comply with ICC A117.1, Section 703.4.

❖ Clocks provided in transportation facilities must be readily distinguishable. When clocks are provided overhead, the numerals must comply with ICC A117.1, Section 703.4.

SECTION E111 QUALIFIED HISTORIC BUILDINGS AND FACILITIES

E111.1 General. Qualified historic buildings and facilities shall comply with Sections E111.2 through E111.5.

❖ Special considerations are given to registered historic buildings (see commentary, Section 3409.8).

E111.2 Qualified historic buildings and facilities. These procedures shall apply to buildings and facilities designated as historic structures that undergo alterations or a change of occupancy.

❖ This section provides a process to evaluate historically registered buildings in the event where accessibility requirements may adversely affect the historical significance of the building.

E111.3 Qualified historic buildings and facilities subject to Section 106 of the National Historic Preservation Act. Where an alteration or change of occupancy is undertaken to a qualified historic building or facility that is subject to Section 106 of the National Historic Preservation Act, the federal agency with jurisdiction over the undertaking shall follow the Section 106 process. Where the State Historic Preservation Officer or Advisory Council on Historic Preservation determines that compliance with the requirements for accessible routes, ramps, entrances or toilet facilities would threaten or destroy the historic significance of the building or facility, the alternative requirements of Section 3409 for that element are permitted.

❖ If a facility is subject to Section 106 of the National Historic Preservation Act, this section outlines the procedure to follow with the federal agency involved to allow the facility to utilize the alternatives offered in Section 3409.8 of the code.

E111.4 Qualified historic buildings and facilities not subject to Section 106 of the National Historic Preservation Act. Where an alteration or change of occupancy is undertaken to a qualified historic building or facility that is not subject to Section 106 of the National Historic Preservation Act, and the entity undertaking the alterations believes that compliance with the requirements for accessible routes, ramps, entrances or toilet facilities would threaten or destroy the historic significance of the building or facility, the entity shall consult with the State Historic Preservation Officer. Where the State Historic Preservation Officer determines that compliance with the accessibility requirements for accessible routes, ramps, entrances or toilet facilities would threaten or destroy the historical significance of the building or facility, the alternative requirements of Section 3409 for that element are permitted.

❖ If a facility is not subject to Section 106 of the National Historic Preservation Act, this section outlines the procedure to follow with the state historic preservation officer to allow the facility to utilize the alternatives offered in Section 3409.8 of the code.

E111.4.1 Consultation with interested persons. Interested persons shall be invited to participate in the consultation process, including state or local accessibility officials, individuals with disabilities and organizations representing individuals with disabilities.

❖ Some type of public forum or meeting must be held to allow for the input of interested parties that may wish to utilize the facilities offered in the historic building.

E111.4.2 Certified local government historic preservation programs. Where the State Historic Preservation Officer has delegated the consultation responsibility for purposes of this section to a local government historic preservation program that has been certified in accordance with Section 101 of the National Historic Preservation Act of 1966 [(16 U.S.C. 470a(c)] and implementing regulations (36 CFR 61.5), the responsibility shall be permitted to be carried out by the appropriate local government body or official.

❖ Where the state has certified a local group in accordance with the National Historic Preservation Act to oversee activity within a historic district or area, that group can take on the role of the state historic preservation officer.

E111.5 Displays. In qualified historic buildings and facilities, where alternative requirements of Section 3409 are permitted, displays and written information shall be located where they can be seen by a seated person. Exhibits and signs displayed horizontally shall be 44 inches (1120 mm) maximum above the floor.

❖ Displays, exhibits and signs in historic buildings must be located so that a person in a wheelchair can see them.

SECTION E112
REFERENCED STANDARDS

DOJ 36Americans with Disabilities ActE109.2.4,
CFR Part 1192(ADA) Accessibility GuidelinesE109.3.2
for Transportation Vehicles
(ADAAG). Washington, D.C.:
Department of Justice, 1991

DOT 49Transportation ServicesE109.3,
CFR Part 37for Individuals withE109.3.2,
Disabilities (ADA),E109.4
Washington, D.C.:
Department of
Transportation, 1999

DOJ 28CFR Part 36, AmericansE109.4
with Disabilities Act (ADA).
Washington, D.C.:
Department of Justice, 1991

ICC/ANSIAccessible and UsableE101.2, et al
A117.1-98Buildings and Facilities

16 USCNational Historic E111.2,
Sec. 470Preservation Act E111.3,
E111.3.2

Y3.H626 2PNational Historic PreservationE111.2,
43/933Act of 1966, as amendedE111.3,
3rd Edition, Washington, D.C.:E111.3.2
U.S. Government Printing
Office, 1993

❖ This section lists the standards that are referenced in Appendix E. They are listed here and not in Chapter 35 of the code, since appendices are not considered part of the code unless specifically referenced in the adopting ordinance.

Bibliography

The following resource materials are referenced in this chapter or are relevant to the subject matter addressed in this chapter.

36 CFR, Parts 1190 and 1191 Draft, *The Americans with Disabilities Act Accessibility Guidelines* (ADAAG). Washington, DC: Architectural and Transportation Barriers Compliance Board, April 2, 2002.

"Accessibility and Egress for People with Physical Disabilities." CABO Board for the Coordination of the Model Codes Report, October 5, 1993.

DOJ 28 CFR, Part 36-91, *Americans with Disabilities Act* (ADA). Washington, DC: Department of Justice, 1991.

DOJ 28 CFR, Part 36-91 (Appendix A), *ADA Accessibility Guidelines for Buildings and Facilities* (ADAAG). Washington, DC: Department of Justice, 1991.

DOJ 36 CFR Part 1192/DOT 49 CFR Part 38, *ADA Accessibility Guidelines for Transportation Vehicles* (ADAAG). Washington, DC: Department of Justice, 1991.

DOT 49 CFR Part 37, *Transportation Services for Individuals with Disabilities* (ADA). Washington, DC: Department of Transportation, 1999.

FED-STD-795-88, *Uniform Federal Accessibility Standards*. Washington, DC: General Services Administration; Department of Defense; Department of Housing and Urban Development; U.S. Postal Service, 1988.

"Final Report, Recommendations for a New ADAAG." ADAAG Review Federal Advisory Committee, September 30, 1996.

Architectural and Transportation Barriers Compliance Board, 36 CFR Parts 1190 and 1191 Draft, *The Americans with Disabilities Act Accessibility Guidelines (ADAAG),* Washington, DC., April 2, 2002

ICC A117.1-98, *Accessible and Usable Buildings and Facilities*. Falls Church, VA: International Code Council, 1998.

"National Historic Preservation Act of 1966, as amended." 3rd ed. Washington, DC: U.S. Government Printing Office, 1993, Y3.H626 2P 43/933.

Appendix F: Rodentproofing

The provisions contained in this appendix are not mandatory unless specifically referenced in the adopting ordinance.

General Comments

The requirements of this appendix apply only where it has been specifically adopted by the governing jurisdiction, as stated in Section 101.2.1.

The provisions of this appendix are minimum mechanical methods to prevent the entry of rodents. If a food supply is available, a rodent problem cannot be eliminated solely by these methods. However, removal of the food supply and general cleanliness and maintenance of the premises will have a significant impact.

Purpose

The purpose of this appendix is to provide regulations for rodentproofing for health purposes.

SECTION F101 GENERAL

F101.1 General. Buildings or structures and the walls enclosing habitable or occupiable rooms and spaces in which persons live, sleep or work, or in which feed, food or foodstuffs are stored, prepared, processed, served or sold, shall be constructed in accordance with the provisions of this section.

❖ In an effort to protect the public health and welfare and to prevent the spread of disease, this section requires all rooms and spaces used for living, sleeping or working and all spaces used for the storage, preparation or serving of food to be protected from rodents in the manner described in the code. Note that a building used for the storage of materials other than foodstuffs is not required to comply with the requirements of this section.

F101.2 Foundation wall ventilation openings. Foundation wall ventilator openings shall be covered for their height and width with perforated sheet metal plates no less than 0.070 inch (1.8 mm) thick, expanded sheet metal plates not less than 0.047 inch (1.2 mm) thick, cast iron grills or grating, extruded aluminum load-bearing vents or with hardware cloth of 0.035 inch (0.89 mm) wire or heavier. The openings therein shall not exceed $^{1}/_{4}$ inch (6.4 mm).

❖ The purpose of this section is to protect ventilation openings in the foundation walls from the entry of rodents. The thickness of the metal shields and the maximum openings that are specified in this section are necessary for the shields to be effective.

F101.3 Foundation and exterior wall sealing. Annular spaces around pipes, electric cables, conduits, or other openings in the walls shall be protected against the passage of rodents by closing such openings with cement mortar, concrete masonry or noncorrosive metal.

❖ During construction, the openings in the foundation and exterior walls that are made for pipes, electrical cables, conduits and the like are significantly larger than the component that passes through the wall. These openings around the component are required to be closed for rodentproofing.

F101.4 Doors. Doors on which metal protection has been applied shall be hinged so as to be free swinging. When closed, the maximum clearance between any door, door jambs and sills shall not be greater than $^{3}/_{8}$ inch (9.5 mm).

❖ This section addresses wood doors located at ground or basement floor levels. Solid sheet metal protection is needed to stop rodents from gnawing their way through the wood into areas where food is located. The limitation on the clearance between the door and the adjacent door jamb and sill is to deter rodent entry.

F101.5 Windows and other openings. Windows and other openings for the purpose of light or ventilation located in exterior walls within 2 feet (610 mm) above the existing ground level immediately below such opening shall be covered for their entire height and width, including frame, with hardware cloth of at least 0.035 inch (0.89 mm) wire or heavier.

❖ The requirement for wire screens or "hardware cloth" for rodent protection applies only to openings 2 feet (610 mm) or less from the ground. This is a minimum requirement and will not keep all rodents out of the building.

F101.5.1 Rodent-accessible openings. Windows and other openings for the purpose of light and ventilation in the exterior walls not covered in this chapter, accessible to rodents by way of exposed pipes, wires, conduits and other appurtenances, shall be covered with wire cloth of at least 0.035 inch (0.89 mm) wire. In lieu of wire cloth covering, said pipes, wires, conduits and other appurtenances shall be blocked from rodent usage by installing solid sheet metal guards 0.024 inch (0.61 mm) thick or heavier. Guards shall be fitted around pipes, wires, conduits or other appurtenances. In addition, they shall be fastened securely to and shall extend perpendicularly from the exterior wall for a minimum distance of 12 inches (305 mm) beyond and on either side of pipes, wires, conduits or appurtenances.

❖ This section addresses screen requirements for windows or other openings that are higher than 2 feet (610 mm) from grade and are made rodent accessible by pipes, wires, conduits and the like. The window or other opening is required to be screened or rodent guards are to be installed on the pipes, wires or conduits so that the window or other opening is not accessible to rodents.

F101.6 Pier and wood construction.

❖ The requirements of Sections F101.6.1 and F101.6.2 apply only if the building does not have a continuous foundation wall of either treated wood, concrete or coated metal. An example of such a building is a post and beam wood-frame building with a slab on grade. Also note that these requirements apply if the building contains habitable or occupiable rooms in which persons live, sleep or work, or in which foodstuffs are stored, prepared, processed, served or sold. Thus, a storage or utility building used for storage of materials other than foodstuffs is not required to meet this section, even though it does not have a continuous foundation wall.

F101.6.1 Sill less than 12 inches above ground. Buildings not provided with a continuous foundation shall be provided with protection against rodents at grade by providing either an apron in accordance with Section F101.6.1.1 or a floor slab in accordance with Section 101.6.1.2.

❖ See the commentary to Section F101.6 for an explanation of when the requirements of this section apply. This section cites two alternative methods that meet the requirements: an apron or a concrete grade floor. Either of these methods provides effective rodent protection.

F101.6.1.1 Apron. Where an apron is provided, the apron shall not be less than 8 inches (203 mm) above, nor less than 24 inches (610 mm) below, grade. The apron shall not terminate below the lower edge of the siding material. The apron shall be constructed of an approved nondecayable, water-resistant rat-proofing material of required strength and shall be installed around the entire perimeter of the building. Where constructed of masonry or concrete materials, the apron shall not be less than 4 inches (102 mm) in thickness.

❖ The apron is usually a concrete, masonry or treated wood wall. Each material meets the requirement of this section. The extension into the ground is to prevent rodents from burrowing underground into the building. Note that a concrete floor slab is not required where the apron is used for rodent protection in accordance with Section F101.6.2.

F101.6.1.2 Grade floors. Where continuous concrete grade floor slabs are provided, open spaces shall not be left between the slab and walls and openings in the slab shall be protected.

❖ This is the second option for rodent protection in Section 101.6.1. When a grade floor that meets the requirements of this section is installed, the apron described in Section F101.6.1.2 is not required.

The grade floor is required to have any openings protected from rodents. For example, if the floor slab has an opening where drain pipes go through it, the area around the pipe is to be sealed with material that will keep rodents out of the building.

F101.6.2 Sill at or above 12 inches above ground. Buildings not provided with a continuous foundation and which have sills 12 or more inches (305 mm) above the ground level shall be provide with protection against rodents at grade in accordance with any of the following:

1. Section F101.6.1.1 or F101.6.1.2.
2. By installing solid sheet metal collars at least 0.024 inch (0.6 mm) thick at the top of each pier or pile and around each pipe, cable, conduit, wire, or other item which provides a continuous pathway from the ground to the floor or
3. By encasing the pipes, cables, conduits or wires in an enclosure constructed in accordance with Section F101.6.1.1.

❖ See the commentary to Section F101.6 for an explanation of when the requirements of this section apply. Where the sill height is 12 inches (305 mm) or more above the ground, this section provides two additional acceptable methods for rodent protection versus methods for instances where the sill height is less than 12 inches (305 mm).

Methods 2 and 3 described in this section provide the rodent barrier directly around the pipes, cables, conduits or wires that would otherwise be a continuous pathway from the ground to the raised floor.

Appendix G: Flood-Resistant Construction

The provisions contained in this appendix are not mandatory unless specifically referenced in the adopting ordinance.

General Comments

The provisions contained in this appendix are not mandatory unless specifically referenced in the adopting ordinance, as stated in Section 101.2.1. This appendix is intended to fulfill the flood-plain management and administrative requirements of the National Flood Insurance Program (NFIP) that are not included in the code. Communities that adopt the code and this appendix without modification will meet the minimum requirements of NFIP as set forth in Title 44 of the Code of Federal Regulations (CFR). This appendix includes administrative requirements of NFIP and requirements concerning modifications to watercourses; permits for flood hazard area development; conditions for the issuance of variances from flood-plain management requirements and site improvements; subdivision planning and installation of manufactured homes; recreation vehicles and tanks in flood hazard areas. It is important to note that many states and communities regulate flood-plain development to higher standards than the minimum requirements of the code and this appendix. Prior to adopting the code or this appendix, communities are advised to consult their state NFIP coordinator or Federal Emergency Management Agency (FEMA) regional office to determine additional actions that may be necessary to provide for continued participation in NFIP.

Purpose

The purpose of this appendix is to provide optional criteria for flood-resistant construction. A jurisdiction that wants to make this appendix mandatory needs to include it in their adoption ordinance. See page v of the code for a sample ordinance for adoption.

SECTION G101 ADMINISTRATION

G101.1 Purpose. The purpose of this appendix is to promote the public health, safety and general welfare and to minimize public and private losses due to flood conditions in specific flood hazard areas through the establishment of comprehensive regulations for management of flood hazard areas designed to:

1. Prevent unnecessary disruption of commerce, access and public service during times of flooding;
2. Manage the alteration of natural flood plains, stream channels and shorelines;
3. Manage filling, grading, dredging and other development which may increase flood damage or erosion potential;
4. Prevent or regulate the construction of flood barriers which will divert floodwaters or which can increase flood hazards; and
5. Contribute to improved construction techniques in the flood plain.

❖ This section sets forth the purposes of this appendix. Communities that administer flood-plain management regulations, such as in this appendix, achieve multiple objectives beyond reducing physical damage to buildings.

G101.2 Objectives. The objectives of this appendix are to protect human life, minimize the expenditure of public money for flood control projects, minimize the need for rescue and relief efforts associated with flooding, minimize prolonged business interruption, minimize damage to public facilities and utilities, help maintain a stable tax base by providing for the sound use and development of flood-prone areas, contribute to improved construction techniques in the flood plain and ensure that potential owners and occupants are notified that property is within flood hazard areas.

❖ This appendix, combined with the provisions in the code, achieve several objectives intended to reduce the impacts of flooding. Management of flood hazard areas in ways that reduce exposure to damage also protects health and safety. Additionally, the provisions of this appendix promote sustainable development in communities subject to flooding. Flooding has both short-term and long-term impacts. Some impacts are obvious, such as damaged homes and businesses. Others are less apparent, including increases in flood levels due to changes in the flood plain; scour and erosion; impaired public and private water and sewage systems and reductions in the natural and beneficial functions of flood plains, including wetland areas.

G101.3 Scope. The provisions of this appendix shall apply to all proposed development in a flood hazard area established in Section 1612 of this code.

❖ In communities that participate in NFIP, its minimum requirements must be adopted and applied to all development in flood hazard areas. The definition of "Development," as stated in Section G201.2, is inclusive: any man-made change to improved or unimproved real estate, including but not limited to buildings or other structures, temporary or permanent storage of materials, mining, dredging, filling, grading, paving, excavations, operations and other land-disturbing activities. This broad definition captures all activities that take place in flood hazard areas that are subject to the requirements of the code or this appendix. Because NFIP requires regulation of all development within special flood hazard areas that could impact on flooding, a code that applies only to buildings or structures does not fulfill the minimum requirements for participation in the program.

G101.4 Violations. Any violation of a provision of this appendix, or failure to comply with a permit or variance issued pursuant to this appendix or any requirement of this appendix, shall be handled in accordance with Section 113.

❖ Violations of the provisions of this appendix are to be corrected or adjudicated under the provisions of Section 113. Communities participating in NFIP enter into an agreement with FEMA: In exchange for the availability of federally backed flood insurance and many forms of federal disaster assistance, communities agree to adopt and effectively enforce the flood-plain management and administrative provisions of NFIP. A significant part of community responsibility takes the form of enforcing flood-plain management and other flood damage reduction ordinances and laws. The procedures set forth in Section 113 are to be followed if activities take place in a flood hazard area without a permit, or if a permit applicant fails to comply with a condition of a permit or variance. If reasonable enforcement actions do not result in compliance, the community is encouraged to contact its state NFIP coordinator or appropriate FEMA regional office for further guidance.

SECTION G102 APPLICABILITY

G102.1 General. This appendix, in conjunction with the *International Building Code*, provides minimum requirements for development located in flood hazard areas, including the subdivision of land, installation of utilities, placement and replacement of manufactured homes, new construction and repair, reconstruction, rehabilitation, or additions to new construction and substantial improvement of existing buildings and structures, including restoration after damage.

❖ The definition of "Development" in Section G201.2 encompasses a wide array of activities that are to be regulated if they are proposed to take place within a designated flood hazard area. Some activities have the potential to increase flooding on other properties. Other activities, if planned in a way that recognizes flood hazards, have the potential to allow reasonable use of land without placing people and site improvements at significant risk.

G102.2 Establishment of flood hazard areas. Flood hazard areas are established in Section 1612.3 of the *International Building Code*, adopted by the governing body on [INSERT DATE].

❖ The flood hazard areas established in Section 1612.3 are subject to the additional provisions of this appendix. The flood hazard map to be adopted is, at a minimum, the flood insurance rate map (FIRM), supported by the flood insurance study prepared by FEMA. A community participating in NFIP may specify another map, provided it designates flood hazard areas that are the same or more extensive and regulatory flood elevations that are the same or higher than those shown on the FIRM. Regardless of which map is specified, flood elevations are the design flood elevations. From time to time, FEMA revises and republishes FIRMs. A formal review process is followed, including community and public comment. When new FIRMs are published, communities are required to use them. The flood hazard areas shown on maps prepared by FEMA are determined using the base flood, which is defined as having a 1-percent chance (one chance in 100) of occurring in any given year. The maps are not intended to show the worst case flood, nor the "flood of record," which usually refers to the most severe flood in the history of the community. Although a 1-percent chance seems fairly remote, larger floods occur regularly throughout the United States. Application of the provisions of this appendix is not intended to prevent or eliminate all future flood damage. These provisions are intended to represent a reasonable balance of the knowledge and awareness of flood hazards, methods to guide development to less hazard-prone locations, methods of design and construction intended to resist flood damage and each community's and landowner's reasonable expectations to use the land. Some coastal communities have FIRMs that show areas that are designated as units of the Coastal Barrier Resource System (CBRS) established by the Coastal Barrier Resource Act (CoBRA) of 1982 and subsequent amendments. Within these designated areas, NFIP is prohibited from offering flood insurance on new or substantially improved buildings. The community is responsible for application of the flood-resistant provisions of the code and this appendix in all designated flood hazard areas, whether or not flood insurance is made available.

SECTION G103
POWERS AND DUTIES

G103.1 Permit applications. The building official shall review all permit applications to determine whether proposed development sites will be reasonably safe from flooding. If a proposed development site is in a flood hazard area, all site development activities, including grading, filling, utility installation and drainage modification, and all new construction and substantial improvements (including the placement of prefabricated buildings and manufactured homes) shall be designed and constructed with methods, practices and materials that minimize flood damage and that are in accordance with this code and ASCE 24.

❖ The building official is empowered to examine all permit applications to determine if the proposed activities will take place in designated flood hazard areas on the community's flood hazard map. Such activities are to be designed and constructed in accordance with both the code and this appendix, as applicable.

G103.2 Other permits. It shall be the responsibility of the building official to assure that approval of a proposed development shall not be given until proof that necessary permits have been granted by federal or state agencies having jurisdiction over such development.

❖ Other federal, state and local regulatory authorities may have jurisdiction over activities within designated flood hazard areas. Examples at the federal level include permitting under Section 404 of the Clean Water Act of 1972 and Section 10 of the Rivers and Harbors Act of 1899 and consultation or permitting under the Endangered Species Act of 1973. State and regional agencies may also regulate activities in flood hazard areas, including activities that impact wetlands; forestry resources; dunes; the coastal zone; subaquatic vegetation; threatened and endangered species; navigation and waterways. The intent is to provide coordination among various levels of government and to avoid added costs to applicants in situations where multiple permits are required. Building officials may satisfy this requirement by either withholding the permit until other permits have been obtained or by issuing the permit contingent on the applicant obtaining other specified permits.

G103.3 Determination of design flood elevations. If design flood elevations are not specified, the building official is authorized to require the applicant to:

1. Obtain, review and reasonably utilize data available from a federal, state or other source, or
2. Determine the design flood elevation in accordance with accepted hydrologic and hydraulic engineering techniques. Such analyses shall be performed and sealed by a registered design professional. Studies, analyses and computations shall be submitted in sufficient detail to allow review and approval by the building official. The accuracy of data submitted for such determination shall be the responsibility of the applicant.

❖ Many FIRMs show flood hazard areas without specifying the base flood elevations (BFE). These areas, often referred to as "unnumbered A or V Zones," are subject to the flood-plain management requirements of the code and the appendix, although the minimum height to which buildings and structures are to be elevated has not been determined by FEMA. An important step in regulating these flood hazard areas is determination of the design flood elevation (DFE). As defined in the code, at a minimum the DFE is the BFE shown on a community's FIRM. In some instances flood elevation information may have been developed by sources other than FEMA, including other federal or state agencies. The building official is to obtain the information or require the permit applicant to do so. If flood elevation information is not available, the building official may require the applicant to develop the DFE in accordance with accepted engineering practices. Local officials unfamiliar with establishing DFEs in unnumbered A and V Zones are encouraged to contact the state NFIP coordinator or appropriate FEMA regional office. For additional guidance, refer to FEMA 265, *Managing Floodplain Development in Approximate Zone A Areas: A Guide for Obtaining and Developing Base (100-Year) Flood Elevations.*

G103.4 Activities in riverine flood hazard areas. In riverine situations, until a regulatory floodway is designated, the building official shall not permit any new construction, substantial improvement or other development, including fill, unless the applicant demonstrates that the cumulative effect of the proposed development, when combined with all other existing and anticipated development, will not increase the design flood elevation more than 1 foot (305 mm) at any point within the community.

❖ Although FEMA has provided floodways along many rivers and streams shown on a community's FIRM, many other riverine flood hazard areas have not had floodways designated. In these areas the potential effects that flood plain activities may have on flood elevations have not been evaluated. If FEMA has not designated a regulatory floodway on a community's FIRM, the community is responsible for regulating development so as not to increase flood elevations by more than 1 foot (305 mm) at any point in the community. In effect, this means a community must either prepare a hydraulic analysis for the proposed activity or require permit applicants to do so. Several states have more restrictive requirements, which can be determined by contacting the state NFIP coordinator.

G103.5 Floodway encroachment. Prior to issuing a permit for any floodway encroachment, including fill, new construction, substantial improvements and other development or land-disturbing activity, the building official shall require submission of a certification, along with supporting technical data, that demonstrates that such development will not cause any increase of the level of the base flood.

❖ The floodway is that part of the riverine flood plain that must be reserved in order to convey the base flood without cumulatively increasing the water surface elevation more than a designated height specified in the flood insurance study provided by FEMA. Generally, this designated height is 1 foot (305 mm), although some states and communities have a more restrictive requirement. Development in the floodway could obstruct flood flows and increase flood levels, causing additional damage. Usually, the floodway is where the water will be deepest and move the fastest. As required in Section G103.5, communities are to prohibit any floodway encroachment, including fill, new construction and substantial improvement, if such activities will cause any increase in flood levels. Permit applications are to include a certification supported with engineering analyses to demonstrate the anticipated impacts. To be acceptable, the certification must demonstrate that the proposed development will not impact base flood elevations, floodway elevations or floodway widths and is to be signed and sealed by a registered professional engineer. The building official is to review the certification and determine that it meets accepted engineering practices. If necessary to make this determination, the building official is advised to seek technical assistance from the community's engineer, the state NFIP coordinator or the FEMA regional office. In limited situations, a community may decide to permit development in the floodway that causes an increase in flood elevations. This may be appropriate for dams or other water resource structures and for bridges and culverts where the most cost-effective alternative results in an increase in flood elevations. Section 103.5 allows for these activities, provided the community or applicant requests and obtains a conditional letter of map revision (CLOMR) and a floodway revision from FEMA. The proposed development may not increase flooding of existing buildings and all impacted property owners must be notified. A permit for the development may be issued after the CLOMR is approved by FEMA and after the community adopts the increased flood elevations. When the project is completed, the community is required to submit as-built certifications to FEMA so that a final map revision can be issued. For guidance on preparing and certifying revisions to NFIP map products, see FEMA Form 81-89 Series, *Revisions to National Flood Insurance Rate Maps: Application/Certification Forms and Instructions for Conditional Letters of Map Revision, Letters of Map Revision and Physical Map Revisions*. Although each floodway proposal is to be reviewed carefully and some states have specific and more restrictive regulations that apply within floodways, there are some activities and uses that may be permitted without extensive engineering analysis. The most important factor is that there is no fill and grading that changes the shape of the land. Possible uses include:

- Agricultural uses not involving buildings or structures;
- At-grade uses such as parking, loading areas and small airport landing strips;
- Passive recreational uses, such as hiking, biking, horse trails, wildlife and nature preserves and hunting and fishing areas;
- Active recreational uses (where improvements are anchored to prevent flotation), such as picnic and playground areas, ball fields, boat launch ramps, swimming areas, target shooting ranges and similar uses and
- Uses incidental to residential buildings, such as lawns, gardens, parking areas, playgrounds and tot lots.

G103.5.1 Floodway revisions. A floodway encroachment that increases the level of the base flood is authorized if the applicant has applied for a conditional Flood Insurance Rate Map (FIRM) revision and has received the approval of the Federal Emergency Management Agency (FEMA).

❖ Under certain circumstances, a permit may be issued for a proposed floodway encroachment that, based on analysis and supporting technical data, will cause an increase in the level of the base flood. The community is to require that the analysis be submitted to FEMA in accordance with the guidance described in Section 103.5 (FEMA Form 81-89). The permit may be issued If FEMA approves the proposal by issuance of a conditional revision to the flood hazard area map. Prior to issuance of a final FIRM revision, FEMA will require that the applicant submit post-construction documentation to demonstrate that the work was conducted as proposed.

G103.6 Watercourse alteration. Prior to issuing a permit for any alteration or relocation of any watercourse, the building official shall require the applicant to provide notification of the proposal to the appropriate authorities of all affected adjacent government jurisdictions, as well as appropriate state agencies. A copy of the notification shall be maintained in the permit records and submitted to FEMA.

❖ Relocation or alteration of the channel of a river or stream can significantly and adversely impact the flood plain. Because such impact may extend a significant distance from a proposed relocation or alteration, NFIP requires notification of adjacent communities, appropriate state agencies and FEMA.

G103.6.1 Engineering analysis. The building official shall require submission of an engineering analysis which demonstrates that the flood-carrying capacity of the altered or relocated portion of the watercourse will not be decreased. Such watercourses shall be maintained in a manner which preserves the channel's flood-carrying capacity.

❖ Prior to issuing a permit that allows modification of a waterway, the applicant is to submit a hydraulic analysis demonstrating that the ability of the altered channel and flood plain to carry flood discharges is not diminished. An important condition of a permit, if issued, is that the applicant or other designated entity is to maintain the altered channel in order to preserve its flood-carrying capacity and to avoid increasing future flood risks.

G103.7 Alterations in coastal areas. Prior to issuing a permit for any alteration of sand dunes and mangrove stands in flood hazard areas subject to high velocity wave action, the building official shall require submission of an engineering analysis which demonstrates that the proposed alteration will not increase the potential for flood damage.

❖ As part of the review of proposed site work in coastal areas, a community is to examine proposals for possible impacts to sand dunes or mangrove stands, which mitigate the effects of coastal flooding. If a proposal is to alter a sand dune or mangrove stand, an engineering analysis of the effect on flood damages is required as part of the documentation. If engineering analysis indicates that the potential for flood damage will be increased due to the proposed alteration, the building official is to require the permit applicant to make necessary modifications to the proposed site work to avoid such increases.

G103.8 Records. The building official shall maintain a permanent record of all permits issued in flood hazard areas, including copies of inspection reports and certifications required in Section 1612.

❖ The building official's responsibility to maintain department records is established in Section 104.7. This section extends that responsibility to require that records for permits in flood hazard areas be maintained permanently. Specifically identified for retention are the inspection reports and certifications required to be submitted in Section 1612. As a condition for participation in NFIP, a community agrees that FEMA or its designee may examine records relating to administration of the community's flood management regulations.

SECTION G104
PERMITS

G104.1 Required. Any person, owner or authorized agent who intends to conduct any development in a flood hazard area shall first make application to the building official and shall obtain the required permit.

❖ The definition of "Development" in Section G201 encompasses a wide array of human activities that may be proposed within designated flood hazard areas. Communities are to require that all such activities are authorized by permits issued in compliance with the applicable flood-plain management requirements of the code and this appendix.

G104.2 Application for permit. The applicant shall file an application in writing on a form furnished by the building official. Such application shall:

1. Identify and describe the development to be covered by the permit.
2. Describe the land on which the proposed development is to be conducted by legal description, street address or similar description that will readily identify and definitely locate the site.
3. Include a site plan showing the delineation of flood hazard areas, floodway boundaries, flood zones, design flood elevations, ground elevations, proposed fill and excavation and drainage patterns and facilities.
4. Indicate the use and occupancy for which the proposed development is intended.
5. Be accompanied by construction documents, grading and filling plans and other information deemed appropriate by the building official.
6. State the valuation of the proposed work.
7. Be signed by the applicant or the applicant's authorized agent.

❖ This section details the information to be shown on or included in an application for a permit to conduct activities within a flood hazard area. The site plan is to show sufficient detail and information about the designated flood hazard area, including floodway, to allow for a complete review of the proposed activities. The horizontal boundary of the designated flood hazard area shown on the community's flood hazard map is an approximate boundary based on topographic information that was available when the map was prepared by FEMA. The determination of the actual boundary of the designated flood hazard is to be based on the best topographic information that is available at the time of the application. All land area below the DFE is subject to regulation even if it is not shown on the community's flood hazard map as being within the designated flood hazard area. Along with necessary flood hazard information, the storm drainage conveyance requirements are to be shown on the site plan.

G104.3 Validity of permit. The issuance of a permit under this appendix shall not be construed to be a permit for, or approval of, any violation of this appendix or any other ordinance of the jurisdiction. The issuance of a permit based on submitted documents and information shall not prevent the building official from requiring the correction of errors. The building official is authorized to prevent occupancy or use of a structure or site which is in violation of this appendix or other ordinances of this jurisdiction.

❖ This section clarifies that it is the owner's responsibility to comply with the code. Permits are approved based on a review of the materials submitted and a determination that the proposed activities are in accordance with the provisions of the code. If errors or deviations are made during construction, or it is discovered that the information and documentation submitted to obtain the permit were in error, the building official has the authority to require correction and compliance.

G104.4 Expiration. A permit shall become invalid if the proposed development is not commenced within 180 days after its issuance, or if the work authorized is suspended or abandoned for a period of 180 days after the work commences. Extensions shall be requested in writing and justifiable cause demonstrated. The building official is authorized to grant, in writing, one or more extensions of time, for periods not more than 180 days each.

❖ Permits issued for activities in a flood hazard area are to expire after 180 days if work either has not commenced during that period of time or is abandoned for any 180-day period after it commences. The building official may grant extensions to the permit for up to 180 days at a time. Section 1612.2 defines "Start of construction" as the date of permit issuance for new construction and substantial improvements to existing structures, provided the actual start of construction, repair, reconstruction, rehabilitation, addition placement or other improvement is within 180 days after the date of issuance. The actual start of construction means the first placement of permanent construction of a building (including a manufactured home) on a site, such as the pouring of a slab or footings; installation of pilings or construction of columns.

Building officials may not grant permit extensions in situations where construction has not begun during the 180-day period if FEMA has issued a revised FIRM that shows increases in flood hazard areas or DFEs at the location of the permitted activity. A new permit is required if the revised FIRM changes the flood risk information at the site of the permit. The new application is to be reviewed based on the revised flood risk information.

G104.5 Suspension or revocation. The building official is authorized to suspend or revoke a permit issued under this appendix wherever the permit is issued in error or on the basis of incorrect, inaccurate or incomplete information, or in violation of any ordinance or code of this jurisdiction.

❖ A permit issued in error or based on incorrect, inaccurate or incomplete information is to be suspended or revoked. This provision does not preclude issuance of a new or revised permit that is based on accurate information. Violation of a permit or permit condition may also prompt suspension or revocation.

SECTION G105
VARIANCES

G105.1 General. The board of appeals established pursuant to Section 112 shall hear and decide requests for variances. The board of appeals shall base its determination on technical justifications, and has the right to attach such conditions to variances as it deems necessary to further the purposes and objectives of this appendix and Section 1612.

❖ Section 112 empowers the board of appeals to hear and decide requests for variances. Variances from the flood-plain management requirements of the code and this appendix may place people and property at significant risk; therefore, communities are cautioned to carefully evaluate the impacts of issuing a variance from the flood-resistant construction provisions of the code and this appendix, particularly the requirements to elevate or floodproof buildings to the DFE. The impacts to be evaluated include impacts on the site, the permit applicant, other parties that may be affected, such as adjacent property owners, and the community. Flood-plain development that is not undertaken in accordance with the flood-resistant construction provisions of the code and this appendix will be exposed to increased flood damages. As a consequence, flood insurance premium rates will be significantly higher. Variance decisions made by the board of appeals are to be based solely on technical justifications outlined in this section, not on the personal circumstances of an owner or applicant.

Applicants sometimes request variances to the minimum elevation requirements for the lowest floor of buildings in flood hazard areas. Such requests may be based on the need to improve access for the disabled and elderly. Generally, variances of this nature are not to be granted since these are personal circumstances that will change as the property changes ownership. Not only would persons of limited mobility be at risk from flooding, but the building would continue to be exposed to flood damage long after the personal need for a variance changes. More appropriate alternatives are to be considered to serve the needs of disabled or elderly persons, such as varying setbacks to allow construction on less flood-prone portions of sites or installing personal elevators.

G105.2 Records. The building official shall maintain a permanent record of all variance actions, including justification for their issuance.

❖ The permanent records of the community that document official actions on permit applications and variance requests concerning flood-plain management requirements are to include the justifications on which variance determinations are based. These records will be examined periodically by FEMA or its designee in an effort to determine if the community is effectively administering its flood management provisions consistent with the minimum requirements of NFIP.

G105.3 Historic structures. A variance is authorized to be issued for the repair or rehabilitation of a historic structure upon a determination that the proposed repair or rehabilitation will not preclude the structure's continued designation as a historic structure, and the variance is the minimum necessary to preserve the historic character and design of the structure.

Exception: Within flood hazard areas, historic structures that are not:

a. Listed or preliminarily determined to be eligible for listing in the National Register of Historic Places; or

b. Determined by the Secretary of the U.S. Department of Interior as contributing to the historical significance of a registered historic district or a district preliminarily determined to qualify as an historic district; or

c. Designated as historic under a state or local historic preservation program that is approved by the Department of Interior.

❖ This provision recognizes the importance and value of historic buildings and structures. If an owner proposes repair or restoration of an historic building or structure that would otherwise be found to be a substantial improvement, the board of appeals may grant a variance to perform the work without requiring full compliance with the flood-resistant provisions of the code and this appendix. This is not to say that work towards flood-resistant provisions should not be performed. It is intended that an historic building should be brought into compliance as much as possible without adversely affecting its historic significance. To qualify for this treatment, the building or structure must be individually listed or eligible to be listed as a historic structure, or be certified as contributing to the historic significance of a historic district. Merely being located in an historic district is insufficient justification for a variance.

A variance may be issued provided the proposed work does not change the historic designation of the building or structure. The board of appeals or the building official may require that the permit applicant obtain a written review and determination to that effect from the appropriate state or local organization responsible for the historic designation.

All variances are to be the minimum necessary to afford relief [see Section G105.7(4)]. There are a number of actions that may be taken to reduce future flood damage to substantially damaged or substantially improved historic buildings or structures, while maintaining their historic and cultural value. For example, if elevation of an historic building or structure would negatively affect its character or cause a loss of its historic designation, a variance from the elevation requirement may be appropriate; however, it may still be possible to require that building utility systems be elevated or otherwise protected from flood damage to the maximum extent possible. A condition of a variance may require that building materials, including interior finishes, be resistant to flood damage (refer to FEMA FIA-TB #2: *Flood Resistant Materials Requirements for Buildings Located in Special Flood Hazard Areas*). Another condition may require that building contents be elevated to the DFE or the maximum extent possible. For further information, contact the state NFIP coordinator, state historic preservation officer or appropriate FEMA regional office.

G105.4 Functionally dependent facilities. A variance is authorized to be issued for the construction or substantial improvement of a functionally dependent facility provided the criteria in Section 1612.1 are met and the variance is the minimum necessary to allow the construction or substantial improvement, and that all due consideration has been given to methods and materials that minimize flood damages during the design flood and create no additional threats to public safety.

❖ Functionally dependent facilities include those that must be located in close proximity to water to fulfill their intended purpose, such as shipbuilding; ship or boat repair and docks and port facilities for loading or unloading cargo or passengers. Some facilities that often are located close to water are specifically excluded from the definition, including long-term storage and manufacturing, sales or service facilities that are associated with water activities. Functionally dependent facilities represent a special case recognized by NFIP as warranting treatment under the variance provisions. Under some circumstances it may impossible for a facility to perform its function unless one or more of the flood-resistant construction provisions of the code or this appendix are varied. For example, temporary storage facilities may need to be at dock level, or a building in a designated flood hazard area subject to high-velocity wave action may need to be elevated on armored fill rather than on piles or columns.

At a minimum, functionally dependent facilities are to be wet floodproofed. This type of flood protection reduces flood damage while allowing floodwaters to enter the building or structure. The keys to successful wet floodproofing are to design the building or structure, to the maximum extent possible, to resist flood and wind loads anticipated during conditions of the design flood; to use flood damage-resistant materials and to elevate or use damage-resistant building utility systems as required by the flood-resistant construction provisions of the code and this appendix. For further guidance, refer to FEMA FIA-TB #1, *Openings in Foundation Walls for Buildings Located in Special Flood Hazard Areas*; FEMA FIA-TB #2, *Flood Resistant Material Requirements for Buildings Located in Special Flood Hazard Areas* and FEMA FIA-TB #7, *Wet Floodproofing Requirements for Structures Located in Special Flood Hazard Areas*.

Within functionally dependent facilities, only those functions that must be carried out below the DFE are to be included within the scope of the variance. For example, a boat repair facility may have areas that are used for a variety of purposes. The actual boat repair activities may need to be conducted at an elevation below the DFE. Other areas that are not functionally dependent, such as employee locker rooms, bathrooms and offices, must either be elevated or dry floodproofed (A Zones only) to or above the DFE as required by the code.

G105.5 Restrictions. The board of appeals shall not issue a variance for any proposed development in a floodway if any increase in flood levels would result during the base flood discharge.

❖ The board of appeals may not issue variances to the requirements in Section G103.5, which addresses activities in floodways that may increase flood elevations. The commentary to Section G103.5 describes submission of engineering analyses and approval by FEMA prior to approving activities that increase flood elevations in designated floodways.

G105.6 Considerations. In reviewing applications for variances, the board of appeals shall consider all technical evaluations, all relevant factors, all other portions of this appendix and the following:

1. The danger that materials and debris may be swept onto other lands resulting in further injury or damage;
2. The danger to life and property due to flooding or erosion damage;
3. The susceptibility of the proposed development, including contents, to flood damage and the effect of such damage on current and future owners;
4. The importance of the services provided by the proposed development to the community;
5. The availability of alternate locations for the proposed development that are not subject to flooding or erosion;
6. The compatibility of the proposed development with existing and anticipated development;
7. The relationship of the proposed development to the comprehensive plan and flood plain management program for that area;
8. The safety of access to the property in times of flood for ordinary and emergency vehicles;
9. The expected heights, velocity, duration, rate of rise and debris and sediment transport of the floodwaters and the effects of wave action, if applicable, expected at the site; and
10. The costs of providing governmental services during and after flood conditions including maintenance and repair of public utilities and facilities such as sewer, gas, electrical and water systems, streets and bridges.

❖ This section provides factors to be taken into account by the board of appeals when it reviews requests for variances from the flood-resistant construction and flood-plain management provisions of the code and this appendix. Section G105.2 requires that documentation be maintained as part of the permanent record to indicate how the board of appeals addressed each consideration.

G105.7 Conditions for issuance. Variances shall only be issued by the board of appeals upon:

1. A technical showing of good and sufficient cause that the unique characteristics of the size, configuration or topography of the site renders the elevation standards inappropriate;
2. A determination that failure to grant the variance would result in exceptional hardship by rendering the lot undevelopable;
3. A determination that the granting of a variance will not result in increased flood heights, additional threats to public safety, extraordinary public expense, nor create nuisances, cause fraud on or victimization of the public or conflict with existing local laws or ordinances;
4. A determination that the variance is the minimum necessary, considering the flood hazard, to afford relief; and
5. Notification to the applicant in writing over the signature of the building official that the issuance of a variance to construct a structure below the base flood level will result in increased premium rates for flood insurance up to amounts as high as $25 for $100 of insurance coverage, and that such construction below the base flood level increases risks to life and property.

❖ Each of the listed conditions for issuance of variances is to be addressed by the board of appeals, especially the requirement that it determine that failure to grant the variance would result in exceptional hardship by rendering the lot undevelopable. By itself, this determination may be insufficient to result in exceptional hardship if other conditions for issuance of a variance cannot be met. The determination of hardship is to be based on the unique characteristics of the site and not the personal circumstances of the applicant.

In guidance materials, FEMA cautions that economic hardship alone is not to be considered an exceptional hardship. Building officials and boards of appeals are cautioned that granting a variance does not affect how the building will be rated for the purposes of NFIP flood insurance. Even if circumstances justify granting a variance to build a floor that is below the design flood elevation, the rate used to calculate the cost of a flood insurance policy will be based on the risk to the building. Flood insurance, required by certain mortgage lenders, may be extremely expensive, costing as much as $25 for each $100 of insurance coverage. Although the applicant may not be required to purchase flood insurance, the requirement may be imposed on subsequent owners. The building official is to provide the applicant with a written notice to this effect, along with the other conditions listed in this section.

SECTION G201 DEFINITIONS

G201.1 General. The following words and terms shall, for the purposes of this appendix, have the meanings shown herein. Refer to Chapter 2 for general definitions.

❖ Definitions of terms can help in the understanding and application of the code requirements. Terms specific to this appendix are defined herein. The use and application of the general definitions of the code are found in Section 202.

G201.2 Definitions.

DEVELOPMENT. Any man-made change to improved or unimproved real estate, including but not limited to, buildings or other structures, temporary or permanent storage of materials, mining, dredging, filling, grading, paving, excavations, operations and other land disturbing activities.

❖ Any activity within a flood hazard area has the potential to damage and thereby alter the characteristics of the area. Some alterations may increase flood frequency or flood levels that in turn may adversely impact adjacent property owners. This term is broadly defined so that communities examine and regulate all activities for potential flood hazard area impacts. Land development activities that are not buildings and structures are also subject to the provisions of this appendix.

FUNCTIONALLY DEPENDENT FACILITY. A facility which cannot be used for its intended purpose unless it is located or carried out in close proximity to water, such as a docking or port facility necessary for the loading or unloading of cargo or passengers, shipbuilding or ship repair. The term does not include long-term storage, manufacture, sales or service facilities.

❖ By the very nature of their use, certain buildings and structures would not be able to fulfill their intended functions if located away from bodies of water. In addition, their functions may be hampered if they are elevated in full compliance with the code. Very specific facilities are defined as functionally dependent facilities, and other facilities that often are located near bodies of water are specifically excluded from the definition.

MANUFACTURED HOME. A structure that is transportable in one or more sections, built on a permanent chassis, designed for use with or without a permanent foundation when attached to the required utilities, and constructed to the Federal Mobile Home Construction and Safety Standards and rules and regulations promulgated by the U.S. Department of Housing and Urban Development. The term also includes mobile homes, park trailers, travel trailers and similar transportable structures that are placed on a site for 180 consecutive days or longer.

❖ Manufactured homes are defined separately from buildings and approvals for them are administered differently by some local jurisdictions. In flood hazard areas, manufactured homes that are not installed according to the provisions of the code will be more vulnerable to flood damage.

MANUFACTURED HOME PARK OR SUBDIVISION. A parcel (or contiguous parcels) of land divided into two or more manufactured home lots for rent or sale.

❖ Typically, manufactured home parks and subdivisions are developed by a single owner and, in most cases, the utilities, pads and foundations are installed before the lots are rented or sold. These development activities are included in the code (see Sections 301.1 and 301.2) so that manufactured homes that are placed or replaced will comply with the requirements for flood hazard areas.

RECREATIONAL VEHICLE. A vehicle that is built on a single chassis, 400 square feet (37.16 m^2) or less when measured at the largest horizontal projection, designed to be self-propelled or permanently towable by a light-duty truck, and designed primarily not for use as a permanent dwelling but as temporary living quarters for recreational, camping, travel or seasonal use. A recreational vehicle is ready for highway use if it is on its wheels or jacking system, is attached to the site only by quick disconnect-type utilities and security devices and has no permanently attached additions.

❖ Recreational vehicles are intended to be used temporarily and, therefore, are typically not placed on permanent foundations that are designed and constructed to resist flood damage. Due to the nature of their placement and susceptibility to movement, the code (see Section G601.1) provides that they not be placed in flood hazard areas subject to high-velocity wave action (V Zones) and floodways.

VARIANCE. A grant of relief from the requirements of this section which permits construction in a manner otherwise prohibited by this section where specific enforcement would result in unnecessary hardship.

❖ Within flood hazard areas, a variance, especially to build below the design flood elevation, exposes people and property to higher risks and increased potential for flood damage. For the purpose of handling requests for variances to the flood-resistant design and construction provisions of the code, specific restrictions, considerations and conditions are set forth in this appendix (see Section G105). Note that repair or rehabilitation of historic structures, as well as construction or substantial improvements of functionally dependent facilities, may be handled by variance.

VIOLATION. A development that is not fully compliant with this appendix or Section 1612, as applicable.

❖ Work that is approved by issuance of a permit is to be conducted in accordance with the code and conditions of the permit. If the work is not in compliance, then it is in violation and is to be handled in accordance with the procedures set forth in the code (see Section 113) and this appendix (see Sections G101.4, G104.3 and G104.5).

SECTION G301 SUBDIVISIONS

G301.1 General. Any subdivision proposal, including proposals for manufactured home parks and subdivisions, or other proposed new development in a flood hazard area shall be reviewed to assure that:

1. All such proposals are consistent with the need to minimize flood damage;
2. All public utilities and facilities, such as sewer, gas, electric and water systems are located and constructed to minimize or eliminate flood damage; and
3. Adequate drainage is provided to reduce exposure to flood hazards.

❖ When land is subdivided, the opportunity arises to recognize flood hazard areas and design the lot layout to reduce future flood damage. It is during this process that many communities elect to guide development to less hazardous locations. At a minimum, public utilities are to be designed and located to avoid or minimize impairment and damage. For additional guidance, refer to APA PAS 473.

G301.2 Subdivision requirements. The following requirements shall apply in the case of any proposed subdivision, including proposals for manufactured home parks and subdivisions, any portion of which lies within a flood hazard area:

1. The flood hazard area, including floodways and areas subject to high velocity wave action, as appropriate, shall be delineated on tentative and final subdivision plats;
2. Design flood elevations shall be shown on tentative and final subdivision plats;
3. Residential building lots shall be provided with adequate buildable area outside the floodway; and
4. The design criteria for utilities and facilities set forth in this appendix and appropriate *International Codes* shall be met.

❖ To facilitate the review of subdivision proposals with respect to flood hazards, tentative and final subdivision plats are to include adequate information. Where the land to be subdivided includes areas that are not within the designated flood hazard area, lots may be laid out to avoid or minimize flood-plain impacts. Residential lots are to be laid out so that building footprints and fills or other obstructions are outside the floodway.

SECTION G401 SITE IMPROVEMENT

❖ The definition of "Development" in Section G201 includes site work necessary to place buildings or structures, as well as site improvements that do not involve buildings or structures. When buildings and structures cannot be located outside of the designated flood hazard area, they are to be elevated so that the lowest floor is at or above the DFE. The most common ways to elevate buildings or structures are on fill, on solid foundation walls surrounding crawl spaces or on posts, pilings or columns. Compacted fill, placed in accordance with the code, may be placed to raise a building pad above the DFE. Note that fill is not to be used to elevate buildings in flood hazard areas subject to high-velocity wave action (V Zones).

Before allowing fill in a designated flood hazard area, the building official is to determine that the fill will not increase flooding or cause drainage problems on neighboring properties. In areas with known drainage and flooding problems, the building official has the authority to require submission of an engineering analysis. Under Section G103.4, this analysis is required when a floodway has not been delineated and, under Sections G103.5 and G401.1, if fill will encroach into a designated floodway. Some states and communities require encroachment analyses for all large fill proposals. In areas that are already developed, increased flood levels resulting from flood hazard area activities may be considered threats to public safety and are to be mitigated to the maximum extent possible by the permit applicant.

Permit applicants may propose placing fill in designated flood hazard areas with the intent of later constructing buildings with excavated basements. The definition of "Basement" in Section 1612.2 is "the portion of a building having its floor subgrade (below ground level) on all sides." When excavated into fill, basements may be subject to damage, especially in designated flood hazard areas where waters remain high for more than a few hours. Fill materials can become saturated and provide inadequate support or water pressure can collapse below-grade walls. Basements below residential buildings are not to be constructed below the DFE, even if excavated into fill that is placed above the DFE. Basements beneath nonresidential buildings are similarly not to be built unless they are designed, constructed and certified to be dry floodproofed in accordance with flood-resistant construction provisions of the code and ASCE 24-98.

If elevated on individual fill pads in designated flood hazard areas, buildings will be surrounded by water during general conditions of flooding. To address emergency management concerns about evacuation and public safety, it may be appropriate to consult with local emergency personnel responsible for evacuations. This could be especially important when reviewing applications for large subdivisions or critical facilities, such as hospitals and care facilities. Many states and communities have provisions that require uninterrupted access to all buildings during flood conditions. These requirements may be administered by agencies responsible for permitting the construction of public rights-of-way.

G401.1 Development in floodways. Development or land disturbing activity shall not be authorized in the floodway unless it has been demonstrated through hydrologic and hydraulic analyses performed in accordance with standard engineering practice that the proposed encroachment will not result in any increase in the level of the base flood.

❖ See the commentary to Section G103.5 for more information.

G401.2 Flood hazard areas subject to high velocity wave action.

1. Development or land disturbing activity shall only be authorized landward of the reach of mean high tide.
2. The use of fill for structural support of buildings is prohibited.

❖ All new construction within flood hazard areas subject to high-velocity wave action (commonly called V Zones) is to be located landward of the reach of mean high tide. The use of fill to provide structural support is prohibited because areas subject to wave action are more likely to experience flood-related erosion that may cause failure of foundations on fill. For additional guidance, refer to FEMA FIA-TB #5, *Free of Obstruction Requirements for Buildings Located in Coastal High Hazard Areas* and FEMA 55, *Coastal Construction Manual.*

G401.3 Sewer facilities. All new or replaced sanitary sewer facilities, private sewage treatment plants (including all pumping stations and collector systems) and on-site waste disposal systems shall be designed in accordance with Chapter 8, ASCE 24, to minimize or eliminate infiltration of floodwaters into the facilities and discharge from the facilities into floodwaters, or impairment of the facilities and systems.

❖ New and replacement sewer facilities are to be designed in accordance with ASCE 24-98, Chapter 8. Adhering to these requirements will minimize damage to such facilities, as well as the likelihood of contamination from such facilities.

G401.4 Water facilities. All new replacement water facilities shall be designed in accordance with the provisions of Chapter 8, ASCE 24, to minimize or eliminate infiltration of floodwaters into the systems.

❖ New and replacement water supply facilities are to be designed in accordance with ASCE 24-98, Chapter 8. Adhering to these requirements will minimize damage to water supply facilities and infiltration of contamination that may be carried by floodwaters.

G401.5 Storm drainage. Storm drainage shall be designed to convey the flow of surface waters to minimize or eliminate damage to persons or property.

❖ In addition to designing a site to minimize damage from flood conditions, storm runoff is to be addressed to minimize or eliminate related damage. Storm drainage conveyance patterns and facilities are to be shown on the site plan described in Section G104.2.

G401.6 Streets and sidewalks. Streets and sidewalks shall be designed to minimize potential for increasing or aggravating flood levels.

❖ The layout of streets and sidewalks may impact both storm runoff and flood flows. The design and construction of streets and roadways, including bridges and culverts, are to reduce or eliminate damage from both storm runoff and flooding. In addition to removing flood storage from designated flood hazard areas, the placement of fill associated with the construction of roads and sidewalks has the potential to alter the flow of both storm runoff and floodwater. Bridges and culverts are to be sized to minimize their effects on flooding. Refer to Section G401.1 if such activities are proposed within floodways.

SECTION G501 MANUFACTURED HOMES

❖ This section requires that all new and substantially improved manufactured homes are to be elevated such that their lowest floors are at or above the DFE. Replacement of an existing manufactured home is considered a new construction and must comply with the provisions of this section. In effect, manufactured homes are to meet the same flood-resistant construction requirements as other buildings and structures. In part, this provision exceeds the minimum requirements of NFIP, which allows, in some limited situations in existing manufactured home parks and subdivisions, replacement manufactured homes to be located below the DFE. Manufactured homes are to be anchored to permanent foundations to resist flotation, collapse and lateral movement. Foundations are to be reinforced and elevated so that the floor of the manufactured home is at or above the DFE. Foundations are to meet all applicable requirements to resist wind, seismic and flood loads prescribed in the code.

Manufactured homes may be placed either on prepared sites in manufactured home parks or subdivisions or on individually owned parcels of land. Manufactured home parks have a single owner, and the pads, foundations and utility hookups are rented to the manufactured home owners. As in a typical subdivision, a manufactured home subdivision normally is developed by a single developer and the lots are sold to individuals for use as manufactured home sites.

G501.1 Elevation. All new and replacement manufactured homes to be placed or substantially improved in a flood hazard area shall be elevated such that the lowest floor of the manufactured home is elevated to or above the design flood elevation.

❖ The elevation requirements set forth in Section 1612 and ASCE 24-98 apply to manufactured homes, thereby treating this type of construction the same as other buildings and structures. The certification of the lowest floor elevation, as required in Section 1612.5, also applies to manufactured homes that are placed in designated flood hazard areas.

G501.2 Foundations. All new and replacement manufactured homes, including substantial improvement of existing manufactured homes, shall be placed on a permanent, reinforced foundation that is designed in accordance with Section 1612.

❖ Flood loads experienced in most designated flood hazard areas include hydrostatic and hydrodynamic components. Impacts from floating debris and waves can impart additional loads. Additionally, in flood hazard areas subject to high-velocity wave action (V Zones), flood and wind loads acting simultaneously, as required in Section 1603, are to be addressed in foundation design. Permanent, reinforced foundations for manufactured homes are necessary to resist anticipated loads. A permanent foundation system for a manufactured home includes the following:

- A below-grade footing capable of providing resistance against overturning;
- Footing depth below the frost line;
- Reinforced piers, driven pilings, embedded posts or poured concrete or reinforced block foundation walls and
- Anchors and connections (see Section G501.3).

G501.3 Anchoring. All new and replacement manufactured homes to be placed or substantially improved in a flood hazard area shall be installed using methods and practices which minimize flood damage. Manufactured homes shall be securely anchored to an adequately anchored foundation system to resist flotation, collapse and lateral movement. Methods of anchoring are authorized to include, but are not limited to, use of over-the-top or frame ties to ground anchors. This requirement is in addition to applicable state and local anchoring requirements for resisting wind forces.

❖ The manufactured home anchoring requirement has two elements:

1. Manufactured homes are to be anchored to a permanent, reinforced foundation to limit movement and resist uplift and overturning due to flood, wind and seismic forces.
2. Manufactured homes are to have ties firmly attached to ground anchors to transfer loads to the ground. Manufacturer specifications for ground anchor depths typically do not include flood loads. Steel strapping, cable, chain or other approved material may be used for ties, which are to be fastened to ground anchors and drawn tight with turnbuckles or other adjustable tensioning devices. Manufacturer specifications for ties typically do not include flood loads. Ties may be over-the-top straps or frame ties, as appropriate to the manufactured home.

SECTION G601
RECREATIONAL VEHICLES

G601.1 Placement prohibited. The placement of recreational vehicles shall not be authorized in flood hazard areas subject to high velocity wave action and in floodways.

❖ By their very nature, recreational vehicles are not normally elevated or permanently attached to a foundation and, therefore, may be very susceptible to flooding. This provision limits the permanent placement of recreational vehicles in designated flood hazard areas where velocities may be high and flood loads may be significant, such as floodways and flood hazard areas subject to high-velocity wave action.

G601.2 Temporary placement. Recreational vehicles in flood hazard areas shall be fully licensed and ready for highway use, and shall be placed on a site for less than 180 consecutive days.

❖ Unless otherwise prohibited under Section 304, recreational vehicles may be temporarily placed in designated flood hazard areas if they are fully licensed, remain highway ready and can be moved when flooding threatens. "Highway ready" means that a recreational vehicle is on its wheels or internal jacking system, is connected to the site only by quick disconnect-type utilities and security devices and has no permanently attached additions.

G601.3 Permanent placement. Recreational vehicles that are not fully licensed and ready for highway use, or that are to be placed on a site for more than 180 consecutive days, shall meet the requirements of Section G501 for manufactured homes.

❖ Recreational vehicles that do not meet the requirements of Section 304.2 are considered permanent buildings and are to be installed in accordance with the requirements for manufactured homes in Section 303.

SECTION 701
TANKS

❖ This section addresses requirements for tanks located in flood hazard areas. Physical damage as well as environment contamination and health risks result when above-ground and underground tanks are not adequately protected and restrained during conditions of flooding. Many hazardous materials, for example gas or oil, are lighter than water and, if released due to damage or unprotected vents and fill openings, contribute to contamination and risks to public health and safety. For additional guidance, reference FEMA 348.

G701.1 Underground tanks. Underground tanks in flood hazard areas shall be anchored to prevent flotation, collapse or lateral movement resulting from hydrostatic loads, including the effects of buoyancy, during conditions of the design flood.

❖ Most tanks will not be entirely filled at all times and will, thus, contain at least some air that contributes to buoyancy. In addition, tanks often contain products that are lighter than water. When the surrounding ground becomes saturated during conditions of flooding, underground tanks may be dislodged or shifted, causing damage and loss of product. Installation designs and methods are to account for buoyancy.

G701.2 Above-ground tanks. Above-ground tanks in flood hazard areas shall be elevated to or above the design flood elevation or shall be anchored or otherwise designed and constructed to prevent flotation, collapse or lateral movement resulting from hydrodynamic and hydrostatic loads, including the effects of buoyancy, during conditions of the design flood.

❖ Above-ground tanks are to be raised above the design flood elevation, which may be accomplished by installation on platforms or other foundations that are designed to resist flood loads. Above-ground tanks that are not elevated are subject to flood loads imposed by moving water and buoyancy and are to be adequately anchored to resist flood damage. Tanks that are exposed to floodwaters may be impacted by floating debris or become dislodged, resulting in loss of product that could contribute to contamination and in becoming floating debris themselves.

G701.3 Tank inlets and vents. In flood hazard areas, tank inlets, fill openings, outlets and vents shall be:

1. At or above the design flood elevation or fitted with covers designed to prevent the inflow of floodwater or outflow of the contents of the tanks during conditions of the design flood.
2. Anchored to prevent lateral movement resulting from hydrodynamic and hydrostatic loads, including the effects of buoyancy, during conditions of the design flood.

❖ Tank inlets and vents are to be protected from the entry of floodwater either by raising them above the design flood elevation or by protective covers or devices. In addition, inlets and vents, and the piping that serve them, are to be anchored and protected from flood damage.

SECTION G801 REFERENCED STANDARDS

ASCE 24–98	Flood Resistance Design and Construction	G103.1, G401.3, G401.4
HUD 24 CFR Part 3280 (1994)	Manufactured Home Construction and Safety Standards	G201
IBC-03	*International Building Code*	G102.2

BIBLIOGRAPHY

The following resource materials are referenced in, or are relevant to the subject matter addressed in, this appendix. See the commentary to Chapter 16 for information on obtaining additional related publications from FEMA.

APA PAS #473, *Subdivision Design in Flood Hazard Areas.* Washington, DC: American Planning Association, 1997.

FEMA 44 CFR Parts 59-73, *National Flood Insurance Program* (NFIP). Washington, DC: Federal Emergency Management Agency.

FEMA 55, *Coastal Construction Manual*. Washington, DC: Federal Emergency Management Agency.

FEMA 265, *Managing Floodplain Development in Approximate Zone A Areas: A Guide for Obtaining and Developing Base (100-Year) Flood Elevations.* Washington, DC: Federal Emergency Management Agency, 1995.

FEMA Form 81-89 Series, *Revisions to National Flood Insurance Rate Maps: Application/Certification Forms and Instructions for Conditional Letters of Map Revision, Letters of Map Revision and Physical Map Revisions.* Washington, DC: Federal Emergency Management Agency, 1999.

FEMA (various dates). *NFIP Technical Bulletin Series.* Washington, DC: National Flood Insurance Program. [Online]. Available: http://www.fema.gov/MIT/techbul.htm

FEMA FIA-TB #1, *Openings in Foundation Walls for Buildings Located in Special Flood Hazard Areas*, 1993.

FEMA FIA-TB #2, *Flood Resistant Material Requirements for Buildings Located in Special Flood Hazard Areas*, 1993.

FEMA FIA-TB #3, *Non-Residential Floodproofing Requirements and Certification for Buildings Located in Special Flood Hazard Areas,* 1993.

FEMA FIA-TB #4, *Elevator Installation for Buildings Located in Special Flood Hazard Areas*, 1993.

FEMA FIA-TB #5, *Free of Obstruction Requirements for Buildings Located in Coastal High Hazard Areas*, 1993.

FEMA FIA-TB #6, *Below Grade Parking Requirements for Buildings Located in Special Flood Hazard Areas*, 1993.

FEMA FIA-TB #7, *Wet Floodproofing Requirements for Structures Located in Special Flood Hazard Areas*, 1993.

FEMA FIA-TB #8, *Corrosion Protection for Metal Connectors In Coastal Areas for Structures Located in Special Flood Hazard Areas*, 1996.

FEMA FIA-TB #9, *Design and Construction Guidance for Breakaway Walls Below Elevated Coastal Buildings*, 1999.

Appendix H: Signs

The provisions contained in this appendix are not mandatory unless specifically referenced in the adopting ordinance.

General Comments

The provisions contained in this appendix are not mandatory unless specifically referenced in the adopting ordinance, as stated in Section 101.2.1.

The regulations contained in this appendix were prompted by improper design, erection, placement, location and maintenance of outdoor signs. Improperly designed and installed signs pose a hazard to people and property. Improperly wired signs or improperly maintained electrical signs can cause fires. The criteria in this chapter are straightforward that, in turn, facilitates straightforward enforcement. Section H102 contains definitions for nine distinct types of signs.

Purpose

The purpose of Appendix H is to locate in one place the provisions that regulate construction or protection requirements for outdoor signs. This appendix also provides the owner, designer, installer and maintainer of outdoor signs with minimum safe practices presented in performance language whenever possible, thus allowing the widest possible application.

SECTION H101 GENERAL

H101.1 General. A sign shall not be erected in a manner that would confuse or obstruct the view of or interfere with exit signs required by Chapter 10 or with official traffic signs, signals or devices. Signs and sign support structures, together with their supports, braces, guys and anchors, shall be kept in repair and in proper state of preservation. The display surfaces of signs shall be kept neatly painted or posted at all times.

❖ The provisions of this appendix are confined to that which is displayed in any manner out of doors for recognized advertising purposes. An orderly procedure is to be followed in erecting and maintaining outdoor signs. Outdoor signs must not block the view of signs regulated in Section 1011 that identify the location and path of travel to exits or street signs regulated by the jurisdiction. Signs must be properly maintained in accordance with this section, including corrosion prevention on all parts of the sign and its supports. Corrosion of a structural member results in loss of cross section, which could render that member incapable of handling the load of the sign or loads (snow, wind, etc.) on the sign. The building official has the authority to order the removal of a sign that is not maintained properly, as it can represent a hazard to the public and, therefore, can be determined to be unsafe (see Section 115).

H101.2 Signs exempt from permits. The following signs are exempt from the requirements to obtain a permit before erection:

1. Painted nonilluminated signs.
2. Temporary signs announcing the sale or rent of property.
3. Signs erected by transportation authorities.
4. Projecting signs not exceeding 2.5 square feet (0.23 m^2).
5. The changing of moveable parts of an approved sign that is designed for such changes, or the repainting or repositioning of display matter shall not be deemed an alteration.

❖ The objective of this section is to describe when a permit is not required to regulate the erection, construction, alteration and maintenance of a new sign. Note that this section applies to new signs only, except for existing signs that are to be altered, and that the owner of an exempt sign is responsible for erecting and maintaining it in a safe manner (see Section H104.1 for a related issue).

A permit is not required for the erection of wall signs (commonly of limited area) without electrical service that, when not maintained, can be painted over or the surface repaired to its original state.

A sale or rental sign is temporary in nature, pertains to the premises on which it is placed and its installation is not likely to represent a hazard; for example, it is substantially fixed in place and without electrical service.

Regulation of street signs in the code would be redundant, since signs erected by the jurisdiction inevitably undergo one or more levels of regulatory review before installation. Directional (informational) signs in conjunction with transit lines, such as subways, trains, buses, ferries and airports, are exempt from permit requirements since it is generally not necessary for a jurisdiction to take out a permit for an activity it is performing.

Projecting signs that do not exceed an area of 2.5 square feet (.23 m^2) do not require a permit, as the size limitation allows access for maintenance. The sign will likely be constructed so as not to create a hazard, for example, lack of headroom for pedestrians, protrusions into a walking surface and electrical supply disrepair.

When an existing sign is altered (i.e., enlarged, relocated, etc.), the alterations must comply with the requirements for new signs. A permit is required for such alterations. The exceptions to the requirements for alterations apply to a sign that has been designed and previously authorized (through the permit process) to have changeable or movable parts. Repainting or reposting of display matter (lettering) is not considered an alteration. If a sign is enlarged or relocated without a permit and, in turn, without a field inspection, it could add excessive loads to a roof, interfere with the exhaust or intake ventilation facilities from a building or create a potential hazard to the public.

SECTION H102 DEFINITIONS

H102.1 General. Unless otherwise expressly stated, the following words and terms shall, for the purposes of this appendix, have the meanings shown herein. Refer to Chapter 2 of the *International Building Code* for general definitions.

❖ Definitions of terms that are associated with the content of this section are contained herein. These definitions can help in the understanding and application of the code requirements. The use and application of all defined terms are set forth in Section 202 of the code.

COMBINATION SIGN. A sign incorporating any combination of the features of pole, projecting and roof signs.

❖ The purpose of this definition is to identify which combination of features constitutes a combination sign. These signs are characterized by being simultaneously supported partly by one or more poles, partly by the wall, partly by the roof of a structure or any combination thereof.

DISPLAY SIGN. The area made available by the sign structure for the purpose of displaying the advertising message.

❖ A display sign constitutes the surface of a sign where the alphabetic or pictographic components of a message are arranged for display.

ELECTRIC SIGN. A sign containing electrical wiring, but not including signs illuminated by an exterior light source.

❖ This definition states that an electrical sign is any sign activated or illuminated by means of electrical energy. Electric signs are characterized by the use of artificial light projecting through, but not reflecting off, its surface(s).

GROUND SIGN. A billboard or similar type of sign which is supported by one or more uprights, poles or braces in or upon the ground other than a combination sign or pole sign, as defined by this code.

❖ This definition identifies those signs that are not part of other buildings or structures. The structural supports of these signs are such that the wind loading is directly transferred into its own foundation system rather than into other buildings or structures. Section H109.1 contains the specific material and height requirements of ground signs.

POLE SIGN. A sign wholly supported by a sign structure in the ground.

❖ Pole signs are those where the structural supports are constructed so that loads are directly transferred to the ground.

PORTABLE DISPLAY SURFACE. A display surface temporarily fixed to a standardized advertising structure which is regularly moved from structure to structure at periodic intervals.

❖ This definition identifies the surfaces available for displaying commercial or noncommercial advertising messages that are not permanently attached to a structure and are periodically placed at different locations by means of manual or remote input.

PROJECTING SIGN. A sign other than a wall sign, which projects from and is supported by a wall of a building or structure.

❖ A projecting sign is the opposite of a wall sign in that it is projecting from the wall that provides its support. Section H112 contains requirements for projecting signs to regulate their materials, maximum projections, clearances and structural capabilities.

ROOF SIGN. A sign erected upon or above a roof or parapet of a building or structure.

❖ This definition identifies that any exterior sign supported on the roof of a building or structure is considered a roof sign. Such a sign directly impacts the design of the roof because of increased wind and snow drifting loads. Section H110.1 contains explicit material and clearance requirements along with height limitations for the erection of roof signs.

SIGN. Any letter, figure, character, mark, plane, point, marquee sign, design, poster, pictorial, picture, stroke, stripe, line, trademark, reading matter, or illuminated service, which shall be constructed, placed, attached, painted, erected, fastened, or manufactured in any manner whatsoever, so that the same shall be used for the attraction of the public to any place, subject, person, firm, corporation, public performance, article, machine, or merchandise, whatsoever, which is displayed in any manner whatsoever outdoors. Every sign shall be classified and conform to the requirements of that classification as set forth in this chapter.

❖ This definition provides the basis for determining which building components are considered as signs, and thus must comply with the provisions of Appendix H. For the purposes of the code, signs are always considered as exterior elements, since the code addresses the hazards associated with exterior exposure. Interior signs must meet the specific code requirements for the materials involved, along with the code provisions for interior finishes and trim (see Chapter 8). The code further subdivides signs into different groups, with specific requirements for each group.

SIGN STRUCTURE. Any structure which supports or is capable of supporting a sign as defined in this code. A sign structure is permitted to be a single pole and is not required to be an integral part of the building.

❖ Structures supporting signs, which may be affixed to the ground and include one or more columns, poles or braces placed in or upon the ground, are not required to be part of the building structural system.

WALL SIGN. Any sign attached to or erected against the wall of a building or structure, with the exposed face of the sign in a plane parallel to the plane of said wall.

❖ This definition details those signs that are part of the wall surfaces of a building or structure, or which are separate signs and are then fastened to the wall surface. Wall signs are separately fastened signs of limited extension from the wall surface; otherwise, they become projecting signs. The code allows certain-sized wall signs to be exempt from permit requirements (see Section H101.2). Section H111.1 specifies the material requirements and extension limitations of wall signs.

SECTION H103 LOCATION

H103.1 Location restrictions. Signs shall not be erected, constructed or maintained so as to obstruct any fire escape or any window or door or opening used as a means of egress or so as to prevent free passage from one part of a roof to any other part thereof. A sign shall not be attached in any form, shape or manner to a fire escape, nor be placed in such manner as to interfere with any opening required for ventilation.

❖ The intent of this section is to prohibit the installation of signs at locations that may interfere with either the evacuation process during an emergency situation or free movement throughout the roof of a building. Fire escapes and openings that are part of the means of egress or required for ventilation are to be free from obstructions that may affect their intended use.

SECTION H104 IDENTIFICATION

H104.1 Identification. Every outdoor advertising display sign hereafter erected, constructed or maintained, for which a permit is required shall be plainly marked with the name of the person, firm or corporation erecting and maintaining such sign and shall have affixed on the front thereof the permit number issued for said sign or other method of identification approved by the building official.

❖ All signs must bear the name of the owner to identify the party responsible for installation and maintenance. The construction documents on file must show where such identification is located on each sign. If the sign is erected improperly or becomes damaged, the field inspector will be able to identify the responsible person to whom any notification is to be directed.

SECTION H105 DESIGN AND CONSTRUCTION

H105.1 General requirements. Signs shall be designed and constructed to comply with the provisions of this code for use of materials, loads and stresses.

❖ This section addresses the general requirements pertaining to the construction, design loads and working stresses of signs. The design and construction of signs must be in accordance with all of the pertinent requirements of the code. In many instances, structural calculations and the resulting design must be submitted showing the type of materials used, their strength capabilities and the loads and resultant stresses placed on the sign and its supports (see Chapter 16).

H105.2 Permits, drawings and specifications. Where a permit is required, as provided in Chapter 1, construction documents shall be required. These documents shall show the dimensions, material and required details of construction, including loads, stresses and anchors.

❖ The intent of this section is to require construction documents when a permit is required for the erection of a sign. In order to grant a permit to the owner, the sign is to be delineated on construction documents, such as plans and specifications, in sufficient detail to allow the determination of code compliance. Section H104.1 affects this section in that it explains that the owner of a sign (new or altered) is not relieved from obtaining the necessary permit, even if he or she holds a building or other type of permit at the time the sign is under construction. The sign permit is usually a separate permit and may also require other related permits (e.g., electrical).

H105.3 Wind load. Signs shall be designed and constructed to withstand wind pressure as provided for in Chapter 16.

❖ A reference is made to Section 1609, which relates to wind loads. Although a reference is not made to snow loads (Section 1608), consideration should also be given thereto, especially for signs with a significant horizontal surface that will have snow accumulation, such as marquee signs.

H105.4 Seismic load. Signs designed to withstand wind pressures shall be considered capable of withstanding earthquake loads, except as provided for in Chapter 16.

❖ The designer and plan reviewer are to consider the provisions set forth in Section 1622 before the sign and its construction can be considered in compliance with the code.

H105.5 Working stresses. In outdoor advertising display signs, the allowable working stresses shall conform to the requirements of Chapter 16. The working stresses of wire rope and its fastenings shall not exceed 25 percent of the ultimate strength of the rope or fasteners.

Exceptions:

1. The allowable working stresses for steel and wood shall be in accordance with the provisions of Chapter 22 and Chapter 23.
2. The working strength of chains, cables, guys or steel rods shall not exceed one-fifth of the ultimate strength of such chains, cables, guys or steel.

❖ Consideration is to be given to the imposed loads and resultant working stresses in the design of signs. The allowable working stresses are to be evaluated in accordance with the provisions of Chapter 16. Structural members are to be of the appropriate size to support the loads applied to the sign (snow, wind, etc.) or the load of the sign itself. All members and connections are to be sufficient to carry all design loads of Chapter 16 without exceeding the specified design values set forth in the referenced chapter.

An exception to this requirement applies only to steel and wood allowable working stresses, which are to comply with the provisions of Chapters 22 and 23. A limitation is placed on wire rope and fastening utilized as structural members supporting signs, where the working stresses are to be below 25 percent of the ultimate strength of the member. An exception to this limitation applies to chains, cables, guys or steel rods where the working stresses of such members may not exceed one-fifth of the ultimate strength of the members.

H105.6 Attachment. Signs attached to masonry, concrete or steel shall be safely and securely fastened by means of metal anchors, bolts or approved expansion screws of sufficient size and anchorage to safely support the loads applied.

❖ In accordance with this section, metal connectors are the prescriptive requirement for the support and transfer of loads when signs are bearing on walls, floors and roofs of masonry, concrete or steel construction. Metal connectors provide points of anchorage at intersecting components and are also used as bonding elements. Section 2103.11 contains the standards and material requirements for joint reinforcement, wire fabric, wire accessories, metal anchors, ties and accessories and the required corrosion protection of these components.

SECTION H106
ELECTRICAL

H106.1 Illumination. A sign shall not be illuminated by other than electrical means, and electrical devices and wiring shall be installed in accordance with the requirements of the ICC *Electrical Code.* Any open spark or flame shall not be used for display purposes unless specifically approved.

❖ The reference to the ICC *Electrical Code*® (ICC EC™) when dealing with electrical aspects is common throughout the *International Codes*® (see commentary, Chapter 27). Improperly installed lighting and wiring are a potential fire or electrocution hazard. Other illumination, utilizing such means as flame or sparks, must be approved by the building official so that such an arrangement does not induce a fire hazard for which commensurate safety precautions are not considered.

H106.1.1 Internally illuminated signs. Except as provided for in Sections 402.14 and 2611, where internally illuminated signs have sign facings of wood or approved plastic, the area of such facing section shall not be more than 120 square feet (11.16 m^2) and the wiring for electric lighting shall be entirely enclosed in the sign cabinet with a clearance of not less than 2 inches (51 mm) from the facing material. The dimensional limitation of 120 square feet (11.16 m^2) shall not apply to sign facing sections made from flame-resistant-coated fabric (ordinarily known as "flexible sign face plastic") that weighs less than 20 ounces per square yard (678 g/m^2) and which, when tested in accordance with NFPA 701, meets the requirements of both the small-scale test and the large-scale test, or which, when tested in accordance with an approved test method, exhibits an average burn time for ten specimens of 2 seconds or less and a burning extent of 15 centimeters or less.

❖ This section applies to outdoor signs that are internally illuminated. The exception is a reminder that the area limitations of this section are not applicable to internally illuminated signs located within a building. Wood and approved plastic are limited because of the need to control the amount of combustibles incorporated into the sign. Once the amount of combustibles is limited, maintenance of a 2-inch (51 mm) clearance is still necessary between this material and electrical wiring. Square-foot limitations are not imposed on flame-resistant controlled fabrics that weigh less than 20 ounces per square yard (678 g/m^2), provided that the material meets the specific criteria listed. This is because of the low fuel content of such fabrics and the resulting lower potential for fire propagation. One standard that is referenced, NFPA 701, deals with not only average flame time and average length of char, but also outdoor conditions and weathering.

H106.2 Electrical service. Signs that require electrical service shall comply with the ICC *Electrical Code*.

❖ This section is concerned with sign wiring being installed in a safe and protected manner so as not to create a potential tripping or electrical shock hazard. Electrical safety is as important as structural safety when dealing with illuminated signs. This section emphasizes the need for control by the building official so that the referenced ICC EC is utilized properly.

SECTION H107 COMBUSTIBLE MATERIALS

H107.1 Use of combustibles. Wood, approved plastic or plastic veneer panels as provided for in Chapter 26, or other materials of combustible characteristics similar to wood, used for moldings, cappings, nailing blocks, letters and latticing, shall comply with Section H109.1, and shall not be used for other ornamental features of signs, unless approved.

❖ The use of combustibles for ornamentation and sign facings is addressed in Sections H107.1 and H107.1.2. The control of combustible signs and portions of signs is necessary to limit the fuel load that the combustible features add to a structure. The use of the term "approved" means approved by the building official in accordance with the provisions of Chapter 26 and other applicable code sections. The reference to Section H109.1 controls the height at which combustibles may be used (see Section H109.1 for details).

H107.1.1 Plastic materials. Notwithstanding any other provisions of this code, plastic materials which burn at a rate no faster than 2.5 inches per minute (64 mm/s) when tested in accordance with ASTM D 635 shall be deemed approved plastics and can be used as the display surface material and for the letters, decorations and facings on signs and outdoor display structures.

❖ These provisions specify those plastic materials, mostly consisting of acrylics, that burn at a faster rate when tested in accordance with ASTM D 635 are to be considered approved plastics for the purpose of installation as display surface materials and alphabetic characters, ornaments and coverings on signs. ASTM D 635 measures the burning rate of a small sample held horizontally and exposed to a Bunsen burner flame to compare the relative linear rate of burning, extent and time of burning or both of plastics in the horizontal position.

H107.1.2 Electric sign faces. Individual plastic facings of electric signs shall not exceed 200 square feet (18.6 m^2) in area.

❖ Consistent with other provisions of the code to limit the spread of fire, this section places restrictions on the size of plastic facings utilized in signs activated by electrical means. This limitation is intended to control the amount of combustible materials that potentially soften or melt when subject to heat.

H107.1.3 Area limitation. If the area of a display surface exceeds 200 square feet (18.6 m^2), the area occupied or covered by approved plastics shall be limited to 200 square feet (18.6 m^2) plus 50% of the difference between 200 square feet (18.6 m^2) and the area of display surface. The area of plastic on a display surface shall not in any case exceed 1,100 square feet (102 m^2).

❖ This section provides requirements for how to determine the permissible amount of combustible materials when the area of a display surface exceeds the 200-square-foot (18.6 m^2) threshold. As an additional safeguard against exterior fire spread, the aggregate area of combustible materials covering a display surface should not exceed the sum of 200 square feet (18.6 m^2) and 50 percent of the difference between the limitation and the area of the display surface. However, at no time can the aggregate area of plastic in display surfaces occupy more than 1,100 square feet (102 m^2).

H107.1.4 Plastic appurtenances. Letters and decorations mounted on an approved plastic facing or display surface can be made of approved plastics.

❖ Under the provisions of this section, combustible lettering and ornaments are permitted to be mounted on display surfaces constructed of approved plastic materials.

SECTION H108 ANIMATED DEVICES

H108.1 Fail-safe device. Signs that contain moving sections or ornaments shall have fail-safe provisions to prevent the section or ornament from releasing and falling or shifting its center of gravity more than 15 inches (381 mm). The fail-safe device shall be in addition to the mechanism and the mechanism's housing which operate the movable section or ornament. The fail-safe device shall be capable of supporting the full deal weight of the section or ornament when the moving mechanism releases.

❖ Animated signs may be operated for many hours at a time, perhaps even continuously. Thus, it is imperative that if the operating mechanism malfunctions, the animated part of the sign will not release or fall, creating a hazard or damaging property.

SECTION H109 GROUND SIGNS

H109.1 Height restrictions. The structural frame of ground signs shall not be erected of combustible materials to a height of more than 35 feet (10668 mm) above the ground. Ground signs constructed entirely of noncombustible material shall not be erected to a height of greater than 100 feet (30 480 mm) above the ground. Greater heights are permitted where approved and located so as not to create a hazard or danger to the public.

❖ Combustible materials exceeding 35 feet (10 668 mm) in height are prohibited in a sign's structural frame. This correlates with the height limitation for structures of Type V` construction in Table 503. The primary concern is the sign and its structure's potential involvement in a fire conflagration. The height of a noncombustible sign structure is limited to 100 feet (30 480 mm) so as to limit reasonably the area exposed to a sign or the sign's structural failure. The building official is permitted to approve signs that exceed 100 feet (30 480 mm) in height, provided that a public hazard is not created. An example of this is an area remote from other buildings and occupied areas, where structural failure would not jeopardize the public.

H109.2 Required clearance. The bottom coping of every ground sign shall be not less than 3 feet (914 mm) above the ground or street level, which space can be filled with platform decorative trim or light wooden construction.

❖ This section regulates the bottom clearance of ground signs. This clear space of 3 feet (914 mm) is permitted to be decorated with nonstructural or decorative trim or light wooden construction, provided this arrangement will not create a hazard.

H109.3 Wood anchors and supports. Where wood anchors or supports are embedded in the soil, the wood shall be pressure treated with an approved preservative.

❖ The intent of this section is to require wood support members in direct contact with the earth to be preservative treated. This provision is consistent with the requirements of Section 2304.11.4. Typically, wood species that are untreated are susceptible to decay. However, when preservative treated with chemicals in accordance with standardized procedures, wood becomes less susceptible to failure due to decay and rot.

SECTION H110 ROOF SIGNS

H110.1 General. Roof signs shall be constructed entirely of metal or other approved noncombustible material except as provided for in Sections H106.1.1 and H107.1. Provisions shall be made for electric grounding of metallic parts. Where combustible materials are permitted in letters or other ornamental features, wiring and tubing shall be kept free and insulated therefrom. Roof signs shall be so constructed as to leave a clear space of not less than 6 feet (1829 mm) between the roof level and the lowest part of the sign and shall have at least 5 feet (1524 mm) clearance between the vertical supports thereof. No portion of any roof sign structure shall project beyond an exterior wall.

> **Exception:** Signs on flat roofs with every part of the roof accessible.

❖ Section H110.1 controls the materials, bottom clearances and arrangements of roof signs (refer to definition of "Sign" in the Section H102.1 for special sign descriptions). Noncombustible materials are required for roof signs. This arrangement will not add to the fuel load present on the building (see Sections H106.1.1 and H107.1 for exceptions that deal with sign facings, electrical wiring and electrical clearances). Grounding of metal signs for lightning control and preventing electrical shock and fire is mandatory. A 6-foot (1829 mm) clear space is to be maintained between the bottom of the sign and the roof. This arrangement will enable wind to travel under the sign, thereby minimizing the collection of snow and debris. It will also facilitate roof repair. An exception to these requirements applies to signs constructed on flat roofs where every portion of the roof is available for access.

H110.2 Bearing plates. The bearing plates of roof signs shall distribute the load directly to or upon masonry walls, steel roof girders, columns or beams. The building shall be designed to avoid over stress of these members.

❖ The provisions of this section require bearing plates of roof signs to transfer the loads directly to masonry or steel structural members. The intent is to minimize the possibility of imposing additional loads on bearing plates not prepared to carry them, which may affect the structural stability of the roof sign. The structural design is to consider the load-carrying capability of the bearing plates to avoid distress of these elements.

H110.3 Height of solid signs. A roof sign having a solid surface shall not exceed, at any point, a height of 24 feet (7315 mm) measured from the roof surface.

❖ This section specifies a height limitation on roof signs of solid construction. The 24-foot (7315 mm) threshold provides a conservative criteria where solid signs are to remain stable against the forces of wind.

H110.4 Height of open signs. Open roof signs in which the uniform open area is not less than 40% of total gross area shall not exceed a height of 75 feet (22 860 mm) on buildings of Type 1 or Type 2 construction. On buildings of other construction types, the height shall not exceed 40 feet (12 192 mm). Such signs shall be thoroughly secured to the building upon which they are installed, erected or constructed by iron, metal anchors, bolts, supports, chains, stranded cables, steel rods or braces and they shall be maintained in good condition.

❖ An open sign is the opposite of a "closed sign" (see Section H110.5) in that the membranes and edges are substantially open, thus allowing wind loads to pass through without any significant structural effects. The code allows increased heights for open signs located on the roofs of buildings based on the type of construction. On buildings of combustible construction, the height is limited to 40 feet (12.2 m), whereas on buildings of noncombustible construction, the height is permitted to be 75 feet (22.9 m). Consistent with other provisions of the code, the use of noncombustible materials will not add to the fuel load present in the building. Open signs are to be properly anchored to their supporting structure and structural supports are to be kept in proper state of preservation and free from rust to avoid failure of the members.

H110.5 Height of closed signs. A closed roof sign shall not be erected to a height greater than 50 feet (15 240 mm) above the roof of buildings of Types 1 and 2 construction, nor more than 35 feet (10 668 mm) above the roof of buildings of Types 3, 4 and 5 construction.

❖ As in the previous section, there are different height limitations for closed signs located on the roofs of buildings. A closed sign usually consists of membranes containing the lettering or design fastened to the structural framework with its edges also enclosed. The 50-foot (15 240 mm) threshold identifies the point at which wind loading on the sign becomes significant and a fire could go undetected for some time.

SECTION H111 WALL SIGNS

H111.1 Materials. Wall signs which have an area exceeding 40 square feet (3.72 m^2) shall be constructed of metal or other approved noncombustible material, except for nailing rails and as provided for in Sections H106.1.1 and H107.1.

❖ This section regulates the materials used and the arrangement of wall signs (see the definition of "Wall sign"). Signs in excess of 40 square feet (3.72 m^2) must be constructed of noncombustible materials. Nailing rails and the exceptions regarding sign facings, electrical wiring and clearances to electrical wires, as stipulated in Section H107.1, are permitted. Signs of less than 40 square feet (3.72 m^2) may be of combustible construction. To date, experience has shown that fire exposure from a combustible sign of this size is minimal.

H111.2 Exterior wall mounting details. Wall signs attached to exterior walls of solid masonry, concrete or stone, shall be safely and securely attached by means of metal anchors, bolts or expansion screws of not less than $^3/_8$-inch (9.5 mm) diameter and shall be embedded at least 5 inches (127 mm). Wood blocks shall not be used for anchorage, except in the case of wall signs attached to buildings with walls of wood. A wall sign shall not be supported by anchorages secured to an unbraced parapet wall.

❖ This section indicates the size, type and location of masonry anchors intended to be used for the attachment of signs to exterior walls of masonry construction. Note that the anchorage required is a minimum $^3/_8$-inch-diameter (9.5 mm) metal anchor, bolt or expansion screw embedded a minimum of 5 inches (127 mm) in solid masonry, concrete or stone construction. The use of wood blocking for attachment of wall signs is prohibited except in buildings where the exterior walls are constructed of wood. The anchors of wall signs are not permitted to be supported by unbraced parapet walls that are not providing adequate structural support for load transfer.

H111.3 Extension. Wall signs shall not extend above the top of the wall, nor extend beyond the ends of the wall to which the signs are attached unless such signs conform to the requirements for roof signs, projecting signs or ground signs.

❖ Signs cannot extend above the top of the wall (access to the roof must be maintained for repair and fire fighting) or beyond the ends of the wall. The signs must be kept on the property, unless exceptions for other types of signs apply.

SECTION H112 PROJECTING SIGNS

H112.1 General. Projecting signs shall be constructed entirely of metal or other noncombustible material and securely attached to a building or structure by metal supports such as bolts, anchors, supports, chains, guys or steel rods. Staples or nails shall not be used to secure any projecting sign to any building or structure. The dead load of projecting signs not parallel to the building or structure and the load due to wind pressure shall be supported with chains, guys or steel rods having net cross-sectional dimension of not less than $^3/_8$-inch (9.5 mm) diameter. Such supports shall be erected or maintained at an angle of at least 45 percent (0.78 rad) with the horizontal to resist the dead load and at angle of 45 percent (0.78 rad) or more with the face of the sign to resist the specified wind pressure. If such projecting sign exceeds 30 square feet (2.8 m^2) in one facial area, there shall be provided at least two such supports on each side not more than 8 feet (2438 mm) apart to resist the wind pressure.

❖ This section establishes explicit provisions for the construction and anchorage of projecting signs (see the definition of "Projecting sign"). Projecting signs are required to be constructed of metal or approved noncombustible materials and are to be anchored to a building or structure only by metal bolts, anchors, supports, chains, guys or steel rods. Metal chains, guys or steel bolts with a minimum $^3/_8$-inch (9.5 mm) diameter are specifically prescribed to support the dead loads and wind pressures imposed on projecting signs perpendicular to the building or structure. These anchorage requirements are based on common construction techniques and are considered to provide a minimum resistance to pressures caused by wind loads. To resist the dead loads, these structural supports are to be located at a minimum angle of 45 degrees (0.78 rad). However, to resist wind pressures, these supports are to be at an angle of more than 45 degrees (0.78 rad) with the face of the sign. The use of staples or nails is prohibited for the attachment of projecting signs to buildings or structures. These limitations recognize that nails, when subjected to wind forces, may not provide adequate connection strength and structural resistance. A minimum of two supports is to be provided with a space between them of at least 8 feet (2438 mm) when the facial area of projecting signs exceeds 30 square feet (2.8 m^2). The intent is to provide a minimum number of structural elements so that adequate load transfer is provided.

H112.2 Attachment of supports. Supports shall be secured to a bolt or expansion screw that will develop the strength of the supporting chains, guys or steel rods, with a minimum $^5/_8$-inch (15.9 mm) bolt or lag screw, by an expansion shield. Turnbuckles shall be placed in chains, guys or steel rods supporting projecting signs.

❖ This section stipulates that structural supports of projecting signs are to be secured to a bolt or lag screw with a diameter of at least $^5/_8$ inch (15.9 mm). The bolt or expansion screw is to develop the strength of the supporting members. Structural supports are to be provided with turnbuckles to adjust the tension of the members.

H112.3 Wall mounting details. Chains, cables, guys or steel rods used to support the live or dead load of projecting signs are permitted to be fastened to solid masonry walls with expansion bolts or by machine screws in iron supports, but such supports shall not be attached to an unbraced parapet wall. Where the supports must be fastened to walls made of wood, the supporting anchor bolts must go through the wall and be plated or fastened on the inside in a secure manner.

❖ This section specifies the use of expansion bolts or machine screws in iron supports to fasten tension members required to support the live or dead loads of projecting signs at all locations of solid masonry walls, except on unbraced parapets. Where walls constructed of wood are provided for fastening the supports, anchor bolts are required to penetrate the wall and be securely plated or fastened on the inside face of the wall.

H112.4 Height limitation. A projecting sign shall not be erected on the wall of any building so as to project above the roof or cornice wall or above the roof level where there is no cornice wall; except that a sign erected at a right angle to the building, the horizontal width of which sign is perpendicular to such a wall and does not exceed 18 inches (457 mm), is permitted to be erected to a height not exceeding 2 feet (610 mm) above the roof or cornice wall or above the roof level where there is no cornice wall. A sign attached to a corner of a building and parallel to the vertical line of such corner shall be deemed to be erected at a right angle to the building wall.

❖ Consistent with other provisions of the code, projecting signs cannot extend beyond the ends of the wall or above the roof or cornice wall of a building. An exception to this limitation is projecting signs constructed perpendicular to the building, with a horizontal width of 18 inches (457 mm) and a height not exceeding 2 feet (610 mm) above the top of the roof or cornice wall or roof level without a cornice wall. In accordance with this section, signs attached to a corner of a building and parallel to the corner's vertical line are considered projecting signs constructed at a right angle to the building wall. Thus, their extension beyond the end of the wall is permitted within the limitations prescribed in this section.

H112.5 Additional loads. Projecting sign structures which will be used to support an individual on a ladder or other servicing device, whether or not specifically designed for the servicing device, shall be capable of supporting the anticipated additional load, but not less than a 100-pound (445 N) concentrated horizontal load and a 300-pound (1334 N) concentrated vertical load applied at the point of assumed or most eccentric loading. The building component to which the projecting sign is attached shall also be designed to support the additional loads.

❖ The intent of this section is to require additional structural capability in projecting signs that may be subjected to loads during maintenance procedures as shown in Figure H112.5. The face of the sign and the supports, therefore, must be capable of withstanding not only an additional concentrated horizontal load anywhere on the face of the sign, but also a 300-pound (1334 N) concentrated load acting vertically down at the point most eccentric from the sign fasteners. The building or building component to which the sign is attached must be capable of withstanding the resultant load.

SECTION H113 MARQUEE SIGNS

H113.1 Materials. Marquee signs shall be constructed entirely of metal or other approved noncombustible material except as provided for in Sections H106.1.1 and H107.1.

❖ A marquee sign is always supported by the building or structure and projects from the building into surrounding spaces. Marquee signs are usually located on canopies or other extensions of the building or structure to obtain a greater visual effect. The code has specific provisions for marquee signs concerning allowable materials. Marquee signs must be of metal or approved noncombustible materials. Sections H113.2 through H113.4 regulate the attachment, dimensions and height limitations of a marquee sign. Section H107.1 describes exceptions regarding sign facings, electrical wiring and clearances to electrical wires.

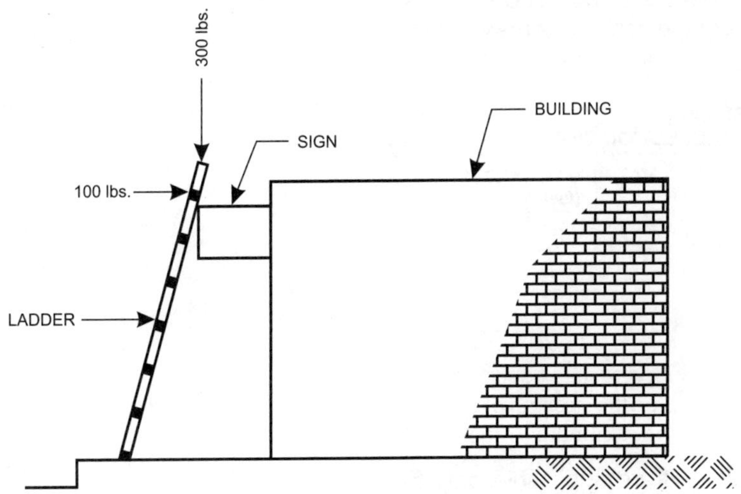

For SI: 1 pound = 4.4 N.

Figure H112.5
ADDITIONAL SIGN LOADS

H113.2 Attachment. Marquee signs shall be attached to approved marquees that are constructed in accordance with Section 3106.

❖ Marquee signs are attached to a structure called a "marquee." The marquee is to be designed and constructed in accordance with Section H113.1. A marquee sign should not be confused with a projecting sign (see Section H112.1).

H113.3 Dimensions. Marquee signs, whether on the front or side, shall not project beyond the perimeter of the marquee.

❖ Marquees accommodate a specially designed marquee sign arrangement. In other words, the sign fits into the marquee. A projecting sign cannot be placed on a marquee because the sign would project beyond the perimeter of the marquee.

H113.4 Height limitation. Marquee signs shall not extend more than 6 feet (1829 mm) above, nor 1 foot (305 mm) below such marquee, but under no circumstances shall the sign or signs have a vertical dimension greater than 8 feet (2438 mm).

❖ This section requires marquee signs to project less than 6 feet (1829 mm) above and not more than 1 foot (305 mm) below the marquee structure. The total vertical dimension of the marquee sign is limited to 8 feet (2438 mm). The intent of this section is to place a limitation on structures with a higher fuel load due to decorations incorporated on the sign that are subject to outdoor conditions and weathering.

SECTION H114 PORTABLE SIGNS

H114.1 General. Portable signs shall conform to requirements for ground, roof, projecting, flat and temporary signs where such signs are used in a similar capacity. The requirements of this section shall not be construed to require portable signs to have connections to surfaces, tie-downs or foundations where provisions are made by temporary means or configuration of the structure to provide stability for the expected duration of the installation.

❖ Portable signs must conform to the general regulations as set forth in this appendix for all signs, except that temporary support or stability methods may be utilized when approved by the building official. Portable signs must also conform to the requirements of this section commensurate to the manner in which the sign is used. This section establishes criteria for those types of signs that are neither mounted on buildings nor provided with a permanent foundation support. A typical example of a portable sign is a trailer-mounted illuminated sign. Such signs are used temporarily at locations and do not require any permanent support to resist wind overturning loads because of their limited size. The code contains electrical supply requirements for portable signs.

TABLE 4-A
SIZE, THICKNESS AND TYPE OF GLASS PANELS IN SIGNS

MAXIMUM SIZE OF EXPOSED PANEL		MINIMUM THICKNESS OF GLASS (inches)	TYPE OF GLASS
Any dimension (inches)	Area (square inches)		
30	500	$^1/_8$	Plain, plate or wired
45	700	$^3/_{16}$	Plain, plate or wired
144	3,600	$^1/_4$	Plain, plate or wired
> 144	> 3,600	$^1/_4$	Wired glass

For SI: 1 inch = 25.4 mm, 1 square inch = 645 mm^2.

❖ Glazing materials installed in all types of signs are required to conform to the limitations of Table 4-A. This table details the maximum allowable size, thickness limitations and type of glass panels to be used in signs. Glass subject to unusual loading conditions may require special engineering analysis to determine its size.

TABLE 4-B
THICKNESS OF PROJECTION SIGN

PROJECTION (feet)	MAXIMUM THICKNESS (feet)
5	2
4	2.5
3	3
2	3.5
1	4

For SI: 1 foot = 304.8 mm.

❖ The requirements of Table 4-B are applicable only for projecting signs constructed in accordance with Section H112.

SECTION H115
REFERENCED STANDARDS

ASTM D635-98	Test for Rate of Burning and/or Extent and Time of Burning of Self-Supporting Plasters in a Horizontal Position.	H107.1.1
ICC EC-2003	ICC *Electrical Code*	H106.1, H106.2
NFPA 701-96	Methods of Fire Test for Flame Resistant Textiles and Films.	H106.1.1

❖ See the commentary to Chapter 35.

Appendix I: Patio Covers

The provisions contained in this appendix are not mandatory unless specifically referenced in the adopting ordinance.

General Comments

These provisions are optional. They apply only if specifically adopted by the jurisdiction as stated in Section 101.2.1. The requirements are intended to apply only to patio covers associated with residential dwelling units.

Purpose

These provisions for patio covers are intended to simplify the requirements for residential patio installations.

SECTION I101 GENERAL

I101.1 General. Patio covers shall be permitted to be detached from or attached to dwelling units. Patio covers shall be used only for recreational, outdoor living purposes and not as carports, garages, storage rooms or habitable rooms. Openings shall be permitted to be enclosed with insect screening, approved translucent or transparent plastic not more that 0.125 inch (3.2 mm) in thickness, glass conforming to the provisions of Chapter 24 or any combination of the foregoing.

❖ The provisions of this appendix are intended to be used in conjunction with Groups R-1, R-2, R-3 and U with attached or detached patio covers. Patio covers must be used for recreational or outdoor living purposes only and not for the uses generally ascribed to Group U and R occupancies. If used for storage or as a garage, an entire new fuel load is introduced, thereby qualifying the patio cover as a private garage governed by Sections 312 and 406.

SECTION I102 DEFINITIONS

I102.1 General. The following word and term shall, for the purposes of this appendix have the meaning shown herein.

❖ Definitions of terms that are associated with the content of this appendix are contained herein. This definition can help in the understanding and application of the code requirements. It is important to emphasize that this term is not exclusively related to this appendix, but is applicable everywhere the term is used throughout the code. The purpose for including this definition within this appendix is to provide more convenient access to it rather than referring back to Chapter 2.

PATIO COVERS. One story structures not exceeding 12 feet (3657 mm) in height. Enclosure walls shall be permitted to be of any configuration, provided the open or glazed area of the longer wall and one additional wall is equal to at least 65 percent of the area below a minimum of 6 feet 8 inches (2032 mm) of each wall, measured from the floor.

❖ A patio cover is defined in order to make a distinction between carports, garages and covered awnings. The code does not consider a patio a hazardous condition with a high fuel load; therefore, special construction or consideration is not necessary. The maximum height a patio cover is allowed without being considered a separate structure is 12 feet (3658 mm) (see Figure I102.1).

SECTION I103 EXTERIOR OPENINGS

I103.1 Light, ventilation and emergency egress. Exterior openings required for light and ventilation shall be permitted to open into a patio structure. However, the patio structure shall be unenclosed if such openings are serving as emergency egress or rescue openings from sleeping rooms. Where such exterior openings serve as an exit from the dwelling unit, the patio struc-

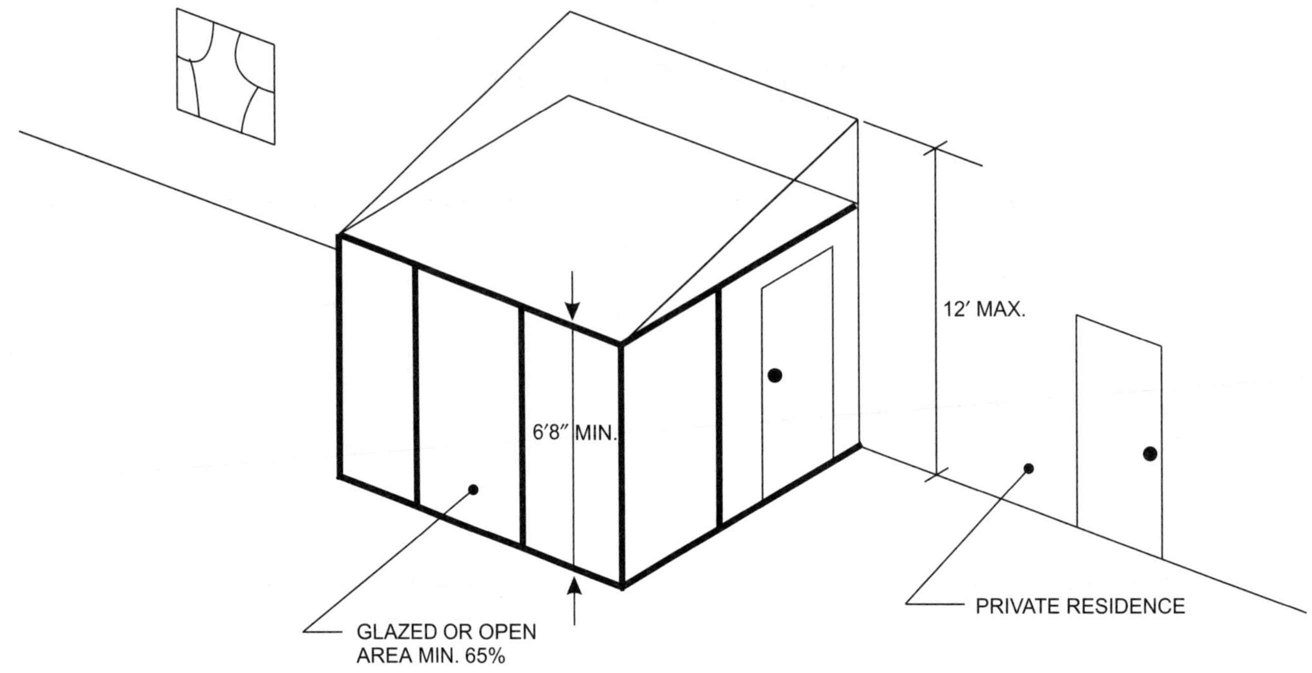

Figure I102.1
PATIO COVERS

ture, unless unenclosed, shall be provided with exits conforming to the provision of Chapter 10.

❖ A patio is allowed to be placed over an opening that is to be used as a means of natural light or ventilation. Once the patio covers an opening used for egress by the resident, the patio must provide an appropriate means of egress conforming with Chapter 10.

SECTION I104 STRUCTURAL PROVISIONS

I104.1 Design loads. Patio covers shall be designed and constructed to sustain, within the stress limits of this code, all dead loads plus a minimum vertical live load of 10 pounds per square foot (0.48 kN/m^2) except that snow loads shall be used where such snow loads exceed this minimum. Such patio covers shall be designed to resist the minimum wind and seismic loads set forth in this code.

❖ The minimum value for the live load of the patio is to be either 10 pounds per square foot (psf) (0.48 kN/m^2) or the roof snow load, whichever is greater. Additionally, where required, the patio must be constructed to be stabilized against the forces of wind and seismic loading conditions.

I104.2 Footings. In areas with a frost depth of zero, a patio cover shall be permitted to be supported on a concrete slab on grade without footings, provided the slab conforms to the provisions of Chapter 19 of this code, is not less than $3^1/_2$ inches (89 mm) thick and further provided that the columns do not support loads in excess of 750 pounds (3.36 kN) per column.

❖ This allowance, which permits foundations to be constructed without footings, is due to the minor value of these structures and the lack of any substantive safety issue involved. This exception addresses the exemption of footings in regard to frost penetration; it does not address allowable load-bearing capacity of the soil.

Appendix J: Grading

The provisions contained in this appendix are not mandatory unless specifically referenced in the adopting ordinance.

General Comments

This appendix contains grading provisions that address soil-related hazards such as slope failure, landslides and erosion. Jurisdictions that include mountainous or hilly terrain often encounter grading work in connection with large commercial developments or the development of residential subdivisions that can pose grading problems that are not addressed by any other section of the code. Note that only excavation, grading and fill in connection with foundation construction is regulated by Section 1803.

In western states the *Uniform Building Code*™ (UBC) appendix chapter on excavating and grading has found very widespread usage in many areas where topography is an issue. While the need for this appendix is based on the extensive use of that UBC appendix, which was originally developed in the 1960s, suggestions from numerous individuals and groups involved in the code development process have been incorporated into this grading appendix to make it compatible for use with the IBC.

Purpose

This appendix provides a set of grading requirements that can be utilized in lieu of (or in addition to) developing local regulations. Where grading is an important consideration, a comprehensive code is available simply by adopting this appendix. It is intended to provide consistent and uniform code requirements anywhere grading is considered an issue.

SECTION J101 GENERAL

J101.1 Scope. The provisions of this chapter apply to grading, excavation and earthwork construction, including fills and embankments. Where conflicts occur between the technical requirements of this chapter and the soils report, the soils report shall govern.

❖ This section states the scope of the chapter and also states that the soils report, which is required by Section J104.3, would supercede the requirements of this chapter when a conflict exists. This gives the professional responsible for the report much latitude to provide recommendations that are appropriate for each individual site.

J101.2 Flood hazard areas. The provisions of this chapter shall not apply to grading, excavation and earthwork construction, including fills and embankments, in floodways within flood hazard areas established in Section 1612.3 unless it has been demonstrated through hydrologic and hydraulic analyses performed in accordance with standard engineering practice that the proposed work will not result in any increase in the level of the base flood.

❖ As the definition in Section 1612.2 indicates, a floodway is that portion of a flood hazard area that is reserved for the discharge of the design flood event. The National Flood Insurance Program (NFIP) requires that the impact of development or encroachment on the floodway be considered. The intent of this section is to permit grading in a floodway only if is demonstrated that this activity will not adversely affect surrounding areas by increasing the base flood elevation.

SECTION J102 DEFINITIONS

J102.1 Definitions. For the purposes of this appendix chapter, the terms, phrases and words listed in this section and their derivatives shall have the indicated meanings.

❖ This section defines terms used in this appendix for uniformity of meaning to facilitate administration of these grading provisions.

BENCH. A relatively level step excavated into earth material on which fill is to be placed.

❖ This term is used to identify the steps that are cut into an existing slope prior to placing fill over that slope (see Figure J107.3).

COMPACTION. The densification of a fill by mechanical means.

❖ This definition aides the application of Section J107.5, which is applicable to all fill material.

CUT. See Excavation.

❖ Earthwork is typically referred to as "cut" and "fill." The term "cut" is synonymous with excavation.

DOWN DRAIN. A device for collecting water from a swale or ditch located on or above a slope, and safely delivering it to an approved drainage facility.

❖ The inclusion of a definition for "Down drain" helps to clarify the term and make the requirements of Section J109.2 enforceable.

EROSION. The wearing away of the ground surface as a result of the movement of wind, water or ice.

❖ This definition describes what is a naturally occurring process of wearing away the ground surface. The provisions of this chapter regulate grading with the intent to minimize the susceptibility to erosion of man-made slopes primarily due to surface water.

EXCAVATION. The removal of earth material by artificial means, also referred to as a cut.

❖ Excavations are regulated by Section J106.

FILL. Deposition of earth materials by artificial means.

❖ Earthwork is typically referred to as "cut" and "fill." Fill is regulated by Section J107

GRADE. The vertical location of the ground surface.

❖ The elevation of the existing ground surface is usually established by topographic surveys. This must be done in order to methodically evaluate a site to determine the need for regrading. The process of developing an appropriate grading plan leads to the required finished grade that must be achieved either by excavating where the existing grade is above the desired finished grade or by filling where the existing grade is below the desire finished grade.

GRADE, EXISTING. The grade prior to grading.

❖ See the definition of "Grade."

GRADE, FINISHED. The grade of the site at the conclusion of all grading efforts.

❖ See the definition of "Grade."

GRADING. An excavation or fill or combination thereof.

❖ This definition clarifies that the term "grading" refers to all earthwork.

KEY. A compacted fill placed in a trench excavated in earth material beneath the toe of a slope.

❖ This term refers to the level area that creates a notch at the bottom of a slope requiring benching in accordance with Section J107.3 (see Figure J107.3).

All slope references in the chapter have been modified to show the horizontal:vertical relationship.

SLOPE. An inclined surface, the inclination of which is expressed as a ratio of horizontal distance to vertical distance.

❖ This definition establishes that slopes in this chapter are expressed in the form of the horizontal to vertical distance ratio.

TERRACE. A relatively level step constructed in the face of a graded slope for drainage and maintenance purposes.

❖ This definition helps to clarify the term "terrace" and make the requirements of Section J109.2 enforceable.

SECTION J103
PERMITS REQUIRED

J103.1 Permits required. Except as exempted in Section J103.2, no grading shall be performed without first having obtained a permit therefor from the building official. A grading permit does not include the construction of retaining walls or other structures.

❖ A grading permit does not include the construction of retaining walls or other structures, since that is covered by Section 1803.

J103.2 Exemptions. A grading permit shall not be required for the following:

1. Grading in an isolated, self-contained area, provided there is no danger to the public, and that such grading will not adversely affect adjoining properties.
2. Excavation for construction of a structure permitted under this code.
3. Cemetery graves.
4. Refuse disposal sites controlled by other regulations.
5. Excavations for wells, or trenches for utilities.
6. Mining, quarrying, excavating, processing or stockpiling rock, sand, gravel, aggregate or clay controlled by other regulations, provided such operations do not affect the lateral support of, or significantly increase stresses in, soil on adjoining properties.
7. Exploratory excavations performed under the direction of a registered design professional This phrase was added to assure that the "exploratory excavation" is not to begin construction of a building prior to receiving a permit for the sole purpose of preparing a soils report.

Exemption from the permit requirements of this appendix shall not be deemed to grant authorization for any work to be done in any manner in violation of the provisions of this code or any other laws or ordinances of this jurisdiction.

❖ The exemptions listed are for grading work that is relatively minor in nature (e.g., graves) or grading work that is normally regulated by other laws (e.g., utility trenching, mining, etc.).

SECTION J104
PERMIT APPLICATION AND SUBMITTALS

J104.1 Submittal requirements. In addition to the provisions of Section 105.3, the applicant shall state the estimated quantities of excavation and fill.

❖ Section 105.3 lists the information that is required for a building permit application, which also applies for a grading permit. Additionally, the quantities of earthwork, commonly referred to as cut and fill, must be furnished.

J104.2 Site plan requirements. In addition to the provisions of Section 106, a grading plan shall show the existing grade and finished grade in contour intervals of sufficient clarity to indicate the nature and extent of the work and show in detail that it complies with the requirements of this code. Drafting requirements were deleted here.The plans shall show the existing grade on adjoining properties in sufficient detail to identify how grade changes will conform to the requirements of this code.

❖ It is necessary for the permit applicant to furnish all applicable construction documents that are required in Section 106. Additionally, grading plans must provide sufficient grading information so that compliance with these requirements is verifiable.

J104.3 Soils report. A soils report prepared by registered design professionals shall be provided which shall identify the nature and distribution of existing soils; conclusions and recommendations for grading procedures; soil design criteria for any structures or embankments required to accomplish the proposed grading; and, where necessary, slope stability studies, and recommendations and conclusions regarding site geology.

Exception: A soils report is not required where the building official determines that the nature of the work applied for is such that a report is not necessary.

❖ This section requires the applicant to furnish a soils report unless the building official determines that it is not necessary because the proposed grading is of relatively minor consequence. The intent is that the appropriate professional provides a report that assesses the hazards of a particular site and includes recommendations for accomplishing the proposed grading. Note that Section J101.1 states that the soils report would supercede the requirements of this chapter when a conflict exists, giving great weight to the professional's recommendations.

J104.4 Liquefaction study. For sites with mapped maximum considered earthquake spectral response accelerations at short periods (S_s) greater than 0.5g as determined by Section 1615, a study of the liquefaction potential of the site shall be provided, and the recommendations incorporated in the plans.

Exception: A liquefaction study is not required where the building official determines from established local data that the liquefaction potential is low.

❖ This section intends to provide a guideline for the building official to determine if a liquefaction study is necessary. The exception provides the latitude to waive this requirement where it is evidenced that liquefaction is a low risk.

The need for a liquefaction study is triggered by the mapped spectral acceleration at short periods (see Section 1615.1) rather than seismic design category (see Section 1616.3), which is based on the nature of an occupancy in addition to the mapped spectral accelerations modified for the site soil classification. While seismic design category is an entirely appropriate criteria for such a requirement in connection with a building foundation under Section 1802, the mapped spectral acceleration is more direct and appropriate for the application of this grading appendix, since occupancy is most likely not applicable. While the mapped acceleration is used as the trigger for a study, it is ultimately the soil profile that determines the susceptibility to liquefaction.

SECTION J105
INSPECTIONS

J105.1 General. Most of this section was deleted or simplified. Inspections shall be governed by Section 109 of this code.

❖ A general reference is made to the inspection requirements in Section 109 of the code. The type and frequency of inspections specific to grading are not enumerated and are, therefore, left to the judgement of the building official.

J105.2 Special inspections. The special inspection requirements of Section 1704.7 shall apply to work performed under a grading permit where required by the building official.

❖ The special inspection requirements of Section 1704.7 that are referred to concern site preparation and fill placement.

SECTION J106
EXCAVATIONS

J106.1 Maximum slope. The slope of cut surfaces shall be no steeper than is safe for the intended use, and shall be no steeper than 2 horizontal to 1 vertical (50 percent) unless the applicant furnishes a soils report justifying a steeper slope.

Exceptions:

1. A cut surface may be at a slope of 1.5 horizontal to 1 vertical (67 percent) provided that all the following are met:
   1.1. It is not intended to support structures or surcharges.
   1.2. It is adequately protected against erosion.
   1.3. It is no more than 8 feet (2438 mm) in height.
   1.4. It is approved by the building official.
2. A cut surface in bedrock shall be permitted to be at a slope of 1 horizontal to 1 vertical (100 percent).

❖ The code establishes a maximum 2 horizontal to 1 vertical (50 percent) slope even though many professionals experienced in hillside urban development are likely to consider a 1.5 horizontal to 1 vertical (67 percent) slope to be adequate for stability and erosion control. The 2 horizontal to 1 vertical (50 percent) slope gradient is required to provide an extra margin of safety and easier maintenance capability, due to the fact that slope failures of cuts (as well as fills) with 1.5 horizontal to 1 vertical (67 percent) slopes have been observed in the past. The cause of these failures was frequently traced to a lack of geotechnical inspection during grading or of proper maintenance after they were constructed.

The building official may allow steeper cut slopes if they are justified by the soils report. Exception 1 allows the steeper cut surface slope (discussed above) of 1.5 horizontal to 1 vertical (67 percent) where the listed conditions are met, indicating a lower slope failure hazard. A steeper cut surface slope in bedrock is permitted because it is inherently more stable.

SECTION J107 FILLS

J107.1 General. Unless otherwise recommended in the soils report, fills shall conform to provisions of this section.

❖ These provisions govern all fills. As is indicated in J101.1, the soils report takes precedence in the event of conflicts with this chapter.

J107.2 Surface preparation. The ground surface shall be prepared to receive fill by removing vegetation, topsoil and other unsuitable materials, and scarifying the ground to provide a bond with the fill material.

❖ Fill must only be placed onto sound bedrock or other competent material to minimize subsidence and/or settlement; therefore, the removal of all unsuitable materials is necessary.

J107.3 Benching. Where existing grade is at a slope steeper than 5 horizontal to 1 vertical (20 percent) and the depth of the fill exceeds 5 feet (1524 mm) benching shall be provided in accordance with Figure J107.3. A key shall be provided which is at least 10 feet (3048 mm) in width and 2 feet (610 mm) in depth.

❖ Benching requirements are illustrated in Figure J107.3. The fill over an existing slope is potentially unstable unless constructed properly. Benching creates a series of (temporary) level steps on which the fill can be reliably placed and compacted. It is required for fills over 5 feet (1524 mm) deep where the existing slope is steeper than 5 horizontal to 1 vertical (20 percent). The toe-of-fill bench is referred to as a key and the code specifies a minimum depth and width. The area of ground surface left at the planned toe-of-fill slope should be graded to drain away from the fill (see Section J108.3).

FIGURE J107.3. See page J-5.

❖ This figure illustrates the intention of the benching requirements. It shows the key that is located at the toe-of-slope and the series of level benches that must be established in order to accomplish the proposed earthwork.

J107.4 Fill material. Fill material shall not include organic, frozen or other deleterious materials. No rock or similar irreducible material greater than 12 inches (305 mm) in any dimension shall be included in fills.

❖ Organic material should not be allowed in fills because they decompose over time, causing settlement or creating planes of weakness. Rocks of limited size are permitted in fills. Where doing so, they should be placed to assure that the fill material can be placed in the voids around the rocks and compacted as required.

J107.5 Compaction. All fill material shall be compacted to 90 percent of maximum density as determined by ASTM D1557, Modified Proctor, in lifts not exceeding 12 inches (305 mm) in depth.

❖ This section specifies a minimum soil density to be achieved uniformly throughout the fill as well as the test to be used. This sets minimum acceptable levels of fill quality to achieve performance that is consistent with the purpose of this chapter.

J107.6 Maximum slope. The slope of fill surfaces shall be no steeper than is safe for the intended use. Fill slopes steeper than 2 horizontal to 1 vertical (50 percent) shall be justified by soils reports or engineering data.

❖ The maximum gradient of 2 horizontal to 1 vertical (50 percent) for fill slopes minimizes the risk of slope failures (see commentary, Section J106.1). Compaction of fill can be accomplished more readily on slopes of 2 horizontal to 1 vertical (50 percent) or less. Also, such slopes are more accessible, increasing the likelihood they will be maintained. Greater slopes are only permitted if substantiated by the soils report or other engineering data.

SECTION J108 SETBACKS

J108.1 General. Cut and fill slopes shall be set back from the property lines in accordance with this section. Setback dimensions shall be measured perpendicular to the property line and shall be as shown in Figure J108.1, unless substantiating data is submitted justifying reduced setbacks.

❖ This section establishes minimum setbacks from property lines between the property being graded and adjacent properties. It is intended to apply at the exterior boundary of a tract and should not be applied, for instance, at the interior lot lines of a proposed development.

FIGURE J108.1. See page J-5.

❖ The figure specifies the setback requirements at the top and bottom of graded slopes.

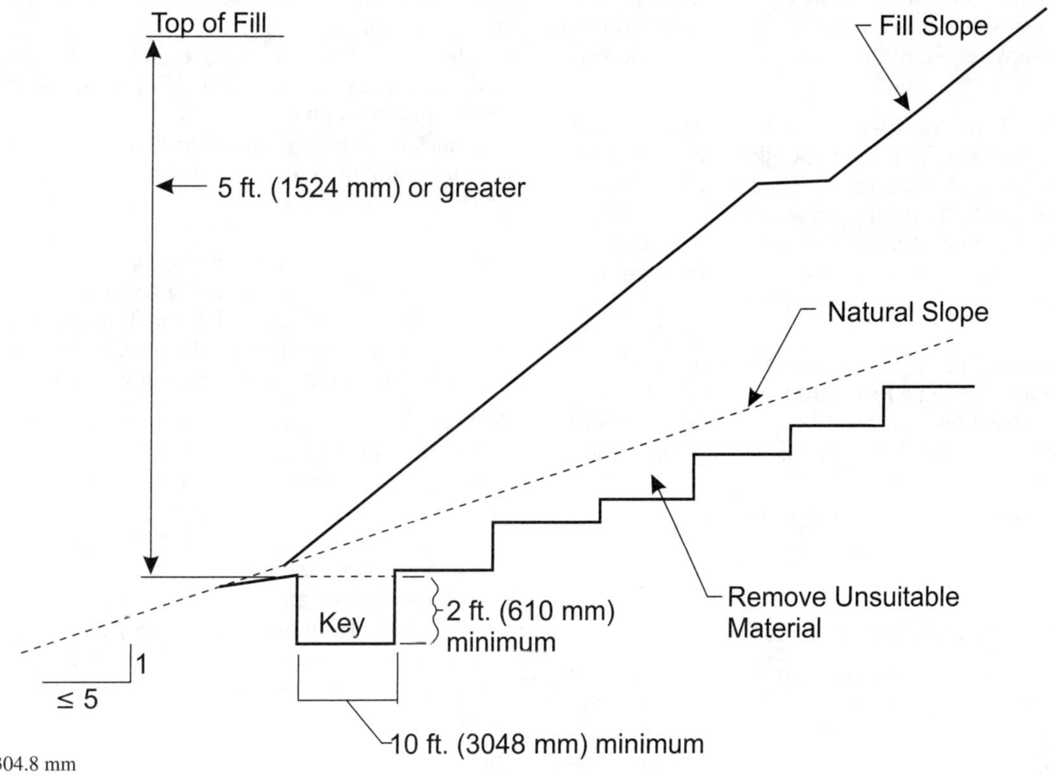

For SI: 1 foot = 304.8 mm

FIGURE J107.3
BENCHING DETAILS

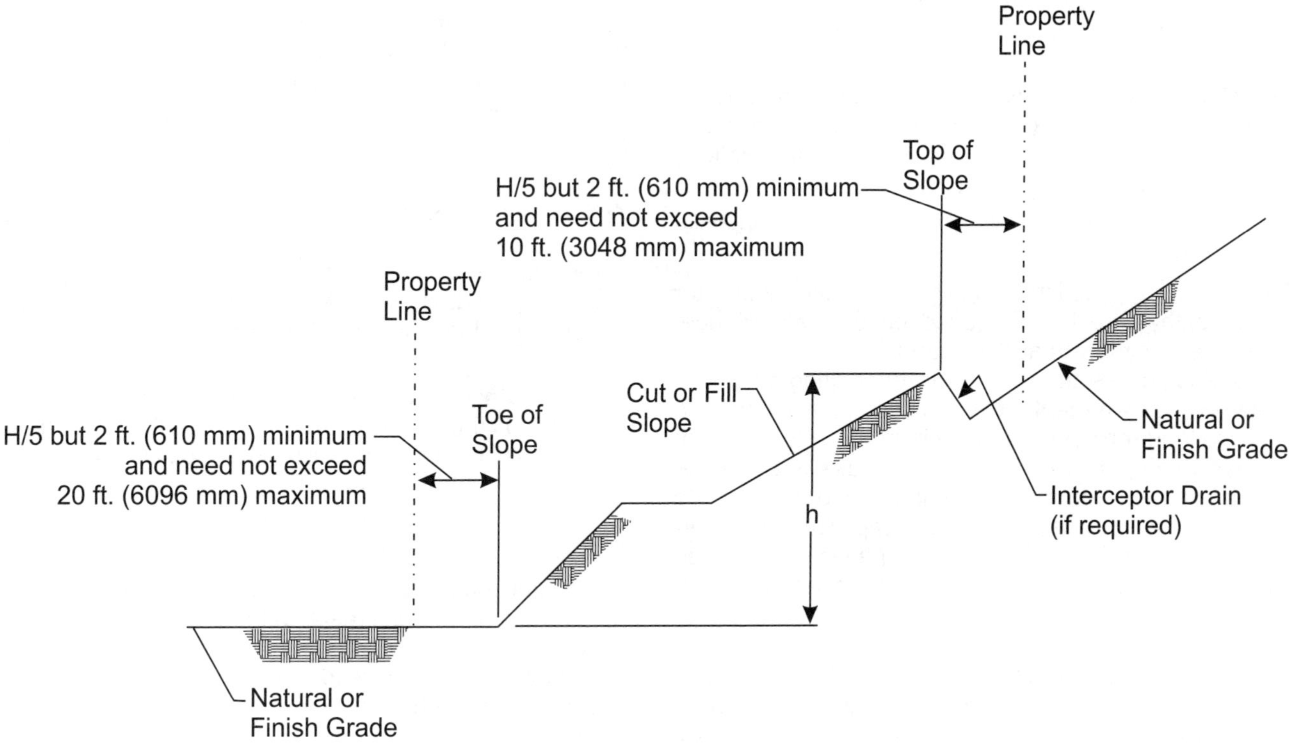

For SI: 1 foot = 304.8 mm

FIGURE J108.1
DRAINAGE DIMENSIONS

J108.2 Top of slope. The setback at the top of a cut slope shall not be less than that shown in Figure J108.1, or than is required to accommodate any required interceptor drains, whichever is greater.

❖ The intention of the top-of-slope setback is to protect the adjacent property. The setback allows for some soil erosion without losing lateral support of the adjacent property, thus protecting structures on that property. It also provides some space for interceptor drain if needed (see Section J109.3). The top-of-slope setback requirements are stated in Figure J108.1.

J108.3 Slope protection. Where required to protect adjacent properties at the toe of a slope from adverse effects of the grading, additional protection, approved by the building official, shall be included. Such protection may include but shall not be limited to:

1. Setbacks greater than those required by Figure J108.1.
2. Provisions for retaining walls or similar construction.
3. Erosion protection of the fill slopes.
4. Provision for the control of surface waters.

❖ The minimum toe-of-fill setback from the adjacent property line is specified in Figure J108.1. Additional clearance must be provided as necessary, for example, to allow for the construction of retaining walls. For example, it may also be necessary to provide an area to collect slope runoff along the toe of slope and to construct a drain if needed.

SECTION J109
DRAINAGE AND TERRACING

J109.1 General. Unless otherwise recommended by a registered design professional, drainage facilities and terracing shall be provided in accordance with the requirements of this section.

Exception: Drainage facilities and terracing need not be provided where the ground slope is not steeper than 3 horizontal to 1 vertical (33 percent).

❖ This section establishes minimum requirements for drainage that are largely based on past experience. They are minimum standards and the design professional should use them as such, evaluating each site based on its own merits. Consideration also should be given to long-term performance in addition to the likelihood of inadequate maintenance of slopes in residential developments. Controlling drainage onto and off the proposed site is one of the most important considerations in grading design, since even a well-compacted fill or graded cut slope can be compromised by inadequate drainage.

J109.2 Terraces. Terraces at least 6 feet (1829 mm) in width shall be established at not more than 30-foot (9144 mm) vertical intervals on all cut or fill slopes to control surface drainage and debris. Suitable access shall be provided to allow for cleaning and maintenance.

Where more than two terraces are required, one terrace, located at approximately mid-height, shall be at least 12 feet (3658 mm) in width.

Swales or ditches shall be provided on terraces. They shall have a minimum gradient of 20 horizontal to 1 vertical (5 percent) and shall be paved with concrete not less than 3 inches (76 mm) in thickness, or with other materials suitable to the application. They shall have a minimum depth of 12 inches (305 mm) and a minimum width of 5 feet (1524 mm).

A single run of swale or ditch shall not collect runoff from a tributary area exceeding 13,500 square feet (1256 m^2) (projected) without discharging into a down drain.

❖ Terraces are required where a graded slope has a height over 30 feet (9144 mm). They provide level surfaces along the graded slope to allow access for slope maintenance and accommodate drainage.

The terrace ditch gradient of 5 percent helps make the drain more self-cleansing and recognizes a typical lack of maintenance, since most of the hillside dwellers will not climb up or down the slope to keep a drain clear of debris. Swales serving larger drainage areas must discharge into a down drain, which is defined in Section J102.1.

J109.3 Interceptor drains. Interceptor drains shall be installed along the top of cut slopes receiving drainage from a tributary width greater than 40 feet, measured horizontally. They shall have a minimum depth of 1 foot (305 mm) and a minimum width of 3 feet (915 mm). The slope shall be approved by the building official, but shall not be less than 50 horizontal to 1 vertical (2 percent). The drain shall be paved with concrete not less than 3 inches (76 mm) in thickness, or by other materials suitable to the application. Discharge from the drain shall be accomplished in a manner to prevent erosion and shall be approved by the building official.

❖ Interceptor drains are important in protecting the face of cut slopes from excessive erosion and are, therefore, required under certain conditions. The discharge of an interceptor drain must not contribute to erosion and requires the building official's approval.

J109.4 Drainage across property lines. Drainage across property lines shall not exceed that which existed prior to grading. Excess or concentrated drainage shall be contained on site or directed to an approved drainage facility. Erosion of the ground in the area of discharge shall be prevented by installation of nonerosive down drains or other devices.

❖ This section protects adjacent properties from exposure to added runoff as a result of the grading as well as any concentrated runoff created by the grading. It requires collection of any excess or concentrated drainage and allows it to be held on site or discharged by approved methods.

SECTION J110
EROSION CONTROL

J110.1 General. The faces of cut and fill slopes shall be prepared and maintained to control erosion. This control shall be permitted to consist of effective planting.

> **Exception:** Erosion control measures need not be provided on cut slopes not subject to erosion due to the erosion-resistant character of the materials.

Erosion control for the slopes shall be installed as soon as practicable and prior to calling for final inspection.

❖ Adequate provision should be made to prevent surface water from damaging the face of a graded slope. Effective planting is a lasting method of minimizing erosion. Other measures may include berms, interceptor drains and terrace drains.

J110.2 Other devices. Where necessary, check dams, cribbing, riprap or other devices or methods shall be employed to control erosion and provide safety.

❖ This requires use of the alternative means of providing erosion control where necessary.

SECTION J111
REFERENCED STANDARDS

This section lists the standards referenced by this appendix chapter. Since appendices are not considered part of the code unless specifically referred to in the adopting ordinance, the standards that are referenced in this chapter are listed here rather than in Chapter 35.

Bibliography

ASTM D 1557-00, *Test Method for Laboratory Compaction Characteristics of Soil Using Modified Effort (56,000ft-lb/ ft).* West Conshohocken, PA: ASTM International, 2000.

Scullin, C. Michael. *Excavation and Grading Code Administration, Inspection and Enforcement.* Englewood Cliffs, NJ: Prentice-Hall, Inc, 1983.

UBC-97, *Uniform Building Code.* Whittier, CA: International Conference of Building Officials, 1997.

INDEX

Note: This is taken from the index for the 2003 International Building Code. Volume II of the IBC Commentary only includes Chapters 16 through 35.

A

B

C

F

G

I

N

R

S